JOHN GERARD

THE HERBAL
OR
GENERAL HISTORY OF PLANTS

The Complete 1633 Edition
as Revised and Enlarged by
THOMAS JOHNSON

Dover Publications, Inc.
NEW YORK

Published in Canada by General Publishing Company, Ltd., 30 Lesmill Road, Don Mills, Toronto, Ontario.
Published in the United Kingdom by Constable and Company, Ltd., 10 Orange Street, London WC 2.

This Dover edition, first published in 1975, is an unabridged republication of the work originally printed by Adam Islip Joice Norton and Richard Whitakers, London, in 1633. For completeness, a few pages have been reproduced from the second edition of 1636. Errors in folios and chapter numbers have been tacitly corrected.
A new Publisher's Note has been written for the present edition.

International Standard Book Number: 0-486-23147-X
Library of Congress Catalog Card Number: 74-18719

Manufactured in the United States of America
Dover Publications, Inc.
180 Varick Street
New York, N.Y. 10014

PUBLISHER'S NOTE

From its inception in ancient Greece until the seventeenth century and beyond, botany was largely an adjunct of medicine and pharmacy. Plants were studied chiefly for their curative powers, and this knowledge was transmitted in books known as herbals, which also contained rudimentary physical descriptions and illustrations of the plants for purposes of identification.

Medieval herbals were generally unoriginal commentaries on Theophrastus and Dioscorides, with large admixtures of folklore. Their illustrators were more concerned with abstract beauty of design than with the accurate rendering of plant anatomy. The same is true of the earliest printed herbals, which form a significant part of the incunabular repertoire.

With the spread of the Renaissance to northern Europe in the sixteenth century, new life was breathed into the study of plants. Otto Brunsfels' Latin herbal (Strasbourg, 1530), with illustrations by Hans Weiditz, deliberately featured natural renderings from life. Before long, important new herbals were available in the vernacular languages: in German, those of Hieronymus Bock (scholarly name: Tragus), 1539, and of Leonhart Fuchs (Fuchsius), 1543 (the illustrations to Fuchs, by Albrecht Meyer, were the most influential and most frequently copied and reused of the century); in Italian, that of Pierandrea Mattioli, 1544; in Flemish/Dutch, that of Rembert Dodoens (Dodonaeus), 1554, which was quickly translated into French by Charles de l'Ecluse and from French into English by Henry Lyte; in English, especially that of the adventurous scholar, physician, theologian and "father of English botany," William Turner, publication of which began in 1551.

The second half of the century was a golden age of early botany. Not only was the scholarship more original and observation more direct, but explorers were bringing specimens of exotic plants from the New World and the Near and Far East. The herbarists themselves were no longer sedentary clerics, but roamed the remote corners of Europe, collecting and trading plants and visiting foreign colleagues. Moreover, European potentates and grandees, from the Holy Roman Emperor on down, as well as learned institutions, maintained extensive gardens and subsidized publications. The major books of this period (all in Latin) were those of Mathias de l'Obel (Lobelius), a Fleming active in England, 1570, of Dodoens (a larger work than that of 1554, the *Stirpium historiae pemptades sex,* 1583) and of Charles de l'Ecluse (Clusius), 1601, the last-named being no longer an herbal, but a work of systematic botany and probably the best-illustrated plant book of the period.

In England, by far the most popular herbal for generations (to the extent that the author's name is inseparable from the concept of herbal to English speakers) was that of John Gerard, published in London in December 1597. Gerard, born in Nantwick, Cheshire, in 1545, became a barber-surgeon in London, slowly rising to the top of his profession. He was superintendent of the gardens owned by Lord Burghley in London and in Hertfordshire. William Cecil, first baron of Burghley, was chief minister to Queen Elizabeth. (In the 1540s he had been secretary to the Duke of Somerset, whose garden was superintended by the above-mentioned William Turner.) Gerard also had a garden of his own in Holborn (London), and published a complete catalogue of its contents in 1596, the first publication of that type in history. His major work, the *Herball or Generall Historie of Plantes,* appeared in the following year (printed by John Norton). (In later years, Gerard was herbarist to James I; he died in London in February 1612.)

The text of Gerard's *Herball* was based largely on a pre-existent translation into English by a Dr. Priest of Dodoens' Latin herbal of 1583. Although Gerard is said to have traveled in Russia and Scandinavia, his text was quite unoriginal and his scholarship was suspect even to contemporaries. Of the 1800 illustrations in the 1597 edition, only a handful were specially drawn, the bulk being derived from a somewhat earlier book; this was the 1588 Frankfurt herbal (*Neuw Kreuterbuch*) of Jacob Dietrich of Bergzabern (Tabernaemontanus), of which the illustrations only were reissued in 1590 as *Eicones plantarum;* Dietrich, a pupil of Brunfels and Bock, had himself utilized illustrations from Bock, Fuchs, Mattioli, Dodoens, de l'Ecluse and de l'Obel.

With all its faults, however, Gerard's *Herball* was England's herbal, and its popularity was unrivaled until John Parkinson appeared on the horizon in the 1620s. Upon the announcement of a major herbal by Parkinson (it did not appear until 1640, and was disappointing), those Londoners who had a vested interest in the Gerard work commissioned the apothecary-botanist Thomas Johnson to prepare a revised and enlarged edition in the short space of one year. The result was the 1633 Gerard-Johnson *Herball,* which is here reprinted in its entirety. (A practically unaltered second edition of this revision appeared in 1636.)

Thomas Johnson was born in Selby, Yorkshire, probably early in the seventeenth century, and was active in London by 1626. He undertook several botanical excursions in England and Wales and made significant contributions to local botany. In the civil war he was a militant Royalist, rising to the rank of lieutenant-colonel. He received a special M.D. degree from Oxford in 1643, and was mortally wounded in battle in the following year.

Johnson appears to have been a more capable and scholarly botanist than Gerard. In his new edition, he corrected errors in the old text, commented skeptically on Gerard's gullibility (the goose tree) or downright dishonesty, and added much material from his own observation and from Continental herbals published since 1597, notably that of de l'Ecluse. The 1633 Gerard-Johnson describes about 2850 plants, some 800 more than the 1597 Gerard, and about 700 illustrations were added, chiefly from the stock of the Antwerp publisher Plantin, who had brought out Dodoens, de l'Obel and de l'Ecluse. Thus, the 1633 edition reprinted here represents, in both text and pictures, a conspectus and summation of the finest research of its age. Written in clear and charming English, it is a lasting monument of Renaissance botany.

The preliminary matter of the book consists of:

Title page (engraved by John Payne; all the other illustrations are woodcuts).

Johnson's dedication (in Latin) to Richard Edwards (Master of the Company of Apothecaries), Edward Cooke, Leonard Stone and all the other members of the Company.

Gerard's dedication to Lord Burghley (in English).

Epistle (in Latin) from Lancelot Browne,

physician of Queen Elizabeth, to Gerard (dated from the Court at Westminster, Dec. 1, 1597).

Epistle (in Latin) from de l'Obel to Gerard (dated London, Dec. 1, 1597); de l'Obel states that Gerard has included many English plants for the first time.

Various Latin poems in Gerard's honor.

Foreword (in English) by the physician Stephen Bredwell, who states that Gerard has here "accommodated" Priest's translation of Dodoens.

Foreword (in English) by George Baker, Master of the London surgeons.

Foreword (in English) by Gerard (dated Holborn, Dec. 1, 1597), who states that Priest's translation of Dodoens "perished."

Foreword (in English) by Johnson (dated Snow Hill [London], Oct. 22, 1633). This long piece, which describes in detail how the 1633 edition is to be understood and used, should be read by all serious owners of the present volume; it also includes a long history of botany and an enlightening, uncomplimentary critique of Gerard's procedures and accomplishments.

"A Catalogue of Additions," listing everything added by Johnson in his edition to both the text and the illustrative matter.

The body of the work follows Gerard's original division into three books (Johnson's additions being inserted in their proper place), the first book containing grasses, rushes, grains, reeds, irises and bulbs; the second, food plants, medicinal plants and sweet-smelling plants; the third, roses, trees, shrubs, bushes, fruit-bearing (food) plants, rosin- and gum-producing plants, heaths, mosses and fungi (as well as corals and sponges, then thought to be plants). For each plant (with appropriate exceptions and modifications) there is a picture, Latin and English names, and text under the following headings: Kinds (rudimentary classifications), Description (physical appearance), Place (where found, with many highly localized English habitats), Time (of flowering, sowing, etc., as the case may be), Names (a variety of English names, as well as names in several Continental languages, "low Dutch" meaning Flemish/Dutch and "high Dutch" meaning German), Nature or Temper (whether the plant is "hot" or "cold," "dry" or "moist"—according to the traditional medical theory of "temperaments" or "humors") and Virtues (the medical properties, harmful or otherwise; use as food; recipes for draughts, poultices, etc.; sometimes there is a special heading "The Danger").

The backmatter of the book consists of:

An "Appendix" (supplement) of still more plant descriptions and figures added by Johnson.

An index of Latin plant names.

Nominum et opinionum harmonia (a glossary/concordance of Latin plant names as used by various cited authors).

A "Table" (index) of English plant names.

"A Supplement . . . unto the generall Table" (a glossary of obsolete or little used English plant names).

"A Catalogue of the Brittish Names of Plants" (a glossary of Welsh names).

An index of the uses and properties ("vertues") of the plants.

Errata.

NOTE TO THE READER ON PAGINATION

With the desire to make this edition of Johnson's Gerard as like the original as possible, the faulty page numbers that occur from time to time have been retained (for instance, between pages 29 and 30 there are four pages marked 30, 29, 30, 29). The reader is assured that the text itself is in proper sequence, with nothing omitted or duplicated.

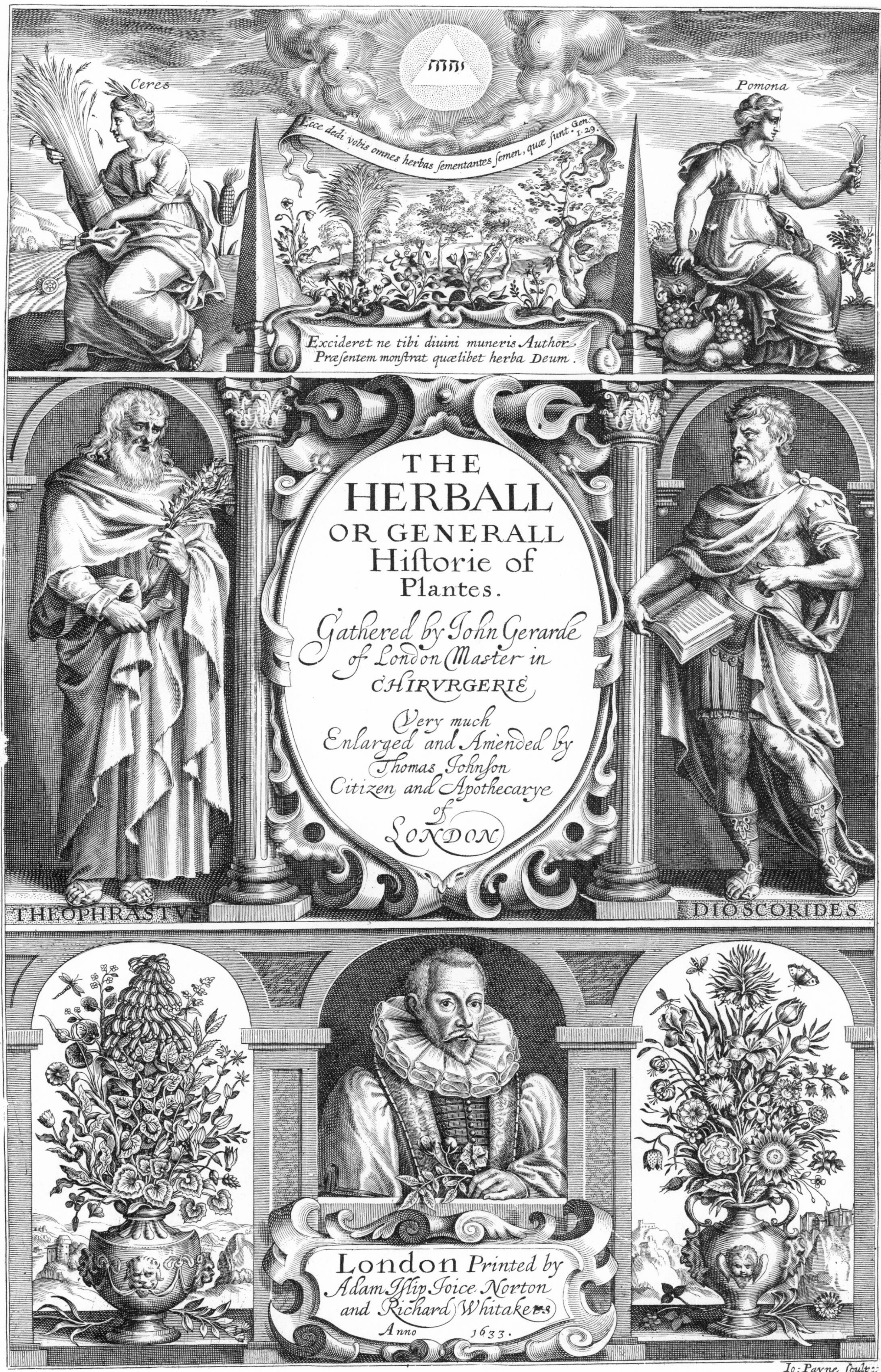

יהוה
Ceres
Pomona
Ecce dedi vobis omnes herbas sementantes semen, quæ sunt Gen: I. 29.
Excideret ne tibi diuini muneris Author
Præsentem monstrat quælibet herba Deum.
THE
HERBALL
OR GENERALL
Historie of
Plantes.
Gathered by John Gerarde
of London Master in
CHIRVRGERIE
Very much
Enlarged and Amended by
Thomas Johnson
Citizen and Apothecarye
of
LONDON
THEOPHRASTVS
DIOSCORIDES
London Printed by
Adam Islip Joice Norton
and Richard Whitakers
Anno 1633.
Io: Payne sculp:

VIRIS

PRVDENTIA, VIRTVTE,
ARTE, RERVMQVE VSV SPECTATISSIMIS,
DIGNISSIMIS

RICHARDO EDWARDS

RECTORI, SIVE MAGISTRO;

EDWARDO COOKE, LEONARDO
STONE GVARDIANIS,

CÆTERISQVE CLARISS. SOCIET.
PHARMACEVT. LOND. SOCIIS,

HOS SVOS IN EMA-
CVLANDO, AVGENDOQVE
HANC PLANTARVM
HISTORIAM,

LABORES, STVDIORVM BOTANICORVM
SPECIMEN, AMORIS SYMBOLVM,
EX ANIMO

D. D.

VESTRÆ, PVBLICÆQVE VTILI-
TATIS STVDIOSISSIMVS

THOM. IOHNSON.

TO THE RIGHT HONORABLE HIS SINGVLAR GOOD LORD AND MASTER, SIR WILLIAM CECIL KNIGHT, BARON OF Burghley, Master of the Court of Wards and Liueries, Chancellor of the Vniuersitie of Cambridge, Knight of the most noble Order of the Garter, one of the Lords of her Maiesties most honorable Priuy Councell, and Lord high Treasurer of *England*.

Mong the manifold creatures of God (right Honorable, and my singular good Lord) that haue all in all ages diuersly entertained many excellent wits, and drawne them to the contemplation of the diuine wisdome, none haue prouoked mens studies more, or satisfied their desires so much as Plants haue done, and that vpon iust and worthy causes: For if delight may prouoke mens labor, what greater delight is there than to behold the earth apparelled with plants, as with a robe of embroidered worke, set with Orient pearles, and garnished with great diuersitie of rare and costly iewels? If this varietie and perfection of colours may affect the eye, it is such in herbs and floures, that no *Apelles*, no *Zeuxis* euer could by any art expresse the like: if odours or if taste may worke satisfaction, they are both so soueraigne in plants, and so comfortable, that no confection of the Apothecaries can equall their excellent vertue. But these delights are in the outward sences: the principall delight is in the minde, singularly enriched with the knowledge of these visible things, setting forth to vs the inuisible wisedome and admirable workmanship of almighty God. The delight is great, but the vse greater, and ioyned often with necessity. In the first ages of the world they were the ordinarie meate of men, and haue continued euer since of necessarie vse both for meates to maintaine life, and for medicine to recouer health. The hidden vertue of them is such, that (as *Pliny* noteth) the very bruite beasts haue found it out: and (which is another vse that he obserues) from thence the Dyars tooke the beginning of their Art. Pli.li.8.ca.27. Ibid.li.22.c.2.

Furthermore, the necessary vse of these fruits of the earth doth plainly appeare by the great charge and care of almost all men in planting and maintaining of gardens, not as ornaments onely, but as a necessarie prouision also to their houses. And here beside the fruit, to speake againe in a word of delight, gardens, especially such as your Honor hath, furnished with many rare Simples, do singularly delight, when in them a man doth behold a flourishing shew of Sommer beauties in the midst of Winters force, and a goodly spring of floures, when abroad a leafe is not to be seene. Besides these and other causes, there are many examples of those that haue honored this science: for to passe by a multitude of the Philosophers, it may please your Honor to call to remembrance that which you know of some noble Princes, that haue ioyned this study with their most important mat-

. de Discr. .l & amic. .in. lib. 25. .ap. 2.

ters of state: *Mithridates* the great was famous for his knowledge herein, as *Plutarch* noteth. *Euax* also King of Arabia, the happy garden of the world for principall Simples, wrot of this argument, as *Pliny* sheweth. *Diocletian* likewise, might haue had his praise, had he not drowned all his honour in the bloud of his persecution. To conclude this point, the example of *Solomon* is before the rest, and greater, whose wisedome and knowledge was such, that hee was able to set out the nature of all plants from the highest Cedar to the lowest Mosse. But my very good Lord, that which sometime was the study of great Philosophers and mightie Princes, is now neglected, except it be of some few, whose spirit and wisdome hath carried them among other parts of wisedome and counsell, to a care and studie of speciall herbes, both for the furnishing of their gardens, and furtherance of their knowledge: among whom I may iustly affirme and publish your Honor to be one, being my selfe one of your seruants, and a long time witnesse thereof: for vnder your Lordship I haue serued, and that way employed my principall study and almost all my time, now by the space of twenty yeares. To the large and singular furniture of this noble Island I haue added from forreine places all the varietie of herbes and floures that I might any way obtaine, I haue laboured with the soile to make it fit for plants, and with the plants, that they might delight in the soile, that so they might liue and prosper vnder our clymat, as in their natiue and proper countrey: what my successe hath beene, and what my furniture is, I leaue to the report of them that haue seene your Lordships gardens, and the little plot of myne owne especiall care and husbandry. But because gardens are priuat, and many times finding an ignorant or a negligent successor, come soone to ruine, there be that haue sollicited me, first by my pen, and after by the Presse to make my Labors common, and to free them from the danger whereunto a garden is subiect: wherein when I was ouercome, and had brought this History or report of the nature of Plants to a iust volume, and had made it (as the Reader may by comparison see) richer than former Herbals, I found it no question vnto whom I might dedicate my Labors; for considering your good Lordship, I found none of whose fauor and goodnesse I might sooner presume, seeing I haue found you euer my very good Lord and Master. Again, considering my duty and your Honors merits, to whom may I better recommend my Labors, than to him vnto whom I owe my selfe, and all that I am able in any seruice or deuotion to performe? Therefore vnder hope of your Honorable and accustomed fauor I present this Herball to your Lordships protection; and not as an exquisite Worke (for I know my meannesse) but as the greatest gift and chiefest argument of duty that my labour and seruice can affoord: whereof if there be no other fruit, yet this is of some vse, that I haue ministred Matter for Men of riper wits and deeper iudgements to polish, and to adde to my large additions where any thing is defectiue, that in time the Worke may be perfect. Thus I humbly take my leaue, beseeching God to grant you yet many dayes to liue to his glory, to the support of this State vnder her Maiestie our dread Soueraigne, and that with great encrease of honor in this world, and all fulnesse of glory in the world to come.

Your Lordships most humble

and obedient Seruant,

IOHN GERARD.

LANCELOTVS BRVNIVS MEDICVS REGINEVS
IOHANNI GERARDO *Chirurgo peritiſsimo,*
& rei Herbariæ callentiſsimo S. P. D.

VM ſingularum medicinæ partium cognitio atque intelligentia libero homine digna conſenda eſt; tum earum nulla vel antiquitate, vel dignitate, vel vtilitate, vel denique iucunditate, cum ſtirpium cognitione iure comparari debet. Antiquiſſimam eam eſſe ex eo liquet, quòd quum ceteræ medicinæ partes (ſicut reliquæ etiam artes) ab ipſis hominibus (prout eos dura preſſit neceſſitas) primum excogitatæ & inuentæ fuerunt: ſola herbarum arborumque cognitio ante hominem formatum condita, eidemque mòx creato ab ipſo mundi archetecto donata videri poteſt. Cuius tanta apud antiqua ſecula exiſtimatio ac dignitas erat, vt & ipſius inuentionem ſapientiſſimo Deorum Apollini veteres tribuerint, & reges celeberrimi in ſtirpium viribus indagandis ſtudium laboremque ſuum conſumere, ſummæ ſibi apud poſteros laudi honorique futurum cenſuerint. Iam verò plantarum vtilitas, atq; etiam neceſſitas, adeò latè patet, vt eius immenſitatem nullius vel acutiſſimi hominis animus capere, nedum meus calamus exprimere queat. Stirpium enim complurimæ nobis in cibos, alimentumque cedunt: innumeræ aduerſus morbos remedia ſuppeditant: ex alijs domos, naues, inſtrumenta tam bellica quam ruſtica fabricamus: aliquot etiam earum veſtes noſtris corporibus ſubminiſtrant. In quibus ſingulis recenſendis diutiùs perſiſtere, hominis eſſet intemperantèr abutentis & otio & literis. Quantas autem, & quam varias voluptates ex ſtirpium ſiue amœnitate oculis capiamus, ſiue fragrantia naribus hauriamus, ſine ſumma inearum conditorem impietate inficiari non poſſumus. Adeò vt abſque ſtirpium ope & ſubſidio vita nobis ne vitalis quidem haberi debeat.

Quum igitur res plantaria reliquis omnibus medicinæ partibus antiquitate antecedat; dignitate, nulli cedat; vtilitate inſuper oblectationeque cæteras longè ſuperet, quis futurus eſt, adeo, aut inſenſatus vt non exploratum habeat, aut ingratus, ut non ingenuè agnoſcat, quanta vniuerſis Anglis commoda, quantaſque voluptates tuus mi *Gerarde* in ſtirpium inueſtigatione & cultu labor indefeſſus, ſtudium inexhauſtum, immenſique ſumptus hoc de ſtirpibus edito libro allaturi ſunt. Macte itaque iſta tua virtute, iſtoque de republica benè merendi ſtudio, & quod inſigni tua cum laude ingreſſus es virtutis gloriæque curriculum, eidem inſiſte animosè & gnauitèr, neq; à re plantaria promouenda prius deſiſte, quam eam à te ad vmbilicum iam fermè productam ipſe plenè abſoluas atque perficias. Sic enim & tibi adhùc ſuperſtiti gloriam paries immortalem, & poſt obitum tantam tui nominis celebritatem relinques, vt tuarum laudum poſteros noſtros nulla vnquam captura ſit obliuio. Bene vale. Ex Aula Reginea Weſtm. ipſis Cal. Decemb. 1597.

MATTHIAS DE L'OBEL
IOHANNI GERARDO
felicitatem.

QVum Londinum appuli, in sinu gauisus sum Gerarde *amicißime, dum typographo formis excudenda Plantarum collectanea tua commissa vidi, de quibus summas, nulla die perituras laudes Anglia tibi Rei-herbariæ familiam vniuersam, medicatricis artis partem, antiquißimum, iucundißimum & vtilißimum studium, retegere cupido, debet. Priscorum enim* Theophrasti, Dioscoridis, Plinij, & Galeni *scripta, paßim toto orbe peruulgata, tanquam fontes; Neotericorum autem, seu riuulos,* Brunfelsij, Fuchsij, Tragi, Ruellij, Matthioli, Dodonæi, Turneri, Clusij, Dalescampij, Camerarij, Tabernæmontani, Penæ, *nostramque nouam methodum & ordinem, à Gramine & notioribus ad Triticea, generatim & speciatim, materno idiomate, Anglicæ genti tuæ cultißimæ, Reipublicæ voluptabili commodo, recludis; quò ipsa stimulata, herbarum delitias & hortorum suauißimum & amœnißimum cultum amplectetur, maximorum Imperatorum, Regum & Heroum tam priscorum quam nuperorum exemplo. Nec satis hoc tibi fuit; sed multò magis insuper præstitisti, quòd copiam multarum elegantißimarum plantarum in Anglia sponte nascentium ab alijs hactenus prætermissarum, historiam descripsisti, magna hoc studio captorum vtilitate & oblectamento: Singulas enim regiones peculiares quasdam plantas, quas in alijs non facilè reperias, gignere certum. Neque magni tibi fuit hac inspectione & è viuis Naturæ typis nosse; quippe qui diu herbas indigenas, inquilinas & peregrinas cum nuperrimè solo erumpentes & pululantes, tum adultas, semineque prægnantes, hortulo tuo suburbano aluisti & fouisti: Exactum enim cognoscendarum ex figura aut facie superficiaria herbarum studium generatim consistit (*Dioscoride *teste) in frequenti & aßidua, temporis omnis, inspectione. Sed alia est interioris & substantialis formæ plantarum, quæ oculis cerni non potest, solers cognitio; quam etiam, quantum potes percunctando, seniorum Græcorum Medicorum more, aperire conaris. Solebant autem antiqui suorum Medicaminum experimenta, in Reipublicæ vtilitatem, scriptis tabellis dare, quibus apud Ephesios templi syluaticæ Dianæ parietes vestiebantur. Compertum etiam est Hippocratem discendi cupidum, permultis regionibus peragratis, idem præstitisse, & in methodum commemorabiliorem restituisse & illustrasse. Melius enim est Reipublicæ quam nostris commodis prospicere. Non est igitur quod huius inuidiosæ procacis ætatis conuiciatores maledici Zoili scripta tua obtrectent: dedisti enim gratis quod potuisti, cætera doctioribus iudicijs relinquens; exortiuis & exoticis incompertarum penè adhuc virium mangonizatis & lenocinijs allectis Floristarum floribus à Flora Dea meretrice nobili dictis, valetudini & vtilitati potius consulens, quam voluptati, valeri iußis. Nonnulli siquidem ex alijs libris herbarum transcriptores rapsodi, ignotis sibi viuis plantis ad medendum maximè necessarijs, aßignant incertis, dubijs & suppositityjs stirpibus aut simplicibus facultates legitimi simplicis medicamenti, maximo errore & summa periclitatione (vnum enim sæpe simplex compositionem ineptam reddit peruertit aut deprauat) quibus nec tutò nec temerè credendum; multoque etiam minus multis herbarum experimentis fallacibus, quibus etiam neque nisi notißimis morbis simplicibus, compositis & implicatis, eorundemque sæuißimis symptomatibus, vtendum, ne inoportunus earum vsus sæpius venenum quam remedium sit. Summo enim ægrotantium dispendio & exercitatißimorum Medicorum tædio periclitatores procaces, contemptis & neglectis artis institutionibus,* Hippocratis *&* Galeni *præceptis, per salutis discrimina & hominum strages medentum tentamenta agunt. Omitto, breuitatis ergô, vulgi opifices, textores sellularios, sordidißimos fabros, interpolatores, circulatores forenses & veteratores scutica dignos, qui professionibus & mechanicis artibus suis fastiditis, scelerato insaniæ lucro, se Medicos* Theophrasteos, *quem vix vnquam summis labris degustarunt, profitentur. Non inuenustè* Syluius *in huiusmodi hominis inuehit, dum ait,* Quam quisque nouit artem, hanc exerceat vnam, atque excolat, & totus in ea versetur, &c. *Et sub finem præfationis rursus ait,* Faxit Deus vt quisque quam exercet Artem, pernoscat, & Medicus nihil eorum quæ ad morbos citò & tutò curandos vtilia vel necessaria esse consueuerunt, ignoret. Præualet Medicus vbi Pharmacopœi fides suspecta est, qui ipse simplicia & composita pernoscit; imò quam infamiæ notam imprudens inurit, dum ignarus horum simplicium medicamentorum, tanquam asinus quidam ad omnia Pharmacopœi rogata, auribus motis, velut annuit: quid quod illi sæpe etiam volens Pharmacopœus illudit. Absurdissimus est ac sæpè ridiculus qui medicinam facit, harum rerum ignarus; & Pharmacopœo ignorantiæ suspectum meritò se reddit. *Plura si quis require apud* Syluium, *ibidem loci.*

Authoris necessaria diligentia in stirpium siue Materiæ Medicæ cognitione commendatur.

Præstigiosas popularium medicastrorũ fallacias detegimus & inueteratos depulimus errores.

Initio prologi Pharmac. Præparand.

Medico

Medico quam plurima perscrutanda, vt satis superq; ad artem medicatricem perdiscendam, annos paucos haudquaquam sufficere, testantur ipsius experientissimi & Diuini senis verba vbi inquit; Ego enim ad finem Medicinæ non perueni, etiamsi iam senex sim. *Et statim per initia Aphorismorum vitam breuem & artem longam pronunciauit. Quomodo ergo tuto medebuntur multi laruati Medici aut Medicastri tam repente creati, nulla Medicinæ parte, Medicamentorumve facultatibus perspectis? Huiusmodi adulatores, assentatores, dubitatores, rixatores, periclitatores & Gnathonicos parasistratos histrionibus qui in tragœdijs introducuntur similimos fecit Hippocrates.* Quemadmodum enim illi (inquit) figuram quidem & habitum ac personam eorum quos referunt habent, illi ipsi autem vere non sunt: Sic & Medici fama quidem & nomine multi, " re autem & opere valde pauci. *Itaque cum paulo ante Medicinam omnium artium præclarissimam esse dixerit:* Verum propter ignorantiam eorum qui eam exercent, & ob vulgi ruditatem, qui tales pro Medicis iudicat & habet; iam eo res deuenisse, vt omnium artium longe vilissima censeatur. At vero hoc peccatum ob hanc potissimum causam committi videtur; soli namque Medicinæ nulla pæna in rebus-publicis statuta est, præterquam ignominiæ. *Ne animam & famam læderit, aut illi insignis ignominia inureretur ob huiusmodi ardua & noxia discrimina, bonus ille & syncerus Dodonæus (quamvis multas herbas ex alijs & Fuchsio transcripserit, cuius methodo vsus est, quemque inchoauerat, vt ipsemet mihi retulit, vernacula Germanica inferiori lingua vertere) vulgatissimis, notissimis ijsque paucis ex tot herbarum millibus, quinquagenis aut septuagenis herbis quibus vtebatur, potius contentus fuit, quam innumeris sibi ignotis periclitari: melius enim omnino medicamento carere, abstinere, & naturæ committere, quam abuti. Vtinam huius nostræ ætatis quamplures auso potiti, medicinam factitantes, eo studio, candore & voto mederentur: Illis id forsitan nequaquam euenerit, quod Philosophis (Hippocrate defuncto) discipulis suis inexpertis & parum adhuc exercitatis medendo, id est necando (vt memoriæ traditum est) contingit: quamobrem ars Medica Athenis, Romæ & per vniuersam Græciam centum & septuaginta annis, interdicta et exul fuit. Merito igitur caute et tute agendum: Opiatis et Diagrediatis, Colocynthide, Tithymalis, Esula, Lathyride, Mercurio, Stibio, & similibus molestissimis simplicibus cum cautione vtendum: optimis ducibus & experientissimis senioribus præceptoribus adhærendum, quorũ sub vexillis fidissime & tutissime rara & præclara, ob barbariem fere extincta, patrum & auorum remedia, maximo et pristino artis ornamento et proximi vtilitate renouantur, et in vsum reuocantur; neglectis, spretis, et exclusis Empiricis verbosis, inuidiosis, suspensis, ambagiosis et exitiosis opinionibus, quibus Mundus immundus regitur et labitur; qui cum decipi velit, decipiatur: in cuius fallacias perapposite finxit et cecinit olim hos versiculos eruditissimus collega D. Iacobus Paradisus nobilis Gandauensis alludens ad nomen tanti versutissimi herois Nostradami Salonensis Gallo-prouinciæ,*

Epist. ad Democritum.

" Multi malunt videri quam esse.

Nostra-damus, cum verba damus, quia fallere nostrum;
Et cum verba damus, nil nisi Nostra-damus.

Vale. Londini ipsis Calendis Decemb. 1597.

In GERARDI Botanologian
φερώνυμον.

VLtimus ecce Gerardus : *at edit an optimus herbas ?*
Quid ni ? non notas ſed dedit ille novas.
Ergo ne inuideas, videas cum nomen & omen
φυτολόγε, *mirum eſt ardua quanta gerit.*
Οὗτος ἄναντα, κάταντα, πάραντα, τε δόχμιά τ' ἦλθε :
Sic liber eſt promus, condus vt hortus erat.
Et γαῖαν ἄρδων *cælumque ſolumq; ſubegit.*
τῷ γ' ἄρα θεσπεσίας ἁρμόζει ἐπὶ γέρας.

ANTONIVS HVNTONVS.
Medicinæ candidatus.

Ad Iohannem Gerardum *Chirurgum Herbariumq; peritiſſimum.*

NVlla oculos hominum ſpecies magis allicit illa,
Quam præſtante manu duxit generoſus Apelles.
Nulla aures animoſque magis facundia, quam quæ
Se fuſam loquitur Ciceronis ab ore diſerti:
Hæc eadem hunc librum commendat cauſa, *Gerarde,*
Cui pro laude ſatis tali natum eſſe parente,
Artifices cui inter dextras pro numine, nomen
Nobilius reliquis herbæ, plantæq; magiſtris.
Illi etenim Europæ ſuccos, Aſiæque liquores
Quæque arente ſolo ſitiens parit Africa, tractant:
Tu veterum inuentis noua conſuis omnia, ſi qua
Indus vterque dedit noſtram factura ſalutem,
Sive aliunde vehit noſtras mercator ad oras,
Hoc ipſo vtilius. Quia quæ ſunt credita ſcriptis,
Illa manu expertus medico, & bene diues ab horto
Explorata diu multumq; emittis in auras,
Quæ curent hominum languentia corpora, multi
Præſtanteſq; viri docuere fideliter artem.
Sed ſi ſuſtuleris plantas, quem verba iuvabunt
Sic animo, ſic fronte minax. In prælia miles
Proſilit, at ſtricto cedit victoria ferro
Quæ tibi pro tanto cedit victoria ferro
Præmia perſoluet, Myrti lauriq; coronas ?
Iſtam nouit edax mercedem abolere vetuſtas.
At tibi pro ſtudio impenſiſque laboribus iſtis,
Queis hominum curas ſertam tectamq; ſalutem,
Ille opifex rerum, cuſtoſq; authorq; ſalutis
Æterna ſtatuit frontem redimire corona.

G. Launæus Medicus.

In hiſtoriam plantarum, Io. Gerardi *ciuis & Chirurgi Londinenſis,* M. Iacobi Ihonſtonij *Scoti Ballincriſæ Regij pagi portionarij* Epigramma.

DEſine, quæ vaſtis pomaria montibus Atlas
 Clauſerat (Heſperij munera rara ſoli)
Auratis folijs auratos deſine ramos
 Mirari, & ramis pendula poma ſuis.
Singula cum Domino periere, & Gorgone viſo
 In montis riguit veſcera verſus Atlas.
Alcinoi perijt qui, cedat penſilis hortus,
 Quem celebrat priſci temporis aura fugax:
Vna Gerardini ſpecies durabilis horti
 Æterno famæ marmore ſculpta manet.
Hic quicquid Zephyrus produxit, quicquid & Eurus,
 Antiquus quicquid & novus orbis habet,
Intulit in patriam naturamq; exprimit arte:
 Sic nullo cedit terra Britanna ſolo.

Quod

Quod magis est Graium & Latium concludit in vno
Margine, & Anglorum iam facit ore loqui:
Sic erit æternum hinc vt viuas, horte *Gerardi*,
Cultoris studio nobilitate tui.

In Plantarum historiam, a solertissimo viro, Reiq; Herbariæ peritissimo, D. Iohanne Gerardo, *Anglice editam*
Epigramma.

EGregiam certè laudem, decus immortale refertis
Tu, socijq; tui, magnum & memorabile nomen
(Illustris DEVORAX) raptoribus orbis IBERIS
Deuictis classe Anglorum; Tuque (Dicasta
Maxime EGERTONE) veterem superans Rhadamanthum,
HEROVM merito ἡμιθεῶν censendus in albo.
Nec laus vestra minor (sacræ pietatis alumni)
Qui mentes hominum diuina pascitis esca.
Ornatis Patriam cuncti, nomenq; Britannum
Augetis, vobisq; viam munitis ad astra.
Quin agite, in partem saltem permittite honoris
Phœbei veniant Vates, qui pellere gnari
Agmina morborum, humanæ insidiantia vitæ.
Huius & ingentes, serena fronte labores
ANGLO-DIOSCORIDIS, Patriæ, vestræq; saluti
Excipite exhaustos: paulum huc diuortite in HORTOS
Quos CHORTEIA colit, quos Flora exornat, & omnes
Naiades, & Dryades, Charites, Nymphæq; Britannæ.
Corporibus hic grata salus, animisq; voluptas.
Hic laxate animos: HABITAVIT NVMEN IN HORTIS.

Fran. Hering Med. D.

Thomas Newtonus, Cestreshyrius, D. Io. Gerardo, *amico non vulgari*, S.

POst tot ab ingenuis conscripta volumina mystis,
Herbarum vires qui reserare docent,
Tu tandem prodis Spartamq; hanc gnauiter ornas,
Dum reliquis palmam præripuisse studes.
Nec facis hoc, rutilo vt possis ditarier auro,
Nec tibi vt accrescat grandis aceruus opum;
Sed prodesse volens, vestitos gramine colles
Perlustras, & agros, frondiferumq; nemus.
Indeq; Pæonias (apis instar) colligis herbas,
Inq; tuum stirpes congeris alueolum.
Mille tibi species plantarum, milleq; notæ;
Hortulus indicio est, quem colis ipse domi.
Pampineæ vites, redolens cèdrus, innuba laurus,
Nota tibi, nota est pinguis oliua tibi.
Balsama, narcyssus, rhododaphne, nardus, amomum,
Saluia, dictamnus, galbana nota tibi.
Quid multis? radix, stirps, flos, cum cortice ramus,
Spicaq; cum siliquis est bene nota tibi.
Gratulor ergo tibi, cunctisq; (*Gerarde*) Britannis,
*N*amptwicoq; tuo gratulor, atq; meo.
Nam Cestreshyrij te ac me genuere parentes,
Tu meliore tamen sydere natus eras.
Macte animo, pergasq; precor, cœptumq; laborem
Vrge etiam vlterius. Viuitur ingenio.
Aurum habeant alij, gemmas, nitidosq; pyropos,
Plantas tu & flores scribe *Gerarde*. Vale.

Vere & ex animo tuus, Tho. Newton, *Ilfordensis* ἱερεύς.

To

To the well affected Reader and peruser of this *Booke, St. Bredwell Physition,* greeting.

Plin.Iun. in pan.

Pen is the campe of glorie and honour for all men, saith the younger Pliny: *not onely men of great birth and dignitie, or men of office endued with publique charge and titles, are seene therein, and haue the garland of praise and preferment waiting to crowne their merits, but euen the common souldier likewise: so as he, whose name and note was erst all obscure, may by egregious acts of valour obtaine a place among the noble. The schoole of science keepeth semblable proportion: whose amplitude, as not alwaies, nor onely, men of great titles and degrees, labour to illustrate; so whosoeuer doth, may confidently account of, at the least, his name to be immortall. What is he then that will denie his voice of gracious commendation to the Authors of this Booke: to euery one, no doubt, there is due a condigne measure.*

Turnerus. Dodonæus. Pena. L Obelius. Tabernamontanus.

The first gatherers out of the Antients, and augmentors by their owne paines, haue alreadie spread the odour of their good names, through all the Lands of learned habitations. D. Priest, *for his translation of so much as* Dodonæus, *hath thereby left a tombe for his honorable sepulture.* M. Gerard *comming last, but not the least, hath many waies accommodated the whole worke vnto our English Nation: for this Historie of Plants, as it is richly replenished by those fiue mens labours laied together, so yet could it full ill haue wanted that new accession he hath made vnto it. Many things hath he nourished in his garden, and obserued in our English fields, that neuer came into their pennes to write of. Againe, the greatest number of these plants, hauing neuer been written of in the English tongue, would haue wanted names for the vulgar sort to call them by: in which defect he hath bin curiously carefull, touching both old and new names to make supply. And lest the Reader should too often languish with frustrate desire, to finde some plant he readeth, of rare vertue, he spareth not to tell (if himselfe haue seene it in England) in what wood, pasture or ditch the same may be seene and gathered. Which when I thinke of, and therewithall remember, with what cheerefull alacritie, and resolute attendance he hath many yeares tilled this ground, and now brought forth the fruit of it, whether I should more commend his great diligence to attaine this skill, or his large beneuolence in bestowing it on his countrie, I cannot easily determine. This booke-birth thus brought forth by* Gerard, *as it is in forme and disposition faire and comely, euery* species *being referred to his likeliest* genus, *of whose stocke it came: so is it accomplished with surpassing varietie, vnto such spreading growth and strength of euery lim, as that it may seeme some heroicall Impe of illustrious race, able to draw the eies and expectation of euery man vnto it. Somewhat rare it will be here for a man to moue a question of this nature, and depart againe without some good satisfaction. Manifold will be the vse both to the Physition and others: for euery man delighteth in knowledge naturally, which (as*

Laert.l.5. cap.1.

Aristotle *said) is in prosperitie an ornament, in aduersitie a refuge. But this booke aboue many others will sute with the most, because it both plenteously ministreth knowledge, which is the food of the minde, and doth it also with a familiar and pleasing taste to euery capacitie. Now as this commoditie is communicated to all, and many shall receiue much fruit thereof, so I wish some may haue the minde to returne a benefit againe; that it might not be true in all that* Iuvenall *saith,*

Iuuenal.7. Sat.

Scire volunt omnes, mercedem soluere nemo: (i.) All desire to know, none to yeeld reward. *Let men think, that the perfection of this knowledge is the high aduancement of the health of man: that perfection is not to be attained, but by strong indeuor: neither can strong indeuor be accomplished without free maintenance. This hath not he, who is forced to labour for his daily bread: but if hee, who from the short houres of his daily and necessarie trauell, stealing as it were some, for the publike behoofe, and setting at length those peeces together, can bring forth so comely a garment as this, meet to couer or put away the ignorance of many: what may be thought he would do, if publicke maintenance did free him from that priuate care, and vnite his thoughts to be wholly intent to the generall good. O Reader, if such men as this sticke not to rob themselues of such wealth as thou haste to inrich thee, with that substance thou wantest, detract not to share out of thine aboundance to merit and encourage their paines: that so fluxible riches, and permanent sciences, may the one become a prop vnto the other.*

Cic.Off.c.1.

Although praise and reward ioined as companions to fruitfull endeuors, are (in part) desired of all men, that vndertake losses, labours, or dangers for the publique behoofe: because they

Simplic. comm. in Epict.

adde sinewes (as it were) vnto reason, and able her more and more to refine her selfe: yet doe they not imbrace that honour in respect of it selfe, nor in respect of those that conferred it vpon them, but as as hauing thereby an argument in themselues, that there is something in them worthy estimation among men: which then doubleth their diligence to deserue it more abundantly. Admirable and for the imitation of Princes, was that act of Alexander, who setting Aristotle to compile commentaries of the bruit creatures, allowed him for the better performance thereof, certaine thousands of men, in all Asia and Greece, most skilfull obseruers of such things, to giue him information touching all beasts, fishes, foules, serpents, and flies. What came of it? A booke written, wherein all learned men in all ages since do exercise themselues principally, for the knowledge of the creatures. Great is the number of those that of their owne priuate haue laboured in the same matter, from his age downe to our present time, which all do not in comparison satisfie vs. Whereas if in those ensuing ages there had risen still new Alexanders, there (certainely) would not haue wanted Aristotles to haue made the euidence of those things an hundred fold more cleered vnto vs, than now they be. Whereby you may perceiue the vnequall effects that follow those vnsutable causes of publike and priuate maintenances vnto labours and studies. Now that I might not dispaire in this my exhortation, I see examples of this munificence in our age to giue me comfort: Ferdinand the Emperor and Cosmus Medices Prince of Tuscane are herein registred for furthering this science of plants, in following of it themselues and becomming skilfull therein: which course of theirs could not be holden without the supporting and aduancing of such as were studious to excell in this kinde. Bellonius likewise (whom for honours cause I name) a man of high attempts in naturall science, greatly extolleth his Kings liberalitie, which endued him with free leisure to follow the studie of plants, seconded also herein by Montmorencie the Constable, the Cardinals Castilion and Loraine, with Oliuerius the Chancellor; by whose meanes he was enabled to performe those his notable peregrinations in Italy, Africa and Asia: the sweet fruit whereof, as we haue receiued some taste by his obseruations, so we should plenteously haue been filled with them, if violent death by most accursed robbers had not cut him off. And as I finde these examples of comfort in forreine nations, so we are (I confesse) much to be thankfull to God, for the experience we haue of the like things at home. If (neuerthelesse) vnto that Physicke lecture lately so well erected, men who haue this worlds goods shall haue hearts also of that spirit, to adde some ingenious labourer in the skill of simples, they shall mightily augment and adorne the whole science of Physicke. But if to that likewise they ioine a third, namely the art of Chimicall preparation; that out of those good creatures which God hath giuen man for his health, pure substances may be procured for those that be sicke, (I feare not to say it, though I see how Momus scorneth) this present generation would purchase more to the perfection of Physicke, than all the generations past since Galens time haue done: that I say, nothing of this one fruit that would grow thereof, to wit, the discouering and abolishing of these pernitious impostures and sophistications, which mount-promising Paracelsians euery where obtrude, through want of a true and constant light among vs to discerne them by. In which behalfe, remembring the mournfull speech of graue Hippocrates; The art of Physicke truly excelleth all arts, howbeit, through the ignorance partly of those that exercise it, and partly of those that iudge rashly of Physitions, it is accounted of all arts the most inferiour: I say in like manner, the art of Chimistrie is in it selfe the most noble instrument of naturall knowledges; but through the ignorance & impiety, partly of those that most audaciously professe it without skill, and partly of them that impudently condemne that they know not, it is of all others most basely despised and scornfully rejected. A principall remedy to remoue such contumelious disgrace from these two pure virgins of one stocke and linage, is this that I haue now insinuated, euen by erecting the laboratory of an industrious Chimist, by the sweet garden of flourishing simples. The Physicke reader by their meanes shall not onely come furnished with authorities of the Ancients, and sensible probabilities for that he teacheth, but with reall demonstrations also in many things, which the reason of man without the light of the fornace would neuer haue reached vnto. I haue vttered my hearts desire, for promoting first the perfection of my profession, and next by necessary consequence, the healthie liues of men. If God open mens hearts to prouide for the former, it cannot be but that the happy fruits shall be seene in the later. Let the ingenious learned iudge whether I haue reason on my side; the partiall addicted sect I shun, as men that neuer meane good to posteritie.

Plin. lib. 8. cap. 16.

Gryllus in orat. de peregr. studij medic.

Bellon. de negl. stirp. cul. prob. 9.

Hipp. de Lege.

George

George Baker, *one of her Maiesties chiefe Chirurgions in ordinarie, and M. of the Chirurgions of the Citie of London, to the Reader.*

ARistotle, a Prince amongst the Philosophers, writing in his Metaphysicks of the nature of mankind, saith, that man is naturally inclined and desirous of science. The which sentence doth teach vs, that all creatures (being vertuously giuen) doe striue to attain to perfection, and draw neare in what they can to the Creator; and this knowledge is one of the principall parts which doth concerne the perfection of vnderstanding: for of the same doth follow, that all such are generally inclined to know the meanes by the which they may conserue their life, health, and reputation. And although it be necessarie for man to learne and know all sciences, yet neuerthelesse the knowledge of naturall philosophie ought to be preferred, as being the most necessarie; and moreouer it doth bring with it a singular pleasure and contentment. The first inuentor of this knowledge was *Chiron* the Centaure, of great renowne, sonne to *Saturne* and *Phillyre*: and others say that it was inuented of *Apollo*: & others of *Esculape* his son; esteeming that so excellent a science could neuer proceed but from the gods immortall, and that it was impossible for man to finde out the nature of Plants, if the great worker, which is God, had not first instructed and taught him. For, as *Pliny* saith, if any thinke that these things haue bin inuented by man, he is vngratefull for the workes of God. The first that we can learn of among the Greekes that haue diligently written of herbes, haue bin *Orpheus*, *Musæus*, and *Hesiode*, hauing bin taught by the Ægyptians: then *Pythagoras* of great renowne for his wisedom, which did write bookes of the nature of Plants, and did acknowledge to learne the same from *Apollo* and *Esculape*. *Democrite* also did compose bookes of Plants, hauing first trauelled ouer all Persia, Arabia, Ethiopia, and Egypt. Many other excellent spirits haue taken great pleasure in this science, which to accomplish haue hazarded their liues in passing many vnknowne regions, to learne the true knowledge of *Elleborus*, and other Medicaments: of which number were *Hippocrates*, *Crateua*, *Aristotle*, *Theophrast*, *Diocles Caristius*, *Pamphylus*, *Montius*, *Hierophile*, *Dioscorides*, *Galen*, *Pliny*, and many others, which I leaue to name, fearing to be too long. And if I may speake without partialitie of the Author of this book, his great paines, his no lesse expences in trauelling far and neere for the attaining of his skill haue bin extraordinarie. For he was neuer content with the knowledge of those simples which grow in those parts, but vpon his propet cost and charges hath had out of all parts of the world all the rare simples which by any means he could attaine vnto, not onely to haue them brought, but hath procured by his excellent knowledge to haue them growing in his garden, which as the time of the yeare doth serue may be seene: for there shall you see all manner of strange trees, herbes, roots, plants, floures, and other such rare things, that it would make a man wonder, how one of his degree, not hauing the purse of a number, could euer accomplish the same. I protest vpon my conscience, I do not think for the knowledge of Plants, that he is inferiour to any: for I did once see him tried with one of the best strangers that euer came into England, and was accounted in Paris the onely man, being recommended vnto me by that famous man Master *Amb.*

Pareus,

Pareus; and he being here was desirous to goe abroad with some of our Herbarists, for the which I was the meane to bring them together, and one whole day we spent therein, searching the rarest Simples: but when it came to the triall, my French man did not know one to his foure. What doth this man deserue that hath taken so much paines for his countrey, in setting out a booke, that to this day neuer any in what language soeuer did the like? First for correcting their faults in so many hundred places, being falsly named, mistaken the one for the other; and then the pictures of a great number of plants now newly cut. If this man had taken this paines in Italy and Germany, where *Matthiolus* did write, he should haue sped as well as he did: For (saith he) I had so great a desire euer to finish my Booke, that I neuer regarded any thing in respect of the publique good, not so much as to thinke how I should finish so great a charge, which I had neuer carried out, but that by Gods stirring vp of the renowned Emperour *Ferdinando* of famous memorie, and the excellent Princes had not helped mee with great sums of money, so that the Commonwealth may say, That this blessing doth rather proceed of them than from me. There haue been also other Princes of Almaine which haue bin liberal in the preferring of this Book, and the most excellent Elector of the Empire the Duke of Saxonie, which sent me by his Post much mony toward my charges: the liberalitie of the which and the magnificence toward me I cannot commend sufficiently. They which followed in their liberalitie were the excellent *Fredericke* Count Palatine of the Rhine, and the excellent *Ioachim* Marques of Brandeburg, which much supplied my wants: and the like did the reuerend Cardinall and Prince of Trent, and the Excellent Archbishop of Saltzperg, the Excellent Dukes of Bauare and Cleues, the duke of Megapolencis Prince of Vandalis, the State Republique of Noremberg, the liberalitie of whom ought to be celebrated for euer: and it doth much reioice me that I had the helpe and reward of Emperors, Kings, Electors of the Roman Empire, archdukes, Cardinalls, Bishops, Dukes and Princes, for it giveth more credit to our Labors than any thing that can be said. Thus far *Matthiolus* his owne writing of the liberalitie of Princes towards him. What age do we liue in here that wil suffer all vertue to go vnrewarded? Master *Gerard* hath taken more pains than euer *Matthiolus* did in his Commentaries, and hath corrected a number of faults that he passed ouer; and I dare affirme (in reuerence be it spoken to that Excellent man) that Master *Gerard* doth know a great number of Simples that were not knowne in his time: and yet I doubt whether he shall taste of the liberalitie of either Princ, Duke, Earle, Bishop, or publique Estate. Let a man excell neuer so much in any excellent knowledge, neuertheles many times he is not so much regarded as a Iester, a Boaster, a Quacksaluer or Mountebanke: for such kinde of men can flatter, dissemble, make of trifles great matters, in praising of this rare secret, or that excellent spirit, or this Elixer or Quintessence; which when it shall come to the triall, nothing shal be found but boasting words,

VALE.

¶¶ To

To the courteous and well willing Readers.

ALthough my paines haue not been ſpent (curteous Reader) in the gracious diſcouerie of golden Mines, nor in the tracing after ſiluer veines, whereby my natiue country might be enriched with ſuch merchandiſe as it hath moſt in requeſt and admiration; yet hath my labour (I truſt) been otherwiſe profitably employed, in deſcrying of ſuch a harmleſſe treaſure of herbes, trees, and plants, as the earth frankely without violence offereth vnto our moſt neceſſarie vſes. Harmleſſe I call them, becauſe they were ſuch delights as man in the perfecteſt ſtate of his innocencie did erſt inioy: and treaſure I may well terme them, ſeeing both Kings and Princes haue eſteemed them as Iewels; ſith wiſe men haue made their whole life as a pilgrimage to attaine to the knowledge of them: by the which they haue gained the hearts of all, and opened the mouthes of many, in commendation of thoſe rare vertues which are contained in theſe terreſtriall creatures. I conf ſſe blind Pluto is now adayes more ſought after than quicke ſighted Phœbus: and yet this duſty mettall, or excrement of the earth (which was firſt deepely buried leaſt it ſhould be an eye-ſore to grieue the corrupt heart of man) by forcible entry made into the bowels of the earth, is rather ſnatched at of man to his owne deſtruction, than directly ſent of God, to the comfort of this life. And yet behold in the compaſsing of this worldly droſſe, what care, what coſt, what aduentures, what myſticall proofes, and chymicall trials are ſet abroach; when as notwithſtanding the chiefeſt end is but vncertaine wealth. Contrariwiſe, in the expert knowledge of herbes, what pleaſures ſtill renewed with varietie? what ſmall expence? what ſecurity? and yet what an apt and ordinary meanes to conduct man to that moſt deſired benefit of health? Which as I deuoutly wiſh vnto my natiue countrey, and to the carefull nurſing mother of the ſame; ſo hauing bent my labours to the benefiting of ſuch as are ſtudiouſly practiſed in the conſeruation thereof, I thought it a chiefe point of my duty, thus out of my poore ſtore to offer vp theſe my far fetched experiments, together with mine owne countries vnknowne treaſure, combined in this compendious Herball (not vnprofitable though vnpoliſhed) vnto your wiſe conſtructions and courteous conſiderations. The drift whereof is a ready introduction to that excellent art of Simpling, which is neither ſo baſe nor contemptible as perhaps the Engliſh name may ſeeme to intimate: but ſuch it is, as altogether hath been a ſtudy for the wiſeſt, an exerciſe for the nobleſt, a paſtime for the beſt. From whence there ſpring floures not onely to adorne the garlands of the Muſes, to decke the boſomes of the beautifull, to paint the gardens of the curious, to garniſh the glorious crownes of Kings; but alſo ſuch fruit as learned Dioſcorides long trauelled for; and princely Mithridates reſerued as precious in his owne cloſet: Mithridates I meane, better knowne by his ſoueraigne Mithridate, than by his ſomtime ſpeaking two and twenty languages. But what this famous Prince did by tradition, Euax king of the Arabians did deliuer in a diſcourſe written of the vertues of herbes, and dedicated it vnto the Emperor Nero. Euery greene Herbariſt can make mention of the herbe Lyſimachia, whoſe vertues were found out by King Lyſimachus, and his vertues no leſſe eterniſed in the ſelfe ſame plant, than the name of Phydias, queintly beaten into the ſhield of Pallas, or the firſt letters of Ajax or Hyacinthus (whether you pleaſe) regiſtred in that beloued floure of Apollo. As for Artemiſia, firſt called Παρθενις, whether the title thereof ſprang from Ἄρτεμις, Diana her ſelfe, or from the renowned Queene of Caria, which diſcloſed the vſe thereof vnto poſteritie, it ſuruiueth as a monument to reuiue the memories of them both for euer. What ſhould we ſpeake of Gentiana, bearing ſtill the cogniſance of Gentius? or of diuers other herbes taking their denominations of their princely Inuentors? What ſhould I ſay of thoſe royall perſonages, Iuba, Attalus, Climenus, Achilles, Cyrus, Maſyniſſa, Semyramis, Diocleſian? but onely thus, to beſpeake their princely loues to Herbariſme, and their euerlaſting honors (which neither old Plinius dead, nor yong Lipſius liuing will permit to die?) Creſcent herbæ, creſcetis amores: creſcent herbæ, creſcetis honores. But had this wonted facultie wanted the authoriſement of ſuch a royall companie, King Solomon, excelling all the reſt for wiſdome, of greater royaltie than they all (though the Lillies of the field out-braued him) he only (I ſay) might yeeld hereunto ſufficient countenance and commendation, in that his lofty wiſedome thought no ſcorne to ſtoupe vnto the lowly plants. I liſt not ſeeeke the

common

common colours of antiquitie, when notwithstanding the world can brag of no more antient Monument than Paradise and the garden of Eden: and the fruits of the earth may contend for seigniori-tie, seeing their mother was the first Creature that conceiued, and they themselues the first fruit she brought forth. Talke of perfect happinesse or pleasure, and what place was so fit for that as the garden place where Adam *was set to be the Herbarist? Whither did the Poets hunt for their sincere delights, but into the gardens of* Alcinous, *of* Adonis, *and the Orchards of* Hesperides? *Where did they dreame that Heauen should be, but in the pleasant garden of* Elysium? *Whither do all men walke for their honest recreation, but thither where the earth hath most beneficially painted her face with flourishing colours? And what season of the yeare more longed for than the Spring, whose gentle breath enticeth forth the kindely sweets, and makes them yeeld their fragrant smells? who would therefore looke dangerously vp at Planets, that might safely looke downe at Plants? And if true be the old prouerbe,* Quæ supra nos, nihil ad nos; *I suppose this new saying cannot be false,* Quæ infra nos, ea maximè ad nos. *Easie therefore is this treasure to be gained, and yet pretious. The science is nobly supported by wise and Kingly Fauorites: the subiect thereof so necessary and delectable, that nothing can be confected either delicate for the taste, daintie for smell, pleasant for sight, wholesome for body, conseruatiue or restoratiue for health, but it borroweth the relish of an herbe, the sauour of a floure, the colour of a leafe, the iuice of a plant, or the decoction of a root. And such is the treasure that this my Treatise is furnished withall, wherein though myne Art be not able to counteruaile Nature in her liuely portraitures; yet haue I counterfeited likenes for life, shapes and shadowes for substance, being ready with the bad Painter to explaine the imperfections of my pensill with my pen, chusing rather to score vpon my pictures such rude marks as may describe my meaning, than to let the beholder to guesse at randome and misse. I haue here therefore set downe not onely the names of sundry Plants, but also their natures, their proportions and properties, their affects and effects, their increase and decrease, their flourishing and fading, their distinct varieties and seuerall qualities, as well of those which our owne Countrey yeeldeth, as of others which I haue fetched further, or drawne out by perusing diuers Herbals set forth in other languages, wherein none of my country-men hath to my knowledge taken any paines, since that excellent Worke of Master Doctor* Turner. *After which time Master* Lyte *a Worshipfull Gentleman translated* Dodonæus *out of French into English: and since that, Doctor* Priest, *one of our London Colledge, hath (as I heard) translated the last Edition of* Dodonæus, *and meant to publish the same; but being preuented by death, his translation likewise perished. Lastly my selfe, one of the least among many, haue presumed to set forth vnto the view of the world, the first fruits of these myne owne Labours, which if they be such as may content the Reader, I shall thinke my selfe well rewarded, otherwise there is no man to be blamed but my selfe, being a worke I confesse for greater Clerkes to vndertake: yet may my blunt attempt serue as a whetstone to set an edge vpon some sharper wits, by whom I wish this my course Discourse might be both fined and refined. Faults I confesse haue escaped, some by the Printers ouersight, some through defects in my selfe to performe so great a Worke, and some by meanes of the greatnesse of the Labour, and that I was constrained to seeke after my liuing, being void of friends to beare some part of the burthen. The rather therefore accept this at my hands (louing Countrey-men) as a token of my good will; and I trust that the best and well minded wil not rashly condemne me, although some thing haue passed worthy reprehension. But as for the slanderer or Enuious I passe not for them, but returne vpon themselues any thing they shall without cause either murmure in corners, or iangle in secret. Farewell.*

From my House in Holborn within the Suburbs
of London, this first of December, 1597.

Thy sincere and vnfeigned Friend,

IOHN GERARD.

TO THE READER.

Courteous READER,

THere are many things which I thinke needfull to impart vnto thee, both concerning the knowledge of plants in generall, as alſo for the better explaining of ſome things pertinent to this preſent Hiſtorie, which I haue here ſet forth much amended and enlarged. For the generall differences, affectionss, &c. of Plant, I hold it not now ſo fitting nor neceſſarie for me to inſiſt vpon them; neither doe I intend *in* any large diſcourſe to ſet forth their many and great vſes and vertues: giue me leaue onely to tell you, That God of his infinit goodneſſe and bountie hath by the *medium* of Plants, beſtowed almoſt all food, clothing, and medicine vpon man. And to this off-ſpring we alſo owe (for the moſt part) our houſes, ſhipping, and infinite other things, though ſome of them *Proteus* like haue run through diuers ſhapes, as this paper wereon I write, that firſt from ſeed became Flax; then after much vexation thred, then cloath, where it was cut and mangled to ſerue the Faſhions of the time: but afterwards rejected and caſt aſide, yet vnwilling ſo to forſake the ſeruice of man for which God had created it, againe it comes (as I may terme it) to the Hammer, from whence it takes a more noble forme and aptitude to be imployed to Sacred, Ciuill, Forreine and Domeſticke vſes. I will not ſpeake of the many and various obiects of delight that theſe preſent to the ſenſes, nor of ſundry other things, which I could plentifully in this kinde deliuer: but rather acquaint you from what Fountaines this Knowledge may be drawne, by ſhewing what Authours haue deliuered to vs the Hiſtorie of Plants, and after what manner they haue done it; and this will be a meanes that many controuerſies may be the more eaſily vnderſtood by the leſſe learned and judicious Reader.

Solomon.

He whoſe name we firſt finde vpon record (though doubtleſſe ſome had treated therof before) that largely writ of Plants, was the wiſeſt of men, euen King *Solomon*, who certainely would not haue medled with this ſubiect, if he in his wiſedome had not knowne it worthy himſelfe, and exceeding fitting: Firſt for the honour of his Creator, whoſe gifts and bleſſings theſe are: Secondly for the good of his Subiects, whereof without doubt, he in this worke had a ſpeciall regard in the curing of their diſeaſes and infirmities. But this kingly worke being loſt, I will not inſiſt vpon it, but come to ſuch as are yet extant, of which (following the courſe of antiquitie) that of *Theophraſtus* firſt takes place.

Theophraſtus.

Now *Theophraſtus* ſucceeded *Ariſtotle* in the gouernement of the Schoole at Athens, about the 114 Olymp. which was ſome 322 yeares before Chriſt. He among many other things writ a Hiſtorie of Plants in ten bookes, and of the cauſes of them, eight bookes; of the former ten there are nine come to our times reaſonable perfect; but there now remain but ſix of the eight of the cauſes of Plants. Some looking vpon the Catalogue of the bookes of *Theophraſtus* his writing, ſet forth in his life, written by *Diogenes Laertius*, may wonder that they finde no mention of theſe bookes of Plants, amongſt theſe he reckons vp, and indeed I thought it ſomewhat ſtrange, and ſo much the more, becauſe this his life is ſet forth by *Daniel Heinſius* before his * Edition of *Theophraſtus*, and there alſo no mention neither in the Greeke nor Latine of thoſe workes. Conſidering this, I thinking to haue ſaid ſomething therof, I found the doubt was long ſince cleared by the learned *Cauſabone* in his notes vpon "*Laertius*, where *pag.* 331. for Περὶ φυτικῶν ἱστορία, and φυτικῶν αἰτιῶν, hee wiſhes you to reade Περὶ φυτικῶν ἱστοριῶν and αἰτιῶν. Thus being certaine of the Authour, let mee ſay ſomewhat of the work, which though by the iniurie of time it hath ſuffered much, yet is it one of the chiefe pieces of Antiquitie, from whence the knowledge of Plants is

Lugd. Batau. 1613.

Excuſ. ab Henr. Stepb. 1593.

to

to be drawne. *Theophrastus* as he followed *Aristotle* in the Schoole, so also in his manner of writing, for according as *Aristotle* hath deliuered his *Historia Animalium*, so hath hee set forth this of Plants, not by writing of each *species* in particular, but of *their differences and nature, by their parts, affections, generations and life.* Which how hard a thing it was, hee tells you in his second Chapter, and renders you this reason, *Because there is nothing common to all Plants, as the mouth and belly is to other liuing creatures*, &c. Now by this manner of writing you may learne the generall differences and affections of Plants, but cannot come to the particular knowledge of any without much labour: for you must goe to many places to gather vp the description of one Plant: neither doth hee (nor is it necessarie for any writing in this manner) make mention of any great number, and of many it may bee but once. His workes being in Greeke were translated into Latine by *Theodore Gaza*, who did them but *Græca fide*, for he omitted some things, otherwhiles rendred them contrary to the minde of the Author: but aboue all, he tooke to himselfe too much libertie in giuing of names in imitation of the Greeke, or of his owne inuention, when it had beene better by much for his Reader to haue had them in the Greeke, as when he renders Ἐλατήριον, *Agitatorium*, ἡλιοτρόπιον, *Solaris*, &c. The learned *Iulius Scaliger* hath set forth *Animaduersiones* vpon these bookes, wherein he hath both much explained the minde of *Theophrastus*, and shewed the errours of *Gaza*. Some since his time haue promised to do something to this Author, as *Daniel Heinsius*, and *Spigelius*, but twentie yeares are past since, and I haue not yet heard of any thing done in this kinde by either of them. Thus much for *Theophrastus*.

Theoph. Hist. pl. l. 1. cap. 1.

Σημεῖον δὲ τὸ μηδὲν εἶναι κοινὸν λαβεῖν ὃ πᾶσιν ὑπάρχει, &c.

Aristotle.

Let me not passe ouer *Aristotle* in silence, though his bookes writ of this subiect were but two, and these according to the coniecture of *Iulius Scaliger* (who hath made a large and curious examination of them) haue either perished, or come to vs not as they were originally written by *Aristotle*, but as they haue been by some later man put into Greeke. Amongst other things *Scaliger* hath these concerning those two bookes: *Reor è textrina Theophrasti detracta fila quædam, iisq; clavos additos, tametsi neque aureos, neque purpureos Quod si protinus autorem tibi dari vis ad Arabum diligentiam propius accedit* And afterwards thus: *Attribuere viri docti, alius alij, at quidem qui aliorum viderem nihil Planudem autorem facienti malim assentiri; extant enim illius alijs in libris similis vestigia semilatinietatis*, &c Thus much for *Aristotle*, whom as you see I haue placed after his Scholler, because there is such doubt of these bookes carried about in his name, and for that *Scaliger* as you see thinks them rather taken out of *Theophrastus*, than written by his Master.

Dioscorides.

The next that orderly followes is *Pedacius Dioscorides Anazarbeus*, who liued (according to *Suidas*) in the time of *Cleopatra*, which was some few yeares before the birth of our Sauiour. Now *Suidas* hath confounded * *Dioscorides Anazarbeus* with *Dioscorides Phacas*, but by some places in *Galen* you may see they were different men: for our Anazarbean *Dioscorides* was of the Empericke sect, but the other was a follower of *Herophylus* and of the Rationall sect. He writ not only of Plants, but *de tota materia medica*; to which studie hee was addicted euen from his childe-hood, which made him trauell much ground, and leade a militarie life, the better to accomplish his ends: and in this he attained to that perfection, that few or none since his time haue attained to, of the excellencie of his worke, which is as it were the foundation and ground-worke of all that hath been since deliuered in this nature. Heare what *Galen* one of the excellentest of Physitions, and one who spent no smal time in this study, affirmes: But, saith he, the Anazarbean *Dioscorides* in fiue bookes hath written of the necessarie matter of medicine, not onely making mention of herbes, but also of trees, fruits, " liquours and iuices, as also of all mineralls, and of the parts of liuing creatures: and in mine opinion he hath with the greatest perfection performed this worke of the matter of Medicine: for although many before him haue written well vpon this subiect, yet none haue writ so well of all. Now *Dioscorides* followes not the method of *Theophrastus*, but treats of each kinde of herbe in particular, first giuing the names, then the description, and then the place where they vsually grow, and lastly their vertues. Yet of some, which then were as frequently knowne with them, as Sage, Rosemary, an Ash or Oke tree are with vs, he hath omitted the descriptions, as not necessarie, as indeed at that time when they were so vulgarly knowne, they might seeme so to be: but now wee know the least of these, and haue no certaintie, but some probable coniectures do direct vs to the knowledge of them. He was not curious about his words nor method, but plainely and truly deliuered that whereof he had certaine and experimentall knowledge, concerning the description and nature of Plants. But the generall method he obserued you may finde set forth by *Bauhine* in his Edition of *Matthiolus*, immediatly after the preface of the first booke, whereto I refer the curious, being too long for me in this place to insist vpon. His

Διοσκορίδης Ἀναζαρβεὺς ἰατρὸς ὁ ἐπικληθεὶς Φακᾶς &c. *Suid.*

De simpl. med. facult. lib. 6. proem.

χυμῶν καὶ ὀπῶν.

workes that haue come to vs are fiue bookes *de materia Medica*. One *de letalibus venenis, eorumq; præcautione et curatione* : another *de Cane rabido, deq; notis quæ morſus ictuſve animalium venenum relinquentium ſequuntur* : a third *De eorum curatione*. Theſe eight bookes within theſe two laſt centuries of yeares haue been tranſlated out of Greeke into Latine, and commented vpon by diuers, as *Hermolaus Barbarus, Iohannes Ruellius, Marcellus Virgilius*, &c. But of theſe and the reſt, as they offer themſelues, I ſhall ſay ſomewhat hereafter. There is alſo another worke which goes vnder his name, and may well be his. It is Περὶ εὐπορίστων *ſiue de facile parabilibus*, diuided into two bookes, tranſlated and confirmed with the conſent of other Greeke Phyſitions, by the great labour of *Iohn Moibane* a Phyſition of Auſpurge, who liued not to finiſh it, but left it to bee perfected and ſet forth by *Conrade Geſner*.

Pliny. The next that takes place is the laborious *Caius Plinius ſecundus*, who liued in the time of *Veſpaſian*, and was ſuffocated by the ſulphureous vapours that came from mount Veſuvius, falling at that time on fire; he through ouermuch curioſitie to ſee and finde out the cauſe thereof approching too nigh, and this was *Anno Domini*, 79. He read and writ exceeding much, though by the iniurie of time wee haue no more of his than 37. books *de Hiſtoria Mundi*, which alſo haue receiued ſuch wounds, as haue tried the beſt skill of our Critickes, and yet in my opinion in ſome places require *medicas manus*. From the twelfth to the end of the twentie ſeuenth of theſe bookes he treats of Plants, more from what he found written in other Authors, than from any certaine knowledge of his owne, in many places following the method and giuing the words of *Theophraſtus*, and in other places thoſe of *Dioſcorides*, though he neuer make mention of the later of them : he alſo mentions, and no queſtion followed many other Authors, whoſe writings haue long ſince periſhed. Sometimes he is pretty large, and otherwhiles ſo briefe, that ſcarce any thing can thence be gathered. From the ſeuenteenth vnto the twentie ſeuenth he variouſly handles them, what method you may quickly ſee by his *Elenchus*, contained in his firſt book, but in the twenty ſeuenth hee handles thoſe whereof hee had made no, or not ſufficient mention, after an Alphabeticall order, beginning with *Æthyopis, Ageratum, Aloe*, &c. ſo going on to the reſt.

Galen. Paulus. Aetius. I muſt not paſſe ouer in ſilence, neither need I long inſiſt vpon *Galen*, *Paulus Ægineta*, and *Aetius*, for they haue only alphabetically named Plants and other ſimple Medicines, briefely mentioning their temperature and faculties, without deſcriptions (ſome very few, and thoſe briefe ones, excepted) and other things pertinent to their hiſtorie.

Macer. The next that preſent themſelues are two counterfeits, who abuſe the World vnder feined titles, and their names haue much more antiquitie than the works themſelues: the firſt goes vnder the title of *Æmilius Macer* a famous Poet, of whom *Ouid* makes mention in theſe verſes;

> *Sæpe ſuas volucres legit mihi grandior ævo,*
> *Quæq; nocet Serpens, quæ iuuat herba Macer.*

Pliny alſo makes mention of this *Macer* : hee in his Poems imitated *Nicander*, but this worke that now is carried about vnder his name, is written in a rude, and ſomewhat barbarous verſe, far different from the ſtile of thoſe times wherein *Macer* liued, and no way in the ſubiect immitating *Nicander*. It ſeemes to haue beene written about 400 or 500 yeares agoe.

Apuleius. The other alſo is of an vnknowne Author, to whom the Printers haue giuen the title of *Apuleius Madaurenſis*, and ſome haue been ſo abſurdly bold of late, as to put it vnto the workes of *Apuleius*; yet the vncurious ſtile and method of the whole booke will conuince them of errour, if there were no other argument. I haue ſeene ſome foure manuſcripts of this Authour, and heard of a fifth, and all of them ſeeme to bee of good Antiquitie : the figures of them all for the moſt part haue ſome reſemblance each of other : the firſt of theſe I ſaw ſome nine yeares agoe with that worthy louer and ſtorer of Antiquities, Sir *Robert Cotton* : it was in a faire Saxon hand, and as I remember in the Saxon tongue; but what title it carried, I at that time was not curious to obſerue. I ſaw alſo another after that, which ſeemed not to be of any ſmall ſtanding, but careleſly obſerued not the title. But ſince I being informed by my friend Maſter *Goodyer* (as you may finde in the Chapter of Saxifrage of the Antients) that his Manuſcript which was very antient, acknowledged no ſuch Author as *Apuleius*, I begunne a little to examine ſome other Manuſcripts, ſo I procured a very faire one of my much honored friend S[r]. *Theod. Mayern*: in the verie beginning of this is writ, *In hoc continentur libri quatuor medicinæ Ypocratis, Platonis Apolienſis vrbis de diuerſis herbis; Sexti Papiri placiti ex Animalibus*, &c. A little after in in the same page at the beginning of a table which is of the vertues, are these words, *In primo libro sunt herbæ descriptæ, quas Apoliensis Plato descripsit, &c.* and thus also he is named in the title of the Epistle or Proeme; but at the end of the worke is *explicit liber Platonis de herbis masculinis,* &c. With this in all things agrees that of Mr. *Goodyer*, as he hath affirmed to me. Besides these, I found one with Mr. *Iohn Tradescant*, which was written in a more ignorant and barbarous time, as one may coniecture by the title, which is thus at the very beginning. *In nomine domini incipit Herboralium Apulei Platonis quod accepit a Scolapio, & Chirone Centauro magistro.* Then followes (as also in the former, and in the printed bookes) the tract ascribed to *Antonius Musa, de herba Betonica*: after that are these words, *Liber Medicinæ Platonis herbaticus explicit.* By this it seemes the Author of this worke either was named, or else called himselfe *Plato*, a thing not without example in these times. This worke was first printed at Basill, 1528. amongst some other workes of Physicke, and one *Albanus Torinus* set it forth by the helpe of many Manuscripts, of whose imperfections he much complaines, and I thinke not without cause: after this, *Gabriel Humelbergius* of Rauenspurge in Germany set it forth with a Comment vpon it, who also complaines o the imperfections of his copies, and thinkes the worke not perfect: indeed both the editions are faultie in many places: and by the help of these Manuscripts I haue seen they might be mended (if any thought it worth their labour) in some things, as I obserued in cursorily looking ouer them. One thing I much maruell at, which is, that I finde not this Author mentioned in any Writer of the middle times, as *Platearius, Bartholomæus Anglus,* &c. Now I coniecture this worke was originally written in Greeke, for these reasons: first, because it hath the Greeke names in such plenty, and many of them proper, significant, and in the first place: Secondly some are onely named in Greeke, as *Hierobulbon, Artemisia Leptophyllos,* and *Artemisia tagantes, Batrachion, Gryas* (which I iudge rather Greeke than Latine) &c. Besides in both the written bookes in very many places amongst the names I finde this word *Omòeos*, but diuersly written; for I coniecture the Greeke names were written in the Greeke character, and ὁμοίως amongst them; and then also when the rest of the worke was translated, which afterwards made the transcribers who vnderstood it not to write it variously, for in the one booke it is alwaies written *Amoeos*, and in the other *Omoeos*, and somtimes *Omeos*, as in the Chapter of *Brittanica*, the one hath it thus, *Nomen herbæ istius Britanica, Amoeos dicunt eam Damasinium,* &c. The other thus: *Nomen herbæ Brittanica, Omeos Damasinius,* &c. & in the chap of *Althæa* the one hath it thus: *Nomen huius herbæ Altea Amoeos vocant hanc herbam Moloche,* &c. The other *Nomen herbæ Ibiscus omoeos Moloce,* &c. If it be certaine which *Philip Ferrarius* affirmes in his *Lexicon Geographicum*, that the citie Apoley is Constantinople, then haue I found *Apoliensis vrbs*, of which I can finde no mention in any antient or moderne Geographer besides; and then it is more than probable that this was written in Greeke, and it may be thought differently translated, which occasions such diuersitie in the copies, as you shall finde in some places. Now I coniecture this worke was written about some 600. yeares agoe.

From these Antients haue sprung all, or the greatest part of the knowledge, that the middle or later times haue had of Plants; and all the controuersies that of late haue so stuffed the bookes of such as haue writ of this subiect, had their beginning by reason that the carelesnesse of the middle times were such, that they knew little but what they transcribed out of these Antients, neuer endeuouring to acquire any perfect knowledge of the things themselues: so that when as learning (after a long Winter) began to spring vp againe, men began to be somwhat more curious, and by the notes and descriptions in these antient Authors they haue laboured to restore this lost knowledge; making inquirie, first whether it were knowne by *Theophrastus, Dioscorides*, or any of the Antients, then by what name. But to returne to my Authors.

About *An. Dom.* 1100. or a little after, liued the Arabians *Auicen, Auerrhoes, Mesve, Rhasis* and *Serapio*; most of these writ but briefely of this subiect; neither haue we their works in the Arabicke wherein they were written, but barbarously translated into Latine, and most part of these workes were by them taken out of the Greekes, especially *Dioscorides* and *Galen*; yet so as they added somewhat of their own, and otherwhiles confounded other things with those mentioned by the Greekes, because they did not well know the things whereof they writ. *Auicen, Auerrhoes,* and *Rhasis* alphabetically and briefly (following the method of *Galen*) giue the names, temperature, and vertues, of the chiefest simple medicines. But *Serapio* after a particular tract of the temperature and qualities of simple medicines in generall, comes to treat of them in particular, and therein followes chiefely *Dioscorides, Galen,* and *Paulus*, and diuers Arabians that went before him. This is the chiefe worke in this kinde of the Arabians, which haue come to vs; he himselfe tells vs his methode

The Arabians. *Auicen.* *Auerrhoes.* *Rhases.* *Serapio.*

thod in his preface, which is (when he comes to particulars) firſt of medicines temperate, then of thoſe that are hot and drie in the firſt degree; then thoſe cold and drie in the ſame degree: after that, thoſe hot and dry in the ſecond degree, &c. and in each of theſe tracts he followes the order of the Arabicke Alphabet.

In or after the times of the Arabians vntill about the yeare 1400. There were diuers obſcure and barbarous writers, who by ſight knew little whereof they writ, but tooke out of the Greekes, Arabians, and one another, all that they writ, giuing commonly rude figures, ſeldome ſetting downe any deſcriptions: I will only name the chiefe of them that I haue ſeene, and as neare as I can gueſſe in that order that one of them ſucceeded another. For the particular times of their liuing is ſomewhat difficult to be found out. One *Iſodore.* of the ancienteſt of them ſeemes to be *Iſidore*; then *Platearius* whoſe worke is Alphabeti- *Platearius.* call and intituled *Circa inſtans*: the next *Matthæus Syluaticus*, who flouriſhed about the yere *Barthol. Angl.* 1319. his worke is called *Pandectæ*: a little after him was *Bartholomæus Anglus*, whoſe workes (as that of *Iſodore*, and moſt of the reſt of thoſe times) treat of diuers other things beſides Plants, as Beaſts, Birds, Fiſhes, &c. His worke is called *De proprietatibus rerum*: the Authors name was *Bartholmew Glanuill*, who was deſcended of the Noble Family of the Earles of Suffolke; and he wrote this worke in *Edward* the thirds time, about the yeare *Hortus ſanitat.* of our Lord, 1397. After all theſe, and much like them is the *Hortus ſanitatis* whoſe Author I know not. But to leaue theſe obſcure men and their writings, et me reckon ſome of later time, who with much more learning and iudgement haue endeuoured to illuſtrat this part of Phyſicke.

About ſome 200 yeare agoe learning againe beginning to flouriſh, diuers begunne to leaue and loath the confuſed and barbarous writings of the middle times, and to haue recourſe to the Antients, from whence together with puritie of language, they might acquite a more certaine knowledge of the things treated of, which was wanting in the o- *Hermol. Barb.* ther. One of the firſt that tooke paines in this kinde was *Hermolaus Barbarus* Patriarch of Aquileia, who not onely tranſlated *Dioſcorides*, but writ a Commentarie vpon him in fiue bookes, which he calls his *Corollarium*; in this worke hee hath ſhewed himſelfe both iudicious and learned.

Merc. Virg. After him *Marcellus Virgilius* Secretarie to the State of Florence, a man of no leſſe learning and indgement than the former, ſet forth *Dioſcorides* in Greeke and Latine with a Comment vpon him.

Iohn. Ruellius. Much about their time alſo *Iohn Ruellius* a French Phyſition, who flouriſhed in the yere 1480, tranſlated *Dioſcorides* into Latine, whoſe tranſlation hath been the moſt followed of all the reſt. Moreouer he ſet forth a large worke, *De natura Stirpium*, diuided into three bookes, wherein he hath accurately gathered all things out of ſundry writers, eſpecially the Greekes and Latines; for firſt hauing (after the manner of *Theophraſtus*) deliuered ſome common precepts and Aduertiſements pertaining to the forme, life, generation, ordering, and other ſuch accidents of plants; he then comes to the particular handling of each *ſpecies*.

Much about this time, the Germanes began to beautifie this ſo neceſſary part of *Otho Brunfel.* Phyſicke; and amongſt them *Otho Brunfelſius*, a Phyſition of good account, writ of plants, and was the firſt that gaue the liuely figures of them; but he treated not in all of aboue 288 Plants. He commonly obſerues this method in his particular chapters: Firſt the figure (yet he giues not the figures of all he writes of) then the Greeke, Latine, and Germane names; after that, the deſcription and hiſtorie out of moſt former Authors; then the temperature and vertues, and laſtly, the Authours names that had treated of them. His worke is in three parts or tomes, the firſt was printed in 1530. the ſecond in 1531, and the third in 1536.

Hieron. Tragus. Next after him was *Hieronymus Tragus* a learned, ingenious, and honeſt writer, who ſet forth his workes in the German tongue, which were ſhortly after tranſlated into Latine by *David Kiber*. He treats of moſt of the Plants commonly growing in Germany, & I can obſerue no generall method he keepes, but his particular one is commonly this: hee firſt giues the figure with the Latine and high Dutch name; then commonly a good deſcription; after that the names, then the temperature, and laſtly the vertues, firſt inwardly, then outwardly vſed. He hath figured ſome 567, and deſcribed ſome 800. his figures are good, (and ſo are moſt of the reſt that follow.) His workes were ſet forth in Latine, *An.* 1552.

Leonhar. Fuch. In his time liued *Leonhartus Fuchſius*, a German Phyſition, being alſo a learned and diligent writer, but he hath taken many of his deſcriptions as alſo vertues word for word out of the Antients, and to them hath put figures; his generall method is after the Greek Alphabet, and his particular one thus: Firſt the names in Greeke and Latine, together oft-

times

times with their Etymologies,as also the German and French names,then the kinds,after that the forme,the place,time,temperature, then the vertues : first out of the Antients,as *Dioscorides,Galen,Pliny,&c.*and sometimes from the late Writers,whom he doth not particularize,but expresses in generall *ex recentioribus*. His worke was set forth at Basil, 1542, in *Fol.*containing,516 figures;also they were set forth in *Octavo*,the historie first,with all the figures by themselues together at the end with the Latine and high Ducth names.

About this time,and a little after,flourished *Conrade Gesner* also a German Physition, who set forth diuers things of this nature,but yet liued not to finish the great and general worke of Plants,which he for many yeres intended,and about which he had taken a great deale of paines,as may be gathered by his Epistles. He was a very learned,painfull,honest and iudicious writer,as may appeare by his many & great workes;wherof those of Plants were first a briefe Alphabeticall Historie of plants without figures, gathered out of *Dioscorides,Theophrastus,Pliny,*&c with the vertues briefely,and for the most part taken out of *Paulus Ægineta*,with their names in Greek and French put in the margent:this was printed at Venice,1541,in a small forme. He set forth a catalogue of Plants,in Latin,Greeke, high-Dutch and French,printed at Zurich,1542. Also another tract *De Lunarÿs & noctu lucentibus cum montis fracti,siue Pilati Lucernatum descriptione, An.*1552.in *quarto*. He also set forth the foure Books of *Valerius Cordus*(who died in his time)and his *Sylua obseruationum* at Strausburgh,1561.in *fol.* and to these he added a Catalogue of the Germane Gardens with an Appendix and *Corollarium* to *Cordus* his Historie. Also another treatise of his *De stirpium collectione*,was set forth at Zurich by *Wolphius, An.* 1587,in *Octavo*. *Conrade Gesner.*

At the same time liued *Adam Lonicerus* a Physition of Frankeford, whose naturall historie was there printed,*An* 1551,and the first part thereof is of Plants;and foure yeres after he added another part thereto,treating also of Plants. I finde no generall method obserued by him,but his particular method vsually is this:first he giues the figure, then the names in Latine and Dutch,then the temperature,&c.as in *Tragus*,from whom & *Cordus*, he borrows the most part of his first tome,as he doth the 2. from *Matth.*& *Amatanus Lusit.* *Lonicerus.*

In his time the Italian Physition *Petrus Andreas Matthiolus* set forth his Commentaries vpon *Dioscorides*,first in Italian with 957 large and very faire figures, and then afterwards in Latine at Venice,with the same figures,*An.*1568. After this he set forth his Epitome in *Quarto*,with 921 smaller figures. Now these his Commentaries are very large; and he hath in them deliuered the historie of many Plants not mentioned by *Dioscorides*; but he is iustly reprehended by some, for that he euery where taxes and notes other Writers,when as he himselfe runs into many errours,and some of them wilfull ones, as when he giues figures framed by his owne fancie,as that of *Dracontium maius, Rhabarbarum,*&c. and falsified othersome in part,the better to make them agree with *Dioscori.* his description,as when he pictures *Arbor Iudæ* with prickles,and giues it for the true *Acatia*: and he oft-times giues bare figures without description of his owne,but saith,it is that described by *Dioscorides, Nullis reclamantibus notis*,for which the Authors of the *Aduersaria* much declaime against him. It had bin fit for him,or any one that takes such a worke in hand, to haue shewed by describing the plant he giues, and conferring it with the description of his Author,that there is not any one note wanting in the description,vertues,or other particulars which his Author sets downe; and if hee can shew that his is such, then will the contrary opinions of all others fall of themselues,and need no confutation. *P.And. Matthiolus.*

Amatus Lusitanus also about the same time set forth Commentaries vpon *Dioscorides*, adding the names in diuers Languages but without figures,at Strausbourgh, *An.*1554.in *Quarto*:he dissented from *Matthiolus* in many things;whereupon *Matthiolus* writ an Apologie against him. He hath performed no great matter in his Enarrations vpon *Dioscorides*, but was an Author of the honestie of *Matthiolus*,for as the one deceiued the world with counterfeit figures, so the other by feined cures to strengthen his opinion,as *Crato* iudges of his *Curationes Medicinales*(another worke of his)which hee thinkes, *potius fictæ,quam factæ*. *Amatus Lusitanus.*

Rembertus Dodonæus a Physition borne at Mechlin in Brabant, about this time begun to write of Plants. Hee first set foorth a Historie in Dutch, which by *Clusius* was turned into French, with some additions, *Anno Domini*, 1560. And this was translated out of French into English by Master *Henry Lite*,and set forth with figures, *Anno Dom.*1578.and diuers times since printed,but without Figures. In the yeare 1552, *Dodonæus* set forth in Latine his *Frugum Historia*,and within a while after his *Florum,purgantium, & deleteriorum Historia*. Afterwards hee put them all together, his former, and those his later Workes, and diuided them into thirtie Bookes, and set them forth with 1305 figures, in *fol.An.*1583. This edition was also translated into English,which became the foundation *Remb. Dodon.*

of

of this present Worke, as I shall shew hereafter. It hath since beene printed in Latine, with the addition of some few new figures: and of late in Dutch, *Anno* 1618. with the addition of the same figures; and most of these in the *Exoticks* of *Clusius*, and great store of other additions. His generall method is this: first he diuides his Works into six Pemptades or fiues: the fifth Pemptas or fiue bookes of these containe Plants in an Alphabeticall order, yet so as that other Plants that haue affinitie with them are comprehended with them, though they fall not into the order of the Alphabet. The second Pempt. containes *Flores Coronarij, Plantæ odoratæ & vmbelliferæ*. The third is *De Radicibus, Purgantibus Herbis, convolvulis, deleterijs ac perniciosis Plantis, Filicibus, Muscis & Fungis*. The fourth is *De Frumentis, Leguminibus, palustribus & aquatilibus*. The fifth, *De Oleribus & Carduis*. The sixth, *de Fruticibus & Arboribus*. The particular method is the same vsed by our Author.

Peter Pena. *Matth. Lobel.* In the yeare 1570, *Peter Pena* and *Matthias Lobel* did here at London set forth a Worke, entituled *Stirpium Aduersaria noua*; the chiefe end and intention whereof being to find out the *Materia medica* of the Antients. The generall method is the same with that of our Author, which is, putting things together as they haue most resemblance one with another in externall forme, beginning with Grasses, Cornes, &c. They giue few figures, but sometimes refer you to *Fuchsius, Dodonæus*, and *Matthiolus*: but where the figure was not giuen by former Authors, then they commonly giue it; yet most part of these figures are very small and vnperfect, by reason (as I coniecture) they were taken from dried plants. In this Worke they insist little vpon the vertues of Plants, but succinctly handle controuersies, and giue their opinions of Plants, together with their descriptions and names, which sometimes are in all these languages, Greeke, Latine, French, high and low Dutch, and English: otherwhiles in but one or two of them. Some Writers for this Work call them *Doctißimi Angli*; yet neither of them were borne here, for *Pena* (as I take it) was a French man, and *Lobel* was borne at Ryssele in Flanders, yet liued most part of his later time in this Kingdome, and here also ended his dayes. In the yeare 1576 he set forth his Obseruations, and ioyned them with the *Aduersaria*, by them two to make one entire Worke: for in his Obseruations he giues most part of the figures and vertues belonging to those herbes formerly described onely in the *Aduersaria*; and to these also adds some new ones not mentioned in the former Worke. After which he set forth an Herball in Dutch, wherein he comprehended all those Plants that were in the two former Workes, and added diuers other to them, the Worke containing some 2116 figures; which were printed afterwards in a longish forme, with the Latine names, and references to the Latine and Dutch bookes. After all these, at London, *Anno* 1605, he againe set forth the *Aduersaria*, together with the second part thereof, wherein is contained some fourty figures, being most of them of Grasses and Floures; but the descriptions were of some 100 plants, varieties and all. To this he added a Treatise of Balsam (which also was set forth alone in Quarto, *Anno* 1598.) and the *Pharmacopæa* of *Rondeletius*, with Annotations vpon it. He intended another great Worke, whose title should haue beene *Stirpium Illustrationes*, but was preuented by death.

Carol. Clusius. Some six yeares after the Edition of the *Aduersaria, Anno* 1576, that learned, diligent, and laborious Herbarist *Carol. Clusius* set forth his Spanish Obseruations, hauing to this purpose trauelled ouer a great part of Spaine; and being afterwards called to the Imperiall Court by *Maximilian* the second, he viewed Austria and the adiacent prouinces, and set forth his there Obseruation, *Anno* 1583. He also translated out of Spanish the Works of *Garcias ab Orta* and *Christopher Acosta*, treating of the simple medicines of the East Indies, and *Nicolas Monardus*, who writ of those of the West Indies. After this he put into one body both his Spanish and Pannonicke Obseruations, with somoother, and those he comprehends in six bookes, entituled *Rariorum Plantarum Historia*: whereto he also addes *Honor. Bellus.* an Appendix, a treatise of Mushroms, six Epistles treating of Plants, from *Honorius Bellus* an Italian Physition liuing at Cydonia in Candy; as also the description of mount Baldus, being a Catalogue with the description and figures of some rare and not before writ*Iohn Pona.* ten of Plants there growing, written by *Iohn Pona* an Apothecarie of Verona (This Description of *Pona's* was afterwards with some new descriptions and thirty six figures set forth alone in Quarto, *An.* 1608.) This first Volume of *Clusius* was printed in Antwerp, *Anno* 1601, in Folio: and in the yeare 1605 he also in Folio set forth in another volume six bookes of *Exoticks* containing various matter, as plants, or some particles of them, as Fruits, Woods, Barks, &c. as also the forenamed translations of *Garcias, Acosta*, and *Monardus*: Three Tracts besides of the same *Monardus*; the first, *De lapide Bezaar, & Herba Scorsonera*. The second, *De Ferro & eius facultatibus*: The third, *De Niue & eius commodis*.

To

To these he also added *Bellonius* his Obseruations or Singularities, and a tract of the same Author, *De neglecta Stirpium cultura*, both formerly translated out of French into Latine by him. He was borne at Atrebas or Arras, the chiefe city of Artois, *Anno* 1526. and died at Leyden, *Ann.* 1609. After his death, by *Euerard Vorstius, Peter Paw*, or some others, were set forth some additions and emendations of his former Works, together with his funerall Oration made by *Vorstius*, his Epitaph, &c. in Quarto, *Anno* 1611, by the name of his *Curæ Posteriores.*

In the yeare 1583, *Andreas Cæsalpinus* an Italian Physition, and Professor at Pisa, set forth an historie of Plants, comprehended in sixteene bookes: his Worke is without figures, and he oft times giues the Tuscane names for Latine; wherefore his worke is the more difficult to be vnderstood, vnlesse it be by such as haue been in Tuscanie, or else are already well exercised in this study. He commonly in his owne words diligently for the most part describes each Plant, and then makes enquirie whether they were knowne by the Antients. He seldome sets downe their faculties, vnlesse of some, to which former Writers haue put downe none. In the first booke he treats of Plants in generall, according as *Theophrastus* doth: but in the following bookes hee handles them in particular: he maketh the chiefe affinity of Plants to consist in the similitude of their seeds and seed vessels. *Andr.Cæsalp.*

Ioachimus Camerarius a Physition of Noremberg flourished about this time: Hee set forth the Epitome of *Mattheolus*, with some additions and accurate figures, in Quarto, at Frankfort, 1586: in the end of which Worke (as also in that set forth by *Matthiolus* himselfe) is *Iter Baldi*, or a journey from Verona to mount Baldus, written by *Francis Calceolarius* an Apothecarie of Verona. Another Worke of *Camerarius* was his *Hortus Medicus*, being an Alphabeticall enumeration of Plants, wherein is set forth many things concerning the names, ordering, vertues, &c. of Plants. To this he anexed *Hyrcinia Saxonothuringica Iohannis Thalij*, or an alphabeticall Catalogue written by *Iohn Thalius*, of such Plants as grew in Harkwald a part of Germanie betweene Saxony and Durengen. This was printed also at Frankfort in Quarto, *An.* 1588. *Ioach. Camer.* *Fr.Calceolarius.* *Ioh.Thalius.*

In the yeare 1587 came forth the great Historie of Plants printed at Lyons, which is therefore vulgarly termed *Historia Lugdunensis*: it was begun by *Dalechampius*: but hee dying before the finishing thereof, one *Iohn Molinæus* set it forth, but put not his name thereto. It was intended to comprehend all that had written before, and so it doth, but with a great deale of confusion; which occasioned *Bauhine* to write a treatise of the errors committed therein, in which he shewes there are about foure hundred figures twice or thrice ouer. The whole number of the figures in this Worke are 2686. This History is diuided into eighteene bookes, and the Plants in each booke are put together either by the places of their growings, as in Woods, copses, mountaines, waterie places, &c. or by their externall shape, as vmbelliferous, bulbous, &c. or by their qualities, as purging, poysonous, &c. Herein are many places of *Theophrastus* and other antient Writers explained. He commonly in each chapter giues the names, place, forme, vertue, as most other do. And at the end thereof there is an Appendix containing some Indian plants, for the most part out of *Acosta*; as also diuers Syrian and Egyptian plants described by *Reinold Rawolfe* a Physition of Ausburgh. *Hist.Lugd.* *Leon.Rawolf.*

At this time, to wit *Anno* 1588, *Iacobus Theodorus Tabernamontanus* set forth an History of Plants in the Germane tongue, and some twelue yeares after his Figures being in all 2087, were set forth in a long forme, with the Latine and high-Dutch names put vnto them; and with these same Figures was this Worke of our Author formerly printed. *Tabernamont.*

Prosper Alpinus a Physition of Padua in Italy, in the yeare 1592 set forth a Treatise of some Egyptian Plants, with large yet not very accurate figures: he there treats of some 46 plants, and at the end thereof is a Dialogue or Treatise of Balsam. Some six yeares agone, *Anno* 1627, his Son set forth two bookes of his fathers, *De Plantis Exoticis*, with the figures cut in Brasse: this Worke containes some 136 Plants. *Prosp.Alpinus.*

Fabius Columna a gentleman of Naples, of the house of *Columna* of Rome, *An.* 1592 set forth a Treatise called *Phytobasanos*, or an Examination of Plants; for therein he examines and asserts some plants to be such and such of the Antients: and in the end of this worke he giues also the historie of some not formerly described plants. Hee also set forth two other bookes, *De minus cognitis*, or of lesse knowne Plants: the first of which was printed at Rome, *Anno* 1606; and the other 1616. He in these works, which in all contain little aboue two hundred thirty six plants, shewes himselfe a man of an exquisit iudgment, and very learned and diligent, duely examining and weighing each circumstance in the writings of the Antients. *Fab.Columna.*

Caspar

Casp. Bauhine. *Caspar Bauhine*, a Physition and Professor of Basil, besides his Anatomicall Works, set forth diuers of Plants. *Anno* 1596 he set forth his *Phytopinax*, or Index of Plants, wherein he followes the best method that any yet found: for according to *Lobels* method (which our Author followed) he begins with Grasses, Rushes, &c. but then he briefely giues the Etymologie of the name in Greeke and Latine, if any such be, and tells you who of the Antients writ thereof, and in what part of their Works: and lastly (which I chiefly commend him for) he giues the *Synonima's* or seuerall names of each plant giuen by each late Writer, and quoteth the pages. Now there is nothing more troubles such as newly enter into this study, than the diuersitie of names, which sometimes for the same plant are different in each Author; some of them not knowing that the plant they mention was formerly written of, name it as a new thing; others knowing it writ of, yet not approuing of the name. In this Worke he went but through some halfe of the historie of Plants. After this, *Anno* 1598, he set forth *Matthiolus* his Commentaries vpon *Dioscorides*, adding to them 330 Figures, and the descriptions of fifty new ones not formerly described by any; together with the *Synonima's* of all such as were described in the Worke. He also *Anno* 1613 set forth *Tabernamontanus* in Dutch, with some addition of historie and figures. In *Anno* 1620 he set forth the *Prodromus*, or fore-runner of his *Theatrum Botanicum*, wherein he giues a hundred and forty new figures, and describes some six hundred plants, the most not described by others. After this, *Anno* 1623, he set forth his *Pinax Theatri Botanici*, whose method is the same with his *Phytopinax*, but the quotations of the pages in the seuerall Authors are omitted. This is indeed the Index and summe of his great and generall Worke, which should containe about six thousand plants, and was a Worke of forty yeares: but he is dead some nine yeares agone, and yet this his great worke is not in the Presse, that I can heare of.

Basil Besler. *Basil Besler* an Apothecarie of Noremberg, *Anno* 1613 set forth the garden of the Bishop of Eystet in Bauaria, the figures being very large, and all curiously cut in brasse, and printed vpon the largest paper: he onely giues the *Synonima's* and descriptions, and diuideth the worke first into foure parts, according to the foure seasons of the yeare; and then againe he subdiuides them, each into three, so that they agree with the moneths, putting in each Classis the plants that flourish at that time.

These are the chiefe and greatest part of those that either in Greeke or Latine (whose Works haue come to our hands) haue deliuered to vs the history of Plants; yet there are some who haue vsed great diligence to helpe forward this knowledge, whose names I wil not passe ouer in silence. The first and antientest of these was *Aloysius Anguillara* a physition of Padua, and President of the publique Garden there: his opinions of some plants were set forth in Italian at Venice, 1561. (Aloys. Anguill.)

Melchior Guillandinus. *Melchior Guillandinus*, who succeeded *Anguillara* in the garden at Padua, writ an Apologie against *Matthiolus*, some Epistles of plants, and a Commentarie vpon three Chapters of *Pliny*, *De Papyro*.

Fer. Imperato. *Ferantes Imperatus* an Apothecary of Naples also set forth a Naturall Historie diuided into twenty eight bookes, printed at Naples *Anno* 1599. In this there is something of Plants: but I haue not yet seene the opinions of *Anguillara*, nor this Naturall Historie: yet you shall find frequent mention of both these in most of the forementioned Authors that writ in their time, or since, wherefore I could not omit them.

Let me now at last looke home, and see who we haue had that haue taken pains in this kinde. The first that I finde worthy of mention is D[r]. *William Turner*, the first of whose works that I haue seene, was a little booke of the names of herbes, in Greeke, Latine, English, Dutch, and French, &c. printed at London *Anno* 1548. In the yeare 1551 he set forth his Herbal or Historie of Plants, where he giues the figures of *Fuchsius*, for the most part: he giues the Names in Latine, Greeke, Dutch, and French: he did not treat of many Plants; his method was according to the Latine alphabet. He was a man of good iudgment and learning, and wel performed what he tooke in hand. (Will. Turner.)

Hen. Lyte. After this, *Dodonæus* was translated into English by M[r]. *Lyte*, as I formerly mentioned. And some yeares after, our Author set forth this Worke, whereof I will presently treat, hauing first made mention of a Worke set forth betweene that former Edition, and this I now present you withall.

Ioh. Parkinson. M[r]. *Iohn Parkinson* an Apothecarie of this city (yet liuing and labouring for the common good) in the yeare 1629 set forth a Worke by the name of *Paradisus terrestris*, wherein he giues the figures of all such plants as are preserued in gardens, for the beauty of their floures, for vse in meats or sauces; and also an Orchard of all trees bearing fruit, and such shrubs as for their raritie or beauty are kept in Orchards and gardens, with the ordering, planting,

planting and preſeruing of all theſe. In this Worke he hath not ſuperficially handled theſe things, but accurately deſcended to the very varieties in each ſpecies: wherefore I haue now and then referred my Reader addicted to theſe delights, to this worke eſpecially in floures and fruits, wherein I was loth to ſpend too much time, eſpecially ſeeing I could adde nothing to what he had done vpon that ſubiect before. He alſo there promiſed another worke, the which I thinke by this time is fit for the Preſſe.

Now am I at length come to this preſent Worke, whereof I know you will expect I ſhould ſay ſomewhat; and I will not fruſtrate your expectation, but labour to ſatisfie you in all I may, beginning with the Author, then his worke, what it was, and laſtly what it now is.

For the Author Mr. *Iohn Gerard* I can ſay little, but what you alſo may gather out of this worke; which is, he was borne in the yeare 1545. in Cheſhire, at Namptwich, from whence hee came to this city, and betooke himſelfe to Surgerie, wherein his endeauours were ſuch, as he therein attained to be a Maſter of that worthy profeſſion: he liued ſome ten yeares after the publiſhing of this worke, and died about the yeare 1607. His chiefe commendation is, that he out of a propenſe good will to the publique aduancement of this knowledge, endeauoured to performe therein more than he could well accompliſh; which was partly through want of ſufficient learning, as (beſides that which he himſelfe ſaith of himſelfe in the chapter of Water Docke) may be gathered by the tranſlating of diuers places out of the *Aduerſaria*; as this for one in the deſcription of * *Aſter Atticus, Caules pedales terni aut quaterni* which is rendred, A ſtalke foure or fiue foot long. He alſo by the ſame defect called burnt Barley, * *Hordeum diſtichon*; and diuided the titles of honour from the name of the perſon whereto they did belong, making two names thereof, beginning one clauſe with * *Iulius Alexandrinus* ſaith, &c. and the next with, *Cæſarius Archiater* ſaith. He alſo was very little conuerſant in the writings of the Antients, neither, as it may ſeeme by diuers paſſages, could hee well diſtinguiſh betweene the antient and moderne writers: for he in one place ſaith, [* Neither by *Dioſcorides, Fuchſius*, or any other antient writer once remembred.] Diuers ſuch there are, which I had rather paſſe ouer in ſilence, than here ſet downe: neither ſhould I willingly haue touched hereon, but that I haue met with ſome that haue too much admired him, as the only learned and iudicious writer. But let none blame him for theſe defects, ſeeing he was neither wanting in pains nor good will, to performe what he intended; and there are none ſo ſimple but know, that heauy burthens are with moſt paines vndergone by the weakeſt men: and although there were many faults in the worke, yet iudge well of the Author; for as a late writer well ſaith, *Falli & hallucinari humanum eſt; ſolitudinem quærat oportet, qui vult cum perfectis viuere. Penſanda vitijs bona cuiuſque ſunt, & quà maior pars ingenij ſtetit, ea iudicandum de homine eſt.*

Iohn Gerard.

See the former Edition in the places heremention ned.
**pag.*391.
*p.*66.
*p.*147.
*p.*518.
*Cun. li.*3.*ca.*3. *de Rep. Heb.*

Now let me acquaint you how this Worke was made vp. *Dodonæus* his Pemptades comming forth *Anno* 1583, were ſhortly after tranſlated into Engliſh by Dr. *Prieſt* a phyſition of London, who died either immediately before or after the finiſhing of this tranſlation. This I had firſt by the relation of one who knew Dr. *Prieſt* and Mr. *Gerard*: and it is apparant by the worke it ſelfe, which you ſhall finde to containe the Pemptades of *Dodonæus* tranſlated, ſo that diuers chapters haue ſcarce a word more or leſſe than what is in him. But I cannot commend my Author for endeauouring to hide this thing from vs, cauilling (though commonly vniuſtly) with *Dodonæus*, whereſoeuer he names him, making it a thing of heare-ſay, * that Dr. *Prieſt* tranſlated *Dodonæus*: when in the Epiſtle of his friend Mr. *Bredwell*, prefixed before this worke, are theſe words: [The firſt gatherers out of the Antients, and augmenters by their owne paines, haue already ſpred the odour of their good names through all the lands of learned habitations: Dr. *Prieſt* for tranſlating ſo much as *Dodonæus*, hath hereby left a tombe for his honorable ſepulture. Mr. *Gerard* comming laſt, but not the leaſt, hath many waies accommodated the whole worke vnto our Engliſh Nation, &c.] But that which may ſerue to cleare all doubts, if any can be in a thing ſo manifeſt, is a place in *Lobels* Annotations vpon *Rondeletius* his *Pharmacopeia*, where *pag.*59. he findes fault with *Dodonæus*, for vſing barbarouſly the word *Seta* for *Sericum*: and with Dr. *Prieſt*, who (ſaith he) at the charges of Mr. *Norton* tranſlated *Dodonæus*, and deceiued by this word *Seta*, committed an abſurd errour in tranſlating it a briſtle, when as it ſhould haue been ſilke. This place ſo tranſlated is to be ſeen in the chapter of the Skarlet Oke, at the letter F. And *Lobel* well knew that it was Dr. *Prieſt* that committed this error, and therefore blames not Mr. *Gerard*, to whom hee made ſhew of friendſhip, and who was yet liuing: but yet he couertly gaue vs to vnderſtand, that the worke wherein that error was committed, was a tranſlation of *Dodonæus*, and that made by

See his Epiſtle to the Reader.

D[r]. *Priest*, and set forth by M[r]. *Norton*. Now this translation became the ground-worke whereupon M[r]. *Gerard* built vp this Worke: but that it might not appeare a translation, he changes the generall method of *Dodonæus*, into that of *Lobel*, and therein almost all ouer followes his *Icones* both in method and names, as you may plainly see in the Grasses and *Orchides* To this translation he also added some plants out of *Clusius*, and other-some out of the *Aduersaria*, and some fourteene of his owne not before mentioned. Now to this historie figures were wanting, which also M[r]. *Norton* procured from Frankfort, being the same wherewith the Works of *Tabernamontanus* were printed in Dutch: but this fell crosse for my Author, who (as it seemes) hauing no great iudgement in them, frequently put one for another: and besides, there were many plants in those Authors which he followed, which were not in *Tabernamontanus*, and diuers in him which they wanted, yet he put them all together, and one for another; and oft times by this meanes so confounded all, that none could possibly haue set them right, vnlesse they knew this occasion of these errors. By this meanes, and after this manner was the Worke of my Author made vp, which was printed at the charges of M[r]. *Norton*, *An*. 1597.

Now it remaines I acquaint you with what I haue performed in this Edition, which is either by mending what was amisse, or by adding such as formerly were wanting: some places I helped by putting out, as the Kindes in the Chapter of Stonecrop, where there was but one mentioned. I haue also put out the Kindes in diuers places else where they were not very necessarie, by this meanes to get more roome for things more necessarie: as also diuers figures and descriptions which were put in two or three places, I haue put them out in all but one, yet so, as that I alwaies giue you notice where they were, and of what. Some words or passages are also put out here and there, which I thinke needlesse to mention. Sometimes I mended what was amisse or defectiue, by altering or adding one or more words, as you may frequently obserue if you compare the former edition with this, in some few chapters almost in any place. But I thinke I shall best satisfie you if I briefely specifie what is done in each particular, hauing first acquainted you with what my generall intention was: I determined, as wel as the shortnesse of my time would giue me leaue, to retaine and set forth whatsoeuer was formerly in the booke described, or figured without descriptions (some varieties that were not necessarie excepted) and to these I intended to adde whatsoeuer was figured by *Lobel*, *Dodonæus*, or *Clusius*, whose figures we made vse of; as also such plants as grow either wilde, or vsually in the gardens of this kingdome, which were not mentioned by any of the forenamed Authors; for I neither thought it fit nor requisite for me, ambitiously to aime at all that *Bauhine* in his *Pinax* reckons vp, or the Exotickes of *Prosper Alpinus* containe, not mentioned in the former. This was my generall intention. Now come I to particulars, and first of figures: I haue, as I said, made vse of those wherewith the Workes of *Dodonæus*, *Lobel*, and *Clusius* were formerly printed, which, though some of them be not so sightly, yet are they generally as truly exprest, and sometimes more. When figures not agreeable to the descriptions were formerly in any place, I giue you notice thereof with a marke of alteration before the title, as also in the end of the Chapter; and if they were not formerly in the booke, then I giue you them with a marke of addition. Such as were formerly figured in the booke, though put for other things, and so hauing no description therein, I haue caused to be new cut and put into their fit places, with descriptions to them, and only a marke of alteration. The next are the descriptions, which I haue in some places lightly amended, without giuing any notice thereof; but when it is much altered, then giue I you this marke † at the beginning thereof; but if it were such as that I could not helpe it but by writing a new one, then shall you finde it with this marke ‡ at the beginning and end thereof, as also whatsoeuer is added in the whole booke, either in description or otherwise. The next is the Place, which I haue seldome altered, yet in some places supplied, and in others I haue put doubts, & do suspect othersome to be false, which because I had not yet viewed, I left as I found. The Time was a thing of no such moment, for any matter worth mentioning to be performed vpon, wherefore I will not insist vpon it. Names are of great importance, and in them I should haue been a little more curious if I had had more time, as you may see I at the first haue beene; but finding it a troublesome worke, I haue onely afterwards where I iudged it most needfull insisted vpon it: *Bauhinus* his *Pinax* may supply what you in this kinde finde wanting. In many places of this worke you shall finde large discourses and sometimes controuersies handled by our Authour in the names; these are for the most part out of *Dodonæus*, & some of them were so abbreuiated, and by that meanes confounded, that I thought it not worth my paines to mend them, so I haue put them out in some few places, and referred you to the places in *Dodonæus* out of which

which they were taken, as in the chapter of Alehoofe: it may be they are not ſo perfect as they ſhould be in ſome very few other places, (for I could not compare all) but if you ſuſpect any ſuch thing, haue recourſe to that Author, and you ſhall finde full ſatisfaction.

Now come I to the Temper and Vertues. Theſe commonly were taken forth of the fore-mentioned Author, and here and there out of *Lobels* Obſeruations, and *Camerarius* his *Hortus medicus.* To theſe he alſo adddedſome few Receipts of his owne: theſe I haue not altered, but here and there ſhewed to which they did moſt properly belong; as alſo if I found them otherwiſe than they ought, I noted it; or if in vnfit places, I haue tranſferred them to the right place, and in diuers things whereof our Author hath bin ſilent, I haue ſupplied that defect.

For my additions I will here ſay nothing, but refer you to the immediate enſuing Catalogue, which will enforme you what is added onely in figure, or deſcription, or in both, by which, and theſe two formerly mentioned marks, you may ſee what is much altered or added in the Work; for this marke † put either to figure, or before any clauſe, ſhews it to haue bin otherwiſe put before; or that clauſe whether it be in deſcription, Place, Time, Names, or Vertues to be much altered. This other marke ‡ put to a figure ſhewes it not to haue been formerly in the worke, but now added; and put in any other place it ſhewes all is added vntill you come to another of the ſame marks. But becauſe it is ſomtimes omitted, I will therefore giue notice in the *Errata* where it ſhould be put, in thoſe places where I obſerue either the former or later of them to be wanting.

Further, I muſt acquaint you how there were the deſcriptions of a few plants here and there put in vnfitting places, which made me deſcribe them as new added, as *Saxifraga maior Matthioli*, *Perſicaria ſiliquoſa*, of which in the chapter of *Perſicaria* there was an ill deſcription, but a reaſonable good one in the chapter of *Aſtrantia nigra*. *Papauer ſpinoſum*, was figured and deſcribed amongſt the *Cardui*; now all theſe (as I ſaid) I added as new in the moſt fitting places: yet found them afterwards deſcribed, but put them out all, except the laſt, whoſe hiſtorie I ſtill retaining, with a reference to the preceding figure and Hiſtorie. Note alſo, whereſoeuer my Author formerly mentioned *Cluſius*, according to his Spaniſh or Pannonicke Obſeruations, I haue made it, according to his Hiſtorie, which containes them both with additions.

Alſo I muſt certifie you, (becauſe I know it is a thing that ſome will thinke ſtrange, that the number of the pages in this booke do no more exceed that of the former, conſidering there is ſuch a large acceſſion of matter and figures) the cauſe hereof is, each page containes diuers lines more than the former, the lines themſelues alſo being longer; and by the omiſſion of deſcriptions and figures put twice or thrice ouer, and the Kindes, vnneceſſarily put in ſome places, I gained as much as conueniently I could, beeing deſirous that it might be bound together in one volume.

Thus haue I ſhewed what I haue performed in this Worke, entreating you to take this my Labor in good part; and if there be any defect therein (as needs there muſt in all humane works) aſcribe it in part to my haſte and many buſineſſes, and in ſome places to the want of ſufficient information, eſpecially in Exoticke things; and in other ſome, to the little conuerſation I formerly had with this Author, before ſuch time as (ouercome by the importunitie of ſome friends, and the generall want of ſuch a Worke) I tooke this taske vpon me. Furthermore I deſire, that none would raſhly cenſure me for that which I haue here done; but they that know in what time I did it, and who themſelues are able to do as much as I haue here performed, for to ſuch alone I ſhall giue free libertie, and will be as ready to yeeld further ſatisfaction if they deſire it, concerning any thing I haue here aſſerted, as I ſhall be apt to neglect and ſcorne the cenſure of the Ignorant and Vnlearned, who I know are ſtill forward to verifie our Engliſh prouerbe * *A fooles bolt is ſoone ſhot.*

I muſt not in ſilence paſſe ouer thoſe from whom I haue receiued any fauour or incouragement, whereby I might be the better enabled to performe this Taske. In the firſt place let me remember the onely Aſſiſtant I had in this Worke, which was M[r]. *Iohn Goodyer* of Maple Durham in Hampſhire, from whom I receiued many accurate deſcriptions, and ſome other obſeruations concerning plants; the which (deſirous to giue euery man his due) I haue cauſed to be ſo printed, as they may be diſtinguiſhed from the reſt: and thus you ſhall know them; in the beginning is the name of the plant in Latine in a line by it ſelfe, and at the end his name is inſerted; ſo that the Reader may eaſily finde thoſe things that I had from him, and I hope together with me will be thankfull to him, that he would ſo readily impart them for the further increaſe of this knowledge.

M[r]. *George Bowles* of Chiſſelhurſt in Kent muſt not here be forgot, for by his trauells and induſtry I haue had knowledge of diuers plants, which were not thought nor formerly knowne to grow wilde in this kingdome, as you ſhall finde by diuers places in this book. My louing friends and fellow Trauellers in this ſtudy, and of the ſame profeſſion, whoſe companie I haue formerly enioyed in ſearching ouer a great part of Kent, and who are ſtill ready to do the like in other places, are here alſo to be remembred, and that the rather, becauſe this Knowledge amongſt vs in this city was almoſt loſt, or at leaſt too much neglected, eſpecially by thoſe to whom it did chiefely belong, and who ought to be aſhamed of ignorance, eſpecially in a thing ſo abſolutely neceſſarie to their profeſſion. They ſhould indeed know them as workemen do their tooles, that is readily to cal them by their names, know where to fetch, and whence to procure the beſt of each kinde; and laſtly, how to handle them.

Thomas Hickes
Iohn Buggs.
William Broad.
Job Weale.
Leonard Buckner.
Iames Clarke.
Robert Lorkin.

I haue already much exceeded the bounds of an Epiſtle, yet haue omitted many things of which I could further haue informed thee Reader, but I will leaue them vntill ſuch time as I finde a gratefull acceptance, or ſome other occaſion that may againe inuite me to ſet Pen to Paper; which, That it may be for my Countreyes good and Gods glory, ſhall euer be the prayers and Endeauours of thy Well-Wiſher

From my houſe on Snow-hill,
Octob. 22. 1633.

THOMAS IOHNSON

A Catalogue of Additions.

Because the markes were not so carefully and right put to these Figures, which were not formerly in the booke, I haue thought good to giue you the names of all such as are added, either in figure or description, or both: together with the booke, chapter, and number or place they hold in each chapter. *F* stands for figure, *D* for Description, and where both are added, you shall finde both these letters; and where the letter *C* is put, the Historie of the whole Chapter is added.

Booke. I.

Chap. 2. 1. *Gram.min.rub.siue Xerampelinum*, f.
Chap. 5. 3. *Gram.arund.minus Disc.*
Chap. 6. 1 *Gram.toment.arundin.* f.
2. *Gram. pan. elegans* d.
Chap. 8 3. *Gram.typhoides spica longis.* d.
Chap. 13. 3. *Gram. Panic. spic. simp.* d.
Chap. 14. 1 *Gram.pal.echin.* f.
3. *Gram.capit.glob.* d.
4. *Gram.mont.echin.* d.
Chap. 16. 8. *Gram.cyper.spic.* d.
Chap. 20, 3. *Gram.dactyloides.* f. d.
C. { Chap 21. 1. *Gram.Cyp.ang.mai.* f d
2. *Pseudocyperus.* f. d.
3. *Cyperus long.inod.* f. d.
4. *Cyperus rot.inod.* f d
5 *Cyper. Gram.mil.* f. d. }
C. { Chap. 22. 1 *Gram.mont. auen.* f. d.
2. *Gram.muror spic.long.* f. d.
3. *Gram.cristatum.* f. d.
4. *Gram.spica secal.* d.
5. *Gram.spica. Brizæ* d.
6. *Gram. lanatum* d.
7. *Gram.iunc.leucanth.* d
8. *Gram.Loliac.min.* d.
9. *Gram.lol.* d.
10. *Gram. sparteum min.* d
11. *Gram.alopecur.sp.aspera.* d.
12. *Gram.scoparium.* d. }
Chap. 24. 3. *Cyperus rotund. Syriacus.* d.
4. *Cyp.min.Cret.* d.
5. *Cyp.rotund.inodorus.* f. d.
Chap. 25. *Cyp.esculentus.* d. C.
Chap. 26. *Galanga maior.* C.
Galanga minor. C.
Chap. 27. *Cyperus Indicus.* C.
Chap. 28. *Zedoaria.* C.
Chap. 29. 5. *Iuncus cap.Equis.* f. d.
Chap. 34. 5. *Spartum nost.par.* f. d.
6. *Spart. Austriacum.* f. d.
Chap. 39. 4 *Phalangium antiq.* f. d.
5. *Phalang.Virgin.* f d
Chap. 42. 4. *Iris Byzantina*, d.
7. *Iris flo.cerul.obsol.* f-d.
8. *Chamæiris niuea*, f. d.
9. *Chamæir.lat.fl.rub.* f. d.
10. *Chamæir.lut.* f. d.
11 *Cham.variegat.* f. d.
Chap. 45. 3. *Calamus aromat.* f. d.
Chap. 63. 3. *Panicum Americanum*, f. d.
Chap 64. 3. *Phalar.prat.altera.* f d
Chap. 65. 2. *Alopecuros Anglica.pal.* d.
Chap. 68. 2 *Melampyrum purp.* f d
3 *Melampyr.cærul.* f d
4 *Melampyr.Lut.* f d
Chap. 70. 5 *Asphodelus minimus*, f d
Chap. 71 3 *Asphod.Lanc.ver.* f d
Chap. 74. 1 *Iris bulbosa Lut.* f d
4 *Iris bulb.versicol*, f d
6 *Iris bulb.flo.cin.* f d
7 *Iris bulb.flo.alb.* f d
Chap. 75 2 *Sisynrichium minus*, f
Chap. 76 4 *Gladiolus lacustris.* f d
Chap. 77 2 *Hyacinthus stel.albicans*, f
3 *Hyacinthus stel.bifol.* f d
6 *Hyac.stel.Byzant.* f d
8 *Hyac.stel. Som.* f d
9 *Hyac.stel.æst.mai.* f d
10 *Hyac.stel.æst.min.* f d
12 *Hyac.Peruv.* f d
Chap. 78 3 *Hyac.stel.ver.* d.
Chap. 79 6 *Hyac.Or.polyanth.* f d
7 *Hyac.Or.purp.* f d
8 *Hyac.Or.alb.* f d
9 *Hyac.Brumalus*, f d
10 *Hyac.Or.caule folioso*, f d
11 *Hyac. Or.flo.pleno*, f d
12 *Hyac. Or.flo.cærul.pleno.* f d
13 *Hyac.Or.flo.cand.plen.* f d
14 *Hyac. obsolet.flo.Hisp.* f d
15 *Hyac.min.Hisp.* f d
16 *Hyac.Ind.tuber.* f d
Chap. 80 3 *Hyacinthus com.Byzant.* f
4 *Hyacinth.com.ramosus.* d
5 *Hyacinth.com.ram.eleg.* f d

Chap.

Chap.84.4 *Narc ss. medio-croceus serot. polyanth.* f d
5 *Narc. medio purp. flo. plen.* f d
11 *Narc. flore pleno albo,* f d
12 *Narc. flo. plen. medio lut.* f d
13 *Narc. flore plen. med. versic.* f d
17 *Narc. iuncifol. roseolut.* f d
18 *Narc. iuncifol. amplo cal.* f d
19 *Narc. iuncifol. reflex. flo. alb.* f d
19 *Narc. iuncifol. reflex. min.* f d
20 *Narc. iuncifol. mult.* f d
23. *Narc. vernus præc. flau.* f d
Chap.85.3 *Pseudonarcissus Hisp.* f
4 *Pseudonarc. min. hispan.* f d
5 *Pseudonarc. albo flore* f d
C. { Chap.86.1 *Narcis. omnium maximus* f d
2 *Narcis. mult. Robini.* d
3 *Pseudonarc. flo. pleno* f d
4 *Narcis. Wilmot.* d
Narc. Tradescant. d
Narc. Parkinson. d
5 *Narcis. Iacob. Indicus* f d
6 *Narcis. iuncifol. mont. min.* f d
7 *Narc. mont. iuncifol. flore fimbriato* f d
8 *Narc. omnium. min. mont. alb.* f d }
Chap. 87. *Tuliparum fig.* 23.
Chap. 88.2 *Leucoium bulbosum præcox Byzant.* f d
5 *Leucoium bulb. Aut. min.* f d
6 *Leuc. bulb. vern. min.* d
Chap.89.3 *Frittillaria Aquitan. min. flo. obsol.* f d.
9 *Frittil. alb. præcox* f d
Cum nominibus & notis nouem aliarum varietatum.
Chap.90.3 *Crocus vernus flo. lut.* f
4 *Crocus vern. flo. albo* f
5 *Crocus vern. flo. purp.* f
6 *Croc. mont. Autum.* f
7 *Croc. mon. Autum. flo. mai. albid. cærul.* f d
8 *Crocus Autum. flo. alb.* f
9 *Crocus vernus angustifol. flo. viol.* f d
10 *Crocus vernus lat. flo. flau. striis viol.* f d.
11 *Croc. vern. lat. striat. flo. dupl.* d
Chap. 91 5 *Colchicum montanum min. versic. flo.* f d
9 *Colchicum latifolium.* f d
10 *Colchicum versic. flore* f d
11 *Colchicum. flo. pleno.* f d
12 *Colchicum biflor.* f d
13 *Colchicum vernum* f d
14 *Colchicum variegat. Chiense.* f d
15 *Hermodactyli offic.* f d.
Chap.92. 4 *Ornithogalum hispan. minus.* f d
5 *Ornithogalum luteum minus.* f d
6 *Bulbus vnifolius* f d
7 *Ornithogalum maius Arabicum.* f d
8 *Ornithogalum spicatum* f d
9 *Ornithogalum Neapolitanum* f d
Chap.93.3 *Cepa Hispanica oblonga* f d
4 *Ascalonitides.* f d.
Chap.94 1 *Scilla Hispanica vulg.* f d
6 *Narcissus tertius Matthioli.* f d
Chap.95 *Porrum sectivum aut tonsile* f d
Chap.96 3 *Ampeloprason, sive Porrum syl.* f d
Chap.97. 2 *Allium syl. rubent. nul.* f d
Chap.99.2 *Scorodoprasum primum Clus.* f
3 *Scorodoprasum minus* f d
4 *Ophioscoridon.* f d
Chap.100.6 *Moly latifolium flore flauo.* f d.
7 *Moly minus flo. albo.* f d

C. { Chap. 101. 1 *Moly Narcissinis foliis prim.* f d
2. *Moly Narc. foliis secund.* f d
3 *Moly Narc. fol. tertium* f d
4 *Moly mont. latifol.* 1 *Clus.* f d
5 *Moly mont. secund. Clus.* f d
6 *Moly mont.* 3 *Clus.* f d
7 *Moly mont.* 4 *spec.* 1 *Clus.* f d
8 *Moly mont.* 4 *spec.* 2. *Clus.* f d
9 *Moly mont.* 5 *Clus.* f d. }
Chap.103.2 *Lilium rubrum,* f d
4 *Lilium cruentum bulbif.* f
5 *Lil. cruent. secundum caul. bulb. don.* f.
6 *Lilium purpureum minus* f d
Chap.105 2 *Lilium Byzantinum flo. purpur. sang.* f d
3 *Lilium Byzant. flo. dilute rubent.* f d
4 *Lilium Byzant. miniatum polyanth,* f d
Chap.106.1 *Lilium rubrum angustif.* f d
2 *Lilium rubrum præcox.* d
3 *Lilium mont. Flo. Flav. punct.* f d
4 *Lilium mont. flore flavo non punct.* f d
Chap. 110.5 *Cynosorchis minor. Pannon.* f d
Chap. 111.3 *Cynosorchis morio minor.* f d
Chap.112.4 *Tragorchis minor Batauica* f d
Chap. 113.15 *Orchis trifol. minor.* f d
16 *Orchis angustifol.* d
Chap.115.3 *Orchis palmata Pannon.* 8. *Clus.* f d
Chap.116.9 *Serapias Batrachites altera* f d
Chap.117.5 *Palma Christi max.* f d
Chap.118.2 *Nidus auis flore & caule violat.* d

LIB 2.

Chap.1. 2 *Rapum radice oblonga* f d
Chap.2.2 *Rapistrum arvense alt.* f d
Chap.3.1 *Bunias,* f
2 *Bunias syl. Lobelii* f
Chap.5.1 *Raphanus sativus* f
2 *Radicula sat. min.* f
Chap.7.3 *Lepidium annuum* f d
Chap. 9.1 *Sinapi sativum,* f
2 *Sinapi vulgare* d
3 *Sinapi sat. alt.* f
4 *Sinapi album,* f d
5 *Sinapi syl. minus* f
Chap. 10. 1 *Eruca sativa,* f
3 *Eruca syl. angustifolia* f
4 *Eruca Nasturtio cogn. tenuifol.* f
5 *Eruca marina* f
6 *Eruca aquatica,* d
Chap. 15. 2 *Erysimum alterum Italicum,* f
Chap. 17.2 *Sium maius angustifol.* d
3 *Sium vmbellatum repens* d
4 *Sium alterum Olusatri facie,* f d
6 *Sium Matthioli & Ital.* f d
Chap. 18.3 *Cardamine altera flo. pleno* f d
7 *Sium minus impatiens,* f d
8 *Cardamine pumila Bellidis fol.* f d
Chap. 19.6 *Thlaspi amarum* d
Chap.20 *Thlaspi Cand. flo, alb.* f d
Chap.21. 8 *Thlaspi petræum minus,* f d
Chap. 22.4 *Thlaspi fruticosum folio Leuc. mar.* f
5 *Thlaspi hederaceum,* f d
Chap.23.2 *Turritis maior* f d
C. { Chap. 24.1 *Draba Dioscoridis* f d
2 *Draba prima repens* f d
3 *Draba altera repens* f d
4 *Draba sive Arabis quorundam.* d }
Chap.27.3 *Erigeron tomentosum alterum,* f d

Chap.

Additions.

Chap.28.2 *Iacobæa angustifolia*, f d
3 *Iacobæa latifolia*. f d
Chap.29.5 *Cichorium spinosum*. f d
Chap.31. 3 *Chondrilla lutea*, f
4 *Chondrilla Hispanica*, f
9 *Cichorium verrucarium*. f d
Chap.32.3 *Dens leonis bulbosus*. f d
Chap.33.2 *Sonchus asperior*. f d
5 *Sonchus læuis muralis*, f d
6 *Sonchus læuis angustif*. f d
9 *Sonchus arborescens alt*. f d
10 *Sonchus syluaticus* d
Chap.34.9 *Hieracium falcatum alterum*. f d
Chap.35. 5 *Hieracium parvum Creticum*. f d
6 *Hieracium Dentis Leonis folio hirs*. f d
C. { Chap.36.1 *Pulmon. Gallica, siue aurea Lat*. f d
2 *Pulm. Gal. siue aur. angustif*. f d
3 *Hierac. hort. latif*. f d }
C. { Chap.38. 1 *Lactuca syl. mai*. f d
2 *Lactuca syl. minor fol. integris*. d
3 *Lactuca syl. fol. dissect*. f d }
Chap.42.3 *Beta rubra Romana*. f
Chap.43. 1 *Blitum maius alb*. f
3 *Blitum minus alb*. f
4 *Blitum minus rubrum*. f
Chap.44.5 *Amaranthus pannicula incurva holo-serica*. f d
Chap.45.3 *Atriplex syl. siue Polyspermon*. d
5 *Atrip. syl. vulgaris*. f d
6 *Atrip. syl. altera*. f d
7 *Atrip. syl. angustifol*. f d
8 *Atrip. baccifera*. f d
Chap.47. 1 *Atrip. syl. latif*. f
2 *Atrip syl. latif. alt*. f d
Chap.52. 1 *Cynocrambe*. f
3 *Phyllon thelygon*. d
Chap.54.2 *Scorpioides Matth*. f
Chap.55.2 *Solanum somniferum*. f d
Chap.58.2 *Mirabilia peruv. flo. alb*. f
Chap.66.3 *Hyoscyamus alb. min*. f d
4 *Hyoscyamus alb. Cret*. f d
5 *Hyoscyamus flo. rub*. f d
Chap.68 3 *Tabacum min*. f d
Chap.71.2 *Capsicum rotundior. filiq*. f d
cum 12. *aliis varietatibus*.
Chap.72. 3 *Papauer corniculatum phœnic. glab*. f d
Chap.73.3 *Papauer fimbriatum album*. f d
5 *Papaver syl*. f d
Chap. 76. 7 *Anemone latifolia duplo flauo flore* f d
C. { Chap.77.1 *Anemone latif. flo. coccin*. f d
2 *Anem. latif. flo. mag. coccin*. f d
3 *Anem. latif. Byzant* f d
4 *Anem. ten. flo. ampl. sang*. f d
5 *Anem ten. flo. Coccin*. f d
6 *Anem. ten. flo. dilute purp*. f d
7 *Anem. ten. flo. exalb*. f d
8 *Anem. ten. flo. cerul. striat*. f d
9 *Anem. ten. flo plen. coccin*. f d
10 *Anem. ten. flo. atropurp*. f d }
Chap.78. 3 *Anem. nemorum. flo. pleno alb*. f
4 *Anem. nem. flo. pleno. purp*. f d
Chap. 79. 4 *Pulsatilla Flore minore*. f d
5 *Pulsatil. flo. lut*. d
Chap.81.3. *Lapathum. syl. fol. min. acuto*. f d
Chap.82.5 *Lapathum sativum. sang*. f
Chap. 83.2 *Rha capitatum angustif*. f d
3 *Rha verum antiq*. f d.
Chap.87. 3 *Ophris bifolia bulb*. f d
Chap.88.2 *Ophioglosson abortivum*. f
Chap.89. 3 *Lunaria min. ramosa*, f d
Chap.90. 2. *Pyrola 2 tenerior. Clus*. f d
3 *Pyrola 3 fruticans Clus*. f d
4 *Pyrola 4 min. Clus*. f d
Chap.92.3 *Limonium folio sinuato*. f d
4 *Limonio congener Clus*. f d
Chap.93.2 *Tripolium vulg. min*. f
Chap.96.2 *Plantago aquat. minor stel*. f d
Chap.97.6 *Plantago panniculis sparsis*. f d
Chap.99.4 *Holosteum siue Leontopod. Cret*. f d
5 *Holost. siue Leontod. Cret. alt* f d
Chap.104 *Herba Dorea Lobel*. f
Herba Dorea altera d
Chap.105. 2 *Gentiana min. purp*. f d
G. { Chap.107. 1 *Gentianella verna maior*. f d
2 *Gent. Alpin. vern*. d
3 *Gent. fugax min*. d
4 *Gent. fugax mai*. d. }
Chap.109.2 *Speculum Veneris minus*. f d
Chap.112.2 *Calceolus Mariæ alter*. f d
Chap.114. 3 *Persicaria pusilla repens*. f d
4 *Persicaria siliquosa*. f d
Chap.116.5 *Trachelium maius petræum* f d
Chap.117.7 *Campanula Cymbalariæ fol*. d
Chap.118, 4 *Rapunculus Alp. Cornicul*. f d
5 *Rapunculus cornic. mont*. f d
6 *Rapunculus saxatilis*, f d
Chap. 119. 4 *Leucoium syl*. d
5 *Leuc. luteum flo. ampl*. d
6 *Leucoinm Flo. albo*- d
Chap.120.2 *Leucoium flo. multipl*. f d
3 *Leucoium spinosum Cret*. f d
Chap 121.2 *Leucoium marinum purp*. d
Chap.122. 2 *Viola matronalis flore. multipl*. d
3. 4 *Viola Mat. Flo. obsol. siue Leucoium Melancholicum*. f d
Chap.124.2. *Alyssum Dioscoridis* f
Chap 126.3 *Lychnis Coronaria mult*. f d
Chap.127.1 *Lychnis syl. rub. Flo*. f
Chap.127.7 *Lychnis caliculis striatis* 2. *Clus*. d
8 *Lych. syl. alb* 9. *Clus*. d
C. { Chap.128. 1 *Lychnis syl. multipl. purp*. f d
2 *Lychnis syl. alba mult*. f d
3 *Lychnis abortiva flo. mult. virid*. f d
4 *Lych. syl. lat. Clus*. f d
5 *Lych. mont. repens* f d }
Chap.129.2 *Lysimachia lutea min*. f d
3 *Lysimachia lut. Flo. glob*. f d
4 *Lysimachia lut. virgin*. f d
7 *Chamænerion*. f
8 *Chamænerion alt. angustif*. f d
9 *Lysimachia cærulea*. f
10 *Lysimachia galericulata*. f
11 *Lysimachia purp. minor*. f d
C. { Chap.131. 1 *Coniza maior*. f d
2 *Coniza min*. f d
3 *Coniza media*. d
4 *Coniza minim*. d
5 *Coniza folijs lacin*. f d
6 *Coniza palustris serratifol*. d
7 *Coniza Aust. Clus*. f d }

8 *Coniza*

8 *Coniza incana*, f.d.
9 *Coniza Alpina pilosis*. f.d.
10 *Coniza cærulea acris*, f.d.
Ch. 132. 2 *Aster Ital*. f.d.
5 *Aster Conizoides Gesn*. f.
6 *Aster lut. sup. Clus*. f.
7 *Aster lut. fol. succisæ*, f.
8 *Aster salicis folio*, f.
9 *Aster Austriacus* 5. *Clus*. f.
10 *Aster* 6. *Clus*. f.
11 *Aster* 7. *Clus*. f.
12 *Aster Virginian. fruticos*. d.
13 *Aster fruticosus minor*, d
Ch. 133. *Glastum sylvestre*, f.
Ch. 135. 2 *Sesamoides Salamanticum parvum*, d.
3 *Sesamoides parvum Matth*. d.
Ch. 139. 10 *Tithymalus characias angustifol*. f.d.
11 *Tithymalus characias serratifol*. f.d.
12 *Tithymalus dendroides ex cod. Cæs*. f.
17 *Esula exigua Tragi*, f.d.
23 *Apios radice oblonga*, f.d.
Ch. 141. 1 *Aloe vulgaris*, f.
Ch. 142. 2 *Sedum maius arborescens*, f.
5 *Sedum maius angustifol*. f.
Ch. 143. 3 *Sedum minus æstivum*, f.
4 *Sedum minus flo. amplo*. f.
5 *Sedum medium teretifolium*, f.d.
6 *Aizoon Scorpioides*, f.
7 *Sedum Portlandicum*, f.
8 *Sedum petræum*, f.
C. { Ch. 144. 1 *Sedum minus palustre*, f.d.
2 *Sedum Alpinum* 1 *Clus*. f.d.
3 *Sedum Alpinum* 3 *Clus*. f.d.
4 *Sedum Alpinum* 4 *Clus*. f.d.
5 *Sedum petræum Bupleuri folio*, f.d. }
Ch. 147. 3 *Telephium legitimum Imperati*, f.d.
Cha. 149. 1 *Halimus latifolius*, f.
2 *Halimus angustifol. procumbens*, f.
3 *Halimus vulgaris*, d.
4 *Vermicularis frutex minor*, f.
5 *Vermicularis frutex maior*, f.
Ch. 150. 5 *Chamæpitys spuria alt. Dod*. f.
6 *Chamæpitys Austr*. f.
Ch. 151. 2 *Umbelicus ven. sive Cotyl. al*. f.
3 *Umbelicus ven. min*. d.
4 *Cotyledon min. mont. alt*. f.d.
6 *Cymbalaria Italica*, f.d.
Ch. 155. 2 *Kali maius sem. cochleato*, f.
3 *Kali minus*, f.
Ch. 157. 2 *Cerinthe asperior flore flavo*, f.d.
Ch. 158. 3 *Hypericum tomentosum Lob*. f.d.
4 *Hypericum supinum glabrum*, f.d.
5 *Hypericum pulchrum Tragi*, d.
Chap. 159. 2 *Ascyron supinum palustre*, d.
Ch. 160. 2 *Androsæmum hypericoides*, f.d.
C. { Ch. 161. 1 *Coris Matth*. f.d.
2 *Coris cœrulea Monspel*. f.d. }
Ch. 162. 2 *Centaurium maius alt*. f.
Ch. 164. 5 *Antirrhinum min. repens*, f.d.
Ch. 165. 3 *Linaria purp. alt*. f.
4 *Linaria Valentina Clus*. f.
7 *Osyris flava syl*. f.d.
8 *Linaria quadrifol. supina*, d
12 *Passerina linariæ folio*, f.
13 *Passerina altera*, d.
14 *Linaria adulterina*, d.
Ch. 166. *Linum sativum*, f.
Ch. 167. 3 *Linum syl. latifol*. f.
5 *Linum syl. catharticum*, f.d.
6 *Linum syl. latifol*. 3 *Clus*. d.
7 *Linum marinum lut*. f.d.
Ch. 170. 3 *Polygonum marinum max*. d.
Ch. 171. 2 *Anthyllis Valentina Clus*. f.
3 *Polygonum serpillifolium*, f.d.
5 *Saxifraga Anglicana alsinefolia*, d.
6 *Saxifraga palustris alsinefolia*, f.d.
Cha. 172. 2 *Millegrana minima*, f.
Ch. 173. 7 *Serpillum citratum*, f.
8 *Serpillum hirsutum*, f.d.
Ch. 175. 4 *Satureia Cretica*, f.d.
Ch. 177. 5 *Hyssopus parva angust. folijs*, f.d.
Ch. 178. 2 *Gratiola angustifolia*, f.d.
Ch. 180. 4 *Stæchas summis caulic. nudis*, f.d.
Ch. 182. *Caryophyll. fig*. 4.
Ch. 183. *Caryophyll. plum. albus odorat*. f.d.
8 *Caryophyll. pumil. Alpinus*, f.d.
11 *Caryophyll. prat*. f.
13 *Caryophyll. mont. hum. lat*. f.
14 *Caryophyll. mont. alb*. f.d.
17 *Caryophyll. hum. flore cand. amæno*, f.d.
Ch. 184. 5 *Armeria prolifera, Lob*. d.
Ch. 185. 3 *Armeria prat. flo. pleno*, f.
Ch. 186. 3 *Muscipula angustifol*. f.d.
C. { Ch. 188. 1 *Saxifrag. mag. Mat*. f.d.
2 *Saxifrag. antiq. Lob*. f.d. }
Ch. 189. 4 *Ptarmica Imperati*, d.
Ch. 191. 3 *Lithospermum Anchusæ fac*. f.
4 *Anchusa degener*, f.
Ch. 192 11 *Alsine rotundifolia*, f.d.
12 *Alsine palust. serpillifol*. f.d.
13 *Alsine baccifera*, f.d.
Ch. 194. 3 *Anagallis tenuifol*. f.d.
Ch. 195. 3 *Anagallis aquat. rotundifol*. f.d
4 *Anagallis aquat*. 4. *Lob*. f.d.
5 *Cepæa*, f.d.
Ch. 196. 1 *Anthyllis lentifolia*, f.
2 *Anthyllis marina incana*, f.
3 *Anthyllis altera Italorum*, d.
Ch. 197. 5 *Veronica fruticans serpilifol*. f.d.
7 *Veron. spicata lat*. d.
8 *Veronica supina*, f.
Ch. 198. 3 *Nummularia flo. purp*. f.d.
Ch. 205. 8. *Gnaphalium Americanum*. f.
13 *Gnaphalium oblongo folio*, f.d.
14 *Gnaphalium minus lat. fol*, f.d.
Ch. 207. 1 *Stæchas citrina*, f.
2 *Amaranthus luteus latifol*. d.
Ch. 208. 3 *Ageratum folijs non serratis*, f.
4 *Ageratum floribus albis*, f.d,
Ch. 209. 4 *Tanacetum in odor. maius*, f.d.
Ch. 210. 3 *Matricaria Alpina Clus*. f.
Ch. 211. 5 *Polium lavandulæ folio*, f.d.
Ch. 213. 3 *Teucrium maius Pann*. f.d.
4 *Teucrium petræum pumil*. f.d.
Ch. 215. *Scorodonia*, f.
Ch. 219. 3 *Tragoriganum Cretense*, f.d.
Ch. 221. 1 *Pulegium regium*, f.
2 *Pulegium mas*, f.
Ch. 222. 4 *Ocimum Indicum*, f.d.
Ch. 223. 3 *Corchorus*, f.
4 *Acinos Anglicum Clusij*, d.
5 *Clinopodium Austr*. f.d.
6 *Clinopodium Alpinum*, f.d.
7 *Acinos odoratiss*. d.

Cll.

Additions.

Ch. 225. 4. *Mentha cardiaca*, f.
Mentha spicata alt. f.d.
Ch. 227. 3 *Mentastrum*, f.d.
4 *Mentastr. niv. Angl.* f.d.
5 *Mentastrum minus*, f.d.
6 *Mentastr. mont.* 1 *Clus.* f.d.
7 *Mentastrum tuberos. rad. Clus.* f.d.
Ch. 229. 3 *Melissa Fuch. flo. alb. & purp.* f.2.
4 *Herba Iudaica Lob.* f.
Ch. 231. 3 *Stachys spinosa Cretica*, f.d.
4 *Stachis Lusitan.* f.d.
5 *Sideritis scordioides*, f.
6 *Sideritis Alpina Hyssopifolia*, f.
C. Ch. 232. 1 *Sideritis vulgaris*, f.d.
2 *Sideritis angustifol.* f.d.
3 *Sideritis procumb. ramosa*, f.d.
4 *Sideritis procumbens non ramosa*, f.d
5 *Sideritis humilis lato obtuso folio*, d.
6 *Sideritis latifolia glabra*, f. d.
7. *Sideritis arvensis flo. rub.* d.
Ch. 233. *Marrubium aquat.* f.
Ch. 234. 2 *Marrubium nigrum longifol.* f.d.
Ch. 235. 4 *Lanium Pannon.* f.
5 *Galeopsis vera*, f.d.
6 *Lamium Pannon.* 3 *Clus.* f.d.
Ch. 238. 2 *Cannabis fœm.* f.
Ch. 239. 2 *Cannabis spuria alt.* f.
3 *Cannabis spuria tert.* f.
Ch. 240. 2 *Eupat. Cannabinum mas*, f.
Ch. 245. 5 *Scrophularia Ind.* f.
3 *Scrophularia flo. lut.* f.d.
Ch. 247. 2 *Scabiosa rubra Austr.* d.
8 *Scabiosa mont. alb.* f. d,
13 *Scabiosa min. Bellidis fol.* f. d.
14 *Scabiosæ flo. pall.* d.
15 *Scabiosa prolifera*, f.d.
16 *Scabiosa rubra Indica*, f.d.
17 *Scabiosa æstivalis Clus.* f.d.
Ch. 249. 7 *Iacea Austr. villosa*, f.d.
8 *Iacea capitulis hirsut.* d
Ch. 250. 4 *Stæbe Rosmarini fol.* f.d.
5 *Stæbe ex Cod. Cæsar.* f.d.
Ch. 251. 9 *Cyanus repens latifol.* f.d.
10 *Cyanus repens angustifol.* f.d.
Ch. 253. 4 *Viperaria angustifol. elatior*, f.
5 *Viper. Pannon angust.* d.
Ch. 256. 1 *Chrysanthemum segetum*, f.
3 *Chrysanth. Alp.* 1 *Clus* f.d.
4 *Chrysanth. Alp.* 2 *Clus.* f.d.
5 *Chrysanth. Cret.* f.d.
6 *Chrysanth. Bæticum Boelij*, d.
7 *Chrysanth. tenuifol. Bæt. Boel.* d.
Ch. 260. *Flos solis pyramidalis*, f.d.C.
Ch. 262. 3. *Leucanthemum Alpinum Clus.* f.d.
Ch. 264. 5 *Doronicum angustifol Austr.* f.d.
6 *Doronicum Stiriacum flo. amp.* f.d.
7 *Doronicum maximum*, f.d.
Ch. 205. 7 *Saluia absinthites*, d.
8 *Saluia Cret. pomifera & non pomif.* f.2.d.
Ch. 266. 2 *Verbascum angustis saluiæ fol.* f.
3 *Phlomos Lychnites Syr.* f.d.
Ch. 267. 3 *Colus Iouis*, f.
Ch. 268. 3 *Horminum syl. latifol.* f.d.
4 *Horminum syl. flo. alb.* f.d.
5 *Horminum Syl. flo. rub.* f.d.
Ch. 271. 3 *Blattaria flo. viridi*, f.
4 *Blattaria flo. ex vir. purpurasc.* f.
5 *Blattaria flo. albo*, f.d.
6 *Blattaria flo. amplo*, f.d.
7 *Blattaria flo. lut.* f.d.
Ch. 273. 8 *Primula veris Hesketbi*, f.
Ch. 277. 3 *Digitalis lutea*, f
4 *Digitalis ferruginea* f.
5 *Digitalis ferrug. minor*, d.
Ch. 278. *Bacchar. Monspel.* f.
Ch. 283. 3. *Buglossa syl. min.* f. d.
Ch. 284. 2 *Anchusa lutea*, f.
3 *Anchusa minor*, f.
Ch. 283. 2 *Echium vulgare*, f.
3 *Echium pullo flore*, f.d.
4 *Echium rubro flo.* f.d.
Ch. 286. 2 *Cynoglossum Cret.* f.
Cynogloss. Cret. alt. f.d.
Cynogloss. minus fol. virente, f.
Ch. 287 3 *Symphytum tuberosum*, f.
4 *Symphytum par. Borag. fac.* f.d.
Ch. 290. 2 *Tussilago Alpina*, f.d.
C. Ch. 292. 1 *Cacalia incano folio*, f.d.
2 *Cacalia folio glabro*, f.d.
Ch. 297. 2 *Potamogeiton angust.* d.
3 *Potamogeiton* 3 *Dod.* f.
4 *Potamogeiton long. acut. folijs*, f.d.
Ch. 298. 2 *Tribulus aquat. min. quer. flo.* f.d.
3 *Tribulus aquat. min Muscat. flo.* f.d.
Ch. 300. 4 *Millefolium tenuifol.* f.
5 *Millefol. palustr. galeric.* f.
6 *Myriophyllon aquat. minus*, d.
Ch. 302. 3 *Stellaria aquatica*, f.
Ch. 304. 2 *Arum Ægyptiacum*, f.
Ch. 307. 2 *Soldanella Alp. maior.* f.
3 *Soldanella Alp. minor*, f.d.
Ch. 308. 2 *Gramen Parnassi flo. dupl.* f.
Ch. 309. *Saxifraga alba petræa*, f.d.
Ch. 310. 3 *Cyclamen vernum*, f.
4 *Cyclamen vernum album*, f.d.
5 *An Cyclaminos alt.* f.
Ch. 311. 4 *Aristolochia Saracenica*, f.
5 *Pistolochia*, f.
6 *Pist. Cret. siue Virginiana*, f.d.
Ch. 314. 2 *Hedera saxatilis*, f.d.
Ch. 315. 3 *Hedera Virginiana*, d.
Ch. 317. 4 *Convolvulus argenteus*, d.
Ch. 318. 2 *Conuolvulus cær. fol. rot.* f.d.
3 *Convolv. cærul. min.* f.d.
Ch. 319. 3 *Scammonium Monspel.* f.
Ch. 321. 3 *Bryonia nigra tantum florens*, d.
Ch. 322. *Ialapium*, d.
Ch. 326. 3 *Clematis cær. flo. pleno*, f.d.
Ch. 327. *Clematis cruciata Alpina*, f.d.
Ch. 330. 2 *Clematis Daphnoides maior*, f.
Ch. 334. *Apocynum Syr. Clus.* f
Ch. 336. 2 *Periploca latifolia*, f.
Ch. 337. 6 *Polygonatum Virginianum*, d.
Ch. 342. 2 *Citrullus minor*, f.
Ch. 345. Macocks *Virginiani*, d.
Melones *aquat. edules*, *Virg.* d.
Ch. 352. 5 *Malua æstiua Hispanica*, f.d.
Ch 353. 5 *Alcea fruticosa cannab.* f.
Ch. 355. 3 *Alcea Ægypt.* f.d.
Ch. 356. 2 *Geranium colum. maius dissect. fol.* d.
3 *Geran. saxatile*, d.
Ch. 360. 2 *Geranium batrachioides alt.* f.d.
3 *Geran. Batrachioides pullo Fl.* f.d.
4 *Geran. batrach. long. rad.* f.d.

Ch. 363.

Additions.

C. Ch.363. 1 *Geranium bulb. Pen.* f.d.
2 *Geran. nodosum Plateau*, f.d.
3 *Geran. argent. Alp.* f.d.
4 *Geran. batrach. flo. var.* d.
5 *Geran. Ind. flo. maculato*, d.
Ch. 367. 13 *Ranunculus hirsut. Alp. flo alb.* f.d.
14 *Ran. mont. hirsut. purp.* f.d.
Ch. 369. 2 *Ranunculus Asiat. Flo. plen. min.* f. d.
3 *Ran. Asiat. Flo. plen. prolifero*, f.d.
5 *Ran. grum. rad. ramosus*, f.d.
6 *Ran. grum. rad. Flo. alb.* f.d.
7 *Ran. Asiat. grum. rad. Flo. flav. var.* f.d.
C. Ch.371. 1 *Ran. Cret latifol.* f.d.
2 *Ran. folio plant.* f.d.
3 *Ran. mont. Flo. min.* f.d.
4 *Ran. mont. Flo. maj.* f. d.
5 *Ran. præcox rut fol.* f.d.
6 *Ran. præcox Thalietrifol.* f.d.
7 *Ran. parvus echinat.* f. d.
Ch. 376. 4 *Aconitum lycoct. ex cod. Cæs.* f.
5 *Acon. lycoct. hirsut.* f.d.
6 *Acon. Violaceum*, f.d.
7 *Aconitum purp. Neuberg.* f.d.
8 *Aconitum max. Iudenberg.* f.d.
9 *Acon. max. nutant. coma.* f d.
Ch. 380. 5 *Pœonia promiscua*, f.
6 *Pœonia fœmina pumila*, f.
7 *Pœonia Byzant.* f.d.
Ch. 381. 5 *Pœonia Pentaphyll. alt.* f.
Ch. 382. 6 *Pentaphyllon sup. Torm. fac.* d.
9 *Pentaphyll. incanum minus rep.* d.
11 *Quinquefol. syl. minus* f.d.
12 *Quinquefol. min. flo. aur.* f. d.
13 *Pentaphyllum fragif.* f.d.
Ch. 385. 4 *Caryophyll. mont. purp.* f.d.
5 *Caryophyll. Alp. min.* f.d.
Ch. 386 *Fragaria fructu hispido*, d.
Ch. 387. 3 *Archangelica*, f.
Ch. 391. *Laserpitium*, f.
Ch. 395. 2 *Coriandrum alt. min. od.* f.
Ch. 396. 3 *Apium, siue Petrosel. Virgin.* d.
Ch. 398. Oreoselinum, f.
Ch. 399. 1 *Petroselinum Macedon. Fuch.* f.
Ch. 400. *Selinum Siȷ fol.* f.d.C.
Ch. 402. *Apium syl. siue Thesselium Plin.* f.
Ch. 403. 2 *Caucalis Apij fol.* f.
4 *Caucalis maior*, f d.
5 *Caucalis minor flosc. rub.* f.d.
6 *Caucalis nodo. echinato sem.* f.d.
Ch. 407. 2 *Pastinaca satiua atrorubens*, f.
Ch. 414. 2 *Anisum Ind. stell.* f.d.
Ch. 415. 2 *Ammi Creticum*, f.
3 *Ammi perpusil.* f.
Ch. 416. 2 *Cerefolium syl.* d.
4 *Myrrhis altera parua*, f.d.
5 *Myrrhis Æquicolor. noua*, d.
6 *Cicutaria alba*, d.
Ch. 417. 2 *Anthriscus*, f.d.
Ch. 419. 2 *Barba Capri Tragi*, d.
Ch. 421. 3 *Pimpinella sanguisorba max.* d.
Ch. 422. 1 *Saxifraga Ang. fac. seseli prat.* f.
2 *Saxifraga Pannon.* f.d.
Chap. 424. 2 *Seseli Cretic. maj.* f.
3 *Seseli montanum maius*, f.
4 *Seseli Massiliense*, f.
Ch. 425. 2 *Meum alt. Ital.* f.
2 *Ferulago*, f.
Ch. 427. 3 *Panax Asclepium*, f.d.
Ch. 435. 2 *Chelidon. mai. fol. mag. dissecto*, f.d.
Ch. 440. 6 *Valeriana Mexican.* f.
8 *Valeriana annua Clus.* f.d.
9 *Valer. Alp. lat.* f.d.
10 *Valer. Alp. angust.* f.d.
Ch. 442. 3 *Consolida reg. Flo. dup.* f.d.
4 *Consol reg. elat. Flo. plen.* f.d.
Ch. 443. 4 *Melanthium Dam. Flo. plen.* f.d.
6 *Nigella Hisp. Flo. amp.* f. d.
Ch. 447. 4 *Aquilegia var.* f.d.
5 *Aquil. Flo. inuerso rubro*, f.d.
6 *Aquil. Flo. inuerso albo*, f.d.
7 *Aquil. Flo. ros.* f.d.
8 *Aquil. degener*, f. d.
Ch. 457. 6 *Drypis*, f.
Ch. 460. 4 *Rubia spicata Cretica*, f.d.
5 *Rubia synanchica*, d.
6 *Rubia minima*, f.d.
Ch. 462. 2 *Rubia cruciata læuis*, f.d.
Ch. 463. 2 *Asperula Flo. cærul.* f.
3 *Saginæ spergula*, f.
4 *Spergula marina*, d.
5 *Spergula rubra*, d.
Ch. 565. *Filicis maris variet.* f.d.
Ch. 467. 3 *Polypodium Ind.* f.d.
Ch. 468. *Dryopteris Adu.* f.d.
Ch. 70. 3 *Hemionitis maior*, f.
4 *Hem. minor*, f.
5 *Hem. peregrina*, f.
Ch. 472. 4 *Chamæfilix mar. Angl.* f
Ch. 475. 2 *Acanthus syl. aculeatus*, f.
Ch. 478. 2 *Carduus globos. acut.* f.
3 *Carduus glob. min.* f.
5 *Carduus glob. cap. latiore*, f.
6 *Carduus eriocephalus*, f.
Ch. 481. 1 *Carlina caulescens*, f.
3 *Carlina acaulos min. Flo. purp.* f.d.
Ch. 485. 5 *Eryngium pusil. plan.* f.d.
Ch. 487. 3 *Dipsacus minor*, f.
Ch. 488. 1 *Carthamus*, f.
Ch. 490. 9. *Picnomos*, d.
C. Ch.493. 1 *Cirsium max. Asphod. rad.* f.d.
2 *Cirsium mai. alter.* f.d.
3 *Cirsium fol. non hirsut.* f.d.
4 *Cirs. mont. cap. parv.* f.d.
5 *Cirs. mont. Angl.* f.3.d.
6 *Carduus mollis fol. dissect.* f.d.
7 *Card. moll. fol. Lapathi*, f.d.
Ch. 494. 3 *Trifol. mai. Flo. alb.* f.
4 *Trifol. mai. Flor. purp.* f.
5 *Trifol. lut. lupul.* f.
6 *Trifol. lut. min.* f.
Ch. 496. 6. *Coronopus ex cod. Cæs.* f.d.
8 *Trifol. lut. sil. corn.* f.d.
Ch. 497. 1 *Lagopus max.* f.
2 *Lagop. mai. spica long.* f.d.
3 *Lagop. angust. Hisp.* f.d.
Ch. 500. *Fœnum-Græcum syl.* f.
Ch. 501. *Lotus siliqua quad.* f.d.
Ch. 502 *Medica fruct. cochleat. spin.* f.d. 4 *var.*
C. Ch.506. 1 *Trifol. siliqua lun.* f.d.
2 *Trifol. ang. Alp.* f.d.
3 *Trifol. spin. Cret.* f.d.
4 *Trifol. fragif.* f.d.
5 *Trifol. stell. hirsut.* d.
6 *Trif. stell. glab.* d.

Ch. 507.

Ch.507. *Faba vulgaris*, d.
Ch.508. 5,6,7, 8,9, *Phaseolor.pegrin. var.* 9. *cum fig*.3.d.9.
Ch.509. 4 *Lupinus mai. flo. car*. f.d.
Ch.515. 1 *Vicia*, f.
2 *Vicia max. dumet*. d.
3 *Vicia syl. flo. alb*. f.d
5 *Vicia syl. siue Cracca min*. f.d.
Ch.516. 1 *Lathyrus mai. latifol*. f.
2 *Lath. ang. flo. alb*. f.
3 *Lath. angust. flo. purp*. f.d.
4 *Lath. Ægypt*. f.d.
5 *Lath. ann. sil. Orobi*, f.d.
6 *Lath. syl, flo. lut*. f.
Ch.518. 2 *Hedysarum glycyrrhizatum*, f.
3 *Hedysar. mai. siliquis artic*. f.
4 *Securidaca min. pal. carul*. f.
5 *Secur. min. lut*. f.d.
6 *Secur. sil. plan. dent*. f.d.
7 *Hedysar. clyp*. f.
Ch.519. 2 *Astragalus syl*. f.d.
Ch.520. 3 *Astragalus Matth*. f.
4 *Astragaloides*, f.
Ch.521. 3 *Ornithopodium mai*. f.
4 *Ornithopod. min*. f.
5 *Scorpioides leguminosa*, f.d.
C. { Ch.526 1 *Orobus venet*. f.d.
2 *Orobus syl. vernus*, f.d.
3 *Orob. mont. flo. alb*. f.d.
4 *Orob. mont. angust*. f.d.
C. { Ch.527. 1 *Ochrus siue Eruilia*, f.d.
2 *Eruum sylv*. f.d.
3 *Aphaca*, f.d.
4 *Legumen mar. long. rad*. d.
Ch.528. 3 *Talictrum mai. Hispan*. d.
Ch.531. 6 *Ruta canin*. f.d.

LIB. 3.

CH.2. 6 *Rosa lut. multipl*. f.d.
8 *Rosa Cinnam. flo. simpl*. f.
Ch.3. 2 *Rosa syl. odor. flo. dup*. f.
Ch.4. 2 *Rubus repens fructu casio*, d.
Ch.5. 19 *Cistus ann. flo. mac*. f.d.
20 *Cistus folio sampsuch*. f.d.
Ch.6. 7 *Chamacistus serpillifol*. f.d.
8 *Chamacistus Fris*. f.d.
Ch.7. 15 *Cistus Ledon folijs Rosm*. f.d.
Ch.12. *Glycyrrhiza vulg*. f.
Ch.17. *Orobanches triplex var*. f.3.
Ch.20. 5 *Genista spinosa humilis*, d.
Ch.25. 2 *Tragacantha min. icon accur*.
3 *Poterion Lob*. f.
Ch.26. 1 *Acacia Diosc*. f.
Ch.27. 2 *Lycium Hisp*. f.
Ch.28. 1 *Rhamnus flo. alb*. f.
Rhamn. alt. flo. purp. f.d.
2 *Rhamnus* 2 *Clus*. f.d.
3 *Rhamnus* 3 *Clus*. d.
Ch.30. 1 *Rhamnus solut*. f.
2 *Rhamn. sol. min*. f.d.
3 *Rham. sol. pumil*. d.
Ch.34. *Ilicis ramus flor*. f.
Ch.35. *Cerri minoris ram. cum flo*. f.
Ch.37. 2 *Galla maior alt*. f.d.
Ch.40. 2 *Picea pumila*, f.
Ch.42. 8 *Pinaster Austr*. f.d.
9 *Pinaster mar. min*. f.d.
Ch.43. 2 *Abies mas*, f.
Abietis ramus cum Iulis, f.

Ch.47. *Taxus glandif. & baccif*. d.
Taxus tant. flor. d.
Ch.48. 3 *Iuniperus Alp. min*. f.d.
Ch.49. 3 *Cedrus lycia alt*. f.d.
Ch.50. 3 *Sabina bacc. alt*. f.d.
Ch.52. 3 *Erica mai. flo. alb*. d.
9 *Erica baccif. procumbens*, f.
10 *Erica baccif. ten*. d.
11 *Erica pum*. 3. *Dod*. f.d.
12 *Erica ternis per interualla ramis*, f.d.
13 *Erica peregrin. Lob*. f.d.
14 *Erica coris folio* 7 *Clus*. f.d.
15 *Erica Coris fol*. 9 *Clus*. f.d.
Ch.54. 2 *Vitex lat. serat. folio*, f.d.
Ch.55. 8 *Salix hum. repens*, f.
Ch.61. 3 *Syringa Arabica*, f.d.
Ch.71. 2 *Myrtus Batica lat*. f.
3 *Myrt. exot*. f.
4 *Myrt. fruct. alb*. f.
5 *Myrtus min*. f.
6 *Myrt. Batica syl*. f.d.
Ch.73. 6 *Vitis Idaa fol. subrotund. mai*. d.
Ch.77. 2 *Sambucus fructu alb*. f.
Ch.89. *Auellana pum. Byz*. f.d.
Ch.91. 3 *Castanea Peru. fruct*. f.d.
Ch.94. 5 *Persica flo. pleno*, d.
Ch.98. 2 *Mespilus sativa alt*. f.d.
4 *Chamamespilus*, f.
Ch.113. 2 *Alnus hirsut*. f.d.
Ch.116. 1 *Vlmus vulgat. fol. lato scabro*, d.
2 *Vlmus min. fol. angusto scabro*, f.d.
3 *Vlmus fol. latiss. scab*. f.d.
4 *Vlmus fol. glab*. d.
Ch.118. 1 *Acer mai*. f.
Ch.119. 5 *Populus alba folijs minor*, f.
Ch.122. 2 *Zizypha Cappadocica*, f.
Ch.124. *Guaiacum Patau. angust*. d.
Ch.133. 2 *Chamaficus*, f.
Ch.136 *Musa fructus exact. icon*, f.d.
Ch.145. 3 *Balsamum Alp*. f.d.
Ch.146. 2 *Molle arboris adulta ramus*, f.
Ch.153. 5 *Piper caudatum*, f.
C. { Ch.159. *Fructus Indici & exotic. quorum fig. ad*.26. *descr*.35.
Ch.162. 6 *Muscus Pyxidatus*, f.
12 *Musc. cleuat. fol. Cypr*. d.
14 *Musc. parv. stell*. f.d.
Ch.164. 3 *Lichen mar. rotund*. f.d.
4 *Quernus mar. var*. f.d.
5 *Quern. mar. secund*. f.d.
6 *Quern. mar. tertia*, f.d.
7 *Quern. mar. quarta*, f.d.
8 *Alga*, f.d.
9 *Fucus phasganoides & polys*. f.d.
10 *Fucus spong. nod*. f.d.
11 *Conferua*, f.d.
Ch.165. 7 *Fucus ferul*. f.d.
8 *Fucus tenuifol. alt*. f.d.
9 *Muscus mar. Clus*. f.d.
10 *Muscus mar. tertius Dod*. f.d.
11 *Abies mar. Belg. Clus*. f.d.
Ch.166. 5 *Coraloides alb*. f.d.
6 *Coral. rub*. f.d.
8 *Spong. infundibuli forma*, f.d.
9 *Spongia ramosa*, f.d.
Ch.167. *Fungorum fig*. 14.
The Appendix containes fig. 46. *descrip*. 72.

THE

THE FIRST BOOKE OF THE HISTORIE OF PLANTS:

Containing Graßes, Rushes, Reeds, Corne, Flags, and Bulbous, or Onion-rooted Plants.

N this Historie of Plants it would be tedious to vse by way of introduction, any curious discourse vpon the generall diuision of Plants, contained in Latine vnder *Arbor, Frutex, Suffrutex, Herba*: or to speake of the differing names of their seuerall parts, more in Latine than our vulgar tongue can well expresse. Or to go about to teach thee, or rather to beguile thee by the smell or taste, to guesse at the temperature of Plants: when as all and euery of these in their place shall haue their true face and note, whereby thou maist both know and vse them.

In three bookes therefore, as in three gardens, all our Plants are bestowed; sorted as neere as might be in kindred & neighbourhood.

The first booke hath Grasses, Rushes, Corne, Reeds, Flags, Bulbous or Onion-rooted Plants.

The second, most sorts of herbes vsed for meate, medicine, or sweet smelling.

The third hath Trees, Shrubs, Bushes, Fruit-bearing Plants, Rosins, Gummes, Roses, Heathes, Mosses, Mushroms, Corall, and their seuerall kindes.

Each booke hath chapters, as for each herbe a bed: and euery Plant presents thee with the Latine and English name in the title, placed ouer the picture of the Plant.

Then followes the kindes, description, place, time, names, natures, and vertues, agreeing with the best receiued opinions.

Last of all thou hast a generall Index, as well in Latine as English, with a carefull supply likewise of an *Index bilinguis*, of barbarous names.

And thus hauing giuen thee a generall view of this garden, now with our friendly labours wee will accompany thee, and leade thee through a Grasse-plot, little or nothing of many Herbarists heretofore touched; and begin with the most common or best knowne Grasse, which is called in Latine, *Gramen pratense*: and then by little and little conduct thee through most pleasant gardens and other delightfull places, where any herbe or plant may be found fit for meate or medicine.

CHAP. 1. *Of Medow-Grasse.*

THere be sundry and infinite kindes of Grasses not mentioned by the Antients, either as vnnecessarie to be set downe, or vnknowne to them: onely they make mention of some few, whose wants we meane to supply, in such as haue come to our knowledge, referring the rest to the curious searcher of Simples.

¶ *The Description.*

1 COmmon Medow Grasse hath very small tufts or roots, with thicke hairy threds depending vpon the highest turfe, matting and creeping on the ground with a most thicke and apparant shew of wheaten leaues, lifting vp long thinne ioynted and light stalks, a foot or a cubit high, growing small and sharpe at the top, with a loose eare hanging downward, like the tuft or top of the common Reed.

2 Small

2 Small medow Graſſe differeth from the former in varietie of the ſoile; for as the firſt kind groweth in medowes, ſo doth this ſmall graſſe clothe the hilly and more dry grounds vntilled, and barren by nature; a Graſſe more fit for ſheepe than for greater cattell. And becauſe the kindes of Graſſe do differ apparantly in root, tuft, ſtalke, leafe, ſheath, eare, or creſt, we may aſſure our ſelues that they are endowed with ſeuerall vertues, formed by the Creator for the vſe of man, although they haue been by a common negligence hidden and vnknowne. And therefore in this our Labor we haue placed each of them in their ſeuerall bed, where the diligent ſearcher of Nature may, if ſo he pleaſe, place his learned obſeruations.

1 *Gramen pratenſe.*
Medow Graſſe.

2 *Gramen pratenſe minus.*
Small Medow-graſſe.

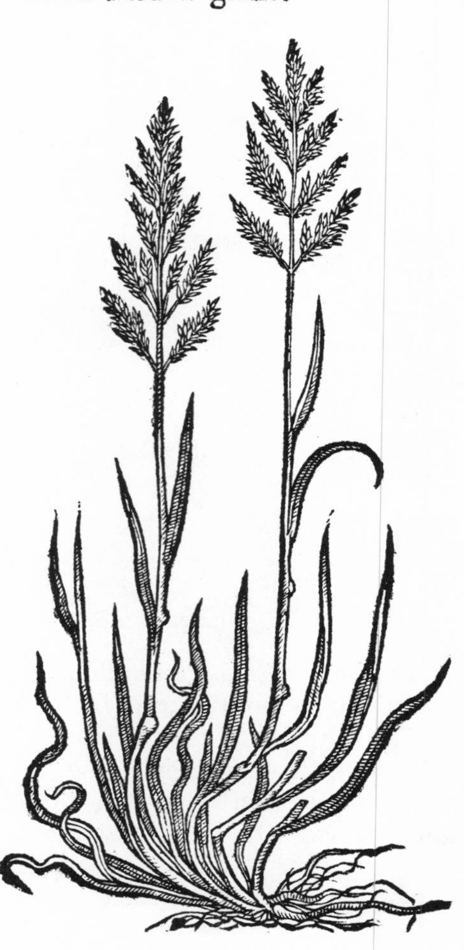

¶ *The Place.*

Common Medow-graſſe groweth of it ſelfe vnſet or vnſowen, euery where, but the ſmall medow graſſe for the moſt part groweth vpon dry and barren grounds, as partly wee haue touched in the deſcription.

¶ *The Time.*

Concerning the time when Graſſe ſpringeth and ſeedeth, I ſuppoſe there is none ſo ſimple but knoweth it, and that it continueth all the whole yeare, ſeeding in Iune and Iuly. Neither needeth it any propagation or replanting by ſeed or otherwiſe; no not ſo much as the watery Graſſes, but that they recouer themſelues againe, although they haue beene drowned in water all the Winter long, as may appeare in the wilde fennes in Lincolnſhire and ſuch like places.

¶ *The Names.*

Graſſe is called in Greeke, ἄγρωστις: in Latine, *Gramen,* as it is thought, *à gradiendo, quod geniculatis internodijs ſerpat crebreque nouas ſpargat radices:* for it groweth, goeth, or ſpreadeth it ſelfe vnſet or vnſowen, naturally ouer all fields or grounds, cloathing them with a faire and perfect greene. It is yearely mowed, in ſome places twice, and in ſome rare places thrice; then is it dried and withered by the heate of the Sunne, with often turning it; and then is it called *Fœnum, neſcio an à fœnore aut fœtu.* In Engliſh, Hay: in French, *Le herbe du praiz.*

¶ *The Nature.*

The roots and ſeeds of Graſſe are of more vſe in phyſicke than the herbe, and are accounted of all Writers moderately to open obſtructions, and prouoke vrine.

¶ The

¶ *The Vertues.*

The decoction of Graſſe with the roots of Parſley drunke, helpeth the diſſurie, and prouoketh vrine. A

The roots of Graſſe, according to *Galen*, doe glew and conſolidate together new and bleeding wounds. B

The iuyce of Graſſe mixed with honey and the pouder of Sothernwood taken in drinke, killeth wormes in children; but if the childe be young, or tender of nature, it ſhall ſuffice to mixe the iuyce of Graſſe, and the gall of an Oxe or Bull together, and therewith anoint the childes belly, and lay a clout wet therein vpon the nauell. C

Fernelius ſaith, that graſſe doth helpe the obſtructions of the liuer, reines and kidnies and the inflammation of the raines called *Nephritis*. D

Hay ſodden in water till it be tender, and applied hot to the chaps of beaſts that be chap-fallen, through long ſtanding in pound or ſtable without meate, is a preſent remedie. E

Chap. 2. *Of Red Dwarfe-Graſſe.*

¶ *The Deſcription.*

1 Dwarfe Graſſe is one of the leaſt of Graſſes. The root conſiſts of many little bulbes, couered with a reddiſh filme or ſkinne, with very many ſmal hairy and white ſtrings: the tuft or eare is of a reddiſh colour, and not much differing from the graſſe called *Iſchæmon*, though the eare be ſofter, broader, and more beautifull.

† 1 *Gramen minimum rubrum, ſiue Xerampelinum.*
Red Dwarfe-graſſe.

2 *Gramen minimum album.*
White Dwarfe-graſſe.

† 2 This kinde of Graſſe hath ſmall hairy roots; the leaues are ſmall and ſhort, as alſo the ſtalke, which on the top thereof beares a pannicle not much vnlike the ſmall medow Graſſe, but leſſe: the colour thereof is ſometimes white, and otherwhiles reddiſh; whence ſome haue giuen two figures, which I thinking needleſſe, haue onely retained the later, and for the former giuen the figure of another Graſſe, intended by our Author to be comprehended in this Chapter.

3 Small hard Grasse hath small roots compact of little strings or threds, from which come forth many soure rushy leaues of the length of an inch and a halfe: the tuft or eare is compact of many pannicles or very little eares, which to your feeling are very hard or harsh. This Grasse is vnpleasant, and no wholesome food for cattell.

4 Rush-grasse is a small plant some handfull high, hauing many small rushy leaues tough and pliant, as are the common Rushes: whereupon do grow small scaly or chaffie huskes, in stead of floures, like those of Rushes, but smaller. The root is threddy like the former. ‡ There is a varietie of this to be found in bogs, with the seeds bigger, and the leaues and whole plant lesser. ‡

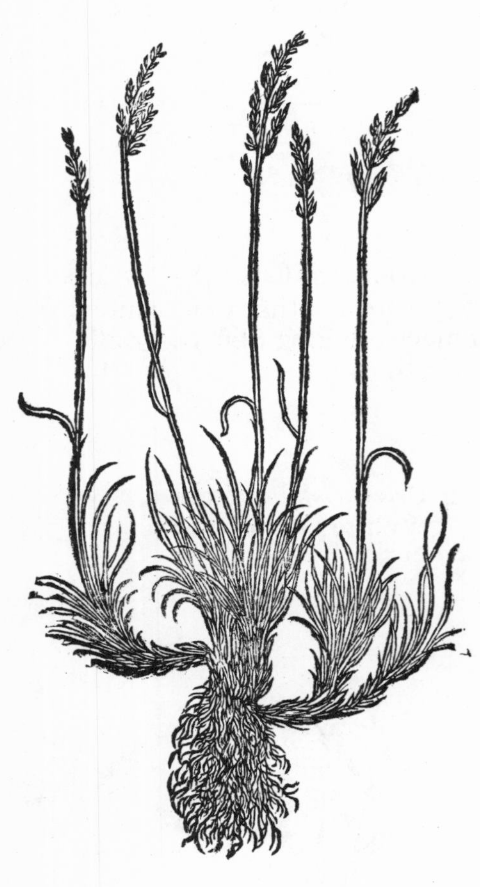

3 *Gramen minus duriusculum.*
Small hard Grasse.

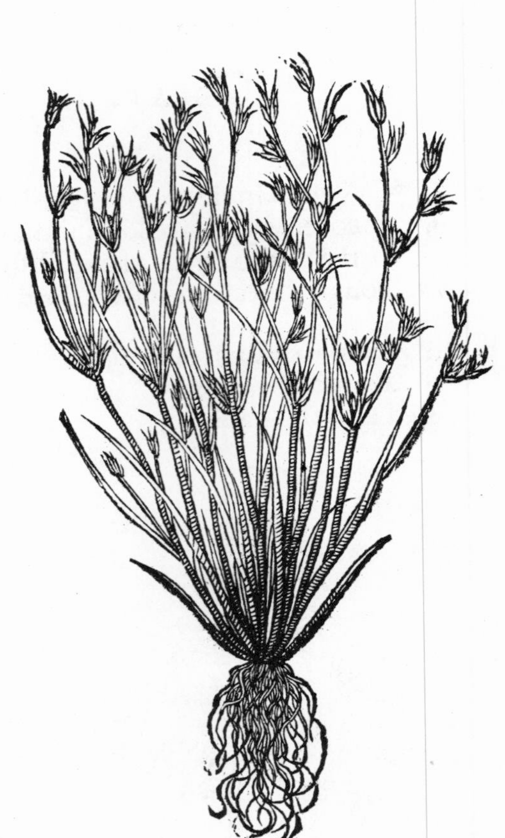

4 *Gramen junceum.*
Rush-grasse, or Toad-grasse.

¶ *The Place.*

The Dwarfe-grasse doth grow on heathy rough and dry barren grounds in most places of England. ‡ That which I haue giuen you I haue not as yet obserued growing in any part of England. ‡

The white Dwarfe-grasse is not so common as the former, yet doth it grow very plentifully among the Hop gardens in Essex and many other places.

Small Hard-grasse groweth in moist fresh marishes, and such like places.

Rush-grasse groweth in salt marishes neere vnto the sea, where the marishes haue beene ouerflowne with salt water. ‡ It also groweth in many wet woods, lanes, and such places, as in the lane going by Totenham Court towards Hampstead. The lesser varietie hereof growes on the bogges vpon Hampstead heath. ‡

¶ *The Time.*

These kindes of Grasses do grow, floure, and flourish when the common Medow grasse doth.

¶ *The Names.*

It sufficeth what hath beene said of the names in the description, as well in English as Latine; onely that some haue deemed White Dwarfe-grasse to be called *Xerampelinum.*

Rush-grasse hath been taken for *Holosteum Matthioli.*

‡ ¶ *The Names in particular.*

1 This I here giue you in the first place is the *Gramen minimum Xerampelinum* of *Lobel:* it is the *Gramen*

Gramen of *Matthiolus*, and *Gramen bulboſum* of *Daleſchampius*. Our Author did not vnderſtand what *Xerampelinus* ſignified, when as he ſaid the white Dwarfe-graſſe was ſo termed; for the word imports red, or murrey, ſuch a colour as the withered leaues of Vines are of. 2. *Tabern* calls this, *Gramen panniculatum minus*. 3. *Lobel* calls this, *Exile Gramen durius*. 4. This by *Matthiolus* was called *Holoſtium*: by *Thalius*, *Gramen epigonatocaulon*: by *Tabernamontanus*, *Gra. Bufonium*, that is, Toad-graſſe. ‡

¶ *The Nature and Vertues.*

Theſe kindes of Graſſes doe agree as it is thought with the common Medow-graſſe, in nature and vertues, notwithſtanding they haue not beene vſed in phyſicke as yet, that I can reade of.

† The firſt figure was onely a varietie of the ſecond, according to *Bauhinus*; yet in my iudgement it was the ſame with the third, which is *Gramen minus duriuſculum*.

CHAP. 3. *Of Corne-Graſſe.*

¶ *The Deſcription.*

1 Corne-graſſe hath many graſſie leaues reſembling thoſe of Rie, or rather Otes, amongſt the which commeth vp ſlender benty ſtalkes, kneed or ioynted like thoſe of corne, whereupon groweth a faire tuft or pannicle not much vnlike to the feather-like tuft of common Reed, but rounder compact together like vnto Millet. The root is threddy like thoſe of Otes.

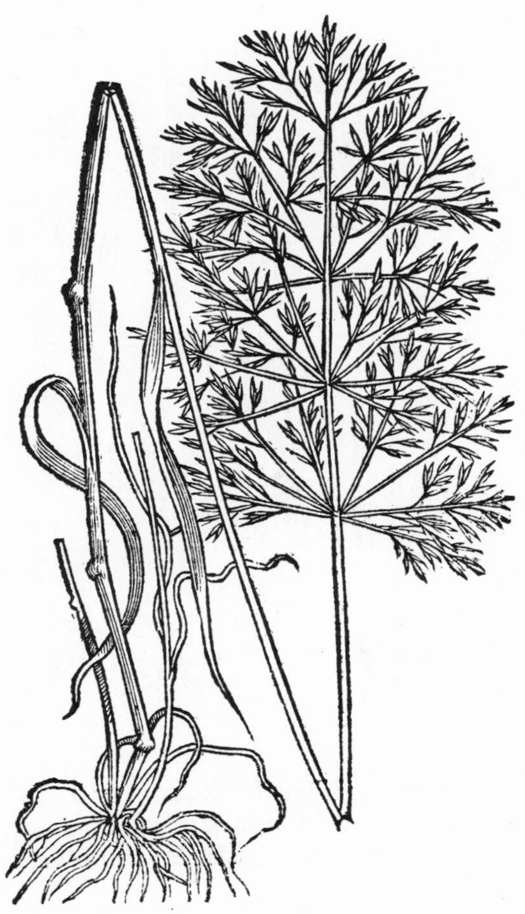

1 *Gramen ſegetale.*
Corne-graſſe.

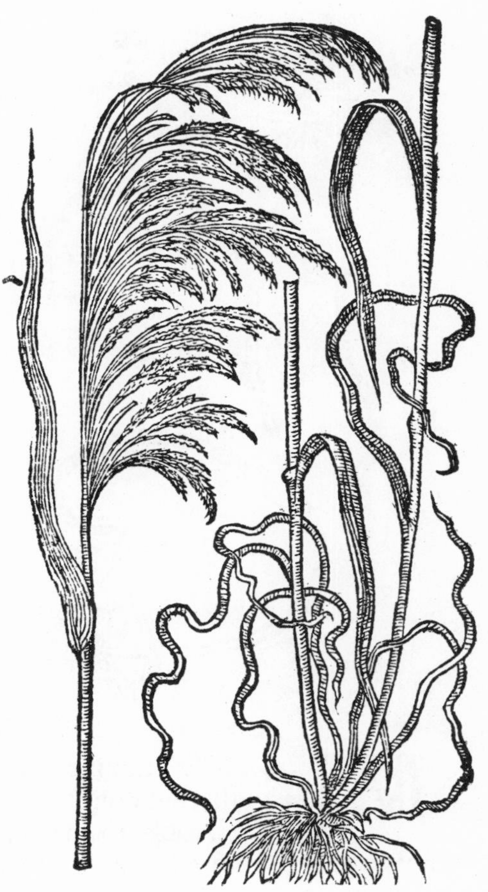

2 *Gramen harundinaceum.*
Reed-graſſe, or Bent.

2 Reed-graſſe hath many thin graſſie leaues like the former: the buſhy top, with his long feather-like pannicles do reſemble the common Reed, which is lightly ſhaken with the winde, branched vpon a long ſlender reeden ſtalke, kneed or ioynted like corne. The root is ſmall and fibrous.

¶ *The Place and Time.*

Theſe kindes of Graſſes grow for the moſt part neere hedges, & in fallow fields in moſt places. Their time of ſpringing, flouring, and fading may be referred to the common Medow-graſſe.

The Names.

† The firſt is called in Engliſh, Corne-graſſe. *Lobelius* calls this, *Segetum gramen pannicula ſpecioſa latiore*: others terme it *Gramen ſegetale*, for that it vſually groweth among corne; the which I haue not as yet ſeene.

The ſecond is called in Engliſh, Reed-graſſe: of *Lobelius* in Latine, *Gramen agrorum latiore, arundinacea, & comoſa pannicula*, for that his tuft or pannicles do reſemble the Reed: and *Spica venti agrorum*, by reaſon of his feather-top, which is eaſily ſhaken with the wind. ‡ Some in Engliſh, much agreeable to the Latine name, call theſe, Windle-ſtrawes. Now I take this laſt to be the Graſſe with which we in London do vſually adorne our chimneys in Sommer time: and we commonly call the bundle of it handſomely made vp for our vſe, by the name of Bents. ‡

¶ *The Temperature and Vertues.*

Theſe Graſſes are thought to agree with common Graſſe, as well in temperature as vertues, although not vſed in phyſicke.

CHAP. 4. *Of Millet Graſſe.*

1 *Gramen Miliaceum.*
Millet Graſſe.

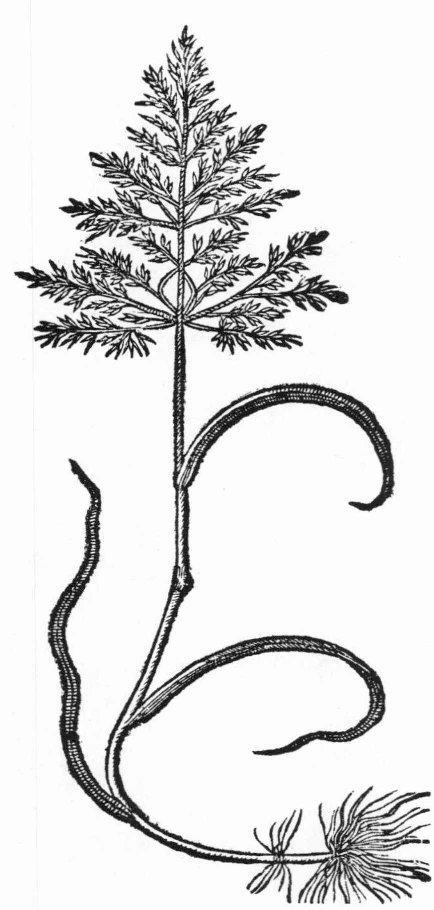

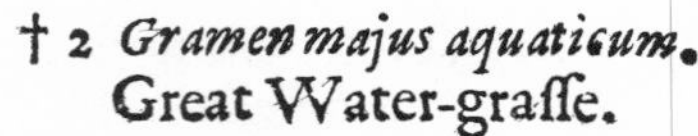

† 2 *Gramen majus aquaticum.*
Great Water-graſſe.

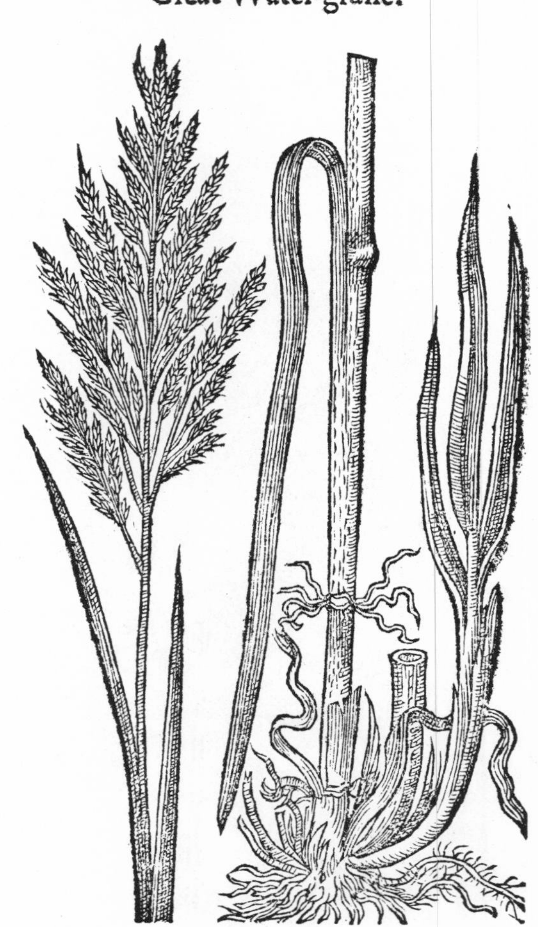

¶ *The Deſcription.*

1 Millet Graſſe is but a ſlender Graſſe, bearing a tuft or eare like vnto the common Medow-graſſe, but conſiſting of ſmall ſeeds or chaffie heads like to *Milium*, or Millet, whereof it tooke the name. The ſtalke or leaues do reſemble the Bent, wherewith countrey people do trimme their houſes.

2 The great Water-graſſe in root, leafe, tuft, and reeden ſtalke doth very well reſemble the Graſſe called in Latine, *Gramen ſulcatum*, or *Pictum*; and by our Engliſh women, Lady-laces, becauſe it is ſtript or furrowed with white and greene ſtreakes like ſilke laces; but yet differs from that, that this Water graſſe doth get vnto it ſelfe ſome new roots from the middle of the ſtalks and ioynts, which the other doth not. ‡ This is a large Graſſe, hauing ſtalkes almoſt as thicke as ones little finger, with the leaues anſwerable vnto them, and a little rougiſh: the tuft is ſomewhat like a reed, but leſſe, and whitiſh coloured. ‡

¶ The Place, Names, Nature, and Vertues.

The former growes in medowes, and about hedges, and the later is to be found in moſt fenny and watery places, and haue their vertues and natures common with the other Graſſes, for any thing that wee can finde in writing. The reaſon of their names may be gathered out of the deſcription.

† This which I giue you in the ſecond place is not of the ſame plant that was figured in the former edition; for that picture was of *Gramen aquaticum harundinaceum panniculatum* of *Taber.* which hath a running root and large ſpecious pannicle like to a Reed, of a browne colour. But it is moſt apparant that our Authour meant this, and framed his deſcription by looking vpon this figure, eſpecially the later part thereof. The true figure of this was in the ſecond place in the next Chapter.

CHAP. 5.

Of Darnell Graſſe.

¶ The Deſcription.

1 DArnell Graſſe, or *Gramen Sorghinum*, as *Lobel* hath very properly termed it, hath a browniſh ſtalke thicke and knotty, ſet with long ſharpe leaues like vnto the common Dogs Graſſe: at the top whereof groweth a tuft or eare of a grayiſh colour, ſomwhat like *Sorghum*, whereof it tooke his name.

1 *Gramen Sorghinum.*
Darnell Graſſe.

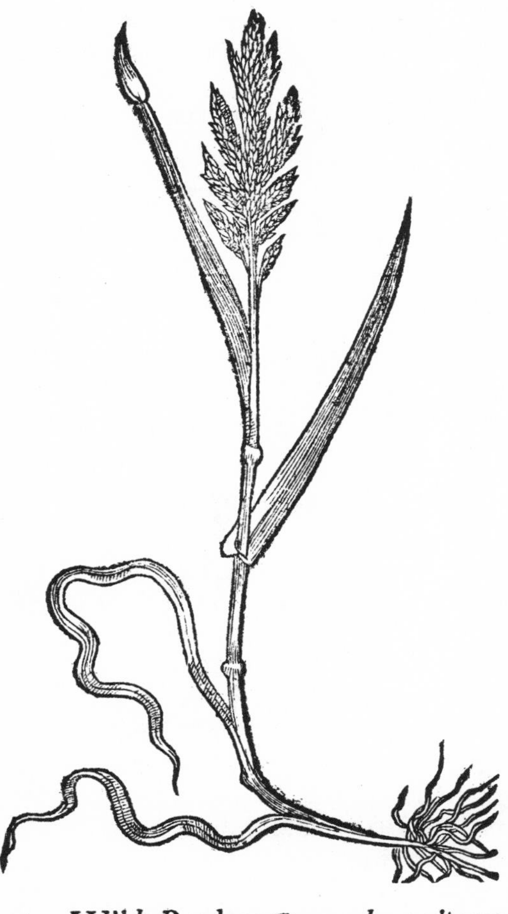

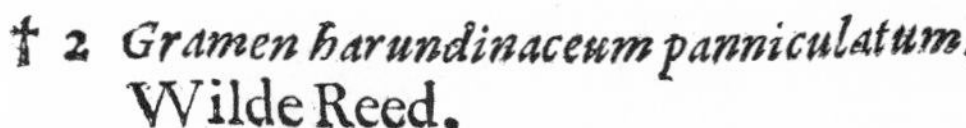

† 2 *Gramen harundinaceum panniculatum.*
Wilde Reed.

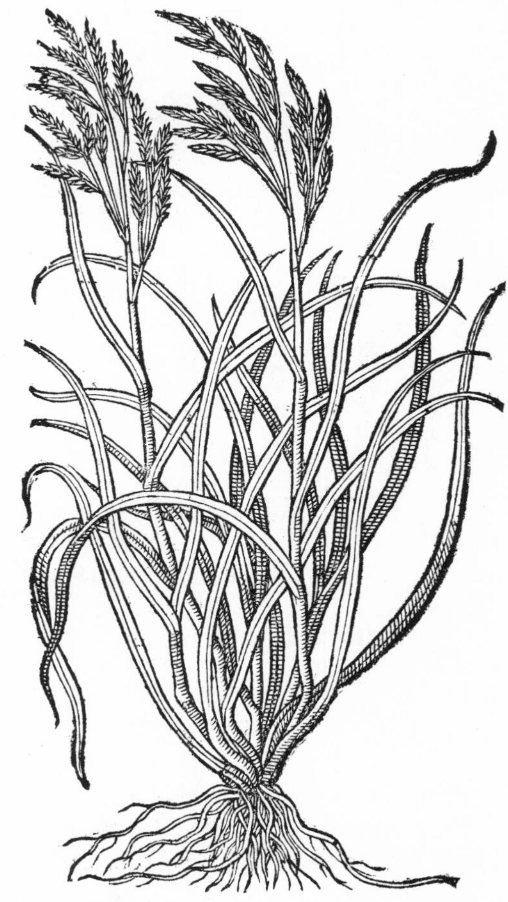

2 Wilde Reed, or *Gramen harundinaceum panniculatum*, called alſo *Calamogroſtis*, is far bigger than Couch graſſe, or Dogs graſſe, and in ſtalkes and leaues more rough, rugged, and cutting. It is bad food for cattell, though they want, or be very hungry; and deadly to Sheepe, becauſe that, as the Husbandman ſaith, it is a cauſe of leanneſſe in them, thirſt, and conſumption: it cutteth their

 tongue,

‡ 3 *Gramen arundinaceum minus.*
The lesser Reed-Grasse.

tongue, straitneth the gullet or throat, and draweth downe bloud into the stomacke or maw; whereof ensueth inflammation, and death for the most part. And not onely this *Calamogrostis* is hurtfull, but also all other kindes of shearing leaued reeds, flagges, sedge, or the like, which haue as it were edges; and cut on both sides like kniues as well mens fingers, as cattels mouthes. This herbe is in a meane between reed & grasse. The root is white, creeping downwards very deepe. The spike or eare is like vnto the reed, being soft and cottony, somewhat resembling Pannicke.

‡ 3 This in root, stalkes, and leaues is like to the last described, but that they are lesser: the top or head is a long single spike or eare, not seuered or parted into many eares like the top of the precedent, and by this and the magnitude it may chiefely be distinguished from it. This was in the twelfth place in the sixteenth chapter, vnder the title of *Gramen harundinaceum minus*: and the *Calamogrostis* but now described, was also there againe in the eleuenth place. ‡

¶ *The Place.*

The first growes in fields and orchards almost euery where; the other grow in fenny waterish places.

¶ *The Names.*

2 This in Lincolneshire is called Sheere-grasse, or Henne: in other parts of England, wild Reed: in Latine, *Calamogrostis* out of the Greeke, καλαμόγρωστις. As for their natures and vertues we doe not finde any great vse of them worth the setting downe.

† The figure that was in the second place was of *Gramen maius aquaticum*, being the second of the precedent Chapter. The true figure of this was page 21. vnder the title of *Gramen harundinaceum maius*. The third being there also, as I haue touched in the description.

Chap. 6. *Of Feather-top, Ferne, and Wood-grasse.*

¶ *The Description.*

‡ 1 This might fitly haue beene put to those mentioned in the foregoing chapter, but that our Author determined it for this, as may appeare by the mention made of it in the names, as also by the description hereof, framed from the figure we here giue you. ‡ This Grasse is garnished with chaffie and downie tufts, set vpon a long benty stalke of two cubits high or somewhat more, naked without any blades or leaues, for the most part. His root is tough and hard. ‡ The top is commonly of a red or murrey colour, and the leaues soft and downy. ‡

‡ 2 This, whose figure was formerly by our Author giuen for the last described, though verie much different from it, is a very pretty and elegant grasse: it in roots and leaues is not vnlike to the vsuall medow Grasse; the stalke riseth to the height of a foot, and at the top thereof it beareth a beautifull pannicle, (whence the French and Spanish Nations call it *Amourettes*, that is, the Louely Grasse.) This head consists of many little eares, shaped much like those of the ordinarie Quaking Grasse, longer and flatter, being composed of more scales, so that each of them somewhat resembles the leafe of a small Ferne, whence I haue called it Ferne-Grasse. These tops when they are ripe are white, and are gathered where they grow naturally to beautifie garlands. ‡

3 Wood-grasse hath many small and threddy roots, compact together in manner of a tuft; from which spring immediately out of the earth many grassy leaues, among the which are sundrie benty

‡ 1 *Gramen tomentosum arundinaceum.*
Feather-top, or Woolly Reed-grasse.

2 *Gramen panniculatum elegasis.*
Ferne-grasse.

3 *Gramen sylvaticum majus.*
The greater Wood-grasse.

benty stalkes, naked and without leaues or blades like the former, bearing at the top a soft spikie tuft or eare much like vnto a Fox-taile, of a brownish colour.

‡ 4 This in leaues, stalks, roots, manner and place of growing is like the last described: the onely difference betweene them is, That this hath much lesse, yet sharper or rougher eares or tufts. The figure and description of this was formerly giuen by our Author in the sixteenth chapter, and ninth place, vnder the title of *Gramen sylvaticum minus*. But because the difference between the last described and this is so small, we haue spared the figure, to make roome for otners more different and note-worthy.

¶ *The Time and Place.*

1 This kinde of Grasse growes in fertil fields and pastures.

2 The second growes in diuers places of Spaine and France.

The other two grow in Woods.

¶ *The Names.*

1 *Lobelius* in Latine calls this *Gramen tomentosum & Acerosum*. Some haue taken it for the second kinde of *Calamogrostis*; but most commonly it

it is called *Gramen plumosum* : and in English, a Bent, or Feather-top Grasse.

2 *Gramen panniculatum* is called by some *Heragrostis* in Greeke. *Lobel* calls this *Gramen panniculosum phalaroides*. And it is named in the *Hist. Lugd. Gramen filiceum, seu polyanthos* : that is, Ferne, or many-floured Grasse. ‡

3 *Gramen syluaticum*, or as it pleaseth others, *Gramen nemorosum*, is called in our tongue, wood Grasse, or shadow Grasse.

Chap. 7. *Of great Fox-taile Grasse.*

¶ *The Description.*

1 The great Fox-taile Grasse hath many threddy roots like the common Medow grasse; and the stalke riseth immediatly from the root, in fashion like vnto Barley, with two or three leaues or blades like Otes; but is nothing rough in handling, but soft and downie, and somewhat hoarie, bearing one eare or tuft on the top, and neuer more; fashioned like a Fox-taile, whereof it tooke his name. At the approch of Winter it dieth, and recouereth it selfe the next yeare by falling of his seed.

1 *Gramen Alopecuroides majus.*
Great Fox-taile Grasse.

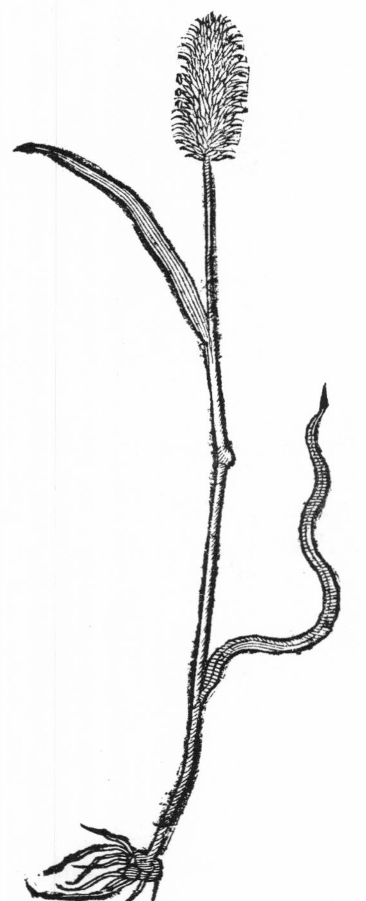

† 2 *Gramen Alopecuroides minus.*
Small Fox-taile Grasse.

2 The lesser Fox-taile Grasse hath a tuffe and hard root compact of many small strings, yeelding a strawie stalke like the former, though somwhat lesser, with the like top or crest, but of a whitish colour.

3 Great bastard Fox-taile Grasse hath a strawie stalke or stemme, which riseth to the height of a cubit and an halfe, hauing a small root consisting of many fibres. His leafe is small and grassie, and hath on his top one tuft or spike, or eare of a hard chaffie substance, some three inches long, composed of longish seeds, each hauing a little beard or awne.

4 Small bastard Fox-taile Grasse doth resemble the former, sauing that this kinde doth not send

send forth such large stalkes and eares as the other, but smaller, and not so close packed together, neither hauing so long beards or awnes.

† 3 *Gramen Alopecurinum majus.*
Great bastard Fox-taile Grasse.

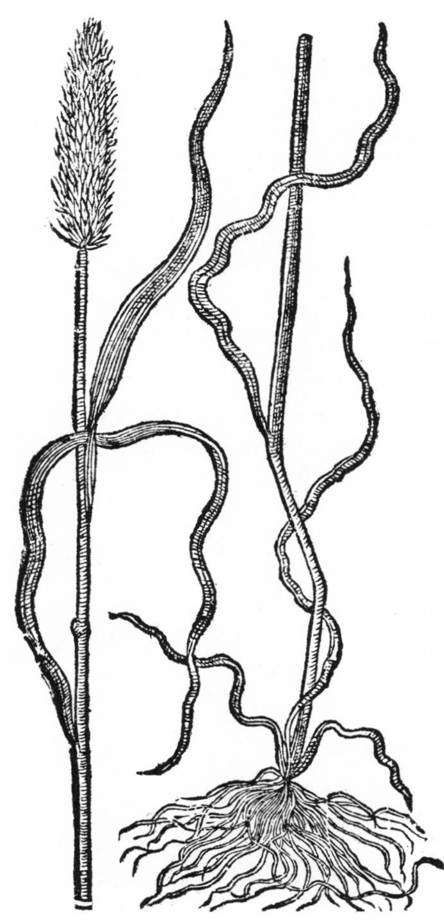

4 *Gramen Alopecurinum minus.*
Small bastard Fox-taile Grasse.

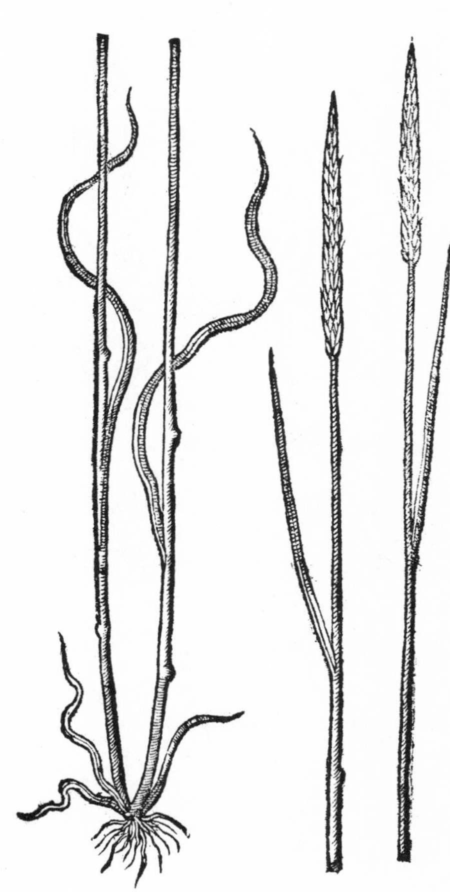

¶ *The Place and Time.*

These wilde bastard Fox-taile Grasses doe grow in the moist furrowes of fertile fields, towards the later end of Sommer.

¶ *The Names.*

‡ The first by *Lobel* and *Tabern.* is called *Gramen phalaroides.* The other *Lobel* calleth 2 *Gramen Alopecuroides.* 3. *minus.* 4. *minus alterum.*

Chap. 8. *Of Great Cats-taile Grasse.*

¶ *The Description.*

1 Great Cats-taile Grasse hath very small roots, compact of many small skins or threds, which may easily be taken from the whole root. The stalke riseth vp in the middest, and is somewhat like vnto wilde Barley, kneed and ioynted like corne, of a foot high or thereabout; bearing at the top a handsome round close compact eare resembling the Cats-taile.

2 The small Cats-taile grasse is like vnto the other, differing chiefely in that it is lesser than it. The root is thicke and cloued like those of Rush Onions, or Ciues, with many small strings or hairie threads annexed vnto it.

‡ 3 There is another that growes plentifully in many places about London, the which may fitly be referred to this Classis. The root thereof is a little bulbe, from whence ariseth a stalke some two foot or better high, set at each ioynt with long grassie leaues: the spike or eare is commonly

Gramen Typhinum minus.
Small Cats-taile Grasse.

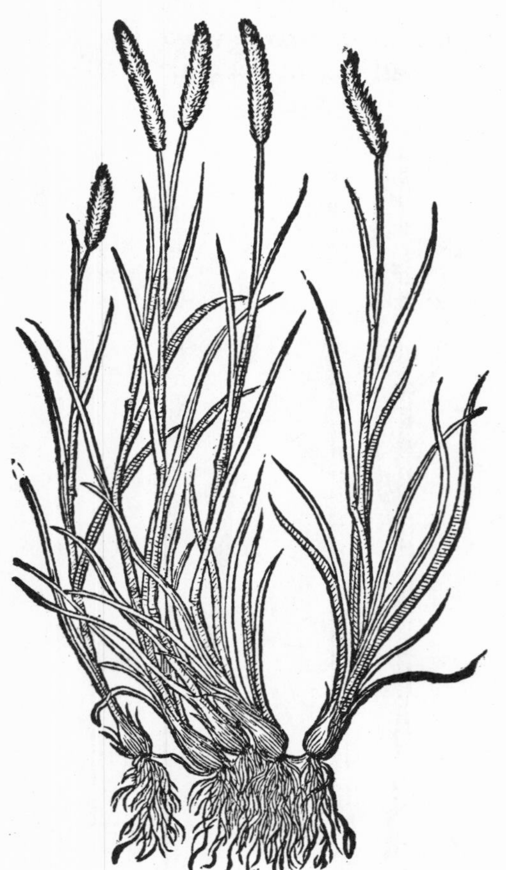

monly foure or fiue inches long, closely and handsomely made in the fashion of the precedent, which in the shape it doth very much resemble. ‡

¶ *The Place and Time.*

These kindes of Grasses do grow very well neere waterie places, as *Gramen Cyperoides* doth, and flourish at the same time that all the others doe.

‡ The latter may be found by the bridge entring into Chelsey field, as one goeth from Saint *Iames* to little Chelsey. ‡

¶ *The Names.*

The Latines borrow these names of the Greekes, and call it *Gramen Typhinum*, of *Typha*, a Cats taile: and it may in English as wel be called round Bent-grasse, as Cats-taile Grasse.

‡ The last described is by *Bauhine*, who first gaue the figure and description thereof in his *Prodomus, pag.* 10. called *Gramen Typhoides maximum spica longissima*; that is, The largest Foxetaile Grasse with a very long eare. ‡

Chap. 9. *Of Cyperus Grasse.*

1 *Gramen Cyperoides.*
Cyperus Grasse.

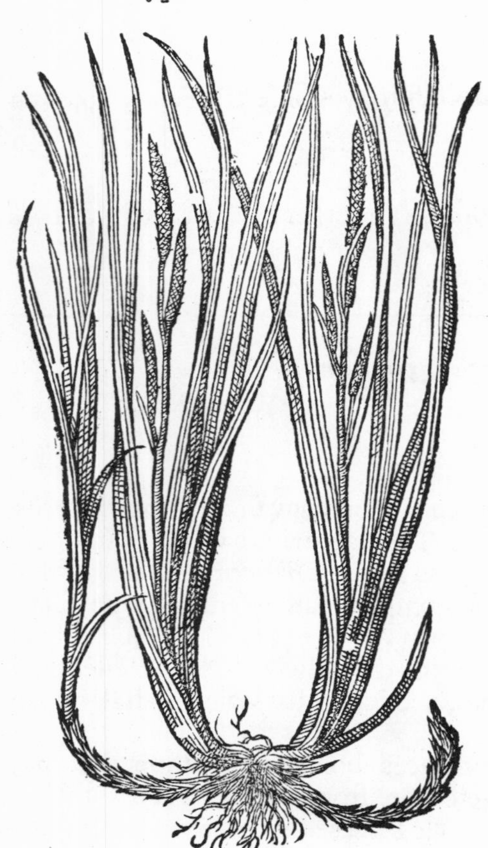

2 *Gramen Iunceum aquaticum.*
Rushy Water-Grasse.

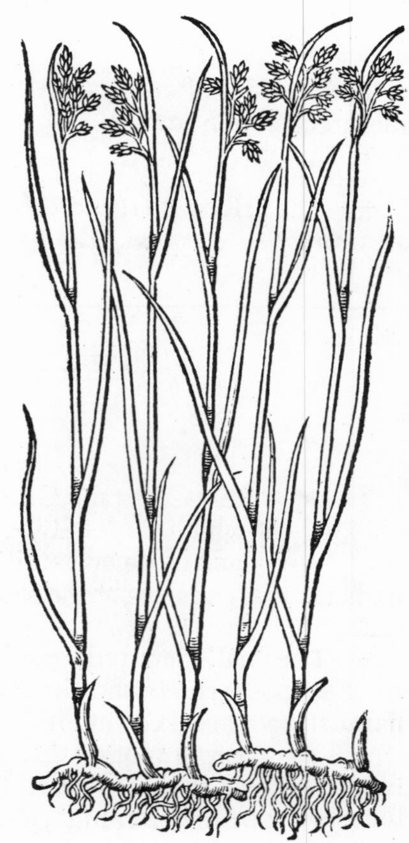

¶ *The Description.*

1 CYperus Grasse hath roots somewhat like Cyperus, whereof it tooke his name: his leaues are long and large like vnto the common reed: the stalke doth grow to the height of a cubit in some places, vpon which groweth little scaly knobs or eares, spike fashion, somewhat like vnto Cats-taile, or Reed-mace, very chaffie, rough, and rugged.

2 Rushy Water-grasse hath his roots like the former, with many fibres or strings hanging at them; and creepeth along vpon the vppermost face of the earth, or rather mud, wherein it groweth, bearing at each ioynt one slender benty stalke, set with a few small grassie blades or leaues, bringing forth at the top in little hoods, small feather-like tufts or eares.

¶ *The Place, Time, and Names.*

They grow, as I haue insinuated, in myrie and muddy grounds, in the same season that others do. And concerning their names there hath been said enough in their titles.

CHAP. 10. *Of Water-Grasse.*

1 *Gramen aquaticum.*
Water-grasse.

2 *Gramen aquaticum spicatum.*
Spiked Water-grasse.

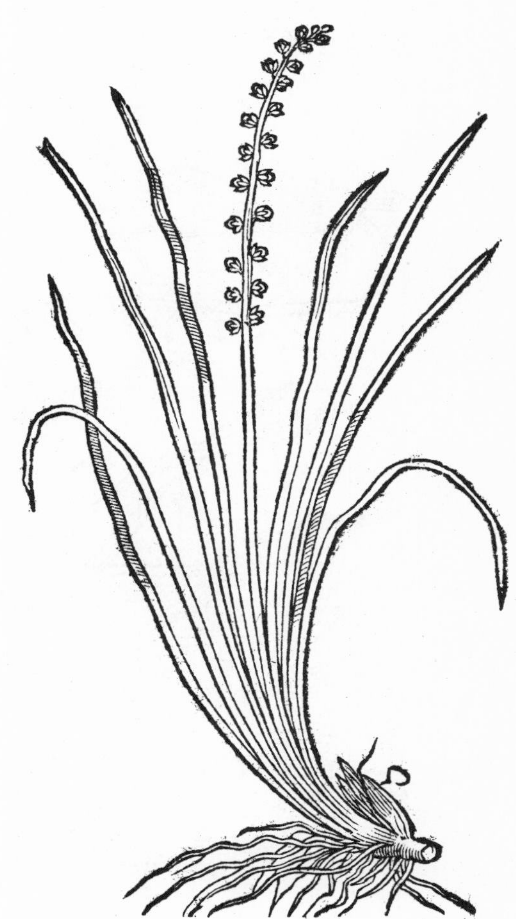

¶ *The Description.*

† 1 WAter-grasse, or as we terme it, Water Burre-grasse, hath a few long narrow slender and ioynted leaues: among which riseth vp a stalke of two foot high, bearing vpon his small and tender branches certaine little rough knobs, or brownish sharpe pointed seeds made vp into cornered heads: his root is small and threddy.

‡ The figure of this plant is not well exprest, for it should haue had the leaues made narrower, and ioynts exprest in them, like as you may see in the *Gramen junceum syluaticum*, which is the ninth in the sixteenth chapter; for that and this are so like, that I know no other difference betweene them, but that this hath leaues longer and narrower than that, and the heads smaller and whiter. There is a reasonable good figure of this in the *Historia Lugd.p.*1001. vnder the name of *Arundo minima.* ‡

2 Spiked

2 Spiked Water-graſſe hath long narrow leaues: the ſtalke is ſmall, ſingle, and naked, without leaues or blades, bearing alongſt the ſame toward the top an eare or ſpike made of certaine ſmall buttons, reſembling the buttonie floures of Sea Worme-wood. His root is thick & tough, full of fibres or threds.

¶ *The Place and Time.*

They differ not from the former kindes of Graſſes in place and time: and their names are manifeſt.

¶ *The Nature and Vertues.*

Their nature and vertues are referred vnto Dogs Graſſe, whereof we will ſpeake hereafter.

CHAP. 11. *Of Flote-Graſſe.*

1 *Gramen fluviatile.* Flote-graſſe.

2 *Gramen fluviatile ſpicatum.* Spiked Flote-graſſe.

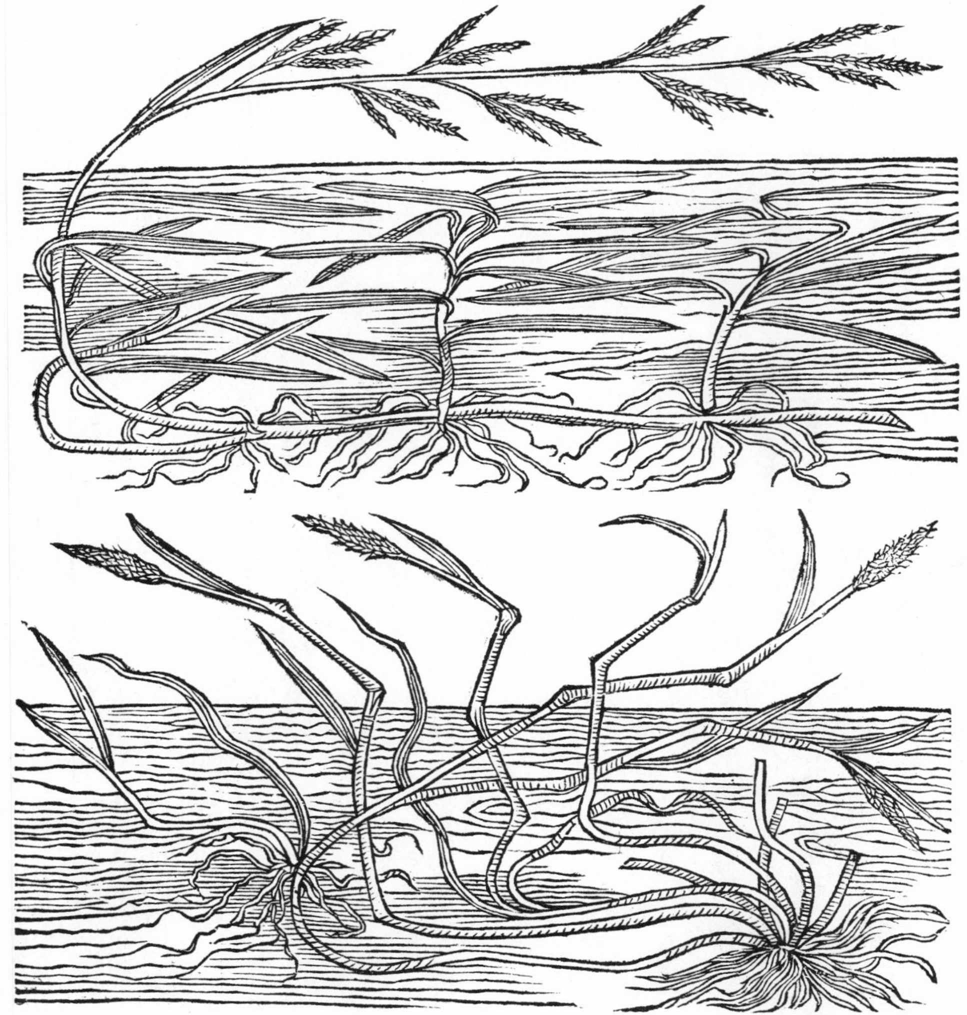

¶ *The Deſcription.*

† 1 Flote-graſſe hath a long and round root ſomewhat thicke, like vnto Dogs-graſſe, ſet on euen ioynts with ſmall ſtrings or threds; from the which riſe vp long and crooked ſtalkes, croſſing, winding, and folding one within another with many flaggie leaues, which horſes eate greedily of. At the top of theſe ſtalks, and ſomewhat lower, there come forth very many little eares of a whitiſh colour, compoſed of two ranks of little chaffie ſeeds ſet alternately, each of theſe ſmall eares being almoſt an inch in length.

2 Spike Flote-Graſſe, or ſpiked Flote-graſſe beareth at the top of each ſlender creeping ſtalke one ſpiked eare and no more, and the other many, which maketh a difference betwixt them; otherwiſe they are one like the other. His root is compact, tufted, and made of many thrummie threds.

¶ *The Place.*

The firſt of theſe growes euery where in waters. The ſecond is harder to be found.

¶ The

¶ *The Names.*

The first is called *Gramen fluviatile*, and also *Gramen aquis innatans*: in English, Flote-grasse. *Tragus* calls it, *Gramen Anatum*, Ducks-grasse.

The second is called *Gramen fluviatile spicatum*, and *fluviatile album* by *Tabernamontanus*. Likewise in English it is called Flote-grasse, and Floter-grasse, because they swimme and flote in the water.

CHAP. 12. *Of Kneed-Grasse.*

¶ *The Description.*

1 KNeed-grasse hath straight and vpright strawie stalkes, with ioynts like to the straw of corne, and beareth small grassie leaues or blades spiked at the top like vnto Pannick, with a rough eare of a darke browne colour. His roots are hairy and threddy, and the ioynts of the straw are very large and conspicuous.

1 *Gramen geniculatum.*
Kneed-grasse.

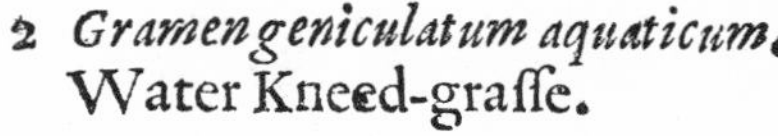

2 *Gramen geniculatum aquaticum.*
Water Kneed-grasse.

2 Water Kneed-grasse hath many long and slender stemmes, ioynted with many knobby and gouty knees like vnto Reed, set with broad flaggy leaues somewhat sharpe pointed; bearing at the top a tuft or pannicle diuided into sundry small branches, of a duskish colour. His root is thred-die like the other.

¶ *The Place, Time, and Names.*

These Grasses do grow in fertile moist medowes, not differing in time from others. And they are called *Geniculata*, because they haue large ioynts like as it were knees.

We haue nothing deliuered vs of their nature and properties.

Chap. 13. *Of Bearded Panicke Grasse.*

1 *Gramen Paniceum.*
Bearded Panick Grasse.

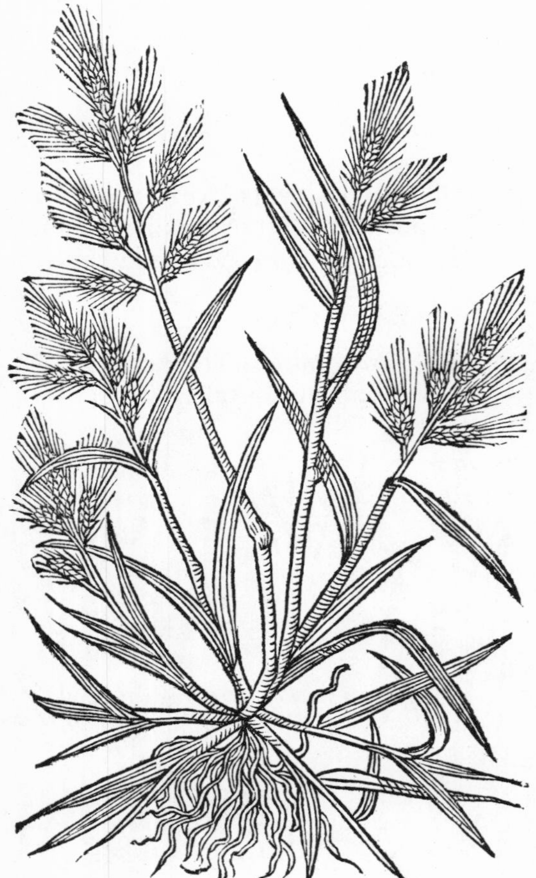

¶ *The Description.*

1 BEarded Panicke grasse hath broad and large leaues like barly, somwhat hoarie, or of an oner-worne russet colour. The stalkes haue two or three ioynts at the most, and many eares on the top, without order; vpon some stalkes more eares; on others fewer, much like vnto the eare of wilde Panicke, but that this hath many beards or awnes, which the other wants.

2 Small Pannicke Grasse, as *Lobelius* writeth, in roots, leaues, ioynts, and stalkes is like the former, sauing that the eare is much lesse, consisting of fewer rowes of seed, contained in small chaffie blackish huskes. This, as the former, hath many eares vpon one stalke.

‡ 3 This small Pannicke Grasse from a threddy root sendeth forth many little stalkes, whereof some are one handfull, other-some little more than an inch high; and each of these stalkes on the top sustaines one single eare, in shape very like vnto the eare of wilde Pannicke, but about halfe the length. The stalkes of this are commonly crooked, and set with grassie leaues like to the rest of this kinde. The figure hereof was vnfitly placed by our Author in the sixteenth place in the eighth chapter, vnder the title of *Gramen Cyperoides spicatum.*

2 *Gramen paniceum parvum.*
Small Panicke Grasse.

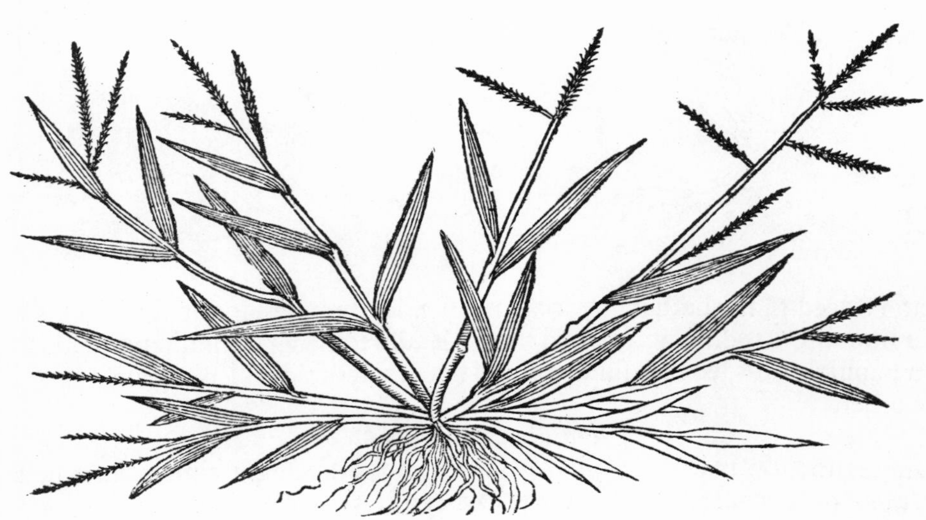

¶ *The Place and Time.*

The first of these two doth grow neere vnto mud walls, or such like places not manured, yet fertile or fruitfull.

The

The ſecond groweth in ſhallow waterie plaſhes of paſtures, and at the ſame time with others.
‡ I haue not as yet obſerued any of theſe three growing wilde. ‡

† 3 *Gramen Pannici effigie ſpica ſimplici.*
Single eared Pannicke Graſſe.

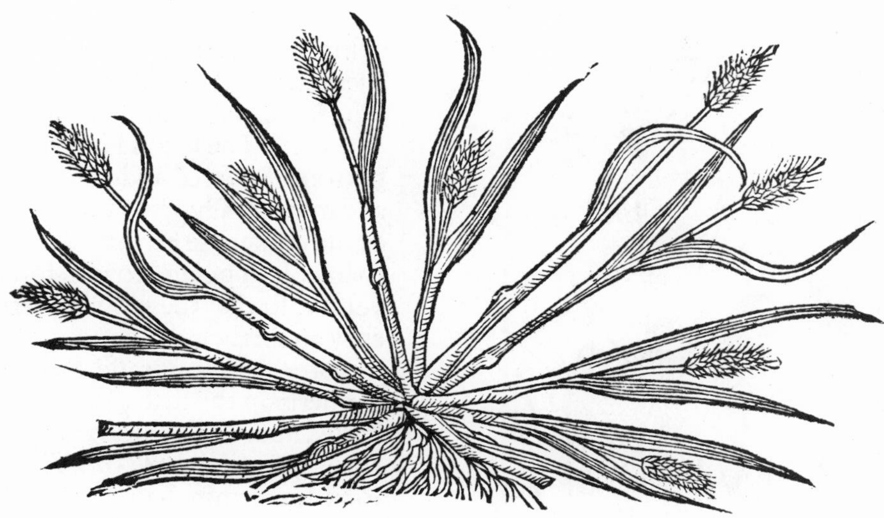

¶ *The Names and Vertues.*

They are called Panicke Graſſes, becauſe they are like the Italian corne called Panicke. Their nature and vertues are not knowne.

CHAP. 14. *Of Hedge-hog Graſſe.*

† 1 *Gramen paluſtre Echinatum.*
Hedge-hog Graſſe.

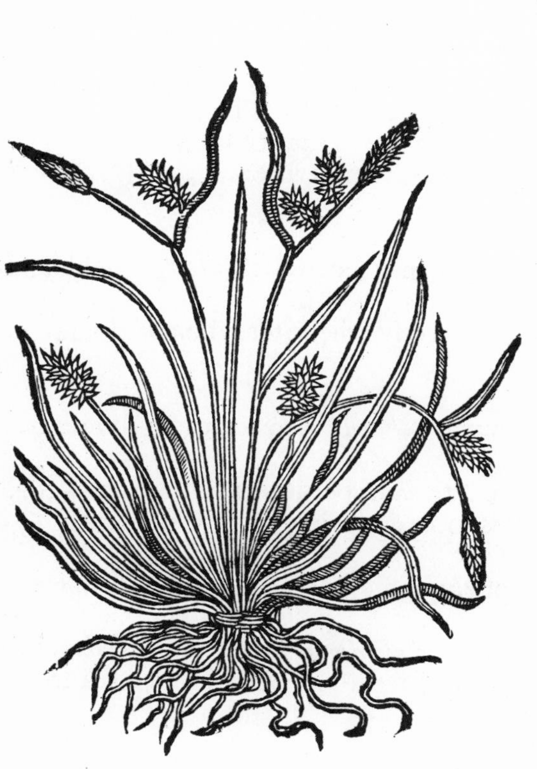

2 *Gramen exile Hirſutum.*
Hairy-graſſe.

‡ 3 *Gramen Capitulis globosis.*
Round headed Siluer-grasse.

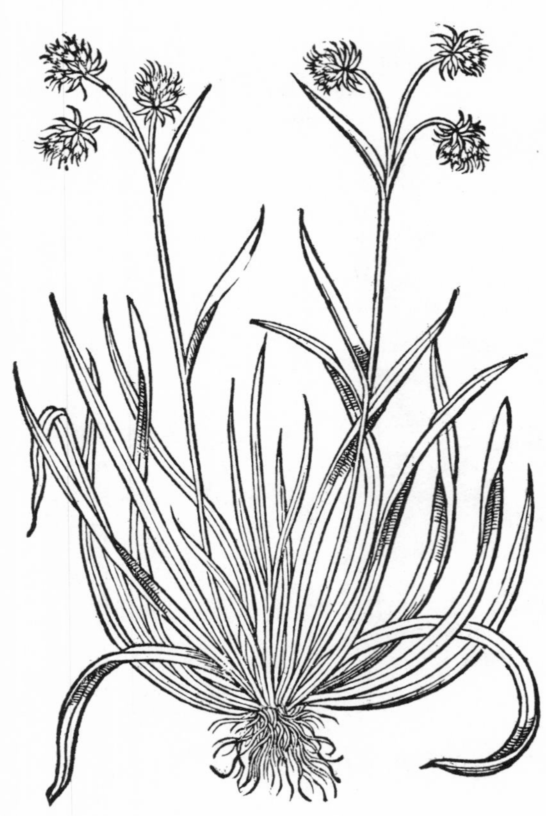

¶ *The Description.*

1 Hedge-hog Grasse hath long stiffe flaggy leaues with diuers stalkes proceeding from a thicke spreading root; and at the top of euery stalke growe certaine round and pricking knobs fashioned like an hedge-hog.

† 2 The second is rough and hairie: his roots do spred and creep vnder the mud and myre as Cyperus doth; and at the top of the stalkes are certaine round soft heads, their colour being browne, intermixed with yellow, so that they looke prettily when as they are in their prime.

‡ 3 This Grasse (whose figure was formerly in the first place in this Chapter) hath a small and fibrous root, from which rise leaues like those of Wheat, but with some long white hairs vpon them like those of the last described: at the tops of the stalks (which are some foot or better high) there grow two or three round heads consisting of soft and white downie threds. These heads are said to shine in the night, and therefore they in Italy call it (according to *Cæsalpinus*) *Luciola, quia noctu lucet.*

4 To this I may adde another growing also in Italy, and first described by *Fabius Columna*. It hath small creeping ioynted roots, out of which come small fibres, and leaues little and very narrow at the first, but those that are vpon the stalkes are as long againe, incompassing the stalks, as in Wheat, Dogs-grasse, and the like. These leaues are crested all along, and a little forked at the end: the straw or stalke is very slender, at the top whereof growes a sharpe prickly round head, much after the manner of the last described: each of the seed-vessels whereof this head consists ends in a prickly stalke hauing fiue or seuen points, whereof the vppermost that is in the middle is the longest. The seed that is contained in these prickly vessels is little and transparent, like in colour to that of Cow-wheat. The floures (as in others of this kinde) hang trembling vpon yellowish small threds. ‡

¶ *The Place and Time.*

† 1 2 They grow in watery medows and fields, as you may see in Saint *Georges* fields and such like places.

3 4 Both these grow in diuers mountainous places of Italy; the later whereof floures in May.

¶ *The Names.*

The first is called Hedge-hog Grasse, and in Latine, *Gramen Echinatum*, by reason of those prickles which are like vnto a hedge-hog.

The second hairy Grasse is called *Gramen exile hirsutum Cyperoides*, because it is small and little, and rough or hairy like a Goat: and *Cyperoides*, because his roots do spring and creepe like the *Cyperus*.

‡ 3 This by *Anguillara* is thought to be *Combretum* of *Pliny*; it is *Gram. lucidum* of *Tabernamontanus*; and *Gramen hirsutum capitulo globoso*, of *Bauhine*, *Pin. pag.* 7.

4 *Fabius Columna* calls this, *Gramen montanum Echinatum tribuloides capitatum*: and *Bauhine* nameth it, *Gramen spica subrotunda echinata*. Wee may call it in English, Round headed Caltrope Grasse.

¶ *The Vertues.*

3 The heade of this (which I haue thought good to call Siluer-grasse) is very good to be applied to greene wounds, and effectuall to stay bleeding, *Cæsalp.* ‡

† It is euident by the name and description, that our Author meant this which we here giue you in the first place; yet his figure was of another Grasse somwhat like the second, which figure and description you may finde here exprest in the third place.

Chap.

CHAP. 15. *Of Hairy Wood-Graſſe.*

¶ *The Deſcription.*

1 HAiry Wood-graſſe hath broad rough leaues ſomewhat like the precedent, but much longer, and they proceed from a threddy root, which is very thicke, and ful of ſtrings, as the common Graſſe, with ſmall ſtalkes riſing vp from the ſame roots; but the top of theſe ſtalkes is diuided into a number of little branches, and on the end of euery one of them ſtandeth a little floure or huske like the top of *Allium Vrſinum*, or common Ramſons, wherein the ſeed is contained when the floure is fallen.

2 Cyperus Wood-graſſe hath many ſheary graſſie leaues, proceeding from a root made of many hairy ſtrings or threds: among which there riſeth vp ſundry ſtraight and vpright ſtalkes, on whoſe tops are certaine ſcaly and chaffie huskes, or rather ſpikie blackiſh eares, not much vnlike the catkins or tags which grow on Nut-trees, or Aller trees.

1 *Gramen hirſutum nemoroſum.*
Hairy Wood-graſſe.

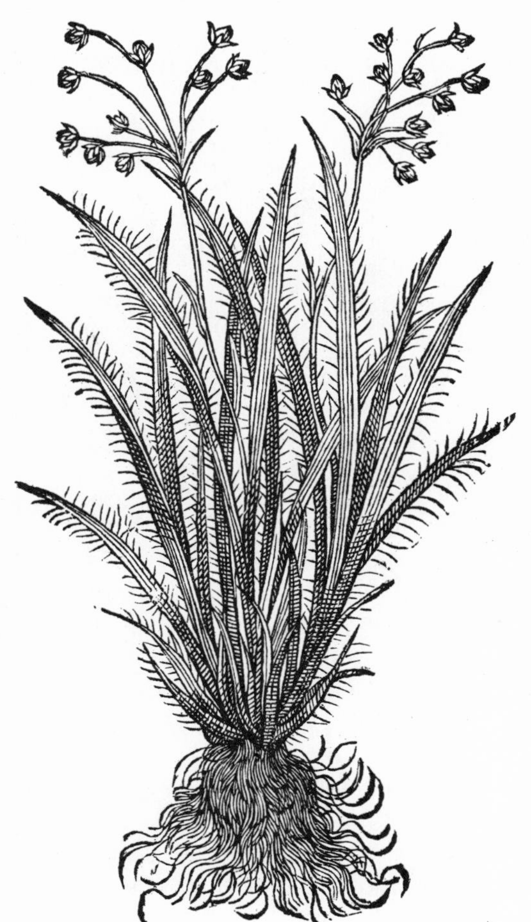

2 *Gramen Cyperinum nemoroſum.*
Cyperus Wood-graſſe.

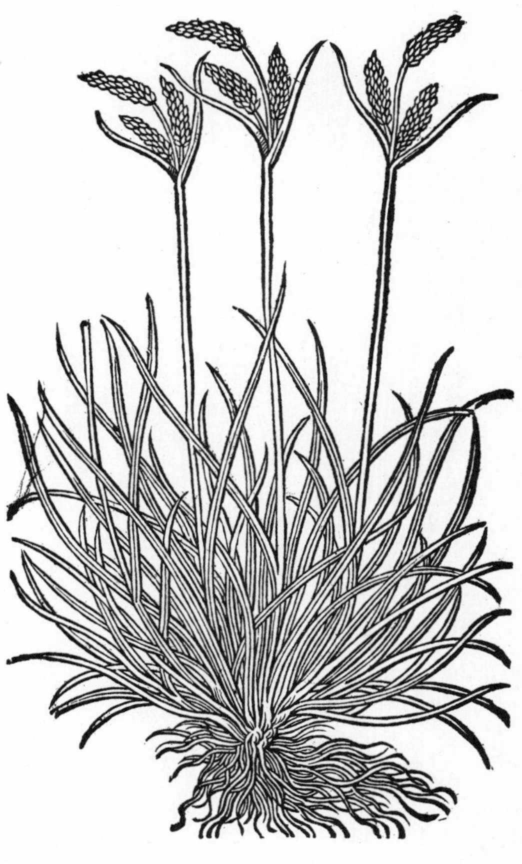

¶ *The Place, Time, and Names.*

Theſe two grow in woods or ſhadowie places, and may in Engliſh be called Wood-graſſes. Their time is common with the reſt.

¶ *Their Nature and Vertues.*

There is nothing to be ſaid of their nature and vertues, being as vnknowne as moſt of the former.

Chap. 16. *Of Sea Spike-Graſſe.*

¶ *The Deſcription.*

† 1 Sea Spike-graſſe hath many ſmall hollow round leaues about ſix inches long, riſing from a buſhy threddy white fibrous root, which are very ſoft and ſmooth in handling. Among theſe leaues there doe ſpring vp many ſmall ruſhy ſtalkes; alongſt which are at the firſt diuers ſmall flouring round buttons; the ſides whereof falling away, the middle part growes into a longiſh ſeed-veſſell ſtanding vpright.

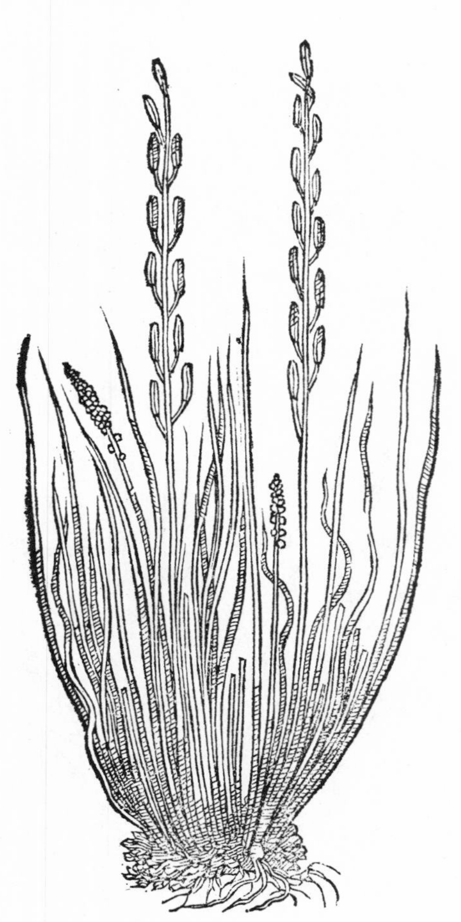

1 *Gramen marinum ſpicatum.*
Sea Spike-graſſe.

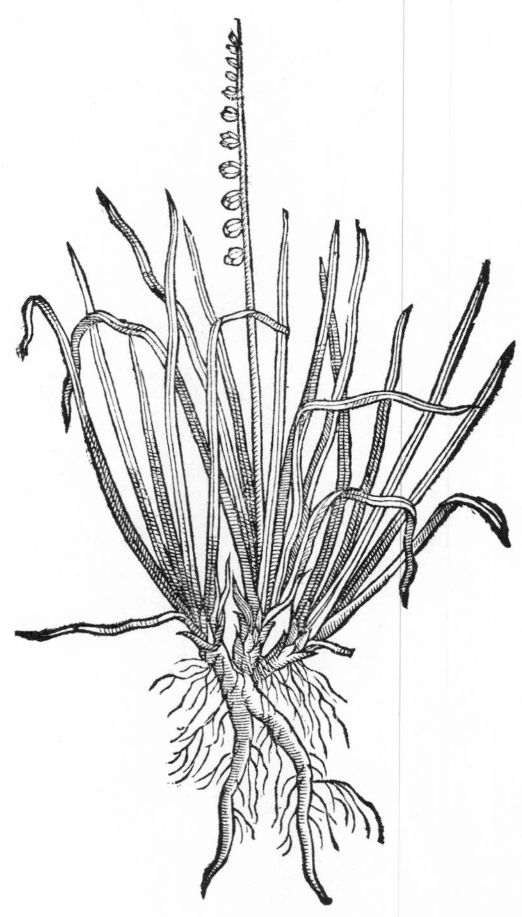

2 *Gramen ſpicatum alterum.*
Salt marſh Spike graſſe.

† 2 Salt-marſh Spike-graſſe hath a woody tough thicke root with ſome ſmall hairy threds faſtned thereunto; out of which ariſe long and thicke leaues very like thoſe of that Sea-graſſe we vulgarly call Thrift. And amongſt theſe leaues grow vp ſlender naked ruſhy ſtalkes which haue on one ſide ſmall knobs or buttons of a greeniſh colour hanging on them.

3 The third hath many ruſhy leaues tough and hard, of a browne colour, well reſembling Ruſhes: his root is compact of many ſmall tough and long ſtrings. His ſtalke is bare and naked of leaues vnto the top, on which it hath many ſmall pretty chaffie buttons or heads.

4 The fourth is like the third, ſauing that it is larger; the ſtalke alſo is thicker and taller than that of the former, bearing at the top ſuch huskes as are in Ruſhes.

5 Great Cypreſſe Graſſe hath diuers long three-ſquare ſtalkes proceeding from a root compact of many long and tough ſtrings or threds. The leaues are long and broad, like vnto the ſedge called *Carex*. The ſpike or eare of it is like the head of Plantaine, and very prickly, and commonly of a yellowiſh greene colour.

6 Small Cypreſſe Graſſe is like vnto the other in root and leaues, ſauing that it is ſmaller. His ſtalke is ſmooth and plaine, bearing at the top certaine tufts or pannicles, like to the laſt deſcribed in roughneſſe and colour.

3 *Gramen junceum marinum.*
Sea Ruſh-graſſe.

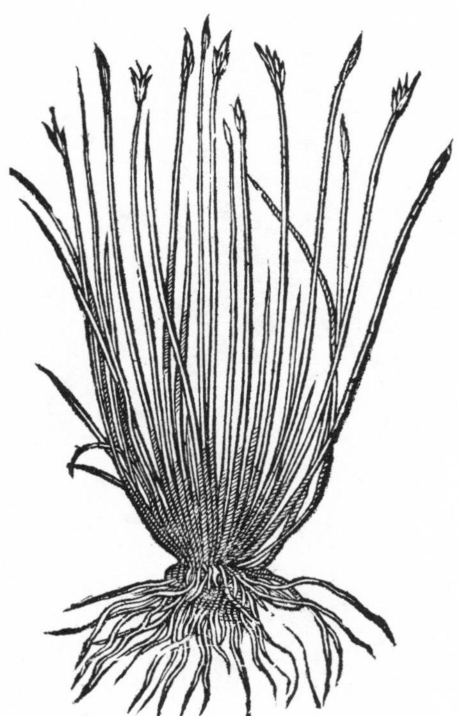

4 *Gramen junceum maritimum.*
Mariſh Ruſh-graſſe.

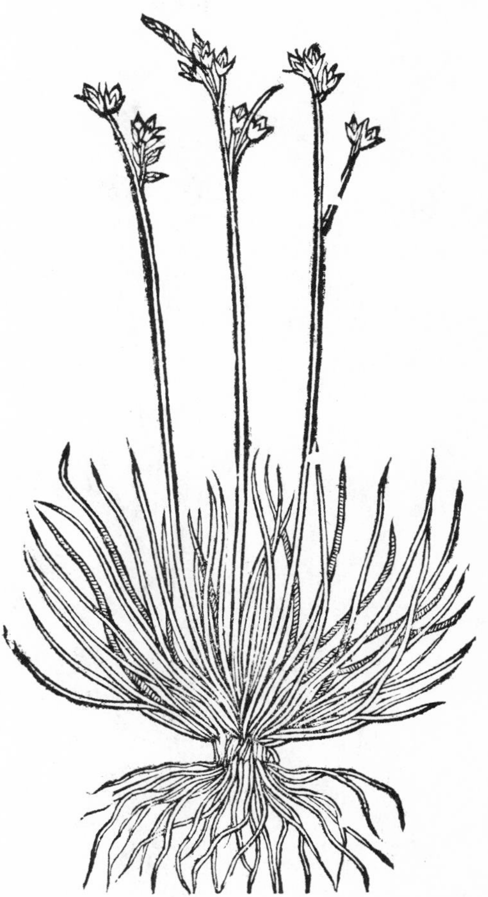

5 *Gramen paluſtris Cyperoides.*
Great Cypreſſe Graſſe.

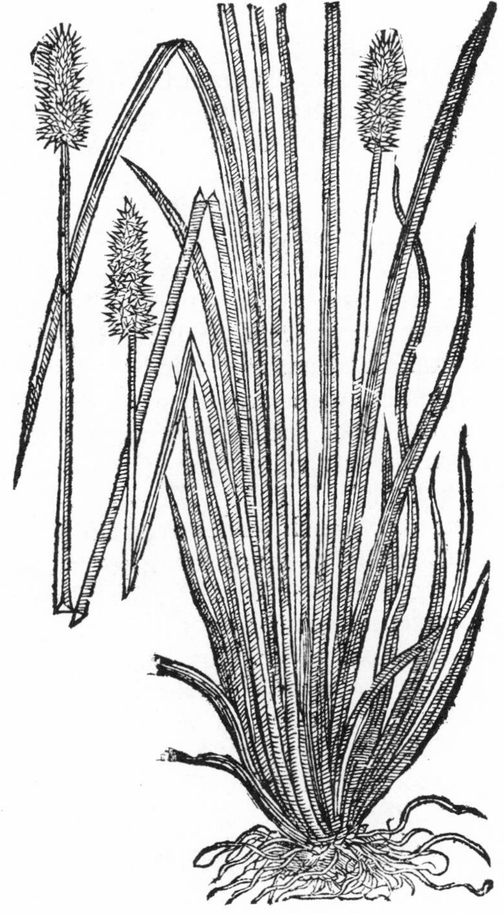

6 *Gramen Cyperoides parvum.*
Small Cypreſſe Graſſe.

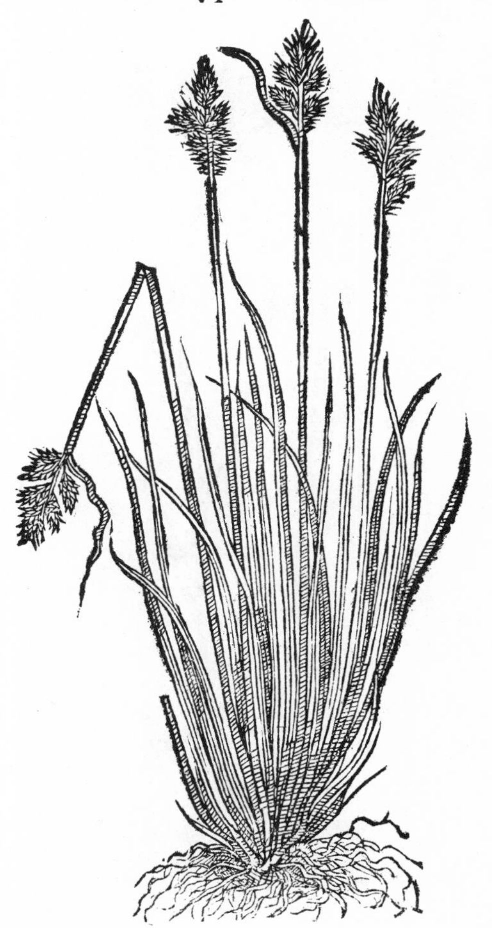

7 *Gramen aquaticum Cyperoides vulgatius.*
Water Cypreſſe Graſſe.

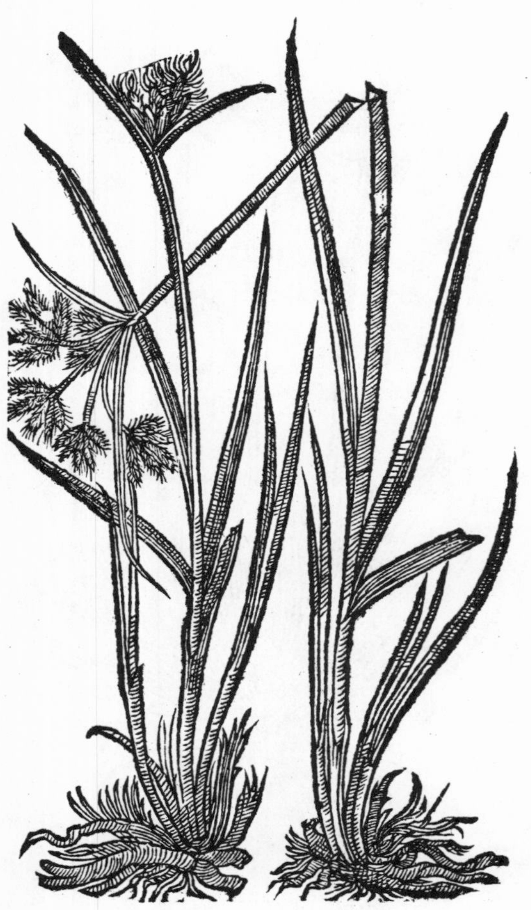

‡ 8 *Gramen Cyperoides ſpicatum.* Spike Cypreſſe Graſſe.

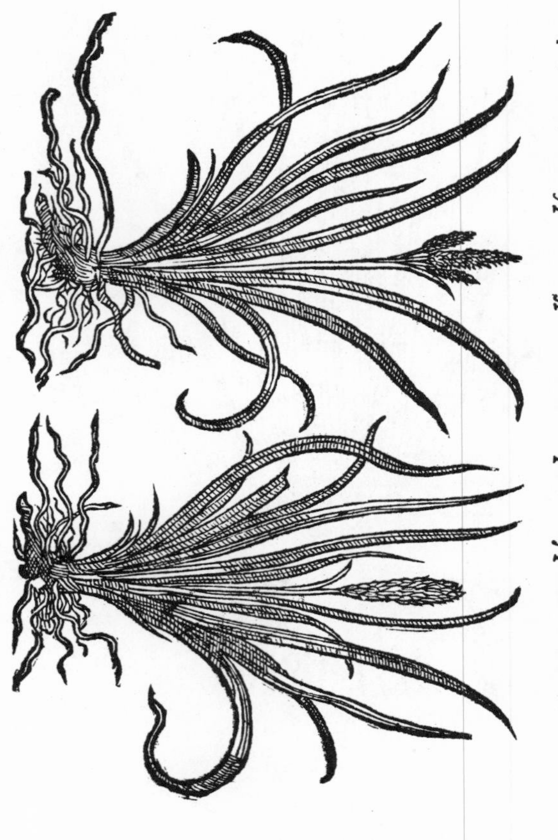

9 *Gramen junceum ſyluaticum.*
Wood Ruſhy-graſſe.

7 The firſt of theſe two kindes hath many crooked and crambling roots of a woody ſubſtance, very like vnto the right Cyperus, differing from it onely in ſmell, becauſe the right Cyperus roots haue a fragrant ſmell, and theſe none at all. His leaues are long and broad, rough, ſharp or cutting at the edges like ſedge. His ſtalke is long, big, and three ſquare, like to Cyperus, and on his top a chaffie vmbel or tuft like vnto the true Cyperus.

‡ 8 The ſecond kinde hath many broad leaues like vnto thoſe of Gillouers, but of a freſher greene: amongſt the which riſeth vp a ſhort ſtalke ſome handful or two high, bearing at the top three or foure ſhort eares of a reddiſh murrey colour, and theſe eares grow commonly together at the top of the ſtalk, and not one vnder another. There is alſo another leſſer ſort hereof, with leaues and roots like the former, but the ſtalke is commonly ſhorter, and it hath but one ſingle eare at the top thereof. You haue the figures of both theſe expreſt in the ſame table or piece. This kinde of Graſſe is the *Gramen ſpicatum folijs Vetonicæ* of *Lobel*. ‡

9 This hath long tough and hairy ſtrings growing deepe in the earth like a turfe, which make the root; from which riſe many crooked tough and ruſhy ſtalks, hauing toward the top ſcaly and chaffie knobs or buttons. ‡ This

grows

growes some halfe yard high, with round brownish heads, and the leaues are ioynted as you see them expressed in the figure we here giue you. ‡

¶ *The Place, Time, Names, Nature, and Vertues.*

All the Grasses which we haue described in this chapter doe grow in marish and watery places neere to the sea, or other fenny grounds, or by muddy and myrie ditches, at the same time that the others do grow and flourish. Their names are easily gathered of the places they grow in, or by their Descriptions, and are of no vertue nor propertie in medicine, or any other necessarie vse as yet knowne.

† Formerly in the eighth place (but very vnfitly) was the figure of *Gramen pannici effigie spica simp.* being the third in the thirteenth chapter. The ninth also is restored to his due place, being the fourth in the sixth chapter. The two Reed-grasses that were in the eleuenth and twelfth places are also before in the fifth Chapter.

CHAP. 17. *Of Couch-Grasse, or Dogs-grasse.*

1 *Gramen Caninum.*
Couch-grasse, or Dogs-grasse.

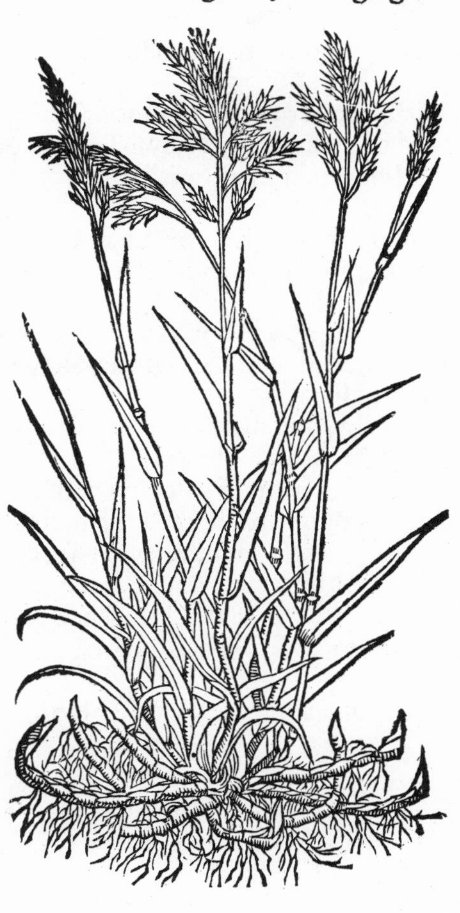

2 *Gramen Caninum nodosum.*
Knotty Dogs-grasse.

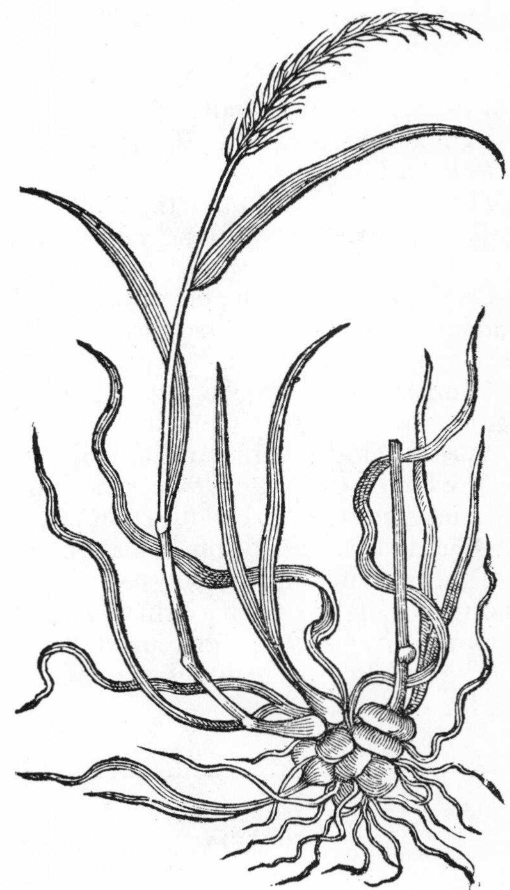

¶ *The Description.*

† 1 THe common or best knowne Dogs-grasse, or Couch-grasse hath long leaues of a whitish greene colour: the stalke is a cubit and a halfe high, with ioynts or knees like wheaten straw, but these ioynts are couered with a little short down or woollinesse. The plume or tuft is like the reed, but smaller and more chaffie, and of a grayish colour: it creepeth in the ground hither and thither with long white roots, ioynted at certaine distances, hauing a pleasant sweet taste, and are platted or wrapped one within another very intricately, insomuch as where it hapneth in gardens amongst pot-herbes, great labour must be taken before it can be destroyed, each piece being apt to grow, and euery way to dilate it selfe.

† 2 Knotty Dogs graſſe is like vnto the former in ſtalke and leafe,but that they are of a deeper colour; alſo the ſpike or eare is greener, and about ſome two handfulls long, much in ſhape reſembling an Oate, yet far ſmaller, and is much more diſperſed than the figure preſents to you. The roots of this are ſomewhat knotty and tuberous, but that is chiefely about the Spring of the yeare, for afterwards they become leſſe and leſſe vntill the end of Summer. And theſe bulbes do grow confuſedly together, not retaining auy certaine ſhape or number.

¶ *The Place.*

1 The firſt growes in gardens and arable lands,as an infirmitie or plague of the fields, nothing pleaſing to Husbandmen; for after that the field is plowed, they are conſtrained to gather the roots together with harrowes and rakes; and being ſo gathered and laid vpon heapes, they ſet them on fire leſt they ſhould grow againe.

2 The ſecond growes in plowed fields and ſuch like places, but not euery where as the other. I haue found of theſe in great plenty,both growing,and plucked vp with harrowes, as before is rehearſed, in the fields next to S.*Iames* wall as ye go to Chelſey,and in the fields as ye go from the Tower-hill of London to Radcliffe.

¶ *The Time.*

Theſe Graſſes ſeldome come to ſhew their eare before Iuly.

¶ *The Names.*

It is called *Gramen Caninum*, or *Sanguinale*, and *Vniola*. The Countreymen of Brabant name it **Peen**: others, **Ledt graſſe**: of the Grecians, ἄγρωστις: of the Latines,by the common name,*Gramen*. It is of ſome named ἀγκοὶ: in Engliſh,Couch-graſſe,Quitch-Graſſe,and Dogs-graſſe.

Gramen Caninum bulboſum, or *nodoſum*, is called in Engliſh, Knobby,or Knotty Couch-graſſe.

¶ *The Nature.*

The nature of Couch-graſſe,eſpecially the roots, agreeth with the nature of common Graſſe: although that Couch-graſſe be an vnwelcome gueſt to fields and gardens, yet his phyſicke vertues do recompence thoſe hurts; for it openeth the ſtoppings of the liuer and reines, without any manifeſt heate.

The learned Phyſitions of the Colledge and Societie of London do hold this bulbous Couch graſſe in temperature agreeing with the common Couch-graſſe, but in vertues more effectuall.

¶ *The Vertues.*

A Couch-graſſe healeth greene wounds. The decoction of the root is good for the kidneys and bladder: it prouoketh vrine gently, and driueth forth grauell. *Dioſcorides* and *Galen* do agree,that the root ſtamped and laid vpon greene wounds doth heale them ſpeedily.

B The decoction thereof ſerueth againſt griping paines of the belly, and difficultie of making water.

C *Marcellus* an old Author maketh mention in his 26 chapter, That ſeuen and twenty knots of the herbe which is called *Gramen*, or Graſſe, boiled in wine till halfe be conſumed, preſſed forth, ſtrained, and giuen to drinke to him that is troubled with the ſtrangurie,hath ſo great vertue,that after the Patient hath once begun to make water without paine, it may not be giuen any more. But it muſt be giuen with water onely to ſuch as haue a Feuer. By which words it appeareth, That this knotted Graſſe was taken for that which is properly called *Gramen*, or *Agroſtis*; and hath bin alſo commended againſt the ſtone and diſeaſes of the bladder.

D The later Phyſitions doe vſe the roots ſometimes of this, and ſometimes of the other indifferently.

Chap. 18. *Of Sea Dogs-Graſſe.*

¶ *The Deſcription.*

1 The Sea Dogs-graſſe is very like vnto the other before named: his leaues are long and ſlender, and very thicke compact together, ſet vpon a knotty ſtalke ſpiked at the top like the former. Alſo the root crambleth and creepeth hither and thither vnder the earth, occupying much ground by reaſon of his great encreaſe of roots.

‡ This Graſſe (whereof *Lobel* gaue the firſt figure and deſcription, vnder the name of *Gramen geniculatum Caninum marinum*) I coniecture to be that which growes plentifully vpon the banks in the ſalt mariſhes by Dartford in Kent,and moſt other ſalt places by the ſea; as alſo in many banks and orchards about London, and moſt other places farre from the ſea. Now *Lobels* figure being not good, and the deſcription not extant in any of his Latine Workes; I cannot certainly affirme any thing. Yet I thinke it fit to giue you an exact deſcription of that I do probably iudge to be it; and

and not onely so, but I iudge it to be the same Grasse that **Bauhine** in his *Prodromus* hath set forth, *pag.* 17. vnder the name of *Gramen latifolium spica triticea compacta*. This is a very tall Grasse; for it sends forth a stalke commonly in good ground to the height of a yard and an halfe: the leaues are large, stiffe, and greene, almost as big as those of white Wheat; the which it also very much resembles in the eare, which vsually is some handfull and an halfe long, little spokes standing by course with their flat sides towards the straw. About the beginning of Iuly it is hung with little

1 *Gramen Caninum marinum.*
Sea Dogs-grasse.

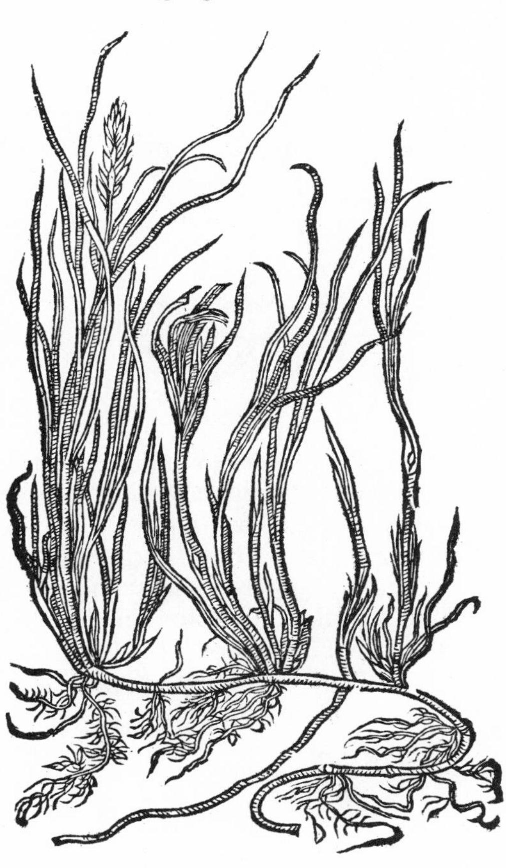

2 *Gramen Caninum marinum alterum.*
Sea Couch-grasse.

whitish yellow floures such as Wheat hath. The roots of this are like those of the first described. This sometimes varies in the largenesse of the whole Plant, as also in the greatnesse, sparsednesse, and compactnesse of the eare. ‡

2 The second Sea Dogs-grasse is according vnto *Lobel* somewhat like the former: his roots are more spreading and longer, dispersing themselues vnder the ground farther than any of the rest. The leaues are like the former, thicke bushed at the top, with a cluster or bush of short thick leaues one folded within another. The stalke and tuft is of a middle kinde, betweene *Ischæmon* and the common Couch-grasse.

¶ *The Place, Time, Names, Nature, and Vertues.*

They grow on the sea shore at the same time that others do; and are so called because they grow neere the sea side. Their nature and vertues are to be referred vnto Dogs-grasse.

CHAP. 19. *Of vpright Dogs-Grasse.*

¶ *The Description.*

1 VPright Dogs-grasse, or Quich-grasse, by reason of his long spreading ioynted roots is like vnto the former, and hath at euery knot in the root sundry strings of hairie substance, shooting into the ground at euery ioint as it spreadeth: the stalks ly creeping, or rise but a little from the ground, and at their tops haue spokie pannicles farre smaller than the common

common Couch-graſſe. By which notes of difference it may eaſily be diſcerned from the other kindes of Dogs-graſſe.

1 *Gramen Caninum ſupinum.*
Vpright Dogs-graſſe.

2 Ladies Laces hath leaues like vnto Millet in faſhion, rough and ſharpe pointed like to the Reed, with many white vaines or ribs, and ſiluer ſtreakes running along through the midſt of the leaues, faſhioning the ſame like to laces or ribbons wouen of white and greene ſilke, very beautifull and faire to behold : it groweth vnto the height of wilde Pannicke, with a ſpoky top not very much vnlike, but more compact, ſoft, white, and chaffie. The root is ſmall and hairie, and white of colour like vnto the Medow-graſſe.

2 *Gramen ſtriatum.*
Lady-lace Graſſe.

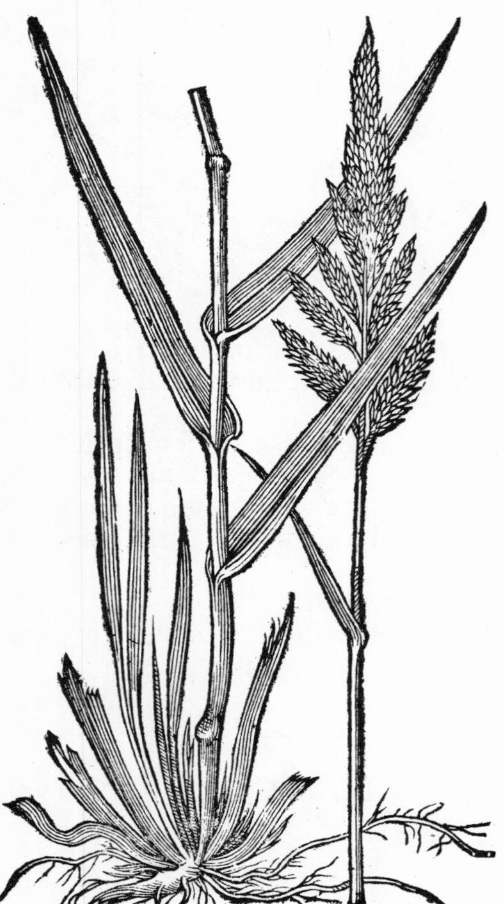

¶ *The Place.*

1 Vpright Dogs-graſſe groweth in dunged grounds and fertile fields.

2 Lady-laces growes naturally in woody and hilly places of Sauoy, and anſwers common Graſſe in his time of feeding.

It is kept and maintained in our Engliſh gardens, rather for pleaſure than vertue, which is yet knowne.

¶ *The Names.*

Lobelius calleth the later, *Gramen ſulcatum*, and *ſtriatum*, or *Gramen pictum* : in Engliſh, the Furrowed Graſſe, the white Chamelion Graſſe, or ſtreaked Graſſe ; and vſually of our Engliſh women it is called Lady-laces, or painted Graſſe : in French, *Aiguillettes d'armes*.

¶ *The Nature and Vertues.*

The vertues are referred vnto the Dogs-graſſes.

CHAP. 20. *Of Dew-Grasse.*

¶ *The Description.*

1 DEw-grasse hath very hard and tough roots long and fibrous: the stalkes are great, of three or foure cubits high, very rough and hairy, ioynted and kneed like the common Reed: the leaues are large and broad like vnto corne. The tuft or eare is diuided into sundry branches, chaffie, and of a purple colour; wherein is contained seed like *Milium*, wherewith the Germanes do make pottage and such like meat, as we in England do with Otemeale; and it is sent into Middleborough and other townes of the Low-countries, in great quantitie for the same purpose, as *Lobel* hath told me.

2 The second kinde of Dew-grasse or *Ischæmon* is somewhat like the first kinde of Medow-grasse, resembling one the other in leaues and stalkes, sauing that the crest or tuft is spred or stretched out abroad like a Cocks foot set downe vpon the ground, whereupon it was called *Galli crus*, by *Apuleius*. These tops are cleere and vpright, of a glistering purple colour, or rather violet; and it is diuided into foure or fiue branches like the former Dew-grasse. The root consists of a great many small fibres.

‡ 3 To these may fitly be added another Grasse, which *Clusius* hath iudged to be the medicinall Grasse of the Antients: and *Lobel* referres it to the Dogs grasses, because it hath a root iointed thicke, and creeping like as the Dogs-grasses: the stalkes are some foot high, round, and of a purplish colour: but the top is very like to that of the last described, of a darke purple colour.

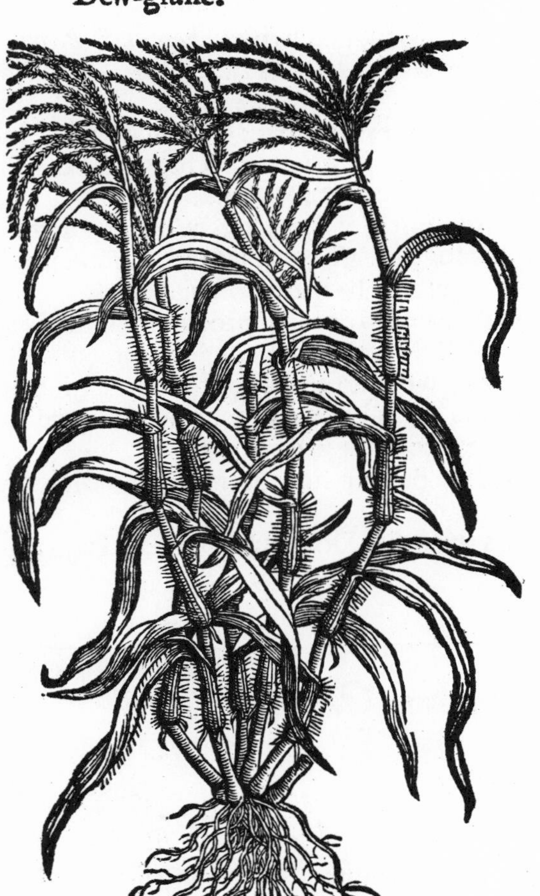

1 *Gramen Mannæ esculentum.*
Dew-grasse.

2 *Ischæmon vulgare.*
Cocks-foot grasse.

¶ *The Place and Time.*

1 The first groweth naturally in Germanie, Bohemia, Italy, and in the territories of Goritia and Carinthia, as *Matthiolus* reporteth.

2 The second groweth neere vnto rough bankes of fields, as I haue seene in the hilly bankes neere Greenhithe in Kent. It differeth not in time from those we haue spoken of.

‡ 3 This

‡ 3 *Gramen dactiloides radice repente.*
Cocks-foot Grasse with creeping roots.

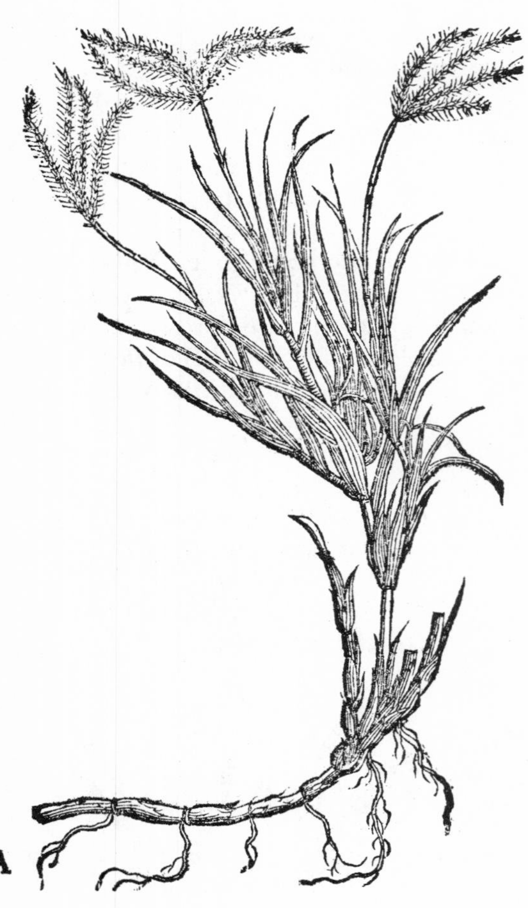

A

‡ 3 This groweth plentifully in most parts of Spaine and France; and it is probable, that this was the grasse that our Author found neere Greenhithe in Kent.

¶ *The Names.*

1 The Germanes call it **Himeldau**: That is to say, *Cœli ros*; whereupon it was called *Gramen Mannæ*: it seemeth to be *Milij sylvestris spurium quoddam genus*, a certaine wilde or bastard kinde of Millet. *Leonicenus* and *Ruellius* name it *Capriola*, and *Sanguinaria*: some would haue it to be *Gramen aculeatum Plinij*, but because the description thereof is very short, nothing can be certainly affirmed. But they are far deceiued who thinke it be *Coronopus*, as some very learned haue set downe: but euery one in these dayes is able to controll that errour. *Lobel* calleth it *Gramen Mannæ esculentum*, for that in Germany and other parts, as Bohemia and Italy, they vse to eate the same as a kind of bread-corne, and also make pottage therewith as wee do with Otemeale; for the which purpose it is there sowen as Corne, and sent into the Low-countries, and there sold by the pound. In English it may be called Manna-grasse, or Dew-grasse; but more fitly Rice-grasse.

2 This is iudged to be *Ischæmon* of *Pliny*; and *Galli crus* of *Apuleius*.

¶ *The Nature.*

These Grasses are astringent and drying, in taste sweet like the common Dogs-grasse.

¶ *The Vertues.*

Apuleius saith, if a plaister be made of this Grasse, Hogs grease, and leuen of household bread, it cureth the biting of mad dogs.

B As in the description I told you, this plant in his tuft or eare is diuided into sundry branches, some tuft into three, some foure, and some fiue clouen parts like Cocks toes. *Apuleius* reporteth, If ye take that eare which is diuided onely into three parts, it wonderfully helpeth the running or dropping of the eyes, and those that begin to be bleare eyed, being bound about the necke, and so vsed for certaine dayes together, it turneth the humors away from the weake part.

C ‡ Manna Grasse, or Rice-grasse is said to be very good to be put into pultesses, to discusse hard swellings in womens brests.

D The Cocks-foot Dogs-grasse is very good in all cases, as the other Dogs-grasses are, and equally as effectuall. ‡

‡ Chap. 21. *Of diuers Cyperus Grasses.*

¶ *The Description.*

‡ 1 THe first of these hath reasonable strong fibrous roots, from whence rise stiffe long and narrow leaues like those of other Cyperus Grasses: the stalkes also (as it is proper to all the plants of this kindred) are three square, bearing at their tops some three brownish eares soft and chaffie like the rest of this kinde, and standing vpright, and not hanging downe as some others do.

2 This hath pretty thicke creeping blacke roots, from whence arise three square stalkes set with leaues shorter, yet broader than those of the last described; and from the top of the stalke come forth three or foure foot-stalkes, whereupon doe hang longish rough scaly and yellowish heads.

3 The roots of this are blacke, without smell, and somewhat larger than those of the last described:

‡ 1 *Gramen Cyperoides anguſtifolium majus.*
Great narrow leaued Cyperus Graſſe.

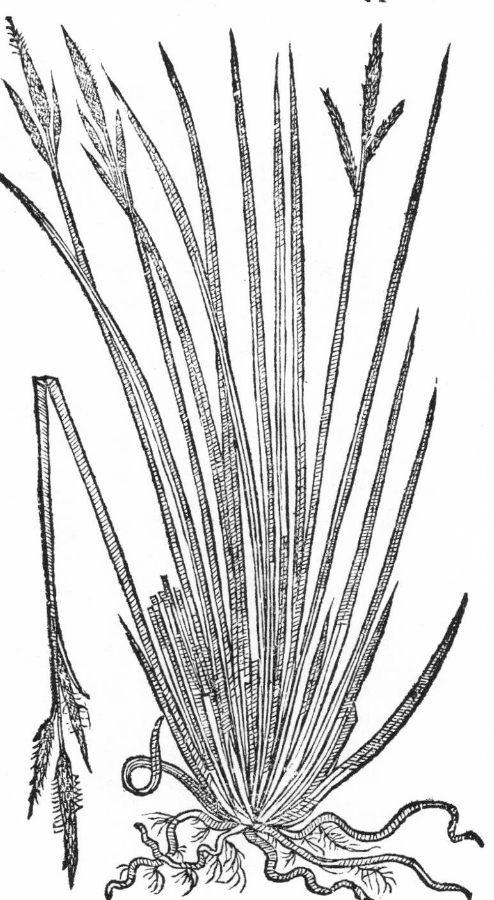

‡ 2 *Pſeudocyperus.*
Baſtard Cyperus.

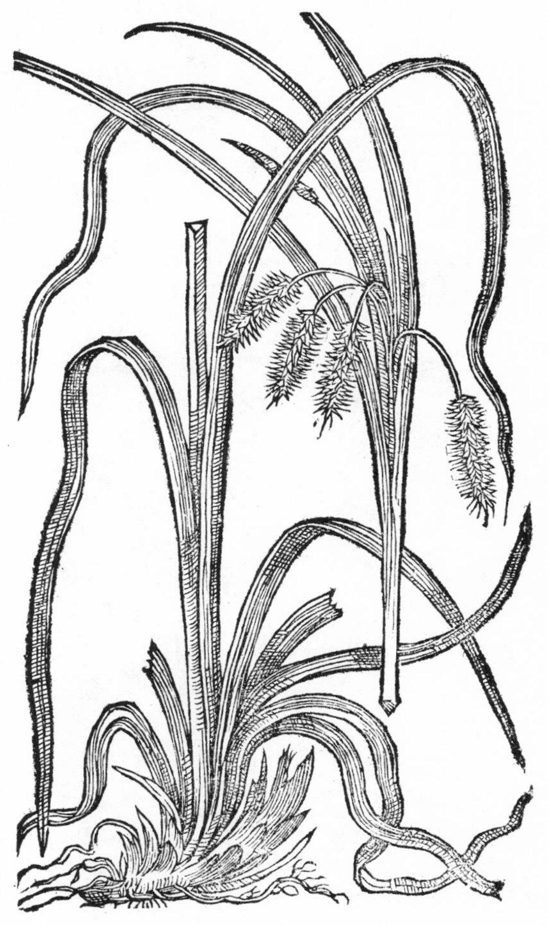

‡ 3 *Cyperus longus inodorus ſylueſtris.*
Long Baſtard Cyperus.

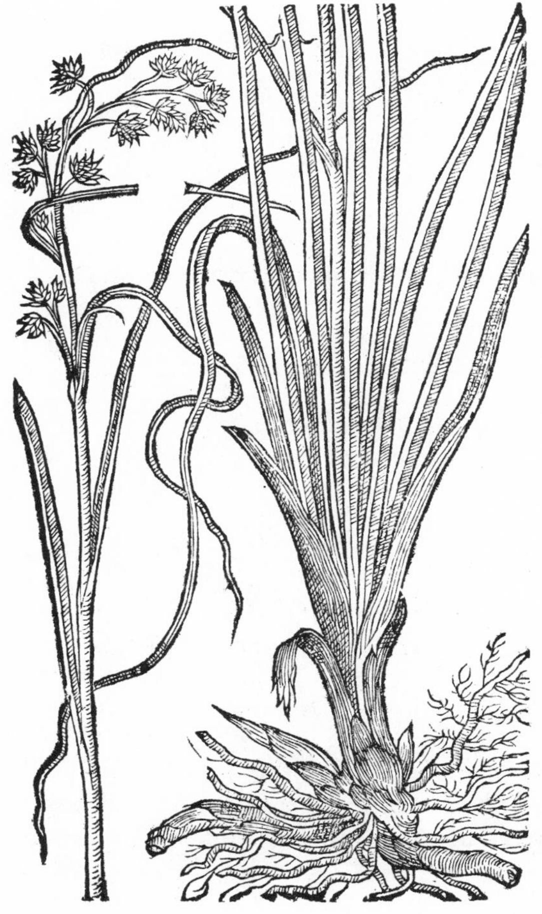

deſcribed: the 3 ſquare ſtalke alſo is ſome two cubits high, bearing at the top diſperſedly round ſcaly heads ſomewhat like thoſe of the wood Ruſh-graſſe: the leaues are ſomewhat ſharpe and triangular like thoſe of the other Cyperus.

4 This Cyperus hath creeping blacke roots, hauing here and there knotty tuberous heads for the moſt part, putting vp leaues like thoſe of the laſt deſcribed, as alſo a ſtalke bearing at the top long chaffy eares like to ſome others of this kinde.

5 This Cyperus Graſſe hath pretty thicke fibrous and blacke roots, from whence ariſeth a ſtalke ſome cubit high, pretty ſtiffe, triangular, ioynted, ſet at each ioynt with a large greene leafe which at the bottome incompaſſes the ſtalke, which is omitted in the figure. At the top of the ſtalke, as in the true Cyperus, come forth two or three pretty large leaues, betweene which riſe vp many ſmall foot-ſtalkes very much branched, and bearing many blacke ſeeds ſomewhat like Millet or ruſhes.

¶ *The Place and Time.*

All theſe grow in ditches and waterie places,

places, and are to be found with their heads about the middle of Sommer, and ſome of them ſooner.

¶ *The Names.*

The firſt of theſe by *Lobel* is called *Gramen paluſtre majus*.

2 This by *Geſner*, *Lobel*, and *Dodonæus* is called *Pſeudocyperus*.

3 *Lobel* names this, *Cyperus longus inodorus ſylueſtris*.

4 He alſo calls this, *Cyperus aquaticus ſeptentrionalis*.

5 This is the *Cyperus graminea miliacea* of *Lobel* and *Pena*: the *Iuncus latus* in the *Hiſtor. Lugd. pag.* 988. and the *Pſeudocyperus polycarpos* of *Thalius*.

‡ 4 *Cyperus rotundus inodorus ſylueſtris.*
Round Baſtard Cyperus.

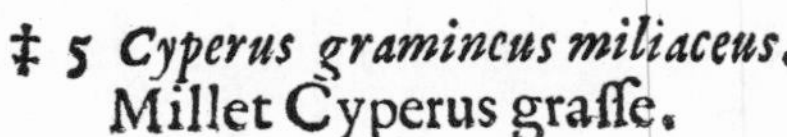

‡ 5 *Cyperus gramineus miliaceus.*
Millet Cyperus graſſe.

¶ *The Temper and Vertue.*

None of theſe are made vſe of in phyſicke; but by their taſte they ſeeme to be of a cold and aſtringent qualitie. ‡

‡ Chap. 22. *Of diuers other Graſſes.*

¶ *The Deſcription.*

‡ 1 This Ote or Hauer-graſſe, deſcribed by *Cluſius*, hath ſmall creeping roots: the ſtalks are ſome cubit high, ſlender ioynted, and ſet with ſhort narrow leaues: at the top of the ſtalke growes the eare, long, ſlender, and bending, compoſed of downy huskes containing a ſeed like to a naked Ote. The ſeed is ripe in Iuly. It growes in the mountainous and ſhadowie woods of Hungary, Auſtria, and Bohemia. Our Author miſtaking himſelfe in the figure, and as much in the title, gaue the figure of this for Burnt Barley, with this title, *Hordeum Diſtichon*. See the former edition, *pag.66.*

2 I cannot omit this elegant Graſſe, found by M. *Goodyer* vpon the wals of the antient city of Wincheſter, and not deſcribed as yet by any that I know of. It hath a fibrous and ſtringy root, from which ariſe leaues long and narrow, which growing old become round as thoſe of *Spartum* or

Mat-

Mat-weed: amongſt theſe graſſie leaues there growes vp a ſlender ſtalke ſome two foot long, ſcarſe ſtanding vpright, but oft times hanging down the head or top of the eare: it hath ſome two ioints, and at each of theſe a pretty graſſy leafe. The eare is almoſt a foot in length, compoſed of many ſmall and ſlender hairy tufts, which when they come to maturitie looke of a grayiſh or whitiſh colour, and do very well reſemble a Capons taile; whence my friend, the firſt obſeruer thereof, gaue it the title of *Gramen* ἀλεκτρυόνειος, or Capons-taile Graſſe: by which name I receiued the ſeed thereof, which ſowen, tooke root, and flouriſhes.

‡ 1 *Gra. montanum avenaceum.*
Mountaine Hauer-graſſe.

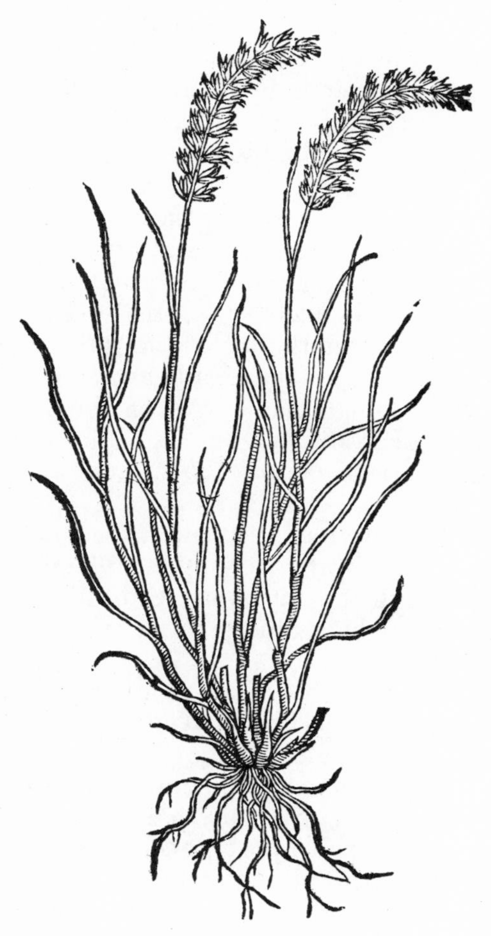

‡ 2 *Gramen murorum ſpica longiſſima.*
Capon-taile Graſſe.

3 Next to this I thinke fit to place the *Gramen Criſtatum*, or Cocks-combe graſſe of *Bauhinus*. This Graſſe hath for the root many white fibrous threds thicke packt together; the leaues are but ſhort, about the bigneſſe of the ordinarie medow graſſe; the ſtalks are ſome cubit and halfe high, with ſome two or three knots a piece: the leaues of the ſtalke are ſome foure or fiue inches long: the eare is ſmall, longiſh, of a pale greene colour, ſomewhat bending, ſo that in ſome ſort it reſembles the combe of a Cocke, or the ſeed-veſſell of that plant which is called *Caput Gallinaceum*. This is ordinarily to be found in moſt medowes about Mid-ſummer.

4 There is alſo commonly about the ſame time in our medowes to be found a Graſſe growing to ſome cubit high, hauing a ſmall ſtalke, at the top whereof there growes an eare ſome inch and an halfe, or two inches long, conſiſting as it were of two rankes of corne: it very much reſembles Rie both in ſhape and colour, and in his ſhort bearded awnes, wherefore it may very fitly be termed *Gramen ſecalinum*, or Rie-graſſe. Yet is it not *Gramen ſpica ſecalina* which *Bauhine* deſcribes in the fifty ſeuenth place, in his *Prodromus, pag*. 18. for that is much taller, and the eare much larger than this of my deſcription.

5 In diuers places about hedges, in Iuly and Auguſt is to be found a fine large tall Graſſe, which *Bauhine* (who alſo firſt deſcribed it) hath vnder the name of *Gramen ſpica Brizæ majus*. This hath ſtalkes as tall as Rie, but not ſo thicke, neither are the leaues ſo broad: at the top of the ſtalk grow diuers pretty little flattiſh eares conſiſting of two rankes of chaffie huskes or ſeed-veſſells, which haue yellowiſh little floures like to thoſe of Wheat.

6 There is alſo commonly to be found about May or the beginning of Iune, in medowes and ſuch

ſuch places that graſſe which in the *Hiſtoria Lugdun.* is ſet forth vnder the name of *Gramen Lanatum Daleſchampij:* the ſtalkes and leaues are much like the common medow graſſe, but that they are more whitiſh and hairy; the head or panicle is alſo ſoft and woolly, and it is commonly of a gray, or elſe a murrie colour.

7 There is to be found in ſome bogs in Summer time about the end of Iuly a pretty ruſhie graſſe ſome foote or better in height, the ſtalke is hard and ruſhie, hauing ſome three ioints, at each whereof there comes forth a leafe as in other graſſes, and out of the boſome of the two vppermoſt of theſe leaues comes out a ſlender ſtalke being ſome 2 or 3 inches high, and at the top thereof growes as in a little vmble a prety white chaffie floure; and at, or nigh to the top of the maine ſtalke there grow three or foure ſuch floures cluſtering together vpon little ſhort and ſlender foot ſtalkes: the leaues are but ſmall, and ſome handfull or better long, the roote I did not obſerue. This ſeemes to haue ſome affinitie with the *Gramen junceum aquaticum*, formerly deſcribed in the ninth chapter. I neuer found this but once, and that was in the companie of M. *Thomas Smith*, and M. *Iames Clarke*, Apothecaries of London; we riding into Windſore Foreſt vpon the ſearch of rare plants, and we found this vpon a bogge neere the high way ſide at the corner of the great parke. I thinke it may very fitly be called *Gramen junceum leucanthemum:* White floured ruſh-graſſe.

8 The laſt yeare at Margate in the Iſle of Tenet, neere to the ſea ſide and by the chalky cliffe I obſerued a pretty litle graſſe which from a ſmall white fibrous roote ſent vp a number of ſtalkes of an vnequall height; for the longeſt, which were thoſe that lay partly ſpred vpon the ground, were ſome handfull high, the other that grew ſtraight vp were not ſo much; and of this, one inch and halfe was taken vp in the ſpike or eare, which was no thicker than the reſt of the ſtalke, and ſeemed nothing elſe but a plaine ſmooth ſtalke, vnleſſe you looked vpon it earneſtly, and then you might perceiue it to be like Darnell graſſe: wherefore in the Iournall that I wrot of this Simpling voyage, I called it *pag.* 3. *Gramen parvum marinum ſpica Loliacea.* I iudge it to be the ſame that *Bauhine* in his *Prodromus, pag.* 19 hath ſet forth vnder the name of *Gramen Loliaceum minus ſpica ſimplici.* It may be called in Engliſh, Dwarfe Darnell Graſſe.

9 The Darnell graſſe that I compared the eare of this laſt deſcribed vnto, is not the *Gramen ſorhginum* (which our Author called Darnel-graſſe) but another graſſe growing in moſt places with ſtalkes about ſome ſpan high, but they ſeldome ſtand vpright, the eare is made iuſt like that which hereafter *chap.* 58. is called *Lolium rubrum*, Red Darnell, of which I iudge this a variety, differing little therefrom but in ſmallneſſe of growth.

10 Vpon Hampſted heath I haue often obſerued a ſmall graſſe whoſe longeſt leaues are ſeldome aboue two or three inches high, and theſe leaues are very greene, ſmall, and perfectly round like the *Spartum Auſtriacum*, or Feather-graſſe: I could neuer finde any ſtalke or eare vpon it: wherefore I haue brought it into the Garden to obſerue it better. In the forementioned Iournall, *pag.* 33. you may finde it vnder the name of *Gramen Spartium capillaceo folio minimum.* It may be this is that graſſe which *Bauhine* ſet forth in his *Prodromus, pag.* 11. vnder the title of *Gramen ſparteum Monſpeliacum capillaceo folio minimum.* I haue thought good in this place to explaine my meaning by theſe two names to ſuch as are ſtudious of plants, which may happen to light by chance (for they were not intended for publicke) vpon our Iournall, that they need not doubt of my meaning.

11 I muſt not paſſe ouer in ſilence two other Graſſes, which for any thing that I know are ſtrangers with vs, the one I haue ſeene whith M. *Parkinſon*, and it is ſet forth by *Bauhine, pag.* 30. of his *Prodromus*. The other by *Lobell* in the ſecond part of his *Adverſaria, pag.* 468. The firſt (which *Bauhine* fitly calls *Gramen alopecuriodes ſpica aſpera*, and thinkes it to be *Gram. Echinatum Daleſchampij*, deſcribed *Hiſt. Lugd. pag.* 432.) hath a fibrous and white root, from which ariſes a ſtiffe ſtalke diuided by many knots, or knees: the leaues are like to the other fox-taile graſſes, but greener: the eare is rough, of ſome inch in length, and growes as it were vpon one ſide of the ſtalke: the eare at firſt is greene, and ſhewes yellowiſh little flowers in Auguſt.

12 This other Graſſe which *Lobell* in the quoted place figures and deſcribes by the name of *Gramen Scoparium Iſchæmi panniculis Gallicum*, hath rootes ſome cubit long, ſlender, and very ſtiffe, (for of theſe are made the head bruſhes which are vulgarly vſed) the ſtraw is ſlender, and ſome cubit high, being heere and there ioynted like to other Graſſes: the top hath foure or fiue eares ſtanding after the manner of Cocks foot Graſſe, whereof it is a kinde. It growes naturally about Orleance, and may be called in Engliſh, Bruſh-graſſe. ‡

CHAP. 23. *Of Cotton Graſſe.*

¶ *The deſcription.*

1 This ſtrange Cotton graſſe, which *L'Obelius* hath comprehended vnder the kindes of Ruſhes; notwithſtanding that it may paſſe with the Ruſhes, yet I finde in mine owne experience, that it doth rather reſemble graſſe than ruſhes, and may indifferently be taken for either, for that it doth participate of both. The ſtalke is ſmall and ruſhy, garniſhed with many graſſy leaues alongſt the ſame, bearing at the top a buſh or tuft of moſt pleaſant downe or cotton like vnto the moſt fine and ſoft white ſilke. The root is very tough, ſmall and threddy.

2 This Water Gladiole, or graſſy Ruſh, of all others is the faireſt and moſt pleaſant to behold, and ſerueth very well for the decking and trimming vp of houſes, becauſe of the beauty and brauerie thereof: conſiſting of ſundry ſmall leaues, of a white colour mixed with carnation, growing at the top of a bare and naked ſtalke, fiue or ſix foot long, and ſometime more. The leaues are long and flaggy, not much vnlike the common reed. The root is threddy, and not long.

1 *Gramen Tomentarium.*
Cotton Graſſe.

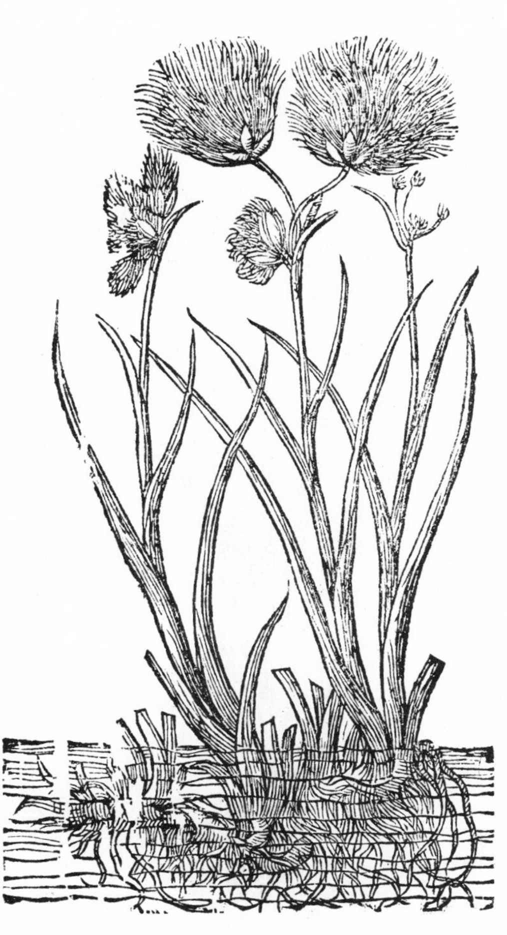

2 *Gladiolus paluſtris Cordi.*
Water Gladiole.

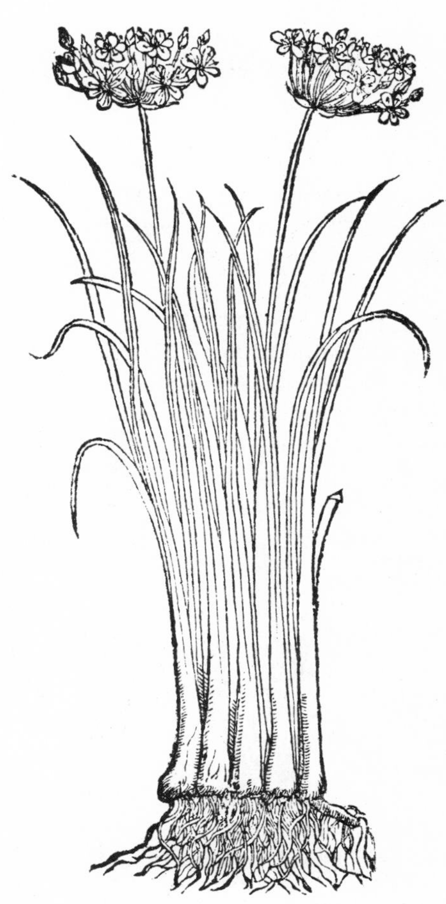

¶ *The place and time.*

1 Cotton graſſe groweth vpon bogs and ſuch like mooriſh places, and it is to be ſeene vpon the bogs on Hampſted heath. It groweth likewiſe in Highgate parke neere London.

2 Water Gladiole groweth in ſtanding pooles, motes, and water ditches. I found it in great plenty being in company with a Worſhipfull Gentleman Maſter *Robert Wilbraham*, at a Village fifteene miles from London called Buſhey. It groweth likewiſe neere Redriffe by London, and many other places: the ſeaſon anſwereth all others.

¶ *The names.*

1 *Gramen Tomentoſum* is called likewiſe *Iuncus bombicinus*: of *Cordus*, *Linum pratenſe*, and *Gnaphalium Hieronymi Bockij*. In Engliſh Cotton graſſe.

2 Water

2 Water Gladiole is called of *L'Obelius*, *Iuncus Cyperoides floridus paludoſus*, Flowring Cypreſſe Ruſh: *Iuncus*, for that his ſtalke is like the ruſh: *Cyperoides*, becauſe his leaues reſemble *Cyperus*: *Floridus*, becauſe it hath on the top of euery ſtalke a fine vmble or tuft of ſmall flowers, in faſhion of the Lilly of Alexandria, the which it is very like, and therefore I had rather call it Lilly graſſe.

The nature and vertues.

A *Cordus* ſaith, That *Iuncus bombicinus* ſodden in wine, and ſo taken, helpeth the throwes and gripings of the belly, that women haue in their childing.

There be alſo ſundry kinds of Graſſes wholly vnknowne, or at the leaſt not remembred of the old Writers, whereof ſome few are touched in name onely by the late and new Writers: now for as much as they haue onely named them, I will referre the better conſideration of them to the induſtrie and diligence of painefull ſearchers of nature, and proſecute my purpoſed labour, to vnfold the diuers ſorts and manifold kindes of *Cyperus*, Flags, and Ruſhes: and becauſe that there is added vnto many of the Graſſes before mentioned, this difference, *Cyperoides*, that is to ſay, reſembling *Cyperus*, I thought it therefore expedient to ioyne next vnto the hiſtory of graſſes, the diſcourſe of *Cyperus*, and his kindes, which are as follow.

Chap. 24. *Of Engliſh Galingale.*

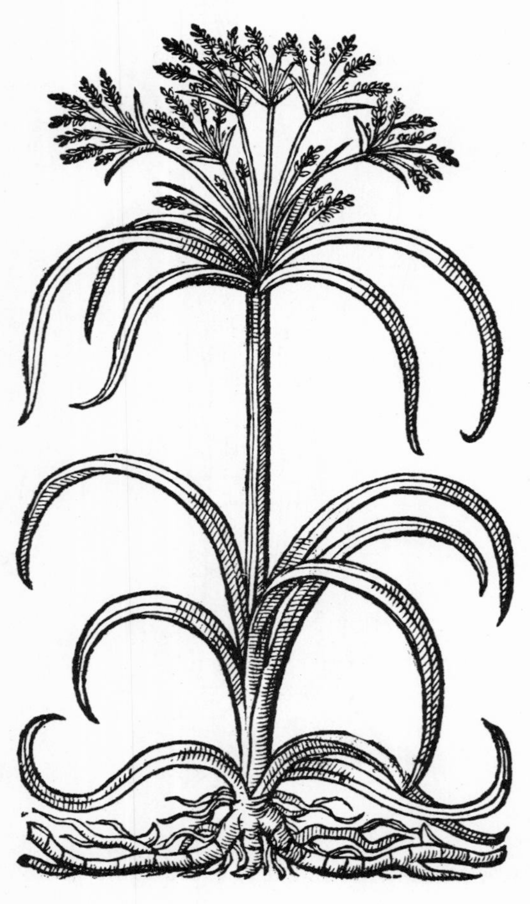

1 *Cyperus longus.*
Engliſh Galingale

2 *Cyperus rotundus vulgaris.*
Round Galingale.

¶ *The deſcription.*

Engliſh Galingale hath leaues like vnto the common Reed, but leſſer and ſhorter. His ſtalke is three ſquare, two cubits high: vpon whoſe top ſtand ſundry branches, euery little branch bearing many ſmall chaffy ſpikes. The root is blacke and very long, creeping hither and thither, occupying much ground by reaſon of his ſpreading: it is of a moſt ſweet and pleaſant ſmell when it is broken.

2 The

2 The common round *Cyperus* is like the former in leaues and tops, but the roots are here and there knotty and round, and not altogether so well smelling as the former.

‡ 3 There is also another *Cyperus* which growes in Syria and Ægypt, whose roots are round, blackish, and large, many hanging vpon one string, and hauing a quicke and aromaticke smell: the leaues and spokyn-tufts resemble the former.

4 There is said to be another kinde of this last described, which is lesser, and the roots are blacker, and it growes in Creet, now called Candy.

5 There is also another round *Cyperus* which growes about ditches and the bankes of Riuers whereas the salt water sometimes comes: the roots of this are hard and blacke without smell, many hanging sometimes vpon one string: the stalke and leaues are much like the former, but the heads vnlike, for they are rough and blackish, about the bignesse of a filbert, and hang some six or seuen at the top of the stalke. It floures in Iuly and August. ‡

5 *Cyperus rotundus littoreus.*
Round Salt-marsh *Cyperus.*

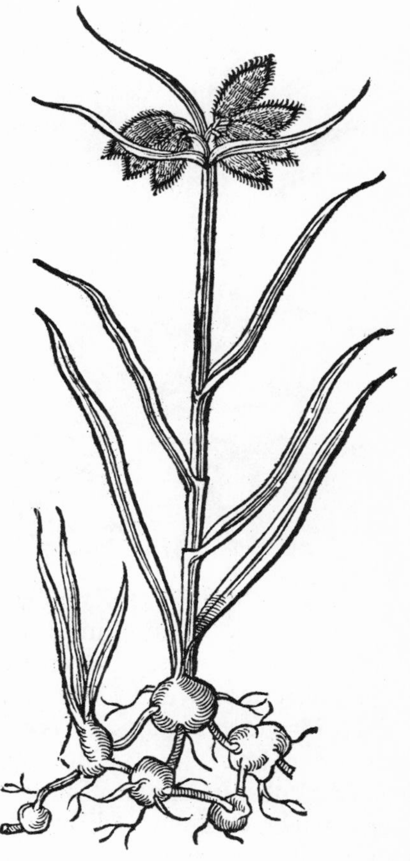

¶ *The place and time.*

1 2 The first and second of these grow naturally in fenny grounds, yet will they prosper exceedingly in gardens, as experience hath taught vs.

3 4 The former of these growes naturally in Syria and Ægypt, the later in Candy.

5 This growes plentifully in the Marishes below Grauesend, in Shipey, Tenet, and other places.

¶ *The name in generall.*

Cyperus is called in Greeke, κύπειρος, or κύπειρις: of the Latines as well *Cypirus* as *Cyperus*: of some *Iuncus quadratus*: of *Pliny Iuncus Angulosus*, and *Triangularis*: of others *Aspalathum* and *Erysisceptron*: in French *Souchet*: in Dutch **Galgan**: in Spanish *Iunco odorosa*: By vs *Cyperus* and English Galangall.

‡ ¶ *The names in particular.*

1 This is called *Cyperus longus*, and *Cyperus longus Oderatior*: in English, Common *Cyperus*, and English Gallingall. 2 This is called *Cyperus rotundus vulgaris*, Round English Galangall. 3 *Cyperus rotundus Cyriacus*, or *Ægyptiacus*, Syrian or Ægyptian round *Cyperus*. 4 *Cyperus minor Creticus*, Candy round *Cyperus*. 5 *Cyperus rotundus inodorus Littoreus*, Round Salt-marsh *Cyperus*, or Galingale. ‡

¶ *The nature.*

Dioscorides saith, That *Cyperus* hath an heating qualitie. *Galen* saith, The roots are most effectual in medicine, and are of an heating and drying qualitie: and some doe reckon it to be hot and dry in the second degree.

¶ *The vertues.*

It maketh a most profitable drinke to breake and expell grauell, and helpeth the dropsie. A

If it be boyled in wine, and drunke, it prouoketh vrine, driueth forth the stone, and bringeth downe the naturall sicknesse of women. B

The same taken as aforesaid, is a remedie against the stinging and poyson of Serpents. C

Fernelius saith, The root of *Cyperus* vsed in Baths helpeth the coldnesse and stopping of the matrix, and prouoketh the termes. D

He writeth also, that it increaseth bloud by warming the body, and maketh good digestion; wonderfully refreshing the spirits, and exhilarating the minde, comforting the senses, and encreasing their liuelinesse, restoring the colour decayed, and making a sweet breath. E

The powder of *Cyperus* doth not onely dry vp all moist vlcers either of the mouth, priuy members, and fundament, but stayeth the humor and healeth them, though they be maligne and virulent, according to the iudgement of *Fernelius*. F

‡ CHAP. 25. *Of Italian Traſi, or Spaniſh Galingale.*

1 *Cyperus Eſculentus ſine Caule & flore.*
Italian Traſi, or Spaniſh Galingall, without ſtalke and floure.

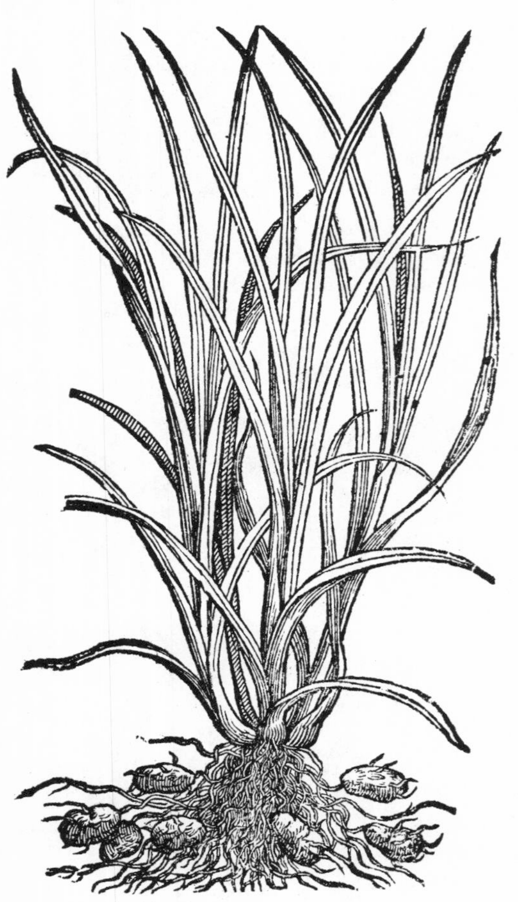

2 *Cyperus Eſculentus, ſiue Traſi Italorum.*
Italian Traſi, or Spaniſh Galingall.

‡ 1 THe Italian Traſi, which is here termed Spaniſh Galingale, is a plant that hath many ſmall roots, hanging at ſtringy fibers like as our ordinary Dropwort roots do, but they are of the bigneſſe of a little Medlar, and haue one end flat and as it were crowned like as a Medlar, and it hath alſo ſundry ſtreakes or lines ſeeming to diuide it into ſeueral parts: it is of a browniſh colour without, and white within; the taſte thereof is ſweet almoſt like a Cheſnut. The leaues are very like thoſe of the garden *Cyperus*, and neuer exceed a cubit in length. Stalkes, flowers, or ſeed it hath none, as *Iohn Pona* an Apothecary of Verona, who diligently obſerued it nigh to that city whereas it naturally growes, affirmes; but he ſaith there growes with it much wild *Cyperus*, which as he judges hath giuen occaſion of their error who giue it the ſtalkes and flowers of *Cyperus*, or Engliſh Galingale, as *Matthiolus* and others haue done. It is encreaſed by ſetting the roots firſt ſteeped in water, at the beginning of Nouember. I haue here giuen you the figure of it without the ſtalke, according to *Pona*, and with the ſtalke, according to *Matthiolus* and others.

¶ *The names.*

The Italian Traſi is called in Greeke by *Theophraſtus* μελιναθάλλη, *Hiſt. plant.* 4. *cap.* 10. as *Fabius Columna* hath proued at large: *Pliny* termes it *Anthalium*: the later writers *Cyperus Eſculentus*, and *Dulcichinum*: The Italians, *Traſi*, and *Dolzolini*, by which names in Italy they are cryed vp and downe the ſtreets, as Oranges and Lemmons are here.

¶ *The temper and vertues.*

A The milke or creame of theſe Bulbous rootes being drunke, mundifies the breſt and lungs, wherefore it is very good for ſuch as are troubled with coughs. Now you muſt beat theſe roots, and macerate them in broth, and then preſſe out the creame through a linnen cloath, which by ſome late Writers is commended alſo to be vſed in venereous potions.

B The ſame creame is alſo good to be drunke againſt the heate and ſharpneſſe of the vrine, eſpecially if you in making it do adde thereto the ſeeds of Pompions, Gourds, and Cucumbers. The Citiſens of Verona eate them for dainties, but they are ſomewhat windy. ‡

‡ CHAP. 26. *Of the true Galingale, the greater and the leſſer.*

‡ 1 *Galanga major.*
The greater Galingale.

‡ 2 *Galanga minor.*
The leſſer Galingale.

THe affinitie of name and nature hath induced me in this place to inſert theſe two, the bigger and the leſſer Galingale; firſt therefore of the greater.

¶ *The deſcription.*

1 The great Galingale, whoſe root onely is in vſe, and brought to vs from Iava in the Eaſt Indies, hath flaggy leaues ſome two cubits high, like theſe of Catſ-taile or Reed-mace: the root is thicke and knotty, reſembling thoſe of our ordinary flagges, but that they are of a more whitiſh colour on the inſide, and not ſo large. Their taſt is very hot and biting, and they are ſomwhat reddiſh on the outſide.

2 The leſſer growing in China, and commonly in ſhops called Galingale, without any addition, is a ſmall root of a browniſh red colour both within and without; the taſte is hot and biting, the ſmell aromaticall, the leaues (if we may beleeue *Garcias ab Horto*) are like thoſe of Myrtles.

¶ *The names.*

1 The firſt is called by *Matthiolus, Lobell*, and others, *Galanga major*. Some thinke it to be the *Acorus* of the Ancients: and *Pena* and *Lobell* in their *Stirp. Aduerſ.* queſtion whither it be not the *Acorus Galaticus* of *Dioſcorides*. But howſoeuer, it is the *Acorus* of the ſhops, and by many vſed in Mithridate in ſtead of the true. The Indians call it *Lancuaz*.

2 The leſſer is called *Galanga*, and *Galanga minor*, to diſtinguiſh it from the precedent. The Chinois call it *Lauandon*: the Indians *Lancuaz*: we in England terme it Galingale, without any addition.

¶ *Their temper and vertue.*

Theſe roots are hot and dry in the third degree, but the leſſer are ſomewhat the hotter.

They ſtrengthen the ſtomacke, and mitigate the paines thereof ariſing from cold and flatulencies. A

The ſmell, eſpecially of the leſſer, comforts the too cold braine; the ſubſtance thereof being chewed ſweetens the breath. It is good alſo againſt the beating of the heart. B

They are vſefull againſt the Collicke proceeding of flatulencies, and the flatulent affects of the wombe; they conduce to venery, and heate the too cold reines. To conclude, they are good againſt all cold diſeaſes. ‡ C

‡ CHAP. 27. *Of Turmericke.*

THis alſo challengeth the next place, as belonging to this Tribe, according to *Dioſcorides*; yet the root, which onely is brought vs, and in vſe, doth more on the outſide reſemble Ginger, but that it is yellower, and not ſo flat, but rounder. The inſide thereof is of a Saffron colour, the taſte hot and bitteriſh; it is ſaid to haue leaues larger than thoſe of Millet, and a leafie ſtalke. There is ſome varietie of theſe roots, for ſome are longer, and others rounder, and the later are the hotter, and they are brought ouer oft times together with Ginger.

¶ *The place.*

It growes naturally in the Eaſt-Indies about Calecut, as alſo at Goa.

¶ *The names.*

This without doubt is the *Cyperus Indicus* of *Dioſcorides, Lib.* 1. *Cap.* 4. It is now vulgarly by most Writers, and in shops, called by the name of *Terra merita*, and *Curcuma*: yet some terme it *Crocus Indicus*, and we in English call it Turmericke.

¶ *The temperature and vertues.*

A This root is certainly hot in the third degree, and hath a qualitie to open obstructions, and it is vsed with good successe in medicines against the yellow Iaundise, and against the cold distempers of the liuer and spleene.

Chap. 28. *Of Zedoarie.*

‡ Zedoarie is also a root growing naturally in the woods of Malavar about Calecut and Cananor in the Indies; the leaues thereof are larger than Ginger, and much like them; the root is also as large, but consisting of parts of different figures, some long and small, others round; their colour is white, and oft times brownish on the inside, and they haue many fibers comming out of them, but they are taken away together with the outward rinde before they come to vs. These roots haue a strong medicine-like smell, and somewhat an vngratefull taste.

‡ *Zerumbeth, siue Zedoaria rotunda.*
Round Zedoarie.

¶ *The names.*

Some call the long parts of these roots *Zedoaria*, and the round (whose figure we here giue you) *Zerumbeth*, and make them different, whenas indeed they are but parts of the same root, as *Lobell* and others haue well obserued. Some make *Zedoaria* and *Zerumbeth* different, as *Auicen*: others confound them and make them one, as *Rhases* and *Serapio*. Some thinke it to be Ἀρνάβω of Ægineta: but that is not so; for he saith, τῶν ἀρωματιζόντων ἐστὶν, ὅθεν τοῖς μύροις μάλιστα μίγνυται; It is an Aromaticke, and therefore chiefely mixed in ointments: which is as much as if he should haue said, That it was put into ointments for the smells sake, which in this is no wayes gratefull, but rather the contrarie.

¶ *The temperature and vertues.*

A It is hot and dry in the second degree; it discusses flatulencies, and fattens by a certaine hidden qualitie. It also dissipates and amends the vngratefull smell which Garlicke, Onions, or too much wine infect the breath withall, if it be eaten after them. It cures the bites and stings of venomous creatures, stops laskes, resolues the Abscesses of the wombe, stayes vomiting, helpes the Collicke, as also the paine of the stomacke.

B It kills all sorts of wormes, and is much vsed in Antidotes against the plague, and such like contagious diseases. ‡

Chap. 29. *Of Rushes.*

‡ I Do not here intend to trouble you with an accurate distinction and enumeration of Rushes; for if I should, it would be tedious to you, laborious to me, and beneficiall to neither. Therefore I will onely describe and reckon vp the chiefe and more note-worthy of them, beginning with the most vsuall and common. ‡

¶ *The description.*

1 The roots of our common Rushes are long and hairy, spreading largely in the ground, from which, as from one entire tuft, proceed a great company of small rushes; so exceedingly well knowne, that I shall not need to spend much time about the description thereof.

2 There be sundry sorts of Rushes besides the former, whose pictures are not here exprest, and the rather, for that the generall description of Rushes, as also their common vse and seruice, are sufficient to leade vs to the knowledge of them. This great Water-Grasse or Bul-Rush, in stead of leaues bringeth forth many strait twiggie shoots or springs, which be round, smooth, sharpe pointed, and without knots. Their tuft or flower breaketh forth a little beneath the top, vpon the one side of the Rush, growing vpon little short stems like Grape clusters, wherein is contained the seed after the fashion of a speares point. The roots be slender and full of strings. *Pliny*, and *Theophrastus* before him, affirme that the roots of the Rush do die euery yeare, and that it groweth

it groweth againe of the ſeed. And they affirme likewiſe that the male is barren, and groweth againe of the yong ſhoots; yet I could neuer obſerue any ſuch thing.

‡ 3 There growes a Ruſh to the thicknes of a Reed, and to ſome two yards and an halfe, or three yards high, in diuers fenny grounds in this kingdome; it is very porous and light, and they vſually make mats, and bottom chaires therewith. The ſeeds are contained in reddiſh tufts, breaking out at the top thereof. The roots are large and ioynted, and it grows not vnleſſe in waters. ‡

4 *Iuncus acutus*, or the ſharpe Ruſh, is likewiſe common and well knowne; not much differing from *Iuncus læuis*, but harder, rougher, and ſharper pointed, fitter to ſtraw houſes and chambers than any of the reſt; for the others are ſo ſoft and pithy, that they turne to duſt and filth with much treading; where contrariwiſe this ruſh is ſo hard that it will laſt ſound much longer.

‡ 5 There is alſo another pretty ſmall kinde of Ruſh growing to ſome foot in heigth, hauing ſmooth ſtalkes which end in a head like to that of the ordinary Horſe-taile. This ruſh hath alſo one little ioynt towards the bottome thereof. It growes in watery places, but not ſo frequently as the former. ‡

1 *Iuncus læuis.* Common Ruſhes.

4 *Iuncus acutus.* Sharpe Ruſh, or hard Ruſh.

3 *Iuncus aquaticus maximus.* Great Water-Ruſh, or Bul-Ruſh.

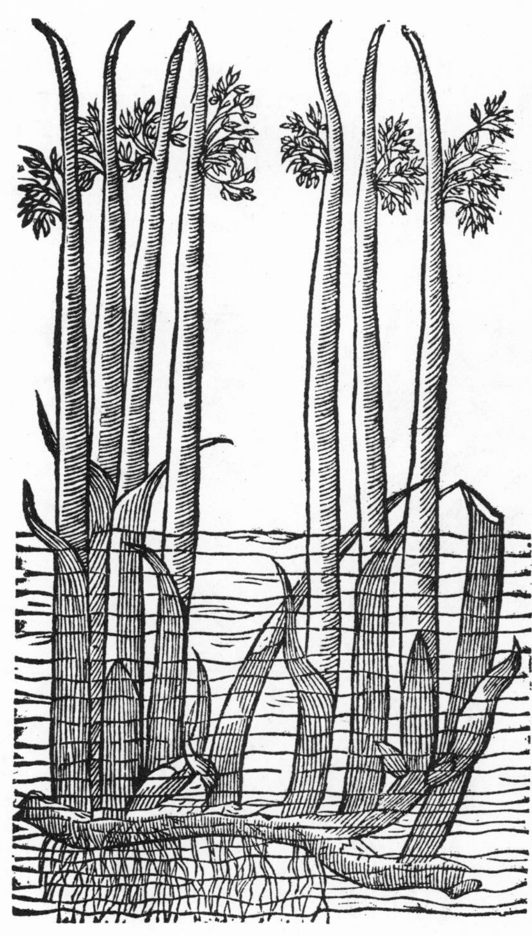

¶ *The place.*

1 *Iuncus læuis* groweth in fertile fields, and meadowes that are ſomewhat moiſt.

2 3 5 Grow in ſtanding pooles, and by riuers ſides in ſundry places.

4 *Iuncus acutus* groweth vpon dry and barren grounds, eſpecially neere the furrows of plowed land. I need not ſpeake of their time of growing, they being ſo common as they are.

¶ *The names.*

The Ruſh is called in Greeke σχοῖνος: in Latine *Iuncus*: in high Dutch **Binken**: in low Dutch **Bieſen**: in Italian *Giunco*: in Spaniſh *Iunco*: in French *Ionc*: in Engliſh Ruſhes.

2 3 The Grecians haue called the Bull-Ruſh ὁλοσχοῖνος. The greater are commonly in many places termed Bumbles.

1 *Iuncus læuis* is that Ruſh which *Dioſcorides* called σχοῖνος λεία.

4 *Iuncus acutus* is called in Greeke ὀξυσχοῖνος: In Dutch **Pieren Bieſen.**

5 This is called by *Lobell*, *Iuncus aquaticus minor Capitulis Equiſeti*: By *Daleſchampius*, *Iuncus clauatus*, or Club-Ruſh.

¶ *The*

¶ *The nature and vertues.*

These Rushes are of a dry nature.

A The seed of Rushes dried at the fire, and drunke with wine alayed with water, stayeth the laske and the ouermuch flowing of womens termes.

B *Galen* yeeldeth this reason thereof, because that their temperature consisteth of an earthy essence, moderately cold and watery, and meanly hot, and therefore doth the more easily drie vp the lower parts, and by little and little send vp the cold humours to the head, whereby it prouoketh drowsinesse and desire to sleepe, but causeth the head-ache; whereof *Galen* yeeldeth the reason as before.

C The tender leaues that be next the root make a conuenient ointment against the bitings of the Spider called *Phalangium*.

D The seed of the Bull-Rush is most soporiferous, and therefore the greater care must be had in the administration thereof, lest in prouoking sleepe you induce a drowsinesse or dead sleepe.

Chap. 30. Of Reeds.

¶ *The kindes.*

OF Reeds the Ancients haue set downe many sorts. *Theophrastus* hath brought them all first into two principall kindes, and those hath he diuided againe into moe sorts. The two principall are these, *Auleticæ*, or *Tibiales Arundines*, and *Arundo vallatoria*. Of these and the rest we will speake in their proper places.

1 *Arundo vallatoria.*
Common Reed.

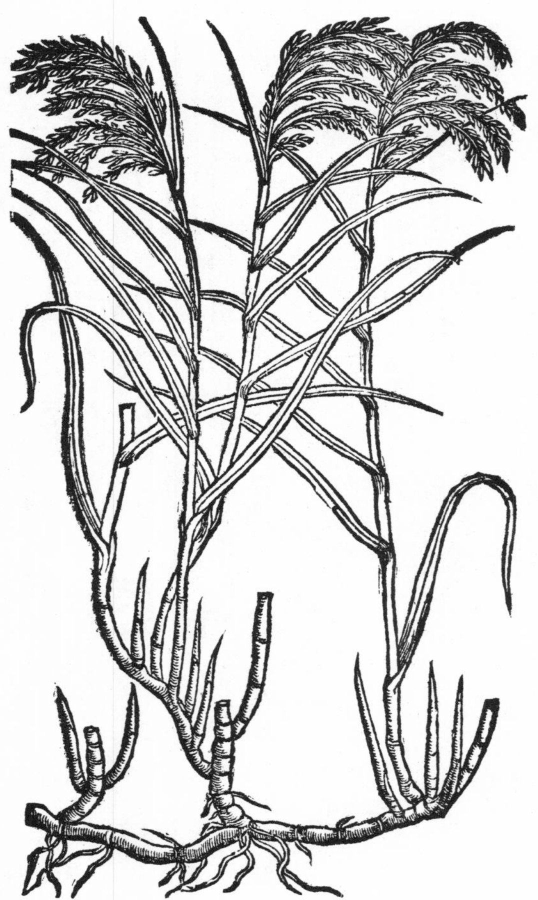

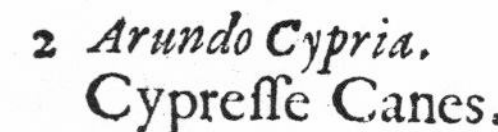

2 *Arundo Cypria.*
Cypresse Canes.

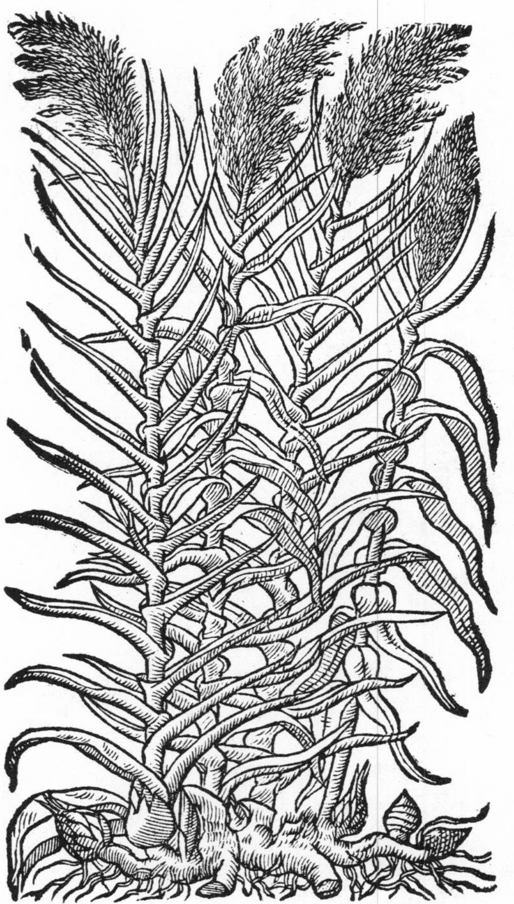

¶ *The description.*

1 THe common Reed hath long strawie stalkes full of knotty joints or knees like vnto corne, whereupon do grow very long rough flaggy leaues. The tuft or spoky eare doth grow at the top of the stalkes, browne of colour, barren and without seed, and doth resemble a bush of feathers, which turneth into fine downe or cotton which is carried away with the winde. The root is thicke, long, and full of strings, dispersing themselues farre abroad, whereby

wherby it doth greatly increase. ‡ *Bauhinus* reports, That he receiued from D.*Cargill* a Scottish-man a Reed whose leaues were a cubit long, and two or three inches broad, with some nerues apparantly running alongst the leafe; these leaues at the top were diuided into two, three, or foure points or parts; as yet I haue not obserued it. *Bauhine* termes it *Arundo Anglica folijs in summitate dissectis*. ‡

1 The Cypresse Reed is a great Reed hauing stalkes exceeding long, sometimes twenty or thirty foot high, of a woody substance, set with very great leaues like those of Turky wheate. It carrieth at the top the like downie tuft that the former doth.

3 *Arundo farcta.*
Stuffed Canes.

4 *Calamus sagittalis Lobelij.*
Small stuffed Reed.

5 *Nastos Clusij.*
Turky walking staues.

6 *Arundo scriptoria.*
Turky writing Reeds.

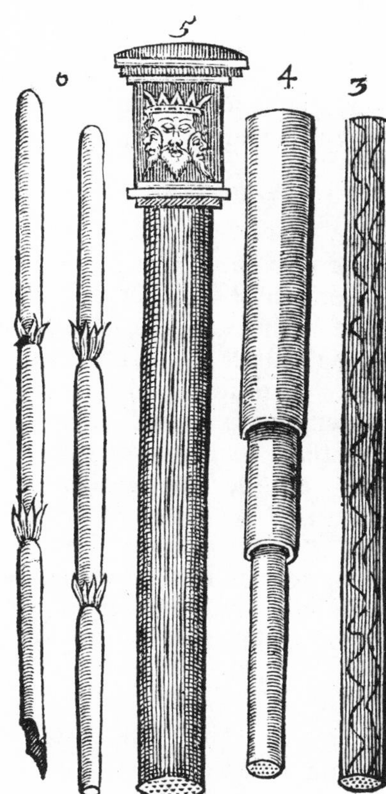

3 These Reeds *Lobelius* hath seene in the Low countries brought from Constantinople, where, as it is said, the people of that countrey haue procured them from the parts of the Adriaticke sea side where they do grow. They are full stuft with a spongeous substance, so that there is no hollownesse in the same, as in Canes & other Reeds, except here and there certaine small pores or passages of the bignesse of a pinnes point; in manner such a pith as is to be found in the Bull-Rush, but more firme and solid.

4 The second differeth in smalnesse, and that it will winde open in fleakes, otherwise they are very like, and are vsed for darts, arrowes, and such like.

5 This great sort of Reeds or Canes hath no particular description to answer your expectation, for that as yet there is not any man which hath written thereof, especially of the manner of growing of them, either of his owne knowledge or report from others: so that it shall suffice that yee know that that great cane is vsed especially in Constantinople and thereabout, of aged and wealthy Citisens, and also Noblemen and such great personages, to make them walking staues of, caruing them at the top with sundry Scutchions, and pretty toyes of imagerie for the beautifying of them; and so they of the better sort do garnish them both with siluer and gold, as the figure doth most liuely set forth vnto you.

6 In like manner the smaller sort hath not as yet beene seene growing of any that haue beene curious in herbarisme, whereby they might set downe any certaintie thereof; onely it hath beene vsed in Constantinople and thereabout, euen to this day, to make writing pens withall, for the which it doth very fitly serue, as also to make pipes, and such like things of pleasure.

¶ *The place.*

The common Reed groweth in standing waters and in the edges and borders of riuers almost euery where: and the other being the angling Cane for Fishers groweth in Spaine and those hot Regions.

¶ *The time.*

They flourish and flower from April to the end of September, at what time they are cut down for the vse of man, as all do know.

¶ *The names.*

The common Reed is called *Arundo* and *Harundo vallatoria*: in French *Roseau*: in Dutch **Riet**: in Italian *Canne a far siepo*: of *Diosc. Phragmitis*: in English, Reed.

Arundo Cypria; or after *Lobelius*, *Arundo Donax*: in French *Canne*: in Spanish *Cana*: in Italian, *Calami a far Connochia*: In English, Pole reed, and Cane, or Canes.

¶ *The nature.*

Reeds are hot and dry in the second degree, as *Galen* saith.

¶ *The vertues.*

The roots of reed stamped smal draw forth thorns and splinters fixed in any part of mans body. A

The same stamped with vineger ease all luxations and members out of ioynt. B

And likewise stamped they heale hot and sharpe inflammations. The ashes of them mixed with vineger helpeth the scales and scurfe of the head, and helpeth the falling of the haire. C

The

D The great Reed or Cane is not vsed in physicke,but is esteemed to make slears for Weauers, sundry sorts of pipes, as also to light candles that stand before Images, and to make hedges and pales, as we do of laths and such like; and also to make certaine diuisions in ships to diuide the sweet oranges from the sowre, the pomecitron and lemmons likewise in sunder, and many other purposes.

Chap. 31. *Of Sugar Cane.*

¶ *The description.*

1 SVgar Cane is a pleasant and profitable Reed, hauing long stalkes seuen or eight foot high, ioynted or kneed like vnto the great Cane; the leaues come forth of euerie ioynt on euery side of the stalke one, like vnto wings, long, narrow, and sharpe pointed. The Cane it selfe, or stalke is not hollow as other Canes or Reeds are, but full, and stuffed with a spongeous substance in taste exceeding sweet. The root is great and long, creeping along within the vpper crust of the earth, which is likewise sweet and pleasant,but lesse hard or woody than other Canes or Reeds; from the which there doth shoot forth many yong siens, which are cut away from the maine or mother plant, because they should not draw away the nourishment from the old stocke, and so get vnto themselues a little moisture,or else some substance not much worth, and cause the stocke to be barren, and themselues little the better; which shoots do serue for plants to set abroad for encrease.

Arundo Saccharina.
Sugar Cane.

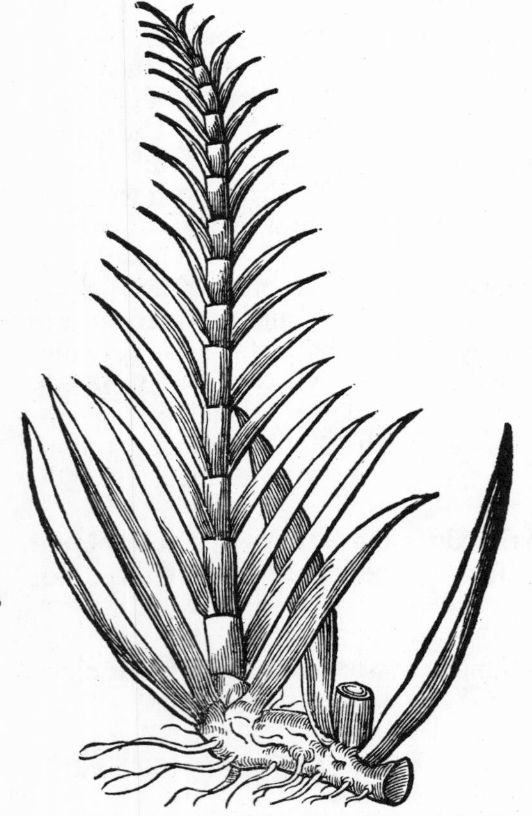

¶ *The place.*

The Sugar Cane groweth in many parts of Europe at this day, as in Spaine, Portugal, Olbia, and in Prouence. It groweth also in Barbarie, generally almost euery where in the Canarie Islands, and in those of Madera, in the East and West Indies, and many other places. My selfe did plant some shoots thereof in my garden, and some in Flanders did the like: but the coldnesse of our clymate made an end of mine,and I thinke the Flemings will haue the like profit of their labour.

¶ *The time.*

This Cane is planted at any time of the yeare in those hot countries where it doth naturally grow,by reason they feare no frosts to hurt the yong shoots at their first planting.

¶ *The names.*

The Latines haue called this plant *Arundo Saccharina,* with this additament,*Indica,*because it was first knowne or brought from India. Of some it is called *Calamus Saccharatus*: in English Sugar Cane: in Dutch **Suyckerriedt.**

¶ *The nature and vertues.*

A The Sugar or juice of this Reed is of a temperate qualitie; it drieth and cleanseth the stomacke, maketh smooth the roughnesse of the brest and lungs, cleareth the voice, and putteth away hoarsenesse, the cough,and all sourenesse and bitternesse,as *Isaac* saith in *Dictis*.

¶ *The vse.*

Of the iuyce of this Reed is made the most pleasant and profitable sweet, called Sugar, whereof is made infinite confections, confectures, syrups, and such like, as also preseruing and conseruing of sundry fruits, herbes, and flowers, as Roses, Violets, Rosemary flowers, and such like, which still retaine with them the name of Sugar, as Sugar Roset, Sugar violet, &c. The which to write of would require a peculiar volume, and not pertinent vnto this historie, for that it is not my purpose to make of my booke a Confectionarie, a Sugar Bakers furnace, a Gentlewomans preseruing pan, nor yet an Apothecaries shop or Dispensatorie; but onely to touch the chiefest matter that I purposed to handle in the beginning,that is, the nature,properties, and descriptions of plants. Notwithstanding I thinke it not amisse to shew vnto you the ordering of these reeds when

when they be new gathered, as I receiued it from the mouth of an Indian my seruant: he saith, They cut them in small pieces, and put them into a trough made of one whole tree, wherein they put a great stone in manner of a mill-stone, whereunto they tie a horse, buffle, or some other beast which draweth it round: in which trough they put those pieces of Canes, and so crush and grind them as we do the barkes of trees for Tanners, or apples for Cyder. But in some places they vse a great wheele, wherein slaues do tread and walke as dogs do in turning the spit: and some others do feed as it were the bottome of the said wheele, wherein are some sharpe or hard things which do cut and crush the Canes into powder. And some likewise haue found the inuention to turne the wheele with water workes, as we do our iron mills. The Canes being thus brought into dust or powder, they put them into great cauldrons with a little water, where they boyle vntill there be no more sweetnesse left in the crushed reeds. Then doe they straine them through mats and such like things, and put the liquor to boyle againe vnto the consistence of honey, which being cold is like vnto sand both in shew and handling, but somewhat softer; and so afterward it is carried into all parts of Europe, where it is by the Sugar Bakers artificially purged and refined to that whitenesse as we see.

CHAP. 32. *Of Flowring Reed.*

Arundo florida.
Flowring Reed.

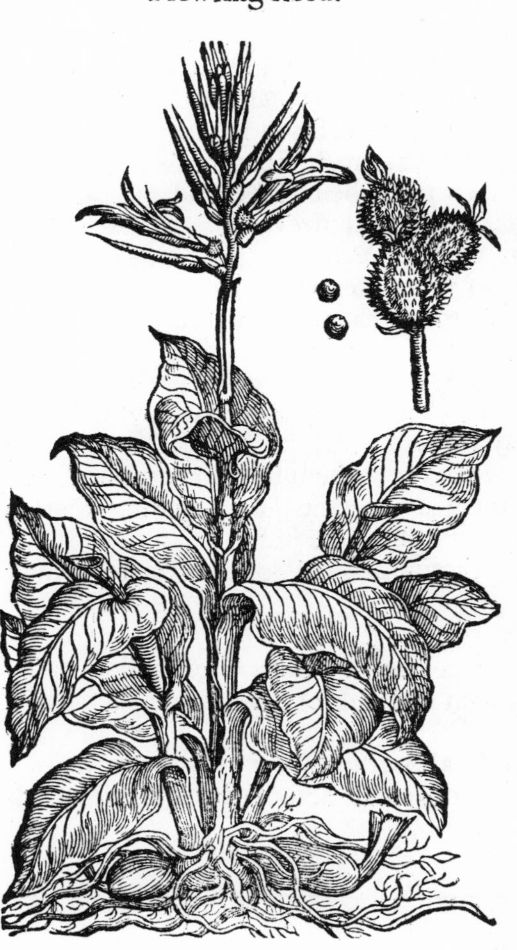

¶ *The description.*

FLourishing Reed hath a thicke and fat stalke of foure or fiue foot high, great below neere the ground, and smaller toward the top, taper-wise: whereupon do grow very faire broad leaues ful of ribs or sinewes like vnto Plantaine, in shape representing the leaues of white Hellebor, or the great Gentian, but much broader and larger euery way: at the top of which stalkes do grow phantasticke flowers of a red or vermilion colour; which being faded, there follow round, rough, and prickly knobs, like those of *Sparganium*, or water-Burre, of a browne colour, and from the middle of those knobs three small leaues. The seed contained in those knobs is exceeding black, of a perfect roundnesse, of the bignesse of the smallest pease. The root is thicke, knobby, and tuberous, with certain small threds fixed thereto. ‡ There is a variety of this, hauing floures of a yellow or Saffron colour, with red spots. ‡

¶ *The place.*

It groweth in Italy in the garden of Padua, and many other places of those hot regions. My selfe haue planted it in my garden diuers times, but it neuer came to flowring or seeding, for that it is very impatient to endure the injurie of our cold clymate. It is a natiue of the West Indies.

¶ *The time.*

It must be set or sowen in the beginning of Aprill, in a pot with fine earth, or in a bed made with horse-dung, and some earth strawed thereon, in such manner as Cucumbers and Muske-Melons are.

¶ *The names.*

The name *Arundo Indica* is diuersly attributed to sundrie of the Reeds, but principally vnto this, called of *Lobelius, Cannacorus*: of others, *Arundo florida*, and *Harundo florida*: in English, the Flowring Reed.

¶ *The nature and vertues.*

There is not any thing set downe as touching the temperature and vertues of this Flourishing Reed, either of the Ancients, or of the new or later Writers.

Chap. 33. *Of Paper Reed.*

PAper Reed hath many large flaggie leaues ſomewhat triangular and ſmooth, not much vnlike thoſe of Catſ-taile, riſing immediatly from a tuft of roots compact of many ſtrings, among ſt the which it ſhooteth vp two or three naked ſtalkes, ſquare, and riſing ſome ſix or ſeuen cubits high aboue the water; at the top whereof there ſtands a tuft or bundle of chaffie threds ſet in comely order, reſembling a tuft of flowers, but barren and void of ſeed.

Papyrus Nilotica.
Paper Reed.

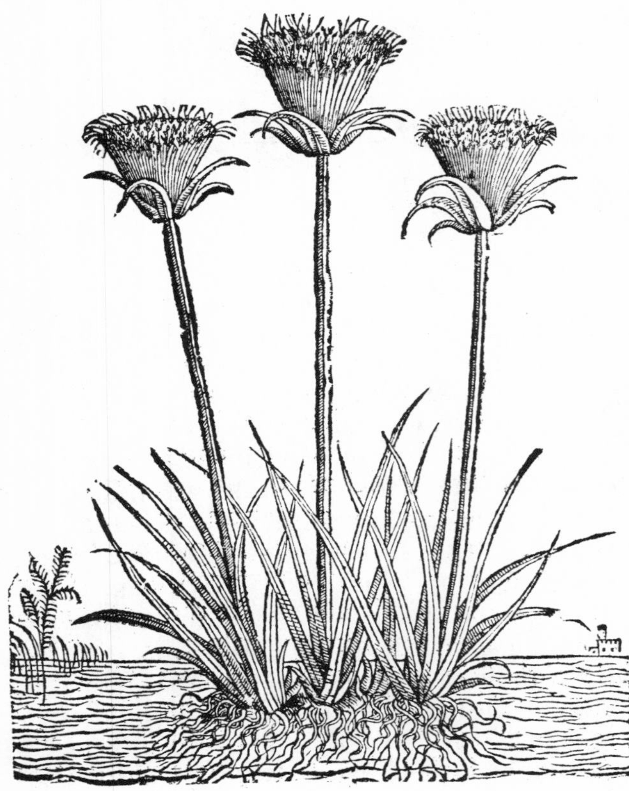

¶ *The place.*

This kinde of Reed growes in the Riuers about Babylon, and neere the city Alcaire, in the riuer Nilus, and ſuch other places of thoſe countries.

¶ *The time.*

The time of ſpringing and flouriſhing anſwereth that of the common Reed.

¶ *The names.*

This kinde of Reed which I haue Engliſhed Paper Reed, or Paper plant, is the ſame (as I do reade) that Paper was made of in Ægypt, before the inuention of paper made of linnen clouts was found out. It is thought by men of great learning and vnderſtanding in the Scriptures, and ſet downe by them for truth, that this plant is the ſame Reed mentioned in the ſecond chapter of *Exodus*; whereof was made that basket or cradle, which was dawbed within and without with ſlime of that countrey, called *Bitumen Iudaicum*, wherein *Moſes* was put being committed to the water, when *Pharaoh* gaue commandement that all the male children of the Hebrewes ſhould be drowned.

¶ *The nature, vertues, and vſe.*

A The roots of Paper Reed doe nouriſh, as may appeare by the people of Ægypt, which do vſe to chew them in their mouthes, and ſwallow downe the juice, finding therein great delight and comfort.

B The aſhes burned aſſwage and conſume hard apoſtumes, tumors, and corraſiue vlcers in any part of the body, but chiefely in the mouth.

C The burnt paper made hereof doth performe thoſe effects more forcibly.

D The ſtalkes hereof haue a ſingular vſe and priuiledge in opening the chanels or hollow paſſages of a Fiſtula, being put therein; for they do ſwell as doth the pith of Elder, or a tent made of a ſponge.

E The people about Nilus do vſe to burne the leaues and ſtalkes, but eſpecially the roots.

F The frailes wherein they put Raiſins and Figs are ſometimes made hereof; but generally with the herbe *Spartum*, deſcribed in the next Chapter.

Chap. 34. *Of Mat-Weed.*

¶ *The kindes.*

There be diuers kindes of Mat-Weeds, as ſhall be declared in their ſeuerall deſcriptions.

¶ *The deſcription.*

THe herbe *Spartum*, as *Pliny* ſaith, groweth of it ſelfe, and ſendeth forth from the root a multitude of ſlender ruſhie leaues of a cubit high, or higher, tough and pliable, of a whitiſh colour, which in time draweth narrow together, making the flat leafe to become round, as is the Ruſh. The ſtub or ſtalke thereof beareth at the top certaine feather-like tufts comming forth

forth of a ſheath or huske, among the which chaffie huskes is contained the ſeed, long and chaffie. The root conſiſterh of many ſtrings folding one within another, by meanes whereof it commeth to the forme of a turfe or haſſocke.

1 *Spartum Plinij Cluſio.*
Plinies Mat-Weed.

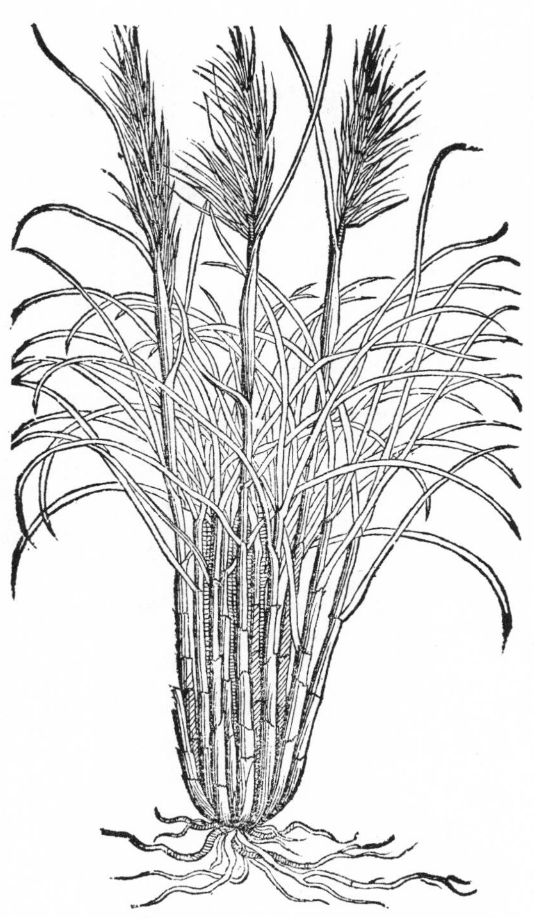

2 *Spartum alterum Plinij.*
Hooded Mat-Weed.

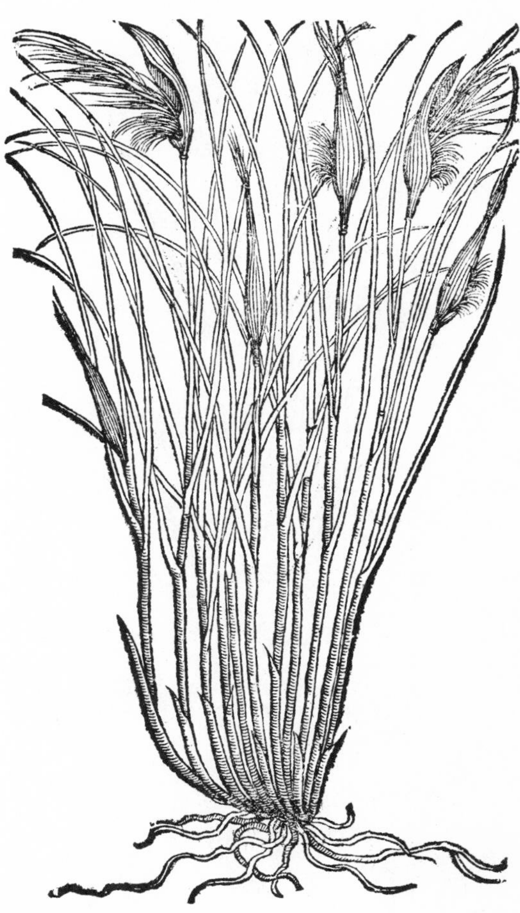

2 The ſecond likewiſe *Pliny* deſcribeth to haue a long ſtalke not much vnlike to Reed, but leſſer, whereupon do grow many graſſie leaues, rough and pliant, hard in handling as are the Ruſhes. A ſpokie chaffie tuft groweth at the top of the ſtalke, comming forth of a hood or ſinewie ſheath, ſuch as encloſeth the flowers of Onions, Leekes, Narciſſus, and ſuch like, before they come to flowring, with ſeed and roots like the precedent.

3 Engliſh Mat-weed hath a ruſhie root, deepely creeping and growing in heapes of ſand and grauell, from the which ariſe ſtiffe and ſharpe pointed leaues a foot and a halfe long, of a whitiſh colour, very much reſembling thoſe of Camels hay. The ſtalke groweth to the height of a cubit or more, whereupon doth grow a ſpike ‡ or eare of ſome fiue or ſix inches long, ſomwhat reſembling Rie; it is the thickneſſe of a finger in the midſt, and ſmaller towards both the ends. The ſeed is browne, as ſmall as Canarie ſeed, but round, and ſomewhat ſharpe at the one end ‡. Of this plant neither Sheepe nor any other Cattle will taſte or eate.

4 The other Engliſh Mat-Weed is like vnto the former, ſauing that the roots of this are long, not vnlike to Dogs Graſſe, but do not thruſt deepe into the ground, but creepe onely vnder the vpper cruſt of the earth. The tuft or eare is ſhorter, and more reſembling the head of Canary ſeed than that of Rie.

‡ 5 *Lobell* giues a figure of another ſmaller Ruſh, leaued *Spartum*, with ſmall heads, but hee hath not deſcribed it in his Latine Workes, ſo that I can ſay nothing certainly of it.

6 To this kindred muſt be added the Feathered Graſſe, though not partaking with the former in place of growth. Now it hath many ſmall leaues of a foots length round, green, and ſharp pointed, not much in forme vnlike the firſt deſcribed Mat-weed, but much leſſe: amongſt theſe leaues riſe vp many ſmall ſtalkes not exceeding the height of the leaues, which beare a ſpike vnlike the forementioned Mat-weeds, hauing 3 or foure ſeeds ending in, or ſending vp very fine white Feathers, reſembling the ſmaller ſort of feathers of the wings of the Bird of Paradiſe. The root conſiſts of many ſmall graſſie fibres.

3 *Spartum Anglicanum.*
English Mat-Weed, or Helme.

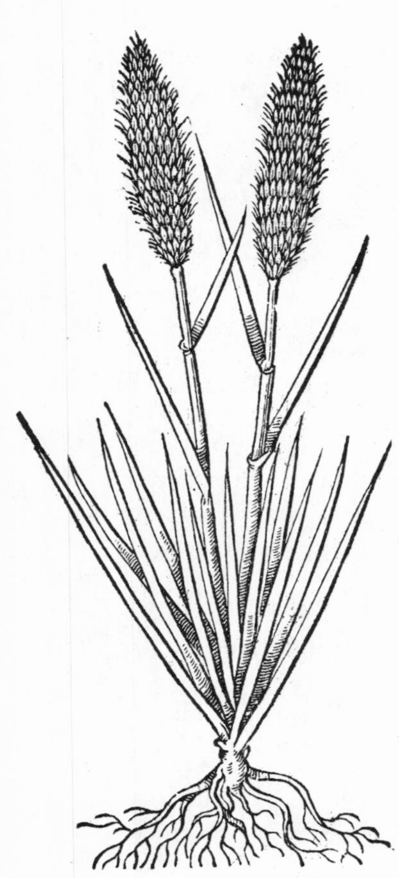

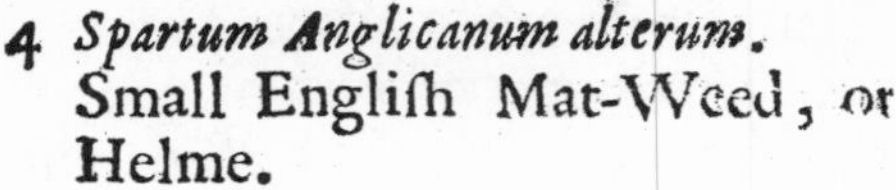

4 *Spartum Anglicanum alterum.*
Small English Mat-Weed, or Helme.

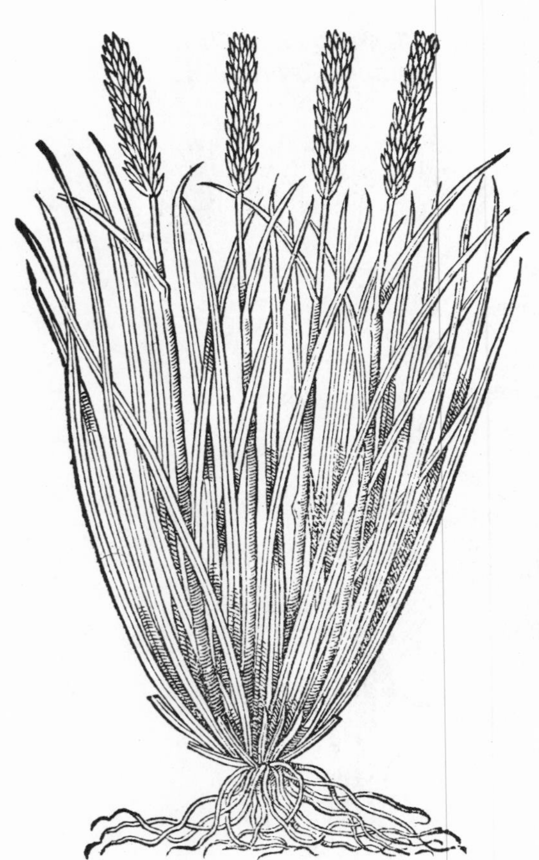

‡ 6 *Spartum Austriacum.*
Feather-Grasse.

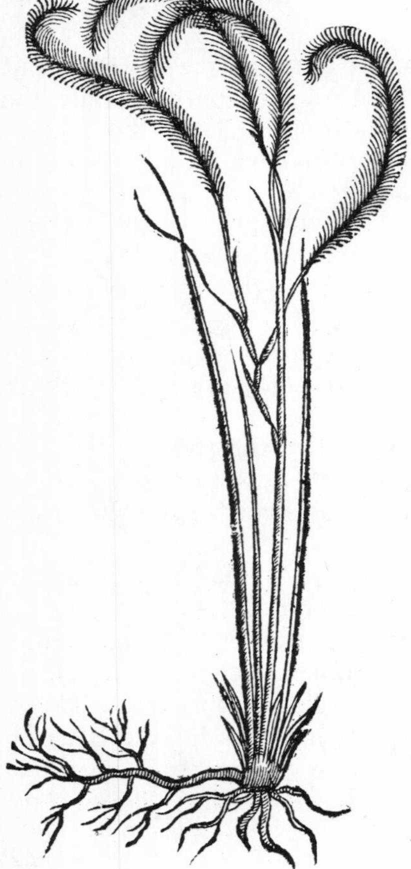

¶ *The place.*

1 2 These two grow in diuers places of Spaine.

3 I being in company with M. *Tho. Hicks, William Broad*, and three other London Apothecaries besides, in August, *1632*, to finde out rare plants in the Island of Tenet, found this bigger English one in great plentie, as soone as we came to the sea side, going betweene Margate and Sandwich.

4 5 These it may be grow also vpon our Coasts; howeuer they grow neere the sea side in diuers parts of the Low-Countries.

6 This elegant Plant *Clusius* first obserued to grow naturally in the mountaines nigh to the Bathes of Baden in Germany, and in diuers places of Austria and Hungarie. It is nourished for the beautie in sundrie of our English gardens.

¶ *The time.*

These beare their heads in the middle, and some in the later end of Sommer.

¶ *The names.*

1 This is called *Spartum primum Plinij*; that is, the first Mat-Weed described by *Pliny*: in Spaine they call it *Sparto*: the French in Prouence terme it *Olpho*.

2 This is *Spartum alterum Plinij*, *Plinie* his second Mat-Weed, or Hooded Mat-weed, it is called *Albardin* in Spaine.

3 This is *Spartum tertium* of *Clusius*, and *Gramen Sparteum secundum Schœnanthinum* of *Taber*. Our Author gaue

gaue *Cluſius* his figure for his firſt, and *Tabernamontanus* figure for the ſecond *Spartum Anglicanum*; but I will thinke them both of one plant (though *Bauhine* diſtinguiſh them) vntill ſome ſhall make the contrary manifeſt. This the Dutch call **Halme**; and our Engliſh in Tenet, Helme. *Turner* calls it Sea-Bent.

4 This is *Spartum herba* 4 *Batavicum* of *Cluſius*; *Gramen Sparteum*, or *Iunci Spartium* of *Tabern.* and our Author gaue *Tabern.* figure in the 23 Chapter of this Booke vnder the title of *Iuncus marinus gramineus*; *Lobell* calls it *Spartum noſtras alterum*.

5 *Lobell* calls this *Spartum noſtras parvum*.

6 *Cluſius* calls this *Spartum Auſtriacum*; *Daleſchampius*, *Gramen pinnatum*; we in England call it *Gramen plumoſum*, or Feathered Graſſe. ‡

¶ *The temperature, vertues, and vſe.*

These kindes of graſſie or rather ruſhie Reed haue no vſe in phyſicke, but ſerue to make Mats, and hangings for chambers, frailes, baskets, and ſuch like. The people of the Countries where they grow do make beds of them, ſtraw their houſes and chambers in ſtead of Ruſhes, for which they do excell, as my ſelfe haue ſeene. *Turner* affirmeth, That they made hats of the Engliſh one in Northumberland in his time. A

They do likewiſe in ſundry places of the Iſlands of Madera, Canaria, Saint *Thomas*, and other of the Iſlands in the tract vnto the Weſt Indies, make of them their boots, ſhooes, Herd-mens Coats, fires, and lights. It is very hurtfull for cattell, as Sheere-graſſe is. B

The Feather-Graſſe is worne by ſundry Ladies and Gentlewomen in ſtead of a Feather, the which it exquiſitely reſembles.

CHAP. 35. *Of Camels Hay.*

1 *Scœnanthum*
Camels Hay.

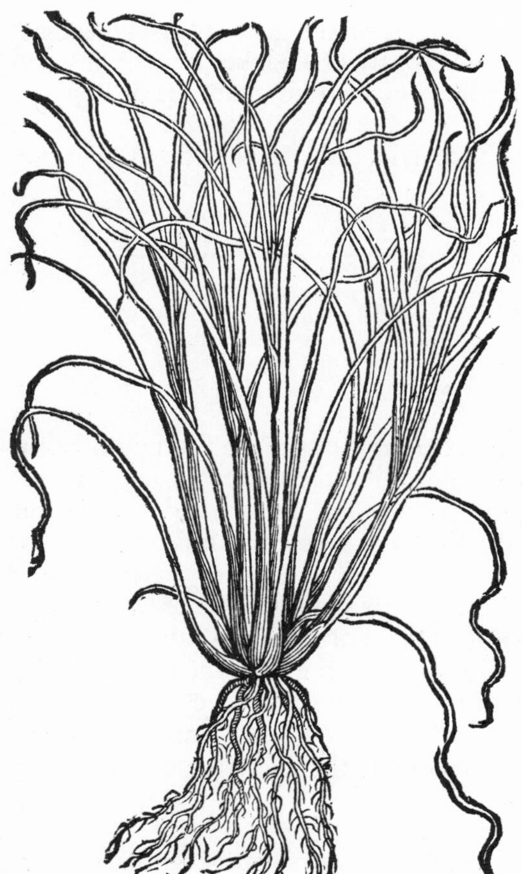

2 *Scœnanthum adulterinum.*
Baſtard Camels Hay.

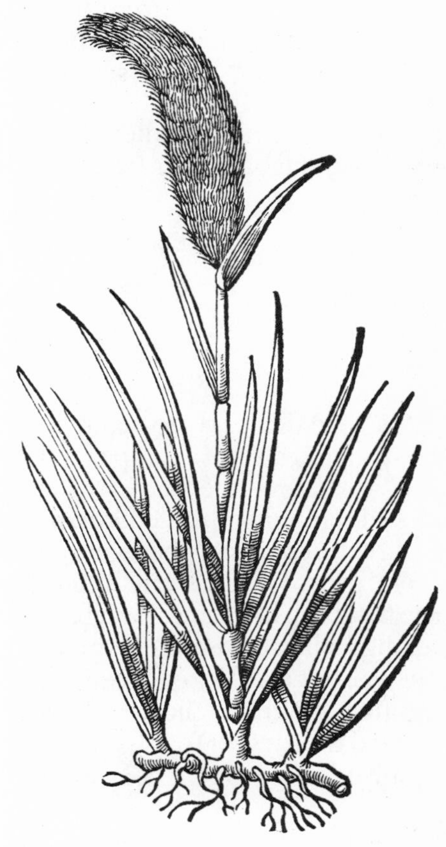

¶ *The deſcription.*

1 Camels Hay hath leaues very like vnto Mat-Weed or Helme; his roots are many, in quantitie meane, full of ſmall haires or threds proceeding from the bigger Root deeply growing in the ground, hauing diuers long ſtalkes like Cyperus Graſſe, ſet with

with some smaller leaues euen vnto the top, where do grow many small chaffie tufts or pannicles like vnto those of the wilde Oats, of a reasonable good smell and sauour, when they are broken, like vnto a Rose, with a certaine biting and nipping of the tongue.

† 2 *Francis Penny*, of famous memory, a good Physitian and skilfull Herbarist, gathered on the coast of the Mediterranean sea, between Aigues Mortes and Pescaire, this beautifull plant, whose roots are creeping, and stalkes and leaues resemble Squinanth. The flowers are soft, pappous, and thicke compact, and some fiue or six inches in length, like to Fox-taile; they in colour resemble white silke or siluer. Thus much *Lobell*. Our Author described this in the first place, *Ch.*23. vnder *Iuncus Marinus Gramineus*, for so *Lobell* also calls it. †

¶ *The place.*

1 This growes in Africa, Nabathæa, and Arabia, and is a stranger in these Northerne Regions.

2 The place of the second is mentioned in the description.

¶ *The time.*

Their time answereth the other Reeds and Flags.

¶ *The names.*

1 Camels Hay is called in Greeke χοῖνος ἀρωματικος: in Latine, *Iuncus odoratus*, and *Scœnanthum*: in shops *Squinanthum*, that is, *Flos Iunci*: in French, *Pasteur de Chammeau*: in English, Camels Hay, and Squinanth.

2 This *Lobell* calls *Iuncus marinus gramineus*, and *Pseudoschœnanthum*: We call it Bastard Squinanth, and Fox-taile Squinanth.

¶ *The temper.*

This plant is indifferently hot, and a little astrictiue.

¶ *The vertues.*

A Camels Hay prouoketh vrine, moueth the termes, and breaketh winde about the stomacke.

B It causeth aking and heauinesse of the head, *Galen* yeeldeth this reason thereof, because it heateth moderately, and bindeth with tenuitie of parts.

C According to *Dioscorides*, it dissolues, digests, and opens the passages of the veines.

D The floures or chaffie tufts are profitable in drinke for them that pisse bloud any wayes: It is giuen in medicines that are ministred to cure the paines and griefes of the guts, stomacke, lungs, liuer, and reines, the fulnesse, loathsomenesse, and other defects of the stomacke, the dropsie, conuulsions, or shrinking of sinews, giuen in the quantitie of a dram, with a like quantitie of Pepper, for some few dayes.

E The same boyled in wine helpeth the inflammation of the matrix, if the woman do sit ouer the fume thereof, and bathe her selfe often with it also.

Chap. 36. *Of Burre-Reed.*

¶ *The description.*

1 The first of these plants hath long leaues, which are double edged, or sharpe on both sides, with a sharpe crest in the middle, in such manner raised vp that it seemeth to be triangle or three square. The stalkes grow among the leaues, and are two or three foot long, being diuided into many branches, garnished with many prickly huskes or knops of the bignesse of a nut. The root is full of hairy strings.

2 The great Water Burre differeth not in any thing from the first kind in roots or leaues, saue that the first hath his leaues rising immediately from the tuft or knop of the root; but this kinde hath a long stalke comming from the root, whereupon, a little aboue the root, the leaues shoot out round about the stalke successiuely, some leaues still growing aboue others, euen to the top of the stalke, and from the top thereof downeward by certaine distances. It is garnished with many round wharles, or rough coronets, hauing here and there among the said wharles one single short leafe of a pale greene colour.

¶ *The place.*

Both these are very common, and grow in moist medowes, and neere vnto water-courses. They plentifully grow in the fenny grounds of Lincolnshire, and such like places; in the ditches about S. *George* his fields, and in the ditch right against the place of execution, at the end of Southwark, called S. *Thomas* Waterings.

¶ *The time.*

They bring forth their burry bullets or seedy knots in August.

1 *Sparganium Ramosum.*
Branched Burre-Reed.

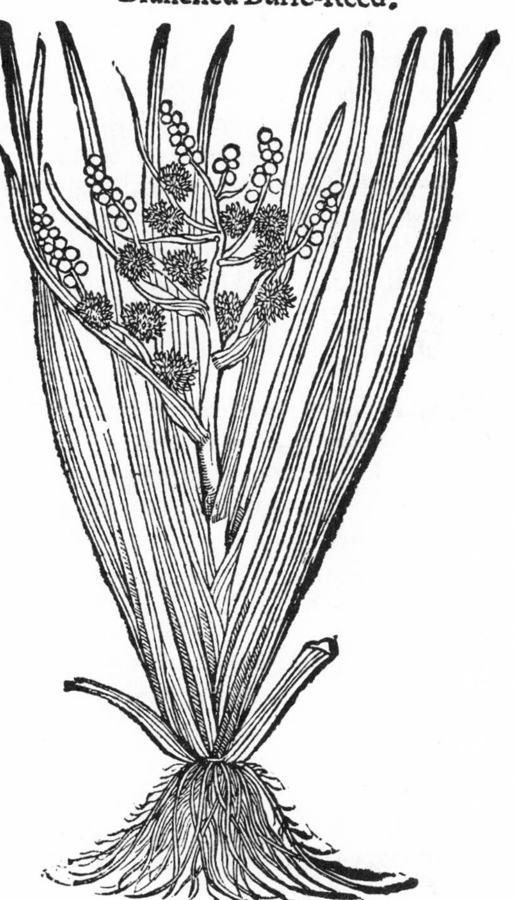

2 *Sparganium latifolium.*
Great Water-Burre.

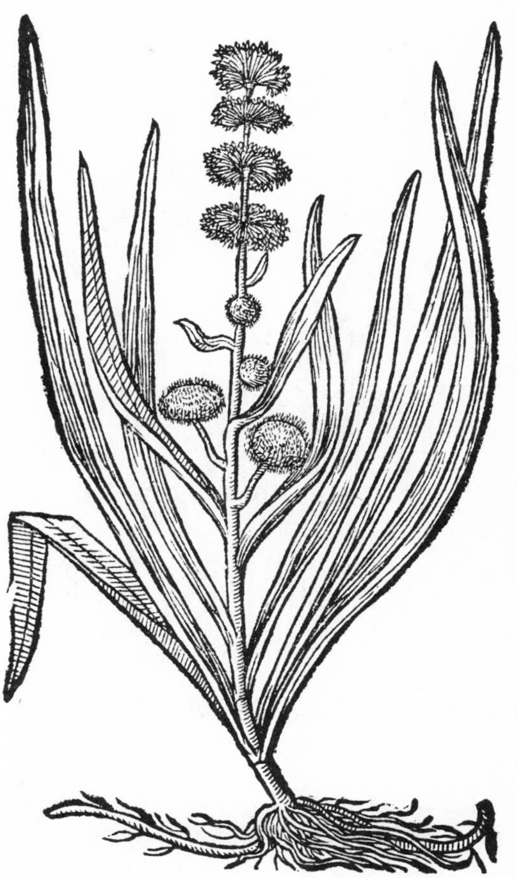

¶ *The names.*

These Plants of some are called *Sparganium* : *Theophrastus* in his fourth Booke and eighteenth Chapter calleth them *Butomus* : of some, *Platanaria* : I call them Burre-Reed : in the Arabian tongue they are called *Safarhe Bamon* : in Italian *Sparganio* : of *Dodoneus*, *Carex* : Some call the first *Sparganium ramosum*, or Branched Burre-Reed. The second, *Sparganium non ramosum*, Not-branching Burre-Reed.

¶ *The temperature.*

They are cold and dry of complexion.

¶ *The vertues.*

Some write, that the knops or rough burres of these plants boyled in wine, are good against the bitings of venomous beasts, if either it be drunke, or the wound washed therewith. A

Chap. 37. *Of Cats Taile.*

¶ *The description.*

Cats Taile hath long and flaggy leaues, full of a spongeous matter, or pith, among which leaues groweth vp a long smooth naked stalke, without knot, fashioned like a speare, of a firme or solid substance, hauing at the top a browne knop or eare, soft, thicke and smooth, seeming to be nothing else but a deale of flockes thicke set and thrust together, which being ripe turneth into a downe, and is carried away with the winde. The Roots be hard, thicke, and white, full of strings, and good to burne, where there is plenty thereof to be had.

¶ *The place.*

It groweth in pooles and such like standing waters, and sometimes in running streames.

I haue found a smaller kinde hereof growing in the ditches and marshie grounds in the Isle of Shepey, going from Sherland house to Feuersham.

¶ *The time.*

They floure and beare their mace or torch in Iuly and August.

Typha.
Cats Taile.

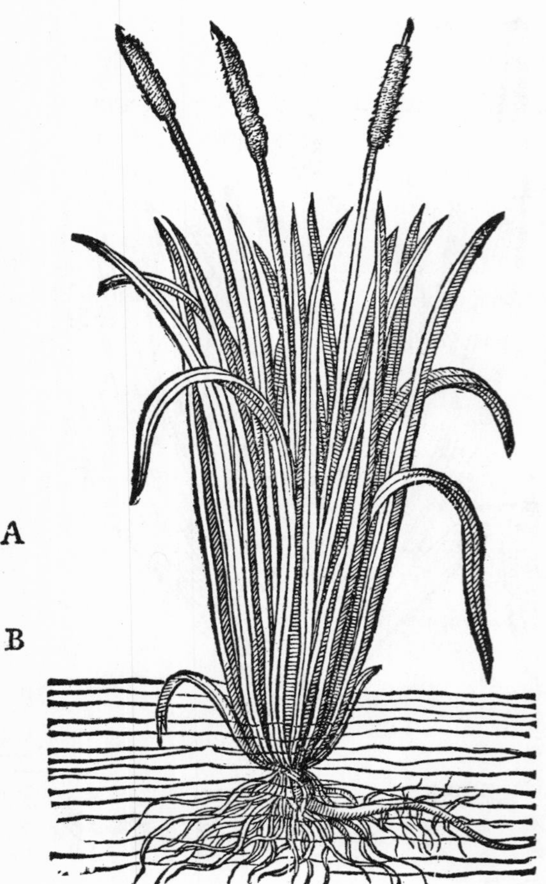

¶ *The names.*

They are called in Greeke τύφη: in Latine *Typha:* of ſome *Ceſtrum Morionis:* in French *Marteau Maſſes:* in Dutch, **Liſchdoden**, and **Donſen:** In Italian *Mazza ſorda:* in Spaniſh *Behordo*, and *Iunco amacorodato:* In Engliſh, Cats Taile, and Reed-Mace. Of this Cats Taile *Ariſtophanes* maketh mention in his Comedy of Frogs, where he bringeth them forth one talking with another, being very glad that they had ſpent the whole day in skipping and leaping *inter Cyperum & Phleum*, among Galingale and Cats Taile. *Ouid* ſeemeth to name this plant *Scirpus*; for he termeth the mats made of the leaues, Catſ-taile Mats, as in his ſixth Booke *Faſtorum*,

At Dominus, diſcedite, ait, plauſtróque morantes
Suſtulit, in plauſtro ſcirpea matta fuit.

¶ *The nature.*

It is cold and dry of complexion.

¶ *The vertues.*

A The ſoft Downe ſtamped with ſwines greaſe well waſhed, healeth burnings or ſcaldings with fire or water.

B Some practitioners by their experience haue found, That the Downe of the Cats taile beaten with the leaues of Betony, the roots of Gladiole, and the leaues of *Hippogloſſon* into powder, and mixed with the yelks of egges hard ſodden, and ſo eaten, is a moſt perfect medicine againſt the diſeaſe in children called in Greeke Εντεροκήλη, which is, when the gut called *Inteſtinum cæcum* is fallen into the cods. This medicine muſt be miniſtred euery day faſting for the ſpace of thirtie dayes, the quantitie thereof to be miniſtred at one time is 1. ʒ. This being vſed as before is ſpecified doth not onely helpe children and ſtriplings, but growne men alſo, if in time of their cure they vſe conuenient ligature or truſſings, and fit conſounding plaiſters vpon the grieued place, according to art appointed for that purpoſe in Chirurgerie.

C This Downe in ſome places of the Iſle of Elie, and the low countries adioyning thereto, is gathered and well ſold to make mattreſſes of, for plowmen and poore people.

D It hath beene alſo often proued to heale kibed or humbled heeles (as they are termed) being applied to them, either before or after the skinne is broken.

Chap. 38. *Of Stitchwort.*

¶ *The deſcription.*

1 Stitchwort, or as *Ruellius* termeth it *Holoſteum*, is of two kindes, and hath round tender ſtalkes full of joints leaning toward the ground; at euery ioynt grow two leaues one againſt another. The flowers be white, conſiſting of many ſmall leaues ſet in the manner of a ſtarre. The roots are ſmall, jointed, and threddy. The ſeed is contained in ſmall heads ſomewhat long, and ſharpe at the vpper end, and when it is ripe it is very ſmall and browne.

2 The ſecond is like the former in ſhape of leaues and flowers, which are ſet in forme of a ſtarre; but the leaues are orderly placed, and in good proportion, by couples two together, being of a whitiſh colour. When the flowers be vaded then follow the ſeeds, which are incloſed in bullets like the ſeed of flax, but not ſo round. The chiues or threds in the middle of the floure are ſometimes of a reddiſh, or of a blackiſh colour. ‡ There are more differences of this plant, or rather varieties, as differing little but in the largeneſſe of the leaues, floures, or ſtalkes. ‡

¶ *The place.*

They grow in the borders of fields vpon banke ſides and hedges, almoſt euery where.

¶ *The time.*

They flouriſh all the Sommer, eſpecially in May and Iune.

¶ *The*

Gramen Leucanthemum.
Stitchwort.

¶ *The names.*

Some (as *Ruellius* for one) haue thought this to be the plant which the Grecians call ὀλόστεον: in Latine, *Tota oſſea*: in Engliſh, All-Bones; whereof I ſee no reaſon, except it be by the figure *Antonomia*; as when we ſay in Engliſh, He is an honeſt man, our meaning is that he is a knaue: for this is a tender herbe hauing no ſuch bony ſubſtance. ‡ *Dodonæus* queſtions, whether this plant be not *Cratæogonon*; and he calls it *Gramen Leucanthemum*, or White-floured Graſſe. The qualitie here noted with B. is by *Dioſcorides* giuen to *Cratæogonon*; but it is with his ἱστορεῖται ὑπό τινων, (that is) Some ſay or report ſo much: which phraſe of ſpeech hee often vſeth when as he writes faculties by heare-ſay and doubts himſelfe of the truth of them. ‡

¶ *The nature.*

The ſeed of Stitchwort, as *Galen* writeth, is ſharpe and biting to him that taſteth it, and to him that vſeth it very like to Mill.

¶ *The vertues.*

They are wont to drinke it in Wine with the powder of Acornes, againſt the paine in the ſide, ſtitches, and ſuch like. A

Diuers report, ſaith *Dioſcorides*, That the Seed of Stitchwort being drunke cauſeth a woman to bring forth a man childe, if after the purgation of her Sickneſſe, before ſhe conceiue, ſhe do drinke it faſting thrice in a day, halfe a dram at a time, in three ounces of water many dayes together. B

Chap. 39. *Of Spiderwort.*

¶ *The deſcription.*

1 The obſcure deſcription which *Dioſcorides* and *Pliny* haue ſet downe for *Phalangium*, hath bred much contention among late Writers. This plant *Phalangium* hath leaues much like Couch Graſſe, but they are ſomewhat thicker and fatter, and of a more whitiſh greene colour. The ſtalkes grow to the height of a cubit. The top of the ſtalke is beſet with ſmall branches, garniſhed with many little white flowers, compact of ſix little leaues. The threds or thrums in the middle are whitiſh, mixed with a faire yellow, which being fallen, there follow blacke ſeeds, incloſed in ſmall round knobs, which be three cornered. The roots are many, tough, and white of colour.

2 The ſecond is like the firſt, but that his ſtalke is not branched as the firſt, and floureth a moneth before the other.

3 The third kinde of Spiderwort, which *Carolus Cluſius* nameth *Aſphodelus minor*, hath a root of many threddy ſtrings, from the which immediately riſe vp graſſie leaues, narrow and ſharpe pointed: among the which come forth diuers naked ſtrait ſtalkes diuided towards the top into ſundry branches, garniſhed on euery ſide with faire ſtarre-like flowers, of colour white, with a purple veine diuiding each leafe in the middeſt: they haue alſo certaine chiues or threds in them. The ſeed followeth incloſed in three ſquare heads like vnto the kindes of Aſphodils.

‡ 4 This Spiderwort hath a root conſiſting of many thicke, long, and white fibers, not much vnlike the precedent, out of which it ſends forth ſome fiue or ſix greene and firme leaues, ſomewhat hollowed in the middle, and mutually inuoluing each other at the root: amongſt theſe there riſeth vp a round greene ſtalke, bearing at the top thereof ſome nine or ten floures, more or leſſe; theſe conſiſt of ſix leaues apiece, of colour white (the three innermoſt leaues are the broader, and more curled, and the three outmoſt are tipt with greene at the tops.) The whole floure much

reſembles

resembles a white Lilly, but much smaller. Three square heads, containing a dusky and vnequall seed, follow after the floure.

1 *Phalangium Ramosum.*
Branched Spiderwort.

2 *Phalangium non ramosum.*
Vnbranched Spiderwort.

† 3 *Phalangium Cretæ.*
Candy Spiderwort.

‡ 4 *Phalangium Antiquorum.*
The true Spiderwort of the Ancients.

5 *Phalangium*

‡ 5 *Phalangium Virginianum Tradescanti.*
Tradescants Virginian Spiderwort.

5 This plant in my iudgement cannot be fitlier ranked with any than these last described; therefore I haue here giuen him the fifth place, as the last commer. This plant hath many creeping stringy roots, which here and there put vp greene leaues, in shape resembling those of the last described: amongst these there riseth vp a pretty stiffe stalke jointed, and hauing at each joint one leafe incompassing the stalke, and out of whose bosome oft times little branches arise: now the stalke at the top vsually diuides it selfe into two leaues, much after the manner of *Cyperus*; between which there come forth many floures consisting of three pretty large leaues a piece, of colour deepe blew, with reddish chiues tipt with yellow standing in their middle. These fading (as vsually they doe the same day they shew themselues) there succeed little heads couered with the three little leaues that sustained the floure. In these heads there is contained a long blackish seed.

¶ *The place.*

1. 2. 3. These grow only in gardens with vs, and that very rarely. 4 This growes naturally in some places of Sauoy. 5 This Virginian is in many of our English gardens, as with M. *Parkinson*, M. *Tradescant*, and others.

¶ *The time.*

1. 4. 5. These floure in Iune: the second about the beginning of May: and the third about August.

¶ *The names.*

The first is called *Phalangium ramosum*, Branched Spiderwort. 2 *Phalangium non ramosum*, Vnbranched Spiderwort. *Cordus* calls it *Liliago*. 3 This, *Clusius* calls *Asphodelus minor*: *Lobell*, *Phalangium Cretæ*, Candy Spiderwort. 4 This is thought to be the *Phalangium* of the Ancients, and that of *Matthiolus*: it is *Phalangium Allobrogicum* of *Clusius*, Sauoy Spiderwort. 5 This by M. *Parkinson* (who first hath in writing giuen the figure and description thereof) is aptly termed *Phalangium Ephemerum Virginianum*, Soone-fading Spiderwort of Virginia, or *Tradescants* Spiderwort, for that M. *Iohn Tradescant* first procured it from Virginia. *Bauhine* hath described it at the end of his *Pinax*, and very vnfitly termed it *Allium, siue Moly Virginianum*. ‡

¶ *The nature.*

Galen saith, *Phalangium* is of a drying qualitie, by reason of the tenuitie of parts.

¶ *The vertues.*

Dioscorides saith, That the leaues, seed, and floures, or any of them drunke in Wine, preuaileth against the bitings of Scorpions, and against the stinging and biting of the Spider called *Phalangium*, and all other venomous beasts. A

The roots tunned vp in new ale, and drunke for a moneth together, expelleth poyson, yea although it haue vniuersally spred it selfe through the body. B

CHAP. 40. *Of the Floure de-luce.*

¶ *The kindes.*

There be many kindes of Iris or Floure de-luce, whereof some are tall and great, some little, small, and low; some smell exceeding sweet in the root, some haue no smell at all: some floures are sweet in smell, and some without; some of one colour, some of many colours mixed: vertues attributed to some, others not remembred: some haue tuberous or knobby roots, others bulbous or Onion roots, some haue leaues like flags, others like grasse or rushes.

¶ *The*

¶ *The deſcription.*

1 THe common Floure de-luce hath long and large flaggy leaues like the blade of a ſword, with two edges, amongſt which ſpring vp ſmooth and plaine ſtalkes two foot long, bearing floures toward the top, compact of ſix leaues ioyned together, whereof three that ſtand vpright are bent inward one toward another; and in thoſe leaues that hang downward there are certaine rough or hairie welts, growing or riſing from the nether part of the leafe vpward, almoſt of a yellow colour. The roots be thicke, long, and knobby, with many hairy threds hanging thereat.

2 The water Floure de-luce, or Water flag, or Baſtard *Acorus*, is like vnto the garden Floure de-luce in roots, leaues, and ſtalkes, but the leaues are much longer, ſometimes of the height of foure cubits, and altogether narrower. The floure is of a perfect yellow colour, and the Root knobby like the other; but being cut, it ſeemeth to be of the colour of raw fleſh.

1 *Iris vulgaris.*
Floure de-luce.

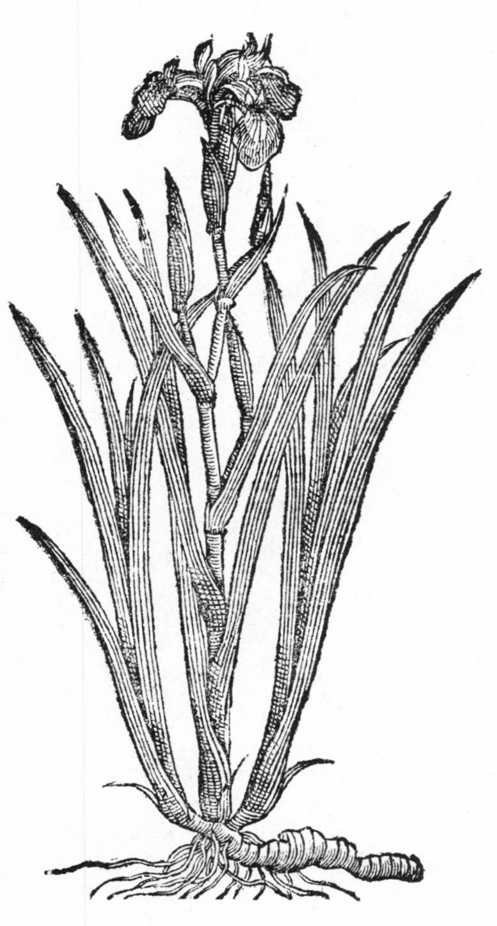

2 *Iris paluſtris lutea.*
Water-flags, or Floure de-luce.

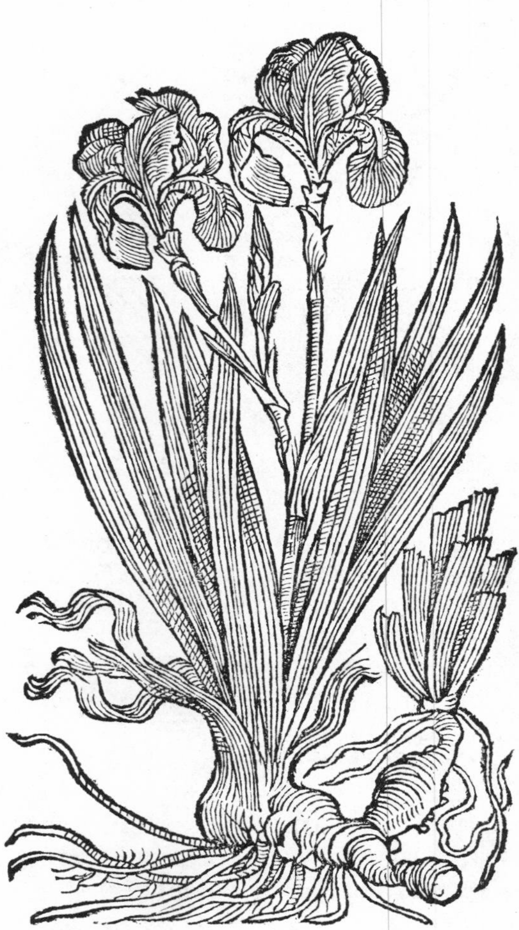

¶ *The place.*

The Water Floure de-luce or yellow flag proſpereth well in moiſt medows, and in the borders and brinks of riuers, ponds, and ſtanding lakes. And although it be a water plant of nature, yet being planted in gardens it proſpereth well.

¶ *The names.*

Floure de-luce is called in Greeke ἶρις: *Athenæus* and *Theophraſtus* reade ἱρις: as though they ſhould ſay, *Conſecratrix*; by which name it is alſo called of the Latines *Radix Marica*, or rather *Radix Naronica*, of the riuer Naron, by which the beſt and greateſt ſtore do grow. Whereupon *Nicander* in his Treacles commendeth it thus:

Iridem quam aluit Drilon, & Naronis ripa.

Which may thus be Engliſhed:

Iris, which *Drilon* water feeds,
And *Narons* bankes with other weeds.

The Italians, *Giglio azurro*: in Spaniſh, *Lilio Cardeno*: in French, *Flambe*: The Germanes, **Gilgen, Schwertel**: in Dutch, **Lisch**.

The ſecond is called in Latine, *Iris paluſtris lutea, Pſeudoacorus*, and *Acorus paluſtris*: in Engliſh, Water-

Waterflags, Bastard Floure de-luce, or Water Floure de-luce: and in the North they call them Seggs.

¶ *The nature.*

1 The roots of the Floure de-luce being as yet fresh and greene, and full of juyce, are hot almost in the fourth degree. The dried roots are hot and dry in the third degree, burning the throat and mouth of such as taste them.

2 The bastard Floure de-luce his root is cold and dry in the third degree, and of an astringent or binding facultie.

¶ *The vertues.*

The root of the common Floure de-luce cleane washed, and stamped with a few drops of Rose water, and laid plaister-wise vpon the face of man or woman, doth in two dayes at the most take away the blacknesse or blewnesse of any stroke or bruse: so that if the skinne of the same woman or any other person be very tender and delicate, it shall be needfull that ye lay a piece of silke, sindall, or a piece of fine laune betweene the plaister and the skinne; for otherwise in such tender bodies it often causeth heate and inflammation. A

The iuyce of the same doth not onely mightily and vehemently draw forth choler, but most especially watery humors, and is a speciall and singular purgation for them that haue the Dropsie, if it be drunke in whay or some other liquor that may somewhat temper and alay his heate. B

The dry roots attenuate or make thinne thicke and tough humours, which are hardly and with difficultie purged away. C

They are good in a loch or licking medicine for shortnesse of breath, an old cough, and all infirmities of the chest which rise hereupon. D

They remedie those that haue euill spleenes, and those that are troubled with convulsions or cramps, biting of serpents, and the running of the reines, being drunke with vinegre, as saith *Dioscorides*; and drunke with wine it bringeth downe the monethly courses of women. E

The decoction is good in womens baths, for it mollifieth and openeth the matrix. F

Being boyled very soft, and laid to plaister-wise it mollifieth or softneth the kings euil, and old hard swellings. G

‡ The roots of our ordinary flags are not (as before is deliuered) cold and dry in the third degree, nor yet in the second, as *Dodonæus* affirmes; but hot and dry, and that at the least in the second degree, as any that throughly tastes them will confesse. Neither are the faculties and vse (as some would persuade vs) to be neglected; for as *Pena* and *Lobell* affirme, though it haue no smell, nor great heat, yet by reason of other faculties it is much to be preferred before the *Galanga major*, or forreigne *Acorus* of shops, in many diseases; for it imparts more heate and strength to the stomacke and neighbouring parts than the other, which rather preyes vpon and dissipates the innate heate and implanted strength of those parts. It bindes, strengthens, and condenses: it is good in bloudy flixes, and stayes the Courses. ‡ H

CHAP. 40. *Of Floure de-luce of Florence.*

¶ *The description.*

1 THe Floure de-luce of Florence, whose roots in shops and generally euery where are called *Ireos*, or *Orice* (whereof sweet waters, sweet pouders, and such like are made) is altogether like vnto the common Floure de-luce, sauing that the flowers of the *Ireos* is of a white colour, and the roots exceeding sweet of smell, and the other of no smell at all.

2 The white Floure de-luce is like vnto the Florentine Floure de-luce in roots, flaggy leaues, and stalkes; but they differ in that, that this *Iris* hath his flower of a bleake white colour declining to yellownesse; and the roots haue not any smell at all; but the other is very sweet, as we haue said.

3 The great Floure de-luce of Dalmatia hath leaues much broader, thicker, and more closely compact together than any of the other, and set in order like wings or the fins of a Whale fish, greene toward the top, and of a shining purple colour toward the bottome, euen to the ground: amongst which riseth vp a stalke of foure foot high, as my selfe did measure oft times in my garden: whereupon doth grow faire large floures of a light blew, or as we terme it, a watchet colour. The floures do smell exceeding sweet, much like the Orenge floure. The seeds are contained in square cods, wherein are packed together many flat seeds like the former. The root hath no smell at all.

1 *Iris Florentina.*
Floure de-luce of Florence.

2 *Iris alba.*
White Floure de-luce.

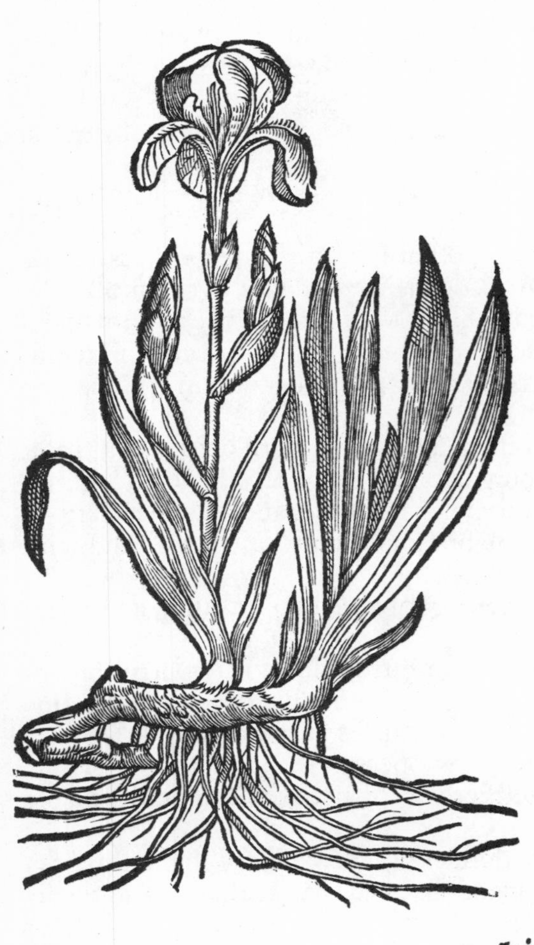

3 *Iris Dalmatica major.*
Great Flourede-luce of Dalmatia.

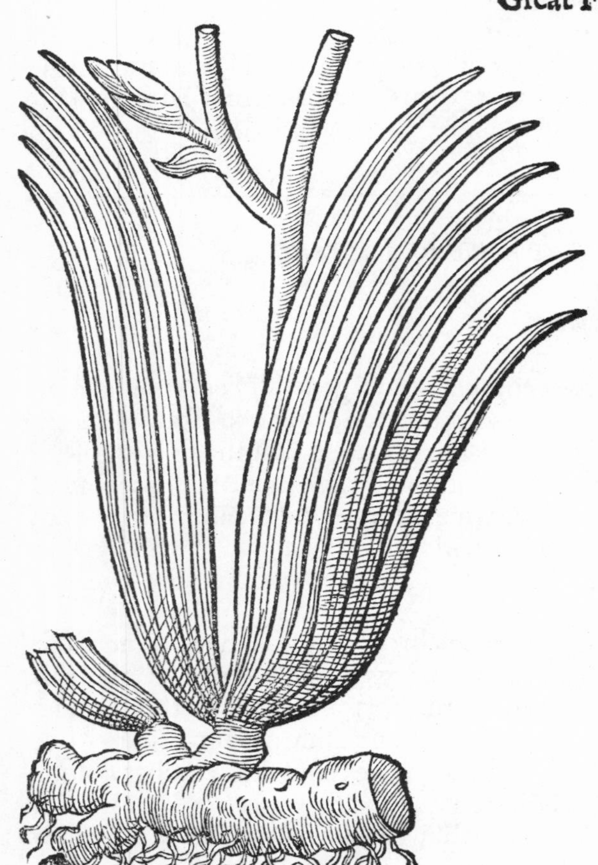

4 *Iris*

4 *Iris Dalmatica minor.*
Small Dalmatian *Iris.*

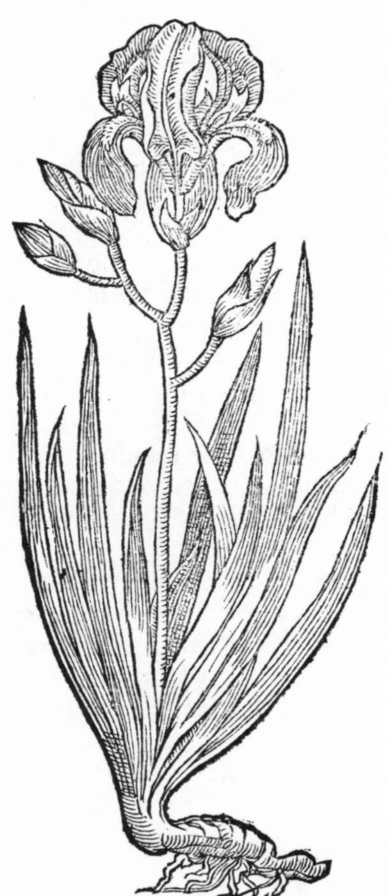

5 *Iris Biflora.*
Twice-flouring Floure de-luce.

6 *Iris Violacea.*
Violet Floure de-luce.

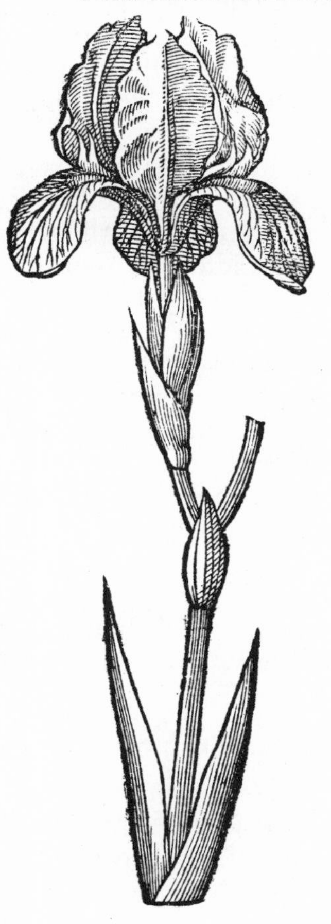

7 *Iris Pannonica.*
Auſtrian Floure de luce.

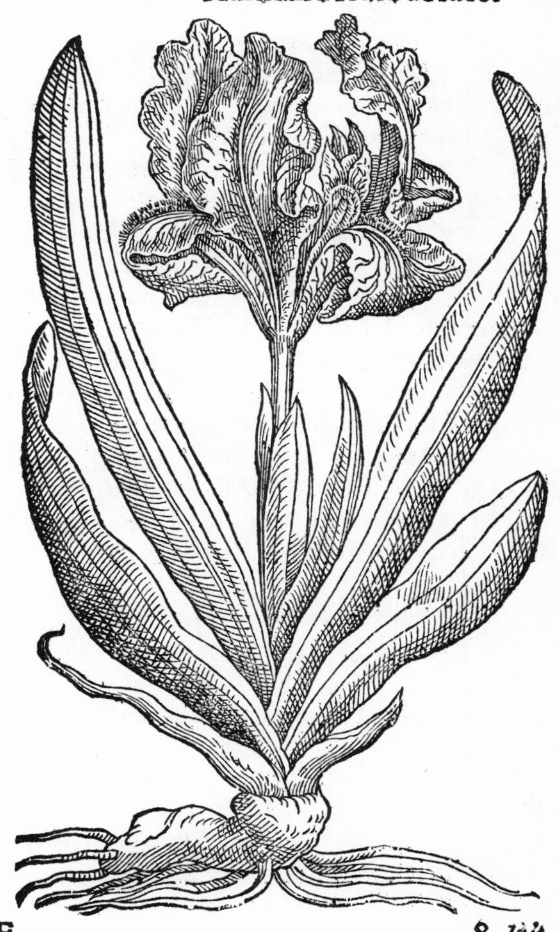

† 8 *Iris Camerarij.*
Germane Floure de-luce.

4 The small Floure de-luce of Dalmatia is in shew like to the precedent, but rather resembling *Iris biflora,* being both of one stature, small and dwarfe plants in respect of the greater. The floures be of a more blew colour. They flower likewise in May as the others do; but beware that ye neuer cast any cold water vpon them presently taken out of a Wel; for their tendernesse is such, that they wither immediatly, and rot away, as I my selfe haue proued: but those which I left vnwatred at the same time liue and prosper to this day.

5 This kinde of Floure de-luce came first from Portugal to vs. It bringeth forth in the Spring time floures of a purple or violet colour, smelling like a violet, with a white hairy welt downe the middle. The root is thick and short, stubborne or hard to breake. In leaues and shew it is like to the lesser Floure de-luce of Dalmatia, but the leaues be more spred abroad, and it commonly hath but one stalke, which in Autumne floureth againe, and bringeth forth the like floures; for which cause it was called *Iris biflora.*

6 *Iris violacea* is like vnto the former, but much smaller, and the floure is of a more deepe violet colour.

7 *Carolus Clusius,* that excellent and learned Father of Herbarists, hath set forth in his Pannonicke Obseruations the picture of this beautifull Floure de-luce, with great broad leaues, thicke and fat, of a purple colour neere vnto the ground, like the great Dalmatian Floure de-luce, which it doth very well resemble. The root is very sweet when it is dry, and striueth with the Florentine *Iris* in sweetnesse. The floure is of all the other most confusedly mixed with sundry colours, insomuch that my pen cannot set downe euery line or streake, as it deserueth. The three leaues that stand vpright do claspe or embrace one another, and are of a yellow colour. The leaues that looke downward, about the edges are of a pale colour, the middle part of white, mixed with a line of purple, and hath many small purple lines stripped ouer the said white floure, euen to the brim of the pale coloured edge. It smelleth like the Hauthorne floures being lightly smelled vnto.

8 The Germane Floure de-luce, which *Camerarius* hath set forth in his Booke named *Hortus Medicus,* hath great thicke and knobby roots: the stalke is thicke and full of iuyce: the leaues be very broad in respect of all the rest of the Floure de-luces. The floure groweth at the top of the stalke, consisting of six great leaues blew of colour, welted downe the middle, with white tending to yellow; at the bottome next the stalke it is white of colour, with some yellownesse fringed about the said white, as also about the brims or edges, which greatly setteth forth his beautie; the which *Ioachimus Camerarius,* the sonne of old *Camerarius* of Noremberg, had sent him out of Hungarie, and did communicate one of the plants thereof to *Clusius*; whose figure he hath most liuely set forth with this description, differing somewhat from that which *Ioachimus* himselfe did giue vnto me at his being in London. The leaues, saith he, are very large, twice so broad as any of the others. The stalke is single and smooth; the floure groweth at the top, of a most bright shining blew colour, the middle rib tending to whitenesse, the three vpper leaues somewhat yellowish. The root is likewise sweet as *Ireos.*

¶ *The place.*

These kindes of Floure de-luces do grow wilde in Dalmatia, Goritia, and Piedmont; notwithstanding our London gardens are very well stored with euery one of them.

¶ *The time.*

Their time of flouring answereth the other Floure de-luces.

¶ *The names.*

The Dalmatian Floure de-luce is called in Greeke of *Athenæus* and *Theophrastes* ἶρις: it is named also οὐρανία, of the heauenly Bow or Rainbow: vpon the same occasion θαυμαστός, or Admirable: for the Poets sometime do call the Rainbow θαυμαντίας: in Latine *Iris,* and in English Floure de-luce. Their seuerall titles do sufficiently distinguish them, whereby they may be knowne one from another.

¶ *The*

¶ *The nature.*

The nature of theſe Floure de-luces are anſwerable to thoſe of the common kinde; that is to ſay, the dry roots are hot and dry in the latter end of the ſecond degree.

¶ *The vertues.*

The iuyce of theſe Floure de-luces doth not onely mightilyand vehemently draw forth choler, but moſt eſpecially waterie humors,and is a ſingular good purgation for them that haue the Dropſie, if it be drunke in ſweet wort or whay.

The ſame are good for them that haue euill ſpleenes, or that are troubled with cramps or conuulſions, and for ſuch as are bit with Serpents. It profiteth alſo much thoſe that haue the Gonorrhea, or running of the reines, being drunke with Vineger, as *Dioſc.* ſaith; and drunke with Wine they bring downe the monethly termes.

CHAP. 42. *Of Variable Floure de-luces.*

1 *Iris lutea variegata.*
Variable Flourede-luce.

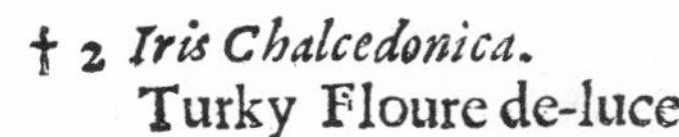

† 2 *Iris Chalcedonica.*
Turky Floure de-luce.

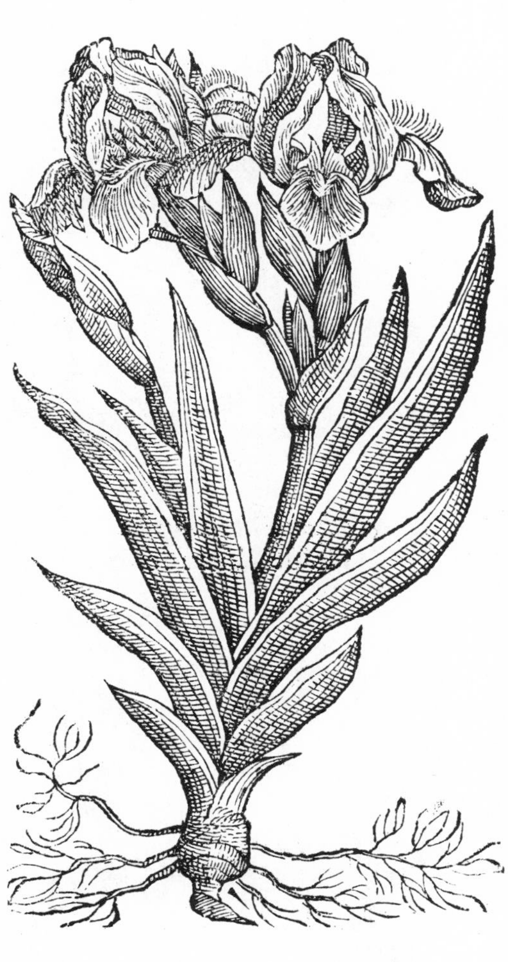

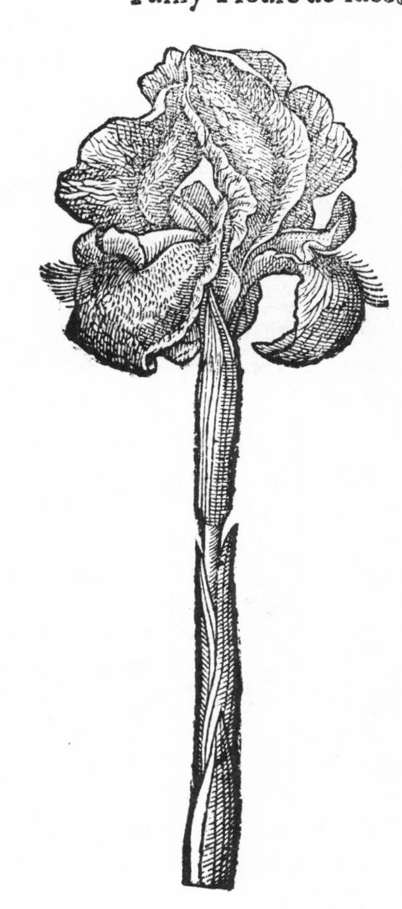

¶ *The deſcription.*

1 THat which is called the Floure de-luce of many colours loſeth his leaues in Winter, and in the Spring time recouereth them anew. I am not able to expreſſe the ſundrie colours and mixtures contained in this floure: it is mixed with purple, yellow, blacke, white, and a fringe or blacke thrum downe the middle of the lower leaues, of a whitiſh yellow, tipped or frized, and as it were a little raiſed vp; of a deep purple colour neere the ground.

2 The ſecond kinde hath long and narrow leaues of a blackiſh greene, like the ſtinking Gladdon; among which riſe vp ſtalkes two foot long, bearing at the top of euery ſtalke one floure compact of ſix great leaues: the three that ſtand vpright are confuſedly and very ſtrangely ſtripped, mixed with white and a duskiſh blacke colour. The three leaues that hang downeward are like a gaping hood, and are mixed in like manner, (but the white is nothing ſo bright as of the other) and are as it were ſhadowed ouer with a darke purple colour ſomewhat ſhining: ſo that according

cording to my iudgement the whole floure is of the colour of a Ginny hen: a rare and beautifull floure to behold.

‡ 3 *Iris maritima Narbonensis.* The Sea Floure de-luce.

4 *Iris sylvestris Bizantina.* Wilde Bizantine Floure de-luce.

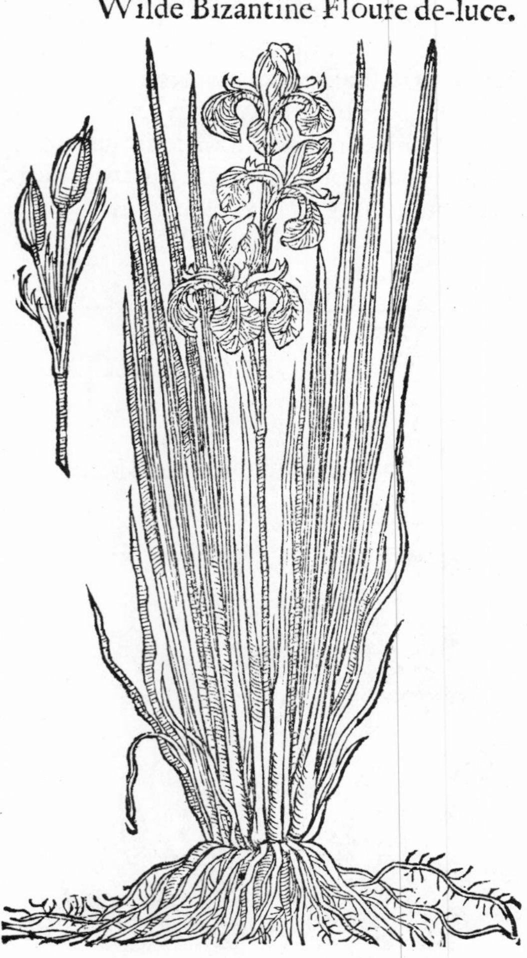

5 *Chamæiris Angustifolia.* Narrow leafed Floure de-luce.

6 *Chamæiris tenuifolia.* Grasse Floure de-luce.

‡ 7 *Iris*

‡ 7 *Iris flore cæruleo obſoleto polyanthos.*
Narrow-leafed many-floured *Iris.*

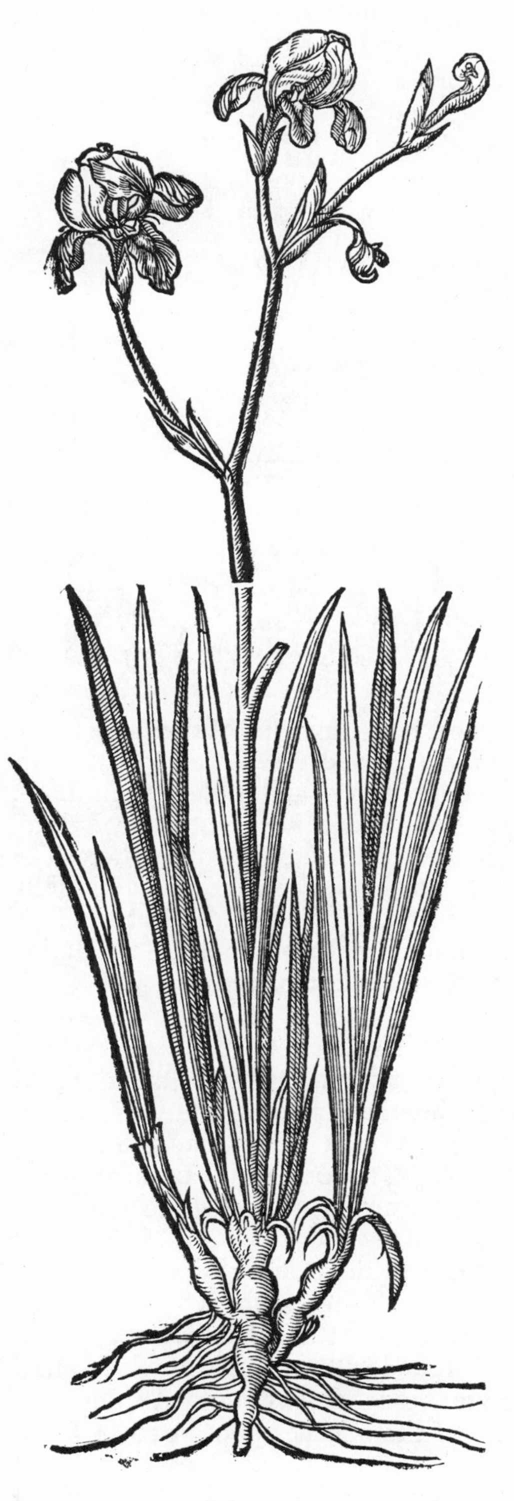

‡ 8 *Chamæiris nivea aut Candida.*
White Dwarfe *Iris.*

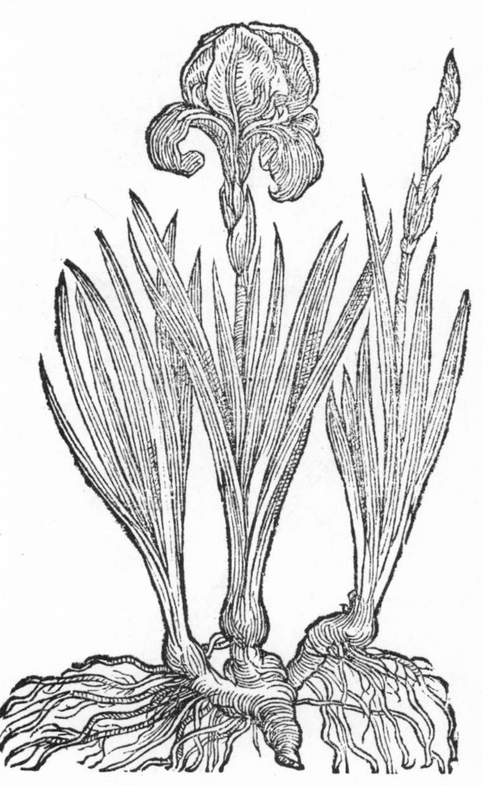

‡ 9 *Chamæiris latifolia flore rubello.*
Red floured Dwarfe *Iris.*

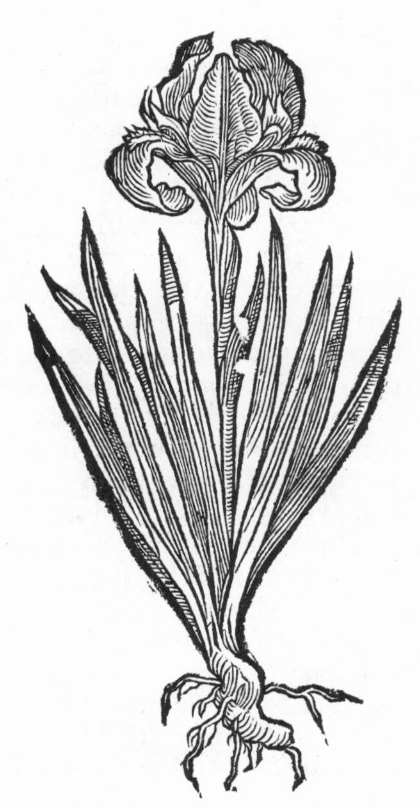

 ‡ 10 *Cha-*

‡ 10 *Chamæiris Lutea.*
Yellow Dwarfe *Iris.*

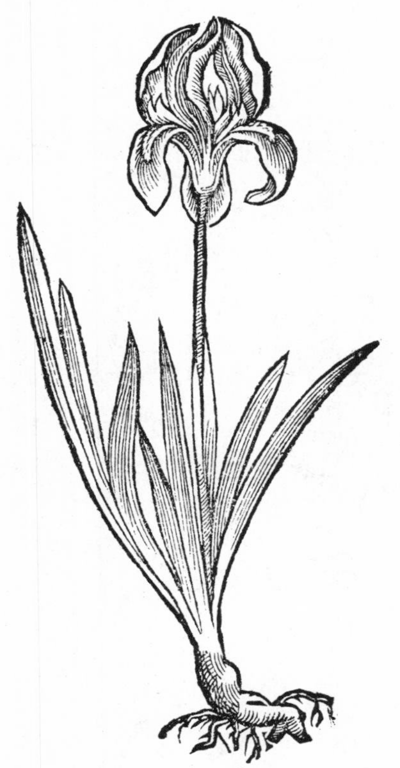

‡ 11 *Chamæiris variegata.*
Varigated Dwarfe *Iris.*

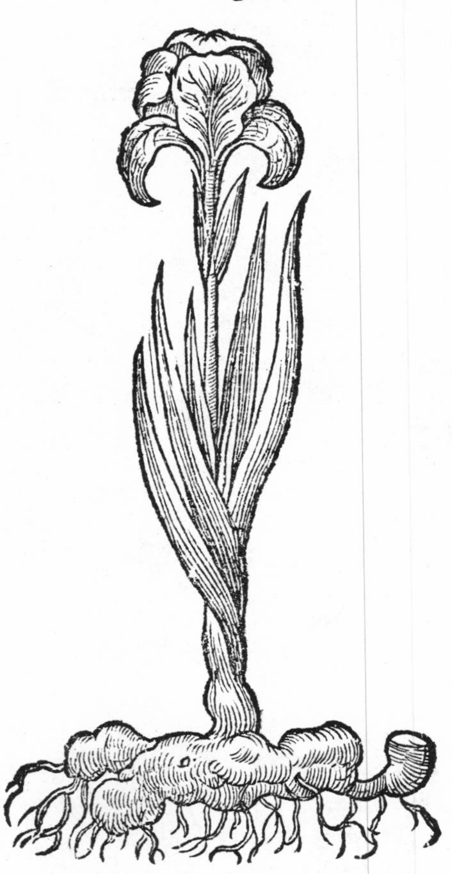

3 The French, or rather Sea Floure de-luce (whereof there is also another of the same kinde altogether lesser) haue their roots without any sauour. In shew they differ little from the garden Floure de-luce, but that the leaues of these are altogether slenderer, and vnpleasant in smell, growing plentifully in the rough crags of the rocks vnder the Alpes, and neere vnto the sea side. The which *Pena* found in the grassie grauelly grounds of the sea coast neere to Montpellier. The learned Doctor *Assatius* a long time supposed it to be *Medium Diosc.* *Matthiolus* deceiued himselfe and others, in that he said, That the root of this plant hath the sent of the peach: but my selfe haue proued it to be without sauour at all. It yeeldeth his floures in Iune, which are of all the rest most like vnto the grasse Floure de-luce. The taste of his root is hot, bitter, and with much tenuitie of parts, as hath been found by physicall proofe.

‡ 4 This *Iris Bizantina* hath long narrow leaues like those of the last described; very narrow, sharpe pointed, hauing no vngratefull smell; the stalks are some cubit and an halfe in length, and somtimes more; at the top they are diuided into 2 or 3 branches that haue 2 or 3 floures a piece, like in shape to the floures of the broad leafed variegated bulbous *Iris*; they haue also a good smell: the ends of the hanging-downe leaues are of a darke colour; the other parts of them are variegated with white, purple, or violet colour. The three other leaues that stand vp are of a deepe violet or purple colour. The root is blackish, slender, hard, knotty. ‡

5 Narrow leafed Floure de-luce hath an infinite number of grassie leaues much like vnto Reed, among which rise vp many stalkes: on the ends of the same spring forth two, sometimes three right sweet and pleasant floures, compact of nine leaues. Those three that hang downward are greater than the rest, of a purple colour, stripped with white and yellow; but those three small leaues that appeare next, are of a purple colour without mixture: those three that stand vpright are of an horse-flesh colour, tipped with purple, and vnder each of these leaues appeare three small browne aglets like the tongue of a small bird.

6 The small grassie Floure de-luce differeth from the former in smalnesse and in thinnesse of leaues, and in that the stalkes are lower than the leaues, and the floures in shape and colour are like those of the stinking Gladdon, but much lesse.

‡ There are many other varieties of the broad leafed Floure de-luces besides these mentioned by our Authour; as also of the narrow leafed, which here wee doe not intend to insist vpon, but referre such as are desirous to trouble themselues with these nicities, to *Clusius* and others.

Not-

Notwithſtanding I judge it not amiſſe to giue the figures and briefe deſcriptions of ſome more of the Dwarfe Floure de-luces, as alſo of one of the narrower leafed.

7 This therefore which we giue you in the ſeuenth place is *Iris flore cæruleo obſoleto,&c. Lobelij*. The leaues of this are ſmall and long like thoſe of the wild *Bizantine* Floure de-luce; the root (which is not very big) hath many ſtrong threds or fibres comming out of it: the ſtalke (which is ſomewhat tall) diuides it ſelfe into two or three branches, whereon grow floures in ſhape like thoſe of the other Floure de-luces, but their colour is of an ouer-worne blew, or Aſh colour.

8 Many are the differences of the *Chamæirides latifoliæ*, or Broad leafed Dwarfe Floure de-luces, but their principall diſtinction is in their floures; for ſome haue flowers of violet or purple colour, ſome of white, other ſome are variegated with yellow and purple, &c. Therefore I will onely name the colour, and giue you their figure, becauſe their ſhapes differ little. This eighth therefore is *Chamæiris nivea aut Candida*, White Dwarfe *Iris*: The ninth, *Chamæiris latifolia flore rubello*, Red floured Dwarfe *Iris*: The tenth, *Camæiris lutea*, Yellow Dwarfe *Iris*: The eleuenth, *Chamæiris variegata*, Variegated Dwarfe *Iris*. The leaues and ſtalkes of theſe plants are vſually about a foot high; the floures, for the bigneſſe of the plants, large, and they floure betimes, as in April. And thus much I thinke may ſuffice for the names and deſcriptions of theſe Dwarfe varieties of Floure de-luces. ‡

¶ *The place.*

Theſe plants do grow in the gardens of London, amongſt Herbariſts and other Louers of Plants.

¶ *The nature.*

They floure from the end of March to the beginning of May.

¶ *The names.*

The Turky Floure de-luce is called in the Turkiſh tongue *Alaia Suſiani*, with this additament from the Italians, *Fiore Belle pintate*; in Engliſh, Floure de-luce. The reſt of the names haue bin touched in their titles and hiſtorie.

¶ *Their nature and vertues.*

The faculties and temperature of theſe rare and beautifull floures are referred to the other ſorts of Floure de-luces, whereunto they do very well accord.

There is an excellent oyle made of the floures and roots of Floure de-luce, of each a like quantitie, called *Oleum Irinum*, made after the ſame manner that oyle of Roſes, Lillies, and ſuch like be made: which oyle profiteth much to ſtrengthen the ſinewes and joints, helpeth the cramp proceeding of repletion, and the diſeaſe called in Greeke *Peripneumonia*. A

The floures of French Floure de-luce diſtilled with *Diatrion ſandalon*, and Cinnamon, and the water drunke, preuaileth greatly againſt the Dropſie, as *Hollerius* and *Geſner* teſtifie. B

Chap. 43. *Of ſtinking Gladdon.*

¶ *The deſcription.*

Stinking Gladdon hath long narrow leaues like *Iris*, but ſmaller, of a darke greene colour, and being rubbed, of a ſtinking ſmell very lothſome. The ſtalkes are many in number, and round toward the top, out of which do grow floures like the Floure de-luce, of an ouer-worne blew colour, or rather purple, with ſome yellow and red ſtreakes in the midſt. After the floures be vaded there come great huskes or cods, wherein is contained a red berry or ſeed as bigge as a peaſe. The root is long, and threddy vnderneath.

¶ *The place.*

Gladdon groweth in many gardens: I haue ſeene it wilde in many places, as in woods and ſhadowie places neere the ſea.

¶ *The time.*

The ſtinking Gladdon floureth in Auguſt, the ſeed whereof is ripe in September.

¶ *The names.*

Stinking Gladdon is called in Greeke ξυρὶς, by *Dioſcorides*; and ἶρις ἀγρία by *Theophraſtus*, according to *Pena*: in Latine *Spatula fœtida* among the Apothecaries: it is called alſo *Xyris*: in Engliſh, ſtinking Gladdon, and Spurgewort.

¶ *The nature.*

Gladdon is hot and dry in the third degree.

¶ *The vertues.*

Such is the facultie of the roots of all the Irides before named, that being pounding they prouoke ſneeſing, and purge the head: generally all the kinds haue a heating & extenuating quality. A

They

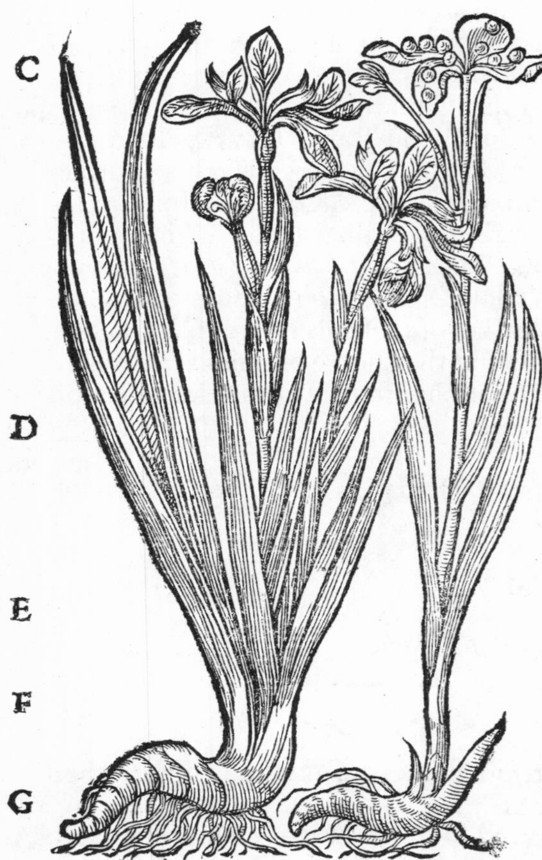

Xyris.
Stinking Gladdon.

B They are effectuall againſt the cough; they eaſily digeſt and conſume the groſſe humors which are hardly concocted: they purge choler and tough flegme: they procure ſleepe, and helpe the gripings within the belly.

C It helpeth the Kings Euill, and Buboes in the groine, as *Pliny* ſaith. If it be drunke in Wine it prouoketh the termes, and being put in Baths for women to ſit ouer, it prouoketh the like effects moſt exquiſitly. The root put in manner of a peſſarie haſtneth the birth. They couer with fleſh bones that be bare,being vſed in plaiſters. The roots boyled ſoft,and vſed plaiſterwiſe,ſoften all old hard tumours, and the ſwellings of the throat called *Strumæ*, that is, the Kings Euill; and emplaiſtered with honey it draweth out broken bones.

D The meale thereof healeth all the rifts of the fundament, and the infirmities thereof called *Condilomata*; and openeth Hemorrhoides. The juice ſniffed or drawne vp into the noſe, prouoketh ſneeſing,and draweth downe by the noſe great ſtore of filthy excrements, which would fall into other parts by ſecret and hidden E waies,and conueiances of the channels.

It profiteth being vſed in a peſſarie, to prouoke the termes, and will cauſe abortion.

F It preuaileth much againſt all euill affections of the breſt and lungs, being taken in a little ſweet wine, with ſome Spiknard; or in Whay with a little Maſticke.

G The Root of *Xyris* or Gladdon is of great force againſt wounds and fractures of the head; for it draweth out all thornes, ſtubs, prickes, and arrow-heads, without griefe; which qualitie it effecteth (as *Galen* ſaith) by reaſon of his tenuitie of parts, and of his attracting, drying, and digeſting facultie, which chiefely conſiſteth in the ſeed or fruit, which mightily prouoketh vrine.

H The root giuen in Wine, called in phyſicke *Paſſum*, profiteth much againſt Conuulſions, Ruptures, the paine of the huckle bones, the ſtrangury, and the flux of the belly. Where note, That whereas it is ſaid that the potion aboue named ſtayeth the flux of the belly, hauing a purging qualitie; it muſt be vnderſtood that it worketh in that manner as *Rhabarbarum* and *Aſarum* do, in that they concoct and take away the cauſe of the laske; otherwiſe no doubt it moueth vnto the ſtoole, as *Rheubarb*, *Aſarum*, and the other Irides do. Hereof the Countrey people of Somerſetſhire haue good experience, who vſe to drinke the decoction of this Root. Others do take the infuſion thereof in ale or ſuch like, wherewith they purge themſelues, and that vnto very good purpoſe and effect.

I The ſeed thereof mightily purgeth by vrine, as *Galen* ſaith, and the country people haue found it true.

Chap. 44. *Of Ginger.*

¶ *The deſcription.*

1 Ginger is moſt impatient of the coldneſſe of theſe our Northerne Regions, as my ſelfe haue found by proofe, for that there haue beene brought vnto me at ſeuerall times ſundry plants thereof, freſh, greene, and full of juyce, as well from the Weſt Indies, as from Barbary and other places; which haue ſprouted and budded forth greene leaues in my garden in the heate of Sommer, but as ſoone as it hath been but touched with the firſt ſharp blaſt of Winter, it hath preſently periſhed both blade and root. The true forme or picture hath not before this time beene ſet forth by any that hath written; but the World hath beene deceiued by a counterfeit figure, which the reuerend and learned Herbariſt *Matthias Lobell* did ſet forth in his Obſeruations. The forme whereof notwithſtanding I haue here expreſſed, with the true and vndoubted

doubted picture also, which I receiued from *Lobelius* his owne hands at the impression hereof. The cause of whose former errour, as also the meanes whereby he got the knowledge of the true Ginger, may appeare by his owne words sent vnto me in Latine, which I haue here inserted. His words are these:

How hard and vncertaine it is to describe in words the true proportion of Plants, (hauing no other guide than skilfull, but yet deceitfull formes of them, sent from friends, or other meanes) they best do know who haue deepliest waded in this sea of Simples. About thirty yeares past or more, an honest and expert Apothecarie *William Dries*, to satisfie my desire, sent me from Antwerpe to London the picture of Ginger, which he held to be truly and liuely drawne: I my selfe gaue him credit easily, because I was not ignorant, that there had bin often Ginger roots brought greene, new, and full of juice, from the Indies to Antwerpe; and further, that the same had budded and growne in the said *Dries* Garden. But not many yeares after, I perceiued that the picture which was sent me by my Friend was a counterfeit, and before that time had been drawne and set forth by an old Dutch Herbarist. Therefore not suffering this error any further to spred abroad, (which I discouered not many yeares past at Flushing in Zeeland, in the Garden of *William* of Nassau Prince of Orange, of famous memorie, through the means of a worthy person, if my memorie faile me not, called *Vander Mill*; at what time he opened, and loosed his first young buds and shoots about the end of Sommer, resembling in leaues, and stalkes of a foot high, the young and tender shoots of the common Reed, called *Harundo vallatoria*) I thought it conuenient to impart thus much vnto Master *Iohn Gerard*, an expert Herbarist, and Master of happy successe in Surgerie; to the end he might let posteritie know thus much, in the painefull and long laboured trauels which now he hath in hand, to the great good and benefit of his Countrey. The plant it selfe brought me to Middleborrough, and set in my Garden, perished through the hardnesse of the Winter.

Thus much haue I set downe, truly translated out of his owne words in Latine; though too fauourably by him done to the commendation of my meane skill.

1 *Zinziberis ficta Icon.*
The feigned figure of Ginger.

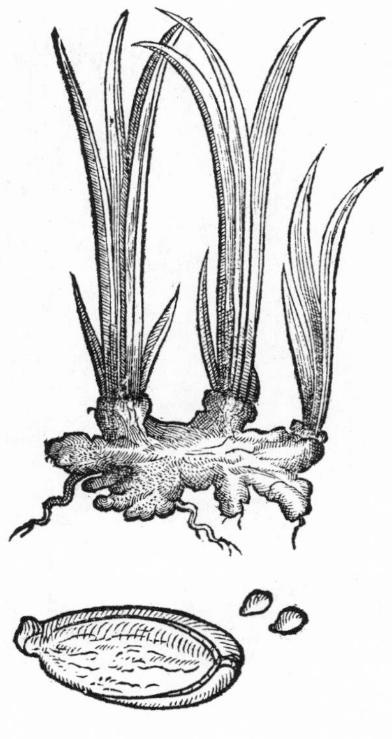

1 *Zinziberis verior Icon.*
The true figure of Ginger.

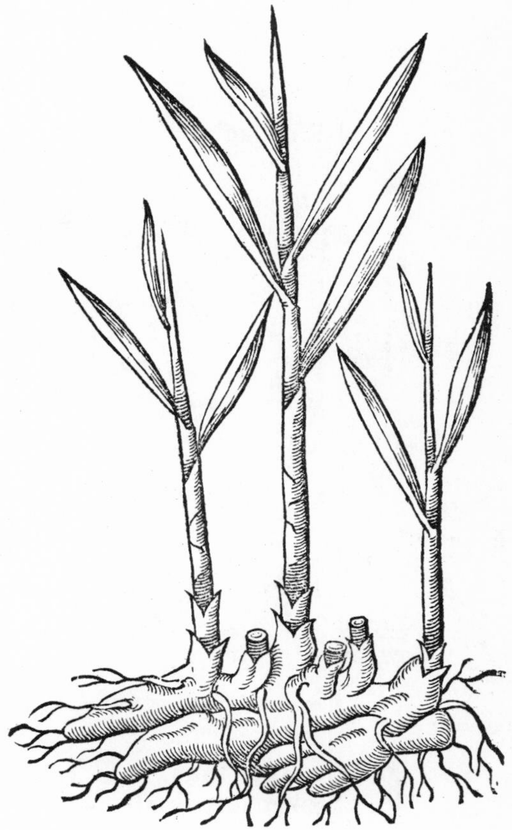

¶ *The place.*

Ginger groweth in Spaine, Barbary, in the Canary Islands, and the Azores. Our men which sacked Domingo in the Indies, digged it vp there in sundry places wilde.

¶ *The*

¶ *The time.*

Ginger flouriſheth in the hot time of Sommer, and loſeth his leaues in Winter.

¶ *The names.*

Ginger is called in Latine *Zinziber* and *Gingiber* : in Greeke, ζιγγίβερις and γιγγίβερι : In French, *Gigembre.*

¶ *The nature.*

Ginger heateth and drieth in the third degree.

¶ *The vertues.*

A Ginger, as *Dioſcorides* reporteth, is right good with meate in ſauces, or otherwiſe in conditures: for it is of an heating and digeſting qualitie; it gently looſeth the belly, and is profitable for the ſtomacke, and effectually oppoſeth it ſelfe againſt all darkneſſe of the ſight; anſwering the qualities and effects of Pepper. It is to be conſidered, That canded, greene or condited Ginger is hot and moiſt in qualitie, prouoking Venerie: and being dried, it heateth and drieth in the third degree.

CHAP. 45. *Of Aromaticall Reeds.*

1 *Acorus verus officinis falsò Calamus, cum julo.*
The true *Acorus* with his floure.

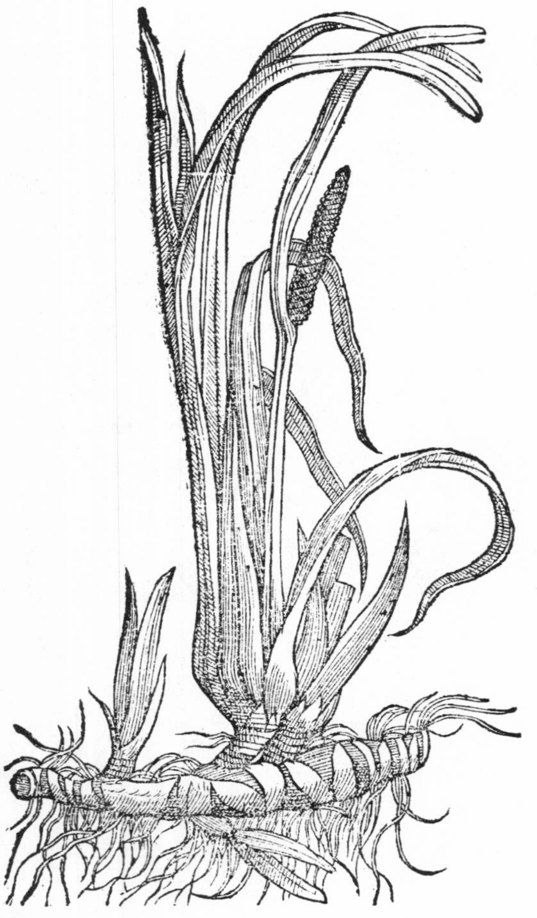

Acorus verus ſine julo.
The true *Acorus* without the floure.

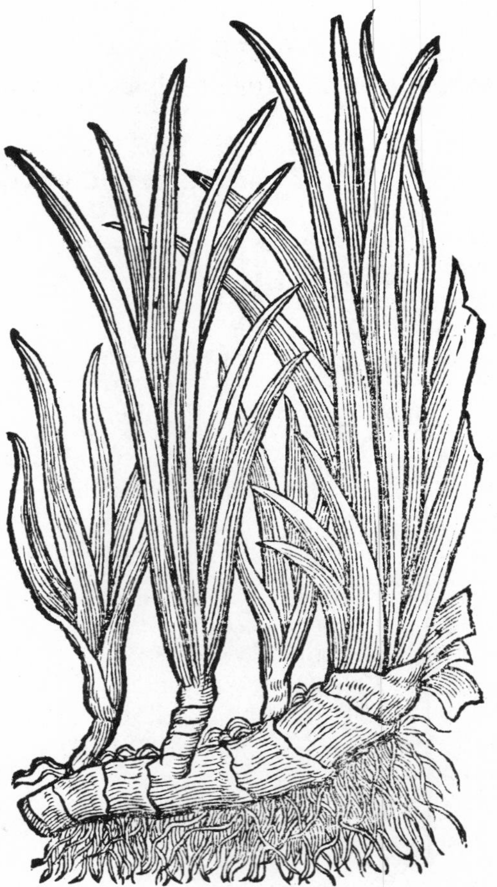

¶ *The deſcription.*

1 THis ſweet-ſmelling Reed is of a darke dun colour, full of joints and knees, eaſie to be broken into ſmall ſplinters, hollow, and full of a certaine pith cobweb-wiſe, ſomewhat gummy in eating, and hanging in the teeth, and of a ſharpe bitter taſte. It is of the thickneſſe of the little finger, as *Lobelius* affirmeth of ſome which he had ſeene in Venice.

2 Baſtard Calamus hath flaggy leaues like vnto the Water floure de-luce or flagge, but narrower, three foot long; of a freſh greene colour, and aromaticke ſmell, which they keepe a long time, although they be dried. Now the ſtalke which beares the floure or fruit is much like another

ther leafe, but onely from the fruit downwards, whereas it is somewhat thicker, and not so broad, but almost triangular. The floure is a long thing resembling the Cats-tailes which grow on Hasels; it is about the thicknesse of an ordinarie Reed, some inch and halfe long, of a greenish yellow colour, curiously chequered, as if it were wrought with a needle with greene and yellow silke intermixt †. I haue not as yet seene it beare his tuft in my garden, and haue read that it is barren, and by proofe haue seene it so: yet for all that I beleeue *Clusius*, who saith hee hath seene it beare his floure in that place where it doth grow naturally, although in England it is altogether barren. The root is sweet in smell, and bitter in taste, and like vnto the common Flagge, but smaller, and not so red.

3 *Calamus Aromaticus Antiquorum.*
The true Aromaticall Reed of the Antients.

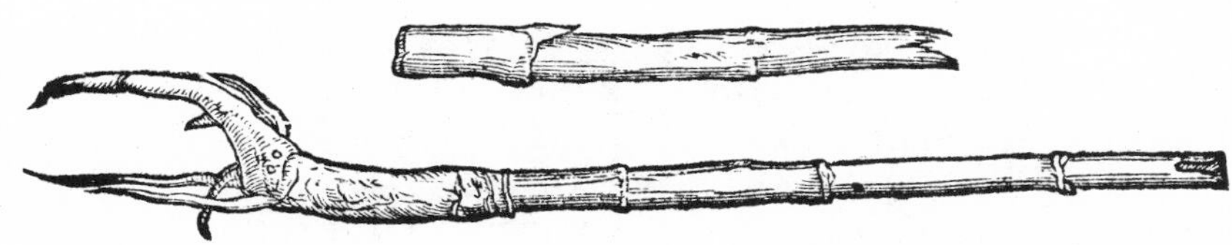

‡ 3 I thinke it very fitting in this place to acquaint you with a Plant, which by the conjecture of the most learned (and that not without good reason) is iudged to be the true *Calamus* of the Ancients. *Clusius* giues vs the historie thereof in his Notes vpon *Garcias ab Horto, lib.1. ca.32.* in these words: When as (saith he) this Historie was to be the third time printed, I very opportunely came to the knowledge of the true *Calamus Aromaticus*; the which the learned *Bernard Paludanus* the Frisian, returning from Syria and Ægypt, freely bestowed vpon me, together with the fruit Habhel, and many other rare seeds, about the beginning of the yeare 1579. Now wee haue caused a figure to be exactly drawne by the fragments thereof (for that it seemes so exquisitly to accord with *Dioscorides* his description.) In myne opinion it is rather to be iudged an vmbellliferous plant than a reedy; for it hath a straight stalke parted with many knots or ioynts, otherwise smooth, hollow within, and inuested on the inside with a slender filme like as a Reed, and it breaketh into shiuers or splinters, as *Dioscorides* hath written: it hath a smell sufficiently strong, and the taste is gratefull, yet bitter, and pertaking of some astriction: The leaues, as by remaines of them might appeare, seeme by couples at euery ioynt to engirt the stalke: the root at the top is somewhat tuberous, and then ends in fibres. Twenty fiue yeares after *Paludanus* gaue me this *Calamus*, the learned *Anthony Coline* the Apothecarie (who lately translated into French these Commentaries the fourth time set forth, *Anno* 1593) sent me from Lyons pieces of the like Reed, certifying me withall, That he had made vse thereof in his Composition of Treacle. Now these pieces, though in forme they resembled those I had from *Paludanus*, yet had they a more bitter taste than his, nether did they partake of any astriction; which peraduenture was to be attributed to the age of one of the two. Thus much *Clusius*. ‡

¶ *The place.*

The true *Calamus Aromaticus* groweth in Arabia, and likewise in Syria, especially in the moorish grounds betweene the foot of Libanus † and another little hill, not the mountaine Antilibanus, as some haue thought, in a small valley neere to a lake, whose plashes are dry in Sommer *Pliny* 12. 22. †

Bastard or false *Calamus* growes naturally at the foot of a hill neere to Prusa a city of Bithynia, not far from a great lake. It prospereth exceeding well in my garden, but as yet it beareth neither floures nor stalke. It groweth also in Candia, as *Pliny* reporteth: in Galatia likewise, and in many other places.

¶ *The time.*

They lose their leaues in the beginning of Winter, and do recouer them againe in the Spring of the yeare. ‡ In May this yeare 1632, I receiued from the Worshipfull Gentleman M. *Thomas Glynn* of Glynnlhivon in Carnaruanshire, my very good friend, the pretty *Iulus*, or floure of this plant; which I could neuer see here about London, though it groweth with vs in many Gardens, and that in great plenty. ‡

¶ *The names.*

‡ The want of the true *Calamus* being supplied by *Acorus* as a *succedaneum*, was the cause (as *Pena* and *Lobell* probably coniecture) that of a substitute it tooke the prime place vpon it; and being as it were made a Vice-Roy, would needs be King. But the falsenesse of the title was disco-

uered

uered by *Matthiolus*, and others, and so it is sent backe to its due place againe; though notwithstanding it yet in shops retaines the title of *Calamus*.

1 The figure that by our Author was giuen for this, is supposed, and that (as I thinke truly) to be but a counterfeit, of *Matthiolus* his inuention; who therein hath beene followed (according to the custome of the world) by diuers others. The description is of a small Reed called *Calamus oderatus Libani*, by *Lobell* in his Obseruations, and figured in his *Irones*, *p*. 54.

2 This is called Ἄκορος and Ἄκορον by the Greekes: by some, according to *Apuleius*, Ἀφροδισίας; and in Latine it is called *Acorus* and *Acorum*; and in shops, as I haue formerly said, *Calamus Aromaticus*: for they vsually take *Galanga major*, (described by me, *Chap*. 26.) for *Acorus*. It may besides the former names be fitly called in English, The sweet Garden Flag.

3 This is iudged to be the κάλαμος ἀρωματικὸς of *Dioscorides*; the κάλαμος εὐώδης of *Theophrastus*; that is, the true *Calamus Aromaticus* that should be vsed in Compositions. ‡

¶ *The nature of the true Acorus, or our sweet garden Flag.*

Dioscorides saith, the roots haue an heating facultie: *Galen* and *Pliny* do affirme, that they haue thin and subtill parts, both hot and dry.

¶ *The vertues of the same.*

A The decoction of the root of *Calamus* drunke prouoketh vrine, helpeth the paine in the side, liuer, spleene, and brest; conuulsions, gripings, and burstings; it easeth and helpeth the pissing by drops.

B It is of great effect, being put in broth, or taken in fumes through a close stoole, to prouoke womens naturall accidents.

C The iuyce strained with a little honey, taketh away the dimnes of the eyes, and helpeth much against poyson, the hardnesse of the spleene, and all infirmities of the bloud.

D The root boyled in wine, stamped and applied plaisterwise vnto the cods, doth wonderfully abate the swelling of the same, and helpeth all hardnesse and collections of humors.

E The quantitie of two scruples and an halfe of the root drunke in foure ounces of Muskadel, helpeth them that be bruised with grieuous beating, or falls.

F The root is with good successe mixed in counterpoysons. In our age it is put into Eclegma's, that is, medicines for the lungs, and especially when the lungs and chest are opprest with raw and cold humors.

G ‡ The root of this preserued is very pleasant to the taste, and comfortable to the stomacke and heart; so that the Turks at Constantinople take it fasting in the morning, against the contagion of the corrupt aire. And the Tartars haue it in such esteeme, that they will not drinke Water (which is their vsuall drinke) vnlesse they haue first steeped some of this root therein. ‡

¶ *The choice.*

The best *Acorus*, as *Dioscorides* saith, is that which is substantiall, and well compact, white within, not rotten, full, and well smelling.

Pliny writeth, That those which grow in Candia are better than those of Pontus, and yet those of Candia worse than those of the Easterne countries, or those of England, although we haue no great quantitie thereof.

¶ *The faculties of the true Calamus out of Dioscorides.*

H ‡ It being taken in drinke moueth vrine; wherefore boyled with the roots of grasse or Smallage seeds, it helpeth such as are hydropick, nephritick, troubled with the strangurie, or bruised.

I It moues the Courses, either drunke or otherwise applied. Also the fume thereof taken by the mouth in a pipe, either alone or with dried Turpentine, helpes coughs.

K It is boyled also in baths for women, and decoctions for Glysters, and it enters into plaisters and perfumes for the smells sake. ‡

Chap. 46. *Of Corne.*

THus farre haue I discoursed vpon Grasses, Rushes, Spartum, Flags, and Floure deluces: my next labour is to set downe for your better instruction, the historie of Corne, and the kindes thereof, vnder the name of Graine; which the Latines call *Cerialia semina*, or Bread-corne; the Grecians, σιτηρὰ and δημητρεία σπέρματα; of which wee purpose to discourse. There belong to the historie of Graine all such things as be made of Corne, as *Far*, *Condrus*, *Alica*, *Tragus*, *Amylum*, *Ptisana*, *Polenta*, *Maza*, *Byne* or Malt, *Zythum*, and whatsoeuer are of that sort. There be also ioyned vnto them many seeds, which *Theophrastus* in his eighth booke placeth among the graines; as Millet, Sorgum, Panicke, Indian wheat, and such like. *Galen* in his first booke of the Faculties of nourishments, reckoneth

vp

vp the diſeaſes of Graine, as well thoſe that come of the graine it ſelfe degenerating, or that are changed into ſome other kinde, and made worſe through the fault of the weather, or of the ſoile; as alſo ſuch as be cumberſome by growing among them, doe likewiſe fitly ſucceed the graines. And beginning with corne, we will firſt ſpeake of Wheat, and deſcribe it in the firſt place, becauſe it is preferred before all other corne.

1 *Triticum ſpica mutica.*
White Wheate.

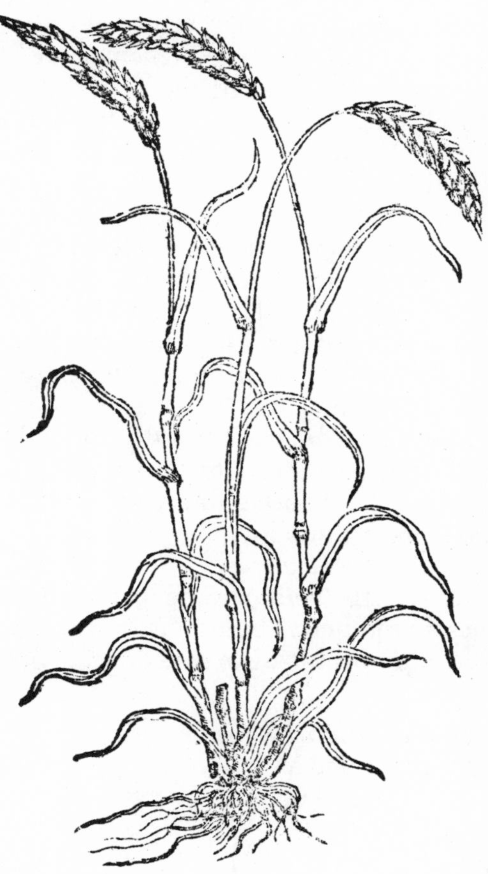

¶ *The deſcription.*

1 THis kinde of Wheate which *Lobelius*, diſtinguiſhing it by the eare, calleth *Spica Mutica*, is the moſt principal of all other, whoſe eares are altogether bare or naked, without awnes or chaffie beards. The ſtalke riſeth from a threddy root, compact of many ſtrings, joynted or kneed at ſundry diſtances; from whence ſhoot forth graſſie blades and leaues like vnto Rie, but broader. The plant is ſo well knowne to many, and ſo profitable to all, that the meaneſt and moſt ignorant need no larger deſcription to know the ſame by.

2 The ſecond kinde of Wheat, in root, ſtalkes, joints, and blades, is like the precedent, differing onely in eare, and number of graines, whereof this kinde doth abound, hauing an eare conſiſting of many ranks, which ſeemeth to make the eare double or ſquare. The root and graine is like the other, but not bare and naked, but briſtled or bearded, with many ſmall and ſharpe eiles or awnes, not vnlike to thoſe of Barley.

3 Flat Wheat is like vnto the other kindes of Wheat in leaues, ſtalkes, and roots, but is bearded and bordered with rough and ſharpe ailes, wherein conſiſts the difference. ‡ I know not what our Author means by this flat Wheat, but I conjecture it to be the long rough eared Wheat, which hath blewiſh eares when as it is ripe, in other things reſembling the ordinary red wheat. ‡

4 The fourth kinde is like the laſt deſcribed, and thus differeth from it, in that, that this kind hath many ſmal ears comming forth of one great eare, & the beards hereof be ſhorter than of the former kind.

5 Bright wheate is like the ſecond before deſcribed, and differeth from it in that, that this kind is foure ſquare, ſomewhat bright and ſhining, the other not.

‡ I thinke it a very fit thing to adde in this place a rare obſeruation, of the tranſmutation of one ſpecies into another, in plants; which though it haue beene obſerued of ancient times, as by *Theophraſtus, de cauſ. plant. lib. 3. cap. 6.* whereas amongſt others hee mentioneth the change of Ζειᾶ πρὸς τὸ βρόμον, Spelt into oates: and by *Virgill* in theſe verſes;

Grandia ſæpe quibus mandauimus Hordea ſulcis,
Infœlix Lolium, & ſteriles dominantur avenæ.

That is;

In furrowes where great Barley we did ſow,
Nothing but Darnel and poore Oats do grow;

yet none that I haue read haue obſerued, that two ſeuerall graines, perfect in each reſpect, did grow at any time in one eare: the which I ſaw this yeare 1632, in an eare of white Wheat, which was found by my very good Friend Maſter *Iohn Goodyer*, a man ſecond to none in his induſtrie and ſearching of plants, nor in his iudgement or knowledge of them. This eare of wheat was as large and faire as moſt are, and about the middle thereof grew three or foure perfect Oats in all reſpects: which being hard to be found, I held very worthy of ſetting downe, for ſome reaſons not to be inſiſted vpon in this place. ‡

¶ *The place.*

Wheat groweth almoſt in all the countries of the world that are inhabited and mannured, and requireth a fruitfull and fat ſoile, and rather Sunny and dry, than watery grounds and ſhadowie: for in a dry ground (as *Columella* reporteth) it groweth harder and better compact: in a moiſt and darke ſoile it degenerateth ſometime to be of another kinde.

2 *Triticum aristis circumvallatum.*
Bearded Wheat, or Red-Wheat.

¶ *The time.*

They are most commonly sowen in the fall of the leafe, or Autumne: somtime in the Spring.

¶ *The names.*

Wheat is called of the Grecians πυρὸς: of the Latines, *Triticum*, and the white Wheate *Siligo*. *Triticum* doth generally signifie any kinde of Corne which is threshed out of the eares, and made clean by fanning or such ordinary meanes. The Germans call it **Weusen**: in low Dutch, **Terwe**: in Italian, *Grano*: the Spaniards, *Trigo*: the French men, *Bled, ou Fourment*: in England we call the first, White-Wheat, and Flaxen Wheat. *Triticum Lucidum* is called Bright Wheat: Red Wheat is called in Kent, Duck-bill Wheate, and Normandy Wheat.

¶ *The nature.*

Wheat (saith *Galen*) is very much vsed of men, and with greatest profit. Those Wheats do nourish most which be hard, and haue their whole substance so closely compact as they can scarcely be bit asunder; for such doe nourish very much: and the contrary but little.

Wheat, as it is a medicine outwardly applied, is hot in the first degree, yet can it not manifestly either dry or moisten. It hath also a certaine clamminesse and stopping qualitie.

¶ *The vertues.*

A Raw Wheat, saith *Dioscorides*, being eaten, breedeth wormes in the belly: being chewed and applied, it doth cnre the biting of mad dogs.

3 *Triticum Typhinum.*
Flat Wheat.

4 *Triticum multiplici spica.*
Double eared Wheat.

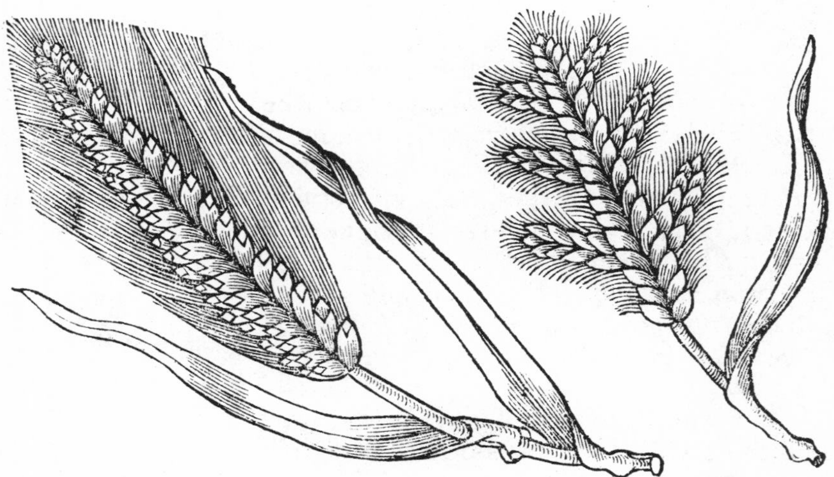

B The floure of wheat being boyled with honey and water, or with oyle and water, taketh away all inflammations, or hot swellings.

C The bran of Wheat boyled in strong Vineger, clenseth away scurfe and dry scales, and dissolueth the beginning of all hot swellings, if it be laid vnto them. And boyled with the decoction of Rue, it slaketh the swellings in womens brests.

D The graines of white Wheat, as *Pliny* writeth in his two and twentieth booke, and seuenth chapter, being dried brown, but not burnt, and the pouder thereof mixed with white wine is good for watering eyes, if it be laid thereto.

E The dried pouder of red Wheat boyled with vineger, helpeth the shrinking of sinewes.

F The meale of Wheat mingled with the juice of Henbane, and plaisterwise applied, appeaseth inflam-

5 *Triticum lucidum.*
Bright Wheat.

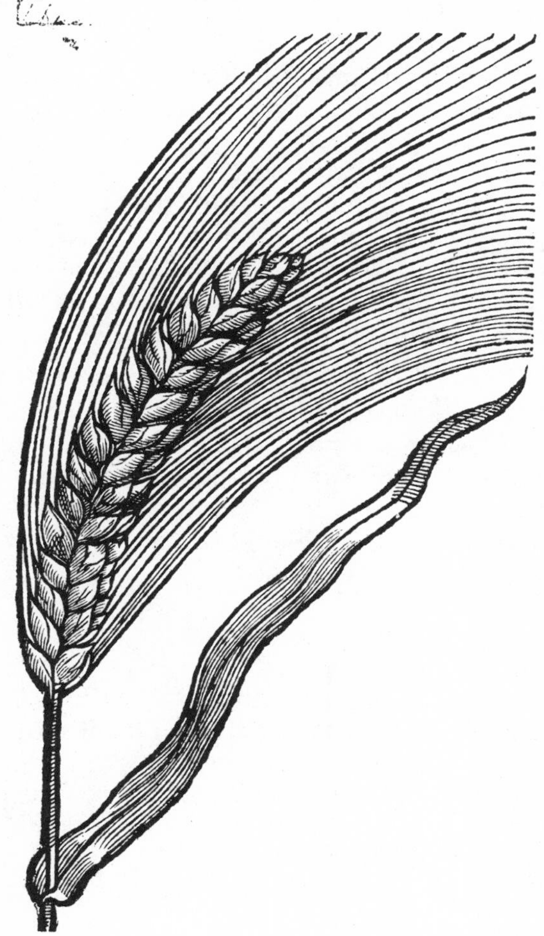

inflammations, as *Ignis ſacer*, or Saint *Anthonies* Fire, and ſuch like, ſtaying the flux of humors to the ioynts, which the Grecians call *Rheumatiſmata*. Paſte made of fine meale, ſuch as Booke-binders vſe, helpeth ſuch as doe ſpit bloud, taken warme one ſpoonfull at once. The bran of wheat boiled in ſharpe vineger, and rubbed vpon them that be ſcuruie and mangie, eaſeth the party very much.

G The leauen made of Wheat hath vertue to heate and draw outward, it reſolueth, concocteth, and openeth all ſwellings, bunches, tumors, and felons, being mixed with ſalt.

H The fine floure mixed with the yolke of an egge, honey, and a little ſaffron, doth draw and heale byles and ſuch like ſores, in children and in old people, very well and quickely. Take crums of wheaten bread one pound and an halfe, barley meale ℥ ij. Fennigreeke and Lineſeed of each an ounce, the leaues of Mallowes, Violets, Dwale, Sengreene, and Cotyledon, *ana* one handfull: boyle them in water and oyle vntill they be tender: then ſtampe them very ſmall in a ſtone morter, and adde thereto the yolks of three egges, oyle of Roſes, and oyle of Violets, *ana* ℥ ij. Incorporate them altogether; but if the inflammation grow to an Eryſipelas, then adde thereto the juice of Nightſhade, Plantaine, and Henbane, *ana* ℥ ij. it eaſeth an Eryſipelas, or Saint *Anthonies* fire, and all inflammations very ſpeedily.

I Slices of fine white bread laid to infuſe or ſteepe in Roſe water, and ſo applied vnto ſore eyes which haue many hot humors falling into them, doth eaſily defend the humour, and ceaſe the paine.

K The oyle of wheat preſſed forth betweene two plates of hot iron, healeth the chaps and chinks of the hands, feet, and fundament, which come of cold, making ſmooth the hands, face, or any other part of the body.

L The ſame vſed as a Balſame doth excellently heale wounds, and being put among ſalues or vnguents, it cauſeth them to worke more effectually, eſpecially in old vlcers.

CHAP. 47. *Of Rie.*

¶ *The deſcription.*

THe leafe of Rie when it firſt commeth vp, is ſomewhat reddiſh, afterward greene, as be the other graines. It groweth vp with many ſtalks, ſlenderer than thoſe of wheat, and longer, with knees or ioynts by certaine diſtances like vnto Wheat: the eares are orderly framed vp in rankes, and compaſſed about with ſhort beards, not ſharpe but blunt, which when it floureth ſtandeth vpright, and when it is filled vp with ſeed it leaneth and hangeth downward. The ſeed is long, blackiſh, ſlender, and naked, which eaſily falleth out of the huskes of it ſelfe. The roots be many, ſlender, and full of ſtrings.

¶ *The place.*

Rie groweth very plentifully in the moſt places of Germany and Polonia, as appeareth by the great quantitie brought into England in times of dearth, and ſcarcitie of corne, as hapned in the yeare 1596, and at other times, when there was a generall want of corne, by reaſon of the aboundance of raine that fell the yeare before; whereby great penutrie enſued, as well of cattell and all other victuals, as of all manner of graine. It groweth likewiſe very wel in moſt places of England, eſpecially towards the North.

Secale.
Rie.

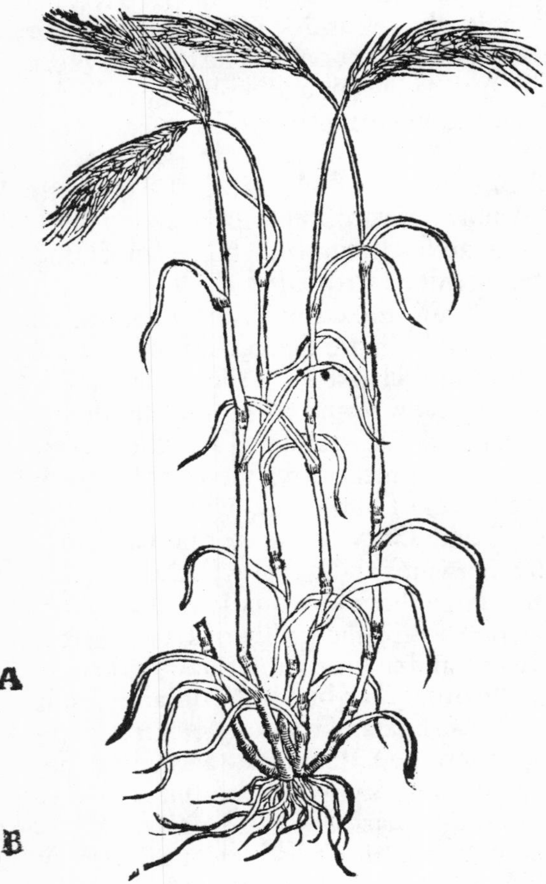

¶ *The time.*

It is for the most part sowen in Autumne, and sometimes in the Spring, which proueth to be a Graine more subiect to putrifaction than that which was sowen in the fall of the leafe, by reason the Winter doth ouertake it before it can attaine to his perfect maturitie and ripenesse.

¶ *The names.*

Rie is called in high Dutch, **Rocken:** in Low-Dutch, **Rogge:** in Spanish, *Centeno:* in Italian, *Segala:* in French, *Seigle:* which soundeth after the old Latine name which in *Pliny* is *Secale* and *Farrago, lib.* 18. *cap.* 16.

¶ *The temperature.*

Rie as a medicine is hotter than wheat, and more forcible in heating, wasting, and consuming away that whereto it is applied. It is of a more clammy and obstructing nature than Wheat, and harder to digest, yet to rusticke bodies that can well digest it, it yeelds good nourishment.

¶ *The vertues.*

A Bread, or the leauen of Rie, as the Belgian Physitians affirme vpon their practise, doth more forcibly digest, draw, ripen, and breake all Apostumes, Botches, and Byles, than the leuen of Wheat.

B Rie Meale bound to the head in a Linnen Cloath, doth asswage the long continuing paines thereof.

Chap. 48. *Of Spelt Corne.*

¶ *The description.*

Spelt is like to Wheat in stalkes and eare: it groweth vp with a multitude of stalks which are kneed and joynted higher than those of Barley: it bringeth forth a disordered eare, for the most part without beards. The cornes be wrapped in certaine dry huskes, from which they cannot easily be purged, and are joyned together by couples in two chaffie huskes, out of which when they be taken they are like vnto wheat cornes: it hath also many roots as wheat hath, whereof it is a kinde.

¶ *The place.*

It groweth in fat and fertile moist ground.

¶ *The time.*

It is altered and changed into Wheat it selfe, as degenerating from bad to better, contrary to all other that do alter or change; especially (as *Theophrastus* saith) if it be clensed, and so sowen, but that not forthwith, but in the third yeare.

¶ *The names.*

The Grecians haue called it ζειὰ and ζέα: the Latines *Spelta:* in the Germane tongue **Speltz**, and **Dinkel:** in low Dutch, **Spelte:** in French, *Espeautre:* of most Italians, *Pirra, Farra:* of the Tuscans, *Biada:* of the Millanois, *Alga:* in English, Spelt Corne *Dioscorides* maketh mention of two kindes of Spelt: one of which he names ἁπλῆ, or single: another, δίκοκκος, which brings forth two cornes ioyned together in a couple of huskes, as before in the description is mentioned. That Spelt which *Dioscorides* calls *Dicoccos,* is the same that *Theophr.* and *Galen* do name *Zea.* The most ancient Latines haue called *Zea* or *Spelta* by the name of *Far,* as *Dionysius Halicarnassæus* doth sufficiently testifie: The old Romans (saith he) did call sacred marriages by the word φαῤῥάκεια, because the

Zea siue Spelta.
Spelt Corne.

the Bride and Bridegroome did eate of that *Far* which the Grecians do call ζέα. The same thing *Asclepiades* affirmeth in *Galen*, in his ninth Booke according to the places affected writing thus; *Farris quod Zea appellant*: that is to say, *Far* which is called *Zea*, &c. And this *Far* is also named of the Latines, *Ador*, *Adoreum*, and *Semen adoreum*.

¶ *The temper.*

Spelt, as *Dioscorides* reporteth, nourisheth more than Barley. *Galen* writeth in his Bookes of the Faculties of simple Medicines, Spelt is in all his temperature in a meane betweene Wheat and Barley, and may in vertue be referred to the kindes of Barley and Wheat, being indifferent to them both.

¶ *The vertues.*

The floure or meale of Spelt corne boyled in water with the pouder of red Saunders, and a little oyle of Roses and Lillies, vnto the forme of a Pultesse, and applied hot, taketh away the swelling of the legs gotten by cold and long standing. A

‡ Spelt (saith *Turner*) is common about Weisenburgh in high Almanie, eight Dutch miles on this side Strausbourgh: and there all men vse it for wheat; for there groweth no wheat at all: yet I neuer saw fairer and pleasanter bread in any place in all my life, than I haue eaten there, made onely of this Spelt. The Corne is much lesse than Wheat, and somewhat shorter than Rie, but nothing so blacke. ‡ B

CHAP. 49. *Of Starch Corne.*

Triticum Amyleum.
Starch Corne.

¶ *The description.*

THis other kind of *Spelta* or *Zea* is called of the Germane Herbarists *Amyleum Frumentum*, or Starch corne; and is a kinde of grain sowen to that end, or a three moneths graine, and is very like vnto wheat in stalke and seed; but the eare thereof is set round about, and made vp with two ranks, with certaine beards, almost after the manner of Barley, and the seed is closed vp in chaffie huskes, and is sowen in the Spring.

¶ *The place.*

Amil corne, or Starch corne is sowen in Germanie, Polonia, Denmarke, and other those Easterne Regions, as well to feed their cattel and pullen with, as also to make starch; for the which purpose it doth very fitly serue.

¶ *The time.*

It is sowen in Autumne, or the fall of the leafe, and oftentimes in the Spring; and for that cause hath beene called *Trimestre*, or three months grain: it bringeth his seed to ripenesse in the beginning of August, and is sowen in the Low-Countries in the Spring of the yeare.

¶ *The names.*

Because the Germanes haue great vse of it to make starch with, they do call it **Amelcorn**: Wee thinke

thinke good to name it in Latine *Amyleum frumentum*: in English it may be called Amelcorne, after the Germane word; and may likewise be called Starch Corne. *Tragus* and *Fuchsius* tooke it to be *Triticum trimestre*, or three moneths wheat; but it may rather be referred to the *Farra*: for *Columella* speaketh of a graine called *Far Halicastrum*, which is sowen in the Spring; and for that cause it is named *Trimestre*, or three moneths *Far*. If any be desirous to learne the making of Starch, let them reade *Dodoneus* last edition, where they shall be fully taught; my selfe not willing to spend time about so vaine a thing, and not pertinent to the story. It is vsed onely to feed cattell, pullen, and make starch, and is in nature somewhat like to wheat or Barley.

Chap. 50. *Of Barley.*

¶ *The description.*

Barley hath an helme or straw which is shorter and more brittle than that of Wheat, and hath more joints; the leaues are broader and rougher; the eare is armed with long, rough, and prickly beards or ailes, and set about with sundry rankes, sometimes two, otherwhiles three, foure, or six at the most, according to *Theophrastus*; but eight according to *Tragus*. The graine is included in a long chaffie huske: the roots be slender, and grow thicke together. Barley, as *Pliny* writeth, is of all graine the softest, and least subiect to casualtie, yeelding fruit very quickely and profitably.

1 *Hordeum Distichon.*
Common Barley.

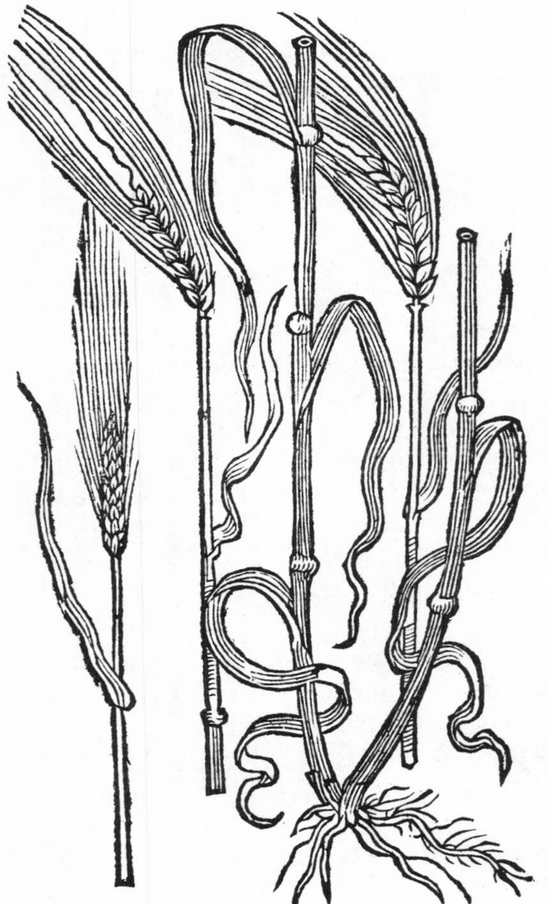

2 *Hordeum Polystichum vernum.*
Beare Barley, or Barley Big.

1 The most vsuall Barley is that which hath but two rowes of Corne in the eare, each graine set iust opposite to other, and hauing his long awne at his end, is couered with a huske sticking close thereto.

2 This which commonly hath foure rowes of corne in the eare, and sometimes more, as wee haue formerly deliuered, is not so vsually sowen with vs; the eare is commonly shorter than the former, but the graine very like; so that none who knowes the former, but may easily know the later at the first sight.

¶ *The*

¶ *The place.*

They are ſowen, as *Columella* teacheth, in looſe and dry ground, and are well knowne all Europe through.

2 The ſecond is ſowen commonly in ſome parts of Yorke ſhire and the Biſhopricke of Durham.

¶ *The names.*

1 The firſt is called of the Grecians κριθὴ: in high Dutch, **Gersten:** in Low Dutch, **Gerst:** in Italian, *Orzo:* in Spaniſh, *Ceuada:* in French, *Orge:* in Engliſh, Barley.

2 The ſecond is called of the Grecians πολυστιχὸν, and alſo ἑξαστιχὸν: *Columella* calleth it *Galaticum*; and *Hippocrates*, ἀχιλλαὶς κριθὴ: of our Engliſh Northerne people, Big, and Big Barley. *Crimmon* (ſaith *Galen* in his Commentaries vpon the ſecond booke of *Hippocrates* his Prognoſticks) is the groſſer part of Barley meale being groſſely ground. Malt is well knowne in England, inſomuch that the word needeth no interpretation; notwithſtanding becauſe theſe Workes may chance into the hands of Strangers, that neuer heard of ſuch a word, or ſuch a thing, by reaſon it is not euerie where made; I thought good to lay downe a word of the making thereof. Firſt, it is ſteeped in water vntill it ſwell; then is it taken from the water, and laid (as they terme it) in a Couch; that is, ſpred vpon an euen floore the thickneſſe of ſome foot and an halfe; and thus is it kept vntill it Come, that is, vntill it ſend forth two or three little ſtrings or fangs at the end of each Corne: then it is ſpred vſually twice a day, each day thinner than other, for ſome eight or ten daies ſpace, vntill it be pretty dry, and then it is dried vp with the heate of the fire, and ſo vſed. It is called in high Dutch, **Maltz:** in low Dutch, **Mout:** in Latine of later time, *Maltum:* which name is borrowed of the Germanes. *Aetius* a Greeke Phyſitian nameth Barley thus prepared, βύνη, or *Bine:* The which Author affirmeth, That a plaiſter of the meale of Malt is profitably laid vpon the ſwellings of the Dropſie. *Zythum*, as *Diodorus Siculus* affirmeth, is not onely made in Ægypt, but alſo in Galatia. The aire is ſo cold (ſaith he, writing of Galatia) that the country bringeth forth neither wine nor oyle; and therefore men are compelled to make a compound drinke of Barley, which they call *Zythum*. *Dioſcorides* nameth one kinde of Barley drinke *Zythum*; another, *Curmi*. *Simeon Zethi* a later Grecian calleth this kind of drinke by an Arabicke name, φούκας: in Engliſh we call it Beere and Ale which is made of Barley Malt.

¶ *The temperature.*

Barley, as *Galen* writeth in his booke of the Faculties of nouriſhments, is not of the ſame temperature that Wheat is; for Wheat doth manifeſtly heate, but contrariwiſe what medicine or bread ſoeuer is made of Barley, is found to haue a certaine force to coole and drye in the firſt degree, according to *Galen* in his booke of the faculties of Simples. It hath alſo a little abſterſiue or cleanſing qualitie, and doth dry ſomewhat more than Beane meale.

¶ *The vertues.*

Barley, ſaith *Dioſcorides*, doth cleanſe, prouoke vrine, breedeth windineſſe, and is an enemie to the ſtomacke. A

Barley meale boyled in an honied water with figges, taketh away inflammations: with Pitch, Roſin, and Pigeons dung, it ſoftneth and ripeneth hard ſwellings. B

With Melilot and Poppy ſeeds it taketh away the paine in the ſides: it is a remedy againſt windineſſe in the guts, being applied with Lineſeed, Fœnugreeke, and Rue: with tarre, wax, oyle, and the vrine of a yong boy, it doth digeſt, ſoften, and ripe hard ſwellings in the throat, called the Kings Euill. C

Boyled with wine, myrtles, the barke of the pomegranate, wilde peares, and the leaues of brambles, it ſtoppeth the laske. D

Further, it ſerueth for *Ptiſana, Polenta, Maza,* Malt, ale, and Beere. The making whereof if any be deſirous to learne, let them reade *Lobelius Aduerſaria,* in the chapter of Barley. But I thinke our London Beere-Brewers would ſcorne to learne to make beere of either French or Dutch, much leſſe of me that can ſay nothing therein of mine owne experience more than by the Writings of others. But I may deliuer vnto you a Confection made thereof (as *Columella* did concerning ſweet wine ſodden to the halfe) which is this; Boyle ſtrong ale till it come to the thickeneſſe of hony, or the forme of an vnguent or ſalue, which applied to the paines of the ſinewes and joints (as hauing the propertie to abate aches and paines) may for want of better remedies be vſed for old and new ſores, if it be made after this manner. E

Take ſtrong ale two pound, one Oxe gall, and boyle them to one pound with a ſoft fire, continually ſtirring it; adding thereto of Vineger one pound, of *Olibanum* one ounce, floures of Camomil and melilot of each ℥ i. Rue in fine pouder ℥s. a little hony, and a ſmall quantitie of the pouder of Comin ſeed; boyle them all together to the forme of an vnguent, and ſo apply it. There be ſundry ſorts of Confections made of Barley, as *Polenta, Ptiſana,* made of water and husked or hulled barley, and ſuch like. *Polenta* is the meate made of parched Barley, which the Grecians doe properly F perly call ἄλφιτον. *Maza* is made of parched Barley tempered with water, after *Hippocrates* and *Xenophon* : *Cyrus* hauing called his souldiers together, exhorteth them to drinke water wherein parched Barley hath beene steeped, calling it by the same name, *Maza*. *Hesychius* doth interpret μάζα to be Barley meale mixed with water and oyle.

Barley meale boyled in water with garden Nightshade, the leaues of garden Poppie, the pouder of Fœnugreeke and Lineseed, and a little Hogs grease, is good against all hot and burning swellings, and preuaileth against the Dropsie, being applied vpon the belly.

Chap. 51. *Of Naked Barley.*

Hordeum nudum.
Naked Barley.

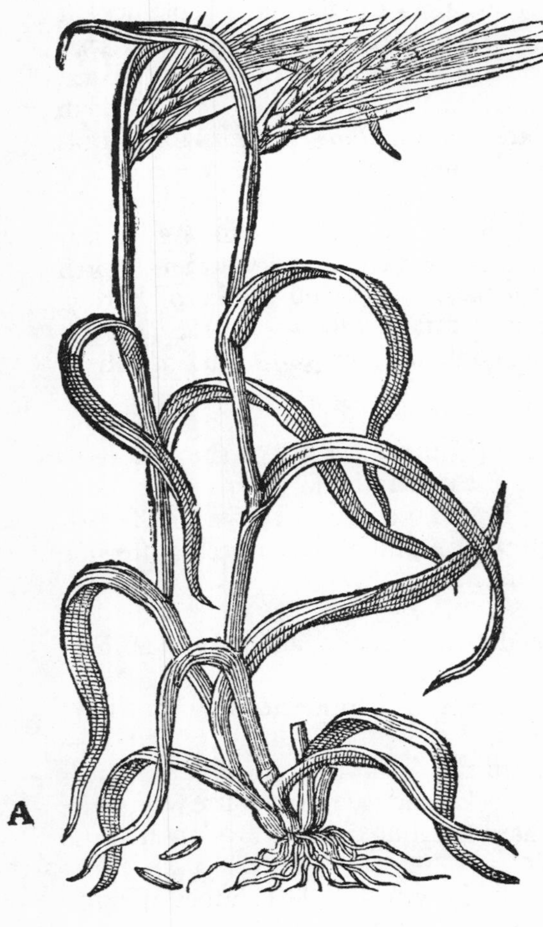

¶ *The description.*

Hordeum nudum is called *Zeopyrum*, and *Tritico-Speltum*, because it is like to *Zea*, otherwise called *Spelta*, and is like to that which is called French Barley, whereof is made that noble drinke for sicke Folkes, called *Ptisana*. The plant is altogether like vnto Spelt, sauing that the eares are rounder, the eiles or beards rougher and longer, and the seed or graine naked without huskes, like to wheat, the which in it's yellowish colour it somewhat resembles.

¶ *The place.*

‡ It is sowne in sundry places of Germany, for the same vses as Barley is.

¶ *The names.*

It is called *Hordeum Nudum*, for that the Corne is without huske, and resembleth Barley. In Greeke it is called ζείπυρον, because it participateth in similitude and nature with *Zea*, that is, Spelt, and *Puros*, (that is) wheat. ‡

¶ *The vertues.*

A This Barley boyled in water cooleth vnnaturall and hot burning choler. In vehement feuers you may adde thereto the seeds of white Poppie and Lettuse, not onely to coole, but also to prouoke sleepe.

B Against the shortnesse of the breath, and paines of the brest, may be added to all the foresaid, figs, raisins of the Sunne, liquorice, and Annise seed.

C Being boyled in the Whay of Milke, with the leaues of Sorrell, Marigolds, and Scabious, it quencheth thirst, and cooleth the heate of the inflamed Liuer, being drunke first in the morning, and last to bedward.

Chap.

Hordeum Spurium.
Wall Barley.

CHAP. 52.

Of Wall Barley.

¶ *The description.*

THis kinde of wilde Barley, called of the Latines *Hordeum Spurium*; is called of *Pliny*, *Holcus*; in English, Wall Barley, Way Barley, or after old English Writers, Way Bennet. It groweth vpon mud walls and stony places by the wayes sides; very well resembling Selfe-sowed Barley, yet the blades are rather like grasse than Barley. ‡ This groweth some foot and better in height, with grassie leaues, the eare is very like that of Rie, and the corne both in colour and shape absolutely resembles it; so that it cannot be fitlier named than by calling it wilde Rie, or Rie grasse. ‡

¶ *The vertues.*

A This Bastard wilde Barley stamped and applied vnto places wanting haire, doth cause it to grow and come forth, whereupon in old time it was called *Ristida*.

CHAP. 53. *Of Saint Peters Corne.*

1 *Briza monococcos.* S. *Peters* Corne.

2 *Festuca Italica.* Hauer Grasse.

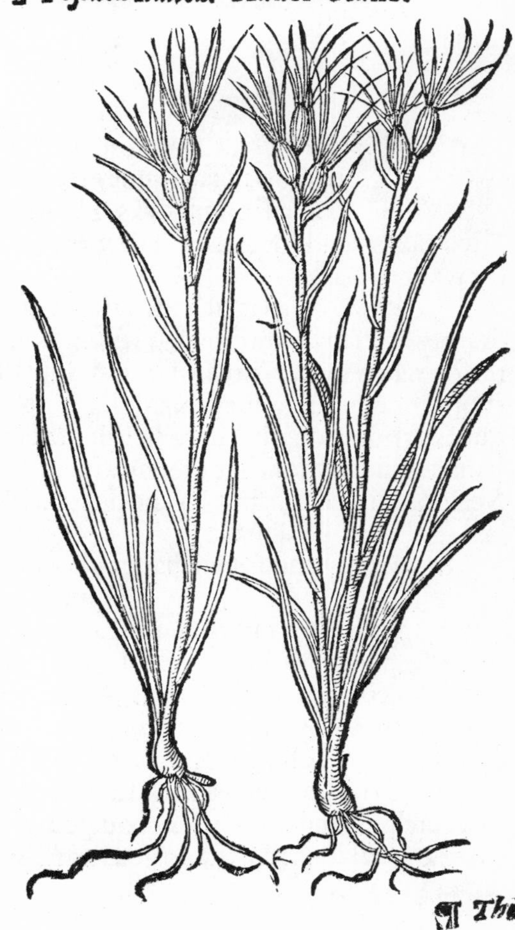

¶ The

¶ *The description.*

† 1 BRiza is a Corne whose leaues, stalkes, and eares are lesse than Spelt; the eare resembles our ordinary Barley, the corne growing in two rowes, with awnes at the top, and huskes vpon it not easily to be gotten off. In colour it much resembles barley; yet *Tragus* saith it is of a blackish red colour.

2 This *Ægilops* in leaues and stalkes resembles wheat or barley, and it growes some two handfuls high, hauing a little eare or two at the top of the stalke, wherein are inclosed two or three seeds a little smaller than Barley, hauing each of them his awne at his end. These seeds are wrapped in a crested filme or skinne, out of which the awnes put themselues forth.

Matthiolus saith, That he by his owne triall hath found this to be true, That as *Lolium*, which is our common Darnel, is certainly knowne to be a seed degenerate from wheat, being found for the most part among wheat, or where wheat hath been: so is *Festuca* a seed or grain degenerating from barley, and is found among Barley, or where barley hath beene.

‡ ¶ *The place.*

1 Briza is sowen in some parts of Germany and France; and my memorie deceiues me if I haue not often times found many eares thereof amongst ordinarie barley, when as I liued in the further side of Lincolneshire, and they there called it Brant Barley.

2 This *Ægilops* growes commonly amongst their Barley in Italy and other hot countries. ‡

¶ *The names.*

1 *Briza Monococcos*, after *Lobelius*, is called by *Tabernamontanus*, *Zea Monococcos*: in English, Saint *Peters* Corne, or Brant Barley.

2 *Festuca* of Narbone in France is called αἰγίλωψ: in Latine, *Ægilops Narbonensis*, according to the Greeke: in English, Hauer-grasse.

¶ *The nature.*

They are of qualitie somewhat sharpe, hauing facultie to digest.

¶ *The vertues.*

A The iuice of *Festuca* mixed with Barley meale dried, and at times of need moistned with Rose water, applied plaisterwise, healeth the disease called *Ægilops*, or Fistula in the corner of the eye: it mollifieth and disperseth hard lumps, and asswageth the swellings in the joynts.

Chap. 54. *Of Otes.*

¶ *The description.*

1 A*Vena Vesca*, Common Otes, is called *Vesca*, *à Vescendo*, because it is vsed in many countries to make sundry sorts of bread, as in Lancashire, where it is their chiefest bread corne for Iannocks, Hauer cakes, Tharffe cakes, and those which are called generally Oten cakes; and for the most part they call the graine Hauer, whereof they do likewise make drink for want of Barley.

2 *Auena Nuda* is like vnto the common Otes; differing in that, that these naked Otes immediately as they be threshed, without helpe of a Mill become Otemeale fit for our vse. In consideration whereof in Northfolke and Southfolke they are called vnhulled and naked Otes. Some of those good house-wiues that delight not to haue any thing but from hand to mouth, according to our English prouerbe, may (whiles their pot doth seeth) go to the barne, and rub forth with their hands sufficient for that present time, not willing to prouide for to morrow, according as the Scripture speaketh, but let the next day bring with it.

¶ *The nature.*

Otes are dry and somewhat cold of temperature, as *Galen* saith.

¶ *The vertues.*

A Common Otes put into a linnen bag, with a little bay salt quilted handsomely for the same purpose, and made hot in a frying pan, and applied very hot, easeth the paine in the side called the stitch, or collicke in the belly.

B If Otes be boyled in water, and the hands or feet of such as haue the *Serpigo* or *Impetigo*, that is, certaine chaps, chinks, or rifts in the palmes of the hands or feet (a disease of great affinitie with the pocks) be holden ouer the fume or smoke thereof in some bowle or other vessell wherein the Otes are put, and the Patient couered with blankets to sweat, being first annointed with that ointment or vnction vsually applied *contra Morbum Gallicum*: it doth perfectly cure the same in sixe times so annointing and sweating.

Otemeale is good for to make a faire and wel coloured maid to looke like a cake of tallow, especially if she take next her stomacke a good draught of strong vineger after it. C

Otemeale vsed as a Cataplasme dries and moderately discusses, and that without biting; for it hath somewhat a coole temper, with some astriction, so that it is good against scourings. D

1 *Auena Vesca.*
Common Otes.

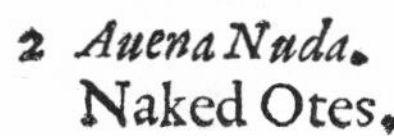

2 *Auena Nuda.*
Naked Otes.

Chap. 55. *Of Wilde Otes.*

The description.

1 Bromos *sterilis*, called likewise *Auena fatua*, which the Italians do call by a very apt name *Vena vana*, and *Auena Cassa*, (in English, Barren Otes, or wilde Otes) hath like leaues and stalkes as our Common Otes; but the heads are rougher, sharpe, many little sharpe huskes making each eare.

† 2 There is also another kinde of *Bromos* or wilde Otes, which *Dodoneus* calleth *Festuca altera*, not differing from the former wilde Otes in stalkes and leaues, but the heads are thicker, and more compact, each particular eare (as I may terme it) consisting of two rowes of seed handsomly compact and ioyned together; being broader next the straw, and narrower as it comes to an end.

‡ ¶ *The time and place.*

‡ The first in Iuly and August may be found almost in euery hedge; the later is to be found in great plenty in most Rie.

¶ *The names.*

1 This is called in Greeke βρῶμος πόα: in Latine, *Bromos stirilis* by *Lobell*: *Ægylops prima* by *Matthiolus*: in English, Wilde-Otes, or Hedge-Otes.

2 *Lobell* calls this *Bromos sterilis altera*: *Dodonæus* termes it *Festuca altera*: in Brabant they call it **Drauich**: in English, Drauke.

1 *Bromos*

1 *Bromos sterilis.*
Wilde Otes.

2 *Bromos altera.*
Drauke, or small wilde Otes.

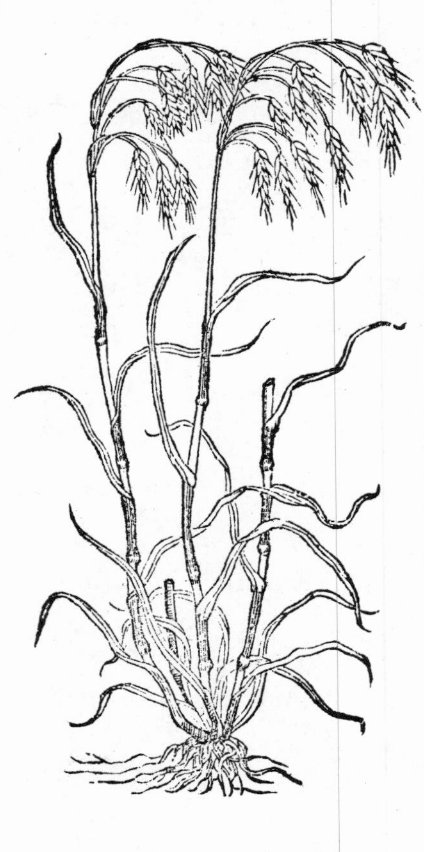

¶ *The nature and vertues.*

A 1 It hath a drying facultie (as *Dioscorides* saith.) Boile it in water together with the roots vntill two parts of three be consumed; then straine it out, and adde to the decoction a quantitie of honey equall thereto: so boile it vntill it acquire the thicknesse of thin honey. This medicine is good against the *Ozæna* and filthy vlcers of the nose, dipping a linnen cloth therein, and putting it vp into the nosthrils; some adde thereto Aloes finely poudred, and so vse it.

B Also boiled in Wine with dried Rose leaues, it is good against a stinking breath. ‡

Chap. 56. *Of Bearded Wilde Otes.*

¶ *The description.*

Ægylops Bromoides Belgarum is a Plant indifferently partaking of the nature of *Ægilops* and *Bromos*. It is in shew like to the naked Otes. The seed is sharpe, hairy, and somewhat long, and of a reddish colour, inclosed in yellowish chaffie huskes like as Otes, and may be Englished, Crested or bearded Otes. I haue found it often among Barley and Rie in sundry grounds. This is likewise vnprofitable and hurtfull to Corne; whereof is no mention made by the Antients worthy the noting.

1 *Ægylops*

† Ægilops Bromoides.
Bearded Wilde Otes.

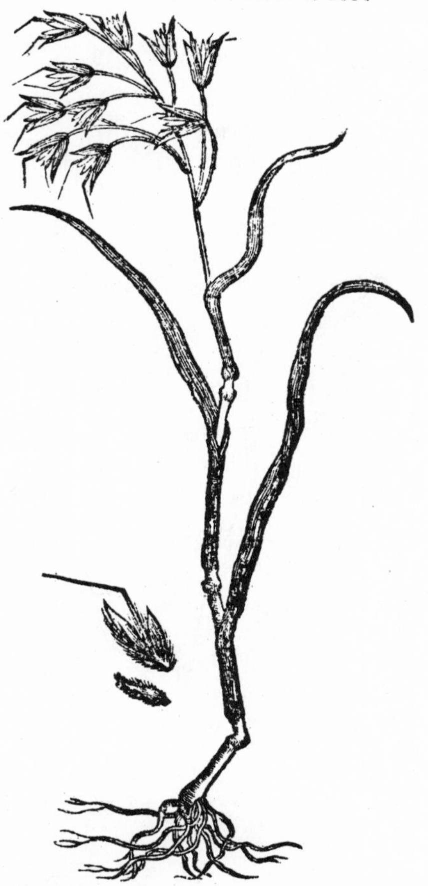

CHAP. 57. Of Burnt Corne.

¶ *The description.*

1 Hordeum vstum, or *Vstilago Hordei*, is that burnt or blasted Barley which is altogether vnprofitable and good for nothing, an enemy vnto corne; for that in stead of an eare with corne, there is nothing else but blacke dust, which spoileth bread, or whatsoeuer is made thereof.

2 Burnt Otes, or *Vstilago Auenæ*, or *Auenacea*, is likewise an vnprofitable Plant, degenerating from Otes, as the other from Barley, Rie, and Wheat. It were in vaine to make a long haruest of such euill corne, considering it is not possessed with one good qualitie. And therefore thus much shall suffice for the description.

3 Burnt Rie hath no one good property in phisicke, appropriate either to man, birds, or beast, and is a hurtfull maladie to all corne where it groweth, hauing an eare in shape like to corne, but in stead of graine it doth yeeld a blacke pouder or dust, which causeth bread to looke blacke, and to haue an euill taste: and that corne where it is, is called smootie corne, and the thing it selfe Burnt Corne, or blasted corne.

1 *Hordeum vstum, siue vstilago hordei.* Burnt Barley. 2 *Vstilago Auenacea.* Burnt Otes.

3 *Vstilago Secalina.* Burnt Rie.

Chap. 58. *Of Darnell.*

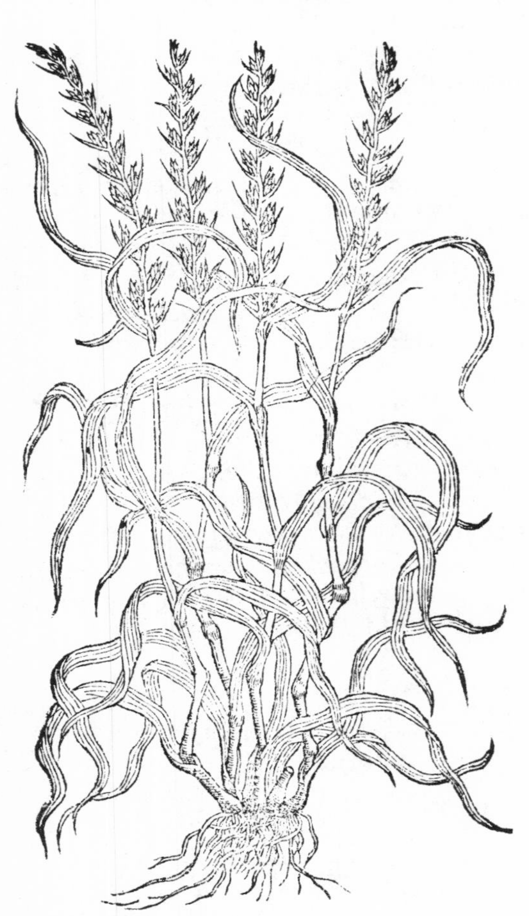

1 *Lolium album.*
White Darnell.

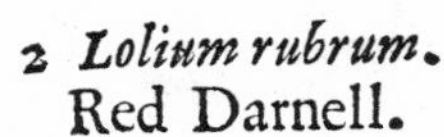

2 *Lolium rubrum.*
Red Darnell.

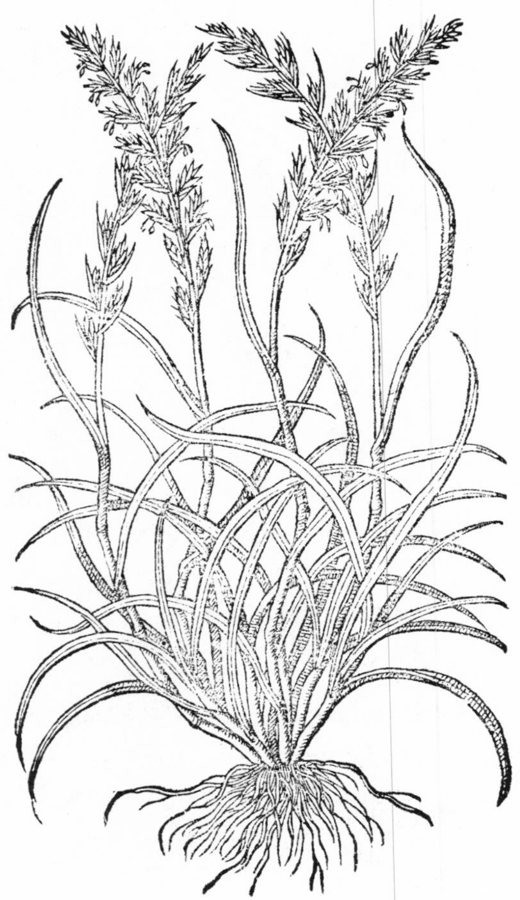

¶ *The description.*

1 Among the hurtfull weeds Darnell is the first. It bringeth forth leaues or stalkes like those of wheat or barley, yet rougher, with a long eare made vp of many little ones, euery particular one whereof containeth two or three graines lesser than those of wheat, scarcely any chaffie huske to couer them with; by reason whereof they are easily shaken out and scattered abroad.

2 Red Darnell is likewise an vnprofitable corne or grasse, hauing leaues like barly. The joints of the straw or stalke are sometimes of a reddish colour, bearing at the top a small and tender eare, flat, and much in forme resembling the former.

¶ *The place.*

They grow in fields among wheat and barley, of the corrupt and bad seed, as *Galen* saith, especially in a moist and dankish soile.

¶ *The time.*

They spring and flourish with the corne, and in August the seed is ripe.

¶ *The names.*

1 Darnell is called in Greeke, αἶρα: in the Arabian Tongue, *Zizania* and *Sceylen:* In French, *Yuray:* in Italian, *Loglio:* in low Dutch, **Dolick:** in English, Darnell: of some, Iuray, and Raye: and of some of the Latines, *Triticum temulentum.*

2 Red Darnell is called in Greeke φοῖνιξ, or *Phœnix,* because of the crimson colour: in Latine, *Lolium Rubrum,* and *Lolium Murinum:* of some, *Hordeum Murinum,* and *Triticum Murinum:* in Dutch, **Muyse coren:** in English, Red Darnell, or great Darnell Grasse.

¶ *The temperature.*

Darnell is hot in the third degree, and dry in the second. Red Darnell drieth without sharpenesse, as *Galen* saith.

¶ *The*

¶ *The vertues.*

The seed of Darnell, Pigeons dung, oile Oliue, and pouder of Lineseed, boiled to the forme of a plaister, consume wennes, hard lumpes, and such like excrescenses in any part of the body. A

The new bread wherein Darnel is, eaten hot, causeth drunkennesse: in like manner doth beere or ale wherein the seed is fallen, or put into the Malt. B

Darnell taken with red wine stayeth the flux of the belly, and the ouermuch flowing of womens termes. C

Dioscorides saith, That Darnell meale doth stay and keepe backe eating sores, Gangrenes, and putrified vlcers; and being boyled with Radish roots, salt, brimstone, and vineger, it cureth spreading scabs, and dangerous tetters, called in Greeke, λειχῆνες, and leprous or naughty scurfe. D

The seed of Darnell giuen in white or Rhenish wine, prouoketh the flowers or menses. E

A fume made thereof with parched barly meale, myrrh, saffron, and frankinsence, made in form of a pultesse, and applied vpon the belly, helps conception, and causeth easie deliuerance of child-bearing. F

Red Darnell (as *Dioscorides* writeth) being drunke in sowre or harsh red Wine, stoppeth the laske, and the ouermuch flowing of the flowers or menses, and is a remedie for those that pisse in bed. G

¶ *The danger.*

Darnell hurteth the eyes, and maketh them dim, if it happen in corne either for bread or drinke; which thing *Ouid* in his first booke *Fastorum* hath mentioned, in this verse:

Et careant loliis oculos vitiantibus agri.

And hereupon it seemeth that the old prouerbe came, That such as are dimme sighted should be said, *Lolio victitare.*

CHAP. 59. *Of Rice.*

Oryza.
Rice.

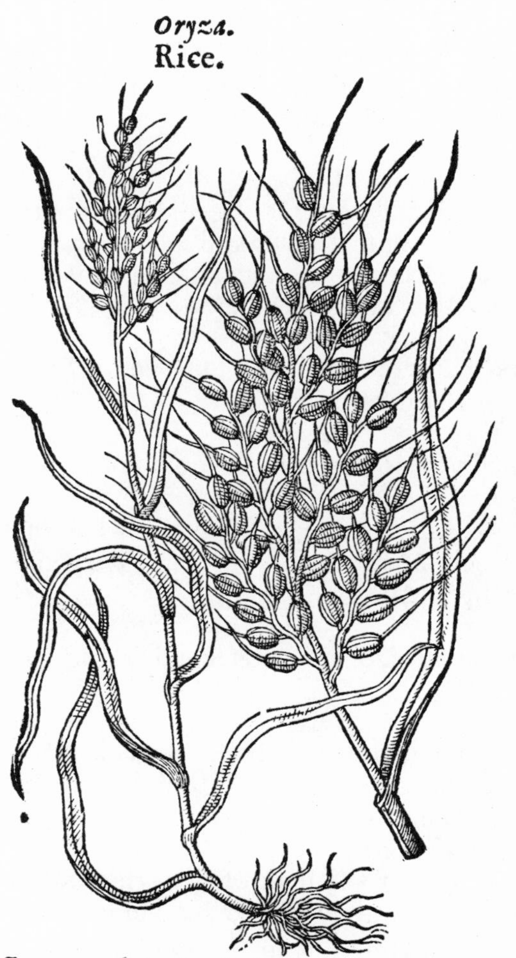

¶ *The description.*

RIce is like vnto Darnell in shew, as *Theophrastus* saith: it bringeth not forth an eare, like corne, but a certaine mane or plume, as Mill, or Millet, or rather like Panick. The leaues, as *Pliny* writeth, are fat and full of substance, like to the blades of leeks, but broader: but (if neither the soile nor climate did alter the same) the plants of Rice that did grow in my garden had leaues soft and grassie like barly. The floure did not shew it selfe with me, by reason of the iniurie of our vnseasonable yere 1596. *Theophrastus* concludeth, that it hath a floure of a purple colour. But, saith my Author, Rice hath leaues like vnto Dogs grasse or Barley, a small straw or stem full of ioynts like corne: at the top whereof groweth a bush or tuft farre vnlike to barley or Darnell, garnished with round knobs like small gooseberries, wherein the seed or graine is contained: euery such round knob hath one small rough aile, taile, or beard like vnto barley hanging thereat. *Aristobulus*, as *Strabo* reporteth, sheweth, That Rice growes in water in Bactria, and neere Babylon, and is two yards high, and hath many eares, and bringeth forth plenty of seed. It is reaped at the setting of the seuen starres, and purged as Spelt and Otemeale, or hulled as French Barley.

¶ *The place.*

It groweth in the territories of the Bactrians, in Babylon, in Susium, and in the lower part of Syria. It groweth in those dayes not onely in those countries before named, but also in the fortunate Islands, and in Spaine, from whence it is brought vnto vs, purged and prepared as we see, after the manner of French Barley. It prospereth best in fenny and waterish places.

¶ *The time.*

It is sowen in the Spring in India, as *Eratosthenes* witnesseth, when it is moistned with Sommer showers.

¶ *The names.*

The Grecians call it ὄρυζα, or as *Theophrastus* saith, ὄρυζον: the Latines keepe the Greeke word *Oryza*: in French it is called *Riz*: in the Germane tongue, **Reiß**, and **Rys**: in English, Rice.

¶ *The temperature and vertues.*

Galen saith, That all men vse to stay the belly with this graine, being boiled after the same manner that *Chondrus* is. In England we vse to make with milke and Rice a certaine food or pottage, which doth both meanly binde the belly, and also nourish. Many other good kindes of food is made with this graine, as those that are skilfull in cookerie can tell.

Chap. 60. Of Millet.

Milium.
Mill, or Millet.

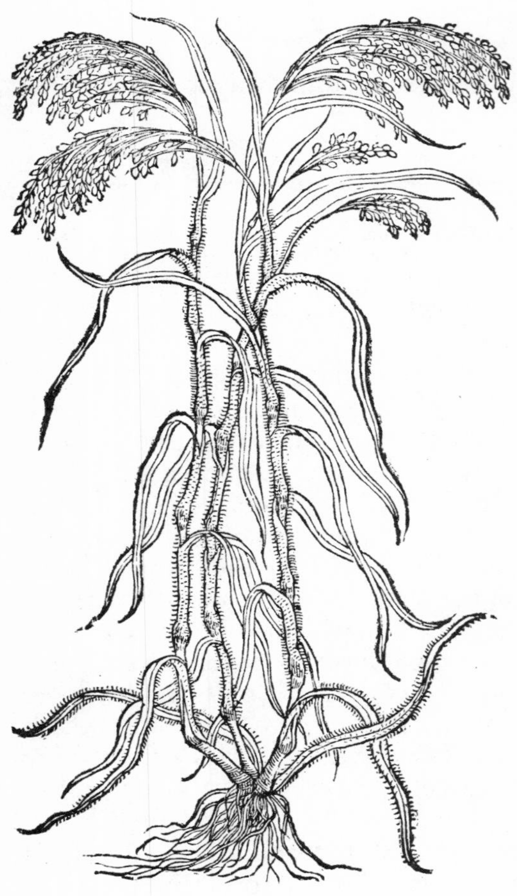

¶ *The description.*

Milium riseth vp with many hairy stalkes knotted or jointed like wheat. The leaues are long, and like the leaues of the Common Reed. It bringeth forth on the top of the stalke a spoky bush or mane, called in Greeke φόβη, like the plume or feather of the Pole reed, hanging downewards, of colour for the most part yellow or white; in which groweth the seed, small, hard, and glistering, couered with a few thinne huskes, out of which it easily falleth. The roots be many, and grow deep in the ground.

2 *Milium nigrum* is like vnto the former, sauing that the eare or plume of this plant is more loose and large, and the seed somewhat bigger, of a shining blacke colour.

¶ *The place.*

It loueth light and loose mould, and prospereth best in a moist and rainy time. And after *Columella*, it groweth in greatest aboundance in Campania. I haue of it yearely in my garden.

¶ *The time.*

It is to be sowen in Aprill and May, and not before, for it ioyeth in warme weather.

¶ *The names.*

It is named of the Grecians, κέγχρος: of some, κεγχρίς: and of *Hippocrates*, *Paspale*, as *Hermolaus* saith: In Spanish, *Mijo*: in Italian, *Miglio*: in High-Dutch, **Hirß**: in French, *Millet*: in Low-Dutch, **Hirs**: in English, Mill, or Millet.

¶ *The temper.*

It is cold in the first degree, as *Galen* writeth, and dry in the third, or in the later end of the second, and is of a thinne substance.

¶ *The vertues.*

A The meale of Mill mixed with tarre is laid to the bitings of serpents, and all venomous beasts.

B There is a drinke made hereof bearing the name of *Sirupus Ambrosij*, or *Ambrose* his syrup, which procureth sweat, and quencheth thirst, vsed in the city of Milan in Tertian agues. The receit whereof *Henricus Rantzonius* in his booke of the gouernment of health setteth downe in this manner: Take (saith he) of vnhusked Mill a sufficient quantitie, boile it till it be broken; then take fiue ounces of the hot decoction, and adde thereto two ounces of the best white wine, and so giue it hot vnto the patient, being well couered with clothes, and then he will sweat throughly. This is likewise commended by *Iohannes Heurneus*, in his booke of Practise.

C Millet parched, and so put hot into a linnen bag, and applied, helpes the griping paines of the belly, or any other paine occasioned by cold.

Chap.

CHAP. 61. Of Turkie Corne.

1 *Frumentum Aſiaticum.*
Corne of Aſia.

2 *Frumentum Turcicum.*
Turkie Corne.

¶ *The kindes.*

OF Turkie cornes there be diuers ſorts, notwithſtanding of one ſtocke or kindred, conſiſting of ſundry coloured graines, wherein the difference is eaſie to be diſcerned, and for the better explanation of the ſame, I haue ſet forth to your view certaine eares of different colours, in their full and perfect ripeneſſe, and ſuch as they ſhew themſelues to be when their ſkinne or filme doth open it ſelfe in the time of gathering.

The forme of the eares of Turky Wheat.

3 *Frumenti Indici ſpica.*
Turkie wheat in the huske, as alſo naked or bare.

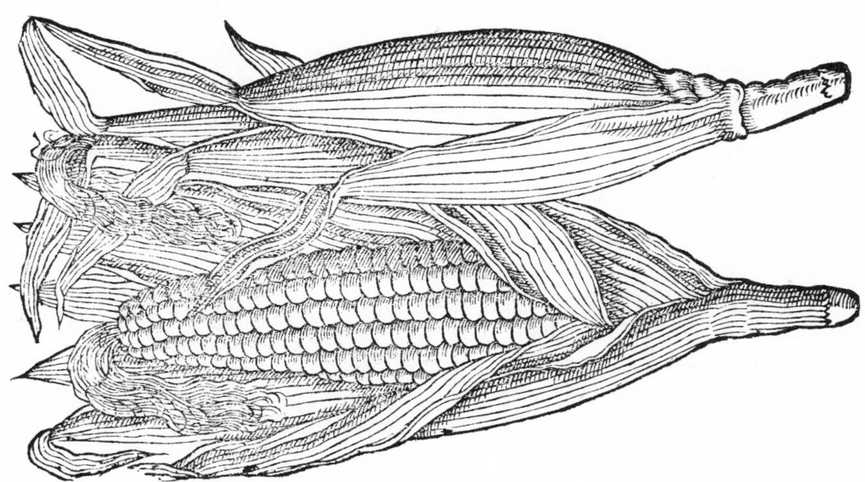

¶ *The description.*

1 Corne of Asia beareth a long great stem or stalke, couered with great leaues like the great Cane reed, but much broader, and of a darke brownish colour towards the bottome: at the top of the stalkes grow idle or barren tufts like the common Reed, somtimes of one colour, and sometimes of another. Those eares which are fruitfull do grow vpon the sides of the stalkes, among the leaues, which are thicke and great, so couered with skins or filmes, that a man cannot see them vntill ripenesse haue discouered them. The graine is of sundrie colours, sometimes red, and sometimes white, and yellow, as my selfe haue seene in myne owne garden, where it hath come to ripenesse.

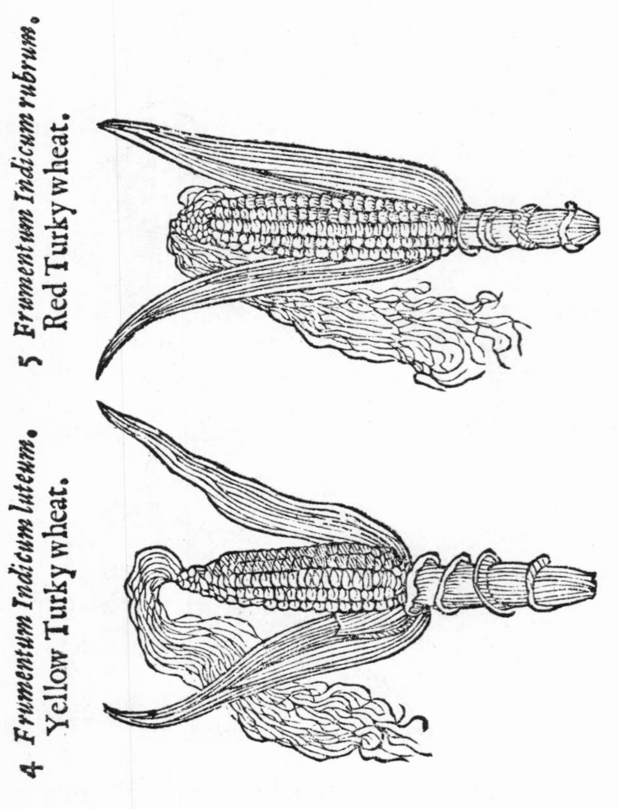

5 *Frumentum Indicum rubrum.* Red Turky wheat.

4 *Frumentum Indicum luteum.* Yellow Turky wheat.

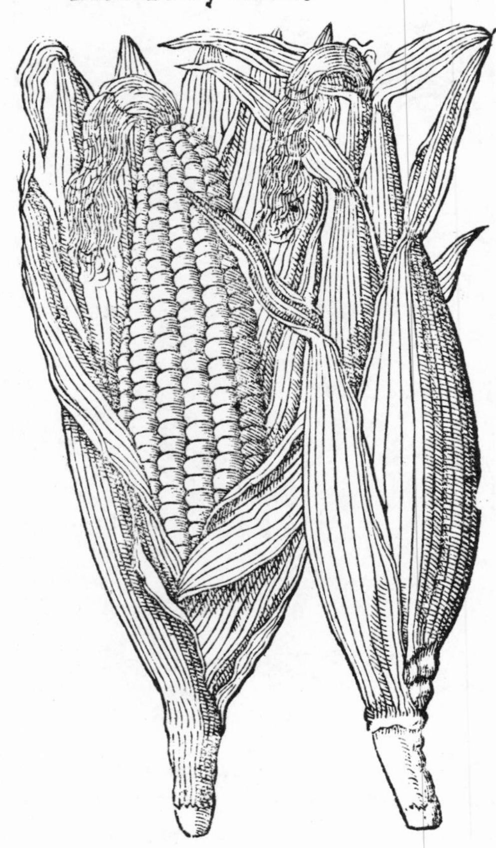

6 *Frumentum Indicum cæruleum.* Blew Turky wheat.

2 The stalke of Turky Wheat is like that of the Reed, full of spongie pith, set with many ioynts, fiue or six foot high, bigge beneath, and now and then of a purple colour, and by little and little small aboue: the leaues are broad, long, set with vaines like those of the Reed. The eares on the top of the stalke be a spanne long, like vnto the feather top of the common Reed, diuided into many plumes hanging downward, empty and barren without seed, yet blooming as Rie doth. The floure is either white, yellow, or purple, that is to say, euen as the fruit will be. The Fruit is contained in very bigge eares, which grow out of the ioynts of the stalke, three or foure from one stalke, orderly placed one aboue another, couered with cotes or filmes like huskes and leaues, as if it were a certaine sheath; out of which do stand long and slender beards, soft and tender, like those laces that grow vpon Sauorie, but greater and longer, euery one fastned vpon his owne seed. The seeds are great, of the bignesse of common peason, cornered in that part whereby they are fastned to the eare, and in the outward part round: being of colour sometimes white, now and then yellow, purple, or red; of taste sweet and pleasant, very closely ioyned together in eight or tenne orders or rankes. This graine hath many roots, strong, and full of strings.

¶ *The place.*

These kindes of graine were first brought into Spaine, and then into other prouinces of Europe: not (as some suppose) out of Asia *minor*, which is the Turks Dominions, but out of America and the Islands adioyning, as out of Florida and Virginia, or Norembega, where they vse to sow or set it, and to make bread of it, where it groweth much higher than in other countries. It is planted in the gardens of these Northerne regions, where it commeth to ripenesse when the sommer falleth out to be faire and hot, as my selfe haue seene by proofe in myne owne garden.

¶ The

¶ *The time.*

It is ſowen in theſe countries in March and Aprill, and the fruit is ripe in September.

¶ *The names.*

† Turky wheat is called of ſome *Frumentum Turcicum*, and *Milium Indicum*, as alſo *MaiZum*, and *Maiz*, or *Mays*. It in all probabilitie was vnknowne to the antient both Greeke and Latine Authors. In Engliſh it is called Turky corne, and Turky wheat. The Inhabitants of America and the Iſlands adioyning, as alſo of the Eaſt and Weſt Indies, do call it *Mais*: the Virginians, *Pagatowr*.

¶ *The temperature and vertues.*

Turky wheat doth nouriſh far leſſe than either wheat, rie, barley, or otes. The bread which is made thereof is meanly white, without bran: it is hard and dry as Bisket is, and hath in it no clammineſſe at all; for which cauſe it is of hard digeſtion, and yeeldeth to the body little or no nouriſhment; it ſlowly deſcendeth, and bindeth the belly, as that doth which is made of Mill or Panick. We haue as yet no certaine proofe or experience concerning the vertues of this kinde of Corne; although the barbarous Indians, which know no better, are conſtrained to make a vertue of neceſſitie, and thinke it a good food: whereas we may eaſily iudge, that it nouriſheth but little, and is of hard and euill digeſtion, a more conuenient food for ſwine than for men.

CHAP. 62. *Of Turkie Millet.*

Sorghum.
Turky Millet.

¶ *The deſcription.*

TVrky Millet is a ſtranger in England. It hath many high ſtalkes, thicke, and jointed commonly with ſome nine ioynts, beſet with many long and broad leaues like Turky Wheat: at the top whereof groweth a great and large tuft or eare like the great Reed. The ſeed is round and ſharpe pointed, of the bigneſſe of a Lentill, ſometimes red, and now and then of a fuller blacke colour. It is faſtned with a multitude of ſtrong ſlender roots like vnto threds: the whole plant hath the forme of a Reed: the ſtalkes and eares when the ſeed is ripe are red.

¶ *The place.*

It ioyeth in a fat and moiſt ground: it groweth in Italy, Spaine, and other hot regions.

¶ *The time.*

This is one of the Sommer graines, and is ripe in Autumne.

¶ *The names.*

The Millanois and other people of Lombardy call it *Melegua*, and *Melega*: in Latine, *Melica*: in Hetruria, *Saggina*: in other places of Italy, *Sorgho*: in Portugal, *Milium Saburrum*: in Engliſh, Turky Mill, or Turky Hirſſe.

‡ This ſeemes to be the *Milium* which was brought into Italy out of India, in the reigne of the Emperour *Nero*: the which is deſcribed by *Pliny*, *lib.* 18. *cap.* 7. ‡

¶ *The temperature and vertues.*

The ſeed of Turky Mill is like vnto Panicke in taſte and temperature. The country People ſometimes make bread hereof, but it is brittle, and of little nouriſhment, and for the moſt part it ſerueth to fatten hens and pigeons with.

CHAP.

Chap. 63. Of Panick.

1 *Panicum Indicum*
Indian Panick.

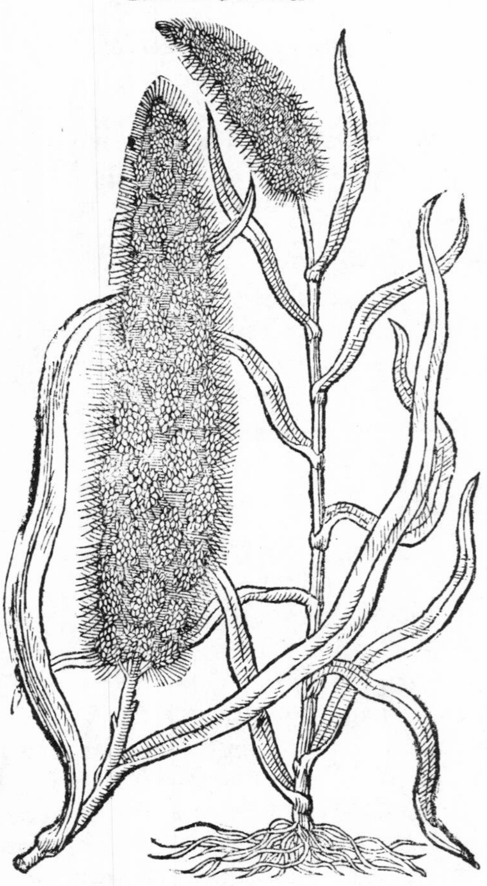

2 *Panicum Cæruleum.* Blew Panicke.

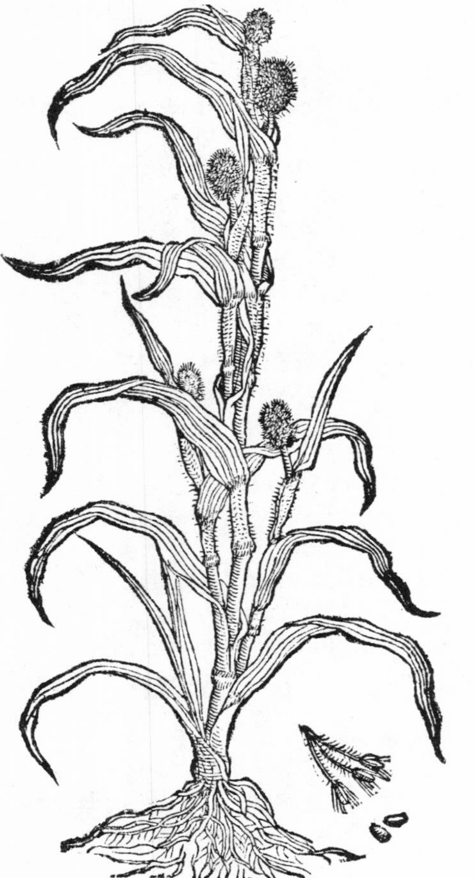

¶ *The kindes.*

There be sundry sorts of Panicke, although of the Antients there haue beene set downe but two, that is to say, the wilde or field Panicke, and the garden or manured Panicke.

¶ *The descrip tin.*

1 The Panick of India groweth vp like Millet, whose straw is knotty, or full of ioynts; the ears be round, and hanging downward, in which is contained a white or yellowish seed, like Canarie seed, or *Alpisti*.

2 Blew Panick hath a reddish stalke like to Sugar cane, as tall as a man, thicker than a finger, full of a fungous pith, of a pale colour: the stalkes be vpright and knotty; these that grow neere the root are of a purple colour: on the top of the stalk commeth forth a spike or eare like the water Cats Taile, but of a blew or purple colour. The Seed is like to naked Otes: The Roots are very small, in respect of the other parts of the plant.

‡ 3 *Panicum Americanum spica longissimo.*
West-Indian Panicke with a very long eare.

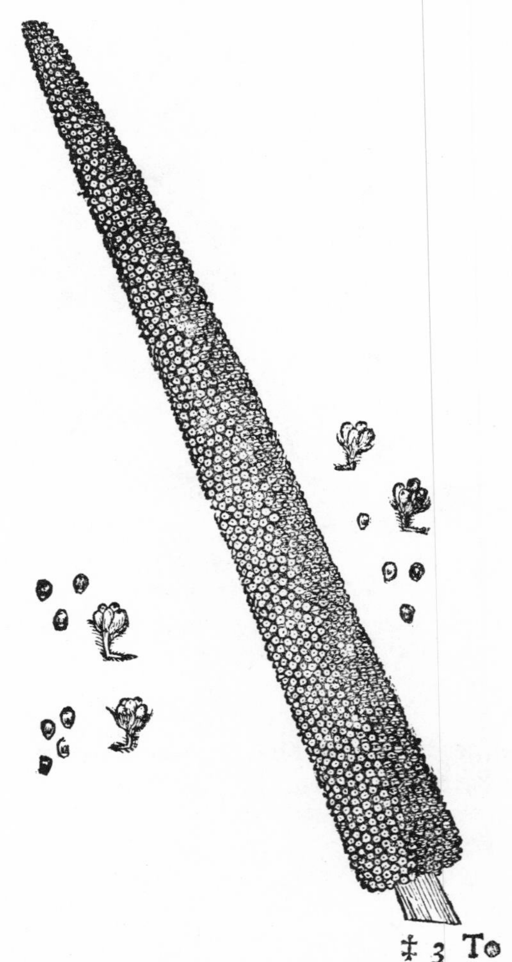

‡ 3 To

‡ 3 To theſe may be added another Weſt-Indian Panicke, ſent to *Cluſius* from M. *Iames Garet* of London. The eare hereof was thicke, cloſe, compact and made Taper-faſhion, ſmaller at the one end than at the other; the length thereof was more than a foot & halfe. The ſhape of the ſeed is much like the laſt deſcribed, but that many of them together are contained in one hairie huske, which is faſtned to a very ſhort ſtalke, as you may ſee repreſented apart by the ſide of the figure ‡

4 *Panicum vulgare.*
Common or Germane Panicke.

5 *Panicum ſylueſtre.*
Wild Panicke.

4 Germane Panicke hath many hairy roots growing thicke together like vnto wheat, as is all the reſt of the plant, as well leaues or blades, as ſtraw or ſtalke. The eare groweth at the top ſingle, not vnlike to Indian Panicke, but much leſſer. The graines are contained in chaffie ſcales, red declining to tawny.

5 The wilde Panicke groweth vp with long reeden ſtalkes, full of ioynts, ſet with long leaues like thoſe of *Sorghum*, or Indian Panicke: the tuft or feather-like top is like vnto the common reed, or the eare of the graſſe called *Iſchæmon*, or *Manna* graſſe. The root is ſmall and threddy.

¶ *The place and time.*

The kindes of Panick are ſowen in the Spring, and are ripe in the beginning of Auguſt. They proſper beſt in hot and dry Regions, and wither for the moſt part with much watering, as doth Mil and Turky wheat: they quickly come to ripeneſſe, and may be kept good a long time.

¶ *The names.*

Panick is called in Greeke ἔλυμος, and μελίνη. *Diocles* the Phyſition nameth it *Mel Frugum*: the Spaniards, *Panizo*: the Latines, *Panicum*, of *Pannicula*: in Engliſh, Indian Panicke, or Otemeale.

¶ *The temperature.*

Panicks nouriſh little, and are driers, as *Galen* ſaith.

¶ *The vertues.*

Panicke ſtoppeth the laske, as Millet doth, being boyled (as *Pliny* reporteth) in Goats milke, and drunke twice in a day. Outwardly in Pulteſſes or otherwiſe, it dries and cooles. A

Bread made of Panick nouriſheth little, and is cold and dry, very brittle, hauing in it neither clammineſſe nor fatneſſe; and therefore it drieth a moiſt belly. B

Chap.

Chap. 64. *Of Canary seed, or Pety Panicke.*

1 *Phalaris.*
Canarie seed.

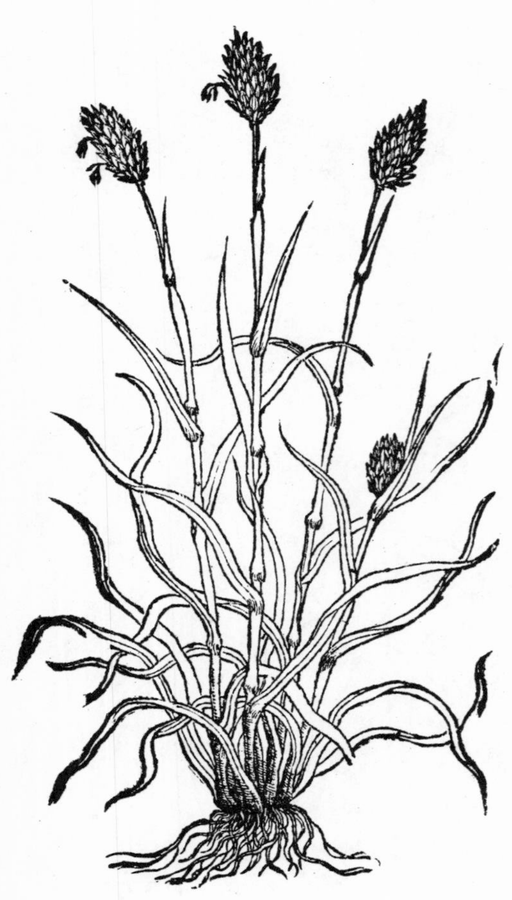

2 *Phalaris pratensis.*
Quaking grasse.

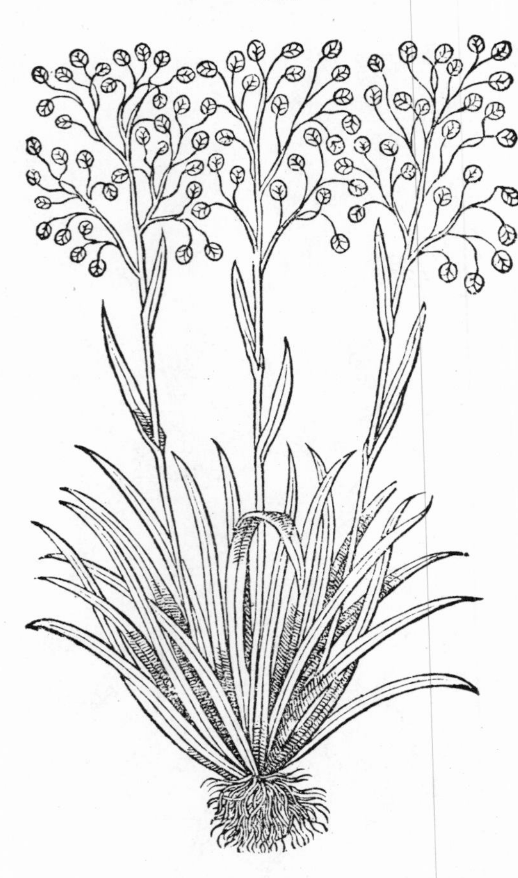

¶ *The description.*

1 CAnarie seed, or Canarie grasse after some, hath many small hairy roots, from which arise small strawie stalkes ioynted like corne, whereupon do grow leaues like those of Barley, which the whole plant doth very well resemble. The small chaffie eare groweth at the top of the stalke, wherein is contained small seeds like those of Panicke, of a yellowish colour, and shining.

2 Shakers, or Quaking Grasse groweth to the height of halfe a foot, and sometimes higher, when it groweth in fertile medowes. The stalke is very small and benty, set with many grassie leaues like the common medow grasse, bearing at the top a bush or tuft of flat scaly pouches, like those of Shepheards purse, but thicker, of a browne colour, set vpon the most small and weake hairy foot stalkes that may be found, whereupon those small pouches do hang: by meanes of which small hairy strings, the knaps which are the floures do continually tremble and shake, in such sort that it is not possible with the most stedfast hand to hold it from shaking.

‡ 3 There is also another Grassie plant which may fitly be referred to these: the leaues and stalkes resemble the last described, but the heads are about the length and bredth of a small Hop, and handsomely compact of light scaly filmes much like thereto; whence some haue termed it *Gramen Lupuli glumis*. The colour of this pretty head when it commeth to ripenesse is white. ‡

¶ *The place.*

1 Canarie seed groweth naturally in Spaine, and also in the Fortunate or Canary Islands, and doth grow in England or any other of these cold Regions, if it be sowen therein.

2 Quaking

3 *Phalaris pratensis altera.*
Pearle Grasse.

2 Quaking *Phalaris* groweth in fertile pastures, and in dry medowes.

3 This growes naturally in some parts of Spaine, and it is sowen yearely in many of our London Gardens.

¶ *The time.*

1 3 These Canarie seeds are sowen in May, and are ripe in August.

¶ *The names.*

1 Canary seed, or Canarie corne is called of the Grecians, Φαλαρὶς: the Latines retaining the same name *Phalaris*: in the Islands of Canarie, *Alpisti*: in English, Cana rie seed, and Canary grasse.

2 *Phalaris pratensis* is called also *Gramen tremulum*: in Cheshire about Nantwich, Quakers and Shakers: in some places, Cow-quakes.

3 This by some is termed *Phalaris altera*: *Clusius* calleth it *Gramen Amourettes majus*: *Bauhine*, *Gramen tremulum maximum*: In English they call it Pearle-Grasse, and Garden-Quakers.

¶ *The nature and vertues.*

I finde not any thing set downe as touching the temperature of *Phalaris*, notwithstanding it is thought to be of the nature of Millet.

A The iuyce and seed, as *Galen* saith, are thought to be profitably drunke against the paines of the bladder. Apothecaries for want of Millet doe vse the same with good successe in fomentations; for in dry fomentations it serueth in stead thereof, and is his *succedaneum*, or *quid pro quo*. We vse it in England also to feed the Canarie Birds.

Alopecuros.
Fox-taile.

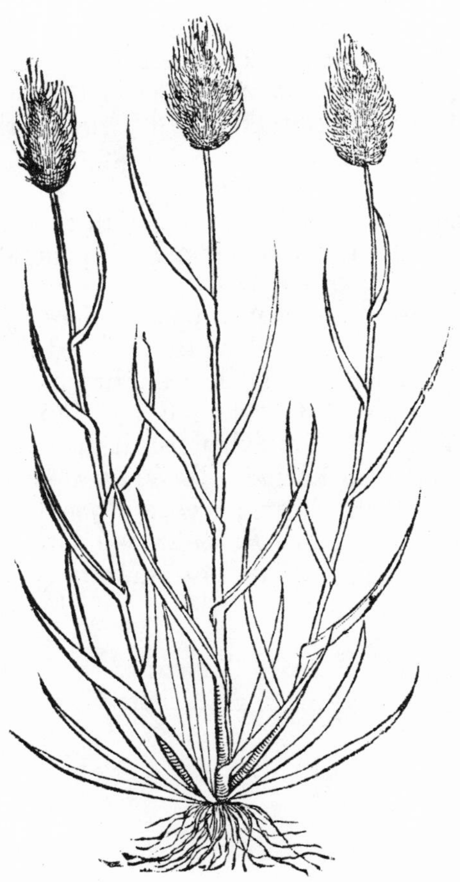

CHAP. 65.

Of Fox-Taile.

¶ *The description.*

1 FOx-taile hath many grassie leaues or blades, rough and hairy, like vnto those of Barley, but lesse and shorter. The stalke is likewise soft and hairy; whereupon doth grow a small spike or eare, soft, and very downy, bristled with very small haires in shape, like vnto a Fox-taile, whereof it tooke his name, which dieth at the approch of Winter, and recouereth it selfe the next yeare by falling of his seed.

‡ There is one or two varieties of this Plant in the largenesse and smalnesse of the eare.

2 Besides these forementioned strangers, there is also another which growes naturally in many watry Salt places of this kingdome, as in Kent by Dartford, in Essex, &c. The stalkes of this plant are grassy, and some two foot high, with leaues like Wheat or Dogs Grasse. The eare is very large, being commonly foure or fiue inches long, downy, soft like silke, and of a brownish colour. ‡

¶ The

¶ *The place.*

1 This kinde of Fox-taile Grasse groweth in England, onely in gardens.

¶ *The time.*

1 This springeth vp in May, of the seed that was scattered the yere before, and beareth his taile with his seed in Iune.

2 This beares his head in Iuly.

¶ *The names.*

1 There hath not beene more said of the antient or later writers, as touching the name, than is set downe, by which they called it in Greeke *Alopecuros*; that is in Latine, *Cauda vulpis*: in English, Fox-taile.

2 This by *Lobell* is called *Alopecuros altera maxima Anglica paludosa*; that is, The large English Marsh Fox-taile.

¶ *The temperature and vertues.*

I finde not any thing extant worthy the memorie, either of his nature or vertues.

Chap. 66. *Of Iobs Teares.*

Lachrima Iob.
Iobs Teares.

¶ *The description.*

Iobs Teares hath many knotty stalks, proceeding from a tuft of threddy roots, two foot high, set with great broad leaues like vnto those of reed, amongst which leaues come forth many small branches like straw of corne: on the end whereof doth grow a gray shining seed or graine hard to breake, and like in shape to the seeds of Gromell, but greater, and of the same colour, whereof I hold it a kinde: euery of which grains are bored through the middest like a bead, and out of the hole commeth a small idle or barren chaffie eare like vnto that of Darnell.

¶ *The place.*

It is brought from Italy and the countries adjoyning, into these countries, where it doth grow very well, but seldome commeth to ripenesse; yet my selfe had ripe seed thereof in my garden, the Sommer being very hot.

¶ *The time.*

It is sowen early in the Spring, or else the winter will ouertake it before it come to ripenesse.

¶ *The names.*

Diuers haue thought it to be *Lithospermi species*, or a kinde of Gromell, which the seed doth very notably resemble, and doth not much differ from *Dioscorides* his Gromell. Some thinke it *Plinies Lithospermum*; and therefore it may verie aptly be called in Latine, *Arundo Lithospermos*, that is in English, Gromell reed, as *Gesner* saith. It is generally called *Lachrima Iob*, and *Lachrima Iobi*: of some it is called *Diospiros*: in English it is called *Iobs* Teares, or *Iobs* Drops, for that euery graine resembleth the drop or teare that falleth from the eye.

¶ *The nature and vertues.*

There is no mention made of this herbe for the vse of physicke: onely in France and those places (where it is plentifully growing) they do make beads, bracelets, and chaines thereof, as we do with pomander and such like.

Chap.

CHAP. 67. *Of Buck-wheat.*

Tragopyron.
Buckwheat, or Bucke.

¶ *The deſcription.*

BVck-wheat may very well be placed among the kinds of graine or corne, for that oftentimes in time of neceſſitie bread is made thereof, mixed among other graine. It hath round fat ſtalkes ſomewhat creſted, ſmooth and reddiſh, which is diuided in many armes or branches, whereupon do grow ſmooth and ſoft leaues in ſhape like thoſe of Iuie or one of the Bindeweeds, not much vnlike Baſil, wherof *Tabernamontanus* called it *Ocymum Cereale*: The floures be ſmall, white, and cluſtred together in one or moe tufts or vmbels, ſlightly daſht ouer here & there with a flouriſh of light Carnation colour. The ſeeds are of a darke blackiſh colour, triangle, or three ſquare like the ſeed of blacke Bindeweed, The root is ſmall and threddy.

¶ *The place.*

It proſpereth very wel in any ground, be it neuer ſo dry or barren, where it is commonly ſowen to ſerue as it were in ſtead of a dunging. It quickly commeth vp, and is very ſoone ripe: it is verie common in and about the Namptwich in Cheſhire, where they ſow it as well for food for their cattell, pullen, and ſuch like, as to the vſe aforeſaid. It groweth likewiſe in Lancaſhire, and in ſome parts of our South country, about London in Middleſex, as alſo in Kent and Eſſex.

¶ *The time.*

This baſe kinde of graine is ſowen in Aprill and the beginning of May, and is ripe in the beginning of Auguſt.

¶ *The names.*

Buck-wheat is called of the high Almaines, **Heydencorn**: of the baſe Almaines, **Buckenweidt**; that is to ſay, *Hirci triticum*, or Goats wheat: of ſome, *Fagi triticum*, Beech Wheat: In Greeke, ἐρύσιμον, by *Theophraſtus*; and by late Writers, τραγόπυρον: in Latine, *Fago triticum*, taken from the faſhion of the ſeed or fruit of the Beech tree. It is called alſo *Fegopyrum*, and *Tragopyron*: In Engliſh, French wheat, Bullimong, and Buck-wheat: In French, *Dragee aux cheuaux*.

¶ *The temper.*

Buck-wheat nouriſheth leſſe than Wheat, Rie, Barley, or Otes; yet more than either Mill or Panicke.

¶ *The vertues.*

Bread made of the meale of Buck-wheat is of eaſie digeſtion, and ſpeedily paſſeth through the belly, but yeeldeth little nouriſhment.

Chap. 68. *Of Cow Wheat.*

1 *Melampyrum album.*
White Cow-wheat.

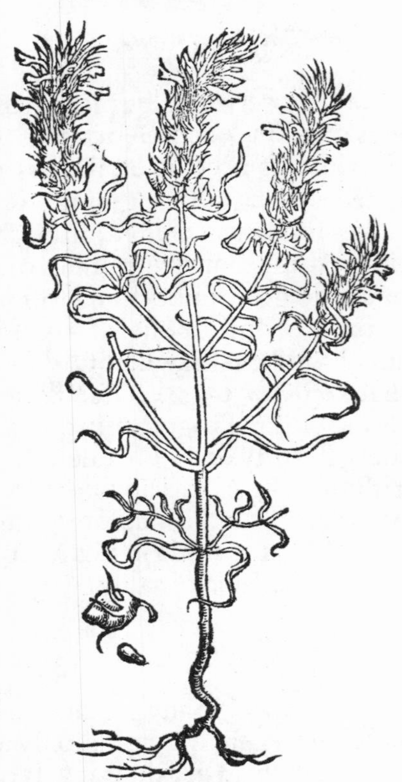

‡ 2 *Melampyrum purpureum.*
Purple Cow-wheat.

‡ 3 *Melampyrum cæruleum.*
Blew Cow-wheat.

‡ 4 *Melampyrum luteum.*
Yellow Cow-wheat.

¶ *The description.*

1 MElampyrum growes vpright, with a straight stalke, hauing other small stalkes comming from the same, of a foot long. The leaues are long and narrow, and of a darke colour. On the top of the branches grow bushy or spikie eares full of floures and small leaues mixed together, and much jagged, the whole eare resembling a Foxe-taile. This eare

eare beginneth to floure below, and so vpward by little and little vnto the top: the small leaues before the opening of the floures, and likewise the buds of the floures, are white of colour. Then come vp broad husks, wherein are enclosed two seeds somewhat like wheat, but smaller and browner. The root is of a woody substance.

‡ 2 3 These two are like the former in stalkes and leaues, but different in the colour of their floures, the which in the one are purple, and in the other blew. *Clusius* calls these, as also the *Cratæogonon* treated of in the next Chapter, by the names of *Parietariæ sylvestres*. ‡

4 Of this kinde there is another called *Melampyrum luteum*, which groweth neere vnto the ground, with leaues not much vnlike Harts horne, among which riseth vp a small straw with an eare at the top like *Alopecuros*, the common Fox-taile, but of a yellow colour.

¶ *The place.*

1 The first groweth among corne, and in pasture grounds that be fruitfull: it groweth plentifully in the pastures about London.

The rest are strangers in England.

¶ *The time.*

They floure in Iune and Iuly.

¶ *The names.*

Melampyrum is called of some *Triticum vaccinium*: in English, Cow-wheat, and Horse-floure: in Greeke, μελάμπυρον: The fourth is called *Melampyrum luteum*: in English, Yellow Cow-wheat.

¶ *The danger.*

The seed of Cow Wheat raiseth vp fumes, and is hot and dry of nature, which being taken in meats and drinks in the manner of Darnell, troubleth the braine, causing drunkennesse and head-ache.

Chap. 69. *Of Wilde Cow-Wheat.*

1 *Cratæogonon album.*
Wilde Cow-wheat.

¶ *The description.*

1 The first kinde of wilde Cow-Wheat *Clusius* in his Pannonick history calls *Parietaria sylvestris*, or wilde Pellitorie: which name, according to his owne words, if it do not fitly answer the Plant, hee knoweth not what to cal it, for that the Latines haue not giuen any name thereunto: yet because some haue so called it, he retaineth the same name. Notwithstanding he referreth it vnto the kindes of *Melampyrum*, or Cow-wheat, or vnto *Cratæogonon*, the wilde Cow-wheat, which it doth very wel answer in diuers points. It hath an hairy foure square stalke, very tender, weake, and easie to breake, not able to stand vpright without the helpe of his neighbours that dwell about him, a foot high or more; whereupon do grow long thin leaues, sharp pointed, and oftentimes lightly snipt about the edges, of a darke purplish colour, sometimes greenish, set by couples one opposite against the other; among the which come forth two floures at one ioynt, long and hollow, somewhat gaping like the floures of a dead nettle, at the first of a pale yellow, and after of a bright golden colour; which do floure by degrees, first a few, and then more, by meanes whereof it is long in flouring. Which being past, there succeed small cups or seed vessels, wherein is contained browne seed not vnlike to wheat. The whole plant is hairy, not differing from the plant Stichwort.

2 Red leafed wilde Cow-wheat is like vnto the former, sauing that the leaues be narrower, and the tuft of leaues more iagged. The stalkes and leaues are of a reddish horse-flesh colour. The

floures in forme are like the other, but in colour differing; for that the hollow part of the floure with the heele or ſpurre is of a purple colour, the reſt of the floure yellow. The ſeed and veſſels are like the precedent.

3 *Cratæogonon Euphroſine.*
Eyebright Cow-wheat.

¶ *The deſcription.*

3 This kinde of wilde Cow-wheat *Tabernamontanus* hath ſet forth vnder the title of *Odontites*: others haue taken it to be a kinde of *Euphraſia* or Eyebright, becauſe it doth in ſome ſort reſemble it, eſpecially in his floures. The ſtalks of this plant are ſmall, woody, rough, and ſquare. The leaues are indented about the edges, ſharpe pointed, and in moſt points reſembling the former Cow-wheat; ſo that of neceſſitie it muſt be of the ſame kinde, and not a kinde of Eyebright, as hath beene ſet downe by ſome.

¶ *The place.*

Theſe wilde kindes of Cow-wheat doe grow commonly in fertile paſtures, and buſhy Copſes, or low woods, and among buſhes vpon barren heaths and ſuch like places.

The two firſt doe grow vpon Hampſted heath neere London, among the Iuniper buſhes and bilberry buſhes in all the parts of the ſaid heath, and in euery part of England where I haue traueled.

¶ *The time.*

They floure from the beginning of May, to the end of Auguſt.

¶ *The names.*

1 The firſt is called of *Lobelius, Cratæogonon*: and of *Tabernamontanus*, *Milium Syluaticum*, or Wood Millet, and *Alſine ſyluatica*, or Wood-Chickweed.

2 The ſecond hath the ſame titles: in Engliſh, Wilde Cow-wheat.

3 The laſt is called by *Tabernamontanus*, *Odontites*: of *Dodonæus*, *Euphraſia altera*, and *Euphroſine*. *Hippocrates* called the wilde Cow-wheat, *Polycarpum*, and *Polycritum*.

¶ *The nature and vertues.*

There is not much ſet downe either of the nature or vertues of theſe plants: onely it is reported that the ſeeds do cauſe giddineſſe and drunkenneſſe as Darnell doth.

The ſeed of *Cratæogonon* made in fine pouder, and giuen in broth or otherwiſe, mightily prouoketh Venerie.

Some write, that it will likewiſe cauſe women to bring forth male children.

† See the vertues attributed to *Cratæogonon* by *Dioſcorides* before, Chap. 38. B.

Chap. 70. *Of White Aſphodill.*

¶ *The kindes.*

Hauing finiſhed the kindes of corne, it followeth to ſhew vnto you the ſundry ſorts of Aſphodils, whereof ſome haue bulbous roots, other tuberous or knobby roots, ſome of yellow colour, and ſome of mixt colours: notwithſtanding *Dioſcorides* maketh mention but of one Aſphodill, but *Pliny* ſetteth downe two; which *Dionyſius* confirmeth, ſaying, That there is the male and female Aſphodil. The latter age hath obſerued many more beſides the bulbed one, of which *Galen* maketh mention.

1 *Aſphodelus*

1 *Aſphodelus* [illegible] *ramoſus.*
White Aſph[illegible]ll.

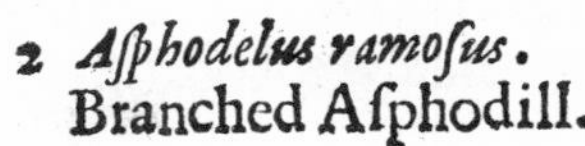
2 *Aſphodelus ramoſus.*
Branched Aſphodill.

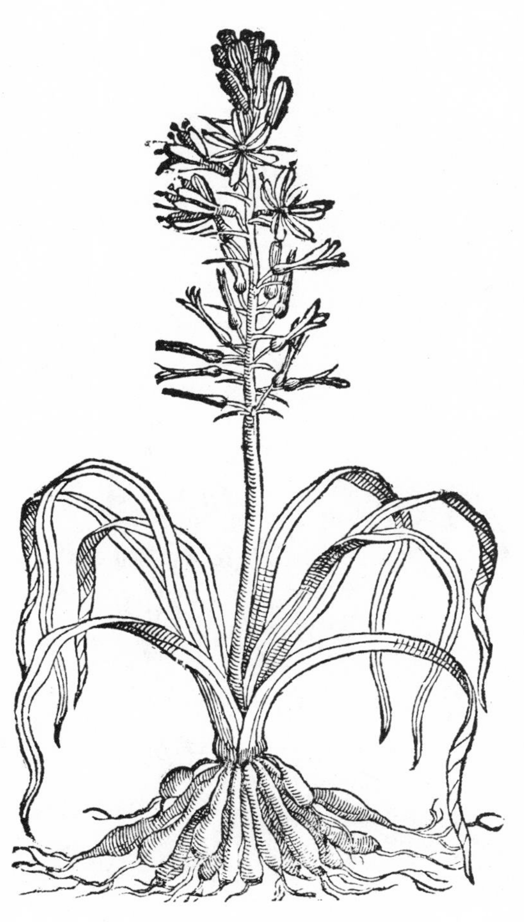

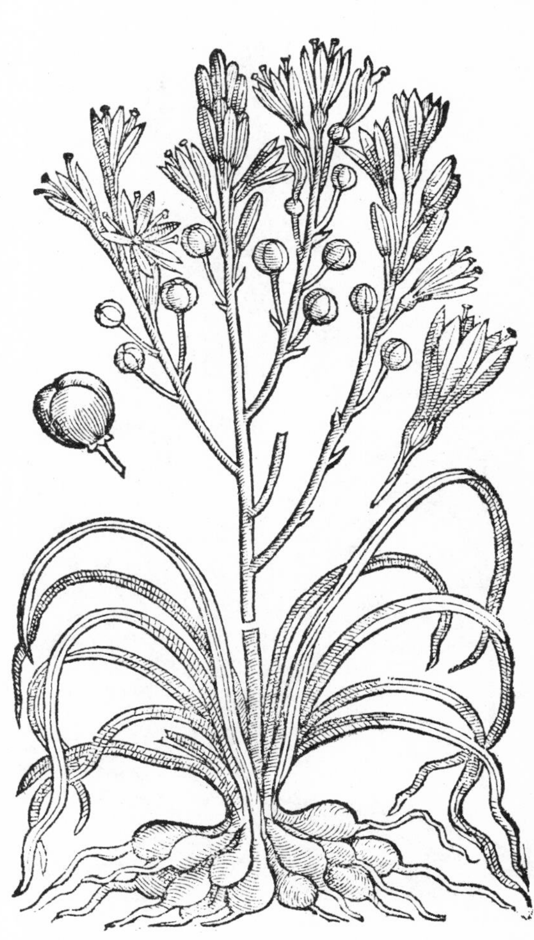

¶ *The deſcription.*

1 THe white Aſphodill hath many long and narrow leaues like thoſe of leeks, ſharpe pointed. The ſtalke is round, ſmooth, naked, and without leaues, two cubits high, garniſhed from the middle vpward with a number of floures ſtarre-faſhion, made of fiue leaues apiece; the colour white, with ſome darke purple ſtreakes drawne downe the backe-ſide. Within the floures be certaine ſmall chiues. The floures being paſt, there ſpring vp little round heads, wherein are contained hard, blacke, and 3 ſquare ſeeds like thoſe of Buck-wheat or Staueſ-acre. The toot is compact of many knobby roots growing out of one head, like thoſe of the Peonie, full of juyce, with a ſmall bitterneſſe and binding taſte.

2 Branched Aſphodill agreeth well with the former deſcription, ſauing that this hath many branches or armes growing out of the ſtalke, whereon the floures do grow, and the other hath not any branch at all, wherein conſiſteth the difference.

3 Aſphodill with the reddiſh floure groweth vp in roots, ſtalke, leafe, and manner of growing like the precedent, ſauing that the floures of this be of a dark red color, & the others white, which ſetteth forth the difference, if there be any ſuch difference, or any ſuch plant at all: for I haue conferred with many moſt excellent men in the knowledge of plants, but none of them can giue mee certaine knowledge of any ſuch, but tell me they haue heard it reported that ſuch a one there is, and ſo haue I alſo, but certainly I cannot ſet downe any thing of this plant vntill I heare more certaintie: for as yet I giue no credit to my Authour, which for reuerence of his perſon I forbeare to name.

4 The yellow Aſphodill hath many roots growing out of one head, made of ſundry tough, fat, and oleous yellow ſprigs, or groſſe ſtrings, from the which riſe vp many graſſy leaues, thick and groſſe, tending to ſquareneſſe; among the which commeth vp a ſtrong thicke ſtalke ſet with the like leaues euen to the floures, but leſſe: vpon the which do grow ſtarre-like yellow floures, otherwiſe like the white Aſphodill.

3 *Asphodelus flore rubente.*
Red Asphodill.

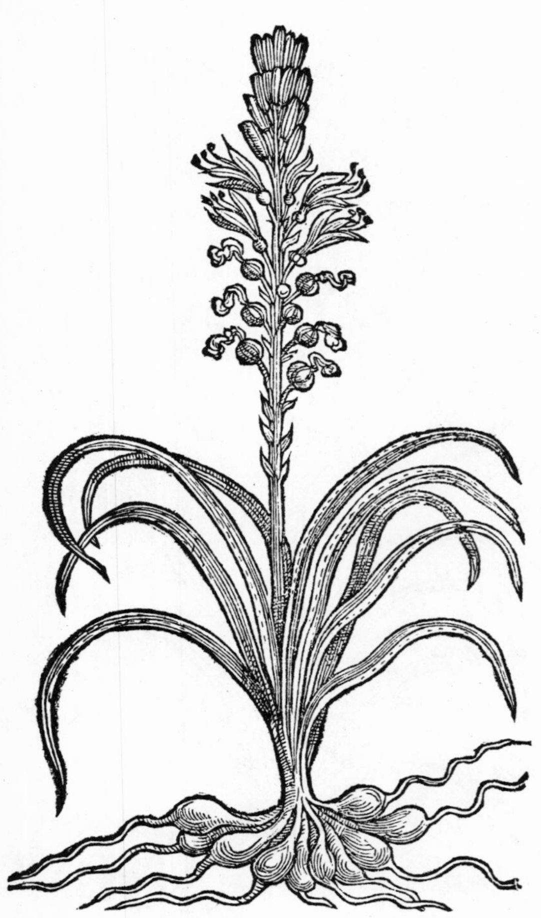

4 *Asphodelus luteus.*
Yellow Asphodill.

‡ 5 *Asphodelus minimus.* Dwarfe Asphodil.

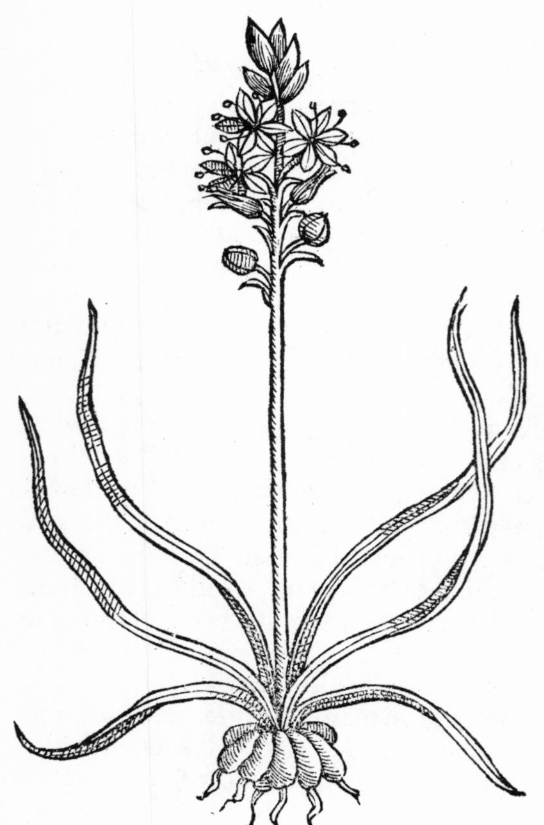

‡ 5 Besides these there is an Asphodill which *Clusius* for the smalnesse calls *Asphodelus minimus*. The roots thereof are knotty and tuberous, resembling those of the formerly described, but lesse: from these arise fiue or sixe very narrow and long leaues; in the middest of which growes vp a stalk of the height of a foot, round and without branches, bearing at the top thereof a spoke of floures, consisting of six white leaues a piece, each of which hath a streake running alongst it, both on the inside and outside, like as the first described. It floures in the beginning of Iuly, when as the rest are past their floures. It loseth the leaues in Winter, and gets new ones againe in the beginning of Aprill. ‡

¶ *The time and place.*

They floure in May and Iune, beginning below, and so flouring vpward: and they grow naturally in France, Italy, Spaine, and most of them in our London Gardens.

¶ *The names.*

Asphodill is called in Latine, *Asphodelus, Albucum, albucus*, and *Hastula Regia*: in Greeke, ἀσφόδελος: in English, Asphodill, not Daffodil; for Daffodill is *Narcissus*, another plant differing from Asphodill. *Pliny* writeth, That the stalke with the floures is called *Anthericos*; and the root, that is to say, the bulbs *Asphodelus*.

Of

Of this Asphodill *Hesiod* maketh mention in his Works, where he saith, that fooles know not how much good there is in the Mallow and in the Asphodill, because the roots of Asphodill are good to be eaten. Yet *Galen* doth not beleeue that he meant of this Asphodill, but of that bulbed one, whereof we will make mention hereafter. And he himselfe testifieth, that the bulbes thereof are not to be eaten without very long seething; and therefore it is not like that *Hesiod* hath commended any such: for he seemeth to vnderstand by the Mallow and the Asphodil, such kinde of food as is easily prepared, and soone made ready.

¶ *The nature.*

These kindes of Asphodils be hot and dry almost in the third degree.

¶ *The vertues.*

After the opinion of *Dioscorides* and *Aetius*, the roots of Asphodill eaten, prouoke vrine and the termes effectually, especially being stamped and strained with wine, and drunke. A

One dram thereof taken in wine in manner before rehearsed, helpeth the paine in the sides, ruptures, convulsions, and the old cough. B

The roots boiled in dregs of wine cure foule eating vlcers, all inflammations of the dugges or stones, and easeth the felon, being put thereto as a pultesse. C

The iuyce of the root boyled in old sweet Wine, together with a little myrrh and saffron, maketh an excellent Collyrie profitable for the eyes. D

Galen saith, the roots burnt to ashes, and mixed with the grease of a ducke, helpeth the *Alopecia*, and bringeth haire againe that was fallen by that disease. E

The weight of a dram thereof taken with wine helpeth the drawing together of sinews, cramps, and burstings, F

The like quantitie taken in broth prouoketh vomit, and helpeth those that are bitten with any venomous beasts. G

The iuyce of the root cleanseth and taketh away the white morphew, if the face be annointed therewith; but first the place must be chafed and wel rubbed with a course linnen cloath. H

CHAP. 71. *Of the Kings Speare.*

1 *Asphodelus luteus minor.* The Kings Speare.

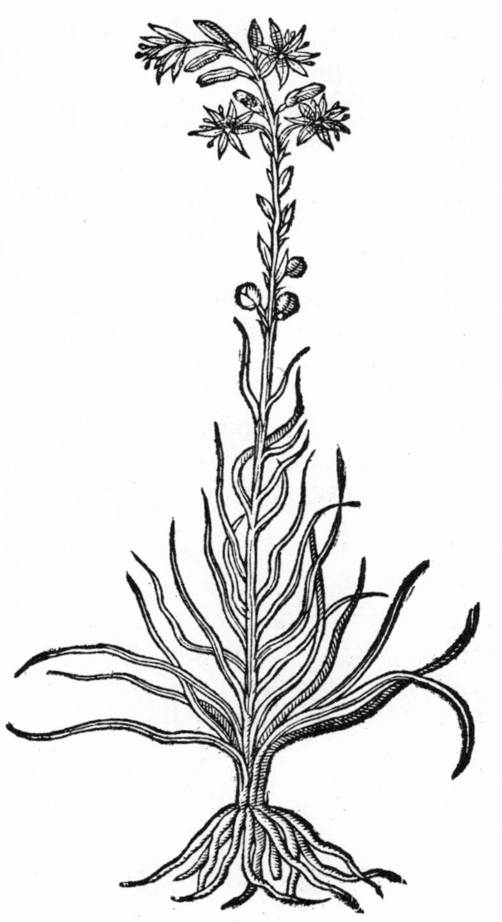

2 *Asphodelus Lancastriæ.* Lancashire Asphodil.

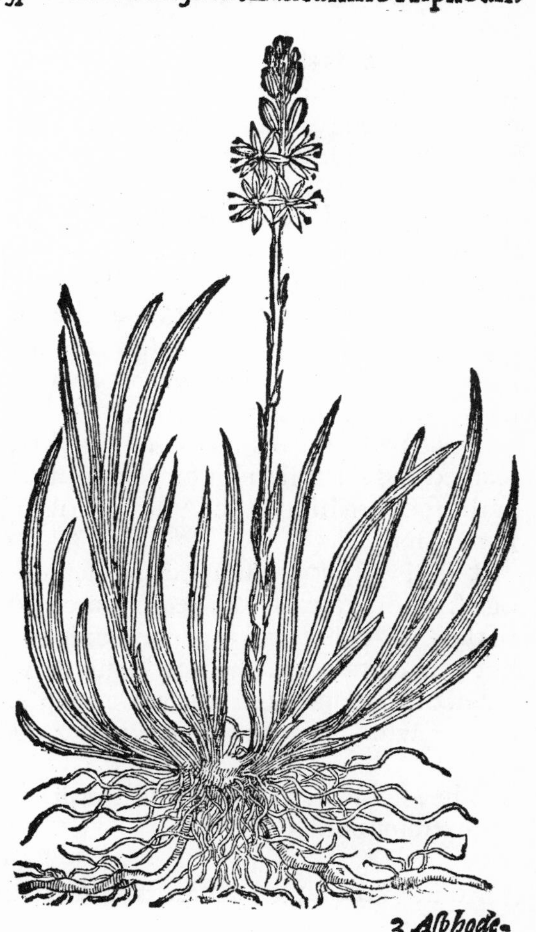

3 *Asphode-*

‡ 3 *Aſphodelus Lancaſtriæ verus.*
The true Lancaſhire Aſphodil.

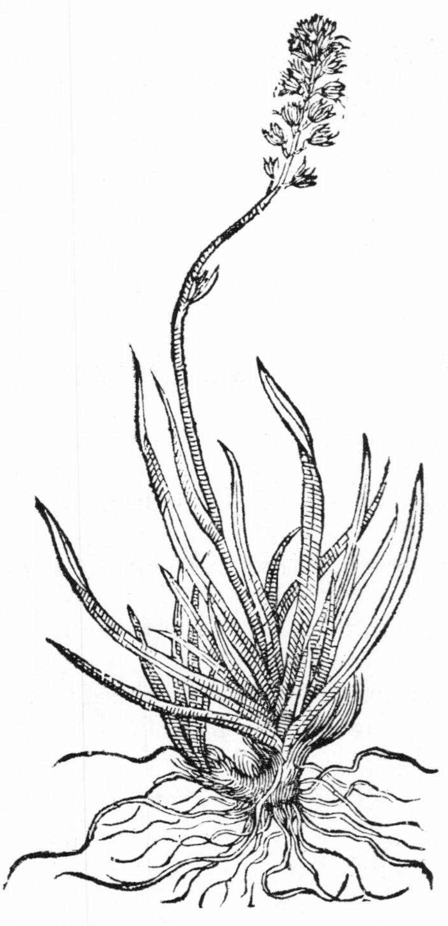

¶ *The deſcription.*

1 The leaues of the Kings Speare are long, narrow, and chamfered or furrowed, of a blewiſh greene colour. The ſtalk is round, of a cubit high. The floures which grow thereon from the middle to the top are very many, in ſhape like to the floures of the other; which being paſt, there come in place thereof little round heads or ſeed-veſſels, wherein the ſeed is contained. The roots in like manner are very many, long, and ſlender, ſmaller than thoſe of the other yellow ſort. Vpon the ſides whereof grow forth certaine ſtrings, by which the plant it ſelfe is eaſily encreaſed and multiplied.

2 There is found in theſe dayes a certaine waterie or mariſh Aſphodill like vnto this laſt deſcribed, in ſtalke and floures, without any difference at all. It bringeth forth leaues of a beautifull greene ſomwhat chamfered, like to thoſe of the Floure de-luce, or corne-flag, but narrower, not full a ſpan long. The ſtalke is ſtrait, a foot high, whereupon grow the floures, conſiſting of ſixe ſmall leaues: in the middle whereof come forth ſmall yellow chiues or threds. The ſeed is very ſmall, contained in long ſharpe pointed cods. The root is long, ioynted, and creepeth as graſſe doth, with many ſmall ſtrings.

‡ 3 Beſides the laſt deſcribed (which our Author I feare miſtaking, termed *Aſphodelus Lancaſtriæ*) there is another water Aſphodill, which growes in many rotten mooriſh grounds in this kingdome, and in Lancaſhire is vſed by women to die their haire of a yellowiſh colour, and therefore by them it is termed Maiden-haire, if we may beleeue *Lobell*.) This plant hath leaues of ſome two inches and an halfe, or three inches long, being ſomewhat broad at the bottome, and ſo ſharper towards their ends. The ſtalke ſeldome attaines to the height of a foot, and it is ſmooth without any leaues thereon; the top thereof is adorned with pretty yellow ſtar-like floures, wherto ſucceed longiſh little cods, vſually three, yet ſometimes foure or fiue ſquare, and in theſe there is contained a ſmall red ſeed. The root conſiſts onely of a few ſmall ſtrings. ‡

¶ *The place.*

1 The ſmall yellow Aſphodill groweth not of it ſelfe wilde in theſe parts, notwithſtanding we haue great plenty thereof in our London gardens.

2 The Lancaſhire Aſphodill groweth in moiſt and mariſh places neere vnto the Towne of Lancaſter, in the mooriſh grounds there, as alſo neere vnto Maudſley and Martom, two Villages not farre from thence; where it was found by a Worſhipfull and learned Gentleman, a diligent ſearcher of ſimples, and feruent louer of plants, M. *Thomas Hesket*, who brought the plants thereof vnto me for the encreaſe of my garden.

I receiued ſome plants thereof likewiſe from Maſter *Thomas Edwards*, Apothecarie in Exceſter, learned and skilfull in his profeſſion, as alſo in the knowledge of plants. He found this Aſphodill at the foot of a hill in the Weſt part of England, called Bagſhot hill, neere vnto a village of the ſame name.

‡ This Aſphodill figured and deſcribed out of *Dodonæus*, and called *Aſphodelus Lancaſtriæ* by our Author, growes in an heath ſome two miles from Bruges in Flanders, and diuers other places of the Low-countries; but whether it grow in Lancaſhire or no, I can ſay nothing of certaintie: but I am certaine, that which I haue deſcribed in the third place growes in many places of the Weſt of England; and this yeare 1632, my kinde friend M. *George Bowles* ſent mee ſome plants thereof, which I keepe yet growing. *Lobell* alſo affirmes this to be the Lancaſhire Aſphodill.

¶ *The time.*

They floure in May and Iune: moſt of the leaues thereof remaine greene in the Winter, if it be not extreme cold.

¶ *The names.*

Some of the later Herbariſts thinke this yellow Aſphodill to be *Iphyon* of *Theophraſtus*, and others

others iudge it to be *Erizambac* of the Arabians. In Latine it is called *Asphodelus luteus*: of some it is called *Hastula Regia*. We haue Englished it, the Speare for a King, or small yellow Asphodill.

2 The Lancashire Asphodill is called in Latine, *Asphodelus Lancastriæ*; and may likewise be called *Asphodelus palustris*, or *Pseudoasphodelus luteus*, or the Bastard yellow Asphodill.

‡ 3 This is *Asphodelus minimus luteus palustris Scoticus & Lancastriensis*, of *Lobell*; and the *Pseudoasphodelus pumilio folys Iridis*, of *Clusius*, as farre as I can iudge; although *Bauhine* distinguisheth them. ‡

¶ *The temperature and vertues.*

It is not yet found out what vse there is of any of them in nourishment or medicines.

CHAP. 72. *Of Onion Asphodill.*

Asphodelus Bulbosus.
Onion Asphodill.

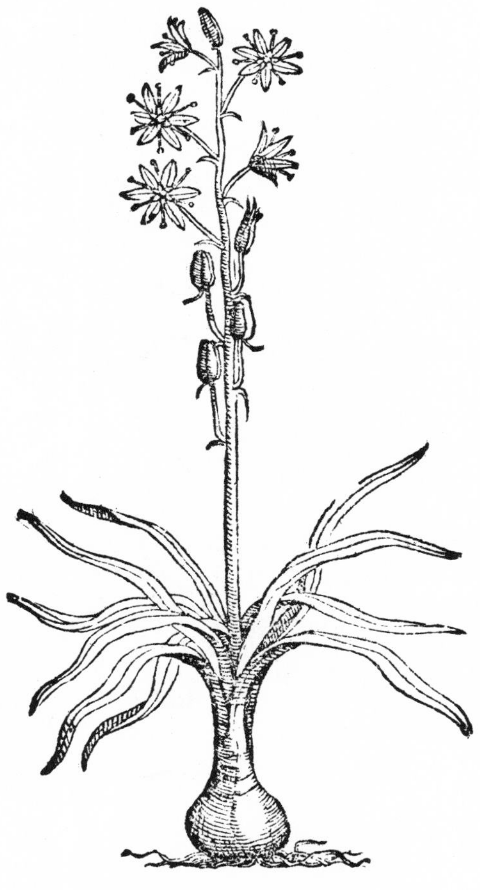

¶ *The description.*

THe bulbed Asphodill hath a round bulbus or Onion root, with some fibres hanging thereat; from the which come vp many grassie leaues, very well resembling the Leeke; among the which leaues there riseth vp a naked or smooth stem, garnished toward the top with many star-like floures, of a whitish greene on the inside, and wholly greene without, consisting of six little leaues sharpe pointed, with certaine chiues or threads in the middle. After the floure is past there succeedeth a small knop or head three square, wherein lieth the seed.

¶ *The place.*

It groweth in the gardens of Herbarists in London, and not elsewhere that I know of, for it is not very common.

¶ *The time.*

It floureth in Iune and Iuly, and somewhat after.

¶ *The names.*

The stalke and floures being like to those of the Asphodill before mentioned, do shew it to be *Asphodeli species*, or a kinde of Asphodill; for which cause also it seemeth to be that Asphodil of which *Galen* hath made mention in his second book of the Faculties of nourishments, in these words; The root of Asphodill is in a manner like to the root of Squill, or Sea Onion, as well in shape as bitternes. Notwithstanding, saith *Galen*, my selfe haue known certaine countrymen, who in time of famine could not with many boilings and steepings make it fit to be eaten. It is called of *Dodonæus*, *Asphodelus fœmina*, and *Asphodelus Bulbosus*, *Hyacintho-Asphodelus*, and *Asphodelus Hyacinthinus* by *Lobell*, and that rightly; for that the root is like the Hyacinth, and the floures like the Asphodill: and therefore as it doth participate of both kindes, so likewise doth the name: in English we may call it Bulbed Asphodill. *Clusius* calls it *Ornithogalum majus*, and that fitly.

¶ *The nature.*

The round rooted Asphodil, according to *Galen*, hath the same temperature and vertue that *Aron*, *Arisarum*, and *Dracontium* haue, namely an absterfiue and cleansing qualitie.

¶ *The vertues.*

A The yong sprouts or springs thereof is a singular medicine against the yellow Iaundise, for that the root is of power to make thin and open.

B *Galen* saith, that the ashes of this Bulbe mixed with oile or hens grease cureth the falling of the haire in an *Alopecia* or scalld head.

Chap.

Chap. 73. *Of Yellow Lillies.*

¶ *The kindes.*

Because we shall haue occasion hereafter to speake of certaine Cloued or Bulbed Lillies, wee will in this chapter entreat onely of another kinde not bulbed, which likewise is of two sorts, differing principally in their roots; for in floures they are Lillies, but in roots Asphodils, participating as it were of both, though neerer approching vnto Asphodils than Lillies.

1 *Lilium non bulbosum.*
The yellow Lillie.

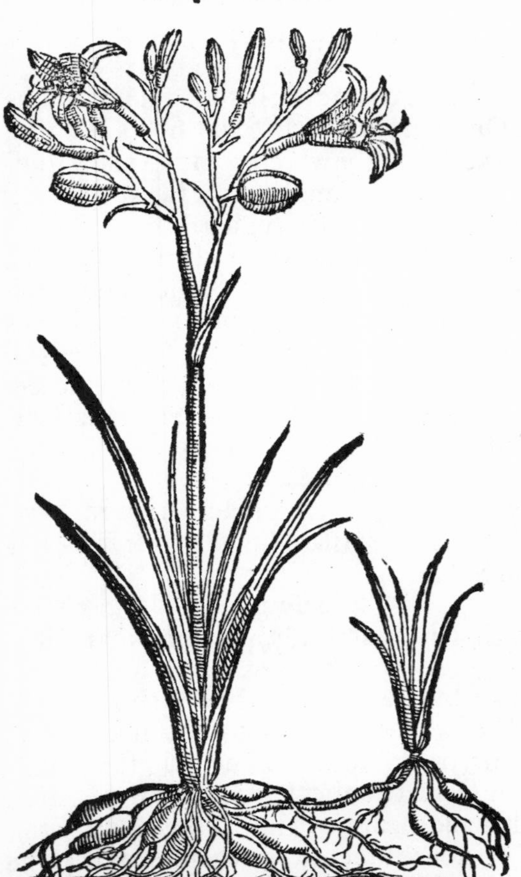

2 *Lilium non bulbosum Phœniceum.*
The Day-Lillie.

¶ *The description.*

1 The yellow Lillie hath very long flaggie leaues, chamfered or channelled, hollow in the middest like a gutter, among the which riseth vp a naked or bare stalke, two cubits high, branched toward the top, with sundry brittle armes or branches, whereon do grow many goodly floures like vnto those of the common white Lillie in shape and proportion, of a shining yellow colour; which being past, there succeed three cornered huskes or cods, full of blacke shining seeds like those of the Peonie. The root consisteth of many knobs or tuberous clogs, proceeding from one head, like those of the white Asphodill or Peonie.

2 The Day-Lillie hath stalkes and leaues like the former. The floures be like the white Lillie in shape, of an Orenge tawny colour: of which floures much might be said which I omit. But in briefe, this plant bringeth forth in the morning his bud, which at noone is full blowne, or spred abroad, and the same day in the euening it shuts it selfe, and in a short time after becomes as rotten and stinking as if it had beene trodden in a dunghill a moneth together, in foule and rainie weather: which is the cause that the seed seldome followes, as in the other of his kinde, not bringing forth any at all that I could euer obserue; according to the old prouerbe, Soone ripe, soone rotten. His roots are like the former.

¶ The

¶ *The place.*

These Lillies do grow in my garden, as also in the gardens of Herbarists, and louers of fine and rare plants; but not wilde in England, as in other countries.

¶ *The time.*

These Lillies do floure somewhat before the other Lillies, and the yellow Lillie the soonest.

¶ *The names.*

Diuers do call this kinde of Lillie, *Liliasphodelus, Liliago*, and also *Liliastrum*, but most commonly *Lilium non bulbosum*: In English, Liriconfancie, and yellow Lillie. The old Herbarists name it *Hemerocallis*: for they haue two kindes of *Hemerocallis*; the one a shrub or woody plant, as witnesseth *Theophrastus*, in his sixth booke of the historie of Plants. *Pliny* setteth downe the same shrub among those plants, the leaues whereof onely do serue for garlands.

The other *Hemerocallis* which they set downe, is a Floure which perisheth at night, and buddeth at the Sunne rising, according to *Athenæus*; and therefore it is fitly called ἡμεροκαλλὶς; that is, Faire or beautifull for a day: and so we in English may rightly terme it the Day-Lillie, or Lillie for a day.

¶ *The nature.*

The nature is rather referred to the Asphodils than to Lillies.

¶ *The vertues.*

Dioscorides saith, That the root stamped with honey, and a mother pessarie made thereof with wooll, and put vp, bringeth forth water and bloud. A

The leaues stamped and applied do allay hot swellings in the dugges, after womens trauell in childe-bearing, and likewise taketh away the inflammation of the eyes. B

The roots and the leaues be laid with good successe vpon burnings and scaldings. C

Chap. 73. *Of Bulbed Floure de-Luce.*

‡ 1 *Iris Bulbosa Latifolia.*
Broad leaued Bulbous Floure de-luce.

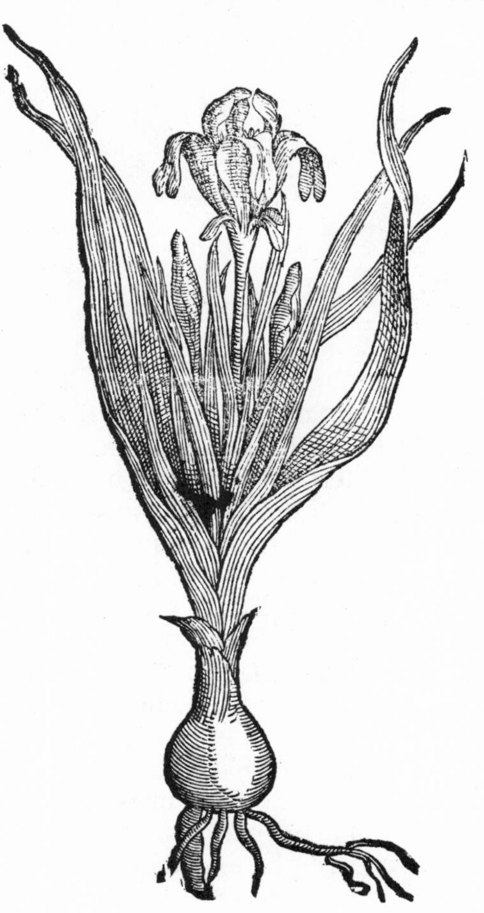

2 *Iris Bulbosa Anglica.*
Onion Floure de-Luce.

¶ The

¶ *The kindes.*

LIke as we haue ſet downe ſundry ſorts of Floure de-luces, with flaggy leaues, and tuberous or knobby roots, varying very notably in ſundry reſpects, which we haue diſtinguiſhed in their proper Chapters: it reſteth that in like manner we ſet forth vnto your view certaine bulbous or Onion-rooted Floure de-luces, which in this place do offer themſelues vnto our conſideration; whereof there be alſo ſundry ſorts, ſorted into one chapter as followeth.

3 *Iris Bulboſa flore vario.*
Changeable Floure de-luce.

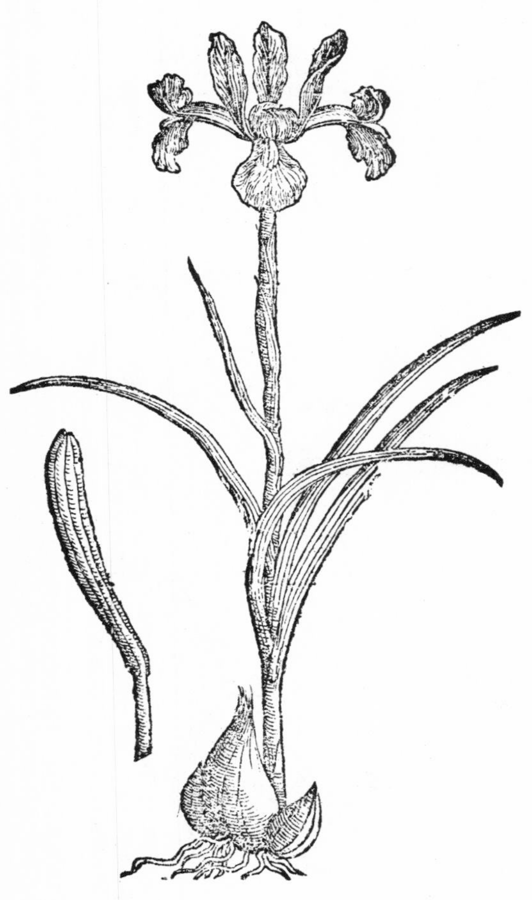

‡ 4 *Iris Bulboſa verſicolor Polyclonos.*
Many branched changeable Floure de-luce.

¶ *The deſcription.*

‡ 1 THe firſt of theſe, whoſe figure here we giue you vnder the name of *Iris Bulboſa Latifolia,* hath leaues ſomewhat like thoſe of the Day-Lillie, ſoft, and ſomewhat paliſh greene, with the vnder ſides ſomewhat whiter; amongſt which there riſeth vp a ſtalk bearing at the top thereof a Floure a little in ſhape different from the formerly deſcribed Floure de-luces. The colour thereof is blew; the number of the leaues whereof it conſiſts, nine: three of theſe are little, and come out at the bottome of the Floure as ſoone as it is opened; three more are large, and being narrow at their bottome, become broader by little and little, vntill they come to turne downwards, whereas then they are ſhapen ſomewhat roundiſh or obtuſe. In the middeſt of theſe there runnes vp a yellow variegated line to the place whereas they bend backe. The three other leaues are arched like as in other Floures of this kinde, and diuided at their vpper end, and containe in them three threads of a whitiſh blew colour.

This is called *Iris Bulboſa Latifolia,* by *Cluſius*; and *Hyacinthus Poetarum Latifolius,* by *Lobell.*

It floures in Ianuarie and Februarie, whereas it growes naturally, as it doth in diuers places of Portugall and Spaine. It is a tender plant, and ſeldome thriues well in our gardens. ‡

2 Onion Floure de-luce hath long narrow blades or leaues, creſted, chamfered, or ſtreaked on the backe ſide as it were welted; below ſomewhat round, opening it ſelfe toward the top, yet remaining as it were halfe round, whereby it reſembleth an hollow trough or gutter. In the bottome of the hollowneſſe it tendeth to whiteneſſe; and among theſe leaues do riſe vp a ſtalke of a cubit high; at the top whereof groweth a faire blew Floure, not differing in ſhape from the common

mon Floure de-luce: the which being past, there come in the place thereof long thicke cods or seed-vessels, wherein is contained yellowish seed of the bignesse of a tare or fitch. The root is round like an Onion, couered ouer with certaine browne skinnes or filmes. Of this kind there are some fiue or six varieties, caused by the various colours of the Floures.

5 *Iris Bulbosa Flore luteo cum flore & semine.*
Yellow bulbed Floure de-luce in floure and seed.

3 Changeable Floure de-luce hath leaues, stalkes, and Roots like the former, but lesser. The Floure hath likewise the forme of the Floure de-luce, that is to say, it consisteth of sixe greater leaues, and three lesser; the greater leaues fold backward and hang downward, the lesser stand vpright; and in the middle of the leaues there riseth vp a yellow welt, white about the brimmes, and shadowed all ouer with a wash of thinne blew tending to a Watchet colour. Toward the stalke they are stripped ouer with a light purple colour, and likewise amongst the hollow places of those that stand vpright (which cannot be expressed in the figure) there is the same faire purple colour; the smell and sauour very sweet and pleasant. The root is Onion fashion, or bulbous like the other.

‡ 4 There is also another variegated Floure de-luce, much like this last described, in the colour of the Floure; but each plant produceth more branches and Floures, whence it is termed *Iris Bulbosa versicolor polyclonos*, Many-branched changeable Floure de-luce. ‡

5 Of which kinde or sort there is another in my Garden, which I receiued from my Brother *James Garret* Apothecarie, far more beautifull than the last described; the which is dasht ouer, in stead of the blew or watchet colour, with a most pleasant gold yellow colour, of smell exceeding sweet, with bulbed roots like those of the other sort.

6 It is reported, that there is in the garden of the Prince Elector the Lantgraue of Hessen, one of this sort or kinde, with white Floures, the which as yet I haue not seene.

‡ Besides these sorts mentioned by our Author, there are of the narrow leaued bulbous Floure de-luces, some twenty foure or more varieties, which in shape of roots, leaues, and Floures differ very little, or almost nothing at all; so that he which knows one of these may presently know the rest. Wherefore because it is a thing no more pertinent to a generall historie of Plants, to insist vpon these accidentall nicities, than for him that writes a historie of Beasts to describe all the colours, and their mixtures, in Horses, Dogs, and the like; I refer such as are desirous to informe

themſelues of thoſe varieties,to ſuch as haue onely and purpoſely treated of Floures and their diuerſities, as *De-Bry*, *Swerts*, and our Countreyman M. *Parkinſon*, who in his *Paradiſus terreſtris*, ſet forth in Engliſh, *Anno* 1629. hath iudiciouſly and exactly comprehended all that hath beene deliuered by others in this nature. ‡

‡ 6 *Iris Bulboſa flore cinereo.*
Aſh-coloured Floure de-luce.

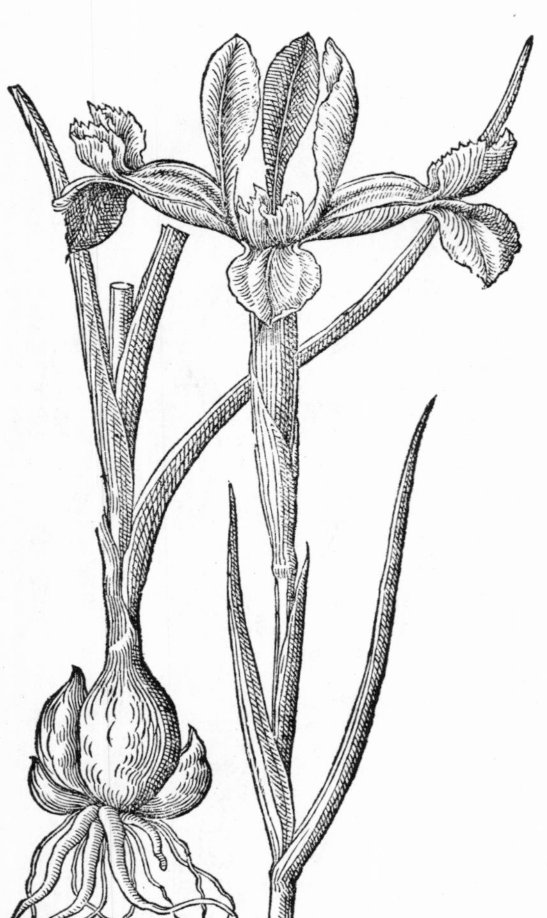

‡ 7 *Iris Bulboſa flore albido.*
Whitiſh Floure de-luce.

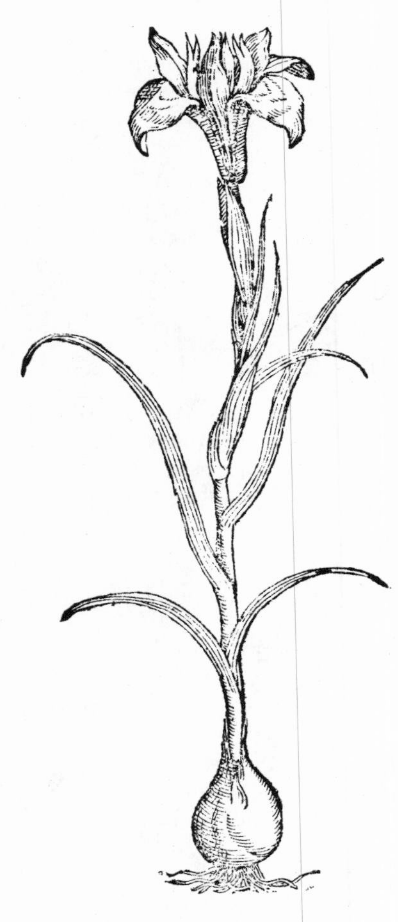

¶ *The place.*

The ſecond of theſe bulbed Floure de-luces growes wilde, or of it ſelfe, in the corne fields of the Weſt parts of England, as about Bathe and Wells, and thoſe places adiacent; from whence they were firſt brought into London, where they be naturaliſed, and encreaſe in great plenty in our London gardens.

The other ſorts do grow naturally in Spaine and Italy wilde, from whence we haue had Plants for our London gardens, whereof they do greatly abound.

¶ *The time.*

They floure in Iune and Iuly, and ſeldome after.

¶ *The names.*

The Bulbed Floure de-luce is called of *Lobelius*, *Iris Bulboſa*, and alſo *Hyacinthus flore Iridis*: of ſome, *Hyacinthus Poetarum*; and peraduenture it is the ſame that *Apuleius* mentioneth in the one and twentieth Chapter, ſaying, That *Iris*, named among the old Writers *Hieris*, may alſo be called, and not vnproperly, *Hierobulbus*, or *Hieribulbus*: as though you ſhould ſay, *Iris Bulboſa*, or Bulbed Ireos; vnleſſe you would haue ἱερόβολβος, called a greater or larger Bulbe: for it is certaine, that great and huge things were called of the Antients, ἱερὰ, or *Sacra*: in Engliſh, Holy.

¶ *The nature.*

The nature of theſe Bulbed Floure de-luces are referred to the kindes of Aſphodils.

¶ *The vertues.*

Take, ſaith *Apuleius*, of the herbe *Hierobulbus* ſix ʒ. Goats ſuet as much, Oile of Alcanna one
A pound; mix them together, being firſt ſtamped in a ſtone morter, it taketh away the paine of the Gout.

Moreouer, if a woman do vſe to waſh her face with the decoction of the root, mixed with the
B meale of Lupines, it forthwith cleanſeth away the freckles & morphew, and ſuch like deformities.

Chap.

CHAP. 75. Of Spaniſh Nut.

1 *Siſynrichium majus.*
Spaniſh Nut.

‡ 2 *Siſynrichium minus.*
Small Spaniſh Nut.

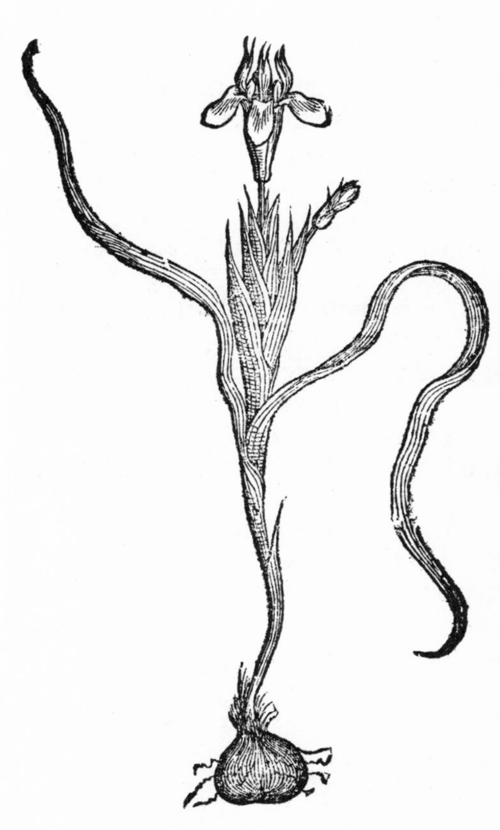

3 *Iris Tuberoſa.* Veluet Floure de-luce.

¶ *The deſcription.*

1 Spaniſh Nut hath ſmall graſſie leaues like thoſe of the Starres of Bethlem, or *Ornithogalum*; among which riſeth vp a ſmall ſtalke of halfe a foot high, garniſhed with the like leaues, but ſhorter. The Floures grow at the top, of a skie colour, in ſhape reſembling the Floure de-luce, or common *Iris*; but the leaues that turne downe are each of them marked with a yellowiſh ſpot: they fade quickely, and being paſt, there ſucceed ſmall cods with ſeeds as ſmall as thoſe of Turneps. The root is round, compoſed of two bulbes, the one lying vpon the other as thoſe of the Corne flag vſually do; and they are couered with a skinne or filme in ſhape like a Net. The Bulbe is ſweet in taſte, and may be eaten before any other bulbed Root.

2 There is ſet forth another of this kinde, ſomewhat leſſer, with Floures that ſmell ſweeter than the former.

3 Veluet Floure de-luce hath many long ſquare leaues, ſpongeous or full of pith, trailing vpon the ground, in ſhape like to the leaues of Ruſhes: among which riſeth vp a ſtalke of a foot

high, bearing at the top a Floure like the Floure de-luce. The lower leaues that turne downward are of a perfect blacke colour, soft and smooth as is blacke Veluet; the blacknesse is welted about with greenish yellow, or as wee terme it a Goose-turd greene; of which colour the vppermost leaues do consist: which being past, there followeth a great knob or crested seed vessell of the bignesse of a mans thumbe, wherein is contained round white seed as bigge as the Fetch or tare. The root consisteth of many knobby bunches like fingers.

¶ *The place.*

These bastard kindes of Floure de-luces are strangers in England, except it be among some few diligent Herbarists in London, who haue them in their gardens, where they increase exceedingly; especially the last described, which is said to grow wilde about Constantinople, Morea, and Greece: from whence it hath beene transported into Italy, where it hath beene taken for *Hermodactylus*, and by some exprest or set forth in writing vnder the title *Hermodactylus*; whereas in truth it hath no semblance at all with *Hermodactylus*.

¶ *The time.*

The wilde or Bastard Floure de-luces do floure from May to the end of Iune.

¶ *The names.*

1 2 These bulbed bastard Floure de-luces, which we haue Englished Spanish Nuts, are called in Spaine, *Nozelhas*; that is, little Nuts: the lesser sort *Parua Nozelha*, and *Macuca*: wee take it to be that kinde of nourishing Bulbe which is named in Greeke, σισύριγχιον: of *Pliny*, *Sisyrichium*.

‡ 3 Some, as *Vlysses Aldroandus*, would haue this to be *Louchitis Prior*, of *Dioscor.* *Matthiolus* makes it *Hermodactylus verus*, or the true Hermodactill: *Dodonæus* and *Lobell* more fitly refer it to the Floure de-luces, and call it *Iris tuberosa*. ‡

¶ *The nature and vertues.*

Of these kindes of Floure de-luces there hath beene little or nothing at all left in writing concerning their natures or vertues; only the Spanish nut is eaten at the tables of rich and delicious, nay vitious persons, in sallads or otherwise, to procure lust and lecherie.

Chap. 76. *Of Corne-Flagge.*

1 *Gladiolus Narbonensis.*
French Corne-Flag, or Sword-Flag.

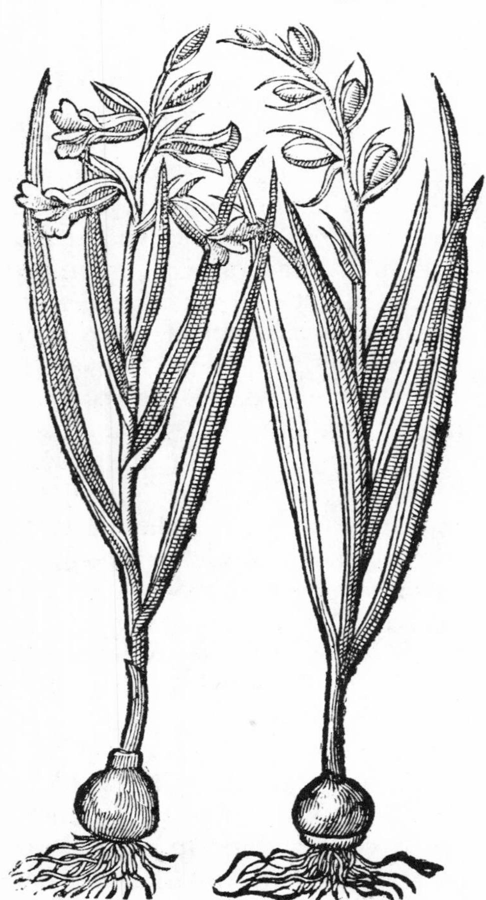

2 *Gladiolus Italicus.*
Italian Corne-Flag, or Sword-Flag.

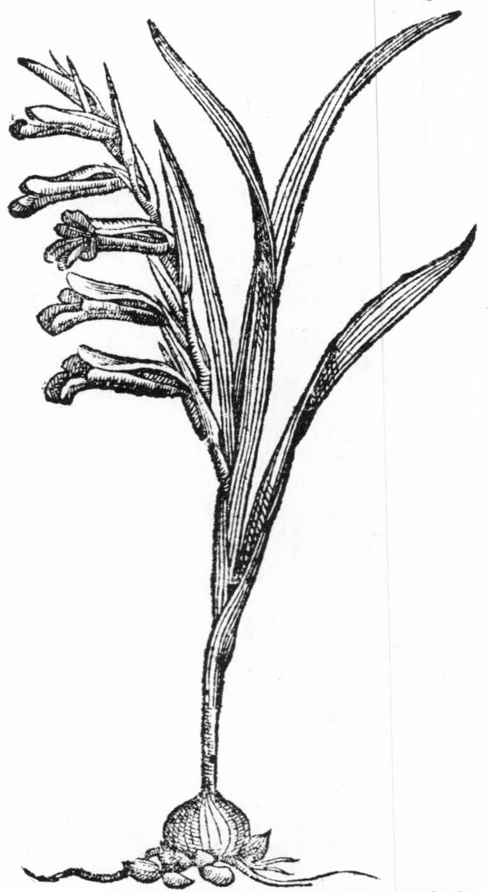

‡ 4 *Gladi-*

‡ 4 *Gladiolus Lacustris.*
Water Sword-Flag.

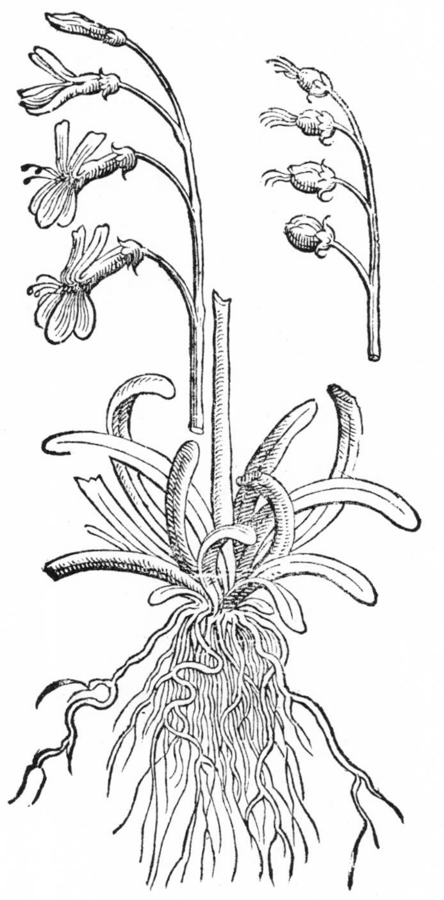

¶ *The description.*

1 FRench Corne-Flagge hath small stiffe leaues, ribbed or chamfered with long nerues or sinewes running through the same, in shape like those of the small Floure de-luce, or the blade of a sword, sharpe pointed, of an ouer-worne greene colour, among the which riseth vp a stiffe brittle stalke two cubits high, whereupon doe grow in comely order many faire purple Floures, gaping like those of Snapdragon, or not much differing from the Fox-Gloue, called in Latine *Digitalis*. After them come round knobbie seed-vessels, full of chaffie seed, very light, of a browne reddish colour. The root consisteth of two Bulbes, one set vpon the other; the vppermost whereof in the beginning of the Spring is lesser, and more ful of juice; the lower greater, but more loose and lithie, which a little while after perisheth.

2 Italian Corn-Flag hath long narrow leaues with many ribbes or nerues running through the same: the stalke is stiffe and brittle, whereupon do grow Floures orderly placed vpon one side of the stalke, whereas the precedent hath his floures placed on both the sides of the stalke, in shape and colour like the former, as are also the roots, but seldome seene one aboue another, as in the former.

3 There is a third sort of Corne-Flag which agreeth with the last described in euerie point, sauing that the Floures of this are of a pale colour, as it were betweene white, and that which we call Maidens Blush.

‡ 4 This Water Sword-Flag, described by *Clusius* in his *Cur. Post.* hath leaues about a span long, thicke and hollow, with a partition in their middles, like as wee see in the cods of Stocke-Gillouers, and the like: their colour is greene, and taste sweet, so that they are an acceptable food to the wilde Ducks ducking downe to the bottome of the water; for they sometimes lie some ells vnder water: which notwithstanding is ouer-topt by the stalke, which springs vp from among these leaues, and beares Floures of colour white, larger than those of Stock-Gillouers, but in that hollow part that is next the stalke they are of a blewish colour, almost in shape resembling the Floures of the Corne-Flag, yet not absolutely like them. They consist of fiue leaues, whereof the two vppermost are reflected towards the stalke; the three other being broader hang downewards. After the floures there follow round pointed vessels filled with red seed. It floures at the end of Iuly.

It was found in some places of West-Friseland, by *Iohn Dortman* a learned Apothecary of Groningen. It growes in waters which haue pure grauell at the bottome, and that bring forth no plant besides.

Clusius, and *Dortman* who sent it him, call it *Gladiolus Lacustris*, or *Stagnalis*. ‡

¶ *The place.*

These kindes of Corne-Flags grow in medowes, and in earable grounds among corne, in many places of Italy, as also in the parts of France bordering thereunto. Neither are the fields of Austria and Morauia without them, as *Cordus* writeth. We haue great plenty of them in our London Gardens, especially for the garnishing and decking them vp with their seemly Floures.

¶ *The time.*

They floure from May to the end of Iuly.

¶ *The names.*

Corne-Flag is called in Greeke ξίφιον: in Latine, *Gladiolus*; and of some, *Ensis*: of others, φάσγανον, and *Gladiolus Segetalis*. *Theophrastus* in his discourse of *Phasganum* maketh it the same with *Xiphion*. *Valerius Cordus* calleth Corne-Flag *Victorialis fœmina*: others, *Victorialis rotunda*: in the Germane

Tongue, Seigwurtz: yet we must make a difference betweene *Gladiolus* and *Victorialis longa*; for that is a kinde of Garlicke found vpon the highest Alpish mountaines, which is likewise called of the Germanes Seigwurtz. The Floures of Corne-Flag are called of the Italians, *Monacuccio*: in English, Corne-Flag, Corne-Sedge, Sword-Flag, Corne Gladin: in French, *Glais*.

¶ *The nature.*

The root of Corne-Flag, as *Galen* saith, is of force to draw, waste, or consume away, and dry, as also of a subtill and digesting qualitie.

¶ *The vertues.*

A The root stamped with the pouder of Frankincense and wine, and applied, draweth forth splinters and thornes that sticke fast in the flesh.

B Being stamped with the meale of Darnell and honied water, doth waste and make subtill hard lumps, nodes, and swellings, being emplaistred.

C Some affirme, that the vpper root prouoketh bodily lust, and the lower causeth barrennesse.

D The vpper root drunke in water is profitable against that kinde of bursting in children called *Enterocele*.

E The root of Corne-Flag stamped with hogs grease and wheaten meale, hath been found by late Practitioners in physicke and Surgerie, to be a certaine and approued remedie against the *Strumæ Scrophulæ*, and such like swellings in the throat.

F The cods with the seed dried and beaten into pouder, and drunk in Goats milke or Asses milke, presently taketh away the paine of the Collicke.

Chap. 77. *Of Starry Hyacinths and their kindes.*

1 *Hyacinthus stellatus Fuchsij.*
Starry Iacinth.

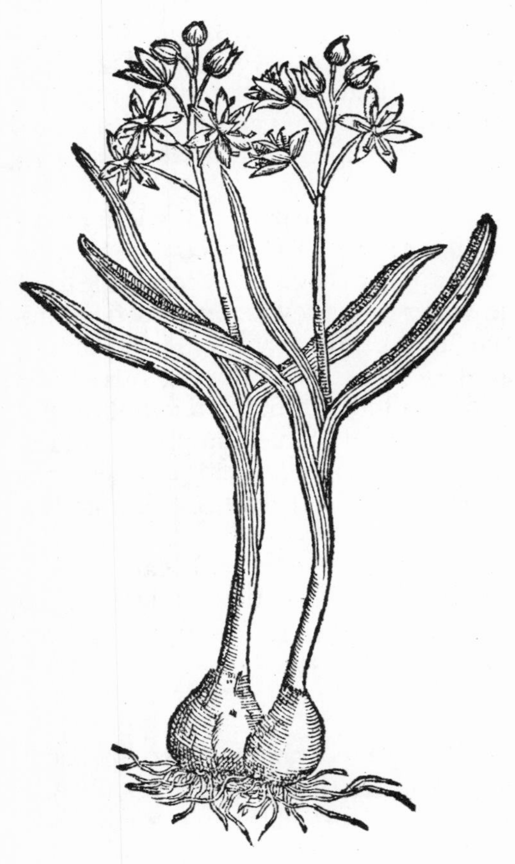

‡ 2 *Hyacinthus stellaris albicans.*
The white floured starry Iacinth.

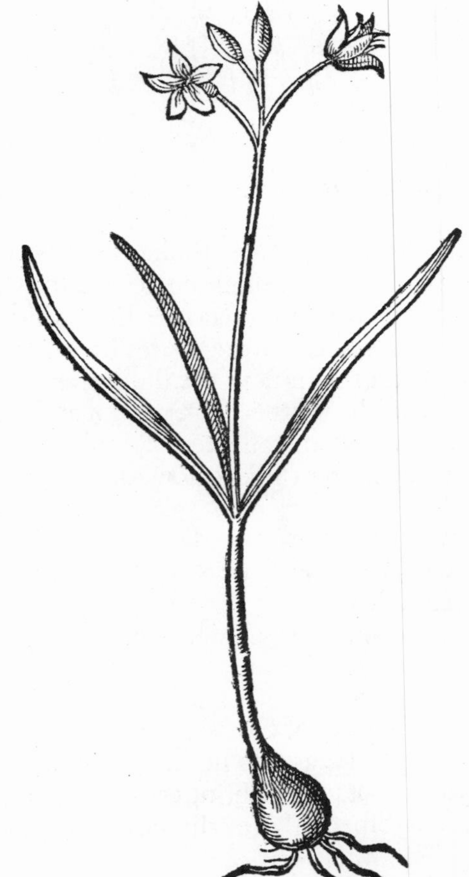

¶ *The kindes.*

THere be likewise bulbous or Onion rooted plants that do orderly succeed, whereof some are to be eaten, as Onions, Garlicke, Leekes, and Ciues; notwithstanding I am first to entreat of of those bulbed roots, whose faire and beautifull Floures are receiued for their grace and ornament in gardens and garlands: the first are the Hyacinths, whereof there is found at this day diuers sorts, differing very notably in many points, as shall be declared in their seueral descriptions.

‡ 3 *Hyacinthus stellatus bifolius.*
Two-leaued starry Iacinth.

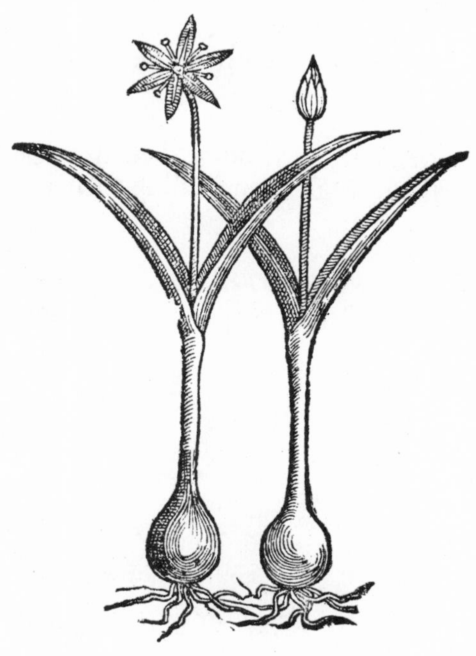

4 *Hyacinthus stellatus Lilifolius cum flore & semine.*
The Lilly leaued starry Iacinth in floure and seed.

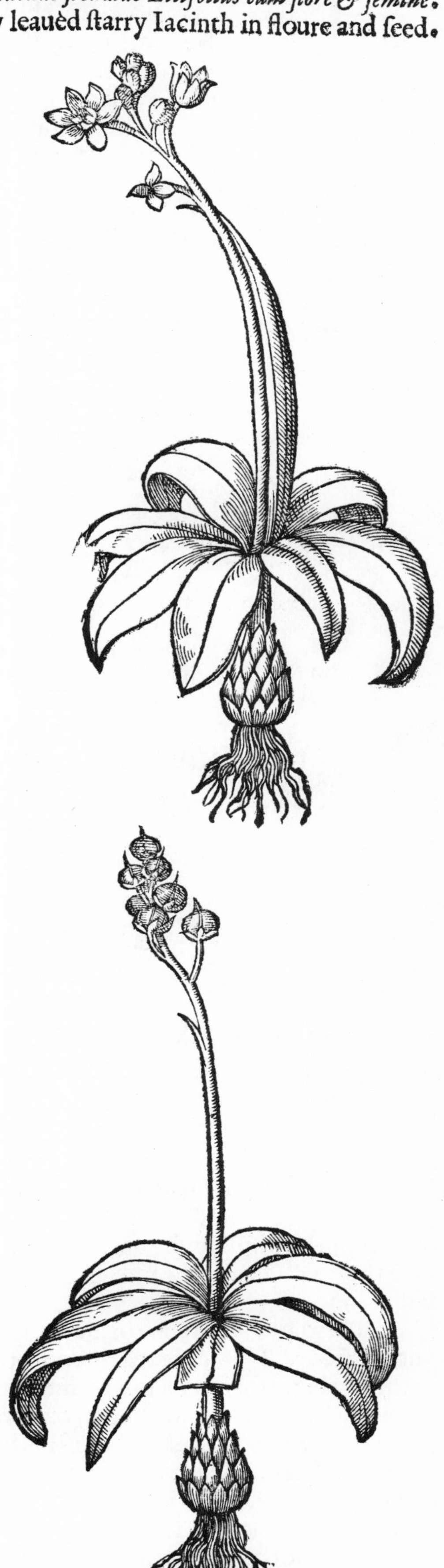

‡ 6 *Hyacinthus stellaris Byzantinus.*
The starry Iacinth of Constantinople.

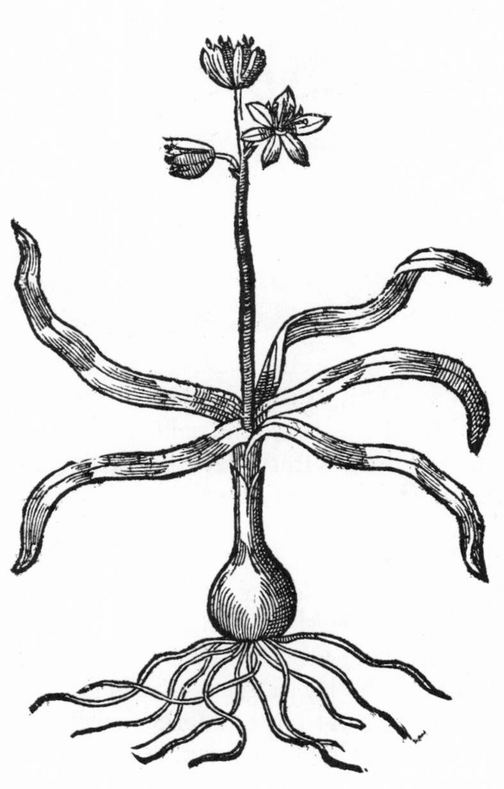

¶ The

¶ *The deſcription.*

1 The firſt kinde of Iacinth hath three very fat thicke browne leaues, hollow like a little trough, very brittle, of the length of a finger: among which ſhoot vp fat, thick browniſh ſtalkes, ſoft and very tender, and full of juyce; whereupon do grow many ſmall blew Floures conſiſting of ſix little leaues ſpred abroad like a ſtarre. The ſeed is contained in ſmall round bullets, which are ſo ponderous or heauy that they lie trailing vpon the ground. The root is bulbous or Onion faſhion, couered with browniſh ſcales or filmes.

2 There is alſo a white floured one of this kinde.

3 There is found another of this kinde which ſeldome or neuer hath more than two leaues. The roots are bulbed like the other. The Floures be whitiſh, ſtarre-faſhion, tending to blewneſſe, which I receiued of *Robinus* of Paris.

‡ 8 *Hyacinthus ſtellaris* Someri.
Somers ſtarry Iacinth.

‡ 9 *Hyacinthus ſtellatus æſtiuus major.*
The greater ſtarry Summer Iacinth.

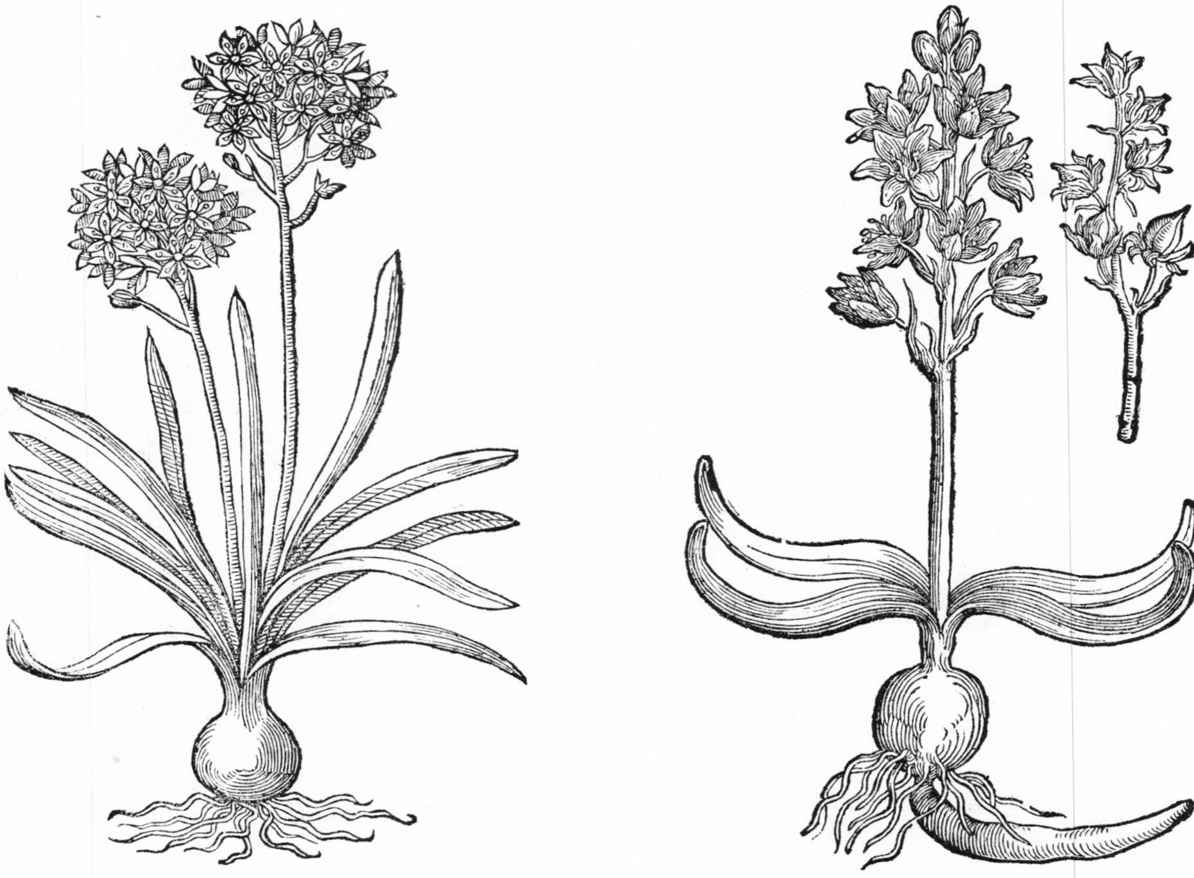

4 This kinde of Hyacinth hath many broad leaues ſpread vpon the ground, like vnto thoſe of Garden Lilly, but ſhorter. The ſtalkes do riſe out of the middeſt thereof bare, naked, and very ſmooth, an handfull high; at the top whereof do grow ſmall blew floures ſtarre-faſhion, very like vnto the precedent. The root is thicke and full of juyce, compact of many ſcaly cloues of a yellow colour.

‡ There are ſome tenne or eleuen varieties of ſtarry Iacinths, beſides theſe two mentioned by our Authour. They differ each from other either in the time of flouring (ſome of them flouring in the Spring, other ſome in Sommer) in their bigneſſe, or the colours of their floures. The leaues of moſt of them are much like to our ordinarie Iacinth, or Hare-bels, and lie ſpread vpon the ground. Their floures in ſhape reſemble the laſt deſcribed, but are vſually more in number, and ſomewhat larger. The colour of moſt of them are blew or purple, one of them excepted, which is of an Aſh colour, and is knowne by the name of *Somers* his Iacinth. I thinke it not amiſſe to giue you their vſuall names, together with ſome of their figures; for ſo you may eaſily impoſe them truly vpon the things themſelues whenſoeuer you ſhall ſee them.

5 *Hya-*

5 *Hyacinthus ſtellaris ByZantinus nigra radice, flore cæruleo.*

The blew ſtarry Iacinth of Conſtantinople, with the blacke root.

6 *Hyacinthus ſtellatus ByZantinus major flore cæruleo.*

The greater blew ſtarry Iacinth of Conſtantiple.

7 *Hyacinthus ſtellatus Byzantinus alter flore boraginis.*

The other blew ſtarry Iacinth of Conſtantinople, with Floures ſomewhat reſembling Borage.

8 *Hyacinthus ſtellaris æſtivus, ſiue exoticus* Someri *flore cinereo.*

Aſh coloured ſtarry Iacinth, or *Somers* Iacinth.

9 *Hyacinthus ſtellatus æſtivus major.*

The greater ſtarry Sommer Iacinth.

10 *Hyacinthus ſtellatus æſtivus minor.*

The leſſer ſtarry Summer Iacinth.

11 *Hyacinthus ſtellaris* Poreti *flore cæruleo ſtrijs purpureis.*

Porets ſtarry Iacinth with blew Floures, hauing purple ſtreakes alongſt their middles.

12 *Hyacinthus Hiſpanicus ſtellaris flore ſaturè cæruleo.*

The Spaniſh ſtarry Iacinth with deepe blew floures.

13 There is another ſtarry Iacinth more large and beautifull than any of theſe before mentioned. The leaues are broad and not very long, ſpread vpon the ground, and in the midſt of them there riſeth vp a ſtalke which at the top beareth a great ſpoke of faire ſtarry floures, which firſt begin to open themſelues below, and ſo ſhew themſelues by little and little to the top of the ſtalke. The vſuall ſort hereof hath blew or purple floures. There is alſo a ſort hereof which hath fleſh-coloured floures, and another with white Floures: This is called *Hyacinthus ſtellatus Peruanus,* The ſtarry Iacinth of Peru.

10 *Hyacinthus ſtellatus æſtivus minor.*
The leſſer ſtarry Summer Iacinth.

13 *Hyacinthus Peruanus.*
Hyacinth of Peru.

Thoſe who are ſtudious in varieties of Floures, and require larger deſcriptions of theſe, may haue recourſe to the Workes of the learned *Carolus Cluſius* in Latine, or to M. *Parkinſons* Worke in Engliſh, where they may haue full ſatisfaction. ‡

¶ *The place.*

The three firſt mentioned Plants grow in many places of Germany in woods and mountaines, as *Fuchſius* and *Geſner* do teſtifie: In Bohemia alſo vpon diuers bankes that are full of Herbes. In England

England we cherish most of these mentioned in this place, in our gardens, onely for the beauty of their floures.

¶ *The time.*

The three first begin to floure in the midst of Ianuarie, and bring forth their seed in May. The other floures in the Spring.

¶ *The names.*

1 The first of these Hyacinths is called *Hyacinthus stellatus*, or *Stellaris Fuchsij*, of the starre-like Floures: *Narcissus cæruleus Bockij*: of some, *Flos Martius stellatus*.

3 This by *Lobell* is thought to be *Hyacinthus Bifolius*, of *Theophrastus*: *Tragus* calls it *Narcissus cæruleus*: and *Fuchsius*, *Hyacinthus cæruleus minor mas*. Wee may call it in English, The small two leaued starrie Iacinth.

4 The Lilly Hyacinth is called *Hyacinthus Germanicus Liliflorus*, or Germane Hyacinth, taken from the countrey where it naturally groweth wilde.

‡ ¶ *The vertues.*

‡ The faculties of the starry Hyacinths are not written of by any. But the Lilly leaued Iacinth, (which growes naturally in a hill in Aquitaine called *Hos*, where the Herdmen call it *Sarahug*) is said by them to cause the heads of such cattell as feed thereon to swell exceedingly, and then kils them: which shewes it hath a maligne and poysonous qualitie. *Clus.* ‡

Chap. 78. *Of Autumne Hyacinths.*

1 *Hyacinthus Autumnalis minor.*
Small Autumne Iacinth.

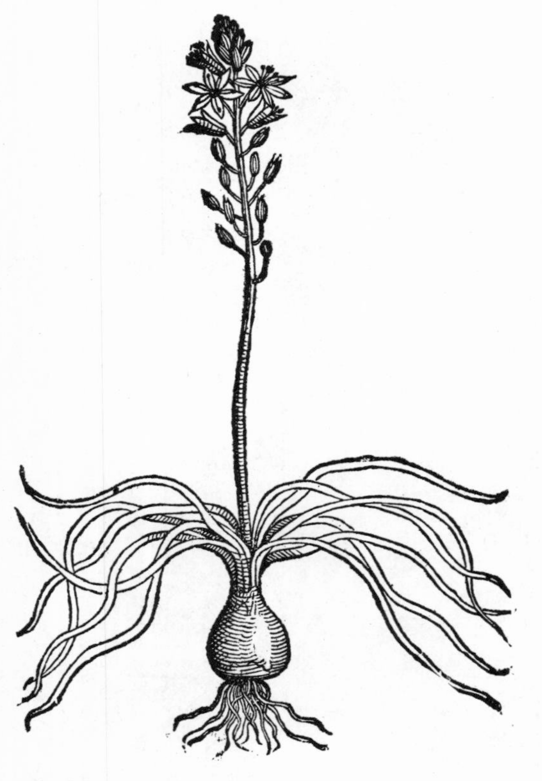

2 *Hyacinthus Autumnalis major.*
Great Autumne Iacinth.

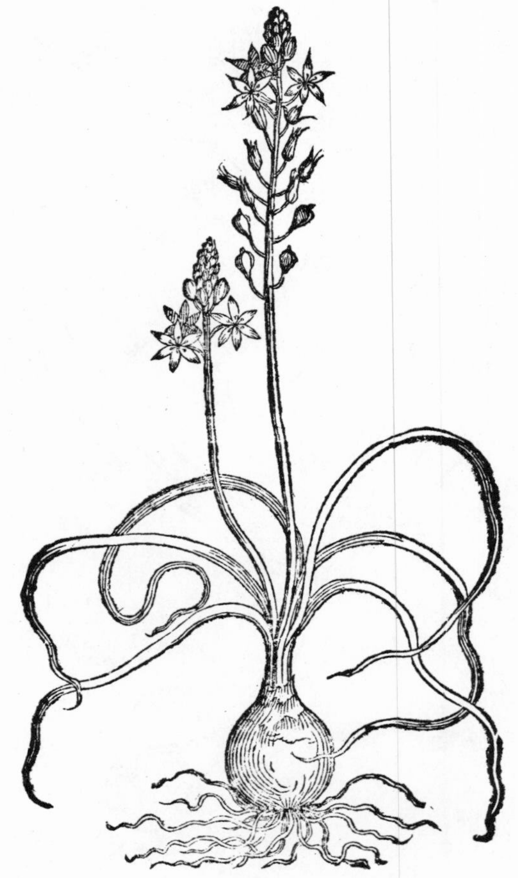

¶ *The description.*

1 Autumne Iacinth is the least of all the Iacinths: it hath small narrow grassy leaues spread abroad vpon the ground; in the middest whereof springeth vp a small naked stalke an handfull high, set from the middle to the top with many small starre-like blew floures, hauing certaine small loose chiues in the middle. The seed is blacke contained in small huskes: the root is bulbous.

2 The

2 The great Winter Iacinth is like vnto the precedent, in leaues, ſtalkes, and floures, not differing in any one point but in greatneſſe.

‡ 3 To theſe I thinke it not amiſſe to adde another ſmall Hyacinth, more different from theſe laſt deſcribed in the time of the flouring, than in ſhape. The root of it is little, ſmall, white, longiſh, with a few fibres at the bottome; the leaues are ſmall and long like the laſt deſcribed. The ſtalke, which is ſcarce an handfull high, is adorned at the top with three or foure ſtarry floures of a blewiſh Aſh colour, each floure conſiſting of ſix little leaues, with ſix chiues and their pointals, of a darke blew, and a peſtill in the middeſt. It floures in Aprill. ‡

¶ *The place.*

† The greater Autumne Iacinth growes not wilde in England, but it is to be found in ſome gardens.

The firſt or leſſer growes wilde in diuers places of England, as vpon a banke by the Thames ſide betweene Chelſey and London. †

¶ *The time.*

They floure in the end of Auguſt, and in September, and ſometimes after.

¶ *The names.*

1 The firſt is called *Hyacinthus Autumnalis minor*, or the leſſer Autumne Iacinth, and Winter Iacinth.

2 The ſecond, *Hyacinthus Autumnalis major*, the great Autumne Iacinth, or Winter Iacinth.

3 This is called by *Lobell*, *Hyacinthus parvulus ſtellaris vernus*, The ſmall ſtarry Spring Iacinth.

CHAP. 79. *Of the Engliſh Iacinth, or Hare-Bels.*

1 *Hyacinthus Anglicus.* Engliſh Hare-bels.

2 *Hyacinthus albus Anglicus.* White Engliſh Hare-bels

¶ *The deſcription.*

1 The blew Hare-bels or Engliſh Iacinth is very common throughout all England. It hath long narrow leaues leaning towards the ground, among the which ſpring vp naked

or

or bare stalkes loden with many hollow blew Floures, of a strong sweet smell, somewhat stuffing the head: after which come the coddes or round knobs, containing a great quantitie of small blacke shining seed. The root is bulbous, full of a slimy glewish juyce, which wil serue to set feathers vpon arrowes in stead of glew, or to paste bookes with: whereof is made the best starch next vnto that of Wake-robin roots.

4 *Hyacinthus Orientalis cæruleus.*
The blew Orientall Iacinth.

5 *Hyacinthus Orientalis Polyanthos.*
Many floured Orientall Iacinth.

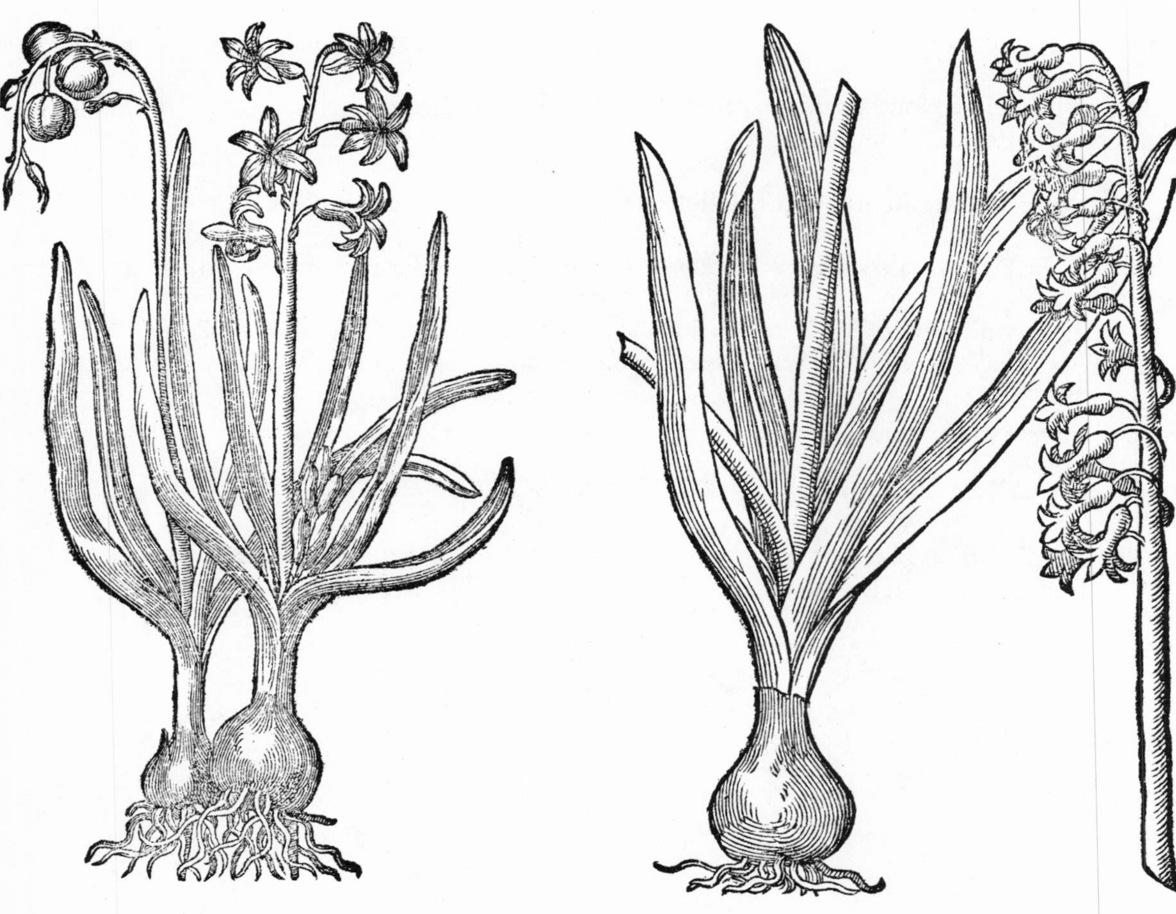

2 The white English Iacinth is altogether like vnto the precedent, sauing that the leaues hereof are somewhat broader, the Floures more open, and very white of colour.

3 There is found wilde in many places of England, another sort, which hath Floures of a faire carnation colour, which maketh a difference from the other.

‡ There are also sundry other varieties of this sort, but I thinke it vnnecessarie to insist vpon them, their difference is so little, consisting not in their shape, but in the colour of their Floures. ‡

The blew Hare-bels grow wilde in woods, copses, and in the borders of fields euery where thorow England.

The other two are not so common, yet do they grow in the woods by Colchester in Essex, in the fields and woods by South-fleet, neere vnto Graues-end in Kent, as also in a piece of ground by Canturbury called the Clapper, in the fields by Bathe, about the woods by Warrington in Lancashire, and other places.

¶ *The time.*

They floure from the beginning of May vnto the end of Iune.

¶ *The names.*

1 The first of our English Hyacinths is called *Hyacinthus Anglicus*, for that it is thought to grow more plentifully in England than elsewhere; of *Dodonæus*, *Hyacinthus non scriptus*, or the vnwritten Iacinth.

2 The second, *Hyacinthus Belgicus candidus*, or the Low-Countrey Hyacinth with white Floures.

‡ 3 This third is called *Hyacinthus Anglicus*, *aut Belgicus Flore incarnato*, Carnation Harebels.

‡ 6 *Hyacinthus*

‡ 6 *Hyacinthus Orientalis polyanthos alter.*
The other many-Floured Oriental Iacinth.

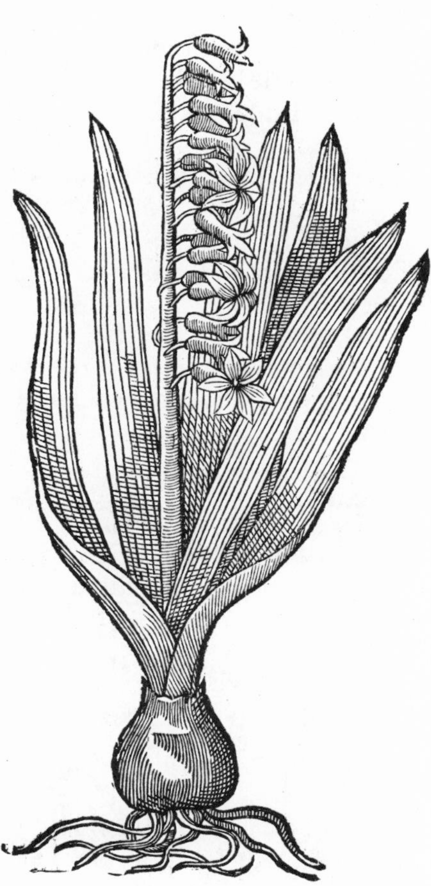

‡ 7 *Hyacinthus Orientalis purpuro rubeus.*
Reddiſh purple Oriental Iacinth.

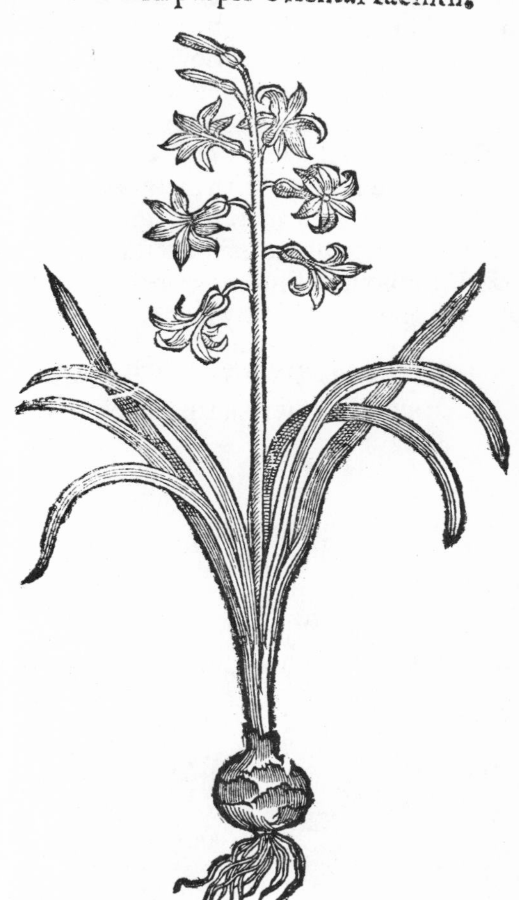

‡ 8 *Hyacinthus Orientalis albus.*
White Oriental Iacinth.

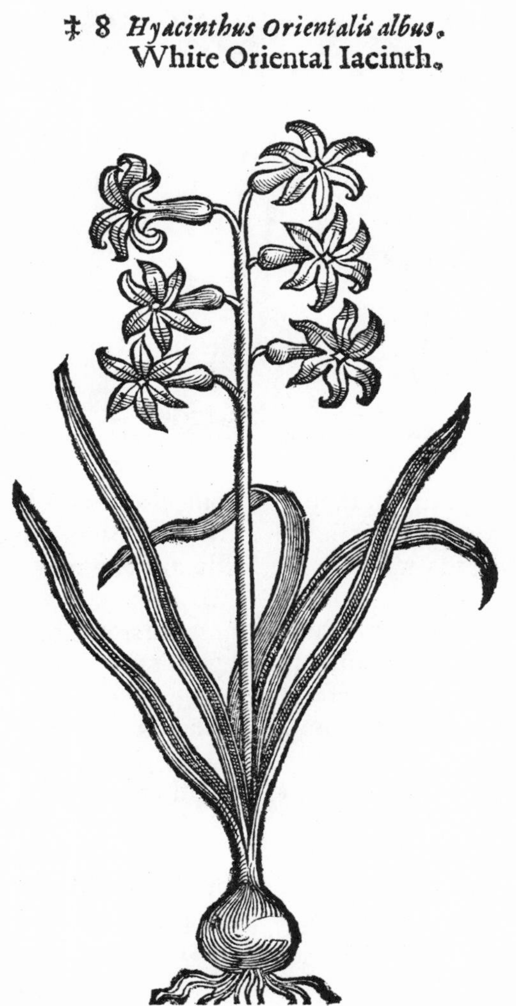

‡ 9 *Hyacinthus Brumalis.*
Winter Iacinth.

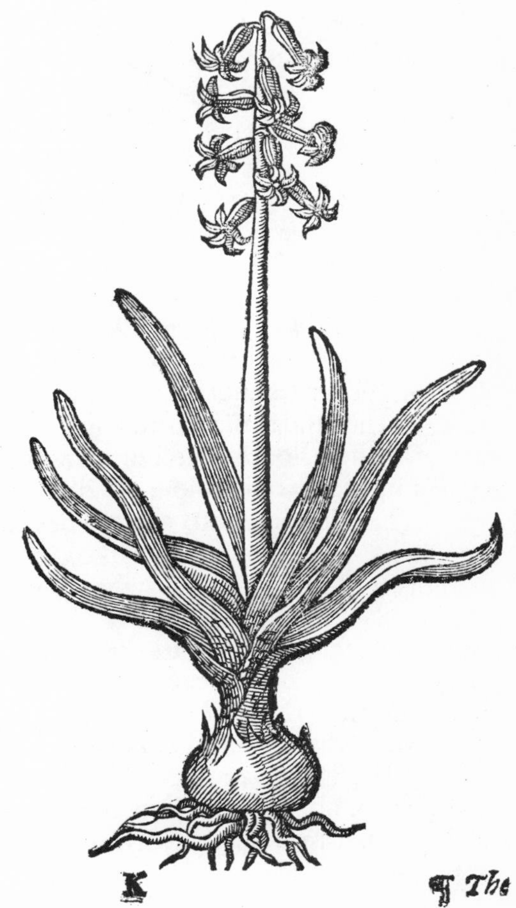

¶ *The Description.*

4 The Orientall Iacinth hath great leaues, thicke, fat, and full of juyce, deepely hollowed in the middle like a trough: from the middle of those leaues riseth vp a stalke two hands high, bare without leaues, very smooth, soft, and full of juice, loden toward the top with many faire blew Floures, hollow like a bell, greater than the English Iacinth, but otherwise like them. The root is great, bulbous, or Onion fashion, couered with many scaly reddish filmes or pillings, such as couer Onions.

5 The Iacinth with many Floures (for so doth the word *Polyanthos* import) hath very many large and broad leaues, short and very thicke, fat, or full of slimy juyce: from the middle whereof rise vp strong thicke grosse stalkes, bare and naked, set from the middle to the top with many blew or skie coloured Floures growing for the most part vpon one side of the stalke. The root is great, thicke, and full of slimy juyce.

‡ 10 *Hyacinthus Orientalis caule folioso*
Orientall Iacinth with leaues on the stalke.

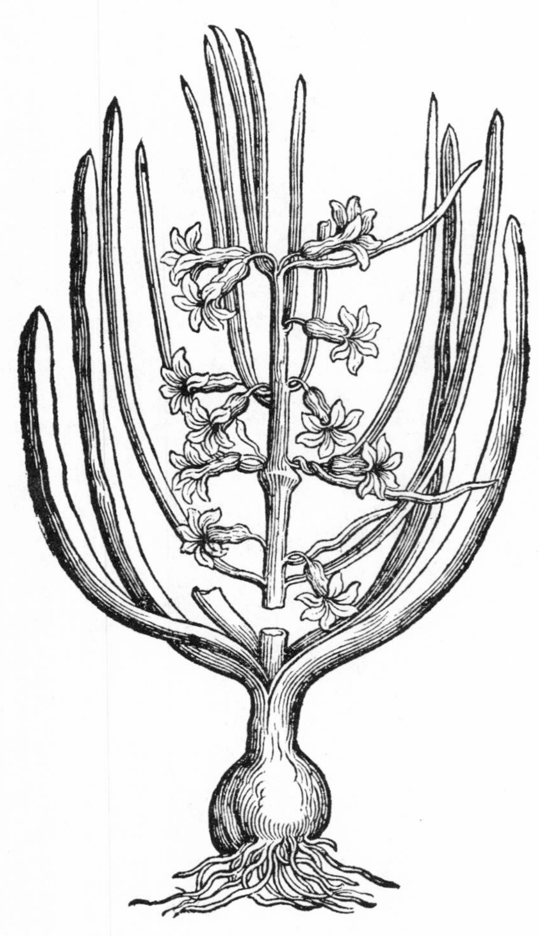

‡ 11 *Hyacinthus Orientalis flore pleno.*
The double floured Oriental Iacinth.

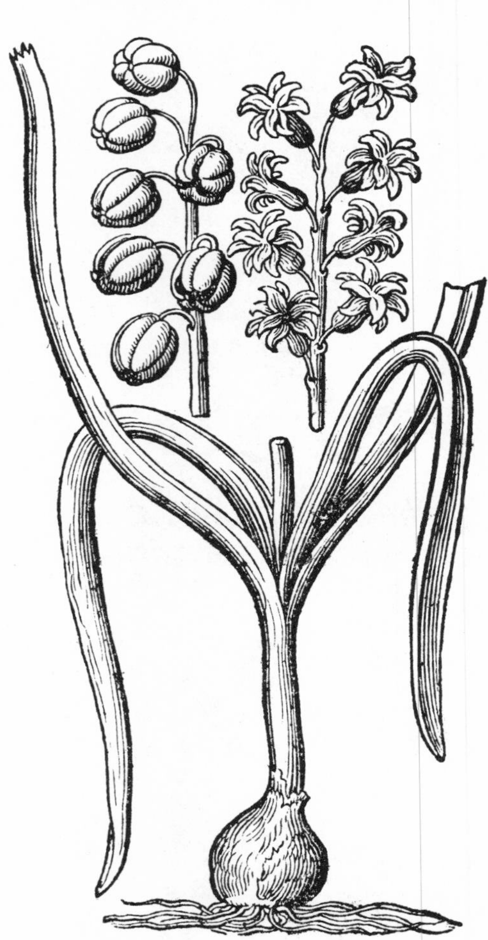

‡ *6* There is another like the former in each respect, sauing that the floures are wholly white on the inside, and white also on the outside, but three of the out-leaues are of a pale whitish yellow. These floures smell sweet as the former, and the heads wherein the seeds are contained are of a lighter greene colour. ‡

7 There is come vnto vs from beyond the seas diuers other sorts, whose figures are not extant with vs; of which there is one like vnto the first of these Oriental Iacinths, sauing that the floures thereof are purple coloured; whence it is termed *Hyacinthus purpuro rubeus*.

8 Likewise there is another called *Orientalis albus*, differing also from the others in colour of the floures, for that these are very white, and the others blew.

9 There is another called *Hyacinthus Brumalis*, or winter Iacinth: it is like the others in shape, but differeth in the time of flouring.

‡ 10 There is another Hyacinth belonging rather to this place than any other, for that in root, leaues, floures, and seeds it resembles the first described Oriental Iacinth; but in one respect it differs not onely from them, but also from all other Iacinths: which is, it hath a leauie stalke, hauing sometimes one, and otherwhiles two narrow long leaues comming forth at the bottome of the

‡ 14 *Hyacinthus obsoleto flore Hispanicus major.*
The greater dusky floured Spanish Iacinth.

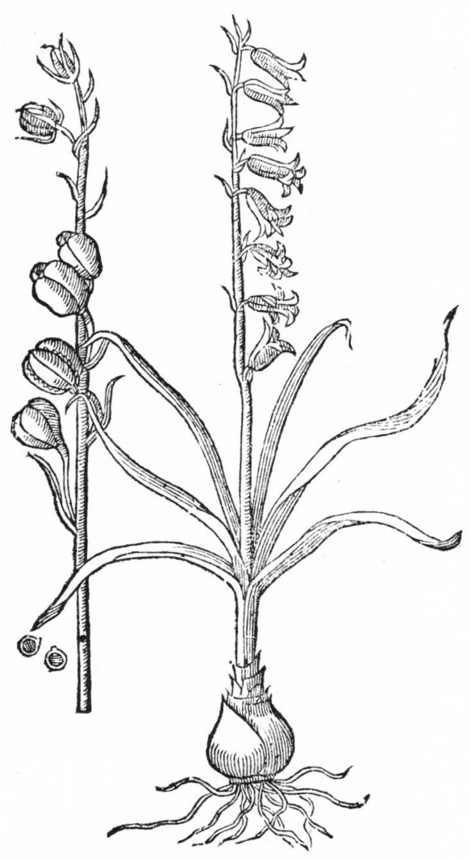

‡ 15 *Hyacinthus minor Hispanicus.*
The lesser Spanish Iacinth.

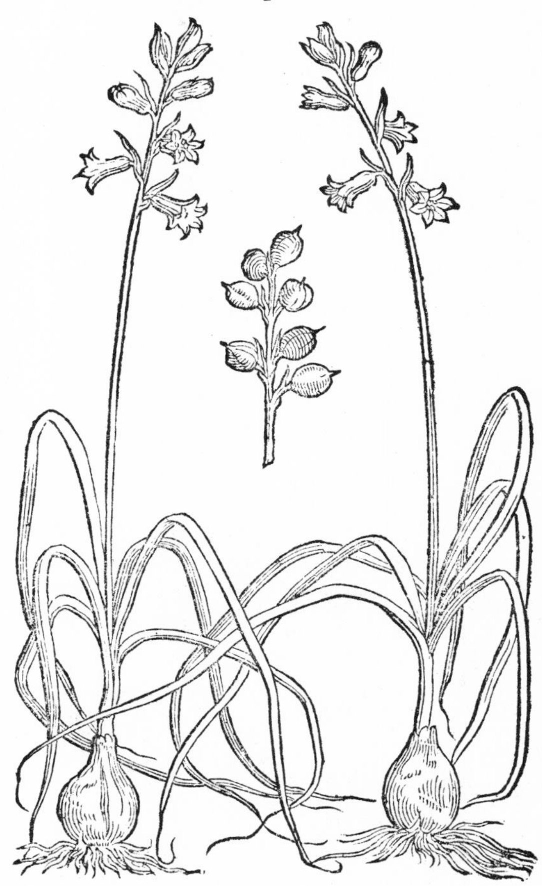

‡ 16 *Hyacinthus Indicus tuberosus.*
The tuberous rooted Indian Iacinth.

the setting on of the floure. Whereupon *Clusius* calls it *Hyacinthus Orientalis caule folioso:* That is, the Oriental Hyacinth with leaues on the stalke.

¶ *Of double floured Oriental Hyacinths.*

Of this kindred there are two or three more varieties, whereof I wil giue you the description of the most notable, and the names of the other two; which, with that I shall deliuer of this, may serue for sufficient description. The first of these (which *Clusius* calls *Hyacinthus Orientalis subvirescente flore,* or, the greenish floured double Orientall Iacinth) hath leaues, roots, and seeds like vnto the formerly described Oriental Iacinths; but the floures (wherin the difference consists) are at the first, before they be open, greene, and then on the out side next to the stalke of a whitish blew; and they consist of six leaues whose tips are whitish, yet retaining some manifest greenes: then out of the midst of the floure comes forth another floure consisting of three leaues, whitish on their inner side, yet keeping the great veine or streake vpon the outer side, each floure hauing in the middle a few chiues with blackish pendants. It floures in Aprill.

 12 This

12 This varietie of the last described is called *Hyacinthus Orientalis flore cæruleo pleno*, The double blew Orientall Iacinth

13 This, *Hyacinthus Orientalis candidissimus flore pleno*, The milke-white double Orientall Iacinth.

14 This, which *Clusius* calls *Hyacinthus obsoletior Hispanicus*, hath leaues somewhat narrower, and more flexible than the *Muscari*, with a white veine running alongst the inside of them: among these leaues there riseth vp a stalke of some foot high, bearing some fifteene or sixteene floures, more or lesse, in shape much like the ordinarie English, consisting of six leaues, three standing much out. and the other three little or nothing. These floures are of a very dusky colour, as it were mixt with purple, yellow, and greene: they haue no smell. The seed, which is contained in triangular heads, is smooth, blacke, scaly, and round. It floures in Iune.

15 The lesser Spanish Hyacinth hath leaues like the Grape-floure, and small floures shaped like the Orientall Iacinth, some are of colour blew, and other some white. The seeds are contained in three cornered seed-vessels. I haue giuen the figure of the white and blew together, with their seed-vessels.

16 This Indian Iacinth with the tuberous root (saith *Clusius*) hath many long narrow sharpe pointed leaues spread vpon the ground, being somewhat like to those of Garlicke, and in the middest of these rise vp many round firme stalkes of some two cubits high, and oft times higher, sometimes exceeding the thicknesse of ones little finger; which is the reason that oftentimes, vnlesse they be borne vp by something, they lie along vpon the ground. These stalkes are at certaine spaces ingirt with leaues which end in sharpe points. The tops of these stalkes are adorned with many white floures, somewhat in shape resembling those of the Orientall Iacinth. The roots are knotty or tuberous, with diuers fibres comming out of them. ‡

¶ *The place.*

These kindes of Iacinths haue beene brought from beyond the Seas, some out of one countrey, and some out of others, especially from the East countries, whereof they tooke their names *Orientalis*.

¶ *The time.*

They floure from the end of Ianuarie vnto the end of Aprill.

¶ *The nature.*

The Hyacinths mentioned in this Chapter do lightly cleanse and binde; the seeds are dry in the third degree; but the roots are dry in the first degree, and cold in the second.

¶ *The vertues.*

A The Root of Hyacinth boyled in Wine and drunke, stoppeth the belly, prouoketh vrine, and helpeth against the venomous bitings of the field Spider.

B The seed is of the same vertue, and is of greater force in stopping the laske and bloudy flix. Being drunke in wine it preuaileth against the falling sicknesse.

C The roots, after the opinion of *Dioscorides*, being beaten and applied with white Wine, hinder or keepe backe the growth of haires.

D ‡ The seed giuen with Southerne-wood in Wine is good against the Iaundice. ‡

Chap. 80. *Of Faire haired Iacinth.*

¶ *The Description.*

1 The Faire haired Iacinth hath long fat leaues, hollowed alongst the inside, trough fashion, as are most of the Hyacinths, of a darke greene colour tending to rednesse. The stalke riseth out of the middest of the leaues, bare and naked, soft and full of slimie juyce, which are beset round about with many small floures of an ouerworne purple colour: The top of the spike consisteth of a number of faire shining purple floures, in manner of a tuft or bush of haires, whereof it tooke his name *Comosus*, or faire haired. The seed is contained in small bullets, of a shining blacke colour, as are most of those of the Hyacinths. The roots are bulbous or Onion fashion, full of slimy juyce, with some hairy threads fastned vnto their bottome.

2 White haired Iacinth differeth not from the precedent in roots, stalkes, leaues, or seed. The floures hereof are of a darke white colour, with some blacknesse in the hollow part of them, which setteth forth the difference.

3 Of this kinde I receiued another sort from Constantinople, resembling the first hairy Hyacinth very notably: but differeth in that, that this is altogether greater, as well in leaues, roots, and floures, as also is of greater beauty without all comparison.

1 *Hyacinthus comosus.*
Faire haired Iacinth.

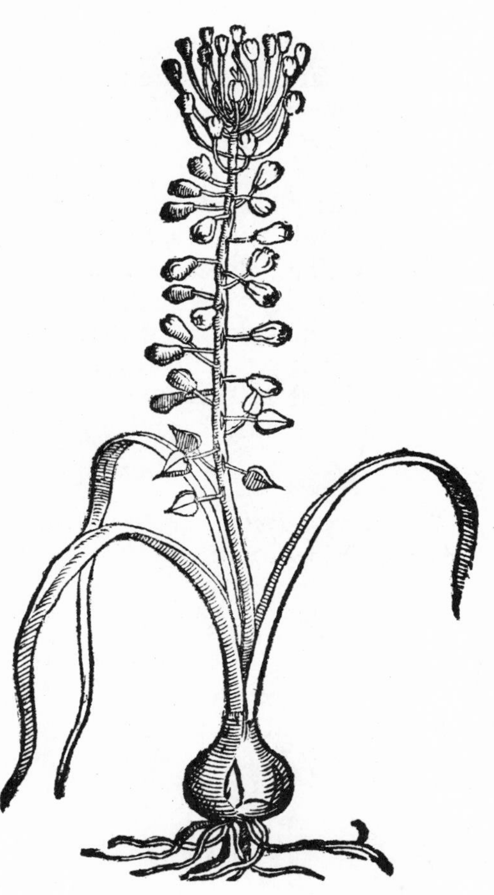

2 *Hyacinthus comosus albus.*
White haired Iacinth.

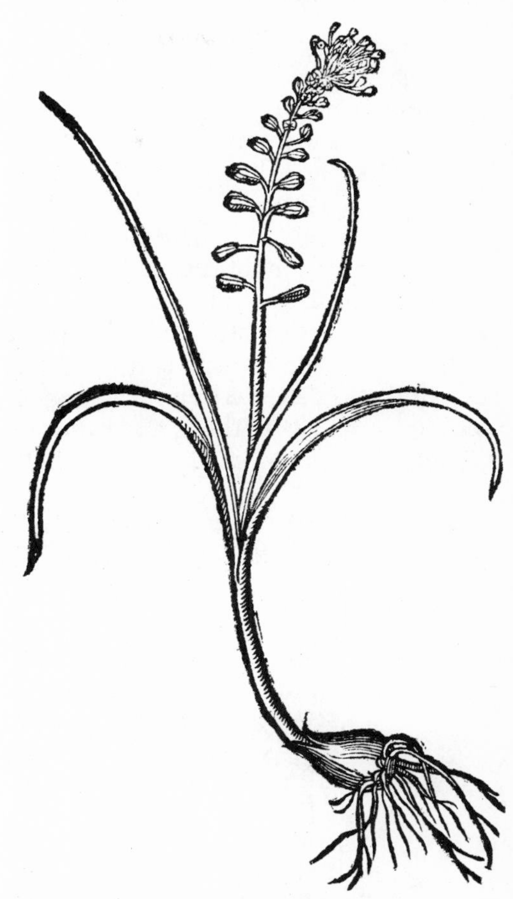

‡ 3 *Hyacinthus comosus Bizantinus.*
Faire-haired Iacinth of Constantinople.

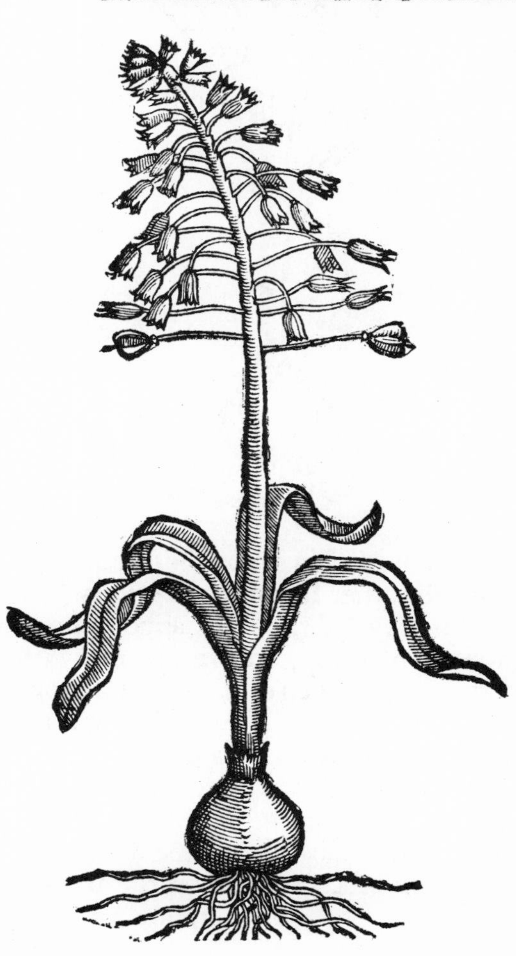

‡ 5 *Hyacinthus comosus ramosus elegantior.*
Faire curld-haired branched Iacinth.

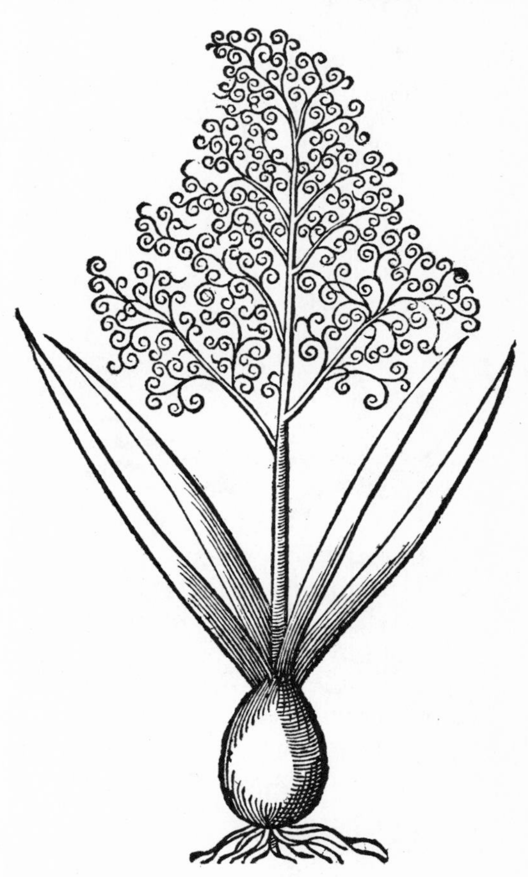

‡ 4 There are two other more beautifull haired Iacinths nouriſhed in the gardens of our prime Floriſts. The firſt of theſe hath roots and leaues reſembling the laſt deſcribed: the ſtalke commonly riſeth to the height of a foot, and it is diuided into many branches on euery ſide, which are ſmall and threddy; and then at the end as it were of theſe threddy branches there come forth many ſmaller threds of a darke purple colour, and theſe ſpread and diuaricate themſelues diuers wayes, much after the manner of the next deſcribed; yet the threds are neither of ſo pleaſing a colour, neither ſo many in number, nor ſo finely curled. This is called *Hyacinthus comoſus ramoſus purpureus*, The faire haired branched Iacinth.

5 This is a moſt beautiful and elegant plant, and in his leaues and roots he differs little from the laſt deſcribed; but his ſtalke, which is as high as the former, is diuided into very many ſlender branches, which ſubdiuided into great plenty of curled threads variouſly ſpread abroad, make a very pleaſant ſhew. The colour alſo is a light blew, and the floures vſually grow ſo, that they are moſt dilated at the bottome, and ſo ſtraiten by little and little after the manner of a Pyramide. Theſe floures keepe their beautie long, but are ſucceeded by no ſeeds that yet could be obſerned. This by *Fabius Columna* (who firſt made mention hereof in writing) is called *Hyacinthus Sanneſius panniculoſa coma*: By others, *Hyacinthus comoſus ramoſus elegantior*, The faire curld-haire Iacinth.

Theſe floure in May. ‡

6 *Hyacinthus botryoides cæruleus.*
Blew Grape-floure.

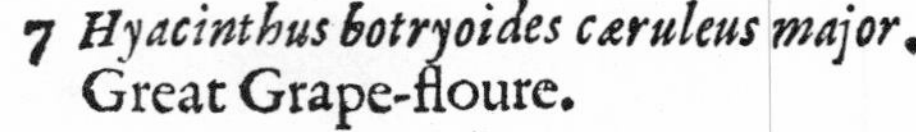

7 *Hyacinthus botryoides cæruleus major.*
Great Grape-floure.

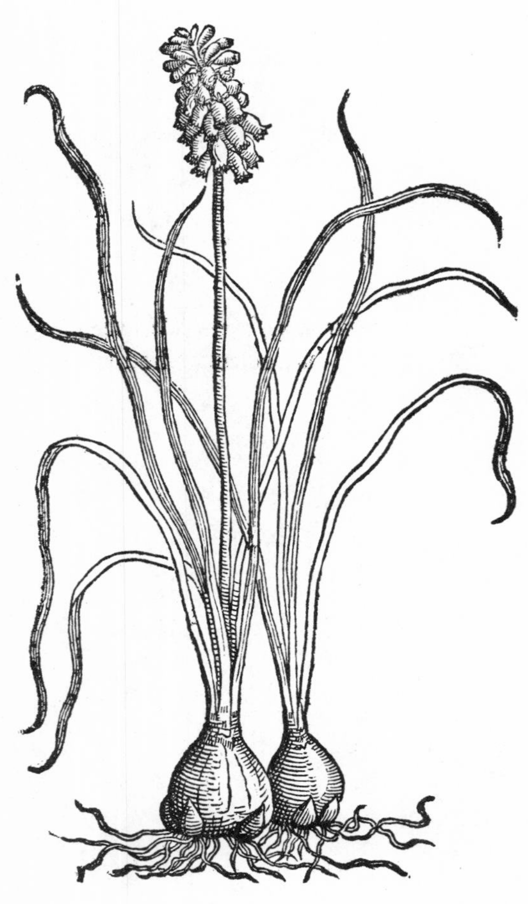

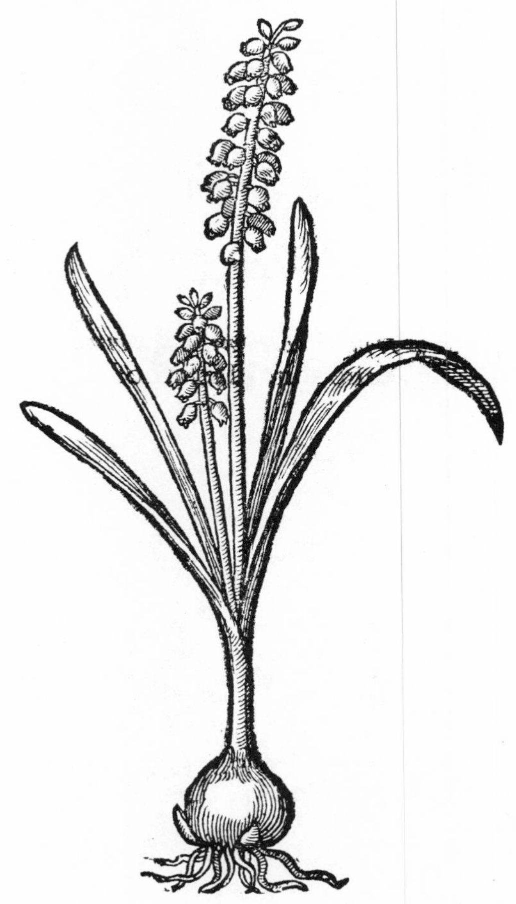

6 The ſmall Grape floure hath many long fat and weake leaues trailing vpon the ground, hollow in the middle like a little trough, full of ſlimie juyce like the other Iacinths; amongſt which come forth thicke ſoft ſmooth and weake ſtalkes, leaning this way and that way, as not able to ſtand vpright by reaſon it is ſurcharged with very heauy floures on his top, conſiſting of many little bottle-like blew floures, cloſely thruſt or packed together like a bunch of grapes, of a ſtrong ſmell, yet not vnpleaſant, ſomewhat reſembling the ſauour of the Orange. The root is round and bulbous, ſet about with infinite young cloues or roots, whereby it greatly increaſeth.

7 The great Grape-floure is very like vnto the ſmaller of his kinde. The difference conſiſteth, in that this plant is altogether greater, but the leaues are not ſo long.

8 The sky-coloured Grape-floure hath a few leaues in reſpect of the other Grape-floures, the which are ſhorter, fuller of juyce, ſtiffe and vpright, whereas the others traile vpon the ground.

The

The floures grow at the top, thruſt or packt together like a bunch of Grapes, of a pleaſant bright sky colour, euery little bottle-like floure ſet about the hollow entrance with ſmall white ſpots not eaſie to be perceiued. The roots are like the former.

8 *Hyacinthus Botryoides cæruleus major.*
Great Grape-floure.

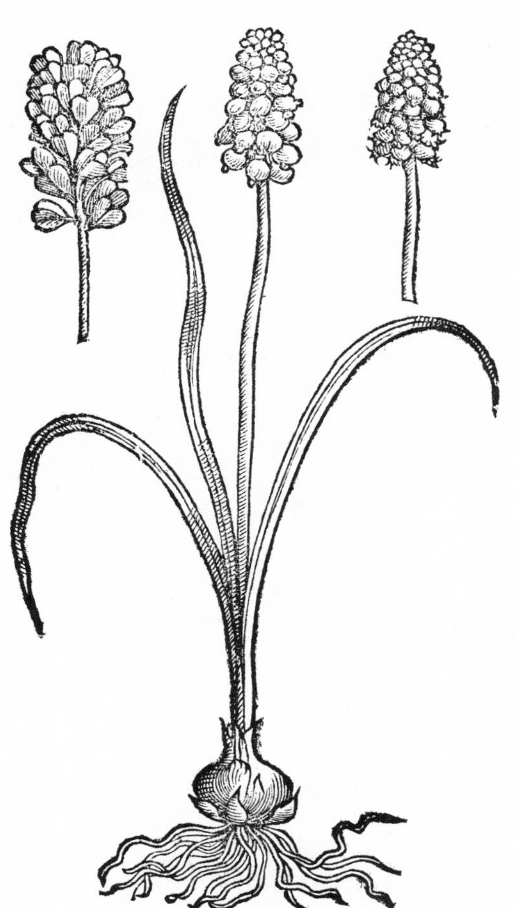

9 The white Grape-floure differeth not from the sky-coloured Iacinth, but in colour of the floure: for this Iacinth is of a pleaſant white colour tending to yellownes, tipped about the hollow part with White, whiter than White it ſelfe, otherwiſe there is no difference.

¶ *The Place.*

Theſe plants are kept in gardens for the beautie of their floures, wherewith our London gardens do abound.

¶ *The Time.*

They floure from Februarie to the end of May.

¶ *The Names.*

The Grape-floure is called *Hyacinthus Botryoides*, and *Hyacinthus Neotricorum Dodonæi*: of ſome, *Bulbus Eſculentus*, *Hyacinthus ſylueſtris cordi*, *Hyacinthus exiguus Tragi*. Some iudge them to be *Bulbinæ*, of *Pliny*.

† *The faire haired Iacinth deſcribed in the firſt place is the* Hyacinthus of *Dioſcorides and the Antients*. †

¶ *The Nature and Vertues.*

† *The vertues ſet downe in the precedent Chapter properly belong to that kinde of Hyacinth which is deſcribed in the firſt place in this Chapter.*

CHAP. 81. *Of Muſcari, or Musked Grape-floure.*

¶ *The Deſcription.*

1 YEllow Muſcarie hath fiue or ſix long leaues ſpread vpon the ground, thicke, fat, and full of ſlimie juyce, turning and winding themſelues crookedly this way & that way, hollowed alongſt the middle like a trough, as are thoſe of faire haired Iacinth, which at the firſt budding or ſpringing vp are of a purpliſh colour, but being growne to perfection, become of a darke greene colour: amongſt the which leaues riſe vp naked, thicke, and fat ſtalkes, infirme and weake in reſpect of the thickneſſe and greatneſſe thereof, lying alſo vpon the ground as do the leaues; ſet from the middle to the top on euery ſide with many yellow floures, euerie one made like a ſmall pitcher or little box, with a narrow mouth, exceeding ſweet of ſmell like the ſauor of muske, whereof it tooke the name *Muſcari*. The ſeed is incloſed in puffed or blowne vp cods, confuſedly made without order, of a fat and ſpongeous ſubſtance, wherein is contained round blacke ſeed. The root is bulbous or onion faſhion, whereunto are annexed certaine fat and thicke ſtrings like thoſe of Dogs graſſe.

2 Aſh-coloured *Muſcari* or grape-floure, hath large and fat leaues like the precedent, not differing in any point, ſauing that theſe leaues at their firſt ſpringing vp are of a pale dusky colour like aſhes. The floures are likewiſe ſweet, but of a pale bleake colour, wherein conſiſteth the difference.

¶ *The*

1 *Muscari flavum.*
Yellow musked Grape-floure.

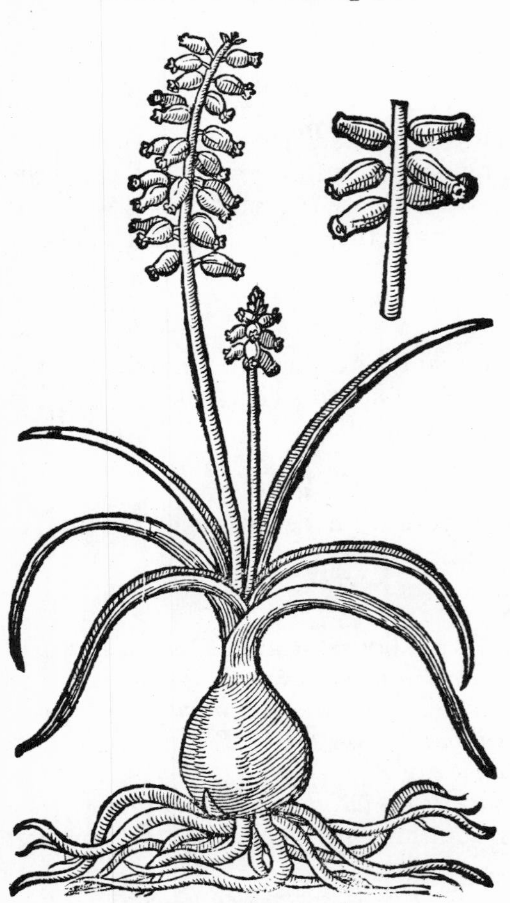

2 *Muscari Clusij.*
Ash-coloured Grape-floure.

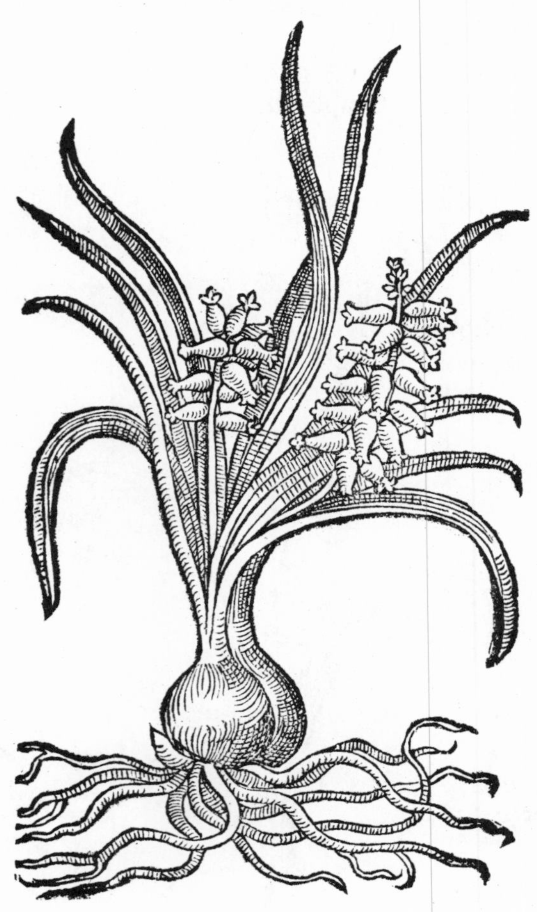

Muscari caulis siliquis onustus.
The stalke of Muscari hanged with the seed-vessels.

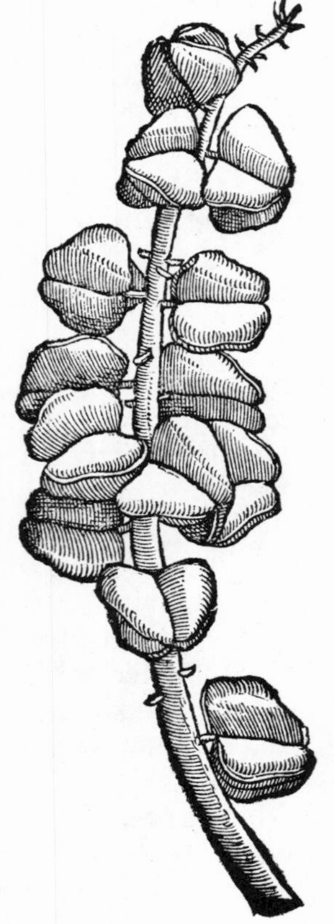

¶ *The Place.*

These Plants came from beyond the Thracian Bosphorus, out of Asia, and from about Constantinople, and by the meanes of Friends haue been brought into these parts of Europe, whereof our London gardens are possessed.

¶ *The Time.*

They floure in March and Aprill, and sometimes after.

¶ *The Names.*

They are called generally *Muscari*: In the Turky Tongue, *Muschoromi, Muscurimi, Tipcadi,* and *Dipcadi,* of their pleasant sweet smell: Of *Matthiolus, Bulbus Vomitorius.* These plants may be referred vnto the Iacinths, whereof vndoubtedly they be kindes.

¶ *The Nature and Vertues.*

There hath not as yet any thing beene touched concerning the nature or vertues of these Plants, onely they are kept and maintained in gardens for the pleasant smell of their floures, but not for their beauty, for that many stinking field floures do in beautie farre surpasse them. But it should seem that *Matthiolus* called them *Vomitorius,* in that he supposed they procure vomiting; which of other Authors hath not bin remembred.

Chap.

CHAP. 82. *Of Woolly Bulbus.*

Bulbus Eriophorus.
Woolly Iacinth.

¶ *The Description.*

THere hath fallen out to be here inserted a bulbous plant consisting of many Bulbes, which hath passed currant amongst all our late Writers. The which I am to set forth to the view of our Nation, as others haue done in sundry languages to theirs, as a kind of the Iacinths, which in roots and leaues it doth very wel resemble; called of the Grecians, Ἐριόφορος: in Latine, *Laniferus*, because of his aboundance of Wooll-resembling substance, wherewith the whole Plant is in euery part full fraught, as well roots, leaues, as stalkes. The leaues are broad, thicke, fat full of juyce, and of a spider-like web when they be broken. Among these leaues riseth vp a stalke two cubits high, much like vnto the stalke of Squilla or Sea-Onion; and from the middle to the top it is beset round about with many small starre-like blew floures without smell, very like to the floures of Asphodill; beginning to floure at the bottome, and so vpward by degrees, whereby it is long before it hath done flouring: which floures the learned Physitian of Vienna, *Iohannes Aicholzius*, desired long to see; who brought it first from Constantinople, and planted it in his Garden, where he nourished it tenne yeares with great curiositie: which time being expired, thinking it to be a barren plant, he sent it to *Carolus Clusius*, with whom in some few yeres it did beare such floures as before described, but neuer since to this day. This painefull Herbarist would gladly haue seene the seed that should succeed these floures; but they being of a nature quickly subiect to perish, decay, and fade, began presently to pine away, leauing onely a few chaffie and idle seed-vessels without fruit. My selfe hath beene possessed with this plant at the least twelue yeares, whereof I haue yearely great encrease of new roots, but I did neuer see any token of budding or flouring to this day: notwithstanding I shall be content to suffer it in some base place or other of my garden, to stand as the cipher o at the end of the figures, to attend his time and leisure, as those men of famous memorie haue done. Of whose temperature and vertues there hath not any thing beene said, but kept in gardens to the end aforesaid.

CHAP. 83. *Of two feigned Plants.*

¶ *The Description.*

1 I Haue thought it conuenient to conclude this historie of the Hyacinths with these two bulbous Plants, receiued by tradition from others, though generally holden for feigned and adulterine. Their pictures I could willingly haue omitted in this historie, if the curious eye could elsewhere haue found them drawne and described in our English Tongue: but because I finde them in none, I will lay them downe here, to the end that it may serue for excuse to others who shall come after, which list not to describe them, being as I said condemned for feined and adulterine, nakedly drawne onely. And the first of them is called *Bulbus* ἐριόφορος: by others, *Bulbus Bombicinus Commentitius*. The description consisteth of these points, *viz.* The floures (saith the Author) are no lesse strange than wonderfull. The leaues and roots are like to those of Hyacinths,

cinths, which hath cauſed it to occupie this place. The floures reſemble the Daffodils or Narciſſus. The whole plant conſiſteth of a woolly or flockie matter: which deſcription with the Picture was ſent vnto *Dodonæus* by *Iohannes Aicholzius*. It may be that *Aicholzius* receiued inſtructions from the Indies, of a plant called in Greeke τειζώδης, which groweth in India, whereof *Theophraſtus* and *Athenæus* do write in this manner, ſaying, The floure is like the *Narciſſus*, conſiſting of a flockie or woolly ſubſtance, which by him ſeemeth to be the deſcription of our bombaſt Iacinth.

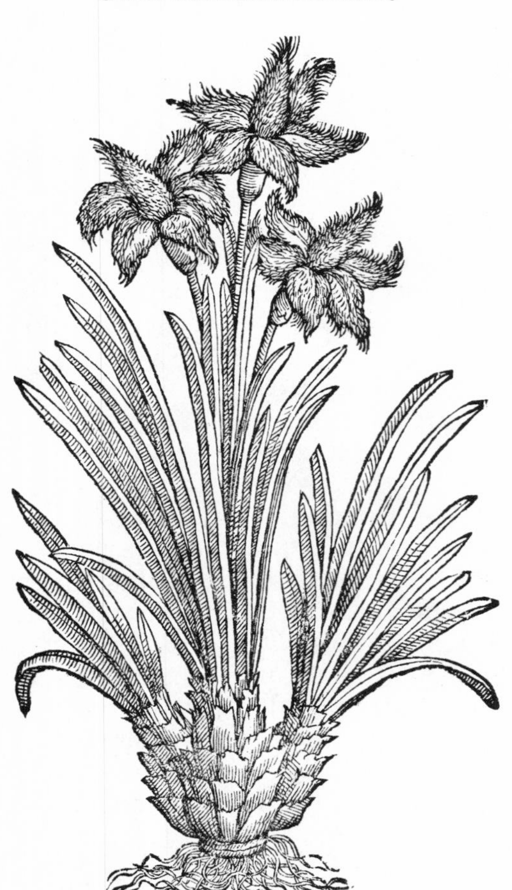

1 *Bulbus Bombicinus Commentitius.*
Falſe bumbaſte Iacinth.

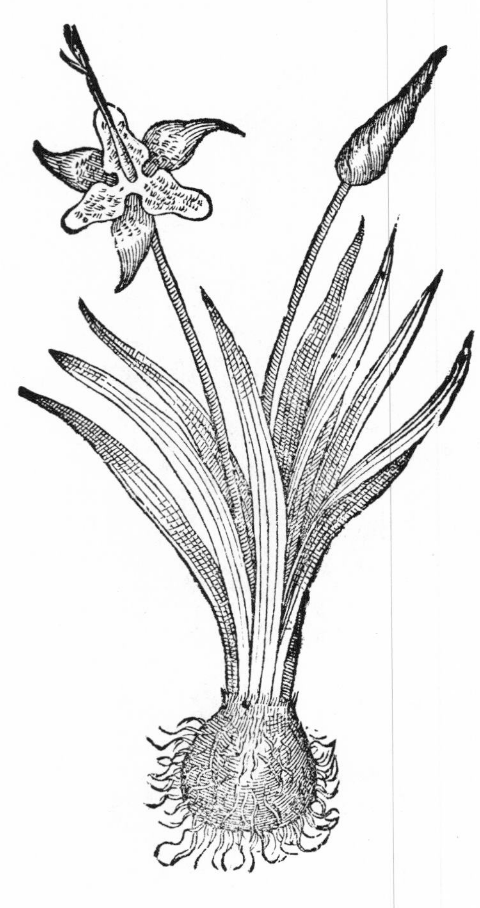

2 *Tigridis flos.*
The floure of Tygris.

2 The ſecond feigned picture hath beene taken of the Diſcouerer and others of later time, to be a kinde of Dragons not ſeene of any that haue written thereof; which hath moued them to thinke it a feigned picture likewiſe; notwithſtanding you ſhall receiue the deſcription thereof as it hath come to my hands. The root (ſaith my Author) is bulbous or Onion faſhion, outwardly blacke; from the which ſpring vp long leaues, ſharpe pointed, narrow, and of a freſh greene colour: in the middeſt of which leaues riſe vp naked or bare ſtalkes, at the top whereof groweth a pleaſant yellow floure, ſtained with many ſmall red ſpots here and there confuſedly caſt abroad: and in the middeſt of the floure thruſteth forth a long red tongue or ſtile, which in time groweth to be the cod or ſeed-veſſell, crooked or wreathed, wherein is the ſeed. The vertues and temperature are not to be ſpoken of, conſidering that we aſſuredly perſuade our ſelues that there are no ſuch plants, but meere fictions and deuices, as we terme them, to giue his friend a gudgeon.

‡ Though theſe two haue beene thought commentitious or feigned, yet *Bauhinus* ſeemeth to vindicate the latter, and *Iohn Theodore de Bry* in his *Florilegium* hath ſet it forth. He giues two Figures thereof, this which we here giue you being the one; but the other is farre more elegant, and better reſembles a naturall plant. The leaues (as *Bauhine* ſaith) are like the ſword-flag, the root like a leeke, the floures (according to *De Bries* Figure) grow ſometimes two or three of a ſtalke: the floure conſiſts of two leaues, and a long ſtile or peſtill: each of theſe leaues is diuided into three parts, the vttermoſt being broad and large, and the innermoſt much narrower and ſharper: the tongue or ſtile that comes forth of the midſt of the floure is long, and at the end diuided into three crooked forked points. All that *De Bry* ſaith thereof is this; *Flos Tigridis rubet egregiè circa medium tamen pallet, albuſque eſt & maculatus; ex Mexico à Caſparo Bauhino*. That is; *Flos Tigridis* is wondrous red, yet is it pale and whitiſh about the middle, and alſo ſpotted; it came from about Mexico, I had it from *Caſpar Bauhine*. ‡

Chap.

CHAP. 84. *Of Daffodils.*

¶ *The Kindes.*

DAffodill, or *Narcissus*, according to *Dioscorides*, is of two sorts: the floures of both are white, the one hauing in the middle a purple circle or coronet; the other with a yellow cup circle or coronet. Since whose time there hath been sundry others described, as shall be set forth in their proper places.

1 *Narcissus medio purpureus.*
Purple circled Daffodill.

‡ 4 *Narcissus medio croceus serotinus Polyanthos.*
The late many floured Daffodill with the Saffron-coloured middle.

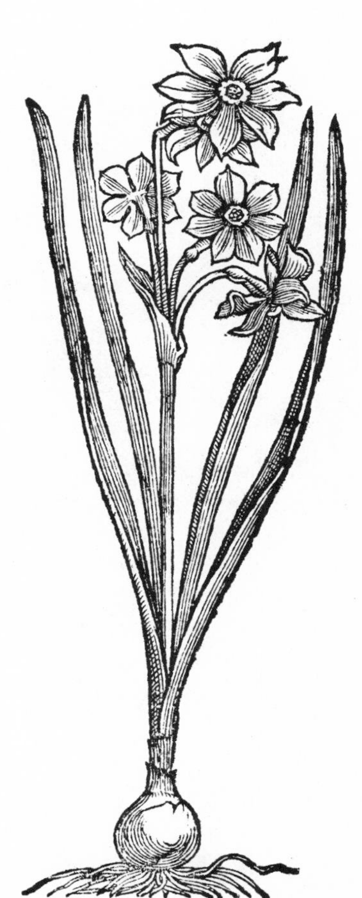

¶ *The Description.*

1 THe first of the Daffodils is that with the purple crowne or circle, hauing small narrow leaues, thicke, fat, and full of slimie juyce; among the which riseth vp a naked stalke, smooth and hollow, of a foot high, bearing at the top a faire milk-white floure growing forth of a hood or thinne filme, such as the floures of onions are wrapped in: in the middest of which floure is a round circle or small coronet of a yellowish colour, purfled or bordered about the edge of the said ring or circle with a pleasant purple colour; which beeing past, there followeth a thicke knob or button, wherein is contained blacke round seed. The root is white, bulbous or Onion fashion.

2 The second kinde of Daffodill agreeth with the precedent in euery respect, sauing that this Daffodill floureth in the beginning of Februarie, and the other not vntill Aprill, and is somewhat lesser. It is called *Narcissus medio purpureus præcox*; That is, Timely purple ringed Daffodill. The next may haue the addition *præcocior*, More timely: and the last in place, but first in time, *præcocissimus*, Most timely, or very early flouring Daffodill.

‡ 5 *Narciſſus medio-purpureus flore pleno.*
Double floured purple circled Daffodill.

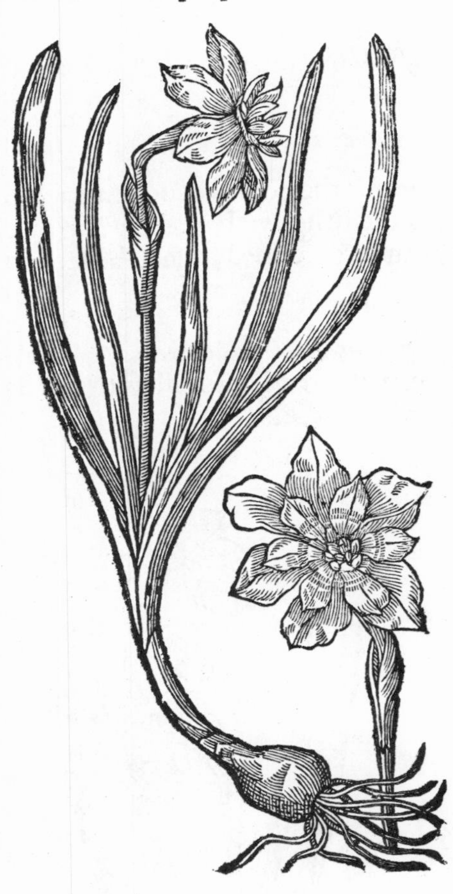

6 *Narciſſus minor ſerotinus.*
The late flouring ſmall Daffodill.

7 *Narciſſus medioluteus.*
Primroſe Pearles, or the common white Daffodill.

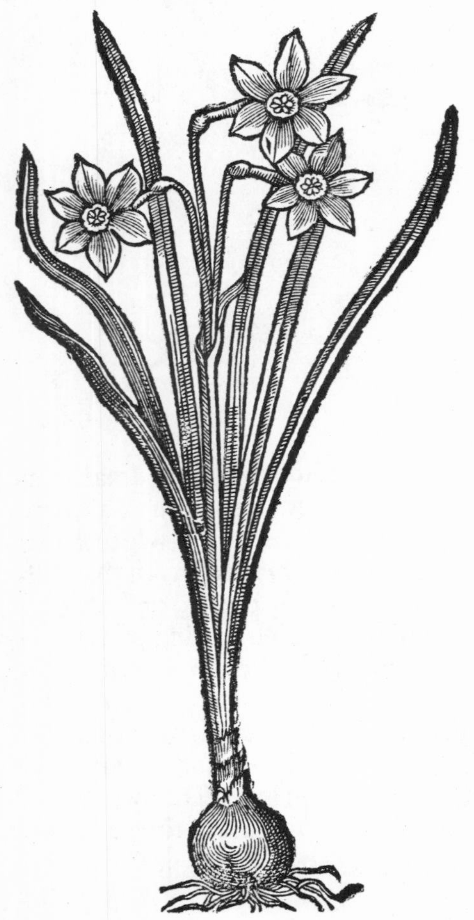

8 *Narciſſus medioluteus polyanthos.*
French Daffodill.

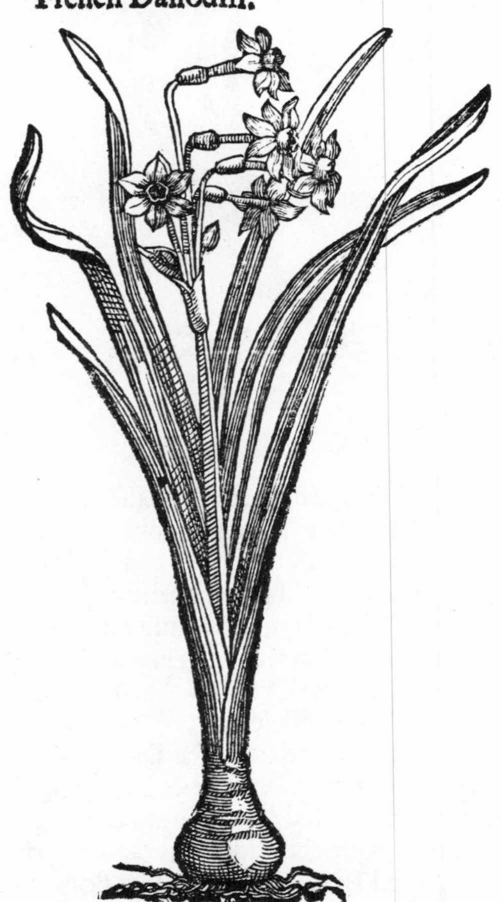

9 Nar-

9 *Narcissus Pisanus.*
Italian Daffodill.

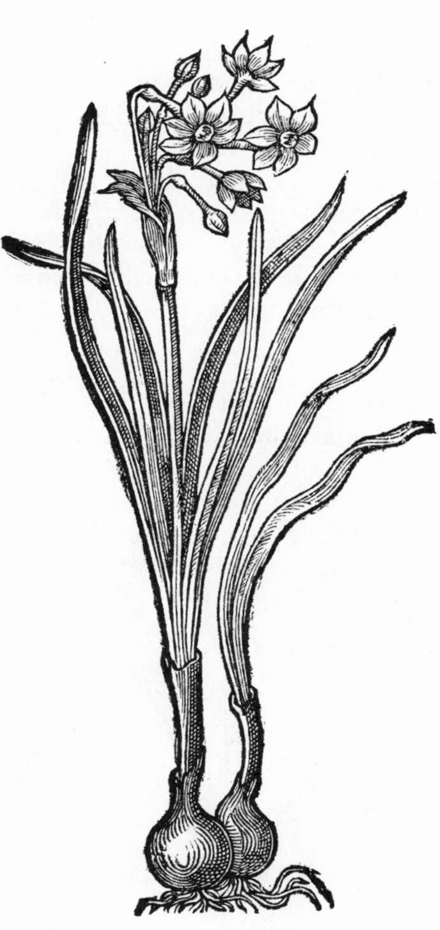

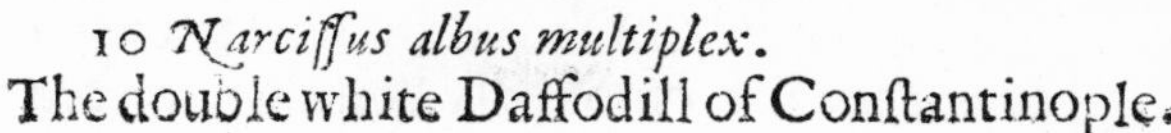

10 *Narcissus albus multiplex.*
The double white Daffodill of Constantinople.

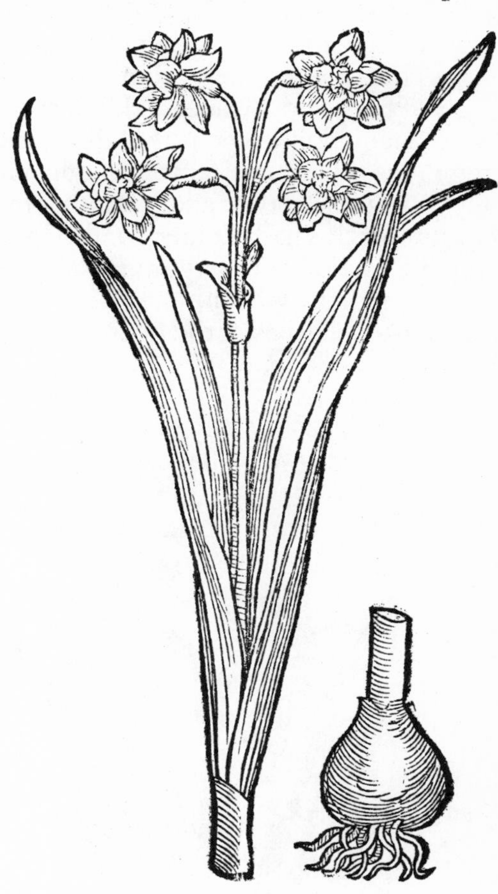

‡ 11 *Narcissus flore pleno albo.*
The other double white Daffodill.

‡ 12 *Narcissus flore pleno, medio luteo.*
Double white Daffodil with the middle yellow.

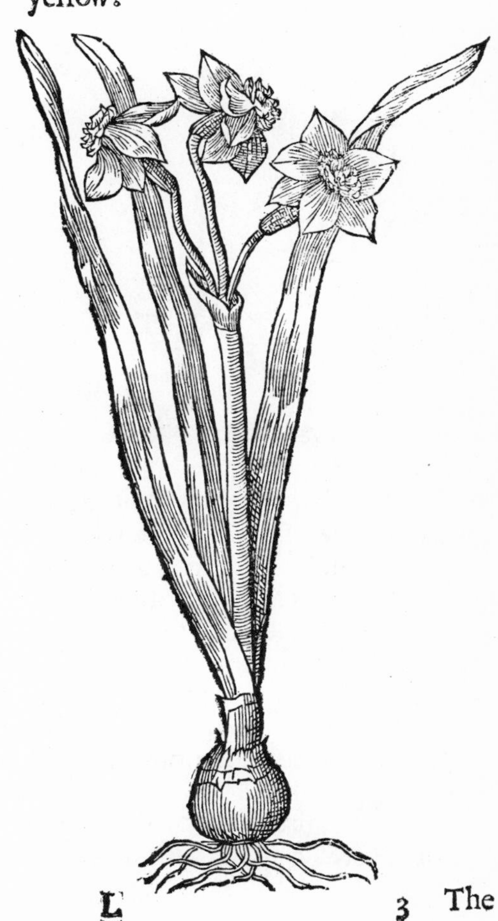

3 The third kind of Daffodil with the pnrple ring or circle in the middle, hath many ſmall narrow leaues, very flat, crookedly bending toward the top; among which riſeth vp a ſlender bare ſtalke, at whoſe top doth grow a faire and pleaſant floure, like vnto thoſe before deſcribed, but leſſer, and floureth ſooner, wherein conſiſteth the difference.

‡ There is alſo another ſomewhat leſſe, and flouring ſomewhat earlier than the laſt deſcribed.

4 This in roots, leaues, and ſtalkes differeth very little from the laſt mentioned kindes; but it beares many floures vpon one ſtalke, the out-leaues being like the former, white, but the cup or ring in the middle of a ſaffron colour, with diuers yellow threds contained therein.

5 To theſe may be added another mentioned by *Cluſius*, which differs from theſe onely in the floures: for this hath floures conſiſting of ſix large leaues fairely ſpread abroad, within which are other ſix leaues not ſo large as the former, and then many other little leaues mixed with threds comming forth of the middle. Now there are purple welts which runne betweene the firſt and ſecond ranke of leaues, in the floure, and ſo in the reſt. This floures in May; and it is *Narciſſus pleno flore quintus*, of *Cluſius*. ‡

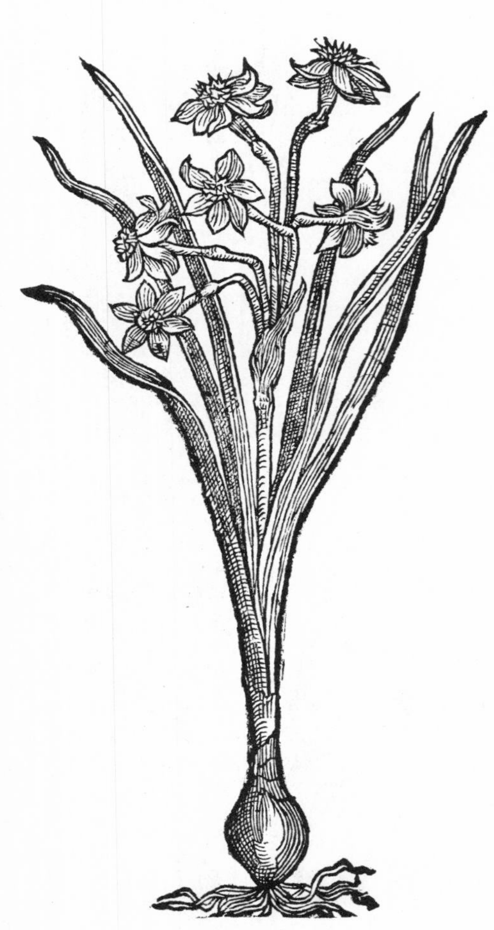

‡ 13 *Narciſſus flore pleno, medio verſicolore.*
Double Daffodill with a diuers coloured middle.

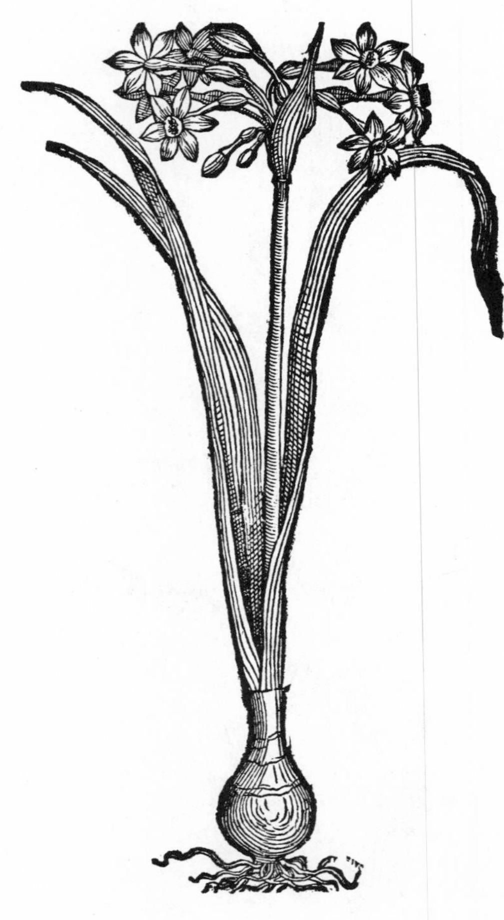

14 *Narciſſus totus albus.*
Milke white Daffodill.

6 This late flouring Daffodill hath many fat thicke leaues, full of juice, among the which riſeth vp a naked ſtalke, on the top whereof groweth a faire white floure, hauing in the middle a ring or yellow circle. The ſeed groweth in knobby ſeed veſſels. The root is bulbous or Onion faſhion. It floureth later than the others before deſcribed, that is to ſay, in Aprill and May.

7 The ſeuenth kinde of Daffodill is that ſort of *Narciſſus* or Primeroſe peereleſſe that is moſt common in our countrey gardens, generally knowne euery where. It hath long fat and thicke leaues, full of a ſlimie juice; among which riſeth vp a bare thicke ſtalke, hollow within and full of juice. The floure groweth at the top, of a yellowiſh white colour, with a yellow crowne or circle in the middle; and floureth in the moneth of Aprill, and ſometimes ſooner. The root is bulbous faſhion.

8 The eighth Daffodill hath many broad and thicke leaues, fat and full of juice, hollow and ſpongeous. The ſtalkes, floures, and roots are like the former, and differeth in that, that this plant bringeth

bringeth forth many floures vpon one ſtalk, and the other fewer, and not of ſo perfect a ſweet ſmel, but more offenſiue and ſtuffing the head. It hath this addition, *Polyanthos*, that is, of many floures, wherein eſpecially conſiſteth the difference.

9 The Italian Daffodill is very like the former, the which to diſtinguiſh in words, that they may be knowne one from another, is impoſſible. Their floures, leaues, and roots are like, ſauing that the floures of this are ſweeter and more in number.

15 *Narciſſus Iuncifolius præcox.*
Ruſh Daffodill, or *Iunquilia.*

16 *Narciſſus Iuncifolius ſerotinus.*
Late flouring Ruſh Daffodill.

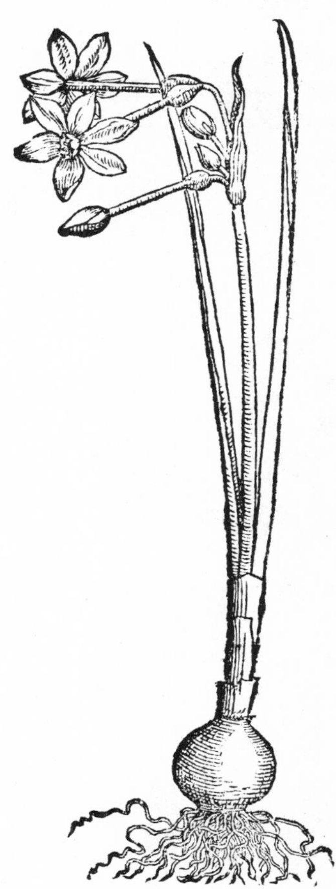

10 The double white Daffodill of Conſtantinople was ſent into England vnto the right honourable the Lord Treaſurer, among other bulbed floures: whoſe roots when they were planted in our London gardens, did bring forth beautifull floures, very white and double, with ſome yellowneſſe mixed in the middle leaues, pleaſant and ſweet in ſmell, but ſince that time we neuer could by any induſtrie or manuring bring them vnto flouring againe. So that it ſhould appeare, when they were diſcharged of that birth or burthen which they had begotten in their owne country, and not finding that matter, ſoile, or clymate to beget more floures, they remaine euer ſince barren and fruitleſſe. Beſides, we found by experience, that thoſe plants which in Autumne did ſhoot forth leaues, did bring forth no floures at all; and the others that appeared not vntill the Spring, did flouriſh and beare their floures. The ſtalks, leaues, and roots are like vnto the other kindes of Daffodils. It is called of the Turks, *Giul Catamer lale*; That is, *Narciſſus* with double floures. Notwithſtanding we haue receiued from beyond the ſeas, as well from the Low Countries, as alſo from France, another ſort of greater beautie, which from yeare to yeare doth yeeld forth moſt pleaſant double floures, and great encreaſe of roots, very like as well in ſtalkes as other parts of the plant, vnto the other ſorts of Daffodils. It differeth onely in the floures, which are very double and thicke thruſt together, as are the floures of our double Primroſe, hauing in the middle of the floure ſome few chiues or welts of a bright purple colour, and the other mixed with yellow as aforeſaid.

‡ 11 This alſo with double white floures, which *Cluſius* ſets forth in the ſixth place, is of the ſame kinde with the laſt deſcribed; but it beares but one or two floures vpon a ſtalke, whereas the other hath many.

12 This, which is *Cluſius* his *Narciſſus flore pleno* 2. is in roots, leaues, and ſtalkes very like the precedent.

precedent; but the floures are composed of six large white out-leaues; but the middle is filled with many faire yellow little leaues much like to the double yellow wall-floure. They smel sweet like as the last mentioned.

13 This differs from the last mentioned onely in that it is lesse, and that the middle of the floure within the yellow cup is filled with longish narrow little leaues, as it were crossing each other. Their colour is white, but mixed with some greene on the outside, and yellow on the inside. ‡

14 The milke white Daffodill differeth not from the common white Daffodill, or Primrose peerelesse, in leaues, stalkes, roots, or floures, sauing that the floures of this plant hath not any other colour in the floure but white, whereas all the others are mixed with one colour or other.

‡ 17 *Narcissus juncifolius Roseoluteus.*
Rose or round floured *Iunquilia.*

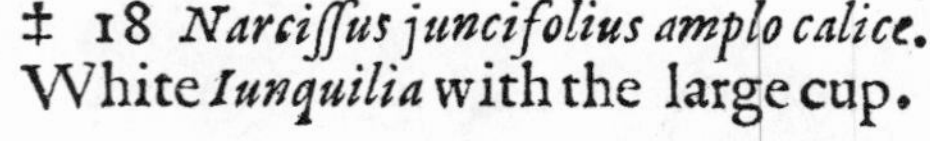

‡ 18 *Narcissus juncifolius amplo calice.*
White *Iunquilia* with the large cup.

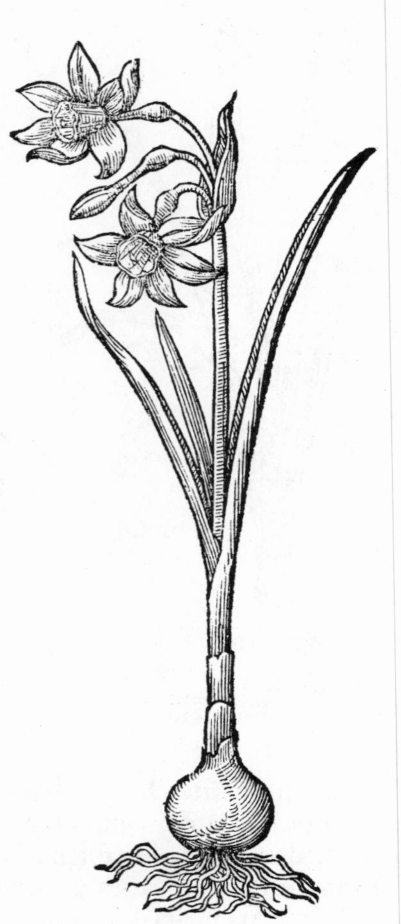

‡ 19 *Narcissus juncifolius reflexus flore albo.*
The white reflex *Iunquilia.*

15 The Rush Daffodill hath long, narrow, and thicke leaues, very smooth and flexible, almost round like Rushes, whereof it tooke his syrname *Iuncifolius* or Rushie. It springeth vp in the beginning of Ianuarie, at which time also the floures doe shoot forth their buds at the top of small rushy stalkes, sometimes two, and often more vpon one stalke, made of six small yellow leaues. The cup or crowne in the middle is likewise yellow, in shape resembling the other Daffodills, but smaller, and of a strong sweet smell. The root is bulbed, white within, and couered with a blacke skin or filme.

16 This Rush Daffodil is like vnto the precedent in each respect, sauing that it is altogether lesser, and longer before it come to flouring. There is also a white floured one of this kinde.

‡ 17 There

‡ 17 There is also another Rush Daffodill or *Iunquilia*, with floures not sharpe pointed, but round with a little cup in the middle: the colour is yellow or else white. This is *Lobels Narcissus juncifolius flore rotundæ circinitatis roseo.*

18 There is also another *Iunquilia* whose leaues and stalkes are like those of the first described Rushy Daffodill, but the cup in the middest of the floure is much larger. The colour of the floure is commonly white. *Clusius* calls this *Narcissus 1 Iuncifolius amplo calice.*

19 There are three or foure reflex *Iunquilia's*, whose cups hang downe, and the sixe incompassing leaues turne vp or backe, whence they take their names. The floures of the first are yellow; those of the second all white, the cup of the third is yellow, and the reflex leaues white. The fourth hath a white cup, and yellow reflex leaues. This seemes to be *Lobels Narcissus montanus minimus coronatus.*

20 This is like to the ordinarie lesser *Iunquilia*, but that the floures are very double, consisting of many long and large leaues mixed together; the shorter leaues are obtuse, as if they were clipt off. They are wholly yellow. ‡

‡ 19 *Narcissus Iuncifolius reflexus minor.*
The lesser reflex *Iunquilia.*

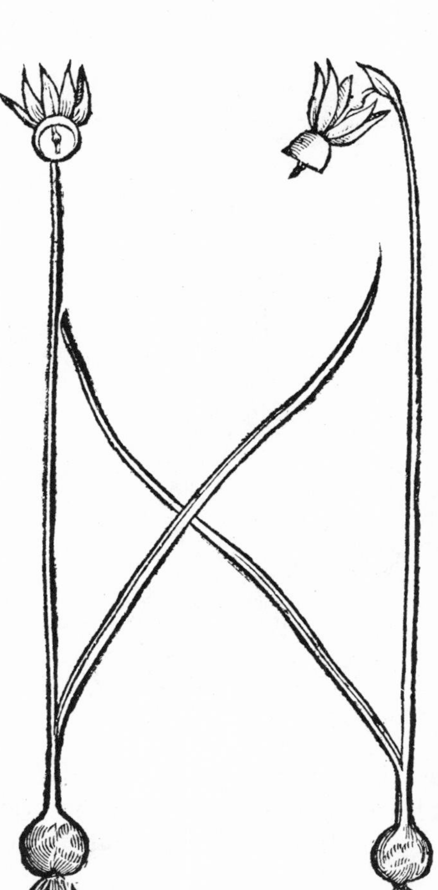

‡ 20 *Narcissus juncifolius multiplex.*
The double *Iunquilia.*

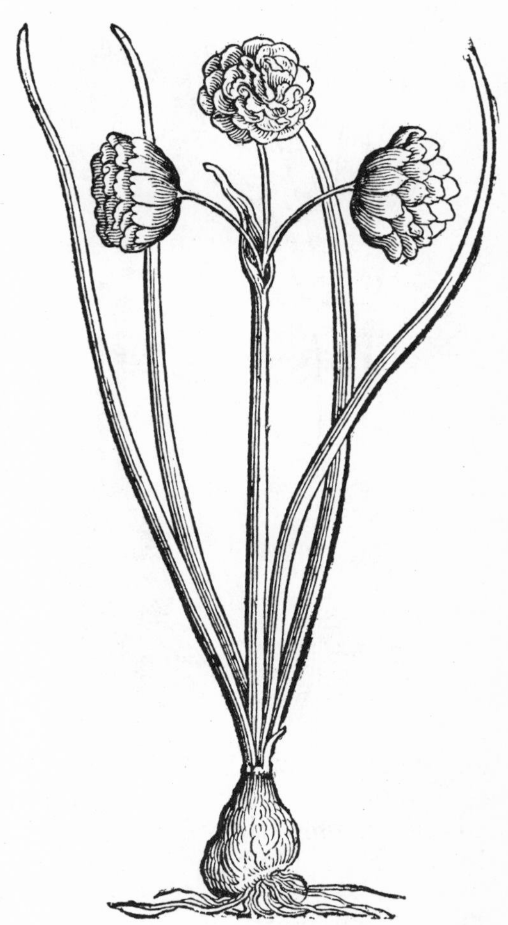

21 The Persian Daffodill hath no stalke at all, but onely a small and tender foot stalke of an inch high, such as the Saffron floure hath: vpon which short and tender stalk doth stand a yellowish floure consisting of six small leaues; of which the three innermost are narrower than those on the out side. In the middle of the floure doth grow forth a long stile or pointall, set about with many small chiues or threds. The whole floure is of an vnpleasant smel, much like to Poppy. The leaues rise vp a little before the floure, long, smooth, and shining. The root is bulbed, thicke, and grosse, blackish on the out side, and pale within, with some threds hanging at the lower part.

22 The Autumne Daffodill bringeth forth long smooth, glittering leaues, of a deepe greene colour: among which riseth vp a short stalke, bearing at the top one floure and no more, resembling the floure of Mead Saffron or common Saffron, consisting of six leaues of a bright shining yellow colour; in the middle whereof stand six threds or chiues, and also a pestell or clapper yellow likewise. The root is thicke and grosse like vnto the precedent.

‡ 23 To this last may be adioyned another which in shape somewhat resembles it. The leaues

leaues are ſmooth,greene, growing ſtraight vp,and almoſt a fingers breadth; among which riſeth vp a ſtalke a little more than halfe a foot in height, at the top of which groweth forth a yellow floure not much vnlike that of the laſt deſcribed Autumne Narciſſe: it conſiſteth of ſixe leaues ſome inch and halfe in length, and ſome halfe inch broad, ſharpe pointed, the three inner leaues being ſomewhat longer than the outer. There grow forth out of the middeſt of the floure three whitiſh chiues,tipt with yellow, and a peſtell in the midſt of them longer than any of them. The root conſiſts of many coats, with fibres comming forth of the bottome thereof like others of this kinde. It floures in Februarie. ‡

21 *Narciſſus Perſicus.*
The Perſian Daffodill.

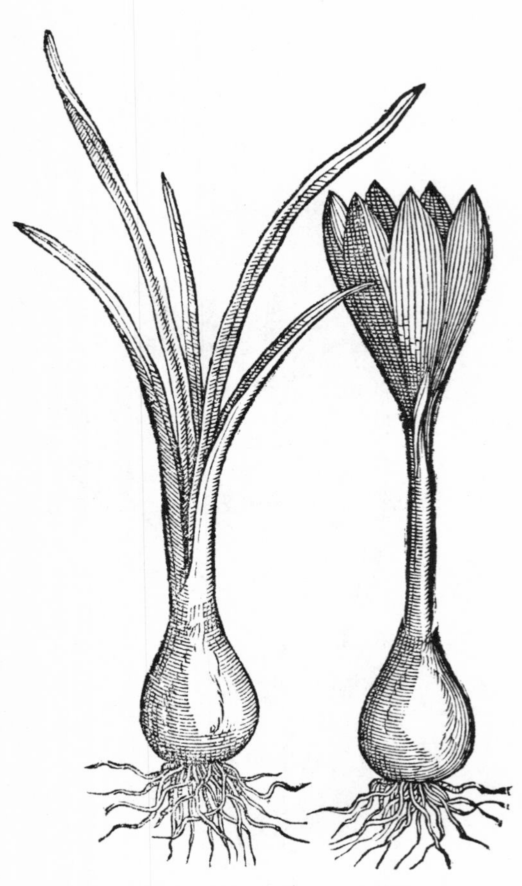

22 *Narciſſus Autumnalis major.*
The great Winter Daffodill.

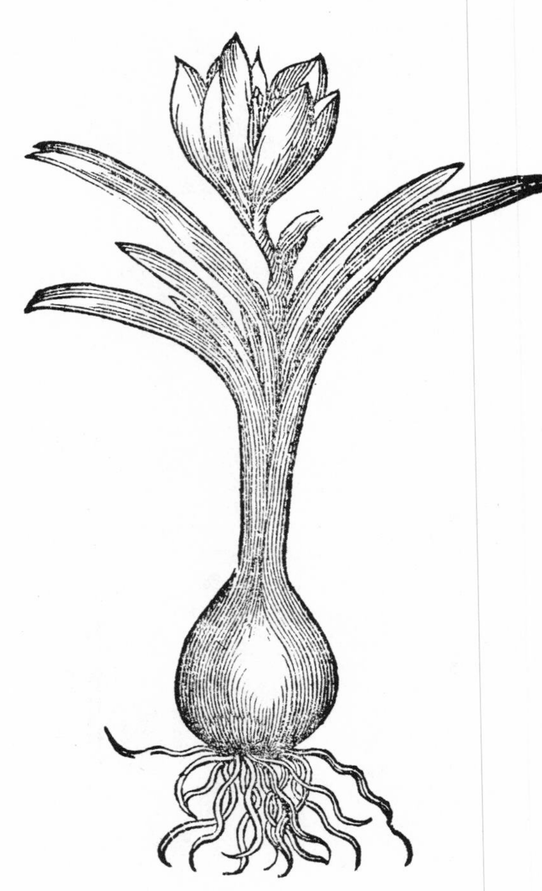

24 Small Winter Daffodill hath a bulbous root, much like vnto the root of Ruſh Daffodil, but leſſer: from the which riſeth vp a naked ſtalke without leaues, on the top whereof groweth a ſmall white floure with a yellow circle in the middle, ſweet in ſmell, ſomething ſtuffing the head as do the other Daffodils.

¶ *The Place.*

The Daffodils with purple coronets do grow wilde in ſundry places of France,chiefly in Bourgondie, and in Suitzerland in medowes.

The Ruſh Daffodill groweth wilde in ſundry places of Spaine, among graſſe and other herbes. *Dioſcorides* ſaith, That they be eſpecially found vpon mountaines. *Theocritus* affirmeth the Daffodils to grow in medowes, in his nineteenth *Eidyl.* or twentieth,according to ſome editions: where he writeth, That the faire Ladie *Europa* entring with her Nymphs into the medowes, did gather the ſweet ſmelling Daffodils; in theſe Verſes:

Αἱ δ', ἐπεὶ οὖν, &c.

Which we may Engliſh thus:

But when the Girles were come into
The medowes flouring all in ſight,
That Wench with theſe, this Wench with thoſe
Trim floures, themſelues did all delight:
She with the Narciſſe *good in ſent,*
And ſhe with Hyacinths *content.*

But

But it is not greatly to our purpose particularly to seeke out their places of growing wilde, seeing that we haue them all and euery of them in our London gardens, in great aboundance. The common white Daffodill groweth wilde in fields and sides of Woods in the West parts of England.

¶ *The Time.*

They floure for the most part in the Spring, that is, from the beginning of Februarie vnto the end of Aprill.

The Persian and Winter Daffodils do floure in September and October.

‡ 23 *Narcissus vernus præcocior flauo flore.*
The timely Spring yellow Daffodill.

24 *Narcissus Autumnalis minor*
Small Winter Daffodill.

¶ *The Names.*

Although their names be set forth in their seuerall titles, which may serue for their appellations and distinctions; notwithstanding it shall not be impertinent to adde a supply of names, as also the cause why they are so called.

The Persian Daffodill is called in the Sclauonian or Turkish tongue, *Zaremcada Persiana*, and *Zaremcatta*, as for the most part all other sorts of Daffodils are. Notwithstanding the double floured Daffodill they name *Giul catamer lale*. Which name they generally giue vnto all double floures.

The common white Daffodil with the yellow circle they call *Serin Cade*, that is to say, the kings Chalice; and *Deue bohini*, which is to say, Camels necke, or as we do say of a thing with long spindle shinnes, Long-shankes, vrging it from the long necke of the floure.

The Rush Daffodill is called of some *Ionquillias*, of the similitude the leaues haue with Rushes. Of *Dioscorides*, *Bulbus Vomitorius*, or Vomiting Bulbe, according to *Dodonæus*.

Generally all the kindes are comprehended vnder this name *Narcissus*, called of the Grecians Νάρκισσος: in Dutch, **Narcissen**: in Spanish, *Iennetten*: in English, Daffodilly, Daffodowndilly, and Primerose peerelesse.

Sophocles nameth them the garland of the infernal gods, because they that are departed and dulled with death, should worthily be crowned with a dulling floure.

Of the first and second Daffodill *Ouid* hath made mention in the third booke of his *Metamorphosis.*

phoſis, where hee deſcribeth the transformation of rhe faire boy *Narciſſus* into a floure of his own name; ſaying,

> *Nuſquam corpus erat, croceum pro corpore florem*
> *Inueniunt, folijs medium cingentibus albis.*
> But as for body none remain'd; in ſtead whereof they found
> A yellow floure, with milke white leaues ingirting of it round.

Pliny and *Plutarch* affirme, as partly hath been touched before, that their narcoticke quality was the very cauſe of the name *Narciſſus*, that is, a qualitie cauſing ſleepineſſe; which in Greekes is νάρκωσις: or of the fiſh Torpedo, called in Greeke νάρκη, which benummes the hands of them that touch him, as being hurtfull to the ſinewes; and bringeth dulneſſe to the head, which properly belongeth to the Narciſſes, whoſe ſmell cauſeth drowſineſſe.

¶ *The Nature.*

The roots of Narciſſus are hot and dry in the ſecond degree.

¶ *The Vertues.*

A *Galen* ſaith, That the roots of *Narciſſus* haue ſuch wonderfull qualities in drying, that they conſound and glew together very great wounds, yea and ſuch gaſhes or cuts as happen about the veins, ſinewes, and tendons. They haue alſo a certaine cleanſing and attracting facultie.

B The roots of *Narciſſus* ſtamped with honey, and applied plaiſter-wiſe, helpeth them that are burned with fire, and ioyneth together ſinewes that are cut in ſunder.

C Being vſed in manner aforeſaid, it helpeth the great wrenches of the ankles, the aches and pains of the ioynts.

D The ſame applied with hony and nettle ſeed helpeth Sun burning and the morphew.

E The ſame ſtamped with barrowes greaſe and leuen of rie bread, haſtneth to maturation hard impoſtumes, which are not eaſily brought to ripeneſſe.

F Being ſtamped with the meale of Darnell and honey, it draweth forth thornes and ſtubs out of any part of the body.

G The root, by the experiment of *Apuleius*, ſtamped and ſtrained, and giuen in drinke, helpeth the cough and collicke, and thoſe that be entred into a ptiſicke.

H The roots whether they be eaten or drunken, do moue vomit; and being mingled with Vineger and nettle ſeed, taketh away lentiles and ſpots in the face.

Chap. 85. *Of the Baſtard Daffodill.*

¶ *The Deſcription.*

1 THe double yellow Daffodill hath ſmall ſmooth narrow leaues, of a darke greene colour; among which riſeth vp a naked hollow ſtalke of two hands high, bearing at the top a faire and beautifull yellow floure, of a pleaſant ſweet ſmell: it ſheddeth his floure, but there followeth no ſeed at all, as it hapneth in many other double floures. The root is ſmall, bulbous, or onion faſhion, like vnto the other Daffodils, but much ſmaller.

2 The common yellow Daffodill or Daffodowndilly is ſo well knowne to all that it needeth no deſcription.

3 We haue in our London gardens another ſort of this common kind, which naturally groweth in Spaine, very like vnto our beſt knowne Daffodill in ſhape and proportion, but altogether fairer, greater, and laſteth longer before the floure doth fall or fade.

‡ 4 This hath leaues and roots like the laſt deſcribed, but ſomewhat leſſe; the floure alſo is in ſhape not vnlike that of the precedent, but leſſe, growing vpon a weake ſlender greene ſtalke, of ſome fingers length: the ſeed is contained in three cornered, yet almoſt round heads. The root is ſmall, bulbous, and blacke on the outſide.

5 This hath a longiſh bulbous root, ſomwhat blacke on the outſide, from which riſe vp leaues not ſo long nor broad as thoſe of the laſt deſcribed: in the midſt of theſe leaues ſprings vp a ſtalk, ſlender, and ſome halfe foot in height; at the top of which, forth of a whitiſh filme, breakes forth a floure like in ſhape to the common Daffodill, but leſſe, and wholly white, with the brim of the cup welted about. It floures in Aprill, and ripens the ſeeds in Iune. ‡

¶ *The Place.*

The double yellow Daffodill I receiued from *Robinus* of Paris, which he procured by meanes of friends from Orleance and other parts of France.

The

1 *Pseudonarcissus luteus multiplex.*
Double yellow Daffodill.

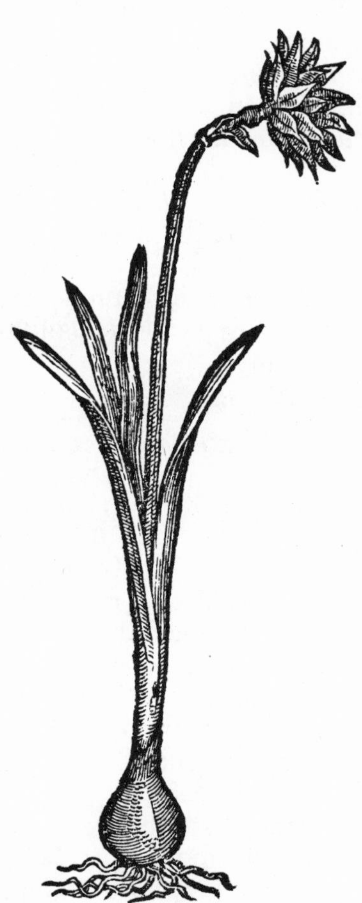

2 *Pseudonarcissus Anglicus.*
Common yellow Daffodill.

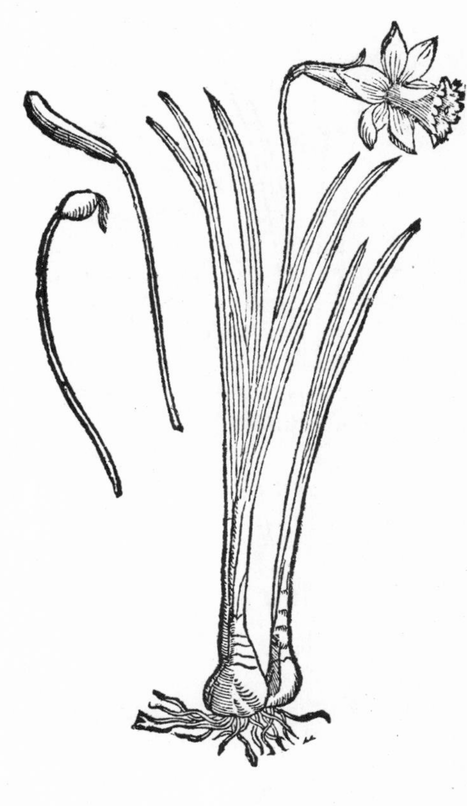

‡ 3 *Pseudonarcissus Hispanicus.*
The Spanish yellow Daffodill.

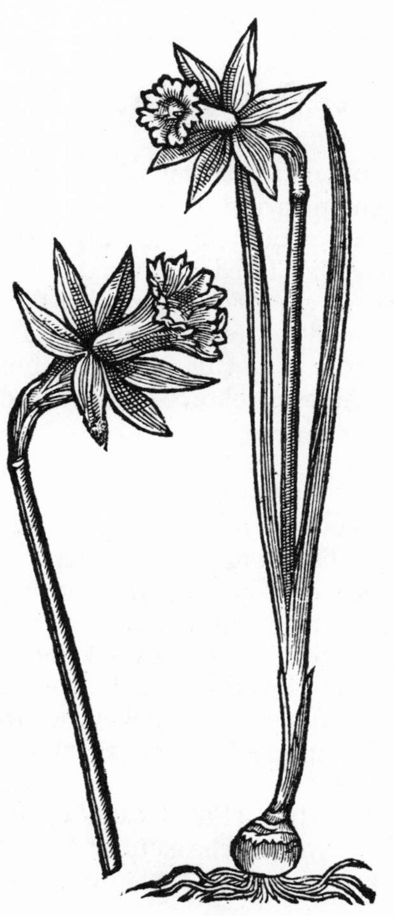

‡ 4 *Pseudonarcissus minor Hispanicus.*
The lesser Spanish Daffodill.

The

‡ 5 *Pseudonarcissus albo flore.*
White Bastard Daffodill.

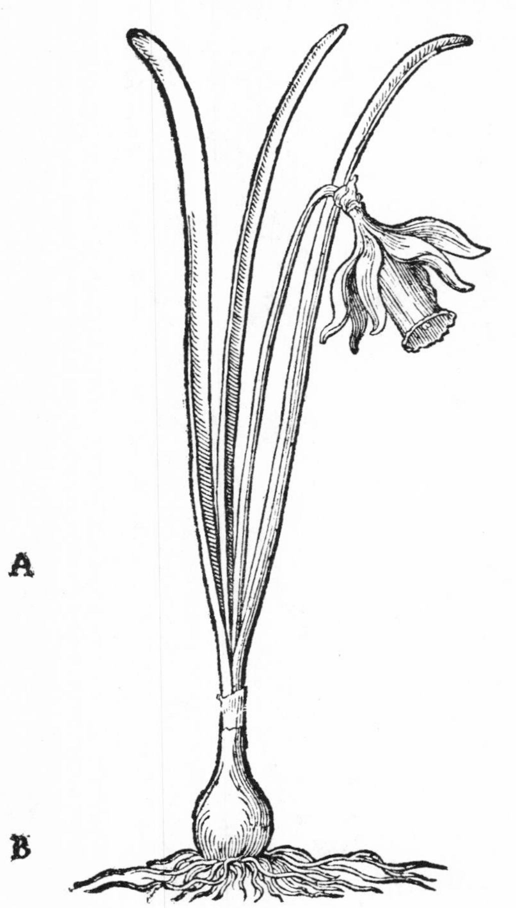

The yellow English Daffodill groweth almost euerie where through England. The yellow Spanish Daffodill doth likewise decke vp our London Gardens, where they increase infinitely.

¶ *The time.*

The double Daffodill sendeth forth his leaues in the beginning of Februarie, and his floures in Aprill.

¶ *The Names.*

The first is called *Pseudonarcissus multiplex*, and *Narcissus luteus Polyanthos*: in English, the double yellow Daffodill, or *Narcissus*.

The common sort are called in Dutch, **Geel Sporckel bloemen**: in English, yellow Daffodill, Daffodilly, and Daffodowndilly.

¶ *The Temperature.*

The temperature is referred vnto the kindes of *Narcissus*.

¶ *The Vertues.*

A Touching the vertues hereof, it is found out by experiment of some of the later Physitians, that the decoction of the roots of this yellow Daffodill do purge by siege tough and flegmaticke humors, and also waterish, and is good for them that are full of raw humors, especially if there be added thereto a little anise seed and ginger, which will correct the churlish hardnesse of the working.

B The distilled water of Daffodils doth cure the Palsie, if the Patient be bathed and rubbed with the sayd liquor by the fire. It hath beene proued by an especiall and trusty Friend of myne, a man learned, and a diligent searcher of nature, M. *Nicholas Belson*, sometimes of Kings Colledge in Cambridge.

Chap. 86. *Of diuers other Daffodils or Narcisses.*

‡ There are besides the forementioned sorts of Daffodils, sundry others, some of which may be referred to them; other some not. I do not intend an exact enumeration of them, it being a thing not so fitting for a historie of Plants, as for a Florilegie, or booke of floures. Now those that require all their figures, and more exact descriptions, may finde satisfaction in the late Worke of my kinde friend M. *Iohn Parkinson*, which is intitled *Paradisus terrestris*: for in other Florilegies, as in that of *De Bry*, *Swertz*, &c. you haue barely the names and figures, but in this are both figures, and an exact historie or declaration of them. Therefore I in this place will but onely briefely describe and name some of the rarest that are preserued in our choice gardens, and a few others whereof yet they are not possest.

¶ *The Descriptions.*

1 The first of these, which for the largenesse is called *Nonpareille*, hath long broad leaues and roots like the other Daffodils. The floure consists of six very large leaues of a pale yellow colour, with a very large cup, but not very long: this cup is yellower than the incompassing leaues, narrower also at the bottome than at the top, and vneuenly cut about the edges. This is called *Narcissus omnium maximus*, or *Non pareille*; the figure well expresseth the floure, but that it is somewhat too little. There is a varietie of this with the open leaues & cup both yellow, which makes the difference. There is also another *Non pareille*, whose floures are all white, and the six leaues that stand spred abroad are vsually a little folded, or turned in at their ends.

2 Besides these former there are foure or fiue double yellow Daffodils, which I cannot passe ouer in silence; the first is that, which is vulgarly amongst Florists knowne by the name of *Robines* Narcisse

Narcisse; and it may be was the same our Author in the precedent chapter mentions he receiued from *Robine*; but he giuing the figure of another, and a description not well fitting this, I can affirme nothing of certaintie. This double Narcisse of *Robine* growes with a stalke some foot in height, and the floure is very double, of a pale yellow colour, and it seemes commonly to diuide it selfe into some six partitions, the leaues of the floure lying one vpon another euen to the middle of the floure. This may be called *Narcissus pallidus multiplex Robini*, *Robines* double pale Narcisse.

‡ 1 *Narcissus omnium maximus.*
The *Nonpareille* Daffodill.

‡ 3 *Pseudonarcissus flore pleno.*
The double yellow Daffodill.

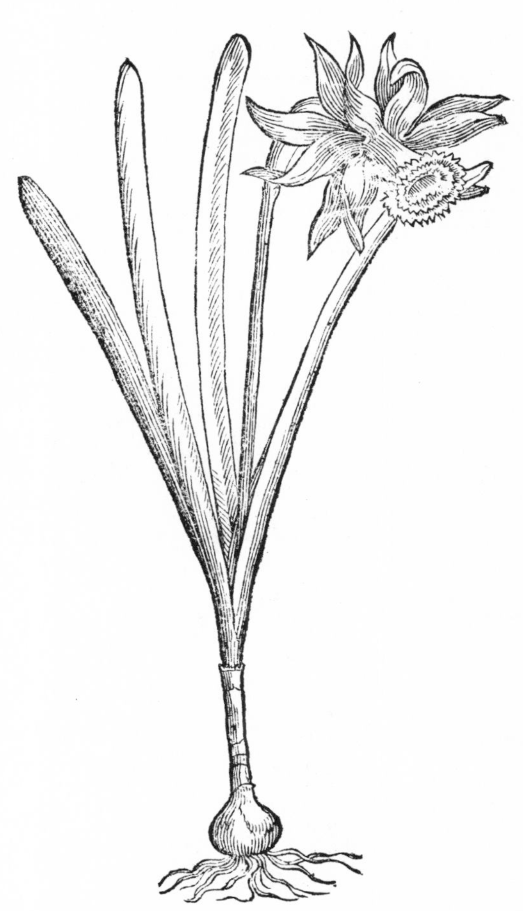

3 The next to this is that which from our Author, the first obseruer thereof, is vulgarly called *Gerrards* Narcisse: the leaues and root do not much differ from the ordinarie Daffodill; the stalk is scarce a foot high, bearing at the top thereof a floure very double; the sixe outmost leaues are of the same yellow colour as the ordinarie one is; those that are next are commonly as deepe as the tube or trunke of the single one, and amongst them are mixed also other paler coloured leaues, with some green stripes here & there among those leaues: these floures are somtimes all contained in a trunk like that of the single one, the sixe out-leaues excepted: other whiles this inclosure is is broke, and then the floure stands faire open like as that of the last described. *Lobel* in the second part of his *Aduersaria* tells, That our Author Master *Gerrard* found this in Wiltshire, growing in the garden of a poore old woman; in which place formerly a Cunning man (as they vulgarly terme him) had dwelt.

This may be called in Latine, according to the English, *Narcissus multiplex Gerardi*, *Gerrards* double Narcisse.

The figure we here giue you is expressed somewhat too tall, and the floure is not altogether so double as it ought to be.

4 There are also two or three double yellow Daffodils yet remaining. The first of these is called *Wilmots* Narcisse, (from Master *Wilmot*, late of Bow) and this hath a very faire double & large yellow floure composed of deeper and paler yellow leaues orderly mixed.

The second (which is called *Tradescants* Narcisse, from Master *Iohn Tradescant* of South-Lambeth) is the largest and stateliest of all the rest; in the largenesse of the floures it exceeds *Wilmots*, which otherwise it much resembles; some of the leaues whereof the floure consists are sharp poin-

This

ted, and theſe are of a paler colour; other ſome are much more obtuſe, and theſe are of a deeper and fairer yellow.

This may be called *Narciſſus Roſeus* Tradeſcanti, *Tradeſcants* Roſe Daffodill.

The third M. *Parkinſon* challengeth to himſelfe; which is a floure to be reſpected, not ſo much for the beautie, as for the various compoſure thereof, for ſome of the leaues are long and ſharpe pointed, others obtuſe and curled, a third ſort long and narrow, and vſually ſome few hollow, and in ſhape reſembling a horne; the vtmoſt leaues are commonly ſtreaked, and of a yellowiſh green; the next to them fold themſelues vp ronnd, and are vſually yellow, yet ſometimes they are edged with greene. There is a deepe yellow peſtill diuided into three parts, vſually in the midſt of this floure. It floures in the end of March. I vſually (before M. *Parkinſon* ſet forth his Florilegie, or garden of floures) called this floure *Narciſſus* πλύμορφος, by reaſon of its various ſhape and colour: but ſince I thinke it fitter to giue it to the Author, and terme it *Narciſſus multiplex varius* Parkinſoni, *Parkinſons* various double Narciſſe.

‡ 5 *Narciſſus Iacobæus Indicus.*
The Indian or Iacobæan Narciſſe.

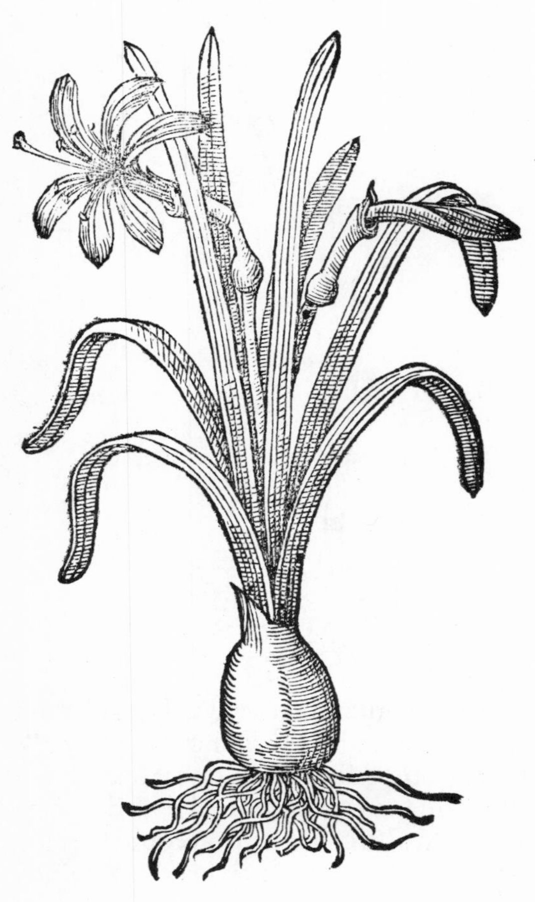

‡ 6 *Narciſſus juncifolius montanus minimus.*
The leaſt Ruſh-leaued Mountaine Narciſſe.

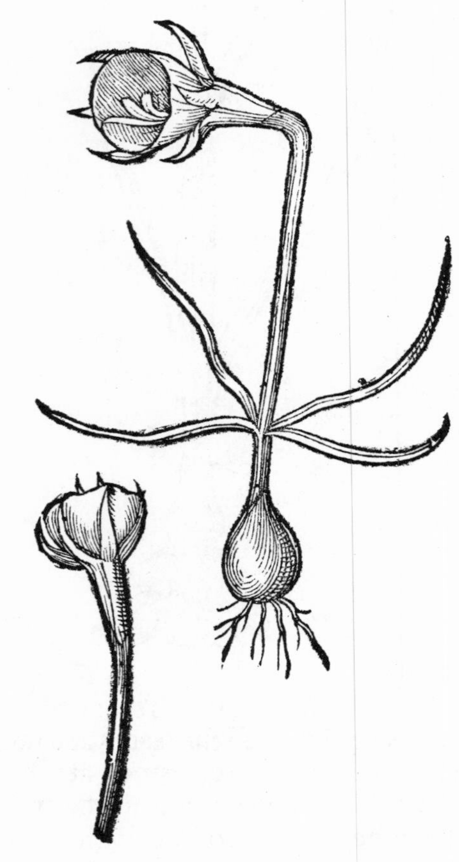

5 Now come I to treat of ſome more rarely to be found in our gardens, if at all. That which takes the firſt place is by *Cluſius* called *Narciſſus Iacobæus Indicus*, the Indian or Iacobæan Narciſſe. The root hereof is much like to an ordinarie onion, the leaues are broad like the other Narciſſes, the ſtalke is ſmooth, round, hollow, and without knots, at the top whereof, out of a certaine skinny huske comes forth a faire red floure like that of the flouring Indian reed, but that the leaues of this are ſomewhat larger, and it hath ſix chiues or threds in the middle thereof of the ſame colour as the floure, and they are adorned with browniſh pendants; in the midſt of theſe there ſtands a little farther out than the reſt, a three forked ſtile, vnder which ſucceeds a triangular head, after the falling of the floure.

This giues his floure in Iune or Iuly.

6 This *Lobell* calls *Narciſſus montanus juncifolius minimus*, The leaſt Ruſh-leaued mountaine Narciſſe. The leaues of this are like the *Iunquilia*; the ſtalke is ſhort, the floure yellow, with the ſix winged leaues ſmall and paler coloured, the cup open and large to the bigneſſe of the floure.

7 This

7 This alſo is much like the former; but the ſix incompaſſing leaues are of a greeniſh faint yellow colour; the cup is indented, or vnequally curled about the edges, but yellow like the precedent. *Lobell* calls this *Narciſſus montanus juncifolius flore fimbriato,* The mountaine Ruſh-leaued Narciſſe with an indented or curled cup.

‡ 7 *Narciſſus montanus juncifolius flore fimbriato.*
The mountaine Ruſh leaued Narciſſe with an indented or curled cup.

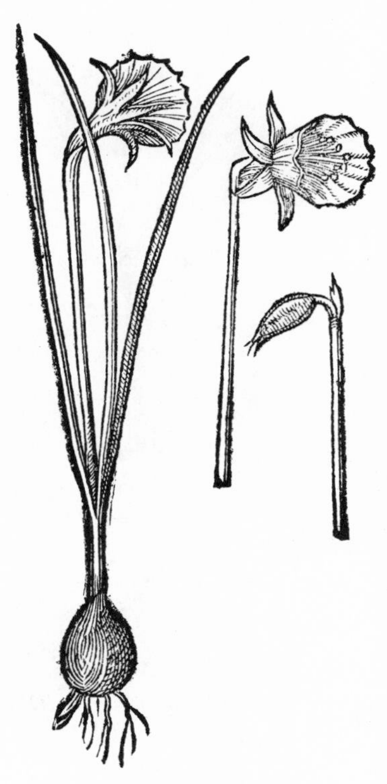

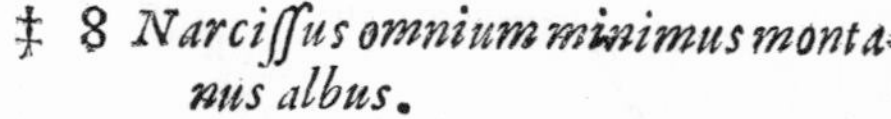

‡ 8 *Narciſſus omnium minimus montanus albus.*
The leaſt mountaine white Narciſſe.

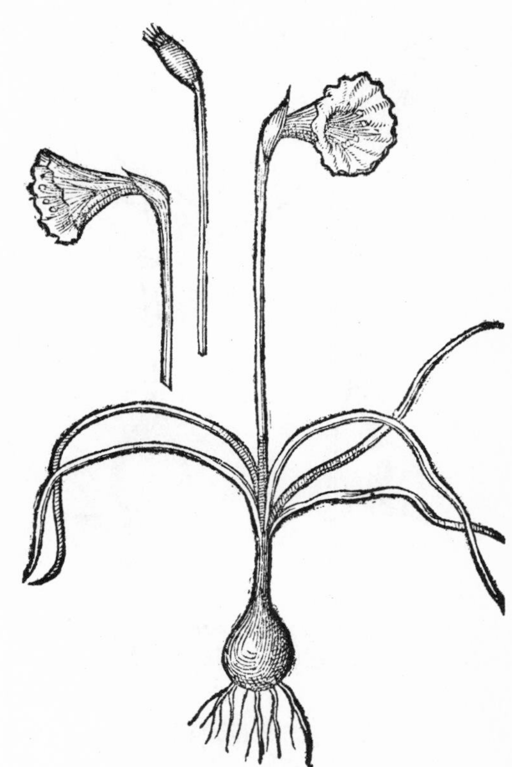

8 The leaues of this are as ſmall as the Autumne Iacinth, the ſtalke ſome handfull high, and the floure like the laſt deſcribed, but it is of a whitiſh colour. *Lobell* calls this laſt deſcribed, *Narciſſus omnium minimus montanus albus,* The leaſt mountaine white Narciſſe. Theſe three laſt vſually floure in Februarie. ‡

Chap. 87. *Of Tulipa, or the Dalmatian Cap.*

¶ *The Kindes.*

T*Vlipa,* or the Dalmatian Cap is a ſtrange and forreine floure, one of the number of the bulbed floures, whereof there be ſundry ſorts, ſome greater, ſome leſſer, with which all ſtudious and painefull Herbariſts deſire to be better acquainted, becauſe of that excellent diuerſitie of moſt braue floures which it beareth. Of this there be two chiefe and generall kindes, *viz. Præcox* and *Serotina*; the one doth beare his floures timely, the other later. To theſe two we will adde another ſort called *Media,* flouring betweene both the others. And from theſe three ſorts, as from their heads, all other kindes do proceed, which are almoſt infinite in number. Notwithſtanding, my louing friend M. *Iames Garret,* a curious ſearcher of Simples, and learned Apothecary of London, hath vndertaken to finde out, if it were poſſible, the infinite ſorts, by diligent ſowing of their ſeeds, and by planting thoſe of his owne propagation, and by others receiued from his Friends

1 *Tulipa Bononiensis.*
Italian Tulipa.

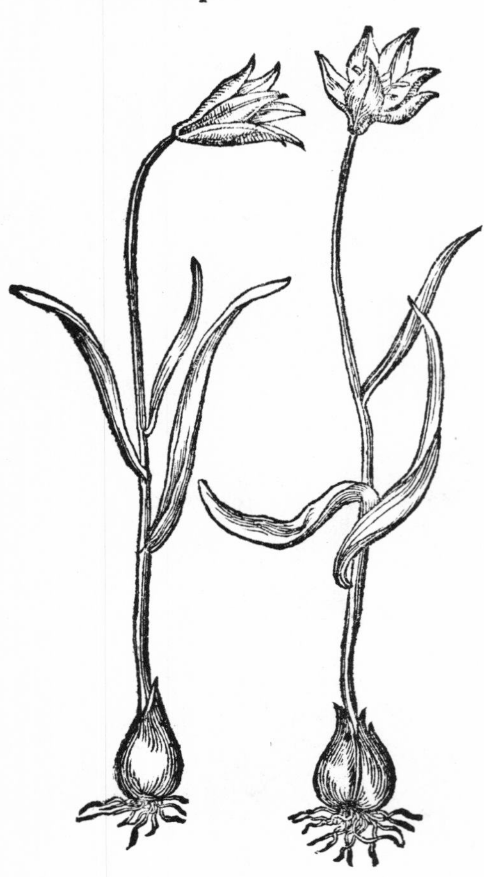

2 *Tulipa Narbonensis.*
French Tulipa.

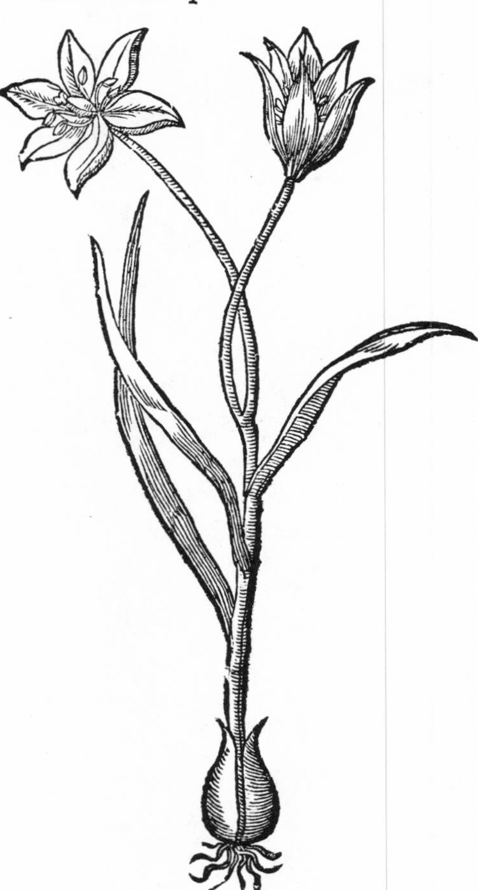

3 *Tulipa præcox tota lutea.*
Timely flouring Tulipa.

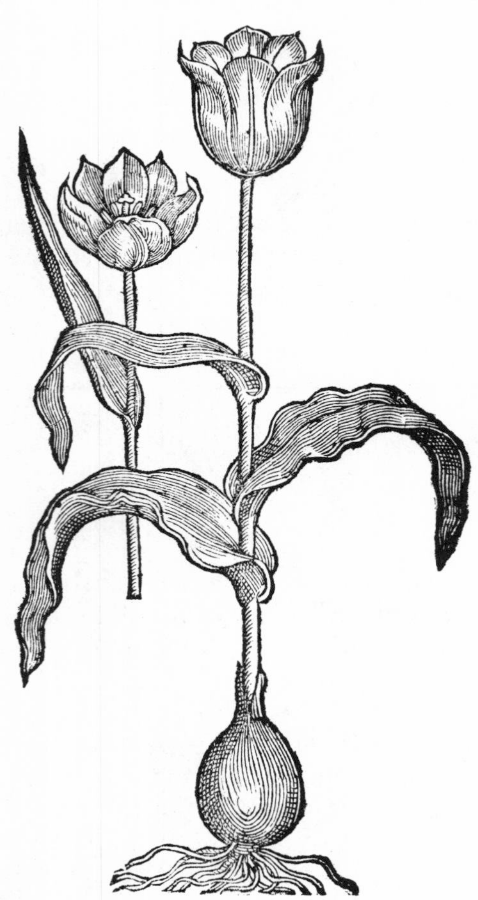

4 *Tulipa Coccinea serotina.*
Late flouring Tulipa.

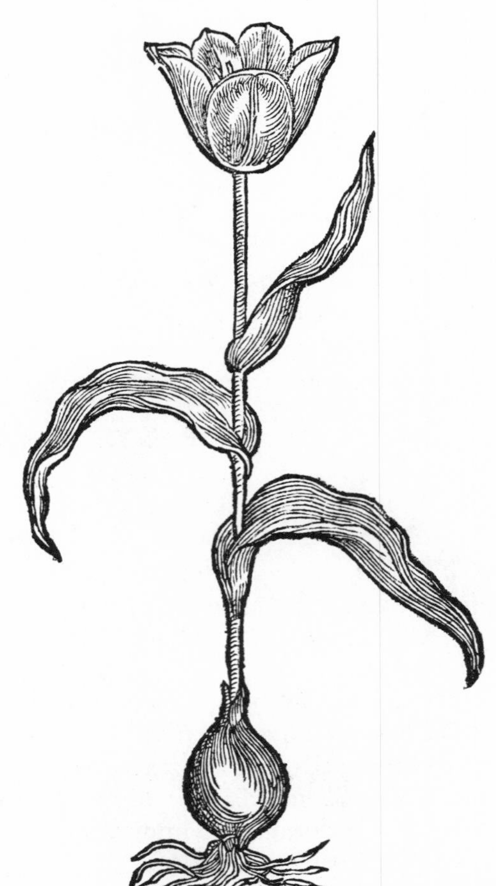

5 *Tulipa*

5 *Tulipa media ſanguinea albis oris.*
Apple bloome Tulipa.

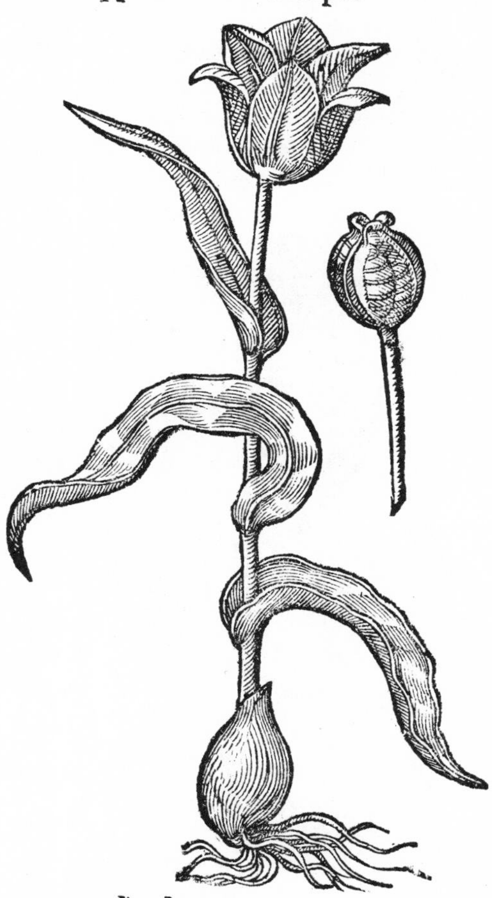

6 *Tulipa Candida ſuaue rubentibus oris.*
Bluſh coloured Tulipa.

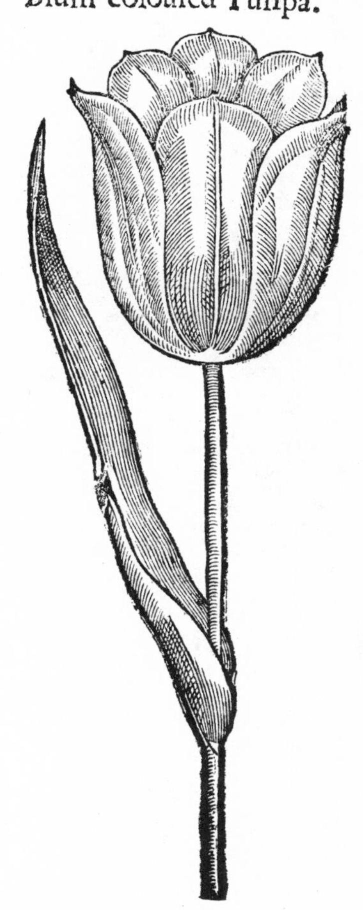

7 *Tulipa bulbifera.*
Bulbous ſtalked Tulipa.

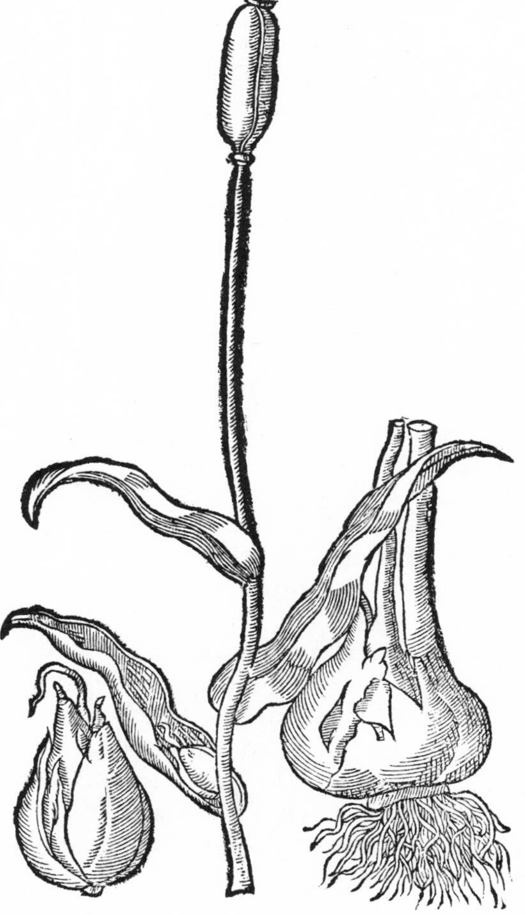

‡ 8 *Tulipa ſanguinea luteo fundo.*
The bloud-red Tulip with a yellow bottome.

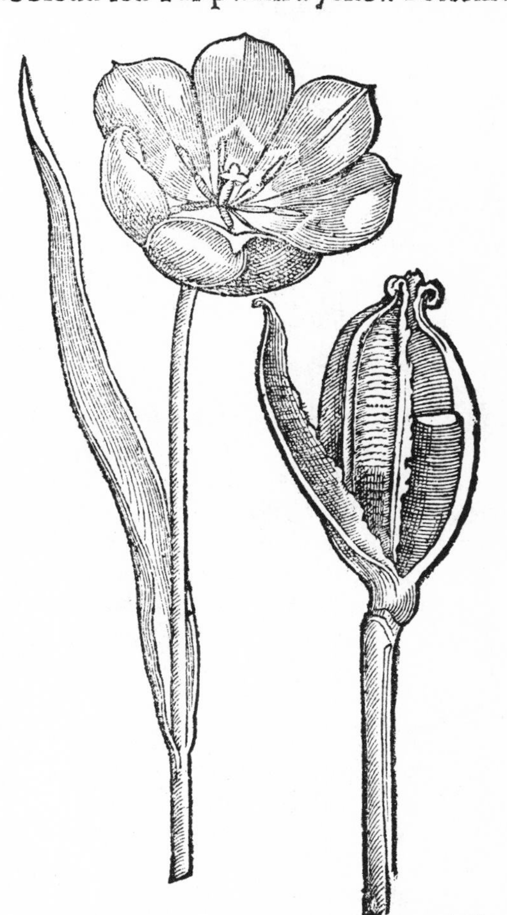

beyond the seas for the space of twenty yeares, not being yet able to attaine to the end of his trauell, for that each new yeare bringeth forth new plants of sundry colours, not before seene : all which to describe particularly were to roll *Sisiphus* stone, or number the sands. So that it shall suffice to speake of and describe a few, referring the rest to some that meane to write of *Tulipa* a particular volume.

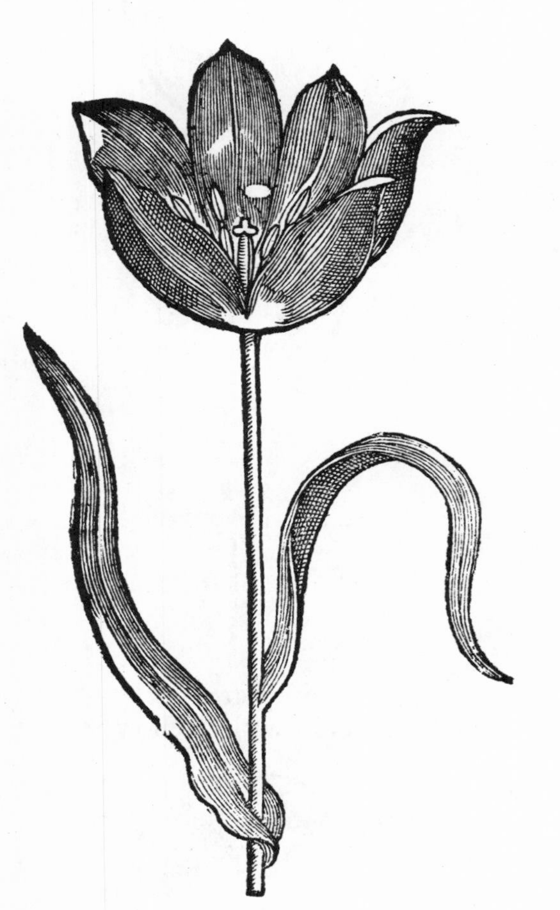

‡ 9 *Tulipa purpurea.*
The purple Tulip.

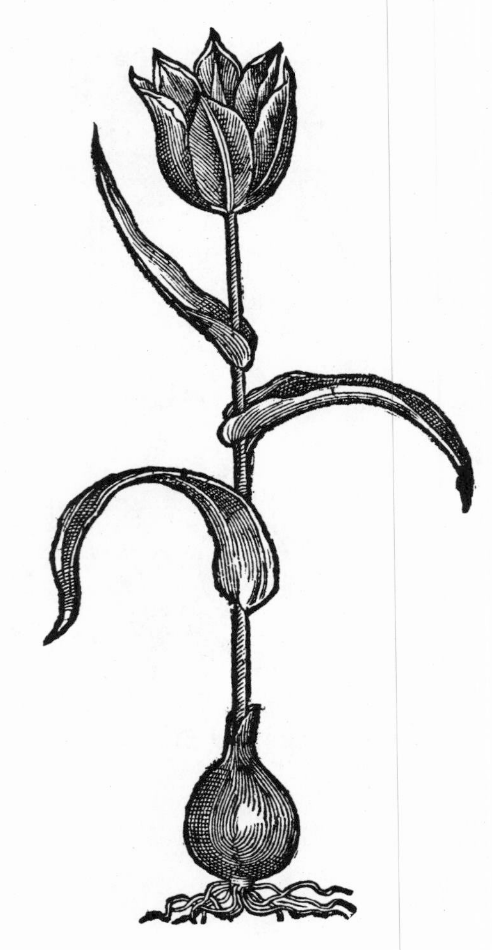

‡ 10 *Tulipa rubra amethistina.*
The bright red Tulip.

¶ *The Description.*

1 THe *Tulipa* of Bolonia hath fat, thicke, and grosse leaues, hollow, furrowed or chanelled, bending a little backward, and as it were folded together : which at their first comming vp seeme to be of a reddish colour, and being throughly growne turne into a whitish greene. In the middest of those leaues riseth vp a naked fat stalke a foot high, or something more, on the top whereof standeth one or two yellow floures, sometimes three or more, consisting of six small leaues, after a sort like to a deepe wide open cup, narrow aboue, and wide in the bottome. After it hath beene some few dayes floured, the points and brims of the floure turne backward, like a Dalmatian or Turkish cap, called *Tulipan, Tolepan, Turban*, and *Turfan*, whereof it tooke his name. The chiues or threads in the middle of the floures be sometimes yellow, otherwhiles blackish or purplish, but commonly of one ouer-worne colour or other, Nature seeming to play more with this floure than with any other that I do know. This floure is of a reasonable pleasant smell, and the other of his kinde haue little or no smell at all. The seed is flat, smooth, shining, and of a gristly substance. The root is bulbous, and very like to a common onion of Saint *Omers*.

2 The French Tulipa agreeth with the former, except in the blacke bottome which this hath in the middle of the floure, and is not so sweet of smell, which setteth forth the difference.

3 The yellow Tulipa that floureth timely hath thicke and grosse leaues full of iuyce, long, hollow, or gutter fashion, set about a tender stalke, at the top whereof doth grow a faire and pleasant shining yellow floure, consisting of six small leaues without smell. The root is bulbous or like an onion.

4 The

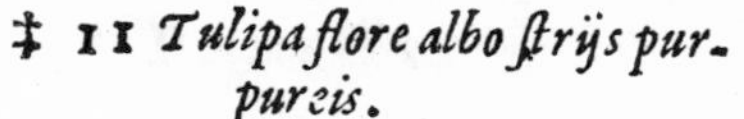

‡ 11 *Tulipa flore albo ſtrijs purpureis.*
The white Tulip with purple ſtreakes.

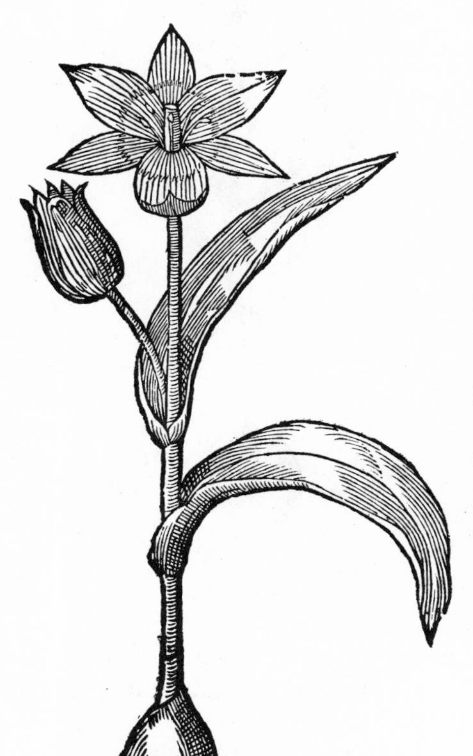

‡ 12 *Tulipa flore albo oris dilute rubentibus.*
The white Tulip with light red edges.
‡ 13 *Tulipa flore pallido.* The ſtraw-coloured Tulip.
‡ 14 *Tulipa flammea ſtrijs flaueſcentibus.*
The flame coloured Tulip with yellowiſh ſtreakes.

‡ 16 *Tulipa ſerotina polyclados major flo. flauo fundo nigro,* Cluſij.
Cluſius his greater many branched Tulip with a yellow floure, and blacke bottome.

‡ 15 *Tulipa polyclonos minor ſerotina flore rubro vel flauo,* Cluſij.
The leſſer many-branched late Tulip of *Cluſius,* with red, or elſe yellow floures.

‡ 17 *Tulipa pumilio obscure rubeus oris virentibus.*
The dwarfe Tulip with darke red floures edged with greene.
‡ 18 *Tulipa pumilio flore purpurascenti intus candido.*
The Dwarfe Tulip with a purplish floure, white within.

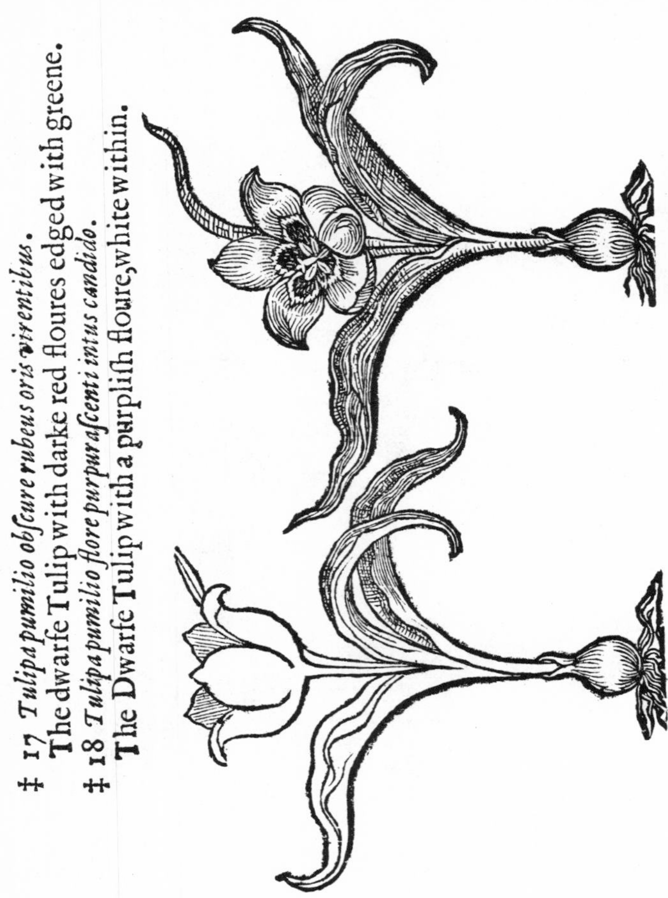

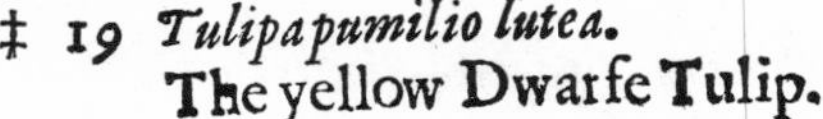

‡ 19 *Tulipa pumilio lutea.*
The yellow Dwarfe Tulip.

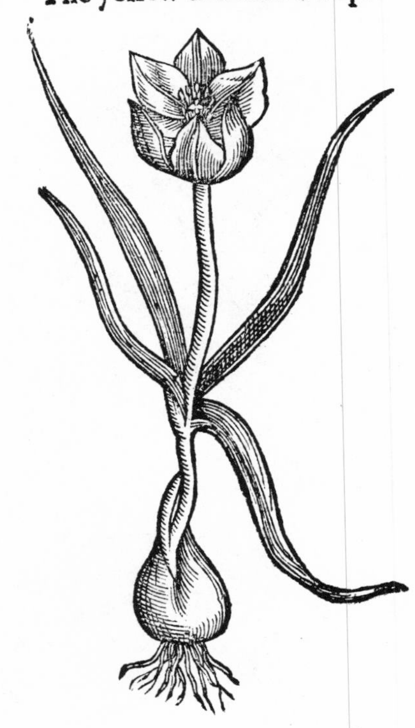

‡ 21 *Tulipa aurea oris rubentibus.*
The gold yellow with red edges.

‡ 20 *Tulipa Persica flore rubro, oris albidis elegans.*
The pretty Persian Tulip hauing a red floure with whitish edges.

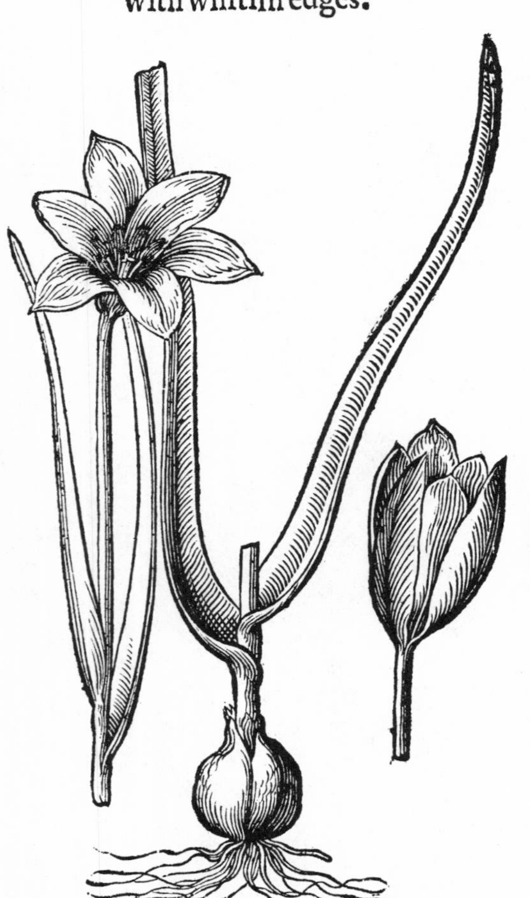

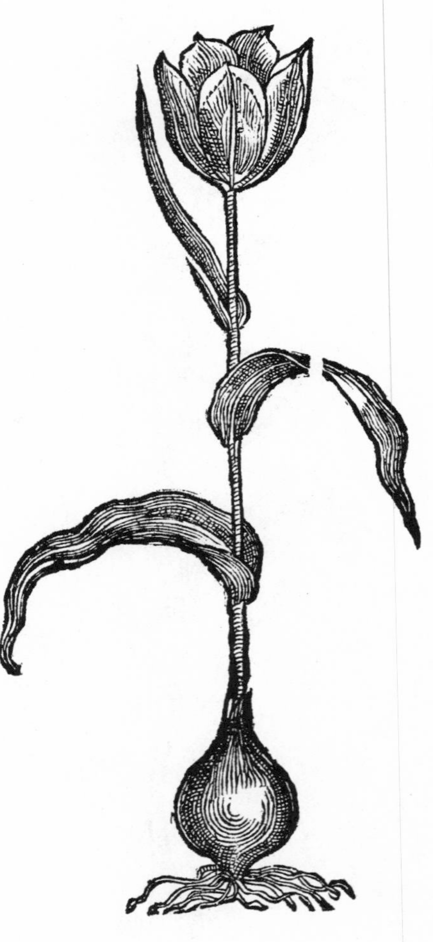

4 The fourth kinde of Tulipa, that floureth later, hath leaues, stalks, and roots like vnto the precedent. The floures hereof be of a skarlet colour, welted or bordered about the edges with red. The middle part is like vnto a hart tending to whitenesse, spotted in the same whitenes with red speckles or spots. The seed is contained in square cods, flat, tough, and sinewie.

‡ 22 *Tulipa miniata.*
The Vermilion Tulip.

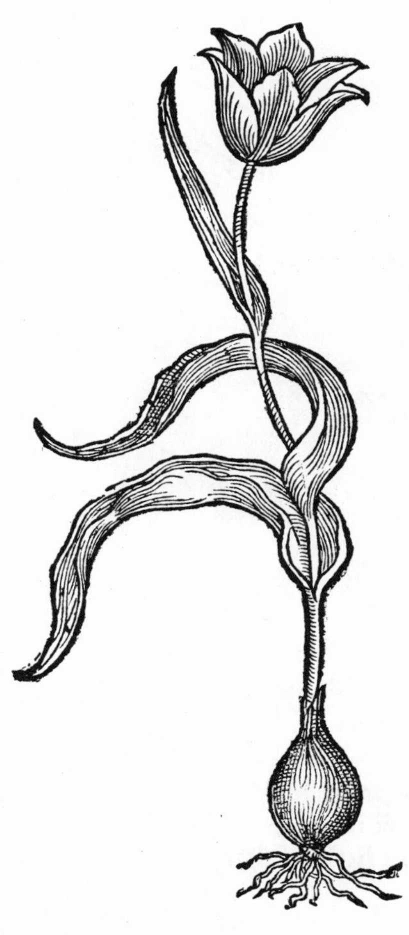

‡ 23 *Tulipa albo & rubro striatus.*
The white and red striped Tulip.

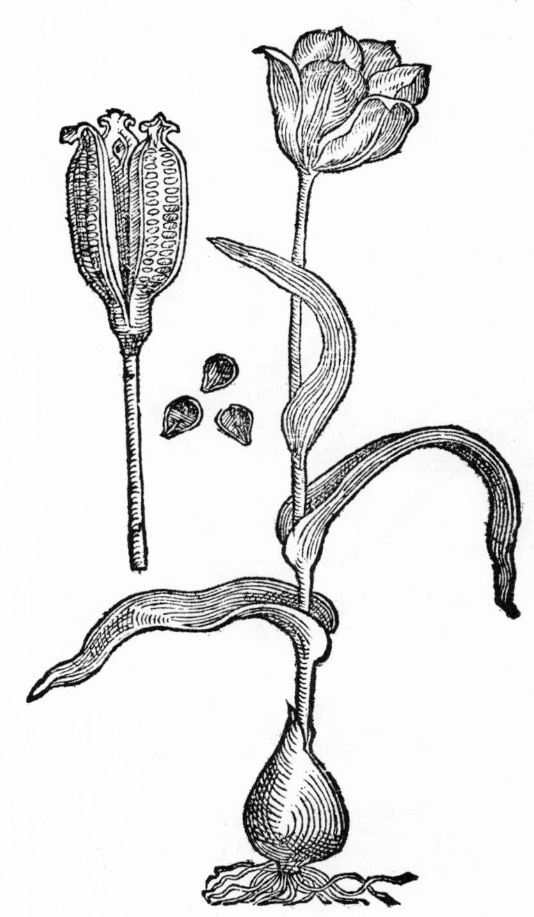

5 The fift sort of Tulipa, which is neither of the timely ones, nor of the later flouring sort, but one that buddeth forth his most beautifull floures betweene both. It agreeth with the last described Tulipa, in leaues, stalkes, roots, and seed, but differeth in floures. The floure consisteth of six small leaues ioyned together at the bottome: the middle of which leaues are of a pleasant bloudy colour, the edges be bordered with white, and the bottome next vnto the stalke is likewise white; the whole floure resembling in colour the blossomes of an Apple tree.

6 The sixth hath leaues, roots, stalkes, and seed like vnto the former, but much greater in euery point. The floures hereof are white, dasht about the brimmes or edges with a red or blush colour. The middle part is stripped confusedly with the same mixture, wherein is the difference.

7 *Carolus Clusius* setteth forth in his Pannonicke historie a kinde of Tulipa that beareth faire red floures, blacke in the bottome, with a pestell in the middle of an ouer-worne greenish colour; of which sort there happeneth some to haue yellow floures, agreeing with the others before touched: but this bringeth forth encrease of root in the bosome of his lowest leafe next to the stalke, contrarie to all the other kindes of Tulipa.

8 *Lobelius* in his learned Obseruations hath set forth many other sorts; one he calleth *Tulipa Chalcedonica*, or the Turky Tulipa, saying it is the least of the small kindes or Dwarfe Tulipa's, whose floure is of a sanguine red colour, vpon a yellow ground, agreeing with the others in roote, leafe, and stalke.

9 He hath likewise set forth another; his floure is like the Lilly in proportion, but in colour of a fine purple.

10 We may also behold another sort altogether greater than any of the rest, whose floure is in colour like the stone called *Amethist*, not vnlike to the floures of Peonie.

11 We haue likewise another of greater beauty, and very much desired of all, with white floures dasht on the backside, with a light wash of watchet colour.

12 There

‡ 24 *Tulipa luteo & rubro ſtriatus.*
The red and yellow Fooles coat.

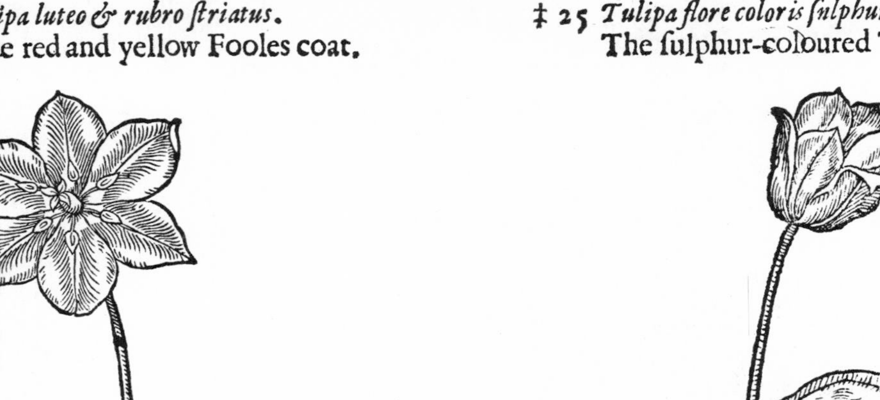

‡ 25 *Tulipa flore coloris ſulphurei*
The ſulphur-coloured Tulip.

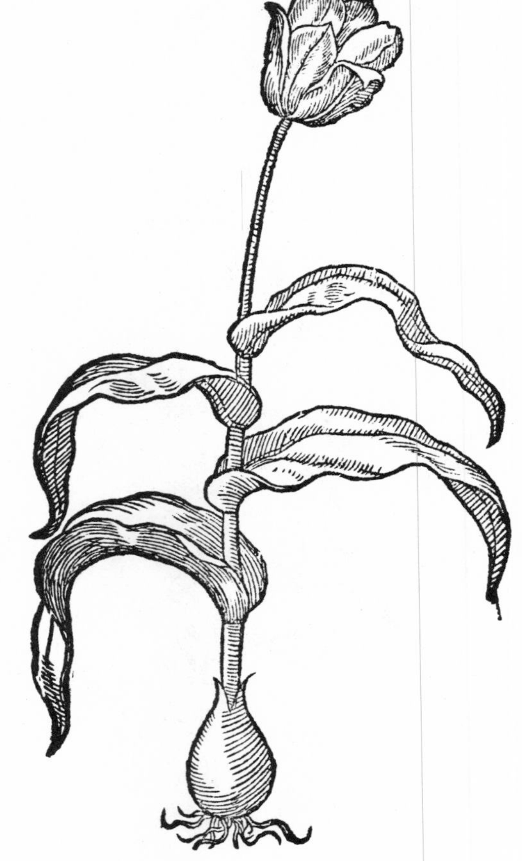

‡ 26 *Tulipa rubra oris pallidis.*
The red Tulip with pale edges.

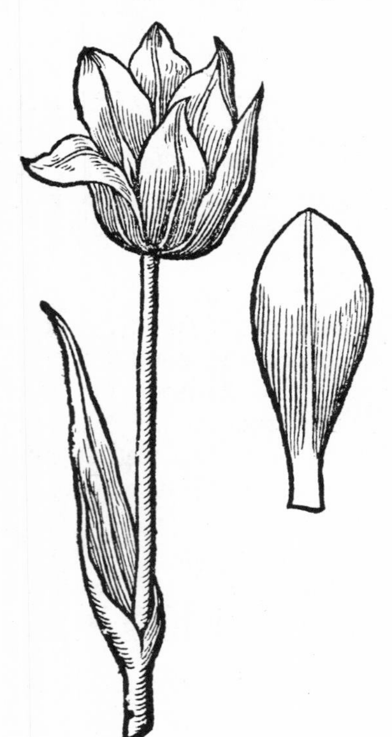

12 There is another alſo in our London gardens, of a ſnow white colour; the edges ſlightly waſht ouer with a little of that we call bluſh colour.

13 We haue another like the former, ſauing that his floure is of a ſtraw colour.

14 There is another to be ſeene with a floure mixed with ſtreaks of red and yellow, reſembling a flame of fire, wherupon we haue called it Flambant.

There be likewiſe ſo many more differing ſo notably in colour of their floures, although in leaues, ſtalke, and roots for the moſt part one like another, that (as I ſaid before) to ſpeake of them ſeuerally would require a peculiar volume.

‡ Therefore not to trouble you any further, I haue giuen you onely the figures and names of the notableſt differences which are in ſhape; as, the dwarfe Tulipa's, and the branched ones, together with the colour of their floures, contained in their titles, that you need not far to ſeeke it. ‡

There be a ſort greater than the reſt, which in forme are like; the leaues whereof are thicke, long, broad, now and then ſomewhat folded in the edges; in the middeſt whereof doth riſe vp a ſtalk a foot high, or ſomthing higher, vpon which ſtandeth onely one floure bolt vpright, conſiſting of ſix leaues, after a ſort like to a deepe wide cup of this forme, *viz.* the bottome turned vpwards, with threads threds or chiues in the middle, of the colour of Saffron. The colour of the floure is ſometimes yellow, ſometimes white, now and then as it were of a light purple, and many times red; and in this there is no ſmall varieties of colours, for the edges of the leaues, and oftentimes the nailes or lower part of the leaues are now & then otherwiſe coloured than the leaues themſelues, and many times there doth runne all along theſe ſtreakes ſome other colours. They haue no ſmell at all that can be perceiued. The roots of theſe are likewiſe bulbed, or Onion faſhion; euery of the which to ſet forth ſeuerally would trouble the writer, and wearie the Reader; ſo that, what hath bin ſaid ſhall ſuffice touching the deſcription of Tulipa's. ‡ True it is that our Author here affirmes, The varieties of theſe floures are ſo infinite, that it would both tyre the Writer and Reader to recount them. Yet for that ſome are more in loue with floures than with Plants in generall, I haue thought good to direct them where they may finde ſomewhat more at large of this Plant: Let ſuch therefore as deſire further ſatisfaction herein haue recourſe to the Florilegies of *De Bry*, *Swerts*, *Robin*, or to M. *Parkinſon*, who hath not onely largely treated of the floures in particular, but alſo of the ordering of them. ‡

‡ 27 *Tulipa lutea ſerotina.*
The late flouring yellow Tulip.

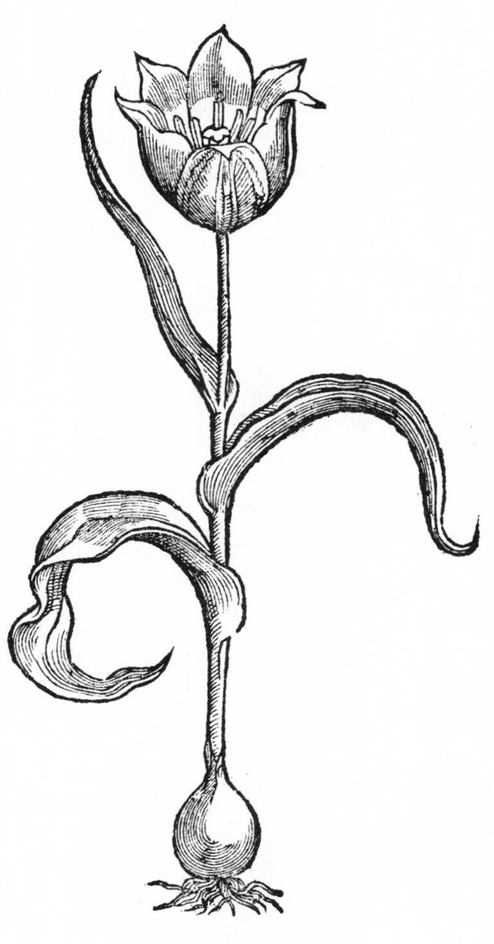

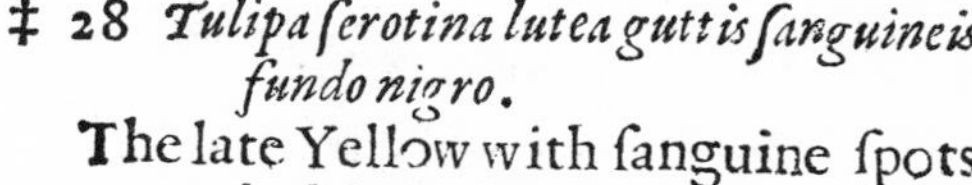
‡ 28 *Tulipa ſerotina lutea guttis ſanguineis fundo nigro.*
The late Yellow with ſanguine ſpots and a blacke bottome.

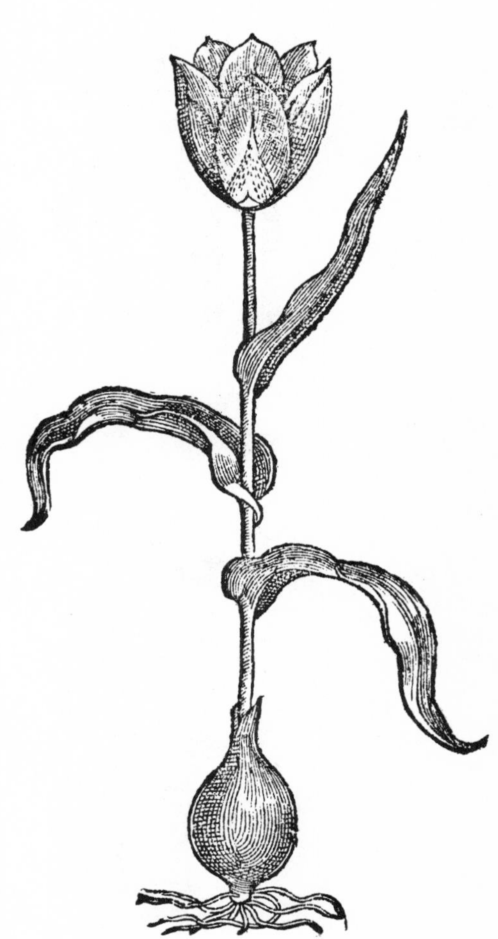

¶ *The Place.*

Tulipa groweth wilde in Thracia, Cappadocia, and Italy; in Bizantia about Conſtantinople, at Tripolis and Alepo in Syria. They are now common in all the gardens of ſuch as affect floures, all ouer England.

¶ *The Time.*

They floure from the end of Februarie vnto the beginning of May, and ſomewhat after; although *Augerius Busbequius* in his journey to Conſtantinople, ſaw between Hadrianople and Conſtantinople, great aboundance of them in floure euery where, euen in the middeſt of Winter, in the moneth of Ianuarie, which that warme and temperate climate may ſeeme to performe.

¶ *The*

The Names.

The later Herbariſts by a Turkiſh and ſtrange name call it *Tulipa,* of the Dalmatian Cap called Tulipa, the forme whereof, the floure when it is open ſeemeth to repreſent.

It is called in Engliſh after the Turkiſh name Tulipa, or it may be called Dalmatian Cap, or the Turkes Cap. What name the antient Writers gaue it is not certainly knowne. A man might ſuſpect it to be τοπιών, if it were a Bulbe that might be eaten, and were of force to make milke cruddy, for *Theophraſtus* reckoneth it among thoſe Bulbes that may be eaten: and it is an herbe, as *Heſychius* ſaith, wherewith milke is crudded. *Conradus Geſnerus* and diuers others haue taken *Tulipa* to be that *Satyrium* which is ſyrnamed *Erythronium,* becauſe one kinde hath a red floure; or altogether a certaine kinde of *Satyrium*: with which it doth agree reaſonable well, if in *Dioſcorides* his deſcription we may in ſtead of λινοσπέρμω, reade κρινοσπέρμῳ or λειριοσπέρμω; for ſuch miſtakes are frequent in antient and moderne Authors, both in writing and printing. In the Turky Tongue it is called *Café lalé, Cauále lalé,* and likewiſe *Turban* and *Turfan,* of the Turks Cap ſo called, as beforeſaid of *Lobelius.*

‡ 29 *Tulipa Holias alba ſtrijs & punctis ſanguineis.*
The white Holias with ſanguine ſpots and ſtreakes.

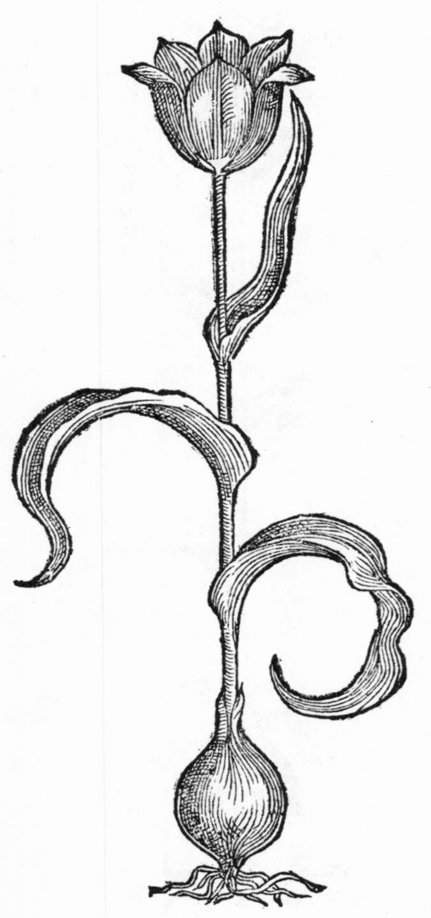

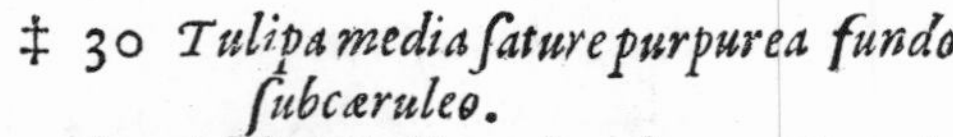
‡ 30 *Tulipa media ſature purpurea fundo ſubcæruleo.*
A middle Tulip of a deepe Purple colour with a blewiſh bottome.

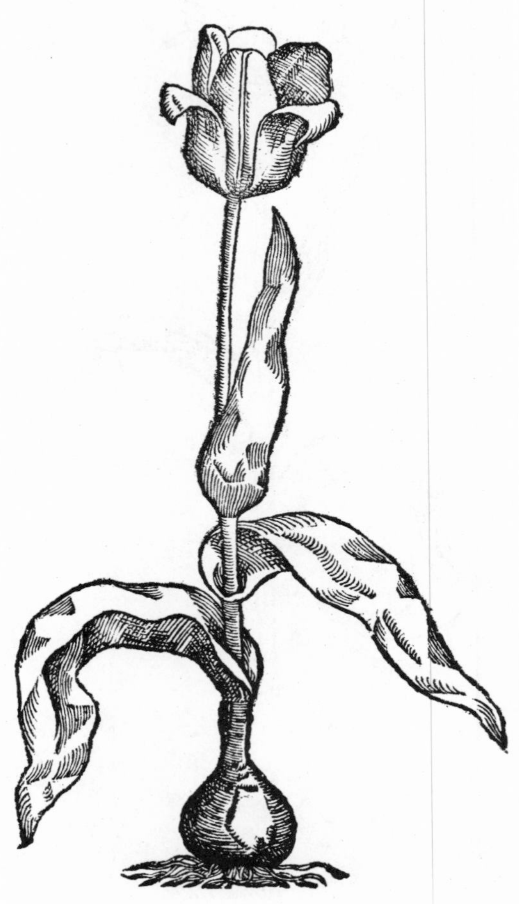

‡ I do verily thinke that theſe are the κρίνα τοῦ ἀγροῦ, the Lillies of the field mentioned by our Sauiour, *Mat.6.28,29.* for he ſaith, That *Solomon* in all his royaltie was not arayed like one of theſe. The reaſons that induce me to thinke thus are theſe: Firſt, their ſhape; for their floures reſemble Lillies, and in theſe places whereas our Sauiour was conuerſant they grow wilde in the fields. Secondly, the infinite varietie of colour, which is to be found more in this than any other ſort of floure: and thirdly, the wondrous beautie and mixtures of theſe floures. This is my opinion, and theſe my reaſons, which any may either approue of or gainſay as he ſhall thinke good. ‡

¶ *The Temperature and Vertues.*

There hath not beene any thing ſet downe of the antient or later Writers as touching the Nature or Vertues of the Tulipa's, but they are eſteemed eſpecially for the beauty of their floures.

‡ The

‡ The roots preserued with sugar, or otherwise dressed, may be eaten, and are no vnpleasant nor any way offensiue meat, but rather good and nourishing. ‡ A

CHAP. 88. *Of Bulbous Violets.*

¶ *The Kindes.*

THeophrastus hath mentioned one kinde of bulbous *Leucoion*, which *Gaza* translates *Viola alba*, or the white Violet. Of this *Viola Theophrasti*, or *Theophrastus* his Violet, we haue obserued three sorts, whereof some bring forth many floures and leaues, others fewer; some floure very early, and others later, as shall be declared.

1 *Leucoium bulbosum præcox minus.*
Timely flouring bulbous Violet.

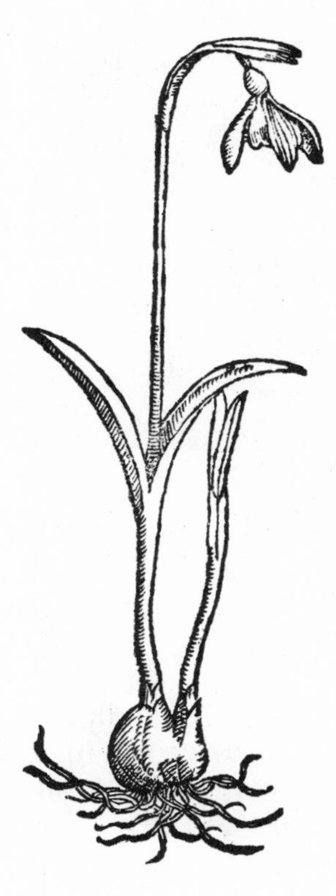

‡ 2 *Leucoium bulbosum præcox Byzantinum.*
The Byzantine early bulbous Violet.

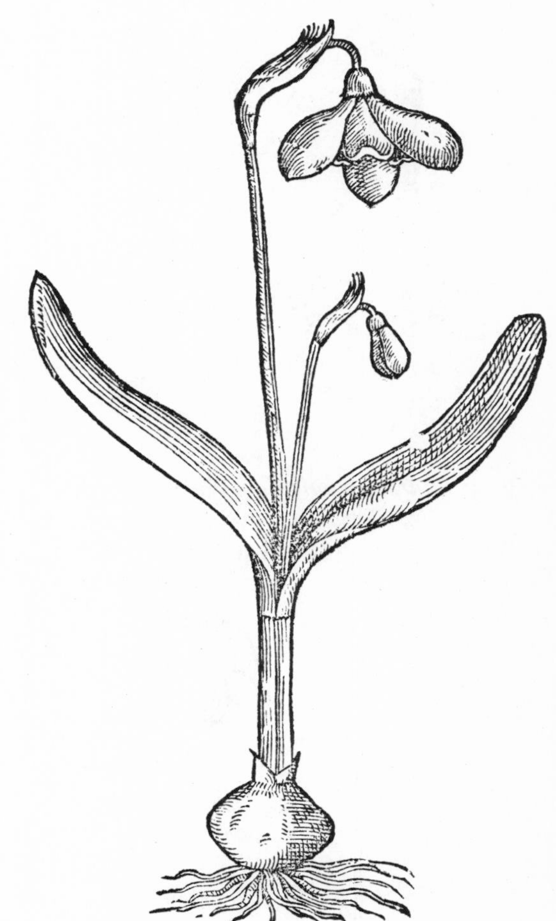

¶ *The Description.*

1 THe first of these bulbous Violets riseth out of the ground, with two small leaues flat and crested, of an ouerworne greene colour, betweene the which riseth vp a small and tender stalke of two hands high; at the top whereof commeth forth of a skinny hood a small white floure of the bignesse of a Violet, compact of six leaues, three bigger, and three lesser, tipped at the points with a light greene: the smaller are fashioned into the vulgar forme of a heart, and pretily edged about with greene; the other three leaues are longer, and sharpe pointed. The whole floure hangeth downe his head, by reason of the weake foot stalke whereon it groweth. The root is small, white, and bulbous.

‡ 2 There are two varieties of this kind which differ little in shape, but the first hath a floure as bigge againe as the ordinarie one, and *Clusius* calls it *Leucoium bulbosum præcox Byzantinum*, The greater early Constantinopolitan bulbous Violet. The other is mentioned by *Lobel*, and differs onely in colour of floures; wherefore he calls it *Leucoium triphyllum flore cæruleo*, The blew floured bulbous Violet.

3 The

3 *Leucoium bulbosum serotinum.*
Late flouring bulbous Violet.

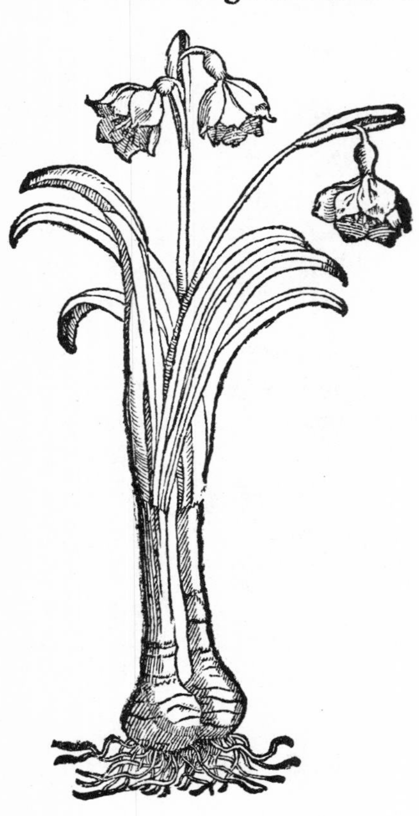

4 *Leucoium bulbosum majus polyanthemum.*
The many floured great bulbous violet.

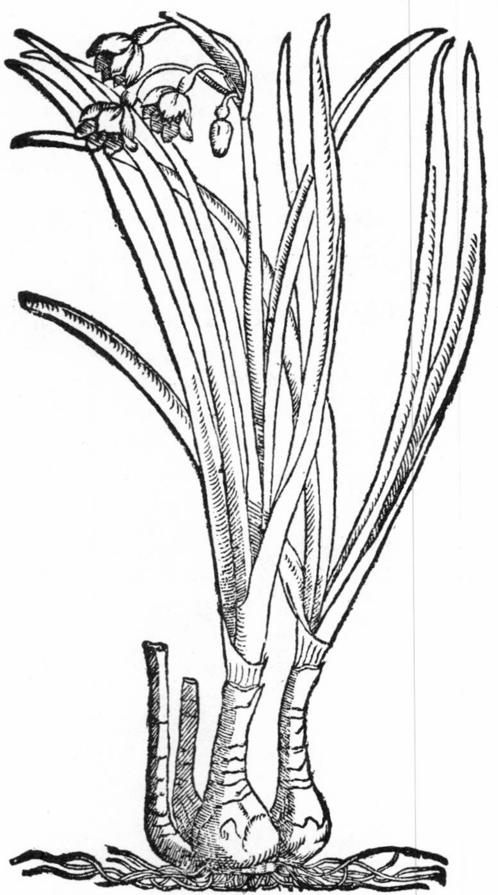

‡ 5 *Leucoium bulbosum Autumnale minimum.*
The least Autumne bulbous Violet.

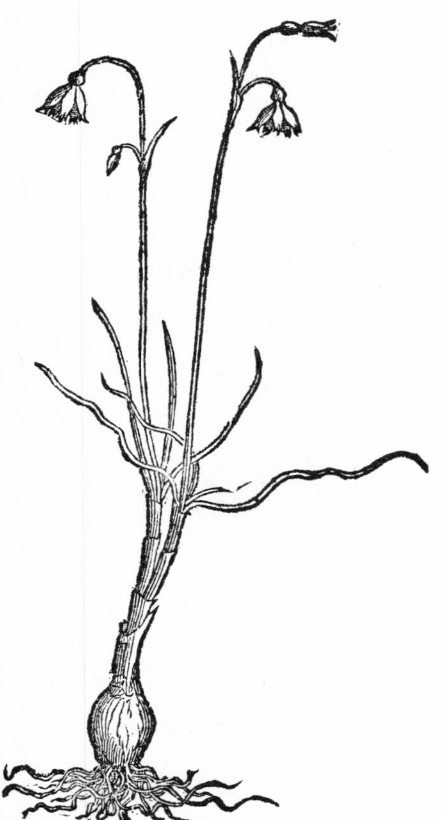

3 The third sort of bulbed Violets hath narrow leaues like those of the leeke, but lesser and smoother, not vnlike to the leaues of the bastard Daffodill. The stalks be slender and naked, two hands high, whereupon doe grow faire white floures, tipped with a yellowish greene colour, with many small chiues or threds in the middest of the floure. The seed is contained in smal round buttons. The root is white and bulbous.

4 The great bulbed Violet is like vnto the third in stalke and leaues, yet greater and higher. It bringeth forth on euery stalke not one floure onely, but fiue or six, blowing or flouring one after another, altogether like the other floures in forme and bignesse.

‡ 5 This small bulbous plant may be annexed to the former, the root is small, compact of many coats: the leaues are also small, and the stalke an handfull high, at the top whereof there hang downe one or two small white floures consisting of six leaues a piece, much resembling the last described, but farre lesse. It floures in Autumne.

6 Besides these, *Clusius* makes mention of a small one much like this, and it floures in the Spring, and the floures are somewhat reddish nigh rhe stalke, and smell sweet. *Clusius* cals this, *Leucoium bulbosum vernum minimum*, The smallest Spring bulbous Violet. ‡

¶ *The*

¶ *The Place.*

Theſe plants do grow wilde in Italy and the places adiacent. Notwithſtanding our London gardens haue taken poſſeſſion of moſt of them many yeares paſt.

¶ *The Time.*

The firſt floureth in the beginning of Ianuary; the ſecond in September; and the third in May; the reſt at their ſeaſons mentioned in their deſcriptions.

¶ *The Names.*

† The firſt is called of *Theophraſtus*, Λευκόιον; which *Gaza* renders *Viola alba*, and *Viola Bulboſa*, or Bulbed Violet. *Lobelius* hath from the colour and ſhape called it *Leuconarciſſolirion*, and that very properly, conſidering how it doth as it were participate of two ſundry plants, that is to ſay, the root of the *Narciſſus*, the leaues of the ſmall Lilly, and the white colour; taking the firſt part *Leuco*, of his whiteneſſe; *Narciſſo*, of the likeneſſe the roots haue vnto *Narciſſus*; and *Lirium*, of the leaues of Lillies, as aforeſaid. In Engliſh we may call it the bulbous Violet; or after the Dutch name, **Somer ſotteken**s; that is, Sommer fooles, and **Druyfkens**. Some call them alſo Snow drops. This name *Leucoium*, without his Epithite *Bulboſum*, is taken for the Wall-floure, and ſtock Gilloſloure, by all moderne Writers.

¶ *The Nature and Vertues.*

Touching the faculties of theſe bulbous Violets we haue nothing to ſay, ſeeing that nothing is ſet downe hereof by the antient Writers, nor any thing obſerued by the modderne, only they are maintained and cheriſhed in gardens for the beautie and rareneſſe of the floures, and ſweetneſſe of their ſmell.

CHAP. 89. *Of Turkie or Ginny-hen Floure.*

1 *Frittillaria.*
Checquered Daffodill.

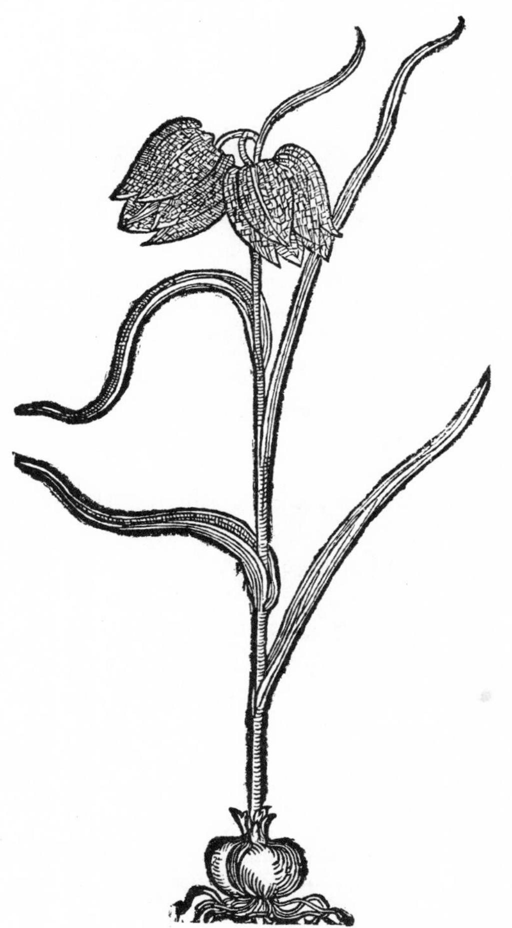

2 *Frittillaria variegata.*
Changeable Checquered Daffodil.

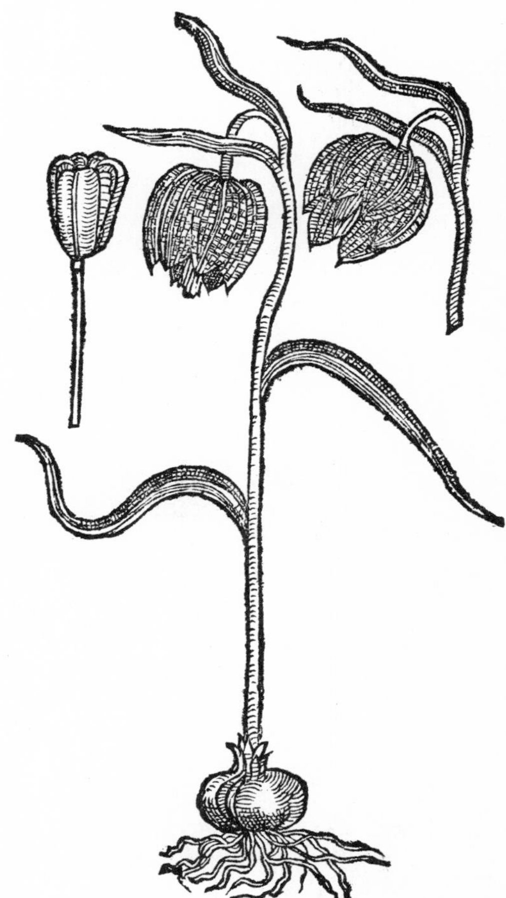

¶ *The*

¶ *The Description.*

1 THe Checquered Daffodill, or Ginny-hen Floure, hath small narrow grassie leaues; among which there riseth vp a stalke three hands high, hauing at the top one or two floures, and sometimes three, which consisteth of six small leaues checquered most strangely: wherein Nature, or rather the Creator of all things, hath kept a very wonderfull order, surpassing (as in all other things) the curiousest painting that Art can set downe. One square is of a greenish yellow colour, the other purple, keeping the same order as well on the backside of the floure, as on the inside, although they are blackish in one square, and of a Violet colour in an other; insomuch that euery leafe seemeth to be the feather of a Ginny hen, whereof it tooke his name. The root is small, white, and of the bignesse of halfe a garden beane.

2 The second kinde of Checquered Daffodill is like vnto the former in each respect, sauing that this hath his floure dasht ouer with a light purple, and is somewhat greater than the other, wherein consisteth the difference.

‡ 3 *Frittillaria Aquitanica minor flore luteo obsoleto.*
The lesser darke yellow Fritillarie.

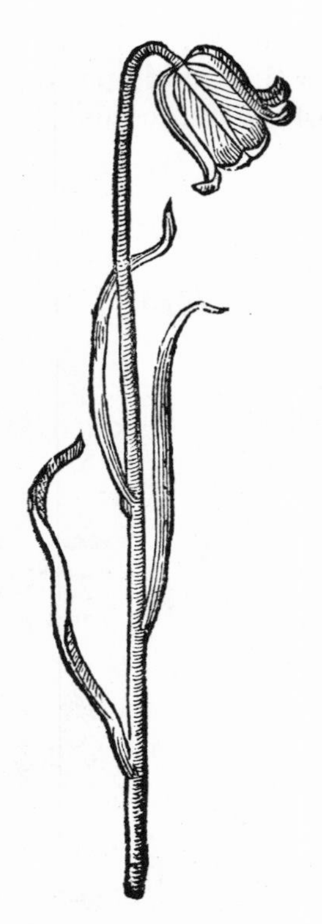

‡ 9 *Frittillaria alba præcox.*
The early white Fritillarie.

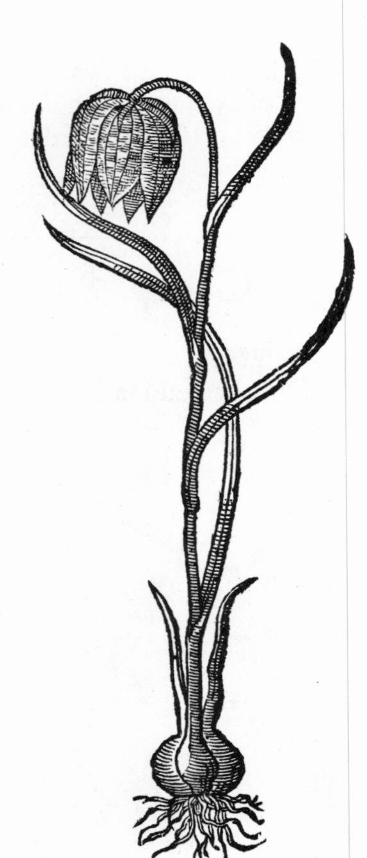

‡ There are sundry differences and varieties of this floure, taken from the colour, largenes, doublenesse, earlinesse and latenes of flouring, as also from the many or few branches bearing floures. We will onely specifie their varieties by their names, seeing their forme differs little from those you haue here described.

4 *Fritillaria maxima ramosa purpurea.* The greatest branched purple checquered Daffodill.

5 *Fritillaria flore purpureo pleno.* The double purple floured checquered Daffodill.

6 *Fritillaria polyanthos flauoviridis.* The yellowish greene many floured checquered Daffodill.

7 *Fritillaria lutea* Someri. *Somers* his yellow Checquered Daffodill.

8 *Fritillaria alba purpureo tessulata.* The white Fritillarie checquered with purple.

9 *Fritillaria alba præcox.* The early white Fritillarie or Checquered Daffodill.

10 *Fritillaria minor flore luteo absoleto.* The lesser darke yellow Fritillarie.

11 *Fritillaria angustifolia lutea variegata paruo flore, & altera flore majore.* Narrow leaued yellow variegated Fritillarie with small floures; and another with a larger floure.

12 *Fritillaria minima pluribus floribus.* The least Fritillarie with many floures.

13 *Fritil*

Fritillaria Hiſpanica vmbellifera. The Spaniſh Fritillarie with the floures ſtanding as it were in an vmbell. ‡

¶ *The Names.*

The Ginny hen floure is called of *Dodonæus*, *Flos Meleagris* : of *Lobelius*, *Lilio-narciſſus variegata*, for that it hath the floure of a Lilly, and the root of *Narciſſus*: it hath beene called *Fritillaria*, of the table or boord vpon which men play at Cheſſe, which ſquare checkers the floure doth very much reſemble; ſome thinking that it was named *Fritillus* : whereof there is no certaintie; for *Martialis* ſeemeth to call *Fritillus*, *Abacus*, or the Tables whereat men play at Dice, in the fifth Booke of his Epigrams, writing to *Galla*.

Iam triſtis, nucibus puer relictis,
Clamoſo reuocatur à magiſtro :
Et blando malè proditus Fritillo
Arcana modò raptus è popina
Ædilem rogat vdus aleator. &c.

The ſad Boy now his nuts caſt by,
Call'd vnto Schole by Maſters cry :
And the drunke Dicer now betray'd
By flattring Tables as he play'd,
Is from his ſecret tipling houſe drawne out,
Although the Officer he much beſought. &c.

In Engliſh we may call it Turky-hen or Ginny-hen Floure, and alſo Checquered Daffodill, and Fritillarie, according to the Latine.

¶ *The Temperature and Vertues.*

Of the facultie of theſe pleaſant floures there is nothing ſet downe in the antient or later Writer, but are greatly eſteemed for the beautifying of our gardens, and the boſoms of the beautifull.

CHAP. 90. *Of true Saffron, and the wilde or Spring Saffrons.*

Crocus florens & ſine flore. Saffron with and without floure.

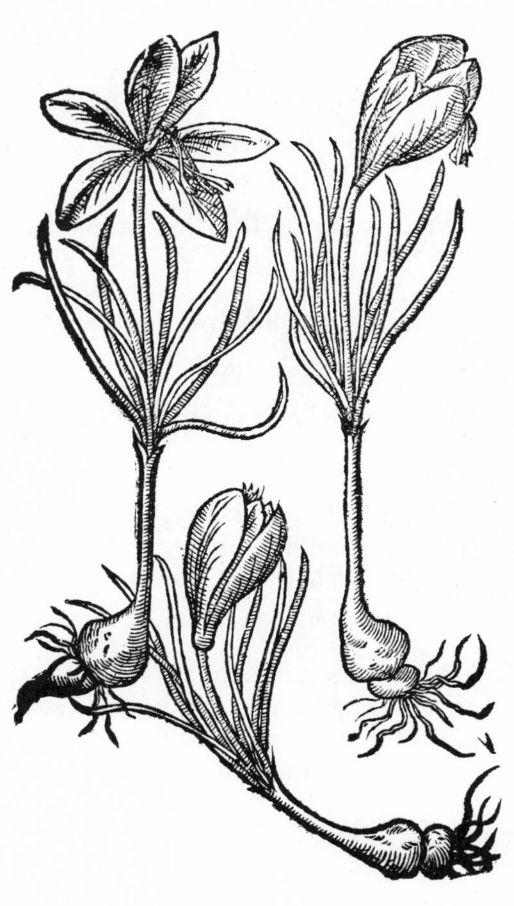

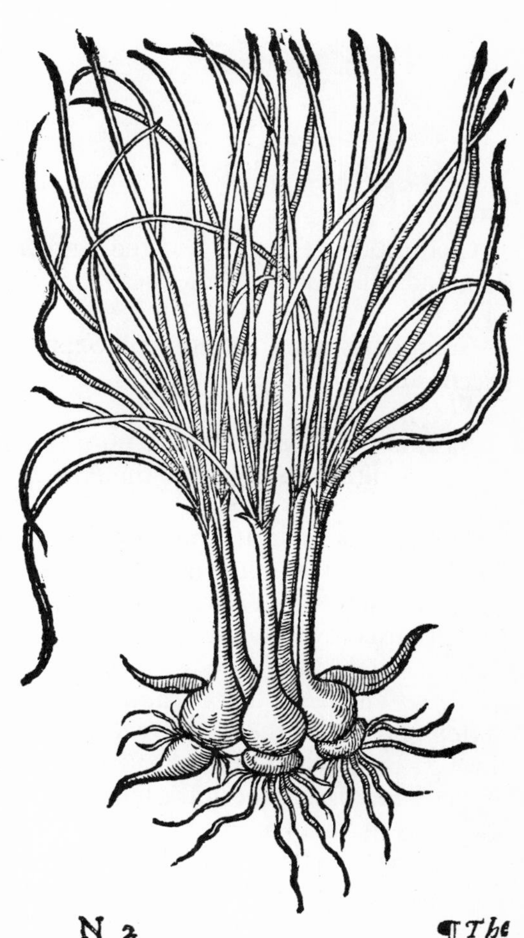

The Deſcription.

ALthough I haue expreſſed two pictures of Saffrons,as you ſee, yet are you to vnderſtand that theſe two do but ſet forth one kinde of plant,which could not ſo eaſily be perceiued by one picture as by two,becauſe his floure doth firſt riſe out of the ground nakedly in September, and his long ſmal graſſy leaues ſhortly after the floure,neuer bearing floure and leafe at once. The which to expreſſe,I thought it conuenient to ſet downe two pictures before you,with this deſcription, *viz.* The root is ſmall, round, and bulbous. The floure conſiſteth of ſixe ſmall blew leaues tending to purple, hauing in the middle many ſmall yellow ſtrings or threds ; among which are two,three, or more thicke fat chiues of a fierie colour ſomewhat reddiſh, of a ſtrong ſmell when they be dried, which doth ſtuffe and trouble the head. The firſt picture ſetteth forth the Plant when it beareth floures, and the other expreſſeth nothing but leaues.

1 *Crocus vernus.*
Early flouring wilde Saffron.

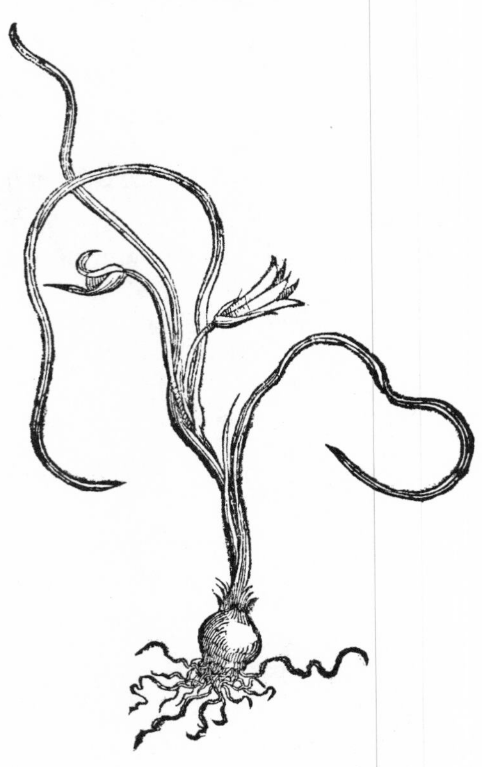

2 *Crocus vernus minor.*
Small wilde Saffron.

¶ *The Place.*

Common, or the beſt knowne Saffron groweth plentifully in Cambridge-ſhire, Saffron-Walden, and other places thereabout, as corne in the fields.

¶ *The Time.*

Saffron beginneth to floure in September, and preſently after ſpring vp the leaues,and remaine greene all the Winter long.

¶ *The Names.*

Saffron is called in Greeke, Κρόκος : in Latine, *Crocus :* in Mauritania, *Saffaran :* in Spaniſh, *Açafron :* in Engliſh, Saffron : in the Arabicke tongue, *Zahafaran.*

¶ *The Temperature.*

Saffron is a little aſtringent or binding, but his hot qualitie doth ſo ouer-rule in it,that in the whole eſſence it is in the number of thoſe herbes which are hot in the ſecond degree, and drie in the firſt : therefore it alſo hath a certaine force to concoct, which is furthered by the ſmall aſtriction that is in it, as *Galen* ſaith.

¶ *The Vertues.*

A *Auicen* affirmeth that it cauſeth head-ache, and is hurtfull to the braine, which it cannot do by taking it now and then, but by too much vſing of it : for too much vſing of it cutteth off ſleepe, through want whereof the head and ſences are out of frame. But the moderate vſe of it is good for the head, and maketh the ſences more quicke and liuely,ſhaketh off heauy and drowſie ſleepe, and maketh a man merry.

B Alſo Saffron ſtrengthneth the heart, concocteth crude and raw humors of the cheſt, openeth the lungs, and remoueth obſtructions.

It is

‡ 3 *Crocus vernus flore luteo.*
Yellow Spring Saffron.

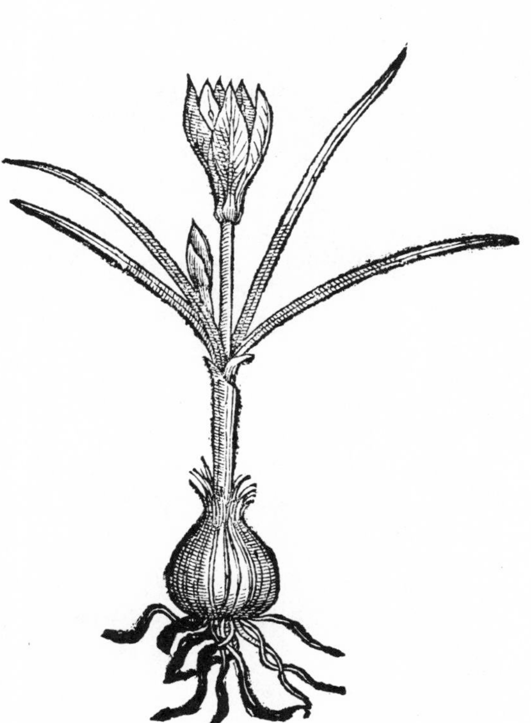

‡ 4 *Crocus vernus flore albo.*
White Spring Saffron.

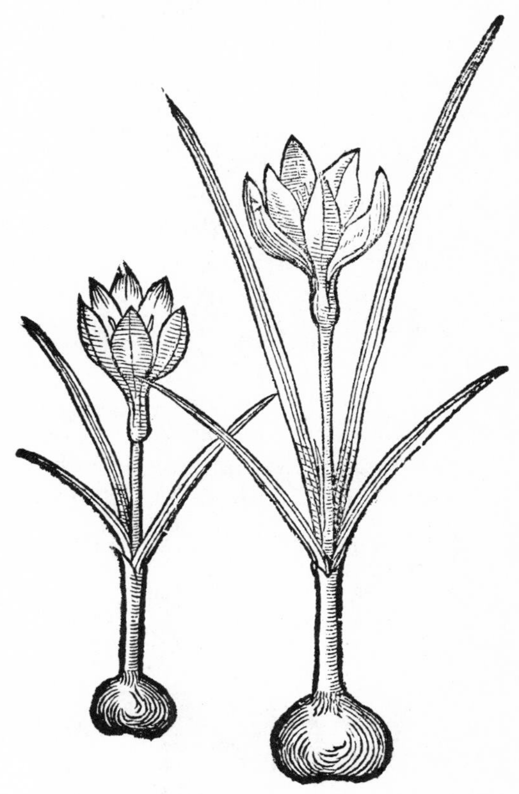

‡ 5 *Crocus vernus flore purpureo.*
Purple Spring Saffron.

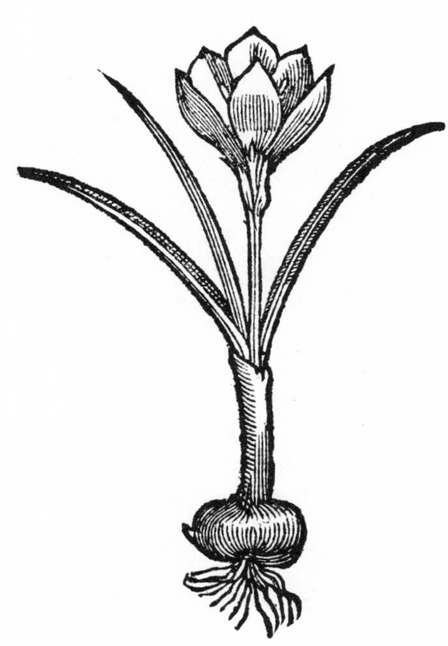

‡ 6 *Crocus montanus Autumnalis.*
Autumne mountaine Saffron.

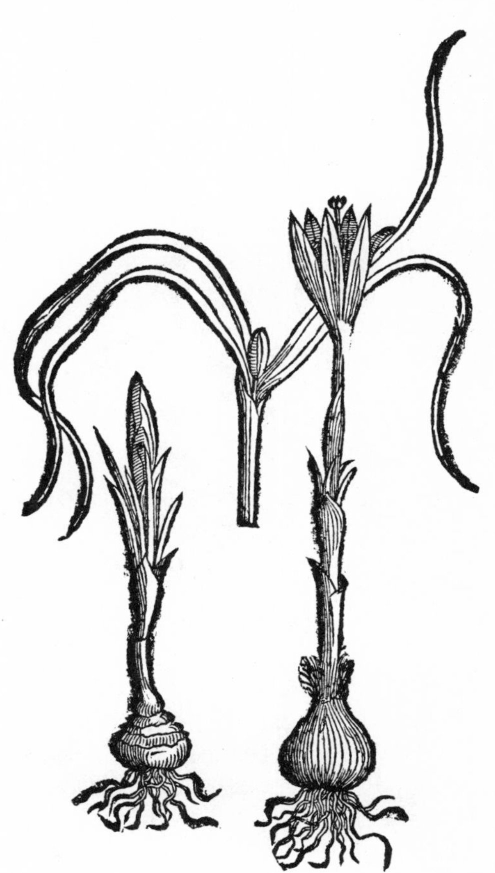

C It is also such a speciall remedie for those that haue consumption of the lungs, and are, as wee terme it, at deaths doore, and almost past breathing, that it bringeth breath again, and prolongeth life for certaine dayes, if ten, or twentie graines at the most be giuen with new or sweet Wine. For we haue found by often experience, that being taken in that sort, it presently and in a moment remoueth away difficultie of breathing, which most dangerously and suddenly hapneth.

D *Dioscorides* teacheth, That being giuen in the same sort it is also good against a surfet.

E It is commended against the stoppings of the liuer and gall, and against the yellow Iaundise: And hereupon *Dioscorides* writeth, That it maketh a man well coloured. It is put into all drinkes that are made to helpe the diseases of the intrailes, as the same Authour affirmeth, and into those especially which bring downe the floures, the birth, and the after burthen. It prouoketh vrine, stirreth fleshly lust, and is vsed in Cataplasmes and pultesses for the matrix and fundament, and also in plaisters and seare-cloaths which serue for old swellings and aches, and likewise for hot swellings that haue also in them S. Anthonies fire.

‡ 7 *Crocus montanus Autumnalis flore majore albido cæruleo.*
Autumne mountaine Saffron with a large whitish blew floure.

‡ 8 *Crocus Autumnalis flore albo.*
White Autumne Saffron.

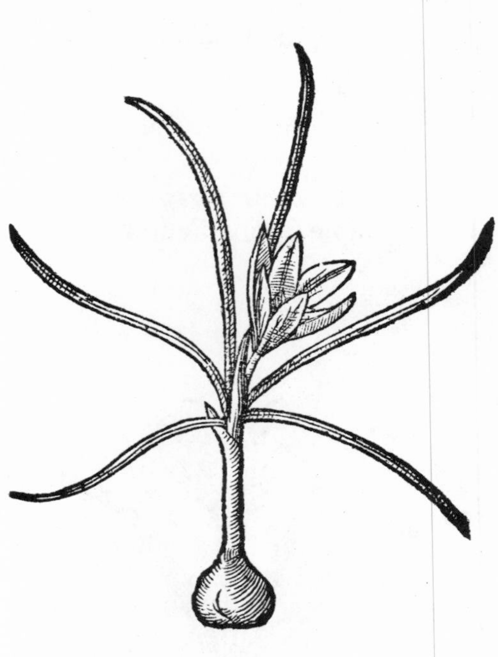

F It is with good successe put into compositions for infirmities of the eares.

G The eyes being annointed with the same dissolued in milke, or fennell or rose water, are preserued from being hurt by the small pox and measels, and are defended thereby from humours that would fall into them.

H The chiues steeped in water, serue to illumine or (as we say) limne pictures and imagerie, as also to colour sundry meats and confections. It is with good successe giuen to procure bodily lust. The confections called *Crocomagna, Oxycroceum,* and *Diacurcuma,* with diuers other emplaisters and electuaries cannot be made without this Saffron.

I The weight of tenne graines of Saffron, the kernels of Wall-nuts two ounces, Figges two ounces, Mithridate one dram, and a few sage leaues, stamped together with a sufficient quantitie of Pimpernell water, and made into a masse or lumpe, and kept in a glasse for your vse, and thereof twelue graines giuen in the morning fasting, preserueth from the Pestilence, and expelleth it from those that are infected.

¶ *The*

‡ 9 *Crocus vernus angustifolius flore violaceo.*
Narrow leaued Spring Saffron
with a violet floure.

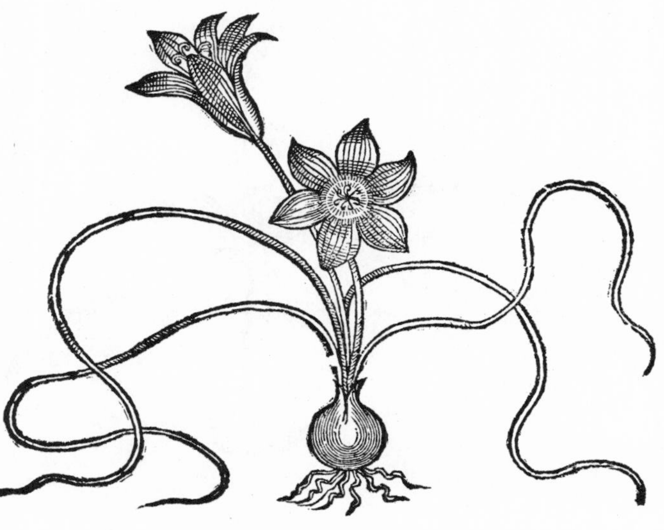

‡ 10 *Crocus vernus latifolius flore flauo strijs violaceis.*
Broad leaued Spring Saffron with
a yellow floure & purple streaks.

‡ 11 *Crocus vernus latifolius striatus flore duplici.*
Double floured streaked Spring
Saffron.

¶ *The Kindes of Spring Saffron*

OF wilde Saffrons there be sundry sorts, differing as well in the colour of the floures, as also in the time of their flouring. Of which, most of the figures shall be set forth vnto you.

¶ *The Description of wilde Saffron.*

1 THe first kind of wilde Saffron hath small short grassie leaues, furrowed or chanelled downe the midst with a white line or streake: among the leaues rise vp small floures in shape like vnto the common Saffron, but differing in colour; for this hath floures of mixt colours; that is to say, the ground of the floure is white, stripped vpon the backe with purple, and dasht ouer on the inside with a bright shining murrey colour; the other not. In the middle of the floures come forth many yellowish chiues, without any smel of saffron at all. The root is small, round, and couered with a browne skinne or filme like vnto the roots of common Saffron.

2 The second wilde Saffron in leaues, roots, and floures is like vnto the precedent, but altogether lesser, and the floures of this are of a purple violet colour.

3 We haue likewise in our London gardens another sort, like vnto the other wilde Saffrons in

‡ 12 *Crocus vernus latifolius flore purpureo.*
Broad leaued Spring Saffron with the purple floure.

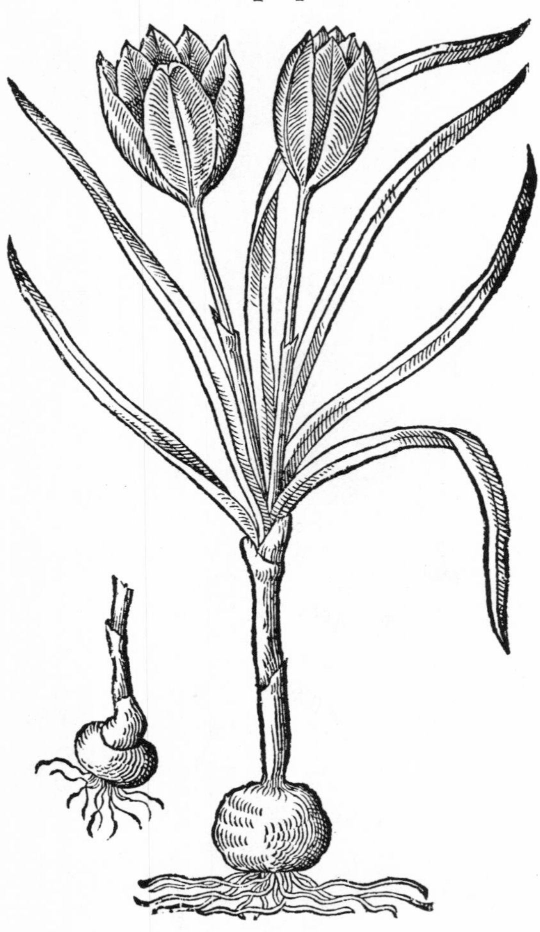

‡ 13 *Crocus vernus flore cinereo striato.*
Spring Saffron with an Ash-coloured streaked floure.

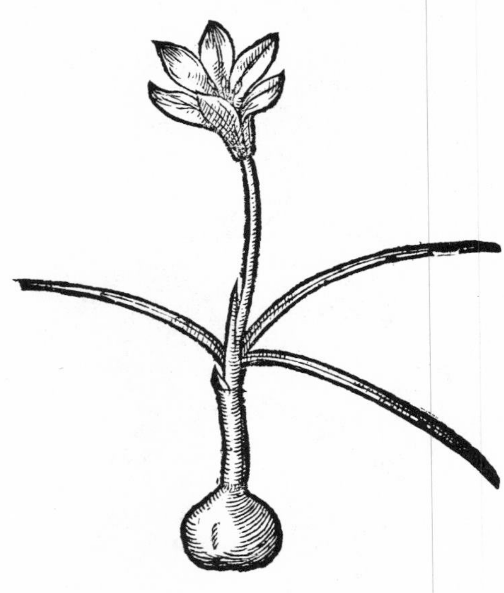

‡ 14 *Crocus vernus latifolius flore flauo-vario duplici.*
Broad leaued Spring Saffron with a double floure yellow & streaked.

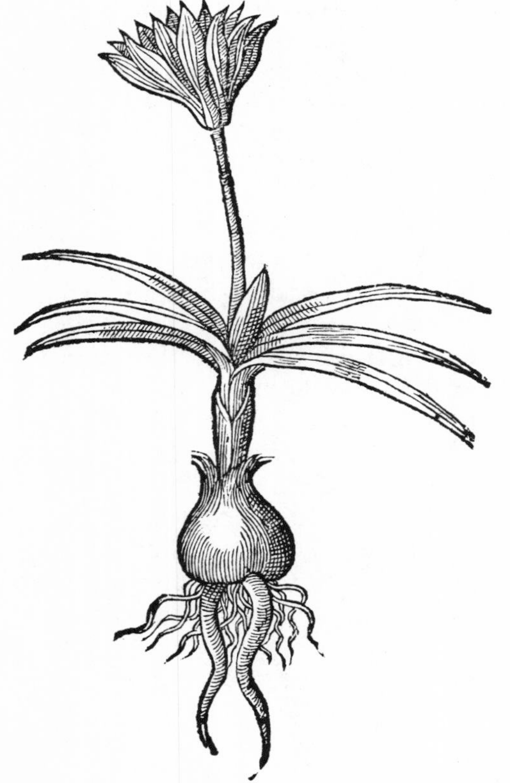

in euery point, sauing that this hath floures of a most perfect shining yellow colour, seeming a far off to be a hot glowing cole of fire, which maketh the difference.

4 There is found among Herbarists another sort, not differing from the others, sauing that this hath white floures, contrarie to all the rest.

5 Louers of Plants haue gotten into their gardens one sort hereof with purple or Violet coloured floures, in other respects like vnto the other.

6 Of these we haue another that floureth in the fall of the leafe, with floures like to the common Saffron, but destitute of those chiues which yeeld the colour, smell, or taste that the right manured Saffron hath.

‡ 7 And of this last kinde there is another with broader leaues, and the floure also is larger, with the leaues thereof not so sharpe pointed, but more round; the colour being at the first whitish, but afterwards intermixt with some blewnesse. ‡

8 There is also another of Autumne wild Saffrons with white floures, which sets forth the distinction.

Many sorts there are in our gardens besides those before specified, which I thought needlesse to entreat of, because their vse is not great. ‡ Therefore I will only giue the figures and names of some of the chiefe of them, and refer such as delight to see or please themselues with the varieties (for they are no specificke differences) of these plants, to the gardens and the bookes of Florists, who are onely the preseruers and admirers of these varieties, not sought after for any vse but delight. ‡

¶ *The*

¶ *The Place.*

All these wilde Saffrons we haue growing in our London Gardens. Those which doe floure in Autumne do grow vpon certaine craggy rockes in Portugall, not far from the sea side. The other haue been sent ouer vnto vs, some out of Italy, and some out of Spaine, by the labour and diligence of that notable learned Herbarist *Carolus Clusius*; out of whose Obseruations, and partly by seeing them in our owne gardens, we haue set downe their descriptions.

That pleasant plant that bringeth forth yellow floures was sent vnto me from *Robinus* of Paris, that painfull and most curious searcher of Simples.

¶ *The Time.*

They floure for the most part in Ianuarie and Februarie; that of the mountain excepted, which floureth in September.

¶ *The Names.*

All these Saffrons are vnprofitable, and therefore they be truly said to be *Croci syluestres*, or wild Saffrons: in English, Spring Saffrons, and vernall Saffrons.

¶ *The Temperature and Vertues.*

Of the faculties of these we haue nothing to set downe, for that as yet there is no knowne vse of them in Physicke.

CHAP. 91. *Of Medow Saffron.*

¶ *The Kindes.*

THere be sundry sorts of Medow Saffrons differing very notably as well in the colour of their floures, as also in stature and Countrey, from whence they had their being, as shall be declared.

1 *Colchicum Anglicum Purpureum.*
Purple English Medow Saffron.

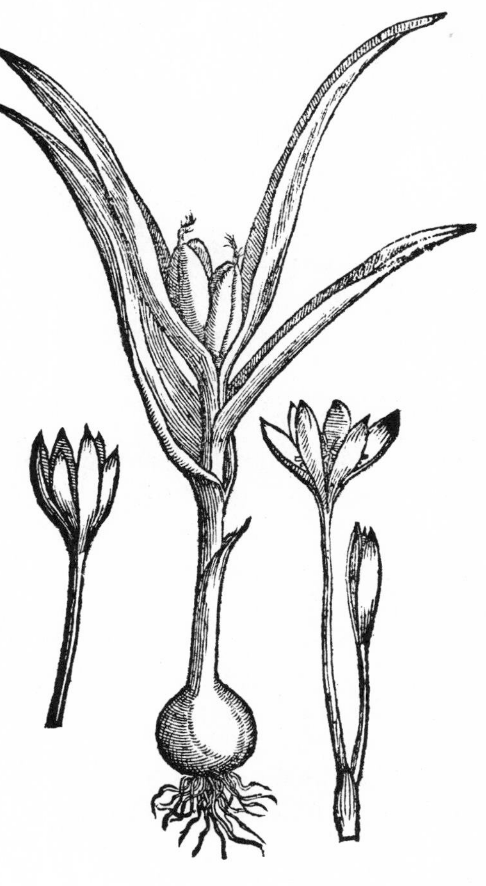

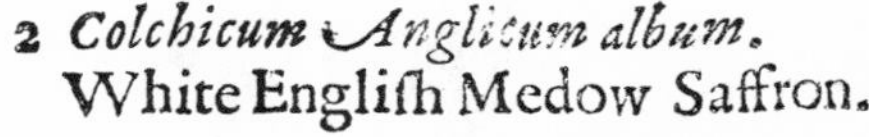

2 *Colchicum Anglicum album.*
White English Medow Saffron.

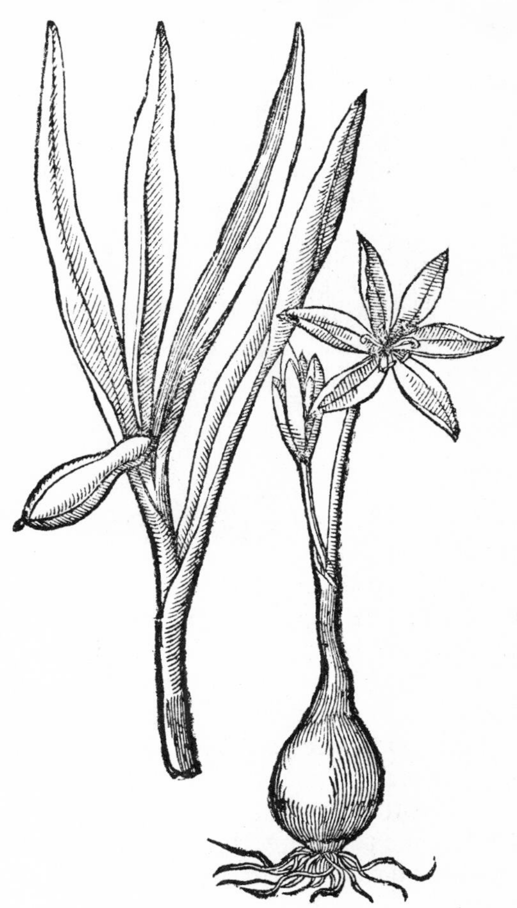

¶ The

¶ *The Description.*

1 MEdow Saffron hath three or foure leaues rising immediately forth of the ground, long, broad, smooth, fat, much like to the leaues of the white Lilly in forme and smoothnesse: in the middle whereof spring vp three or foure thicke cods of the bignesse of a small Wall-nut, standing vpon short tender foot-stalkes three square, and opening themselues when they be ripe, full of seed something round, and of a blackish red colour: and when this seed is ripe, the leaues together with the stalkes doe fade and fall away. In September the floures bud forth, before any leaues appeare, standing vpon short tender and whitish stemmes, like in forme and colour to the floures of Saffron, hauing in the middle small chiues or threads of a pale yellow colour, altogether vnfit for meat or medicine. The root is round or bulbous, sharper at the one end than at the other, flat on the one side, hauing a deepe clift or furrow in the same flat side when it floureth, and not at any time else: it is couered with blackish coats or filmes, it sendeth downe vnto the lowest part certaine strings or threds. The root it selfe is full of a white substance, yeelding a juyce like milke, whilest it is greene and newly digged out of the earth. It is in taste sweet, with a little bitternesse following, which draweth water out of the mouth.

3 *Colchicum Pannonicum florens & sine flore.*
Hungary mede Saffron with and without Floure.

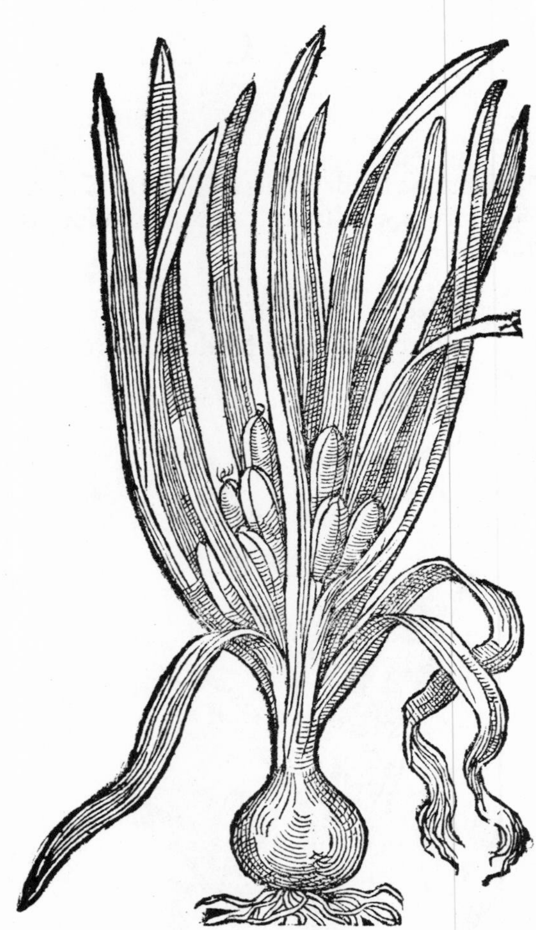

2 The second kinde of Mede Saffron is like the precedent, differing onely in the colour of the floures, for that this plant doth bring forth white leaues, which of some hath beene taken for the true *Hermodactylus*; but in so doing they haue committed the greater error.

3 These two figures expresse both but one and the selfe same plant, which is distinguished because it neuer beareth floures and leaues both at one time. So that the first figure sets it forth when it is in leaues and seed, and the other when it floureth; and therefore one description shall suffice for them both. In the Spring of the yeare it bringeth forth his leaues, thicke, fat, shining, and smooth, not vnlike the leaues of Lillies, which do continue greene vnto the end of Iune; at which time the leaues do wither away, but in the beginning of September there shooteth forth of the ground naked milke white floures without any greene leafe at all: but so soone as the Plant hath done bearing of floures, the root remaines in the ground, not sending forth any thing vntill Februarie in the yeare following.

It

‡ It beares plentifull store of reddish seed in loose triangular heads. The root hereof is bigger than that of the last described. ‡

† 4 The small medow Saffron hath three or foure thicke fat leaues narrower than any of the rest. The floure appeareth in the fall of the leafe, in shape, colour, and manner of growing like the common mede Saffron, but of a more reddish purple colour, and altogether lesser. The leaues in this, contrarie to the nature of these plants, presently follow after the floure, and so continue all the Winter and Spring, euen vntill May or Iune. The root is bulbous, and not great; it is couered with many blackish red coats, and is white within.

‡ 5 This medow Saffron hath roots and leaues like to those of the last described, but the leaues of the floure are longer and narrower, and the colour of them is white on the inside, greene on the middle of the backe part, and the rest thereof of a certaine flesh colour.

4 *Colchicum montanum minus Hispanicum cum flore & semine.*
Small Spanish medow Saffron in floure and seed.

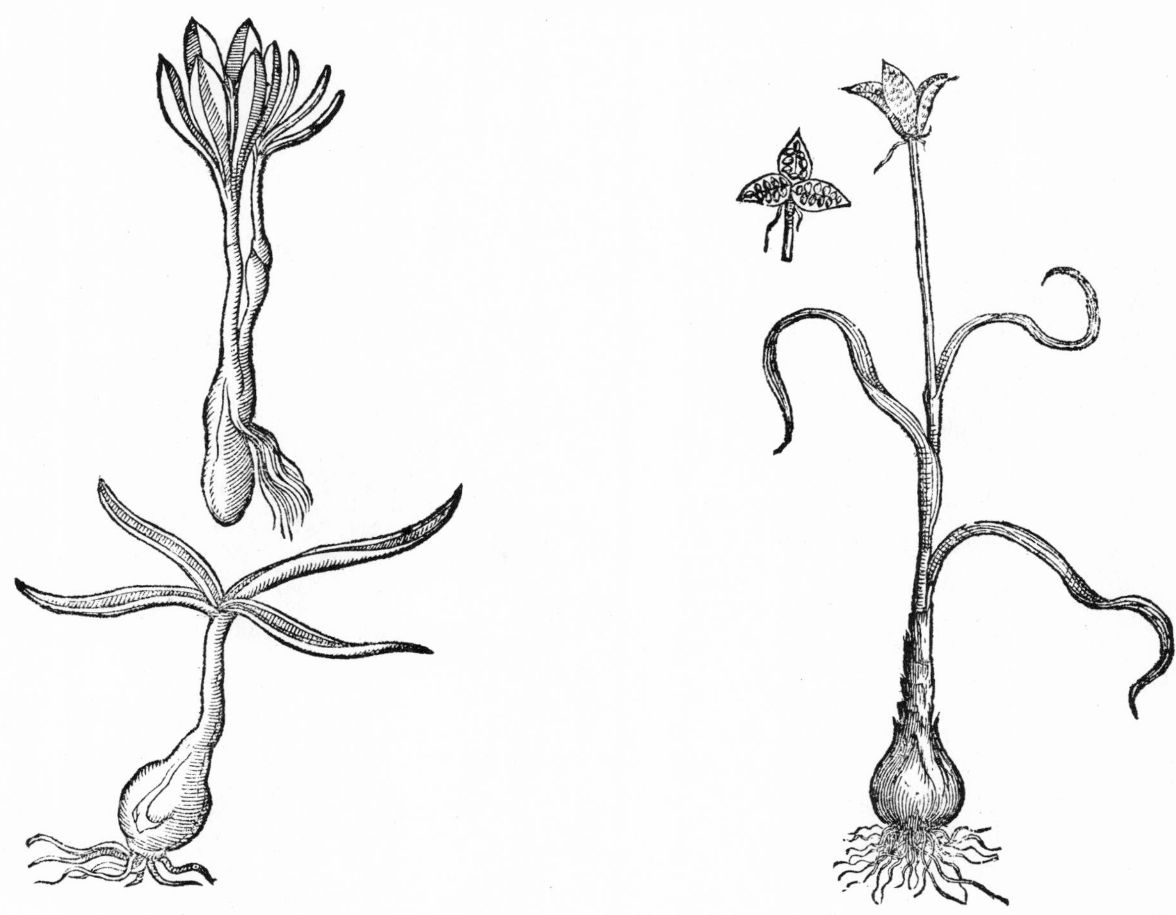

6 The medow Saffron of Illyria hath a great thicke and bulbous root, full of substance: from which riseth vp a fat, thicke, and grosse stalke, set about from the lower part to the top by equall distances, with long, thicke, and grosse leaues, sharpe pointed, not vnlike to the leaues of leekes; among which leaues do grow yellowish floures like vnto the English medow Saffron, but smaller

7 The Assyrian medow Saffron hath a bulbous root, made as it were of two pieces; from the middle cleft whereof riseth vp a soft and tender stalke set with faire broad leaues from the middle to the top: among which commeth forth one single floure like vnto the common medow Saffron, or the white Anemone of *Matthiolus* description.

8 The mountaine wilde Saffron is a base and low plant, but in shape altogether like the common medow Saffron, but much lesser. The floures are smaller, and of a yellow colour, which setteth forth the difference. ‡ The leaues and roots (as *Clusius* affirmes) are more like to the Narcisses; and therefore he calls this *Narcissus Autumnalis minor*, The lesser Autumne Narcisse. ‡

‡ *9* This, whose figure we here giue you, is by *Clusius* called *Colchicum Byzantinum latifolium*, The broad leaued *Colchicum* of Constantinople. The leaues of this are not in forme and magnitude much vnlike to those of the white Hellebor, neither lesse neruous, yet more greene. It beares many floures in Autumne, so that there come sometimes twenty from one root. Their forme and colour are much like the ordinarie sort, but that these are larger, and haue thicker stalkes. They are

are of a lighter purple without, and of a deeper on the inſide, and they are marked with certaine veines running alongſt theſe leaues. The roots and ſeeds of this plant are thrice as large as thoſe of the common kinde.

10 This hath roots and leaues like to the firſt deſcribed, but the floure is ſhorter, and growes vpon a ſhorter ſtalke, ſo that it riſes but little aboue the earth: the three inner leaues are of a reddiſh purple; the three out leaues are either wholly white, or purpliſh on the middle in the inſide, or ſtreaked with faire purple veins, or ſpotted with ſuch coloured ſpots: all the leaues of the floure are blunter and rounder than in the common kinde.

11 This in leaues, roots, manner and time of growing, as alſo in the colour of the floures, differs not from the firſt deſcribed, but the floures, as you may perceiue by the figure here expreſſed, are very double, and conſiſt of many leaues.

‡ 5 *Colchicum montanum minus verſicolore flore.*
The leſſer mountaine Saffron with a various coloured floure.

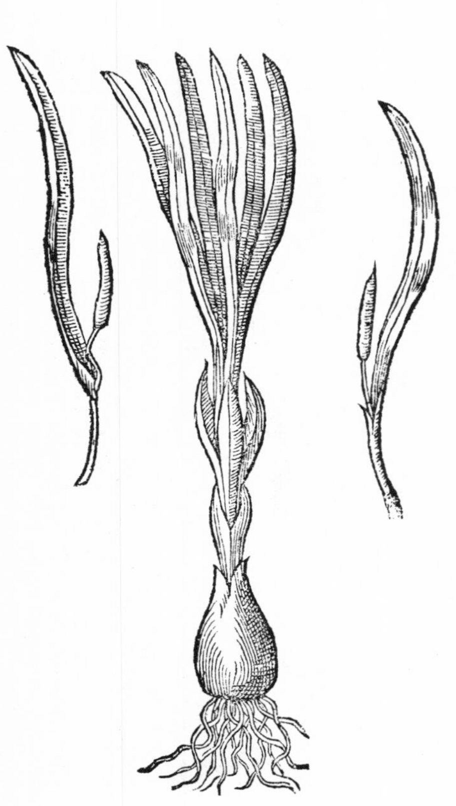

6 *Colchicum Illyricum.*
Greeke medow Saffron.

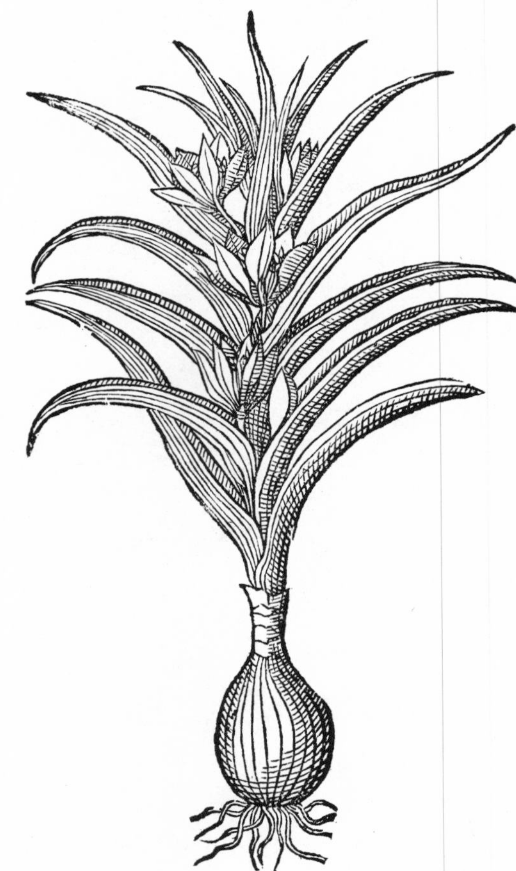

12 This *Colchicum* differs little from the firſt ordinarie one, butthat the floures are ſomewhat leſſe, and the three out-leaues are ſomwhat bigger than the three inner leaues; the colour is a little deeper alſo than that of the common one; but that wherein the principall difference conſiſts, is, That this floures twice in a yeare, to wit, in the Spring and Autumne: and hence *Cluſius* hath called it *Colchicum biflorum*, Twice-flouring Mede Saffron.

13 This alſo in the ſhape of the root and leaues is not much different from the ordinary, but the leaues of the floure are longer and narrower, the colour alſo when they begin to open and ſhew themſelues, is white, but ſhortly after they are changed into a light purple: each leafe of the floure hath a white thread tipt with yellow growing out of it, and in the middle ſtands a white three forked one longer than the reſt. The floure growes vp between three or foure leaues narrower than thoſe of the ordinarie one, and broader than thoſe of the ſmall Spaniſh kinde. *Cluſius*, to whom we are beholden for this, as alſo for moſt of the reſt, calls it *Colchicum vernum*, or Spring Mede-Saffron, becauſe it then floures together with the Spring Saffrons and Dogs Tooth.

14 There are other Mede-Saffrons beſides theſe I haue mentioned, but becauſe they may be referred

7 *Colchicum Syriacum Alexandrinum.*
Assyrian Mede Saffron.

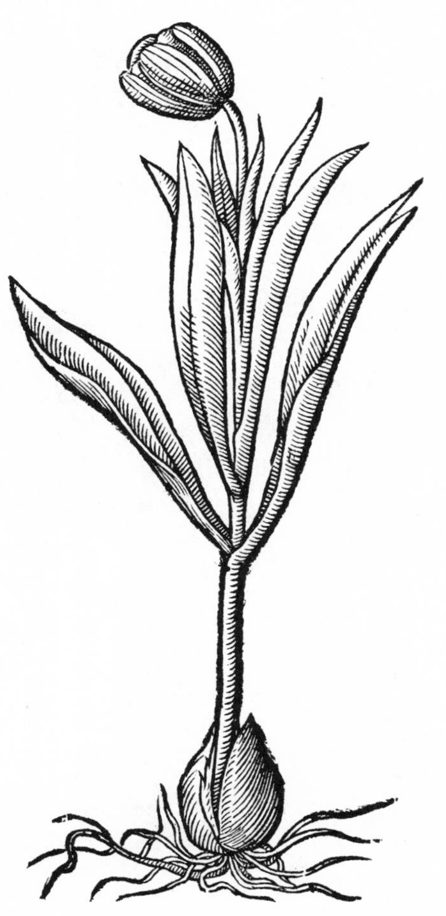

referred easily to some of these, for that their difference chiefely consists either in the doublenesse or colour of the floures, whereof some are striped, some fraided, others variegated, I will not insist vpon them, but referre such as desire their further acquaintance to look into the gardens of our Florists, as M. *Parkinsons*, M. *Tuggies*, *&c.* or else into the booke of floures set forth not long since by M. *Parkinson*, where they shall finde them largely treated of. Yet I cannot passe ouer in silence that curious *Colchicum* which is called by some, *Colchicum variegatum Chiense*. The floure thereof is very beautiful, consisting of six pretty broad and sharp pointed leaues, all curiously checkered ouer with deepe blew or purple, the rest of the floure being of a light whitish colour: the leaues, that rise vp in the Spring, are not very long, but somewhat broad and sharpe pointed; the root is like others of this kinde. I haue giuen you an exact and large figure of this, as I tooke it from the growing floure some three yeares agone, it being at that time amongst her Maiesties floures kept at Edgcombe in Surry, in the garden of my much honoured friend Sir *Iohn Tunstall*, Gentleman Vsher vnto her Maiestie.

15 I giue you here in this place the true Hermodactill of the shops, which probably by all is adiudged to this Tribe, though none can certainly say what floures or leaues it beares: the Roots are onely brought to vs, and from what place I cannot tell; yet I coniecture from some part of Syria or the adiacent countries. Now how hard it is to iudge of Plants by one part or particle, I shall shew you more at large when I come to treat of *Pistolochia*, wherefore I will say nothing thereof in this place. These roots, which wanting the maligne qualitie of *Colchicum*, either of their owne nature, or by drinesse, are commonly about the bignesse of a Chesnut, smooth, flattish, and sharpe at the one end, but somewhat full at the other, and on the one side there is a little channell or hollownesse, as is in the roots of Mede-Saffron where the stalke of the floure comes vp. Their colour is either white, browne, or blackish on the outside, and very white within, but those are the best that are white both without and within, and may easily be made into a fine white meale or pouder. ‡

8 *Colchicum parvum montanum luteum.*
Yellow mountaine Saffron.

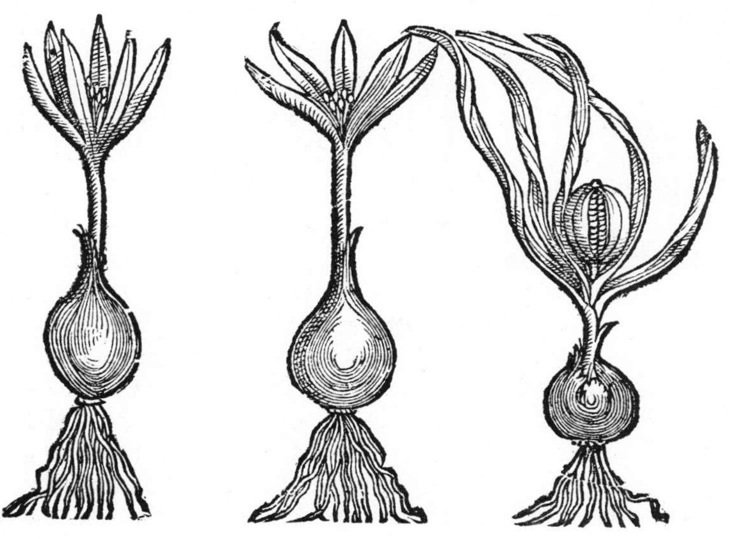

¶ *The Place.*

Medow Saffron, or *Colchicum*, groweth in Meſſinia, and in the Iſle of Colchis, whereof it tooke his name. The titles of the reſt do ſet forth their natiue countries; notwithſtanding our London gardens are poſſeſſed with the moſt part of them.

The two firſt do grow in England in great aboundance, in fat and fertile medowes, as about Vilford and Bathe, as alſo in the medowes neere to a ſmall village in the Weſt part of England, called Shepton Mallet, in the medowes about Briſtoll, in Kingſtroppe medow neere vnto a Water-mill as you go from Northampton to Holmeby Houſe, vpon the right hand of the way, and likewiſe in great plenty in Nobottle wood two miles from the ſaid towne of Northampton, and many other places. ‡ The reſt for the moſt part may be found in the gardens of the Floriſts among vs. ‡

‡ 9 *Colchicum latifolium.* Broad leaued Mede Saffron.

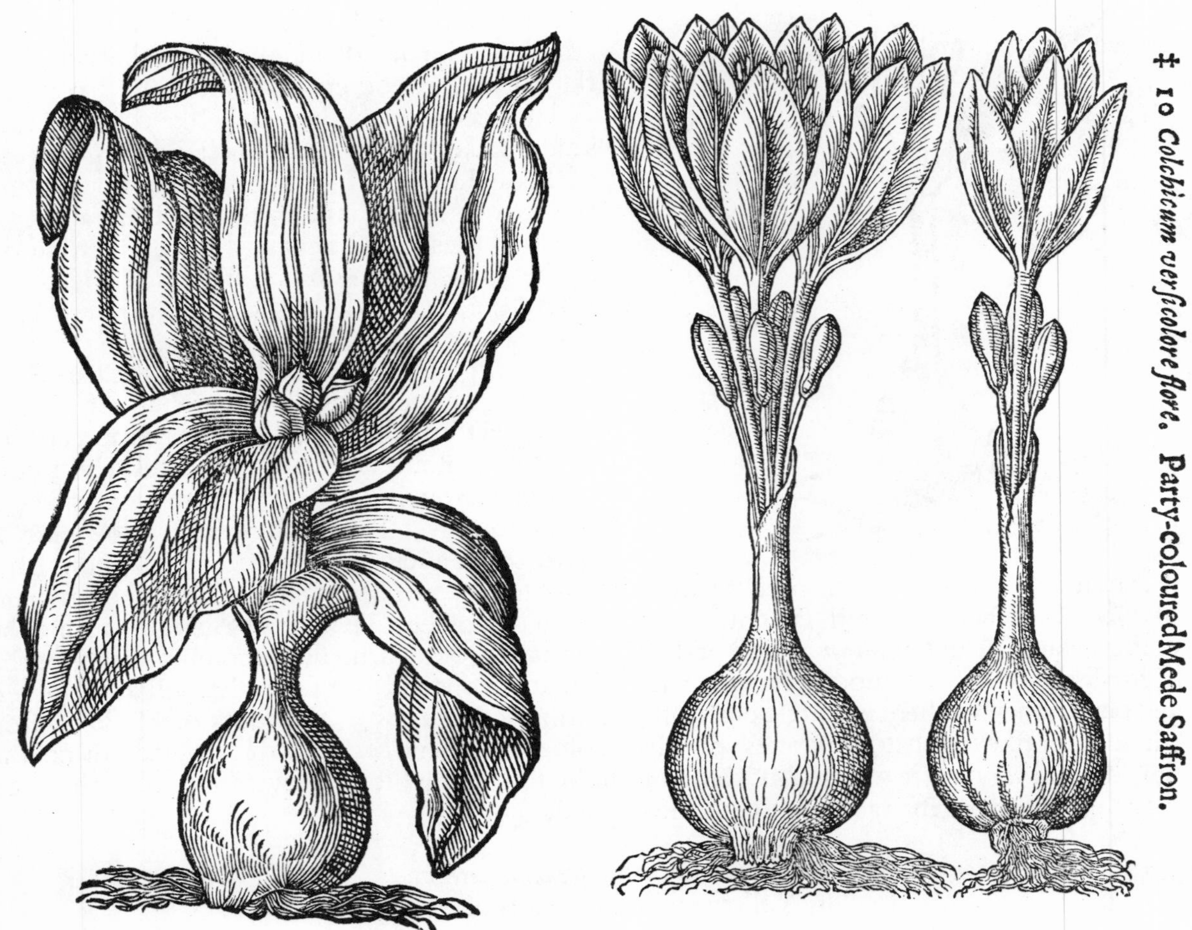

‡ 10 *Colchicum verſicolore flore.* Party-coloured Mede Saffron.

¶ *The Time.*

The leaues of all the kindes of Mede-Saffron do begin to ſhew themſelues in Februarie; The ſeed is ripe in Iune. The leaues, ſtalkes, and ſeed do periſh in Iuly, and their pleaſant floures doe come forth of the ground in September.

¶ *The Names.*

Dioſcorides calleth Medow Saffron Κολχικόν: ſome, Ἐφήμερον: notwithſtanding there is another *Ephemeron* which is not deadly. Diuers name it in Latine *Bulbus agreſtis*, or wild Bulbe: in high Dutch it is called **Zeitloosen**: in low Dutch, **Tijteloosen**: in French, *Mort au Chien.* Some haue taken it to be the true Hermodactyl, yet falſely. Other ſome call it *Filius ante Patrem*, although there is a kinde of *Lyſimachia* or Looſe-ſtrife ſo called, becauſe it firſt bringeth forth his long cods with ſeed, and then the floure after, or at the ſame time at the end of the ſaid cod. But in this Mede-Saffron it is far otherwiſe, becauſe it bringeth forth leaues in Februarie, ſeed in May, and floures in September, which is a thing cleane contrarie to all other plants whatſoeuer, for that they do firſt floure, and after ſeed; but this Saffron ſeedeth firſt, and foure moneths after brings forth floures: and therefore ſome haue thought this a fit name for it, *Filius ante Patrem*: and we accordingly may call

‡ 11 *Colchicum flore pleno.*
Double floured Mede-Saffron.

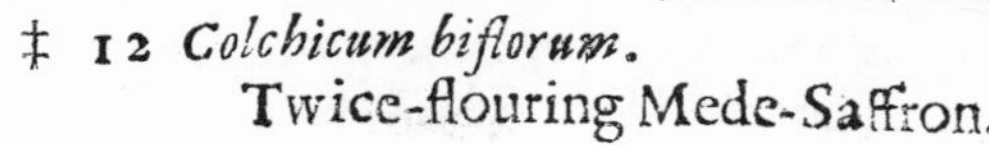

‡ 12 *Colchicum biflorum.*
Twice-flouring Mede-Saffron.

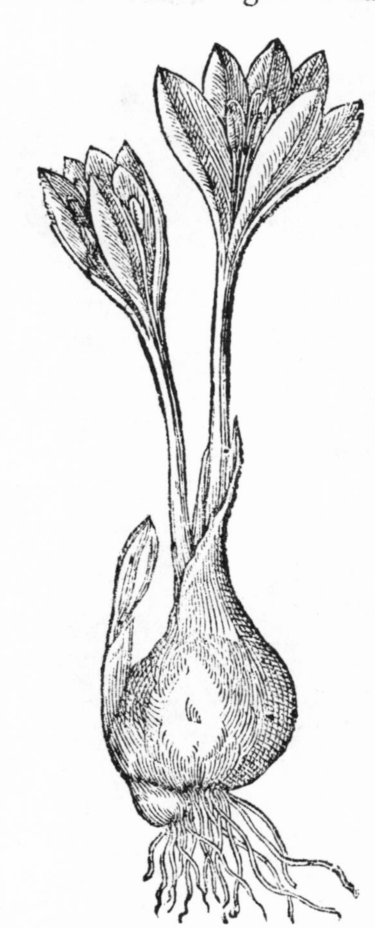

‡ 13 *Colchicum vernum.*
Spring Mede-Saffron.

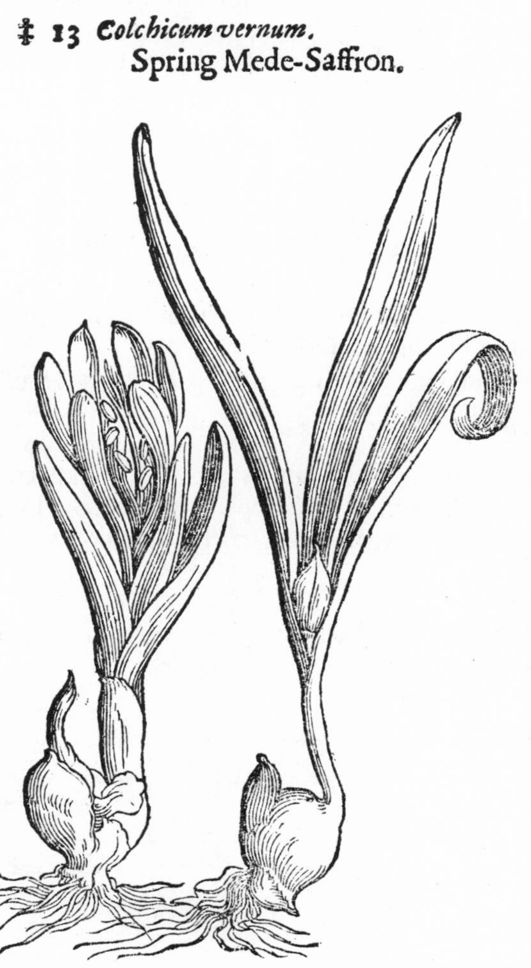

‡ 14 *Colchicum variegatum Chienſe.*
Checquered Mede Saffron of Chio.

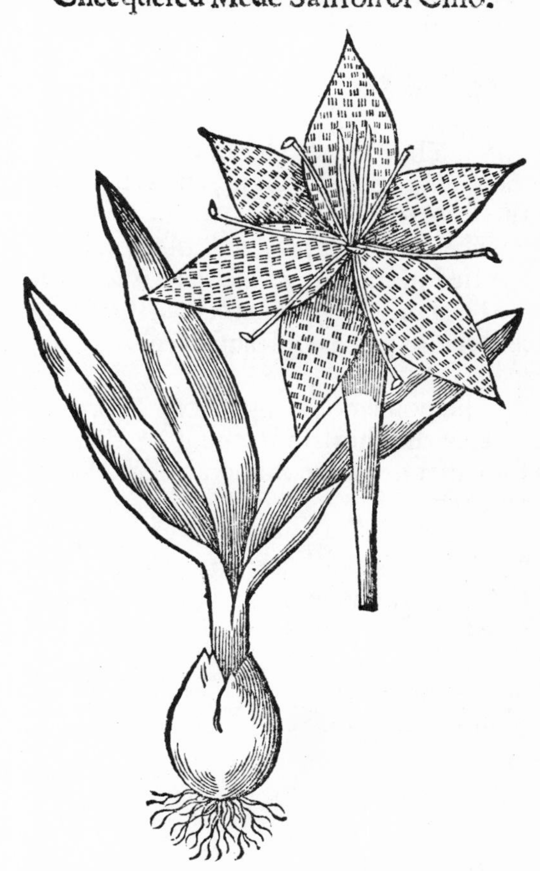

‡ 15 *Hermodactyli Officinarum.*
The true Hermodactyls of the ſhops.

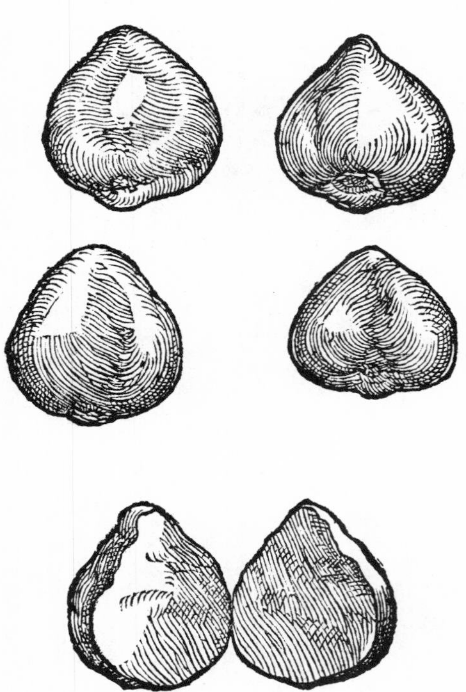

call it, The Sonne before the Father.

‡ Our Author in this chapter was of many mindes; for firſt, in the deſcription of *Colchicum Anglicum*, being the ſecond, hee reproues ſuch as make that white floured *Colchicum* the true Hermodactyl. Then in the deſcription of the eighth he hath theſe words, which being omitted in that place I here ſet downe. *Of all theſe kindes* (ſaith he) *of Medow Saffrons it hath not beene certainly knowne which hath been the true Hermodactyll; notwithſtanding wee haue certaine knowledge that the Illyrian Colchicum is the Phyſicall Hermodactyll.* Yet when he comes to ſpeake of the names, after that out of *Dodonæus* he had ſet downe the truth in theſe words; *But notwithſtanding that Hermodactyll which we do vſe in compound medicines, differeth from this* (to wit, *Colchicum*) *in many notable points, for that the true Hermodactyll hath a bulbe or round root, which being dried continueth very white within, and without not wrinkled at all, but full and ſmooth, of a meane hardneſſe*; and that he had out of the ſame Authour alledged the words of *Valerius Cordus* and *Auicen*, (which are here omitted) he concludes contrarie to the truth, his firſt admonition, and ſecond aſſertion, That the white Medow Saffron which we haue in the Weſt part of England, growing eſpecially about Shepton Mallet, is the Hermodactyll vſed in ſhops.

Thoſe we haue in ſhops ſeeme to be the Hermodactyls of *Paulus Ægineta*; yet not thoſe of *Nicholaus* and *Actuarius*, which were cordial, and increaſers of ſperme; the which the Authors of the *Aduerſaria, pag.* 55. thinke to be the *Behen album & rubrum* of the Arabians. And to theſe vnknowne ones are the vertues ſet downe by our Author in the third place vnder C, to be referred. ‡

¶ *The Temperature.*

Medow Saffron is hot and dry in the ſecond degree.

¶ *The Vertues of Hermodactyls.*

A † The roots of Hermodactyls are of force to purge, and are properly giuen (ſaith *Paulus*) to thoſe that haue the Gout, euen then when the humors are in flowing. And they are alſo hurtful to the ſtomacke.

B The ſame ſtamped, and mixed with the whites of egges, barley meale, and crums of bread, and applied plaiſterwiſe, eaſe the paine of the Gout, ſwellings and aches about the ioynts.

C The ſame ſtrengthneth, nouriſheth, and maketh good iuyce, encreaſeth ſperme or naturall ſeed, and is alſo good to cleanſe vlcers or rotten ſores.

¶ *The correction.*

The pouder of Ginger, long Pepper, Anniſe ſeed or Cumine ſeed, and a little Maſticke, correcteth the churliſh working of that Hermodactyll which is vſed in Shops. But thoſe which haue eaten of the common medow Saffron muſt drinke the milke of a cow, or elſe death preſently enſueth.

¶ *The Danger.*

The roots of all the ſorts of Mede Saffrons are very hurtfull to the ſtomacke, and being eaten they kill by choaking, as Muſhromes do, according vnto *Dioſcorides*; whereupon ſome haue called it *Colchicum ſtrangulatorium.*

† That which was ſet forth by our Author in the fourth place, vnder the title of *Colchicum montanum minus*, was nothing but the former *Colchicum minus* expreſſed in ſeed. The ninth and tenth were the ſame with the firſt and ſecond. The ſixth and ſeuenth, which are *Colchicum Illyricum* and *Syriacum* I haue left with their figures and hiſtorie, though they be ſuſpected to be counterfeits; and *Cluſius* probably geſſes, that the latter is the Apennine Tulip, the Painter making the leaues of the floure too round, and thoſe of the plant too broad and ſhort. †

CHAP. 92. *Of Starre of Bethlem.*

¶ *The Kindes.*

THere be ſundry ſorts of wilde field Onions called Starres of Bethlehem, differing in ſtature, taſte, and ſmell, as ſhall be declared.

1 *Ornithogalum.*
Star of Bethlehem.

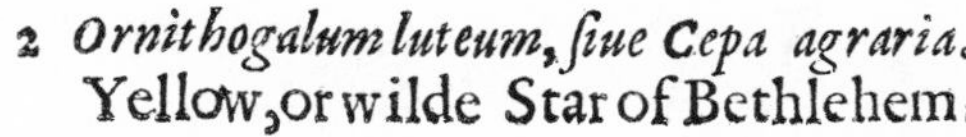

2 *Ornithogalum luteum, ſiue Cepa agraria.*
Yellow, or wilde Star of Bethlehem.

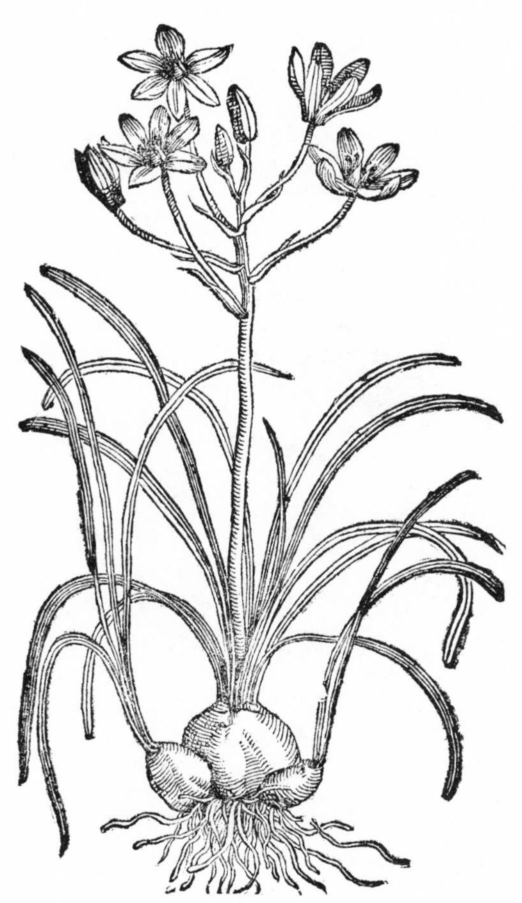

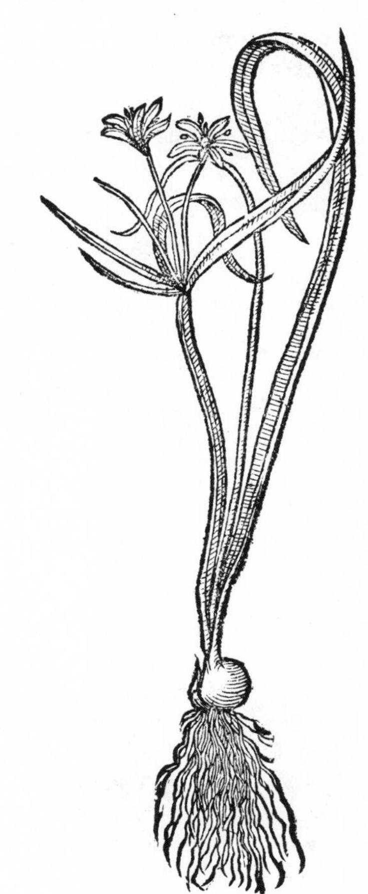

¶ *The Deſcription.*

1 OVr common Starre of Bethlehem hath many narrow leaues, thicke, fat, full of iuyce, and of a very greene colour, with a white ſtreake downe the middle of each leafe: among the which riſe vp ſmall naked ſtalkes, at the top whereof grow floures compact of ſix little leaues, ſtripped on the backeſide with lines of greene, the inſide being milke-white. Theſe floures open themſelues at the riſing of the Sunne, and ſhut againe at the Sun ſetting; whereupon this Plant hath beene calle'd by ſome, *Bulbus Solſequius.* The floures being paſt, the ſeed doth follow incloſed in three cornered husks. The root is bulbous, white both within and without.

† 2 The ſecond ſort hath two or three graſſy leaues proceeding from a clouen bulbous root. The ſtalke riſeth vp in the middeſt naked, but toward the top there doe thruſt forth more leaues like vnto the other, but ſmaller and ſhorter; among which leaues do ſtep forth very ſmall, weake, and tender foot-ſtalkes. The floures of this are on the backeſide of a pale yellow ſtripped with greene, on the inſide of a bright ſhining yellow colour, with Saffron coloured threds in their middles. The ſeed is contained in triangular veſſels.

† 3 This Star of Hungarie, contrarie to the cuſtome of other plants of this kinde, ſendeth forth before Winter fiue or ſix leaues ſpread vpon the ground, narrow, and of ſome fingers length, ſomewhat whitiſh greene, and much reſembling the leaues of Gillofloures, but ſomewhat roughiſh. In Aprill the leaues beginning to decay, amongſt them riſes vp a ſtalke bearing at the top a

spoke of floures, which consisting of six leaues apiece shew themselues open in May; they in colour are like the first described, as also in the greene streake on the lower side of each leafe. The seed is blacke, round, and contained in triangular heads. The root is bulbous, long, and white. †

‡ 4 This fourth, which is the *Ornithogalum Hispanicum minus* of *Clusius*, hath a little white root which sends forth leaues like the common one, but narrower, and destitute of the white line wherwith the other are marked. The stalke is some two handfulls high, bearing at the top thereof some seuen or eight floures growing each aboue other, yet so, as that they seeme to make an vmbell: each of these floures hath six leaues of a whitish blew colour, with so many white chiues or threds, and a little blewish vmbone in the midst. This floures in Aprill.

5 This fifth first sends vp one onely leafe two or three inches long, narrow, and of a whitish colour, and of an acide taste: nigh whereto riseth vp a small stalke some inch or two high, hauing one or two leaues thereon, betweene which come forth small star-floures, yellow within, and of a greenish purple without. The seed, which is reddish and small, is contained in triangular heads. The root is white, round, and couered with an Ash-coloured filme.

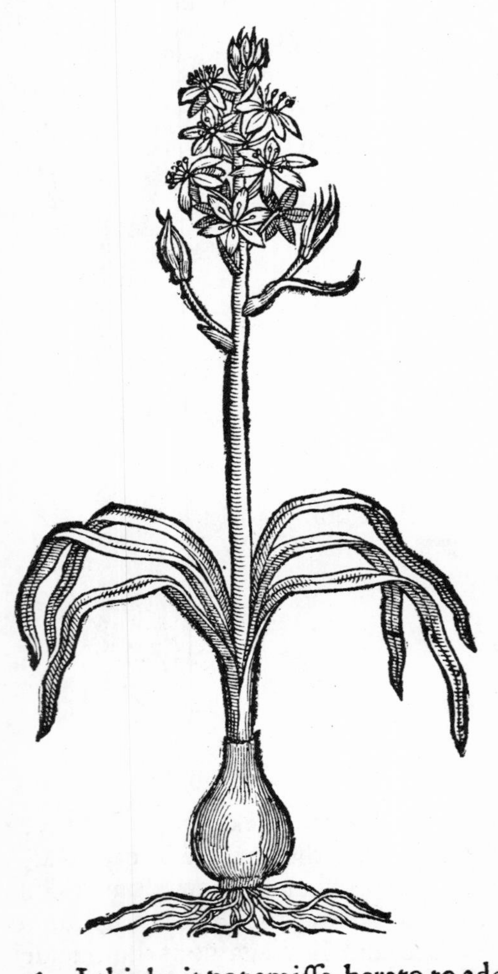

3 *Ornithogalum Pannonicum.*
Star of Hungary.

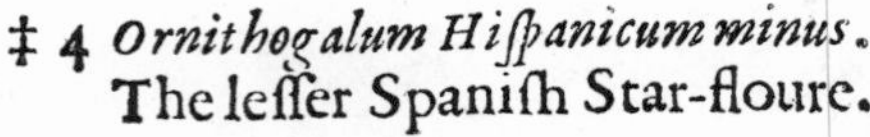

‡ 4 *Ornithogalum Hispanicum minus.*
The lesser Spanish Star-floure.

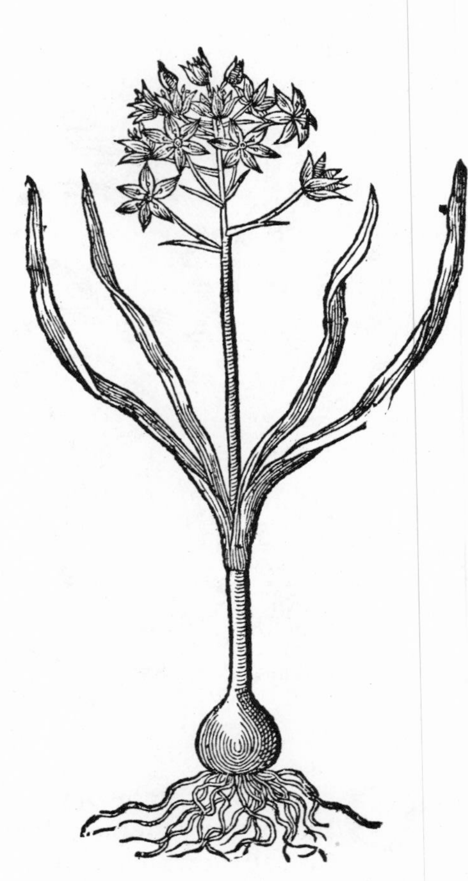

6 I thinke it not amisse, hereto to adde another small bulbous plant, which *Clusius* calls *Bulbus μονόφυλλος*, The one leaued Bulbe. This from a small root sends forth one rush-leafe of some foot in length, which about two inches aboue the earth being somewhat broader than in the other places, and guttered, sends forth a little stalke some three inches long, whose top is set with three little floures, each standing aboue other, about the bignesse here presented vnto your view in the figure: each of those consisteth of six very white leaues, and are not much vnlike the floures of the grasse of Parnassus, but yet without leaues to sustaine the floure, as it hath: six white threds tipt with yellow, and a three square head with a white pointall possesse the middest of the floure; the smell thereof is somewhat like that of the floures of the Haw-thorne. It floures in the midst of Iune.

7 Hauing done with these two small plants, I must acquaint you with three or foure larger, belonging also to this Classis. The first of these is that which *Dodonæus* calls *Ornithogalum majus*, and *Clusius*, *Ornithogalum Arabicum*: This by *Lobel* and some others is called *Lilium Alexandrinum*, or

‡ 5 *Ornithogalum luteum parvum.*
Dwarfe yellow Star of Bethlehem.

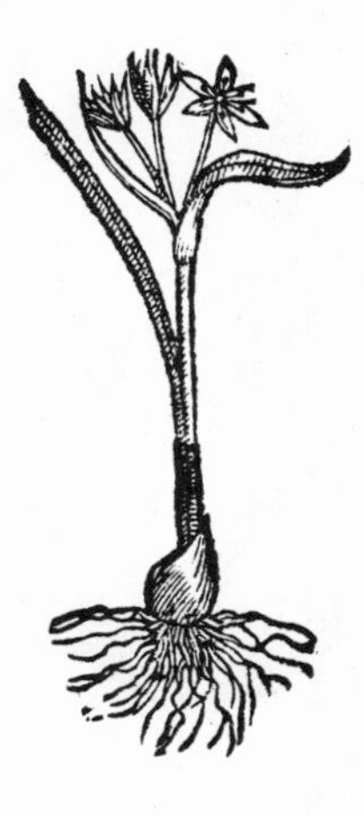

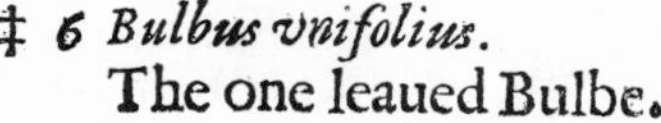

‡ 6 *Bulbus vnifolius.*
The one leaued Bulbe.

‡ 7 *Ornithogalum majus Arabicum.*
The great Arabicke Star-floure.

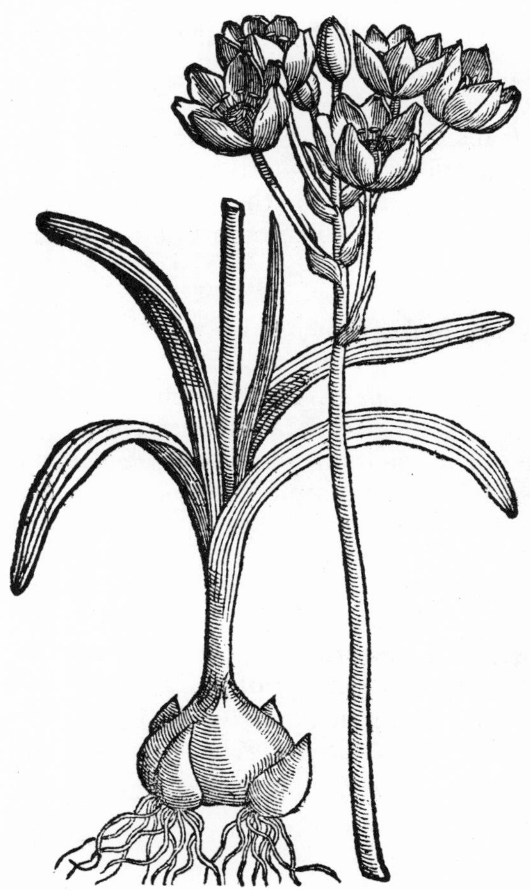

or the Lilly of Alexandria, as our Author calls it in the chapter of Cotton-grasse. This faire, but tender plant, hath broad greene leaues comming from a large white flat bottomed root; amongst which riseth vp a stalke some cubit high, whose top is garnished with sundry pretty large floures made of sixe very white leaues, with a shining blackish head, ingirt with six white threds tipt with yellow. This floures in May.

8 This, which is commonly called *Ornithogalum spicatum,* hath large leaues and roots, and the stalke growes some cubit or more high, wheron grow many starre-floures in shape and colour like those of the ordinarie, but larger, and they begin to floure below, and floure vpwards to the top. There is a larger sort of this *Spicatum,* whose floures are not streaked with greene on their backes. There is also a lesser, differing from the first of these onely in bignesse.

9 This Neapolitan hath three or foure long leaues not much vnlike those of the Hyacinths, but narrower, the stalke is pretty thicke, some foot high, and hath vsually growing theron some fiue or six floures hanging one way, though their stalkes grow alternately out of each side of the maine stemme. These floures are composed of six leaues, being about an inch long, and some quarter of an inch broad, white within, and of an Ash-coloured greene without, with white edges, the

the middle of the floure is poſſeſſed by another little floure, conſiſting alſo of ſix little leaues, hauing in them ſix threads headed with yellow, and a white pointall. A blacke wrinkled ſeed is contained in three cornered heads, which by reaſon of their bigneſſe weigh downe the ſtalke. This floures in Aprill. ‡

‡ 8 *Ornithogalum ſpicatum.*
Spike faſhioned Star-floure.

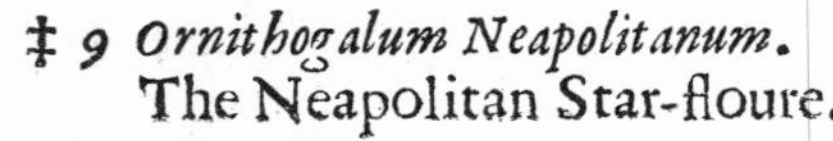

‡ 9 *Ornithogalum Neapolitanum.*
The Neapolitan Star-floure.

¶ *The Place.*

Stars of Bethlehem, or Star-floures, eſpecially the firſt and ſecond, grow in ſundry places that lie open to the aire, not onely in Germany and the Low-countries, but alſo in England, and in our gardens very common. The yellow kinde *Lobell* found in Somerſet-ſhire in the corne fields. The reſt are ſtrangers in England; yet we haue moſt of them, as the third, fourth, eighth, and ninth, in ſome of our choice gardens.

¶ *The Time.*

Theſe kindes of bulbed plants do floure from Aprill to the end of May.

¶ *The Names.*

Touching the names, *Dioſcorides* calls it Ὀρνιθόγαλον: *Pliny*, *Ornithogale*: in high Dutch it is called **Feldz wibel, Ackerz wibel**: as you ſhould ſay, *Cepa agraria*: in Engliſh, Stars of Bethlehem.

‡ The reſt are named in their titles and hiſtory; but *Cluſius* queſtions whether the *Bulbus vnifolius* be not *Bulbine* of *Theophraſtus*, 7. *hiſt.* 13. *Bauhinus* ſeemes to affirme the *Spicatum* to be *Moly* of *Dioſcorides* and *Theophraſtus*, and *Epimedium* of *Pliny*.

¶ *The Nature.*

Theſe are temperate in heate and drineſſe.

¶ *The Vertues.*

A The vertues of moſt of them are vnknowne; yet *Hieronymus Tragus* writeth, That the root of the Star of Bethlehem roſted in hot embers, and applied with honey in manner of a Cataplaſme or pulteſſe, healeth old eating vlcers, and ſoftens and diſcuſſes hard tumors.

The roots, ſaith *Dioſcorides*, are eaten both raw and boyled.

† That which was the ſecond of our Author, vnder the title of *Cepa agraria*, and the third vnder *Ornithogalum luteum* were figures of the ſame plant, but in the later, as *Bauhine* obſerues, the bottome leaues are omitted, becauſe they fall away when as it is growne vp to floure. †

Chap.

Chap. 93. *Of Onions.*

¶ *The Kindes.*

THere be, ſaith *Theophraſtus,* diuers ſorts of Onions, which haue their ſyr-names of the places where they grow: ſome alſo leſſer, others greater; ſome be round, and diuers others long; but none wilde, as *Pliny* writeth.

1 *Cepa alba.*
White Onions.

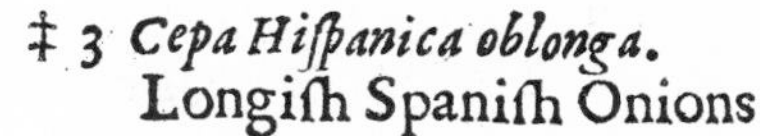
‡ 3 *Cepa Hiſpanica oblonga.*
Longiſh Spaniſh Onions.

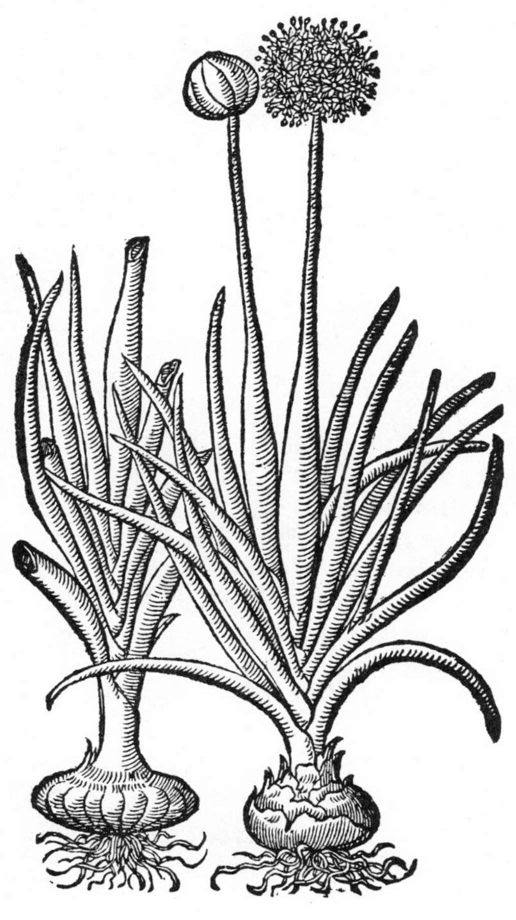

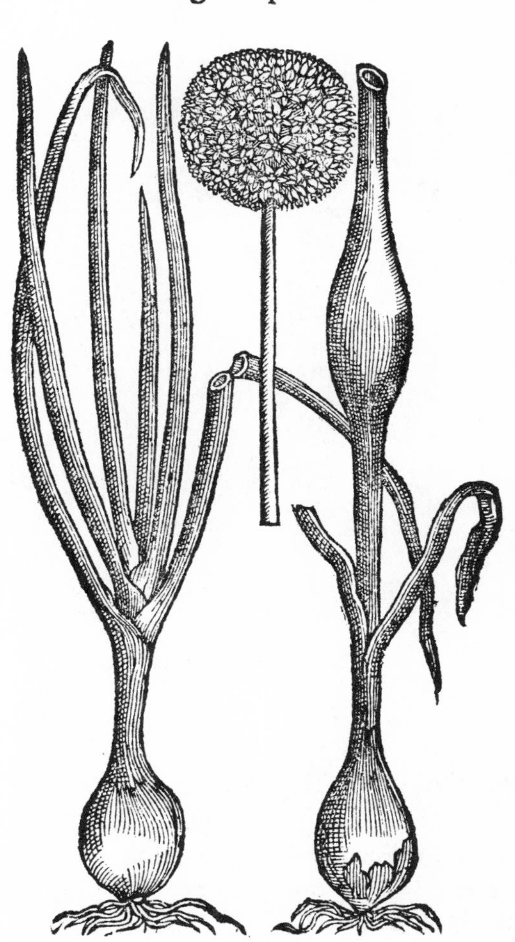

¶ *The Deſcription.*

1 THe Onion hath narrow leaues, and hollow within; the ſtalke is ſingle, round, biggeſt in the middle, on the top whereof groweth a round head couered with a thinne skin or filme, which being broken, there appeare little white floures made vp in forme of a ball, and afterward blacke ſeed three cornered, wrapped in thinne white skinnes. In ſtead of the root there is a bulbe or round head compact of many coats, which oftentimes becommeth great in manner of a Turnep, many times long like an egge. To be briefe, it is couered with very fine skinnes for the moſt part of a whitiſh colour.

2 The red Onion differeth not from the former but in ſharpneſſe and redneſſe of the roots, in other reſpects there is no difference at all.

‡ 3 There is alſo a Spaniſh kinde, whoſe root is longer than the other, but in other reſpects very little different.

‡ 4 There is alſo another ſmall kinde of Onion, called by *Lobel, Aſcalonitis Antiquorum,* or Scallions; this hath but ſmall roots, growing many together: the leaues are like to Onions, but leſſe. It ſeldome beares either ſtalke, floure, or ſeed. It is vſed to be eaten in ſallads.

¶ *The*

¶ *The Place.*

The Onion requireth a fat ground well digged and dunged, as *Palladius* saith. It is cherished euery where in kitchen gardens: it is now and then in beds sowne alone, and many times mixed with other herbes, as with Lettuce, Parseneps, and Carrets. *Palladius* liketh well that it should be sowne with Sauory, because, saith *Pliny*, it prospereth the better, and is more wholesome.

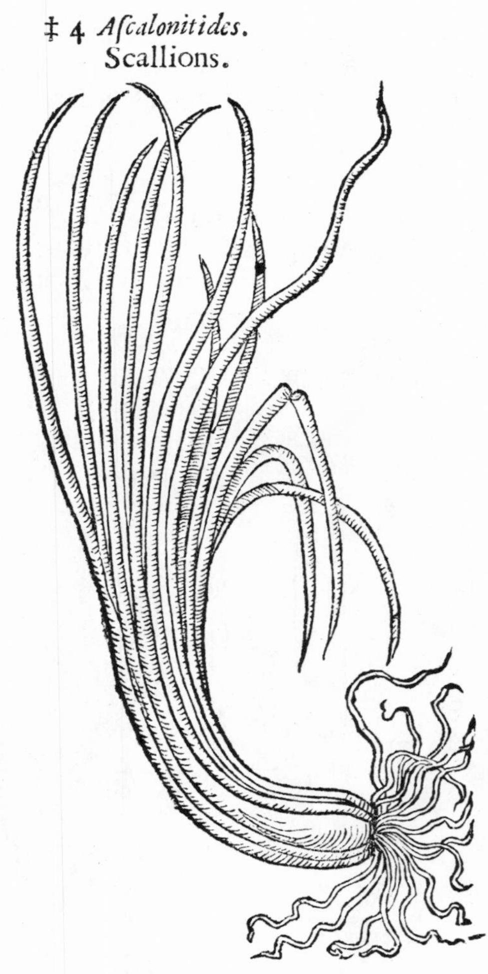

‡ 4 *Ascalonitides.* Scallions.

¶ *The Time.*

It is sowne in March or Aprill, and somtimes in September.

¶ *The Names.*

The Onion is called in Greeke, Κρόμμυον: in Latine, *Cepa*, and many times *Cepe* in the neuter gender: the shops keepe that name. The old Writers haue giuen vnto this many syr-names of the places where they grow, for some are named *Cipriæ*, *Sardiæ*, *Creticæ*, *Samothraciæ*, *Ascaloniæ*, of a towne in Iudea, otherwise called *Pompeiana*: in English, Onions. Moreouer, there is one named *Marisca*, which the Countrey-men call *Vnio* saith *Columella*; and thereupon it commeth that the French men call it *Oignon*, as *Ruellius* thinketh: and peraduenture the Low-Dutch men name it **Aiuein**, of the French word corrupted: they are called *Setaniæ* which are very little and sweet; and these are thought to be those which *Palladius* nameth *Cepullæ*, as though he called them *parvæ Cepæ*, or little Onions.

There is an Onion which is without an head or bulbe, and hath as it were a long necke, and spends it selfe wholly in the leaues, and it is often cropped or cut for the pot like the Leekes. This *Theophrastus* names Γήθυον: of this *Pliny* also writeth, in his nineteenth booke, and sixt chapter. There is with vs two principall sorts of Onions, the one seruing for a sauce, or to season meate with, which some call *Gethyon*, and others *Pallacana*: and the other is the headed or common Onion, which the Germanes call **Onion zwibel**: the Italians, *Cipolla*: the Spaniards, *Cebolla*, *Ceba*, and *Cebola*.

¶ *The Temperature.*

All Onions are sharpe, and moue teares by the smell. They be hot and dry, as *Galen* saith, in the fourth degree, but not so extreme hot as Garlick. The iuyce is of a thin waterie and airy substance: the rest is of thicke parts.

¶ *The Vertues.*

A The Onions do bite, attenuate, or make thinne, and cause drinesse: being boyled they doe lose their sharpenesse, especially if the water be twice or thrice changed, and yet for all that they doe not lose their attenuating qualitie.

B they also breake winde, prouoke vrine, and be more soluble boyled than raw; and raw they nourish not at all, and but a little though they be boyled.

C They be naught for those that are cholericke, but good for such as are replete with raw and flegmaticke humors; and for women that haue their termes stayed vpon a cold cause, by reason they open the passages that are stopped.

D *Galen* writeth, That they prouoke the Hemorrhoides to bleed if they be laid vnto them, either by themselues, or stamped with vineger.

E The iuyce of Onions sniffed vp into the nose, purgeth the head, and draweth forth raw flegmaticke humors.

F Stamped with salt, rew, and honey, and so applied, they are good against the biting of a mad Dog.

G Rosted in the embers, and applied, they ripen and breake cold Apostumes, Biles, and such like.

The

The iuyce of Onions mixed with the decoction of Pennirial, and annointed vpon the goutie member with a feather, or a cloath wet therein, and applied, easeth the same very much. H

The iuice annointed vpon a pild or bald head in the sunne, bringing againe the haire very speedily. I

The iuyce taketh away the heate of scalding with water or oyle, as also burning with fire and gun-pouder, as is set forth by a very skilfull Chirurgion named Master *William Clowes*, one of the Queens Chirurgions; and before him by *Ambrose Parey*, in his Treatise of wounds made by gun shot. K

Onions sliced, and dipped in the iuyce of Sorrell, and giuen vnto the sicke of a tertian Ague, to eate, take away the fit in once or twice so taking them. L

¶ *The Hurts.*

The Onion being eaten, yea though it be boyled, causeth head-ache, hurteth the eyes, and maketh a man dimme sighted, dulleth the sences, ingendreth windinesse, and prouoketh ouermuch sleepe, especially being eaten raw.

CHAP. 94. *Of Squils, or Sea-Onions.*

‡ 1 *Scilla Hispanica vulgaris.* The common Spanish Squill.

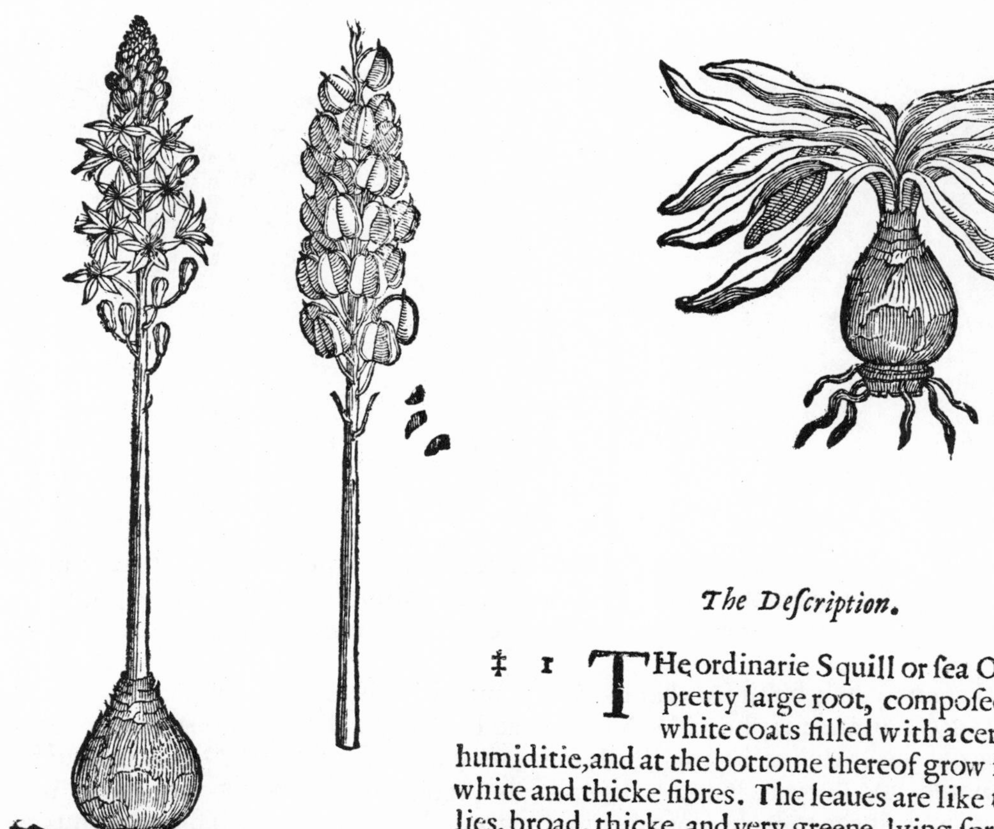

The Description.

‡ 1 THe ordinarie Squill or sea Onion hath a pretty large root, composed of sundrie white coats filled with a certain viscous humiditie, and at the bottome thereof grow forth sundry white and thicke fibres. The leaues are like those of Lillies, broad, thicke, and very greene, lying spred vpon the ground, and turned vp on the sides. The stalke groweth some cubit or more high, straight, naked without leaues, beautified at the top with many starre-fashioned floures, very like those of the bigger *Ornithogalum*. The seed is contained in chaffie three cornered seed-vessels, being it selfe also black, smooth, and chaffie. It floures in August and September, and the seed is ripe in October. The leaues spring vp in Nouember and December, after that the seed is ripe, and stalke decayed. ‡

2 The great Sea Onion, which *Clusius* hath set forth in his Spanish historie, hath very great and broad leaues, as *Dioscorides* saith, longer than those of the Lilly, but narrower. The bulbe or headed root is very great, consisting of many coats or scaly filmes of a reddish colour. The floure is sometimes yellow, sometimes purple, and sometimes of a light blew. ‡ *Clusius* saith it is like that of the former, I thinke he meanes both in shape and colour. ‡

3 The sea-Onion of Valentia, or rather the sea Daffodill, hath many long and fat leaues, and narrow like those of Narcissus, but smoother and weaker, lying vpon the ground; among which riseth vp a stalke a foot high, bare and naked, bearing at the top a tuft of white floures, in shape like vnto

vnto our common yellow Daffodil. The ſeed is incloſed in thicke knobby huskes,blacke,flat,and thicke,very ſoft,in ſhape like vnto the ſeeds of *Ariſtolochia longa,* or long Birth-wort. The root is great,white,long, and bulbous.

4 Red floured Sea Daffodill, or ſea Onion, hath a great bulbe or root like to the precedent; the leaues long, fat, and ſharpe pointed, the ſtalke bare and naked,bearing at the top ſundry faire red floures in ſhape like to the laſt deſcribed.

2 *Pancratium Cluſij.*
Great Squill, or Sea Onion.

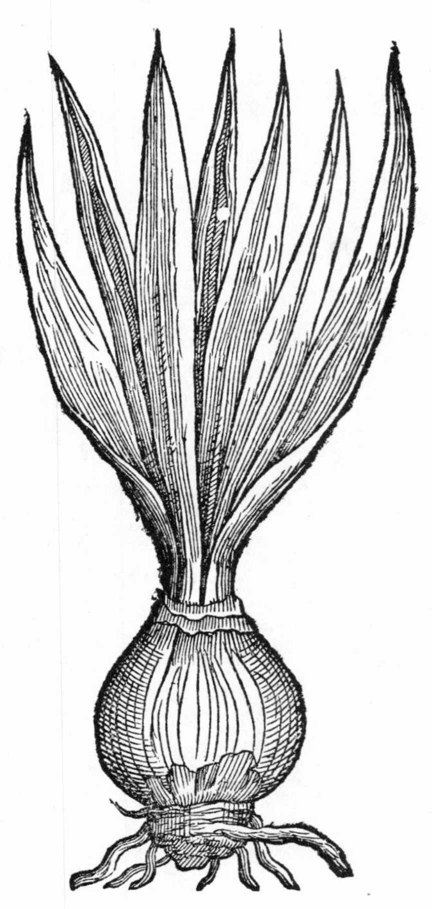

3 *Pancratium Marinum.*
Sea Onion of Valentia.

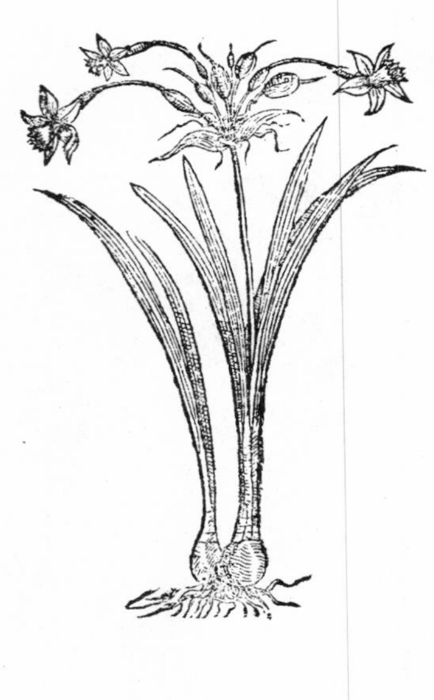

5 The yellow floured ſea Daffodill, or ſea Onion, hath many thicke fat leaues like vnto the common Squill or ſea Onion,among which riſeth vp a tender ſtraight ſtalke full of iuyce,bearing at the top many floures like the common yellow Daffodill. The ſeed and root is like the precedent.

‡ 6 To theſe may fitly be added that elegant plant which is knowne by the name of *Narciſſus tertius* of *Matthiolus*, and may be called White Sea Daffodill. This plant hath large roots,as bigge ſometimes as the ordinarie Squill; the leaues are like thoſe of other Daffodils, but broader, rounder pointed, and not very long. The ſtalke is pretty thicke, being ſometimes round, otherwhiles cornered, at the top whereof grow many large white floures: each floure is thus compoſed; it hath ſix long white leaues, in the midſt growes forth a white pointall which is incompaſſed by a welt or cap diuided into ſix parts,which ſix are againe by threes diuided into eighteen iagges or diuiſions, a white thred tipt with greene,of an inch long,comming forth of the middle of each diuiſion. This floureth in the end of May. It is ſaid to grow naturally about the ſea coaſt of Illyria. ‡

¶ *The Place.*

The firſt is found in Spaine and Italy, not far from the ſea ſide.

The ſecond alſo neere vnto the ſea, in Italy, Spaine, and Valentia. I haue had plants of them brought me from ſundry parts of the Mediterranean ſea ſide, as alſo from Conſtantinople, where it is numbred among the kindes of Narciſſus.

The third groweth in the ſands of the ſea, in moſt places of the coaſt of Narbone, and about Montpellier.

The fourth groweth plentifully about the coaſts of Tripolis and Aleppo, neere to the ſea, and alſo in the ſalt marſhes that are ſandie and lie open to the aire.

¶ *The*

¶ *The Time.*

They floure from May to the end of Iuly, and their seed is ripe in the end of August.

¶ *The Names.*

The first is called of the Grecians, σκίλλα: and of the Latines also *Scilla*: the Apothecaries name it *Squilla*: Diuers, *Cepa muris*: the Germanes, **Meer zwibel**: the Spaniards, *Cebolla albarrana*: the French-men, *Oignon de mer*: in English, Squill, and Sea Onion.

‡ The second is called Παγκράτιον, and *Scilla rubra major*.

3, 4, 5. These are all figures of the same plant, but the least (which is the worst) is the figure of the *Aduersaria*, where it is called *Pancratium marinum*: *Dodonæus* calls it *Narcissus marinus*: and *Clusius*, *Hemerocallis Valentina*; and it is iudged to be the Ημεροκαλλὶς of *Theoporastus*, *Lib. 6. Hist. cap. 1.* The Spaniards call this *Amores mios*: the Turkes, *Con zambach*: the Italians, *Giglio marino*. These three (as I said) differ no otherwise than in the colour of their floures.

The sixth is *Narcissus tertius*, or *Constantinopolitanus*, of *Matthiolus*: *Clusius* calls it *Lilionarcissus Hemerocallidis facie*. ‡

4 *Pancratium floribus rubris.*
Red floured sea Daffodill.

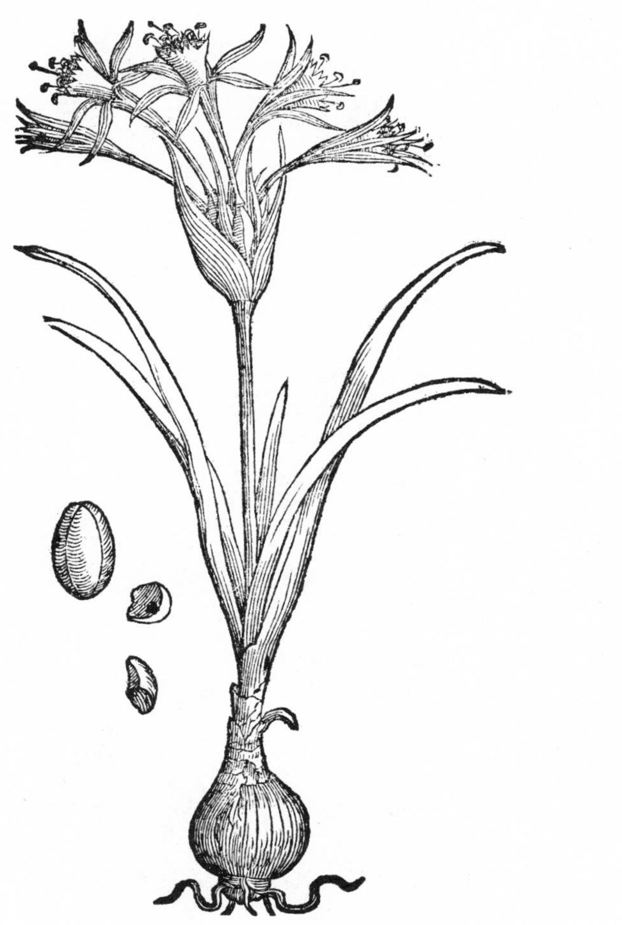

‡ 6 *Narcissus tertius Matthioli.*
The white sea Daffodill.

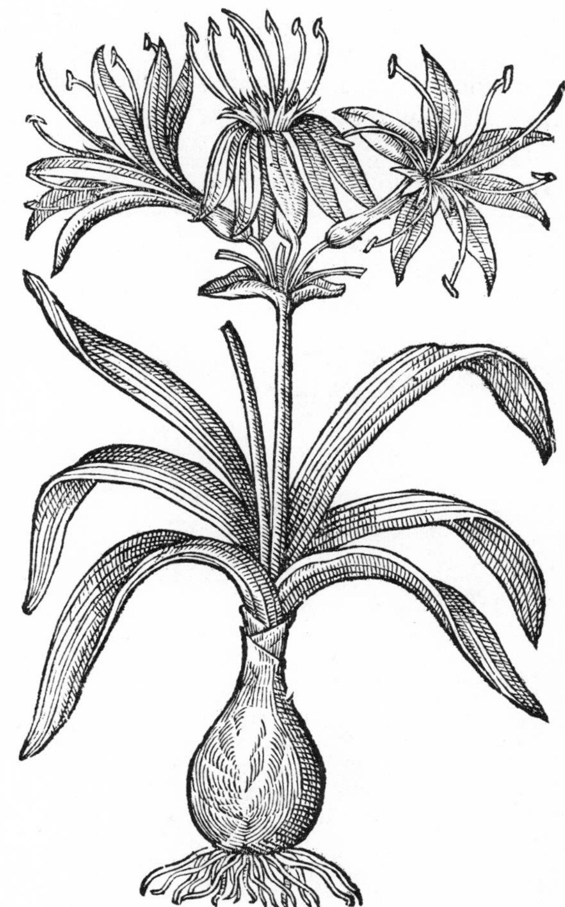

¶ *The Temperature.*

The sea Onion is hot in the second degree, and cutteth very much, as *Galen* saith. It is best when it is taken baked or rosted, for so the vehemencie of it is taken away.

¶ *The Vertues of Squills.*

A The root is to be couered with paste or clay, (as *Dioscorides* teacheth) and then put into an ouen to be baked, or else buried in hot embers till such time as it be throughly rosted: for not being so baked or rosted it is very hurtfull to the inner parts.

B It is likewise baked in an earthen pot close couered and set in an ouen. That is to be taken especially which is in the midst, which being cut in pieces must be boyled, but the water is still to be changed, till such time as it is neither bitter nor sharpe: then must the pieces be hanged on a thread, and dried in the shadow, so that no one piece touch another. ‡ Thus vsed it loseth most of the strength; therefore it is better to vse it lightly dried, without any other preparation. ‡

P These

C These slices of the Squill are vsed to make oyle, wine, or vineger of Squill. Of this vineger of Squill is made an Oxymel. The vse whereof is to cut thicke, tough, and clammy humors, as also to be vsed in vomits.

D This Onion rosted or baked is mixed with potions and other medicines which prouoke vrine, and open the stoppings of the liuer and spleene, and is also put into treacles. It is giuen to those that haue the Dropsie, the yellow Iaundise, and to such as are tormented with the gripings of the belly, and is vsed in a licking medicine against an old rotten cough, and for shortnesse of breath.

E One part of this Onion being mixed with eight parts of salt, and taken in the morning fasting to the quantitie of a spoonefull or two, looseth the belly.

F The inner part of Squilla boyled with oyle and turpentine, is with great profit applied to the chaps or chil-blanes of the feet or heeles.

G It driueth forth long and round wormes if it be giuen with honey and oyle.

‡ The *Pancratium marinum*, or *Hemerocallis Valentina* (saith *Clusius*) when as I liued with *Rondeletius*, at Montpellier, was called *Scilla*; and the Apothecaries thereof made the trochisces for the composition of Treacle: afterwards it began to be called *Pancratium flore Lilij*. *Rondeletius* also was wont to tell this following story concerning the poysonous and maligne qualitie thereof. There were two Fishermen, whereof the one lent vnto the other (whom he hated) his knife, poysoned with the iuyce of this Hemerocallis, for to cut his meate withall; he suspecting no treachery cut his victuals therewith, and so eat them, the other abstaining therefrom, and saying that he had no stomacke. Some few dayes after, he that did eate the victuals died; which shewed the strong and deadly qualitie of this plant: which therefore (as *Clusius* saith) cannot be the *Scilla Epimenidia* of *Pliny*, which was eatable, and without malignitie ‡

Chap. 95. Of Leekes.

1 *Porrum capitatum.*
Headed, or set Leeke.

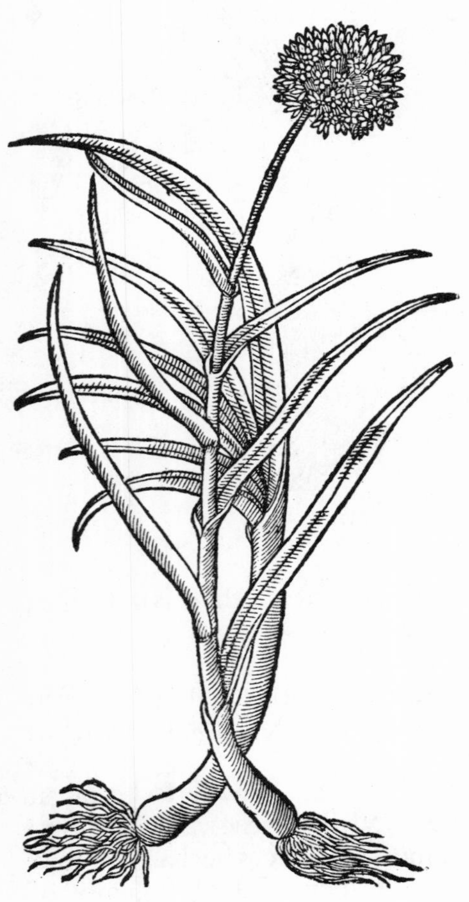

‡ 2 *Porrum sectivum aut tonsile.*
Cut, or vnset Leeke.

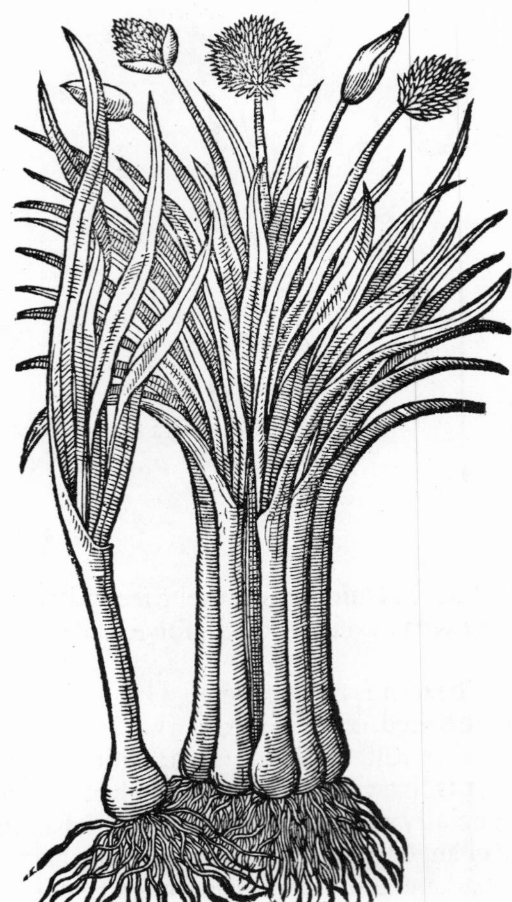

¶ The

¶ *The Deſcription.*

1 THe leaues or the blades of the Leeke be long, ſomewhat broad, and very many, hauing a keele or creſt in the backſide, in ſmell and taſte like to the Onion. The ſtalks, if the blades be not often cut, do in the ſecond or third yeare grow vp round, bringing forth on the top floures made vp in a round head or ball as doth the Onion. The ſeeds are like. The bulbe or root is long and ſlender, eſpecially of the vnſet Leeke. That of the other Leeke is thicker and greater.

‡ 2 Moſt Writers diſtinguiſh the common Leeke into *Porrum capitatum & ſectivum*; and *Lobel* giues theſe two figures wherewith we here preſent you. Now both theſe grow of the ſame ſeed, and they differ onely in culture; for that which is often cut for the vſe of the kitchen is called *Sectivum*: the other, which is headed, is not cut, but ſpared, and remoued in Autumne. ‡

¶ *The Place.*

It requireth a meane earth, fat, well dunged and digged. It is very common euery where in other countries, as well as in England.

¶ *The Time.*

It may be ſowne in March or Aprill, and it to be remoued in September or October.

¶ *The Names.*

The Grecians call it πράσον: the Latines, *Porrum*. The Emperour *Nero* had great pleaſure in this root, and therefore he was called in ſcorne, *Porrophagus*. But *Palladius* in the maſculine gender called it *Porrus*: the Germanes, **Lauch**: the Brabanders, **Poreue**: the Spaniards, *Puerro*: the French, *Porreau*: the Engliſh-men, Leeke, or Leekes.

¶ *The Temperature.*

The Leeke is hot and dry, and doth attenuate or make thinne as doth the Onion.

¶ *The Vertues.*

Being boyled it is leſſe hurtfull, by reaſon that it loſeth a great part of his ſharpeneſſe: and yet being ſo vſed it yeeldeth no good iuyce. But being taken with cold herbes his too hot quality is tempered.

Being boyled and eaten with Ptiſana or barley creame, it concocteth and bringeth vp raw humors that lie in the cheſt. Some affirme it to be good in a loch or licking medicine, to clenſe the pipes of the lungs. A

The iuyce drunke with honey is profitable againſt the bitings of venomous beaſts, and likewiſe the leaues ſtamped and laid thereupon. B

The ſame iuyce, with vineger, frankincenſe, and milke, or oyle of roſes, dropped into the eares, mitigateth their paine, and is good for the noyſe in them. C

Two drams of the ſeed, with the like weight of myrtill berries drunk, ſtop the ſpitting of bloud which hath continued a long time. The ſame ingredients put into Wine keepe it from ſouring, and being alreadie ſoure, amend the ſame, as diuers write. It cutteth and attenuateth groſſe and tough humors. D

‡ *Lobel* commends the following Loch as very effectuall againſt phlegmatick Squinances, and other cold catarrhes which are like to cauſe ſuffocation. This is the deſcription thereof; Take blanched almonds three ounces, foure figges, ſoft *Bdellium* halfe an ounce, iuyce of Liquorice, two ounces, of ſugar candy diſſolued in a ſufficient quantitie of iuyce of Leekes, and boyled in *Balneo* to the height of a Syrup, as much as ſhall be requiſit to make the reſt into the forme of an *Eclegma*. ‡ E

¶ *The Hurts.*

It heateth the body, ingendreth naughty bloud, cauſeth troubleſome and terrible dreames, offendeth the eyes, dulleth the ſight, hurteth thoſe that are by nature hot and cholericke, and is noyſome to the ſtomacke, and breedeth windineſſe.

Chap. 96.

Of Ciues or Chiues, and wilde Leekes.

¶ *The Kindes.*

THere be diuers kindes of Leekes, ſome wilde, and ſome of the garden, as ſhall be declared, Thoſe called Ciues haue beene taken of ſome for a kinde of wilde Onion: but all the Authors that I haue beene acquainted with, do accord that there is not any wild Onion.

1 *Schœnoprason.*
Ciues or Chiues.

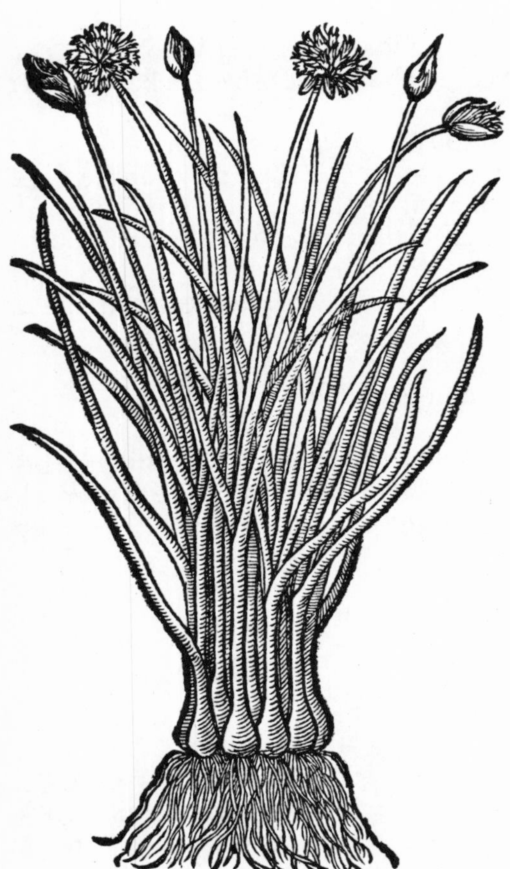

2 *Porrum vitigineum.*
French Leekes, or Vine Leekes.

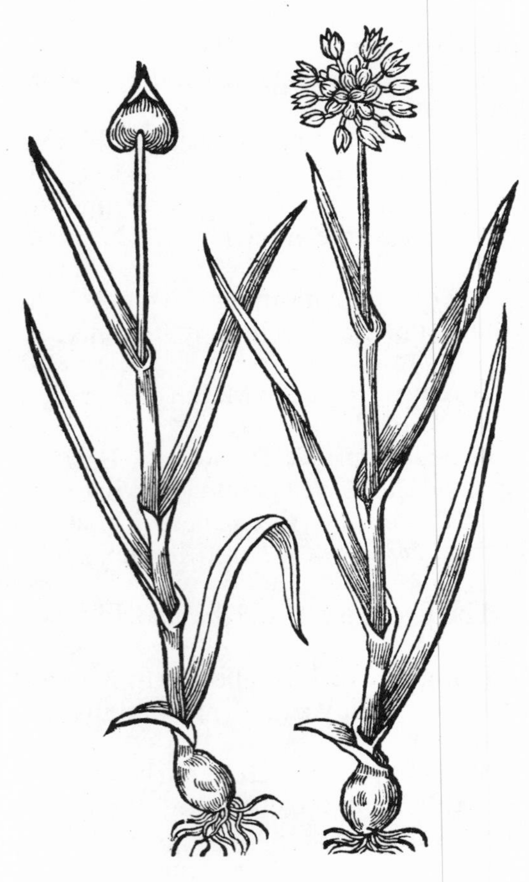

3 *Ampeloprason siue porrum siluestre.*
Wilde Leeke.

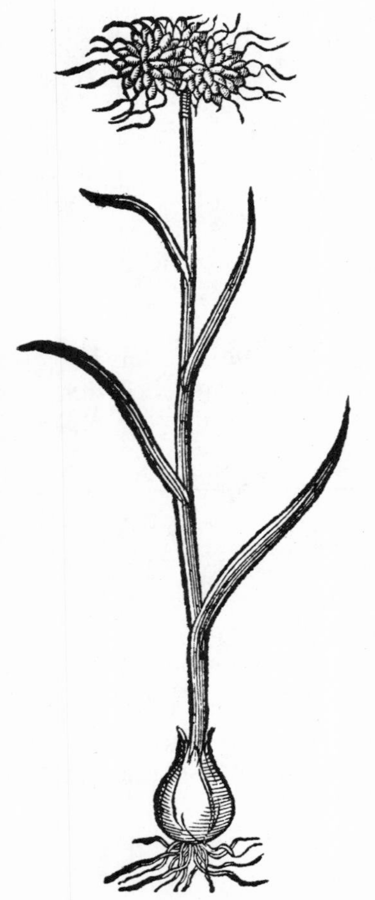

¶ *The Description.*

1 Ciues bring forth many leaues about a hand-full high, long, slender, round, like to little rushes; amongst which grow vp small and tender stalkes, sending forth certaine knops with floures like those of the Onion, but much lesser. They haue many little bulbes or headed roots fastned together: out of which grow downe into the earth a great number of little strings, and it hath both the smell and taste of the Onion and Leeke, as it were participating of both.

2 The Vine Leeke or French Leeke groweth vp with blades like those of Leekes: the stalke is a cubit high, on the top whereof standeth a round head or button, couered at the first with a thinne skinne, which being broken, the floures and seeds come forth like those of the Onion. The bulbe or headed root is round, hard, and sound, which is quickly multiplied by sending forth many bulbes.

‡ 3 The wilde Leeke hath leaues much like vnto those of Crow-garlicke, but larger, and more acride. The floures and seeds also resemble those of the Crow-garlicke, the seeds being about the bignesse of cornes of wheat, with smal strings comming forth at their ends. ‡

¶ *The*

¶ *The Time and Place.*

1 Ciues are ſet in gardens, they flouriſh long, and continue many yeares, they ſuffer the cold of Winter. They are cut and polled often, as is the vnſet Leeke.

2 The Vine-leeke groweth of it ſelfe in Vineyards, and neere vnto Vines in hot regions, wherof it both tooke the name Vine-Leeke, and French Leeke. It beareth his greene leaues in Winter, and withereth away in the Sommer. It groweth in moſt gardens of England.

‡ Thus farre our Author deſcribes and intimates to you a garden Leeke, much like the ordinarie in all reſpects, but ſomewhat larger. But the following names belong to the wilde Leeke, which here we giue you in the third place. ‡

¶ *The Names.*

Ciu ; are called in Greeke, σχοινόπρασον, *Shœnopraſum* : in Dutch, **Bieſloack**, as though you ſhould ſay, *Iunceum Porrum*, or Ruſh Leeke : in Engliſh, Ciues, Chiues, Ciuet and Sweth : in French, *Brelles*.

† 2 The Vine-leeke, or rather wild Leeke, is called in Greeke, Ἀμπελόπρασον, of the place where it naturally groweth : it may be called in Latine, *Porrum Vitium*, or *Vitigineum Porrum* : in Engliſh, after the Greeke and Latine, Vine Leeke, or French Leeke.

¶ *The Temperature.*

Ciues are like in facultie vnto the Leeke, hot and dry. The Vine leeke heateth more than doth the other Leeke.

¶ *The Vertues.*

Ciues attenuate or make thinne, open, prouoke vrine, ingender hot and groſſe vapours, and are hurtfull to the eyes and braine. They cauſe troubleſome dreames, and worke all the effects that the Leeke doth. A

The Vine-leeke, or Ampelopraſon, prouoketh vrine mightily, and bringeth downe the floures. It cureth the bitings of venomous beaſts, as *Dioſcorides* writeth. B

† The figure of *Ampelopraſum* was in the first place, in the Chapter next but one, by the name of *Allium ſylueſtre*.

CHAP. 97. *Of Garlicke.*

¶ *The Deſcription.*

1 THe bulbe or head of Garlicke is couered with moſt thinne skinnes or filmes of a very light white purple colour, conſiſting of many cloues ſeuered one from another, vnder which in the ground below groweth a taſſell of threddy fibres : it hath long greene leaues like thoſe of the Leeke, among which riſeth vp a ſtalke at the end of the ſecond or third yeare, whereupon doth grow a tuft of floures couered with a white skinne, in which, being broken when it is ripe, appeareth round blacke ſeeds.

‡ 2 There is alſo another Garlicke which growes wilde in ſome places of Germanie and France, which in ſhape much reſembles the ordinarie, but the cloues of the roots are ſmaller and redder. The floure is alſo of a more duskie and darke colour than the ordinarie. ‡

¶ *The Place and Times.*

Garlick is ſeldome ſowne of ſeed, but planted in gardens of the ſmall cloues in Nouember and December, and ſometimes in Februarie and March.

¶ *The Names.*

It is called in Latine, *Allium* : in Greeke, σκόροδον : The Apothecaries keepe the Latine name: the Germanes call it **Knoblauch** : the Low Dutch, **Look** : the Spaniards, *Aios*, *Alho* : the Italians, *Aglio* : the French, *Ail* or *Aux* : the Bohemians, *Czeſnek* : the Engliſh, Garlicke, and poore mans Treacle.

¶ *The Temperature.*

Garlicke is very ſharpe, hot, and dry, as *Galen* ſaith, in the fourth degree, and exulcerateth the skinne by raiſing bliſters.

¶ *The Vertues.*

Being eaten, it heateth the body extremely, attenuateth and maketh thinne thicke and groſſe humors ; cutteth ſuch as are tough and clammy, digeſteth and conſumeth them ; alſo openeth obſtructions, is an enemie to all cold poyſons, and to the bitings of venomous beaſts : and therefore *Galen* nameth it *Theriaca Ruſticorum*, or the husbandmans Treacle. A

It yeeldeth to the body no nouriſhment at all, it ingendreth naughty and ſharpe bloud. Therefore B

fore

fore such as are of a hot complexion must especially abstaine from it. But if it be boyled in water vntill such time as it hath lost his sharpenesse, it is the lesse forcible, and retaineth no longer his euill iuyce, as *Galen* saith.

C It taketh away the roughnesse of the throat, it helpeth an old cough, it prouoketh vrine, it breaketh and consumeth winde, and is also a remedie for the Dropsie which procceedeth of a cold cause.

D It killeth wormes in the belly, and driueth them forth. The milke also wherein it hath beene sodden is giuen to yong children with good successe against the wormes.

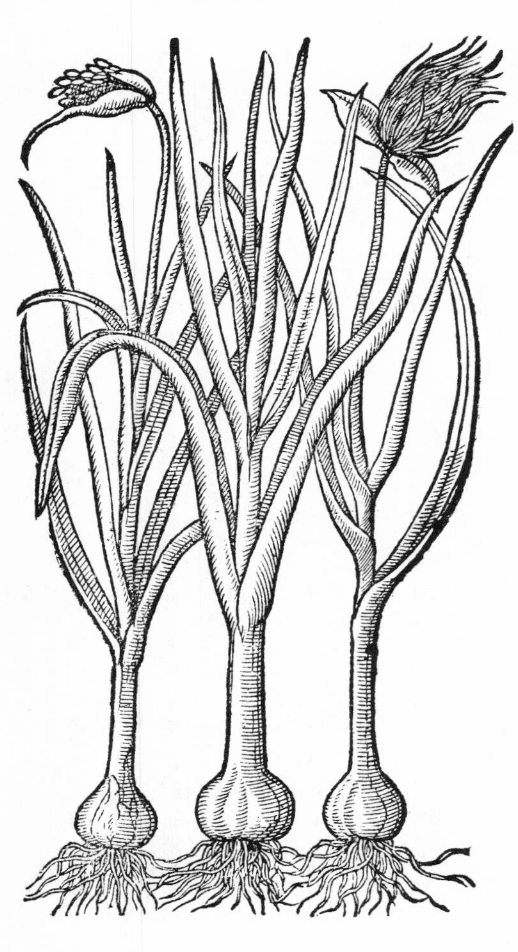

1 *Allium.*
Garlicke.

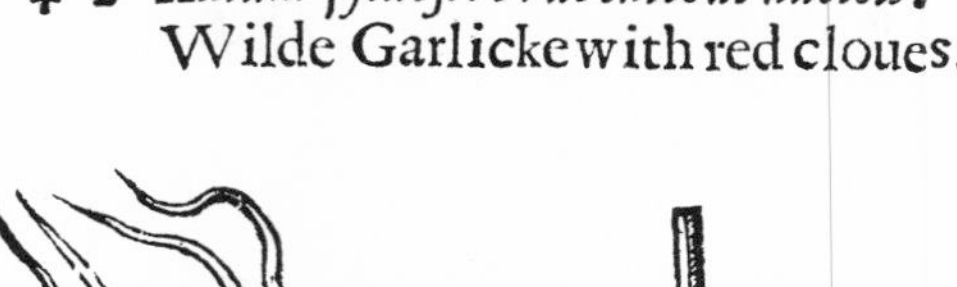

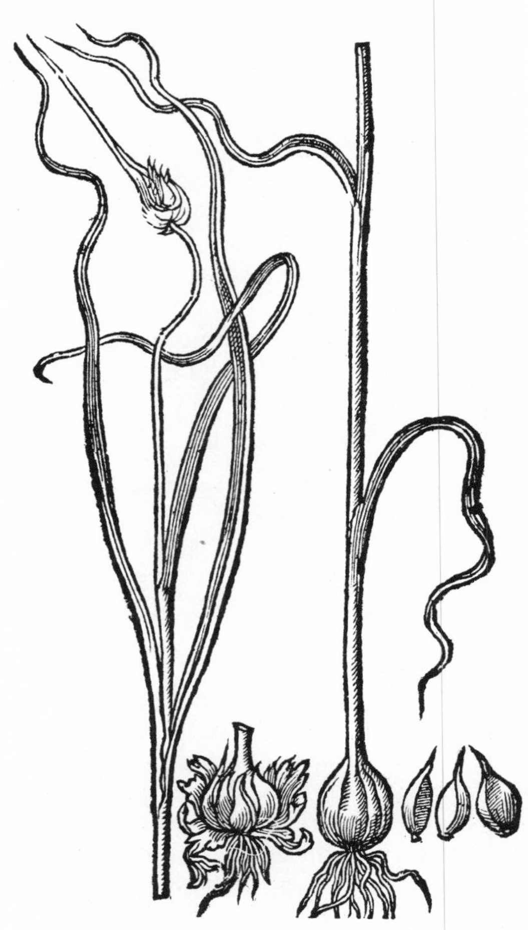

‡ 2 *Allium syluestre rubentibus nucleis.*
Wilde Garlicke with red cloues.

E It helpeth a very cold stomacke, and is a preseruatiue against the contagious and pestilent aire.

F The decoction of Garlick vsed for a bath to sit ouer, bringeth downe the floures and secondines or after-burthen, as *Dioscorides* saith.

G It taketh away the morphew, tetters, or ring-wormes, scabbed heads in children, dandraffe and scurfe, tempered with honey, and the parts anointed therewith.

H With Fig leaues and Cumin it is laid on against the bitings of the Mouse called in Greeke, μυγάλη: in English, a Shrew.

Chap.

CHAP. 98. *Of Crow-Garlicke and Ramſons.*

¶ *The Deſcription.*

1 THe wilde Garlicke or Crow-garlicke hath ſmall tough leaues like vnto ruſhes, ſmooth and hollow within; among which groweth vp a naked ſtalke, round, ſlipperie, hard and ſound: on the top whereof, after the floures be gone, grow little ſeeds made vp in a round cluſter like ſmall kernels, hauing the ſmell and taſte of Garlick. In ſtead of a root there is a bulbe or round head without any cloues at all.

2 Ramſons do ſend forth two or three broad longiſh leaues ſharpe pointed, ſmooth, and of a light greene colour. The ſtalke is a ſpan high, ſmooth and ſlender, bearing at the top a cluſter of white ſtar-faſhioned floures. In ſtead of a root it hath a long ſlender bulbe, which ſendeth downe a multitude of ſtrings, and is couered with skinnes or thicke coats.

† 1 *Allium ſylueſtre.*
Crow Garlicke.

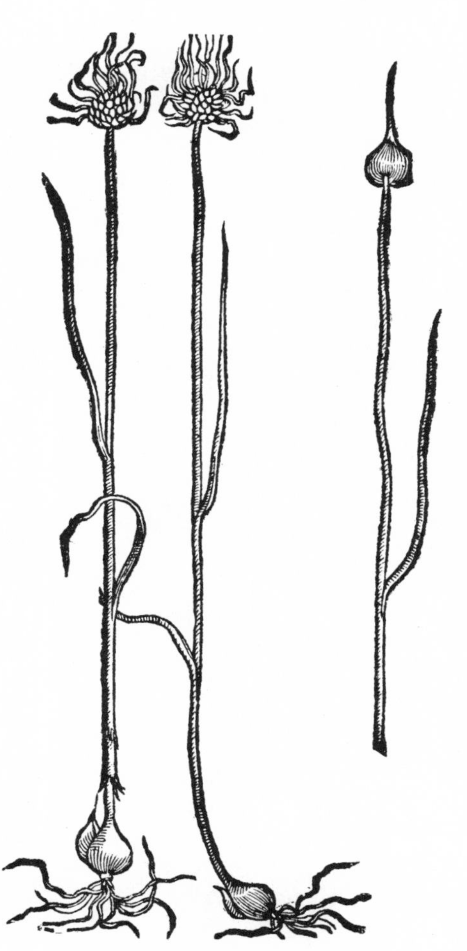

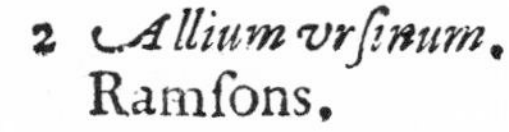

2 *Allium vrſinum.*
Ramſons.

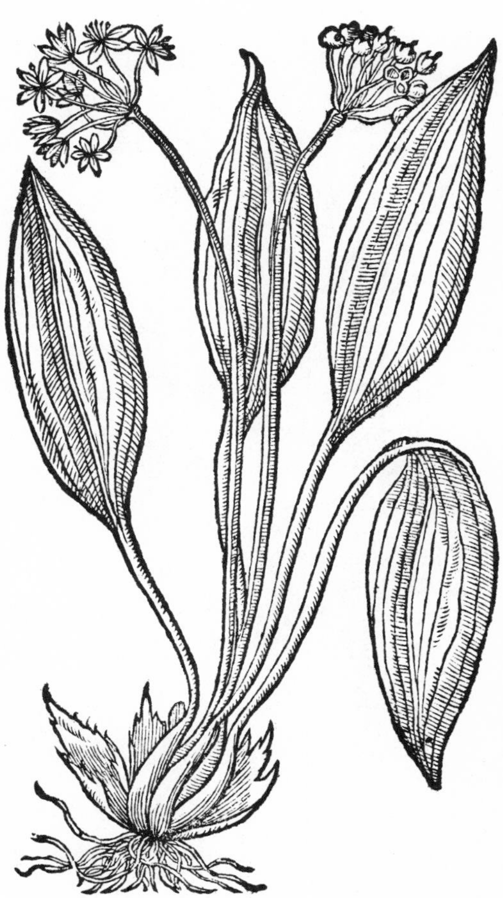

¶ *The Time.*

They ſpring vp in Aprill and May. Their ſeed is ripe in Auguſt.

¶ *The Place.*

The Crow Garlicke groweth in fertile paſtures in all parts of England. I found it in great plentie in the fields called the Mantels, on the backſide of Iſlington by London.

Ramſons grow in the Woods and borders of fields vnder hedges, among the buſhes. I found it in the next field vnto Boobies barne, vnder that hedge that bordereth vpon the lane; and alſo vpon the left hand, vnder an hedge adioyning to a lane that leadeth to Hampſted, both places neere London.

¶ *The*

¶ *The Names.*

Both of them be wilde Garlicke, and may be called in Latine, *Allium syluestria*: in Greeke, σκόροδα ἄγρια: The first, by *Dodonæus* and *Lobell* is called *Allium syluestre tenuifolium.*

Ramsons are named of the later practioners, *Allium Vrsinum*, or Beares Garlicke: *Allium latifolium*, and *Moly Hippocraticum*: in English, Ramsons, Ramsies, and Buckrams.

¶ *The Nature.*

The temperatures of these wilde Garlickes are referred vnto those of the gardens.

¶ *The Vertues.*

A Wilde Garlicke, or Crow-Garlicke, as *Galen* saith, is stronger and of more force than the garden Garlicke.

B The leaues of Ramsons be stamped and eaten of diuers in the Low-countries, with fish for a sauce, euen as we do eate greene-sauce made with sorrell.

C The same leaues may very well be eaten in April and May with butter, of such as are of a strong constitution, and labouring men.

D The distilled water drunke breaketh the stone, and driueth it forth, and prouoketh vrine.

Chap. 99. *Of Mountaine Garlicks.*

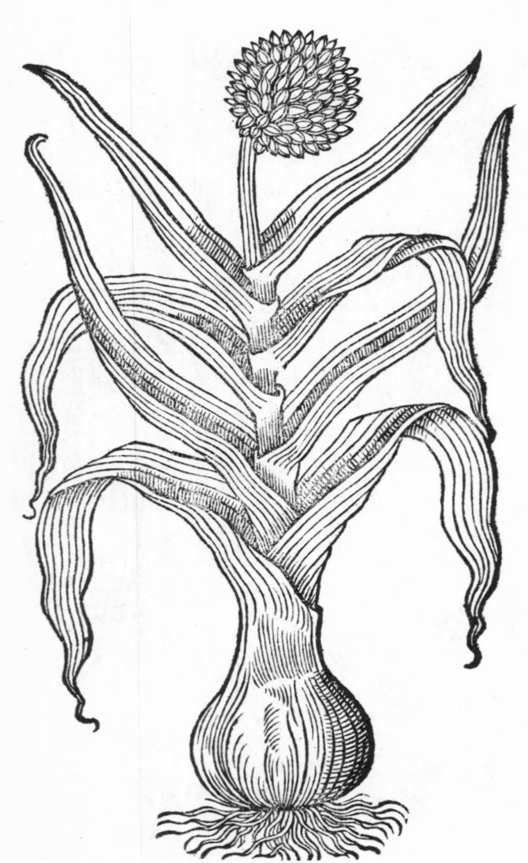

1 *Scorodoprasum.*
Great mountaine Garlicke.

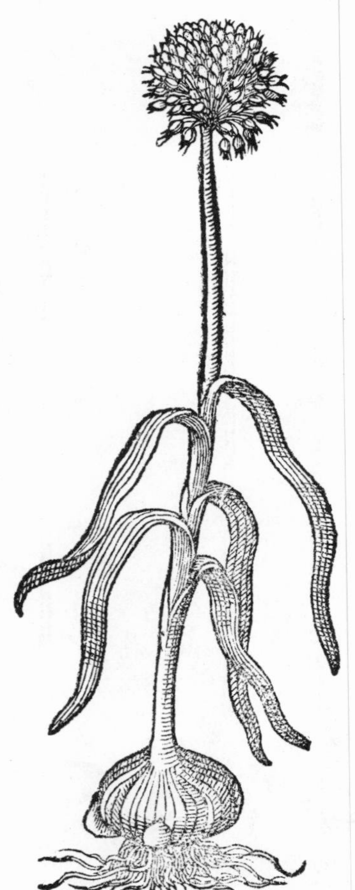

‡ 2 *Scorodoprasum primum Clusij.*
Clusius his great mountaine Garlicke.

¶ *The Description.*

1 2 The great Mountaine Garlicke hath long and broad leaues like those of Leekes, but much greater and longer, embracing or clasping about a great thicke stalke, soft and full of juyce, bigger than a mans finger, and bare toward the top; vpon which is set a great head bigger than a tennise ball, couered with a skinne after the manner of an Onion. The skinne when it commeth to perfection breaketh, and discouereth a great multitude of whitish floures; which being past, blacke seeds follow, inclosed in a three cornered huske. The root is bulbous, of the bignesse of a great Onion. The whole plant smelleth very strong like vnto Garlick,

Garlicke, and is in shew a Leeke, whereupon it was called *Scorodoprasum*, as if we should say, Garlicke Leeke; participating of the Leeke and Garlicke, or rather a degenerate Garlicke growne monstrous.

‡ I cannot certainely determine what difference there may be betweene the plants expressed by the first figure, which is our Authors, and the second figure which is taken out of *Clusius*. Now the historie which *Clusius* giues vs to the second, the same is (out of him) giuen by our Author to the first: so that by this reason they are of one and the same plant. To the which opinion I rather incline, than affirme the contrarie with *Bauhine*, who distinguishing them, puts the first amongst the Leekes, vnder the name of *Porrum folio latissimo*: following *Tabernamontanus*, who first gaue this figure, vnder the name of *Porrum Syriacum*.

3 This plant is lesser in all the parts than the former; the root is set about with longer and slenderer bulbes wrapped in brownish skinnes; the floures and leaues are like, yet smaller than Garlicke.

‡ 3 *Scorodoprasum minus.*
The lesser leeke-leaued Garlicke.

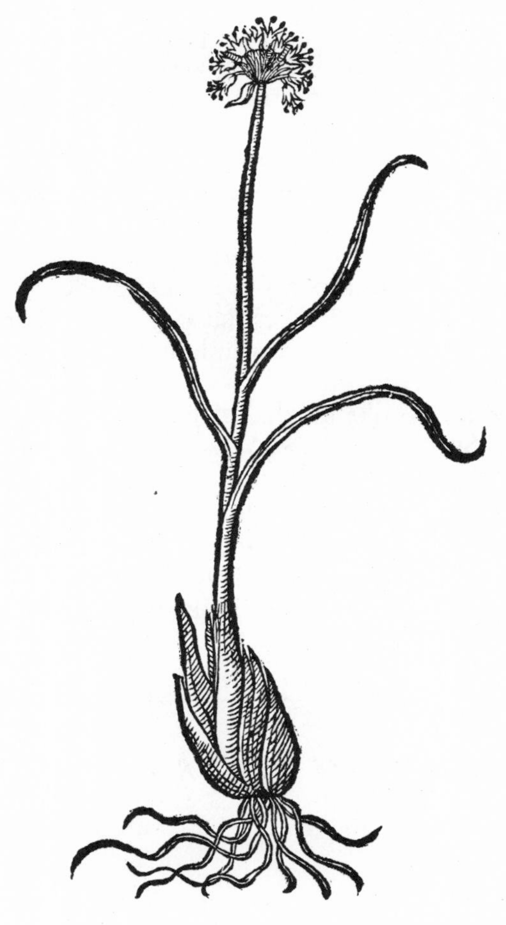

‡ 4 *Ophioscoridon.*
Vipers Garlike.

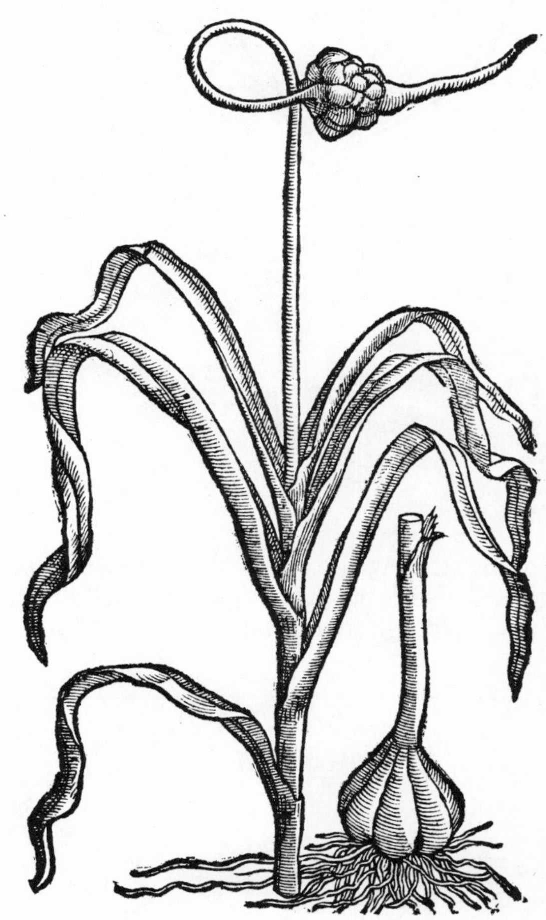

4 The third, which *Clusius* makes his second *Scorodoprasum*, hath stalkes some two cubits high, hauing many leaues like those of Leekes from the bottome of the stalke to the middle thereof; their smell is betweene that of Leekes and Garlicke; the rest of the stalke is naked, green, smooth, sustaining at the top a head composed of many bulbes, couered with a whitish skinne ending in a long greene point; which skinne by the growth of the bulbes being broken, they shew themselues, being first of a purplish, and afterwards of a whitish colour, amongst which are some floures. The top of the stalke at first twines it selfe, so that it in some sort represents a serpent; then by little it vntwines againe, and beares the head straight vp. The root consists of many cloues much like that of Garlicke. ‡

5 The broad leaued Mountaine Garlicke, or rather the Mountaine Ramsons, riseth vp with a stalke a cubit high, a finger thicke, yet very weake, full of a spongeous substance, neere to the bottome of a purplish colour, and greene aboue, bearing at the top a multitude of small whitish floures, somewhat gaping, star-fashion. The leaues are three or foure, broad ribbed like the leaues of great Gentian, resembling those of Ramsons, but greater. The root is great and long, couered with many scaly coats and hairy strings.

¶ The

5 *Allium Alpinum latifolium, seu Victorialis.*
Broad leaued Mountaine Garlicke.

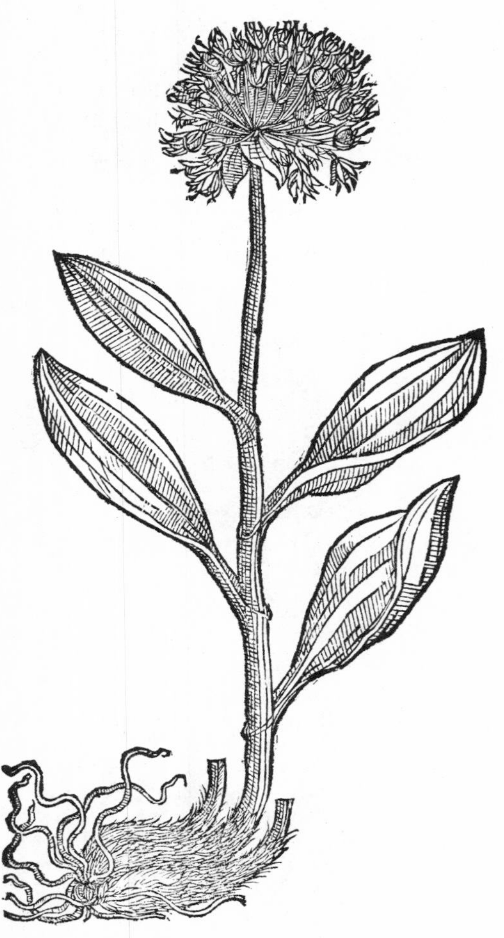

¶ *The Place.*

The great mountaine Garlicke growes about Constantinople, as saith *Clusius*. I receiued a plant of it from M. *Thomas Edwards* Apothecary of Excester, who found it growing in the West parts of England.

Victorialis groweth in the mountaines of Germany, as saith *Carolus Clusius*, and is yet a stranger in England for any thing that I do know.

‡ ¶ *The Time.*

‡ Most of these plants floure in the months of Iune and Iuly.

¶ *The Names.*

Of the first and second I haue spoken already. The third is *Scorodoprassum minus* of *Lobell*. The fourth is *Allium satiuum secundum* of *Dodonæus*, and *Scorodoprasum secundum* of *Clusius*. The fifth is *Allium anguinum* of *Matthiolus*; *Ophioscoridon* of *Lobell*, and *Victorialis* of *Clusius* and others, as also *Allium Alpinum*. The Germanes call it **Seigwurtz.**

¶ *The Temper.*

They are of a middle temper between Leekes and Garlicke.

¶ *Their Vertues.*

Scorodoprasum, as it partakes of the temper, so also of the vertues of Leekes and Garlicke; that is, it attenuates grosse and tough matter, helpes expectoration, &c.

Victorialis is like Garlicke in the operation thereof. Some (as *Camerarius* writeth) hang the root thereof about the necks of their cattell being falne blinde, by what occasion soeuer it happen, and persuade themselues that by this meanes they will recouer their sight. Those that worke in the mines in Germany affirme, That they find this root very powerfull in defending them from the assaults of impure spirits or diuels, which often in such places are troublesome vnto them. *Clus.* ‡

Chap. 100. *Of Moly, or the Sorcerers Garlicke.*

¶ *The Description.*

1 THe first kinde of Moly hath for his root a little whitish bulbe somewhat long, not vnlike to the root of the vnset Leeke, which sendeth forth leaues like the blades of corne or grasse: among which doth rise vp a slender weake stalke, fat and full of iuyce, at the top whereof commeth forth of a skinny filme a bundle of milke-white floures, not vnlike to those of Ramsons. The whole plant hath the smell and taste of Garlicke, whereof no doubt it is a kinde.

2 Serpents Moly hath likewise a small bulbous root with some fibres fastned to the bottom, from which rise vp weake grassie leaues of a shining greene colour, crookedly winding and turning themselues toward the point like the taile of a Serpent, whereof it tooke his name: the stalke is tough, thicke, and full of iuyce, at the top whereof standeth a cluster of small red bulbes, like vnto the smallest cloue of Garlicke, before they be pilled from their skinne. And among those bulbes there do thrust forth small and weake foot-stalkes, euery one bearing at the end one small white floure tending to a purple colour: which being past, the bulbes do fall downe vpon the ground, where they without helpe do take hold and root, and thereby greatly encrease, as also by the infinite bulbes that the root doth cast off: all the whole plant doth smell and taste of Garlick, whereof it is also a kinde.

3 *Homers* Moly hath very thicke leaues, broad toward the bottome, sharpe at the point, and hollowed

1 *Moly Dioſcorideum.*
Dioſcorides his Moly.

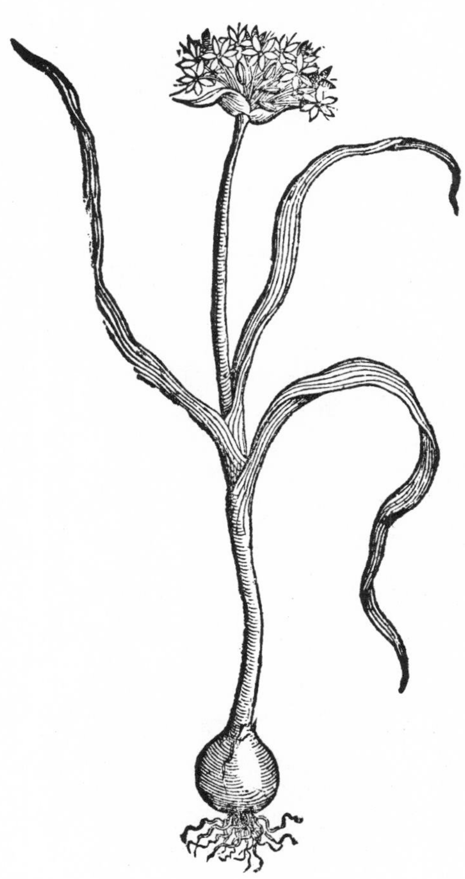

2 *Moly Serpentinum.*
Serpents Moly.

3 *Moly Homericum.*
Homers Moly.

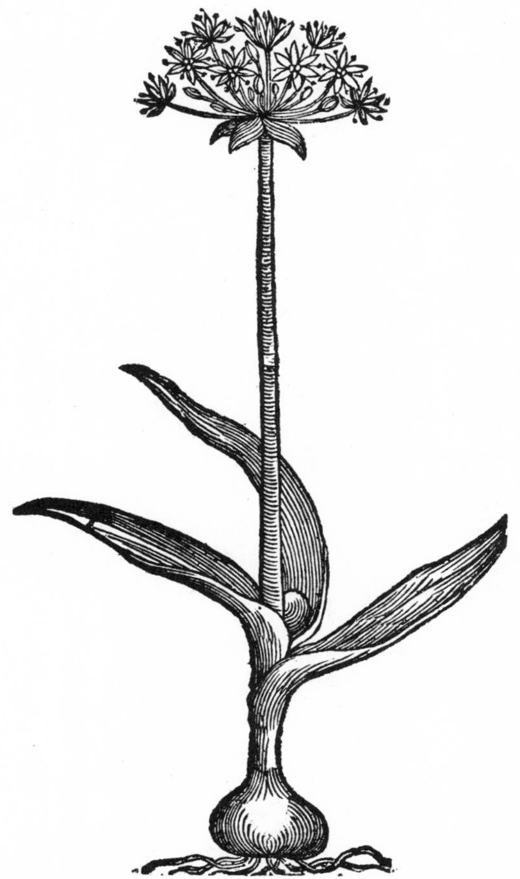

hollowed like a trough or gutter, in the boſome of which leaues neere vnto the bottome commeth forth a certaine round bulbe or ball of a gooſe-turd greene colour: which being ripe and ſet in the ground groweth and becommeth a faire plant ſuch as is the mother. Among thoſe leaues riſeth vp a naked ſmooth thicke ſtalke, of two cubits high, as ſtrong as a ſmall walking ſtaffe: at the top of the ſtalke ſtandeth a bundle of faire whitiſh floures, daſhed ouer with a waſh of purple colour, ſmelling like the floures of Onions. When they be ripe there appeareth a blacke ſeed wrapped in a white skinne or huske. The root is great and bulbous, couered with a blackiſh skinne on the outſide, and white within, and of the bigneſſe of a great Onion.

4 Indian Moly hath very thicke fat ſhort leaues, and ſharpe pointed; in the boſome wherof commeth forth a thicke knobby bulbe like that of *Homers* Moly. The ſtalke is alſo like the precedent, bearing at the top a cluſter of ſcaly bulbes included in a large thinne skin or filme. The root is great, bulbous faſhion, and full of iuyce.

5 *Caucaſon*, or withering Moly, hath a very great bulbous root, greater than that of *Homers* Moly, and fuller of a ſlimie iuyce; from which do ariſe three or foure great thicke and broad leaues withered alwaies at the point; wherein

wherein consisteth the difference betweene these leaues and those of *Homers* Moly, which are not so. In the middle of the leaues riseth vp a bunch of smooth greenish bulbes set vpon a tender footstalke, in shape and bignesse like to a great garden Worme, which being ripe and planted in the earth, do also grow vnto a faire plant like vnto their mother.

‡ These two last mentioned (according to *Bauhine*, and I thinke the truth) are but figures of one and the same plant; the later whereof is the better, and more agreeing to the growing of the plant.

6 To these may be fitly added two other Molyes: the first of these, which is the yellow Moly, hath roots whitish and round, commonly two of them growing together; the leaues which it sends forth are long and broad, and somwhat resemble those of the Tulip, and vsually are but two in number; betweene which rises vp a stalke some foot high, bearing at the top an vmbell of faire yellow star-like floures tipt on their lower sides with a little greene. The whole plant smelleth of Garlicke.

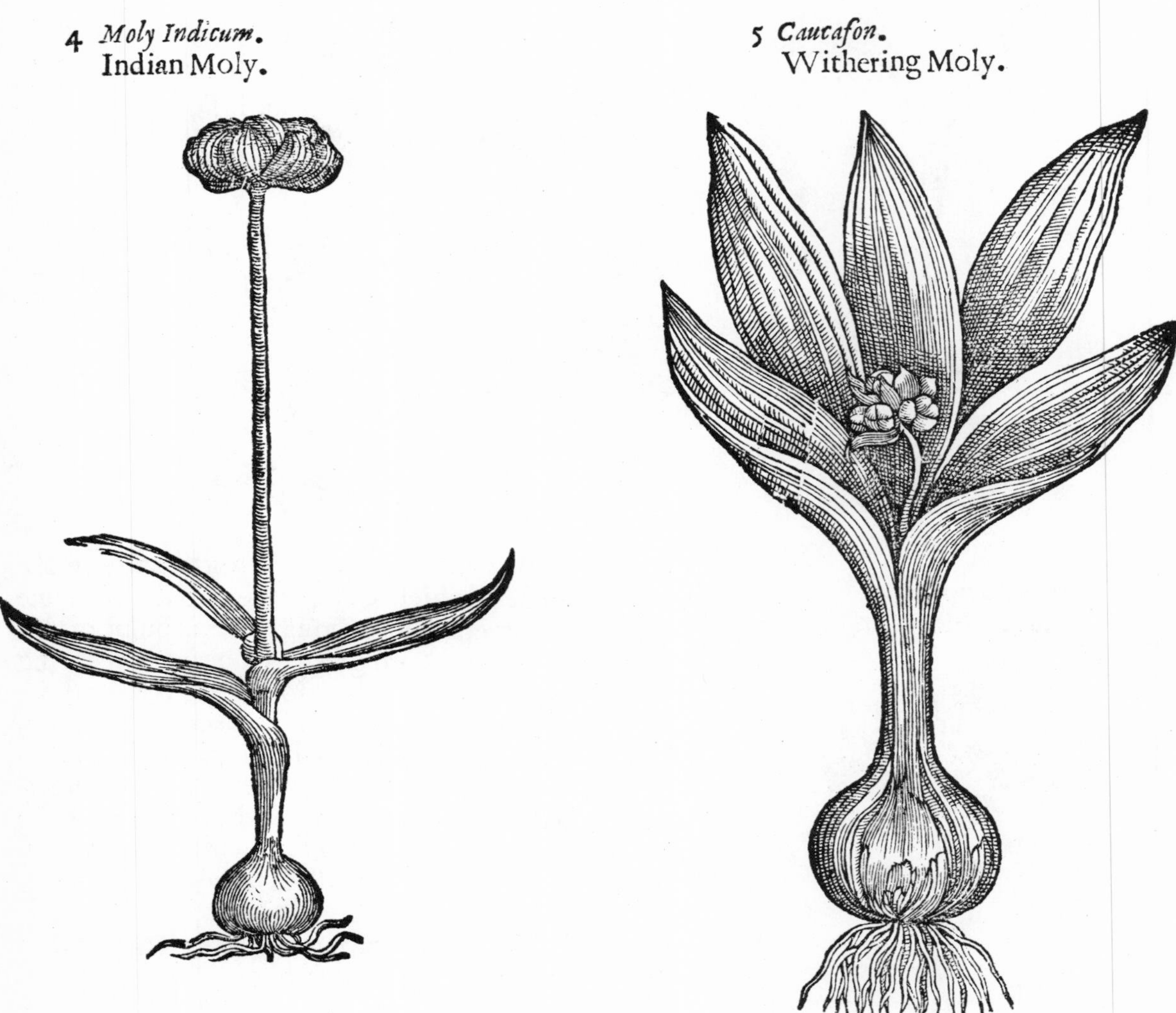

4 *Moly Indicum.*
Indian Moly.

5 *Caucason.*
Withering Moly.

7 This little Moly hath a root about the bignesse of an Hasell nut, white, with some fibres hanging thereat; the stalke is of an handfull or little more in height, the top thereof is adorned with an vmbel of ten or twelue white floures, each of which consists of six leaues, not sharpe pointed, but turned round, and pretty large, considering the bignesse of the plant. This plant hath also vsually but two leaues, and those like those of Leekes, but far lesse. ‡

¶ *The Place.*

† These plants grow in the garden of M. *Iohn Parkinson* Apothecarie, and with M. *Iohn Tradescant* and some others, studious in the knowledge of plants.

¶ *The Time.*

They spring forth of the ground in Februarie, and bring forth their floures, fruit, and seed in the end of August.

¶ *The Names.*

† Some haue deriued the name *Moly* from these Greeke words, Μωλύειν τὰς νόσους: that is, to driue away diseases. It may probably be argued to belong to a certaine bulbous plant, and that a kind of

of Garlicke, by the words μώλυζα, and μῶλυξ. The former, *Galen* in his *Lexicon* of some of the difficulter words vsed by *Hippocrates*, thus expounds: Σκόροδον ἁπλῆν τὴν κεφαλὴν ἔχον, καὶ μὴ διαλυομένην εἰς ἀγλῖθας· τινὲς δὲ τὸ μῶλυ. That is, *Moliza* is a Garlicke hauing a simple or single head, and not to be parted or distinguished into cloues: some terme it *Moly*. *Erotianus* in his *Lexicon* expounds the later thus: μῶλυξ (saith hee) Σκορόδου κεφαλὴ ϲωλοειδής, &c. That is; *Molyx* is a head of Garlicke, round, and not to be parted into cloues. ‡

¶ *The Names in particular.*

‡ 1 This is called *Moly* by *Matthiolus*; *Moly Angustifolium* by *Dodonæus*; *Moly Dioscorideum* by *Lobel* and *Clusius*.

2 This, *Moly Serpentinum vocatum*, by *Lobel* and the Author of the *Hist. Lugd.*

3 This same is thought to be the *Moly* of *Theophrastus* and *Pliny*, by *Dodonæus*, *Clusius*, &c. and some also would haue it to be that of *Homer*, mentioned in his twentieth *Odyss.* *Lobel* calleth it *Moly Liliflorum*.

4 5 The fourth and fifth being one, are called *Caucason*, and *Moly Indicum* by *Lobel*, *Clusius*, and others.

6 This is *Moly Montanum latifolium flauo flore* of *Clusius*, and *Moly luteum* of *Lobel*, *Aduersar. par. 2.*

7 This same is *Moly minus* of *Clusius*. ‡

‡ 6 *Moly latifolium flore flauo.*
Broad leaued Moly with the yellow floure.

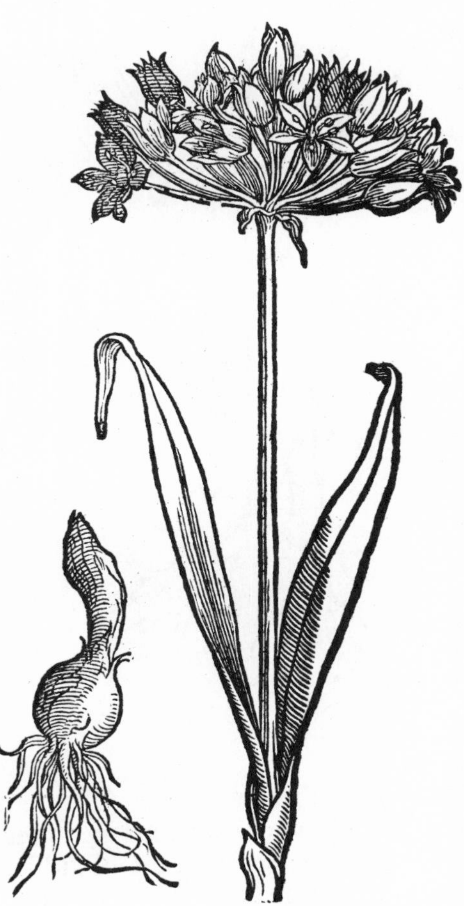

‡ 7 *Moly minus flore albo.*
Dwarfe white floured Moly.

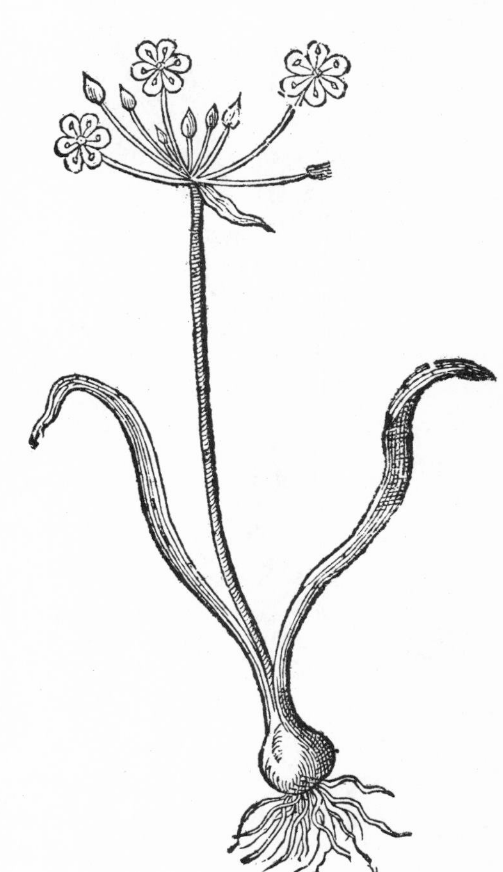

¶ *The Temperature and Vertues.*

These Molyes are very hot, approching to the nature of Garlicke, and I doubt not but in time some excellent man or other will find out as many good vertues of them, as their stately and comely proportion should seeme to be possessed with. But for my part, I haue neither proued, nor heard of others, nor found in the writings of the Antients, any thing touching their faculties. Only *Dioscorides* reporteth, That they are of maruellous efficacie to bring downe the termes, if one of them be stamped with oyle of Floure de-luce according to art, and vsed in manner of a pessarie or mother suppositorie.

‡ Chap. 79. *Of diuers other Molyes.*

‡ BEsides the Garlickes and Molyes formerly mentioned by our Author, and those I haue in this Edition added, there are diuers others, which, mentioned by *Clusius*, and belonging vnto this Tribe, I haue thought good in this place to set forth. Now for that they are more than conueniently could be added to the former chapters, (which are sufficiently large) I thought it not amisse to allot them a place by themselues.

‡ 1 *Moly Narcissinis folijs primum.*
The first Narcisse-leaued Moly.

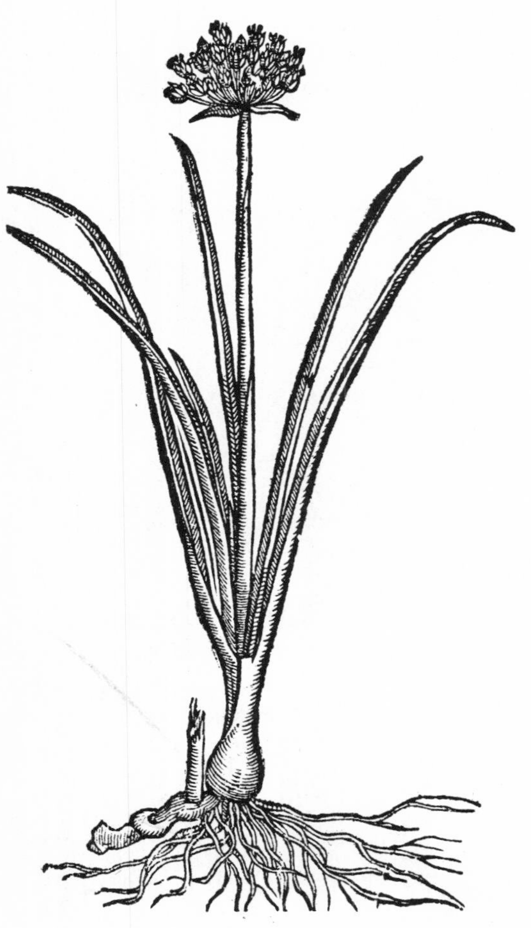

‡ 2 *Moly Narcissinis folijs secundum.*
The second Narcisse-leaued Moly.

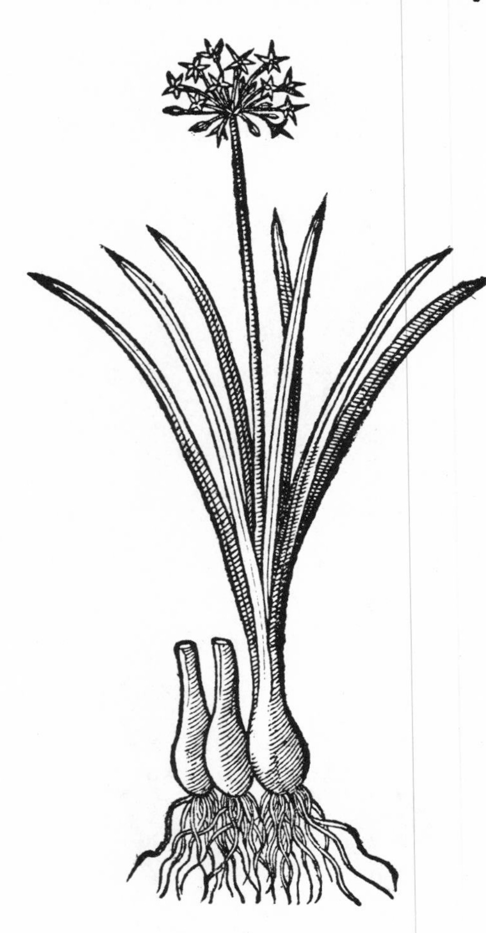

¶ *The Description.*

‡ 1 THis, which in face nighest represents the Molyes described in the last Chapter, hath a root made of many scales, like as an Onion in the vpper part, but the lower part is knotty, and runnes in the ground like as *Solomons* Seale; the Onion-like part hath many fibres hanging thereat; the leaues are like those of the white Narcisse, very greene and shining, amongst which riseth vp a stalke of a cubit high, naked, firme, greene, and crested; at the top come forth many floures consisting of six purplish leaues, with as many chiues on their insides: after which follow three square heads, opening when they are ripe, and containing a round blacke seed.

2 This other being of the same kinde, and but a varietie of the former, hath softer and more Ash-coloured leaues with the floures of a lighter colour. Both these floure at the end of Iune, or in Iuly.

3 This hath fiue or six leaues equally as broad as those of the last described, but not so long, being somewhat twined, greene, and shining. The stalke is some foot in length, smaller than that of the former, but not lesse stiffe, crested, and bearing in a round head many floures, in manner of growing and shape like those of the former, but of a more elegant purple colour. In seed and root it

‡ 3 *Moly Narciſsinis folijs tertium.*
The third Narciſſe-leaued Moly.

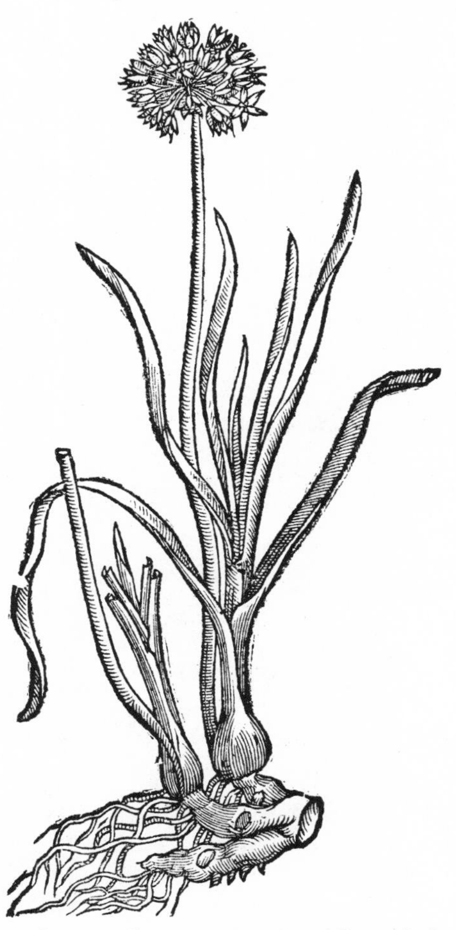

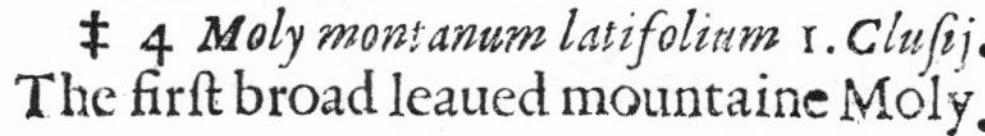

‡ 4 *Moly montanum latifolium* 1. *Cluſij.*
The firſt broad leaued mountaine Moly.

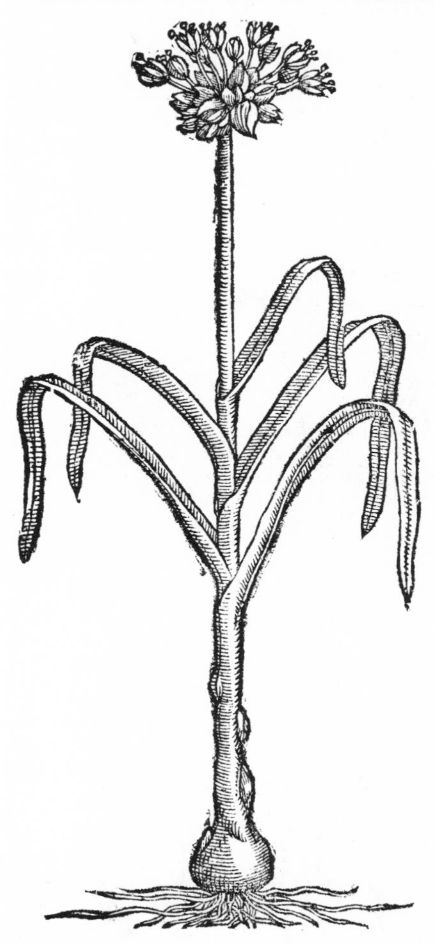

‡ 5 *Moly montanum ſecundum Cluſij.*
The ſecond mountaine Moly.

it reſembles the precedent. There is alſo a varietie of this kinde, with leaues longer and narrower, neither ſo much twined, the ſtalks weaker, and floures much lighter coloured.

This floures later than the former, to wit, in Iuly and Auguſt.

All theſe plants grow naturally in Leitenberg and other hills neere to Vienna in Auſtria, where they were firſt found and obſerued by *Carolus Cluſius.*

4 This hath a ſtalke ſome two cubits high, which euen to the middle is incompaſſed with leaues much longer and broader than thoſe of Garlicke, and very like thoſe of the Leeke : on the top of the ſmooth and ruſh-like ſtalke groweth a tuft conſiſting of many darke purple coloured bulbs growing cloſe together, from amongſt which come forth pretty long ſtalkes bearing light purple ſtarre-faſhioned floures, which are ſucceeded by three cornered ſeed-veſſels. The root is bulbous, large, conſiſting of many cloues, and hauing many white fibres growing forth thereof. Moreouer, there grow out certain round bulbes about the root, almoſt like thoſe which grow in the head, and being planted apart, they produce plants of the ſame kinde. This is *Allium, ſiue Moly montanum latifolium* 1. *Cluſij.*

5 This hath a ſmooth round greene ſtalke ſome cubit high, whereon doe grow moſt commonly

monly three leaues narrower than thoſe of the former, and as it were graſſy. The top of the ſtalke ſuſtaines a head wrapped in two lax filmes, each of them running out with a ſharpe point like two hornes, which opening themſelues, there appeare many ſmall bulbes heaped together, amongſt which are floures compoſed of ſix purpliſh little leaues, and faſtned to long ſtalkes. The root is round and white, with many long white fibres hanging thereat. *Cluſius* calls this, *Allium, ſiue Moly montanum ſecundum*. And this is *Lobels Ampelopraſon proliferum*.

6 Like to the laſt deſcribed is this in height and ſhape of the ſtalke and leaues, as alſo in the forked or horned skinne inuoluing the head, which conſiſteth of many ſmall bulbes of a reddiſh greene colour, and ending in a long greene point; amongſt which, vpon long and ſlender ſtalkes hang downe floures like in forme and magnitude to the former, but of a whitiſh colour, with a darke purple ſtreake alongſt the middle, and vpon the edges of each leafe. The root is round and white, like that of the laſt deſcribed. This *Cluſius* giues vnder the title of *Allium ſiue Moly montanum tertium*.

‡ 6 *Moly montanum* 3. *Cluſ.*
The third mountaine Moly.

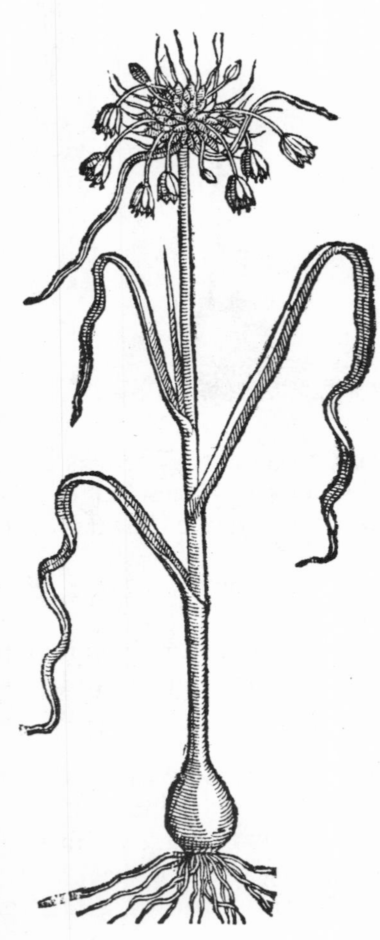

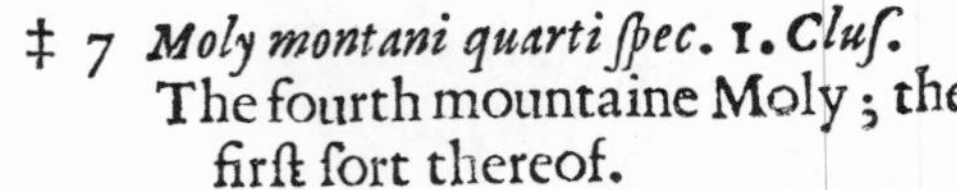
‡ 7 *Moly montani quarti ſpec.* 1. *Cluſ.*
The fourth mountaine Moly; the firſt ſort thereof.

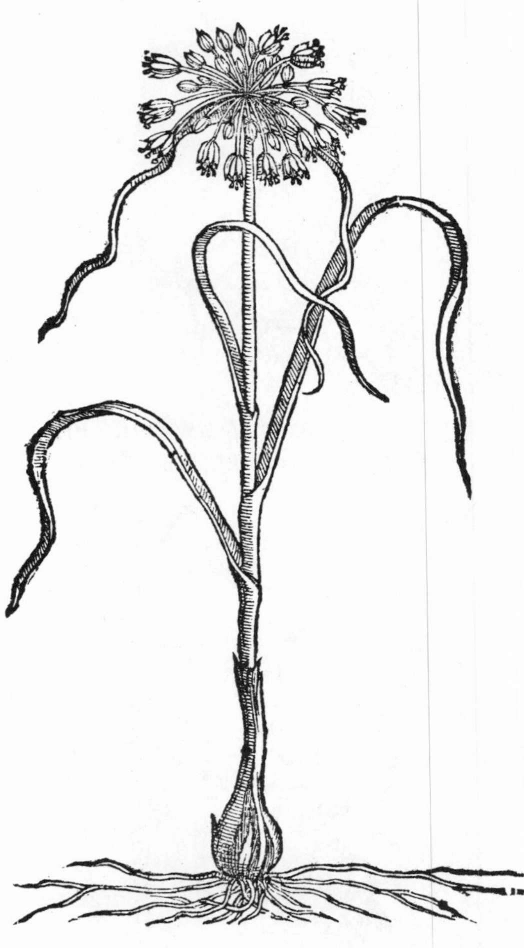

7 This alſo hath three ruſhy leaues, with a round ſtalke of ſome cubit high, whoſe top is likewiſe adorned with a forked membrane, containing many pale coloured floures hanging vpon long ſtalkes, each floure conſiſting of ſix little leaues, with the like number of chiues, and a peſtil in the midſt. This tuft of floures cut off with the top of the ſtalke, and carried into a chamber, wil yeeld a pleaſant ſmell (like that which is found in the floures in the earlier *Cyclamen*) but it will quickly decay. After theſe floures are paſt ſucceed three cornered heads containing a blacke ſmall ſeed, not much vnlike Gillofloure ſeed. The root is round like the former, ſometimes yeelding off-ſets. This is *Alij montani* 4. *ſpecies* 1. of *Cluſius*.

8 There is another kinde of this laſt deſcribed, which growes to almoſt the ſame height, and hath like leaues, and the head ingirt with the like skinny long pointed huskes; but the floures of this are of a very darke colour, The roots are like the former, with off-ſets by their ſide. This is

Cluſius

Cluſius his *Moly montani quarti ſpecies ſecunda*. The roots of the three laſt deſcribed ſmell of garlick, but the leaues haue rather an hearby or graſſe-like ſmell.

The fifth and ſixth of theſe grow naturally in the Styrian and Auſtrian Alpes. The ſeuenth growes about Presburg in Hungarie, about Niclaſpurg in Morauia, but moſt aboundantly about the Baths in Baden.

‡ 8 *Moly montani quarti ſpecies ſecunda Cluſij*.
The ſecond kinde of the fourth mountaine Moly.

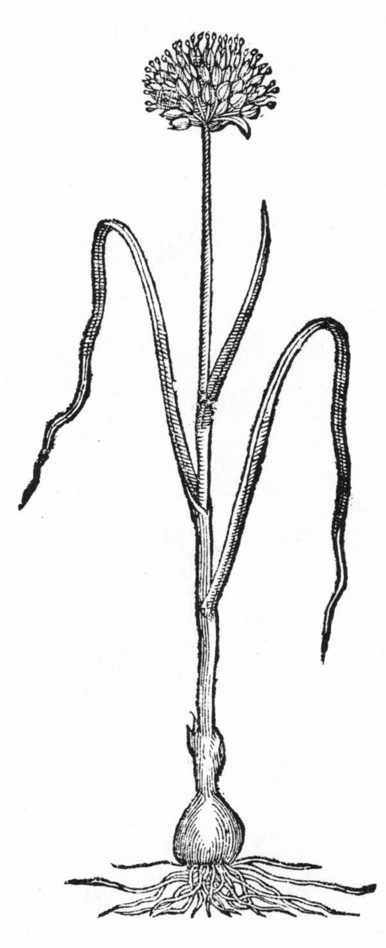

‡ 9 *Moly montanum quintum Cluſij*.
The fifth mountaine Moly.

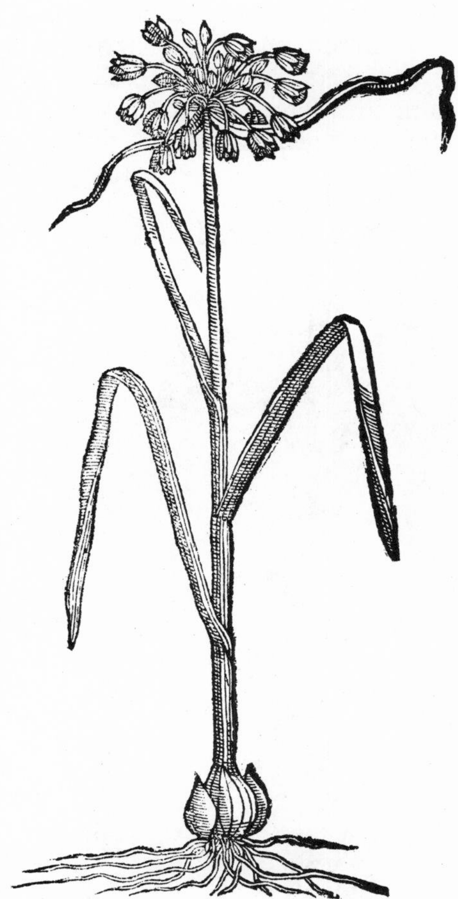

9 This growes to the like height as the former, with a greene ſtalke, hauing few leaues thereupon, and naked at the top, where it carieth a round head conſiſting of many ſtar-like ſmall floures, of a faire purple colour, faſtned to ſhort ſtalkes, each floure being compoſed of ſixe little leaues, with as many chiues, and a peſtill in the middle. The root is bulbous and white, hauing ſomtimes his off-ſets by his ſides. The ſmell of it is like Garlicke. This groweth alſo about Presburgh in Hungarie, and was there obſerued by *Cluſius* to beare his floure in May and Iune. He calleth this *Allium, ſeu Moly montanum quintum*. ‡

CHAP. 102. *Of White Lillies.*

¶ *The Kindes.*

THere be ſundry ſorts of Lillies, whereof ſome be wilde, or of the field; others tame, or of the garden; ſome white, others red; ſome of our owne countries growing, others from beyond the ſeas: and becauſe of the variable ſorts we will diuide them into chapters, beginning with the two white Lillies, which differ little but in the natiue place of growing.

¶ *The Deſcription.*

1 THe white Lillie hath long, ſmooth, and full bodied leaues, of a graſſie or light greene colour. The ſtalkes be two cubits high, and ſometimes more, ſet or garniſhed with the like leaues, but growing ſmaller and ſmaller toward the top; and vpon them doe grow faire white floures ſtrong of ſmell, narrow toward the foot of the ſtalke whereon they doe grow, wide or open in the mouth like a bell. In the middle part of them doe grow ſmall tender pointals tipped with a duſty yellow colour, ribbed or chamfered on the backe ſide, conſiſting of ſix ſmall leaues thicke and fat. The root is a bulbe made of ſcaly cloues, full of tough and clammie iuyce, wherewith the whole plant doth greatly abound.

2 The white Lilly of Conſtantinople hath very large and fat leaues like the former, but narrower and leſſer. The ſtalke riſeth vp to the height of three cubits, ſet and garniſhed with leaues alſo like the precedent, but much leſſe. Which ſtalke oftentimes doth alter and degenerate from his naturall roundneſſe to a flat forme, as it were a lath of wood furrowed or chanelled alongſt the ſame, as it were ribs or welts. The floures grow at the top like the former, ſauing that the leaues do turne themſelues more backward like the Turkes cap, and beareth many more floures than our Engliſh white Lilly doth.

1 *Lilium album.*
The white Lilly.

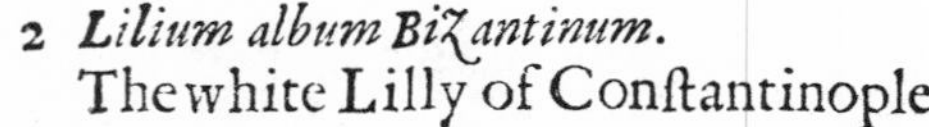
2 *Lilium album Bizantinum.*
The white Lilly of Conſtantinople.

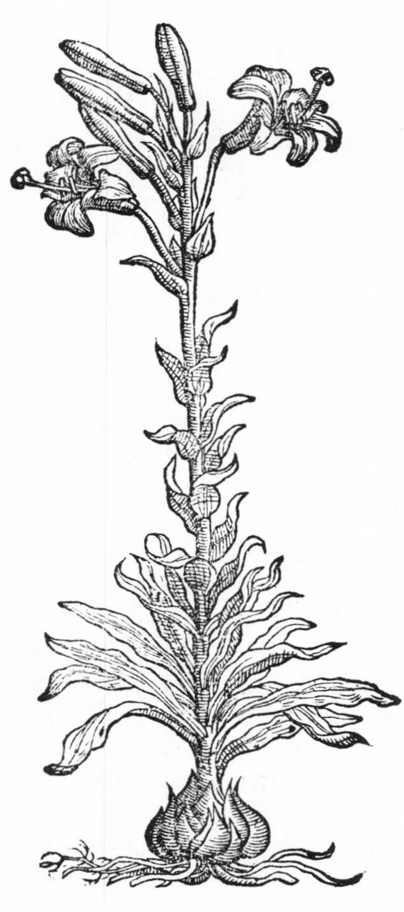

¶ *The Place.*

Our Engliſh white Lilly groweth in moſt gardens of England. The other groweth naturally in Conſtantinople and the parts adiacent, from whence wee had plants for our Engliſh gardens, where they flouriſh as in their owne countrey.

¶ *The Time.*

The Lillies floure from May to the end of Iune.

The Names.

The Lillie is called in Greeke κρίνον: in Latine, *Lilium*, and alſo *Roſa Iunonis*, or *Iuno's* Roſe, becauſe as it is reported it came vp of her milke that fell vpon the ground. For the Poets feigne, That *Hercules*, who *Iupiter* had by *Alcumena*, was put to *Iuno's* breaſts whileſt ſhee was aſleepe; and after the ſucking there fell away aboundance of milke, and that one part was ſpilt in the heauens, and the other on the earth; and that of this ſprang the Lilly, and of the other the circle in the heauens called

called *Lacteus Circulus*, or the milky way, or otherwise in English Watling street. Saint *Basill* in the explication of the 44. Psalme saith, That no floure so liuely sets forth the frailty of mans life as the Lilly. It is called in high Dutch, **Weisz Gilgen**: in low Dutch, **Witte Lilien**: in Italian, *Giglio*: in Spanish, *Lirio blanco*: in French, *Lys blanc*: in English, the white Lilly.

The other is called *Lilium album Bizantinum*, and also *Martagon album Bizantinum*: in English, the white Lilly of Constantinople: of the Turkes themselues, *Sultan Zambach*, with this addition, (that they might be the better knowne which kinde of Lilly they meant when they sent roots of them into these countries) *Fa fiora grandi Bianchi*, so that *Sultan Zambach fa fiora grandi Bianchi*, is as much to say as, Sultans great Lilly with white floures.

¶ *The Nature.*

The white Lilly is hot, and partly of a subtill substance. But if you regard the root, it is dry in the first degree, and hot in the second.

¶ *The Vertues.*

The root of the garden Lilly stamped with honey gleweth together sinewes that be cut in sunder. It consumeth or scoureth away the vlcers of the head called Achores, and likewise all scuruinesse of the beard and face. A

The Root stamped with Vineger, the leaues of Henbane, or the meale of Barley, cureth the tumours and apostumes of the priuy members. It bringeth the haire againe vpon places which haue beene burned or scalded, if it be mingled with oyle or grease, and the place anointed therewith. B

The same root rosted in the embers, and stamped with some leauen of Rie bread and Hogges grease, breaketh pestilentiall botches. It ripeneth Apostumes in the flankes, comming of venery and such like. C

The floures steeped in Oyle Oliue, and shifted two or three times during Sommer, and set in the Sunne in a strong glasse, is good to soften the hardnesse of sinewes, and the hardnesse of the matrix. D

Florentinus a writer of husbandry saith, That if the root be curiously opened, and therein be put some red, blew, or yellow colour that hath no causticke or burning qualitie, it will cause the floure to be of the same colour. E

Iulius Alexandrinus the Emperors Physitian saith, That the water thereof distilled and drunke causeth easie and speedy deliuerance, and expelleth the secondine or after-burthen in most speedy manner. F

He also saith, the leaues boyled in red Wine, and applied to old wounds or vlcers, doe much good, and forward the cure, according to the doctrine of *Galen* in his seuenth booke *de simpl. med. facultat.* G

The root of a white Lilly stamped and strained with wine, and giuen to drinke for two or three dayes together, expelleth the poyson of the pestilence, and causeth it to breake forth in blisters in the outward part of the skinne, according to the experience of a learned Gentleman M. *William Godorus*, Sergeant Surgeon to the Queenes Maiestie: who also hath cured many of the Dropsey with the iuyce thereof, tempered with Barley meale, and baked in cakes, and so eaten ordinarily for some moneth or six weekes together with meate, but no other bread during that time. H

Chap. 103. *Of Red Lillies.*

¶ *The Kindes.*

There be likewise sundry sorts of Lillies, which we do comprehend vnder one generall name in English, Red Lillies, whereof some are of our owne countries growing, and others of beyond the seas, the which shall be distinguished seuerally in this chapter that followeth.

¶ *The Description.*

1 The gold-red Lilly groweth to the height of two, and sometimes three cubits, and often higher than those of the common white Lilly. The leaues be blacker and narrower, set very thicke about the stalke. The floures in the top be many, from ten to thirty floures, according to the age of the plant, and fertilitie of the soile, like in forme and greatnesse to those of the white Lilly, but of a red colour tending to a Saffron, sprinkled or poudred with many little blacke specks, like to rude vnperfect draughts of certaine letters. The roots be great bulbes, consisting of many cloues, as those of the white Lilly.

‡ 2 In ſtead of the Plantaine leaued red Lilly, deſcribed and figured in this ſecond place by our Author out of *Tabernamontanus*, for that I iudge both the figure and deſcription counterfeit, I haue omitted them, and here giue you the many-floured red Lilly in his ſtead. This hath a root like that of the laſt deſcribed, as alſo leaues and ſtalkes; the floure alſo in ſhape is like that of the former, but of a more light red colour, and in number of floures it exceedeth the precedent, for ſometimes it beares ſixty floures vpon one ſtalke. ‡

† 3 This red Lilly is like vnto the former, but not ſo tall; the leaues be fewer in number, broader, and downy towards the top of the ſtalke, where it beares ſome bulbes. The floures in ſhape be like the former, ſauing that the colour hereof is more red, and thicke daſht with blacke ſpecks. The root is ſcaly like the former.

4 There is another red Lilly which hath many leaues ſomewhat ribbed, broader than the laſt mentioned, but ſhorter, and not ſo many in number. The ſtalke groweth to the height of two cubits, and ſometimes higher, whereupon do grow floures like the former: among the foot-ſtalks of which floures come forth certaine bulbes or cloued roots, browne of colour, tending vnto redneſſe; which do fall in the end of Auguſt vpon the ground, taking root and growing in the ſame place, whereby it greatly encreaſeth, for ſeldome or neuer it bringeth forth ſeed for his propagation.

1 *Lilium aureum.*
Gold-red Lilly.

† 2 *Lilium rubrum.*
The red Lilly.

5 There is another ſort of red Lillie hauing a faire ſcaly or cloued root, yellow aboue, and browne toward the bottome; from which riſeth vp a faire ſtiffe ſtalke creſted or furrowed, of an ouer-worne browne colour, ſet from the lower part to the branches, whereon the floures doe grow with many leaues, confuſedly placed without order. Among the branches cloſe by the ſtem grow forth certaine cloues or roots of a reddiſh colour, like vnto the cloues of Garlicke before they are pilled: which being fallen vpon the ground at their time of ripeneſſe, do ſhoot forth certaine tender ſtrings or roots that do take hold of the ground, whereby it greatly encreaſeth. The floures are in ſhape like the other red Lillies, but of a darke Orange colour, reſembling a flame of fire ſpotted with blacke ſpots.

‡ 6 This hath a much ſhorter ſtalke, being but a cubit or leſſe in height, with leaues blackiſh

kiſh, and narrower than thoſe afore going. The floures, as in the reſt, grow out of the top of the ſtalke, and are of a purpliſh Saffron colour, with ſome blackiſh ſpots. The root in ſhape is like the precedent. ‡

¶ *The Place.*

Theſe Lillies do grow wilde in the plowed fields of Italy and Languedocke, in the mountaines and vallies of Hetruria and thoſe places adiacent. They are common in our Engliſh gardens, as alſo in Germany.

¶ *The Time.*

Theſe red Lillies do floure commonly a little before the white Lillies, and ſometimes together with them.

3 *Lilium cruentum latifolium.*
The fierie red Lilly.

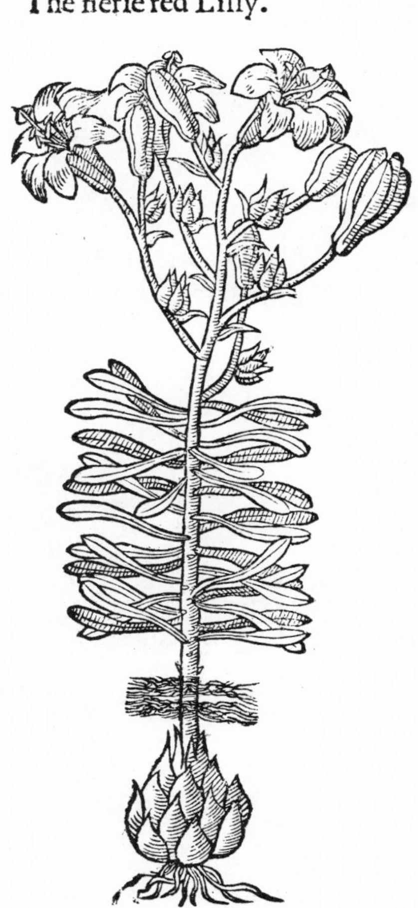

‡ 4 *Lilium cruentum bulbiferum.*
Red bulbe-bearing Lilly.

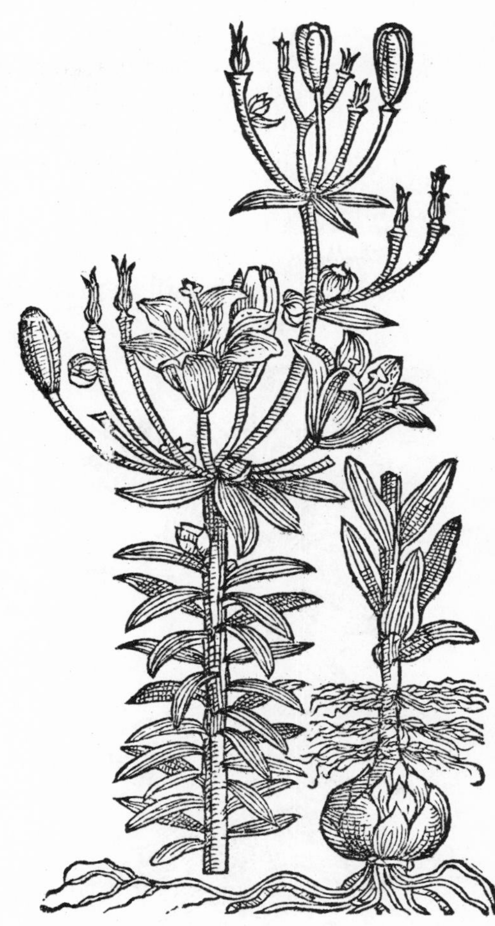

¶ *The Names.*

‡ 1 The firſt of theſe is thought by ſome to be the *Bulbus cruentus* of *Hippocrates*; as alſo the *Lilium purpureum* of *Dioſcorides:* Yet *Matthiolus* and ſome others would haue it his *Hemerocallis*. *Dodonæus* and *Bapt. Porta* thinke it the *Hyacinthus* and *Coſmoſandalos* of the Poets, of which you ſhall finde more hereafter. It is the *Martagon Chymiſtarum* of *Lobell*, and the *Lilium aureum majus* of *Tabernamontanus*.

2 This is *Martagon Chymiſtarum alterum* of *Lobell.* 3 This is *Cluſius* his *Martagon bulbiferum ſecundum.* 4 *Martagon bulbiferum primum* of *Cluſius.* 5 This *Dodonæus* calls *Lilium purpureum tertium*, and it is *Martagon bulbiferum tertium* of *Cluſius.* 6 This laſt *Lobell* and *Dodonæus* call *Lilium purpureum minus*.

‡ I haue thought good here alſo to giue you that diſcourſe touching the Poets Hyacinth, which being tranſlated out of *Dodonæus*, was formerly vnfitly put into the chapter of Hyacinths; which therefore I there omitted, and haue here reſtored to his due place, as you may ſee by *Dodonæus*, *Pempt.2.lib.2.cap.2.* ‡

† There is a Lilly which *Ouid*, *Metamorph.lib.* 10. calls *Hyacinthus*, of the boy *Hyacinthus*, of whoſe bloud he feigneth that this floure ſprang, when he periſhed as he was playing with *Apollo*, for

forwhose sake, he saith, that *Apollo* did print certaine letters and notes of his mourning. These are his words:

Ecce cruor, qui fusus humo signauerat herbas,
Desinit esse cruor, Tyrioque nitentior ostro
Flos oritur, formamque capit, quam Lilia, si non
Purpureus color his argenteus esset in illis.
Non satis hoc Phœbo est, (is enim fuit auctor honoris)
Ipse suos gemitus folijs inscribit, & ai ai,
Flos habet inscriptum, funestaque litera ducta est.

Which lately were elegantly thus rendred in English by M. *Sands*:

Behold! the bloud which late the grasse had dy'de
Was now no bloud: from thence a floure full blowne,
Far brighter than the Tyrian scarlet shone:
Which seem'd the same, or did resemble right
A Lilly, changing but the red to white.
Nor so contented, (for the Youth receiu'd
That grace from *Phœbus*) in the leaues he weau'd
The sad impression of his sighs, Ai, Ai,
They now in funerall characters display, &c.

‡ 5 *Lilium cruentum secundum caulem bulbulis donatum.*
Red Lilly with bulbes growing alongst the stalke.

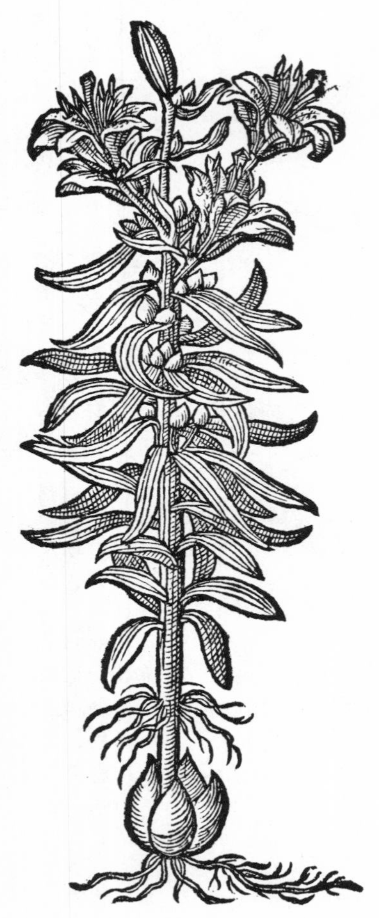

‡ 6 *Lilium purpureum minus.*
The small red Lilly.

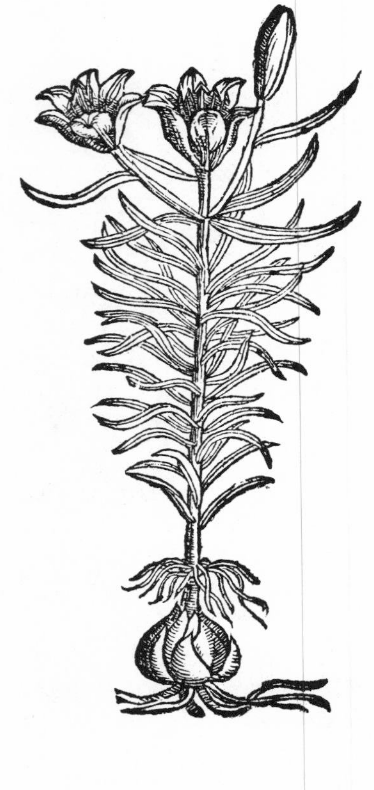

Theocritus also hath made mention of this Hyacinth, in *Bions* Epitaph, in the 19. *Eidyl.* which *Eidyl* by some is attributed to *Moschus*, and made his third. The words are these:

Νῦν ὑάκινθε λαλεῖ τὰ σὰ γράμματα ϗ πλέον αἶ αἶ.
Λάμβανε σοῖς πετάλοισι.

In English thus:

Now Iacinth speake thy letters, and once more
Imprint thy leaues with Ai, Ai, as before.

Likewise

Likewise *Virgill* hath written hereof in the third *Eclog* of his *Bucolicks*.

Et me Phœbus amat, Phœbo sua semper apud me
Munera sunt, lauri & suaue rubens Hyacinthus.
Phœbus loues me, his gifts I alwayes haue,
The e're greene Laurel, and the Iacinth braue.

In like manner also *Nemesianus* in his second *Eclog* of his *Bucolicks*:

Te sine me, misero mihi Lilia nigra videntur,
Pallentesque Rosæ, nec dulce rubens Hyacinthus:
At si tu venias, & candida Lilia fient
Purpureæque Rosæ, & dulce rubens Hyacinthus.
Without thee, Loue, the Lillies blacke do seeme;
The Roses pale, and Hyacinths I deeme
Not louely red. But if thou com'st to me,
Lillies are white, red Rose and Iacinths be.

The Hyacinths are said to be red which *Ouid* calleth purple; for the red colour is somtimes termed purple. Now it is thought this *Hyacinthus* is called *Ferrugineus*, for that it is red of a rusty iron colour: for as the putrifaction of brasse is named *Ærugo*; so the corruption of iron is called *Ferrugo*, which from the reddish colour is stiled also *Rubigo*. And certainly they are not a few that would haue *Color ferrugineus* to be so called from the rust which they thinke *Ferrugo*. Yet this opinion is not allowed of by all men; for some iudge, that *Color ferrugineus* is inclining to a blew, for that when the best iron is heated and wrought, when as it is cold againe it is of a colour neere vnto blew, which from *Ferrum* (or iron) is called *Ferrugineus*. These latter ground themselues vpon *Virgils* authoritie, who in the sixth of his *Æneidos* describeth *Charons* ferrugineous barge or boat, and presently calleth the same blew. His words are these:

Ipse ratem conto subigit velisque ministrat,
Et ferruginea subuectat corpora Cymba,
He thrusting with a pole, and setting sailes at large,
Bodies transports in ferrugineous barge.

And then a little after he addes;

Cæruleam aduertit puppim, ripæque propinquat.
He then turnes in his blew Barge, and the shore
Approches nigh to.

And *Claudius* also, in his second booke of the carrying away of *Proserpina*, doth not a little confirme their opinions; who writeth, That the Violets are painted, *ferrugine dulci*, with a sweet iron colour.

Sanguineo splendore rosas, vaccinea nigro
Induit, & dulci violas ferrugine pingit.
He trimmes the Rose with bloudy bright,
And Prime-tree berries blacke he makes,
And decks the Violet with a sweet
Darke iron colour which it takes.

But let vs returne to the proper names from which we haue digressed. Most of the later Herbarists do call this Plant *Hyacinthus Poeticus*, or the Poets Hyacinth. *Pausanias* in his second booke of his Corinthiackes hath made mention of *Hyacinthus* called of the Hermonians, *Comosandalos*, setting downe the ceremonies done by them on their festiuall dayes, in honour of the goddesse *Chthonia*. The Priests (saith he) and the Magistrates for that yeare being, doe leade the troupe of the pompe; the women and men follow after; the boyes solemnly leade forth the goddesse with a stately shew: they go in white vestures, with garlands on their heads made of a floure which the Inhabitants call *Comosandalos*, which is the blew or sky-coloured Hyacinth, hauing the marks and letters of mourning as aforesaid.

¶ *The Nature.*

The floure of the red Lilly (as *Galen* saith) is of a mixt temperature, partly of thinne, and partly of an earthly essence. The root and leaues do dry and cleanse, and moderately digest, or waste and consume away.

¶ *The Vertues.*

The leaues of the herbe applied are good against the stinging of Serpents. A

The same boiled and tempered with vineger are good against burnings, and heale green wounds and Vlcers. B

The root rosted in the embers, and pounded with oyle of Roses cureth burnings, and softneth hardnesse of the matrix. C

D The same stamped with honey cureth the wounded sinewes and members out of ioynt. It takes away the morphew, wrinkles, and deformitie of the face.

E Stamped with Vineger, the leaues of Henbane, and wheat meale, it remoueth hot swellings of the stones, the yard, and matrix.

F The roots boyled in Wine (saith *Pliny*) causeth the cornes of the feet to fall away within few dayes, with remouing the medicine vntill it haue wrought his effect.

G Being drunke in honied water, they driue out by siege vnprofitable bloud.

Chap. 104. *Of Mountaine Lillies.*

¶ *The Description.*

1 THe great mountaine Lilly hath a cloued bulbe or scaly root like to those of the Red Lilly, yellow of colour, very small in respect of the greatnesse of the plant: From the which riseth vp a stalke, sometimes two or three, according to the age of the plant; whereof the middle stalke commonly turneth from his roundnesse into a flat forme, as those of the white Lilly of Constantinople. Vpon these stalkes do grow faire leaues of a blackish greene colour, in roundles and spaces as the leaues of Woodroofe, not vnlike to the leaues of white Lillie, but smaller at the top of the stalkes. The floures be in number infinite, or at the least hard to be counted, very thicke set or thrust together, of an ouerworne purple, spotted on the inside with many small specks of the colour of rusty iron. The whole floure doth turne it selfe backeward at such time as the Sunne hath cast his beames vpon it, like vnto the Tulipa or Turkes Cap, as the Lilly or Martagon of Constantinople doth; from the middle whereof doe come forth tender pointalls with small dangling pendants hanging thereat, of the colour the floure is spotted with.

1 *Lilium montanum majus.*
The great mountaine Lilly.

2 *Lilium montanum minus.*
Small Mountaine Lilly.

2 The

2 The ſmall mountaine Lilly is very like vnto the former in root, leafe, ſtalke, and floures: differing in theſe points; The whole plant is leſſer, the ſtalke neuer leaueth his round forme, and beareth fewer floures.

‡ There are two or three more varieties of theſe plants mentioned by *Cluſius*; the one of this leſſer kinde, with floures on the outſide of a fleſh colour, and on the inſide white, with blackiſh ſpots; as alſo another wholly white without ſpots. The third varietie is like the firſt, but differs in that the floures blow later, and ſmell ſweet.

Theſe plants grow in the woody mountaines of Styria and Hungarie, and alſo in ſuch like places on the North of Francfort, vpon the Mœne. ‡

The ſmall ſort I haue had many yeares growing in my garden; but the greater I haue not had till of late, giuen me by my louing friend M. *Iames Garret* Apothecarie of London.

¶ *The Time.*

Theſe Lillies of the mountaine floure at ſuch time as the common white Lilly doth, and ſometimes ſooner.

¶ *The Names.*

The great mountaine Lilly is called of *Tabernamontanus*, *Lilium Saracenicum*, receiued by Maſter *Garret* aforeſaid from Liſle in Flanders, by the name of *Martagon Imperiale*: of ſome, *Lilium Saracenicum mas*: It is *Hemerocallis flore rubello*, of *Lobel*.

The ſmall mountaine Lilly is called in Latine, *Lilium montanum*, and *Lilium ſylueſtre*: of *Dodonæus*, *Hemerocallis*: of others, *Martagon*: but neither truly; for that there is of either, other Plants properly called by the ſame names. In high Dutch it is called **Goldwurtz**, from the yellowneſſe of the roots: in low Dutch, **Lilikens van Caluarien**: in Spaniſh, *Lirio Amarillo*: in French, *Lys Sauvage*: in Engliſh, Mountaine Lilly.

¶ *The Nature and Vertues.*

There hath not beene any thing left in writing either of the nature or vertues of theſe plants: notwithſtanding we may deeme, that God which gaue them ſuch ſeemely and beautifull ſhape, hath not left them without their peculiar vertues; the finding out whereof we leaue to the learned and induſtrious Searcher of Nature.

Chap. 105. *Of the Red Lillie of Conſtantinople.*

1 *Lilium Bizantinum.*
The red Lilly of Conſtantinople.

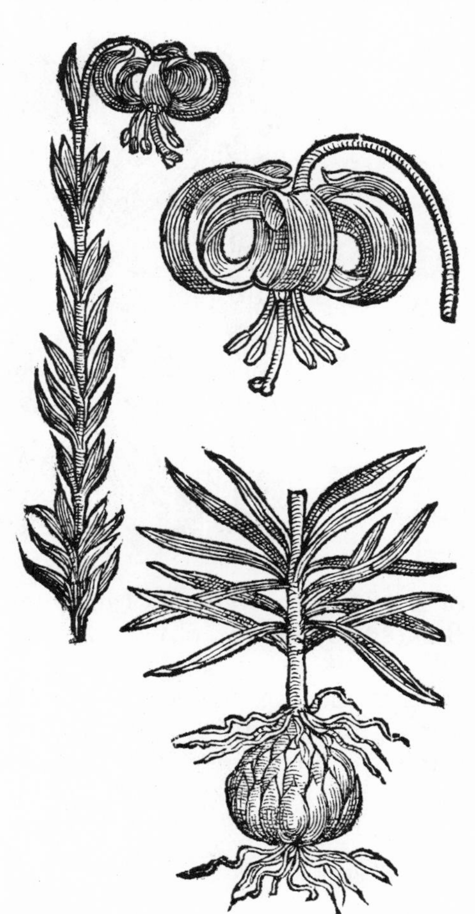

‡ 2 *Lilium Byzantinum flo. purpuro ſanguineo.*
The Byzantine purpliſh ſanguine-coloured Lilly.

¶ *The Description.*

1 THe red Lilly of Constantinople hath a yellow scaly or cloued Root like vnto the Mountaine Lilly, but greater: from the which ariseth vp a faire fat stalke a finger thicke, of a darke purplish colour toward the top; which sometimes doth turne from his naturall roundnesse into a flat forme, like as doth the great mountaine Lilly: vpon which stalk grow sundry faire and most beautifull floures, in shape like those of the mountaine Lilly, but of greater beauty, seeming as it were framed of red wax, tending to a red leade colour. From the middle of the floure commeth forth a tender pointall or pestell, and likewise many small chiues tipped with loose pendants. The floure is of a reasonable pleasant sauour. The leaues are confusedly set about the stalke like those of the white Lilly, but broader and shorter.

‡ 2 This hath a large Lilly-like root, from which ariseth a stalke some cubit or more in height, set confusedly with leaues like the precedent. The floures also resemble those of the last described, but vsually are more in number, and they are of a purplish sanguine colour.

‡ 3 *Lilium Byzantinum flo. dilute rubente.*
The light red Byzantine Lilly.

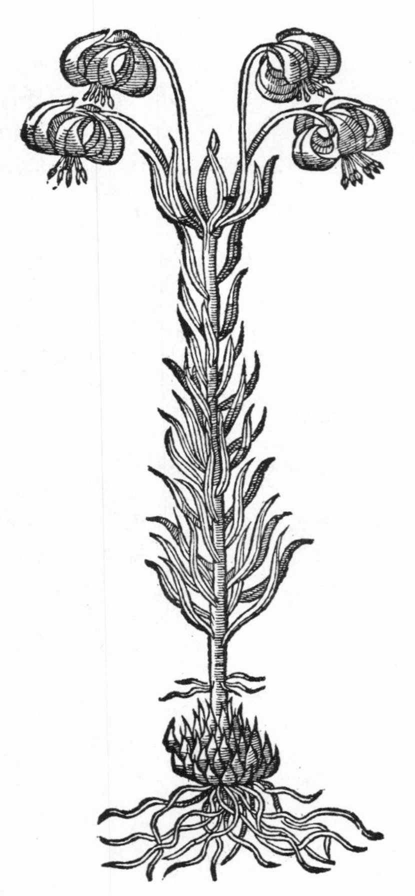

‡ 4 *Lilium Byzantinum miniatum polyanthos.*
The Vermilion Byzantine many-floured Lilly.

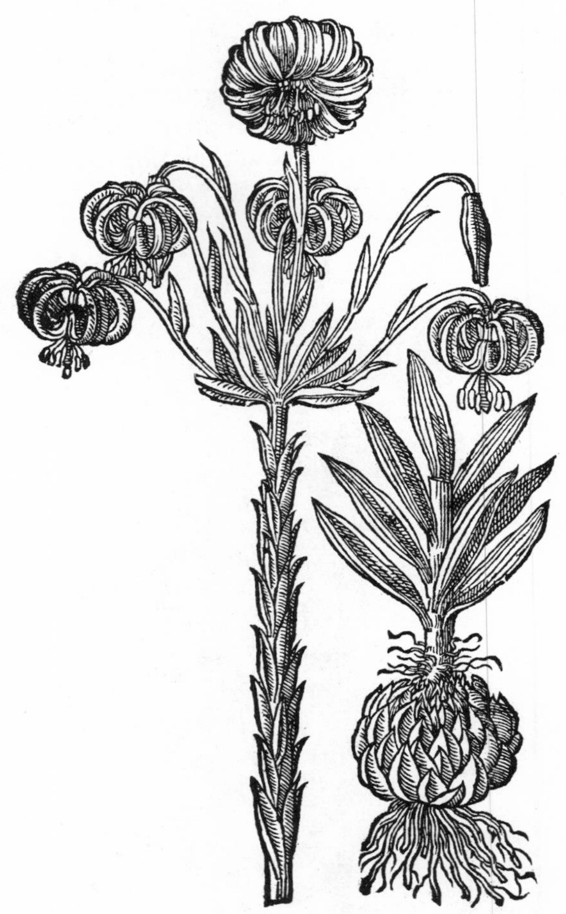

3 This differs little from the last, but in the colour of the floures, which are of a lighter red colour than those of the first described. The leaues and stalkes also, as *Clusius* obserueth, are of a lighter greene.

4 This may also more fitly be termed a varietie from the former, than otherwise: for according to *Clusius*, the difference is onely in this, that the floures grow equally from the top of the stalke, and the middle floure rises higher than any of the rest, and sometimes consists of twelue leaues as it were a twinne, as you may perceiue by the figure. ‡

¶ *The Time.*

They floure and flourish with the other Lillies.

¶ *The*

¶ *The Names.*

The Lilly of Conſtantinople is called likewiſe in England, Martagon of Conſtantinople: of *Lobel*, *Hemerocallis Chalcedonica*, and likewiſe *Lilium Bizantinum*: of the Turks it is called *Zuſiniare*: of the Venetians, *Marocali*.

¶ *The Nature and Vertues.*

Of the nature or vertues there is not any thing as yet ſet down, but it is eſteemed eſpecially for the beautie and rareneſſe of the floure; referring what may be gathered hereof to a further conſideration.

‡ CHAP. 106.

Of the narrow leaued reflex Lillies.

¶ *The Deſcription.*

‡ 1 THe root of this is not much vnlike that of other Lillies; the ſtalke is ſome cubit high, or better; the leaues are many and narrow, and of a darker green than thoſe of the ordinarie Lilly; the floures are reflex, like thoſe treated of in the laſt chap. of a red or Vermilion colour. This floures in the end of May: wherefore *Cluſius* calls it *Lilium rubrum præcox*, The early red Lilly.

‡ 1 *Lilium rubrum anguſtifolium.*
The red narrow leaued Lilly.

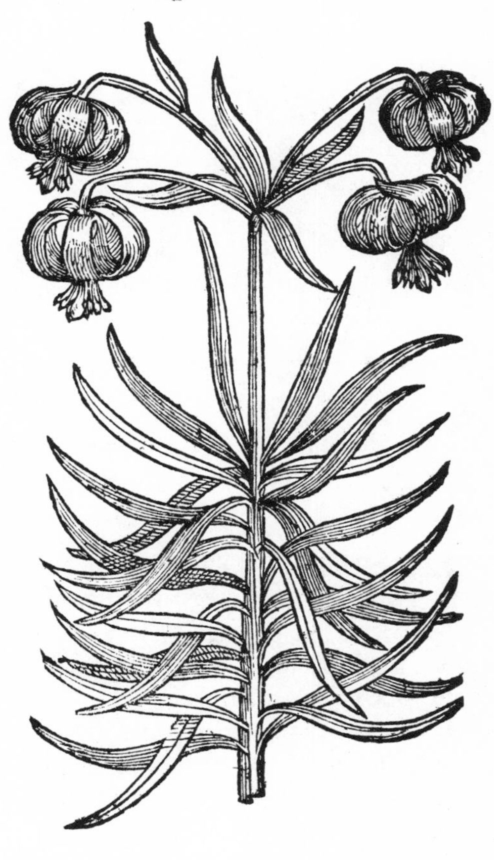

‡ 3 *Lilium mont. flore flauo punctato.*
The yellow mountaine Lilly with the ſpotted floure.

2 This Plant is much more beautifull than the laſt deſcribed; the roots are like thoſe of Lillies, the ſtalke ſome cubit and an halfe in height, being thicke ſet with ſmall graſſie leaues. The floures grow out one aboue another, in ſhape and colour like thoſe of the laſt deſcribed, but oft-

times are more in number, so that some one stalke hath borne some 48 floures. The root is much like the former.

‡ 4 *Lilium mont. flore flauo non punctato.*
The yellow Mountaine Lilly with the vnspotted floure.

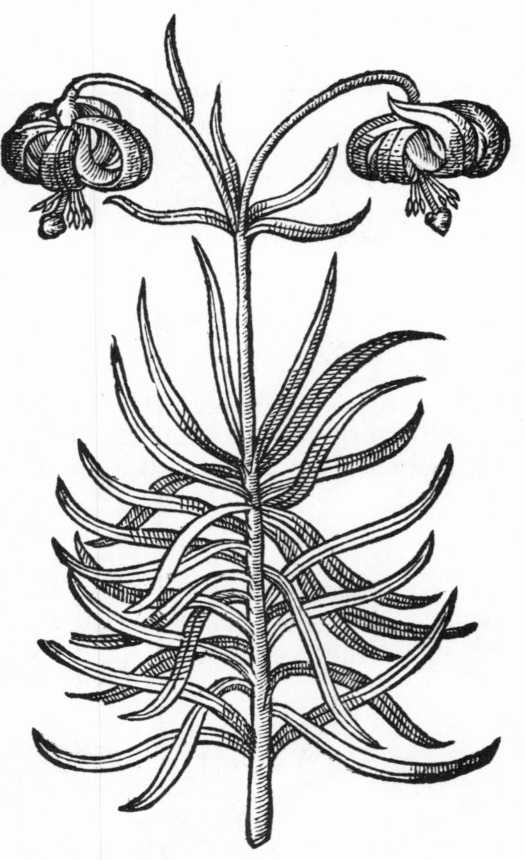

3 This in roots is like those afore described, the stalke is some 2 cubits high, set confusedly with long narrow leaues, with three conspicuous nerues running alongst them. The floures are at first pale coloured, afterwards yellow, consisting of six leaues bended backe to their stalkes, & marked with blackish purple spots.

4 There is also another differing from the last described onely in that the floure is not spotted, as that of the former.

¶ *The Place.*

These Lillies are thought Natiues of the Pyrenean mountaines, and of late yeares are become Denizons in some of our English gardens.

¶ *The Time.*

The first (as I haue said) floures in the end of May: the rest in Iune.

¶ *The Names.*

1 This is called by *Clusius*, *Lilium rubrum præcox*.

2 *Clusius* names this, *Lilium rubrum præcox 3. angustifolium*. *Lobel* stiles it, *Hemerocallis Macedonica*, and *Martagon Pomponeum*.

3 This is *Lilium flauo flore maculis distinctum* of *Clusius*, and *Lilium montanum flauo flo.* of *Lobel*.

4 This being a varietie of the last, is called by *Clusius*, *Lilium flauo flore maculis non distinctum*.

¶ *The Temper and Vertues.*

These in all likelihood cannot much differ from the temper and vertues of other Lillies, which in all their parts they so much resemble. ‡

Chap. 107. *Of the Persian Lilly.*

¶ *The Description.*

The Persian Lilly hath for his root a great white bulbe, differing in shape from the other Lillies, hauing one great bulbe firme or solid, full of juyce, which commonly each yeare setteth off or encreaseth one other bulbe, and sometimes more, which the next yeare after is taken from the mother root, and so bringeth forth such floures as the old plant did. From this root riseth vp a fat thicke and straight stemme of two cubits high, whereupon is placed long narrow leaues of a greene colour, declining to blewnes as doth those of the woade. The floures grow alongst the naked part of the stalke like little bels, of an ouer-worne purple colour, hanging down their heads, euery one hauing his owne foot-stalke of two inches long, as also his pestell or clapper from the middle part of the floure; which being past and withered, there is not found any seed at all, as in other plants, but is increased onely in his root.

¶ *The Place.*

This Persian Lilly groweth naturally in Persia and those places adiacent, whereof it tooke his name, and is now (by the industrie of Trauellers into those countries, louers of Plants) made a Denizon in some few of our London gardens.

¶ The

¶ *The Time.*

This plant floureth from the beginning of May, to the end of Iune.

¶ *The Names.*

This Perſian Lilly is called in Latine, *Lilium Perſicum, Lilium Suſianum, Pennaciò Perſiano*, and *Pannaco Perſiano*, either by the Turks themſelues, or by ſuch as out of thoſe parts brought them into England; but which of both is vncertaine. *Alphonſus Pancius*, Phyſition to the Duke of Ferrara, when as he ſent the figure of this Plant vnto *Carolus Cluſius*, added this title, *Pennacio Perſiano è Pianta belliſsima & è ſpecie di Giglio ô Martagon, diuerſo della corona Imperiale*: That is in Engliſh, This moſt elegant plant *Pennacio* of Perſia is a kinde of Lilly or Martagon, differing from the floure called the Crowne Imperiall.

Lilium Perſicum.
The Perſian Lilly.

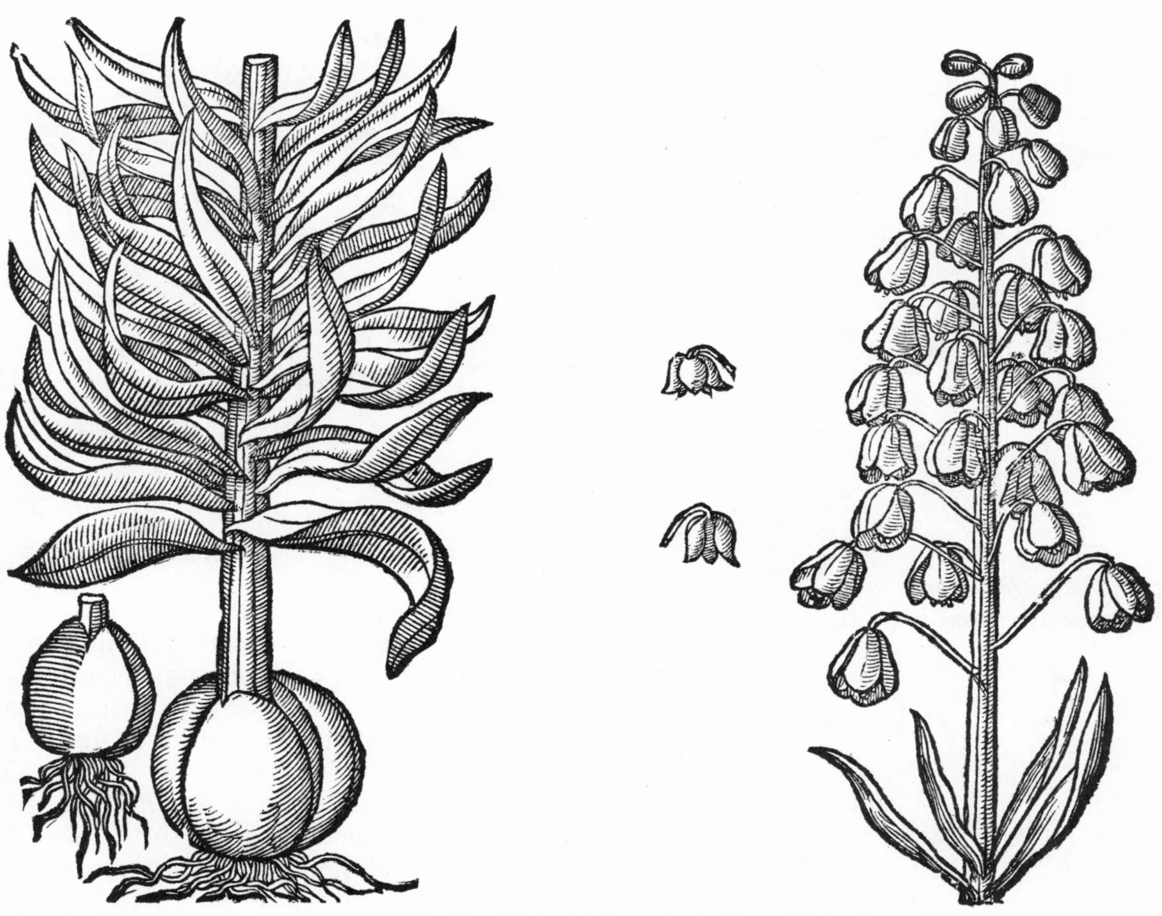

¶ *The Nature and Vertues.*

There is not any thing knowne of the nature or vertues of this Perſian Lilly, eſteemed as yet for his rareneſſe and comely proportion; although (if I might be ſo bold with a ſtranger that hath vouchſafed to trauell ſo many hundreds of miles for our acquaintance) we haue in our Engliſh fields many ſcores of floures in beauty far excelling it.

CHAP. 108. *Of the Crowne Imperiall.*

¶ *The Deſcription.*

THE Crowne Imperial hath for his root a thicke firme and ſolid bulbe, couered with a yellowiſh filme or skinne, from the which riſeth vp a great thicke fat ſtalke two cubits high, in the bare and naked part of a darke ouerworne dusky purple colour. The leaues grow confuſedly about the ſtalke like thoſe of the white Lilly, but narrower: the floures grow at the top of the ſtalke, incompaſſing it round in forme of an Imperiall crowne, (whereof it tooke his name) hanging their

 heads

Corona Imperialis. The Crowne Imperiall.

Corona Imperialis duplici corona.
The double Crowne Imperiall.

Corona Imperialis cum ſemine.
Crowne Imperiall with the ſeed.

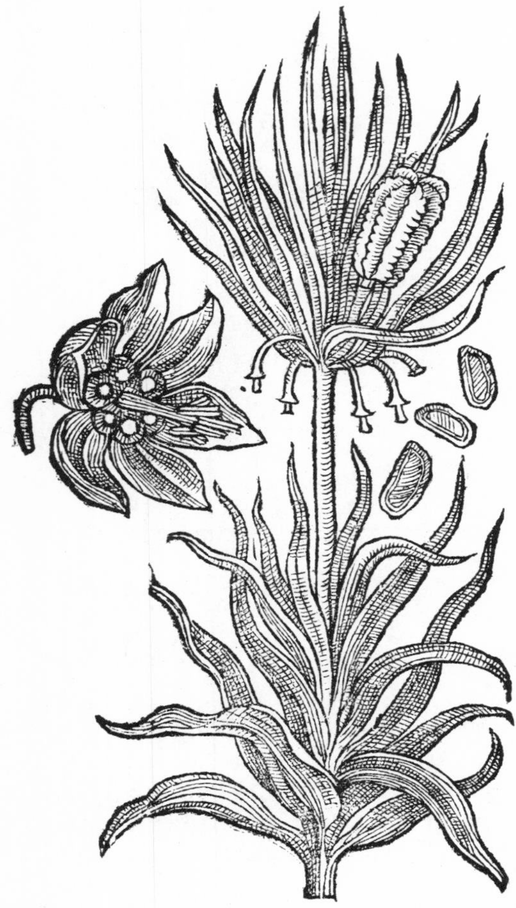

heads downward as it were bels: in colour it is yellowiſh; or to giue you the true colour, which by words otherwiſe cannot be expreſſed, if you lay ſap berries in ſteepe in faire water for the ſpace of two houres, and mix a little Saffron with that infuſion, and lay it vpon paper, it ſheweth the perfect colour to limne or illumine the floure withall. The backſide of the ſaid floure is ſtreaked with purpliſh lines, which doth greatly ſet forth the beauty thereof. In the bottome of each of theſe bells there is placed ſix drops of moſt cleere ſhining ſweet water, in taſt like ſugar, reſembling in ſhew faire Orient pearles; the which drops if you take away, there do immediately appeare the like: notwithſtanding if they may be ſuffered to ſtand ſtill in the floure according to his owne nature, they wil neuer fall away, no not if you ſtrike the plant vntill it be broken. Amongſt theſe drops there ſtandeth out a certaine peſtell, as alſo ſundry ſmal chiues tipped with ſmall pendants like thoſe of the Lilly: aboue the whole floures there growes a tuft of green leaues like thoſe vpon the ſtalke, but ſmaller. After the floures be faded, there follow cods or ſeed-veſſels ſix ſquare, wherein

is

is contained flat seeds, tough and limmer, of the colour of Mace. The whole plant, as well roots as floures, do sauour or smell very like a Fox. As the plant groweth old, so doth it wax rich, bringing forth a Crowne of floures amongst the vppermost greene leaues, which some make a second kinde, although in truth they are but one and the selfe same, which in time is thought to grow to a triple crowne, which hapneth by the age of the root, and fertilitie of the soile; whose figure or tipe I haue thought good to adioyne with that picture also which in the time of his infancie it had.

¶ *The Place.*

This plant likewise hath been brought from Constantinople amongst other bulbous roots, and made Denizons in our London gardens, whereof I haue great plenty.

¶ *The Time.*

It floureth in Aprill, and sometimes in March, when as the weather is warme and pleasant. The seed is ripe in Iune.

¶ *The Names.*

This rare & strange Plant is called in Latine, *Corona Imperialis*, and *Lilium Byzantinum:* the Turks doe call it *Cauale lale*, and *Tusai*. And as diuers haue sent into these parts of these roots at sundry times, so haue they likewise sent them by sundry names; some by the name *Tusai*; others, *Tousai*, and *Tuyschiachi*, and likewise *Tursani* and *Tursanda*. ‡ *Clusius*, and that not without good reason, iudgeth this to be the *Hemerocallis* of *Dioscorides*, mentioned *lib.3. cap.*120.

¶ *The Nature and Vertues.*

The vertue of this admirable plant is not yet knowne, neither his faculties or temperature in working.

† If this be the *Hemerocallis* of *Dioscorides*, you may finde the vertues thereof specified *pag.* 99. of this Worke; where in my iudgement they are not so fitly placed as they might haue beene here: yet we at this day haue no knowledge of the physicall operation of either of those plants mentioned in that place, or of this treated of in this chapter.

CHAP. 109. *Of Dogs Tooth.*

¶ *The Description.*

1 THere hath not long since beene found out a goodly bulbous rooted plant, and termed Satyrion, which was supposed to be the true Satyrion of *Dioscorides*, after that it was cherished, and the vertues thereof found out by the studious searchers of nature. Little difference hath bin found betwixt that plant of *Dioscorides* and this *Dens caninus*, except in the colour, which (as you know) doth commonly vary according to the diuersitie of places where they grow, as it falleth out in Squilla, Onions, and the other kindes of bulbous plants. It hath most commonly two leaues, very seldome three; which leafe in shape is very like to *Allium Vrsinum*, or Ramsons, though farre lesse. The leaues turne downe to the groundward; the stalke is tender and flexible like to *Cyclamen*, or Sow-bread, about an handfull high, bare and without leaues to the root. The proportion of the floure is like that of Saffron or the Lilly floure, full of streames of a purplish white colour. The root is bigge, and like vnto a date, with some fibres growing from it: vnto the said root is a small flat halfe round bulbe adioyning, like vnto *Gladiolus*, or Corn-flag.

2 The second kinde is farre greater and larger than the first, in bulbe, stalke, leaues, floure, and cod. It yeeldeth two leaues for the most part, which do close one within another, and at the first they doe hide the floure (for so long as it brings not out his floure) it seemes to haue but one leafe like the Tulipa's, and like the Lillies, though shorter, and for the most part broader; wherefore I haue placed it and his kindes next vnto the Lillies before the kinds of *Orchis* or stones. The leaues which it beareth are spotted with many great spots of a darke purple colour, and narrow below, but by little and little toward the top wax broad, and after that grow to be sharpe pointed, in form somewhat neere Ramsons, but thicker and more oleous. When the leaues be wide opened the floure sheweth it selfe vpon his long weake naked stalke, bowing toward the earth-ward, which floure consisteth of six very long leaues of a fine delayed purple colour, which with the heat of the Sunne openeth it selfe, and bendeth his leaues backe againe after the manner of the Cyclamen floure, within which there are six purple chiues, and a white three forked stile or pestell. This floure is of no pleasant smell, but commendable for the beauty: when the floure is faded, there succeedeth a three square huske or head, wherein are the seeds, which are very like them of *Leucoium bulbosum præcox*; but longer, slenderer, and of a yellow colour. The root is long, thicker below than aboue, set with many white fibres, waxing very tender in the vpper part, hauing one or more off-sets, or young shoots, from which the stalke ariseth out of the ground (as hath been said) bringing forth two leaues, and not three, or onely one, saue when it will not floure.

3 The third kinde is in all things like the former, saue in the leaues, which are narrower, and in the colour of the floure, which is altogether white, or consisting of a colour mixt of purple and white. Wherefore sith there is no other difference, it shall suffice to haue said thus much for the description.

¶ *The Place.*

These three plants grow plentifully at the foot of certain hills in the greene and moist grounds of Germanie and Italy, in Styria not far from Gratz, as also in Modena and Bononia in Italy, and likewise in some of the choice gardens of this countrey.

¶ *The Time.*

They floure in Aprill, and sometimes sooner, as in the middle of March.

1 *Dens caninus.*
Dogs tooth.

2 *Dens caninus flore albo angustioribus folijs.*
White Dogs tooth.

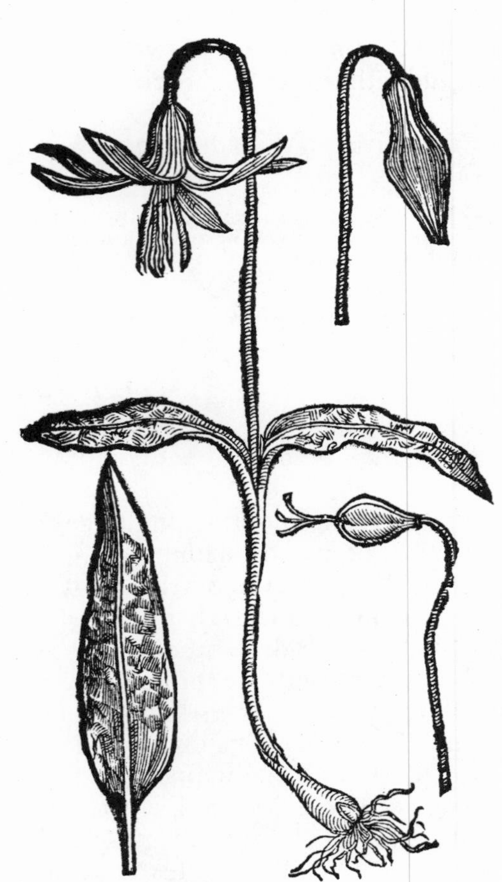

¶ *The Names.*

This plant is called in Latine, *Dens caninus*; and some haue iudged it *Satyrium Erythronium*. *Matthiolus* calls it *Pseudohermodactylus*. The men of the countrey where it groweth call it **Schostwurtz**: and the Physitians about Styria call it *Dentali*. The secoud may for distinctions sake be termed *Dens caninus flore albo, angustioribus folijs*; that is, Dogs tooth with the white floure and narrow leaues.

¶ *The Nature.*

These are of a very hot temperament, windie, and of an excrementitious nature, as may appeare by the vertues.

¶ *The Vertues.*

A The Women that dwell about the place where these grew, and do grow, haue with great profit put the dried meale or pouder of it in their childrens pottage, against the wormes of the belly.

B Being drunke with Wine it hath been proued maruellously to asswage the Collicke passion.

C It strengthneth and nourisheth the body in great measure, and being drunke with water it cureth children of the falling sicknesse.

Chap.

CHAP. 110. *Of Dogs stones.*

¶ *The Kindes.*

STones or Testicles, as *Dioscorides* saith, are of two sorts, one named *Cynosorchis*, or Dogs stones, the other *Orchis Serapias*, or Serapias his stones. But because there be many and sundry other sorts differing one from another, I see not how they may be contained vnder these two kinds onely: therefore I haue thought good to diuide them as followeth. The first kind we haue named *Cynosorchis*, or Dogs stones: the second, *Testiculus Morionis*, or Fooles stones: the third, *Tragorchis*, or Goats stones: the fourth, *Orchis Serapias*, or Serapia's stones: the fifth, *Testiculus odoratus*, or sweet smelling stones, or after *Cordus*, *Testiculus Pumilio*, or Dwarfe stones.

† 1 *Cynosorchis maior.*
Great Dogs stones.

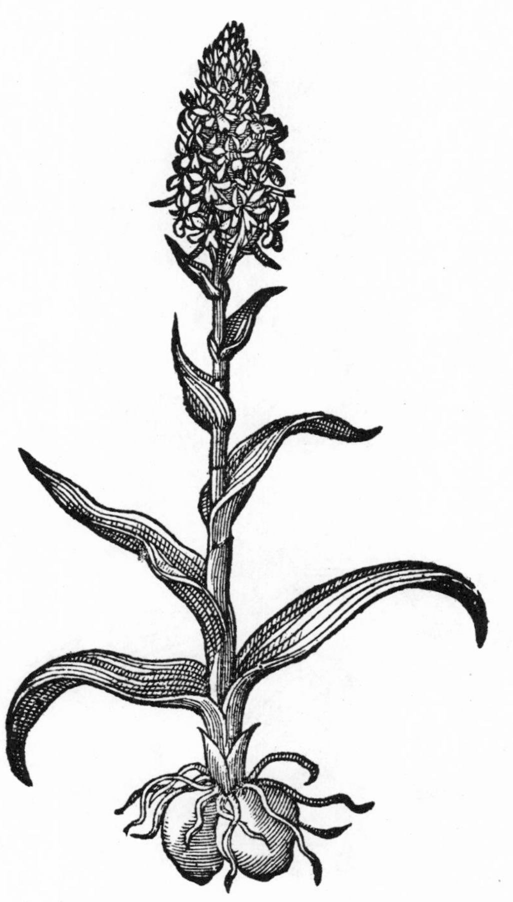

† 2 *Cynosorchis major altera.*
White Dogs stones.

¶ *The Description.*

1 GReat Dogs stones hath foure, and sometimes fiue, great broad thicke leaues, somwhat like those of the garden Lilly, but smaller. The stalke riseth vp a foot or more in height; at the top whereof doth grow a thicke tuft of carnation or horse-flesh coloured floures, thick and close thrust together, made of many small floures spotted with purple spots, in shape like to an open hood or helmet. And from the hollow place there hangeth forth a certain ragged chiue or tassell, in shape like to the skinne of a Dog, or some such other foure footed beast. The roots be round like vnto the stones of a Dog, or two oliues, one hanging somewhat shorter than the other, whereof the highest or vppermost is the smaller, but fuller and harder. The lowermost is the greatest, lightest, and most wrinkled or shriueled, not good for any thing.

2 Whitish Dogs stones hath likewise smooth, long broad leaues, but lesser and narrower than those of the first kinde. The stalke is a span long, set with fiue or six leaues clasping or embracing the same round about. His spikie floure is short, thicke, bushy, compact of many small whitish purple

purple coloured floures, ſpotted on the inſide with many ſmall purple ſpots and little lines or ſtreakes. The ſmall floures are like an open hood or helmet, hauing hanging out of euery one as it were the body of a little man without a head, with armes ſtretched out, and thighes ſtradling abroad, after the ſame manner almoſt that the little boyes are wont to be pictured hanging out of *Saturnes* mouth. The roots be like the former.

3 Spotted Dogs ſtones bring forth narrow leaues, ribbed in ſome ſort like vnto the leaues of narrow Plaintaine or Rib-wort, daſht with many blacke ſtreakes and ſpots. The ſtalke is a cubit and more high: at the top whereof doth grow a tuft or eare of violet-coloured floures, mixed with a darke purple, but in the hollowneſſe thereof whitiſh, not of the ſame forme or ſhape that the others are of, but leſſer, and as it were reſembling ſomewhat the floures of Larkes-ſpur. The roots be like the former.

4 Mariſh Dogs ſtones haue many thicke blunt leaues next the root, thick ſtreaked with lines or nerues like thoſe of Plantaine. The floure is of a whitiſh red or carnation: the ſtalk and roots be like the former.

† 3 *Cynoſorchis maculata.*
Spotted Dogs ſtones.

4 *Cynoſorchis paluſtris.*
Mariſh Dogs ſtones.

‡ 5 This hath fiue or ſix little leaues; the ſtalke is ſome handfull or better in height, ſet about with ſomewhat leſſe leaues: the tuft of floures at the top of the ſtalke are of a purple colour, ſmall, with a white lip diuided into foure partitions hanging downe, which alſo is lightly ſpotted with purple; it hath a little ſpurre hanging downe on the hinder part of each floure. The ſeed is ſmall, and contained in ſuch twined heads as in other plants of this kinde. The roots are like the former, but much leſſe. ‡

¶ *The Place.*

Theſe kindes of Dogs ſtones do grow in moiſt and fertile medowes. The mariſh Dogs ſtones grow for the moſt part in moiſt and wateriſh woods, and alſo in mariſh grounds. ‡ The 5 growes in many hilly places of Auſtria and Germanie. ‡

¶ *The Time.*

They floure from the beginning of May to the midſt of Auguſt.

¶ *The*

¶ *The Names.*

The firſt and ſecond are of that kinde which *Dioſcorides* calleth *Cynoſorchos* ; that is in Engliſh, Dogs ſtones, after the common or vulgar ſpeech ; the one the greater, the other the leſſer.

‡ 1 This is *Cynoſorchis prior* of *Dodonæus* , *Cynoſorchis noſtra major* of *Lobel.*

2 *Dodonæus* names this *Cynoſorchis altera. Lobel, Cynoſorchis majoris ſecunda ſpecies.*

3 This *Lobel* calls *Cynoſorchis Delphinia,&c. Tabern. Cynoſorchis maculata.*

4 *Dodonæus* calls this, *Cynoſorchis tertia : Lobel, Cynoſorchis major altera noſtras : Tabernam. Cynoſorchis major quarta.*

5 This is *Cluſius* his *Orchis Pannonica quarta.*

¶ *The Temperature.*

Theſe kindes of Dogs ſtones be of temperature hot and moiſt ; but the greater or fuller ſtone ſeemeth to haue much ſuperfluous windineſſe, and therefore being drunke it ſtirreth vp fleſhly luſt.

The ſecond, which is leſſer, is quite contrarie in nature, tending to a hot and dry temperature ; therefore his root is ſo far from mouing venerie, that contrariwiſe it ſtaieth and keepeth it backe, as *Galen* teacheth.

‡ 5 *Cynoſorchis minor Pannnonica.*
The leſſer Auſtrian Dogs ſtones.

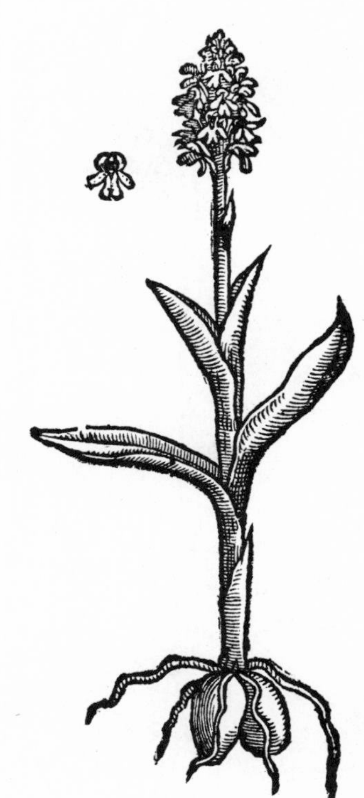

He alſo affirmeth, that Serapias ſtones are of a more dry facultie, and doe not ſo much preuaile to ſtirre vp the luſt of the fleſh.

¶ *The Vertues.*

Dioſcorides writeth that it is reported, That if men doe eate of the great full or fat roots of theſe kindes of Dogs ſtones, they cauſe them to beget male children ; and if women eate of the leſſer dry or barren root which is withered or ſhriueled, they ſhall bring forth females. Theſe are ſome Doctors opinions onely. A

It is further reported, That in Theſſalia the women giue the tender full root to be drunke in Goats milke, to moue bodily luſt, and the dry to reſtraine the ſame. B

¶ *The Choice.*

Our age vſeth all the kindes of ſtones to ſtirre vp venery, and the Apothecaries mix any of them indifferently with compoſitions ſeruing for that purpoſe. But the beſt and moſt effectuall are theſe Dogs ſtones, as moſt haue deemed: yet both the bulbes or ſtones are not to be taken indifferently, but the harder and fuller, and that which containes moſt quantity of iuyce, for that which is wrinkled is leſſe profitable, or not fit at all to be vſed in medicine. And the fuller root is not alwaies the greater, but often the leſſer, eſpecially if the roots be gathered before the plant hath ſhed his floure, or when the ſtalke firſt commeth vp ; for that which is fuller of iuyce is not the greateſt before the ſeed be perfectly ripe. For ſeeing that euery other yeare by courſe one ſtone or bulbe waxeth full, the other empty and periſheth, it cannot be that the harder and fuller of iuyce ſhould be alwaies the greater ; for at ſuch time as the leaues come forth, the fuller then beginns to encreaſe, and whilſt the ſame by little & little encreaſeth, the other doth decreaſe and wither till the ſeed be ripe : then the whole plant, together with the leaues and ſtalkes doth forthwith fall away and periſh, and that which in the meane time encreaſed, remaineth ſtill freſh and full vnto the next yeare.

† The figures of the firſt and ſecond were tranſpoſed in the former Edition : the third was of the *Cynoſorchis morio mas*, following in the next chapter.

CHAP.

CHAP. 111. *Of Fooles Stones.*

¶ *The Deſcription.*

1 THe male Foole ſtones hath fiue, ſometimes ſix long broad and ſmooth leaues, not vnlike to thoſe of the Lilly, ſauing that they are daſht and ſpotted in ſundry places with blacke ſpots and ſtreakes. The floures grow at the top, tuft or ſpike faſhion, ſomewhat like the former, but thruſt more thicke together, in ſhape like to a fooles hood, or cocks combe, wide open, or gaping before, and as it were creſted aboue, with certaine eares ſtanding vp by euery ſide, and a ſmall taile or ſpur hanging downe, the backeſide declining to a violet colour, of a pleaſant ſauour or ſmell.

† 1 *Cynoſorchis Morio mas.*
The male Foole ſtones.

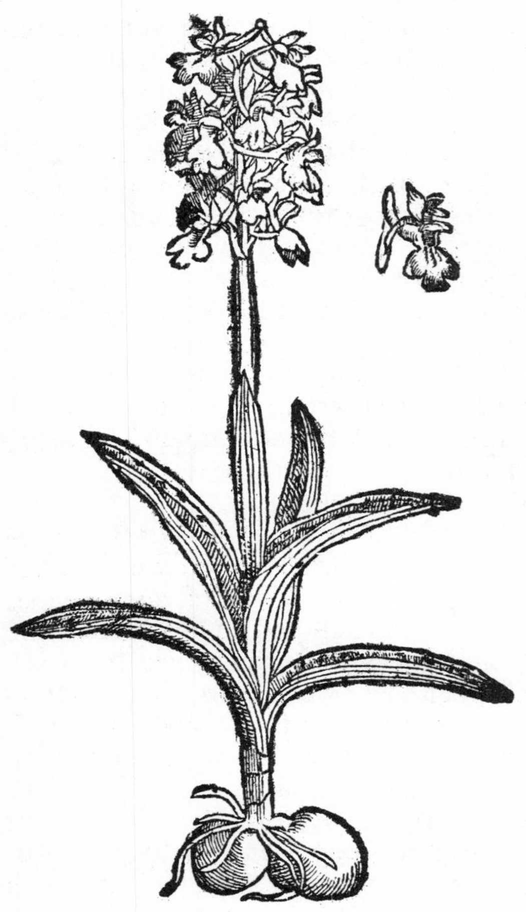

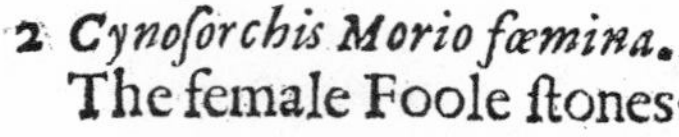

2 *Cynoſorchis Morio fœmina.*
The female Foole ſtones.

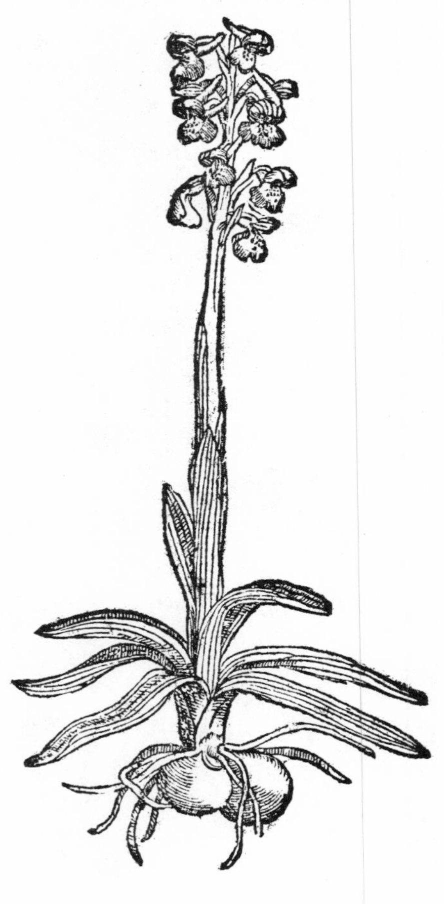

2 The Female Fooles ſtones haue alſo ſmooth narrow leaues, ribbed with nerues like thoſe of Plantaine. The floures be likewiſe gaping, and like the former, as it were open hoods, with a little horne or heele hanging behinde euerie one of them, and ſmall greene leaues ſorted or mixed among them, reſembling cockes combes, with little eares, not ſtanding ſtraight vp, but lying flat vpon the hooded floure, in ſuch ſort, that they cannot at the ſudden view be perceiued. The roots are a paire of ſmall ſtones like the former. The floures of this ſort doe varie infinitely in colour, according to the ſoile or countrey where they do grow: ſome bring forth their floures of a deepe violet colour, ſome as white as ſnow; ſome of a fleſh colour, and ſome garniſhed with ſpots of diuers colours, which are not poſſible to be diſtinguiſhed.

‡ 3 This hath narrow ſpotted leaues, with a ſtalke ſome foot or more high, at the top whereof groweth a tuft of purple floures in ſhape much like thoſe of the laſt deſcribed, each floure conſiſting of a little hood, two ſmall wings or ſide leaues, and a broad lippe or leafe hanging downe. ‡

¶ *The*

‡ 3 *Cynoſorchis Morio minor.*
The leſſer ſpotted Fooles ſtones.

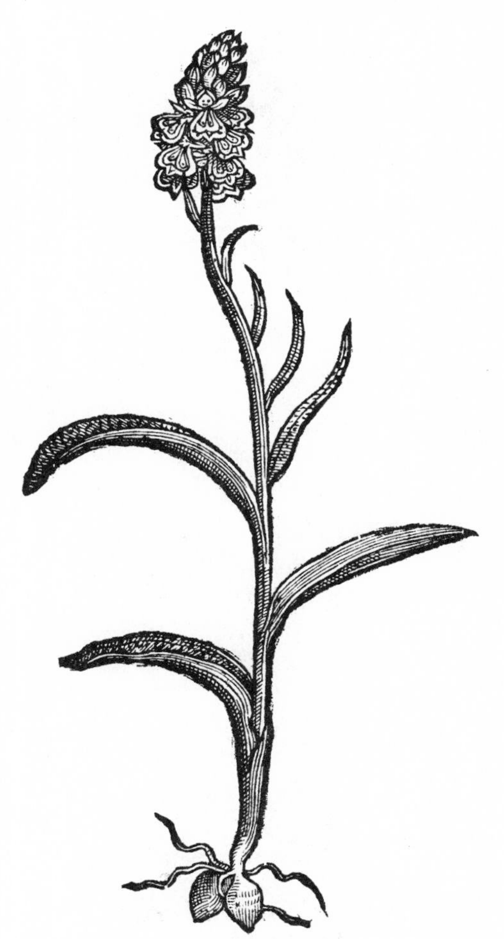

† The firſt was of *Cynoſorchis maculata*, being the third in the former chapter.

¶ *The Place.*

Theſe kindes of Fooles ſtones do grow naturally to their beſt liking in paſtures and fields that ſeldome or neuer are dunged or manured.

¶ *The Time.*

They floure in May and Iune. Their ſtones are to be gathered for medicine in September, as are thoſe of the Dogs ſtones.

¶ *The Names.*

The firſt is called *Cynoſorchis Morio* : of *Fuchſius*, *Orchis mas anguſtifolia* · of *Apuleius*, *Satyrion*: and alſo it is the *Orchis Delphinia* of *Cornelius Gemma*.

‡ The ſecond is *Cynoſorchis morio fœmina* of *Lobel* : *Orchis anguſtifolia fœmin.* of *Fuchſius* : *Teſticulus Morionis fœmina* of *Dodonæus*.

3 This is *Cynoſorchis minimis & ſecundum caulem, &c. maculoſis folijs*, of *Lobel*. ‡

¶ *The Temperature.*

Fooles ſtones both male and female are hot and moiſt of nature.

¶ *The Vertues.*

Theſe Fooles ſtones are thought to haue the vertues of Dogs ſtones, whereunto they are referred.

CHAP. 112. *Of Goats Stones.*

¶ *The Deſcription.*

1 THe greateſt of the Goats ſtones bringeth forth broad leaues, ribbed in ſome ſort like vnto the broad leaued Plantaine, but larger : the ſtalke groweth to the height of a cubit, ſet with ſuch great leaues euen to the top of the ſtalke by equall diſtances. The tuft or buſh of floures is ſmall and flat open, with many tender ſtrings or laces comming from the middle part of thoſe ſmall floures, crookedly tangling one with another, like to the ſmall tendrels of the Vine, or rather the laces or ſtrings that grow vpon the herbe Sauorie. The whole floure conſiſteth of a purple colour. The roots are like the reſt of the Orchides, but greater.

2 The male Goats ſtones haue leaues like to thoſe of the garden Lilly, with a ſtalke a foot long, wrapped about euen to the tuft of the floure with thoſe his leaues. The floures which grow in this buſh or tuft be very ſmall, in forme like vnto a Lizard, becauſe of the twiſted or writhen tailes, and ſpotted heads. Euery of theſe ſmall floures is at the firſt like a round cloſe huske, of the bigneſſe of a peaſe, which when it openeth there commeth out of it a little long and tender ſpurre or taile, white toward the ſetting of it to the floure; the reſt ſpotted with red daſhes, hauing vpon each ſide a ſmall thing adioyning vnto it, like to a little leg or foot; the reſt of the ſaid taile is twiſted crookedly about, and hangeth downward. The whole plant hath a ranke or ſtinking ſmel or ſauour like the ſmell of a Goat, whereof it tooke his name.

3 The female Goats ſtones haue leaues like the male kinde, ſauing that they be much ſmaller, hauing many floures on the tuft reſembling the flies that feed vpon fleſh, or rather ticks. The ſtones or roots, as alſo the ſmell are like the former.

1 *Tragorchis maximus.*
The greatest Goat stones.

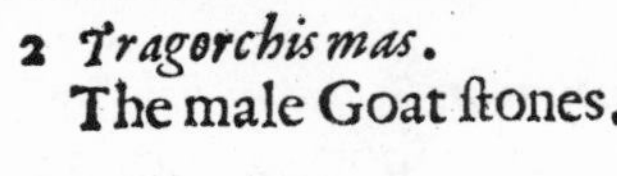

2 *Tragorchis mas.*
The male Goat stones.

3 *Tragorchis fœmina.*
The female Goats stones.

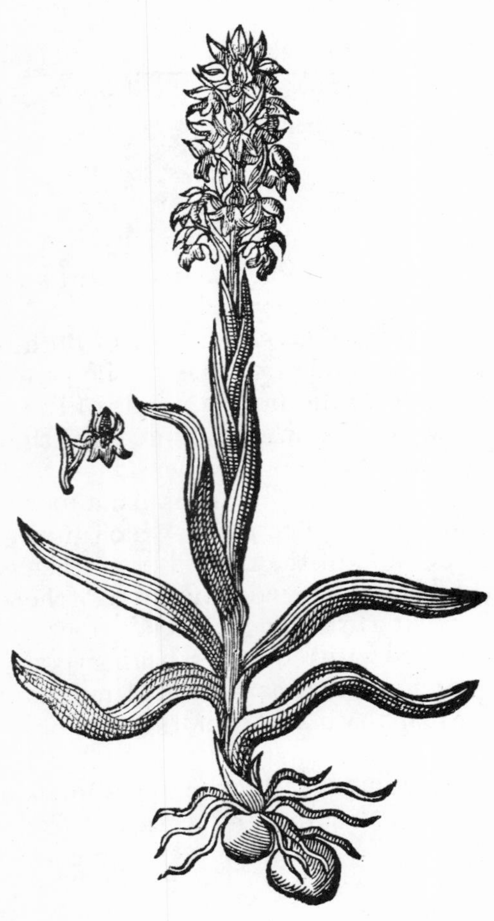

‡ 4 *Tragorchis minor Batauica.*
The small Goat stones of Holland.

‡ 4 This alſo becauſe of the vnpleaſant ſmell may fitly be referred to this Claſſis. The roots hereof are ſmall, and from them ariſe a ſtalke ſome halfe a foot high, beſet with three or foure narrow leaues: the tuft of floures which groweth on the top of this ſtalke is ſmall, and the colour of them is red without, but ſomewhat paler within; each floure hanging downe a lippe parted in three. ‡

¶ The Place.

1. 2. 3. Theſe kindes of Goats ſtones delight to grow in fat clay grounds, and ſeldome in any other ſoile to be found.

‡ 4 This growes vpon the ſea bankes in Holland, and alſo in ſome places neere vnto the Hage. ‡

¶ The Time.

They floure in May and Iune with the other kindes of Orchis.

¶ The Names.

† 1 Some haue named this kind of Goats ſtones in Greeke, τραγόρχις: in Latine, *Teſticulus Hircinus*, and alſo *Orchis Saurodes*, or *Scincophora*, by reaſon that the floures reſemble Lizards.

The ſecond may be called *Tragorchis mas*, male Goats ſtones; and *Orchis Saurodes*, or *Scincophora*, as well as the former.

The third, *Tragorchis fœmina*, as alſo *Corioſmites*, and *Coriophora*, for that the floures in ſhape and their vngratefull ſmell reſemble Ticks, called in Greeke κόρεις: †

¶ The Nature and Vertues.

The temperature and vertues of theſe are referred to the Fooles ſtones, notwithſtanding they are ſeldome or neuer vſed in phyſick, in regard of the ſtinking and loathſome ſmell and ſauor they are poſſeſſed with.

Chap. 113. *Of Fox Stones.*

1 *Orchis Hermaphroditica.*
Butter-fly Satyrion.

† 2 *Teſticulus pſycodes.*
Gnat Satyrion.

 ¶ The

¶ *The Kindes.*

THere be diuers kindes of Fox-stones, differing very much in shape of their leaues, as also in floures: some haue floures wherein is to be seene the shape of sundry sorts of liuing creatures; some the shape and proportion of flies, in other gnats, some humble bees, others like vnto honey Bees; some like Butter-flies, and others like Waspes that be dead; some yellow of colour, others white; some purple mixed with red, others of a browne ouer-worne colour: the which seuerally to distinguish, as well those here set downe, as also those that offer themselues dayly to our view and consideration, would require a particular volume; for there is not any plant which doth offer such varietie vnto vs as these kindes of Stones, except the Tulipa's, which go beyond all account: for that the most singular Simplest that euer was in these later ages, *Carolus Clusius* (who for his singular industrie and knowledge herein is worthy triple honour) hath spent at the least fiue and thirty yeares, sowing the seeds of Tulipa's from yeare to yeare, and to this day he could neuer attaine to the end or certaintie of their seuerall kindes of colours. The greatest reason whereof that I can yeeld is this; that if you take the seeds of a Tulipa that bare white floures, and sow them in some pan or tub with earth, you shall receiue from that seed plants of infinite colours: contrariwise, if you sow the seeds of a plant that beareth floures of variable colours, the most of those plants will be nothing like the plant from whence the seed was taken. It shall be sufficient therefore to set downe most of the varieties, and comprehend them in this chapter.

¶ *The Description.*

1 BVtter-fly Orchis, or Satyrion, beareth next the root two very broad leaues like those of the Lilly, seldome three: the floures be white of colour, resembling the shape of a Butter-fly: the stalke is a foot high; the root is two stones like the other kindes of stones or Cullions, but somewhat sharper pointed.

† 3 *Testiculus Vulpinus* 2. *sphegodes.*
Humble Bee Orchis.

4 *Testiculus Vulpinus major sphegodes.*
Waspe Orchis.

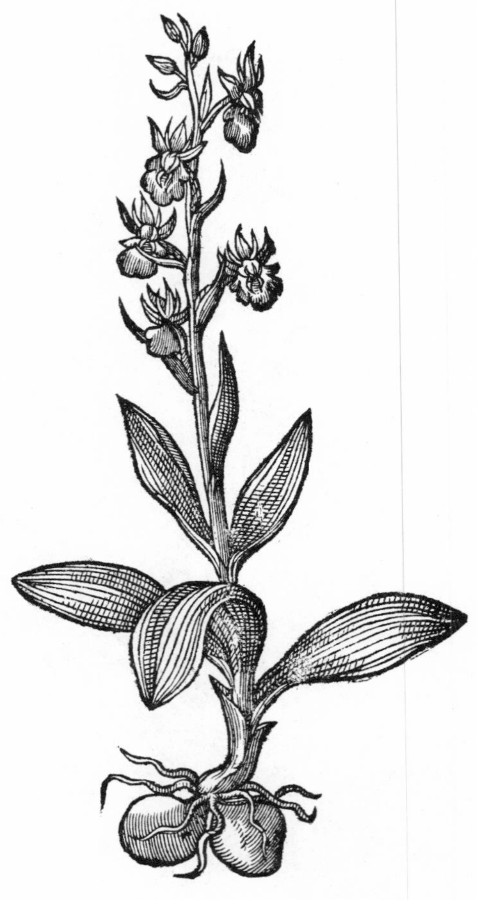

2 Waspe Satyrion commeth forth of the ground, bearing two, somtimes three leaues like the former, but much smaller. The stalke groweth to the height of an hand, whereon are placed very orderly small floures like in shape to Gnats, and of the same colour. The root is like the former.

3 The

3 The Humble Bee Orchis hath a few ſmall weake and ſhort leaues, which grow ſcatteringly about the ſtalke: the floures grow at the top among the ſmall leaues, reſembling in ſhape the humble Bee. The root conſiſteth of two ſtones or bulbes, with ſome few threds annexed thereunto.

4 The Waſpe Satyrion groweth out of the ground, hauing ſtalkes ſmall and tender. The leaues are like the former, but ſomewhat greater, declining to a browne or darke colour. The floures be ſmall, of the colour of a dry Oken leafe, in ſhape reſembling the great Bee, called in Engliſh an Hornet or drone Bee. The root is like the other.

5 The leaues of Bee Satyrion are longer than the laſt before mentioned, narrower, turning themſelues againſt the Sun as it were round. The ſtalke is round, tender, and very fragile. At the top grow the floures, reſembling the ſhape of the dead carkaſſe of a Bee. The ſtones or bulbes of the roots be ſmaller and rounder than the laſt deſcribed.

6 The Fly Satyrion is in his leaues like the other, ſauing that they be not of ſo dark a colour: the floures be ſmaller and more plentifully growing about the ſtalke, in ſhape like vnto flies, of a greeniſh colour.

† 5 *Orchis Melittias.*
Bee Orchis.

† 6 *Orchis Myodes.*
Fly Satyrion.

7 Yellow Orchis riſeth out of the ground with browne leaues, ſmaller than the laſt before mentioned: the ſtalk is tender and crooked. The floures grow at the top yellow of colour, in ſhape reſembling the yellow flies bred in the dung of Kine after raine.

8 The ſmall yellow Satyrion hath leaues ſpread vpon the ground, at the firſt comming vp; the ſlender ſtalke riſeth vp in the midſt, of halfe a hand high. The floures grow ſcatteringly toward the top, reſembling the flies laſt before mentioned, darke or ruſty of colour. The ſtones or bulbes are very round.

9 Birds Orchis hath many large ribbed leaues, ſpread vpon the ground like vnto thoſe of Plantaine; among the which riſe vp tender ſtalkes couered euen to the tuft of the floures with the like leaues, but leſſer, in ſuch ſort that the ſtalkes cannot be ſeene for the leaues. The floures grow at the top, not ſo thicke ſet or thruſt together as the others, purple of colour, like in ſhape vnto little Birds, with their wings ſpread abroad ready to fly. The roots be like the former.

10 Spotted Birds Satyrion hath leaues like vnto the former, ſauing that they be daſhed or

ſpotted here and there with darke ſpots or ſtreakes, hauing a ſtalke couered with the like leaues, ſo that the plants differ not in any point, except the blacke ſpots which this kinde is daſht with.

11 White Birds Satyrion hath leaues riſing immediately forth of the ground like vnto the blades or leaues of Leekes, but ſhorter; among the which riſeth vp a ſlender naked ſtalke two handfulls high; on the top whereof be white floures reſembling the ſhape or forme of a ſmall bird ready to fly, or a white Butter-fly with her wings ſpread abroad. The roots are round, and ſmaller than any of the former.

12 Souldiers Satyrion bringeth forth many broad large and ribbed leaues, ſpred vpon the ground like vnto thoſe of the great plantaine: among the which riſeth vp a fat ſtalke full of ſap or iuyce, cloathed or wrapped in the like leaues euen to the tuft of floures, whereupon do grow little floures reſembling a little man hauing a helmet vpon his head, his hands and legges cut off, white vpon the inſide, ſpotted with many purple ſpots, and the backe part of the floure of a deeper colour tending to redneſſe. The roots be greater than any of the other kindes of Satyrions.

† 7 *Orchis Myodes Lutea.*
Yellow Satyrion.

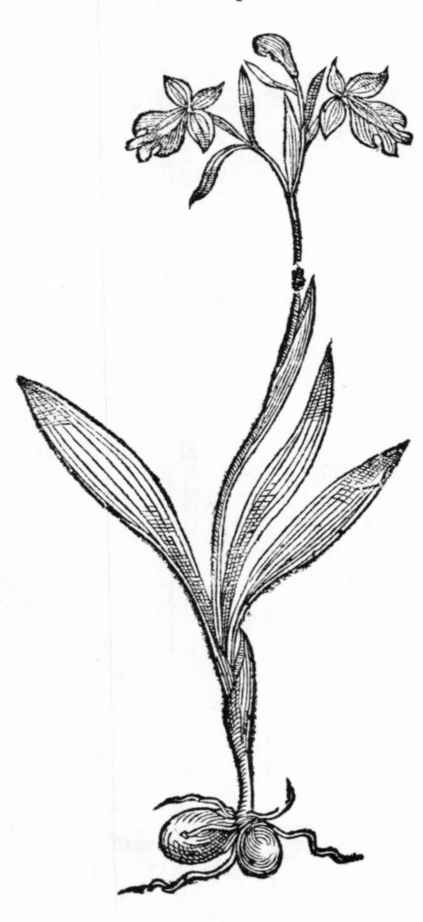

† 9 *Orchis Myodes minor.*
Small yellow Satyrion.

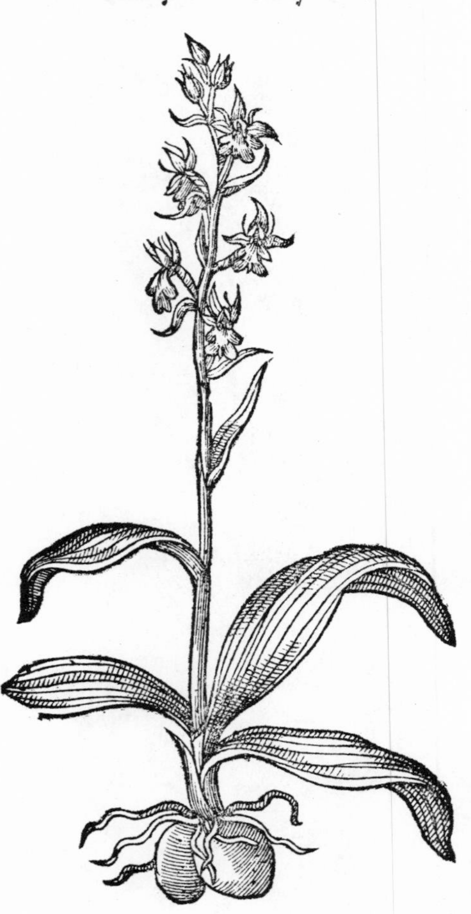

13 Souldiers Cullions hath many leaues ſpread vpon the ground, but leſſer than the ſoldiers Satyrion, as is the whole plant. The backſide of the floures are ſomwhat mixed with whiteneſſe, and ſometimes are aſh coloured: the inſide of the floure is ſpotted with white likewiſe.

14 Spider Satyrion hath many thinne leaues like vnto thoſe of the Lilly, ſcatteringly ſet vpon a weake and feeble ſtalk, whereupon doth grow ſmall floures, reſembling as well in ſhape as colour the body of a dead humble Bee, ‡ or rather of a Spider; and therefore I thinke *Lobel*, who was the Author of this name, would haue ſaid *Arachnitis*, of Ἀράχνης, a Spider. ‡

‡ 15 This by right ſhould haue beene put next the Gnat Satyrion, deſcribed in the ſecond place. It hath ſhort, yet pretty broad leaues, and thoſe commonly three in number, beſides thoſe ſmall ones ſet vpon the ſtemme. The floures are ſmall, and much like thoſe of the ſecond formerly deſcribed.

‡ 16 Our Author gaue you this figure in the fourteenth place, vnder the title of *Orchis Andrachnitis*; but it is of the *Orchis* 16. *minor* of *Tabernam.* or *Orchis Anguſtifolia* of *Bauhinus*. This Orchis is of the kinde of the *Myodes*, or Fly Satyrions, but his leaues are farre longer and narrower than

† 9 *Orchis Ornithophora.*
Birds Satyrion.

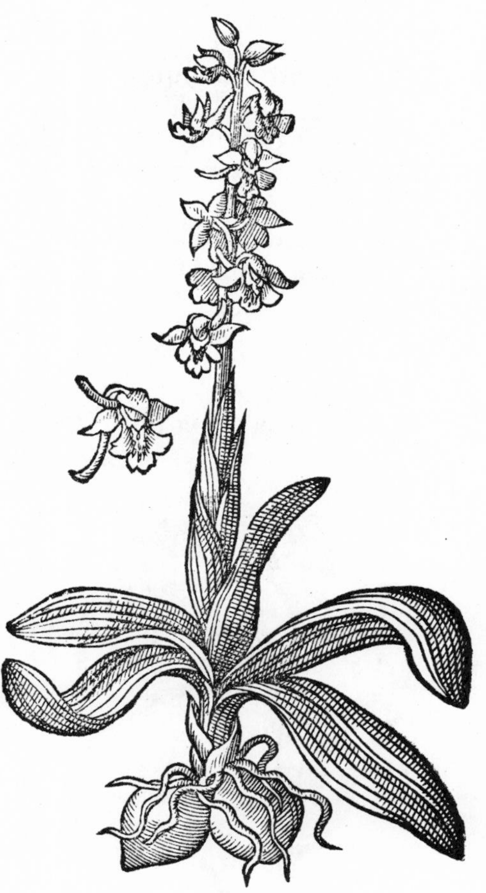

† 10 *Orchis Ornithophora folio maculoso.*
Spotted Birds Orchis.

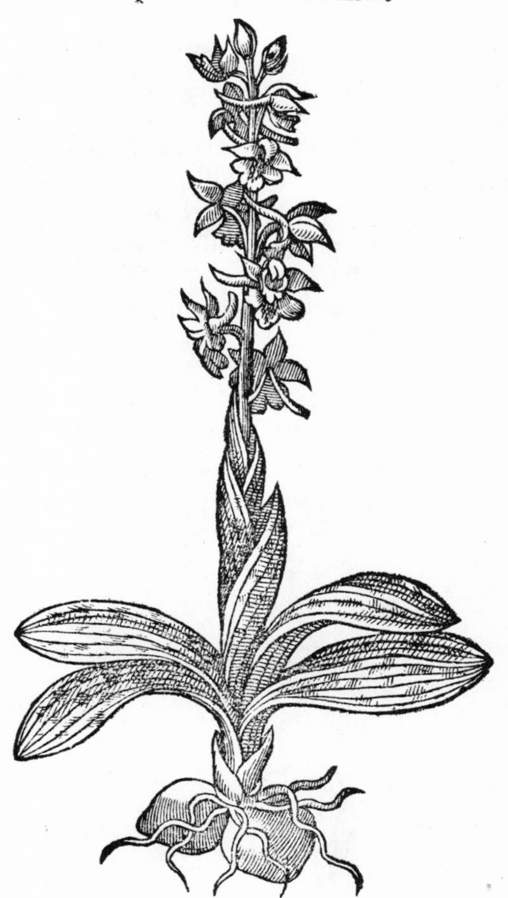

† 11 *Orchis Ornithophora candida.*
White Birds Orchis.

† 12 *Orchis Strateumatica.*
Souldiers Satyrion.

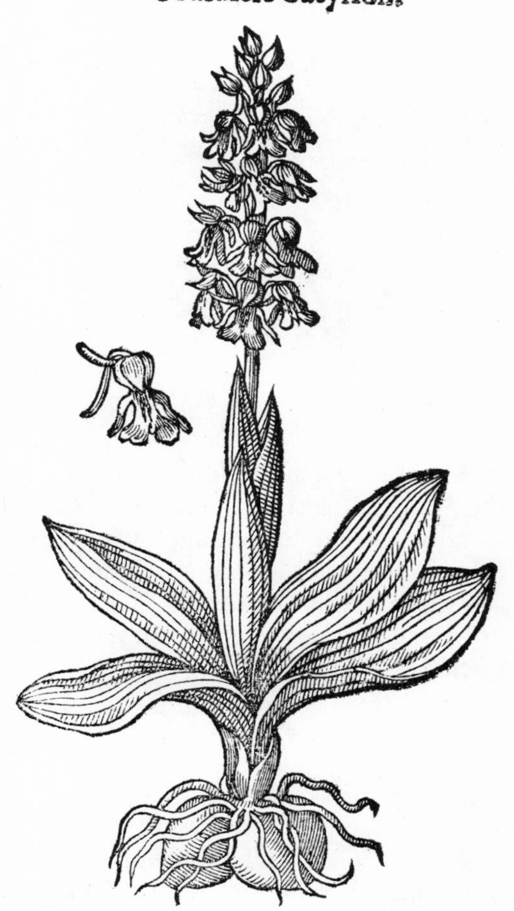

† 13 *Orchis*

than any of the reſt of that kinde, and therein conſiſts the onely and chiefeſt difference. ‡

¶ *The Place.*

Theſe kindes of Orchis grow for the moſt part in moiſt medowes and fertile paſtures, as alſo in moiſt woods.

The Bee, the Fly, and the Butter-fly Satyrions do grow vpon barren chalkie hills and heathie grounds, vpon the hils adioyning to a village in Kent named Green-hithe, vpon long field downes by South-fleet, two miles from the ſame place, and in many other places of Kent: likewiſe in a field adioyning to a ſmall groue of trees, halfe a mile from Saint Albons, at the South end thereof. They grow likewiſe at Hatfield neere S. Albons, by the relation of a learned Preacher there dwelling, M. *Robert Abot*, an excellent and diligent Herbariſt.

† 13 *Orchis Strateumatica minor.*
Souldiers Cullions.

† 14 *Orchis Andrachnitis.*
Spider Satyrion.

That kinde which reſembleth the white Butter-fly groweth vpon the declining of the hill at the North end of Hampſted heath, neere vnto a ſmall cottage there in the way ſide, as yee go from London to Henden a village thereby. It groweth in the fields adioyning to the pound or pinnefold without the gate, at the Village called High-gate, neere London: and likewiſe in the wood belonging to a Worſhipfull Gentleman of Kent named Maſter *Sidley*, of South-fleet; where doe grow likewiſe many other rare and daintie Simples, that are not to be found elſewhere in a great circuit.

¶ *The Time.*

They floure for the moſt part from May to the end of Auguſt, and ſome of them ſooner.

¶ *The Names.*

Theſe kindes of Orchis haue not bin much written of by the Antients, neither by the late writers to any purpoſe, ſo that it may content you for this time to receiue the names ſet down in their ſeuerall

ſeuerall titles, reſeruing what elſe might be ſaid as touching t he Greeke, French, or Dutch names or any generall definition vntill a further conſideration.

‡ 15 *Orchis trifolia minor.*
Small Gnat Satyrion.

‡ 16 *Orchis anguſtifolia.*
Narrow leaued Satyrion.

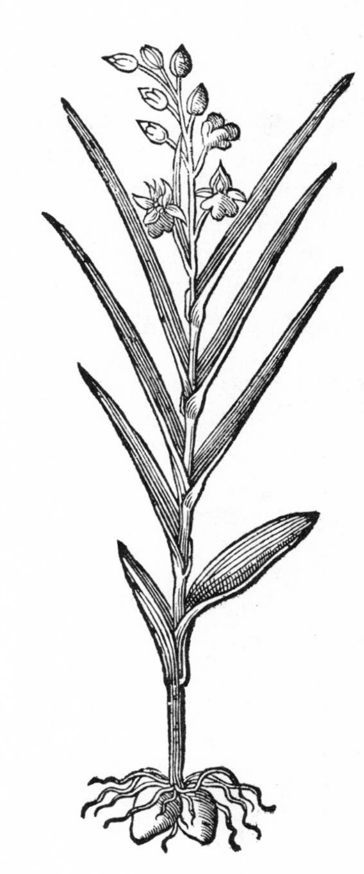

¶ *The Nature and Vertues.*

The nature and vertues of theſe kindes of Orchis are referred vnto the others, namely to thoſe of the Fox ſtones; notwithſtanding there is no great vſe of theſe in phyſicke, but they are chiefly regarded for the pleaſant and beautifull floures, wherewith Nature hath ſeemed to play and diſport her ſelfe.

† Theſe Figures in this Chapter were formerly much miſplaced: as thus; The ſecond was of *Orchis Ornith. fol. macul.* being the tenth. The third was of *Triorchis mas minor* of *Tabern.* being a varietie of *Cynoſorchis morio fœmina.* The fifth was of *Orchis Batrachitis.* The ſixth, of *Orchis Melittias.* The ſeuenth and eighth were onely tranſpoſed, or put the one for the other. The ninth was of the ſecond, called formerly *Teſticulus ſphegodes.* The tenth was of the third, called *Teſticulus Vulpinus.* The eleuenth was of *Strateumatica.* The twelfth was of *Strateumatica minor.* The thirteenth was a varietie of the fourth. The fourteenth was of *Orchis Anguſtifolia,* which we here giue you in the ſixteenth place.

Chap. 114. *Of Sweet Cullions.*

¶ *The Kindes.*

There be ſundry ſorts of ſweet ſmelling Teſticles or Stones, whereof the firſt is moſt ſweet and pleaſant in ſmell, the others of leſſe ſmell or ſauour, differing in floure and roots. Some haue white floures, others yellow; ſome fleſh coloured; ſome daſht vpon white with a little reddiſh waſh: ſome haue two ſtones, others three, and ſome foure, wherein their difference conſiſteth.

¶ *The Deſcription.*

1 The firſt kinde of Sweet ſtones is a ſmall baſe and low plant in reſpect of all the reſt: The leaues be ſmall, narrow, and ſhort, growing flat vpon the ground; amongſt the which riſeth vp a ſmall weake and tender ſtalke of a finger long, whereupon doe grow

ſmall

ſmall white floures ſpike faſhion, of a pleaſant ſweet ſmel. The roots are two ſmall ſtones in ſhape like the other.

2 Triple Orchis hath commonly three, yet ſometimes foure bulbes or tuberous roots, ſomewhat long, ſet with many ſmall fibres or ſhort threads; from the which roots riſe immediately many flat and plaine leaues, ribbed with nerues alongſt them like thoſe of Plantaine: among the which come forth naked ſtalkes, ſmall and tender, whereupon are placed certaine ſmall white floures, trace faſhion, not ſo ſweet as the former in ſmell and ſauour. ‡ The top of the ſtalke whereon the floures do grow, is commonly as if it were twiſted or writhen about. ‡

3 Frieſeland Lady traces hath two ſmall round ſtones or bulbes, of the bigneſſe of the peaſe that we call Rouncifalls; from the which riſe vp a few hairy leaues, leſſer than thoſe of the triple ſtones, ribbed as the ſmall leafed Plantaine: among the which commeth forth a ſmall naked ſtalk, ſet round about with little yellow floures, not trace faſhion as the former.

4 Liege Lady traces hath for his roots two greater ſtones, and two ſmaller; from the which come vp two and ſometimes more leaues, furrowed or made hollow in the midſt like to a trough, from the which riſeth vp a ſlender naked ſtalke, ſet with ſuch floures as the laſt deſcribed, ſauing that they be of an ouerworne yellow colour.

1 *Teſticulus odoratus.*
Lady Traces.

2 *Triorchis.*
Triple Lady Traces.

¶ *The Place.*

Theſe kindes of Stones or Cullions do grow in dry paſtures and heaths, and likewiſe vpon chalkie hills, the which I haue found growing plentifully in ſundry places, as in the field by Iſlington, neere London, where there is a bowling place vnder a few old ſhrubby Okes. They grow likewiſe vpon the heath at Barne-elmes, neere vnto the head of a conduit that ſendeth water to the houſe belonging to the late Sir *Francis Walſingham*. They grow in the field next vnto a Village called Thiſtleworth, as you go from Branford to her Maieſties houſe at Richmond; alſo vpon a common Heath by a Village neere London called Stepney, by the relation of a learned merchant of London, named M. *Iames Cole*, exceedingly well experienced in the knowledge of Simples.

The yellow kindes grow in barren paſtures and borders of fields about Ouenden and Clare in Eſſex

Eſſex. Likewiſe neere vnto Muche Dunmow in Eſſex, where they were ſhewed me by a learned Gentleman Maſter *Iames Twaights*, excellently well ſeene in the knowledge of plants.

‡ I receiued ſome roots of the ſecond from my kinde friend M. *Thomas Wallis* of Weſtminſter, the which he gathered at Dartford in Kent, vpon a piece of ground commonly called the Brimth: but I could not long get them to grow in a garden, neither do any of the other Satyrions loue to be pent vp in ſuch ſtraight bounds. ‡

3 *Orchis Friſia lutea.*
Frieſeland Lady-traces.

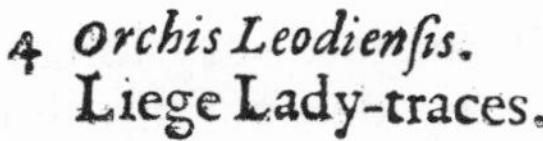

4 *Orchis Leodienſis.*
Liege Lady-traces.

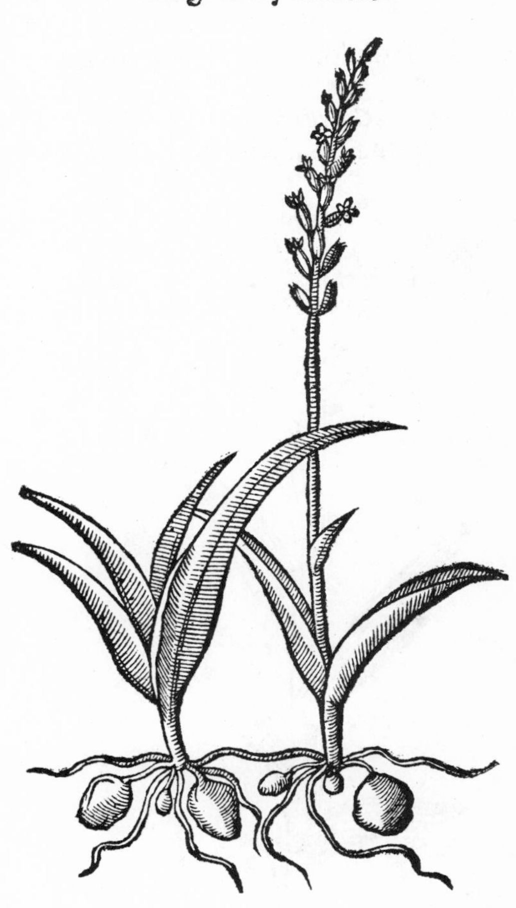

¶ *The Time.*

Theſe kindes of ſtones do floure from Auguſt to the end of September.

¶ *The Names.*

The firſt is called in Latine *Teſticulus Odoratus*: in Engliſh, Sweet ſmelling Teſticles or ſtones, not of the ſweetneſſe of the roots, but of the floures. It is called alſo *Orchis ſpiralis*, or *Autumnalis*, for that this (as alſo that which is ſet forth in the next place) hath the top of the ſtalke as it were twiſted or twined ſpire faſhion, and for that it commeth to flouring in Autumne: of our Engliſh women they be called Lady-traces; but euery countrey hath a ſeuerall name; for ſome call them Sweet Ballocks, ſweet Cods, ſweet Cullions, and Stander-graſſe.. In Dutch, **Knabenkraut**, and **Stondelcraut**: In French, *Satyrion*.

The ſecond ſort is called *Triorchis*, and alſo *Tetrorchis*: in Engliſh, Triple Lady-traces, or white Orchis.

The third is called *Orchis Friſia*: in Engliſh Frieſeland Orchis.

The laſt of theſe kindes of Teſticles or Stones is called of ſome in Latine, *Orchis Leodienſis*, and *Orchis Lutea*, as alſo *Baſilica minor Serapias*, and *Triorchis Æginetæ*: In Engliſh, Yellow Lady-traces.

¶ *The Temperature.*

Theſe kindes of ſweet Cullions are of nature and temperature like the Dogs ſtones, although not vſed in Phyſicke in times paſt; notwithſtanding later Writers haue attributed ſome vertues vnto them as followeth.

¶ *The Vertues.*

The full and ſappy roots of Lady-traces eaten or boyled in milke, and drunke, prouoke venery, A
nouriſh and ſtrengthen the body, and be good for ſuch as be fallen into a Conſumption or Feuer Hectique.

Chap.

Chap. 115. *Of Satyrion Royall.*

¶ *The Deſcription.*

1 THe male Satyrion royal hath large roots,knobbed,not bulbed as the others,but branched or cut into ſundry ſections like an hand, from the which come vp thick and fat ſtalkes ſet with large leaues like thoſe of Lillies,but leſſe; at the top whereof groweth a tuft of floures, ſpotted with a deepe purple colour.

1 *Palma Chriſti mas.*
The male Satyrion Royall.

2 *Palma Chriſti fœmina.*
The female Satyrion Royall.

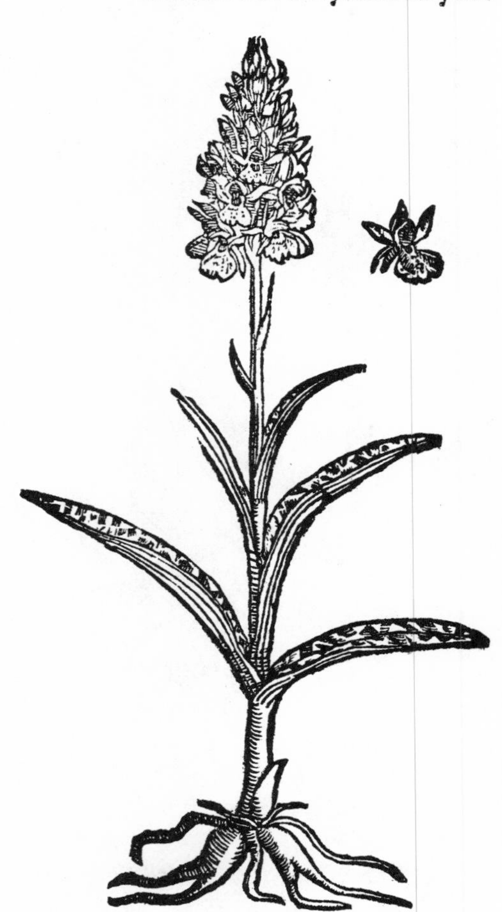

2 The female Satyrion hath clouen or forked roots,with ſome fibres ioyned thereto. The leaues be like the former, but ſmaller and narrower, and confuſedly daſhed or ſpotted with black ſpots: from the which ſpringeth vp a tender ſtalke, at the top whereof doth grow a tuft of purple floures, in faſhion like vnto a Friers hood, changing or varying according to the ſoile and clymat, ſometimes red,ſometimes white, and ſometimes light carnation or fleſh colour.

‡ 3 This in roots and leaues is like the former,but that the leaues want the black ſpots,the ſtalke is but low, and the top thereof hath floures of a whitiſh colour, not ſpotted: they on the foreſide reſemble gaping hoods,with eares on each ſide,and a broad lip hanging down; the backe part ends in a broad obtuſe ſpur. Theſe floures ſmell like Elder bloſſomes. ‡

¶ *The Place.*

The royall Satyrions grow for the moſt part in moiſt and fenny grounds, medowes,and Woods that are very moiſt and ſhadowie. I haue found them in many places, eſpecially in the midſt of a wood in Kent called Swaineſcombe wood neere to Graueſend,by the village Swaineſcombe, and likewiſe in Hampſted wood foure miles from London.

¶ *The Time.*

They floure in May and Iune, but ſeldome later.

¶ *The*

‡ 3 *Orchis Palmata Pannonica 8. Cluſ.*
The Auſtrian handed Satyrion.

¶ *The Names.*

† Royal Satyrion,or finger Orchis is called in Latine, *Palma Chriſti* ; notwithſtanding there is another herbe or plant called by the ſame name, which otherwiſe is called *Ricinus*. This plant is called likewiſe of ſome, *Satyrium Baſilicum*, or *Satyrium regium*. Some would haue it to be *Buzeiden*, or *Buzidan Arabum*, but *Auicen* ſaith *Buzeiden* is a woody Indian medicine : and *Serapio* ſaith, *Buzeiden* be hard white roots like thoſe of *Behen album*, and that it is an Indian drug:but contrariwiſe the roots of *Palma Chriſti* are nothing leſſe than woody, ſo that it cannot be the ſame. *Matthiolus* would haue Satyrion royall to be the *Digiti Citrini* of *Auicen* ; finding fault with the Monkes which ſet forth Commentaries vpon *Meſues* Compoſitions, for doubting and leauing it to the iudgement of the diſcreet Reader. Yet do we better allow of the Monkes doubt, than of *Matthiolus* his aſſertion. For *Auicens* words be theſe ; What is *Aſabaſafra*, or *Digiti Citrini ?* and anſwering the doubt himſelfe, he ſaith, It is in figure or ſhape like the palme of a mans hand, of a mixt colour betweene yellow and white, and it is hard, in which there is a little ſweetneſſe, and there is a Citrine ſort duſty and without ſweetneſſe. *Rhaſis* alſo in the laſt booke of his Continent calls theſe, *Digiti Crocei*, or Saffron fingers ; and he ſaith it is a gumme or veine for Dyars. Now theſe roots are nothing leſſe than of a Saffron colour, and wholly vnfit for Dying. Wherefore without doubt theſe words of *Auicen* and *Rhaſis*, in the eares of men of iudgment do confirme, That Satyrion Royall, or *Palma Chriſti*, are not thoſe *Digiti Citrini*. The Germans call it **Creutſblum**: the low Dutch, **Handekens cruyt**: the French, *Satyrion royal*.

¶ *The Temperature and Vertues.*

The Roots of Satyrion royall are like to *Cynoſorchis* or Dogs ſtones, both in ſauour and taſte, and therefore are thought by ſome to be of like faculties. Yet *Nicolaus Nicolus*, in the chapter of the cure of a Quartaine Ague, ſaith, That the roots of *Palma Chriſti* are of force to purge vpward and downward ; and that a piece of the root as long as ones thumbe ſtamped and giuen with wine before the fit commeth, is a good remedie againſt old Quartaines after purgation : and reporteth, That one *Baliolus*, after he had endured 44 fits, was cured therewith.

† This facultie of purging and vomiting, which our Author out of *Dodonæus*, and he out of *Nicolus*, giue to the root of *Palma Chriſti*, I doubt is miſtaken and put in the wrong place : for I iudge it to belong to the *Ricinus*, which alſo is called *Palma Chriſti* ; for that *Nicolus* ſaith, a piece of root muſt be taken as long as ones thumbe ; now the whole root of this plant is not ſo long. And beſides, *Ricinus* is knowne to haue a vomitorie or purging facultie.

Chap. 116. *Of Serapia's Stones.*

¶ *The Kindes.*

THere be ſundry ſorts of Serapias ſtones, whereof ſome be male, others female ; ſome great, and ſome of a ſmaller kinde ; varying likewiſe in colour of the floures, whereof ſome be white, others purple ; altering according to the ſoile or clymate, as the greateſt part of bulbous roots do. Moreouer, ſome grow in marſhie and fenny grounds, and ſome in fertile paſtures, lying open to the Sun, varying likewiſe in the ſhape of their floures ; retaining the forme of flies, Butter-flies, and Gnats, like thoſe of the Fox ſtones.

1 *Serapias Candido flore.*
White handed Orchis.

2 *Serapias minor, nitente flore.*
Red handed Orchis.

3 *Serapias palustris latifolia.*
Marish Satyrion.

4 *Serapias palustris leptophylla.*
Fenny Satyrion.

† 5 *Serapias Montana.*
Mountaine Satyrion.

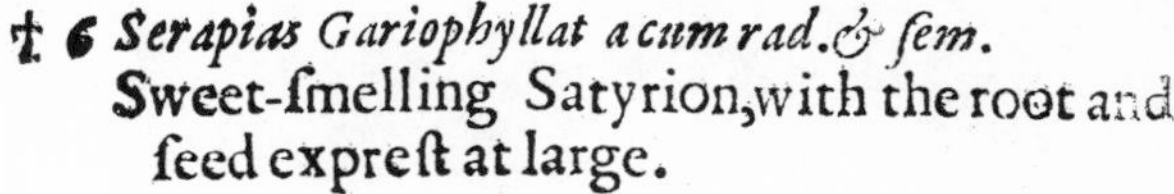

† 6 *Serapias Gariophyllat a cum rad. & ſem.*
Sweet-ſmelling Satyrion, with the root and ſeed expreſt at large.

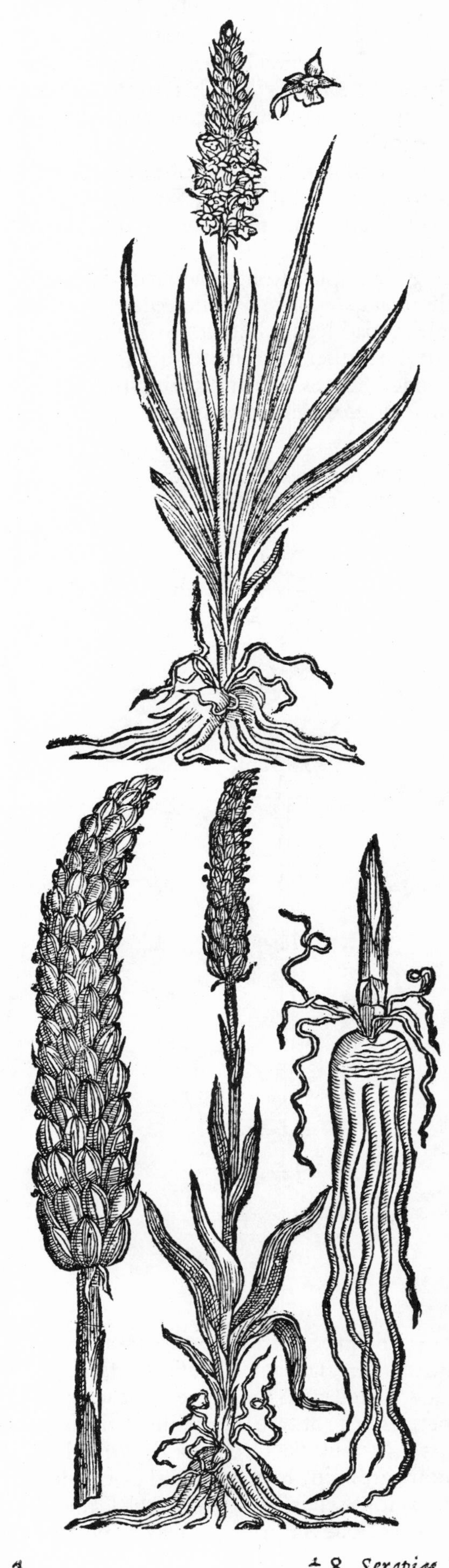

7 *Serapias Caſtrata.*
Gelded Satyrion.

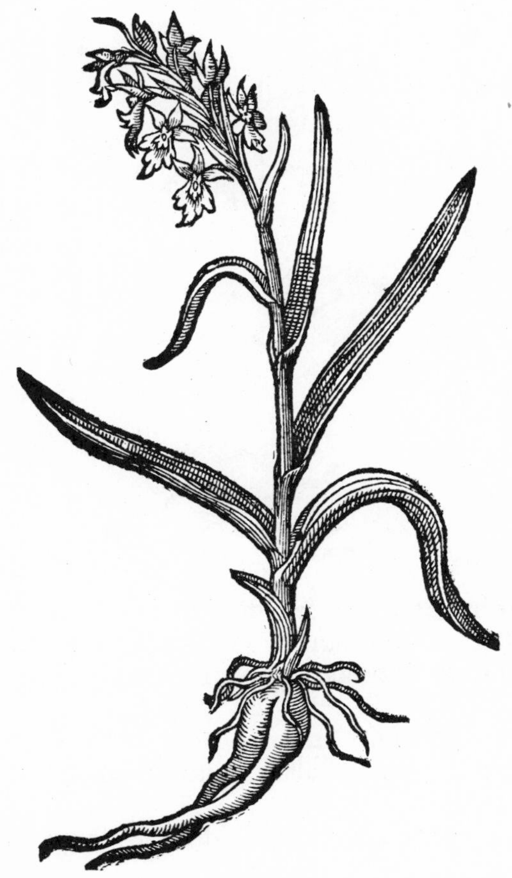

¶ *The Deſcription.*

1 THe white handed Orchis or Satyrion hath long and large leaues, ſpotted and daſhed with blacke ſpots, from the which doth riſe vp a ſmall fragile or brittle ſtalke of two hands high, hauing at the top a buſh or ſpoky tuft of white floures, like in ſhape to thoſe of *Palma Chriſti*, whereof this is a kinde. The root is thicke, fat, and full of iuyce, faſhioned like the hand and fingers of a man, with ſome tough and fat ſtrings faſtned to the vpper part thereof.

2 Red handed Satyrion is a ſmall low and baſe herbe, hauing a ſmall tender ſtalke ſet with two or three ſmall leaues, like vnto thoſe of the Leeke, but ſhorter. The floure groweth at the top tuft faſhion, of a gliſtering red colour, with a root faſhioned like an hand, but leſſer than the former.

3 Serapia's ſtones, or mariſh Satyrion hath a thicke knobby root, diuided into fingers like thoſe of *Palma Chriſti*, whereof it is a kinde: from which riſe thicke fat and ſpongeous ſtalkes, ſet with broad leaues like thoſe of Plantaine, but much longer, euen to the top of the tuft of floures; but the higher they riſe toward the top the ſmaller they are. The floure conſiſteth of many ſmall hooded floures ſomewhat whitiſh, ſpotted within with deepe purple ſpots; the backſide of theſe little floures are Violet mixed with purple.

† 8 *Serapias Batrachites.*
Frog Satyrion.

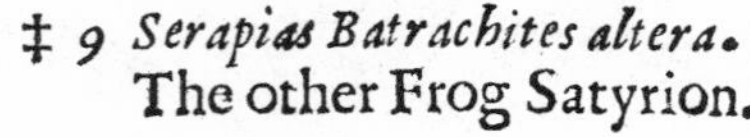

‡ 9 *Serapias Batrachites altera.*
The other Frog Satyrion.

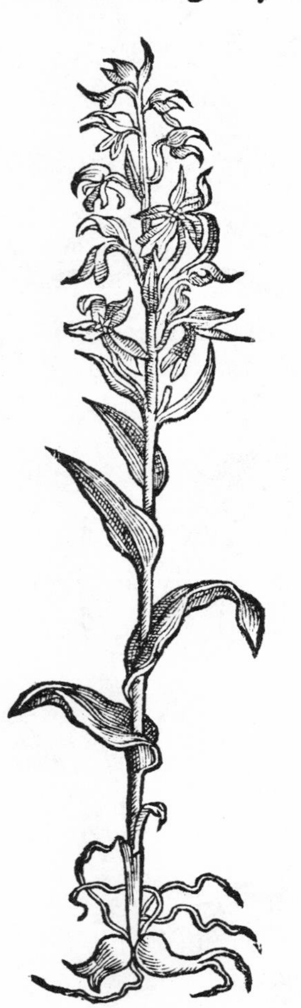

4 Fenny Satyrion (or Serapia's ſtones) differeth little from the former, ſauing that the leaues are ſmaller, and ſomewhat ſpotted, and the tuft of floures hath not ſo many greene leaues, nor ſo long, mixed with the floures, neither are they altogether of ſo darke or purpliſh a colour as the former. The roots are like thoſe of the laſt deſcribed.

5 Mountaine Orchis or Satyrion hath thicke fat and knobby roots, the one of them for the moſt part being handed, and the other long. It growes like the former in ſtalkes, leaues, and floures, but is ſomewhat bigger, with the leaues ſmoother, and more ſhining.

6 Cloue Satyrion, or ſweet ſmelling Orchis, hath flat and thicke roots diuided into fingers

like

like thoſe of *Palma Chriſti*, ſauing that the fingers are longer, ſmaller, and more in number; from the which riſe vp long and narrow leaues like thoſe of Narciſſus or Daffodill: among which commeth forth a ſmall tender ſtalke, at the top whereof groweth a purple tuft compact of many ſmall floures reſembling Flies, but in ſauour and ſmell like the Cloue, or Cloue Gillo-floure; but farre ſweeter and pleaſanter, as my ſelfe with many others can witneſſe now liuing, that haue both ſeene and ſmelt them in my garden. ‡ After the floure is paſt, come many ſeed veſſels filled with a ſmall ſeed, and growing after the manner as you ſee them here at large expreſſed in a figure, together with the root alſo ſet forth at full. ‡

7 Gelded Satyrion hath leaues with nerues and ſinewes like to thoſe of Daffodill, ſet vpon a weake and tender ſtalke, with floures at the top white of colour, ſpotted within the floure, and in ſhape they are like Gnats and little Flies. The ſtalke is gelded as it were, or the ſtones and hands cut off, leauing for the root two long legges or fingers, with many ſtrings faſtned vnto the top.

8 Frog Satyrion hath ſmall flat leaues ſet vpon a ſlender weake ſtem; at the top wherof growes a tuft of floures compact of ſundry ſmall floures, which in ſhape do reſemble little frogges, whereof it tooke his name. The root is likewiſe gelded, onely reſerued two ſmall miſhapen lumps with certaine fibres annexed thereto.

‡ 9 This alſo may fitly be added to the laſt deſcribed, the root ſhewing it to be of a kinde betweene the Serapia's and Orchis. It groweth to the height of the former, with ſhort leaues engirting the ſtalke at their ſetting on. The floures on the top reſemble a Frogge, with their long leaues; and if you looke vpon them in another poſture, they will ſomewhat reſemble little Flies: wherefore *Lobel* calls it as well *Myoides*, as *Batrachites*. ‡

¶ *The Time.*

Theſe Plants flouriſh in the moneth of May and Iune, but ſeldome after, except ſome degenerate kinde, or that it hath had ſome impediment in the time when it ſhould haue floured, as often hapneth.

¶ *The Names.*

We haue called theſe kindes, Serapia's ſtones, or Serapiades, eſpecially for that ſundry of them do bring forth floures reſembling Flies and ſuch like fruitfull and laſciuious inſects, as taking their name from *Serapias* the god of the citiſens of Alexandria in Ægypt, who had a moſt famous Temple at Canopus, where he was worſhipped with all kinde of laſciuious wantonneſſe, ſongs, and dances, as we may reade in *Strabo*, in his ſeuenteenth Booke. *Apuleius* confounds the Orchides and Serapiades, vnder the name of both the Satyrions; and withall ſaith it is called *Entaticos*, *Panion*, and of the Latines, *Teſticulus Leporinus*. In Engliſh we may call them Satyrions, and finger Orchis, and Hares ſtones.

¶ *The Nature and Vertues.*

Serapia's ſtones are thought to be in nature, temperature, and Vertues, like vnto the Satyrion Royall; and although not ſo much vſed in phyſicke, yet doubtleſſe they worke the effect of the other Stones.

† The fifth was the figure of *Satyrium trifolium* of Tabern. and is a kinde of *Teſticulus pſycodes*. 6 In this place formerly was the figure of the laſt before, to wit *Serapias montana*. 8 Here was the figure of *Orchis Myodes*, which ſhould haue beene in the ſixth place in the 101 Chapter of the former Edition, being the 113 of this.

CHAP. 117.

Of Fenny Stones.

¶ *The Deſcription.*

† 1 THis hath cleft or diuided roots like fingers, much like vnto the Roots of other *Palma Chriſti's*; whereof this is a kinde: from the which riſeth vp a ſtalke of a foot high, ſet here and there with very faire Lilly-like leaues, of colour red, the which do clip or embrace the ſtalkes almoſt round about, like the leaues of Thorow-wax. At the

top of the stalke groweth a faire bush of very red floures, among the which floures do grow many small sharpe pointed leaues. The seed I could neuer obserue, being a thing like dust that flieth in the winde.

2 The other Marish handed Satyrion differeth little from the precedent, but in the leaues and floures, for that the leaues are smaller and narrower, and the floures are faire white, gaping wide open; in the hollownesse whereof appeare certaine things obscurely hidden, resembling little helmets, which setteth forth the difference.

† 1 *Serapias Dracontias palustris.*
Marish Dragon Satyrion.

† 2 *Serapias palustris leptophylla altera.*
The other Marish handed Satyrion.

3 This third handed Satyrion hath roots fashioned like an hand, with some strings fastned to the vpper part of them; from which riseth vp a faire stiffe stalke armed with large leaues, very notably dasht with blackish spots, clipping or embracing the stalke round about: at the top of the stalke standeth a faire tuft of purple floures, with many greene leaues mingled amongst the same, which maketh the bush or tuft much greater. The seed is nothing else but as it were dust like the other of his kinde: ‡ and it is contained in such twined vessels as you see exprest apart by the side of the figure; which vessels are not peculiar to this, but common to most part of the other Satyrions. ‡

4 The creeping rooted Orchis or Satyrion without testicles, hath many long roots dispersing themselues, or creeping far abroad in the ground, contrarie to all the rest of the Orchides: which Roots are of the bignesse of strawes, in substance like those of Sopewort; from the which immediately doth rise foure or fiue broad smooth leaues like vnto the small Plantaine, from the which shooteth vp a small and tender stalke, at the top whereof groweth a pleasant spikie eare of a whitish colour, spotted on the inside with little speckes of a bloudie colour. The seed also is very small.

‡ 5 This from handed roots like others of this kinde sends vp a large stalke, sometimes attaining to the height of two cubits; the leaues are much like to those of the marish Satyrions; the floures are of an elegant purple, with little hoods like the top of an helmet (whence *Gemma* termed

3 *Palma Christi palustris.*
The third handed marish Satyrion.

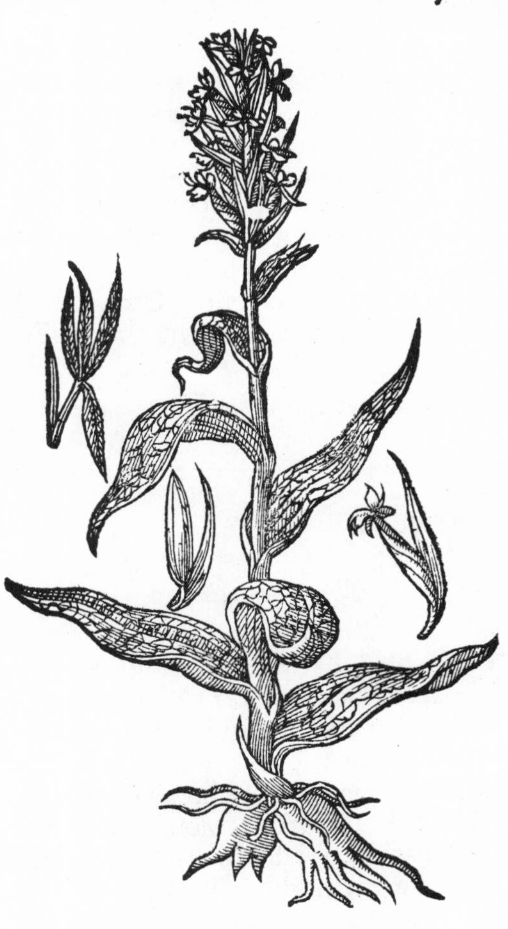

4 *Palma Christi, radice repente.*
Creeping Satyrion.

‡ 5 *Palma Christi maxima.*
The greatest handed Satyrion.

termed the plant, *Cynosorch. conopsæa*; and from the height he called it *Macrocaulos*.) These floures smel sweet, and are succeeded by seeds like those of the rest of this kindred.

It delights to grow in grounds of an indifferent temper, not too moist nor too dry. It floures from mid-May to mid-Iune. ‡

The Place.

They grow in marish and fenny grounds, and in shadowie woods that are very moist.

The fourth was found by a learned Preacher called Master *Robert Abbot*, of Bishops Hatfield, in a boggy groue where a Conduit head doth stand, that sendeth water to the Queenes house in the same towne.

‡ It growes also plentifully in Hampshire, within a mile of a market Towne called Peters-field, in a moist medow named Wood-mead, neere the path leading from Peters-field, towards Beryton. ‡

¶ *The Time.*

They floure and flourish about May and Iune.

‡ ¶ *The Names.*

‡ 1 This is *Cynosorchis Dracuntias* of *Lobell* and *Gemma*.

2 This

2 This is *Cynoſorchis paluſtris altera Leptaphylla*, of *Lobell*; *Teſticulus Galericulatus*, of *Tabernamontanus*.

3 *Lobell* and *Gemma* terme this, *Cynoſorchis paluſtris altera Lophodes, vel nephelodes*.

4 This is *Orchis minor radice repente*, of *Camerarius*.

5 This by *Lobell* and *Gemma* is called *Cynoſorchis macrocaulos, ſiue Conopſæa*.

¶ *The Temperature and Vertues.*

There is little vſe of theſe in phyſicke; onely they are referred vnto the handed Satyrions, whereof they are kindes: notwitſtanding *Daleſcampius* hath written in his great Volume, that the Mariſh Orchis is of greater force than any of the Dogs ſtones in procuring of luſt.

Camerarius of Noremberg, who was the firſt that deſcribed this kinde of creeping Orchis, hath ſet it forth with a bare deſcription onely; and I am likewiſe conſtrained to do the like, becauſe as yet I haue had no triall thereof.

† The firſt of theſe was the third in the former Chapter; in lieu whereof I giue you the *Dracuntias* of *Lobel*, whoſe figure was here in the ſecond place.

Chap. 118. Of Birds neſt.

1 *Satyrium abortinum, ſiue Nidus auis.*
Birds neſt.

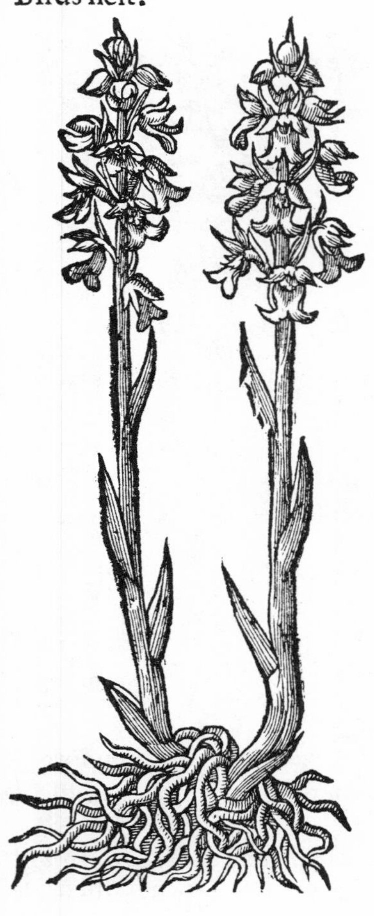

¶ *The Deſcription.*

1 Birds Neſt hath many tangling roots platted or croſſed one ouer another very intricately, which reſembleth a Crowes neſt made of ſtickes; from which riſeth vp a thicke ſoft groſſe ſtalk of a browne colour, ſet with ſmall ſhort leaues of the colour of a dry Oken leafe that hath lien vnder the tree all the winter long. On the top of the ſtalke groweth a ſpikie eare or tuft of floures, in ſhape like vnto Maimed Satyrion, whereof doubtleſſe it is a kinde. The whole plant, as well ſticks, leaues, and floures, are of a parched browne colour.

‡ I receiued out of Hampſhire from my often remembred friend Maſter *Goodyer* this following deſcription of a *Nidus auis* found by him the twenty ninth of Iune, 1621.

¶ *Nidus avis flore & caule violaceo purpureo colore; an Pſeudoleimodoron Cluſ. Hiſt. Rar. plant. pag. 270.*

This riſeth vp with a ſtalke about nine inches high, with a few ſmal narrow ſharpe pointed ſhort skinny leaues, ſet without order, very little or nothing at all wrapping or incloſing the ſtalke; hauing a ſpike of floures like thoſe of *Orobanche*, without tailes or leaues growing amongſt them: which fallen, there ſucceed ſmall ſeed-veſſels. The lower part of the ſtalke within the ground is not round like *Orobanche*, but ſlender or long, and of a yellowiſh white colour, with many ſmall brittle roots growing vnderneath confuſedly, wrapt or folded together like thoſe of the common *Nidus auis*. The whole plant as it appeareth aboue ground, both ſtalkes, leaues, and floures, is of a violet or deepe purple colour. This I found wilde in the border of a field called Marborne, neere Habridge in Haliborne, a mile from a towne called Alton in Hampſhire, being the land of one *William Balden*. In this place alſo groweth wilde the thiſtle called *Corona fratrum*. *Ioh. Goodyer*.

¶ *The Place.*

This baſtard or vnkindely Satyrion is very ſeldome ſeene in theſe Southerly parts of England.

It

It is reported, That it groweth in the North parts of England, neere vnto a village called Knaesborough. I found it growing in the middle of a Wood in Kent two miles from Grauesend, neere vnto a worshipfull Gentlemans house called Master *William Swan*, of Howcke Greene. The wood belongeth to one Master *Iohn Sidley* : which plant I did neuer see elsewhere ; and because it is very rare, I am the more willing to giue you all the markes in the wood for the better finding it, because it doth grow but in one piece of the Wood : that is to say, The ground is couered all ouer in the same place neere about it with the herbe Sanycle, and also with the kinde of Orchis called *Hermaphroditica*, or Butter-fly Satyrion.

¶ *The Time.*

It floureth and flourisheth in Iune and August. The dusty or mealy seed (if it may be called seed) falleth in the end of August ; but in my iudgement it is an vnprofitable or barren dust, and not any seed at all.

¶ *The Names.*

It is called *Satyrium abortirum* : of some, *Nidus auis* : in French *Nid d' oiseau* : in English, Birds nest, or Goose-nest : in Low-Dutch, **Vogels nest** : in High-Dutch, **Margen dzehen.**

¶ *The Temperature and Vertues.*

It is not vsed in Physicke that I can finde in any authoritie either of the antient or later Writers, but is esteemed as a degenerate kinde of Orchis, and therefore not vsed.

THE

THE SECOND BOOKE OF THE HISTORIE OF PLANTS:

Containing the deſcription, place, time, names, nature, and vertues of all ſorts of Herbes for meate, medicine, or ſweet ſmelling vſe, &c.

WE haue in our firſt booke ſufficiently deſcribed the Graſſes, Ruſhes, Flags, Corne, and bulbous rooted Plants, which for the moſt part are ſuch as with their braue and gallant floures decke and beautifie Gardens, and feed rather the eyes than the belly. Now there remaine certaine other bulbes, whereof the moſt (though not all) ſerue for food: of which we will alſo diſcourſe in the firſt place in this booke, diuiding them in ſuch ſort, that thoſe of one kinde ſhall be ſeparated from another. ‡ In handling theſe and ſuch as next ſucceed them, we ſhall treat of diuers, yea the moſt part of thoſe Herbes that the Greekes call by a generall name Λάχανα: and the Latines, *Olera*: and we in Engliſh, Sallet-herbes. When we haue paſt ouer theſe, we ſhall ſpeake of other plants, as they ſhall haue reſemblance each to other in their externall forme. ‡

CHAP. 1. *Of Turneps.*

¶ *The Kindes.*

THere be ſundry ſorts of Turneps; ſome wilde; ſome of the garden; ſome with round roots globe faſhion; other ouall or peare faſhion; and another ſort longiſh or ſomwhat like a Radiſh: and of all theſe there are ſundry varieties, ſome being great, and ſome of a ſmaller ſort.

¶ *The Deſcription.*

1 THe Turnep hath long rough and greene leaues, cut or ſnipt about the edges with deepe gaſhes. The ſtalke diuideth it ſelfe into ſundry branches or armes, bearing at the top ſmall floures of a yellow colour, and ſometimes of a light purple: which being paſt, there do ſucceed long cods full of ſmall blackiſh ſeed like rape ſeed. The root is round like a bowle, and ſometimes a little ſtretched out in length, growing very ſhallow in the ground, and often ſhewing it ſelfe aboue the face of the earth.

‡ 2 This is like the precedent in each reſpect, but that the root is not made ſo globous or bowle-faſhioned as the former, but ſlenderer, and much longer, as you may perceiue by the figure wee here giue you. ‡

3 The ſmall Turnep is like vnto the firſt deſcribed, ſauing that it is leſſer. The root is much ſweeter in taſte, as my ſelfe hath often proued.

4 There is another ſort of ſmall Turnep ſaid to haue red roots; ‡ and there are other-ſome whoſe roots are yellow both within and without; ſome alſo are greene on the outſide, and other-ſome blackiſh. ‡

¶ *The Place.*

The Turnep proſpereth wel in a light, looſe, and fat earth, and ſo looſe, as *Petrus Creſcentius* ſaith, that

that it may be turned almost into dust. It groweth in fields and diuers vineyards or Hop gardens in most places of England.

The small Turnep groweth by Hackney, in a sandy ground; and those that are brought to Cheape-side market from that Village are the best that euer I tasted.

¶ *The Time.*

Turneps are sowne in the spring, as also in the end of August. They floure and seed the second yeare after they are sowen: for those which floure the same yeare that they are sowen are a degenerate kinde, called in Cheshire about the Namptwitch, Mad neeps, of their euill qualitie in causing frensie and giddinesse of the braine for a season.

1 *Rapum majus.*
Great Turnep.

‡ 2 *Rapum radice oblonga.*
Longish rooted Turnep.

¶ *The Names.*

The Turnep is called in Latine, *Rapum*: in Greeke, γογγύλη: the name commonly vsed in shops and euery where is *Rapa*. The Lacedemonians call it γαστήρ: the Boetians, ζακελτίς, as *Athenæus* reporteth: in high Dutch, **Ruben**: in low Dutch, **Rapen**: in French, *Naueau rond*: in Spanish, *Nabo*: in English, Turnep, and Rape.

¶ *The Temperature and Vertues.*

A The bulbous or knobbed root, which is properly called *Rapum* or Turnep, and hath giuen the name to the plant, is many times eaten raw, especially of the poore people in Wales, but most commonly boiled. The raw root is windy, and engendreth grosse and cold bloud; the boyled doth coole lesse, and so little, that it cannot be perceiued to coole at all, yet it is moist and windy.

B It auaileth not a little after what manner it is prepared; for being boyled in water, or in a certaine broth, it is more moist, and sooner descendeth, and maketh the body more soluble; but being rosted or baked it drieth, and ingendreth lesse winde, and yet it is not altogether without winde. But howsoeuer they be dressed, they yeeld more plenty of nourishment than those that are eaten raw: they do increase milke in womens brests, and naturall seed, and prouoke vrine.

C The decoction of Turneps is good against the cough and hoarsenesse of the voice, being drunke in the euening with a little sugar, or a quantitie of clarified honey.

D *Dioscorides* writeth, That the Turnep it selfe being stamped, is with good successe applied vpon mouldie

mouldie or kibed heeles, and that also oile of roses boiled in a hollow turnep vnder the hot embers doth cure the same.

E The young and tender shootes or springs of Turneps at their first comming forth of the ground, boiled and eaten as a sallade, prouoke vrine.

F The seed is mixed with counterpoisons and treacles: and being drunke it is a remedie against poisons.

G They of the lowe countries doe giue the oile which is pressed out of the seed, against the after throwes of women newly brought to bed, and also minister it to young children against the wormes, which it both killeth and driueth forth.

H The oile washed with water doth allaie the feruent heat and ruggednesse of the skin.

CHAP. 2. *Of wilde Turneps.*

¶ *The Kindes.*

THere be three sorts of wilde Turneps; one our common Rape which beareth the seed whereof is made rape oile, and feedeth singing birds. the other the common enemy to corne, which call Charlock; whereof there be two kindes, one with a yellow, or els purple floure, the other with a white floure: there is also another of the water and marish grounds.

1 *Rapum sylvestre.*
Wilde Turneps.

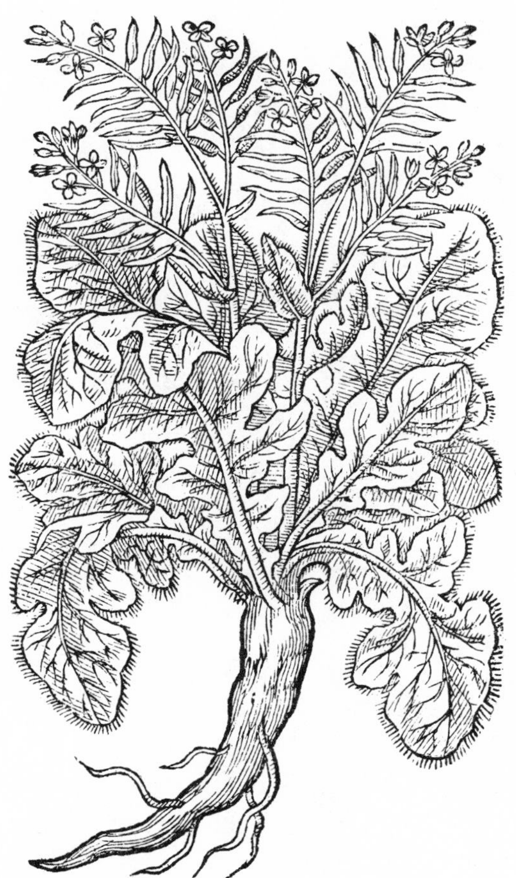

2 *Rapistrum aruorum.*
Charlocke or Chadlocke.

¶ *The Description.*

1 WIlde Turneps or Rapes, haue long, broad, and rough leaues like those of Turneps, but not so deeply gashed in the edges. The stalkes are slender and brittle, somewhat hairie, of two cubits high, diuiding themselues at the top into many armes or branches, whereon doe grow little yellowish flowers: which being past, there doe succeed small long cods which containe the seed like that of the Turnep, but smaller, somewhat reddish, and of a firie hot

 and

and biting taſte as is the muſtard, but bitterer. The root is ſmall, and periſheth when the ſeed is ripe.

2 Charlocke, or the wilde rape, hath leaues like vnto the former, but leſſer, the ſtalke and leaues being alſo rough. The ſtalkes bee of a cubite high, ſlender, and branched; the floures are ſometimes purpliſh, but more often yellow. The rootes are ſlender, with certaine threds or ſtrings hanging on them.

‡ There is alſo another varietie hereof with the leaues leſſe diuided, and much ſmoother than the two laſt deſcribed, hauing yellow floures and cods not ſo deeply joynted as the laſt deſcribed: this is that, which is ſet forth by *Matthiolus* vnder the name of *Lampſana*.

3 Water Chadlock groweth vp to the height of three foot or ſomewhat more, with branches ſlender and ſmooth in reſpect of any of the reſt of his kinde, ſet with rough ribbed leaues, deeply indented about the lower part of the leafe. The floures grow at the top of the branches, vmble or tuft faſhion, ſometimes of one colour, and ſometimes of another. ‡ The root is long, tough, and full of ſtrings, creeping and putting forth many ſtalkes: the ſeed veſſells are ſhort and ſmall. *Bauhine* hath this vnder the title of *Raphanus aquaticus alter*. ‡

2 *Rapiſtrum aruenſe alterum.*
Another wilde Charlocke.

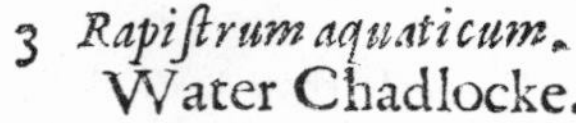

3 *Rapiſtrum aquaticum.*
Water Chadlocke.

¶ *The Place.*

Wilde Turneps or Rapes, doe grow of themſelues in fallow fields, and likewiſe by high wayes neere vnto old walls, vpon ditch-bankes, and neere vnto townes and villages, and in other vntoiled and rough places.

The Chadlocke groweth for the moſt part among corne in barraine grounds, and often by the borders of fields and ſuch like places.

Water Chadlocke groweth in moiſt medowes and mariſh grounds, as alſo in water ditches, and ſuch like places.

¶ *The Time.*

Theſe doe floure from March, till Summer be farre ſpent, and in the meane ſeaſon the ſeed is ripe.

¶ *The*

¶ *The Names.*

Wilde Turnep is called in Latine *Rapiſtrum*, *Rapum ſylueſtre*, and of ſome, *Sinapi ſylueſtre*, or wild muſtard: in high Dutch, **Hederich:** in low Dutch, **Herick:** in French, *Velar*: in Engliſh, Rape, and Rape ſeed. *Rapiſtrum aruorum* is called Charlock, and Carlock.

¶ *The Temperature.*

The ſeed of theſe wild kindes of Turneps as alſo the water Chadlock, are hot and drie as muſtard ſeed is. Some haue thought that Carlock hath a drying and clenſing qualitie, and ſomewhat digeſting.

¶ *The Vertues.*

Diuers vſe the ſeed of Rape in ſteed of muſtard ſeed, who either make hereof a ſauce bearing the name of muſtard, or elſe mixe it with muſtard ſeed: but this kinde of ſauce is not ſo pleaſant to the taſte, becauſe it is bitter. A

Galen writeth that theſe being eaten engender euill blood: yet *Dioſcorides* ſaith, they warme the ſtomacke and nouriſh ſomewhat. B

CHAP. 3. *Of Nauewes.*

¶ *The Kindes.*

THere be ſundrie kindes of Nape or Nauewes degenerating from the kindes of Turnep; of which ſome are of the garden; and other wilde, or of the field.

¶ *The Deſcription.*

1 NAuew gentle is like vnto Turneps in ſtalkes, floures, and ſeed, as alſo in the ſhape of the leaues, but thoſe of the Nauew are much ſmoother; it alſo differeth in the root: the Turnep is round like a globe, the Nauew root is ſomewhat ſtretched forth in length.

† 1 *Bunias.*
Nauew Gentle.

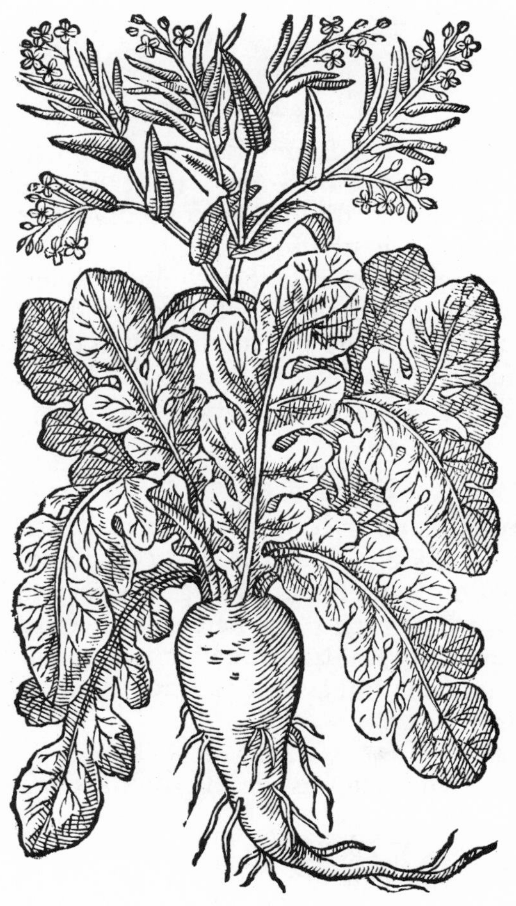

† 2 *Bunias ſylueſtris L'Obelij.*
Wilde Nauew.

The ſmall or wilde Nauew is like vnto the former, ſauing that it is altogether leſſer. The root is ſmall, ſomewhat long, with threads long and tough at the end thereof.

¶ *The Place.*

Nauew-gentle requireth a looſe and yellow mould euen as doth the Turnep, and proſpereth in a fruitfull ſoile: he is ſowen in France, Bauaria, and other places in the fields for the ſeeds ſake, as is likewiſe that wild Colewort called of the old writers *Crambe*: for the plentifull increaſe of the ſeeds bringeth no ſmall gaine to the husbandmen of that countrey, becauſe that being preſſed they yeeld an oile which is vſed not onely in lampes, but alſo in the making of ſope; for of this oile and a lie made of certaine aſhes, is boiled a ſope which is vſed in the Lowe-countries euery where to ſcoure and waſh linnen clothes. I haue heard it reported that it is at this day ſowen in England for the ſame purpoſe.

The wilde Nauew groweth vpon ditch bankes neere vnto villages and good townes, as alſo vpon freſh marſhie bankes in moſt places.

¶ *The Time.*

The Nauew is ſowen, floureth and ſeedeth at the ſame time that the Turnep doth.

¶ *The Names.*

The Nauew is called in Latine *Napus*, and alſo *Bunias*: in Greeke Βυνιάς: the Germaines call it **Steckruben**: the Brabanders, **Steckropen**: in Spaniſh, *Naps*: in Italian, *Nauo*: the Frenchmen, *Naueau*: in Engliſh, Nauew-gentle, or French Naueau. The other is called *Napus ſylueſtris*, or wild Nauew.

¶ *The Temperature and Vertues.*

The Nauew and the Turnep are all one in temperature and vertues, yet ſome ſuppoſe that the Nauew is a little drier, and not ſo ſoone concocted, nor paſſeth downe ſo eaſily, and doth withall ingender leſſe winde. In the reſt it is anſwerable to the Turnep.

A ‡ The ſeeds of theſe taken in drinke or broth are good againſt poyſon, and are vſually put into Antidotes for the ſame purpoſe. ‡.

† The figure that was in the firſt place is a kinde of the long Turnep deſcribed by me in the ſecond place of the firſt chapter of this ſecond booke. And that in the ſecond place was a leſſer kinde of the ſame.

Chap. 4. *Of Lyons Turnep, or Lyons leafe.*

Leontopetalon.
Lyons leafe.

¶ *The Deſcription.*

Lyons Turnep or Lyons leafe, hath broad leaues like vnto Coleworts, or rather like the pionyes cut and diuided into ſundry great gaſhes: the ſtalke is two foot long, thicke, and full of iuyce, diuiding it ſelfe into diuers branches or wings; in the tops whereof ſtand red floures: afterward there appeareth long cods in which lie the ſeeds like vnto tares, or wilde chichs. The root is great, bumped like a Turnep, and blacke without.

¶ *The Place.*

It groweth among corne in diuers places of Italy, in Candie alſo, and in other Prouinces towards the South and Eaſt. The right honorable Lord *Zouch* brought a plant hereof from Italy at his returne into England, the which was planted in his garden. But as farre as I doe know, it periſhed.

¶ *The Time.*

It floureth in winter, as witneſſeth *Petrus Belloniis*.

¶ *The Names.*

The Grecians call it Λεοντοπέταλον, that is, *Leonis folium*, or Lyons leafe: *Plinie* doth call it alſo *Leontopetalon*: *Apuleius*, *Leontopodion*: yet there is another plant called by the ſame name. There bee many baſtard Names giuen vnto it, as

Rapeium,

Rapeium, Papauerculum, Semen Leoninum, Pes Leoninus, and *Brumaria*: in Engliſh Lyons leafe, and Lyons Turnep.

¶ *The Temperature.*

Lyons Turnep is of force to digeſt; it is hot and drie in the third degree, as *Galen* teacheth.

¶ *The Vertues.*

The root (ſaith *Dioſcorides*) taken in wine doth helpe them that are bitten of Serpents, and it A
doth moſt ſpeedily alay the paine. It is put into gliſters which are made for them that bee tormented with the Sciatica.

Chap. 5. *Of Radiſh.*

¶ *The Kindes.*

There be ſundrie ſorts of Radiſh, whereof ſome be long and white; others long and blacke; ſome round and white; others round, or of the forme of a peare, and blacke of colour; ſome wilde, or of the field; and ſome tame, or of the garden, whereof we will intreat in this preſent chapter.

† 1 *Raphanus ſatiuus.* Garden Radiſh.

† 2 *Radicula ſatiua minor.* Small garden Radiſh.

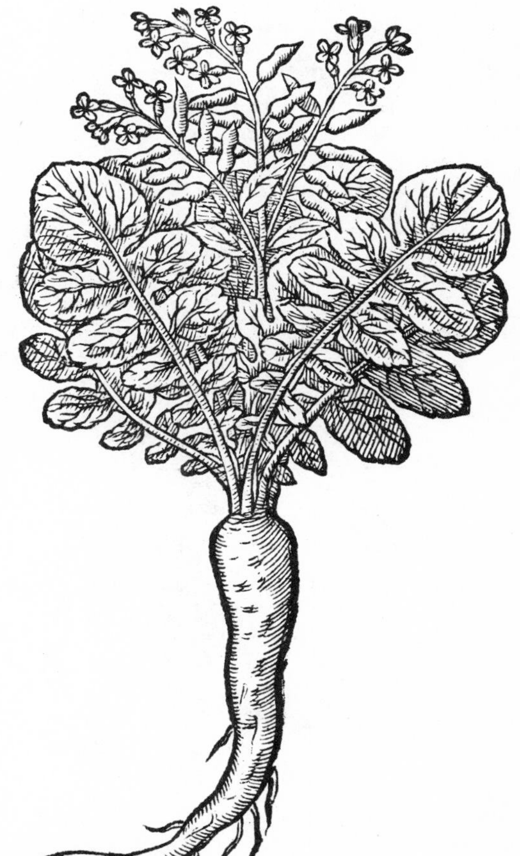

¶ *The Deſcription.*

1 The garden Radiſh ſendeth forth great and large leaues, greene, rough, cut on both ſides with deepe gaſhes, not vnlike to the garden Turnep, but greater. The ſtalkes bee round and parted into many branches; out of which ſpring ſmall floures of a light purple colour, made of foure little leaues: and when they be paſt, there doe come in place ſharpe pointed cods huft or blowne vp toward the ſtalke, full of ſpungious ſubſtance, wherein is contained the ſeed, of a light browne colour, ſomewhat greater than the ſeeds of Turneps or Coleworts. The root is groſſe, long, and white both without and within, and of a ſharpe taſte.

2 The small garden Radish hath leaues like the former, but smaller, and more brittle in handling. The stalke of two cubits high, whereon be the floures like the former. The seed is smaller, and not so sharpe in taste. The root is small, long, white both within and without, except a little that sheweth it selfe aboue the ground of a reddish colour.

3 Radish with a round root hath leaues like the garden Turnep: among which leaues springeth vp a round and smooth stalke, diuiding it selfe toward the top into two or three branches, whereon doe grow small purplish floures made of foure leaues apeece: which being past, there doe come in place small long cods puft vp or bunched in two, and sometimes three places, full of pith as the common Radish; wherein is contained the seed, somewhat smaller than the Colewort seed, but of a hotter taste. The root is round and firme, nothing waterish like the common Radish, more pleasant in taste, wholsomer, not causing such stinking belchings as the garden Radish doth.

4 The Radish with a root fashioned like a peare, groweth to the height of three or foure cubits, of a bright reddish colour. The leaues are deeply cut or iagged like those of the Turnep, somewhat rough. The floures are made of foure leaues, of a light carnation or fleshie colour. The seed is contained in small bunched cods like the former. The root is fashioned like a peare or long Turnep, blacke without and white within, of a firme and solide substance. The taste is quicke and sharpe, biting the tongue as the other kindes of Radish, but more strongly.

3 *Rhaphanus orbiculatus.*
Round Radish.

4 *Raphanus pyriformis, siue radice nigra.*
The blacke, or Peare-fashion Radish.

¶ *The Place.*

All the kindes of Radish require a loose ground which hath beene long manured and is somewhat fat. They prosper well in sandie ground, where they are not so subiect to wormes, as in other grounds.

¶ *The Time.*

These kindes of Radish are most fitly sowen after the Summer Solstice in Iune or Iulie: for being sowen betimes in the spring they yeeld not their roots so kindly nor profitably, for then they doe for the most part quickly run vp to stalke and seed, where otherwise they doe not floure and seed till the next spring following. They may be sowen ten moneths in the yeere, but as I said before, the best time is in Iune and Iulie.

¶ *The*

¶ *The Names.*

Radiſh is called in Greeke of *Theophraſtus, Dioſcorides, Galen*, and other old writers ῥαφανίς. in ſhops, *Raphanus*, and *Satiua Radicula* : in high Dutch, **Rettich** : in low Dutch, **Radys** : in French, *Raifort* : in Italian, *Raphano* : in Spaniſh, *Rauano* : in Engliſh, Radiſh, and Rabone: in the Bohemian tongue **Rzedkew.** *Cælius* affirmeth that the ſeed of Radiſh is called of *Marcellus Empericus, Bacanon*; and ſo likewiſe of *Aëtius* in the ſecond chapter of the ſecond booke of his Tetrabible : yet *Cornarius* doth not reade *Bacanon*, but *Cacanon* : The name of *Bacanum* is alſo found in *N. Myrepſus*, in the 255. Compoſition of his firſt booke.

¶ *The Temperature.*

Radiſh doth manifeſtly heat and drie, open and make thin by reaſon of the biting quality that ruleth in it. *Galen* maketh them hot in the third degree, and drie in the ſecond, and ſheweth that it is rather a ſauce than a nouriſhment.

¶ *The Vertues.*

A Radiſh are eaten raw with bread in ſtead of other food ; but being eaten after that manner, they yeeld very little nouriſhment, and that faultie and ill. But for the moſt part, they are vſed as ſauce with meates to procure appetite, and in that ſort they ingender blood leſſe faulty, than eaten alone or with bread onely : but ſeeing they be of a harder digeſtion than meates, they are alſo many times troubleſome to the ſtomacke ; neuertheleſſe, they ſerue to diſtribute and diſperſe the nouriſhment, eſpecially being taken after meat ; and taken before meat, they cauſe belchings, and ouerthrow the ſtomacke.

B Before meate they cauſe vomiting, and eſpecially the rinde: the which as it is more biting than the inner ſubſtance, ſo doth it with more force cauſe that effect if it be giuen with Oximel, which is a ſyrupe made with vineger and hony.

C Moreouer, Radiſh prouoketh vrine, and diſſolueth cluttered ſand, and driueth it forth, if a good draught of the decoction thereof be drunke in the morning. *Pliny* writeth, and *Dioſcorides* likwiſe, that it is good againſt an old cough ; and to make thin, thicke and groſſe flegme which ſticketh in the cheſt.

D In ſtead hereof the Phiſitions of our age doe vſe water diſtilled thereof: which likewiſe procureth vrine mightily, and driueth forth ſtones in the kidnies.

E The root ſliced and laid ouer night in white or Rheniſh wine, and drunke in the morning, driueth out vrine and grauell mightily, but in taſte and ſmell it is very lothſome.

F The root ſtamped with hony and the powder of a ſheepes heart dried, cauſeth haire to grow in ſhort ſpace.

G The ſeed cauſeth vomite, prouoketh vrine : and being drunke with honied vineger, it killeth and driueth forth wormes.

H The root ſtamped with the meale of Darnell and a little white wine vineger, taketh away all blew and blacke ſpots, and bruſed blemiſhes of the face.

I The root boiled in broth, and the decoction drunke, is good againſt an old cough : it moueth womens ſickneſſe, and cauſeth much milke.

† Thoſe figures that were in the firſt and ſecond place, were varietyes of the long Turnep deſcribed in the ſecond place, in the firſt Chapter of this ſecond booke.

CHAP: 6. Of wilde Radiſh.

¶ *The Deſcription.*

1 Wilde Radiſh hath a ſhorter narrower leafe than the common Radiſh, and more deeply cut or iagged, almoſt like the leaues of Rocket, but much greater. The ſtalke is ſlender and rough, of two cubits high, diuided toward the top into many branches. The floures are ſmall and white : the cod is long, ſlender, aud ioynted, wherein is the ſeed. The root is of the bigneſſe of the finger, white within and without, of a ſharpe and biting taſte.

2 The water Radiſh hath long and broad leaues, deeply indented or cut euen to the middle rib. The ſtalke is long, weake, and leaneth this way and that way, being not able to ſtand vpright without a prop, in ſo much that yee ſhall neuer find it, no not when it is very young, but leaning down vpon the mud or mire where it groweth. The floures grow at the top made of foure ſmall yellow leaues. The root is long, ſet in ſundrie ſpaces with ſmall fibres or threds like the rowell of a ſpur, hot and burning in taſte more than any of the garden Radiſhes.

¶ *The Place.*

The firſt growes vpon the borders of bankes and ditches caſt vp, and in the borders of fields.

The

The second growes in ditches, standing waters, and riuers; as on the stone wall that bordereth vpon the riuer Thames by the Sauoy in London.

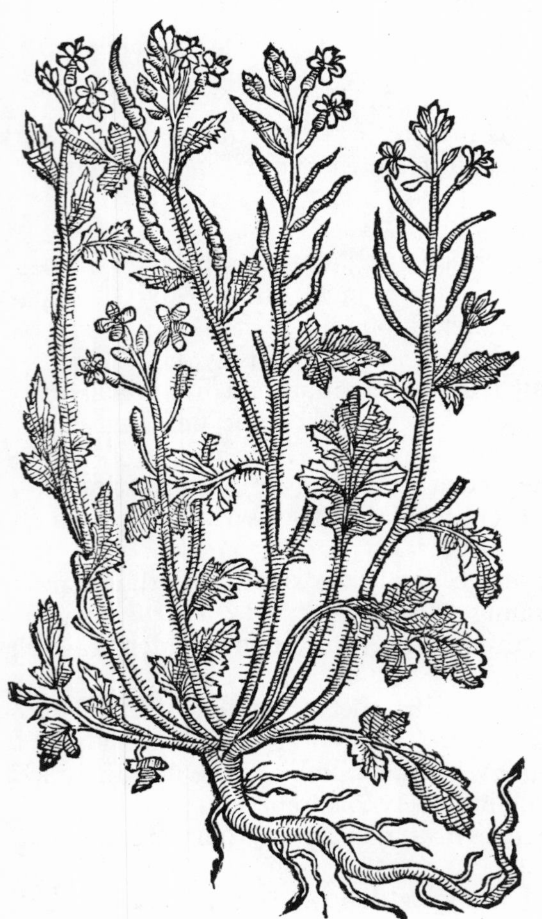
1 *Raphanus syluestris.*
Wilde Radish.

2 *Raphanus aquaticus.*
Water Radish.

¶ *The Time.*

They floure in Iune, and the seed is ripe in August.

¶ *The Names.*

† The first of these is *Rapistrum flore albo Erucæ folijs*, of *Lobell*: *Armoratia*, or *Rapistrum album* of *Tabernamontanus*: and *Raphanus sylvestris*, of our Author: in English, wilde Radish.

The second is *Radicula sylvestris* of *Dodonæus*: and *Rhaphanus aquaticus*, or *palustris* of others: in English, water Radish.

¶ *The Temperature.*

The wilde Radishes are of like temperature with the garden Radish, but hotter and drier.

¶ *The Vertues.*

A *Dioscorides* writeth, that the leaues are receiued among the pot herbes, and likewise the boiled root, which as he saith, doth heate, and prouoke vrine.

Chap. 7. *Of Horse Radish.*

¶ *The Description.*

1 Horse Radish bringeth forth great leaues, long, broad, sharpe pointed and snipped about the edges, of a deepe greene colour like those of the great garden Docke, called, of some Monkes Rubarbe, of others Patience, but longer and rougher. The stalke is slender and brittle, bearing at the top small white floures: which being past, there follow small cods, wherein is the seed. The root is long and thicke, white of colour, in taste sharpe, and very much biting the tongue like mustard.

2 Dittander or pepperwort, hath broad leaues, long, and sharpe pointed, of a blewish greene colour like woad, somewhat snipt or cut about the edges like a sawe. The stalke is round and tough:

tough: vpon the branches whereof grow little white floures. The root is long and hard, creeping farre abroad in the ground, in such sort that when it is once taken in a ground, it is not possible to root it out, for it will vnder the ground creepe and shoot vp and bud forth in many places farre abroad. The root also is sharp and biteth the tongue like pepper, whereof it tooke the name pepperwort.

‡ 3 This which we giue you in the third place hath a small fibrous root, the stalke growes vp to the height of two cubits, and it is diuided into many branches furnished with white floures, after which follow seeds like in shape and taste to Thlaspi, or Treacle mustard. The leaues are somewhat like those of Woad. This is nourished in some Gardens of the Low Countryes, and *Lobell* was the first that gaue the figure hereof, and that vnder the same title as wee here giue you it. ‡.

1 *Raphanus rusticanus.*
Horse Radish.

2 *Raphanus sylvestris Offic. Lepidium Æginetæ Lob.*
Dittander, and Pepperwort.

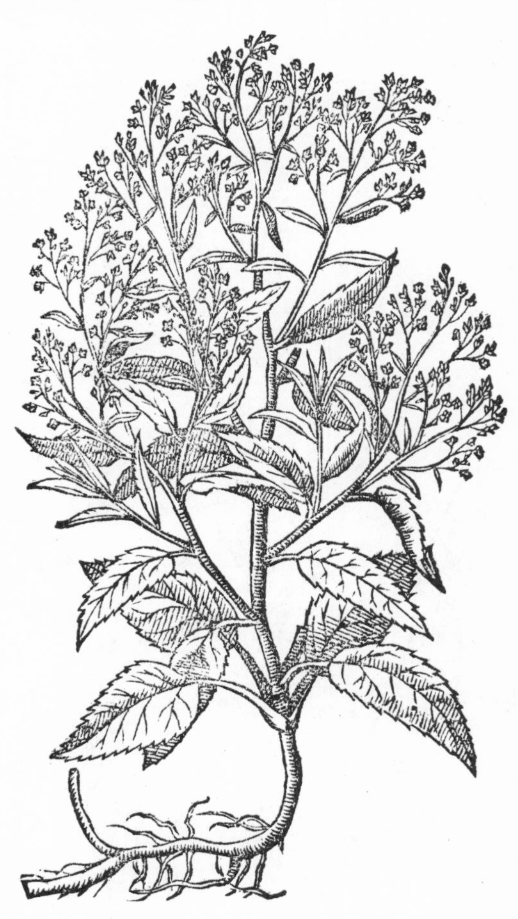

¶ *The Place.*

Horse Radish for the most part groweth and is planted in gardens, yet haue I found it wilde in sundrie places, as at Namptwich in Cheshire, in a place called the Milne eye, and also at a small village neere London called Hogsdon, in the field next vnto a farme house leading to Kings-land, where my very good friend master *Bredwell* practitioner in Phisick, a learned and diligent searcher of Simples, and master *William Martin* one of the fellowship of Barbers and Chirurgians, my deere aud louing friend, in company with him found it, and gaue me knowledge of the place, where it flourisheth to this day.

Dittander is planted in gardens, and is to be found wild also in England in sundry places, as at Clare by Ouenden in Essex, at the Hall of Brinne in Lancashire, and neere vnto Excester in the West parts of England. It delighteth to grow in sandie and shadowie places somewhat moist.

¶ *The Time.*

Horse Radish for the most part floureth in Aprill or May, and the seed is ripe in August, and that so rare or seldome seene, as that *Petrus Placentius* hath written, that it bringeth forth no seed at all. Dittander floures in Iune and Iuly.

¶ *The Names.*

Horse Radish is commonly called *Raphanus rusticanus*, or *Magnus*, and of diuers simply *Raphanus sylvestris:*

syluestris: of the high Dutch men, **Merrettich, Krain or Kren**: in French, *Grand raifort*: of the low Germaines, **Merradus**: in English, mountaine radish, Great Raifort, and Horse Radish. It is called in the North part of England, Redcole.

Diuers thinke that this Horse Radish is an enemy to Vines, and that the hatred betweene them is so great, that if the roots hereof be planted neere to the Vine it bendeth backward from it, as not willing to haue fellowship with it.

It is also reported that the root hereof stamped, and cast into good and pleasant wine, doth forthwith turne it into vineger: but the old writers doe ascribe this enmity to the vine and *Brassica*, our coleworts, which the most ancients haue named ῥάφανος.

Dittander is described of *Pliny* by the name of *Lepidium* in his 19. booke, 9. Chapter: likewise *Ægineta* maketh mention of this plant, by the name *Lepidium*: in shops, *Raphanus syluestris*, and *Piperitis*: the Germans call it, **Pfefferkraut**: the lowe Dutch men, **Pepper cruyt**: the English men, Dittander, Dittany, and Pepperwort.

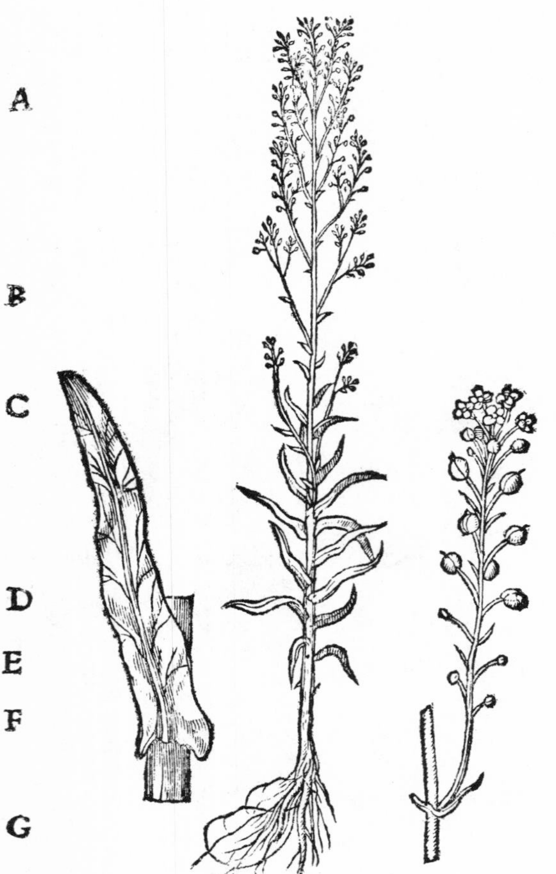

3 *Lepidium Annum.*
Annuall Dittander.

¶ *The Temperature.*

These kindes of wilde Radishes, are hot and drie in the third degree: they haue a drying and clensing quality, and somewhat digesting.

¶ *The Vertues.*

A Horse Radish stamped with a little vineger put thereto, is commonly vsed among the Germanes for sauce to eate fish with, and such like meates, as we doe mustard; but this kinde of sauce doth heate the stomacke better, and causeth better digestion than mustard.

B Oximel or syrupe made with vineger and honie, in which the rindes of Horse radish haue beene infused three dayes, causeth vomit, and is commended against the quartaine ague.

C The leaues boiled in wine, and a little oile oliue added thereto and laid vpon the grieued parts in manner of a Pultis, doe mollifie and take away the hard swellings of the liuer and milte; and being applied to the bottome of the belly is a remedie for the strangurie.

D It profiteth much in the expulsion of the secondine or after-birth.

E It mittigateth and asswageth the paine of the hip or haunch, commonly called Sciatica.

F It profiteth much against the collicke, strangurie, and difficultie of making water, vsed in stead of mustard as aforesaid.

G The root stamped and giuen to drinke, killeth the wormes in children: the iuyce giuen doth the same: an ointment made thereof, doth the like, being annointed vpon the belly of the child.

H The leaues of Pepperwort but especially the rootes, be extreame hot, for they haue a burning, and bitter taste. It is of the number of scorching and blistring simples, saith *Pliny* in his 20. booke, the 17. chap and therefore by his hot qualitie, it mendeth the skin in the face, and taketh away scabs, scarres, and manginesse, if any thing remaine after the healing of vlcers and such like.

Chap. 8. *Of Winter Cresses.*

¶ *The Description.*

The Winter Cresses hath many greene, broad, smoothe and flat leaues like vnto the common turneps, whose stalkes be round, and full of branches, bringing forth at the top small yeilow floures: after them doe follow small cods, wherein is conteined small reddish seed.

¶ The

1 *Barbarea.*
Winter Cresses.

¶ *The Place.*

It groweth in gardens among pot herbes, and very common in the fields, neere to pathes and high wayes, almost euery where.

¶ *The Time.*

This herbe is green all winter long, it floureth in May, and seedeth in Iune.

¶ *The Names.*

Winter Cresse is called of the Latines, *Cardamum*, or *Nasturtium Hibernum*, of some, *Barbarea*, and *Pseudobunium*: the Germanes call it **S. Barberen kraut**: in lowe Dutch, **Winter kersse.**

It seemeth to be *Dioscorides* his ψευδοβούνιον, that is to say, false or bastard *Bunium*: in English, winter Cresses, or herbe Saint Barbara.

¶ *The Nature.*

This herbe is hot and drie in the second degree.

¶ *The Vertues.*

The seed of winter Cresse causeth one to make water, and driueth forth grauell, and helpeth the strangurie. A

The iuyce thereof mundifieth corrupt and filthy vlcers, being made in forme of an vnguent with waxe, oyle, and turpentine. B

In winter when salad herbes bee scarce, this herbe is thought to be equall with Cresses of the garden, or Rocket. C

This herbe helpeth the scuruie, being boiled among scuruie grasse, called in Latine *Cochlearia*, causing it to worke the more effectually. D

CHAP. 9. *Of Mustard.*

¶ *The Description.*

1 THe tame or garden Mustard, hath great rough leaues like to those of the Turnep, but rougher and lesser. The stalke is round, rough, and hairie, of three cubits high, diuided into many branches, whereon doe grow small yellow floures, and after them long cods, slender and rough, wherein is contained round seed bigger then Rape seed, of colour yellow, of taste sharpe, and biting the tongue as doth our common field mustard.

‡ 2 Our ordinary Mustard hath leaues like Turneps, but not so rough, the stalkes are smooth, and grow sometimes to three, foure, or fiue cubits high, they haue many branches, and the leaues vpon these branches, especially the vppermost, are long and narrow, and hang downeward on small stalkes; the cods are short, and lie flat and close to the branches, and are somewhat square; the seed is reddish or yellow. ‡

3 The other tame Mustard is like to the former in leaues, and branched stalkes, but lesser, and they are more whitish and rough. The floures are likewise yellow, and the seed browne like the Rape seed, which is also not a little sharpe or byting.

‡ 4 This which I giue you bere being the *Sinapi satiuum alterum*, of *Lobel*; and the *Sinapi album* of the shops, growes but low, and it hath rough crooked cods, and whitish seeds; the stalks, floures, and leaues, are mnch like the first described. ‡.

5 The wilde Mustard hath leaues like those of shepheards purse, but larger, and more deeply indented, with a stalke growing to the height of two foot, bearing at the top small yellow floures made of foure leaues: the cods be small and slender, wherein is contained reddish seed, much smaller than any of the others, but not so sharpe or biting.

† 1 *Sinapi*

† 1 *Sinapi ſatiuum.*
Garden Muſtard.

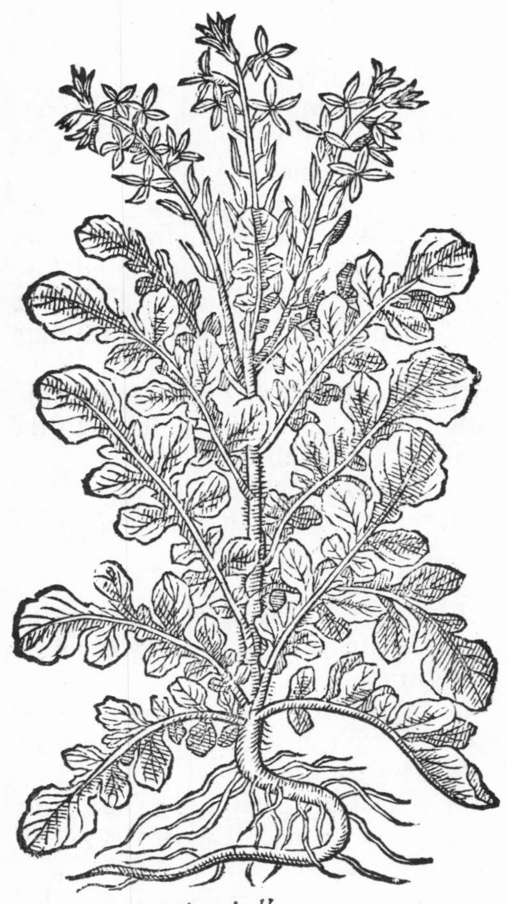

‡ 3 *Sinapi ſatiuum alterum, Dod.*
Field Muſtard.

‡ 4 *Sinapi album.*
White Muſtard.

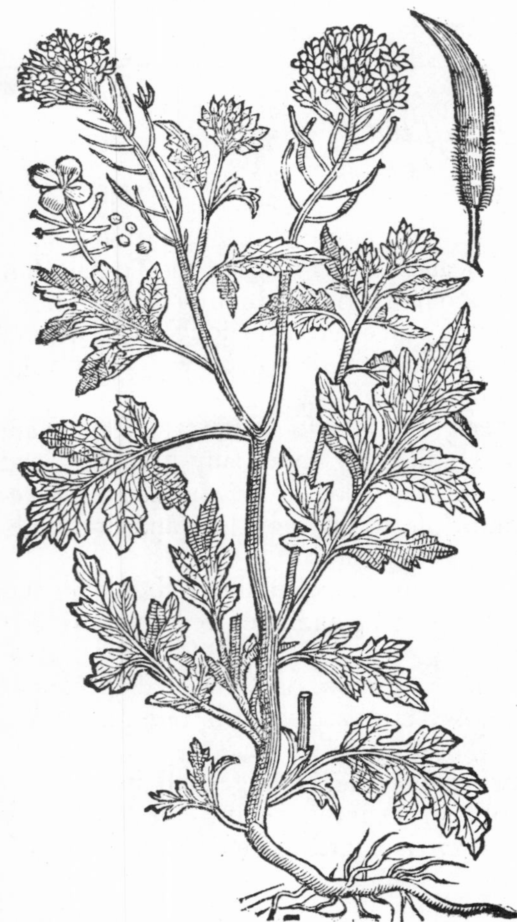

† 5 *Sinapi ſylueſtre minus.*
Small wilde Muſtard.

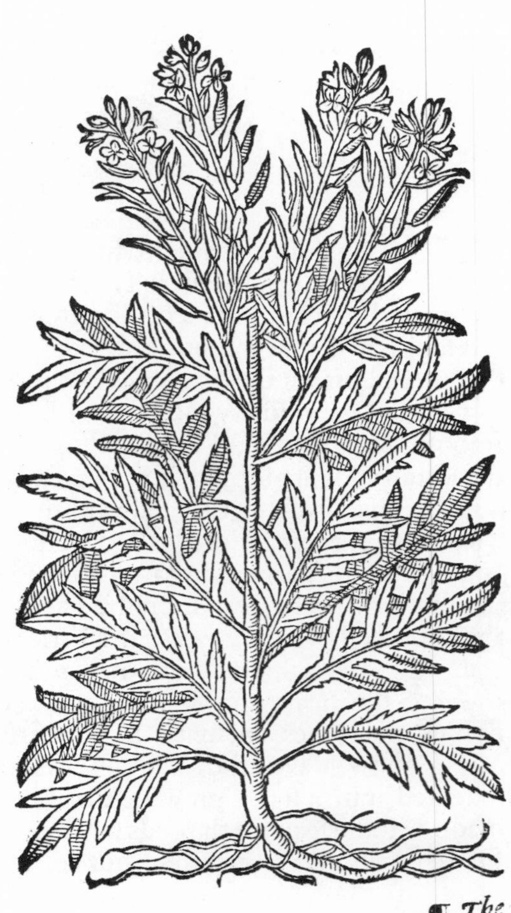

¶ *The*

¶ The Place.

‡ Our ordinarie Muſtard (whoſe deſcription I haue added) as alſo the wilde and ſmall grow wilde in many places of this kingdome, and may all three be found on the bankes about the backe of Old-ſtreet, and in the way to Iſlington. ‡

¶ The Time.

Muſtard may be ſowen in the beginning of the Spring: the ſeed is ripe in Iuly or Auguſt: It commeth to perfection the ſame yeare that it is ſowen.

¶ The Names.

The Greekes call Muſtard, σίναπι: the Athenians called it νᾶπυ: the Latines, *Sinapi*: the rude and barbarous, *Sinapium*: the Germanes, **Senff**: the French, *Seneue* and *Mouſtarde*: the low-Dutchmen, **Mostaert saet**: the Spaniards, *Moſtaza*, and *Moſtilla*: the Bohemians, *Horcice*: *Pliny* calls it *Thlaſpi*, whereof doubtleſſe it is a kinde: and ſome haue called it *Saurion*.

‡ Theſe kindes of Muſtard haue beene ſo briefely treated of by all Writers, that it is hard to giue the right diſtinctions of them, and a matter of more difficultie than is expected in a thing ſo vulgarly knowne and vſed: I will therefore endeauour in a few words to diſtinguiſh thoſe kindes of muſtard which are vulgarly written of.

1 The firſt is *Sinapi primum* of *Matthiolus* and *Dodonæus*; and *Sinapi ſatiuum Erucæ aut Rapi folio* of *Lobel*.

2 The ſecond I cannot iuſtly referre to any of thoſe which are written of by Authours; for it hath not a cod like Rape, as *Pena* and *Lobel* deſcribe it; nor a ſeed bigger than it, as *Dodonæus* affirmeth; yet I ſuſpect, and almoſt dare affirme that it is the ſame with the former mentioned by them, though much differing from their figures and deſcription.

3 The third (which alſo I ſuſpect is the ſame with the fourth) is *Sinapi alterum* of *Matthiolus*, and *Sinapi agreſte Apij, aut potius Laueris folio*, of *Lobel*: and *Sinapi ſatiuum alterum* of *Dodonæus*.

4 The fourth is by *Lobel* called *Sinapi alterum ſatiuum*; and this is *Sinapi album Officinarum*, as *Pena* and *Lobel* affirme, *Aduerſ. pag.* 68.

5 The fifth is *Sinapi ſylueſtre* of *Dodonæus*: and *Sinapi ſylueſtre minus Burſæ paſtoris folio*, of *Lobel*. It is much like Rocket, and therefore *Bauhine* fitly calls it *Sinapi Erucæ folio*: in Engliſh it may be called Small wilde Muſtard. ‡

¶ The Temperature.

The ſeed of Muſtard, eſpecially that which we chiefely vſe, doth heat and make thinne, and alſo draweth forth. It is hot and dry in the fourth degree, according to *Galen*.

¶ The Vertues.

A The ſeed of Muſtard pounded with vineger, is an excellent ſauce, good to be eaten with any groſſe meates either fiſh or fleſh, becauſe it doth helpe digeſtion, warmeth the ſtomacke, and prouoketh appetite.

B It is giuen with good ſucceſſe in like manner to ſuch as be ſhort winded, and are ſtopped in the breaſt with tough flegme from the head and braine.

C It appeaſeth the tooth-ache being chewed in the mouth.

D They vſe to make a gargariſme with honey, vineger, and muſtard ſeed, againſt the tumours and ſwellings of the Vuula, and the almonds about the throat and root of the tongue.

E Muſtard drunke with water and honey prouoketh the termes and vrine.

F The ſeed of muſtard beaten and put into the noſthrils, cauſeth ſneeſing, and raiſeth women ſicke of the mother out of their fits.

G It is good againſt the falling ſickeneſſe, and ſuch as haue the Lithargie, if it be laid plaiſterwiſe vpon the head (after ſhauing) being tempered with figs.

H It helpeth the Sciatica, or ache in the hip or huckle bone: it alſo cureth all manner of paines proceeding of a cold cauſe.

I It is mixed with good ſucceſſe with drawing plaiſters, and with ſuch as waſte and conſume nodes and hard ſwellings.

K It helpeth thoſe that haue their haire pulled off; it taketh away the blew and blacke marks that come of bruiſings.

L ‡ The ſeed of the white Muſtard is vſed in ſome Antidotes, as *Electuarium de ouo, &c.*

† The three figures in the former edition were all falſe: The firſt was of *Barbarea*, deſcribed in the precedent chapter: The ſecond, of *Eruca aquatica maior* of *Tabern.* The third, of *Eruca aquat. minor, Tab.*

Chap. 10. *Of Rocket.*

¶ *The Kindes.*

THere be sundry kindes of Rocket; some tame, or of the garden; some wilde, or of the field; some of the water, and of the sea.

† 1 *Eruca satiua.*
Garden Rocket.

2 *Eruca syluestris.*
Wilde Rocket.

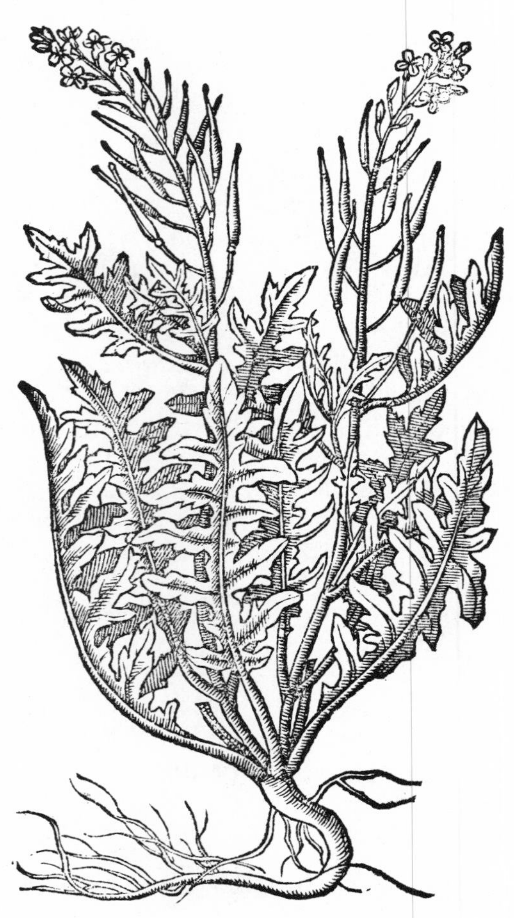

¶ *The Description.*

1 GArden Rocket, or Rocket gentle, hath leaues like those of Turneps, but not neere so great nor rough. The stalks rise vp of a cubit, & somtimes two cubits high, weak and brittle; at the top whereof grow the floures of a whitish colour, and sometimes yellowish; which being past, there do succeed long cods, which containe the seed, not vnlike to rape seed, but smaller.

2 The common Rocket, which some keepe in Gardens, and which is vsually called the wilde Rocket, is lesser than the Romane Rocket, or Rocket-gentle, the leaues and stalkes narrower, and more iagged. The floures be yellow, the cods also slenderer, the seed thereof is reddish, and biteth the tongue.

3 This kinde of Rocket hath long narrow leaues almost such as those of Tarragon, but thicker and fatter, resembling rather the leaues of Myagrum, altogether vnlike any of the rest of the Rockets, sauing that the branch, floure, and seed are like the garden Rocket.

4 There is another kinde of Rocket, thought by that reuerend and excellent Herbarist *Carolus Clusius* to be a kinde of Cresses; if not Cresses it selfe, yet cousine germane at the least. Vnto whose censure *Lobelius* is indifferent, whether to call it Rocket with thinne and narrow leaues, or to call it Cousine to the kindes of Cresses, hauing the taste of the one, and the shape of the other. The leaues are much diuided, and the floures yellow.

5 There is is a wild kind of Sea-Rocket which hath long weake and tender branches trailing vpon

vpon the ground, with long leaues like vnto common Rocket, or rather Groundſwell, hauing ſmall and whitiſh blew floures; in whoſe place commeth ſmall cods, wherein is contained ſeed like that of Barley.

‡ 6 Beſides theſe there is another plant, whoſe figure which here I giue was by our Author formerly ſet forth in the precedent chapter, vnder the title of *Sinapi ſylueſtre*; together with a large kinde thereof, vnder the name of *Sinapi ſatiuum alterum*. Now I will onely deſcribe the later, which I haue ſometimes found in wet places: The root is woody: the ſtalke ſome foot long, creſted, and hauing many branches, lying on the ground: the leafe is much diuided, and that after the manner of the wilde Rocket: the floures are of a bright yellow, and are ſucceeded by ſhort crooked cods, wherein is contained a yellowiſh ſeed. ‡

† 3 *Eruca ſylueſtris anguſtifolia.*
Narrow leaued wilde Rocket.

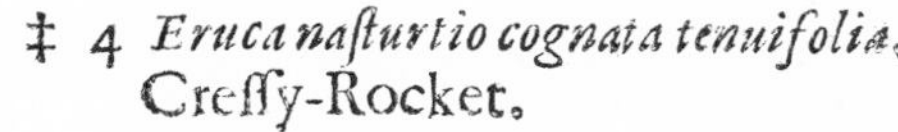

‡ 4 *Eruca naſturtio cognata tenuifolia.*
Creſſy-Rocket.

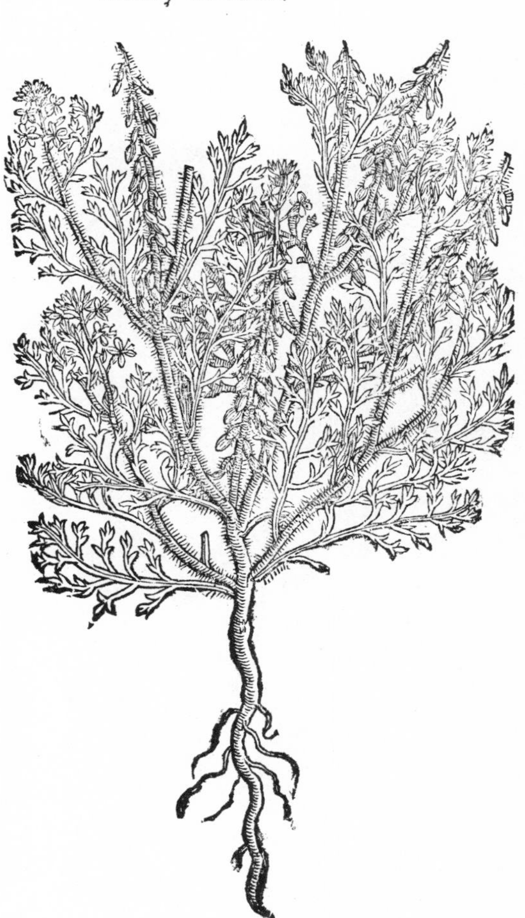

¶ *The Place.*

Romane Rocket is cheriſhed in Gardens.

Common or wilde Rocket groweth in moſt gardens of it ſelfe: you may ſee moſt bricke and ſtone walls about London and elſewhere couered with it.

The narrow leaued Rocket groweth neere vnto water ſides, in the chinkes and creuiſes of ſtone walls among the morter. I found it as ye go from Lambeth bridge to the village of Lambeth, vnder a ſmall bridge that you muſt paſſe ouer hard by the Thames ſide.

I found Sea Rocket growing vpon the ſands neere vnto the ſea in the Iſle of Thanet, hard by a houſe wherein Sir *Henry Criſpe* did ſometimes dwell, called Queakes houſe.

¶ *The Time.*

Theſe Kindes of Rocket floure in the moneths of Iune and Iuly, and the ſeed is ripe in September.

The Romane Rocket dieth euery yeare, and recouereth it ſelfe againe by the falling of his owne ſeed.

¶ *The Names.*

Rocket is called in Greeke, εὔζωμον : in Latine, *Eruca* : in high Dutch, **Rauckenkraut** : in French, *Roquette* : in Low-Dutch, **Rakette** : in Italian, *Ruchetta* : in Spaniſh, *Oruga*, in Engliſh, Rocket, and Racket. The Poets do oft times name it *Herba ſalax* : *Eruca* doth ſignifie likewiſe a certaine canker worme, which is an enemie to pot-herbes, but eſpecially to Coleworts.

‡ The firſt is called *Eruca ſatiua*, or *Hortenſis major* : Great Garden Rocket.

2 The ſecond, *Eruca ſylueſtris* : Wilde Rocket.

3 This third is by *Lobel* called *Eruca ſylueſtris anguſtifolia* : Narrow leaued wilde Rocket.

4 *Cluſius* fitly calls this, *Naſturtium ſylueſtre* : and he reprehendeth *Lobel* for altering the name into *Eruca Naſturtio cognata tenuifolia* : Creſſy-Rocket.

5 The fifth is *Eruca marina*, (thought by *Lobel* and others to be *Cakile Serapionis*,) Sea Rocket.

6 *Eruca aquatica* : Water Rocket.

‡ 5 *Eruca marina.*
Sea Rocket.

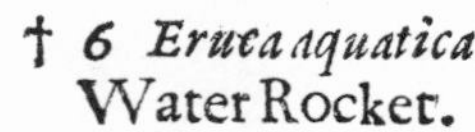

† 6 *Eruca aquatica.*
Water Rocket.

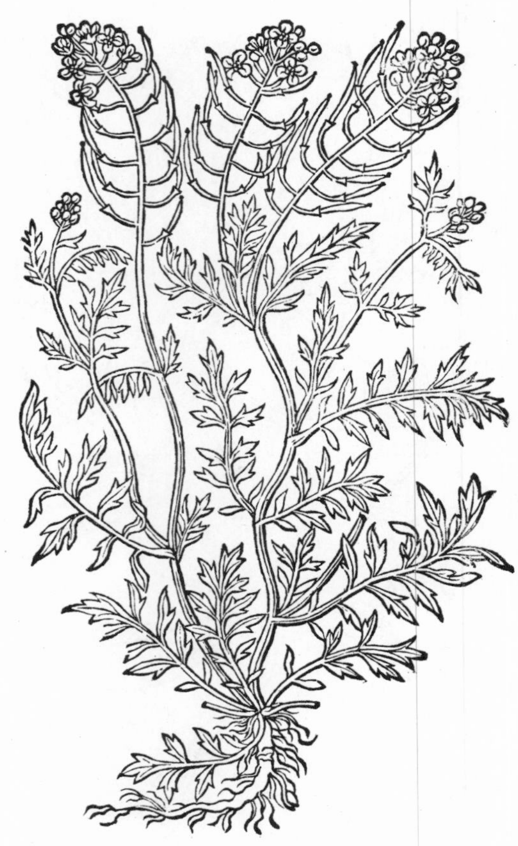

¶ *The Temperature.*

Rocket is hot and dry in the third degree, therefore ſaith *Galen* it is not fit nor accuſtomed to be eaten alone.

¶ *The Vertues.*

A Rocket is a good ſallet herbe, if it be eaten with Lettuce, Purſlane, and ſuch cold herbes; for being ſo eaten it is good and wholeſome for the ſtomacke, and cauſeth that ſuch cold herbes do not ouer-coole the ſame : otherwiſe, to be eaten alone, it cauſeth head-ache, and heateth too much.

B The vſe of Rocket ſtirreth vp bodily luſt, eſpecially the ſeed.

C It prouoketh vrine, and cauſeth good digeſtion.

D *Pliny* reporteth, That whoſoeuer taketh the ſeed of Rocket before he be whipt, ſhall be ſo hardened, that he ſhall eaſily endure the paines.

E The root and ſeed ſtamped, and mixed with Vineger and the gall of an Oxe, taketh away freckles, lentiles, blacke and blew ſpots, and all ſuch deformities of the face.

† The figure that was in the third place, vnder the title of *Eruca ſyl. anguſtifolia*, is of the ſame plant that in the Chapter of *Turritis* is called *Camelina*, where you ſhall finde it treated of at large. And that in the firſt place is *Eryſimum ſecundum* of *Tabern.* and I queſtion whither it be not of *Sinapi ſylueſtre minus*.

CHAP. 11. *Of Tarragon.*

Draco herba.
Tarragon.

¶ *The deſcription.*

TArragon the ſallade herbe hath long and narrow leaues of a deepe greene colour, greater and longer than thoſe of common Hyſſope, with ſlender brittle round ſtalkes two cubites high: about the branches whereof hang little round flowers, neuer perfectly opened, of a yellow colour mixed with blacke, like thoſe of common VVormewood. The root is long and fibrous, creeping farre abroad vnder the earth, as doe the rootes of Couch-graſſe, by which ſprouting forth it increaſeth, yeelding no ſeede at all, but as it were a certaine chaffie or duſtie matter that flieth away with the winde.

¶ *The place.*

Tarragon is cheriſhed in gardens, and is encreaſed by the young ſhootes: *Ruellius* and ſuch others haue reported many ſtrange tales hereof ſcarſe worth the noting, ſaying, that the ſeed of flaxe put into a radiſh roote or ſea Onion, and ſo ſet, doth bring forth this herbe Tarragon.

¶ *The time.*

It is greene all Summer long, and a great part of Autumne, and floureth in Iuly.

The names.

It is called in Latine, *Draco, Dracunculus hortenſis*, and *Tragum vulgare* by *Cluſius*; Of the Italians, *Dragoncellum*; in French, *Dragon*, in Engliſh, Tarragon.

It is thought to be that *Tarchon* which *Auicen* mentioneth in his *686.* chapter: but he writeth ſo little thereof, as that nothing can certainly be affirmed of it. *Simeon Sethi* the Greeke alſo maketh mention of *Tarchon*.

¶ *The temperature and vertues.*

Tarragon is hot and drie in the third degree, and not to be eaten alone in ſallades, but ioyned with other herbes, as Lettuce, Purſlain, and ſuch like, that it may alſo temper the coldnes of them, like as Rocket doth, neither doe we know what other vſe this herbe hath.

CHAP. 12. *Of garden Creſſes.*

¶ *The deſcription.*

1 GArden Creſſes or Towne Creſſes hath ſmall narrow iagged leaues, ſharpe and burning in taſte. The ſtalks be round, a cubite high, which bring forth many ſmall white flowers, and after little flat huskes or ſeede veſſels, like to thoſe of ſhe pheards purſe, wherein are contained ſeeds of a browne reddiſh colour. The roote dieth when the ſeede is ripe.

2 There is another kinde in taſte like the former, but in leaues farre different, which I recouered of ſeedes, ſent me from *Robinus* dwelling in Paris. The ſtalkes riſe vp to the height of a foot, garniſhed with many broad leaues deepely cut or indented about the edges: the middle of the leafe is deckt and garniſhed with many little ſmall leaues or rather ſhreds of leaues, which make the ſame like a curlde fanne of feathers. The ſeede is like the former in ſhape.

3 Spaniſh Creſſes riſeth forth of the ground like vnto Baſill, afterwards the leaues grow larger and broader, like thoſe of Marigolds; among the which riſeth vp a crooked lymmer ſtalke, where-

whereupon do grow small tufts or spokie rundles of white flowers. The seede followeth, browne of colour, and bitter in taste. The whole plant is of a loathsome smell and sauour.

4 Stone-Cresse groweth flat vpon the ground, with leaues iagged and cut about the edges like the oake leafe, resembling well the leaues of shepheardes purse. I haue not seene the flowers, and therefore they be not exprest in the figure; notwithstanding it is reported vnto me, that they bee small and white of colour, as are those of the garden Cresses. The seed is contained in small pouches or seede vessels, like those of Treacle mustard or Thlaspi.

¶ *The Place.*

Cresses are sowne in gardens, it skils not what soile it be; for that they like any ground, especially if it be well watered. ‡ M. *Bowles* found the fourth growing in Shropshire in the fields about Birch in the parish of Elesmere, in the grounds belonging to M. *Richard Herbert*, and that in great plenty. ‡

¶ *The Time.*

It may be sowne at any time of the yeere, vnlesse it be in Winter; it groweth vp quickly, and bringeth forth betimes both stalke and seede: it dieth euery yeere, and recouereth it selfe of the fallen or shaken seed.

1 *Nasturtium hortense.*
Garden Cresses.

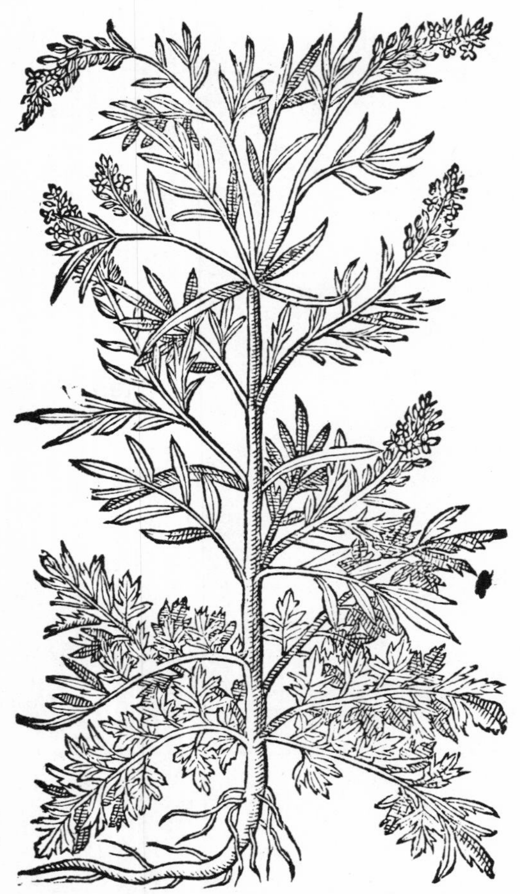

¶ *The Names.*

Cresses is called in Greeke κάρδαμον: in Latine *Nasturtium*; in English Cresses: the Germaines call it **Kersse**: and in French, *Cresson*: the Italians, *Nasturtio*, and *Agretto*: of some, towne Cresses, and garden Karsse. It is called *Nasturtium*, as *Varro* and *Plinie* thinke *à narribus torquendis*, that is to say, of writhing the nosthrils, which also by the loathsome smell and sharpnesse of the seede doth cause sneesing. ‡ The first is called *Nasturtium hortense*, Garden Cresses. 2 *Nasturtium hortense crispum*, Garden Cresses with crispe, or curled leaues. 3 *Nasturtium Hispanicum*, or *Latifolium*; Spanish Cresses, or Broad-leaued Cresses. 4 This is *Nasturtium petræum* of *Tabernamontanus* (and not of *Lobell*, as our Author termed it.) Stone Cresses. ‡

¶ *The Temperature.*

The herbe of garden Cresses is sharpe and biting the tongue; and therefore it is very hot and drie, but lesse hot whilest it is yong and tender, by reason of the waterie moisture mixed therewith, by which the sharpenesse is somewhat allaied.

The seede is much more biting then the herbe, and is hot and drie almost in the fourth degree.

¶ *The Vertues.*

A *Galen* saith that the Cresses may be eaten with bread *Veluti obsonium*, and so the Antient Spartanes vsually did; and the low-Countrie men many times doe, who commonly vse to feed of Cresses with bread and butter. It is eaten with other sallade hearbes, as Tarragon and Rocket: and for this cause it is chiefely sowen.

B It is good against the disease which the Germaines call **Scorbuch** and **Scorbuye**: in Latine, *Scorbutus*: which we in England call the Scuruie, and Scurby, and vpon the seas the Skyrby: it is as good and as effectuall as the Scuruie grasse, or water Cresses.

C *Dioscorides* saith, if the seed be stamped and mixed with hony, it cureth the hardnesse of the milt: with Vineger and Barley meale parched it is a remedie against the Sciatica, and taketh away hard swellings and inflammations. It scoureth away tetters, mixed with brine: it ripeneth felons, called in Greeke, δοθιῆνες: it forcibly cutteth and raiseth vp thicke and tough humors of the chest, if it be mixed with things proper against the stuffing of the lungs.

Dioscorides saith it is hurtfull to the stomacke, and troubleth the belly.

It

3 *Nasturtium Hispanicum.*
Spanish Cresses.

4 *Nasturtium Petreum.*
Stone Cresses.

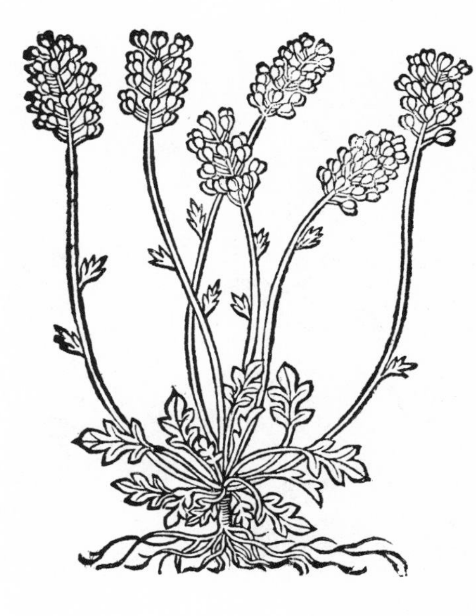

It driueth forth wormes, bringeth downe the floures, killeth the child in the mothers womb, D
and prouoketh bodily lust.

Being inwardly taken, it is good for such as haue fallen from high places: it dissolueth clutte- E
red bloud, and preuenteth the same that it do not congeale and thicken in any part of the body: it procureth sweat, as the later Physitions haue found and tried by experience.

CHAP. 13. *Of Indian Cresses.*

¶ *The Description.*

CResses of India haue many weake and feeble branches, rising immediately from the ground, dispersing themselues far abroade; by meanes whereof one plant doth occupie a great circuit of ground, as doth the great Bindeweede. The tender stalkes diuide themselues into sundry branches, trailing likewise vpon the ground, somewhat bunched or swollen vp at euery ioint or knee, which are in colour of a light red, but the spaces betweene the ioints are greene. The leaues are round like wall peniwort, called Cotyledon, the footestalke of the leafe commeth forth on the backeside almost in the middest of the leafe, as those of Frogbit, in taste and smell like the garden Cresses. The flowers are dispersed throughout the whole plant, of colour yellow, with a crossed starre ouerthwart the inside, of a deepe orange colour; vnto the backe part of the same doth hang a taile or spurre, such as hath the Larkes heele, called in Latine *Consolida Regalis*, but greater, and the spurre or heele longer; which being past there succeed bunched and knobbed cods or seede vessels, wherein is contained the seede, rough, browne of colour, and like vnto the seedes of the beete, but smaller.

¶ *The Place.*

The seedes of this rare and faire plant came first from the Indies into Spaine, and thence into France and Flanders, from whence I receiued seede that bore with me both flowers and seede, especially those I receiued from my louing friend *Iohn Robin* of Paris.

¶ *The Time.*

The seedes must be sowen in the beginning of Aprill, vpon a bed of hot horse dung, and some fine

fine sifted earth cast thereon of an handfull thicke. The bed must be couered in sundry places with hoopes or poles, to sustaine the mat or such like thing that it must be couered with in the night, and layd open to the Sunne in the day time. The which being sprung vp, and hauing gotten three leaues, you must replant them abroad in the hottest place of the garden, and most fine and fertile mold. Thus may you do with Muske-Melons, Cucumbers, and all cold fruits that require haste; for that otherwise the frost will ouertake them before they come to fruit-bearing.

‡ They may also be sowen in good mold like as other seeds, and vsually are. ‡

Nasturtium Indicum cum flore & semine.
Indian Cresses with floure and seed.

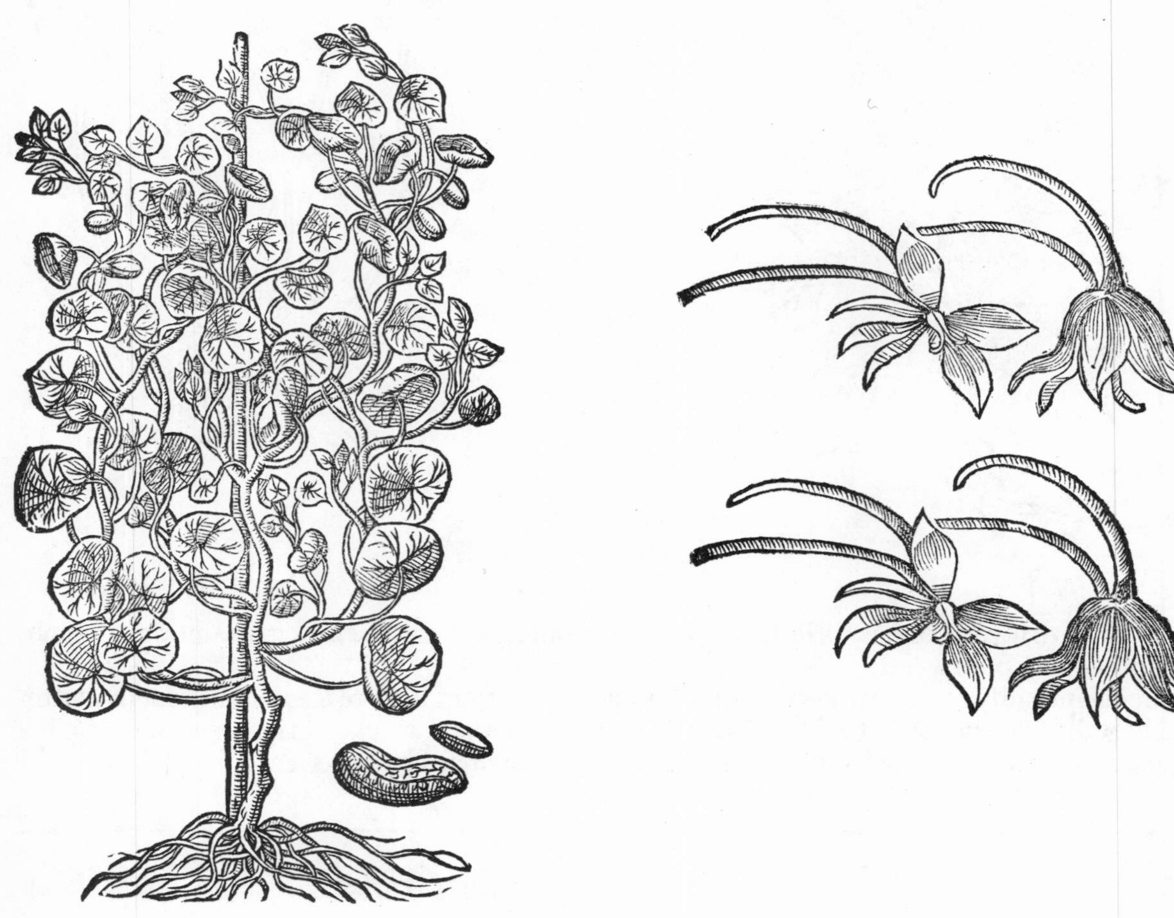

¶ *The Names.*

This beautifull plant is called in Latine, *Nasturtium Indicum*: in English, Indian Cresses. Although some haue deemed it a kinde of *Conuolvulus*, or Binde-weed; yet I am well contented that it retaine the former name, for that the smell and taste shew it to be a kinde of Cresses.

¶ *The Nature and Vertues.*

We haue no certain knowledge of his nature or vertues, but are content to refer it to the kindes of Cresses, or to a further consideration.

Chap. 14. *Of Sciatica Cresses.*

¶ *The Description.*

1 Sciatica Cresses hath many slender branches growing from a stalke of a cubit high, with small long and narrow leaues like those of Garden Cresses. The floures be very small, and yellow of colour; the seed-vessels be little flat chaffie huskes, wherein is the seed of a reddish gold colour, sharpe and very bitter in taste. The root is small, tough, white within and without, and of a biting taste.

‡ The plant whose figure I here giue you in stead of that with the narrower leaues of our Author, hath leaues somewhat like Rocket, but not so deepe cut in, being only snipt about the edges: the

the vpper leaues are not ſnipt, nor diuided at all, and are narrower. The floures decking the tops of the branches are ſmall and white, the ſeed veſſels are leſſe then thoſe of Creſſes, and the ſeed it ſelfe exceeding ſmall, and of a blackiſh colour; the root is woody, ſometimes ſingle, otherwhiles diuided into two branches. ‡

¶ *The Place.*

It groweth vpon old wals and rough places by high waies ſides, and ſuch like: I haue found it in corne fields about Southfleete neere to Grauesend in Kent.

Iberis Cardamantica.
Sciatica Creſſes.

¶ *The Time.*

It floureth according to the late or earely ſowing of it in the fields, in Iune and Iuly.

¶ *The Names.*

Sciatica Creſſes is called in Greeke ἰβηρὶς, and καρδαμαντική: in Latine *Iberis*: of *Plinie*, *Heberis*, and *Naſturtium ſylueſtre*, and in like manner alſo *Lepidium*. There is another *Lepidium* of *Plinie*: in Engliſh, Sciatica Creſſe. ‡ The firſt deſcribed may be called *Iberis Cardamantica tenuifolia*, Small leaued Sciatica Creſſes. The ſecond, *Iberis latiore folio*, broad leaued Sciatica Creſſes. ‡

¶ *The Nature.*

Sciatica Creſſe is hot in the fourth degree, and like to garden Creſſes both in ſmell and in taſte.

¶ *The Vertues.*

The rootes gathered in Autumne, ſaith *Dioſcorides*, doe heate and burne, and are with good ſucceſſe with ſwines greaſe made vp in manner of a plaiſter, and put vpon ſuch as are tormented with the Sciatica: it is to lie on the grieued place but foure hours at the moſt, and then taken away, and the patient bathed with warme water, and the place afterwards anointed with oile, and wooll laid on it; which things *Galen* in his ninth booke of medicines, according to the place greeued, citeth out of *Democrates*, in certaine verſes tending to that effect. A

CHAP. 15. *Of Banke Creſſes.*

¶ *The Deſcription.*

1 BAnke Creſſes hath long leaues, deepely cut or jagged vpon both ſides, not vnlike to thoſe of Rocket, or wilde muſtard. The ſtalkes be ſmal, limber or pliant, yet very tough, and wil twiſt and writhe as doth the Ozier or water willow, wherupon do grow ſmall yellow flowers, which being paſt there do ſucceed little ſlender cods, full of ſmall ſeedes, in taſte ſharpe and biting the tongue as thoſe of Creſſes.

2 The ſecond kinde of banke Creſſes hath leaues like vnto thoſe of Dandelion, ſomewhat reſembling Spinach. The branches be long, tough, and pliant like the other. The flowers be yellowiſh, which are ſucceeded by ſmal long cods, hauing leaues growing amongſt them: in theſe cods is contained ſmall biting ſeed like the other of this kinde. The ſmell of this plant is very vngratefull.

¶ *The Place.*

Banke Creſſes is found in ſtonie places among rubbiſh, by path waies, vpon earth or mudde walls, and in other vntoiled places.

The

The ſecond kinde of banke Creſſes groweth in ſuch places as the former doth: I found it growing at a place by Chelmes forde in Eſſex called little Baddowe, and in ſundrie other places.

‡ If our Author meant this which I haue deſcribed and giuen you the figure of, (as it is probable he did) I doubt he ſcarce found it wilde: I haue ſeene it in the garden of Maſter *Parkinſon*, and it groweth wilde in many places of Italy. ‡

¶ *The Time.*

They flower in Iune and Iuly, and the ſeed is ripe in Auguſt and September.

¶ *The Names.*

Banke Creſſes is called in Latine *Irio* and *Eryſimum*: in Greeke ἐρύσιμον, and of ſome, χαμαίπλιον, according to *Dioſcorides*: *Theophraſtus* hath an other *Eryſimum*. ‡ The firſt is called *Irio*, or *Eryſimum* by *Matthiolus Dodonæus*, and others. *Turner*, *Fuchſius* and *Tragus* call it *Verbena fœmina*, or *recta*. The ſecond is *Irio alter* of *Matthiolus*, and *Saxifraga Romanorum*; *Lugd*. It may be called Italian Banke Creſſed: or Roman Saxifrage. ‡

1 *Eryſimum Dioſcoridis, Lobelij.*
Bancke Creſſes.

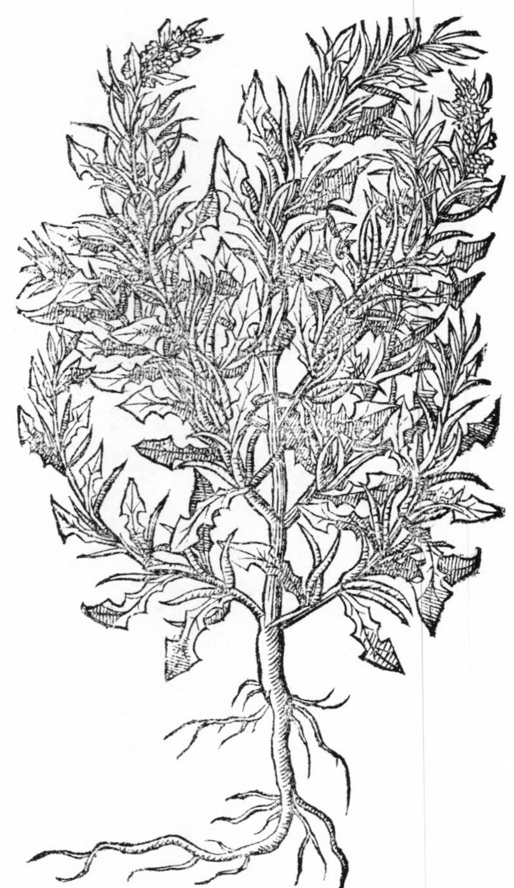

† 2 *Eryſimum alterum Italicum.*
Italian bancke Creſſes.

¶ *The Nature.*

The ſeed of bancke Creſſes is like in taſte to garden Creſſes, and is as *Galen* ſaith of a fierie temperature, and doth extreamely attenuate or make thinne.

¶ *The Vertues.*

A The ſeed of bancke Creſſes is good againſt the rheume that falleth into the cheſt, by rotting the ſame.

B It remedieth the cough, the yellow jaundiſe, and the Sciatica or ache of the hucklebones, if it be taken with hony in manner of a lohoc and often licked.

C It is alſo drunke againſt deadly poiſons, as *Dioſcorides* addeth: and being made vp in a plaiſter with water and hony and applied, it is a remedie againſt hidden cankrous apoſtumes behind the eares hard ſwellings and inflammations of the pappes and ſtones.

D ‡ The ſeeds of the Italian Banke Creſſes, or Roman Saxifrage taken in the weight of a dram, in in a decoction of Graſſe roots, effectually cleanſe the reines, and expell the ſtone, as the Authour of the *hiſt. Lugd.* affirmes. ‡

† The figure that was here in the ſecond place was of the *Sonchus ſyluaticus*, or *Libanotis Theophraſti ſterilis* of *Tabernamontanus*. You ſhall finde mention of it amongſt the *Sonchi*, or *Sow-thiſtles*.

CHAP. 16. *Of Docke Creſſes.*

† *Lampſana.*
Docke Creſſes.

¶ *The Deſcription.*

† Docke-Creſſes is a wilde Wort or pot-herbe hauing roughiſh hairy leaues of an ouerworne greene colour, deepely cut or indented vpon both ſides like the leaues of ſmall Turneps. The ſtalkes grow to the height of two or three cubits, and ſometimes higher, diuiding themſelues toward the top into ſundry little branches, whereon do grow many ſmall yellow floures like thoſe of *Hieracium*, or Hawke-weed; which decaying, are ſucceeded by little creſted heads containing a longiſh ſmall ſeed ſomewhat like Lettice ſeed, but of a yellowiſh colour: the plant is alſo milkie, the ſtalke woody, and the root ſmall, fibrous, and white.

¶ *The Place.*

Dock-Creſſes grow euery where by High-waies, vpon walls made of mud or earth, and in ſtony places.

¶ *The Time.*

It floureth from May to the end of Auguſt: the ſeed is ripe in September.

¶ *The Names.*

Docke-Creſſes are called in Greeke, Λαμψάνη: in Latine, *Lampſana*, and *Napium*, by *Dodonæus*: *Tabernamontanus* calleth this, *Sonchus ſyluaticus*: *Camerarius* affirmes, That in Pruſſia they call it *Papillaris*.

¶ *The Nature.*

Docke-Creſſes are of nature hot, and ſomewhat abſterſiue or cleanſing.

¶ *The Vertues.*

Taken in meate, as *Galen* and *Dioſcorides* affirme, it ingendreth euill iuyce and naughtie nouriſhment. A

‡ *Camerarius* affirmeth, That it is vſed with good ſucceſſe in Pruſſia againſt vlcerated or ſore breaſts. ‡ B

† The figure that was here, was of the *Rapiſtrum aruorum*, deſcribed in the ſecond chapter of this booke; and the true figure of this plant here deſcribed was pag. 231. vnder the name of *Sonchus ſyluaticus*.

CHAP. 17.
Of Water-Parſenep, and Water-Creſſes.

1 Great Water-Parſenep groweth vpright, aud is deſcribed to haue leaues of a pleaſant ſauour, fat and full of iuyce as thoſe of Alexanders, but ſomewhat leſſer, reſembling the Garden Parſenep: the ſtalke is round, ſmooth, and hollow, like to Kexe or Cashes: the root conſiſteth of many ſmall ſtrings or threds faſtned vnto the ſtalke within the water

or

1 *Sium majus latifolium.*
Great Water Parsenep.

† *Sium majus angustifolium.*
The lesser water Parsenep.

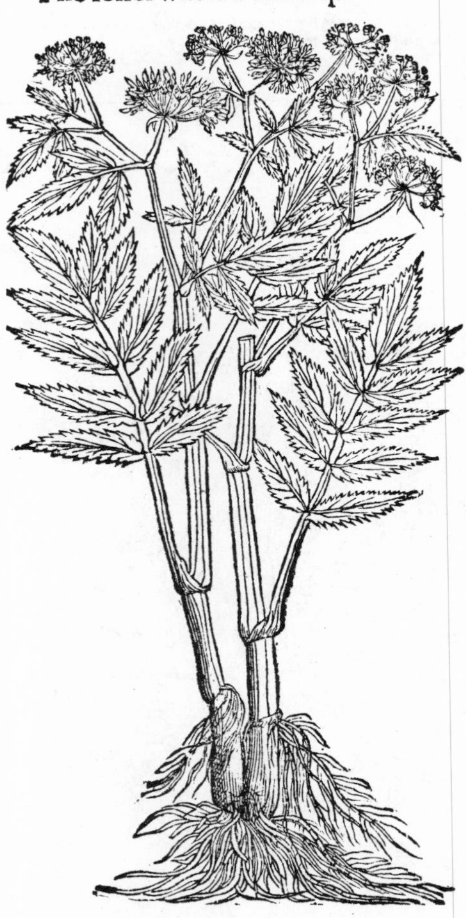

‡ 4 *Sium alterum Olusatri facie.*
Long leaued water-Cresses.

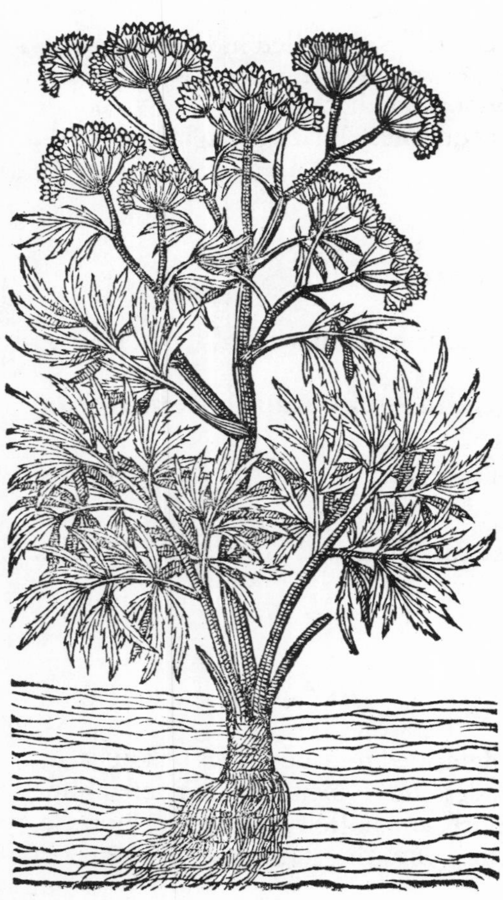

or miry ground: at the top go grow many white floures, in spoky roundles like fennell; which being bruised do yeeld a very strong sauour, smelling like *Petroleum*, as doth the rest of the plant.

‡ 2 This plant much resembles the last described, and growes vp some cubit and a halfe high, with many leaues finely snipt about the edges, growing vpon one rib, and commonly they stand bolt vpright. The vmbell consists of little white floures: the root is smal, and consisteth of many strings.

‡ 3 There is another very like this, but they thus differ: the stalkes and leaues of this later are lesse than those of the precedent, and not so many vpon one rib; the other growes vpright, to some yard or more high: this neuer growes vp, but alwaies creepes, and almost at euerie ioynt puts forth an vmbel of floures.

4 To these may be added another, whose root consists of aboundance of writhen and small blacke fibres; the stalkes are like Hemlock, some three cubits high; the leaues are long, narrow, and snipped about

about the edges, growing commonly two or three together: the vmbel of floures is commonly of a yellowiſh greene: the ſeed is like parſley ſeed, but in taſte ſomewhat reſembles *Cumine*, *Daucus Creticus*, and the rinde of a Citron, yet ſeemes ſomewhat hotter. ‡

5 Water-Creſſe hath many fat and weake hollow branches trailing vpon the grauell and earth where it groweth, taking hold in ſundry places as it creepeth; by meanes whereof the plant ſpreadeth ouer a great compaſſe of ground. The leaues are likewiſe compact and winged with many ſmall leaues ſet vpon a middle rib one againſt another, except the point leafe, which ſtands by it ſelfe, as doth that of the aſh, if it grow in his naturall place, which is in a grauelly ſpring. The vpper face of the whole plant is of a browne colour, and greene vnder the leaues, which is a perfect marke to know the phyſicall kinde from the others. The white floures grow alongſt the ſtalkes, and are ſucceeded by cods wherein the ſeed is contained. The root is nothing elſe but as it were a thrumme or bundle of threds.

† 5 *Naſturtium aquaticum, ſiue Crateuæ Sium.*
Common Water-Creſſes.

‡ 6 *Sium Matthioli & Italorum.*
Italian Water-Creſſe.

6 There is alſo another kinde hereof, hauing leaues growing many on one ſtalke, ſnipt about the edges, being in ſhape betweene the garden Creſſes and Cuckow-floures: the ſtalke is creſted, and diuided into many branches, the floures white, and are ſucceeded by cods like thoſe of our ordinarie Water-Creſſe laſt deſcribed.

¶ *The Place.*

‡ 1 The firſt of theſe I haue not found growing, nor as yet heard of within this kingdome.

2 The ſecond I firſt found in the company of M. *Robert Larkin*, going betweene Redriffe and Deptford, in a rotten boggy place on the right hand of the way.

3 The third growes almoſt in euery watery place about London.

4 This is more rare, and was found by Mr. *Goodyer* in the ponds about Moore Parke; and by M. *George Bowles* in the ditches about Elleſmere, and in diuers ponds in Flint-ſhire.

5 The fifth is as frequent as the third, and commonly they grow neere together.

6 This *Lobel* ſaith he found in Piemont, in riuelets amongſt the hills: I haue not yet heard that it growes with vs. ‡

¶ *The Time.*

They spring and wax greene in Aprill, and floure in Iuly.

The water Cresse to be eaten in sallads sheweth it selfe in March, when it is best, and floureth in Summer with the rest.

¶ *The Names.*

‡ 1 The first of these is *Sium maius latifolium* of *Tabernamontanus*.

2 This is *Sion odoratum Tragi* : *Sium*, of *Matthiolus*, *Dodonæus*, and others : it is taken to be *Sium*, or *Lauer*, of *Dioscorides*. *Lobel* calls it also *Pastinaca aquatica*, or water Parsenep.

3 This may be called *Sium vmbellatum repens*, Creeping water Parsenep. Of this there is a reasonable good figure in the *Historia Lugdunensis*, *pag.* 1092. vnder the title of *Sium verum Matthioli*; but the description is of that we here giue you in the sixth place.

4 This is *Sium alterum* of *Dodonæus* : and *Sium alterum olusatri facie* of *Lobel*.

5 Many iudge this to be the *Sisymbrium alterum*, or *Cardamine* of *Dioscorides* : as also the *Sion* of *Crateuas* : and therefore *Lobel* termes it *Sion Crateuæ erucæ folium*. It is called by *Dodonæus*, and vulgarly in shops knowne by the name of *Nasturtium aquaticum*, or water Cresses.

6 This is called *Sium vulgare* by *Matthiolus* : *Lobel* also termes it *Sium Matthioli & Italorum*. This was thought by our Countrey-man Doctor *Turner* to be no other than the second here described : of which opinion I must confesse I also was ; but vpon better consideration of that which *Lobel* and *Bauhine* haue written, I haue changed my minde. ‡

¶ *The Temperature.*

Water-Cresse is euidently hot and dry.

¶ *The Vertues.*

A Water-Cresse being boyled in Wine or Milke, and drunke for certaine dayes together, is very good against the Scuruy or Scorbute.

B Being chopped or boyled in the broth of flesh, and eaten for thirty dayes together, at morning, noone, and night, it prouoketh vrine, wasts the stone, and driueth it forth. Taken in the same maner, it doth cure yong maidens of the green sicknesse, bringeth downe the termes, and sendeth into the face their accustomed liuely colour, lost by the stopping of their *Menstrua*.

Chap. 18. *Of wilde Water-Cresses, or Cuckow Floures.*

¶ *The Description.*

1 THe first of the Cuckow floures hath leaues at his springing vp somwhat round, and those that spring afterward grow iagged like the leaues of Greeke Valerian : among which riseth vp a stalke a foot long, set with the like leaues, but smaller, and more iagged, resembling those of Rocket. The floures grow at the top in small bundles, white of colour, hollow in the middle, resembling the white sweet-Iohn : after which do come small chaffie husks or seed vessels, wherein the seed is contained. The root is small and threddy.

2 The second sort of Cuckow floures hath small iagged leaues like those of small water Valerian, agreeing with the former in stalkes and roots : the floures be white, ouerdasht or declining toward a light carnation.

‡ 3 The leaues and stalks of this are like those of the last described ; neither are the floures which first shew themselues much vnlike them ; but when as they begin to faile, in their middle rise vp heads of pretty double floures made of many leaues, like in colour to these of the single. ‡

4 The fourth sort of Cuckow Floures groweth creeping vpon the ground, with small threddy stalkes, whereon do grow leaues like those of the field Clauer, or three leaued Grasse : amongst which do come vp small and tender stalkes two handfulls high, hauing floures at the top in greater quantitie than any of the rest, of colour white ; and after them follow cods containing a small seed. The root is nothing else but as it were a bundle of thrums or threds.

5 Milke white Lady-smocke hath stalkes rising immediately from the root, diuiding themselues into sundry small twiggy and hard branches, set with leaues like those of *Serpillum*. The floures grow at the top, made of foure leaues of a yellowish colour : the root is tough and woody, with some fibres annexed thereto. ‡ This is no other than the first described, differing onely therefrom in that the floures are milke white, as our Author truly in the title of his figure made them ; yet forgetting himselfe in his description, he makes them yellowish, contrarie to himselfe and the truth. ‡

6 Mountaine

1 *Cardamine.* Cuckow floures.

2 *Cardamine altera.* Ladies-smocks.

‡ 3 *Cardamine altera flore pleno.*
Double floured Lady-smocke.

4 *Cardamine Trifolia.*
Three leaued Lady-smocke.

6 Cardamine Alpina.
Mountaine Lady-Smocke.

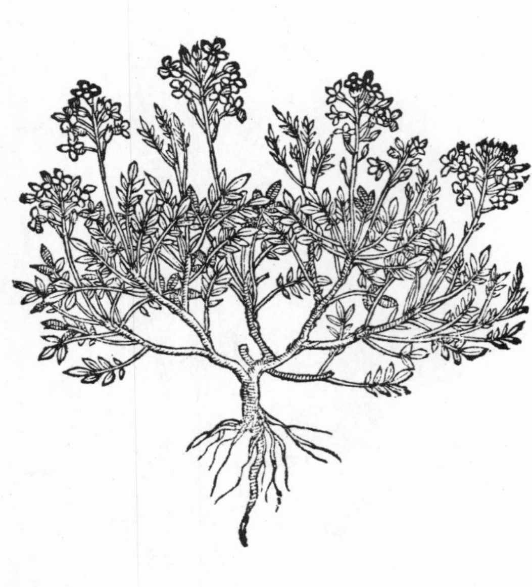

‡ *7 Sium minus impatiens.*
The impatient Lady-ſmocke.

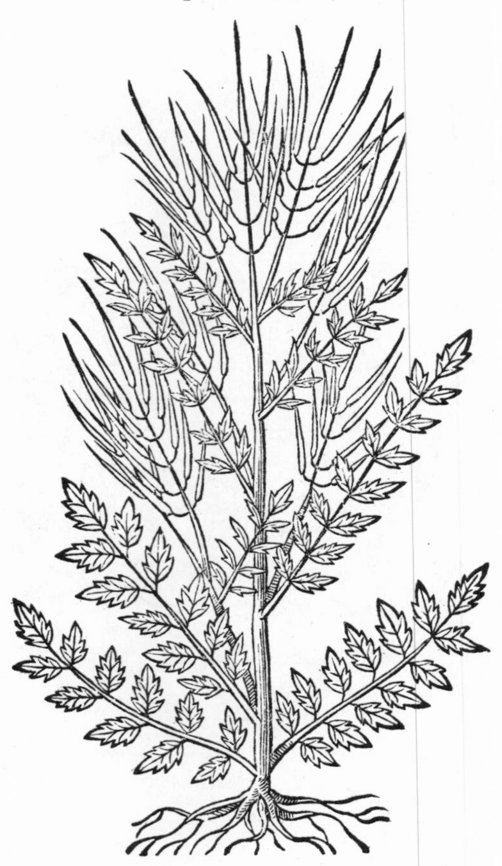

8 Cardamine pumila Bellidis folio Alpina.
The Dwarfe Daſie-leaued Lady-ſmocke of the Alpes.

6 Mountaine Lady-ſmocke hath many roots, nothing elſe but as it were a bundle of threddy ſtrings, from the which do come forth three or foure ſmall weak or tender leaues made of ſundry ſmall leaues, in ſhew like to thoſe of ſmall water Valerian. The ſtalkes be ſmall and brittle, whereupon doe grow ſmall floures like the firſt kinde.

‡ 7 I ſhould be blame worthy if in this place I omitted that pretty conditioned *Sium* which is kept in diuers of our London gardens, and was firſt brought hither by that great Treaſurer of Natures rarieties, M. *Iohn Tradeſcant*. This plant hath leaues ſet many vpon a rib, like as the other *Sium* deſcribed in the ſecond place hath; but they are cut in with two or three prety deep gaſhes: the ſtalk is ſome cubit high, & diuided into many branches, which haue many ſmall white floures growing vpon them: after theſe floures are paſt there follow ſmall long cods containing a ſmall white ſeed. Now the nature of this plant is ſuch, that if you touch but the cods when as the ſeed is ripe, though you do it neuer ſo gently, yet will the ſeed fly all abroad with violence, as diſdaining to be touched: whence they vſually call it *Noli me tangere*; as they for the like qualitie name the *Perſicaria ſiliquoſa*. The nature of this plant is ſomewhat admirable, for if the ſeeds (as I ſaid) be fully ripe,

ripe,though you put but your hand neere them,as profering to touch them,though you doe it not, yet will they fly out vpon you,and if you expect no fuch thing,perhaps make you affraid by reafon of the fuddenneffe thereof. This herbe is written of onely by *Profper Alpinus*, vnder the title of *Sium Minimum*: and it may be called in Englifh,Impatient Lady-fmocke, or Cuckow floure. It is an annuall, and yeerely fowes it felfe by the falling feeds. ‡

‡ 8 The leaues of this fomewhat refemble thofe of Dafyes,but leffe,and lie fpread vpon the ground,amongft which rifes vp a weake and flender ftalke fet with 3 or 4 leaues at certaine diftances,it being fome handful high,the top is adorned with fmal white floures confifting of foure leaues apeece,after which follow large and long cods, confidering the fmallnes of the plant; within thefe in a double order is conteined a fmall reddifh feed,of fomewhat a biting tafte. The root creepes vpon the top of the ground,putting vp new buds in diuers places.*Clufius* found this growing vpon the rockes on the Etfcherian mountaine in Auftria,and hath giuen vs the hiftory and figure thereof vnder the name of *Plantula Cardamines emula*,and *Sinapi pumulum Alpinum*.

¶ *The Time and Place.*

That of the Alpifh mountaines is a ftranger in thefe cold Countries: the reft are to be found euery where, as aforefaid, efpecially in the caftle ditch at Clare in Effex. ‡ The feuenth growes naturally in fome places of Italy.

Thefe flower for the moft part in Aprill and May, when the Cuckowe doth begin to fing her pleafant notes with out ftammering.

¶ *The Names.*

They are commonly called in Latine, *Flos Cuculi*, by *Brunfelfius* and *Dodonæus*, for the reafon aforefaid; and alfo fome call them *Nafturtium aquaticum minus*, or leffer water Creffe: of fome, *Cardamine*, and *Sifymbrium alterum* of *Diofcorides*: it is called in the Germane tongue, Wilderkreſz: in French, *Pafferage fauuage*. in Englifh, Cuckowe flowers: in Northfolke, Canterbury bells: at the Namptwich in Chefhire, where I had my beginning, Ladie fmockes, which hath giuen me caufe to Chriften it after my Country fafhion.

¶ *The Nature and Vertues.*

Thefe herbes be hot and drie in the fecond degree: we haue no certaine proofe or authority of their vertues, but furely from the kindes of water Creffe they cannot much differ, and therefore to them they may be referred in their vertues.

† The figure that was in the fourth place; being of the fame plant that is defcribed in the firft place; the counterfeit ftalkes and heades being taken away, as *Bauhine* rightly ha h obferued; as alfo the defcription thereof, which (as many other) our Author frames by looking vpon the figure, and the ftrength of his owne fancie: I haue omitted as impertinent.

Chap. 19. *Of Treacle Muftard.*

¶ *The Defcription.*

1 Treacle muftard hath long broad leaues,efpecially thofe next the ground, the others leffer, flightly indented about the edges like thofe of Dandelion. The ftalkes be long and brittle,diuided into many branches euen from the ground to the top,where grow many fmall idle flowers tuft fafhion,after which fucceed large, flat, thin, chaffie huskes or feed veffels heart fafhion,wherein are conteined browne flat feeds, fharpe in tafte, burning the tongue as doth muftard feed,leauing a tafte or fauour of Garlicke behinde for a farewell.

2 Mithridate Muftard hath long narrow leaues like thofe of Woad,or rather Cow Bafil. The ftalkes be inclofed with fmall fnipt leaues euen to the branches, Pyramidis fafhion, that is to fay,fmaller and fmaller toward the top, where it is diuided into fundrie branches, whereon doe grow fmall flowers: which being paft, the cods, or rather thinne chaffie huskes do appeare full of fharpe feed,like the former. The roote is long and flender.

3 The third kinde of Treacle Muftard,named Knaues Muftard,(for that it is too bad for honeft men) hath long, fat,and broad leaues,like thofe of Dwale or deadly Nightfhade: in tafte like thofe of Vuluaria or ftinching Orach, fet vpon a round ftalke two cubits high,diuided at the top into fmall armes or branches,whereon do grow fmall foolifh white fpokie flowers. The feed is conteined in flat pouches like thofe of Shepheards purfe,brown,fharpe in tafte,and of an ill fauor.

4 Bowyers Muftard hath the lower leaues refembling the ordinary Thlafpi,but the vpper are very fmall like tode flaxe but fmaller. The ftalkes be fmall, flender,and many. The flowers be fmall,and white,each confifting of foure leaues. The feeds be placed vpon the branches from the loweft part of them to the top,exceeding fharpe and hot in tafte,and of a yellowifh colour. The roote is fmall and woody.

5 Grecian muftard hath many leaues fpred vpon the ground, like thofe of the common Daifie,of a darke greenifh colour: from the midft whereof fpring vp ftalkes two foote long, diuided into

1 *Thlaſpi Dioſcoridis.*
Treacle Muſtatd.

2 *Thlaſpi Vulgatiſsimum.*
Mithridate Muſtard.

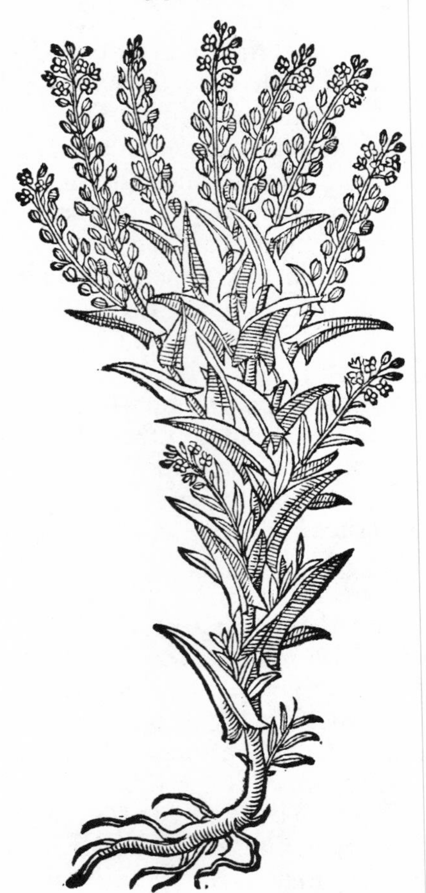

3 *Thlaſpi maius.*
Knaues Muſtard.

4 *Thlaſpi minus.*
Bowyers Muſtard.

5 *Thlaſpi*

5 *Thlaſpi Græcum.*
Grecian Muſtard.

6 *Thlaſpi amarum.*
Clownes Muſtard.

7 *Thlaſpi Clypeatum Lobelij.*
Buckler Muſtard.

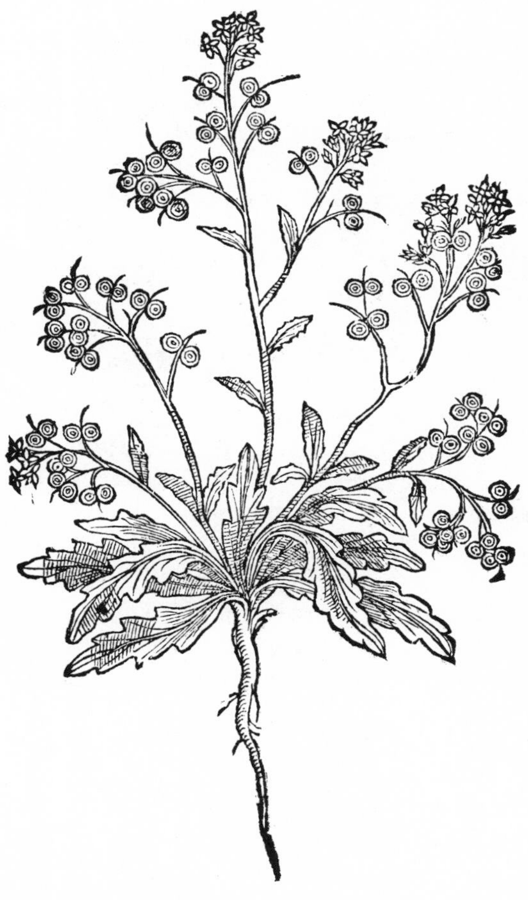

8 *Thlaſpi minus Clypeatum.*
Small Buckler Muſtard.

into many ſmall branches, whereupon grow ſmall white flowers compoſed of foure leaues, after which ſucceed round flat huskes or ſeed veſſels, ſet vpon the ſtalke by couples, as it were ſundry paires of ſpectacles, wherein the ſeed is contained, ſharpe and biting as the other. This is ſometimes ſeen with yellow flowers.

† 6 Clownes muſtard hath a ſhort white fibrous root,from whence ariſeth vp a ſtalke of the height of a foot,which a little aboue the root diuides it ſelfe into ſome foure or fiue branches,and theſe againe are ſubdiuided into other, ſmaller ſo that it reſembles a little ſhrub: longiſh narrow leaues notched after the manee of Sciatica Creſſes by turees garniſh theſe branches,and theſe leaues are as bitter as the ſmaller Centaury. The flowers ſtand thicke together at the tops of theſe branches in manner of little vmbels,and are commonly of a light blew and white mixed together (being ſeldome onely white,or yellow.) After the flowers ſucceed ſeed veſſels after the manner of the other plants of this kinde,and in them is conteined a ſmall hot ſeed †

7 Buckler muſtard hath many large leaues,ſpread vpon the ground like *Hieracium* or Hawkeweede, ſomewhat more toothed or ſnipt about the edges: among which comes vp ſtalkes ſmall and brittle,a cubit high, garniſhed with many ſmall pale yellowiſh flowers: in whoſe place ſucceed many round flat cods or pouches,buckler faſhion,conteining a ſeed like vnto the others.

8 Small Buckler Muſtard,is a very ſmall, baſe,or low plant,hauing whitiſh leaues like thoſe of wild Time,ſet vpon ſmall,weake and tender branches. The flowers grow at the top like the other buckler Muſtard. The ſeed veſſels are like, but not ſo round, ſomewhat ſharpe pointed, ſharp in taste,& burning the tongue.The whole plant lieth flat vpon the ground,like wild Tyme.

¶ *The Place.*

Treacle or rather Mithridate Muſtard growes wild in ſundry places in corn fields,ditch banks, and in ſandy,drie,and barren ground.I haue found it in corne fie'ds betweene Croydon & Godsſtone in Surrey,at South-fleete in Kent,by the path that leadeth from Harnſey (a ſmall village by London) vnto Waltham croſſe, and in many other places.

The other do grow vnder hedges, oftentimes in fields and in ſtonie and vntoiled places; they grow plentifully in Bohemia and Germany: they are ſeene likewiſe on the ſtonie bankes of the riuer Rhene. They are likewiſe to be found in England in ſundrie places wilde,the which I haue gathered into my garden. ‡ I haue found none but the firſt and ſecond growing wilde in any part of England as yet;but I deny not,but that ſome of the other may be found, though not all. ‡

¶ *The Time.*

Theſe treacle Muſtards are found with their flowers from May to Iuly,and the ſeed is ripe in the end of Auguſt.

¶ *The Names.*

The Grecians call theſe kindes of herbes θλάσπι, θλασπίδιον, or Σίναπι ἄγριον, of the huske or ſeed veſſell, which is like a little ſhield. They haue alſo other names which be found among the baſtard words: as *Scandulaceum, Capſella, Pes gallinaceus*. Neither be the later writers without their names, as *Naſturtium tectorum*,and *Sinapi ruſticum*: it is called in Dutch, **Wilde Kerſe**: in French,*Seneue ſauuage*: in Engliſh,Treacle Muſtard, diſh Muſtard, Bowyers Muſtard:of ſome, *Thlaſpi*, after the Greeke name,Churles muſtard,and wilde Creſſes.

‡ 1 This is *Thlaſpi Dioſcoridis Drabæ,aut Chamelinæ folio* of *Lobell*: *Thlaſpi Latius* of *Dodonæus*: and the ſecond *Thlaſpi* of *Matthiolus*.

2 This,*Thlaſpi Vulgatiſſimum Vaccariæ folio* of *Lobell*: the firſt *Thlaſpi* of *Matthiolus*,and ſecond of *Dodonæus*; and this is that *Thlaſpi* whoſe ſeed is vſed in ſhops.

3 This is *Thlaſpi majus* of *Tabernamontanus*.

4 This is *Thlaſpi minus* of *Dodonæus*: *Thlaſpi: anguſtifolium* of *Fuchſius*: *Thlaſpi minus hortenſe Oſyridis folio,&c.* of *Lobell*: and *Naſturtium ſylveſtre* of *Thalius*.

5 This is *Alyſſon* of *Matthiolus*: *Thlaſpi Græcum Polygonati folio*, of *Lobell* and *Tabern*.

6 This the Author of the *Hiſt. Lug.* calls *Naſturtium ſylveſtre*; *Tabern.* calls it *Thlaſpi amarum*.

7 *Lobell* termes this *Thlaſpi parvum Hieracifolium*,and *Lunaria Lutea Monſpelienſium*.

8 This is *Thlaſpi minus clypeatum Serpillifolio* of *Lobell*. ‡

† The figures of theſe two laſt mentioned were tranſpoſed in the former Edition.

¶ *The Temperature.*

The ſeed of theſe kindes of Treacle Muſtards be hot and drie in the end of the third degree.

¶ *The Vertues.*

The ſeed of Thlaſpi or treacle Muſtard eaten, purgeth colour both vpward and downeward, prouoketh flowers, and breaketh inward apoſthumes.

The ſame vſed in clyſters, helpeth the ſciatica,and is good vnto thoſe purpoſes for which Muſtard ſeed ſerueth.

¶ *The Danger.*

The ſeed of theſe herbes be ſo extreame hot and vehement in working,that being taken in too great

great a quantitie, purgeth and scoureth euen vnto bloud,and is hurtfull to women with child,and therefore great care is to be had in giuing them inwardly in any great quantitie.

CHAP. 20. *Of Candie Mustard.*

¶ *The Description.*

CAndie mustard excelleth all the rest,as well for the comely floures that it bringeth forth for the decking vp of gardens and houses,as also for that it goeth beyond the rest in his physicall vertues. It riseth vp with a very brittle stalke of a cubit high,which diuideth it selfe into sundry bowes or branches, set with leaues like those of stocke gillifloures, of a gray or ouerworne greene colour.The floures grow at the top of thes stalke roundt,hicke clustering together, like those of Scabious or diuels bit,sometimes blew,often purple,carnation or horse flesh,but seldome white for any thing that I haue seen;varying according to the soile or Clymate. The seed is reddish,sharpe,and biting the tongue,wrapped in little huskes fashioned like an heart. ‡ There is a lesser variety of this with white well smelling flowers,in other respects little differing from the ordinary. ‡

Thlaspi Candiæ.
Candie Mustard.

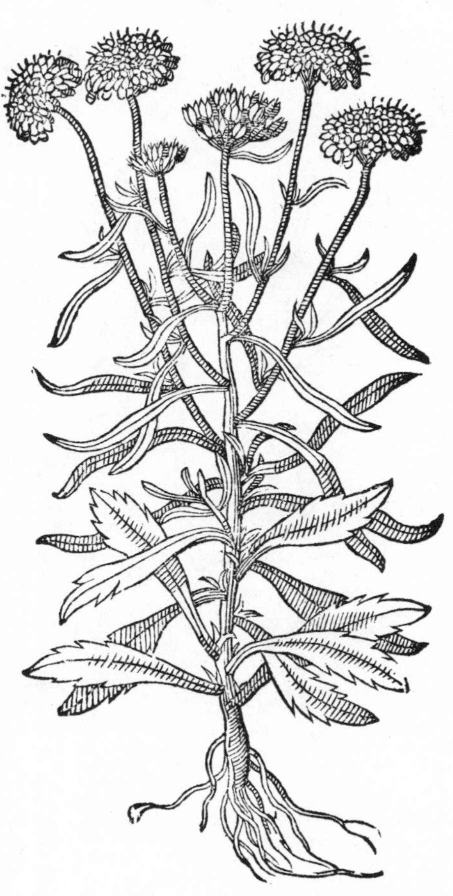

‡ *Thlaspi Candiæ parvum flo. albo.*
Small Candy mustard with a white floure.

¶ *The Place.*

This growes naturally in some places of Austria,as also in Candy, Spaine,& Italy,from whence I receiued seeds by the liberality of the right Honorable the Lord *Edward Zouch*, at his returne into England from those parts. ‡ *Clusius* found the later as he trauelled through Switzerland into Germany. ‡

¶ *The Time.*

It floureth from the beginning of May vnto the end of September, at which time you shall haue floures and seeds vpon one branch, some ripe,ond some that will not ripen at all.

¶ *The Name.*

† This plant is called by *Dodonæus* (but not rightly) *Arabis* and *Draba*: as also *Thlaspi Candiæ:* which last name is reteined by most writers : in English,Candy Thlaspi,or Candy Mustard. †

¶ *The Temperature.*

The seed of Candie Mustard is hot and drie at the end of the third degree,as is that called *Scorodothlaspi*, or treacle mustard.

Chap.

CHAP. 21. *Of Treacle Muſtard.*

¶ *The Deſcription.*

1 ROund leaued Muſtard hath many large leaues laid flat vpon the ground like the leaues of the wilde Cabbage, and of the ſame colour; among which riſe vp many ſlender ſtalkes of ſome two handfulls high or thereabouts, which are ſet with leaues far vnlike to thoſe next the ground, encloſing or embracing the ſtalkes as do the leaues of *Perfoliatum*, or Thorow-wax. The floures grow at the top of the branches, white of colour; which being paſt, there do ſucceed flat huskes or pouches like vnto thoſe of Shepheards purſe, with hot ſeed biting the tongue.

1 *Thlaſpi rotundifolium.*
Round leaued Muſtard.

2 *Thlaſpi Pannonicum Cluſij.*
Hungary Muſtard.

2 Hungary Muſtard bringeth forth ſlender ſtalkes of one cubit high: the leaues which firſt appeare are flat, ſomewhat round like thoſe of the wilde Beet; but thoſe leaues which after doe garniſh the ſtalkes are long and broad like thoſe of the garden Colewort, but leſſer and ſofter, greene on the vpper ſide, and vnder declining to whiteneſſe, ſmelling like Garlicke. The floures be ſmall and white, conſiſting of foure ſmall leaues, which in a great tuft or vmbel do grow thick thruſt together: which being paſt, there followeth in euery ſmall huske one duskiſh ſeed and no more, bitter and ſharpe in taſte. The root is white and ſmall, creeping vnder the ground far abroad like the roots of Couch-graſſe; preparing new ſhoots and branches for the yeare following, contrarie to all the reſt of his kinde, which are encreaſed by ſeed, and not otherwiſe.

3 Churles Muſtard hath many ſmall twiggy ſtalkes, ſlender, tough, and pliant, ſet with ſmall leaues like thoſe of Cudweed, or Lauander, with ſmall white floures: the huskes and ſeeds are ſmall, few, ſharpe, bitter, and vnſauorie: the whole plant is of a whitiſh colour.

4 Peaſants Muſtard hath many pretty large branches, with thin and iagged leaues like thoſe of Creſſes, but ſmaller, in ſauor and taſte like to the ordinarie *Thlaſpi*: the floures be whitiſh, and grow in a ſmall ſpoky tuft. The ſeed in taſte and ſauor is equall with the other of his kinde and countrey, or rather exceeds them in ſharpneſſe.

3 *Thlaſpi Narbonenſe Lobelij.*
Churles Muſtard.

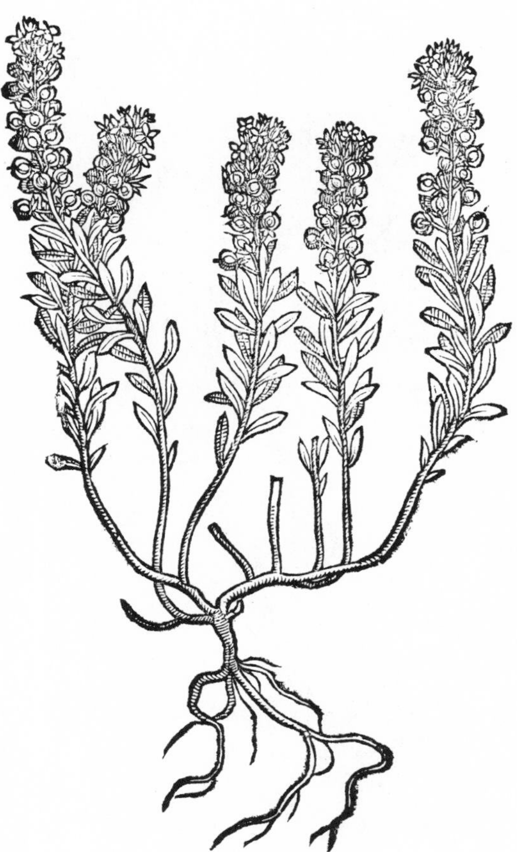

4 *Thlaſpi vmbellatum Narbonenſe.*
Peaſants Muſtard of Narbone.

† 5 *Thlaſpi ſupinum luteum.*
Yellow Muſtard.

5 Yellow Muſtard hath an exceeding number of whitiſh leaues ſpred vpon the ground in manner of a turfe or haſſocke, from the midſt whereof riſeth vp an vpright ſtalke of three foot high, putting forth many ſmall branches or armes: at the top whereof grow many ſmall yellow floures like thoſe of the wall-floure, but much leſſer: which being paſt, the husks appeare flat, pouch-faſhion, wherein is the ſeed like Treacle Muſtard, ſharp alſo and biting.

6 White Treacle Muſtard hath leaues ſpred vpon the ground like the other, but ſmaller: the ſtalkes riſe vp from the middeſt thereof, branched, ſet with leaues ſmaller than thoſe that lie vpon the ground euen to the top, where doth grow a tuft of white floures in faſhion like to thoſe of the other Thlaſpies: the ſeed is like the other: ‡ The cods of this are ſometimes flat, and otherwhiles round: the floures alſo grow ſometimes ſpike-faſhion, otherwhiles in an vmbell. I haue giuen you two figures expreſſing both theſe varieties. ‡

7 The

6 Thlaspi album supinum, & eius varietas.
White Treacle Mustard.

7 Thlaspi minus Clusij.
Clusius his small Mustard.

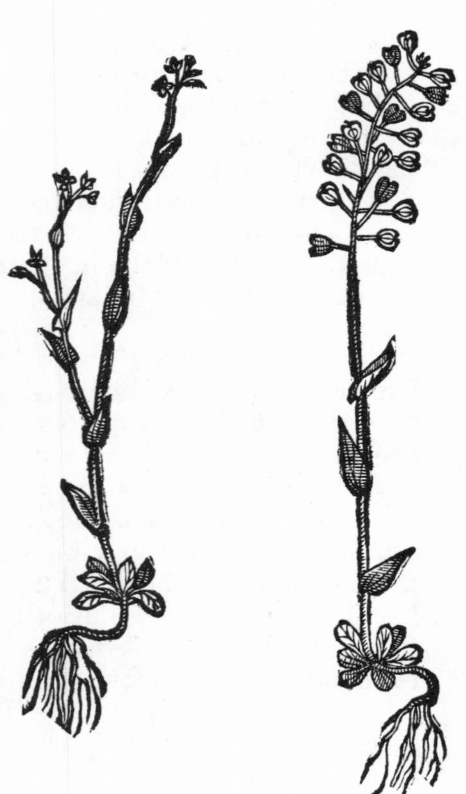

‡ 8 *Thlaspi petræum minus.*
Small Rocke Mustard.

7 This ſmall kinde of Muſtard hath a few ſmall leaues ſpread vpon the ground like thoſe of the leſſer Daſie, but of a blewiſher greene colour; from which riſe vp ſmall tender ſtalks ſet with three, and ſometimes foure ſmall ſharpe pointed leaues: the floures grow at the top, ſmall and white; the cods are flat, pouch-faſhion, like thoſe of Shepheards purſe, and in each of them there is contained two or three yellowiſh ſeeds.

‡ 8. To theſe we may fitly adde a other ſmall mountaine Thlaſpi, firſt deſcribed by that diligent and learned Apothecarie *Iohn Iona* of Verona, in his deſcription of Mount Baldus. This from a threddy root brings forth many ſmall whitiſh leaues lying ſpred vpon the ground, and a little nicked about their edges: among theſe riſeth vp a ſtalke ſome two or three handfulls high, diuaricated toward the top into diuers ſmall branches, vpon which grow white little floures conſiſting of foure leaues apiece: which fading, there follow round ſeed-veſſels, like to thoſe of *Myagrum*: whence *Pona*, the firſt deſcriber thereof, calls it *Thlaſpi petræum myagrodes*. The ſeed is as ſharpe and biting as any of the other Thlaſpies. This growes naturally in the chinkes of the rocks, in that part of Baldus that is termed *Vallis frigida*, or, The cold Valley. ‡

¶ *The Place.*

Theſe kindes of Treacle Muſtard grow vpon hills and mountaines in corne fields, in ſtony barren and grauelly grounds.

¶ *The Time.*

Theſe floure in May, Iune, and Iuly: the ſeed is ripe in September.

¶ *The Names.*

‡ 1 This is *Thlaſpi oleraceum* of *Tabernamontanus*: *Thlaſpi primum*, of *Daleſchampius*: *Thlaſpi mitius rotundifolium* of *Columna*. Our Author confounded it with that whoſe figure is the firſt in the enſuing Chapter, and called it *Thlaſpi incanum*.

2 *Thlaſpi montanum peltatum* of *Cluſius*: and *Thlaſpi Pannonicum* of *Lobel* and *Taber.*

3 *Thlaſpi Narbonenſe centunculi anguſtifolio*, of *Lobel*: and *Thlaſpi maritimum* of *Daleſchampius.*

4 *Thlaſpi vmbellatum Naſturtij hortenſis folio Narbonenſe*, of *Lobel*. The figures of this and the precedent were tranſpoſed in the former edition.

5 *Thlaſpi ſupinum luteum* of *Lobel*. Our Authors figure was a varietie of the next following.

6 *Thlaſpi album ſupinum* of *Lobel*: *Thlaſpi montanum ſecundum* of *Cluſius.*

7 *Thlaſpi pumilum* of *Cluſius*: *Thlaſpi minimum* of *Tabernamontanus.*

8 *Thlaſpi petræum myagrodes* of *Pona*: *Thlaſpi tertium ſaxatile* of *Camerarius*, in his *Epit.* of *Matthiolus.* ‡

¶ *The Temperature and Vertues.*

The ſeeds of theſe churliſh kindes of Treacle Muſtard haue a ſharpe or biting qnalitie, breake inward apoſtumes, bring downe the floures, kill the birth, and helpeth the Sciatica or pain in the hip. They purge choler vpward and downeward, if you take two ounces and a halfe of them, as *Dioſcorides* writeth. They are mixed in counterpoyſons, as Treacle, Mythridate, and ſuch like Compoſitions. A

CHAP. 22.

Of Woodդy Muſtard.

¶ *The Deſcription.*

1 WOoddy Muſtard hath long narrow leaues declining to whiteneſſe, like thoſe of the ſtocke Gilloflower, but ſmaller, very like the leaues of Roſemary, but ſomewhat broader, with rough ſtalks very tough and pliant, being of the ſubſtance of wood: the floures grow at the top, white of colour: the ſeeds do follow, in taſte ſharpe and biting. The huskes or ſeed-veſſels are round and ſomewhat longiſh.

2 Small wooddy Muſtard groweth to the height of two cubits, with many ſtalkes ſet with ſmall narrow leaues like thoſe of Hyſſop, but rougher; and at the top grow floures like thoſe of Treacle Muſtard, or Thlaſpi. The whole plant groweth as a ſhrub or hedge-buſh.

3 Thorny Muſtard groweth vp to the height of foure cubits, of a wooddy ſubſtance, like vnto a hedge-buſh, or wilde ſhrub, with ſtalkes beſet with leaues, floures, and ſeeds like the laſt before mentioned; agreeing in all points, ſauing in the cruell pricking ſharpe thornes wherewith this plant is armed; the other not. The root is tough, wooddy, and ſome ſtrings or fibres annexed thereto.

1 *Thlaspi fruticosum incanum.*
Hoary wooddy Mustard.

2 *Thlaspi fruticosum minus.*
Small wooddy Mustard.

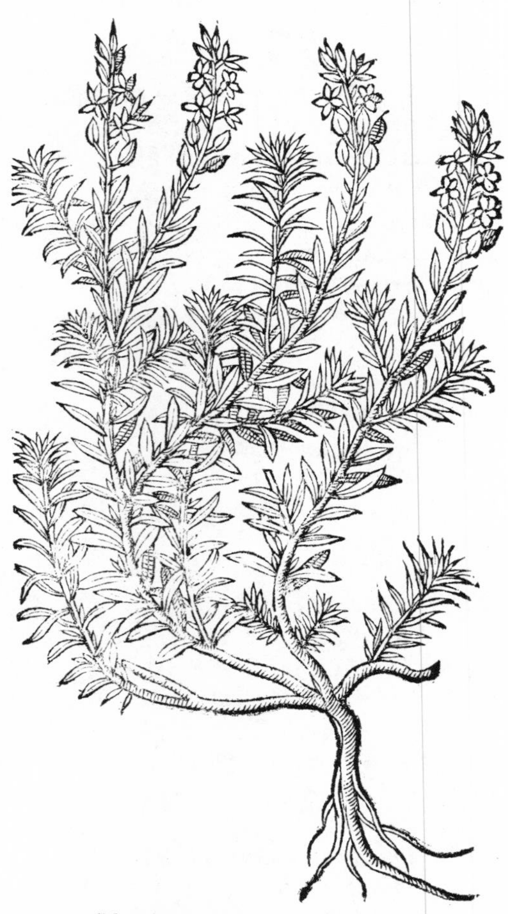

3 *Thlaspi spinosum.*
Thorny Mustard.

‡ 4 *Thlaspi fruticosum folio Leucoij marini.*
Bushy Mustard.

‡ 5 *Thlaspi*

‡ 5 *Thlaspi hederacium.*
Iuy Mustard.

4 There is another sort of wooddy Mustard growing in shadowie and obscure mountaines, and rough stony places resembling the last described; sauing that this plant hath no pricks at all, but many small branches set thick with leaues, resembling those of the lesser sea *Leucoion*: the floures are many and white; the seed like the other Thlaspies: the root is wooddy and fibrous.

‡ 5 There is (saith *Lobel*) in Portland and about Plimouth, and vpon other rockes on the sea coast of England, a creeping little herbe hauing small red crested stalkes about a spanne high: the leaues are thicke and fashioned like Iuy; the white floures and small seeds do in taste and shape resemble the Thlaspies. ‡

¶ *The Place.*

‡ 1 The first of these groweth about Mechline.

2. 3. 4. These plants grow vpon the Alpish and Pyrene mountaines: in Piemont and in Italy, in stony and rockie grounds.

¶ *The Time.*

They floure when the other kindes of Thlaspies do; that is, from May to the end of August.

¶ *The Names.*

‡ 1 This *Clusius* and *Lobel* call *Thlaspi incanum Mechliniense*. *Bauhine* thinks it to be the *Iberis prima* of *Tabernamontanus*, whose figure retained this place in the former edition.

2 This is *Thlaspi fruticosum alterum* of *Lobel*: *Thlaspi* 5. *Hispanicum* of *Clusius*.

3 *Lobel* calls this, *Thlaspi fruticosum spinosum*.

4 *Camerarius* calls this, *Thlaspi sempervirens bisflorum folio Leucoij, &c.* *Lobel*, *Thlaspi fruticosum folio Leucoij, &c.*

5 This *Lobel* calls *Thlaspi hederaceum*. ‡

¶ *The Nature and Vertues.*

I finde nothing extant of their nature or vertues, but they may be referred to the kinds of Thlaspies, whereof no doubt they are of kindred and affinitie, as well in facultie as forme.

CHAP. 23. *Of Towers Mustard.*

¶ *The Description.*

1 Towers Mustard hath beene taken of some for a kinde of Cresses, and referred by them to it: of some, for one of the Mustards, and so placed among the Thlaspies as a kinde thereof; and therefore my selfe must needs bestow it somewhere with others. Therefore I haue with *Clusius* and *Lobel* placed it among the Thlaspies, as a kinde thereof. It commeth out of the ground with many long and large rough leaues, like those of Hounds-tongue, especially those next the ground: amongst which riseth vp a long stalke of a cubit or more high, set abont with sharpe pointed leaues like those of Woad. The floures grow at the top, if I may terme them floures, but they are as it were a little dusty chaffe driuen vpon the leaues and branches with the winde: after which come very small cods, wherein is small reddish seed like that of Cameline or English Worm-seed, with a root made of a tuft full of innumerable threds or strings.

 ‡ 2 This

‡ 2 This ſecond kinde hath a thicker and harder root than the precedent,hauing alſo fewer fibers;the leaues are bigger than thoſe of the laſt deſcribed,ſomewhat curled or ſinuated,yet leſſe, rough,and of a lighter greene;in the middeſt of theſe there riſe vp one or two ſtalkes or more, vſually ſome two cubits high,diuided into ſome branches,which are adorned with leaues almoſt ingirting them round at there ſetting on. The floures are like thoſe of the former, but ſomewhat larger,and the colour is either white, or a pale yellow : after theſe ſucceed many long cods filled with a ſeed ſomewhat larger than the laſt deſcribed. ‡

3 Gold of pleaſure is an herbe with many branches ſet vpon a ſtraight ſtalke,round,and diuided into ſundry wings,in height two cubits. The leaues be long,broad,and ſharpe pointed,ſomewhat ſnipt or indented about the edges like thoſe of Sow-thiſtles. The flowers along the ſtalkes are white; the ſeed contained in round little veſſels is fat and oily.

1 *Turritis.*
Towers Muſtard.

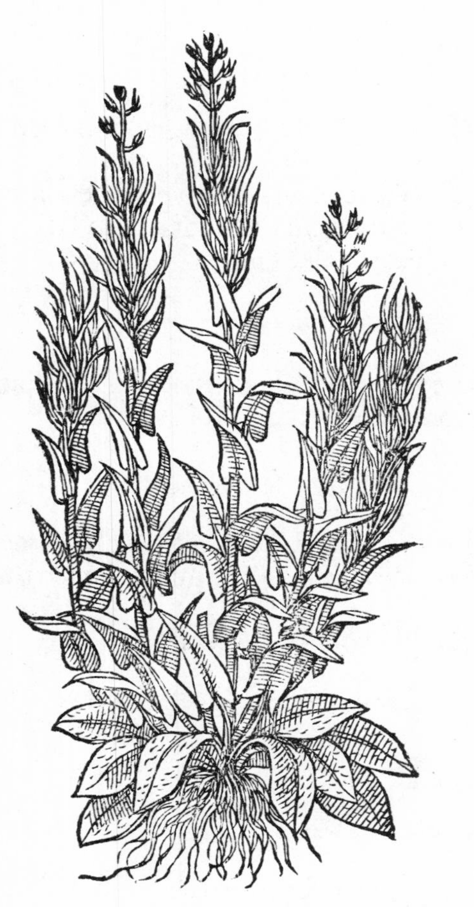

‡ 2 *Turritis major.*
Great Tower Muſtard.

4 Treacle Wormeſeed riſeth vp with tough and pliant branches,whereupou do grow many ſmall yellow flowers ; after which come long ſlender cods like Flixe-weed, or Sophia, wherein is conteined ſmall yellowiſh ſeed,bitter as Wormeſeed or Coliquintida. The leaues are ſmall and darke of colour,in ſhape like thoſe of the wilde ſtocke Gillofloures, but not ſo thicke,nor fat. The root is ſmall and ſingle.

¶ *The Place.*

Towers Treacle groweth in the Weſt part of England,vpon dunghils and ſuch like places. I haue likewiſe ſeen it in ſundrie other places,as at Pyms by a village called Edmonton neere London,by the Citie wals of Weſt-cheſter in corne fields, and where flaxe did grow about Cambridge. ‡ The ſecond is a ſtranger with vs; yet I am deceiued if I haue not ſeene it growing in M. *Parkinſons* garden. ‡

The other grow in the territorie of Leiden in Zeeland,and many places of the Low-countries; and likewiſe wilde in ſundrie places of England.

¶ *The Time.*

Theſe herbes doe floure in May and Iune,and their ſeed is ripe in September.

¶ *The Names.*

‡ 1 This is *Turritis* of *Lobell*: *Turrita Vulgatior* of *Cluſius*.

2 This is *Turrita maior*, of *Cluſius*,who thinkes it to be *Braſſica Virgata* of *Cordus*.

3 *Mat-*

3 *Matthiolus* calls this, *Pseudomyagrum* : *Tragus* calls it, *Sesamum* : *Dodonæus*, *Lobel*, and others call it *Myagrum*.

4 This *Lobel* calls *Myagrum thlaspi effigie*. *Tabernamontanus* hath it twice; first vnder the name of *Erysimum tertium* : secondly, of *Myagrum secundum*. And so also our Authour (as I formerly noted) had it before vnder the name of *Eruca syluestris angustifolia*; and here vnder the name of *Camelina*. ‡

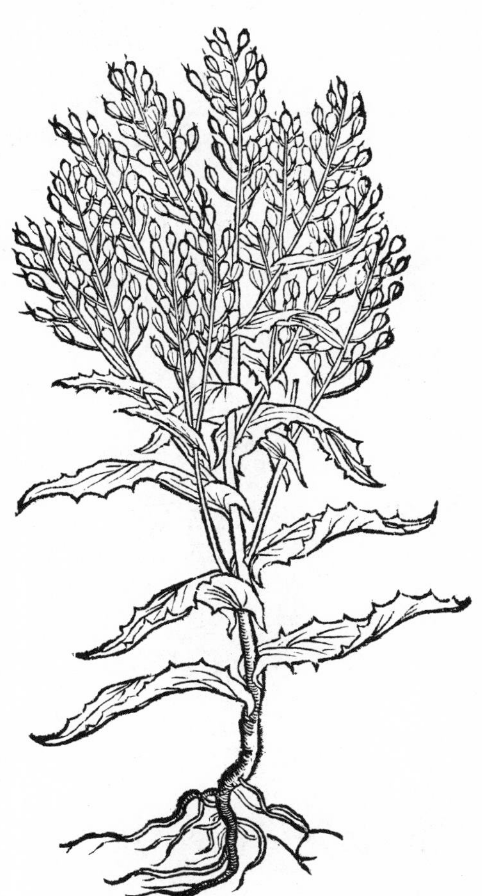

3 *Myagrum*.
Gold of pleasure.

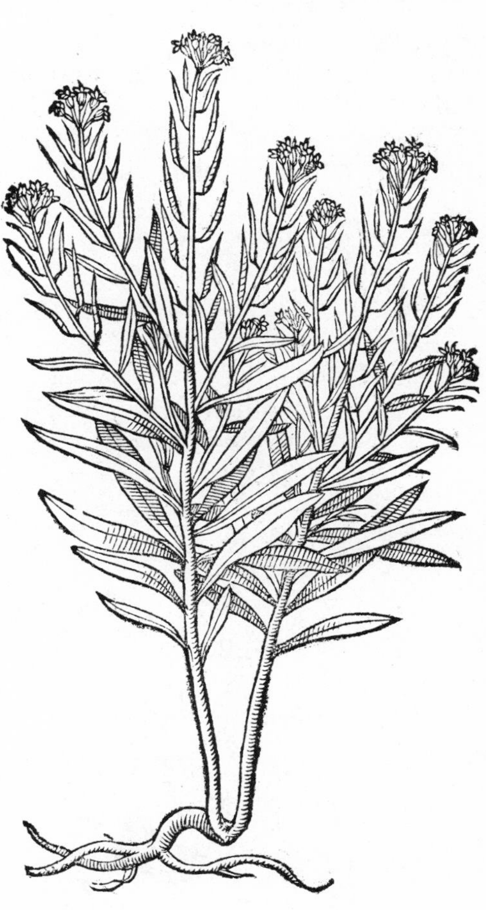

4 *Camelina*.
Treacle Worm-seed.

¶ *The Temperature*.

These Plants be hot and dry in the third degree.

¶ *The Vertues*.

It is thought, saith *Dioscorides*, That the roughnesse of the skinne is polished and made smooth with the oylie fatnesse of the seed of *Myagrum*. A

Ruellius teacheth, That the iuyce of the herbe healeth vlcers of the mouth; and that the poore peasant doth vse the oile in banquets, and the rich in their lampes. B

The seed of *Camelina* stamped, and giuen children to drinke, killeth the wormes, and driueth them forth both by siege and vomit. C

† The two Drabaes here omitted are treated of at large in the following Chapter.

‡ Chap. 24. *Of Turky Cresses*.

‡ Our Author did briefely in the precedent Chapter make mention of the two plants wee first mention in this Chapter; but that so briefely, that I thought it conuenient to discourse more largely of them, as also to adde to them other two, being by most Writers adiudged to be of the same Tribe or kindred. The vertues of the first were by our Author out of *Dodonæus* formerly put to the *Thlaspi Candiæ*, Chapter 20. from whence I haue brought them to their proper place, in the end of this present Chapter.

¶ *The Description.*

† 1 The first hath crested slender, yet firme stalkes of some foot long, which are set with leaues of some inch in length, broad at the setting on, sinuated about the edges, and sharpe pointed; their colour is a whitish greene, and taste acride; the leaues that are at the bottome of the stalke are many, and larger. The tops of the stalkes are diuided into many branches of an vnequall length, and sustain many floures; each whereof consists of foure litle white leaues, so that together they much resemble the vmbell of the Elder when it is in floure. Little swolne seed vessels diuided into two cells follow the fading floures: the seed is whitish, about the bignesse of millet, the root also is white, slender and creeping.

† 2 This hath creeping roots, from which arise many branches lying vpon the ground here and there, taking root also; the leaues, which vpon the lower branches are many, are in forme and colour much like those of the last described, but lesse, and somewhat snipt about the edges. The stalkes are about a handfull high, or somewhat more, round, greene, and hairy, hauing some leaues growing vpon them. The floures grow spoke fashion at the top of the stalkes, white, and consisting of foure leaues; which fallen, there follow cods conteining a small red seed.

1 *Draba Dioscoridis.*
Turkie Cresses.

‡ 2 *Draba prima repens.*
The first creeping Cresse.

3 From a small and creeping root rise vp many shootes, which while they are young haue many thicke juicy and darke greene leaues rose fashion adorning their tops, out of the middest of which spring out many slender stalkes of some foot high, which at certain spaces are encompassed (as it were) with leaues somewhat lesser then the former, yet broader at the bottome: the floures, cods, and seed are like the last mentioned.

4 There is a plant also by some refer'd to this Classis; and I for some reasons thinke good to make mention thereof in this place. It hath a strong and very long root of colour whitish, and of as sharpe a taste as Cresses; the stalkes are many, and oft times exceed the height of a man, yet slender, and towards their tops diuided into some branches, which make no vmbell, but carry their floures dispersed; which consist of foure small yellow leaues: after the floure is past there follow long slender cods conteining a small, yellowish, acride seed. The leaues which adorne this plant are long, sharpe pointed, and snipt about the edges, somewhat like those of Saracens Consound, but that these towards the top are more vnequally cut in.

‡ 3 *Draba altera repens.*
The other creeping Cresse.

¶ *The Time.*

The first of these floures in May and the beginning of Iune. The 2 and 3 in Aprill. The fourth in Iune and Iuly.

¶ *The Place.*

None of these (that I know of) are found naturally growing in this kingdome; the last excepted, which I thinke may be found in some places.

¶ *The Names.*

1 This by a generall consent of *Matthiolus, Anguillara, Lobell, &c.* is iudged to be the *Arabis*, or *Draba* of the Ancients.

2 *Draba altera* of *Clusius*.

3 *Draba tertia succulento folio*, of *Clusius*. *Eruca Muralis* of *Daleschampius*.

4 This by *Camerarius* is set forth vnder the name of *Arabis quorundam*, and he affirmes in his *Hor. Med.* that he had it out of England vnder the name of *Solidago*; The which is very likely, for without doubt this is the very plant that our Author mistooke for *Solidago Sarracenica*, for he bewraies himselfe in the Chapter of *Epimedium*, whereas he saith it hath cods like *Sarraccens Consound*; when as both he, and all other giue no cods at all to *Sarracens Consound*. My very good friend Mr. *Iohn Goodyer* was the first, I thinke, that obserued this mistake in our Author; for which his obseruation, together with some others formerly and hereafter to be remembred, I acknowledge my selfe beholden to him.

¶ *The Vertues, attributed to the first.*

1 *Dioscorides* saith, that they vse to eate the dryed seed of this herbe with meate, as we do pepper especially in Cappadocia. A

They vse likewise to boyle the herbe with the decoction of barly, called Ptisana; which being so boiled, concocteth and bringeth forth of the chest tough and raw flegme which sticketh therein. B

The rest are hot, and come neere to the vertues of the precedent. ‡ C

CHAP. 25. *Of Shepheards-purse.*

¶ *The Description.*

1 THe leaues of Shepheards purse grow vp at the first long, gashed in the edges like those of Rocket, spred vpon the ground: from these spring vp very many little weake stalks diuided into sundry branches, with like leaues growing on them, but lesser; at the top whereof are orderly placed small white floures: after these come vp little seed vessels, flat, and cornered, narrow at the stem like to a certaine little pouch or purse, in which lieth the seed. The root is white not without strings. ‡ There in another of this kinde with leaues not sinuated, or cut in. ‡

2 The small Shepheards purse commeth forth of the ground like the Cuckow floure, which I haue Englished Ladie-smockes, hauing small leaues deepely indented about the edges; among which rise vp many small tender stalkes with floures at the top, as it were chaffe. The huskes and seed is like the other before mentioned.

¶ *The Place.*

These herbes do grow of themselues for the most part, neere common high waies, in desert and vntilled places, among rubbish and old walls.

¶ *The*

1 *Bursa Pastoris.*
Shepheards purse.

2 *Bursa Pastoria minima.*
Small Shepheards purse.

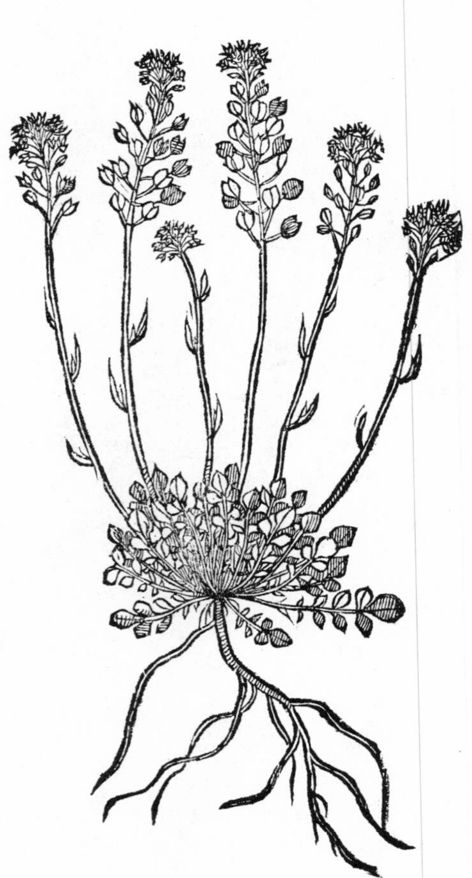

¶ *The Time.*

They floure, flourish, and seed all the Sommer long.

¶ *The Names.*

Shepheards purse is called in Latine, *Pastoris bursa*, or *Pera pastoris*: in high Dutch, **Seckel**: in low-Dutch, **Borsekens cruyt**: in French, *Bourse de pasteur ou Curé*: in English, Shepheards purse or scrip: of some, Shepheards pouch, and poore mans Parmacetie: and in the North part of England, Toy-wort, Pick-purse, and Case-weed.

¶ *The Temperature.*

They are of temperature cold and dry, and very much binding, after the opinion of *Ruellius*, *Matthiolus*, and *Dodonæus*; but *Lobel* and *Pena* hold them to be hot and dry, iudging the same by their sharpe taste: which hath caused me to insert them here among the kindes of Thlaspi, considering the fashion of the leaues, cods, seed, and taste thereof: which do so wel agree together, that I might very well haue placed them as kindes thereof. But rather willing to content others that haue written before, than to please my selfe, I haue followed their order in marshalling them in this place, where they may stand for cousine germanes.

¶ *The Vertues.*

A Shepheards purse stayeth bleeding in any part of the body, whether the iuyce or the decoction thereof be drunke, or whether it be vsed pultesse-wise, or in bath, or any other way else.

B In a Clyster it cureth the bloudy flix: it healeth greene and bleeding wounds: it is maruellous good for inflammations new begun, and for all diseases which must be checked backe and cooled.

C The decoction doth stop the laske, the spitting and pissing of bloud, and all other fluxes of bloud.

Chap. 26. *Of Italian Rocket.*

¶ *The Description.*

1 ITalian Rocket hath long leaues cut into many parts or diuisions like those of the Ash tree, resembling *Ruellius* his Bucks-horne : among which rise vp stalks weake and tender, but thicke and grosse, two foot high, garnished with many small yellowish floures like the middle part of Tansie floures, of a naughty sauor or smell. The seed is small like sand or dust, in taste like Rocket seed, whereof in truth wee suspect it to be a kinde. The root is long and wooddy.

1 *Rheseda Plinij.*
Italian Rocket.

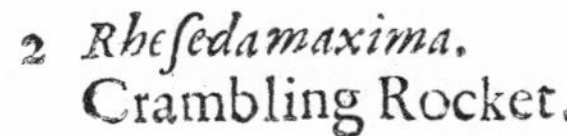

2 *Rhesedamaxima.*
Crambling Rocket.

2 Crambling Rocket hath many large leaues cut into sundry sections, deeply diuided to the middle rib, branched like the hornes of a stag or hart : among which there do rise vp long fat and fleshy stalkes two cubits high, lying flat vpon the ground by reason of his weake and feeble branches. The floures grow at the top, clustering thicke together, white of colour, with brownish threds in them. The seed is like the former. ‡ *Lobel* affirmes it growes in the Low-country gardens with writhen stalkes, sometimes ten or twelue cubits high, with leaues much diuided. ‡

¶ *The Place.*

These Plants grow in sandy, stony, grauelly, and chalky barren grounds. I haue found them in sundry places of Kent, as at South-fleet, vpon Long-field downes, which is a chalkie and hilly ground very barren. They grow at Greenhithe vpon the hills, and in other places of Kent. ‡ The first growes also vpon the Wolds in Yorke-shire. The second I haue not seene growing except in gardens, and much doubt whether it grow wilde with vs or no. ‡

¶ *The Time.*

These Plants do flourish in Iune, Iuly, and August.

¶ The

¶ *The Names.*

The first is called of *Pliny, Reseda, Eruca peregrina,* & *Eruca Cantabrica:* in English, Italian Rocket.

The second is called *Reseda maxima :* of *Anguillara, Pignocomon,* whereof I finde nothing extant worthy the memorie, either of temperature or vertues.

Chap. 27. *Of Groundsell.*

¶ *The Description.*

1 The stalke of Groundsell is round, chamfered and diuided into many branches: the leaues be greene, long, and cut in the edges almost like those of Succorie, but lesser, like in a manner to the leaues of Rocket. The floures be yellow, and turne to downe, that is carried away with the winde. The root is full of strings and threds.

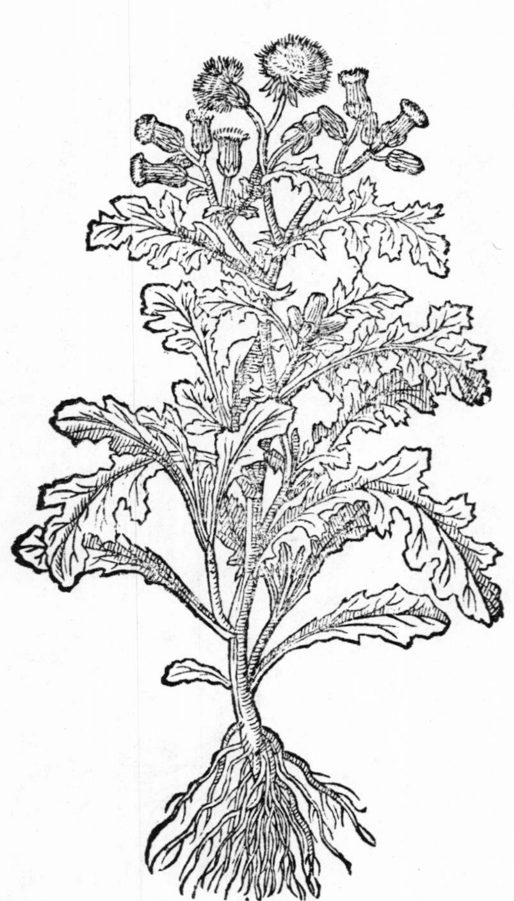

1 *Erigerum.*
Groundsell.

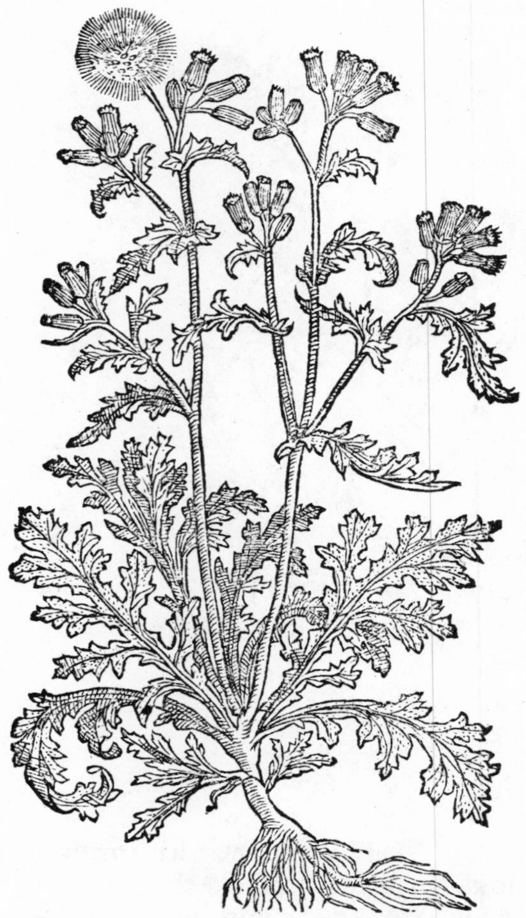

2 *Erigerum Tomentosum.*
Cotton Groundsell.

2 Cotton Groundsel hath a straight stalke of a browne purple colour, couered with a fine cotton or downy haire, of the height of two cubits. The leaues are like those of S. Iames Wort, or Rag-wort; and at the top of the stalke grow small knops, from which come floures of a pale yellow colour; which are no sooner opened and spred abroad, but they change into downe like that of the Thistle, euen the same houre of his flouring, and is carried away with the winde: the root is small and tender.

‡ 3 There is another with leaues more iagged, and finelier cut than the last mentioned, soft also and downie: the floures are fewer, lesse and paler than in the ordinarie, but turne speedily into downe like as the former. ‡

¶ *The Place.*

These herbes are very common throughout England, and do grow almost euery where.

¶ *The Time.*

They flourish almost euery moneth of the yeare.

‡ 2 *Erigeron tomentoſum alterum.*
The other Cotton Groundſell.

¶ *The Names.*

Groundſel is called in Greek ἠριγέρων: in Latine, *Senecio*, becauſe it waxeth old quickely: by a baſtard name *Herbutum*: in Germany, **Creutzwurtz**: in low-Dutch, **Cruys cruyt**, and **Cruyſken cruyt**: in Spaniſh, *Yerua cana*: in Italian, *Cardoncello*, *Speliciosa*: in Engliſh, Groundſel.

Cotton Groundſell ſeemeth to be all one with *Theophraſtus* his *Aphace*; hee maketh mention of *Aphace* in his ſeuenth booke, which is not onely a kinde of pulſe, but an herbe alſo, vnto which this kinde of Groundſell is very like. For as *Theophraſtus* ſaith, The herbe *Aphace* is one of the pot-herbs and kindes of Succorie: adding further, That it floureth in haſte, but yet ſoone is old, and turneth into down; and ſuch a one is this kind of Groundſell. But *Theophraſtus* ſaith further, That it floureth all the winter long, and ſo long as the Spring laſteth, as my ſelfe haue often ſeene this Groundſell do.

¶ *The Temperature.*

Groundſell hath mixt faculties; it cooleth, and withall digeſteth, as *Paulus Ægineta* writeth.

¶ *The Vertues.*

The leaues of Groundſell boyled in wine or water, and drunke, healeth the paine and ache of the ſtomacke that proceedeth of choler. A

The leaues and floures ſtamped with a little Hogs greaſe ceaſeth the burning heat of the ſtones and fundament. By adding to a little ſaffron or ſalt it helpeth the *Struma* or Kings Euill. B

The leaues ſtamped and ſtrained into milke and drunke, helpeth the red gummes and frets in children. C

Dioſcorides ſaith, That with the fine pouder of Frankinſence it healeth wounds in the ſinewes. The like operation hath the downe of the floures mixed with vineger. D

Boyled in Ale with a little honey and vineger, it prouoketh vomit, eſpecially if you adde thereto a few roots of *Aſſarabacca*. E

CHAP. 28. *Of Saint Iames his Wort.*

¶ *The Kindes.*

THe herbe called Saint Iames his wort is not without cauſe thought to be a kinde of Groundſel: of which there be ſundry ſorts; ſome of the paſture, and one of the ſea; ſome ſweet-ſmelling, and ſome of a loathſome ſauor. All which kindes I will ſet downe.

¶ *The Deſcription.*

1 SAint Iames his wort or Rag-wort is very well knowne euery where, and bringeth forth at the firſt broad leaues, gaſhed round about like to the leaues of common Wormewood, but broader, thicker, not whitiſh or ſoft, of a deepe greene colour, with a ſtalke which riſeth vp aboue a cubit high, chamfered, blackiſh, and ſomewhat red withall. The armes or wings are ſet with leſſer leaues like thoſe of Groundſell or of wilde Rocket. The floures at the top be of a yellow colour like Marigolds, as well the middle button as the ſmall floures that ſtand in a pale round about, which turne into downe as doth Groundſell. The root is thready.

‡ 2 This hath ſtalkes ſome cubit high, creſted, and ſet with long whitiſh leaues; the lower leaues are the ſhorter; but the vpper leaues the longer, yet the narrower: at the top of the ſtalke grow ſome foure or fiue floures as in an vmbell, which are of a darke red colour before they open themſelues,

1 *Iacobæa.* Rag-wort.

‡ 2 *Iacobæa anguſtifolia.* Narrow leaued Rag-weed.

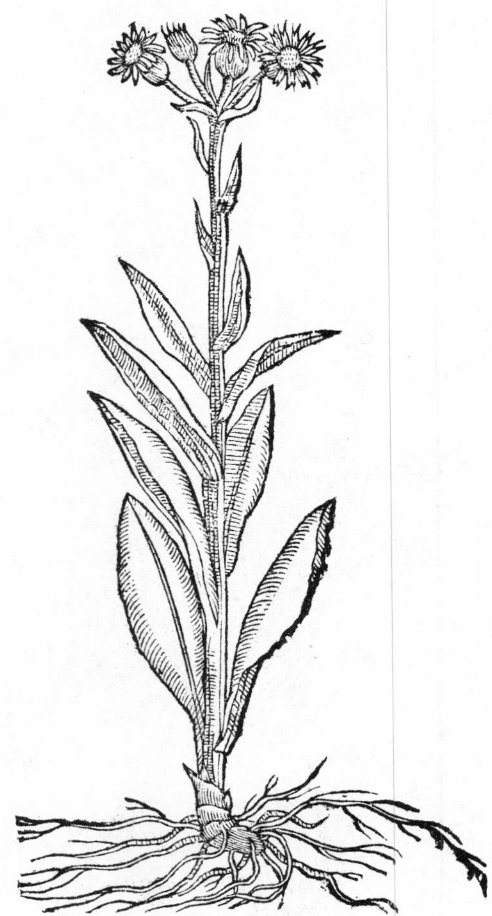

‡ 3 *Iacobæa latifolia.*
Broad leaued Rag-weed.

4 *Iacobæa marina.*
Sea Rag-weed.

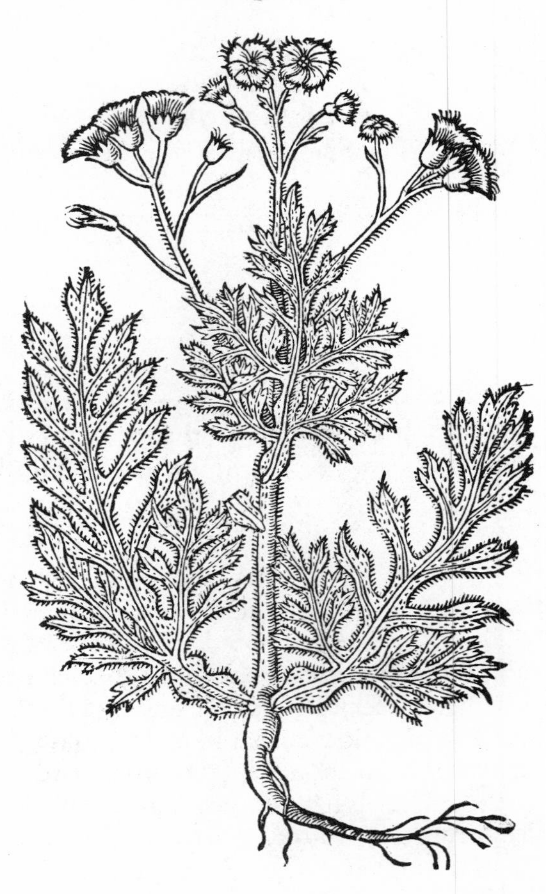

themselues ; but opened, of a bright golden colour, and those are ingirt by fifteene or more little leaues, which are of a flame colour aboue, and red vnderneath. The floures fly away in downe, and the seed is blackish, and like that of the former. The roots are made of many strings like those of the precedent.

3 This broad leaued Rag-weed hath stiffe crested stalkes, which are set with broad wrinckled sharpe pointed leaues, of a greene colour: the bottome leaues are the larger and rounder, the top leaues the lesse, and more diuided. The floures grow at the top of the stalkes, in shape and colour like those of the common Rag-weed, but much bigger: They also turne into Downe as the former. ‡

4 Sea Rag-wort groweth to the height of two cubits: the stalkes be not reddish as the other, but contrariwise Ash-coloured, gray and hoary: the leaues be greater and broader than the other: the floures grow at the top, of a pale yellow colour, couered on the cup or huske of the floure, as also the leaues, with a certaine soft white Downe or freese: the floures vanish into Downe, and fly away with the winde.

¶ *The Place.*

Land Rag-wort groweth euery where in vntilled pastures and fields, which are somewhat moist especially, and neere vnto the borders of fields.

‡ 2 3 These grow vpon the Austrian and Heluetian Alpes. ‡

The fourth kinde of Rag-wort groweth neere the sea side in sundry places: I haue seene it in the field by Margate, by Queakes house, and by Byrchenton in the Isle of Tenet: likewise it groweth neere the Kings ferry in the Isle of Shepey, in the way leading to Sherland house, where S. *Edward Hobby* dwelleth: and likewise at Queenborough castle in the same Isle; and in other places. ‡ I haue been at the former and later of these places to finde out plants, yet could I not see this plant. It growes in the garden of Mr. *Ralph Tuggy*; but I feare hardly wilde in this kingdome. ‡

¶ *The Time.*

They floure in Iuly and August, at which time they are carried away with the Downe.

¶ *The Names.*

The first is called in Latine, *Herba S. Iacobi*, or *S. Iacobi flos*, and *Iacobæa*: in high-Dutch, **Sant Jacobs bloumen**: in low-Dutch, **Sant Jacobs cruyt**: in French, *Fluer de S. Iacques*: in English, S. Iames his Wort: the countrey people do call it Stagger-wort, and Staner-wort, and also Rag-wort, ‡ and Rag-weed. In Holdernesse in Yorke-shire they call it Seggrum.

The second is *Iacobæa Pannonica* 2. of *Clusius*.

The third is his *Iacobæa latifolia*. *Gesner* calls it *Coniza montana*. ‡

The fourth is named *Cineraria*, or Ash-coloured S. *Iames* Wort: some call it *Erigeron marinum*, or Sea Groundsell: of some, *Artemisia marina*. ‡ And by *Prosper Alpinus*, *Artemisia alba*. ‡

¶ *The Temperature.*

S. Iames wort is hot and dry in the second degree, and also cleansing, by reason of the bitternesse which it hath.

¶ *The Vertues.*

It is commended by the later Physitions to be good for greene wounds, and old filthy Vlcers which are not scoured, mundified, and made cleane; it also healeth them, with the iuyce hereof tempered with honey and May butter, and boyled together to the forme of an Vnguent or salue. A

It is much commended, and not without cause, to helpe old aches and pains in the armes, hips, and legs, boyled in hogs grease to the forme of an ointment. B

Moreouer, the decoction hereof gargarised is much set by as a remedie against swellings and impostumations of the throat, which it wasteth away and throughly healeth. C

The leaues stamped very small, and boyled with some hogs grease vnto the consumption of the iuyce, adding thereto in the end of the boyling a little Masticke and Olibanum, and then strained, taketh away the old ache in the huckle-bones called Sciatica. D

‡ The Egyptians (saith *Prosper Alpinus*) vse the Sea Rag-wort, for many things: for they commend the decoction made with the leaues thereof against the stone in the kidnies and bladder, as also to helpe the old obstructions of the inward parts, but principally those of the wombe; as also the coldnesse, strangulation, barrennesse, inflation thereof, and it also brings downe the intercepted courses: wherefore women troubled with the mother are much eased by baths made of the leaues and floures hereof. ‡ E

Chap. 29. *Of Garden Succorie.*

¶ *The Kindes.*

THere be ſundry ſorts of plants comprehended vnder the title of *Cichoracea,* that is to ſay, Cichorie, Endiue, Dandelion, &c. differing not ſo much in operation and working, as in ſhape and forme, which hath cauſed many to deeme them diuers, who haue diſtinguiſhed them vnder the titles aforeſaid: of euery which kinde there be diuers ſorts, the which ſhall be diuided in their ſeuerall chapters, wherein the differences ſhall be expreſt.

¶ *The Deſcription.*

1 GArden Succory is of two ſorts, one with broad leaues, and the other with narrow, deeply cut and gaſhed on both ſides The firſt hath broad leaues ſomewhat hairie, not much vnlike to Endiue, but narrower; amongſt which doe riſe vp ſtalkes, whereon are placed the like leaues, but ſmaller. The ſtalke diuideth it ſelfe toward the top into many branches, whereon doe grow little blew floures conſiſting of many ſmall leaues, after which followeth white ſeed. The root is tough, long, and white of colour, continuing many yeeres; from the which as from euery part of the plant doth iſſue forth bitter and milkie juice. The whole plant is of a bitter taſte likewiſe.

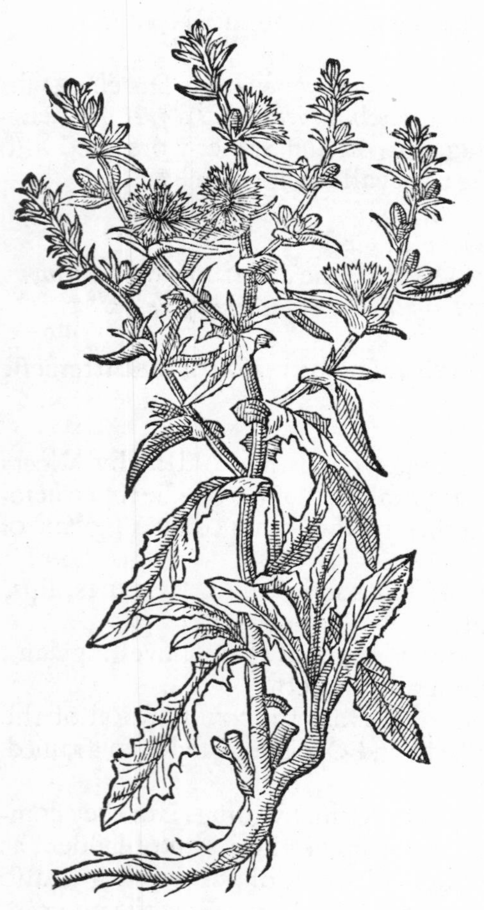

2 *Cichorium ſatiuum.*
Garden Succorie.

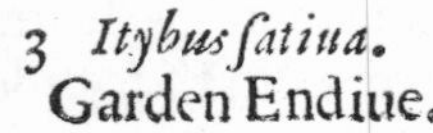

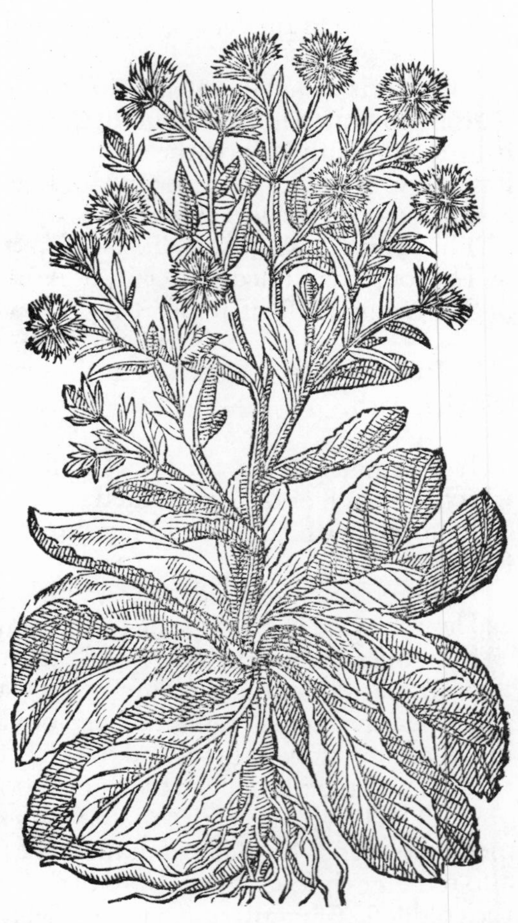

3 *Itybus ſatiua.*
Garden Endiue.

2 The ſecond kinde of Succorie is like vnto the former, but greater in euery point. That which cauſeth the difference is, that this beareth floures white of colour tending to blewneſſe; the others blew, as I haue ſaid.

3 Garden Endiue bringeth forth long leaues, broad, ſmooth, more greene than white, like almoſt to thoſe of lettuce, ſomething nicked in the edges. The ſtalke groweth vp among the leaues, being round and hollow, diuided into branches; out of which being broken or cut there iſſueth a juice like milke, ſomewhat bitter: the floures vpon the branches conſiſt of many leaues, in colour com-

‡ 5 *Cichorium ſpinoſum.*
Thornie Succory.

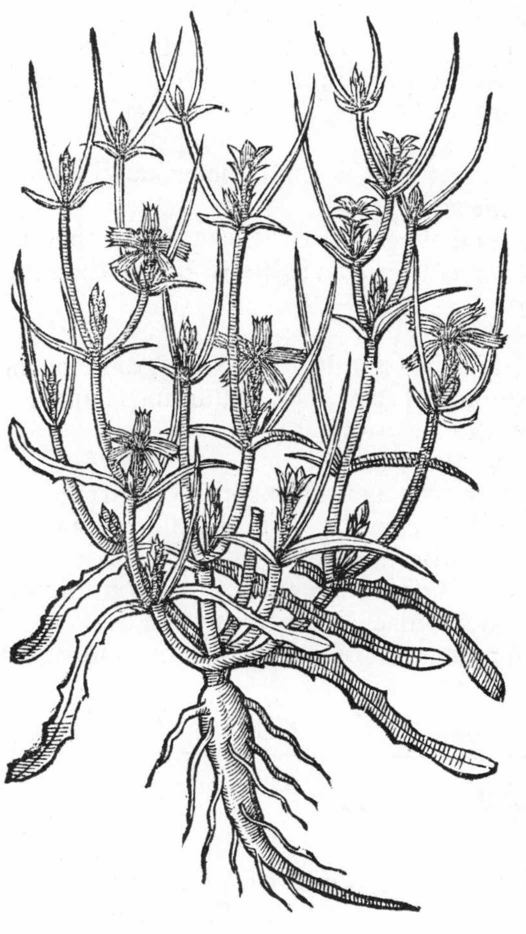

commonly blew, ſeldome white. The root is long,white, with ſtrings growing thereat, which withereth after the ſeed is ripe.

4 Curled Endiue hath leaues not vnlike to thoſe of the curled or Cabbage lettuce, but much greater; among which riſe vp ſtrong and thicke ſtalkes,ſet with the like leaues,but leſſer, and not ſo notably curled or criſped. The flours grow at the top,blew of colour. The root periſheth,as doth the whole plant, when it hath brought forth his ripe ſeed.

‡ 5 To theſe may fitly be added the thorny or prickly Succory of Candy, being of this kindred, and there vſed in defect of the true Succorie,in ſtead thereof. The root is pretty long, white, with few fibers hanging thereat; the ſtalke is hard,wooddy,and diuaricated into many branches,which commonly end in two or three prickes like hornes: The leaues are bitter, long, narrow, and ſharpe pointed, and lie ſpread vpon the ground,and are a little ſinuated, or cut about the edges: The flours,which vſually grow vpon little footſtalkes at the diuiſions of the branches,are much like thoſe of the ordinary Succory, yet much leſſe, conſiſting of fiue blew leaues, with yellow chiues in the middle. The ſeed is like thoſe of the common Succory. It floures in Iuly and Auguſt. ‡

¶ *The Place and Time.*

This Succory, and theſe Endiues are only ſowne in gardens.

Endiue being ſowen in the ſpring quickly commeth vp to floure, which ſeedeth in harueſt,and afterward dieth. But being ſowen in Iuly it remaineth till winter, at which time it is taken vp by the roots,and laid in the ſunne or aire for the ſpace of two houres; then will the leaues be tough,and eaſily endure to be wrapped vpon an heape, and buried in the earth with the roots vpward,where no earth can get within it(which if it did,would cauſe rottenneſſe)the which ſo couered may be taken vp at times conuenient, and vſed in ſallades all the winter,as in London and other places is to be ſeen;and then it is called white Endiue, whereof *Pliny* ſeemeth not to be ignorant,ſpeaking to the ſame purpoſe in his 20.booke and 8.chapter.

¶ *The Names.*

Theſe herbes be called by one name in Greek Σέριδες: notwithſtanding for diſtinctions ſake they called the garden Succory, Σέρις ἥμερος, and the wilde Succory, σέρις ἀγρία; *Pliny* nameth the Succory *Hedypnois*: and the bitterer *Dioſcorides* calleth πικρίς: in Latine,*Intybum ſylueſtre*, *Intybum agreſte*, *Intybum erraticum*,and *Cichorium*: in ſhops it is called *Cichorea*, which name is not onely allowed of the later Phyſitions, but alſo of the Poet *Horace* in the 31.Ode of his firſt booke,

Me paſcunt oliuæ,
Me Cichorea,leueſque maluæ.

With vs,ſaith *Pliny* in his 20.booke,8.chapter, they haue called *Intybum erraticum*, or wilde Endiue, *Ambugia* (others reade *Ambubeia*:)and ſome there be that name it *Roſtrum porcinum*: and others, as *Guilielmus Placentinus*, and *Petrus Creſcentius*, terme it *Sponſa ſolis*: the Germanes call it **Wegwarten**,which is as much to ſay,as the keeper of the waies: the Italians, *Cichorea*: the Spaniards, *Almerones*: the Engliſh-men, Cicorie and Succory: the Bohemians, *Czakanka*.

Endiue is named in Greeke Σέρις ἥμερος: in Latine,*Intybum ſatiuum*: of ſome, *Endiuia*: of *Auicen* and *Serapio*, *Taraxacon*: of the Italians, *Scariola*, which name remaineth in moſt ſhops; alſo *Seriola*, as though they ſhould fitly call it *Seris*, but not ſo well *Serriola*, with a double *r*: for *Serriola* is *Lactuca ſylueſtris*,or wilde lettuce: it is called in Spaniſh, *Serraya Enuide*: in Engliſh, Endiue,and Scariole: and when it hath been in the earth buried as aforeſaid, then it is called white Endiue.

‡ 5 This was first set forth by *Clusius* vnder this name, *Chondrillæ genus elegans cæruleo flore:* since, by *Pona* and *Bauhine*, by the title we giue you, to wit, *Cichorium spinosum. Honorius Bellus* writes that in Candy where as it naturally growes, they vulgarly terme it σταμναγκαθι, that is, *Hydriæ spina*, the Pitcher Thorne; because the people fetch all their water in stone pots or Pitchers, which they stop with this plant, to keepe mice and other such things from creeping into them: and it growes so round, that it seems by nature to be prouided for that purpose. ‡

¶ *The Nature.*

Endiue and Succorie are cold and drie in the second degree, and withall somewhat binding: and because they be something bitter, they doe also clense and open.

Garden Endiue is colder, and not so drie or clensing, and by reason of these qualities they are thought to be excellent medicines for a hot liuer, as *Galen* hath written in his 8. book of the compositions of medicines according to the places affected.

¶ *The Vertues.*

A These herbs when they be greene haue vertue to coole the hot burning of the liuer, to helpe the stopping of the gall, yellow jaundice, lacke of sleepe, stopping of vrine, and hot burning feauers.

B A syrup made thereof and sugar is very good for the diseases aforesaid.

C The distilled water is good in potions, cooling and purging drinkes.

D The distilled water of Endiue, Plantaine, and roses, profiteth against excoriations in the conduit of the yard to be iniected with a syringe, whether the hurt came by vncleanenesse or by small stones and grauell issuing forth with the vrine; as often hath been seene.

E These herbes eaten in sallades or otherwise, especially the white Endiue, doth comfort the weake and feeble stomacke, and cooleth and refresheth the stomacke ouermuch heated.

F The leaues of Succorie brused are good against inflammation of the eyes, being outwardly applied to the grieued place.

Chap. 30. *Of wilde Succorie.*

† 1 *Cichorium syluestre.*
Wilde Succorie.

† 2 *Cichorium luteum.*
Yellow Succorie.

¶ *The*

¶ *The Kindes.*

IN like manner as there be ſundrie ſorts of Succories and Endiues, ſo is there wilde kindes of either of them.

¶ *The Deſcription.*

1 WIlde Succorie hath long leaues, ſomewhat ſnipt about the edges like the leaues of Sow-thiſtle, with a ſtalke growing to the height of two cubits, which is diuided towards the top into many branches. The floures grow at the top blew of colour: the root is tough, and wooddie, with many ſtrings faſtned thereto.

2 Yellow Succorie hath long and large leaues, deepely cut about the edges like thoſe of the Hawkeweed. The ſtalke is branched into ſundry arms, wheron do grow yellow flours very double, reſembling the floures of Dandelion, or Piſſe-abed; the which being withered, it flieth away in downe with euery blaſt of winde.

3 *Intybum ſylueſtre.*
Wilde Endiue.

3 Wilde Endiue hath long ſmooth leaues ſlightly ſnipt about the edges. The ſtalke is brittle and full of milkie juice, as is all the reſt of the plant: the floures grow at the top, of a blew or skie colour: the root is tough and threddie.

4 Medow Endiue, or Endiue with broad leaues, hath a thicke, tough, and wooddie root with many ſtrings faſtened thereto, from which riſe vp many broad leaues ſpread vpon the ground like thoſe of garden Endiue, but leſſer, and ſomewhat rougher, among which riſe vp many ſtalkes immediately from the root; euery of them are deuided into ſundrie branches, whereupon doe grow many floures like thoſe of the former, but ſmaller.

¶ *The Place.*

Theſe plants doe grow wilde in ſundrie places in England, vpon wilde and vntilled barren grounds, eſpecially in chalkie and ſtonie places.

¶ *The Time.*

They floure from the middeſt to the end of Auguſt.

¶ *The Names.*

‡ The firſt of theſe is *Seris Picris* of *Lobell*, or *Cichorium ſylueſtre*: or *Intybus erratica* of *Tabernamontanus*. ‡

Yellow Succorie is not without cauſe thought to be *Hyoſiris*, or (as ſome copies haue it) *Hyoſciris*, of which *Pliny* in his 20. booke and 8. chapter writeth; *Hyoſiris* (ſaith he) is like to Endiue, but leſſer and rougher: it is called of *Lobelius*, *Hedypnois*: the reſt of the names ſet forth in their ſeuerall titles ſhall be ſufficient for this time.

¶ *The Temperature.*

They agree in temperature with the garden Succorie, or Endiue.

¶ *The Vertues.*

The leaues of theſe wilde herbes are boiled in pottage or brothes, for ſicke, and feeble perſons that haue hot, weake, and feeble ſtomackes, to ſtrengthen the ſame. A

They are iudged to haue the ſame vertues with thoſe of the garden, if not of more force in working. B

† The firſt figure was of *Cichoreum album ſatiuum* of *Tabernamontanus*. The ſecond is *Cichoreum luteum*. But the true figures of thoſe our Author meant were vnder theſe titles. The firſt, of *Hieracium Latifolium*. The ſecond, *Dens Leonis Cichoriſata*; for that is *Lobells Hedypnois*.

Chap. 31. *Of Gumme Succorie.*

¶ *The Deſcription.*

1 Gvmme Succorie with blew floures hath a thicke and tough root, with ſome ſtrings annexed thereto, full of a milkie iuyce, as is all the reſt of the plant, the floures excepted. The leaues are great and long, in ſhape like to thoſe of garden Succorie, but deeplier cut or iagged, ſomewhat after the manner of wilde Rocket: among which riſe tender ſtalkes very eaſie to be broken, branched toward the top in two or ſometimes three branches, bearing very pleaſant floures of an azure colour or deepe blew; which being paſt, the ſeed flieth away in downe with the winde.

1 *Chondrilla cærulea.*
Blew Gum Succorie.

2 *Chondrilla cærulea latifolia.*
Robinus Gum Succory.

2 Gum Succorie with broad leaues, which I haue named *Robinus* Gum Succorie (for that he was the firſt that made any mention of a ſecond kind, which he ſent me as a great dainty, as indeed I confeſſe it) in roots is like the former: the leaues be greater, not vnlike to thoſe of Endiue, but cut more deeply euen to the middle rib: the ſtalkes grow to the height of two foot: the floures likewiſe are of an azure colour, but ſprinckled ouer as it were with ſiluer ſand; which addeth vnto the floure great grace and beauty.

3 Yellow gum Succorie hath long leaues like in forme and diuiſion of the cut leaues to thoſe of wild Succorie, but leſſer, couered all ouer with a hoarie down. The ſtalke is two foot high, white and downie alſo, diuided into ſundry branches, whereupon doe grow torne floures like thoſe of Succorie, but in colour yellow, which are turned into downe that is caried away with the winde. The root is long, and of a meane thickneſſe, from which, as from all the reſt of the plant, doth iſſue forth a milky iuyce, which being dried is of a yellowiſh red. ſharp, or biting the tongue. There is found vpon the branches hereof a gum, as *Dioſcorides* ſaith, which is vſed at this day in phyſicke in the Iſle Lemnos, as *Bellonius* witneſſeth.

4 Spaniſh

4 Spanish Gum Succorie hath many leaues spred vpon the ground, in shape like those of Groundsell, but much more diuided, and not so thicke nor fat: amongst which rise vp branched stalkes set with leaues like those of *Stœbe salamantica minor*, or Siluer-weed, whereof this is a kinde. The floures grow at the top, of an ouerworne purple colour, which seldome shew themselues abroad blowne: ‡ The seed is like that of *Carthamus* in shape, but blacke and shining. ‡

† 3 *Chondrilla lutea.*
Yellow Gum Succorie.

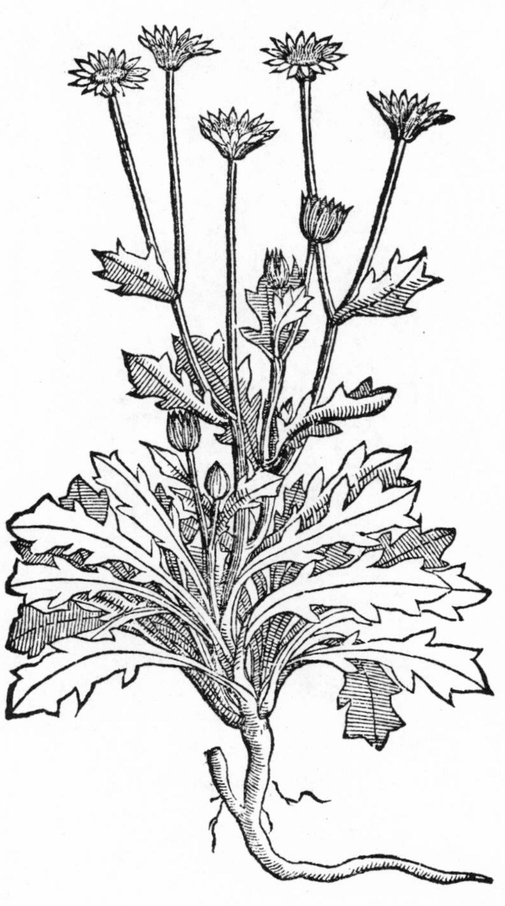

† 4 *Chondrilla Hispanica.*
Spanish Gum Succorie.

5 Rushy Gum Succorie hath a tough and hard root, with a few short threds fastned thereto; from the which rise vp a few iagged leaues like those of Succorie, but much more diuided: The stalke groweth vp to the height of two foot, tough and limmer like vnto rushes, whereon are set many narrow leaues. The floures be yellow, single, and small; which being faded doe fly away with the winde: the whole plant hauing milky iuyce like vnto the other of his kinde.

‡ There is another sort of this plant to be found in some places of this kingdome, and it is mentioned by *Bauhinus* vnder the name of *Chondrilla viscosa humilis.*

† 6 Sea Gum Succorie hath many knobby or tuberous roots full of iuyce, of a whitish purple colour, with long strings fastned to them; from which immediately rise vp a few small thinne leaues fashioned like those of Succory, narrower below, and somewhat larger towards their ends; among which spring vp small tender stalkes, naked, smooth, hollow, round, of some foot high, or thereabout: each of these stalkes haue one floure, in shape like that of the Dandelion, but lesser. The whole plant is whitish or hoary, as are many of the sea plants. †

7 Swines Succorie hath white small and tender roots, from the which rise many indented leaues like those of Dandelion, but much lesse, spred or laid flat vpon the ground; from the midst whereof rise vp small soft and tender stalkes, bearing at the top double yellow floures like those of Dandelion or Pisse-abed, but smaller: the seed with the downy tuft flieth away with the wind.

8 The male Swines Succorie hath a long and slender root, with some few threds or strings fastned thereto; from which spring vp small tender leaues about the bignesse of those of Dasies, spred vpon the ground, cut or snipt about the edges confusedly, of an ouerworne colour, full of a milky iuyce: among which rise vp diuers small tender naked stalkes, bearing at the top of euery stalke one floure and no more, of a faint yellow colour, and something double: which being ripe, do

5 *Chondrilla juncea.*
Ruſhy Gum Succorie.

6 *Chondrilla marina Lobelij.*
Sea Gum Succorie.

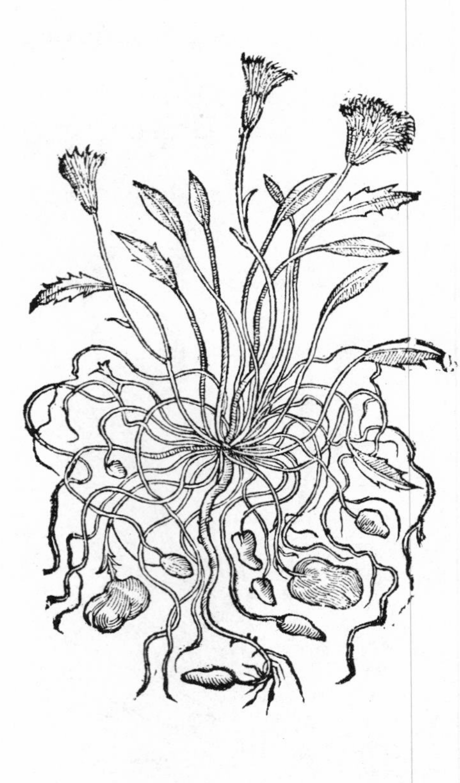

7 *Hypochæris, Porcellia.*
Swines Succorie.

8 *Hyoſeris maſcula.*
Male Swines Succorie.

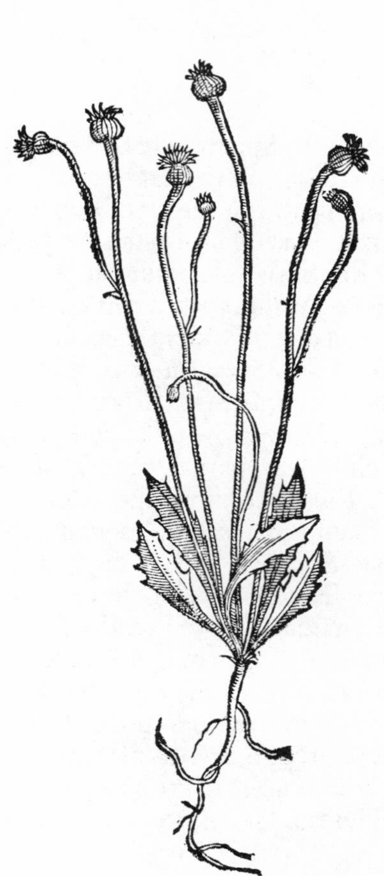

doe turne into downe that is carried away with the winde: the ſeed likewiſe cleaueth vnto the ſaid downe, and is alſo carried away with the winde. The whole plant periſheth when it hath perfected his ſeed, and recouereth it ſelfe againe by the falling thereof.

‡ 9 *Cichorium verracarium.*
Wart-Succorie.

‡ 9 I thinke it expedient in this place to deliuer vnto you the hiſtorie of the *Cichorium verrucarium*, or *Zacintha* of *Matthiolus*; of which our Author maketh mention in his Names and Vertues, although he neither gaue figure, nor the leaſt deſcription thereof. This Wart-Succory (for ſo I will call it) hath leaues almoſt like Endiue, greene, with pretty deepe gaſhes on their ſides; the ſtalkes are much creſted, and at the top diuided into many branches; betweene which, and at their ſides grow many ſhort ſtalkes with yellow floures like thoſe of Succorie, but that theſe turne not into Downe, but into cornered and hard heads, moſt commonly diuided into eight cels or parts, wherein the ſeed is contained. ‡

¶ *The Place.*

† Theſe plants are found only in gardens in this country; the ſeuenth & eighth excepted, which peraduenture may be found to grow in vntilled places, vpon ditches bankes and the borders of fields, or the like.

¶ *The Time.*

They do floure from May to the end of Auguſt.

¶ *The Names.*

Gum Succorie hath beene called of the Grecians, χονδρίλη: of the Latines, *Condrilla*, and *Chonarilla*: *Dioſcorides* and *Pliny* call it *Cichorion*, and *Seris*, by reaſon of ſome likeneſſe they haue with Succorie, eſpecially the two firſt, which haue blew floures as thoſe of the Succories. *Lobelius* maketh *Cichorea verrucaria* to be *Zacintha* of *Matthiolus*.

‡ ¶ *Names in particular.*

‡ 1 This is called *Chondrilla cærulea Belgarum*, of *Lobel*: *Apate*, of *Daleſchampius*.

2 *Condrilla* 2. of *Matthiolus*: *Chondrilla latifolia cærulea*, of *Tabernamontanus*.

3 *Chondrilla prior Dioſcoridis*, of *Cluſius* and *Lobel*.

4 *Chondrilla rara purpurea, &c.* of *Lobel*: *Chondrilla Hiſpanica Narbonenſis*, of *Tabern. Seneciocarduus Apulus*, of *Columna*.

5 *Chondrilla prima Dioſcoridis*, of *Columna* and *Bauhinus*: *Viminea, viſcoſa*, of *Lobel* and *Cluſius*.

6 *Chondrilla altera Dioſcoridis*, of *Columna*: ſome thinke it to be ἀφάκη of *Theophraſtus*: *Lobell* calls it, *Chondrilla puſilla marina lutea bulboſa*.

7 *Hypochæris, porcellia*, of *Tabernamontanus*.

8 *Hieracium minimum* 9. of *Cluſius*: *Hyoſeris latifolia*, of *Tabern.* The two laſt ſhould haue bin put among the *Hieracia*.

9 *Cichorium verrucarium*, and *Zacinthus* of *Matthiolus* and *Cluſius*. ‡

¶ *The Nature and Vertues.*

Theſe kinds of gum Succorie are like in temperature to the common Succory, but drier.

The root and leaues tempered with hony, and made into Trochiskes, or little flat cakes, with niter or ſalt-peter added to them, cleanſe away the morphew, ſun-burnings, and all ſpots of the face. A

The gum which is gathered from the branches, whereof it tooke his name, layeth downe the ſtairing haires of the eye-browes and ſuch like places: and in ſome places it is vſed for Maſtick, as *Bellonius* obſerues. B

The gum poudered with myrrh, and put into a linnen cloath, and a peſſarie made thereof like a finger, and put vp, bringeth downe the termes in yong Wenches and ſuch like. C

The

D The ſeedes of *Zazintha* beate to powder, and giuen in the decreaſing of the Moone to the quantitie of a ſpoonefull, taketh away warts, and ſuch like excreſcence, in what part of the body ſoeuer they be; the which medicine a certaine Chirurgion of Padua did much vſe, whereby he gained great ſums of mony, as reporteth that ancient Phyſition *Ioachimus Camerarius* of Noremberg a famous citie in Germanie. And *Matthiolus* affirmes that he hath knowne ſome helped of warts, by once eating the leaues hereof in a Sallade.

† The figure of the third was of the ſame plant as the firſt, and was *Chondrilla alba* of *Taber*. The fourth was of *Hieracium montanum maius Latifolium* of *Tabern.* which you ſhall finde in the tenth place in the foure and thirtieth Chapter.

Chap. 32. *Of Dandelion.*

¶ *The Deſcription.*

1 The herbe which is commonly called Dandelion doth ſend forth from the root long leaues deepely cut and gaſhed in the edges like thoſe of wilde Succorie, but ſmoother: vpon euery ſtalke ſtandeth a floure greater than that of Succorie, but double, and thicke ſet together, of colour yellow, and ſweet in ſmell, which is turned into a round downie blowball, that is carried away with the winde. The root is long, ſlender, and full of milkie juice when any part of it is broken, as is the Endiue or Succorie, but bitterer in taſte than Succorie.

‡ There are diuers varieties of this plant, conſiſting in the largeneſſe, ſmallneſſe, deepeneſſe, or ſhallowneſſe of the diuiſions of the leafe, as alſo in the ſmoothneſſe and roughneſſe thereof. ‡

1 *Dens Leonis.*
Dandelion.

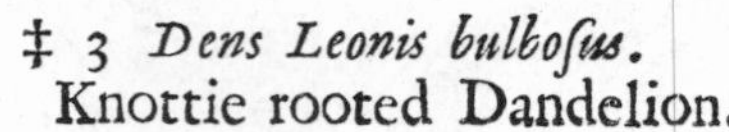

‡ 3 *Dens Leonis bulboſus.*
Knottie rooted Dandelion.

2 There is alſo another kinde of Succorie which may be referred heereunto, whoſe leaues are long, cut like thoſe of broad leafed Succorie: the ſtalkes are not vnlike, being diuided into branches as thoſe of Dandelion, but leſſer, which alſo vaniſheth into downe when the ſeed is ripe, hauing a long and white root.

‡ 3 There is another *Dens Leonis*, or Dandelion, which hath many knotty, and tuberous roots like

like thoſe of the Aſphodil; the leaues are not ſo deeply cut in as thoſe of the common Dandelion, but larger, and ſomewhat more hairy: the floures are alſo larger, and of a paler yellow, which flie away in ſuch downe as the ordinary. ‡

¶ *The Place.*

They are found often in medowes neere vnto water ditches, as alſo in gardens and high waies much troden.

¶ *The Time.*

They floure moſt times in the yeere, eſpecially if the winter be not extreame cold.

¶ *The Names.*

Theſe plants belong to the Succory which *Theophraſtus*, & *Pliny* call *Aphaca*, or *Aphace Leonardus: Fuchſius* thinketh that Dandelion is *Hedypnois Plinij*, of which he writeth in his 20. booke, and eighth chapter, affirming it to be a wilde kinde of broad leafed Succorie, and that Dandelion is *Taraxacon*: but *Taraxacon*, as *Auicen* teacheth in his 692. chapter, is garden Endiue, as *Serapio* mentioneth in his 143. chapter; who citing *Paulus* for a witneſſe concerning the faculties, ſetteth down theſe words which *Paulus* writeth of Endiue and Succorie. Diuers of the later Phyſitions do alſo call it *Dens Leonis*, or Dandelion: it is called in high Dutch, **Kolkraut:** in low-Dutch, **Papencruit:** in French, *Piſſenlit ou couronne de preſtre*, or *Dent de lyon*: in Engliſh, Dandelion: and of diuers, Piſſeabed. The firſt is alſo called of ſome, and in ſhops *Taraxacon*, *Caput monachi*, *Roſtrum porcinum*, and *Vrinaria*. The other is *Dens Leonis Monſpelienſium* of *Lobell*, and *Cichoreum Conſtantinopolitanum*, of *Matthiolus*.

¶ *The Temperature and Vertues.*

A Dandelion is like in temperature to Succorie, that is to ſay, to wilde Endiue. It is cold, but it drieth more, and doth withall clenſe, and open by reaſon of the bitterneſſe which it hath ioyned with it: and therefore it is good for thoſe things for which Succory is. ‡ Boiled, it ſtrengthens the weake ſtomacke, and eaten raw it ſtops the bellie, and helpes the Dyſentery, eſpecially being boyled with Lentiles; The juice drunke is good againſt the vnuoluntary effuſion of ſeed; boyled in vineger, it is good againſt the paine that troubles ſome in making of water; A decoction made of the whole plant helpes the yellow jaundice. ‡

† The figure which was in the 2 place was of the *Cich. Luteum*, where you may find it, but to what plant the deſcription may be referred, I cannot yet determine.

Chap. 20. *Of Sow-thiſtle.*

† 1 *Sonchus aſper.* Prickly Sow-thiſtle.

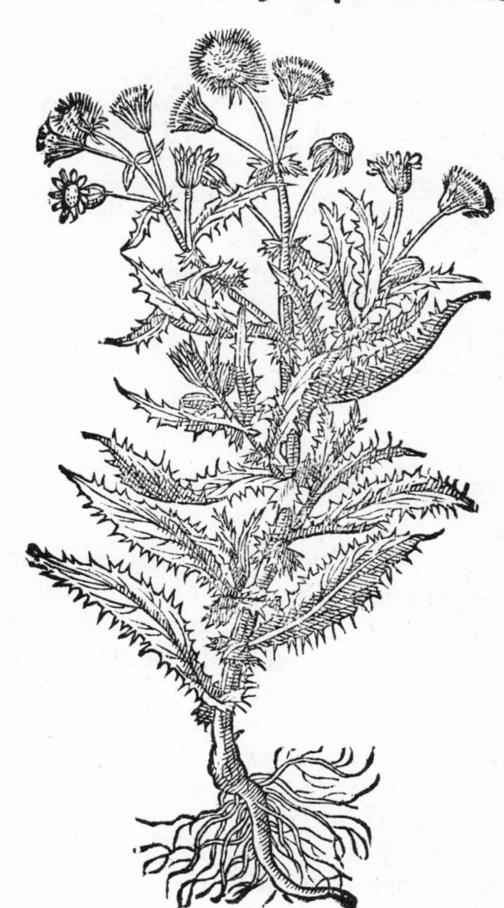

‡ 2 *Sonchus aſperior.* The more prickly Sow-thiſtle.

¶ The

¶ *The Kindes.*

There be two chiefe kindes of Sow-thistles; one tenderer and softer; the other more pricking and wilder: but of these there be sundry sorts more found by the diligence of the later Writers; all which shall be comprehended in this chapter, and euery one be distinguished with a seuerall description.

¶ *The Description.*

1 The prickly Sow-thistle hath long broad leaues cut very little in, but full of smal prickles round about the edges something hard and sharpe, with a rough and hollow stalke: the floures stand on the tops of the branches, consisting of many small leaues, single, and yellow of colour; and when the seed is ripe it turneth into downe, and is carried away with the winde. The whole plant is full of a white milky iuyce.

‡ 2 There is another kinde of this, whose leaues are sometimes prettily deepe cut in like as those of the ordinarie Sow-thistle; but the stalkes are commonly higher than those of the last described, and the leaues more rough and prickly; but in other respects not differing from the rest of this kinde. It is also sometimes to be found with the leaues lesse diuided. ‡

† 3 *Sonchus Læuis.*
Hares Lettuce.

4 *Sonchus lænis latifolius.*
Broad leaued Sow-thistle.

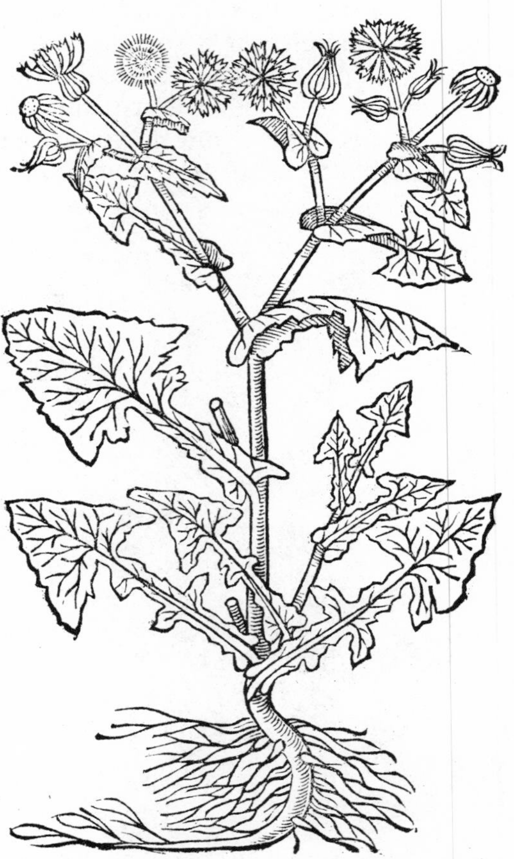

3 The stalke of Hares Lettuce, or smooth-Thistle is oftentimes a cubit high, edged and hollow, of a pale colour, and sometimes reddish: the leaues be greene, broad, set round about with deepe cuts or gashes, smooth, and without prickles. The floures stand at the top of the branches, yellow of colour, which are caried away with the winde when the seed is ripe. ‡ This is sometimes found with whitish, and with snow-white floures, but yet seldome: whence our Authour made two kindes more, which were the fourth and fifth; calling the one, The white floured Sow-thistle; and the other, The snow-white Sow-thistle. Both these I haue omitted as impertinent, and giue you others in their stead. ‡

4 Broad leaued Sow-thistle hath a long thicke and milky root, as is all the rest of the Plant, with many strings or fibres; from the which commeth forth a hollow stalke branched or diuided into sundry sections. The leaues be great, smooth, sharpe pointed, and greene of colour: the floures be

be white, in ſhape like the former. ‡ The floures of this are for the moſt part yellow like as the former. ‡

‡ 5 Wall Sow-thiſtle hath a fibrous wooddy root, from which riſes vp a round ſtalke not creſted: the leaues are much like to thoſe of the other Sow-thiſtles, broad at the ſetting on, then narrower, and after much broader, and ſharpe pointed, ſo that the end of the leafe much reſembles the ſhape of an iuy leafe; theſe leaues are very tender, and of ſomewhat a whitiſh colour on the vnder ſide: the top of the ſtalke is diuided into many ſmall branches, which beare little yellow floures that fly away in downe.

6 This hath longiſh narrow leaues ſoft and whitiſh, vnequally diuided about the edges. The ſtalkes grow ſome foot high, hauing few branches, and thoſe ſet with few leaues, broad at their ſetting on, and ending in a ſharpe point: the floures are pretty large like to the great Hawk-weed, and fly away in downe: the root is long, white, and laſting. It floures moſt part of Summer; and in Tuſcany, where it plentifully growes, it is much eaten in ſallets, with oile and vineger, it hauing a ſweetiſh and ſomewhat aſtringent taſte. ‡

‡ 5 *Sonchus læuis muralis.*
Wall (or Iuy-leaued) Sow-thiſtle.

‡ 6 *Sonchus læuis anguſtifolius.*
Narrow leaued Sow-thiſtle.

† 7 This blew floured Sow-thiſtle is the greateſt of all the reſt of the kindes, ſomewhat reſembling the laſt deſcribed in leaues; but thoſe of this are ſomewhat rough or hairy on the vnder ſide: the floures are in ſhape like thoſe of the ordinarie Sow-thiſtle, but of a faire blew colour; which fading, flie away in Downe that carries with it a ſmall aſh-coloured ſeed. The whole plant yeeldeth milke as all the reſt do. †

8 Tree Sow thiſtle hath a very great thicke and hard root ſet with a few hairy threds; from which ariſeth a ſtrong and great ſtalke of a wooddy ſubſtance, ſet with long leaues not vnlike to Languedebeefe, but more deepely cut in about the edges, and not ſo rough: vpon which do grow faire double yellow floures, which turne into Downe, and are caried away with the winde. The whole plant is poſſeſt with ſuch a milky iuyce as are the tender and hearby Sow-thiſtles; which certainly ſheweth it to be a kinde thereof: otherwiſe it might be referred to the Hawke-weeds, whereunto in face and ſhew it is like. ‡ This hath a running root, and the heads and tops of the ſtalkes are very rough and hairy. ‡

7 *Sonchus flore cæruleo.*
Blew-floured Sow-thiſtle.

8 *Sonchus Arboreſcens.*
Tree Sow-thiſtle.

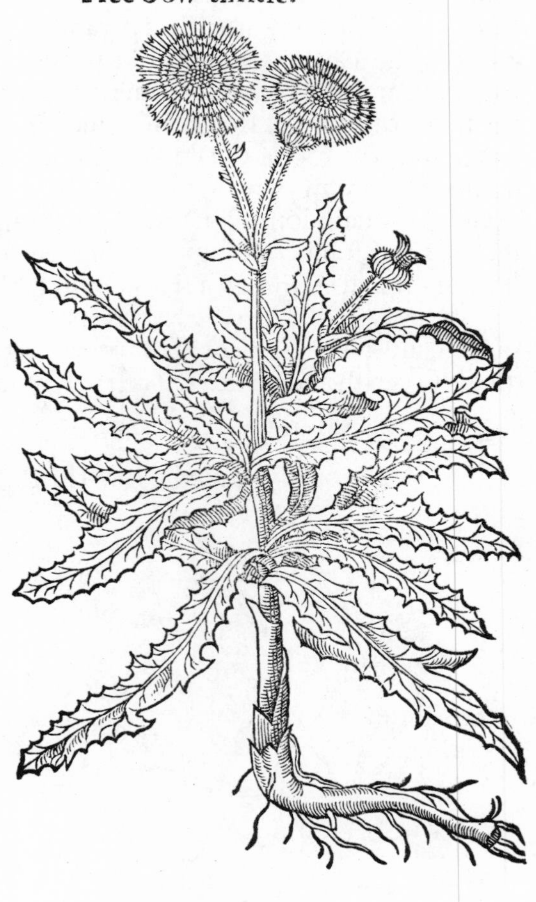

‡ 9 *Sonchus arboreſcens alter.*
The other Tree Sow-thiſtle.

‡ 10 *Sonchus ſyluaticus.*
Wood Sow-thiſtle.

‡ 9 This other Tree Sow-thistle growes to a mans height or more, hauing a firme crested stalke, smooth, without any prickles, and set with many leaues incompassing the stalke at their setting on, and afterwards cut in with foure, or sometimes with two gashes only: the vpper leaues are not diuided at all: the colour of these leaues is green on the vpper side, and grayish vnderneath: the top of the stalke is hairy, and diuided into many branches, which beare the floures in an equall height, as it were in an vmbell: the floures are not great, considering the largenesse of the plant, but vsually as big as those of the common Sow-thistle, and yellow, hauing a hairy head or cap: the seed is crested, longish, and ash-coloured, and flies away with the downe: the root is thicke, whitish, hauing many fibres, putting out new shoots, and spreading euery yeare. *Bauhine* maketh this all one with the other, according to *Clusius* his description: but in my opinion there is some difference betweene them, which chiefely consists, in that the former hath larger and fewer floures; the plant also not growing to so great a height. ‡

‡ 10 This plant (whose figure our Author formerly gaue, *pag.* 148. vnder the title of *Erysimum sylvestre*) hath long knotty creeping roots, from whence ariseth a round slender stalke some two foot high, set at first with little leaues, which grow bigger and bigger as they come neerer the middle of the stalke, being pretty broad at their setting on, then somewhat narrower, and so broader againe, and sharpe pointed, being of the colour of the Wall (or Iuy-leaued) Sow-Thistle. The top is diuided into many small branches, which end in small scaly heads like those of the wilde Lettuce, containing floures consisting of foure blewish purple leaues, turned backe and snipped at their ends; there are also some threds in the midle of the floure, which turning into Downe, carry away with them the seed, which is small, and of an Ash-colour. *Bauhine* makes a bigger and a lesser of these, distinguishing betweene that of *Clusius* (whose figure I here giue you) and that of *Columna*; yet *Fabius Columna* himselfe could finde no difference, but that *Clusius* his plant had fiue leaues in the floure, and his but foure: which indeed *Clusius* in his description affirmes; yet his figure (as you may see) expresses but foure: adding, That the root is not well expressed; which notwithstanding *Clusius* describes according to *Columna's* expression. ‡

¶ *The Place*

The first foure grow wilde in pastures, medowes, woods, and marishes neere the sea, and among pot-herbes.

The fifth growes vpon walls, and in wooddy mountainous places.

The Tree Sow-thistle growes amongst corne in waterie places.

The sixth, seuenth and tenth are strangers in England.

¶ *The Time.*

They floure in Iune, Iuly, August, and sometimes later.

¶ *The Names.*

Sow-thistle is called in Greeke, σόγχος: in Latine, *Sonchus*: of diuers, *Cicerbita, lactucella*, and *Lacterones*. *Apuleius* calleth it *Lactuca Leporina*, or Hares-thistle: of some, *Brassica Leporina*, or Hares Colewort. The English names are sufficiently touched in their seuerall titles: In Dutch it is called **Hasen Latouwe**: the French, *Palays de lieure*.

‡ ¶ *Names in particular.*

1 This is *Sonchus asper major* of *Cordus*: *Sonchus tenerior aculeis asperior* of *Lobel*: *Sonchus* 3. *asperior* of *Dodonæus*.

2 This is *Sonchus asper*, of *Matthiolus*, *Fuchsius*, and others.

3 This, *Matthiolus*, *Dodonæus*, *Lobel*, and others call *Sonchus læuis*: *Tragus* calls it *Intybus erratica tertia*.

4 This *Tabernamontanus* onely giues, vnder the title as you haue it here.

5 *Matthiolus* stiles this, *Sonchus læuis alter*: *Cæsalpinus* calls it *Lactuca murorum*: and *Tabern. Sonchus sylvaticus quartus*: *Lobel*, *Sonchus alter folio sinuato hederaceo*.

6 *Lobel* calls this, *Sonchus læuis Matthioli*: it is *Terracrepulus* of *Cæsalpinus*: and *Crepis* of *Dalechampius*.

7 *Clusius* and *Camerarius* giue vs this vnder the title of *Sonchus cœruleus*.

8 Onely *Tabern.* hath this figure, vnder the title our Author giues it: *Bauhine* puts it amongst the *Hieracia*, calling it *Hieracium arborescens palustre*.

9 This *Bauhine* also makes an *Hieracium*, and would persuade vs that *Clusius* his description belongs to the last mentioned, and the figure to this: to which opinion I cannot consent. *Clusius* giueth it vnder the name of *Sonchus* 3 *læuis altissimus*.

10 This *Clusius* giues vnder the name of *Sonchus læuior Pannonicus* 4. *flore purp.* *Tabern.* calls it *Libanotis Theophrasti sterilis*: *Columna* hath it by the name of *Sonchus montanus purpureus* πρώτιστον: *Cordus*, *Gesner*, *Thalius*, and *Bauhine* refer it to the *Lactucæ sylvestres*: the last of them terming it, *Lactuca montana purpuro-cœrulea*. ‡

¶ *The Temperature.*

The Sow-thiſtles, as *Galen* writeth, are of a mixt temperature; for they conſiſt of a watery and earthy ſubſtance, cold, and likewiſe binding.

¶ *The Vertues.*

A Whileſt they are yet yong and tender they are eaten as other pot-herbes are; but whether they be eaten, or outwardly applied in manner of a pulteſſe, they do euidently coole: therefore they be good for all inflammations or hot ſwellings, if they be laid thereon.

B Sow-thiſtle giuen in broth taketh away the gnawings of the ſtomacke proceeding of an hot cauſe, and increaſe milke in the breaſts of Nurſes, cauſing the children whom they nurſe to haue a good colour: and of the ſame vertue is the broth if it be drunken.

C The iuyce of theſe herbes doth coole and temper the heate of the fundament and priuy parts.

CHAP. 34. *Of Hawke-weed.*

¶ *The Kindes.*

HAwke-weed is alſo a kinde of Succorie; of which *Dioſcorides* maketh two ſorts, and the later Writers more: the which ſhall be deſcribed in this chapter following, where they ſhall be diſtinguiſhed as well with ſeuerall titles as ſundry deſcriptions.

† 1 *Hieracium majus Dioſcoridis.*
Great Hawke-weed.

2 *Hieracium minus, ſiue Leporinum.*
Small Hares Hawk-weed, or Yellow Diuels-bit.

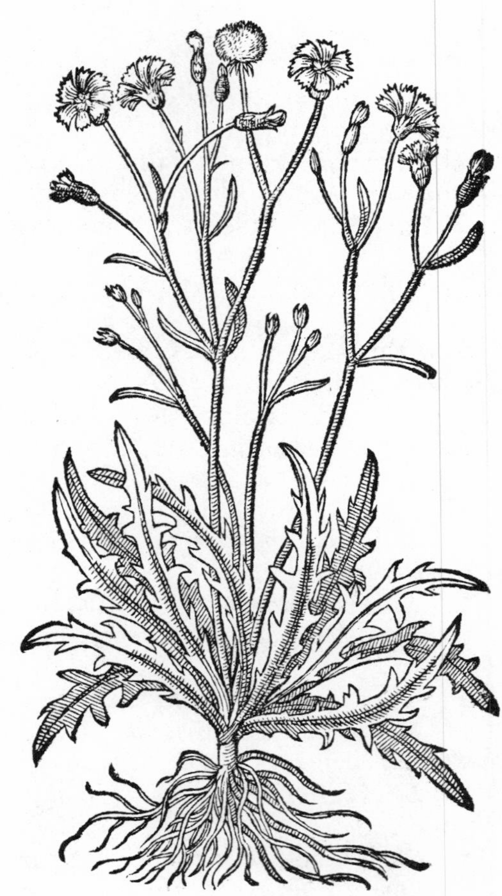

¶ *The Deſcription.*

1 THe great Hawke-weed hath large and long leaues ſpred vpon the ground, in ſhape like thoſe of the Sow-thiſtle: the ſtalk groweth to the height of two cubits, branched into ſundry armes or diuiſions, hollow within as the yong Kexe, reddiſh of colour; whereupon do grow yellow floures thicke and double, which turne into Downe that flieth away with the winde when the ſeed is ripe. The root is thicke, tough and threddy.

2 The

2 The ſmall Hawke-weed, which of moſt writers hath been taken for yellow Diuels-bit, hath long leaues deepely cut about the edges, with ſome ſharpe roughneſſe thereon like vnto Sow-thiſtle. The ſtalkes and floures are like the former: the root is compact of many ſmall ſtrings, with a ſmall knob, or as it were the ſtumpe of an old root in the middle of thoſe ſtrings, cut or bitten off; whereupon it tooke his name Diuels bit.

3 Blacke Hawke-weed hath very many long iagged leaues, not much vnlike to thoſe of Bucks horne, ſpred flat and farre abroad vpon the ground, which the picture cannot expreſſe as is requiſite, in ſo little roome: among which riſe vp many ſtalkes ſlender and weake, the floures growing at the top yellow and very double: it hath alſo a threddy root.

‡ Our Author formerly gaue three figures, and ſo many deſcriptions of this ſmall *Hieracium*, which I haue contracted into two; for the onely difference that I can finde is, that the one hath the root as it were bitten off, with the leaues leſſe cut in; the other hath a root ſomewhat longer, and fibrous as the former; the leaues alſo in this are much more finely and deepe cut in: in other reſpects there is no difference. ‡

3 *Hieracium nigrum.*
Blacke Hawke-weed.

4 *Hieracium Aphacoides.*
Succory Hawke-weed.

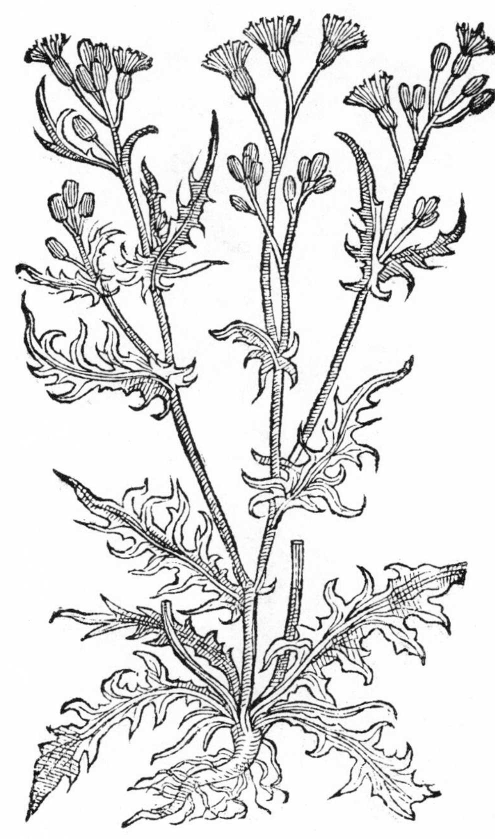

4 Succory Hawke-weed hath many long and large leaues ſpred vpon the ground, deepely cut on both ſides almoſt to the middle rib; from which riſe vp ſmall ſtalkes and floures like thoſe of the leſſe Dandelion, but leſſer. The root conſiſteth of many ſmall threddy ſtrings.

5 Endiue Hawke-weed hath many broad leaues, indented about the edges very like vnto Garden Endiue, but narrower; among which riſe vp ſtalkes a foot or more high, ſlender, hairy, and brittle: the floures are yellow, and grow at the top double, and thick ſet in a ſcaly huske like the Knapweed or Iacea, hauing great thicke and threddy roots. ‡ This hath a ſtalke ſometimes more, and otherwhiles leſſe rough, with the leaues ſomtimes more cut in, more long and narrow, and againe otherwhiles more ſhort and broad. ‡

6 Long rooted Hawk-weed hath many broad leaues ſpread vpon the ground, ſleightly & confuſedly indented about the edges, with ſomewhat a bluntiſh point; among which leaues ſpring vp ſtrong and tough ſtalks a foot and halfe high, ſet on the top with faire double yellow floures much like vnto a Piſſe-abed. The root is very long, white and tough.

7 Sharpe Hawk-weed hath leaues like thoſe of Languebeefe or Ox-tongue, but much narrower, ſharpe about the edges, and rough in the middle: the ſtalks be long and ſlender, ſet with the like leaues, but leſſer: the floures grow at the top, double and yellow: the root is tough & threddy.

5 *Hieracium intybaceum.*
Endiues Hawke-weed.

6 *Hieracium longius radicatum.*
Long rooted Hawke-weed.

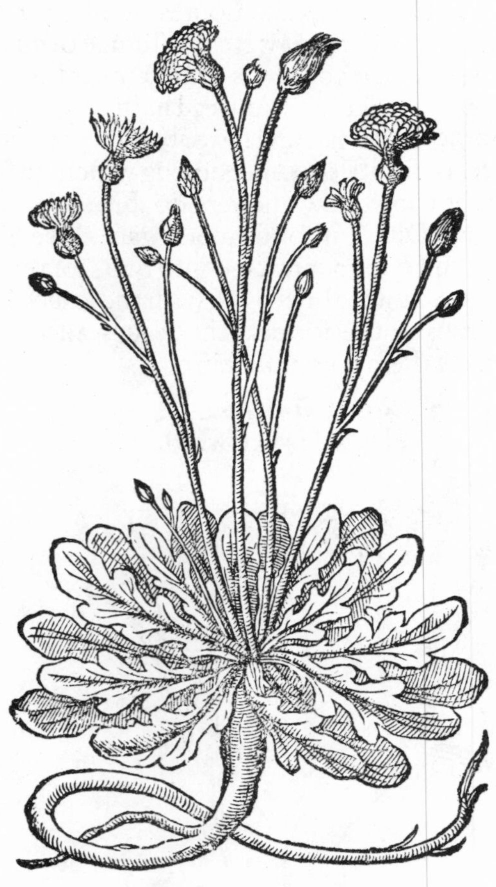

7 *Hieracium asperum.*
Sharpe Hawke-weed.

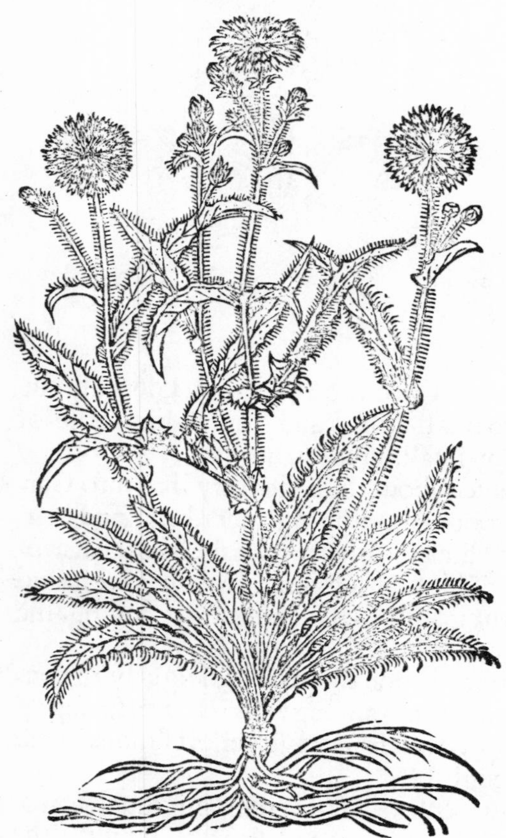

8 *Hieracium falcatum Lobelij.*
Crooked Hawke-weed.

‡ 9 *Hieracium.*

† 8 Crooked or falked Hawkeweed hath leaues like vnto the garden Succory, yet much ſmaller, and leſſe diuided, ſlightly indented on both ſides, with tender, weake, and crooked ſtalkes; whereupon doe grow floures like thoſe of *Lampſana*, of a blacke, or pale yellow colour, and the roote ſmall and threddy. The ſeedes are long, and falcated, or crooked, ſo that they ſomewhat reſemble the foot or clawes of a bird, and from theſe ſeeds the plant hath this Epithite, *Falcatum*, or crooked in maner of a Sicle or Sithe.

‡ 9 This in leaues is not much vnlike the laſt deſcribed, but that they are ſomewhat broader, and leſſe cut in, hauing little or no bitterneſſe nor milkineſſe, the ſtalkes are ſome foot high commonly bending, or falling vpon the ground; the floures are ſmall and yellow, and ſeeme to grow out of the middeſt of the ſeed, whenas indeed they grow at the top of them, the reſt being but an empty huske which is falcated like that of the laſt deſcribed. This figure we giue you was taken before the floures were blowne, ſo that by that meanes the falcated or crooked ſeed veſſels are not expreſt in this, but you may ſee there manner of growing by the former. ‡

‡ 9 *Hieracium falcatum alterum.*
The other crooked Hawkeweed.

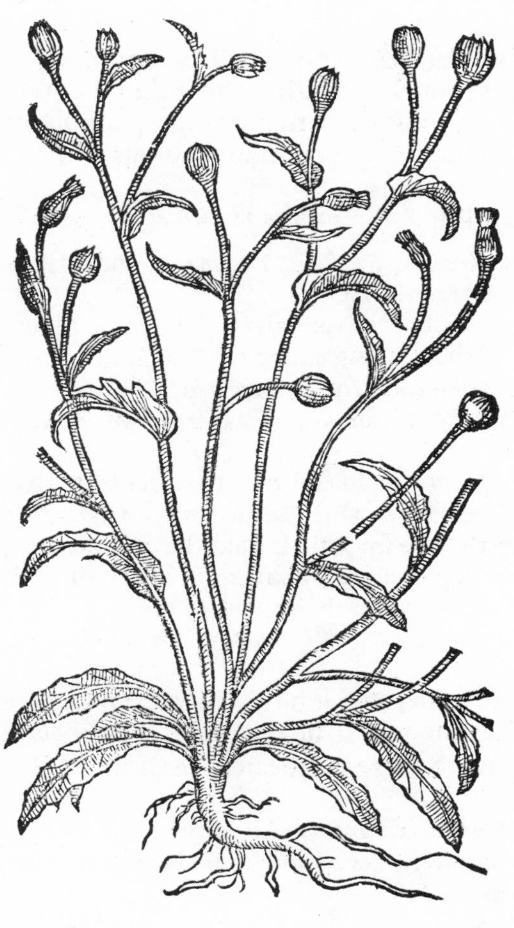

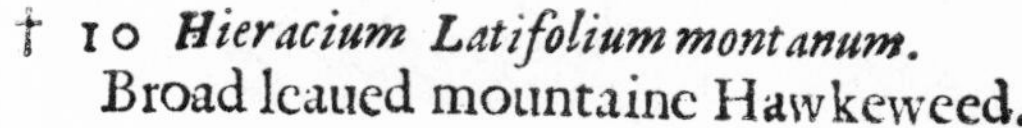

† 10 *Hieracium Latifolium montanum.*
Broad leaued mountaine Hawkeweed.

10 The broad leaued mountaine Hawkeweed hath broad, long, ſmooth leaues, deepely indented toward the ſtalke, reſembling the leaues of the greateſt Sowthiſtle. The ſtalke is hollow, and ſpungious, full of a milkie iuice, as is the reſt of the plant, as alſo all the other of his kinde: the floures grow at the top of the ſtalkes, double and yellow.

11 The narrow leaued mountaine Hawkeweed hath leaues like thoſe of the laſt deſcribed, but narrower. The ſtalkes be fat, hollow, and full of milke: the floures grow at the top double, and yellow of colour. The root is ſmall and threddy.

There is a ſmall mountaine Hawkeweed hauing leaues like vnto the former, but more deepely cut about the edges, and ſharper pointed, the ſtalkes are tender and weake; the floures be double and yellow like thoſe of Piloſella, or great Mouſe-eare; the root is ſmall and threddy.

¶ *The Place.*

Theſe kindes of herbes doe grow in vntoiled places neere vnto the borders of corne fields, in medowes, high-waies, woode, mountaines, and hillie places, and neere to the brinks of ditches. ‡ The two falcated Hawkeweeds grow onely in ſome few gardens. ‡

¶ *The*

11 *Hieracium montanum Latifolium minus.*
The lesser broad leaued mountaine Hawke-weed.

¶ *The Time.*

They floure for the most part all the summer long, some sooner, and others later.

¶ *The Names in generall.*

These plants are all conteined vnder the name of *Hieracium*: which is called in Greeke also ἱεράκιον: diuers name it in Latine, *Accipitrina*, which is termed in French, *Cichoree iaulne*: in English, Hawkeweed. These herbes tooke there name from a Hawke, which is called in Latine *Accipiter*, and in Greeke, ἱέραξ, for they are reported to cleere their sight by conueying the juice heereof into their eyes. *Gaza* calleth it *Porcellia* for it is numbred among the Succories, they are called also *Lampuca*.

Yellow Hawkeweed is called of some *Morsus diaboli*, or yellow Diuels bit, for that the root doth very well resemble the bitten or cropt root of the common Diuels bit, being like Scabious.

‡ ¶ *The Names in particular.*

1 *Matthiolus*, *Fuchsius*, *Dodonæus*, and others call this *Hieracium maius*.

2 3 These are varieties of the same plant, the first of them being called by *Fuchsius*, *Dodonæus*, and *Matthiolus*, *Hieracium minus*, *Lobell* calls it, *Hieracium minus præmorsa radice*. That sort of this with more cut leaues is by *Tabernamontanus* called, *Hieracium nigrum*.

4 *Lobell* calls this *Hieracium folijs & facie Chondrillæ*; *Bauhinus* makes this to differ from that which our Author gaue in this 4. place out of *Tabern.* for he termes this *Hieracium Chondrillæ folio hirsutum*, and the other, *Hieracium Chondrillæ folio Glabrum*; the one smooth leaued, the other rough; yet that which growes frequently with vs, and is very well represented by this figure, hath smooth leaues, as he also obserued it to haue in Italy and about Mountpelier in France.

5 This is *Hieracium alterum grandius*, and *Hieracium montanum angustifolium primum* of *Tabernamontanus*.

6 *Lobell* calls this from the length of the root (though sometimes it be not so long) *Hieracium Longius radicatum*; as also *Taber. Hieracium macrorhizon*, it is thought to be the *Apargia* of *Theophrastus*, by *Daleschampius* in the *Hist. Lugd. pag.* 562. but the figure there that beares the title is of *Hieracium minus*.

7 *Tabernamontanus* first gaue this vnder the name of *Hieracium intybaceum asperum*: *Bauhine* refers it to the wilde yellow Succories, and calls it *Cichoreum montanum angustifolium hirsutie asperum*.

8 This *Lobell* calls *Hieracium Narbonense falcata siliqua*.

9 He calls this *Hieracium facie Hedypnois*: and *Cæsalpinus* termes this *Rhagadiolus*; and the last mentioned, *Rhagadiolus alter*.

10 This by *Tabernamontanus* is called *Hieracium montanum majus Latifolium*; The figure of this was giuen by our Author, *chap.* 30. vnder the title of *Chondrilla Hispanica*.

11 *Tabernamontanus* also stiles this *Hieracium montanum Latifolium minus.* ‡

¶ *The Nature.*

The kindes of Hawkeweed are cold and dr ie, and somewhat binding.

¶ *The Vertues.*

A They are in vertue and operation like to *Sonchus* or Sowthistle, and being vsed after the same manner, be as good to all purposes that it doth serue vnto.

B They be good for the eie-sight, if the juice of them be dropped into the eyes, especially that which is called Diuels bit, which is thought to be the best, and of greatest force.

Therefore

Therefore as *Dioſcorides* writeth,it is good for an hot ſtomacke, and for inflammations if it be laid vpon them. C

The herbe and root being ſtamped and applied,is a remedie for thoſe that be ſtung of the ſcorpion;which effect not onely the greater Hawkeweeds, but the leſſer ones alſo doe performe. D

CHAP. 35. *Of Cluſius Hawkeweed.*

¶ *The Kindes.*

THere be likewiſe other ſorts of Hawkeweeds, which *Carolus Cluſius* hath ſet forth in his Pannonicke obſeruations,the which likewiſe require a particular chapter, for that they do differ in forme very notably.

1 *Hieracium primum latifolium Cluſij.* The firſt Hawkeweed of *Cluſius.*

2 *Hieracium* 5. *Cluſij.* *Cluſius* his 5.kinde of Hawkeweed.

¶ *The Deſcription.*

1 THe firſt of *Cluſius* his Hawkeweeds haue great broad leaues ſpred vpon the ground, ſomewhat hairie about the edges, oftentimes a little iagged, alſo ſoft as is the leafe of Mullen,or Higtaper,and ſometimes daſht here and there with ſome blacke ſpots, in ſhape like the garden Endiue, full of a milkie juice: among which riſeth vp a thicke hollow ſtalke of a cubit high,diuiding it ſelfe at the top into two or three branches,whereupon do grow ſweete ſmelling floures not vnlike to thoſe of yellow Succorie, ſet or placed in a blacke hoarie and woollie cup or huske, of a pale bleake yellow colour,which turneth into a downie blowball that is caried away with the winde: the root entereth deepely into the ground, of the bigneſſe of a finger,full of milke,and couered with a thicke blacke barke.

2 The ſecond ſort of great Hawkeweed according to my computation,and the 5. of *Cluſius*, hath leaues like the former, that is to ſay ſoft,and hoarie, and as it were couered with a kinde of white

white woollineſſe or hairineſſe, bitter in taſte, of an inche broad. The ſtalke is a foot high, at the top whereof doth grow one yellow floure like that of the great Hawkeweed, which is caried away with the winde when the ſeed is ripe. The root is blacke and full of milkie juice, and hath certaine white ſtrings annexed thereto.

3 This kinde of Hawkeweed hath blacke roots a finger thicke, full of milkie juice, deepely thruſt into the ground, with ſome ſmall fibers belonging thereto: from which come vp many long leaues halfe an inch or more broad, couered with a ſoft downe or hairineſſe, of an ouerworne ruſſet colour: and amongſt the leaues come vp naked and hard ſtalkes, whereupon doe grow yellow floures ſet in a woollie cup or chalice, which is turned into downe, and caried away with his ſeed by the winde.

4 The fourth Hawkeweed hath a thicke root aboue a finger long, blackiſh, creeping vpon the top of the ground, and putting out ſome fibres, and it is diuided into ſome heads, each whereof at the top of the earth putteth out ſome ſix or ſeuen longiſh leaues ſome halfe an inche broad, and ſomewhat hoarie, hairie, and ſoft as are the others precedent, and theſe leaues are ſnipt about the edges, but the deepeſt gaſhes are neereſt the ſtalkes, where they are cut in euen to the middle rib, which is ſtrong and large. The ſtalke is ſmooth, naked, and ſomewhat high: the floures be yellow and double as the other.

3 *Hieracium 6. Cluſij.*
Cluſius his 6. Hawkeweed.

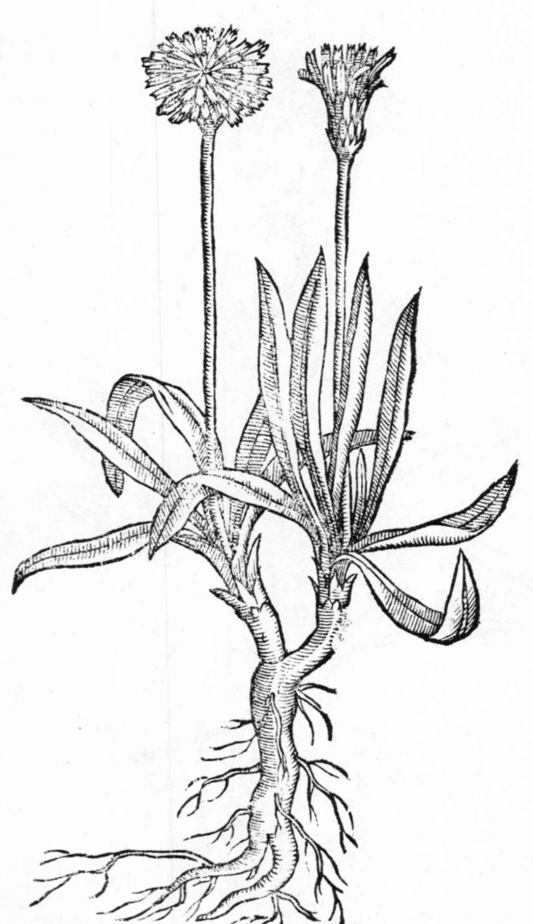

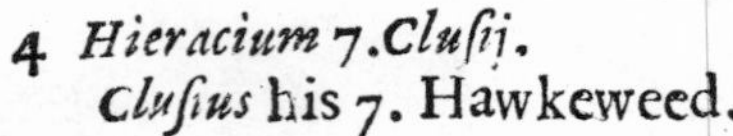

4 *Hieracium 7. Cluſij.*
Cluſius his 7. Hawkeweed.

‡ 5 The ſame Author hath alſo ſet forth another *Hieracium*, vnder the name of *Hieracium paruum Creticum*, which he thus deſcribes; this is an elegant little plant ſpreading ſome ſix, or more leaues vpon the top of the ground, being narrower at that part whereas they adhere to the root, and broader at the other end, and cut about the edges, hauing the middle rib of a purple colour; amongſt theſe riſe vp two or three little ſtalkes about a foot high, without knot vntill you come almoſt to the top, whereas they are diuided into two little branches, at which place growes forth leaues much diuided; the floures grow at the top of a ſufficient bigneſſe, conſidering the magnitude of the plant, and they conſiſt of many little leaues lying one vpon another, on the vpper ſide wholly white, and on the vnder ſide of a fleſh colour. The root is ſingle, longiſh, growing ſmall towards

towards the end, and putting forth stringy fibres on the sides. Thus much *Clusius*, who receiued this figure and description from his friend *Iaques Plateau* of Tournay. I coniecture this to be the same plant that *Bauhine* hath somewhat more accurately figured and described in his *Prod. pag*. 68. vnder the title of *Chondrilla purpurascens fœtida*: which plant being an annuall, I haue seen growing some yeares since with Mr. *Tuggy* at Westminster; and the last Summer with an honest and skilfull Apothecarie one Mr. *Nicholas Swayton* of Feuersham in Kent: but I must confesse I did not compare it with *Clusius*; yet now I am of opinion, that both these figures and descriptions are of one and the same plant. It floures in Iuly and August, at the later end of which moneth the seeds also come to ripenesse.

6 This other (not described by *Clusius*, but by *Lobel*) hath long rough leaues cut in and toothed like to Dandelion, with naked hairy stalkes, bearing at their tops faire large and very double yellow floures, which fading fly away in downe. It growes in some medowes.

‡ 5 *Hieracium parvum Creticum.*
Small Candy Hawk-weed.

‡ 6 *Hieracium Dentis leonis folio hirsutum.*
Dandelion Hawk-weed.

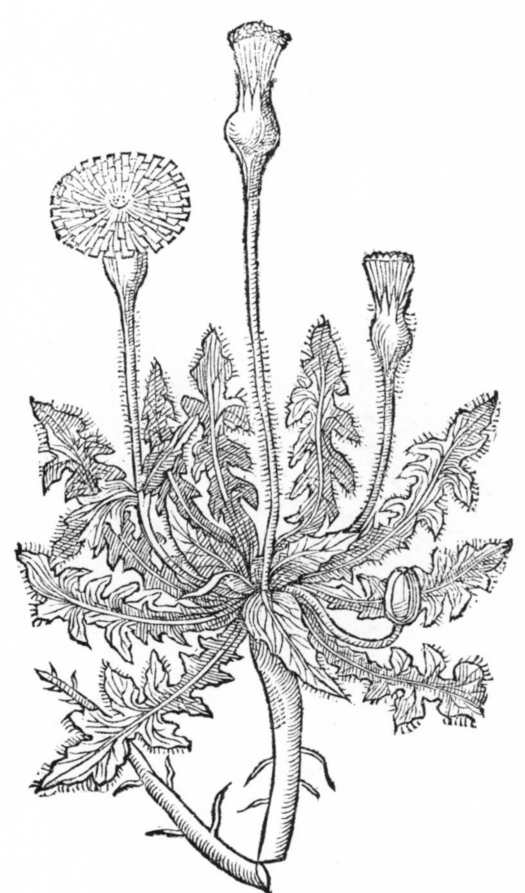

¶ *The Place.*

These kinds of Hawke-weeds, according to the report of *Clusius*, do grow in Hungarie and Austria, and in the grassy dry hills, and herby and barren Alpish mountaines, and such like places: notwithstanding if my memorie faile me not I haue seene them growing in sundry places in England; which I meane, God willing, better to obserue hereafter, as opportunitie shall serue me.

¶ *The Time.*

He saith they floure from May to August, at what time the seed is ripe.

¶ *The Names.*

The Author himselfe hath not said more than here is set downe as touching the names, so that it shall suffice what hath now been said, referring the handling thereof to a further consideration.

¶ *The Nature and Vertues.*

I finde not any thing at all set downe either of their nature or vertues, and therefore I forbeare to say any thing else of them, as a thing not necessarie to write of their faculties vpon my owne conceit and imagination.

Chap.

Chap. 36.

‡ *Of French or Golden Lung-wort.*

‡ 1 *Pulmonaria Gallica ſiue aurea latifolia.*
Broad-leaued French or golden Lung-wort.

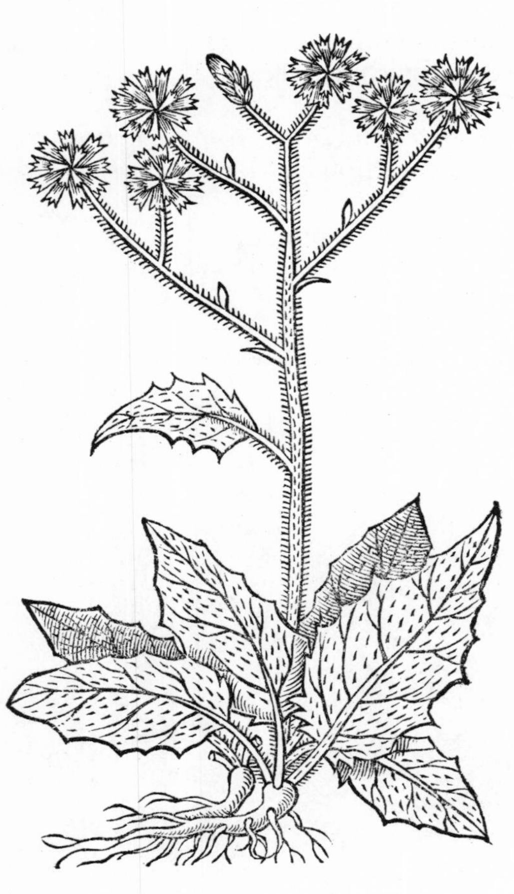

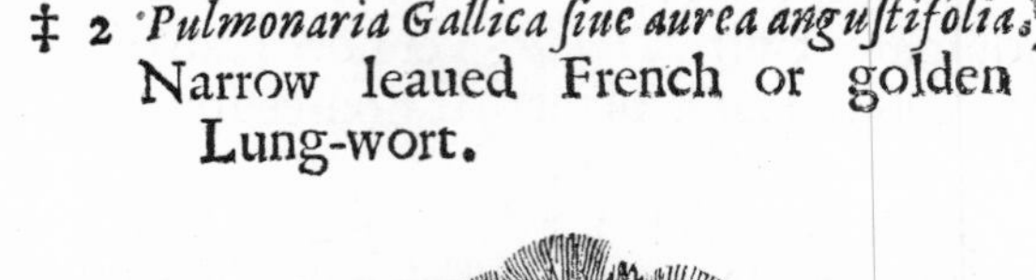

‡ 2 *Pulmonaria Gallica ſiue aurea anguſtifolia.*
Narrow leaued French or golden Lung-wort.

¶ *The Deſcription.*

‡ 1 This which I here giue you in the firſt place, as alſo the other two, are of the kinds of Hawke-weed, or *Hieracium*; wherefore I thought it moſt fit to treat of them in this place, and not to handle them with the *Pulmonaria maculoſa*, or Sage of Ieruſalem: whereas our Author gaue the name *Pulmonaria Gallorum*, and pointed at the deſcription; but his figure being falſe, and the deſcription imperfect, I iudged it the beſt to handle it here next to thoſe plants which both in ſhape and qualities it much reſembles. This firſt hath a pretty large yet fibrous and ſtringy root; from the which ariſe many longiſh leaues, hairy, ſoft, and vnequally diuided, and commonly cut in the deepeſt neereſt the ſtalke; they are of a darke green colour, and they are ſometimes broader and ſhorter, and otherwhiles narrower and longer (whence *Tabernamontanus* makes three ſorts of this, yet are they nothing but varieties of this ſame plant.) Amongſt theſe leaues grow vp one or two naked ſtalks, commonly hauing no more than one leafe apiece, and that about the middle of the ſtalke; theſe ſtalks are alſo hairy, and about a cubit high, diuided at their tops into ſundry branches, which beare double yellow floures of an indifferent bigneſſe, which fading and turning into downe, are together with the ſeed carried away with the winde. This whole plant is milky like as the other Hawk-weeds.

2 This Plant (though confounded by ſome with the former) is much different from the laſt deſcribed; for the root is ſmall and fibrous; the leaues alſo are ſmall, of the bigneſſe, and ſomewhat of the ſhape (though otherwiſe indented) of Daſie leaues, whitiſh and hoarie; the ſtalke is not aboue an handfull high, creſted, hoary, and ſet with many longiſh narrow leaues; and at the top on ſhort foot-ſtalkes it beares foure or fiue floures of a bright yellow colour, and pretty large, conſidering

‡ 3 *Hieracium hortense latifolium, siue Pilosella major.*
Golden Mouse-eare, or Grimme the Colliar.

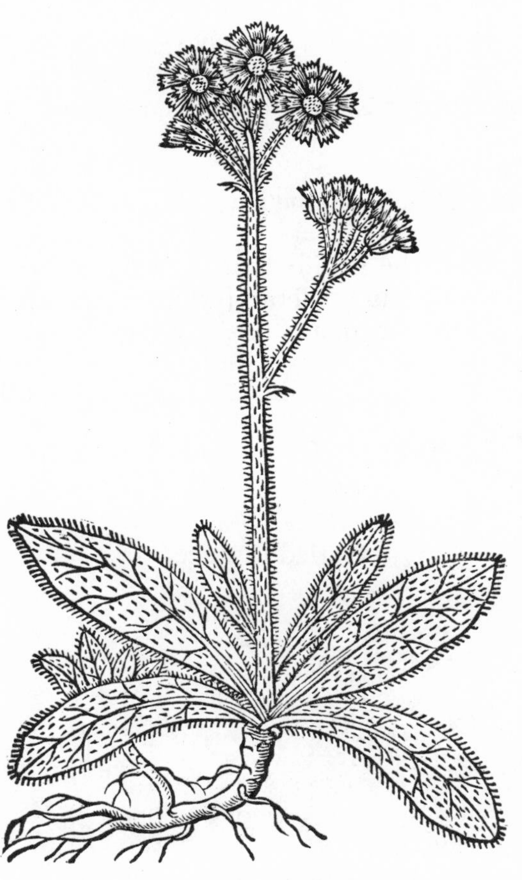

considering the smallnesse of the plant. The floures, like as others of this kinde, fly away in downe, and carry the seeds with them.

3 This plant (which some also haue confounded with the first described) hath a root at the top, of a reddish or brownish colour, but whitish within the earth, & on the lower side sending forth whitish fibres: it bringeth forth in good and fruitfull grounds leaues about a foot long, and two or three inches broad, of a darke greene colour, and hairy, little or nothing at all cut in about the edges; amongst these leaues riseth vp a stalke some cubit high, round, hollow, and naked, but that it sometimes hath a leafe or two toward the bottome, and towards the top it puts forth a branch or two. The floures grow at the top as it were in an vmbell, and are of the bignes of the ordinarie Mouse-eare, and of an orange colour. The seeds are round, & blackish, and are caried away with the downe by the wind. The stalkes and cups of the floures are all set thicke with a blackish downe or hairinesse as it were the dust of coles; whence the women, who keep in it gardens for noueltie sake, haue named it Grim the Colliar.

¶ *The Time.*

All these floure in Iune, Iuly, and August, about the later part of which moneth they ripen their seed.

¶ *The Place.*

1 I receiued some plants of this from Mr. *Iohn Goodyer*, who first found it May 27, 1631. in floure; and the 3 of the following May, not yet flouring, in a copse in Godlemen in Surrey, adioyning to the orchard of the Inne whose signe is the Antilope.

2 This I had from my kinde friend Mr. *William Coote*, who wrot to mee, That he found them growing on a hill in the Lady *Bridget Kingsmills* ground, in an old Romane campe, close by the *Decumane* port, on the quarter that regards the West-South-West, vpon the skirts of the hill.

3 This is a stranger, and onely to be found in some few gardens.

¶ *The Names.*

1 This was first set forth by *Tragus*, vnder the name of *Auricula muris major*: and by *Tabern.* (who gaue three figures expressing the seuerall varieties thereof) by the name of *Pulmonaria Gallica siue aurea*: *Dalechampius* hath it vnder the name of *Corchorus*.

2 This was by *Lobel* (who first set it forth) confounded with the former, as you may see by the title ouer the figure in his Obseruations, *pag.* 317. yet his figure doth much differ from that of *Tragus*, who neither in his figure nor description allowes so much as one leafe vpon the stalke; and *Tabernamontanus* allowes but one, which it seldome wants. Now this by *Lobels* figure hath many narrow leaues; and by the Description, *Aduers. pag.* 253. it is no more than an handfull, or handfull and halfe high: which very well agrees with the plant wee heere giue you, and by no meanes with the former, whose naked stalkes are at least a cubit high. So it is manifest that this plant I haue described is different from the former, and is that which *Pena* and *Lobel* gaue vs vnder the title of *Pulmonaria Gallorum flore Hieracij*. *Bauhine* also confounds this with the former.

3 *Basil Besler* in his *Hortus Eystettensis* hath well exprest this plant vnder the title of *Hieracium latifolium peregrinum Phlomoides*: *Bauhinus* calls it *Hieracium hortense floribus atropurpurascentibus*; and saith that some call it *Pilosella major*: and I iudge it to be the *Hieracium Germanicum* of *Fabius Columna*. This also seemes rather to be the herbe *Costa* of *Camerarius*, than the first described; and I dare almost be bold to affirme it the same: for he saith that it hath fat leaues lying flat vpon the ground, and as much as he could discerne by the figure, agreed with the *Hieracium latifolium* of *Clusius*: to which indeed in the leaues it is very like, as you may see by the figure which is in the first place in the foregoing chapter, which very well resembles this plant, if it had more and smaller floures.

¶ *The Temper and Vertues.*

I iudge these to be temperate in qualitie, and endued with a light astriction.

A 1 The decoction or the distilled water of this herbe taken inwardly, or outwardly applied, conduce much to the mundifying and healing of greene wounds; for some boyle the herb in wine, and so giue it to the wounded Patient; and also apply it outwardly.

B It also is good against the internall inflammations and hot distempers of the heart, stomacke, and liuer.

C The iuyce of this herbe is with good successe dropped into the eares when they are troubled with any pricking or shooting paine or noyse.

D Lastly, The water hath the same qualitie as that of Succorie. *Tragus.*

E 2 *Pena* and *Lobel* affirme this to be commended against whitlowes, and in the diseases of the lungs.

F 3 This (if it be the *Costa* of *Camerarius*) is of singular vse in the Pthisis, that is, the vlceration or consumption of the lungs: whereupon in Misnia they giue the conserue, syrrup, and pouder thereof for the same purpose: and they also vse it in broths and otherwise. *Cam.* ‡

CHAP. 37. *Of Lettuce.*

1 *Lactuca satiua.*
Garden Lettuce.

2 *Lactuca crispa.*
Curled Lettuce.

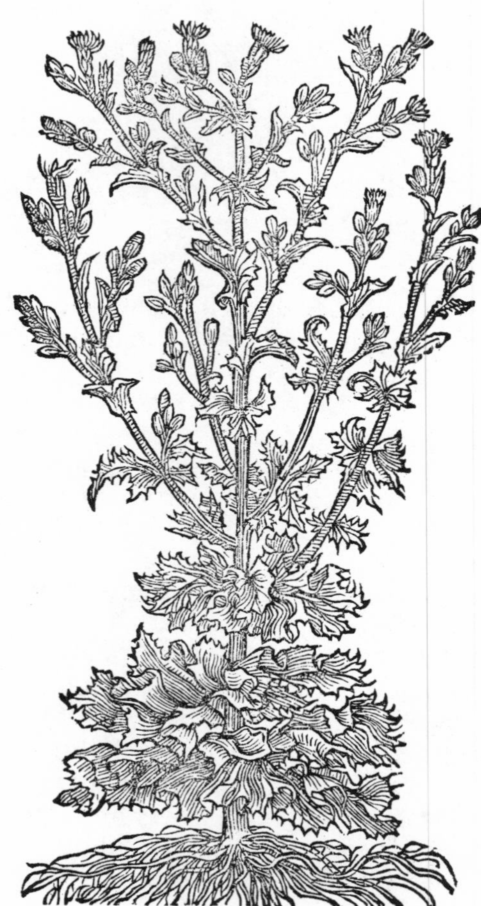

¶ *The Kindes.*

THere be according to the opinion of the Antients, of Lettuce two sorts; the one wilde, or of the field; the other tame, or of the Garden: but time, with the industrie of later Writers haue found out others both wilde and tame, as also artificiall, which I purpose to lay downe.

¶ *The*

¶ *The Deſcription.*

1 GArden Lettuce hath a long broad leafe,ſmooth,and of a light green colour : the ſtalke is round, thicke ſet with leaues full of milky iuyce, buſhed or branched at the top: whereupon do grow yellowiſh floures, which turne into downe that is carried away with the winde. The ſeed ſticketh faſt vnto the cottony downe,and flieth away likewiſe, white of colour,and ſomewhat long : the root hath hanging on it many long tough ſtrings,which being cut or broken, do yeeld forth in like manner as doth the ſtalke and leaues, a iuyce like to milke. And this is the true deſcription of the naturall Lettuce, and not of the artificiall ; for by manuring, tranſplanting, and hauing a regard to the Moone and other circumſtances, the leaues of the artificiall Lettuce are oftentimes transformed into another ſhape : for either they are curled, or elſe ſo drawne together,as they ſeeme to be like a Cabbage or headed Colewort, and the leaues which be within and in the middeſt are ſomething white, tending to a very light yellow.

5 *Lactuca capitata.*
Cabbage Lettuce.

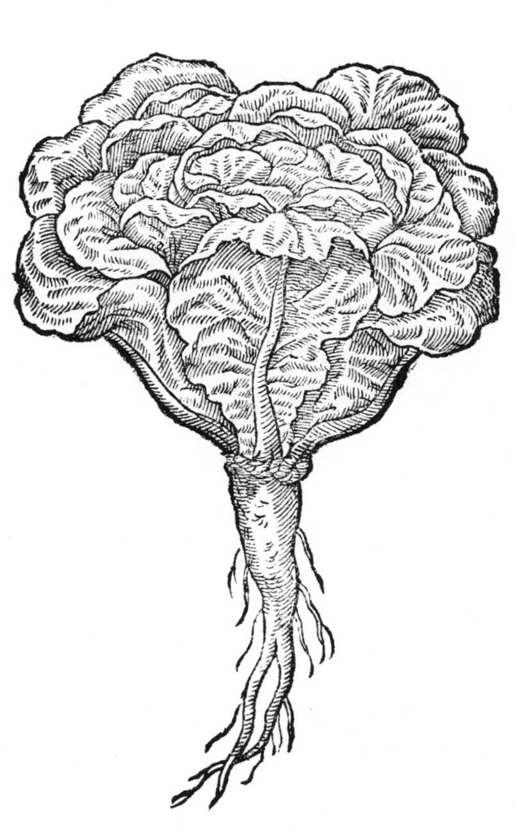

6 *Lactuca intybacea.*
Lumbard Lettuce.

2 The curled Lettuce hath great and large leaues deeply cut or gaſhed on both the ſides,not plaine or ſmooth as the former, but intricately curled and cut into many ſections. The floures are ſmall, of a bleake colour, the which do turne into downe, and is carried away with the winde. The ſeed is like the former, ſauing that it changeth ſometime into blackneſſe, with a root like vnto the former.

3 This ſmall ſort of curled Lettuce hath many leaues hackt and torne in pieces very confuſedly, and withall curled in ſuch an admirable ſort, that euery great leafe ſeemeth to be made of many ſmall leaues ſet vpon one middle rib, reſembling a fan of curled feathers vſed among Gentlewomen : the floures,roots,and ſeeds agree with the former.

4 The Sauoy Lettuce hath very large leaues ſpred vpon the ground, at the firſt comming vp broad, cut or gaſht about the edges, criſping or curling lightly this or that way, not vnlike to the leaues of Garden Endiue, with ſtalkes, floures,and ſeeds like the former, as well in ſhape, as yeelding that milky iuyce wherewith they do all abound.

5 Cabbage Lettuce hath many plaine and ſmooth leaues at his firſt growing vp, which for the moſt part lie flat ſtill vpon the ground : the next that do appeare are thoſe leaues in the midſt, which turn themſelues together,embracing each other ſo cloſely,that it is formed into that globe or

or round head, whereof the ſimpleſt is not ignorant. The ſeed hereof is blacke, contrary to all the reſt; which may be as it were a rule whereby ye may know the ſeed of Cabbage Lettuce from the other ſorts.

6 The Lumbard Lettuce hath many great leaues ſpred vpon the ground like vnto thoſe of the garden Endiue, but leſſer. The ſtalkes riſe vp to the height of three foot: the floures be yellowiſh, which turne into downe and flie away with the winde: the ſeed is white as ſnow.

¶ *The Place.*

Lettuce delighteth to grow, as *Palladius* ſaith, in a mannured, fat, moiſt, and dunged ground: it muſt be ſowen in faire weather in places where there is plenty of water, as *Columella* ſaith, and proſpereth beſt if it be ſowen very thin.

¶ *The Time.*

It is certaine, ſaith *Palladius*, that Lettuce may well be ſowen at any time of the yeare, but eſpecially at euery firſt ſpring, and ſo ſoone as winter is done, till ſummer be well nigh ſpent.

¶ *The Names.*

Garden Lettuce is called in Latine, *Lactuca ſatiua*: *Galen* names it θριδακίνη: the Pythagorians εὐνοῦχον: ſome iudge it to be *Lactuca, à Lacteo ſucco*, called of the milkie iuice which iſſueth forth of the wounded ſtalkes and rootes: the Germanes name it **Lattich**: the low Duch, **Latouwe**: the Spaniards, *Lechuga*, and *Alface*: the Engliſh, Lettuce: and the French, *Laictue*. When the leaues of this kinde are curled or crompled, it is named of *Pliny*, *Lactuca criſpa*: and of *Columella*, *Lactuca Ceciliana*: in Engliſh, curl'd or crompled Lettuce.

The Cabbage Lettuce is commonly called *Lactuca capitata*, and *Lactuca ſeſſilis*: *Pliny* nameth it *Lactuca Laconica*: *Columella*, *Lactuca Bœtica*: *Petrus Creſcentius*, *Lactuca Romana*: in Engliſh, Cabbage Lettuce, and Loued Lettuce.

There is another ſort with reddiſh leaues, called of *Columella*, *Lactuca Cypria*: in Engliſh, red Lettuce.

¶ *The Temperature.*

Lettuce is a cold and moiſt pot-herbe, yet not in the extreame degree of cold or moiſture, but altogether moderately; for otherwiſe it were not to be eaten.

¶ *The Vertues.*

A Lettuce cooleth the heate of the ſtomacke, called the heart-burning; and helpeth it when it is troubled with choller: it quencheth thirſt, cauſeth ſleepe, maketh plenty of milke in nurſes, who through heate and drineſſe grow barren and drie of milke: for it breedeth milke by tempering the drieneſſe and heate. But in bodies that be naturally cold, it doth not ingender milke at all, but is rather an hinderance thereunto.

B Lettuce maketh a pleaſant ſallad, being eaten raw with vineger, oyle, and a little ſalt: but if it be boyled it is ſooner digeſted, and nouriſheth more.

C It is ſerued in theſe dayes, and in theſe countries in the beginning of ſupper, and eaten firſt before any other meate: which alſo *Martiall* teſtifieth to be done in his time, maruelling why ſome did vſe it for a ſeruice at the end of ſupper, in theſe verſes.

> *Claudere quæ cœnas Lactuca ſolebat auorum,*
> *Dic mihi, cur noſtras incohat illa dapes?*
> Tell me why Lettuce, which our Grandſires laſt did eate,
> Is now of late become, to be the firſt of meate?

D Notwithſtanding it may now and then be eaten at both thoſe times to the health of the body: for being taken before meat it doth many times ſtir vp appetite: and eaten after ſupper it keepeth away drunkenneſſe which commeth by the wine; and that is by reaſon that it ſtayeth the vapors from riſing vp into the head.

F The iuice which is made in the veines by Lettuce is moiſt and cold, yet not ill, nor much in quantitie: *Galen* affirmeth that it doth neither binde the belly nor looſe it, for it hath in it no harſhnes nor ſtiptike qualitie by which the belly is ſtayed, neither is there in it any ſharpe or biting facultie, which ſcoureth and prouoketh to the ſtoole.

F But howſoeuer *Galen* writeth this, and howſoeuer the ſame wanteth theſe qualities, yet it is found by experience, that it maketh the body ſoluble, eſpecially if it be boyled; for by moiſtning of the belly it maketh it the more ſlippery: which *Martial* very well knew, writing in his 11. booke of Epigrams in this manner: *Prima tibi dabitur, ventri Lactuca mouendo Vtilis.*

G Lettuce being outwardly applied mitigateth all inflammations; it is good for burnings and ſcaldings, if it be laid thereon with ſalt before the bliſters doe appeare, as *Plinie* writeth.

H The iuice of Lettuce cooleth and quencheth the naturall ſeed if it be too much vſed, but procureth ſleepe.

‡ Chap.

‡ Chap. 38. *Of Wilde Lettuce.*

¶ The Deſcription.

‡ There are three ſorts of wilde Lettuce growing wilde here with vs in England, yet I know not any that haue mentioned more than two; yet I thinke all three of them haue beene written of, though two of them be confounded together and made but one (a thing often happening in the hiſtory of Plants) and vnleſſe I had ſeene three diſtinct ones, I ſhould my ſelfe haue beene of the ſame opinion.

1 The firſt and rareſt of theſe hath long and broad leaues, not cut in, but only ſnipt about the edges, and thoſe leaues are they that are on the lower part of the ſtalke almoſt to the midle thereof: then come leaues from thence to the top, which are deepely diuided with large gaſhes: the ſtalke if it grow in good grounds exceeds the height of a man, (for I haue ſeene it grow in a garden to the height of eight or nine foot) it is large, round, and ſmooth, and towards the top diuided into many branches which beare yellow floures ſomewhat like to the garden Lettuce, after which alſo ſucceed blackiſh ſeeds like to other plants of this kinde. The whole plant is full of a clammy milky iuice, which hath a very ſtrong and grieuous ſmell of *Opium*.

‡ 1 *Lactuca ſyl. maior odore Opij.*
The greater wilde Lettuce ſmelling of *Opium.*

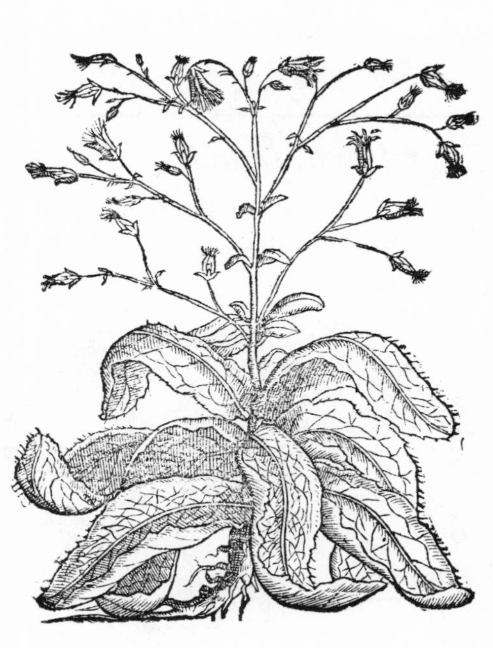

‡ 3 *Lactuca ſylveſtris folijs diſſectis.*
The wilde Lettuce with the diuided Leafe.

2 This hath broad leaues only cut about the edges, but not altogether ſo large as thoſe of the laſt deſcribed: the ſtalke, which commonly is two cubits or better high, is alſo ſmooth, and diuided into many branches, bearing ſuch floures and ſeeds as the laſt deſcribed; and this alſo hath a milky iuice of the ſame ſmell as the laſt deſcribed, from which it differs only in the magnitude, and that this hath all the leaues whole, and not ſome whole and ſome diuided, as the former.

3 This in ſtalkes, floures and ſeedes is like to the laſt deſcribed, but the leaues are much different, for they are all deeply gaſhed or cut in like as the leaues of Succory, or Dandelion. This alſo is full of a milky iuice, but hath not altogether ſo ſtrong a ſent of *Opium* as the two former, though it partake much thereof. The ſtalke of this is ſometimes a little prickly, and ſo alſo is the middle rib vpon the backeſide of the leafe. All theſe three haue woddy roots which die euery yeare, and ſo they come vp againe of the ſcattered ſeed.

¶ The Place.

The firſt of theſe was found in Hampſhire by Mr. *Goodyer* and the ſeeds hereof ſent to Mr. *Parkinſon* in in whose garden I saw it growing some two yeares agoe. The other grow plentifully between London and Pancridge Church, about the ditches and highway side.

¶ *The Time.*

They come vp in the Spring, and sometimes sooner, and ripen their seed in Iuly and August.

¶ *The Names.*

1 I take the first of these to be the *Lactuca Sylvestris* of *Dioscorides* and the Ancients, and that which the Authours of the *Adversaria* gaue vs vnder the title of *Lactuca agrestis scariolæ hortensis folio, Lactucæ flore, Opij odore vehementi, soporifero & viroso.*

2 This is the *Endiuia* of *Tragus, pag.* 268. and the *Thesion* of *Daleschampius, pag.* 564. *Bauhine* confounds this with the former.

3 This is the *Lactuca Sylvestris prior*, of *Tragus*: the *Lactuca Sylvestris* of *Matthiolus*, *Fuchsius*, *Dodonæus*, and others: it is the *Seris Domestica* of *Lobell.*

The Temper.

These certainly, especially the two first, are cold, and that in the later end of the third or beginning of the fourth degree (if *Opium* be cold in the fourth.)

The Vertues.

A Some (saith *Dioscorides*) mix the milkie iuice hereof with *Opium*; (for his *Meconium* is our *Opium*.) in the making thereof.

B He also saith, that the iuice hereof drunke in Oxycrate in the quantity of 2 *obuli*, (which make some one scruple) purgeth watrish humors by stoole; it also clenseth the little vlcer in the eye called *Argemon* in Greeke, as also the mystines or darknesse of sight.

C Also beaten and applied with womans milke it is good against burnes and scaldes.

D Lastly, it procures sleepe, asswages paine, moues the courses in women, and is drunke against the stingings of scorpions, and bitings of spiders.

E The seed taken in drinke, like as the Garden Lettuce, hindreth generation of seed and venereous imaginations. ‡

Chap. 39. *Of Lambs Lettuce, or Corne sallad.*

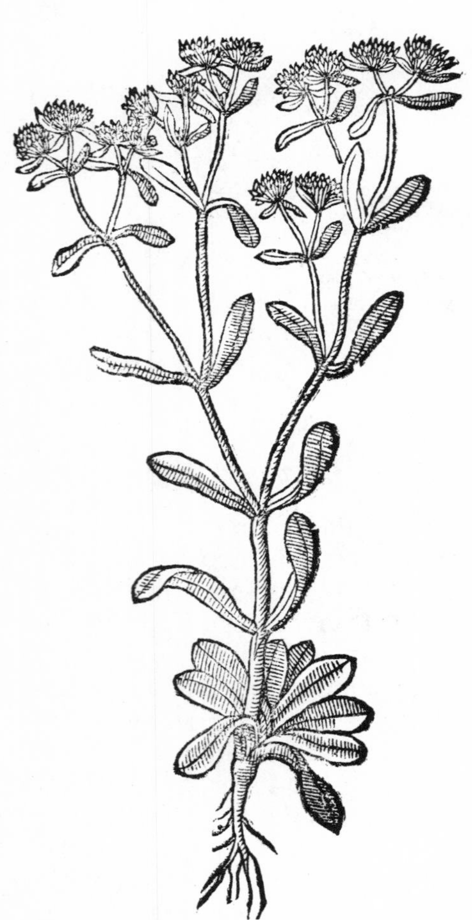

1 *Lactuca Agnina.*
Lambes Lettuce.

2 *Lactuca Agnina latifolia.*
Corne sallade.

¶ The

¶ *The Description.*

1 THe plant which is commonly called *Olus album*, or the white pot-herbe (which of some hath been set out for a kinde of Valerian, but vnproperly, for that it doth very notably resemble the Lettuce, as well in forme, as in meate to be eaten, which propertie is not to be found in Valerian, and therefore by reason and authoritie I place it as a kinde of Lettuce) hath many slender weake stalkes trailing vpon the ground, with certaine edges a foot high when it growes in most fertile ground; otherwise a hand or two high, with sundry ioynts or knees: out of euery one whereof grow a couple of leaues narrow and long, not vnlike to Lettuce at the first comming vp, as well in tendernesse as taste in eating; and on the top of the stalkes stand vpon a broad tuft as it were certaine white floures that be maruellous little, which can scarsely be known to be floures, sauing that they grow many together like a tuft or vmbel: it hath in stead of roots a few slender threads like vnto haires.

2 The other kind of Lettuce, which *Dodonæus* in his last edition setteth forth vnder the name of *Album olus*: the Low-countrey men call it **Weitmoes**, and vse it for their meate called Wermose; with vs, Loblollie. This plant hath small long leaues a finger broad, of a pale green colour; among which shooteth vp a small cornered and slender stem halfe a foot high, ioynted with two or three ioynts or knees, out of which proceed two leaues longer than the first, bearing at the top of the branches tufts of very smal white floures closely compact together, with a root like the former.

‡ Both these are of one plant, differing in the bignesse and broadnesse of the leafe and the whole plant besides. ‡

¶ *The Place.*

These herbes grow wilde in the corne fields; and since it hath growne in vse among the French and Dutch strangers in England, it hath beene sowen in gardens as a sallad herbe.

¶ *The Time.*

They are found greene almost all Winter and Sommer.

¶ *The Names.*

The Dutch-men do call it **Weymoes**; that is to say, *Album olus*: of some it is called **Veltcrop**: the French terme it *Sallade de Chanoine*: it may be called in Greeke, λευκολάχανον: in English, The White Pot-herbe; but commonly, Corne sallad.

¶ *The Temperature and Vertues.*

This herbe is cold and something moist, and not vnlike in facultie and temperature to the garden Lettuce; in stead whereof, in Winter and in the first moneths of the Spring it serues for a sallad herbe, and is with pleasure eaten with vineger, salt and oile, as other sallads be; among which it is none of the worst.

CHAP. 40. *Of Coleworts.*

¶ *The Kindes.*

DIoscorides maketh two kindes of Coleworts; the tame and the wilde: but *Theophrastus* makes more kindes hereof; the ruffed or curled Cole, the smooth Cole, and the wilde Cole. *Cato* imitating *Theophrastus*, setteth downe also three Coleworts: the first hee describeth to be smooth, great, broad leaued, with a big stalke; the second ruffed; the third with little stalks, tender, and very much biting. The same distinction also *Pliny* maketh, in his twentieth booke, and ninth chapter; where he saith, That the most ancient Romanes haue diuided it into three kindes; the first roughed, the second smooth, and the third which is properly called κράμβη, or Colewort. And in his nineteenth booke he hath also added to these, other moe kindes; that is to say, *Tritianum, Cumanum, Pompeianum, Brutianum, Sabellium*, and *Lacuturrium*.

The Herbarists of our time haue likewise obserued many sorts, differing either in colour or else in forme; other headed with the leaues drawne together; most of them white, some of a deepe greene, some smooth leaued, and others curled or ruffed; differing likewise in their stalkes, as shall be expressed in their seuerall descriptions.

¶ The

1 *Braßica vulgaris sativa.*
Garden Colewort.

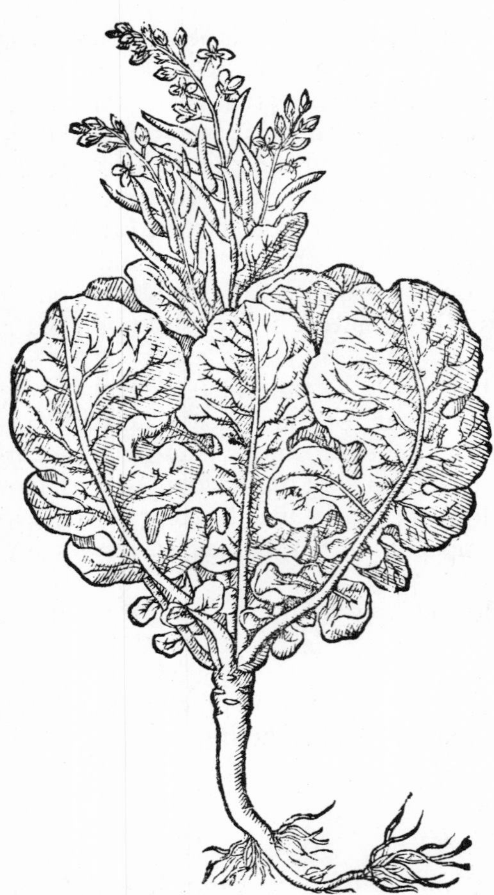

2 *Braßica sativa crispa.*
Curled Garden Cole.

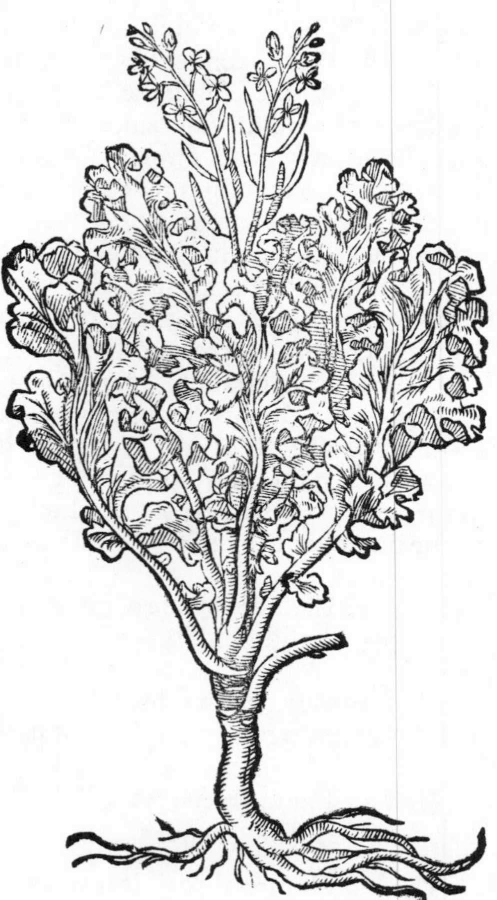

3 *Braßica rubra.*
Red Colewort.

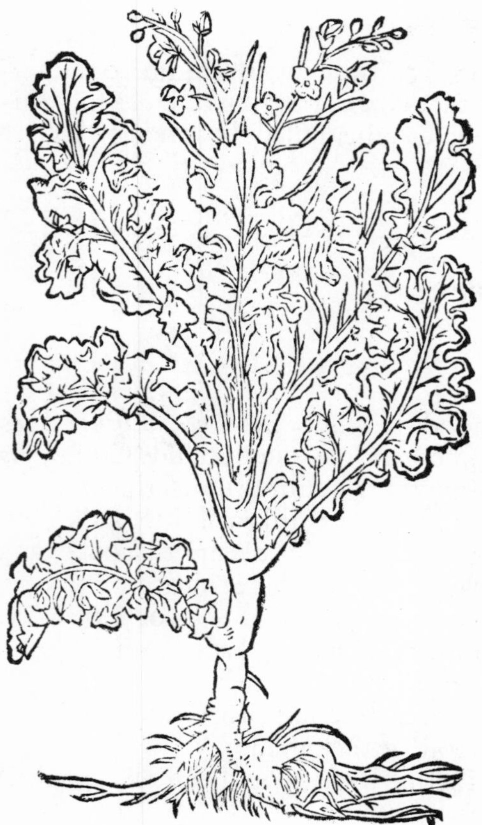

4 *Braßica capitata alba.*
White Cabbage Cole.

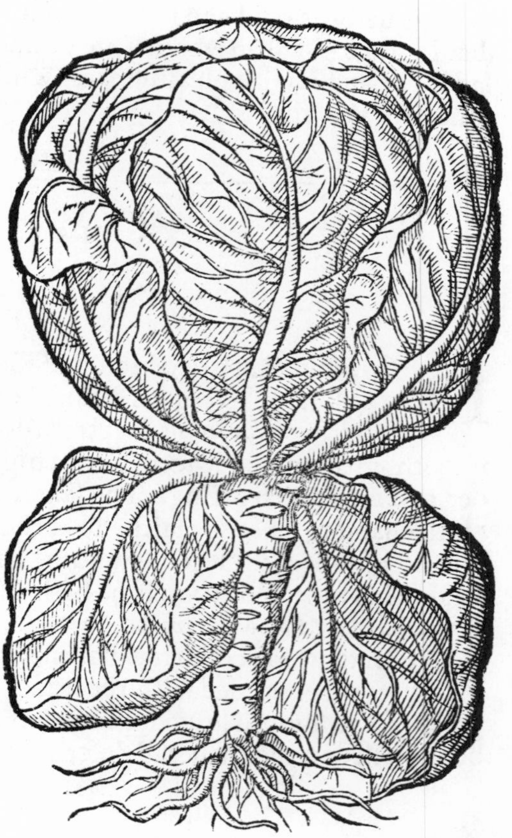

5 *Braßica*

¶ *The Description.*

1 THe Garden Colewort hath many great broad leaues of a deepe blacke greene colour, mixed with ribs and lines of reddish and white colours: the stalke groweth out of the middest from among the leaues, branched with sundry armes bearing at the top little yellow floures: and after they be past, there do succeed long cods full of round seed like those of the Turnep, but smaller, with a wooddy root hauing many strings or threds fastned thereto.

2 There is another lesser sort than the former, with many deepe cuts on both sides euen to the middest of the rib, and very much curled and roughed in the edges; in other things it differeth not.

3 The red kinde of Colewort is likewise a Colewort of the garden, and differeth from the common in the colour of his leaues, which tend vnto rednesse; otherwise very like.

4 There is also found a certaine kinde hereof with the leaues wrapped together into a round head or globe, whose head is white of colour, especially toward Winter when it is ripe. The root is hard, and the stalkes of a wooddy substance. ‡ This is the great ordinarie Cabbage knowne euery where, and as commonly eaten all ouer this kingdome. ‡

5 *Brassica capitata rubra.*
Red Cabbage Cole.

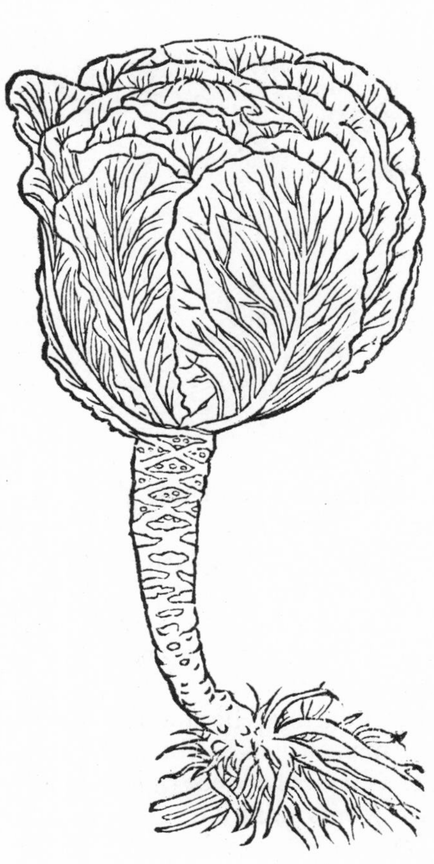

6 *Brassica patula*
Open Cabbage Cole.

5 There is another sort of Cabbage or loued Colewort which hath his leaues wrapped together into a round head or globe, yet lesser than that of the white Cabbage, and the colour of the leaues of a lighter red than those of the former.

6 The open loued Colewort hath a very great hard or wooddy stalke, whereupon do grow very large leaues of a white greene colour, and set with thicke white ribs, and gathereth the rest of the leaues closely together, which be lesser than those next the ground; yet when it commeth to the shutting vp or closing together, it rather dilateth it selfe abroad, than closeth all together.

7 Double Colewort hath many great and large leaues, whereupon doe grow here and there other small iagged leaues, as it were made of ragged shreds and iagges set vpon the smooth leafe, which giueth shew of a plume or fan of feathers. In stalke, root, and euery other part besides it doth agree with the Garden Colewort.

8 The

8 The double criſpe or curled Colewoort agreeth with the laſt before deſcribed in euery reſpect, onely it differeth in the leaues, which are ſo intricately curled, and ſo thick ſet ouer with other ſmall cut leaues, that it is hard to ſee any part of the leafe it ſelfe, except ye take and put aſide ſome of thoſe iagges and ragged leaues with your hand.

9 Braßica florida.
Cole-Florie.

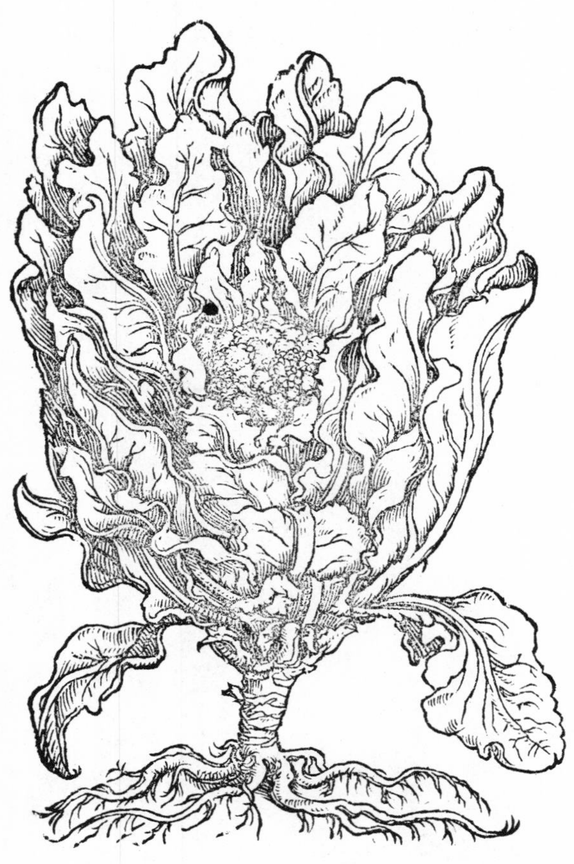

10 Braßica Tophoſa.
Swollen Colewoort.

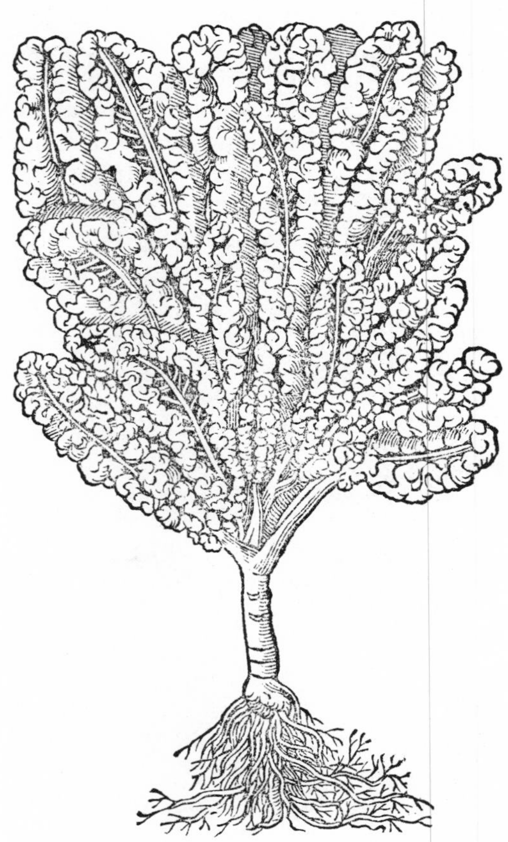

9 Cole flore, or after ſome Colieflore, hath many large leaues ſleightly indented about the edges, of a whitiſh greene colour, narrower and ſharper pointed than Cabbage: in the middeſt of which leaues riſeth vp a great white head of hard floures cloſely thruſt together, with a root full of ſtrings; in other parts like vnto the Coleworts.

10 The ſwollen Colewort of all other is the ſtrangeſt, which I receiued from a worſhipfull merchant of London maſter *Nicholas Lete*, who brought the ſeed thereof out of France; who is greatly in loue with rare and faire floures & plants, for which he doth carefully ſend into Syria, hauing a ſeruant there at Aleppo, and in many other countries, for the which my ſelfe and likewiſe the whole land are much bound vnto him. This goodly Colewort hath many leaues of a blewiſh green, or of the colour of Woade, bunched or ſwollen vp about the edges as it were a peece of leather wet and broiled on a gridiron, in ſuch ſtrange ſort that I cannot with words deſcribe it to the full. The floures grow at the top of the ſtalkes, of a bleake yellow colour. The root is thicke and ſtrong like to the other kindes of Coleworts.

11 Sauoy Cole is alſo numbred among the headed Colewoorts or Cabbages. The leaues are great and large very like to thoſe of the great Cabbage, which turne themſelues vpwards as though they would embrace one another to make a loued Cabbage, but when they come to the ſhutting vp they ſtand at a ſtay, and rather ſhew themſelues wider open, than ſhut any neerer together; in other reſpects it is like vnto the Cabbage.

12 The curled Sauoy Cole in euery reſpect is like the precedent, ſauing that the leaues hereof doe ſomewhat curle or criſpe about the midle of the plant: which plant if it be opened in the ſpring time, as ſometimes it is, it ſendeth forth branched ſtalks, with many ſmall white floures at the top, which being paſt their follow long cods and ſeeds like the common or firſt kinde deſcribed.

13 This kinde of Colewoort hath very large leaues deepely iagged euen to the middle rib, in face reſembling great and ranke parſley. It hath a great and thicke ſtalke of three cubits high, whereupon doe grow floures, cods, and ſeed like the other Colewoorts.

14 The

11 *Brassica Sabauda.*
Sauoy Cole.

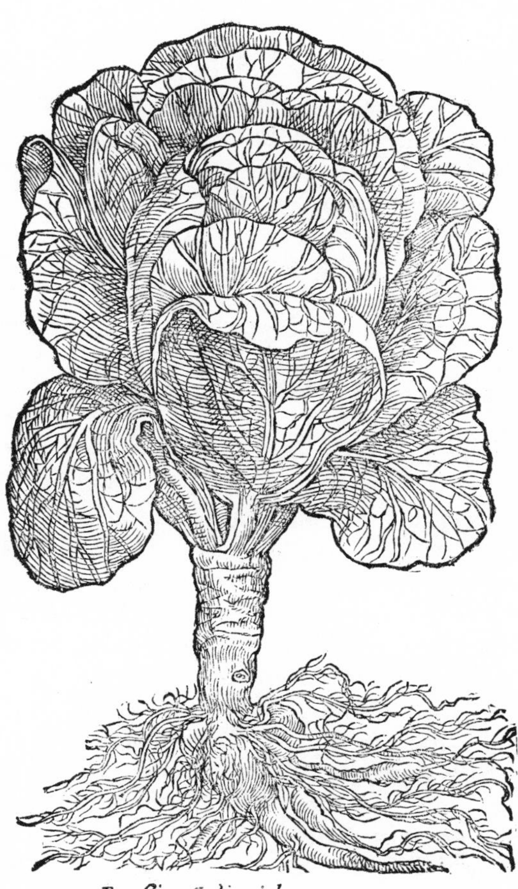

12 *Brassica Sabauda crispa.*
Curled Sauoy Cole.

13 *Brassica Selinoides.*
Parseley Colewoort.

† 15 *Brassica marina Anglica.*
English sea Colewoorts.

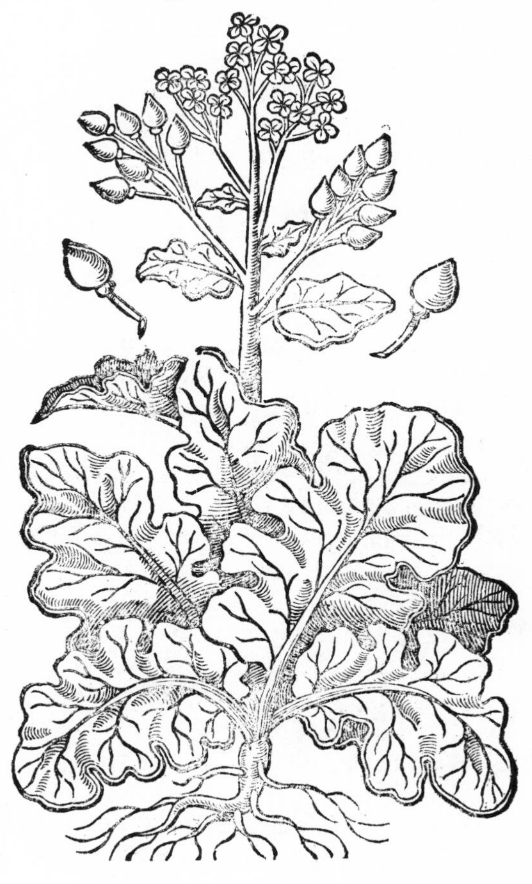

14 The

14 The ſmall cut Colewoort hath very large leaues, wonderfully cut, hackt and hewen euen to the middle rib, reſembling a kinde of curled parſley, that ſhall be deſcribed in his place, (which is not common nor hath not beene knowne nor deſcribed vntill this time) very well agreeing with the laſt before mentioned, but differeth in the curious cutting and iagging of the leaues: in ſtalke floures and ſeed not vnlike.

† 16 *Braſsica ſylueſtris.*
Wilde Colewoorts.

15 Sea Colewoort hath large and broad leaues very thicke and curled, and ſo brittle that they cannot be handled without breaking, of an ouerworne greene colour, tending to grayneſſe: among which riſe vp ſtalkes two cubits high, bearing ſmall pale floures at the top; which being paſt their follow round knobs wherein is contained one round ſeed and no more, blacke of colour, of the bigneſſe of a tare and a fetch: ‡ And therefore *Pena* and *Lobell* called it *Braſsica marina monoſpermos*. ‡

16 The wilde Colewoort hath long broad leaues not vnlike to the tame Colewoort, but leſſer, as is all the reſt of the plant, and is of his owne nature wilde, and therefore not ſought after as a meate, but is ſowen and husbanded up on ditch bankes and ſuch like places for the ſeeds ſake, by which oftentimes great gaine is gotten.

¶ *The Place.*

The greateſt ſort of Colewoorts doe grow in gardens, and doe loue a ſoile which is fat and throughly dunged and well manured: they doe beſt proſper when they be remooued, and every of them grow in our Engliſh gardens, except the wilde, which groweth in fields and new digged ditch banks.

The ſea Colewoort groweth naturally vpon the bayche and brims of the ſea, where there is no earth to bee ſeene, but ſand and rowling pibble ſtones, which thoſe that dwell neere the ſea doe call Bayche: I found it growing betweene Whytſtable and the Ile of Thanet neere the brinke of the ſea, and in many places neere to Colcheſter and elſewhere by the ſea ſide.

¶ *The Time.*

Petrus Creſcentius ſaith that the Colewoort may bee ſowen and remooued at any time of the yeere; whoſe opinion I altogether miſlike. It is ſowen in the ſpring, as March, Aprill, and oftentimes in May, and ſometimes in Auguſt, but the ſpeciall time is about the beginning of September.

The Colewoort, ſaith *Columella*, muſt be remoued when it attaineth to ſix leaues, after it is come vp from ſeed; the which muſt be done, in Aprill or May, eſpecially thoſe that were ſowne in Autumne; which afterwards flouriſh in the winter moneths, at what time, they are fitteſt for meate.

But the Sauoy-Cole, and the Cole florey, muſt be ſowne in Aprill, in a bed of hot horſedung, and couered with ſtraw or ſuch like, to keepe it from the cold, and froſty mornings; and when it hath gotten ſix leaues after this ſort, then ſhall you remoue him as aforeſaid, otherwiſe if you tarry for temperate weather before you ſow, the yeare will be ſpent, before it come to ripeneſſe.

¶ *The Names.*

Euery of the Colewoorts, is called in Greeke by *Dioſcorides* and *Galen* κράμβη: it is alſo called ἀμέθυσος: ſo named, not only becauſe it driueth away drunkenneſſe, but alſo for that it is like in colour to the precious ſtone called the Amethyſt: which is meant by the firſt and garden Colewoort. The Apothecaries and the common Herbariſts doe call it *Caulis*, of the goodneſſe of the ſtalke: in the Germane tongue it is called **Koole kraut**: in French, *des Choux*: in Engliſh, Colewoorts.

Cole-florey is called in Latine *Braſsica Cypria*, and *Cauliflora*: in Italian, *Caulifiore*: it ſeemeth to agree with *Braſsica pompeiana* of *Pliny*, whereof he writeth in his 19. booke, and 8. chapter.

¶ *The*

¶ *The Temperature.*

All the Colewoorts haue a drying and binding facultie, with a certaine nitrous or ſalt quality, whereby they mightily cleanſe, either in the iuice, or in the broth. The whole ſubſtance or body of the Colewoort is of a binding and drying faculty, becauſe it leaueth in the decoction this ſalt quality; which lieth in the iuyce and watry part thereof: the water wherein it is firſt boyled, draweth to it ſelfe all the quality; for which cauſe the decoction thereof looſeth the belly, as doth alſo the iuyce of it, if it be drunke: but if the firſt broth in which it was boyled be caſt away, then doth the Colewoort dry and binde the belly. But it yeeldeth to the body ſmall nouriſhment, and doth not ingender good, but a groſſe and Melancholicke bloud. The white Cabbage is beſt next vnto the Cole-florey; yet *Cato* doth chiefly commend the ruſſet Cole: but he knew neither the white ones, nor the Cole-florey; for if he had, his cenſure had beene otherwiſe.

¶ *The Vertues.*

Dioſcorides teacheth, that the Colewoort being eaten is good for them that haue dim eyes, and that are troubled with the ſhaking palſie. A

The ſame author affirmeth, that if it be boiled and eaten with vineger, it is a remedie for thoſe that be troubled with the ſpleene. B

It is reported, that the raw Colewoort being eaten before meate, doth preſerue a man from drunkenneſſe: the reaſon is yeelded, for that there is a naturall enmity betweene it and the vine, which is ſuch, as if it grow neere vnto it, forthwith the vine periſheth and withereth away: yea, if wine be poured vnto it while it is in boyling, it will not be any more boiled, and the colour thereof quite altered, as *Caſsius* and *Dionyſius Vticenſis* doe write in their bookes of tillage: yet doth not *Athenæus* aſcribe that vertue of driuing away drunkenneſſe to the leaues, but to the ſeeds of Colewoort. C

Moreouer, the leaues of Colewoorts are good againſt all inflammations, and hot ſwellings; being ſtamped with barley and meale, and laid vpon them with ſalt: and alſo to breake carbuncles. D

The iuyce of Colewoorts, as *Dioſcorides* writeth, being taken with floure-deluce and niter, doth make the belly ſoluble: aud being drunke with wine, it is a remedie againſt the bitings of venomous beaſts. E

The ſame being applyed with the powder of Fennugreeke, taketh away the paine of the gout, and alſo cureth old and foule vlcers. F

Being conueied into the noſthrils, it purgeth the head: being put vp with barley meale it bringeth downe the floures. G

Pliny writeth, that the iuyce mixed with wine, and dropped into the eares, is a remedie againſt deafeneſſe. H

The ſeed, as *Galen* ſaith, driueth forth wormes, taketh away freckles of the face, ſun-burning, and what thing ſoeuer that need to be gently ſcoured or clenſed away. I

They ſay that the broth wherein the herbe hath beene ſodden is maruellous good for the ſinewes and ioynts, and likewiſe for Cankers in the eies, claled in Greeke *Carcinomata*, which cannot be healed by any other meanes, if they be waſhed therewith. K

† The fifteenth and ſixteenth figures were formerly tranſported.

CHAP. 41. *Of Rape-Cole.*

¶ *The Deſcription.*

1 THe firſt kinde of Rape Cole hath one ſingle long root, garniſhed with many threddy ſtrings: from which riſeth vp a great thicke ſtalke, bigger than a great Cucumber or great Turnep: at the top whereof ſhooteth forth great broad leaues, like vnto thoſe of Cabbage Cole. The floures grow at the top on ſlender ſtalkes, compact of foure ſmall yellow floures: which being paſt the ſeed followeth incloſed in litle long cods, like the ſeed of Muſtard.

2 The ſecond hath a long fibrous root like vnto the precedent; the tuberous ſtalke is very great and long, thruſting forth in ſome few places here and there, ſmall footſtalkes; whereupon doe grow ſmooth leaues, ſleightly indented about the edges: on the top of the long Turnep ſtalke grow leane ſtalkes and floures like the former. ‡ This ſecond differs from the former onely in the length of the ſwolne ſtalke, whence they call it *Caulorapum longum*, or Long Rape Cole. ‡

¶ *The Place.*

They grow in Italy, Spaine, and ſome places of Germanie, from whence I haue receiued ſeedes for my garden, as alſo from an honeſt and curious friend of mine called maſter *Goodman*, at the Minories neere London.

1 *Caulorapum rotundum.*
Round rape Cole.

¶ *The Time.*

They floure and flourish when the other Colewoorts doe, whereof no doubt they are kinds, and must be carefully set and sowne, as muske Melons and Cucumbers are.

¶ *The Names.*

They are called in Latine, *Caulorapum*, and *Rapocaulis*, bearing for their stalkes, as it were Rapes and Turneps, participating of two plants, the Colewort and Turnep; whereof they tooke their names.

¶ *The Temperature and Vertues.*

There is nothing set downe of the faculties of these plants, but are accounted for daintie meate, contending with the Cabbage Cole in goodnesse and pleasant taste.

Chap. 42. *Of Beets.*

¶ *The Description.*

1 THe common white Beet hath great broad leaues, smooth, and plain: from which rise thicke crested or chamfered stalks: the floures grow along the stalks clustering together, in shape like little starres; which being past, there succeed round and vneuen prikly seed. The root is thicke, hard, and great.

1 *Beta alba.* White Beets.

2 *Beta rubra.* Red Beets.

2 There

‡ 3 *Beta rubra Romana.*
Red Roman Beet.

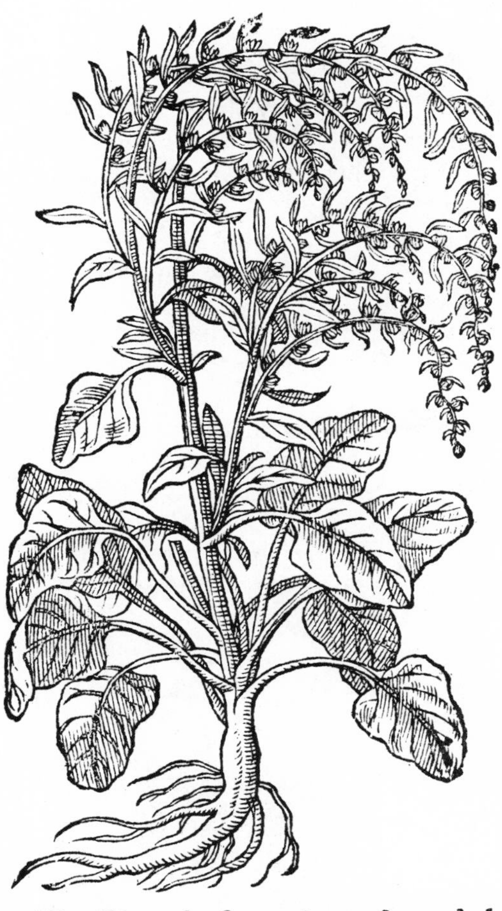

2 There is another ſort like in ſhape and proportion to the former, ſauing that the leaues of this be ſtreaked with red here and there confuſedly, which ſetteth forth the difference.

3 There is likewiſe another ſort hereof, that was brought vnto me from beyond the ſeas, by that courteous merchant maſter *Lete*, before remembred, the which hath leaues very great, and red of colour, as is all the reſt of the plant, as well root, as ſtalke, and floures, full of a perfect purple iuyce tending to redneſſe: the middle rib of which leaues are for the moſt part very broad and thicke, like the middle part of the Cabbage leafe, which is equall in goodneſſe with the leaues of Cabbage being boyled. It grew with me 1596. to the height of viij. cubits, and did bring forth his rough and vneuen ſeed very plentifully: with which plant nature doth ſeeme to play and ſport herſelfe: for the ſeeds taken from that plant, which was altogether of one colour and ſowen, doth bring forth plants of many and variable colours, as the worſhipfull gentleman maſter *Iohn Norden* can very well teſtifie, vnto whom I gaue ſome of the ſeeds aforeſaid, which in his garden bruoght forth many other of beautifull colours.

¶ *The Place.*

The Beete is ſowen in gardens: it loueth to grow in a moiſt and fertile ground. ‡ The ordinary white Beet growes wilde vpon the ſea-coaſt of Tenet and diuers other places by the Sea, for this is not a different kind as ſome would haue it. ‡

¶ *The Time.*

The fitteſt time to ſow it is in the ſpring: it flouriſheth and is greene all ſommer long, and likewiſe in winter, and bringeth forth his ſeed the next yeare following.

¶ *The Names.*

The Grecians haue named it σεῦτλον, τεῦτλον: the Latines, *Beta:* the Germanes, Mangolt: the Spaniards, *Aſelgas*: the French, *de la Porée, des Iotes*, and *Beets*: *Theophraſtus* ſaith, that the white Beete is ſurnamed σικελικὴ, that is to ſay, *Sicula*, or of Sicilia: hereof commeth the name *Sicla*, by which the Barbarians, and ſome Apothecaries did call the Beet; the which word we in England doe vſe, taken for the ſame.

¶ *The Nature.*

The white Beets are in moiſture and heate temperate, but the other kinds are drie, and all of them abſterſiue: ſo that the white Beete is a cold and moiſt pot-herbe, which hath ioyned with it a certaine ſalt and nitrous quality, by reaſon whereof it clenſeth and draweth flegme out of the noſthrils.

¶ *The Vertues.*

Being eaten when it is boyled, it quickly deſcendeth, looſeth the belly, and prouoketh to the ſtoole, eſpecially being taken with the broth wherein it is ſodden: it nouriſheth little or nothing, and is not ſo wholeſome as Lettuce. A

The iuyce conueied vp into the noſthrils doth gently draw forth flegme, and purgeth the head. B

The great and beautifull Beet laſt deſcribed may be vſed in winter for a ſallad herbe, with vineger, oyle, and ſalt, and is not onely pleaſant to the taſte, but alſo delightfull to the eye. C

The greater red Beet or Roman Beet, boyled and eaten with oyle, vineger and pepper, is a moſt excellent and delicate ſallad: but what might be made of the red and beautifull root (which is to be preferred before the leaues, as well in beauty as in goodneſſe) I refer vnto the curious and cunning cooke, who no doubt when he hath had the view thereof, and is aſſured that it is both good and wholeſome, will make thereof many and diuers diſhes, both faire and good. D

Chap. 43. *Of Blites.*

¶ *The Deſcription.*

1 THe great white Blite groweth three or foure foot high, with grayiſh or white round ſtalkes: the leaues are plaine and ſmooth, almoſt like to thoſe of the white Orach, but not ſo ſoft nor mealy: the floures grow thruſt together like thoſe of Orach: after that commeth the ſeed incloſed in little round flat husky skinnes.

2 There is likewiſe another ſort of Blites very ſmooth and flexible like the former, ſauing that the leaues are reddiſh, mixed with a darke greene colour, as is the ſtalke and alſo the reſt of the plant.

3 There is likewiſe found a third ſort very like vnto the other, ſauing that the ſtalkes, branches, leaues, and the plant is altogether of a greene colour. But this growes vpright, and creepes not at all.

4 There is likewiſe another in our gardens very like the former, ſauing that the whole Plant traileth vpon the ground: the ſtalks, branches, and leaues are reddiſh: the ſeed is ſmall, and cluſtering together, greene of colour, and like vnto thoſe of *Ruellius* his *Coronopus*, or Bucks-horne.

‡ 1 *Blitum majus album.*
The great white Blite.

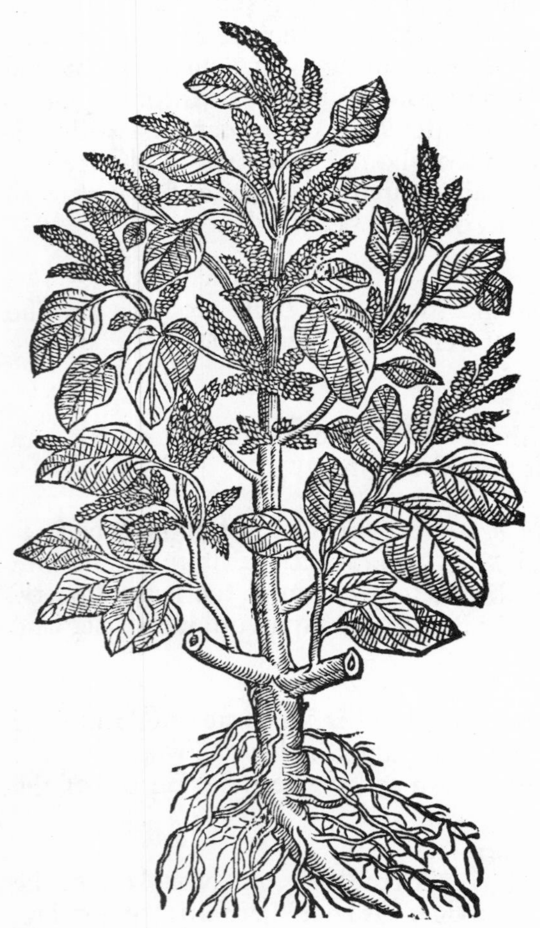

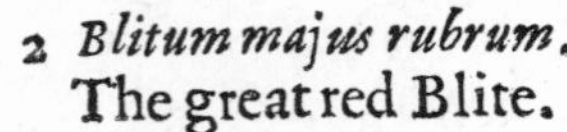

2 *Blitum majus rubrum.*
The great red Blite.

¶ *The Place.*

The Blites grow in Gardens for the moſt part, although there be found of them wilde many times.

¶ *The Time.*

They flouriſh all the Summer long, and grow very greene in Winter likewiſe.

¶ *The Names.*

It is called in Greeke, βλίτον: in Latine, *Blitum*: in Engliſh, Blite, and Blites: in French, *Blites*, or *Blitres*.

¶ The

‡ 3 *Blitum minus album.*
The small white Blite.

‡ 4 *Blitum minus rubrum.*
The small red Blite.

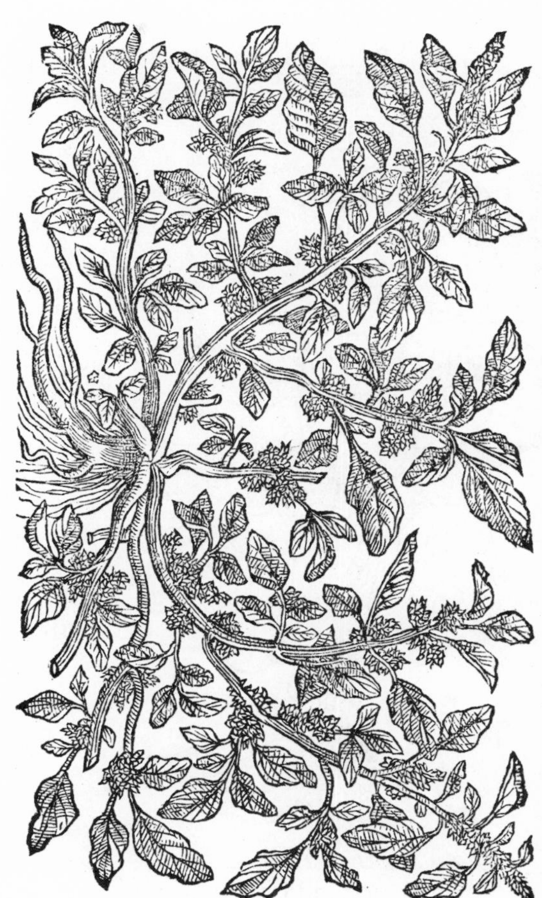

¶ *The Nature.*

The Blite (saith *Galen* in his sixth booke of the faculties of simple medicines) is a pot-hearbe which serueth for meate, being of a cold moist temperature, and that chiefely in the second degree. It yeeldeth to the body small nourishment, as in his second booke of the faculties of nourishments he plainly shewes; for it is one of the pot-herbes that be vnsauoury or without taste, whose substance is waterish.

¶ *The Vertues.*

A The Blite doth nourish little, and yet is fit to make the belly soluble, though not vehemently, seeing it hath no nitrous or sharpe qualitie whereby the belly should be prouoked. I haue heard many old wiues say to their seruants, Gather no Blites to put into my pottage, for they are not good for the eye-sight: whence they had those words I know not, it may be of some Doctor that neuer went to schoole, for that I can finde no such thing vpon record, either among the old or later Writers.

CHAP. 44. *Of Floure-Gentle.*

¶ *The Kindes.*

THere be diuers sorts of floure-Gentle, differing in many points very notably; as in greatnesse and smallnesse; some purple, and others of a skarlet colour; and one aboue the rest wherewith Nature hath seemed to delight her selfe, especially in the leaues, which in variable colours do striue with the Parats feathers for beautie.

1 *Amaranthus purpureus.*
Purple Floure-Gentle.

2 *Amaranthus coccineus.*
Scarlet Floure-Gentle.

3 *Amaranthus tricolor.*
Floramor and Passeuelours.

4 *Amaranthus Pannicula sparsa.*
Branched Floure-Gentle.

5 *Amaranthus*

¶ *The Description.*

1 PVrple floure Gentle riseth vp with a stalke a cubit high, and sometime higher, streaked or chamfered alongst the same, often reddish toward the root, and very smooth: which diuideth it selfe toward the top into small branches, about which stand long leaues, broad, sharpe pointed, soft, slippery, of a greene colour, and sometimes tending to a reddish: in stead of floures, come vp eares or spokie tufts, very braue to looke vpon, but without smell; of a shining light purple, with a glosse like veluet, but far passing it: which when they are bruised, doe yeeld a iuyce almost of the same colour, and being gathered, doe keepe their beauty a long time after; insomuch that being set in water, it will reuiue againe as at the time of his gathering, and it remaineth so, many yeares, whereupon likewise it hath taken its name. The seed standeth in the ripe eares, of colour blacke, and much glittering: the root is short, and full of strings.

‡ 5 *Amaranthus pannicula incurua holoserica.*
Veluet Floures Gentle.

2 The second sort of floure Gentle hath leaues like vnto the former: the stalke is vpright with a few small slender leaues set vpon it: among which doe grow small clusters of scaly floures, of an ouerworne scarlet colour. The seed is like the former.

3 It far exceedeth my skill to describe the beauty and excellency of this rare plant called *Floramor*; and I thinke the pensill of the most curious painter wil be at a stay, when he shall come to set him downe in his liuely colours: but to colour it after my best manner this I say: *Floramor* hath a thicke knobby root, whereupon doe grow many threddy strings: from which riseth a thicke stalke, but tender and soft, which beginneth to deuide himselfe into sundry branches at the ground and so vpward, whereupon doe grow many leaues, wherein doth consist his beauty: for in few words, euery leafe doth resemble in colours the most faire and beautifull feather of a Parrat, especially those feathers that are mixed with most sundry colours, as a stripe of red, and a line of yellow, a dash of white, and a rib of green colour, which I canot with words set forth, such are the sundry mixtures of colours that nature hath bestowed in her greatest iollitie vpon this floure: the floures doe grow betweene the foot-stalkes of those leaues and the body of the stalke or trunke, base, and of no moment in respect of the leaues, being as it were little chaffie husks of an ouerworne tawnie colour: the seed is blacke, and shining like burnished horne. ‡ I haue not seene this thus variegated as our Author mentions, but the leaues are commonly of three colours; the lower part, or that next to the stalke is greene; the middle red, and the end yellow; or else the end red, the middle yellow, and the bottome greene. ‡

4 This plant hath a great many of threds and strings, of which his roots doe consist. From which doe rise vp very thicke fat stalkes, crested and streaked, exceeding smooth, and of a shining red colour, which begin at the ground to diuide themselues into branches; whereupon doe grow many great and large leaues of a darke greene colour tending to rednesse, in shew like those of the red Beet, streaked and dasht here and there with red, mixed with greene. The floures grow alongst the stalkes, from the middest thereof euen to the top, in shape like *Panicum*, that is, a great number of chaffie confused eares thrust hard together, of a deepe purple colour. I can compare the shape thereof to nothing so fitly as to the veluet head of a Stag, compact of such soft matter as is the same: wherein is the seed, in colour white, round, and bored through the middle.

‡ 5 This in stalkes and leaues is much like the purple floure Gentle, but the heads are larger, bended round, and laced, or as it were wouen one with another looking very beautifully like to Crimson veluet: this is seldome to be found with vs; but for the beauties sake is kept in the Gardens of Italy, whereas the women esteemed it not only for the comelinesse and beautious aspect, but

1 *Atriplex satiua alba.*
White Orach.

† 2 *Atriplex satiua purpurea.*
Purple Orach.

3 *Atriplex syluestris, siue Polyspermon.*
Wilde Orach, or All-seed.

† 4 *Atriplex marina.*
Sea Orach.

but alſo for the efficacy thereof againſt the bloudy iſſues, and ſanious vlcers of the wombe and kidneyes, as the Authors of the *Aduerſaria* affirme. ‡

¶ *The Place and Time.*

Theſe pleaſant floures are ſowen in gardens, eſpecially for their great beauty.

They floure in Auguſt, and continue flouriſhing till the froſt ouertake them, at what time they periſh. But the Floramor would be ſowne in a bed of hot horſe-dung, with ſome earth ſtrewed thereon in the end of March, and ordered as we doe muske Melons, and the like.

¶ *The Names.*

This plant is called in Greeke Ἀμάραντος, becauſe it doth not wither and wax old: in Latin, *Amaranthus purpureus*: in Duch, **Samatbluomen**: in Italian, *Fior velluto*: in French, *Paſſe velours*: in Engliſh, floure Gentle, purple Veluet floure, Floramor; and of ſome floure Velure.

¶ *The Temperature, and Vertues.*

Moſt attribute to floure Gentle a binding faculty, with a cold and dry temperature.

A It is reported they ſtop all kinds of bleeding; which is not manifeſt by any apparant quality in them, except peraduenture by the colour only that the red eares haue: for ſome are of opinion, that all red things ſtanch bleeding in any part of the body: becauſe ſome things, as *Bole armoniacke, ſanguis Draconis, terra Sigillata*, and ſuch like of red colour doe ſtop bloud: But *Galen, lib. 2. & 4. de ſimp. facult.* plainly ſheweth, that there can be no certainty gathered from the colours, touching the vertues of ſimple and compound medicines: wherefore they are ill perſuaded, that thinke the floure Gentle to ſtanch bleeding, to ſtop the laske or bloody flix, becauſe of the colour only, if they had no other reaſon to induce them thereto.

Chap. 45. *Of Orach.*

¶ *The Deſcription.*

1 The Garden white Orach hath an high and vpright ſtalke, with broad ſharpe pointed leaues like thoſe of Blite, yet ſmoother and ſofter. The floures are ſmall and yellow, growing in cluſters: the ſeed round, and like a leafe couered with a thin skin, or filme, and groweth in cluſters. The root is wooddy and fibrous: the leaues and ſtalkes at the firſt are of a glittering gray colour, and ſprinkled as it were with a meale or floure.

2 This differs from the former, only in that it is of an ouerworne purple colour.

‡ 3 This might more fitly haue beene placed amongſt the Blites, yet finding the figure here (though a contrary diſcription) I haue let it inioy the place. It hath a white and ſlender root, and it is ſomewhat like, yet leſſe then the Blite, with narrow leaues ſomewhat reſembling Baſill: it hath aboundance of ſmall floures, which are ſucceeded by a numerous ſort of ſeeds, which are blacke and ſhining. ‡

4 There is a wilde kinde growing neere the ſea, which hath pretty broad leaues, cut deepely about the edges, ſharpe pointed, and couered ouer with a certaine mealineſſe, ſo that the whole plant as well leaues, as ſtalkes and floures, looke of an hoary or gray colour. The ſtalks lye ſpred, on the ſhore or Beach, whereas it vſually growes.

‡ 5 The common wilde Orach hath leaues vnequally ſinuated, or cut in ſomewhat after the manner of an oaken leafe, and commonly of an ouerworne grayiſh colour: the floures and ſeeds are much like thoſe of the garden, but much leſſe.

6 This is like the laſt deſcribed, but the leaues are leſſer and not ſo much diuided, the ſeeds grow alſo in the ſame manner as thoſe of the precedent.

7 This alſo in the face and manner of growing is like thoſe already deſcribed, but the leaues are long and narrow, ſometimes a little notched: and from the ſhape of the leafe *Lobell* called it *Atriplex Sylueſtris polygoni, aut Helxines folio.*

8 This elegant Orach hath a ſingle and ſmall root, putting forth a few fibers, the ſtalkes are ſome foot high, diuided into many branches, and lying along vpon the ground; and vpon theſe grow leaues at certaine ſpaces whitiſh and vnequally diuided, ſomewhat after the manner of the wilde Orach; about the ſtalke or ſetting on of the leaues grow as it were little berries, ſomewhat like a little mulberry, and when theſe come to ripeneſſe, they are of an elegant red colour, and make a fine ſhew. The ſeed is ſmall round and aſh coloured. ‡

¶ *The Place.*

The Garden Oraches grow in moſt gardens. The wilde Oraches grow neere paths-wayes and ditch ſides; but moſt commonly about dung-hils and ſuch fat places. Sea Orach I haue found at Queeneborough, as alſo at Margate in the Ile of Thanet: and moſt places about the ſea ſide. ‡ The eighth groweth only in ſome choice gardens, I haue ſeen it diuers times with Mr. *Parkinſon*. ‡

¶ *The*

‡ 5 *Atriplex sylvestris vulgaris.*
Common wilde Orach.

‡ 6 *Atriplex sylvestris altera.*
The other wilde Orach.

‡ 7 *Atriplex sylvestris angustifolia.*
Narrow leaued wilde Orach.

‡ 8 *Atriplex baccifera.*
Berry-bearing Orach.

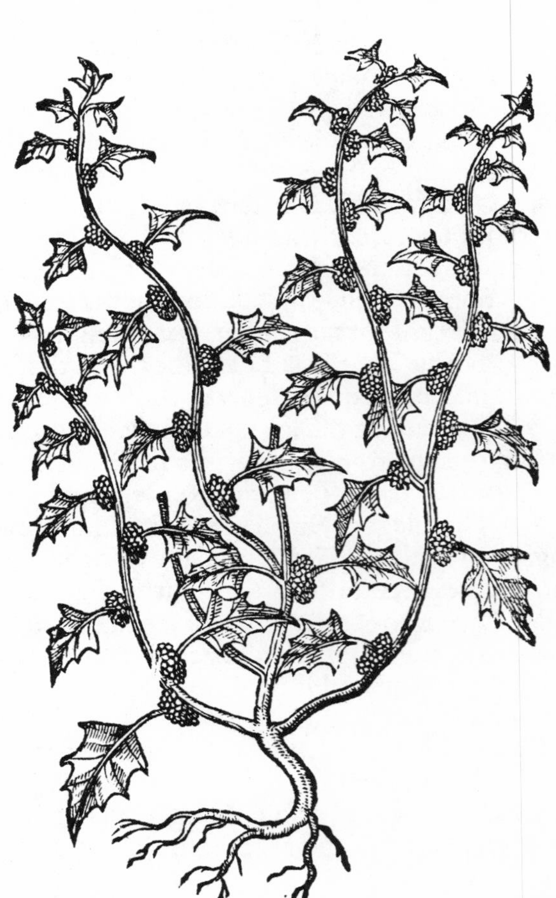

¶ *The Time.*

They floure and ſeed from Iune to the end of Auguſt.

¶ *The Names.*

Garden Orach is called in Greeke, ἀτράφαξις: in Latine, *Atriplex*, and *Aureum olus*: in Dutch, **Melde**: in French, *Arrouches ou bonnes dames*: in Engliſh, Orach, and Orage: in the Bohemian tongue, *Leboda*: *Pliny* hath made ſome difference betweene *Atriplex* and *Chryſolachanum*, as though they differed one from another; for of *Atriplex* he writeth in his twentieth booke; and of *Chryſolachanum* in his twenty eighth booke, and eighth chapter: where hee writeth thus, *Chryſolachanum*, ſaith he, groweth in Pinetum like Lettuce: it healeth cut ſinewes if it be forthwith applied.

3 This wilde Orach hath beene called of *Lobel, Polyſpermon Caſſani Baſſi*, or All ſeed.

¶ *The Temperature.*

Orach, ſaith *Galen*, is of temperature moiſt in the ſecond degree, and cold in the firſt.

¶ *The Vertues.*

Dioſcorides writeth, That the garden Orach is both moiſt and cold, and that it is eaten boyled as other ſallad herbes are, and that it ſoftneth and looſeth the belly. A

It conſumeth away the ſwellings of the throat, whether it be laid on raw or ſodden. B

The ſeed being drunke with meade or honied water, is a remedie againſt the yellow jaundice. C

Galen thinketh, that for that cauſe it hath a clenſing qualitie, and may open the ſtoppings of the liuer. D

† The figure which was in the ſecond place was of *Pes Anſerinus* 2. of *Taber*. The figure in the fourth place was of the wild Orach, that I haue deſcribed in the fifth place.

CHAP. 46. *Of Stinking Orach.*

Atriplex olida.
Stinking Orach.

¶ *The Deſcription.*

STinking Orach growes flat vpon the ground and is a baſe and low plant with many weak and feeble branches, whereupon doe grow ſmall leaues of a grayiſh colour, ſprinkled ouer with a certaine kinde of duſty mealineſſe, in ſhape like the leaues of Baſill: amongſt which leaues here and there confuſedly be the ſeeds diſperſed, as it were nothing but duſt or aſhes. The whole plant is of a moſt loathſome ſauour or ſmel; vpon which plant if any ſhould chance to reſt and ſleepe, he might very well report to his friends, that he had repoſed himſelfe among the chiefe of *Scoggins* heires.

¶ *The Place.*

It groweth vpon dunghills, and in the moſt filthy places that may be found, as alſo about the common piſſing places of great princes and Noblemens houſes. Sometime it is found in places neere bricke kilns and old walls, which doth ſomewhat alter his ſmell, which is like toſted cheeſe: but that which groweth in his naturall place ſmells like ſtinking ſalt-fiſh, whereof it tooke his name *Garoſmus*.

¶ *The Time.*

It is an herbe for a yeare, which ſpringeth vp, and when the ſeed is ripe it periſheth, and recouereth it ſelfe againe of his owne ſeed; ſo that if it be gotten into a ground, it cannot be deſtroyed.

¶ *The Names.*

Stinking Orach is called of *Cordus, Garoſmus*, becauſe it ſmelleth like ſtinking fiſh: it is likewiſe called

called *Tragium Germanicum*, and *Atriplex fœtida garum olens*, by *Pena* and *Lobel*: for it smelleth more stinking than the rammish male Goat: whereupon some by a figure haue called it *Vulvaria*: and it may be called in English, stinking Mother-wort.

¶ *The Nature and Vertues.*

A There hath been little or nothing set down by the Antients, either of his nature or vertues, notwithstanding it hath beene thought profitable, by reason of his stinking smell, for such as are troubled with the mother: for as *Hyppocrates* saith, when the mother doth stifle or strangle, such things are to be applied vnto the nose as haue a ranke and stinking smell.

Chap. 47. *Of Goose-foot.*

¶ *The Description.*

1 Goose-foot is a common herbe, and thought to be a kinde of Orach: it riseth vp with a stalke a cubit high or higher, somewhat chamfered and branched; the leaues be broad, smooth, sharpe pointed, shining, hauing certaine deepe cuts about the edges, and resembling the foot of a goose: the floures be small, something red: the seed standeth in clusters vpon the top of the branches, being very like the seed of wilde Orach, and the root is diuided into sundry strings.

‡ 2 This differs from the last described, in that the leaues are sharper cut, and more diuided, the seed somewhat smaller, and the colour of the whole plant is a deeper or darker greene.

‡ 1 *Atriplex syluestris latifolia, siue Pes Anserinus.* Goose-foot.

‡ 2 *Atriplex syluestris latifolia altera.* The other Goose-foot.

¶ *The Place.*

It growes plentifully in obscure places neere old walls and high-waies, and in desart places.

¶ *The Time.*

It flourisheth when the Orach doth, whereof this is a wilde kinde.

¶ *The Names.*

The later Herbarists haue called it *Pes anserinus*, and *Chenopodium*, of the likenesse the leaues haue with the foot of a Goose: in English, Goose-foot, and wilde Orach.

¶ The

¶ *The Temperature.*

This herbe is cold and moiſt, and that no leſſer than Orach, but as it appeareth more cold.

¶ *The Vertues.*

It is reported that it killeth ſwine if they do eate thereof: it is not vſed in Phyſicke: and much leſſe as a ſallade herbe.

Chap. 48. *Of Engliſh Mercurie.*

Bonus Henricus.
Engliſh Mercurie, or good Henrie.

¶ *The Deſcription.*

Good Henrie called *Tota bona*, ſo named of the later Herbariſts, is accounted of them to be one of the Dockes, but not properly. This bringeth forth very many thicke ſtalkes, ſet with leaues two foot high; on the branches wherof towards the top ſtand greene floures in cluſters, thicke thruſt together. The ſeed is flat like that of the Orach, whereof this is a kinde. The leaues be faſtened to long foote-ſtalkes, broad behinde, and ſharpe pointed, faſhioned like the leaues of Aron, or Wake-robin, white, or grayiſh of colour, and as it were couered ouer with a fine meale: in handling it is fat and olious, with a very thicke root, and parted into many diuiſions, of a yellow colour within, like the ſharpe pointed Docke.

¶ *The Place.*

It is commonly found in vntilled places, and among rubbiſh neere common waies, old walls, and by hedges in fields.

¶ *The Time.*

It floureth in Iune and Iuly eſpecially.

¶ *The Names.*

It is called of ſome *Pes Anſerinus*, and *Tota bona*: in Engliſh, All-Good, and Good Henrie: in Cambridgſhire it is called Good king Harry: the Germanes call it **Guter Heinrick**, of a certaine good qualitie it hath, as they alſo name a certaine pernicious herbe, *Malus Henricus*, or bad Henry. It is taken for a kinde of Mercurie, but vnproperly, for that it hath no participation with Mercurie, either in forme or quality, except yee will call euery herbe Mercurie which hath power to looſe the belly.

¶ *The Temperature.*

Bonus Henricus, or Good Henrie is moderately hot and dry, clenſing and ſcouring withall.

¶ *The Vertues.*

The leaues boiled with other pot-herbes and eaten, maketh the body ſoluble. A

The ſame bruſed and laid vpon greene wounds, or foule and old vlcers, doth ſcoure, mundifie and heale them. B

Chap. 49. *Of Spinach.*

Spinacia.
Spinach.

¶ *The Deſcription.*

1 SPinach is a kinde of Blite, after ſome; notwithſtanding I rather take it for a kinde of Orach. It bringeth forth ſoft and tender leaues of a darke greene colour, full of juice, ſharpe pointed, and in the largeſt part or neather end ſquare; parted oftentimes with a deepe gaſh on either ſide next to the ſtemme or foot-ſtalke: the ſtalke is round, a foot high, hollow within: on the tops of the branches ſtand litle floures in cluſters, in whoſe places doth grow a prickly ſeed. The root conſiſteth of many ſmall threds.

2 There is another ſort found in our gardens like vnto the former in goodneſſe, as alſo in ſhape, ſauing that the leaues are not ſo great, nor ſo deepely gaſht or indented: and the ſeed hath no prickles at all, for which cauſe it is called round Spinach.

¶ *The Place.*

It is ſowne in gardens without any great labor or induſtrie, and forſaketh not any ground being but indifferent fertill.

¶ *The Time.*

It may be ſowne almoſt at any time of the yeere, but being ſowne in the ſpring it quickly groweth vp, and commeth to perfection within two moneths: but that which is ſowne in the fall of the leafe groweth not ſo ſoone to perfection, yet continueth all the win terand ſeedeth preſently vpon the firſt ſpring.

¶ *The Names.*

It is called in theſe daies *Spinachia:* of ſome, *Spinacheum olus*: of others, *Hiſpanicum olus*: *Fuchſius* nameth it Σπινάχια: the Arabians and *Serapio* call it *Hiſpane*: the Germanes, Spinet: in Engliſh, Spinage and Spinach: in French, *Eſpinas.*

¶ *The Nature.*

Spinach is euidently cold and moiſt almoſt in the ſecond degree, but rather moiſt. It is one of the pot-herbes whoſe ſubſtance is waterie, and almoſt without taſte, and therefore quickly deſcendeth and looſeth the bellie.

¶ *The Vertues.*

A It is eaten boiled, but it yeeldeth little or no nouriſhment at all: it is ſomething windie, and eaſily cauſeth a deſire to vomit: it is vſed in ſallades when it is young and tender.

B This herbe of all other pot-herbes and ſallade herbes maketh the greateſt diuerſitie of meates and ſallades.

Chap. 50. *Of Pellitorie of the wall.*

¶ *The Deſcription.*

PEllitorie of the wall hath round tender ſtalkes ſomewhat browne or reddiſh of colour and ſomewhat ſhining: the leaues be rough like to the leaues of Mercurie, nothing ſnipt about the edges. The floures be ſmall, growing cloſe to the ſtemmes: the ſeed is blacke and very ſmall, couered with a rough huſke which hangeth faſt vpon garments: the root is ſomewhat reddiſh.

¶ The

Parietaria.
Pellitorie of the wall.

¶ *The Place.*

It groweth neere to old walls in the moiſt corners of Churches and ſtone buildings, among rubbiſh and ſuch like places.

¶ *The Time.*

It commeth vp in May: it ſeedeth in Iuly and Auguſt: the root onely continueth and is to be found in Winter.

¶ *The Names.*

It is commonly called *Parietaria*, or by a corrupt word *Paritaria*, becauſe it groweth neere to walls: and for the ſame cauſe it is named of diuers *Muralis*: alſo *Muralium* of *Pliny* and *Celſus*: of the Grecians ἑλξίνη. There is alſo another *Helxine* ſyrnamed *Ciſſampelos*: ſome call it *Perdicium*, of Partridges which ſomtimes feed hereon: ſome, *Vrceolaris*, and *Vitraria*, becauſe it ſerueth to ſcoure glaſſes, pipkins, and ſuch like: it is called in high-Dutch, **Tag vnd nacht**: in Spaniſh, *Yerua del muro*: in Engliſh, Pellitorie of the wall: in French, *Parietaire.*

¶ *The Temperature.*

Pellitorie of the wall (as *Galen* ſaith) hath force to ſcoure, and is ſomething cold and moiſt.

¶ *The Vertues.*

Pellitory of the wall boyled, and the decoction of it drunken, helpeth ſuch as are vexed with an old cough, the grauell and ſtone, and is good againſt the difficultie of making water, and ſtopping of the ſame, not onely inwardly, but alſo outwardly applied vpon the region of the bladder, in manner of a fomentation or warme bathing, with ſpunges or double clouts, or ſuch like. A

Dioſcorides ſaith, That the iuyce tempered with Ceruſe or white leade maketh a good ointment againſt Saint Anthonies fire and the Shingles: and mixed with the Cerot of Alcanna, or with the male Goats tallow, it helpeth the gout in the feet: which *Pliny* alſo affirmeth, *Lib.22.cap.17.* B

It is applied (ſaith he) to paines of the feet with Goats ſuet and wax of Cyprus; where in ſtead of wax of Cyprus there muſt be put the Cerot of Alcanna. C

Dioſcorides addeth, That the iuyce hereof is a remedy for old coughs, and taketh away hot ſwellings of the almonds in the throat, if it be vſed in a gargariſme, or otherwiſe applied: it mitigateth alſo the paines of the eares, being poured in with oile of Roſes mixed therewith. D

It is affirmed, That if three ounces of the iuyce be drunke it prouoketh vrine out of hand.

The leaues tempered with oyle of ſweet almonds in manner of a pulteſſe, and laid to the pained parts, is a remedie for them that be troubled with the ſtone, and that can hardly make water.

Chap. 51. *Of French Mercurie.*

¶ *The Kindes.*

There be two kindes of Mercury reckoned for good, and yet both ſomtimes wilde; beſides two wilde neuer found in gardens, vnleſſe they be brought thither.

¶ *The Deſcription.*

1 The male garden Mercurie hath tender ſtalks full of ioints and branches, whereupon do grow greene leaues like Pellitorie of the wall, but ſnipt about the edges: amongſt which come forth two hairy bullets round, and ioyned together like thoſe of Gooſegraſſe or Cleuers, each containing in it ſelfe one ſmall round ſeed: the root is tender, and full of white hairy ſtrings.

2 The female is like vnto the former in leaues, ſtalks, and manner of growing, differing but in

the floures and ſeed : for this kinde hath a greater quantitie of floures and ſeed growing together like little cluſters of grapes, of a yellowiſh colour. The ſeed for the moſt part is loſt before it can be gathered.

1 *Mercurialis mas.*
Male Mercurie.

2 *Mercurialis fœmina.*
Female Mercurie.

¶ *The Place.*

French Mercurie is ſowen in Kitchen gardens among pot-herbes; in Vineyards, and in moiſt ſhadowie places : I found it vnder the dropping of the Biſhops houſe at Rocheſter; from whence I brought a plant or two into my garden, ſince which time I cannot rid my garden from it.

¶ *The Time.*

They floure and flouriſh all the Sommer long.

¶ *The Names.*

It is called in Greeke, λινόζωστις, and ἑρμῦ βοτάνιον, or Mercurie his herbe; whereupon the Latines call it *Mercurialis* : it is called in Italian, *Mercorella* : in Engliſh, French Mercurie : in French, *Mercuriale, Vignoble*, and *Foirelle, quia Fluidam laxamue alvum reddit, Gallobelgæ enim foiZe & foizeus, ventris Fluorem vocant.*

¶ *The Temperature.*

Mercury is hot and dry, yet not aboue the ſecond degree : it hath a cleanſing facultie, and (as *Galen* writerh) a digeſting qualitie alſo.

¶ *The Vertues.*

A It is vſed in our age in cliſters, and thought very good to clenſe and ſcoure away the excrements and other filth contained in the guts. It ſerueth to purge the belly, being eaten or otherwiſe taken, voiding out of the belly not only the excrements, but alſo phlegme and choler. *Dioſcorides* reporteth, that the decoction hereof purgeth wateriſh humors.

B The leaues ſtamped with butter, and applied to the fundament, prouoketh to the ſtoole; and the herbe bruiſed and made vp in manner of a peſſary, cleanſeth the mother, and helpeth conception.

C *Coſtæus* in his booke of the nature of plants ſaith, that the iuyce of Mercurie, Hollihocks, & purſlane mixed together, and the hands bathed therein, defendeth them from burning, if they be thruſt into boyling leade.

Chap.

CHAP. 52. *Of Wilde Mercurie.*

‡ 1 *Cynocrambe.*
Dogs Mercury.

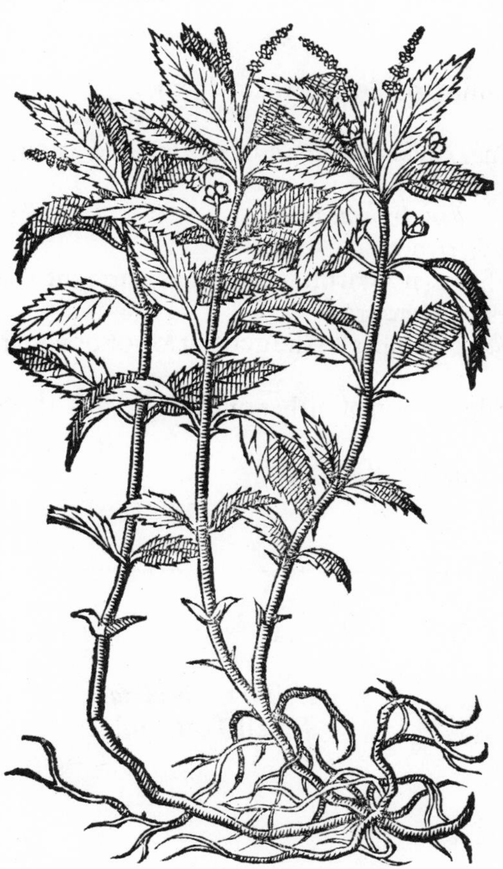

† 2 *Phyllon arrhenogonon, ſiue marificum.*
Male childrens Mercury.

3 *Phyllon Thelygonon, ſiue Fœminificum.*
Childrens Mercurie, the female.

¶ *The Deſcription.*

1 Dogs Mercurie is ſomewhat like vnto the garden Mercury, ſauing the leaues hereof are greater, and the ſtalke not ſo tender, and yet very brittle, growing to the height of a cubit, without any branches at all, with ſmal yellow floures. The ſeed is like the female Mercurie. ‡ It is alſo found like the male Mercurie, as you ſee them both expreſt in the figure; and ſo there is both male & female of this Mercury alſo. ‡

2 Male childrens Mercury hath three or foure ſtalkes, or moe: the leaues be ſomwhat long, not much vnlike the leaues of the oliue tree, couered ouer with a ſoft downe or wooll gray of colour; and the ſeed alſo like thoſe of Spurge, growing two together, being firſt of an aſh-colour, but after turne to a blew.

‡ 3 This is much in ſhape like to the laſt deſcribed, but the ſtalkes are weaker, and haue more leaues vpon them; the floures alſo are ſmall and moſſy, and they grow vpon long ſtalks, whereas the ſeeds of the other are faſtned to very ſhort ones: the ſeed is contained in round little heads, being ſometimes two, otherwhiles three or more in a cluſter. ‡

¶ The Place.

They grow in woods and copses, in the borders of fields, and among bushes and hedges. ‡ But the two last described are not in England, for any thing that I know. ‡

The Dogs Mercurie I haue found in many places about Green-hithe, Swaines-combe village, Grauesend, and South-fleet in Kent; in Hampsted wood, and all the villages thereabout, foure miles from London.

¶ The Time.

These flourish all the Sommer long, vntill the extreame frost do pull them downe.

¶ The Names.

Dogs Mercurie is called in Greeke, κυνοκράμβη: in Latine, *Canina*, and *Brassica Canina*, and *Mercurialis syluestris*: in English, Dogs Cole, and Dogs Mercury.

Childrens Mercury is called *Phyllon thelygonon*, and *Phyllon Arrhenogonon*.

¶ The Temperature and Vertues.

These wilde kindes of Mercurie are not vsed in physicke; notwithstanding it is thought they agree as well in nature as qualitie with the other kindes of Mercury.

A ‡ It is reported by the Antients, that the male *Phyllon* conduces to the generation of boyes, and the female to girles.

B At Salamantica they giue and much commend the decoction of either of these against the bitings of a mad dog.

C The Moores at Granado vse them frequently in womens diseases. ‡

† The figure of the *Cynocrambe* was omitted, and in stead thereof was put the figure of *Phyllon marificum*.

Chap. 53. *Of Torne-sole.*

1 *Heliotropium maius.*
Great Torne-sole.

† 2 *Heliotropium minus.*
Small Torne-sole.

¶ The Kindes.

There be foure sorts of Torne-sole, differing one from another in many notable points, as in greatnesse and smallnesse, in colour of floures, in forme and shape.

¶ The

The deſcription.

1 The great Torneſole hath great ſtraight ſtalks couered with a white hairy cotton,eſpecially about the top; the leaues are ſoft and hairy in handling,in ſhape like the leaues of Baſill: the floures grow at the top of the branches, in colour white, thicke together in rowes vpon one ſide of the ſtalke, which ſtalke doth bend or turne backward like the taile of a ſcorpion: the root is ſmall and hard.

2 The ſmall Torneſole hath many little and weake branches trailing vpon the ground,whereupon doe grow ſmall leaues, like thoſe of the leſſer Baſill. The floures doe grow without any certaine order, amongſt the leaues and tender branches, gray of colour, with a little ſpot of yellow in the middeſt, the which turne into crooked tailes like thoſe of the precedent,but not altogether ſo much.

† 3 *Heliotropium ſupinum Cluſij & L'obelij.* Hairie Torneſole.

4 *Heliotropium Tricoccum.* Widowwaile Tornſole.

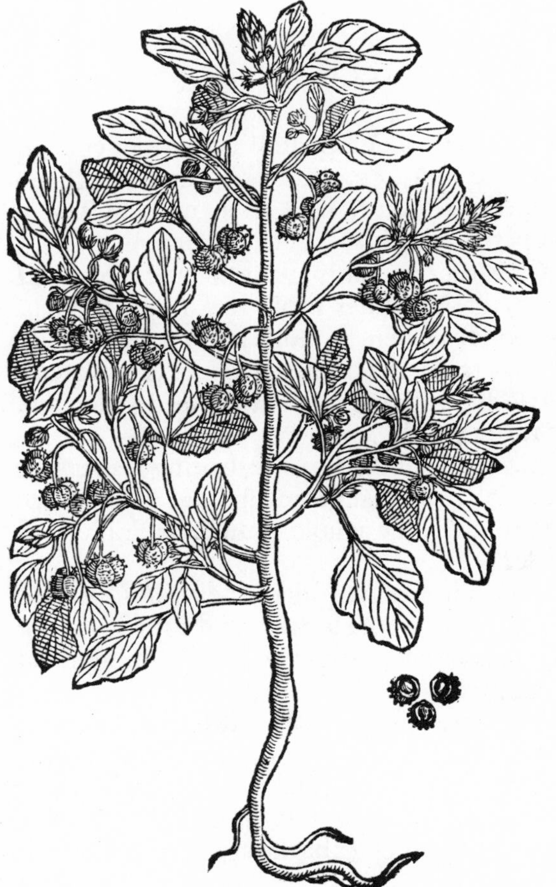

3 Hairy Torneſole hath many feeble and weake branches trailing vpon the ground, ſet with ſmall leaues,leſſer than the great Tornſole, of which it is a kinde, hauing the ſeed in ſmall chaffie husks,which do turne back like the taile of a ſcorpion, iuſt after the manner of the firſt deſcribed.

4 This kinde of Torneſole hath leaues very like to thoſe of the great Tornſole, but of a blacker greene colour:the floures be yellow,and vnprofitable; for they are not ſucceeded by the fruit, but after them commeth out the fruit hanging vpon ſmall foot-ſtalks three ſquare,and in euery corner there is a ſmall ſeed like to thoſe of the Tythimales;the root is ſmall and threddy.

¶ *The Place.*

Tornſole, as *Dioſcorides* ſaith, doth grow in fennie grounds and neere vnto pooles and lakes. They are ſtrangers in England as yet: It doth grow about Montpelier in Languedock, where it is had in great vſe to ſtaine and die clouts withall, wherewith through Europe meat is coloured.

¶ *The Time.*

They flouriſh eſpecially in the Sommer ſolſtice, or about the time when the ſun entreth into Cancer.

¶ *The Names.*

The Græcians call it *Heliotropium*: the Latines keepe theſe names, *Heliotropium magnum*, and

and *Scorpiurum*: of *Ruellius*, *Herba Cancri* · it is named *Heliotropium*, not because is is turned about at the daily motion of the sun, but by reason it flowreth in the sommer solstice, at which time the sun being farthest gone from the Æquinoctiall circle, returneth to the same: and *Scorpiurum* of the twiggie tops, that bow backeward like a scorpions taile: of the Italians, *Tornesole bobo*; in French, *Tournsol*: some thinke it to be *Herba Clytiæ*, into which the Poets feigne *Clytia* to be metamorphosed; whence one hath these verses:

Herba velut Clytiæ semper petit obuia solem,
Sic pia mens Christum, quo prece spectet, habet.

¶ *The Nature.*

Tornsole, as *Paulus Ægineta* writeth, is hot and dry, and of a binding faculty.

¶ *The Vertues.*

A A good handfull of great Tornsole boyled in wine, and drunke, doth gently purge the body of hot cholericke humours and tough clammie or slimie flegme.

B The same boyled in wine and drunke is good against the stingings of Scorpions, or other venomous beasts, and is very good to be applyed outwardly vpon the griefe or wound.

C The seed stamped and layd vpon warts and such like excrescences, or superfluous out-growings, causeth them to fall away.

D The small Tornesole and his seed boyled with Hyssope, Cresses, and salt-peter and drunke, driueth forth flat and round wormes.

E With the small Tornsole they in France doe die linnen rags and clouts into a perfect purple colour, wherewith cookes and confectioners doe colour iellies, wines, meates, and sundry confectures: which clouts in shops be called Tornsole, after the name of the herbe.

† The second and third figures were formerly transposed: the fourth was the figure of the hairy Scorpion-grasse, described in the fourth place, in the following Chapter.

Chap. 54. *Of Scorpion Grasse.*

¶ *The Description.*

1 Scorpion grasse hath many smooth, plaine, euen leaues, of a darke greene colour; stalks small, feeble and weake, trailing vpon the ground, and occupying a great circuit in respect of the plant. The floures grow vpon long and slender foot-stalks, of colour yellow, in shape like to the floures of broome; after which succeed long, crooked, rough cods, in shape and colour like vnto a Caterpiller; wherein is contained yellowish seed like vnto a kidney in forme. The roote is small and tender · the whole plant perisheth when the seed is ripe.

2 There is another Scorpion grasse, found among (or rather resembling) pease and tares, and thereupon called *Scorpioides Leguminosa*, which hath small and tender roots like small threds: branches many, weake and tender, trailing vpon the ground, if there be nothing to take hold vpon with his clasping and crooked seed vessels; otherwise it rampeth vpon whatsoeuer is neere vnto it. The leaues be fewe and small: the floures very little and yellow of colour: the seed followeth, little and blackish, conteined in little cods, like vnto the taile of a Scorpion.

3 There is another sort almost in euery shallow grauelly running streame, hauing leaues like to *Becabunga* or Brooklime. The floures grow at the top of tender fat greene stalkes, blew of colour, and sometimes with a spot of yellow among the blew; the whole branch of floures doe turne themselues likewise round like the scorpions taile.

There is also another growing in watrie places, with leaues like innto *Anagallis aquatica*, or water Chickweed, hauing like slender stalkes and branches as the former, and the floures not vnlike, sauing that the floures of this are of a light blew or watched colour, somewhat bigger, and layd more open, whereby the yellow spot is better seene.

4 There is likewise another sort growing vpon most dry grauelly and barren ditch bankes, with leaues like those of or Mouse-eare: this is called *Myosotis scorpioides*; it hath rough and hairy leaues, of an ouerworne russet colour: the floures doe grow vpon weake, feeble, and rough branches, as is all the rest of the plant. They likewise grow for the mostpart vpon one side of the stalke, blew of colour, with a like little spot of yellow as the others, turning themselues backe againe like the taile of a Scorpion.

There

There is another of the land called *Myosotis Scorpioides repens*, like the former: but the floures are thicker thrust together, and doe not grow all vpon one side as the other, and part of the floures are blew, and part purple, confusedly mixt together.

¶ *The Place.*

1, 2 These Scorpion grasses grow not wilde in England, notwithstanding I haue receiued seed of the first from beyond the seas, and haue dispersed them through England, which are esteemed of gentlewomen for the beauty and strangenesse of the crooked cods resembling Caterpillers.

The others doe grow in waters and streames, as also on drie and barren bankes.

¶ *The Time.*

The first floureth from May to the end of August: the others I haue found all the sommer long.

¶ *The Names.*

‡ 1 *Fabius Columna* iudges this to be the *Clymenon* of *Dioscorides*: others call it *Scorpioides*, and *Scorpioides buplevri folio*.

2 This is the *Scorpioides* of *Matthiolus*, *Dod. Lobell*, and others; and I iudge it was this plant our Author in this place intended, and not the *Scorpioides Leguminosa* of the *Aduersaria*, for that hath not a few leaues, but many vpon one rib; and besides, *Dodonæus*, whom in descriptions & history our Author chiefely followes, describes this immediatly after the other: *Guillandinus*, *Cæsalpinus*, and *Bauhine* iudge it to be the *Telephium* of *Dioscorides*.

3 This and the next want no names, for almost euery writer hath giuen them seuerall ones: *Brunfelsius* called it *Cynoglossa minor*: *Tragus*, *Tabernamontanus*, and our Author (*page* 537. of the former edition) haue it vnder the name of *Euphrasia Cærulea*: *Dodonæus* cals it *Scorpioides fœmina*: *Lonicerus*, *Leontopodium*; *Cæsalpinus*, *Heliotropium minus in paluſtribus*: *Cordus* and *Thalius*, *Echium palustre*.

4 This is *Auricula muris minor tertia*, *Euphrasia quarta*, and *Pilosella sylvestris of Tragus*: *Scorpioides mas* of *Dodonæus*; *Alsine Myosotis*: and *Myosotis hirsuta repens* of *Lobell*; *Heliotropium minus alterum* of *Cæsalpinus*; *Echium minimum* of *Columna*; and *Echium palustre alterum* of *Thalius*: our Authour had it thrise: first in the precedent chapter, by the name of *Heliotropium rectum*, with a figure: secondly in this present chapter, without a figure: and thirdly *pag.* 514, also with a figure vnder the name of *Pilosella flore cæruleo*. ‡

¶ *The Nature and Vertues.*

There is not any thing remembred of the temperature: yet *Dioscorides* saith, that the laeues of Scorpion grasse applyed to the place, is a present remedy against the stinging of Scorpions: and likewise boyled in wine and drunke, preuaileth against the said bitings, as also of adders, snakes, and such venomous beasts: being made in an vnguent with oile, wax, and a little gum *Elemni*, is profitable against such hurts as require a healing medicine. A

CHAP. 55. *Of Nightshade.*

¶ *The Kindes.*

THere be diuers Nightshades, whereof some are of the garden; and some that loue the fields, and yet euery of them found wilde; whereof some cause sleepinesse euen vnto death: others cause sleepinesse, and yet Physicall: and others very profitable vnto the health of man, as shall be declared in their seuerall vertues.

¶ *The Description.*

1 GArden Nightshade hath round stalkes a foot high, and full of branches, whereon are set leaues of a blackish colour, soft and full of iuice, in shape like to leaues of Basill, but much greater: among which doe grow small white floures with yellow pointals in the middle; which being past, there succeed round berries, greene at the first, and blacke when they be ripe, like those of Iuy: the root is white, and full of hairy strings.

‡ 2 The root of this is long, pretty thicke and hard, being couered with a brownish skin; from this root grow vp many smal stalks of the height of a cubit and better, somewhat thick withall: the leaues that grow alongst the stalke are like those of the Quince tree, thicke, white, soft and downye. The floures grow about the stalke at the setting on of the leafe, somewhat long and of a pale colour, diuided into foure parts, which are succeeded by seeds contained in hairy or woolly receptacles: which when they come to ripenesse are red, or of a reddish saffron colour. ‡

¶ *The Place.*

This Nightshade commeth vp in many places, and not only in gardens, of which not withstanding

1 *Scorpioides Bupleuri folio, Penæ & L'Obely.*
Scorpion graſſe, or Caterpillers.

‡ 2 *Scorpioides Matthioli.*
Matthiolus his Scorpion graſſe.

‡ 3 *Myoſotis ſcorpioides paluſtris.*
Water Scorpion graſſe.

‡ 4 *Myoſotis ſcorpioides aruenſis hirſuta.*
Mouſe-eare Scorpion graſſe.

It

ding it hath taken his surname, and in which it is often found growing with other herbes; but also neere common high waies, the borders of fields, by old walls and ruinous places.

‡ 2 This growes not with vs, but in hotter Countries *Clusius* found it growing among rubbish at Malago in Spaine.‡

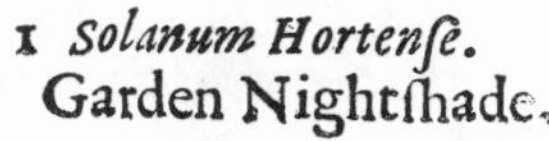

1 *Solanum Hortense.*
Garden Nightshade.

‡ 2 *Solanum Somniferum.*
Sleepie Nightshade.

¶ *The Time.*

It flowreth in Sommer, and oftentimes till Autumne be well spent; and then the fruit commeth to ripenesse.

‡ 2 This *Clusius* found in flower and with the seede ripe in Februarie; for it liueth many yeares in hot Countries, but in cold it is but an annuall.‡

¶ *The Names.*

It is called of the Græcians Στρύχνος: of the Latines, *Solanum*, and *Solanum Hortense*: in shops, *Solatrum*: of some, *Morella*, *Vua Lupina*, and *Vua Vulpis*: in Spanish likewise, *Morella*, and *Yerua Mora*: *Marcellus* an old Phisicke writer, and diuers others of his time called it *Strumum*: *Plinie* in his 27. booke chap. 8. sheweth that it is called *Cucubalus*: both these words are likewise extant in *Apuleius* among the confused names of Nightshade; who comprehending all the kindes of Nightshade together in one chapter, being so many, hath strangely & absurdly confounded their names. In English it is called Garden Nightshade, Morel, and Petie Morell: in French, *Morelle*, *Gallobelgis*: *seu ardent: quia medetur igni sacro.*

¶ *The Temperature.*

Nightshade (as *Galen* saith in his booke of the faculties of simple medicines) is vsed for those infirmities that haue need of cooling and binding; for these two qualities it hath in the second degree: which thing also hee affirmeth in his booke of the faculties of nourishments, where hee saith that there is no pot-herbe which wee vse to eat that hath so great astriction or binding as Nightshade hath, and therefore Physitions do worthily vse it, and that seldome as a nourishment, but alwaies as a medicine.

¶ *The Vertues.*

1 *Dioscorides* writeth, that Nightshade is good against S. Anthonies fire, the shingles, paine of the head, the heart burning or heat of the stomacke, and other like accidents proceeding of sharp and biting humours. Notwithstanding that it hath these vertues, yet it is not alwaies good that it should be applied vnto those infirmities, for that many times there hapneth more dangers by applying A

plying of theſe remedies, than of the diſeaſe it ſelfe: for as *Hippocrates* writeth in the 6. booke of his Aphoriſmes, the 25. particular, that it is not good, that S. Anthonies fire ſhould be driuen from the outward parts to the inward. And likewiſe in his Prognoſticks he ſaith, that it is neceſſary that S. Anthonies fire ſhould breake forth, and that it is death to haue it driuen in; which is to be vnderſtood not onely of S. Anthonies fire, but alſo of other like burſtings out procured by nature. For by vſing of theſe kindes of cooling and repelling medicines, the bad, corrupt, and ſharpe humors are driuen backe inwardly to the chiefe and principall parts, which cannot be done without great danger and hazard of life. And therefore we muſt not vnaduiſedly, lightly, or raſhly miniſter ſuch kinde of medicines vpon the comming out of Saint Anthonies fire, the ſhingles, or ſuch hot pimples and blemiſhes of the skinne.

B The iuice of the greene leaues of Garden Nightſhade mixed with Barley meale, is very profitably applied vnto Saint Anthonies fire, and to all hot inflammations.

C The iuice mixed with oile of Roſes, Ceruſe, and Littarge of gold, and applied, is more proper and effectuall to the purpoſes before ſet downe.

D † Neither the iuice heereof, nor any other part is vſually giuen inwardly, yet it may without any danger.

E The leaues ſtamped are profitably put into the ointment of Popler buds, called *Vnguentum populeon*, and it is good in all other ointments made for the ſame purpoſe.

F ‡ 2 The barke of the root of Sleepie Nightſhade, taken in the weight of ʒ 1. hath a ſomniferous qualitie; yet is it milder then *Opium*, and the fruit thereof vehemently prouokes vrine. But (as *Pliny* ſaith) the remedies hereof are not of ſuch eſteeme that we ſhould long inſiſt vpon them, eſpecially ſeeing wee are furniſhed with ſuch ſtore of medicines leſſe harmefull, yet ſeruing for the ſame purpoſe. ‡

† The Figure in the ſecond place was of the *Solanum Pomiferum*, or *Mala Æthiopica*, treated of at large in the 61. Chap. of this Booke, and therefore it is omitted here: and in ſtead thereof another put in the place.

Chap. 56. *Of ſleepy Nightſhade.*

Solanum Lethale.
Dwale, or deadly Nightſhade.

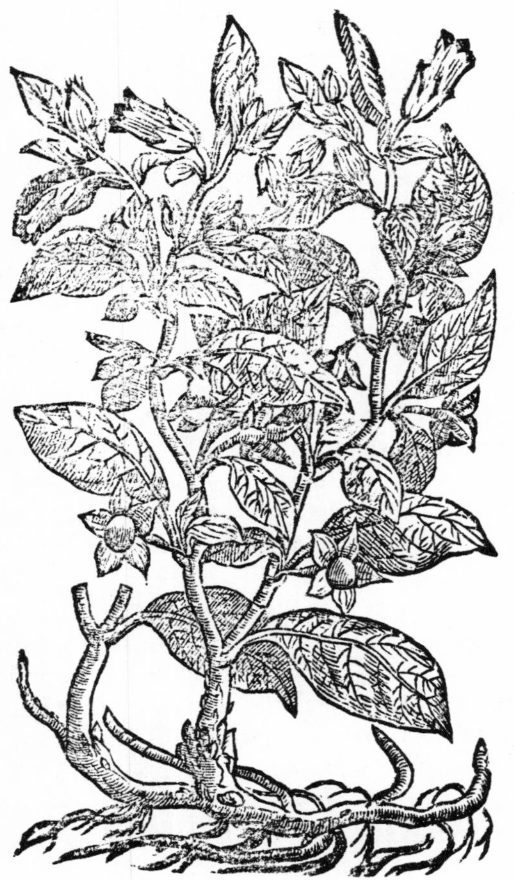

¶ *The Deſcription.*

Dwale or ſleeping Nightſhade hath round blackiſh ſtalkes ſix foot high, wherupon do grow great broad leaues of a darke greene colour; among which doe grow ſmall hollow flowers bel faſhion, of an ouerworne purple colour; in the place wherof come forth great round berries of the bigneſſe of the blacke cherry, greene at the firſt, but when they be ripe of the color of black iette or burniſhed horne, ſoft and full of purple iuice: among which iuice lie the ſeeds like the berries of Iuy: the root is very great, thicke, and long laſting.

The Place.

It groweth in vntoiled places neere vnto high waies and the ſea marſhes, and ſuch like places.

It groweth very plentifully in Holland in Lincolnſhire, and in the Ile of Ely at a place called Walſoken, neere vnto Wisbitch.

I found it growing without the gate of Highgate neere vnto a pound or pinfold on the left hand.

The Time.

This flouriſheth all the Sommer and Spring, beareth his ſeed and flower in Iuly and Auguſt.

¶ *The Names.*

It is called of *Dioſcorides*, στρύχνος ὑπνωτικός: of *Theophraſtus*, στρύχνος ὑπνώδης: of the Latines, *Solanum ſomniferum*,

ſomniferum, or ſleeping Nightſhade; and *Solanum lethale*, or deadly Nightſhade; and *Solanum manicum*, raging Nightſhade: of ſome, *Apollinaris minor vlticana*, and *Herba Opſago*: in Engliſh, Dwale, or ſleeping Nightſhade: the Venetians and Italians call it *Bella dona*: the Germanes, **Dollwurtz**: the low Dutch, **Dulle beſſen**: in French, *Morelle mortelle*: it commeth very neere vnto *Theophraſtus* his *Mandragoras*, (which differeth from *Dioſcorides* his *Mandragoras*.)

¶ *The Nature.*

It is cold euen in the fourth degree.

¶ *The Vertues.*

This kinde of Nightſhade cauſeth ſleep, troubleth the minde, bringeth madneſſe if a few of the berries be inwardly taken, but if moe be giuen they alſo kill and bring preſent death. *Theophraſtus* in his *6.* booke doth likewiſe write of Mandrake in this manner; Mandrake cauſeth ſleepe, and if alſo much of it be taken it bringeth death. A

The greene leaues of deadly Nightſhade may with great aduice be vſed in ſuch caſes at Petti-mortell: but if you will follow my counſell, deale not with the ſame in any caſe, and baniſh it from your gardens and the vſe of it alſo, being a plant ſo furious and deadly: for it bringeth ſuch as haue eaten thereof into a dead ſleepe wherein many haue died, as hath been often ſeen and proued by experience both in England and elſewhere. But to giue you an example heereof it ſhall not be amiſſe: It came to paſſe that three boyes of Wisbich in the Ile of Ely did eate of the pleaſant & beautifull fruite hereof, two whereof died in leſſe than eight houres after that they had eaten of them. The third child had a quantitie of hony and water mixed together giuen him to drinke, cauſing him to vomit often: God bleſſed this meanes and the child recouered. Baniſh therefore theſe pernicious plants out of your gardens, and all places neere to your houſes, where children or women with child do reſort, which do oftentimes long and luſt after things moſt vile and filthie; and much more after a berry of a bright ſhining blacke colour, and of ſuch great beautie, as it were able to allure any ſuch to eate thereof. B

The leaues heereof laid vnto the temples cauſe ſleepe, eſpecially if they be imbibed or moiſtened in wine vineger. It eaſeth the intollerable paines of the head-ache proceeding of heate in furious agues, cauſing reſt being applied as aforeſaid. C

CHAP. 57. *Of winter Cherries.*

¶ *The Deſcription.*

1 THe red winter Cherrie bringeth forth ſtalkes a cubit long, round, ſlender, ſmooth, and ſomewhat reddiſh, reeling this way and that way by reaſon of his weakeneſſe, not able to ſtand vpright without a ſupporter: whereupon do grow leaues not vnlike to thoſe of common Nightſhade, but greater; among which leaues come forth white floures, conſiſting of fiue ſmall leaues: in the middle of which leaues ſtandeth out a berry, greene at the firſt, and red when it is ripe, in colour of our common Cherry and of the ſame bigneſſe, incloſed in a thinne huske or little bladder, it is of a pale reddiſh colour, in which berrie is conteined many ſmall flat ſeeds of a pale colour. The rootes be long, not vnlike to the rootes of Couch-graſſe, ramping and creeping within the vpper cruſt of the earth farre abroad, whereby it encreaſeth greatly.

2 The blacke winter Cherrie hath weake and ſlender ſtalkes ſomewhat creſted, and like vnto the tendrels of the vine, caſting it ſelfe all about, and taketh hold of ſuch things as are next vnto it: whereupon are ſet jagged leaues deepely indented or cut about the edges almoſt to the middle ribbe. The floures be very ſmall and white ſtanding vpon long foote-ſtalkes or ſtemmes. The skinnie bladders ſucceed the floures, parted into three ſells or chambers, euery of the which conteineth one ſeed and no more, of the bigneſſe of a ſmall peaſe, and blacke of colour, hauing a marke of white colour vpon each berrie, in proportion of an heart. The roote is very ſmall and threddie.

¶ *The Place.*

The red winter Cherrie groweth vpon old broken walls, about the borders of fieldes, and in moiſt ſhadowie places, and in moſt gardens, where ſome cheriſh it for the beautie of the berries, and others for the great and worthy vertues thereof.

2 The blacke winter Cherrie is brought out of Spaine and Italy, or other hot regions, from whence I haue had of thoſe blacke ſeeds marked with the ſhape of a mans hart, white, as aforeſaid: and haue planted them in my garden where they haue borne floures, but haue periſhed before the fruit could grow to maturitie, by reaſon of thoſe vnſeaſonable yeeres, 1594. 95. 96.

¶ *The Time.*

The red winter Cherrie beareth his floures and fruite in August.

The blacke beareth them at the same time, where it doth naturally grow.

¶ *The Names.*

The red winter Cherrie is called in Greeke, Στρύχνος : in Latine, *Vesicaria*, and *Solanum Vesicarium* : in shops, *Alkekengi* : *Plinie* in his 21. booke nameth it *Halicacabus*, and *Vesicaria*, of the little bladders : or as the same Author writeth, because it is good for the bladder and the stone : it is called in Spanish, *Vexiga de porro* : in French, *Alquequenges*, *Bagenauldes*, and *Cerises d'outre mer* : in English, red Nightshade, Winter Cherries, and Alkakengie.

1 *Solanum Halicacabum.*
Red winter Cherries.

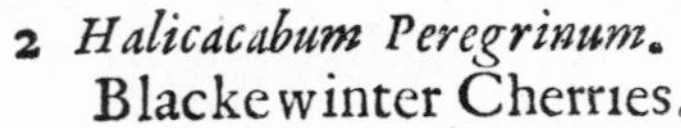

2 *Halicacabum Peregrinum.*
Blacke winter Cherries.

The blacke winter Cherrie is called *Halacacabum Peregrinum*, *Vesicaria Peregrina*, or strange winter Cherrie : of *Pena* and *Lobel* it is called, *Cor Indum*, *Cor Indicum* : of others, *Pisum Cordatum* . in English, the Indian heart, or heart pease: some haue taken it to be *Dorycnion*, but they are greatly deceiued, being in truth not any of the Nightshades; it rather seemeth to agree with the graine named of *Serapio*, *Abrong*, or *Abrugi*, of which he writeth in his 153. chapter in these words : It is a little graine spotted with blacke and white, round, and like the graine Maiz, with which notes this doth agree.

¶ *The Temperature.*

The red winter Cherrie is thought to be cold and drie, and of subtile parts.

The leaues differ not from the temperature of the garden Nightshade, as *Galen* saith.

¶ *The Vertues.*

A The fruite brused and put to infuse or steepe in white wine two or three houres, and after boiled two or three bublings, straining it, and putting to the decoction a little sugar and cinnamon, and drunke, preuaileth very mightily against the stopping of vrine, the stone and grauell, the difficultie and sharpenes of making water, and such like diseases : if the griefe be old, the greater quantity must be taken ; if new and not great, the lesse : it scoureth away the yellow jaundise also, as some write.

CHAP. 58. *Of the Maruell of the World.*

Mirabilia Peruuiana flore luteo.
The maruell of Peru with yellowiſh floures.

‡ *Mirabilia Peruuiana flore albo.*
The maruell of Peru with white floures.

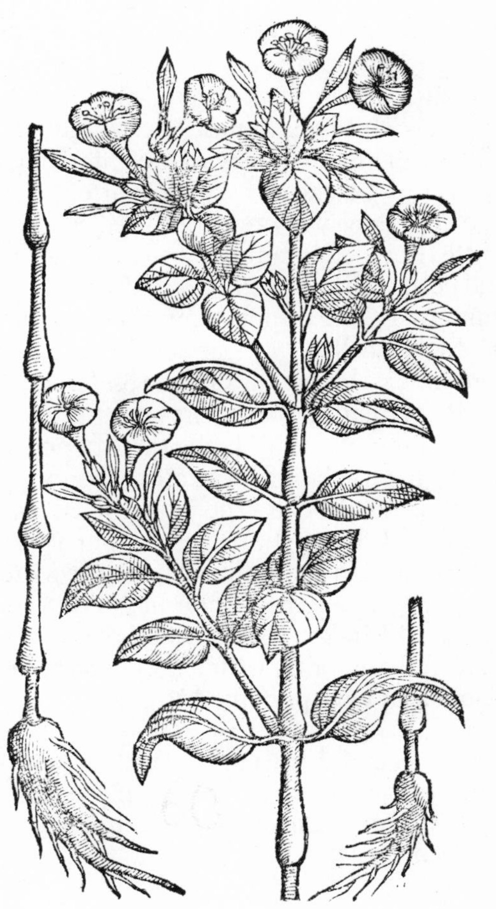

The deſcription.

THis admirable plant called the maruell of Peru, or the maruell of the World, ſpringeth forth of the ground like vnto Baſill in leaues ; amongſt which it ſendeth out a ſtalke two cubits and a halfe high, of the thickeneſſe of a finger ; full of iuice, very firme, and of a yellowiſh greene colour, knotted or kneed with ioints ſomewhat bunching forth, of purpliſh color, as in the female Balſamina : which ſtalke diuideth it ſelfe into ſundrie branches or boughes, and thoſe alſo knottie like the ſtalke. His branches are decked with leaues growing by couples at the ioints like the leaues of wilde Peaſcods, greene, fleſhie, and full of ioints; which beeing rubbed doe yeeld the like vnpleaſant ſmell as wilde Peaſcods doe, and are in taſte alſo verie vnſauorie, yet in the latter end they leaue a taſte and ſharpe ſmacke of Tabaco. The ſtalkes towards the top are garniſhed with long hollow ſingle flowers, folded, as it were, into fiue parts before they be opened ; but being fully blowne doe reſemble the flowers of Tabaco, not ending into ſharpe corners, but blunt and round as the flowers of Bindeweede, and larger than the flowers of Tabaco, glittering oftentimes with a fine purple or Crimſon colour ; many times of an horſe-fleſh ; ſometime yellow ; ſometime pale, and ſometime reſembling an old red or yellow colour ; ſometime whitiſh, and moſt commonly two colours occupying halfe the flower, or intercourſing the whole flower with ſtreakes and orderly ſtreames, now yellow, now purple, diuided through the whole ; hauing ſometime great, ſometime little ſpots of a purple colour, ſprinkled and ſcattered in a moſt variable order, and braue mixture. The ground or field of the whole flower is either pale, red, yellow, or white, containing in the middle of the hollowneſſe a pricke or pointell ſet round about with ſixe ſmall ſtrings or chiues. The flowers are verie ſweet and pleaſant, reſembling the Narciſſe or white Daffodill, and are very ſuddenly fading ; for at night they are flowred wide open, and ſo continue vntill eight of the clocke the next morning : at which time they beginne to cloſe or ſhut vp (after the manner of the Bindeweede) eſpecially if the weather be very hot : but if the aire be more temperate they remaine open the whole day, and are cloſed onely at night, and ſo periſh, one flower laſting

sting but onely one day, like the true Ephemerum or Hemerocallis. This maruellous varietie doth not without cause bring admiration to all that obserue it. For if the flowers be gathered and reserued in seuerall papers, and compared with those flowers that will spring and flourish the next day, you shall easily perceiue that one is not like another in colour, though you should compare one hundreth which flower one day, and another hundred which you gathered the next day; and so from day to day during the time of their flowring. The cups and huskes which containe and embrace the flowers are diuided into fiue pointed sections, which are greene, and, as it were, consisting of skinnes, wherein is contained one seede and no more, couered with a blackish skinne, hauing a blunt point whereon the flower groweth; but on the end next the cup or huske it is adorned with a little fiue cornered crowne. The seed is as bigge as a pepper corne, which of it selfe fadeth with any light motion. Within this seede is contained a white kernell, which being bruised, resolueth into a very white pulpe like starch. The root is thicke and like vnto a great radish, outwardly blacke, and within white, sharpe in taste, wherewith is mingled a superficiall sweetnes. It bringeth new floures from Iuly vnto October in infinite number, yea euen vntill the frosts doe cause the whole plant to perish: notwithstanding it may be reserued in pots, and set in chambers and cellars that are warme, and so defended from the iniurie of our cold climate; prouided alwaies that there be not any water cast vpon the pot, or set forth to take any moisture in the aire vntill March following; at which time it must bee taken forth of the pot and replanted in the garden. By this meanes I haue preserued many (though to small purpose) because I haue sowne seeds that haue borne floures in as ample manner and in as good time as those reserued plants.

Of this wonderfull herbe there be other sorts, but not so amiable or so full of varietie, and for the most part their floures are all of one color. But I haue since by practise found out another way to keepe the roots for the yeare following with very little difficultie, which neuer faileth. At the first frost I dig vp the rootes and put vp or rather hide the roots in a butter ferkin, or such like vessell, filled with the sand of a riuer, the which I suffer still to stand in some corner of a house where it neuer receiueth moisture vntill Aprill or the midst of March, if the weather be warme; at which time I take it from the sand and plant it in the garden, where it doth flourish exceeding well and increaseth by roots; which that doth not which was either sowne of seed the same yeare, nor those plants that were preserued after the other manner.

¶ *The Place*

The seed of this strange plant was brought first into Spaine, from Peru, whereof it tooke his name *Mirabilia Peruana*, or *Peruuiana*: and since dispersed into all the parts of Europe: the which my selfe haue planted many yeares, and haue in some temperate yeares receiued both floures and ripe seed.

¶ *The Time.*

It is sowne in the midst of Aprill, and bringeth forth his variable floures in September, and perisheth with the first frost, except it be kept as aforesaid.

¶ *The Names.*

It is called in Peru of those Indians there, *Hachal*. Of others after their name *Hachal Indi*: of the high and low Dutch, *Solanum Odoriferum*: of some, *Iasminum mexicanum*: and of *Carolus Clusius*, *Admirabilia Peruuiana*: in English rather the Maruell of the World, than of Peru alone.

¶ *The Nature and Vertues.*

We haue not as yet any instructions from the people of India concerning the nature or vertues of this plant: the which is esteemed as yet rather for his rarenesse, beautie, and sweetnesse of his floures, than for any vertues knowne; but it is a pleasant plant to decke the gardens of the curious. Howbeit *Iacobus Antonius Cortusus* of Padua hath by experience found out, that two drams of the root thereof taken inwardly doth very notably purge waterish humours.

Chap. 59. *Of Madde Apples.*

¶ *The Description.*

Raging Apples hath a round stalke of two foot high, diuided into sundry branches, set with broad leaues somewhat indented about the edges, not vnlike the leaues of white Henbane, of a darke browne greene colour, somewhat rough. Among the which come the floures of of a white colour, and ſome times changing into purple, made of ſix parts, wide open like a ſtarre with certaine yellow chiues or thrums in the middle; which beeing paſt the fruit commeth in place, ſet in a cornered cup or huske after the manner of the great Nightſhade, great and ſomewhat long, of the bigneſſe of a ſwans egge, and ſomtimes much greater, of a white color, ſometimes yellow, and often browne, wherein is contained ſmall flat ſeed of a yellow colour. The root is thicke, with many threds faſtned thereto.

Mala inſana.
Madde or raging Apples.

¶ *The Place.*

This plant groweth in Egypt almoſt euery where in ſandie fields euen of it ſelfe, bringing forth fruit of the bigneſſe of a great Cucumber, as *Petrus Bellonius* reporteth in the ſecond booke of his ſingular obſeruations.

Wee had the ſame in our London Gardens, where it hath borne floures; but the Winter approching before the time of ripening, it periſhed: notwithſtanding it came to beare fruit of the bigneſſe of a gooſe egge one extraordinarie temperate yeare, as I did ſee in the garden of a worſhipfull Merchant Mr. *Haruie* in Limeſtreet, but neuer to the full ripeneſſe.

¶ *The Time.*

This herbe muſt be ſowne in Aprill in a bed of hot horſe doung, as Muske-Melons are, and floureth in Auguſt.

¶ *The Names.*

Petrus Bellonius hath iudged it to bee *Malinathalla Theophraſti.* In the Dukedome of Millaine it is called *Melongena*: and of ſome, *Melanzana*: in Latine, *Mala inſana*: and in Engliſh, Mad Apples: in the Germaine tongue, **Dollopffel**: In Spaniſh, *Verangenes.*

¶ *The Nature.*

The hearbe is cold almoſt in the fourth degree.

¶ *The vſe and danger.*

A The people of Tolledo do eat them with great deuotion being boiled with fat fleſh, putting thereto ſome ſcraped cheeſe, which they do keepe in vineger, honie, or ſalt pickell all Winter to procure luſt.

B *Petrus Bellonius*, and *Hermolaus Barbarus*, report that in Egypt and Barbary they vſe to eat the fruit of *Mala inſana* boiled or roſted vnder aſhes, with oile, vineger, & pepper, as people vſe to eat Muſhroms. But I rather wiſh Engliſh men to content themſelues with the meat and ſauce of our owne Countrey, than with fruit and ſauce eaten with ſuch perill: for doubtleſſe theſe apples haue a miſchieuous qualitie, the vſe whereof is vtterly to be forſaken. And as wee ſee and know many haue eaten and doe eat Muſhroms more for wantonneſſe than for need: for there are two kindes thereof venemous and deadly, which being in the handling of an vnskilful cooke, may procure vntimely death. Therefore it is better to eſteeme this plant and haue him in the Garden for your pleaſure and the rareneſſe threof, than for any vertue or good qualities yet knowne.

CHAP. 60. *Of Apples of Loue.*

¶ *The Deſcription.*

The Apple of Loue bringeth forth very long round ſtalkes or branches, fat and full of iuice, trailing vpon the ground, not able to ſuſtaine himſelfe vpright by reaſon of the tenderneſſe of the ſtalkes, and alſo the great weight of the leaues and fruit wherewith it is ſurcharged. the leaues are great and deeply cut or iagged about the edges, not vnlike to the leaues of Agrimony, but greater, and of a whiter greene colour: among which come forth yellow floures growing vpon ſhort ſtems or foot ſtalks,cluſtering together in bunches : which being fallen,there do come in place faire and goodly apples, chamfered,vneuen, and bunched out in many places ; of a bright ſhining red colour,and the bigneſſe of a gooſe egge or a large pippin.The pulpe or meat is verie full of moiſture, ſoft, reddiſh, and of the ſubſtance of a wheat plumme.The ſeed is ſmall,flat and rough : the root ſmall and threddie : the whole Plant is of a ranke and ſtinking ſauour.

There hath happened vnto my hands another ſort, agreeing very notably with the former,as well in leaues and ſtalkes as alſo in floures and roots, onely the fruit hereof was yellow of colour, wherein conſiſted the difference.

Poma Amoris.
Apples of Loue.

¶ *The Place.*

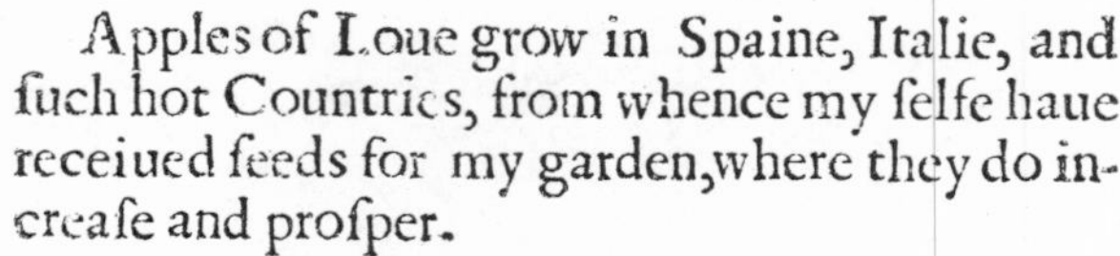

Apples of Loue grow in Spaine, Italie, and ſuch hot Countries, from whence my ſelfe haue receiued ſeeds for my garden,where they do increaſe and proſper.

¶ *The Time.*

It is ſowne in the beginning of Aprill in a bed of hot horſe dung,after the maner of musk Melons and ſuch like cold fruits.

¶ *The Names.*

The Apple of Loue is called in Latine *Pomum Aureum*, *Poma Amoris*, and *Lycoperſicum* : of ſome, *Glaucium* : in Engliſh, Apples of Loue, and Golden Apples:in French,*Pommes d'amours*. Howbeit there be other golden Apples whereof the Poets doe fable,growing in the Gardens of the daughters of *Heſperus*, which a Dragon was appointed to keepe,who,as they fable,was killed by *Hercules*.

¶ *The Temperature.*

The Golden Apple, with the whole herbe it ſelfe is cold, yet not fully ſo cold as Mandrake, after the opinion of *Dodonæus*.But in my iudgement it is very cold, yea perhaps in the higheſt degree of coldneſſe:my reaſon is,becauſe I haue in the hotteſt time of Sommer cut away the ſuperfluous branches from the mother root, and caſt them away careleſly in the allies of my garden,the which (notwithſtanding the extreme heat of the Sun,the hardneſſe of the trodden allies, and at that time when no raine at all did fall) haue growne as freſh where I caſt them, as before I did cut them off ; which argueth the great coldneſſe contained therein. True it is,that it doth argue alſo a great moiſture wherewith the plant is poſſeſſed, but as I haue ſaid,not without great cold,which I leaue to euery mans cenſure.

¶ *The Vertues.*

A In Spaine and thoſe hot Regions they vſe to eat the Apples prepared and boiled with pepper, ſalt,and oile : but they yeeld very little nouriſhment to the bodie, and the ſame nought and corrupt.

B Likewiſe they doe eat the Apples with oile, vineger and pepper mixed together for ſauce to their meat,euen as we in theſe cold Countries doe Muſtard.

Chap. 61. *Of the Æthiopian Apple.*

¶ *The Deſcription.*

THe Apple of Æthiopia hath large leaues of a whitiſh greene colour, deepely indented about the edges, almoſt to the middle rib ; the which middle rib is armed with a few ſharpe prickles. The floures be white,conſiſting of ſix ſmal leaues,with a certain yellow pointel in the midſt. The

Mala Æthiopica.
Apples of Æthiopia.

The fruit is round, and bunched with vneuen lobes or bankes lesser than the golden Apple, of colour red, and of a firme and sollid substance; wherein are contained small flat seeds. The root is small and threddy.

¶ *The Place.*

The seeds of this plant haue beene brought vnto vs out of Spaine, and also sent into France and Flanders: but to what perfection it hath come vnto in those parts I am ignorant, but mine perished at the first approch of Winter. His first original was from Æthiopia, whereof it tooke his name.

¶ *The Time.*

This Plant must be sowen as Muske-Melons, and at the same time. They floure in Iuly, and the fruit is ripe in September.

¶ *The Names.*

In English wee haue thought good to call it the Æthiopian Apple, for the reason before alledged: in Latine, *Mala Æthiopica*: of some it hath been thought to be *Malinathalla*. ‡ This is the *Solanum pomiferum* of *Lobel* and others; by which name our Author also formerly had it, in the fiftieth chapter of the former edition. ‡

¶ *The Nature.*

The temperature agreeth with the Apple of Loue.

¶ *The Vertues.*

These Apples are not vsed in physicke that I can reade of; onely they are vsed for a sauce and seruice vnto rich mens tables to be eaten, being first boyled in the broth of fat flesh with pepper and salt, and haue a lesse hurtfull iuyce than either mad apples or golden Apples.

Chap. 62. *Of Thornie-Apples.*

¶ *The Description.*

1 THe stalkes of Thorny-apples are oftentimes aboue a cubit and a halfe high, seldome higher, an inch thicke, vpright and straight, hauing very few branches, sometimes none at all, but one vpright stemme; whereupon doe grow leaues smooth and euen, little or nothing indented about the edges, longer and broader than the leaues of Night-shade, or of the mad Apples. The floures come forth of long toothed cups, great, white of the forme of a bell, or like the floures of the great Withwinde that rampeth in hedges; but altogether greater and wider at the mouth, sharpe cornered at the brimmes, with certaine white chiues or threds in the middest, of a strong ponticke sauour, offending the head when it is smelled vnto: in the place of the floure commeth vp round fruit full of short and blunt prickles, of the bignesse of a greene Wall-nut when it is at the biggest, in which are the seeds of the bignesse of tares or of Mandrakes, and of the same forme. The herbe it selfe is of a strong sauor, and doth stuffe the head, and causeth drowsinesse. The root is small and threddy.

2 There is another kinde hereof altogether greater than the former, whose seeds I receiued of the right honorable the Lord *Edward Zouch*; which he brought from Constantinople, and of his liberalitie did bestow them vpon me, as also many other rare & strange seeds; and it is that Thorn-apple that I haue dispersed through this land, whereof at this present I haue great vse in Surgery, as well in burnings and scaldings, as also in virulent and maligne vlcers, apostumes, and such like. The which plant hath a very great stalke in fertile ground, bigger than a mans arme, smooth, and greene of colour, which a little aboue the ground diuideth it selfe into sundry branches or armes, in manner of an hedge tree; whereupon are placed many great leaues cut and indented deepely about

about the edges, with many vneuen ſharpe corners: among theſe leaues come white round floures made of one piece in manner of a bell, ſhutting it ſelfe vp cloſe toward night, as do the floures of the great Binde-weed, whereunto it is very like, of a ſweet ſmell, but ſo ſtrong, that it offends the ſences. The fruit followeth round, ſometimes of the faſhion of an egge, ſet about on euerie part with moſt ſharpe prickles; wherein is contained very much ſeed of the bigneſſe of tares, and of the ſame faſhion. The root is thicke, made of great and ſmall ſtrings: the whole plant is ſowen, beareth his fruit, and periſheth the ſame yeare. ‡ There are are ſome varieties of this plant, in the colour and doubleneſſe of the floures. ‡

1 *Stramonium Peregrinum.*
The Apple of Peru.

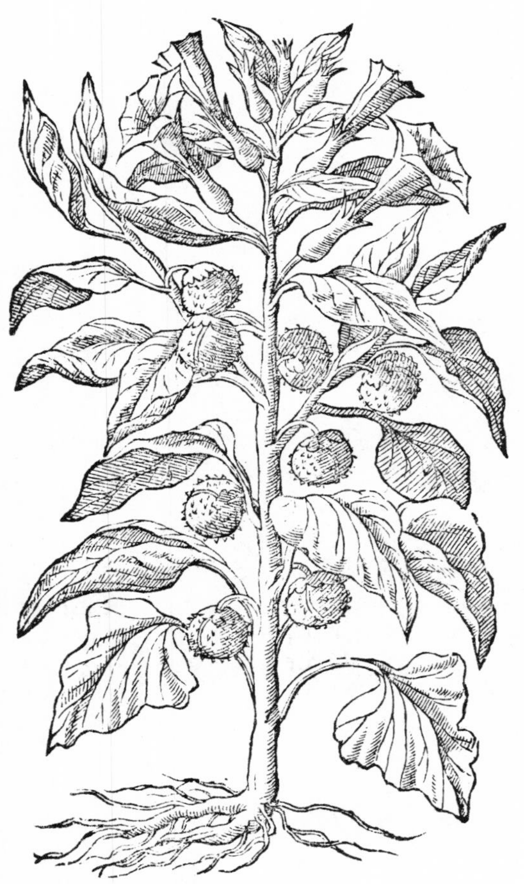

2 *Stramonium ſpinoſum.*
Thorny Apples of Peru.

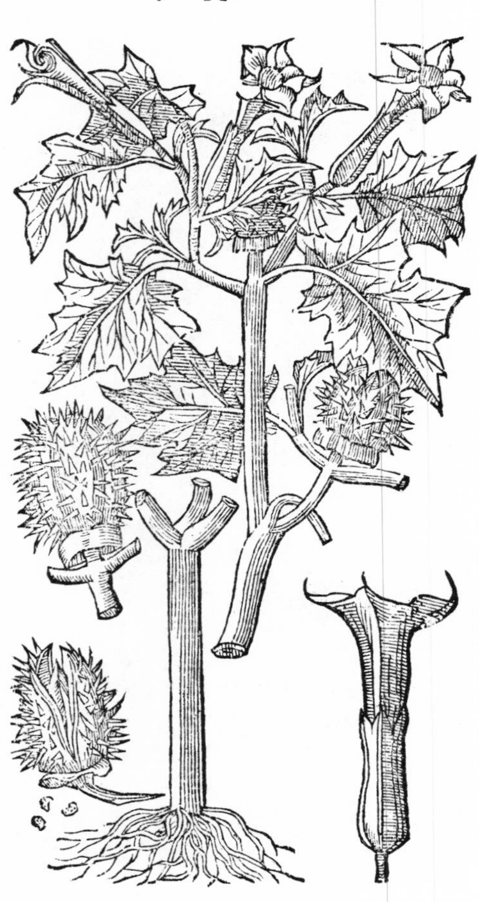

¶ The Place.

1 This plant is rare and ſtrange as yet in England: I receiued ſeeds thereof from *Iohn Robin* of Paris, an excellent Herbariſt; which did grow and bare floures, but periſhed before the fruit came to ripeneſſe.

2 The Thorne-apple was brought in ſeed from Conſtantinople by the right honourable the Lord *Edward Zouch*, and giuen vnto me, and beareth fruit and ripe ſeed.

¶ The Time.

The firſt is to be ſowen in a bed of horſe-dung, as we do cucumbers and Muske-melons.

The other may be ſowen in March or Aprill, as other ſeeds are.

¶ The Names.

The firſt of theſe Thorne-apples may be called in Latine, *Stramonia,* and *Pomum,* or *Malum ſpinoſum*: of ſome, *Corona regia,* and *Meloſpinum*: The Grecians of our time name it πτεγκόκκαλον, or rather Καρυοκόκκαλον; as though they ſhould ſay, a nut ſtuffing, and cauſing drowſineſſe and diſquiet ſleepe: the Italians, *Paracoculi*: it ſeemeth to *Valerius Cordus* to be *Hyoſcyamus Peruvianus,* or Henbane of Peru: *Cardanus* doubteth whether it ſhould be inſerted among the Night-ſhades as a kinde thereof: of *Matthiolus* and others it is thought to be *Nux methel*: *Serapio, cap.* 375. ſaith, That *Nux methel* is like vnto *Nux vomica*; the ſeed whereof is like that of Mandrake: the huske is rough or full of prickles; the taſte pleaſing and ſtrong: the qualitie thereof is cold in the fourth degree. Which deſcription agreeth herewith, except in the forme or ſhape it ſhould haue with *Nux vomica*: *Anguillara* ſuſpecteth it to be *Hippomanes* which *Theocritus* mentioneth, wherewith in his ſecond *Eclog*

he

he ſheweth that horſes are made mad: for *Crateuas*, whom *Theocritus* his Scholiaſt doth cite, writeth, That the plant of *Hippomanes* hath a fruit full of prickles, as hath the fruit of wilde Cucumbers. In Engliſh it may be called Thorne-apple, or the Apple of Peru.

‡ The words of *Theocritus, Eidyll.* 2. are theſe:

Ἱππομανὲς φυτόν ἐστι παρ᾽ Ἀρκάσι, &c.

Which is thus in Engliſh:

Hippomanes 'mongſt the Arcadians ſprings, by which euen all
The Colts and agile Mares in mountaines mad do fall.

Now in the Greeke *Scholia* amongſt the Expoſitions there is this: κρατεύας φησι, &c. That is; *Crateuas* ſaith, That the plant hath a fruit like the wilde Cucumber, but blacker; the leaues are like a poppie, but thorny or prickly. Thus I expound theſe words of the Greeke Scholiaſt, being *pag.* 51. of the edition ſet forth by *Dan. Heinſius, Ann. Dom.* 1603. *Iulius Scaliger* blames *Theocritus*, becauſe he calls *Hippomanes* φυτὸν, a Plant: but *Heinſius*, as you may ſee in his notes vpon *Theocritus, pag.* 120, probably iudges, that φυτὸν in this place ſignifies nothing but χρῆμα, a Thing [*growing*.] Such as are curious may haue recourſe to the places quoted, where they may finde it more largely handled than is fit for me in this place to inſiſt vpon. There is no plant at this day knowne, in mine opinion, whereto *Crateuas* his deſcription may be more fitly referred, than to the *Papauer ſpinoſum*, or *ficus infernalis*, which we ſhall hereafter deſcribe. ‡

¶ *The Nature.*

The whole plant is cold in the fourth degree, and of a drowſie and numming qualitie, not inferior to Mandrake.

¶ *The Vertues.*

The iuyce of Thorne-apples boiled with hogs greaſe to the forme of an vnguent or ſalue, cureth all inflammations whatſoeuer, all manner of burnings or ſcaldings, as well of fire, water, boyling leade, gun-pouder, as that which comes by lightning, and that in very ſhort time, as my ſelf haue found by my dayly practiſe, to my great credit and profit. The firſt experience came from Colcheſter, where Miſtreſſe *Lobel* a Merchants wife there being moſt grieuouſly burned with lightning, and not finding eaſe or cure in any other thing, by this found helpe when all hope was paſt, by the report of Mr. *William Ramme*, publique Notarie of the ſaid towne, was perfectly cured. A

The leaues ſtamped ſmall, and boiled with oyle Oliue vntill the herbes be as it were burnt, then ſtrained and ſet to the fire againe with ſome wax, roſin, and a little Turpentine, and made into a ſalue, doth moſt ſpeedily cure old vlcers, new and freſh wounds, vlcers vpon the glandulous part of the yard, and other ſores of hard curation. B

CHAP. 63.

Of Bitter-ſweet, or Wooddy Nightſhade.

¶ *The Deſcription.*

Bitter-ſweet bringeth forth wooddy ſtalkes as doth the Vine, parted into many ſlender creeping branches, by which it climeth and taketh hold of hedges and ſhrubs next vnto it. The barke of the oldeſt ſtalkes are rough and whitiſh, of the colour of aſhes, with the outward rinde of a bright greene colour, but the yonger branches are greene as are the leaues: the wood brittle, hauing in it a ſpongie pith; it is clad with long leaues, ſmooth, ſharpe pointed, leſſer than thoſe of the Binde-weed. At the lower part of the ſame leaues doth grow on either ſide one ſmall or leſſer leafe like vnto two eares. The floures be ſmall, and ſomewhat cluſtered together, conſiſting of fiue little leaues apiece, of a perfect blew colour, with a certaine pricke or yellow pointall in the middle: which being paſt, there do come in place faire berries, more long than round, at the firſt green, but very red when they be ripe; of a ſweet taſte at the firſt, but after very vnpleaſant, of a ſtrong ſauour, growing together in cluſters like burniſhed coral. The root is of a meane bigneſſe, and full of ſtrings.

I haue found another ſort which bringeth forth moſt pleaſant white floures with yellow pointals in the middle, in other reſpects agreeing with the former.

¶ *The Place.*

Bitter-ſweet doth grow in moiſt places about ditches, riuers, and hedges, almoſt euery where.

The

Amara-dulcis.
Bitter-ſweet.

The other ſort with the white floures I found in a ditch ſide againſt the right honorable the Earle of Suſſex his garden wall at his houſe in Bermonſey ſtreet by London, as you go from the court which is full of trees, vnto a farme houſe neere thereunto.

¶ *The Time.*

The leaues come forth in the Spring, the floures in Iuly, the berries are ripe in Auguſt.

¶ *The Names.*

The later Herbariſts haue named this plant *Dulcamara, Amarodulcis,* and *Amaradulcis*; that is in Greeke, γλυκύπικρον: they call it alſo *Solanum lignoſum*, and *Siliquaſtrum*: *Pliny* calleth it *Melortum*: *Theophraſtus, Vitis ſylueſtris*: in Engliſh we call it Bitter-ſweet, and Wooddy Nightſhade. But euery Author muſt for his credit ſay ſomthing, although to ſmall purpoſe; for *Vitis ſylueſtris* is that which we call our Ladies Seale, which is no kinde of Nightſhade: for *Tamus* and *Vitis ſylueſtris* are both one; as likewiſe *Solanum lignoſum* or *Fruticoſum*; and alſo *Solanum rubrum*: whereas indeed it is no ſuch plant, nor any of the Nightſhades, although I haue followed others in placing it here. Therfore thoſe that vſe to mixe the berries thereof in compoſitions of diuers cooling ointments, in ſtead of the berries of Nightſhade haue committed the greater errour; for the fruit of this is not cold at all, but hot, as forthwith ſhall be ſhewed. *Dioſcorides* ſaith it is *Cyclaminus altera*; deſcribing it by the deſcription of thoſe with white floures aforeſaid, whereunto it doth very well agree. ‡ *Dioſcorides* deſcribeth his *Muſcoſo flore* with a moſſy floure, that is, ſuch an one as conſiſts of ſmall chiues or threds, which can by no meanes be agreeable to the floure of this plant. ‡

¶ *The Temperature.*

The leaues and fruit of Bitter-ſweet are in temperature hot and dry, clenſing and waſting away.

¶ *The Vertues.*

A The decoction of the leaues is reported to remoue the ſtoppings of the liuer and gall; and to be drunke with good ſucceſſe againſt the yellow jaundice.

B The iuyce is good for thoſe that haue fallen from high places, and haue beene thereby bruiſed, or dry beaten: for it is thought to diſſolue bloud congealed or cluttered any where in the intrals, and to heale the hurt places.

C *Hieronymus Tragus* teacheth to make a decoction of Wine with the wood finely ſliced and cut into ſmall pieces; which he reporteth to purge gently both by vrine and ſiege thoſe that haue the dropſie or jaundice.

D *Dioſcorides* doth aſcribe vnto *Cyclaminus altera*, or Bitter-ſweet with white floures as I conceiue it, the like faculties.

E The fruit (ſaith he) being drunke in the weight of one dram, with three ounces of white wine, for forty dayes together helpeth the ſpleene.

F It is drunke againſt difficultie of breathing: it throughly cleanſeth women that are newly brought a bed.

Chap. 64. *Of Binde-weed Nightſhade.*

¶ *The Deſcription.*

Inchanters Night-ſhade hath leaues like to Peti-morel, ſharpe at the point like vnto Spinage: the ſtalke is ſtraight and vpright, very brittle, two foot high: the floures are white tending to carnation, with certaine ſmall browne chiues in the midſt: the ſeed is contained in ſmall round bullets

Circæa Lutetiana.
Inchanters Night-ſhade.

bullets, rough and very hairy. The roots are tough, and many in number, thruſting themſelues deep into the ground, and diſper ng far abroad; whereby it doth greatly increaſe, inſomuch that when it hath once taken faſt rooting, it can hardly with great labour be rooted out or deſtroyed.

¶ *The Place.*

It groweth in obſcure and darke places, about dung-hills, and in vntoiled grounds, by path-wayes and ſuch like.

¶ *The Time.*

It flouriſheth from Iune to the end of September.

¶ *The Names.*

It is called of *Lobel*, *Circæa Lutetiana*: in Engliſh, Inchanters Night-ſhade, or Binde-weed Nightſhade.

¶ *The Nature and Vertues.*

There is no vſe of this herbe either in phyſicke or Surgerie that I can reade of; which hath happened by the corruption of time and the errour of ſome who haue taken *Mandragoras* for *Circæa*; in which errour they haue ſtill perſiſted vnto this day, attributing vnto *Circæa* the vertues of *Mandragoras*; by which means there hath not any thing been ſaid of the true *Circæa*, by reaſon, as I haue ſaid, that *Mandragoras* hath been called *Circæa*: but doubtleſſe it hath the vertue of Garden Night-ſhade, and may ſerue in ſtead thereof without error.

CHAP. 65. *Of Mandrake.*

¶ *The Deſcription.*

THe male Mandrake hath great broad long ſmooth leaues of a darke greene colour, flat ſpred vpon the ground: among which come vp the floures of a pale whitiſh colour, ſtanding euery one vpon a ſingle ſmall and weake foot-ſtalke of a whitiſh greene colour: in their places grow round Apples of a yellowiſh colour, ſmooth, ſoft, and glittering, of a ſtrong ſmell; in which are contained flat and ſmooth ſeeds in faſhion of a little kidney, like thoſe of the Thorne-apple. The root is long, thicke, whitiſh, diuided many times into two or three parts reſembling the legs of a man, with other parts of his body adioyning thereto, as the priuy part, as it hath beene reported; whereas in truth it is no otherwiſe than in the roots of carrots, parſeneps, and ſuch like, forked or diuided into two or more parts, which Nature taketh no account of. There hath been many ridiculous tales brought vp of this plant, whether of old wiues, or ſome runnagate Surgeons or phyſicke-mongers I know not, (a title bad enough for them) but ſure ſome one or moe that ſought to make themſelues famous and skilfull aboue others, were the firſt brochers of that errour I ſpeake of. They adde further, That it is neuer or very ſeldome to be found growing naturally but vnder a gallowes, where the matter that hath fallen from the dead body hath giuen it the ſhape of a man; and the matter of a woman, the ſubſtance of a female plant, with many other ſuch doltiſh dreams. They fable further and affirme, That he who would take vp a plant thereof muſt tie a dog therunto to pull it vp, which will giue a great ſhreeke at the digging vp; otherwiſe if a man ſhould do it, he ſhould ſurely die in ſhort ſpace after. Beſides many fables of louing matters, too full of ſcurrilitie to ſet forth in print, which I forbeare to ſpeake of. All which dreames and old wiues tales you ſhall from henceforth caſt out of your books and memory; knowing this, that they are all and euerie part of them falſe and moſt vntrue: for I my ſelfe and my ſeruants alſo haue digged vp, planted, and replanted very many, and yet neuer could either perceiue ſhape of man or woman, but ſometimes one ſtraight root, ſometimes two, and often ſix or ſeuen branches comming from the maine

great

great root, euen as Nature liſt to beſtow vpon it, as to other plants. But the idle drones that haue little or nothing to do but eate and drinke, haue beſtowed ſome of their time in caruing the roots of Brionie, forming them to the ſhape of men & women : which falſifying practiſe hath confirmed the errour amongſt the ſimple and vnlearned people, who haue taken them vpon their report to be the true Mandrakes.

The female Mandrake is like vnto the male, ſauing that the leaues hereof be of a more ſwart or darke greene colour ; and the fruit is long like a peare, and the other is round like an apple.

Mandragoras mas & fœmina.
The male and female Mandrake.

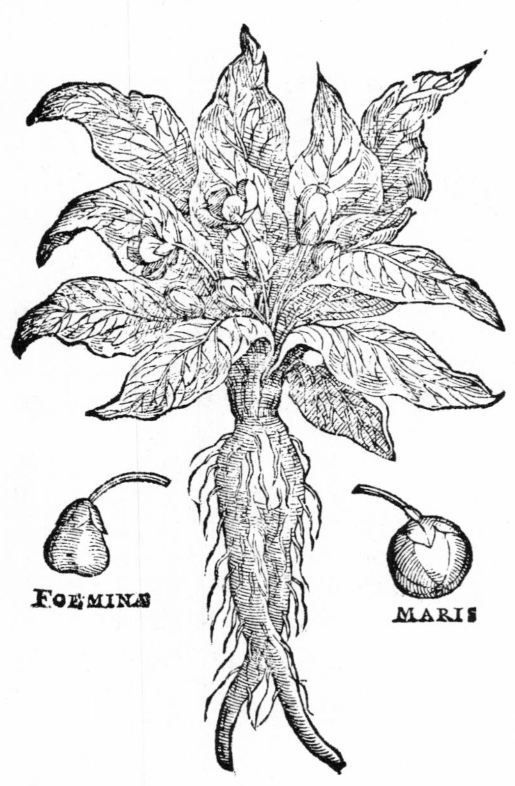

¶ *The Place.*

Mandrake groweth in hot Regions, in woods and mountaines, as in mount Garganus in Apulia, and ſuch like places; we haue them onely planted in gardens, and are not elſewhere to be found in England.

¶ *The Time.*

They ſpring vp with their leaues in March, and floure in the end of Aprill : the fruit is ripe in Auguſt.

¶ *The Names.*

Mandrake is called of the Grecians Μανδραγόρας : of diuers, Κιρκαία, and *Circæa*, of *Circe* the witch, who by art could procure loue : for it hath beene thought that the Root hereof ſerueth to win loue : of ſome, ἀντίμηλον, *Anthropomorphos*, and *Morion* : ſome of the Latines haue called it *Terræ malum*, and *Terreſtre malum*, and *Canina malus* : Shops, and alſo other Nations doe receiue the Greeke name. *Dioſcorides* ſaith, That the male is called of diuers *Morion* : and deſcribeth alſo another Mandrake by the name of *Morion*, which, as much as can be gathered by the deſcription, is like the male, but leſſe in all parts : in Engliſh we cal it Mandrake, Mandrage, and Mandragon.

¶ *The Temperature.*

Mandrake hath a predominate cold facultie, as *Galen* ſaith, that is to ſay cold in the third degree : but the root is cold in the fourth degree.

¶ *The Vertues.*

A *Dioſcorides* doth particularly ſet downe many faculties hereof; of which notwithſtanding there be none proper vnto it, ſauing thoſe that depend vpon the drowſie and ſleeping power thereof: which qualitie conſiſteth more in the root than in any other part.

B The Apples are milder, and are reported that they may be eaten, being boyled with pepper and other hot ſpices.

C *Galen* ſaith that the Apples are ſomething cold and moiſt, and that the barke of the root is of greateſt ſtrength, and doth not onely coole, but alſo dry.

D The iuyce of the leaues is very profitably put into the ointment called *Populeon*, and all cooling ointments.

E The iuyce drawne forth of the roots dried, and taken in ſmall quantitie, purgeth the belly exceedingly from flegme and melancholike humors.

F It is good to be put into medicines and colliries that do mitigate the paine of the eyes; and put vnder a peſſarie it draweth forth the dead childe and ſecondine.

G The greene leaues ſtamped with barrowes greaſe and barley meale, coole all hot ſwellings and inflammations ; and they haue vertue to conſume apoſtumes and hot vlcers, being bruiſed and applied thereon.

H A ſuppoſitorie made with the ſame iuyce, and put into the fundament cauſeth ſleepe.

I The wine wherein the root hath beene boyled or infuſed prouoketh ſleepe and aſſwageth paine.

K The ſmell of the Apples moueth to ſleepe likewiſe ; but the iuyce worketh more effectually if you take it in ſmall quantitie.

Great

Great and ſtrange effects are ſuppoſed to be in the Mandrakes, to cauſe women to be fruitfull L
and beare children, if they ſhall but carry the ſame neere vnto their bodies. Some doe from hence ground it, for that *Rahel* deſired to haue her ſiſters Mandrakes (as the text is tranſlated) but if we looke well into the circumſtances which there we ſhall finde, we may rather deeme otherwiſe. Young *Ruben* brought home amiable and ſweet ſmelling floures (for ſo ſignifieth the Hebrew word, which is vſed *Cantic.* 7. 13. in the ſame ſenſe:) and the lad brought them home, rather for their beauty and ſmell, than for their vertue. Now in the floures of Mandrake there is no ſuch delectable or amiable ſmell as was in theſe amiable floures which *Ruben* brought home. Beſides, wee read not that *Rahel* conceiued hereupon, for *Leah Iacobs* wife had foure children before God granted that bleſſing of fruitfulneſſe vnto *Rahel.* And laſt of all (which is my chiefeſt reaſon) *Iacob* was angry with *Rahel* when ſhe ſaid, Giue me children elſe I die: and demanded of her, whether he were in the ſtead of God or no, who had withheld from her the fruit of her body. And we know that the Prophet *Dauid* ſaith, Children and the fruit of the wombe are the inheritance, that commeth from the Lord, *Pſalm.* 127.

Serapio, Auicen, and *Paulus Ægineta* doe write, that the ſeed and fruit of *Mandragoras* taken M
in drinke, doe clenſe the matrix or mother, and *Dioſcorides* wrote the ſame long before them.

He that would know more hereof, may reade that chapter of doctor *Turner* his booke, concerning this matter, where he hath written largely and learnedly of this Simple.

CHAP. 66. *Of Henbane.*

1 *Hyoſcyamus Niger.*
Blacke Henbane.

2 *Hyoſcyamus Albus.*
White Henbane.

¶ *The Deſcription.*

1 THe common blacke Henbane hath great and ſoft ſtalkes: leaues very broad, ſoft, and woolly, ſomewhat iagged, eſpecially thoſe that grow neere vnto the ground, and thoſe that grow vpon the ſtalke, narrower, ſmaller, and ſharper. The floures are bel faſhion, of a faint yellowiſh white and browne within towards the bottome; when the floures are

gone, there commeth hard knobby huskes, like small cups or boxes, wherein are small browne seeds.

2 The White Henbane is not much vnlike to the blacke, sauing that his leaues are smaller, whiter and more woolly, and the floures also whiter. The cods are like the other, but without pricks; it dieth in winter, and must likewise be sowne againe the next yeere.

‡ 3 *Hyoscyamus albus minor.*
The lesser White Henbane.

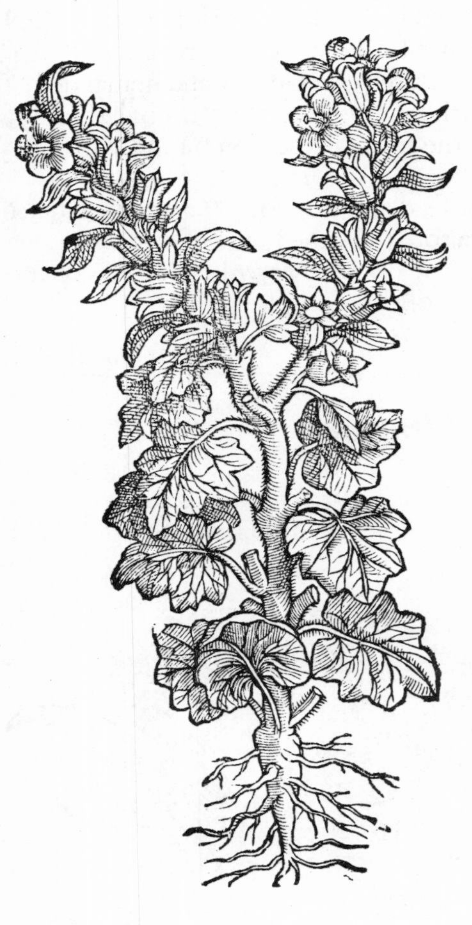

‡ 4 *Hyoscyamus albus Creticus.*
White Henbane of Candy.

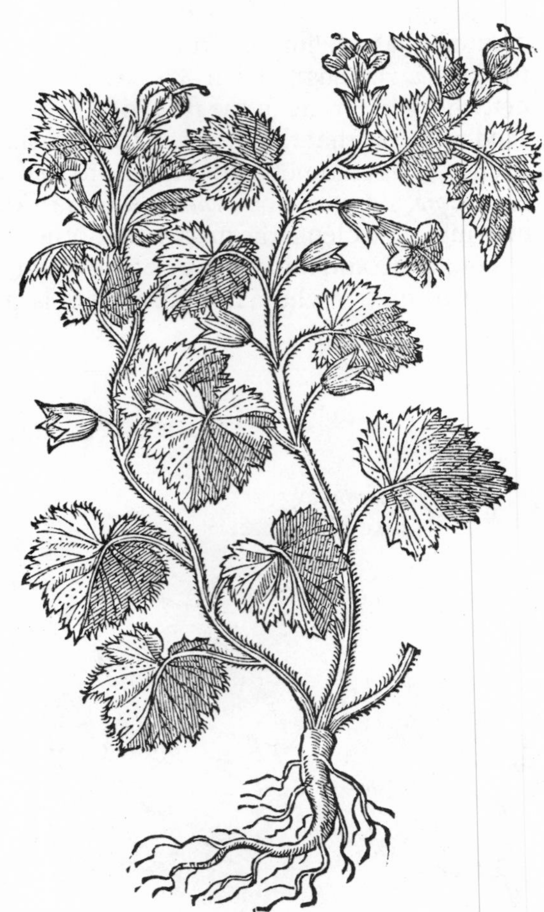

‡ 3 This other white Henbane is much like the last described, but that it is lesser: the leaues smaller and rounder, hanging vpon pretty long stalkes, the floures and seed vessels are like those of the last mentioned.

4 This is softer and tenderer than the last described, the leaues also hang vpon long foot-stalkes, and are couered ouer with a soft downines: and they are somewhat broader, yet thinner and more sinuated than those of the white, and somewhat resemble the forme of a vine leafe, being snipt about the edges; the stalkes are also couered with a white downe. The floures are of a gold yellow, with a veluet coloured circle in their middles: the root is sufficiently thicke and large: *Clusius* had the figure and description of this from his friend *Iaques Plateau*, who had the plant growing of seed receiued from Candy.

5 The stalke of this growes some cubit high, being pretty stiffe, about the thickenesse of ones little finger, and couered ouer with a soft and white downe: the leaues grow dispersed vpon the stalk, not much vnlike those of the common kinde, but lesser and more diuided, and white (while they are young) couered with a slender and long downinesse: the top of the stalke is diuided into certaine branches that bend or hang downe their heads, which alternately amongst narrower, lesser and vndiuided leaues carry cups like as the common one, ending in fiue pretty stiffe points, in which are contained floures at first somewhat like the common kinde, but afterwards, as they grow bigger, they change into an elegant red purplish colour, with deepe coloured veines: neither is the ring or middle part purple as in the common kinde, but whitish, hauing a purplish pointall, and fiue threds in the middle: the seeds and seed vessels are like those of the common kind. *Clusius* receiued the seed hereof from *Paludanus* returning from his trauailes into Syria and Egypt, wherefore he calls it *Hyoscyamus Ægyptius*, Egyptian Henbane. ‡

¶ *The*

‡ 5 *Hyoſcyamus flore rubello.*
Henbane with a reddiſh floure.

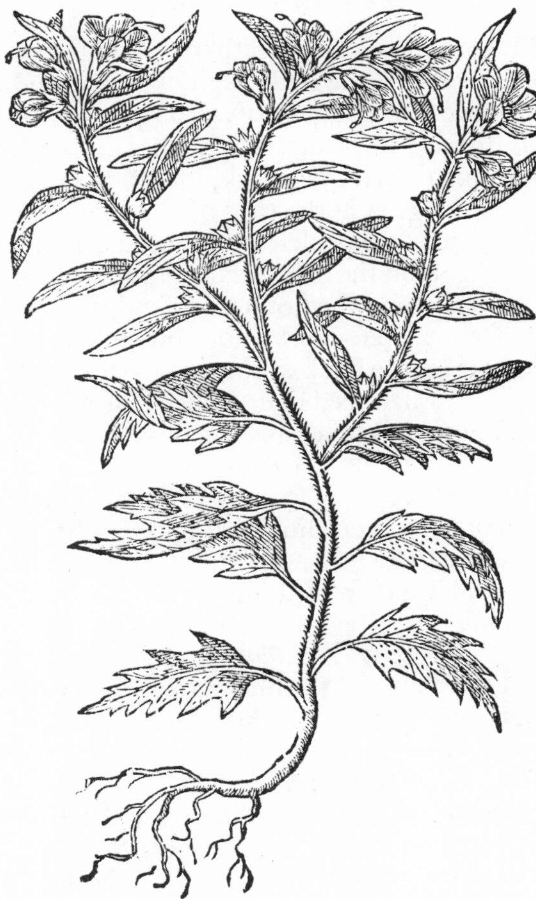

¶ *The Place.*

Blacke Henbane grows almoſt euerie where by high-wayes, in the borders of fields, about dung-hills and vntoiled places; the white Henbane is not found but in the gardens of thoſe that loue phyſicall plants: the which groweth in my garden, and doth ſow it ſelfe from yeare to yeare.

¶ *The Time.*

They ſpring out of the ground in May, bring forth their floures in Auguſt, and the ſeed is ripe in October.

¶ *The Names.*

Henbane is called of the Grecians, ὑοσκύαμος: of the Latines, *Apollinaris*, and *Faba ſuilla*: of the Arabians, as *Pliny* ſaith, *Altercum*: of ſome, *Faba Iouis*, or *Iupiters* beame: of *Pythagoras*, *Zoroaſtes*, and *Apuleius Inſana*, *Alterculum*, *Symphoniaca*, and *Calicularis*: of the Tuſcanes, *Faulonia*, and *Faba lupina*: of *Matthæus ſyluaticus*, *Dens Caballinus*, *Milimandrum*, *Caßilago*: of *Iacobus a Manlijs*, *Herbapinnula*: in ſhops it is called *Inſquiamus*, and *Hyoſcyamus*: in Engliſh, Henbane in Italian, *Hyoſquiamo*: in Spaniſh, *Veleno*: in high Dutch, **Bilſen kraut**: in French, *Hannebane*, *Endormie*: the other is called *Hyoſcyamus albus*, or white Henbane.

¶ *The Temperature.*

Theſe kindes of Henbane are cold in the fourth degree.

¶ *The Vertues.*

Henbane cauſeth drowſineſſe, and mitigateth all kinde of paine: it is good againſt hot and ſharpe diſtillations of the eyes and other parts: it ſtayeth bleeding and the diſeaſe in women: it is applied to inflammations of the ſtones and other ſecret parts. A

The leaues ſtamped with the ointment *Populeon*, made of poplar buds, aſſwageth the paine of the gout, and the ſwellings of the ſtones, aud the tumors of womens breſts, and are good to be put into the ſame ointment, but in ſmall quantitie B

To waſh the feet in th e decoction of Henbane cauſeth ſleepe; or giuen in a cliſter it doth the ſame; and alſo the often ſmelling to the floures. C

The leaues, ſeed, and iuyce taken inwardly cauſeth an vnquiet ſleepe like vnto the ſleep of drunkenneſſe, which continueth long, and is deadly to the party. D

The ſeed of white Henbane is good againſt the cough, the falling of waterie humours into the eyes or breſt; againſt the inordinate flux of womens iſſues, and all other iſſues of bloud, taken in the weight of ten graines, with water wherein honey hath beene ſodden. E

The root boyled with vineger, and the ſame holden hot in the mouth, eaſeth the paine of the teeth. The ſeed is vſed by Mountibanke Tooth-drawers which runne about the countrie, for to cauſe wormes come forth of mens teeth, by burning it in a chafing-diſh with coles, the party holding his mouth ouer the fume thereof: but ſome crafty companions to gaine mony conuey ſmall lute ſtring into the water, perſuading the patient that thoſe ſmall creeping beaſts came out of his mouth or other parts which he intended to eaſe. F

Chap. 67. *Of yellow Henbane, or English Tabaco.*

Hyoſcyamus luteus.
Yellow Henbane.

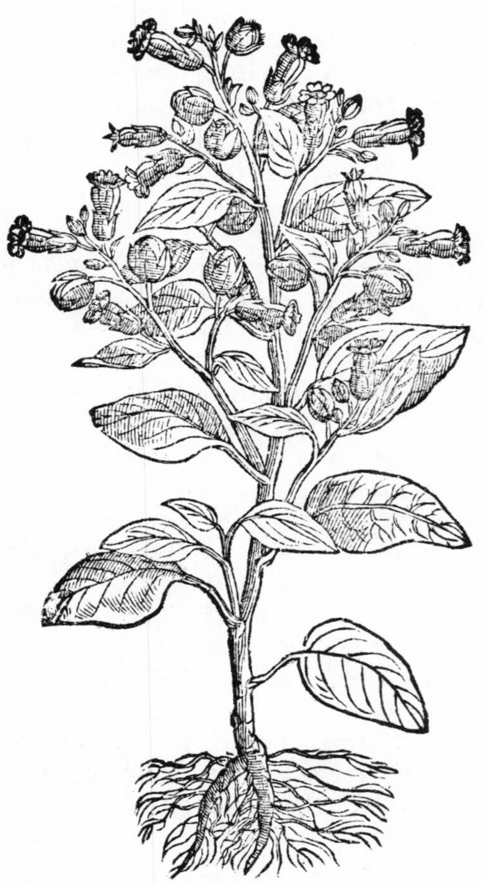

¶ *The Deſcription.*

Yellow Henbane groweth to the height of two cubits: the ſtalke is thicke, fat, and greene of colour, full of a ſpongeous pith, and is diuided into ſundry branches ſet with ſmooth and euen leaues, thick, and ful of juice. The floures grow at the tops of the branches, orderly placed, of a pale yellow colour, ſomething leſſer than thoſe of the blacke Henbane. The cups wherein the floures do ſtand are like, but leſſer, tenderer, and without ſharpe points, wherein is ſet the husk or cod ſomwhat round, full of very ſmall ſeed like the ſeed of Marjerome. The root is ſmall and threddy.

¶ *The Place.*

Yellow Henbane is ſowen in gardens, where it doth proſper exceedingly, inſomuch that it cannot be deſtroyed where it hath once ſowen it ſelfe, and it is diſperſed into the moſt parts of England.

¶ *The Time.*

It floureth in the Sommer moneths, and oftentimes till Autumne be farre ſpent, in which time the ſeed commeth to perfection.

¶ *The Names.*

Yellow Henbane is called *Hyoſcyamus luteus*: of ſome, *Petum*, and *Petun*: of others, *Nicoſiana*, of *Nicot* a Frenchman that brought the ſeeds from the Indies, as alſo the ſeeds of the true Tabaco, whereof this hath beene taken for a kinde; inſomuch that *Lobel* hath called it *Dubius Hyoſcyamus*, or doubtfull Henbane, as a plant participating of Henbane and Tabaco: and it is vſed of diuers in ſtead of Tabaco, and called by the ſame name, for that it hath beene brought from Trinidada, a place ſo called in the Indies, as alſo from Virginia and Norembega, for Tabaco, which doubtleſſe taken in ſmoke worketh the ſame kinde of drunkenneſſe that the right Tabaco doth. ‡ Some vſe to call this Nicotian, in Engliſh, being a name taken from the Latine. ‡

¶ *The Nature.*

This kinde of Henbane is thought of ſome to be cold and moiſt; but after *Lobel* it rather heateth than cooles at all, becauſe of the biting taſte, as alſo that roſenninesſe or gummines it is poſſeſſed of; which is euidently perceiued both in handling and chewing it in the mouth.

¶ *The Vertues.*

A This herbe auaileth againſt all apoſtumes, tumors, inueterate vlcers, botches, and ſuch like, being made into an vnguent or ſalue as followeth: Take of the greene leaues three pounds and an halfe, ſtampe them very ſmall in a ſtone morter; of Oyle Oliue one quart; ſet them to boyle in a braſſe pan or ſuch like, vpon a gentle fire, continually ſtirring it vntill the herbes ſeem blacke, and will not boyle or bubble any more: then ſhall you haue an excellent greene oyle; which being ſtrained from the feces or droſſe, put the cleare and ſtrained oyle to the fire againe; adding thereto of wax halfe a pound, of roſen foure ounces, and of good Turpentine two ounces: melt them all together, and keepe it in pots for your vſe, to cure inueterate vlcers, apoſtumes, burnings, greene wounds, and all cuts and hurts in the head; wherewith I haue gotten both crownes and credit.

B It is vſed of ſome in ſtead of Tabaco, but to ſmall purpoſe or profit, although it do ſtupifie and dull the ſences, and cauſe that kinde of giddineſſe that Tabaco doth, and likewiſe ſpitting; which any other herbe of hot temperature will do, as Roſemary, Time, winter Sauorie, ſweet Marjerome, and ſuch like: any of the which I like better to be taken in ſmoke than this kinde of doubtfull henbane.

CHAP. 68.

Of Tabaco, or Henbane of Peru.

¶ *The Kindes.*

THere be two ſorts or kindes of Tabaco; one greater, the other leſſer: the greater was brought into Europe out of the prouinces of America, which we call the Weſt Indies; the other from Trinidada, an Iſland neere vnto the continent of the ſame Indies. Some haue added a third ſort: and others make the yellow Henbane a kinde thereof.

† 1 *Hyoſcyamus Peruvianus.*
Tabaco or Henbane of Peru.

† 2 *Sana Sancta Indorum.*
Tabaco of Trinidada.

¶ *The Deſcription.*

1 TAbaco, or Henbane of Peru hath very great ſtalkes of the bigneſſe of a childes arme, growing in fertile and well dunged ground of ſeuen or eight foot high, diuiding it ſelfe into ſundry branches of great length; whereon are placed in moſt comely order very faire long leaues, broad, ſmooth, and ſharpe pointed, ſoft, and of a light greene colour, ſo faſtned about the ſtalke, that they ſeeme to embrace and compaſſe it about. The floures grow at the top of the ſtalkes, in ſhape like a bell-floure, ſomewhat long and cornered, hollow within, of a light carnation colour, tending to whiteneſſe toward the brimmes. The ſeed is contained in long ſharpe pointed cods or ſeed-veſſels like vnto the ſeed of yellow Henbane, but ſomewhat ſmaller and browner of colour. The root is great, thicke, and of a wooddy ſubſtance, with ſome threddie ſtrings anexed thereunto.

2 Trinidada Tabaco hath a thicke tough and fibrous root, from which immediately riſe vp long broad leaues and ſmooth, of a greeniſh colour, leſſer than thoſe of Peru: among which riſeth vp a ſtalke diuiding it ſelfe at the ground into diuers branches, whereon are ſet confuſedly the like leaues, but leſſer: at the top of the ſtalks ſtand vp long necked hollow floures of a pale purple tending to a bluſh colour: after which ſucceed the cods or ſeed-veſſels, including many ſmall ſeeds like vnto the ſeed of Marjerome. The whole plant periſheth at the firſt approch of Winter.

 ‡ 3 This

‡ 3 *Tabacum minimum.*
Dwarfe Tabaco.

‡ 3 This third is an herbe some spanne or better long, not in face vnlike the precedent, neither defectiue in the hot and burning taste. The floures are much lesse than those of the yellow Henbane, & of a greenish yellow. The leaues are small, and narrower those of Sage of Ierusalem. The root is small and fibrous. ‡

¶ *The Place.*

These were first brought into Europe out of America, which is called the West Indies, in which is the prouince or countrey of Peru: but being now planted in the gardens of Europe it prospereth very well, and commeth from seed in one yeare to beare both floures and seed. The which I take to be better for the constitution of our bodies than that which is brought from India; and that growing in the Indies better for the people of the same Countrey: notwithstanding it is not so thought, nor receiued of our Tabaconists; for according to the English Prouerbe, Far fetcht and deare bought is best for Ladies.

¶ *The Time.*

Tabaco must be sowen in the most fruitfull ground that may be found, carlesly cast abroad in the sowing, without raking it into the ground or any such paine or industrie taken as is requisite in the sowing of other seeds, as my self haue found by proofe, who haue experimented euery way to cause it quickly to grow: for I haue committed some to the earth in the end of March, some in Aprill, and some in the beginning of May, because I durst not hasard all my seed at one time, lest some vnkindely blast should happen after the sowing, which might be a great enemie thereunto.

¶ *The Names.*

The people of America call it *Petun*: Some, as *Lobel* and *Pena*, haue giuen it these Latine names, *Sacra herba*, *Sancta herba*, and *Sana sancta Indorum*: and other, as *Dodonæus*, call it *Hyoscyamus Peruvianus*, or Henbane of Peru: *Nicolaus Monardus* names it *Tabacum*. That it is *Hyoscyami species*, or a kind of Henbane, not onely the forme being like to yellow Henbane, but the qualitie also doth declare; for it bringeth drowsinesse, troubleth the sences, and maketh a man as it were drunke by taking of the fume onely; as *Andrew Theuet* testifieth, (and common experience sheweth:) of some it is called *Nicotiana*: the which I refer to the yellow Henbane, for distinctions sake.

¶ *The Temperature.*

It is hot and dry, and that in the second degree, as *Monardis* thinketh, and is withall of power to discusse or resolue, and to cleanse away filthy humors, hauing also a small astriction, and a stupifying or benumming qualitie, and it purgeth by the stoole: and *Monardis* writeth that it hath a certaine power to resist poyson. And to proue it to be of an hot temperatute, the biting qualitie of the leaues doth shew, which is easily perceiued by taste: also the greene leaues laid vpon vlcers in sinewie parts may serue for a proofe of heate in this plant; because they do draw out filth and corrupted matter, which a cold Simple would neuer do. The leaues likewise being chewed draw forth flegme and water, as doth also the fume taken when the leaues are dried: which things declare that this is not a little hot; for what things soeuer, that being chewed or held in the mouth bring forth flegme and water, the same be all accounted hot; as the root of Pellitorie of Spaine, of Saxifrage, and other things of like power. Moreouer, the benumming qualitie hereof is not hard to be perceiued, for vpon the taking of the fume at the mouth there followeth an infirmitie like vnto drunkennesse, and many times sleepe; as after the taking of *Opium*: which also sheweth in the taste a biting qualitie, and therefore is not without heate; which when it is chewed

and

and inwardly taken, it doth forthwith shew, causing a certaine heat in the chest, and yet withall troubling the wits, as *Petrus Bellonius* in his third Booke of Singularities doth declare; where also hee sheweth, that the Turkes oftentimes doe vse *Opium*, and take one dramme and a halfe thereof at one time; without any other hurt following, sauing that they are thereupon (as it were) taken with a certaine light drunkennesse. So also this Tabaco being in taste biting, and in temperature hot, hath notwithstanding a benumming qualitie. Hereupon it seemeth to follow, that not onely this Henbane of Peru, but also the iuice of poppie otherwise called *Opium*, consisteth of diuers parts; some biting and hot, and others extreame cold, that is to say, stupifying or benumming: if so bee that this benumming qualitie proceed of extreme cold (as *Galen* and all the old Physitions doe hold opinion) Then should this bee cold; but if the benumming facultie doth not depend of an extreme cold qualitie, but proceedeth of the essence of the substance; then Tabaco is not cold and benumming; but hot and benumming, and the latter not so much by reason of his temperature, as through the propertie of his substance; no otherwise than a purging medicine, which hath his force not from the temperature, but from the essence of the whole substance.

¶ *The Vertues.*

Nicolaus Monardis saith, that the leaues hereof are a remedy for the paine in the head called the Megram or Migraime that hath beene of long continuance: and also for a cold stomacke, especially in children; and that it is good against the paines in the kidneies. A

It is a present remedie for the fits of the Mother: it mitigateth the paine of the gout if it bee rosted in hot embers and applied to the grieued part. B

It is likewise a remedie for the tooth-ache, if the teeth and gums be rubbed with a linnen cloth dipped in the iuice; and afterward a round ball of the leaues laid vnto the place. C

The iuice boiled with Sugar in forme of a sirrup and inwardly taken, driueth forth wormes of the bellie; if withall a leafe be laid to the Nauell. D

The same doth likewise scoure and clense old and rotten vlcers, and bringeth them to perfect digestion as the same Author affirmeth. E

In the Low Countreyes it is vsed against scabbes and filthinesse of the skinne, and for the cure of wounds: but some hold opinion that it is to bee vsed but onely to hot and strong bodies: for they say that the vse is not safe in weake and old folkes: and for this cause, as it seemeth, the women in America (as *Theuet* sayth) abstayne from the hearbe *Petun* or Tabaco, and doe in no wise vse it. F

The weight of foure ounces of the iuice heereof drunke purgeth both vpwards and downewards, and procureth after, a long and sound sleepe, as wee haue learned of a friend by obseruation, affirming that a strong Countreyman of a middle age, hauing a dropsie, tooke of it, and being wakened out of his sleepe, called for meat and drinke, and after that became perfectly whole. G

Moreouer the same man reported, that he had cured many countriemen of agues with the distilled water of the leaues drunke a little while before the fit. H

Likewise there is an oile to be taken out of the leaues that healeth merry-gals, kibed heels and such like. I

It is good against poison, and taketh away the malignitie thereof, if the iuice be giuen to drink or the wounds made by venemous beasts be washed therewith. K

The drie leaues are vsed to be taken in a pipe set on fire and suckt into the stomacke, and thrust forth againe at the nosthrils against the pains of the head, rheumes, aches in any part of the body whereof soeuer the originall proceed, whether from France, Italy, Spaine, Indies, or from our familiar and best knowne diseases: those leaues doe palliate or ease for a time, but neuer performe any cure absolutely: for although they emptie the body of humours, yet the cause of the griefe cannot be so taken away. But some haue learned this principle, that repletion requireth euacuation; that is, fulnesse craueth emptinesse, and by euacuation assure themselues of health: But this doth not take away so much with it this day, but the next bringeth with it more: As for example, a Well doth neuer yeeld such store of water as when it is most drawne and emptied. My selfe speake by proofe, who haue cured of that infectious disease a great many; diuers of which had couered or kept vnder the sickenesse by the helpe of Tabaco as they thought, yet in the end haue beene constrained to haue vnto such an hard knot, a crabbed wedge, or else had vtterly perished. L

Some vse to drinke it (as it is tearmed) for wantonnesse or rather custome, and cannot forbeare it, no not in the midst of their dinner, which kind of taking is vnwholesome and very dangerous: although to take it seldome and that Physically is to be tolerated and may do some good: but I commend the syrrup aboue this fume or smokie medicine. M

It

N It is taken of some physically in a pipe for that purpose once in a day at the most, and that in the morning fasting against paines in the head, stomacke, and griefe in the brest and lungs: against catarrhes and rheumes, and such as haue gotten cold and hoarsenesse.

O Some haue reported that it little preuaileth against an hot disease, and that it profiteth an hot complexion nothing at all: but experience hath not shewed it to bee iniurious vnto either.

P They that haue seene the proofe hereof haue credibly reported, that when the Moores and Indians haue fainted either for want of food or rest, this hath beene a present remedie vnto them to supplie the one, and to helpe them to the other.

Q The priests and Inchanters of the hot countries do take the fume thereof vntill they be drunke, that after they haue lien for dead three or foure houres, they may tell the people what wonders, visions, or illusions they haue seene, and so giue them a propheticall direction or foretelling (if we may trust the Diuell) of the successe of their businesse.

R The iuice or distilled water of the first kind is very good against catarrhes, the dizzinesse of the head, and rheumes that fall downe the eies, against the paine called the Megram, if either you applie it vnto the temples, or take one or two greene leaues, or a dry leafe moistned in wine, and dried cunningly vpon the embers and laid thereto.

S It cleereth the sight and taketh away the webs and spots thereof, being annointed with the iuice bloud warme.

T The oile or iuice dropped into the eares is good against deafenesse; a cloth dipped in the same and laid vpon the face, taketh away the lentils, rednesse, and spots thereof.

V Many notable medicines are made hereof against the old and inueterate cough, against asthmaticall or pectorall griefes, which if I should set downe at large, would require a peculiar Volume.

X It is also giuen to such as are accustomed to swoune, and are troubled with the Collicke and windinesse, against the Dropsie, the Wormes in children, the Piles and the Sciatica.

Y It is vsed in outward medicines either the herbe boiled with oile, waxe, rosin and turpentine, as before is set downe in yellow Henbane, or the extraction thereof with salt, oile, balsame, the distilled water and such like, against tumours, apostumes, old vlcers, of hard curation, botches, scabbes, stinging with nettles, carbuncles, poisoned arrowes, and wounds made with gunnes or any other weapon.

Z It is excellent good in burnings and scaldings with fire, water, oile, lightning, or such like, boiled with Hogges greace in forme of an Ointment, which I haue often prooued, and found most true, adding a little of the iuice of thorne apple leaues, spreading it vpon a cloth and so applying it.

A I doe make hereof an excellent balsame to cure deepe wounds and punctures, made by some narrow sharpe pointed weapon. Which balsame doth bring vp the flesh from the bottome verie speedily, and also heale simple cuts in the flesh according to the first intention, that is, to glew or soder the lips of the wound together, not procuring matter or corruption vnto it, as is commonly seene in the healing of wounds. The receit is this: Take oile of roses, oile of S. Iohns wort, of either one pinte, the leaues of Tabaco stamped small in a stone morter two pounds, boile them together to the consumption of the iuice, straine it and put it to the fire againe, adding thereto of Venice Turpentine two ounces, of Olibanum and masticke of either halfe an ounce, in most fine and subtill pouder, the which you may at all times make an vnguent or salue by putting thereto wax and rosin to giue vnto it a stiffe body, which worketh exceeding well in maligne and virulent vlcers, as in wounds and punctures. I send this iewell vnto you women of all sorts, especially to such as cure and helpe the poore and impotent of your Countrey without reward. But vnto the beggerly rabble of witches, charmers, and such like couseners, that regard more to get money, than to helpe for charitie, I wish these few medicines far from their vnderstanding, and from those deceiuers whom I wish to be ignorant herein. But courteous gentlewomen, I may not for the malice that I doe beare vnto such, hide any thing from you of such importance: and therefore take one more that followeth, wherewith I haue done very many and good cures, although of small cost, but

B regard it not the lesse for that cause. Take the leaues of Tabaco two pound, hogges grease one pound, stampe the herbe small in a stone morter, putting thereto a small cup full of red or claret wine, stir them well together, couer the morter from filth and so let it rest vntill morning; then put it to the fire and let it boile gently, continually stirring it vntill the consumption of the wine; straine it, and set it to the fire againe, putting thereto the iuice of the herbe one pound, of Venice turpentine foure ounces; boile them togethet to the consumption of the iuice, then adde therto of the

the roots of round *Ariſtolochia* or Birthwoort in moſt fine pouder two ounces, ſufficient waxe to giue it a body, the which keep for thy wounded poore neighbour, as alſo the old and filthy vlcers of the legs and other parts of ſuch as haue need of helpe.

† The figures were formerly tranſpoſed.

Chap. 69. *Of Tree Nightſhade.*

Amomum Plinij.
Tree Nightſhade.

¶ *The Deſcription.*

This rare and pleaſant Plant, called tree Nightſhade, is taken of ſome to be a kinde of Ginnie pepper, but not rightly; of others for a kinde of Nightſhade, whoſe iudgement and cenſure I gladly admit; for that it doth more fitly anſwer it both in the forme and nature. It groweth vp like vnto a ſmall ſhrubbe or wooddy hedge buſh, two or three cubits high, couered with a greeniſh barke ſet with many ſmall twiggie branches, and garniſhed with many long leaues very greene, like vnto thoſe of the Peach tree. The floures are white, with a certaine yellow pricke or pointell in the middle, like vnto the floures of garden Nightſhade. After which ſucceede ſmall round berries verie red of colour, and of the ſame ſubſtance with Winter Cherries, wherein are contained little flat yellow ſeeds. The root is compact of many ſmall hairie yellow ſtrings.

¶ *The Place.*

It groweth not wilde in theſe cold regions, but we haue them in our gardens, rather for pleaſure than profit, or any good qualitie as yet knowne.

¶ *The Time.*

It is kept in pots and tubs with earth and ſuch like in houſes during the extremity of Winter, becauſe it cannot indure the coldneſſe of our colde climate, and is ſet abroad into the Garden in March or Aprill: it floureth in May, and the fruit is ripe in September.

¶ *The Names.*

Tree Nightſhade is called in Latine *Solanum Arboreſcens*: of ſome, *Strychnodendron*: and ſome iudge it to be *Amomum* of *Plinie*: it is *Pſeudocapſicum* of *Dodonæus*.

¶ *The Nature and Vertues.*

We haue not as yet any thing ſet downe as touching the temperature or vertues of this Plant, but it is referred of ſome to the kindes of Ginnie pepper, but without any reaſon at all; for Ginny pepper though it bring forth fruit very like in ſhape vnto this plant, yet in taſte moſt vnlike, for that *Capſicum* or Ginny pepper is more ſharpe in taſte than our common pepper, and the other hath no taſte of biting at all, but is like vnto the Berries of Garden Nightſhade in taſte, although they differ in colour: which hath moued ſome to call this plant red Nightſhade, of the colour of the berries: and Tree Nightſhade, of the wooddy ſubſtance which doth continue and grow from yeare to yeare: and Ginnie pepper dieth at the firſt approch of Winter.

Chap. 70. *Of Balme Apple, or Apple of Hierusalem.*

1 *Balsamina mas.*
The male Balsam Apple.

2 *Balsamina fœmina.*
The female Balsam Apple.

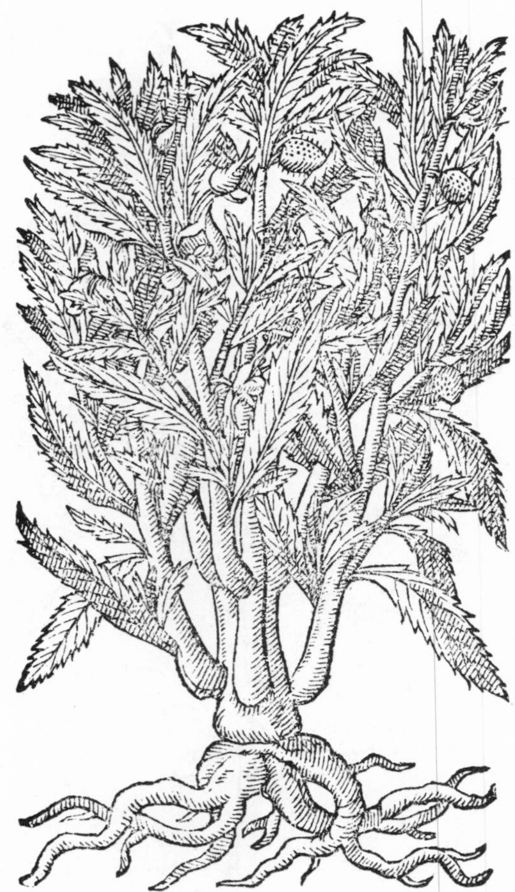

The Description.

1 THe male Balme Apple hath long, small, and tender branches, set with leaues like those of the vine; and the like small clasping tendrels wherewith it catcheth hold of such things as do grow neere vnto it, not able by reason of his weakenesse to stand vpright without some pole or other thing to support it. The floures consist of fiue small leaues of a meane bignesse, and are of a faint yellow colour: which being past, there doe come in place long Apples, something sharpe toward the point almost like an egge, rough all ouer as it were with small harmelesse prickles, red both within and without when they be ripe, and cleaue in sunder of themselues: in the Apple lieth great broad flat seeds, like those of Pompion or Citrull, but something blacke when they be withered. The root is threddie, and disperseth it selfe far abroad in the ground.

2 The female Balm Apple doth not a little differ from the former: it bringeth forth stalks not running or climing like the other, but a most thicke and fat truncke or stocke full of iuice, in substance like the stalks of Purslane, of a reddish color and somewhat shining. The leaues be long and narrow, in shape like those of Willow or the Peach tree, somewhat toothed or notched about the edges: among which grow the floures of an incarnate colour tending to blewnesse, hauing a small spur or taile annexed thereto as hath the Larks heele, of a faire light crimson colour: in their places come vp the fruit or Apples rough and hairy, but lesser than those of the former, yellow when they be ripe, which likewise cleaue asunder of themselues and cast abroad their seedes much like vnto Lentils, saith mine Author. But those which I haue from yeare to yeare in my Garden bring forth seed like the Cole-florey or Mustard seed; whether they be of two kindes, or the climate doe alter the shape, it resteth disputable.

¶ The

¶ *The Place.*

Theſe plants do proſper beſt in hot Regions : they are ſtrangers in England,and doe with great labour and induſtrie grow in theſe cold Countries.

¶ *The Time.*

They muſt be ſowne in the beginning of Aprill in a bed of hot horſe dung, euen as Muske-Melons,Cucumbers, and ſuch like cold fruits are; and replanted abroad from the ſaid bed into the moſt hot and fertile place of the Garden at ſuch time as they haue gotten three leaues a peece.

¶ *The Names.*

Diuerſly hath this plant been named; ſome calling it by one name,and ſome by another,euery one as it ſeemed good to his fancie. *Baptiſta Sardus* calleth it *Balſamina Cucumerina*: others, *Viticella*,and *Charantia*,as alſo *Pomum Hieroſolymitanum*, or Apples of Hieruſalem : in Engliſh, Balme Apple:in Italian,*Caranza* : in the Germane tongue,**Balſam opffel** : in French,*Merueille* · ſome of the Latines haue called it *Pomum mirabile*, or maruellous Apples.It is thought to be named *Balſamina*, becauſe the oile wherein the ripe Apples be ſteeped or infuſed,is taken to bee profitable for many things, as is *Opobalſamum*,or the liquour of the plant *Balſamum*.

The female Balſam Apple is likewiſe called *Balſamina*, and oftentimes in the Neuter Gender *Balſaminum* . *Geſner* chooſeth rather to name it *Balſamina amygdaloides* : *Valerius Cordus*,*Balſamella*: others,*Balſamina fœmina* : in Engliſh, the Female Balme Apples.

¶ *The Nature.*

The fruit or apples hereof, as alſo the leaues, doe notably drie,hauing withall a certaine moderate coldneſſe very neere to a meane temperature, that is after ſome hot,in the firſt,and drie in the ſecond degree.

¶ *The Vertues.*

The leaues are reported to heale greene wounds if they be bruiſed and laid thereon; and taken with wine they are ſaid to be a remedie for the collicke; and an effectuall medicine for burſtings and conuulſions or crampes. A

The leaues of the male *Balſamina* dried in the ſhadow , and beaten into pouder and giuen in wine vnto thoſe that are mortally wounded in the body,doth cure them inwardly,and helpeth alſo the Collicke. B

The oile which is drawne forth of the fruit doth cure all greene and freſh wounds as the true naturall Balſam : it helpeth the crampes and conuulſions,and the ſhrinking of ſinewes,being annointed therewith. C

It profiteth women that are in great extremitie of childe-birth in taking away the paine of the matrix, cauſing eaſie deliuerance beeing applied to the place, and annointed vpon their bellies,or caſt into the matrix with a ſyring,and eaſeth the dolour of the inward parts. D

It cureth the Hemorrhoides and all other paines of the fundament,being thereto applied with lint of old clouts. E

The leaues drunken in wine, heale ruptures. F

I finde little or nothing written of the property or vertues of the female kinde, but that it is thought to draw neere vnto the firſt in temperament and vertue. G

Oile oliue in which the fruit (the ſeede taken forth) is either ſet in the Sun,as we do when wee make oile of roſes,or boiled in a double glaſſe ſet in hot water,or elſe buried in hot horſe dung,taketh away inflammations that are in wounds. It doth alſo eaſily and in ſhort time conſolidate or glew them together,and perfectly cure them. H

It cureth the vlcers of the dugs or paps, the head of the yard or matrix,as alſo the inflammation thereof being iniected or conueied into the place with a ſyringe or mother peſſarie. I

This apple is with good ſucceſſe applied vnto wounds,prickes and hurts of the ſinewes.It hath great force to cure ſcaldings and burnings : it taketh away ſcarres and blemiſhes,if in the meane time the pouder of the leaues be taken for certaine daies together. K

It is reported that ſuch as be barren are made fruitfull herewith, if the woman firſt be bathed in a fit and conuenient bath for the purpoſe,& the parts about the ſhare and matrix annointed herewith,and the woman preſently haue the company of her husband. L

Chap. 71. *Of Ginnie or Indian Pepper.*

1 *Capsicum longioribus siliquis.*
Long codded Ginnie Pepper.

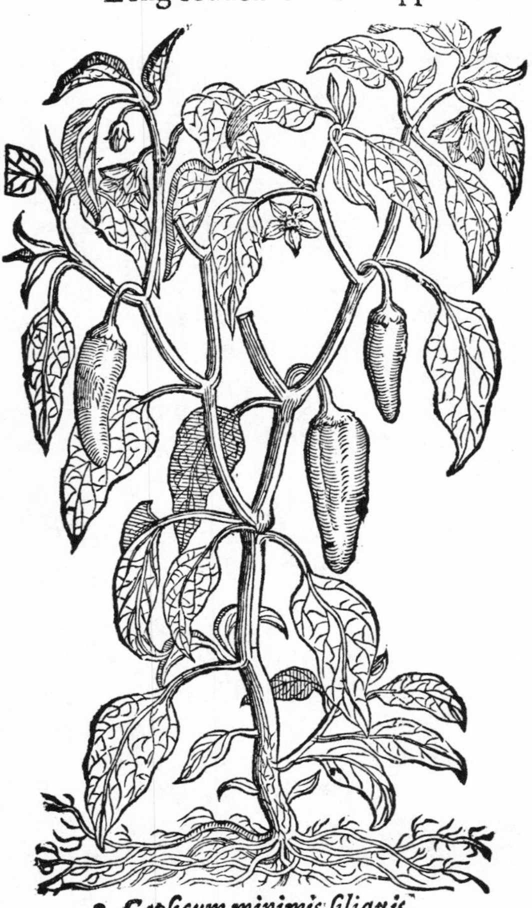

‡ 2 *Capsicum rotundioribus siliquis.*
Round codded Ginnie Pepper.

3 *Capsicum minimis siliquis.*
Small codded Ginnie Pepper.

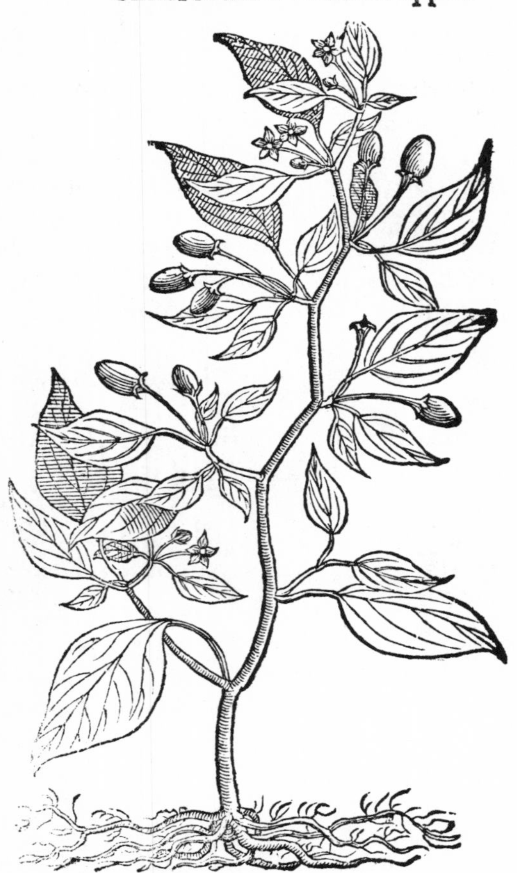

‡ *Capsici siliquæ variæ.*
Varieties of the cods of Ginnie Pepper.

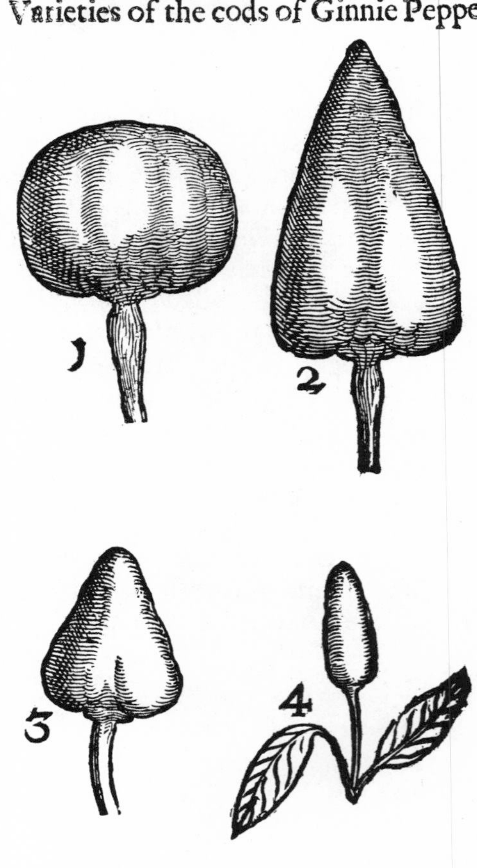

¶ The

¶ *The Description.*

1 THe first of these plants hath square stalkes a foot high or somewhat more, set with many thicke and fat leaues, not vnlike to those of garden Nightshade, but narrower and sharper pointed, of a darke greene colour. The floures grow alongst the stalkes, out of the wings of the leaues, of a white colour, hauing for the most part fiue small leaues blasing out like a star, with a greene button in the middle. After them grow the cods, greene at the first, and when they be ripe of a braue colour glittering like red corall, in which is contained little flat seeds, of a light yellow colour, of a hot biting taste like common pepper, as is also the cod it selfe: which is long, and as big as a finger, and sharpe pointed.

‡ 2 The difference that is betweene this and the last described is small, for it consists in nothing but that the cods are pretty large and round, after the fashion of cherries, and not so long as those of the former. ‡

3 The third kinde of Ginnie pepper is like vnto the precedent in leaues, floures, and stalkes. The cods hereof are small, round, and red, very like to the berries of *Dulcamara* or wooddy Nightshade, both in bignesse, colour, and substance, wherein consisteth the difference: notwithstanding the seed and cods are very sharpe and biting, as those of the first kinde.

‡ *Capsici siliquæ variæ.*
Varieties of the cods of Ginnie pepper.

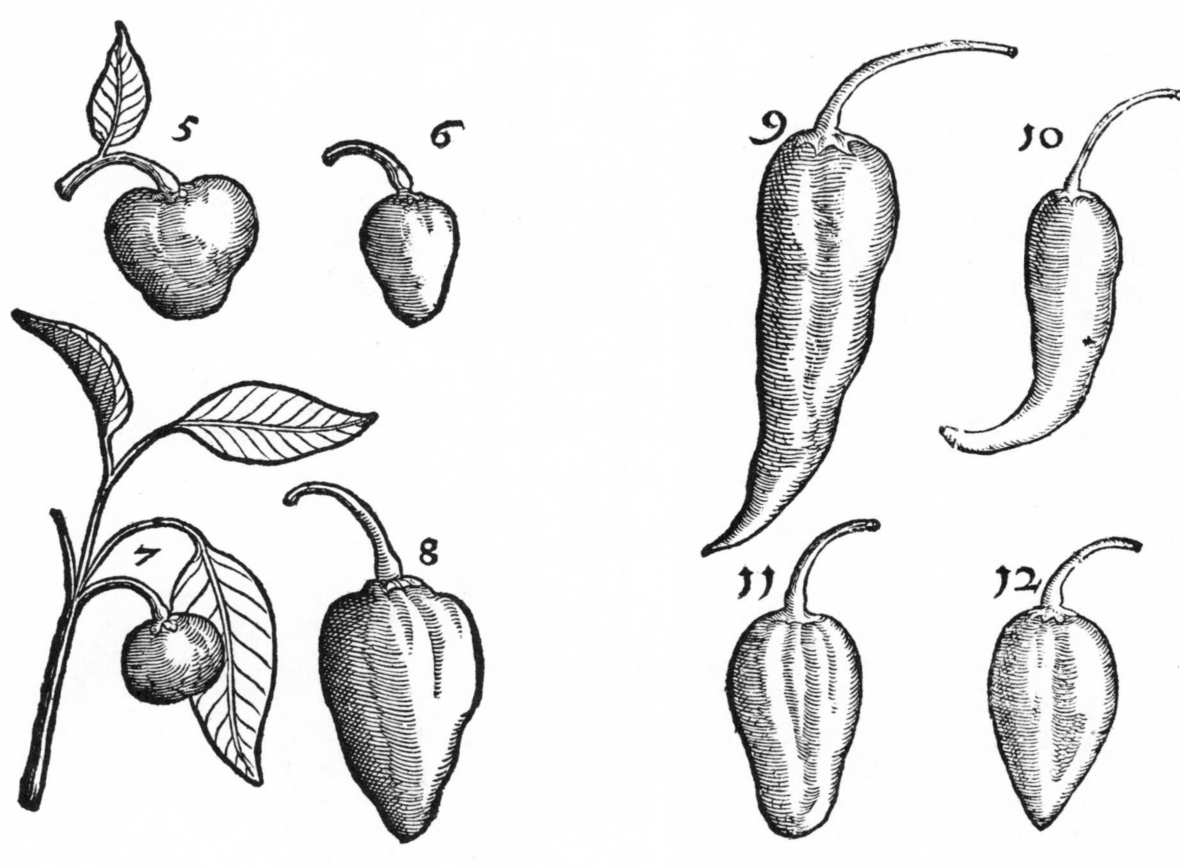

‡ There are many other varieties of Ginnie pepper, which chiefly consist in the shape and colour of the cods: wherefore I thought good (and that chiefely because it is a plant that will hardly brooke our climate) only to present you with the figures of their seuerall shapes, whereof the cods of some stand or grow vpright, and other some hang downe: such as desire further information of this plant, may be aboundantly satisfied in *Clusius* his *Curæposter*. from *pag*. 95. to *pag*. 108. where they shall finde these treated of at large in a treatise written in Italian by *Gregory de Regio*, a Capuchine Fryer, and sent to *Clusius*, who translating it into Latine, left it to be set forth with other his obseruations, whith was performed 2. yeares after his death, to wit *Anno Domini* 1611. The figures we here giue are the same which are in that tractate. ‡

¶ *The Place.*

These plants are brought from forrein countries, as Ginnie, India, and those parts, into Spaine and Italy: from whence we haue receiued seed for our English gardens, where they come to fruitbearing: but the cod doth not come to that bright red colour which naturally it is possessed with, which hath happened by reason of these vnkindly yeeres that are past: but we expect better when God shall send vs a hot and temperate yeere.

¶ *The Time.*

The seeds hereof must be sowen in a bed of hot horse-dung, as muske-Melons are, and remooued into a pot when they haue gotten three or foure leaues, that it may the more conueniently be caried from place to place to receiue the heate of the sunne: and are toward Autumne to be caried into some house, to auoide the iniurie of the cold nights of that time of the yeere, when it is to beare his fruite.

¶ *The Names.*

Actuarius calleth it in Greeke καψικόν: in Latine, *Capsicum*: and it is thought to be that which *Auicen* nameth *Zinziber caninum*, or dogs Ginger: and *Pliny*, *Siliquastrum*, which is more like in taste to pepper than is *Panax*, and it is therefore called *Piperitis*, as he hath written in his 19. booke, 12. chap. *Panax* (saith he) hath the tast of pepper and *Siliquastrum*, for which cause it is called *Piperitis*. The later Herbarists do oftentimes call it *Piper Indianum*, or *Indicum*, sometimes *Piper Calicuthium*, or *Piper Hispanicum*: in English it is called Ginnie pepper, and Indian pepper: in the Germane tongue, **Indianischer Pfeffer**: in low Dutch, **Brestlie Peper**: in French, *Poiure d' Inde*, verie well knowne in the shops at Billingsgate by the name of Ginnie pepper, where it is vsually to be bought.

¶ *The Temperature.*

Ginnie pepper is extreame hot and drie euen in the fourth degree: that is to say, far hotter and drier then *Auicen* sheweth dogs ginger to be.

¶ *The Vertues.*

A Ginnie pepper hath the taste of pepper, but not the power or vertue, notwithstanding in Spaine and sundrie parts of the Indies they do vse to dresse their meate therewith, as we doe with Calecute pepper: but (saith my Authour) it hath in it a malicious qualitie, whereby it is an enemy to the liuer and other of the entrails. *Auicen* writeth that it killeth dogs.

B It is said to die or colour like Saffron; and being receiued in such sort as Saffron is vsually taken, it warmeth the stomacke, and helpeth greatly the digestion of meates.

C It dissolueth the swellings about the throat called the Kings Euill, as kernels and cold swellings; and taketh away spots and lentiles from the face, being applied thereto with honie.

Chap. 72. *Of horned Poppie.*

¶ *The Description.*

1 The yellow horned Poppie hath whitish leaues very much cut or jagged, somewhat like the leaues of garden Poppie, but rougher and more hairie. The stalks be long, round, and brittle. The floures be large and yellow, consisting of foure leaues; which being past, there come long huskes or cods, crooked like an horne or cornet, wherein is conteined small blacke seede. The roote is great, thicke, scalie, and rough, continuing long.

2 The second kinde of horned Poppie is much slenderer and lesser than the precedent, and hath leaues with like deepe cuts as Rocket hath, and something hairie. The stalks be very slender, brittle, and branched into diuers armes or wings, the floures small, made of foure little leaues, of a red colour, with a small strake of blacke toward the bottome, after which commeth the seed, inclosed in slender, long, crooked cods full of blackish seed. The root is small and single, and dieth euery yeere.

‡ 3 This is much like the last described, and according ro *Clusius*, rather a variety than difference. It is distinguished from the last mentioned by the smoothnes of the leaues, and the colour of the floures, which are of a pale yellowish red, both which accidents *Clusius* affirmes happen to the former, towards the later end of sommer. ‡

4 There is another sort of horned Poppie altogether lesser than the last described, hauing tenderer leaues, cut into fine little parcels: the floure is likewise lesser, of a blew purple colour like the double Violets

¶ The

1 *Papauer cornutum flore luteo.*
Yellow horned Poppie.

2 *Papauer cornutum flore rubro.*
Red horned Poppie.

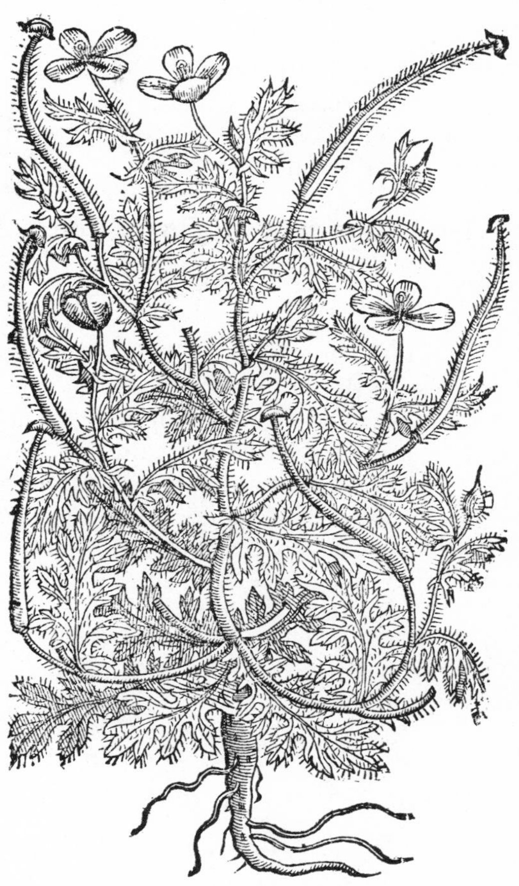

‡ 3 *Papauer corniculatum phœniceum glabrum.*
Red horned Poppie with ſmooth leaues.

4 *Papauer cornutum flore violaceo.*
Violet coloured horned Poppie.

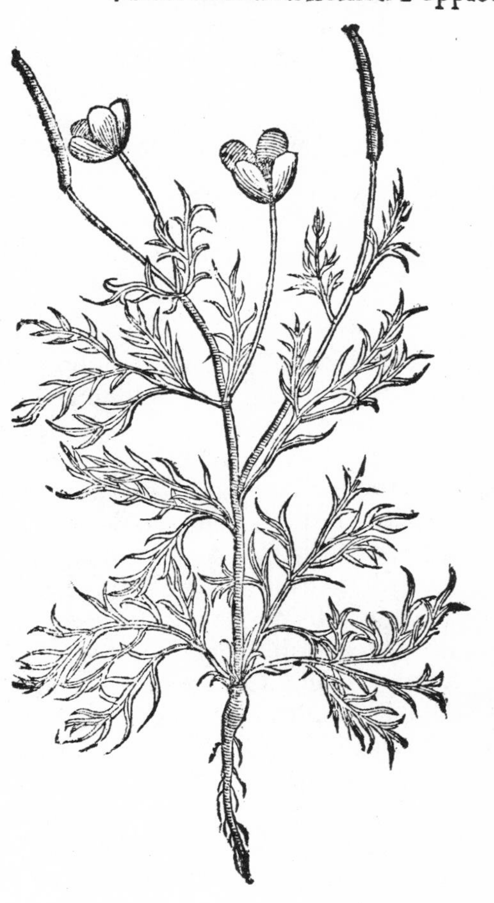

¶ *The Place.*

The yellow horned Poppie groweth vpon the ſands and banks of the ſea: I haue found it growing neere vnto Rie in Kent, in the Iles of Shepey and Thanet, at Lee in Eſſex, at Harwich, at Whiteſtable, and many other places alongſt the Engliſh coaſt.

The ſecond groweth not wilde in England. *Angelus Palea*, and *Bartholomæus ab Vrbe-veterum*, who haue commented vpon *Meſue*, write that they found this red horned Poppie in the kingdomes of Arragon and Caſtile in Spaine, and the fields neere vnto common paths. They doe grow in my Garden very plentifully.

¶ *The Time.*

They floure from May to the end of Auguſt.

¶ *The Names.*

Moſt Writers haue taken horned Poppie, eſpecially that with red floures to be *Glaucium*: neither is this their opinion altogether vnprobable; for as *Dioſcorides* ſaith, *Glaucium* hath leaues like thoſe of horned Poppey, but λιπαρώτερα, that is to ſay fatter, χαμαίζηλα, low, or lying on the ground, of a ſtrong ſmell and of a bitter taſte; the iuice alſo is much like in colour to Saffron. Now *Lobel* and *Pena* witneſſe, that this horned Poppie hath the ſame kinde of iuice, as my ſelfe likewiſe can teſtifie. *Dioſcorides* ſaith that *Glaucium* groweth about Hierapolis, a citie in Syria; but what hindereth that it ſhould not bee found alſo ſomewhere elſe? Theſe things ſhew it hath a great affinity with *Glaucium*, if it be not the true and legitimate *Glaucium* of *Dioſcorides*. Howbeit the firſt is the *Mecon Ceratites*, or *Papauer corniculatum* of the Antients, by the common conſent of all late Writers: in Engliſh, Sea Poppie and Horned Poppie: in Dutch, **Geelheul** and **Hozne Heule**: in the Germane Tongue, **Gelbomag**: in French, *Pauot Cornu*: in Spaniſh, *Dormidera marina*.

¶ *The Nature.*

Horned Poppies are hot and drie in the third degree.

¶ *The Vertues.*

A The root of horned Poppie boiled in water vnto the conſumption of the one halfe, and drunke, prouoketh vrine, and openeth the ſtopping of the liuer.

B The ſeed taken in the quantitie of a ſpoonefull looſeth the belly gently.

C The iuice mixed with meale and honie, mundifieth old rotten and filthie vlcers.

D The leaues and floures put into vnguents or ſalues appropriate for greene wounds, digeſt them that is, bring them to white matter, with perfect quitture or ſanies.

† The figure that formerly was in the fourth place of this chap. vnder the title of *Papauer cornutum luteum minus*, was of a Bindeweed called by *Cluſius*, *Conuolulus, fol. Althea*. You ſhall finde it hereafter in the due place. The Deſcription as far as I can iudge was of the *Cuminum corniculatum* which was pag. 909.

Chap. 73. *Of Garden Poppies.*

¶ *The Deſcription.*

1 The leaues of white Poppie are long, broad, ſmooth, longer than the leaues of Lettuce, whiter, and cut in the edges: the ſtem or ſtalke is ſtraight and brittle, oftentimes a yard and a halfe high: on the top whereof grow white floures, in which at the very beginning appeareth a ſmall head, accompanied with a number of threds or chiues, which being full growne is round, and yet ſomething long withall, and hath a couer or crownet vpon the top; it is with many filmes or thin skins diuided into coffers or ſeuerall partitions, in which is contained abundance of ſmall round and whitiſh ſeed. The root groweth deepe, and is of no eſtimation nor continuance.

2 Like vnto this is the blacke garden Poppie, ſauing that the floures are not ſo white and ſhining, but vſually red, or at leaſt ſpotted or ſtraked with ſome lines of purple. The leaues are greater, more iagged, and ſharper pointed. The ſeed is likewiſe blacker, which maketh the difference.

‡ 3 There is alſo another garden Poppie whoſe leaues are much more ſinuated, or creſted, and the floure alſo is all iagged or finely cut about the edges, and of this ſort there is alſo both blacke and white. The floures of the blacke are red, and the ſeed blacke; and the other hath both the floures and ſeed white.

4 There are diuers varieties of double Poppies of both theſe kindes, and their colours are commonly either white, red, darke purple, ſcarlet, or mixt of ſome of theſe. They differ from the former onely in the doubleneſſe of their floures.

1 Papauer

1 *Papauer satiuum album.*
White garden Poppie.

2 *Papauer satiuum nigrum.*
Blacke Garden Poppie.

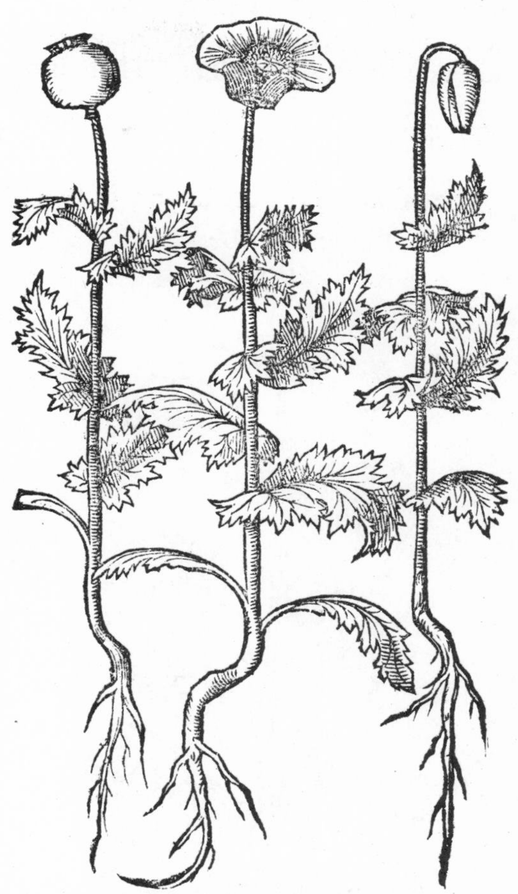

‡ 3 *Papauer fimbriatum album.*
White iagged Poppie.

4 *Papauer flo. multipl. albo & nigro.*
The double white and blacke Poppie.

5 There

5 There is also another kinde of Poppie which oft times is found wilde, the stalkes, leaues floures, and heads are like, but lesse than those of the precedent: the floures are of an ouerworn blewish purple color; after which follow heads short and round, which vnder their couer or crownet haue little holes by which the seed may fall out; contrarie to the heads of the former, which are close and open not of themselues. There is also a double one of this kinde. ‡

¶ *The Place.*

These kinde of Poppies are sowne in gardens, & do afterward come of the fallings of their seed.

¶ *The Time.*

They floure most commonly in Iune. The seed is perfected in Iuly and August.

5 *Papauer syluestre.*
Wilde Poppie.

¶ *The Names.*

Poppie is called of the Græcians μήκων: of the Latines, *Papauer*: the shops keepe the Latine name: it is called in high Dutch, **Magsamen**: in low Dutch, **Huel** and **Mancop**: in English, Poppie & Cheesebowls: in French, *Pauot*, and *Oliette*, by the Wallons.

The garden Poppie which hath blacke seeds, is surnamed of *Dioscorides* ἄγριον, or wilde, and is as hee saith called ῥοιάς, because *Opium* flowes from it: of *Pliny* and of the Latines, *Papauer nigrum*. whereof there be many variable colours, and of great beautie, although of euill smell, whereupon our gentlewomen doe call it Ione Siluer pin.

¶ *The Temperature.*

All the Poppies are cold, as *Galen* testifieth in his booke of the Faculties of simple medicines.

¶ *The Vertues.*

A This seed, as *Galen* saith in his booke of the Faculties of nourishments, is good to season bread with; but the white is better than the black. He also addeth, that the same is cold and causeth sleepe, and yeeldeth no commendable nourishment to the body; it is often vsed in comfits, serued at the table with other iunketting dishes.

B The oile which is pressed out of it is pleasant and delightfull to be eaten, and is taken with bread or any other waies in meat, without any sence of cooling.

C A greater force is in the knobs or heads, which doe specially preuaile to mooue sleepe, and to stay and represse distillations or rheums, and come neere in force to *Opium*, but more gentle. *Opium*, or the condensed iuice of Poppie heads is strongest of all: *Meconium* (which is the iuice of the heads and leaues) is weaker. Both of them any waies taken either inwardly, or outwardly applied to the head, prouoke sleepe. *Opium* somewhat too plentifully taken doth also bring death, as *Plinie* truely writeth.

D It mitigateth all kinde of paines: but it leaueth behinde it oftentimes a mischiefe worse than the disease it selfe, and that hard to be cured, as a dead palsie and such like.

E The vse of it, as *Galen* in his 11. booke of medicines according to the places affected, saith, is so offensiue to the firme and solide parts of the body, as that they had need afterwards to be restored.

F So also collirries or eie medicines made with *Opium* haue beene hurtfull to many; insomuch that they haue weakned the eies and dulled the sight of those that haue vsed it: whatsoeuer is compounded of *Opium* to mittigate the extreeme paines of the eares bringeth hardnesse of hearing. Wherefore all those medicines and compounds are to bee shunned that are to be made of *Opium*, and are not to be vsed but in extreme necessitie; and that it is, when no other mitigater or asswager of paine doth any thing preuaile, as *Galen* in his third booke of Medicines, according to the places affected, doth euidently declare.

G The leaues of poppie boiled in water with a little sugar and drunke, causeth sleep: or if it be boiled without sugar, and the head, feet, and temples bathed therewith, it doth effect the same.

H The heads of Poppie boiled in water with sugar to a sirrup causeth sleepe, and is good against rheumes and catarrhes that distill & fal downe from the brain into the lungs, & easeth the cough.

I The greene knops of Poppie stamped with barley meale, and a little barrowes grease, helpeth S. Anthonies fire, called *Ignis sacer*.

The

The leaues, knops and ſeed ſtamped with vineger, womans milke, and ſaffron, cureth an *Eryſipe-* K
las, (another kinde of S. Anthonies fire, and eaſeth the gout mightily, and put in the fundament as a cliſter cauſeth ſleepe.

The ſeed of black Poppy drunke in wine ſtoppeth the flux of the belly, and the ouermuch flow- L
ing of womens ſickneſſe.

A Caudle made of the ſeeds of white poppy, or made into Almond milk, and ſo giuen cauſeth M
ſleepe.

† It is manifeſt that this wilde Poppy (which I haue deſcribed in the fifth place) is that of N
which the compoſition *Diacodium* is to be made, as *Galen* hath at large treated in his ſeuenth booke of Medicines, according to the places affected. *Crito* alſo, and after him *Themiſon* and *Democrates* do appoint ἄγριον, or the wilde Poppy, to be in the ſame compoſition; and euen that ſame *Democritus* addeth, that it ſhould be that which is not ſowen: and ſuch an one is this, which groweth without ſowing. *Dod.*

CHAP. 74. *Of Corne-Roſe, or wilde Poppy.*

1 *Papauer Rhœas.*
Red Poppy, or Corne-roſe.

‡ 4 *Papauer ſpinoſum.*
Prickly Poppy.

¶ *The Deſcription.*

1 THe ſtalkes of red Poppy be blacke, tender, and brittle, ſomewhat hairy: the leaues are cut round about with deepe gaſhes like thoſe of Succory or wilde Rocket: the floures grow forth at the tops of the ſtalks, being of a beautifull and gallant red colour, with blackiſh threds compaſſing about the middle part of the head: which being fully growne, is leſſer than that of the garden Poppy: the ſeed is ſmall and blacke.

† 2 There is alſo a kinde hereof in all points agreeing with the former, ſauing that the floures of this are very double and beautifull, and therein only conſiſts the difference. †

‡ 3 There

‡ 3 There is a ſmall kinde of red Poppy growing commonly wilde together with the firſt deſcribed, which is leſſer in all parts, and the floures are of a fainter or ouerworne red, inclining ſomewhat to orange.

‡ 4 Beſides theſe there is another rare plant, which all men, and that very fitly, haue referred to the kindes of Poppy. This hath a ſlender long and fibrous root, from which ariſes a ſtalke ſome cubit high, diuided into ſundry branches, round, creſted, prickly, and full of a white pith. The leaues are diuided after the maner of horned poppy, ſmooth, with white veins & prickly edges: the floure is yellow, and conſiſts of foure or fiue leaues; after which ſucceeds a longiſh head, being either foure, fiue, or ſix cornered, hauing many yellow threds incompaſſing it: the head whileſt it is tender is reddiſh at the top, but being ripe it is blacke, and it is ſet with many and ſtiffe pricks. The ſeed is round, blacke, and pointed, being ſix times as big as that of the ordinary Poppy. ‡

¶ *The Place.*

They grow in earable grounds, among wheat, ſpelt, rie, barley, otes, and other graine, and in the borders of fields. ‡ The double red, and prickly Poppy are not to be found in this kingdome, vnleſſe in the gardens of ſome prime herbariſts. ‡

¶ *The Time.*

The fields are garniſhed and ouerſpred with theſe wilde poppies in Iune and Auguſt.

¶ *The Names.*

† Wilde Poppy is called in Greeke of *Dioſcorides*, μήκων ῥοιὰς: in Latine, *Papauer erraticum: Gaza* according to the Greeke nameth it *Papauer fluidum*: as alſo *Lobel*, who cals it *Pap. Rhœas*, becauſe the floure thereof ſoone falleth away. Which name *Rhœas* may for the ſame cauſe be common, not onely to theſe, but alſo to the others, if it be ſo called of the ſpeedy falling of the floures: but if it be ſyrnamed *Rhœas* of the falling away of the ſeed (as it appeareth) then ſhall it be proper to that which is deſcribed in the fifth place in the foregoing chapter, out of whoſe heads the ſeed eaſily and quickly falls; as it doth alſo out of this, yet leſſe manifeſtly. They name it in French *Cocquelicot, Conſanons, Pauot ſauvage*: in Dutch, **Collen bloemen, Coren roſen**: in high Dutch, **Klapper Roſſen**: in Engliſh, Red Poppy, and Corne-roſe.

‡ 4 Some haue called this *Ficus infernalis*, from the Italian name *Figo del inferno*. But *Cluſius* and *Bauhine* haue termed it *Papauer ſpinoſum*: and the later of them would haue it (and that not without good reaſon) to be *Glaucium* of *Dioſcorides, lib. 3. cap. 100.* And I alſo probably coniecture it to be the *Hippomanes* of *Crateuas*, mentioned by the Greeke Scholiaſt of *Theocritus*, as I haue formerly briefely declared *Chap. 62.* ‡

¶ *The Nature.*

The facultie of the wilde poppies is like to that of the other poppies; that is to ſay cold, and cauſing ſleepe.

¶ *The Vertues.*

A Moſt men being led rather by falſe experiments than reaſon, commend the floures againſt the Pleuriſie, giuing to drinke as ſoone as the paine commeth, either the diſtilled water, or ſyrrup made by often infuſing the leaues. And yet many times it happeneth that the paine ceaſeth by that meanes, though hardly ſometimes, by reaſon that the ſpittle commeth vp hardly, and with more difficultie, eſpecially in thoſe that are weake, and haue not a ſtrong conſtitution of body. *Baptiſta Sardus* might be counted the Author of this error; who hath written, That moſt men haue giuen the floures of this poppy againſt the paine of the ſides, and that it is good againſt the ſpitting of bloud.

Chap. 75. *Of Baſtard wilde Poppy.*

¶ *The Deſcription.*

The firſt of theſe baſtard wilde Poppies hath ſlender weake ſtemmes a foot high, rough and hairy, ſet with leaues not vnlike to thoſe of Rocket, made of many ſmall leaues deeply cut or iagged about the edges. The floures grow at the top of the ſtalkes, of a red colour, with ſome ſmall blackneſſe toward the bottome. The ſeed is ſmall, contained in little round knobs. The ſeed is ſmall and threddy.

2 The ſecond is like the firſt, ſauing that the cods hereof be long, and the other more round, wherein the difference doth conſiſt.

¶ *The Place.*

Theſe plants do grow in the corne fields in Somerſetſhire, and by the hedges and high-wayes, as ye trauell from London to Bathe. *Lobel* found it growing in the next field vnto a village in Kent called

called Southfleet, my felfe being in his company, of purpofe to difcouer fome ftrange plants not hitherto written of.

‡ Mr. *Robert Lorkin* and I found both thefe growing in Chelfey fields, as alfo in thofe belonging to Hamerfmith: but the fhorter headed one is a floure of a more elegant colour, and not fo plentifull as the other. ‡

1 *Argemone capitulo torulo.*
Baftard wilde Poppy.

2 *Argemone capitulo longiore.*
Long codded wilde Poppy.

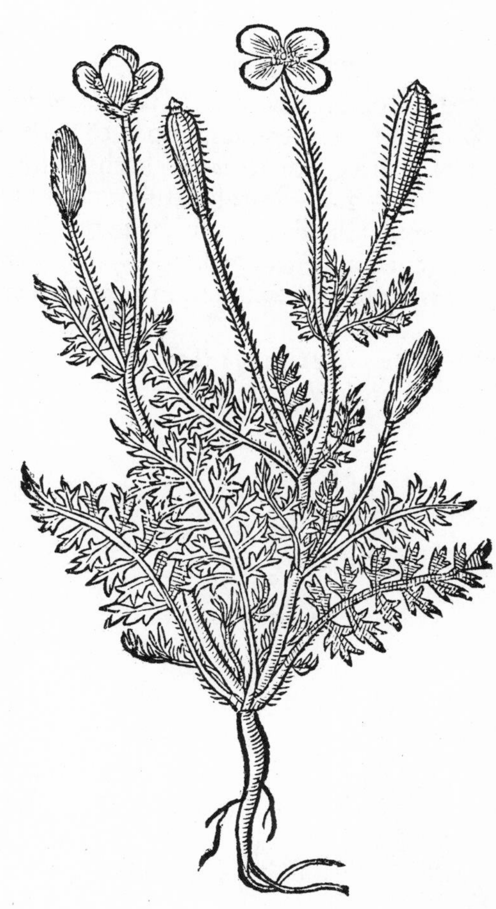

¶ *The Time.*

They floure in the beginning of Auguft, and their feed is ripe at the end thereof.

¶ *The Names.*

The baftard wilde Poppy is called in Greeke Ἀργεμώνη: in Latine, *Argemone, Argemonia, Concordia, Concordalis*, and *Herba liburnica*: of fome, *Pergalium, Arfela*, and *Sacrocolla Herba*: in Englifh, Windrofe, and baftard wilde poppy.

¶ *The Temperature.*

They are hot and dry in the third degree.

¶ *The Vertues.*

The leaues ftamped, and the iuyce dropped into the eyes eafeth the inflammation thereof; and cureth the difeafe of the eye called *Argema*, whereof it tooke his name: which difeafe when it hapneth on the blacke of the eye it appeares white; and contrariwife when it is in the white then it appeareth blacke of colour. A

The leaues ftamped and bound vnto the eyes or face that are blacke or blew by meanes of fome blow or ftripe, doth perfectly take it away. The dry herbe fteeped in warme water worketh the like effect. B

The leaues and roots ftamped, and the iuyce giuen in drinke, helpeth the wringings or gripings of the belly. The dry herbe infufed in warme water doth the fame effectually. C

The herbe ftamped, cureth any wound, vlcer, canker, or fiftula, being made vp into an vnguent or falue, with oile, wax, and a little turpentine. D

The iuyce taken in the weight of two drammes, with wine, mightily expelleth poyfon or venome. E

The

F The iuyce taketh away warts if they be rubbed therewith; and being taken in meate it helpes the milt or spleene if it be wasted.

Chap. 76.
Of Winde-floures.

¶ *The Kindes.*

The stocke or kindred of the *Anemones* or Winde-floures, especially in their varieties of colours, are without number, or at the least not sufficiently knowne vnto any one that hath written of plants. For *Dodonæus* hath set forth fiue sorts; *Lobel* eight; *Tabernamontanus* ten: My selfe haue in my garden twelue different sorts: and yet I do heare of diuers more differing very notably from any of these; which I haue briefely touched, though not figured, euery new yeare bringing with it new and strange kindes; and euery countrey his peculiar plants of this sort, which are sent vnto vs from far countries, in hope to receiue from vs such as our countrey yeeldeth.

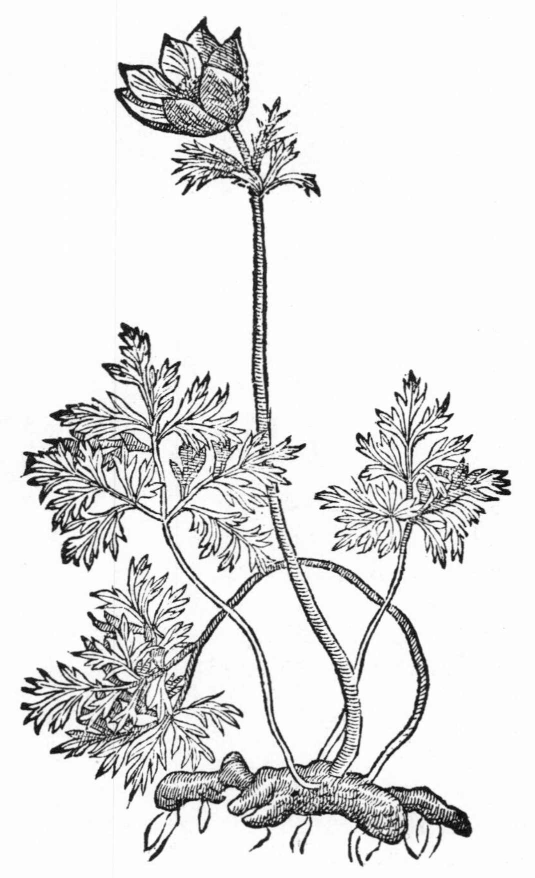

1 *Anemone tuberosa radice.*
Purple Winde-floure.

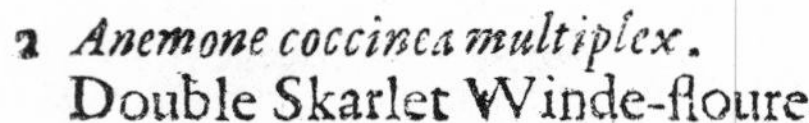

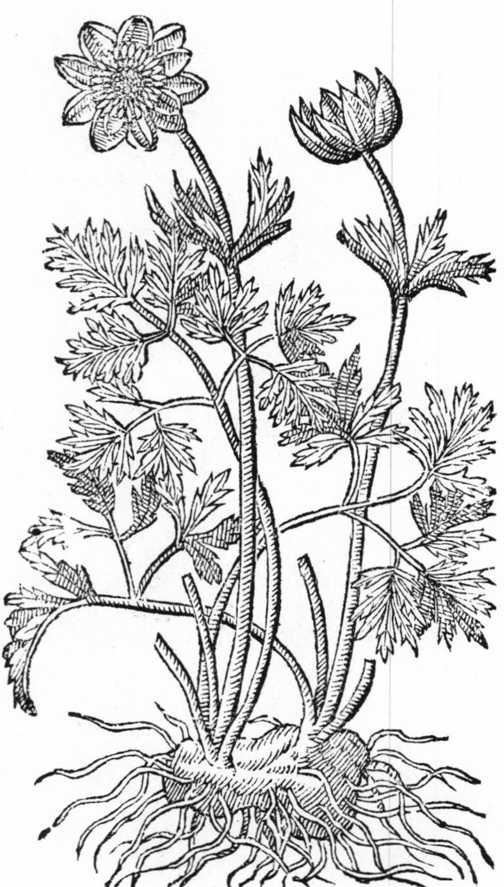

2 *Anemone coccinea multiplex.*
Double Skarlet Winde-floure.

¶ *The Description.*

1 The first kinde of *Anemone* or Winde-floure hath small leaues very much snipt or iagged almost like vnto Camomile, or Adonis floure: among which riseth vp a stalke bare or naked almost vnto the top; at which place is set two or three leaues like the other: and at the top of the stalke commeth forth a faire and beautifull floure compact of seuen leaues, and sometimes eight, of a violet colour tending to purple. It is impossible to describe the colour in his full perfection, considering the variable mixtures. The root is tuberous or knobby, and very brittle.

The

3 *Anemone maxima Chalcedonica polyanthos.*
The great double Winde-floure of Bithynia.

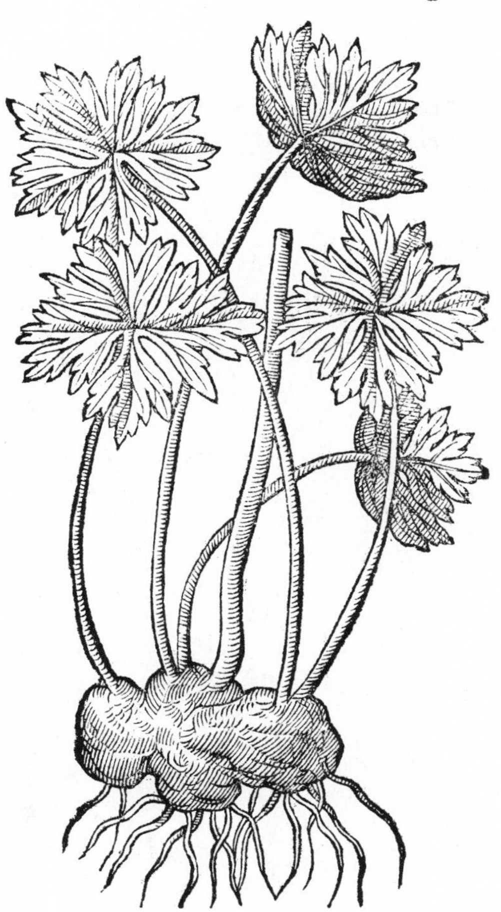

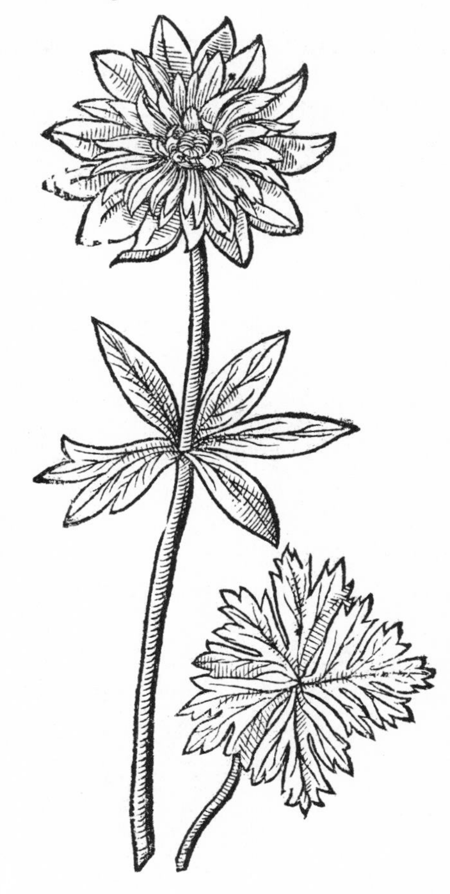

4 *Anemone Chalcedonica simplici flore.*
The single Winde-floure of Bithynia.

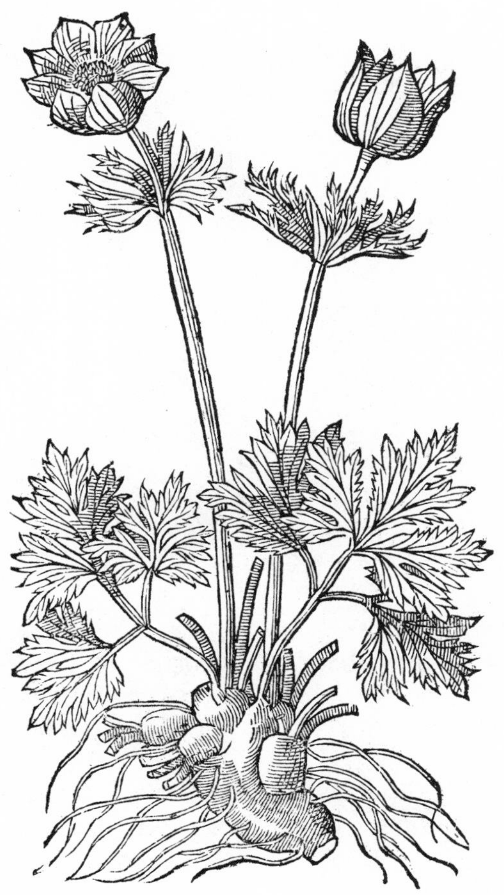

5 *Anemone Bulbocastani radice.*
Chesnut Winde-floure.

2 The second kind of *Anemone* hath leaues like to the precedent, insomuch that it is hard to distinguish the one from the other but by the floures onely: for those of this plant are of a most bright and faire skarlet colour, and as double as the Marigold; and the other not so. The root is knobby and very brittle, as is the former.

3 The great *Anemone* hath double floures, vsually called the *Anemone* of Chalcedon (which is a city in Bithynia) and great broad leaues deeply cut in the edges, not vnlike to those of the field Crow-foot, of an ouerworne greene colour: amongst which riseth vp a naked bare stalke almost vnto the top, where there stand two or three leaues in shape like the others, but lesser; sometimes changed into reddish stripes, confusedly mixed here and there in the said leaues. On the top of the stalke standeth a most gallant floure very double, of a perfect red colour, the which is sometimes striped amongst the red with a little line or two of yellow in the middle; from which middle commeth forth many blackish thrums. The seed is not to be found that I could euer obserue, but is carried away with the winde. The root is thicke and knobby.

4 The fourth agreeth with the first kind of *Anemone*, in roots, leaues, stalks, and shape of floures, differing in that, that this plant bringeth forth faire single red floures, and the other of a violet colour, as aforesaid.

5 The fifth sort of *Anemone* hath many small iagged leaues like those of Coriander, proceeding from a knobby root resembling the root of *Bulbocastanum* or earth Chesnut. The stalke rises vp amongst the leaues of two hands high, bearing at the top a single floure, consisting of a pale or border of little purple leaues, somtimes red, and often of a white colour set about a blackish pointall, thrummed ouer with many small blackish haires.

6 *Anemone latifolia Clusij.*
Broad leaued Winde-floure.

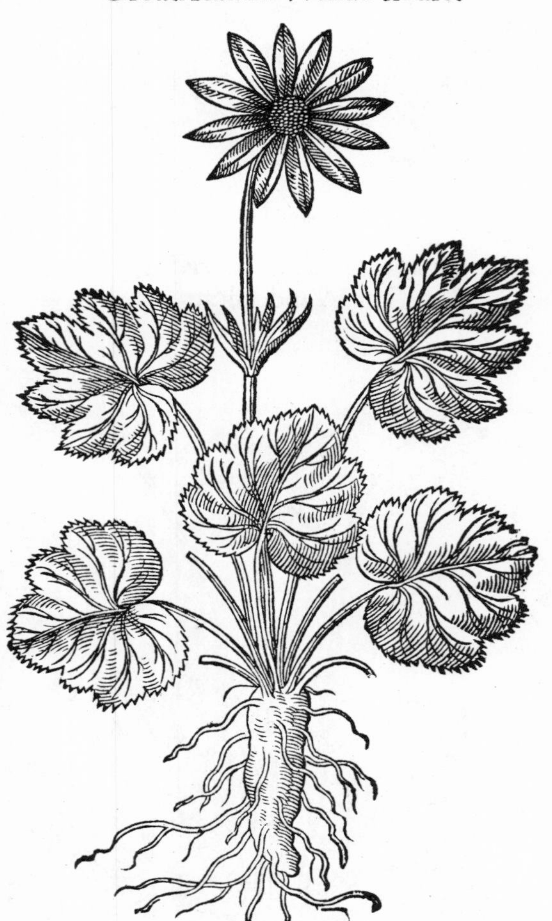

‡ 7 *Anemone latifolia duploflauo flore.*
The double yellow wind-floure.

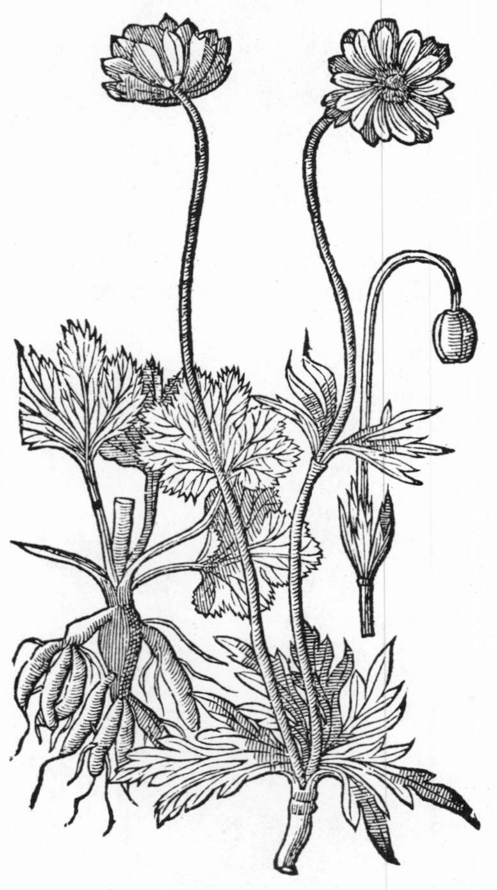

6 The sixt hath very broad leaues in respect of all the rest of the *Anemones*, not vnlike to those of the common Mallow, but greene on the vpper part, and tending to rednesse vnderneath, like the leaues of Sow-bread. The stalke is like that of the last described, on the top whereof growes a faire yellow star-floure, with a head ingirt with yellow thrums. The root (saith my Author) is a finger long, thicke and knobby.

‡ 7 There is also another whose lower leaues resemble those of the last described, yet those which grow next aboue them are more diuided or cut in: amongst these leaues riseth vp a stalke

8 *Anemone Geranifolia.*
Storkes bill Winde-floure.

9 *Anemone Matthioli.*
Matthiolus white Winde-floure.

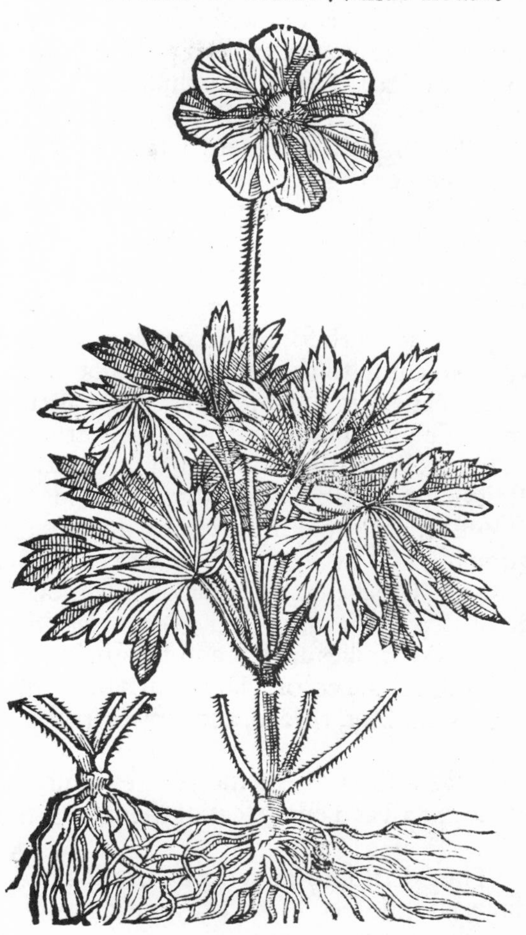

10 *Anemone trifolia.*
Three leaued Winde-floure.

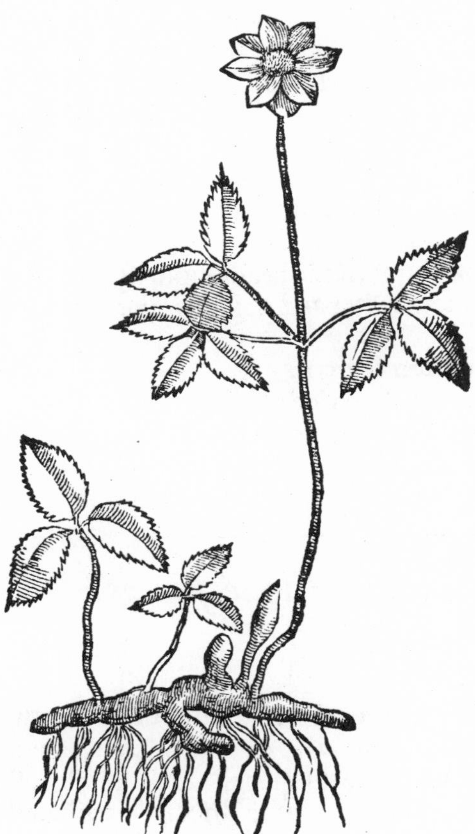

11 *Anemone Papaueracea.*
Poppy Winde-floure.

ſome foot high. the top whereof is adorned with a floure conſiſting of two ranks of leaues, whereof thoſe on the outſide are larger, rounder pointed, and ſometimes ſnipt in a little; the reſt are narrower and ſharper pointed: the colour of theſe leaues is yellow, deeper on the inſide, and on the outſide there are ſome ſmall purple veines running alongſt theſe leaues of the floure. The root is ſome two inches long, the thickeneſſe of ones little finger, with ſome tuberous knobs hanging thereat ‡

8 The eighth hath many large leaues deeply cut or iagged, in ſhape like thoſe of the Storks bil or Pinke-needle; among which riſeth vp a naked ſtalke, ſet about toward the top with the like leaues, but ſmaller and more finely cut, bearing at the top of the ſtalke a ſingle floure conſiſting of many ſmall blew leaues, which do change ſometimes into purple, and oftentimes into white, ſet about a blackiſh pointall, with ſome ſmall threds like vnto a pale or border. The root is thick and knobby.

9 The ninth ſort of Anemone hath leaues like vnto the garden Crow-foot: the ſtalke riſeth vp from amongſt the leaues, of a foot high, bearing at the top faire white floures made of fiue ſmall leaues; in the middle whereof are many little yellow chiues or threds. The root is made of many ſlender threds or ſtrings, contrarie to all the reſt of the Winde-floures.

10 The tenth ſort of Anemone hath many leaues like vnto the common medow Trefoile, ſleightly ſnipt about the edges like a ſaw: on the top of the ſlender ſtalkes ſtandeth a ſingle white floure tending to purple, conſiſting of eight ſmall leaues, reſembling in ſhape the floures of common field Crow-foot. The root is knobby, with certaine ſtrings faſtned thereto.

11 The eleuenth kinde of Anemone hath many iagged leaues cut euen to the middle rib, reſembling the leaues of *Geranium Columbinum*, or Doues foot. The leaues that do embrace the tender weake ſtalkes are flat and ſleightly cut: the floures grow at the top of the ſtalkes, of a bright ſhining purple colour, ſet about a blackiſh pointall, with ſmall thrums or chiues like a pale. The root is knobby, thicke, and very brittle, as are moſt of thoſe of the Anemones.

¶ *The Place.*

All the ſorts of Anemones are ſtrangers, and not found growing wilde in England; notwithſtanding all and euery ſort of them do grow in my garden very plentifully.

¶ *The Time.*

They do floure from the beginning of Ianuarie to the end of Aprill, at what time the floures do fade, and the ſeed flieth away with the winde, if there be any ſeed at all; the which I could neuer as yet obſerue.

¶ *The Names.*

Anemone, or Winde floure is ſo called, ἀπὸ τῶ ἀνέμου; that is to ſay, of the winde; for the floure doth neuer open it ſelfe but when the winde doth blow, as *Pliny* writeth: wherupon alſo it is named of diuers *Herba venti*: in Engliſh, Winde-floure.

Thoſe with double floures are called in the Turky tongue *Giul*, and *Gul Catamer*: and thoſe with ſmall iagged leaues and double floures are called *Lalé benzede*, and *Galipoli lalé*. They do call thoſe with ſmall iagged leaues and ſingle floures *Biniẓate & binizade*, and *Biniẓante*.

¶ *The Temperature.*

All the kindes of Anemones are ſharpe, biting the tongue, and of a binding qualitie.

¶ *The Vertues.*

A The leaues ſtamped, and the iuyce ſniffed vp into the noſe purgeth the head mightily.

B The root champed or chewed procureth ſpitting, and cauſeth water and flegme to run forth out of the mouth, as Pellitorie of Spaine doth.

C It profiteth in collyries for the eyes, to ceaſe the inflammation thereof.

D The iuyce mundifieth and clenſeth maligne, virulent, and corroſiue vlcers.

E The leaues and ſtalkes boyled and eaten of Nurſes cauſe them to haue much milke: it prouoketh the termes, and eaſeth the leproſie, being bathed therewith.

‡ Chap. 77. *Of diuers other Anemones, or Winde-floures.*

¶ *The Kindes.*

‡ These floures which are in ſuch eſteeme for their beauty may well be diuided into two ſorts, that is, the *Latifolia*, or broad leaued, and the *Tenuifolia*, or narrow leaued: now of each of theſe ſorts there are infinite varieties, which conſiſt in the ſingleneſſe and doubleneſſe of the floures, and in their diuerſitie of colours; which would aske a large diſcourſe to handle exactly. Wherefore I onely intend (beſides thoſe ſet downe by our Authour) to giue you the figures

figures of some few others, with their description, briefly taken out of the Workes of the learned and diligent Herbarist *Carolus Clusius*; where such as desire further discourse vpon this subiect may be aboundantly satisfied: and such as do not vnderstand Latine may finde as large satisfaction in the late Worke of Mr. *Iohn Parkinson*; whereas they shall not onely haue their historie at large, but also learne the way to raise them of seed, which hath been a thing not long knowne (except to some few;) and thence hath risen this great varietie of these floures, wherewith some gardens so much abound.

¶ *The Description.*

1 THe root of this is like to that of the great double red *Anemone* described in the third place of the precedent chapter; and the leaues also are like, but lesser and deeper coloured. The stalke growes some foot high, slender and greene, at the top whereof groweth a single floure, consisting of eight leaues of a bright shining skarlet colour on the inside, with a paler coloured ring incompassing a hairy head set about with purple thrums: the outside of the floure is hairy or downie. This is *Anem. latifol. simpl. flo.* 16. of *Clusius*.

‡ 1 *Anemone latifolia flore coccineo.*
The broad leaued skarlet Anemone.

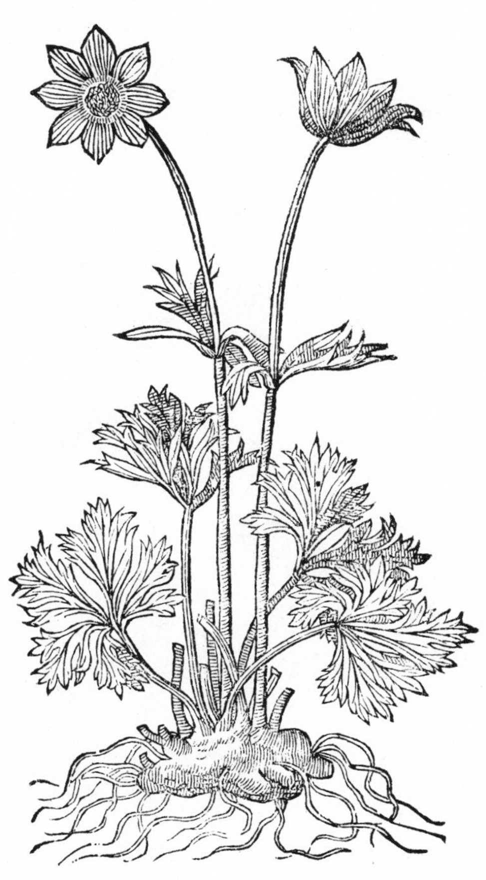

‡ 2 *Anemone latifolia flore magno coccineo.*
The skarlet Anemone with the large floure.

2 This in shape of roots & leaues is like the former, but the leaues are blacker, and more shining on their vpper sides: the stalke also is like to others of this kinde, and at the top carrieth a large floure consisting of eight broad leaues, being on the inside of a bright skarlet colour, without any circle; and the thrums that ingirt the hairy head are of a sanguine colour. This head (as in others of this kindred) growes larger after the falling of the floure, and at length turnes into a downie substance, wherein a smooth blacke seed is inclosed like as in other Anemones; which sowen as soone as it is ripe vsually comes vp before winter. This is *Anem. latifol. simpl. flore* 17. of *Clusius*.

3 This differs not from the former but in floures, which are of an orange-tawny colour, like that of Corne-rose, or red Poppy; and the bottomes of the leaues of the floures are of a paler colour, which make a ring or circle about the hairy head. This is the eighteenth of *Clusius*.

Besides these varieties here mentioned, there are many others, which in the colour of the leaues of the floure, or the nailes which make a circle at the bottome thereof, doe differ each from other. Now let vs come to the narrow leaued ones, which also differ little but in colour of their floures.

‡ 3 *Anemone latifolia ByZantina.*
The broad leaued Anemone of Conſtantinople.

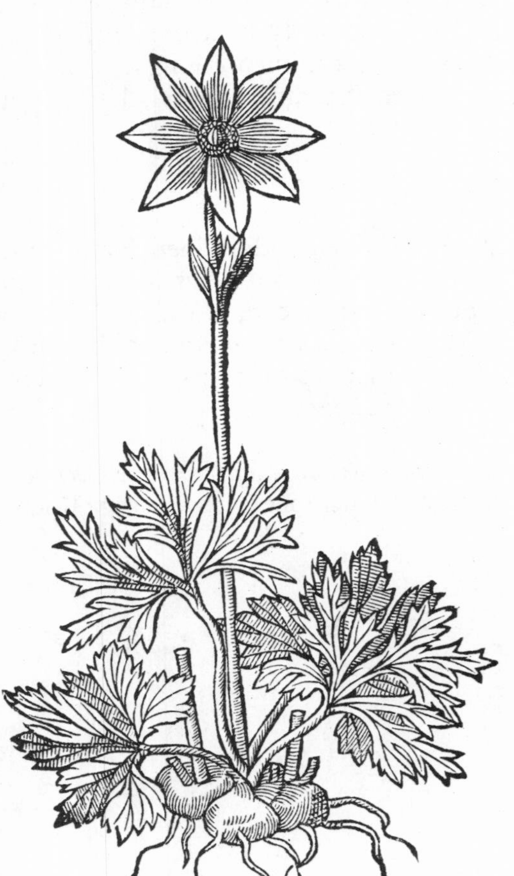

‡ 4 *Anemone tenuifolia flore amplo ſanguineo.*
Small leaued Anemone with the ſanguine floure.

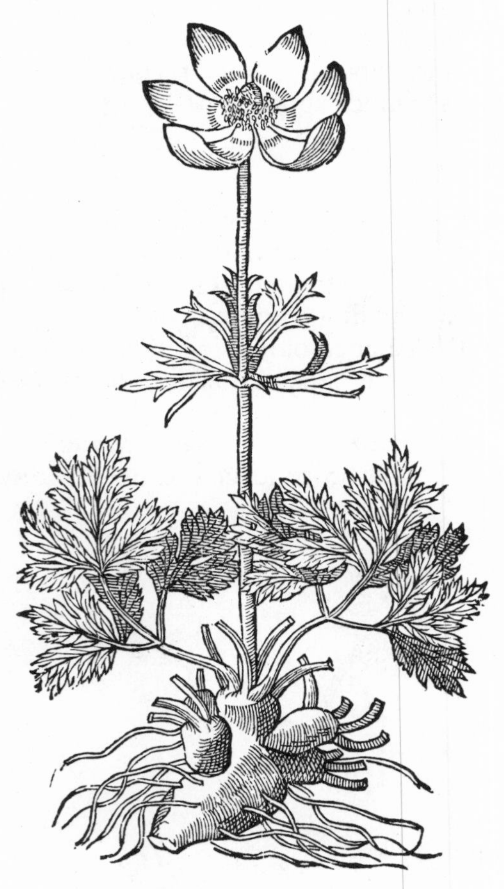

‡ 5 *Anemone tenuifolia flore coccineo.*
The ſmall leaued skarlet Anemone.

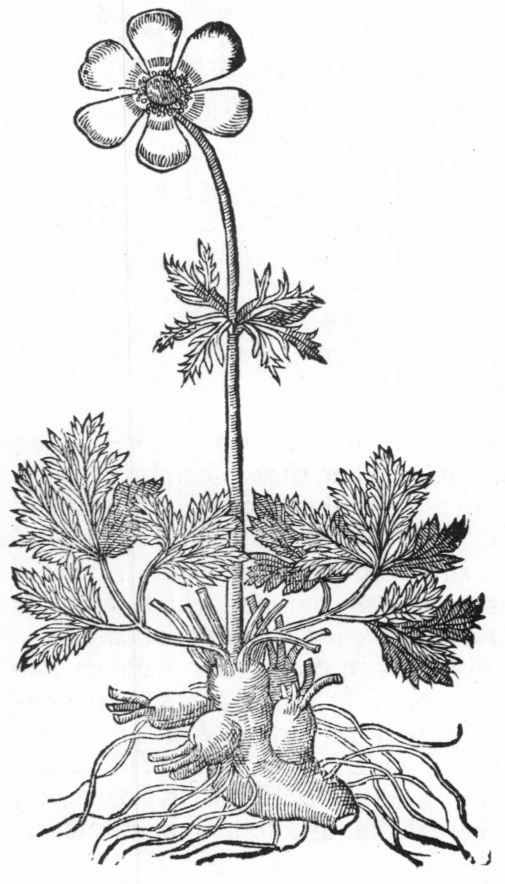

‡ 6 *Anemone tenuifol. flo. dilute purpureo.*
The light purple ſmall leaued Anemone.

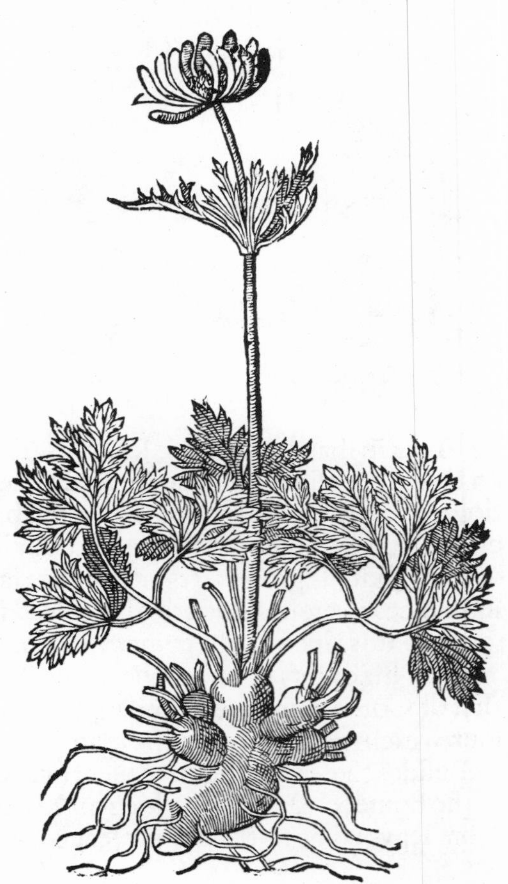

‡ 7 *Anemone tenuifol. flo. exalbido.*
The whitiſh ſmall leaued Anemone.

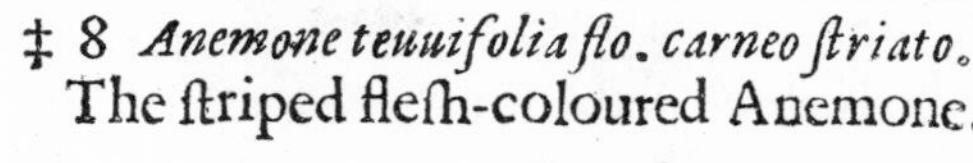

‡ 8 *Anemone tenuifolia flo. carneo ſtriato.*
The ſtriped fleſh-coloured Anemone.

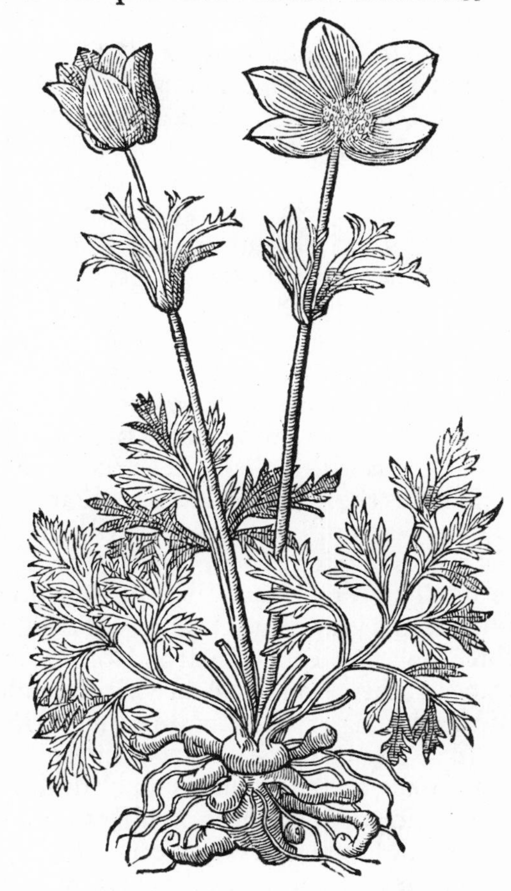

‡ 9 *Anemone tenuifol. flo. pleno coccin.*
The ſmall leaued double crimſon Anemone.

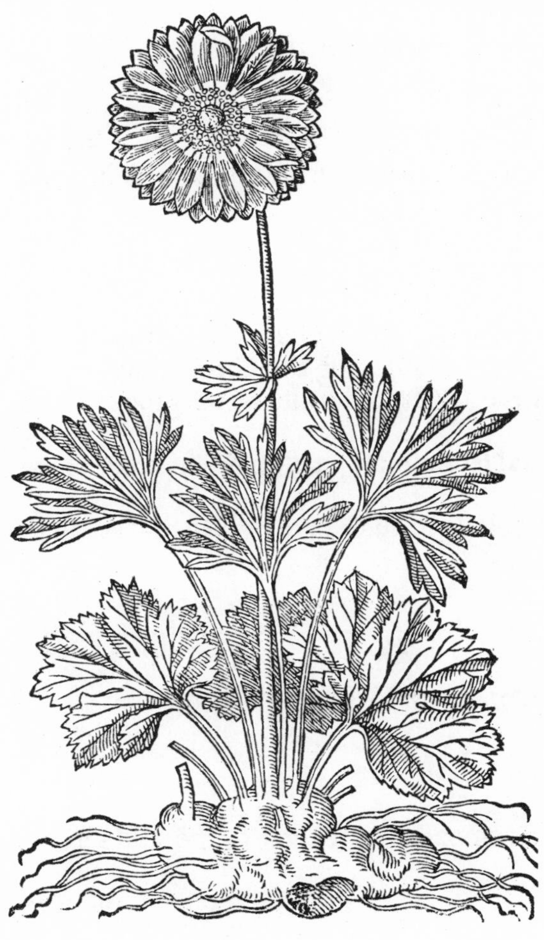

‡ 10 *Anemone tenuifol. flo. pleno atropurpuraſcente.*
The double darke purple Anemone.

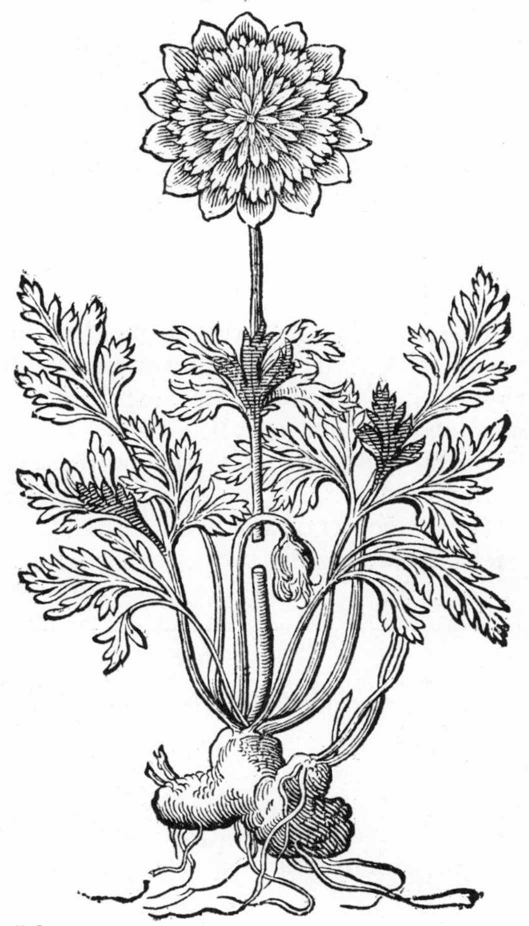

4 The root of this is knotty and tuberous like thoſe of other Anemones, and the leaues are much diuided and cut in like to thoſe of the firſt deſcribed in the former Chapter: the ſtalke (which hath three or foure leaues ingirting it, as in all other Anemones) at the top ſuſtaineth a faire ſanguine floure conſiſting of ſix large leaues with great white nailes. The ſeeds are contained in downie heads like as thoſe of the former. This is *Anem. tenuifol. ſimpl. flo. 6.* of *Cluſius*.

5 This differs from the former in the floure, which conſiſts of ſix leaues made ſomwhat rounder than thoſe of the precedent: their colour is betweene a skarlet and ſanguine. And there is a varietie hereof alſo of a bricke colour. This is the eighth of *Cluſius*.

6 This differs from the reſt, in that the floure is compoſed of ſome fourteene or more leaues, and theſe of a light purple, or fleſh-colour. This is the ninth of *Cluſius*.

7 The floure of this is large, conſiſting of ſix leaues, being at the firſt of a whitiſh greene, and then tending to a fleſh colour, with their nailes greene on the outſide, and white within, and the threds in the middle of a fleſh colour. There is a leſſer of this kinde, with the floure of a fleſh colour, and white on the outſide, and wholly white within, with the nailes greeniſh. Theſe are the tenth and eleuenth of *Cluſius*.

8 This floure alſo conſiſts of ſix leaues of a fleſh colour, with whitiſh edges on the outſide; the inſide is whitiſh, with fleſh coloured veines running to the middeſt thereof.

Beſides theſe ſingle kindes there are diuers double both of the broad and narrow leaued Anemones, whereof I will only deſcribe and figure two, and refer you to the forementioned Authors for the reſt, which differ from theſe onely in colour.

9 This broad leaued double Anemonie hath roots, ſtalkes, and leaues like thoſe of the ſingle ones of this kinde, and at the top of the ſtalke there ſtands a faire large floure compoſed of two or three rankes of leaues, ſmall and long, being of a kinde of skarlet or orange-tawny colour; the bottomes of theſe leaues make a whitiſh circle, which giues a great beauty to the floure, and the downie head is ingirt with ſanguine threds tipt with blew. This is the *Pauo major* 1. of *Cluſius*.

10 This in ſhape of roots, leaues, and ſtalkes reſembles the formerly deſcribed narrow leaued Anemones, but the floure is much different from them; for it conſiſts firſt of diuers broad leaues, which incompaſſe a great number of ſmaller narrow leaues, which together make a very faire and beautifull floure: the outer leaues hereof are red, and the inner leaues of a purple Veluet colour.

Of this kinde there are diuers varieties, as the double white, crimſon, bluſh, purple, blew, carnation, roſe-coloured, &c.

¶ *The Place and Time.*

Theſe are onely to be found in gardens, and bring forth their floures in the Spring.

¶ *Their Names.*

I iudge it no waies pertinent to ſet downe more of the names than is already deliuered in their ſeuerall titles and deſcriptions.

¶ *Their Temper and Vertues.*

A Theſe are of a hot and biting facultie, and not (that I know of) at this day vſed in medicines, vnleſſe in ſome one or two ointments: yet they were of more vſe amongſt the Greeke Phyſitions, who much commend the iuyce of them for taking away the ſcares and ſcales which grow on the eyes; and by them are called ἕλκη, and λευκώματα.

B *Trallianus* alſo ſaith, That the floures beaten in oyle, and ſo anointed, cauſe haire to grow where it is deficient.

The vertues ſet downe in the former Chapter do alſo belong to theſe here treated of, as theſe here deliuered are alſo proper to them. ‡

Chap. 78. *Of wilde Anemones, or Winde-floures.*

¶ *The Kindes.*

Like as there be many and diuers ſorts of the garden Anemones, ſo are there of the wild kindes alſo, which do vary eſpecially in their floures.

¶ *The*

1 *Anemone nemorum lutea.*
Yellow wilde Winde floure.

2 *Anemone nemorum alba.*
White winde floure.

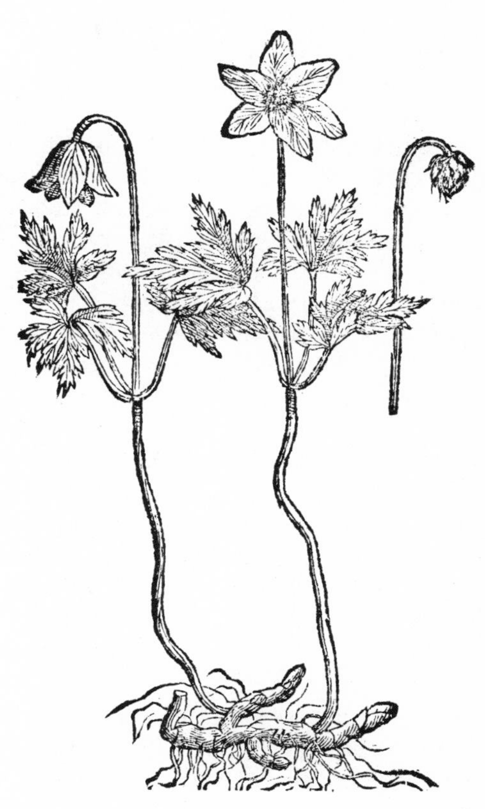

‡ 3 *Anemone nemorum flo. pleno albo.*
The double white wood Anemone.

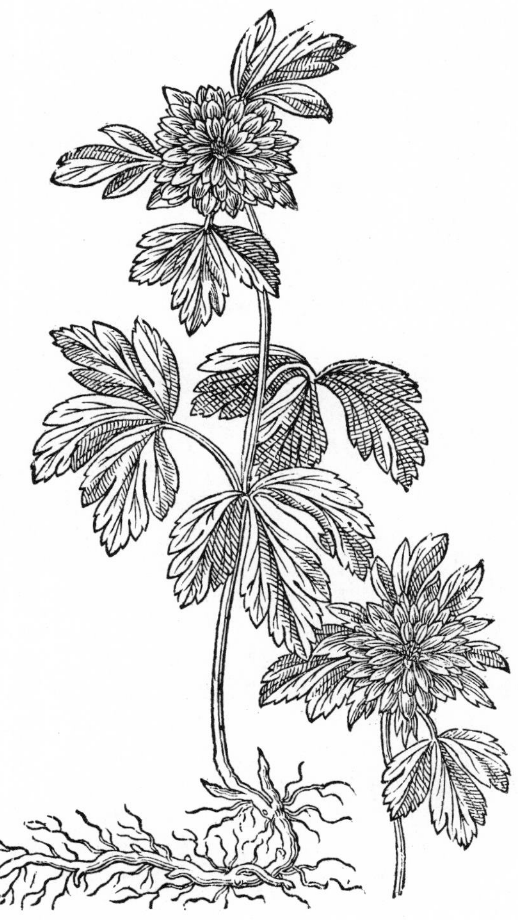

‡ 4 *Anemone nemorum flo. pleno purpuraſcente.*
The double purpliſh wood Anemone.

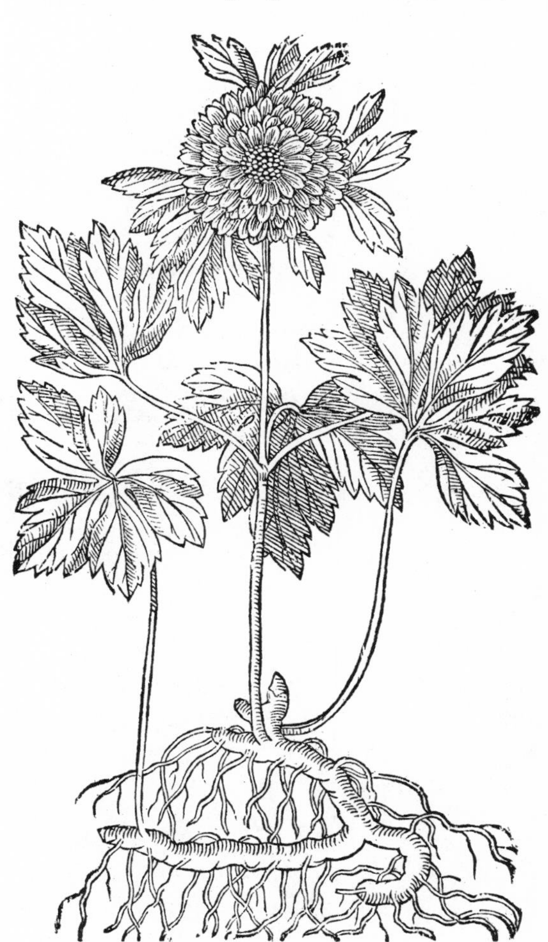

¶ *The description.*

1 THe first of these wilde *Anemones* hath iagged leaues deepely cut or indented, which do grow vpon the middle part of a weake and tender stalke: at the top whereof doth stand a prettie yellow floure made of six small leaues, and in the middle of the floure there is a little blackish pointell, and certaine slender chiues or threds. The root is small, somewhat knottie and very brittle.

2 The second hath iagged leaues, not vnlike to water Crowfoot or mountaine Crowfoot. The flower groweth at the top of the stalke not vnlike to the precedent in shape, sauing that this is of a milke white colour, the root is like the other.

‡ There is also of this single kinde two other varieties, the one with a purple floure, which wee may therefore call *Anemone nemorum purpurea*, the wilde purple Winde-floure. And the other with a Scarlet (or rather a Blush) coloured floure, which we may terme *Anemone nemorum coccinia*, The wilde Scarlet wind floure. These two differ not in other respects from the white wind floure. ‡

3 There is in some choice gardens one of this kinde with white floures very double, as is that of the Scarlet *Anemone*, and I had one of them giuen mee by a worshipfull Merchant of London, called Mr. *Iohn Franqueuille*, my very good friend.

‡ 4 This in roots and stalkes is like the last described wood *Anemones*, or winde floures. But this and the last mentioned double one haue leaues on two places of their stalks; whereas the single ones haue them but in one, and that is about the middle of the stalkes. The floure of this double one consists of some fortie or more little leaues, whereof the outermost are the biggest; the bottomes or nailes of these leaues are of a deepe purple, but the other parts of a lighter blush colour. ‡

¶ *The Place.*

All these wilde single *Anemones* grow in most woods and copses through England, except that with the yellow floure, which as yet I haue not seene: notwithstanding I haue one of the greater kindes which beareth yellow floures, whose figure is not expressed nor yet described, for that it doth very notably resemble those with single floures, but is of small moment, either in beautie of the floure, or otherwise. ‡ The double ones grow onely in some few gardens. ‡

¶ *The Time.*

They floure from the middest of Februarie vnto the end of Aprill, or the midst of May.

¶ *The Names.*

‡ The first of these by most Writers is referred to the *Ranunculi*, or Crowfeet; and *Lobel* cals it fitly *Ranunculus nemorosus luteus*: only *Dodonæus*, *Cæsalpinus*, and our Authour haue made it an *Anemone*.

2 This with the varieties also, by *Tragus*, *Fuchsius*, *Cordus*, *Gesner*, *Lobell*, and others, is made a *Ranunculus*: yet *Dodonæus*, *Cæsalpinus*, and our Authour haue referred it to the *Anemones*. *Clusius* thinkes this to be *Anemone*, Λειμώνια of *Theophrastus*.

3 *Clusius* calls this *Anemone Limonia*, or *Ranunculus syluarum flo. pleno albo*.

4 And he stiles this *Anem. limonia*, or *Ranunc. syl. flore pleno purpurascente*. ‡

¶ *The Temperature and Vertues.*

The faculties and temperature of these plants are referred to the garden sorts of *Anemones*.

Chap. 79. *Of Bastard Anemones, or Pasque floures.*

¶ *The Description.*

1 THe first of these Pasque floures hath many small leaues finely cut or iagged, like those of Carrots: among which rise vp naked stalkes, rough and hairie; whereupon doe grow beautifull floures bell fashion, of a bright delaied purple colour: in the bottome whereof groweth a tuft of yellow thrums, and in the middle of the thrums it thrusteth forth a small purple pointell: when the whole floure is past there succeedeth an head or knop compact of many gray hairy lockes, and in the solide parts of the knops lieth the seed flat and hoarie, euery seed hauing his owne small haire hanging at it. The root is thicke and knobby, of a finger long, running right downe, and therefore not like vnto those of the *Anemone*, which it doth in all other parts very notably resemble, and whereof no doubt this is a kinde.

2 There is no difference at all in the leaues, roots, or seedes, betweene this red Pasque floure and the precedent, nor in any other point, but in the colour of the floures: for whereas the other

are of a purple colour, these are of a bright red, which setteth forth the difference.

3 The white Passe floures hath many fine iagged leaues, closely couched or thrust together, which resemble an Holi-water sprinckle, agreeing with the others in rootes, seedes, and shape of floures, sauing that these are of a white colour, wherein chiefely consisteth the difference.

‡ 4 This also in shape of roots and leaues little differs from the precedent, but the floures are lesser, of a darker purple colour, and seldome open or shew themselues so much abroad as the other of the first described, to which in all other respects it is very like.

5 There is also another kinde with leaues lesse diuided, but in other parts like those already described, sauing that the floure is of a yellow colour something inclining to a red. ‡

1 *Pulsatilla vulgaris.*
Purple Passe floure.

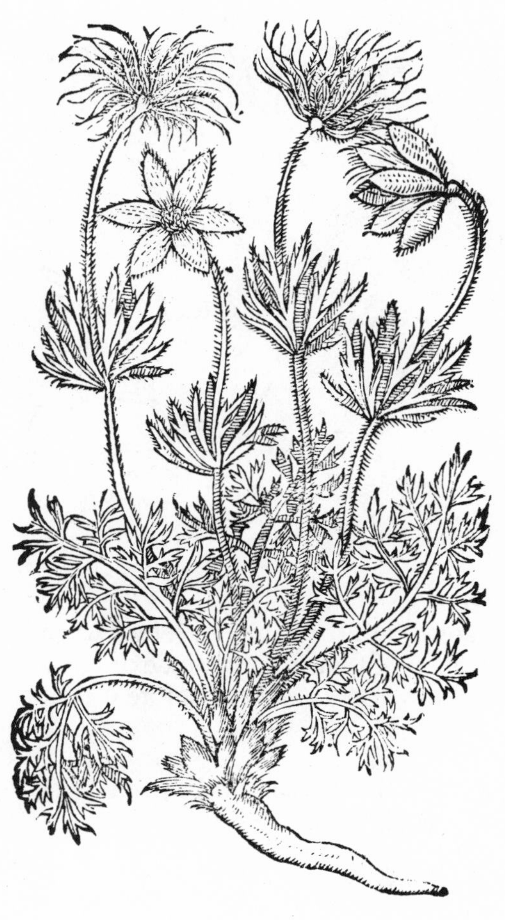

2 *Pulsatilla rubra.*
Red Passe floure.

¶ *The Place.*

Ruellius writeth, that the Passe floure groweth in France in vntoiled places: in Germanie they grow in rough and stonie places, and oftentimes on rockes.

Those with purple floures doe grow verie plentifully in the pasture or close belonging to the parsonage house of a small village six miles from Cambridge, called Hildersham: the Parsons name that liued at the impression hereof was Mr. *Fuller*, a very kind and louing man, and willing to shew vnto any man the said close, who desired the same.

¶ *The Time.*

They floure for the most part about Easter, which hath mooued mee to name it *Pasque Floure*, or Easter floure: and often they doe floure againe in September. ‡ The yellow kinde floures in May. ‡

¶ *The Names.*

† Passe floure is called commonly in Latine *Pulsatilla*: and of some, *Apium risus*, *& herba venti*. *Daleschampius* would haue it to be *Anemone Limonia* & *Samolus* of *Pliny*: in French, *Coquelourdes*: in Dutch, **Kneckenschell**: in English, Pasque floure, or Passe floure, and after the Latine name *Pulsatill*, or Flaw floure: in Cambridge-shire where they grow, they are named Couentrie bels.

¶ *The*

3 *Pulsatilla flore albo.*
White Passe floure.

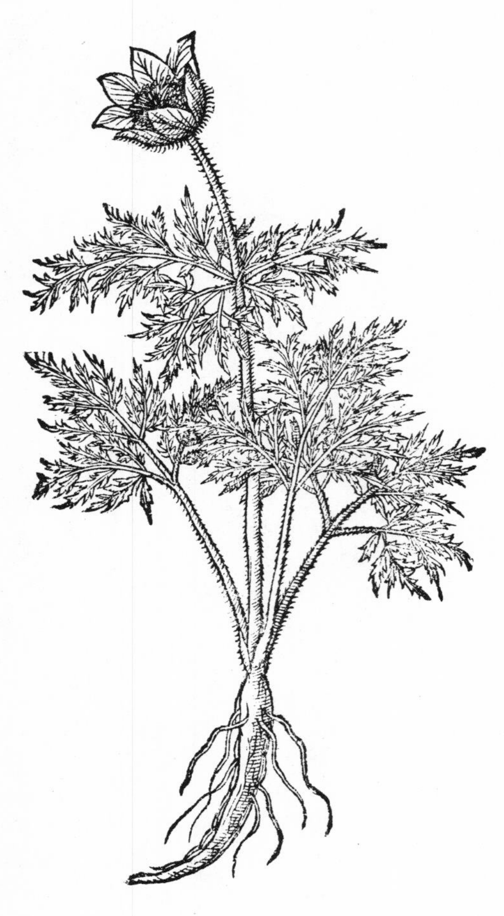

‡ 4 *Pulsatilla flore minore.*
The lesser purple Passe floure.

¶ *The Temperature.*

Passe floure doth extremely bite, and exulcerateth and eateth into the skinne if it be stamped and applied to any part of the body; whereupon it hath been taken of some to be a kinde of Crowfoot, and not without reason, for that it is not inferiour to the Crowfoots : and therefore it is hot and drie.

¶ *The Vertues.*

There is nothing extant in writing among Authours of any peculiar vertue, but they serue onely for the adorning of gardens and garlands, being floures of great beautie.

Chap. 80. *Of Adonis floure.*

¶ *The Description.*

1 THe first hath very many slender weake stalkes, trailing or leaning to the ground, set on euerie part with fine iagged leaues very deepely cut like those of Camomill, or rather those of May-weed : vpon which stalkes do grow small red floures, in shape like the field Crowfoot, with a blackish greene pointell in the middle, which being growne to maturitie turneth into a small greenish bunch of seeds, in shape like a little bunch of grapes. The root is small and threddie.

2 The second differeth not from the precedent in any one point, but in the colour of the floures, which are of a perfect yellow colour, wherein consisterh the difference.

¶ *The Place.*

The red floure of Adonis groweth wilde in the West parts of England among their corne, euen as May-weed doth in other parts, and is likewise an enemie to corne as May-weed is : from thence I brought the seed, and haue sowne it in my garden for the beautie of the floures sake. That with the yellow floure is a stranger in England.

¶ *The*

1 *Flos Adonis flore rubro.*
Adonis,with red floures.

¶ *The Time.*

They floure in the Sommer moneths, May, Iune,and Iuly,and ſometimes later.

¶ *The Names.*

Adonis floure is called in Latine *Flos Adonis*,and *Adonidis*: of the Dutchmen, Feldroſzlin: in Engliſh wee may call it Red Maythes, by which name it is called of them that dwell where it groweth naturally,and generally Red Camomill: in Greeke, ἠράνθεμον, & *Eranthemum*: our London women doe call it Roſe-a-rubie.

¶ *The Temperature.*

There hath not beene any that hath written of the Temperature hereof; notwithſtanding, ſo farre as the taſte thereof ſheweth, it is ſomething hot,but not much.

¶ *The Vertues.*

The ſeed of Adonis flower is thought to bee good againſt the ſtone: amongſt the Ancients it was not knowne to haue any other facultie: albeit experience hath of late taught vs,that the ſeed ſtamped, and the pouder giuen in wine,ale, or beere to drinke, doth wonderfully and with great effect helpe the collicke. A

CHAP. 81. *Of Dockes.*

¶ *The Kindes.*

D*Ioſcorides* ſetteth forth foure kindes of Dockes; wilde or ſharpe pointed Docke; Garden Docke; round leafed Docke; and the Soure Docke called Sorrell: beſides theſe the later Herbariſts haue added certaine other Dockes alſo, which I purpoſe to make mention of.

¶ *The Deſcription.*

1 THat which among the Latines ſignifieth to ſoften, eaſe, or purge the bellie, the ſame ſignification hath λαπάσσειν, among the Græcians: whereof *Lapathum* and ἀλάπαθον (as ſome do reade) tooke their names for herbes which are vſed in pottage and medicine, very well knowne to haue the power of cleanſing: of theſe there be many kindes and differences, great ſtore euery where growing, among whom is that which is now called ſharpe pointed Docke,or ſharpe leafed Docke. It groweth in moſt medowes and by running ſtreames,hauing long narrow leaues ſharpe and hard pointed: among the which commeth vp round hollow ſtalks of a browne colour, hauing ioynts like knees, garniſhed with ſuch like leaues, but ſmaller: at the end whereof grow many floures of a pale colour, one aboue another; and after them commeth a browniſh three ſquare ſeede, lapped in browne chaffie huskes like Patience. The roote is great, long, and yellowe within.

‡ There is a varietie of this with criſped or curled leaues whoſe figure was by our Authour giuen in the ſecond place in the following chapter, vnder the Title of *Hydrolapathum minus*. ‡

2 The ſecond kind of ſharpe pointed Docke is like the firſt,but much ſmaller,and doth beare his ſeed in rundles about his branches in chaffie huskes,like Sorrell,not ſo much in vſe as the former,called alſo ſharpe pointed Docke.

‡ 3 This in roots,ſtalkes,and ſeeds is like to the precedent; but the leaues are ſhorter,and rounder than thoſe of the firſt deſcribed,& therin conſiſts the chiefe difference betwixt this & it.‡

¶ *The Place.*

Theſe kindes of Docks do grow,as is before ſaid,in medowes and by riuers ſides.

¶ The

† 1 *Lapathum acutum.*
Sharpe pointed Docke.

2 *Lapathum acutum minimum.*
Small sharpe Docke.

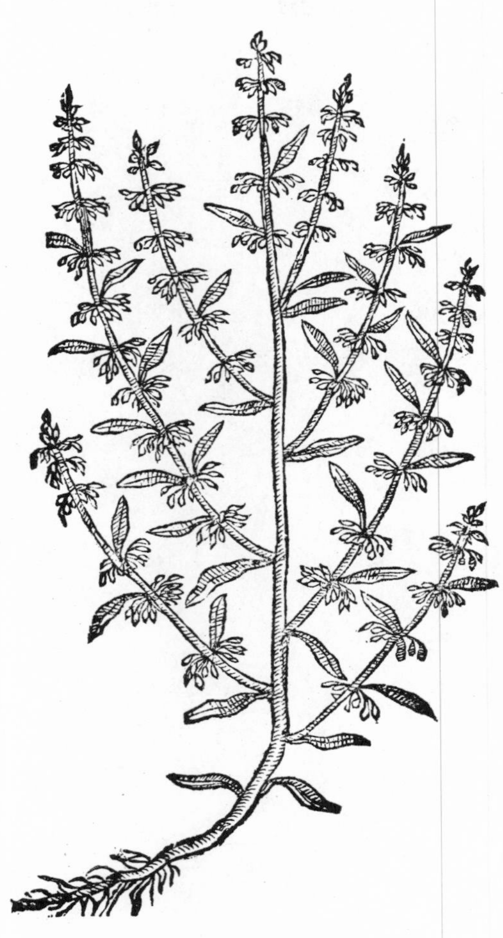

‡ 3 *Lapathum sylvestre fol. minus acuto.*
The roundish leaued wilde Docke.

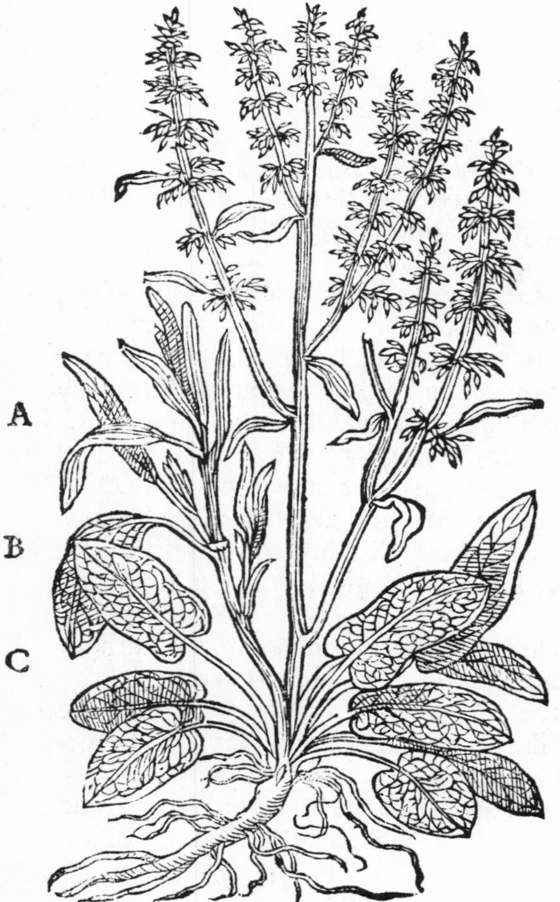

¶ *The Time.*

They floure in Iune and Iuly.

¶ *The Names.*

They are called in Latine *Lapathum acutum, Rumex, Lapatium,* & *Lapathium:* of some, *Oxylapathum:* in English, Docke, and sharpe pointed Docke, the greater and the lesser: of the Græcians, ὀξυλάπαθον: in high Dutch, **Mengelwurtz, Streiffwurtz:** in Italian, *Rombice:* in Spanish, *Romaza, Paradella,* in Low Dutch, **Patich** (which word is deriued of *Lapathum*) and also **Peerdick:** in French, *Pareille.*

‡ The third is *Lapathum folio retuso,* or *minus acuto* of *Lobell;* and *Hippolapathum sylvest.* of *Tabern.* ‡

¶ *The Nature and Vertues.*

These herbes are of a mixture betweene cold and heat, and almost drie in the third degree, especially the seed which is very astringent.

The pouder of any of the kinds of Docks drunk in wine, stoppeth the laske and bloudie flixe, and easeth the pains of the stomacke.

The roots boiled til they be very soft, and stamped with barrowes grease, and made into an ointment helpeth the itch and all scuruie scabs and mangines. And for the same purpose it shall bee necessarie to boile them in water, as aforesaid, and the partie to be bathed and rubbed therewith.

† The first figure in the former edition was of *Hydrolapathum magnum,* being the first in the next chapter; and the figure of that we giue you in the third place of this chapt. was that in the first place of the following chap. vnder the forementioned title.

CHAP.

CHAP. 82. *Of Water Dockes.*

† 1 *Hydrolapathum magnum.*
Great Water Docke.

† 2 *Hydrolapathum minus.*
Small Water Docke.

† 3 *Hippolapathum sativum*
Patience, or Munkes Rubarb.

4 *Hippolapathum rotundifolium.*
Bastard Rubarb.

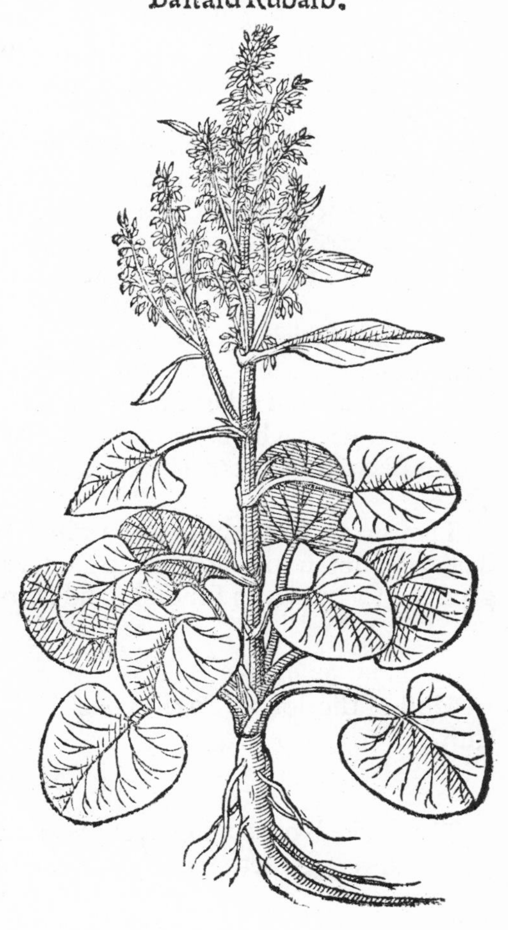

‡ 5 *Lapathum sativum sanguineum.*
Bloudwoort.

The Description.

1 The Great water Docke hath very long and great leaues, stiffe, and hard, not vnlike to the Garden Patience, but much longer. The stalke riseth vp to a great height, oftentimes to the height of fiue foot or more. The floure groweth at the top of the stalke in spokie tufts, brown of colour. The seed is contained in chaffie huskes, three square, of a shining pale colour. The root is very great, thicke, browne without, and yellowish within.

2 The small water Docke hath short narrow leaues, set vpon a stiffe stalke. The floures grow from the middle of the stalke vpward in spokie rundles, set in spaces by certaine distances round about the stalke, as are the floures of Horehound: Which Docke is of all the kindes most common and of lesse vse, and taketh no pleasure or delight in any one soile or dwellingplace, but is found almost euery where, as well vpon the land as in waterie places, but especially in gardens among good and holesome pot-herbes, being there better known than welcome or desired: wherefore I intend not to spend further time about his description.

3 The Garden Patience hath very strong stalks, furrowed or chamfered, of eight or nine foot high when it groweth in fertile ground, set about with great large leaues like to those of the water Docke, hauing alongst the stalkes toward the top floures of a light purple colour declining to browneneſſe. The seed is three square, contained in thin chaffie huskes, like those of the common Docke. The root is verie great, browne without, and yellow within, in colour and taste like the true Rubarb.

4 Bastard Rubarb hath great broad round leaues, in shape like those of the great Bur-docke. The stalke and seeds are so like vnto the precedent, that the one cannot be knowne from the other, sauing that the seeds of this are somewhat lesser. The root is exceeding great and thicke, very like vnto the Rha of Barbarie, as well in proportion as in colour and taste; and purgeth after the same manner, but must be taken in greater quantitie, as witnesseth that famous learned Physition now liuing, Mr. Doctor *Bright*, and others, who haue experimented the same.

5 This fifth kinde of Docke is best knowne vnto all, of the stocke or kindred of Dockes; it hath long thin leaues, sometimes red in euery part thereof, and often stripped here and there with lines and strakes of a darke red colour; among which rise vp stiffe brittle stalkes of the same colour: on the top whereof come forth such floures and seed as the common wilde docke hath. The root is likewise red, or of a bloudie colour.

¶ *The Place.*

They do grow for the most part in ditches and water-courses, very common through England. The two last saue one do grow in gardens; my selfe and others in London and elswhere haue them growing for our vse in Physicke and chirurgerie. The last is sowne for a pot-herbe in most gardens.

¶ *The Time.*

Most of the dockes do rise vp in the Spring of the yeare, and their seed is ripe in Iune and August.

¶ *The Names.*

The docke is called in Greeke λάπαθον: in Latine, *Rumex*, and *Lapathum*; yet *Pliny* in his 19 Booke, 12. Chapter, seemeth to attribute the name of *Rumex* onely to the garden docke.

The

The Monkes Rubarbe is called in Latine *Rumex ſativus*, and *Patientia*, or Patience, which word is borrowed of the French, who call this herbe *Patience*: after whom the Dutch men name this pot herbe alſo **Patientie**: of ſome, *Rhabarbarum Monachorum*, or Monkes Rubarbe; becauſe as it ſhould ſeeme ſome Monke or other haue vſed the root hereof in ſtead of Rubarbe.

Bloudwoort, or bloudy Patience, is called in Latine *Lapathum ſanguineum*: of ſome, *Sanguis Draconis*, of the bloudie colour wherewith the whole plant is poſſeſt, and is of pot-herbes the chiefe or principall, hauing the propertie of the baſtard Rubarbe; but of leſſe force in his purging quality.

¶ *The Temperature.*

Generally all the Dockes are cold, ſome little and moderately, and ſome more: they doe all of them drie, but not all after one manner: notwithſtanding ſome are of opinion that they are dry almoſt in the third degree.

¶ *The Vertues.*

The leaues of the Garden Docke or Patience may be eaten, and are ſomewhat colde, but more moiſt, and haue withall a certaine clammineſſe; by reaſon whereof they eaſily and quickely paſſe through the belly when they be eaten: and *Dioſcorides* writeth, that all the Dockes beeing boiled doe mollifie the bellie: which thing alſo *Horace* hath noted in his ſecond booke of Sermons, the fourth Satyre, writing thus, A

——*Si dura morabitur alvus*
Mugilus, & viles pellent obſtantia conchæ,
Et lapathi breuis herba.

He calleth it a ſhort herbe, being gathered before the ſtalke be growne vp; at which time it is fitteſt to be eaten. B

And being ſodden, it is not ſo pleaſant to bee eaten as either Beetes or Spinage: it ingendreth moiſt bloud of a meane thickneſſe, and which nouriſheth little. C

The leaues of the ſharpe pointed Dockes are cold and drie: but the ſeed of Patience, and the water Docke doe coole, with a certaine thinneſſe of ſubſtance. D

The decoction of the roots of Monkes Rubarbe is drunke againſt the bloudy flix, the laske, the wambling of the ſtomacke which commeth of choler: and alſo againſt the ſtinging of ſerpents, as *Dioſcorides* writeth. E

It is alſo good againſt the ſpitting of bloud, being taken with Acacia (or his *ſuccedaneum*, the dried iuice of ſloes) as *Plinie* writeth. F

Monkes Rubarb or Patience is an excellent wholeſome pot-herbe; for being put into the pottage in ſome reaſonable quantitie, it doth looſen the belly, helpeth the iaunders, the timpany and ſuch like diſeaſes, proceeding of cold cauſes. G

If you take the roots of Monkes Rubarb, and red Madder, of each halfe a pound; Sena foure ounces, anniſe ſeed and licorice, of each two ounces; Scabiouſe and Agrimonie, of each one handfull; ſlice the roots of the Rubarb, bruiſe the anniſe ſeed and licorice, breake the herbes with your hands, and put them into a ſtone pot called a ſteane, with foure gallons of ſtrong ale to ſteepe or infuſe the ſpace of three daies; and then drinke this liquour as your ordinarie drinke for three weekes together at the leaſt, though the longer you take it, ſo much the better; prouiding in a readineſſe another ſteane ſo prepared that you may haue one vnder another, being alwaies carefull to keepe a good diet: it cureth the dropſie, the yellow iaunders, all manner of itch, ſcabbes, breaking out, and mangineſſe of the whole body: it purifieth the bloud from all corruption; preuaileth againſt the greene ſickneſſe very greatly, and all oppilations or ſtoppings: maketh young wenches to looke faire and cherrie like, and bringeth downe their tearmes, the ſtopping whereof hath cauſed the ſame. H

The ſeed of baſtard Rubarb is of a manifeſt aſtringent nature, inſomuch that it cureth the bloudy flix, mixed with the ſeed of Sorrell, and giuen to drinke in red wine. I

There haue not beene any other faculties attributed to this plant either of the antient or later writers, but generally of all it hath beene referred to the other Docks or Monks Rubarb, of which number I aſſure my ſelfe this is the beſt, and doth approch neereſt vnto the true Rubarb. Manie reaſons induce me ſo to thinke and ſay, firſt this hath the ſhape and proportion of Rubarbe, the ſame colour, both within and without, without any difference. They agree as well in taſte as ſmell: it coloureth the ſpittle of a yellow colour when it is chewed, as Rubarb doth; and laſtly it purgeth the belly after the ſame gentle manner that the right Rubarb doth, onely herein it differeth, that this muſt be giuen in three times the quantitie of the other. Other diſtinctions and differences, with the temperature and euery other circumſtance, I leaue to the learned Phyſitions of our London colledge (who are very well able to ſearch this matter) as a thing farre aboue my reach, being K no graduate, but a Countrey Scholler, as the whole framing of this Historie doth well declare: but I hope my good meaning will be well taken, considering I doe my best; not doubting but some of greater learning will perfect that which I haue begun according to my small skill, especially the ice being broken vnto him, and the wood rough hewed to his hands. Notwithstanding I thinke it good to say thus much more in mine owne defence, that although there bee many wants and defects in me, that were requisite to performe such a worke; yet may my long experience by chance happen vpon some one thing or other that may do the learned good: considering what a notable experiment I learned of one *Iohn Bennet* a Chirurgion of Maidstone in Kent, a man as slenderly learned as my selfe, which he practised vpon a Butchers boy of the same towne, as himselfe reported vnto me; his practise was this: Being desired to cure the foresaid lad of an ague, which did grieuously vex him, he promised him a medicine, & for want of one for the present (for a shift as himselfe confessed vnto me) he tooke out of his garden three or foure leaues of this plant of Rubarb, which my selfe had among other simples giuen him, which he stamped & strained with a draught of ale, and gaue it the lad in the morning to drinke: it wrought extremely downeward and vpward within one houre after, and neuer ceased vntill night. In the end the strength of the boy ouercame the force of the Physicke, it gaue ouer working, and the lad lost his ague; since which time (as hee saith) he hath cured with the same medicine many of the like maladie, hauing euer great regard vnto the quantitie, which was the cause of the violent working in the first cure. By reason of which accident, that thing hath been reuealed vnto posteritie, which heretofore was not so much as dreamed of. Whose blunt attempt may set an edge vpon some sharper wit, and greater iudgement in the faculties of plants, to seeke farther into their nature than any of the Antients haue done: and none fitter than the learned Physitions of the Colledge of London; where are many singularly wel learned and experienced in naturall things.

L The roots sliced and boiled in the water of *Carduus Benedictus* to the consumption of the third part, adding thereto a little honie, of the which decoction eight or ten spoonfuls drunke before the fit, cureth the ague in two or three times so taking it at the most: vnto robustous or strong bodies twelue spoonfuls may be giuen. This experiment was practised by a worshipfull Gentlewoman mistresse *Anne Wylbraham*, vpon diuers of her poore Neighbours with good successe.

† That figure that was in the first place was of the *Lapathum fol minus acuto*, described by me in the third place of the preceding chapter. The second was of *Lapathum acutum crispum* of *Tabernamontanus*. The third was of *Hydrolapathum minus*.

Chap. 83. *Of Rubarb.*

‡ IT hath happened in this as in many other forreine medicines or simples, which though they be of great and frequent vse, as Hermodactyls, Muske, Turbeth, &c. yet haue we no certaine knowledge of the very place which produces them, nor of their exact manner of growing, which hath giuen occasion to diuers to thinke diuersly, and some haue been so bold as to counterfeit figures out of their owne fancies, as *Matthiolus*: so that this saying of *Pliny* is found to be very true, *Nulla medicinæ pars magis incerta, quam quæ ab alio quam nostro orbe petitur*. But we will endeauour to shew you more certaintie of this here treated of than was knowne vntill of very late yeres. ‡

¶ *The Description.*

1 THis kinde of Rubarb hath very great leaues, somewhat snipt or indented about the edges like the teeth of a Saw, not vnlike the leaues of *Enula campana*, called by the vulgar sort Elecampane, but greater: among which riseth vp a straight stalke of two cubits high, bearing at the top a scalie head like those of Knappe-weed, or *Iacea maior*: in the middle of which knap or head thrusteth forth a faire floure consisting of many purple threds like those of the Artichoke; which being past, there followeth a great quantitie of downe, wherein is wrapped long seede like vnto the great Centorie, which the whole plant doth very well resemble. The root is long and thicke, blackish without, and of a pale colour within: which being chewed maketh the spittle very yellow, as doth the Rubarb of Barbarie.

‡ 2 This other bastard Rha, which is also of *Lobels* description, hath a root like that of the last described: but the leaues are narrower almost like those of the common Docke, but hoarie on the other side: the stalke growes vp straight, and beareth such heads and floures as the precedent.

‡ 3 I haue thought good here to omit the counterfeit figure of *Matthiolus*, giuen vs in this place by our Authour; as also the Historie, which was not much pertinent, and in lieu of them to present you with a perfect figure and description of the true *Rha Ponticum* of the Antients, which

1 *Rha Capitatum L'obelij.*
Turkie Rubarbe.

‡ 2 *Rha Capitatum angustifolium.*
The other bastard Rubarbe.

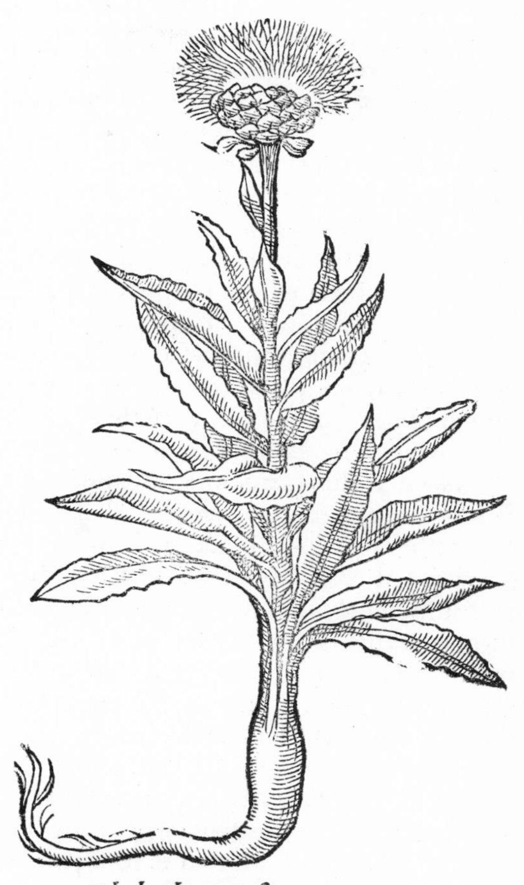

‡ 3 *Rha verum antiquorum.*
The true Rubarbe of the Antients.

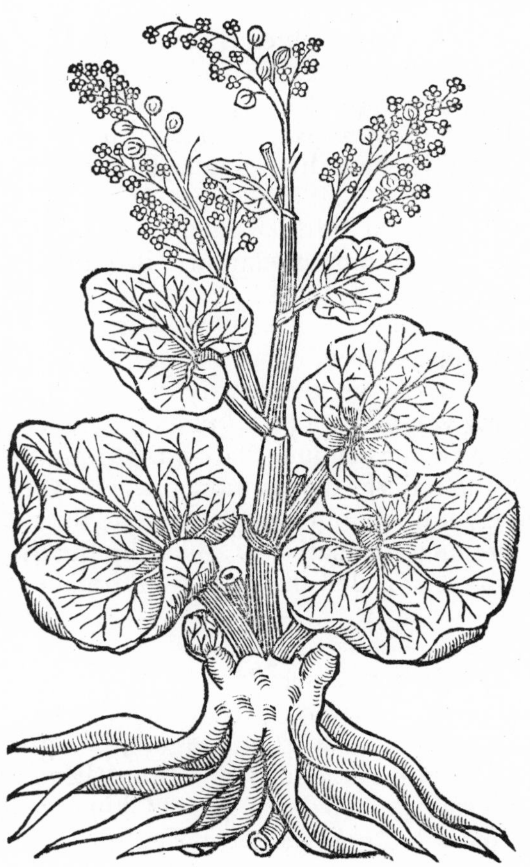

Rhabarbarum siccatum.
The drie roots of Rubarbe.

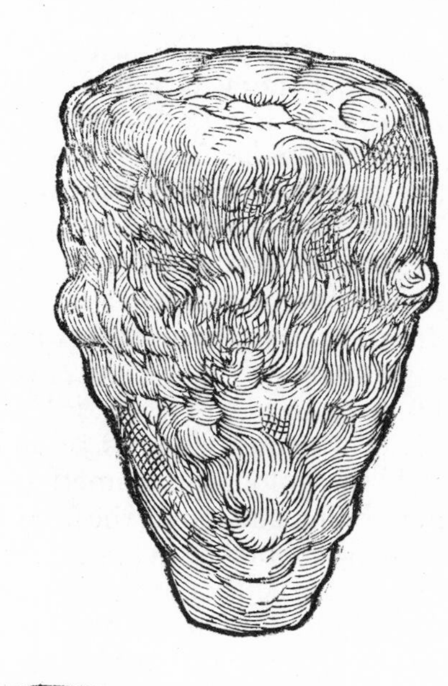

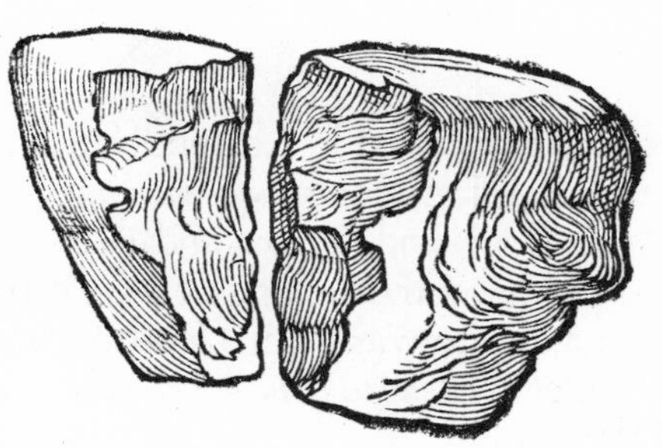

was first of late discouered by the learned *Prosper Alpinus*, who writ a peculiar tract thereof, and it is also againe figured and described in his worke *de Plantis exoticis*. Our Countryman M[r] *Iohn Parkinson* hath also set forth very well both the figure and description hereof, in his *Paradisus terrestris*. This plant hath many large roots diuersly spreading in the ground, of a yellow colour, from which grow vp many very great leaues like those of the Butter-burre, but of a fresh greene colour, with great and manifest veines dispersed ouer them. The stalke also is large and crested, sending forth sundry branches bearing many small white floures, which are succeeded by seeds three square and brownish, like as those of other Docks. D[r]. *Lister* one of his Maiesties Physitions was the first that enricht this kingdome with this elegant and vsefull plant, by sending the seedes thereof to M[r]. *Parkinson*. *Prosper Alpinus* proues this to be the true *Rha* of the Antients, described by *Dioscorides*, *Lib*. 3. *cap*. 2. yet neither he nor any other (that I know of) haue obserued a fault, which I more than probably suspect to bee in the text of *Dioscorides* in that place, which is in the word μέλανα, which I iudge should be μήλινα, that is, yellow, and not blacke, as *Ruellius* and others haue translated it: now μήλινος is a word frequently vsed by *Dioscorides*, as may appeare by the Chapters of *Hieracium magnum & parvum*, *Conyra*, *Peucedanum*, *Ranunculus*, and diuers others, and I suspect the like fault may bee found in some other places of the same Authour. But I will no further insist vpon this, seeing the thing it selfe in all other respects, as also in yellownesse shewes it selfe to be that described by *Dioscorides*, and that my coniecture must therefore be true. And besides, the root wherto he compares it is ὑπέρυθρος, that is *Rubescens*, or rather *ex flauo rubescens*, as any versed in reading *Dioscorides* may easily gather by diuers places in him. Now I here omit his words, because they are in the next description alledged by our Authour, as also the description of our ordinarily vsed Rubarb, for that it is sufficiently described vnder the following title of the choise thereof. M[r]. *Parkinson* is of opinion that this is the true Rubarbe vsed in shops, onely lesse heauy, bitter, and strong in working, by reason of the diuersity of our climat from that whereas the dried Rubarb brought vs vsually grows. This his opinion is very probable; and if you compare the roots together, you may easily bee induced to be of the same beleefe. ‡

† 4 The Ponticke Rubarbe is lesser and slenderer than that of Barbarie. Touching Pontick Rubarbe *Dioscorides* writeth thus: Rha that diuers call Rheon, which groweth in those places that are beyond Bosphorus, from whence it is brought, hath yellow roots like to the great Centorie, but lesser and redder, ἄοσμος, that is to say, without smell (*Dodonæus* thinkes it should bee εὔοσμος, that is, well smelling) spongie, and something light. That is the best which is not worme-eaten, and tasted is somewhat viscide with a light astriction, and chewed becomes of a yellow or Saffron colour.

¶ *The Place.*

It is brought out of the Countrey of Sina (commonly called China) which is toward the East in the vpper part of India, and that India which is without the riuer Ganges: and not at all *Ex Scenitarum prouincia*, (as many do vnaduisedly thinke) which is in Arabia the Happie, and far from China: it groweth on the sides of the riuer Rha now called Volga, as *Amianus Marcellus* saith, which riuer springeth out of the Hyperborean mountaines, and running through Muscouia, falleth into the Caspian or Hircan sea.

‡ The Rha of the Antients growes naturally, as *Alpinus* saith, vpon the hill Rhodope in Thrace, now called Romania. It growes also as I haue been informed vpon some mountaines in Hungarie. It is also to be found growing in some of our choice gardens. ‡

The choice of Rubarbe.

The best Rubarbe is that which is brought from China fresh and new, of a light purplish red, with certaine veines and branches, of an vncertaine varietie of colour, commonly whitish: but when it is old the colour becommeth ill fauored by turning yellowish or pale, but more, if it bee worme eaten: being chewed in the mouth it is somewhat gluie and clammie, and of a saffron colour, which being rubbed vpon paper or some white thing sheweth the colour more plainely: the substance thereof is neither hard or closely compacted, nor yet heauy; but something light, and as it were in a middle betweene hard and loose and something spungie: it hath also a pleasing smell. The second in goodnesse is that which commeth from Barbarie. The last and worst from Bosphorus and Pontus.

¶ *The Names.*

It is commonly called in Latine *Rha Barbarum*, or *Rha Barbaricum*: of diuers, *Rheu Barbarum*: the Moores and Arabians doe more truely name it *Raued Seni, a Sinensi prouincia*; from whence it is brought into Persia and Arabia, and afterwards into Europe: and likewise from Tanguth, through the land of Cataia into the land of the Persians, whereof the Sophie is the ruler, and from thence into Ægypt, and afterwards into Europe. It is called of the Arabians and the people of China, and the parts adiacent, *Rauend Cini*, *Raued Seni*, and *Raued Sceni*: in shops, *Rhabarbarum*: in English, Rubarb, and Rewbarbe.

¶ *The*

4 *Rha Ponticum Siccatum.*
Rubarb of Pontus dried.

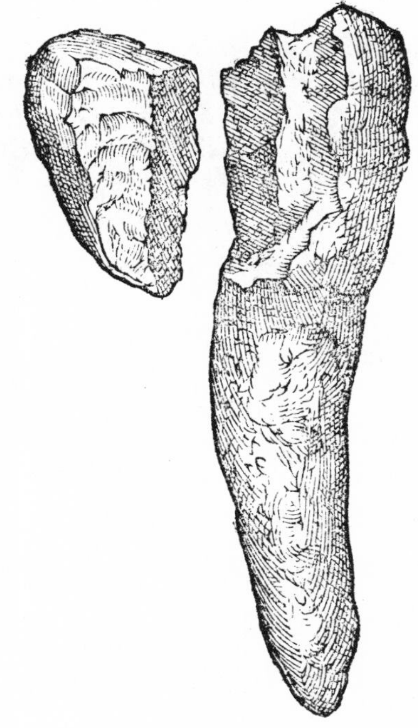

¶ *The Temperature.*

Rubarb is of a mixt substance, temperature and faculties: some of the parts thereof are earthy, binding and drying: others thin, airious, hot, and purging.

¶ *The Vertues.*

A Rubarb is commended by *Dioscorides* against windinesse, weaknesse of the stomack, and all griefes thereof, convulsions, diseases of the spleene, liuer, and kidnies, gripings and inward gnawings of the guts, infirmities of the bladder and chest, swelling about the heart, diseases of the matrix, paine in the huckle bones, spitting of bloud, shortnesse of breath, yexing, or the hicket, the bloudie flix, the laske proceeding of raw humors, fits in Agues, and against the bitings of venomous beasts.

B Moreouer he saith, that it taketh away blacke and blew spots, and tetters or Ringwormes, if it be mixed with vineger, and the place anointed therewith.

C *Galen* affirmes it to be good for burstings, cramps, and convulsions, and for those that are short winded, and that spit bloud.

D But touching the purging facultie neither *Dioscorides* nor *Galen* hath written any thing, because it was not vsed in those daies to purge with. *Galen* held opinion, that the thinne airious parts doe make the binding qualitie of more force; not because it doth resist the cold and earthy substance, but by reason that it carrieth the same, and maketh it deeply to pierce, and thereby to worke the greater effect; the dry and thinne essence containing in it selfe a purging force and qualitie to open obstructions, but helped and made more facile by the subtil and airious parts. *Paulus Ægineta* seemeth to be the first that made triall of the purging facultie of Rubarb; for in his first booke, Chap. 43. he maketh mention thereof, where he reckoneth vp Turpentine among those medicines which make the bodies of such as are in health soluble: But when we purpose, saith he, to make the turpentine more strong, we adde vnto it a little Rubarb. The Arabians that followed him brought it to a further vse in physicke, as chiefely purging downward choler, and oftentimes flegme.

E The purgation which is made with Rubarb is profitable and fit for all such as be troubled with choler, and for those that are sicke of sharpe and tertian feuers, or haue the yellow jaundice, or bad liuers.

F It is a good medicine against the pleurisie, inflammation of the lungs, the squinancie or Squincie, madnesse, frensie, inflammation of the kidnies, bladder, and all the inward parts, and especially against S. Anthonies fire, as well outwardly as inwardly taken.

G Rubarb is vndoubtedly an especiall good medicine for the liuer and infirmities of the gall; for besides that it purgeth forth cholericke and naughty humors, it remoueth stoppings out of the conduits.

H It also mightily strengthneth the intrals themselues: insomuch as Rubarb is iustly termed of diuers the life of the liuer; for *Galen* in his eleuenth booke of the method or manner of curing, affirmeth that such kinde of medicines are most fit and profitable for the liuer, as haue ioyned with a purging and opening qualitie an astringent or binding power. The quantitie that is to be giuen is from one dram to two; and the infusion from one and a halfe to three.

I It is giuen or steeped, and that in hot diseases, with the infusion or distilled water of Succory, Endiue, or some other of the like nature; and likewise in Whay; and if there be no heate it may be giuen in Wine.

It

K It is also oftentimes giuen being dried at the fire, but so, that the least or no part thereof at all be burned; and being so vsed it is a remedie for the bloudy flix, and for all kindes of laskes: for it both purgeth away naughty and corrupt humors, and likewise withall stoppeth the belly.

L The same being dried after the same manner doth also stay the ouermuch flowing of the monethly sicknesse, and stoppeth bloud in any part of the body, especially that which commeth thorow the bladder; but it should be giuen in a little quantitie, and mixed with some other binding thing.

M *Mesues* saith, That Rubarb is an harmelesse medicine, and good at all times, and for all ages, and likewise for children and women with childe.

‡ My friend Mr. *Sampson Iohnson* Fellow of *Magdalen* Colledge in Oxford assures me, That the Physitions of Vienna in Austria vse scarce any other at this day than the Rubarb of the Antients, which grows in Hungary not far from thence: and they prefer it before the dried Rubarb brought out of Persia and the East Indies, because it hath not so strong a binding facultie as it, neither doth it heate so much; onely it must be vsed in somewhat a larger quantitie. ‡

Chap. 84.

Of Sorrell.

¶ *The Kindes.*

THere be diuers kindes of Sorrell, differing in many points, some of the garden, others wilde; some great, and some lesser.

1 *Oxalis, siue Acetosa.*
Sorrell.

2 *Oxalis tuberosa.*
Knobbed Sorrell.

¶ *The Description.*

THough *Dioscorides* hath not expressed the *Oxalides* by that name, yet none ought to doubt but that they were taken and accounted as the fourth kinde of *Lapathum*. For though some like

like it not well that the seed should be said to be *Drimus*; yet that is to be vnderstood according to the common phrase, when acride things are confounded with those which be sharpe and soure; else we might accuse him of such ignorance as is not amongst the simplest women. Moreouer, the word *Oxys* doth not onely signifie the leafe, but the sauour and tartnesse, which by a figure drawne from the sharpnesse of kniues edges is therefore called sharpe: for ὀξὺς χύμος signifieth a sharpe or soure iuyce which pierceth the tongue like a sharpe knife: whereupon also *Lapathum* may be called *Oxalis*, as it is indeed. The leaues of this are thinner, tenderer, and more vnctuous than those of *Lapatium acutum*, broader next to the stem, horned and crested like Spinage and *Atriplex*. The stalke is much streaked, reddish, and full of iuyce: the root is yellow and fibrous; the seed sharpe, cornered and shining, growing in chaffie huskes like the other Docks.

2 The second kinde of *Oxalis* or Sorrell hath large leaues like Patience, confusedly growing together vpon a great tall stalke, at the top whereof grow tufts of a chaffie substance. The root is tuberous, much like the Peonie, or rather Filipendula, fastned to the lower part of the stem with small long strings and laces.

3 The third kinde of Sorrell groweth very small, branching hither and thither, taking hold (by new shoots) of the ground where it groweth, whereby it dispersethit selfe far abroad. The leaues are little and thin, hauing two small leaues like eares fastned thereto, in shew like the herbe *Sagittaria*: the seed in taste is like the other of his kinde.

4 The fourth kinde of Sorrell hath leaues somewhat round and cornered, of a whiter colour than the ordinarie, and hauing two short eares anexed vnto the same. The seed and root in taste is like the other Sorrels.

3 *Oxalis tenuifolia.*
Sheepes Sorrell.

4 *Oxalis Franca seu Romana.*
Round leaued, or French Sorrel.

5 This kinde of curled Sorrell is a stranger in England, and hath very long leaues, in shape like the garden Sorrell, but curled and crumpled about the edges as is the curled Colewort. The stalke riseth vp among the leaues, set here and there with the like leaues, but lesser. The floures, seeds, and roots are like the common Sorrell or soure Docke.

6 The small Sorrell that groweth vpon dry barren sandy ditch-banks, hath small grassy leaues somewhat forked or crossed ouer like the crosse hilt of a rapier. The stalkes rise vp amongst the leaues, small, weake, and tender, of the same soure taste that the leaues are of. The floure, seed, and root is like the other Sorrels, but altogether lesser.

7 The

6 *Oxalis minor.*
Small Sorrell.

7 The smallest sort of Sorrell is like vnto the precedent, sauing that the lowest leaues that ly vpon the ground be somewhat round, and without the little eares that the other hath, which setteth forth the difference.

‡ 8 There is also kept in some gardens a verie large sorrel, hauing leaues thicke, whitish, and as large as an ordinarie Docke, yet shaped like Sorrell, and of the same acide taste. The stalkes and seed are like those of the ordinary, yet whiter coloured. ‡

¶ *The Place.*

† The common Sorrell groweth for the most part in moist medowes and gardens. The second by waters sides, but not in this kingdome that I know of. The fourth also is a garden plant with vs, as also the fifth: but the third and last grow vpon grauelly and sandie barren ground and ditch bankes. †

¶ *The Time.*

They flourish at that time when as the other kinds of Docks do floure.

¶ *The Names.*

Garden Sorrell is called in Greeke ὀξαλὶς, and ἀνάξυρις: of *Galen*, ὀξυλάπαθον: that is to say, *Acidum lapathum*, or *Acidus rumex*, soure Docke: and in shops commonly *Acetosa*: in the Germane Tongue, **Sawrampffer**: in low-Dutch, **Surckele**, and **Surinck**: the Spaniards, *Azederas*, *Agrelles*, and *Azedas*: in French, *Ozeille*, and *Surelle*, *Aigrette*: in English, Garden Sorrell.

The second is called of the later Herbarists *Tuberosa acetosa*, and *Tuberosum lapathum*: in English Bunched or Knobbed Sorrell.

The third is called in English Sheepes Sorrell: in Dutch, **Schap Surkel.**

The fourth, Romane Sorrell, or round leaued Sorrell.

The fifth, Curled Sorrell.

The sixth and seuenth, Barren Sorrell, or Dwarfe Sheepes Sorrell.

‡ The eighth is called *Oxalis*, or *Acetosa maxima latifolia*, Great broad leaued Sorrell. ‡

¶ *The Nature.*

The Sorrels are moderately cold and dry.

¶ *The Vertues.*

A Sorrell doth vndoutedly coole and mightily dry; but because it is soure it likewise cutteth tough humors.

B The iuyce hereof in Sommer time is a profitable sauce in many meats, and pleasant to the taste: it cooleth an hot stomacke, moueth appetite to meate, tempereth the heate of the liuer, and openeth the stoppings thereof.

C The leaues are with good successe added to decoctions which are vsed in Agues.

D The leaues of Sorrell taken in good quantitie, stamped and strained into some Ale, and a posset made thereof, cooleth the sicke body, quencheth the thirst, and allayeth the heate of such as are troubled with a pestilent feuer, hot ague, or any great inflammation within.

E The leaues sodden, and eaten in manner of a Spinach tart, or eaten as meate, softneth and loosneth the belly, and doth attemper and coole the bloud exceedingly,

F The seed of Sorrell drunke in grosse red wine stoppeth the laske and bloudy flix.

Chap.

CHAP. 85. *Of Biſtort or Snake-weed.*

¶ *The Deſcription.*

1 THe great Biſtort hath long leaues much like Patience,but ſmaller,and more wrinkled or crumpled, on the vpper ſide of a darke greene, and vnderneath of a blewiſh greene colour, much like Woad. The ſtalke is long, ſmooth, and tender, hauing at the top a ſpiked knap or eare, ſet full of ſmall whitiſh floures declining to carnation. The root is all in a lumpe, without faſhion ; within of a reddiſh colour like vnto fleſh, in taſte like the kernell of an Acorne.

2 The ſmall Biſtort hath leaues about three inches long, and of the bredth of a mans naile ; the vpper ſide is of a greene colour, and vnderneath of an ouerworne greeniſh colour : amongſt the which riſeth vp a ſtalke of the height of a ſpanne, full of ioynts or knees, bearing at the top ſuch floures as the great Biſtort beareth ; which being fallen, the ſeeds appeare of the bignes of a tare, reddiſh of colour, euery ſeed hauing one ſmall greene leafe faſtned thereunto, with many ſuch leaues thruſt in among the whole bunch of floures and ſeed. The root is tuberous like the other, but ſmaller, and not ſo much crooked.

1 *Biſtorta major.*
Snake-weed.

2 *Biſtorta minor.*
Small Snake-weed.

3 Broad leaued Snake-weed hath many large vneuen leaues,ſmooth and very greene , among which riſe vp ſmall brittle ſtalkes of two hands high,bearing at the top a faire ſpike of floures like vnto the great Biſtort. The root is knobby or bunched,crookedly turned or wrythed this way and that way, whereof it tooke his name *Biſtorta*. ‡ It differs from the firſt onely in that the root is ſomewhat more twined in, and the leaues broader and more crumpled. ‡

¶ *The Place.*

1 The great Biſtort groweth in moiſt and waterie places, and in the darke ſhadowie Woods, and is very common in moſt gardens.

2 The

2 The small Bistort groweth in great aboundance in Westmerland, at Crosby, Rauensswaith, at the head of a Parke belonging to one Mr. *Pickering*: from whence it hath beene dispersed into many gardens; as also sent vnto me from thence for my garden.

¶ *The Time.*

They floure in May, and the seed is ripe in Iune.

¶ *The Names.*

Bistorta is called in English Snake-weed: in some places, Oisterloit: in Cheshire, Passions, and Snake-weed, and there vsed for an excellent Pot-herbe. It is called *Bistorta* of his wrythed roots, and also *Colubrina, Serpentaria, Brittanica; Dracontion, Plinij; Dracunculus, Dodonæi*; and *Limonium Gesneri*.

¶ *The Nature.*

Bistort doth coole and dry in the third degree.

¶ *The Vertues.*

A The iuyce of Bistort put into the nose preuaileth much against the Disease called *Polypus*, and the biting of Serpents or any venomous beast, being drunke in Wine or the water of Angelica.

B The root boyled in wine and drunke, stoppeth the laske and bloudy flix; it stayeth also the ouermuch flowing of womens monethly sicknesses.

C The root taken as aforesaid stayeth vomiting, and healeth the inflammation and sorenesse of the mouth and throat: it likewise fastneth loose teeth, being holden in the mouth for a certaine space, and at sundry times.

Chap. 86. *Of Scuruy-Grasse, or Spoon-wort.*

¶ *The Description.*

1 Round leaued Scuruy-Grasse is a low or base herbe: it bringeth forth leaues vpon smal stems or foot-stalks of a meane length, comming immediately from the root, very many in number, of a shining greene colour, somewhat broad, thicke, hollow like a little spoone, but of no great depth, vneuen, or cornered about the edges: among which leaues spring vp small stalkes of a spanne high, whereon doe grow many little white floures: after which commeth the seed, small and reddish, contained in little round pouches or seed-vessels: the roots be small, white, and threddy. The whole plant is of a hot and spicie taste.

2 The common Scuruy-grasse or Spoone-wort hath leaues somewhat like a spoone, hollow in the middle, but altogether vnlike the former: the leaues hereof are bluntly toothed about the edges, sharpe pointed, and somewhat long: the stalkes rise vp among the leaues, of the length of halfe a foot; whereon do grow white floures with some yellownesse in the middle: which being past, there succeed small seed-vessels like vnto a pouch, not vnlike to those of Shepheards purse, greene at the first, next yellowish, and lastly when they be ripe, of a browne colour, or like a filberd nut. The root is small and tender, compact of a number of threddy strings very thicke thrust together in manner of a little turfe.

¶ *The Place.*

The first groweth by the sea side at Hull, at Boston, and Lynne, and in many other places of Lincolnshire neere vnto the sea, as in Whaploade and Holbecke Marshes in Holland in the same County. It hath beene found of late growing many miles from the sea side, vpon a great hill in Lancashire called Ingleborough hill; which may seeme strange vnto those that do not know that it will be content with any soile, place, or clyme whatsoeuer: for proofe whereof, my selfe haue sowen the seeds of it in my garden, and giuen them vnto others, with whom they floure, flourish, and bring forth their seed, as naturally as by the sea side; and likewise retaine the same hot spicie taste: which proueth that they refuse no culture, contrary to many other sea-plants.

The second, which is our common scuruie grasse, groweth in diuers places vpon the brimmes of the famous riuer Thames, as at Woolwich, Erith, Greenhithe, Grauesend, as well on the Essex shore as the Kentish; at Portsmouth, Bristow, and many other places alongst the Western coast: but toward the North I haue not heard that any of this kinde hath growne.

¶ The

¶ *The Time.*

It floureth and flouriſheth in May. The ſeed is ripe in Iune.

2 *Cochlearia rotundifolia.*
Round leafed Scuruie graſſe.

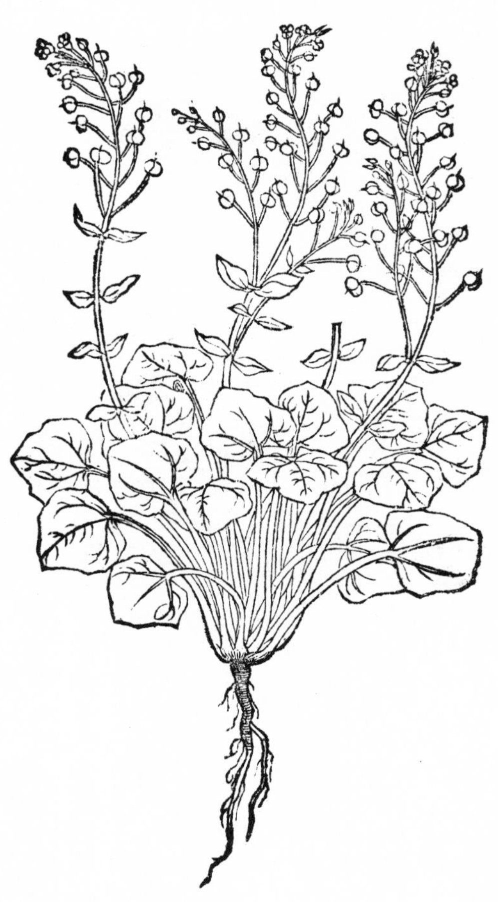

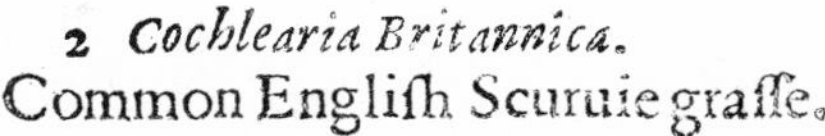

2 *Cochlearia Britannica.*
Common Engliſh Scuruie graſſe.

¶ *The Names.*

† We are not ignorant that in low Germany, this hath ſeemed to ſome of the beſt learned to be the true *Britannica,* and namely to thoſe next the Ocean in Frieſland and Holland. The Germanes call it **Leffelkraut**: that is, *Cochlearia* or Spoonwort, by reaſon of the compaſſed roundnes and hollownes of the leaues, like a ſpoone; and haue thought it to be *Plinie's Britannica,* becauſe they finde it in the ſame place growing, and endued with the ſame qualities. Which excellent plant *Cæſars* ſoldiers (when they remooued their camps beyond the Rhene) found to preuaile (as the Friſians had taught it them) againſt that plague and hurtfull diſeaſe of the teeth, gums, and ſinewes, called the Scuruie, being a depriuation of all good bloud and moiſture, in the whole bodie, called *Scorbutum*; in Engliſh, the Scuruie, and Skyrby, a diſeaſe happening at the ſea among Fiſhermen, and freſh-water ſouldiers, and ſuch as delight to ſit ſtill without labour and exerciſe of their bodies; and eſpecially aboue the reſt of the cauſes, when they make not cleane their biſket bread from the floure or mealines that is vpon the ſame, which doth ſpoile many. But ſith this agrees not with *Plinies* deſcription, and that there be many other water plants as *Naſturtium, Sium, Cardamine,* and ſuch others, like in taſte, and not vnlike in proportion and vertues, which are remedies againſt the diſeaſes aforeſaid, there can be no certaine argument drawne therefrom to prooue it to be *Britannica.* For the leaues at their firſt comming forth are ſomewhat long like *Pyrola* or Adders tongue, ſoone after ſomewhat thicker, and hollow like a nauell, after the manner of Sun-dew, but in greatneſſe like *Soldanella,* in the compaſſe ſomewhat cornered, in faſhion ſomewhat like a ſpoone: the floures white, and in ſhape like the Cuckow floures: the ſeed reddiſh, like the ſeed of *Thlaſpi,* which is not to be ſeen in *Britannica,* which is rather holden to be Biſtort or garden Patience, than Scuruie graſſe. In Engliſh it is called Spoonewort, Scruby graſſe, and Scuruie graſſe.

¶ *The Temperature.*

Scuruie graſſe is euidently hot and drie, very like in taſte and qualitie to the garden Creſſes, of an aromaticke or ſpicie taſte.

¶ *The Vertues.*

A The juice of Spoonewoort giuen to drinke in Ale or Beere, is a ſingular medicine againſt the corrupt and rotten vlcers, and ſtench of the mouth: it perfectly cureth the diſeaſe called of *Hippocrates, Volulus Hematites*: of *Pliny, Stomacace*: of *Marcellus, Oſcedo*: and of the later writers, *Scorbutum*: of the Hollanders and Friſians, Scuerbuyck: in Engliſh, the Scuruie: either giuing the juice in drinke as aforeſaid, or putting ſix great handfuls to ſteepe, with long pepper, graines, aniſe-ſeede, and liquorice, of each one ounce, the ſpices being braied, and the herbes bruſed with your hands, and ſo put into a pot, ſuch as is before mentioned in the chapter of baſtard Rubarbe, and vſed in like maner; or boiled in milke or wine and drunke for certaine daies together it worketh the like effect.

B The juice drunke once in a day faſting in any liquor, ale, beere, or wine, doth cauſe the foreſaid medicine more ſpeedily to worke his effect in curing this filthy, lothſome, heauy, and dull diſeaſe, which is very troubleſome, and of long continuance. The gums are looſed, ſwolne, and exulcerate; the mouth greeuouſly ſtinking; the thighes and legs are withall very often full of blew ſpots, not much vnlike thoſe that come of bruſes: the face and the reſt of the body is oftentimes of a pale colour: and the feet are ſwolne, as in a dropſie.

C There is a diſeaſe (ſaith *Olaus magnus* in his hiſtorie of the Northerne regions) haunting the campes, which vexe them that are beſieged and pinned vp: and it ſeemeth to come by eating of ſalt meates, which is increaſed and cheriſhed with the cold vapors of the ſtone walls. The Germanes call this diſeaſe (as we haue ſaid) Scorbuck, the ſymptome or paſſion which hapneth to the mouth, is called of *Pliny* στομακάκη: *Stomacace*: and that which belongeth to the thighes σκελοτύρβη: *Marcellus* an old writer nameth the infirmities of the mouth *Oſcedo*: which diſeaſe commeth of a groſſe cold and tough bloud, ſuch as malancholy juice is, not by aduſtion, but of ſuch a bloud as is the feculent or droſſie part thereof: which is gathered in the body by ill diet, ſlothfulneſſe to worke, laiſineſſe (as we terme it) much ſleepe and reſt on ſhip boord, and not looking to make cleane the biſquet from the mealineſſe, and vncleane keeping their bodies, which are the cauſes of this diſeaſe called the ſcuruie or ſcyrby; which diſeaſe doth not onely touch the outward parts, but the inward alſo: for the liuer oftentimes, but moſt commonly the ſpleene, is filled with this kinde of thicke, cold and tough juice, and is ſwolne by reaſon that the ſubſtance thereof is ſlacke, ſpungie and porous, very apt to receiue ſuch kinde of thick and cold humors. Which thing alſo *Hippocrates* hath written of in the ſecond booke of his Prorrhetikes: their gums (ſaith he) are infected, and their mouthes ſtinke that haue great ſpleenes or milts: and whoſoeuer haue great milts and vſe not to bleed, can hardly be cured of this malladie, eſpecially of the vlcers in the legs, and blacke ſpots. The ſame is affirmed by *Paulus Ægineta* in his third booke, 49. chapter, where you may eaſily ſee the difference between this diſeaſe and the black jaunders; which many times are ſo confounded together, that the diſtinction or difference is hard to be known, but by the expert chirurgion: who oftentimes ſeruing in the ſhips, as wel her Maieſties as merchants, are greatly peſtered with the curing thereof: it ſhall be requiſite to carrie with them the herbe dried: the water diſtilled, and the juice put into a bottle with a narrow mouth, full almoſt to the necke, and the reſt filled vp with oile oliue, to keep it from putrifaction: the which preparations diſcreetly vſed, will ſtand them in great ſtead for the diſeaſe aforeſaid.

D The herbe ſtamped and laid vpon ſpots and blemiſhes of the face, will take them away within ſix houres, but the place muſt be waſhed after with water wherein bran hath been ſodden.

Chap. 87. *Of Twayblade, or herbe Bifoile.*

¶ *The Deſcription.*

1 Herbe Byfoile hath many ſmall fibres or threddy ſtrings, faſtened vnto a ſmall knot or root, from which riſeth vp a ſlender ſtem or ſtalke, tender, fat, and full of juice; in the middle whereof are placed in comely order two broad leaues, ribbed and chamfered, in ſhape like the leaues of Plantaine: vpon the top of the ſtalke groweth a ſlender greeniſh ſpike made of many ſmall floures, each little floure reſembling a gnat, or little goſling newly hatched, very like thoſe of the third ſort of Serapias ſtones.

2 *Ophris Trifolia*, or Trefoile Twaiblade, hath roots, tender ſtalkes, and a buſh of flours like the precedent; but differeth in that, that this plant hath three leaues which do clip or embrace the ſtalke

ſtalke about; and the other hath but two, and neuer more, wherein eſpecially conſiſteth the difference: although in truth I thinke it a degenerate kinde, and hath gotten a third leafe *per accidens*, as doth ſometimes chance vnto the Adders Tongue, as ſhall be declared in the Chapter that followeth.

‡ 3 This kinde of Twaibald, firſt deſcribed in the laſt edition of *Dodonæus*, hath leaues, floures, and ſtalkes like to the ordinarie; but at the bottome of the ſtalke aboue the fibrous roots it hath a bulbe greeniſh within, and couered with two or three skins: it growes in moiſt and wet low places of Holland. ‡

1 *Ophris bifolia.*
Twaibald.

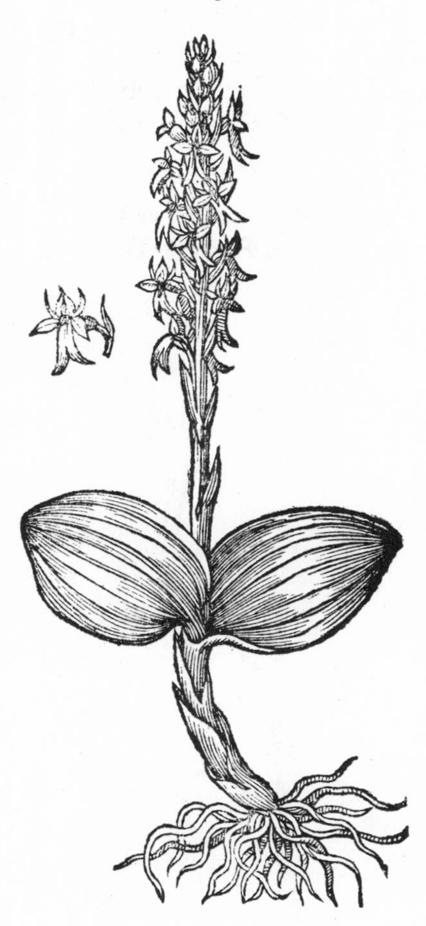

‡ 3 *Ophris bifolia bulboſa.*
Bulbous Twaibald.

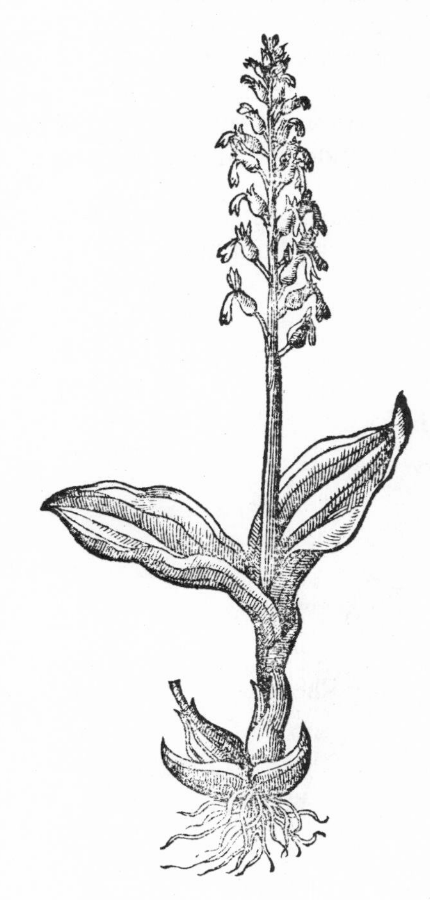

¶ *The Place.*

The firſt groweth in moiſt medowes, fenny grounds, and ſhadowie places. I haue fonnd it in many places, as at Southfleet in Kent, in a Wood of Maſter *Sidleys* by Long-field Downes, in a Wood by London called Hampſtead Wood, in the fields by High-gate, in the Woods by Ouenden neere to Clare in Eſſex, and in the Woods by Dunmow in Eſſex. The ſecond ſort is ſeldome ſeene.

¶ *The Time.*

They floure in May and Iune.

¶ *The Names.*

It is called of the later Herbariſts, *Bifolium*, and *Ophris*.

¶ *The Nature and Vertues.*

A These are reported of the Herbariſts of our time to be good for greene wounds, burſtings, and ruptures; whereof I haue in my vnguents and Balſams for greene wounds had great experience, and good ſucceſſe.

Chap. 88. Of Adders-Tongue.

¶ *The Description.*

1 O*Phioglosson*, or *Lingua Serpentis* (called in English Adders tongue; of some, Adders Grasse, though vnproperly) riseth forth of the ground, hauing one leafe and no more, fat or oleous in substance, of a finger long, and very like the yong and tender leaues of Marigolds: from the bottome of which leafe springeth out a small and tender stalke one finger and a halfe long, on the end whereof doth grow a long small tongue not vnlike the tongue of a serpent, whereof it tooke the name.

2 I haue seene another like the former in root, stalke, and leafe; and differeth, in that this plant hath two, and sometimes more crooked tongues, yet of the same fashion, which if my iudgment faile not chanceth *per accidens*, euen as we see children borne with two thumbes vpon one hand: which moueth me so to thinke, for that in gathering twenty bushels of the leaues a man shall hardly finde one of this fashion.

1 *Ophioglosson.*
Adders-Tongue.

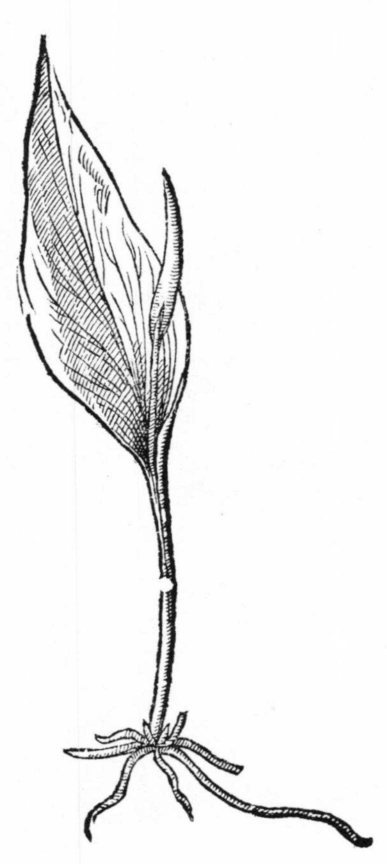

‡ 2 *Ophioglosson abortivum.*
Mis-shapen Adders-Tongue.

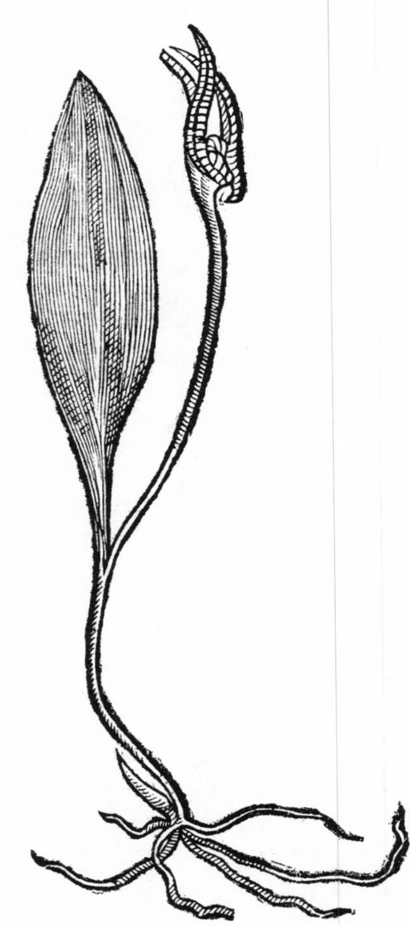

¶ *The Place.*

Adders-Tongue groweth in moist medowes throughout most parts of England; as in a Meadow neere the preaching Spittle adioyning to London; in the Mantels by London, in the medowes by Cole-brooke, in the fields in Waltham Forrest, and many other places.

¶ *The Time.*

They are to be found in Aprill and May; but in Iune they are quite vanished and gone.

¶ *The Names.*

Ophioglossum is called in shops *Lingua serpentis*, *Linguace*, and *Lingualace*: it is also called *Lancea Christi*, *Enephyllon*, and *Lingua vulneraria*: in English, Adders tongue, or Serpents tongue: in Dutch, **Natertonguen**: of the Germanes, **Nater zungelin**.

¶ The

¶ *The Nature.*

Adders-tongue is dry in the third degree.

¶ *The Vertues.*

The leaues of Adders tongue ſtamped in a ſtone morter, and boyled in Oile Oliue vnto the conſumption of the iuyce, and vntill the herbes be dry and partched, and then ſtrained, will yeeld a moſt excellent greene oyle, or rather a balſam for greene wounds, comparable vnto oyle of S. *Iohns* wort, if it do not farre ſurpaſſe it by many degrees: whoſe beauty is ſuch, that very many Artiſts haue thought the ſame to be mixed with Verdigreaſe. A

CHAP. 89.

Of One-berry, or Herbe True-loue, and Moone-wort.

1 *Herba Paris.*
One-Berry, or Herbe True-loue.

2 *Lunaria minor.*
Small Moone-wort.

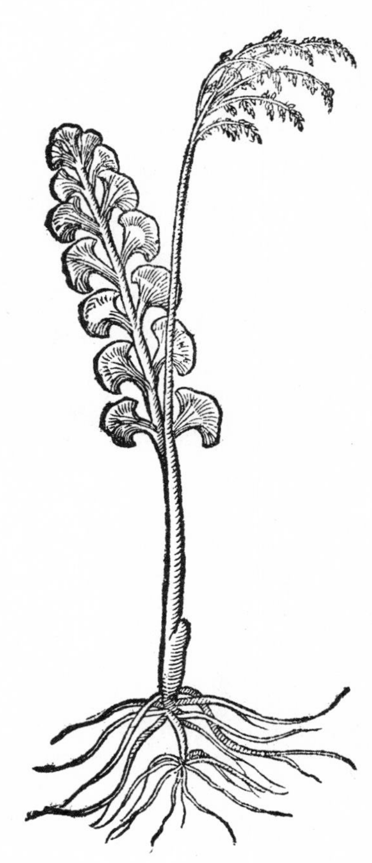

¶ *The Deſcription.*

1 HErbe Paris riſeth vp with one ſmall tender ſtalke two hands high; at the very top whereof come forth foure leaues directly ſet one againſt another in manner of a Burgundian Croſſe or True-loue knot: for which cauſe among the Antients it hath bin called Herbe True-loue. In the midſt of the ſaid leafe comes forth a ſtar-like floure of an herby or graſſie colour; out of the middeſt whereof there ariſeth vp a blackiſh browne berrie: the root is long and tender, creeping vnder the earth, and diſperſing it ſelfe hither and thither.

2 The ſmall Lunary ſpringeth forth of the ground with one leafe like Adders-tongue, iagged or cut on both ſides into fiue or ſix deepe cuts or notches, not much vnlike the leaues of *Scolopendria*, or *Ceterach*, of a greene colour; whereupon doth grow a ſmall naked ſtem of a finger long, bearing at the top many little ſeeds cluſtering together; which being gathered and laid in a platter or ſuch like thing for the ſpace of three weekes, there will fall from the ſame a fine duſt or meale of a whitiſh colour, which is the ſeed if it bring forth any. The root is ſlender, and compact of many ſmall threddy ſtrings.

‡ In England (ſaith *Camerarius*) there growes a certaine kinde of *Lunaria*, which hath many leaues, and ſometimes alſo ſundry branches; which therefore I haue cauſed to be delineated, that other Herbariſts might alſo take notice hereof. Thus much *Camerarius, Epit. Mat. p. 644.* where he giues an elegant figure of a varietie hauing more leaues and branches than the ordinary, otherwiſe not differing from it.

3 Beſides this varietie there is another kinde ſet forth by *Cluſius*; whoſe figure and deſcription I thinke good here to ſet downe. This hath a root conſiſting of many fibres ſomewhat thicker than thoſe of the common kinde: from which ariſe one or two winged leaues, that is, many leaues ſet to one ſtalke; and theſe are like the leaues of the other *Lunaria*, but that they are longer, thicker, and more diuided, and of a yellowiſh greene colour. Amongſt theſe leaues there comes vp a ſtalke fat and juycie, bearing a greater tuſt of floures or ſeeds (for I know not whether to cal them) than the ordinarie, but otherwiſe very like thereto. It groweth in the mountaines of Sileſia, and in ſome places of Auſtria. ‡

‡ 3 *Lunaria minor ramoſa.*
Small branched Moon-wort.

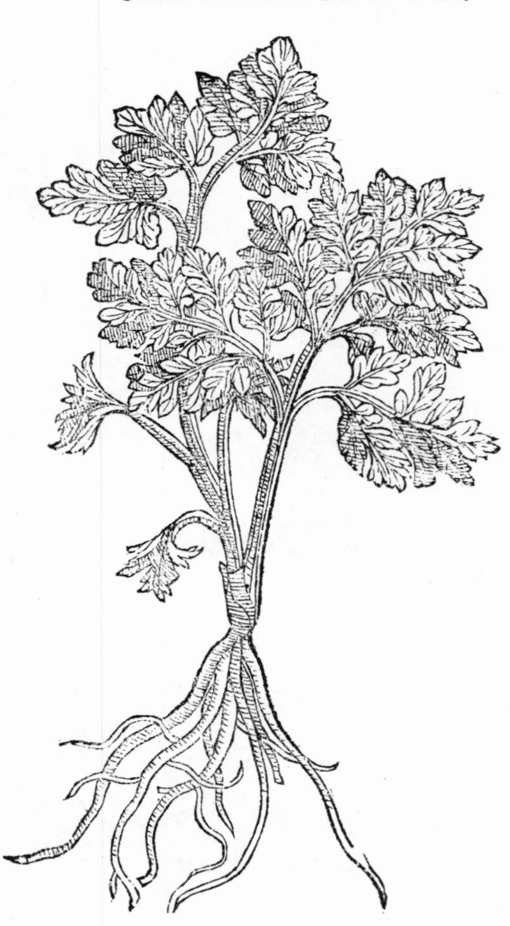

¶ *The Place.*

Herba Paris groweth plentifully in all theſe places following; that is to ſay, in Chalkney wood neere to wakes Coulne, ſeuen miles from Colcheſter in Eſſex, and in the wood by Robinhoods well, neere to Nottingham; in the parſonage orchard at Radwinter in Eſſex, neere to Saffron Walden; in Blackburne at a place called Merton in Lancaſhire; in the Moore by Canturbury called the Clapper; in Dingley wood, ſix miles from Preſton in Aunderneſſe; in Bocking parke by Braintree in Eſſex; at Heſſet in Lancaſhire, and in Cotting wood in the North of England; as that excellent painefull and diligent Phyſition Mr. Doctor *Turner* of late memorie doth record in his Herbal.

Lunaria or ſmall Moone-wort groweth vpon dry and barren mountaines and heaths. I haue found it growing in theſe places following; that is to ſay, about Bathe in Somerſetſhire in many places, eſpecially at a place called Carey, two miles from Bruton, in the next Cloſe vnto the Church-yard; on Cockes Heath betweene Lowſe and Linton, three miles from Maidſtone in Kent: it groweth alſo in the ruines of an old bricke-kilne by Colcheſter, in the ground of Mr. *George Sayer*, called Miles end: it groweth likewiſe vpon the ſide of Blacke-heath, neere vnto the ſtile that leadeth vnto Eltham houſe, about an hundred paces from the ſtile: alſo in Lancaſhire neere vnto a Wood called Faireſt, by Latham: moreouer, in Nottinghamſhire by the Weſt wood at Gringley, and at Weſton in the Ley field by the Weſt ſide of the towne; and in the Biſhops field at Yorke, neere vnto Wakefield, in the Cloſe where Sir *George Sauill* his houſe ſtandeth, called the Heath Hall, by the relation of a learned Doctor in Phyſicke called Mr. *Iohn Merſhe* of Cambridge, and many other places.

¶ *The Time.*

Herba Paris floureth in Aprill, and the berry is ripe in the end of May.

Lunaria or ſmall Moone-wort is to be ſeene in the moneth of May.

¶ *The Names.*

One-berry is alſo called Herbe True-loue, and Herbe Paris: in Latine, *Herba Paris*, and *Solanum tetraphyllum* by *Geſner* and *Lobel*.

Lunaria minor is called in Engliſh Small Lunarie, and Moon-wort.

¶ *The Nature.*

Herbe Paris is exceeding cold; whereby it repreſſes the rage and force of poiſon.

Lunaria minor is cold and dry of temperature.

¶ *The*

¶ *The Vertues.*

A The berries of Herbe Paris giuen by the space of twentie daies, are excellent good against poison,or the pouder of the herbe drunke in like manner halfe a spoonfull at a time in the morning fasting.

B The same is ministred with great successe vnto such as are become peeuish, or without vnderstanding,being ministred as is aforesaid,euery morning by the space of twentie daies, as *Baptista Sardus*, and *Matthiolus* haue recorded. Since which time there hath been further experience made thereof against poison,and put in practice in the citie of Paris,in Louaine,and at the baths in Heluetia,by the right excellent Herbarists *Matthias de L'obel*, and *Petrus Pena*, who hauing often read, that it was one of the Aconites,called *Pardalianches*,and so by consequence of a poisoning quality, they gaue it vnto dogs and lambes, who receiued no hurt by the same: wherefore they further prosecuted the experience thereof,and gaue vnto two dogs fast bound or coupled together,a dram of Arsenicke,and one dram of Mercurie sublimate mixed with flesh (‡ in the *Aduersaria* it is but of each halfe a dram, and there *pag.* 105. you may finde this Historie more largely set downe. ‡) which the dogs would not willingly eat, and therefore they had it crammed downe their throats: vnto one of these dogs they gaue this Antidote following in a little red wine, whereby he recouered his former health againe within a few houres: but the other dog which had none of the medicine,died incontinently.

This is the receit.

R. *vtriusque Angelicæ (innuit) domesticam,& syluestrem,Vicetoxici,Valerianæ domesticæ,Polipodij querni,radicum Altheæ,& Vrticæ,ana ℥.iij, Corticis Mezerei Germanici, ℥.ij. granorum herbæ Paridis, N.24. foliorum eiusdem cumtoto, Num.36. Ex maceratis in aceto radicibus, & siccatis fit omnium pulvis.*

C The people in Germany do vse the leaues of Herbe Paris in greene wounds, for the which it is very good,as *Ioachimus Camerarius* reporteth;who likewise saith,that the pouder of the roots giuen to drink,doth speedily cease the gripings and paine of the Collicke.

D Small Moonewoort is singular to heale greene and fresh wounds:it staieth the bloudy flix. It hath beene vsed among the Alchymistes and witches to doe wonders withall,who say,that it will loose lockes,and make them to fall from the feet of horses that grase where it doth grow,and hath beene called of them *Martagon*,whereas in truth they are all but drowsie dreames and illusions;but it is singular for wounds as aforesaid.

CHAP. 90. *Of Winter-Greene.*

¶ *The Description.*

1 *Pyrola* hath many tender and verie greene leaues,almost like the leaues of Beete, but rather in my opinion like to the leaues of a Peare-tree, whereof it tooke his name *Pyrola*, for that it is *Pyriformis*. Among these leaues commeth vp a stalke garnished with prettie white floures, of a verie pleasant sweet smell, like *Lillium Conuallium*, or the Lillie of the Valley. The root is small and threddie, creeping farre abroad vnder the ground.

‡ 2 This differs from the last described in the slendernesse of the stalkes, and smalnesse of the leaues and floures: for the leaues of this are not so thicke and substantiall, but very thinne, sharpe pointed,and very finely snipt about the edges, blacker, and resembling a Peare-tree leafe. The floures are like those of the former,yet smaller and more in number:to which succeed fiue cornered seed vessels with a long pointell as in the precedent: the root also creepes no lesse than that of the former,and here and there puts vp new stalkes vnder the mosse. It growes vpon the Austrian and Styrian Alpes,and floures in Iune and Iuly.

3 This is an elegant plant,and sometimes becomes shrubbie, for the new and short branches growing vp each yeare,doe remaine firme and greene for some yeares,and grow straight vp, vntill at length borne downe by their owne weight they fall downe and hide themselues in the mosse. It hath commonly at each place where new branches growe forth, two, three, or foure thicke verie greene and shining leaues,almost in forme and magnitude like to the leaues of *Laureola*,yet snipt about the edges,of a very drying taste,and then bitterish. From among these leaues at the Spring of the yeare new branches shoot vp,hauing small leaues like scailes vpon them,and at their toppes grow

1 *Pyrola.*
Winter Greene.

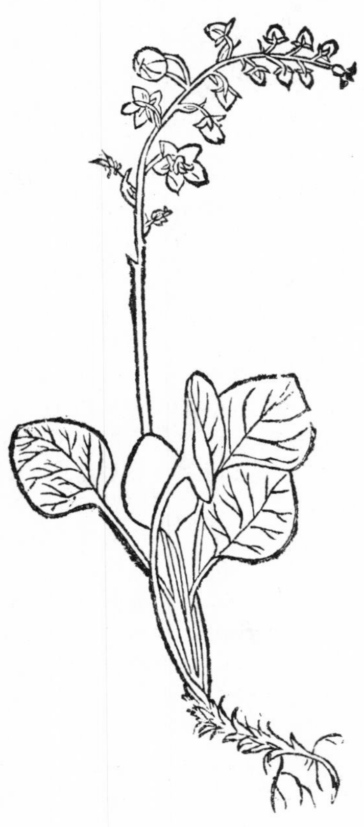

‡ 2 *Pyrola 2 tenerior Cluſ.*
The ſmaller Winter-Greene.

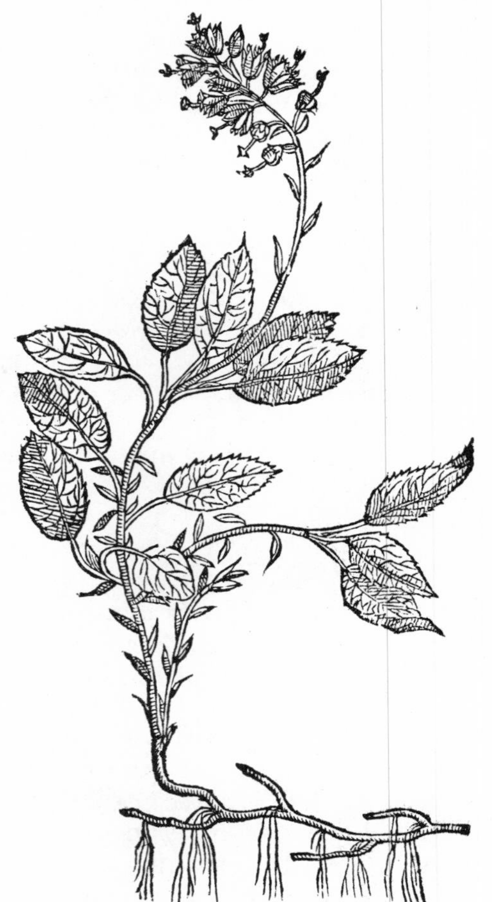

‡ 3 *Pyrola 3. fruticans Cluſ.*
Shrubby Winter-Greene.

‡ 4 *Pyrola 4. minima Cluſ.*
Round leaued Winter Greene.

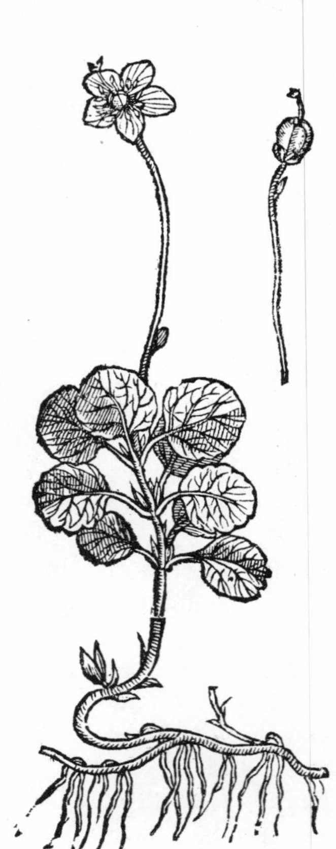

5 *Monophyllon.*
One Blade.

grow floures like to thoſe of the firſt deſcribed, yet ſomewhat larger, of a whitiſh purple colour; which fading, are ſucceeded by fiue cornered ſeed veſſels containing a very ſmall ſeed; the roots are long & creeping. It growes a little from Vienna in Auſtria in the woods of Entzeſtorf, and in diuers places of Bohemia and Sileſia.

4 This from creeping roots ſends vp ſhort ſtalkes, ſet at certaine ſpaces with ſmall, round, and thin leaues, alſo ſnipt about the edges, amongſt which vpon a naked ſtem growes a floure of a pretty bigneſ, conſiſting of fiue white ſharpiſh pointed leaues with ten threds, and a long pointell in the midſt. The ſeed is contained in ſuch heads as the former, and it is very ſmall. This growes in the ſhadowie places of the Alpes of Sneberge, Hochbergerin, Durrenſtaine, towards the roots of theſe great mountaines. *Cluſ.* ‡

5 *Monophyllon*, or *Vnifolium*, hath a leafe not much vnlike the greateſt leafe of Iuie, with many ribs or ſinewes like the Plantaine leafe; which ſingle leafe doth alwaies ſpring forth of the earth alone, but when the ſtalke riſeth vp, it bringeth vpon his ſides two leaues, in faſhion like the former; at the top of which ſlender ſtalke come forth fine ſmall floures like *Pyrola*, which being vaded, there ſucceed ſmall red berries. The roote is ſmall, tender, and creeping farre abroad vnder the vpper face of the earth.

¶ *The Place.*

1 *Pyrola* groweth in Lanſdale, and Crauen, in the North part of England, eſpecially in a cloſe called Crag-cloſe.

2 *Monophyllon* groweth in Lancaſhire in Dingley wood, ſix miles from Preſton in Aunderneſſe; and in Harwood, neere to Blackburne likewiſe.

¶ *The Time.*

1 *Pyrola* floureth in Iune and Iuly, and groweth winter and ſommer.

2 *Monophyllon* floureth in May, and the fruit is ripe in September.

¶ *The Names.*

1 *Pyrola* is called in Engliſh Winter-greene: it hath beene called *Limonium* of diuers, but vntruly.

2 *Monophyllon*, according to the etymologie of the word, is called in Latine *Vnifolium*: in Engliſh, One-blade, or One-leafe.

¶ *The Nature.*

1 *Pyrola* is cold in the ſecond degree, and drie in the third.

2 *Monophyllon* is hot and dry of complexion.

¶ *The Vertues.*

Pyrola is a moſt ſingular wound-hearbe, either giuen inwardly, or applied outwardly: the leaues whereof ſtamped and ſtrained, and the iuice made into an vnguent, or healing ſalue, with waxe, oile, and turpentine, doth cure wounds, vlcers, and fiſtulaes, that are mundified from the callous & tough matter, which keepeth the ſame from healing. A

The decoction hereof made with wine, is commended to cloſe vp and heale wounds of the entrailes, and inward parts: it is alſo good for vlcers of the kidneies, eſpecially made with water, and the roots of Comfrey added thereto. B

The leaues of *Monophyllon*, or *Vnifolium*, are of the ſame force in wounds with *Pyrola*, eſpecially in wounds among the nerues and ſinewes. Moreouer, it is eſteemed of ſome late writers a moſt perfect medicine againſt the peſtilence, and all poiſons, if a dram of the root be giuen in vineger mixed with wine or water, and the ſicke go to bed and ſweat vpon it. C

CHAP.

Chap. 91. *Of Lilly in the valley, or May Lilly.*

1 *Lilium conuallium.*
Conuall Lillies.

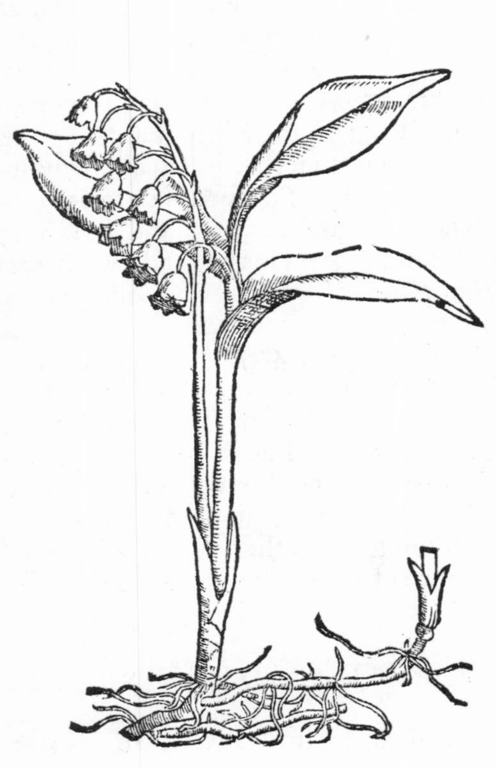

2 *Lilium conuallium floribus suaue-rubentibus.*
Red Conuall Lillies.

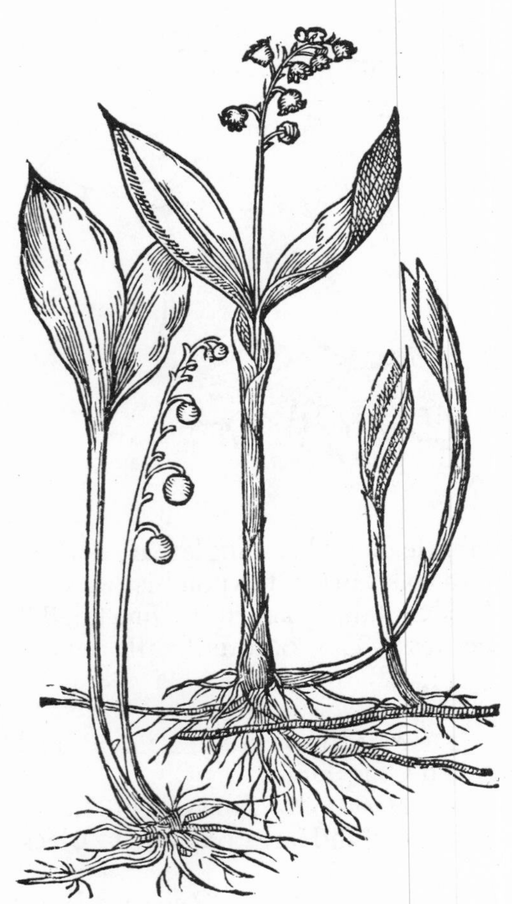

¶ *The Description.*

1 THe Conuall Lillie, or Lilly of the Vally, hath many leaues like the smallest leaues of Water Plantaine; among which riseth vp a naked stalke halfe a foot high, garnished with many white floures like little bels, with blunt and turned edges, of a strong sauour, yet pleasant enough; which being past, there come small red berries, much like the berries of *Asparagus*, wherein the seed is contained. The root is small and slender, creeping far abroad in the ground.

2 The second kinde of May Lillies, is like the former in euery respect; and herein varieth or differeth, in that this kinde hath reddish floures, and is thought to haue the sweeter smell.

¶ *The Place.*

1 The first groweth on Hampsted heath, foure miles from London, in great abundance: neere to Lee in Essex, and vpon Bushie heath, thirteene miles from London, and many other places.

2 That other kind with the red floure is a stranger in England: howbeit I haue the same growing in my garden.

¶ *The Time.*

They floure in May, and their fruit is ripe in September.

¶ *The Names.*

The Latines haue named it *Lilium Conuallium*: *Gesner* doth thinke it to be *Callionymum*: in the Germane tongue, **Meyen blumlen**: the low Dutch, **Meyen bloemkens**: in French, *Muguet*: yet there is likewise another herbe which they call *Muguet*, commonly named in English, Woodroof. It is called in English Lillie of the Valley, or the Conuall Lillie, and May Lillies, and in some places Liriconfancie.

¶ *The Nature.*

They are hot and drie of complexion.

¶ The

¶ *The Vertues.*

The floures of the Valley Lillie diſtilled with wine, and drunke the quantitie of a ſpoonfull, reſtoreth ſpeech vnto thoſe that haue the dum palſie and that are falne into the Apoplexie, and is good againſt the gout, and comforteth the heart. A

The water aforeſaid doth ſtrengthen the memorie that is weakened and diminiſhed; it helpeth alſo the inflammation of the eies, being dropped thereinto. B

The floures of May Lillies put into a glaſſe, and ſet in a hill of antes cloſe ſtopped for the ſpace of a moneth and then taken out, therein you ſhall find a liquour, that appeaſeth the paine & griefe of the gout, being outwardly applied; which is commended to be moſt excellent. C

CHAP. 92. *Of Sea Lauander.*

1 *Limonium.*
Sea Lauander.

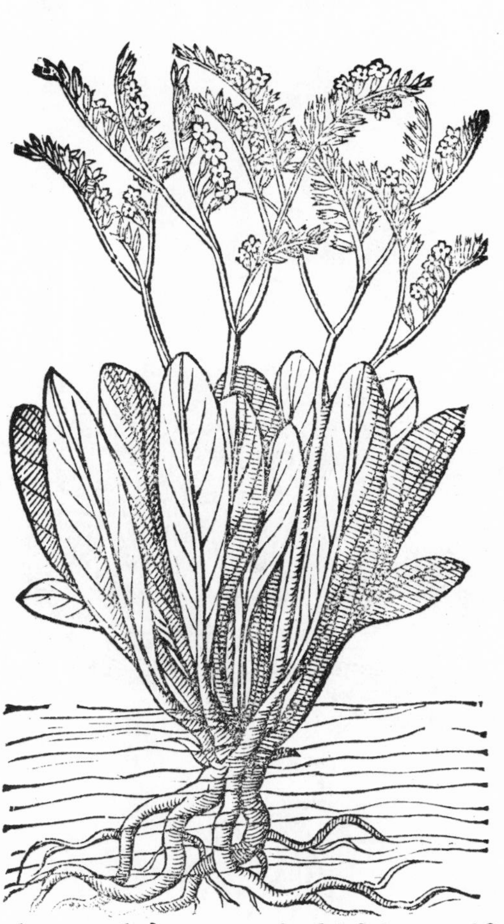

2 *Limonium paruum.*
Rocke Lauander.

¶ *The Deſcription.*

1 There hath beene among writers from time to time, great contention about this plant *Limonium*, no one authour agreeing with another: for ſome haue called this herbe *Limonium*; ſome another herb by this name; & ſome in remouing the rock, haue mired themſelues in the mud, as *Matthiolus*, who deſcribed two kindes, but made no diſtinction of them, nor yet expreſſed which was the true *Limonium*; but as a man heerein ignorant, hee ſpeakes not a word of them. Now then to leaue controuerſies and cauilling, the true *Limonium* is that which hath faire leaues, like the Limon or Orenge tree, but of a darke greene colour, ſomewhat fatter, and a little crumpled: amongſt which leaues riſeth vp an hard and brittle naked ſtalke of a foot high, diuided at the top into ſundry other ſmall branches, which grow for the moſt part vpon the one ſide, full of little blewiſh floures, in ſhew like Lauander, with long red ſeed, and a thicke root like vnto the ſmall Docke.

2 There is a kinde of *Limonium* like the firſt in each reſpect, but leſſer, which groweth vpon rockes and chalkie cliffes.

‡ 3 Beſides theſe two here deſcribed, there is another elegant Plant by *Cluſius* and others referred to this kindred: the deſcription thereof is thus; from a long ſlender root come forth long greene leaues lying ſpred vpon the ground, being alſo deepely ſinuated on both ſides, and ſomewhat roughiſh. Amongſt theſe leaues grow vp the ſtalkes welted with ſlender indented skinnes, and towards their tops they are diuided into ſundry branches after the manner of the ordinarie one; but theſe branches are alſo winged, and at their tops they carry floures ſome foure or fiue clu-

clustering together, consisting of one thin crispe or crumpled leafe of a light blew colour (which continues long, if you gather them in their perfect vigour, and so drie them) and in the middest of this blew comes vp little white floures, consisting of fiue little round leaues with some white threds in their middles. This plant was first obserued by *Rauwolfius* at Ioppa in Syria : but it growes also vpon the coasts of Barbarie, and at Malacca and Cadiz in Spaine : I haue seene it growing with many other rare plants, in the Garden of my kinde friend M^r^. *Iohn Tradescant* at South Lambeth.

4 *Clusius* in the end of his fourth Booke *Historiæ Plantarum*, sets forth this, and saith, hee receiued this figure with one dryed leafe of the plant sent him from Paris from *Claude Gonier* an Apothecarie of that citie, who receiued it (as you see it here exprest) from Lisbone. Now *Clusius* describes the leafe that it was hard, and as if it had been a piece of leather, open on the vpper side, and distinguished with many large purple veines on the inside, &c. for the rest of his description was onely taken from the figure (as he himselfe saith) which I hold impertinent to set downe, seeing I heere giue you the same figure, which by no meanes I could omit, for the strangenesse thereof, but hope that some or other that trauell into forraine parts may finde this elegant plant, and know it by this small expression, and bring it home with them, that so we may come to a perfecter knowledge thereof. ‡

‡ 3 *Limonium folio sinuato.*
Sea-Lauander with the indented leafe.

‡ 4 *Limonio congener, Clus.*
Hollow leaued Sea-Lauander.

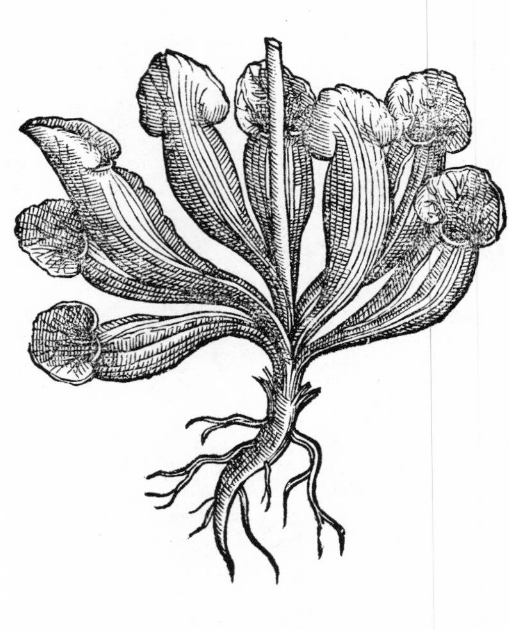

¶ *The Place.*

1 The first groweth in great plentie vpon the walls of the fort against Grauesend: but abundantly on the bankes of the Riuer below the same towne, as also below the Kings Store-house at Chattam : and fast by the Kings Ferrey going into the Isle of Shepey : in the salt marshes by Lee in Essex : in the Marsh by Harwich, and many other places.

2 The

The ſmall kinde I could neuer finde in any other place but vpon the chalky cliffe going from the towne of Margate downe to the ſea ſide, vpon the left hand.

¶ *The Time.*

They floure in Iune and Iuly.

¶ *The Names.*

It ſhall be needleſſe to trouble you with any other Latine name than is expreſt in their titles: the people neere the ſea ſide where it groweth do call it Marſh Lauander, and ſea Lauander.

‡ This cannot be the *Limonium* of *Dioſcorides*, for the leaues are not longer than a Beet, nor the ſtalke ſo tall as that of a Lillie, but you ſhall finde more hereafter concerning this in the Chapter of water Plantaine. I cannot better refer this to any plant deſcribed by the Antients than to *Britannica* deſcribed by *Dioſcorides, lib.* 4. *cap.* 2. ‡

¶ *The Nature.*

The ſeed of *Limonium* is very aſtringent or binding.

¶ *The Vertues.*

The ſeed beaten into pouder, and drunke in wine, helpeth the collicke, ſtrangurie, and Dyſenteria. A

The ſeed taken as aforeſaid, ſtaieth the ouermuch flowing of womens termes, and all other fluxes of bloud. B

Chap. 93. *Of Serapias Turbith, or Sea Starwort.*

1 *Tripolium vulgare majus.*
Great Sea Starwort.

‡ 2 *Tripolium vulgare minus.*
Small Sea Starwort.

¶ *The Deſcription.*

1 The firſt kinde of *Tripolium* hath long and large leaues ſomewhat hollow or furrowed, of a ſhining greene colour declining to blewneſſe, like the leaues of Woade: among which riſeth vp a ſtalke of two cubits high, and more, which toward the top is diuided into many ſmall branches garniſhed with many floures like Camomill, yellow in the middle, ſet about

or bordered with small blewish leaues,like a pale,as in the floures of Camomill, which grow into a whitish rough downe, that flieth away with the winde. The root is long and thready.

2 There is another kinde of *Tripolium* like the first, but much smaller, wherein consisteth the difference.

¶ *The Place.*

These herbs grow plentifully alongst the English coasts in many places,as by the fort against Grauesend, in the Ile of Shepey in sundry places, in a marsh which is vnder the towne walls of Harwich, in the marsh by Lee in Essex, in a marsh which is betweene the Ile of Shepey and Sandwich, especially where it ebbeth and floweth: being brought into gardens, it flourisheth a long time, but there it waxeth huge,great, and ranke; and changeth the great roots into strings.

¶ *The Time.*

These herbs do floure in May and Iune.

¶ *The Names.*

It is reported by men of great fame and learning, that this plant was called *Tripolium*, because it doth change the colour of his floures thrice in a day. This rumour we may beleeue,and it may be true, for that we see and perceiue things of as great and greater wonder to proceed out of the earth. This herbe I planted in my garden, whither (in his season) I did repaire to finde out the truth hereof, but I could not espie any such variablenesse herein; yet thus much I may say, that as the heate of the sunne doth change the colour of diuers floures, so it fell out with this, which in the morning was very faire, but afterward of a pale or wan colour. Which prooueth that to be but a fable which *Dioscorides* saith is reported by some, that in one day it changeth the colour of his floures thrice: that is to say, in the morning it is white, at noone purple, and in the euening φοινικοῦν: or crimson. But it is not vntrue, that there may be found three colours of the floures in one day, by reason that the floures are not all perfected together (as before I partly touched) but one after another by little and little. And there may easily be obserued three colours in them, which is to be vnderstood of them that are beginning to floure, that are perfectly floured, and those that are falling away. For they that are blowing and be not wide open and perfect, are of a purplish colour, and those that are perfect and wide open, of a whitish blew; and such as haue fallen away haue a white down: which changing hapneth vnto sundry other plants. This herbe is called of *Serapio*, *Turbith*: women that dwell by the sea side, call it in English, blew Daisies, or blew Camomill; and about Harwich it is called Hogs beanes, for that the swine do greatly desire to feed thereon: as also for that the knobs about the roots doe somewhat resemble the Garden Beane. It is called in Greeke τριπόλιον: and diuers others τριπόλιον: it may be fitly called *Aster Marinus*, or *Amellus Marinus*: in English, Sea Starwort, Serapio's Turbith: of some, Blew Daisies. The Arabian *Serapio*, doth call Sea Starwort, Turbith, and after him, *Auicen*: yet *Actuarius* the Grecian doth thinke that Turbith is the root of *Alypum*: *Mesues* iudged it to be the root of an herbe like fennell. The Historie of Turbith of the shops shall be discoursed vpon in his proper place.

¶ *The Nature.*

Tripolium is hot in the third degree, as *Galen* saith.

¶ *The Vertues.*

A The root of *Tripolium* taken in wine by the quantitie of two drams, driueth forth by siege waterish and grosse humors, for which cause it is often giuen to them that haue the dropsie.

B It is an excellent herbe against poison, and comparable with *Pyrola*, if not of greater efficacy in healing of wounds either outward or inward.

Chap. 94. Of Turbith of Antioch.

¶ *The Description.*

Garcias a Portugal Physition saith that Turbith is a plant hauing a root which is neither great nor long: the stalke is of two spans long, sometimes much longer, a finger thicke, which creepeth in the ground like Iuie, and bringeth forth leaues like those of the marish Mallow. The floures be also like those of the Mallow, of a reddish white colour: the lower part of the stalke only, which is next to the root and gummie, is that which is profitable in medicine, and is the same that is vsed in shops: they chuse that for the best which is hollow, and round like a reed, brittle, and with a smooth barke, as also that whereunto doth cleaue a congealed gum, which is said to be *gummosum*, or gummy, and somewhat white. But, as *Garcias* saith, it is not alwaies gummie

gummie of his owne nature, but the Indians because they see that our merchants note the best Turbith by the gumminesse, are wont before they gather the same, either to writhe or else lightly to bruse them, that the sap or liquor may issue out; which root being once hardned, they picke out from the rest to sell at a greater price. It is likewise made white, as the said Author sheweth, being dried in the sunne: for if it be dried in the shadow it waxeth blacke, which notwithstanding may be as good as the white which is dried in the Sunne.

Turbith Alexandrinum officinarum.
Turpetum, or Turbith o fthe shops.

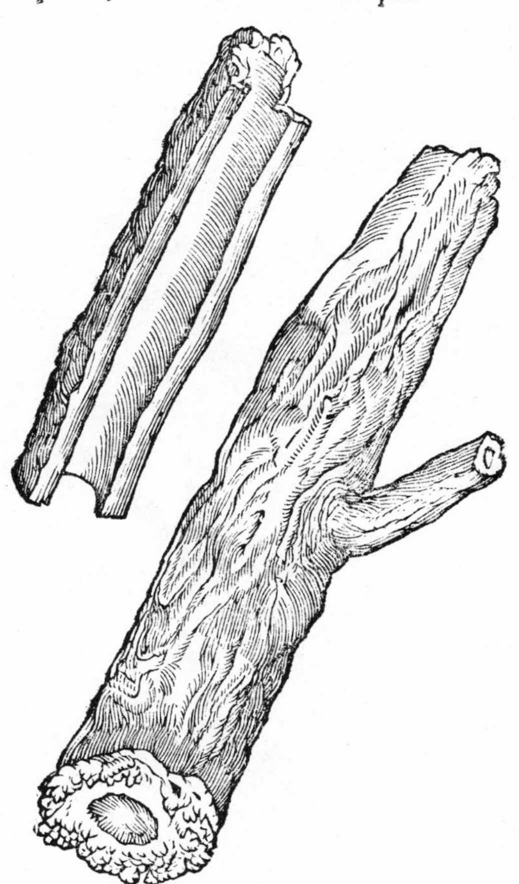

¶ *The Place.*

It groweth by the sea side, but yet not so neere that the wash or water of the sea may come to it, but neere about, and that for two or three miles in vntilled grounds, rather moist than drie. It is found in Cambaya, Surrate, in the Ile Dion, Bazaim, and in places hard adioining; also in Guzarate, where it groweth plentifully, from whence great abundance of it is brought into Persia, Arabia, Asia the lesse, and also into Portingale and other parts of Europe: but that is preferred which groweth in Cambaya.

¶ *The Names.*

It is called of the Arabians, Persians, and Turkes *Turbith*: and in Guzarata *Barcaman*: in the prouince Canara, in which is the city Goa, *Tiguar*: likewise in Europe the learned call it diuersly, according to their seuerall fancies, which hath bred sundry controuersies, as it hath fallen out aswell in Hermodactyls, as in Turbith; the vse and possession of which we cannot seeme to want: but which plant is the true Turbith, we haue great cause to doubt; Some haue thought our *Tripolium marinum*, described in the former chapter, to be Turbith: others haue supposed it to be one of the *Tithymales*, but which kinde they know not: *Guillandinus* saith, that the root of *Tithymalus myrsinitis* is the true Turbith; which caused *Lobelius* and *Pena* to plucke vp by the roots all the kindes of *Tithymales*, and drie them very curiously; which when they had beheld, and throughly tried, they found it nothing so. The Arabians and halfe Moores that dwell in the East parts haue giuen diuers names vnto this plant: and as their words are diuers, so haue they diuers significatious; but this name Turbith they seeme to interpret to be any milky root which doth strongly purge flegme, as this plant doth. So that as men haue thought good, pleasing themselues, they haue made many and diuers constuctions which haue troubled many excellent learned men to know what root is the true Turbirh. But briefly to set downe my opinion, not varying from the iudgment of men which are of great experience; I thinke assuredly that the root of Scammony of Antioch is the true and vndoubted Turbith, one reason especially that moueth me so to thinke is, for that I haue taken vp the roots of Scammony which grew in my garden, and compared them with the roots of Turbith, between which I found little ot no difference at all

‡ Through all Spain (as *Clusius* in his notes vpon *Garcias* testifies) they vse the roots of *Thapsia* for Turbith which also haue been brought hither, and I keepe some of them by me, but they purge little or nothing at all being drie, though it may be the green root or juice may haue some purging faculty. ‡

¶ *The Temperature and Vertues.*

The Indian physitions vse it to purge flegme, to which if there be no feuer they adde ginger, otherwise they giue it without in the broth of a chicken, and sometimes in faire water. A

Mesues writeth, that Turbith is hot in the third degree; and that it voideth thicke tough flegme out of the stomacke, chest, sinewes, and out of the furthermost parts of the body: but (as he saith) it is slow in working, and troubleth and ouerturneth the stomacke: and therefore ginger, masticke, and other spices are to be mixed with it; also oile of sweet almondes, or almondes themselues, or sugar, least the body with the vse herof should pine and fall away. Others temper B

per

per it with Dates, ſweet Almonds, and certaine other things, making thereof a compoſition (that the Apothecaries call an Electuarie) which is named διαφοινικῶν: common in ſhops, and in continuall vſe among expert Phyſitions.

C There is giuen at one time of this Turbith one dram (more or leſſe) two at the moſt: but in the decoction, or in the infuſion three or foure.

Chap. 95. *Of Arrow-head, or Water-archer.*

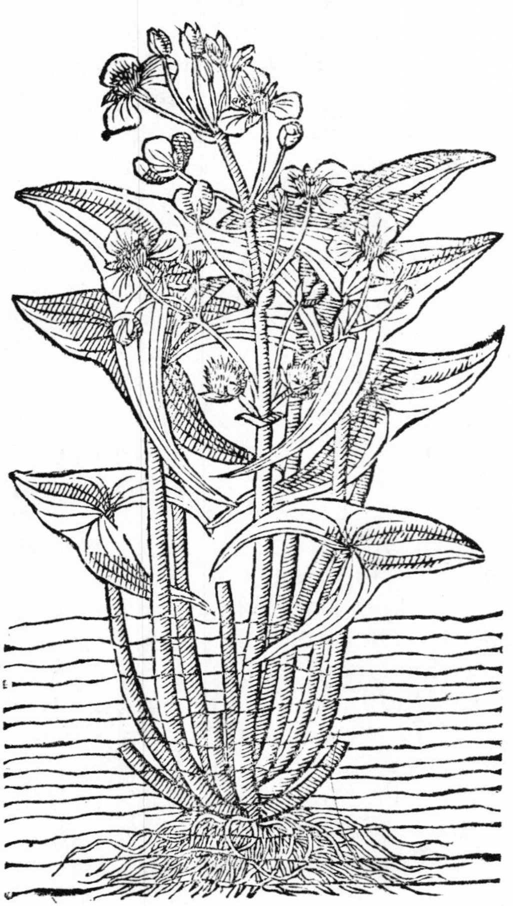

1 *Sagittaria maior.* Great Arrow-head.

2 *Sagittaria minor.* Small Arrow-head.

¶ *The Deſcription.*

1 The firſt kinde of Water-archer or Arrow-head, hath large and long leaues, in ſhape like the ſigne *Sagittarius*, or rather like a bearded broad Arrow head. Among which riſeth vp a fat and thicke ſtalke, two or three foot long, hauing at the top many prettie white floures, declining to a light carnation, compact of three ſmall leaues: which being paſt, there come after great rough knops or burres wherein is the ſeed. The root conſiſteth of many ſtrings.

2 The ſecond is like the firſt, and differeth in that this kinde hath ſmaller leaues and floures, and greater burres and roots.

3 The third kinde of Arrow-head hath leaues in ſhape like the broad Arrow-head, ſtanding vpon the ends of tender foot ſtalkes a cubit long: among which riſe vp long naked ſmooth ſtalks of a greeniſh colour, from the middle whereof to the top doe grow floures like to the precedent. The root is ſmall and threddie.

¶ *The Place.*

Theſe herbes doe grow in the watrie ditches by Saint George his field neere vnto London; in the Tower ditch at London; in the ditches neere the wals of Oxford; by Chelmesford in Eſſex, and many other places, as namely in the ditch neere the place of execution, called Saint *Thomas Waterings* not far from London.

¶ *The Time.*

They floure in May and Iune.

¶ *The*

¶ *The Names.*

Sagittaria, may be called in Engliſh the Water-archer, or Arrow-head. ‡ Some would haue it the *Phleum* of *Theophraſtus*; and it is the *Piſtana Magonis*, and *Sagitta* of *Pliny*, *lib.* 21. *cap.* 17.

¶ *The Nature and Vertues.*

I finde not any thing extant in writing either concerning their vertues or temperament, but doubtleſſe they are cold and drie in qualitie, and are like Plantaine in facultie and temperament.

CHAP. 96. *Of Water Plantaine.*

1 *Plantago aquatica maior.*
Great Water Plantaine.

2 *Plantago aquatica minor ſtellata.*
Starry headed ſmall Water Plantaine.

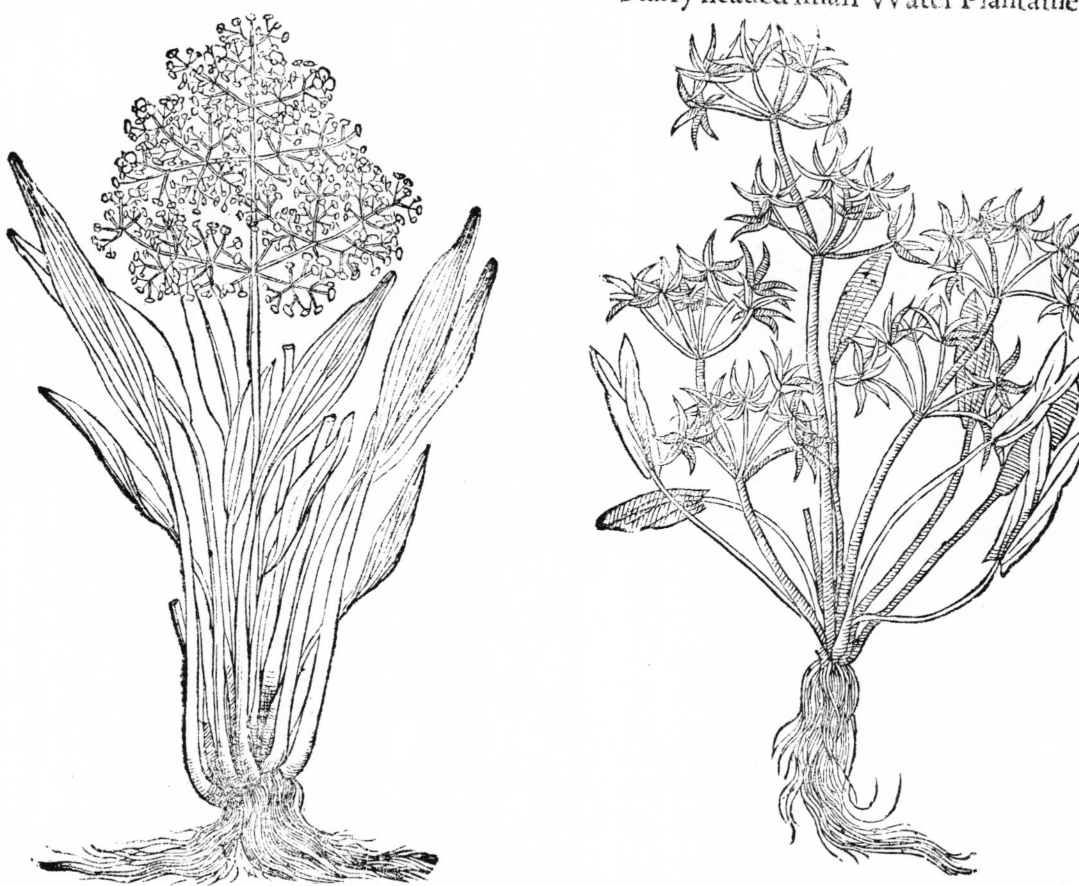

3 *Plantago aquatica humilis.* Dwarfe water Plantaine.

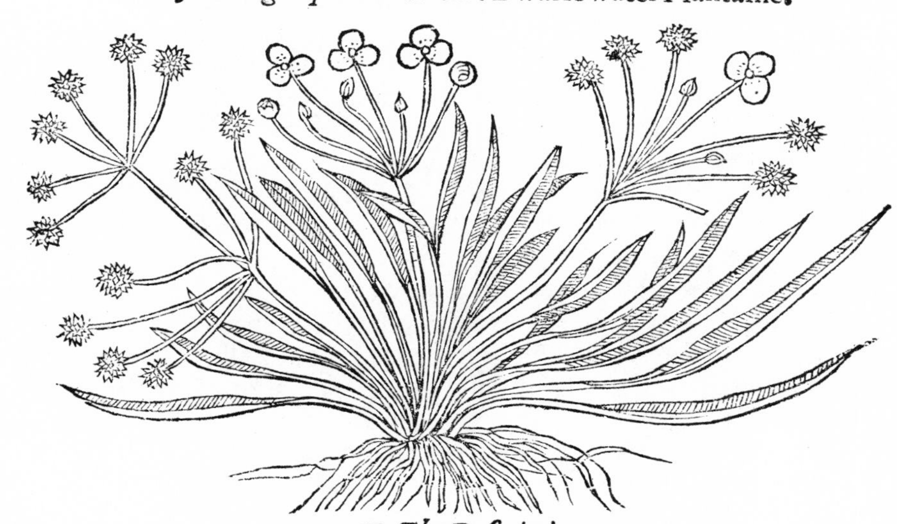

¶ *The Deſcription.*

1 THe firſt kinde of water Plantaine hath faire great large leaues like the land Plaintaine, but ſmoother, and full of ribs or ſinewes: among which riſeth vp a tall ſtemme foure foot high, diuiding it ſelfe into many ſlender branches, garniſhed with infinit ſmall white floures,

which being past there appeare triangle huskes or buttons wherein is the seed. The root is as it were a great tuft of threds or thrums.

‡ 2 This plant in his roots and leaues is like the last described, as also in the stalke, but much lesse in each of them, the stalke being about some foot high; at the top whereof stand many pretty starre-like skinny seed-vessels, containing a yellowish seed. ‡

3 The second kinde hath long, little, and narrow leaues, much like the Plantaine called Ribwoort: among which rise vp small and feeble stalks branched at the top, whereon are placed white floures, consisting of three slender leaues; which being fallen, there come to your view round knobs, or rough burs: the root is thready.

¶ *The Place.*

1 This herbe growes about the brinkes of riuers, ponds and ditches almost euery where.

‡ 2 3 These are more rare. I found the second a little beyond Ilford, in the way to Rumford, and Mr. *Goodyer* found it also growing vpon Hounslow heath. I found the third in the Company of Mr. *William Broad*, and Mr. *Leonard Buckner*, in a ditch on this side Margate in the Isle of Tenet. ‡

¶ *The Time.*

They floure from Iune till August.

¶ *The Names.*

The first kinde is called *Plantago aquatica*, that is, water Plantaine. ‡ The second *Lobell* calls *Alisma pusillum Angustifolium muricatum*, and in the *Hist. Lugd.* it is called *Damasonium stellatum.* ‡ The third is named *Plantago aquatica humilis*, that is, the low water Plantaine.

‡ I thinke it fit here to restore this plant to his antient dignitie, that is, his names and titles wherewith he was anciently dignified by *Dioscorides* and *Pliny*. The former whereof calls it by sundry names, and all very significant and proper, as λειμώνιον, ποταμογείτων, νευροειδὲς, λογχῖτις: thus many are Greek, and therefore ought not to be reiected, as they haue been by some without either reason or authoritie. For the barbarous names we can say nothing; now it is said to be called *Limonium*, because ἐν λειμῶσι φύεται: it growes in wet or ouerflowen medowes: it is called *Neuroides*, because the leafe is composed of diuers strings or fibres running from the one end thereof to the other, as in Plantain, which therfore by *Dioscorides* is termed by the same reason πολύνευρος: Also it may be as fitly termed *Lonchitis* for the similitude which the leafe hath to the top or head of a lance which λόγχη properly signifies, as that other plant described by *Dios. lib. 3. cap. 161.* for that the seed (a lesse eminent part) resembles the same thing. And for *Potamogeiton* which signifies a neighbour to the Riuer or water, I thinke it loues the water aswell, and is as neere a neighbour to it as that which takes it's name from thence, and is described by *Dioscorides, lib. 4. cap. 101.* Now to come to *Pliny, lib. 20. cap. 8.* he calls it, *Beta siuestris, Limonion*, and *Neuroides*: the two later names are out of *Dioscorides*, and I shall shew you where also you shall finde the former in him. Thus much I thinke might serue for the vindication of my assertion, for I dare boldly affirme that no late writer can fit all these names to any other plant, and that makes me more to wonder that all our late Herbarists as *Matthiolus, Dodonæus, Fuchsius, Cæsalpinus, Daleschampius*, but aboue all *Pena* and *Lobell*, who *Aduers. pag. 126.* call it to question, should not allow this plant to be *Limonium*, especially seing that *Anguillara* had before or in their time asserted it so to be; but whether he gaue any reasons or no for his assertion, I cannot tell, because I could neuer by any meanes get his Opinions, but only finde by *Bauhine* his *Pinax* that such was his opinion hereof. But to returne from whence I digrest, I will giue you *Dioscorides* his description, and a briefe explanation thereof, and so desist; it is thus: It hath leaues like a Beet, thinner and larger, 10. or more; a stalke slender, straight, and as tall as that of a Lilly, and full of seeds of an astringent taste. The leaues of this you see are larger than those of a Beet, and thin, and as I formerly told you in the names, neruous; which to be so may be plainely gathered by *Dioscorides* his words in the description of white Hellebore, whose leaues he compares to the leaues of Plantaine and the wilde Beet: now there is no wild Beet mentioned by any of the Antients, but only this by *Pliny* in the place formerly quoted, nor no leafe more fit to compare those of white Hellebore to, than those of water Plantaine, especially for the nerues and fibres that run alongst the leaues, the stalke also of this is but slender considering the height, and it growes straight, and as high as that of a Lilly, with the top plentirifully stored with astringent seed; so that no one note is wanting in this, nor scarse any to be found in the other plants that many haue of late set forth for *Limonium.* ‡

¶ *The Nature.*

Water Plantaine is cold and dry of temperature.

¶ The

¶ *The Vertues.*

The leaues of water Plantaine, as ſome Authors report, are good to be laid vpon the legs of ſuch as are troubled with the Dropſie, and hath the ſame propertie that the land Plantaine hath. A

‡ *Dioſcorides* and *Galen* commend the ſeed hereof giuen in Wine, againſt Fluxes, Dyſenteries, the ſpitting of bloud, and ouermuch flowing of womens termes. B

Pliny ſaith, the leaues are good againſt burnes. ‡ C

CHAP. 97. *Of Land Plantaine.*

1 *Plantago latifolia.*
Broad leaued Plantaine.

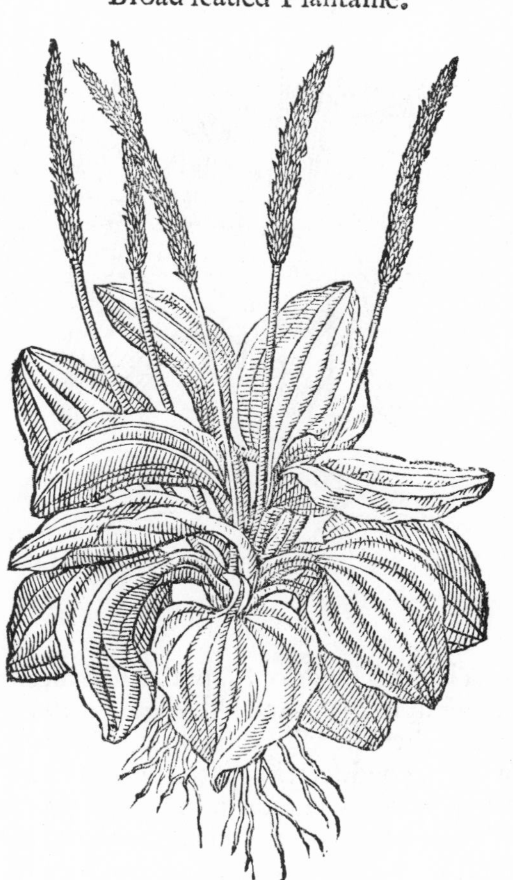

2 *Plantago incana.*
Hoarie Plantaine.

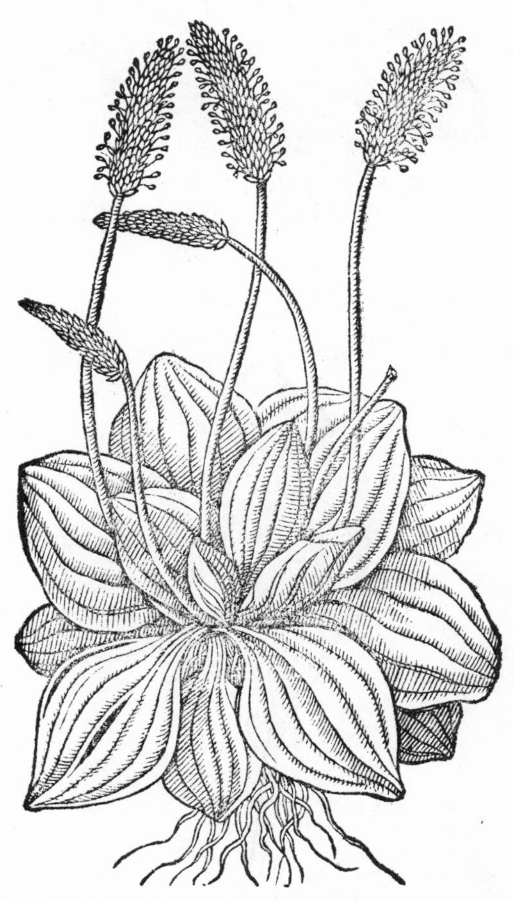

¶ *The Deſcription.*

1 AS the Greekes haue called ſome kindes of Herbes Serpents tongue, Dogs tongue, and Oxe tongue; ſo haue they termed a kind of Plantaine *Arnogloſſon*, which is as if you ſhould ſay Lambes tongue, very well knowne vnto all, by reaſon of the great commoditie and plenty thereof growing euery where; and therefore it is needleſſe to ſpend time about them. The greatneſſe and faſhion of the leaues hath been the cauſe of the varieties and diuerſities of their names

2 The ſecond is like the firſt kinde, and differeth in that, that this kinde of Plantaine hath greater, but ſhorter ſpikes or knaps: and the leaues are of an hoarie or ouerworne greene colour: the ſtalkes are likewiſe hoary and hairy.

3 The ſmall Plantaine hath many tender leaues ribbed like vnto the great Plantaine, and is very like in each reſpect vnto it, ſauing that it is altogether leſſer.

4 The ſpiked Roſe Plantaine hath very few leaues, narrower than the leaues of the ſecond kinde of Plantaine, ſharper at the ends, and further growing one from another. It beareth a very double floure vpon a ſhort ſtem like a roſe, of a greeniſh colour tending to yellowneſſe. The ſeed groweth vpon a ſpikie tuft aboue the higheſt part of the plant; notwithſtanding it is but very low in reſpect of the other Plantaines aboue mentioned.

5 The

4 *Plantago Roſea ſpicata.*
Spiked Roſe Plantaine.

5 *Plantago Roſea exotica.*
Strange Roſe Plantaine.

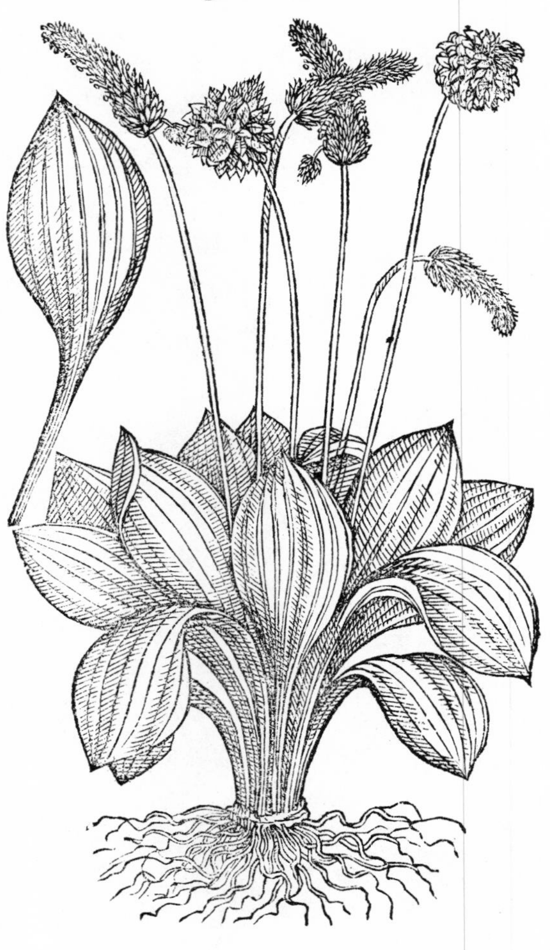

‡ 6 *Plantago panniculis ſparſis.*
Plantaine with ſpoky tufts.

5 The fifth kinde of Plantaine hath beene a ſtranger in England and elſewhere, vntill the impreſſion hereof. The cauſe why I ſay ſo is, the want of conſideration of the beauty which is in this plant, wherein it excelleth all the other. Moreouer, becauſe that it hath not bin written of or recorded before this preſent time, though plants of leſſer moment haue beene very curiouſly ſet forth. This plant hath leaues like vnto them of the former, and more orderly ſpred vpon the ground like a Roſe: among which riſe vp many ſmall ſtalks like the other plantaines, hauing at the top of euery one a fine double Roſe altogether vnlike the former, of an hoary or ruſty greene colour.

‡ I take this ſet forth by our Author to be the ſame with that which *Cluſius* receiued from *Iames Garret* the yonger, from London; and therefore I giue you the figure thereof in this place, together with this addition to the hiſtorie out of *Cluſius*: That ſome of the heads are like thoſe of the former Roſe Plantaine; other ſome are ſpike faſhion, and ſome haue a ſpike growing as it were out of the midſt of the Roſe, and ſome heads are otherwiſe ſhaped: alſo the whole plant is more hoary than the common Roſe Plantaine.

6 This plantain muſt not here be forgot, though it be ſomwhat hard to be found: his leaues, roots, and ſtalkes are like thoſe of the ordinarie, but in ſtead of a compact ſpike it hath one much diuided after the manner as you ſee it here expreſſed in the figure, and the colour thereof is greeniſh. ‡

¶ *The*

¶ *The Place.*

The greater Plantaines do grow almost euery where.

The lesser Plantaine is found on the sea coasts and bankes of great riuers, which are sometimes washed with brackish water.

‡ The Rose Plantaines grow with vs in gardens; and the sixth with spokie tufts groweth in some places in the Isle of Tenet, where I first found it, being in company with Mr. *Thomas Hickes*, Mr. *Leonard Buckner*, and other London Apothecaries, *Anno* 1632. ‡

¶ *The Time.*

They are to be seene from Aprill vnto September.

¶ *The Names.*

Plantaine is called in Latine *Plantago*, and in Greeke ἀρνόγλωσσος, and *Arnoglossa*; that is to say, Lambes tongue: the Apothecaries keepe the Latine name: in Italian, *Piantagine*, and *Plantagine*: in Spanish, *Lhantem*: the Germanes, Wegerich: in Low-Dutch, Weechbre: in English, Plantain, and Weybred: in French, *Plantain*.

¶ *The Temperature.*

Plantaine (as *Galen* saith) is of a mixt temperature; for it hath in it a certaine waterie coldnesse, with a little harshnesse, earthy, dry, and cold: therefore they are cold and dry in the second degree. To be briefe, they are dry without biting, and cold without benumming. The root is of like temperature, but drier, and not so cold. The seed is of subtill parts, and of temperature lesse cold.

¶ *The Vertues.*

Plantaine is good for vlcers that are of hard curation, for fluxes, issues, rheumes, and rottennesse, and for the bloudy flix: it stayeth bleeding, it heales vp hollow sores and vlcers, as well old as new. Of all the Plantaines the greatest is the best, and excelleth the rest in facultie and vertue. A

The iuyce or decoction of Plantaine drunken stoppeth the bloudy flix and all other fluxes of the belly, stoppeth the pissing of bloud, spitting of bloud, and all other issues of bloud in man or woman, and the desire to vomit. B

Plantaine leaues stamped and made into a Tansie, with the yelkes of egges, stayeth the inordinate flux of the termes, although it haue continued many yeares. C

The root of Plantaine with the seed boyled in white Wine and drunke, openeth the conduits or passages of the liuer and kidnies, cures the jaundice, and vlcerations of the kidnies and bladder. D

The juyce dropped in the eyes doth coole the heat and inflammation thereof. I finde in ancient Writers many good-morrowes, which I thinke not meet to bring into your memorie againe; as that three roots will cure one griefe, foure another disease, six hanged about the necke are good for another maladie, &c. all which are but ridiculous toyes. E

The leaues are singular good to make a water to wash a sore throat or mouth, or the priuy parts of a man or woman. F

The leaues of Plantaine stamped and put into Oyle Oliue, and set in the hot Sun for a moneth together, and after boyled in a kettle of seething water (which we doe call *Balneum Mariæ*) and then strained, preuaileth against the paines in the eares, the yard, or matrix, (being dropped into the eares, or cast with a syringe into the other parts before rehearsed) or the paines of the fundament; proued by a learned Gentleman Mr. *William Godowrus* Sergeant Surgeon to the Queenes Maiestie. G

CHAP. 98. *Of Rib-wort.*

¶ *The Description.*

1 RIb-wort or small Plantaine hath many leaues flat spred vpon the ground, narrow, sharp pointed, and ribbed for the most part with fiue nerues or sinewes, and therefore it was called *Quinque-neruia*; in the middle of which leaues riseth vp a crested or ribbed stalke, bearing at the top a darke or dusky knap, set with a few such white floures as are the floures of wheat. The root and other parts are like the other Plantaines.

‡ There is another lesse kinde of this Rib-wort, which differs not from the last mentioned in any thing but the smallnesse thereof. ‡

2 Rose Rib-wort hath many broad and long leaues of a darke greene colour, sharpe pointed, and ribbed with fiue nerues or sinewes like the common Rib-wort; amongst which rise vp naked stalkes furrowed, chamfered, or crested with certaine sharpe edges: at the top whereof groweth a great and large tuft of such leaues as those are that grow next the ground, making one entire tuft of

or vmbel, in shape resembling the rose (wherof I thought good to giue it his syrname Rose) which is from his floure.

‡ This also I think differs not from that of *Clusius*; wherefore I giue his figure in the place of that set forth by our Author. ‡

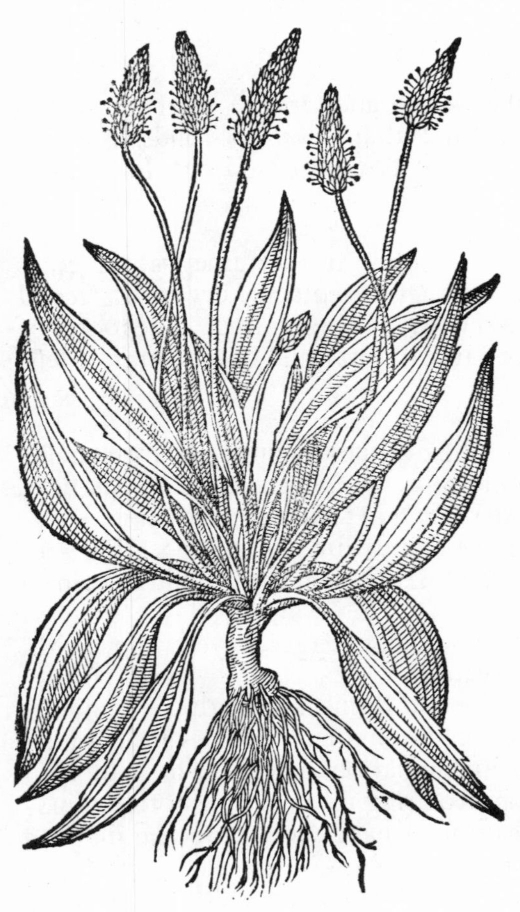

1 *Plantago quinqueneruia.*
Ribwort Plantaine.

2 *Plantago quinqueneruia rosea.*
Rose Ribwort.

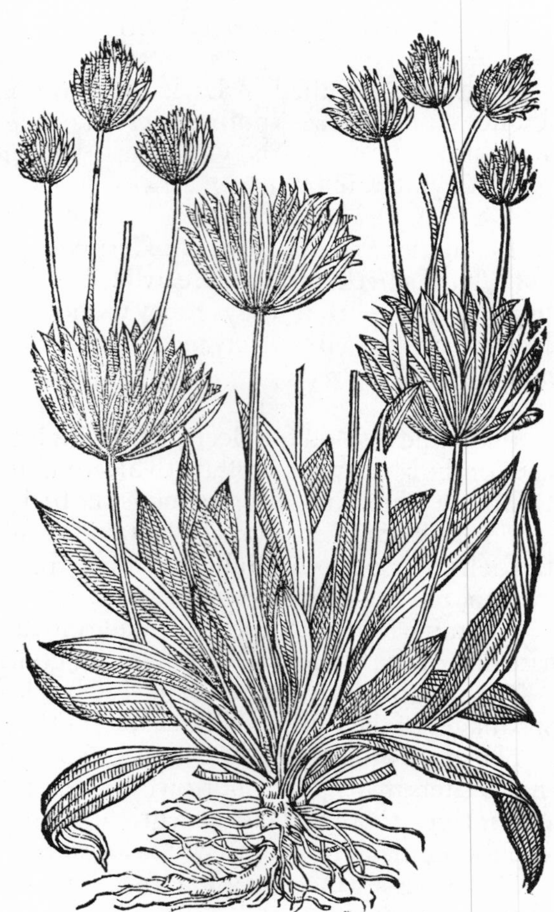

¶ *The Place.*

Ribwort groweth almost euery where in the borders of path-wayes and fertile fields.

Rose Ribwort is not very common in any place, notwithstanding it groweth in my garden, and wilde also in the North parts of England; and in a field neere London by a village called Hogsdon, found by a learned merchant of London Mr. *Iames Cole*, a louer of plants, and very skilfull in the knowledge of them.

¶ *The Time.*

They floure and flourish when the other Plantaines do.

¶ *The Names.*

Ribwort is called in Greeke, Αρνογλωσσον μικρὸν: and of some, πεντάνευρον: in Latine, *Plantago minor*, *Quinqueneruia*, and *Lanceola*, or *Lanceolata*: in high Dutch, **Spitziger wegrich**: in French, *Lanceole*: in Low-Dutch, **Hondts ribbe**; that is to say in Latine, *Costa canina*, or Dogs rib: in English, Ribwort, and Ribwort Plantaine.

The second I haue thought meet to cal Rose Ribwort in English, and *Quinqueneruia rosea* in Latine.

¶ *The Temperature.*

Ribwort is cold and dry in the second degree, as are the Plantains.

¶ *The Vertues.*

The vertues are referred to the kindes of Plantaines.

Chap.

Chap. 99. *Of Sea Plantaines.*

1 *Holosteum Salamanticum.*
Flouring sea Plantaine.

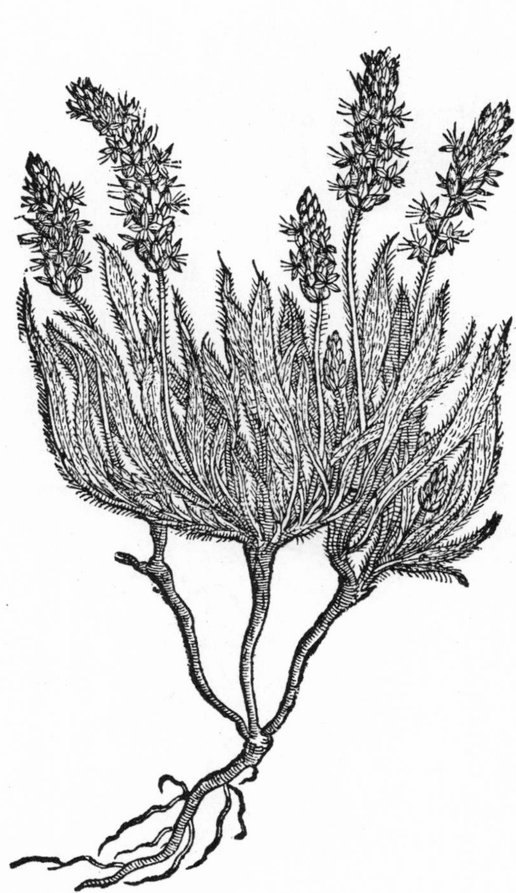

2 *Holosteum parvum.*
Small sea Plantaine.

3 *Plantago marina.*
Sea Plantaine.

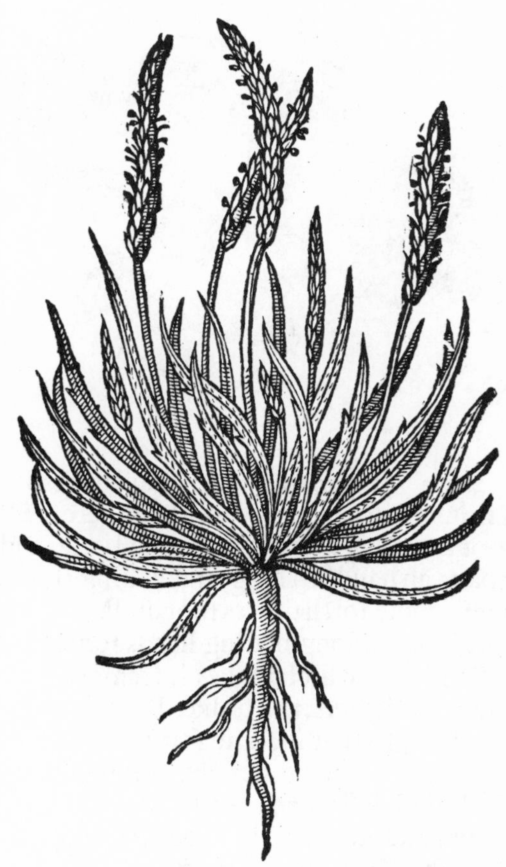

¶ *The Description.*

1 *Carolus Clusius* that excellent Herbarist hath referred these two sorts of *Holosteum* vnto the kindes of Sea Plantaine. The first hath long leaues like the common Rib-wort, but narrower, couered with some hairinesse or wollinesse; among which there riseth vp a stalke, bearing at the top a spike like the kindes of Plantaine, beset with many small floures of an herby colour, declining to whitenesse. The seed is like that of the Plantaine: the root is long and wooddy. This floures in Aprill or May.

2 The second is like the former, but smaller, and not so gray or hoary: the floures are like to *Coronopus*, or the lesser Ribwort. This floures at the same time as the former.

3 The third kinde, which is the sea Plantaine, hath small and narrow leaues like Bucks-horn, but without any manifest incisure, cuttings or notches vpon the one side: among which riseth vp a spikie stalke, like the common kinde, but smaller.

‡ 4 *Holosteum, siue Leontopodium Creticum.*
Candy Lyons foot.

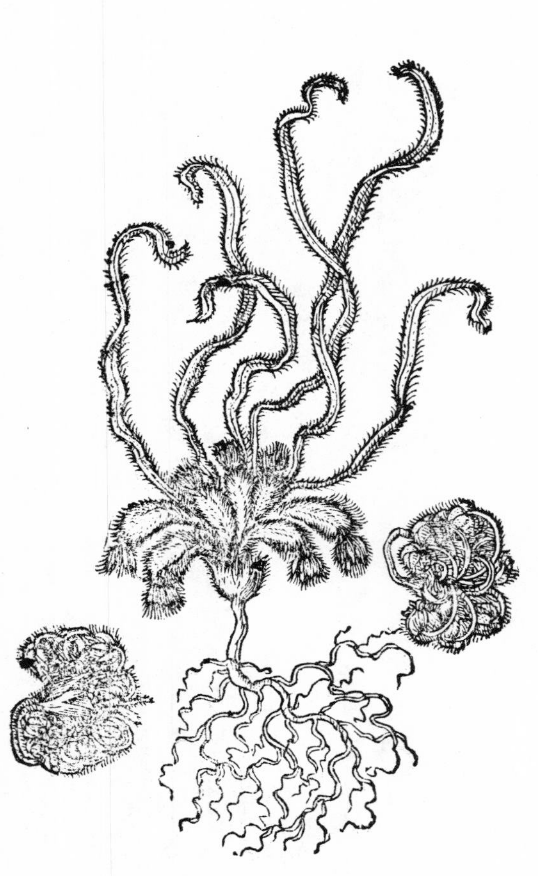

‡ 5 *Holosteum, siue Leontopod. Cret. alterum.*
The other Candy Lyons foot.

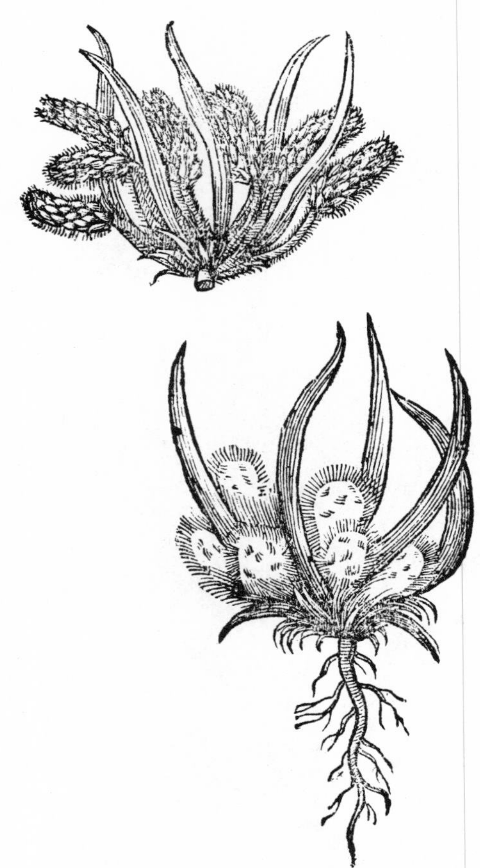

‡ 4 These two following Plants are by *Clusius* and *Bauhine* referred to this Tribe; wherefore I thinke it fitting to place them here. The former of them from a reddish, and as it were scaly root growing lesse by little and little, and diuided into fibres, sends forth many leaues, narrow, hoary, an handfull long, and hauing three nerues or ribbes running alongst each of them: amongst these come forth diuers foot-stalkes, couered with a soft reddish downe, and being some two or three inches long, hauing heads somewhat thicke and reddish: the floures are whitish, with a blackish middle, which makes it seeme as if it were perforated or holed. Now when the plant growes old, and withers, the stalkes becomming more thicke and stiffe, bend downe their heads towards the root, so that in some sort they resemble the foot of a Lyon.

5 This Plant which is figured in the vpper place (for I take the lower to be an exacter figure of the last described) hath leaues like to the small sea Plantaine, but tenderer, and standing vpright; and amongst these on little foot-stalkes grow heads like those of *Psyllium*, but prettier, and of a whitish red colour. ‡

¶ *The Place.*

The two first grow in most of the kingdomes of Spaine. *Carolus Clusius* writeth, that hee neuer saw greater or whiter than neere to Valentia a city of Spaine, by the high-waies. Since, they haue beene found at Bastable in the isle of Wight, and in the isles of Gernsey and Iarsey.

The third doth grow neere vnto the sea in all the places about England where I haue trauelled, especially by the forts on both the sides of the water at Grauesend; at Erith neere London; at Lee in Essex; at Rie in Kent; at West-Chester, and at Bristow.

‡ The fourth and fifth grow in Candy, from whence they haue been sent to Padua and diuers other places. ‡

¶ *The Names.*

Holoſteum is alſo called by *Dodonæus*, *Plantago anguſtifolia albida*, or *Plantago Hiſpanienſis*: in Engliſh, Spaniſh hairy ſmall Plantaine, or flouring ſea Plantaine.

‡ The fourth is called by *Cluſius*, *Leontopodium Creticum*: by ſome it hath beene thought to be *Catanance* of *Dioſcorides*: the which *Honorius Bellus* will not allow of: *Bauhine* calls it *Holoſteum, ſiue Leontopodium Creticum.*

The fifth is *Leontopodium Creticum alterum* of *Cluſius*; the *Habbures* of *Camerarius*; and the *Holoſteum Creticum alterum* of *Bauhine*. ‡

¶ *The Temperature and Vertues.*

Galen ſaith, That *Holoſteum* is of a binding and drying facultie.

A *Galen*, *Dioſcorides*, and *Pliny* haue proued it to be ſuch an excellent wound herbe, that it preſently cloſeth or ſhutteth vp a wound, though it be very great and large: and by the ſame authority I ſpeake it, that if it be put into a pot where many pieces of fleſh are boyling, it will ſoder them together.

Theſe herbes haue the ſame faculties and vertues that the other Plantains haue, and are thought to be the beſt of all the kindes.

† That which was formerly in the fourth place of this chapter, vnder the name of *Holoſteum petræum*, you ſhall finde hereafter vnder the title of *Muſcus corniculatus*; for vnder that name our Author alſo gaue another figure thereof, with a deſcription; and I iudge it more fitly placed in that place, than here amongſt the Plantaines.

CHAP. 100. *Of Sea Buck-horne Plantaines.*

1 *Coronopus.*
Sea Buck-horne.

2 *Coronopus, ſiue Serpentina minor.*
Small Sea Buck-horne.

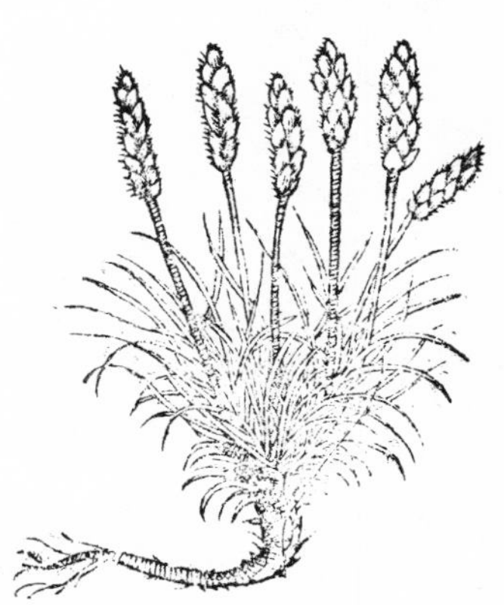

¶ *The Deſcription.*

1 THe new Writers following as it were by tradition thoſe that haue written long agone, haue beene content to heare themſelues ſpeake and ſet downe certainties by vncertaine ſpeeches; which hath wrought ſuch confuſion and corruption of writings, that ſo many Writers, ſo many ſeuerall opinions; as may moſt euidently appeare in theſe plants and in others: And my ſelfe am content rather to ſuffer this ſcar to paſſe, than by correcting the error, to renew the old wound. But for mine owne opinion thus I thinke, the plant which is reckoned for a kinde of *Coronopus* is doubtleſſe a kinde of *Holoſteum*: my reaſon is, becauſe it hath graſſie leaues, or rather leaues like *Vetonica ſylueſtris* or wilde Pinks, a root like thoſe of *Garyophyllata* or Auens, and the ſpikie eare of *Holoſteum* or Sea Plantaine: which are certaine arguments that theſe writers haue neuer ſeene the Plant, but onely the picture thereof, and ſo haue ſet downe their opinions by heare-ſay.

 This

This plant likewiſe hath beene altogether vnknowne vnto the old Writers. It groweth moſt plentifully vpon the cliffes and rocks and the tops of the barren mountains of Auergne in France, and in many places of Italy.

2 The ſecond ſort of wilde ſea Plantaine or *Serpentina* differeth not from the former but onely in quantitie and ſlenderneſſe of his ſtalkes, and the ſmallneſſe of his leaues, which exceed not the height of two inches. It groweth on the hills and rockes neere the waſhings of the ſea at Maſſilia in great plenty almoſt euery where among the *Tragacanthum*, hauing a moſt thicke and ſpreading cluſter of leaues after the manner of *Sedum minimum ſaxeum montanum*, ſomewhat like *Pinaſter*, or the wilde Pine, as well in manner of growing, as ſtiffeneſſe, and great increaſe of his ſlender branches. It hath the ſmall ſeed of Plantaine, or *Serpentina vulgaris*, contained within his ſpiky eares. The root is ſomewhat long, wooddy, and thicke, in taſte ſomewhat hot and aromaticall.

3 *Coronopus ſiue Serpentina minima.*
Small Buck-horne Plantaine.

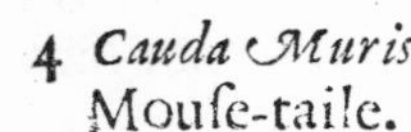

4 *Cauda Muris.*
Mouſe-taile.

3 This ſmall ſea plant is likewiſe one of the kindes of ſea Plantaine, participating as well of Buck-horne as of *Holoſtium*, being as it were a degenerate kinde of ſea Plantaine. It hath many graſſie leaues very like vnto the herbe Thrift, but much ſmaller; among which come forth little tender foot-ſtalkes, whereon do grow ſmall ſpikie knops like thoſe of ſea Plantaine. The root is tough and threddy.

4 Mouſe-taile or *Cauda muris* reſembleth the laſt kinde of wilde *Coronopus* or ſea Plantaine, in ſmall ſpikie knops, leaues, and ſtalkes, that I know no reaſon to the contrarie, but that I may as well place this ſmall herbe among the kindes of *Coronopus* or Bucks horne, as other Writers haue placed kindes of *Holoſtium* in the ſame ſection: and if that be pardonable in them, I truſt this may be tolerated in me, conſidering that without controuerſie this little and baſe herbe is a kinde of *Holoſtium*, hauing many ſmall ſhort graſſie leaues ſpred on the ground, an inch long or ſomewhat more: among which do riſe ſmall tender naked ſtalkes of two inches long, bearing at the top a little blackiſh torch or ſpikie knop in ſhape like that of the Plantaines, reſembling very notably the taile of a Mouſe, whereof it tooke his name. The root is ſmall and threddy.

¶ *The Place.*

The firſt and ſecond of theſe plants are ſtrangers in England; notwithſtanding I haue heard ſay that they grow vpon the rocks in Silley, Garnſey, and the Iſle of man.

Mouſe-taile groweth vpon a barren ditch banke neere vnto a gate leading into a paſture on the right hand of the way, as ye go from London to a village called Hampſtead; in a field as you goe from Edmonton (a village neere London) vnto a houſe thereby called Pims, by the foot-paths ſides; in Woodford Row in Waltham Forreſt, and in the Orchard belonging to M[r]. *Francis Whetſtone* in Eſſex, and in other places.

¶ *The*

¶ *The Time.*

They floure and flouriſh in May and Iune.

¶ *The Names.*

Matthiolus writeth, That the people of Goritia do commonly call theſe two former plants *Serpentaria* and *Serpentina*; but vnproperly, for that there be other plants which may better be called *Serpentina* than theſe two: we may cal them in Engliſh wild ſea Plantaine, whereof doubtleſſe they are kindes.

Mouſe taile is called in Latine *Cauda muris*, and *Cauda murina*: in Greeke, μυόσουρος, or μυὸς οὐρά. *Myoſuros* is called of the French-men *Queue de ſouris*: in Engliſh, Bloud-ſtrange, and Mouſe-taile.

¶ *The Temperature.*

Coronopus is cold and dry much like vnto the Plantaine. Mouſe-taile is cold and ſomthing drying, with a kinde of aſtriction or binding qualitie.

¶ *The Vertues.*

Their faculties in working are referred vnto the Plantaines and Harts-horne.

CHAP. 101.

Of Bucke-horne Plantaines, or Harts-horne.

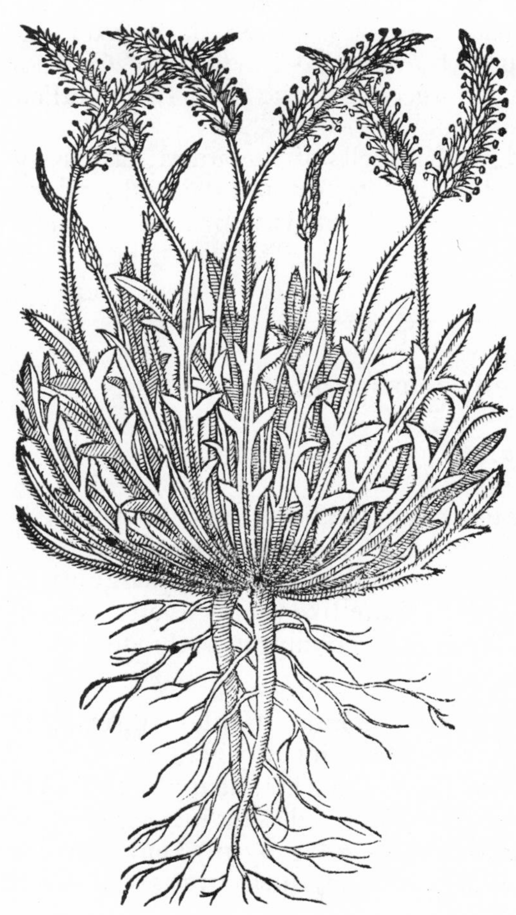

1 *Cornu Ceruinum.*
Harts-horne.

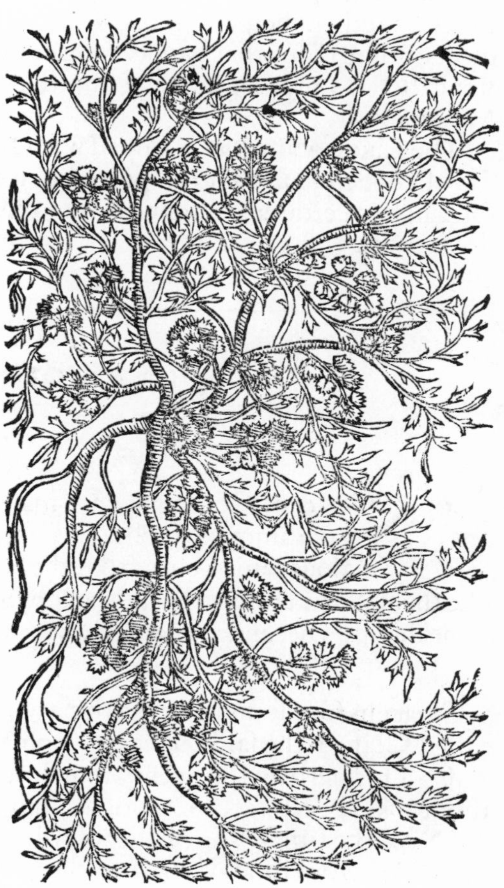

2 *Coronopus Ruellij.*
Swines Creſſes, or Bucks-horne.

¶ *The Deſcription.*

1 BVcks-horne or Harts-horne hath long narrow hoary leaues, cut on both the ſides with three or foure ſhort ſtarts or knags, reſembling the branches of a harts horne, ſpreading it ſelfe on the ground like a ſtar: from the middle whereof ſpring vp ſmall round naked hairy ſtalks; at the top whereof do grow little knops or ſpikie torches like thoſe of the ſmal Plantaines. The root is ſlender and threddy.

2 *Ruellius* Bucks-horne or Swines Cresses hath many smal and weake stragling branches, trailing here and there vpon the ground, set with many small cut or iagged leaues, somewhat like the former, but smaller, and nothing at all hairy as is the other. The floures grow among the leaues, in small rough clusters, of an herby greenish colour: which being past, there come in place little flat pouches broad and rough, in which the seed is contained. The root is white, threddy, and in taste like the garden Cresses.

¶ *The Place.*

They grow in barren plaines, and vntilled places, and sandy grounds; as in Touthill field neere vnto Westminster, at Waltham twelue miles from London, and vpon Blacke-heath also neere London.

¶ *The Time.*

They floure and flourish when the Plantaines doe, whereof these haue beene taken to be kindes.

¶ *The Names.*

Bucks-horne is called in Latine *Cornu Ceruinum*, or Harts-horne: diuers name it *Herba stella*, or *Stellaria*, although there be another herbe so called: in low-Dutch, **Hertshooren**: in Spanish, *Guiabella*: in French, *Corne de Cerf*: It is thought to *Dioscorides* his κορωνόποδα, which doth signifie *cornicis pedem*, a Crowes foot. It is called also by certaine bastard names, as *Harenarea, Sanguinaria*: and of many, Herbe Iuy, or herbe Eue.

¶ *The Temperature.*

Bucks-horne is like in temperature to the common Plantaine, in that it bindeth, cooleth, and drieth.

¶ *The Vertues.*

A The leaues of Buckes-horne boyled in drinke, and giuen morning and euening for certaine dayes together, helpeth most wonderfully those that haue sore eyes, waterie or blasted, and most of the griefes that happen vnto the eyes; experimented by a learned Physition of Colchester called Master *Duke*; and the like by an excellent Apothecarie of the same Towne called M[r]. *Buckstone*.

B The leaues and roots stamped with Bay salt, and tied to the wrests of the armes, take away fits of the Ague: and it is reported to worke the like effect being hanged about the necke of the Patient in a certaine number; as vnto men nine plants, roots and all; and vnto women and children seuen.

Chap. 102. *Of Saracens Consound.*

¶ *The Description.*

1 Saracens Consound hath many long narrow leaues cut or sleightly snipt about the edges: among which rise vp faire browne hollow stalkes of the height of foure cubits; along which euen from the bottome to the top it is set with long and prety large leaues like them of the Peach tree: at the top of the stalkes grow faire starre-like yellow floures, which turne into downe, and are carried away with the winde. The root is very fibrous or threddy.

¶ *The Place.*

Saracens Consound groweth by a wood as ye ride from great Dunmow in Essex, vnto a place called Clare in the said countrey; from whence I brought some plants into my garden.

‡ I formerly in the twenty fourth Chapter of this second booke told you what plant our Author tooke for Saracens Consound, and (as I haue been credibly informed) kept in his garden for it. Now the true *Solidago* here described and figured was found *Anno* 1632, by my kinde Friends M[r]. *George Bowles* and M[r]. *William Coot*, in Shropshire in Wales, in a hedge in the way as one goeth from Dudson in the parish of Cherbery to Guarthlow. ‡

¶ *The Time.*

It floureth in Iuly, and the seed is ripe in August.

¶ *The Names.*

Saracens Consound is called in Latine *Solidago Saracenica*, or Saracens Comfrey, and *Consolida Saracenica*: in Dutch, **Heidnisch Wundtkraut**: of some, *Herba fortis*: in English, Saracens Consound, or Saracens Wound-wort.

¶ *The*

† *Solidago Saracenica.*
Saracens Conſound.

¶ *The Nature.*

Saracens Conſound is dry in the third degree, with ſome manifeſt heate.

¶ *The Vertues.*

A Saracens Conſound is not inferiour to any of the wound-herbes whatſoeuer, being inwardly miniſtred, or outwardly applied in ointments or oyles. With it I cured Maſter *Cartwright* a Gentleman of Grayes Inne, who was grieuouſly wounded into the lungs, and that by Gods permiſſion in ſhort ſpace.

B The leaues boyled in water and drunke, doth reſtraine and ſtay the waſting of the liuer, taketh away the oppilation and ſtopping of the ſame, and profiteth againſt the Iaundice and Feuers of long continuance.

C The decoction of the leaues made in water is excellent againſt the ſoreneſſe of the throat, if it be therewith gargariſed: it increaſeth alſo the vertue and force of lotion or waſhing waters, appropriate for priuy maimes, ſore mouthes, and ſuch like, if it be mixed therewith.

† The figure that was formerly in this place was of *Conſolida paluſtris* of *Tabernamontanus*; and the true figure belonging to this hiſtorie was in the next chapter ſaue one, vnder the title of *Herba Dorea Lobelii.*

Chap. 103. *Of Golden Rod.*

¶ *The Deſcription.*

1 Golden Rod hath long broad leaues ſomwhat hoary and ſharpe pointed; among which riſe vp browne ſtalkes two foot high, diuiding themſelues toward the top into ſundry branches, charged or loden with ſmall yellow floures; which when they be ripe turne into downe which is carried away with the winde. The root is threddy and browne of colour. ‡ *Lobel* makes this with vnſnipt leaues to be that of *Arnoldus de villa noua.* ‡

2 The ſecond ſort of Golden Rod hath ſmall thin leaues broader than thoſe of the firſt deſcribed, ſmooth, with ſome few cuts or nickes about the edges, and ſharpe pointed, of a hot and harſh taſte in the throat being chewed; which leaues are ſet vpon a faire reddiſh ſtalke. It tooke his name from the floures which grow at the top of a gold yellow colour: which floures turne into Downe, which is carried away with the winde, as is the former. The root is ſmall, compact of many ſtrings or threds.

¶ *The Place.*

They both grow plentifully in Hampſtead Wood, neere vnto the gate that leadeth out of the wood vnto a Village called Kentiſh towne, not far from London; in a wood by Rayleigh in Eſſex, hard by a Gentlemans houſe called M[r]. *Leonard*, dwelling vpon Dawes heath; in Southfleet and in Swaineſcombe wood alſo, neere vnto Graueſend.

¶ *The Time.*

They floure and flouriſh in the end of Auguſt.

¶ *The Names.*

It is called in Engliſh Golden Rod: in Latine, *Virga aurea*, becauſe the branches are like a golden rod: in Dutch, **Gulden roede**: in French, *Verge d' or.*

1 *Virga aurea.*
Golden Rod.

2 *Virga aurea Arnoldi Villanouani.*
Arnold of the new towne his Golden rod.

¶ *The Temperature.*

Golden Rod is hot and dry in the second degree: it clenseth, with a certaine astriction or binding qualitie.

¶ *The Vertues.*

A Golden Rod prouoketh vrine, wasteth away the stones in the kidnies, and expelleth them, and withall bringeth downe tough and raw flegmatick humors sticking in the vrine vessels, which now and then do hinder the comming away of the stones, and causeth the grauell or sand which is brittle to be gathered together into one stone. And therefore *Arnoldus Villanouanus* by good reason hath commended it against the stone and paine of the kidnies.

B It is of the number of those plants that serue for wound-drinks, and is reported that it can fully performe all those things that Saracens Consound can; and in my practise shall be placed in the formost ranke.

C *Arnoldus* writeth, That the distilled water drunke with wine for some few dayes together, worketh the same effect, that is, for the stone and grauell in the kidnies.

D It is extolled aboue all other herbes for the stopping of bloud in sanguinolent vlcers and bleeding wounds; and hath in times past beene had in greater estimation and regard than in these dayes: for in my remembrance I haue knowne the dry herbe which came from beyond the sea sold in Bucklers Bury in London for halfe a crowne an ounce. But since it was found in Hampstead wood, euen as it were at our townes end, no man will giue halfe a crowne for an hundred weight of it: which plainly setteth forth our inconstancie and sudden mutabilitie, esteeming no longer of any thing, how pretious soeuer it be, than whilest it is strange and rare. This verifieth our English prouerbe, Far fetcht and deare bought is best for Ladies. Yet it may be more truely said of phantasticall Physitions, who when they haue found an approued medicine and perfect remedie neere home against any disease; yet not content therewith, they wil seeke for a new farther off, and by that meanes many times hurt more than they helpe. Thus much I haue spoken to bring these new fangled fellowes backe againe to esteeme better of this admirable plant than they haue done, which no doubt hath the same vertue now that then it had, although it growes so neere our owne homes in neuer so great quantitie.

CHAP. 103. *Of Captaine Andreas Dorias his Wound-woort.*

† *Herba Doria L'obelij.*
Dorias Woundwoort.

¶ *The Deſcription.*

THis plant hath long and large thicke and fat leaues, ſharp pointed, of a blewiſh greene like vnto Woad, which being broken with the hands hath a prettie ſpicie ſmell. Among theſe leaues riſeth vp a ſtalk of the height of a tal man, diuided at the top into many other branches, whereupon grow ſmall yellowiſh floures, which turneth into downe that flieth away with the wind. The root is thick almoſt like *Helleborus albus*.

Of which kinde there is another like the former, but that the leawes are rougher, ſomewhat bluntly indented at the edges, and not ſo fat and groſſe.

‡ *Herba Doria altera.*

This herbe growes vp with a green round brittle ſtalke, very much champhered, ſinewed, or furrowed, about foure or fiue foot high, full of white pith like that of Elder, and ſendeth forth ſmall branches: the leaues grow on the ſtalk out of order, & are ſmooth, ſharpe pointed, in ſhape like thoſe of *Herba Doria*, but much ſhorter & narrower, the broadeſt and longeſt ſeldome being aboue ten or eleuen inches long, and ſcarce two inches broad, and are more finely and ſmally nickt or indented about the edges; their ſmell being nothing pleaſant, but rather when together with the ſtalke they are broken and rubbed yeeld forth a ſmell hauing a ſmall touch of the ſmell of Hemlocke. Out of the boſomes of theſe leaues ſpring other ſmaller leaues or branches. The floures are many, and grow on ſmall branches at the tops of the ſtalkes like thoſe of *Herba Doria*, but more like thoſe of *Iacobæa*, of a yellow colour, as well the middle button, as the ſmall leaues that ſtand round about, euery floure hauing commonly eight of thoſe ſmall leaues. Which beeing paſt the button turneth into downe and containeth very ſmall long ſeedes which flie away with the winde. The root is nothing elſe but an infinite of ſmall ſtrings which moſt hurtfully ſpread in the ground, and by their infinite increaſing deſtroyeth and ſtarueth other herbes that grow neere it. Its naturall place of growing I know not, for I had it from Mr. *Iohn Coys*, and yet keep it growing in my garden. *Iohn Goodyer.* ‡

¶ *The Place.*

Theſe plants grow naturally about the borders or brinkes of riuers neere to Narbone in France, from whence they were brought into England, and are contented to be made denizons in my garden, where they flouriſh to the height aforeſaid.

¶ *The Time.*

They floured in my garden about the twelfth of Iune.

¶ *The Nature.*

The roots are ſweet in ſmell, and hot in the third degree.

¶ *The Vertues.*

Two drams of the roots of *Herba Doria* boiled in wine and giuen to drinke, draweth downe wateriſh humors, and prouoketh vrine. A

The ſame is with good ſucceſſe vſed in medicines that expell poiſon. B

‡ All

‡ All these Plants mentioned in the three last Chapters, to wit, *Solidago, Virga aurea* and this *Herba Dorea*, are by *Bauhine* fitly comprehended vnder the title of *Virga aurea*; because they are much alike in shape, and for that they are all of the same facultie in medicine. ‡

† The figure that was here was of *Solidago Saracenica*.

Chap. 105. *Of Felwoort, or Baldmoney.*

¶ *The Kindes.*

There be diuers sorts of Gentians or Felwoorts, whereof some be of our owne countrey; others more strange and brought further off: and also some not before this time remembred, either of the antient or later writers, as shall be set forth in this present chapter.

¶ *The Description.*

The first kinde of Felwoort hath great large leaues, not vnlike to those of Plantaine, very well resembling the leaues of the white Hellebore: among which riseth vp a round hollow stalke as thicke as a mans thumbe, full of ioints or knees, with two leaues at each of them, and towards the top euery ioint or knot is set round about with small yellow starre-like floures, like a coronet or garland: at the bottome of the plant next the ground the leaues do spread them selues abroad, embracing or clipping the stalke in that place round about, set together by couples one opposite against another. The seede is small, browne, flat, and smooth like the seeds of the Stocke Gillo-floure. The roote is a finger thicke. The whole Plant is of a bitter taste.

1 *Gentiana maior.*
Great Felwoort.

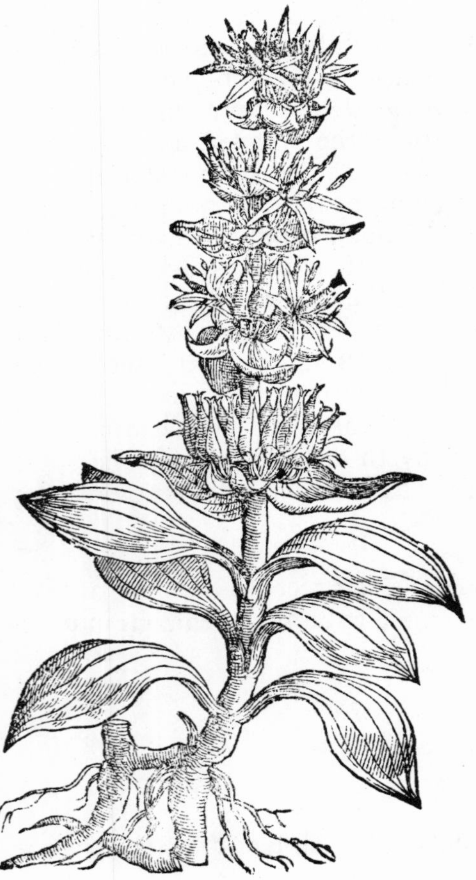

‡ 2 *Gentiana maior purpurea, 1. Clusij.*
Great Purple Felwoort.

3 *Gen-*

3 *Gentiana maior ij. cæruleo flore Cluſij.*
Blew floured Felwoort.

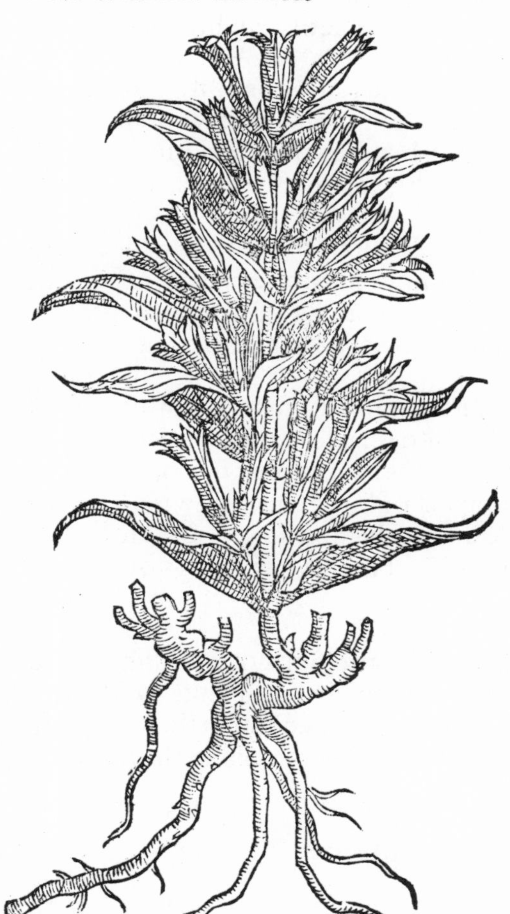

4 *Gentiana minor Cruciata.*
Croſſewoort Gentian.

5 *Gentiana Pennei minor.*
Spotted Gentian of Dr. *Pennie.*

‡ 2 This deſcribed by *Cluſius*, hath leaues and ſtalkes like the precedent; theſe ſtalkes are ſome cubite and halfe or two cubits high, and towards the toppes they are ingirt with two or three coroners of faire purple floures, which are not ſtar-faſhioned, like thoſe of the former, but long and hollow, diuided as it were into ſome fiue or ſix parts or leaues, which towards the bottome on the inſide are ſpotted with deepe purple ſpots: theſe floures are without ſmell, & haue ſo many chiues as they haue iagges, and theſe chiues compaſſe the head, which is parted into two cells, and containes ſtore of a ſmooth, chaffie, reddiſh ſeed. The root is large, yellow on the outſide, and white within, very bitter, & it ſends forth euery yere new ſhoots. It growes in diuers places of the Alps, it floures in Auguſt, and the ſeeds are ripe in September. ‡

3 *Carolus Cluſius* alſo ſetteth forth another ſort of a great Gentian, riſing forth of the ground with a ſtiffe, firme or ſolide ſtalke, ſet with leaues like vnto *Aſclepias*, by couples one oppoſite againſt another, euen from the bottome to the top in certaine diſtances: from the boſome of the leaues

leaues there ſhoot forth ſet vpon ſlender foot-ſtalkes certaine long hollow floures like bels, the mouth whereof endeth in fiue ſharpe corners. The whole floure changeth many times his colour according to the ſoile and climate; now and then purple or blew, ſometimes whitiſh, and often of an aſhe colour. The root and ſeed is like the precedent.

4 Croſſe-woort Gentian hath many ribbed leaues ſpred vpon the ground, like vnto the leaues of Sopewoort, but of a blacker greene colour: among which riſe vp weake iointed ſtalkes trailing or leaning toward the ground. The floures grow at the top in bundles thicke thruſt together, like thoſe of ſweet Williams, of a light blew colour. The root is thicke, and creepeth in the ground far abroad, whereby it greatly increaſeth.

5 *Carolus Cluſius* hath ſet forth in his Pannonicke hiſtorie a kinde of Gentian, which he receiued from M^r^. *Thomas Pennie* of London, D^r^. in Phiſicke, of famous memorie, and a ſecond *Dioſcorides* for his ſingular knowledge in Plants: which *Tabernamontanus* hath ſet forth in his Dutch booke for the ſeuenth of *Cluſius*, wherein he greatly deceiued himſelfe, and hath with a falſe deſcription wronged others.

This twelfth ſort or kinde of Gentian after *Cluſius*, hath a round ſtiffe ſtalke, firme and ſolide, ſomewhat reddiſh at the bottome, iointed or kneed like vnto Croſſewoort Gentian. The leaues are broad, ſmooth, full of ribbes or ſinewes, ſet about the ſtalkes by couples, one oppoſite againſt another. The floures grow vpon ſmall tender ſtalkes, compact of fiue ſlender blewiſh leaues, ſpotted very curiouſly with many blacke ſpots and little lines; hauing in the middle fiue yellow chiues. The ſeed is ſmall like ſand: the root is little, garniſhed with a few ſtrings of a yellowiſh colour.

¶ *The Place.*

Gentian groweth in ſhadowie woods, and the mountains of Italie, Sclauonia, Germany, France, and Burgundie; from whence M^r^. *Iſaac de Laune* a learned Phiſition ſent me plants for the increaſe of my garden. Croſſewoort Gentian groweth in a paſture at the Weſt end of little Rayne in Eſſex on the North ſide of the way leading from Braintree to Much-Dunmow; and in the horſe way by the ſame cloſe.

¶ *The Time.*

They floure and flouriſh in Auguſt, and the ſeed is ripe in September.

¶ *The Names.*

Gentius King of Illyria was the firſt finder of this herbe, and the firſt that vſed it in medicine, for which cauſe it was called Gentian after his owne name: in Greeke γεντιανὴ: which name alſo the Apothecaries retaine vnto this day, and call it *Gentiana*: it is named in Engliſh Felwoort Gentian, Bitterwoort; Baldmoyne, and Baldmoney.

1 This by moſt Writers is called *Gentiana*, and *Gentiana maior Lutea*.

2 *Geſner* calleth this *Gentiana punicea*; *Cluſius*, *Gentiana maior flore purpureo*.

3 This is *Gentiana folijs hirundinariæ* of *Geſner*: and *Gentiana Aſclepiadis folio* of *Cluſius*.

4 This, *Cruciata*, or *Gentiana Cruciata*, of *Tragus*, *Fuchſius*, *Dodon.* *Geſner* and others: it is the *Gentiana minor* of *Matthiolus*.

5 *Cluſius* calls this *Gentiana maior pallida punctis diſtincta*.

¶ *The Temperature.*

The root of Felwoort is hot, as *Dioſcorides* ſaith, clenſing or ſcouring: diuers copies haue, that it is likewiſe binding, and of a bitter taſte.

¶ *The Vertues.*

A It is excellent good, as *Galen* ſaith, when there is need of attenuating, purging, clenſing, and remouing of obſtructions, which qualitie it taketh of his extreme bitterneſſe.

B It is reported to be good for thoſe that are troubled with crampes and conuulſions; for ſuch as are burſt, or haue falne from ſome high place: for ſuch as haue euill liuers and bad ſtomacks. It is put into Counterpoiſons, as into the compoſition named *Theriaca diateſſaron*: which *Aetius* calleth *Myſterium*, a myſterie or hid ſecret.

C This is of ſuch force and vertue, ſaith *Pliny*, that it helpeth cattell which are not onely troubled with the cough, but are alſo broken winded.

D The root of Gentian giuen in powder the quantitie of a dramme, with a little pepper and herbe Grace mixed therewith, is profitable for them that are bitten or ſtung with any manner of venomous beaſt or mad dog: or for any that hath taken poiſon.

E The decoction drunke is good againſt the ſtoppings of the liuer, and cruditie of the ſtomacke, helpeth digeſtion, diſſolueth and ſcattereth congealed bloud, and is good againſt all cold diſeaſes of the inward parts.

Chap.

CHAP. 106. *Of Engliſh Felwoort.*

¶ *The Deſcription.*

Hollow leafed Felwoort or Engliſh Gentian hath many long tough roots, diſperſed hither and thither within the vpper cruſt of the earth; from which immediatly riſeth a fat thicke ſtalke, iointed or kneed by certaine diſtances, ſet at euery knot with one leafe, and ſometimes moe, keeping no certaine number: which leaues doe at the firſt incloſe the ſtalkes round about, being one whole and entire leafe without any inciſure at all, as it were a hollow trunke; which after it is growne to his fulneſſe, breaketh in one ſide or other, and becommeth a flat ribbed leafe, like vnto the great Gentian or Plantaine. The floures come forth of the boſome of the vpper leaues, ſet vpon tender foot ſtalkes, in ſhape like thoſe of the ſmall Bindweed, or rather the floures of Sopewoort, of a whitiſh colour, waſht about the brims with a little light carnation. Then followeth the ſeed, which as yet I haue not obſerued.

Gentiana concaua.
Hollow Felwoort.

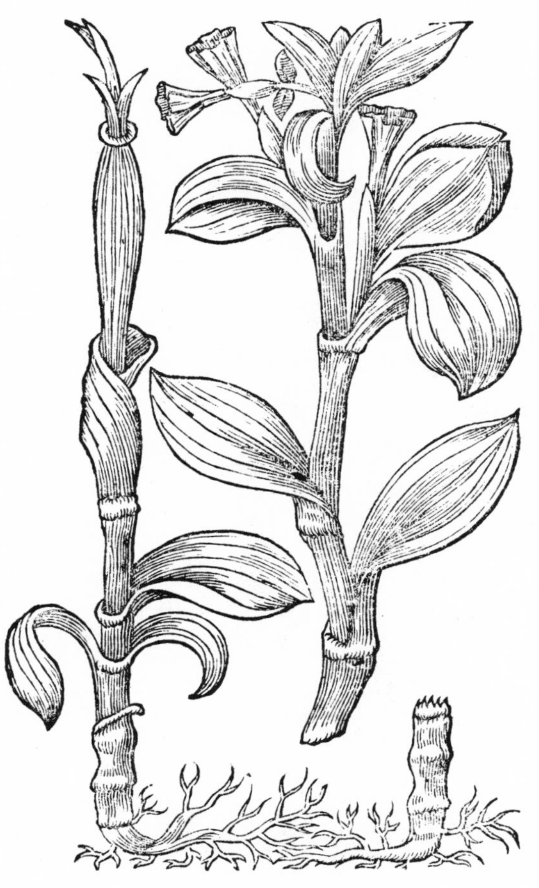

¶ *The Place.*

I found this ſtrange kind of Gentian in a ſmall groue of a wood called the Spinie, neere vnto a ſmall village in Northampton ſhire called Lichbarrow: elſewhere I haue not heard of it.

¶ *The Time.*

It ſpringeth forth of the ground in Aprill, and bringeth forth his floures and ſeed in the end of Auguſt.

¶ *The Names.*

I haue thought good to giue vnto this plant, in Engliſh, the name Gentian, being doubtleſſe a kinde therof. The which hath not been ſet forth, nor remembred by any that haue written of plants vntil this time. In Latine we may call it *Gentiana concaua*, of the hollow leaues. It may be called alſo hollow leaued Felwoort.

¶ *The Temperature and Vertues.*

Of the faculties of this plant as yet I can ſay nothing, referring it vnto the other Gentians, vntill time ſhall diſcloſe that which yet is ſecret and vnknowne.

‡ *Bauhine* receiued this plant with the figure thereof from Doctor *Liſter* one of his Maieſties Phyſitions, and he referres it vnto *Saponaria*, calling it *Saponaria concaua Anglica*; and (as farre as I can coniecture) hath a good deſcription thereof in his *Prodrom. pag.* 103. Now both by our Authour and *Bauhines* Deſcription, I gather, that the roote in this Figure is not rightly expreſſed, for that it ſhould bee long, thicke, and creeping, with few fibers adhering thereunto; when as this figure expreſſeth an annuall wooddy root. But not hauing as yet ſeene the plant, I can affirme nothing of certaintie. ‡

CHAP.

‡ CHAP. 107. Of Baſtard Felwoort.

¶ The Deſcription.

‡ OVr Authour in this Chapter ſo confounded all, that I knew not well how, handſomely to ſet all right; for his deſcriptions they were ſo barren, that little might be gathered by them, and the figures agreed with their titles, but the place contradicts all; for the firſt figured is found in England; and the ſecond is not that euer I could learne: alſo the ſecond floures in the ſpring, according to *Cluſius* and all others that haue written thereof, and alſo by our Authours owne title, truely put ouer the figure: yet he ſaid they both floure and flouriſh from Auguſt to the end of September. Theſe things conſidered, I thought it fitter both for the Readers benefit, and my owne credit to giue you this chapter wholly new with additions, rather than mangled and confuſed, as otherwiſe of neceſſitie it muſt haue beene. ‡

‡ 1 This elegant *Gentianella* hath a ſmall yellowiſh creeping root, from which ariſe many greene ſmooth thicke hard and ſharpe pointed leaues like thoſe of the broad leaued Myrtle, yet larger, and hauing the veines running alongſt the leaues as in Plantaine. Amongſt the leaues come vp ſhort ſtalkes, bearing very large floures one vpon a ſtalke; and theſe floures are hollow like a Bel-floure, and end in fiue ſharpe points with two little eares betweene each diuiſion, and their colour is an exquiſite blew. After the floure is paſt there followes a ſharpe pointed longiſh veſſell, which opening it ſelfe into two equall parts, ſhewes a ſmall creſted darke coloured ſeede.

‡ 1 *Gentianella verna maior.*
Spring large floured Gentian.

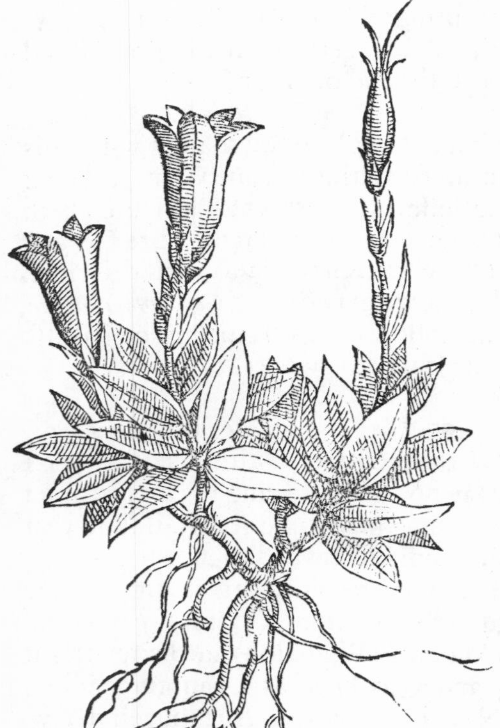

2 *Gentianella Alpina verna.*
Alpes Felwoort of the ſpring time.

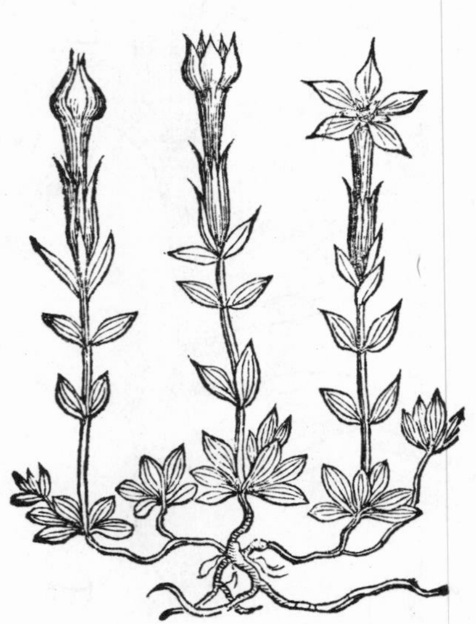

2 This ſecond riſes vp with a ſingle ſlender and purpliſh ſtalke, ſet at certaine ſpaces with ſix or eight little ribbed leaues, ſtanding by couples one againſt another. At the top ſtands a cup, out whereof comes one long floure without ſmell, and as it were diuided at the top into fiue parts; and it is of ſo elegant a colour, that it ſeemes to exceed blewneſſe it ſelfe; each of the foldes or little leaues of the floure hath a whitiſh line at the ſide, and other fiue as it were pointed leaues or appendices ſet betweene them: and in the middeſt of the floure are certaine pale coloured chiues: a longiſh ſharpe pointed veſſell ſucceeds the floure which contains a ſmall hard round ſeed. The root is ſmall, yellowiſh and creeping, putting vp here and there ſtalkes bearing floures, and in other places onely leaues lying orderly ſpred vpon the ground.

3 *Gentianella fugax minor.*
Bastard or Dwarfe Felwoort.

3 Besides these two whose roots last long and increase euery yeare, there are diuers other Dwarfe or Bastard Gentians which are annuall, and wholly perish euery yeare assoone as they haue perfected their seed; and therefore by *Clusius* they are fitly called *Gentianæ fugaces*. Of these I haue onely obserued two kindes (or rather varieties) in this Kingdome, which I wil here describe vnto you. The first of these, which is the lesser, & whose figure we here giue you, is a proper plant some two or three inches high, diuided immediatly from the root into three or foure or more branches, set at certaine spaces with little longish leaues, being broadest at the setting on, and so growing narrower or sharper pointed. The tops of these stalkes are beautified with long, hollow, and pretty large floures, considering the magnitude of the plant, and these floures are of a darke purplish colour, and at their tops diuided into fiue parts. The root is yellowish, small, and wooddy. The seede which is small and round is contained in longish vessels. The stalkes and leaues are commonly of a darke green, or else of a brownish colour.

4 This from a root like, yet a little larger than the former, sends vp a pretty stiffe round stalke of some span high; which at certaine spaces is set with such leaues as the last described, but larger: and out of the bosomes of these leaues from the bottome to the top of the stalke come forth little foot stalkes, which vsually carry three floures a piece; two set one against another, and the third vpon a stalke somewhat higher; and sometimes there comes forth a single floure at the root of these foot stalkes. The floures in their shape, magnitude and colour, are like those of the last mentioned, and also the seed and seed vessels. The manner of growing of this is very well presented by the figure of the third Gentian, formerly described in the Chapter last saue one aforegoing.

¶ *The Place.*

1 2 These grow not wilde in England that I know of, but the former is to bee found in most of our choice Gardens. As with M^r. *Parkinson*, Master *Tradescant*, and Master *Tuggye*, &c.

3 4 These are found in diuers places, as in the Chalke-dale at Dartford in Kent, and according to our Authour (for I know he meant these) in Waterdowne Forest in Sussex, in the way that leadeth from Charlwoods lodge, vnto the house of the Lord of Abergauenie, called Eridge house by a brooke side there, especially vpon a Heath by Colbrooke neere London: on the Plain of Salisburie, hard by the turning from the said Plaine, vnto the right Honourable the Lord of Pembrooks house at Wilton, and vpon a Chalkie banke in the high way betweene Saint Albons and Goramberrie.

¶ *The Time.*

1 2 These two floure in Aprill and May. The other from August vnto the end of October.

¶ *The Names.*

1 This is the *Gentiana* 4. of *Tragus*. The *Gentianella Alpina* of *Gesner*. *Gentianella campanulæ flore* and *Heluetica* of *Lobel*; the *Gentiana* 5. or *Gentianella maior verna* of *Clusius*.

2 *Gesner* called this *Calathiana verna*: *Lobel*, *Gentianella Alpina*: and *Clusius*, *Gentiana* 6. and *Gentianella minor verna*.

3 This is the *Calathiana vera* of *Daleschampius*: and the *Gentiana fugax* 5. or *Gentiana* 11. *minima* of *Clusius*.

4 I take this to bee *Clusius* his *Gentiana fugax* 4. or *Gentiana* 10. We may call this in English, Small Autumne Gentian.

¶ *Their Temperature and Vertues.*

Theſe by their taſte and forme ſhould be much like to the greater Gentians in their operation and working, yet not altogether ſo effectuall. ‡

Chap. 106. *Of Calathian Violet, or Autumne Bel-floure.*

¶ *The Deſcription.*

Among the number of the baſe Gentians there is a ſmal plant, which is late before it commeth vp, hauing ſtalks a ſpan high, and ſometimes higher, narrow leaues like vnto Time, ſet by couples about the ſtalkes by certaine diſtances: long hollow floures growing at the top of the ſtalks, like a cup called a Beaker, wide at the top, and narrower toward the bottome, of a deepe blew colour tending to purple, with certain white threds or chiues in the bottome: the floure at the mouth or brim is fiue cornered before it be opened, but when it is opened it appeareth with fiue clifts or pleats. The whole plant is of a bitter taſte, which plainly ſheweth it to be a kinde of wilde Gentian. The root is ſmall, and periſheth when it hath perfected his ſeed, and recouereth it ſelfe by falling of the ſame.

Pneumonanthe.
Calathian Violet.

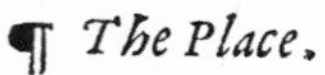

¶ *The Place.*

It is found ſometimes in Meadowes, oftentimes in vntilled places. It groweth vpon Long-field downes in Kent, neere vnto a village called Longfield by Grauesend, vpon the chalkie cliffes neere Greene-Hythe and Cobham in Kent, and many other places. It likewiſe groweth as you ride from Sugar-loafe hill vnto Bathe, in the Weſt countrey.

‡ This plant I neuer found but once, and that was on a wet Mooriſh ground in Lincolnſhire, 2. or 3. miles on this ſide Caſter, and as I remember, the place is called Netleton Moore. Now I ſuſpect that our Authour knew it not; firſt, becauſe he deſcribes it with leaues like vnto Time, when as this hath long narrow leaues more like to Hyſſop or Roſemary. Secondly, for that he ſaith the root is ſmall & periſheth when as it hath perfected the ſeed: whereas this hath a liuing, ſtringie and creeping root. Beſides, this ſeldome or neuer growes on chalkie cliffes, but on wet Mooriſh grounds and Heaths: wherefore I ſuſpect our Authour tooke the ſmall Autumne Gentian (deſcribed by me in the fourth place of the laſt Chapter) for this here treated of. ‡

¶ *The Time.*

The gallant floures hereof be in their brauerie about the end of Auguſt, and in September.

¶ *The Names.*

‡ This is thought to be *Viola Calathiana* of *Ruellius*, yet not that of *Pliny*; and thoſe that deſire to know more of this may haue recourſe to the twelfth chapter of the firſt booke of the 2. *Pempt.* of *Dodon.* his Latine Herball, whence our Authour tooke thoſe words that were formerly in this place, though he did not well vnderſtand nor expreſſe them ‡. It is called *Viola Autumnalis*, or Autumne Violet, and ſeemeth to bee the ſame that *Valerius Cordus* doth call *Pneumonanthe*, which he ſaith is named in the Germane tongue **Lungenblumen**, or Lung-floure: in Engliſh, Autumne Bel-floures, Calathian Violets, and of ſome, Haruest-bels.

¶ *The Temperature.*

This wilde Felwoort or Violet is in Temperature hot, ſomewhat like in facultie to Gentian, whereof it is a kinde, but far weaker in operation.

¶ *The Vertues.*

A The latter Phyſitions hold it to be effectuall againſt peſtilent diſeaſes, and the bitings & ſtingings of venomous beaſts.

Chap.

Chap. 109. *Of Venus Looking-glaſſe.*

¶ *The Deſcription.*

1 BEſides the former Bel-floures, there is likewiſe a certaine other, which is low and little, the ſtalkes whereof are tender, two ſpans long, diuided into many branches moſt commonly lying vpon the ground. The leaues about the ſtalks are little, ſleightly nicked in the edges. The floures are ſmall, of a bright purple colour tending to blewnes, very beautifull, with wide mouths like broad bels, hauing a white chiue or thred in the middle. The floures in the day time are wide open, and about the ſetting of the Sun are ſhut vp and cloſed faſt together, in fiue corners, as they are before their firſt opening, and as the other Bel-floures are. The roots be verie ſlender, and periſh when they haue perfected their ſeed.

‡ 2 There is another which from a ſmall and wooddy root ſendeth vp a ſtraight ſtalk, ſometimes but two or three inches, yet otherwhiles a foot high, when as it lights into good ground. This ſtalke is creſted and hollow, hauing little longiſh leaues crumpled or ſinuated about the edges ſet thereon: and out of the boſomes of thoſe leaues towards the top of the ſtalke and ſometimes lower, come little branches bearing little winged cods, at the tops of which in the middeſt of fiue little greene leaues ſtand ſmall purple floures, of little or no beauty; which being paſt the cods become much larger, and containe in them a ſmall yellowiſh ſeed, and they ſtill retaine at their tops the fiue longiſh greene leaues that incompaſſed the floure. This plant is an annuall like as the former. ‡

1 *Speculum Veneris.*
Venus Looking-glaſſe.

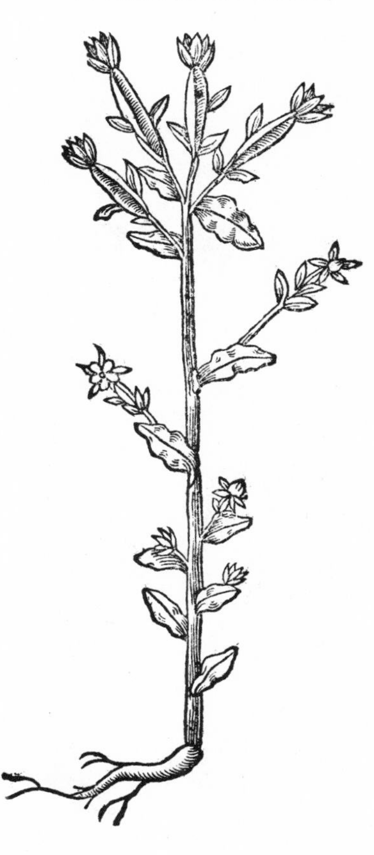

‡ 2 *Speculum Veneris minus.*
Codded corne violet.

¶ *The Place.*

It groweth in ploughed fields among the corne, in a plentifull and fruitfull ſoile. I found it in a field among the corne by Greene-hithe, as I went from thence toward Dartford in Kent, and in many other places thereabout, but not elſwhere: from whence I brought of the ſeeds for my Garden, where they come vp of themſelues from yeare to yeare by falling of the ſeed.

‡ That which is here figured and deſcribed in the firſt place I neuer found growing in England

land, I haue seene only some branches of it brought from Leiden by my friend Mr. *William Parker*. The other of my description I haue diuers times found growing among the corn in Chelsey field, and also haue had it brought me from other places by Mr. *George Bowls*, & M. *Leonard Buckner*. ‡

¶ *The Time.*

It floureth in Iune and Iuly, and the seed is ripe in the end of August.

¶ *The Names.*

It is called *Campana Aruensis*, and of some *Onobrychis*, but vnproperly, of other *Cariophyllus segetum*, or corne Gillofloure, or Corne pinke, and *Speculum Veneris*, or Ladies glasse The Brabanders in their tongue call it **Urowen Spiegel**.

‡ *Tabernamontanus* hath two figures thereof, the one vnder the name of *Viola aruensis*, and the other by the title of *Viola Pentagonia*, because the floure hath fiue folds or corners. 2 This of my description is not mentioned by any Authour; wherefore I am content to follow that name which is giuen to the former, and terme it in Latine *Speculum Veneris minus*: and from the colour of the floure and codded seed vessell, to call it in English, Codded Corne Violet.

¶ *The Temperature and Vertues.*

We haue not found any thing written either of his vertue or temperature, of the antient or late Writers.

Chap. 110. *Of Neesing root, or Neesewoort.*

1 *Helleborus albus.*
White Hellebor.

2 *Helleborus albus præcox.*
Timely white Hellebor.

¶ *The Description.*

1 The first kinde of white Hellebor hath leaues like vnto great Gentian, but much broader, and not vnlike the leaues of the great Plantaine, folded into pleats like a garment pleated to bee laied vp in a chest; amongst these leaues riseth vp a stalke a cubite long, set towards

towards the top full of little ſtarre-like floures, of an herbie green colour tending to whitenesse; which being paſt there come ſmall huskes containing the ſeed. The root is great and thicke, with many ſmall threds hanging thereat.

2 The ſecond kinde is very like the firſt, and differeth in that, that this hath blacke reddiſh floures, and commeth to flouring before the other kinde, and ſeldome in my garden commeth to ſeeding.

¶ *The Place.*

The white Hellebor groweth on the Alps, and ſuch like mountains where Gentian doth grow. It was reported vnto me by the biſhop of Norwich, that white Hellebor groweth in a wood of his owne neere to his houſe at Norwich. Some ſay likewiſe that it doth grow vpon the Mountaines of Wales. I ſpeake this vpon report, yet I thinke not, but that it may be true. Howbeit I dare aſſure you, that they grow in my garden at London, where the firſt kinde floureth and ſeedeth very well.

¶ *The Time.*

The firſt floureth in Iune, and the ſecond in May.

¶ *The Names.*

Neeſewoort is called in Greeke ἐλλέβορος λευκὸς: in Latine, *Veratrum Album*, *Helleborus albus*, and *Sanguis Herculeus*. The Germans call it **Weiſz nieſwurt**: the Dutchmen, **Nieſwortel**: the Italians, *Elleboro bianco*: The Spaniards, *Verde gambre blanco*: the French, *Ellebore blanche*: and we of England call it white Hellebor, Nieſwoort, Lingwoort, and the root Neeſing pouder.

¶ *The Temperature.*

The root of white Hellebor, is hot and drie in the third degree.

¶ *The Vertues.*

The root of white Hellebor procureth vomite mightily, wherein conſiſteth his chiefe vertue, and by that means voideth all ſuperfluous ſlime and naughtie humors. It is good againſt the falling ſickneſſe, phrenſies, ſciatica, dropſies, poiſon, and againſt all cold diſeaſes that bee of hard curation, and will not yeeld to any gentle medicine. A

This ſtrong medicine made of white Hellebor, ought not to be giuen inwardly vnto delicate bodies without great correction, but it may more ſafely be giuen vnto Country people which feed groſſely, and haue hard, tough, and ſtrong bodies. B

The root of Hellebor cut in ſmall pieces, ſuch as may aptly and conueniently be conueied into the Fiſtulaes doth mundifie them, and taketh away the callous matter which hindereth curation, and afterward they may be healed vp with ſome incarnatiue vnguent, fit for the purpoſe. ‡ This facultie by *Dioſcorides* is attributed to the blacke Hellebor, and not to this. ‡ C

The pouder drawne vp into the noſe cauſeth ſneeſing, and purgeth the braine from groſſe and ſlimie humours. D

The root giuen to drinke in the weight of two pence, taketh away the fits of agues, killeth Mice and rats being made vp with honie and floure of wheat: *Pliny* addeth that it is a medicine againſt the Louſie euill. E

CHAP. III. *Of Wilde white Hellebor.*

¶ *The Deſcription.*

1 Helleborine is like vnto white Hellebor, and for that cauſe we haue giuen it the name of *Helleborine*. It hath a ſtraight ſtalke of a foot high, ſet from the bottome to the tuft of floures, with faire leaues, ribbed and chamfered like thoſe of white Hellebor, but nothing neere ſo large, of a darke greene colour. The floures bee orderly placed from the middle to the top of the ſtalke, hollow within, and white of colour, ſtraked here and there with a daſh of purple, in ſhape like the floures of Satyrion. The ſeed is ſmall like duſt or motes in the Sun. The root is ſmall, full of iuice, and bitter in taſte.

2 The ſecond is like vnto the firſt, but altogether greater, and the floures white, without any mixture at all, wherein conſiſteth the difference.

3 The third kind of *Helleborine*, being the 6. after *Cluſius* account, hath leaues like the firſt deſcribed,

bed, but smaller and narrower. The stalke riseth vp to the height of two spans; at the top whereof grow faire shining purple coloured floures, consistiug of six little leaues, within or among which lieth hid things like small helmets. The plant in proportion is like the othet of this kinde. The The root is small, and creepeth in the ground.

1 *Helleborine.*
Wilde white Hellebore.

3 *Helleborine angustifolia 6. Clusij.*
Narrow leafed wilde Neesewoort.

¶ *The Place.*

They bee found in dankish and shadowie places; the first was found growing in the woods by Digges well pastures, halfe a mile from Welwen in Hartfordshire: it groweth in a wood fiue miles from London, neere vnto a bridge called Lockbridge: by Nottingham neere Robinhoods well, where my friend M[r]. *Steuen Bredwell* a learned Physition found the same: in the woods by Dunmowe in Essex: by Southfleet in Kent; in a little groue of Iuniper, and in a wood by Clare in Essex.

¶ *The Time.*

They floure in May and Iune, and perfect their seed in August.

¶ *The Names.*

The likenesse that it hath with white Hellebor, doth shew it may not vnproperly bee named *Helleborine*, or wilde white Hellebor, which is also called of *Dioscorides* and *Pliny* ἐπιπακτὶς, or *Epipactis*; But from whence that name came it is not apparant: it is also named [illegible].

¶ *The Temperature.*

They are thought to be hot and drie of nature.

¶ *The Vertues.*

A The faculties of these wilde Hellebors are referred vnto the white Neesewoort, whereof they are kindes.

B It is reported that the decoction of wilde Hellebor drunken, openeth the stoppings of the Liuer, and helpeth any imperfections of the same.

Chap.

Chap. 112. *Of our Ladies Slipper.*

¶ *The Deſcription.*

1 OVr Ladies Shoo or Slipper hath a thicke knobbed root, with certaine marks or notes vpon the ſame, ſuch as the roots of Solomons Seale haue, but much leſſer, creeping within the vpper cruſt of the earth: from which riſeth vp a ſtiffe and hairy ſtalke a foot high, ſet by certaine ſpaces with faire broad leaues, ribbed with the like ſinewes or nerues as thoſe of the Plantaine. At the top of the ſtalke groweth one ſingle floure, ſeldome two, faſhioned on the one ſide like an egge; on the other ſide it is open, empty, and hollow, and of the forme of a ſhoo or ſlipper, whereof it tooke his name; of a yellow colour on the outſide, and of a ſhining deepe yellow on the inſide. The middle part is compaſſed about with foure leaues of a bright purple colour, often of a light red or obſcure crimſon, and ſometimes yellow as in the middle part, which in ſhape is like an egge, as aforeſaid.

‡ 2 This other differs not from the former, vnleſſe in the colour of the floure; which in this hath the foure long leaues white, and the hollow leafe or ſlipper of a purple colour. ‡

1 *Calceolus Mariæ.*
Our Ladies Slipper.

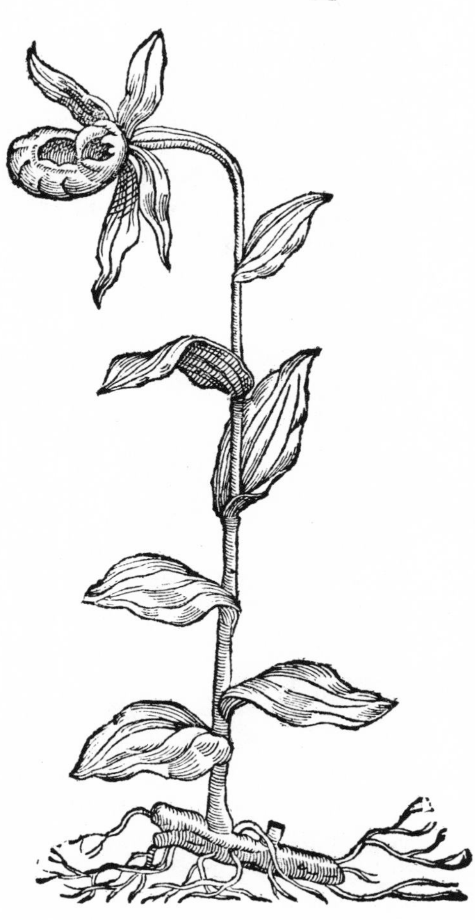

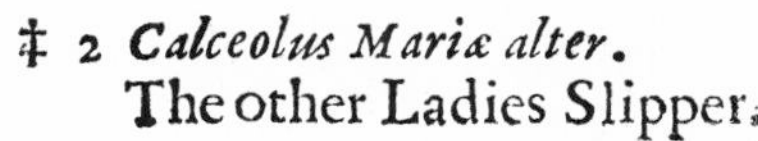

‡ 2 *Calceolus Mariæ alter.*
The other Ladies Slipper.

¶ *The Place.*

Ladies Slipper groweth vpon the mountains of Germany, Hungary, and Poland. I haue a plant thereof in my garden, which I receiued from M[r]. *Garret* Apothecary, my very good friend.

‡ It is alſo reported to grow in the North parts of this kingdome; and I ſaw it in floure with M[r]. *Tradeſcant* the laſt Sommer. ‡

¶ *The Time.*

It floureth about the midſt of Iune.

¶ *The Names.*

It is commonly called *Calceolus D. Mariæ*, and *Marianus*: of ſome, *Calceolus Sacerdotis*: of ſome, *Aliſma*, but vnproperly: in Engliſh, Our Ladies ſhoo or ſlipper: in the Germane tongue, **Pfaffen Schuech, Papen ſcoen**: and of ſome, *Damaſonium nothum*.

¶ *The*

¶ *The Temperature and Vertues.*

Touching the faculties of our Ladies Shoo we haue nothing to write, it being not sufficiently knowne to the old Writers, no nor to the new.

Chap. 113. *Of Sope-wort.*

¶ *The Description.*

The stalkes of Sope-wort are slipperie, slender, round, ioynted, a cubit high or higher : the leaues are broad, set with veines very like broad leaued Plantaine, but yet lesser, standing out of euery ioynt by couples for the most part, and especially those that are the neerest the roots bowing backwards. The floures in the top of the stalkes and about the vppermost ioynts are many, well smelling, sometimes of a beautifull red colour like a Rose; other-while of a light purple or white, which grow out of long cups consisting of fiue leaues, in the middle of which are certaine little threds. The roots are thicke, long, creeping aslope, hauing certaine strings hanging out of them like to the roots of blacke Hellebor : and if they haue once taken good and sure rooting in any ground it is impossible to destroy them.

‡ There is kept in some of our gardens a varietie of this, which differs from it in that the floures are double and somewhat larger : in other respects it is altogether like the precedent. ‡

1 *Saponaria.*
Sope-wort, or Bruse-wort.

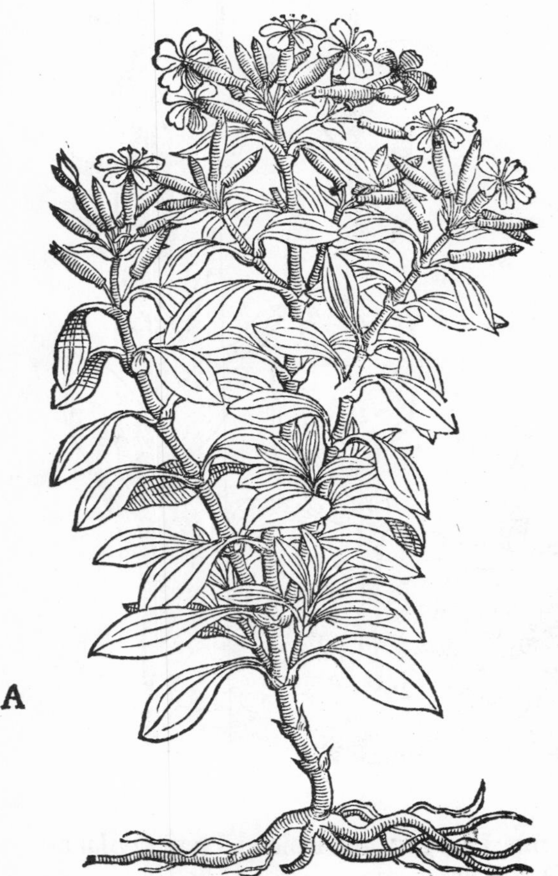

¶ *The Place.*

It is planted in gardens for the floures sake, to the decking vp of houses, for the which purpose it chiefely serueth. It groweth wild of it selfe neere to riuers and running brookes in sunny places.

¶ *The Time.*

It floureth in Iune and Iuly.

¶ *The Names.*

It is commonly called *Saponaria*, of the great scouring qualitie that the leaues haue: for they yeeld out of themselues a certaine iuyce when they are bruised, which scoureth almost as well as Sope: although *Ruellius* describe a certaine other Sopewort. Of some it is called *Alisma*, or *Damasonium* of others, *Saponaria Gentiana*, whereof doubtlesse it is a kinde : in English it is called Sopewort, and of some Bruisewort.

¶ *The Temperature and Vertues.*

A It is hot and dry, and not a little scouring withall, hauing no vse in physicke set downe by any Author of credit.

‡ Although our Authour and such as before him haue written of Plants were ignorant of the facultie of this herbe, yet hath the industrie of some later men found out the vertue thereof: and *Septalius* reports that it was one *Zapata* a Spanish Empericke. Since whose time it hath beene written of by *Rudius*, *lib.* 5. *de morbis occult. & venenat. cap.* 18. And by *Cæsar Claudinus*, *de ingressu ad infirmos*, *pag.* 411. & *pag.* 417. But principally by *Ludouicus Septalius*, *Animaduers. med. lib.* 7. *num.* 214. where treating of decoctions in vse against the French Poxes, he mentions the singular effect of this herb against that filthy disease. His words are these : I must not in this place omit the vse of another Alexipharmicall decoction, being very effectuall and vsefull for the poorer sort; namely that which is made of Sope-wort, an herbe common and knowne to all. Moreouer, I haue sometimes vsed it with happy successe in the most contumacious disease : but it is of somewhat an vngratefull taste, and

and therefore it muſt be reſerued for the poorer ſort. The decoction is thus made: R. *Saponariæ virid. M. ij. infundantur per noctem in lib. viij. aquæ mox excoquantur ad cocturam Saponariæ: deinde libra vna cum dimidia aquæ cum herba iam cocta excoletur cum expreßione, quæ reſeruetur pro potione matutina ad ſudores proliciendos ſumendo ℥ vij. aut viij, quod vero ſupereſt dulcoretur cum paſſulis aut ſaccaro pro potu cum cibis: æſtate & bilioſis naturis addi poterit aut Sonchi, aut Cymbalariæ M. j. Valet & pro mulieribus ad menſtrua alba abſumenda cum M. ß. Cymbalariæ, & addito tantundem Philipendulæ.* Thus much *Septalius*, who ſaith that he had vſed it *ſæpè ac ſæpius*, often and often againe.

Some haue commended it to be very good to be applied to greene wounds, to hinder inflammation, and ſpeedily to heale them. ‡ B

CHAP. 114. *Of Arſmart or Water-Pepper.*

¶ *The Deſcription.*

1 ARſmart bringeth forth ſtalkes a cubit high, round, ſmooth, ioynted or kneed, diuiding themſelues into ſundry branches; whereon grow leaues like thoſe of the Peach or of the Sallow tree. The floures grow in cluſters vpon long ſtems, out of the boſome of the branches and leaues, and likewiſe vpon the ſtalkes themſelues, of a white colour tending to a bright purple: after which commeth forth little ſeeds ſomewhat broad, of a reddiſh yellow, and ſometimes blackiſh, of an hot and biting taſte, as is all the reſt of the Plant, and like vnto pepper, whereof it tooke his name; yet hath it no ſmell at all.

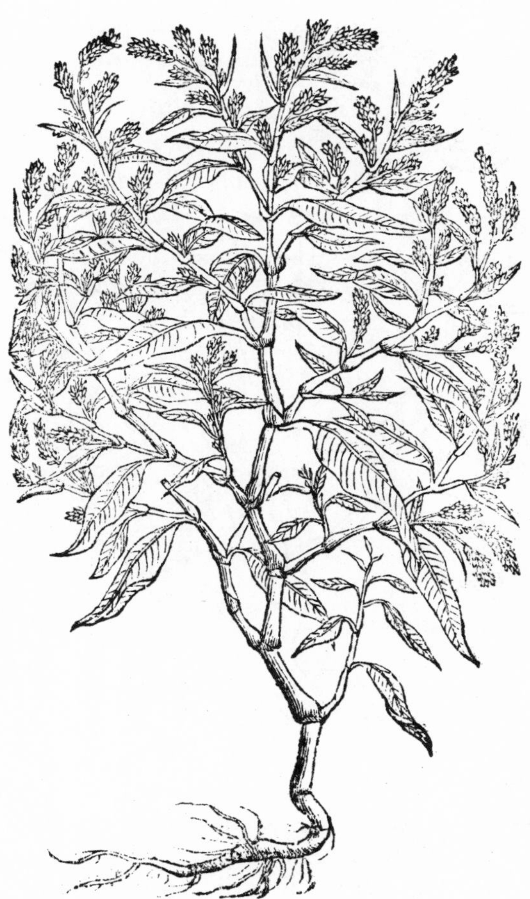

1 *Hydropiper.*
Arſmart.

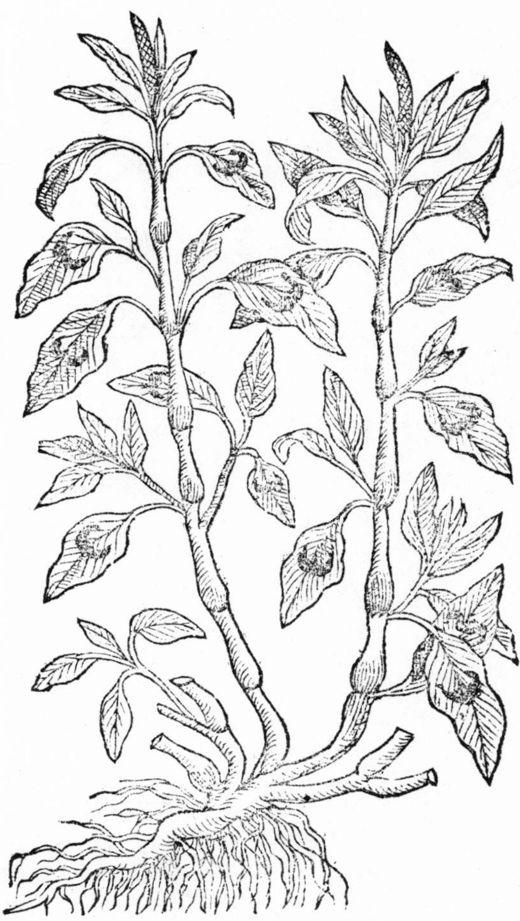

2 *Perſicaria maculoſa.*
Dead or ſpotted Arſmart.

2 Dead Arſmart is like vnto the precedent in ſtalkes, cluſtering floures, roots and ſeed, and differeth in that, that this plant hath certaine ſpots or marks vpon the leaues, in faſhion of a halfe moone, of a darke blackiſh colour. The whole plant hath no ſharpe or biting taſte, as the other hath, but as it were a little ſoure ſmacke vpon the tongue. The root is likewiſe full of ſtrings or threds, creeping vp and downe in the ground.

‡ 3 This

‡ 3 This in roots, leaues, and manner of growing is very like the first described, but lesser by much in all these parts: the floures also are of a whitish, and sometimes of a purplish colour: it growes in barren grauelly and wet places.

4 I haue thought good to omit the impertinent description of our Author fitted to this plant, and to giue one somewhat more to the purpose: the stalkes of this are some two foot high, tender, greene, and sometimes purplish, hollow, smooth, succulent and transparent, with large and eminent ioynts, from whence proceed leaues like those of French Mercurie, a little bigger, and broader toward their stalkes, and thereabout also cut in with deeper notches: from the bosomes of each of these leaues come forth long stalkes hanging downewards, and diuided into three or foure branches; vpon which hang floures yellow, and much gaping, with crooked spurs or heeles, and spotted also with red or sanguine spots: after these are past succeed the cods, which containe the seed, and they are commonly two inches long, slender, knotted, and of a whitish greene colour, crested with greenish lines; and as soone as the seed begins to be ripe, they are so impatient that they will by no meanes be touched, but presently the seed will fly out of them into your face. And this is the cause that *Lobel* and others haue called this Plant *Noli me tangere*. As for the like reason some of late haue imposed the same name vpon the *Sium minimum* of *Alpinus*, formerly described by me in the seuenth place of the eighteenth chapter of this booke, *pag.* 260. ‡

‡ 3 *Persicaria pusilla repens.*
Small creeping Arsmart.

4 *Persicaria siliquosa.*
Codded Arsmart.

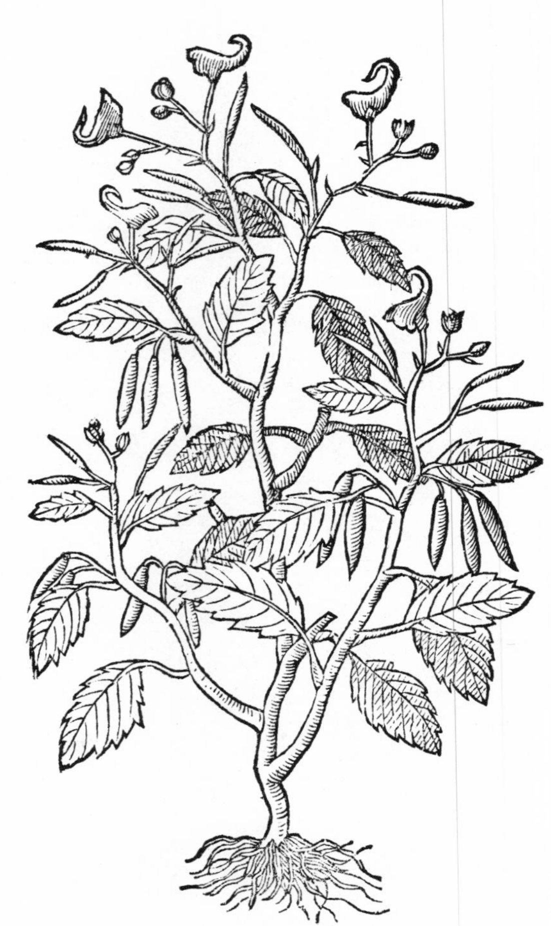

¶ *The Place and Time.*

They grow very common almost euery where in moist and waterish plashes, and neere vnto the brims of riuers, ditches, and running brookes. They floure from Iune to August.

‡ The codded or impatient Arsmart was first found to grow in this kingdome by the industrie of my good friend Mr. *George Bowles*, who found it at these places: first in Shropshire, on the bankes of the riuer Kemlet at Marington in the parish of Cherberry, vnder a Gentlemans house called Mr. *LLoyd*; but especially at Guerndee in the parish of Cherstocke, halfe a mile from the foresaid Riuer, amongst great Alder trees in the highway. ‡

¶ *The Names.*

1 Arsmart is called in Greeke ὑδροπέπερι: of the Latines, *Hydropiper*, or *Piper aquaticum*, or *Aquatile*, or water Pepper: in high-Dutch, **Wasser Pfeffer**: in low-Dutch, **Water Peper**: in French, *Curage*,

Curage, or *Culrage* : in Spanish, *Pimenta aquatica* : in English, Water-Pepper, Culrage, and Arsesmart, according to the operation and effect when it is vsed in the absterfion of that part.

2 Dead Arsmart is called *Persicaria*, or Peach-wort, of the likenesse that the leaues haue with those of the Peach tree. It hath beene called *Plumbago* of the leaden coloured markes which are seene vpon it : but *Pliny* would haue *Plumbago* not to be so called of the colour, but rather of the effect, by reason that it helpeth the infirmitie of the eyes called *Plumbum*. Yet there is another *Plumbago* which is rather thought to be that of *Plinies* description, as shal be shewed in his proper place. In English we may call it Peach-wort, and dead Arsmart, because it doth not bite those places as the other doth.

‡ 3 This is by *Lobel* set forth, and called *Persicaria pusilla repens* : of *Tabernamontanus*, *Persicaria pumila*.

4 No plant I thinke hath found more varietie of names than this : for *Tragus* calls it *Mercurialis syluestris altera* ; and he also calls it *Esula* : *Leonicerus* calls it *Tithymalus syluestris* : *Gesner*, *Camerarius*, and others, *Noli me tangere* : *Dodonæus*, *Impatiens herba* : *Cæsalpinus*, *Catanance altera* : in the *Hist. Lugd.* (where it is some three times ouer) it is called besides the names giuen it by others, *Chrysæa* : *Lobel*, *Thalius*, and others call it *Persicaria siliquosa* : yet none of these well pleasing *Columna*, he hath accurately described and figured it by the name of *Balsamita altera* : and since him *Bauhine* hath named it *Balsamina lutea* : yet both these and most of the other keepe the title of *Noli me tangere*. ‡

¶ *The Temperature.*

Arsmart is hot and dry, yet not so hot as Pepper, according to *Galen*.

Dead Arsmart is of temperature cold, and something dry.

¶ *The Vertues.*

The leaues and seed of Arsmart do waste and consume all cold swellings, dissolue and scatter congealed bloud that commeth of bruisings or stripes. A

The same bruised and bound vpon an impostume in the ioynts of the fingers (called among the vulgar sort a fellon or vncome) for the space of an houre, taketh away the paine : but (saith the Author) it must be first buried vnder a stone before it be applied ; which doth somewhat discredit the medicine. B

The leaues rubbed vpon a tyred jades backe, and a good handfull or two laid vnder the saddle, and the same set on againe, wonderfully refresheth the wearied horse, and causeth him to trauell much the better. C

It is reported that Dead Arsmart is good against inflammations and hot swellings, being applied in the beginning : and for greene wounds, if it be stamped and boyled with oyle Oliue, waxe, and Turpentine. D

‡ The faculties of the fourth are not yet knowne. *Lobel* saith it hath a venenate qualitie : and *Tragus* saith a vomitorie : yet neither of them seemes to affirme any thing of certaintie, but rather by heare-say. ‡ E

CHAP. 115. *Of Bell-Floures.*

¶ *The Description.*

1 Couentry-Bells haue broad leaues rough and hairy, not vnlike to those of the Garden Buglosse, of a swart greene colour : among which do rise vp stiffe hairie stalks the second yeare after the sowing of the seed : which stalkes diuide themselues into sundry branches, whereupon grow many faire and pleasant bell-floures, long, hollow, and cut on the brim with fiue sleight gashes, ending in fiue corners toward night, when the floure shutteth it selfe vp, as do most of the Bell-floures : in the middle of the floures be three or foure whitish chiues, as also much downy haire, such as is in the eares of a Dog or such like beast. The whole floure is of a blew purple colour : which being past, there succeed great square or cornered seed-vessels, diuided on the inside into diuers cels or chambers, wherein doe lie scatteringly many small browne flat seeds. The root is long and great like a Parsenep, garnished with many threddy strings, which perisheth when it hath perfected his seed, which is in the second yeare after his sowing, and recouereth it selfe againe by the falling of the seed.

2 The second agreeth with the first in each respect, as well in leaues, stalkes, or roots, and differeth in that, that this plant bringeth forth milke-white floures, and the other not so.

¶ *The*

Viola Mariana. Blew Couentry-Bells.

A

¶ *The Place and Time.*

They grow in woods, mountaines, and darke vallies, & vnder hedges among the bushes, especially about Couentry, where they grow very plentifully abroad in the fields, & are there called Couentry-bels; and of some about London Canturbury-bels, but vnproperly, for that there is another kinde of Bell-floure growing in Kent about Canturbury, which may more fitly be called Canturbury-bells, because they grow there more plentifully than in any other Country. These pleasant Bel-floures we haue in our London gardens especially for the beauty of their floure, although they be kindes of Rampions, and the roots eaten as Rampions are.

They floure in Iune, Iuly, and August; the seed waxeth ripe in the meane time; for these plants bring not forth their floures all at once; but when one floureth another seedeth.

¶ *The Names.*

Couentry bels are called in Latine *Viola Mariana:* in English, Mercuries violets, or Couentry Rapes; and of some, Mariets. It hath bin taken to be *Medium,* but vnfitly: of some it is called *Rapū syluestre:* which the Greeks cal γογγύλην ἀγρίαν.

¶ *The Temperature and Vertues.*

The root is cold and somewhat binding, and not vsed in physicke, but only for a sallet root boyled and eaten with oyle, vineger, and pepper.

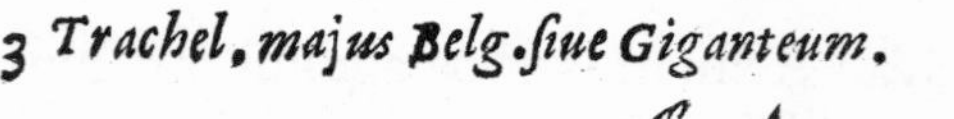

Chap. 116. *Of Throat-wort, or Canturbury-Bells.*

1 *Trachelium majus.* Blew Canturbury-Bels.

3 *Trachel. majus Belg. siue Giganteum.* Gyant Throatwort.

¶ *The Deſcription.*

1 The firſt of the Canterbury bells hath rough and hairy brittle ſtalkes, creſted into a certaine ſquareneſſe, diuiding themſelues into diuers branches, whereupon do grow very rough ſharpe pointed leaues, cut about the edges like the teeth of a ſawe; and ſo like the leaues of nettles, that it is hard to know the one from the other, but by touching them. The floures are hollow, hairy within, and of a perfect blew colour, bell faſhion, not vnlike to the Couentry bells. The root is white, thicke, and long laſting. ‡ There is alſo in ſome Gardens kept a variety hereof hauing double floures. ‡

2 The white Canterbury bells are ſo like the precedent, that it is not poſſible to diſtinguiſh them, but by the colour of the floures; which of this plant is a milke white colour, and of the other a blew, which ſetteth forth the difference.

4 *Trachelium minus.*
Small Canterbury bells.

‡ 5 *Trachelium majus petræum.*
Great Stone Throtewort.

‡ Our Author much miſtaking in this place (as in many other) did againe figure and deſcribe the third and fourth, and of them made a fift and ſixt, calling the firſt *Trachelium Giganteum*, and the next *Viola Calathiana*; yet the figures were ſuch as *Bauhine* could not coniecture what was meant by them, and therefore in his *Pinax*, he ſaith, *Trachelium Giganteum, & Viola Calathiana apud Gerardum, quid?* but the deſcriptions were better, wherefore I haue omitted the former deſcription and here giuen you the later. ‡

3 Giants Throtewort hath very large leaues of an ouerworne greene colour, hollowed in the middle like the Moſcouites ſpoone, and very rough, ſlightly indented about the edges. The ſtalke is two cubits high, whereon thoſe leaues are ſet from the bottome to the top; from the boſome of each leafe commeth forth one ſlender footeſtalke, whereon doth grow a faire and large floure faſhioned like a bell, of a whitiſh colour tending to purple. The pointed corners of each floure turne themſelues backe like a ſcrole, or the Dalmatian cap; in the middle whereof commeth forth a ſharpe ſtile or clapper of a yellow colour. The root is thicke, with certaine ſtrings annexed thereto.

4 The ſmaller kinde of Throtewort hath ſtalkes and leaues very like vnto the great Throte-

woort, but altogether leſſer, and not ſo hairy: from the boſome of which leaues ſhoot forth very beautifull floures bell faſhion, of a bright purple colour, with a ſmall peſtle or clapper in the middle, and in other reſpects is like the precedent.

‡ 5 This from a wooddy and wrinkled root of a pale purple colour ſends forth many rough creſted ſtalkes of ſome cubit high, which are vnorderly ſet with leaues, long, rough, and ſnipt lightly about their edges, being of a darke colour on the vpper ſide, and of a whitiſh on their vnder part. At the tops of the ſtalkes grow the floures, being many, and thicke thruſt together, white of colour, and diuided into fiue or ſeuen parts, each floure hauing yellowiſh threds, and a pointall in their middles. It floures in Auguſt, and was firſt ſet forth and deſcribed by *Pona* in his deſcription of Mount Baldus. ‡

¶ *The Place.*

The firſt deſcribed and ſometimes the ſecond growes very plentifully in the low woods and hedge-rowes of Kent, about Canterbury, Sittingborne, Graueſend, Southfleet, and Greenehyth, eſpecially vnder Cobham Parke-pale in the way leading from Southfleet to Rocheſter, at Eltham about the parke there not farre from Greenwich; in moſt of the paſtures about Watford and Buſhey, fifteene miles from London.

‡ 3 The third was kept by our Author in his Garden, as it is alſo at this day preſerued in the garden of Mr. *Parkinſon*: yet in the yeere *1626* I found it in great plenty growing wilde vpon the bankes of the Riuer Ouſe in Yorkſhire, as I went from Yorke to viſite Selby the place whereas I was borne, being ten miles from thence. ‡

The fourth groweth in the medow next vnto Ditton ferrie as you goe to Windſore, vpon the chalky hills about Greenehithe in Kent; and in a field by the high way as you go from thence to Dartford; in Henningham parke in Eſſex; and in Sion medow neere to Brandford, eight miles from London.

The fifth growes on Mount Baldus in Italy.

¶ *The Time.*

All the kindes of bell floures do floure and flouriſh from May vntill the beginning of Auguſt, except the laſt, which is the plant that hath been taken generally for the Calathian violet, which floureth in the later end of September; notwithſtanding the Calathian violet or Autumne violet is of a moſt bright and pleaſant blew or azure colour, as thoſe are of this kinde, although this plant ſometimes changeth his colour from blew to whiteneſſe by ſome one accident or other.

¶ *The Names.*

1 2 Throtewoort is called in Latine *Ceruicaria*, and *Ceruicaria major*: in Greeke, τραχήλιον: of moſt, *Vuularia*: of *Fuchſius*, *Campanula*: in Dutch, **Halſcruyt**: in Engliſh, Canterburie bells, Haskewoorte, Throtewoort, or *Vuula* woort, of the vertue it hath againſt the paine and ſwelling thereof.

‡ 3 This is the *Trachelium majus Belgarum* of *Lobell*, and the ſame (as I before noted) that our Author formerly ſet forth by the name of *Trachelium Giganteum*, ſo that I haue put them, as you may ſee, together in the title of the plant.

4 This is the *Trachelium maius* of *Dodonæus*, *Lobell*, and others: the *Ceruicaria minor* of *Tabernamontanus*; and *Vuularia exigua* of *Tragus*: Our Author gaue this alſo another figure and deſcription by the name of *Viola Calathiana*, not knowing that it was the laſt ſaue one which he had deſcribed by the name of *Trachelium minus*. ‡

¶ *The Temperature.*

Theſe plants are cold and dry, as are moſt of the Bell floures.

¶ *The Vertues.*

A The Antients for any thing that we know haue not mentioned, and therefore not ſet downe any thing concerning the vertues of theſe Bell floures: notwithſtanding we haue found in the later writers, as alſo of our owne experience, that they are excellent good againſt the inflammation of the throte and *Vuula* or almonds, and all manner of cankers and vlcerations in the mouth, if the mouth and throte be gargarized and waſhed with the decoction of them: and they are of all other herbes the chiefe and principall to be put into lotions or waſhing waters, to inject into the priuy parts of man or woman being boiled with hony and Allom in water, with ſome white wine.

Chap.

CHAP. 117. *Of Peach-bells and Steeple-bells.*

¶ *The Deſcription.*

1 THe Peach-leaued Bell-floure hath a great number of ſmall and long leaues, riſing in a great buſh out of the ground, like the leaues of the Peach tree: among which riſeth vp a ſtalke two cubits high: alongſt the ſtalke grow many floures like bells, ſomtime white, and for the moſt part of a faire blew colour; but the bells are nothing ſo deepe as they of the other kindes; and theſe alſo are more dilated or ſpred abroad than any of the reſt. The ſeed is ſmall like Rampions, and the root a tuft of laces or ſmall ſtrings.

2 The ſecond kinde of Bell-floure hath a great number of faire blewiſh or Watchet floures, like the other laſt before mentioned, growing vpon goodly tall ſtems two cubits and a halfe high, which are garniſhed from the top of the plant vnto the ground with leaues like Beets, diſorderly placed. This whole plant is exceeding full of milke, inſomuch as if you do but breake one leafe of the plant, many drops of a milky iuyce will fall vpon the ground. The root is very great, and full of milk alſo: likewiſe the knops wherein the ſeed ſhould be are empty and void of ſeed, ſo that the whole plant is altogether barren, and muſt be increaſed with ſlipping of his root.

1 *Campanula perſicifolia.*
Peach-leaued Bell-floure.

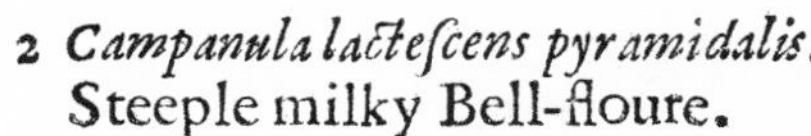
2 *Campanula lacteſcens pyramidalis.*
Steeple milky Bell-floure.

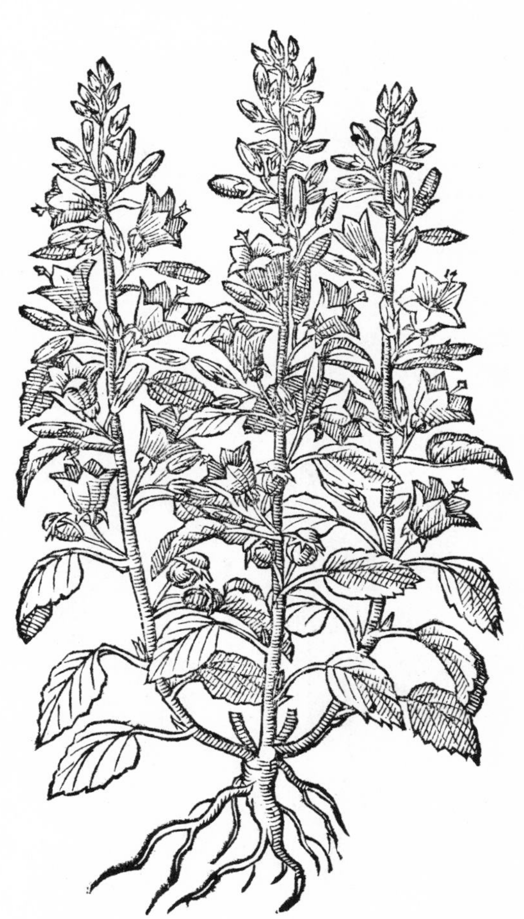

3 The ſmall Bell-floure hath many round leaues very like thoſe of the common field Violet, ſpred vpon the ground; among which riſe vp ſmall ſlender ſtems, diſorderly ſet with many graſſie narrow leaues like thoſe of flax. The ſmall ſtem is diuided at the top into ſundry little branches, whereon do grow pretty blew floures bell-faſhion. The root is ſmall and threddy.

4 The yellow Bell-floure is a very beautifull plant of an handfull high, bearing at the top of his weake and tender ſtalkes moſt pleaſant floures bel-faſhion, of a faire and bright yellow colour. The leaues and roots are like the precedent, ſauing that the leaues that grow next to the ground of this plant are not ſo round as the former. ‡ Certainly our Author in this place meant to ſet forth the *Campanula lutea linifolia flore volubilis*, deſcribed in the *Aduerſ. pag.* 177. and therefore I haue giuen you the figure thereof. ‡

5 The

3 *Campanula rotundifolia.*
Round leaued Bell-floure.

† 4 *Campanula lutea linifolia.*
Yellow Bell-floure.

5 *Campanula minor alba, ſiue purpurea.*
Little white or purple Bel-floure.

5 The little white Bell-floure is a kinde of wilde Rampions, as is that which followeth, and alſo the laſt ſaue one before deſcribed. This ſmall plant hath a ſlender root of the bigneſſe of a ſmall ſtraw, with ſome few ſtrings anexed thereto The leaues are ſomwhat long, ſmooth, and of a perfect greene colour, lying flat vpon the ground: from thence riſe vp ſmall tender ſtalkes, ſet heere and there with a few leaues. The floures grow at the top, of a milke white colour.

6 The other ſmall Bell-floure or wilde Rampion differeth not from the precedent but onely in colour of the floures; for as the others are white, theſe are of a bright purple colour, which ſets forth the difference.

‡ 7 Beſides theſe here deſcribed, there is another very ſmall and rare Bell-floure, which hath not beene ſet forth by any but onely by *Bauhine*, in his *Prodrom.* vnder the title of *Campanula Cymbalariæ folijs*, and that fitly, for it hath thinne and ſmall cornered leaues much after the maner of *Cymbalaria*, and theſe are ſet without order on very ſmall weake and tender ſtalkes ſome handfull long; and at the tops of the branches grow little ſmall and tender Bell-floures of a blew colour. The root, like as the whole plant, is very ſmall and threddy. This pretty plant was firſt diſcouered to grow in England by Maſter *George Bowles*, *Anno* 1632. who found it in Montgomerie ſhire, on the dry bankes in the high-way as one rideth from Dolgeogg a Worſhipfull Gentlemans houſe called Mr. *Francis Herbert*, vnto a market towne called Mahuntleth, and in all the way from thence to the ſea ſide. It may be called in Engliſh, The tender Bell-floure. ‡

¶ *The*

¶ The Place.

The two firſt grow in our London gardens, and not wilde in England.

The reſt, except that ſmall one with yellow floures, do grow wilde in moſt places of England, eſpecially vpon barren ſandy heaths and ſuch like grounds.

¶ The Time.

Theſe Bell-floures do flouriſh from May vnto Auguſt.

¶ The Names.

Their ſeuerall titles ſet forth their names in Engliſh and Latine, which is as much as hath been ſaid of them.

¶ The Temperature and Vertues.

Theſe Bell-floures, eſpecially the foure laſt mentioned, are cold and dry, and of the nature of Rampions, whereof they be kindes.

† The figure in the fourth place was of *Rapunculus nemoroſus* 3. *of Tabern.* whereof you ſhall finde mention in the following chapter.

CHAP. 118. *Of Rampions, or wilde Bell-floures.*

1 *Rapuntium majus.*
Great Rampion.

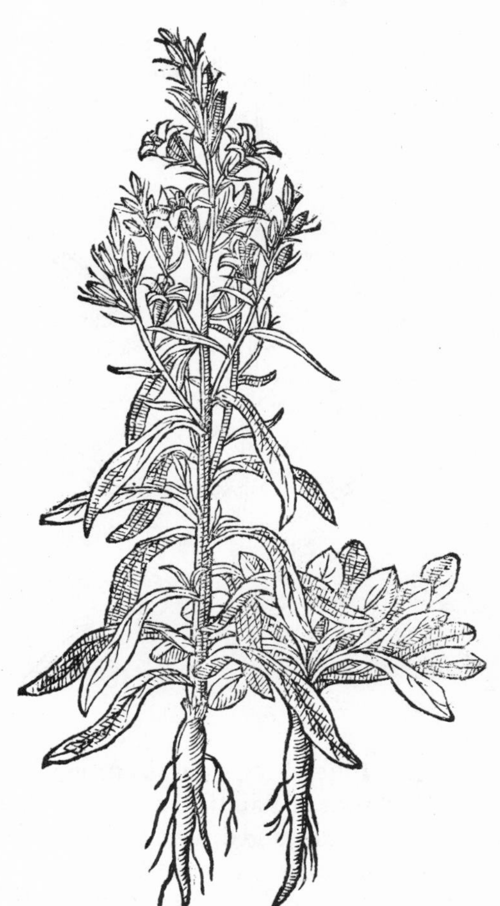

2 *Rapuntium parvum.*
Small Rampion.

¶ The Deſcription.

1 THe great Rampion being one of the Bell-floures, hath leaues which appeare or come forth at the beginning ſomewhat large and broad, ſmooth and plaine, not vnlike to the leaues of the ſmalleſt Beet. Among which riſe vp ſtemmes one cubit high, ſet with ſuch like leaues as thoſe are of the firſt ſpringing vp, but ſmaller, bearing at the top of the ſtalke a great thicke buſhy eare full of little long floures cloſely thruſt together like a Fox-taile: which ſmall floures before their opening are like little crooked hornes, and being wide opened they are ſmall blew-bells, ſometimes white, or ſometimes purple. The root is white, and as thicke as a mans thumbe.

2 The second kind being likewise one of the bel-floures, and yet a wild kind of Rampion, hath leaues at his first comming vp like vnto the garden Bell-floure. The leaues which spring vp afterward for the decking vp of the stalke are somewhat longer and narrower. The floures grow at the top of tender and brittle stalkes like vnto little bells, of a bright blew colour, sometimes white or purple. The root is small, long, and somewhat thicke.

3 This is a wilde Rampion that growes in woods: it hath small leaues spred vpon the ground, bluntly indented about the edges: among which riseth vp a straight stem of the height of a cubit, set from the bottome to the top with longer and narrower leaues than those next the ground: at the top of the stalkes grow small Bell-floures of a watchet blewish colour. The root is thicke and tough, with some few strings anexed thereto.

‡ There is another varietie of this, whose figure was formerly by our Author set forth in the fourth place of the last chapter: it differs from this last onely in that the floures and other parts of the plant are lesser a little than those of the last described. ‡

3 *Rapunculus nemorosus.*
Wood Rampions.

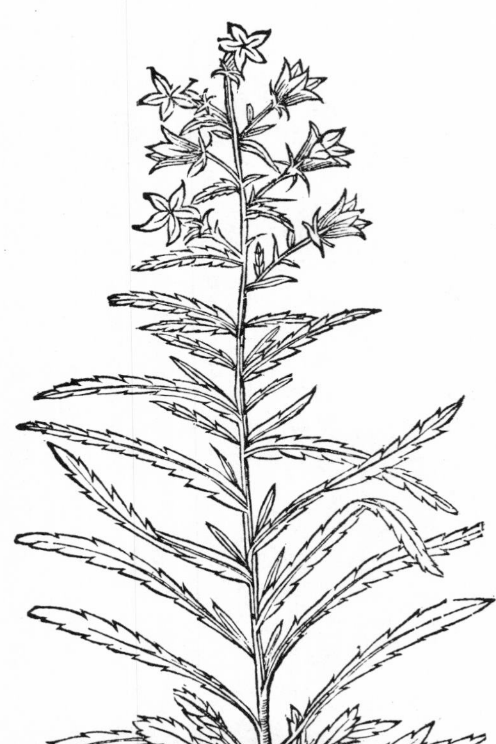

‡ 4 *Rapunculus Alpinus Corniculatus.*
Horned Rampions of the Alpes.

‡ 4 This which growes amongst the rockes in the highest Alpes hath a wooddy and verie wrinckled root an handfull and halfe long, from which arise many leaues set on pretty long stalks, somewhat round, and diuided with reasonable deepe gashes, hauing many veines, and being of a darke greene colour: amongst these grow vp little stalkes, hauing one leafe about their middles, and three or foure set about the floure, being narrower and longer than the bottome leaues. The floures grow as in an vmbell, and are shaped like that Chymicall vessell we vsually call a Retort, being big at their bottomes, and so becomming smaller towards their tops, and hauing many threds in them, whereof one is longer than the rest, and comes forth in the middle of the floure: it floures in August. *Pona* was the first that described this, vnder the name of *Trachelium petraum minus*.

5 The roots of this other kinde of horned Rampion grow after an vnusuall manner; for first or lowermost is a root like to that of a Rampion, but slenderer, and from the top of that commeth forth as it were another root or two, being smallest about that place whereas they are fastned to the vnder root, and all these haue small fibres comming from them. The leaues which first grow vp are smooth, and almost like those of a Rampion, yet rounder, and made somwhat after the maner of a violet leafe, but nothing so big: at the bottome of the stalk come forth 7 or eight long narrow

leaues ſnipt about the edges,and ſharpe pointed,and vpon the reſt of the ſtalke grow alſo three or foure narrow ſharp pointed leaues. The floures which are of a purple colour,at firſt reſemble thoſe of the laſt deſcribed ; but afterwards parte themſelues into fiue ſlender ſtrings with threds in the middles ; which decaying,they are ſucceeded by little cups ending in fiue little pointels,and containing a ſmall yellow ſeed. This is deſcribed by *Fabius Columna*, vnder the name of *Rapuntium Corniculatum montanum :* And I receiued ſeeds and roots hereof from Mr. *Goodyer*, who found it growing plentifully wilde in the incloſed chalkie hilly grounds by Maple-Durham neere Peterſfield in Hampſhire.

6 This which is deſcribed in *Cluſius* his *Curæ poſter*. by the name of *Pyramidalis,* and was firſt found and ſent to him by *Gregory de Reggio* a Capuchine Frier, is alſo of this kindred ; wherefore I will giue you a briefe deſcription thereof. The root is white, and long laſting; from which come diuers round hairie and writhen ſtalkes,about a ſpan long more or leſſe. At the top of theſe ſtalks and all amongſt the leaues, grow many elegant blew floures,which are ſucceeded by ſeed veſſels like thoſe of the leſſer *Trachelium*, being full of a ſmall ſeed. The whole plant yeelds milke like as the reſt of this kinde, and the leaues as well in ſhape as hoarineſſe on their vnder ſides, well reſemble thoſe of the ſecond French or Golden Lungwoort of my deſcription. It was firſt found growing in the chinkes of hard rockes about the mouthes of Caues, in the mountaines of Breſcia in Italy by the foreſaid Frier. ‡

‡ 5 *Rapunculus Corniculatus montanus.*
Mountaine horned Rampions.

‡ 6 *Rapunculus ſaxatilis,ſiue Pyramidalis alter.*
Rocke Rampion.

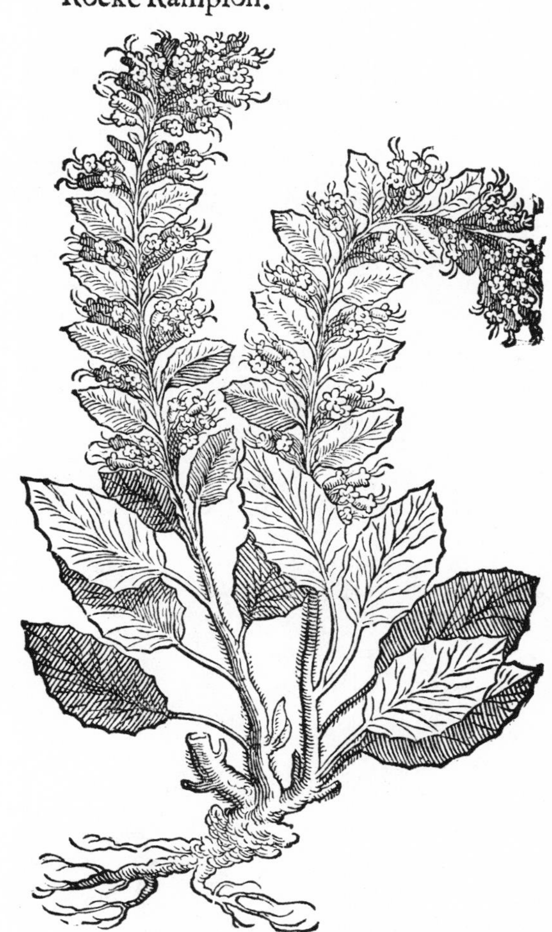

¶ *The Place.*

The firſt is ſowne and ſet in Gardens,eſpecially becauſe the rootes are eaten in Sallads.

The ſecond groweth in woods and ſhadowie places,in fat and clayie ſoiles.

¶ *The Time.*

They floure in May,Iune,and Iuly.

¶ *The Names.*

Rampions by a generall name are called *Rapuntium* and *Rapunculus* ; and the firſt by reaſon of the long ſpokie tufte of floures is called *Rapuntium maius Alopecuri comoſo flore* by *Lobell* and *Pena*: *Rapunculum ſylueſtre,* and *Rapunculus ſylueſtris ſpicatus* by others. The ſecond,which is is the ordinary Rampion is called *Rapunculus*, and *Rapuntium minus*; *Lobell* thinkes it the *Pes Locuſtæ* of *Auicen*; and *Columna* iudges it to be *Erinus* of *Nicander* and *Dioſcorides*. The third is the *Rapunculus nemoroſus ſecundus* of *Tabernamontanus*; & the varietie of it is *Rapunc. nemor. tertius*. The names of the reſt are ſhewen in their deſcriptions. ‡

¶ *The Temperature.*

The roots of theſe are of a cold temperature, and ſomething binding.

¶ *The Vertues.*

A The roots are eſpecially vſed in ſallads, being boiled and eaten with oile, vineger, and pepper.

B Some affirme, that the decoction of the roots are good for all inflammations of the mouth, and Almonds of the throte, and other diſeaſes happening in the mouth and throte, as the other Throte-woorts.

Chap. 119. *Of Wall-floures, or yellow Stocke-Gillo-floures.*

¶ *The Kindes.*

‡ These plants which wee terme commonly in Engliſh, Wal-floures and Stocke Gillo-floures are comprehended vnder one generall name of *Leucoion*, (1) *Viola alba*, White Violet, λευκὸς ſignifying white, and ἴον a Violet, which as ſome would haue it is not from the whiteneſſe of the floure, for that the moſt and moſt vſuall of them are of other colours, but from the whitenes or hoarineſſe of the leaues, which is proper rather to the Stocke Gillouers than to the wal-floures, I therefore thinke it fit to diſtinguiſh them into *Leucoia folijs viridibus*, that is VVal-floures; and *Leucoia foliis incanis*, Stocke Gillouers. Now theſe againe are diſtinguiſhed into ſeuerall ſpecies, as you may finde by the following Chapters Moreouer you muſt remember there is another *Viola alba* or *Leucoion* (which is thought to be that of *Theophraſtus* and whereof we haue treated in the firſt booke) which is far different from this, and for diſtinction ſake called *Leucoium bulboſum*. ‡

1 *Viola Lutea.*
VVal-floure.

2 *Viola lutea multiplex.*
Double VVal-floure.

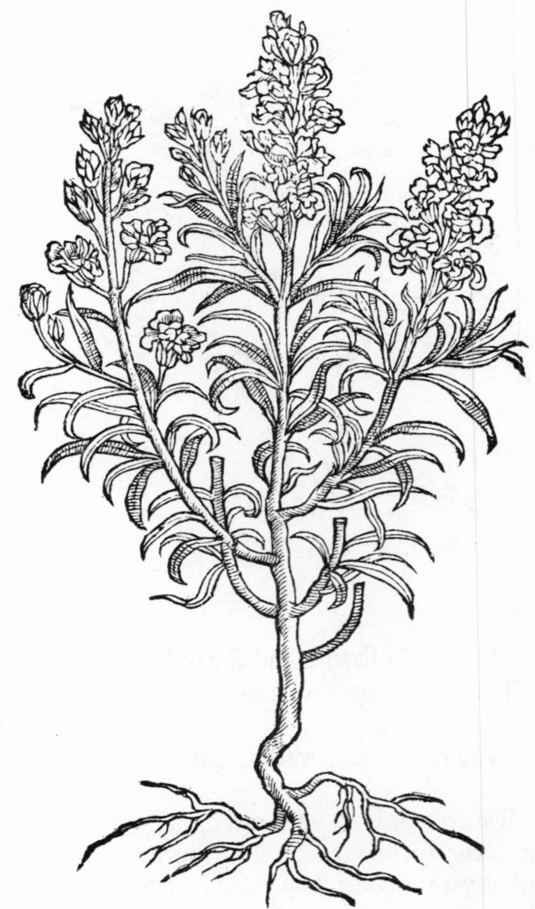

¶ *The*

† 4 *Leucoium syluestre.*
Wilde wall floure.

¶ *The Description.*

1 THe stalks of the Wall floure are full of greene branches, the leaues are long, narrow, smooth, slippery, of a blackish greene colour, and lesser than the leaues of stocke Gillofloures. The floures are small, yellow, very sweete of smell, and made of foure little leaues; which being past, there succeed long slender cods, in which is contained flat reddish seed. The whole plant is shrubby, of a wooddie substance, and can easily endure the colde of winter.

2 The double Wall floure hath long leaues greene aud smooth, set vpon stiffe branches, of a wooddie substance: whereupon do grow most pleasant sweet yellow flours very double, which plant is so well knowne to all, that it shall be needlesse to spend much time about the description.

3 Of this double kinde we haue another sort that bringeth his floures open all at once, whereas the other doth floure by degrees, by meanes whereof it is long in flouring.

‡ 4 This plant which was formerly seated in the fourth place of the following chapter, I haue brought to enjoy the same place in this, for that by reason of the greenesse of his leaues and other things he comes nearest to these here described, also I wil describe it anew, because the former was almost wholly false: It hath many greene leaues at the top of the root like to these of the wall floure, but narrower, and bitter of taste, among which rise vp one or more stalks of a foot or more in height, crested and set with carinated leaues. The floures grow at the tops of the stalkes many together, consisting of foure yellow leaues a piece, lesser than those of the ordinary wall floures; there floures are succeeded by long cods containing a flat seed. The root is long and whitish, with many fibres.

5 Besides these, there is in some gardens kept another wall-floure differing from the first in the bignesse of the whole plant, but especially of the floure, which is yellow and single, yet very large and beautifull.

6 Also there is another with very greene leaues, and pure white and well smelling floures. ‡

¶ *The Place.*

The first groweth vpon bricke and stone walls, in the corners of churches euery where, as also among rubbish and such other stony places.

The double Wall-floure groweth in most gardens of England.

¶ *The Time.*

They floure for the most part all the yeere long, but especially in winter, whereupon the people in Cheshire do call them Winter-Gillofloures.

¶ *The Names.*

The Wallfloure is called in Greeke λευκόϊον: in Latine, *Viola lutea*, and *Leucoium luteum*: in the Arabicke tongue *Keyri*: in Spanish, *Violettas Amarillas*: in Dutch, **Violieren**: in French, *Girofflees iaulnes, Violieres des murailles*: in English, Wall-Gillofloure, Wall-floure, yellow stocke Gillofloure, and Winter-Gillofloure.

¶ *The Temperature.*

All the whole shrub of Wall-Gillofloures, as *Galen* saith, is of a clensing faculty, and of thinne parts.

¶ *The Vertues.*

A *Dioscorides* writeth that the yellow Wall-floure is most vsed in physicke, and more than the rest of stocke-Gillofloures, whereof this is holden to be a kinde: which hath mooued me to preferre it vnto the first place. He saith, that the juice mixed with some vnctious or oilie thing, and boiled to the forme of a lyniment, helpeth the chops or rifts of the fundament.

The

B The herbe boiled with white wine, honie, and a little allom, doth cure hot vlcers, and cankers of the mouth.

C The leaues ſtamped with a little bay ſalt, and bound about the wriſts of the hands, taketh away the ſhaking fits of the Ague.

D ‡ A decoction of the floures together with the leaues, is vſed with good ſucceſſe to mollifie Schirrous tumors.

E The oile alſo made with theſe is good to be vſed to anoint a Paralyticke, as alſo a goutie part to mitigate paine.

F Alſo a ſtrong decoction of the floures drunke, moueth the Courſes, and expelleth the dead childe. ‡

Chap. 120. *Of Stocke Gillofloures.*

1 *Leucoium album, ſiue purpureum, ſiue violaceum.*
White, purple, or Violet coloured Stocke Gilloflоure.

‡ 2 *Leucoium flore multiplici.*
Double Stocke Gilloflоure.

¶ *The Deſcription.*

1 The ſtalke of the great ſtocke Gilloflоure is two foot high or higher, round, and parted into diuers branches. The leaues are long, white, ſoft, and hauing vpon them as it were a downe like vnto the leaues of willowe, but ſofter: the floures conſiſt of foure little leaues growing all along the vpper part of the branches, of a white colour, exceeding ſweet of ſmell: in their places come vp long and narrow cods, in which is contained broad, flat, and round ſeed. The root is of a wooddy ſubſtance, as is the ſtalke alſo.

The purple ſtocke Gilloflоure is like the precedent in each reſpect, ſauing that the floures of this plant are of a pleaſant purple colour, and the others white, which ſetteth forth the difference: of which kinde we haue ſome that beare double floures, which are of diuers colours, greatly eſteemed for the beautie of their floures, and pleaſant ſweet ſmell.

This

3 *Leucoium ſpinoſum Creticum.*
Thornie Stocke Gillouers.

This kinde of Stocke Gilloflouré that beareth floures of the colour of a Violet, that is to ſay of a blew tending to a purple colour, which ſetteth forth the difference betwixt this plant & the other ſtocke Gillofloures, in euery other reſpect is like the precedent.

2 ‡ There were formerly 3 figures of the ſingle Stocks, which differ in nothing but the colour of their floures; wherefore we haue made them content with one, & haue giuen (which was formerly wanting) a figure of the double Stock, of which there are many and prettie varieties kept in the garden of my kinde friend Mr *Ralph Tuggye* at VVeſtminſter, and ſet forth in the bookes of ſuch as purpoſely treat of floures and their varieties. ‡

‡ 3 To theſe I thinke it not amiſſe to adde that plant which *Cluſius* hath ſet forth vnder the name of *Leucoium ſpinoſum Creticum*. It growes ſome foot or more high, bringing forth many ſtalkes which are of a grayiſh colour, and armed at the top with many and ſtrong thorny prickles: the leaues which adorne theſe ſtalkes are like thoſe of the ſtocke Gillouer, yet leſſe and ſomewhat hoary; the floures are like thoſe of Mulleine, of a whitiſh yellow colour, with ſome purple threds in their middles; the cods which ſucceede the floures are ſmall and round, containing a little ſeed in them. They vſe, ſaith *Honorius Bellus*, to heat ovens therewith in Candy, where it plentifully growes; and by reaſon of the ſimilitude which the prickles hereof haue with *Stœbe* and the white colour, they cal it *Gala Stivida*, or *Galaſtivida*, and not becauſe it yeelds milke, which *Gala* ſignifies.

¶ *The Place.*

1. 2. Theſe kindes of Stocke Gillofloures do grow in moſt Gardens throughout England.

¶ *The Time.*

They floure in the beginning of the Spring, and continue flouring all the Sommer long.

¶ *The Names.*

The Stocke Gilloflourе is called in Greeke λευκόϊον: in Latine, *Viola alba*: in Italian, *Viola bianca*: in Spaniſh, *Violettas blanquas*: in Engliſh, Stocke Gilloflоure, Garnſey Violet, and Caſtle Gillofloure.

¶ *The Temperature and Vertues.*

A They are referred vnto the VVal-floure, although in vertue much inferiour, yet are they not vſed in Phyſicke, except amongſt certaine Empericks and Quackſaluers, about loue and luſt matters, which for modeſtie I omit.

B *Ioachimus Camerarius* reporteth, that a conſerue made of the floures of Stocke Gilloflоure, and often giuen with the diſtilled water thereof, preſerueth from the Apoplexy, and helpeth the palſie.

CHAP. 121. *Of Sea Stocke Gillofloures.*

¶ *The Kindes.*

OF Stocke Gillofloures that grow neere vnto the Sea there bee diuers and ſundrie ſorts, differing as well in leaues as floures, which ſhall bee comprehended in this Chapter next following.

1 *Leucoium marinum flore candido L'obelij.*
White Sea Stocke Gillofloures.

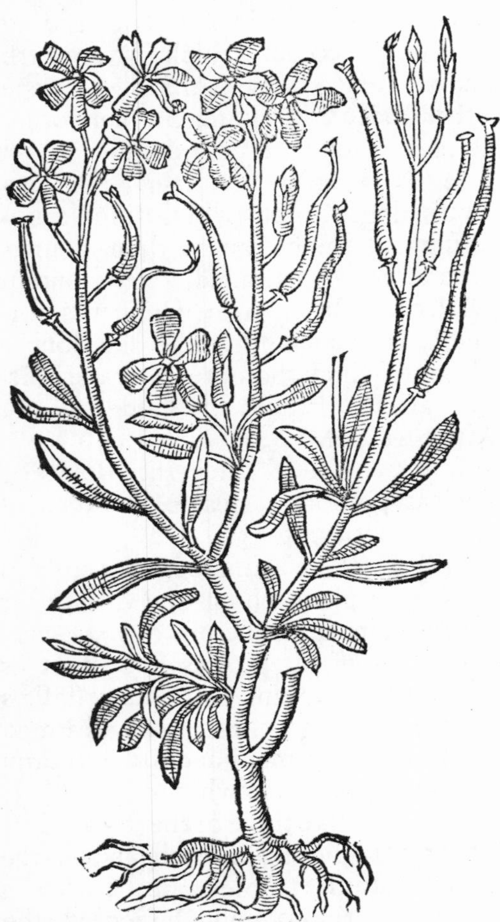

§ *Leucoium marinum latifolium.*
Broad leafed ſea ſtocke Gillofloure.

2 *Leucoium marinum purpureum L'obelij.*
Purple ſea Stocke Gillofloures.

¶ *The Deſcription.*

1 THe Sea ſtocke Gillofloure hath a ſmall wooddy root very threddie; from which riſeth vp an hoarie white ſtalke of two foot high, diuided into diuers ſmall branches, whereon are placed confuſedly many narrow leaues of a ſoft hoarie ſubſtance. The floures grow at the top of the branches, of a whitiſh colour, made of foure little leaues; which being paſt, there follow long coddes and ſeed, like vnto the garden ſtocke Gillofloure.

‡ 2 The purple ſtocke Gillofloure hath a very long tough root, thruſting it ſelfe deepe into the ground; from which riſe vp thicke, fat, ſoft, and hoarie ſtalkes. The leaues come forth of the ſtalkes next the ground, long, ſoft, thicke, full of iuice, couered ouer with a certaine downie hoarineſſe, and ſinuated ſomewhat deepe on both ſides, after the manner you may ſee expreſt in the figure of the fourth deſcribed in this Chapter. The ſtalke is ſet here and there with the like leaues, but leſſer. The floures grow at the top of the ſtalkes, compact of foure ſmall leaues, of a light purple colour. The ſeede is contained in long crooked cods like the garden ſtocke Gillofloure.

‡ The figure of *Lobels* which here we giue you was taken of a dried plant, and therefore the leaues are not expreſt ſo ſinuate as they ſhould be. ‡

3 This ſea ſtock Gillofloure hath many broad leaues ſpred vpon the ground, ſomwhat ſnipt or cut on the edges; amongſt which riſe vp ſmall naked ſtalkes, bearing at the top many little floures of a blew colour tending to a purple. The ſeede is in long cods like the others of his kinde.

4 The

4 The great Sea stock Gillofloure hath many broad leaues, growing in a great tuft, sleightly indented about the edges. The floures grow at the top of the stalkes, of a gold yellow colour. The root is small and single.

5 The small yellow Sea stocke Gillofloure hath many smooth, hoary, and soft leaues, set vpon a branched stalke: on the top whereof grow pretty sweet smelling yellow floures, bringing his seed in little long cods. The root is small and threddy. ‡ The Floures of this are sometimes of a red, or purplish colour. ‡

4 *Leucoium marinum luteum maius Clusij & L'obelij.*
The yellow Sea stocke Gillofloure.

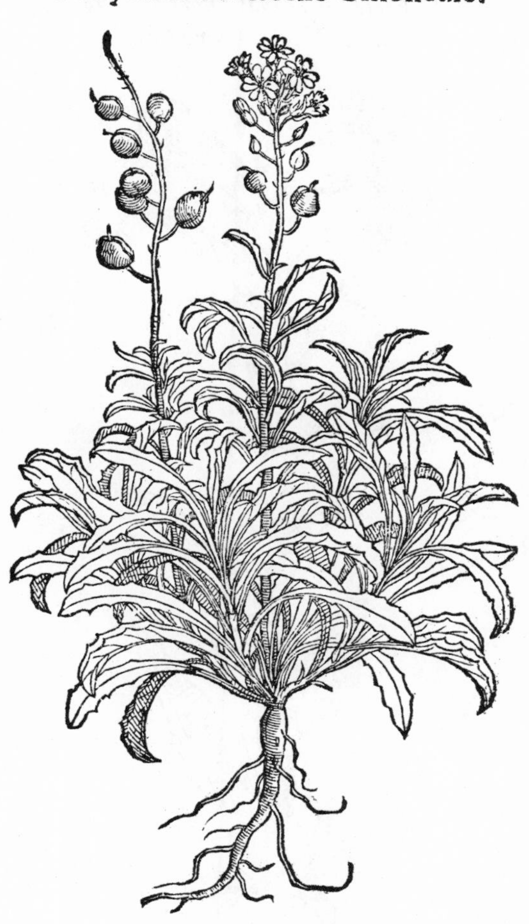

5 *Leucoium marinum minus L'obelij & Clus.*
Small yellow Sea stocke Gillofloure.

¶ *The Place.*

These plants do grow neere vnto the sea side, about Colchester, in the Isle of Man, neere Preston in Aundernesse, and about Westchester.

‡ I haue not hard of any of these wilde on our coasts but onely the second, which it may bee growes in these places here set downe; for it was gathered by Mr. *George Bowles* vpon the Rocks at Aberdovye in Merioneth shire. ‡

¶ *The Time.*

They flourish from Aprill to the end of August.

¶ *The Names.*

There is little to bee said as touching the names, more than hath been touched in their seuerall titles.

¶ *The Temperature and Vertues.*

There is no vse of these in Physicke, but they are esteemed for the beauty of their floures.

CHAP. 122. *Of Dames Violets, or Queenes Gillofloures.*

¶ *The Description.*

1 DAmes Violets or Queenes Gillofloures, haue great large leaues of a darke greene colour, somewhat snipt about the edges: among which spring vp stalkes of the height of

two cubits, set with such like leaues: the floures come forth at the tops of the branches, of a faire purple colour, verie like those of the stocke Gillofloures, of a very sweet smell, after which come vp long cods, wherein is contained small long blackish seed. The root is slender and threddie.

The Queenes white Gillofloures are like the last before remembred, sauing that this plant bringeth forth faire white floures, and the other purple.

‡ 2 By the industrie of some of our Florists, within this two or three yeares hath beene brought to our knowledge a very beautifull kinde of these Dame Violets, hauing very faire double white floures, the leaues, stalks and roots, are like to the other plants before described. ‡

1 *Viola Matronalis flore purpureo, siue albo.*
Purple, or white Dames Violets.

‡ 3 *Viola matronalis flore obsoleto.*
Russet Dames Violets.

‡ 3 This plant hath a stalke a cubit high, and is diuided into many branches, vpon which in a confused order grow leaues like those of the Dame Violet, yet a little broader and thicker, being first of somewhat an acide, and afterwards of an acride taste; at the tops of the branches in long cups grow floures like those of the Dames violet, consisting of foure leaues, which stand not faire open, but are twined aside, and are of a ouerworn russet colour, composed as it were of a yellow and browne with a number of blacke purple veines diuaricated ouer them. Their smell on the day time is little or none, but in the euening very pleasing and sweet. The floures are succeeded by long, and here and there swolne cods, which are almost quadrangular and containe a reddish seed like that of the common kinde. The root is fibrous, and vsually liues not aboue two yeares, for after it hath borne seed it dies; yet if you cut it downe and keepe it from seeding, it sometimes puts forth shouts whereby it may bee increased. I very much suspect that this figure and description which I here giue you taken out of *Clusius*, is no other plant than that which is kept in some of our gardens, and set foorth in the *Hortus Eystettensis* by the name of *Leucoium Melancholicum*: now I iudge the occasion of this error to haue come from the figure of *Clusius* which we here present you with, for it is in many particulars different from the description: first in that it expresses not many branches: secondly, in that the leaues are not snipt & diuided: thirdly, in that the Floures are not exprest wrested or twined: fourthly, the veins are not rightly exprest in the floure; & lastly, the cods are omitted. Now the *Leucoium melancholicum* hath a hairy stalke diuided into sundry branches of the height formerly mentioned, and the leaues about the middle of the stalke are somewhat sinuated or deepely or vnequally cut in; the shape and colour of the floure is the same with that now described,

‡ 4 *Leucoium melancholicum.*
The Melancholly floure.

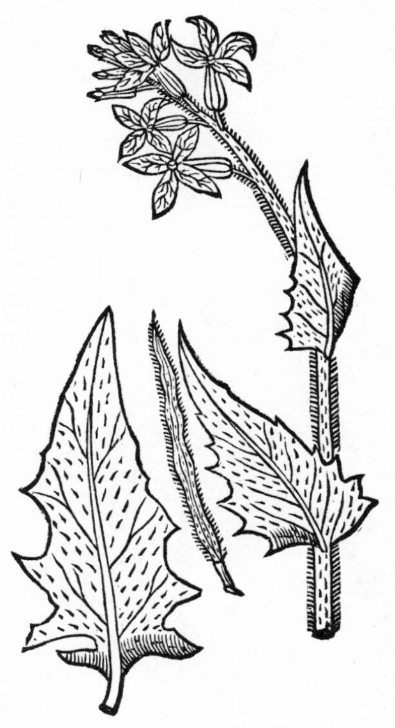

described, and the seed vessels the same, as far as I remember: for I must confesse, I did not in writing take any particular note of them though I haue diuers times seene them, neither did I euer compare them with this description of *Clusius*; onely I tooke some yeares agone an exact figure of a branch with the vpper leaues and floures, whereof one is exprest as they vsually grow twining backe, and the rest faire open, the better to set forth the veines that are spred ouer it. There are also exprest a cod or seede vessell, and one of the leaues that grow about the middle of the stalke; all which are agreeable to *Clusius* description in mine opinion; wherefore I onely giue you the figure that I then drew, with he title that I had it by. ‡

¶ *The Place.*

They are sown in gardens for the beauty of their floures.

¶ *The Time.*

They especially floure in Maie and Iune, the second yeare after they are sowne.

¶ *The Names.*

Dames Violet is called in Latine *Viola matronalis*, and *Viola Hyemalis*, or Winter Violets, and *Viola Damascena*: It is thought to be the *Hesperis* of *Pliny, lib.* 21. *cap.* 7. so called, for that it smels more, & more pleasantly in the euening or night, than at any other time. They are called in French *Violettes des Dames, & de domas*, and *Giroffleés des dames*, or *Matrones Violettes*. in English, Damaske Violets, winter Gillofloures, Rogues Gillofloures, and close Sciences.

¶ *The Temperature.*

The leaues of Dames Violets are in taste sharpe and hot, very like in taste and facultie to *Eruca* or Rocket, and seemeth to be a kinde thereof.

¶ *The Vertues.*

The distilled water of the floures hereof is counted to be a most effectuall thing to procure sweat. A

CHAP. 123. *Of White Sattin floure.*

¶ *The Description.*

1 Bolbonac or the Sattin floure hath hard and round stalkes, diuiding themselues into many other small branches, beset with leaues like Dames Violets, or Queenes Gillofloures, somewhat broad, and snipt about the edges, and in fashion almost like Sauce alone, or Iacke by the hedge, but that they are longer and sharper pointed. The stalkes are charged or loden with many floures like the common stocke Gillofloure, of a purple colour, which being falne, the seed

seed commeth forth contained in a flat thin cod, with a sharp point or prick at one end, in fashion of the Moone, and somewhat blackish. This cod is composed of three filmes or skins, whereof the two outmost are of an ouerworne ash colour, and the innermost, or that in the middle, whereon the seed doth hang or cleaue, is thin and cleere shining, like a shred of white Sattin newly cut from the peece. The whole plant dieth the same yeare that it hath borne seed, & must be sowne yearely. The root is compact of many tuberous parts like key clogs, or like the great Asphodill.

2 The second kind of *Bolbonac* or white Sattin hath many great and broad leaues, almost like those of the great burre Docke: among which riseth vp a very tall stem of the height of foure cubits, stiffe, and of a whitish greene colour, set with the like leaues, but smaller. The floures grow vpon the slender branches, of a purple colour, compact of foure small leaues like those of the stocke Gillofloure; after which come thin long cods of the same substance and colour of the former. The root is thicke, whereunto are fastened an infinite number of long threddie strings: which roote dieth not euery yeare as the other doth, but multiplieth it selfe as well by falling of the seede, as by new shoots of the root.

1 *Viola Lunaris siue Bolbonac.*
White Sattin.

2 *Viola Lunaris longioribus siliquis.*
Long codded white Sattin.

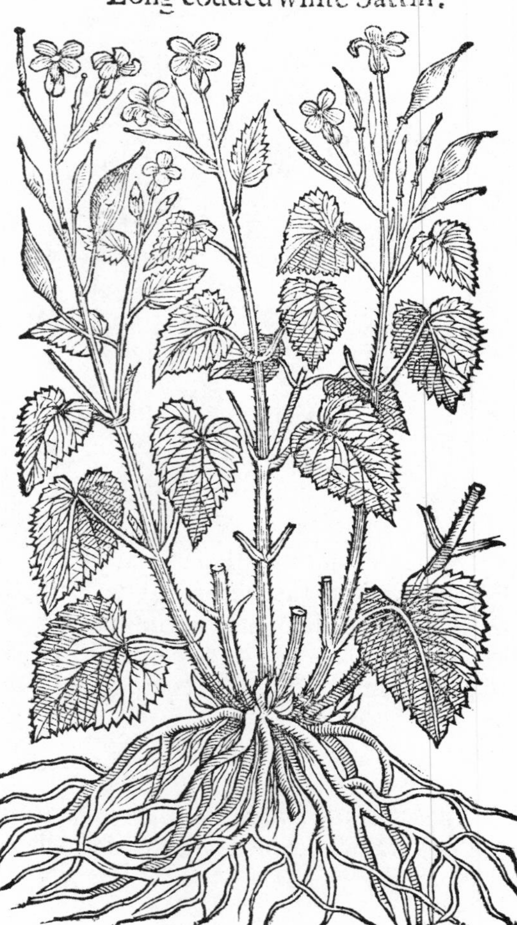

¶ *The Place.*

These plants are set and sowne in gardens; notwithstanding the first hath been found wilde in the woods about Pinner, and Harrow on the hill, twelue miles from London; and in Essex likewise about Horn-church.

The second groweth about Watford, fifteene miles from London.

¶ *The Time.*

They floure in Aprill the next yeare after they be sowne.

¶ *The Names.*

They are commonly called *Bolbonac* by a barbarous name: we had rather call it with *Dodonæus* & *Clusius*, *Viola latifolia*, and *Viola lunaris*, or as it pleaseth most Herbarists, *Viola peregrina*: the Brabanders name it **Penninck bloemen**, of the fashion of the coddes, like after a sort to a groat or testerne, and **Paesch bloemen**, because it alwaies floureth neere about the Feast of Easter: most of the later Herbarists doe call it *Lunaria*: Others, *Lunaria Græca*, either of the fashion of the seed, or of the siluer brightnesse that it hath, or of the middle skinne of the cods, when the two outtermost skinnes or huskes and seedes likewise are falne away. We call this herbe in English Penny floure, or Money floure, Siluer Plate, Pricke-songwoort; in Norfolke, Sattin, and White Sattin,

Sattin,and among our women it is called Honeſtie: it ſeemeth to be the old Herbariſts *Thlaſpi alterum*,or ſecond Treacle muſtard,and that which *Crateuas* deſcribeth, called of diuers *Sinapi Perſicum*; for as *Dioſcorides* ſaith,*Crateuas* maketh mention of a certaine *Thlaſpi* or Treacle Muſtard, with broad leaues and bigge roots,and ſuch this Violet hath,which we ſurname *Latifolia* or broad leafed: generally taken of all to be the great *Lunaria*, or Moonwoort.

¶ *Their Temperature and Vertues.*

The ſeed of Bolbonac is of Temperature hot and drie,and ſharpe of taſte,and is like in taſte and force to the ſeed of Treacle Muſtard;the roots likewiſe are ſomewhat of a biting qualitie, but not much: they are eaten with ſallads as certaine other roots are. A

A certaine Chirurgian of the Heluetians compoſed a moſt ſingular vnguent for wounds of the leaues of Bolbonac and Sanicle ſtamped together,adding thereto oile and wax.The ſeed is greatly commended againſt the falling ſickneſſe. B

CHAP. 124. *Of* Galen *and* Dioſcorides *Moonwoorts or Madwoorts.*

1 *Alyßum Galeni.*
Galens Madwoort.

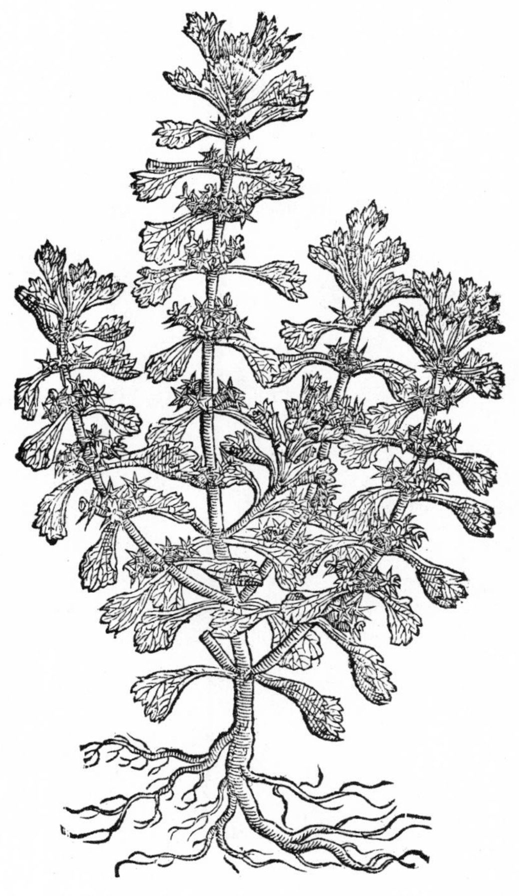

† 2 *Alyſſum Dioſcoridis.*
Dioſcorides Moonwoort or Madwoort.

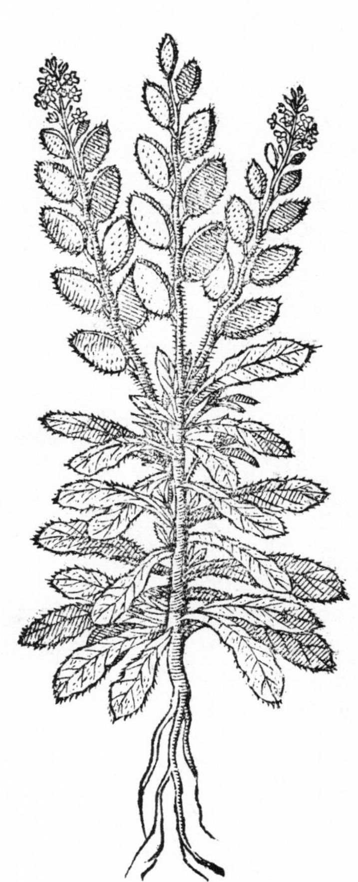

¶ *The Deſcription.*

1 THis might be one of the number of the Horehounds,but that *Galen* vſed it not for a kind thereof,but for *Alyſſon*,or Madwoort: it is like in forme and ſhew vnto Horehound,and alſo in the number of the ſtalks,but the leaues thereof are leſſer,more curled,more hoary, & whiter, without any manifeſt ſmell at all. The little coronets or ſpokie whurles that compaſſe the ſtalkes round about are full of ſharpe prickles: out of which grow floures of a blewiſh purple colour like to thoſe of Horehound. The root is hard,woody,and diuerſly parted.

2 I haue one growing in my garden,which is thought to be the true & right Lunary or Moonwoort of *Dioſcorides* deſcription,hauing his firſt leaues ſomewhat round,and afterward more long, whitiſh, and rough, or ſomewhat woolly in handling: among which riſe vp rough brittle ſtalkes, ſome cubite high, diuided into many branches, whereupon doe growe many little yellow floures:

floures; the which being past,there follow flat and rough huskes,of a whitish colour,in sh ape like little targets or bucklers,wherein is contained flat seed,like to the seeds of stock Gillofloures, but bigger. The whole huske is of the same substance, fashion,and colour that those are of the white Sattin.

¶ *The Place.*

These Plants are sowne now and then in Gardens, especially for the rarenesse of the m ; the seede beeing brought out of Spaine and Italy, from whence I receiued some for my Garden.

¶ *The Time.*

They floure and flourish in May ; the seede is ripe in August, the second yeare after their sowing.

¶ *The Names.*

Madwoort,or Moonwoort is called of the Græcians ἄλυσσος or ἄλυσσον : of the Latines *Alyssum*: in English, *Galens* Madwoort: of some,Heale-dog: and it hath the name thereof,because it is a present remedy for them that are bitten of a mad dogge, as *Galen* writeth;who in his second booke *De Antidotis*, in *Antoninus Cous* his composition describeth it in these words: Madwoort is an herbe very like to Horehound,but rougher, and more full of prickles about the floures:it beareth a floure tending to blew.

‡ 2 The second by *Dodonæus*,*Lobell*,*Camerarius* and others, is reputed to bee the *Alysson* of *Dioscorides*; *Gesner* mames it *Lunaria aspera* ;and *Columna*,*Leucoium Montanum Lunatum*. ‡

¶ *The Temperature and Vertues.*

Galen saith it is giuen vnto such as are inraged by the biting of a mad dogge,which thereby are
A perfectly cured, as is knowne by experience,without any artificiall application or method at all. The which experiment if any shall proue,he shall finde in the working thereof. It is of temperature meanly drie,digesteth and something scoureth withall:for this cause it taketh away the morphew and Sun-burning,as the same Authour affirmeth.

† That which was formerly described in the second place, being a Kinde of *Sideritis*, I haue here omitted,that I may giue you it more fitly amongst the rest of that name and kindred hereafter.

Chap. 125. *Of Rose Campion.*

Lychnis Chalcedonica.
Floure of Constantinople.

¶ *The Kindes.*

There be diuers sorts of Rose Campions; some of the Garden,and others of the Field : the which shal be diuided into seuerall chapters:and first of the Campion of Constantinople.

¶ *The Description.*

The Campion of Constantinople hath sundry vpright stalks,two cubits high and ful of ioynts, with a certaine roughnesse ; and at euery ioynt two large leaues,of a browne greene colour. The floures grow at the top like Sweet-Williams,or rather like Dames violets,of the colour of red lead, or Orenge tawny. The root is somewhat sharpe in taste.

‡ There are diuers varieties of this, as with white and blush coloured floures,as also a double kinde with very large,double and beautiful floures of a Vermelion colour like as the single one here described. ‡

¶ *The Place.*

The floure of Constantinople is planted in Gardens, and is very common almost euerie where.

‡ The white and blush single, and the double one are more rare,and not to be found but in the Gardens of our prime Florists. ‡

¶ *The Time.*

It floureth in Iune and Iuly, the second yeare after it is planted, and many yeares after ; for it consisteth

conſiſteth of a root full of life; and endureth long, and can away with the cold of our clymate.

¶ *The Names.*

It is called *Conſtantinopolitanus flos*, and *Lychnis Chalcedonica*: of *Aldrouandus*, *Flos Creticus*, or Floure of Candy: of the Germans, *flos Hieroſolymitanus*, or Floure of Ieruſalem: in Engliſh, Floure of Conſtantinople; of ſome, Floure of Briſtow, or None-ſuch.

¶ *The Temperature and Vertues.*

Floure of Conſtantinople, beſides that grace and beauty which it hath in gardens and garlands, is, for ought we know, of no vſe, the vertues thereof being not as yet found out.

CHAP. 126. *Of Roſe-Campion.*

1 *Lychnis Coronaria rubra.*
Red Roſe Campion.

2 *Lychnis Coronaria alba.*
White Roſe Campion.

¶ *The Deſcription.*

1 THe firſt kinde of Roſe-Campion hath round ſtalks very knotty and woolly, and at euerie knot or ioynt there do ſtand two woolly ſoft leaues like Mulleine, but leſſer, and much narrower. The floures grow at the top of the ſtalke, of a perfect red colour; which being paſt, there follow round cods full of blackiſh ſeed. The root is long and threddy.

2 The ſecond Roſe Campion differs not from the precedent in ſtalkes, leaues, or faſhion of the floures: the onely difference conſiſteth in the colour; for the floures of this plant are of a milke white colour, and the other red.

‡ 3 This alſo in ſtalks, roots, leaues, and manner of growing differs not from the former; but the floures are much more beautifull, being compoſed of ſome three or foure rankes or orders of leaues lying each aboue other. ‡

¶ *The*

‡ 3 *Lychnis coronaria multiplex.*
Double Rose Campion.

¶ *The Place.*

The Rose Campion growes plentifully in most gardens.

¶ *The Time.*

They floure from Iune to the end of August.

¶ *The Names.*

The Rose Campion is called in Latine *Dominarum Rosa, Mariana Rosa, Cœli Rosa, Cœli flos:* of *Dioscorides*, λυχνὶς στεφανωματικὴ: that is, *Lychnis Coronaria*, or *Satiua: Gaza* translateth λυχνίδα, *Lucernula*, because the leaues thereof be soft, and fit to make weekes for candles, acccording to the testimonie of *Dioscorides*: it was called *Lychnis*, or *Lychnides*, that is, a torch, or such like light, according to the signification of the word, cleere, bright, and light-giuing floures: and therefore they were called the Gardners Delight, or the Gardeners Eye: in Dutch, **Christes eie:** in French, *Oeillets*, & *Oeilets Dieu:* in high-Dutch, **Marien roszlin**, and **Himmel roszlin.**

¶ *The Temperature.*

The seed of Rose-Campion, saith *Galen*, is hot and dry after a sort in the second degree.

¶ *The Vertues.*

A The seed drunken in wine is a remedie for them that are stung with a Scorpion, as *Dioscorides* testifieth.

Chap. 127. *Of wilde Rose-Campions.*

¶ *The Description.*

1 The wilde Rose-Campion hath many rough broad leaues somewhat hoary and woolly; among which rise vp long soft and hairy stalkes branched into many armes, set with the like leaues, but lesser. The floures grow at the top of the stalkes, compact of fiue leaues of a reddish colour: the root is thicke and large, with some threds anexed thereto.

‡ There also growes commonly wilde with vs another of this kinde, with white floures, as also another that hath them of a light blush colour. ‡

2 The sea Rose Campion is a small herbe, set about with many greene leaues from the lower part vpward; which leaues are thicke, somewhat lesser and narrower than the leaues of sea Purslane. It hath many crooked stalkes spred vpon the ground, a foot long; in the vpper part whereof there is a small white floure, in fashion and shape like a little cup or box, after the likenesse of *Behen album*, or Spatling Poppy, hauing within the said floure little threds of a blacke colour, in taste salt, yet not vnpleasant.

It is reported vnto me by a Gentleman one M[r]. *Tho Hesket*, that by the sea side in Lancashire, from whence this plant came, there is another sort hereof with red floures.

‡ 3 This brings many stalkes from one root, round, long, and weaker than those of the first described, lying vsually vpon the ground: the leaues grow by couples at each ioynt, long, soft, and hairy; amongst which alternately grow the floures, about the bignesse of those of the first described, and of a blush colour; and they are also succeeded by such seed-vessels, containing a reddish seed. The root is thicke and fibrous, yet commonly outliues not the second yeare.

† 1 *Lychnis*

† 1 *Lychnis sylvestris rubello flore.*
Red wilde Campion.

2 *Lychnis marina Anglica.*
English Sea Campion.

3 *Lychnis sylvestris hirta, 5. Clusij.*
Wilde hairy Campion.

4 *Lychnis sylvestris 8. Clusij.*
Hoary wilde Campion.

5 *Lychnis hirta minima, 6. Clus.*
Small Hairy Campion.

† 6 *Lychnis syluestris incana, Lob.*
Ouerworne Campion.

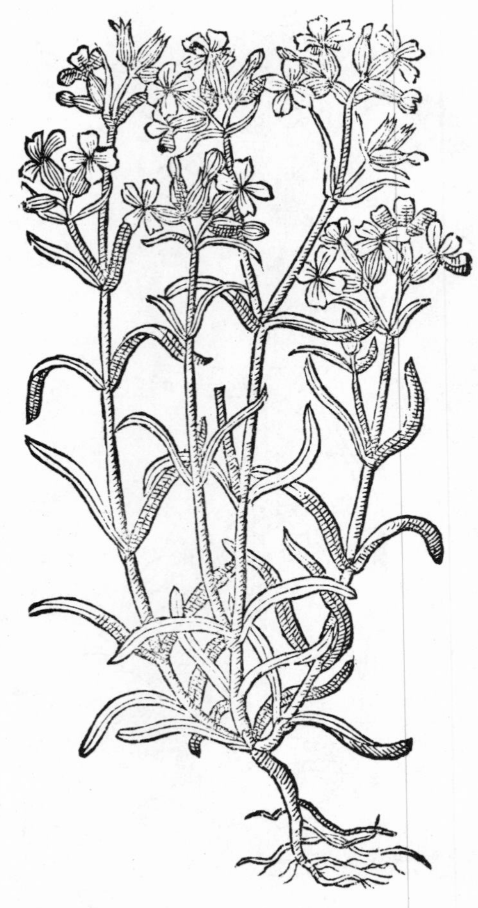

7 *Lychnis caliculis striatis 2. Clusij.*
Spatling Campion.

† 8 *Lychnis syluestris alba 9. Clus.*
Whitewilde Campion.

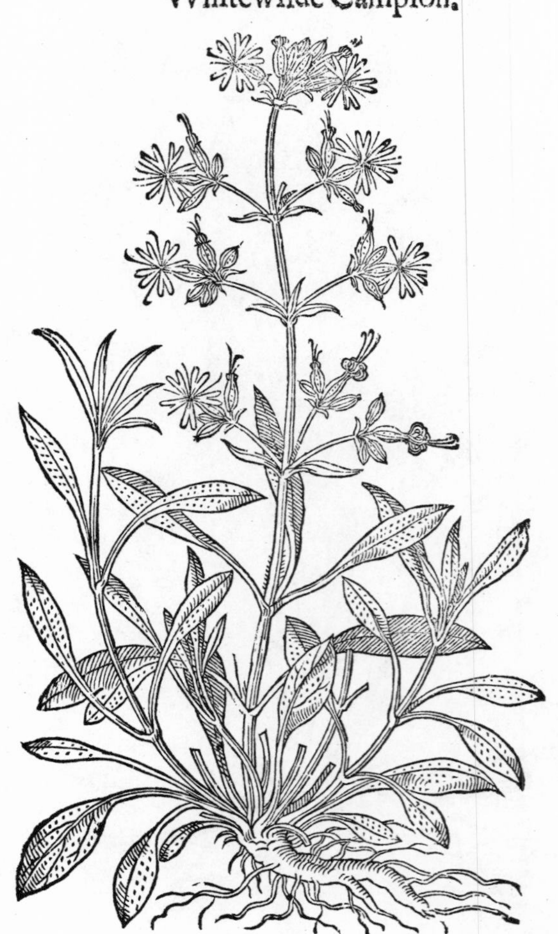

4 The fourth kinde of wilde Campions hath long and ſlender ſtems, diuiding themſelues into ſundry other branches, which are full of ioynts, hauing many ſmall and narrow leaues proceeding from the ſaid ioynts, and thoſe of a whitiſh greene colour. The floures do grow at the top of the ſtalke, of a whitiſh colour on the inner ſide, and purpliſh on the outer ſide, conſiſting of fiue ſmall leaues, euery leafe hauing a cut in the end, which maketh it of the ſhape of a forke: the ſeed is like the wilde Poppy; the root ſomewhat groſſe and thicke, which alſo periſheth the ſecond yeare.

5 The fifth kinde of wilde Campion hath three or foure ſoft leaues ſomewhat downy, lying flat vpon the ground; among which riſeth vp an hairy aſh-coloured ſtalke, diuided into diuers branches; whereupon do grow at certaine ſpaces, euen in the ſetting together of the ſtalke and branches, ſmall and graſſe-like leaues, hairy, and of an ouerworne dusky colour, as is all the reſt of the Plant. The floures grow at the top of the branches, compoſed of fiue ſmall forked leaues of a bright ſhining red colour. The root is ſmall, and of a wooddy ſubſtance.

6 The ſixth kinde of wilde Campion hath many long thicke fat and hoary leaues ſpred vpon the ground, in ſhape and ſubſtance like thoſe of the garden Campion, but of a very duſty ouerworne colour: among which riſe vp ſmall and tender ſtalkes ſet at certaine diſtances by couples, with ſuch like leaues as the other, but ſmaller. The floures do grow at the top of the ſtalks in little tufts like thoſe of ſweet Williams, of a red colour. The root is ſmall, with many threddy ſtrings faſtned to it.

‡ 7 This growes ſome cubit high, with ſtalkes diſtinguiſhed with ſundry ioynts, at each whereof are ſet two leaues, greene, ſharpe pointed, and ſomewhat ſtiffe: the floures grow at the tops of the branches, like to thoſe of *Muſcipula* or Catch-fly, yet ſomewhat bigger, and of a darke red: which paſt, the ſeed (which is aſh-coloured, and ſomewhat large) is contained in great cups or veſſels couered with a hard and very much creſted skin or filme; whence it is called *Lychnis caliculis ſtriatis*, and not *Cauliculis ſtriatis*, as it is falſly printed in *Lobels Icones*, which ſome as fooliſhly haue followed. The root is ſingle, and not large, and dies euery yeare.

8 That which our Author figured in this place had greene leaues and red floures, which no way ſorted with his deſcription: wherefore I haue in lieu thereof giuen you one out of *Cluſius*, which may fitly carry the title. This at the top of the large fibrous and liuing root ſendeth forth many leaues ſomewhat greene, and of ſome fingers length, growing broader by degrees, and at laſt ending againe in a ſharpe point. The ſtalkes are ſome cubit high, ſet at each ioynt with two leaues as it were embracing it with their foot-ſtalkes; which leaues are leſſe and leſſe as they are higher vp, and more ſharpe pointed. At the tops of the branches grow the floures, conſiſting of fiue white leaues deepely cut in almoſt to the middle of the floure, and haue two ſharpe pointed appendices at the bottome of each of them, and fiue chiues or threds come forth of their middles: theſe when they fade contract and twine themſelues vp, and are ſucceeded by thicke and ſharpe pointed ſeed-veſſels, containing a ſmall round Aſh-coloured ſeed. I coniecture that the figure of the *Lychnis plumaria*, which was formerly here in the ninth place out of *Tabern.* might be of this plant, as well as of that which *Bauhine* refers it to, and which you ſhall finde mentioned in the end of the chapter. ‡

¶ *The Place.*

They grow of themſelues neere to the borders of plowed fields, medowes, and ditch banks, common in many places. ‡ I haue obſerued none of theſe, the firſt and ſecond excepted, growing wilde with vs. ‡

The ſea Campion groweth by the ſea ſide in Lancaſhire, at a place called Lytham, fiue miles from Wygan, from whence I had ſeeds ſent me by Mr. *Thomas Hesketh*; who hath heard it reported, that in the ſame place doth grow of the ſame kinde ſome with red floures, which are very rare to be ſeene. ‡ This plant (in my laſt Kentiſh Simpling voyage, 1632, with Mr. *Thomas Hickes*, Mr. *Broad*, &c.) I found growing in great plenty in the low mariſh ground in Tenet that lieth directly oppoſite to the towne of Sandwich. ‡

¶ *The Time.*

They floure and flouriſh moſt part of the Sommer euen vnto Autumne.

¶ *The Names.*

The wilde Campion is called in Greeke Λυχνὶς ἀγρία: in Latine, *Lychnis ſylueſtris*: in Engliſh, wilde Roſe Campion.

¶ *The Temperature.*

The temperature of theſe wilde Campions are referred vnto thoſe of the garden.

¶ *The Vertues.*

The weight of two drammes of the ſeed of Wilde Campion beaten to pouder and drunke, A

doth

doth purge choler by the ſtoole, and it is good for them that are ſtung or bitten of any venomous beaſt.

† The figure that was in the firſt place, and was intended for our ordinary wilde Campion, is that which you ſee here in the eighth place; and thoſe that were in the ſixth and eighth places you ſhall hereafter finde with *Muſcipula* or Catch-fly, whereto they are of affinitie. That figure which was in the ninth place, out of *Tabern.* vnder the title of *Lychnis plumaria*, as alſo the deſcription, I haue omitted as impertinent: for the figure *Bauhine* himſelfe (who corrected and againe ſet forth the Workes of *Tabernamontanus*) could not tell what to make thereof; but queſtions, *Quid ſit? an Muſcipula flore muſcoſo?* Which if it be, you ſhall finde that plant hereafter deſcribed, vnder the title of *Seſamoides magnum Salmanticum*: for our Authors deſcription it is not worth the ſpeaking of, being framed onely from imagination.

‡ Chap. 128. *Of diuers other wilde Campions.*

¶ *The Deſcription.*

‡ 1 THe firſt of theſe which we here giue you is like in leaues, ſtalkes, roots, and manner of growing vnto the ordinarie wilde Campion deſcribed in the firſt place of the precedent Chapter; but the floures are very double, compoſed of a great many red leaues thicke packt together, and they are commonly ſet in a ſhort and broken huske or cod. Now the ſimilitude that theſe floures haue to the iagged cloath buttons anciently worne in this kingdome gaue occaſion to our Gentlewomen and other louers of floures in thoſe times to call them Bachelours Buttons.

2 This differs not in ſhape from the laſt deſcribed, but only in the colour of the floures, which in this plant are white.

‡ 1 *Lychnis ſyl. multiplex purpurea.*
Red Bachelors Buttons.

‡ 2 *Lychnis ſyl. alba multiplex.*
White Bachelors Buttons.

3 Neither in roots, leaues, or ſtalkes is there any difference betweene this either degenerate or accidentall varietie of Bachelors buttons, from the two laſt mentioned; onely the floures hereof are of a greeniſh colour, and ſometimes through the middeſt of them they ſend vp ſtalkes, bearing alſo tufts of the like double floures.

4 This (ſaith *Cluſius*) hath fibrous roots like to thoſe of Primroſes; out of which come leaues of

‡ 3 *Lychnis abortiua flore multiplici viridi.*
Degenerate Bachelors Buttons with greene floures.

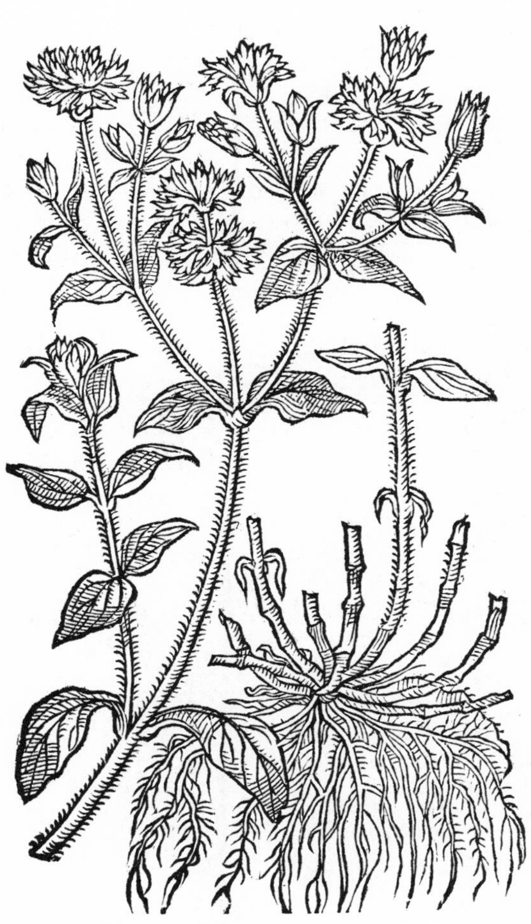

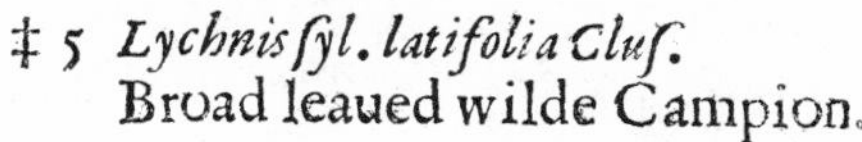

‡ 5 *Lychnis syl. latifolia Clus.*
Broad leaued wilde Campion.

‡ 5 *Lychnis montana repens.*
Creeping mountaine Campion.

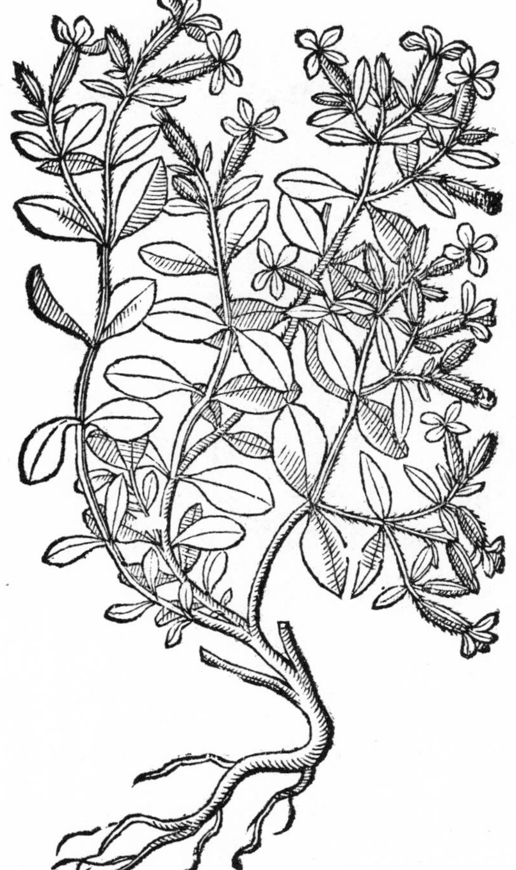

of a fufficient magnitude,not much vnlike thofe of the great yellow Beares-eare,yet whiter,more downy, thicke, and iuycie. The next yeare after the fowing thereof it fends vp a ftalke of two or three cubits high, here and there fending forth a vifcous and glutinous iuyce, which detaines and holds faft flies and fuch infects as do chance to light thereon. At the top of the branches it yeeldeth many floures fet as it were in an vmbel, euen fometimes an hundred; yet fufficiently fmall,confidering the magnitude of the plant; and each of thefe confifts of fiue little yellowifh greene forked leaues.

5 The ftalkes of this are flender, ioynted, and creeping like to thofe of the greater Chickweed, and at each ioynt grow two leaues like thofe of the myrtle, or of Knot-graffe yet fomewhat broader. The floures grow in fuch long cups like as thofe of *Saponaria*, and are much leffe, yet of the fame colour. The root is fmall.

¶ *The Place.*

1. 2. Thefe are kept in many Gardens of this kingdome for their beauty, efpecially the firft,which is the more common.

The fourth growes naturally in Candy; and the fifth by riuelets in the mountainous places of Sauoy.

¶ *The Time.*

Thefe floure in Iune and Iuly with the other wild Campions.

Rr ¶ *The*

¶ *The Names.*

1 The first of these is *Lychnis agrestis multiflora* of *Lobel*; and *Ocymoides flore pleno* of *Camerarius*.

2 The second is by *Pena* and *Lobel* also called *Lychnis syluestris multiflora*: it is the *Ocymastrum multiflorum* of *Tabernamontanus*; by which title our Author also had it in the former edition, *p*. 551.

3 *Lobel* hath this by the name of *Lychnis agrestis abortiua multiplici viride flore*.

4 *Clusius* calls this *Lychnis syluestris latifolia*; and he saith he had the seed from *Ioseph de Casa Bona*, by the name of *Muscipula auriculæ vrsi facie*: *Bauhine* hath it by the name of *Lychnis auriculæ vrsi facie*.

5 This (according to *Bauhine*) was set forth by *Matthiolus*, by the name of *Cneoron aliud Theophrasti*: it is the *Ocimoides repens polygonifolia flore Saponariæ*, in the *Aduersaria*: and *Saponaria minor Daleschampij*, in the *Hist. Lugd*. It is also *Ocimoides Alpinum*, of *Gesner*; and *Ocymoides repens*, of *Camerarius*.

¶ *The Nature and Vertues.*

The natures and vertues of these, as of many others, lie hid as yet, and so may continue, if chance, or a more curious generation than yet is in being do not finde them out. ‡

CHAP. 129. *Of Willow-herbe, or Loose-strife.*

1 *Lysimachia lutea.*
Yellow Willow-herbe.

‡ 2 *Lysimachia lutea minor.*
Small yellow Willow-herbe.

¶ *The Description.*

1 THe first kinde of Willow-herbe hath long and narrow leaues of a grayish greene colour, in shape like the Willow or Sallow leaues, standing three or foure one against another at seuerall distances round about the stalke; which toward the top diuideth it selfe into many other branches, on the tops whereof grow tufts of faire yellow floures, consisting of fiue leaues apiece, without smell: which being past, there commeth forth seed like Coriander. The root is long and slender.

‡ 2 This

‡ 2 This leſſer of *Cluſius* his deſcription hath a ſtalke a cubit high, and ſometimes higher firme, hard, and downy; about which at certaine diſtances grow commonly foure leaues together, yet ſometimes but three, and they are ſoft and ſomewhat downy, leſſer than thoſe of the former, being firſt of an acide taſte, and then of an acride; and they are vſually marked on their lower ſides with blacke ſpots. About the top of the ſtalke, out of the boſomes of each leafe come forth little branches bearing ſome few floures, or elſe foot-ſtalkes carrying ſingle floures, which is more vſuall towards the top of the ſtalke. The floures are yellow, with ſomewhat a ſtrong ſmell, conſiſting of fiue ſharpe pointed yellow leaues, with ſo many yellow threds in their middle. The root is ioynted, or creeping here and there, putting vp new ſhouts.

‡ 3 *Lyſimachia lutea flore globoſo.*
Yellow Willow-herbe with bunched floures.

‡ 4 *Lyſimachia lutea Virginiana.*
Tree Primroſe.

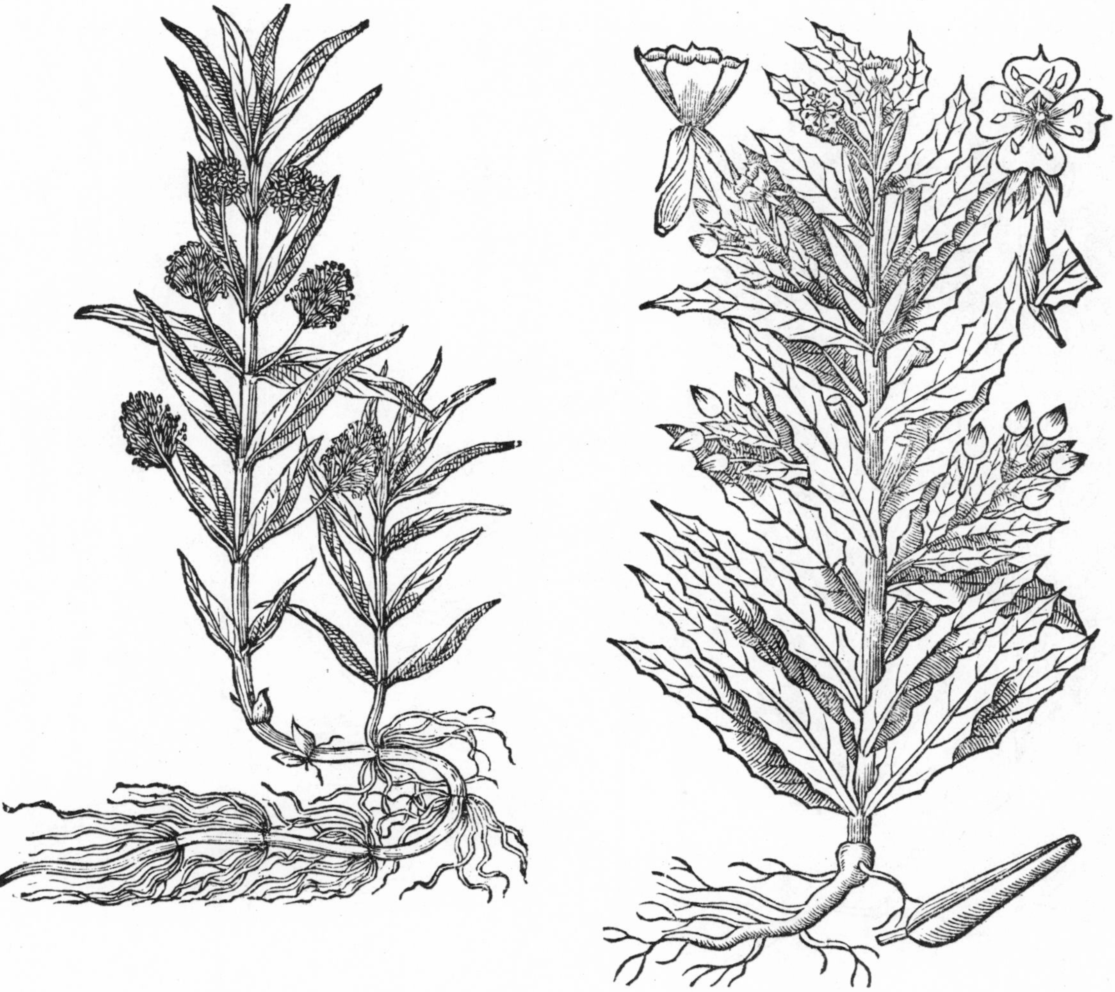

3 This alſo may fitly be referred to the former. The ſtalke is a cubit high, ſtraight, and as it were ioynted, naked oft times below by the falling away of the leaues; but from the middle to the top ſet with two leaues at a ioynt, like thoſe of the former; and out of their boſoms on ſhort ſtalks grow round tufts of ſmall yellow floures as in bunches: the root which creepes ſends forth many ſmall fibres at each ioynt. This was ſet forth by *Lobel* vnder the title of *Lyſimachia lutea altera*, or *Lyſimachia ſalicaria*: *Dodonæus* hath it by the name of *Lyſimachium aquatile*: and *Cluſius* calls it *Lyſimachia lutea tertia, ſiue minor*.

4 This Virginian hath beene deſcribed and figured onely by *Proſper Alpinus*, vnder the title of *Hyoſcyamus Virginianus*: and by Mr. *Parkinſon*, by the name of *Lyſimachia lutea ſiliquoſa Virginiana*: Alſo *Bauhine* in the Appendix of his *Pinax* hath a large deſcription thereof, by the name of *Lyſimachia lutea corniculata*. The root hereof is longiſh, white, about the thickneſſe of ones thumbe, from whence growes vp a tall ſtalke diuided into many branches of an ouerworne colour, and a little hairie: the leaues are like thoſe of the former, but ſomewhat ſinuated alongſt their edges, and hauing their middle veine of a whitiſh colour: toward the tops of the branches amongſt the leaues come vp pretty thicke cods, which growing ſmaller on their tops ſuſtaine pretty large yellow floures conſiſting of foure leaues, with a peſtill in the middle vpon which ſtand foure yellowiſh thrums

in fashion of a crosse; and there are also eight threds with their pointals in the middles of them. These floures haue somewhat the smell of a Primrose (whence M^r. *Parkinson* gaue it the English name, which I haue also here giuen you:) after the floures are fallen, the cods grow to be some two inches long, being thicker below, and sharper at the top, and somwhat twined, which in fine open themselues into foure parts to shatter their seed, which is blacke and small; and sowne, it growes not the first yeare into a stalke, but sends vp many large leaues lying handsomely one vpon another Rose-fashion. It floures in Iune, and ripens the seed in August. ‡

5 The second kinde of Willow-herbe in stalks and leaues is like the first, but that the leaues are longer, narrower, and greener. The floures grow along the stalke toward the top, spike-fashion, of a faire purple colour: which being withered turne into downe, which is carried away with the winde.

5 *Lysimachia purpurea spicata.*
Spiked Willow-herbe.

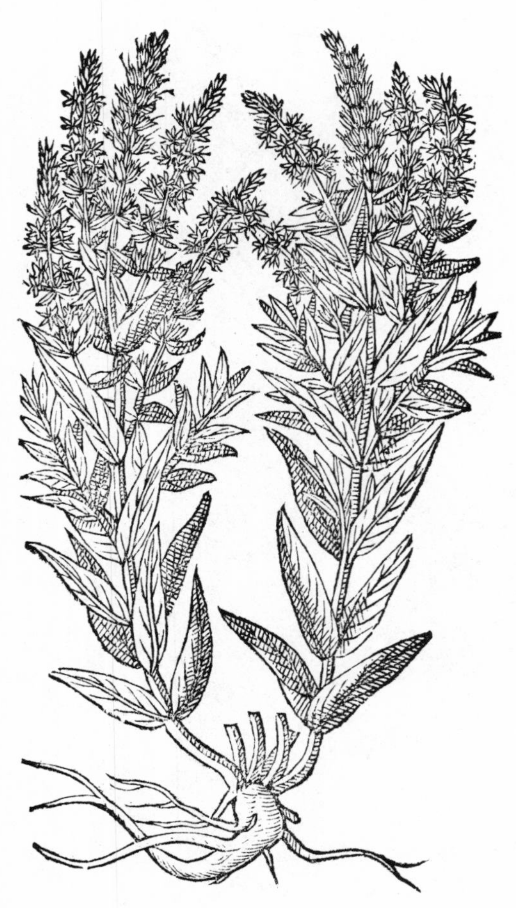

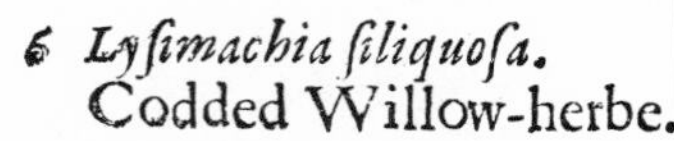

6 *Lysimachia siliquosa.*
Codded Willow-herbe.

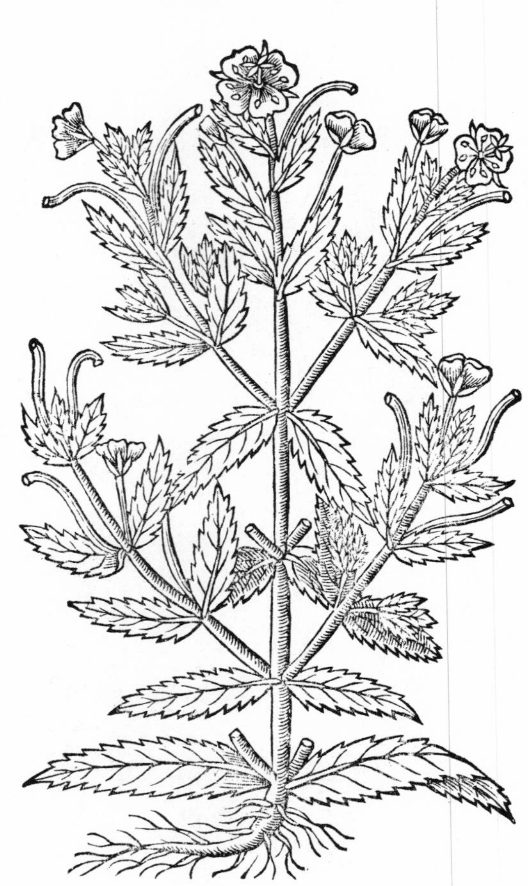

6 This *Lysimachia* hath leaues and stalkes like vnto the former. The floure groweth at the top of the stalke, comming out of the end of a small long cod, of a purple colour, in shape like a stocke Gillofloure, and is called of many *Filius ante Patrem* (that is, The Sonne before the Father) because that the cod commeth forth first, hauing seeds therein, before the floure doth shew it selfe abroad. ‡ The leaues of this are more soft, large, and hairy than any of the former: they are also snipt about the edges, and the floure is large, wherein it differs from the twelfth, hereafter described; and from the eleuenth in the hairinesse of the leaues, and largenesse of the floures also, as you shall finde hereafter. ‡

7 This being thought by some to be a bastard kinde, is (as I do esteeme it) of all the rest the most goodly and stately plant, hauing leaues like the greatest Willow or Ozier. The branches come out of the ground in great numbers, growing to the height of six foot, garnished with braue floures of great beauty, consisting of foure leaues a piece, of an orient purple colour, hauing some threds in the middle of a yellow colour. The cod is long like the last spoken of, and full of downy matter, which flieth away with the winde when the cod is opened.

‡ 8 This also, which is the *Chamænerion* of *Gesner*, as also his *Epilobion, quasi* ἐπὶ λόβου ἴον, a Violet or floure vpon a cod, may iustly challenge the next place. *Dodonæus* calls it *Pseudolysimachium purpureum*

† 7 *Chamænerion.*
Roſe bay Willow-herbe.

‡ 8 *Chamænerion alterum anguſtifolium.*
Narrow leaued Willow-floure.

‡ 9 *Lyſimachia cœrulea.*
Blew Looſe-ſtrife.

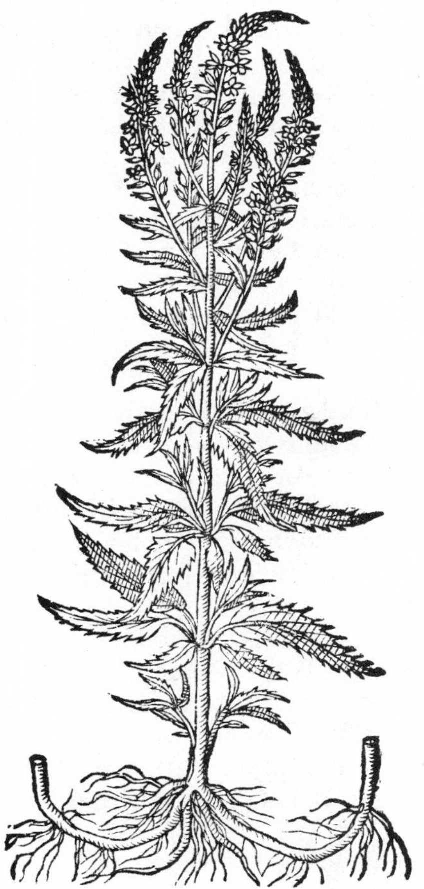

‡ 10 *Lyſimachia galericulata.*
Hooded Looſe-ſtrife.

11 *Lysimachia campestris.*
Wilde Willow-herbe.

purpureum minus: and it is in the *Histor. Lugdun.* vnder the name of *Linaria rubra*. It groweth vp with stalkes some foot high, set with many narrow leaues like those of Toad-flax, of a grayish colour, and the stalke is parted into diuers branches, which at their tops vpon long cods carrie purple floures consisting of foure leaues apiece. The root is long, yellowish, and wooddy. ‡

9 There is another bastard Loose-strife or Willow-herbe hauing stalkes like the other of his kinde. whereon are placed long leaues snipt about the edges, in shape like the great *Veronica* or herbe *Fluellen*. The floures grow along the stalkes, spike-fashion, of a blew colour; after which succeed small cods or pouches. The root is small and fibrous: it may be called *Lysimachia cœrulea*, or blew Willow-herbe.

10 We haue likewise another Willow-herbe that groweth neere vnto the bankes of riuers and water-courses. This I found in a waterie lane leading from the Lord Treasurer his house called Theobalds, vnto the backeside of his slaughter-house, and in other places, as shall be declared hereafter. Which *Lobel* hath called *Lysimachia galericulata*, or hooded Willow-herbe. It hath many small tender stalkes trailing vpon the ground, beset with diuers leaues somwhat snipt about the edges, of a deep green colour, like to the leaues of *Scordium* or water Germander: among which are placed sundrie small blew floures fashioned like a little hood, in shape resembling those of Ale-hoofe. The root is small and fibrous, dispersing it selfe vnder the earth farre abroad, whereby it greatly increaseth.

11 The wilde Willow-Herbe hath fraile and very brittle stalkes, slender, commonly about the height of a cubit, and sometimes higher; whereupon doe grow sharpe pointed leaues somewhat snipt about the edges, and set together by couples. There come forth at the first long slender coddes, wherein is contained small seed, wrapped in a cottony or downy wooll, which is carried away with the winde when the seed is ripe: at the end of which commeth forth a small floure of a purplish colour; whereupon it was called *Filius ante Patrem*, because the floure doth not appeare vntill the cod be filled with his seed. But there is another Sonne before the Father, as hath beene declared in the Chapter of Medow-Saffron. The root is small and thred-die. ‡ This differeth from the sixth onely in that the leaues are lesse, and lesse hairy, and the floure is smaller. ‡

12 The Wood VVillow-hearbe hath a slender stalke diuided into other smaller branches, whereon are set long leaues rough and sharpe pointed, of an ouerworne greene colour. The floures grow at the tops of the branches, consisting of foure or fiue small leaues, of a pale purplish colour tending to whitenesse: after which come long cods, wherein are little seeds wrapped in a certaine white Downe that is carried away with the winde. The root is threddie. ‡ This differs from the sixth in that it hath lesser floures. There is also a lesser sort of this hairie *Lysimachia* with small floures.

There are two more varieties of these codded Willow-herbes; the one of which is of a middle growth, somewhat like to that which is described in the eleuenth place, but lesse, with the leaues also snipped about the edges, smooth, and not hairie: and it may fitly be called *Lysimachia siliquosa glabra media*, or *minor*, The lesser smooth-leaued Willow-herbe. The other is also smooth leaued, but they are lesser and narrower: wherefore it may in Latine be termed, *Lysimachia siliquosa glabra minor angustifolia*: in English, The lesser smooth and narrow leaued Willow-herbe.

‡ 13 This lesser purple Loose-strife of *Clusius*, hath stalkes seldome exceeding the height of a cubit, they are also slender, weake and quadrangular, towards the top, diuided into branches growing

growing one againſt another, the leaues are leſſe and narrower than the common purple kinde, and growing by couples, vnleſſe at the top of the ſtalkes and branches, whereas they keepe no certaine order; and amongſt theſe come here and there cornered cups containing floures compoſed of ſix little red leaues with threds in their middles. The root is hard, woody, and not creeping, as in others of this kinde, yet it endures all the yeere, and ſends forth new ſhoots. It floures in Iune and Iuly, and was found by *Cluſius* in diuers wet medowes in Auſtria. ‡

¶ *The Place.*

The firſt yellow *Lyſimachia* groweth plentifully in moiſt medewos, eſpecially along the medowes as you go from Lambeth to Batterſey neere London, and in many other places throughout England.

‡ 13 *Lyſimachia purpurea minor Cluſ.*
Small purple Willow herbe.

‡ The ſecond and third I haue not yet ſeene.

The fourth groweth in many gardens. ‡

The fift groweth in places of greater moiſture, yea almoſt in the running ſtreames and ſtanding waters, or hard by them. It groweth vnder the Biſhops houſe wall at Lambeth, neere the water of Thames, and in moiſt ditches in moſt places of England.

The ſixth groweth neere the waters (and in the waters) in all places for the moſt part.

The ſeuenth groweth in Yorkſhire in a place called the Hooke, neer vnto a cloſe called a Cow paſture, from whence I had theſe plants, which doe grow in my garden very goodly to behold, for the decking vp of houſes and gardens.

‡ The eighth I haue not yet found growing.

The ninth growes wild in ſome places of this kingdome, but I haue ſeene it only in Gardens.

The tenth growes by the ponds and waters ſides in Saint Iames his Parke, in Tuthill fields and many other places. ‡

The eleuenth groweth hard by the Thames, as you goe from a place called the Diuels Neckerchiefe to Redreſſe, neere vnto a ſtile that ſtandeth in your way vpon the Thames banke, among the plankes that doe hold vp the ſame banke. It groweth alſo in a ditch ſide not farre from the place of execution, called Saint Thomas Waterings.

‡ The other varieties of this grow in wet places, about ditches, and in woods and ſuch like moiſt grounds. ‡

¶ *The Time.*

Theſe herbes floure in Iune and Iuly, and oftentimes vntill Auguſt.

¶ *The Names.*

Lyſimachia, as *Dioſcorides* and *Pliny* write tooke his name of a ſpeciall vertue that it hath in appeaſing the ſtrife and vnrulineſſe which falleth out among oxen at the plough, if it bee put about their yokes: but it rather retaineth and keepeth the name *Lyſimachia*, of King *Lyſimachus* the ſonne of *Agathocles*, the firſt finder out of the nature and vertues of this herb, as *Pliny* ſaith in his 25. book chap. 7. which retaineth the name of him vnto this day, and was made famous by *Eraſiſtratus*. *Ruellius* writeth, that it is called in French *Cornelle* and *Corneola*: in Greeke, λυσιμάχιον: of the Latines, *Lyſimachium*: of *Pliny*, *Lyſimachia*: of the later Writers, *Salicaria*: in high Dutch, **Weiderick**: in Engliſh, Willow herbe, or herbe Willow, and Looſe ſtrife.

Chamænerium is called of *Geſner*, *Epilobion*: in Engliſh, Bay Willow, or bay yellow herbe.

‡ The

‡ The names of ſuch as I haue added haue been ſufficiently ſet forth in their titles and Hiſtories. ‡

¶ *The Nature.*

The yellow *Lyſimachia*, which is the chiefe and beſt for Phyſicke vſes, is cold and drie, and very aſtringent.

¶ *The Vertues.*

A The iuice, according to *Dioſcorides*, is good againſt the bloudy flix, being taken either by potion or Cliſter.

B It is excellent good for greene wounds, and ſtancheth the bloud: being alſo put into the noſthrils, it ſtoppeth the bleeding at the noſe.

C The ſmoke of the burned herbe driueth away ſerpents, and killeth flies and gnats in a houſe; which *Pliny* ſpeaketh of in his 25. book, chap. 8. Snakes, ſaith he, craull away at the ſmell of Looſſtrife. The ſame Authour affirmeth in his 26 booke, laſt chap. that it dieth haire yellow, which is not very vnlike to be done by reaſon the floures are yellow.

D The others haue not been experimented, wherefore vntill ſome matter worthy the noting doth offer it ſelfe vnto our conſideration, I will omit further to diſcourſe hereof.

E The iuice of yellow *Lyſimachia* taken inwardly, ſtoppeth all fluxe of bloud, and the Dyſenteria or bloudy flix.

F The iuice put into the noſe, ſtoppeth the bleeding of the ſame, and the bleeding of wounds, and mightily cloſeth and healeth them, being made into an vnguent or ſalue.

G The ſame taken in a mother ſuppoſitorie of wooll or cotton, bound vp with threds (as the manner thereof is, well knowne to women) ſtaieth the inordinate flux or ouermuch flowing of womens termes.

H It is reported, that the fume or ſmoke of the herbe burned, doth driue away flies and gnats, and all manner of venomous beaſts.

Chap. 130. *Of Barren-woort.*

Epimedium.
Barren Woort.

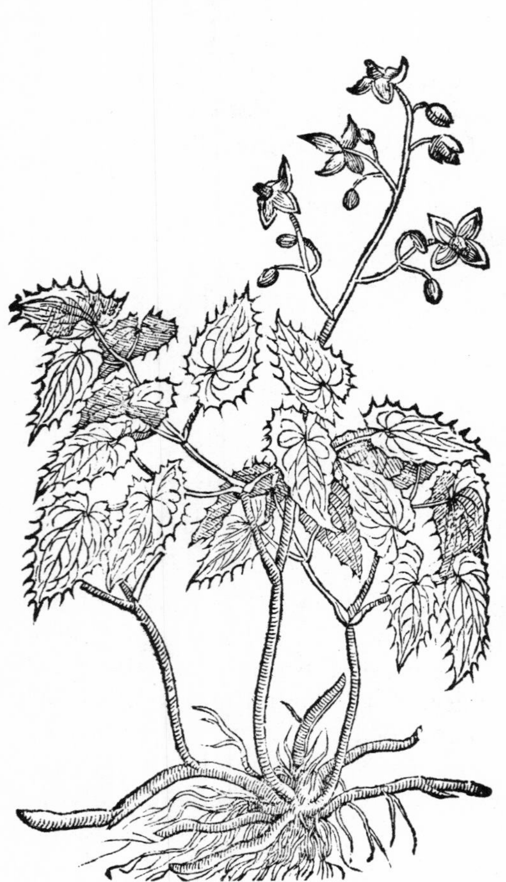

¶ *The Deſcription.*

This rare and ſtrange plant was ſent to me from the French Kings Herbariſt *Robinus*, dwelling in Paris at the ſigne of the blacke head, in the ſtreet called *Du bout du Monde*, in Engliſh, The end of the world. This herbe I planted in my garden, & in the beginning of May it came forth of the ground, with ſmall, hard & woodie crooked ſtalks: whereupon grow rough & ſharpe pointed leaues, almoſt like *Alliaria*, that is to ſay, Sauce alone, or Iacke by the hedge. *Lobel* and *Dod.* ſay, that the leaues are ſomewhat like Iuie; but in my iudgement they are rather like *Alliaria*, ſomwhat ſnipt about the edges, and turning themſelues flat vpright, as a man turneth his hand vpwards when hee receiueth money. Vpon the ſame ſtalkes come forth ſmall floures, conſiſting of foure leaues, whoſe outſides are purple, the edges on the inner ſide red, the bottome yellow, & the middle part of a bright red colour, and the whole floure ſomewhat hollow. The root is ſmal, and creepeth almoſt vpon the vppermoſt face of the earth. It beareth his ſeed in very ſmall cods like Saracens Confound, (‡ to wit that of our Authour formerly

merly deſcribed, *pag.* 274.‡) but ſho ter : which came not to ripeneſſe in my garden,by reaſon that it was dried away with the extreme and vnaccuſtomed heat of the Sun,which happened in the yeare 1590. ſince which time from yeare to yeare it bringeth ſeed to perfection. Further,*Dioſcorides* and *Pliny* do report,that it is without floure or ſeed.

¶ *The Place.*

† It groweth in the moiſt medowes of Italie about Bononia and Vincentia:it groweth in the garden of my friend Mr. *Iohn Milion* in Old-ſtreet,and ſome other gardens about towne.

¶ *The Time.*

It floureth in Aprill and May,when it hath taken faſt hold and ſetled it ſelfe in the earth a yeare before.

¶ *The Names.*

It is called *Epimedium* : I haue thought good to call it Barren woort in Engliſh ; not becauſe that *Dioſcorides* ſaith it is barren both of floures and ſeeds,but becauſe(as ſome authors affirme)being drunke it is an enemie to conception.

¶ *The Temperature and Vertues.*

Galen affirmeth that it is moderately cold,with a waterie moiſture : we haue as yet no vſe hereof in Phyſicke.

‡ CHAP. 131. *Of Fleabane.*

‡ 1 *Conyza maior.*
Great Fleawoort.

‡ 2 *Conyza minor vera.*
Small Fleabane.

‡ THe ſmalneſſe of the number of theſe plants here formerly mentioned, the confuſion notwithſtanding in the figures, their nominations & hiſtorie,not oneagreeing with another,hath cauſed me wholly toomit the deſcriptions of our Authour,and to giue you new, agreeable to the figures ; together with an addition of diuers other plants belonging to this kindred. Beſides there is one thing I muſt aduertiſe you of,which is,that our Authour in the firſt place deſcribed the *Baccharis Monſpelienſium* of *Lobel*,or *Conyza maior* of *Matthiolus*,& it is that which grows in Kent and Eſſex on chalkie hils;yet he gaue no figure of it,but as it were forgetting what he had don,allotted it a particular chap.afterwards,where alſo another figure was put for it, but there you ſhall now finde it,though I muſt confeſſe that this is as fit or a fitter place for it;but I will follow the courſe of my Authour,whoſe matter, not method I indeauour to amend.

¶ The

¶ *The Deſcription.*

1 THis great Fleawoort or Fleabane, from a thick long liuing fibrous root ſends forth many ſtalkes of ſome yard high or more; hard, wooddy, rough, fat, and of an ou erworne colour: the leaues are many, without order, and alternately embrace the ſtalkes, twice as big as thoſe of the Oliue tree, rough and fat, being as it were beſmeared with a gummineſſe or fattineſſe, and of a yellowiſh greene colour: the floures grow after a ſort ſpoke faſhion, ſtanding at the ends of footſtalkes comming out of the boſomes of the leaues, and they are yellow and round almoſt like to Groundſwell, and flie away in downe like as they doe; the ſeed is ſmall and aſh coloured. The whole plant is fattie and glutinous, with a ſtrong, yet not altogether vnpleaſant ſmell. This growes not that I know of in theſe cold Countries, vnleſſe ſowne in gardens. *Cluſius* found it by Lisbone, and in diuers places of Spaine. He, as alſo *Dodonæus*, *Lobel*, and others, call this *Conyza maior*, and it is thought to be the *Conyza mas* of *Theophraſtus*, and *Conyza maior* of *Dioſcorides*.

2 The leſſer ſeldome ſends vp more than one ſtalke, and that of a cubit high, yet vſually not ſo much: it is diuided into little branches, and alſo rough and glutinous as the precedent, but more greene. The leaues are three times leſſe than thoſe of the former, ſomewhat ſhaped like thoſe of Toad-flax, yet hairy and vnctious; the tops of the branches as in the bigger, carrie leſſe, and leſſe ſhining and ſightly floures, vaniſhing in like ſort into downe. The root is ſingle and annuall, and the whole plant more ſmelling than the former. This is iudged the *Conyza fœmina* of *Theophraſtus*; and *Con. minor* of *Dioſcorides*; it is the *Con. minor* of *Geſner*, *Lobel*, *Cluſius* and others. It growes in diuers parts of Spaine and Prouince in France, but not here, vnleſſe in Gardens.

† 3 *Conyza media.*
Middle Fleawoort.

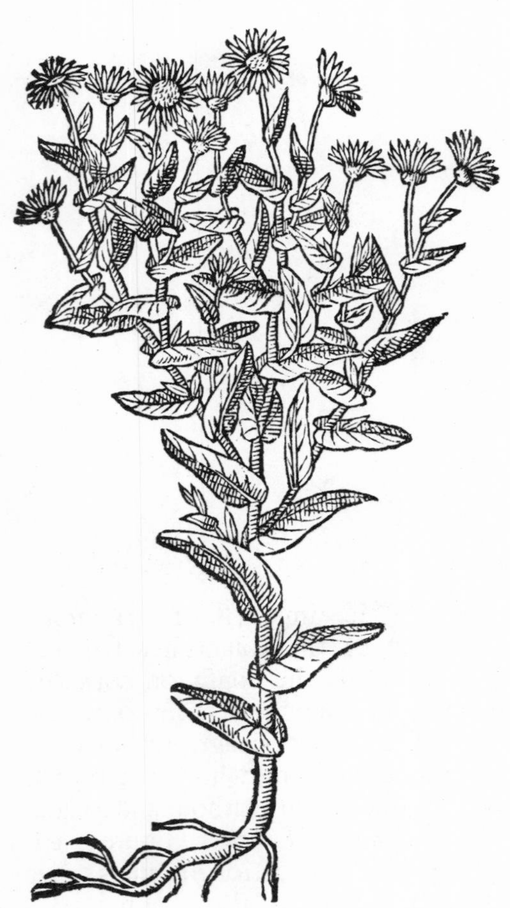

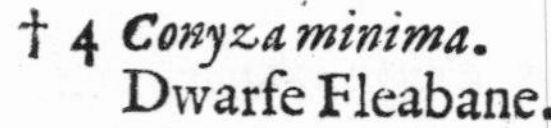

† 4 *Conyza minima.*
Dwarfe Fleabane.

3 The root of this middle kinde is prettie large and fibrous, from whence ariſeth a branched ſtalke of ſome cubite high, engirt at certaine ſpaces with thicke, rough, grayiſh greene leaues: at the tops of the branches grow pretty faire yellow floures of the bignes of a little Marigold; which fading turne to downe, and are carried away with the winde. This floures in Iuly and Auguſt, and may be found growing in moſt places about riuers and pond ſides, as in S. Iames his Parke, Tuthill fields, &c. This is *Conyza media* of *Matthiolus*, *Dodonæus*, and others. Some haue referred it vnto the

Mints

Mints, as *Fuchſius*, who makes it *Calaminthæ 3. genus*; and *Lonicerus*, who calls it *Mentha Lutea*. In Cheape-ſide the herbe-women call it Herbe Chriſtopher, and ſell it to Empericks, who with it (as they ſay) make Medicines for the eyes, but againſt what affect of them, or with what ſucceſſe I know not.

4 In like places, or rather ſuch as are plaſhy in winter this may be plentifully found growing. The roots are ſmall and fibrous; from whence ariſeth a branched ſtalke ſome foot high, ſet with ſmall longiſh leaues ſomewhat roundiſh pointed, ſoft alſo and woolly, with a ſmell not altogether vnpleaſant, like as the laſt deſcribed: the floures are compoſed of many yellowiſh threds like to the middle part of Camomill floures, or thoſe of Tanſey: and as the former, turne into downe, and are carried away with the winde; it floures in Iuly and Auguſt. This is the *Conyza minor* of *Tragus*, *Matthiolus*, and others: *Lobel* and *Dodon.* call it *Conyza minima*.

5 This cut leaued Fleabane hath ſmall fibrous roots, from which ariſe thicke, creſted, & hollow ſtalks, diuided towards the tops into ſundry branches: the leaues that incompaſſe the ſtalke are gaſhed, or elſe onely ſinuated on the edges: the floures are ſtar faſhion and yellow, and alſo flie away in downe; the whole plant is couered ouer with a ſoft and tender downe, and hath ſomewhat the ſmell of Honie. This is a varietie of the third, and is called by *Dodon. Conyzæ mediæ ſpecies altera*. *Lobel* names it *Conyza helenitis folijs laciniatis*.

6 The figure which you haue in this ſixth place was formerly vnfitly giuen by our Authour for *Solidago Saracenica*; it hath a large root which ſends foorth many fibres, and a creſted hollow ſtalke ſome two cubites or more high, which is vnorderly ſet, with long, yet narrow ſnipt leaues ſomewhat hairie and ſharpe pointed: the toppe is diuided into branches, which beare prettie large yellow floures, made after the manner of thoſe of Ragwort, and like as they, are alſo carried away with the winde. This *Thalius* cals *Conyza maxima ſerratifolia*. It is the *Lingua maior* of *Daleſchampius*, and the *Conſolida paluſtris* of *Tabernamontanus*. It groweth neere water ſides, and floures towards the latter end of Sommer: I haue not yet heard that it doth grow wilde amongſt vs.

‡ 5 *Conyza folijs laciniatis.*
Great iagged leaued Fleabane.

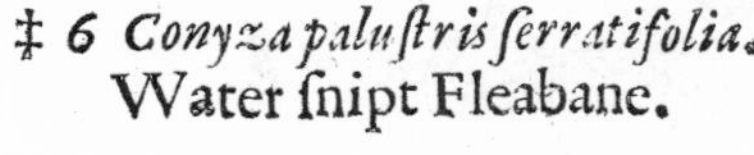

‡ 6 *Conyza paluſtris ſerratifolia.*
Water ſnipt Fleabane.

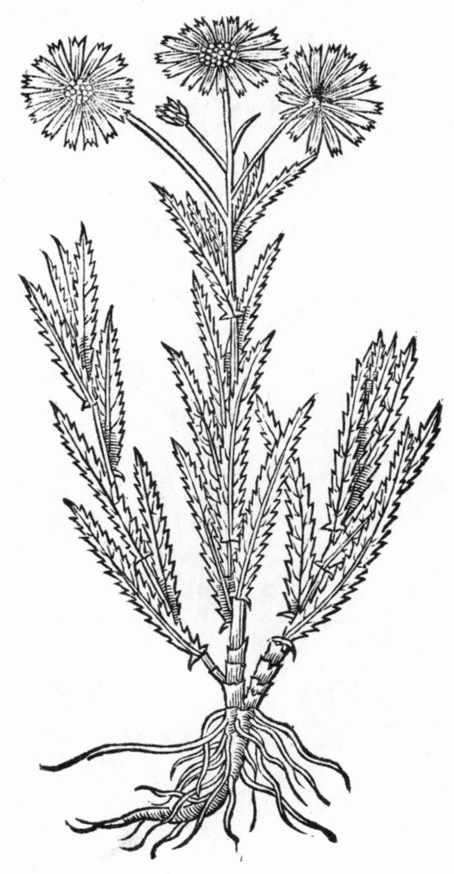

7 The

‡ 7 *Conyza Auſtriaca Cluſij.*
Auſtrian Fleabane.

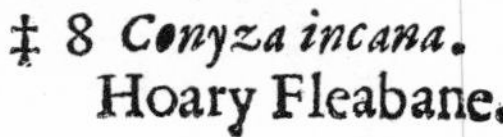

‡ 8 *Conyza incana.*
Hoary Fleabane.

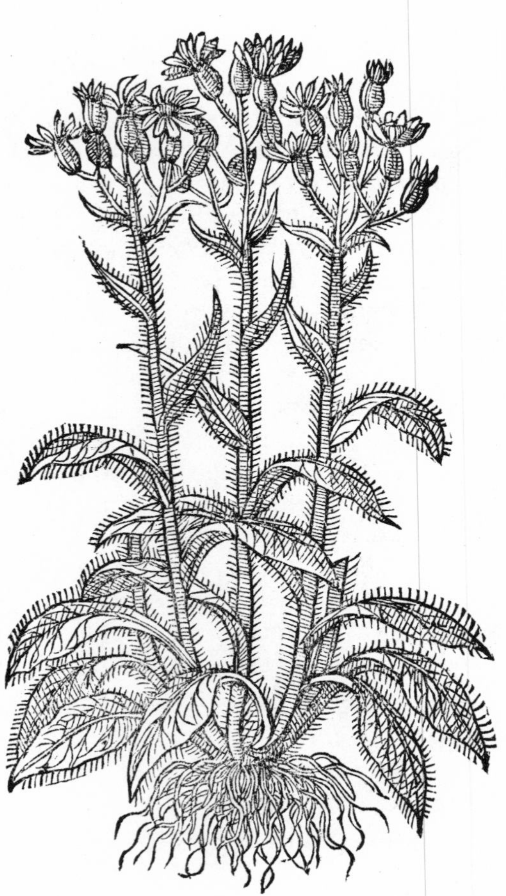

‡ 9 *Conyza Alpina piloſiſsima.*
Hairie Fleabane of the Alpes.

† 10 *Conyza Cærulea acris.*
Blew floured Fleabane.

7 The stalkes of this are about a foot high, straight, stiffe, hard, and couered with a whitish downe: the leaues at the root grow vpon long stalkes, and are soft and hairie; but those which are higher vp, haue a short, or else no stalke at all, and rubbed, they yeeld no vnpleasant smell, and tasted, they are somwhat bitter and acride. The floures that grow vpon the tops of the branches are large, and fashioned like those of Elecampane, and are of the same yellow color. The root is long, slender and blackish, creeping and putting vp new stalkes; it hath many white fibres and a resinous smell. *Clusius* found it growing on dry hilly places in Austria, and calls it *Conyza* 3. *Austriaca*.

8 This which *Lobel* sets forth vnder the title of *Conyza helentis mellita incana*, I take to be the same Plant that I last figured and described out of *Clusius*, onely the root is better exprest in *Clusius* his figure; otherwise by the figures I cannot find any difference, though *Bauhine* reckon it vp in his *Pinax*, as differing therefrom.

9 This also seemes not much to differ from the last mentioned, but onely in the hairinesse of the leaues and stalkes, and that the floures are smaller. This *Lobel* cals *Conyza Helenitis mellita incana*: *Helenitis*, because the floures and leaues haue some semblance of Elecampane; and *Mellita*, for that they smell somewhat like Honie. These last grow vpon mountaines, but none of them with vs in England that I can yet heare of.

10 This hath a small fibrous and yellowish root, of a very hot and biting taste, which sends vp diuers longish leaues about the head thereof; the stalke is some foot and halfe high, and set alternately with twined, longish, narrow and somewhat rough leaues of an ouerworne greene colour; the top of the stalke and branches are adorned with floures set in longish scaly heads like those of *Hieracium*: the outer little leaues are of a faint blew colour, and the inner threds are yellow. It floures in August, and the floures quickly turne into downe, and are carried away with the wind. It grows in many Chalkie hils, and I first obserued it in the company of Mr. *George Bowles*, Mr. *Iohn Bugs* and others, close by Farmingham in Kent; and the last yeare Mr. *William Broad* found it growing at the Blockehouse at Grauesend. *Tragus* calls it *Tinctorius flos alter*: *Dodonæus* because the floure quickly turns to downe makes it *Erigeron quartum*: and *Gesner* for that the root is hot, and drawes rheume like as Pellitorie of Spaine, which therefore is vsed against the Tooth-ache, names it *Dentelaria*: he also cals it *Conyza muralis*, and *Conyzoides Cærulea*: *Tabernamontanus* also calls it *Conyza cærulea*: and lastly, *Fabius Columna* hath it by the name of *Amellus Montanus*, to which kinde it may in mine opinion be as fitly referred, as to these *Conyza's*. Our Authour had the figure hereof in the third place in this Chapter.

¶ *The Place, Time, and Names.*

All these haue beene sufficiently shewne in their particular Titles and Descriptions. ‡

¶ *The Nature.*

Conyza is hot and drie in the third degree.

¶ *The Vertues.*

The leaues and floures be good against the strangurie, the iaundise, and the gnawing or griping of the bellie. A

The same taken with Vineger, helpeth the Epilepsie or falling sicknesse. B

If Women doe sit ouer the decoction thereof, it greatly easeth their paines of the Mother. C

The Herbe burned, where flies, Gnats, fleas, or any venemous things are, doth driue them away. D

† The first was formerly of *Conyza media*; the second was of *Conyza minima*; and the third of *Conyza Cærulea acris*.

CHAP. 132. *Of Starre-woort.*

¶ *The Description.*

1 THe first kinde of *Aster* or *Inguinalis*, hath large broad leaues like *Verbascum Salvifolium* or the great *Conyza*: among which riseth vp a stalke foure or fiue handfuls high, hard, rough and hairie, beset with leaues like Rose Campions, of a darke greene colour. At the top of the said stalkes come forth floures, of a shining and glistering golden colour; and vnderneath about these floures grow fiue or six long leaues, sharpe pointed and rough, not much in

ſhape vnlike the fiſh called *Stella marina*. The floures turne into downe, and are carried away with the winde. The root is fibrous, of a binding and ſharpe taſte.

‡ 2 The ſecond called Italian Starrewoort hath leaues not much vnlike Marigolds,but of a darke greene colour,and rough,and they are ſomewhat round at the vpper end : the ſtalkes are many,and grow ſome cubite high ; and at their tops are diuided into ſundry branches,which beare faire blewiſh purple floures,yellow in their middles, and ſhaped like Marigolds,and almoſt of the ſame bigneſſe, whence ſome haue called them blew Marigolds. ‡

3 The third kinde hath leaues ſo like Italian Starwort, that a man can ſcarcely at the ſudden diſtinguiſh the one from the other. The ſingle ſtalke is a cubit long,vpright and ſlender;on the top whereof grow faire yellow floures,like thoſe of *Enula Campana*, and they fly away in downe : the root is ſmall and threddie.

4 The fourth kinde in talneſſe and floure is not much vnlike that laſt before ſpecified, but in ſtalke and leaues more hairie, and longer,ſomewhat like our ſmall Houndſ-tongue;and the rootes are leſſe fibrous or threddie than the former.

5 There is another ſort that hath a browne ſtalke,with leaues like the ſmall *Coniza*. The floures are of a darke yellow,which turne into downe that flieth away with the wind like *Conyza*. The root is full of threds or ſtrings.

6 There is alſo another that hath leaues like the great Campion,ſomewhat hairie ; amongſt which come vp crooked crambling ſtalkes, leaning lamely many waies. Whereupon doe growe faire yellow floures, Starre-faſhion ; which paſt, the cups become ſo hard,that they will ſcarcely be broken with ones nailes to take forth the ſeed. The root is long and ſtraight as a finger, with ſome few ſtrings annexed vnto the vppermoſt part thereof. It groweth wilde in ſome parts of Spaine.

1 *Aſter Atticus.*
Starrewoort.

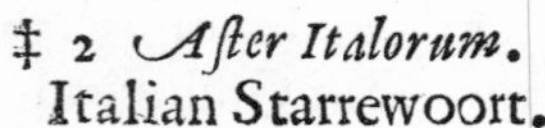
‡ 2 *Aſter Italorum.*
Italian Starrewoort.

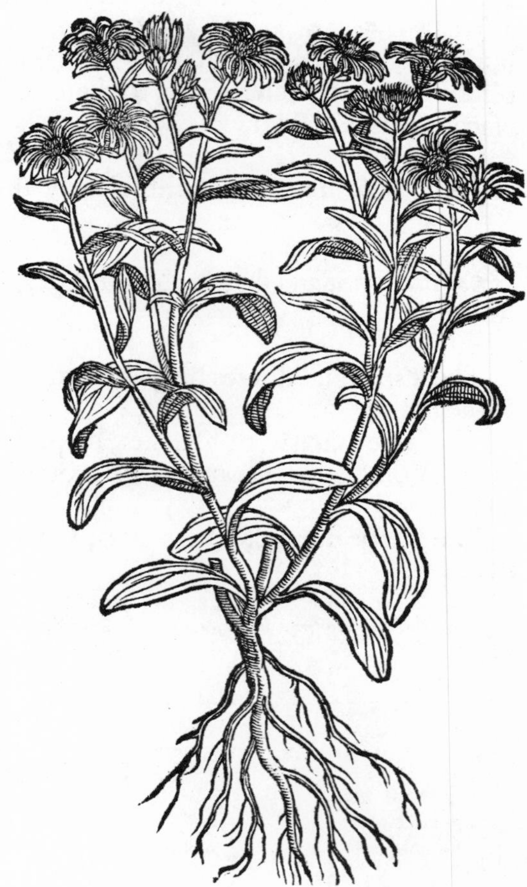

7 There groweth another kinde of Starrewoort, which hath many leaues like Scabious,but thinner,and of a more greene colour, couered with a woollie hairineſſe, ſharpe and bitter in taſte; amongſt which ſpringeth vp a round ſtalke more than a cubite high ; often growing vnto a reddiſh

diſh colour; ſet with the like leaues, but ſmaller and ſharper pointed, diuiding it ſelfe toward the top into ſome few branches; whereon doe grow large yellow floures like *Doronicum* or *Sonchus*. The root is thicke and crooked. ‡ This is *Aſter Pannonicus maior*, *ſive tertius* of *Cluſ.* and his *Auſtriacus primus*.

8 Wee haue ſeene growing vpon wilde Mountaines another ſort, which hath leaues much leſſer than the former, ſomewhat like to the leaues of Willow, of a faire greene colour, which doe adorne and decke vp the ſtalke euen to the top; whereupon doe grow yellow floures ſtarre faſhion, like vnto the former. The root is ſmall and tender, creeping farre abroad, whereby it mightily increaſeth. ‡ This is *Aſter Pannonicus ſalignis folijs: ſive Aſter* 4. *Auſtriacus* 2. of *Cluſius*. It is *Bubonium luteum* of *Tabern*. And our Authour gaue the Figure heereof for *Aſter Italorum*. ‡

9 *Cluſius* hath ſet forth a kinde that hath an vpright ſtalke, ſomewhat hairy, two cubits high, beſet with leaues ſomewhat woollie like to thoſe of the Sallow, hauing at the top of the ſtalke faire yellow floures like *Enula Campana*, which turne into down that is carried away with the wind, the root is thicke, with ſome haires or threds faſtened thereto. ‡ This is *Aſter lanuginoſo folio, ſive* 5. of *Cluſius*. Our Authour gaue the figure hereof vnder the title of *Aſter Hirſutus*: it is *Aſter flore Luteo* of *Taber*.

10 Hee hath likewiſe deſcribed another ſort, that hath leaues, ſtalks, floures, and roots like the ninth, but neuer groweth to the height of one cubite. ‡ It bringeth forth many ſtalkes, and the leaues that grow diſorderly vpon them are narrower, blacker, harder and ſharper pointed than the former, not vnlike thoſe of the common *Ptarmica*, yet not ſnipt about the edges: the floures are yellow and like thoſe of the laſt deſcribed, but leſſe. This is the *Aſter anguſtifolius ſive ſextus* of *Cluſius*. ‡

11 There is likewiſe ſet forth in his Pannonicke obſeruation, a kind of *Aſter* that hath many ſmall hairie leaues like the common great Daiſie: among which riſeth vp an hairy ſtalke of a foot high, hauing at the top faire blew floures inclining to purple, with their middle yellow, which turn (in the time of ſeeding) into a woollie downe, that flieth away with the winde. The whole plant hath a drying, binding, and bitter taſte. The root is threddie like the common Daiſie, or that of Scabious. ‡ This is *Aſper Alpinus cæruleo flore, ſive* 7 of *Cluſius*. ‡

3 *Aſter montanus flore amplo.* Mountaine Starwoort.

4 *Aſter hirſutus.* Hairie Srarwoort.

‡ 5 *Aſter Conyzoides Geſneri.* Fleabane Starrewoort.

‡ 6 *Aſter Luteus ſupinus Cluſij.*
Creeping Starwoort.

‡ 7 *Aſter luteus foliis Succiſæ.*
Scabious leaued Starwoort.

‡ 8 *Aſter Salicis folio.*
Willow leaued Starwoort.

‡ 9 *Aſter Auſtriacus, 5 Cluſ.*
Sallow leaued Starwoort.

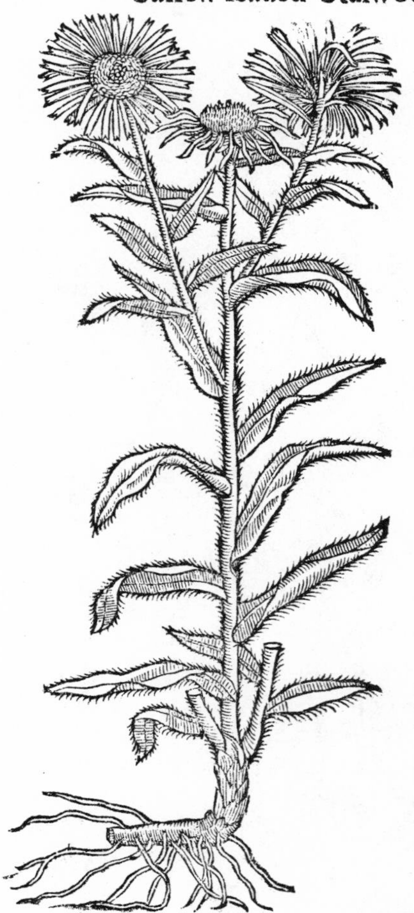

‡ 12 There

‡ 12 There are kept in the Gardens of M^r. *Tradeſcant*, M^r. *Tuggye*, and others, two Starre-woorts different much from all theſe formerly mentioned: the firſt of them is to bee eſteemed, for that it floures in October and Nouember when as few other floures are to be found: the root is large and liuing, which ſends vp many ſmall ſtalks ſome two cubits high, wooddy, ſlender, and not hollow, and towards the top they are diuided into aboundance of ſmall twiggie branches: the leaues that grow alternately vpon the ſtalkes, are long, narrow, and ſharpe pointed, hauing foure or ſixe ſcarce diſcernable nicks on their edges: the floures which plentifully grow on ſmall branches much after the manner of thoſe of *Virga aurea*, conſiſt of twelue white leaues ſet in a ring, with many threds in their middles; which being young are yellow, but becomming elder and larger they are of a reddiſh colour, and at length turne into downe. I haue thought fit to call this plant, not yet deſcribed by any that I know of, being reported to be a Virginian, by the name of *Aſter Virginianus fruticoſus*, Shrubbie Starwoort.

13 This which in gardens floures ſome moneth before the former, growes not ſo high, neither are the ſtalkes ſo ſtraight, but often crooked, yet are they diuided into many branches which beare ſmall blewiſh floures like thoſe of the former: the leaues are longiſh and narrow. This alſo is ſaid to haue come from Canada or Virginia; and it may be called *Aſter fruticoſus minor*, Small ſhrubby Starwoort. ‡

‡ 10 *Aſter 6. Cluſij.*
Narrow leaued Starwoort.

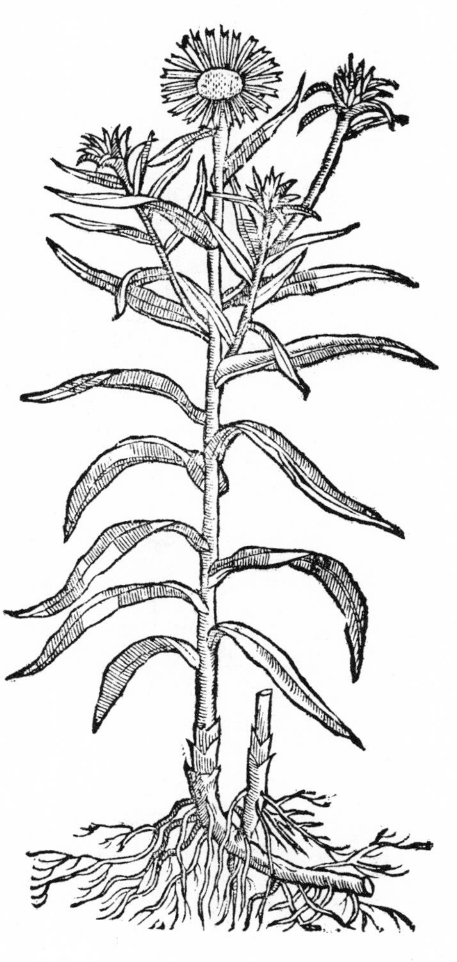

‡ 11 *Aſter 7. Cluſij.*
Dwarfe Daſie leaued Starwoort. ‡

¶ *The Place.*

The kindes of Starwoort grow vpon mountaines and hillie places, and ſometimes in woods and medowes lying by riuers ſides.

The two firſt kindes doe grow vpon Hampſtead heath foure miles from London, in Kent vpon Southfleet Downes, and in many other ſuch downie places. ‡ I could neuer yet finde nor heare of any of theſe Starfloures to grow wilde in this kingdome, but haue often ſeene the Italian Starwort growing in gardens. Theſe two kindes that our Authour mentions to grow on Hampſtead heath and in Kent, are no other than two *Hieracia*, or Hauke-weedes, which are much differing from theſe. ‡

¶ *The Time.*

They floure from Iuly to the end of Auguſt.

¶ *The Names.*

This herbe is called in Greeke ἀστὴρ ἀττικὸς, and also βουβώνιον: in Latine, *Aster Atticus*, *Bubonium*, and *Inguinalis*: of some, *Asterion*, *Asteriscon*, and *Hyophthalmon*: in high Dutch, Megerkraut: in Spanish, *Bobas*: in French, *Estrille*, and *Asper goutte menne*: in English, Starwoort and Sharewoort.

¶ *The Nature.*

It is of a meane temperature in cooling and drying. *Galen* saith it doth moderately waste and consume, especially while it is yet soft and new gathered.

That with the blew floure or purple, is thought to be that, which is of *Virgil* called *flos Amellus*: of which he maketh mention in the fourth booke of his Georgickes.

Est etiam flos in pratis, cui nomen Amello
Fecere agricolæ · facilis quærentibus herba;
Namque vno ingentem tollit de cespite sylvam:
Aureus ipse, sed in folijs, quæ plurima circum
Funduntur, violæ sublucet purpura nigræ.

In English thus.

In Meades there is a floure *Amello* nam'd,
By him that seekes it easie to be found,
For that it seemes by many branches fram'd
Into a little Wood: like gold the ground
Thereof appeares, but leaues that it beset
Shine in the colour of the Violet.

¶ *The Vertues.*

A The leaues of *Aster* or *Inguinalis* stamped, and applied vnto botches, imposthumes, and venereous bubones (which for the most part happen *in Inguine*, that is, the flanke or share) doth mightily maturate and suppurate them, whereof this herbe *Aster* tooke the name *Inguinalis*.

B It helpeth and preuaileth against the inflammation of the fundament, and the falling forth of the gut called *Saccus ventris*.

C The floures are good to be giuen vnto children against the Squinancie, and the falling sicknes.

† That figure which formerly was in the second place vnder the title of *Aster Atticus*, was of the eighth here described; also in the third place formerly were these two figures which we here giue you, whereof the former is of *Aster montanus*, and the latter of *Aster hirsutus*; and that which was vnder the title of *Aster hirsutus* in the fourth place, belongs to the ninth description.

Chap. 133. *Of Woade.*

¶ *The Description.*

1 Glastum or Garden Woad hath long leaues of a blewish greene colour. The stalk groweth two cubits high, set about with a great number of such leaues as come vp first, but smaller, branching it selfe at the top into many little twigs, whereupon do grow many small yellow floures: which being past, the seed commeth forth like little blackish tongues: the root is white and single.

2 There is a wilde kinde of VVoad very like vnto the former in stalks, leaues, and fashion, sauing that the stalke is tenderer, smaller, and browner, and the leaues and little tongues narrower; otherwise there is no difference betwixt them.

¶ *The Place.*

The tame or garden VVoad groweth in fertile fields, where it is sowne: the wilde kind growes where the tame kinde hath been sowne.

¶ *The Time.*

They floure from Iune to September.

¶ *The Names.*

VVoad is called in Greeke ἰσάτις: in Latine, *Isatis*, and *Glastum*: *Cæsar* in his fifth booke of the French wars saith, that all the Brittons do colour themselues with VVoad, which giueth a blew colour: the which thing also *Pliny* in his 22. booke, chap. 1. doth testifie. in France they call it *Glastum* which is like vnto Plantaine, wherewith the Brittish wiues and their daughters are coloured all ouer, and go naked in some kinde of sacrifices. It is likewise called of diuers *Guadum*: of the Italians, *Guado*; a word as it seemeth, wrung out of the word *Glastum*. in Spanish and French, *Pastel*: in Dutch, Weet: in English, VVoad, and VVade.

¶ The

1 *Glaſtum ſativum.*
Garden Woade.

‡ 2 *Glaſtum ſylueſtre.*
Wilde Woade.

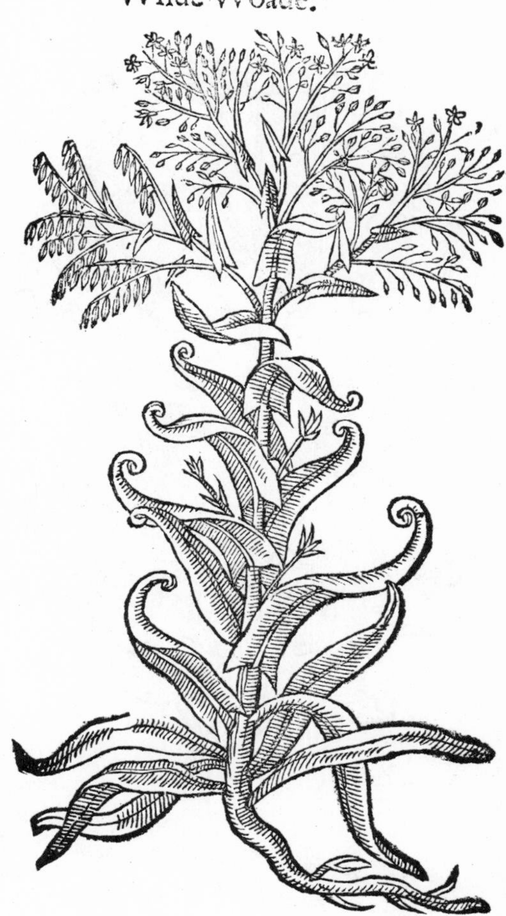

¶ *The Nature.*

Garden Woade is dry without ſharpeneſſe: the wilde Woade drieth more, and is more ſharpe and biting.

¶ *The Vertues.*

The decoction of Woade drunken is good for ſuch as haue any ſtopping or hardneſſe in the milt or ſpleene, and is alſo good for wounds or vlcers in bodies of a ſtrong conſtitution, as of countrey people, and ſuch as are accuſtomed to great labour and hard courſe fare. A

It ſerueth well to dye and colour cloath, profitable to ſome few; and hurtfull to many. B

Chap. 134. *Of Cow-Baſill.*

¶ *The Deſcription.*

1 This kinde of wilde Woade hath fat long leaues like *Valeriana rubra Dodocæi*, or *Behen rubrum*: the ſtalke is ſmall and tender, hauing thereupon little purple floures conſiſting of foure leaues; which being paſt, there come ſquare cornered huskes full of round blacke ſeed like Coleworts. The whole plant is couered ouer with a clammy ſubſtance like Bird-lime, ſo that in hot weather the leaues thereof will take flies by the wings (as *Muſcipula* doth) in ſuch manner that they cannot eſcape away.

2 *Ephemerum Matthioli* hath long fat and large leaues like vnto Woad, but much leſſe; among which riſeth vp a round ſtalke a cubit high, diuiding it ſelfe into many branches at the top, the which are ſet with many ſmall white floures conſiſting of fiue leaues, which being paſt, there follow little round bullets containing the ſeed. The root is ſmall and full of fibres.

¶ *The Place.*

Cow-Baſill groweth in my garden: but *Ephemerum* is a ſtranger as yet in England.

¶ *The Time.*

They floure in May and Iune.

¶ *The*

1 *Vaccaria.*
Cow-Basill.

2 *Ephemerum Matthioli.*
Quicke-fading floure.

‡ ¶ *The Names.*

1 Cow-Basill is by *Cordus* called *Thamecnemon*: by some, according to *Gesner*, *Lychnis & Perfoliata rubra*: *Lobel* termes it *Isatis sylvestris*, and *Vaccaria*: the last of which names is retained by most late Writers.

2 This by *Lobel* is said to be *Ephemerum* of *Matthiolus*; yet I thinke *Matthiolus* his figure, (which was in this place formerly) was but a counterfeit; and so also doe *Columna* and *Bauhinus* judge of it; and *Bauhine* thinkes this of *Lobel* to be some kinde of *Lysimachia*. ‡

¶ *The Nature and Vertues.*

I finde not any thing extant concerning the Nature and Vertues of *Vaccaria* or Cow-Basill.

A *Ephemerum* (as *Dioscorides* writeth) boyled in wine, and the mouth washed with the decoction thereof, taketh away the tooth-ache.

Chap. 135.
Of Sesamoides, or Bastard Weld or Woade.

¶ *The Description.*

1 THe great *Sesamoides* hath very long leaues and many, slender toward the stalk, and broader by degrees toward the end, placed confusedly vpon a thicke stiffe stalke: on the top whereof grow little foolish or idle white floures: which being past, there follow small seeds like vnto Canarie seed that birds are fed withall. The root is thicke, and of a wooddy substance.

‡ 2 This lesser *Sesamoides* of Salamanca, from a long liuing, white, hard, and prettie thicke root sends vp many little stalks set thicke with small leaues like those of Line; and from the middle to the top of the stalke grow many floures, at first of a geeenish purple, and then putting forth yellowish threds; out of the midst of which appeare as it were foure greene graines, which when the floure is fallen grow into little cods full of a small blackish seed. It growes in a stony soile vpon the hills neere Salamanca, where it floures in May, and shortly after perfects his seed. ‡

3 Our

1 *Seſamoides Salamanticum magnum.*
Great baſtard Woade.

2 *Seſamoides Salamanticum parvum.*
Small Baſtard Woade.

3 *Seſamoides parvum Matthioli.*
Bucks-horne Gum-Succorie

‡ 3 Our Author formerly in the Chapter of *Chondrilla* ſpoke (in *Dodonæus* his words) againſt the making of this plant a *Seſamoides* ; for of this plant were the words of *Dodonæus* ; which are theſe: Diuers (ſaith he) haue taken the plant with blew floures to be *Seſamoides parvum*, but without any reaſon ; for that *Seſamoides* hath borrowed his name from the likeneſſe it hath with *Seſamum* : but this herbe is not like to *Seſamum* in any one point, and therefore I thinke it better referred vnto the Gum Succories ; for the floures haue the form and colour of Gum Succory, and it yeeldeth the like milky juyce. Our Authour it ſeemes was either forgetfull or ignorant of what he had ſaid ; for here hee made it one, and deſcribed it meerly by the figure and his fancie. Now I following his tract, haue (though vnfitly) put it here, becauſe there was no hiſtorie nor figure of it formerly there, but both here, though falſe and vnperfect. This plant hath a root ſomewhat like that of Goatſ-beard ; from which ariſe leaues rough and hairy, diuided or cut in on both ſides after the manner of Bucks-horne, and larger than they. The ſtalke is ſome foot high, diuided into branches, which on their tops carry floures of a faire blew colour like thoſe of Succorie, which ſtand in rough ſcaly heads like thoſe of Knapweed. ‡

¶ The

¶ *The Place.*

These do grow in rough and stony places, but are all strangers in England.

¶ *The Time.*

These floure in May and Iune, and shortly after ripen their seed.

‡ ¶ *The Names.*

‡ 1 I thinke none of these to be the *Sesamoides* of the Antients: The first is set forth by *Clusius* vnder the name we here giue you: it is the *Muscipula altera muscoso flore* of *Lobel*: *Viscago maior* of *Camerarius*.

2 This also *Clusius* and *Lobel* haue set forth by the same name as we giue you them.

3 *Matthiolus, Camerarius*, and others haue set this forth for *Sesamoides parvum*: in the *Historia Lugd.* it is called *Catanance quorundam*: but most fitly by *Dodon. Chondrillæ species tertia*, The third kinde of Gum-Succory. ‡

¶ *The Temperature.*

Galen affirmeth that the seed containeth in it selfe a bitter qualitie, and saith that it heateth, breaketh, and scoureth.

¶ *The Vertues.*

A *Dioscorides* affirmeth, that the weight of an halfe-penny of the seed drunke with Meade or honied water purgeth flegme and choler by the stoole.

B The same being applied doth waste hard knots and swellings.

† That which here formerly enioyed the third place, by the title of *Sesamoides maius Scaligeri*, was no other than the plant that is hereafter described by the name of *Tarton-Raire Gallo-provinciæ*, where you may finde both the figure and description.

Chap. 136. *Of Dyers Weed.*

Luteola. Dyers weed or yellow weed.

¶ *The Description.*

Dyers weed hath long narrow and greenish yellow leaues, not much vnlike to woad, but a great deale smaller and narrower; from among which commeth vp a stalke two cubits high, beset with little narrow leaues: euen to the top of the stalke come forth small pale yellow floures, closely clustering together one aboue another, which doe turne into small buttons, cut as it were crosse-wise, wherein the seed is contained. The root is very long and single.

¶ *The Place.*

Dyers weed groweth of it selfe in moist, barren, and vntilled places, in and about Villages almost euery where.

¶ *The Names.*

Pliny, lib. 33. cap. 5. maketh mention by the way of this herbe, and calleth it *Lutea*: *Vitruvius* in his seuenth booke, *Lutum*: it is the *Anticarhinum* of *Tragus*: & *Pseudostruthium* of *Mathiolus*: *Virgill*, in his Bucolickes, Eglog 4. cals it also *Lutum*: in English, Welde, or Dyers weed.

¶ *The Time.*

This herbe flourisheth in Iune and Iuly.

¶ *The Nature.*

It is hot and dry of temperature.

‡ ¶ *The Vertues.*

A The root as also the whole herbe heates and dries in the third degree: it cuts, attenuates, resolueth, opens, digests. Some also commend it against the punctures and bites of venomous creatures,

tures, not onely outwardly applied to the wound, but alſo taken inwardly in drinke.

Alſo it is commended againſt the infection of the Plague: ſome for theſe reaſons terme it *Theriacaria*; *Mat.* ‡ B

Chap. 137. *Of Staues-acre.*

Staphis-agria. Staues-acre.

¶ *The Deſcription.*

Staues-acre hath ſtraight ſtalkes of a browne colour, with leaues clouen or cut into ſundry ſections, almoſt like the leaues of the wilde Vine. The floures do grow vpon ſhort ſtems, faſhioned ſomewhat like vnto our common Monks hood, of a perfect blew colour; which being paſt, there ſucceed welted huskes like thoſe of Wolfsbane, wherein is contained triangular browniſh rough ſeed. The root is of a wooddy ſubſtance, and periſheth when it hath perfected his ſeed.

¶ *The Place.*

It is with great difficultie preſerued in our cold countries, albeit in ſome milde VVinters I haue kept it couered ouer with a little Ferne, to defend it from the iniury of the March winde, which doth more harme vnto plants that come forth of hot Countries, than doth the greateſt froſts.

¶ *The Time.*

It floureth in Iune, and the ſeed is ripe the ſecond yeare of his ſowing.

¶ *The Names.*

It is called in Greeke σταφὶς ἀγρία: in Latine, *Herba Pedicularis*, and *Peduncularia*, as *Marcellus* reporteth. *Pliny* in his 26 Booke, chap. 13. ſeemeth to name it *Vua Taminia*: of ſome, *Pituitaria*, and *Paſſula montana*: in ſhops, *Staphiſ-agria*: in Spaniſh, *Yeruapiolente*: in French, *Herbe aux poulx*: in high-Dutch, **Leus kraut**: in low-Dutch, **Luyscruit**: in Engliſh, Staues-acre, Louſe-wort, and Louſe-pouder.

¶ *The Temperature.*

The ſeeds of Staues-acre are extreame hot, almoſt in the fourth degree, of a biting and burning qualitie

¶ *The Vertues.*

A Fifteene ſeeds of Staues-acre taken with honied water, will cauſe one to vomit groſſe flegme and ſlimie matter, but with great violence, and therefore thoſe that haue taken them ought to walke without ſtaying, and to drinke honied water, becauſe it bringeth danger of choking and burning the throat, as *Dioſcorides* noteth. And for this cauſe they are reiected, and not vſed of the phyſitions, either in prouoking vomit, or elſe in mixing them with other inward medicines.

B The ſeed mingled with oyle or greaſe, driueth away lice from the head, beard, and all other parts of the body, and cureth all ſcuruy itch and mangineſſe.

C The ſame boyled in Vineger, and holden in the mouth, aſſwageth the tooth-ache.

D The ſame chewed in the mouth draweth forth much moiſture from the head, and cleanſeth the braine, eſpecially if a little of the root of Pellitorie of Spaine be added thereto.

E The ſame tempered with vineger is good to be rubbed vpon louſie apparell, to deſtroy and driue away Lice.

F The ſeeds hereof are perillous to be taken inwardly without good aduice, and correction of the ſame: and therefore I aduiſe the ignorant not to be ouer-bold to meddle with it, ſith it is ſo dangerous that many times death enſueth vpon the taking of it.

Chap.

Chap. 138. *Of Palma Christi.*

¶ *The Description.*

1 Ricinus, *Palma Christi*, or *Kik* hath a great round hollow stalke fiue cubits high, of a browne colour, died with a blewish purple vpon greene. The leaues are great and large, parted into sundry sections or diuisions, fashioned like the leaues of a fig-tree, but greater, spred or wide open like the hand of a man; and hath toward the top a bunch of floures clustering together like a bunch of grapes, whereof the lowest are of a pale yellow colour, and wither away without bearing any fruit; and the vppermost are reddish, bringing forth three cornered huskes which containe seed as big as a kidney beane, of the colour and shape of a certaine vermine which haunteth cattell, called a Tik.

2 This *Palma Christi* of America growes vp to the height and bignesse of a small tree or hedge shrub, of a wooddy substance, whose fruit is expressed by the figure, being of the bignes of a great beane, somewhat long, and of a blackish colour, rough and scaly.

1 *Ricinus.*
Palma Christi.

2 *Ricinus Americanus.*
Palma Christi of America.

¶ *The Place.*

The first kinde of *Ricinus* or Palma Christi groweth in my garden, and in many other gardens likewise.

¶ *The Time.*

Ricinus or Kik is sowne in Aprill, and the seed is ripe in the end of August.

¶ *The Name, and cause thereof.*

Ricinus (whereof mention is made in the fourth chapter and sixt verse of the prophecie of *Ionas*) was

was called of the Talmudiſts, קיק *Kik*, for in the Talmud we reade thus, ולא בשמן קיק *Velo beſchemen Kik*. that is, in Engliſh, And not with the oile of *Kik*: which oile is called in the Arabian tongue, *Alkerua*, as *Rabbi Samuel* the ſonne of *Hophni* teſtifieth. Moreouer, a certaine Rab bine moueth a queſtion, ſaying, what is *Kik?* Hereunto *Reſch Lachiſh* maketh anſwer in Ghemara, ſaying, Kik is nothing elſe but *Ionas* his Kikaijon. And that this is true, it appeareth by that name κίκι: which the ancient Greeke Phyſitions, and the Ægiptians vſed; which Greeke word commeth of the Hebrew word *Kik*. Hereby it appeareth, that the old writers long agoe called this plant by the true and proper name. But the old Latine writers knew it by the name *Cucurbita*, which euidently is manifeſted by an hiſtory which Saint *Auguſtine* recordeth in his Epiſtle to Saint *Ierome*, where in effect he writeth thus; That name *Kikaïjon* is of ſmall moment, yet ſo ſmall a matter cauſed a great tumult in Africa. For on a time a certaine Biſhop hauing an occaſion to intreat of this which is mentioned in the fourth chapter of *Ionas* his prophecie (in a collation or ſermon, which he made in his cathedrall church or place of aſſembly) ſaid, that this plant was called *Cucurbita*, a Gourde, becauſe it encreaſed vnto ſo great a quantitie, in ſo ſhort a ſpace, or elſe (ſaith he) it is called *Hedera*. Vpon the nouelty and vntruth of this his doctrine, the people were greatly offended, and thereof ſuddenly aroſe a tumult and hurly burly; ſo that the Biſhop was inforced to goe to the Iewes, to aske their iudgment as touching the name of this plant. And when he had receiued of them the true name, which was *Kikaïjon*: he made his open recantation, and confeſſed his error, & was iuſtly accuſed for a falſifier of the holy ſcripture. ‡ The Greeks called this plant alſo κρότων: *i. Ricinus*, by reaſon of the ſimilitude that the ſeed hath with that inſect, to wit, a Tik. ‡

¶ *The Nature.*

The ſeed of Palma Chriſti, or rather *Kik*, is hot and dry in the third degree.

¶ *The Vertues.*

Ricinus his ſeed taken inwardly, openeth the belly, and cauſeth vomit, drawing ſlimy flegme and choler from the places poſſeſſed therewith. A

The broth of the meate ſupped vp, wherin the ſeed hath been ſodden, is good for the collicke and the gout, and againſt the paine in the hips called *Sciatica*: it preuaileth alſo againſt the jaundiſe and dropſie. B

The oile that is made or drawne from the ſeed is called *Oleum Cicinum*: in ſhops it is called *Oleum de Cherua*. it heateth and drieth, as was ſaid before, and is good to anoint and rub all rough hardneſſe and ſcuruineſſe gotten by itch. C

This oile, as *Rabbi Dauid Chimchi* writeth, is good againſt extreme coldneſſe of the body. D

Chap. 139. *Of Spurge.*

¶ *The Deſcription.*

1 The firſt kinde of Sea Spurge riſeth forth of the ſands, or baich of the ſea, with ſundry reddiſh ſtems or ſtalkes growing vpon one ſingle roote, of a wooddy ſubſtance: and the ſtalkes are beſet with ſmall, fat, and narrow leaues like vnto the leaues of Flaxe. The floures are yellowiſh, and grow out of little diſhes or ſaucers like the common kinde of Spurge. After the floures come triangle ſeeds, as in the other Tithymales.

2 The ſecond kinde (called *Helioſcopius*, or *Soliſequius*: and in Engliſh, according to his Greek name, Sunne Spurge, or time Tithymale, of turning or keeping time with the ſunne) hath ſundry reddiſh ſtalkes of a foot high: the leaues are like vnto Purſlane, not ſo great nor thicke, but ſnipt about the edges: the floures are yellowiſh, and growing in little platters.

3 The third kinde hath thicke, fat, and ſlender branches trailing vpon the ground, beſet with leaues like Knee-holme, or the great Myrtle tree. The ſeed and floures are like vnto the other of his kinde.

4 The fourth is like the laſt before mentioned, but it is altogether leſſer, and the leaues are narrower; it groweth more vpright, otherwiſe alike.

5 Cypres Tithymale hath round reddiſh ſtalkes a foot high, long and narrow like thoſe of Flaxe, and growing buſhie, thicke together like as thoſe of the Cypres tree. The floures, ſeed, and root, are like the former, ſometimes yellow, oftentimes red.

6 The ſixt is like the former, in floures, ſtalkes, rootes, and ſeedes, and differeth in that, this kinde hath leaues narrower, and much ſmaller, growing after the faſhion of thoſe of the Pine tree, otherwiſe it is like.

7 There is another kinde that groweth to the height of a man; the ſtalke is like the laſt mentioned

1 *Tithymalus paralius.*
Sea Spurge.

2 *Tithymalus Helioſcopius.*
Sunne Spurge.

3 *Tithymalus Myrtifolius latifolius.*
Broad leaued Myrtle Spurge.

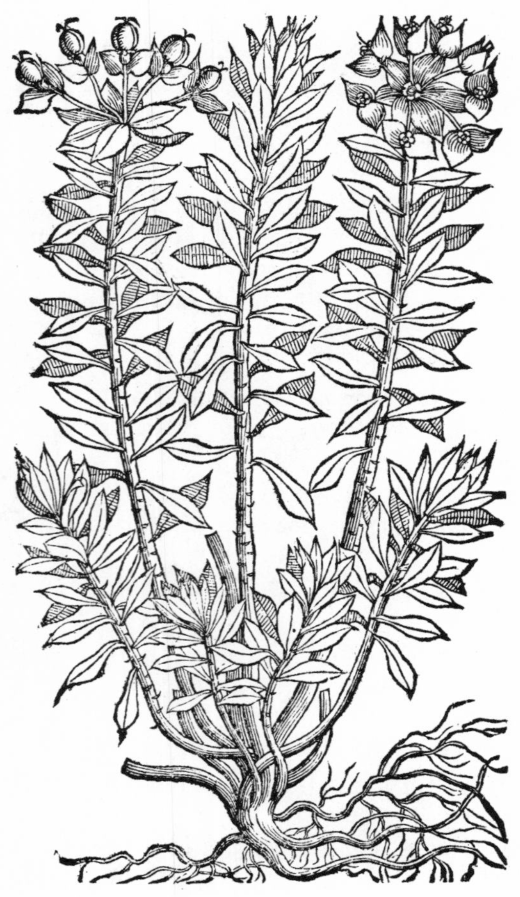

4 *Tithymalus Myrſinitis anguſtifolius.*
Narrow leaued myrtle Spurge.

5 *Tithymalus Cupressinus.*
Cypresse Spurge.

6 *Tithymalus Pineus.*
Pine Spurge.

† 7 *Tithymalus Myrsinitis arborescens.*
Tree Myrtle Spurge.

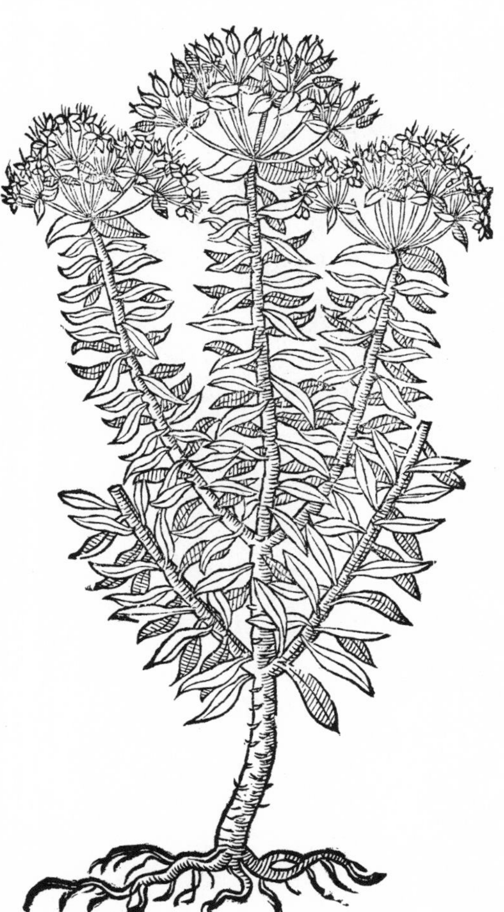

† 8 *Tithymalus Characias Monspell.*
Sweet wood Spurge.

† 9 *Tithymalus Characias Amygdaloides.*
Vnſauorie Wood-ſpurge.

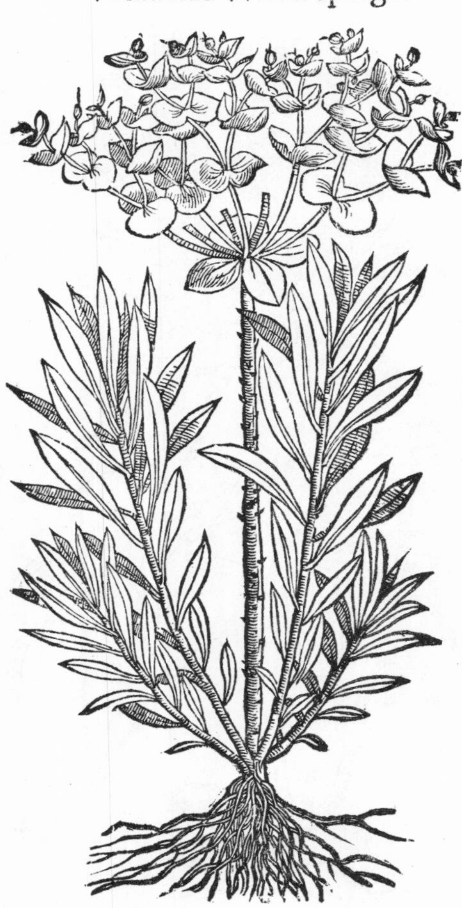

‡ 10 *Tithymalus Characias anguſtifolius.*
Narrow leaued Wood-ſpurge.

‡ 11 *Tithymalus Characias ſerratifolius.*
Cut leaued Wood-ſpurge.

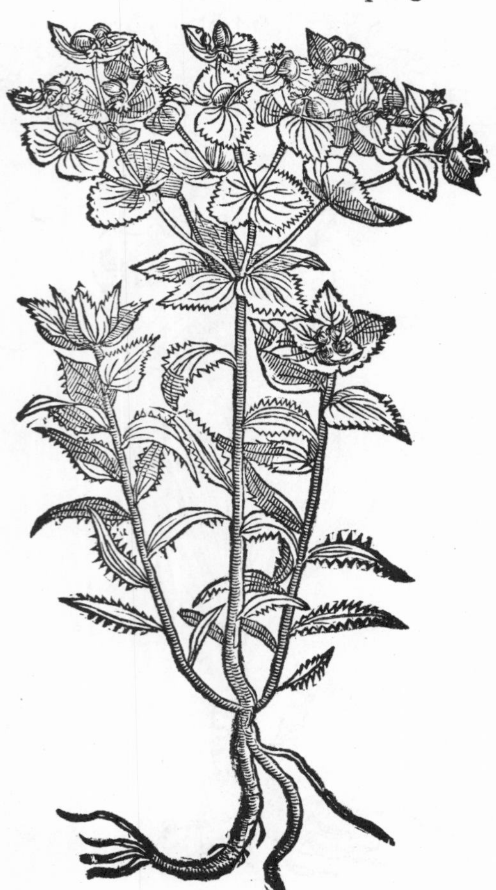

12 *Tithymalus platyphyllos.*
Broad leaued Spurge.

mentioned, but diuided into sundry branches a finger thicke, and somewhat hairy, not red as the others, but white: the leaues be long and narrow, whitish, and a little downy: the floures are yellow, but in other points like to the rest of this kinde.

8 The eighth kinde riseth vp with one round reddish stalke two cubits high, set about with long thin and broad leaues like the leaues of the Almond tree: the floures come forth at the top like the others, and of a yellow colour. The seed and root resemble the other of his kinde.

9 The ninth (which is the common kinde growing in most woods) is like the former, but his leaues be shorter and lesse, yet like to the leaues of an Almond tree: the floures are also yellow, and the seed contained in three cornered seed-vessels.

‡ 10 This fourth kinde of *Tithymalus Characias*, or Valley Tithymale (for so the name imports) hath long, yet somewhat narrower leaues than the former, whitish also, yet not hoary; the vmbels or tufts of floures are of a greenish yellow, which before they be opened do represent the shape of a longish fruit, as an Almond, yet in colour it is like the rest of the leaues: the floures and seeds are like those of the former, and the root descends deepe into the ground.

11 The fifth *Characias* hath also long leaues sharpe pointed, and broader at their setting on, and of a light greene colour, and snipt or cut about the edges like the teeth of a saw. The vmbels are smaller, yet carry such floures and seeds as the former. ‡

12 This kinde hath great broad leaues like the yong leaues of Woad, set round about a stalk of a foot high, in good order: on the top whereof grow the floures in small platters like the common kinde, of a yellow colour declining to purple. The whole plant is full of milke, as are all the rest before specified.

‡ 13 *Tithymalus Dendroides ex Cod. Cæsareo.*
Great Tree Tithymal.

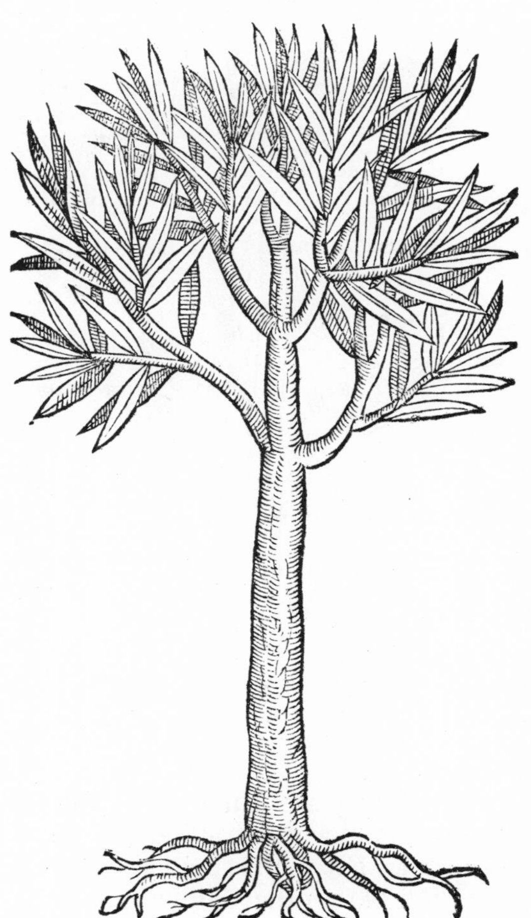

14 *Esula maior Germanica.*
Quacksaluers Turbith.

13 There is another kinde of Tithymale, whose figure was taken forth of a Manuscript of the Emperors by *Dodonæus*, that hath a stalke of the bignesse of a mans thigh, growing like a tree vnto the height of two tall men, diuiding it selfe into sundry armes or branches toward the top, of a red colour. The leaues are small and tender, much like vnto the leaues of *Myrtus*: the seed is like vnto that of wood Tithymale, or *Characias*, according to the authority of *Peter Bellone*.

14 There is a kinde of Tithymale called *Esula maior*, which *Martinus Rulandus* had in great veneration,

veneration, as by his extraction which he vsed for many infirmities may and doth appeare at large, in his bookes entituled *Centuriæ curationum Empiricarum*, dedicated vnto the duke of Bauaria. This plant of *Rulandus* hath very great and many roots couered ouer with a thicke barke, plaited as it were with many surculous sprigs; from which arise sundry strong and large stemmes of a fingers thicknesse, in height two cubits, beset with many pretty large and long leaues like *Lathyris*, but that they are not so thicke: the seed and floure are not vnlike the other Tithymales.

15 This is like the fifth, saue that it hath smaller and more feeble branches; and the whole plant is altogether lesser, growing but some span or foot high; and the floures are of a red or else a greene colour.

16 There is another rare and strange kinde of *Esula*, in alliance and likenesse neere vnto *Esula minor*, that is the small *Esula* or *Pityusa* vsed among the Physitions and Apothecaries of Venice as a kind of *Esula*, in the Confection of their *Benedicta* and Catharticke pills, in stead of the true *Esula*: It yeeldeth a fungous, rough, and browne stalke two cubits high, diuiding it selfe into sundry branches, furnished with stiffe and fat leaues like Licorice, growing together by couples. The floures are pendulous, hanging downe their heads like small bells, of a purple colour, and within they are of a darke colour like *Aristolochia rotunda*.

† 15 *Esula minor, seu Pityusa.*
Small Esula.

‡ 16 *Esula Veneta maritima.*
Venetian Sea-Spurge.

‡ 17 There growes in many chalkie grounds and such dry hilly places, among corne, a small Spurge which seldome growes to two handfuls high; the root is small, and such also are the stalks and leaues, which grow pretty thicke thereon; which oft times are not sharpe, but flat pointed: the seed-vessels and floures are very small, yet fashioned like those of the other Tithymales. It is to be found in corne fields in Iuly and August. ‡

18 The bigger *Cataputia* or the common garden Spurge is best knowne of all the rest, and most vsed; wherefore I will not spend time about his description.

The small kinde of *Cataputia* is like vnto the former, but lesser, whereby it may easily be distinguished; being likewise so well knowne vnto all, that I shall not need to describe it.

‡ These two (I meane the bigger and lesser *Cataputia* of our Author) differ not but by reason of their age, and the fertilenesse and barrennesse of the soile, whence the leaues are somtime broader, and otherwhiles narrower. ‡

19 The

‡ 17 *Esula exigua Tragi.*
Dwarfe Esula.

18 *Lathyris seu Cataputia minor.*
Garden Spurge.

19 *Peplus, sive Esula rotunda.*
Pettie Spurge.

20 *Peplis.*
Isope Spurge.

21 *Chamæſyce.*
Spurge Time.

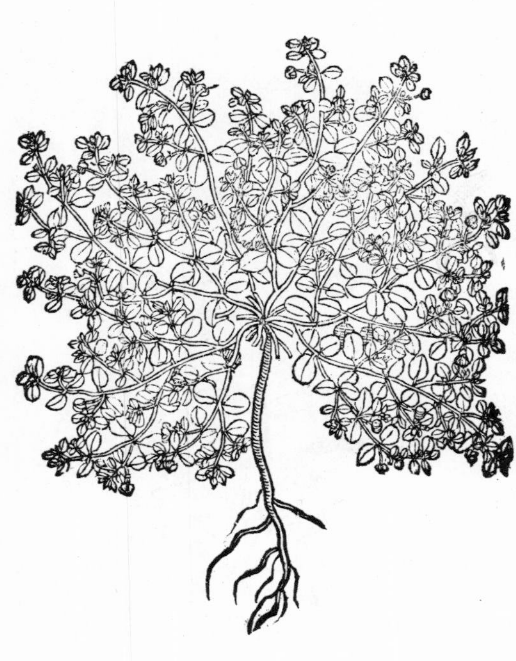

22 *Apios vera.*
Knobbed Spurge.

‡ 23 *Apios radice oblonga.*
Long knottie rooted Spurge.

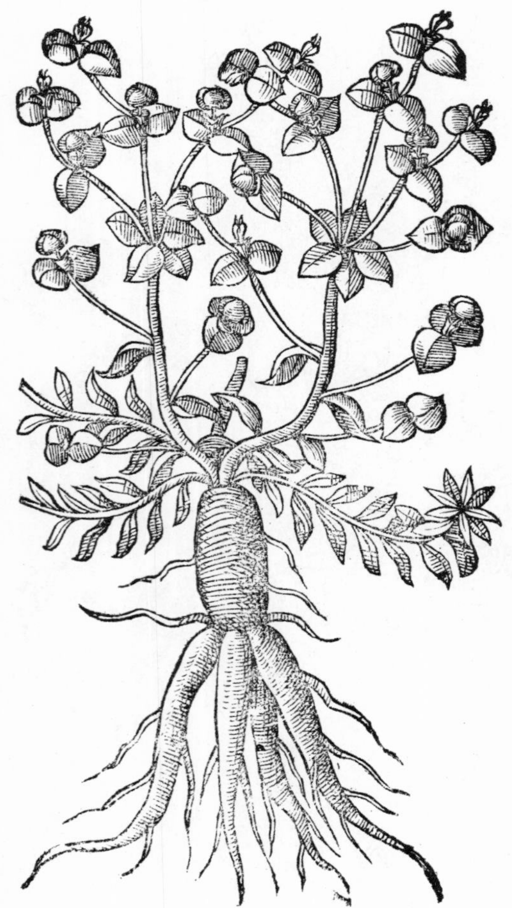

19 The fifteenth kinde called *Peplus*, hath a ſmall, and fibrous root, bringing forth many fruitfull branches two handfuls long, but little and tender, with leaues like the Sun Tithymale, but rounder and much ſmaller: it hath alſo ſmall yellow floures: which being paſt there appeareth a ſlender pouchet, three cornered like the other Tithymales, hauing within it a very medullous whitiſh ſeed like Poppie, the whole plant yeelding a milky iuice, which argueth it to be a kinde of Tithymale.

20 As in name ſo in ſhape this twentieth reſembleth *Peplus*, and commeth in likelihood neerer the ſignification of *Peplum*, or *Flammeolum* than the other; therefore *Dioſcorides* affirmeth it to be *Thamnos amphilaphes*, for that it bringeth foorth a greater plentie of branches, more cloſely knit and wound together, with ſhining twiſts and claſpers an handfull and a halfe long. The leaues are leſſer than thoſe of *Peplus*, of an indifferent likeneſſe and reſemblance betweene *Chamæſyce* and wilde Purſlane. The ſeed is great, and like that of *Peplus*: the root is ſmall and ſingle.

21 The one and twentieth kinde may be eaſily knowne from the two laſt before mentioned, although they be verie like. It hath

hath many branches and leaues creeping on the ground of a pale greene colour, not vnlike to *Herniaria*, but giuing milke as all the other Tithymales doe, bearing the like seed, pouch, and floures, but smaller in each respect.

22 The two and twentieth kinde of Tithymale hath a round root like a small Turnep, as euery Authour doth report: yet my selfe haue the same plant in my garden which doth greatly increase, of which I haue giuen diuers vnto my friends, whereby I haue often viewed the roots, which do appeare vnto me somewhat tuberous, and therein nothing answering the descriptions which *Dioscorides, Pena*, and others haue expressed and set forth. This argueth, that either they were deceiued, and described the same by heare-say, or else the plant doth degenerate being brought from his natiue soile. The leaues are set all alongst a small rib like *Fraxinella*, somewhat round, greene aboue, and reddish vnderneath. The seed groweth among the leaues like the seed of *Peplus*. The whole plant is full of milke like the other Tithymales.

‡ Our Authour here wrongfully taxes other Writers of plants, & *Dioscorides* & *Pena* by name, which shewes that he either neuer read, or else vnderstood not what they writ, for neither of them (nor any other that I know of) resembles the root of this to a Turnep, but say it hath a tuberous peare fashioned root, &c as you may see in *Diosc. lib.* 4. *cap.* 177. and in the *Aduersaria, pag.* 204. The leaues also grow not by couples one against another, as in *Fraxinella*, but rather alternately, or else without any certaine order, as in other Tithymales. ‡

‡ 23 This, saith *Clusius*, hath also a tuberous root, but not peare fashioned like as the former, but almost euery where of an equall thicknesse, beeing about an inch and sometimes two inches long, and the lower part thereof is diuided into foure other roots, or thicke fibers, growing smaller by little and little, and sending forth some few fibers: it is blacke without, and white within, & full of a milkie iuice: the stalkes are short and weake, set with little leaues like those of the former: the floures are of a yellowish red colour, and the seede is contained in such vessels as the other Tithymales. This is *Tithymalus tuberosus*, or *Ischas altera* of *Clusius*. ‡

¶ *The Place.*

The first kinde of Spurge groweth by the Sea side vpon the rowling Sand and Baich, as at Lee in Essex, at Lang-tree point right against Harwich, at Whitstable in Kent, and in many other places.

The second groweth in grounds that lie waste, and in barren earable soile, almost euery where.

The third and fourth, as also the foureteenth and eighteenth, grow in gardens, but not wilde in England.

The ninth Spurge called *Characias* groweth in most VVoods of England that are drie and warme.

The eighteenth and nineteenth grow in salt marshes neere the sea, as in the isle of Thanet by the sea side, betwixt Reculvers and Margate in great plentie.

¶ *The Time.*

These plants floure from Iune to the end of Iuly.

¶ *The Names.*

Sea Spurge is called in Latine *Tithymalus paralius*: in Spanish, *Leche tresua*: in high Dutch, **Wolfer milch**, that is to say *Lupinum lac*, or Wolfes milke. Wood Spurge is called *Tithymalus characias*. The first is called in English Sea Spurge, or Sea VVartwoort. The second, Sun Spurge; the third and fourth, Mirtle Spurge: the fifth Cypresse Spurge; or among women, VVelcome to our house; the sixth Pine Spurge; the seuenth shrub Spurge, and tree Mirtle Spurge; the eighth and ninth VVood Spurge; the twelfth Broad leafed Spurge: the thirteenth Great Tree Spurge: the foureteenth and fifteenth Quacksaluers Spurge; the sixteenth Venice Spurge, the seuenteenth Dwarfe Spurge; the eighteenth common Spurge; the nineteenth and twentieth Petie Spurge; the one and twentieth Spurge Time: The two and twentieth, True *Apios* or the knobbed Spuige.

¶ *The Temperature.*

All the kinds of Tithymales or Spurges are hot and drie almost in the fourth degree, of a sharp and biting qualitie, fretting or consuming. First the milke and sap is in speciall vse, then the fruit and leaues, but the root is of least strength. The strongest kinde of Tithymale, and of greatest force is that of the sea. A

Some write by report of others, that it inflameth exceedingly, but my selfe speak by experience; for walking along the sea coast at Lee in Essex, with a Gentleman called Mr. *Rich*, dwelling in the same towne, I tooke but one drop of it into my mouth; which neuerthelesse did so inflame and swell in my throte that I hardly escaped with my life. And in like case was the gentleman, which caused vs to take our horses, and poste for our liues vnto the next farme house to drinke some milk to quench the extremitie of our heate, which then ceased. B

¶ The

¶ *The Vertues.*

A The iuice of Tithymale, I do not meane ſea Tithymale, is a ſtrong medicine to open the bellie, and cauſing vomite, bringeth vp tough flegme and cholericke humours. Like vertue is in the ſeed and root, which is good for ſuch as fall into the dropſie, being miniſtred with diſcretion and good aduice of ſome excellent Phyſition, and prepared with his Correctories by ſome honeſt Apothecarie.

B The iuice mixed with honie, cauſeth haire to fall from that place which is anointed therewith, if it be done in the Sun.

C The iuice or milke is good to ſtop hollow teeth, being put into them warily, ſo that you touch neither the gums, nor any of the other teeth in the mouth with the ſaid medicine.

D The ſame cureth all roughneſſe of the skin, mangineſſe, leprie, ſcurfe, and running ſcabs, and the white ſcurfe of the head. It taketh away all manner of warts, knobs, and the hard callouſneſſe of Fiſtulaes, hot ſwellings and Carbuncles.

E It killeth fiſh, being mixed with any thing that they will eat.

F Theſe herbes by mine aduiſe would not be receiued into the bodie, conſidering that there be ſo many other good and wholeſome potions to be made with other herbes, that may bee taken without perill.

† The ſeuenth figure was formerly of *Tithymalus myrſinites 3. anguſtifolius* of *Tabernamontanus*: The 8. and 9. were both of the ſame plant: the 12. was the figure of the *Eſula exigua Tragi*, whoſe hiſtorie I haue giuen you in the 17. place.

Chap. 140. *Of Herbe Terrible.*

1 *Alypum montis Ceti.*
Herbe Terrible.

2 *Tarton-Raire Gallo-Prouincia.*
Gutwoort.

¶ *The Deſcription.*

1 Herbe Terrible is a ſmall ſhrub two or three cubits high, branched with many ſmall twigges, hauing a thin rinde firſt browne, then purple, with many little and thinne leaues like Myrtle. The floures are rough like the middle of Scabious floures, of a blew purple colour. The root is two fingers thicke, browne of colour, and of a wooddie ſubſtance: the whole plant very bitter, and of an vnpleaſant taſte like *Chamelæa*, yea ſomewhat ſtronger.

2 Tartonraire, called in Engliſh Gutwoort, groweth by the ſea, and is Cathartical, and a ſtranger with vs. In the mother tongue of the Maſſilians, it is called Tartonraire, of that abundant and

and vnbridled facultie of purging, which many times doth cause *Dysenteria*, and such like immoderate fluxes, especially when one not skilfull in the vse thereof shall administer the pouder of the leaues, mixed with any liquor. This plant groweth in manner of a shrub, like *Chamelæa*, and bringeth forth many smal, tough, and pliant twigs, set about with a thin and cottony hairinesse, & hath many leaues of a glistering siluer colour, growing from the lowest part euen to the top, altogether like *Alypum* before mentioned: and vpon these tough and thick branches (if my memory faile not) do grow small floures, first white, afterward of a pale yellow: the seed is of a russet colour: the root hard and wooddy, not very hot in the mouth, leauing vpon the tongue some of his inbred heat and taste, somewhat resembling common Turbith, and altogether without milke.

¶ *The Place.*

These plants do grow vpon the mountains in France, and other places in the grauelly grounds, and are as yet strangers in England.

¶ *The Time.*

They flourish in August and September. ‡ The first *Clusius* found flouring in diuers parts of Spaine, in Februarie and March; and I coniecture the other floures about the same time, yet I can finde nothing said thereof in such as haue deliuered the historie of it. ‡

¶ *The Names.*

There are not any other names appropriate to these plants more than are set forth in the titles.

‡ The first of these is the *Alypum montis Ceti*, & *Herba terribilis* of *Lobel*; *Clus.* calls it *Hippoglossum Valentinum*, & in *Hist. Lugd.* it is named *Alypum Penæ*, & *Empetrum Phacoides*. The second is the *Tartonraire Galloprouinciæ Massiliensium*, in the *Aduersaria*; *Sesamoides maius multorum* of *Dalesc.* & the *Sesamoides maius Scalegeri* of *Tabern.* by which title our Author also gaue his figure, in the 397. *pag.* of the former Edition. ‡

¶ *The Temperature and Vertues.*

There is nothing either of their nature or vertues, more than is set forth in the Descriptions.

‡ Both these plants haue a strong purging faculty like as the Tithymales; but the latter is far more powerfull, and comes neere to the qualitie of *Mezereon*; wherefore the vse of it is dangerous, by reason of the violence and great heat thereof. ‡

CHAP. 141. *Of Herbe Aloe, or Sea Housleeke.*

‡ 1 *Aloe vulgaris, sive Sempervivum marinum.*
Common *Aloe*, or Sea-Housleeke.

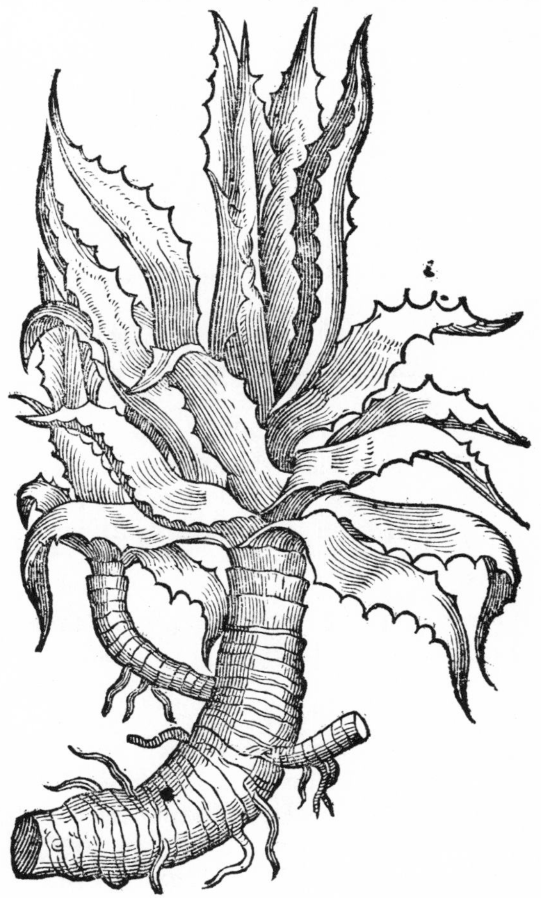

2 *Aloe folio mucronato.*
Prickly herbe Aloe, or Sea Housleeke.

¶ *The*

¶ *The Deſcription.*

1 HEarbe Alloe hath leaues like thoſe of ſea Onion, very long, broad, ſmooth, thick, bending backewards, notched in the edges, ſet with certaine little blunt prickles, full of tough and clammie iuice like the leaues of Houſleeke. The ſtalke, as *Dioſcorides* ſaith, is like to the ſtalke of Affodill: the floure is whitiſh; the ſeed like that of Affodill; the root is ſingle, of the faſhion of a thicke pile thruſt into the ground. The whole herbe is extreme bitter, ſo is the iuice alſo that is gathered thereof.

† 2 There is another herbe Aloe that groweth likewiſe in diuers prouinces of America; the leaues are two cubits long, alſo thicker, broader, greater, and ſharper pointed than the former, and it hath on the edges far harder prickles. The ſtalke is three cubits high, and a finger thicke, the which in long cups beares violet coloured floures. †

¶ *The Place.*

This plant groweth very plentifully in India, and in Arabia, Cœloſyria, & Egypt, from whence the iuice put into skins is brought into Europe. It groweth alſo, as *Dioſcorides* writeth, in Aſia, on the ſea coaſts, and in Andros, but not verie fit for iuice to be drawne out. It is likewiſe found in Apulia, and in diuers places of Granado and Andaluſia, in Spaine, but not far from the ſea: the iuice of this is alſo vnprofitable.

¶ *The Time.*

The herbe is alwaies greene, and likewiſe ſendeth forth branches, though it remaine out of the earth, eſpecially if the root be couered with lome, and now and then watered: for ſo being hanged on the ſeelings and vpper poſts of dining roomes, it doth not onely continue a long time greene, but it alſo groweth and bringeth forth new leaues: for it muſt haue a warme place in winter time, by reaſon it pineth away if it be frozen.

¶ *The Names.*

The herbe is called in Greeke ἀλόη: in Latine, and in ſhops alſo, *Aloe*: and ſo is likewiſe the iuice. The plant alſo is named ἀμφίβιον, ἠρίγγιον, ἔρμινον, τραγωκέρος: but they are baſtard words: it is called ἀμφίβιον becauſe it liueth not onely in the earth, but alſo out of the earth. It is named in French, *Poroquet*: in Spaniſh, *Azeuar*, and *Yerua bauoſa*: in Engliſh, *Aloes*; herbe *Aloes*, Sea Houſeleeke, Sea Aigrene.

The hearbe is called of the latter Herbariſts oftentimes *Semperviuum*, and *Semperviuum Marinum*, becauſe it laſteth long after the manner of Houſe-leeke. It ſeemeth alſo that *Columella* in his tenth booke nameth it *Sedum*, where he ſetteth downe remedies againſt the canker-wormes in trees.

Profuit & plantis latices infundere amaros
Marrubij, multoque Sedi contingere ſucco.

In Engliſh thus:

Liquours of Horehound profit much b'ing pour'd on trees:
The ſame effect Sea Houſleeke works as well as theſe.

For he reciteth the iuice of *Sedum* or Houſeleeke among the bitter iuices, and there is none of the Houſleekes bitter but this.

The Temperature.

Aloë, that is to ſay, the iuice which is vſed in Phyſicke, is good for many things. It is hot, and that in the firſt or ſecond degree, but drie in the third, extreme bitter, yet without biting. It is alſo of an emplaiſticke or clammie qualitie, and ſomething binding, externally applied.

¶ *The Vertues.*

A It purgeth the belly, and is withall a wholeſome and conuenient medicine for the ſtomacke, if any at all bee wholeſome. For as *Paulus Aegineta* writeth, when all purging medicines are hurtfull to the ſtomacke, *Aloës* onely is comfortable. And it purgeth more effectually if it be not waſhed: and if it be, it then ſtrengtheneth the ſtomacke the more.

B It bringeth forth choler, but eſpecially it purgeth ſuch excrements as be in the ſtomacke, the firſt veines, and in the neereſt paſſages. For it is of the number of thoſe medicines, which the Græcians call ἐκκοπρωτικά, of the voiding away of the Ordure; and of ſuch whoſe purging force paſſeth not far beyond the ſtomacke. Furthermore *Aloës* is an enemie to all kindes of putrefactions; and defendeth the body from all manner of corruption. It alſo preſerueth dead carkaſes from putrifying;

it

it killeth and purgeth away all manner of wormes of the belly. It is good againſt a ſtinking breath
proceeding from the imperfection of the ſtomacke: it openeth the piles or hemorrhoides of the
fundament; and being taken in a ſmall quantity, it bringeth down the monthly courſe: it is thought
to be good and profitable againſt obſtructions and ſtoppings in the reſt of the intrals. Yet ſome
there be who thinke, that it is not conuenient for the liuer.

One dramme thereof giuen, is ſufficient to purge. Now and then halfe a dramme or little more C
is enough.

It healeth vp greene wounds and deepe ſores, clenſeth vlcers, and cureth ſuch ſores as are hard- D
ly to bee helped, eſpecially in the fundament and ſecret parts. It is with good ſucceſſe mixed with
ἐναίμοις, or medicines which ſtanch bleeding, and with plaiſters that be applied to bloudy wounds;
for it helpeth them by reaſon of his emplaiſticke qualitie and ſubſtance. It is profitably put into
medicines for the eies forasmuch as it clenſeth and drieth without biting.

Dioſcorides ſaith, that it muſt be torrified or parched at the fire, in a cleane and red hot veſſell, E
and continually ſtirred with a *Spatula*, or Iron Ladle, till it bee torrified in all the parts alike:
and that it muſt alſo bee waſhed, to the end that the vnprofitable and ſandie droſſe may
ſinke downe vnto the bottome, and that which is ſmooth and moſt perfect bee taken and reſerued.

The ſame Authour alſo teacheth, that mixed with honie it taketh away blacke and blew ſpots, F
which come of ſtripes: that it helpeth the inward ruggedneſſe of the eye-lids, and itching in the
corners of the eies: it remedieth the head-ache, if the temples and forehead bee annointed therewith, being mixed with vineger and oile of Roſes: being tempered with wine, it ſtaieth the falling
off of the haire, if the head be waſhed therewith: and mixed with wine and honie, it is a remedie
for the ſwelling of the Vuula, and ſwelling of the Almonds of the throte, for the gums & all vlcers
of the mouth.

The iuice of this herbe *Aloe*: (whereof is made that excellent and moſt familiar purger, called G
Aloe Succotrina, the beſt is that which is cleere and ſhining, of a browne yellowiſh colour) it openeth the bellie, purging cold, flegmaticke, and cholericke humours, eſpecially in thoſe bodies that
are ſurcharged with ſurfetting, either of meat or drinke, and whoſe bodies are fully repleat with
humours, fairing daintily, and wanting exerciſe. This *Aloes* I ſay, taken in a ſmall quantitie after ſupper (or rather before) in a ſtewed prune, or in water the quantitie of two drammes in the
morning, is a moſt ſoueraigne medicine to comfort the ſtomacke, and to clenſe and driue foorth
all ſuperfluous humours. Some vſe to mixe the ſame with Cinnamon, Ginger, and Mace, for
the purpoſe aboue ſaid; and for the Iaundies, ſpitting of bloud, and all extraordinarie iſſues of
bloud.

The ſame vſed in vlcers, eſpecially thoſe of the ſecret parts or fundament, or made into pouder, H
and ſtrawed on freſh wounds, ſtaieth the bloud, and healeth the ſame, as thoſe vlcers before ſpoken of.

The ſame taken inwardly cauſeth the Hemorrhoids to bleed, and being laid thereon it cauſeth I
them to ceaſe bleeding.

CHAP. 142. *Of Houſleeke or Sengreene.*

¶ *The Kindes.*

Sengreene, as *Dioſcorides* writeth, is of three ſorts, the one is great, the other ſmall, and the third is that which is called *Illecebra*, biting Stone-crop, or VVall pepper.

¶ *The Deſcription.*

1 The great Sengreene, which in Latine is commonly called *Iouis Barba*, Iupiters beard, bringeth forth leaues hard adioyning to the ground and root, thicke, fat, full of tough iuice, ſharpe pointed, growing cloſe and hard together, ſet in a circle in faſhion of an eye, and bringing forth very many ſuch circles, ſpreading it ſelfe out all abroad: it oftentimes alſo ſendeth forth ſmall ſtrings, by which it ſpreadeth farther, and maketh new circles; there riſeth vp oftentimes in the middle of theſe an vpright ſtalke about a foot high, couered with leaues growing leſſe and leſſe toward the points, parted at the top into certaine wings or branches, about which are floures orderly placed, of a darke purpliſh colour: the root is all of ſtrings.

Vu 2 There

2 There is alſo another great Houſleek or Sengreen (ſyrnamed tree Houſleeke) that bringeth forth a ſtalke a cubit high, ſometimes higher, and often two; which is thicke, hard, woody, tough, and that can hardly be broken, parted into diuers branches, and couered with a thicke groſſe barke, which in the lower part reſerueth certaine prints or impreſſed markes of the leaues that are fallen away. The leaues are fat, well bodied, full of juice, an inch long and ſomewhat more, like little tongues, very curiouſly minced in the edges, ſtanding vpon the tops of the branches, hauing in them the ſhape of an eye. The floures grow out of the branches, which are diuided into many ſprings; which floures are ſlender, yellow, and ſpred like a ſtar; in their places commeth vp very fine ſeed, the ſprings withering away: the root is parted into many off-ſprings. This plant is alwaies greene, neither is it hurt by the cold in winter, growing in his natiue ſoile; whereupon it is named ἀείζωον, and *Semperuiuum*, or Sengreene.

1 *Semperuiuum maius.*
Great Houſleeke.

‡ 2 *Sedum maius arboreſcens.*
Tree Houſleeke.

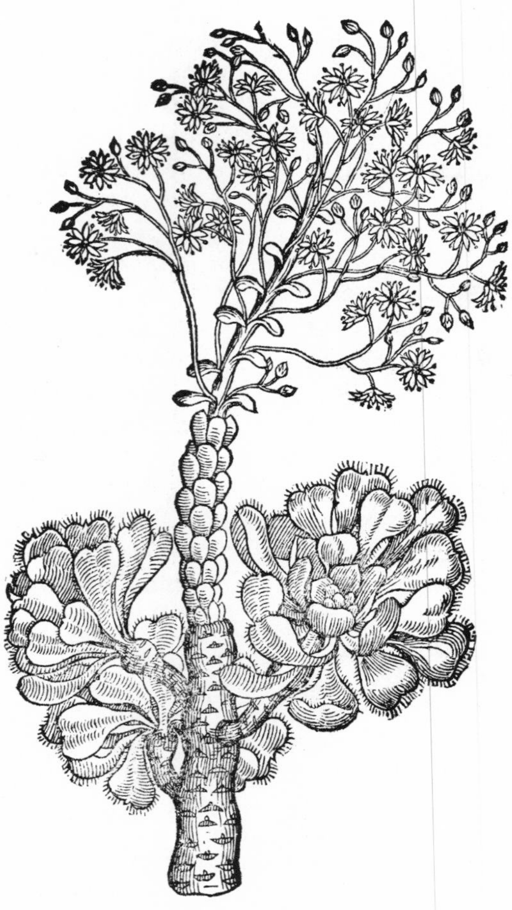

3 There is alſo another of this kinde, the circles whereof are anſwerable in bigneſſe to thoſe of the former, but with leſſer leaues, moe in number, and cloſeſly ſet, hauing ſtanding on the edges very fine haires as it were like ſoft prickles. This is ſomewhat of a deeper greene: the ſtalke is ſhorter, and the floures are of a pale yellow. ‡ This is the third of *Dodonæus* deſcription, *Pemptad.* 1. *lib.* 5. *cap.* 8. ‡

4 There is likewiſe a third to be referred hereunto: the leaues hereof be of a whitiſh greene, and are very curiouſly nicked round about. ‡ The floure is great, conſiſting of ſix white leaues; This is that deſcribed by *Dodonæus* in the 4. place: and it is the *Cotyledon altera ſecunda* of *Cluſius*. ‡

5 There is alſo a fourth, the circles whereof are leſſer, the leaues ſharpe pointed, very cloſely ſet, of a darke red colour on the top, and hairy in the edges: the floures on the ſprigs are of a gallant purple colour. ‡ This is the fift of *Dodonæus*; and the *Cotyledon altera tertia* of *Cluſius*. ‡

¶ *The Place.*

1 The great Sengreen is well knowne not onely in Italy, but alſo in France, Germany, Bohemia, and the Lowe-Countries. It groweth on ſtones in mountaines, vpon old walls, and ancient buildings, eſpecially vpon the tops of houſes. The forme hereof doth differ according to the nature of the ſoile; for in ſome places the leaues are narrower and leſſer, but mo in number, and haue one onely circle; in ſome they are fewer, thicker, and broader: they are greene, and of a deeper greene

greene in ſome places; and in others of a lighter greene: for thoſe which we haue deſcribed grow not in one place, but in diuers and ſundry.

‡ 5 *Sedum maius anguſtifolium.*
Great narrow leaued Houſleeke.

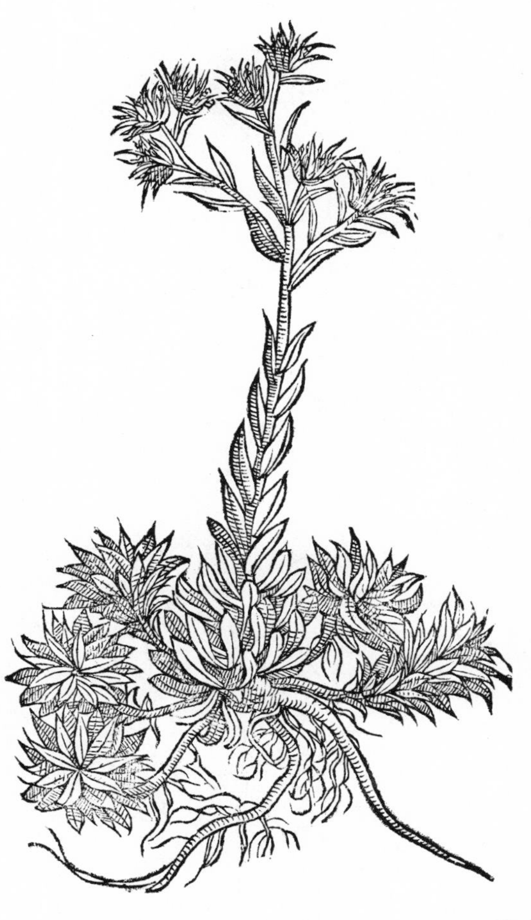

2 Great Sengreene is found growing of it ſelfe on the tops of houſes, old walls, and ſuch like places in very many prouinces of the Eaſt, and of Greece: and alſo in the Iſlands of the Mediterranian ſea; as in Crete, which now is called Candy, Rhodes, Zant, & others; neither is Spaine without it: for (as *Carolus Cluſius* witneſſeth) it groweth in many places of Portingale; otherwiſe it is cheriſhed in earthen pots. In cold countries, and ſuch as lie Northward, as in both the Germanies, it neither groweth of it ſelfe, nor yet laſteth long, though it be carefully planted, and diligently looked vnto, but through the extremitie of the weather, and the ouermuch cold of winter it periſheth.

¶ *The Time.*

The ſtalke of the firſt doth at length floure after the Summer Solſtice, which is in Iune about Saint *Barnabies* day, and now and than in the moneth of Auguſt; but in Aprill, that is to ſay, after the æquinoctiall in the ſpring, which is about a moneth after the ſpring is begun, there grow out of this among the leaues ſmall ſtrings, which are the groundwork of the circles, by which being at length full growne, it ſpreadeth it ſelfe into very many circles.

2 Houſleeke that groweth like a tree, doth floure in Portingale at the beginning of the yeere preſently after the winter Solſtice, which is December, about S. *Lucies* day.

¶ *The Names.*

The firſt is commonly called *Iouis barba*, or Iupiters beard, and alſo *Sedum maius vulgare*: the Germanes call it **Hanſzwurtz, Groſz Donderbaer**: they of the Low-countries, **Donderbaert**: the Hollanders, **Huyſlook**: the French-men, *Ioubarbe*: the Italians, *Sempreuiuo maggiore*: the Spaniards, *Siempreuiua, yerua pentera*: the Engliſh-men, Houſleeke, and Sengreene, and Aygreene: of ſome, Iupiters eie, Bullocks eie, and Iupiters beard: of the Bohemians, *Netreske*. Many take it to be *Cotyledon altera Diſocoridis*; but we had rather haue it one of the Sengreenes: for it is continually greene, and alwaies flouriſheth, and is hardly hurt by the extremity of winter.

The other without doubt is *Dioſcorides* his ἀείζωον μέγα: that is, *Semperuivum magnum*, or *Sedum majus*, great Houſleek, or Sengreen: *Apuleius* calleth it *Vitalis*, and *Semperflorium*: it is alſo named ζωόφθαλμον, στεργηθρόν, αιθαλής.

¶ *The Temperature.*

The great Houſleeks are cold in the third degree: they are alſo dry, but not much, by reaſon of the watery eſſence that is in them.

¶ *The Vertues.*

They are good againſt Saint Anthonies fire, the ſhingles, and other creeping vlcers and inflammations, as *Galen* ſaith, that proceed of rheumes and fluxes: and as *Dioſcorides* teacheth, againſt the inflammations or fiery heate in the eyes: the leaues, ſaith *Pliny*, being applied, or the juice laid on, are a remedy for rheumatike and watering eies. A

They take away the fire in burnings and ſcaldings; and being applied with Barly meale dried, do take away the paine of the gout. B

Dioſcorides teacheth, that they are giuen to them that are troubled with a hot laske: that they likewiſe driue forth wormes of the belly if they be drunke with wine. C

The juice put vp in a peſſary do ſtay the fluxes in women, proceeding of a hot cauſe: the leaues held in the mouth do quench thirſt in hot burning feauers. D

The juice mixed with Barly meale and vineger preuaileth againſt S. Anthonies fire, all hot burning and fretting vlcers, and againſt ſcaldings, burnings, and all inflammations, and alſo the gout comming of an hot cauſe. E

A The iuice of Housleeke, Garden Nightshade, and the buds of Poplar boiled in *Axungia porci*, or hogs grease, maketh the most singular Populeon that euer was vsed in Chirurgerie.

B The iuice hereof taketh away cornes from the toes and feet, if they be washed and bathed therwith, and euery day and night as it were implaistered with the skin of the same Housleeke, which certainly taketh them away without incision or such like, as hath beene experimented by my very good friend M. *Nicholas Belson*, a man painefull and curious in searching forth the secrets of Nature.

C The decoction of Housleeke, or the iuice thereof drunke, is good against the bloudie flixe, and cooleth the inflammation of the eies being dropped thereinto, and the bruised hearbe layed vpon them.

Chap. 143. *Of the Lesser Housleekes or Prickmadams.*

1 *Sedum minus hæmatoides.*
Pricke-madame.

2 *Sedum minus Officinarum.*
White floured Prickmadam.

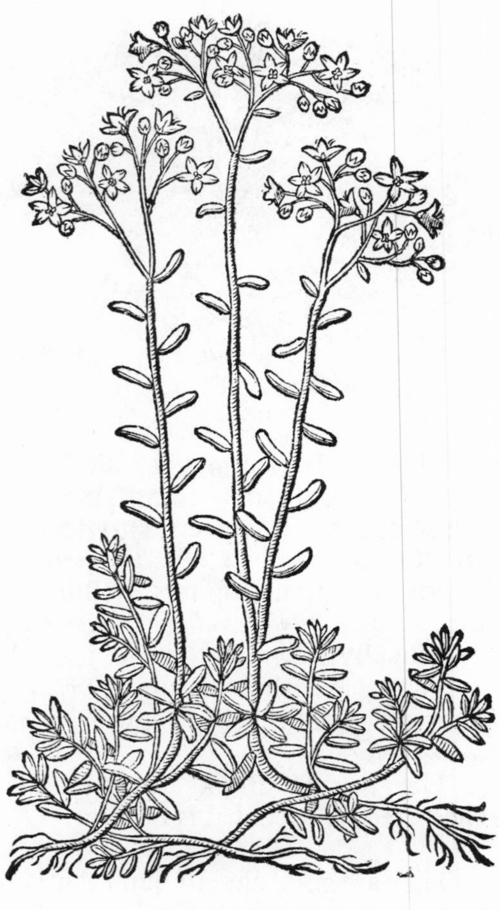

¶ *The Description.*

1 The first of these is a very little herb, creeping vpon the ground with many slender stalks, which are compassed about with a great number of leaues, that are thicke, ful of ioints, little, long, sharpe pointed, inclining to a greene blew. There rise vp among these, little stalkes, a handful high, bringing forth at the top, as it were a shadowie tuft; and in these fine yellow floures: the root is full of strings.

2 The other little Sengreene is also a small herbe, bringing forth many slender stalkes, seldome aboue a span high; on the tops whereof stand little floures like those of the other, in small loose

‡ 3 *Sedum minus æſtivum.*
Small Sommer Sengreene.

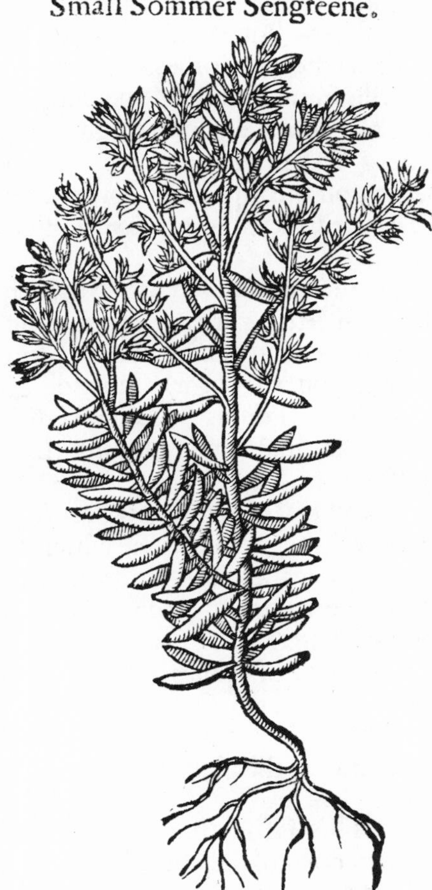

‡ 4 *Sedum minus flore amplo.*
Small large floured Sengreene.

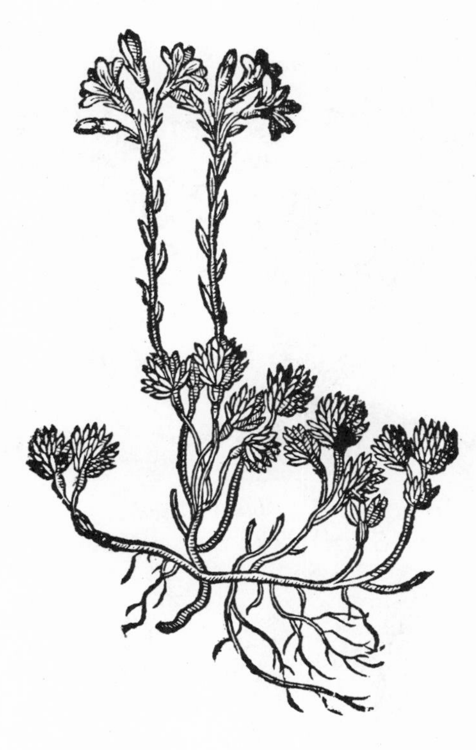

‡ 5 *Sedum medium teretifolium.*
Small Prickmadam.

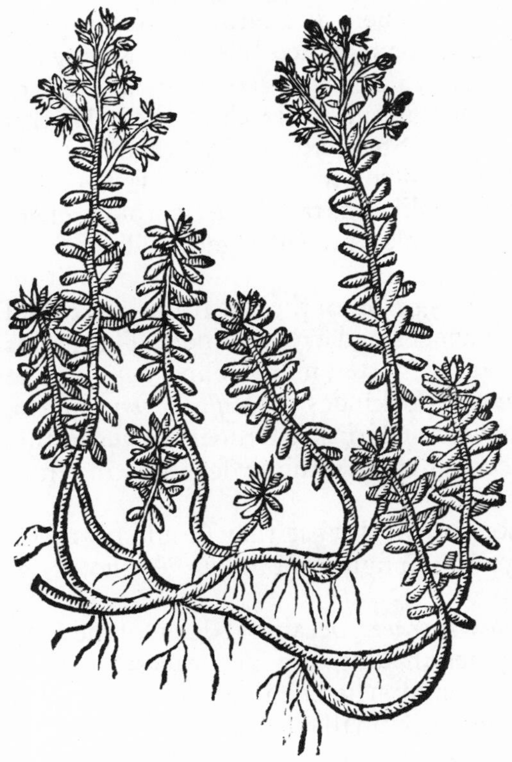

‡ 6 *Aizoon Scorpioides.*
Scorpion Sengreene.

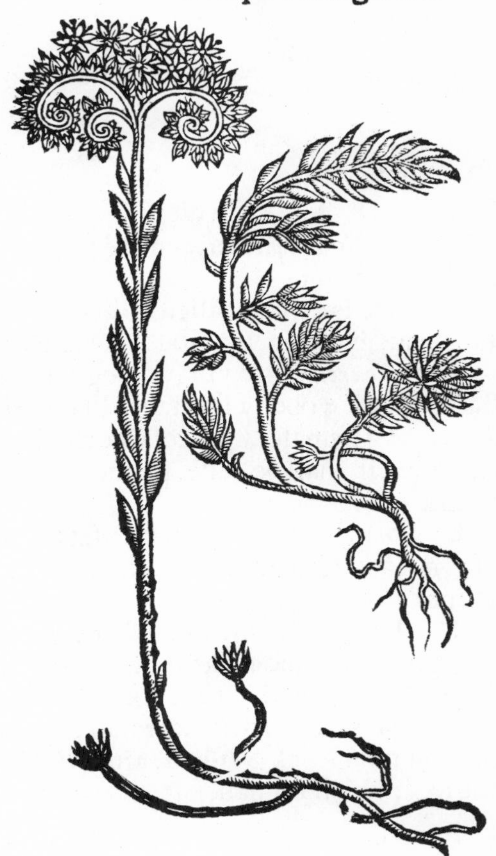

‡ 7 *Sedum Portlandicum.*
Portland Sengreene.

‡ 8 *Sedum petræum.*
Small rocke Sengreene.

loose tufts; but they are white and something lesser: the leaues about the stalkes are few and little, but long, blunt, and round, bigger than wheaten cornes, something lesser than the kernels of the Pine Apples, otherwise not vnlike; which oftentimes are something red, stalkes and all: the roote creepeth vpon the superficiall or vppermost part of the earth, sending downe slender threds.

3 There is a small kinde of Stonecrop, which hath little narrow leaues, thicke, sharpe pointed, and tender stalkes, full of fattie iuice; on the top whereof doe grow small yellowe floures, Starre fashion. The roote is small, and running by the ground.

4 There is likewise another Stonecrop called Frog Stonecrop, which hath little tufts of leaues rising from small and and threddie rootes, creeping vpon the ground like vnto *Kali* or Frog-grasse; from the which tufts of leaues riseth a slender stalke, set with a few such like leaues, hauing at the top prettie large yellow floures, the smalnesse of the plant beeing considered.

‡ 5 This is like that which is described in the second place, but that the stalks are lesser, and not so tall, and the floures of this are star fashioned, and of a golden yellow colour. ‡

6 There is another Stonecrop, or Prickmadam called *Aizoon Scorpioides*, which is altogether like the great kinde of Stonecrop, and differeth in that, that this kinde of Stonecrop or Prickmadam hath his tuft of yellow floures turning again, not much vnlike the taile of a Scorpion, resembling *Myositis Scorpioides*, and the leaues somewhat thicker, and closer thrust together. The root is small and tender.

7 There is a plant called *Sedum Portlandicũ*, or Portland Stonecrop, of the English Island called Portland, lying in the South coast, which hath goodly branches and a rough rinde. The leaues imitate *Laureola*, growing among the Tithymales, but thicker, shorter, more fat and tender. The stalke is of a wooddy substance like *Laureola*, participating of the kindes of *Crassula, Sempervivum*, and the Tithymales, whereof wee thinke it to bee a kinde; yet not daring to deliuer any vncertaine sentence, it shall be lesse preiudiciall to the truth, to account it as a shrubbe, degenerating from both kindes.

‡ *Pena* and *Lobel*, who first set this foorth knewe, not verie well what they should say thereof; nor any since them: wherefore I haue onely giuen you their figure put to our Authours description. ‡

8 There is a plant which hath receiued his name *Sedum Petræum*, because it doth for the most part grow vpon the rocks, mountains, & such like stonie places, hauing very smal leaues, comming forth of the ground in tufts like *Pseudo-Moly*; that is, our common herbe called Thrift: amongst the leaues come forth slender stalkes an handfull high, loden with small yellow floures like vnto the common Prick-Madam: after which come little thicke sharpe pointed cods, which containe the seed, which is small, flat, and yellowish.

¶ The

¶ *The Place.*

The former of theſe groweth in gardens in the Low-countries : in other places vpon ſtone wals and tops of houſes in England almoſt euery where.

The other groweth about rubbiſh in the borders of fields, and in other places that lye open to the Sunne.

¶ *The Time.*

They floure in the Sommer moneths.

¶ *The Names.*

The leſſer kinde is called in Greeke ἀείζωον μικρὸν : in Latine, *Sedum*, and *Semperuiuum minus* : of the Germanes, **Kleyn Donderbaer**, and **Kleyn Hauſzwurtz** : of the Italians, *Semperuiuo minore* : of the Frenchmen, *Tricque-madame* : of the Engliſh men, Pricke-Madam, Dwarfe Houſe-leeke, and ſmall Sengreene.

The ſecond kinde is named in ſhops *Craſſula minor*; and they ſyrname it *minor*, for difference betweene it and the other *Craſſula*, which is a kinde of Orpin: it is alſo called *Vermicularis* : in Italian, *Pignola, Granelloſa*, and *Graſella* : in low-Dutch, **Blader looſen** : in Engliſh, Wilde Pricke-Madam, Great Stone-crop, or Worme-graſſe. ‡ That which is vulgarly knowne and called by the name of Stone-crop is the *Illecebra* deſcribed in the following chapter, and ſuch as grow commonly with vs of theſe ſmall Houſeleekes mentioned in this chapter are generally named Pricke-Madames : but our Author hath confounded them in this and the next chapter; which I would not alter, thinking it ſufficient to giue you notice thereof. ‡

¶ *The Temperature and Vertues.*

All theſe ſmall Sengreens are of a cooling nature like vnto the great ones, and are good for thoſe things that the others be. The former of theſe is vſed in many places in ſallads, in which it hath a fine reliſh, and a pleaſant taſte : it is good for the heart-burne.

---

‡ CHAP. 144. *Of diuers other ſmall Sengreenes.*

¶ *The Deſcription.*

‡ 1 THe ſtalke of this ſmall water Sengreene is ſome ſpanne long, reddiſh, ſucculent, and weake : the leaues are longiſh, a little rough, and full of iuyce : the floures grow vpon the tops of the ſtalkes, conſiſting of ſix purple or elſe fleſh-coloured leaues; which are ſucceeded by as many little cods containing a ſmall ſeed : the root is ſmall and threddy, and the whole plant hath an inſipide or wateriſh taſte. This was found by *Cluſius* in ſome waterie places of Germany about the end of Iune; and he calls it *Sedum minus* 3 *ſiue paluſtre*.

2 This ſecond from ſmall fibrous and creeping roots ſends vp ſundry little ſtalkes ſet with leaues like thoſe of the ordinary Pricke-Madam, yet leſſe, thicke, and flatter, and of a more aſtringent taſte : the floures, which are pretty large, grow at the tops of the branches, and conſiſt of fiue pale yellowiſh leaues. It growes in diuers places of the Alps, and floures about the end of Iuly, and in Auguſt. This is the *Sedum minus* 6. or *Alpinum* 1. of *Cluſius*.

3 This hath ſmall little and thicke leaues, lying bedded, or compact cloſe together, and are of an Aſh colour inclining to blew : the ſtalkes are ſome two inches long, ſlender, and almoſt naked; vpon which grow commonly ſome three floures conſiſting of fiue white leaues apiece, with ſome yellow threds in the middle. This mightily encreaſes, and will mat and couer the ground for a good ſpace together. It floures in Auguſt, and growes vpon the craggy places of the Alpes. *Cluſius* calls it *Sedum minus nonum, ſiue Alpinum* 3.

4 The leaues of this are ſomewhat larger and longer, yet thicke, and ſomewhat hairy about their edges; at firſt alſo of an acide taſte, but afterwards bitteriſh and hot : it alſo ſendeth forth ſhoots, and in the middeſt of the leaues it puts forth ſtalkes ſome two inches high, which at the top as in an vmbel carry ſome ſix little floures conſiſting of fiue leaues apiece, hauing their bottomes of a yellowiſh colour. It is found in the like places, and floures at the ſame time as the former. *Cluſius* makes it his *Sedum minus* 10. *Alpinum* 4. and in the *Hiſt. Lugd.* It is called *Iaſme montana*.

5 For

‡ 1 *Sedum minus palustre.*
Small water Sengreene.

‡ 2 *Sedum Alpinum* 1. *Clusij.*
Small Sengreene of the Alps.

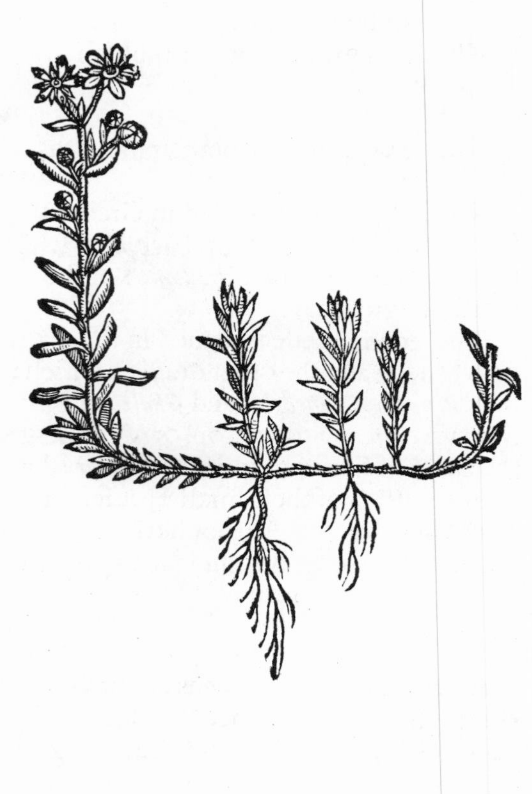

‡ *Sedum Alpinum* 3. *Clusij.*
White Sengreene of the Alpes.

‡ 4 *Sedum Alpinum* 4. *Clusij.*
Hairy Sengreene of the Alpes.

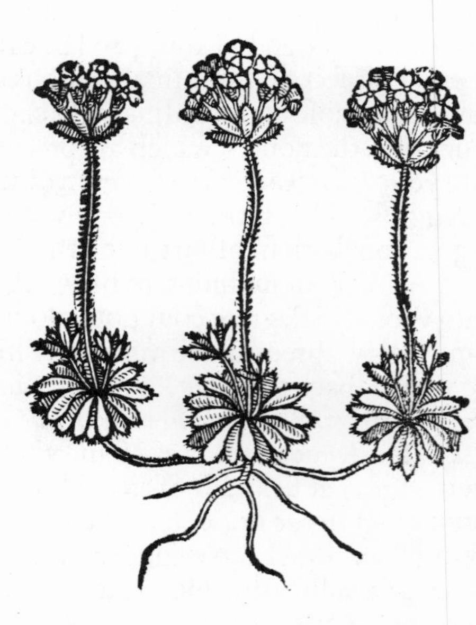

‡ 5 *Sedum petræum Bupleurifolio.*
Long leaued Rocke Sengreene.

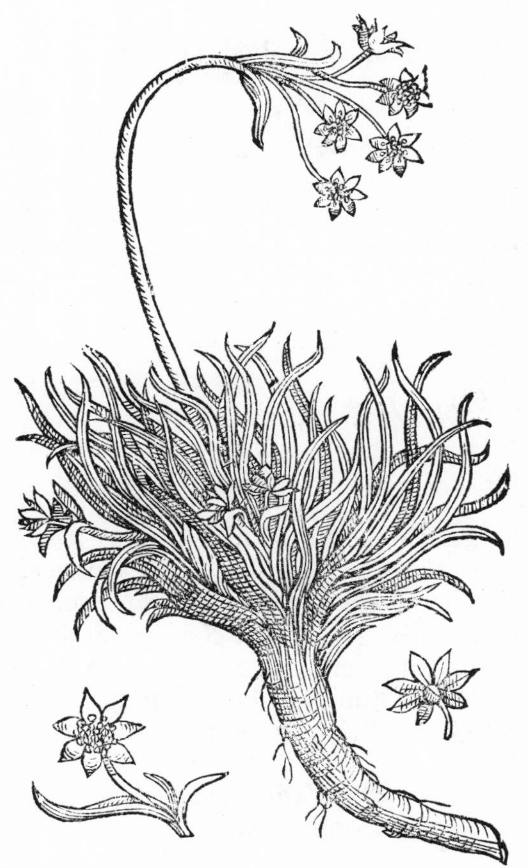

5 For these foure last described we are beholden to *Clusius*; and for this fifth to *Pona*, who thus describes it: It hath one thicke and large root with few or no fibres, but som knots bunching out here and there: it is couered with a thicke barke, and is of a blackish red colour on the outside: the leaues are many, long and narrow, lying spred vpon the ground; the stalke grows some foot high, and is round and naked, and at the top carries floures consisting of 7 sharpe pointed pale yellow leaues; which are succeeded by seeds like those of *Bupleurum*, and of a strong smell. It floures about the middle of Iuly, and the seed is ripe about the middle of August. *Pona*, who first obserued this growing vpon Mount Baldus in Italy, sets it forth by the name of *Sedum petræum Bupleurifolio*. *Bauhine* hath it by the name of *Perfoliata Alpina Gramineo folio*, and *Bupleuron angustifolium Alpinum.*

¶ *The Temper and Vertues.*

The three first described without doubt are cold, and partake in vertues with the other small Sengreenes; but the two last are rather of an hot and attenuating facultie. None of them are commonly knowne or vsed in Physicke. ‡ A

Vermicularis siue Illecebra minor acris.
Wall-Pepper, or Stone-crop.

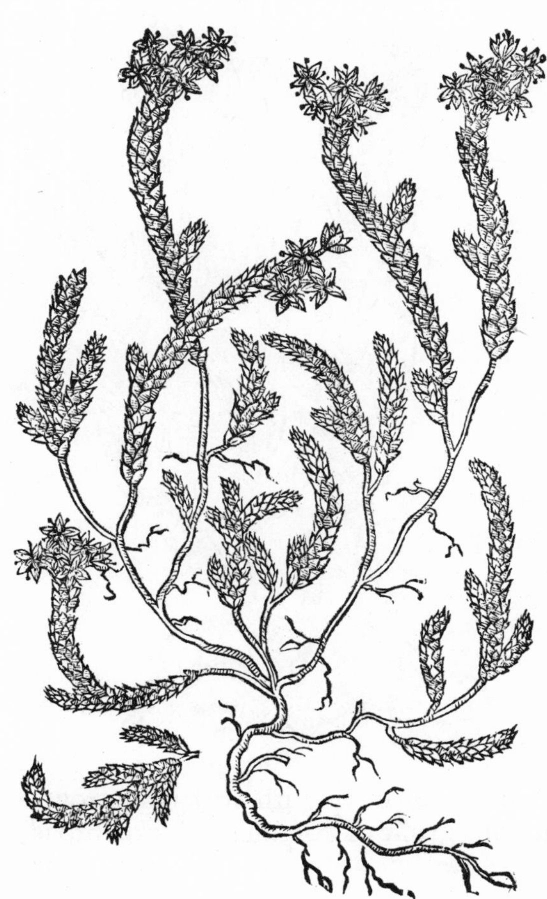

CHAP. 145.

Of Stone-crop, called Wall-pepper.

¶ *The Description.*

THis is a low and little herbe: the stalks be slender and short: the leaues about these stand very thicke, and small in growth, full bodied, sharpe pointed, and full of iuyce: the floures stand on the top, and are maruellous little, of colour yellow, and of a sharp biting taste: the root is nothing but strings.

¶ *The Place.*

It groweth euery where in stony and dry places, and in chinks and crannies of old wals, and on the tops of houses: it is alwaies green, and therefore it is very fitly placed among the Sengreenes.

¶ *The Time.*

It floureth in the Sommer moneths.

¶ *The Names.*

This is *Tertium sempervivum Dioscoridis*, or *Dioscorides* his third Sengreene, which he saith is called of the Grecians ἀίζωον ἄγρια, and τηλέφιον: and of the Romanes, *Illecebra*. *Pliny* also witnesseth, that the Latines name it *Illecebra*. Yet there is another ἀίζωον ἄγρια, and another τηλέφιον: the Germanes call this herbe **Mauerpfeffer**, and **Katzen treuble**: the French men, *Pain d'oiseau*:

oiseau: the Low-countrey men, **Muer Pepper**: the English men, Stone-crop, and Stone-hore, little Stone-crop, Pricket, Mouse-taile, Wall-Pepper, Countrey Pepper, and Iacke of the Butterie.

¶ *The Temperature.*

This little herbe is sharpe and biting, and very hot. Being outwardly applied it raiseth blisters, and at length exulcerateth.

¶ *The Vertues.*

A It wasteth away hard kernels, and the Kings Euill, if it be layd vnto them, as *Dioscorides* writes.

B The iuyce hereof extracted or drawne forth, and taken with vineger or other liquor, procureth vomit, and bringeth vp grosse and flegmaticke humors, and also cholericke; and doth thereby oftentimes cure the Quartan Ague and other Agues of long continuance: and giuen in this manner it is a remedy against poysons inwardly taken.

Chap. 146. *Of Orpyne.*

¶ *The Description.*

1 The Spanish Orpyne sendeth forth round stalkes, thicke, slipperie, hauing as it were little ioynts, somewhat red now and then about the root: the leaues in like manner be thicke, smooth, grosse, full of tough iuyce, sometimes sleightly nicked in the edges, broader leafed, and greater than those of Purslane; otherwise not much vnlike; which by couples are set opposit one against another vpon euery joint, couering the stalke in order by two and two: the floures in the round tufts are of a pale yellow: the root groweth full of bumpes like vnto long kernels, waxing sharpe toward the point: these ketnels be white, and haue strings growing forth of them.

1 *Crassula major Hispanica.*
Spanish Orpyne.

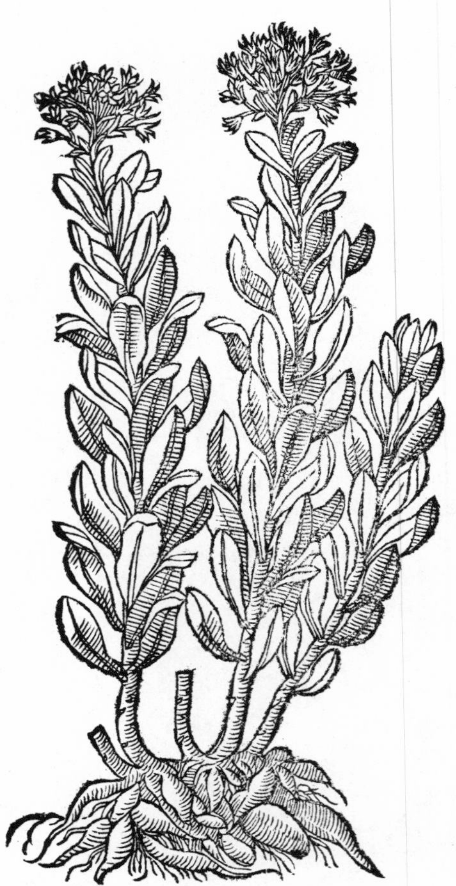

2 *Crassula siue faba inuersa.*
Common Orpyne.

2 The second, which is our common Orpyne, doth likewise rise vp with very many round stalkes that are smooth, but not ioynted at all: the leaues are grosse or corpulent, thicke, broad, and

and oftentimes ſomewhat nicked in the edges, leſſer than thoſe of the former, placed out of order. The floures be either red or yellow, or elſe whitiſh: the root is white, well bodied, and full of kernels. This plant is very full of life: the ſtalkes ſet onely in clay continue greene a long time; and if they be now and then watered they alſo grow. We haue a wilde kinde of Orpyne growing in corne fields and ſhadowy woods in moſt places of England, in each reſpect like that of the garden, ſauing that it is altogether leſſer.

¶ *The Place.*

They proſper beſt in ſhadowie and ſtony places, in old walls made of lome or ſtone. *Oribaſius* ſaith, That they grow in Vineyards and tilled places. The firſt groweth in gardens; the other euerie where: the firſt is much found in Spaine and Hungarie; neither is Germanie without it; for it groweth vpon the bankes of the riuer of Rhene neere the Vineyards, in rough and ſtony places, nothing at all differing from that which is found in Spaine.

The ſecond groweth plentifully both in Germany, France, Bohemia, England, and in other countries among vines, in old lomie daubed and ſtony walls.

¶ *The Time.*

The Orpynes floure about Auguſt or before.

¶ *The Names.*

The firſt is that which is called of the Grecians τηλέφιον, and ἀείζωον ἄγριον: of the Latines, *Telephium*, and *Sempervivum ſylueſtre*, and *Illecebra*: but *Illecebra* by reaſon of his ſharpe and biting qualitie doth much differ from it, as we haue declared in the former Chapter. Some there be that name it ἀνδράχνη, or *Portulaca ſylueſtris*: yet there is another *Portulaca ſylueſtris*, or wilde Purſlane, like to that which groweth in gardens, but leſſer: we may call this in Engliſh, Spaniſh Orpyne, Orpyne of Hungarie, or ioynted Orpyne.

The ſecond kinde of Orpyne is called in ſhops *Craſſula*, and *Craſſula Fabaria*, and *Craſſula maior*, that it may differ from that which is deſcribed in the chapter of little Houſleeke: it is named alſo *Fabaria*: in high-Dutch, **Mundkraut, Knauenkraut, Fortzwang**, and **Fotzweyn**: in Italian, *Faba graſſa*: in French, *Ioubarbe des vignes*, *Feue eſpeſſe*: in low-Dutch, **Smer wortele**, and **Hemel Sluetel**: in Engliſh, Orpyne; alſo Liblong, or Liue-long.

¶ *The Temperature.*

The Orpyns be cold and dry, and of thin or ſubtill parts

¶ *The Vertues.*

Dioſcorides ſaith, That being laid on with Vineger it taketh away the white morphew: *Galen* ſaith the blacke alſo; which thing it doth by reaſon of the ſcouring or cleanſing qualitie that it hath. Whereupon *Galen* attributeth vnto it an hot facultie, though the taſte ſheweth the contrarie: which aforeſaid ſcouring facultie declareth, That the other two alſo be likewiſe cold. But cold things may as well cleanſe, if drineſſe of temperature and thinneſſe of eſſence be ioyned together in them. A

CHAP. 147. *Of the ſmaller Orpyns.*

¶ *The Deſcription.*

1 THe Orpyn with purple floures is lower and leſſer than the common Orpyn: the ſtalkes be ſlenderer, and for the moſt part lie along vpon the ground. The leaues are a ſo thinner and longer, and of a more blew greene, yet well bodied, ſtanding thicker below than aboue, confuſedly ſet together without order: the floures in the tufts at the tops of the ſtalks be of a pale blew tending to purple. The roots be not ſet with lumpes or knobbed kernels, but with a multitude of hairy ſtrings.

2 This ſecond Orpyn, as it is knowne to few, ſo hath it found no name, but that ſome Herbariſts do call it *Telephium ſempervivum* or *virens*: for the ſtalkes of the other do wither in winter, the root remaineth greene; but the ſtalkes and leaues of this endure alſo the ſharpeneſſe of Winter; and therefore we may call it in Engliſh, Orpyn euerlaſting, or Neuer-dying Orpyn. This hath leſſer and rounder leaues than any of the former: the floures are red, and the root fibrous.

‡ 3 *Cluſius* receiued the ſeeds of this from *Ferranto Imperato* of Naples, vnder the name of *Telephium*

1 *Telephium floribus purpureis.*
Purple Orpyn.

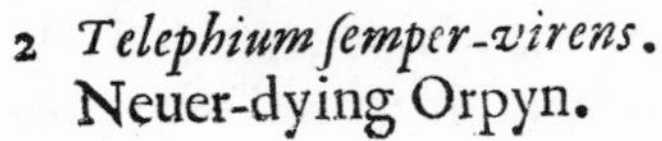
2 *Telephium ſemper-virens.*
Neuer-dying Orpyn.

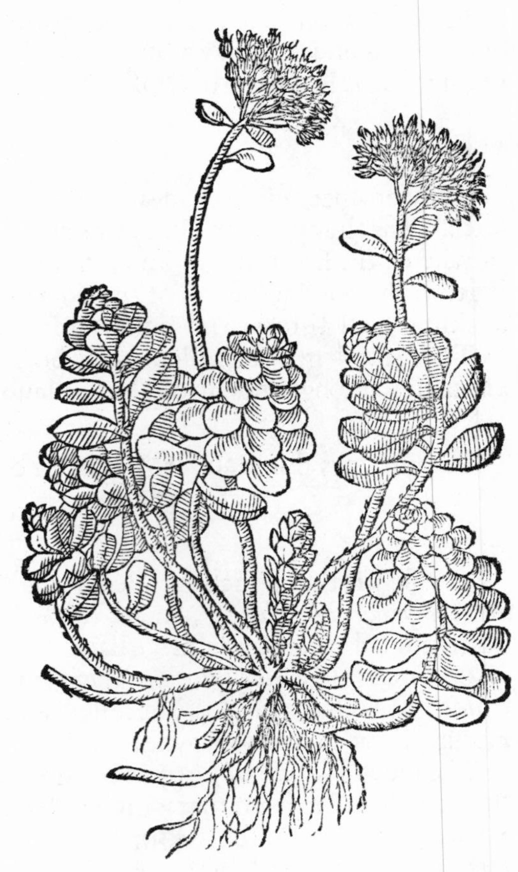

‡ 2 *Telephium legitimum Imperati.* Creeping Orpyn.

Telephium legitimum; and he hath thus giuen vs the hiſtory thereof: It produces from the top of the root many branches ſpred vpon the ground, which are about a foot long, ſet with many leaues, eſpecially ſuch as are not come to floure; for the other haue fewer: theſe leaues are ſmaller, leſſe thicke alſo and ſucculent than thoſe of the former kindes, neither are they ſo brittle: their colour is green, inclining a little to blew: the tops of the branches are plentifully ſtored with little floures growing thicke together, and compoſed of fiue little white leaues apiece: which fading, there ſucceed cornered ſeed-veſſels full of a browniſh ſeed. The root is ſometimes as thicke as ones little finger, tough, white, diuided into ſome branches, and liuing many yeares. ‡

¶ *The Place, Time, Names, Temperature, and Vertues.*

The firſt growes not in England. The ſecond flouriſhes in my garden. ‡ The third is a ſtranger with vs ‡. They floure when the common Orpyn doth. Their names are ſpecified in their ſeuerall deſcriptions: and their temperature and faculties in working are referred to the common Orpyn.

Chap.

CHAP. 148. *Of Purſlane.*

¶ *The Deſcription.*

1 THe ſtalkes of the great Purſlane be round, thicke, ſomewhat red, full of juice, ſmooth, glittering, and parted into certaine branches trailing vpon the ground: the leaues be an inch long, ſomething broad, thicke, fat, glib, ſomewhat greene, whiter on the neither ſide: the floures are little, of a faint yellow, and grow out at the bottome of the leaues. After them ſpringeth vp a little huske of a greene colour, of the bigneſſe almoſt of halfe a barly corne, in which is ſmall blacke ſeed: the root hath many ſtrings.

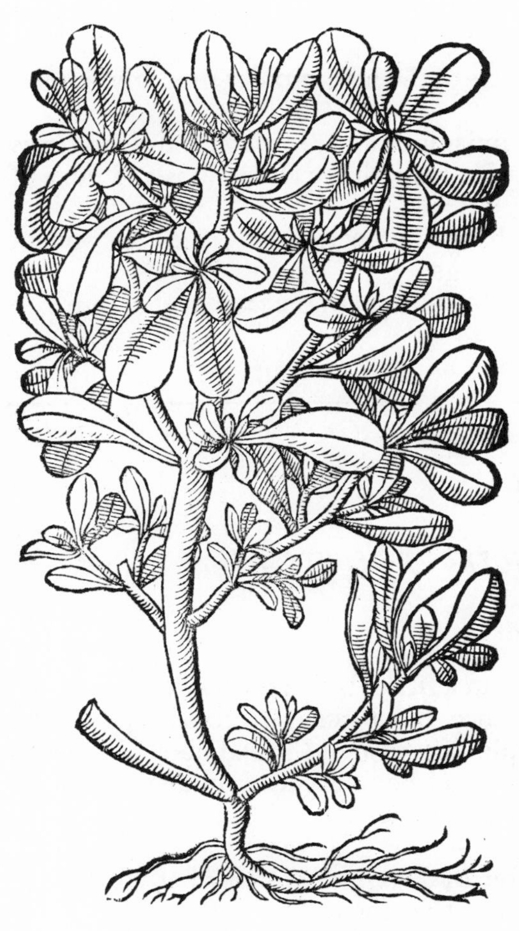

1 *Portulaca domeſtica.*
Garden Purſlane.

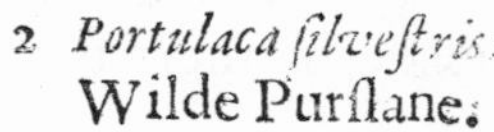

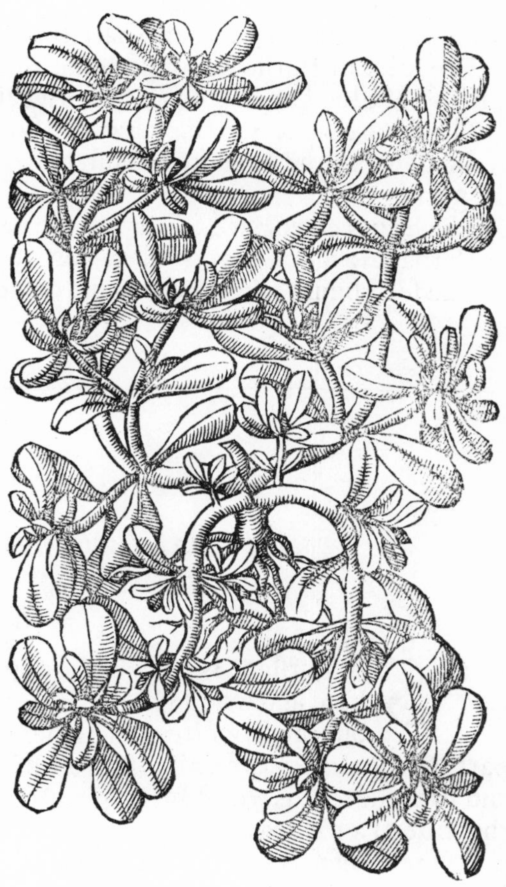

2 *Portulaca ſilveſtris.*
Wilde Purſlane.

2 The other is leſſer and hath like ſtalkes, but ſmaller, and it ſpreadeth on the ground: the leaues be like the former in faſhion, ſmoothneſſe, and thickneſſe, but farre leſſer.

¶ *The Place.*

The former is fitly ſowne in gardens, and in the waies and allies thereof being digged and dunged; it delighteth to grow in a fruitfull and fat ſoile not dry.

The other commeth vp of his owne accord in allies of gardens and vineyardes, and oftentimes vpon rocks: this alſo is delighted with watery places being once ſowne, if it be let alone till the ſeed be ripe it doth eaſily ſpring vp afreſh for certaine yeeres after.

¶ *The Time.*

It may be ſowne in March or Aprill; it flouriſheth and is greene in Iune, and afterwards euen vntill winter.

¶ *The Names.*

Purſlane is called in Greeke, ἀνδράχνη: in Latine, *Portulaca*: in high Dutch, **Burkelkraut**: in French, *Poupier*: in Italian, *Prochaccia*: in Spaniſh, *Verdolagas*: in Engliſh, Purſlane, and Porcelane.

¶ *The Temperature.*

Purslane is cold,and that in the third degree,and moist in the second : but wilde Purslane is not so moist.

¶ *The Vertues.*

A Rawe Purslane is much vsed in sallades,with oile, salt, and vineger : it cooleth an hot stomacke, and prouoketh appetite ; but the nourishment which commeth thereof is little, bad, cold, grosse, and moist : being chewed it is good for teeth that are set on edge or astonied ; the juice doth the same being held in the mouth,and also the distilled water.

B Purslane is likewise commended against wormes in young children, and is singular good,especially if they be feuerish withall, for it both allaies the ouermuch heate, and killeth the wormes : which thing is done through the saltnes mixed therewith,which is not only an enemy to wormes, but also to putrifaction.

C The leaues of Purslane either rawe, or boiled, and eaten as sallades, are good for those that haue great heate in their stomackes and inward parts, and doe coole and temper the inflamed bloud.

D The same taken in like manner is good for the bladder and kidnies, and allaieth the outragious lust of the body : the juice also hath the same vertue.

E The juice of Purslane stoppeth the bloudy fluxe, the fluxe of the hemorroides,monthly termes, spitting of bloud,and all other fluxes whatsoeuer.

F The same thrown vp with a mother syringe,cureth the inflammations,frettings,and vlcerations of the matrix ; and put into the fundament with a clister pipe,helpeth the vlcerations and fluxe of the guts.

G The leaues eaten rawe,take away the paine of the teeth, and fasteneth them ; and are good for teeth that are set on edge with eating of sharpe or soure things.

H The seed being taken, killeth and driueth forth wormes, and stoppeth the laske.

Chap. 149.

Of sea Purslane, and of the shrubby Sengreens.

¶ *The Description.*

1 Sea Purslane is not a herbe as garden Purslane,but a little shrub : the stalkes whereof be hard and woddy : the leaues fat,full of substance, like in forme to common Purslane, but much whiter and harder : the mossie purple floures stand round about the vpper parts of the stalkes,as do almost those of Blyte,or of Orach : neither is the seed vnlike,being broad and flat : the root is wooddy, long lasting,as is also the plant, which beareth out the winter with the losse of a few leaues.

† 2 There is another sea Purslane or *Halimus*, or after *Dodonæus*,*Portulaca marina,* which hath leaues like the former,but not altogether so white, yet are they somewhat longer and narrower, not much vnlike the leaues of the Oliue tree. The slender branches are not aboue a cubit or cubit and halfe long,and commonly lie spred vpon the ground,and the floures are of a deepe ouerworne herby colour, and after them follow seedes like those of the former,but smaller.

‡ 3 Our ordinary *Halimus* or sea Purslane hath small branches some foot or better long, lying commonly spred vpon the ground, of an ouerworne grayish colour,and sometimes purple;the leaues are like those of the last mentioned,but more fat and thicke, yet lesse hoary. The floures grow on the tops of the branches,of an herby purple colour,which is succeeded by small seeds like to that of the second kinde. ‡

4 There is found another wilde sea Purslane,whereof I haue thought good to make mention; which doth resemble the kindes of Aizoons. The first kinde groweth vpright,with a trunke like a small tree or shrub,hauing many vpright woody branches,of an ashe colour, with many thicke, darke,greene leaues like the small Stone crop,called *Vermicularis* : the floures are of an herby yellowish greene colour : the root is very hard and fibrous : the whole plant is of a salt tang taste, and the juice like that of Kaly.

5 There is another kinde like the former,and differeth in that,this strange plant is greater,the leaues more sharpe and narrower,and the whole plant more wooddy, and commeth neere to the forme of a tree. The floures are of a greenish colour.

¶ The

‡ 1 *Halimus latifolius.*
Tree Sea Purſlane.

‡ 2 *Halimus anguſtifolius procumbens.*
Creeping Sea Purſlane.

† 3 *Halimus vulgaris, ſiue Portulaca marina.*
Common Sea Purſlane.

‡ 4 *Vermicularis frutex minor.*
The leſſer ſhrubby Sengreen.

‡ 5 *Vermicularis frutex major.*
The greater Tree Stone-crop.

¶ *The Place.*

‡ The first and second grow vpon the Sea coasts of Spaine and other hot countries ‡ : and the third groweth in the salt marishes neere the sea side, as you passe ouer the Kings ferrey vnto the isle of Shepey, going to Sherland house (belonging sometime vnto the Lord *Cheiny*, and in the yeare 1590, vnto the Worshipfull S^r^. *Edward Hobby*) fast by the ditches sides of the same marish : it groweth plentifully in the isle of Thanet as you go from Margate to Sandwich, and in many other places along the coast. The other sorts grow vpon bankes and heapes of sand on the Sea coasts of Zeeland, Flanders, Holland, and in like places in other countries, as besides the Isle of Purbecke in England ; and on Rauen-spurne in Holdernesse, as I my selfe haue seene.

¶ *The Time.*

These flourish and floure especially in Iuly and August.

¶ *The Names.*

Sea Purslane is called *Portulaca Marina* : In Greeke, ἅλιμος : it is also called in Latine *Halimus* : in Dutch, **Zee Porceleijne** : in English, Sea Purslane.

The bastard ground Pines are called of some, *Chamepitys virmiculata* : in English, Sea ground Pine : ‡ or more fitly, Tree Ston-crop, or Pricket, or Shrubby Sengreene. ‡

¶ *The Temperature.*

Sea Purslane is (as *Galen* saith) of vnlike parts, but the greater part thereof is hot in a meane, with a moisture vnconcocted, and somewhat windie.

¶ *The Vertues.*

A The leaues (saith *Dioscorides*) are boyled to be eaten : a dram weight of the root being drunke with meade or honied water, is good against crampes and drawings awrie of sinewes, burstings, and gnawings of the belly : it also causeth Nurses to haue store of milke. The leaues be in the Low-countries preserued in salt or pickle as capers are, and be serued and eaten at mens tables in stead of them, and that without any mislike of taste, to which it is pleasant. *Galen* doth also report, that the yong and tender buds are wont in Cilicia to be eaten, and also laid vp in store for vse.

B ‡ *Clusius* saith, That the learned Portugal Knight *Damianus a Goes* assured him, That the leaues of the first described boyled with bran, and so applied, mitigate the paines of the Gout proceeding of an hot cause. ‡

† The figure that was formerly giuen by our Author by the title of *Portulaca marina*, and is set forth by *Tabern.* vnder the same name, is either of none of these plants, or else it is vnperfect. *Bauhine* knowes not what to make of it, but questions, *Quid sit?*

Chap. 150. *Of Herbe-Iuy, or Ground-Pine.*

¶ *The Description.*

1 THe common kinde of *Chamæpitys* or Ground-Pine is a small herbe and very tender, creeping vpon the ground, hauing small and crooked branches trailing about. The leaues be small, narrow and hairy, in sauour like the Firre or Pine tree ; but if my sence of smelling be perfect, me thinkes it is rather like vnto the smell of hempe. The floures be little, of a pale yellow colour, and somtimes white : the root is small and single, and of a wooddy substance.

† 2 The second hath pretty strong foure square ioynted stalkes, browne and hairy ; from which grow pretty large hairy leaues much clouen or cut : the floures are of a purple colour, and grow about the stalks in roundles like the dead Nettle : the seed is black and round, and the whole plant sauoureth like the former : ‡ which sheweth this to be fitly referred to the *Chamæpytis*, and not to be well called *Chamædrys fœmina*, or Iagged Germander, as some haue named it. ‡

3 This

1 *Chamæpitys mas.*
The male ground Pine.

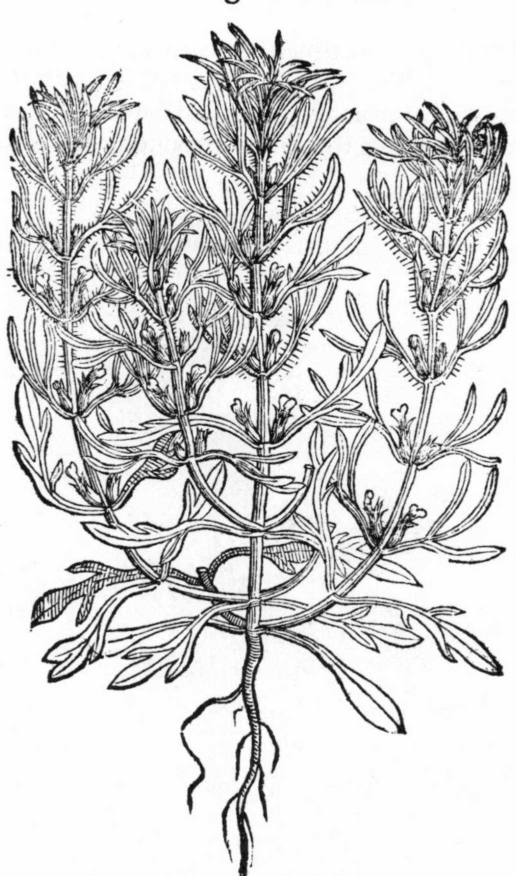

2 *Chamæpitys fœmina.*
The female ground-Pine.

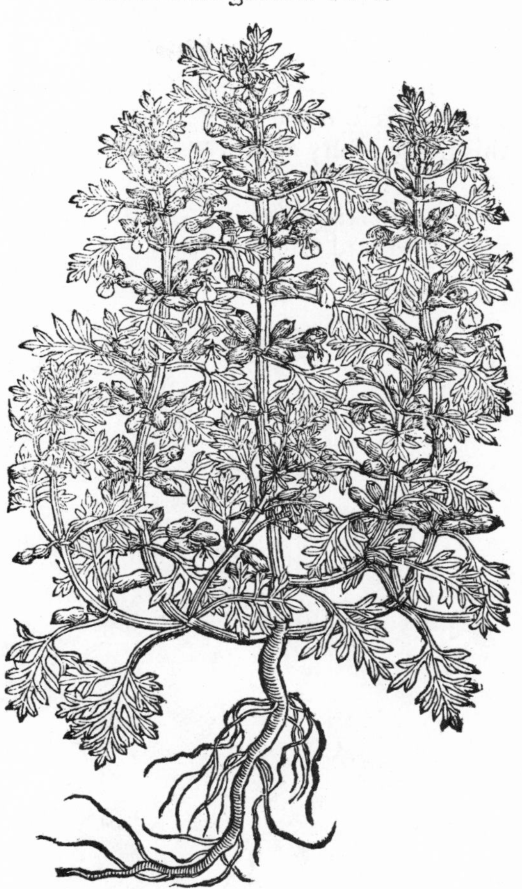

3 *Chamæpitys 3. Dodon.*
Small Ground-Pine.

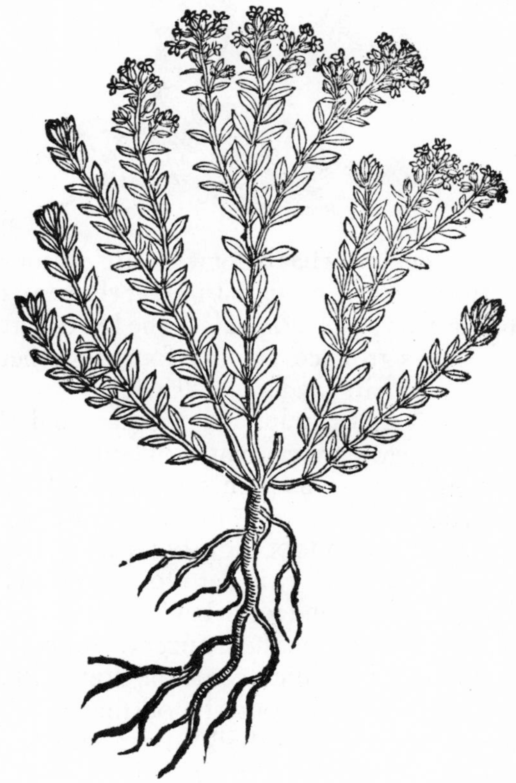

4 *Iua muscata Monspeliaca.*
French Herbe-Iuy or Ground-Pine.

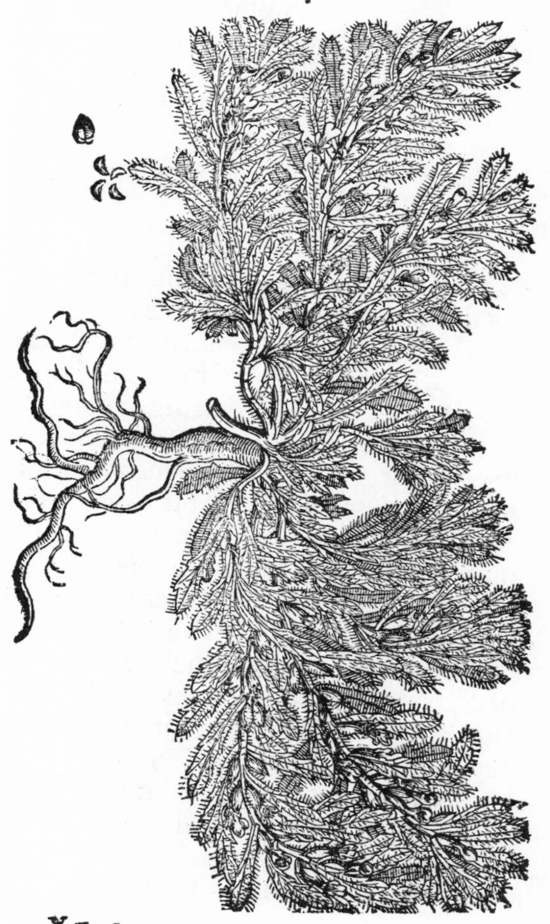

3 This kinde of Herb-Iuy, growing for the most part about Montpelier in France, is the least of all his kind, hauing smal white and yellow floures, in smell and proportion like vnto the others, but much smaller.

† 4 There is a wilde or bastard kinde of *Chamæpitys*, or ground-Pine, that hath leaues somewhat like vnto the second kinde, but not iagged in that manner, but onely snipt about the edges. The root is somewhat bigger, wooddy, whitish, and bitter, and like vnto the root of Succorie. All this herbe is very rough, and hath a strong vnpleasant smell, not like that of the ground-Pines.

‡ 5 *Chamæpitys spuria altera Dodon.*
Bastard Ground-Pine.

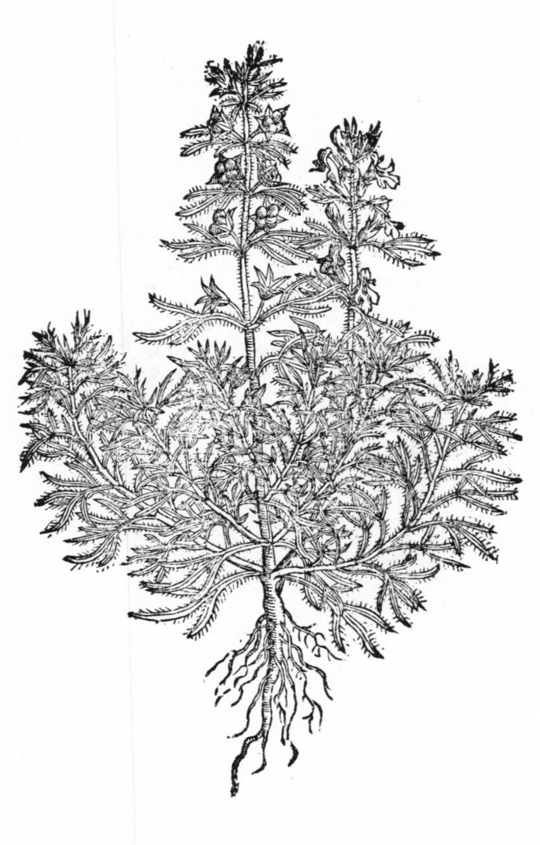

‡ 6 *Chamæpitys Austriaca.*
Austrian Ground-Pine.

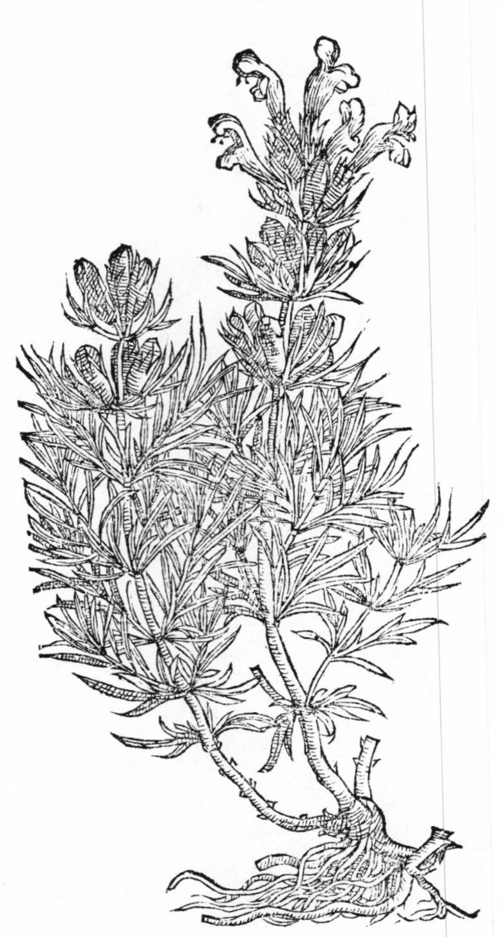

† 5 There is another kind that hath many small and tender branches beset with little leaues for the most part three together, almost like the leaues of the ordinarie ground-Pine: at the top of which branches grow slender white floures; which being turned vpside downe, or the lower part vpward, do somewhat resemble the floures of *Lamium*: the seeds grow commonly foure together in a cup, and are somewhat big and round: the root is thicke, whitish, and long lasting.

6 There groweth in Austria a kinde of *Chamæpitys*, which is a most braue and rare plant, and of great beautie, yet not once remembred either of the ancient or new Writers, vntill of late that famous *Carolus Clusius* had set it forth in his Pannonicke Obseruations; who for his singular skil and industrie hath woon the garland from all that haue written before his time. This rare and strange plant I haue in my garden, growing with many square stalkes of halfe a foot high, beset euen from the bottome to the top with leaues so like our common Rosemary, that it is hard for him which doth not know it exactly to finde the difference; being greene aboue, and somwhat hairy and hoarie vnderneath: among which come forth round about the stalkes (after the manner of roundles or coronets) certain small cups or chalices of a reddish colour; out of which come the floures like vnto Archangell in shape, but of a most excellent and stately mixed colour, the outside purple declining to blewnesse, and sometimes of a violet colour. The floure gapeth like the mouth of a beast, and hath as it were a white tongue; the lower and vpper iawes are white likewise, spotted with many bloudy spots: which being past, the seeds appeare very long, of a shining blacke colour, set in order in the small huskes as the *Chamæpitys spuria*. The root is blacke and hard, with manie hairy strings fastned thereto.

¶ The

¶ *The Place.*

These kindes of *Chamæpitys* (except the two last) grow very plentifully in Kent, especially about Grauesend, Cobham, Southfleet, Horton, Dartford, and Sutton, and not in any other shire in England that euer I could finde.

‡ None of these, except the first, for any thing I know, or can learne, grow wilde in England; the second I haue often seene in Gardens. ‡

¶ *The Time.*

They floure in Iune, and often in August.

¶ *The Names.*

Ground Pine is called in Greeke χαμαίπιτυς: in Latine, *Ibiga*, *Aiuga*, and *Abiga*: in shops, *Iua Arthritica* and *Iua moschata*: in Italian, *Iua*: in Spanish, *Chamæpiteos*: in High Dutch, **Vergiss mich nicht**: in low Dutch, **Veldt Cipres**: in French, *Iue moschate*: In English, Herbe Iuie, Forget me not, Ground Pine, and field Cypresse.

‡ 1 The first of these is the *Chamæpitys prima*, of *Matthiolus*, *Dodonæus* and others, and is that which is commonly vsed in shops and in Physicke.

2 This *Matthiolus* cals *Chamædrys altera*: *Lobel*, *Chamædrys Laciniatis folijs*: *Lonicerus*, *Trixago vera*; *Tabernamontanus*, *Iva moschata*; and *Dodon.* (whom in this Chapter we chiefely follow) *Chamæpitys altera.*

3 Thirdly, this is the *Chamæpitys* 1. of *Fuchsius* and others; the *Chamæpitys* 1. *Dioscoridis odoratior* of *Lobel*; and the *Chamæpitys* 3. of *Matthiolus* and *Dodon.*

4 *Gesner* cals this *Chamæpitys species Monspellij*: *Clusius*, *Dodon.* *Anthyllis altera*; and *Lobel*, *Anthyllis Chamepityides minor*; and *Tabern.* *Iua Moschata Monspeliensium.*

5 This is *Chamæpitys adulterina* of *Lobel*: *Pseudochamæpitys* and *Aiuga adulterina* of *Clusius*: and *Chamæpitys spuria altera* of *Dodon.*

6 This is *Chamæpitys Austriaca* of *Clusius*; and *Chamæpitys cærulea* of *Camerarius*. ‡

¶ *The Nature.*

These herbes are hot in the second degree, and drie in the third.

¶ *The Vertues.*

The leaues of *Chamæpytis* tunned vp in Ale, or infused in wine, or sodden with hony, and drunke by the space of eight or ten daies, cureth the iaundies, the Sciatica, the stoppings of the liuer, the difficultie of making water, the stoppings of the spleene, and causeth women to haue their natural sicknesse. A

Chamæpytis stamped greene with honie cureth wounds, malignant and rebellious vlcers, and dissolueth the hardnesse of womens brests or paps, and profitably helpeth against poison, or biting of any venomous beast. B

The decoction drunke, dissolueth congealed bloud, and drunke with vineger, driueth forth the dead childe. C

It clenseth the intrals: it helpeth the infirmities of the liuer and kidneies; it cureth the yellow iaundies being drunke in wine: it bringeth downe the desired sicknesse, and prouoketh vrine: being boiled in Mead or honied water and drunke, it helpeth the Sciatica in fortie daies. The people of Heraclea in Pontus do vse it against Wolfes bane in stead of a counterpoison. D

The pouder hereof taken in pils with a fig, mollifieth the bellie: it wasteth away the hardnesse of the paps: it healeth wounds, it cureth putrified vlcers being applied with hony: and these things the first ground Pine doth performe, so doth the other two: but not so effectually, as witnesseth *Dioscorides.* E

Clusius of whom mention was made, hath not said any thing of the Vertues of *Chamæpytis Austriaca*: but verily I thinke it better by many degrees for the purposes aforesaid: my coniecture I take from the taste, smell, and comely proportion of this Hearbe, which is more pleasing and familiar vnto the nature of man, than those which wee haue plentifully in our owne Countrey growing. F

CHAP. 152. *Of Nauelwoort, or Penniwoort of the Wall.*

¶ *The Description.*

1 THe great Nauelwoort hath round and thicke leaues, somewhat bluntly indented about the edges, and somewhat hollow in the midst on the vpper part, hauing a short tender stem

ſtemme faſtened to the middeſt of the leafe, on the lower ſide vnderneath the ſtalke whereon the floures doe grow,is ſmall and hollow,an handfull high and more,beſet with many ſmall floures of an ouerworne incarnate colour. The root is round like an oliue,of a white colour.

‡ The root is not well expreſt in the figure,for it ſhould haue been more vnequall or tuberous, with the fibers not at the bottome but top thereof. ‡

2 The ſecond kinde of Wall Penniwoort or Nauelwoort hath broad thicke leaues ſomewhat deepely indented about the edges : and are not ſo round as the leaues of the former, but ſomewhat long towards the ſetting on,ſpred vpon the ground in manner of a tuft,ſet about the tender ſtalke, like to Sengreene or Houſleeke : among which riſeth vp a tender ſtalke whereon do grow the like leaues. The floures ſtand on the top conſiſting of fiue ſmall leaues of a white colour, with red ſpots in them. The root is ſmall and threddie. ‡ This by ſome is called *Sedum Serratum*. ‡

‡ 3 This third kinde hath long thicke narrow leaues, very finely ſnipt or nickt on the edges, which lie ſpred very orderly vpon the ground; and in the midſt of them riſes vp a ſtalke ſome foot high, which beares at the top thereof vpon three or foure little branches,diuers white floures conſiſting of fiue leaues apiece.

4 The leaues of this are long and thicke,yet not ſo finely ſnipt about the edges,nor ſo narrow as thoſe of the former : the ſtalke is a foot high, ſet here and there with ſomewhat ſhorter and rounder leaues than thoſe below ; and towards the top thereof,out of the boſſomes of theſe leaues come ſundry little foot-ſtalkes,bearing on their tops pretty large floures of colour white,and ſpotted with red ſpots. The rootes are ſmall, and here and there put vp new tufts of leaues,like as the common Houſleeke. ‡

5 There is a kinde of Nauelwoort that groweth in waterie places,which is called of the huſbandmen Sheeps bane,becauſe it killeth ſheepe that do eat thereof : it is not much vnlike the precedent,but the round edges of the leaues are not ſo euen as the other ; and this creepeth vpon the ground,and the other vpon the ſtone walls.

1 *Vmbilicus Veneris.*
Wall Penniwoort.

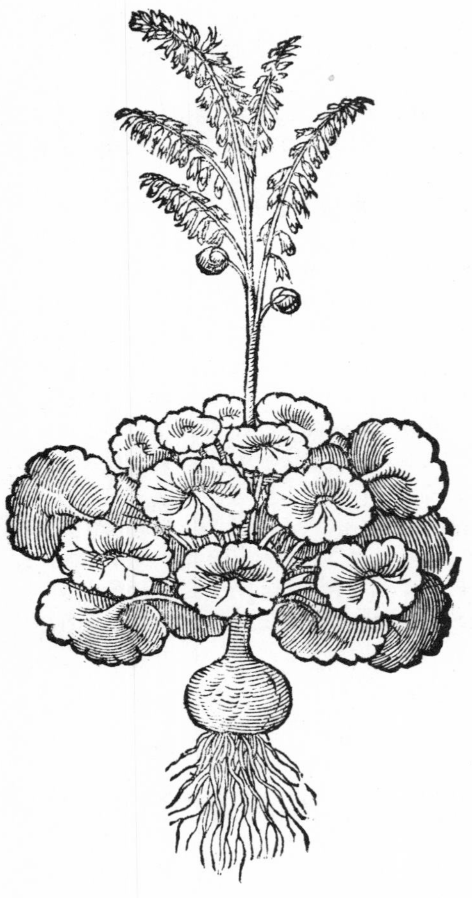

‡ 2 *Vmbilicus Ven.ſive Cotyledon altera.*
Iagged or Roſe Penniwoort.

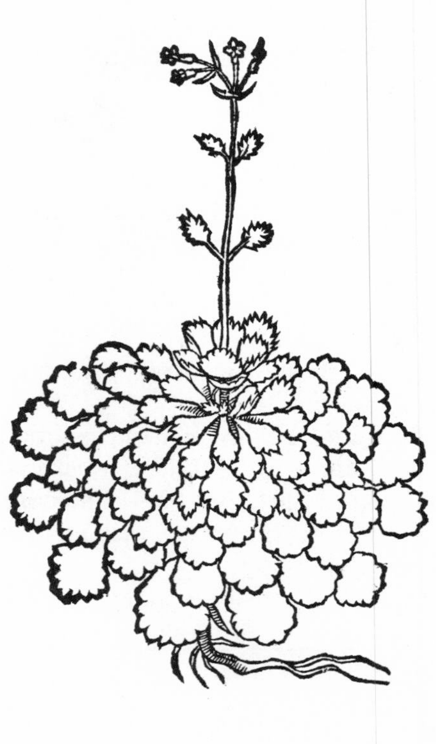

‡ 6 Becauſe ſome in Italy haue vſed this for *Vmbilicus Veneris*,and otherſome haue ſo called it,I thought it not amiſſe to follow *Matthiolus*,and giue you the hiſtory thereof in this place,rather than to omit it, or giue it in another which may be perhaps as vnfit, for indeed I cannot fitly ranke it

it with any other plant. *Bauhine* ſets it betweene *Hedera Terreſtris,* and *Naſturtium Indicum* : and *Columna* refers it to the *Linaria's*, but I muſt confeſſe I cannot referre it to any ; wherefore I thinke it as proper to giue it here as in any other place. The branches of this are many, long, ſlender, and creeping, vpon which grow without any certaine order many little ſmooth thicke leaues faſhioned like thoſe of Ivie, and faſtened to ſtalkes of ſome inch long : and together with theſe ſtalkes come forth others of the ſame length, that carry ſpur-faſhioned floures, of the ſhape and bigneſſe of thoſe of the female Fluellen : their outſide is purple, their inſide blew, with a ſpot of yellow in the opening. The root is ſmall, creeping, and threddie. It floures toward the end of Sommer, and growes wilde vpon walls in Italie, but in gardens with vs. *Matthiolus* calls it *Cymbalaria* (to which *Lobel* addes) *Italica Hederaceo folio* : *Lonicerus* termes it *Vmbilicus Veneris Officinarum* : and laſtly *Columna* cals it *Linaria hederæ folio.* ‡

¶ *The Place.*

The firſt kind of Penniwoort groweth plentifully in Northampton vpon euery ſtone wall about the towne, at Briſtow, Bathe, Wells, and moſt places of the Weſt Countrie vpon ſtone walls. It groweth vpon Weſtminſter Abbey, ouer the doore that leadeth from *Chaucers* tombe to the old palace. ‡ In this laſt place it is not now to be found. ‡

The ſecond, third, and fourth grow vpon the Alpes neere Piedmont, and Bauier, and vpon the mountaines of Germanie : I found the third growing vpon Bieſton Caſtle in Cheſhire.

‡ The fifth growes vpon the Bogges vpon Hampſtead Heath, and many ſuch rotten grounds in other places. ‡

¶ *The Time.*

They are greene and flouriſh eſpecially in VVinter : They floure alſo in the beginning of Sommer.

¶ *The Names.*

Nauelwoort is called in Greeke κοτυληδὼν : in Latine, *Vmbilicus Veneris*, and *Acetabulum* : of diuers, *Herba Coxendicum* · *Iacobus Manlius* nameth it *Scatum Cœli*, and *Scatellum* : in Dutch, **Nauelcruyt** : in Italian, *Cupertoiule* : in French, *Eſcuelles* : in Spaniſh, *Capadella* : of ſome, *Hortus Veneris*, or Venus garden, and *Terræ vmbilicus*, or the Nauel of the earth : in Engliſh, Penniwoort, Wall-penniwoort, Ladies nauell, Hipwoort and Kidney-woort.

VVater Penniwoort is called in Latine *Cotyledon paluſtris*. in Engliſh, Sheepe-killing Pennigraſſe, Penny-rot, and in the North Countrey VVhite-rot : for there is alſo Red-rot, which is *Roſa ſolis* : in Northfolke it is called Flowkwoort. ‡ *Columna* and *Bauhine* fitly refer this to the *Ranunculi*, or Crowfeet ; for it hath no affinitie at all with the Cotyledons (but onely in the roundneſſe of the leafe) the former of them cals it *Ranunculus aquaticus vmbilicato folio*, and the later, *Ranunculus aquat. Cotyledonis folio.*

¶ *The Temperature.*

Nauelwoort is of a moiſt ſubſtance and ſomewhat cold, and of a certaine obſcure binding qualitie : it cooleth, repelleth, or drieth backe, ſcoureth, and conſumeth, or waſteth away, as *Galen* teſtifieth.

‡ The VVater Pennywoort is of an hot and vlcerating qualitie, like to the Crowfeet, whereof it is a kinde. The baſtard Italian Nauelwoort ſeemes to partake with the true in cold and moiſture. ‡

¶ *The Vertues.*

The iuice of VVall Pennywoort is a ſingular remedie againſt all inflammations and hot tumors, A
as Eryſipelas, Saint Anthonies fire, and ſuch like : and is good for kibed heeles, being bathed therwith, and one or more of the leaues laid vpon the heele.

The leaues and rootes eaten doe breake the ſtone, prouoke vrine, and preuaile much againſt the B
dropſie.

The ignorant Apothecaries doe vſe the VVater Pennywoort in ſtead of this of the wall, which C
they cannot doe without great error, and much danger to the patient : for husbandmen know well, that it is noiſome vnto Sheepe, and other cattell that feed thereon, and for the moſt part bringeth death vnto them, much more to men by a ſtronger reaſon.

3 *Vmbilicus Veneris minor.*
Small Nauelwoort.

‡ 4 *Cotyledon minor montana altera.*
The other ſmall mountaine Nauelwoort.

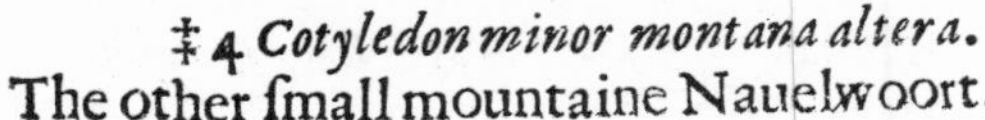

5 *Cotyledon paluſtris.*
Water Penniwoort.

‡ 6 *Cymbalaria Italica.*
Italian Baſtard Nauelwoort.

CHAP. 152. *Of Sea Pennywoort.*

1 *Androſace Matthioli.*
Sea Nauel-woort.

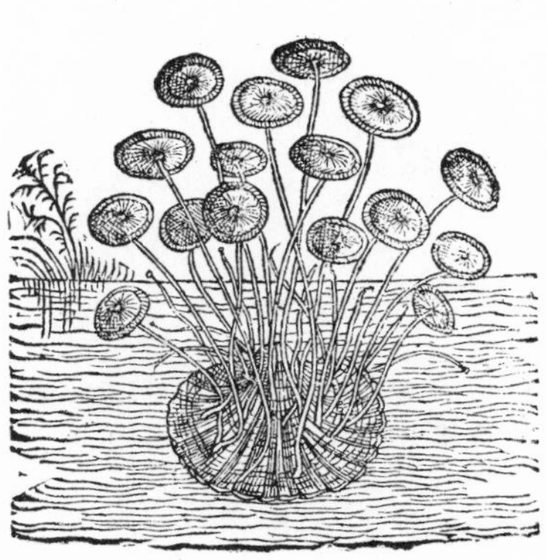

2 *Androſace annua ſpuria.*
One Sommers Nauell-woort.

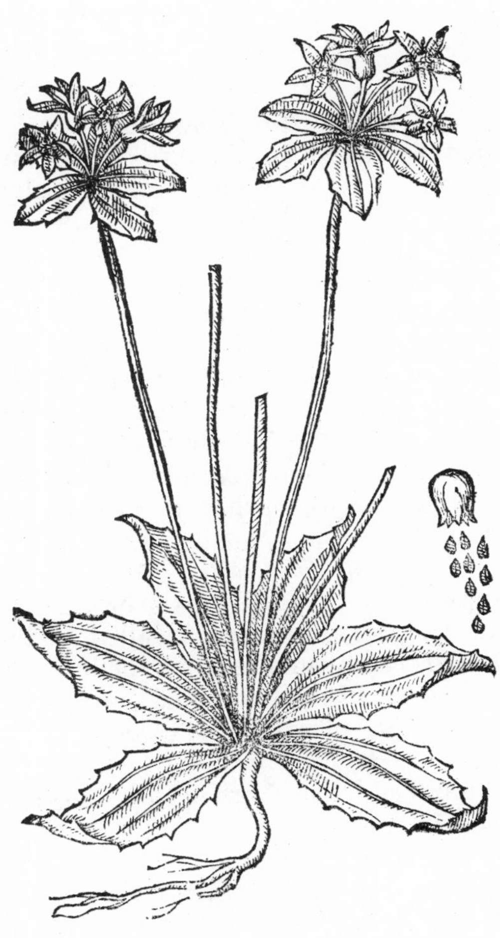

¶ *The Deſcription.*

1 THe Sea Nauel-woort hath many round thicke leaues like vnto little ſaucers, ſet vpon ſmall & tender ſtalks, bright, ſhining, and ſmooth, of two inches long, for the moſt part growing vpon the furrowed ſhels of cockles or the like, euery ſmall ſtem bearing vpon the end or point, one little buckler and no more, reſembling a nauell; the ſtalke and leafe ſet together in the middle of the ſame. Whereupon the Herbariſts of Montpelier haue called it *Vmbilicus Marinus*, or ſea Nauel. The leaues and ſtalkes of this plant, whileſt they are yet in the water, are of a pale aſh colour, but being taken forth, they preſently waxe white, as Sea Moſſe, called *Corallina*, or the ſhel of a Cockle. It is thought to be barren of ſeed, and is in taſte ſaltiſh.

2 The ſecond *Androſace* hath little ſmooth leaues, ſpred vpon the ground like vnto the leaues of ſmall Chickweed or Henbit, whereof doubtles it is a kind: among which riſeth vp a ſlender ſtem, hauing at the top certaine little chaffie floures of a purpliſh colour. The ſeed is contained in ſmall ſcaly husks, of a reddiſh colour, & a bitter taſte. The whole plant periſheth when it hath perfected his ſeede, and muſt be ſowne againe the next yeare: which plant was giuen to *Mathiolus* by *Cortuſus*, who (as he affirmeth) receiued it from Syria; but I thinke hee ſaid ſo to make *Matthiolus* more ioyfull: but ſurely I ſurmiſe he picked it out of one old wal or other, where it doth grow euen as the ſmall Chickweed, or Nailewoort of the wall do.

‡ The figure that was here was that vnperfect one of *Matthiolus*; and the deſcription of our Authour was framed by it, vnleſſe the laſt part therof, which was taken out of the *Aduerſaria pag. 166.* to amend both theſe, we here preſent you with the true figure and deſcription, taken out of the workes of the iudicious and painfull Herbariſt *Carolus Cluſius*. It hath (ſaith he) many leaues lying flat vpon the ground like to thoſe of Plantaine, but leſſer and of a pale greene colour, and toothed about the edges, ſoft alſo and iuicie, and of ſomewhat a biting taſte. Amongſt theſe leaues riſe vp fiue or ſix ſtalkes of an handfull high, commonly of a green, yet ſometimes of a purple colour, naked and ſomewhat hairy, which at their tops carry in a circle fiue roundiſh leaues alſo a little toothed and hairy; from the midſt of which ariſe fiue or more footſtalks, each bearing a greeniſh rough or hairie, cup & parted alſo into fiue little leaues or iags, in the middeſt

midſt of which ſtands a little white floure parted alſo into fiue; after which ſucceed pretty large ſeed veſſels which containe an vnequall red ſeed like that of Primroſes, but bigger : the root is ſingle and ſlender, and dies as ſoon as the ſeed is perfected. It growes naturally in diuers places of Auſtria, and amongſt the corne about the Bathes of Baden ; whereas it floures in Aprill, and ripens the ſeed in May and Iune. ‡

¶ *The Place.*

Androſace will not grow any where but in water: great ſtore of it is about Frontignan by Montpellier in Languedoc, where euery fiſher-man doth know it.

The ſecond groweth vpon old ſtone and mud walls : notwithſtanding I haue (the more to grace *Matthiolus* great iewell) planted it in my garden.

¶ *The Time.*

The baſtard *Androſace* floureth in Iuly, and the ſeed is ripe in Auguſt.

¶ *The Names.*

Androſace is of ſome called *Vmbilicus marinus*, or ſea Nauell.

‡ The ſecond is knowne and called by the name of *Androſace altera Matthioli*. ‡

¶ *The Temperature.*

The ſea Nauell is of a diureticke qualitie, and more drie than *Galen* thought it to be, and leſſe hot than others haue deemed it : there can no moiſture be found in it.

¶ *The Vertues.*

A Sea Nauelwoort prouoketh vrine, and digeſteth the filthineſſe and ſlimineſſe gathered in the ioints.

B Two drams of it, as *Dioſcorides* ſaith, drunke in wine, bringeth downe great ſtore of vrine out of their bodies that haue the dropſie, and maketh a good plaiſter to ceaſe the paine of the gout.

CHAP. 153. *Of Roſe-woort, or Roſeroot.*

Rhodia radix, Roſe-root.

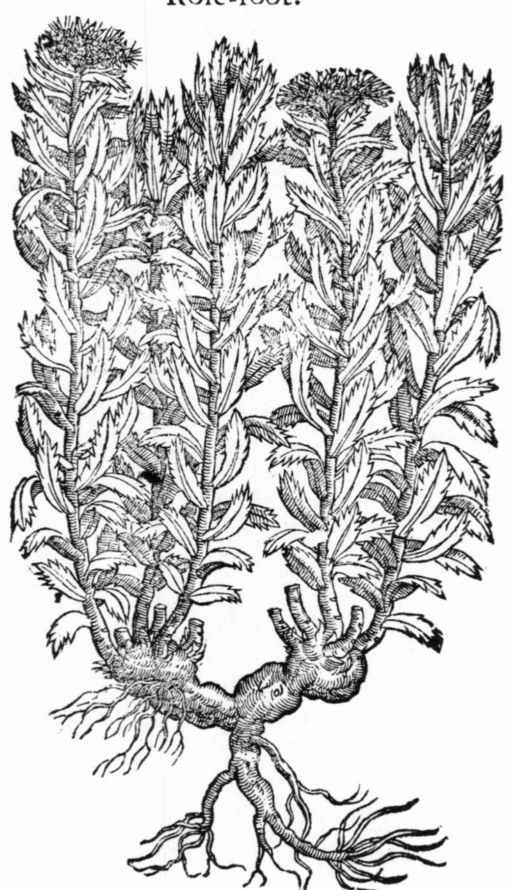

¶ *The Deſcription.*

Roſewoort hath many ſmall, thicke, and fat ſtems, growing from a thicke and knobby root: the vpper end of it for the moſt part ſtandeth out of the ground, and is there of a purpliſh colour, bunched & knobbed like the root of Orpin, with many hairy ſtrings hanging therat, of a pleaſant ſmell when it is broken, like the damaske roſe, whereof it tooke his name. The leaues are ſet round about the ſtalks, euen from the bottome to the top, like thoſe of the field Orpin, but narrower, and more ſnipt about the edges. The floures grow at the top of a faint yellow colour.

¶ *The Place.*

It groweth very plentifully in the North part of England, eſpecially in a place called Ingleborough Fels, neere vnto the brookes ſides, and not elſewhere that I can as yet finde out, from whence I haue had plants for my garden.

¶ *The Time.*

It floureth and flouriſheth in Iuly, and the ſeed is ripe in Auguſt.

¶ *The Names.*

Some haue thought it hath taken the name *Rhodia* of the Iſland in the Mediterranean ſea, called Rhodes : but doubtleſſe it took his name *Rhodia radix*, of the root which ſmelleth like a roſe: in Engliſh, Roſe-root, and Roſe-woort.

¶ *The Vertues.*

There is little extant in writing of the faculties of Roſewoort : but this I haue found, that if the root be ſtamped with oile of Roſes and laid to the temples of the head, it eaſeth the paine of the head.

Chap. 144. *Of Sampier.*

1 *Crithmum marinum.*
Rocke Sampier.

2 *Crithmum Spinosum.*
Thornie Sampier.

3 *Crithmum chrysanthemum.*
Golden Sampier.

¶ *The Description.*

1 Rocke Sampier hath many fat and thicke leaues, somewhat like those of the lesser Purslane, of a spicy taste with a certaine saltnesse; amongst which riseth vp a stalke, diuided into many smal spraies or sprigs; on the top wherof doe grow spokie tufts of white floures, like the tufts of Fenell or Dill; after that commeth the seed like the seed of Fenell, but greater. The root is thicke and knobbie, being of smell delightfull & pleasant

2 The second Sampier called *Pastinaca marina*, or Sea Parsnep, hath long fat leaues, very much iagged or cut euen to the middle rib, sharpe or prickley pointed, which are set vpon large fat iointed stalks; on the top whereof do grow tuftes of whitish, or els reddish floures. The seed is wrapped in thornie huskes. The root is thick and long, not vnlike to the Parsnep, very good and wholesome to be eaten.

3 Golden Sampier bringeth forth many stalks from one root, compassed about with a multitude of long fat leaues, set together by equall distances; at the top whereof come yellow floures. The seed is like those of the Rocke Sampier.

¶ *The Place.*

Rocke Sampier groweth on the rockie cliffes at Douer, VVinchelsey, by Rie, about Southhampton, the Isle of VVight, and most rockes about the West and North-west parts about England.

The second groweth neere the Sea vpon the sands, and Bayche between Whitstable and the Isle of Thanet, by Sandwich, and by the sea neere Westchester.

The third groweth in the myrie marsh in the Isle of Shepey, as you go from the Kings Ferrie, to Sherland house.

¶ *The Time.*

Rocke Sampier flourisheth in May and Iune, and must be gathered to be kept in pickle in the beginning of August.

¶ *The Names.*

Rocke Sampier is called in Greeke κρίθμον: in Latine, *Crithmum*: and of diuers, *Bati* in some shops, *Creta marina*: of *Petrus Crescentius*, *Cretamum*, and *Rincum marinum*: in high Dutch, **Meerfenchel**: which is in Latine, *Fœniculum marinum*, or Sea Fenell: in Italian, *Fenocchio marino*, *Herba di San Pietro*; and hereupon diuers name it *Sampetra*: in Spanish, *Perexil de la mer*, *Hinoio marino*, *Fenolmarin*: in English, Sampier, and Rocke Sampier, and of some, Crestmarine; and these bee the names of the Sampier generally eaten in sallads.

The other two be also *Crithma* or Sampiers, but most of the later writers would draw them to some other plant: for one calleth the second *Pastinaca marina*, or sea Parsnep, and the third *Aster atticus marinus*; and *Lobel* names it *Chrysanthemum Littoreum*: but we had rather entertaine them as *Matthiolus* doth, among the kindes of *Crithmum*, or Sampier.

¶ *The Temperature.*

Sampier doth drie, warme, and scoure, as *Galen* saith.

¶ *The Vertues.*

A The leaues, seeds, and roots, as *Dioscorides* saith, boiled in wine and drunke, prouoke vrine, and womens sicknesse, and preuaile against the iaundies.

B The leaues kept in pickle, and eaten in sallads with oile and vineger, is a pleasant sauce for meat, wholesome for the stoppings of the liuer, milt, kidneies and bladder: it prouoketh vrine gently; it openeth the stoppings of the intrals, and stirreth vp an appetite to meat.

C It is the pleasantest sauce, most familiar, and best agreeing with mans body, both for digestion of meats, breaking of the stone, and voiding of grauell in the reines and bladder.

Chap. 155. *Of Glasse Saltwoort.*

¶ *The Description.*

1 Glassewoort hath many grosse, thicke and round stalkes a foot high, full of fat and thicke sprigges, set with many knots or ioints, without any leaues at all, of a reddish greene colour. The whole Plant resembleth a branch of Corall. The root is very small and single.

2 There is another kinde of Saltwoort, which hath been taken among the antient Herbarists for a kinde of Sampier. It hath a little tender stalke a cubite high, diuided into many small branches, set full of little thicke leaues very narrow, somewhat long and sharpe pointed, yet not pricking; amongst which commeth forth small seed, wrapped in a crooked huske, turned round like a crooked perwinkle. The stalkes are of a reddish colour. The whole plant is of a salt and biting taste. The root is small and threddie.

† 3 There is likewise another kinde of *Kali*, whereof *Lobel* maketh mention vnder the name of *Kali minus*, which is like to the last before remembred, but altogether lesser, ‡ hauing many slender weake branches lying commonly spred vpon the ground, and set with many small round long sharpe pointed leaues, of a whitish green colour: the seed is small and shining, not much vnlike that of Sorrell: the root is slender with many fibers; the whole plant hath a saltish taste like as the former. *Dodon.* cals this *Kali album.* ‡

¶ The

1 *Salicornia, siue Kali geniculatum.*
Glassewoort, Saltwoort, or Sea-grape.

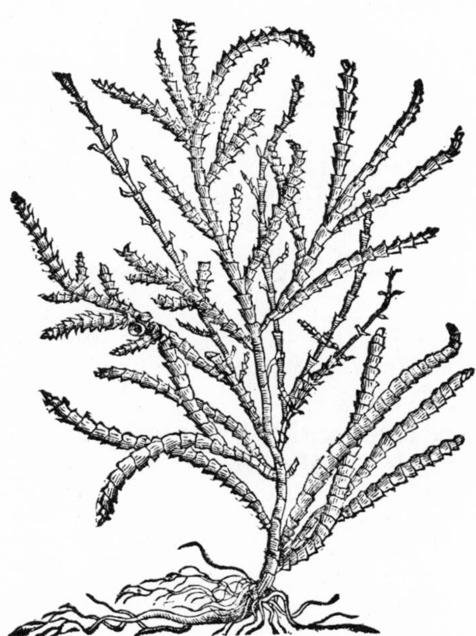

‡ 2 *Kali maius semine cochleato.*
Snaile Glassewoort.

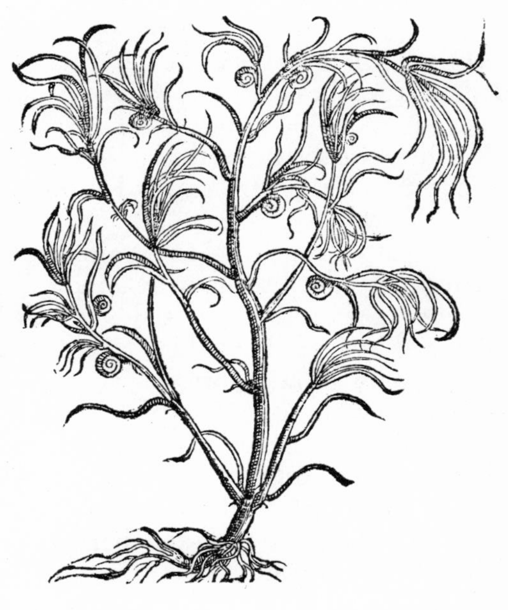

‡ 3 *Kali minus.*
Small Glassewoort,

¶ *The Place.*

These plants are to be found in salt marshes almost euery where.

‡ The second excepted, which growes not here, but vpon the coasts of the Mediterranean sea. ‡

¶ *The Time.*

They floure and flourish in the Sommer moneths.

¶ *The Names.*

Saltwoort is called of the Arabians *Kali*, and *Alkali*. *Auicen*, chap. 724. describeth them vnder the name of *Vsnen*, which differeth from *Vsnee*: for *Vsnee* is that which the Græcians call βρύον: and the Latines, *Muscus*, or Mosse of some, as *Baptista Montanus*: it hath bin iudged to be *Empetron*.

The axen or ashes hereof are named of *Matthæus Siluaticus*, *Soda*: of most, *Sal Alkali*: diuers call it *Alumen catinum*. Others make this kind of difference betweene *Sal Kali*, and *Alumen catinum*, that *Alumen catinum* is the ashes it self: and that the salt that is made of the ashes is *Sal Alkali*.

Stones are beaten to pouder, & mixed with ashes, which beeing melted together become the matter wherof glasses are made. VVhich while it is made red hot in the furnace, and is melted, becomming liquide and fit to work vpon, doth yeeld as it were a fat floting aloft; which, when it is cold, waxeth as hard as a stone, yet it is brittle, and quickely broken. This is commonly called *Axungia vitri*. In English, Sandeuer: in French, *Suin de Voirre*: in Italian, *Fior de cristalo*, (i) Floure of Christall. The Herbe is also called of of diuers *Kali articulatum*, or iointed Glasse-woort: and in English, Crabbe-grasse, and Frogge-grasse.

¶ *The Temperature.*

Glasse-woort is hot and drie: the ashes are both drier and hotter, and that euen to the fourth degrée: the ashes haue a causticke or burning qualitie.

¶ *The Vertues.*

A A little quantitie of the herbe taken inwardly, doth not onely mightily prouoke vrine, but in like sort casteth forth the dead childe. It draweth forth by siege waterish humours, and purgeth away the dropsie.

B A great quantitie taken is mischieuous and deadly. The smell and smoke also of this hearbe being burnt doth driue away serpents.

C The ashes are likewise tempered with those medicines that serue to take away scabs and filth off the skin: it easily consumeth proud and superfluous flesh that groweth in poisonsome vlcers, as *Auicen* and *Serapio* report.

D Wee read in the copies of *Serapio*, that Glasse-woort is a tree so great, that a man may stand vnder the shadow thereof. but it is very like, that this errour proceedeth rather from the interpreter, than from the Authour himselfe.

E The floure of Christall, or (as they commonly terme it, Sandeuer) doth wonderfully drie. It easily taketh away scabbes and manginesse, if the foule parts be washed and bathed with the water wherein it is boiled.

Chap. 156. *Of Thorow Waxe.*

1 *Perfoliata vulgaris.*
Common Thorow-waxe.

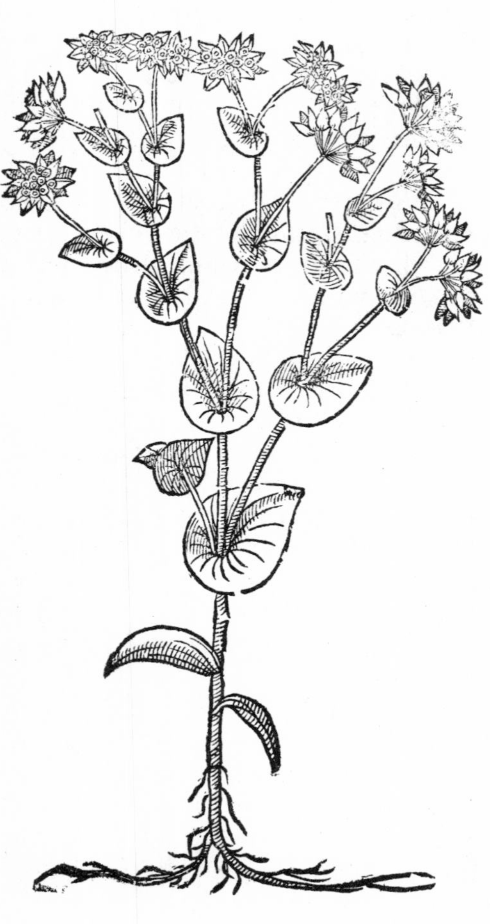

2 *Perfoliata siliquosa.*
Codded Thorow-waxe

¶ *The*

¶ *The Deſcription.*

1 THorow-wax or Thorow-leafe,hath a round,ſlender,and brittle ſtalk,diuided into many ſmall branches, which paſſe or goe thorow the leaues, as though they had beene drawne or thruſt thorow,and to make it more plaine, euery branch doth grow thorow euery leafe,making them like hollow cups or ſaucers. The ſeed groweth in ſpokie tufts or rundles like Dill,long and blackiſh.The floures are of a faint yellow colour.The root is ſingle,white and threddie.

2 Codded Thorow-wax reckoned by *Dodonæus* among the Braſſickes or Colewoorts,and making it a kinde thereof,and calling it *Braßica ſylueſtris perfoliata*: though in mine opinion without reaſon,ſith it hath neither ſhape,affinitie,nor likeneſſe with any of the Colewoorts,but altogether moſt vnlike,reſembling very well the common Thorow-wax, whereunto I rather refer it. It hath ſmall,tender,and brittle ſtalkes two foot high, bearing leaues, which wrap and incloſe themſelues round about,although they do not run thorow as the other do,yet they grow in ſuch manner, that vpon the ſudden view thereof, they ſeeme to paſſe thorow as the other:vpon the ſmall branches do grow little white floures: which being paſt,there ſucceed ſlender and long cods like thoſe of Turneps or Nauewes, whoſe leaues and cods do ſomewhat reſemble the ſame,from whence it hath the name *Napifolia*,that is,Thorow-wax with leaues like vnto the Nauew.The root is long and ſingle, and dieth when it hath brought forth his ſeed.

There is a wilde kinde hereof growing in Kent,in many places among the corne,like to the former in each reſpect,but altogether leſſer:the which no doubt brought into the garden would proue the very ſame.

¶ *The Place.*

‡ The firſt deſcribed growes plentifully in many places about Kent, and betweene Farningham and Ainsford it growes in ſuch quantitie(as I haue been informed by M[r]. *Bowles*) in the corne fields on the tops of the hils,that it may well be termed the infirmitie of them.

The later growes not wilde with vs that euer I could finde,though *Lobell* ſeemes to affirme the contrary. ‡

They grow in the gardens of Herbariſts, and in my garden likewiſe.

¶ *The Time.*

They floure in May and Iune, and their ſeed is ripe in Auguſt.

¶ *The Names.*

1 It hath beene called from the beginning *Perfoliata*,becauſe the ſtalke doth paſſe thorow the leafe, following the ſignification of the ſame: wee call it in Engliſh,Thorow-waxe, or Thorow-leafe.

‡ 2 This by the moſt and beſt part of VVriters(though our Authour be of another opinion) is very fitly referred to the wilde Cole-woorts, and called *Braſſica campeſtris* by *Cluſius* and by *Camerarius*; *Braſſica agreſtis* by *Tragus*: yet *Lobel* calleth it *Perfoliata Napifolia Anglorum ſiliquoſa*. ‡

The Temperature.

Thorow-waxe is of a dry complexion.

¶ *The Vertues.*

The decoction of Thorow-wax made of water or wine,healeth wounds. The iuice is excellent for wounds made either into an oile or vnguent. A

The greene leaues ſtamped,boiled with wax,oile,roſine and turpentine,maketh an excellent vnguent or ſalue to incarnate,or bring vp fleſh in deepe wounds. B

CHAP. 157. *Of Honie-woort.*

¶ *The Deſcription.*

1 C*Erinthe* or Honie-woort riſeth forth of the ground after the ſowing of his ſeed, with two ſmall leaues like thoſe of Baſil;betweene the which leaues commeth forth a thick fat,ſmooth,tender,and brittle ſtalk full of iuice,that diuideth it ſelfe into many other branches;which alſo are diuided in ſundry other armes or branches likewiſe,crambling or leaning toward the ground, being not able without props to ſuſtaine it ſelfe,by reaſon of the great weight

1 *Cerinthe maior.*
Great Honie-woort.

‡ 2 *Cerinthe aſperior flore flavo.*
Rough Honie-woort.

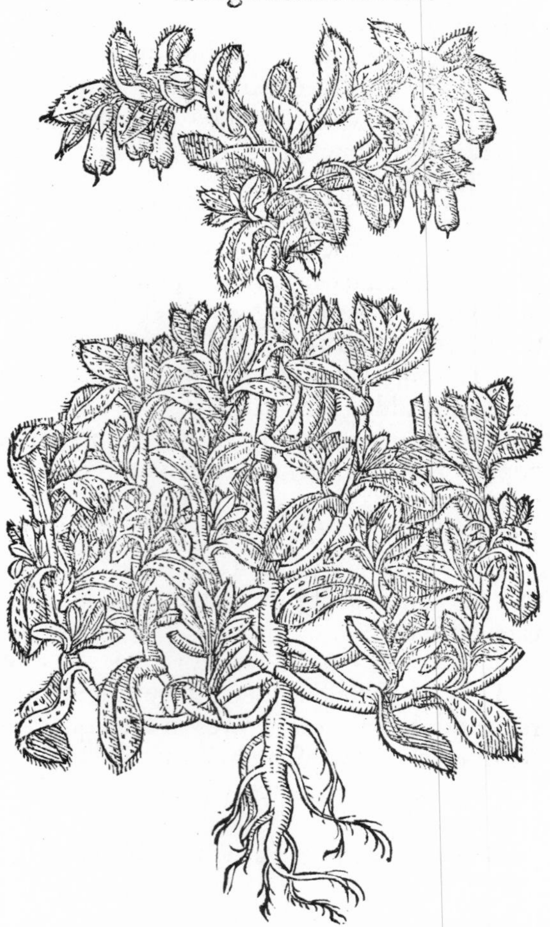

3 *Cerinthe minor.*
Small Honie-woort.

of leaues, branches, & much iuice, the whole plant is ſurcharged with; vpon which branches are placed many thicke rough leaues, ſet with very ſharpe prickles like the rough skinne of a Thornebacke, of a blewiſh green colour, ſpotted very notably with white ſtrakes and ſpots, like thoſe leaues of the true *Pulmonaria* or Cowſlips of Ieruſalem, and in ſhape like thoſe of the codded Thorow-waxe, which leaues do clip or embrace the ſtalkes round about: from the boſome whereof come forth ſmall cluſters of yellow floures, with a hoope or band of bright purple round about the middle of the yellow floure. The floure is hollow, faſhioned like a little boxe, of the taſte of honie when it is ſucked, in the hollowneſſe whereof are many ſmal chiues or threds; which being paſt, ther ſucceed round blacke ſeed, contained in ſoft skinnie husks. The root periſheth at the firſt approch of Winter. ‡ This varies in the colour of the floures, which are yellow, or purple, and ſometimes of both commixt. ‡

‡ 2 The leaues of this other great Honiewort (of *Cluſ.* deſcription) are ſhaped like thoſe laſt deſcribed, but that they are narrower at their ſetting on, and rougher; the floures are alſo yellow of color, but in ſhape & magnitude like the former, as it is alſo in the ſeeds, & all the other parts thereof. ‡

3 This

3 This other *Cerinth* or Honywort hath small long and slender branches, reeling this way and that way, as not able to sustaine it selfe, very brittle, beset with leaues not much vnlike the precedent, but lesser, neither so rough nor spotted, of a blewish greene colour. The floures be small, hollow, and yellow. The seed is small, round, and as blacke as Iet: the root is white, with some fibres, the which dieth as the former. There is a taste as it were of new wax in the floures or leaues chewed, as the name doth seeme to import.

¶ *The Place.*

These plants do not grow wilde in England, yet I haue them in my garden; the seeds wherof I receiued from the right honorable the Lord *Zouch*, my honorable good friend.

¶ *The Time.*

They floure from May to August, and perish at the first approch of Winter, and must be sowen againe the next Spring.

‡ ¶ *The Names.*

‡ 1 The first of these by *Gesner* is called *Cynoglossa montana* and *Cerinthe*: *Dodonæus* calleth it *Maru herba*: and *Lobel* and others, *Cerinthe major*.

2 The second is *Cerinthe quorundam major flauo flore* of *Clusius*.

3 The third by *Dodonæus* is called *Maru herba minor*: and by *Clusius*, *Cerinthe quorundam minor flauo flore*: *Lobel* also calls it *Cerinthe minor*. ‡

¶ *The Temperature and Vertues.*

Pliny and *Auicen* seeme to agree, that these herbes are of a cold complexion; notwithstanding there is not any experiment of their vertues worth the writing.

Chap. 158. *Of S. Iohns wort.*

1 *Hypericum.*
S. Iohns wort.

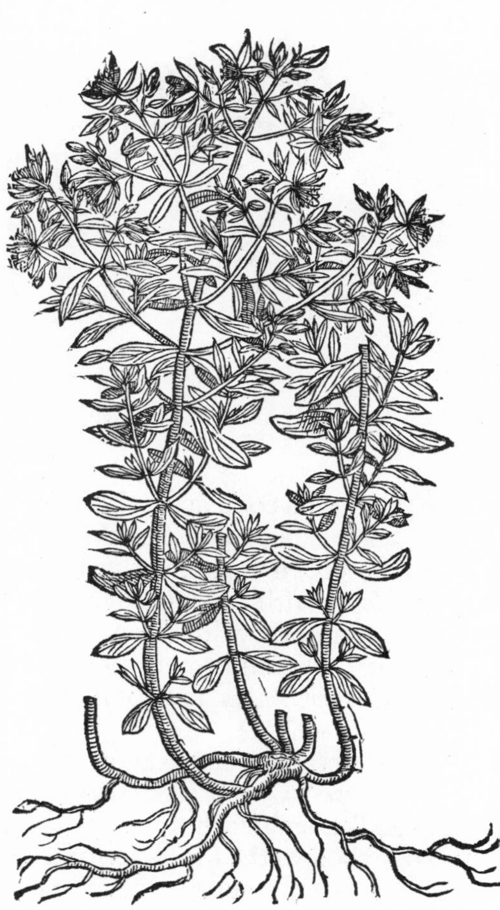

2 *Hypericum Syriacum.*
Rew S. Iohns wort

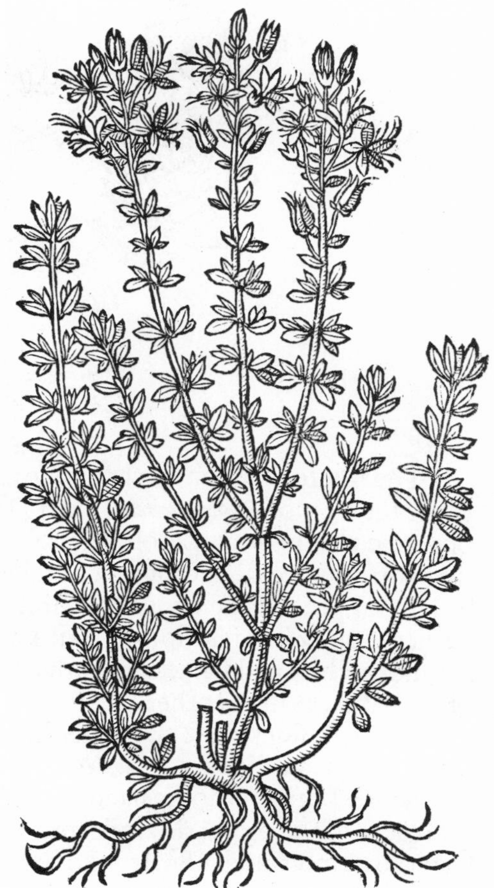

¶ *The*

¶ *The Description.*

1 Saint Iohns wort hath brownish stalkes beset with many small and narrow leaues, which if you behold betwixt your eyes and the light do appeare as it were bored or thrust thorow in an infinite number of places with pinnes points. The branches diuide themselues into sundry small twigs, at the top whereof doe grow many yellow floures, which with the leaues bruised do yeeld a reddish iuyce of the colour of bloud. The seed is contained in little sharpe pointed huskes, blacke of colour, and smelling like Rosin. The root is long, yellow, and of a wooddy substance.

2 The second kinde of S. Iohns wort named *Syriacum*, of those that haue not seene the fruitfull and plentifull fields of England, wherein it groweth aboundantly, hauing small leaues almost like to Rew or Herbe-Grace: wherein *Dodonæus* hath failed, entituling the true *Androsæmum* by the name of *Ruta syluestris*; whereas indeed it is no more like Rew than an Apple to an Oister. This plant is altogether like the precedent, but smaller, wherein consisteth the difference. ‡ It had beene fitter for our Author to haue giuen vs a better and perfecter description of this plant (which as he saith growes so aboundantly with vs) than so absurdly to cauill with *Dodonæus*, for calling, as he saith, the true *Androsæmum*, *Ruta syluestris*: for if that be the true *Androsæmum* which *Dodonæus* made mention of by the foresaid name, why did not our Author figure and describe it in the next chapter saue one, for *Androsæmum*, but followed *Dodonæus* in figuring and describing *Tutsan* for it? See more hereof in the chapter of *Tutsan*. I cannot say I haue seene this plant; but *Lobel* the Author and setter forth thereof thus briefely describes it: the leaues are foure times lesser than those of ours, which grow thicke together as in rundles vpon stalkes, being a cubit high. The floures are yellow, and like those of our common kinde. ‡

3 Woolly S. Iohns wort hath many small weake branches trailing vpon the ground, beset with many little leaues, couered ouer with a certaine soft kinde of downinesse: among which commeth forth weake and tender branches charged with small pale yellow floures. The seeds and roots are like vnto the true S. Iohns wort.

‡ 3 *Hypericum tomentosum Lobelij.* *Lobels* woolly S. Iohns wort.

‡ The figure that our Author gaue was of that which I here giue you second in the third place, vnder the title of *Hyper. toment. Clusij*; for *Clusius* saith it was his, and blames *Lobel* for making it all one with that he found about Montpelier; whose figure also I giue you first in the third place, that you may see what difference you can obserue by them: for *Clusius* saith *Lobels* is but an handfull high; yet tells he not vs how high his growes, neither instances how they differ, neither can I gather it by *Lobels* description: but I coniecture it is thus; That of *Clusius* his description is taller, more white and hairy, and hath the floures growing along little foot-stalks, and not in manner of an vmbel, as in the other.

‡ 4 Besides these two creeping hoary S. Iohns worts here described, there is another small kinde which is called by *Dodonæus*, *Hypericum minus*; and by *Lobel*, *Hypericum minimum supinum Septentrionale*. It growes some handfull or more high, with weake and slender branches set with leaues like those of the ordinarie kinde, but lesse: the floures are also like those of the first described, but fewer in number, and lesse. It is to be found in dry and barren grounds, and floures at the same time as the the former.

5 I haue obserued growing in S. Iohns wood and other places, that kinde of S. Iohn Wort which

which by *Tragus* is called *Hypericum pulchrum*; and both by him and *Lonicerus* is thought to be *Dioſcorides* his *Androſæmum*; the which we in Engliſh may for diſtinctions ſake call Vpright Saint Iohns wort. It hath roots like thoſe of the ordinarie kinde; from which ariſe ſtraight ſlender ſtalks ſome cubit high, ſet at equall ſpaces with pretty ſmooth leaues, broad, and almoſt incompaſſing the ſtalke at their ſetting on, and being ſometimes of a green, and otherwhiles of a reddiſh colour: towards the top they are parted into ſome few branches, which beare ſuch yellow floures as the common kinde, but ſomewhat ſmaller. It floures about the ſame time as the former, or a little after. ‡

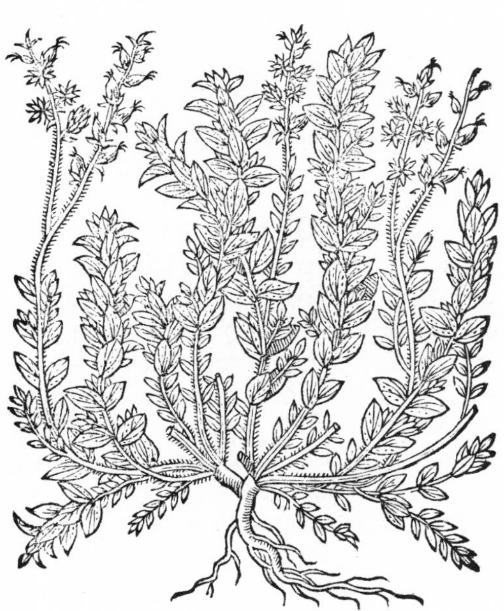

3 *Hypericum tomentoſum Cluſij.*
Woolly S. Iohns wort of *Cluſius.*

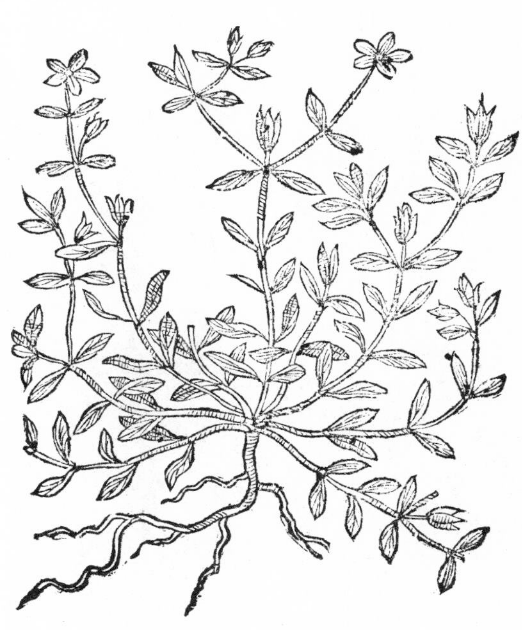

‡ 4 *Hypericum ſupinum glabrum.*
Small creeping S. Iohns wort.

¶ *The Place.*

They grow very plentifully in the paſtures in euery countrey.

¶ *The Time.*

They floure and flouriſh for the moſt part in Iuly and Auguſt.

¶ *The Names.*

S. Iohns wort is called in Greeke ὑπερικὸν: in Latine, *Hypericum*: in ſhops, *Perforata*: of diuers, *Fuga dæmonum*: in Dutch, **San Johans kraut**: in Italian, *Hyperico*: in Spaniſh, *Caraconzillo*: in French, *Mille Pertuys*: in Engliſh, S. Iohns wort, or S. Iohns graſſe.

¶ *The Temperature.*

S. Iohns wort (as *Galen* teacheth) is hot and dry, being of ſubſtance thinne.

¶ *The Vertues.*

S. Iohns wort with his floures and ſeed boyled and drunken, prouoketh vrine, and is right good against the ſtone in the bladder, and ſtoppeth the laske. The leaues ſtamped are good to be layd vpon burnings, ſcaldings, and all wounds; and alſo for rotten and filthy vlcers. A

The leaues, floures, and ſeeds ſtamped, and put into a glaſſe with oyle Oliue, and ſet in the hot Sunne for certaine weekes together, and then ſtrained from thoſe herbes, and the like quantitie of new put in, and ſunned in like manner, doth make an oyle of the colour of bloud, which is a moſt precious remedy for deepe wounds, and thoſe that are thorow the body, for ſinewes that are prickt, or any wound made with a venomed weapon. I am accuſtomed to make a compound oyle hereof; the making of which ye ſhall receiue at my hands, becauſe that I know in the world there is not a better, no not naturall balſam it ſelfe; for I dare vndertake to cure any ſuch wound as abſolutely in each reſpect, if not ſooner and better, as any man whatſoeuer ſhall or may with naturall balſam. B

Take white wine two pintes, oyle oliue foure pounds, oile of Turpentine two pounds, the leaues, floures, and ſeeds of S. Iohns wort, of each two great handfulls gently bruiſed; put them all together into a great double glaſſe, and ſet it in the Sunne eight or ten dayes; then boyle them in the ſame glaſſe *per balneum Mariæ*, that is, in a kettle of water with ſome ſtraw in the bottome, wherein the glaſſe muſt ſtand to boyle: which done, ſtraine the liquor from the herbes, and do as you did before, putting in the like quantitie of herbes, floures, and ſeeds, but not any more wine. And ſo haue you a great ſecret for the purpoſes aforeſaid. C

Dioſcorides

E *Dioscorides* saith, That the seed drunke for the space of fourty dayes together, cureth the *Sciatica*, and all aches that happen in the hips.

F The same Author saith, That being taken with Wine it taketh away Tertian and Quartane Agues.

Chap. 159.

Of Saint Peters wort, or square S. Johns Grasse.

1 *Ascyron.*
S. Peters wort.

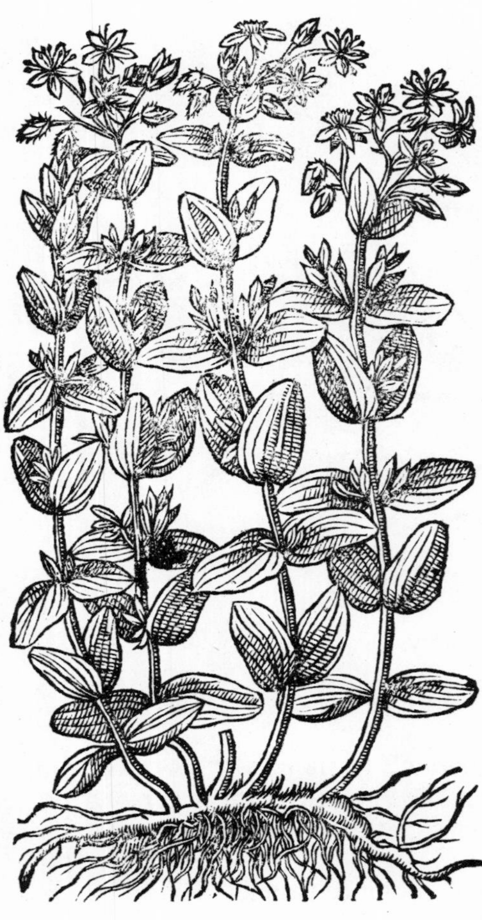

¶ *The Description.*

1 Saint Peters wort groweth to the height of a cubit and a halfe, hauing a straight vpright stalke somewhat browne, set by couples at certaine distances, with leaues much like those of S. Iohns wort, but greater, rougher, and rounder pointed: from the bosome of which leaues come forth many smaller leaues, the which are not bored through, as those of S. Iohns wort are; yet sometime there be some few so bored through. The floures grow at the top of the branches of a yellow colour: the leaues and floures when they are bruised do yeeld forth a bloudy iuyce as doth S. Iohns wort, whereof this is a kinde. The root is tough, and of a wooddy substance.

‡ 2 Vpon diuers boggy grounds of this kingdome is to be found growing that S. Peters Wort which *Clusius* describes in his *Auctarium*, by the name of *Ascyrum supinum* ἑλώδες. This sends forth diuers round hairy creeping stalkes, which heere and there put out new fibres or roots; and these are set at certaine spaces with very round and hairie leaues of a whitish colour, two at a ioynt, and on the tops of these stalkes grow a few small yellow floures which consist of fiue leaues a piece; these stalks seldome send forth branches, vnlesse it be one or two at the tops. It may well be called in English, Round leaued S. Peters wort. ‡

¶ *The Place.*

S. Peters wort, or S. Iohns Grasse groweth plentifully in the North part of England, especially in Landsdale and Crauen: I haue found it in many places of Kent, especially in a copse by Master *Sidleys* house neere Southfleet.

¶ *The Time.*

It floureth and flourisheth when S. Iohns wort doth.

¶ *The Names.*

It is called in Greeke ἀσκυρον: the Latines haue no other name but this Greeke name *Ascyron*. It is called of some *Androsæmum*: *Galen* maketh it both a kinde of Tutsan, and S. Iohns Wort: and saith it is named *Ascyron*, and *Ascyroides*: in English, S. Peters wort, Square or great S. Iohns grasse: and of some, Hardhay. Few know it from S. Iohns wort.

¶ *The Temperature.*

This herbe is of temperature hot and dry.

¶ *The Vertues.*

A It is endued with the same vertues that S. Iohns wort is endued withall. The seed, saith *Dioscorides*, being drunke in foure ounces and a halfe of Meade, doth plentifully purge by siege cholericke excrements. *Galen* doth likewise affirme the same.

Chap

CHAP. 160. *Of Tutſan or Parke-leaues.*

¶ *The Deſcription.*

1 THe ſtalkes of Tutſan be ſtraight, round, chamfered or creſted, hard and wooddy, being for the moſt part two foot high. The leaues are three or foure times bigger than thoſe of S. Iohns wort, which be at the firſt greene; afterwards, and in the end of Sommer of a dark red colour: out of which is preſſed a iuyce not like blacke bloud, but Claret or Gaſcoigne wine. The floures are yellow, and greater than thoſe of S. Peters wort; after which riſeth vp a little round head or berry, firſt greene, afterwards red, laſt of all blacke, wherein is contained yellowiſh red ſeed. The root is hard, wooddy, and of long continuance.

‡ 2 This (which *Dodonæus* did not vnfitly call *Ruta ſylueſtris Hypericoides*, and which others haue ſet forth for *Androſæmum*, and our Author the laſt chapter ſaue one affirmed to be the true *Androſæmum*, though here it ſeemes he had either altered his minde, or forgot what he formerly wrot) may fitly ſtand in competition with the laſt deſcribed, which may paſſe in the firſt place for the *Androſæmum* of the Antients, for *adhuc ſub iudice lis eſt*. I will not here inſiſt vpon the point of controuerſie, but giue you a deſcription of the plant, which is this: It ſends vp round ſlender reddiſh ſtalkes ſome two cubits high, ſet with fewer yet bigger leaues than the ordinarie S. Iohns Wort, and theſe alſo more hairy: the floures and ſeeds are like thoſe of the common S. Iohns wort, but ſomewhat larger. It growes in ſome mountainous and wooddy places, and in the *Aduerſaria* it is called *Androſæmum excellentius, ſeu magnum*: and by *Dodonæus* (as we but now noted) *Ruta ſylueſtris Hypericoides*, thinking it to be the *Ruta ſylueſtris* which is deſcribed by *Dioſcorides, lib. 3. cap.* 48. in the old Greeke edition of *Manutius*, Κεφ. υμζ. And in that of *Marcellus Virgilius* his Interpretation, in the chapter and booke but now mentioned; but reiected amongſt the *Notha* in the Paris Edition *Anno* 1549. You may finde the deſcription alſo in *Dodonæus, Pempt. primæ, lib.* 3. *cap.* 25. whither I refer the curious, being loath here to inſiſt further vpon it. ‡

1 *Clymenon Italorum.*
Tutſan, or Parke leaues.

‡ 2 *Androſæmum Hypericoides.*
Tutſan S. Iohns wort.

¶ The

¶ *The Place.*

Tutsan groweth in woods and by hedges, especially in Hampsted wood, where the Golden rod doth grow; in a wood by Railie in Essex, and many other places.

¶ *The Time.*

It floureth in Iuly and August: the seed in the meane time waxeth ripe. The leaues become red in Autumne; at that time is very easily pressed forth his winie iuyce.

¶ *The Names.*

It is called in Greeke ἀνδρόσαιμον: and the Latines also *Androsæmon*: it is likewise called *Dionysias*, as *Galen* witnesseth. They are farre from the truth that take it to be *Clymenum*, and it is needlesse to finde fault with their error. It is also called *Siciliana*, and *Herba Siciliana*: in English, Tutsan, and Parke-leaues.

¶ *The Temperature*

The faculties are such as S. Peters wort, which doth sufficiently declare it to be hot and dry.

¶ *The Vertues.*

A The seed hereof beaten to pouder, and drunke to the weight of two drams, doth purge cholericke excrements, as *Dioscorides* writeth; and is a singular remedie for the Sciatica, prouided that the Patient do drinke water for a day or two after purging.

B The herbe cureth burnings, and applied vpon new wounds it stancheth the bloud, and healeth them.

C The leaues laid vpon broken shins and scabbed legs healeth them, and many other hurts and griefes, whereof it tooke his name Tout-saine, or Tutsane, of healing all things.

‡ CHAP. 161. *Of Bastard S. Johnswort.*

‡ 1 *Coris Matthioli.*
Matthiolus his bastard S. Iohns wort.

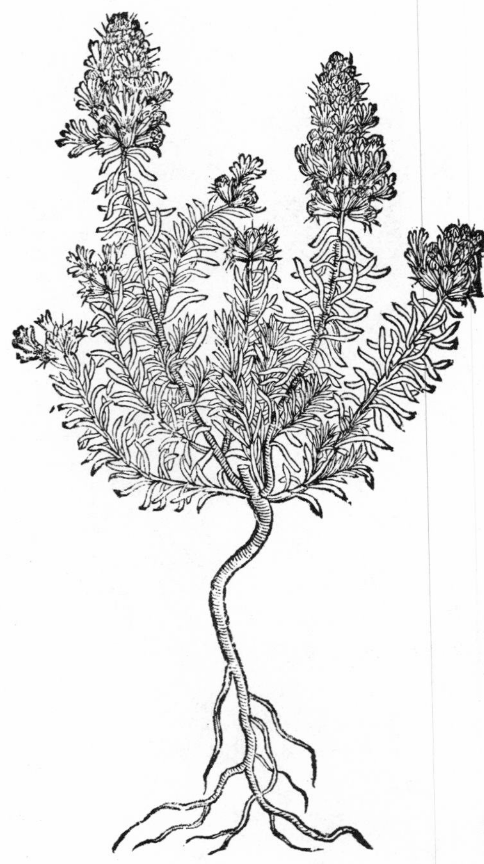

‡ 2 *Coris cærulea Monspeliaca.*
French bastard S. Iohns wort.

‡ THe diligence of these later times hath beene such to finde out the *Materia medica* of the Antients, that there is scarse any plant described by them, but by some or other of late there haue been two or more seuerall plants referred thereto: and thus it hath happened vnto that which

which *Dioſcorides lib.* 3. *cap.* 174. hath ſet forth by the name of *Coris*; and preſently deſcribes after the kindes of *Hypericon*, and that with theſe words; οἱ δὲ καὶ τοῦτο ὑπερικὸν καλοῦσι. Some alſo call this *Hypericon*; to which *Matthiolus* and others haue fitted a plant, which is indeed a kinde of *Hypericon*, as you may perceiue by the figure and deſcription which I giue you in the firſt place. Some (as *Heſychius*) referre it to *Chamæpytis*, (and indeed by *Dioſcorides* it is placed betweene *Androſæmon* and *Chamæpytis*) and to this that which is deſcribed by *Pena* and *Lobel* in the *Aduerſ.* and by *Cluſius* in his Hiſtorie, may fitly be referred: this I giue you in the ſecond place.

¶ *The Deſcription.*

1 THe firſt hath a wooddy thicke and long laſting root, which ſendeth vp many branches ſome foot or more high, and it is ſet at certaine ſpaces with round leaues like thoſe of the ſmall Glaſſe-wort or Sea-Spurry, but ſhorter: the tops of the ſtalkes are diuided into ſundrie branches, which carry floures like thoſe of S. Iohns wort, of a whitiſh red colour, with threds in their middles hauing little yellow pendants. It growes in Italy and other hot countries, in places not far from the ſea ſide. This is thought to be the true *Coris*, by *Matthiolus*, *Geſner*, *Lonicerus*, *Lacuna*, *Bellus*, *Pona*, and others.

2 This from a thicke root red on the outſide ſendeth vp ſundry ſtalkes, ſome but an handfull, other ſome a foot or more long, ſtiffe, round, purpliſh, ſet thicke with leaues like thoſe of Heath, but thicker, more ſucculent and bitter, which ſometimes grow orderly, and otherwhiles out of order. The ſpikes or heads grow on the tops of the branches, conſiſting of a number of little cups, diuided into fiue ſharpe points, and marked with a blacke ſpot in each diuiſion: out of theſe cups comes a floure of a blew purple colour, of a moſt elegant and not fading colour; and it is compoſed of foure little bifide leaues, whereof the two vppermoſt are the larger: the ſeed, which is round and blackiſh, is contained in ſeed-veſſels hauing points ſomewhat ſharpe or prickly. It floures in Aprill and May, and is to be found growing in many places of Spaine, as alſo about Mompelier in France; whence *Pena* and *Lobel* called it *Coris Monſpeliaca*; and *Cluſius*, *Coris quorundam Gallorum & Hiſpanorum*.

¶ *The Temperature.*

Theſe Plants ſeeme to be hot in the ſecond or third degree.

¶ *The Vertues.*

Dioſcorides ſaith, That the ſeed of *Coris* drunke moue the courſes and vrine, are good againſt A
the biting of the Spider *Phalangium*, the Sciatica; and drunke in Wine, againſt that kinde of Convulſion which the Greekes call *Opiſthotonos*, (which is when the body is drawne backwards) as alſo againſt the cold fits in Agues. It is alſo good anointed with oyle, againſt the aforeſaid Convulſion. ‡

CHAP. 162. *Of the great Centorie.*

¶ *The Deſcription.*

1 THe great Centory bringeth forth round ſmooth ſtalkes three cubits high: the leaues are long, diuided as it were into many parcels like to thoſe of the Walnut tree, and of an ouerworne grayiſh colour, ſomewhat ſnipt about the edges like the teeth of a ſaw. The floures grow at the top of the ſtalks in ſcaly knaps like the great Knapweed, the middle thrums whereof are of a light blew or sky colour: when the ſeed is ripe the whole knap or head turneth into a downy ſubſtance like the head of an Artichoke, wherein is found a long ſmooth ſeed, bearded at one end like thoſe of Baſtard Saffron, called *Cartamus*, or the ſeed of *Cardus Benedictus*. The root is great, long, blacke on the outſide, and of a ſanguine colour on the inſide, ſomewhat ſweet in taſte, and biting the tongue.

2 There is likewiſe another ſort, hauing great and large leaues like thoſe of the water Docke, ſomewhat ſnipt or toothed about the edges. The ſtalke is ſhorter than the other, but the root is more oleous or fuller of iuyce, otherwiſe like. The floure is of a pale yellow purpliſh colour, and the ſeed like that of the former.

1 *Centaurium magnum.*
Great Centorie.

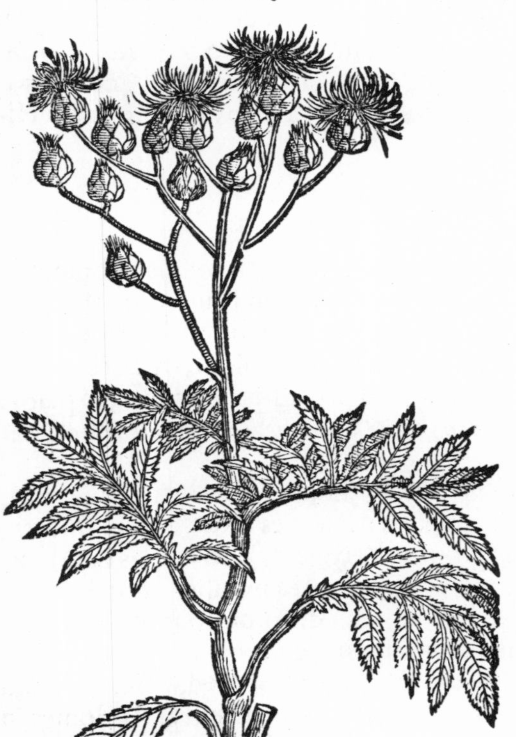

‡ 2 *Centaurium maius alterum.*
Whole leaued great Centorie.

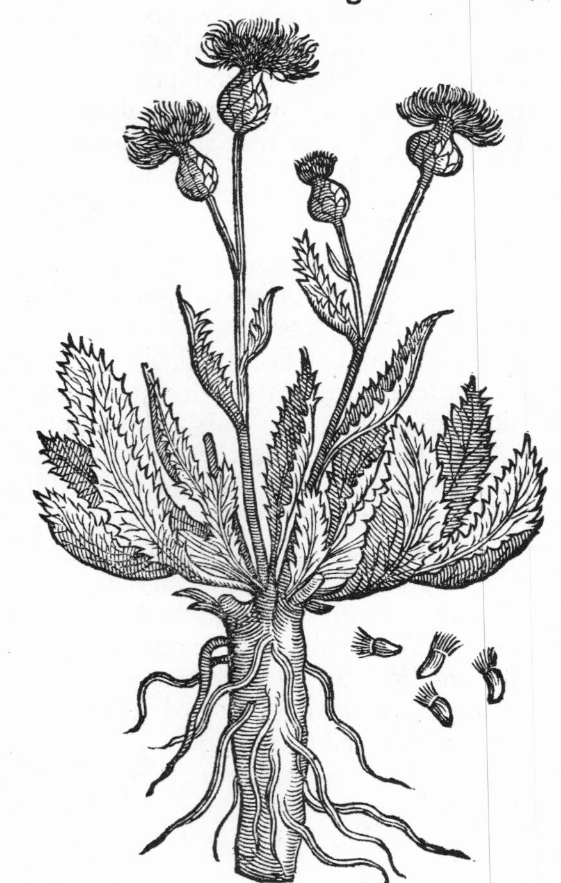

¶ *The Place.*

The great Centorie ioyeth in a fat and fruitfull soile, and in Sunny bankes full of Grasse and herbes. It groweth very plentifully, saith *Dioscorides*, in Lycia, Peloponnesus, Arcadia, and Morea: and it is also to be found vpon Baldus a mountaine in the territories of Verona, and likewise in my garden.

¶ *The Time.*

It floureth in Sommer, and the roots may be gathered in Autumne.

¶ *The Names.*

It is called in Greeke Κενταύρειον τὸ μέγα: of *Theophrastus* also *Centauris*: in diuers shops falsly *Rha Ponticum*: for *Rha Ponticum* is *Rha* growing in the countries of Pontus; a plant differing from great Centorie. *Theophrastus* and *Pliny* set downe among the kindes of *Panaces* or All-heales, this great Centorie, and also the lesser, whereof we will write in the next chapter following. *Pliny* reciting the words of *Theophrastus*, doth in his twenty fifth booke and fourth chapter write, that they were found out by *Chiron* the Centaure, and syrnamed *Centauria*. Also affirming the same thing in his sixth chapter (where he more largely expoundeth both the Centauries) hee repeateth them to be found out by *Chiron*: and thereupon he addeth, that both of them are named *Chironia*. Of some it is reported, That the said *Chiron* was cured therewith of a wound in his foot, that was made with an arrow that fell vpon it when he was entertaining *Hercules* into his house; whereupon it was called *Chironium*: or of the curing of the wounds of his souldiers, for the which purpose it is most excellent.

¶ *The Temperature.*

It is hot and dry in the third degree. *Galen* saith, by the taste of the root it sheweth contrarie qualities, so in the vse it performeth contrarie effects.

¶ *The Vertues.*

A The root taken in the quantitie of two drams is good for them that be bursten, or spit bloud; against the crampe and shrinking of sinewes, the shortnesse of wind or difficultie of breathing, the cough and gripings of the belly.

B There is not any part of the herbe but it rather worketh miracles than ordinarie cures in greene wounds; for it ioyneth together the lips of simple wounds in the flesh, according to the first intention, that is, glewing the lips together, not drawing to the place any matter at all.

The

The root of this Plant (ſaith *Dioſcorides*) is a remedie for ruptures, conuulſions, and cramps, taken in the weight of two drams, to be giuen with wine to thoſe that are without a feuer, and vnto thoſe that haue, with water. A

Galen ſaith, that the iuyce of the leaues thereof performeth thoſe things that the root doth; which is alſo vſed in ſtead of *Lycium*, a kinde of hard iuyce of a ſharpe taſte. B

CHAP. 163. *Of Small Centorie.*

¶ *The Deſcription.*

1 THe leſſer Centorie is a little herbe: it groweth vp with a cornered ſtalke halfe a foot high, with leaues in forme and bigneſſe of S. Iohns wort: the floures grow at the top in a ſpoky buſh or rundle, of a red colour tending to purple; which in the day time and after the Sun is vp do open themſelues, but towards euening ſhut vp againe: after them come forth ſmall ſeed-veſſels, of the ſhape of wheat cornes, in which are contained very little ſeeds. The root is ſlender, hard, and ſoone fading.

2 The yellow Centorie hath leaues, ſtalkes, and ſeed like the other, and is in each reſpect alike, ſauing that the floures hereof are of a perfect yellow colour, which ſetteth forth the difference.

‡ This is of two ſorts; the one with broad leaues through which the ſtalkes paſſe; and the other hath narrow leaues like thoſe of the common Centorie. ‡

1 *Centaurium paruum.*
Small Centorie.

2 *Centaurium paruum luteum Lobelij.*
Yellow Centorie.

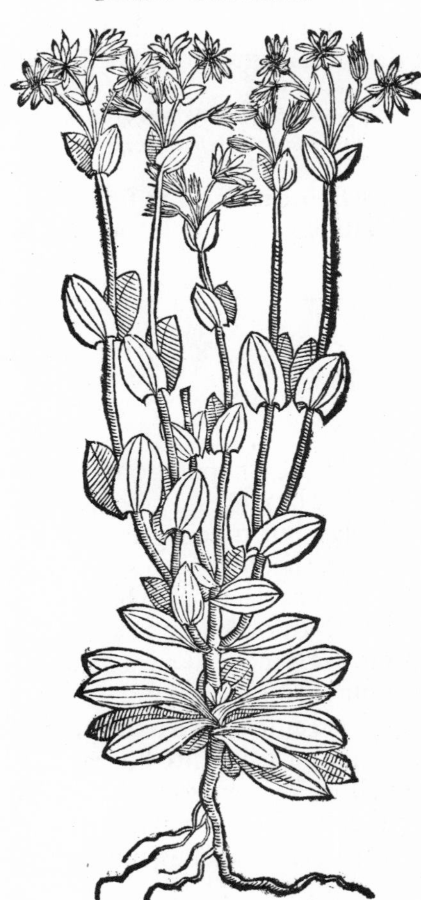

¶ *The Place.*

1 The firſt is growing in great plenty throughout all England, in moſt paſtures and graſſie fields.

2 The yellow doth grow vpon the chalkie cliffes of Greenhithe in Kent, and ſuch like places.

¶ *The Time.*

They are to be gathered in their flouring time, that is in Iuly and August: of some that gather them superstitiously they are gathered betweene the two Lady dayes.

¶ *The Names.*

The Greekes call this Κενταύρειον μικρόν: in Latine it is called *Centaurium minus*; yet *Pliny* nameth it *Libadion*, and by reason of his great bitternesse, *Fel terræ*. The Italians in Hetruria call it *Biondella*: in Spanish, *Centoria*: in low-Dutch, **Centorye**: in English, Small, little, or common Centorie: in French, *Centoire*.

¶ *The Temperature.*

The small Centorie is of a bitter qualitie, and of temperature hot and dry in the second degree; and the yellow Centorie is hot and dry in the third degree.

¶ *The Vertues.*

A Being boyled in water and drunke it openeth the stoppings of the liuer, gall, and spleene, it helpeth the yellow iaundice, and likewise long and lingering agues: it killeth the wormes in the bellie; to be briefe, it cleanseth, scoureth, and maketh thinne humors that are thicke, and doth effectually performe whatsoeuer bitter things can.

B *Dioscorides*, and *Galen* after him report, that the decoction draweth downe by siege choler and thicke humors, and helpeth the Sciatica; but though wee haue vsed this often and luckily, yet could we not perceiue euidently that it purges by the stoole any thing at all, and yet it hath performed the effects aforesaid.

C This Centorie being stamped and laid on whilest it is fresh and greene, doth heale and close vp greene wounds, cleanseth old vlcers, and perfectly cureth them.

D The iuyce is good in medicines for the eyes; mixed with honey it cleanseth away such things as hinder the sight; and being drunke it hath a peculiar vertue against the infirmities of the sinues, as *Dioscorides* teacheth.

E The Italian Physitions do giue the pouder of the leaues of yellow Centorie once in three daies in the quantitie of a dram, with annise or caraway seeds, in wine or other liquor, which preuaileth against the dropsie and greene sicknesse. Of the red floured, *Ioannes Postius* hath thus written:

Flos mihi suaue rubet, sed inest quoque succus amarus,
Qui juvat obsessum bile, aperitque jecur.

My floure is sweet in smell, bitter my iuyce in taste,
Which purge choler, and helps liuer, that else would waste.

Chap. 164. *Of Calues snout, or Snapdragon.*

¶ *The Description.*

1 The purple Snapdragon hath great and brittle stalks, which diuideth it selfe into many fragile branches, whereupon do grow long leaues sharpe pointed, very greene, like vnto those of wilde flax, but much greater, set by couples one opposite against another. The floures grow at the top of the stalkes, of a purple colour, fashioned like a frogs mouth, or rather a dragons mouth, from whence the women haue taken the name Snapdragon. The seed is blacke, contained in round huskes fashioned like a calues snout, (whereupon some haue called it Calues snout) or in mine opinion it is more like vnto the bones of a sheeps head that hath beene long in the water, the flesh consumed cleane away.

2 The second agreeth with the precedent in euery part, except in the colour of the floures, for this plant bringeth forth white floures, and the other purple, wherein consists the difference.

3 The yellow Snapdragon hath a long thicke woody root, with certain strings fastned thereto; from which riseth vp a brittle stalke of two cubits and a halfe high, diuided from the bottome to the top into diuers branches, whereupon doe grow long greene leaues like those of the former, but greater and longer. The floures grow at the top of the maine branches, of a pleasant yellow colour, in shape like vnto the precedent.

4 The small or wilde Snapdragon differeth not from the others but in stature: the leaues are lesser and narrower: the floures purple, but altogether smaller: the heads or seed-vessels are also like those of the former.

‡ 5 There is another kinde hereof which hath many slender branches lying oft times vpon the ground: the leaues are much smaller than these of the last described: the floures and seed-vessels are also like, but much lesser, and herein consists the onely difference. ‡

¶ The

1. 2. *Antirrhinum purpureum ſiue album.*
Purple or white floured Snapdragon.

3 *Antirrhinum luteum.*
Yellow Snapdragon.

4 *Antirrhinum minus.*
Small Snapdragon.

‡ 5 *Antirrhinum minimum repens.*
Small creeping Snapdragon.

¶ *The Place.*

The three firſt grow in moſt gardens; but the yellow kinde groweth not common, except in the gardens of curious Herbariſts.

‡ The fourth and fifth grow wilde amongſt corne in diuers places. ‡

¶ *The Time.*

That which hath continued the whole Winter doth floure in May, and the reſt of Sommer afterwards; and that which is planted later, and in the end of Sommer, floureth in the ſpring of the following yeare: they do hardly endure the iniurie of our cold Winter.

¶ *The Names.*

Snapdragon is called in Greeke ἀντίῤῥινον: in Latine alſo *Antirrhinum*: of *Apuleius*, *Canis cerebrum*, *Herba Simiana*, *Venuſta minor*, *Opalis grata*, and *Orontium*: it is thought to be *Leo herba*, which *Columella*, *lib*. 10. reckons among the floures: yet *Geſner* hath thought that this *Leo* is *Columbine*, which for the ſame cauſe he hath called *Leontoſtomium*: but this name ſeemeth to vs to agree better with Calues ſnout than with Columbine; for the gaping floure of Calues ſnout is more like to Lyons ſnap than the floure of Columbine: it is called in Dutch **Orant**: in Spaniſh, *Cabeza de ternera*: in Engliſh, Calues ſnout, Snapdragon, and Lyons ſnap: in French, *Teſte de chien*, and *Teſte de Veau*.

¶ *The Temperature.*

They are hot and dry, and of ſubtill parts.

¶ *The Vertues.*

A The ſeed of Snapdragon (as *Galen* ſaith) is good for nothing in the vſe of phyſicke; and the herb it ſelfe is of like facultie with *Bubonium* or Star-wort, but not ſo effectuall.

B They report (ſaith *Dioſcorides*) that the herbe being hanged about one preſerueth a man from being bewitched, and that it maketh a man gracious in the ſight of people.

C *Apuleius* writeth, that the diſtilled water, or the decoction of the herbe and root made in water, is a ſpeedy remedy for the watering of eyes proceeding of a hot cauſe, if they be bathed therewith.

Chap. 165. *Of Tode-Flax.*

1 *Linaria vulgaris lutea.*
Great Tode-flax.

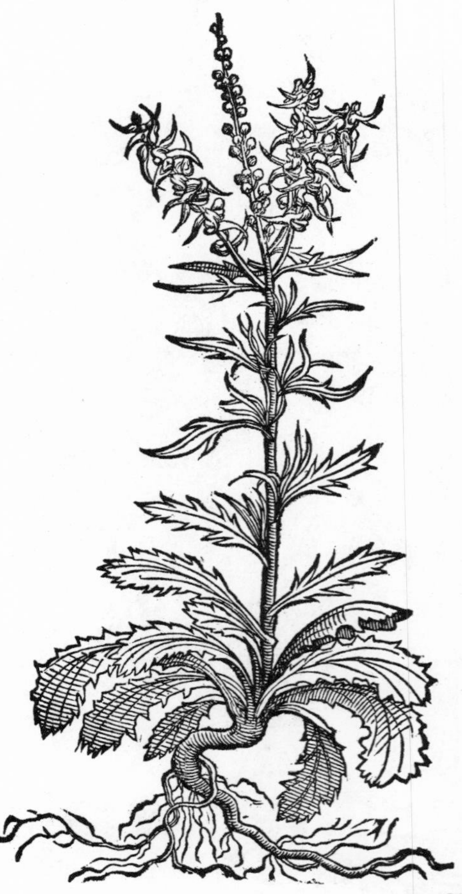

2 *Linaria purpurea odorata.*
Sweet purple Tode-flax.

¶ *The Description.*

1 L*Inaria* being a kinde of *Antyrrhinum*, hath small, slender, blackish stalkes; from which do grow many long narrow leaues like flax. The floures be yellow, with a spur hanging at the same like vnto a Larkes spur, hauing a mouth like vnto a frogs mouth, euen such as is to bee seene in the common Snapdragon; the whole plant before it come to floure so much resembleth *Esula minor*, that the one is hardly knowne from the other, but by this old verse:

Esula lactescit, sine lacte Linaria crescit;

‡ *Esula* with milke doth flow,
Toad-flax without milke doth grow. ‡

2 The second kinde of Tode-flax hath leaues like vnto *Bellis maior*, or the great Dasie, but not so broad, and somewhat iagged about the edges. The stalke is small and tender, of a cubit high, beset with many purple floures like vnto the former in shape. The root is long, with many threds hanging thereat, the floures are of a reasonable sweet sauour.

3 The third, being likewise a kinde of Tode-flax, hath small and narrow leaues like vnto the first kinde of *Linaria*: the stalke is a cubit high, beset with floures of a purple colour, in fashion like *Linaria*, but that it wanteth the taile or spurre at the end of the floure which the other hath. The root is small and threddie.

† 4 *Linaria Valentina* hath leaues like the lesser Centorye, growing at the bottome of the stalke by three and three, but higher vp towards the top, without any certaine order: the stalkes are of a foot high; and it is called by *Clusius*, *Valentina*, for that it was found by himselfe in *Agro Valentino*, about Valentia in Spaine, where it beareth yellow floures about the top of the stalke like common *Linaria*, but the mouth of the floure is downie, or mossie, and the taile of a purple colour. It floureth at Valentia in March, and groweth in the medowes there, and hath not as yet been seene in these Northerne parts.

5 *Osyris alba* hath great, thick, and long roots, with some threds or strings hanging at the same, from which rise vp many branches very tough and pliant, beset towards the top with floures not much vnlike the common Toad-flaxe, but of a pale whitish colour, and the inner part of the mouth somewhat more wide and open, and the leaues like the common Tode-flax.

† 3 *Linaria purpurea altera.*
Variable Tode-flax.

† 4 *Linaria Valentia Clus.*
Tode-flax of Valentia.

† 6 *Osyris*

† 5 *Oſyris alba, Lob.*
White Tode-flax.

6 *Oſyris purpurocærulea* is a kinde of Tode-flax that hath many ſmall and weake branches, trailing vpon the ground, beſet with many little leaues like flaxe. The floures grow at the top of the ſtalke like vnto the common kinde, but of a purple colour declining to blewneſſe. The root is ſmall and threddie.

‡ 7 This hath many ſmall creeping branches ſome handfull or better high, and hath ſuch leaues, floures, and ſeed, as the common kinde, but all of them much leſſe, and therein conſiſteth the difference. It growes naturally in the dry fields about Salamanca in Spaine, and floures all Sommer long. *Lobel* calls it *Oſyris flava ſylueſtris*; and *Cluſius, Linaria Hiſpanica.*

8 The branches of this eight kind are ſpred vpon the ground, and of the length of thoſe of the laſt deſcribed: the leaues are leſſer than thoſe of the common Tode-flax, thicke, iuicie, and of a whitiſh greene colour, and they grow not diſorderly vpon the ſtalks, but at certaine ſpaces ſometimes three, but moſt vſually foure together: the floures in ſhape are like thoſe of the ordinarie kinde, but of a moſt perfect Violet colour, and the lower lip where it gapes of a golden yellow, the taſte is bitter. After the floures are paſt come veſſels round & thick, which contain a flat black ſeed in two partitions or cells: the root is ſlender, white, and long laſting, and it floures vnto the end of Autumne. It grows naturally vpon the higheſt Alps. *Geſner* cals it *Linaria Alpina:* and *Cluſius, Linaria tertia Styriaca.* ‡

† 6 *Oſyris Purpurocærulea repens.* Purple Tode-flaxe.

† 9 Foraſmuch as this plant is ſtalked and leafed like common Flaxe, and thought by ſome to be *Oſyris*; the new writers haue called it *Linoſyris*: it hath ſtalkes very ſtiffe and wooddie, beſet with leaues like the common *Linaria,* with floures at the top of the ſtalkes of a faint ſhining yellow colour, in forme and ſhape ſomewhat like vnto *Conyza maior.* The whole plant groweth to the height of two cubits, and is in taſte ſharpe and clammie, or glutinous, and ſomewhat bitter. The root is compact of many ſtrings, intangled one within another.

† 10 *Guillandinus* calleth this plant *Hyſſopus vmbellifera Dioſcoridis,* that is, *Dioſcorides* his Hyſope,

ſope, which beareth a tuft in all points like *Linoſyris*, whereof it is a kinde, not differing from it in ſhew & leaues. The ſtalks are a cubit high, diuided aboue into many ſmall branches, the tops wherof are garniſhed with tufts of ſmall floures, each little floure being parted into fiue parts with a little thred or peſtell in the middle, ſo that it ſeemes full of many golden haires or thrums. The ſeed is long and blackiſh, and is carried away with the winde. ‡ *Bauhine* in his *Pinax* makes this all one with the former, but vnfitly, eſpecially if you marke the deſcriptions of their floures which are far vnlike. *Fabius Columna* hath proued this to be the *Chryſocome* deſcribed by *Dioſc. lib. 4. cap.* 55. ‡

‡ 7 *Oſyris flava ſylueſtris.* Creeping yellow Tode-flax.

‡ 8 *Linaria quadrifolia ſupina.*
Foure leaued creeping Tode-flax.

† 9 *Linoſyris Nuperorum, Lob.*
Golden Star-floured Tode-flax.

† 11 *Sco-*

10 *Linaria aurea Tragi.*
Golden Tode-flax.

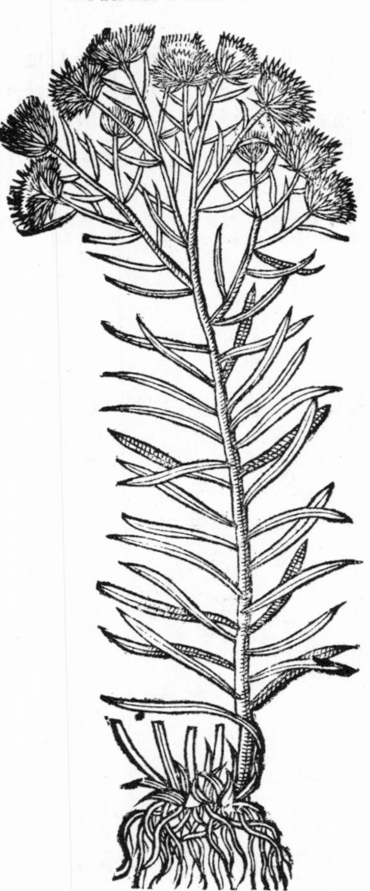

11 *Scoparia siue Osyris Græcorum.*
Bushieor Besome Tode-flax.

† 12 *Passerina linariæ folio, Lob.*
Sparrowes Tode-flax.

† 13 *Passerina altera.*
Sparrow-tongue.

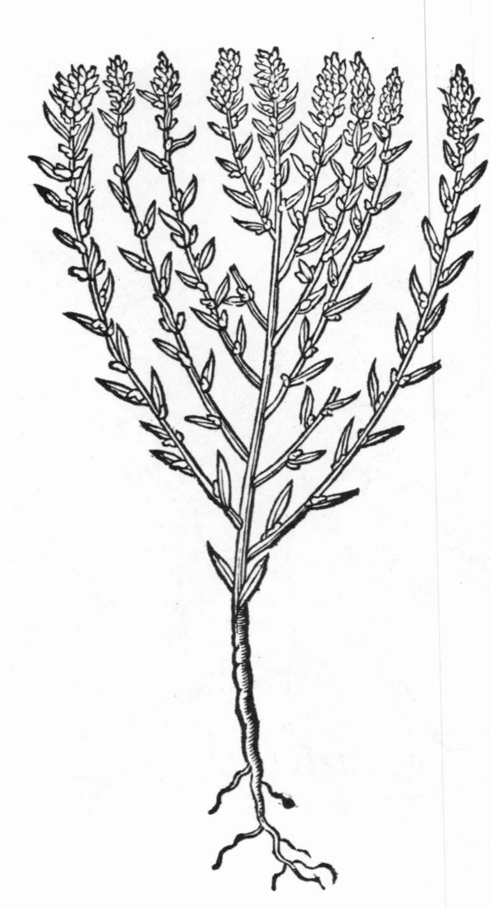

‡ 14 *Linaria adulterina.*
Bastard Tode-flax.

† 11 *Scoparia*, or after *Dodonæus*, *Osyris*, which the Italians cal *Belvidere*, hath very many shoots or sprigs rising from one smal stalk, making the whole plant to resemble a Cypres tree, the branches grow so handsomely: now it growes some three foot high, and very thick and bushie, so that in some places where it naturally groweth they make besomes of it, whereof it tooke the name *Scoparia*. The leaues be small and narrow, almost like to the leaues of flax. The floures be small, and of an hearbie colour, growing among the leaues, which keep greene all the Winter. ‡ I neuer knew it here to ripen the seed, nor to out-liue the first frost. ‡

12 This plant also for resemblance sake is referred to the Linaries, because his leaues be like *Linaria*. At the top of the small branched stalks do grow little yellowish floures, pale of colour, somewhat like the tops of *Chrysocome*. *Iohn Mouton* of Turnay taketh it to be *Chrysocome altera*. And because there hath bin no concordance among Writers, it's sufficient to set forth his description with his name *Passerina*. ‡ *Bauhine* refers it to the *Gromills*, and calls it *Lithospermum Linariæ folio Monspeliacum*.

‡ 13 This which *Tabern.* calls *Lingua Passerina*, and whose figure was giuen by our Authour for the former, hath a small single whitish root, from which it sends vp a slender stalke some cubit and halfe high, naked on the lower part, but diuided into little branches on the vpper, which branches are set thicke with little narrow leaues like those of Winter Sauorie or Tyme: amongst which grow many little longish seeds of the bignesse and taste of Millet, but somewhat hotter and bitterer. The floures consist of foure small yellow leaues. *Tragus* calls this *Passerina*; *Dodonæus* makes it *Lithospermum minus*: and *Columna* hath set it forth by the name of *Linaria altera botryodes montana*. ‡

‡ 14 This which *Clusius*, hath set forth by the name of *Anonymos*, or Namelesse, is called in the *Hist. Lugd. pag.* 1150. *Anthyllis montana*; and by *Tabern. Linaria adulterina*. It hath many hard pale greene branches of some foot high; and vpon these without any order grow many hard narrow long leaues like those of flaxe, at first of a very tart, and afterwards of a bitterish taste: the tops of the stalkes are branched into sundry foot-stalkes, which carry little white floures consisting of fiue small leaues lying starre-fashion, with some threds in their middles: after which at length come single seeds fiue cornered, containing a white pith in a hard filme or skin. The root is white, diuided into sundry branches, and liues long, euery yeare sending vp many stalkes, and sometimes creeping like that of Tode-flax. It floures in May, and grows vpon mountainous places of Germany; Mr. *Goodyer* found it growing wilde on the side of a chalkie hill in an inclosure on the right hand of the way, as you goe from Droxford to Poppie hill in Hampshire. ‡

¶ *The Place.*

The kindes of Tode-flax grow wilde in many places, as vpon stone walls, grauelly grounds, barren medowes, and along by hedges.

‡ I do not remember that I haue seene any of these growing wilde with vs, vnlesse the first ordinary kinde, which is euery where common. ‡

¶ *The Time.*

They floure from Iune to the end of August.

¶ *The Names.*

† Tode-flax is called of the Herbarists of our time, *Linaria*, or Flax-weed, and *Vrinalis*: of some, *Osyris*, in high Dutch, **Lynkraut**, and **Unser frawen flasch**: low Dutch, **Wilt Vlas**: in English, Wild-flax, Tode-flax, and Flax-weed: the eleuenth is called in Italian, *Bel-videre*, or Faire in sight. The same plant is also called *Scoparia*, and *Herba studiosorum*, because it is a fit thing to make brooms of,

of,wherewith ſchollers and ſtudents may ſweepe their owne ſtudies and cloſets. The particular names are expreſſed both in Latine and Engliſh in their ſeuerall titles, whereby they may be diſtinguiſhed.‡ It is thought by moſt that this *Belvidere*,or *Scoparia* is the *Oſyris* deſcribed by *Dioſcorides lib*.4.*cap*.143. For beſides the notes,it hath agreeing with the deſcription:it is at this day by the Greeks called ἀξύρις. ‡

¶ *The Temperature.*

The kindes of Tode-flax are of the ſame temperature with wilde Snap-dragons,whereof they are kindes.

¶ *The Vertues.*

A The decoction of Tode-flax taketh away the yellowneſſe and deformitie of the skinne, beeing waſhed and bathed therewith.

B The ſame drunken, openeth the ſtoppings of the Liuer and ſpleene,and is ſingular good againſt the iaundiſe which is of long continuance.

C The ſame decoction doth alſo prouoke vrine,in thoſe that piſſe drop after drop,vnſtoppeth the kidneies and bladder.

† The figure in this chapter were moſt of them falſe placed,as thus: The third was of *Linaria Panon*.1.of *Cluſius*,being the *Linaria alba* of *Lobel*,deſcribed in the fifth place. The fourth was of the *Oſyris flava ſyl.* of *Lobel*,deſcribed here by me in the ſeuenth place. The fifth was of *Linaria* 3,*Stiriaca* of *Cluſius*,which you may find deſcribed by me in the eighth place. The ſixth was of *Linaria aurea minor* of *Tabern*. being onely a varietie of the *Linaria aurea* ſet forth in the tenth place. The ſeuenth was of the *Linaria Adulterina*, whoſe hiſtorie I haue giuen you in the fourteenth place. That which was formerly vnder the title of *Paſſerina Linaria* is with a hiſtorie fitted thereto in the thirteenth place.

Chap. 166. *Of Garden flaxe.*

† 1 *Linum ſativum.*
Garden flax.

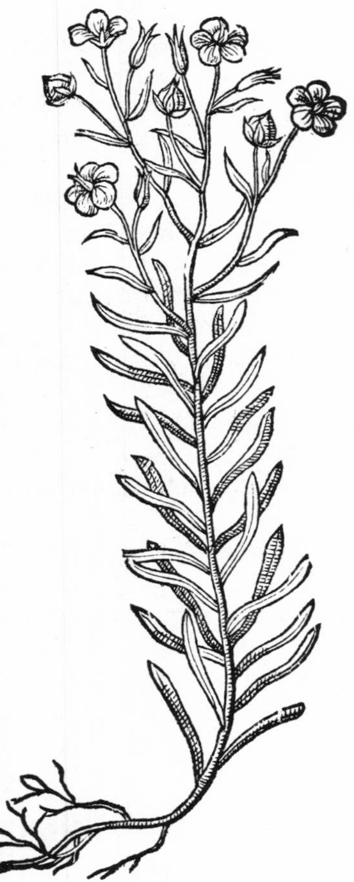

The Deſcription.

Flaxe riſeth vp with ſlender and round ſtalks The leaues thereof bee long, narrow, and ſharpe pointed : on the tops of the ſprigs are faire blew floures, after which ſpring vp little round knops or buttons,in which is contained the ſeed,in forme ſomewhat long,ſmooth,glib or ſlipperie,of a dark colour. The roots be ſmal and threddie.

¶ *The Place.*

It proſpereth beſt in a fat and fruitfull ſoile, in moiſt and not drie places; for it requireth as *Columella* ſaith a very fat ground,and ſomewhat moiſt. Some,ſaith *Palladius*,do ſow it thicke in a leane ground,& by that means the flax groweth fine. *Pliny* ſaith that it is to be ſowne in grauelly places , eſpecially in furrowes : *Nec magis feſtinare aliud* · and that it burneth the ground,and maketh it worſer:which thing alſo *Virgil* teſtifieth in his Georgickes.

Vrit lini campum ſeges,vrit Auena.
Vrunt lethæo perfuſa papauera ſomno.

In Engliſh thus:

Flaxe and Otes ſowne conſume
 The moiſture of a fertile field:
The ſame worketh Poppie,whoſe
 Iuice a deadly ſleepe doth yeeld.

¶ The

¶ *The Time.*

Flaxe is ſowne in the ſpring, it floureth in Iune and Iuly. After it is cut downe (as *Pliny* in his 19. booke, firſt chapter ſaith) the ſtalkes are put into the water ſubject to the heate of the ſunne, and ſome weight laid on them to be ſteeped therein; the looſenes of the rinde is a ſigne when it is well ſteeped: then is it taken vp and dried in the ſunne, and after vſed as moſt huſwiues can tell better than my ſelfe.

¶ *The Names.*

It is called both in Greeke and Laine λίνον: *Linum*: in high Dutch, **Flachſz**: in Italian and Spaniſh, *Lino*: in French, *Dulin*: in low Dutch, **Vlas**: in Engliſh, Flaxe, and Lyne.

¶ *The Temperature and Vertues.*

A *Galen* in his firſt booke of the faculties of nouriſhments ſaith, that diuers vſe the ſeed hereof parched as a ſuſtenacne with *Garum*; no otherwiſe than made ſalt.

B They alſo vſe it mixed with hony, ſome likewiſe put it among bread but it is hurtfull to the ſtomacke, and hard of digeſtion, and yeeldeth to the body but little nouriſhment: but touching the quality which maketh the belly ſoluble, neither will I praiſe or diſpraiſe it; yet that it hath ſome force to prouoke vrine, is more apparant when it is parched: but then it alſo ſtayeth the belly more.

C The ſame author in his bookes of faculties of ſimple medicines ſaith, that Lineſeed being eaten is windy although it be parched, ſo full is it of ſuperfluous moiſture: and it is alſo after a ſort hot in the firſt degree, and in a meane betweene moiſt and dry. But how windy the ſeed is, and how full of ſuperfluous moiſture it is in euery part, might very well haue been perceiued a few yeeres ſince as at Middleborough in Zeland, where for want of graine and other corne, moſt of the Citizens were faine to eate bread and cakes made hereof with hony and oile, who were in ſhort time after ſwolne in the belly below the ſhort ribs, faces, & other parts of their bodies in ſuch ſort, that a great number were brought to their graues thereby: for theſe ſymptomes or accidents came no otherwiſe than by the ſuperfluous moiſture of the ſeed, which cauſeth windineſſe.

D Lineſeed as *Dioſcorides* hath written, hath the ſame properties that Fenugreeke hath: it waſteth away and mollifieth all inflammations or hot ſwellings, as well inward as outward, if it be boiled with hony, oile, and a little faire water, and made vp with clarified hony; it taketh away blemiſhes of the face, and the ſunne burning, being raw and vnboiled; and alſo foule ſpots, if it be mixed with ſalt-peter and figs: it cauſeth rugged and ill fauoured nailes to fall off, mixed with hony and water Creſſes.

E It draweth forth of the cheſt corrupted flegme and other filthy humors, if a compoſition with hony be made thereof to licke on, and eaſeth the cough.

F Being taken largely with pepper and hony made into a cake, it ſtirreth vp luſt.

G The oile which is preſſed out of the ſeed, is profitable for many purpoſes in phyſicke and ſurgery; and is vſed of painters, picture makers, and other artificers.

H It ſofteneth all hard ſwellings; it ſtretcheth forth the ſinewes that are ſhrunke and drawne together, mitigateth paine, being applied in maner of an ointment.

I Some alſo giue it to drinke to ſuch as are troubled with paine in the ſide and collicke; but it muſt be freſh and newly drawne: for if it be old and ranke, it cauſeth aptneſſe to vomit, and withall it ouermuch heateth.

K Lineſeed boiled in water with a little oile, and a quantity of Anniſe-ſeed impoudered and implaiſtered vpon an *angina*, or any ſwelling in the throat, helpeth the ſame.

L It is with good ſucceſſe vſed plaiſterwiſe, boiled in vineger, vpon the diſeaſes called *Coliaca* and *Dyſenteria*, which are bloudy fluxes and paines of the belly.

M The ſeeds ſtamped with the roots of wilde Cucumbers, draweth forth ſplinters, thornes, broken bones, or any other thing fixed in any part of the body.

N The decoction is an excellent bath for women to ſit ouer for the inflammation of the ſecret parts, becauſe it ſofteneth the hardneſſe thereof, and eaſeth paine and aking.

O The ſeed of Line and Fenugreek made into powder, boiled with Mallowes, violet leaues, Smallage, and Chickweed, vntill the herbs be ſoft; then ſtamped in a ſtone morter with a little hogs greaſe to the forme of a cataplaſme or pulteſſe, appeaſeth all maner of paine, ſoftneth all cold tumors or ſwellings, mollifieth and bringeth to ſuppuration all apoſtumes; defendeth wounded members from ſwellings and rankling, and when they be already rankled, it taketh the ſame away being applied very warme euening and morning.

† The figure that was formerly in this place for the ordinary flaxe was of *Linum ſylueſtre latifolium* 3. of *Cluſius*, which is deſcribed by me in the ſixth place in the enſuing Chapter.

Chap. 167. *Of Wilde Flaxe.*

¶ *The Description.*

1 THis Wilde kinde of Line or Flaxe hath leaues like those of garden Flaxe, but narrower, growing vpon round bright and shining sprigs, a foot long, and floures like the manured flaxe, but of a white colour. The root is tough and small, with some fibres annexed thereto. ‡ This is sometimes found with deep blew floures, with violet coloured floures, and sometimes with white, streaked with purple lines. ‡

1 *Linum sylvestre floribus albis.*
Wilde white flaxe.

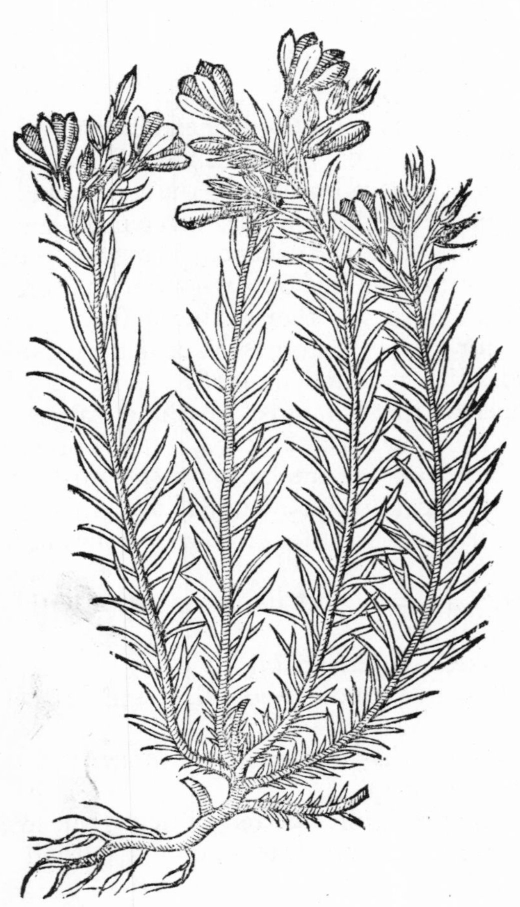

2 *Linum sylvestre tenuifolium.*
Thin leaued wilde flaxe.

2 The narrow and thinne leafed kinde of Line is very like to the common flaxe, but in all points lesser. The floures consist of fiue leaues, which do soone fade and fall away, hauing many stalkes proceeding from one root, of a cubit high, beset with small leaues, yea lesser than those of *Linaria purpurea*.

‡ Our Author in the former edition gaue two figures vnder this one title of *Linum sylvestre tenuifolium*, making them the second and third; but the description of the third was of the Rough broad leaued wilde flaxe, whose figure therefore we haue put in that place. Now the two whose figures were formerly here are but varieties of one species, and differ thus; the former of them (whose figure we haue omitted as impertinent) hath fewer leaues, which therefore stand thinner vpon the stalke, and the floures are either blew or else white. The later, whose figure you may finde here set forth, hath more leaues, and these growing thicker together: the floure is of a light purple or flesh colour. ‡

3 There is a kinde of wilde flaxe which hath many hairy branches, rising vp from a very small root, which doth continue many yeeres without sowing, increasing by roots into many other plants, with stalkes amounting to the height of one cubite, beset with many rough and hairy broad leaues. at the top of the stalkes do grow many blew floures, compact of fiue leaues, much greater and fairer than common Line or flaxe; which being past, there succeed small sharpe pointed heads full of seeds, like Lineseed, but of a blackish shining colour.

4 *Chamælinum*

4 *Chamælinum* (of some called *Linum syluestre perpusillum*, and may be called in English very low or Dwarfe wilde flaxe; for this word *Chamæ* ioined to any simple, doth signifie, that it is a low or dwarfe kinde thereof) beeing scarce an handfull high, hath pale yellow floures: but as it is in all things like vnto flaxe, so the floures, leaues, and stalkes, and all other parts thereof, are foure times lesser than *Linum*.

‡ 5 There is also growing wilde in this kingdome a small kinde of wilde flaxe, which I take to be the *Linocarpos* described by *Thalius*, and mentioned by *Camerarius*, by the name of *Linum syluestre pusillum candicantibus floribus*. *Anno* 1629, when as I first found it, in a Iournall (written of such plants as we gathered) I set downe this by the name of *Linum syluestre pusillum candidis floribus*, which my friend M[r]. *Iohn Goodyer* seeing, he told me he had long knowne the plant, and refer'd it to *Lines*. but there were some which called it in English, Mil-mountaine, and vsed it to purge, and of late he hath sent me this historie of it, which you shall haue as I receiued it from him.

Linum syluestre catharticum. Mil-mountaine.

It riseth vp from a small white threddy crooked root, sometime with one, but most commonly with fiue or six or more round stalks, about a foot or nine inches high, of a browne or reddish color, euery stalk diuiding it selfe neere the top, or from the middle vpward into many parts or branches of a greener colour than the lower part of the stalke: the leaues are small, smooth, of colour green, of the bignes of Lentill leaues, and haue in the middle one rib or sinew, and no more that may bee perceiued, & grow alongst the stalke in very good order by couples, one opposite against the other: at the tops of the small branches grow the floures, of a white colour, consisting of fiue small leaues apiece, the nailes whereof are yellow: in the inside are placed small short chiues also of a yellow colour, after which come vp little knobs or buttons, the top whereof when the seede is ripe diuideth it selfe into fiue parts; wherein is contained small, smooth, flat, slippery, yellow seed: when the seed is ripe the herbe perisheth: the whole herbe is of a bitter taste, and herby smell. It groweth plentifully in the vnmanured inclosures of Hampshire, on chalkie downs, & on Purfleet hils in Essex, and in many other places. It riseth forth of the ground at the beginning of the Spring, and floureth all the Sommer.

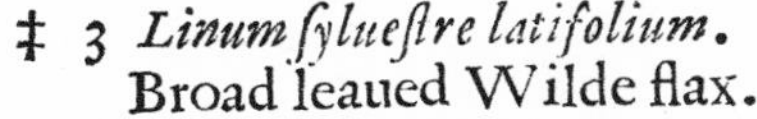
‡ 3 *Linum syluestre latifolium*.
Broad leaued Wilde flax.

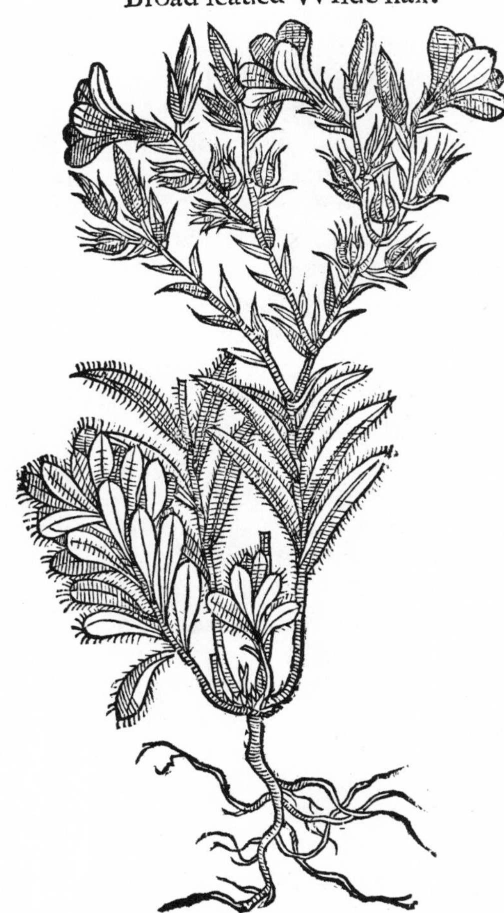

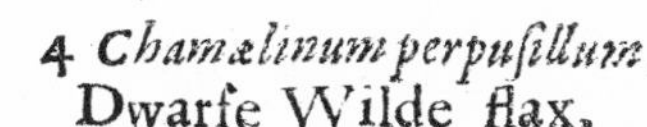
4 *Chamælinum perpusillum*.
Dwarfe Wilde flax.

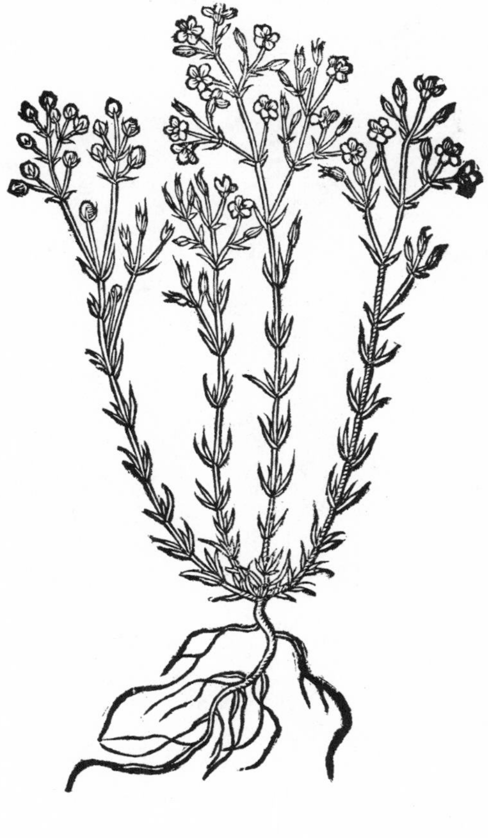

‡ 5 *Linum ſyl. catharticum.*
Mil-mountaine.

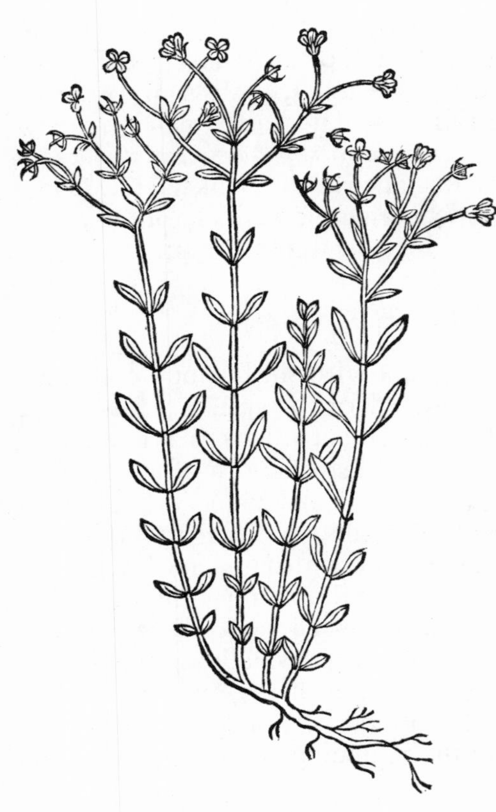

‡ 6 *Linum ſyl. latifolium* 3. *Cluſ.*
The third broad leaued Wilde flax.

‡ 7 *Linum marinum Luteum.*
Yellow floured Wilde flax.

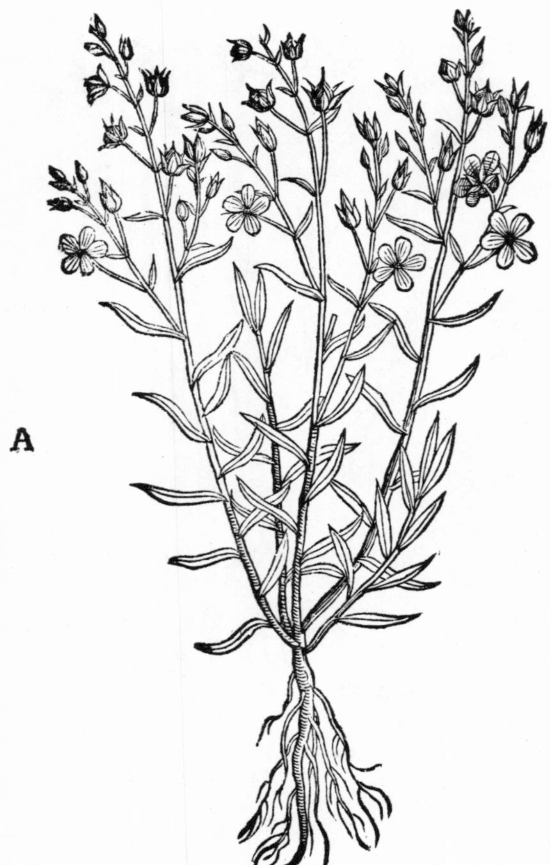

I came to know this herbe by the name of Mil-mountaine, and his vertue by this meanes. On the ſecond of October 1617. going by M[r]. *Colſons* ſhop an Apothecary of VVinchester in Hampſhire, I ſaw this herbe lying on his ſtall, which I had ſeene growing long before: I deſired of him to know the name of it, he told me that it was called Mill-mountain, and he alſo told me that beeing at Doctour *Lake* his houſe at Saint Croſſe a mile from VVincheſter, ſeeing a man of his haue this hearbe in his hand, he deſired the name; hee told him as before, and alſo the vſe of it, which is this.

A Take a handfull of Mill mountaine, the whole plant, leaues, ſeedes, floures and all, bruiſe it and put it in a ſmall tunne or pipkin of a pinte filled with white VVine, and ſet in on the embers to infuſe all night, and drinke that wine in the morning faſting, and hee ſaid it would giue eight or tenne ſtooles. This Doctour *Lake* was afterward made Biſhop of Bath and VVells, who alwaies vſed this hearbe for his purge, after the ſaid manner, as his man affirmed, Iuly 20. 1619. *Iohn Goodyer.*

I haue not as yet made tryall hereof, but ſince in *Geſn. de Lunarijs pag.* 34. I haue found the

I haue found the like or a more purging facultie attributed to this herbe,as I thinke(for I cannot referre it to any other) where hee would haue it to be *Helleborine* of the Antients ; I thinke it not amiſſe here to ſet downe his words,becauſe the booke is not commonly to be had, being ſet forth *Anno* 1555. *Ante annos* 15.*aut circiter cum Anglus quidam, ex Italia rediens,me ſalutaret* (Turnerus *is fuerit, vir excellentis tum in re medica tum alijs pleriſque diſciplinis doctrinæ, aut alius quiſpiam,vix ſatis memini) inter alias rariorum ſtirpium icones quas depingendas commodabat, Elleborinem quoque oſtendebat pictam,herbulam fruticoſam, pluribus ab vna radice cauliculis quinque fere digitorum proceritate erectis, foliolis perexiguis, binis per intervalla (ciuſmodi vt ex aſpectu genus quoddam Alſinæ exiguum videretur) vaſculis in ſummo exiguis, rotundis tanquam lini. Hanc ajebat creſcere in pratis ſiccis, vel clivis Montium; inutili radice, ſubamara, purgare vtrinque & in Anglia vulgo vſurpari a ruſticis.* Thus much for *Geſner*.

6 *Cluſius* amongſt other wilde Lines or Flaxes hath ſet forth this,which from a liuing,thicke, writhen root, ſendeth vp many ſtalkes almoſt a cubite high, ſomewhat red and ſtiffe,ſet with prettie large and thicke leaues not rough and hairie, but ſmooth and hard ; the floures grow plentifully on the tops of the ſtalkes, being large,and compoſed of fiue leaues of a faire yellow colour,with fiue threds comming forth of their middles, with as many ſmaller and ſhorter haires. The ſeed is contained in flatter heads than thoſe of the firſt deſcribed, containing a blacke, but not ſhining ſeed. It floures in Iune and Iuly, and ripens the ſeed in Auguſt. It growes naturally vpon diuers hils in Germany.

7 *Matthiolus* and *Dodonæus* haue vnder the name of *Linum ſylueſtre* ; and *Lobel* by the name of *Linum marinum luteum Narbonenſe*,ſet forth another yellow floured wilde flaxe. This growes with ſlender ſtalkes ſome cubite high,ſet with leaues like thoſe of flaxe, but ſomewhat leſſer, and fewer in number : at the tops of the ſtalkes grow floures ſmaller than thoſe of the common Line, and yellow of colour.It growes naturally vpon the coaſts of France that lie towards the Mediterranean ſea,but not in England that I haue heard of. ‡

¶ *The Place.*

They grow generally in grauelly grounds. The firſt groweth in well mannured places,as in gardens and ſuch like ſoiles. The ſecond groweth by the ſea ſide. The third and fourth grow vpon rockes and cliffes neere to the ſea ſide I haue ſeene them grow vpon the ſea bankes by Lee in Eſſex,and in many places of the Iſle of Shepey. They grow alſo betweene Quinborow and Sherland houſe.

‡ I haue not ſeene any of theſe growing wilde, but onely the fifth of my deſcription. ‡

¶ *The Time.*

They floure from May to the middeſt of Auguſt.

¶ *The Names.*

Their names are ſufficiently expreſt in their ſeuerall titles.

¶ *The Nature and Vertues.*

The faculties of theſe kindes of Wilde-flax are referred vnto the manured flax,but they are ſeldome vſed either in Phyſicke or chirurgerie.

CHAP. 168. *Of Blacke Saltwoort.*

¶ *The Deſcription.*

IN old time,ſay the Authours of the *Aduerſaria*, this plant was vſed for meat, and receiued among the *Legumina*. It was called *Glaux*,by reaſon of the colour of the leaues, which are of a blewiſh gray colour,called in Latine,*Glaucus color*,ſuch as is in the Sallow leafe : of others it is called *Galax* or *Glax* . and *Eugalacton, quaſi lactea* or *lactifica*,becauſe it is good to increaſe milke in the breſts of women,if it be much vſed. *Ruellius* and others haue ſet downe *Galega, Securidica, Polygala*, and many other plants for the true *Glaux*,which hath bred a confuſion. The true *Glaux* of *Dioſcorides* hath many ſmall branches,ſome creeping on the ground,and ſome ſtanding vpright,tender,and ſmall,beſet with many little fat leaues like *Tribulus terreſtris*,or *Herniaria*,growing along the ſtalks by couples ; betweene whom grow ſmall purple floures; which being paſt,there ſucceed certaine little bullets or ſeed veſſels. The root is very ſmall and threddy,and taking hold of the vpper face of the earth,as it doth run abroad,by which meanes it doth mightily increaſe.

Glaux exigua maritima.
Blacke Saltwoort.

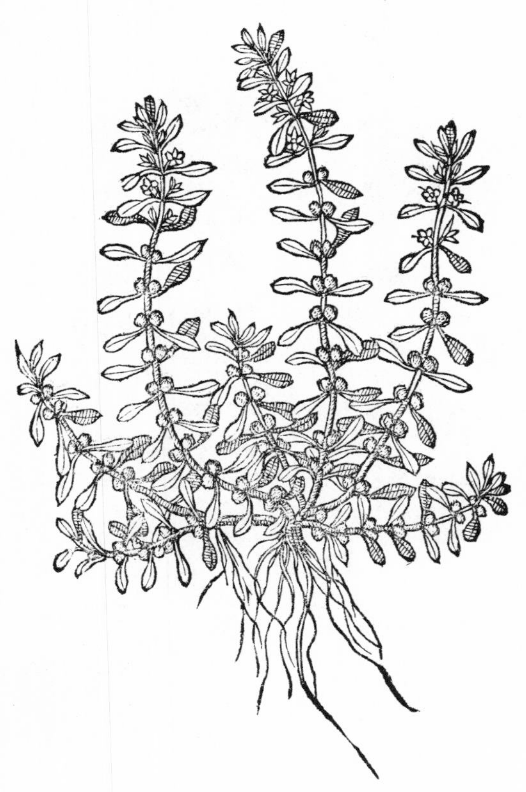

¶ *The Place.*

The true *Glaux* or Milkwoort groweth very plentifully in salt places & marshes neere the sea, from whence I haue brought it into my garden, where it prospereth as well as in his natiue soile. I found it especially betweene Whitstable and the Isle of Thanet in Kent, and by Grauesend in the same countie, by Tilbery Block-house in Essex, and in the isle of Shepey, going from Kings ferry to Sherland house.

¶ *The Time.*

It floureth in May, and the seed is ripe in Iune.

¶ *The Names.*

The names haue beene sufficiently spoken of in the description. It shall suffice to call it in English, Sea Milkwoort.

¶ *The Nature.*

Paulus Aegineta saith, it is hot and moist of temperature.

¶ *The Vertues.*

This Milkwoort taken with milke, drinke, or pottage, ingendereth store of milke, and therefore it is good to be vsed by nurses that want the same.

Chap. 169. *Of Milke-woort.*

¶ *The Description.*

1 There haue beene many plants neerely resembling *Polygala*, and yet not the same indeed, which doth verifie the Latine saying, *Nullum simile est idem.* This neere resemblance doth rather hinder those that haue spent much time in the knowledge of simples, than increase their knowledge. And this also hath been an occasion that many haue imagined a sundry *Polygala* vnto themselues, and so of other plants. Of which number, this (whereof I speake) is one, obtaining this name of the best writers and herbarists of our time, describing it thus. It hath many thick spreading branches, creeping on the ground: bearing leaues like them of *Herniaria,* standing in rowes like the Sea Milkwoort; among which do grow small whorles, or crownets of white floures, the root being exceeding small and threddie.

2 The second kinde of *Polygala* is a small herbe with pliant slender stems, of a wooddie substance, an handfull long, creeping by the ground; the leaues be small & narrow like to Lintels, or little Hyssope. The floures grow at the top, of a blew color, fashioned like a little bird, with wings, taile, and bodie, easie to be discerned by them that doe obserue the same: which being past, there succeed small pouches, like those of *Bursa pastoris,* but lesser. The root is small and wooddie

3 This third kinde of *Polygala* or Milkewoort, hath leaues and stalkes like the last before mentioned, and differeth from it herein, that this kinde hath smaller branches, and the leaues are not so thicke thrust together, and the floures are like the other, but that they bee of a red or purple colour.

4 The fourth kinde is like the last spoken of in euery respect, but that it hath white floures; otherwise it is very like.

5 Purple Milkewoort differeth from the others in the colour of the floures; it bringeth foorth moe branches than the precedent, and the floures are of a purple colour; wherein especially consisteth the difference.

6 The

1 *Polygala repens.*
Creeping Milke-wort.

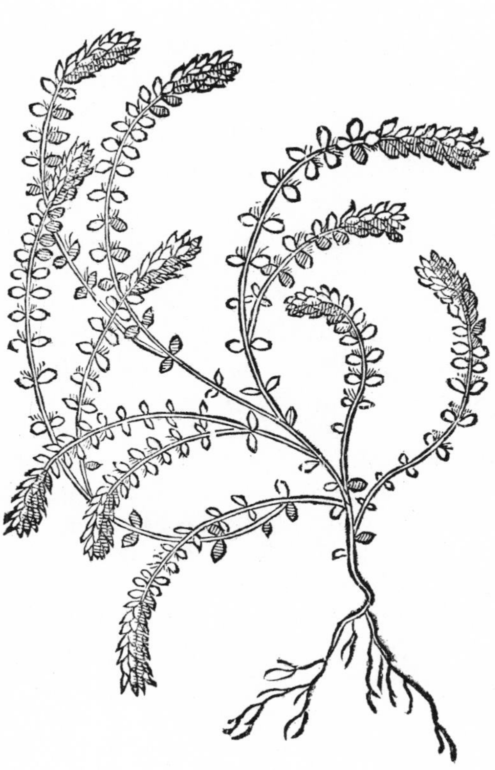

2 *Polygala flore cæruleo.*
Blew Milke-wort.

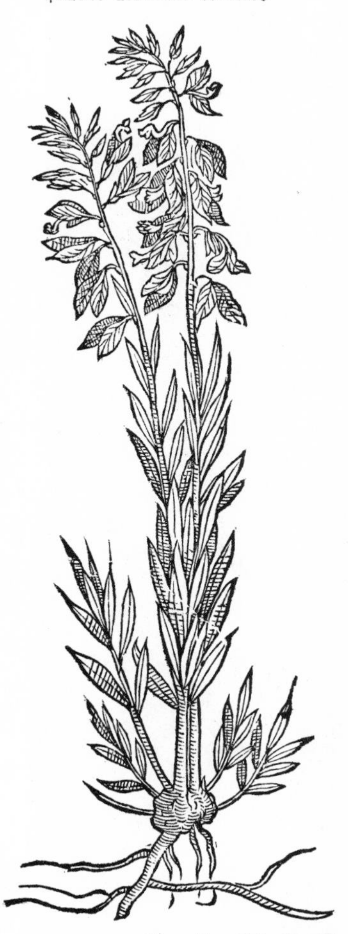

3 *Polygala rubris floribus.*
Red Milkewort.

4 *Polygala albis floribus.*
White Milkewort.

6 The sixth Milke-wort is like vnto the rest in each respect, sauing that the floures are of an ouerworne ill fauoured colour, which maketh it to differ from all the other of his kinde.

5 *Polygala purpurea.*
Purple Milke-wort.

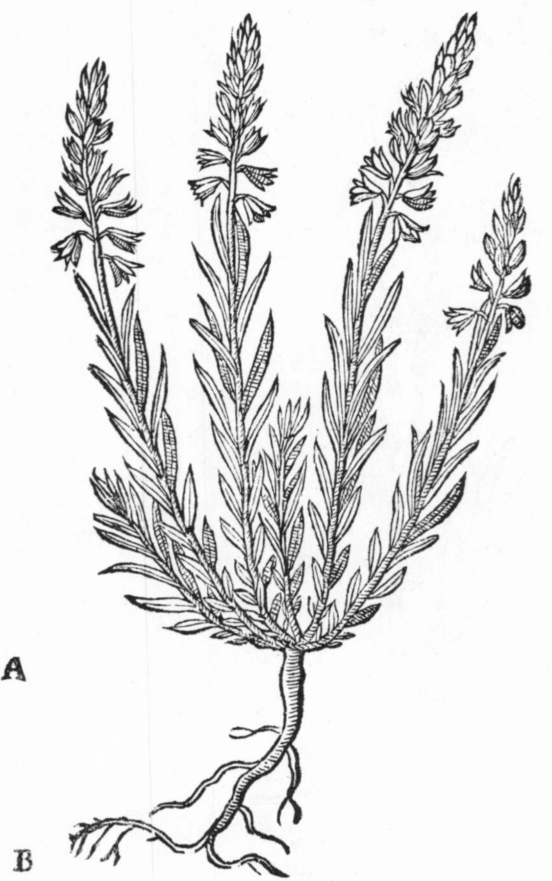

¶ *The Place.*

These plants or Milke-worts grow commonly in euery wood or fertil pasture wheresoeuer I haue trauelled.

¶ *The Time.*

They floure from May to August.

¶ *The Names.*

Milke-wort is called by *Dodonæus*, *Flos Ambarualis*; so called because it doth especially flourish in the Crosse or Gang weeke, or Rogation weeke; of which floures the maidens which vse in the countries to walke the Procession doe make themselues garlands and nosegaies: in English we may call it Crosse-floure, Procession-floure, Gang-floure, Rogation-floure, and Milkewort, of their vertues in procuring milke in the brests of nurses. *Hieronymus Tragus*, as also *Dioscorides* calleth it *Polygalon*. ‡ *Gesner* calls this *Crucis flos*; and in his Epistles he nameth it *Amarella*: it is vulgarly knowne in Cheapside to the herbe-women by the name of Hedge-Hyssop; for they take it for *Gratiola*, or Hedge-Hyssop, and sell it to such as are ignorant for the same. ‡

¶ *The Vertues.*

A *Galen*, *Dioscorides*, and *Theophrastus* doe account these for Milkeworts, and that they may without error be vsed for those purposes whereunto *Glaux* serueth.

B ‡ I doubt that this is not the *Polygalon* of *Dioscorides*; for *Gesner* affirmes that an handful hereof steeped all night in wine, and drunke in the morning, will purge choler effectually by stoole without any danger, as he himselfe had tried. ‡

CHAP. 170. *Of Knot-Grasse.*

¶ *The Description.*

1 THe common male Knot-grasse creepes along vpon the ground, with long slender weake branches full of knots or ioynts, whereof it tooke his name. The leaues grow vpon the weake branches, like those of small S. Iohns wort, but longer and narrower. The floures are maruellous little, and grow out of the knots, of an herby colour; in their places come vp triangular seed. The root is long, slender, and full of strings.

2 The second differeth not from the former, but onely that it is altogether lesser, wherein especially consisteth the difference. ‡ Because the difference is no otherwise, I hane thought good to omit the figure.

3 The Authors of the *Aduersaria* mention another larger Knot-grasse which growes in diuers places of the coast of the Mediterranean sea, hauing longer and larger branches and leaues, and those of a white shining colour. The seeds grow at the ioynts in chaffie white huskes; and the whole plant is of a salt and astringent taste. They call it *Polygonum marinum maximum*. ‡

¶ *The Place.*

These Knot-grasses do grow in barren and stony places almost euery where.

¶ *The Time.*

They are in floure and seed all the Sommer long.

¶ The

1 *Polygonum mas vulgare.*
Common Knot-grasse.

¶ *The Names.*

Knot-grasse is called of the Grecians, πολύγονον ἄῤῥεν: that is to say, *Polygonum mas*, or Male Knot-grasse: in Latine, *Seminalis, Sanguinaria*: of *Columella, Sanguinalis*: in shops, *Centumnodia*, and *Corrigiola*: of *Apuleius, Proserpinaca*: in high-Dutch, **Wegdrzyt**: in low-Dutch, **Verkens gras**, and **Duysent knop**: in Italian, *Polygono*: in Spanish, *Corriola*: in French, *Renouee*: in Wallon, *Mariolaine de Cure*: in English, Knot-grasse, and Swines grasse: In the North, Birds tongue.

¶ *The Temperature.*

Knot-grasse (as *Galen* teacheth) is of a binding qualitie, yet is it cold in the second, if not in the beginning of the third degree.

¶ *The Vertues.*

The iuyce of Knot-grasse is good against the spitting of bloud, the pissing of bloud, and all other issues or fluxes of bloud, as *Brasanolus* reporteth: and *Camerarius* saith he hath cured many with the iuyce thereof, that haue vomited bloud, giuen in a little stipticke Wine. It greatly preuaileth against the *Gonorrhæa*, that is, the running of the reines, and the weaknesse of the backe comming by meanes thereof, being shred and made in tansie with egges and eaten.

The decoction of it cures the disease aforesaid in as ample manner as the iuyce: or giuen in pouder in a reare egge it helpeth the backe very much. B

The herbe boyled in wine and hony cureth the vlcers and inflammations of the secret parts of man or woman, adding thereto a little allom, and the parts washed therewith. C

Dioscorides saith that it prouoketh vrine, and helpeth such as do pisse drop after drop, when the vrine is hot and sharpe. D

It is giuen vnto Swine with good successe, when they are sicke and will not eate their meate; whereupon the countrey people do call it Swines grasse and Swines skir. E

CHAP. 171. *Of sundry sorts of Knot-Grasses.*

¶ *The Description.*

1 THe snowie white and least kinde of *Polygonum* or Knot-grasse, called of *Clusius, Paronychia Hispanica*, is a strange and worthy plant to behold, handle, and consider, although it be but small. It is seldome aboue a foot long, hauing small branches, thicke, tough, hard, and full of ioynts; out of which the leaues come forth like small teeth, lesser than the leaues of *Herniaria*, or *Thymum tenuifolium*. At the top of the stalkes stand most delicate floures framed by nature as it were, with fine parchment leaues about them, standing in their singular whitenesse and snowie colour, resembling the perfect white silke, so many in number at the top, and so thicke, that they ouershadow the rest of the plant beneath. The root is slender, and of a wooddy substance. The seed is couered as it were with chaffe, as it were with chaffe, and is as small as dust, or the motes in the Sunne.

2 *Anthyllis* of Valentia, being likewise a kinde of Knot-grasse, hath small leaues like *Glaux exigua*, or rather like *Chamæsyce*, set orderly by couples at the ioynts: among which come floures consisting of foure little whitish purple leaues, and other small leaues like the first but altogether lesser. The root is small, blacke, and long, and of a wooddy substance.

‡ Our Author, though he meant to haue giuen vs the figure of Knawell in the third place, as may be perceiued by the title, yet he described it in the fourth, and in the third place went about to

1 *Polygonum montanum.* Mountaine Knot-graſſe.

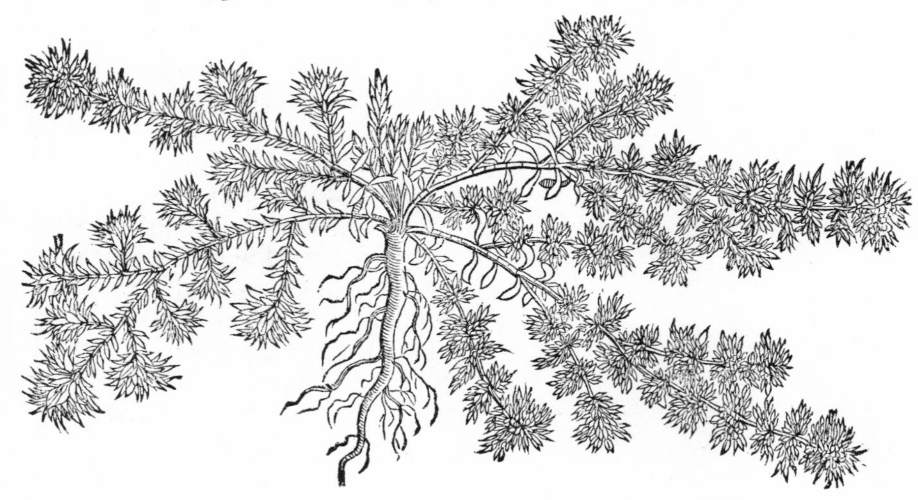

‡ 2 *Anthyllis Valentina Cluſij.* Valentia Knot-graſſe.

‡ 3 *Polygonum ſerpillifolium.*
Small round leaued Knot-graſſe.

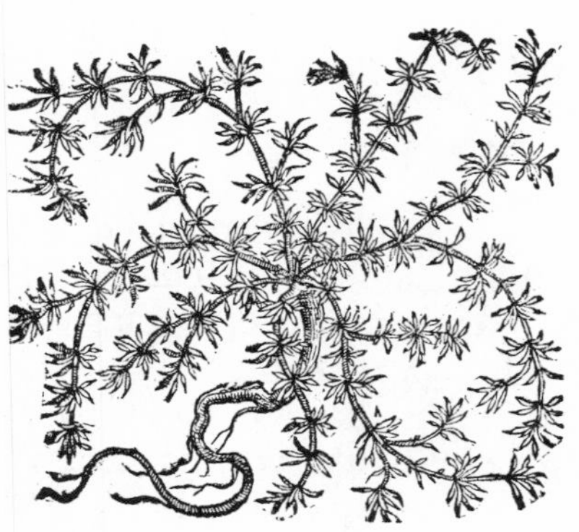

† 4 *Polygonum Selinoides, ſiue Knawel.*
Parſley Piert.

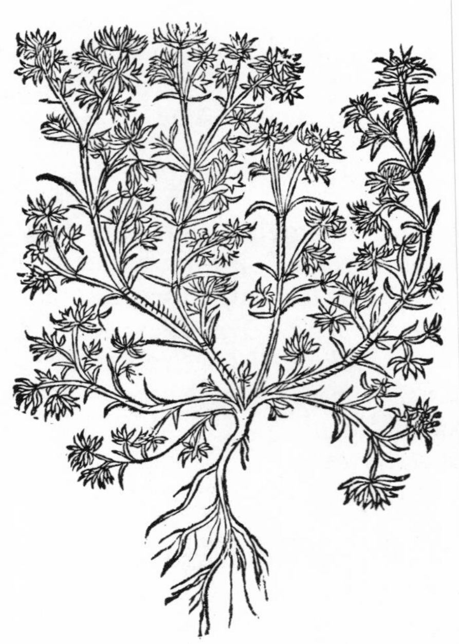

to describe *Polygonum Serpillifolio* of *Pena*; as may be gathered by the description which should haue stood, but that I opportunely receiued a better from my oft mentioned friend Mr. *Goodyer*, which therefore I thought good to impart vnto you.

Polygonum alterum pusillo vermiculato Serpilli foliolo Penæ.

This hath many small round smooth wooddy branches, somewhat reddish, trailing vpon the ground, nine inches or a foot long; whereon by small distances on short ioynts grow tufts of very small short blunt topped smooth greene leaues, in a maner round, like those of the smallest Time, but much smaller, and without smell, diuiding themselues at the bosomes of those leaues into small branches; at the tops of which branches grow small floures, one floure on a branch, and no more, consisting of foure little round topped leaues apiece of a faint or pale purplish colour: I obserued no seed. The root is wooddy, blackish without, very bitter, with some taste of heate, and groweth deepe into the ground. The leaues are nothing so full of iuyce as Aizoon. I found it flouring the third day of September, 1621, on the ditch bankes at Burseldon ferrey by the sea side in Hampshire. *Io. Goodyer.* ‡

4 Among the Knot grasses may well be suted this small plant, but lately written of, and not so commonly knowne as growing in England, being about an handfull high, and putting out from a fibrous root sundry slender stalkes full of little branches and ioynts: about which grow confusedly many narrow leaues, for the most part of an vnequall quantitie, yet here and there two longer than the rest, and much alike in greatnesse: at the outmost parts of the branches and stalks (where it hath thickest tufts) appeare out of the middest of the leaues little floures of an herby colour, which are succeeded by seed-vessels ending in fiue sharpe points: the whole plant is of a whitish colour. If my memorie faile me not, *Pena* means this herbe where he speaketh of *Saxifr. Angl.* in his *Aduer. p.* 103. and also reporteth that he found this plant by the way side as he rode from London to Bristow, on a little hill not far from Chipnam: his picture doth very well resemble the kinde of Knot-grasse called among the Germanes **Knawel**: and calling it *Saxifraga Anglicana* causeth me to thinke, that some in the West parts where he found it do call it Saxifrage, as we do call sundry other herbes, especially if they serue for the stone. My friend Mr. *Stephen Bredwel*, Practitioner of physicke in those parts, heard of a simple man who did much good with a medicine that he made with Parsley Piert against the stone, which he ministred vnto all sorts of people. This my friend requested the poore man to shew him the herbe called Parsley Piert; who frankly promised it him, and the next morning brought him an handfull of the herbe, and told him the composition of his medicine withall, which you shall finde set downe in the vertues, and proued by sundry of good account to be a singular remedie for the same.

† 5 *Saxifraga Anglicana alsinefolia.*
Chick-weed Breake-stone.

‡ 6 *Saxifraga palustris alsinefolia.*
Small water Saxifrage.

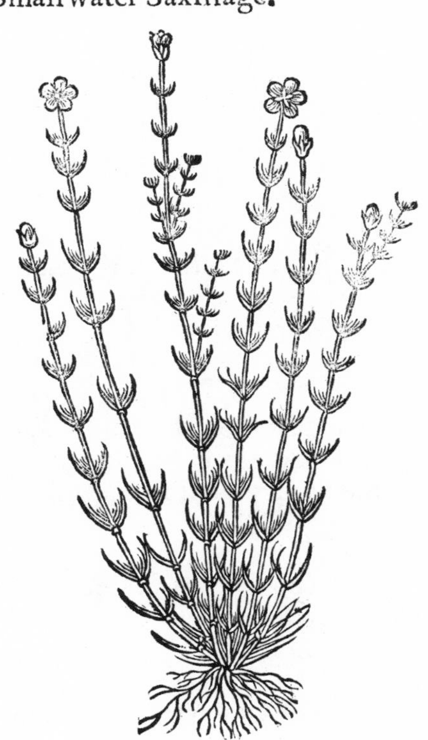

‡ 5 Our

‡ 4 Our Author here in the fourth place described the Knawel, and he figured it in the second place, vnder the title of *Anthyllis Valentina Clusij*: for the figure which was in the third place we here giue you in the fifth; and I coniecture it is not of Knawel, but of *Saxifraga Anglicana* of the *Aduers*. So that our Authors words are true, if he meant of the picture which he set forth by the name of *Polygonum selynoides siue Knawel*; but false if of the plant which he described. But if the coniecture of *Pena* and *Lobel* be true, who iudge their *Saxifraga Anglicana* to be *Synanchice Daleschampij*, then it is neither of *Knawel*, as our Author would haue it, nor of this which I here giue, but of a small plant which you shall finde amongst the *Rubia's*. Now this plant that I take to be the *Saxifraga Anglicana* of *Pena* and *Lobel*, is a small little herbe growing thicke, with very many branches some two or three inches high, with some stalkes standing vpright, and other some creeping: at each ioynt grow two short narrow sharpe pointed greene leaues, out of whose bosomes come diuers lesser leaues: at the tops of the branches vpon pretty long stalkes grow vpon each stalke one round whitish scaly head, consisting commonly of foure vnder greenish leaues which make the cup, and foure grayish or whitish leaues which are the floure. Now after these come to some maturitie they appeare all of a whitish colour, and through the thinne filmes of these heads appeares the seed, which at the first view seems to be pretty large and blacke; for it lies all clustering together; but if you rub it out you shall finde it as small as sand, and of a darke reddish colour. The taste of this plant is very hot and piercing, like that of Golden rod or our common Saxifrage, and without doubt it is more effectuall to moue vrine than the former Knawel. I haue found it growing in many places about bricke and stone walls, and vpon chalky barren grounds. I called this in my Iournall *Anno* 1632. *Saxifraga minor altera flosculis albis semine nigro*; and questioned whether it were not *Alsine Sassifraga angustifolia minima montana* of *Columna*. But now I thinke it rather (if the number of leaues in the floure did not disagree) the other which is described in the next place, of which I since that time haue receiued both the figure and description, as also a dry plant from Mr. *Goodyer*. He coniectures it may be this plant which I haue here described, that is set forth in the *Historia Lugd. pag.* 1235. by the name of *Alsine muscosa*.

Alsine palustris folijs tenuissimis: siue Saxifraga palustris alsinefolia.

6 This hath a great number of very small grasse-like leaues, growing from the root, about an inch long, a great deale smaller and slenderer than small pinnes; amongst which spring vp many small slender round smooth firme branches some handfull or handfull and halfe high, from which sometimes grow a few other smaller branches, whereon at certaine ioynts grow leaues like the former, and those set by couples with other shorter comming forth of their bosomes; and so by degrees they become shorter and shorter towards the top, so that toward the top this plant somwhat resembleth *Thymum durius*. The floures are great for the slendernesse of the plant, growing at the tops of the branches, each floure consisting of fiue smal blunt roundish topped white floures, with white chiues in the middest. The seed I obserued not. The root is small, growing in the myre with a few strings. This groweth plentifully on the boggy ground below the red Well of Wellingborough in Northampton shire. This hath not beene described that I finde. I obserued it at the place aforesaid, August 12. 1626. *Iohn Goodyer*. ‡

¶ *The Place.*

† The first and second are strangers in England: the rest grow in places mentioned in their descriptions.

¶ *The Time.*

These floure for the most part from May to September.

¶ *The Names.*

That which hath beene said of their names in their seuerall descriptions shall suffice.

¶ *The Nature.*

They are cold in the second degree, and dry in the third, astringent and making thicke.

‡ These, especially the three last, are hot in the second or third degree, and of subtill parts; but the Parsley Piert seemes not to be so hot as the other two. ‡

¶ *The Vertues.*

Here according to my promise I haue thought good to insert this medicin made with Knawel;
A which herbe is called (as I said before) Parsley Piert, but if I might without offence it should be called *Petra pungens*: for that barbarous word Parsley Piert was giuen by some simple man (‡ as also the other, that sauors of as much simplicitie ‡) who had not wel learned the true terme. The composition which followeth must be giuen in warme white wine, halfe a dram, two scruples, or more, according to the constitution of the body which is to receiue it.

The

The leaues of Parsley Piert, Mous-eare, of each one ounce when the herbes be dried, bay berryes, Turmericke, Cloues, the seeds of the great Burre, the feeds in the berryes of Hippes, or Briertree, Fenugreeke, of each one ounce, the stone in the oxe gall, the weight of 24. Barley cornes, or halfe a dram, made together into a most fine and subtill pouder, taken and drunke in maner aforesaid hath been proued most singular for the disease aforesaid. E

‡ The fifth and sixth are of the same faculty, and may be vsed in the like cases. ‡

† The figure that formerly was in the second place was of Knawell, and that in the third place of *polygonum minus polycarpon* of Tabern.

CHAP. 172. *Of Rupture woort.*

1 *Herniaria.* Rupture woort.

‡ 2 *Millegrana minima.* Dwarfe Allseed.

¶ *The Description.*

1 There is also a kinde of Knot grasse commonly called in Latine *Herniaria:* in English, Rupture woort, or Rupture grasse. It is a base and low creeping herbe, hauing many small slender branches trailing vpon the ground, yet very tough, and full of little knots somewhat reddish, whereupon do grow very many small leaues like those of Time; among which come forth little yellowish floures which turne into very small seed, and great quantity thereof, considering the smallnesse of the plant, growing thicke clustering together by certaine spaces. The whole plant is of a yellowish greene colour. The root is very slender and single.

2 There is another kinde of *Herniaria*, called *Mille grana* or All-seed, that groweth vpright a handfull high, with many small and tender branches, set with leaues like the former, but few in number, hauing as it were two smal leaues & no more. The whole plant seemeth as it were couered ouer with seeds or graines, like the seed of Panicke, but much lesser. ‡ I haue not seen many plants of this, but all that euer I yet saw neuer attained to the height of two inches. ‡

¶ *The Place.*

1 It ioyeth in barren and sandy grounds, and is likewise found in dankish places that lie wide open to the sunne: it doth grow and prosper in my garden exceedingly. ‡ 2 I found this in Kent on a Heath not farre from Chiste-hurst, being in company with Mr. *Bowles* and diuers others, in Iuly, 1630.

¶ *The Time.*

It floureth and flourisheth in May, Iune, Iuly, and August.

¶ *The Names.*

It is called of the later Herbarists *Herniaria* and *Herniola*; taken from the effect in curing the disease *Hernia*: of diuers, *Herba Turca*, and *Empetron*; in French, *Boutonet*: in English, Rupture woort, and Burstwoort.

¶ *The Temperature and Vertue.*

A Rupture woort doth notably drie, and throughly cloſeth vp together and faſteneth. It is reported that being drunke it is ſingular good for Ruptures, and that very many that haue been burſten, were reſtored to health by the vſe of this herbe; alſo the pouder hereof taken with wine, doth make a man to piſſe that hath his water ſtopt; it alſo waſteth away the ſtones in the kidnies, and expelleth them.

Chap. 173. *Of wilde Time.*

1 *Serpillum vulgre.*
Wilde Time.

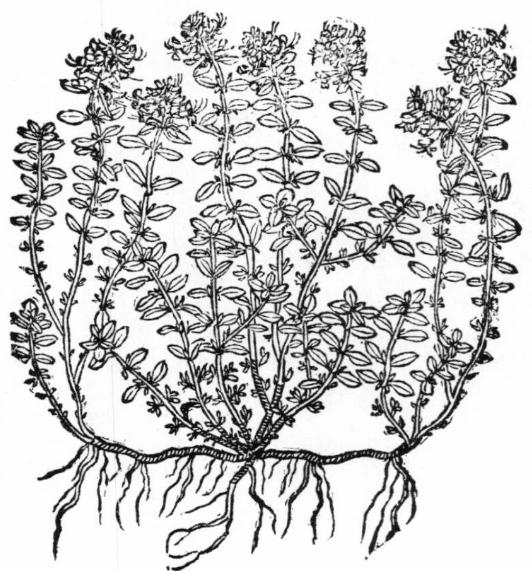

3 *Serpillum majus flore purpureo.*
Great purple wilde Time.

¶ *The Deſcription.*

1 Both *Dioſcorides* and *Pliny* make two kindes of *Serpillum*, that is, of creeping or wilde Time; whereof the firſt is our common creeping Time, which is ſo well knowne, that it needeth no deſcription; yet this ye ſhall vnderſtand, that it beareth floures of a purple colour, as euery body knoweth. Of which kinde I found another ſort, with floures as white as ſnow, and haue planted it in my garden, where it becommeth an herbe of great beauty:

2 This wilde Time that bringeth forth white floures differeth not from the other, but only in the colour of the floures, whence it may be called *Serpillum vulgare flore albo.* White floured Wilde Time.

There is another kinde of *Serpillum*, which groweth in gardens, in ſmell and ſauour reſembling Marjerome. It hath leaues like Organy, or wilde Marjerome, but ſomewhat whiter, putting forth many ſmall ſtalkes, ſet full of leaues like Rue, but longer, narrower, and harder. The floures are of a biting taſte, and pleaſant ſmell. The whole plant groweth vpright, whereas the other creepeth along vpon the earth, catching hold where it growes, & ſpreading it ſelfe far abroad.

3 This great wilde Time creepeth not as the others doe, but ſtandeth vpright, and bringeth forth little ſlender branches full of leaues like thoſe of Rue; yet narrower, longer, and harder. The flours be of a purple colour, and of a twinging biting taſte: it groweth vpon rocks, and is hotter than any of the others.

4 This other great one with white floures differeth not from the precedent, hauing many knaps or heads, of a milke white colour, which ſetteth forth the difference; and it may be called *Serpillum maius flore albo.* Great white floured wilde Time.

5 This wilde Time creepeth vpon the ground, ſet with many leaues by couples like thoſe of Marjerom, but leſſer, of the ſame ſmel: the flours are of a reddiſh color. The root is very threddy.

6 Wilde Time of Candy is like vnto the other wild Times, ſauing that his leaues are narrower and longer and more in number at each joint. The ſmell is more aromaticall than any of the others, wherein is the difference.

7 There is a kinde of wilde Time growing vpon the mountaines of Italy, called *Serpillum Citratum*,

5 *Serpillum folijs amaraci.*
Marjerome Time.

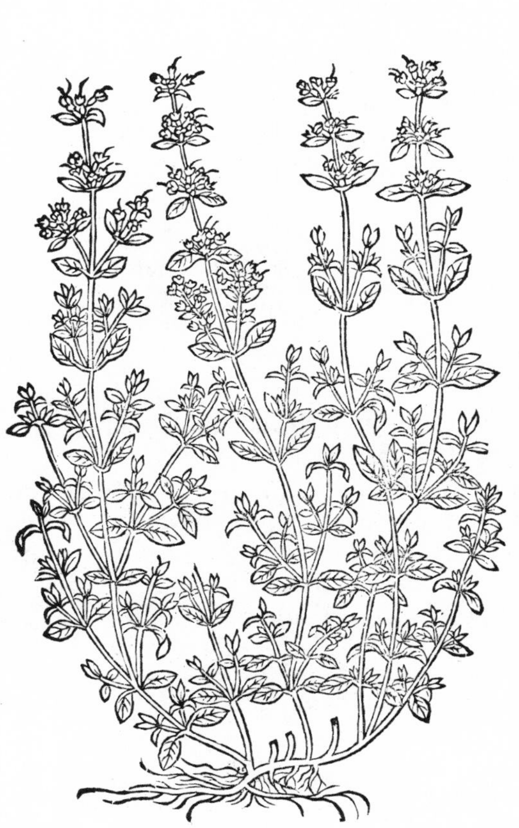

6 *Serpillum Creticum.*
Wilde Time of Candy.

‡ 7 *Serpillum citratum.*
Limon Time.

‡ 8 *Serpillum hirsutum.*
Hoary wilde Time.

Citratum, that is, hauing the smel of a Pome Citron, or a limon, which giueth it the difference from the other wilde times. ‡ It growes in many gardens also, and (as I haue been told) wilde in diuers places of Wales.

8 This (which is the *Serpillum Pannonicum* 3. of *Clusius*) runnes or spreds it selfe far vpon the ground. For though it haue a hard and woody root like as the former kindes, yet the branches which lie spred round about here and there take root, which in time become as hard and woody as the former. The leaues and stalkes are like those of the last described, but rough and hoarie: the floures also are not vnlike those of the common kind. The whole plant hath a kinde of resinous smell. It floures in Iune with the rest, and growes vpon the like mountainous places; bnt whether with vs in England or no I cannot yet affirme any thing of certaintie. ‡

¶ *The Place.*

The first groweth vpon barren hills and vntoiled places: the second groweth in gardens. The white kinde I found at Southfleet in Kent, in a barren field belonging to one M[r]. *William Swan.*

¶ *The Time.*

They floure from May to the end of Sommer.

¶ *The Names.*

Wild Time is called in Latine *Serpillum, à serpendo*, of creeping: in high and low-Dutch, **Quendel**, and **wilden Thymus**, and also **Onser Vrouwen bedstroo:** in Spanish, *Serpoll:* in Italian, *Serpillo:* in French, *Pillolet:* in English, Wilde Time, Puliall Mountaine, Pella mountaine, running Time, creeping Time, Mother of Time: in shops it is called *Serpyllum*; yet some call it *Pulegium montanum:* and it is euery where (saith *Dodonæus*) thought to be the *Serpyllum* of the Antients. Notwithstanding it answereth not so wel to the wilde Times as to *Dioscorides* his *Saxifranga*; for if it be diligently compared with the description of both the *Serpilla* and the *Saxifranga*, it shal be found to be little like the wilde Times, but very much like the *Saxifranga:* for (saith *Dioscorides*) *Saxifranga* is an herbe like Time, growing on rockes, where our common wilde Time is oftentimes found.

Ælianus in his ninth booke of his sundry Histories seemeth to number wilde Time among the floures. *Dionysius Iunior* (saith he) comming into the city Locris in Italy, possessed most of the houses of the city, and did strew them with roses, wild Time, and other such kindes of floures. Yet *Virgil* in the second Eclog of his Bucolicks doth most manifestly testifie that wilde Time is an herbe, in these words:

Thestylis & rapido fessis messoribus æstu
Allia, serpillumque, herbas contundit olentes.

Thestilis for mowers tyr'd with parching heate,
Garlicke, wilde Time, strong smelling herbes doth beate.

Out of which place it may be gathered, that common wilde time is the true and right *Serpillum*, or wilde Time, which the Grecians call ἕρπυλλος. *Marcellus* an old antient Author among the Frenchmen saith it is called *Gilarum*; as *Plinius Valerianus* saith it is called of the same, *Laurio.*

¶ *The Temperature.*

Wilde Time is of temperature hot and dry in the third degree: it is of thin and subtill parts, cutting and much biting.

¶ *The Vertues.*

A It bringeth downe the desired sicknesse, prouoketh vrine, applied in bathes and fomentations it procureth sweat: being boyled in wine, it helpeth the ague, it easeth the strangurie, it stayeth the hicket, it breaketh the stones in the bladder, it helpeth the Lethargie, frensie, and madnesse, and stayeth the vomiting of bloud.

B Wilde Time boyled in wine and drunke, is good against the wambling and gripings of the bellie, ruptures, conuulsions, and inflammations of the liuer.

C It helpeth against the bitings of any venomous beast, either taken in drinke, or outwardly applied.

D *Aetius* writeth, That *Serpillum* infused well in Vineger, and then sodden and mingled with rose water, is a right singular remedie to cure them that haue had a long phrensie or lethargie.

E *Galen* prescribeth one dram of the iuyce to be giuen in vineger against the vomiting of bloud, and helpeth such as are grieued with the spleene.

CHAP. 174. *Of Garden Time.*

¶ *The Description.*

1 THe first kinde of Time is so well knowne that it needeth no description; because there is not any which are ignorant what *Thymum durius* is, I meane our common garden Time.

2 The second kinde of Time with broad leaues hath many wooddy branches rising from a threddy root, beset with leaues like *Myrtus*. The floures are set in rundles about the stalke like Horehound. The whole plant is like the common Time in taste and smell.

1 *Thymum durius.*
Hard Time.

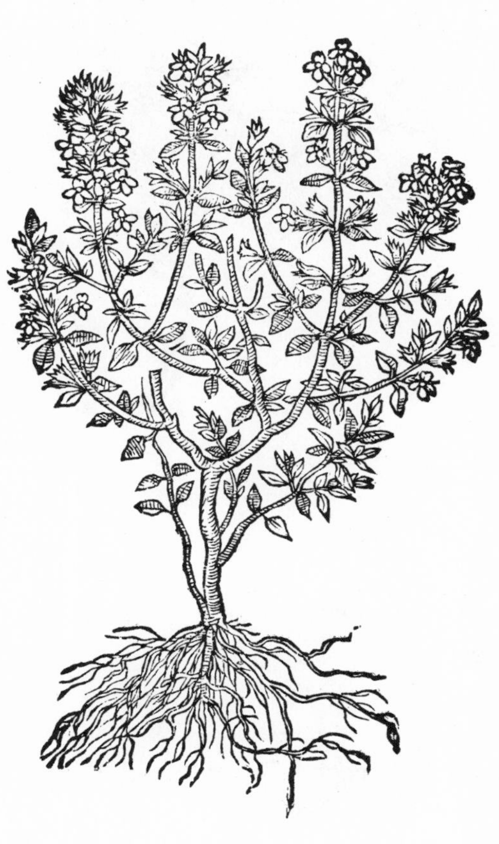

† 2 *Thymum latifolium.*
Great or broad leaued Time.

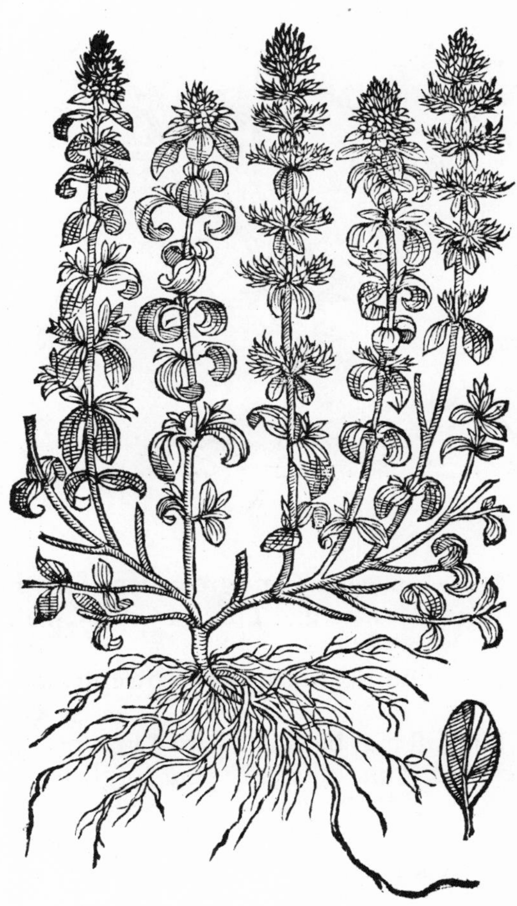

3 Time of Candy is in all respects like vnto common Time, but differeth in that, that this kinde hath certaine knoppy tufts not much vnlike the spikes or knots of *Stœcados*, but much lesser, beset with slender floures of a purple colour. The whole plant is of a more gracious smell than any of the other Times, and of another kinde of taste, as it were sauouring like spice. The root is brittle, and of a wooddy substance.

‡ 4 Doubtlesse that kinde of Time whereon *Epithymum* doth grow, and is called for that cause *Epithymum*, and vsed in shops, is nothing else than Dodder that growes vpon Time; and is all one with ours, though *Matthiolus* makes a controuersie and difference thereof: for *Pena* trauelling ouer the hills in Narbone neere the sea, hath seene not onely the garden Time, but the wilde Time also loden and garnished with this *Epithymum*. So that by his sight and mine owne knowledge I am assured, that it is not another kinde of Time that beareth *Epithymum*, but is common Time: for I haue often found the same in England, not onely vpon our Time, but vpon Sauorie, and other herbes also: notwithstanding thus much I may coniecture, that the clymate of those Countries doth yeeld the same forth in greater aboundance than ours, by reason of the intemperance of cold, whereunto our countrey is subiect.

† 3 *Thymum Creticum.*
Time of Candy.

4 *Epithymum Græcorum.*
Laced Time.

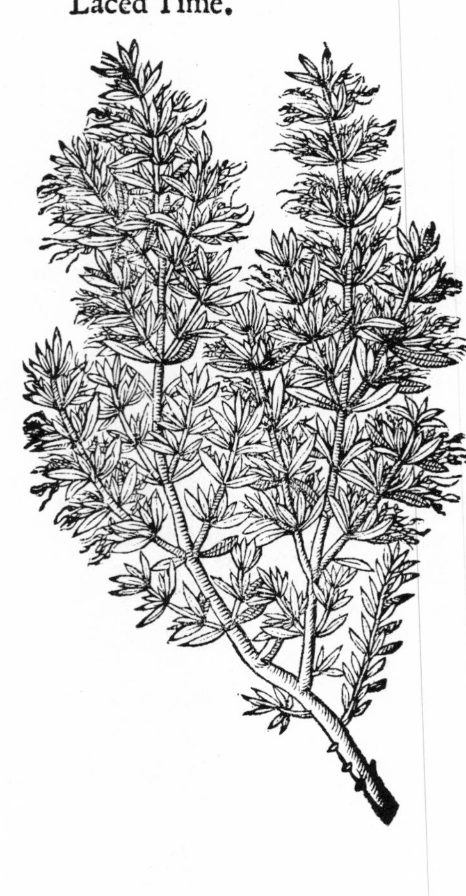

¶ *The Place.*

These kindes of Time grow plentifully in England in moſt gardens euery where, except that with broad leaues, and Time of Candy, which I haue in my garden.

¶ *The Time.*

They flouriſh from May vnto September.

¶ *The Names.*

The firſt may be called hard Time, or common garden Time: the ſecond, Broad leaued Time: the third, Time of Candy; our Engliſh women call it Muske Time: the laſt may be called Dodder Time.

¶ *The Temperature.*

Theſe kindes of Time are hot and dry in the third degree.

¶ *The Vertues.*

A Time boyled in water and hony drunken, is good againſt the cough and ſhortnes of the breath; it prouoketh vrine, expelleth the ſecondine or after-birth, and the dead childe, and diſſolues clotted or congealed bloud in the body.

B The ſame drunke with vineger and ſalt purgeth flegme: and boyled in Mede or Methegline, it cleanſeth the breaſt, lungs, reines, and matrix, and killeth wormes.

C Made into pouder, and taken in the weight of three drams with Mede or honied vineger, called Oxymel, and a little ſalt, purgeth by ſtoole tough and clammie flegme, ſharpe and cholericke humors, and all corruption of bloud.

D The ſame taken in like ſort, is good againſt the Sciatica, the paine in the ſide and breſt, againſt the winde in the ſide and belly, and is profitable alſo for ſuch as are fearefull, melancholike, and troubled in minde.

E It is good to be giuen vnto thoſe that haue the falling ſickneſſe to ſmell vnto.

F *Epithymum*, after *Galen*, is of more effectuall operation in phyſicke than Time, being hot and dry in the third degree, more mightily cleanſing, heating, drying, and opening than *Cuſcuta*, hauing right good effect to eradicat melancholy, or any other humor in the ſpleen, or other diſeaſe, ſprung by occaſion of the ſpleene.

It

It helpeth the long continued paines of the head, and beſides his ſingular effects about ſplene- G
ticall matters, it helpeth the lepry, or any diſeaſe of melancholy; all quartaine agues, and ſuch like griefes proceeding from the ſpleene.

Dioſcorides ſaith, *Epithymum* drunke with honied water, expelleth by ſiege, ſlegme, and melan- H
choly.

Of his natiue propertie it relieueth them which be melancholicke, ſwolne in the face and other I
parts, if you pound *Epithymum*, and take the fine pouder thereof in the quantity of foure ſcruples in the liquour which the Apothecaries call *Paſſum*, or with Oxymell and ſalt, which taketh away all flatuous humours and ventoſities.

† The ſecond figure was of *Serpillum Citratum* deſcribed in the ſeuenth place of the foregoing chapter; the third was of *Marum Matthioli, Tabern.* being the *Tragoriganum alterum* of *Lobel*.

CHAP. 175. *Of Sauorie.*

¶ *The Kindes.*

THere be two kindes of Sauorie, the one that indureth VVinter, and is of long continuance. the other an annuall or yearely plant, that periſheth at the time when it hath perfected his ſeed, and muſt be ſowne againe the next yeare; which we call Sommer Sauorie, or Sauorie of a yeare. There is likewiſe another, which is a ſtranger in England, called of *Lobel Thymbra S. Iuliani*, denying it to be the right *Satureia*, or Sauorie: whether that of *Lobel*, or that we haue in our Engliſh gardens be the true winter Sauorie, is yet diſputable; for we thinke that of S. Iulians rocke to be rather a wilde kinde than otherwiſe. ‡ *Pena* and *Lobel* do not denie, but affirme it in theſe words, *Nullus non fatetur Satureiam veram*; that is, which none can denie to be the true *Satureia* or Sauorie. *Vid. Aduerſar. pag.* 182. ‡

1 *Satureia hortenſis.*
VVinter Sauorie.

2 *Satureia hortenſis æſtiva.*
Sommer Sauorie.

¶ *The*

¶ *The Description.*

1 Winter Sauorie is a plant resembling Hyssope, but lower,more tender and brittle: it bringeth forth very many branches, compassed on euery side with narrow and sharpe pointed leaues,longer than those of Time;among which grow the floures from the bottome to the top, out of small husks, of colour white,tending to a light purple. The root is hard and wooddie,as is the rest of the plant.

2 Sommer Sauorie groweth vp with a slender brittle stalke of a foot high, diuided into little branches: the leaues are narrow, lesser than those of Hysope,like the leaues of winter Sauorie,but thinner set vpon the branches. The floures stand hard to the branches,of a light purple tending to whitenesse. The root is small,full of strings, and perisheth when it hath perfected his seed.

3 *Satureia Sancti Iuliani.*
Rocke Sauorie.

‡ 4 *Satureia Cretica.*
Candie Sauorie.

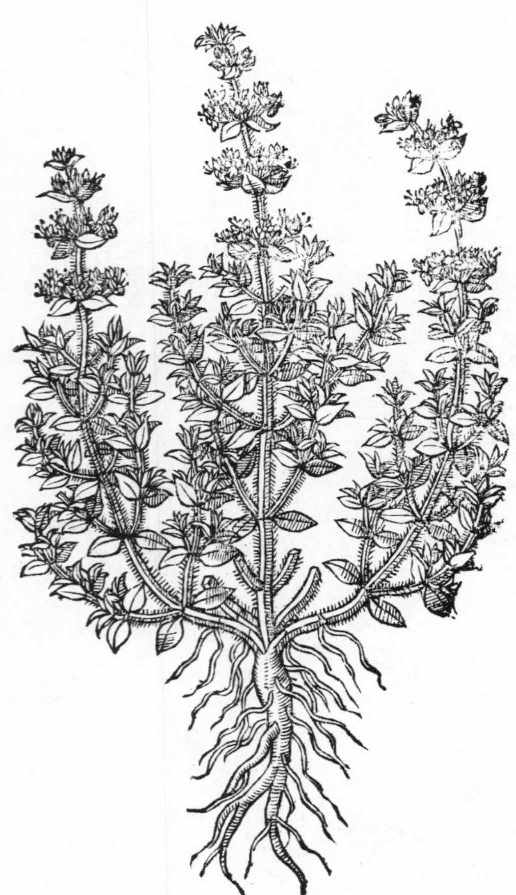

3 This small kinde of Sauorie, which *Lobel* hath set forth vnder the title of *Thymbra S. Iuliani*, because it groweth plentifully vpon the rough cliffes of the Tyrrhenian sea in Italie,called Saint Iulians rocke,hath tender twiggie branches an handfull high, of a wooddie substance,set ful of leaues from the bottome to the top, very thicke thrust together like vnto those of Time,sauing that they be smaller & narrower, bringing forth at the top of the sprigs a round spikie tuft of small purplish floures. The whole plant is whitish, tending to a bleake colour, and of a verie hot and sharpe taste, and also well smelling.

‡ 4 This in the opinion of *Honorius Bellus*,*Clusius*, and *Pona*, is thought, and not without good reason,to be the true *Thymbra*, or *Satureia* of *Dioscorides* and the Antients,for (besides that it agrees with their description, it is to this day called in Candie θυμβρι and θρυβι,) *Clusius* describes it thus:It sends forth many branches immediately from the roote like as Tyme,and those quadrangular,rough, and of a purplish colour: vpon these growe alternately little roughish leaues much like those of the true Tyme; and out of their bosomes come little branches set with the like, but lesser leaues. The toppes of the branches are compassed with a rundle made of manie little leaues, whereout come floures of a fine purple colour, and like the floures of Tyme, beeing diuided into foure parts, whereof the lower is the broader, and hangs downe: The vpper is also broad but shorter, and the other two lesse. Out of the middle of the floure come fiue whitish threds, pointed with browne,and a forked stile. The seed is small and blacke like that of Tyme. The root hard and wooddie. It floured with *Clusius* (who receiued the seedes out of Candie from *Honorius Bellus*) in October and Nouember. ‡

¶ *The Place.*

They are sowne in Gardens, and bring foorth their floures the first yeare of their sowing.

¶ The

¶ *The Time.*

They floure in Iuly and Auguſt.

¶ *The Names.*

Sauorie is called in Greeke θύμβρα, neither hath it any other true name in Latine than *Thymbra*. The Interpreters would haue it called *Satureia*, wherein they are repugnant to *Columella* a Latine Writer, who doth ſhew a manifeſt difference betweene *Thymbra* and *Satureia*, in his tenth booke, where hee writeth, that Sauorie hath the taſte of Tyme, and of *Thymbra* or the Winter Sauorie.

Et Satureia Thymi referens Thymbræq; ſaporem.

† Notwithſtanding this aſſertion of *Columella*, *Pliny lib.* 19. *cap* 8. makes *Satureia*, or Sauorie, to be that *Thymbra* which is called alſo *Cunila*. Sauorie in High Dutch is called **Kunel Saturey**, and **Sadaney**: in Low Dutch, **Ceulen**: which name, as it ſeemeth, is drawne out *Cunila*: in Italian, *Sauoreggia*: in Spaniſh, *Axedrea*, and *Sagorida*: in French, *Sarriette*: in Engliſh, Sauorie, Winter Sauorie, and Sommer Sauorie.

¶ *The Temperature and Vertues.*

Winter Sauorie is of temperature hot and drie in the third degree, it maketh thin, cutteth, it clenſeth the paſſages: to be briefe, it is altogether of like vertue with Time. A

Sommer Sauorie is not full ſo hot as winter Sauorie, and therefore ſaith *Dioſcorides*, more fit to be vſed in medicine: it maketh thin, and doth maruellouſly preuaile againſt winde: therefore it is with good ſucceſſe boiled and eaten with beanes, peaſon, and other windie pulſes, yea if it be applied to the belly in a fomentation, it forthwith helpeth the affects of the mother proceeding from winde. B

CHAP. 176. *Of Dodder.*

Cuſcuta ſiue Caſſutha.
Dodder.

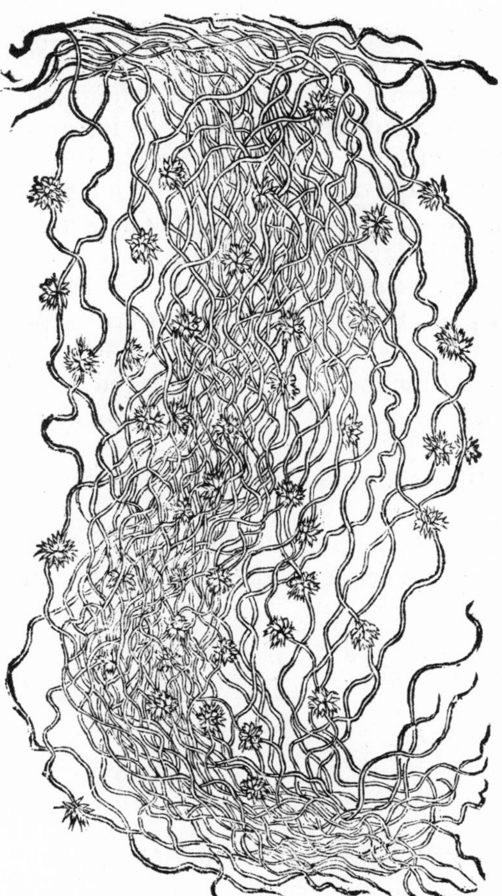

¶ *The Deſcription.*

Cuſcuta, or Dodder, is a ſtrange herbe, altogether without leaues or root, like vnto threds very much ſnarled or wrapped together, confuſedly winding it ſelfe about buſhes and hedges, and ſundry kindes of herbes. The threds are ſomewhat red: vpon which grow here & there little round heads or knops, bringing forth at the firſt ſlender white floures, afterward a ſmall ſeed.

¶ *The Place.*

This herbe groweth vpon ſundry kindes of herbes, as vpon Tyme, VVinter Sauorie, Germander, and ſuch like, taking his name from the herbe whereupon it doth grow, as that vpon Tyme is called *Epithymum*, vpon Line or flaxe *Epilinum*: and ſo of others, as *Dodonæus* ſetteth forth at large: yet hath he forgotten one among the reſt, which groweth very plentifully in Sommerſetſhire vpon nettles: neither is it the leaſt among many, either in beautie or operation, but comparable to the beſt *Epithymum*. following therefore the example of *Dioſcorides*, I haue thought good to call it *Epiurtica*, or rather Επακαλύφη, and ſo of the reſt according to the herbes whereon they do grow.

¶ *The Names.*

The greateſt is called in ſhops euery where *Cuſcuta*: and of diuers becauſe it groweth vpon Flaxe

flaxe or Lyne, *Podagra Lini*; the better learned do name it *Cassutha*, or *Cassytha*: and *Gesnerus*, λινόδεσμον: the Arabians, *Kessuth* and *Chasuth*: in Dutch, **Schorfte**, and **Wranghe**: in High Dutch, **Filkraut**: in French, *Goute d'Lin*, and *Tigne de Lin*: in English, Dodder.

The lesser and slenderer which wrappeth it selfe vpon Time and Sauorie, is called of *Dioscorides* ἐπίθυμον: the Apothecaries keep the name *Epithymum*. others, among whom is *Actuarius*, name that *Epithymum* which groweth vpon Tyme onely, and that which groweth on Sauorie *Epithymbrum*, and that also which hangeth vpon *Stæbe*, they terme *Epistæbe*, giuing a peculiar name to euery kind.

¶ *The Nature.*

The nature of this herb changeth and altereth, according to the nature and qualitie of the herbs whereupon it groweth: so that by searching of the nature of the plant you may easily finde out the temperament of the laces growing vpon the same. But more particularly: it is of temperature somewhat more drie than hot, and that in the second degree: it also clenseth with a certaine astrictiue or binding qualitie, and especially that which is found growing vpon the bramble: for it also receiueth a certaine nature from his parents on which it groweth; for when it groweth vpon the hotter herbes, as Tyme and Sauory, it becommeth hotter and drier, and of thinner parts: that which commeth of Broome prouoketh vrine more forcibly, and maketh the belly more soluble: and that is moister which groweth vpon flaxe: that which is found vpon the bramble hath ioined with it as wee haue said a binding qualitie, which by reason of this facultie ioyned with it is good to cure the infirmities of the Liuer and Milt· for seeing that it hath both a purging and binding facultie vnited to it, it is most singular good for the entrals: for *Galen* in his thirteenth Booke of the Methode of curing, doth at large declare that such Medicines are fittest of all for the Liuer and Milt.

¶ *The Vertues.*

A Dodder remoooeth the stoppings of the liuer and of the milt or spleene, it disburdeneth the veines of flegmaticke, cholericke, corrupt and superfluous humours: prouoketh vrine gently, and in a meane openeth the kidneies, cureth the yellow iaundise which are ioyned with the stopping of the liuer and gall: it is a remedie against lingring agues, bastard and long tertians, quartains also, and properly agues in infants and young children, as *Mesues* saith in *Scrapio*; who also teacheth, that the nature of Dodder is to purge choler by the stoole, and that more effectually if it haue Wormewood ioined with it; but too much vsing of it is hurtfull to the stomacke: yet *Auicen* writeth that it doth not hurt it, but strengtheneth a weake or feeble stomacke; which opinion also we do better allow of.

B *Epithymum*, or the Dodder which groweth vpon Tyme, is hotter and drier than the Dodder that groweth vpon flax, that is to say euen in the third degree, as *Galen* saith. It helpeth all the infirmities of the milt: it is a remedy against obstructions and hard swellings. It taketh away old headaches, the falling sicknesse, madnesse that commeth of Melancholy, and especially that which proceedeth from the spleene and parts thereabout: it is good for those that haue the French disease, and such as be troubled with contagious vlcers, the leprosie, and the scabbie euill.

C It purgeth downewards blacke and Melancholicke humours, as *Aetius*, *Actuarius*, and *Mesue* write, and also flegme, as *Dioscorides* noteth: that likewise purgeth by stoole which groweth vpon Sauorie and Scabious, but more weakly, as *Actuarius* saith.

D *Cuscuta*, or Dodder that groweth vpon flax, boiled in water or wine and drunke, openeth the stoppings of the liuer, the bladder, the gall, the milt, the kidneies and veines, and purgeth both by siege and vrine cholericke humours.

E It is good against the ague which hath continued a long time, and against the iaundise, I meane that Dodder especially that groweth vpon brambles.

F *Epiurtica* or Dodder growing vpon nettles, is a most singular and effectuall medicine to prouoke vrine, and to loose the obstructions of the body, and is proued oftentimes in the West parts with good successe against many maladies.

Chap. 177. *Of Hyssope.*

¶ *The Description.*

1 Dioscorides that gaue so many rules for the knowledge of simples, hath left Hyssope altogether without description, as beeing a plant so well knowne that it needed none: whose example I follow not onely in this plant, but in many others which bee common, to auoid tediousnesse to the Reader.

2 The

1 *Hyſſopus Arabum.*
Hyſſope with blew floures.

2 *Hyſſopus Arabum flore rubro.*
Hyſſope with reddiſh floures.

† 3 *Hyſſopus albis floribus.*
VVhite floured Hyſſope.

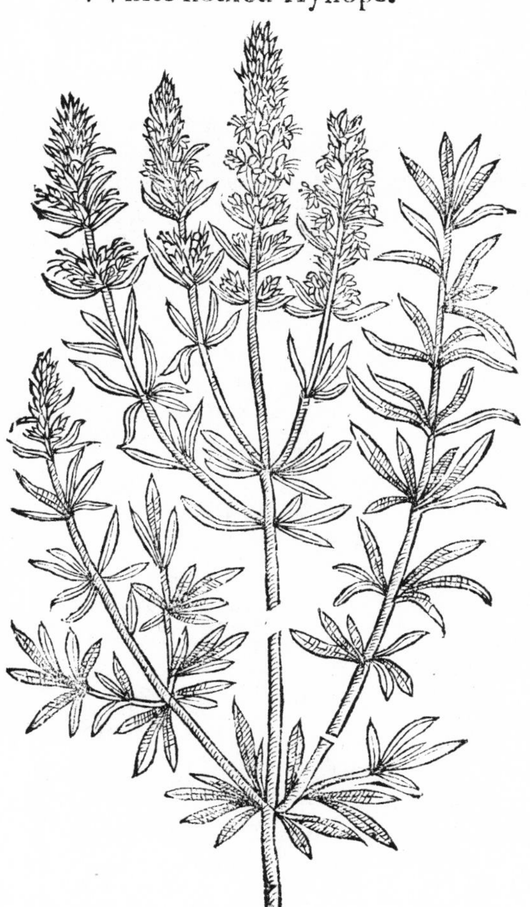

4 *Hyſſopus tenuifolia.*
Thinne leafed Hyſſope.

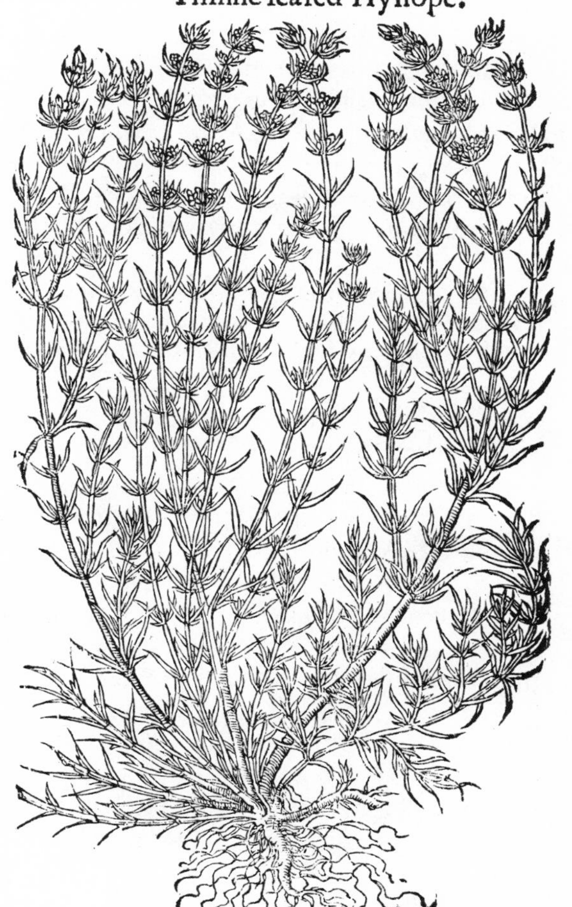

‡ 5 *Hyssopus parva angustis folijs.*
Dwarfe narrow leaued Hyssope.

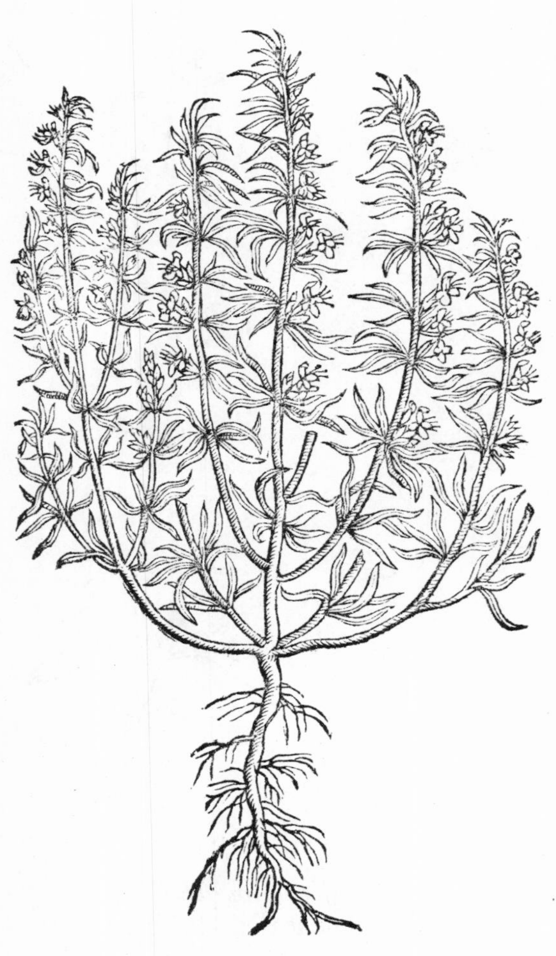

2 The second kind of Hyssope is like the former, which is our common Hyssope, and differeth in that, that this Hyssope hath his small and slender branches decked with faire red floures.

3 The third kinde of Hyssop hath leaues stalkes, branches, seed, and root, like the common Hyssope, and differeth in the floures only, which are as white as snow.

4 This kinde of Hyssope of all the rest is of the greatest beauty; it hath a wooddie root tough, and full of strings; from which rise vp small, tough, and slender flexible stalkes, wherupon do grow infinite numbers of small Fennel-like leaues, much resembling those of the smallest grasse, of a pleasant sweet smel, & aromatick taste, like vnto the rest of the Hyssops but much sweeter; at the top of the stalks do grow amongst the leaues smal hollow floures, of a blewish colour tending to purple. The seeds as yet I could neuer obserue.

‡ 5 This differs from the first described, in that the stalkes are weaker and shorter, the leaues also narrower, and of a darker colour: the floures grow after the same manner, & are of the same colour as those of the common kinde. ‡

We haue in England in our gardens another kinde, whose picture it shall be needlesse to expresse, considering that in few words it may be deliuered. It is like vnto the former, but the leaues are some of them white, some greene, as the other; and some green and white mixed and spotted, very goodly to behold.

Of which kinde we haue in our gardens moreouer another sort, whose leaues are wonderfully curled, rough, and hairie, growing thicke thrust together, making as it were a tuft of leaues; in taste and smell, and in all other things like vnto the common Hyssope.

I haue likewise in my garden another sort of Hyssope, growing to the forme of a small wooddie shrub, hauing very faire broad leaues like vnto those of *Numularia*, or Monywoort, but thicker, fuller of iuice, and of a darker greene colour; in taste and smell like the common Hyssope.

¶ *The Place.*

All these kindes of Hyssope do grow in my garden, and in some others also.

¶ *The Time.*

They floure from Iune to the end of August.

¶ *The Names.*

Hyssope is called in Latine *Hyssopus:* the which name is likewise retained among the Germans, Brabanders, French-men, Italians, and Spaniards. Therefore that shall suffice which hath been set downe in their seuerall titles.

‡ This is by most Writers iudged to be Hyssope vsed by the Arabian Physitions, but not that of the Greekes, which is neerer to *Origanum* and Maricrome, as this is to *Satureia* or Sauorie. ‡

¶ *The Temperature and Vertues.*

A A decoction of Hyssope made with figs, and gargled in the mouth and throte, ripeneth & breaketh the tumors and imposthumes of the mouth and throte, and easeth the difficultie of swallowing, comming by cold rheumes.

B The same made with figges, water, honie, and rue, and drunken, helpeth the inflammation of the lungs, the old cough, and shortnesse of breath, and the obstructions or stoppings of the breast.

C The sirrup or iuice of Hyssope taken with the sirrup of vineger, purgeth by stoole tough and clammie flegme, and driueth forth wormes if it be eaten with figges.

D The distilled water drunke, is good for those diseases before named, but not with that speed and force.

† That figure in the third place was of the *Satureia Romana.* 2. of *Tabernamontanus.*

CHAP.

CHAP. 178. *Of Hedge Hyffope.*

¶ *The Defcription.*

1 HEdge Hyffope is a low plant or herbe about a fpan long, very like vnto the common Hyffope, with many fquare ftalkes or flender branches, befet with leaues fomewhat larger than Hyffope, but very like. The floures grow betwixt the leaues vpon fhort ftems, of a white colour declining to blewneffe. All the herbe is of a moft bitter tafte, like the fmall Centory. The root is little and threddy, dilating it felfe farre abroad; by which meanes it multiplieth greatly, and occupieth much ground where it groweth.

1 *Gratiola.* Hedge Hyffope. ‡ 2 *Gratiola anguftifolia.* Graffe Poley. 3 *Gratiola latifolia.* Broad leaued Hedge Hyffope.

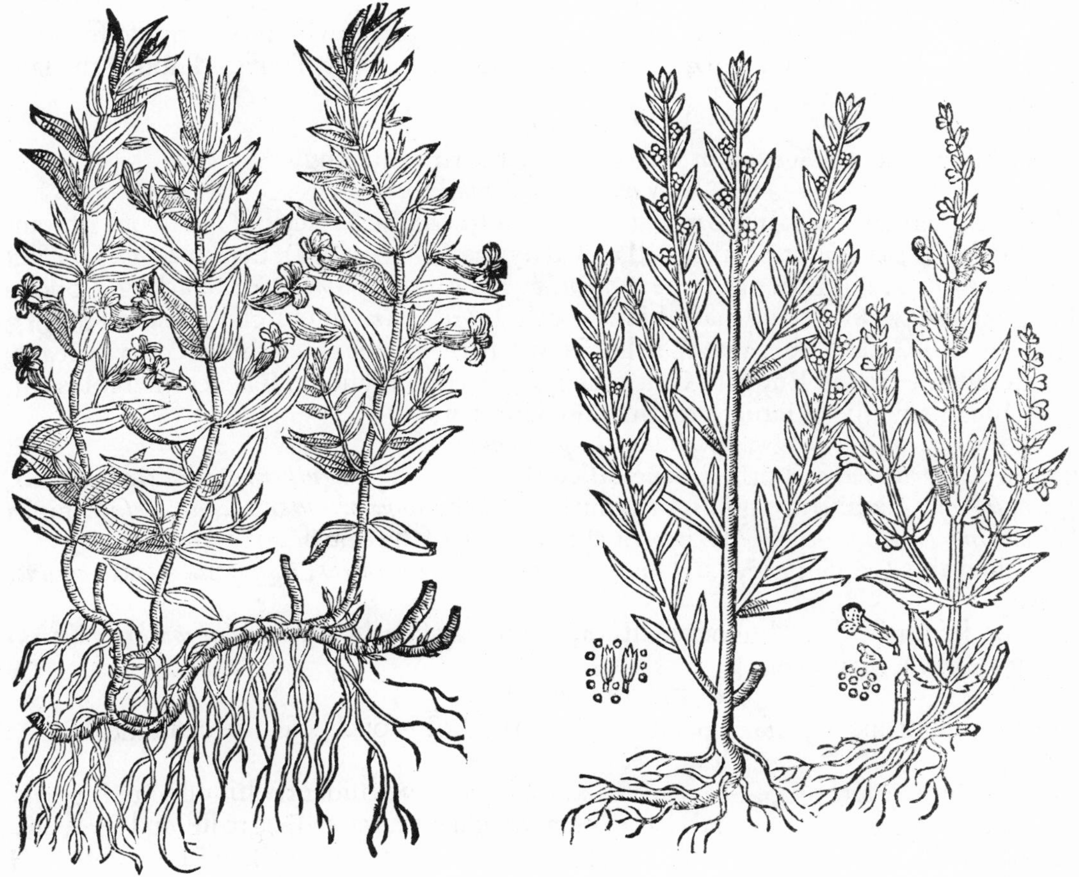

‡ 2 Narrow leaued Hedge Hyffope from a fmall fibrous white root fends vp a reddifh round crefted ftalke diuided into fundry branches, which are fet with leaues like thofe of knot graffe of a pale greene colour, and without any ftalkes: out of the bofome of thefe come floures fet in long cups compofed of foure leaues of a pleafing blew colour, which are fucceeded by longifh feed-veffells conteyning a fmall dusky feed. The whole plant is without fmell, neither hath it any bitterneffe or other manifeft tafte. It varies in leaues, fometimes broader, and otherwhiles narrower, the plant growing fometimes but an handfull, and otherwhiles a foot high. *Gefner* called this *Gratiola minor:* and *Camerarius*, *Hyffopoides:* and *Bauhine* onely hath figured it, and that by the name of *Hyffopifolia fiue Gratiola minor*. *Cordus* firft mentioned it, and that by the Dutch name of Graffe Poley, which name we may alfo very fitly retaine in Englifh. ‡

3 Broad leaued hedge Hyffope hath many fmall and tender branches, foure fquare, and fomewhat hollow or furrowed, befet with leaues by couples one oppofite againft another, like vnto the former, but fomewhat fhorter, and much broader: among which grow the floures of a purple colour,

colour, spotted on the inside with white, and of a brighter purple than the rest of the floure, fashioned like the smallest *Antirrhinum*, or least Snapdragon; which being past there succeed little seed vessels, fashioned like the nut of a crossebow, which containe small yellowish seed, extreame bitter of taste. The whole plant is likewise bitter, as the common or well knowne *Gratiola*. The root is compact of a great number of whitish strings, entangled one within another, which mightily encreaseth and spreadeth abroad.

‡ This plant is onely a lesser kinde of the *Lysimachia galericulata* of *Lobell*, which some haue called *Gratiola latifolia*. our Authors figure was very ill, wherefore I haue endeauoured by the helpe of some dried plants and my memory to present you with a better expression thereof. ‡

¶ *The Place.*

The first groweth in low and moist places naturally, which I haue planted in my Garden. ‡ The second was found growing by my oft mentioned friend M[r]. *Bowles* at Dorchester in Oxfordshire, at the backe side of the enclosed grounds on the left hand of the towne, if you would ride from thence to Oxford in the grassie places of the Champion corne fields. ‡ The third groweth likewise in moist places. I found it growing vpon the bog or marrish ground at the further end of Hampstead heath, and vpon the same heath towards London, neere vnto the head of the springs that were digged for water to be conueied to London, 1590. attempted by that carefull citizen *Iohn Hart* Knight, Lord Major of the City of London: at which time my selfe was in his Lordships company, and viewing for my pleasure the same goodly springs, I found the said plant, not heretofore remembred.

¶ *The Time.*

The first floureth in May: the second in Iune and Iuly: the third in August.

¶ *The Names in generall.*

Hedge Hyssope is called in Latine *Gratiola*, and *Gratia Dei*, or the Grace of God; notwithstanding there is a kind of *Geranium*, or Storkes bill, called by the later name. Of *Cordus*, *Limnesium*, and *Centauroides*: of *Anguillaria* it is thought to be *Dioscorides* his *Papauer spumeum*, or Spatling Poppy: but some think *Papauer spumeum* to be that which we call *Behen album*: in Dutch it is called **Godts gratie**: in Italian, *Stanca cauallo*, because that horses when they haue eaten thereof do wax leane, and languish thereupon: and in English, Gratia Dei, and Hedge Hyssope. The seed hereof is called *Gelbenech*, which name the Arabians retaine vnto this day.

‡ ¶ *Names in particular.*

‡ 1 *Matthiolus*, *Dodonæus* and others haue called this *Gratiola*; *Anguillara*, *Gratia Dei*; *Cordus*, *Limnesium*, *Centauroides*; he also thought it but vnfitly to be the *Eupatoreum* of *Mesue*: *Gesner* thinks it may be *Polemonium palustre amarum* of *Hippocrates*, that write of the diseases of cattell. ‡

2 *Cordus* called this Grasse Poley; *Gesner*, *Gratiola minor*; *Camerarius*, *Hyssopoides*: and *Bauhine*, *Hyssopifolia*.

3 This is not set forth by any but our Author, and it may fitly be named *Lysimachia galericulata minor*, as I haue formerly noted. ‡

¶ *The Temperature.*

Hedge Hyssope is hot and dry of temperature. And the first is onely vsed in medicine.

¶ *The Vertues.*

A Who so taketh but one scruple of *Gratiola* brused, shall perceiue euidently his effectuall operation and vertue, in purging mightely, and that in great abundance, waterish, grosse, and slimy humors. *Conradus Gesnerus* experimented this, and found it to be true, and so haue I my selfe, and many others.

B *Gratiola* boiled, and the decoction drunke or eaten with any kinde of meate, in manner of a sallade, openeth the belly, and causeth notable loosenes, and to scoure freely, and by that meanes purgeth grosse flegme and cholericke humors.

C *Gratiola* or Hedge Hyssope boiled in wine and giuen to drinke, helpeth feuers of what sort soeuer, and is most excellent in dropsies, and such like diseases proceeding of cold and watery causes.

D The extraction giuen with the powder of cinamon and a little of the juice of Calamint, preuaileth against tertian and quotidian feuers, set downe for most certaine by the learned *Ioachimus Camerarius*.

Chap. 179. *Of Lauander Spike.*

¶ *The Deſcription.*

1 Lauander Spike hath many ſtiffe branches of a wooddie ſubſtance, growing vp in the manner of a ſhrubbe, ſet with many long hoarie leaues, by couples for the moſt part; of a ſtrong ſmell, and yet pleaſant enough to ſuch as doe loue ſtrong ſauours. The floures grow at the top of the branches ſpike faſhion, of a blew colour. The roote is hard and wooddie.

2 The ſecond differeth not from the precedent, but in the colour of the floures: For this Plant bringeth milke white floures; and the other blew, wherein eſpecially conſiſteth the difference.

3 Wee haue in our Engliſh gardens a ſmall kinde of Lauander, which is altogether leſſer than the other, ‡ and the floures are of a more purple colour and grow in much leſſe and ſhorter heads, yet haue they a farre more gratefull ſmell: the leaues are alſo leſſe and whiter than thoſe of the ordinarie ſort. This did, and I thinke yet doth grow in great plentie, in his Maieſties priuate Garden at White-hall. And this is called Spike, without addition, and ſometimes Lauander Spike: and of this by diſtillation is made that vulgarly known and vſed oile which is termed *Oleum ſpicæ*, or oile of Spike. ‡

1 *Lavandula flore cæruleo.*
Common Lauander.

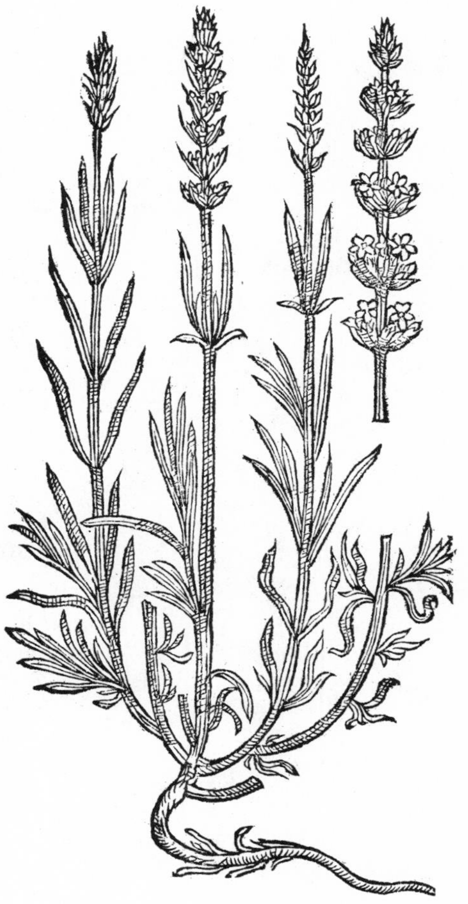

2 *Lavandula flore albo.*
VVhite floured Lauander.

¶ *The Place.*

In Spaine and Languedocke in France, moſt of the mountaines and deſert fields, are as it were coured

3 *Lavendula minor, sive Spica.*
Lauander Spike.

couered ouer with Lauander. In these cold countries they are planted in gardens.

¶ *The Time.*

They floure and flourish in Iune and Iuly.

¶ *The Names.*

Lauander Spike is called in Latine *Lauendula*, and *Spica*: in Spanish, *Spigo*, and *Languda*. The first is the male, and the second the female. It is thought of some to be that sweet herbe *Casia*, whereof Virgil maketh mention in the second Eclog of his Bucolicks:

Tum Casia atque aliis intexens suavibus herbis,
Mollia luteola pingit vacinia Caltha.

(infold)
And then shee'l Spike and such sweet herbs
And paint the Iacinth with the Marygold.

And likewise in the fourth of his Georgicks, where hee intreateth of choosing of seats and places for Bees, and for the ordering thereof, he saith thus:

Hæc circum Casiæ virides, & olentia late
Serpilla, & grauiter spirantis copia Thymbræ
Floreat, &c. ——, ———

About them let fresh Lauander and store
Of wild Time with strong Sauorie to floure.

Yet there is another *Casia* called in shops *Casia Lignea*, as also *Casia nigra*, which is named *Casia fistula*; and another a small shrubbie plant extant among the shrubs or hedge bushes, which some thinke to be the *Casia Poetica*, mentioned in the precedent verses.

¶ *The Temperature.*

Lauander is hot and drie, and that in the third degree, and is of a thin substance, consisting of many airie and spirituall parts. Therefore it is good to be giuen any way against the cold diseases of the head, and especially those which haue their originall or beginning not of abundance of humours, but chiefely of a cold quality onely.

¶ *The Vertues.*

A The distilled water of Lauander smelt vnto, or the temples and forehead bathed therewith, is a refreshing to them that haue the Catalepsie, a light Migram, & to them that haue the falling sicknesse, and that vse to swoune much. But when there is abundance of humours, especially mixt with bloud, it is not then to be vsed safely, neither is the composition to be taken which is made of distilled wine: in which such kinde of herbes, floures, or seeds, and certaine spices are infused or steeped, though most men do rashly and at aduenture giue them without making any difference at all For by vsing such hot things that fill and stuffe the head, both the disease is made greater, and the sicke man also brought into danger, especially when letting of bloud, or purging haue not gone before. Thus much by way of admonition, because that euery where some vnlearned Physitions and diuers rash and ouerbold Apothecaries, and other foolish women, do by and by giue such compositions, and others of the like kinde, not only to those that haue the Apoplexy; but also to those that are taken, or haue the Catuche or Catalepsis with a Feuer; to whom they can giue nothing worse, seeing those things do very much hurt, and oftentimes bring death it selfe.

B The floures of Lauander picked from the knaps, I meane the blew part and not the huske, mixed with Cinamon, Nutmegs, and Cloues, made into pouder, and giuen to drinke in the distilled water thereof, doth helpe the panting and passion of the heart, preuaileth against giddinesse, turning, or swimming of the braine, and members subiect to the palsie.

C Conserue made of the floures with sugar, profiteth much against the diseases aforesaid, if the quantitie of a beane be taken thereof in the morning fasting.

D It profiteth them much that haue the palsie, if they bee washed with the distilled water of the floures,

floures, or annointed with the oile made of the floures, and oile oliue, in ſuch manner as oile of roſes is, which ſhall be expreſſed in the treatiſe of Roſes.

Chap. 180. *Of French Lauander, or Stickeadoue.*

¶ *The Deſcription.*

1 French Lauander hath a bodie like Lauander, ſhort, and of a wooddie ſubſtance, but ſlenderer, beſet with long narrow leaues, of a whitiſh colour, leſſer than thoſe of Lauander: it hath in the top buſhy or ſpikie heads, well compact or thruſt together; out of the which grow forth ſmall purple floures, of a pleaſant ſmell. The ſeede is ſmall and blackiſh: the roote is hard and woodd ie.

2 This iagged Sticadoue hath many ſmall ſtiffe ſtalks of a wooddy ſubſtance; whereupon do grow iagged leaues in ſhape like vnto the leaues of Dill, but of an hoarie colour: on the top of the ſtalkes do grow ſpike floures of a blewiſh colour; and like vnto the common Lauander Spike: the root is likewiſe wooddie. ‡ This by *Cluſius* who first deſcribed it, as alſo by *Lobel*, is called *Lavendula multifido folio*, or Lauander with the diuided leafe; the plant more reſembling Lauander than Sticadoue. ‡

3 There is alſo a certaine kind e hereof, differing in ſmalneſſe of the leaues onely, which are round about the edges nicked or toothed like a ſaw, reſembling thoſe of Lauander cotton. The root is likewiſe wooddie.

‡ 4 There is alſo another kinde of *Stœchas* which differs from the firſt or ordinarie kind, in that the tops of the ſtalkes are not ſet with leaues almoſt cloſe to the head as in the common kinde, but are naked and wholly without leaues: alſo at the tops of the ſpike or floures (as it were to recompence their defect below) there growe larger and fairer leaues than in the other ſorts. The other parts of the plant differ not from the common *Stœchas*. ‡

†1 *Stœchas ſiue ſpica hortulana.*
Sticadoue and Sticados.

2 *Stœchas multifida.*
Iagged Sticados.

¶ *The*

3 *Stœchas folio serrato.*
Toothed Sticadoue.

‡ 4 *Stœchas summis cauliculis nudis.*
Naked Sticadoue.

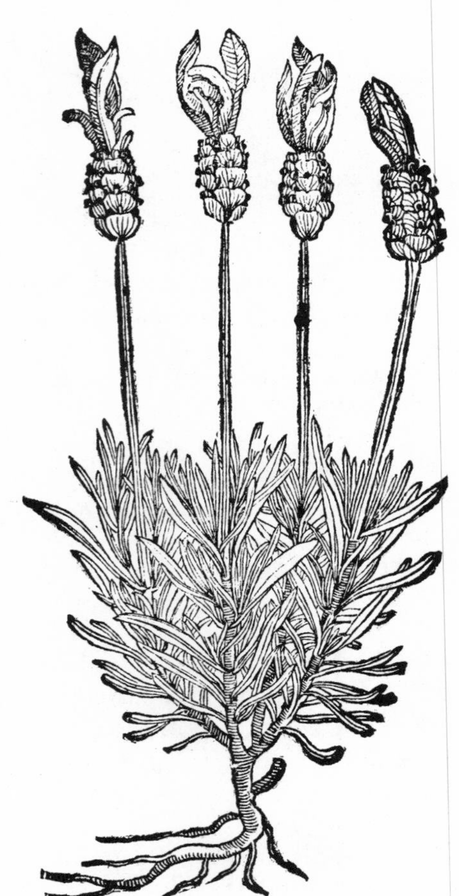

¶ *The Place.*

These herbes do grow wilde in Spaine, in Languedocke in France, and the Islands called Stœchades ouer against Massilia: we haue them in our gardens, and keepe them with great diligence from the iniurie of our cold clymate.

¶ *The Time.*

They are sowne of seed in the end of Aprill, and couered in the Winter from the cold, or els set in pots or tubs with earth, and carried into houses.

¶ *The Names.*

The Apothecaries call the floure *Stœcados*: *Dioscorides*, στιχὰς: *Galen*, στοιχὰς, by the dipthong οι in the first syllable: in Latine, *Stœchas*: in High Dutch, **Stichas kraut**: in Spanish, *Thomani*, and *Cantuesso*: in English, French Lauander, Steckado, Stickadoue, Cassidonie, and some simple people imitating the same name do call it Cast me downe.

¶ *The Temperature.*

French Lauander saith *Galen* is of temperature compounded of a little cold earthie substance, by reason whereof it bindeth: it is of force to take away obstructions, to extenuate or make thinne, to scoure and clense, and to strengthen not onely all the entrails, but the whole bodie also.

¶ *The Vertues.*

A *Dioscorides* teacheth that the decoction hereof doth helpe the diseases of the chest, and is with good successe mixed with counterpoisons.

B The later Physitions affirme, that *Stœchas*, and especially the floures of it, are most effectuall against paines of the head, and all diseases thereof proceeding of cold causes, and therefore they be mixed in all compositions almost which are made against head-ache of long continuance, the Apoplexie, the falling sicknesse, and such like diseases.

C The decoction of the husks and floures drunke, openeth the stoppings of the liuer, the lungs, the milt, the mother, the bladder, and in one word all other inward parts, clensing and driuing forth all euill and corrupt humours, and procuring vrine.

Chap.

CHAP. 181. *Of Flea-wort.*

¶ *The Description.*

1 PSyllium, or the common Flea-wort hath many round and tender branches, set full of long and narrow leaues somewhat hairy. The top of the stalkes are garnished with sundrie round chaffie knops, beset with small yellow floures: which being ripe containe many little shining seeds, in proportion, colour, and bignesse like vnto fleas.

2 The second kinde of *Psyllium* or Flea-wort hath long and tough branches, of a wooddy substance like the precedent, but longer and harder, with leaues resembling the former, but much longer and narrower. The chaffie tuft which containeth the seed is like the other, but more like the eare of *Phalaris*, which is the eare of *Alpisti*, the Canarie seed which is meate for birds that come from the Islands of Canarie. The root hereof lasteth all the Winter, and likewise keepeth his greene leaues, whereof it tooke this addition of *Sempervirens*.

1 *Psyllium siue pulicaris herba.*
Flea-wort.

2 *Psyllium sempervirens Lobelij.*
Neuer dying Flea-wort.

¶ *The Place.*

These plants are not growing in our fields of England, as they doe in France and Spaine, yet I haue them growing in my garden.

¶ *The Time.*

They floure in Iune and Iuly.

¶ *The Names.*

Flea-wort is called in Greeke ψύλλιον: in Latine, *Pulicaria*, and *Herba Pulicaris*: in shops, *Psyllium*: in English, Flea-wort; not because it killeth fleas, but because the seeds are like fleas: of some, Flea-bane, but vnproperly: in Spanish, *Zargatona*: in French, *L'herbe aus pulces*: in Dutch, **Vuyls vloye-cruyt.**

¶ *The Temperature.*

Galen and *Serapio* record, that the seed of *Psyllium* (which is chiefely vsed in medicine) is co'd in the second degree, and temperate in moisture and drinesse.

¶ The

¶ *The Vertues.*

A The seed of Flea-wort boyled in water or infused, and the decoction or infusion drunke, purgeth downewards adust and cholericke humors, cooleth the heate of the inward parts, hot feauers, burning agues, and such like diseases proceeding of heate, and quencheth drought and thirst.

B The seed stamped, and boyled in water to the forme of a plaister, and applied, taketh away all swellings of the ioynts, especially if you boyle the same with vineger and oyle of Roses, and apply it as aforesaid.

C The same applied in manner aforesaid vnto any burning heate, called S. Anthonies fire, or any hot and violent impostume, asswageth the same, and bringeth it to ripenesse.

D Some hold that the herbe strowed in the chamber where many fleas be, will driue them away; for which cause it tooke the name Flea-wort: but I thinke it is rather because the seed doth resemble a flea so much, that it is hard to discerne the one from the other.

¶ *The Danger.*

Too much Flea-wort seed taken inwardly is very hurtful to mans nature: so that I wish you not to follow the minde of *Galen* and *Dioscorides* in this point, being a medicine rather bringing a maladie, than taking away the griefe: remembring the old prouerbe, A man may buy gold too deare; and the hony is too deare that is lickt from thornes.

‡ *Dioscorides* nor *Galen* mention no vse of this inwardly; but on the contrarie, *Dioscorides* in his sixth booke, which treats wholly of the curing and preuenting of poysons, mentions this in the tenth chapter for a poyson, and there sets downe the symptomes which it causes, and refers you to the foregoing chapter for the remedies. ‡

Chap. 185. *Of Cloue Gillofloures.*

Caryophyllus maximus multiplex.
The great double Carnation.

2 *Caryophyllus multiplex.*
The double Cloue Gillofloure.

¶ *The Kindes.*

There are at this day vnder the name of *Caryophyllus* comprehended diuers and sundry sorts of plants, of such various colours, and also seuerall shapes, that a great and large volume would not

not suffice to write of euery one at large in particular; considering how infinite they are, and how euery yeare euery clymate and countrey bringeth forth new sorts, such as haue not heretofore bin written of; some whereof are called Carnations, others Cloue Gillofloures, some Sops in wine, some Pagiants, or Pagion colour, Horse-flesh, blunket, purple, white, double and single Gillofloures, as also a Gillofloure with yellow floures: the which a worshipfull Merchant of London Mr. *Nicolas Lete* procured from Poland, and gaue me thereof for my garden, which before that time was neuer seene nor heard of in these countries. Likewise there be sundry sorts of Pinkes comprehended vnder the same title, which shall be described in a seuerall chapter. There be vnder the name of Gillofloures also those floures which wee call Sweet-Iohns and Sweet-Williams. And first of the great Carnation and Cloue Gillefloure.

‡ There are very many kindes both of Gillofloures, Pinkes, and the like, which differ very little in their roots, leaues, seeds, or manner of growing, though much in the colour, shape, and magnitude of their floures; wherof some are of one colour, other some of more; and of them some are striped, others spotted, &c. Now I (holding it a thing not so fit for me to insist vpon these accidentall differences of plants, hauing specifique differences enough to treat of) refer such as are addicted to these commendable and harmelesse delights to suruey the late and oft mentioned Worke of my friend Mr. *Iohn Parkinson*, who hath accurately and plentifully treated of these varieties; and if they require further satisfaction, let them at the time of the yeare repaire to the garden of Mistresse *Tuggy* (the wife of my late deceased friend Mr. *Ralph Tuggy*) in Westminster, which in the excellencie and varietie of these delights exceedeth all that I haue seene: as also hee himselfe whilest he liued exceeded most, if not all of his time, in his care, industry, and skill in raising, encreasing, and preseruing of these plants and some others; whose losse therefore is the more to be lamented by all those that are louers of plants. I will onely giue you the figures of some three or foure more, whereof one is of the single one, which therefore some terme a Pinke, though in mine opinion vnfitly, for that it is produced by the seed of most of the double ones, and is of different colour and shape as they are, varying from them onely in the singlenesse of the floures. ‡

‡ *Caryophyllus maior & minor, rubro & albo variegati.*
The white Carnation, and Pageant.

‡ *Caryophyllus purpureus profunde laciniatus.*
The blew, or deep purple Gillofloure.

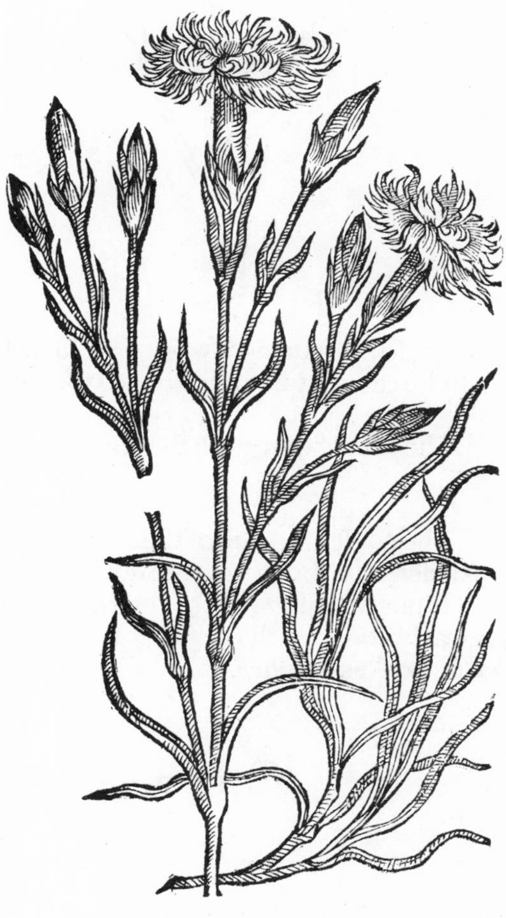

¶ The

¶ The Description.

1 THe great Carnation Gillow-floure hath a thicke round wooddy root, from which riseth vp many strong ioynted stalkes set with long greene leaues by couples: on the top of the stalkes do grow very faire floures of an excellent sweet smell, and pleasant Carnation colour, whereof it tooke his name.

2 The Cloue Gillofloure differeth not from the Carnation but in greatnesse as well of the floures as leaues. The floure is exceeding well knowne, as also the Pinks and other Gillofloures; wherefore I will not stand long vpon the description.

‡ *Caryophyllus simplex maior.*
The single Gillofloure or Pinke.

¶ *The Place.*

These Gillofloures, especially the Carnations, are kept in pots from the extremity of our cold Winters. The Cloue Gillofloure endureth better the cold, and therefore is planted in gardens.

¶ *The Time.*

They flourish and floure most part of the Sommer.

¶ *The Names.*

The Cloue Gillofloure is called of the later Herbarists *Caryophylleus flos*, of the smell of cloues wherewith it is possessed: in Italian, *Garofoli*: in Spanish, *Clauel*: in French, *Oeilletz*: in low-Dutch, **Ginoffelbloemen**: in Latine of most; *Ocellus Damascenus, Ocellus Barbaricus*, and *Barbarica*: in English, Carnations, and Cloue Gillofloures. Of some it is called *Vetonica*, and *Herba Tunica*. The which *Bernardus Gordonius* hath set downe for *Dioscorides* his *Polemonium*.

That worthy Herbarist and learned Physition of late memorie Mr. Doctor *Turner* maketh *Caryophyllus* to be *Cantabrica*; which *Pliny, lib.* 23. *cap.* 8. writeth to haue beene found out in Spaine about *Augustus* time, and that by those of Biscay.

Iohannes Ruellius saith, That the Gillofloure was vnknowne to the old writers: whose iudgment is very good, especially because this herbe is not like to that of *Vetonica* or *Cantabrica*. It is maruell, saith he, that such a famous floure, so pleasant and sweet, should lie hid, and not be made knowne by the old Writers: which may be thought not inferiour to the rose in beauty, smell, and varietie.

¶ *The Temperature.*

The Gillofloure with the leaues and roots for the most part are temperate in heate and drinesse.

¶ *The Vertues.*

A The conserue made of the floures of the Cloue Gillofloure and sugar, is exceeding cordial, and wonderfully aboue measure doth comfort the heart, being eaten now and then.

B It preuaileth against hot pestilentiall feuers, expelleth the poyson and furie of the disease, and greatly comforteth the sicke, as hath of late beene found out by a learned Gentleman of Lee in Essex, called Mr. *Rich.*

Chap. 183. *Of Pinks, or wilde Gillofloures.*

¶ *The Description.*

1 THe double purple Pinke hath many grassie leaues set vpon small ioynted stalkes by couples, one opposite against another, whereupon doe grow pleasant double purple floures,

1 *Caryophyllus ſylueſtris ſimplex.*
Single purple Pinks.

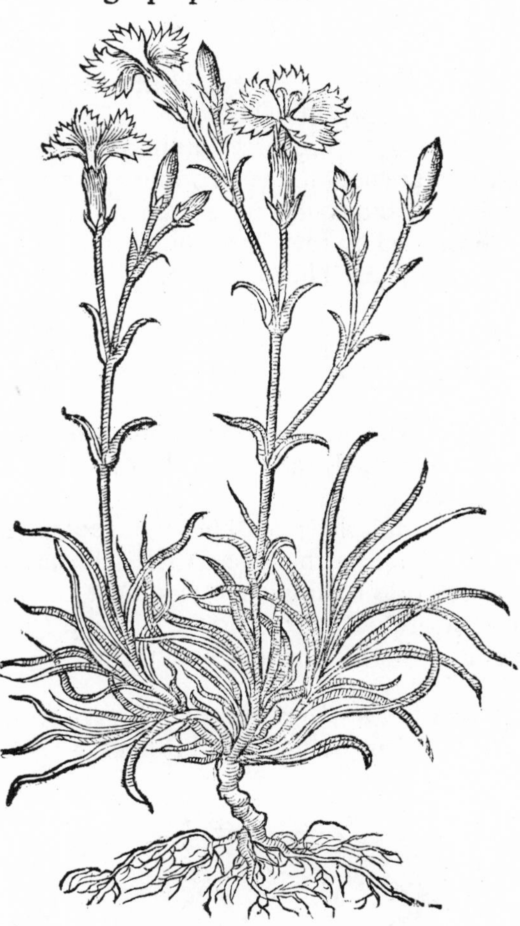

2 *Caryophyllus ſylueſtris ſimplex, ſuaue rubens.*
Single red Pinks.

3 *Caryophyllus plumarius albus.*
White jagged Pinks.

‡ *Caryophyllus plumarius albus odoratior.*
Large white jagged Pinks.

floures of a moſt fragrant ſmell, not inferiour to the Cloue Gilloſloure. The root is ſmall and wooddy.

‡ There is alſo a ſingle one of this kinde, whoſe figure I here giue you in ſtead of the double one of our Author. ‡

2 The ſingle red Pinke hath likewiſe many ſmall graſſie leaues leſſer than the former: The floures grow at the top of the ſmall ſtalkes ſingle, and of a ſweet bright red colour.

3 The white iagged Pinke hath a tough wooddy root: from which riſe immediately many graſſie leaues, ſet vpon a ſmall ſtalke full of ioynts or knees, at euery ioynt two one againſt another euen to the top; whereupon do grow faire double purple floures of a ſweet and ſpicie ſmell, conſiſting of fiue leaues, ſometimes more, cut or deepely iagged on the edges, reſembling a feather: whereupon I gaue it the name *Plumarius*, or feathered Pinke. The ſeed is ſoft, blackiſh, and like vnto Onion ſeed.

‡ There is another varietie of this, with the leaues ſomewhat larger and greener than the laſt mentioned: the floures alſo are ſomewhat bigger, more cut in or diuided, and of a much ſweeter ſmell. ‡

4 This purple coloured Pinke is very like the precedent in ſtalkes, roots, and leaues. The floures grow at the top of the branches leſſer than the laſt deſcribed, and not ſo deepely iagged; of a purple colour tending to blewneſſe, wherein conſiſteth the difference.

There be diuers ſorts of Pinks more, whereof to write particularly were to ſmall purpoſe, conſidering they are all well knowne to the moſt, if not to all. Therefore theſe few ſhall ſerue at this time for thoſe that we do keepe in our gardens: notwithſtanding I thinke it conuenient to place theſe wilder ſorts in this ſame chapter, conſidering their nature and vertues doe agree, and few or none of them be vſed in phyſicke, beſides their neereneſſe in kindred and neighbourhood.

4 *Caryophyllus plumarius purpureus.*
Purple jagged Pinkes.

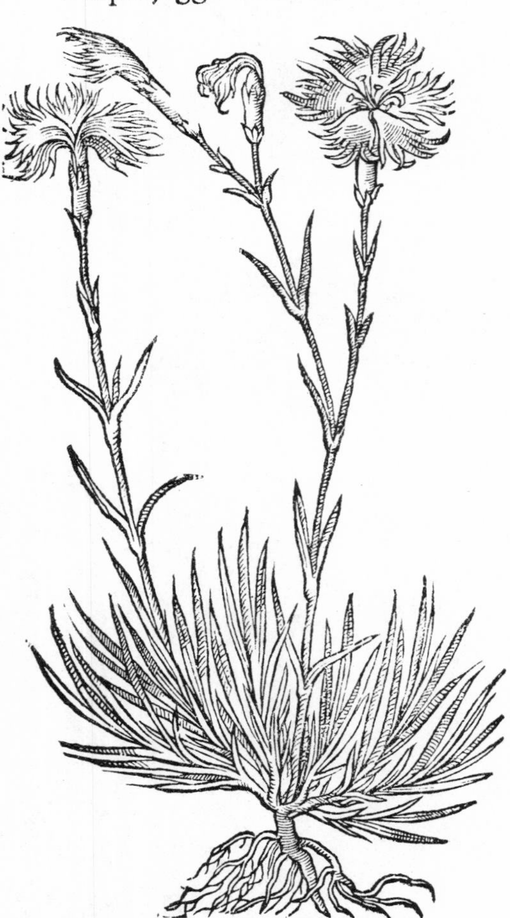

5 *Cariophyllus plumarius ſylueſtris albus.*
White wilde jagged Pinkes.

5 This wilde iagged Pinke hath leaues, ſtalkes, and floures like vnto the white iagged Pinke of the garden, but altogether leſſer, wherein they eſpecially differ.

6 The purple mountaine or wilde Pinke hath many ſmall graſſie leaues: among which riſe vp ſlender ſtalks ſet with the like leaues, but leſſer; on the top whereof do grow ſmall purple floures, ſpotted finely with white or elſe yellowiſh ſpots, and much leſſer than any of the others before deſcribed.

6 *Caryophyllus montanus purpureus.*
Wilde Purple iagged Pinke.

7 *Caryophyllus montanus Clusij.*
Clusius mountaine Pinke.

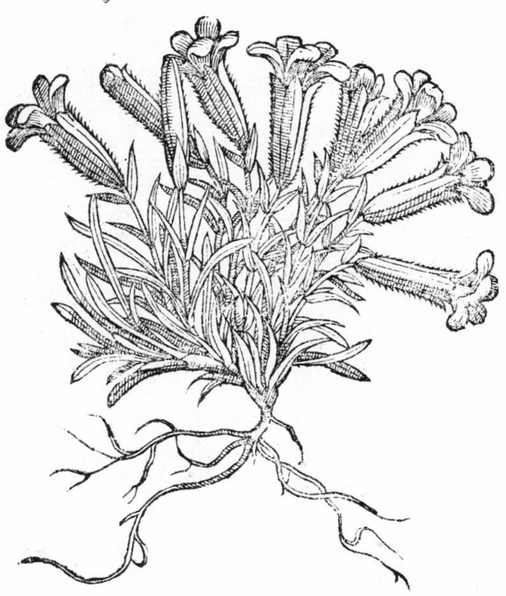

‡ 8 *Caryophyllus pumilio Alpinus.*
Dwarfe Mountaine Pinke.

9 *Caryophyllus cæruleus siue Aphyllanthos.*
Leafeles Pinke, or rushy Pinke.

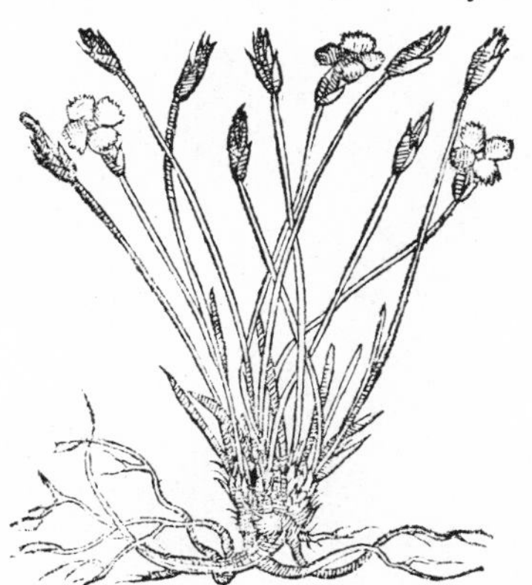

7 The mountaine Pinke of *Clusius* his description hath many leaues growing into a tuft like vnto those of Thrift, and of a bitter taste: amongst which rise vp small slender foot-stalkes, rather than stalkes or stems themselues, of the height of two inches; whereupon do grow such leaues as those that were next the ground, but lesser, set by couples one opposite to another: at the top of each small foot-stalke doth stand one red floure without smell, consisting of fiue little leaues set in a rough hairy huske or hose fiue cornered, of a greenish colour tending to purple. The root is tough and thicke, casting abroad many shoots, whereby it greatly encreaseth.

‡ 8 This for his stature may iustly take the next place; for the stalke is some inch high, set with little sharpe pointed greene grassie leaues: the floures which grow vpon these stalks are composed of fiue little flesh-coloured leaues a little diuided in their vpper parts: the seed is contained in blacke shining heads, and it is small and reddish, and shaped somwhat like the fashion of a kidney, whereby it comes neerer to the *Lychnides*, than to the *Caryophylli* or Pinkes. The root is long, blacke, and much spreading, whereby this little plant couers the ground a good space together like as a mosse, and makes a curious shew when the floures are blowne, which is commonly in Iune. It

10 *Caryophyllus montanus albus.*
White mountaine Pinke.

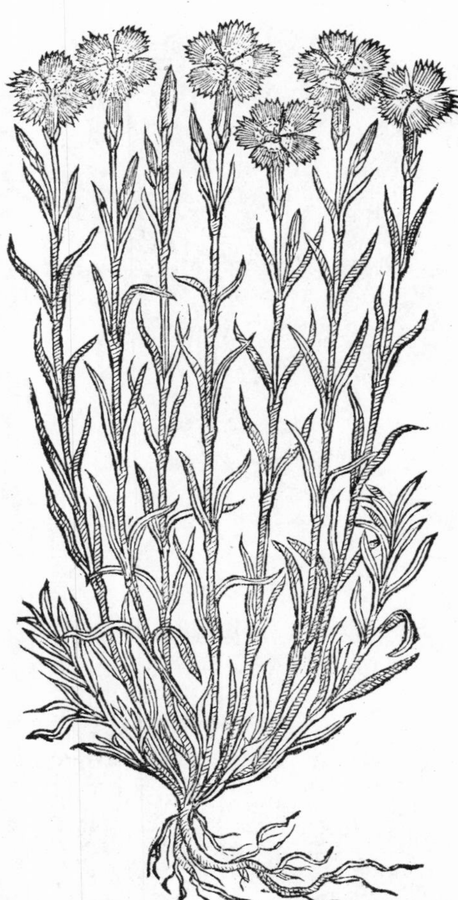

‡ 11 *Caryophyllus pratensis.*
Deptford Pinke.

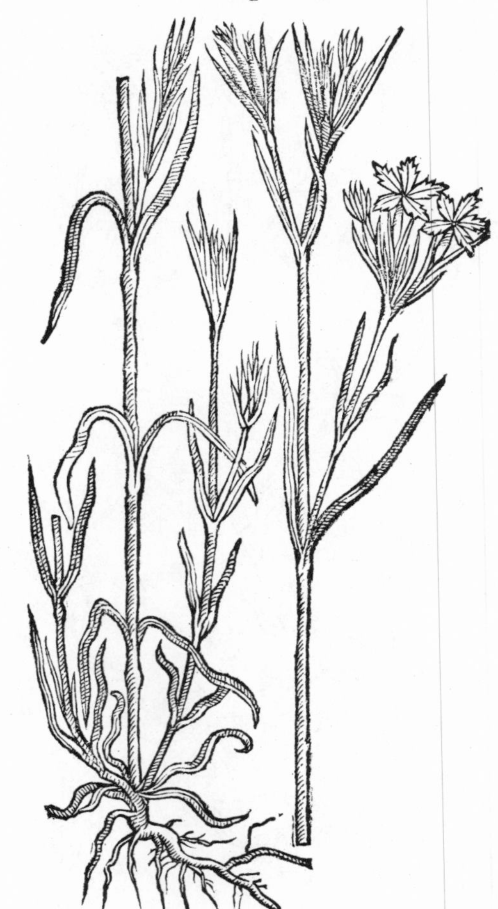

12 *Caryophyllus Virgineus.*
Maidenly Pinkes.

‡ 13 *Caryophyllus montanus humilis latifolius.*
Small mountaine broad leaued Pink.

‡ 14 *Caryophyllus montanus albus.*
White mountaine Pinke.

15 *Caryophyllus Holostius.*
Wilde Sea Pinke.

16 *Caryophyllus Holostius aruensis.*
Broad leafed wilde Pinke.

‡ 17 *Caryophyl. humilis flor cand. amœno.*
White Campion Pinke.

It growes naturally on diuers places of the Alpes. *Gesner* called it *Muscus floridus: Pona, Ocimoides Muscosus:* and *Clusius, Caryophyllus pumilio Alpinus 9.* ‡

9 This leafe-lesse Pinke (as the Greeke word doth seeme to import) hath many small rushy or benty leaues rising immediately from a tough rushy root: among which rise vp stalkes like vnto rushes, of a span high, without any ioynt at all, but smooth and plaine; on the top whereof groweth a small floure of a blewish or sky colour, consisting of foure little leaues somewhat iagged in the edges, not vnlike those of wilde flax. The whole plant is very bitter, and of a hot taste.

10 The white mountaine Pinke hath a great thicke and wooddy root; from the which immediately rise vp very many small and narrow leaues, finer and lesser than grasse, not vnlike to the smallest rush: among which rise vp little tender stalkes, ioynted or kneed by certaine distances, set with the like leaues euen to the top by couples, one opposite against another: at the top whereof grow pretty sweet smelling floures composed of fiue little white leaues. The seed is small and blackish.

11 There is a wilde creeping Pinke which groweth in our pastures neere about London, and in other places, but especially in the great field next to Detford, by the path side as you goe from Redriffe to Greenwich; which hath many small tender leaues shorter than any of the other wilde Pinkes, set vpon little tender stalkes which lie flat vpon the ground, taking hold of the same in sundry places, whereby it greatly encreaseth; whereupon grow little reddish floures. The root is small, tough, and long lasting.

12 This Virgin-like Pinke is like vnto the rest of the garden Pinkes in stalkes, leaues, and roots. The floures are of a blush colour, whereof it tooke his name, which sheweth the difference from the other.

‡ This whose figure I giue you for that small leaued one that was formerly in this place, hath slender stalkes some spanne high, set with two long narrow hard sharpe pointed leaues at each ioynt. The floures (which grow commonly but one on a stalke) consist of fiue little snipt leaues of a light purple colour, rough, and deeper coloured about their middles, with two little crooked threds or hornes: the seed is chaffie and blacke: the root long, and creeping: it floures in Aprill and May, and is the *Flos caryophylleus syluestris* 1. of *Clusius*. ‡

13 *Clusius* mentions also another whose stalkes are some three inches high: the leaues broader, softer, and greener than the former: the floures also that grow vpon the top of the stalkes are larger than the former, and also consist of fiue leaues of a deeper purple than the former, with longer haires finely intermixt with purple and white.

‡ 14 This from a hard wooddy root sends vp such stalks as the former, which are set at the ioynts with short narrower and darker greene leaues: the floures are white, sweet-smelling, consisting of fiue much diuided leaues, hauing two threds or hornes in their middle. It floures in May, and it is the *Caryophyllus syluestris quintus* of *Clusius*. ‡

15 This wilde sea Pinke hath diuers small tender weake branches trailing vpon the ground, whereupon are set leaues like those of our smallest garden Pinke, but of an old hoary colour tending to whitenesse, as are most of the sea Plants. The floures grow at the top of the stalks in shape like those of Stitch-wort, and of a whitish colour. Neither the seeds nor seed-vessels haue I as yet obserued: the root is tough and single.

16 There is another of these wilde Pinkes which is found growing in ploughed fields, yet in such as are neere vnto the sea: it hath very many leaues spred vpon the ground of a fresh green colour; amongst which rise vp tender stalkes of the height of a foot, set with the like leaues by couples at certaine distances. The floures grow at the top many together, in manner of the Sweet-William, of a white, or sometimes a light red colour. The root is small, tough, and long lasting. ‡ This is a kinde of *Gramen Leucanthemum*, or *Holosteum Ruellij*, described in the 38. Chapter of the first booke.

17 *Clusius* makes this a *Lychnis*: and *Lobel* (whom I here follow) a Pinke, calling it *Caryophyllus minimus humilis alter exoticus flore candido amæno.* This from creeping roots sendeth vp euery yeare many branches some handfull and better high, set with two long narrow greene leaues at each ioynt: the floures which grow on the tops of the branches are of a pleasing white colour, composed of fiue iagged leaues without smell. After the floures are gone there succeed round blunt pointed vessels, containing a small blackish flat seed like to that of the other Pinks. This hath a viscous or clammy iuyce like as that of the *Muscipula's* or Catch-flies. *Clusius* makes this his *Lychnis syluestris decima.* ‡

¶ *The Place.*

These kindes of Pinkes do grow for the most part in gardens, and likewise many other sorts, the which were ouer long to write of particularly. Those that be wilde doe grow vpon mountaines, stony rockes, and desart places. The rest are specified in their descriptions.

¶ *The*

¶ *The Time.*

They floure with the Cloue Gilloſloure, and often after.

¶ *The Names.*

The Pinke is called of *Pliny* and *Turner*, *Cantabrica* and *Stactice* : of *Fuchſius* and *Dodonæus*, *Vetonica altera*, and *Vetonica altilis* : of *Lobelius* and *Fuchſius*, *Superba* : in French, *Gyroffiees*, *Oeilletz*, and *Violettes herbues* : in Italian, *Garofoli*, and *Garoni* ; in Spaniſh, *Clauis* : in Engliſh, Pinkes, and Small Honeſties.

¶ *The Temperature.*

The temperature of the Pinkes is referred vnto the Cloue Gillofloures.

¶ *The Vertues.*

These are not vſed in Phyſicke, but eſteemed for their vſe in Garlands and Noſegaies. They are good to be put into Vineger, to giue it a pleaſant taſte and gallant colour, as *Ruellius* writeth. *Fuchſius* ſaith, that the roots are commended againſt the infection of the plague; and that the iuice thereof is profitable to waſte away the ſtone, and to driue it forth : and likewiſe to cure them that haue the falling ſickneſſe. A

CHAP. 184. *Of Sweet Saint Johns and Sweet Williams.*

1 *Armeria alba.*
White Iohns.

2 *Armeria alba & rubra multiplex.*
Double white and red Iohns.

¶ *The Deſcription.*

1 SWeet Iohns haue round ſtalkes as haue the Gillofloures, (whereof they are a kinde) a cubit high, whereupon do grow long leaues broader than thoſe of the Gillofloure, of a greene graſſie colour : the floures grow at the top of the ſtalkes, very like vnto Pinks, of a perfect white colour.

2 The ſecond differeth not from the other but in that, that this plant hath red floures, and the other white.

We haue in our London gardens a kinde hereof bearing moſt fine and pleaſant white floures, ſpotted very confuſedly with reddiſh ſpots, which ſetteth forth the beauty thereof; and hath bin taken of ſome (but not rightly) to be the plant called of the later Writers *Superba Auſtriaca*, or the Pride of Auſtria. ‡ It is now commonly in moſt places called London-Pride. ‡

† Wee haue likewiſe of the ſame kinde bringing forth moſt double floures, and theſe either very white, or elſe of a deepe purple colour.

3 *Armeria rubra latifolia.*
Broad leaued Sweet-Williams.

4 *Armeria ſuaue rubens.*
Narrow leaued Sweet-Williams.

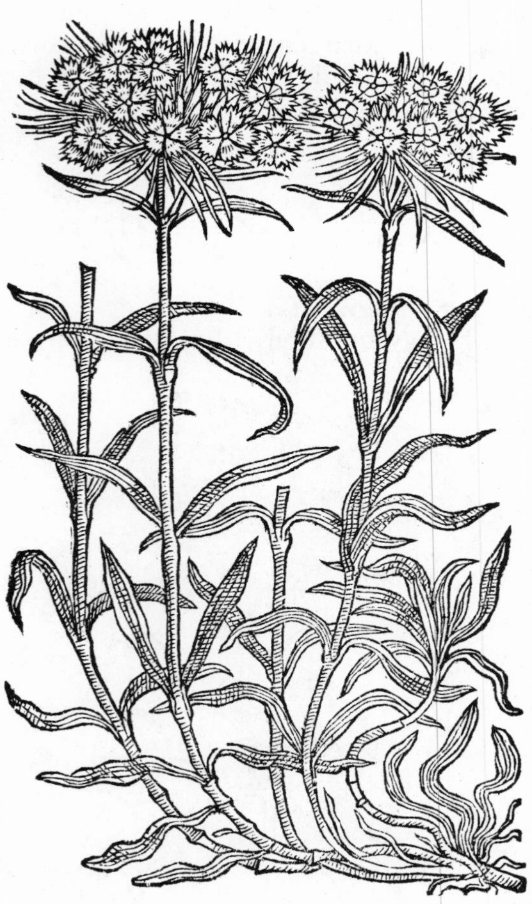

3 The great Sweet-William hath round ioynted ſtalkes thicke and fat, ſomewhat reddiſh about the lower ioynts, a cubit high, with long broad and ribbed leaues like as thoſe of the Plantaine, of a greene graſſie colour. The floures at the top of the ſtalkes are very like to the ſmall Pinkes, many ioyned together in one tuft or ſpoky vmbel, of a deepe red colour: the root is thick and wooddy.

4 The narrow leaued Sweet-William groweth vp to the height of two cubits, very wel reſembling the former, but leſſer, and the leaues narrower: the floures are of a bright red colour, with many ſmall ſharpe pointed graſſie leaues ſtanding vp amongſt them, wherein eſpecially conſiſteth the difference.

‡ 5 This little fruitfull Pinke (whoſe figure our Author formerly gaue in the firſt place of the next chapter ſaue one) hath a ſmall whitiſh wooddy root, which ſends forth little ſtalks ſome handfull and better high; and theſe at each ioynt are ſet with two thinne narrow little leaues: at the top of each of theſe ſtalkes growes a ſingle skinny ſmooth ſhining huske, out of which (as in other Pinkes) growes not one onely floure, but many, one ſtill comming out as another withers; ſo that oft times out of one head come ſeuen, eight, or nine floures one after another, which as they fade leaue behinde them a little pod containing ſmall blacke flattiſh ſeed. The floure is of a light red, and very ſmall, ſtanding with the head ſomewhat far out of the hoſe or huske. ‡

¶ *The Place.*

Theſe plants are kept and maintained in gardens more for to pleaſe the eye, than either the noſe or belly.

‡ 5 *Armeria prolifera, Lob.*
Childing ſweet Williams.

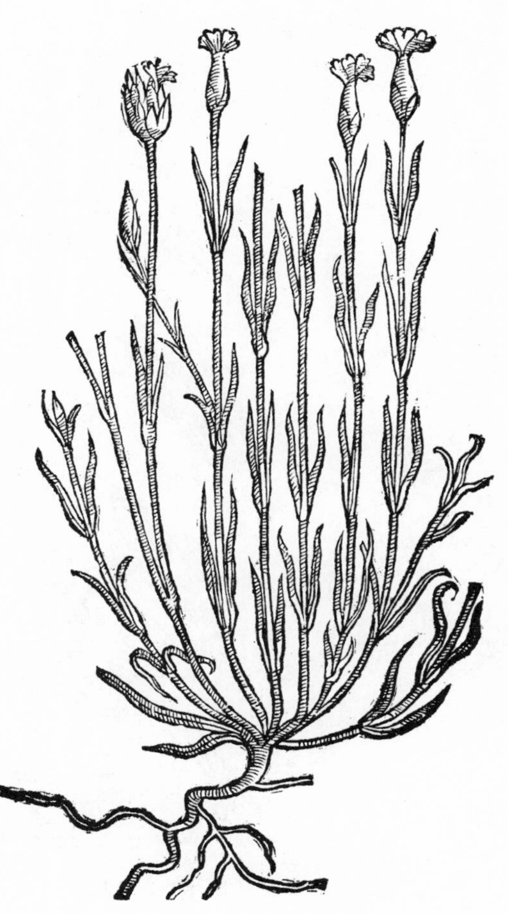

¶ *The Time.*

They flouriſh and bring forth their floures in April and May, ſomewhat before the Gilloſloures, and after beare their floures the whole Sommer.

¶ *The Names.*

The ſweet Iohn, and alſo the ſweet William are both comprehended vnder one title, that is to ſay, *Armeria*: of ſome, *Superba*, and *Caryophyllus ſylueſtris*: of ſome Herbariſts, *Vetonica agreſtis*, or *Sylueſtris*: of ſome, *Herba tunica*: but it doth no more agree her with than the Cloue Gilloſloure doth with *Vetonica altera*, or *Polemonium*. in French, *Armoires*: hereupon *Ruellius* nameth them *Armery Flores*: in Dutch, **Keykens**: as though you ſhould ſay, a bundell or cluſter, for in their vulgar tongue bundles of floures or noſegaies they call **Keykens**: doubtleſſe they are wild kindes of Gilloſloures: In Engliſh the firſt two are called Sweet Iohns; and the two laſt, Sweet Williams, Tolmeiners, and London Tufts.

¶ *The Temperature and Vertues.*

Theſe plants are not vſed either in meat or medicine, but eſteemed for their beauty to decke vp gardens, the boſomes of the beautifull, garlands and crownes for pleaſure.

CHAP. 185. *Of Crow floures, or Wilde Williams.*

¶ *The Deſcription.*

1 BEſides theſe kindes of Pinkes before deſcribed, there is a certaine other kinde, either of the Gilloſloures or elſe of the Sweete Williams, altogether and euery where wilde, which of ſome hath beene inſerted amongſt the wilde Campions; of others taken to be the true *Flos Cuculi*. Notwithſtanding I am not of any of their mindes, but doe hold it for neither: but rather a degenerate kinde of wilde Gilloſloure. The Cuckow floure I haue comprehended vnder the title of *Siſimbrium*: Engliſhed, Ladies ſmocks; which plant hath been generally taken for *Flos Cuculi*. It hath ſtalks of a ſpan or a foot high, wherupon the leaues do ſtand by couples out of euery ioint; they are ſmall and bluntly pointed, very rough and hairy. The floures are placed on the tops of the ſtalkes, many in one tuft, finely and curiouſly ſnipt in the edges, leſſer than thoſe of Gilloſloures, very well reſembling the Sweet VVilliam (whereof no doubt it is a kinde) of a light red or Scarlet colour.

2 This female Crow-floure differeth not from the male, ſauing that this plant is leſſer, and the floures more finely iagged like the feathered Pinke, whereof it is a kinde.

3 Of theſe Crow-floures we haue in our gardens one that doth not differ from the former of the field, ſauing that the plant of the garden hath many faire red double floures, and thoſe of the field ſingle.

¶ *The Place.*

Theſe grow all about in Medowes and paſtures, and dankiſh places.

¶ The

1 *Armoraria pratensis mas.*
The male Crow floure.

‡ 3 *Armoraria pratensis flore pleno.*
The double Crow-floure.

¶ *The Time.*

They begin to floure in May, and end in Iune.

¶ *The Names.*

The Crow floure is called in Latine *Armoraria syluestris,* and *Armoracia :* of some, *Flos Cuculi,* but not properly ; it is also called *Tunix :* of some, *Armeria, Armerius flos primus* of *Dodon.* and likewise *Caryophillus minor syluestris folijs latioribus* : in Dutch, **Craeynbloemkens :** that is to say, *Cornicis flores :* in French, *Cuydrelles.* In English, Crow floures, wilde Williams, marsh Gillofloures, and Cockow Gillofloures.

The Temperatures and Vertues.

These are not vsed either in medicine or in nourishment : but they serue for garlands & crowns, and to decke vp gardens.

Chap. 186. *Of Catch-Flie, or Limewoort.*

¶ *The Description.*

1 This plant, called *Viscaria,* or Lymewoort, is likewise of the stocke and kindred of the wilde Gillofloures : notwithstanding *Clusius* hath ioined it with the wilde Campions, making it a kinde thereof, but not properly. *Lobel* among the Sweet Williams, wherof doubtlesse it is a kinde. It hath many leaues rising immediately from the root like those of the Crowfloure, or wilde sweet VVilliam : among which rise vp many reddish stalkes iointed or kneed at certaine spaces, set with leaues by couples one against another : at the top whereof come foorth prettie red floures ; which being past there commeth in place small blackish seed. The root is large with many fibres. The whole plant, as well leaues and stalkes, as also the floures, are here and there couered ouer with a most thick and clammie matter like vnto Bird-lime, which if you take in your hands,

† 1 *Viscaria, siue Muscipula.*
Limewoort.

2 *Muscipula Lobelij.*
Catch Flie.

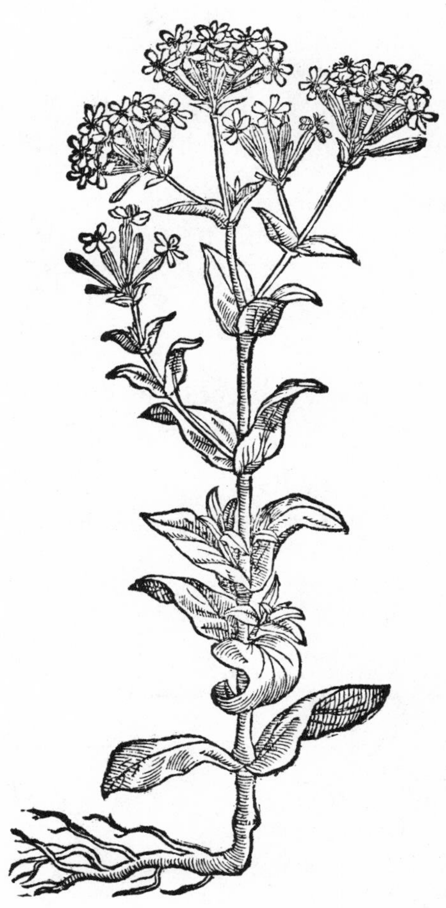

‡ 3 *Muscipula angustifolia.*
Narrow leaued Catch-flie.

hands, the sliminesse is such, that your fingers will stick and cleaue together, as if your hand touched Bird-lime: and furthermore, if flies do light vpon the same, they will be so intangled with the liminesse, that they cannot flie away; insomuch that in some hot day or other you shal see many flies caught by that means. VVhereupon I haue called it Catch Flie, or Limewoort. ‡ This is *Lychnis syl.* 3. of *Clusius*; *Viscago* of *Camerarius*; and *Muscipula siue Viscaria* of *Lobel*. ‡

2 This plant hath many broad leaues like the great sweet VVilliam, but shorter (whereof it is likewise a kinde) set vpon a stiffe and brittle stalk; from the bosom of which leaues, spring forth smaller branches, clothed with the like leaues, but much lesser. The floures grow at the top of the stalkes many together tuft fashion, of a bright red colour. The whole plant is also possessed with the like liminesse as the other is, but lesse in quantitie. ‡ This is *Lychnis syl.* 1. of *Clusius*; and *Muscipula sive Armoraria altera* of *Lobel*: *Dodonæus* calls it *Armerius flos* 3. in his first Edition: but makes it his fourth in the last Edition in *Folio*. ‡

‡ 3 There is also belonging vnto this kindred another plant which *Clusius* makes his *Lychnis syl.* 4. It comes vp commonly with one stalke a foot or more high, of a green purplish

pliſh colour, with two long ſharpe pointed thicke greene leaues, ſet at each ioint: from the middle to the top of the ſtalke grow little branches, which vpon pretty long ſtalkes carry floures conſiſting of fiue little round leaues, yet diuided at the tops; they are of a faire incarnate colour, with a deepe purple ring in their middles, without ſmell: after the floures are paſt ſucceede skinny and hard heads, ſmaller towards the ſtalkes, and thicker aboue; and in theſe are contained verie ſmall darke red ſeeds. The root is thicke and blacke, with many fibers, putting vp new ſhootes and ſtalks after the firſt yeare, and not dying euery yeare like as the two laſt deſcribed.

¶ *The Place.*

Theſe plants do grow wilde in the fields in the VVeſt parts of England, among the corne: wee haue them in our London gardens rather for toyes of pleaſure, than any vertues they are poſſeſſed with, that hath as yet been knowne.

¶ *The Time.*

They floure and flouriſh moſt part of the Sommer.

¶ *The Names.*

Catch Flies hath beene taken for *Behen*, commonly ſo called, for the likeneſſe that it hath with *Behen rubente flore*: or with *Behen* that hath the red floure, called of ſome *Valeriana rubra*, or red Valerian; for it is ſomething like vnto it in iointed ſtalkes and leaues, but more like in colour: of *Lobel, Muſcipula* and *Viſcaria*: of *Dodon. Armerius flos tertius*: of *Cluſius, Lychnis ſylueſtris, Silene Theophraſti*, and *Behen rubrum Salamanticum*: in Engliſh, Catch Flie, and Limewoort.

¶ *The Nature and Vertues.*

The nature and vertues of theſe wilde VVilliams are referred to the Wilde Pinkes and Gillofloures.

† Our Authour certainly intended in this firſt place to figure and deſcribe the *Muſcipula* or *Viſcaria* of *Lobel*, but the figure he here gaue in the firſt place was of that plant which I haue giuen you in the laſt Chapter ſaue one by the name of *Armeria prolifera Lobelii*. The figure which belonged to this place was in the Chapter of wilde Campions, vnder the title of *Lychnis ſylueſtris incana*.

Chap. 187. *Of Thrift, or our Ladies Cuſhion.*

1 *Caryophyllus marinus minimus Lobelij.*
Thrift or Sea Gillofloure.

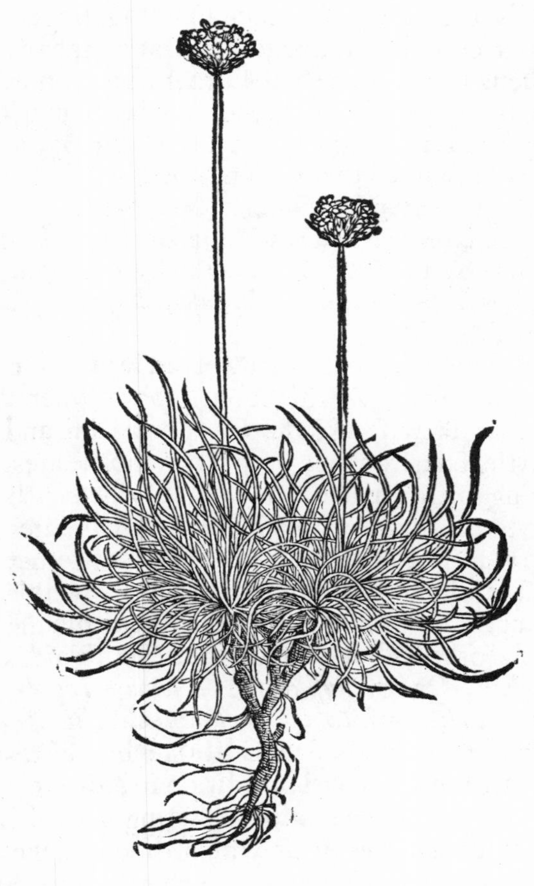

2 *Caryophyllus Mediterraneus.*
Leuant Thrift, or Sea Gillofloure.

¶ The

¶ *The Deſcription.*

1 Thrift is alſo a kind of Gilioflоure,by *Dodonæus* reckoned among graſſes, which brings forth leaues in great tufts, thick thruſt together,ſmaller,ſlenderer,& ſhorter than graſſe: among which riſe vp ſmall tender ſtalkes of a ſpanne high,naked and without leaues; on the tops wherupon ſtand little floures in a ſpokie tuft,of a white colour tending to purple. The root is long and threddie.

The other kinde of Thriſt,found vpon the mountaines neere vnto the Leuant or Mediterranean ſea,differeth not from the precedent in leaues,ſtalkes,or floures, but yet is altogether greater, and the leaues are broader.

¶ *The Place.*

2 The firſt is found in the moſt ſalt marſhes in England,as alſo in Gardens,for the bordering vp of beds and bankes,for the which it ſerueth very fitly. The other is a ſtranger in theſe Northerne Regions.

¶ *The Time.*

They floure from May,till Sommer be far ſpent.

¶ *The Names.*

Thrift is called in Latine *Gramen Polyanthemum*,of the multitude of the floures: of ſome, *Gramen marinum*: of *Lobel*, *Caryophyllus Marinus*: In Engliſh, Thrift, Sea-graſſe,and our Ladies Cuſhion.

¶ *The Temperature and Vertues.*

Their vſe in Phyſicke as yet is not knowne,neither doth any ſeeke into the Nature thereof, but eſteeme them onely for their beautie and pleaſure.

CHAP. 188. *Of the Saxifrage of the Antients, and of that great one of Matthiolus, with that of Pena and Lobel.*

‡ This name *Saxifraga* or Saxifrage, hath of late been impoſed vpon ſundry plants farre different in their ſhapes,places of growing,& temperature,but all agreeing in this one facultie of expelling or driuing the ſtone out of the Kidneies, though not all by one meane or manner of operation. But becauſe almoſt all of them are deſcribed in their fit places by our Authour,I will not inſiſt vpon them: yet I thinke it not amiſſe a little to enquire,whether any *Saxifraga* were knowne to the Antients; and if knowne,to what kinde it may probably be referred. Of the Antients, *Dioſcorides*, *Paulus Ægineta*,and *Apuleius*,ſeeme to mention one *Saxifraga*, but *Pliny*, *lib.* 22. *cap.* 21. by the way,ſhewes that ſome called *Adianthum* by the name of *Saxifragum*: but this is nothing to the former; wherefore I will not inſiſt vpon it,but returne to examine that the other three haue written thereof. *Dioſc. lib.* 4. betweene the Chapters of *Tribulus* and *Limonium*, to wit,in the ſeuenteenth place hath deliuered the Hiſtorie of this plant,both in the Greeke Edition of *Aldus Manutius*,as alſo in that of *Marcellus Virgilius*,yet the whole Chapter in the Paris Edition, 1549, is reiected and put amongſt the *Notha*. The beginning thereof (againſt which they chiefely except) is thus: Σαρξίφαγον, οἱ δὲ σαρξίφραγον, οἱ δὲ ἔμπετρον, Ῥωμαῖοι σαρξιφράγα, (i) *Sarxiphagon, aly vero Sarxifrangon, alij vero Empetron, Romani, Sarxifranga.* The firſt exception of *Marcellus Virgilius* againſt this Chapter is *Peregrina Græcis & aliena vox Saxifraga eſt,&c.* The ſecond is, *Quod multo feliciores in componendis ad certiorem rei alicuius ſignificationem vocibus Græci, quam Latini,&c.* The third is, *Solam in toto hoc opere primam, & a principio propoſitam audiri Romanam vocem, tamque inopes in appellanda hac herba fuiſſe Græcos, vt niſi Romana voce eam indicaſſent, nulla ſibi futura eſſet.* Theſe are the arguments which he vſes againſt this Chapter; yet reiects it not,but by this means hath occaſioned others without ſhewing any reaſon, to doe it: Now I will ſet downe what my opinion is concerning this matter,and ſo leaue it to the iudgement of the Learned. I grant *Marcellus*,that *Saxifraga* is a ſtrange and no Greeke word; but the name in the title, and firſt in the Chapter both in his owne Edition and all the Greeke Editions that I haue yet ſeene is Σαρξίφαγον, which none,no not he himſelfe can denie to haue a Greeke originall ἀπὸ τοῦ τὴν σάρκα φαγεῖν: of eating the fleſh: yet becauſe there is no ſuch facultie as this denomination imports attributed thereto by the Authour, therefore hee will not allow it to be ſo. But you muſt note that many names are impoſed by the vulgar,and the reaſon of the name not alwais explained by thoſe that haue written of them,as in this ſame Author may be ſeene

ſeene in the Chapters of *Catanance*, *Cynosbatos*, *Hemerocallis*, *Crataogonon*, and diuers others, which are or ſeeme to be ſignificant, and to import ſomething by their name; yet he ſaith nothing thereof. It may be that which they would expreſſe by the name, was, that the hearbe had ſo piercing a facultie that it would eat into the very fleſh. The ſecond and third Argument both are anſwered, if this firſt word be Greeke, as I haue alreadie ſhewed it to be, and there are not many words in Greeke that more frequently enter into ſuch compoſition than φαγω: as *Pamphagos*, *Polyphagos*, *Opſiphagos*, and many other may ſhew. Moreouer, it hath beene obſurd from *Dioſcorides*, or any elſe how ſimple ſoeuer they were, if they had knowne the firſt word to haue beene Latine and *Saxifraga*, to ſay againe preſently after that the Romanes called it *Saxifranga*, or *Saxifraga*, for ſo it ſhould be, and not *Sarxifranga*: but I feare that the affinitie of ſounds more than of ſignification hath cauſed this confuſion, eſpecially in the middle times betweene vs and *Dioſcorides*, when learning was at a very low ebbe. The chiefe reaſons that induce mee to thinke this Chapter worthie to keepe his former place in *Dioſcorides*, are theſe: Firſt, the generall conſent of all both Greeke and Latine copies (as *Marcellus* ſaith) how antient ſoeuer they be. Secondly, the mention of this herbe for the ſame effect in ſome Greeke Authours of a reaſonable good antiquitie; for *Paulus Ægineta* teſtifieth that Σαρξίφαγες διουρητικόν τε ἐστὶ καὶ λίθων θρυπτικόν. Then *Trallianus* amongſt other things in a *Conditum Nephriticum* mentions Σαρξίφαγον: but *Nonus* a later Greeke calls it Σαξίφραγος: ſo that it is euident they knew and vſed ſome ſimple medicine that had both the names of *Sarxiphagon* and *Saxiphragos*, which is the Latine *Saxifraga*. Now ſeeing they had, and knew ſuch a ſimple medicine, it remaines we enquire after the ſhape and figure thereof. *Dioſcorides* deſcribes it to be a ſhrubby plant, growing vpon rockes and craggie places, like vnto *Epithymum*: boiled in wine and drunke, it hath the faculty to helpe the Strangurie and Hicket; it alſo breakes the ſtone in the bladder and prouokes vrine. This word *Epithymum* is not found in moſt copies, but a ſpace left for ſome word or words that were wanting: But *Marcellus* ſaith, he found it expreſt in a booke which was *Omnium vetuſtiſſimus & probatiſſimus:* and *Hermolaus Barbarus* ſaith, *Veterem in* Dioſcoride *picturam huius herbæ vidi, non plus folijs quam cirris minutis per ramos ex interuallo conditis, nec frequentibus, in cacumine ſurculorum flocci ſeu arentes potius quam floſculi, ſubrubida radice non ſine fibris.* A figure reaſonable well agreeing with this deſcription of *Hermolaus*, I lately receiued from my friend M^r. *Goodyer*, who writ to me that he had ſought to know what *Saxifraga* (to wit, of the Antients) ſhould be; and finding no antient Authour that had deſcribed it to any purpoſe, he ſought *Apuleius*; which word *Apuleius* (ſaith he) is the printed title: my Manuſcript acknowledgeth no Authour but *Apolienſis Plato*; there is no deſcription neither, but the Manuſcript hath a figure which I haue drawne and ſent you, and all that *verbatim* that hee hath written of it, I ſhould be glad to haue this figure cut and added to your worke, together with his words, becauſe there hath beene ſo little written thereof by the Antients. This his requeſt I thought fit to performe, and haue (for the better ſatisfaction of the Reader) as you ſee made a further enquirie thereof: wherefore I will onely adde this, that the plants here deſcribed, and the *Alſine Saxifraga* of *Colum.* together with the two Chickweed *Saxifrages* formerly deſcribed Chap. 171. come neereſt of any that I know to the figure and deliniation of this of the Antients.

Nomen iſtius herbæ, Saxifraga.

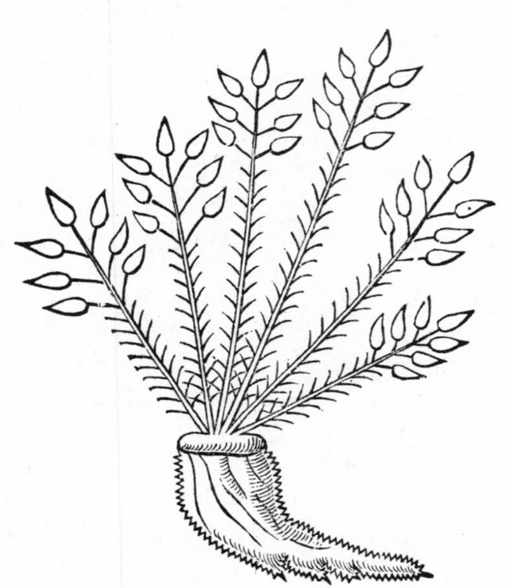

Icon & deſcriptio ex Manuſcripto vetutiſſimo.

Quidam dicunt eam Scolopendriam, alij Scoliomos, alij Vitis canum, quidam vero Brucos. Itali Saxifragam. Egyptij Peperem, alij Lamprocam eam nominant. Naſcitur enim in Montibus & locis ſaxoſis.

Vna cura ipſius ad calculos expellendos.

Herbam iſtam Saxifragam contuſam calculoſo potum dabis in vino. Ipſe vero ſi febricitauerit cum aqua calida, tam preſens effectum ab expertis traditum, vt eodem die perfectis eiectiſque calculis ad ſanitatem vſque producit.

1 This firſt little herb, ſaith *Camerarius*, hath been called *Saxifraga magna*, not from the greatneſſe of his growth, but of his faculties: The ſtalke is woodie, writhen

writhen, and below ſometimes as thicke as ones little finger, from which grow many ſmall & hard branches, and thoſe ſlender ones; the leaues are little, long and ſharpe pointed: the floures are white and ſmall, and grow in cups, which are finely ſnipt at the top in manner of a coronet, wherein is contained a ſmall red ſeed: the rootes grow ſo faſt impact in the Rockes, that it cannot by any meanes be got out. It grows vpon diuers rocks in Italy and Germany; and it is the *Saxifraga magna* of *Matthiolus*, and the Italians.

‡ 1 *Saxifraga magna Matthioli.*
Matthiolus his great Saxifrage.

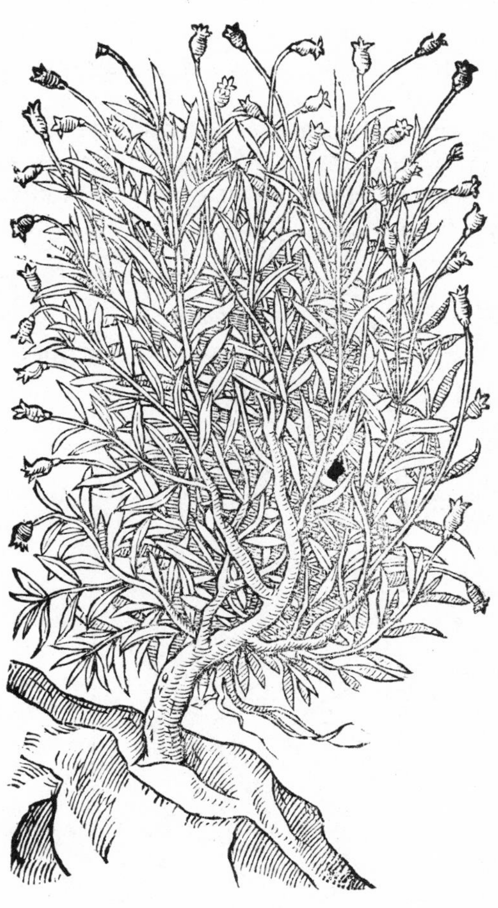

‡ 2 *Saxifraga Antiquorum, Lob.*
Saxifrage of the Antients, according to *Lob.*

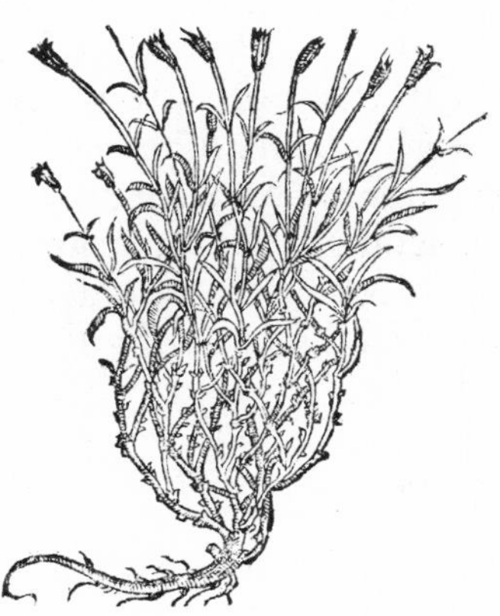

2 *Pena* and *Lobel* ſay, this growes in great plenty in Italie, in Dolphonie in France, and England, hauing many ſmall ſlender branches a foot high, intricately wrapped within one another, where they are ſet with many graſſie ioynts: the roote is ſmall and white with ſome few fibers: the leaues ſtand by couples at the ioynts, beeing long and narrow; of the bigneſſe and ſimilitude of thoſe of the wilde Pinks, or Rocke Sauorie: vpon each wooddie, ſmall, capillarie, ſtraight, and creeping little branch, growes one little floure ſomewhat like a Pinke, beeing finely ſnipt about the edges: and in the head is contained a round ſmall reddiſh ſeed. The foreſaid Authours call this *Saxifragra, ſive Saxifraga Antiquorum.*

The Vertues.

1 *Matthiolus* ſaith, that *Calceolarius* of Verona mightily commended this plant to him, for the ſingular qualitie it had to expell or driue forth the ſtone of the Kidneies, and that I might in verie deed beleeue it, he ſent me abundance of ſtones, whereof diuers exceeded the bigneſſe of a beane, which were voided by drinking of this plant by one onely Citizen of Verona, called *Hieronymo de Tortis*; but this made me moſt to wonder, for that there were ſome ſtones amongſt them, that ſeemed rather to come out of the Bladder, than forth of the Kidneies. A

2 This (ſay the Authours of the *Aduerſ.*) as it is the lateſt receiued in vſe and name for Saxifrage, ſo is it the better & truer, eſpecially ſo thought by the Italians, both for the highly commended facultie, as alſo for the neere affinitie which it ſeemes to haue with *Epithymum*, &c. ‡ B

Chap. 189. *Of Sneeſewoort.*

¶ *The Deſcription.*

1 The ſmall Sneeſe-woort hath many round and brittle branches, beſet with long and narrow leaues, hackt about the edges like a ſaw, at the tops of the ſtalks do grow ſmall ſingle floures like the wilde field Daiſie. The root is tender and full of ſtrings, creeping far abroad in the earth, and in ſhort time occupieth very much ground: the whole plant is ſharpe, biting the tongue and mouth like Pellitorie of Spaine, for which cauſe ſome haue called it wilde Pellitorie. The ſmell of this plant procureth ſneeſing, whereof it tooke the name *Sternutamentoria*, that is the herbe which doth procure ſneeſing, or Neeſewoort.

2 Double floured Sneeſewoort, or *Ptarmica*, is like vnto the former in leaues, ſtalks, and roots, ſo that vnleſſe you behold the floure, you cannot diſcerne the one from the other, and it is exceeding white, and double like vnto double Fetherfew. This plant is of great beautie, and if it be cut downe in the time of his flouring, there will come within a month after a ſupplie or crop of floures fairer than the reſt.

1 *Ptarmica.*
Sneeſewoort.

2 *Ptarmica duplici flore.*
Double floured Sneeſwoort.

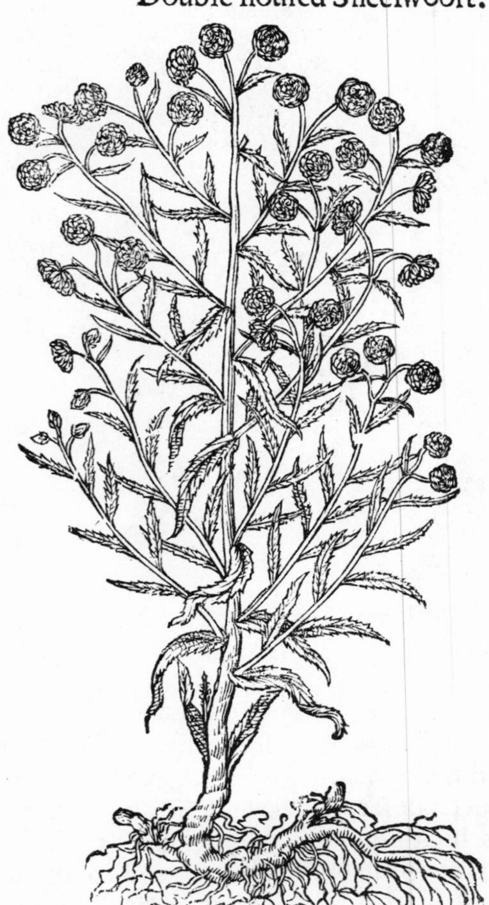

3 There is alſo another kind hereof, of exceeding great beauty, hauing long leaues ſomewhat narrow like thoſe of Oliue tree: the ſtalks are of a cubit high, on the top whereof doe growe verie beautifull floures of the bignes of a ſmall ſingle Marygold, conſiſting of fifteene or ſixteene large leaues, of a bright ſhining red colour tending to purple; ſet about a ball of thrummie ſubſtance, ſuch as is in the middle of the Daiſie, in manner of a pale; which floures ſtand in ſcalie knops like thoſe of Knapweed, or Matfellon. The root is ſtraight, and thruſteth deepe into the ground.

‡ *Ptarmica Imperati; an Ptarmicæ Auſtriacæ ſpecies Cluſ. Cur. poſt. p. 32.*

4 This riſeth vp with a ſmall hard tough cornered whitiſh woolly ſtalke, diuided into many branches,

3 *Ptarmica Austriaca.*
Sneesewoort of Austrich.

branches, and those againe diuided into other branches like those of *Cyanus* about two foot high, wherein grow long narrow whitish Cottonie leaues out of order, of a bitter taste, whiter below than aboue, of the colour of the leaues of Wormwood, hauing but one rib or sinew & that in the middle of the leafe, and commonly turne downewards: on the top of each slender branch groweth one small scalie head or knap, like that of *Cyanus*, which bringeth forth a pale purple floure without smell, containing sixe, seuen, eight, or more, smal hard drie sharp pointed leaues: in the middle whereof groweth many stiffe chiues, their tops being of the colour of the floures: these floures fall not away till the whole hearbe perisheth, but change into a rustie colour: amongst those chiues grow long flat blackish seed, with a little beard at the top. The root is small, whitish, hard and threddie, and perisheth when the seed is ripe, and soone springeth vp by the fall of the seede, and remaineth greene all the Winter, and at the Spring sendeth foorth a stalke as aforesaid. The herbe touched or rubbed sendeth forth a pleasant aromaticall smell. Iuly 26. 1620. *Iohn Goodyer.* ‡

¶ *The Place.*

The first kinde of Sneesewoort grows wilde in drie and barren pastures in many places, and in the three great fieldes next adioyning to a Village neere London called Kentish towne, and in sundry fields in Kent about Southfleet.

† The rest grow onely in gardens.

¶ *The Time.*

They floure from May to the end of September.

¶ *The Names.*

Sneesewoort is called of some *Ptarmica*, and *Pyrethrum syluestre*, and also *Draco syluestris*, or *Tarcon syluestris*: of most, *Sternutamentoria*, taken from his effect, because it procureth sneesing: of *Tragus* & *Tabern. Tanacetum acutum album*: in English, wilde Pellitorie, taking that name from his sharp and biting taste; but it is altogether vnlike in proportion to the true Pellitorie of Spaine.

¶ *The Nature.*

They are hot and drie in the third degree.

¶ *The Vertues.*

The iuice mixed with Vineger and holden in the mouth easeth much the paine of the Toothache. A

The herbe chewed and holden in the mouth, bringeth mightily from the braine slimie flegme, like Pellitorie of Spaine, and therefore from time to time it hath beene taken for a wilde kinde thereof. B

CHAP. 190. *Of Hares Eares.*

¶ *The Description.*

1 Narrow leafed Hares Eares is called in Greeke Βούπλευρον, and is reputed of the late writers to be *Bupleurum Plinij*, from which the name or figure disagreeth not: it hath the long narrow and grassie leaues of *Lachryma Iob*, or *Gladiolus*, streaked or balked as it were with sundry stiffe streakes or ribbes running along euery leafe, as *Plinie* speaketh of his

his *Heptapleurum*. The ſtalkes are a cubite and a halfe long, full of knots or knees, very rough or ſtiffe, ſpreading themſelues into many branches: at the tops whereof grow yellow floures in round tufts or heads like Dill. The root is as big as a finger, and blacke like *Peucedanum*, whereunto it is like in taſte, ſmell, and reſemblance of ſeede, which doth the more perſuade me that it is the true *Bupleurum*, whereof I now ſpeake, and by the authoritie of *Nicander* and *Pliny* confirmed.

1 *Bupleurum anguſtifolium Monſpelienſe.*
Narrow leafed Hares Eare.

2 *Bupleurum latifolium Monſpelienſe.*
Broad leafed Hares Eare.

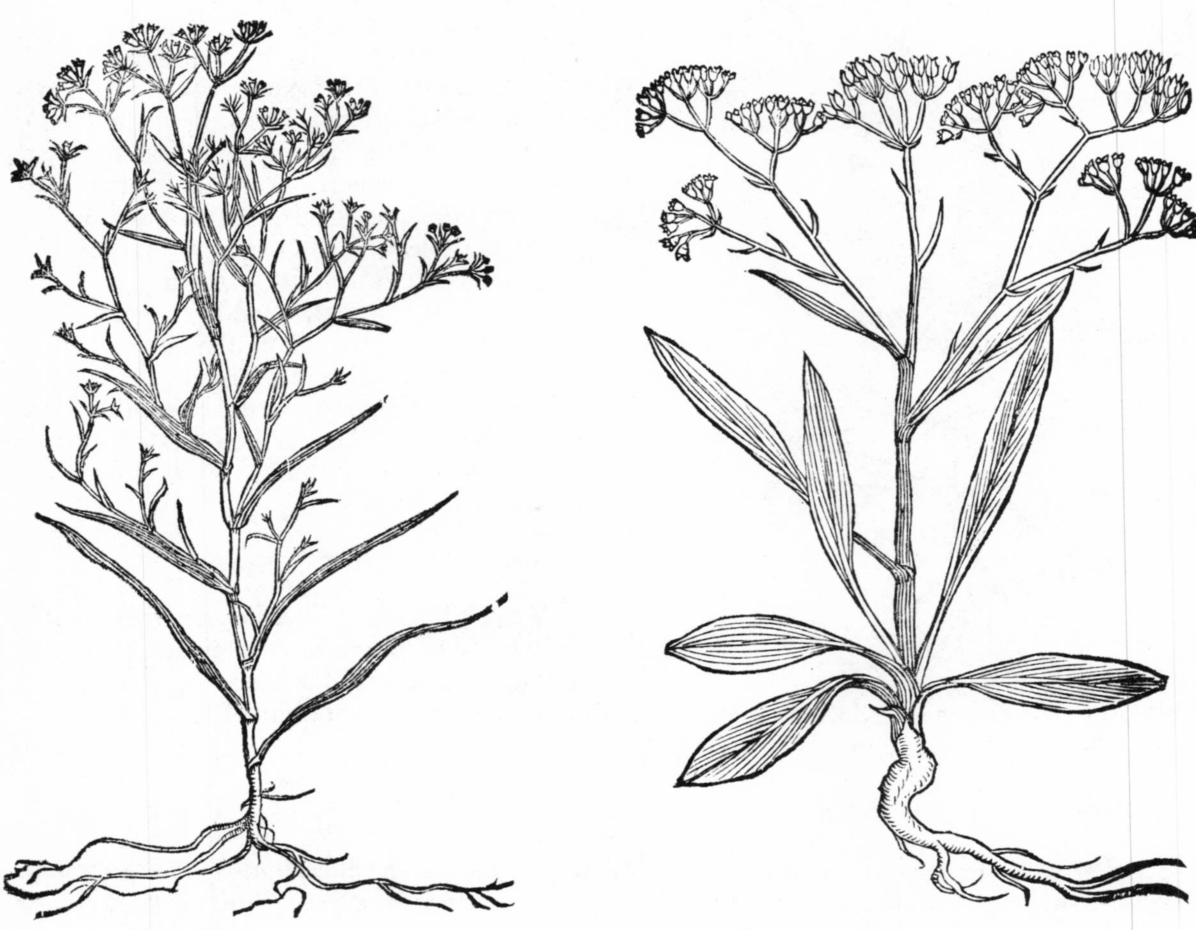

2 The ſecond kinde called broad leafed Hares Eares, in figure, tuftes, and floures, is the very ſame with the former kinde, ſaue that the leaues are broader and ſtiffer, and more hollow in the midſt: which hath cauſed me to call it Hares Eares, hauing in the middle of the leafe ſome hollowneſſe reſembling the ſame. The root is greater and of a wooddie ſubſtance.

¶ *The Place.*

They grow among Oken woods in ſtony and hard grounds in Narbon. I haue found them growing naturally among the buſhes vpon Bieſton caſtle in Cheſhire.

¶ *The Time.*

They floure and bring forth their ſeed in Iuly and Auguſt.

¶ *The Names.*

Hares Eare is called in Latine *Bupleurum*: in Greeke, βούπλευρον: the Apothecaries of Montpelier in France do call it *Auricula leporis*, and therefore I terme it in Engliſh Hares-Eare: *Valerius Cordus* nameth it *Iſophyllon*, but whence he had that name, it is not knowne.

¶ *The Temperature.*

They are temperate in heat and drineſſe.

¶ *The Vertues.*

Hippocrates hath commended it in meats, for ſallads and Pot-hearbs: but by the authoritie of *Glaucon* and *Nicander*, it is effectuall in medicine, hauing the taſte and ſauour of *Hypericon*, ſeruing in the place thereof for wounds, and is taken by *Tragus* for *Panax Chironium*, who doth reckon it *inter Herbas vulnerarias*.

The

The leaues stamped with salt and wine, and applied, doe consume and driue away the swelling of the neck, called the Kings euill, and are vsed against the stone and Grauell. B

CHAP. 191. Of Gromell.

¶ The Description.

1 THe great Gromell hath long, slender and hairie stalkes, beset with long, browne & hoarie leaues; among which grow certaine bearded huskes, bearing at the first small blew floures; which being past, there succeedeth a gray stonie seed somewhat shining. The root is hard, and of a wooddie substance.

2 The second kinde of Gromell hath straight, round, wooddie stalks, full of branches: The leaues long, small, and sharpe, of a darke greene colour; smaller than the leaues of great Gromell: among which come forth little white floures; which being past, there doth follow such seed as the former hath, but smaller.

† 3 There is another kinde of Gromell, which hath leaues and stalkes like the small kinde: the seed is not so white, neither so smooth and plaine, but somewhat shriueled or wrinckled. The leaues are somewhat rough like vnto the common Gromell, but the floures are of a purple colour, and in shape like those of that wilde kinde of Buglosse, called *Anchusa*, for which cause it carrieth that additament *Anchusæ facie*.

4 There is also a degenerate kinde hereof called *Anchusa degener*, being either a kinde of wilde Buglosse, or a kinde of wilde Gromell, or else a kinde of neither of both, but a plant participating of both kindes: it hath the seeds and stalkes of *Milium solis*, or Gromell: the leaues and rootes of *Anchusa*, which is Alkanet, and is altogether of a red colour like the same.

1 *Lithospermum maius.*
Great Gromell.

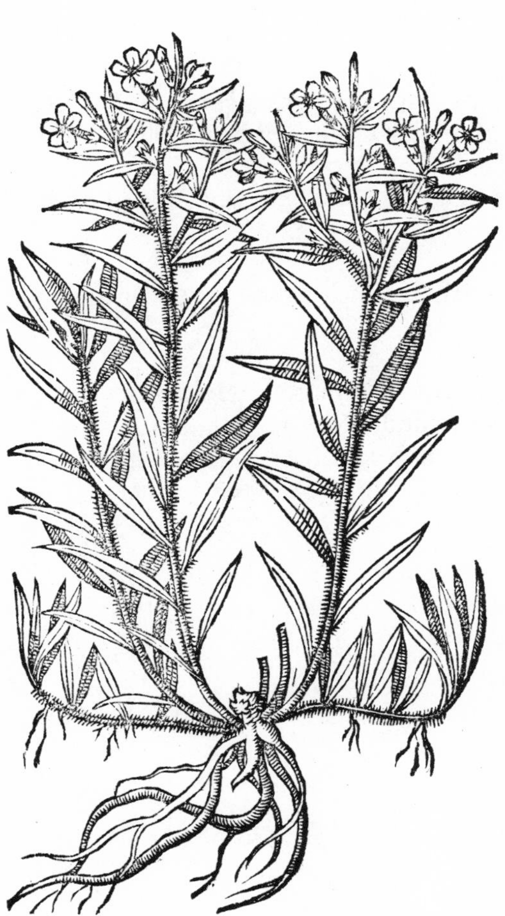

2 *Lithospermum minus.*
Small Gromell.

‡ 3 *Lithospermum Anchusæ facie.*
Purple floured Gromell.

‡ 4 *Anchusa degener facie Milij solis.*
Bastard Gromell.

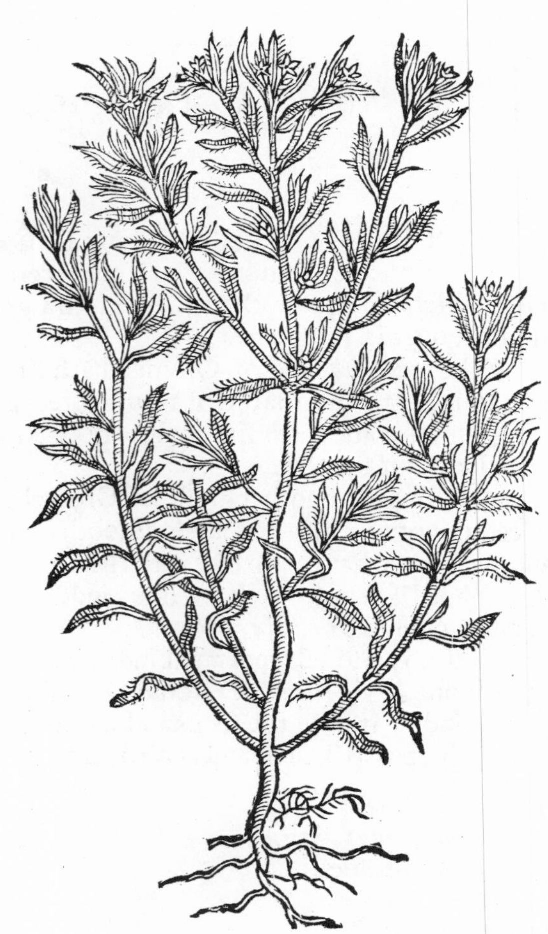

¶ *The Place.*

The two first kindes do grow in vntoiled places, as by the high waies sides, and barren places, in the street at Southfleet in Kent, as you goe from the church vnto an house belonging to a gentleman of worship, called Mr. *William Swan*, and in sundry other places.

The two last kindes grow vpon the sands and Bach of the Sea, in the isle of Thanet neere Reculvers, among the kindes of wilde Buglosse there growing.

¶ *The Time.*

They floure from the Sommer Solstice, or from the twelfth day of Iune euen vnto Autumne, and in the meane season the seed is ripe.

¶ *The Names.*

Gromell is called in Greeke λιθόσπερμον, of the hardnesse of the seed: of diuers, *Gorgonium*: of others, *Aegonychon*, *Leontion*, or *Diosporon*, or *Diospyron*, as *Plinie* readeth it, and also *Heracleos*: of the Arabians, *Milium soler*: in shops, and among the Italians, *Milium solis*: in Spanish, *Mijo del sol*: in French, *Gremill*, and *Herbe aux perles*: in English, Gromell: of some, Pearle plant; and of others, Lichwale.

¶ *The Temperature.*

The seed of Gromell is hot and drie in the second degree.

¶ *The Vertues.*

A The seed of Gromell pound, and drunke in White wine, breaketh, dissolueth, and driueth forth the stone, and prouoketh vrine, and especially breaketh the stone in the bladder.

Chap. 192 *Of Chickeweed.*

¶ *The Description.*

1 The great Chickeweede riseth vp with stalkes a cubit high, and sometime higher, a great many from one roote, long and round, slender, full of ioints, with a couple of leaues growing

growing out of euery knot or ioynt aboue an inch broad, and longer than the leaues of Pellitorie of the wall, whereunto they are very like in ſhape, but ſmooth without haires or downe, and of a light greene colour: the ſtalkes are ſomthing cleere, and as it were tranſparent or thorow-ſhining, and about the ioynts they be oftentimes of a very light red colour, as be thoſe of Pellitorie of the wall: the floures be whitiſh on the top of the branches, like the floures of Stitchwort, but yet leſſer: in whoſe places ſucceed long knops, but not great, wherein the ſeed is contained. The root conſiſteth of fine little ſtrings like haires.

2 The ſecond Chickweed for the moſt part lyeth vpon the ground: the ſtalkes are ſmall, ſlender, long, and round, and alſo ioynted: from which ſlender branches do ſpring leaues reſembling the precedent, but much leſſer, as is likewiſe the whole herbe, which in no reſpect attaineth to the greatnes of the ſame: the floures are in like ſort little and white: the knops or ſeed-heads are like the former: the root is alſo full of little ſtrings.

1 *Alſine maior.*
Great Chickweed.

2 *Alſine minor, ſiue media.*
Middle or ſmall Chickweed.

3 The third is like the ſecond, but farre leſſer: the ſtalkes be moſt tender and fine: the leaues are very ſmall, the floures very little, the root maruellous ſlender.

4 Alſo there is a fourth kinde which groweth by the ſea: this is like to the ſecond, but the ſtemmes are thicker, ſhorter, and fuller of ioynts: the leaues in like ſort be thicker: the knops or ſeed-heads be not long and round, but ſomewhat broad, in which are three or foure ſeeds contained.

5 The vpright Chickweed hath a very ſmall ſingle threddy root, from which riſeth vp a ſlender ſtemme, diuiding it ſelfe into diuers branches euen from the bottome to the top; whereon do grow ſmall leaues, thicke and fat in reſpect of the others, in ſhape like thoſe of Rue or Herbe-Grace. The floures grow at the top of the branches, conſiſting of foure ſmall leaues of a blew colour.

6 The ſtone Chickweed is one of the common Chickweeds, hauing very threddy branches couering the ground farre abroad where it groweth: the leaues be ſet together by couples: the floures be ſmall and very white: the root is tough and very ſlender.

7 Speedwell

3 *Alſine minima.*
Fine Chickweed.

4 *Alſine marina.*
Sea Chickweed.

5 *Alſine recta.*
Right Chickweed.

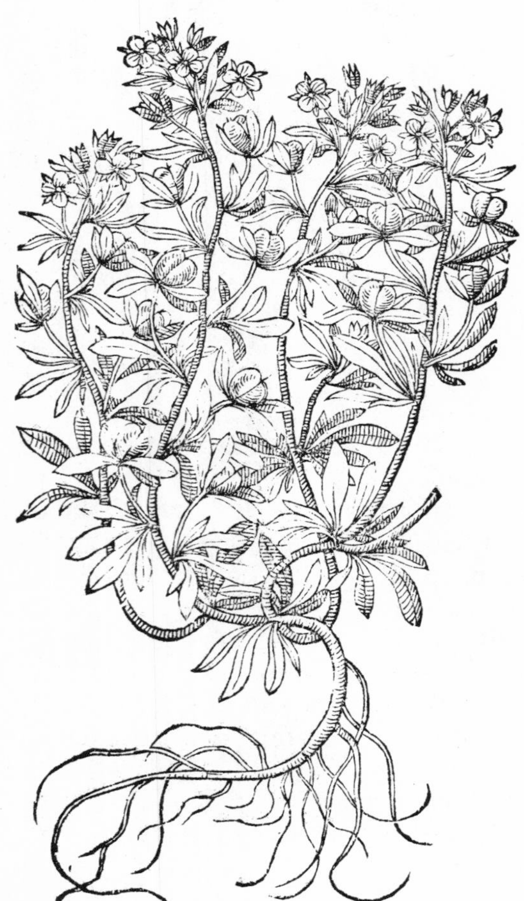

6 *Alſine Petræa.*
Stone Chickweed.

7 *Alsine folijs Veronicæ.*
Speed-well Chickweed.

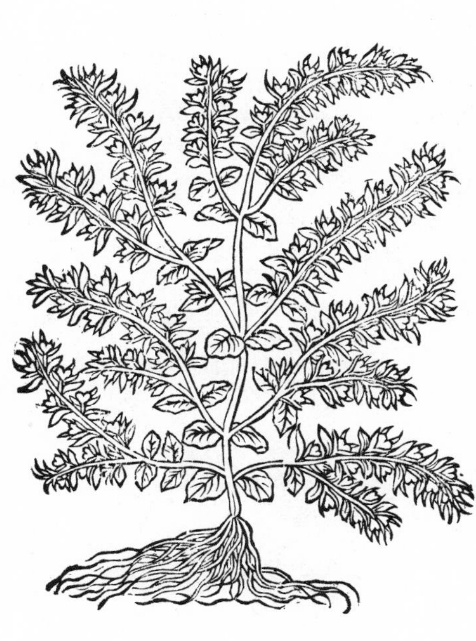

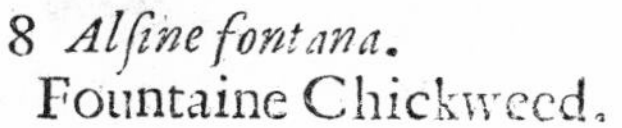

8 *Alsine fontana.*
Fountaine Chickweed.

9 *Alsine fluviatilis.*
Riuer Chickweed.

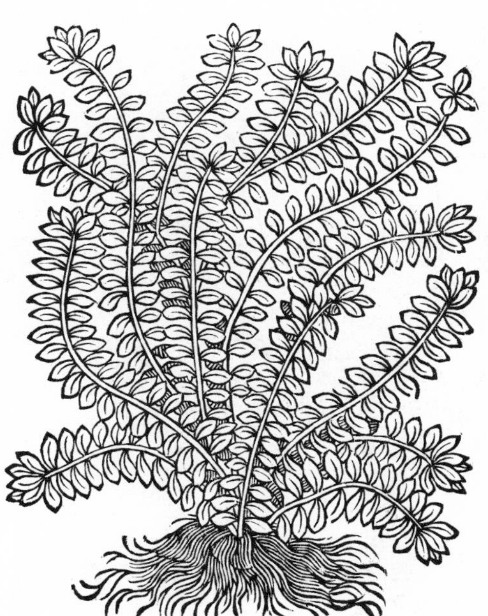

10 *Alsine palustris.*
Marish Chickweed.

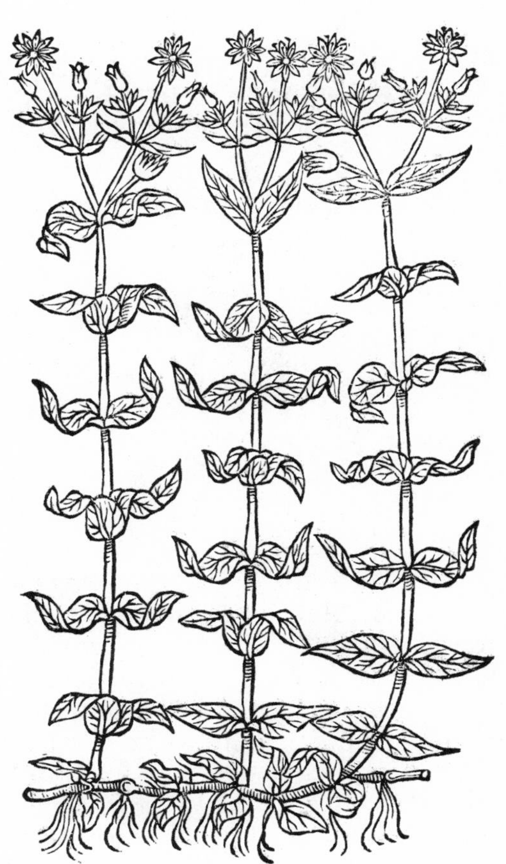

7 Speedwel Chickweed hath a little tender stalk, from which come diuers small armes or branches as it were wings, set together by couples; whereon do grow leaues set likewise by couples, like those of *Veronica*, or herbe Fluellen, whereof it tooke his name. The floures grow along the branches of a blew colour; after which come little pouches wherein is the seed: the root is small, and likewise threddy. This in the *Hist. Lugd.* is called *Elatine polyschides:* and *Fabius Columna* iudgeth it to be the *Alysson* of *Dioscorides*. ‡

8 There is a kind of Chickweed growing in the brinks and borders of Wels, Fountains, & shallow Springs, hauing many threddy roots from which rise vp diuers tender stalks, whereupon doe grow long narrow leaues; from the bosomes of which come forth diuers smaller leaues of a bright greene colour. The floures grow at the top of the stalkes, small, and white of colour.

9 There is likewise another water Chickweed smaller than the last described, hauing for his root a thicke hassocke or tuft of threddy strings: from which rise vp very many tender stems, stretching or trailing along the streame; whereon do grow long leaues set vpon a middle rib, like those of Lentils or wilde Fetch: the floures and seeds are like the precedent, but much smaller.

‡ 11 *Alsine rotundifolia, siue Portulaca aquatica.*
Water Purslane.

‡ 12 *Alsine palustris serpillifolia.*
Creeping water Chickweed.

13 *Alsine baccifera.*
Berry-bearing Chickweed.

10 There growes in the marish or waterish grounds another sort of Chickweed, not much vnlike the rest of the stocke or kindred of Chickweeds. It hath a long root of the bignesse of a wheat straw, with diuers strings hanging thereat, very like the root of Couch-grasse: from the which riseth vp diuers vpright slender stalkes, set with pretty large sharpe pointed leaues standing by couples at certaine distances: on the top of the stalkes grow small white floures like those of Stitchwort, but lesser, and of a white colour.

‡ 11 To these water Chickweeds may fitly be added those two which I mentioned and figured in my last iournall: the former of which, that I haue there called *Alsine aquatica folijs rotundioribus, siue Portulaca aquatica,* (that is) Round leaued Chickweed, or water Purslane, hath a small stringy root which sends forth diuers creeping square branches, which here and there at the ioynts put

put out small fibres, and take root againe: the leaues grow at the ioynts by couples, somwhat longish, and round at the points, resembling those of Purslane, but much smaller, and of a yellowish greene colour: at the bosomes of the leaues come forth little floures, which are succeeded by little round seed-vessels containing a small round seed. *Bauhine* hath set this forth by the name of *Alsine palustris minor folijs oblongis*.

12 The other water Chickweed, which *Iohn Bauhine* hath mentioned by the name of *Serpillifolia*; and *Casper Bauhine* by the title of *Alsine palustris minor Serpillifolia*, hath also weake and tender creeping branches lying spred vpon the ground; set with two narrow sharp pointed leaues at each ioynt, greene aboue, and of a whitish colour below: at the setting on of these leaues grow small vessels parted as it were into two, with a little crest on each side, and in these is contained a verie small seed. Both these may be found in waterie places in Iuly and August, as betweene Clapham heath and Touting, and betweene Kentish towne and Hampstead.

13 This Plant that *Clusius* and others haue called *Alsine repens major*, and some haue thought the *Ciclaminus altera* of *Dioscorides*; and *Cucubalus* of *Pliny*, may fitly be put in this ranke; for it sendeth vp many long weake branches like the great Chickweed, set with two leaues at a ioynt, bigger than those of the greatest Chickweed, yet like them in shape and colour: at the tops of the branches, out of pretty large cups come whitish greene floures, which are succeeded by berries as big as those of Iuniper, at first greene, but afterwards blacke: the seed is small and smooth: the root white, very fibrous, long and woddy, and it endures for many yeares. It floures most part of Sommer, and growes wilde in sundry places of Spaine and Germany, as also in Flanders and England, according to *Pena* and *Lobel*: yet I haue not seene it growing but in the garden of my friend Mr. *Pemble* at Marribone. The Authors last mentioned affirme the berries hereof to haue a poysonous facultie like as those of Dwale or deadly Nightshade. ‡

¶ *The Place.*

Chickweeds, some grow among bushes and briers, old walls, gutters of houses, and shadowie places. The places where the rest grow are set forth in their seuerall descriptions.

¶ *The Time.*

The Chickweeds are greene in Winter, they floure and seed in the Spring.

¶ *The Names.*

Chickweed or Chickenweed is called in Greeke Ἀλσίνη: in Latine it retaineth the same name *Alsine*: of some of the Antients it is called *Hippia*. The rest of the plants are distinguished in their seuerall titles, with proper names which likewise setteth forth the place of their growings.

¶ *The Temperature.*

Chickweed is cold and moist, and of a waterish substance; and therefore it cooleth without astriction or binding, as *Galen* saith.

¶ *The Vertues.*

The leaues of Chickweed boyled in water very soft, adding thereto some hogs grease, the pouder of Fenugreeke and Lineseed, and a few roots of marsh Mallowes, and stamped to the forme of cataplasme or pultesse, taketh away the swellings of the legs or any other part; bringeth to suppuration or matter hot apostumes; dissolueth swellings that wil not willingly yeeld to suppuration; easeth members that are shrunke vp; comforteth wounds in sinewie parts; defendeth foule maligne and virulent vlcers from inflammation during the cure: in a word, it comforteth, digesteth, defendeth, and suppurateth very notably. A

The leaues boyled in Vineger and salt are good against mangines of the hands and legs, if they be bathed therewith. B

Little birds in cadges (especially Linnets) are refreshed with the lesser Chickweed when they loath their meat; whereupon it was called of some *Passerina*. C

CHAP. 193. *Of the bastard Chickweeds.*

¶ *The Description.*

1 GErmander Chickweed hath small tender branches trailing vpon the ground, beset with leaues like vnto those of *Scordium*, or VVater Germander. Among which come forth little blew floures: which being faded, there appeare small flat husks or pouches, wherein lieth the seed. The root is small and threddy; which being once gotten into a garden ground is hard to be destroyed, but naturally commeth vp from yeare to yeare as a noisome weed.

1 *Alsine folijs trissaginis.*
Germander Chickweed.

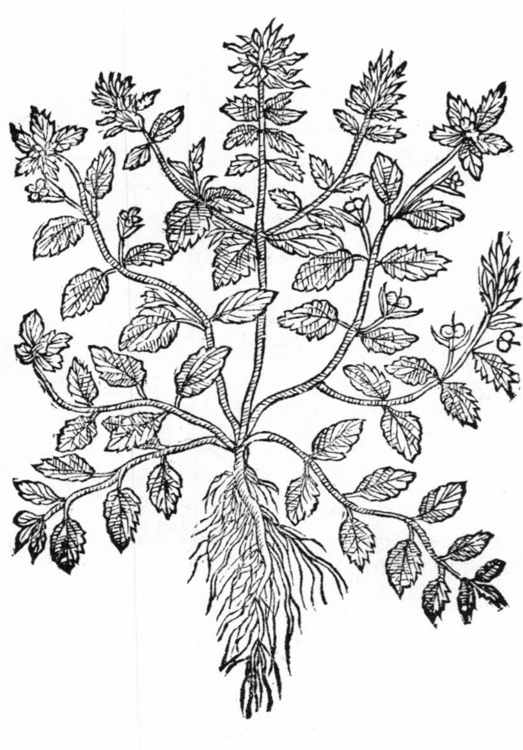

2 *Alsine corniculata Clusij.*
Horned Chickweed.

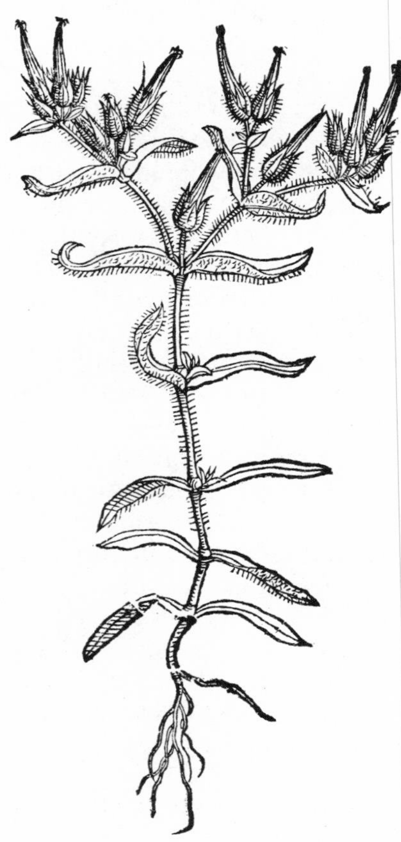

3 *Alsine Hederacea.*
Iuy Chickweed.

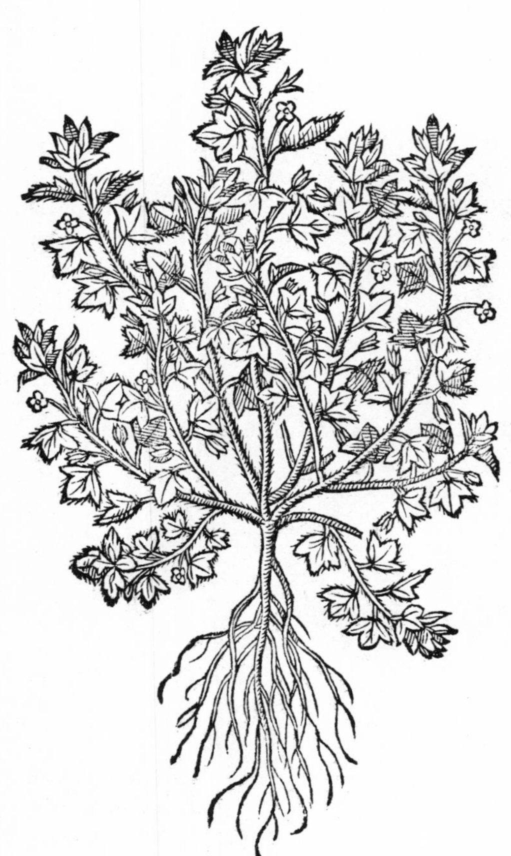

4 *Alsine Hederula altera.*
Great Henne-bit.

2 *Cluſius*, a man ſingular in the knowledge of plants, hath ſet downe this herbe for one of the Chickweeds, which doth very well reſemble the Storks bill, and might haue been there inſerted. But the matter being of ſmall moment I let it paſſe; for doubtleſſe it participateth of both, that is, the head or beake of Storkes bill, and the leaues of Chickweed, which are long and hairy, like thoſe of Scorpion Mouſe-eare. The floures are ſmall, and of an herby colour; after which come long horned cods or ſeed-veſſels, like vnto thoſe of the Storks bill. The root is ſmall and ſingle, with ſtrings faſtened thereto.

3 Iuie Chickeweed or ſmall Henbit, hath thin hairy leaues ſomewhat broad, with two cuts or gaſhes in the ſides, after the maner of thoſe of ground Iuie, whereof it tooke his name, reſembling the backe of a Bee when ſhe flieth. The ſtalkes are ſmall, tender, hairy, and lying flat vpon the ground. The floures are ſlender, and of a blew colour. The root is little and threddy.

4 The great Henbit hath feeble ſtalkes leaning toward the ground, whereupon doe grow at certaine diſtances leaues like thoſe of the dead Nettell; from the boſome whereof come forth ſlender blew floures tending to purple; in ſhape like thoſe of the ſmall dead Nettle. The root is tough, ſingle, and a few ſtrings hanging thereat.

¶ *The Place.*

Theſe Chickweeds are ſowne in gardens among potherbes, in darke ſhadowie places, and in the fields after the corne is reaped.

¶ *The Time.*

They flouriſh and are greene when the other Chickweedes are.

¶ *The Names.*

The firſt and third is called *Morſus Gallinæ*, Hens bit, *Alſine Hederula*, and *Hederacea*: *Lobell* alſo calls the fourth *Morſus Galinæ folio Hederulæ alter*: in high Dutch **Hunerbiſz**: in French, *Morſgelin*, and *Morgeline*: in low Dutch, **Hoenderebeet**: in Engliſh, Henbit the greater and the leſſer.

¶ *The Temperature and Vertues.*

Theſe are thought alſo to be could and moiſt, and like to the other Chickweeds in vertue and operation.

CHAP. 194. *Of Pimpernell.*

1 *Anagallis mas.*
Male Pimpernell.

2 *Anagallis fœmina.*
Female Pimpernell.

¶ *The*

¶ *The Deſcription.*

1 PImpernell is like vnto Chickeweed; the ſtalkes are foure ſquare,trailing here and there vpon the ground,whereupon do grow broad leaues,and ſharpe pointed, ſet together by couples:from the boſome whereof come forth ſlender tendrells, whereupon doe grow ſmall purple floures tending to redneſſe: which being paſt there ſucceed fine round bullets, like vnto the ſeed of Corianders,wherein is conteined ſmall duſtie ſeed. The root conſiſteth of ſlender ſtrings.

2 The female Pimpernell differeth not from the male in any one point,but in the colour of the floures; for like as the former hath reddiſh floures,this plant bringeth forth floures of a moſt perfect blew colour, wherein is the difference.

‡ 3 Of this there is another variety ſet forth by *Cluſius* by the name of *Anagallis tenuifolia Monelli*,becauſe he receiued the figure and Hiſtory thereof from *Iohn Monell* of Tournay in France; it differs thus from the laſt mentioned, the leaues are longer and narrower,ſomewhat like thoſe of *Gratiola*,and they now and then grow three at a joint, and out of the boſomes of the leaues come commonly as many little footſtalkes as there are leaues,which carry floures of a blew colour with the middle purpliſh,and theſe are ſomewhat larger than them of the former,otherwiſe like. ‡

‡ 3 *Anagallis tenuifolia.*
Narrow leaued Pimpernell.

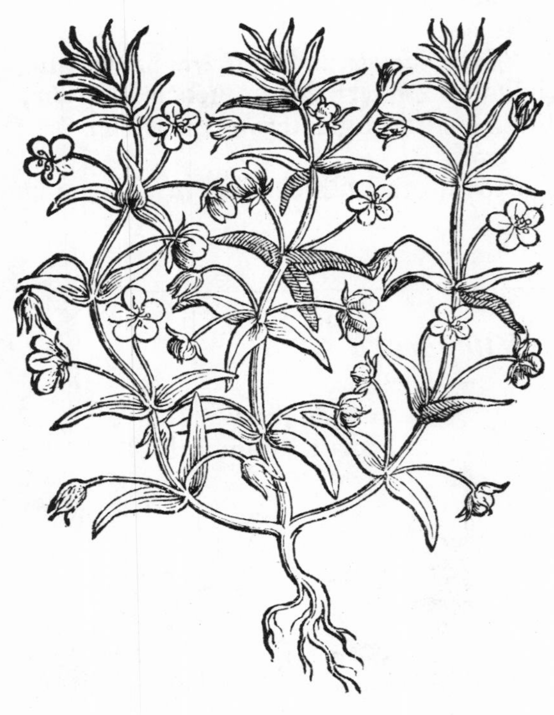

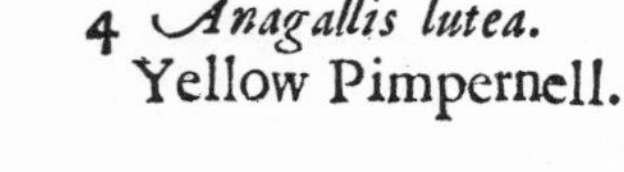

4 *Anagallis lutea.*
Yellow Pimpernell.

4 The yellow Pimpernell hath many weake and feeble branches trailing vpon the ground, beſet with leaues one againſt another like the great Chickweed, not vnlike to *Nummularia*, or Money woort; betweene which and the ſtalkes, come forth two ſingle and ſmall tender footeſtalkes, each bearing at their top one yellow floure and no more. The root is ſmall and threddy

¶ *The Place.*

They grow in plowed fields neere path waies, in gardens and vineyardes, almoſt euery where. I found the female with blew floures in a chalkie corne field in the way from Mr. *William Swaines* houſe of Southfleet to Long field downs, but neuer any where elſe. ‡ I alſo being in Eſſex in the company of my kind friend Mr. *Nathaniel Wright* found this among the corne at Wrightsbridge, being the ſeate of Mr. *Iohn Wright* his brother. ‡ The yellow Pimpernell growes in the woods betweene High-gate and Hampſtead, and in many other woods.

¶ *The Time.*

They floure in Summer,and eſpecially in the moneth of Auguſt,at what time the husbandmen hauing occaſion to go vnto their harueſt worke, will firſt behold the floures of Pimpernell,whereby they know the weather that ſhall follow the next day after: as for example, if the floures be ſhut cloſe vp,it betokeneth raine and foule weather; contrariwiſe, if they be ſpread abroad, faire weather.

¶ *The*

¶ *The Names.*

It is called in Greeke Ἀναγάλλις: in Latine alſo *Anagallis*: of diuers, (as *Pliny* reporteth) *Corchorus*, but vntruly: of *Marcellus* an old Writer, *Macia*; the word is extant in *Dioſcorides* among the baſtard names. That with the crimſon floure, being the male, is named *Phœnicion*, and *Corallion* of this is made the compoſition or receit called *Diacorallion*, that is vſed againſt the gout; which compoſition *Paulus Ægineta* ſetteth downe in his ſeuenth booke. Among the baſtard names it hath beene called *Aetitis*, *Ægitis*, and *Sauritis*: in Engliſh, Red Pimpernell, and blew Pimpernel.

¶ *The Temperature.*

Both the ſorts of Pimpernell are of a drying facultie without biting, and ſomewhat hot, with a certaine drawing quality, inſomuch that it doth draw forth ſplinters and things fixed in the fleſh, as *Galen* writeth.

¶ *The Vertues.*

Dioſcorides writes, That they are of power to mitigate paine, to cure inflammations or hot ſwel- A
lings, to draw out of the body and fleſh thornes, ſplinters, or ſhiuers of wood, and to helpe the Kings Euill.

The iuyce purgeth the head by gargariſing or waſhing the throat therewith; it cures the tooth- B
ache being ſniſt vp into the noſethrils, eſpecially into the contrary noſethrill.

It helpeth thoſe that be dim ſighted: the iuyce mixed with honey cleanſes the vlcers of the eye C
called in Latine *Argema*.

Moreouer he affirmeth, That it is good againſt the ſtinging of Vipers, and other venomous D
beaſts.

It preuaileth againſt the infirmities of the liuer and kidneyes, if the iuyce be drunk with wine. E
He addeth further, how it is reported, That Pimpernel with the blew floure helpeth vp the fundament that is fallen downe; and that red Pimpernell applied, contrariwiſe bringeth it downe.

CHAP. 195. *Of Brooke-lime, or water Pimpernell.*

¶ *The Deſcription.*

1 BRooke-lime or Brooklem hath fat thicke ſtalkes, round, and parted into diuers branches: the leaues be thicke, ſmooth, broad, and of a deepe greene colour. The floures grow vpon ſmall tender foot-ſtalkes, which thruſt forth of the boſome of the leaues, of a perfect blew colour, not vnlike to the floures of land Pimpernell: the root is white, low creeping, with fine ſtrings faſtned thereto: out of the root ſpring many other ſtalkes, whereby it greatly encreaſeth.

‡ There is a leſſer varietie of this, which our Author ſet forth in the fourth place, differing not from this but onely in that it is leſſe in all the parts thereof; wherefore I haue omitted the hiſtorie and figure, to make roome for more conſpicuous differences. ‡

2 The great water Pimpernell is like vnto the precedent, ſauing that this plant hath ſharper pointed or larger leaues, and the floures are of a more whitiſh or a paler blew colour, wherein conſiſteth the difference.

‡ There is alſo a leſſer varietie of this, whoſe figure and deſcription our Authour gaue in the next place; but becauſe the difference is in nothing but the magnitude I haue made bold to omit it alſo.

3 Now that I haue briefely giuen you the hiſtory of the foure formerly deſcribed by our Author, I will acquaint you with two or three more plants which may fitly be here inſerted: The firſt of theſe *Lobel* calls *Anagallis aquatica tertia*; and therefore I haue thought fit to giue you it in the ſame place here. It hath a white and fibrous root; from which ariſeth a round ſmooth ſtalke a foot and more high, (yet I haue ſometimes found it not aboue three or foure inches high:) vpon the ſtalkes grow leaues round, greene, and ſhining, ſtanding not by couples, but one aboue another on all ſides of the ſtalkes. The leaues that lie on the ground are longer than the reſt, and are in ſhape ſomewhat like thoſe of the common Daiſie, but that they are not ſnipped about the edges: the floures are white, conſiſting of one leafe diuided into fiue parts; and they grow at the firſt as it were in an vmbel, but afterwards more ſpike faſhioned. It floures in Iune and Iuly, and groweth in many waterie places, as in the mariſhes of Dartford in Kent, alſo betweene Sandwich and Sandowne caſtle, and in the ditches on this ſide Sandwich. *Bauhine* ſaith, That *Guillandinus* called it ſometimes *Aliſma*, and otherwhiles *Cochlearia*: and others would haue it to be *Samolum* of *Pliny*, *lib.* 25. *cap.* 11. *Bauhine* himſelfe fitly calls it *Anagallis aquatica folio rotundo non crenato.*

1 *Anagallis seu Becabunga.*
Brooke-lime.

2 *Anagallis aquatica maior.*
Great long leaued Brook-lime.

‡ 3 *Anagallis aquatica rotundifolia.*
Round leaued water Pimpernel.

4 I coniecture this figure which we here giue you with the Authors title to be onely the lesser variety of that which our Author describes in the second place; but because I haue no certaintie hereof (for that *Lobel* hath giuen vs no descripti- on thereof in any of his Latine Workes, and also *Bauhinus* hath distinguished them) I am forced to giue you onely the figure thereof; not intending to deceiue my reader by giuing descriptions from my fancie and the figure, as our Author somtimes made bold to do.

5 This which is set forth by most writers for *Cepæa*, and which some may obiect to be more fit to be put next the Purslanes, I will here giue you, hauing forgot to doe it there; and I thinke this place not vnfit, because our Author in the Names in this Chapter takes occasion in *Dodonæus* his words to make mention thereof. It hath a small vnprofitable root, sending vp a stalke some foot high, diuided into many weake branches, which are here and there set with thicke leaues like those of Purslane, but much lesse, and narrower, and sharper pointed: the floures which grow in good plenty vpon the tops of the branches are compo- sed of fiue small white leaues; whereto succeeds small heads, wherein is contained a seed like that of Orpine. This by *Matthiolus* and others is called *Cepæa*: but *Clusius* doubts that it is not the true *Cepæa* of the Antients. ‡

¶ The

‡ 4 *Anagallis aquatica quarta, Lob.*
Lobels fourth water Pimpernel.

‡ 5 *Cepæa.*
Garden Brook-lime.

¶ *The Place.*

They grow by riuers ſides, ſmall running brookes, and waterie ditches. The yellow Pimpernell I found growing in Hampſted wood neere London, and in many other woods and copſes.

¶ *The Time.*

They bring forth their floures and ſeed in Iune, Iuly, and Auguſt.

¶ *The Names.*

Water Pimpernel is called *Anagallis aquatica*: of moſt, *Becabunga*, which is borrowed of the Germane word **Bachpunghen:** in low-Dutch, **Beeckpunghen:** in French, *Berle*; whereupon ſome do call it *Berula*: notwithſtanding *Marcellus* reporteth, That *Berula* is that which the Grecians call καρδάμινη, or rather Creſſes: it is thought to be *Cepæa*; that is to ſay, of the garden; which *Dioſcorides* writeth to be like vnto Purſlane, whereunto this Brook-lime doth very well agree. But if it be therefore ſaid to be κηπαία, becauſe it groweth either onely or for the moſt part in gardens, this Pimpernel or Brook-lime ſhall not be like vnto it, which groweth no where leſſe than in gardens, being altogether of his owne nature wilde, deſiring to grow in waterie places, and ſuch as be continually ouerflowne: in Engliſh the firſt is called Brooklime, and the reſt by no particular names; but we may call them water Pimpernels, or Brook-limes.

¶ *The Temperature.*

Brook-lime is of temperature hot and dry like water Creſſes, yet not ſo much.

¶ *The Vertues.*

Brooke-lime is eaten in ſallads as Water-Creſſes are, and is good againſt that ἐπιχώριον *malum* of ſuch as dwell neere the Germane ſeas, which they call **Scuerbuycke:** or as we terme it, the Scuruie, or Skirby, being vſed after the ſame manner that Water Creſſes and Scuruy graſſe is vſed, yet is it not of ſo great operation and vertue. A

The herbe boyled maketh a good fomentation for ſwollen legs and the dropſie. B

The leaues boyled, ſtrained, and ſtamped in a ſtone morter with the pouder of Fenugreek, Lineſeeds, the roots of mariſh Mallowes, and ſome hogs greaſe, vnto the forme of a cataplaſme or pulteſſe, taketh away any ſwelling in leg or arme; wounds alſo that are ready to fall into apoſtumation it mightily defendeth, that no humor or accident ſhall happen thereunto. C

 The

D The leaues of Brooke-lime stamped, strained, and giuen to drinke in wine, helpeth the strangurie, and griefes of the bladder.

E The leaues of Brook-lime, and the tendrels of *Asparagus*, eaten with oyle, vineger, and Pepper, helpeth the strangurie and stone.

Chap. 196. *Of stinking Ground-Pine.*

¶ *The Kindes.*

‡ *Dioscorides* hath antiently mentioned two sorts of *Anthyllis:* one with leaues like to the Lentill, & the other like to *Chamæpitys*. To the first, some late writers haue referred diuers plants, as the two first described in this Chapter; The *Anthyllis Leguminosa Belgarum* hereafter to be described; the *Anthyllis Valentina Clusij* formerly set forth Chap. 171. To the second are referred the *Iua Moschata Monspeliaca*, described in the fourth place of the 150. Chap. of this booke; the *Linaria adulterina* described formerly chap. 165. in the 14. place, and that which is here described in the third place of this chapter, by the name of *Anthyllis altera Italorum*. ‡

¶ *The Description.*

1 There hath beene much adoe among Writers about the certaine knowledge of the true *Anthyllis* of *Dioscorides :* I will therefore set downe that plant which of all others is found most agreeable thereunto. It hath many small branches full ioynts, not aboue an handfull high, creeping sundry wayes, beset with small thicke leaues of a pale colour, resembling *Lenticula*, or rather *Alsine minor*, the lesser Chickweed. The floures grow at the top of the stalke, starre-fashion, of an herby colour like boxe, or *Sedum minus :* it fostereth his small seeds in a three cornered huske. The root is somewhat long, slender, ioynted, and deepely thrust into the ground like *Soldanella :* all the whole plant is saltish, bitter in taste, and somewhat heating.

‡ 1 *Anthyllis lentifolia, siue Alsine cruciata marina.*
Sea Pimpernell.

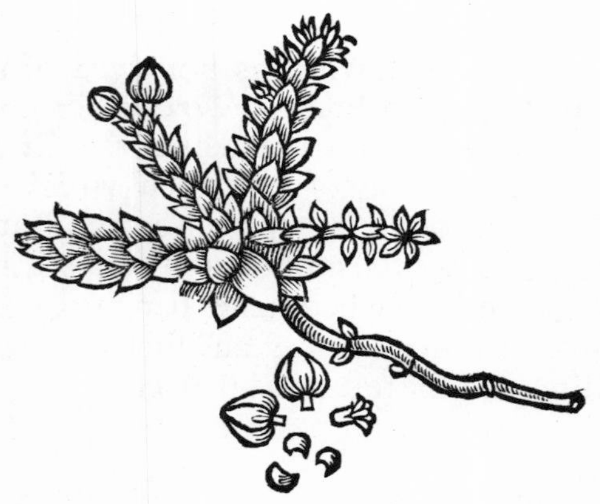

‡ 2 *Anthyllis Marina incana Alsinefolia.*
Many floured Ground-Pine.

‡ This description was taken out of the *Aduersaria, pag.* 195. where it is called *Anthyllis prior lentifolia Peplios effigie maritima :* also *Clusius* hath described it by the name of *Alsines genus pelagicum :* I haue called it in my last iournall by the name of *Alsine cruciata marina,* because the leaues which grow thicke together by couples crosse each other, as it happens in most plants which haue square stalkes with two leaues at each ioynt. I haue Englished it Sea Pimpernell, because the leaues in shape are as like those of Pimpernel as of any other Plant; and also for that our Author hath called another plant by the name of Sea Chickeweed. The figure of the *Aduersaria* was not good, and *Clusius* hath none; which hath caused some to reckon this *Anthyllis* of *Lobel*, and *Alsine* of *Clusius* for two seuerall plants, which indeed are not so. I haue giuen you a figure hereof which I tooke from the growing plant, and which well expresseth the growing thereof. ‡

2 There

3 *Anthyllis altera Italorum.*
Stinking ground Pine.

2 There is likewiſe another ſort of *Anthyllis* or Sea Ground Pine, but in truth nothing els than a kinde of Sea Chickeweed, hauing ſmall branches trailing vpon the ground of two hands high, whereupon do grow little leaues like thoſe of Chickweed, not vnlike thoſe of *Lenticula marina*, or Sea Lentils: on the top of the ſtalks ſtand many ſmall moſſie floures of a white colour. The whole plant is of a bitter and ſaltiſh taſte. ‡ This is the *Marina incana Anthyllis Alſinefolia Narbonenſium* of *Lobel*: it is the *Paronychia altera* of *Matthiolus*. ‡

‡ 3 To this figure (which formerly was giuen for the firſt of theſe by our Authour) I will now giue you a briefe deſcription. This in the branches, leaues, and whole face thereof is very like the French Herbe-Iuie, or Ground Pine, but that it is much leſſe in all the parts thereof, but chiefely in the leaues which alſo are not ſnipt like thoſe of the French Ground Pine, but ſharp pointed: the tops of the branches are downie or woolly, and ſet with little pale yellow floures. ‡

¶ *The Place.*

Theſe do grow in the South Iſles belonging to England, eſpecially in Portland in the grauelly and ſandy foords, which lie low and againſt the ſea; and likewiſe in the iſle of Shepey neere the water ſide. ‡ I haue onely found the firſt deſcribed, and that both in Shepey, as alſo in Weſt-gate bay by Margate in the Iſle of Thanet. ‡

¶ *The Time.*

They floure and flouriſh in Iune and Iuly.

¶ *The Names.*

Their titles and deſcriptions ſufficiently ſet forth their ſeuerall names.

¶ *The Temperature.*

Theſe ſea herbes are of a temperate facultie betweene hot and cold.

The Vertues.

Halfe an ounce of the dried leaues drunke, preuaileth greatly againſt the hot piſſe, the ſtrangurie, or difficultie of making water, and purgeth the reines. A

The ſame taken with Oxymell or honied water is good for the falling ſickneſſe, giuen firſt at morning, and laſt at night. B

† There was formerly three deſcriptions, yet but one figure in this chapter, and that was marked with the figure 1. and called *Anthyllis lentifolia*, but vnfitly: wherefore I haue giuen you the title which *Lobel* the firſt Author thereof puts vpon it, with a deſcription thereto, that it may not ſtand as a cipher, as it formerly did. That deſcription which formerly held the ſecond place was of the *Anthyllis Valentina* of *Cluſius*, deſcribed formerly chap. 171. and therefore I haue omitted it here.

CHAP. 197. *Of Whiteblow, or Whitelow Graſſe.*

¶ *The Kindes.*

1 THe firſt is a very ſlender plant hauing a fewe ſmall leaues like the leaſt Chickeweede, growing in little tufts, from the midſt whereof riſeth vp a ſmall ſtalke, three or foure inches long; on whoſe top do grow very little white floures; which being paſt, there come in place ſmall flat pouches compoſed of three filmes; which being ripe, the two outſides fall away, leauing the middle part ſtanding long time after which is like white Sattin, as is that of *Bolbonac*, which our women call white Sattin, but much ſmaller: the taſte is ſomewhat ſharpe.

2 This kinde of *Paronychia*, hath ſmall thicke and fat leaues, cut into three or more diuiſions, much reſembling the leaues of Rue, but a great deale ſmaller. The ſtalks are like the former, & the

leaues

leaues also; but the cases wherein the seede is contained, are like vnto the seed vessels of *Myositis Scorpioides*, or Mouseare Scorpion grasse. The floures are small and white.

There is another sort of Whitlow grasse or Nailewoort, that is likewise a low or base herbe, hauing a small tough roote, with some threddie strings annexed thereto: from which rise vp diuers slender tough stalkes, set with little narrow leaues confusedly like those of the smallest Chickweed whereof doubtlesse these be kindes·alongst the stalks do grow very little white floures, after which come the seeds in small buttons, of the bignesse of a pins head. ‡ Our Author seemes here to describe the *Paronychia* 2. of *Tabern.* ‡

1 *Paronychia vulgaris.*
Common Whitlow grasse.

2 *Paronychia Rutaceo folio.*
Rew leafed or iagged Whitlow grasse.

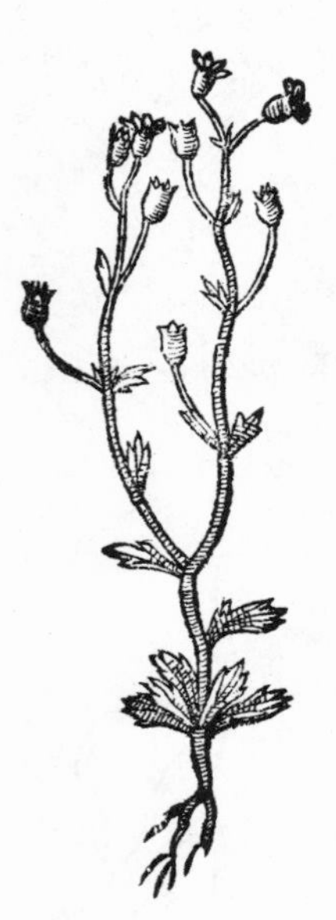

¶ *The Place.*

These small, base and low herbs grow vpon bricke and stone wals, vpon old tiled houses, which are growne to haue much mosse vpon them, and vpon some shadowie, and dry muddy wals. It groweth plentifully vpon the bricke wall in Chancerie Lane, belonging to the Earle of Southampton, in the Suburbs of London, and sundry other places.

¶ *The Time.*

These floure many times in Ianuary and February, and when hot weather approcheth, they are no more to be seen all the yeare after.

¶ *The Names.*

The Græcians haue called these plants παρωνυχία: which *Cicero* calleth *Reduvia*: There be many kindes of plants, called by the said name of *Paronychia*, which hath caused many writers to doubt of the true kinde: but you may very boldly take these plants for the same, vntill time hath reuealed or raised vp some new plant, approching neerer vnto the truth: which I thinke will neuer be, so that we may call them in English, Naile-woort, and Whitelow grasse.

¶ *The Tmperatures and Vertues.*

A As touching the qualitie hereof, we haue nothing to set downe: onely it hath beene taken to heale the disease of the nailes called a Whitlow, whereof it took his name.

† Our Authour here gaue vs two figures, and as many descriptions of both these plants, wherefore I haue omitted 2. of the figures, and the more vnperfect Descriptions.

CHAP. 198. *Of the female Fluellen, or Speedwell.*

¶ *The Description.*

1 THe first kinde of *Elatine*, beeing of *Fuchsius* and *Matthiolus*, called *Veronica fœmina*, or the female Fluellen, shooteth from a small and fibrous root many flexible and tender branches, dispersed flat vpon the ground, ramping & creeping with leaues like *Nummularia*, but that the leaues of *Elatine* are of an hoarie, hairie, and ouerworne greene colour; among which come forth many small floures, of a yellow colour mixed with a little purple, like vnto the small Snapdragon, hauing a certaine taile or Spur fastened vnto euery such floure, like the herbe called Larkes spurre. The lower iaw or chap of the floure is of a purple colour, and the vpper iawe of a faire yellowe; which beeing past, there succeedes a small blacke seede contained in round husks.

2 The second kinde of *Elatine* hath stalkes, branches, floures, and roots, like the first: but the leaues are fashioned like the former, but that they haue two little ears at the lower end, somewhat resembling an arrow head, broad at the setting on: but the spur or taile of the floure is longer, and more purple mixed with the yellow in the floure.

1 *Veronica fœmina Fuchsij, siue Elatine.*
The Female Fluellen.

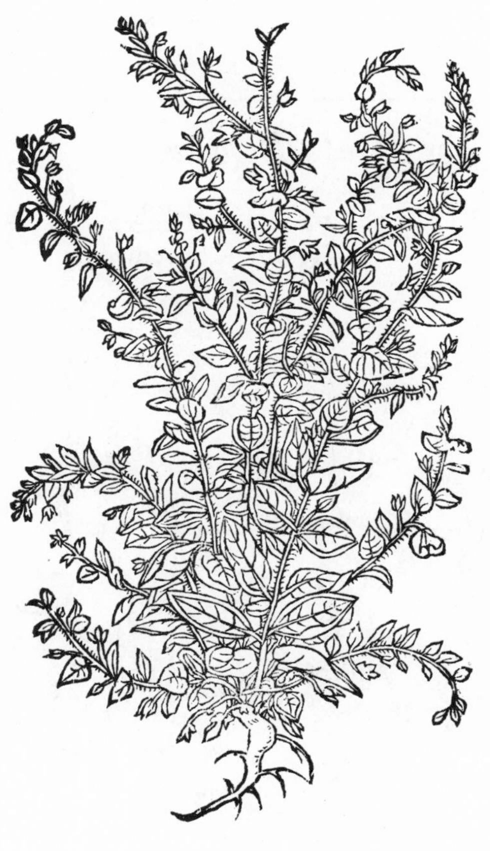

2 *Elatine altera.*
Sharpe pointed Fluellen.

¶ *The Place.*

Both these plants I haue found in sundry places where corne hath growne, especially barley, as in the fields about Southfleet in Kent, where within six miles compasse there is not a field wherein it doth not grow.

Also it groweth in a field next vnto the house sometime belonging to that honourable gentleman Sir *Frances Walsingham*, at Barn-elmes, and in sundry places of Essex; and in the next field vnto the Churchyard at Chiswicke neere London, towards the midst of the field.

¶ *The Time.*

They floure in August and September.

¶ The

¶ *The Names.*

Their ſeuerall titles ſet forth their names as well in Latine as Engliſh.

¶ *The Nature and Vertues.*

A These plants are not onely of a ſingular aſtringent facultie, and thereby helpe them that bee grieued with the Dyſenterie and hot ſwelling;but of ſuch ſingular efficacy to heale ſpreading and eating cankers,and coroſiue vlcers,that their vertue in a manner paſſeth all credit in theſe fretting ſores,vpon ſure proofe done vnto ſundry perſons, and eſpecially vpon a man whom *Pena* reporteth to haue his noſe eaten moſt grieuouſly with a canker or eating ſore,who ſent for the Phyſitions & Chirurgions that were famouſly knowne to be the beſt,and they with one conſent concluded to cut the ſaid noſe off, to preſerue the reſt of his face : among theſe Surgeons and Phyſitions came a poore ſorie Barbar, who had no more skill than he had learned by tradition, and yet vndertooke to cure the patient.This foreſaid Barbar ſtanding in the companie and hearing their determination, deſired that he might make triall of an herbe which he had ſeene his maſter vſe for the ſame purpoſe, which herbe *Elatine*, though he were ignorant of the name whereby it was called,yet hee knew where to fetch it.To be ſhort,this herbe he ſtamped,and gaue the iuice of it vnto the patient to drinke,and outwardly applied the ſame plaiſterwiſe,and in very ſhort ſpace perfectly cured the man,and ſtaied the reſt of his body from further corruption,which was ready to fall into a leproſie, *Aduerſar.pag.197.*

B *Elatine* helpeth the inflammation of the eies,and defendeth humours flowing vnto them,beeing boiled,and as a pultus applied thereto.

C The leaues ſodden in the broth of a hen,or Veale,ſtaieth the dyſenterie.

D The new writers affirme,that the female Fluellen openeth the obſtructions or ſtoppings of the liuer and ſpleen,prouoketh vrine,driueth forth ſtones,and clenſeth the kidneies and bladder,according to *Paulus*.

E The weight of a dram or of a French crowne,of the pouder of the herbe,with the like waight of treacle,is commended againſt peſtilent Feuers.

Chap. 192. *Of Fluellen the male, or Paul's Betonie.*

1 *Veronica vera & maior.*
Fluellen,or Speedwell.

† 2 *Veronica recta mas.*
The male Speedwell.

¶ *The*

¶ *The Deſcription.*

1 THe firſt kinde of *Veronica* is a ſmall herbe, and creepeth by the ground, with little reddiſh and hairy branches. The leafe is ſomething round and hairy, indented or ſnipped round about the edges. The floures are of a light blew colour, declining to purple: the ſeed is contained in little flat pouches: the root is fibrous and hairy.

† 2 The ſecond doth alſo creepe vpon the ground, hauing long ſlender ſtemmes, ſome foot high, and ſomewhat large leaues a little hairy, and pleaſantly ſoft. The floures be blew like as thoſe of the former, but ſomewhat bigger, and of a brighter colour; and they are alſo ſucceeded by round ſeed veſſels.

3 The third kinde of *Veronica* creepeth with branches and leaues like vnto *Serpillum*, for which cauſe it hath beene called *Veronica Serpillifolia*. The floures grow along the ſmall and tender branches, of a whitiſh colour declining to blewneſſe. The root is ſmall and threddie, taking hold vpon the vpper face of the earth, where it ſpreadeth. The ſeed is contained in ſmall pouches like the former.

4 The fourth hath a root ſomewhat wooddie, from the which riſe vp leaues like vnto the former. The ſmall vpright ſtalke is beſet with the like leaues, but leſſer, at the top whereof commeth forth a ſlender ſpike cloſely thruſt together, and full of blewiſh floures, which are ſucceeded by many horned ſeed veſſels.

‡ 5 This hath many wooddie round ſmooth branches, ſome handfull and halfe high or better: the leaues are like thoſe of wilde Tyme, but longer, and of a blacker colour, ſometimes lightly ſnipt: at the tops of the branches grow floures of a whitiſh blew colour, conſiſting of foure, fiue, or elſe ſix e little leaues a piece; which falling, there follow round ſeede veſſels, containing a round ſmall and blacke ſeed. It floures in Auguſt, and growes vpon cold and high mountaines, as the Alpes. *Pona* calls this *Veronica Alpina minima Serpillifolio*: and *Cluſius* hath it by the name of *Veronica 3. fruticans*. ‡

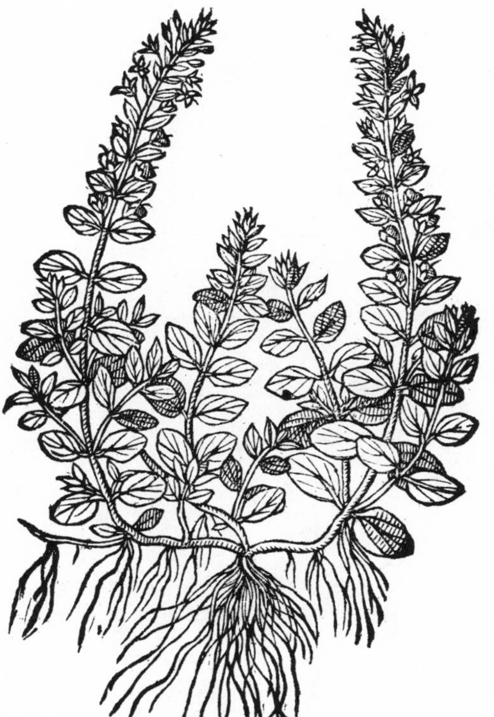

3 *Veronica minor.*
Little Fluellen.

4 *Veronica recta minima.*
The ſmalleſt Fluellen.

‡ 5 *Veronica*

‡ 5 *Veronica fruticans Serpillifolia.*
Shrubbie Fluellen.

6 *Veronica assurgens, siue Spicata.*
Tree Fluellen.

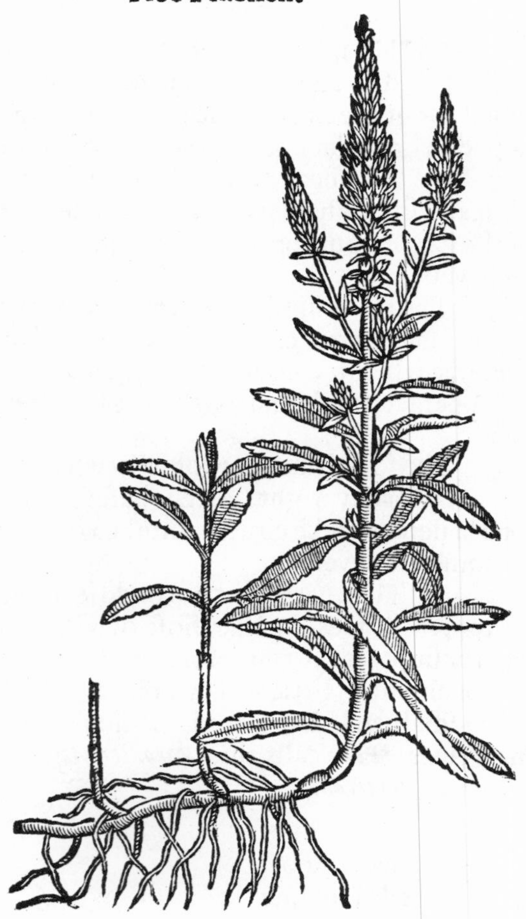

† 7 *Veronica spicata latifolia.*
Vpright Fluellen.

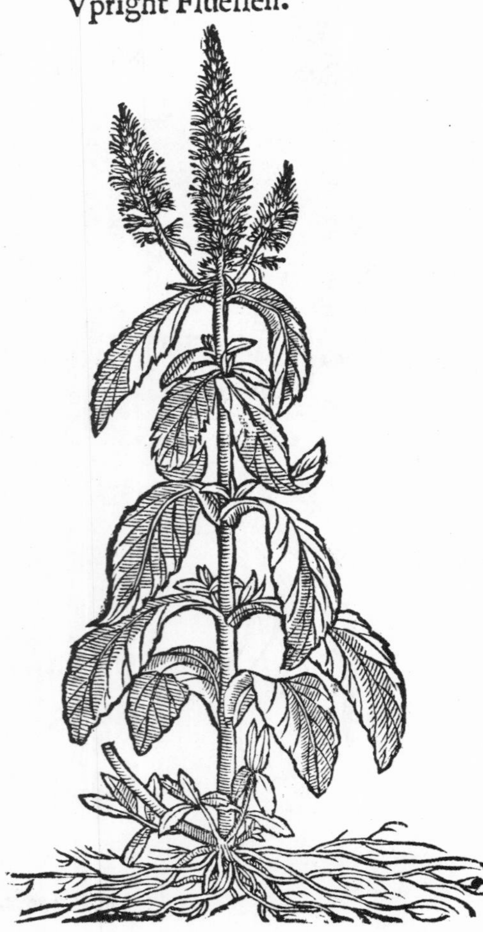

‡ 8 *Veronica supina.*
Leaning Fluellen.

6 The ſixt kinde of *Veronica* hath many vpright branches a foote high and ſometimes more, diuiding themſelues into ſundry other ſmall twigs; at the top whereof do grow faire ſpikie tufts, bearing bright and ſhining blew floures. The leaues are ſomewhat long, indented about the edges like a ſaw: the root is compact of many threds, or ſtrings.

‡ 7 This hath ſtalkes ſome cubit high and ſometimes more, and theſe not very full of branches, yet hauing diuers joints, at each whereof do grow forth two leaues, two or three inches long, and one broad, and theſe leaues are alſo thicke, ſmooth, and ſhining, lightly ſnipt or cut about the edges, and of a very aſtringent and drying taſte, and at laſt ſomewhat biting. At the top of the ſtalkes grow ſpokie tufts or blew floures like thoſe of the laſt mentioned, but of ſomewhat a lighter colour, and they begin firſt to floure or ſhew themſelues below, and ſo go vpwards; the ſeed, which is ſmall and blacke, is conteyned in flat ſeed veſſels: the roote is thicke with many fibres, euery yeere thruſting vp new ſhoots. There is a variety of this with the leaues not ſo blacke and ſhining, but hauing more branches; and another which hath a longer ſpike or tuft of floures. *Cluſius* calls this *Veronica erectior latifolia*. ‡

8 The eighth hauing his ſtalkes leaning vpon the ground looketh with his face vpright, hauing ſundry flexible branches, ſet with leaues like vnto wilde Germander by couples, one right againſt another, deeply jagged about the edges, in reſpect of the other before mentioned. The floures are of a blew colour: the root is long, with ſome threds appendant thereto.

¶ *The Place.*

Veronica groweth vpon bankes, borders of fields, and graſsie mole-hils, in ſandy grounds, and in woods, almoſt euery where.

The fourth kinde, my good friend M[r]. *Stephen Bredwell*, practitioner in phyſicke found and ſhewed it me in the cloſe next adjoining to the houſe of M[r]. *Bele*, chiefe of the clerkes of her Maieſties Counſell, dwelling at Barnes neere London. The ſixth is a ſtranger in England, but I haue it growing in my garden.

¶ *The Time.*

Theſe floure from May to September.

¶ *The Names.*

† Theſe plants are comprehended vnder this generall name *Veronica*; and *Dodonæus* would haue the firſt of them to be the *Betonica* of *Paulus Ægineta*; and *Turner* and *Geſner* the third: we do call them in Engliſh, Pauls Betony, or Speedwell: in Welch it is called Fluellen, and the Welch people do attribute great vertues to the ſame: in high Dutch, **Grondtheill**: in low Dutch, **Eeren prijs**, that is to ſay, Honor and praiſe.

¶ *The Nature.*

Theſe are of a meane temperature, betweene heate and drineſſe.

¶ *The Vertues.*

The decoction of *Veronica* drunke, ſodereth and healeth all freſh and old wounds, clenſeth the bloud from all corruption, and is good to be drunke for the kidnies, and againſt ſcuruineſſe and foule ſpredding tetters, and conſuming and fretting ſores, the ſmall pox and meaſels. A

The water of *Veronica* diſtilled with wine, and re-diſtilled ſo often vntill the liquor wax of a reddiſh colour, preuaileth againſt the old cough, the drineſſe of the lungs, and all vlcers and inflammation of the ſame. B

† The ſecond and third were both figures of that deſcribed in the third place: and thoſe that were formerly in the fifth and ſixth places were alſo of the ſame plant, to wit that which is here deſcribed in the ſixth place and which was formerly in the fifth.

CHAP. 198. *Of herbe Two pence.*

¶ *The Deſcription.*

1 Herbe Two pence hath a ſmall and tender root, ſpreding and diſperſing it ſelfe farre within the ground; from which riſe vp many little, tender, flexible ſtalkes trailing vpon the ground, ſet by couples at certaine ſpaces, with ſmooth greene leaues ſomewhat round, whereof it tooke his name: from the boſome of which leaues ſhoote forth ſmall tender foot-ſtalkes, whereon do grow little yellow floures, like thoſe of Cinkefoile or Tormentill.

2 There is a kinde of Money woort or herbe Two pence, like the other of his kinde in each reſpect, ſauing it is altogether leſſer, wherein they differ.

‡ 3 There is another kinde of Money-woort which hath many very ſlender creeping branches which here and there put forth fibres, and take root againe: the leaues are ſmall and round, ſtanding by couples one againſt another; and out of the boſomes come ſlender foote-ſtalkes

 bearing

bearing prety little whitish purple floures consisting of fiue little leaues standing together in manner of a little bell-floure, and seldome otherwise: the seed is small, and conteined in round heads. This growes in many wet rotten grounds and vpon bogges: I first found it *Anno 1626*, in the Bishopricke of Durham, and in two or three places of Yorkshire, and not thinking any had taken notice thereof, I drew a figure of it & called it *Nummularia pusilla flore ex albo purpurascente*; but since I haue found that *Bauhine* had formerly set it forth in his *Prodromus* by the name of *Nummularia flore purpurascente*. It growes also on the bogges vpon the heath, neare Burnt wood in Essex: it floures in Iuly and August. ‡

1 *Nummularia.*
Herbe Two pence.

‡ 3 *Nummularia flore purpurascente.*
Purple floured Money-woort.

¶ *The Place.*

It groweth neere vnto ditches and streames, and other watery places, and is sometimes found in moist woods: I found it vpon the banke of the riuer of Thames, right against the Queenes pallace of White hall; and almost in euery countrey where I haue trauelled.

¶ *The Time.*

It floureth from May till Summer be well spent.

¶ *The Names.*

Herbe Two pence is called in Latine *Nummularia*, and *Centummorbia*: and of diuers *Serpentaria*. It is reported that if serpents be hurt or wounded, they do heale themselues with this herbe, whereupon came the name *Serpentaria*: it is thought to be called *Centummorbia*, of the wonderfull effect which it hath in curing diseases; and it is called *Nummularia* of the forme of money, whereunto the leaues are like: in Dutch, **Penninckcruyt**: in English, Money woort, Herbe Two pence, and Two penny grasse.

¶ *The Temperature.*

That this herbe is dry, the binding taste thereof doth shew: it is also moderate colde.

¶ *The Vertues.*

A The floures and leaues stamped and laid vpon wounds and vlcers doth cure them: but it worketh most effectually being stamped and boiled in oile oliue, with some rosen, wax, and turpentine added thereto.

The iuice drunke in wine, is good for the bloudie flix, and all other iſſues in bloud of man or woman; the weakeneſſe and looſeneſſe of the belly and laske; it helpeth thoſe that vomite bloud, and the Whites in ſuch as haue them. B

Boiled with wine and honie it cureth the wounds of the inward parts, and vlcers of the lungs, & in a word, there is not a better wound herbe, no not Tabaco it ſelfe, nor any other whatſoeuer. C

The herbe boiled in wine with a little honie, or meade, preuaileth much againſt the cough in children, called the Chinne cough. D

CHAP. 200. *Of Bugle or Middle Comfrey.*

¶ *The Description.*

1 B*Vgula* ſpreadeth and creepeth alongſt the ground like Monie woort; the leaues be long, fat, & oleous, and of a brown colour for the moſt part. The floures grow about the ſtalks in rundles, compaſſing the ſtalke, leauing betweene euery rundle bare or naked ſpaces; and are of a faire blew colour, and often white. I found many plants of it in a moiſt ground vpon Blacke Heath neere London, faſt by a village called Charleton, but the leaues were green, and not browne at all like the other.

1 *Bugula.*
Middle Conſound.

2 *Bugula flore albo, ſiue carneo.*
White or carnation floured Bugle.

2 Bugle with the white floure differeth not from the precedent, in roots, leaues, and ſtalks the onely difference is, that this plant bringeth forth faire milk white floures, and the other thoſe that are blew. ‡ It is alſo found with a fleſh coloured floure, and the leaues are leſſe ſnipt than thoſe of the former. *Bauhine* makes mention of one much leſſe than thoſe, with round ſnipt leaues and a yellow floure, which he ſaith he had out of England, but I haue not as yet ſeene it, nor found any other mention thereof. ‡

¶ *The Place.*

Bugula groweth almoſt in euery wood and copſe, and ſuch like ſhadowie and moiſt places, and is much planted in gardens: the other varieties are ſeldome to be met withall.

¶ *The Time.*

Bugula floureth in Aprill and May.

¶ *The Names.*

Bugle is reckoned among the Conſounds or wound herbes: and it is called of ſome *Conſolida media, Bugula*, and *Buglum*: in High Dutch, **Guntzel**: in Low Dutch **Senegroen**: of *Matthiolus, Herba Laurentina*: in Engliſh, Browne Bugle: of ſome, Sicklewoort, and herbe Carpenter, but not truly.

¶ *The Nature.*

Bugle is of a meane temperature, betweene heat and drineſſe.

¶ *The Vertues.*

A It is commended againſt inward burſtings, and members torne, rent, and bruiſed: and therefore it is put into potions that ſerue for nodes, in which it is of ſuch vertue, that it can diſſolue & waſte away congealed and clotted bloud. *Ruellius* writeth that they commonly ſay in France, how he needeth neither Phyſition nor Surgeon that hath Bugle and Sanickle, for it doth not only cure rotten wounds being inwardly taken, but alſo applied to them outwardly; it is good for the infirmities of the Liuer, it taketh away the obſtructions, and ſtrengthneth it.

B The decoction of Bugle drunken, diſſolueth clotted or congealed bloud within the bodie, healeth and maketh ſound all wounds of the bodie, both inward and outward.

C The ſame openeth the ſtoppings of the Liuer and gall, and is good againſt the iaundiſe and feuers of long continuance.

D The ſame decoction cureth the rotten vlcers and ſores of the mouth and gums.

E *Bugula* is excellent in curing wounds and ſcratches, and the iuice cureth the wounds, vlcers and ſores of the ſecret parts, or the herbe bruiſed and laid thereon.

CHAP. 201 *Of Selfe-heale.*

1 *Prunella.*
Selfe-heale.

2 *Prunella Lobelij.*
The ſecond Selfe-heale.

¶ *The*

3 *Prunella flore albo.*
White floured Selfe-heale.

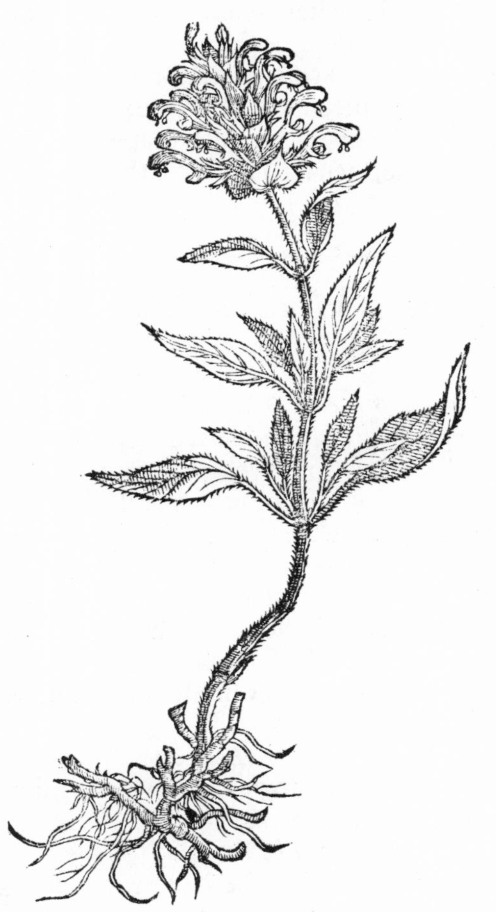

¶ *The Deſcription.*

1 PRunell or Brunel hath ſquare hairy ſtalks of a foot high, beſet with long, hairy and ſharpe pointed leaues, & at the top of the ſtalks grow floures thicke ſet together, like an eare or ſpiky knap, of a browne colour mixed with blew floures, and ſometimes white, of which kinde I found ſome plants in Eſſex neere Henningham caſtle. The root is ſmall and very threddie.

† 2 *Prunella altera*, or after *Lobel* and *Pena*, *Symphytum petræum*, hath leaues like the laſt deſcribed, but ſomewhat narrower, and the leaues that grow commonly towards the tops of the ſtalks, are deeply diuided or cut in, after the manner of the leaues of the ſmall Valerian, and ſometimes the lower leaues are alſo diuided, but that is more ſeldom; the heads and floures are like thoſe of the former, and the colour of the floures is commonly purple yet ſomtimes it is found with fleſh coloured, and otherwhiles with white or aſhe coloured floures.

3 The third ſort of Selfe-heale is like vnto the laſt deſcribed in root, ſtalke, & leaues, & in euery other point, ſauing that the floures hereof are of a perfect white colour, and the others not ſo, which maketh the difference.

‡ The figure which our Authour gaue in this third place, was of the *Prunella ſecunda* of *Tabern.* which I iudge to be all one with the *Prunella* 1. *non vulgaris* of *Cluſius*, and that becauſe the floures in that of *Tabernamontanus* are expreſſed *Ventre laxiore*, which *Cluſius* complaines his drawer did not obſerue; the other parts alſo agree: now this of *Cluſius* hath much larger floures than the ordinary, and thoſe commonly of a deeper purple colour, yet they are ſometimes whitiſh, and otherwhiles of an aſhe colour: the leaues alſo are ſomewhat more hairie, long and ſharpe pointed, than the ordinary, and herein conſiſts the greateſt difference. ‡

¶ *The Place.*

The firſt kinde of Prunell or Brunell groweth verie commonly in all our fieldes throughout England.

The ſecond Brunel or *Symphytum petræum* groweth naturally vpon rocks, ſtonie mountaines, and grauelly grounds.

‡ The third for any thing that I know is a ſtranger with vs: but the firſt common kinde I haue found with white floures. ‡

¶ *The Time.*

Theſe plants floure for the moſt part all Sommer long.

¶ *The Names.*

Brunel is called in Engliſh Prunell, Carpenters herbe, Selfe-heale, and Hooke-heale, and Sicklewoort. It is called of the later Herbariſts *Brunella:* and *Prunella*, of *Matthiolus*, *Conſolida minor*, and *Solidago minor*; but ſaith *Ruellius*, the Daiſie is the right *Conſolida minor*, and alſo the *Solidago minor*.

¶ *The Nature.*

Theſe herbes are of the temperature of *Bugula*, that is to ſay, moderately hot and drie, and ſomething binding.

¶ *The Vertues.*

The decoction of Prunell made with wine or water, doth ioine together and make whole and ſound all wounds, both inward and outward, euen as Bugle doth. A

Prunell bruiſed with oile of Roſes and Vineger, and laied to the forepart of the head, ſwageth and helpeth the paine and aking thereof. B

To bee ſhort, it ſerueth for the ſame that Bugle doth, and in the world there are not two better wound herbs, as hath been often proued. C

D It is commended against the infirmities of the mouth, and especially the ruggednesse, blackenesse, and drinesse of the tongue, with a kinde of swelling in the same. It is an infirmitie amongst souldiers that lie in campe. The Germans call it **De Braun**, which happeneth not without a continuall ague and frensie. The remedie hereof is the decoction of Selfe-heale, with common water, after bloud letting out of the veins of the tongue: and the mouth and tongue must be often washed with the same decoction, and sometimes a little vineger mixed therewith. This disease is thought to be vnknowne to the old writers: but notwithstanding if it be conferred with that which *Paulus Aegineta* calleth *Erysipelas Cerebri*, an inflammation of the braine, then will it not be thought to bee much differing, if it be not the very same.

Chap. 202. *Of the great Daisie, or Maudelen woort.*

1 *Bellis maior.*
The great Daisie.

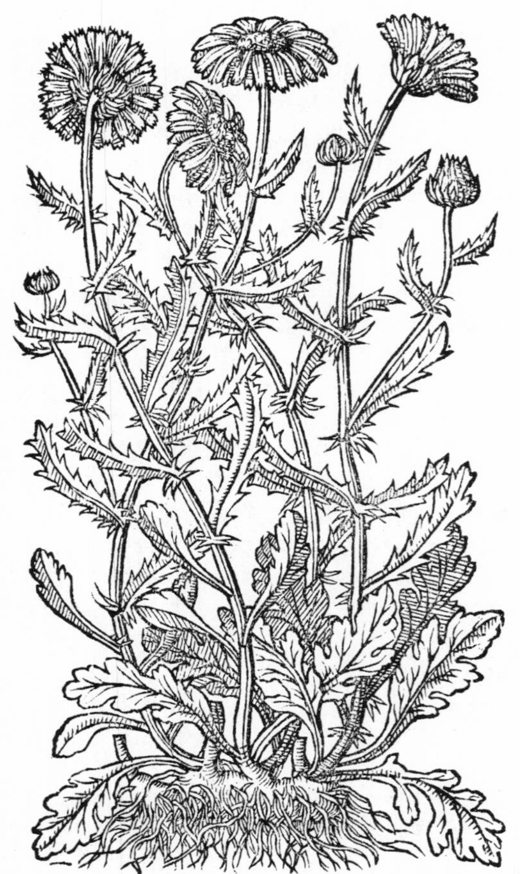

¶ *The Description.*

1 The great Daisie hath very many broad leaues spred vpon the ground, somewhat indented about the edges, of the breadth of a finger, not vnlike those of groundswell: among which rise vp stalkes of the height of a cubit, set with the like leaues, but lesser, in the top whereof do grow large white floures with yellow thrums in the middle like those of the single field Daisy or Mayweed, without any smell at all. The root is full of strings.

¶ *The Place.*

It groweth in Medowes and in the borders of fields almost euery where.

¶ *The Time.*

It floureth and flourisheth in May and Iune.

¶ *The Names.*

It is called (as we haue said) *Bellis maior*, and also *Consolida media vulnerariorum*, to make a difrence betweene it and *Bugula*, which is the true *Consolida media*: notwithstanding this is holden of all to bee *Consolida medij generis*, or a kinde of middle Consound: in High Dutch, as *Fuchsius* reporteth, **Genszblume**: in English, the Great Daisie and Maudelen woort.

¶ *The Temperature.*

This great Daisie is moist in the end of the second degree, and cold in the beginning of the same.

¶ *The Vertues.*

A The leaues of the great Maudleine woort are good against all burning vlcers and apostemes, against the inflammation and running of the eies, being applied thereto.

B The same made vp in an vnguent or salue with wax, oile, and turpentine, is most excellent for wounds, especially those wherein is any inflammation, and will not come to digestion or maturation, as are those weeping wounds made in the knees, elbowes, and other ioints.

C The iuice, decoction, or distilled water, is drunk to very good purpose against the rupture or any inward burstings.

D The herbe is good to be put into Vulnerarie drinks or potions, as one simple belonging thereto most necessarie, to the which effect, the best practised do vse it as a simple in such cases of great effect.

E It likewise asswageth the cruell torments of the gout, vsed with a few Mallows and butter boiled and made to the forme of a pultis.

F The same receipt aforesaid vsed in Clysters, profiteth much against the vehement heat in agues, and ceaseth the torments or wringing of the guts or bowels.

CHAP. 203. *Of little Daiſies.*

¶ *The Deſcription.*

1 THe Daiſie bringeth forth many leaues from a threddy root, ſmooth, fat, long, and ſomwhat round withall, very ſleightly indented about the edges, for the moſt part lying vpon the ground: among which riſe vp the floures, euery one with his owne ſlender ſtem, almoſt like thoſe of Camomill, but leſſer, of a perfect white colour, and very double.

2 The double red Daiſie is like vnto the precedent in euery reſpect, ſauing in the colour of the floures: for this plant bringeth forth floures of a red colour, and the other white as aforeſaid.

‡ Theſe double Daiſies are of two ſorts, that is either ſmaller or larger, and theſe againe either white or red, or of both mixed together: wherefore I haue giuen you in the firſt place the figure of the ſmall, and in the ſecond that of the larger.

3 Furthermore, there is another pretty double daiſie which differs from the firſt deſcribed only in the floure, which at the ſides thereof puts forth many foot-ſtalkes carrying alſo little double floures, being commonly of a red colour; ſo that each ſtalke carries as it were an old one and the brood thereof: whence they haue fitly termed it the childing Daiſie. ‡

1 *Bellis minor multiplex flore albo vel rubro.*
The leſſer double red or white Daiſie.

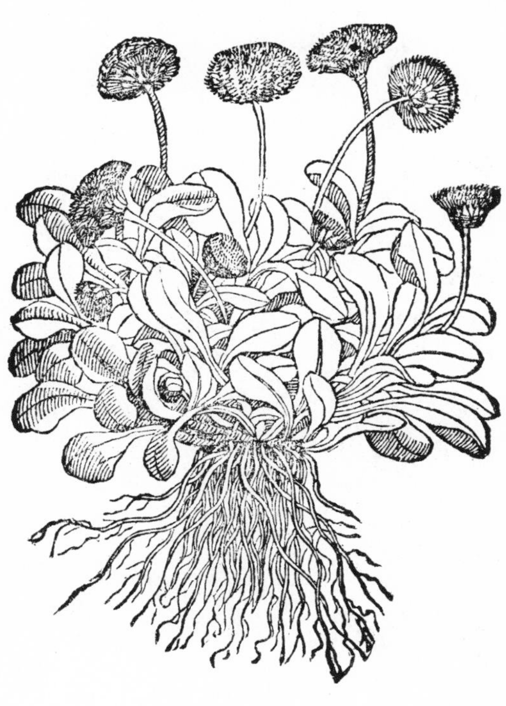

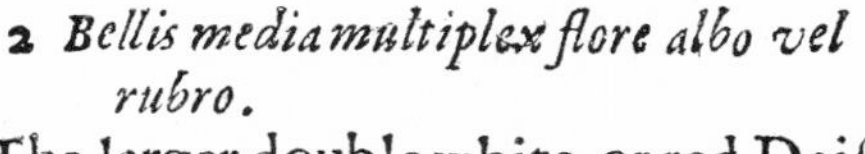

2 *Bellis media multiplex flore albo vel rubro.*
The larger double white or red Daiſie.

4 The wilde field Daiſie hath many leaues ſpred vpon the ground like thoſe of the garden Daiſie: among which riſe vp ſlender ſtems; on the top whereof do grow ſmall ſingle floures like thoſe of Camomill, ſet about a bunch of yellow thrums, with a pale of white leaues, ſometimes white, now and then red, and often of both mixed together. The root is threddy.

5 There doth likewiſe grow in the fields another ſort of wilde Daiſie, agreeing with the former in each reſpect, ſauing that it is ſomewhat greater than the other, and the leaues are ſomwhat more cut in the edges, and larger.

6 The blew Italian Daiſie hath many ſmall threddy roots, from the which riſe vp leaues like

thoſe

‡ 3 *Bellis minor prolifera.*
Childing Daisie.

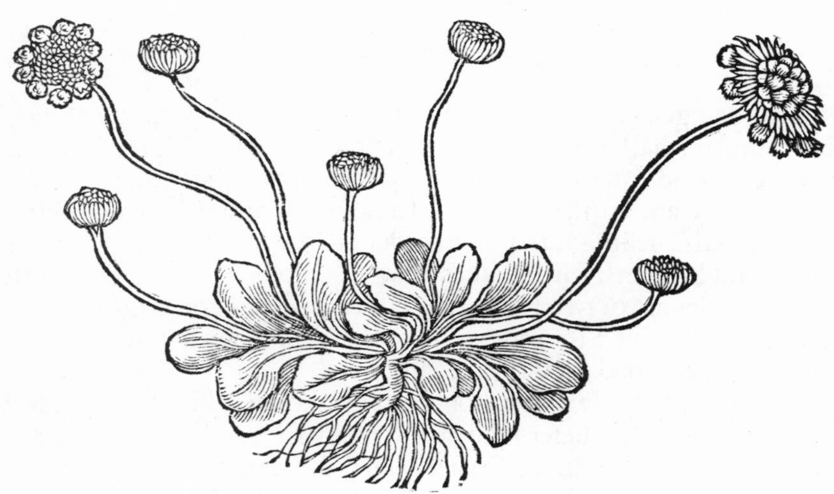

4 *Bellis minor syluestris.*
The small wilde Daisie.

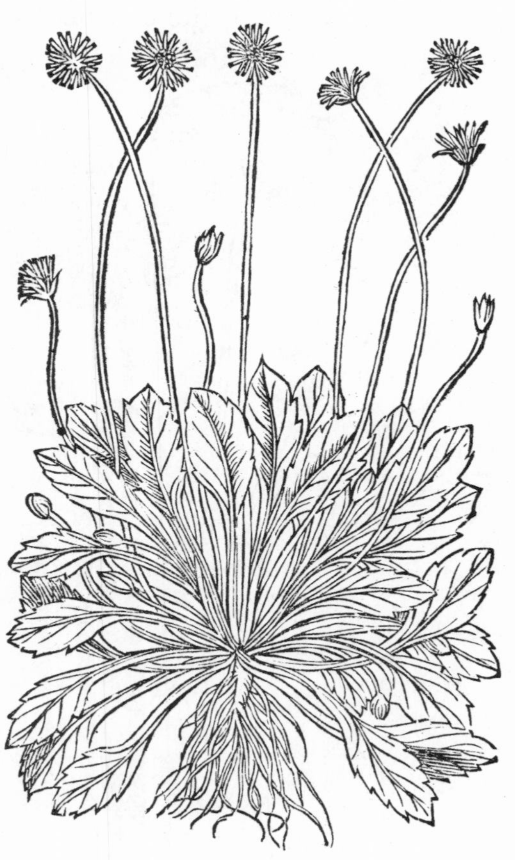

5 *Bellis media syluestris.*
The middle wilde Daisie.

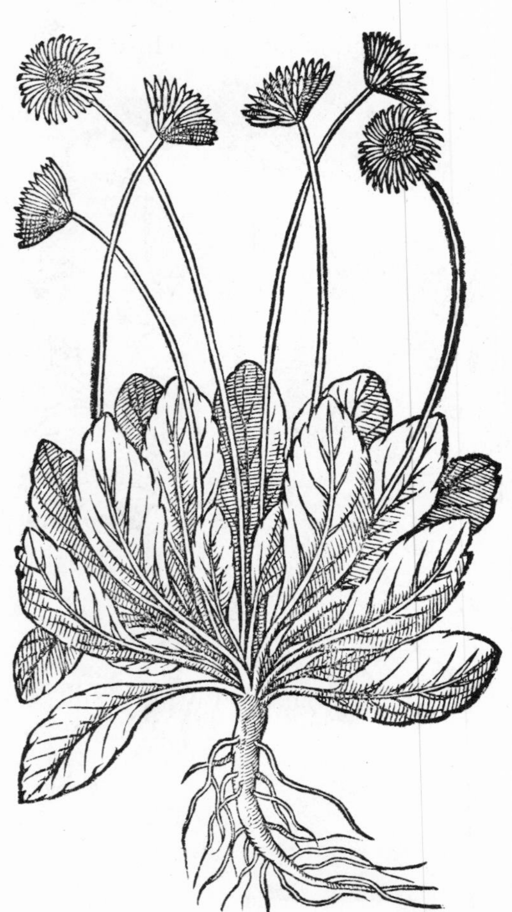

those of the common Daisie, of a darke greene colour: among which commeth vp a fat stemme set round about with the like leaues, but lesser. The floures grow at the top globe-fashion, that is, round like a ball, of a perfect blew colour, verie like vnto the floures of Mountaine Scabious.

7 The French blew Daisie is like vnto the other blew Daisies in each respect, sauing it is altogether lesser, wherein consisteth the difference.

‡ There were formerly three figures and descriptions of this blew Daisie, but one of them might haue serued; for they differ but in the tallnesse of their growth, and in the bredth and narrownesse of their leaues. ‡

¶ *The Place.*

The double Daisies are planted in gardens: the others grow wilde euery where.

The

The blew Daiſies are ſtrangers in England; their naturall place of abode is ſet forth in their ſeuerall titles.

6 *Bellis cœrulea ſiue Globularia Apula.*
The blew Italian Daiſie.

7 *Bellis cœrulea Monſpeliaca.*
Blew French Daiſies.

¶ *The Time.*

The Daiſies do floure moſt part of the Sommer.

¶ *The Names.*

The Daiſie is called in high-Dutch **Maſzlieben**: in low Dutch, **Margrieten**: in Latine, *Bellis minor*, and *Conſolida minor*, or the middle Conſound: of *Tragus*, *Primula veris*; but that name is more proper vnto Primroſe: of ſome, *Herba Margarita*, or Margarites herbe: in French, *Marguerites*, and *Caſſaudes*: in Italian, *Fiori di prima veri gentili*. In Engliſh, Daiſies, and Bruiſewort.

☞ The blew Daiſie is called *Bellis cœrulea*: of ſome, *Globularia*, of the round forme of the floure: it is alſo called *Aphyllanthes*, and *Frondiflora*: in Italian, *Botanaria*: in Engliſh, blew Daiſies, and Globe Daiſie.

¶ *The Temperature.*

The leſſer Daiſies are cold and moiſt, being moiſt in the end of the ſecond degree, and cold in the beginning of the ſame.

¶ *The Vertues.*

A The Daiſies doe mitigate all kinde of paines, but eſpecially of the ioynts, and gout proceeding from an hot and dry humor, if they be ſtamped with new butter vnſalted, and applied vpon the pained place; but they worke more effectually if Mallowes be added thereto.

B The leaues of Daiſies vſed amongſt other Pot-herbes doe make the belly ſoluble; and they are alſo put into Clyſters with good ſucceſſe, in hot burning feuers, and againſt inflammations of the inteſtines.

C The iuyce of the leaues and roots ſnift vp into the noſthrils, purgeth the head mightily of foule and filthy ſlimie humors, and helpeth the megrim.

D The ſame giuen to little dogs with milke keepeth them from growing great.

E The leaues ſtamped taketh away bruiſes and ſwellings proceeding of ſome ſtroke, if they be ſtamped and laid thereon; whereupon it was called in old time Bruiſewort.

F The iuyce put into the eyes cleareth them, and taketh away the watering.

G The decoction of the field Daiſie (which is the beſt for phyſicks vſe) made in water and drunke, is good againſt agues, inflammation of the liuer, and all other the inward parts.

CHAP.

CHAP. 204. *Of Mouse-eare.*

¶ *The Description.*

1 THe great Mouse-eare hath great and large leaues greater than our common *Pylosella*, or Mouse-eare, thicke, and full of substance: the stalkes and leaues be hoarie and white, with a silken mossinesse in handling like silke, pleasant and faire in view: it beareth three or foure quadrangled stalkes, somewhat knotty, a foot long: the roots are hard, wooddy, and full of strings: the floures come forth at the top of the stalke, like vnto the small Pisseabed, or Dandelion, of a bright yellow colour.

2 The second kinde of *Pylosella* is that which we call *Auricula muris*, or Mouse-eare, being a very common herb, but few more worthy of consideration because of his good effect, and yet clean vnremembred of the old Writers. It is called *Pylosella* of the rough hairy and whitish substance growing on the leaues, which are somewhat long like the little Daisie, but that they haue a small hollownesse in them resembling the eare of a Mouse: vpon the which consideration some haue called it *Myosotis*; wherein they were greatly deceiued, for it is nothing like vnto the *Myosotis* of *Dioscorides*: his small stalkes are likewise hairy, slender, and creeping vpon the ground; his floures are double, and of a pale yellow colour, much like vnto *Sonchus*, or *Hieracium*, or Hawkeweed.

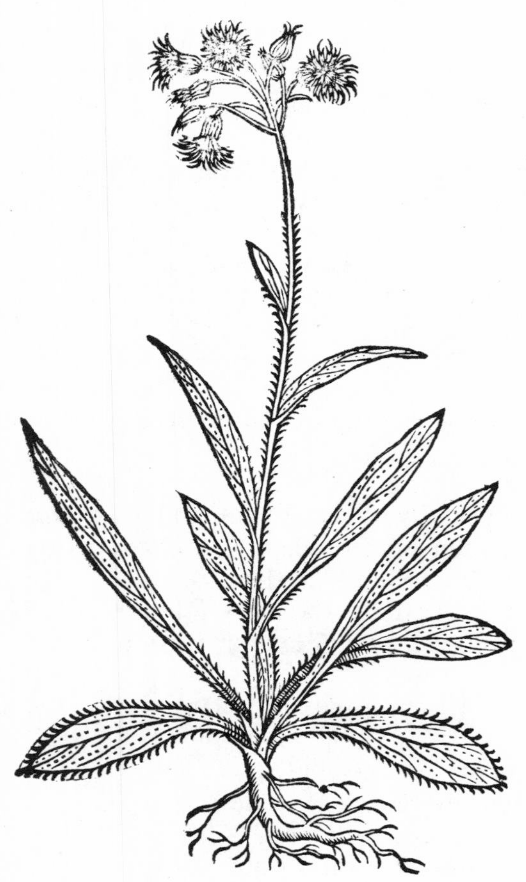

1 *Pylosella major.*
Great Mouse-eare.

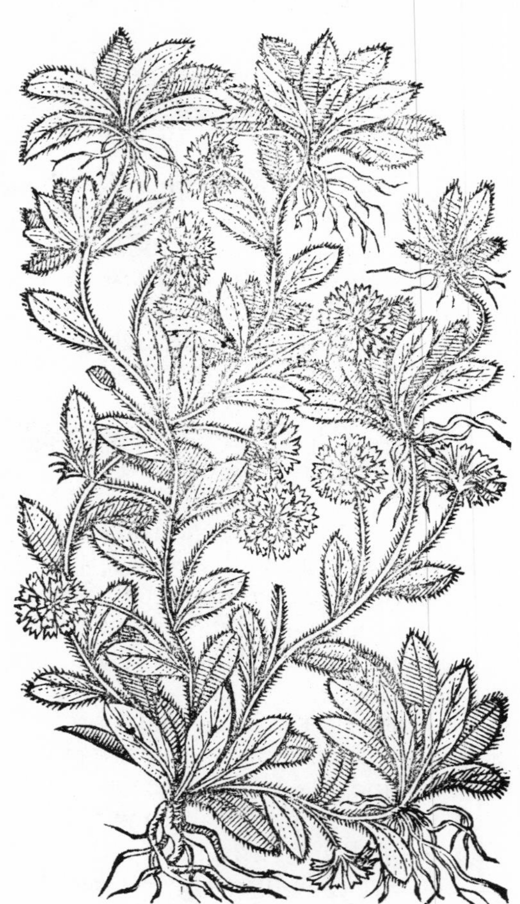

2 *Pylosella repens.*
Creeping Mouse-eare.

3 The small Mouse-eare with broad leaues hath a small tough root, from which rise vp many hairy and hoarie broad leaues spred vpon the ground, among which growes vp a slender stem, at the top whereof stand two or three small yellow floures, which being ripe turne vnto downe that is caried away with the winde.

¶ *The Place.*

These plants do grow vpon sandy bankes and vntoiled places that lie open to the aire.

¶ *The Time.*

They floure in May and Iune.

¶ *The*

¶ The Names.

Great Mouſe-eare is called of the later herbariſts *Pyloſella*: the ſmaller likewiſe *Pyloſella*, and *Auricula muris*: in Dutch, Nagelcruijt, and Muysoor: *Lacuna* thinkes it *Holoſtium*: in French, *Oreille de rat, ou ſouris*: in Italian, *Peloſella*: in Engliſh, Mouſe-eare.

¶ The Temperature.

They are hot and dry of temperature, of an excellent aſtringent facultie, with a certaine hot tenuitie admixed.

¶ The Vertues.

The decoction of *Pyloſella* drunke doth cure and heale all wounds, both inward and outward: it cureth hernies, ruptures, or burſtings. A

The leaues dried and made into pouder, do profit much in healing of wounds, being ſtrewed thereupon. B

The decoction of the iuyce is of ſuch excellencie, that if ſteele-edged tooles red hot be drenched and cooled therein oftentimes, it maketh them ſo hard, that they will cut ſtone or iron, be they neuer ſo hard, without turning the edge or waxing dull. C

This herbe being vſed in gargariſmes cureth the looſeneſſe of the Vuula. D

Being taken in drinke it healeth the fluxes of the wombe, as alſo the diſeaſes called *Dyſenteria* and *Enterocele*: it glueth and conſoundeth wounds, ſtayeth the ſwelling of the ſpleene, and the bloudy excrements procured thereby. E

The Apothecaries of the Low-countries make a ſyrrup of the iuyce of this herb, which they vſe for the cough, conſumption, and ptiſicke. F

† I haue in this chapter omitted two figures and one deſcription: the firſt of the two omitted figures, which ſhould haue beene the third, differs little from the firſt but in the ſmallneſſe of the ſtalke, and fewneſſe of the floures at the top thereof: the other, which was in the fourth place, was figured and deſcribed by me formerly in the fourth place of the 54 chapter of this booke.

CHAP. 205. *Of Cotton-weed or Cud-weed.*

1 *Gnaphalium Anglicum.*
Engliſh Cudweed.

2 *Gnaphalium vulgare.*
Common Cudweed.

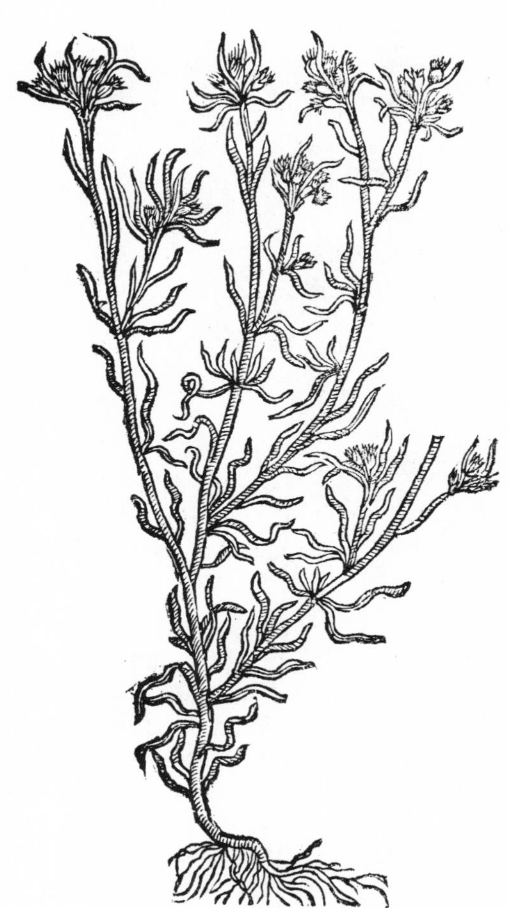

¶ The

¶ *The Defcription.*

1 English Cudweed hath fundry flender and vpright ftalkes diuided into many branches, and groweth as high as common Wormwood, whofe colour and fhape it doth much refemble. The leaues fhoot from the bottome of the turfe ful of haires, in fhape fomewhat like a Willow leafe below, but aboue they be narrower, and like the leaues of *Pfyllium* or Flea-wort: among which do grow fmall pale coloured floures like thofe of the fmall *Coniza* or Flea-bane. The whole plant is of a bitter tafte.

2 The fecond being our common *Gnaphalium* or Cudweed is a bafe or low herbe, nine or ten inches long, hauing many fmall ftalks or tender branches, and little leaues, couered all ouer with a certain white cotton or fine wooll, and very thick: the floures be yellow, and grow like buttons at the top of the ftalkes.

3 The third kinde of Cudweed or Cotton-weed, being of the fea, is like vnto the other Cudweed laft defcribed, but is altogether fmaller and lower, feldome growing much aboue a handfull high: the leaues grow thicke vpon the ftalkes, and are fhort, flat, and very white, foft and woolly. The floures grow at the top of the ftalkes in fmall round buttons, of colour and fafhion like the other Cudweed.

4 The fourth being the Cotton-weed of the hills and ftony mountains, is fo exceeding white and hoary, that one would thinke it to be a plant made of wooll, which may very eafily be known by his picture, without other defcription.

3 *Gnaphalium marinum.*
Sea Cudweed.

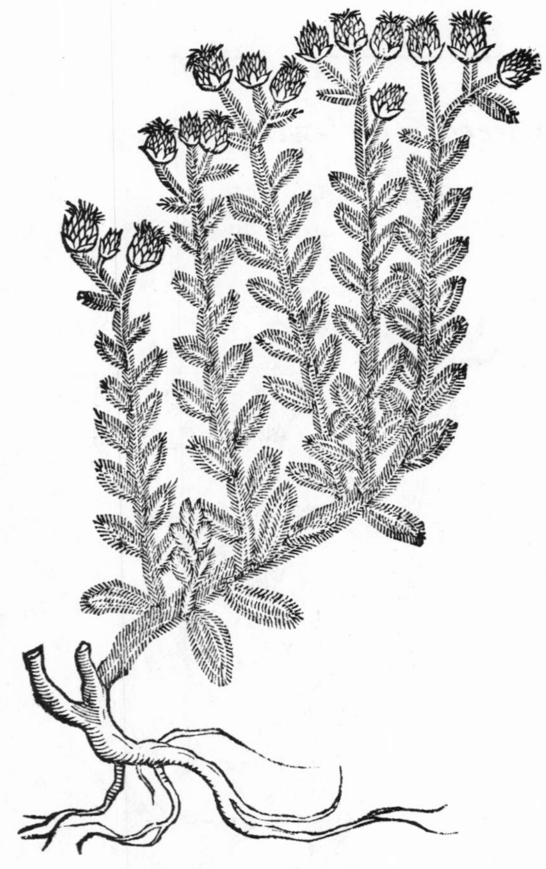

4. 5. *Gnaphalium montanum purpureum & album.*
White and purple mountaine Cotton-weed.

5 The fifth kinde of Cotton-weed hath leaues and ftalkes like the other of his kinde, and differeth in that, that this plant beareth a bufh or tuft of purple floures, otherwife it is very like.

6 The fixth is like vnto the laft recited, but greater: the floures are of an exceeding bright red colour, and of an aromaticall fweet fmell.

7 The feuenth kinde of *Gnaphalium* or Cotton-weed of *Clufius* his defcription, growes nine or ten inches high, hauing little long leaues like the fmall Mouf-eare, woolly within, and of a hoarie colour on the outfide: the ftalkes in like manner are very woolly, at the top whereof commeth forth a faire floure and a ftrange, hauing fuch woolly leaues bordering the floure about, that a man would thinke it to be nothing elfe but wooll it felfe: and in the middeft of the floure come forth fundry

6 *Gnaphalium montanum suaue rubens.*
Bright red mountaine Cotton-weed.

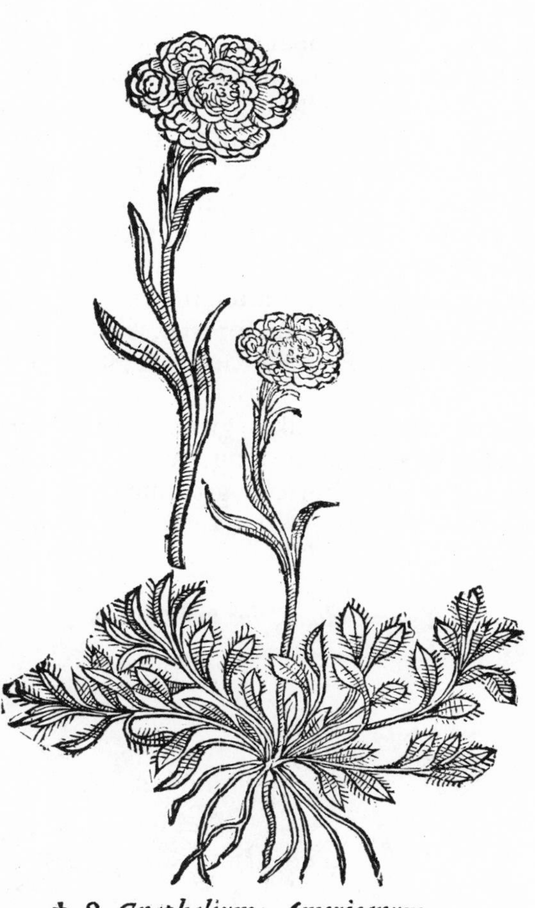

† 7 *Gnaphalium Alpinum.*
Rocke Cotton-weed.

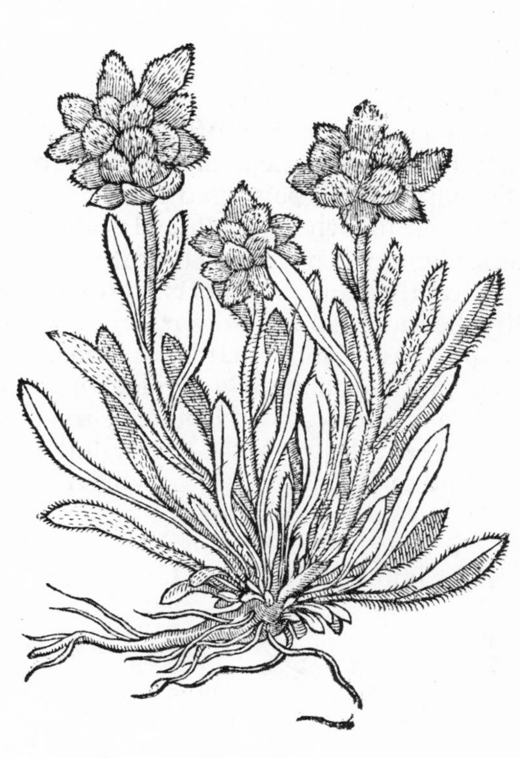

‡ 8 *Gnaphalium Americanum.*
Liue for euer.

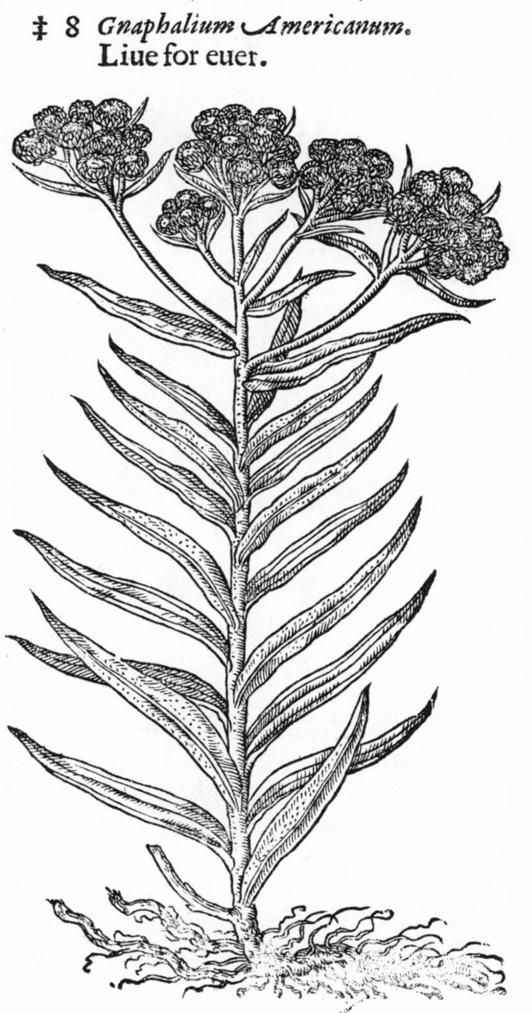

9 *Filago minor.*
Small Cud-weed.

sundry small heads of a pale yellow colour, like vnto the other of this kinde. The root is blacke and somewhat fibrous.

8 There is a kinde of Cotton-weed, being of greater beauty than the rest, that hath strait and vpright stalks 3 foot high or more, couered with a most soft and fine wooll, and in such plentifull manner, that a man may with his hands take it from the stalke in great quantitie: which stalke is beset with many small long and narrow leaues, greene vpon the inner side, and hoary on the other side, fashioned somewhat like the leaues of Rosemary, but greater. The floures do grow at the top of the stalkes in bundles or tufts, consisting of many small floures of a white colour, and very double, compact, or as it were consisting of little siluer scales thrust close together, which doe make the same very double. When the floure hath long flourished, and is waxen old, then comes there in the middest of the floure a certaine browne yellow thrumme, such as is in the midst of the Daisie: which floure being gathered when it is young, may be kept in such manner as it was gathered (I meane in such freshnesse and well liking) by the space of a whole yeare after, in your chest or elsewhere: wherefore our English women haue called it Liue-long, or Liue for euer, which name doth aptly answer his effects. ‡ *Clusius* receiued this plant out of England, and first set it forth by the name of *Gnaphalium Americanum*, or *Argyrocome*. ‡

9 This plant hath three or foure small grayish cottony or woolly stalkes, growing strait from the root, and commonly diuided into many little branches: the leaues be long, narrow, whitish, soft, and woolly, like the other of his kinde: the floures be round like buttons, growing very many together at the top of the stalkes, but nothing so yellow as Mouse-eare, which turne into downe, and are caried away with the winde.

10 *Filago, siue Herba impia.*
Herbe impious, or wicked Cudweed.

11 *Leontopodium, siue Pes Leoninus.*
Lions Cudweed.

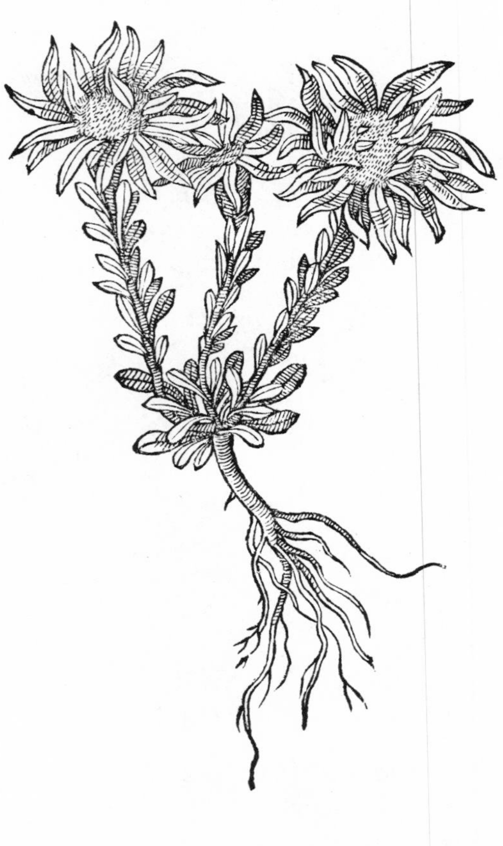

10 The tenth is like vnto the last before mentioned, in stalkes, leaues, and floures, but much larger, and for the most part those floures which appeare first are the lowest and basest, and they are ouertopt by other floures which come on younger branches, and grow higher, as children seeking to ouergrow or ouertop their parents, (as many wicked children do) for which cause it hath beene called *Herba impia,* that is, the wicked Herbe, or Herbe Impious.

11 The eleuenth plant comprehended vnder the title of *Gnaphalium,* (being without doubt a kinde thereof, as may appeare by the shape of his floures and stalks, couered ouer with a soft wool like vnto the other kindes of Cotton-weed) is an handfull high or thereabouts, beset with leaues like

‡ 12 *Leontopodium parvum.*
Small Lyons Cudweed.

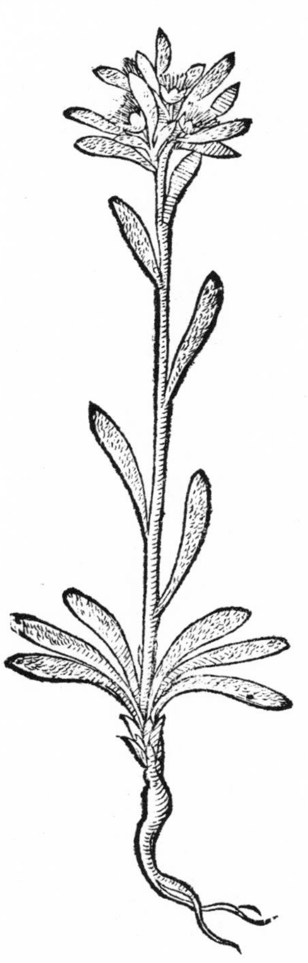

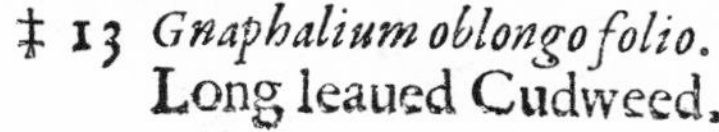
‡ 13 *Gnaphalium oblongo folio.*
Long leaued Cudweed.

‡ 14 *Gnaphalium minus latiore folio.*
Small broad leaued Cudweed.

like *Gnaphalium Anglicum*, but somewhat broader. At the top of the stalke groweth a floure of a blackish brown violet colour, beset about with rough and woolly hairie leaues, which make the whole floure to resemble the rough haired foot of a Lyon, of a Hare, or a Beare, or rather in mine opinion of a rough footed Doue. The heads of these floures when they are spred abroad carry a greater circumference than is required in so small a plant; and when the floure is faded, the seed is wrapped in such a deale of wooll that it is scarsely to be found out.

12 This small kinde of *Leontopodum* being likewise a kind of Cotton-weed, neither by *Dioscorides* or any other antient writer once remembred, hath one single stalke nine inches in height, and the leaues of *Gnaphalium montanum*; which leaues and stalkes are white, with a thicke hoary woollinesse, bearing at the top pale yellow floures like *Gnaphalium montanum*: the root is slender and wooddy.

‡ 13 This, which *Clusius* calls *Gnaphalium Plateau* 2. hath small stalkes some handfull high or somewhat more, of which some

some stand vpright, others lie along vpon the ground, being round, hairy, and vnorderly set with soft hoary leaues ingirting their stalkes at their setting on, and sharpe pointed at their vpper ends. The tops of the stalkes carry many whitish heads full of a yellowish downe: the root is thicke and blackish, with some fibres.

14 This sends vp one stalke parted into seuerall branches set here and there with broad soft and hoarie leaues, and at the diuision of the branches and amongst the leaues grow seuen or eight little heads thicke thrust together, being of a grayish yellow colour, and full of much downe: the root is vnprofitable, and perishes as soone as it hath perfected his seed. *Clusius* calls this *Gnaphalium Plateau* 3. he hauing as it seemes receiued them both from his friend *Iaques Plateau*. ‡

¶ *The Place.*

The first groweth in the darke woods of Hampsted, and in the woods neere vnto Deptford by London. The second groweth vpon dry sandy bankes. The third groweth at a place called Merezey, six miles from Colchester, neere vnto the sea side. ‡ I also had it sent me from my worshipfull friend Mr. *Thomas Glynn*, who gathered it vpon the sea coast of Wales. ‡

The rest grow vpon mountaines, hilly grounds, and barren pastures.

The kinde of *Gnaphalium* newly set forth (to wit *Americanum*) groweth naturally neere vnto the Mediterranean sea, from whence it hath beene brought and planted in our English gardens. ‡ If this be true which our Author here affirmes, it might haue haue had a fitter (at least a neerer) denomination than from America: yet *Bauhine* affirmes that it growes frequently in Brasill, and it is not improbable that both their assertions be true. ‡

¶ *The Time.*

They floure for the most part from Iune to the end of August.

¶ *The Names.*

Cotton-weed is called in Greeke *Gnaphalion*; and it is called *Gnaphalion*, because men vse the tender leaues of it in stead of bombaste or Cotton, as *Paulus Ægineta* writeth. *Pliny* saith it is called *Chamæxylon*, as though he should say Dwarfe Cotton; for it hath a soft and white cotton like vnto bombaste: whereupon also it was called of diuers *Tomentitia*, and *Cotonaria*: of others, *Centunculus*, *Centuncularis*, and *Albinum*; which word is found among the bastard names: but the later word, by reason of the white colour, doth reasonably well agree with it. It is also called *Bombax*, *Humilis filago*, and *Herba Impia*, because the yonger, or those floures that spring vp later, are higher, and ouertop those that come first, as many wicked children do vnto their parents, as before touched in the description: in English, Cotton-weed, Cud-weed, Chaffe-weed, and petty Cotton.

¶ *The Nature.*

These herbes be of an astringent or binding and drying qualitie.

¶ *The Vertues.*

A *Gnaphalium* boyled in strong lee cleanseth the haire from nits and lice: also the herbe being laid in ward-robes and presses keepeth apparell from moths.

B The same boyled in wine and drunken, killeth wormes and bringeth them forth, and preuaileth against the bitings and stingings of venomous beasts.

C The fume or smoke of the herbe dried, and taken with a funnell, being burned therein, and receiued in such manner as we vse to take the fume of Tabaco, that is, with a crooked pipe made for the same purpose by the Potter, preuaileth against the cough of the lungs, the great ache or paine of the head, and clenseth the brest and inward parts.

† The figure that was formerly in the seuenth place should haue beene in the eleuenth; and that in the eleuenth in the seuenth.

Chap. 206.

Of Golden Moth-wort, or Cudweed.

¶ *The Description.*

1 Golden Moth-wort bringeth forth slender stalkes somewhat hard and wooddy, diuided into diuers small branches; whereupon do grow leaues somewhat rough, and of a white colour, very much iagged like Southernwood. The floures stand on the tops of the stalkes, ioyned together in tufts, of a yellow colour glittering like gold, in forme resembling the scaly floures of Tansie, or the middle button of the floures of Camomil; which being gathred before they be ripe or withered, remaine beautifull long time after, as my selfe did see in the hands of Mr. *Wade*, one of the Clerks of her Maiesties Counsell, which were sent him among other things from

from Padua in Italy. For which cause of long lasting, the images and carued gods were wont to weare garlands thereof: whereupon some haue called it Gods floure. For which purpose *Ptolomy* King of Egypt did most diligently obserue them, as *Pliny* writeth.

1 *Elyochryson, siue Coma aurea.*
Golden Moth-wort.

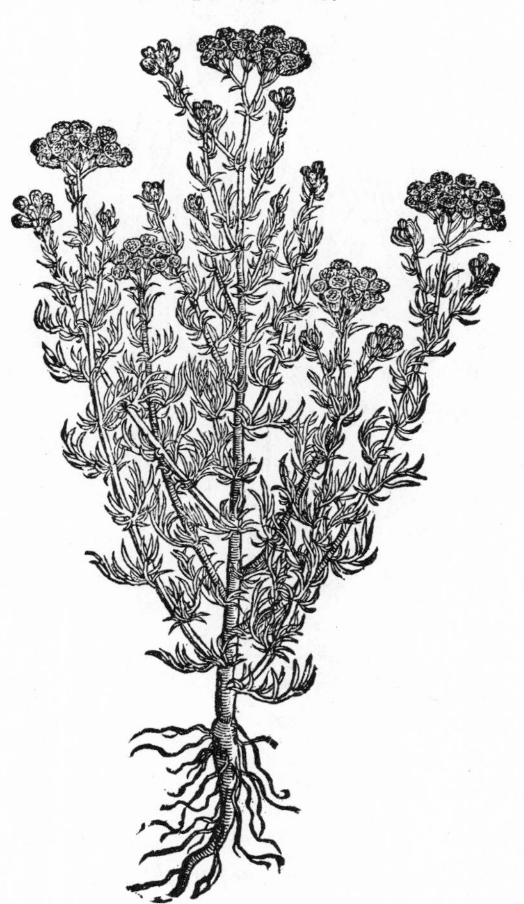

¶ *The Place.*

It growes in most vntilled places of Italy and Spaine, in medowes where the soile is barren, and about the banks of riuers; it is a stranger in England.

¶ *The Time.*

It floures in August and September: notwithstanding *Theophrastus* and *Pliny* reckon it among the floures of the Spring.

¶ *The Names.*

Golden Moth-wort is called of *Dioscorides Elichryson*. *Pliny* and *Theophrastus* call it *Helichryson*: *Gaza* translates it *Aurelia*: in English, Gold-floure, Golden Moth-wort.

¶ *The Temperature.*

It is (saith *Galen*) of power to cut and make thinne.

¶ *The Vertues.*

A *Dioscorides* teacheth, that the tops thereof drunke in wine are good for them that can hardly make water; against stingings of Serpents, paines of the huckle bones: and taken in sweet wine it dissolueth congealed bloud.

B The branches and leaues laid amongst cloathes keepeth them from moths, whereupon it hath beene called of some Moth-weed, or Mothwort.

† Here formerly were two figures and descriptions of the same Plant.

CHAP. 207. *Of Golden Floure-Gentle.*

¶ *The Description.*

1 THis yellow Euerlasting or Floure-Gentle, called of the later Herbarists Yellow Stœcas, is a plant that hath stalkes of a span long, and slender, whereupon do grow narrow leaues white and downie, as are also the stalks. The floures stand on the tops of the stalks, consisting of a scattered or disordered scaly tuft, of a reasonable good smell, of a bright yellow colour; which being gathered before they be ripe, do keep their colour and beauty a long time without withering, as do most of the Cottonweeds or Cudweeds, whereof this is a kinde. The root is blacke and slender. ‡ There is some varietie in the heads of this plant, for they are sometimes very large and longish, as *Camerarius* notes in his Epitome of *Matthiolus*; otherwhiles they are very compact and round, and of the bignesse of the ordinarie.

2 This growes to some foot or more high, and hath rough downie leaues like the former, but broader: the floures are longer, but of the same yellow colour and long continuance as those of the last described. This varies somthing in the bredth and length of the leaues, whence *Tabernamontanus* gaue three figures thereof, and therein was followed by our Author, as you shall finde more particularly specified at the end of the chapter. ‡

3 About Nemausium and Montpelier there growes another kinde of *Chrysocome*, or as *Lobel* termes it, *Stœchas Citrina altera*, but that as this plant is in all points like, so in all points it is lesser and slenderer, blacker, and not of such beauty as the former, growing more neere vnto an ash colour, consisting of many small twigs a foot long. The root is lesser, and hath fewer strings annexed thereto; and it is seldome found but in the cliffes and crags, among rubbish, and on walls of cities. This plant is browne, without sent or sauor like the other: euery branch hath his own bunch of floures comming forth of a scaly or round head, but not a number heaped together as in the first kinde. It prospereth well in our London Gardens.

 4 There

† 1 *Stœchas Citrina, siue Amaranthus luteus.*
Golden Stœchas, or Goldilockes.

† 2 *Amaranthus luteus latifolius.*
Broad leaued Goldilockes.

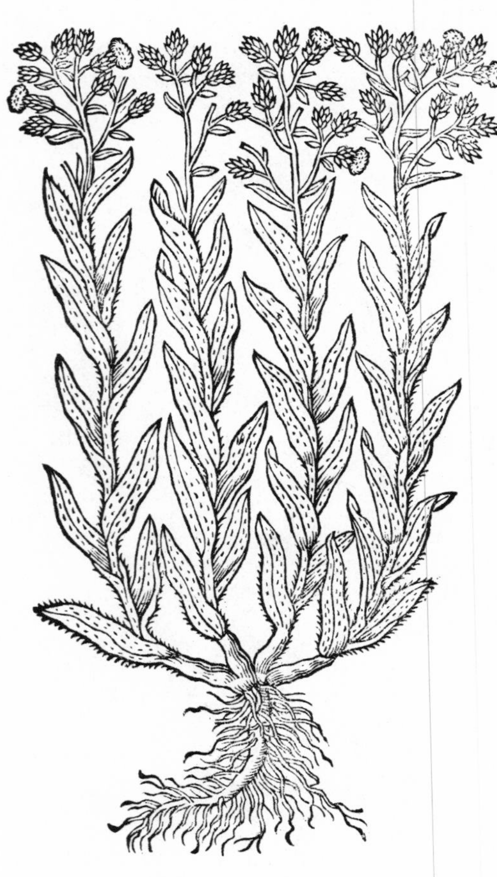

† 3 *Chrysocome capitulis conglobatis.*
Round headed Goldilockes.

† 4 *Amaranthus luteus flore oblongo.*
Golden Cudweed.

4 There is a kinde hereof beeing a very rare plant, and as rare to be found where it naturally groweth, which is in the woods among the Scarlet-Okes betweene Sommieres and Mountpellier. It is a fine and beautifull plant, in shew passing the last described *Stœchas Citrina altera:* but the leaues of this kinde are broad, and somewhat hoarie, as is all the rest of the whole plant; the stalke a foot long, and beareth the very floures of *Stœchas Citrina altera,* but bigger and longer, and somewhat like the floures of *Lactuca agrestis:* the root is like the former, without any manifest smel, little knowne, hard to finde, whose faculties be yet vnknowne.

† 5 This

† 5 *Heliochrysos syluestris*,
Wilde Goldylockes.

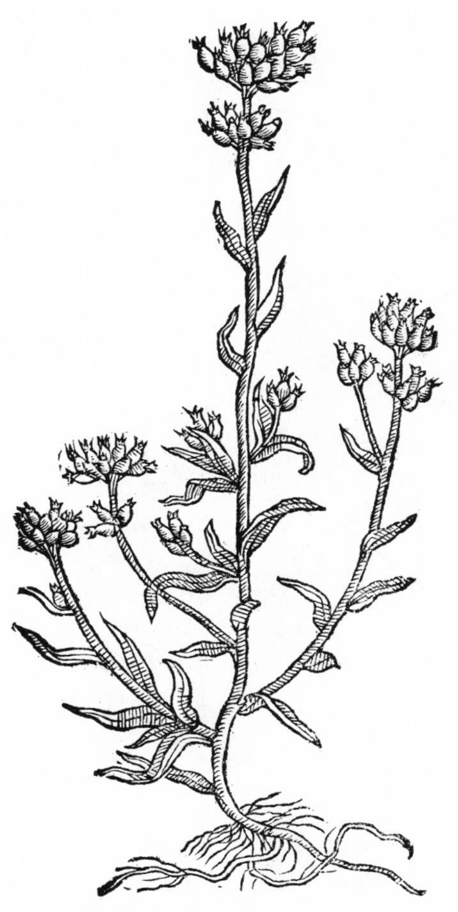

† 5 This is a wilde kinde (which *Lobel* setteth forth) that here may be inserted, called *Eliochrysos syluestris*. The woolly or flockey leafe of this plant resembleth *Gnaphalium vulgare*, but that it is somewhat broader in the middle: the floures grow clustering together vpon the tops of the branches, of a yellow colour, and almost like those of Maudline: the roots are blacke and wooddie.

¶ *The Place.*

The first mentioned growes in Italy, and other hot countries: and the second growes in rough and grauelly places almost euery where neere vnto the Rhene, especially between Spires and Wormes.

¶ *The Time*

They floure in Iune and Iuly.

¶ *The Names.*

Golden floure is called in Latine *Coma aurea*, of his golden locks or beautifull bush, and also *Tineraria*: in shops, *Stœchas citrina*, *Amaranthus luteus*, *Fuchsij*, & *Tragi*: of some, *Linaria aurea*, but not truely: in Greeke, *Chrysocome*: in Dutch, **Reynbloemen**, and **Motten crupt**; in Italian, *Amarantho Giallo*: in English, Gold-floure, Gods floure, Goldilockes, and Golden *Stœchas*.

¶ *The Temperature and Vertues.*

A The floures of Golden Stœchados boiled in wine and drunke, expell worms out of the bellie; and being boiled in Lee made of strong ashes doth kill lice and nits, if they bee bathed therewith. The other faculties are refered to the former plants mentioned in the last chapter.

† There were formerly the same number of figures as are now in this Chapter, but no way agreeing with the descriptions; the first was of *Millefolium Luteum*, being the *Helichrysum Italicum* of *Matthiolus*: The second was of the *Amaranthus primus* of *Tragus*, which stil keeps the 2. place: and the 4. & 5 were onely varieties of this, according to *Bauhine*: but if they be not varieties, but made to expresse the 2. figures of the *Aduersar.* which we here giue, as I coniecture they were, then should the fourth haue beene put in the third place, and the fift in the fourth, & the third should haue been put in the fifth, as you may see now it is.

Chap. 208. *Of Costmarie and Maudelein.*

¶ *The Description.*

1 Costmary groweth vp with round hard stalkes two foot high, bearing long broad leaues finely nicked in the edges, of an ouerworn whitish green colour. The tuft or bundle is of a golden colour, consisting of many little floures like clusters, ioyned together in a rundle after the manner of golden Stœchados. The root is of a woody substance, by nature verie durable, not without a multitude of little strings hanging thereat. The whole plant is of a pleasant smell, sauour, or taste.

2 Maudleine is somewhat like to Costmary (whereof it is a kinde) in colour, smell, taste, and in the golden floures, set vpon the tops of the stalks in round clusters. It bringeth forth a number of stalkes, slender, and round. The leaues are narrow, long, indented, and deepely cut about the edges. The cluster of floures is lesser than that of Costmarie, but of a better smell, and yellower colour. The roots are long lasting and many.

‡ 3 There is another kinde of *Balsamita minor*, or *Ageratum*, which hath leaues lesser and narrower than the former, and those not snipt about the edges: the vmbel or tuft of floures is yellow

1 *Balsamita mas.*
Costmarie.

2 *Balsamita fœmina, siue Ageratum.*
Maudelein.

‡ 3 *Ageratum folijs non serratis.*
Maudelein with vncut leaues.

4 *Ageratum floribus albis.*
White floured Maudlein.

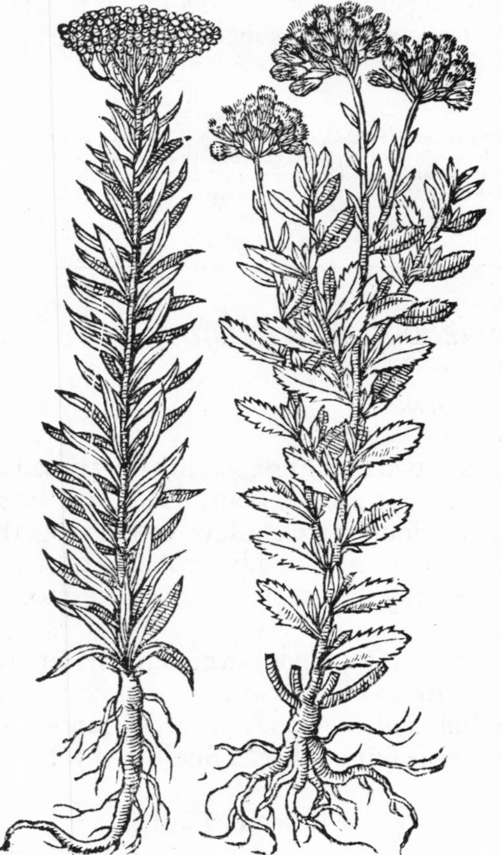

yellow like as the former, and you may call each of these last described at your pleasure, either *Ageratum*, or *Balsamita:* the Græcians call it Ἀγήρατον, which is in Latine *Ageratum, vel non senescens*, called in shops (though vntruly) *Eupatorium Mesuæ*. The floures are of a beautifull and seemely shew, which will not lose their excellencie of grace in growing, vntill they be very old, and therefore called *Ageratum*, or *Non senescens*, as before; and are like in tuft to *Eliochryson*, but of a white colour; and this is thought to be the true and right *Ageratum* of *Dioscorides*, although there hath been great, controuersie which should be the true plant.

‡ 4 This differeth not from the common Maudelein, but in the colour of the floures, which are white, when as those of the ordinarie sort are yellow. ‡

¶ *The Place.*

They grow euery where in gardens, and are cherished for their sweet floures and leaues.

¶ *The Time.*

They bring forth their tufts of yellow floures in the Sommer moneths.

¶ The

¶ *The Names.*

Costmarie is called in Latine *Balsamita maior* or *mas*: of some, *Costus hortorum*: it is also called *Mentha Græca*: and *Saracenica Officinarum*: of *Tragus*, *Alisma*: of *Matthiolus*, *Herba Græca*: of others, *Saluia Romana*, and *Herba lassulata*: of some, *Herba D. Mariæ*: in English, Costmarie, and Ale-coast: in High Dutch, **Frauwenkraut**: in low Dutch, **Heydnisch windkraut**: in French, *Coq*.

Maudlein is without doubt a kinde of Costmarie, called of the Italians *Herba Giulia*: of *Valerius Cordus*, *Mentha Corymbifera minor*: and *Eupatorium Mesue*: It is iudged to be *Dioscorides* his *Ageratum*, and it is the *Costus minor hortensis* of *Gesner*: we call it in English Maudlein.

¶ *The Nature.*

They are hot and drie in the second degree.

¶ *The Vertues.*

These plants are very effectuall, especially Maudlein, taken either inwardly or else outwardly to prouoke vrine; and the fume thereof doth the same, and mollifieth the hardnesse of the Matrix. A

Costmarie is put into Ale to steepe, as also into the barrels and Stands amongst those herbes wherewith they doe make Sage Ale; which drinke is very profitable for the diseases before spoken of. B

The leaues of Maudleine and Adders tongue stamped and boiled in Oile Oliue, adding thereto a little wax, rosin, and a little turpentine, maketh an excellent healing vnguent, or incarnatiue salue to raise or bring vp flesh from a deepe and hollow wound or vlcer, whereof I haue had long experience. C

The Conserue made with the leaues of Costmarie and Sugar, doth warme and drie the braine, and openeth the stoppings of the same: stoppeth all Catarrhes, rheumes and distillations, taken in the quantitie of a beane. D

The leaues of Costmarie boiled in wine and drunken, cureth the griping paine of the bellie, the guts and bowels, and cureth the bloudie flix. E

It is good for them that haue the greene sicknesse, or the dropsie, especially in the beginning; and it helpeth all that haue a weake and cold liuer. F

The seed expelleth all manner of wormes out of the belly, as wormseed doth. G

Chap. 209. *Of Tansie.*

¶ *The Description.*

1 Tansie groweth vp with many stalkes, bearing on the tops of them certaine clustered tufts, with floures like the round buttons of yellow Romane Cammomill, or Feuerfew (without any leaues paled about them) as yellow as gold. The leaues be long, made as it were of a great many set together vpon one stalke, like those of Agrimony, or rather wild Tansie, very like to the female Ferne, but softer and lesser, and euery one of them slashed in the edges as are the leaues of Ferne. The root is tough and of a wooddie substance. The whole plant is bitter in taste, and of a strong smell, but yet pleasant.

2 The double English Tansie hath leaues infinitly iagged and nicked, and curled withall, like vnto a plume of feathers: it is altogether like vnto the other, both in smell and taste, as also in floures, but more pleasantly smelling by many degrees, wherein especially consisteth the difference.

3 The third kinde of Tansie hath leaues, roots, stalkes, and branches like the other, and differeth from them, in that this hath no smell or sauour at all, and the floures are like the common single Fetherfew.

‡ 4 *Clusius* hath described another bigger kind of vnsauorie Tansie, whose figure here we giue you; it grows some cubit and halfe high, with crested stalks, hauing leaues set vpon somwhat longer stalks than those of the last described, otherwise much like them: the floures are much larger, being of the bignesse of the great Daisie, and of the same colour: the seede is long and blacke: The root is of the thicknes of ones finger, running vpon the surface of the ground, & putting forth some fibres, and it lasts diuers yeares, so that the plant may be encreased thereby. This floures in May and Iune, and grows wilde vpon diuers hills in Hungary and Austria. ‡

1 *Tanacetum.*
Tansie.

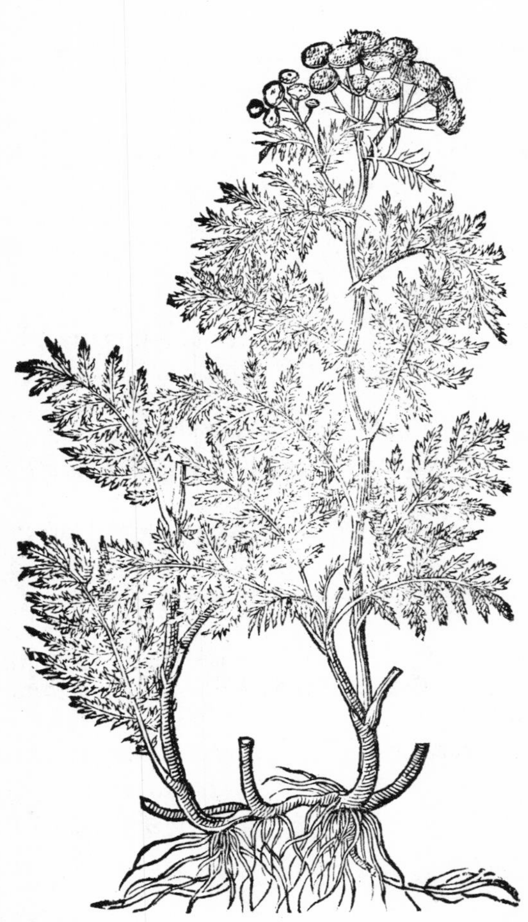

2 *Tanacetum crispum Anglicum.*
Double English Tansie.

3 *Tanacetum non odorum.*
Vnsauorie Tansie.

‡ 4 *Tanacetum inodorum maius.*
Great vnsauorie Tansie.

† 4 *Tanacetum minus album.*
Small white Tanſie.

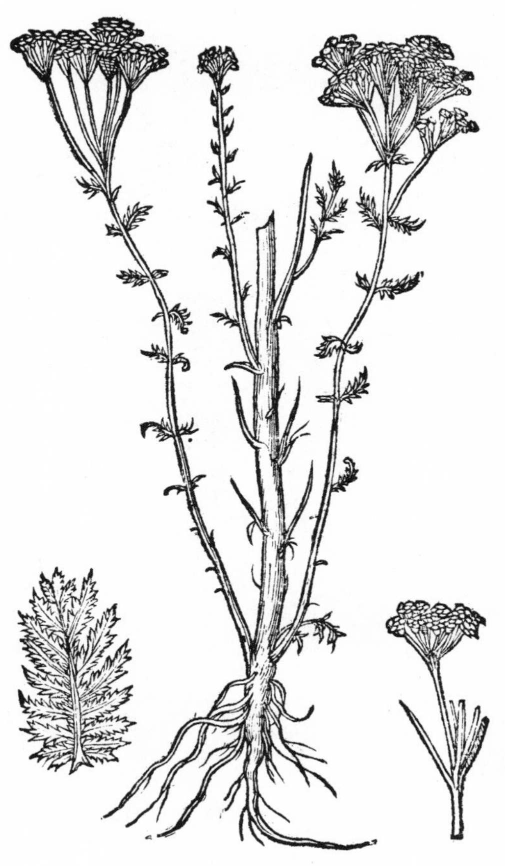

5 The fifth kinde of Tanſie hath broad leaues,much iagged and wel cut,like the leaues of Fetherfew,but ſmaller,and more deeply cut. The ſtalke is ſmall,a foot long,whereupon doe grow little tufts of little white floures,like the tuft of Milfoile or Yarrow. The herbe is in ſmell and ſauour like the common Tanſie, but not altogether ſo ſtrong.

¶ *The Place.*

The firſt groweth wilde in fields as well as in gardens: the others grow in my garden.

¶ *The Time.*

They floure in Iuly and Auguſt.

¶ *The Names.*

The firſt is called Tanſie; the ſecond double Tanſie,the third vnſauory Tanſie,the laſt white Tanſie: in Latine,*Tanacetum*,and *Athanaſia*,as though it were immortall: becauſe the floures do not ſpeedily wither: of ſome, *Artemiſia*, but vntruly.

¶ *The Nature.*

The Tanſies which ſmel ſweet are hot in the ſecond degree,and dry in the third. That without ſmell is hot and drie, and of a meane temperature.

¶ *The Vertues.*

In the Spring time are made with the leaues hereof newly ſprung vp,and with egs, cakes or tanſies, which be pleaſant in taſte,and good for the ſtomacke. For if any bad humours cleaue thereunto,it doth perfectly concoct them, and ſcowre them downewards. The root preſerued with hony or ſugar, is an eſpecial thing againſt the gout,if euery day for a certaine ſpace,a reaſonable quantitie thereof be eaten faſting. A

The ſeed of Tanſie is a ſingular and approoued medicine againſt Wormes,for in what ſort ſoeuer it be taken, it killeth and driueth them forth. B

The ſame pound,and mixed with oile Oliue,is very good againſt the paine and ſhrinking of the ſinewes. C

Alſo being drunke with wine,it is good againſt the paine of the bladder, and when a man cannot piſſe but by drops. D

† The figure that was formerly in the fourth place was onely the varietie of the ordinary Tanſie,hauing a white floure,but that which agreed with the deſcription was *pag.* 915. vnder the title of *Achillea, ſiue Millefolium nobile.*

CHAP. 210. *Of Fetherfew.*

¶ *The Deſcription.*

1 Feuerfew bringeth forth many little round ſtalkes, diuided into certaine branches. The leaues are tender, diuerſly torne and iagged,and nickt on the edges like the firſt and nethermoſt leaues of Coriander, but greater. The floures ſtand on the tops of the branches,with a ſmall pale of white leaues,ſet round about a yellow ball or button, like the wilde field Daiſie. The root is hard and tough: the whole plant is of a light whitiſh greene colour,of a ſtrong ſmell and bitter taſte

2 The ſecond kinde of Feuerfew, *Matricaria*, or *Parthenium*, differeth from the former, in that it hath double floures; otherwiſe in ſmell,leaues,and branches, it is all one with the common Feuerfew.

3 There is a third ſort called Mountaine Feuerfew, of *Carolus Cluſius* his deſcription,that hath ſmall

1 *Matricaria.*
Feuerfew.

2 *Matricaria duplici flore.*
Double Feuerfew.

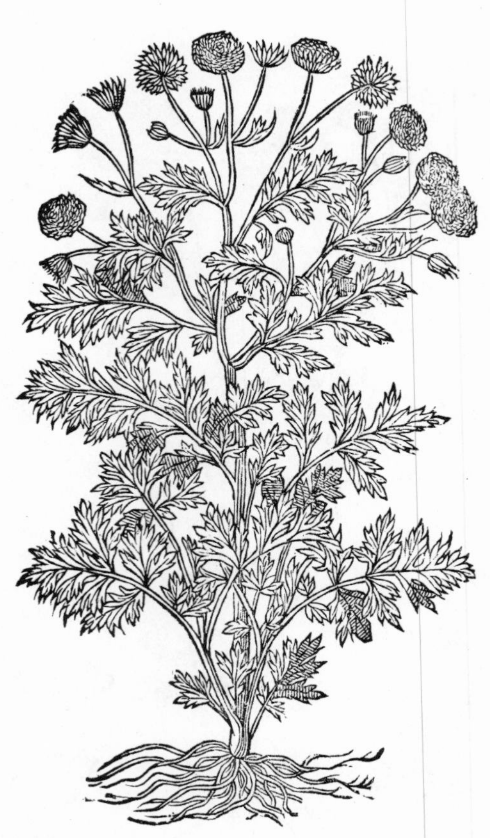

‡ 3 *Matricaria Alpina Clusij.*
Mountaine Feuerfew.

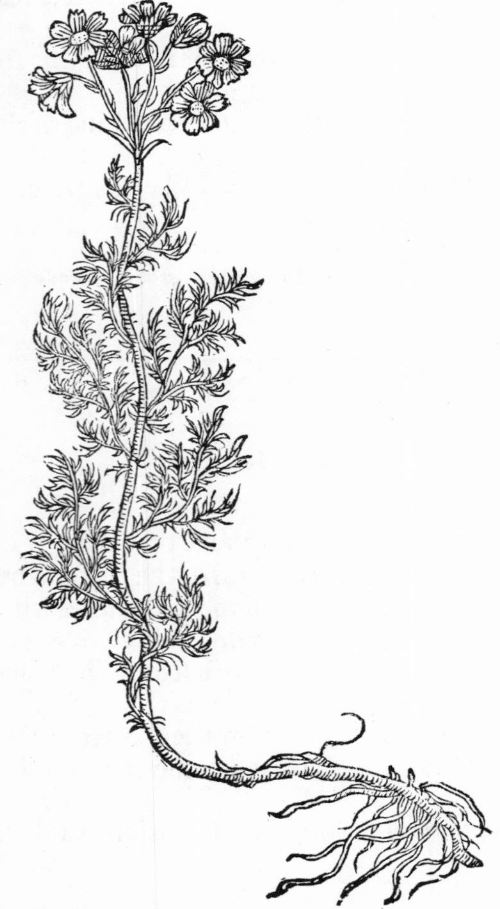

ſmall and fibrous roots; from which proceed ſlender wooddie ſtalks, a foot high and ſomewhat more, beſet or garniſhed about with leaues like Camomill, deepely iagged or cut, of the ſauour or ſmell of Feuerfew, but not ſo ſtrong; in taſte hot, but not vnpleaſant. At the top of the ſtalks there come forth ſmal white floures not like vnto the firſt, but rather like vnto *Abſynthium album*, or White Wormewood.

4 I haue growing in my Garden another ſort, like vnto the firſt kinde, but of a moſt pleaſant ſweet ſauour, in reſpect of any of the reſt. ‡ This ſeemes to be the *Matricaria altera ex Ilua*, mentioned by *Camerarius* in his *Hortus medicus*. ‡

¶ *The Place.*

The common ſingle Feuerfew groweth in hedges, gardens, and about old wals, it ioyeth to grow among rubbiſh. There is oftentimes found when it is digged vp a little cole vnder the ſtrings of the root, and neuer without it, whereof *Cardane* in his booke of Subtilties ſetteth down diuers vaine and trifling things.

¶ *The Time.*

They floure for the moſt part all the Sommer long.

¶ *The Names.*

Feuerfew is called in Greeke of *Dioſcorides* παρθένιον: of *Galen*, and *Paulus* one of his ſect, ἀμάρακος: in Latine, *Parthenium*, *Matricaria*, and *Febrifu-*

Febrifuga, of *Fuchsius*, *Artemisia Tenuifolia* in Italian, *Amarella*. in Dutch, Moeder cruyt: in French, *Espargoute*: in English, Fedderfew and Feuerfew, taken from his force of driuing away Agues.

¶ *The Temperature.*

Feuerfew doth manifestly heat, it is hot in the third degree, and drie in the second; it clenseth, purgeth, or scoureth, openeth and fully performeth all that bitter things can do.

¶ *The Vertues.*

It is a great remedie against the diseases of the matrix; it procureth womens sicknes with speed; it bringeth forth the after birth and the dead childe, whether it bee drunke in a decoction, or boiled in a bath and the woman sit ouer it; or the herbes sodden and applied to the priuie part, in manner of a cataplasme or pultis. A

Dioscorides also teacheth, that it is profitably applied to Saint Anthonies fire, to all hot inflammations, and hot swellings, if it be laid vnto, both leaues and floures. B

The same Author affirmeth, that the pouder of Feuerfew drunke with Oxymell, or syrup of Vineger, or wine for want of the others, draweth away flegme and melancholy, and is good for them that are pursie, and haue their lungs stuffed with flegme, and is profitable likewise to be drunke against the stone, as the same Author saith. C

Feuerfew dried and made into pouder, and two drams of it taken with honie or sweet wine, purgeth by siege melancholy and flegme; wherefore it is very good for them that are giddie in the head, or which haue the turning called *Vertigo*, that is a swimming and turning in the head. Also it is good for such as be melancholike, sad, pensiue, and without speech. D

The herbe is good against the suffocation of the mother, that is, the hardnesse and stopping of the same, being boiled in wine, and applied to the place. E

The decoction of the same is good for women to sit ouer, for the purposes aforesaid. F

It is vsed both in drinks, and bound to the wrests with bay salt, and the pouder of glasse stamped together, as a most singular experiment against the ague. G

CHAP. 211. *Of Poley, or Pellamountaine.*

1 *Polium montanum album.*
White Poley mountaine.

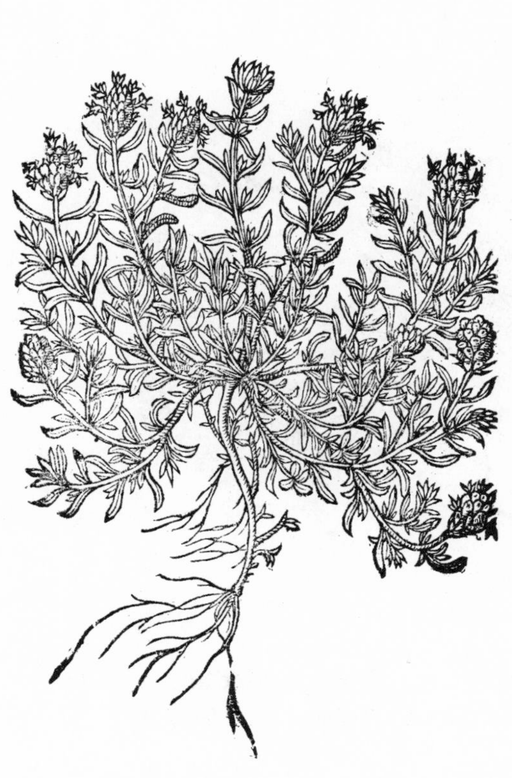

2 *Polium montanum luteum.*
Yellow Poley mountaine.

¶ *The*

¶ *The Deſcription.*

1 THe firſt kinde of *Polium*, or in Engliſh Poley of the mountain, is a little tender and ſweet ſmelling herbe, verie hoarie, whereupon it tooke his name: for it is not onely hoary in part, but his hoarie flockineſſe poſſeſſeth the whole plant, tufts and all, being no leſſe hoarie than *Gnaphalium*, eſpecially where it groweth neere the Sea at the bending of the hils, or neere the ſandie ſhores of the Mediterranean Sea: from his wooddie and ſomewhat threddie root ſhooteth forth ſtraight from the earth a number of ſmall round ſtalkes nine inches long, and by certaine diſtances from the ſtalke proceed ſomewhat long leaues like *Gnaphalium*, which haue light nickes about the edges, that ſtand one againſt another, incloſing the ſtalke: in the toppe of the ſtalkes ſtand ſpokie tufts of floures, white of colour like *Serpillum*. This plant is ſtronger of ſent or ſauour than any of the reſt following, which ſent is ſomewhat ſharp, and affecting the noſe with his ſweetneſſe.

2 The tuftes of the ſecond kinde of *Polium* are longer than the tuftes or floures of the laſt before mentioned, and they are of a yellow colour; the leaues alſo are broader, otherwiſe they are very like.

3 From the wooddie rootes of this third kinde of *Polium* proceed a great number of ſhootes like vnto the laſt rehearſed, lying flat vpright vpon the ground, whoſe ſlender branches take hold vpon the vpper part of the earth where they creepe. The floures are like the other, but of a purple colour.

4 The laſt kinde of *Polium*, and of all the reſt the ſmalleſt, is of an indifferent good ſmell, in all points like vnto the common *Polium*, but that it is foure times leſſer, hauing the leaues not ſnipt, & the floures white.

‡ 5 This ſends vp many branches from one root like to thoſe of the firſt deſcribed, but ſhorter and more ſhrubbie, lying partly vpon the ground; the leaues grow by couples at certain ſpaces, ſomewhat like, but leſſer than thoſe of Roſemarie or Lauander, greene aboue, and whitiſh beneath, not ſnipt about their edges; their taſte is bitter, and ſmell ſomewhat pleaſant: the floures grow plentifully vpon the tops of the branches, white of colour, and in ſhape not vnlike thoſe of the other Poleyes: they grow on a bunch together, and not Spike faſhion: the ſeed is blackiſh and contained in ſmall veſſels: the root is hard and wooddie, with many fibres. *Cluſius* calls this *Polium* 7. *albo flore*. It is the *Polium alterum* of *Matthiolus*, and *Polium recentiorum fœmina Lavandulæ folio* of *Lobel*. I here giue you (as *Cluſius* alſo hath done) two figures to make one good one: the former ſhews the floures and their manner of growing; the other, the ſeede veſſels, and the leaues growing by couples, together with a little better expreſſion of the root. ‡

3 *Polium montanum purpureum.* Purple Poley.

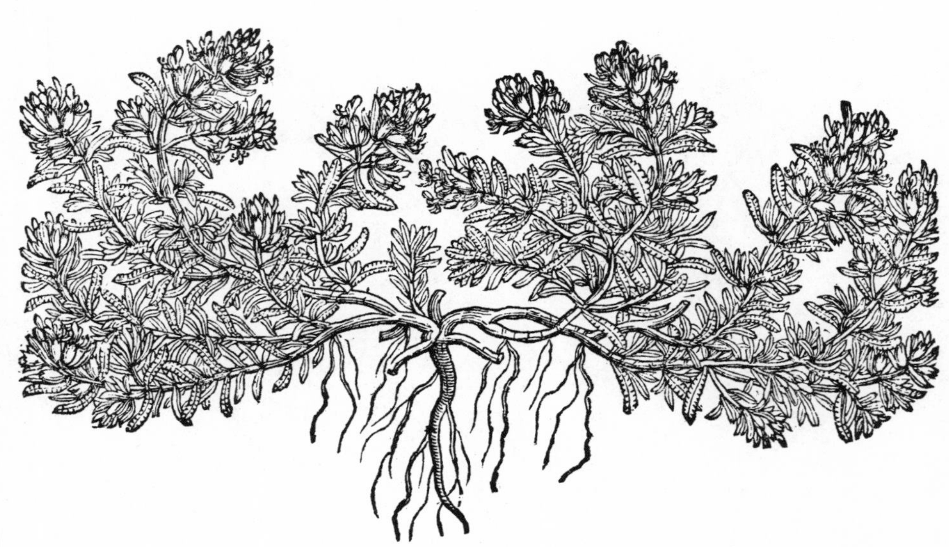

¶ *The Place.*

Theſe plants do grow naturally vpon the mountaines of France, Italie, Spaine, and other hot regions. They are ſtrangers in England, notwithſtanding I haue plants of that Poley with yellow floures by the gift of *Lobel*.

¶ *The Time.*

They floure from the end of May, to the beginning of Auguſt.

¶ *The*

4 *Polium montanum minimum.*
Creeping Poley.

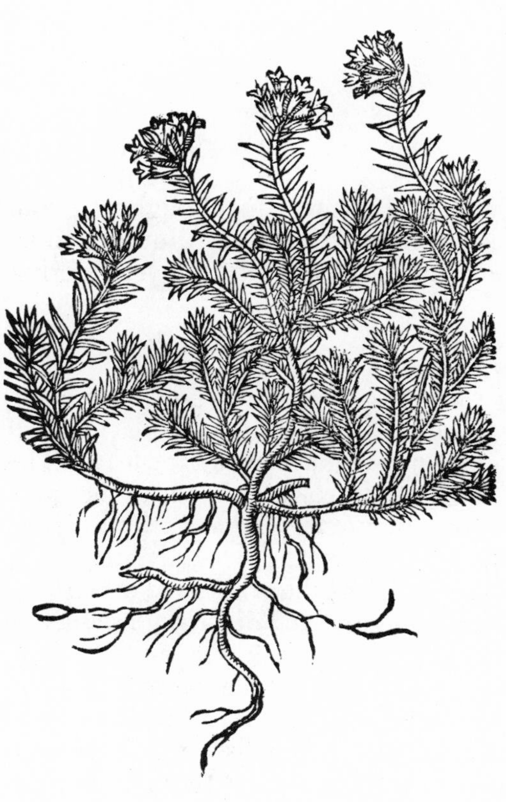

‡ 5 *Polium Lavandulæ folio, flore albo.*
Lavander leaued Poley.

Another figure of the Lauander leaued Poley.

¶ *The Names.*

Poley mountaine is called in Greeke πόλιον, of his hoarineſſe, and in Latine alſo *Polium*. Diuers ſuſpect that *Polium* is *Leucas*, and that *Dioſcorides* hath twiſe intreated of that herbe, vnder diuers names; the kindes, the occaſion of the name, and likewiſe the faculties do agree. There bee two of the *Leucades*, one ὀρεινὴ: that is of the mountaine: the other, ἥμερος, which is that with the broader leafe: it is called *Leucas* of the whitiſh colour, and *Polion* of the hoarineſſe, becauſe it ſeemeth like to a mans hoarie head; for whatſoeuer waxeth hoarie, is ſaid to be white.

¶ *The Temperature.*

Poley is of temperature drie in the third degree, and hot in the end of the ſecond.

¶ *The Vertues.*

A *Dioſcorides* ſaith, it is a remedie for them that haue the dropſie, the yellow iaundice, and that are troubled with the ſpleene.

B It prouoketh vrine, & is put into Mithridate, treacle, and counterpoiſons.

C It profiteth much againſt the bitings of venomous beaſts, and driueth away all venomous beaſts from the place where it is ſtrewed or burnt.

D The ſame drunke with vineger, is good for the

the diseases of the milt and spleene; it troubleth the stomacke, and afflicteth the head, and prouoketh the loosenesse of the bellie.

Chap. 212. *Of Germander.*

¶ *The Kindes.*

THe old writers haue set downe no certaine kinds of Germander, yet we haue thought it good, and not without cause, to intreat of mo sorts than haue been obserued of all, diuiding those vnder the title of *Teucrium* from *Chamædryes* : although they are both of one kind, but yet differing very notably.

¶ *The Description.*

3 THe first Germander groweth lowe, with very many branches lying vpon the ground, tough, hard, and wooddie, spreading it selfe here and there : whereupon are placed small leaues snipt about the edges like the teeth of a saw, resembling the shape of an oken leafe. The floures are of a purple colour, very small, standing close to the leaues toward the top of the branches. The seed is little and blacke. The root slender and full of strings, creeping, and alwaies spreading within the ground, whereby it greatly increaseth. ‡ This is sometimes found with bigger leaues, otherwhiles with lesse; also the floure is sometimes white, and otherwhiles red in the same plant, whence *Tabernam.* gaue two figures, and our Authour two figures and descriptions, whereof I haue omitted the later, and put the two titles into one. ‡

2 The second Germander riseth vp with a little straight stalk a span long, and sometimes longer, wooddie and hard like vnto a little shrub : it is afterwards diuided into very many little small branches. The leaues are indented and nicked about the edges, lesser than the leaues of the former, great creeping Germander : the floures likewise stand neere to the leaues, and on the vpper parts of the sprigs, of colour sometimes purple, and oftentimes tending to blewnesse : the roote is diuersly dispersed with many strings.

1 *Chamædrys maior latifolia.*
Great broad leaued Germander.

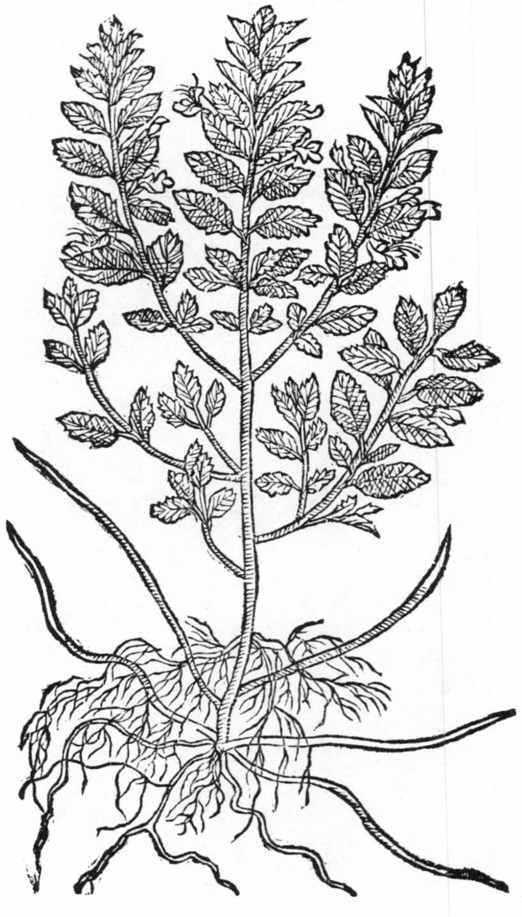

2 *Chamædrys minor.*
Small Germander.

3 Wilde

3 *Chamædrys syluestris.*
Wilde Germander.

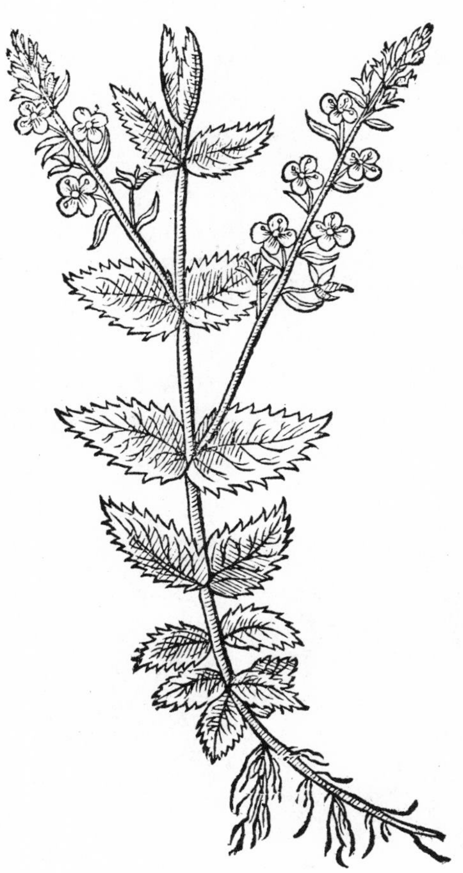

3 Wilde Germander hath little stalkes, weake and feeble, edged or cornered, somewhat hairie, and set as it were with ioints; about the which by certaine distances there come forth at each ioint two leaues something broad, nicked in the edges, and something greater than the leaues of creeping Germander, and softer. The floures be of a gallant blew colour, made of foure small leaues a peece, standing orderly on the tops of the tender spriggie spraies; after which come in place little huskes or seede vessels. The root is small and threddie.

¶ *The Place.*

These plants do grow in rocky and rough grounds, and in gardens they do easily prosper.

The wilde Germander groweth in manie places about London in Medowes and fertil fields, and in euery place wheresoeuer I haue trauelled in England.

¶ *The Time.*

They floure and flourish from the end of May, to the later end of August.

¶ *The Names.*

Garden Germander is called in Greeke, χαμαίδρυς, *Chamædrys*: of some, *Trissago*, & *Trixago*, and likewise *Quercula minor*; notwithstanding most of these names do more properly belong to *Scordium*, or water Germander: in Italian, *Querciuola*: in English, Germander, or English Treacle: in French, *Germandre*. Before creeping Germander was knowne, this wilde kinde bare the name of Germander amongst the Apothecaries, and was vsed for the right Germander in the compositions of Medicines: but after the former were brought to light, this began to be named *Syluestris*, and *Spuria Chamædrys*: that is wilde and bastard Germander: of some, *Teucrium pratense*, and without errour; because all the sorts of plants comprehended vnder the title of *Teucrium*, are doubtlesse kindes of Germander. Of some it hath been thought to be the plant that *Dioscorides* called ἱερὰ βοτάνη, *Hierabotane*; that is to say, the Holie herbe, if so bee that the Holie herbe, and *Verbenaca*, or Veruaine, which is called in Greeke περιστερεών, be sundrie herbes. *Dioscorides* maketh them sundrie herbes, describing them apart, the one after the other: but other Authors, as *Paulus*, *Aetius*, and *Oribasius*, make no mention of *Herba Sacra*, the Holie herbe, but onely of *Peristereon*: and this same is found to be likewise called *Hierabotane*, or the Holie Herb, and therefore it is euident that it is one and the selfe same plant, called by diuers names: the which things considered, if they say so, and say truely, this wilde Germander cannot be *Hierabotane* at all, as diuers haue written and said it to be.

¶ *The Temperature.*

Garden Germander is of thin parts, and hath a cutting facultie, it is hot and drie almost in the third degree, euen as *Galen* doth write of *Teucrium*, or wilde Germandet.

The wilde Germander is likewise hot and drie, and is not altogether without force or power to open and clense: it may be counted among the number of them that do open the liuer and spleen.

¶ *The Vertues.*

A Germander boiled in water and drunk, deliuereth the bodie from all obstructions or stoppings, diuideth and cutteth tough and clammie humors: being receiued as aforesaid, it is good for them that haue the cough, and shortnesse of breath, the strangurie or stopping of vrine, and helpeth those which are entring into a dropsie.

B The leaues stamped with honie and strained, and a drop at sundrie times put into the eies, takes away the web and hawe in the same, or any dimnesse of sight.

C It prouoketh mightily the termes, being boiled in wine, and the decoction drunk; with a fomentation or bath made also thereof, and the secret parts bathed therewith.

Chap. 213. *Of Tree Germander.*

¶ *The Description.*

1 THe first kinde of Tree Germander riseth vp with a little straight stalke a cubite high, wooddie and hard like vnto a small wooddie shrubbe. The stalke diuideth it selfe from the bottome vnto the toppe into diuers branches, whereon are set indented leaues nicked about the edges, in shape not much vnlike the leafe of the common Germander. The floures grow among the leaues of a purple colour. The root is wooddie,as is all the rest of the plant.

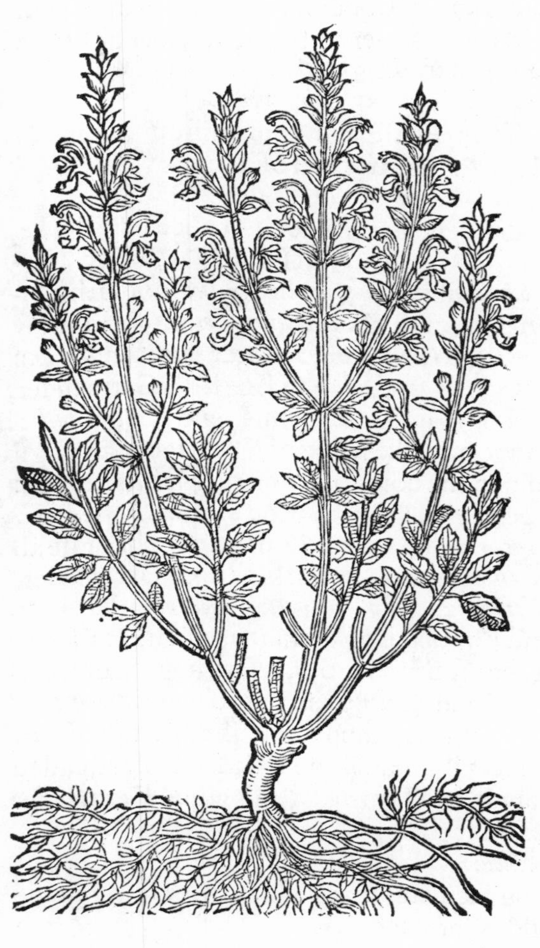
1 *Teucrium latifolium.*
Tree Germander with broad leaues.

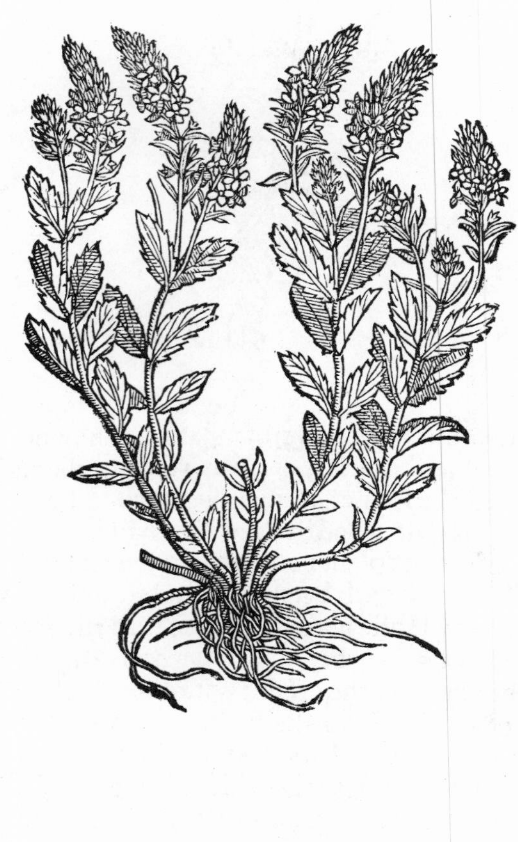
2 *Teucrium Pannonicum.*
Hungarie Germander.

2 The Tree Germander of Hungarie hath many tough threddie roots,from which rise vp diuers weake and feeble stalks,reeling this way and that way;whereupon are set together by couples, long leaues iagged in the edges,not vnlike those of the vpright Fluellen : on the tops of the stalks stand the floures Spike fashion, thicke thrust together, of a purple colour tending towards blewnesse.

‡ 3 This (which is the fourth of *Clusius* description) hath diuers stalkes some cubite high, foure square, rough, and set at certaine spaces with leaues growing by couples like those of the wilde Germander : the tops of the stalkes are diuided into sundry branches, carrying long spokes of blew floures,consisting of foure leaues,whereof the vppermost leafe is the largest,and distinguished with veines : after the floures are past follow such flat seed vessels as in Fluellen : the root is fibrous and liues long,sending forth euery yeare new branches. ‡

4 This

‡ 3 *Teucrium maius Pannonicum.*
Great Austrian Germander.

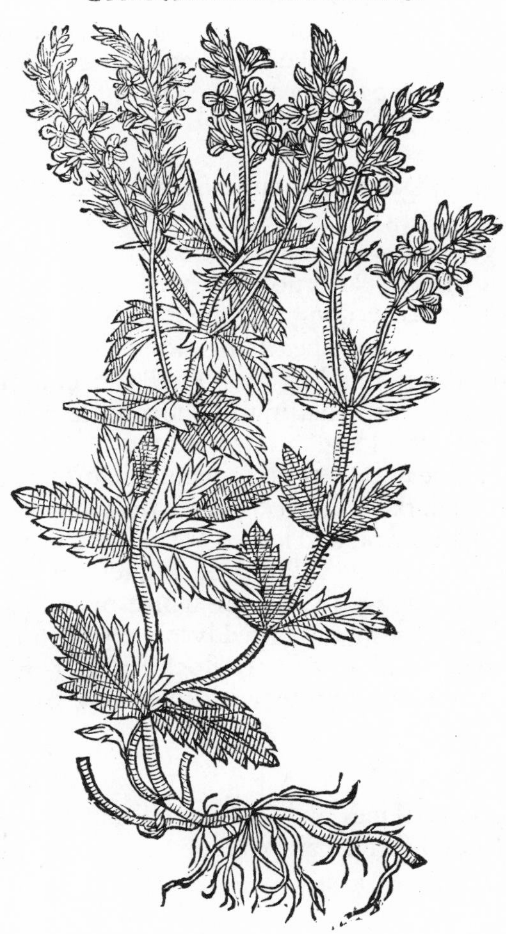

‡ 4 *Teucrium petræum pumilum.*
Dwarfe Rocke Germander.

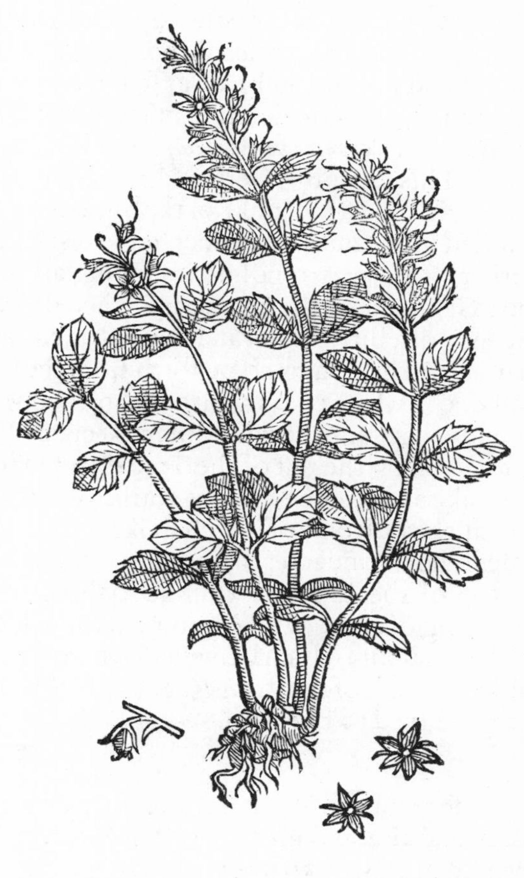

5 *Teucrium Bæticum.*
Spanish Tree Germander.

6 *Teucrium Alpinum Cisti flore.*
Rough headed Tree Germander.

4 This Dwarfe Germander sends vp stalkes some handfull high, round, not branched: the leaues grow vpon these stalkes by couples, thicke, shining, a little hairy and greene on their vpper sides, and whitish below: the tops of the stalkes carry spoky tufts of floures, consisting of foure or fiue blewish leaues; which falling, there followes a seed-vessell, as in the *Veronica's*. The root is knotty and fibrous, and growes so fast amongst the rockes that it cannot easily be got out. It floureth in Iuly. *Clusius* describes this by the name of *Teucrium 6. Pumilum*: and *Pona* sets it forth by the name of *Veronica petræa semper virens*. ‡

5 This Spanish Germander riseth vp oft times to the height of a man, in manner of a hedge bush, with one stiffe stalke of the bignesse of a mans little finger, couered ouer with a whitish bark, diuided sometimes into other branches, which are alwayes placed by couples one right against another, of an ouerworne hoarie colour; and vpon them are placed leaues not much vnlike the common Germander, the vpper parts whereof are of a grayish hoarie colour, and the lower of a deepe greene; of a bitter taste, and somewhat crooked, turning and winding themselues after the manner of a welt. The floures come forth from the bosome of the leaues, standing vpon small tender footstalkes of a white colour, without any helmet or hood on their tops, hauing in the middle many threddy strings. The whole plant keepeth greene all the Winter long.

6 Among the rest of the Tree Germanders this is not of least beauty and account, hauing many weake and feeble branches trailing vpon the ground, of a darke reddish colour, hard and wooddie; at the bottome of which stalks come forth many long broad iagged leaues not vnlike the precedent, hoary vnderneath, and greene aboue, of a binding and drying taste. The floures grow at the top of the stalkes, not vnlike to those of *Cistus fœmina*, or Sage-rose, and are white of colour, consisting of eight or nine leaues, in the middle whereof do grow many threddy chiues without smell or sauour: which being past, there succeedeth a tuft of rough threddy or flocky matter, not vnlike to those of the great Auens or *Pulsatilla*: the root is wooddy, and set with some few hairie strings fastned to the same.

¶ The Place.

These plants do ioy in stony and rough mountaines and dry places, and such as lie open to the Sunne and aire, and prosper well in gardens: and of the second sort I haue receiued one plant for my garden of Mr. *Garret* Apothecarie.

¶ The Time.

They floure, flourish, and seed when the other Germanders do.

¶ The Names.

Tree Germander is called in Greeke χαμαίδρυς, retaining the name of the former *Chamædrys*, and τεύκριον, according to the authoritie of *Dioscorides* and *Pliny*: in Latine *Teucrium*: in English, Great Germander, vpright Germander, and Tree Germander.

¶ The Temperature and Vertues.

Their temperature and faculties are referred vnto the garden Germander, but they are not of such force and working, wherefore they be not much vsed in physicke.

Chap. 214.

Of Water Germander, or Garlicke Germander.

¶ The Description.

1 Scordium or water Germander hath square hairie stalkes creeping by the ground, beset with soft whitish crumpled leaues, nickt and snipt round about the edges like a Saw: among which grow small purple floures like the floures of dead Nettle. The root is small and threddy, creeping in the ground very deepely. The whole plant being bruised smelleth like Garlicke, whereof it tooke that name *Scordium*. ‡ This by reason of goodnesse of soile varieth in the largenesse thereof, whence *Tabernamontanus* and our Author made a bigger and a lesser thereof, but I haue omitted the later as superfluous. ‡

¶ The Place.

Water Germander groweth neere to Oxenford, by Ruley, on both sides of the water, and in a medow

medow by Abington called Nietford, by the relation of a learned Gentleman of S. *Iohns* in the said towne of Oxenford, a diligent φιλοβοτανὸς, my very good friend, called M[r]. *Richard Slater*. Also it groweth in great plenty in the Isle of Elie, and in a medow by Harwood in Lancashire, and diuers other places.

1 *Scordium.*
Water Germander.

¶ *The Time.*

The floures appeare in Iune and Iuly: it is best to gather the herbe in August: it perisheth not in Winter, but onely loseth the stalkes, which come vp againe in Sommer: the root remaineth fresh all the yeare.

¶ *The Names.*

The Grecians call it Σκόρδιον: the Latines do keepe that name *Scordium*: the Apothecaries haue no other name: It is called of some *Trixago Palustris*, *Quercula*, and also *Mithridatium*, of *Mithridates* the finder of it out. It tooke the name *Scordium* from the smel of Garlicke, which the Grecians call σκόροδον, and σκόρδιον, of the ranckness of the smell: in high-Dutch, **Wasser battenig**: in French, *Scordion*: in Italian, *Chalamandrina palustre*: in English, Scordium, Water Germander, and Garlicke Germander.

¶ *The Temperature.*

Water Germander is hot and dry: it hath a certaine bitter taste, harsh and sharpe, as *Galen* witnesseth.

¶ *The Vertues.*

Water Germander cleanseth the intrals, and likewise old vlcers, being mixed with honey according to art: it prouoketh vrine, and bringeth downe the monethly sickenesse: it draweth out of the chest thicke flegme and rotten matter: it is good for an old cough, paine in the sides which commeth of stopping and cold, and for burstings and inward ruptures. A

The decoction made in wine and drunke, is good against the bitings of Serpents, and deadly poysons; and is vsed in antidotes or counterpoysons with good successe. B

It is reported to mitigate the paine of the gout, being stamped and applied with a little vineger and water. C

Some affirme, that raw flesh being laid among the leaues of Scordium, may be preserued a long time from corruption. D

Being drunke with wine it openeth the stoppings of the liuer, the milt, kidnies, bladder, and matrix, prouoketh vrine, helpeth the strangurie, that is, when a man cannot pisse but by drops, and is a most singular cordiall to comfort and make merry the heart. E

The pouder of Scordion taken in the quantitie of two drams in meade or honied water, cureth and stoppeth the bloudy flix, and comforteth the stomacke. Of this Scordium is made a most singular medicine called *Diascordium*, which serueth very notably for all the purposes aforesaid. F

The same medicine made with Scordium is giuen with very good successe vnto children and aged people, that haue the small pockes, measles, or the Purples, or any other pestilent sicknesse whatsoeuer, euen the plague it selfe, giuen before the sicknes haue vniuersally possessed the whole body. G

Chap. 215. *Of Wood Sage, or Garlicke Sage.*

¶ *The Description.*

That which is called Wilde Sage hath stalkes foure square, somewhat hairie, about which are leaues like those of Sage, but shorter, broader, and softer: the floures grow vp all vpon one side of the stalke, open and forked as those of dead Nettle, but lesser, of a pale white colour: then grow the seeds foure together in one huske: the root is full of strings. It is a plant that liueth but a yeare: it smelleth of garlicke when it is bruised, being a kinde of Garlicke Germander, as appeareth by the smell of garlicke wherewith it is possessed.

† *Scorodonia, siue Saluia agrestis.*
Wood Sage, or Garlicke Sage.

¶ *The Place.*

It groweth vpon heaths and barren places: it is also found in Woods, and neere vnto hedge-rowes, and about the borders of fields: it somewhat delighteth in a leane soile, and yet not altogether barren and dry.

¶ *The Time.*

It floureth and seedeth in Iune, Iuly, and August, and it is then to be gathered and laid vp.

¶ *The Names.*

It is called of the later Herbarists *Saluia agrestis*: of diuers also *Ambrosia*; but true *Ambrosia*, which is Oke of Cappadocia, differs from this. *Valerius Cordus* names it *Scordonia*, or *Scorodonia*, and *Scordium alterum*. *Ruellius* saith it is called *Boscisaluia*, or *Saluia Bosci*: in high Dutch, **Waldt salbey**: in English, wilde Sage, wood Sage, and Garlicke Sage.

It seemeth to be *Theophrastus* his σφάκελος, *Sphacelus*, which is also taken for the small Sage, but not rightly.

¶ *The Temperature.*

Wilde Sage is of temperature hot and drie, yet lesse than common Sage, therefore it is hot and dry in the second degree.

¶ *The Vertues.*

A It is commended against burstings, dry beatings, and against wounds: the decoction thereof is giuen to them that fall, and are inwardly bruised: it also prouoketh vrine.

B Some likewise giue the decoction hereof to drinke, with good successe, to them that are infected with the French Pox; for it causeth sweat, drieth vp vlcers, digesteth humors, wasteth away and consumeth swellings, if it be taken thirtie or forty dayes together, or put into the decoction of *Guiacum*, in stead of *Epithymum* and other adiutories belonging to the said decoction.

† The figure which was formerly here was of *Calamintha montana præstantior* of *Lobel*.

Chap. 216. *Of Eye-bright.*

¶ *The Description.*

Evphrasia or Eye-bright is a small low herbe not aboue two handfuls high, full of branches, couered with little blackish leaues dented or snipt about the edges like a saw: the floures are small

ſmall and white, ſprinkled and poudered on the inner ſide, with yellow and purple ſpecks mixed therewith. The root is ſmall and hairie.

Euphraſia.
Eye-bright.

¶ *The Place.*

This plant growes in dry medows, in green and graſſie wayes and paſtures ſtanding againſt the Sunne.

¶ *The Time.*

Eye-bright beginneth to floure in Auguſt and continueth vnto September, and muſt be gathered while it floureth for phyſicks vſe.

¶ *The Names.*

It is commonly called *Euphraſia*, as alſo *Euphroſyne*; notwithſtanding there is another *Euphroſyne, viz*. Bugloſſe: it is called of ſome *Ocularis*, & *Ophthalmica* of the effect: in high-Dutch, **Augen troſt**; in low-Dutch, **Ooghen trooſt**: in Italian, Spaniſh, and French, *Eufraſia*, after the Latine name: in Engliſh, Eye-bright.

¶ *The Nature.*

This herbe is hot and dry, but yet more hot than dry.

¶ *The Vertues.*

It is very much commended for the eyes. Being taken it ſelfe alone, or any way elſe, it preſerues the ſight, and being feeble and loſt it reſtores the ſame: it is giuen moſt fitly being beaten into pouder; oftentimes a like quantitie of Fennel ſeed is added thereto, and a little mace, to the which is put ſo much ſugar as the weight of them all commeth to. A

Eye-bright ſtamped and layd vpon the eyes, or the iuyce thereof mixed with white Wine, and dropped into the eyes, or the diſtilled water, taketh away the darkneſſe and dimneſſe of the eyes, and cleareth the ſight. B

Three parts of the pouder of Eye-bright, and one part of maces mixed therewith, taketh away all hurts from the eyes, comforteth the memorie, and cleareth the ſight, if halfe a ſpoonfull be taken euery morning faſting with a cup of white wine. C

† That which was formerly here ſet forth in the ſecond place vnder the title of *Euphraſia cærulea Tabern.* was deſcribed by our Authour amongſt the Scorpion graſſes, in the third place, Chap. 54. and the figure is pag. 338. vnder the title of *Myoſotis Scorpioides paluſtris.*

CHAP. 217. *Of Marierome.*

¶ *The Deſcription.*

1 SWeet Marjerome is a low and ſhrubby plant, of a whitiſh colour and maruellous ſweet ſmell, a foot or ſomewhat more high. The ſtalkes are ſlender, and parted into diuers branches; about which grow forth little leaues ſoft and hoarie: the floures grow at the top in ſcaly or chaffie ſpiked eares, of a white colour like vnto thoſe of Candy Organy. The root is compact of many ſmall threds. The whole plant and euerie part thereof is of a moſt pleaſant taſte, and aromaticall ſmell, and periſheth at the firſt approch of Winter.

2 Pot Marierome or Winter Maierome hath many threddy tough roots, from which riſe immediately diuers ſmall branches, whereon are placed ſuch leaues as the precedent, but not ſo hoarie, nor yet ſo ſweet of ſmell, bearing at the top of the branches tufts of white floures tending to purple. The whole plant is of long continuance, and keepeth greene all the Winter; whereupon our Engliſh women haue called it, and that very properly, Winter Marierome.

3 Marierome Gentle hath many branches riſing from a threddy root, whereupon do grow ſoft and ſweet ſmelling leaues of an ouerworne ruſſet colour. The floures ſtand at the top of the ſtalks, compact

1 *Mariorana maior.*
Great sweet Marierome.

2 *Mariorana maior Anglica.*
Pot Marierome.

3 *Mariorana tenuifolia.*
Marierome gentle.

compact of diuers small chaffie scales, of a white colour tending to a blush. The whole plant is altogether like the great sweet Marierome, sauing that it is altogether lesser, and far sweeter, wherein especially consisteth the difference.

4 *Epimaiorana* is likewise a kind of Marierome, differing not from the last described, sauing in that, that this plant hath in his naturall country of Candy, and not elsewhere, some laces or threds fastned vnto his branches, such, and after the same manner as those are that doe grow vpon Sauorie, wherein is the difference.

¶ *The Place.*

These plants do grow in Spaine, Italy, Candy, and other Islands thereabout, wilde, and in the fields; from whence wee haue the seeds for the gardens of our cold countries.

¶ *The Time.*

They are sowne in May, and bring forth their scaly or chaffie huskes or eares in August. They are to be watered in the middle of the day, when the Sunne shineth hottest, euen as Basill should be, and not in the euening nor morning, as most Plants are.

¶ *The Names.*

Marierome is called *Mariorana*, and *Amaracus*, and also *Marum* and *Sampsychum* of others: in high-Dutch, **Mayoran**: in Spanish, *Mayorana, Moradux*, and *Almoradux*: in French, *Mariolaine*: in English, Sweet Marierome, Fine Marierome, and Marierome

rome gentle; of the best sort Marjerane. The pot Marjerome is also called Winter Marjerome Some haue made a doubt whether *Maiorana* and *Sampsychum* be all one; which doubt, as I take it, is because that *Galen* maketh a difference betweene them, intreating of them apart, and attributeth to either of them their operations. But *Amaracus Galeni* is *Parthenium*, or Feuerfew. *Dioscorides* likewise witnesseth, that some do call *Amaracus, Parthenium*; and *Galen* in his booke of the faculties of simple medicines, doth in no place make mention of *Parthenium*, but by the name of *Amaracus*. *Pliny* in his 21 booke, chap. 2. witnesseth, that *Diocles* the physition, and they of Sicily did call that *Amaracus*, which the Ægyptians and the Syrians did call *Sampsychum*.

Virgill in the first booke of his *Æneidos* sheweth, that *Amaracus* is a shrub bearing floures, writing thus:

Vbi mollis Amaracus illum
Floribus, & dulci aspirans complectitur vmbra.

Likewise *Catullus* in his *Epithalamium*, or mariage song of *Iulia* and *Mallius* saith,

Cinge tempora floribus
Suaue olentis Amaraci.

Compasse the temples of the head with floures.
Of Amarac affording sweete sauours.

Notwithstanding it may not seeme strange, that Majorane is vsed in stead of *Sampsychum*, seeing that in *Galens* time also *Marum* was in the mixture of the ointment called *Amaracinum vnguentum*, in the place of *Sampsychum*, as he himselfe witnesseth in his first booke of counterpoisons.

¶ *The Temperature.*

They are hot and dry in the second degree; after some copies, hot and dry in the third.

¶ *The Vertues.*

Sweete Marjerome is a remedy against cold diseases of the braine and head, being taken any way to your best liking; put vp into the nosthrils it prouoketh sneesing, and draweth forth much baggage flegme: it easeth the tooth-ache being chewed in the mouth; being drunke it prouoketh vrine, and draweth away waterish humors, and is vsed in medicines against poison. A

The leaues boiled in water, and the decoction drunke, helpeth them that are entering into the dropsie: it easeth them that are troubled with difficultie of making water, and such as are giuen to ouermuch sighing, and easeth the paines of the belly. B

The leaues dried and mingled with hony, and giuen, dissolueth congealed or clotted blood, and putteth away blacke and blew markes after stripes and bruses, being applied thereto. C

The leaues are excellent good to be put into all odoriferous ointments, waters, pouders, broths, and meates. D

The dried leaues poudered, and finely searched, are good to be put into Cerotes, or Cere-cloths and ointments, profitable against colde swellings, and members out of joint. E

There is an excellent oile to be drawne forth of these herbes, good against the shrinking of sinewes, crampes, convulsions, and all aches proceeding of a colde cause. F

CHAP. 218. *Of wilde Marjerome.*

¶ *The Description.*

1 Bastard Marjerome groweth straight vp with little round stalkes of a reddish colour, full of branches, a foot high and sometimes higher. The leaues be broad, more long than round of a whitish greene colour: on the top of the branches stand long spikie scaled eares, out of which shoot forth little white floures like the flouring of wheate. The whole plant is of a sweete smell, and sharpe biting taste.

2 The white Organy, or bastard Marjerome with white floures, differing little from the precedent, but in colour and stature. This plant hath whiter and broader leaues, and also much higher, wherein consisteth the difference.

3 Bastard Marjerome of Candy hath many threddy roots; from which rise vp diuers weake and feeble branches trailing vpon the ground, set with faire greene leaues, not vnlike those of Penny Royall, but broader and shorter: at the top of those branches stand scalie or chaffie eares of a purple colour. The whole plant is of a most pleasant sweet smell. The root endured in my garden

 and

1 *Origanum Heracleoticum.*
Baſtard Marjerome.

† 2 *Origanum album, Tabern.*
White baſtard Marjerome.

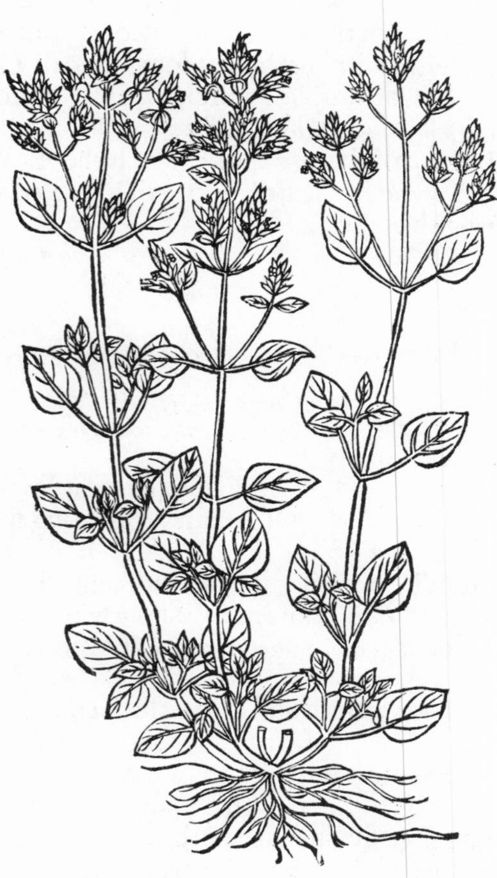

† 3 *Origanum Creticum.*
Wilde Marjerome of Candy.

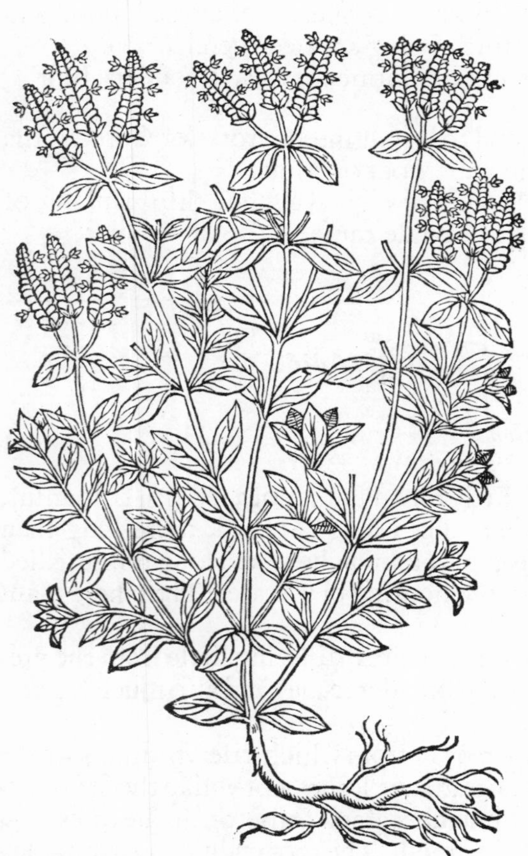

4 *Origanum Anglicum.*
Engliſh wilde Marjerome.

and the leaues also greene all this winter long, 1597 although it hath been said that it doth perish at the first frost, as sweete Marjerome doth.

4 English wilde Marjerome is exceedingly well knowne to all, to haue long, stiffe, and hard stalkes of two cubits high, set with leaues like those of sweet Marjerome, but broader and greater, of a russet greene colour, on the top of the branches stand tufts of purple floures, composed of many small ones set together very closely vmbell fashion. The root creepeth in the ground, and is long lasting.

¶ *The Place.*

These plants do grow wilde in the kingdome of Spaine, Italy, and other of those hot regions. The last of the foure doth grow wilde in the borders of fields, and low copses, in most places of England.

¶ *The Time.*

They floure and flourish in the Sommer moneths, afterward the seed is perfected.

¶ *The Names.*

Bastard Marjerome is called in Greeke, ὀρίγανος; and that which is surnamed *Heracleoticum*, ὀρίγανος ἡρακλεωτικὴ: of diuers it is called *Cunila*: in shops, *Origanum Hispanicum*, Spanish Organy: our English wilde Marjerome is called in Greeke of *Dioscorides*, *Galen*, and *Pliny*, *Onitis*, of some, *Agriorigànum*, or *Sylvestre Origanum*: in Italian, *Origano*: in Spanish *Oregano*: in French, *Mariolaine bastarde*: in English, Organe, bastard Marjerome: and that of ours, wilde Marjerome, and groue Marjerome.

¶ *The Temperature.*

All the Organies do cut, attenuate, or make thin, dry, and heate, and that in the third degree; and *Galen* teacheth that wilde Marjerome is more forceable and of greater strength; notwithstanding Organy of Candy which is brought dry out of Spaine (whereof I haue a plant in my garden) is more biting than any of the rest, and of greatest heate.

¶ *The Vertues.*

Organy giuen in wine is a remedy against the bitings, and stingings of venomous beasts, and cureth them that haue drunke *Opium*, or the juice of blacke poppy, or hemlockes, especially if it be giuen with wine and raisons of the sunne. A

The decoction of Organy prouoketh vrine, bringeth downe the monethly course, and is giuen with good successe to those that haue the dropsie. B

It is profitably vsed in a looch, or a medicine to be licked, against an old cough and the stuffing of the lungs. C

It healeth scabs, itches, and scuruinesse, being vsed in bathes, and it taketh away the bad colour which commeth of the yellow jaundice. D

The weight of a dram taken with meade or honied water, draweth forth by stoole blacke and filthy humors, as *Dioscorides* and *Pliny* write. E

The juice mixed with a little milke, being poured into the eares, mitigateth the paines thereof. F

The same mixed with the oile of *Ireos*, or the rootes of the white Florentine floure de luce, and drawne vp into the nosthrils, draweth downe water and flegme: the herbe strowed vpon the ground drieth away serpents. G

The decoction looseth the belly, and voideth choler; and drunke with vineger helpeth the infirmities of the spleene, and drunke in wine helpeth against all mortall poisons, and for that cause it is put into mithridate and treacles prepared for that purpose. H

These plants are easie to be taken in potions, and therefore to good purpose they may be vsed and ministred vnto such as cannot brooke their meate, and to such as haue a sowre and sqamish and watery stomacke, as also against the swouning of the heart. I

† The second and third figures were formerly transposed.

CHAP. 219. *Of Goates Marjerome, or Organy.*

¶ *The Description.*

1 THe stalkes of Goates Organy are slender, hard and wooddy, of a blackish colour; whereon are set long leaues, greater than those of the wilde Time, sweete of smell, rough, and somewhat hairy. The floures be small, and grow out of little crownes or wharles round about the top of the stalkes, tending to a purple colour. The root is small and threddy.

† 1 *Tragoriganum Dod.*

† *Tragoriganum Lob.*

Goats Marierome.

† 2 *Tragoriganum Clusij.*
Clusius his Goats Marierome.

‡ 3 *Tragoriganum Cretense.*
Candy Goats Marierome.

2 *Carolus Cluſius* hath ſet forth in his Spaniſh Obſeruations another ſort of Goats Marierome growing vp like a ſmall ſhrub: the leaues are longer and more hoarie than wilde Marierome, and alſo narrower, of a hot biting taſte, but of a ſweet ſmell, though not very pleaſant. The floures do ſtand at the top of the ſtalkes in ſpokie rundles, of a white colour. The root is thicke and wooddy.

‡ 3 This differs little in forme and magnitude from the laſt deſcribed, but the branches are of a blacker colour, with rougher and darker coloured leaues: the floures alſo are leſſer, and of a purple colour. Both this and the laſt deſcribed continue alwaies greene, but this laſt is of a much more fragrant ſmell. This floures in March, and was found growing wilde by *Cluſius* in the fields of Valentia: he calls it *Tragoriganum Hiſpanicum tertium*. *Pena* and *Lobel* call it *Tragoriganum Cretenſe apud Venetas*; that is, the Candy Goats Marierome of the Venetians. ‡

¶ *The Place.*

These plants grow wilde in Spaine, Italy, and other hot countries. The firſt of theſe I found growing in diuers barren and chalky fields and high-wayes neere vnto Sittingburne and Rocheſter in Kent, and alſo neere vnto Cobham houſe and Southfleet in the ſame county.

‡ I doubt our Author was miſtaken, for I haue not heard of this growing wilde with vs. ‡

¶ *The Time.*

They floure in the moneth of Auguſt. I remember (ſaith *Dodonæus*) that I haue ſeene *Tragoriganum* in the Low-countries, in the gardens of thoſe that apply their whole ſtudy to the knowledge of plants; or as we may ſay, in the gardens of cunning Herbariſts.

¶ *The Names.*

Goats Organie is called in Greeke τραγορίγανος: in Latine likewiſe *Tragoriganum*: in Engliſh, goats Organie, and Goats Marierome.

¶ *The Temperature.*

Goats Organies are hot and dry in the third degree: They are (ſaith *Galen*) of a binding qualitie.

¶ *The Vertues.*

Tragoriganum or Goats Marierome is very good againſt the wamblings of the ſtomacke, and the ſoure belchings of the ſame, and ſtayeth the deſire to vomit, eſpecially at ſea. A

Theſe baſtard kindes of Organie or wilde Marieromes haue the ſame force and faculties that the other Organies haue for the diſeaſes mentioned in the ſame chapter. B

† There were formerly two figures in this chapter; the firſt whereof was of that which is deſcribed in the ſecond place: the ſecond was of *Tragoriganum* of *Matthiolus*, whereof here is no mention made. The figure of the *Tragoriganum alterum* of *Lobel* (which as I haue formerly ſaid, *Bauhine* would haue all one with that of *Dodonæus*) was formerly vnder the name of *Thymum Creticum*, pag. 459 of the former edition.

CHAP. 220. *Of Herbe Maſticke.*

¶ *The Deſcription.*

1 THe Engliſh and French herbariſts at this day do in their vulgar tongues call this herb Maſticke or Maſtich, taking this name *Marum* of *Maro* King of Thrace; though ſome rather ſuppoſe the name corruptly to be deriued from this word *Amaracus*, the one plant being ſo like the other, that many learned haue taken them to be one and the ſelfe ſame plant: others haue taken *Marum* for *Sampſuchus*, which doubtleſſe is a kinde of Marierome. Some (as *Dodonæus*) haue called this our *Marum* by the name of *Clinopodium*; which name rather belongs to another plant than to Maſticke. ‡ This growes ſome foot high, with little longiſh leaues ſet by couples: at the tops of the ſtalkes amongſt white downie heads come little white floures: the whole plant is of a very ſweet and pleaſing ſmell. ‡

2 If any be deſirous to ſearch for the true *Marum*, let them be aſſured that the plant laſt mentioned is the ſame: but if any do doubt thereof, for nouelties ſake here is preſented vnto your view a plant of the ſame kinde (which cannot be reiected) for a ſpeciall kind thereof, which hath a moſt pleaſant ſent or ſmell, and in ſhew reſembleth Marierome and *Origanum*, conſiſting of ſmal twigs a foot and more long; the heads tufted like the common Marierome; but the leaues are leſſe, and like *Myrtus*: the root is of a wooddy ſubſtance, with many ſtrings hanging thereat.

3 There is another kinde hereof ſet forth by *Lobel*, which I haue not as yet ſeen, nor himſelfe hath well deſcribed, which I leaue to a better conſideration. ‡ Though our Authour knew not how to deſcribe this creeping *Marum* of *Lobel*, yet no queſtion, if he had knowne ſo much, he would haue giuen vs the figure thereof as wel in this place, as in the third place of the next chapter

 for

1 *Marum.*
Herbe Masticke.

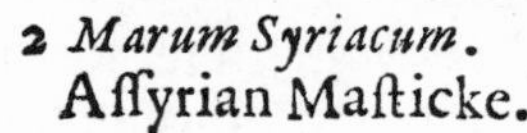
2 *Marum Syriacum.*
Assyrian Masticke.

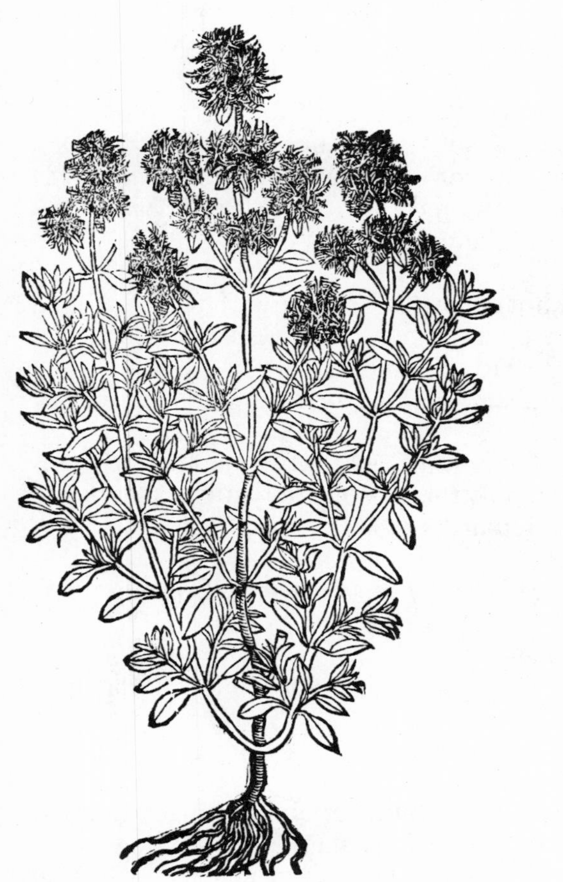

† 3 *Marum supinum Lobelij.*
Creeping Masticke.

for a Penny-Royall; and might as well here as there, and much more fitly haue ventured at a description. But that which is defectiue in him and *Lobel,* I will endeauour to supply out of *Cæsalpinus.* This plant hath many creeping branches like to those of wilde Time, but set with whiter and shorter leaues like to those of the smaller Marjerome, but somewhat narrower: the floures grow in rundles amongst the leaues, as in Calamint, and are of a purple colour: the whole plant is of a strong and sweet smell, and of an hot and bitter taste. *Cæsalpinus* thinkes this to be the *Sampsuchum* of *Dioscorides*: and so also do the Authors of the *Aduersaria. Tabernamontanus* calls it *Marum repens.* ‡

¶ *The Place.*

These plants are set and sowne in the gardens of England, and there maintained with great care and diligence from the iniurie of our cold clymate.

¶ The

¶ *The Time.*

They floure about August, and somewhat later in cold Sommers.

¶ *The Names.*

‡ Masticke is called of the new writers *Marum:* and some, as *Lobel* and *Anguillara* thinke it the *Helenium odorum* of *Theophrastus*. *Dodonæus* iudges it to be the *Clinopodium* of *Dioscorides*. *Clusius* makes it his *Tragoriganum* 1. and saith he receiued the seeds thereof by the name of *Ambra dulcis*. ‡

¶ *The Nature.*

These plants are hot and drie in the third degree.

¶ *The Vertues.*

Dioscorides writeth, that the herbe is drunke, and likewise the decoction thereof, against the bitings of venomous beasts, crampes and conuulsions, burstings and the strangurie. A

The decoction boiled in wine till the third part be consumed, and drunke, stoppeth the laske in them that haue an ague, and vnto others in water. B

† That we here giue you in the third place was formerly vnfitly figured in the third place of the ensuing Chapter by the name of *Pulegium Angustifolium*.

CHAP. 221. *Of Pennie Royall, or pudding grasse.*

† 1 *Pulegium regium.*
Pennie Royall.

† 2 *Pulegium mas.*
Vpright Pennie Royall.

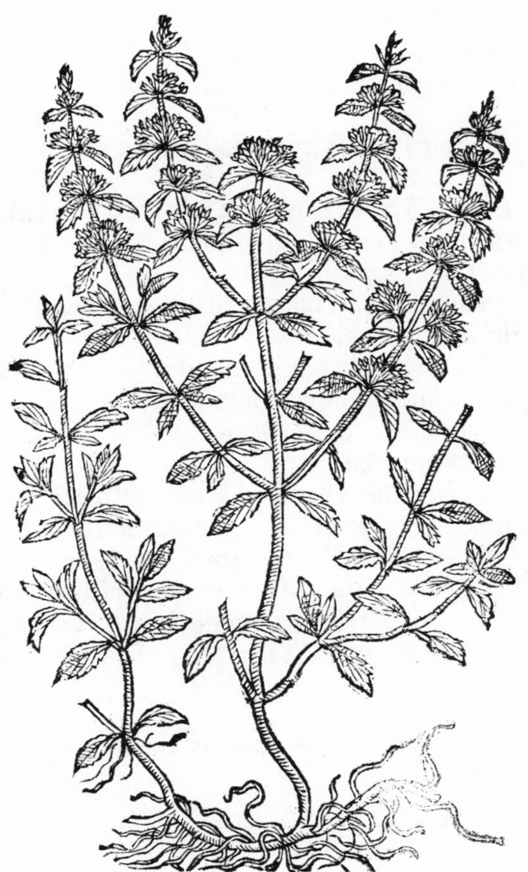

¶ *The Description.*

1 PVlegium regium vulgatum is so exceedingly well knowne to all our English Nation, that it needeth no description, being our common Pennie Royall.

2 The second being the male Pennie Royall is like vnto the former, in leaues, floures and smell, and differeth in that this male kinde groweth vpright of himselfe without creeping, much like in shew vnto wilde Marierome.

3 The

† 3 *Pulegium angustifolium.*
Narrow leafed Pennie Royall.

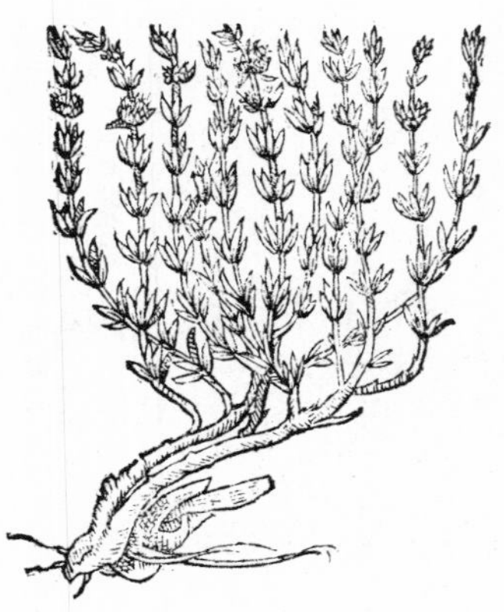

3 The third kinde of Pennie Royall growes like vnto Tyme, and is of a wooddie substance, somewhat like vnto the thinne leafed Hyssope, of the sauour of common Pennie Royall, ‡ but much stronger and more pleasant: the longish narrow leaues stand vpon the stalkes by couples, with little leaues comming forth of their bosomes: and towards the tops of the branches grow rundles of small purple floures. This grows plentifully about Montpellier, and by the Authors of the *Aduersaria*, who first set it forth, it is stiled *Pulegium, angustifol. siue ceruinum Monspeliensium.* ‡

¶ *The Place.*

The first and common Pennie Royall groweth naturally wilde in moist and ouerflown places, as in the Common neere London called Miles end, about the holes & ponds thereof in sundry places, from whence poore women bring plentie to sell in London markets; and it groweth in sundrie other Commons neere London likewise.

The second groweth in my garden: the third I haue not as yet seene.

¶ *The Time.*

They floure from the beginning of Iune to the end of August.

¶ *The Names.*

Pennie Royall is called in Greeke γλήχων, and oftentimes βλήχων: in Latine, *Pulegium*, and *Pulegium regale*, for difference sake betweene it and wilde Tyme, which of some is called *Pulegium montanum*: in Italian, *Pulegio*: in Spanish, *Poleo*: in Dutch, **Poley**: in French, *Pouliot*: in English, Pennie Royall, Pudding grasse, Puliall Royall, and of some Organie.

¶ *The Nature.*

Pennie Royall is hot and drie in the third degree, and of subtill parts, as *Galen* saith.

¶ *The Vertues.*

A Pennie Royall boiled in wine and drunken, prouoketh the monthly termes, bringeth forth the secondine, the dead childe and vnnaturall birth: it prouoketh vrine, and breaketh the stone, especially of the kidneies.

B Pennie Royall taken with honie clenseth the lungs, and cleereth the breast from all grosse and thicke humours.

C The same taken with honie and Aloes, purgeth by stoole melancholie humours; helpeth the crampe and drawing together of sinewes.

D The same taken with water and Vineger asswageth the inordinate desire to vomite, & the pains of the stomacke.

E If you haue when you are at the sea Pennie Royal in great quantitie drie, and cast it into corrupt water, it helpeth it much, neither will it hurt them that drinke thereof.

F A Garland of Pennie royall made and worne about the head is of great force against the swimming in the head, the paines and giddinesse thereof.

G The decoction of Pennie Royall is very good against ventositie, windinesse, or such like, and against the hardnes & stopping of the mother being vsed in a bath or stew for the woman to sit ouer.

† It is apparant by the titles and descriptions that our Authour in this chapter followed *Lobel* but the figures were not agreeable to the historie, for the two first figures were of the *Pulegium Angustifolium* described in the third place; and the third figure was of the *Marum supinum* described in the last place of the foregoing Chapter.

Chap. 222. *Of Basill.*

¶ *The Description.*

1 Garden Basill is of two sorts, differing one from another in bignesse. The first hath broad, thicke, and fat leaues, of a pleasant sweet smell, and of which some one here and there are of a black reddish colour, somewhat snipped about the edges, not vnlike the leaues of French Mercurie. The stalke groweth to the height of halfe a cubite, diuiding it self into diuers branches, whereupon doe stand small and base floures sometimes whitish, and often tending to a darke purple. The root is threddie, and dieth at the approch of Winter.

2 The

1 *Ocimum magnum.*
Great Baſill.

2 *Ocimum medium citratum.*
Citron Baſill.

3 *Ocimum minus Gariophyllatum.*
Buſh Baſill.

‡ 4 *Ocimum Indicum.*
Indian Baſill.

2 The middle Baſill is very like vnto the former, but it is altogether leſſer. The whole plant is of a moſt odoriferous ſmell, not vnlike the ſmell of a Limon, or Citron, whereof it tooke his ſurname.

3 Buſh Baſill, or fine Baſill, is a low and baſe plant, hauing a threddie root, from which riſe vp many ſmall and tender ſtalks, branched into diuers armes or boughes; whereupon are placed many little leaues, leſſer than thoſe of Pennie Royall. The whole plant is of a moſt pleaſing ſweete ſmell.

‡ 4 This which ſome call *Ocimum Indicum*, or rather (as *Camerarius* ſaith) *Hiſpanicum*, ſends vp a ſtalk a foot or more high, foure ſquare, and of a purple colour, ſet at each ioint with two leaues, and out of their boſomes come little branches: the largeſt leaues are ſome two inches broad, and ſome three long; growing vpon long ſtalks, and deepely cut in about their edges, being alſo thicke, fat and iuicie, and either of a darke purple colour, or elſe ſpotted with more or leſſe ſuch coloured ſpots. The tops of the branches end in ſpokie tufts of white floures with purple veines running alongſt them. The ſeede is contained in ſuch ſeed veſſels as that of the other Baſils, and is round, blacke and large. The plant periſhes euery yeare as ſoone as it hath perfected the ſeed. *Cluſius* calls this *Ocimum Indicum*. ‡

¶ *The Place.*

Baſil is ſowne in gardens, and in earthen pots. It commeth vp quickly, and loueth little moiſture except in the middle of the day; otherwiſe if it be ſowne in rainie weather, the ſeed will putrifie, and grow into a iellie or ſlime, and come to nothing.

¶ *The Time.*

Baſill floureth in Iune and Iuly, and that by little and little, whereby it is long a flouring, beginning firſt at the top.

¶ *The Names.*

Baſill is called in Greeke ὄκιμον, and more commonly with ω in the firſt ſyllable ὤκιμον: in Latine, *Ocimum*. It differeth from *Ocymum* which ſome haue called *Cereale* as we (ſaith *Dodonæus*) haue ſhewed in the Hiſtorie of Graine. The later Græcians haue called it βασιλικόν: in ſhops likewiſe *Baſilicum*, and *Regium*: in Spaniſh, *Albahaca*: in French, *Baſilic*: in Engliſh, Baſill, garden Baſill, the greater Baſill Royall, the leſſer Baſill gentle, and buſh Baſill: of ſome, *Baſilicum Gariophyllatum*, or Cloue Baſill.

¶ *The Temperature.*

Baſill, as *Galen* teacheth, is hot in the ſecond degree, but it hath adioined with it a ſuperfluous moiſture, by reaſon whereof he doth not like that it ſhould be taken inwardly; but being applied outwardly, it is good to digeſt or diſtribute, and to concoct.

¶ *The Vertues.*

A *Dioſcorides* ſaith that if Baſil be much eaten, it dulleth the ſight, it mollifieth the belly, breedeth winde, prouoketh vrine, drieth vp milke, and is of a hard digeſtion.

B The iuice mixed with fine meale of parched Barly, oile of roſes and Vineger, is good againſt inflammations, and the ſtinging of venomous beaſts.

C The iuice drunke in wine of *Chios* or ſtrong Sacke, is good againſt head ache.

D The iuice clenſeth away the dimmeneſſe of the eyes, and drieth vp the humour that falleth into them.

E The ſeede drunke is a remedie for melancholicke people, for thoſe that are ſhort winded, and them that can hardly make water.

F If the ſame be ſnift vp in the noſe, it cauſeth often neeſing: alſo the herbe it ſelfe doth the ſame.

G There be that ſhunne Baſill and will not eat thereof, becauſe that if it be chewed and laid in the Sun, it ingendreth wormes.

H They of Africke do alſo affirme, that they who are ſtung of the Scorpion and haue eaten of it, ſhall feele no paine at all.

I The Later writers, among whom *Simeon Zethy* is one, doe teach, that the ſmell of Baſill is good for the heart and for the head. That the ſeede cureth the infirmities of the heart, taketh away ſorrowfulneſſe which commeth of melancholy, and maketh a man merry and glad.

CHAP. 223. *Of wilde Baſill.*

¶ *The Deſcription.*

1 THe wilde Baſil or *Acynos*, called of *Pena*, *Clinopodium vulgare*, hath ſquare hairie ſtems, beſet with little leaues like vnto the ſmall Baſil, but much ſmaller, and more hairie, ſharp pointed, and a little ſnipt towards the end of the leafe, with ſmall floures of a purple colour, faſhioned like vnto the garden Baſill. The root is full of hairie threds, and creepeth along the ground, and ſpringeth vp yearely anew of it ſelfe without ſowing. ‡ This is the *Clinopodium alterum* of *Matthiolus*. ‡

2 This kinde of wilde Baſill called amongſt the Græcians ἄκινος, which by interpretation is *Sine ſemine*, or *Sterilis*, hath cauſed ſundry opinions and great doubts concerning the words of *Plinie* and *Theophraſtus*, affirming that this herbe hath no floures nor ſeeds; which opinions I am ſure of mine owne knowledge to be without reaſon: but to omit controuerſies, this plant beareth purple floures, wharled about ſquare ſtalkes, rough leaues and hairie, verie like in ſhape vnto Baſil: ‡ The ſtalkes are ſome cubite and more high, parted into few branches, and ſet at certaine ſpaces with leaues growing by couples. This is the *Clinopodium vulgare* of *Matthiolus*, and that of *Cordus*, *Geſner*, and others; it is the *Acinos* of *Lobel*. ‡

3 *Serapio* and others haue ſet forth another wilde Baſill vnder the title of *Molochia*; and *Lobel* after the minde of *Iohn Brancion*, calleth it *Corcoros*, which we haue Engliſhed, Fiſh Baſill, the ſeeds whereof the ſaid *Brancion* receiued from Spaine, ſaying that *Corcoros Plinij* hath the leaues of Baſil: the ſtalkes are two handfuls high, the floures yellow, growing cloſe to the ſtalkes, bearing his ſeed in ſmal long cods. The root is compact and made of an innumerable companie of ſtrings, creeping far abroad like running Time. ‡ This figure of *Lobels* which here we giue you is (as *Camerarius* hath obſerued) vnperfect, for it expreſſes not the long cods wherein the ſeed is contained, neither the two little ſtrings or beards that come forth at the ſetting on of each leafe to the ſtalke. ‡

1 *Ocymum ſylueſtre.*
Wilde Baſill.

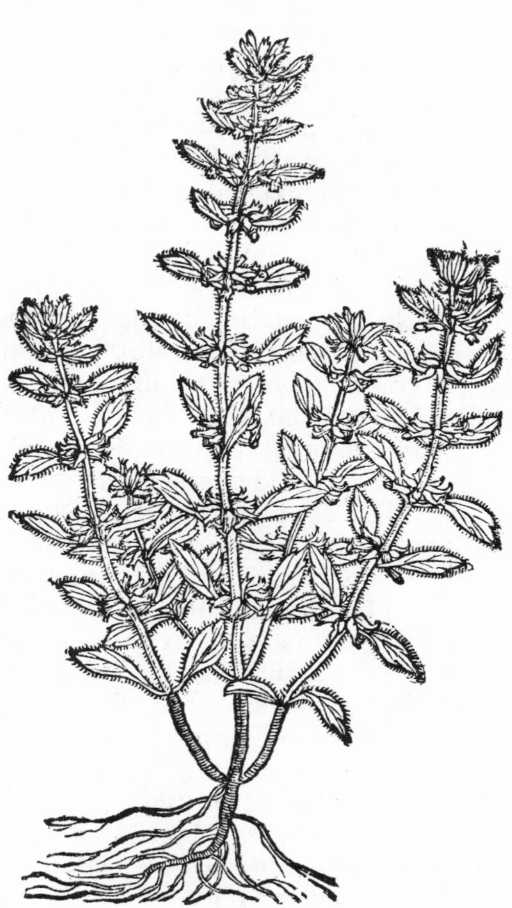

2 *Acynos.*
Stone Baſill.

‡ 3 *Corchoros.*
Fish Basill.

‡ 5 *Clinopodium Austriacum.*
Austrian field Basill.

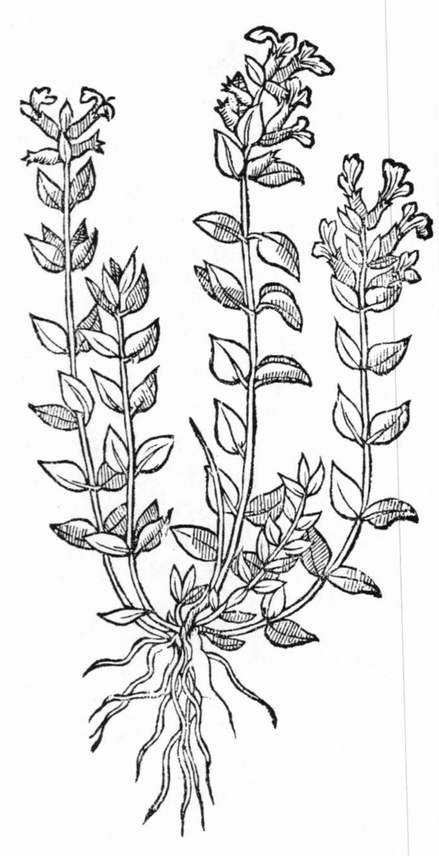

‡ 6 *Clinopodium Alpinum*
Wilde Basill of the Alpes.

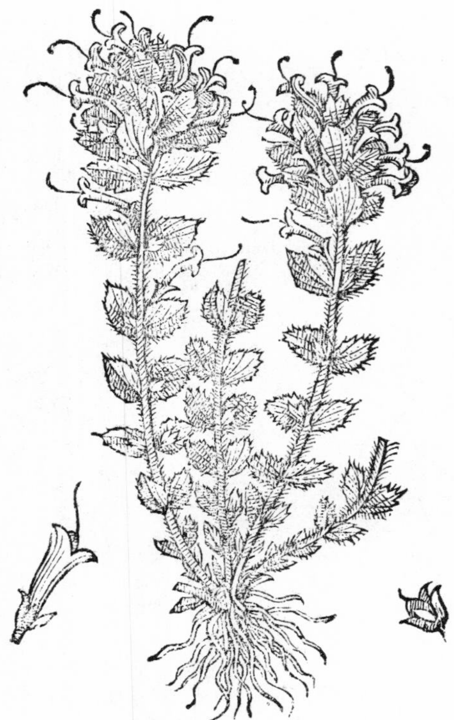

‡ 4 It may be our Authour would haue described this in the first place, as I coniecture by those words which he vsed in mentioning the place of their growing; and [*Clinopodium vulgare* groweth in great plentie vpon Longfield downs in Kent;] but to this neither figure nor description did agree, wherefore I will giue you the Historie therof. It sends vp many little square stalks some handful and an halfe high, seldome diuided into branches: at each ioint stand two smal greenish leaues, little hairy, and not diuided or snipt about the edges, and much like those of the next described, as you see them exprest in the figure: the little hollow and somewhat hooded floures grow in roundles towards the tops of the stalkes, as in the first described, and they are of a blewish violet colour. The seeds I haue not yet obserued: the root is fibrous and wooddie, and lasts for many yeares. The whole plant hath a pretty pleasing but weake smell. It floures in Iuly and August. I first obserued it *Anno* 1626, a little on this side Pomfret in Yorkshire, and since by Datford in Kent, and in the Ile of Tenet. I haue sometimes seene it brought to Cheapside market, where the herbe women called it Poley mountaine, some it may bee that haue taken it for *Polium montanum* misinforming them · *Clusius* first tooke notice of this plant, and called it *Acinos Anglicum*, finding it growing in Kent, *Anno*, 1581. and he thinkes it to be

be the *Acinos* of *Dioſcorides*: now the vertues attributed by *Dioſcorides* to his *Acinos* are ſet downe at the end of the chapter vnder the letter B.

5 This which *Cluſius* hath alſo ſet forth by the name of *Clinopodiū*, or *Acinos Auſtriacum*, doth not much differ from the laſt deſcribed, for it hath tender ſquare hard ſtalkes like thoſe of the laſt deſcribed, ſet alſo with two leaues at each joint, heere and there a little ſnipt (which is omitted in the figure) the floures grow onely at the tops of the ſtalkes, and theſe pretty large, and of a violet colour (yet they are ſometimes found white:) they hang commonly forward, and at is were with there vpper parts turned downe. The ſeed veſſels are like thoſe of the firſt deſcribed, and containe each of them foure little blacke ſeeds: This floures in May, and the ſeed is ripe in Iune: It growes about the bathes of Badon and in diuers places of Auſtria.

6 *Pena* alſo hath giuen vs knowledge of another, that from a fibrous root ſends vp many quadrangular rough branches, of the height of the two former, ſet alſo with two leaues at each joint, and theſe rough and lightly ſnipt about the edges; the floures grow thicke together at the tops of the ſtalkes of a darke red colour, and in ſhape like thoſe of the mountaine Calaminte. It floures in the beginning of Iuly, and growes vpon mount Baldus in Italy; *Pona* ſets it forth by the name of *Clinopodium Alpinum*.

7 To theſe I thinke fit to adde another, whoſe deſcription was ſent me by Mr. *Goodyer*, and I queſtion whether it may not be the plant which *Fabius Columna Phytobaſani, pag. 23.* ſets forth by the name of *Acinos Dioſcoridis*; for he makes his to be endued *odore fragrantiſſimo*: but to the purpoſe.

Acinos odoratiſſimum.

This herbe hath foure, fiue, or more, foure ſquare hard wooddy ſtalkes growing from one root, diuided into many branches, couered with a ſoft white hairineſſe, two or three foot long or longer, not growing vpright, but trailing vpon the ground; the leaues grow on little-ſhort footſtalkes by couples of a light greene colour, ſomewhat like the leaues of Baſill, very like the leaues of *Acinos Lobelij*, but ſmaller, about three quarters of an inch broad, and not fully an inch long, ſomewhat ſharpe pointed, lightly notched about the edges, alſo couered with a light ſoft hoary hairineſſe, of a very ſweete ſmell, little inferiour to Garden Marjerome, of a hot biting taſte: out of their boſomes grow other ſmaller leaues, or elſe branches; the floures alſo grow forth of the boſomes of the leaues toward the tops of the ſtalkes and branches, not in whorles like the ſaid *Acinos*, but hauing one little ſhort footſtalke growing forth of the boſome of each leafe, on which is placed three, foure, or more ſmall floures, gaping open, and diuided into foure vnequall parts at the top, like the floures of Baſill, and very neare of the likeneſſe and bigneſſe of the floures of Garden Marjerome, but of a pale blewiſh colour tending towards a purple. The ſeed I neuer obſerued by reaſon it floured late. This plant I firſt found growing in the Garden of Mr. *William Yalden* in Sheete neere Petersfield in Hampſhire, *Anno* 1620. amongſt ſweete Marjerome, and which by chance they bought with the ſeedes thereof. It is to be conſidered whether the ſeedes of ſweete Marjerome degenerate and ſend forth this herbe or not. 11. October, 1621. *Iohn Goodyer.* ‡

¶ *The Place.*

The wilde kindes doe grow vpon grauelly grounds by water ſides, and eſpecially I found the three laſt in the barren plaine by an houſe in Kent two miles from Dartford, called Saint Iones, in a village called Sutton; and *Clinopodium vulgare* groweth in great plentie vpon Long field downes in Kent. ‡ One of the three laſt of our Authors deſcription is omitted, as you may finde noted at the end of the chapter: yet I cannot be perſuaded that euer he found any of the foure he deſcribed euer wilde in this kingdome, vnleſſe the ſecond, which growes plentifully in Autumne almoſt by euery hedge: alſo the fourth being of my deſcription growes neere Dartford and in many ſuch dry barren places in ſundry parts of the kingdome. ‡

¶ *The Time.*

Theſe herbes floure in Iune and Iuly.

¶ *The Names.*

Vnprofitable Baſil, or wilde Baſill is called by ſome *Clinopodium.*

¶ *The Nature.*

The ſeed of theſe herbes are of complexion hot and dry.

¶ *The Vertues.*

A Wilde Baſill pound with wine appeaſeth the paine of the eyes, and the juice doth mundifie the ſame, and putteth away all obſcurity and dimneſſe, all catarrhes and flowing humors that fall into the eies, being often dropped into the ſame.

‡ The

B † The ſtone Baſill howſoeuer it be taken ſtoppeth the laske,and courſes ; and outwardly applied it helpes hot Tumors and inflammations.

‡ Theſe plants are good for all ſuch effects as require moderate heate and aſtriction. ‡

† The figure that was formerly in the third place of this chapter was of the *Calamentha Ocymoides* of *Tabernmontanus*, and it was deſcribed by our Authour in the fourth place of the next chapter ſaue one, and there you ſhall finde it : the deſcription ſeemes to be of the *Ocymoides repens Polygonifolio* of the *Aduerſaria*, formerly deſcribed by me in the fifth place of the 128. chapter of this booke ; if that the place and floures in the omitted deſcription of our Author did not ſeeme to vary : howeuer I iudge it the ſame and therefore haue heere excluded it.

Chap. 224. *Of Baſill Valerian.*

¶ *The Deſcription.*

1 The firſt kinde of *Ocymaſtrum*, called of *Dodonæus*, *Valeriana rubra*, bringeth forth long and brittle ſtalkes two cubits high, full of knots or joints, in which place is joined long leaues much like vnto great Baſill, but greater, broader, and larger, or rather like the leaues of Woade. At the top of the ſtalkes do grow very pleaſant and long red floures, of the faſhion of the floures of Valerian, which hath cauſed *Dodonæus* to call this plant red Valerian ; which being paſt, the ſeedes are caried away with the winde being,few in number, and little in quantity, ſo that without great diligence the ſeed is not to be gathered or preſerued : for my ſelfe haue often indeuoured to ſee it,and yet haue loſt my labour. The roote is very thicke,and of an excellent ſweete ſauour.

1 *Valeriana rubra Dodonæi.*
Red Valerian.

2 *Behen album.*
Spatling poppy.

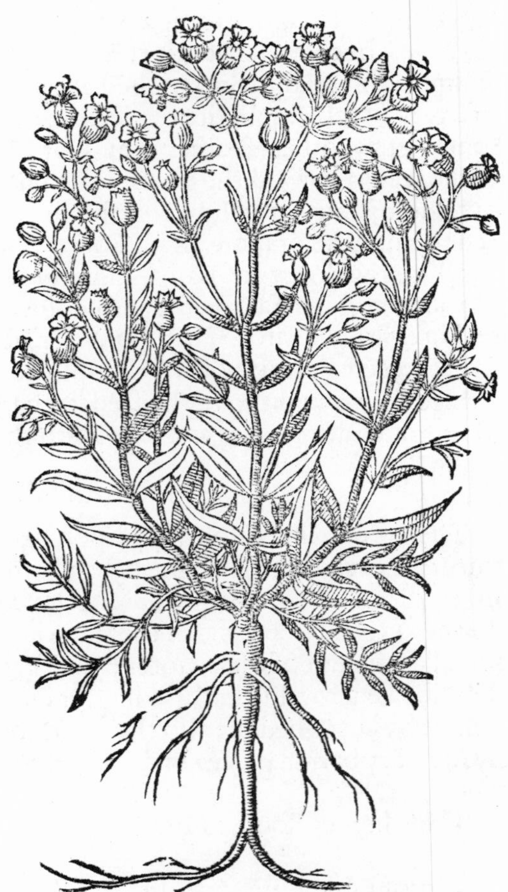

2 The ſecond is taken for *Spumeum papauer*, in reſpect of that kinde of frothy ſpattle, or ſpume, which we call Cuckow ſpittle, that more aboundeth in the boſomes of the leaues of theſe plants, than in any other plant that is knowne : for which cauſe *Pena* calleth it *Papaver ſpumeum*, that is,frothy,or ſpatling Poppy : his floure doth very little reſemble any kinde of Poppy, but onely the ſeede and cod, or bowle wherein the ſeede is contained, otherwiſe it is like the other *Ocymaſtrum* ;

ſtrum. the floures grow at the top of the ſtalkes hanging downewards, of a white colour, and it is taken generally for *Behen album*: the roote is white, plaine, and long, and very tough and hard to breake.

¶ *The Place.*

The firſt groweth plentifully in my garden, being a great ornament to the ſame, and not common in England.

The ſecond groweth almoſt in euery paſture.

¶ *The Time.*

Theſe plants do floure from May to the end of Auguſt.

¶ *The Names.*

Red Valerian hath beene ſo called of the likeneſſe of the floures and ſpoked rundles with Valerian; by which name we had rather haue it called, than raſhly to lay vpon it an vnproper name. There are ſome alſo who would haue it to be a kinde of *Behen* of the later Herbariſts, naming the ſame *Behen rubrum*, for difference between it and the other *Behen album*, that of ſome is called *Ocymaſtrum*, and *Papauer ſpumeum*: which I haue Engliſhed, Spatling Poppie; and is in truth another plant, much differing from *Behen* of the Arabians: it is alſo called *Valerianthon*, *Saponaria altera*, *Struthium Aldroandi*, and *Condurdum*: in Engliſh, red Valerian, and red Cow Baſill.

Spatling Poppie is called *Behen album*, *Ocymaſtrum alterum*; of ſome, *Polemonium*, and *Papauer ſpumeum*: in Engliſh, Spatling Poppie, frothie Poppie, and white Ben.

¶ *The Nature.*

Theſe plants are drie in the ſecond degree.

¶ *The Vertues.*

The root of *Behen Album* drunke in wine, is good againſt the bloudie fluxe: and beeing pound A
leaues and floures, and laid to, cureth the ſtingings of Scorpions and ſuch like venomous beaſts;
inſomuch that who ſo doth hold the ſame in his hand, can receiue no damage or hurt by any venomous beaſt.

The decoction of the root made in water and drunke, prouoketh vrine, it helpeth the ſtrangurie, B
and paines about the backe and Huckle bone.

† That which was formerly here ſet forth in the third place by the name of *Ocymaſtrum multiflorum*, is nothing elſe but the *Lychnis ſylueſtris alba multiplex*, which I haue deſcribed amongſt the reſt of the ſame kinde in the 128. Chapter of this booke.

CHAP. 225. *Of Mints.*

¶ *The Kindes.*

There be diuers ſorts of Mints; ſome of the garden; other wilde, or of the field; and alſo ſome of the water.

¶ *The Deſcription.*

1 The firſt tame or garden Mint commeth vp with ſtalkes foure ſquare, of an obſcure red colour ſomewhat hairie, which are couered with round leaues nicked in the edges like a ſaw, of a deepe greene colour: the floures are little and red, and grow about the ſtalkes circle wiſe, as thoſe of Pennie Royall: the roote creepeth aſlope in the ground, hauing ſome ſtrings on it, and now and then in ſundry places it buddeth out afreſh: the whole herbe is of a pleaſant ſmell, and it rather lieth downe than ſtandeth vp.

2 The ſecond is like to the firſt in hairie ſtalkes ſomething round, in blackiſh leaues, in creeping roots, and alſo in ſmell, but the floures do not at all compaſſe the ſtalke about, but ſtand vp in the tops of the branches being orderly placed in little eares, or rather catkines or aglets.

3 The leaues of Speare-Mint are long like thoſe of the Willow tree, but whiter, ſofter, and more hairie: the floures are orderly placed in the tops of the ſtalks, and in ears like thoſe of the ſecond. The root hereof doth alſo creepe no otherwiſe than doth that of the firſt, vnto which it is like.

4 There is another ſort of Mint which hath long leaues like to the third in ſtalks, yet in leaues and in roots leſſer; but the floures hereof ſtand not in the tops of the branches, but compaſſe the ſtalks about circle-wiſe as do thoſe of the firſt, which be of a light purple colour.

‡ 5 This hath round leaues broader than the common Mint, rounder alſo, and as criſp or curled as thoſe deſcribed in the ſecond place (of which it ſeemes but a larger varietie:) the ſtalkes are

† 1 *Mentha sativa rubra.*
Red Garden Mints.

† 2 *Mentha cruciata, sive crispa.*
Crosse Mint, or curled Mint.

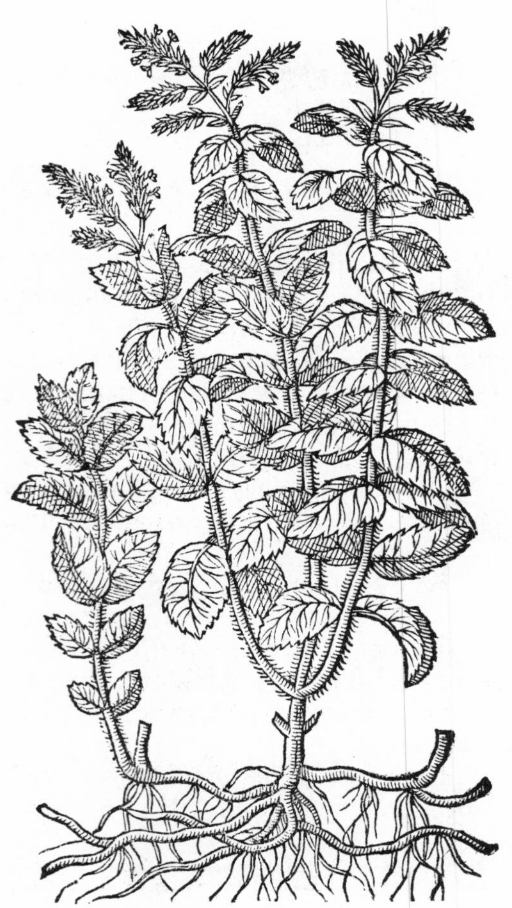

† 3 *Mentha Romana.*
Speare Mint.

‡ 4 *Mentha Cardiaca.*
Heart Mint.

‡ 5 *Mentha ſpicata altera.*
Balſam Mint.

foure ſquare, and the floures grow in eares or ſpokie tufts, like thoſe of the ſecond. ‡

¶ *The Place.*

Moſt vſe to ſet Mints in Gardens almoſt euery where.

¶ *The Time.*

Mints do floure and flouriſh in Sommer, in Winter the roots onely remaine: being once ſet they continue long, and remaine ſure and faſt in the ground.

¶ *The Names.*

Mint is called in Greeke ἡδύοσμος and μίνθη: the ſweet ſmell ſaith *Pliny* in his 19. booke *cap.8.* hath changed the name among the Græcians when as otherwiſe it ſhould be called *Mintha*, from whence our old writers haue deriued the name: for ἡδὺς ſignifieth ſweet, and ὀσμὸς ſmel: The Apothecaries, Italians, and French men, do keepe the Latine name *Mentha*. the Spaniards do call it *Yerua buena*, and *Ortelana*. in High Dutch, **Muntz**: in Low Dutch, **Munte**: in Engliſh, Mint.

The firſt Mint is called in High Dutch, **Diement**: in Low Dutch, **Bruyn heylighe**: he that would tranſlate it into Latin, muſt call it *Sacra nigricans*, or the holy blackiſh mint: in Engliſh, browne Mint, or red Mint.

The ſecond is alſo called in High Dutch **krauſz diement**, **krauſz muntz**, and **krauſz balſam**: that is to ſay, *Mentha cruciata*: in French, *Beaume creſpu*: in Engliſh, Croſſe-Mint, or curled Mint.

The third is called of diuers *Mentha Sarracenica*, *Mentha Romana*: it is called in High Dutch **Balſam muntz**, **Unſer frawen muntz**, **Spitzer muntz**, **Spitzer balſam**: it may be called *Mentha anguſtifolia*: that is to ſay, Mint with the narrow leafe: and in Engliſh, Speare Mint, common garden Mint, our Ladies Mint, browne Mint, and Macrell Mint.

The fourth is called in High Dutch **Hertzkraut**, as though it were to bee named *Cardiaca*, or *Cardiaca Mentha*: in Engliſh, Hart-woort, or Heart-mint ‡ This is the *Siſymbrium ſatiuum* of *Matthiolus*, and *Mentha hortenſis altera* of *Geſner*: the Italians call it *Siſembrio domeſtico*, and *Balſamita*; the Germanes, **Katzenbalſam.** ‡

¶ *The Temperature.*

Mint is hot and drie in the third degree. It is ſaith *Galen*, ſomewhat bitter and harſh, and it is inferiour to Calamint. The ſmell of Mint, ſaith *Pliny* doth ſtir vp the minde, and the taſte to a greedy deſire to meat.

¶ *The Vertues.*

Mint is maruellous wholeſome for the ſtomacke, it ſtaieth the Hicket, parbraking, vomiting & ſcowring in the Cholerike paſſion, if it be taken with the iuice of a ſoure pomegranate. A

It ſtoppeth the caſting vp of bloud, being giuen with water and vineger, as *Galen* teacheth. B

And in broth ſaith *Pliny*, it ſtaieth the floures, and is ſingular good againſt the whites, that is to ſay, that Mint which is deſcribed in the firſt place. For it is found by experience, that many haue had this kinde of flux ſtaied by the continuall vſe of this onely Mint: the ſame being applied to the forehead, or to the temples, as *Pliny* teacheth, doth take away the headache. C

It is good againſt watering eies, and all manner of breakings out in the head, and againſt the infirmities of the fundament, it is a ſure remedie for childrens ſore heads. D

It is poured into the eares with honie water. It is taken inwardly againſt Scolopenders, Beare-wormes, Sea-ſcorpions and ſerpents. E

It is applied with ſalt to the bitings of mad dogs. It will not ſuffer milke to cruddle in the ſtomacke (*Pliny* addeth to wax ſoure.) therefore it is put in Milke that is drunke for feare that thoſe who haue drunke thereof ſhould be ſtrangled. F

It is thought, that by the ſame vertue it is an enemy to generation, by ouerthickning the ſeed. G

H *Dioscorides* teacheth, that being applied to the secret part of a woman before the act, it hindreth conception.

I Garden Mint taken in meat or drinke warmeth and strengtheneth the stomacke, and drieth vp all superfluous humours gathered in the same, and causeth good digestion.

K Mints mingled with the leaues of parched Barly, consumeth tumors and hard swellings.

L The water of Mints is of like operation in diuers medicines, it cureth the trenching and griping paines of the belly and bowels, it appeaseth headach, staieth yexing and vomiting.

M It is singular against the grauell and stone in the kidneies, and against the strangurie, being boiled in wine and drunke.

N They lay it to the stinging of waspes and bees with good successe.

† The figures which were formerly in this Chapter were no way agreeable to the descriptions and names taken forth of *Dodonæus*. The first was of the *Calamintha montana vulgaris* of *Lobel* & *Tab*. The 2. was of that which is described in the third place, the third was of the *Mentha Cattaria angustifolia* described in the third place of the next Chapter. The figure agreeing to the 4. description was in the chapter next saue one afore by the title of *Ocymoides repens*.

Chap. 226. *Of Nep, or Cat Mint.*

¶ *The Description.*

1 CAt Mint or Nep groweth high; it bringeth forth stalks aboue a cubit long, cornered, chamfered, and full of branches: the leaues are broad, nicked in the edges like those of Bawme, or of Horehound, but longer. The floures are of a whitish colour, they partly compasse about the vppermost sprigs, and partly grow on the very top, they are set in a manner like an eare or catkin: the root is diuersly parted, and ful of strings, and endureth a long time. The whole herbe together with the leaues and stalks is soft, and couered with a white downe, but lesser than Horse-mint; it is of a sharpe smel, and pearceth into the head: it hath a hot taste with a certaine bitternesse.

‡ 2 Our Authour figured this and described the next in the second place of this Chapter. This hath pretty large square stalks, set at each ioint with two leaues like those of Costmary, but of a gray or ouerworn colour: the floures grow at the tops of the stalks in long spokie tufts like those of the last described, and of a whitish colour, the smel is pleasanter than that of the last described. ‡

1 *Mentha Felina, seu Cattaria.*
Nep or Cat-mint.

2 *Mentha Cattaria altera.*
Great Cat-mint.

3 There is alſo another kind hereof that hath a longer and narrower leafe, and not of ſo white a colour: the ſtalkes hereof are foure ſquare: the floures be more plentifull, of a red light purple colour inclining to blew, ſprinkled with little fine purple ſpecks; the ſmell hereof is ſtronger, but the taſte is more biting. ‡ The figure of this was formerly in the third place of the laſt chapter. ‡

† 3 *Mentha Cattaria anguſtifolia.*
Small Cat-Mint.

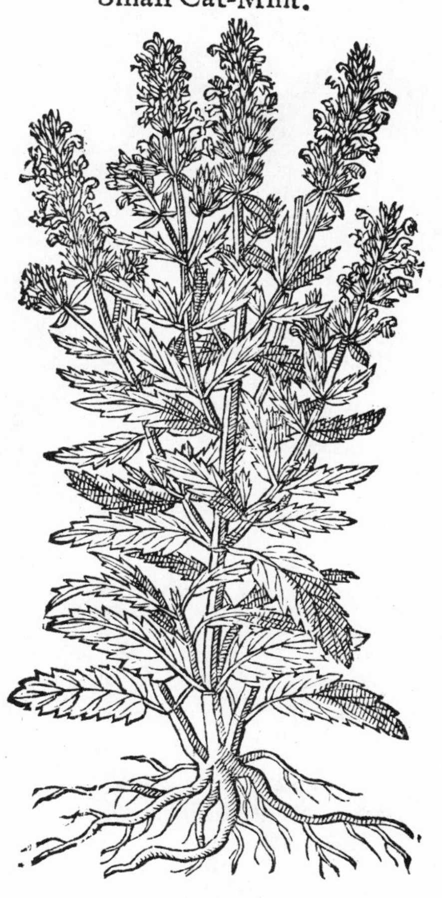

¶ *The Place.*

The firſt growes about the borders of gardens and fields, neere to rough bankes, ditches, and common wayes: it is delighted with moiſt and waterie places: it is brought into gardens.

‡ The other two commonly grow in gardens with vs. ‡

¶ *The Time.*

The Cat-Mints flouriſh by and by after the Spring: they floure in Iuly and Auguſt.

¶ *The Names.*

The later Herbariſts do cal it *Herba Cattaria*, and *Herba Catti*, becauſe the Cats are very much delighted herewith; for the ſmel of it is ſo pleaſant to them, that they rub themſelues vpon it, and wallow or tumble in it, and alſo feed on the branches and leaues very greedily. It is named of the Apothecaries *Nepeta*: but *Nepeta* is properly called (as we haue ſaid) wilde Penny-royall: in high-Dutch, **Katzen Müntz**: in Low-Dutch, **Catte cruijt**: in Italian, *Gattaria*, or *herba Gatta*: in Spaniſh, *Yerua Gatera*: in Engliſh, Cat Mint and Nep. ‡ The true *Nepeta* is *Calamintha Pulegij odore*. ‡

¶ *The Temperature.*

Nep is of temperature hot and dry, and hath the faculties of the Calamints.

¶ *The Vertues.*

It is commended againſt cold paines of the head, ſtomacke, and matrix, and thoſe diſeaſes that grow of flegme and raw humors, and of winde. It is a preſent helpe for them that be burſten inwardly by meanes of ſome fall receiued from an high place, and that are very much bruiſed, if the iuyce be giuen with wine or mede. A

It is vſed in baths and decoctions for women to ſit ouer, to bring downe their ſickneſſe, and to make them fruitfull. B

‡ It is alſo good againſt thoſe diſeaſes for which the ordinarie Mints do ſerue and are vſed. ‡ C

Chap. 227. *Of Horſe-Mint or Water-Mint.*

¶ *The Deſcription.*

1 Water Mint is a kinde of wilde Mint, it is like to the firſt Garden Mint; the leaues thereof are round, the ſtalkes cornered, both the leaues and ſtalkes are of a darke red colour: the roots creepe far abroad, but euery part is greater, and the herbe it ſelfe is of a ſtronger ſmell: the floures in the tops of the branches are gathered together into a round eare, of a purple colour.

† 2 The ſecond kinde of water Mint in each reſpect is like the others, ſauing that the ſame hath a more odoriferous ſauor being lightly touched with the hand: otherwiſe being hardly touched, the ſauour is ouer hot to ſmell vnto: it beareth his floures in ſundry tufts or roundles ingirting the ſtalkes in many places; and they are of a light purple colour: the leaues are alſo leſſe than thoſe of the former, and of an hoary gray colour.

‡ 3 This common Horſe-Mint hath creeping roots like as the other Mints, from which proceed ſtalkes partly leaning, and partly growing vpright: the leaues are pretty large, thicke, wrinkled,

† 1 *Mentha aquatica, siue Sisymbrium.*
Water Mint.

† 2 *Calamintha aquatica.*
Water Calamint.

‡ 3 *Mentastrum.*
Horse-Mint.

‡ 4 *Mentastrum niueum Anglicum.*
Party coloured Horse-Mint.

‡ 5 *Mentaſtrum minus.*
Small Horſe-Mint.

‡ 6 *Mentaſtrum montanum* 1. *Cluſij.*
Mountaine Horſe-Mint.

‡ 7 *Mentaſtrum tuberoſa radice Cluſij.*
Turnep-rooted Horſe-Mint.

wrinkled, hoary and rough both aboue and below, and lightly ſnipped about the edges; the floures grow in thicke compact eares at the tops of the ſtalks, and are like thoſe of common Mint. The whole plant is of a more vnpleaſant ſent than any of the other Mints. It growes in diuers wet and moiſt grounds, and floures in Iune and Iuly. This by moſt writers is called only *Mentaſtrum*, without any other attribute.

4 In ſome of our Engliſh gardens (as *Pena* and *Lobel* obſerued) growes another Horſe-mint, much leſſe, and better ſmelling than the laſt mentioned, hauing the leaues partly greene, and partly milke white; yet ſometimes the leaues are ſome of them wholly white, but more, and more commonly all greene: the ſtalkes, floures, and other parts are like thoſe of the former, but leſſe. This is the *Mentaſtrum niueum Anglicum*, of *Lobel*; and *Mentaſtrum alterum* of *Dodonæus*.

5 This growes in waterie places, hauing a ſtalke of a cubit or cubit and halfe high, ſet with longiſh hoary leaues like thoſe of Horſe-mint: the floures grow in ſpokie tufts at the tops of the ſtalkes, of a duskie purple colour, and in ſhape like thoſe of the common Mint: the ſmell of this comes neere to that of the water Mint. This is the *Mentaſtri folia aquatica hirſuta, ſiue Calamintha* 3. *Dioſcoridis*, of *Lobel*: in the *hiſt Lugd.* it is called *Mentaſtrum minus ſpicatum*.

6 The

6 The ſtalke of this is ſome cubit and halfe high, ſquare, and full of pith: the leaues are like in ſhape to thoſe of Cat-Mint, but not hoarie, but rather greene: the tops of the branches are ſet with roundles of ſuch white floures as thoſe of the Cats-mint: the ſmell of this plant is like to that of the Horſe-Mint; whence *Cluſius* calls it *Mentaſtrum montanum primum*. It floures in Auguſt, and growes in the mountainous places of Auſtria.

7 The ſame Author hath alſo ſet forth another by the name of *Mentaſtrum tuberoſa radice*. It hath roughiſh ſtalkes like the former, and longiſh crumpled leaues ſomewhat ſnipt about the edges like thoſe of the laſt deſcribed: the floures grow in roundles alongſt the tops of rhe branches, and are white of colour, and like thoſe of Cat-Mint. The root of this (which, as alſo the leaues, is not well expreſt in the figure) is like a Radiſh, and blackiſh on the out ſide, ſending forth many ſuccours like to little Turneps, and alſo diuers fibres: theſe ſuccours taken from the maine root will alſo take root and grow. It floures in Iune. *Cluſius* receiued the ſeed of it from Spaine. ‡

¶ *The Place.*

They grow in moiſt and waterie places, as in medowes neere vnto ditches that haue water in them, and by riuers.

¶ *The Time.*

They floure when the other Mints do, and reuiue in the Spring.

¶ *The Names.*

It is called in Greeke Σισύμβριον: in Latine, *Siſymbrium*: in high-Dutch, **Roſzmuntz, Waſſermuntz**: in French, *Menthe ſauuage*: in Engliſh, Water Mint, Fiſh-Mint, Brooke-Mint, and Horſemint.

¶ *The Temperature.*

Water Mint is hot and dry as is the Garden Mint, and is of a ſtronger ſmell and operation.

¶ *The Vertues.*

A It is commended to haue the like vertues that the garden Mint hath; and alſo to be good againſt the ſtinging of Bees and Waſpes, if the place be rubbed therewith.

B The ſauour or ſmell of the Water-Mint reioyceth the heart of man; for which cauſe they vſe to ſtrew it in chambers and places of recreation, pleaſure, and repoſe, and where feaſts and banquets are made.

C There is no vſe hereof in phyſicke whileſt we haue the garden Mint, which is ſweeter, and more agreeing to the nature of man.

† The figure that was in the firſt place was of the Horſe-Mint, and that in the ſecond place ſhould haue beene in the firſt, as now it is.

Chap. 228. Of Mountaine Mint or Calamint.

¶ *The Deſcription.*

1 Mountaine Calamint is a low herbe, ſeldome aboue a foot high, parted into many branches: the ſtalkes are foure ſquare, and haue ioynts as it were, out of euery one whereof grow forth leaues ſomething round, leſſer than thoſe of Baſill, couered with a very thinne hairy downe, as are alſo the ſtalkes, ſomwhat whitiſh, and of a ſweet ſmell: the tops of the branches are gallantly deckt with floures, ſomewhat of a purple colour; then groweth the ſeed which is blacke: the roots are full of ſtrings, and continue.

2 This moſt excellent kinde of Calamint hath vpright ſtalkes a cubit high, couered ouer with a woolly moſſineſſe, beſet with rough leaues like a Nettle, ſomewhat notched about the edges; among the leaues come forth blewiſh or sky-coloured floures: the root is woody, and the whole plant is of a very good ſmell.

3 There is another kinde of Calamint which hath hard ſquare ſtalks, couered in like manner as the other with a certaine hoary or fine cotton. The leaues be in ſhape like Baſill, but that they are rough; and the floures grow in roundles toward the tops of the branches, ſometimes three or foure vpon a ſtemme, of a purpliſh colour. The root is thready, and long laſting.

† 4 There is a kinde of ſtrong ſmelling Calamint that hath alſo ſquare ſtalks couered with ſoft cotton, and almoſt creeping by the ground, hauing euermore two leaues ſtanding one againſt another, ſmall and ſoft, not much vnlike the leaues of Penny-Royall, ſauing that they are larger and whiter: the floures grow about the ſtalks like wharles or garlands, of a blewiſh purple colour; the root is ſmall and threddy: the whole plant hath the ſmell of Penny-Royal; whence it hath the addition of *Pulegij odore*.

¶ The

1 *Calamintha montana vulgaris.*
Calamint, or Mountaine Mint.

† 2 *Calamintha montana præstantior.*
The more excellent Calamint.

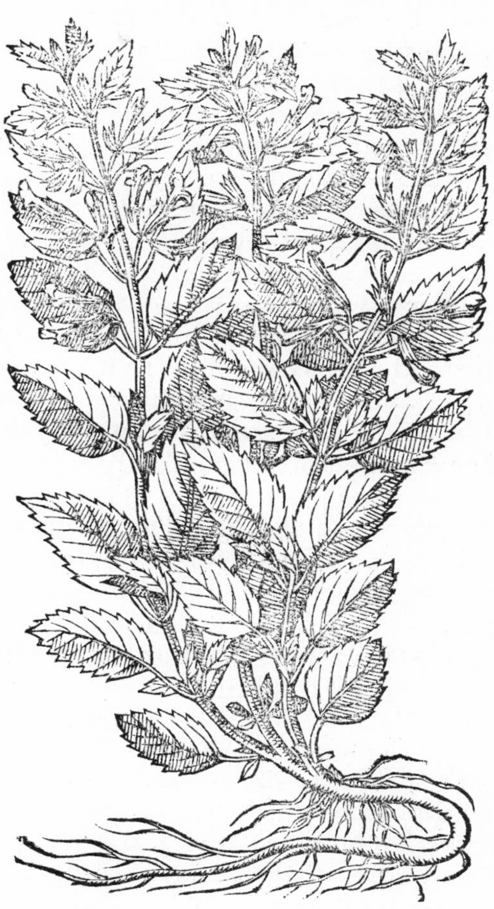

† 3 *Calamintha vulgaris Officinarum.*
Common Calamint.

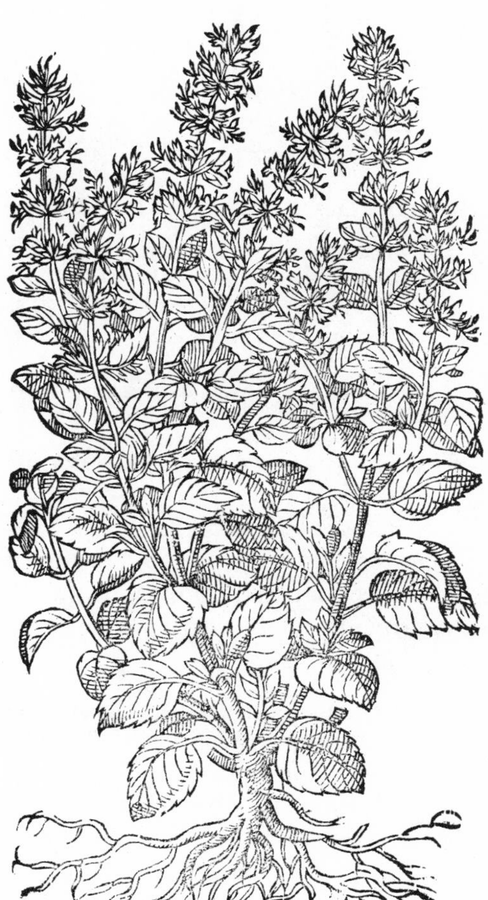

† 3 *Calamintha odore Pulegij.*
Field Calamint.

¶ *The Place.*

It delighteth to grow in mountaines, and in the shadowy and grauelly sides thereof: it is found in many places of Italy and France, and in other countries: it is brought into gardens, where it prospereth maruellous well, and very easily soweth it selfe. I haue found these plants growing vpon the chalkie grounds and highwayes leading from Grauesend vnto Canturbury, in most places, or almost euery where. ‡ I haue onely obserued the third and fourth to grow wilde with vs in England. ‡

¶ *The Time.*

It flourisheth in Sommer, and almost all the yeare thorow: it bringeth forth floures and seed from Iune to Autumne.

¶ *The Names.*

It is called in Greeke Καλαμίνθη, as though you should say, *Elegans aut vtilis Mentha*, a gallant or profitable Mint: the Latines keepe the name *Calamintha*: *Apuleius* also nameth it amisse, *Mentastrum*, and confoundeth the names one with another: the Apothecaries call it *Montana Calamintha, Calamentum*, and sometime *Calamentum montanum*: in French, *Calament*: in English, Mountain Calamint. ‡ The fourth is certainly the second Calamint of *Dioscorides*, and the true *Nepeta* of the Antients. ‡

¶ *The Temperature.*

This Calamint which groweth in mountaines is of a feruent taste, and biting, hot and of a thin substance, and dry after a sort in the third degree, as *Galen* saith: it digesteth or wasteth away thin humors, it cutteth, and maketh thicke humors thin.

¶ *The Vertues.*

A Therefore being inwardly taken by it selfe, and also with meade, or honied water, it doth manifestly heate, prouoketh sweat, and consumeth superfluous humors of the body; it taketh away the shiuerings of Agues that come by fits.

B The same also is performed by the sallet oyle in which it is boyled, if the body be anointed and well rubbed and chafed therewith.

C The decoction thereof drunke prouoketh vrine, bringeth downe the monethly sicknesse, and expelleth the childe, which also it doth being but onely applied.

D It helpeth those that are bruised, such as are troubled with crampes and conuulsions, and that cannot breathe vnlesse they hold their necks vpright (that haue the wheesing of the lungs, saith *Galen*) and it is a remedie saith *Dioscorides* for a cholericke passion, otherwise called the Felony.

E It is good for them that haue the yellow jaundice, for that it remoueth the stoppings of the liuer and gall, and withall clenseth: being taken afore-hand in Wine, it keepeth a man from being poysoned: being inwardly taken, or outwardly applied it cureth them that are bitten of Serpents: being burned or strewed it driues serpents away: it takes away black and blew spots that come by blowes or dry beatings, making the skin faire and white; but for such things (saith *Galen*) it is better to be laid to greene than dry.

F It killeth all manner of wormes of the belly, if it be drunk with salt and honey: the iuyce dropped into the eares doth in like manner kill the wormes thereof.

G *Pliny* saith, that if the iuyce be conueyed vp into the nosthrils it stancheth the bleeding at the nose, and the root (which *Dioscorides* writeth to be good for nothing) helpeth the Squincie, if it be gargarised, or the throat washed therewith, being vsed in Cute, and Myrtle seed withall.

H It is applied to those that haue the Sciatica or ache in the huckle bone, for it drawes the humor from the very bottome, and bringeth a comfortable heat to the whole ioynt: *Paulus Ægineta* saith, that for the paine of the haunches or huckle bones it is to be vsed in Clysters.

I Being much eaten it is good for them that haue the leprosie, so that the patient drinke whay after it, as *Dioscorides* witnesseth.

K *Apuleius* affirmeth, that if the leaues be often eaten, they are a sure and certaine remedy against the leprosie.

L There is made of this an Antidote or composition, which *Galen* in his fourth booke of the Gouernment of health describes by the name of *Diacalaminthos*, that doth not onely notably digest or waste away crudities, but also is maruellous good for young maidens that want their courses, if their bodies be first well purged; for in continuance of time it bringeth them downe very gently without force.

† The figure which formerly was in the second place belonged to the fourth description; and the figure that belonged thereto was before falsly put for the *Scorodonia* or Wood-Sage. As also that which should haue beene put in the fourth place was put in the first place of the last chapter saue two, for the Red Garden Mint.

Chap. 229. *Of Bawme.*

¶ *The Deſcription.*

1 A*Piaſtrum*, or *Meliſſa*, is our common beſt knowne Balme or Bawme, hauing many ſquare ſtalkes and blackiſh leaues like to *Ballote*, or blacke Hore-hound, but larger, of a pleaſant ſmell, drawing neere in ſmell and ſauour vnto a Citron: the floures are of a Carnation colour; the root of a wooddy ſubſtance.

2 The ſecond kinde of Bawme was brought into my garden and others, by his ſeed from the parts of Turky, wherefore we haue called it Turky Balme: it excelleth the reſt of the kinds, if you reſpect the ſweet ſauour and goodly beauty thereof, and deſerueth a more liuely deſcription than my rude pen can deliuer. This rare plant hath ſundry ſmall weake and brittle ſquare ſtalkes and branches, mounting to the height of a cubit and ſomewhat more, beſet with leaues like to Germander or *Scordium*, indented or toothed very bluntly about the edges, but ſomewhat ſharpe pointed at the top. The floures grow in ſmall coronets, of a purpliſh blew colour: the root is ſmall and threddy, and dieth at the firſt approch of Winter, and muſt be ſowne anew in the beginning of May, in good and fertill ground.

1 *Meliſſa.*
Bawme.

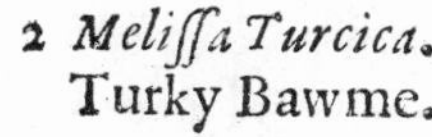

2 *Meliſſa Turcica.*
Turky Bawme.

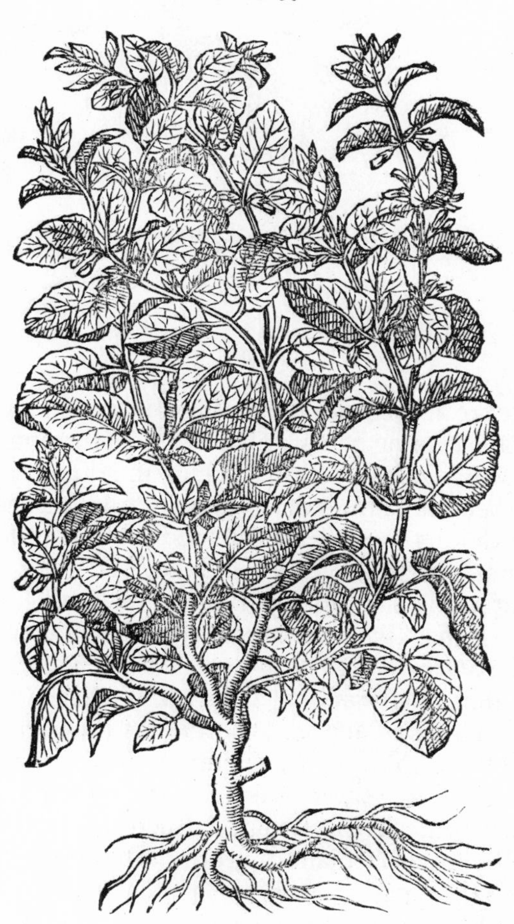

3 *Fuchſius* ſetteth forth a kinde of Bawme hauing a ſquare ſtalke, with leaues like vnto common Bawme, but larger and blacker, and of an euill ſauour; the floures white, and much greater than thoſe of the common Bawme; the root hard, and of a wooddy ſubſtance. ‡ This varies with the leaues ſometimes broader and otherwhiles narrower: alſo the floures are commonly purple, yet ſometimes white, and otherwhiles of diuers colours: the leaues are alſo ſometimes broader, otherwhiles narrower: wherefore I haue giuen you one of the figures of *Cluſius*, and that of *Lobel*, that you may ſee the ſeuerall expreſſions of this plant. *Cluſius*, and after him *Bauhine*, referre it to the *Lamium*, or Arch-angell: and the former calls it *Lamium Pannonicum*: and the later, *Lamium montanum Meliſſæ folio*. ‡

4 There is a kinde of Bawme called *Herba Iudaica*, which *Lobel* calles *Tetrahit*, that hath many

‡ 3 *Melissa Fuchsij flore albo.*
Bastard Bawme with white floures.

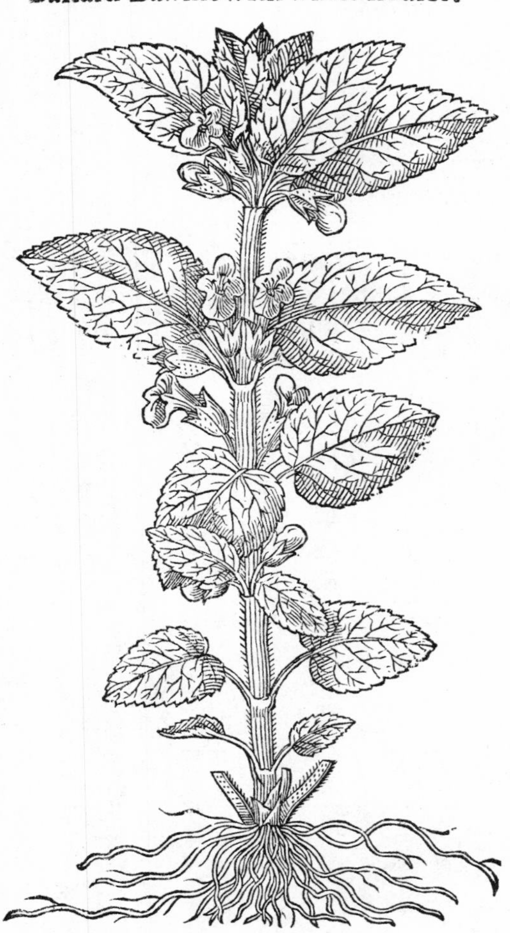

‡ 3 *Melissa Fuchsij flore purpureo.*
Bastard Bawme with purple floures.

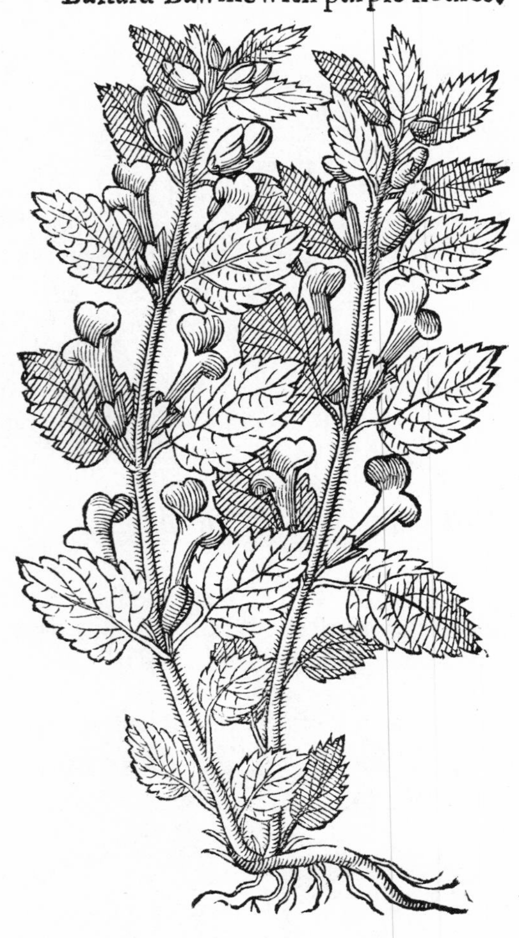

‡ 4 *Herba Iudaica Lobelij.*
Smiths Bawme, or Iewes All-heale.

weake and tender square hairie branches, some leaning backward, and others turning inward, diuiding themselues into sundry other small armes or twigs, which are beset with long rough leaues dented about, and smaller than the leaues of Sage. And growing in another soile or clymat, you shal see the leaues like the oken leaf; in other places like *Marrubium Creticum*, very hoary, which caused *Dioscorides* to describe it with so many shapes, and also the floures, which are sometimes blew and purple, and oftentimes white: the root is small and crooked, with some hairie strings fastned thereto. All the whole plant draweth to the sauour of Balme, called *Melissa*. ‡ This might much more fitly haue beene put to the rest of the *Siderites*, but that our Authour had thrust it as by force into this Chapter. ‡

5 There be also two other plants comprehended vnder the kindes of Balme, the one very like vnto the other, although not knowne to many Herbarists, and haue been of some called by the title of *Cardiaca*: the first kinde *Pena* calleth *Cardiaca Melica*, or *Molucca Syriaca*, so called for that it was first brought out of Syria: it groweth three cubits

cubits high, and yeeldeth many ſhoots from a wooddy root, full of many whitiſh ſtrings; the ſtalkes be round, ſomewhat thicke, and of a reddiſh colour, which are hollow within, with certain obſcure prints or ſmall furrowes along the ſtalkes, with equall ſpaces halfe kneed or knotted, and at euery ſuch knee or ioynt ſtand two leaues one againſt another, tufted like *Meliſſa*, but more rough and deeply indented, yet not ſo deepely as our common *Cardiaca*, called Mother-wort, nor ſo ſharpe pointed: about the knees there come forth ſmall little prickles, with ſix or eight ſmall open wide bells, hauing many corners thinne like parchment, and of the ſame colour, ſomewhat ſtiffe and long; and at the top of the edge of the bell it is cornered and pointed with ſharpe prickles; and out of the middle of this prickly bell riſeth a floure ſomewhat purple tending to whiteneſſe, not vnlike our *Lamium* or *Cardiaca*, which bringeth forth a cornered ſeed, the bottome flat, and ſmaller toward the top like a ſteeple: the ſauour of the plant draweth toward the ſent of *Lamium*.

6 The other kinde of *Melica*, otherwiſe called *Molucca aſperior* (whereof *Pena* writeth) differeth from the laſt before mentioned, in that the cups or bells wherein the floures grow are more prickly than the firſt, and much ſharper, longer, and more in number: the ſtalke of this is foure ſquare, lightly hollowed or furrowed; the ſeed three cornered, ſharpe vpward like a wedge; the tunnels of the floures browniſh, and not ſo white as the firſt.

5 *Meliſſa molucca læuis.*
Smooth Molucca Bawme.

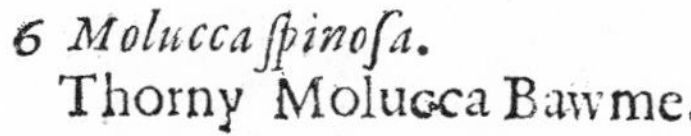

6 *Molucca ſpinoſa.*
Thorny Molucca Bawme.

¶ *The Place.*

Bawme is much ſowen and ſet in gardens, and oftentimes it groweth of it ſelfe in Woods and mountaines, and other wilde places: it is profitably planted in gardens, as *Pliny* writeth, *lib.* 21. *cap.* 12. about places where Bees are kept, becauſe they are delighted with this herbe aboue others, whereupon it hath beene called *Apiaſtrum*: for, ſaith he, when they are ſtrayed away, they do finde their way home againe by it, as *Virgil* writeth in his Georgicks:

Huc tu iuſſos aſperge liquores,
Trita Meliphylla, & Cerinthe nobile gramen.

Vſe here ſuch helpe as husbandry doth vſually preſcribe,
Bawme bruiſed in a mortar, and baſe Hony-wort beſide.
All theſe I haue in my garden from yeare to yeare.

¶ *The Time.*

Bawme floureth in Iune, Iuly, and August: it withereth in the Winter; but the root remaineth, which in the beginning of the Spring bringeth forth fresh leaues and stalkes.

The other sorts do likewise flourish in Iune, Iuly, and August; but they doe perish when they haue perfected their seed.

¶ *The Names.*

Bawme is called in Greeke μελισσόφυλλον: by *Pliny*, *Melitis*: in Latine, *Melissa*, *Apiastrum*, and *Citrago*: of some, *Melissophyllon*, and *Meliphyllon*: in Dutch, **Consille de greyn**: in French, *Poucyrade, ou Melisse*: in Italian, *Cedronella*, and *Arantiata*: in Spanish, *Torongil*: in English, Balme, or Bawme.

¶ *The Temperature.*

Bawme is of temperature hot and dry in the second degree, as *Auicen* saith: *Galen* saith it is like Horehound in facultie.

¶ *The Vertues.*

A Bawme drunke in wine is good against the bitings of venomous beasts, comforts the heart, and driueth away all melancholy and sadnesse.

B Common Bawme is good for women which haue the strangling of the mother, either being eaten or smelled vnto.

C The iuyce thereof glueth together greene wounds, being put into oyle, vnguent, or Balme, for that purpose, and maketh it of greater efficacie.

D The herbe stamped, and infused in *Aqua vitæ*, may be vsed vnto the purposes aforesaid (I meane the liquour and not the herbe) and is a most cordiall liquour against all the diseases before spoken of.

E The hiues of Bees being rubbed with the leaues of Bawme, causeth the Bees to keep togethe and causeth others to come vnto them.

F The later age, together with the Arabians and Mauritanians, affirme Balme to be singular good for the heart, and to be a remedie against the infirmities thereof; for *Auicen* in his booke written of the infirmities of the heart, teacheth that Bawme makes the heart merry and ioyfull, and strengtheneth the vitall spirits.

G *Serapio* affirmeth it to be comfortable for a moist and cold stomacke, to stir vp concoction, to open the stopping of the braine, and to driue away sorrow and care of the minde.

H *Dioscorides* writeth, That the leaues drunke with wine, or applied outwardly, are good against the stingings of venomous beasts, and the bitings of mad dogs: also it helpeth the tooth-ache, the mouth being washed with the decoction, and is likewise good for those that canot take breath vnlesse they hold their necks vpright.

I The leaues being mixed with salt (saith the same Author) helpeth the Kings Euil, or any other hard swellings and kernels, and mitigateth the paine of the Gout.

K Smiths Bawme or Carpenters Bawme is most singular to heale vp greene wounds that are cut with iron; it cureth the rupture in short time; it stayeth the whites. *Dioscorides* and *Pliny* haue attributed like vertues vnto this kinde of Bawme, which they call Iron-wort. The leaues (say they) being applied, close vp wounds without any perill of inflammation. *Pliny* saith that it is of so great vertue, that though it be but tied to his sword that hath giuen the wound, it stancheth the bloud.

CHAP. 230. *Of Horehound.*

¶ *The Description.*

1 WHite Horehound bringeth forth very many stalkes foure square, a cubit high, couered ouer with a thin whitish downinesse: whereupon are placed by couples at certaine distances, thicke whitish leaues somewhat round, wrinkled and nicked on the edges, and couered ouer with the like downinesse; from the bosomes of which leaues come forth small floures of a feint purplish colour, set round about the stalke in round wharles, which turne into sharpe prickly husks after the floures be past. The whole plant is of a strong sauor, but not vnpleasant: the root is threddy.

2 The second kinde of Horehound hath sundry crooked slender stalkes, diuided into many small branches couered ouer with a white hoarinesse or cottony downe. The leaues are likewise hoarie and cottony, longer and narrower than the precedent, lightly indented about the edges, and sharply pointed like the Turky Bawme, and of the same bignesse, hauing small wharles of white floures,

1 *Marrubium album.*
White Horehound.

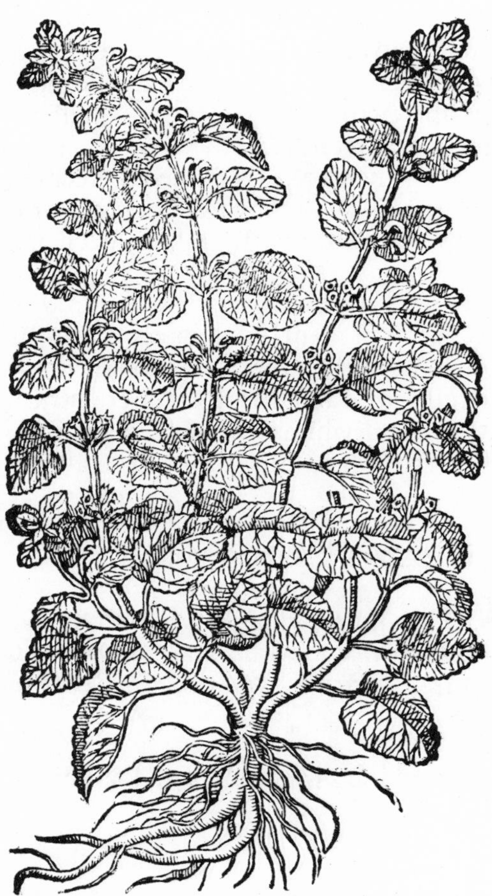

2 *Marrubium candidum.*
Snow white Horehound.

3 *Marrubium Hiſpanicum.*
Spaniſh Horehound.

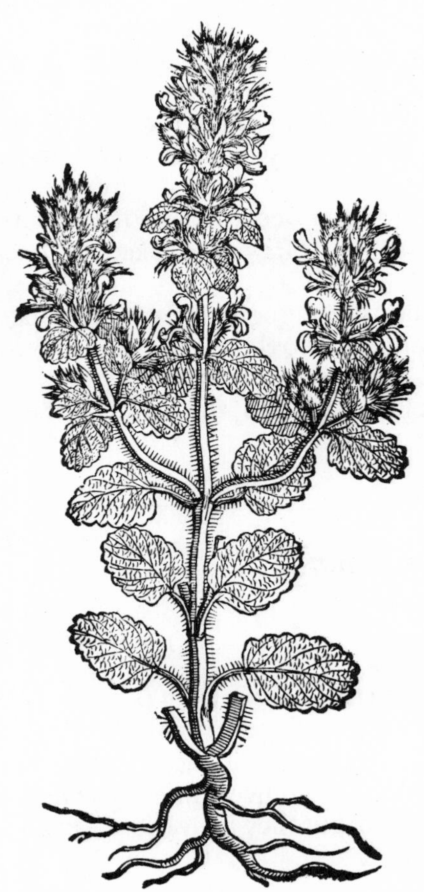

4 *Marrubium Creticum.*
Candy Horehound.

floures, and prickly rundles or seed-vessels set about the stalks by certaine distances. The root is likewise threddy.

3 Spanish Horehound hath a stiffe hoarie and hairy stalke, diuiding it selfe at the bottome into two wings or more armes, and likewise toward the top into two others; whereupon are placed by couples at certaine spaces faire broad leaues, more round than any of the rest, and likewise more woolly and hairy. The floures grow at the top of the stalkes, spike fashion, composed of small gaping floures of a purple colour. The whole plant hath the sauor of Stœchados.

4 Candy Horehound hath a thicke and hard root, with many hairy threds fastned thereunto; from which rise vp immediately rough square stalkes, set confusedly with long leaues of a hoarie colour, of a most pleasant strong smell. The floures grow toward the top of the stalkes in chaffie rundles, of a whitish colour.

¶ *The Place.*

The first of these Horehounds, being the common kinde, groweth plentifully in all places of England, neere vnto old walls, highwayes, and beaten paths, in vntilled places. It groweth in all other countries likewise, where it altereth according to the scituation and nature of the countries; for commonly that which growes in Candy and in Hungary is much whiter, and of a sweeter smel, and the leaues oftentimes narrower and lesser than that which groweth in England and these Northerne Regions.

¶ *The Time.*

They floure in Iuly and August, and that in the second yeare after the sowing of them.

¶ *The Names.*

Horehound is called in Greeke πράσιον: in Latine, *Marrubium*: in shops, *Prassium*, and also *Marrubium*. There be certaine bastard names found in *Apuleius*, as *Melittena, Labeonia*, and *Vlceraria*: in Italian, *Marrubio*: in Spanish, *Marruuio*: in Dutch, **Malrouie**: in French, *Marubin*: in English, Horehound. ‡ *Clusius* calls the third *Ocimastrum Valentinum*. ‡

¶ *The Temperature.*

Horehound (as *Galen* teacheth) is hot in the second degree, and dry in the third, and of a bitter taste.

¶ *The Vertues.*

A Common Horehound boyled in water and drunke, openeth the liuer and spleene, cleanseth the brest and lungs, and preuailes greatly against an old cough, the paine of the side, spitting of bloud, the ptysicke, and vlcerations of the lungs.

B The same boyled in wine and drunke, bringeth downe the termes, expelleth the secondine, after birth, or dead childe, and also easeth those that haue sore and hard labour in childe-bearing.

C Syrrup made of the greene fresh leaues and sugar, is a most singular remedie against the cough and wheesing of the lungs.

D The same syrrup doth wonderfully and aboue credit ease such as haue lien long sicke of any consumption of the lungs, as hath beene often proued by the learned Physitions of our London Colledge.

E It is likewise good for them that haue drunke poyson, or that haue beene bitten of Serpents. The leaues are applied with honey to cleanse foule and filthy vlcers. It stayeth and keepeth back the pearle or web in the eyes.

F The iuyce pressed forth of the leaues, and hardned in the Sun, is very good for the same things, especially if it be mixed with a little wine and honey; and dropped into the eyes, it helps them, and cleereth the sight.

G Being drawne vp into the nosthrils it cleanseth the yellownesse of the eyes, and stayeth the running and watering of them.

Chap. 231. *Of wilde Horehound.*

¶ *The Description.*

1 Wild Horehound is also like to common Horehound: there riseth from the root hereof a great number of stalkes high and ioynted, and out of euery ioynt a couple of leaues opposite, or set one against another, somewhat hard, a little longer than those of common Horehound, and whiter, as also the stalkes are set with soft haires, and of a sweet smell: the floures do compasse the stalke about as those doe of common Horehound, but they are yellow, and the wharles be narrower: the root is wooddy and durable.

1 *Stachys.*
Wilde Hore-hound.

2 *Stachys Fuchſij.*
Wilde ſtinking Horehound.

‡ 3 *Stachys ſpinoſa Cretica.*
Thorny Horehound.

‡ 4 *Stachys Luſitanica.*
Portugall Wilde Horehound.

‡ 5 *Sideritis Scordioides.*
Germander Ironwoort.

‡ 6 *Sideritis Alpina Hyssopifolia.*
Hyssop-leaued Iron-wort.

2 Besides this there is also another described by *Fuchsius*: the stalkes hereof are thicke, foure square, now and then two or three foot long: the leaues be broad, long, hoarie, nicked in the edges, hairie as are also the stalks, and much broader than those of the common Horehound: the floures in the whorles which compasse the stalke about, are of a purple colour; the seede is round and blackish: the root hard & somthing yellow.

‡ 3 This thorny *Stachys* hath leaues before it comes to send forth the stalk, like those of the lesser Sage, but more white & hairie, those that grow vpon the stalkes are much narrower: the stalks are square some foot high: and at the parting of them into branches grow alwaies two leaues one opposit against another: the tops of the branches end in long sharpe thornie prickles: the floures grow about the toppes of the branches like those of Sage, but of somewhat a lighter colour. This grows naturally in Candy, about a Towne called Larda, where *Honorius Bellus* first obserued it, there it is called *Guidarothymo*, or Asses Tyme, though it agree with Tyme in nothing but the place of growth. *Clusius* sets it forth by the name of *Stachys spinosa*.

4 *Lobel* hath giuen vs the figure and first description of this by the name of *Stachys Lusitanica*. It hath creeping and downie stalkes some handfull and halfe high, set with little leaues: amongst which in rundles grow smal floures like those of the other wilde Horehounds; the whole plant is of somewhat a gratefull smell. ‡

5 There is another wilde Horehound of Mountpelier, called *Sideritis Monspelliaca Scordioides, siue Scordij folio*: being that kind of *Sideritis* or wilde Horehound which is like vnto *Scordium*, or water Germander, which groweth to the height of a handfull and a halfe, with many small branches rising vpright, of a wooddie substance, hauing the tops and spokie coronets of Hyssop, but the leaues do resemble *Dioscorides* his *Scordium*, saue that they be somewhat lesser, stiffer, more wrinckled or curled and hairie, than *Tetrahit*, or the Iudaicall herb: the floures do resemble those of the common Sauorie, in taste bitter, and of an aromaticall smell.

6 Mountaine *Sideritis* beeing also of the kindes of Horehound, was first found by *Valerandus Donraz*, in the mountains of Sauoy, resembling very wel the last described, but the leaues are much narrower, and like those of Hyssope: the floures grow in small rough rundlets or tufts, pale of colour like *Marrubium* or *Tetrahit*; the root long and bending, of a wooddie substance, and purple colour, bitter in taste, but not vnpleasant, whose vertue is yet vnknowne.

¶ *The Place.*

These herbes are forreiners, they grow in rough and barren places, notwithstanding I haue them growing in my garden. ‡ My kinde friend Mr. *Buckner* an Apothecary of London the last yeare, beeing being 1632, found the ſecond of theſe growing wilde in Oxfordſhire in the field ioyning to Witney Parke a mile from the Towne. ‡

¶ *The Time.*

They floure in the Sommer moneths, and wither towards winter: the root remaineth aliue a certaine time.

¶ *The Names.*

The former is taken for the right *Stachys*, which is called in Greeke στάχυς: it is knowne in ſhoppes and euery where: we name it in Engliſh yellow Horehound, and wilde Horehoond. ‡ *Lobel* calls it *Stachys Lychnites ſpuria Flandrorum.* ‡

The other wilde Horehound, ſeeing it hath no name, is to be called *Stachys ſpuria*: for it is not the right, neither is it *Sphacelus* (as moſt haue ſuſpected) of which *Theophraſtus* hath made mention: it is called in Engliſh purple Horehound, baſtard wild Horehound, & *Fuchſius* his wild Horehound. ‡ *Fabius Columna* proues the ſecond to be the *Sideritis Heraclia* of *Dioſcorides* and the Antients. ‡

¶ *The Temperature.*

Theſe herbes are of a biting and bitter taſte, and are hot in the third degree according to *Galen*.

‡ The *Stachys Fuchſij* and *Sideritides* ſeem to be hot and drie in the firſt degree. ‡

¶ *The Vertues.*

A The decoction of the leaues drunk doth draw downe the menſes and the ſecondine, as *Dioſcorides* teacheth.

B ‡ 2 This is of ſingular vſe (as moſt of the herbes of this kinde are) to keep wounds from inflammation, and ſpeedily to heale them vp, as alſo to ſtay all fluxes and defluctions, hauing a drying and moderate aſtrictiue facultie.

C *Aetius* and *Aegineta* commend the vſe of it in medicines vſed in the cure of the biting of a mad Dog. ‡

‡ CHAP. 232. *Of the Ironwoorts or Alheales.*

‡ 1 *Sideritis vulgaris.*
Ironwoort, or Alheale.

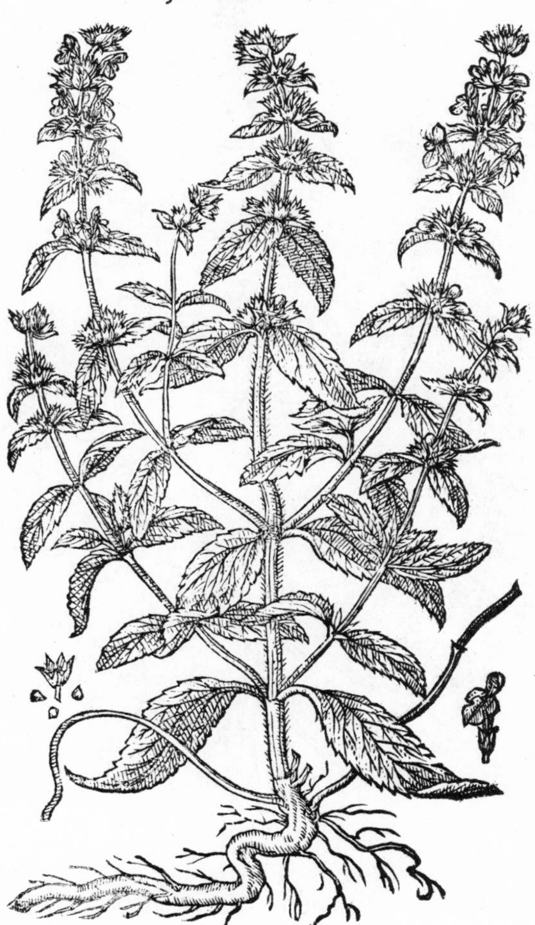

‡ 2 *Sideritis Anguſtifolia.*
Narrow leaued Alheale.

¶ The

¶ *The Kindes.*

‡ There are many plants that belong to this kindred of the *Sideritides*, or Ironwoorts, and some of them are already treated of, though in seuerall places, & that not verie fitly by our Authour; and one of them is also set forth hereafter by the name of Clownes Alheale: these that are formerly handled, and properly belong to this Chapter, are first the *Herba Iudaica Lobelij*, being in the fourth place of the 229. Chapter. Secondly, the *Stachys Fuchsij* (being the first *Sideritis* of *Dioscorides*) described in the second place of the last chapter. Thirdly, the *Sideritis Scordioides* set forth in the fift place; and fourthly the *Sideritis Alpina Hyssopifolia* set forth in the sixt place of the last chapter. Now besides all these, I will in this Chapter giue you the Descriptions of some others like to them in face and Vertues, and all of them may be referred to the first *Sideritis* of *Dioscorides* his description.

¶ *The Description.*

1 This hath square stalkes some cubite high, rough, and iointed with two leaues at each ioint which are wrinkled and hairie, of an indifferent bignesse, snipt about the edges, of a strong smell, and of a bitterish and somewhat hottish taste: almost forth of euery ioint grow branches, set with lesser leaues: the floures which in roundles incompasse the tops of the stalks end in a spike, being somewhat hooded, whitish, well smelling, and marked on the inside with sanguine spots. The seed is rough and blacke, being contained in fiue cornered seed vessels. The root is hard and wooddie, sending forth many stalkes. This is the *Sideritis prima* of *Fuchsius*, *Cordus*, *Clusius*, and others, it hath a very great affinitie with the *Panax Coloni*, or Clownes Al-heale of our Authour, and the difference betweene them certainly is very small.

‡ 3 *Sideritis procumbens ramosa.*
Creeping branched Ironwoort.

‡ 3 *Sideritis procumbens non ramosa.*
Not branched Creeping Ironwoort.

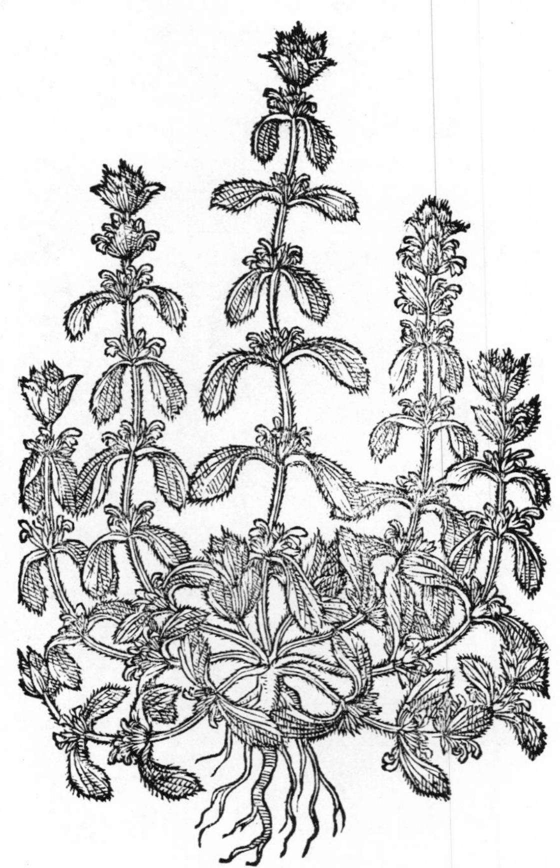

2 The foure square stalke of this plant is not aboue a foot high, and it is presently from the root diuided into diuers branches; the leaues are long and narrow with some nerues or veines running

‡ 6 *Sideritis latifolia glabra.*
Smooth broad leaued Alheale.

ning alongſt them, being alſo very hairie, but not ſnipt about the edges: the floures grow alongſt the branches, and vpon the main ſtalk in roundles like thoſe of the firſt mentioned, but leſſer, and of a darke colour, with a yellowiſh ſpot on their inſides: the ſeed is alſo contained in fiue cornered veſſels like as the former. It floures in Iune and Iuly, and growes amongſt the corne in Hungarie and Auſtria. This is onely ſet forth by *Cluſius*, and that vnder the name of *Sideritis 6. Pannonica.*

3 This hath ſome branches lying along vpon the ground, ſlender, quadrangular & hairie, which at certain ſpaces are ſet with leaues growing by couples, almoſt like thoſe of the firſt, but much leſſe, and ſnipt onely from the middle to the end: the floures grow after the manner of the former, and (as *Cluſius* thinkes) are like them, as is alſo the ſeed. *Cluſius* hath this by the name of *Sideritis* 4.

4 The ſame Authour hath alſo giuen vs another, which from the top of the root ſends foorth many branches, partly lying ſpred on the ground, and partly ſtanding vpright, being hairy, iointed, and ſquare like thoſe of the former, and ſuch alſo are the leaues, but that they are leſſe ſnipt about the edges: and in their boſomes from the bottome of the ſtalkes to the top grow roundles of whitiſh floures ſhaped like others of this kinde. *Cluſius* calls this *Sideritis* 5. He had onely the figures of theſe elegantly drawne by the hand of *Iaques Plateau*, and ſo ſent him.

5 This from a ſmall wooddie root ſends forth a ſquare hairie ſtalke ſome halfe foot high, and ſometimes higher, and this ſtalke moſt commonly ſends forth ſome foure branches, which ſubdiuide themſelues into ſmaller ones, all of them ſometimes lying vpon the ground, and the ſtalke ſtanding vpright; the leaues grow by couples at each ioint, from a broader bottome, ending in an obtuſe point, the lower leaues being ſome inch long, and not much leſſe in breadth: the floures are whitiſh, or light purple, ſmall and hooded, engirting the ſtalkes in roundles, which falling, foure longiſh blacke ſeeds are contained in fiue cornered veſſels. I firſt found it Auguſt 1626 in floure and ſeed amongſt the corne in a field ioining to a wood ſide not far from Greene-hiue in Kent, and I at that time, not finding it to be written of by any, called it *Sideritis humilis lato obtuſo folio*; but ſince I finde that *Bauhine* hath ſet it forth in his *Prodromus* by the name of *Sideritis Alſine Triſſaginis folio.*

6 This (which *Tabernamontanus* calls *Alyſſum Germanicum*, and whoſe figure was formerly giuen with the ſame title by our Authour in the 118 Chapter of the former Edition, with a Deſcription no waies agreeing therewith) grows vp with ſquare ſtalkes ſome cubite high, ſet with pretty large and greene ſmooth leaues ſnipt about the edges: the floures grow in roundles at the tops of the branches, being hooded, and of a pale yellow colour. This grows in the Corne fields in ſome places of Germany and Italy: and it is the *Sideritis* 2. of *Matthiolus* in *Bauhines* opinion, who cals it *Sideritis aruenſis latifolia glabra.*

7 There is another plant that growes frequently in the Corne fields of Kent, and by Purfleet in Eſſex which may fitly be ioined to theſe, for *Camerarius* calls it *Sideritis arvenſis flore rubro*, and in the *Hiſtoria Lugd.* it is named *Tetrahit anguſtifolium*, and thought to be *Ladanum ſegetum* of *Pliny*, mentioned *lib.* 29. *cap.* 8. and *lib.* 26. *cap.* 11. It hath a ſtalke ſome foot or better high, ſet with ſharp pointed longiſh leaues, hauing two or three nickes on their ſides, and growing by couples; at the top of the branches, and alſo the maine ſtalke it ſelfe, ſtand in one or two roundles faire red hooded floures: the root is ſmall and fibrous, dying euery yeare when it hath perfected the ſeed. It floures in Iuly and Auguſt. This is alſo ſometimes found with a white floure.

¶ *The Time, Place, &c.*

All theſe are ſufficiently deliuered in the deſcriptions.

¶ *The*

¶ *The Temperatures and Vertues.*

A These plants are drie with little or no heat, and are endued with an astrictiue faculty. They conduce much to the healing of greene wounds being beaten and applied, or put in vnguents or plaisters made for that purpose.

B They are also good for those things that are mentioned in the last chapter, in B, and C.

C *Clusius* saith, the first and second are vsed in Stiria in fomentations, to bathe the head against the paines or aches thereof, as also against the stiffenesse and wearinesse of the limbs or ioints.

D And the same Author affirmes that he hath knowne the decoction vsed with very good successe in curing the inflammations and vlcerations of the legs. ‡

Chap. 233. *Of Water Horehound.*

‡ 1 *Marubium aquaticum.*
Water Horehound.

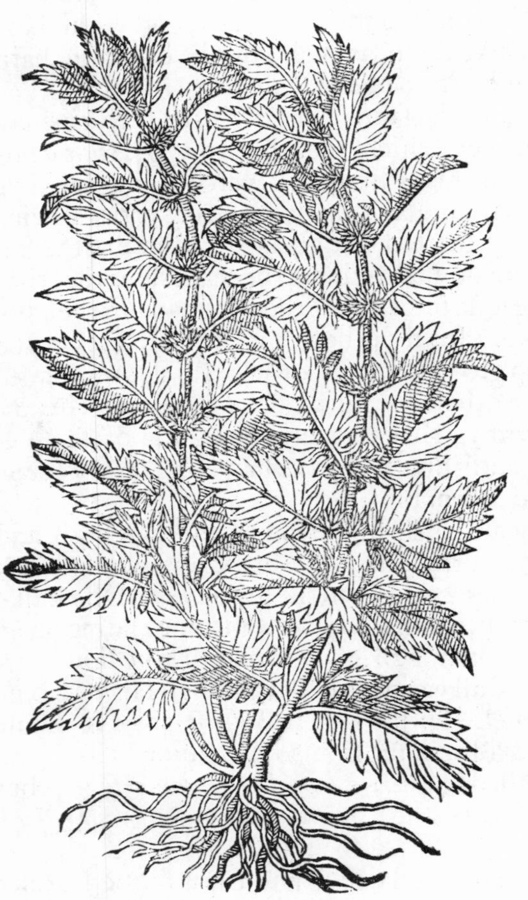

¶ *The Description.*

1 Water Horehound is very like to blacke and stinking Horehound in stalke and floured cups, which are rough, pricking, & compassing the stalks round about like garlands: the leaues thereof be also blacke, but longer, harder, more deeply gashed in the edges than those of stinking Horehound, yet not hairie at all, but wrinkled: the floures be small and whitish: the root is fastened with many blacke strings.

¶ *The Place.*

It growes in Brooks on the brinks of water ditches and neere vnto motes, for it requireth store of water, and groweth not in drie places.

¶ *The Time.*

It flourishes and floures in the Sommer moneths, in Iuly and August.

¶ *The Names.*

It is called *Aquatile*, and *Palustre Marubium*: In English, water Horehound. *Matthiolus* taketh it to be *Species prima Sideritidis*; or a kind of Ironwoort, which *Dioscorides* hath described in the first place; but with this doth better agree that which is called *Herba Iudaica*, or Glidwoort; it much lesse agreeth with *Sideritis secunda*, or the second Ironwoort, which opinion also hath his fauourers, for it is like in leafe to none of the Fernes. Some also thinke good to cal it *Herba Ægyptia*, because they that feine themselues Egyptians (such as many times wander like vagabonds from citie to citie in Germanie and other places) do vse with this herbe to giue themselues a swart colour, such as the Egyptians and the people of Africke are of; for the iuice of this herbe doth die euery thing with this kinde of colour, which also holdeth so fast, as that it cannot be wiped or washed away: insomuch as linnen cloth being died herewith, doth alwaies keepe that colour.

¶ *The Temperature.*

It seemeth to be cold, and withall very astringent or binding.

¶ *The Vertues.*

There is little vse of the water Horehound in Physicke.

† The figure that heretofore was in the first place was of the *Marrubium nigrum* described in the next chapter; and the figure and description that were in the second place by the name of *Marrubium aquaticum acutum*, were of the so much magnified *Panax Coloni* or Clowns Al-heale of our Author, and therefore here omitted to auoid *Tautologie*.

Chap.

Chap. 234. *Of blacke or ſtinking Horehound.*

¶ *The Deſcription.*

1 Blacke Horehound is ſomewhat like vnto the white kinde. The ſtalkes be alſo ſquare and hairie. The leaues ſomewhat larger, of a darke ſwart or blackiſh colour, ſomewhat like the leaues of Nettles, ſnipt about the edges, of an vnpleaſant and ſtinking ſauour. The floures grow about the ſtalks in certain ſpaces, of a purple colour, in ſhape like thoſe of Archangell or dead Nettle. The roote is ſmall and threddie. ‡ I haue found this alſo with white floures.

‡ 2 To this may fitly be referred that plant which ſome haue called *Parietaria, Sideritis*, and *Herba venti*, with the additament of *Monſpelienſium* to each of theſe denominations: but *Bauhine*, who I herein follow, calls it *Marrubium nigrum longifolium*. It is thus deſcribed: the root is thicke and very fibrous, ſending vp many ſquare rough ſtalkes ſome cubite high, ſet at certaine ſpaces with leaues longer and broader than Sage, rough alſo and ſnipt about the edges ÷ and out of their boſomes come floures, hooded, and purple of colour, engirting the ſtalkes as in other plants of this kinde. Some haue thought this to be *Othonna* of the Antients, becauſe the leaues not falling off in Winter, are either eaten by the Wormes, or waſted by the iniurie of the weather to the very nerues or veines that runne ouer them; ſo that by this meanes they are all perforated, and eaſily blowne thorow by each blaſt of winde: which cauſed ſome to giue it alſo the name of *Herba venti*. It grows in the corne fields about Montpelier. ‡

† 1 *Marrubium nigrum.*
Stinking Horehound.

‡ 2 *Marrubium nigrum Longifolium.*
Long leaued Horehound.

¶ *The Place.*

It is found in gardens amongſt pot herbes, and oftentimes amongſt ſtones and rubbiſh in drie ſoiles.

¶ *The Time.*

It floureth and flourisheth when the others do.

¶ *The Names.*

It is called in Greeke βαλλωτὴ, and μέλαν πράσιον, as *Pliny* testifieth in his 27. booke, 8. chapter: of some, *Marrubiastrum*, or *Marrubium spurium*, or bastard Horehound: in shops, *Prasium fœtidum*, and *Ballote*: in Italian, *Marrubiastro*: in Spanish, *Marrauio negro*: in French, *Marubin noir & putant*: in English stinking Horehound.

¶ *The Temperature.*

Stinking Horehound is hot and dry, and as *Paulus Ægineta* teacheth, of a sharpe and clensing faculty.

¶ *The Vertues.*

A Being stamped with salt and applied, it cureth the biting of a mad dogge, against which it is of great efficacy, as *Dioscorides* writeth.

B The leaues rosted in hot embers do waste and consume away hard lumpes or knots in or about the fundament. It also clenseth foule and filthy vlcers, as the same Author teacheth.

† The figure was of *Lamium album*, or Archangell with the white floure; and the figure that should haue beene here was in the former Chapter.

Chap. 235. *Of Archangell, or dead Nettle.*

† 1 *Lamium album.*
White Archangell.

2 *Lamium luteum.*
Yellow Archangell.

¶ *The Description.*

1 WHite Archangell hath foure square stalkes, a cubit high, leaning this way and that way, by reason of the great weight of his ponderous leaues, which are in shape like those of Nettles, nicked round about the edges, yet not stinging at all, but

but ſoft, and as it were downy. The floures compaſſe the ſtalkes round about at certaine diſtances, euen as thoſe of Horehound doe, whereof doubtleſſe this is a kinde, and not of Nettles, as hath been generally holden: which floures are white of colour, faſhioned like to little gaping hoods or helmets. The root is very threddy. ‡ There is alſo a variety of this hauing red or purple floures. ‡

2 Yellow Archangell hath ſquare ſtalkes riſing from a threddy root, ſet with leaues by couples, very much cut or hackt about the edges, and ſharpe pointed. The vppermoſt whereof are oftentimes of a faire purple colour. The floures grow among the ſame leaues, of a gold yellow colour, faſhioned like thoſe of the white Archangell, but greater, and more wider gaping open.

3 Red Archangell being called *Vrtica non mordax*, or dead Nettle, hath many leaues ſpred vpon the ground; among which riſe vp ſtalkes hollow, and ſquare, whereupon do grow rough leaues of an ouer worne colour, among which come forth purple floures, ſet about in round wharles, or rundles. The root is ſmall, and periſheth at the firſt approach of winter.

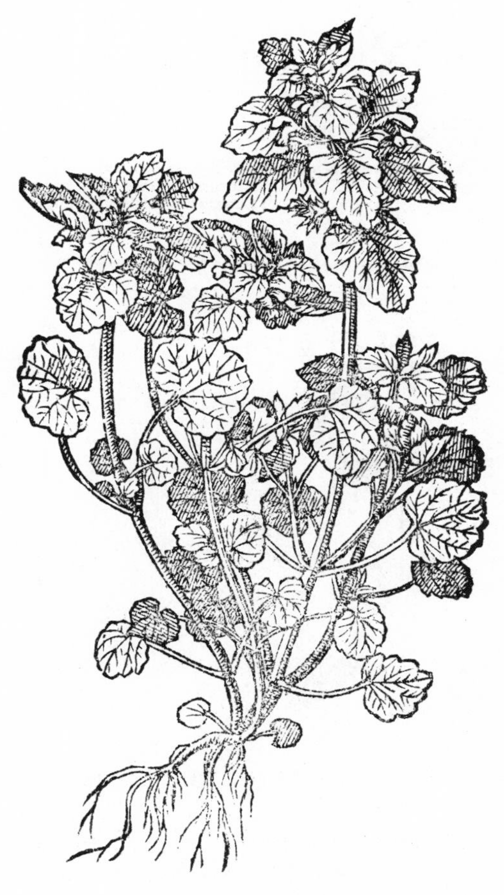

† 3 *Lamium rubrum.*
Red Archangell.

‡ 4 *Lamium Pannonicum, ſiue Galeopſis.*
Hungary dead Nettle.

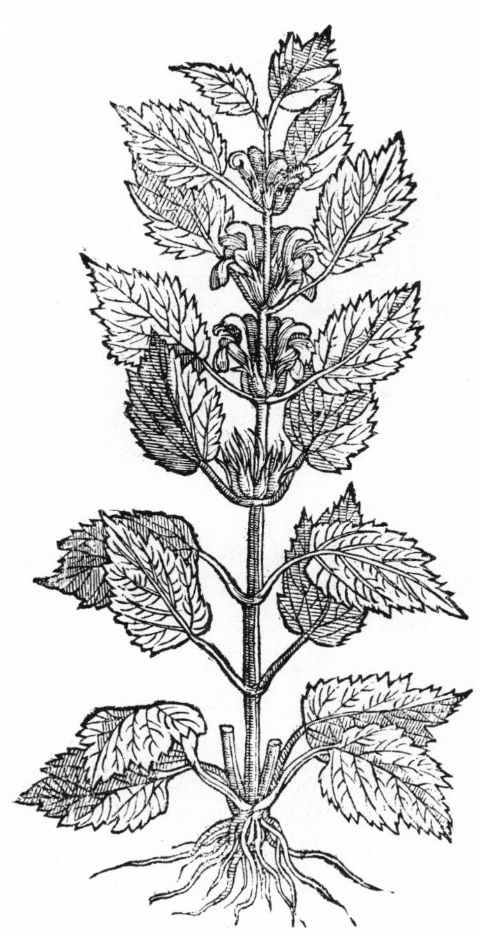

4 Dead Nettle of Hungary hath many large rough leaues very much curled or crumpled like thoſe of the ſtinging nettle, of a darke greene colour, ſnipt about the edges like the teeth of a ſawe, ſet vpon a foure ſquare ſtalke by couples; from the boſome of which leaues come forth the floures cloſe to the ſtalkes, of a perfect purple colour, in ſhape like thoſe of the white Archangell, gaping like a dragons mouth, the lower chap whereof is of a bright purple ſpotted with white, which being paſt, there doth follow ſeed incloſed in rough huskes, with fine ſharpe points ſticking out. The root is thicke, tough, conſiſting of many threds and long ſtrings.

‡ 5 To this of *Cluſius*, we may fitly refer 2. other plants; the firſt of which *Tragus* and others call *Vrtica Heraclea*, or *Herculea*, and *Cluſius* iudges it to be the true *Galeopſis* of *Dioſcorides*, as *Tragus* alſo thought before him. The root hereof is fibrous and creeping, ſending forth many foure ſquare ſtalkes, vpon which at each joint grow two leaues vpon long ſtalkes very like thoſe of Nettles, but more ſoft and hairy, not ſtinging: the tops of the branches end as it were in a ſpike made of ſeuerall roundles of floures like thoſe of Archangell but leſſe, and of a purple colour ſpotted with white on their inſides; the ſeedes are conteined foure in a veſſell, and are blacke when they come to be ripe; It growes about hedges in very many places, and floures in Iune and Iuly.

6 This

‡ 6 This hath roots like thoſe of the laſt deſcribed, ſending vp alſo ſquare ſtalkes a foot high, ſet at each ioint with leaues growing vpon long ſtalkes like thoſe of the ſmall dead Nettle, or rather like thoſe of Alehoofe: out of the boſoms of thoſe come three or foure ſtalks carrying floures like thoſe of Alehoof, gaping, but without a hood, but with a lip turned vp, which is variegated with blew, white, and purple. This hiſtorie *Cluſius* (who did not ſee the plant, but an exact figure thereof in colours) giues vs, and he names it as you finde expreſt in the title. ‡

‡ 5 *Galeopſis vera.*
Hedge Nettle.

6 *Lamium Pannonicum 3. Cluſij.*
Hungary Nettle with the variegated floure.

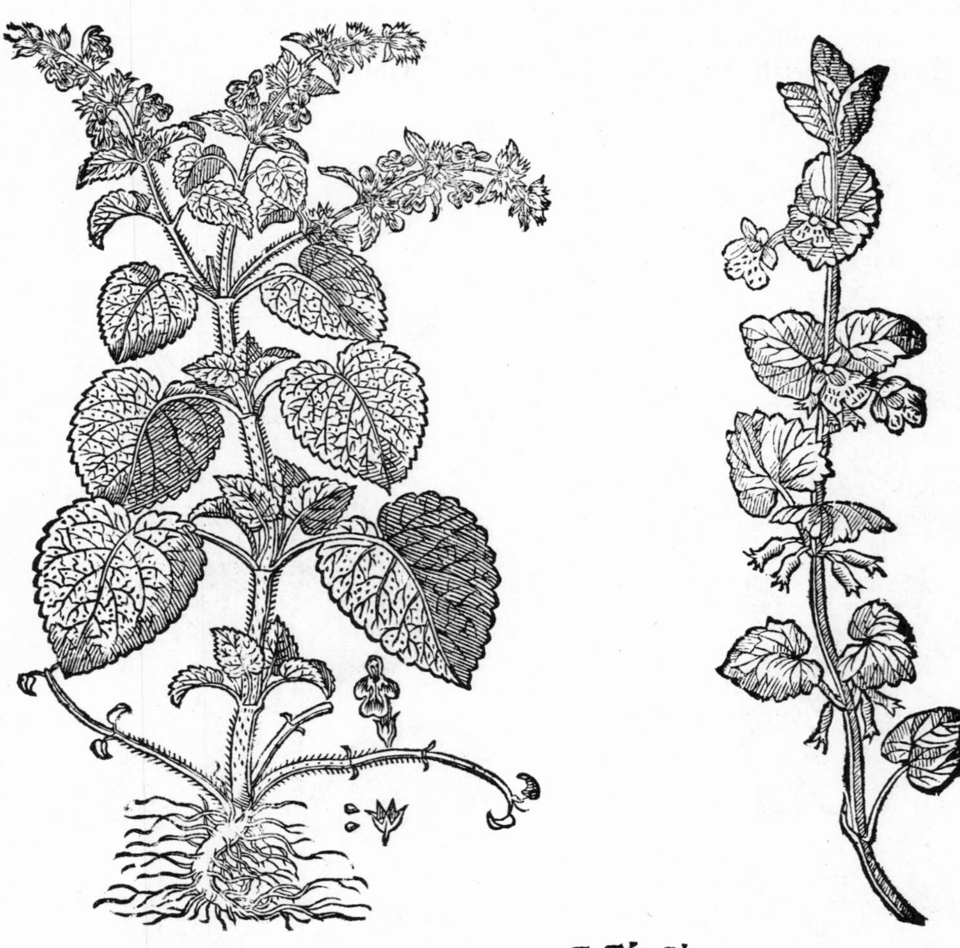

¶ *The Place.*

Theſe plants are found vnder hedges, old wals, common waies, among rubbiſh, in the borders of fieldes, and in earable grounds, oftentimes in gardens ill husbanded.

That with the yellow floure groweth not ſo common as the others. I haue found it vnder the hedge on the left hand as you go from the village of Hampſted neere London to the Church, and in the wood thereby, as alſo in many other copſes about Lee in Eſſex, neere Watford and Buſhie in Middleſex, and in the woods belonging to the Lord Cobham in Kent.

¶ *The Time.*

They floure for the moſt part all Sommer long, but chiefely in the beginning of May.

¶ *The Names.*

Archangell is called of ſome *Vrtica iners*, and *Mortua:* of ſome, *Lamium* : in Engliſh, Archangell, blinde Nettle, and dead Nettle.

¶ *The Temperature.*

They are hotter and drier than Nettle, approching to the temperature of Horehound.

¶ *The Vertues.*

A Archangel [or rather the hedge Nettle] ſtamped with vineger, and applied in manner of a pultis taketh away Wens and hard ſwellings, the Kings euill, inflammation of the kernels vnder the eares and iawes, and alſo hot fierie inflammations of the kernels of the necke, arme-holes and flanks.

B It is good to bathe thoſe parts with the decoction of it, as *Dioſcorides* and *Pliny* ſay.

C The later Phyſitions thinke that the white floures of Archangell doe ſtaie the whites, and for the ſame purpoſe diuers do make of them a Conſerue, as they call it of the floures and ſugar, which they appoint to be taken for certaine daies together.

Th

The floures are baked with ſugar as roſes are, which is called ſugar Roſet: as alſo the diſtilled water of them, which is vſed to make the heart merry; to make a good colour in the face, and to make the vitall ſpirits more freſh and liuely. D

† The firſt figure that was formerly in this Chapter, was of the *Galiopſis* 1. of *Tabern.* being a kinde of dead Nettle that hath the leaues ſpotted with white, & ſomewhat ſmaller than the ordinary one: the figure that ſhould haue been here was in the laſt Chapter; the third was the ſame with the firſt (that ſhould haue bin) differing onely in colour of floures, and that which ſhould haue beene in the third place was in the fourth.

Chap. 236. *Of Motherwoort.*

Cardiaca.
Mother-woort.

¶ *The Deſcription.*

Motherwoort bringeth forth ſtalks foure ſquare, thick, hard, two cubites high, of an obſcure or ouerworn red colour: the leaues are ſomewhat black, like thoſe of Nettles, but greater and broader than the leaues of Horehound, deeply indented or cut on the edges. The husks are hard & pricking, which do compaſſe the ſtalks about like wharles, or little crownets, out of which do grow purpliſh floures, not vnlike to thoſe of dead Nettle, but leſſer: The roote is compact of many ſmall ſtrings, the whole plant is of a very ranke ſmel and bitter taſte.

¶ *The Place.*

It ioieth among rubbiſh, in ſtony and other barren and rough places, eſpecially about Oxford; it profiteth well in gardens.

¶ *The Time.*

It flouriſheth, floureth, and ſeedeth from Iune to September: the leaues and ſtalks periſh in winter, but the root indureth.

¶ *The Names.*

It is called in our age *Cardiaca*: in High Dutch, **Hertzgeſpoꝛt**: in Low Dutch, **Hertegeſpan**: in French, *Agripaulme*: in Engliſh, Motherwoort. Some there be that make it a kinde of Bawme, it ſeemes that it may be alſo referred to *Sideritis Herculana*, or Hercules Ironwoort.

¶ *The Temperature.*

Motherwoort is hot and dry in the ſecond degree, by reaſon of the clenſing and binding quality that it hath.

¶ *The Vertues.*

A Diuers commend it againſt the infirmities of the heart: it is iudged to be ſo forceable, that it is thought it tooke his name *Cardiaca* of the effect.

B It is alſo reported to cure conuulſions and cramps, and likewiſe the palſie: to open the obſtructions or ſtoppings of the intrails: to kill all kindes of Wormes of the bellie.

C The pouder of the herbe giuen in wine, prouoketh not onely vrine, or the monthly courſe, but alſo is good for them that are in hard trauell with childe.

D Moreouer, the ſame is commended for greene wounds: it is alſo a remedie againſt certain diſeaſes in cattell, as the cough and murreine, and for that cauſe diuers husbandmen oftentimes much deſire it.

Chap. 237. *Of stinging Nettle.*

¶ *The Description.*

1 The stalkes of the first be now and then halfe a yard high,round,and hollow wirhin: the leaues are broad,sharp pointed,cut round about like a saw,they be rough on both sides, and couered with a stinging downe,which with a light touch onely causeth a great burning, and raiseth hard knots in the skin like blisters,and sometimes maketh it red. The seed commeth from the roots of the leaues in round pellets bigger than Pease; it is slippery, glittering like Line-seed, but yet lesser and rounder. The roote is set with strings.

1 *Vrtica Romana.*
Romane Nettle.

2 *Vrtica vrens.*
Common stinging Nettle.

2 The second Nettle beeing our common Nettle is like to the former in leaues and stalkes, but yet now and then higher and more full of branches: it is also couered with a downe that stingeth and burneth as well as the other: the seed hereof is small,and groweth not in round bullets, but on long slender strings,as it were in clusters,as those of the female Mercury,which grow along the stalkes and branches aboue the leaues,very many. The root is full of strings; of colour something yellow,and creepeth all about. ‡ This hath the stalkes and rootes sometimes a little reddish,whence *Tabernamontanus* and our Authour gaue another figure thereof by the name of *Vrtica rubra*,Red Nettle. ‡

3 The third is like to the second in stalkes, leaues and seed,that groweth by clusters, but lesser,and commonly more full of branches of a light greene,more burning and stinging; the root is small and not without strings.

¶ *The*

3 *Vrtica minor.*
Small Nettle.

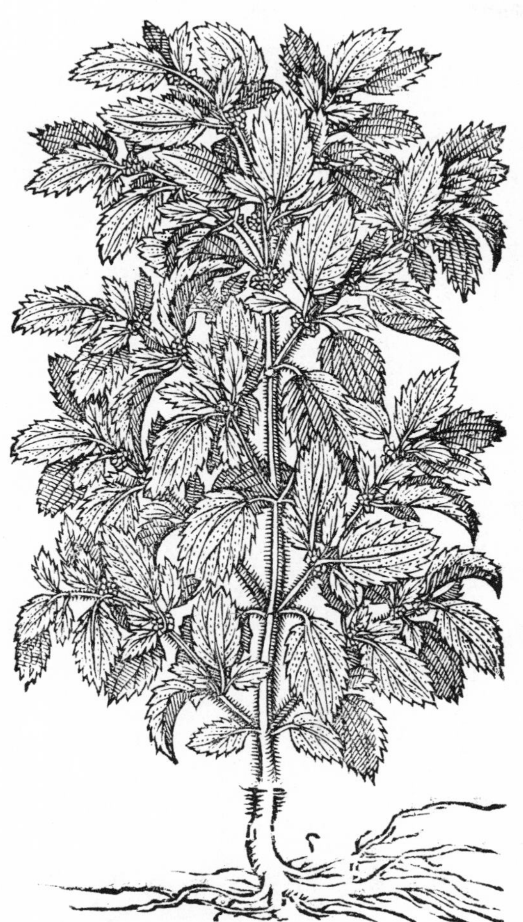

¶ *The Place.*

Nettles grow in vntilled places,and the firſt in thicke woods, and is a ſtranger in England, notwithſtanding it groweth in my garden.

The ſecond is more common, and groweth of it ſelfe neere vnto hedges, buſhes, brambles, and old walls, almoſt euery where.

The third alſo commeth vp in the ſame places, which notwithſtanding groweth in gardens and moiſt earable grounds.

¶ *The Time.*

They all flouriſh in Sommer : the ſecond ſuffereth the winters cold : the ſeed is ripe, and may be gathered in Iuly and Auguſt.

¶ *The Names.*

It is called in Greeke ἀκαλύφη : in Latine, *Vrtica, ab vrendo*, of his burning and ſtinging qualitie : whereupon *Macer* ſaith,

— nec immerito nomen ſumpſiſſe videtur,
Tacta quod exurat digitos vrtica tenentis.

Neither without deſert his name hee ſeemes to git,
As that which quickly burnes the fingers touching it.

And of diuers alſo κνίδη, becauſe it ſtingeth with hurtfull downe : in high-Dutch, **Nessel** : in Italian, *Ortica* : in Spaniſh, *Hortiga* : in French, *Ortie* : in Engliſh, Nettle. The firſt is called in low Dutch **Roomsche Netelen**, that is, *Romana vrtica*, or Roman Nettle : and likewiſe in high-Dutch **Walsche Nesselen**, that is, *Italica vrtica*, Italian Nettle, becauſe it is rare, and groweth but in few places,and the ſeed is ſent from other countries, and ſowne in gardens for his vertues : it is alſo called of diuers *Vrtica mas* : and of *Dioſcorides, Vrtica ſyluestris*, or wilde Nettle, which he ſaith is more rough,with broader and longer leaues, and with the ſeed of Flax, but leſſer. *Pliny* maketh the wilde Nettle the male,and in his 21 booke, chap. 15. ſaith that it is milder and gentler : it is called in Engliſh Romane Nettle, Greeke Nettle, Male Nettle. The ſecond is called *Vrtica fœmina*, and oftentimes *Vrtica maior*, that it may differ from the third Nettle : in Engliſh, Female Nettle, Great Nettle, or common Nettle. The third is named in high-Dutch **Heyter Nessel** : in the Brabanders ſpeech, **Heyte Netelen**, ſo called of the ſtinging qualitie : in Engliſh, Small Nettle, Small burning Nettle : but whether this be that or no which *Pliny* calleth *Cania*, or rather the firſt, let the Students conſider. There is in the wilde Nettle a more ſtinging qualitie, which, ſaith he, is called *Cania*, with a ſtalke more ſtinging, hauing nicked leaues.

¶ *The Temperature.*

Nettle is of temperature dry, a little hot, ſcarſe in the firſt degree : it is of thin and ſubtil parts ; for it doth not therefore burne and ſting by reaſon it is extreme hot, but becauſe the downe of it is ſtiffe and hard, piercing like fine little prickles or ſtings, and entring into the skin : for if it be withered or boyled it ſtingeth not at all, by reaſon that the ſtiffeneſſe of the downe is fallen away.

¶ *The Vertues.*

Being eaten, as *Dioſcorides* ſaith, boyled with Perywinkles, it maketh the body ſoluble,doing it A
by a kinde of cleanſing qualitie : it alſo prouoketh vrine, and expelleth ſtones out of the kidneyes:
being boyled with barley creame it bringeth vp tough humours that ſticke in the cheſt, as it is
thought.

Being ſtamped, and the iuyce put vp into the noſthrils,it ſtoppeth the bleeding of the noſe:the B
iuyce is good againſt the inflammation of the Vuula.

The ſeed of Nettle ſtirreth vp luſt, eſpecially drunke with Cute : for (as *Galen* ſaith) it hath in it C
a certaine windineſſe.

D It concocteth and draweth out of the cheſt raw humors.

E It is good for them that cannot breathe vnleſſe they hold their necks vpright, and for thoſe that haue the pleuriſie, and for ſuch as be ſick of the inflammation of the lungs, if it be taken in a looch or licking medicine, and alſo againſt the troubleſome cough that children haue, called the Chincough.

F *Nicander* affirmeth that it is a remedie againſt the venomous qualitie of Hemlocke, Muſhroms, and Quick-ſiluer.

G And *Apollodoris* ſaith that it is a counterpoyſon for Henbane, Serpents, and Scorpions.

H As *Pliny* witneſſeth, the ſame Author writeth, that the oyle of it takes away the ſtinging which the Nettle it ſelfe maketh.

I The ſame groſſely powned, and drunke in white wine, is a moſt ſingular medicine againſt the ſtone either in the bladder or in the reines, as hath beene often proued, to the great eaſe and comfort of thoſe that haue been grieuouſly tormented with that maladie.

K It expelleth grauell, and cauſeth to make water.

L The leaues of any kinde of Nettle, or the ſeeds, do worke the like effect, but not with that good ſpeed and ſo aſſuredly as the Romane Nettle.

Chap. 238. *Of Hempe.*

1 *Cannabis mas.*
Male or Steele Hempe.

‡ 2 *Cannabis fœmina.*
Femeline, or Female Hempe.

¶ *The Deſcription.*

1 Hempe bringeth forth round ſtalkes, ſtraight, hollow, fiue or ſix foot high, full of branches when it groweth wilde of it ſelfe; but when it is ſowne in fields it hath very few or no branches at all. The leaues thereof be hard, tough, ſomewhat blacke, and if they be bruiſed they be of a ranke ſmell, made vp of diuers little leaues ioyned together, euery particular leafe whereof is narrow, long, ſharpe pointed, and nicked in the edges: the ſeeds come forth from the bottomes of the wings and leaues, being round, ſomewhat hard, full of white ſubſtance. The roots haue many ſtrings.

2 There is another, being the female Hempe, yet barren and without ſeed, contrarie vnto the

nature

nature of that sex; which is very like to the other being the male, and one must be gathered before the other be ripe, else it will wither away, and come to no good purpose.

¶ *The Place.*

Hempe, as *Columella* writeth, delighteth to grow in a fat dunged and waterie soile, or plaine and moist, and deepely digged.

¶ *The Time.*

Hempe is sowne in March and Aprill; the first is ripe in the end of August, the other in Iuly.

¶ *The Names.*

This is named of the Grecians κάνναβις: also of the Latines *Cannabis*: the Apothecaries keep that name: in high-Dutch, **Zamer hanff**: of the Italians *Canape*: of the Spaniards, *Canamo*: in French, *Chanure*: of the Brabanders, **Kemp**: in English, Hempe. The male is called Charle Hempe, and Winter Hempe: the female, Barren Hempe, and Sommer Hempe.

¶ *The Temperature and Vertues.*

A The seed of Hempe, as *Galen* writeth in his bookes of the faculties of simple medicines, is hard of digestion, hurtfull to the stomacke and head, and containeth in it an ill iuyce: notwithstanding some do vse to eate the same parched, *cum aliis tragematis*, with other junkets.

B It consumeth winde, as the said Author saith in his booke of the faculties of medicines, and is so great a drier, as that it drieth vp the seed if too much be eaten of it.

C *Dioscorides* saith, That the iuyce of the herbe dropped into the eares asswageth the paine thereof proceeding (as I take it) of obstruction or stopping, as *Galen* addeth.

D The inner substance or pulpe of the seed pressed out in some kinde of liquor, is giuen to those that haue the yellow jaundice, when the disease first appeares, and oftentimes with good successe, if the disease come of obstruction without an ague; for it openeth the passage of the gall, and dispserseth and concocteth the choler through the whole body.

E *Matthiolus* saith, that the seed giuen to hens causeth them to lay egges more plentifully.

CHAP. 239. *Of wilde Hempe.*

1 *Cannabis Spuria.* Wilde Hempe.

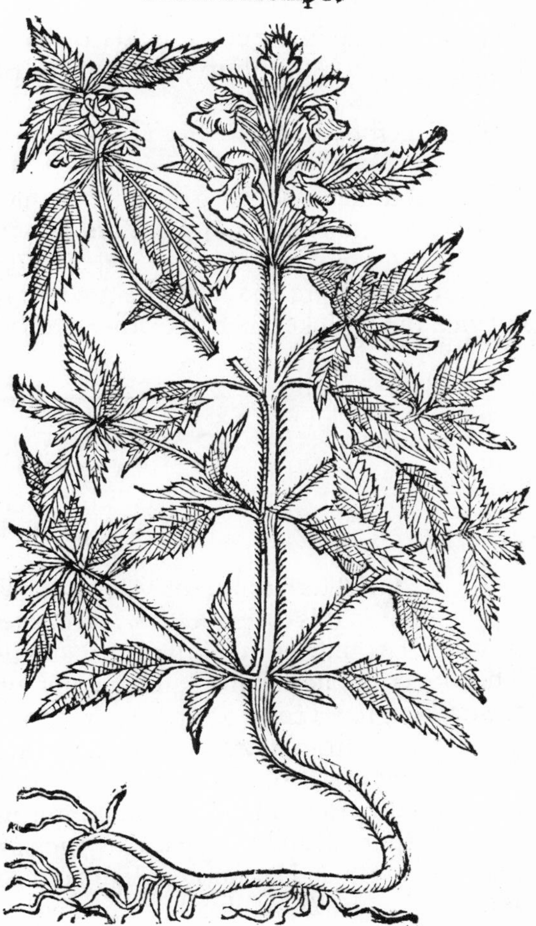

‡ 2 *Cannabis Spuria altera.* Bastard Hempe.

¶ The

‡ 3 *Cannabis Spuria tertia.*
Small Bastard Hempe.

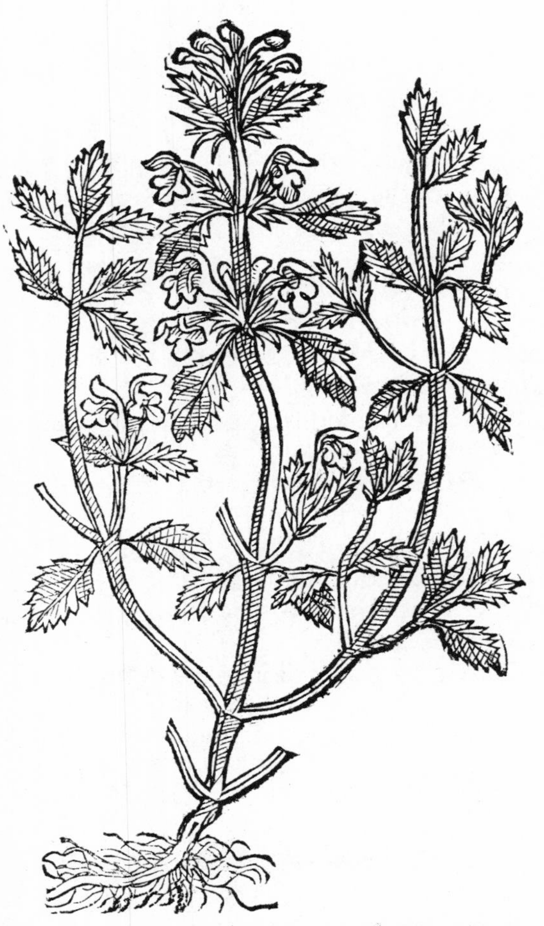

¶ *The Description.*

1 This wilde Hempe, called *Cannabis Spuria*, or Bastard Hempe, hath smal slender hoary and hairie stalkes a foot high, beset at euery ioynt with two leaues, smally indented about the edges somewhat like a Nettle. The floures grow in rundles about the stalkes, of a purple colour, and sometimes also white: the root is little and threddy.

2 There is likewise another kind of wild Hempe which hath hairie stalkes and leaues like the former, but the floures are greater, gaping wide open like the floures of *Lamium*, or dead Nettle, whereof this hath been taken for a kinde: but hee that knoweth any thing may easily discerne the sauor of hempe from the smell of dead Nettle. The floures are of a cleare and light carnation colour, declining to purple.

3 There is also another kinde of wilde Hempe like vnto the last before mentioned, sauing that it is smaller in each respect, and not so hairy. The leafe is somewhat rounder: the root small and threddy: the floure is larger, being purple or white, with a yellow spot in the inside.

¶ *The Place.*

These kinds of wild or bastard Hempe do grow vpon hills and mountaines, and barren hilly grounds, especially in earable land, as I haue often seene in the corne fields of Kent, as about Grauesend, Southfleet, and in all the tract from thence to Canturbury, and in many places about London.

¶ *The Time.*

These herbes do floure from Iuly to the end of August.

¶ *The Names.*

It shall suffice what hath been set downe in the titles for the Latine names: in English, Wilde Hempe, Nettle Hempe, and Bastard Hempe.

¶ *The Temperature and Vertues.*

The temperature and faculties are referred to the manured Hempe, notwithstanding they are not vsed in physicke where the other may be had.

Chap. 240. *Of Water-Hempe.*

¶ *The Description.*

1 Water-Hempe or Water-Agrimony is seldome found in hot regions, for which cause it is called *Eupatorium Cannabinum fœmina Septentrionalium,* and groweth in the cold Northerne countries in moist places, and in the midst of ponds, slow running riuers, and ditches. The root continueth long, hauing many long and slender strings, after the nature of water herbes: the stalkes grow a cubit and a halfe high, of a darke purple colour, with many branches standing by distances one from another. The leaues are more indented and lesse hairy than the male kind: the floures grow at the top, of a browne yellow colour, spotted with blacke spots like *Aster atticus*; which consisteth of such a substance as is in the midst of the Daisie, or the Tansie floure, and is set about with small and sharpe leaues, such as are about the Rose, which causeth the whole floure to resemble a star, and it sauoreth like gum *Elemni, Rosine*, or Cedar wood when it is burned. The seed is long like *Pyrethrum*, closely thrust together, and lightly cleaueth to any woollen garment, that it

toucheth

toucheth by reaſon of his roughneſſe. ‡ This is found with the leaues whole, and alſo with them parted into three parts: the firſt varietie was expreſt by our Authors figure; and the ſecond is expreſt by this we giue you in the place thereof. ‡

2 There is another wilde Hempe growing in the water, whereof there be two ſorts more, delighting to grow in the like ground, in ſhew differing very little. This ſprings vp with long round ſtalkes, and ſomewhat reddiſh, about two cubits high, or ſomething higher: they are beſet with long greene leaues indented about the edges, whereof you ſhal ſee commonly fiue or ſeuen of thoſe leaues hanging vpon one ſtem like the leaues of Hempe, but yet ſofter. The floures are little, of a pale reddiſh colour, conſiſting of ſoft round tufts, and ſtand perting vpon the top of the ſprigges, which at length vaniſh away into downe: the root vnderneath is full of threddy ſtrings of a mean bigneſſe.

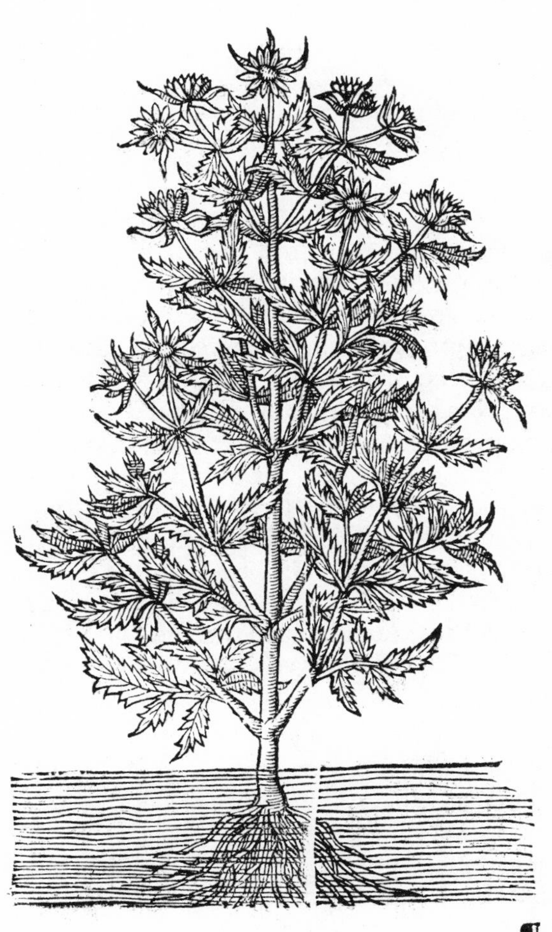

1 *Eupatorium Cannabinum fœmina.*
Water Hempe, or Water Agrimony.

‡ 2 *Eupatorium Cannabinum mas.*
Common Dutch Agrimonie.

¶ *The Place.*

They grow about the brinks of ditches, running waters, and ſtanding pooles, and in watery places almoſt euery where.

¶ *The Time.*

They floure and flouriſh in Iuly and Auguſt: the root continues, but the ſtalkes and leaues wither away in Winter.

¶ *The Names.*

The baſtard or wilde Hempes, eſpecially thoſe of the water, are commonly called *Hepatorium Cannabinum*: of diuers alſo *Eupatorium, Leonhar. Fuchſius* nameth it *Eupatorium Adulterinum*: of moſt, *Cannabina*, of the likeneſſe it hath with the leaues of *Cannabis*, Hempe, and *Eupatorium Auicennæ*. It is thought alſo to be that which *Baptiſta Sardus* doth terme *Terzola*: in high-Dutch, **S. Kunigund kraut**; that is to ſay in Latine, *Sanctæ Cunigundæ herba*, S. Cunigunds herbe: in Low-Dutch, **Boelkens kruit**: in Engliſh, Water Hempe, Baſtard and water Agrimonie. It is called *Hepatorium*, of the facultie, being good for *Hepar*, the liuer. ‡ I haue named the ſecond Common Dutch Agrimonie, becauſe it is commonly vſed for Agrimonie in the ſhops of that countrey. ‡

¶ *The Temperature.*

The leaues and roots of theſe herbes are bitter, alſo hot and dry in the ſecond degree: they haue vertue to ſcoure and open, to attenuate or make thinne thicke and groſſe humours, and to expell or driue them forth by vrine: they clenſe and purifie the bloud.

¶ *The*

¶ *The Vertues, which chiefely belong to the last described.*

A The decoction hereof is profitably giuen to those that be scabbed and haue filthy skinnes ; and likewise to such as haue their spleen and liuer stopped or swolne : for it taketh away the stoppings of both those intrals, and also of the gall : wherefore it is good for them that haue the jaundice, especially somewhat after the beginning.

B The herbe boyled in wine or water is singular good against tertian Feuers.

C The decoction drunke, and the leaues outwardly applied, do heale all wounds both inward and outward.

D ‡ *Fuchsius* saith that the second is very effectuall against poyson. And *Gesner* in his Epistles affirmeth, that he boyled about a *pugil* of the fibres of the root of this plant in wine, and drunke it, which an houre after gaue him one stoole, and afterwards twelue vomits, whereby he cast vp much flegme : so that it workes (saith he) like white Hellebor, but much more easily and safely, and it did me very much good. ‡

Chap. 241. *Of Egrimonie.*

Agrimonia.
Agrimonie.

¶ *The Description.*

The leaues of Agrimonie are long & hairie, greene aboue, and somewhat grayish vnderneath, parted into diuers other small leaues snipt round about the edges, almost like the leaues of hempe : the stalke is two foot and a halfe long, rough & hairy, whereupon grow many small yellow floures one aboue another vpwards toward the top : after the floures come the seeds somewhat long and rough, like to small burs hanging downwards ; which when they be ripe doe catch hold vpon peoples garments that passe by it. The root is great, long, and blacke.

¶ *The Place.*

It growes in barren places by highwayes, inclosures of medowes, and of corne fields, and oftentimes in woods and copses, and almost euery where.

¶ *The Time.*

It floureth in Iune and somwhat later, and seedeth after that a great part of Sommer.

¶ *The Names.*

The Grecians call it εὐπατόριον : and the Latines also *Eupatorium* : *Pliny*, *Eupatoria* : yet there is another *Eupatorium* in *Apuleius*, and that is *Marrubium*, Horehound. In like maner the Apothecaries of Germany haue another *Hepatorium* that is there commonly vsed, being described in the last chapter, and may be named *Hepatorium adulterinum*. Agrimonie is named *Lappa inuersa* : and it is so called, because the seeds which are rough like burres do hang downwards : of some, *Philanthropos*, of the cleauing qualitie of the seeds hanging to mens garments : the Italians and Spaniards call it *Agramonia* : in high Dutch, **Odermeng, Bruckwurtz** : in low-Dutch, in French, and in English, *Agrimonie*, and *Egrimonie* : *Eupatorium* taketh the name of *Eupator*, the finder of it out : and (saith *Pliny*) it hath a royall and princely authoritie.

¶ *The Temperature.*

It is hot, and doth moderately binde, and is of a temperate drinesse. *Galen* saith that Agrimonie is of fine and subtill parts, that it cutteth and scoureth ; therefore, saith he, it remoues obstructions or stoppings out of the liuer, and doth likewise strengthen it by reason of the binding quality that is in it.

¶ The

¶ *The Vertues.*

The decoction of the leaues of Egrimony is good for them that haue naughty liuers, and for ſuch as piſſe bloud vpon the diſeaſes of the kidnies. A

The ſeed being drunke in wine (as *Pliny* affirmeth) doth helpe the bloudy flixe. B

Dioſcorides addeth, that it is a remedy for them that haue bad liuers, and for ſuch as are bitten with ſerpents. C

The leaues being ſtamped with old ſwines greaſe, and applied, cloſeth vp vlcers that be hardly healed, as *Dioſcorides* ſaith. D

‡ Agrimony boiled in wine and drunke, helpes inueterate hepaticke fluxes in old people. ‡ E

Chap. 242. *Of Sawewoort.*

1. 2. *Serratula purpurea, ſiue alba.*
Saw-woort with purple, or white floures.

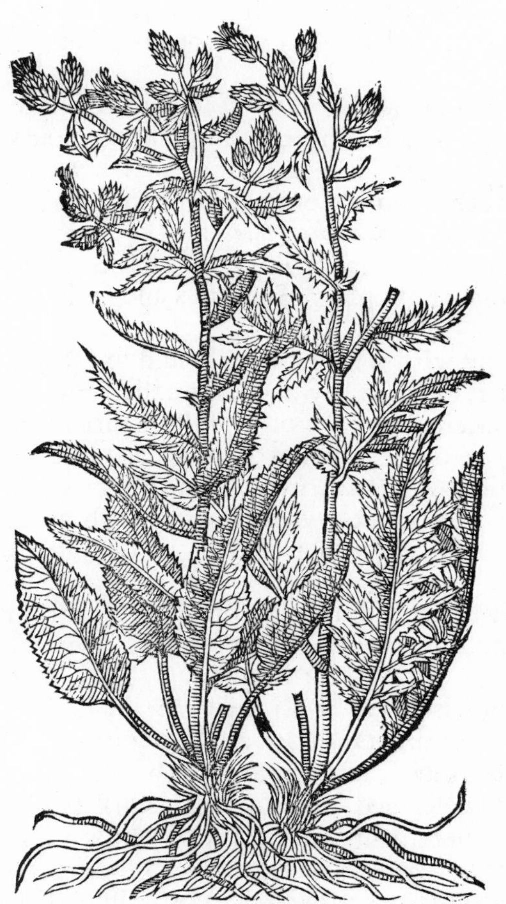

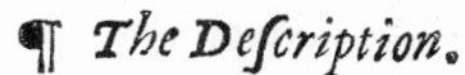

¶ *The Deſcription.*

1 The plant which the new writers haue called *Serratula* differeth from *Betonica*, although the Antients haue ſo called Betony; It hath large leaues ſomewhat ſnipt about the edges like a ſaw (whereof it tooke his name) riſing immediately from the root: among which come vp ſtalkes of a cubite high, beſet with leaues very deepely cut or jagged euen to the middle of the rib, not much vnlike the male Scabious. The ſtalkes towards the top diuide themſelues into other ſmall branches, at the top whereof they beare floures ſomewhat ſcaly, like the Knapweed, but not ſo great nor hard: at the top of the knap commeth forth a buſhie or thrummy floure, of a purple colour. The root is thready, and thereby increaſeth and becommeth of a great quantity.

2 Sawewoort with white floures differeth not from the precedent, but in the colour of the floures: for as the other bringeth forth a buſh of purple floures; in a manner this plant bringeth forth floures of the ſame faſhion, but of a ſnow white colour, wherein conſiſteth the difference.

‡ Our Authour out of *Tabernamontanus* gaue three figures, with as many deſcriptions of this plant, yet made it onely to vary in the colour of the floures, being either purple, white, or red; but he did not touch the difference which *Tabernamontanus* by his figures expreſt, which was, the firſt had all the leaues whole, being only ſnipt about the edges; the lower leaues of the ſecond were moſt of them whole, and thoſe vpon the ſtalkes deepely cut in, or diuided, and the third had the leaues both below and aboue all cut in or deepely diuided. The figure which we here giue you expreſſes the firſt and third varieties, and if you pleaſe, the one may be with white, and the other with red or purple floures. ‡

¶ *The Place.*

Sawe-woort groweth in woods and ſhadowie places, and ſometimes in medowes. They grow in Hampſted wood: likewiſe I haue ſeene it growing in great abundance in the wood adjoining to Iſlington, within halfe a mile from the further end of the towne, and in ſundry places of Eſſex and Suffolke.

¶ *The Time.*

They floure in Iuly and Auguſt.

¶ *The*

¶ *The Names.*

The later age doe call them *Serratula*, and *Serratula tinctoria*, it differeth as we haue said from Betony, which is also called *Serratula*: other names if it haue any we know not: it is called in English Sawewoort. ‡ *Cæsalpinus* calls it *Cerretta* and *Serretta*, and *Thalius*, *Centauroides*, or *Centaurium maius sylvestre Germanicum*. ‡

¶ *The Temperature and Vertues.*

A *Serratula* is wonderfully commended to be most singular for wounds, ruptures, burstings, and such like: and is referred vnto the temperature of Sanicle.

Chap. 243. *Of Betony.*

¶ *The Description.*

1 Betony groweth vp with long leaues and broad, of a darke greene colour, slightly indented about the edges like a saw. The stalke is slender, foure square, somewhat rough, a foote high more or lesse. It beareth eared floures, of a purplish colour, and sometimes reddish; after the floures, commeth in place long cornered seed. The root consisteth of many strings.

2 Betony with white floures is like the precedent in each respect, sauing that the flours of this plant are white, and of greater beautie, and the others purple or red, as aforesaid.

1 *Betonica.*
Betony.

¶ *The Place.*

Betony loues shadowie woods, hedge-rowes, and copses, the borders of pastures, and such like places.

Betony with white floures is seldome seene. I found it in a wood by a village called Hampstead, neere vnto a worshipfull Gentlemans house, one of the Clerkes of the Queenes counsell called Mr. *Wade*, from whence I brought plants for my garden, where they flourish as in their naturall place of growing.

¶ *The Time.*

They floure and flourish for the most part in Iune and Iuly.

¶ *The Names.*

Betony is called in Greeke κέστρον: in Latine, *Betonica*: of diuers *Vetonica*: but vnproperly. There is likewise another *Betonica*, which *Paulus Ægineta* described; and *Galen* in his first booke of the gouernment of health sheweth that it is called κέστρον, that is to say, *Betonica*, Betonie, and also *Sarxiphagon*: *Dioscorides* notwithstanding doth describe another *Sarxiphagon*.

¶ *The Temperature.*

Betony is hot and dry in the second degree: it hath force to cut, as *Galen* saith.

¶ *The Vertues.*

A Betony is good for them that be subject to the falling sickenesse, and for those also that haue ill heads vpon a cold cause.

B It clenseth the lungs and chest, it taketh away obstructions or stoppings of the liuer, milt, and gall: it is good against the yellow jaundise.

It maketh a man to haue a good stomack and appetite to his meate: it preuaileth against sower belchings:

belchings : it maketh a man to piſſe well : it mitigateth paine in the kidnies and bladder : it breaketh ſtones in the kidnies, and driueth them forth.

It is alſo good for ruptures,cramps,and conuulſions : it is a remedie againſt the bitings of mad D
dogs and venomous ſerpents, being drunke, and alſo applied to the hurts, and is moſt ſingular againſt poyſon.

It is commended againſt the paine of the Sciatica, or ache of the huckle bone. E

There is a Conſerue made of the floures and ſugar good for many things,and eſpecially for the F
head-ache. A dram weight of the root of Betonie dried, and taken with meade or honied water, procureth vomit, and bringeth forth groſſe and tough humors, as diuers of our age do report.

The pouder of the dried leaues drunke in wine is good for them that ſpit or piſſe bloud,and cureth all inward wounds, eſpecially the greene leaues boyled in wine and giuen. G

The pouder taken with meate looſeth the belly very gently,and helpeth them that haue the falling ſickneſſe with madneſſe and head-ache. H

It is ſingular againſt all paines of the head : it killeth wormes in the belly ; helpeth the Ague : I
it cleanſeth the mother, and hath great vertue to heale the body, being hurt within by bruiſing or ſuch like.

CHAP. 244. *Of Water-Betony.*

¶ *The Deſcription.*

WAter Betony hath great ſquare hollow and brown ſtalks, whereon are ſet very broad leaues notched about the edges like vnto thoſe of Nettles, of a ſwart greene colour, growing for the moſt part by two and two as it were from one ioynt, oppoſite, or ſtanding one right againſt an other. The floures grow at the top of the branches,of a darke purple colour, in ſhape like to little helmets. The ſeed is ſmall, contained in round bullets or buttons. The root is compact of many and infinite ſtrings.

Betonica aquatica.
Water Betony.

¶ *The Place.*

It groweth by brookes and running waters, by ditch ſides, and by the brinks of riuers, and is ſeldome found in dry places.

¶ *The Time.*

It floureth in Iuly and Auguſt, and from that time the ſeed waxeth ripe.

¶ *The Names.*

Water Betonie is called in Latine *Betonica aquatica* : ſome haue thought it *Dioſcorides* his *Clymenum* : others, his *Galeopſis* : it is *Scrophularia altera* of *Dodonæus* : of *Turner*, *Clymenon* : of ſome, *Seſamoides minus*, but not properly : of others, *Serpentaria* : in Dutch, **S. Antonies cruyd** : in Engliſh, Water Betonie : and by ſome, Browne-wort : in Yorke-ſhire, Biſhops leaues.

¶ *The Temperature.*

Water Betony is hot and dry.

¶ *The Vertues.*

The leaues of Water Betony are of a ſcou- A
ring or cleanſing qualitie, and is very good to mundifie foule and ſtinking vlcers, eſpecially the iuyce boyled with honey.

It is reported, if the face be waſhed with B
the iuyce thereof, it taketh away the redneſſe and deformitie of it.

Chap. 245.

Of Great Figge-wort, or Brownewort.

¶ *The Deſcription.*

1 The great Fig-wort ſpringeth vp with ſtalkes foure ſquare, two cubits high, of a darke purple colour, and hollow within: the leaues grow alwayes by couples, as it were from one ioynt, oppoſite, or ſtanding one right againſt another, broad, ſharpe pointed, ſnipped round about the edges like the leaues of the greater Nettle, but bigger, blacker, and nothing at all ſtinging when they be touched: the floures in the tops of the branches are of a darke purple colour, very like in forme to little helmets: then commeth vp little ſmal ſeed in pretty round buttons, but ſharpe at the end: the root is whitiſh, beſet with little knobs and bunches as it were knots and kernels.

2 There is another Figge-wort called *Scrophularia Indica,* that hath many and great branches trailing here and there vpon the ground, full of leaues, in faſhion like the wilde or common Thiſtle, but altogether without prickes: among the leaues appeare the floures in faſhion like a hood, on the out ſide of a feint colour, and within intermixt with purple; which being fallen and withered, there come in place ſmall knops very hard to breake, and ſharpe at the point as a bodkin: which containeth a ſmall ſeed like vnto Time. The whole plant periſheth at the firſt approch of Winter, and muſt be ſowen againe in Aprill, in good and fertile ground. ‡ This is the *Scrophularia Cretica* 1. of *Cluſius*. ‡

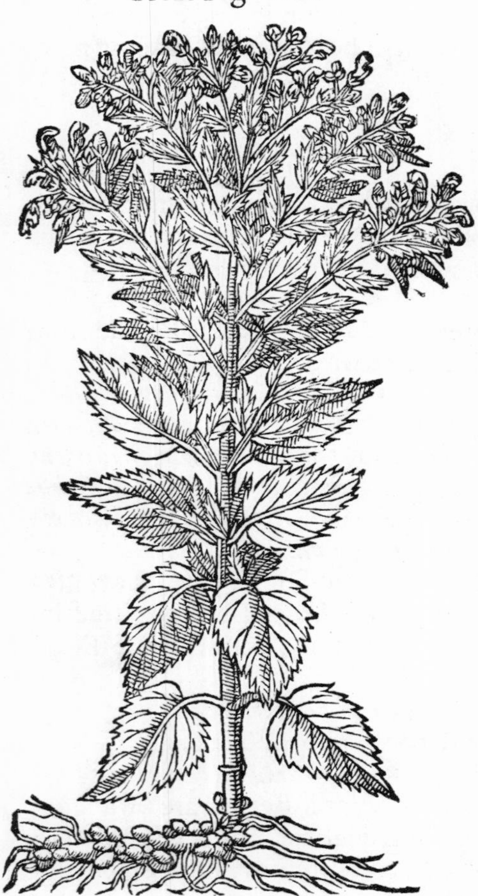

1 *Scrophularia maior.*
Great Fig-wort.

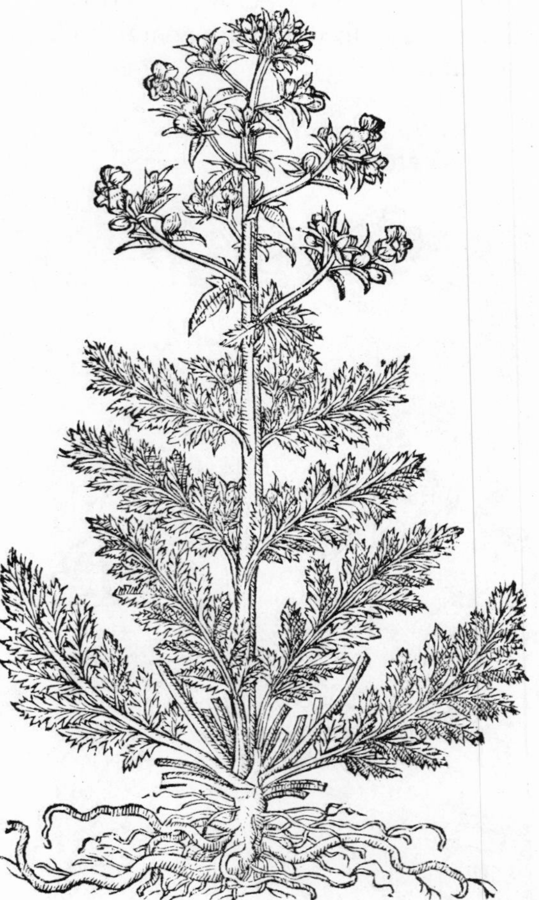

‡ 2 *Scrophularia Indica.*
Indian Fig-wort.

‡ 3 The ſtalke of this is alſo ſquare, and ſome yard high, ſet with leaues like thoſe of the hedge Nettle, but ſomewhat larger and thicker, and a little deeper cut in: out of the boſomes of theſe leaues come little rough foot-ſtalkes ſome inch or two long, carrying ſome foure or fiue hollow round floures of a greeniſh yellow colour, with ſome threds in them, being open at the top, and cut in with fiue little gaſhes: the ſeeds are blacke, and contained in veſſels like thoſe of the firſt

firſt deſcribed: the root is like that of the Nettle, and liues many yeares: it floures in May, and the ſeeds are ripe in Iune. I haue not found nor heard of this wilde with vs, but ſeen it flouriſhing in the garden of my kinde friend Mr. *Iohn Parkinſon*. *Cluſius* calls it *Lamium* 2. *Pannonicum exoticum*: and *Bauhine* hath ſet it forth by the name of *Scrophularia flore luteo*: whom in this I follow. ‡

‡ 3 *Scrophularia flore luteo.*
Yellow floured Fig-wort.

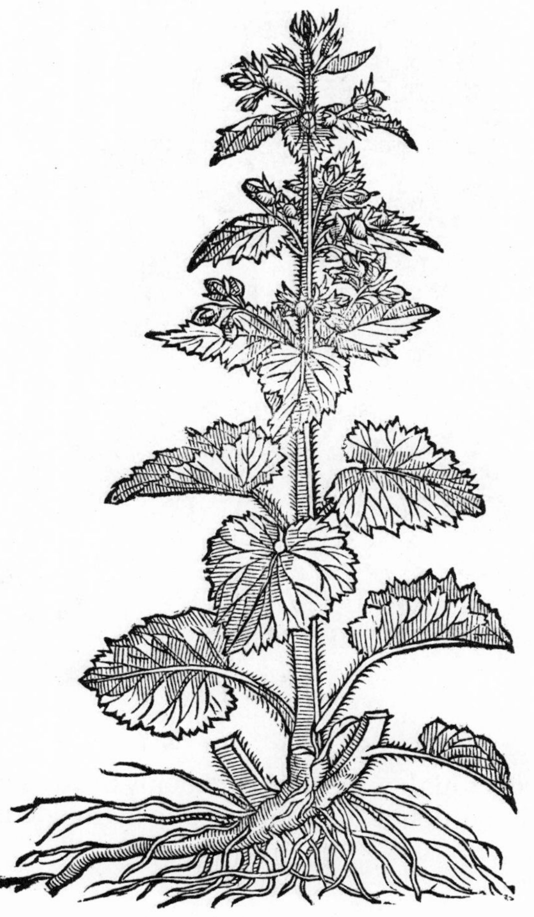

¶ *The Place.*

The great *Scrophularia* groweth plentifully in ſhadowie Woods, and ſometimes in moiſt medowes, eſpecially in greateſt aboundance in a wood as you go from London to Harneſey, and alſo in Stow wood and Shotouer neere Oxford.

The ſtrange Indian figure was ſent me from Paris by *Iohn Robin* the Kings Herbariſt, and it now groweth in my garden.

¶ *The Time.*

They floure in Iune and Iuly.

¶ *The Names.*

Fig-wort or Kernel-wort is called in Latine *Scrophularia maior*, that it might differ from the leſſer Celandine, which is likewiſe called *Scrophularia*, with this addition *minor*, the leſſer: it is called of ſome *Millemorbia*, and *Caſtrangula*: in Engliſh, great Fig-wort, or Kernel-wort, but moſt vſually Brown-wort.

¶ *The Vertues.*

A Fig-wort is good againſt the hard kernells which the Grecians call χοιράδες: the Latines, *Strumas*, and commonly *Scrophulas*, that is, the Kings Euill: and it is reported to be a remedy againſt thoſe diſeaſes whereof it tooke his name, as alſo the painefull piles and ſwelling of the hæmorrhoides.

B Diuers do raſhly teach, that if it be hanged about the necke, or elſe carried about one, it keepeth a man in health.

C Some do ſtampe the root with butter, and ſet it in a moiſt ſhadowie place fifteene dayes together: then they do boyle it, ſtraine it, and keepe it, wherewith they anoint the hard kernels, and the hæmorrhoide veines, or the piles which are in the fundament, and that with good ſucceſſe.

Chap. 246. *Of Veruaine.*

¶ *The Deſcription.*

1 The ſtalke of vpright Veruaine riſeth from the root ſingle, cornered, a foot high, ſeldome aboue a cubite, and afterwards diuided into many branches. The leaues are long, greater than thoſe of the Oke, but with bigger cuts and deeper: the floures along the ſprigs are little, blew, or white, orderly placed: the root is long, with ſtrings growing on it.

2 Creeping Veruaine ſendeth forth ſtalkes like vnto the former, now and then a cubit long, cornered, more ſlender, for the moſt part lying vpon the ground. The leaues are like the former, but with deeper cuts, and more in number. The floures at the tops of the ſprigs are blew, and purple withall, very ſmall as thoſe of the laſt deſcribed, and placed after the ſame manner and order. The root groweth ſtraight downe, being ſlender and long, as is alſo the root of the former.

1 *Verbena communis.*
Common Veruaine.

2 *Verbena sacra.*
Common Veruaine.

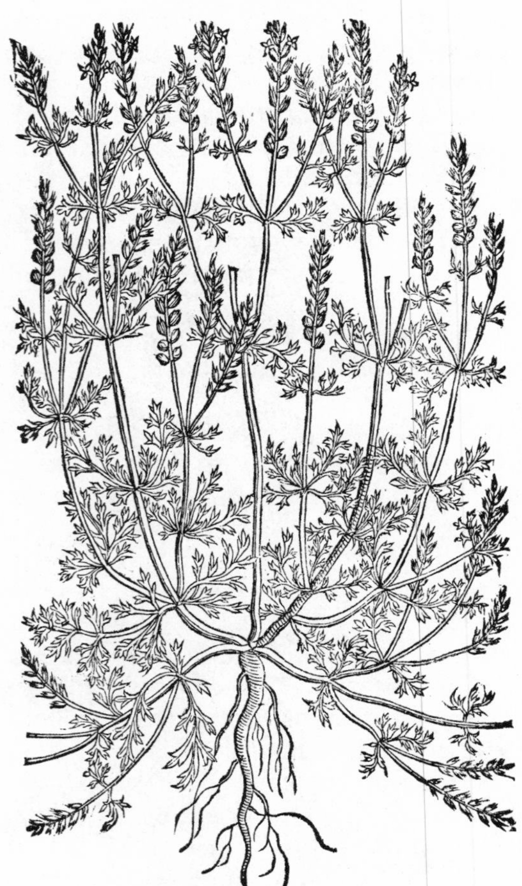

¶ *The Place.*

Both of them grow in vntilled places neere vnto hedges, high-wayes, and commonly by ditches almost euery where. ‡ I haue not seene the second, and doubt it is not to be found wilde in England. ‡

¶ *The Time.*

The Veruaines floure in Iuly and August.

¶ *The Names.*

Veruaine is called in Greeke περιστερεών: in Latine, *Verbena*, and *Verbenaca, Herculania, Ferraria*, and *Exupera*: of some, *Matricalis*, and *Hiera botane*: of others, *Veruena*, and *Sacra herba*: *Verbenæ* are herbes that were taken from the Altar, or from some holy place, which because the Consull or Pretor did cut vp, they were likewise called *Sagmina*, which oftentimes are mentioned in *Liuy* to be grassie herbes cut vp in the Capitoll. *Pliny* also in his two and twentieth booke, and eleuenth Chapter witnesseth, That *Verbenæ* and *Sagmina* be all one: and this is manifest by that which wee reade in *Andræa* in *Terence*: *Ex ara verbenas hinc sume*; Take herbes here from the Altar: in which place *Terence* did not meane Veruaine to be taken from the Altar, but some certaine herbes: for in *Menander*, out of whom this Comedie was translated, is read μυρσίνη, or Myrtle, as *Donatus* saith. In Spanish it is called *Vrgebaom*: in Italian, *Verminacula*: in Dutch, **Iser cruijt**: in French, *Veruaine*: in English, Iuno's teares, Mercuries moist bloud, Holy-herbe; and of some, Pigeons grasse, or Columbine, because Pigeons are delighted to be amongst it, as also to eat thereof, as *Apuleius* writeth.

¶ *The Temperature.*

Both the Veruaines are of temperature very dry, and do meanly binde and coole.

¶ *The Vertues.*

A The leaues of Veruaine pownd with oile of Roses or hogs grease, doth mitigate and appease the paines of the mother, being applied thereto.

B The leaues of Veruaine and Roses stamped with a little new hogs grease, and emplaistered after the manner of a pultesse, doth cease the inflammation and grieuous paines of wounds, and suffereth them not to come to corruption: and the greene leaues stamped with hogs grease takes away the swelling and paine of hot impostumes and tumors, and cleanseth corrupt and rotten vlcers.

C It is reported to be of singular force against the Tertian and Quartane Feuers: but you must obserue

obserue mother *Bombies* rules, to take iust so many knots or sprigs, and no more, lest it fall out so that it do you no good, if you catch no harm by it. Many odde old wiues fables are written of Veruaine tending to witchcraft and sorcerie, which you may read elsewhere, for I am not willing to trouble your eares with reporting such trifles, as honest eares abhorre to heare.

Archigenes maketh a garland of Veruaine for the head-ache, when the cause of the infirmitie D
proceedeth of heat.

The herbe stamped with oile of roses and Vineger, or the decoction of it made in oile of roses, E
keepeth the haires from falling, being bathed or annointed therewith.

It is a remedie against putrified vlcers, it healeth vp wounds, and perfectly cureth Fistulaes, it F
wasteth away old swellings, and taketh away the heat of inflammations.

The decoction of the roots and leaues swageth the tooth-ache, and fasteneth them, and healeth G
the vlcers of the mouth.

They report saith *Pliny*, that if the dining roome be sprinckled with water in which the herbe H
hath beene steeped, the guests will be the merrier, which also *Dioscorides* mentioneth.

Most of the latter Physitions do giue the iuice or decoction hereof to them that hath the plague I
but these men are deceiued, not only in that they looke for some truth from the father of falshood and leasings, but also because in stead of a good and sure remedie they minister no remedy at all for it is reported, that the Diuell did reueale it as a secret and diuine medicine.

CHAP. 247. *Of Scabious.*

† 1 *Scabiosa maior vulgaris.*
Common Scabious.

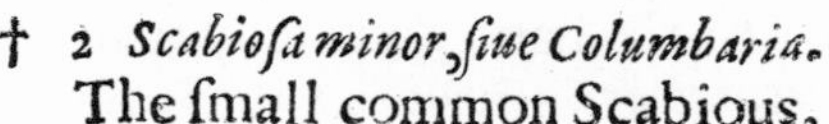

† 2 *Scabiosa minor, siue Columbaria.*
The small common Scabious.

¶ *The Description.*

1 THe first kinde of Scabious being the most common and best known, hath leaues long and broad, of a grayish, hoary, and hairy colour, spred abroad vpon the ground, among which rise vp round and rough stems, beset with hairy iagged leaues, in fashion like great

great Valerian, which we call Setwall. At the top of the stalks grow blew floures in thicke tufts or buttons. The root is white and single.

2 The second is like vnto the former, sauing that his leaues are much cut or iagged, and the whole plant is altogether lesser, scarcely growing to the height of a foot.

3 The third kinde of Scabious is in all things like vnto the second, sauing that the knap or head doth not dilate it selfe so abroad, and is not so thicke or closely thrust together, and the lowest leaues are not so deepely cut or iagged, but the vpper are much smaller, and also the more diuided.

† 4 The fourth groweth with large stalkes, hauing two leaues one set right against another, very much iagged, almost like vnto common Ferne, or rather Ash: and at the top of the stalks there grow larger floures, like vnto the first, but greater, and the roote is also like it, and it differs no waie from the first described, but onely by reason of the soile.

5 Purple floured Scabious hath a rough hairie stalke, whereon doe grow broad leaues deepely cut in the edges, in forme like those of Sowthistle, rough likewise and hairie; the floures grow at the top of the stalks, composed of an innumerable sort of purple thrums: after which come scaly knaps like those of *Iacea*, or Knapweed, wherein is the seed. The root is small and threddie.

‡ 6 The sixth sort of Scabious hath stalks some cubite high, round, and set with leaues not cut and iagged almost to the middle rib, as in the former, yet somewhat rough and hairie, snipt about the edges, and of a light greene colour; amongst which rise vp rough stalkes, on the top whereof do grow faire red floures consisting of a bundle of thrummes. The root is long, tough, and fibrous. ‡

7 The seuenth kinde of Scabious hath sundrie great, rough and round stemmes, as high as a tall man, beset with leaues like the first Scabious, but far greater. The floures grow at the top of the stalkes like vnto the others, but of a faint yellow colour, which fall as soone as it is touched with the hand, whereby it mightily increaseth, notwithstanding the roote endureth for many yeares, and groweth to be wonderfull great: and in my garden it did grow to the bignesse of a mans body.

† 3 *Scabiosa media.*
Middle Scabious.

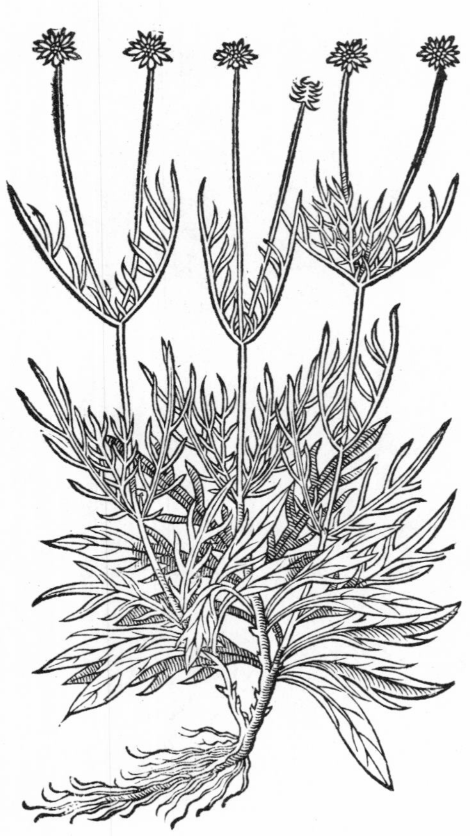

4 *Scabiosa campestris, siue segetum.*
Corne Scabious.

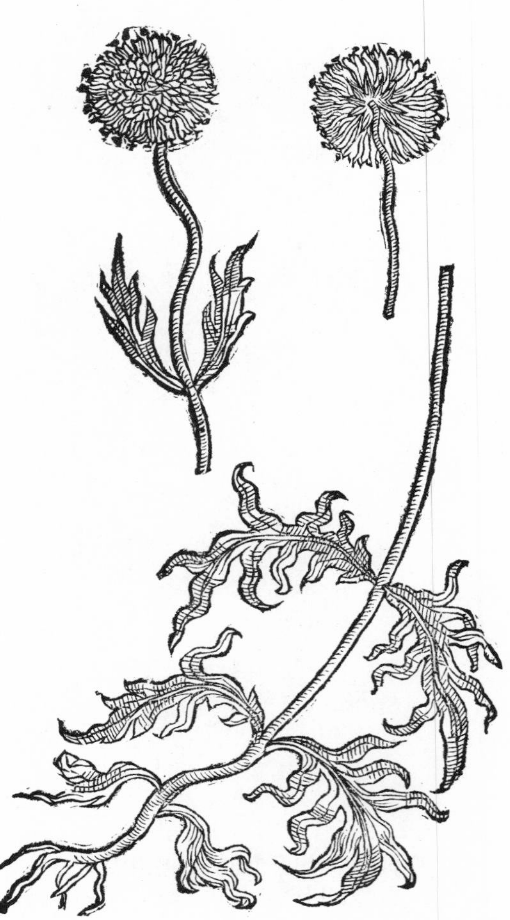

‡ 8 The

5 *Scabioſa flore purpureo.*
Purple floured Scabious.

† 6 *Scabioſa rubra Auſtriaca.*
Red Scabious of Auſtrich.

† 7 *Scabioſa montana maxima.*
Mountaine Scabious.

‡ 8 *Scabioſa montana alba.*
White mountaine Scabious.

6 *Scabiosa maior Hispanica.*
Spanish Scabious.

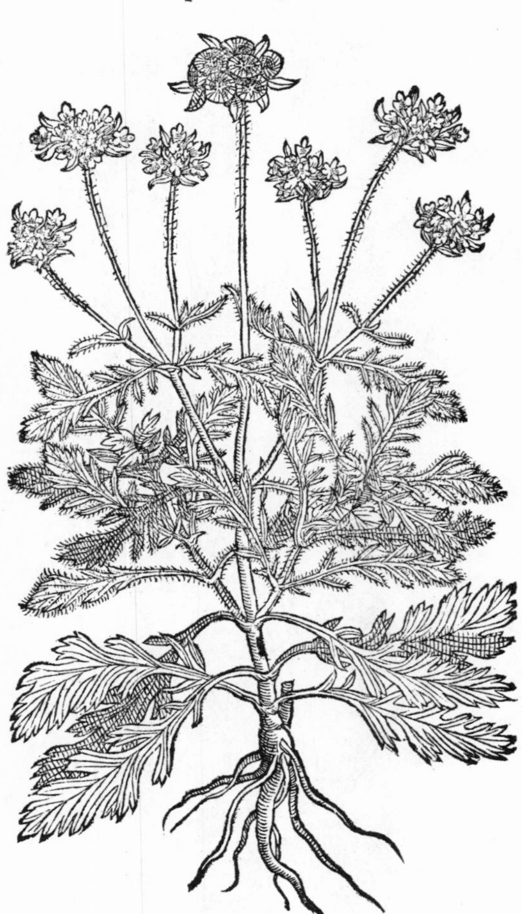

10 *Scabiosa peregrina.*
Strange Scabious.

† 11 *Scabiosa omnium minima.*
Sheepes Scabious.

‡ 8 The white mountaine Scabious hath broad leaues spred vpon the ground, like those of the field Primrose, but greater. Amongst which riseth vp a great stiffe stalke smooth and plain, garnished with leaues not like those next the ground, but lesser, much more diuided, and of a greener colour & harder. The floures are like those of the common Scabious, but white of colour: the root of this perishes euery yeare after the perfecting of the seed. ‡

9 The ninth kinde of Scabious is like vnto the mountaine Scabious, but lower and smaller, hauing sundry large and broad leaues next the ground, snipt confusedly and out of order at the edges like the Oken leafe; among which riseth vp a stem two cubites high, diuiding it selfe into sundry other branches. The floures are set at the top of the naked stalkes, of a whitish colour; which being past, the seed appeareth like a tuft of small bucklers, round, and somewhat hollow within, and made as it were of parchment, very strange to behold: and within the bucklers there are sundry small crosses of blacke fastened to the bottome, as it were the needle in a diall, running vpon the point of a needle. The plant dieth at the beginning of winter, and must be sowne in Aprill in good and fertile ground.

10 The tenth is like vnto the last before mentioned, in stalkes, root, and floures, and differeth that this plant hath leaues altogether without any cuts or iagges about the edges, but is smooth and plaine like the leaues of Marigolds, or Diuels bit, and the floures are like vnto those of the last described.

11 Sheeps Scabious hath small and tender branches trailing vpon the ground, whereupon do grow small leaues very finely iagged or minced euen almost to the middle ribbe, of an ouerworne colour. The floures grow at the top of a blewish colour, consisting of much thrummie matter, hard thrust together like a button: the root is small, and creepeth in the ground.

12 The

12 *Scabioſa minima hirſuta.*
Hairie Sheepes Scabious

‡ 13 *Scabioſa minima Bellidis folio.*
Daiſie leaued Scabious.

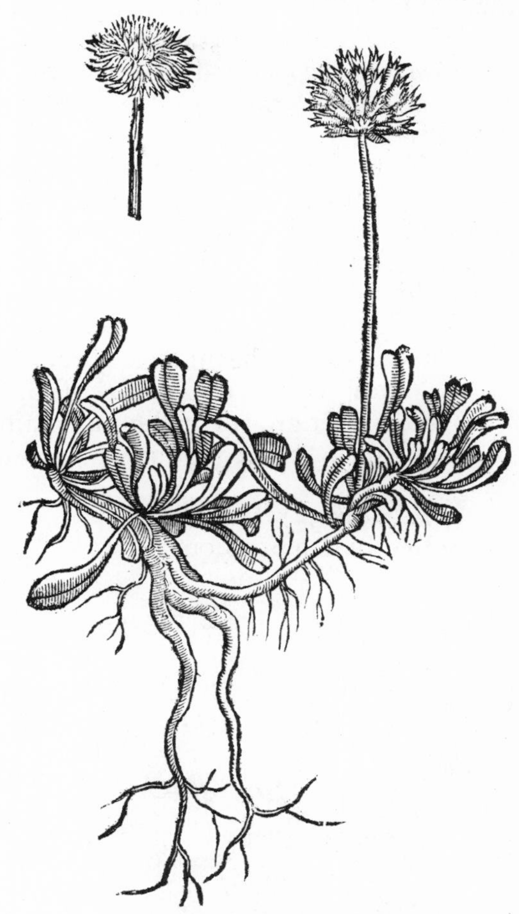

‡ 14 *Scabioſa flore pallido.*
Yellow Scabious.

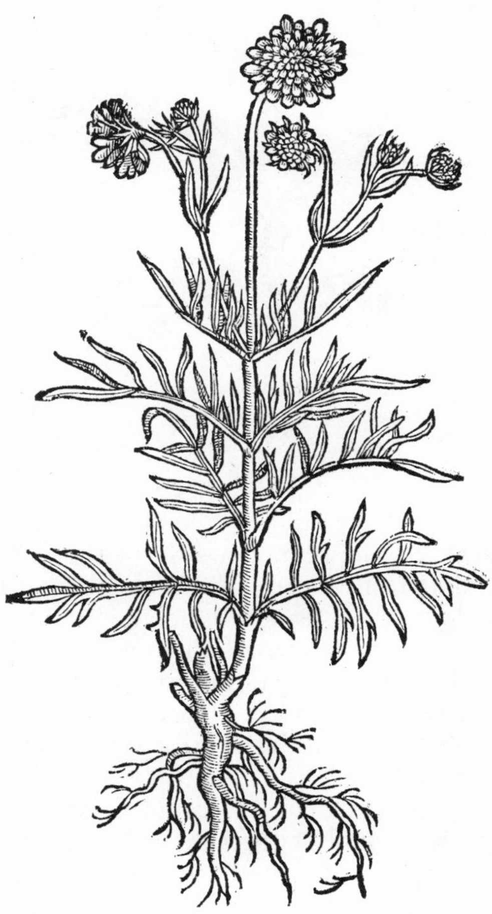

‡ 15 *Scabioſa prolifera.*
Childing Scabious.

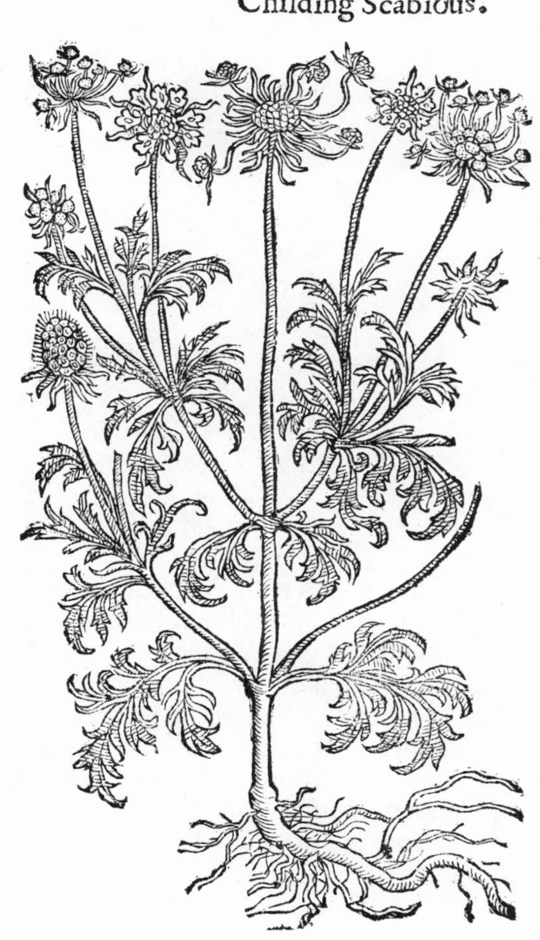

‡ 12 The other Sheeps Scabious of our Author (according to the figure) is greater than the laſt deſcribed, growing ſome foot or better high, with ſlender rough branches ſet with leaues not ſo much diuided, but onely nicked about the edges: the floures are in colour and ſhape like thoſe of the laſt deſcribed, or of the blew daiſie; the root is ſingle, and like that of a Rampion, whence *Fabius Columna* (the ſeed and milkie juice inducing him) hath refer'd this to the Rampions, calling it *Rapuntium montanum capitatum leptophyllon*. *Lobell* calls it *Scabioſa media*: and *Dodonæus*, *Scabioſa minor*.

13 To theſe little plants we may fitly adde another ſmall one refer'd by *Cluſius* to this Claſſis, and called *Scabioſa*. 10. *ſiue repens*: yet *Bauhine* refers it to the Daiſies, and termes it *Bellis cærulea montana fruteſcens*; but it matters not to which we referre it: the deſcription is thus; The root is hard, blacke, and creeping, ſo that it ſpreds much vpon the ſurface of the ground, ſending forth many thicke, ſmooth, greene leaues, like thoſe of the blew Daiſie, not ſharpe pointed, but ending as we vulgarly figure an heart, hauing a certaine graſsie but not vnpleaſant ſmell, and ſomewhat a bitter and hot taſte: out of the middeſt of theſe leaues grow ſlender naked ſtalks ſome hand high, hauing round floures on their tops, like thoſe of Diuells bit, and of the ſame colour, yet ſometimes of a lighter blew. It growes in the mountaines of Hungary and Auſtria. It floures in Aprill and May, and ripens the ſeed in Iuly and Auguſt.

‡ 16 *Scabioſa rubra Indica*.
Red Indian Scabious.

‡ 17 *Scabioſa æſtiualis Cluſij*.
Sommer Scabious.

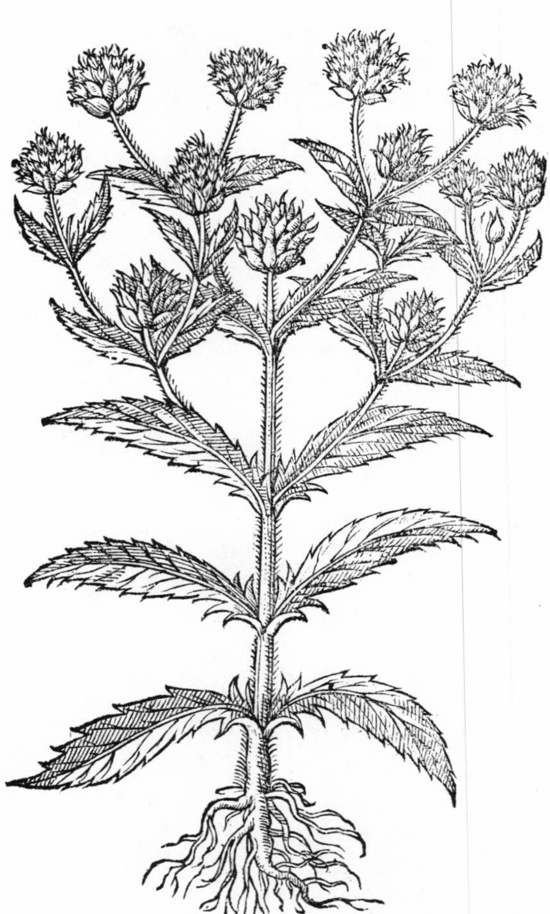

14 This (which is the ſeuenth Scabious of *Cluſius*, and which he termes ὠχρολευκὸς, of the whitiſh yellow colour of the floure) hath round, ſlender, ſtiffe, and greene ſtalkes ſet at each joint with two large and much diuided leaues of a whitiſh greene colour: thoſe leaues that come from the root before the ſtalke grow vp are broader, and leſſe diuided; vpon the tops of the branches and ſtalkes grow floures like thoſe of the common Scabious, being white or rather (before they be throughly open) of a whitiſh yellow colour; which fading, there follow ſeedes like as in the ordinary kinde. This floures in Iune and Iuly, and growes very plentifully in all the hilly grounds and dry Meades of Auſtria and Morauia.

15 There is alſo a kinde of Scabious hauing the leaues much cut and diuided, and the ſtalkes and floures like to the common ſort, of a blewiſh purple colour, but differing in this, that at

the ſides of the floure it puts forth little ſtalkes, bearing ſmaller floures, as is ſeene in ſome other plants, as in Daiſies and Marigolds, which therefore are fitly termed in Latine *Prolifera* or Childing. This growes onely in Gardens, and floures at the ſame time with the former.

16 The ſtalkes of the red Scabious grow ſome cubit or more in height, and are diuided into many very ſlender branches, which at the tops carry floures compoſed after the manner of the other ſorts of Scabious, that is, of many little floures diuided into fiue parts at the top, and theſe are of a perfect red colour, and haue ſmall threds with pendants at them comming forth of the midle of each of theſe little floures, which are of a whitiſh colour, and make a pretty ſhow. The leaues are greene, and very much diuided or cut in. The ſtarry ſeeds grow in long round hairy heads handſomely ſet together. This is an annuall, and periſhes as ſoone as it hath perfected the ſeed. *Cluſius* makes it his ſixt Scabious, and calls it *Scabioſa Indica*. It floures in Iuly, and growes in the Gardens of our prime Herbariſts.

17 The ſame Authour hath alſo giuen vs the figure and deſcription of another Scabious, which ſends vp a ſtalke ſome three cubits or more high, ſet at certaine ſpaces with leaues large, and ſnipt about their edges, and a little cut in neere their ſtalkes. The ſtalkes are diuided into others, which at there tops carry blewiſh floures in long ſcaly heads, which are ſucceeded by long whitiſh ſeed. The roote is whitiſh and fibrous, and dyes euery yeare. This is the *Scabioſa 9, ſiue æſtiualis* of *Cluſius*. ‡

¶ *The Place.*

Theſe kindes of Scabious do grow in paſtures, medowes, corne fields, and barren ſandy grounds almoſt euery where.

The ſtrange ſorts do grow in my garden, yet are they ſtrangers in England.

¶ *The Time.*

They floure and flouriſh in the Sommer moneths.

¶ *The Names.*

Scabious is commonly called *Scabioſa*; diuers thinke it is named ψώρα, which ſignifieth a ſcabbe, and a certaine herbe ſo called by *Aëtius*: I do not know, ſaith *Hermolaus Barbarus*, whether it be Scabious which *Aëtius* doth call *Pſora*, the ſmoake of which being burnt doth kill cankers or little wormes. The Author of the Pandects doth interpret *Scabioſa* to be *Dioſcorides* his *Stæbe*: *Dioſcorides* deſcribeth *Stæbe* by no markes at all, being commonly knowne in his time; and *Galen* in his firſt booke of Antidotes ſaith thus: There is found amongſt vs a certaine ſhrubby herbe, hot, very ſharpe and biting, hauing a little kind of aromaticall or ſpicy ſmell, which the inhabitants do call *Colymbade*, and *Stæbe* ſingular good to keepe and preſerue wine: but it ſeemeth that this *Stæbe* doth differ from that of which he hath made mention in his booke of the faculties of medicines, which agreeth with that of *Dioſcorides*: for he writeth that this is of a binding quality without biting; ſo that it cannot be very ſharpe.

¶ *The Temperature.*

Scabious is hot and dry in the later end of the ſecond degree, or neere hand in the third, and of thin and ſubtile parts: it cutteth, attenuateth, or maketh thin, and throughly concocteth tough and groſſe humours.

¶ *The Vertues.*

Scabious ſcoureth the cheſt and lungs; it is good againſt an old cough, ſhortneſſe of breath, paine in the ſides, and ſuch like infirmities of the cheſt. A

The ſame prouoketh vrine, and purgeth now and then rotten matter by the bladder, which happeneth when an impoſtume hath ſomewhere lien within the body. B

It is reported that it cureth ſcabs, if the decoction thereof be drunke certain daies, and the juice vſed in ointments. C

The later Herbariſts doe alſo affirme that it is a remedy againſt the bitings of Serpents and ſtingings of venomous beaſts, being outwardly applied or inwardly taken. D

The juice being drunke procureth ſweat, eſpecially with Treacle; and it ſpeedily conſumeth plague ſores, if it be giuen in time, and forthwith at the beginning: but it muſt be vſed often. E

It is thought to be forceable, and that againſt all peſtilent feuers. F

† Formerly the 1. 2. 3. 11. figures were all nothing elſe than the varieties of one Plant, being of the 1. 2. 3. 4. *Scabioſa minor* of *Tabern.* they differ onely in the more or leſſe cutting or diuiding of the leaues: I haue of theſe onely reſerued the third, and in other places put ſuch figures as are agreeable to the titles. The figure that was in the ſixt place was of the ordinary firſt deſcribed Scabious; and the figure that ſhould haue beene there was in the eighth place; and that which was in the ſeuenth place belongs to the plant deſcribed by me in the fourteenth place.

Chap. 248. *Of Diuels bit.*

Morsus Diaboli.
Diuels bit.

¶ *The Description.*

Diuels bit hath small vpright round stalks of a cubite high, beset with long leaues somewhat broad, very little or nothing snipt about the edges, somewhat hairie and euen. The floures also are of a darke purple colour, fashioned like the floures of Scabious, which being ripe are carried away with the winde. The root is blacke, thicke, hard and short, with many threddie strings fastened thereto. The great part of the root seemeth to be bitten away: old fantasticke charmers report, that the diuell did bite it for enuie, because it is an herbe that hath so many good vertues, and is so beneficiall to mankinde.

¶ *The Place.*

Diuels bit groweth in drie medows and woods, and about waies sides. I haue found great store of it growing in Hampstead wood neere London, at Lee in Essex, and at Raleigh in Essex, in a wood called Hammerell, and sundrie other places.

¶ *The Time.*

It floureth in August, and is hard to be knowne from Scabious, sauing when it floureth.

¶ *The Names.*

It is commonly called *Morsus Diaboli*, or Diuels bit, of the root (as it seemeth) that is bitten off: for the superstitious people hold opinion, that the diuell for enuie that he beareth to mankinde bit it off, because it would be otherwise good for many vses: it is called of *Fuchsius*, *Succisa*: in High Dutch **Teuffels abbisz**: in Low Dutch, **Duyuelles beet** in French *Mors du Diable*: in English, Diuels bit, and Forebit. ‡ *Fabius Columna* iudgeth it to bee the *Pycnocomon* of *Dioscorides*, described by him *lib.* 4. *cap.* 176. ‡

¶ *The Temperature.*

Diuels bit is something bitter, and of a hot and drie temperature, and that in the later end of the second degree.

¶ *The Vertues.*

A There is no better thing against old swellings of the Almonds, and vpper parts of the throat that be hardly ripened.

B It clenseth away slimie flegme that sticketh in the iawes; it digesteth and consumeth it: and it quickely taketh away the swellings in those parts, if the decoction thereof be often held in the mouth and gargarized, especially if a little quantitie of *Mel Rosarum*, or honie of Roses be put into it.

C It is reported to be good for the infirmities that Scabious serueth for, and to be of no lesse force against the stingings of venomous beasts, poisons, and pestilent diseases, and to consume and waste away plague sores, being stamped and laid vpon them.

D And also to mitigate the paines of the matrix or mother, and to driue forth winde, if the decoction thereof be drunke.

CHAP. 249. *Of Matfellon or Knapweed.*

¶ *The Deſcription.*

1 MAtfellon or blacke Knapweed is doubtleſſe a kinde of Scabious, as all the others are, intituled with the name of *Iacea*; yet for diſtinction I haue thought good to ſet them downe in a ſeuerall Chapter, beginning with that kinde which is called in Engliſh Knapweed and Matfellon, or *Materfilon*. It hath long and narrow leaues, of a blackiſh green colour, in ſhape like Diuels bit, but longer, ſet vpon ſtalks two cubits high, ſomewhat bluntly cut or ſnipt about the edges: the floures do grow at the top of the ſtalks, being firſt ſmall ſcaly knops, like to the knops of Corne floure, or blew bottles, but greater; out of the midſt thereof groweth a purple thrummie or threddie floure. The root is thicke and ſhort.

2 The great Knapweed is very like vnto the former, but that the whole plant is much greater, the leaues bigger, and more deeply cut, euen to the middle rib: the floures come forth of ſuch like ſcaly heads, of an excellent faire purple colour, and much greater.

3 The third kinde of Matfellon, or Knapweed is very like vnto the former great Knapweed laſt before mentioned, ſauing that the floures of this plant are of an excellent faire yellow colour, proceeding forth of a ſcaly head or knop, beſet with moſt ſharp pricks, not to be touched without hurt: the floure is of a pleaſing ſmel, and very ſweet; the root is long and laſting, and creepeth far abroad, by means whereof it greatly increaſeth.

1 *Iacea nigra.*
Blacke Matfellon.

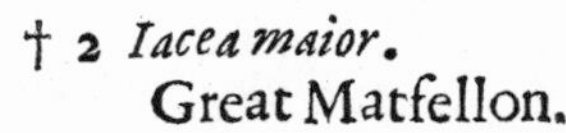

† 2 *Iacea maior.*
Great Matfellon.

4 The mountaine Knapweede of Narbone in France, hath a ſtrong ſtem of two cubits high, and is very plentifull about Couentrie among the hedges and buſhes: the leaues are very much iagged, in forme of *Lonchitis*, or Spleenewoort; the floures are like the reſt of the Knapweeds, of a purple colour.

3 *Iacea maior lutea.*
Yellow Knapweed.

4 *Iacea montana.*
Mountaine Knapweed.

5 *Iacea flore albo.*
White floured Knapweed.

6 *Iacea tuberoſa.*
Knobbed Knapweed.

‡ 7 *Iacea Auſtriaca villoſa.*
Rough headed Knapweed.

‡ 5 The white floured Knapweed hath creeping roots, which ſend vp pretty large whitiſh greene leaues, much diuided or cut in almoſt to the middle rib; from the midſt of which riſes vp a ſtalke ſome two foot high, ſet alſo with the like diuided leaues, but leſſer: the floures are like thoſe of the common ſort, but of a pleaſing white colour. I firſt found this growing wilde in a field nigh Martine Abbey in Surrey, and ſince in the Iſle Tenet. ‡

6 The tuberous or knobbie Knapweed being ſet forth by *Tabernam.* which and is a ſtranger in theſe parts, hath many leaues ſpred vpon the ground, rough, deeply gaſht or hackt about the edges, like thoſe of Sowthiſtle: among which riſeth vp a ſtraight ſtalke, diuiding it ſelfe into other branches, whereon do grow the like leaues, but ſmaller: the knappie floures ſtand on the top of the branches, of a bright red colour, in ſhape like the other Knapweeds. The root is great, thicke and tuberous, conſiſting of many cloggie parcels, like thoſe of the Aſphodill.

‡ 7 This (ſaith *Cluſius*) is a comely plant, hauing broad and long leaues white, ſoft, and lightly ſnipt about the edges: the taſte is gummy, & not a little bitter: it ſends vp many creſted ſtalks from one root, ſome cubit high or more: at the toppes of them grow the heads ſome two or three together, conſiſting of many ſcales, whoſe ends are hairy, and they are ſet ſo orderly, that by this meanes the heads ſeeme as they were incloſed in little nets: the floures are purple, and like thoſe of the firſt deſcribed; the ſeede is ſmall and long, and of an aſh colour. This *Cluſius* calls *Iacea 4. Auſtriaca villoſo capite.*

Iacea capitulis hirſutis Boelij.

8 This hath many ſmall cornered ſtraked hairie trailing branches growing from the root, and thoſe again diuided into many other branches, trailing or ſpreading vpon the ground three or foure foot long, imploying or couering a good plot of ground, whereon grow hairy leaues diuided or iagged into many parts, like the leaues of *Iacea maior*, or Rocket, of a very bitter taſte: at the top of each branch groweth one ſcaly head, each ſcale ending with fiue, ſix, or ſeuen little weake prickles growing orderly like halfe the rowell of a ſpurre, but farre leſſer: the floures grow forth of the heads of a light purple colour, conſiſting of many ſmal floures, like thoſe of the common *Iacea*, the bordering floures, being bigger and larger than thoſe of the middle of the floure, each ſmall floure being diuided into fiue ſmall parts or leaues, not much vnlike thoſe of *Cyanus*: the ſeed is ſmall, and incloſed in downe. The root periſheth when the ſeed is ripe.

This plant hath not been hitherto written of that I can find. Seeds of it I receiued from Mr. *William Coys*, with whom alſo I obſerued the plant, October 10. 1621. he receiued it from *Boelius* a Low countrey man. *Iohn Goodyer.* ‡

¶ *The Place.*

The two firſt grow commonly in euery fertile paſture: the reſt grow in my garden.

¶ *The Time.*

They floure in Iune and Iuly.

¶ *The Names.*

The later age doth call it *Iacea nigra*, putting *nigra* for a difference betweene it and the Hearts-eaſe or Pancie, which is likewiſe called *Iacea*: it is called alſo *Materſillon*, and *Matreſillen*: in Engliſh, Matfellon, Bulweed, and Knapweed.

¶ *The Temperature and Vertues.*

A These plants are of the nature of Scabious, whereof they be kindes, therefore their faculties are like, although not so proper to Physickes vse.

B They be commended against the swellings of the Vvula, as is Diuels bit, but of lesse force and vertue.

† The figure that was formerly in the second place was of the *Iacea tertia* of *Tabern.* which differs from that our Author meant and described, whose figure we haue giuen you in the place thereof.

Chap. 250. *Of Siluer Knapweed.*

¶ *The Description.*

1 The great Siluer Knapweed hath at his first comming vp diuers leaues spred vpon the ground, of a deepe greene colour, cut and iagged as are the other Knapweeds, straked here and there with some siluer lines downe the same, whereof it tooke his surname, *Argentea*: among which leaues riseth vp a straight stalke, of the height of two or three cubits, somwhat rough and brittle, diuiding it selfe toward the top into other twiggie branches: on the tops whereof do grow floures set in scaly heads or knaps like the other Matfellons, of a gallant purple colour, consisting of a number of threds or thrums thicke thrust together: after which the seedes appeare, slipperie, smooth at one end, and bearded with blacke haires at the other end, which maketh it to leap and skip away when a man doth but lightly touch it. The root is small, single, and perisheth when the seed is ripe. ‡ This is not streaked with any lines, as our Author imagined, nor called *Argentea* by any but himselfe, and that very vnfitly. ‡

† 1 *Stæbe argentea maior.*
Great Siluer Knapweed.

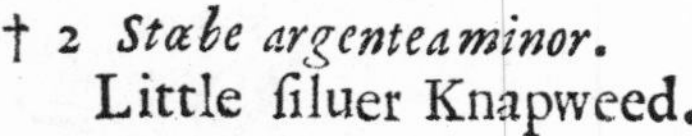

† 2 *Stæbe argentea minor.*
Little siluer Knapweed.

2 The second agreeth with the first in each respect, sauing that the leaues hereof are more iagged, and the siluer lines or strakes are greater, and more in number, wherein consisteth the difference.

‡ 4 *Stœbe Roſmarini folio.*
Narrow leafed Knapweed.

‡ 5 *Stœbe ex Codice Cæſareo.*
Thornie Knapweed.

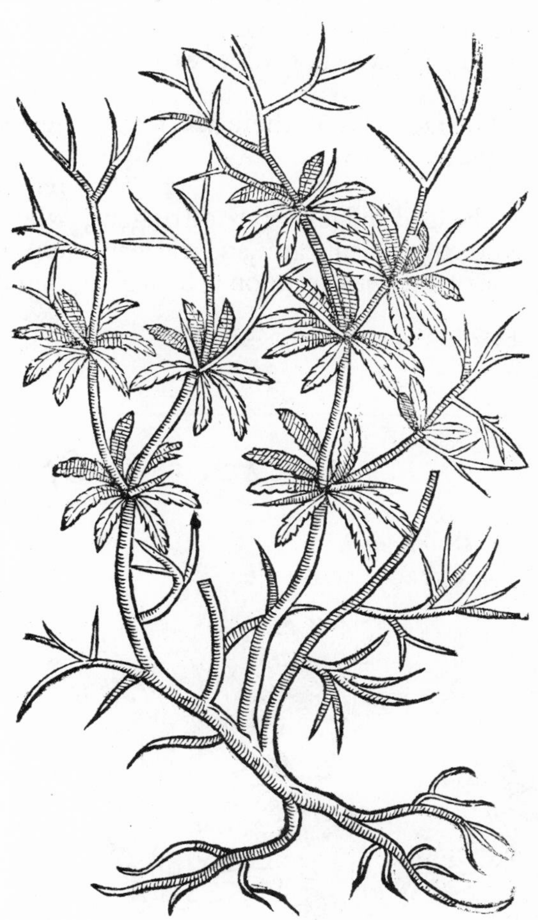

rence. ‡ The leaues of this are very much diuided and hoarie, the ſtalkes ſome two cubites high, ſet alſo with much diuided leaues, that end in ſoft harmleſſe prickles: at the tops of the branches ſtand the heads compoſed as it were of ſiluer ſcales (whence *Lobel* and others haue called this plant *Stœbe argentea*) and out of theſe ſiluer heads come floures like thoſe of the blew bottles, but of a light purple colour, the ſeed is ſmall, blackiſh, and hairy at the tops. ‡

‡ 3 There is another like this in each reſpect, but that the heads haue not ſo white a ſhining ſiluer colour: and this I haue alſo ſeene growing with Maſter *Iohn Tradeſcant* at South Lambeth. ‡

‡ 4 To theſe may be added that plant which *Pona* hath ſet forth by the name of *Stœbe capitata Roſmarini folio.* It hath a whitiſh wooddy root, from whence ariſe diuers branches ſet with long narrow leaues ſomewhat like thoſe of Roſemary, but liker thoſe of the Pine, of a greeniſh colour aboue, and whitiſh below: at the tops of the branches grow ſuch heads as in the firſt deſcribed *Stœbe*, with floures of ſomewhat a deeper purple colour: the ſeed is like that of *Carthamus*, but blackiſh. The root is not annuall, but laſts many yeares. ‡

‡ 5 Though theſe plants haue of late been vulgarly ſet forth by the name of *Stœbe's*, yet are they not iudged to bee the true *Stœbe* of *Dioſcorides* and the Antients, but rather another, whoſe figure which we here giue was by *Dodonæus* taken forth of a manuſcript in the Emperours Library, and he ſaith; *Paludanus* brought home ſome of the ſame out of Cyprus and Morea, as he returned from his journey out of Syria: the bottome leaues are ſaid to be much diuided; thoſe on the ſtalkes long, and onely ſnipt about the edges, and white: the floures white, and contained in ſcaly heads like the Blew-bottles, and the tops of the branches end in ſharpe prickles. ‡

¶ *The Place.*

Theſe doe grow of themſelues in fields neere common high waies, and in vntilled places, but they are ſtrangers in England, neuertheleſſe I haue them in my garden.

¶ The

¶ *The Time.*

They spring vp in April, they floure in August, and the seed is ripe in September.

¶ *The Names.*

Siluer Knapweed is called of *Lobel*, *Stæbe Salamantica* : of **Dodonæus**, *Aphyllanthes*, that is, without leaues, for the floures consist onely of a number of threds without any leaues at all: in English, Siluer Knapweed, or Siluer Scabious, whereof doubtlesse they be kindes.

¶ *The Temperature and Vertues.*

The faculties of these Matfellons are not as yet found out, neither are they vsed for meat or medicine.

‡ *The faculties of Stæbe out of Dioscorides.*

A The seed and leaues are astringent, wherefore the decoction of them is cast vp in Dysenteries, and into purulent eares, and the leaues applied in manner of a pultis are good to hinder the blacknesse of the eies occasioned by a blow, and stop the flowing of bloud. ‡

† The figures were formerly transposed.

Chap. 251. *Of Blew Bottle, or Corne Floure.*

1 *Cyanus maior.*
Great blew Bottle.

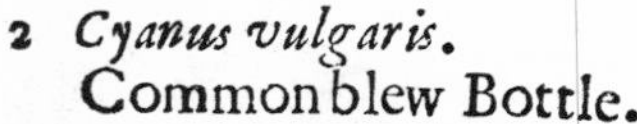

2 *Cyanus vulgaris.*
Common blew Bottle.

¶ *The Description.*

1 The great blew Bottle hath long leaues, smooth, soft, downie, and sharpe pointed: among the leaues rise vp crooked, and prettie thicke branches, chamfered, furrowed, and garnished with such leaues as are next the ground : on the toppes whereof stand faire blew floures tending to purple, consisting of diuers little floures, set in a scaly huske or knap like those of the Knapweeds : the seed is rough or bearded at one end, smooth at the other, and shining. The root is tough, and long lasting (contrary to the rest of the Corne floures) and groweth yearely into new shoots and springs, whereby it greatly increaseth.

2 The

7 *Cyanus cœruleus multiflorus.*
Double Blew-Bottles.

8 *Cyanus purpureus multiflorus.*
Double Purple-Bottles.

‡ 9 *Cyanus repens latifolius.*
Broad leafed creeping Blew-Bottle.

‡ 10 *Cyanus repens anguſtifolius.*
Small creeping Blew-Bottle.

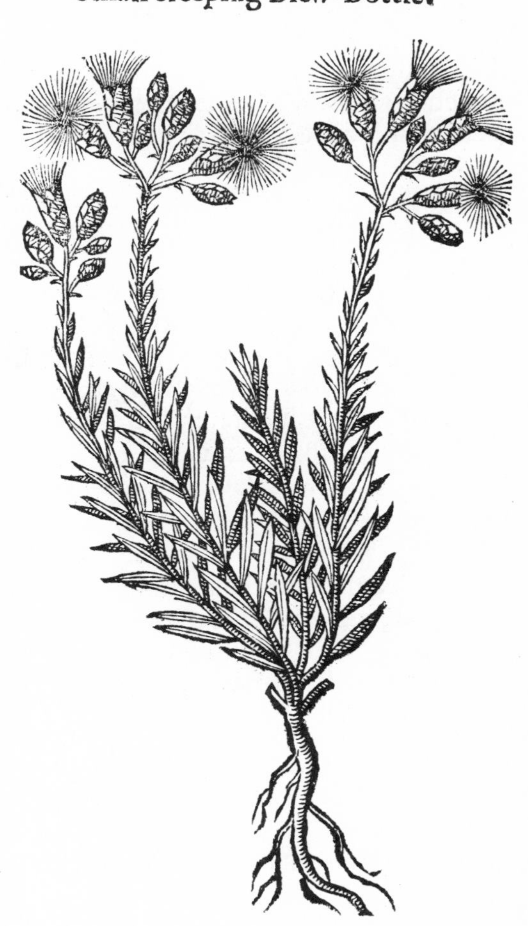

2 The common Corne-floure hath leaues spred vpon the ground, of a whitish greene colour, somewhat hackt or cut in the edges like those of Corne Scabious: among which riseth vp a stalke diuided into diuers small branches, whereon do grow long leaues of an ouerworne greene colour, with few cuts or none at all. The floures grow at the top of the stalkes, of a blew colour, consisting of many small floures set in a scaly or chaffie head like those of the Knapweeds: the seed is smooth, bright shining, and wrapped in a woolly or flocky matter. The root is small and single, and perisheth when it hath perfected his seed.

3 This Bottle is like the last described in each respect, sauing in the colour of the floures, which are purple, wherein consisteth the difference.

4 The fourth Bottle is also like the precedent, not differing in any point but in the floures; for as the last before mentioned are of a purple colour, contrariwise these are of a milke white colour, which setteth forth the difference.

5 The Violet-coloured Bottle or Corne-floure is like the precedent, in stalkes, leaues, seeds, and roots: the onely difference is, that this bringeth floures of a violet colour, and the others not so.

6 Variable Corne-floure is so like the others in stalks, leaues, and proportion, that it cannot be distinguished with words; onely the floures hereof are of two colours mixed together, that is, purple and white, wherein it differeth from the rest.

7 There is no difference to be found in the leaues, stalkes, seed, or rootes of this Corne-floure from the other, but onely that the floures hereof are of a faire blew colour, and very double.

8 The eighth Corne-floure is like vnto the precedent, without any difference at all, sauing in the colour of the floures, the which are of a bright purple colour, that setteth forth the difference.

‡ 9 This from a small root sends vp diuers creeping branches some foot long, set with long hoary narrow leaues: at the tops of the stalkes stand the floures in scaly heads, like as the other Blew-Bottles, but of a darke purple colour. The whole plant is very bitter and vngratefull to the taste. *Lobel* calls this *Cyanus repens*.

10 This is like the last described, but that the leaues are much smaller or narrower, also the scaly heads of this are of a finer white siluer colour: and this plant is not possessed with such bitternesse as the former. *Lobel* calls this *Cyanus minimus repens*. ‡

¶ *The Place.*

The first groweth in my garden, and in the gardens of Herbarists, but not wilde that I know of. The others grow in corne fields among wheat, Rie, Barley, and other graine: it is sowne in gardens, which by cunning to looking doth oft times become of other colours, and some also double, which hath beene touched in their seuerall descriptions. ‡ The two last grow wilde about Montpellier in France. ‡

¶ *The Time.*

They bring forth their floures from the beginning of May vnto the end of haruest.

¶ *The Names.*

The old Herbarists called it *Cyanus flos*, of the blew colour which it naturally hath: most of the later sort following the common Germane name, call it *Flos frumentorum*, for the Germans name it **Corn blumen**: in low-Dutch, **Corn bloemen**: in French, *Blaueole*, and *Bluet*: in Italian, *Fior campese*, and *Bladiseris*, i. *Seris bladi*, and *Battisecula*, or *Baptisecula*, as though it should be called *Blaptisecula*, because it hindereth and annoyeth the Reapers, by dulling and turning the edges of their sickles in reaping of corne: in English it is called Blew-Bottle, Blew-Blow, Corne-floure, & hurt-Sickle. ‡ *Fabius Columna* would haue it to be the *Papauer spumeum*, or *Heracleum* of the Antients. ‡

¶ *The Temperature and Vertues.*

A The faculties of these floures are not yet sufficiently knowne. Sith there is no vse of them in physicke, we will leaue the rest that might be said to a further consideration: notwithstanding some haue thought the common Blew-Bottle to be of temperature something cold, and therefore good against the inflammation of the eyes, as some thinke.

CHAP. 252. *Of Goats Beard, or Go to bed at noone.*

¶ *The Deſcription.*

1 GOats-beard, or Go to bed at noone hoth hollow ſtalks, ſmooth, and of a whitiſh green colour, whereupon do grow long leaues creſted downe the middle with a ſwelling rib, ſharpe pointed, yeelding a milky iuyce when it is broken, in ſhape like thoſe of Garlick; from the boſome of which leaues thruſt forth ſmall tender ſtalks ſet with the like leaues but leſſer: the floures grow at the top of the ſtalkes, conſiſting of a number of purple leaues, daſht ouer as it were with a little yellow duſt, ſet about with nine or ten ſharpe pointed greene leaues: the whole floure reſembles a ſtarre when it is ſpred abroad; for it ſhutteth it ſelfe at twelue of the clocke, and ſheweth not his face open vntill the next dayes Sun doth make it floure anew, whereupon it was called Go to bed at noon: when theſe floures be come to their full maturitie and ripeneſſe, they grow into a downy Blow-ball like thoſe of Dandelion, which is carried away with the winde. The ſeed is long, hauing at the end one piece of that downie matter hanging at it. The root is long and ſingle, with ſome few threds thereto anexed, which periſheth when it hath perfected his ſeed, yeelding much quantitie of a milky iuyce when it is cut or broken, as doth all the reſt of the plant.

2 The yellow Goats beard hath the like leaues, ſtalkes, root, ſeed, and downie blow-balls that the other hath, and alſo yeeldeth the like quantitie of milke, inſomuch that if the pilling while it is greene be pulled from the ſtalkes, the milky iuyce followeth: but when it hath there remained a little while it waxeth yellow. The floures hereof are of a gold yellow colour, and haue not ſuch long greene leaues to garniſh it withall, wherein conſiſteth the difference.

1 *Tragopogon purpureum.*
Purple Goats-beard.

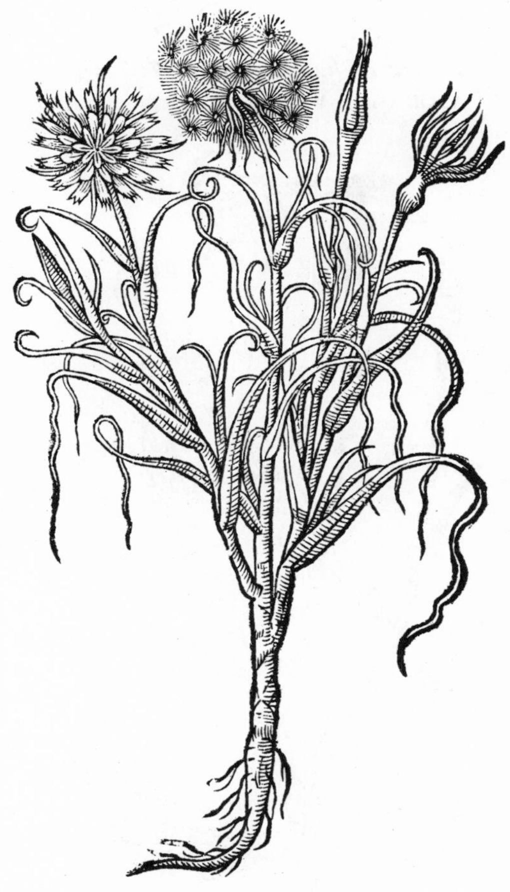

2 *Tragopogon luteum.*
Yellow Goats-beard.

3 There is another ſmall ſort of Goats-beard or Go to bed at noone, which hath a thicke root full of a milky ſap, from which riſe vp many leaues ſpred vpon the ground, very long, narrow, thin, and like vnto thoſe of graſſe, but thicker and groſſer: among which riſe vp tender ſtalkes, on the tops whereof do ſtand faire double yellow floures like the precedent, but leſſer. The whole plant yeeldeth

yeeldeth a milkie ſap or iuyce as the others do: it periſheth like as the other when it hath perfected his ſeed. This may be called *Tragopogon minus anguſtifolium*, Little narrow leaued Goats-beard.

¶ *The Place.*

The firſt growes not wild in England that I could euer ſee or heare of, except in Lancaſhire vpon the banks of the riuer Chalder, neere to my Lady *Heskiths* houſe, two miles from Whawley: it is ſowen in gardens for the beauty of the floures almoſt euery where. The others grow in medows and fertil paſtures in moſt places of England. It growes plentifully in moſt of the fields about London, as at Iſlington, in the medowes by Redriffe, Detford, and Putney, and in diuers other places.

¶ *The Time.*

They floure and flouriſh from the beginning of Iune to the end of Auguſt.

¶ *The Names.*

Goats-beard is called in Greeke τραγοπώγων: in Latine, *Barba hirci*, and alſo *Coma*: in high-Dutch, **Bocxbaert:** in low-Dutch, **Iosephes bloemen:** in French, *Barbe de bouc*, and *Saſsify*: in Italian, *Saſſefrica*: in Spaniſh, *Barba Cabruna*: in Engliſh, Goats-beard, Ioſephs floure, Star of Ieruſalem, Noone-tide, and Go to bed at noone.

¶ *The Temperature.*

Theſe herbes are temperate betweene heate and moiſture.

¶ *The Vertues.*

A The roots of Goats-beard boyled in wine and drunke, aſſwageth the paine and pricking ſtitches of the ſides.

B The ſame boyled in water vntill they be tender, and buttered as parſeneps and carrots, are a moſt pleaſant and wholſome meate, in delicate taſte farre ſurpaſſing either Parſenep or Carrot: which meate procures appetite, warmeth the ſtomacke, preuaileth greatly in conſumptions, and ſtrengthneth thoſe that haue been ſicke of a long lingring diſeaſe.

Chap. 253. *Of Vipers-Graſſe.*

1 *Viperaria, ſiue Scorzonera Hiſpanica.*
Common Vipers Graſſe.

2 *Viperaria humilis.*
Dwarfe Vipers Graſſe.

† 3 *Viperaria Pannonica.*
Auſtrian Vipers graſſe.

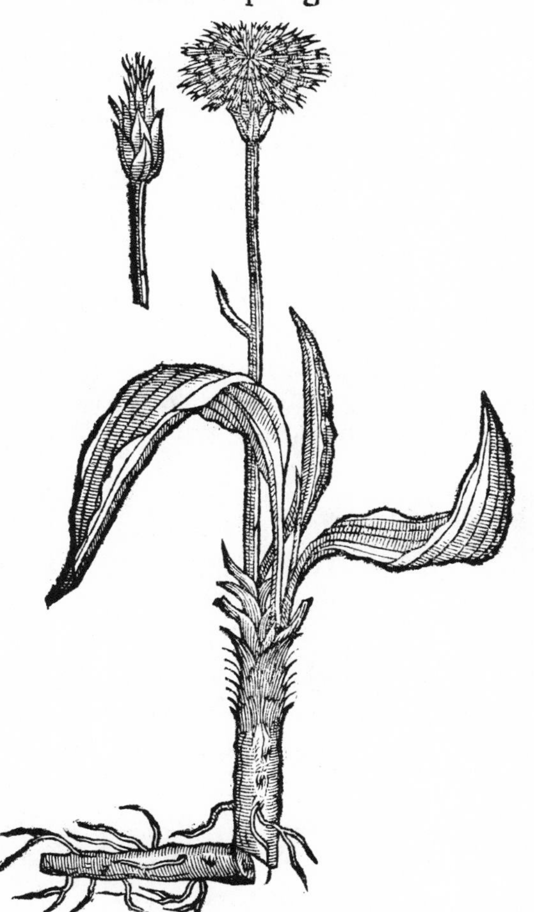

‡ 4 *Viperaria anguſtifolia elatior.*
Hungary Vipers graſſe.

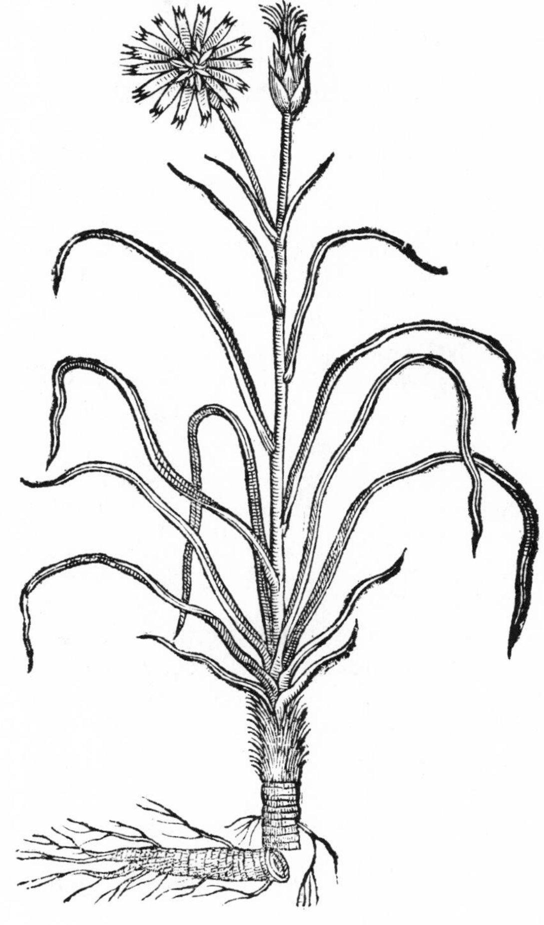

5 *Viperaria Pannonica anguſtifolia.*
Narrow leafed Vipers graſſe.

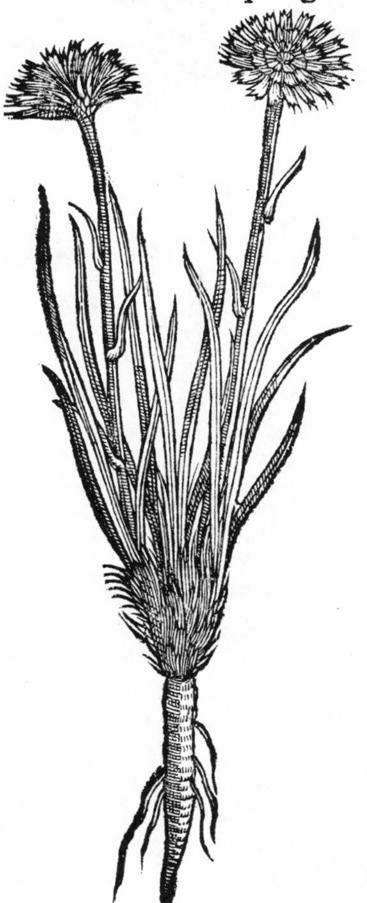

¶ *The Deſcription.*

1 THe firſt of the Viper graſſes hath long broad leaues, fat, or ful bodied, vneuen about the edges, ſharpe pointed, with a high ſwolne ribbe downe the middle, and of an ouer-worne colour, tending to the colour of Woade: among which riſeth vp a ſtiffe ſtalke, ſmooth and plaine, of two cubits high, whereon do grow ſuch leaues as thoſe next the ground. The flours ſtand on the top of the ſtalkes, conſiſting of many ſmall yellow leaues thicke thruſt together, very double, as are thoſe of Goates beard, whereof it is a kinde, as are all the reſt that doe follow in this preſent chapter: the root is long, thicke, very brittle, continuing many yeeres, yeelding great increaſe of roots, blacke without, white within, and yeelding a milkie juice, as doe the leaues alſo, like vnto the Goates beard.

2 The dwarfe Vipers graſſe differeth not from the precedent, ſauing that it is altogether leſſer, wherein eſpecially conſiſteth the difference.

† 3 The broad leafed Auſtrian Vipers-graſſe hath broad leaues ſharpe pointed, vneuen about the edges, of a blewiſh greene colour: the ſtalke riſeth vp to the height of a foot or better; on the top whereof do ſtand faire yellow floures, very double, greater and broader than any of the reſt

of a reſonable good ſmell. The ſeed followeth,long and ſharpe, like vnto thoſe of Goates-beard. The root is thicke,long,and full of a milkie juice, as are the leaues alſo.

4 The narrow leaued Hungary Vipers-graſſe hath long leaues like to thoſe of Goates-beard, but longer and narrower,among which riſeth vp a ſlender hollow ſtalke, ſtiffe and ſmooth, on the top whereof do ſtand faire double floures of a faire blew colour tending to purple, in ſhape like the other of his kinde,of a pleaſant ſweet ſmell, like the ſmell of ſweet balls made of *Benzoin*. The ſeed is conteined in ſmall cups like thoſe of Goates beard, wrapped in a downie matter, that is caried away with the winde. The root is not ſo thicke nor long as the others, very ſingle, bearded at the top,with certain hairy thrums yeelding a milkie juice of a reſinous taſte, and ſomewhat ſharpe withall. It endureth the winter euen as the others do.

‡ 5 This (whoſe figure was by our Authour put to the laſt deſcription) hath leaues like thoſe of Goates-beard,but ſtiffer and ſhorter,amongſt which there growes vp a ſhort hollow ſtalke ſome handfull high,ſet with a few ſhort leaues, bearing a yellow floure at the top,almoſt like that of the laſt ſaue one, but leſſe, the ſeed is conteined in ſuch cups as the common Vipers-graſſe, and being ripe is caried away with the leaſt winde. The root is blacke,with a wrinkled barke,and full of milke,hauing the head hairy,as alſo the laſt deſcribed hath. This by *Cluſius* is called *Scorſonera humilis anguſtifolia Pannonica.* ‡

¶ *The Place and Time.*

Moſt of theſe are ſtrangers in England. The two firſt deſcribed do grow in my garden. The reſt are touched in their ſeuerall titles.

They floure and flouriſh from May to the end of Iuly.

¶ *The Names.*

Vipers-graſſe is called of the Spaniards *Scorzonera*,which ſoundeth in Latine *Viperaria*, or *Viperina*,or *Serpentaria*, ſo called becauſe it is accounted to be of force and efficacy againſt the poiſons of Vipers and ſerpents, for *Vipera* or a viper is called in Spaniſh *Scurzo*: it hath no name either in the high or low Dutch,nor in any other,more than hath been ſaid,that I can reade: in Engliſh we may call it Scorzoner,after the Spaniſh name,or Vipers-graſſe.

¶ *The Temperature.*

They are hot and moiſt as are the Goates-beards.

¶ *The Vertues.*

A It is reported by thoſe of great iudgement, that Vipers-graſſe is moſt excellent againſt the infections of the plague, and all poiſons of venomous beaſts, and eſpecially to cure the bitings of vipers, (of which there be very many in Spaine and other hot countries, yet haue I heard that they haue been ſeen in England) if the juice or herbe be drunke.

B It helpeth the infirmities of the heart,and ſuch as vſe to ſwoune much: it cureth alſo them that haue the falling ſickeneſſe, and ſuch as are troubled with giddineſſe in the head.

C The root being eaten, either roſted in embers, ſodden, or raw, doth make a man merry, and remoueth all ſorrow.

D The root condited with ſugar,as are the roots of *Eringos* and ſuch like, worke the like effects: but more familiarly, being thus dreſſed.

Formerly there were ſix figures in this chapter, whereof the firſt and fourth were both of one plant, and the fifth which was of the *Scorſonera Boemica* of *Matthiolus* did not much differ from them; if it differ at all. In the title and hiſtory of the third there ſhould haue been put *Pannonica* in ſtead of *Hiſpanica*; as now it is.

Chap. 254. *Of Marigolds.*

¶ *The Deſcription.*

1 THe greateſt double Marigold hath many large,fat,broad leaues,ſpringing immediately from a fibrous or threddy root; the vpper ſides of the leaues are of a deepe greene, and the lower ſide of a more light and ſhining greene: among which riſe vp ſtalkes ſomewhat hairie, and alſo ſomewhat jointed, and full of a ſpungious pith. The floures in the top are beautifull, round,very large and double, ſomething ſweet, with a certaine ſtrong ſmell, of a light ſaffron colour, or like pure gold: from the which follow a number of long crooked ſeeds, eſpecially the outmoſt, or thoſe that ſtand about the edges of the floure; which being ſowne commonly bring forth ſingle floures, whereas contrariwiſe thoſe ſeeds in the middle are leſſer, and for the moſt part bring forth ſuch floures as that was from whence it was taken.

2 The common double Marigold hath many fat, thicke, crumpled leaues ſet vpon a groſſe and ſpungious ſtalke: whereupon do grow faire double yellow floures, hauing for the moſt part in the middle a bunch of threddes thicke thruſt together: which being paſt there ſucceed ſuch crooked ſeeds as the firſt deſcribed. The root is thicke and hard, with ſome threds annexed thereto.

3 The

1. 2. *Calendula maior polyanthos.*
The greater double Marigold.

4 *Calendula multiflora orbiculata.*
Double globe Marigold.

6 *Calendula ſimplici flore.*
Single Marigold.

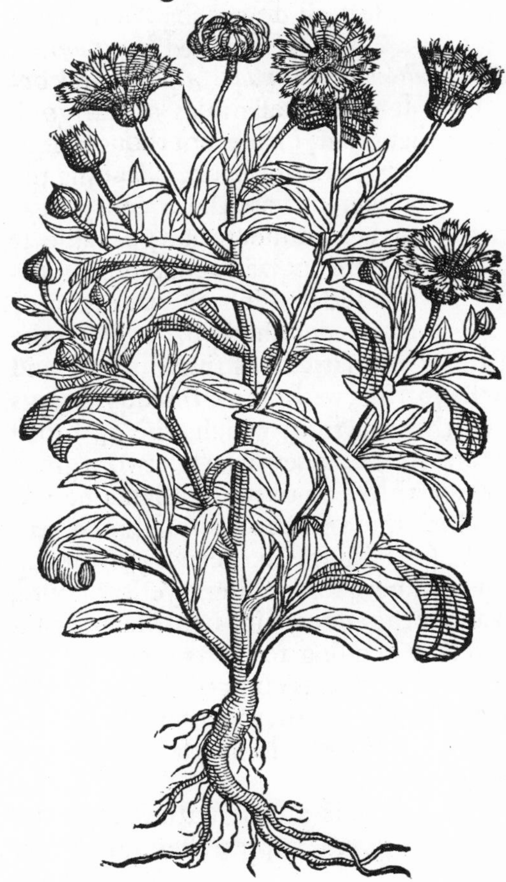

7 *Calendula prolifera.*
Fruitfull Marigold.

3 The ſmaller or finer leafed double Marigold groweth vpright, hauing for the moſt part one ſtem or fat ſpongeous ſtalke, garniſhed with ſmooth and fat leaues confuſedly. The floures grow at the top of the ſmall branches, very double, but leſſer than the other, conſiſting of more fine iaggedneſſe, and of a faire yellow gold colour. The root is like the precedent.

4 The Globe-flouring Marigold hath many large broad leaues riſing immediately forth of the ground; among which riſeth vp a ſtalke of the height of a cubit, diuiding it ſelfe toward the top into other ſmaller branches, ſet or garniſhed with the like leaues, but confuſedly, or without order. The floures grow at the top of the ſtalkes, very double; the ſmall leaues whereof are ſet in comely order by certaine rankes or rowes, as ſundry lines are in a Globe, trauerſing the whole compaſſe of the ſame; whereupon it tooke the name *Orbiculata*.

5 The fifth ſort of double Marigold differeth not from the laſt deſcribed, ſauing in the colour of the floures; for this plant bringeth forth floures of a ſtraw or light yellow colour, and the others not ſo, wherein conſiſteth the difference.

‡ All theſe fiue here deſcribed, and which formerly had ſo many figures, differ nothing but in the bigneſſe and littleneſſe of the plants and floures, and in the intenſeneſſe and remiſneſſe of their colour, which is either orange, yellow, or of a ſtraw colour. ‡

6 The Marigold with ſingle floures differeth not from thoſe with double floures, but in that it conſiſteth of fewer leaues, which we therefore terme Single, in compariſon of the reſt, and that maketh the difference.

7 This fruitfull or much bearing Marigold is likewiſe called of the vulgar ſort of women, Iacke-an-apes on horſebacke: it hath leaues, ſtalkes, and roots like the common ſort of Marigold, differing in the ſhape of his floures, for this plant doth bring forth at the top of the ſtalke one floure like the other Marigolds; from the which ſtart forth ſundry other ſmal floures, yellow likewiſe, and of the ſame faſhion as the firſt, which if I be not deceiued commeth to paſſe *per accidens*, or by chance, as Nature oftentimes liketh to play with other floures, or as children are borne with two thumbes on one hand, and ſuch like, which liuing to be men, do get children like vnto others; euen ſo is the ſeed of this Marigold, which if it be ſowen, it brings forth not one floure in a thouſand like the plant from whence it was taken.

8 The other fruitfull Marigold is doubtleſſe a degenerate kind, comming by chance from the ſeed of the double Marigold, whereas for the moſt part the other commeth of the ſeed of the ſingle floures, wherein conſiſteth the difference. ‡ The floure of this (wherein the onely difference conſiſts) you ſhall finde expreſt at the bottome of the fourth figure. ‡

9 *Calendula Alpina.*
Mountaine Marigold.

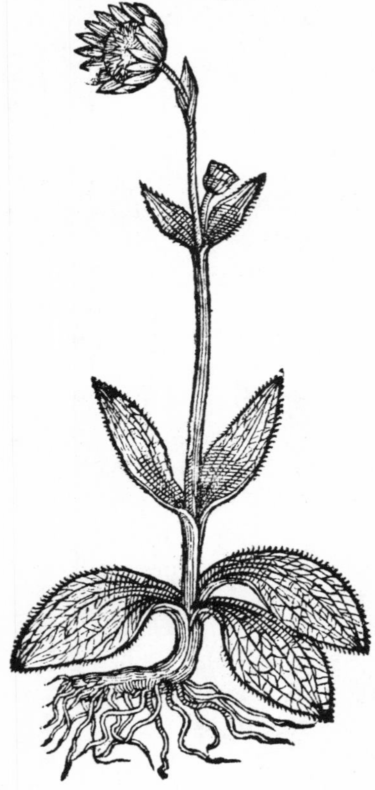

9 The Alpiſh or mountaine Marigold, which *Lobelius* ſetteth downe for *Nardus Celtica*, or *Plantago Alpina*, is called by *Tabernamontanus*, *Caltha*, or *Calendula Alpina*: and becauſe I ſee it rather reſembles a Marigold, than any other plant, I haue not thought it amiſſe to inſert it in this place, leauing the conſideration thereof vnto the friendly Reader, or to a further conſideration, becauſe it is a plant that I am not well acquainted withall; yet I doe reade that it hath a thicke root, growing aſlope vnder the vpper cruſt of the earth, of an aromaticall or ſpicie taſte, and ſomewhat biting, with many threddy ſtrings annexed thereto: from which riſe vp broad thicke and rough leaues of an ouerworn green colour, not vnlike to thoſe of Plantaine: among which there riſeth vp a rough and tender ſtalke ſet with the like leaues; on the top wherof commeth forth a ſingle yellow floure, paled about the edges with ſmall leaues of a light yellow, tending to a ſtraw colour; the middle of the floure is compoſed of a bundle of threds, thicke thruſt together, ſuch as is in the middle of the field Daiſie, of a deepe yellow colour.

‡ This Plant is all one with the two deſcribed in the next Chapter: they vary onely thus;

thus; the ſtalkes and leaues are ſometimes hairy, otherwhiles ſmooth; the floure is yellow, or elſe blew. I hauing three figures ready cut, thinke it not amiſſe to giue you one to expreſſe each varietie. ‡

10 The wilde Marigold is like vnto the ſingle garden Marigold, but altogether leſſer, and the whole plant periſheth at the firſt approch of Winter, and recouereth it ſelfe againe by falling of the ſeed.

¶ *The Place.*

Theſe Marigolds, with double floures eſpecially, are ſet and ſowen in gardens: the reſt, their titles do ſet forth their naturall being.

¶ *The Time.*

The Marigold floureth from Aprill or May euen vntill Winter, and in Winter alſo, if it be warme.

¶ *The Names.*

The Marigold is called *Calendula*: it is to be ſeene in floure in the Calends almoſt of euerie moneth: it is alſo called *Chryſanthemum*, of his golden colour: of ſome, *Caltha*, and *Caltha Poetarum*: whereof *Columella* and *Virgil* doe write, ſaying, That *Caltha* is a floure of a yellow colour: whereof *Virgil* in his Bucolickes, the ſecond Ecloge, writeth thus;

Tum Caſia atque aliȷs intexens ſuauibus herbis
Mollia Luteola pingit vaccinia Caltha.

And then ſhee'l Spike and ſuch ſweet herbes infold,
And paint the Iacinth with the Marigold.

Columella alſo in his tenth booke of Gardens hath theſe words;

Candida Leucoia & flauentia Lumina Calthæ.

Stock-Gillofloures exceeding white,
And Marigolds moſt yellow bright.

It is thought to be *Gromphena Pliniȷ*: in low-Dutch it is called **Goudt bloemen**: in high-Dutch, **Ringleblumen**: in French, *Souſiȷ & Goude*: in Italian, *Fior d' ogni meſe*: in Engliſh, Marigolds, and Ruddes.

¶ *The Temperature and Vertues.*

The floure of the Marigold is of temperature hot, almoſt in the ſecond degree, eſpecially when it is dry: it is thought to ſtrengthen and comfort the heart very much, and alſo to withſtand poyſon, as alſo to be good againſt peſtilent Agues, being taken any way. *Fuchſius* hath written, That being drunke with wine it bringeth downe the termes, and that the fume thereof expelleth the ſecondine or after-birth A

But the leaues of the herbe are hotter; for there is in them a certain biting, but by reaſon of the moiſture ioyned with it, it doth not by and by ſhew it ſelfe; by meanes of which moiſture they mollifie the belly, and procure ſolubleneſſe if it be vſed as a pot-herbe. B

Fuchſius writeth, That if the mouth be waſhed with the iuyce it helpeth the tooth-ache. C

The floures and leaues of Marigolds being diſtilled, and the water dropped into red and watery eyes, ceaſeth the inflammation, and taketh away the paine. D

Conſerue made of the floures and ſugar taken in the morning faſting, cureth the trembling of the heart, and is alſo giuen in time of plague or peſtilence, or corruption of the aire. E

The yellow leaues of the floures are dried and kept throughout Dutchland againſt Winter, to put into broths, in phyſicall potions, and for diuers other purpoſes, in ſuch quantity, that in ſome Grocers or Spice-ſellers houſes are to be found barrels filled with them, and retailed by the penny more or leſſe, inſomuch that no broths are well made without dried Marigolds. F

CHAP. 255. *Of Germane Marigolds.*

¶ *The Deſcription.*

1 GOlden Marigold with the broad leafe doth forthwith bring from the root long leaues ſpred vpon the ground, broad, greene, ſomething rough in the vpper part, vnderneath ſmooth, and of a light greene colour: among which ſpring vp ſlender ſtalks a cubit high,

high, something hoarie, hauing three or foure ioynts, out of euery one whereof grow two leaues, set one right against another, and oftentimes little slender stems; on the tops whereof stand broad round floures like those of Ox-eye, or the corne Marigold, hauing a round ball in the middle (such as is in the middle of those of Camomil) bordered about with a pale of bright yellow leaues. The whole floure turneth into downe that is carried away with the winde; among which down is found long blackish seed. The root consisteth of threddy strings.

† 2 The lesser sort hath foure or fiue leaues spred vpon the ground like vnto those of the last described, but altogether lesser and shorter: among which riseth vp a slender stalke two hands high; on the top whereof stand such floures as the precedent, but not so large, and of a blew colour.

‡ These two here described, and that described in the ninth place of the foregoing Chapter, are all but the varieties of one and the same plant, differing as I haue shewed in the foregoing Chapter. ‡

1 *Chrysanthemum latifolium.*
Golden Marigold with the broad leafe.

2 *Chrysanthemum latifolium minus.*
The lesser Dutch Marigold.

¶ *The Place.*

They be found euery where in vntilled places of Germanie, and in woods, but are strangers in England.

¶ *The Time.*

They are to be seene with their floures in Iune, in the gardens of the Low-countries.

¶ *The Names.*

Golden Marigold is called in high-Dutch **Waldtblume.** There are that would haue it to be *Alisma Dioscoridis*; which is also called *Damasonium*, but vnproperly; therefore we must rather call it *Chrysanthemum latifolium*, than rashly attribute vnto it the name of *Alisma*. ‡ This plant indeed is a *Doronicum*, and the figure in the precedent chapter by *Clusius* is set forth by the name of *Doronicum 6. Pannonicum*: *Matthiolus* calls this plant *Alisma*: *Gesner*, *Caltha Alpina*: *Dodonæus*, *Chrysanthemum latifolium*: *Pena* and *Lobel*, *Nardus Celtica altera*. Now in the *Historia Lugd.* it is set forth in in foure seuerall places by three of the former names; and *pag.* 1169. by the name of *Ptarmica montana Daleschampij.*

¶ *The Temperature.*

It is hot and dry in the second degree being greene, but in the third being dry.

¶ The

¶ *The Vertues.*

The women that liue about the Alps wonderfully commend the root of this plant againſt the ſuffocation of the mother, the ſtoppings of the courſes, and the green ſickneſſe and ſuch like affects in maids. *Hiſtor. Lugd.* ‡

Chap. 256. *Of Corne-Marigold.*

¶ *The Deſcription.*

1 Corne Marigold or golden Corne floure hath a ſoft ſtalke, hollow, and of a greene colour, wherupon do grow great leaues, much hackt and cut into diuers ſections, and placed confuſedly or out of order: vpon the top of the branches ſtand faire ſtarlike floures, yellow in the middle, and ſuch likewiſe is the pale or border of leaues that compaſſeth the ſoft bal in the middle, like that in the middle of Camomill floures, of a reaſonable pleaſant ſmel. The roots are full of ſtrings.

† 1 *Chryſanthemum ſegetum.*
Corne Marigold.

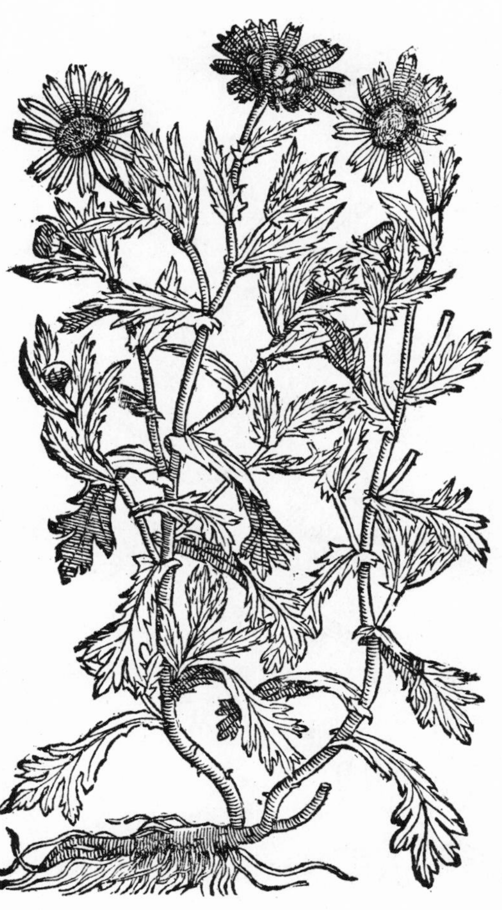

2 *Chryſanthemum Valentinum.*
Corne Marigold of Valentia.

2 The golden floure of Valentia hath a thicke fat ſtalk, rough, vneuen, and ſomewhat crooked, whereupon do grow long leaues, conſiſting of a long middle rib, with diuers little fetherlike leaues ſet thereon without order. The floures grow at the top of the ſtalks, compoſed of a yellow thrummie matter, ſuch as in the middle of the Camomill floures, and is altogether like the Corne Marygold laſt deſcribed, ſauing it doth want that border or pale of little leaues that do compaſſe the ball or head: the root is thicke, tough, and diſperſeth it ſelfe far abroad.

‡ 3 To theſe may be added diuers other, as the *Chryſanthema Alpina*, of *Cluſius*, & his *Chryſanthemum Creticum*, & others. The firſt of theſe ſmal mountain Marigolds of *Cluſius* his deſcription hath leaues like thoſe of white Wormewood, but greener and thicker: the ſtalks grow ſome handfull high, ſet with few and much diuided leaues; and at the tops, as in an vmbell, they carry ſome dozen

zen floures more or leſſe, not much vnlike in ſhape,colour, and ſmell,to thoſe of the common *Iacobæa*,or Ragwoort. The root is ſomewhat thicke, and puts forth many long white fibres. It floures in Iuly and Auguſt,and growes vpon the Alpes of Stiria. *Cluſius* calls it *Chryſanthemum Alpinum.* 1.

4 The ſecond of his deſcription hath many leaues at the root, like to the leaues of the male Sothernwood,but of a lighter and brighter greene,and of no vnpleaſant ſmell, though the taſte be bitteriſh and vngratefull : in the middeſt of the leaues grow vp ſtalkes ſome foot high,diuided at their tops into ſundry branches,which carry each of them two or three floures bigger than,yet like thoſe of the common Cammomill,but without ſmel,and wholly yellow : the root is fibrous,blackiſh,and much ſpreading. It floures in Auguſt,and growes in the like places as the former. *Bauhine* iudges this to be the *Achillæa montana Artemiſiæ tenuifoliæ facie* of the *Aduerſ.* and the *Ageratum ferulaceum* in the *Hiſt. Lugd.* But I cannot be of that opinion; yet I iudge the *Achillæa montana*,and *Ageratum ferulaceum* to be but of the ſame plant. But different from this, & that chiefely in that it hath many more,and thoſe much leſſe floures than thoſe of the plant here figured and deſcribed.

5 Now ſhould I haue giuen you the hiſtorie of the *Chryſanthemum Creticum* of the ſame Author,but that my friend Mr. *Goodyer* hath ſaued me the labour, by ſending an exact deſcription thereof, together with one or two others of this kinde, which I thinke fit here to giue you.

‡ 3 *Chryſanthemum Alpinum 1. Cluſ.*
Small mountaine Marigold.

‡ 4 *Chryſanthemum Alpinum 2. Cluſ.*
The other Alpine Marigold.

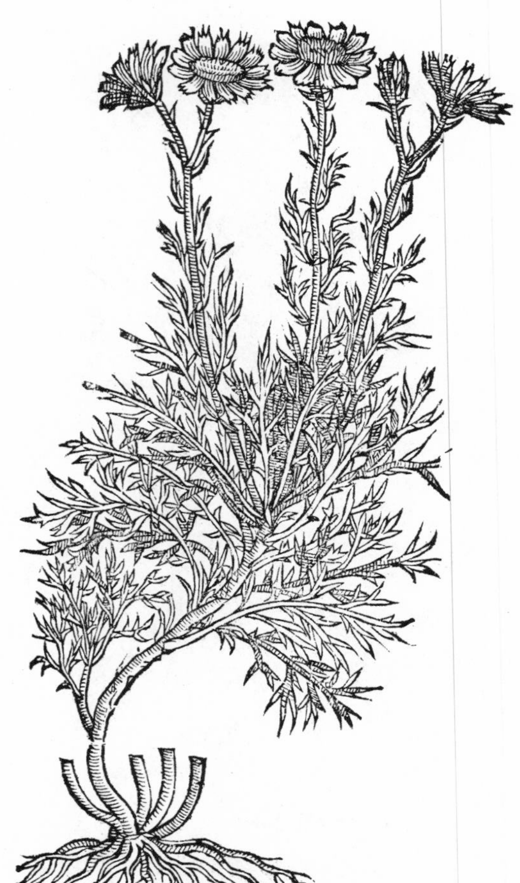

Chryſanthemum Creticum primum Cluſij, pag. 334.

The ſtalkes are round,ſtraked,branched,hard,of a whitiſh greene,with a very little pith within ; neere three foot high : the leaues grow out of order,diuided into many parts,and thoſe again ſnipt or diuided, of the colour of the ſtalkes : at the tops of the ſtalkes and branches grow great floures, bigger than any of the reſt of the Corne-floures,forth of ſcaly heads,conſiſting of twelue or more broad leaues apeece, notched at the top,of a ſhining golden colour at the firſt,which after turne to a pale,whitiſh,or very light yellow,and grow round about a large yellow ball, of ſmell ſomewhat ſweet. The floures paſt,there commeth abundance of ſeed cloſely compact or thruſt together,and it is ſhort,blunt at both ends,ſtraked,of a ſalue colour,ſomwhat flat,& of a reaſonable bignes. The root

‡ 5 *Chryſanthemum Creticum.*
Candy Corne Marigold.

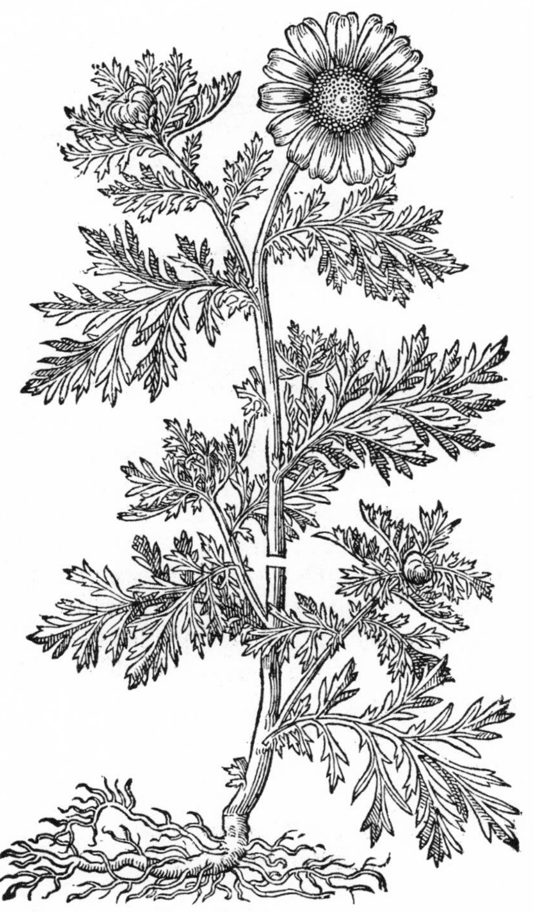

root is whitiſh, neere a fingers bigneſſe, ſhort, with many threds hanging thereat,and periſheth when the ſeede is ripe ; and at the Spring groweth vp againe by the falling of the ſeed.

Chryſanthemum Bæticum Boelij, inſcriptum.

The ſtalks are round,ſtraked,reddiſh brown, diuided into branches,containing a ſpungious white pith within, a cubite high : the leaues grow out of order,without footſtalkes, about three inches long,and an inch broad, notched about the edges, not at all diuided, of a darke greene colour : the floures grow at the tops of the ſtalkes and branches, forth of great ſcaly heads, containing twentie leaues a piece or more, notched at the top, of a ſhining yellow colour,growing about a round yellow ball,of a reaſonable good ſmell, very like thoſe of the common *Chryſanthemum ſegetum* : the ſeede groweth like the other,and is very ſmall,long, round,crooked and whitiſh : the root is ſmall, whitiſh, threddie,and periſheth alſo when the ſeed is ripe.

Chryſanthemum tenuifolium Bæticum Boelij.

The ſtalks are round,ſmall,ſtraked,reddiſh, ſomewhat hairie, branched, a cubit high or higher:the leaues are ſmall,much diuided,iagged, and very like the leaues of *Cotula fætida* : the floures are yellow, ſhining like gold,compoſed of thirteene or fourteene leaues a piece, notched at the top,ſet about a yellow ball,alſo like the common *Chryſanthemum ſegetum* : the ſeed groweth amongſt white flattiſh ſcales,which are cloſely compacted in a round head together,and are ſmall,flat,grayiſh,and broad at the top : the root is ſmall,whitiſh,with a few threds,and dyeth when the ſeed is ripe. Iuly 28. 1621. *Iohn Goodyer.* ‡

¶ *The Place.*

The firſt groweth among corne,and where corne hath been growing : it is found in ſome places with leaues more iagged,and in others leſſe.

The ſecond is a ſtranger in England.

¶ *The Time.*

They floure in Iuly and Auguſt.

¶ *The Names.*

Theſe plants are called by one name in Greeke,of the golden glittering colour, χρυσάνθεμον : in High Dutch, **Sant Iohans blum** : in Low Dutch, **Vokelaer** : in Engliſh, Corne Marigold, yellow Corne floure,and golden Corne floure.

There be diuers other floures called *Chryſanthemum* alſo,as *Batrachion*,a kinde of yellow Crowfoot, *Heliochryſon*,but theſe golden floures differ from them.

¶ *The Temperature.*

They are thought to be of a meane temperature betweene heat and moiſture.

¶ *The Vertues.*

The ſtalks and leaues of Corne Marigold, as *Dioſcorides* ſaith,are eaten as other pot-herbes are. A

The floures mixed with wax,oile,roſine, and frankinſence, and made vp into a ſeare-cloth, waſteth away cold and hard ſwellings. B

The herbe it ſelfe drunke, after the comming forth of the bath,of them that haue the yellow iaundiſe,doth in ſhort time make them well coloured. C

† The figure that was in the firſt place was of the *Chryſanthemum* of *Matthiolus*,which is a ſtranger with vs,and the leaues of it are much like thoſe of Feuerſew,or Mugwoort, the floure is ſomewhat like, but larger than that of Feuerſew,and wholly yellow.

CHAP.

Chap. 257. *Of Oxe-Eie.*

¶ *The Deſcription.*

1 The plant which wee haue called *Buphthalmum*, or Oxe-eie, hath ſlender ſtalks growing from the roots, three, foure, or more, a foot high, or higher, about which be green leaues finely iagged like to the leaues of Fenell, but much leſſer: the floures in the tops of the ſtalks are great, much like to Marigolds, of a light yellow colour, with yellow threds in the middle, after which commeth vp a little head or knap like to that of red Mathes before deſcribed, called *Adonis*, conſiſting of many ſeeds ſet together. The roots are ſlender, and nothing but ſtrings, like to the roots of blacke Ellebor, whereof it hath beene taken to be a kinde.

2 The Oxe-eie which is generally holden to be the true *Buphthalmum* hath many leaues ſpred vpon the ground, of a light greene colour, laied far abroad like wings, conſiſting of very many fine iags, ſet vpon a tender middle rib: among which ſpring vp diuers ſtalks, ſtiffe and brittle, vpon the top whereof do grow faire yellow leaues, ſet about a head or ball of thrummie matter, ſuch as in the middle of Cammomill, like a border or pale. The root is tough and thicke, with certaine ſtrings faſtned thereto.

3 The white Oxe-eie hath ſmall vpright ſtalks of a foot high, whereon do grow long leaues, compoſed of diuers ſmal leaues, and thoſe ſnipt about the edges like the teeth of a ſaw. The floures grow on the tops of the ſtalks, in ſhape like thoſe of the other Oxe-eie; the middle part whereof is likewiſe made of a yellow ſubſtance, but the pale or border of little leaues, are exceeding white, like thoſe of great Daſie, called *Conſolida media vulnerariorum*. The root is long, creeping alongſt vnder the vpper cruſt of the earth, whereby it greatly increaſeth. ‡ This by the common conſent of all writers that haue deliuered the hiſtorie thereof, hath not the pale or out leaues of the floure white, as our Author affirmes, but of a bright and perfect yellow colour. And this is the *Buphthalmum*, of *Tragus*, *Matthiolus*, *Lobel*, *Cluſius* and others. ‡

1 *Buphthalmum ſive Helleborus niger ferulaceus.*
Oxe-eie.

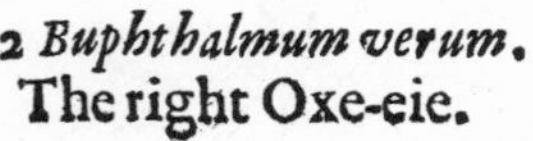
2 *Buphthalmum verum.*
The right Oxe-eie.

¶ The

3 *Buphthalmum vulgare.*
White Oxe-eie.

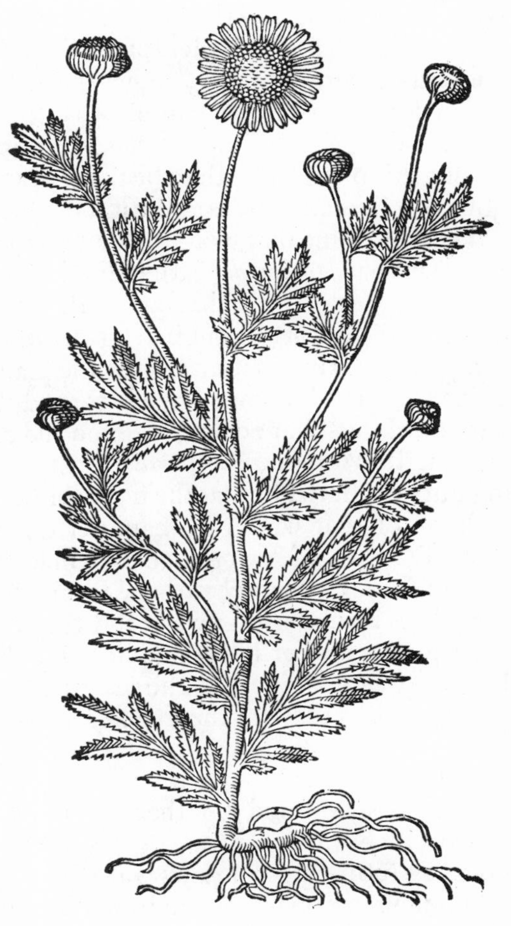

¶ *The Place.*

The two first growe of themselues in Germany, Bohemia, and in the Gardens of the Low-countries; of the first I haue a plant in my garden. The last groweth in barren pastures and fields almost euery where.

‡ The last is also a stranger with vs, for any thing that I know or can learne; neither can I coniecture what our Authour meant here: first in that he said the floures of this were white, and secondly in that it grew in barren pastures and fields almost euery where. ‡

¶ *The Time.*

They floure in May and Iune. The last in August.

¶ *The Names.*

Touching the naming of the first of those plants the late writers are of diuers opinions: some would haue it to be a kind of *Veratrum nigrum*, blacke Hellebor: other some *Consiligo*, or Bearefoot; and againe, others, *Sesamoides*; and some, *Elleborastrum*: But there be found two kindes of blacke Ellebor among the old writers, one with a leafe like vnto Laurel, with the fruit of *Sesamum*: the other with a leafe like that of the Plane tree, with the seed of *Carthamus* or Bastard Saffron. But it is most euident, that this *Buphthalmum*, in English, Oxe-eie, which in this Chapter we in the first place haue described, doth agree with neither of these: what form *Consiligo* is of, we finde not among the old writers. *Pliny 26 cap. 7.* saith, That in his time it was found amongst the *Marsi*, and was a present remedy for the infirmitie of the lungs of swine, and of all kinde of cattell, though it were but drawne thorow the eare. *Columella* in his *6.* booke, Chap. 5. doth also say, that in the mountaines called Marsi there is very great store thereof, and that it is very helpfull to all kind of cattel, and he telleth how and in what manner it must be put into the eare; the roots also of our Oxe-eie are said to cure certain infirmities of cattel, if they be put into the slit or bored eare: but it followeth not that for the same reason it should be *Consiligo*; and it is an ordinary thing to find out plants that are of a like force and qualitie: for *Pliny* doth testifie in his 25 booke, 5 chapter, That the roots also of blacke Ellebor can do the same: it cureth (saith he) the cough in cattel, if it be drawn thorow the eare, and taken out again the next day at the same houre: which is likewise most certaine by experiments of the countrey men of our age; who do cure the diseases of their cattell with the roots of common black Ellebor. The roots of white Ellebor also do the like, as *Absyrtus*, and after him *Hierocles* doth write: who notwithstanding do not thrust the roots of white Ellebor into the eare, but vnder the skin of the brest called the dewlap: after which manner also *Vegetius Renatus* doth vse *Consiligo*, in his first booke of the curing of cattell, chapter 12. intituled, Of the cure of the infirmities vnder the skin: although in his 3 booke, 2. Chapter, *de Malleo*, he writeth, that they also must be fastned thorow the eare: which things do sufficiently declare, that sundry plants haue oftentimes like faculties: and that it doth not at all follow by the same reason, that our Oxe eie is *Consiligo*, because it doth cure diseases in Cattél as wel as *Consiligo* doth. But if we must coniecture by the faculties, *Consiligo* then should bee White Ellebor: for *Vegetius* vseth *Consiligo* in the very same maner that *Absyrtus* and *Hierocles* do vse white Ellebor. This suspition is made the greater, because it is thought that *Vegetius* hath taken this manner of curing from the Grecians; for which cause also most doe take *Consiligo* to be nothing else but white Ellebor: the which if it be so, then shall this present Oxe-eie much differ from *Consiligo*; for it is nothing at all like to white Ellebor.

And that the same is not *Sesamoides*, either the first or the second, it is better knowne, than needfull to be confuted.

This

This same also is vnproperly called *Helleborastrum*; for that may aptly bee called *Helleborastrum* which hath the forme and likenesse of Hellebor: and this Oxe eie is nothing at all like to Ellebore. For all which causes it seemeth that none of these names agree with this plant,but only the name *Buphthalmum*, with whose description which is extant in *Dioscorides* this plant doth most aptly agree. We take it to be the right Oxe eie; for Oxe eie bringeth forth slender soft stalks, and hath leaues of the likenesse or similitude of Fennell leaues: the floure is yellow, bigger than that of Cammomill, euen such an one is this present plant,which doth so exquisitly expresse that form or likenesse of Fennell leaues,both in slendernesse and manifold iaggednesse of the leaues,as no other little leafed herb can do better;so that without all doubt this plant seemeth to be the true and right Oxe eie. Oxe eie is called *Cachla*, or rather *Caltha*;but *Caltha* is *Calendula*, or Marigold,which we said that our Oxe eie in floure did neerest represent. There are some that would haue *Buphthalmum* or Oxe eie to be *Chrysanthemum*,yellow Cammomil,& say that *Dioscorides* hath in sundry places,and by diuers names intreated of this herbe; but if those men had somewhat more diligently weighed *Dioscorides* his words,they would haue bin of another minde: for although descriptions of either of them do in many things agree,yet there is no property wanting that may shew the plants to differ. The leaues of *Chrysanthemum* are said to be diuided and cut into many fine iags: and the leaues of *Buphthalmum* to be like the leaues of Fennell: for all things that be finely iagged and cut into many parts haue not the likenesse of the leaues of Fennell. Moreouer, *Dioscorides* saith, that *Chrysanthemum* doth bring forth a floure much glittering,but he telleth not that the floure of *Buphthalmum*,or Oxe eie is much glittering,neither doth the floure of that which we haue set downe glitter,so that it can or ought not to be said to glitter much. Do not these things declare a manifest difference betweene *Buphthalmum* and *Chrysanthemum*,and confirme that which we haue set down to be the true and right Oxe eie? We are of that minde,let others thinke as they will: and they that would haue *Chrysanthemum* to be *Buphthalmum*,let them seeke out another,if they denie this to bee Oxe eie: for that which we and others haue described for *Chrysanthemum* cannot be the true *Buphthalmum* or Oxe eie; for the leaues of it are not like Fennell,such as those of the true *Buphthalmum* ought to be.

¶ *The Temperature.*

But concerning the faculties *Matthiolus* saith,that all the Physitions and Apothecaries in Bohemia,vse the roots of this Oxe eie in stead of those of blacke Ellebor,namely for diseases in cattel: but he doth not affirme that the roots hereof in medicines are substitutes, or *quid pro quo*; for,saith he,I do remember that I once saw the roots hereof in a sufficient big quantitie put by certain Physitions into decoctions which were made to purge by siege, but they purged no more than if they had not been put in at all: which thing maketh it most plaine,that it cannot be any of the Ellebors, although it hath been vsed to be fastned through the eares of cattell for certaine diseases,and doth cure them as Ellebor doth. The roots of *Gentian* do mightily open the orifices of Fistulaes, which be too narrow,so do the roots of *Aristolochia*,or Birthwoort,or Brionie,or pieces of spunges,which notwithstanding do much differ one from another in other operations: wherefore though the roots of Oxe eie can do something like vnto blacke Ellebor,yet for al that they cannot perform all those things that the same can. We know that thornes,stings,splinters of wood,and such like,bring pain, cause inflammations,draw vnto them humors from the parts neere adioining, if they be fastned in any part of the bodie;no part of the bodie is hurt without pain;the which is increased if any thing be thrust through, or put into the wound: peraduenture also if any other thing beside be put into the slit or bored eare,the same effect would follow which hapned by the root of this plant thrust in; notwithstanding we here affirme nothing,we onely make way for curious men to make more diligent search touching the operations hereof. ‡ *Clusius* affirmes that when hee came to Vienna in Austria,this was vulgarly bought,sold,and vsed for the true blacke Ellebor, the ignorance of the Physitions and Apothecaries in the knowledge of simples was such to make vse of this so far different plant, when as they had the true blacke Hellebor growing plentifully wilde within seuen miles of the citie,the which afterward vpon his admonition, they made vse of. ‡

¶ *The Vertues.*

A *Dioscorides* saith,that the floures of Oxe eie made vp in a seare-cloth doe asswage and waste away cold hard swellings;and it is reported that if they be drunk by and by after bathing,they make them in short time well coloured that haue been troubled with the yellow iaundice.

Chap.

Chap. 258. *Of French Marigold, or African Marigold.*

¶ The Deſcription.

1 THe great double African Marigold hath a great long browne reddiſh ſtalke, creſted, furrowedand ſomewhat knobby, diuiding it ſelfe toward the top into other branches; whereupon do grow leaues compoſed of many ſmall leaues ſet vpon a middle rib by couples,much like vnto the leaues of wilde Valerian, bearing at the top very faire and beautifull double yellow floures, greater and more double than the greateſt Damaske Roſe, of a ſtrong ſmell, but not vnpleaſant. The floures being paſt, there ſucceedeth long blacke flat ſeed: the whole plant periſheth at the firſt approach of winter.

2 There is little difference betweene this and the precedent, or laſt deſcribed, ſauing that this plant is much leſſer, and bringeth forth more ſtore of floures, which maketh the difference. ‡ And we may therefore call it *Flos Aphricanus minor multiflorus*, The ſmall double Africane Marigold. ‡

1 *Flos Aphricanus maior Polyanthos.*
The great African double Marigold.

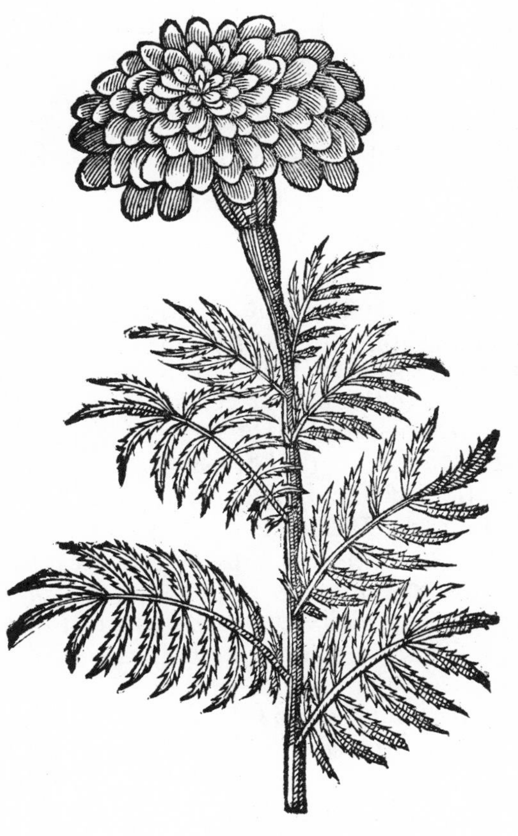

3 *Flos Aphricanus maior ſimplici flore.*
The great ſingle French Marigold.

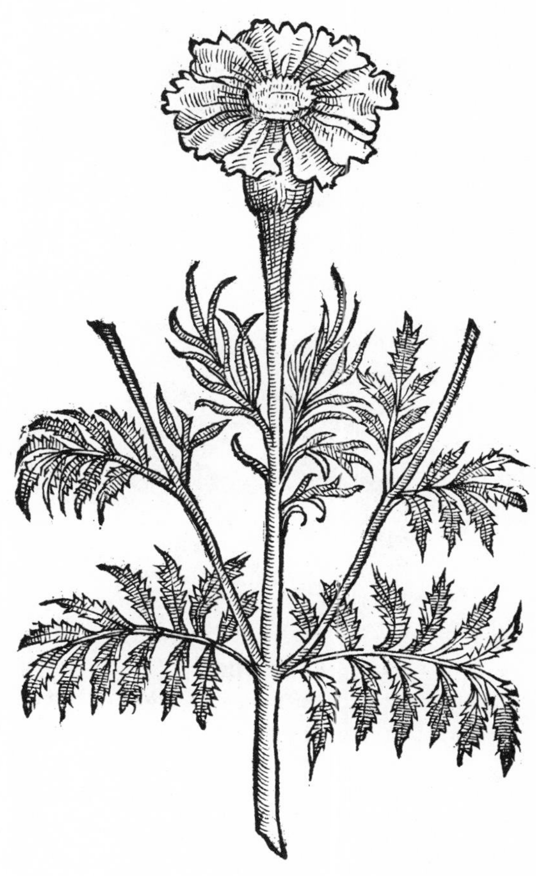

3 The ſingle great Africane Marigold hath a thicke root, with ſome fibres annexed thereto; from which riſeth vp a thicke ſtalke chamfered and furrowed, of the height of two cubits, diuided into other ſmall branches; whereupon are ſet long leaues, compact or compoſed of many little leaues like thoſe or the Aſh tree, of a ſtrong ſmell, yet not very vnpleaſant: on the top of the branches do grow yellow ſingle floures, compoſed in the middle of a bundle of yellow thrummes hard thruſt together, paled about the edges with a border of yellow leaues; after which commeth long blacke ſeed. The whole plant periſheth with the firſt froſt, and muſt be ſowne yeerely as the other ſorts muſt be.

4 The common Africane or as they vulgarly terme it French Marigold hath ſmall weake and tender branches trailing vpon the ground, reeling and leaning this way and that way, beſet with leaues conſiſting of many particular leaues, indented about the edges, which being held vp againſt the ſunne, or to the light, are ſeene to be full of holes like a ſieue, euen as thoſe of Saint Iohns

woort:

woort: The floures stand at the top of the springie branches forth of long cups or huskes, consisting of eight or ten small leaues, yellow vnderneath, on the vpper side of a deeper yellow tending to the colour of a darke crimson veluet, as also soft in handling: but to describe the colour in words, it is not possible, but this way; lay vpon paper with a pensill a yellow colour called Masticot, which being dry, lay the same ouer with a little saffron steeped in water or wine, which setteth forth most liuely the colour. The whole plant is of a most ranke and vnwholesome smell, and perisheth at the first frost.

4 *Flos Aphricanus minor simplici flore.*
The small French Marigold.

¶ *The Place.*

They are cherished and sowne in gardens euery yeere: they grow euery where almost in Africke of themselues, from whence we first had them, and that was when *Charles* the fifth Emperour of Rome made a famous conquest of Tunis; whereupon it was called *Flos Aphricanus*, or *Flos Tunetanus*.

¶ *The Time.*

They are to be sowne in the beginning of Aprill, if the season fall out to be warme, otherwise they must be sowne in a bed of dung, as shall be shewed in the chapter of Cucumbers. They bring forth their pleasant floures very late, and therefore there is the more diligence to be vsed to sow them very earely, because they shall not be ouertaken with the frost before their seed be ripe.

¶ *The Names.*

The Africane or French Marigold is called in Dutch, **Thunis bloemen**: in high Dutch, **Indianisch negelin**, that is, the floure or Gillofloure of India: in Latine, *Cariophillus Indicus*; whereupon the French men call it *Oeilletz d' Inde*. *Cordus* calleth it *Tanacetum Peruvianum*, of the likenesse the leaues haue with Tansie, and of Peru a Prouince of America, from whence hee thought, it may be, it was first brought into Europe. *Gesner* calleth it *Caltha Aphricana*, and saith that it is called in the Carthagenian tongue, *Pedua*: some would haue it to be *Petilius flos Pliny*, but not properly: for *Petilius flos* is an Autumne floure growing among briers and brambles. *Andreas Lacuna* calleth it *Othonna*, which is a certaine herbe of the Troglodytes, growing in that part of Arabia which lieth toward Ægypt, hauing leaues full of holes as though they were eaten with mothes. *Galen* in his first booke of the faculties of Simple medicines, maketh mention of an herbe called *Lycopersicum*, the juice whereof a certain Centurion did cary out of Barbarie all Ægypt ouer with so rancke a smell, and so lothsome, as *Galen* himselfe durst not so much as taste of it, but conjectured it to be deadly; yet that Centurion did vse it against the extreme paines of the joints, and it seemeth to the patients themselues, to be of a very cold temperature; but doubtlesse of a poisonsome quality, very neere to that of hemlockes.

¶ *The Temperature and Vertues.*

A The vnpleasant smell, especiall that common sort with single floures (that stuffeth the head like to that of Hemlocke, such as the juice of *Lycopersium* had) doth shew that is of a poisonsome and cooling qualitie; and also the same is manifested by diuers experiments: for I remember, saith *Dodonæus*, that I did see a boy whose lippes and mouth when he began to chew the floures did swell extremely; as it hath often happened vnto them, that playing or piping with quils or kexes of Hemlockes, do hold them a while between their lippes: likewise he saith, we gaue to a cat the floures with their cups, tempered with fresh cheese, she forthwith mightely swelled, and a little while after died: also mice that haue eaten of the seed thereof haue been found dead. All which things do declare that this herbe is of a venomous and poisonsome facultie; and that they are not

not to be hearkned vnto,that ſuppoſe this herbe to be an harmles plant:ſo to conclude,theſe plants are moſt venomous and full of poiſon, and therefore not to be touched or ſmelled vnto,much leſſe vſed in meat or medicine.

CHAP. 259. *Of the floure of the Sun,or the Marigold of Peru.*

¶ *The Deſcription.*

1 THe Indian Sun or the golden floure of Peru is a plant of ſuch ſtature and talneſſe that in one Sommer being ſowne of a ſeede in Aprill,it hath riſen vp to the height of fourteene foot in my garden,where one floure was in weight three pound and two ounces, and croſſe ouerthwart the floure by meaſure ſixteene inches broad. The ſtalkes are vpright and ſtraight,of the bigneſſe of a ſtrong mans arme, beſet with large leaues euen to the top, like vnto the great Clot Bur: at the top of the ſtalke commeth forth for the moſt part one floure,yet many times there ſpring out ſucking buds,which come to no perfection:this great floure is in ſhape like to the Cammomil floure,beſet round about with a pale or border of goodly yellow leaues,in ſhape like the leaues of the floures of white Lillies: the middle part whereof is made as it were of vnſhorn veluet,or ſome curious cloth wrought with the needle,which braue worke, if you do thorowly view and marke well,it ſeemeth to be an innumerable ſort of ſmall floures, reſembling the noſe or nozell of a candleſticke,broken from the foot thereof: from which ſmall nozell ſweateth forth excellent fine and cleere Turpentine,in ſight,ſubſtance,ſauour and taſte. The whole plant in like manner being broken,ſmelleth of Turpentine: when the plant groweth to maturitie,the floures fal away,in place whereof appeareth the ſeed,blacke, and large, much like the ſeed of Gourds, ſet as though a cunning workeman had of purpoſe placed them in very good order, much like the honiecombes of Bees: the root is white, compact of many ſtrings, which periſh at the firſt approch of winter,and muſt be ſet in moſt perfect dunged ground: the manner how,ſhall be ſhewed when vpon the like occaſion I ſhall ſpeake of Cucumbers and Melons.

1 *Flos Solis maior.*
The greater Sun floure.

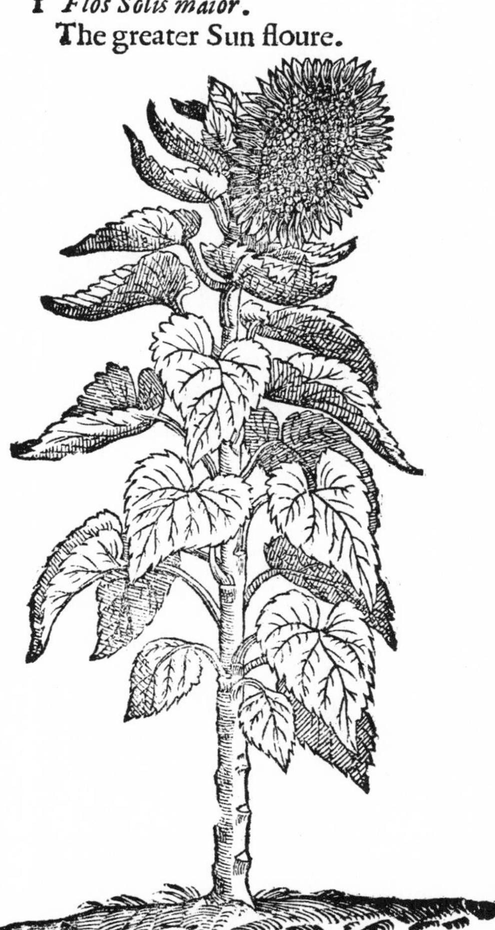

2 *Flos Solis minor.*
The leſſer Sunne floure.

2 The

2 The other golden floure of Peru is like the former,ſauing that it is altogether lower,and the leaues more iagged,and very few in number.

3 The male floure of the Sun of the ſmaller ſort hath a thicke root,hard,and of a wooddy ſubſtance,with many threddie ſtrings annexed thereto, from which riſeth vp a gray or ruſſet ſtalke,to the height of fiue or ſix cubits, of the bigneſſe of ones arme,whereupon are ſet great broad leaues with long foot-ſtalkes,very fragill or eaſie to breake,of an ouerworne greene colour,ſharp pointed, and ſomewhat cut or hackt about the edges like a ſaw: the floure groweth at the top of the ſtalks, bordered about with a pale of yellow leaues:the thrummed middle part is blacker than that of the laſt deſcribed. The whole floure is compaſſed about likewiſe with diuers ſuch ruſſet leaues as thoſe are that do grow lower vpon the ſtalks,but leſſer and narrower. The plant and euery part thereof doth ſmell of Turpentine,and the floure yeeldeth forth moſt cleere Turpentine,as my ſelfe haue noted diuers yeares. The ſeed is alſo long and blacke,with certaine lines or ſtrakes of white running alongſt the ſame. The roote and euery part thereof periſheth when it hath perfected his ſeed.

4 The female or Marigold Sun floure hath a thicke and wooddie root, from which riſeth vp a ſtraight ſtem,diuiding it ſelfe into one or more branches, ſet with ſmooth leaues ſharpe pointed, ſleightly indented about the edges. The floures grow at the top of the branches,of a faint yellow colour,the middle part is of a deeper yellow tending to blackneſſe,of the forme and ſhape of a ſingle Marigold,whereupon I haue named it the Sunne Marigold. The ſeed as yet I haue not obſerued.

¶ *The Place.*

Theſe plants do grow of themſelues without ſetting or ſowing,in Peru,and in diuers other prouinces of America, from whence the ſeeds haue beene brought into theſe parts of Europe. There hath been ſeen in Spaine and other hot regions a plant ſowne and nouriſhed vp from ſeed,to attain to the height of 24.foot in one yeare.

¶ *The Time.*

The ſeed muſt be ſet or ſowne in the beginning of Aprill if the weather be temperate, in the moſt fertile ground that may be,and where the Sun hath moſt power the whole day.

¶ *The Names.*

The floure of the Sun is called in Latine *Flos Solis*, taking that name from thoſe that haue reported it to turne with the Sun, the which I could neuer obſerue, although I haue endeuored to finde out the truth of it; but I rather thinke it was ſo called becauſe it doth reſemble the radiant beames of the Sun,whereupon ſome haue called it *Corona Solis*, and *Sol Indianus*, the Indian Sunne floure: others haue called it *Chryſanthemum Peruuianum*, or the golden floure of Peru: in Engliſh, the floure of the Sun,or the Sun floure.

¶ *The Temperature.*

They are thought to be hot and dry of complexion.

¶ *The Vertues.*

A There hath not any thing been ſet downe either of the antient or later writers concerning the vertues of theſe plants,notwithſtanding we haue found by triall,that the buds before they be floured,boiled and eaten with butter,vineger,and pepper,after the manner of Artichokes, are exceeding pleaſant meat,ſurpaſſing the Artichoke far in procuring bodily luſt.

B The ſame buds with the ſtalks neere vnto the top(the haírineſſe being taken away)broiled vpon a gridiron, and afterward eaten with oile,vineger, and pepper,haue the like property.

Chap. 260. *Of Jeruſalem Artichoke.*

ONe may wel by the Engliſh name of this plant perceiue that thoſe that vulgarly impoſe names vpon plants haue little either iudgement or knowledge of them. For this plant hath no ſimilitude in leafe,ſtalke,root or manner of growing with an Artichoke, but onely a little ſimilitude of taſte in the dreſſed root; neither came it from Ieruſalem or out of Aſia, but out of America, whence *Fabius Columna* one of the firſt ſetters of it forth fitly names it *Aſter Peruuianus tuberoſus*, and *Flos ſolis Farneſianus*,becauſe it ſo much reſembles the *Flos ſolis*,and for that he firſt obſerued it growing in the garden of Cardinall *Farneſius*, who had procured roots thereof from the Weſt Indies. *Pelliterius* calls this *Heliotropium Indicum tuberoſum*; and *Bauhinus* in his *Prodromus* ſets this forth by the name of *Chryſanthemum latifolium Braſilianum*;but in his *Pinax* he hath it by the name of of *Helianthemum Indicum tuberoſum*. Alſo our Countreyman Mr. *Parkinſon* hath exactly deliuered the hiſtory of this by the name of *Battatas de Canada*, Engliſhing it Potatoes of Canada: now all theſe that haue written and mentioned it, bring it from America, but from far different places, as from Peru, Braſil, and Canada: but this is not much material, ſeeing it now grows ſo wel & plentifully in ſo many places of England. I will now deliuer you the Hiſtorie, as I haue receiued it from my oft mentioned friend Mr. *Goodyer*, who, as you may ſee by the date, took it preſently vpon the firſt arriuall into England.

‡ *Flos Solis Pyramidalis.*
Ieruſalem Artichoke.

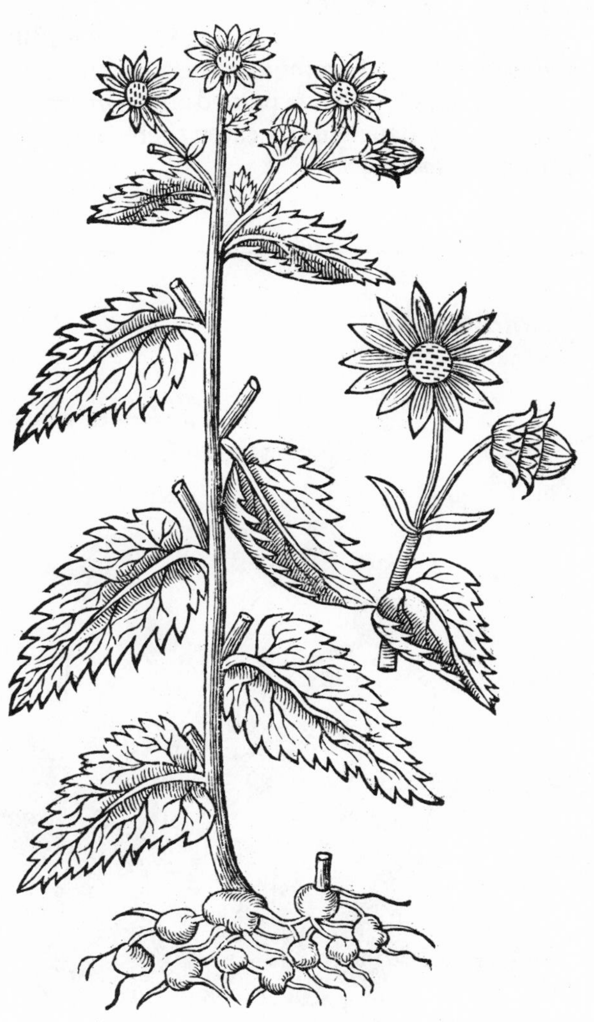

¶ *The Deſcription.*

Flos ſolis Pyramidalis, paruo flore, tuberoſa radice.
Heliotropium Indicum quorundam.

1 THis wonderfull increaſing plant hath growing vp from one root, one, ſometimes two, three or more round green rough hairy ſtraked ſtalks, commonly about twelue foot high, ſometimes fixteene foot high or higher, as big as a childs arme, full of white ſpungious pith within. The leaues grow all alongſt the ſtalkes out of order, of a light green color, rough, ſharp pointed, about eight inches broad, and ten or eleuen inches long, deeply notched or indented about the edges, very like the leaues of the common *flos ſolis Peruanus*, but nothing crompled, and not ſo broad. The ſtalkes diuide themſelues into many long branches euen from the roots to their very tops, bearing leaues ſmaller and ſmaller toward the tops, making the herbe appeare like a little tree, narrower and ſlenderer toward the top, in faſhion of a ſteeple or Pyramide. The floures with vs grow onely at the toppes of the ſtalkes and branches, like thoſe of the ſaid *flos ſolis*, but no bigger than our common ſingle Marigold, conſiſting of twelue or thirteene ſtraked ſharpe pointed bright yellow bordering leaues, growing foorth of a ſcaly ſmall hairie head, with a ſmall yellow thrummie matter within. Theſe floures by reaſon of their late flouring, which is commonly two or three weeks after Michaelmas, neuer bring their ſeed to perfection, & it maketh ſhew of abundance of ſmall heads neere the tops of the ſtalkes and branches forth of the boſomes of the leaues, which neuer open and floure with vs, by reaſon they are deſtroyed with the froſts, which otherwiſe it ſeemes would be a goodly ſpectacle. The ſtalke ſendes foorth many ſmall creeping roots whereby it is fed or nouriſhed, full of hairie threddes euen from the vpper part of the earth, ſpreading farre abroad: amongſt which from the maine root grow forth many tuberous roots, cluſtering together, ſometimes faſtened to the great root it ſelfe, ſometimes growing on long ſtrings a foot or more from the root, raiſing or heauing vp the earth aboue them, and ſometimes appearing aboue the earth, producing from the increaſe of one root, thirty, forty, or fifty in number, or more, making in all vſually aboue a pecke, many times neere halfe a buſhell, if the ſoile be good. Theſe tuberous roots are of a reddiſh colour without, of a ſoft white ſubſtance within, bunched or bumped out many waies, ſometimes as big as a mans fiſt, or not ſo big, with white noſes or peaks where they will ſprout or grow the next yeare. The ſtalkes bowed downe, and ſome part of them couered ouer with earth, ſend forth ſmal creeping threddie roots, and alſo tuberous roots like the former, which I haue found by experience. Theſe tuberous roots will abide aliue in the earth all winter,

 though

though the ſtalkes and rootes by the which they were nouriſhed vtterly rot and periſh away, andwill beginne to ſpring vp againe at the beginning of May,ſeldome ſooner.

¶ *The Place.*

Where this plant groweth naturally I know not, in *Anno* 1617 I receiued two ſmall roots thereof from Maſter *Franqueuill* of London, no bigger than hens egges: the one I planted, and the other I gaue to a friend, mine brought mee a pecke of roots, wherewith I ſtored Hampſhire.

¶ *The Vertues.*

A These rootes are dreſſed diuers waies; ſome boile them in water,and after ſtew them with ſacke and butter,adding a little Ginger: others bake them in pies,putting Marrow,Dates,Ginger, Raiſons of the Sun,Sacke,&c. Others ſome other way,as they are led by their skill in Cookerie. But in my iudgement,which way ſoeuer they be dreſt and eaten they ſtirre and cauſe a filthie loathſome ſtinking windewithin the bodie, thereby cauſing the belly to bee pained and tormented, and are a meat more fit for ſwine,than men: yet ſome ſay they haue vſually eaten them, and haue found no ſuch windie qualitie in them. 17.Octob.1621. *Iohn Goodyer.* ‡

Chap. 261. *Of Cammomill.*

1 *Chamæmelum.*
Cammomill.

2 *Chamæmelum nudum odoratum.*
Sweet naked Cammomill.

¶ *The*

¶ *The Deſcription.*

1 TO diſtinguiſh the kindes of Cammomils with ſundry deſcriptions would be but to enlarge the volume, and ſmall profit would thereby redound to the Reader, conſidering they are ſo well knowne to all: notwithſtanding it ſhall not be amiſſe to ſay ſomething of them, to keepe the order and method of the booke, hitherto obſerued. The common Cammomill hath many weake and feeble branches trailing vpon the ground, taking hold vpon the top of the earth, as it runneth, whereby it greatly encreaſeth. The leaues are very fine, and much iagged or deepely cut, of a ſtrong ſweet ſmell: among which come forth the floures like vnto the field Daiſie, bordered about the edge with a pale of white leaues: the middle part is yellow, compoſed of ſuch thrums cloſe thruſt together, as is that of the Daiſie. The root is very ſmall and threddy.

2 The ſecond kinde of Cammomill hath leaues, roots, ſtalks, and creeping branches like the precedent: the floures grow at the tops of ſmall tender ſtems, which are nothing elſe but ſuch yellow thrummie matter as is in the midſt of the reſt of the Cammomils, without any pale or border of white floures, as the others haue: the whole plant is of a pleaſing ſweet ſmell; whereupon ſome haue giuen it this addition, *Odoratum.*

3 This third Cammomil differeth not from the former, ſauing that the leaues hereof are very much doubled with white leaues, inſomuch that the yellow thrum in the middle is but little ſeen, and the other very ſingle, wherein conſiſteth the difference.

3 *Chamæmelum Anglicum flore multiplici.*
Double floured Cammomill.

4 *Chamæmelum Romanum.*
Romane Cammomil.

4 Romane Cammomill hath many ſlender ſtalkes, yet ſtiffer and ſtronger than any of the others, by reaſon whereof it ſtandeth more vpright, and doth not creepe vpon the earth as the others doe. The leaues are of a more whitiſh colour, tending to the colour of the leaues of Woad. The floures be likewiſe yellow in the middle, and paled about with a border of ſmall white floures.

¶ *The Place.*

Theſe plants are ſet in gardens both for pleaſure and alſo profit.

¶ *The Time.*

They floure moſt part of all the Sommer.

¶ *The*

¶ *The Names.*

Cammomill is called *Chamæmelum*: of some, *Anthemis*, and *Leucanthemis*, and also *Leucanthemon*, especially that double floured Cammomill: which Greeke name is taken from the whitenes of his floure: in English, Cammomill: it is called Cammomil, because the floures haue the smel of μῆλον, an apple, which is plainly perceiued in common Cammomill.

¶ *The Temperature.*

Cammomill, saith *Galen*, is hot and dry in the first degree, and is of thinne parts: it is of force to digest, slacken, and rarifie; also it is thought to be like the Rose in thinnesse of parts, comming to the operation of oyle in heate, which are to man familiar and temperate: wherefore it is a speciall helpe against wearisomenesse; it easeth and mitigateth paine, it mollifieth and suppleth, and all these operations are in our vulgar Cammomill, as common experience teacheth, for it heateth moderately, and drieth little.

¶ *The Vertues.*

A Cammomill is good against the collicke and stone; it prouoketh vrine, and is most singular in Clysters which are made against the foresaid diseases.

B Oile of Cammomill is exceeding good against all manner of ache and paine, bruisings, shrinking of sinewes, hardnesse, and cold swellings.

C The decoction of Cammomill made in wine and drunke, is good against coldnesse in the stomacke, soure belchings, voideth winde, and mightily bringeth downe the monethly courses.

D The Egyptians haue vsed it for a remedie against all cold agues; and they did therefore consecrate it (as *Galen* saith) to their Deities.

E The decoction made in white wine and drunk, expelleth the dead child, and secondine or afterbirth, speedily, and clenseth those parts.

F The herbe boyled in posset Ale, and giuen to drinke, easeth the paine of the chest comming of winde, and expelleth tough and clammy flegme, and helpeth children of the Ague.

G The herbe vsed in baths prouoketh sweat, rarifieth the skinne, and openeth the pores: briefely, it mitigateth gripings and gnawings of the belly; it alayeth the paines of the sides, mollifies hard swellings, and wasteth away raw and vndigested humors.

H The oyle compounded of the floures performeth the same, and is a remedie against all wearisomenesse, and is with good successe mixed with all those things that are applied to mitigate paine.

Chap. 262.

Of May-weed, or wilde Cammomill.

¶ *The Kindes.*

There be three kindes of wilde Cammomill, which are generally called in Latine *Cotula*; one stinking, and two other not stinking: the one hath his floure all white throughout the compasse, and also in the middle; and the other yellow. Besides these there is another with verie faire double floures voyd of smell, which a Kentish Gentleman called Mr. *Bartholmew Lane* found growing wilde in a field in the Isle of Thanet, neere vnto a house called Queakes, sometime the house of Sir *Henry Crispe*. Likewise Mr. *Hesketh*, before remembred, found it in the garden of his Inne at Barnet, if my memorie faile me not, at the signe of the red Lyon, or neere vnto it, and in a poore womans garden as he was riding into Lancashire.

‡ The double floured May-weed, the last yeare, being 1632. I (being in company with Mr. *William Broad*, Mr. *Iames Clarke*, and some other London Apothecaries in the Isle of Thanet) found it growing wild vpon the cliffe side, close by the towne of Margate, and in some other places of the Island. ‡

¶ *The Description.*

1 May-weed bringeth forth round stalkes, greene, brittle, and full of iuyce, parted into many branches thicker and higher than those of Cammomil; the leaues in like maner are broader, and of a blackish greene colour. The floures are like in forme and colour, yet commonly larger, and of a ranke and naughty smell: the root is woody, and perisheth when the seed is ripe. The whole plant stinketh, and giueth a ranke smell.

‡ This

‡ This herbe varies, in that it is found ſometimes with narrower, and otherwhiles with broader leaues; as alſo with a ſtrong vnpleaſant ſmell, or without any ſmell at all: the floures alſo are ſingle, or elſe (which is ſeldome found) very double. ‡

2 The yellow May-weed hath a ſmall and tender root, from which riſeth vp a feeble ſtalke diuiding it ſelfe into many other branches, whereupon do grow leaues not vnlike to Cammomill, but thinner, and fewer in number. The floures grow at the top of the ſtalkes, of a gold yellow colour, ‡ This I take to be no other than the *Buphthalmum verum* of our Author, formerly deſcribed in the ſecond place of the 257. chapter.

3 This mountaine Cammomill hath leaues ſomewhat deepely cut in almoſt to the middle rib, thicke alſo and iuycie, of a bitteriſh taſte, and of no pleaſant ſmell: the ſtalkes are weake, and ſome foot high, carrying at their tops ſingle floures, bigger, yet like thoſe of Cammomill, yellow in the middle, with a border of twenty or more long white leaues, encompaſſing it. It increaſeth much, as Cammomill doth, and hath creeping roots. It is found vpon the Stirian Alpes, and floureth in Iuly and Auguſt. *Cluſius* hath ſet this forth by the name of *Leucanthemum Alpinum*. ‡

1 *Cotula fœtida.*
May-weed.

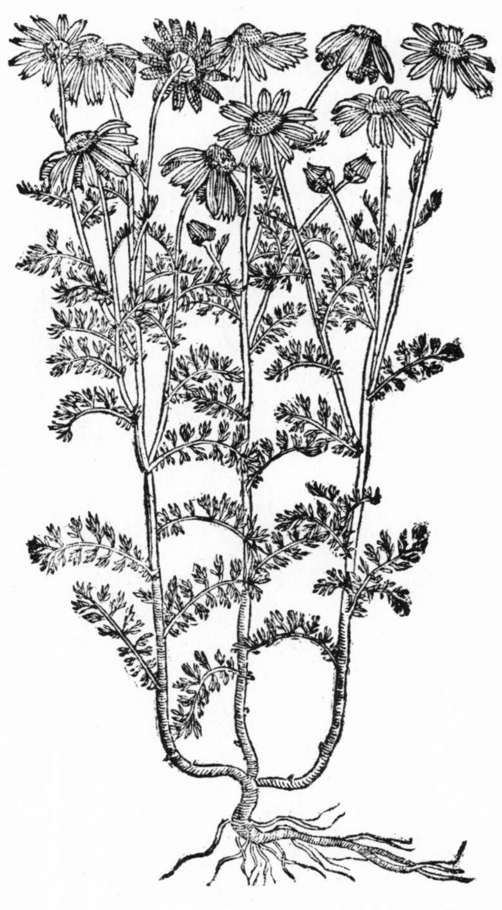

‡ 3 *Leucanthemum Alpinum Cluſij.*
Wilde Mountaine Cammomill.

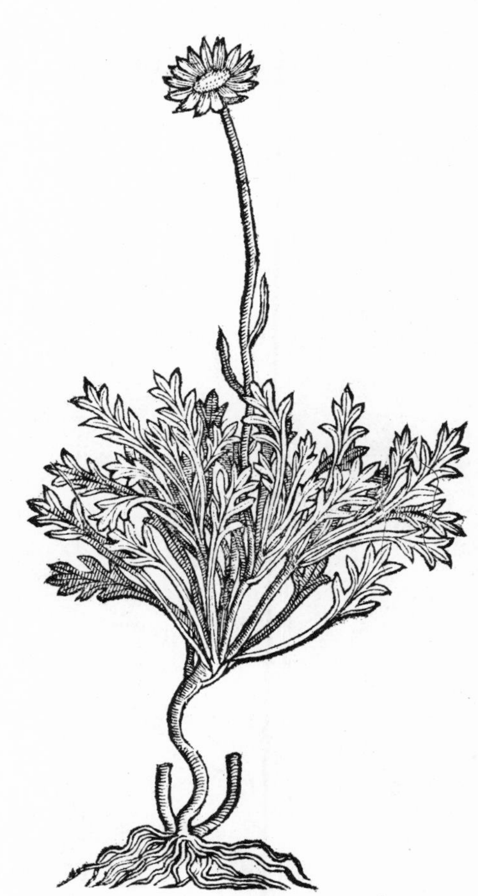

¶ *The Place.*

They grow in Corne fields neere vnto path waves, and in the borders of fields.

¶ *The Time.*

Thee floure in Iuly and Auguſt.

¶ *The Names.*

May-weed is called in ſhops *Cotula fœtida*: of *Leonhartus Fuchſius*, *Parthenium*, and *Virginea*, but not truly: of others, *Κυνάνθεμις*: in high-Dutch, **Krotendill**: in low-Dutch, **Paddebloemen**: in French, *Eſpargoutte*: in Engliſh, May-weed, wilde Cammomill, and ſtinking Mathes.

¶ *The Temperature and Vertues.*

May-weed is not vſed for meate nor medicine, and therefore the faculties are vnknowne; yet all of them are thought to be hot and dry, and like after a ſort in operation to Cammomill, but nothing at all agreeing with mans nature; notwithſtanding it is commended againſt the infirmities of the mother, ſeeing all ſtinking things are good againſt thoſe diſeaſes. A

It is

It is an vnprofitable weed among corne, and raiseth blisters vpon the hands of the weeders and reapers.

Chap. 263. *Of Pellitorie of Spaine.*

¶ *The Description.*

1 Pyrethrum, in English, Pellitorie of Spaine (by the name whereof some doe vnproperly call another plant, which is indeed the true *Imperatoria*, or Master-wort, and not Pellitorie) hath great and fat leaues like vnto Fennell, trailing vpon the ground: amongst which, immediately from the root riseth vp a fat great stem, bearing at the top a goodly floure, fashioned like the great single white Daisie, whose bunch or knob in the midst is yellow like that of the Daisie, and bordered about with a pale of small leaues, exceeding white on the vpper side, and vnder of a faire purple colour: the root is long, of the bignesse of a finger, very hot, and of a burning taste.

2 The wilde Pellitorie groweth vp like vnto wilde Cheruile, resembling the leaues of *Caucalis*, of a quicke and nipping taste, like the leaues of Dittander, or Pepper-wort: the floures grow at the top of slender stalkes, in small tufts or spoky vmbels, of a white colour: the root is tough, and of the bignesse of a little finger, with some threds thereto belonging, and of a quicke biting taste.

1 *Pyrethrum officinarum.*
Pellitorie of Spaine.

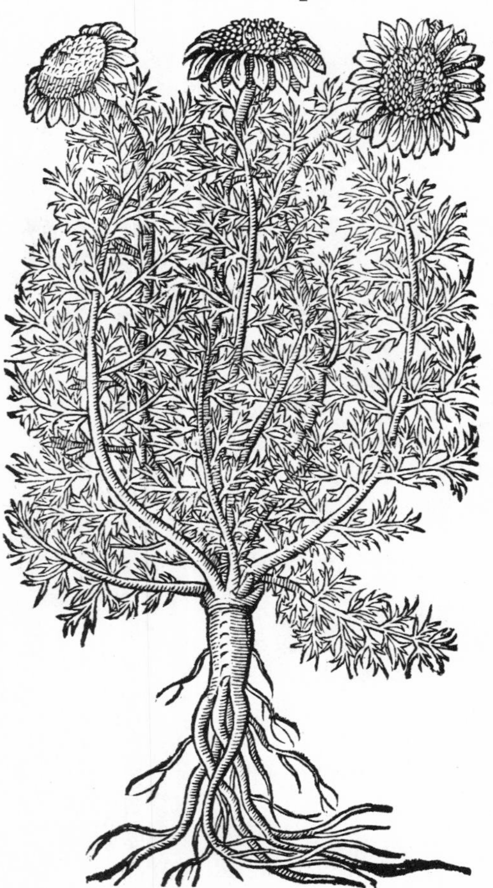

2 *Pyrethrum syluestre.*
Wilde Pellitorie.

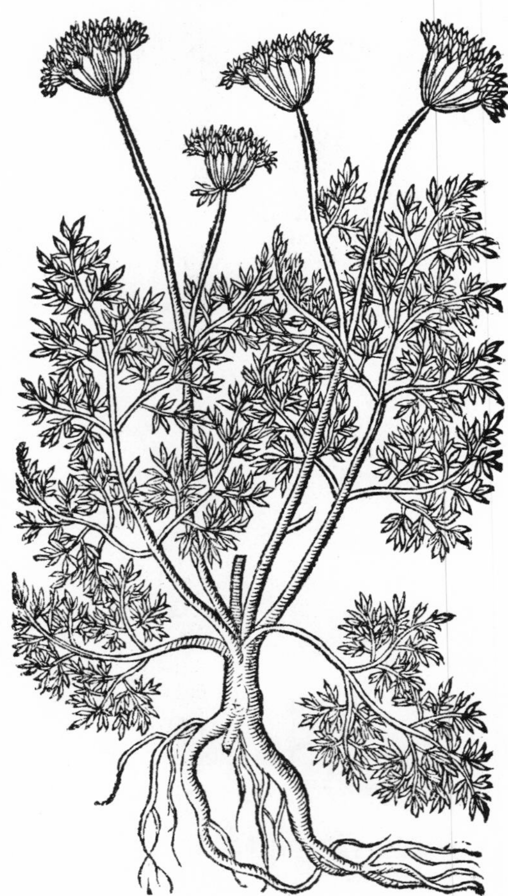

¶ *The Place.*

It groweth in my garden very plentifully.

¶ *The Time.*

It floureth and seedeth in Iuly and August.

¶ *The Names.*

Pellitorie of Spaine is called in Greeke πύρεθρον, by reason of his hot and fierie taste: in shops also *Pyrethrum*: in Latine, *Saliuaris*: in Italian, *Pyrethro*: in Spanish, *Pelitre*: in French, *Pied d'Alexandre*,

Alexandre, that is to ſay, *Pes Alexandrinus*, or Alexanders foot: in high and low Dutch, Bertram: in Engliſh, Pellitorie of Spaine; and of ſome, Bertram, after the Dutch name: and this is the right *Pyrethrum*, or Pellitorie of Spaine; for that which diuers here in England take to be the right, is not ſo, as I haue before noted.

¶ *The Temperature and Vertues.*

The root of Pellitorie of Spaine is very hot and burning, by reaſon whereof it taketh away the cold ſhiuering of Agues, that haue been of long continuance, and is good for thoſe that are taken with a dead palſie, as *Dioſcorides* writeth. A

The ſame is with good ſucceſſe mixed with Antidotes or counterpoyſons which ſerue againſt the megrim or continuall paine of the head, the dizzineſſe called *Vertigo*, the apoplexie, the falling ſickneſſe, the trembling of the ſinewes, and palſies, for it is a ſingular good and effectuall remedy for all cold and continuall infirmities of the head and ſinewes. B

Pyrethrum taken with honey is good againſt all cold diſeaſes of the braine. C

The root chewed in the mouth draweth forth great ſtore of rheume, ſlime, and filthy wateriſh humors, and eaſeth the paine of the teeth, eſpecially if it be ſtamped with a little Staueſ-acre, and tied in a ſmall bag, and put into the mouth, and there ſuffered to remaine a certaine ſpace. D

If it be boyled in Vineger, and kept warme in the mouth it hath the ſame effect. E

The oyle wherein Pellitorie hath been boyled is good to anoint the body to procure ſweating, and is excellent good to anoint any part that is bruiſed and blacke, although the member be declining to mortification: it is good alſo for ſuch as are ſtricken with the palſie. F

It is moſt ſingular for the Surgeons of the Hoſpitals to put into their vnctions *contra Neapolitanum morbum*, and ſuch other diſeaſes that be couſin germanes thereunto. G

Chap. 264. *Of Leopards bane.*

† 1 *Doronicum minus officinarum.*
Small Leopards bane.

† 2 *Doronicum maius Officinarum.*
Great Leopards bane.

¶ *The Deſcription.*

1 Of this Plant *Doronicum* there be ſundry kindes, whereof I will onely touch foure: *Dodonæus* vnproperly calleth it *Aconitum pardalianches*, which hath hapned through the negligence

negligence of *Dioſcorides* and *Theophraſtus*, who in deſcribing *Doronicum*, haue not onely omitted the floures thereof, but haue committed that negligence in many and diuers other plants, leauing out in many plants which they haue deſcribed, the ſpecial accidents; which hath not a little troubled the ſtudy and determination of the beſt herbariſts of late yeares, not knowing certainely what to determine and ſet downe in ſo ambiguous a matter, ſome taking it one way, and ſome another, and ſome eſteeming it to be *Aconitum*. But for the better vnderſtanding hereof, know that this word *Aconitum*, as it is a name attributed to diuers plants, ſo it is to be conſidered, that all plants called by this name are malignant and venomous, as with the iuyce and root whereof ſuch as hunted after wilde and noyſome beaſts were wont to embrue and dip their arrowes, the ſooner and more ſurely to diſpatch and ſlay the beaſt in chaſe. But for the proofe of the goodneſſe of this *Doronicum* and the reſt of his kind, know alſo, That *Lobel* writeth of one called *Iohn de Vroede*, who ate very many of the roots at ſundry times, and found them very pleaſant in taſt, and very comfortable. But to leaue controuerſies, circumſtances, and obiections which here might be brought in and alledged, aſſure your ſelues that this plant *Doronicum minus Officinarum* (whoſe roots *Pena* reporteth to haue found plentifully growing vpon the Pede-mountaine hills and certaine high places in France) hath many leaues ſpred vpon the ground, ſomewhat like Plantaine: among which riſe vp many tender hairy ſtalks ſome handfull and an halfe high, bearing at the top certain ſingle yellow floures, which when they fade change into downe, and are caried away with the winde. The roots are thicke and many, very crookedly croſſing and tangling one within another, reſembling a Scorpion, and in ſome yeares do grow in our Engliſh gardens into infinite numbers.

3 *Doronicum radice repente.*
Cray-fiſh Wolfes bane.

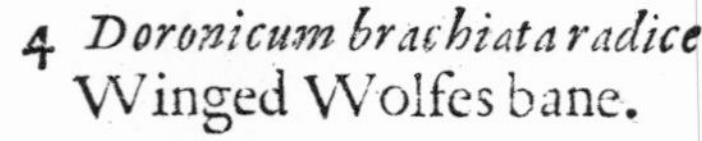

4 *Doronicum brachiata radice.*
Winged Wolfes bane.

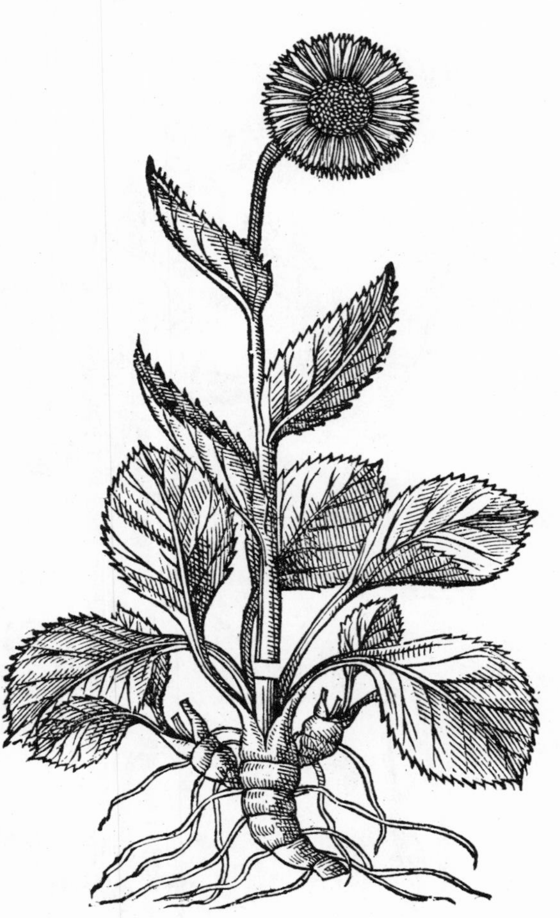

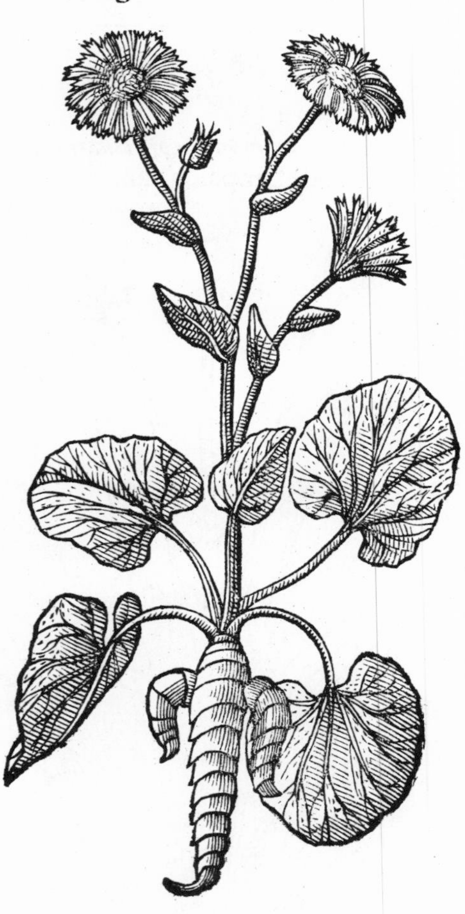

2 The ſecond kinde of *Dorouicum* hath larger leaues than the former, but round, and broader, almoſt like the ſmall leaues of the Clot or Burre; among which riſeth vp a ſtalke ſcarſe a cubit high: the floures are like the former: the root is longer and bigger than the former, barred ouer with many ſcaly barks, in colour white, and ſhining like white marble, hauing on each ſide one arme or finne, not vnlike to the ſea Shrimpe called *Squilla marina*, or rather like the ribbes or ſcales of a Scorpions body, and is ſweet in taſte.

3 The third kinde of *Doronicum*, growing naturally in great aboundance in the mountaines of

of France, is alſo brought into and acquainted with our Engliſh grounds, bearing very large leaues of a light yellowiſh greene, and hairy like *Piloſella*, or *Cucumis agreſtis*. The ſtalkes are a cubit high, hauing at the top yellow floures like *Buphthalmum*, or *Conſolida media vulnerariorum*: all the root is barred and welted ouer with ſcales like the taile of a Scorpion, white of colour, and in taſte ſweet, with ſome bitterneſſe, yeelding forth much clamminеſſe, which is very aſtringent.

4 The fourth kinde hereof is found in the wooddy mountaines about Turin and Sauoy, very like vnto the former, ſauing that the leaues are ſomwhat rougher, the floures greater, and the ſtalks higher. But to be ſhort, each of theſe kindes are ſo like one another, that in ſhew, taſte, ſmell, and manner of growing they ſeeme to be as it were all one: therefore it were ſuperfluous to ſtand vpon their varietie of names, *Pardalianches*, *Myoctonum*, *Thelyphonum*, *Camorum*, and ſuch like, of *Theophraſtus*, *Dioſcorides*, *Pliny*, or any of the new Writers, which names they haue giuen vnto *Doronicum*; for by the opinion of the moſt skilfull in plants, they are but Synonimies of one kinde of plant. And though theſe old writers ſpeake of the hurtfull qualities of theſe plants; yet experience teacheth vs that they haue written what they haue heard and read, and not what they haue knowne and proued; for it is apparant, that *Doronicum* (by the conſent of the old and new writers) is vſed as an antidote or certaine treacle, as well in the confections *de Gemmis Meſuæ*, as in *Electuario Aromatum*. And though *Matthiolus* diſclaimeth againſt the vſe thereof, and calleth it *Pardalianches*, that is, Wolfes bane; yet let the Learned know, that *quantitas, non qualitas, nocet*: for though Saffron be comfortable to the heart, yet if you giue thereof, or of muske, or any ſuch cordial thing, too great a quantitie, it killeth the party which receiueth it.

‡ 5 *Doronicum anguſtifolium Auſtriacum.*
Narrow leaued Wolfes bane.

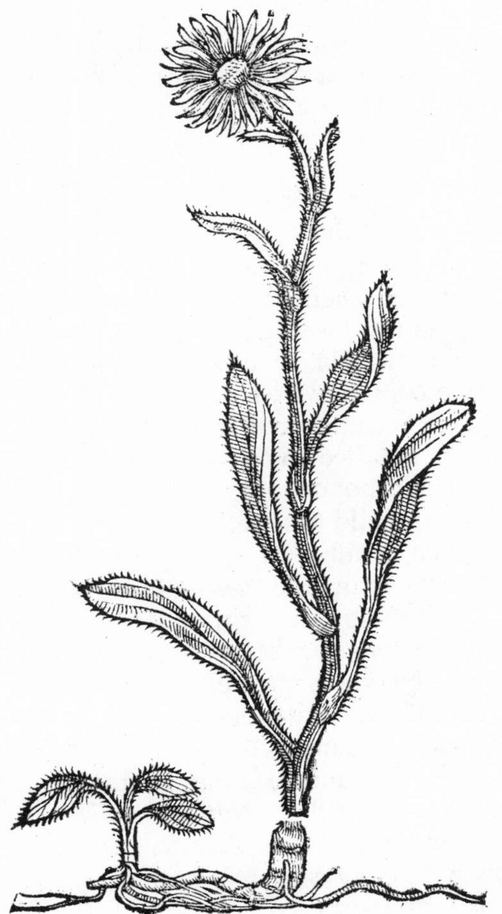

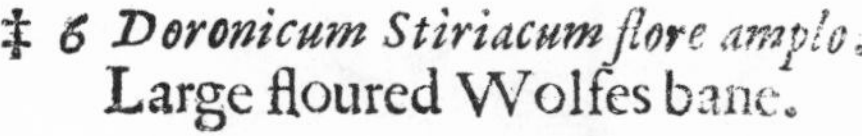

‡ 6 *Doronicum Stiriacum flore amplo.*
Large floured Wolfes bane.

‡ 5 To theſe foure formerly intended by our Author, may we fitly adde ſome others out of *Cluſius*. The firſt of theſe hath a ſtalke ſome foot high, ſoft, rough, and creſted: the leaues are few, thicke, narrow, long, very greene and ſhining, yet hairy on their vpper ſides, but ſmooth on the lower ſides, and of a lighter greene; yet thoſe that adorne the ſtalke are narrower: there groweth commonly at the top of the ſtalke one ſingle floure of the ſhape and bigneſſe of the common *Doronicum* deſcribed in the ſecond place, but of a brighter yellow: the ſeed is little and blackiſh, and is carried away with the winde: the root is ſmall, blackiſh, and ioynted, hauing ſomewhat thicke

 white

white fibres, and an aromaticke taste. This floures in Iuly and August, and growes in rockie places vpon the highest Alpes. *Clusius* (the first and onely describer thereof) calls it *Doronicum* 2. *siue Austriacum* 1.

6 This growes somewhat higher than the last described, and hath much broader and rounder leaues, and those full of veines, and snipt about the edges. The knots and off-sets of the roots descend not down, but run on the surface of the ground, and so send forth fibres on each side, to fasten them and attract nourishment. The floure is like that of the former, but much larger This groweth in the high mountainous places of Stiria, and floures at the same time as the former. *Clusius* calls this *Doronicum* 4. *Stiriacum*.

7 This is the largest of all the rest, and hath a stalke two cubits or more high, of the thicknesse of ones little finger, crested, rough, and towards the top diuided into sundry branches. The leaues next to the root are round, wrinkled, hairy, and fastned to a long stalke: those towards the top of the stalke are longer and narrower, and ingirt the stalke at their setting on. The floures are large and yellow, like to the other plants of this kinde: the seed also is carried away with the winde, and is longish, and of a greenish colour: the root is knotty or ioynted like to a little Shrimpe, and of a whitish greene colour. This floures in Iune or Iuly, and growes vpon the like places as the former. *Clusius* calls this *Doronicum* 7. *Austriacum* 3. ‡

‡ 7 *Doronicum maximum.*
The greatest Wolfe-bane.

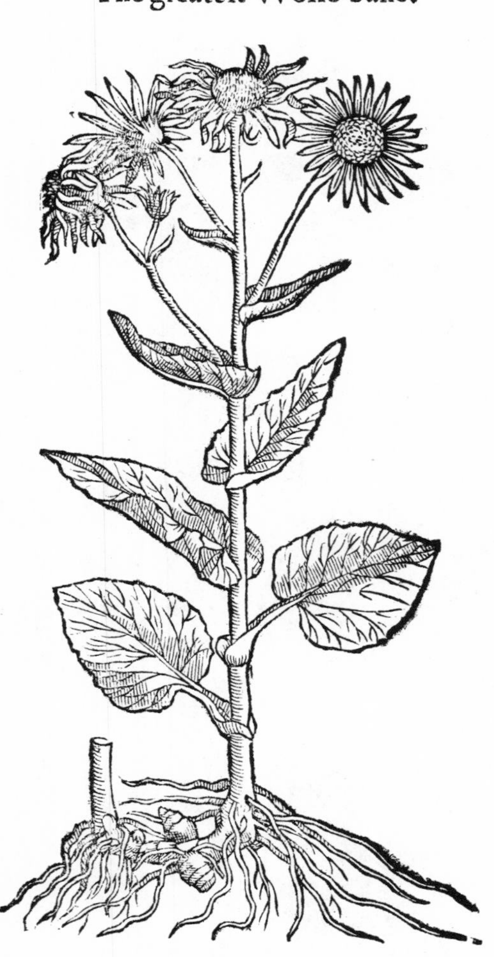

¶ *The Place.*

The place is sufficiently set forth in the description; yet you shall vnderstand, that I haue the two first in my garden; the second hath beene found and gathered in the cold mountaines of Northumberland, by Dr. *Penny* lately of London deceased, a man of much experience and knowledge in Simples, whose death my selfe and many others do greatly bewaile.

¶ *The Time.*

They floure in the months of Iune and Iuly.

¶ *The Names.*

Concerning their names I haue already spoken; yet sith I would be glad that our English women may know how to call it, they may terme *Doronicum* by this name, Cray-fish Pisse-a-bed, because the floure is like Dandelion, which is called Pisse-a-bed.

‡ Our Author certainly at the beginning of this chapter did not well vnderstand what he said, when he affirmes, That the reason of the not wel knowing the *Doronicum* of the Antients was, [through the negligence of *Dioscorides* and *Theophrastus*, who in describing *Doronicum*, &c.] Now it is manifest, that neither of these Authors, nor any of the antient Greekes euer so much as named *Doronicum*: but that which he should haue said, was, That the want of exact describing the *Aconitum thelyphonon* in *Theophrastus*, and *Aconitum Pardalianches* in *Dioscorides*, (which are iudged to be the same plant and all one with our *Doronicum*) hath beene the cause, that the controuersie which *Matthiolus* and others haue of late raised cannot be fully determined; which is, Whether that the vulgar *Doronicum*, vsed in shops, and described in this chapter, be the *Aconitum Pardalianches*? *Matthiolus* affirmes it is, and much and vehemently exclaimes against the vse thereof in cordiall Electuaries, as that which is of a most pernitious and deadly qualitie because that (as he affirmes) it will kill dogs: now *Dodonæus* also seems to incline to his opinion: but others (and not without good reason) deny it; as *Gesner* in his Epistles, who made often triall of it vpon himselfe: part of his words are set downe hereafter by our Author (being translated out of *Dodonæus*) and some part also you shall finde added in the end of the vertues: and these are other some; *Plura alia nunc omitto, quibus ostendere liquido possem, nec Doronicum nostrum, nec Aconitum*

Aconitum vllo modo esse venenatum homini. Canibus autem letiferum esse scio, non solum si drachmarum 4. *sed etiam si vnius pondere sumant.* And before he said, *quasi non alia multa canibus sint venena, quæ homini salubr a sunt; vt de asparago fertur.* Of the same opinion with *Gesner* is *Pena* and *Lobel*, who, *Aduers.p.*290,*&* 291. do largely handle this matter, & exceedingly deride and scoffe at *Matthiolus*, for his vehement declaiming against the vse thereof. Now briefely my opinion is this, That the *Doronicum* here mentioned is not that mentioned and written of by *Serapio* and the Arabians; neither is it the *Aconitum Pardalianches* of *Dioscorides*, nor of so malignant a qualitie as *Matthiolus* would haue it; for I my selfe also haue often eaten of it, and that in a pretty quantitie, without the least offence. ‡

¶ *The Nature and Vertues.*

I haue sufficiently spoken of that for which I haue warrant to write, both touching their natures and vertues; for the matter hath continued so ambiguous and so doubtfull, yea, and so ful of controuersies, that I dare not commit that to the world which I haue read: these few lines therefore shall suffice for this present; the rest which might be said I referre to the great and learned Doctors, and to your owne consideration. A

These herbes are mixed with compound medicines that mitigate the paine of the eyes, and by reason of his cold qualitie, being fresh and greene, it helpeth the inflammation or fierie heate of the eyes. B

It is reported and affirmed, that it killeth Panthers, Swine, Wolues, and all kindes of wilde beasts, being giuen them with flesh. *Theophrastus* saith, That it killeth Cattell, Sheepe, Oxen, and all foure-footed beasts, within the compasse of one day, not by taking it inwardly onely, but if the herbe or root be tied vnto their priuy parts. Yet he writeth further, That the root being drunke is a remedie against the stinging of Scorpions; which sheweth, that this herbe or the root thereof is not deadly to man, but to diuers beasts onely: which thing also is found out by trial and manifest experience; for *Conrade Gesner* (a man in our time singularly learned, and a most diligent searcher of many things) in a certaine Epistle written to *Adolphus Occo*, sheweth, That he himselfe hath oftentimes inwardly taken the root hereof greene, dry, whole, preserued with honey, and also beaten to pouder; and that euen the very same day in which hee wrote these things, hee had drunke with warme water two drams of the roots made into fine pouder, neither felt he any hurt thereby: and that he oftentimes also had giuen the same to his sicke Patients, both by it selfe, and also mixed with other things, and that very luckily. Moreouer, the Apothecaries in stead of *Doronicum* doe vse (though amisse) the roots thereof without any manifest danger. C

That this *Aconite* killeth dogs, it is very certaine, and found out by triall: which thing *Matthiolus* could hardly beleeue, but that at length he found it out to be true by a manifest example, as he confesseth in his Commentaries. D

‡ I haue (saith *Gesner*) oft with very good successe prescribed it to my Patients, both alone, as also mixed with other medicines, especially in the *Vertigo* and falling sicknesse: somtimes also I mix therewith Gentian, the pouder of Misle-toe, and *Astrantia*: thus it workes admirable effects in the Epilepsie, if the vse thereof be continued for some time. ‡ E

† Formerly the figure that was in the first place should haue beene in the second, and the first and second were confounded in the description.

CHAP. 265. *Of Sage.*

¶ *The Description.*

1 THe great Sage is very full of stalkes, foure square, of a wooddy substance, parted into branches, about the which grow broad leaues, long, wrinckled, rough, whitish, verie like to the leaues of wilde Mullein, but rougher, and not so white, like in roughnesse to woollen cloath thread-bare: the floures stand forked in the tops of the branches like those of dead Nettle, or of Clarie, of a purple blew colour, in the place of which doth grow little blackish seeds, in small huskes. The root is hard and wooddy, sending forth a number of little strings.

2 The lesser Sage is also a shrubby plant, spred into branches like to the former, but lesser: the stalkes hereof are tenderer: the leaues be long, lesser, narrower, but not lesse rough; to which there do grow in the place wherein they are fixed to the stalke, two little leaues standing on either side one right against another, somewhat after the manner of finnes or little eares: the floures are

eared blew like thoſe of the former: the root alſo is wooddy: both of them are of a certaine ſtrong ſmell, but nothing at all offenſiue; and that which is the leſſer is the better.

3 This Indian Sage hath diuers branches of a wooddy ſubſtance, whereon doe grow ſmall leaues, long, rough, and narrow, of an ouerworne colour, and of a moſt ſweet and fragrant ſmell. The floures grow alongſt the top of the branches, of a white colour, in forme like the precedent. The root is tough and wooddy.

1 *Saluia maior.*
Great Sage.

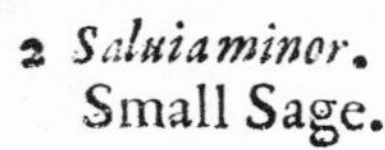

2 *Saluia minor.*
Small Sage.

4 The Mountaine Sage hath an vpright ſtalke ſmooth and plaine, whereupon do grow broad rough and rugged leaues, ſleightly nicked, and vneuenly indented about the edges, of an hoarie colour, ſharpe pointed, and of a ranke ſmell: the floures grow alongſt the top of the ſtalke, in ſhape like thoſe of Roſemarie, of a whitiſh red colour. The root is likewiſe wooddy.

5 We haue in our gardens a kinde of Sage, the leaues whereof are reddiſh; part of thoſe red leaues are ſtriped with white, others mixed with white, greene, and red, euen as Nature liſt to play with ſuch plants. This is an elegant varietie, and is called *Saluia variegata elegans*, Variegated or painted Sage.

6 We haue alſo another, the leaues whereof are for the moſt part white, ſomewhat mixed with greene, often one leafe white, and another greene, euen as Nature liſt, as we haue ſaid. This is not ſo rare as the former, nor neere ſo beautifull, wherefore it may be termed *Saluia variegata vulgaris*, Common painted Sage.

‡ 7 There is kept in ſome of our chiefe gardens a fine Sage, which in ſhape and manner of growing reſembles the ſmaller Sage, but in ſmell and taſte hath ſome affinity with Wormwood; whence it may be termed *Saluia Abſinthites*, or Wormewood Sage. *Bauhine* onely hath mentioned this, and that in the fourth place in his *Pinax*, *pag.* 237. by the name of *Saluia minor altera*: and hee addes, *Hæc odore & ſapore eſt Abſinthij, floreque rubente*: That is, This hath the ſmell and taſte of Wormewood, and a red floure: but ours (if my memorie faile me not) hath a whitiſh floure: it is a tender plant, and muſt be carefully preſerued from the extremitie of Winter. I firſt ſaw this Sage with Mr. *Cannon*, and by him it was communicated to ſome others.

8 This

3 *Saluia Indica.*
Indian Sage.

4 *Saluia Alpina.*
Mountaine Sage.

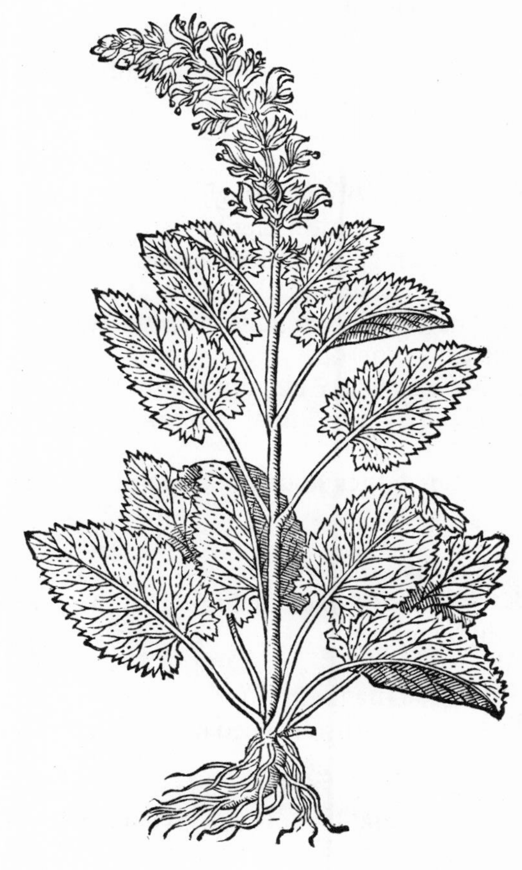

‡ 8 *Saluia Cretica pomifera.*
Apple-bearing Sage of Candy,

‡ 8 *Saluia Cretica non pomifera.*
Candy Sage without Apples.

8 This which we here giue you hath pretty large leaues, and those also very hairy on the vnder side, but rough on the vpper side like as the ordinarie Sage. The stalkes are rough and hairie, foure square below, and round at their tops. The floures in their growing and shape are like those of the ordinarie, but of a whitish purple colour; and fading, they are each of them succeeded by three or foure seeds, which are larger than in other Sages, and so fill their seed-vessels, that they shew like berries. The smell of the whole plant is somewhat more vehement than that of the ordinarie: the leaues also haue sometimes little eares or appendices, as in the smaller or Pig-Sage: and in Candy (the naturall place of the growth) it beares excrescences, or Apples (if we may so terme them) of the bignesse of large Galls, or Oke-Apples: whence *Clusius* hath giuen you two figures by the same titles as I here present the same to your view. *Matthiolus, Dodonæus*, and others also haue made mention hereof. ‡

¶ *The Place.*

These kindes of Sage grow not wilde in England: I haue them all in my garden: most of them are very common.

‡ The fine or elegant painted Sage was first found in a countrey garden, by Mr. *Iohn Tradescant*, and by him imparted to other louers of plants. ‡

¶ *The Time.*

These Sages floure in Iune and Iuly, or later: they are fitly remoued and planted in March.

¶ *The Names.*

Sage is called in Greeke ἐλελίσφακος: the Apothecaries, the Italians, and the Spaniards keepe the Latine name *Saluia*: in high-Dutch, **Salben**: in French, *Sauge*: in low-Dutch, **Sauie**: in English, Sage.

¶ *The Temperature.*

Sage is manifestly hot and dry in the beginning of the third degree, or in the later end of the second; it hath adioyned no little astriction or binding.

¶ *The Vertues.*

A *Agrippa* and likewise *Aetius* haue called it the Holy-herbe, because women with childe if they be like to come before their time, and are troubled with abortments, do eate thereof to their great good; for it closeth the matrix, and maketh them fruitfull, it retaineth the birth, and giueth it life, and if the woman about the fourth day of her going abroad after her childing, shal drink nine ounces of the iuyce of Sage with a little salt, and then vse the companie of her husband, she shall without doubt conceiue and bring forth store of children, which are the blessing of God. Thus far *Agrippa*.

B Sage is singular good for the head and braine; it quickneth the sences and memory, strengthneth the sinewes, restoreth health to those that haue the palsie vpon a moist cause, takes away shaking or trembling of the members; and being put vp into the nosthrils, it draweth thin flegme out of the head.

C It is likewise commended against the spitting of bloud, the cough, and paines of the sides, and bitings of Serpents.

D The iuyce of Sage drunke with honey is good for those that spit and vomit bloud, and stoppeth the flux thereof incontinently, expelleth winde, drieth the dropsie, helpeth the palsie, strengthneth the sinewes, and cleanseth the bloud.

E The leaues sodden in water, with Wood-binde leaues, Plantaine, Rosemary, Honey, Allome, and some white wine, make an excellent water to wash the secret parts of man or woman, and for cankers or other sorenesse in the mouth, especially if you boyle in the same a faire bright shining Sea-cole, which maketh it of greater efficacie.

F No man needs to doubt of the wholesomnesse of Sage Ale, being brewed as it should be, with Sage, Scabious, Betony, Spikenard, Squinanth, and Fennell seeds.

G The leaues of red Sage put into a woodden dish, wherein is put very quicke coles, with some ashes in the bottome of the dish to keepe the same from burning, and a little vineger sprinkled vpon the leaues lying vpon the coles, and so wrapped in a linnen cloath, and holden very hot vnto the side of those that are troubled with a grieuous stitch, taketh away the paine presently: The same helpeth greatly the extremitie of the pleurisie.

Chap. 266. Of French Sage or wooddie Mullein.

1 *Verbascum Matthioli.*
French Sage.

‡ 2 *Verbascum angustis Saluiæ folijs.*
The lesser French Sage.

‡ 3 *Phlomos Lychnites Syriaca.*
Syrian Sage-leaued Mullein.

¶ *The Description.*

1 Wild Mullein, wooddie Mullein, *Matthiolus* his Mullein, or French Sage groweth vp like a small wooddie shrub, hauing many wooddie branches of a woollie and hoarie colour, soft and downie: whereupon are placed thicke hoarie leaues, of a strong ponticke sauour, in shape like the leaues of Sage, whereupon the vulgar people call it French Sage: toward the top of the branches are placed roundles or crownets of yellow gaping floures like those of dead Nettle, but much greater. The root is thicke, tough, and of a wooddie substance, as is all the rest of the plant.

† 2 There is another sort hereof that is very like the other, sauing that the leaues & euery other part of this plant, hath a most sweet and pleasant smell, and the other more strong and offensiue: the leaues also are much lesser and narrower, somewhat resembling those of the lesser Sage.

‡ 3 I thinke it not amisse here to insert this no lesse rare than beautifull plant, which differs from the last described in the manner of growing & shape of the floures, which resemble those of the *Lychnis Chalcedonica*, or None-such, but are of a yellow colour. The leaues are hairy, narrow, and sharp pointed; the stalkes square, and root wooddie. *Lobel*

(to

(to whom we are beholden for this figure and description) calls this, *Phlomos Lychnites altera Syriaca.* ‡

¶ *The Place.*

These wilde Mulleins do grow wilde in diuers Prouinces of Spaine, and also in Languedoc, vpon drie bankes, and stony places: I haue them both in my garden, and many others likewise.

¶ *The Time.*

They floure in Iune and Iuly.

¶ *The Names.*

They are called of the learned men of our time, *Verbasca Syluestria:* the first is called of the Grecians φλόμος or φλογμος: in Latine, *Elychnium*, or after others, *Elychinium*, because of the Cottonie substance thereof, matches, or weeks were made to keep light in lamps: *Verbascum Lychnitis*, as *Dioscorides* himselfe testifieth, is named also *Thryallis* or Rose Campion; but the floure of *Thryallis* is red of colour, as *Nicander* in his Counterpoisons doth shew, but the floures of these are yellow: therefore they are neither *Thryallis* nor *Lychnitis*, but *Syluestre Verbascum*, or wilde Mullein, as we haue already taught in the Chapter of Rose Campion, that *Thryallis* is *Lychnitis satiua*, or Rose Campion. There is nothing to the contrary, but that there may be many plants with soft downie leaues fit to make Candle weeke of: in English it is generally called French Sage: wee may call it Sage Mulleine.

¶ *The Temperature.*

As these be like in vertues to the others going before, so they be likewise drie in temperature.

¶ *The Vertues.*

A *Dioscorides* saith, that the leaues are stamped and laied in manner of a pultis vpon burnings and scaldings.

Chap. 267. Of Clarie.

1 *Gallitricum, siue Horminum.*
Common Clarie.

2 *Gallitricum alterum.*
Small Clarie.

¶ *The*

‡ 3 *Horminum syluestre, Fuchsij.*
Fuchsius his wilde Clarie.

¶ *The Description.*

1 THe first kinde of Clarie which is the right, bringeth forth thick stalks foure square, two foot long, diuided into branches: it hath many leaues growing both from the rootes, and along the stalkes and branches by distances, one against another by two and two, great, a handfull broad or broader, somewhat rough, vnequall, whitish and hairie, as be also the stalkes. The floures are like those of Sage, or of dead Nettle, of colour white, out of a light blew: after which grow vp long toothed huskes in stead of cods, in which is blacke seed. The root is full of strings: the whole herbe yeeldeth forth a rank and strong smell that stuffeth the head: it perisheth after the seed is ripe, which is in the second yeare after it is sowne.

2 The second kinde of Clarie hath likewise stalkes foure square, a foot and a halfe high: the leaues also be rough and rugged, lesser, and not so white. The floures be alike, of colour purple or blew: the rootes bee as those of the former are. This hath not so strong a sent by a great deale.

3 There is a kinde of Clarie which *Fuchsius* pictureth for wilde Clarie, that hath shorter stalkes, hairie, and also foure square: the leaues lesser, long, deeper indented: the floures blew of colour, sweet of smell, but not so sweet as those of

† 4 *Colus Iouis.* Iupiters distaffe.

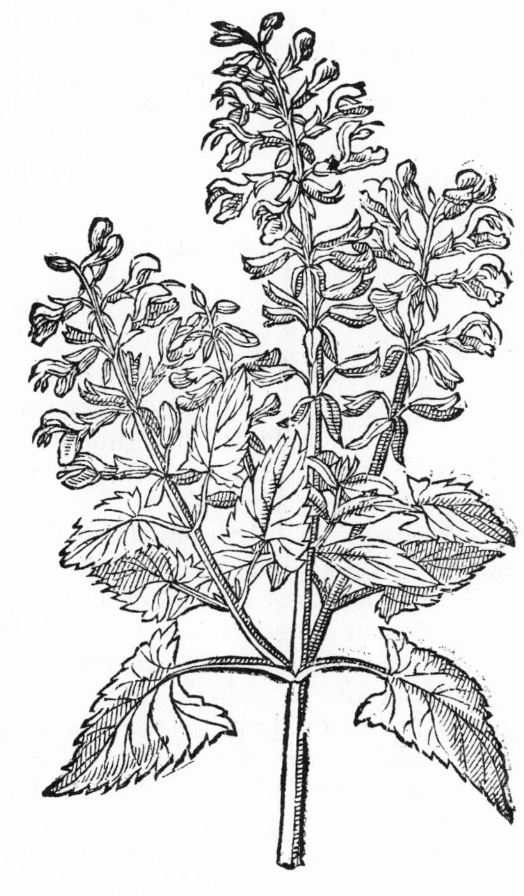

the

the right Clarie: the husks or cods when they are ripe bend downwards: the seed is blackish; the roots in like manner are blacke and full of strings.

4 The fourth kind of *Horminum*, called *Iovis Colus*, representeth in the highest top of the stalke a distaffe, wrapped about with yellow flax, whereof it tooke his name, hauing knobbie roots, with certaine strings annexed thereto like *Galeopsis*, or like vnto the roots of Clarie, which doe yeeld forth sundry foure square rough stalks, two cubits high; whereon do grow leaues like those of the Nettle, rough, sharpe pointed, and of an ouerworne greene colour: the floures do grow alongst the top of the stalks, by certaine spaces, set round about in smal coronets, or wharles, like those of Sage in forme, but of a yellow colour.

¶ *The Place.*

These doe grow wilde in some places, notwithstanding they are manured and planted in Gardens, almost euery where, except Iupiters distaffe, beeing a kinde thereof, which I haue in my Garden.

¶ *The Time.*

They floure in Iune, Iuly, and August.

¶ *The Names.*

Clarie is called of the Apothecaries *Gallitricum*; it is likewise named *Oruala*: of some, *Tota bona*, but not properly: of others, *Scariea*, *Sclarea*, *Centrum Galli*, and *Matrisaluia*. in Italian, *Sciaria*: in French, *Oruale*: in High Dutch, **Scharlach**: in Low Dutch, **Scharleye**: in English, Clarie, or Cleere eie.

Iupiters distaffe is called *Colus Iovis*: of some, *Galeopsis lutea*, but not properly: of diuers, *Horminum luteum*, or yellow Clarie, and *Horminum Tridentinum*, or Clarie of Trent.

¶ *The Temperature.*

Clarie is hot and drie in the third degree.

¶ *The Vertues.*

A The seed of Clarie poudered, finely searced and mixed with hony, taketh away the dimnesse of the eies, and cleereth the sight.

B The same stamped, infused, or laied to steepe in warme water, the mussilag or slimie substance taken and applied plaisterwise, draweth forth splinters of wood, thornes, or any other thing fixed in the bodie: it also scattereth and dissolueth all kindes of swellings, especially in the ioints.

C The seed poudered and drunke with wine, stirreth vp bodily lust.

D The leaues of Clarie taken any manner of way, helpeth the weaknesse of the backe proceeding of the ouermuch flowing of the whites, but most effectually if they be fried with egges in manner of a Tansie, either the leaues whole or stamped.

† The figure which formerly was vnder the title of *Colus Iouis*, was of the *Horminum syluestre* of *Fuchsius*, which is described immediatly before it.

Chap. 268. *Of wilde Clarie, or* Oculus Christi.

¶ *The Description.*

1 Oculus Christi is also a kinde of Clarie, but lesser: the stalkes are many, a cubite high, squared, and somewhat hairie: the leaues be broad, rough, and of a blackish green colour. The floures grow alongst the stalkes, of a blewish colour. The seed is round and blackish, the root is thicke and tough, with some threds annexed thereto. ‡ This is *Hormini syluestris 4. quinta species* of *Clusius*. ‡

2 The purple Clarie hath leaues somewhat round, layd ouer with a hoarie cottony substance, not much vnlike Horehound: among which rise vp small hairie square stalks, set toward the top with little leaues of a purple colour, which appeare at the first view to be floures and yet are nothing else but leaues, turned into an excellent purple colour: and among these beautifull leaues come forth small floures of a blewish or watchet colour, in fashion like vnto the floures of Rosemarie; which being withered, the husks wherein they did grow containe certaine blacke seed, that falleth forth vpon the ground very quickely, because that euery such huske doth turne and hang downe his head toward the ground. The root dieth at the first approch of Winter.

‡ 3 Broad leaued Clarie hath a square stalke some cubite high, hairy, firme, and iointed; the leaues are large, rough, and sharpe pointed, snipt about the edges, wrinckled, and standing by couples at each ioint: vpon the branches in roundles grow purple floures, lesse than those of Clarie, and scarce any bigger than those of Lauander: the seed is smal and blacke: the root is large, hard, black,

† 1 *Horminum ſylueſtre.*
Wilde Clarie, or *Oculus Chriſti.*

2 *Horminum ſylueſtre folijs purpureis.*
Clarie with purple leaues.

‡ 3 *Horminum ſylueſtre latifolium.*
Broad leaued wilde Clarie.

‡ 4 *Horminum ſylueſtre flore albo.*
White floured wilde Clarie.

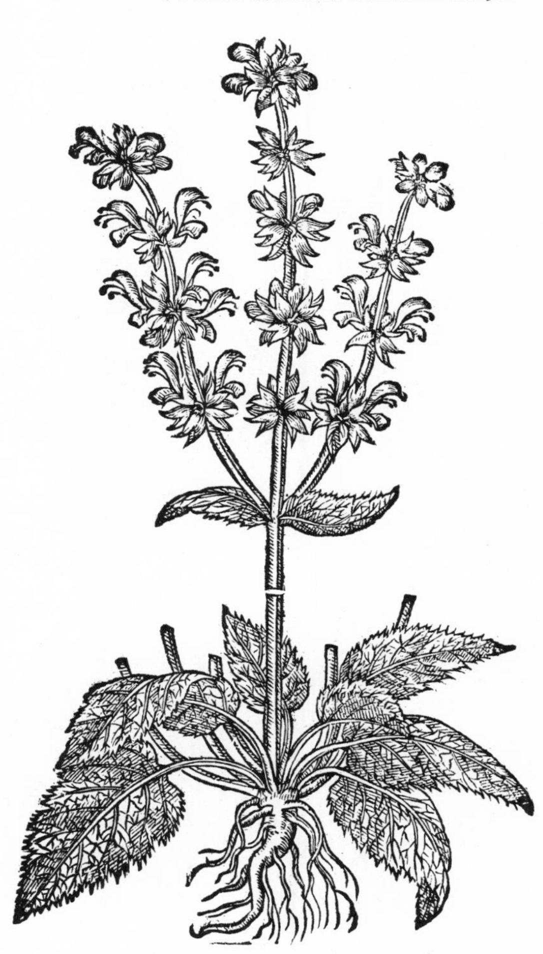

‡ 5 *Horminum syluestre flore rubro.*
Red floured wilde Clarie.

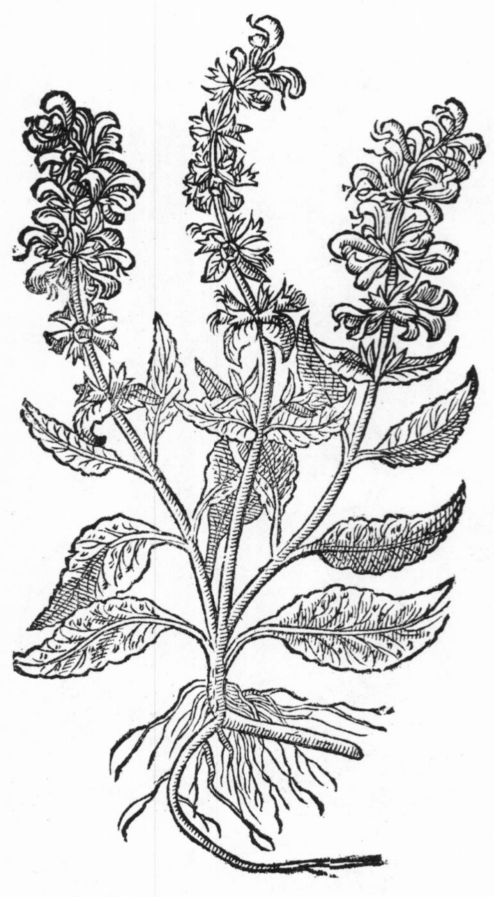

blacke, and liues many yeares. It floures in Iune and Iuly, and growes wilde in many mountainous places of Germany. *Clusius* calls it *Horminum syluestre tertium.*

4 This hath long leaues next vnto the ground, growing vpon prettie long stalkes, broad at their setting on, and so ending by little and little in sharpe points, they are not deeply cut in, but onely lightly snipt about the edges: they are also wrinckled on the vpper side, and whitish, but hairie on the vnder side. The square stalkes are some cubite high, iointed, and set with two leaues at each ioint. The floures grow along st the tops of the branches, and are of a snow white colour. There is a varietie of this with the leaues greener, and the floures of an elegant deepe purple colour. This is the *Horminum syluestris quarti species prima* of *Clusius*, and the varietie with the white floures is his *Hormini syluestris quarti species prima*; and the figure that our Authour gaue in the first place was of these.

5 There is another variety of the last described, which also hath square stalks set with rough snipt leaues, which end in sharp points, but are narrower at the lower end than the former, and they are greene of colour: vpon the tops of the stalkes grow red hooded floures, and those not very large: the seed is small and blacke, and the root liues many yeares. This floures in Iuly. *Clusius* makes this his *Hormini syluestris quarti species quarta.* ‡

¶ *The Place.*

The first groweth wilde in diuers barren places, almost in euery Country, especially in the fields of Holborne neere vnto Grayes Inne, in the high way by the end of a bricke wall: at the end of Chelsey next to London, in the high way as you go from the Queenes pallace of Richmond to the waters side, and in diuers other places.

The other is a stranger in England: it groweth in my garden.

¶ *The Time.*

They floure and flourish from Iune to the end of August.

¶ *The Names.*

Wilde Clarie is called after the Latine name *Oculus Christi*, of his effect in helping the diseases of the eies: in Greeke ὅρμινον: and likewise in Latine, *Horminum*: of some, *Geminalis*: in English, wild Clarie, and *Oculus Christi.*

The second is thought of some to be the right Clarie, and they haue called it *Horminum verum*, but with greater errour: it may be called in Latine *Horminum syluestre folijs & floribus purpureu*, Clarie with leaues and floures of a purple colour.

‡ Our Authour should haue shewn his reasons why this is not the *Horminum verum*, to haue conuinced the errour of *Anguillara*, *Matthiolus*, *Gesner*, *Dodonæus*, *Lobel* and others, who haue accounted it so, as I my selfe must needs do, vntill some reason be shewne to the contrarie, the which I thinke cannot be done, ‡

¶ *The Temperature and Vertues.*

A The temperature and faculties are referred vnto the garden Claries: yet *Paulus Ægineta* saith it is hot and moderately drie, and it also clenseth.

B The seed of wilde Clarie, as *Dioscorides* writeth, being drunke with wine, stirreth vp lust, it clenseth the eies from filmes and other imperfections, being mixed with honie.

The seede put whole into the eies, clenseth and purgeth them exceedingly from waterish humours, rednesse, inflammation, and diuers other maladies, or all that happen vnto the eies, and takes away the paine and smarting thereof, especially being put into the eies one seed at one time, and

no

no more,which is a generall medicine in Cheſhire and other Countries thereabout, knowne of all, and vſed with good ſucceſſe.

The leaues are good to be put into pottage or brothes among other potherbes; for they ſcatter congealed bloud,warme the ſtomacke,and helpe the dimneſſe of the eies. D

† The figure that formerly was in the firſt place,was of that which you may here finde figured and deſcribed in the fourth.

CHAP. 269. *Of Mullein.*

¶ *The Deſcription.*

1 THe male Mullein or Higtaper hath broad leaues, very ſoft,whitiſh and downie;in the midſt of which riſeth vp a ſtalke,ſtraight,ſingle,and the ſame alſo whitiſh all ouer,with a hoarie downe,and couered with the like leaues,but leſſer and leſſer euen to the top:among which taperwiſe are ſet a multitude of yellow floures, conſiſting of fiue leaues apeece: in the places whereof come vp little round veſſels,in which is contained very ſmall ſeed.The root is long,a finger thicke,blacke without,and full of ſtrings.

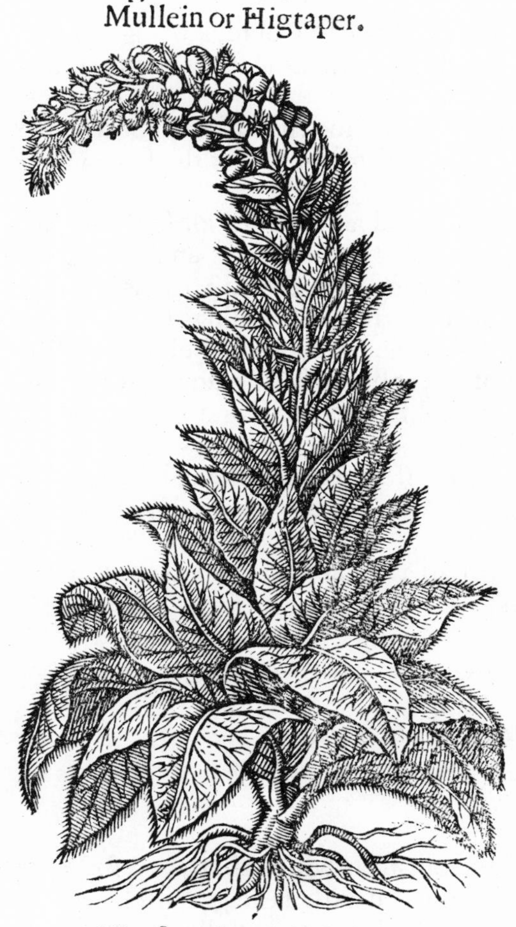

1 *Tapſus Barbatus.*
Mullein or Higtaper.

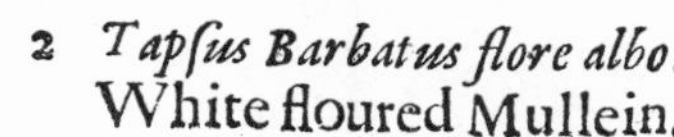

2 *Tapſus Barbatus flore albo.*
White floured Mullein.

2 The female Mullein hath likewiſe many white woolley leaues,ſet vpon an hoarie cottonie vpright ſtalke,of the height of foure or fiue cubits: the top of the ſtalks reſembleth a torch decked with infinite white floures,which is the ſpeciall marke to know it from the male kinde,being like in euery other reſpect.

¶ *The Place.*

Theſe plants do grow of themſelues neere the borders of paſtures, and plowed fields,or cauſies, and drie ſandie ditch banks,and in other vntilled places. They grow in great plentie neere vnto a lyme kill vpon the end of black Heath next to London,as alſo about the Queens houſe at Eltham neere vnto Dartford in Kent: in the high waies about Highgate neere London,and in moſt countries of England that are of a ſandie ſoile.

¶ *The Time.*

They are found with their floure from Iuly to September, and bring forth their seed the second yeare after the seed is sowne.

¶ *The Names.*

Mullein is called in Greeke φλόμος: in shops, *Tapsus Barbatus*: of diuers, *Candela Regia*, *Candelaria*, and *Lanaria*: *Dioscorides*, *Pliny*, and *Galen*, do call it *Verbascum*: in Italian, *Verbasco*, and *Tasso Barbasso*: in Spanish, *Gordolobo*: in High Dutch, **Wullkraut**: in French, *Bouillon*: in English, Mullein, or rather Woollen, Higtaper, Torches, Long-woort, and Bullockes Long-woort; and of some Haresbeard.

¶ *The Temperature.*

Mullein is of temperature drie: the leaues haue also a digesting and clensing qualitie, as *Galen* affirmeth.

¶ *The Vertues.*

A The leaues of Mullein being boiled in water, and laid vpon hard swellings and inflammations of the eies, cureth and ceaseth the paine.

B The root boiled in red wine and drunke, stoppeth the laske and bloudy flix.

C The same boiled in water and drunke, is good for them that are broken and hurt inwardly, and preuaileth much against the old cough.

D A little fine treacle spred vpon a leafe of Mullein, and laied to the piles or Hemorhoides, cureth the same: an ointment also made with the leaues thereof and old hogs grease worketh the same effect.

E The leaues worne vnder the feet day and night, in manner of a shooe sole or sock, bringeth down in yong maidens their desired sicknesse, being kept vnder their feet with some socks or other thing for falling away.

F The Countrey people, especially the husbandmen in Kent, doe giue their cattell the leaues to drinke against the cough of the lungs, being an excellent approued medicine for the same, whereupon they doe call it Bullocks Lung-woort.

G Frankensence and Masticke burned in a chafing dish of coles, and set within a close stoole, and the fume thereof taken vnderneath, doth perfectly cure the piles, hemorrhoids, and all diseases happening in those lower parts, if also there be at euery such fuming (which must bee twice euerie day) a leafe of the herbe bound to the place, and there kept vntill the next dressing.

H There be some who thinke that this herbe being but carried about one, doth helpe the falling sicknesse, especially the leaues of that plant which hath not as yet borne floures, and that is gathered when the Sun is in Virgo, and the Moone in Aries; which thing notwithstanding is vaine and superstitious.

I The later P ysitions commend the yellow floures, beeing steeped in Oile and set in warme doung vntill they bee wasted into the Oile and consumed away, to bee a remedie against the piles.

K The report goeth, saith *Pliny*, that figges do not putrifie at all that are wrapped in the leaues of Mullein: which thing *Dioscorides* also maketh mention of.

Chap. 270. *Of base Mullein.*

¶ *The Description.*

1 The base white Mullein hath a thicke wooddie root, from which riseth vp a stiffe and hairie stalke, of the height of foure cubites, garnished with faire grayish leaues like those of Elecampane, but lesser: the floures grow round about the stalks taper or torch fashion, of a white colour, with certaine golden thrums in the middle: the seed followeth, smal, and of the colour of dust.

2 Blacke Mullein hath long leaues, not downie at all, large and sharp pointed, of an ouerworne blackish green colour, somewhat rough, and strongly smelling: the floures grow at the top of the stalks, of a golden yellow colour, with certaine threds in the middle thereof. The root differeth not from the precedent.

3 Candleweeke Mullein hath large, broad, and woollie leaues, like vnto those of the common Mullein: among which riseth vp a stalke couered with the like leaues, euen to the branches wheron the floures do grow, but lesser and lesser by degrees. The stalke diuideth it selfe toward the top into diuerse branches, whereon is set round about many yellow floures, which oftentimes doe change into white, varying according vnto the soile and clymate. The root is thick and woody.

4 The

1 *Verbascum album.*
Base white Mullein.

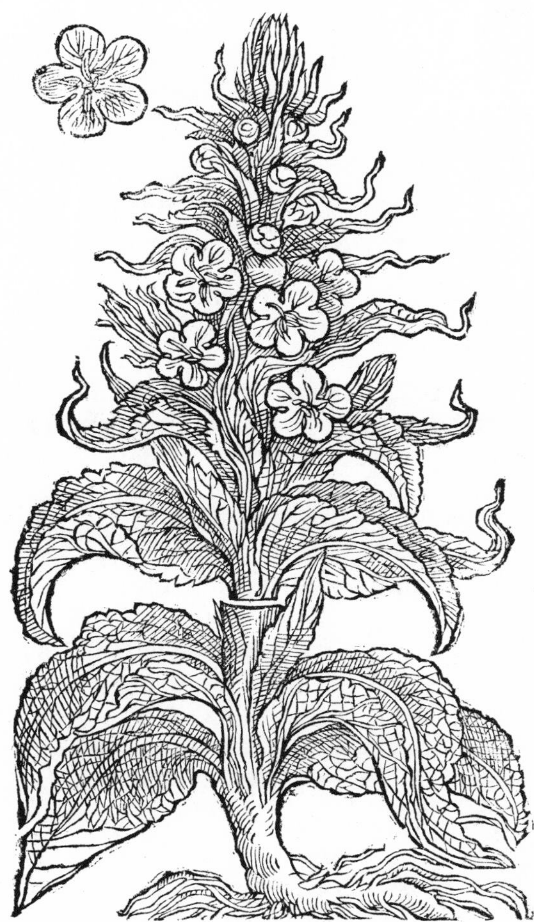

2 *Verbascum nigrum.*
Base blacke Mullein.

3 *Verbascum Lychnite Matthioli.*
Candle-weeke Mullein.

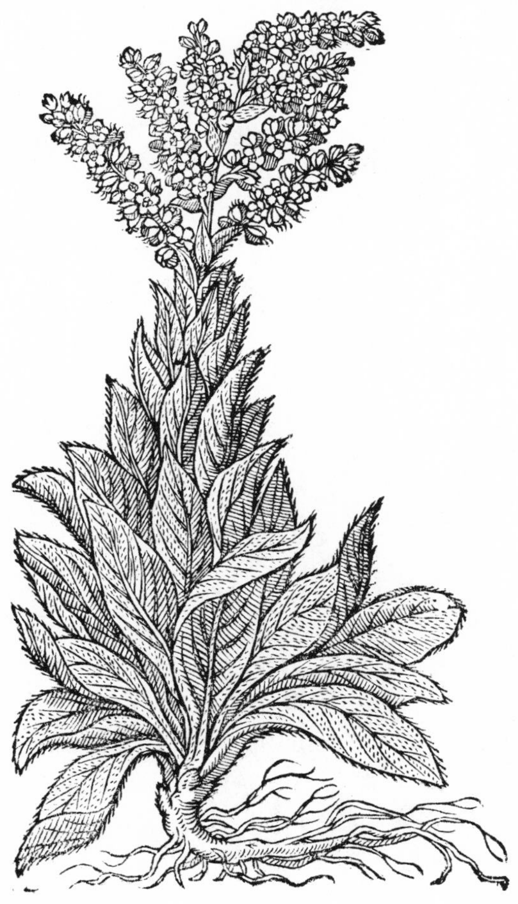

4 *Verbascum Lychnite minus.*
Small Candle-weeke Mullein.

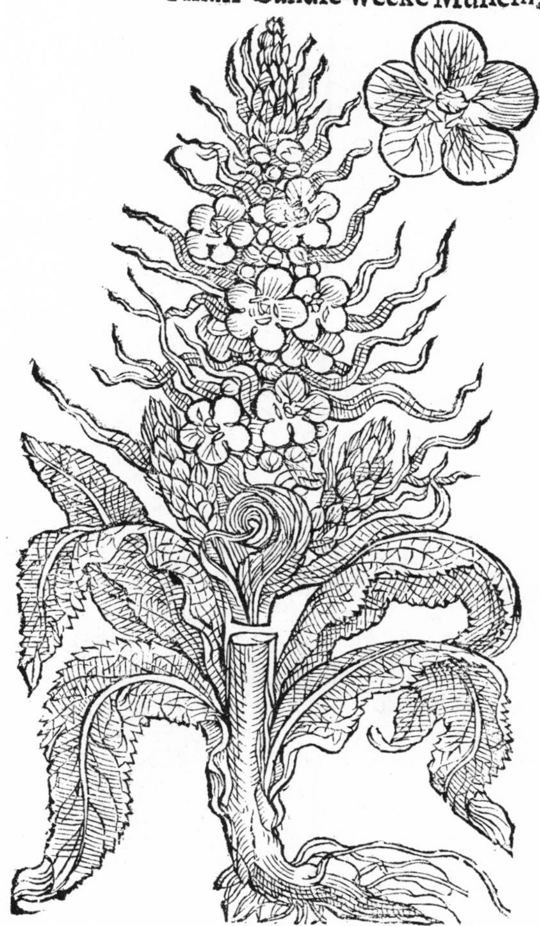

4 The ſmall Candle-weeke Mullein differeth little from the laſt rehearſed, ſauing that the whole plant of this is of a better ſauour, wherein eſpecially conſiſteth the difference. ‡ The floure alſo is much larger, and of a ſtraw, or pale yellow colour. ‡

¶ *The Place.*

Theſe plants do grow where the other Mulleins do, and in the like ſoile.

¶ *The Time.*

The time likewiſe anſwereth their flouring and ſeeding.

¶ *The Names.*

Their capitall names expreſſed in the titles ſhal ſerue for theſe baſe Mulleins, conſidering they are all and euery of them kindes of Mulleins.

¶ *The Temperature.*

Theſe Mulleins are drie without any manifeſt heat, yet doubtleſſe hotter and drier than the common Mullein or Hygtaper.

¶ *The Vertues.*

A The blacke Mullein, with his pleaſant yellow floures, boiled in water or wine and drunken, is good againſt the diſeaſes of the breſt and lungs, and againſt all ſpitting of corrupt rotten matter.

B The leaues boiled in water, ſtamped and applied pultis wiſe vpon cold ſwellings (called *Oedemata*) and alſo vpon the vlcers and inflammations of the eies, cureth the ſame.

C The floures of blacke Mullein are put into lie, which cauſeth the haire of the head to wax yellow, if it be waſhed and combed therewith.

D The leaues are put into cold ointments with good ſucceſſe, againſt ſcaldings and burnings with fire or water.

E *Apuleius* reporteth a tale of *Vlyſſes*, *Mercurie*, and the inchantreſſe *Circe*, and their vſe of theſe herbes in their incantations and witchcrafts.

Chap. 271. *Of Moth Mullein.*

1 *Blattaria Plinij.*
Plinies Moth Mullein.

2 *Blattaria flore purpureo.*
Purple Moth Mullein.

¶ The

¶ *The Deſcription.*

1 PLinie hath ſet forth a kinde of *Blattaria*, which hath long and ſmooth leaues, ſomewhat iagged or ſnipt about the edges: the ſtalke riſeth vp tothe height of three cubits, diuiding it ſelſe toward the top into ſundry armes or branches, beſet with yellow floures like vnto blacke Mullein.

2 *Blattaria* with purple floures hath broad blacke leaues, without any manifeſt ſnips or notches by the ſides, growing flat vpon the ground: among which riſeth vp a ſtalke two cubits high, garniſhed with floures like vnto the common *Blattaria*, but that they are of a purple colour, and thoſe few threds or chiues in the middle of a golden colour: the root is as thick as a mans thumb, with ſome threds hanging thereat, and it indureth from yeare to yeare.

3 There is another kinde like vnto the blacke Mullein, in ſtalks, roots, and leaues, and other reſpects, ſauing that his ſmall floures are of a greene colour.

4 There is another like vnto the laſt before written, ſauing that his leaues are not ſo deepely cut about the edges, and that the ſmall floures haue ſome purple colour mixed with the greenneſſe.

‡ 3 *Blattaria flore viridi.*
Greene Moth Mullein.

‡ 4 *Blattaria flore ex viridi purpuraſcente.*
Moth Mullein with the greeniſh purple coloured floure.

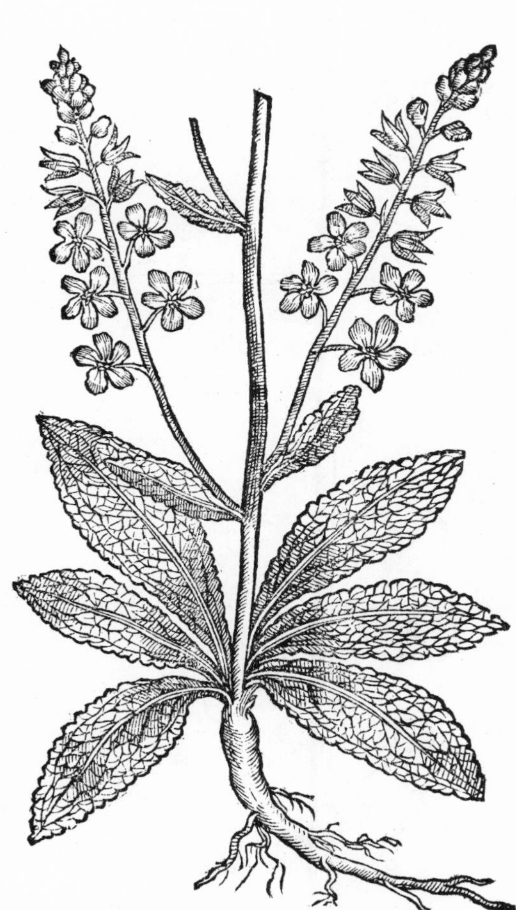

‡ 5 This is ſomewhat like the firſt deſcribed in leaues and ſtalks, but much leſſe, the floures alſo are of a whitiſh or grayiſh colour, and therein conſiſts the chiefeſt difference.

6 There is alſo another varietie of this kinde, which hath very faire and large floures, and theſe either of a bright yellow, or elſe of a purple colour.

7 This hath long narrow leaues like thoſe of the ſecond, ſnipt about the edges, and of a darke greene colour: the ſtalkes grow ſome two cubits high, and ſeldome ſend forth any branches; the floures are large and yellow, with rough threddes in their middles tipt with red, and theſe grow in ſuch an order that they ſomewhat reſemble a flie: the ſeed is ſmall, and contained in round buttons. This is an annuall, and periſheth when the ſeed is ripe. ‡

‡ 5 *Blattaria flore albo.*
White floured Moth Mullein.

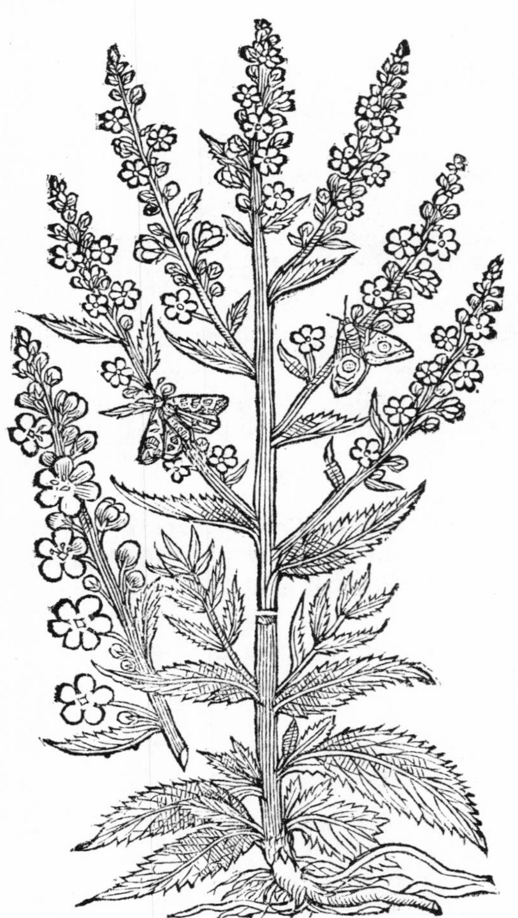

‡ 6 *Blattaria flore amplo.*
Moth Mullein with the great floure.

‡ 7 *Blattaria flore Luteo.*
Yellow Moth Mullein.

¶ *The Place.*

† The firſt and fift of theſe grow wilde in ſundrie places, and the reſt onely in gardens with vs.

¶ *The Time.*

They floure in Iuly and Auguſt.

¶ *The Names.*

The later Herbariſts call Moth Mullein by the name of *Blattaria*, and doe truly take it to bee that which *Plinie* deſcribeth in his 22. booke, *cap.* 9. in theſe words. [There is an herbe like Mullein, or *Verbaſcum nigrum*, which oftentimes deceiueth, being taken for the ſame, with leaues not ſo white, moe ſtalks, and with yellow floures (as wee haue written) which do agree with blacke Mullein, but we haue not as yet learned by obſeruation that they do gather mothes and flies vnto them, as wee haue ſaid.] *Valerius Cordus* names it *Verbaſcum Leptophyllon*, or narrow leafed Mullein: their ſeueral titles ſufficiently ſet forth their Engliſh names.

¶ *The Nature and Vertues.*

Concerning the plants comprehended vnder the titles of *Blattaria*, or Moth Mulleins, I find nothing written of them, ſauing that moths, butterflies, and all manner of ſmall flies and bats do reſort to the place where theſe herbs are laied or ſtrewed.

‡ The decoctio of the floures or leaues of the firſt deſcribed opens the obſtructions of the bowels, as alſo of the Meſeraicke veins, as *Camerar.* affirmes. ‡

CHAP.

CHAP. 272. *Of Mullein of Æthiopia.*

Æthiopis.
Æthiopian Mullein.

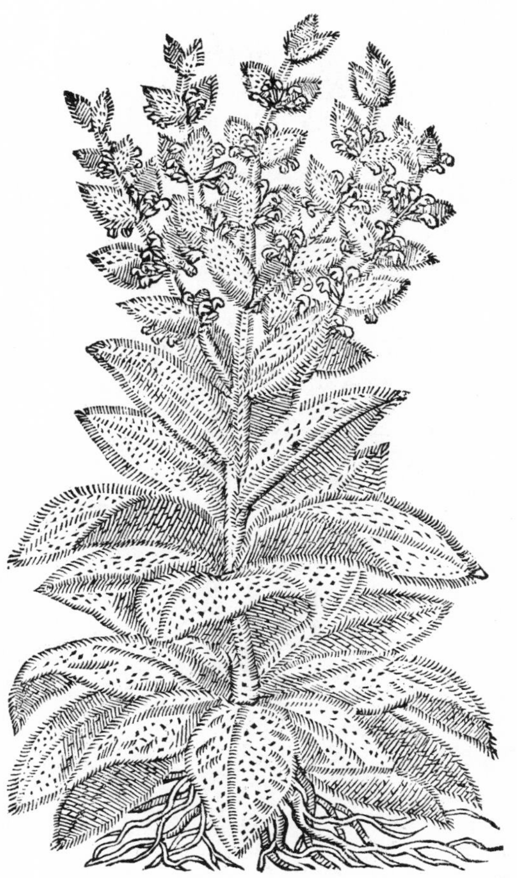

¶ *The Description.*

MVllein of Æthiopia hath many very broad hoary leaues spred vpon the ground, very soft and downy, or rather woolly, like to those of Hygtaper, but farre whiter, softer, thicker, and fuller of woollinesse; which wooll is so long, that one may with his fingers pull the same from the leaues, euen as wooll is pulled from a Sheeps skinne: among which leaues riseth vp a foure square downy stalke, set with the like leaues, but smaller; which stalke is diuided at the top into other branches, set about and orderly placed by certaine distances, hauing many floures like those of Archangell, of a white colour tending to blewnesse: which being past, there succeedeth a three square browne seed: the root is blacke, hard, and of a wooddy substance.

¶ *The Place.*

It groweth naturally in Ethiopia, and in Ida, a hill hard by Troy, and in Messenia a prouince of Morea, as *Pliny* sheweth in his twenty seuenth booke, chap. 4. it also groweth in Meroe, an Island in the riuer Nilus: it likewise groweth in my garden.

¶ *The Time.*

It floureth and flourisheth in Iune, and perfecteth his seed toward the end of August.

¶ *The Names.*

It is called in Greek Αἰθιοπὶς: and in Latine *Æthiopis*, of the countrey; and for that cause it is likewise called *Meroides*, of *Meroë*, as *Pliny* writeth: of some because the Greeke word Αἰθὼν, signifieth in Latine *Fauilla adusta*, or *Cinere aspersa*, or couered with ashes: in English we may cal it Mullein of Æthiopia, or woolly Mullein.

¶ *The Nature.*

Æthiopis is dry without any manifest heate.

¶ *The Vertues.*

A *Æthiopis* is good for those that haue the Pleurisie, and for those that haue their brests charged with corrupt and rotten matter, and for such as are grieued with the asperitie and roughnesse in the throat, and against the Sciatica, if one drinke the decoction of the root thereof.

B For the diseases of the brest and lungs it is good to licke oftentimes of a confection made with the root hereof and honey, and so are the roots condited with sugar, in such manner as they condite the roots of Eringos.

CHAP. 273. *Of Cowslips.*

¶ *The Description.*

1 THose herbes which at this day are called Primroses, Cowslips, and Oxlips, are reckoned among the kindes of Mulleins; notwithstanding for distinctions sake I haue marshalled them in a chapter, comming in the rereward as next neighbors to the Mullens, for that the Antients haue named them *Verbasculi*, that is to say, Small Mullens. The first, which is called in English the field Cowslip, is as common as the rest, therefore I shall not need to spend much time about the description.

2 The second is likewise well knowne by the name of Oxlip, and differeth not from the other, saue

saue that the floures are not so thicke thrust together as the former, and they are fairer, and fewer in number, and do not smell so pleasantly as the other: of which kinde wee haue one lately come into our gardens, whose floures are curled and wrinkled after a most strange manner, which our women haue named Iack-an-apes on horsebacke.

1 *Primula veris maior.*
Field Cowslips.

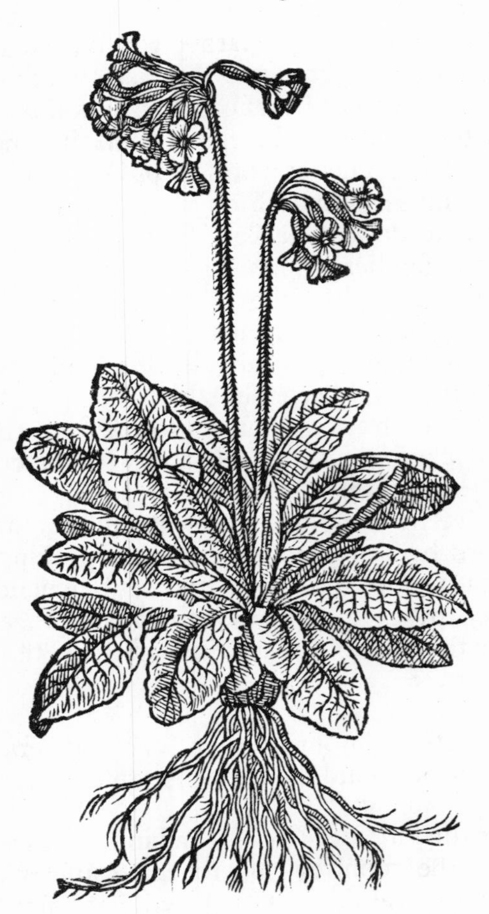

2 *Primula pratensis inodora lutea.*
Field Oxlips.

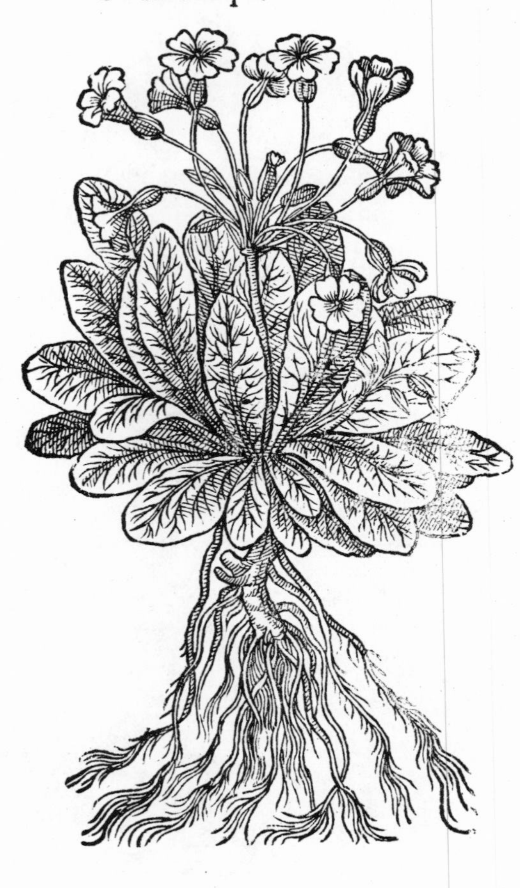

3 Double Paigle, called of *Pena, Primula hortensis Anglica omnium maxima, & serotina floribus plenus*; that is, The greatest English garden Cowslip with double yellow floures, is so commonly knowne that it needeth no description.

4 The fourth is likewise known by the name of double Cowslips, hauing but one floure within another, which maketh the same once double, where the other is many times double, called by *Pena, Geminata,* for the likenesse of the floures, which are brought forth as things against nature, or twinnes.

5 The fifth being the common white field Primrose, needeth no description.

6 The sixth, which is our garden double Primrose, of all the rest is of greatest beauty, the description whereof I refer vnto your owne consideration.

7 The seuenth kinde is also very well knowne, being a Primrose with greenish floures somewhat welted about the edges: for which cause *Pena* hath called it *Siluarum primula, floribus obscure virentibus fimbriatis.*

8 There is a strange Primrose found in a wood in Yorkshire growing wilde, by the trauell and industrie of a learned gentleman of Lancashire called Mr. *Thomas Hesketh,* a diligent searcher of Simples, who hath not onely brought to light this amiable and pleasant kinde of Primrose, but many others likewise, neuer before his time remembred or found out. This kinde of Primrose hath leaues and roots like the wilde field Primrose in each respect: it bringeth forth amongst the leaues a naked stalke of a grayish or ouerworne greenish colour: at the top whereof doth grow in the Winter time one floure and no more, like vnto that single one of the field: but in the Sommer time it bringeth forth a soft russet huske or hose, wherein are contained many small floures, sometimes foure or fiue, and oftentimes more, very thicke thrust together, which maketh one entire floure, seeming to be one of the common double Primroses, whereas indeed it is one double floure made of a number of small single floures, neuer ceasing to beare floures Winter nor Sommer, as before is specified.

‡ Besides

3 *Primula hortenſis Anglica.*
Double Paigles.

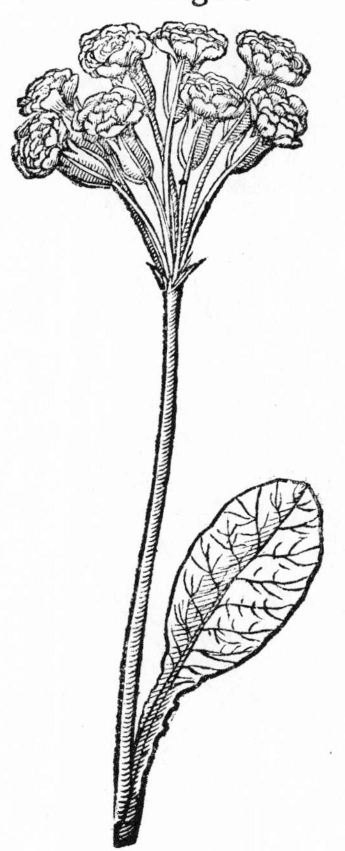

4 *Primula veris flore geminato.*
Cowſlips two in a hoſe.

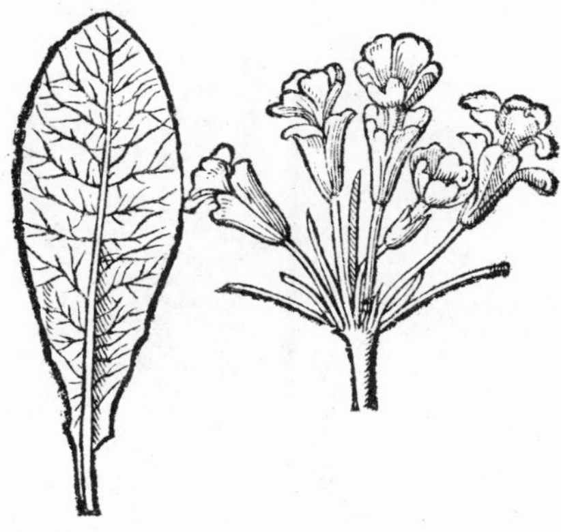

6 *Primula veris flore pleno.*
Double white Primroſe.

5 *Primula veris minor.*
Field Primroſe.

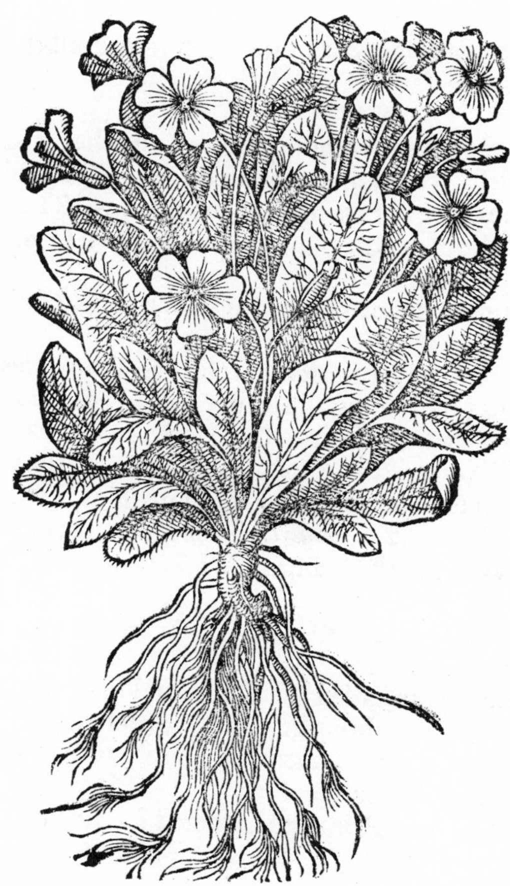

‡ Beſides theſe, there are kept in our gardens, and ſet forth by M^r. *Parkinſon* (to whoſe Worke I referre the curious Reader) two or three more varieties; one a double Cowſlip hoſe in hoſe, naked, without any huske; the other two beare many greene leaues on the tops of the ſtalkes, the one of them hauing yellowiſh floures amongſt the leaues, and the other onely longiſh narrow greene leaues. The firſt of theſe he calls *Paralyſis inodora flore geminato*, Double Oxlips hoſe in hoſe. The ſecond, *Paralyſis fatua*, The fooliſh Cowſlip. And the laſt, *Paralyſis flore viridi roſeo calamiſtrato*, The double greene feathered Cowſlip. ‡

¶ *The*

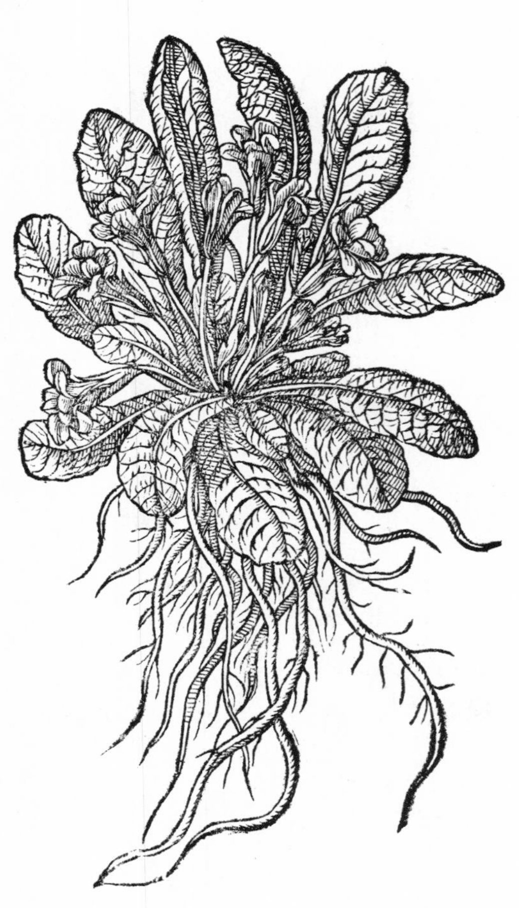
7 *Primula flore viridi.*
Greene Primrose.

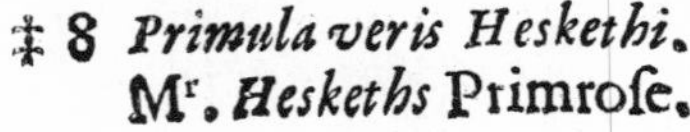
‡ 8 *Primula veris Hesketbi.*
Mr. *Hesketbs* Primrose.

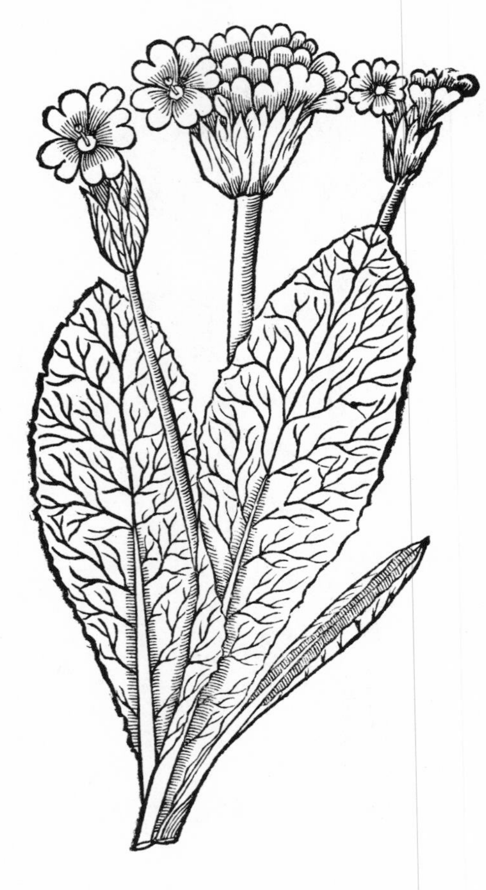

¶ *The Place.*

Cowslips and Primroses ioy in moist and dankish places, but not altogether couered with water; they are found in woods and the borders of fields: the Primrose found by Mr. *Hesketh* growes in a wood called Clap-dale, three miles from a towne in Yorkeshire called Settle.

¶ *The Time.*

They flourish from Aprill to the end of May, and some one or other of them do floure all the Winter long.

¶ *The Names.*

They are commonly called *Primula veris*, because they are the first among those plants that doe floure in the Spring, or because they do floure with the first. They are also named *Arthritica*, and *Herbæ paralysis*, for they are thought to be good against the paines of the ioynts and finewes. They are called in Italian, *Brache cuculi*: in English, Petty Mulleins, or Palsie-worts: of most, Cowslips.

The greater sort, called for the most part Oxlips or Paigles, are named of diuers *Herba S. Petri*. In English, Oxlip, and Paigle.

The common Primrose is vsually called *Primula veris*: most Herbarists do refer the Primroses to the φλομίδες, called in Latine *Verbascula*, or Petty Mulleins; but seeing the leaues be neither woollie nor round, they are hardly drawn vnto them: for *Phlomides* are described by leaues, as *Pliny* hath interpreted it, *Hirsutis & Rotundis*, Hairy and round; which *Pliny*, *lib.* 25. *cap.* 10. translateth thus: *Sunt & Phlomides duæ Hirsutæ, rotundis folijs, humiles*: which is as much to say in English as, There be also two pretty Mulleins, hairy, round leafed, low, or short. ‡ *Fabius Columna* refers these to the *Alisma* of *Dioscor* and calls the Cowslip *Alismapratorum*: and the Primrose, *Alisma syluarum*. ‡

¶ *The Temperature.*

The Cowslips and Primroses are in temperature dry, and a little hot.

¶ *The Vertues.*

A The Cowslips are commended against the paine of the ioynts called the Gout, and slackenesse of the sinewes, which is the palsie. The decoction of the roots is thought to be profitably giuen against the stone in the kidneyes and bladder; and the iuyce of the leaues for members that are loose and out of ioynt, or inward parts that are hurt, rent, or broken.

B A dramme and a halfe of the pouder of the dried roots of field Primrose gathered in Autumne, giuen

giuen to drinke in Ale or Wine purgeth by vomit very forcibly (but ſafely) wateriſh humours, choler, and flegme, in ſuch manner as *Azarum* doth, experimented by a learned and skilfull Apothecarie of Colcheſter Mr. *Thomas Buckſtone*, a man ſingular in the knowledge of Simples.

A conſerue made with the floures of Cowſlips and ſugar preuaileth wonderfully againſt the palſie, convulſions, cramps, and all the diſeaſes of the ſinewes. C

Cowſlips or Paigles do greatly reſtraine or ſtop the belly in the time of a great laske or bloudy flix, if the decoction thereof be drunke warme. D

A practitioner in London, who was famous for curing the frenſie, after that hee had performed his cure by the due obſeruation of phyſicke, accuſtomed euery yeare in the moneth of May to diet his patients after this manner: Take the leaues and floures of Primroſe, boyle them a little in fountaine water, and in ſome Roſe and Betony waters, adding thereto ſugar, pepper, ſalt, and butter, which being ſtrained, he gaue them to drinke thereof firſt and laſt. E

The roots of Primroſe ſtamped and ſtrained, and the iuyce ſniffed into the noſe with a quill or ſuch like, purgeth the braine, and qualifieth the paine of the megrim. F

An vnguent made with the iuyce of Cowſlips and oyle of Linſeed cureth all ſcaldings or burnings with fire, water, or otherwiſe. G

The floures of Primroſes ſodden in vineger and applied, do heale the Kings Euill, as alſo the almonds of the throat and uvula, if you gargariſe the part with the decoction thereof. H

The leaues and floures of Primroſes boyled in wine and drunke, is good againſt all diſeaſes of the breſt and lungs, and draweth forth of the fleſh any thorne or ſplinter, or bone fixed therein. I

CHAP. 274. *Of Birds-eine.*

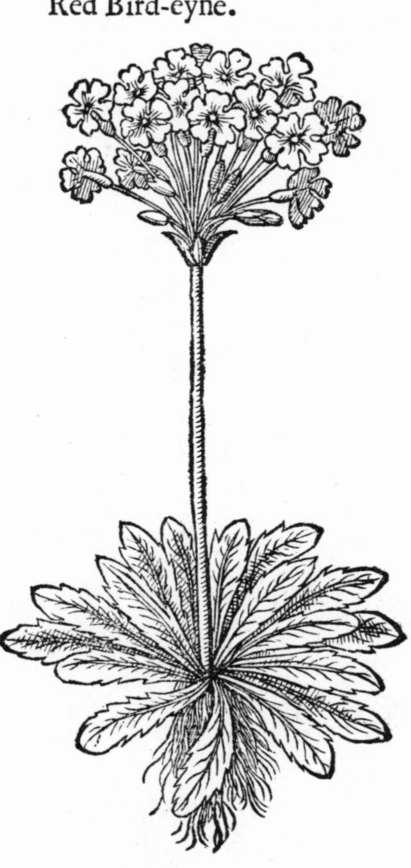

1 *Primula veris flore rubro.*
Red Bird-eyne.

2 *Primula veris flore albo.*
White Bird-eyne.

¶ *The Deſcription.*

1 Some Herbariſts call this plant by the name of *Sanicula anguſtifolia*, making thereof two kinds, and diſtinguiſhing them by theſe termes, *maior & minor, ſiue media*: others cal them *Paralytica alpina*, which without controuerſie are kindes of Cowſlips, agreeing with them as well in ſhape, as in their nature and vertues, hauing leaues much like vnto Cowſlips, but ſmaller, growing

growing flat vpon the ground, of a faint greeniſh colour on the vpper ſide,& vnderneath of a white or mealy colour: among which riſe vp ſmall and tender ſtalkes of a foot high, hauing at the top of euery ſtalke a buſh of ſmall floures in ſhape like the common Oxlip, ſauing that they are of a faire ſtammell colour tending to purple: in the middle of euery ſmall floure appeareth a little yellow ſpot, reſembling the eye of a bird; which hath moued the people of the North parts (where it aboundeth) to call it Birds eyne. The ſeed is ſmall like duſt, and the root white and threddy.

2 The ſecond is like the firſt, ſauing that the whole plant is greater in each reſpect, and that the floures are of a whitiſh colour.

¶ *The Place.*

Theſe plants grow very plentifully in moiſt and ſqually grounds in the North parts of England, as in Harwood neere to Blackburne in Lancaſhire, and ten miles from Preſton in Aunderneſſe; alſo at Crosby, Rauenſwaith, and Crag-Cloſe in Weſtmerland.

They likewiſe grow in the medowes belonging to a village in Lancaſhire neere Maudſley, called Harwood, and at Hesketh not far from thence, and in many other places of Lancaſhire, but not on this ſide Trent, that I could euer haue any certaine knowledge of. *Lobel* reporteth, That doctor *Penny* (a famous Phyſition of our London Colledge) did finde them in theſe Southerne parts.

¶ *The Time.*

They floure and flouriſh from Aprill to the end of May.

¶ *The Names.*

The firſt is called Primroſe with the red floure: the ſecond, Primroſe with the white floure, and Birds eyne.

¶ *The Nature and Vertues.*

The nature and vertues of theſe red and white Primroſes muſt be ſought out amongſt thoſe aboue named.

Chap. 275. *Of Beares eares, or Mountaine Cowſlips.*

1 *Auricula vrſi flore luteo.*
Yellow Beares-eare.

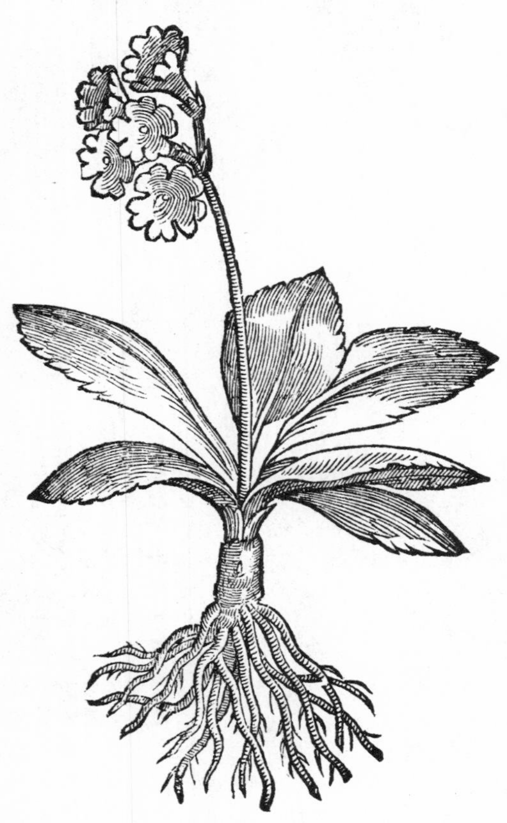

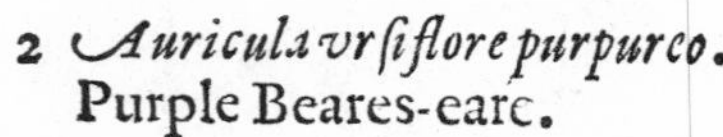

2 *Auricula vrſi flore purpureo.*
Purple Beares-eare.

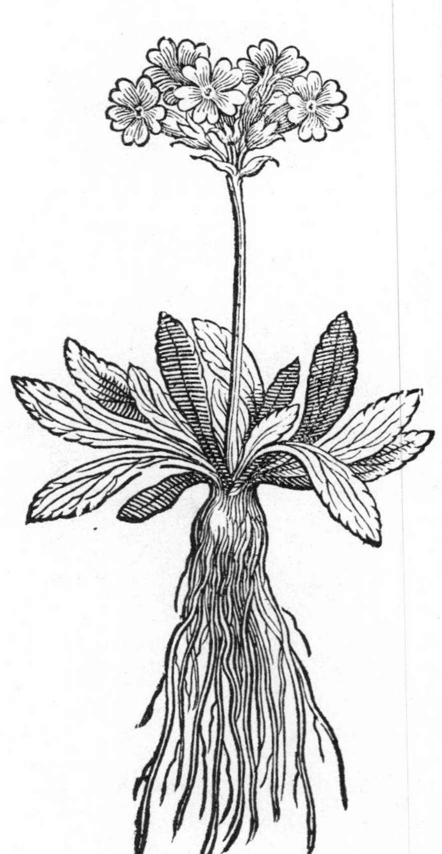

¶ The

¶ *The Kindes.*

THere be diuers ſorts of Mountaine Cowſlips, or Beares-eares, differing eſpecially in the colour of their floures, as ſhall be declared, notwithſtanding it may appeare to the curious, that there is great difference in the roots alſo, conſidering ſome of them haue knobby roots, and others threddy: notwithſtanding there is no difference in the roots at all.

‡ There are diuers varieties of theſe floures, and the chiefe differences ariſe, either from the leaues or floures; from their leaues, which are either ſmooth and greene, or elſe gray and hoary, againe they are ſmooth about the edges, or ſnipt more or leſſe; The floures ſome are fairer then otherſome, and their colours are ſo various, that it is hard to finde words to expreſſe them, but they may be refer'd to whites, reds, yellowes, and purples; for of all the varieties and mixtures of theſe they chiefely conſiſt. The gardens of Mr. *Tradeſcant* and Mr. *Tuggie* are at this preſent furniſhed with very great varieties of theſe floures. ‡

3 *Auricula Vrſi ij. Cluſij.*
Red Beares eare.

4 *Auricula Vrſi iij. Cluſij.*
Scarlet Beares eare.

¶ *The Deſcription.*

1 A*Vricula Vrſi* was called of *Matthiolus*, *Pena*, and other Herbariſts, *Sanicula Alpina*, by reaſon of his ſingular facultie in healing of wounds, both inward and outward. They do all call it *Paralityca*, becauſe of his vertues in curing the palſies, cramps, and conuulſions, and is numbred among the kindes of Cowſlips, whereof no doubt they are kinds, as others are which do hereafter follow vnder the ſame title, although there be ſome difference in the colour of the floures. This beautifull and braue plant hath thicke, greene, and fat leaues, ſomewhat finely ſnipt about the edges, not altogether vnlike thoſe of Cowſlips, but ſmoother, greener, and nothing rough or crumpled: among which riſeth vp a ſlender round ſtem a handfull high, bearing a tuft of floures at the top, of a faire yellow colour, not much vnlike to the floures of Oxe-lips, but more open and conſiſting of one only leafe like Cotiledon: the root is very threddy, and like vnto the Oxe-lip.

2 The leaues of this kinde which beareth the purple floures are not ſo much ſnipt about the edges: theſe ſaid purple floures haue alſo ſome yellowneſſe in the middle, but the floures are not ſo much laid open as the former, otherwiſe in all reſpects they are like.

3 *Carolus Clusius* setteth forth in the booke of his Pannonicke trauels two kindes more, which he hath found in his trauell ouer the Alpes and other mountaines of Germanie and Heluetia, being the third in number, according to my computation: it hath leaues like the former, but longer, smaller, and narrower toward the bottome, greene aboue, and of a pale colour vnderneath. The floures are in fashion like to the former, but of a most shining red colour within, and on the outside of the colour of a mulberry: the middle or eye of the floure is of a whitish pale colour: the root is like the former.

4 The fourth is a smaller plant than any of the foresaid, whose leaues are thicke and fat, nothing at all snipt about the edges, greene aboue, and grayish vnderneath. The floures are like the former, shining about the edges, of an ouerworne colour toward the middle, and in the middle commeth a forke couered with an hairinesse: the root is blacke and threddy.

5 *Auricula Vrsi erubescens.*
Blush coloured Beares eare.

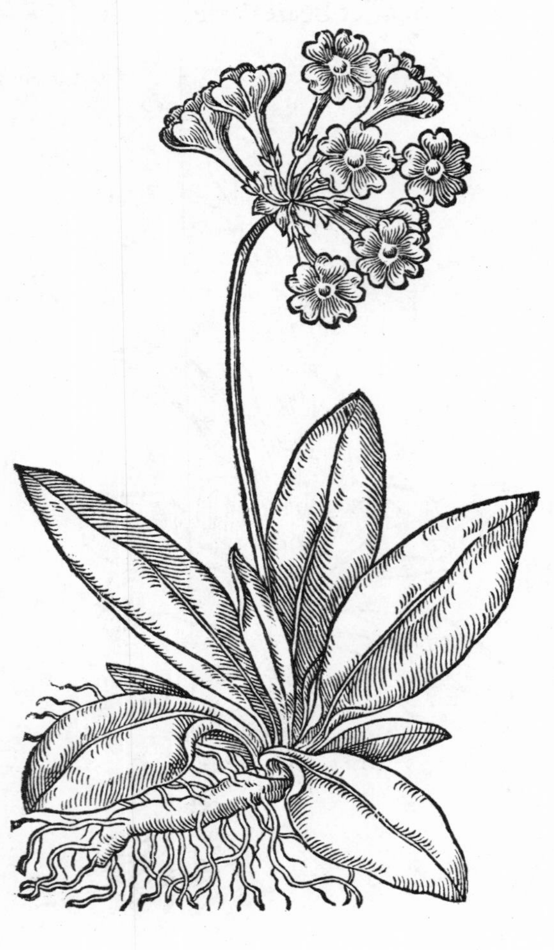

6 *Auricula Vrsi suaue rubens.*
Bright red Beares eare.

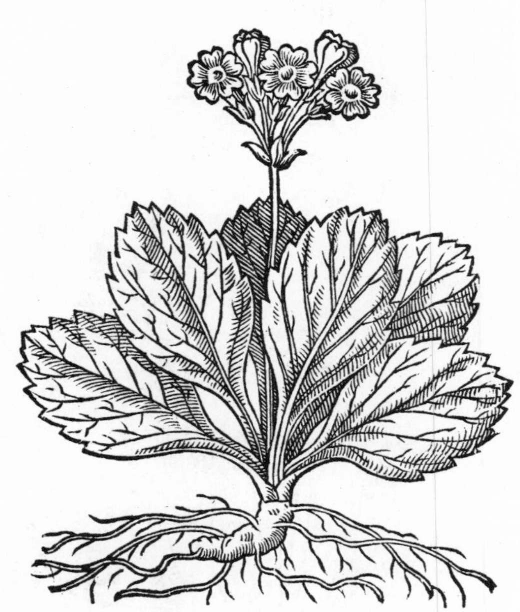

7 *Auricula Vrsi minima.*
Stamell Beares eare.

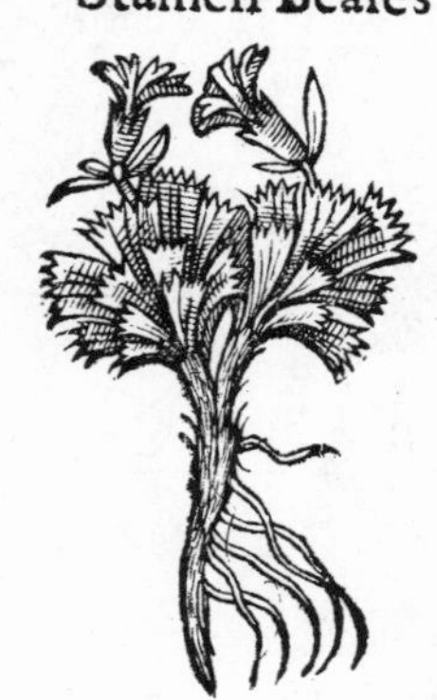

5 The blush-coloured Beares eare hath diuers thicke fat leaues spred vpon the ground, of a whitish green colour, sleightly or not at all indented in the edges: among which riseth vp a naked stalke likewise hairy or whitish, on the top whereof stand very faire floures, in shape like those of the common Cowslip, but of a whitish colour tending to purple, which wee terme blush-colour. The root is tough and threddy, as are all the rest.

6 The bright shining red Beares eare of *Matthiolus* description seemes to late Herbarists to be rather a figure made by conceit or imagination, than by the sight of the plant it self; for doubtlesse we are persuaded that there is no such plant, but onely a figure foisted for ostentations sake, the description whereof we leaue to a further consideration, because we haue not seene any such plant, neither do we beleeue there is any such. ‡ Our Author is here without cause iniurious to *Matthiolus*, for he figures and describes onely the common first described yellow Beares eare: yet if he had said the floures were of a light shining red, he had not erred; for I haue seen these floures of all the reds both bright and darke that one may imagine. ‡

7 *Pena*

7 *Pena* setteth forth a kinde of Beares eare vnder the name of *Sanicula Alpina*, hauing his vppermost leaues an inch long, somewhat iagged and hem'd at the ends, and broad before like a shouel; the lower leaues next the ground are somewhat shorter, but of the same forme; among which riseth a small slender foot-stalke of an inch long, whereon doth stand a small floure, consisting of fiue little leaues of a bright red or stammell colour.

8 The snow white Beares eare differeth not from the last described but in the colour of the floure, for as the others are red, contrarie these are very white, and the whole plant is lesser, wherein consisteth the difference. The root is long, tough, with some fibres thereto belonging. Neither of these two last described will be content to grow in gardens.

¶ *The Place.*

They grow naturally vpon the Alpish and Heluetian mountaines: most of them do grow in our London gardens.

¶ *The Time.*

These herbes do floure in Aprill and May.

¶ *The Names.*

Either the antient writers knew not these plants, or else the names of them were not by them or their successors diligently committed vnto posteritie. *Matthiolus* and other later writers haue giuen names according to the similitude, or of the shape that they beare vnto other plants, according to the likenesse of the qualities and operations: you may call it in English, Beares eare: they that dwell about the Alps doe call it **Drastkrawt**, and **Schwindlekrawt**, by reason of the effects thereof; for the root is amongst them in great request for the strengthning of the head, that when they are on the tops of places that are high, giddinesse and the swimming of the braine may not afflict them: it is there called the Rocke-rose, for that it groweth vpon the rockes, and resembleth the braue colour of the Rose. ‡ *Fabius Columna* proues this to be the *Alisma* or *Damasonium* of *Dioscorides* and the Antients.

¶ *The Nature.*

These herbes are dry and very astringent.

¶ *The Vertues.*

It healeth all outward and inward wounds of the brest, and the enterocele also, if for some reasonable space of time it be put in drinkes, or boyled by it selfe. A

These plants are of the nature and temperature of *Primula veris*, and are reckoned amongst the Sanicles by reason of their vertue. B

Those that hunt in the Alps and high mountaines after Goats and Bucks, do as highly esteeme hereof as of *Doronicum*, by reason of the singular effects that it hath, but (as I said before) one especially, euen in that it preuenteth the losse of their best ioynts (I meane their neckes) if they take the roots hereof before they ascend the rocks or other high places. C

‡ The root of *Damasonium* (according to *Dioscorides*) taken in the weight of one or two drams, helpeth such as haue deuoured the *Lepus marinus* or sea Hare, or haue been bitten by a Toad, or taken too great a quantitie of *Opium*. D

It is also profitably drunke, either by it selfe, or with the like quantitie of *Daucus* seeds, against gripings in the belly, and the bloudy flux. E

Also it is good against conuulsions and the affects of the wombe. F

The herbe stayes the fluxes of the belly, moues the courses, and applied in forme of a pultis asswageth œdematous tumors. ‡ G

CHAP. 276. *Of Mountaine Sanicle.*

¶ *The Kindes.*

THere be sundry sorts of herbes contained vnder the name of Sanicle, and yet not one of them agreeing with our common Sanicle, called *Diapensia*, in any one respect, except in the vertues, whereof no doubt they tooke that name; which number doth dayly increase, by reason that the later writers haue put downe more new plants, not written of before by the Antients, which shall be distinguished in this chapter by seuerall titles.

¶ *The Description.*

1 SPotted Sanicle of the mountaine hath small fat & round leaues, bluntly indented about the edges, and fashioned like vnto the leaues of *Saxifragia aurea*, or rather *Cyclamen folio hederæ*, of a darke greene colour, and somewhat hairy vnderneath: amongst which rise

1 *Sanicula guttata.*
Spotted Sanicle.

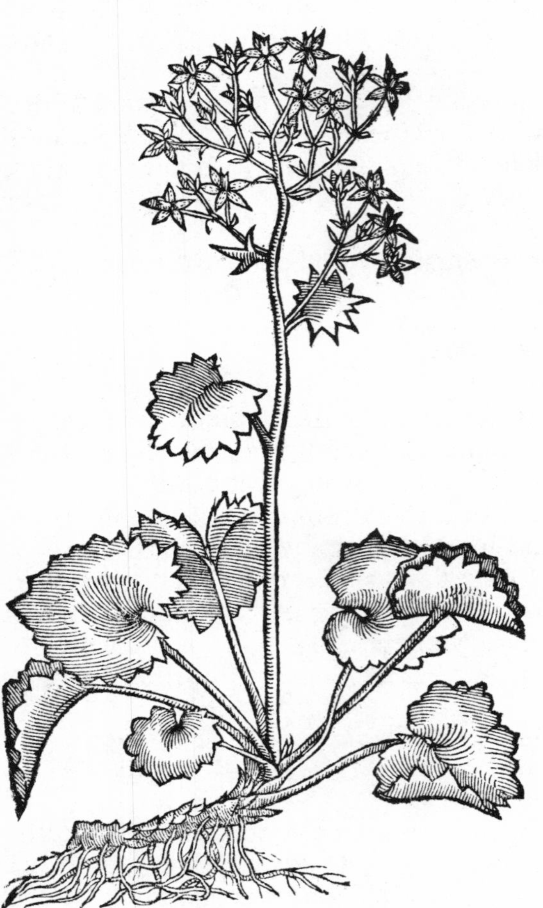

2 *Pinguicula siue Sanicula Eboracensis.*
Butterwort, or Yorkshire Sanicle.

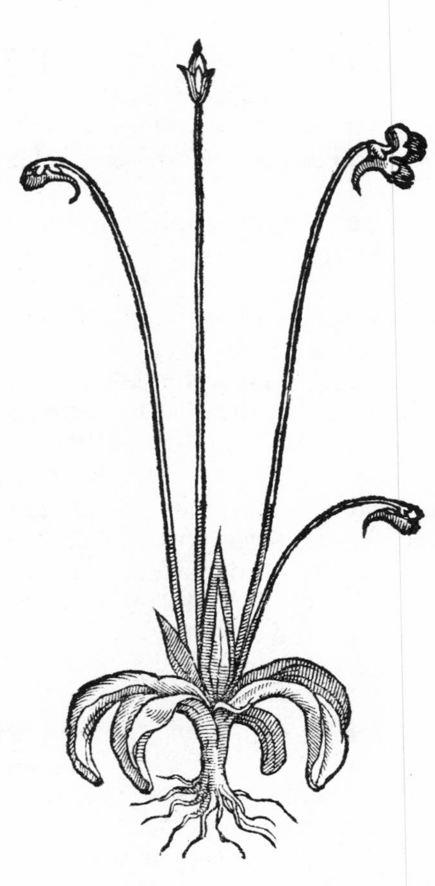

3 *Sanicula Alvina Clusij, siue Cortusa Matthioli.*
Beares eare Sanicle.

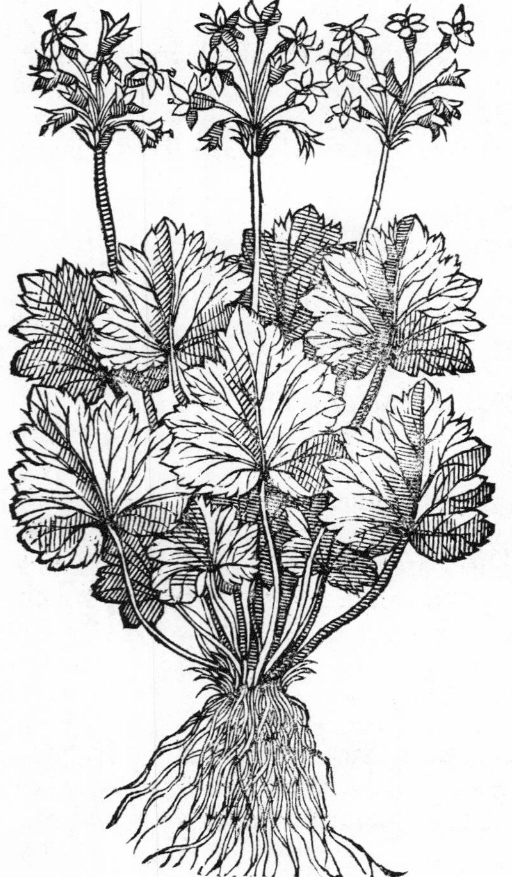

vp sundry stalkes, beset with like leaues, but smaller, and of a cubit high, diuiding themselues into many small armes or branches, bearing diuers little white floures, spotted most curiously with bloudy specks or prickes, insomuch that if you marke the admirable workemanship of the same wrought in such glorious manner, it must needs put euery creature in minde of his Creator: the floures are in smell like the May floures or Hawthorne: the seed is small and blacke, contained in small pointals like vnto white Saxifrage: the root is scaly and full of strings.

2 The second kind of Sanicle, which *Clusius* calleth *Pinguicula,* not before his time remembred, hath small thicke leaues, fat and ful of iuyce, being broad towards the root, and sharpe towards the point, of a faint greene colour, and bitter in taste: out of the middest wherof sprouteth or shooteth vp a naked slender stalke, nine inches long, euery stalke bearing one floure and no more, sometimes white and commonly of a blewish purple colour, fashioned like vnto the common *Consolida regalis,* hauing the like spur or Larks heele anexed thereto.

3 The third kinde of mountaine Sanicle some

ſome Herbariſts haue called *Sanicula alpina flore rubro*: the leaues ſhoot forth in the beginning of the Spring, very thicke and fat, and are like a purſe or round lumpe at their firſt comming out of the ground; and when it is ſpred abroad, the vpper part thereof is full of veines or ſinewes, and houen vp or curled like *Ranunculus Luſitanicus*, or like the crumpling of a cabbage leafe; and are not onely indented about the edges, but each leafe is diuided into ſix or more iagges or cuts, deepely hacked, greeniſh aboue, and of an ouerworne greene colour vnderneath, hot in taſte; from the middle whereof ſhooteth forth a bar or naked ſtalke, ſix inches long, ſomewhat purple in colour, bearing at the top a tuſt of ſmall hollow floures, looking or hanging downewards like little bells, not vnlike in forme to the common Cowſlips, but of a fine deepe red colour tending to purple, hauing in the middle a certaine ring or circle of white, and alſo certaine pointals or ſtrings, which turne into an head wherein is contained ſeed. The whole plant is couered as it were with a rough woollineſſe: the root is fibrous and threddy.

¶ *The Place.*

Theſe plants are ſtrangers in England; their naturall countrey is the Alpiſh mountains of Heluetia: they grow in my garden, where they flouriſh exceedingly, except Butterwort, which groweth in our Engliſh ſqually wet grounds, and will not yeeld to any culturing or tranſplanting: it groweth eſpecially in a field called Crag-Cloſe, and at Crosby, Rauenſwaith, in Weſtmerland, vpon Ingleborow fels twelue miles from Lancaſter, and in Harwood in the ſame countie neere to Blackburne, ten miles from Preſton in Aunderneſſe vpon the bogs and mariſh grounds, and in the boggie medowes about Biſhops Hatfield; and alſo in the fens in the way to Wittles meare from London, in Huntingdonſhire. ‡ It groweth alſo in Hampſhire, and aboundantly in many places of Wales. ‡

¶ *The Time.*

They floureand flouriſh from May to the end of Iuly.

¶ *The Names.*

The firſt is called *Sanicula guttata*, taken from the ſpots wherewith the floures are marked: of *Lobel*, *Geum Alpinum*, making it a kind of Auens: in Engliſh, ſpotted Sanicle: of our London dames, Pratling Parnell.

The ſecond is called *Pinguicula*, of the fatneſſe or fulneſſe of the leafe, or of fatning: in Yorkeſhire, where it doth eſpecially grow, and in greateſt aboundance, it is called Butterworts, Butterroot, and white root; but the laſt name belongeth more properly to Solomons Seale.

¶ *The Temperature and Vertues.*

They are hot and dry in the third degree.

The husbandmens wiues of Yorkſhire do vſe to anoint the dugs of their kine with the fat and A
oilous iuyce of the herbe Butterwort, when they are bitten with any venomous worme, or chapped, rifted, and hurt by any other meanes.

They ſay it rots their ſheepe, when for want of other food they eat rhereof. B

Chap. 277. *Of Fox-Gloues.*

¶ *The Deſcription.*

1 Fox-gloue with the purple floure is moſt common; the leaues whereof are long, nicked in the edges, of a light greene, in manner like thoſe of Mullein, but leſſer, and not ſo downie: the ſtalke is ſtraight, from the middle whereof to the top ſtand the floures, ſet in a courſe one by another vpon one ſide of the ſtalke, hanging downwards with the bottome vpward, in forme long, like almoſt to finger ſtalks, whereof it tooke his name *Digitalis*, of a red purple colour, with certaine white ſpots daſht within the floure; after which come vp round heads, in which lies the ſeed, ſomewhat browne, and as ſmall as that of Time. The roots are many ſlender ſtrings.

2 The Fox-gloue with white floures differs not from the precedent but in the colour of the floures; for as the others were purple, theſe contrariwiſe are of a milke-white colour.

3 We haue in our gardens another ſort hereof, which bringeth forth moſt pleaſant yellow floures, and ſomewhat leſſe than the common kinde, wherein they differ. ‡ This alſo differs from the common kind in that the leaues are much ſmoother, narrower, and greener, hauing the nerues or vrines running alongſt it, neither are the nerues ſnipt, nor ſinuated on their edges. ‡

4 We haue alſo another ſort, which we call *Digitalis ferruginea*, whoſe floures are of the colour of ruſty iron; whereof it tooke his name, and likewiſe maketh the difference. ‡ Of this ſort there is a bigger and a leſſer; the bigger hath the lower leaues ſome foot long, of a darke green colour, with veines running along them; the ſtalks are ſome yard and halfe high: the floures large, and

1 *Digitalis purpurea.*
Purple Fox-gloues.

2 *Digitalis alba.*
White Fox-gloues.

‡ 3 *Digitalis lutea.*
Yellow Fox-gloues.

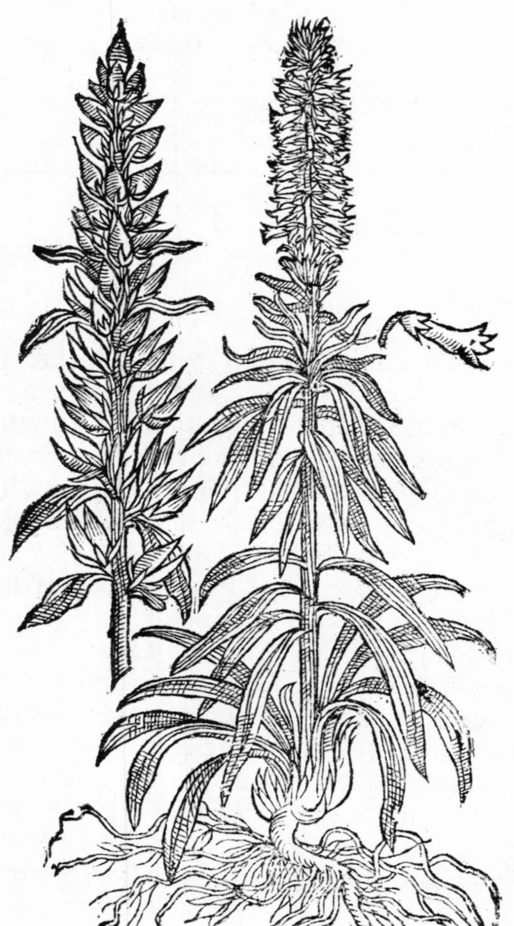

‡ 4 *Digitalis ferruginea.*
Dusky Fox-gloues.

and ending in a sharpe turned vp end as you see in the figure, and they are of a rustie colour, mixed of a yellow and red.

5 The lesser duskie Fox-gloue hath much lesse leaues and those narrow, smooth, and exceeding greene: amongst which comes vp a stalke some foot high, hauing small floures of the colour of the last described. This I obserued the last yeare 1632, in floure with Mr. *Iohn Tradescant* in the middle of Iuly. It may fitly be called *Digitalis ferruginea minor*, Small duskie Fox-gloues. ‡

¶ *The Place.*

Fox-gloue groweth in barren sandie grounds, and vnder hedges almost euery where.

Those with white floures do grow naturally in Landesdale, and Crauen, in a field called Cragge close, in the North of England: likewise by Colchester in Essex; neere Excester in the West parts, and in some few other places. The other two are strangers in England, neuerthelesse they do grow with the others in my garden.

¶ *The Time.*

They floure and flourish in Iune and Iuly.

¶ *The Names.*

Fox-gloues some call in Greeke φλόμος, and make it to be *Verbasci speciem*, or a kinde of Mullein: in Latine, *Digitalis*: in High Dutch, **Fingerhut**, and **Fingher kraut**: in Low Dutch, **Vinger hoet**: in French, *Gantes nostre dame*: in English, Fox-gloues. ‡ *Fabius Columna* thinks it to be that *Ephemerum* of *Dioscorides* described in his fourth booke, and *cap*. 75. ‡

¶ *The Temperature.*

The Fox-gloues in that they are bitter, are hot and drie, with a certaine kinde of clensing qualitie ioined therewith; yet are they of no vse, neither haue they any place amongst medicines, according to the Antients.

¶ *The Vertues.*

Fox-gloue boiled in water or wine, and drunken, doth cut and consume the thicke toughnesse of grosse and slimie flegme and naughtie humours; it openeth also the stopping of the liuer, spleene, and milt, and of other inward parts. A

The same taken in like manner, or boiled with honied water or sugar, doth scoure and clense the brest, ripeneth and bringeth forth tough and clammie flegme. B

They serue for the same purposes whereunto Gentian doth tend, and hath beene vsed in stead thereof, as *Galen* saith. C

‡ Where or by what name *Galen* either mentions, or affirmes this which our Authour cites him for, I must confesse I am ignorant. But I probably coniecture that our Authour would haue said *Fuchsius*: for I onely finde him to haue these words set downe by our Authour, in the end of his Chapter of *Digitalis*. ‡ D

CHAP. 278. *Of Baccharis out of* Dioscorides.

¶ *The Description.*

1 About this plant *Baccharis* there hath beene great contention amongst the old and new writers; *Matthiolus* and *Dodonæus* haue mistaken this plant, for *Coniza maior*, or *Coniza Helenitis Cordi*; *Virgil* and *Athenæus* haue confounded *Baccharis*, and *Azarum* together: but following the antient writers, it hath many blackish rough leaues, somewhat bigger than the leaues of Primrose: amongst which riseth vp a stalke two cubits high, bearing at the top little chaffie or scalie floures in small bunches, of a darke yellowish or purple colour, which turne into downe, and are carried away with the winde, like vnto the kindes of thistles: the root is thick, grosse, and fat, spreading about in the earth, full of strings: the fragrant smell that the root of this plant yeeldeth, may well be compared vnto the sauour of Cinnamon, *Helenium*, or *Enula Campana*, beeing a plant knowne vnto very many or most sorts of people, I meane in most parts of England.

¶ *The Place.*

Baccharis delighteth to grow in rough and craggy places, and in a leane soile where no moisture is:

Baccharis Monſpelienſium.
Plowmans Spikenard.

is: it groweth very plentifully about Montpellier in France, and diuers places in the Weſt parts of England.

¶ *The Time.*

It ſpringeth vp in April, it floureth in Iune, and perfecteth his ſeed in Auguſt.

¶ *The Names.*

The learned Herbariſts of Montpellier haue called this plant *Baccharis*: the Grecians, βακχάρις, or after others, πάγχαρις, by reaſon of that ſweet and aromaticall ſauour which his root containeth and yeeldeth: in Engliſh it may be called the Cinamom root, or Plowmans Spiknard: *Virgill* in his ſeuenth Ecloge of his Bucolicks maketh mention of *Baccharis*, and doth not onely ſhew that it is a Garland plant, but alſo ſuch a one as preuaileth againſt inchantments, ſaying,

——Bacchare frontem
Cingite, ne vati noceat mala lingua futuro.

With Plowmans Nard my forehead girt,
Leſt euill tongue thy Poet hurt.

Baccharis is likewiſe an ointment in *Athenæus*, in his 15 booke, which may take his name of the ſweet herbe *Baccharis*: for as *Pliny* writeth, *Ariſtophanes* of old, being an antient comical Poet witneſſeth, that ointments were wont to bee made of the root thereof: to bee briefe, *Crateuas* his *Aſarum* is the ſame that *Dioſcorides* his *Baccharis* is. ‡ This plant here deſcribed is the *Coniza maior* of *Matthiolus*, *Tragus*, and others. ‡

¶ *The Temperature.*

Baccharis or Plowmans Spiknard is of temperature very aſtringent or binding.

¶ *The Vertues.*

A *Baccharis*, or the decoction of the root, as *Paulus Ægineta* briefely ſetteth downe, doth open the pipes and paſſages that are ſtopped, prouoketh vrine, and bringeth downe the deſired ſickneſſe: the leaues thereof for that they are aſtringent or binding, ſtop the courſe of fluxes and rheumes.

B *Baccharis* is a ſingular remedie to heale inflammations and Saint Anthonies fire, called *Ignis ſacer*, and the ſmell thereof prouoketh ſleepe.

C The decoction of the roots of *Baccharis* helpeth ruptures and convulſions, thoſe alſo that haue falne from an high place, and thoſe that are troubled with the ſhortneſſe of breath.

D It helpeth alſo the old cough, and difficultie to make water.

E When it is boiled in wine it is giuen with great profit againſt the bitings of Scorpions, or any venomous beaſt, being implaiſtered and applied thereto.

F A bath made thereof and put into a cloſe ſtoole, and receiued hot, mightily voideth the birth, and furthereth thoſe that haue extreame labour in their childing, cauſing them to haue eaſie deliuerance.

CHAP. 279: *Of Elecampane.*

¶ *The Deſcription.*

Elecampane bringeth forth preſently from the root great white leaues, ſharpe pointed, almoſt like thoſe of great Comfrey, but ſoft, and couered with a hairie downe, of a whitiſh greene co-

[lour,

Helenium.
Elecampane.

lour, and are more white vnderneath, ſleightly nicked in the edges: the ſtalke is a yard and a halfe long, about a finger thicke, not without downe, diuided at the top into diuers branches, vpon the top of euery ſprig ſtand great floures broad and round, of which not only the long ſmal leaues that compaſſe round about are yellow, but alſo the middle ball or circle, which is filled vp with an infinit number of threds, and at length is turned into fine downe; vnder which is ſlender and long ſeed: the root is vneuen, thicke, and as much as a man may gripe, not long, oftentimes blackiſh without, white within, and full of ſubſtance, ſweet of ſmell, and bitter of taſte.

¶ *The Place.*

It groweth in medowes that are fat and fruitfull: it is alſo oftentimes found vpon mountains, ſhadowie places, that be not altogether drie: it groweth plentifully in the fields on the left hand as you go from Dunſtable to Puddle hill: alſo in an orchard as you go from Colbrook to Ditton ferry, which is the way to Windſor, and in ſundry other places, as at Lidde, and Folkeſtone, neere to Douer by the ſea ſide.

¶ *The Time.*

The floures are in their brauerie in Iune & Iuly: the roots be gathered in Autumne, and oftentimes in Aprill and May.

¶ *The Names.*

That which the Græcians name ἑλένιον, the Latines call *Inula* and *Enula*: in ſhops *Enula campana*: in high Dutch, **Alantwurtz**: in low Dutch, **Alandt wortele**: in Italian, *Enoa*, and *Enola*: in Spaniſh, *Raiz del alla*: in French, *Enula Campane*: in Engliſh, Elecampane, and Scab-woort, and Horſe-heale: ſome report that this plant tooke the name *Helenium* of *Helena* wife to *Menelaus*, who had her hands full of it when *Paris* ſtole her away into Phrygia.

¶ *The Temperature.*

The root of this Elecampane, is maruellous good for many things, being of nature hot and drie in the third degree, eſpecially when it is drie: for beeing greene and as yet full of iuice, it is full of ſuperfluous moiſture, which ſomewhat abateth the hot and drie qualitie thereof.

¶ *The Vertues.*

It is good for ſhortneſſe of breath, and an old cough, and for ſuch as cannot breathe vnleſſe they hold their necks vpright. A

It is of great vertue both giuen in a looch, which is a medicine to be licked on, and likewiſe preſerued, as alſo otherwiſe giuen to purge and void out thicke, tough, and clammie humours, which ſticke in the cheſt and lungs. B

The root preſerued is good and wholeſome for the ſtomack: being taken after ſupper it doth not onely helpe digeſtion, but alſo keepeth the belly ſoluble. C

The iuice of the ſame boiled, driueth forth all kinde of wormes of the belly, as *Pliny* teacheth: who alſo writeth in his twentie booke, and fift chapter, the ſame being chewed faſting, doth faſten the teeth.

The root of Elecampane is with good ſucceſſe mixed with counterpoiſons: it is a remedie againſt the bitings of ſerpents, it reſiſteth poiſon: it is good for them that are burſten, and troubled with cramps and conuulſions. D

Some alſo affirme, that the decoction thereof, and likewiſe the ſame beaten into powder and mixed with honie in manner of an ointment, doth clenſe and heale vp old vlcers. E

Galen ſaith, that herewith the parts are to be made red, which be vexed with long & cold griefs: as are diuers paſſions of the huckle bones, called the Sciatica, and little and continual bunnies and looſeneſſe of certaine ioints, by reaſon of ouermuch moiſture. F

The

H The decoction of *Enula* drunken, prouoketh vrine, and is good for them that are grieued with inward burſtings, or haue any member out of ioint.

I The root taken with honie or ſugar, made in an electuarie, clenſeth the breſt, ripeneth tough flegme, and maketh it eaſie to be ſpet forth, and preuaileth mightily againſt the cough and ſhortneſſe of breath, comforteth the ſtomacke alſo, and helpeth digeſtion.

K The roots condited after the manner of *Eringos* ſerueth for the purpoſes aforeſaid.

L The root of *Enula* boiled very ſoft, and mixed in a morter with freſh butter and the pouder of Ginger, maketh an excellent ointment againſt the itch, ſcabs, mangineſſe, and ſuch like.

M The roots are to be gathered in the end of September, and kept for ſundrie vſes, but it is eſpecially preſerued by thoſe that make Succade and ſuch like.

Chap. 280. *Of Sauce alone, or Iacke by the hedge.*

Alliaria.
Sauce alone.

¶ *The Deſcription.*

Sauce alone hath affinitie with Garlicke in name, not becauſe it is like it in forme, but in ſmell: for if it be bruiſed or ſtamped it ſmelleth altogether like Garlicke: the leaues hereof are broad, of a light green colour, nicked round about, and ſharpe pointed: the ſtalke is ſlender, about a cubit high, about the branches whereof grow little white floures; after which come vp ſlender ſmal and long cods, & in theſe black ſeed: the root is long, ſlender, and ſomething hard.

¶ *The Place.*

It groweth of it ſelfe by garden hedges, by old wals, by highwaies ſides, or oftentimes in the borders of fields.

¶ *The Time.*

It floureth chiefely in Iune and Iuly, the ſeed waxeth ripe in the meane ſeaſon. The leaues are vſed for a ſauce in March or Aprill.

¶ *The Names.*

The later writers call it *Alliaria*, and *Alliaris*: of ſome, *Rima Maria*: it is not *Scordium*, or water Germander, which the apothecaries in times paſt miſtooke for this herbe: neither is it *Scordij ſpecies*, or a kinde of water Germander, whereof wee haue written: it is named of ſome, *Pes Aſininus*: it is called in High Dutch, **Knoblauch kraut Leuchel**, and **Salzkraut**: and in Low Dutch, **Loock ſonder Loock**: you may name it in Latine, *Allium non bulboſum*: in French, *Alliayre*: in Engliſh, Sauce alone, and Iacke of the hedge.

¶ *The Temperature.*

Iacke of the hedge is hot and drie, but much leſſe than Garlicke, that is to ſay, in the end of the ſecond degree, or in the beginning of the third.

¶ *The Vertues.*

A We know not what vſe it hath in medicine: diuers eat the ſtamped leaues hereof with Salt-fiſh, for a ſauce, as they do thoſe of Ramſons.

B Some alſo boile the leaues in cliſters which are vſed againſt the paine of the collicke and ſtone, in which not only winde is notably waſted, but the pain alſo of the ſtone mitigated and very much eaſed.

CHAP. 281. *Of Dittany.*

¶ *The Description.*

1 Dittanie of Crete now called Candie (as *Dioscorides* saith) is a hot and sharpe hearbe, much like vnto Penni-roiall, sauing that his leaues be greater and somewhat hoary, couered ouer with a soft downe or white woollie cotton: at the top of the branches grow small spikie eares or scaly aglets, hanging by little small stemmes, resembling the spiky tufts of Marierome, of a white colour: amongst which scales there doe come forth small floures like the flouring of wheat, of a red purple colour; which being past, the knop is found full of small seed, contrarie to the saying of *Dioscorides*, who saith, it neither beareth floure nor seed, but my selfe haue seene it beare both in my Garden. the whole plant perished in the next VVinter following.

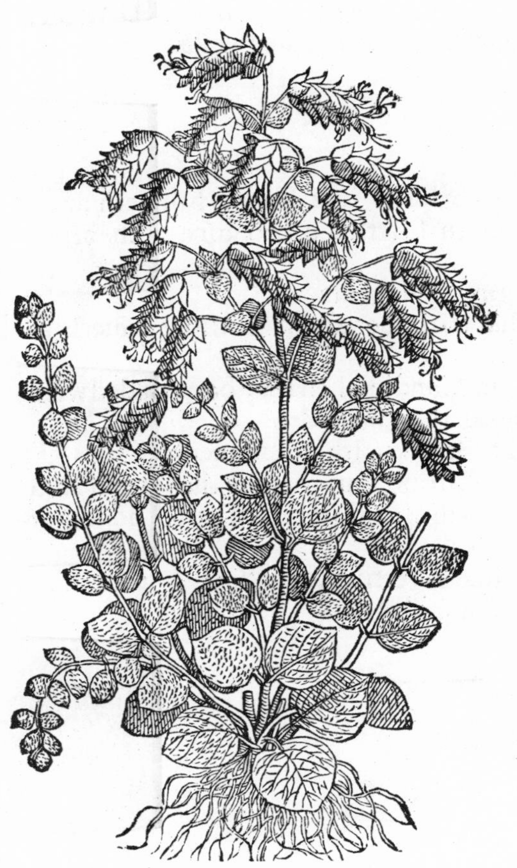

1 *Dictamnum Creticum.*
Dittanie of Candie.

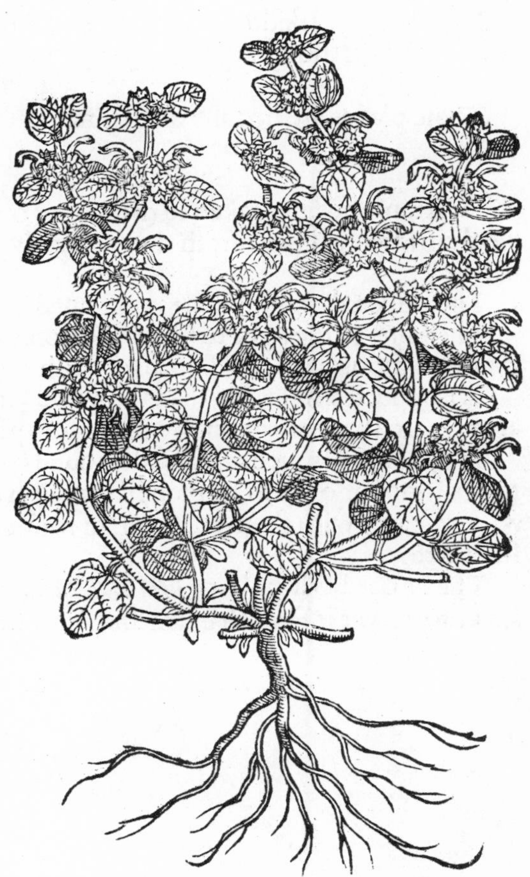

2 *Pseudodictamnum.*
Bastard Dittanie.

2 The second kind called *Pseudodictamnum*, that is, Bastard Dittanie, is much like vnto the first sauing that it is not sweet of smell, neither doth it bite the tongue, hauing round soft woolly stalks with knots and ioints, and at euery knot two leaues somewhat round, soft, woolly, and somewhat bitter: the floures be of a light purple color, compassing the stalks by certain spaces like garlands or wharles, and like the floures of Peni-roiall. The root is of a wooddie substance: the whole plant groweth to the height of a cubite and an halfe, and lasteth long.

¶ *The Place.*

The first Dittanie commeth from Crete, an Iland which we call Candie, where it growes naturally: I haue sowne it in my garden, where it hath floured and borne seed; but it perished by reason of the iniurie of our extraordinarie cold winter that then happened: neuerthelesse *Dioscorides* writeth

writeth againſt all truth, that it neither beareth floures nor ſeed: after *Theophraſtus*, *Virgil* witneſſeth that it doth beare floures in the twelfth of his Æneidos.

Dictamnum genitrix Cretæa carpit ab Ida,
Puberibus caulem folijs, & flore comantem
Purpureo.———

In Engliſh thus:

His mother from the Cretæan Ida crops
Dictamnus hauing ſoft and tender leaues,
And purple floures vpon the bending tops, &c.

¶ *The Time.*

They floure and flouriſh in the Sommer moneths, their ſeed is ripe in September.

¶ *The Names.*

It is called in Greeke δίκταμνος: in Latine, *Dictamnus* and *Dictamnum*: of ſome, *Pulegium ſylueſtre*, or wilde Pennie-roiall: the Apothecaries of Germanie for *Dictamnum* with *c*, in the firſt ſyllable, doe read *Diptamnum* with *p*: but (ſaith *Dodonæus*) this errour might haue beene of ſmall importance, if in ſtead of the leaues of Dittanie, they did not vſe the rootes of *Fraxinella* for Dittany, which they falſely call *Dictamnum*: in Engliſh, Dittanie, and Dittanie of Candie.

The other is called *Pſeudodictamnum*, or baſtard Dittanie, of the likeneſſe it hath with Dittanie, it skilleth not, though the ſhoppes know it not: the reaſon why let the Reader gueſſe.

¶ *The Temperature.*

Theſe plants are hot and drie of nature.

¶ *The Vertues.*

A Dittanie beeing taken in drinke, or put vp in a peſſarie, or vſed in a fume, bringeth away dead children: it procureth the monethly termes, and driueth foorth the ſecondine or the after-birth.

B The iuice taken with wine is a remedie againſt the ſtinging of ſerpents.

C The ſame is thought to be of ſo ſtrong an operation, that with the very ſmell alſo it driueth away venomous beaſts, and doth aſtoniſh them.

D It is reported likewiſe that the wilde Goats and Deere in Candie when they be wounded with arrowes, do ſhake them out by eating of this plant, and heale their wounds.

E It preuaileth much againſt all wounds, and eſpecially thoſe made with inuenomed weapons, arrowes ſhot out of guns, or ſuch like, and is very profitable for Chirurgians that vſe the ſea and land wars, to carry with them and haue in readineſſe: it draweth forth alſo ſplinters of wood, bones, or ſuch like.

F The baſtard Dittanie, or *Pſeudodictamnum*, is ſomewhat like in vertues to the firſt, but not of ſo great force, yet it ſerueth exceeding well for the purpoſes aforeſaid.

CHAP. 282. *Of Borage.*

¶ *The Deſcription.*

1 Borage hath broad leaues, rough, lying flat vpon the ground, of a blacke or ſwart green colour: among which riſeth vp a ſtalke two cubits high, diuided into diuers branches, whereupon do grow gallant blew floures, compoſed of fiue leaues apiece; out of the middle of which grow forth blacke threds ioined in the top, and pointed like a broch or pyramide: the root is threddie, and cannot away with the cold of winter.

2 Borage with white floures is like vnto the precedent, but differeth in the floures, for thoſe of this plant are white, and the others of a perfect blew colour, wherein is the difference.

† 3 Neuer dying Borage hath manie verie broad leaues, rough and hairie, of a blacke darke greene colour: among which riſe vp ſtiffe hairie ſtalkes, whereupon doe grow faire blew floures, ſomewhat rounder pointed than the former: the root is blacke and laſting, hauing leaues both winter and Sommer, and hereupon it was called *Semper virens*, and that very properly, to diſtinguiſh it from the reſt of this kinde, which are but annuall. ‡

4 There

1 *Borago hortensis.*
Garden Borage.

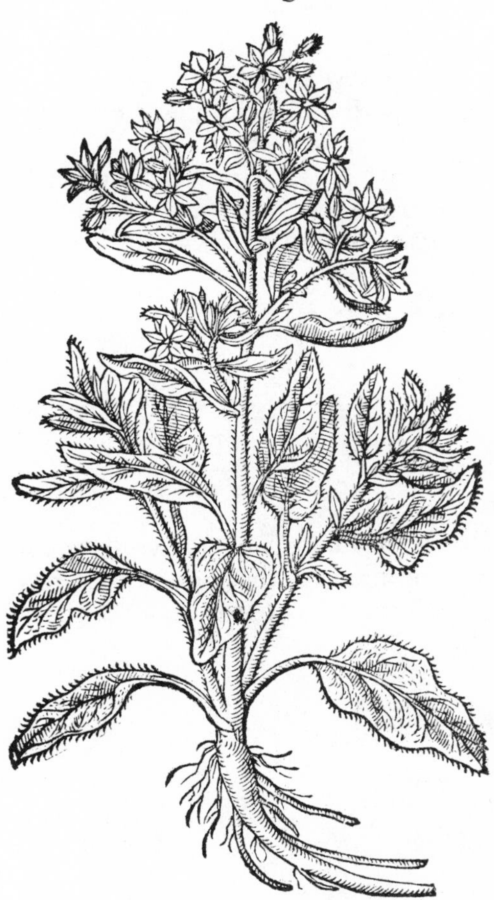

2 *Borago flore albo.*
White floured Borage.

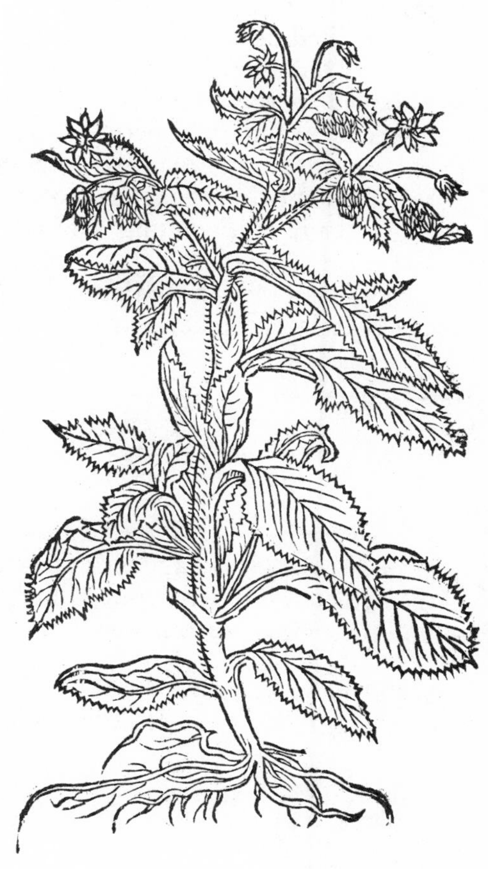

3 *Borago semper virens.*
Neuer dying Borage.

4 There is a fourth sort of Borage that hath leaues like the precedent, but thinner and lesser, rough and hairy, diuiding it selfe into branches at the bottom of the plant, whereupon are placed faire red floures, wherein is the chiefest difference between this and the last described. ‡ The figure which belonged to this description was put hereafter for *Lycopsis Anglica.* ‡

¶ *The Place.*

These grow in my garden, and in others also.

¶ *The Time.*

Borage floures and flourishes most part of all Sommer, and till Autumne be far spent.

¶ *The Names.*

Borage is called in shops *Borago*: of the old Writers, Βούγλωσσον, which is called in Latine *Lingua Bubula*: *Pliny* calleth it *Euphrosinum*, because it maketh a man merry and ioyfull: which thing also the old verse concerning Borage doth testifie:

Ego Borago gaudia semper ago.

I Borage bring alwaies courage.

It is called in high Dutch **Burretsch**: in Italian, *Boragine*: in Spanish, *Boraces*: in low Dutch, **Beruagie**: in English, Borage.

¶ *The Temperature.*

It is euidently moist, and not in like sort hot, but seemes to be in a meane betwixt hot and cold.

¶ *The Vertues.*

A Those of our time do vse the floures in sallads, to exhilerate and make the mind glad. There be also many things made of them, vsed euery where for the comfort of the heart, for the driuing away of sorrow, and encreasing the ioy of the minde.

B The leaues boyled among other pot-herbes do much preuaile in making the belly soluble, they being boyled in honied water be also good against the roughnesse of the throat, and hoarsenesse, as *Galen* teacheth.

C The leaues and floures of Borage put into Wine make men and women glad and merry, ar driue away all sadnesse, dulnesse, and melancholy, as *Dioscorides* and *Pliny* affirme.

D Syrrup made of the floures of Borage comforteth the heart, purgeth melancholy, and quieteth the phrenticke or lunaticke person.

E The floures of Borage made vp with sugar do all the aforesaid with greater force and effect.

F Syrrup made of the iuyce of Borage with sugar, adding thereto pouder of the bone of a Stags heart, is good against swouning, the cardiacke passion of the heart, against melancholy and the falling sicknesse.

G The root is not vsed in medicine: the leaues eaten raw ingender good bloud, especially in those that haue been lately sicke.

Chap. 283. *Of Buglosse.*

¶ *The Kindes.*

Like as there be diuers sorts of Borage, so are there sundry of the Buglosses; notwithstanding after *Dioscorides*, Borage is the true Buglosse: many are of opinion, and that rightly, that they may be both referred to one kinde; yet will we diuide them according to the custome of our time, and their vsuall denominations.

1 *Buglossa vulgaris.*
Common Buglosse, or Garden Buglosse.

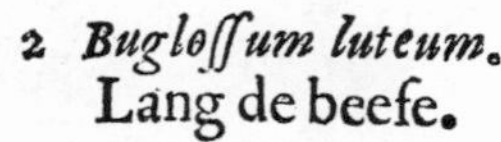
2 *Buglossum luteum.*
Lang de beefe.

¶ *The Description.*

1 That which the Apothecaries call Buglosse bringeth forth leaues longer than those of Borage, sharpe pointed, longer than the leaues of Beets, rough and hairy. The stalke groweth vp to the height of two cubits, parted aboue into sundry branches, whereon are orderly placed blewish floures, tending to a purple colour before they be opened, and afterward more blew. The root is long, thicke, and grosse, and of long continuance.

‡ 3 *Buglossa syluestris minor.*
Small wilde Buglosse.

2 *Lang de Beefe* is a kinde hereof, altogether lesser, but the leaues hereof are rougher, like the rough tongue of an oxe or cow, whereof it tooke his name. ‡ The leaues of *Lang-de-Beefe* are very rough, the stalke some cubit and halfe high, commonly red of colour: the tops of the branches carry floures in scaly rough heads: these floures are composed of many small yellow leaues in manner of those of Dandelion, and flie away in down like as they do: the floures are of a verie bitter taste, whence *Lobel* calls it *Buglossum echioides luteum Hieracio cognatum. Tabernamontanus* hath fitly called it *Hieracium echioides*.

3 There is another wilde Buglosse which *Dodonæus* hath by the name of *Buglossa syluestris*: it hath a small white root, from which arises a slender stalke some foot and halfe high set with smal rough leaues sinuated or cut in on the edges: the stalkes at the top are diuided into three or foure small branches, bearing small blew floures in rough huskes. ‡

¶ *The Place.*

These do grow in gardens euery where. ‡ The *Lang-de-Beefe* growes wilde in many places; as betweene Redriffe and Deptford by the waterie ditch sides. The little wilde Buglosse growes vpon the drie ditch bankes about Pickadilla, and almost euery where. ‡

¶ *The Time.*

They floure from May, or Iune, euen to the end of Sommer. The leaues perish in Winter, and new come vp in the Spring.

¶ *The Names.*

Garden Buglosse is called of the later Herbarists *Buglossa*, and *Buglossa Domestica*: or garden Buglosse.

Lang-de Beefe is called in Latine *Lingua bouis*, and *Buglossum Luteum Hieracio cognatum*, and also *Buglossa syluestris*, or wilde Buglosse.

‡ Small wilde Buglosse is called *Borago syluestris* by *Tragus*; *Echium Germanicum Spinosum* by *Fuchsius*; and *Buglossa syluestris* by *Dodonæus*. ‡

¶ *The Temperature and Vertues.*

The root, saith *Dioscorides*, mixed with oile, cureth greene wounds, and adding thereto a little barley meale, it is a remedie against Saint Anthonies fire. A

It causeth sweat in agues, as *Plinie* saith, if the iuice be mixed with a little *Aqua vitæ*, and the body rubbed therewith. B

The Physitions of the later time vse the leaues, floures, and roots in stead of Borage, and put them both into all kindes of medicines indifferently, which are of force and vertue to driue away sorrow and pensiuenesse of the minde, and to comfort and strengthen the heart. The leaues are of like operation with those of Borage, and are vsed as potherbes for the purposes aforesaid, as wel Buglosse as *Lang-de-Beefe*, and also to keepe the belly soluble. C

Chap. 284. *Of Alkanet or wilde Buglosse.*

¶ *The Description.*

These herbes comprehended vnder the name of *Anchusa*, were so called of the Greeke word ἀγχούσειν: (i) *Illinere succo, vel pigmentis*, that is, to colour or paint any thing: whereupon these plants

† 1 *Anchusa Alcibiadion.*
Red Alkanet.

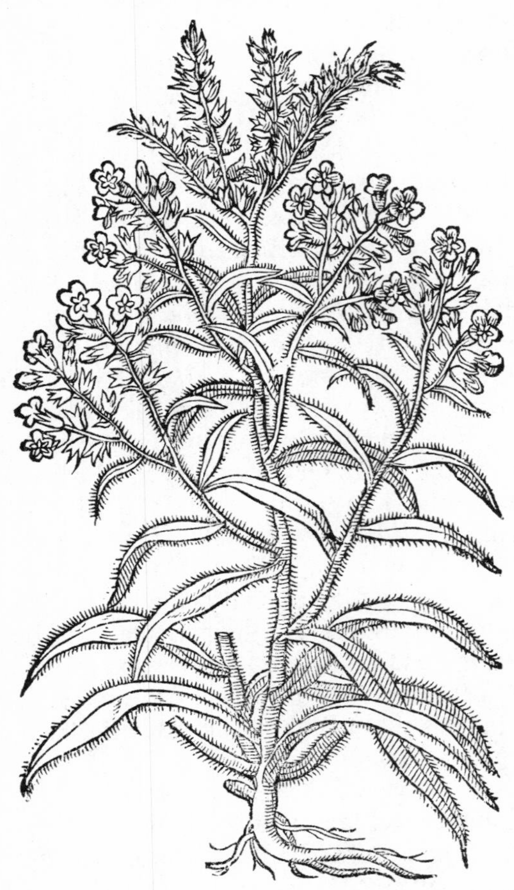

† 2 *Anchusa lutea.*
Yellow Alkanet.

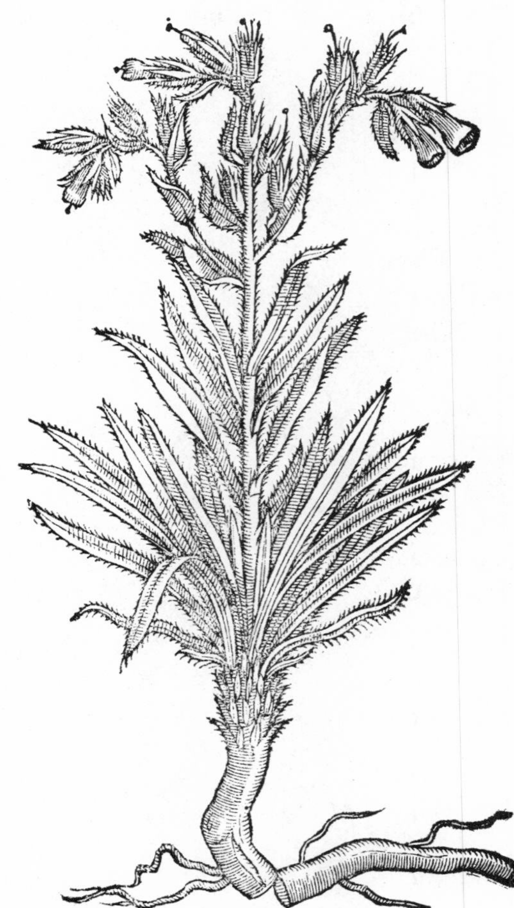

‡ 3 *Anchusa minor.*
Small Alkanet.

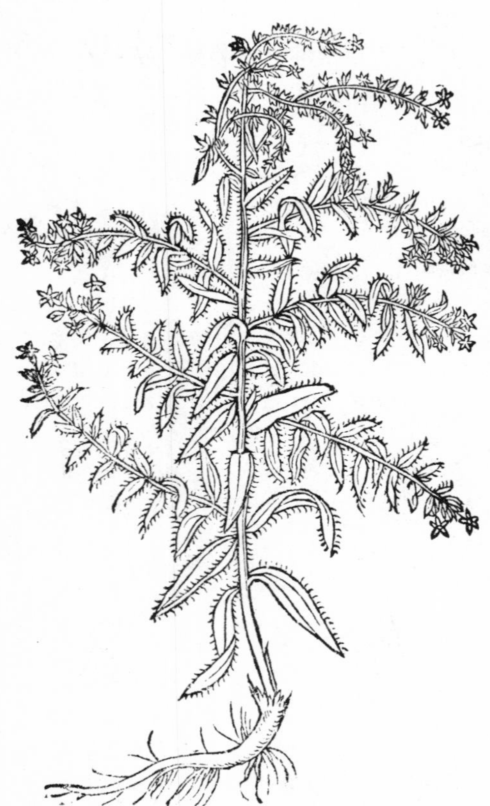

plants were called *Anchusa* of that flourishing and bright red colour which is in the root, euen as red as pure and cleere bloud: for that is the onely marke or note whereby to distinguish these herbes from those which be called *Echum*, *Lycopsis*, and *Buglossa*, whereto they haue a great resemblance: I haue therefore expressed foure differences of this plant *Anchusa* or Alkanet from the other kindes, by the leaues, floures, and bignesse.

1 The first kinde of Alkanet hath many leaues like *Echium*, or small Buglosse, couered ouer with a prickie hoarinesse, hauing commonly but one stalke, which is round, rough, and a cubite high. The cups of the floures are of a skie colour tending to purple, not vnlike the floures of *Echium*; the seed is small, somewhat long, and of a pale colour: the root is a finger thicke, the pith or inner part thereof is of a wooddie substance, dying the hands or whatsoeuer toucheth the same, of a bloudie colour, or of the colour of saunders.

2 The second kinde of *Anchusa* or Alkanet is of greater beautie and estimation than the first; the branches are lesse and more bushie in the toppe: it hath also greater plentie of leaues, and those more woollie or hairie: the stalke groweth to the height of two cubites: at the top grow floures of a yellow colour, far different from the other: the root is more shining, of an excellent delicate purplish colour, and more full of iuice than the first.

3 There

3 There is a small kinde of Alkanet, whose root is greater and more ful of iuice and substance than the roots of the other kindes: in all other respects it is lesse, for the leaues are narrower, smaller, tenderer, and in number more, very greene like vnto Borage, yeelding forth many little tender stalks: the floures are lesse than of the small Buglosse, and red of colour: the seed is of an ashe colour, somewhat long and slender, hauing the taste of Buglosse.

4 There is also another kinde of Alkanet, which is as the others before mentioned, a kinde of wilde Buglosse, notwithstanding for distinctions sake I haue separated and seuered them. This last *Anchusa* hath narrow leaues, much like vnto our common Sommer Sauorie. The stalkes are two handfuls high, bearing very smal floures, and of a blewish or skie colour: the root is of a dark brownish red colour, dying the hands little or nothing at all, and of a woodie substance.

¶ *The Time.*

These plants do grow in the fields of Narbone, and about Montpellier and many other parts of France: I found these plants growing in the Isle of Thanet neere vnto the sea, betwixt the house sometime belonging to Sir *Henrie Crispe*, and Margate; where I found some in their naturall ripenesse, yet scarcely any that were come to that beautifull colour of Alkanet: but such as is sold for very good in our Apothecaries shops I found there in great plentie.

‡ I doubt whether our Authour found any of these in the place heere set downe, for I haue sought it but failed of finding; yet if he found any it was onely the first described, for I think the other three are strangers. ‡

¶ *The Time.*

The Alkanets floure and flourish in the Sommer moneths: the roots doe yeeld their bloudie iuice in haruest time, as *Dioscorides* writeth.

¶ *The Names.*

Alkanet is called in Greeke ἄγχουσα in Latine also *Anchusa*: of diuers, *Fucus herba*, and *Onocleia*, *Buglossa Hispanica*, or Spanish Buglosse: in Spanish, *Soagene*: in French, *Orchanet*: and in English likewise Orchanet and Alkanet.

¶ *The Temperature.*

The roots of Alkanet are cold and drie, as *Galen* writeth, and binding, and because it is bitter it clenseth away cholericke humours: the leaues bee not so forceable, yet doe they likewise binde and drie.

¶ *The Vertues.*

Dioscorides saith, that the root being made vp in a cerote, or seareclоth with oile, is very good for old vlcers; that with parched barley meale it is good for the leprey, and for tetters and ring-worms. A

That being vsed as a pessarie it bringeth forth the dead birth. B

The decoction being inwardly taken with Mead or honied water, cureth the yellow iaundise, diseases of the kidneies, the spleene and agues. C

It is vsed in ointments for womens paintings: and the leaues drunke in wine is good against the laske. D

Diuers of the later Physitions do boile with the root of Alkanet and wine, sweet butter, such as hath in it no salt at all, vntill such time as it becommeth red, which they call red butter, and giue it not onely to those that haue falne from some high place, but also report it to be good to driue forth the measels and small pox, if it be drunke in the beginning with hot beere. E

The roots of these are vsed to color sirrups, waters, gellies, & such like confections as Turnsole is. F

Iohn of *Ardern* hath set down a composition called *Sanguis Veneris*, which is most singular in deep punctures or wounds made with thrusts, as follows: take of oile oliue a pint, the root of Alkanet two ounces, earth worms purged, in number twenty, boile them together & keep it to the vse aforesaid. G

The Gentlewomen of France do paint their faces with these roots, as it is said. H

† The two figures that were formerly here were both of the ordinary Buglosse, whereof the first might well enough serue, but the 2. was much different from that it should haue been.

CHAP. 285. *Of Wall and Vipers Buglosse.*

¶ *The Description.*

1 *Lycopsis Anglica*, or wilde Buglosse, so called for that it doth not grow so commonly elsewhere, hath rough and hairie leaues, somewhat lesser than the garden Buglosse: the floures grow for the most part vpon the side of the slender stalke, in fashion hollow like a little bell, whereof some be blew, and others of a purple colour.

2 There is another kinde of *Echium* that hath rough and hairy leaues likewise, much like vnto the former; the stalke is rough, charged full of little branches, which are laden on euery side with diuers small narrow leaues, sharp pointed, and of a brown colour: among which leaues grow floures, each floure being composed of one leafe diuided into fiue parts at the top, lesse, and not so wide open as that of *Lycopsis*; yet of a sad blew or purple colour at the first, but when they are open they shew to be of an azure colour, long and hollow, hauing certaine smal blew threds in the middle: the seed is small and black, fashioned like the head of a snake or viper: the root is long, and red without.

† 1 *Lycopsis Anglica.*
Wall Buglosse.

‡ 2 *Echium vulgare.*
Vipers Buglosse.

‡ 3 *Echium pullo flore.*
Rough Vipers Buglosse.

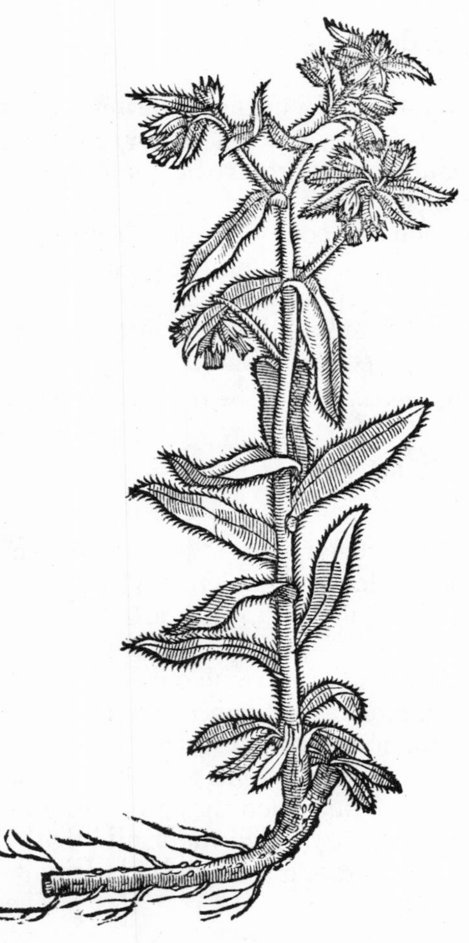

‡ 4 *Echium rubro flore.*
Red floured Vipers Buglosse.

‡ 3 This hath a crested very rough and hairy stalke some foot high; the leaues are like those of Vipers Buglosse, and couered ouer with a soft downinesse, and grow disorderly vpon the stalke, which towards the top is parted into sundry branches, which are diuided into diuers foot-stalkes carrying small hollow floures diuided by fiue little gashes at their tops; and they are of a darke purple colour, and contained in rough cups lying hid vnder the leaues. The seed, as in other plants of this kinde, resembles a Vipers head: the root is long, as thicke as ones little finger, of a dusky colour on the outside, and it liues diuers yeares. This floures in May, and growes in the dry medowes and hilly grounds of Austria. *Clusius* calls it *Echium pullo flore*.

4 This other being also of *Clusius* his description hath long and narrow leaues like those of the common Vipers Buglosse, yet a little broader: the stalkes rise vp some cubit high, firme, crested, and hairy; vpon which grow aboundance of leaues, shorter and narrower than those below; and amongst these towards the top grow many floures vpon short foot-stalks, which twine themselues round like a Scorpions taile: these floures are of an elegant red colour, and in shape somwhat lik those of the common kinde; and such also is the seed, but somewhat lesse: the root is lasting, long also, hard, wooddy, and blacke on the outside, and it sometimes sends vp many, but most vsually but one stalke. It floures in May, and was found in Hungary by *Clusius*, who first set it forth by the name of *Echium rubro flore*. ‡

¶ *The Place.*

Lycopsis groweth vpon stone walls, and vpon dry barren stony grounds.

Echium groweth where Alkanet doth grow, in great aboundance.

¶ *The Time.*

They flourish when the other kindes of Buglosses do floure.

¶ *The Names.*

It is called in Greeke *Echium*, and Ἀλκιβιάδιον, of *Alcibiades* the finder of the vertues thereof: of some it is thought to be *Anchusæ species*, or a kinde of Alkanet: in high-Dutch, wilde **Ochsenzungen**: in Spanish, *Yerua de la Biuora*, or *Chupamel*: in Italian, *Buglossa saluatica*: in French, *Buglosse sauuage*: in English, Vipers Buglosse, Snakes Buglosse; and of some, Vipers herbe, and wilde Buglosse the lesser.

¶ *The Temperature.*

These herbes are cold and dry of complexion.

¶ *The Vertues.*

The root drunke with wine is good for those that be bitten with Serpents, and it keepeth such from being stung as haue drunk of it before: the leaues and seeds do the same, as *Dioscorides* writes. *Nicander* in his book of Treacles makes Vipers Buglosse to be one of those plants which cure the biting of serpents, and especially of the Viper, and that driue serpents away. A

If it be drunke in wine or otherwise it causeth plenty of milke in womens brests. B

The herbe chewed, and the iuyce swallowed downe, is a most singular remedie against poyson and the bitings of any venomous beast; and the root so chewed and layd vpon the sore workes the same effect. C

† That figure which formerly stood in the second place, vnder the title of *Onosma*, and whereof there was no more mention made by our Author, neither in description, name, nor otherwise, I take to be nothing else than the *Lycopsis* which lies with long leaues spred vpon the ground before it comes to send vp the stalke; as you may see it exprest apart by it selfe in the figure we giue you; which is the true figure of that plant our Author described and meant: for the figure which he gaue was nothing but of the common Borage with narrower leaues, which he described in the fourth place of the chapter of Borage, as I haue formerly noted.

Chap. 286. *Of Hounds-tongue.*

¶ *The Description.*

1 The common Hounds tongue hath long leaues much like the garden Buglosse, but broader, and not rough at all, yet hauing some fine hoarinesse or softnesse like veluet. These leaues stinke very filthily, much like to the pisse of dogs; wherefore the Dutch men haue called it **Hounds pisse**, and not Hounds tongue. The stalkes are rough, hard, two cubits high, and of a browne colour, bearing at the top many floures of a darke purple colour: the seed is rough, cleauing to garments like Agrimonie seed: the root is blacke and thicke. ‡ These plants for one yeare after they come vp of seed bring forth onely leaues, and those pretty large; and the second yere they send vp their stalks, bearing both floures and seed, and then vsually the root perisheth. I haue therefore presented you with the figures of it, both when it floures, and when it sendeth forth onely leaues. ‡

1 *Cynogloſſum maius vulgare ſine flore.*
Hounds-tongue without the floure.

1 *Cynogloſſum maius cum flore & ſemine.*
Hounds-tongue with the floure and ſeed.

‡ 2 *Cynogloſſum Creticum* 1.
The firſt Candy Dogs-tongue.

‡ 3 *Cynogloſſum Creticum alterum.*
The other Candy Dogs-tongue.

2 We haue receiued another sort hereof from the parts of Italy, hauing leaues like Woade, somewhat rough, and without any manifest smell, wherein it differeth from the common kinde; the seed hereof came vnder the title *Cynoglossum Creticum*, Hounds-tongue of Candy. ‡ The floures are lesser and of a lighter colour than those of the former; the seeds also are rough, and grow foure together, with a point comming out of the middle of them as in the common kind, but yet lesser; the root is long and whitish. *Clusius* hath this by the name of *Cynoglossum Creticum* 1.

3 This second *Cynoglossum Creticum* of *Clusius* hath leaues some handfull long, and some inch and better broad: among which, the next yeare after the sowing, comes vp a stalke some cubit or more high, crested, stiffe, and straight, and somewhat downy as are also the leaues, which grow vpon the same, being somwhat broad at their setting on, and of a yellowish greene colour. The top of the stalke is diuided into sundry branches, which twine or turne in their tops like as the Scorpion grasse, and carry shorter yet larger floures than the ordinarie kinde, and those of a whitish colour at the first, with many small purplish veines, which after a few dayes become blew. The seeds are like the former in their growing, shape, and roughnesse. ‡

4 We haue another sort of Hounds-tongue like vnto the common kinde, sauing it is altogether lesser: the leaues are of a shining greene colour.

‡ 4 *Cynoglossum minus folio virente.*
Small greene leaued Hound-tongue.

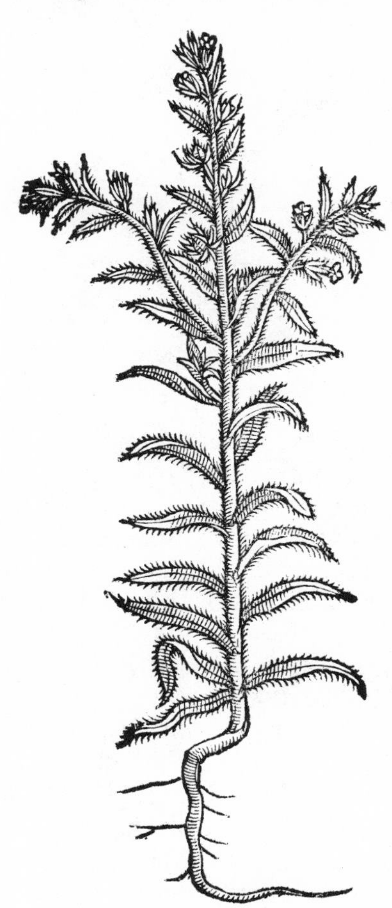

¶ *The Place.*

The great Hounds-tongue growes almost euery where by high-wayes and vntoiled ground: the small Hounds-tongue groweth very plentifully by the waies side as you ride Colchester highway from Londonward, betweene Esterford and Wittam in Essex

¶ *The Time.*

They floure in Iune and Iuly.

¶ *The Names.*

Hounds-tongue is called in Greeke, Κυνόγλωσσον: in Latine, *Lingua canis*: of *Pliny*, *Cynoglossos*; and he sheweth two kinds thereof: in English, Hounds-tongue, or Dogs-tongue, but rather Hounds-pisse, for in the world there is not any thing smelleth so like vnto Dogs-pisse as the leaues of this Plant doe.

¶ *The Nature.*

Hounds-tongue, but especially his root, is cold and dry.

¶ *The Vertues.*

The roots of Hounds-tongue rosted in the embers and layd to the fundament, healeth the hemorrhoides, and the disease called *Ignis sacer*, or wilde-fire. A

The iuyce boiled with honey of roses and Turpentine, to the forme of an vnguent, is most singular in wounds and deepe vlcers. B

Dioscorides saith, That the leaues boyled in wine and drunk, do mollifie the belly, and that the leaues stamped with old swines grease are good against the falling away of the haire of the head, which proceedeth of hot sharpe humors. C

Likewise they are a remedie against scaldings or burnings, and against the biting of dogs, as the same Author addeth. D

CHAP. 287. Of Comfrey, or great Consound.

¶ *The Description.*

1 THe stalke of this Comfrey is cornered, thicke, and hollow like that of Sow-thistle: it groweth two cubits or a yard high: the leaues that spring from the root, and those that grow

1 *Consolida maior flore purpureo.*
Comfrey with purple floures.

3 ‡ *Symphytum tuberosum.*
Comfrey with the knobby root.

‡ 4 *Simphytum parvum Boraginis facie.*
Borage-floured Comfrey.

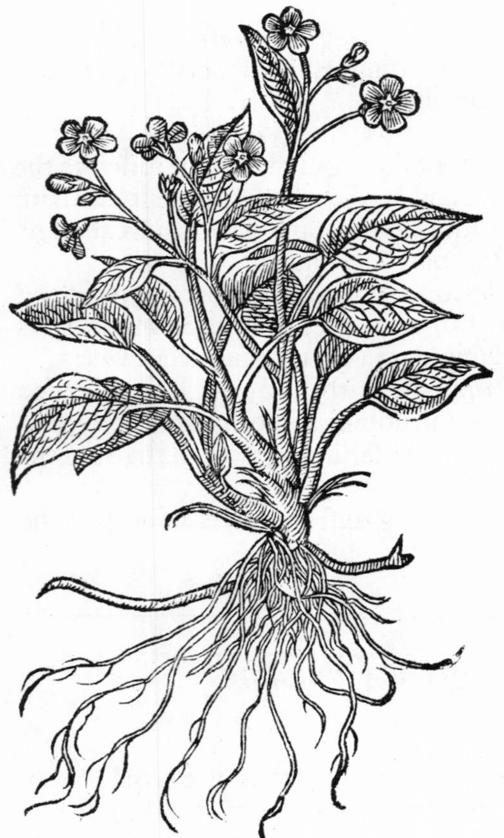

grow vpon the stalkes are long, broad, rough, and pricking withall, something hairie, and being handled make the hands itch; very like in colour and roughnes to those of Borage, but longer, and sharpe pointed, as be the leaues of Elecampane: from out the wings of the stalkes appeare the floures orderly placed, long, hollow within, of a light red colour: after them groweth the seed, which is blacke. The root is long and thick, blacke without, white within, hauing in it a clammy juice, in which root consisteth the vertue.

2 The great Comfrey hath rough hairy stalks, and long rough leaues much like the garden Buglosse, but greater and blacker: the floures be round and hollow like little bells, of a white colour: the root is blacke without, and white within, and very slimy. ‡ This differeth no way from the former but onely in the colour of the floure, which is yellowish or white, when as the other is reddish or purple. ‡

3 There is another kinde of Comfrey which hath leaues like the former, sauing that they be lesser: the stalks are rough and tender: the floures be like the former, but that they be of an ouerworn yellow colour: the roots are thicke, short, blacke without, and tuberous, ‡ which in the figure are not expressed so large and knobby as they ought to haue been. ‡

4 This

‡ 4 This pretty plant hath fibrous and blackiſh roots, from which riſe vp many leaues like thoſe of Borage, or Comfrey, but much ſmaller and greener, the ſtalkes are ſome eight inches high, and on their tops carry pretty floures like thoſe of Borage, but not ſo ſharpe pointed, but of a more pleaſing blew colour. This floures in the ſpring and is kept in ſome choice Gardens. *Lobell* calls it *Symphytum pumilum repens Borraginis facie, ſiue Borrago minima Herbariorum*. ‡

¶ *The Place.*

Comfrey joyeth in watery ditches, in fat and fruitfull medowes; they grow all in my Garden.

¶ *The Time.*

They floure in Iune and Iuly.

¶ *The Names.*

It is called in Greeke σύμφυτον: in Latine *Symphytum*, and *Solidago*: in ſhops, *Conſolida maior*, and *Symphytum maius*: of *Scribonius Largus, Inula ruſtica*, and *Alus Gallica*: of others, *Oſtecollon*: in high Dutch, **Walwurtz**: in low Dutch, **Waelwortele**: in Italian, *Conſolida maggiore*: in Spaniſh, *Suelda maiore*, and *Conſuelda maior*: in French, *Conſire*, and *Oreille d'aſne*: in Engliſh, Comfrey, Comfrey Conſound; of ſome, Knit backe, and Blackewoort.

¶ *The Temperature.*

The root of Comfrey hath a cold quality, but yet not much: it is alſo of a clammie and gluing moiſture, it cauſeth no itch at all, neither is it of a ſharpe or biting taſte, vnſauory, and without any qualitie that may be taſted; ſo far is the tough and gluing moiſture from the ſharpe clammineſſe of the ſea Onion, as that there is no compariſon betweene them. The leaues may cauſe itching not through heate or ſharpeneſſe, but through their ruggedneſſe, as we haue already written, yet leſſe than thoſe of the Nettle.

¶ *The Vertues.*

The rootes of Comfrey ſtamped, and the juice drunke with wine, helpeth thoſe that ſpit bloud, and healeth all inward wounds and burſtings. A

The ſame bruiſed and laid to in manner of a plaiſter, doth heale all freſh and greene woundes, and are ſo glutenatiue, that it will ſodder or glew together meate that is chopt in peeces ſeething in a pot, and make it in one lumpe. B

The rootes boiled and drunke, doe clenſe the breſt from flegme, and cure the griefes of the lungs, eſpecially if they be confect with ſugar and ſyrrup; it preuaileth much againſt ruptures or burſtings. C

The ſlimie ſubſtance of the root made in a poſſet of ale, and giuen to drinke againſt the paine in the backe, gotten by any violent motion, as wraſtling, or ouermuch vſe of women, doth in foure or fiue daies perfectly cure the ſame: although the inuoluntary flowing of the ſeed in men be gotten thereby. D

The roots of Comfrey in number foure, Knotgraſſe and the leaues of Clarie of each an handfull, being ſtamped all together, and ſtrained, and a quart of Muſcadell put thereto, the yolkes of three egges, and the powder of three Nutmegs, drunke firſt and laſt, is a moſt excellent medicine againſt a Gonorrhæa or running of the reines, and all paines and conſumptions of the backe. E

There is likewiſe a ſyrrup made hereof to be vſed in this caſe, which ſtaieth voiding of bloud: tempereth the heate of agues: allaieth the ſharpeneſſe of flowing humors: healeth vp vlcers of the lungs, and helpeth the cough: the receit whereof is this: Take two ounces of the roots of great Comfrey, one ounce of Liquorice; two handfulls of Folefoot, roots and all; one ounce and an halfe of Pine-apple kernells; twenty iuiubes; two drams or a quarter of an ounce of Mallow ſeed; one dram of the heads of Poppy; boile all in a ſufficient quantitie of water, till one pinte remaine, ſtraine it, and and adde to the liquor ſtrained ſix ounces of very white ſugar, and as much of the beſt hony, and make thereof a ſyrrup that muſt be throughly boiled. F

The ſame ſyrrup cureth the vlcers of the kidnies, though they haue been of long continuance; and ſtoppeth the bloud that commeth from thence. G

Moreouer, it ſtaieth the ouermuch flowing of the monethly ſickeneſſe, taken euery day for certaine daies together. H

It is highly commended for woundes or hurts of all the reſt alſo of the intrailes and inward parts, and for burſtings or ruptures. I

The root ſtamped and applied vnto them, taketh away the inflammation of the fundament, and ouermuch flowing of the hemorrhoides. K

CHAP.

Chap. 288. *Of Cowslips of Jerusalem.*

1 *Pulmonaria maculosa.*
Spotted Cowslips of Ierusalem.

2 *Pulmonaria folijs Echij.*
Buglosse Cowslips.

3 *Pulmoria angustifolia ij Clusij.*
Narrow leafed Cowslips of Ierusalem.

¶ *The Description.*

1 Cowslips of Ierusalem, or the true and right Lungwort, hath rough, hairy, and large leaues, of a brown green color, confusedly spotted with diuers spots, or drops of white: amongst which spring vp certaine stalkes, a span long, bearing at the top many fine floures, growing together in bunches like the floures of cowslips, sauing that they be at the first red, or purple, and sometimes blew, and oftentimes al these colours at once. The floures being fallen, there come small buttons full of seed. The root is blacke and threddy. ‡ This is sometimes found with white floures. ‡

2 The second kinde of Lungwort is like vnto the former, but greater in each respect: the leaues bigger than the former, resembling wilde Buglosse, yet spotted with white spots like the former: the floures are like the other, but of an exceeding shining red colour.

3 *Carolus Clusius* setteth forth a third kinde of Lungwoort, which hath rough and hairie leaues, like vnto wilde Buglosse, but narrower: among which rises vp a stalke a foot high, bearing at the top a bundle of blew floures, in fashion like vnto those of Buglosse or the last described.

¶ *The*

¶ *The Place.*

These plants do grow in moist shadowie woods, and are planted almost euery where in gardens. ‡ M^r. *Goodyer* found the *Pulmonaria folijs Echij*, being the second, May 25. *Anno* 1620. flouring in a wood by Holbury house in the New Forrest in Hampshire. ‡

¶ *The Time.*

They floure for the most part in March and Aprill.

¶ *The Names.*

Cowslips of Ierusalem, or Sage of Ierusalem, is called of the Herbarists of our time, *Pulmonaria* and *Pulmonalis*; of *Cordus*, *Symphitum syluestre*, or wilde Comfrey: but seeing the other is also of nature wilde, it may aptly be called *Symphytum maculosum*, or *Maculatum*: in high Dutch, **Lungenkraut**: in low Dutch, **Onser vrouwen melcruijt**: in English, spotted Comfrey, Sage of Ierusalem, Cowslip of Ierusalem, Sage of Bethlem, and of some Lungwort; notwithstanding there is another Lungwort, of which we will intreat among the kindes of Mosses.

¶ *The Temperature.*

Pulmonaria should be of like temperature with the great Comfrey, if the roote of this were clammie: but seeing that it is hard and woody, it is of a more drying quality, and more binding.

¶ *The Vertues.*

The leaues are vsed among pot-herbes. The roots are also thought to be good against the infirmities and vlcers of the lungs, and to be of like force with the great Comfrey.

† The figure which formerly was in the fourth place of this Chapter, was onely of the first described with white floures. But the Title *Pulmonaria Gallorum*, and the description fitted to it (though litle to the purpose, and therefore omitted) were intended for the *Pulmonaria Gallorum siue aurea*, whereof I haue in the due place largely treated, as you may see in this booke, *pag.* 304. *chap*, 36.

CHAP. 289. *Of Clote Burre, or Burre Docke.*

1 *Bardana maior.*
The great Burre Docke.

2 *Bardana minor.*
The lesse Burre Docke.

¶ *The Deſcription.*

1 Clot Burre bringeth forth broad leaues and hairie, far bigger than the leaues of Gourds, and of greater compaſſe, thicker alſo, and blacker, which on the vpper ſide are of a darke greene colour, and on the nether ſide ſomewhat white: the ſtalke is cornered, thicke, beſet with like leaues, but far leſſe, diuided into very many wings and branches, bringing forth great Burres round like bullets or balls, which are rough all ouer, and full of ſharpe crooking prickles, taking hold on mens garments as they paſſe by; out of the tops whereof groweth a floure thrummed, or all of threds, of colour purple: the ſeed is perfected within the round ball or bullet, and this ſeed when the burres open, and the winde bloweth, is caried away with the winde: the root is long, white within, and blacke without.

‡ There is another kinde hereof which hath leſſer and ſofter heads, with weaker prickles; theſe heads are alſo hairy or downy, and the leaues and whole plant ſomewhat leſſe, yet otherwiſe like the fore deſcribed; *Lobell* calls this *Arction montanum*, and *Lappa minor Galeni*: it is alſo the *Lappa minor altera* of *Matthiolus*. *Lobell* found this growing in Somerſetſhire three miles from Bath, neere the houſe of one M^r^. *Iohn Colt*.

2 The leſſer Burre hath leaues farre ſmaller than the former, of a grayiſh ouerworne colour like to thoſe of Orach, nicked round about the edges: the ſtalke is a foot and a halfe high, full of little blacke ſpots, diuiding it ſelfe into many branches: the floures before the Burres come forth do compaſſe the ſmall ſtalkes round about; they are but little, and quickly vade away: then follow the Burres or the fruit out of the boſome of the leaues, in forme long, on the tops of the branches, as big as an Oliue or a Cornell berry, rough like the balles of the Plane tree, and being touched cleaue faſt vnto mens garments: they do not open at all, but being kept cloſe ſhut bring forth long ſeeds. The root is faſtened with very many ſtrings, and groweth not deepe.

¶ *The Place.*

The firſt groweth euery where: the ſecond I found in the high way leading from Draiton to Iuer, two miles from Colbrooke, ſince which time I haue found it in the high way betweene Stanes and Egham ‡ It alſo groweth plentifully in Southwickſheet in Hampſhire, as I haue been enformed by M^r^. *Goodyer*. ‡

¶ *The Time.*

Their ſeaſon is in Iuly and Auguſt.

¶ *The Names.*

The great Burre is called in Greeke ἄρκειον: in Latine, *Perſonata, perſonatia*, and *Arcium*: in ſhops, *Bardana*, and *Lappa maior*: in high Dutch, **Groſzkletten**: in low Dutch, **Groote cliſſen**: in French, *Glouteron*: in Engliſh, Great Burre, Burre Docke, or Clot Burre: *Apuleius* beſides theſe doth alſo ſet downe certaine other names belonging to Clot Burre, as *Dardana, Bacchion, Elephantoſis, Nephelion, Manifolium*.

The leſſer Burre Docke is called of the Græcians ξάνθιον: in Latine, *Xanthium*: in ſhops, *Lappa minor, Lappa inuerſa*, and of diuers, *Strumaria*: *Galen* ſaith it is alſo called, *Phaſganion*, and *Phaſganon*, or herbe victory, being but baſtard names, and therefore not properly ſo called: in Engliſh, Louſe Burre, Ditch Burre, and leſſer Burre Docke: it ſeemeth to be called *Xanthium* of the effect, for the Burre or fruite before it be fully withered, being ſtamped and put into an earthen veſſell, and afterwards when need requireth the weight of two ounces thereof and ſomewhat more, being ſteeped in warme water and rubbed on, maketh the haires of the head red; yet the head is firſt to be dreſſed or rubbed with niter, as *Dioſcorides* writeth.

¶ *The Temperature.*

The leaues of Clot Burre are of temperature moderately dry and waſting; the root is ſomething hot.

The ſeed of the leſſer Burre, as *Galen* ſaith, hath power to digeſt, therefore it is hot and dry.

¶ *The Vertues.*

A The roots being taken with the kernels of Pine Apples, as *Dioſcorides* witneſſeth, are good for them that ſpit bloud and corrupt matter.

B *Apeleius* ſaith that the ſame being ſtamped with a little ſalt, and applied to the biting of a mad dog, cureth the ſame, and ſo ſpeedily ſetteth free the ſicke man.

C He alſo teacheth that the juice of the leaues giuen to drinke with hony, procureth vrine, and taketh away the paines of the bladder; and that the ſame drunke with old wine doth wonderfully helpe againſt the bitings of ſerpents.

D *Columella* declareth, that the herbe beaten with ſalt and laid vpon the ſcarifying, which is made with the launcet or raſer, draweth out the poiſon of the viper: and that alſo the root being ſtamped is more auaileable againſt ſerpents, and that the root in like maner is good againſt the Kings euill.

The

The ſtalke of Clot-burre before the burres come forth,the rinde pilled off,being eaten raw with ſalt and pepper, or boyled in the broth of fat meate, is pleaſant to be eaten: being taken in that manner it increaſeth ſeed and ſtirreth vp luſt. E

Alſo it is a good nouriſhment, eſpecially boyled: if the kernell of the Pine Apple be likewiſe added it is the better, and is no leſſe auailable againſt the vlcer of the lungs,and ſpitting of bloud, than the root is. F

The root ſtamped and ſtrained with a good draught of Ale is a moſt approued medicine for a windie or cold ſtomacke. G

Treacle of Andromachus, and the whites of egges, of each a like quantitie,laboured in a leaden mortar, and ſpred vpon the Burre leafe, and ſo applied to the gout, haue been proued many times moſt miraculouſly to appeaſe the paine thereof. H

Dioſcorides commendeth the decoction of the root of *Arcion,* together with the ſeed,againſt the tooth-ache, if it be holden a while in the mouth: alſo that it is good to foment therewith both burnings and kibed heeles; and affirmeth that it may be drunke in wine againſt the ſtrangury and paine in the hip. I

Dioſcorides reporteth that the fruit is very good to be laid vnto hard ſwellings. K

The root cleane picked, waſhed, ſtamped and ſtrained with Malmeſey, helpeth the running of the reines, the whites in women, and ſtrengthneth the backe, if there be added thereto the yelks of egges, the pouder of acornes and nutmegs brued or mixed together,and drunke firſt and laſt. L

Chap. 290. *Of Colts-foot, or Horſe-foot.*

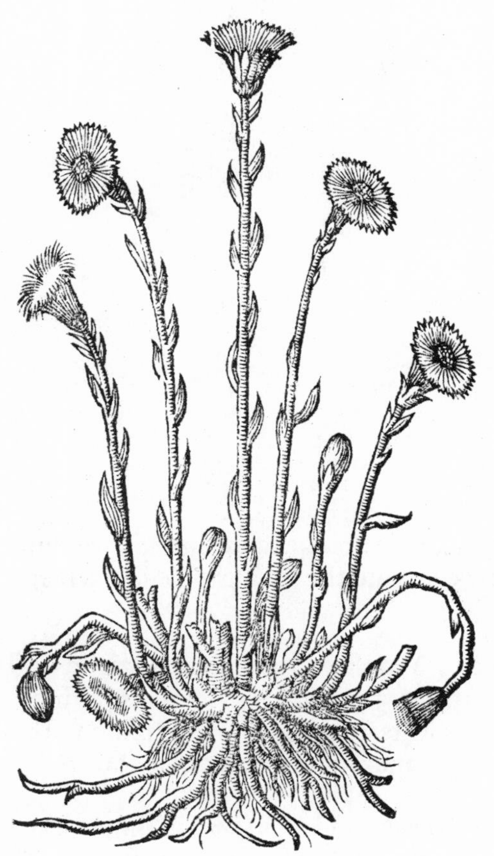

1 *Tuſsilago florens.*
Colts-foot in floure.

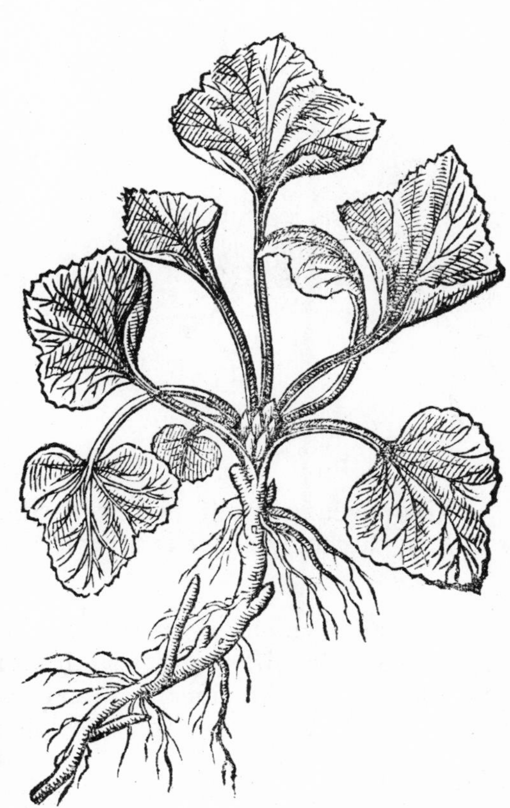

1 *Tuſsilaginis folia.*
The leaues of Colts-foot.

¶ *The Deſcription.*

1 Tvſsilago or Fole-foot hath many white and long creeping roots, ſomewhat fat; from which riſe vp naked ſtalkes (in the beginning of March and Aprill) about a ſpanne long, bearing at the top yellow floures, which change into down,and are caried away with the winde: when the ſtalke and ſeed is periſhed, there appeare ſpringing out of the earth many

many broad leaues,green aboue,and next the ground of a white hoarie or grayiſh colour,faſhioned like an horſe foot; for which cauſe it was called Fole-foot, and Horſe-hoofe: ſeldome or neuer ſhall you find leaues and floures at once, but the floures are paſt before the leaues come out of the ground; as may appeare by the firſt picture,which ſetteth forth the naked ſtalkes and floures; and by the ſecond, which pourtraiteth the leaues onely.

‡ 2 Beſides the commonly growing and deſcribed Colts-foot, there are other two ſmall mountaine Colts-feet deſcribed by *Cluſius*; the firſt whereof I will here preſent you with, but the ſecond you ſhall finde hereafter in the chapter of *Aſarum*, by the name of *Aſarina Matthioli*. This here delineated hath fiue or ſix leaues not much vnlike thoſe of Alehoofe,of a darke ſhining green colour aboue,and very white and downy below: the ſtalke is naked ſome handfull high,hollow and downy,bearing one floure at the top compoſed of purpliſh threds,and flying away in downe: after which the ſtalke falls away, and ſo the leaues onely remaine during the reſt of the yeare: the root is ſmall and creeping. It growes on the tops of the Auſtrian and Stirian mountaines, where it floures in Iune or Auguſt. Brought into gardens it floures in Aprill. *Cluſius* calls it *Tuſsilago Alpina* 1. and he hath giuen two figures thereof, both which I here giue you by the ſame titles as he hath them. ‡

‡ 2 *Tuſsilago Alpina flore aperto.*
Mountaine Colts-foot full in floure.

‡ 2 *Tuſsilago Alpina flore evanido.*
Mountaine Colts-foot with the floure fading.

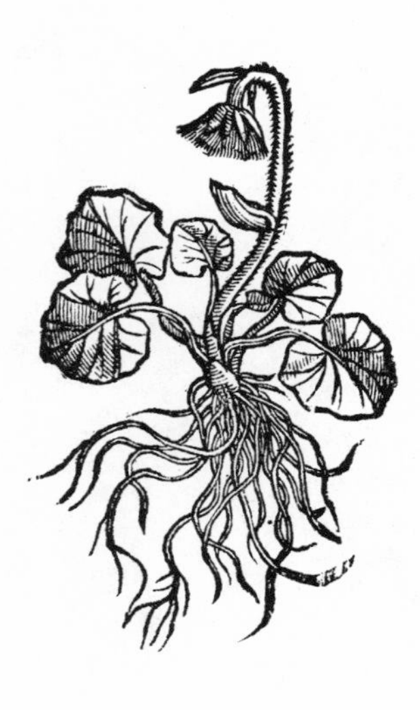

¶ *The Place.*

This groweth of it ſelfe neere vnto Springs, and on the brinkes of brookes and riuers,in wet furrowes, by ditches ſides,and in other moiſt and watery places neere vnto the ſea,almoſt euery where.

¶ *The Time.*

The floures,which quickly fade, are to be ſeene in the end of March,and about the Calends of Aprill, which ſpeedily wither together with the ſtems: after them grow forth the leaues, which remaine greene all Sommer long: and hereupon it came that Colts-foot was thought to be without floures; which thing alſo *Pliny* hath mentioned in his ſixe and twentieth booke, *cap*. 6.

¶ *The Names.*

Folefoot is called in Greeke βήχιον: of the Latines likewiſe *Bechion*, and *Tuſsilago*: in ſhops, *Farfara*, and *Vngula Caballana*: of diuers, *Pata equina*: in Italian, *Vnghia di Cauallo*: in Spaniſh, *vnha d' aſno*: in French, *Pas d' aſne*: in Engliſh, Fole-foot, Colts-foot, Horſe-hoofe, and Bull-foot. The ſame is alſo *Chamæleuce*, which *Pliny* in his twenty eighth booke,and fifteenth chapter reporteth to be likewiſe called *Farfugium*, and *Farranum*, if there be not an error in the copy: which thing alſo *Aëtius* in his firſt booke affirmeth, pretermitting the name of *Bechium*, and attributing vnto it all the vertues and faculties of *Bechium* or Colts-foot. Whoſe opinion *Orabaſius* ſeemeth to be of, in his fifteenth booke of his medicinable Collections, making mention of *Chamæleuce*: only *Pliny* alſo

alſo agreeth with them; ſhewing that ſome thinke, that *Bechium* is called by another name *Chamæleuce*, in his twenty ſixth booke, *cap. 6.* and it may be that *Dioſcorides* hath written of one and the ſelfe ſame herbe in ſundry places, and by diuers names. *Bechium* and *Tuſsilago*, which may alſo be Engliſhed Coughwort, ſo called of the effect, and *Farfara*, of the white Poplar tree, to whoſe leaues it is like; which was named of the Antients *Farfarus*, as *Plautus* writeth in his Comedie called *Pœnulus*:

——— viſcum legioni dedi.
fundaſque eos proſternebam vt folia Farfari.

To the company I gaue both lime buſh and ſling.
That to the ground as Poplar leaues I might them fling.

‡ *Dodonæus* (from whom our Author tooke this) ſets downe this place in *Plautus* as you finde it here, but not well; for the laſt verſe ſhould be *Fundaſque, eo præſternebant folia Farfari.* Thus it is in moſt editions of *Plautus*, and that rightly, as the enſuing words in that place declare. ‡

The white Poplar tree is called in Greeke Λεύκη, and hereupon *Bechion* or Colts-foot was alſo called *Chamæleuce*.

¶ *The Temperature and Vertues.*

The leaues of Colts-foot being freſh and greene are ſomething cold, and haue withall a drying qualitie; they are good for vlcers and inflammations: but the dried leaues are hot and drie, and ſomewhat biting. A

A decoction made of the greene leaues and roots, or elſe a ſyrrup thereof, is good for the cough that proceedeth of a thin rheume. B

The green leaues of Fole-foot pound with hony, do cure and heale the hot inflammation called Saint Anthonies fire, and all other inflammations. C

The fume of the dried leaues taken through a funnell or tunnell, burned vpon coles, effectually helpeth thoſe that are troubled with the ſhortneſſe of breath, and fetch their winde thicke and often, and breaketh without perill the impoſtumes of the breſt. D

Being taken in manner as they take Tobaco, it mightily preuaileth againſt the diſeaſes aforeſaid. E

CHAP. 291. *Of Butter-Burre.*

¶ *The Deſcription.*

1 BVtter-Burre doth in like manner bring forth floures before the leaues, as doth Colts-foot, but they are ſmall, moſſie, tending to a purple colour; which being made vp into a big eare as it were, do quickly (together with the ſtem, which is thicke, full of ſubſtance, and brittle) wither and fall away: the leaues are very great like to a round cap or hat, called in Latine *Petaſus*, of ſuch a wideneſſe, as that of it ſelfe it is big and large enough to keepe a mans head from raine, and from the heate of the Sunne: and therefore they be greater than the leaues of the Clot-burre, of colour ſomewhat white, yet whiter vnderneath: euery ſtem beareth his leafe; the ſtem is oftentimes a cubit long, thicke, full of ſubſtance; vpon which ſtandeth the leafe in the centre or middlemoſt part of the circumference, or very neere, like to one of the greateſt Muſhroms, but that it hath a cleft that ſtandeth about the ſtem, eſpecially when they are in periſhing and withering away: at the firſt the vpper ſuperficiall or outſide of the Muſhroms ſtandeth out, and when they are in withering ſtandeth more in; and euen ſo the leafe of Butter-bur hath on the outſide a certaine ſhallow hollowneſſe: the root is thicke, long, blacke without, white within, of taſte ſomewhat bitter, and is oftentimes worme-eaten.

¶ *The Place.*

This groweth in moiſt places neere vnto riuers ſides, and vpon the brinks and banks of lakes and ponds, almoſt euery where.

¶ *The Time.*

The eare with the floures flouriſh in Aprill or ſooner: then come vp the leaues, which continue till Winter, with new ones ſtill growing vp.

1 *Petasites florens.*
Butter-Burre in floure.

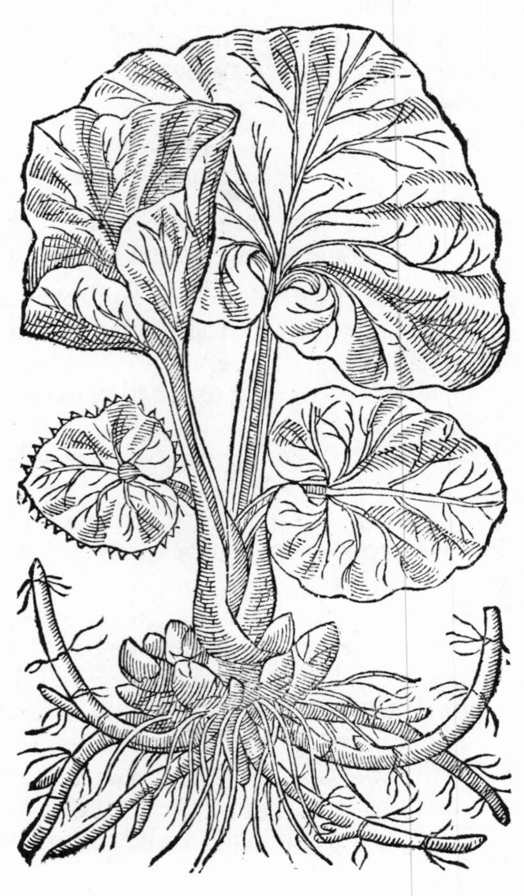

1 *Petasitis fol. a.*
The leaues of Butter-burre.

¶ *The Names.*

Butter-bur is called in Greeke πετασίτης, of the hugenesse of the leafe that is like to πέτασος, or a hat: the Latines call it *Petasites*: in high-Dutch, **Pestilentzwurtz**: in low-Dutch, **Dockebladeren**: in English it is named Butter-Burre: it is very manifest that this is like to Colts-foot, and of the same kinde.

¶ *The Temperature.*

Butter-Burre is hot and dry in the second degree, and of thinne parts.

¶ *The Vertues.*

A The roots of Butter-burre stamped with ale, and giuen to drinke in pestilent and burning Feuers, mightily cooleth and abateth the heate thereof.

B The roots dried and beaten to pouder, and drunke in wine, is a soueraigne medicine against the plague and pestilent feuers, because it prouoketh sweat, and driueth from the heart all venome and ill heate: it killeth wormes, and is of great force against the suffocation of the mother.

C The same cureth all naughty filthy vlcers, if the pouder be strewed therein.

D The same kills wormes in the belly: it prouokes vrine, and brings downe the monthly termes.

‡ CHAP. 292. *Of Mountaine Horse-foot.*

¶ *The Description.*

‡ 1 THis plant (which the moderne Writers haue referred to the *Cacalia* of the Antients, and to the kindes of Colts-foot) I haue thought good to name in English, Horse-foot, for that the leaues exceed Colts-foot in bignesse, yet are like them in shape: and of this plant *Clusius* (whom I here chiefely follow) hath described two sorts: the first of these hath many leaues almost like vnto those of Colts-foot, but larger, very round, and snipt about the edges, of a light greene colour aboue, and hoarie vnderneath, hauing also many veines or nerues running vp and downe them; and these leaues are of an vngratefull taste, and grow vpon long purplish crested stalkes: The stemme is some two cubits high, crested likewise, and of a purplish colour, set also at certaine spaces with leaues very like vnto the other, but lesser than those next

‡ 1 *Cacalia incano folio.*
Hoarie leaued Horſe-foot.

‡ 2 *Cacalia folio glabro.*
Smooth leaued Horſe-foot.

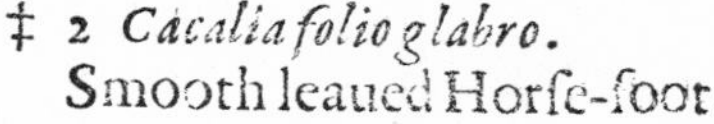

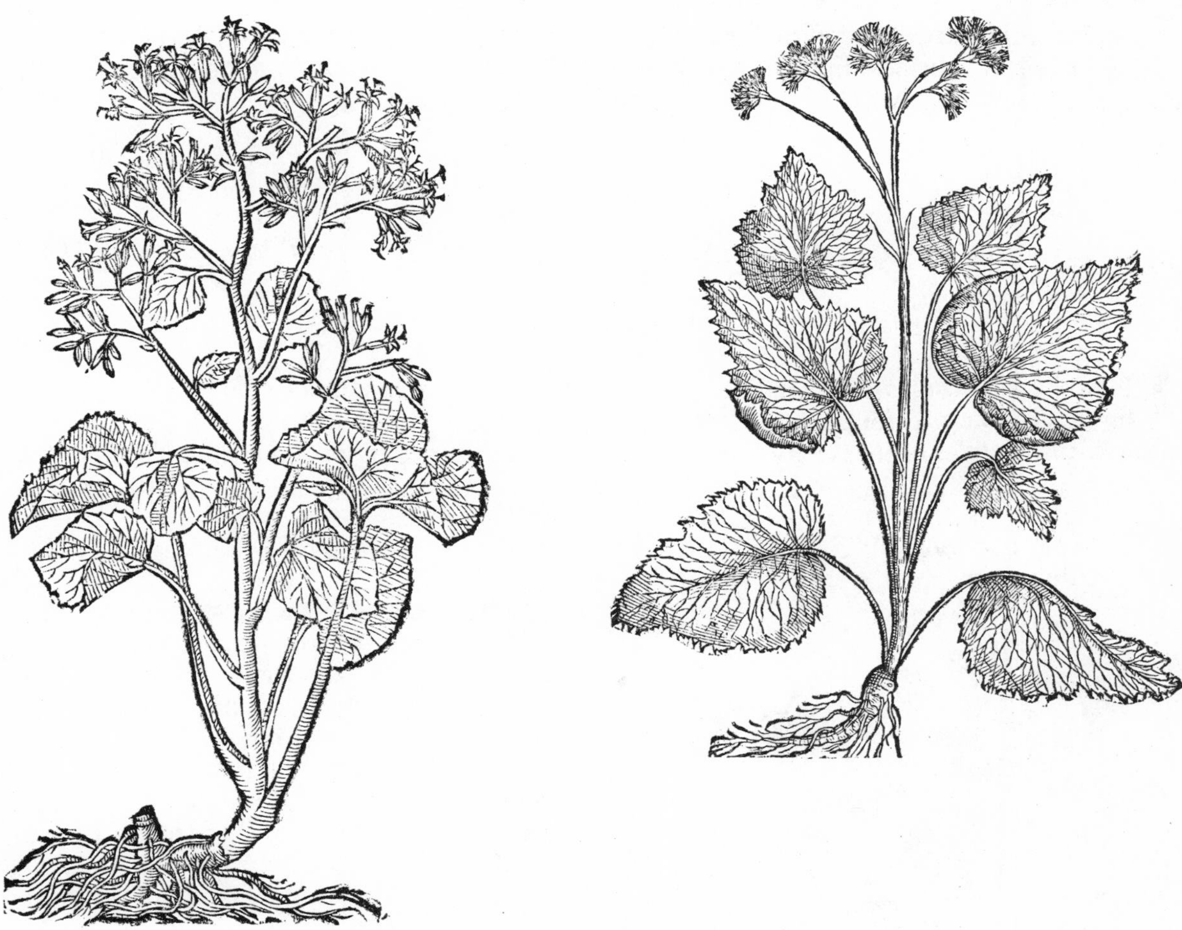

next the ground, and more cornered and ſharper pointed; the tops of the ſtalkes and branches carrie bunches of purple floures, as in an vmbell: and commonly in each bunch there are three little floures conſiſting of foure leaues a peece, and a forked peſtell, and theſe are of a purple colour, and a weake, but not vnpleaſant ſmell, and they at length turne into downe, amongſt which lies hid a longiſh ſeed: the root, if old, ſends forth diuers heads, as alſo ſtore of long whitiſh fibres.

2 The leaues of this are more thin, tough and hard, and of a deeper greene on the vpper ſides, neither are they whitiſh below, nor come ſo round or cloſe whereas they are faſtened to their ſtalks (which are not creſted as thoſe of the other, but round and ſmooth) they are alſo full of veines, and nickt about the edges, and of ſomewhat an vngratefull hot and bitter taſte. The ſtalkes are alſo ſmoother, and the floures of a lighter colour.

¶ *The Place.*

Both theſe grow in the Auſtrian and Stirian Alpes vnder the ſides of woods, among buſhes and ſuch ſhadowie places: but not in England, that I haue yet heard of.

¶ *The Time.*

I find it not ſet downe when theſe floure and ſeed, but iudge it about the ſame time that Coltſ-foot doth.

¶ *The Names.*

This by *Cluſius*, *Lobel* and others, hath beene called *Cacalia*, and referred to that deſcribed by *Dioſcorides*, *lib*. 4. *cap*. 123. which is thought to be that ſet forth by *Galen* by the name of *Cancanus*. In the *Hiſtoria Lugd. pag*. 1052. The later of theſe two here deſcribed is figured by the name of *Tuſſilago Alpina ſive montana*, and the former is there, page 1308, by the name of *Cacalia*, but the floures are not rightly expreſt: and if my iudgement faile me not, the figure which is in the ſeuenteenth page of the *Appendix* of the ſame Authour, by the title of *Aconitum Pardalianches primum*, is of no other than this very plant. But becauſe I haue not as yet ſeene the plant, I will not poſitiuely affirme it: but referre this my opinion to thoſe that are iudicious and curious, to know the plant that raiſed ſuch controuerſie between *Matthiolus* and *Geſner*, and whereof neither *Camerarius* nor *Bauhine*, who haue ſet forth *Matthiolus* his Commentaries, haue giuen vs any certain or probable knowledge.

¶ The

¶ *The Temperature and Vertues, out of the Antients.*

A The root of *Cacalia* is void of any biting qualitie, and moderately dries, and it is of a grosse and emplaisticke substance; wherefore steeped in wine and so taken it helpes the cough, the roughnesse of the Arterie or hoarsnes, like as *Tragacanth*: neither if you chew it and swallow downe the iuice doth it lesse auaile against those effects than the iuice of Liquorice. ‡

Chap. 293. *Of small Celandine or Pile-woort.*

¶ *The Kindes.*

There be two kindes of Celandine, according to the old writers, much differing in forme and figure: the one greater, the other lesser, which I intende to diuide into two distinct chapters, marshalling them as neere as may be with their like, in forme and figure, and first of the small Celandine.

Chelidonium minus.
Pile-woort.

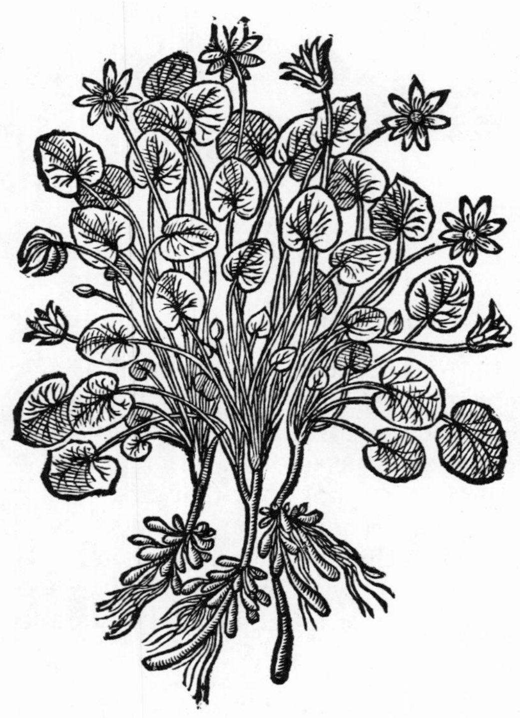

¶ *The Description.*

The lesser Celandine hath greene round leaues, smooth, slipperie, and shining, lesse than the leaues of the Iuie: the stalks are slender, short, and for the most part creeping vpon the ground: they bring forth little yellow flours like those of Crow-foot; and after the floures there springeth vp a little fine knop or head full of seede: the root consisteth of slender strings, on which doe hang as it were certaine graines, of the bignes of wheat cornes, or bigger.

¶ *The Place.*

It groweth in medows, by common waies, by ditches and trenches, and it is common euery where, in moist and dankish places.

¶ *The Time.*

It commeth forth about the Calends of March, and floureth a little after: it beginneth to fade away in Aprill, it is quite gone in May, afterwards it is hard to be found, yea scarcely the root.

¶ *The Names.*

It is called in Greeke χελιδόνιον: of the Latines *Chelidonium minus*, and *Hirundinaria minor*: of diuers, *Scrophularia minor*, *Ficaria minor*: of *Serapio*, *Memiren*: in Italian, *Fauoscello*: in High Dutch, **Feigwurtzenkraut**: in French, *Esclere*, and *Petit Bassinet*: in English, little Celandine, Fig-woort, and Pile-woort.

¶ *The Temperature.*

It is hot and drie, also more biting and hotter than the greater: it commeth neerest in facultie to the Crowfoot.

‡ This which is here, and by most Authours set forth for *Chelidonium minus*, hath no such great heat and Acrimonie as *Dioscorides* and *Galen* affirme to be in theirs; making it hot in the fourth degree, when as this of ours scarce exceedes the first, as farre as wee may coniecture by the taste. ‡

¶ *The Vertues.*

B It presently, as *Galen* and *Dioscorides* affirme, exulcerateth or blistereth the skin: it maketh rough and corrupt nailes to fall away.

B The iuice of the roots mixed with honie, and drawne vp into the nosthrils, purgeth the head of foule and filthie humours.

The

The later age vſe the roots and graines for the piles, which being often bathed with the iuice mixed with wine, or with the ſickmans vrine, are drawne together and dried vp, and the paine quite taken away. C

There be alſo who thinke, that if the herbe be but carried about one that hath the piles, the pain forthwith ceaſeth. D

CHAP. 294. *Of Marſh Marigold.*

¶ *The Deſcription.*

1 MArſh Marigold hath great broad leaues ſomewhat round, ſmooth, of a gallant greene colour, ſleightly indented or purld about the edges: among which riſe vp thicke fat ſtalkes, likewiſe greene; whereupon doe grow goodly yellow floures, glittering like gold, and like to thoſe of Crow-foot, but greater: the root is ſmall, compoſed of verie manie ſtrings.

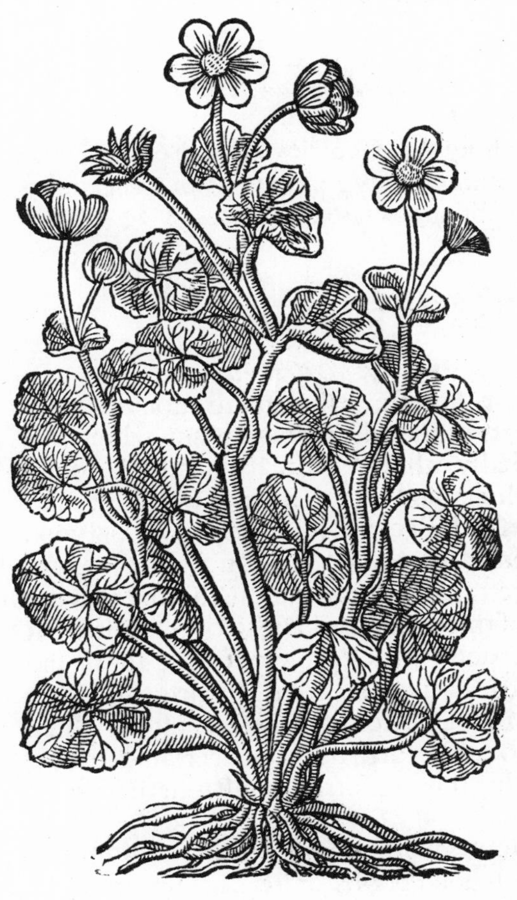

1 *Caltha paluſtris maior.*
The great Marſh Marigold.

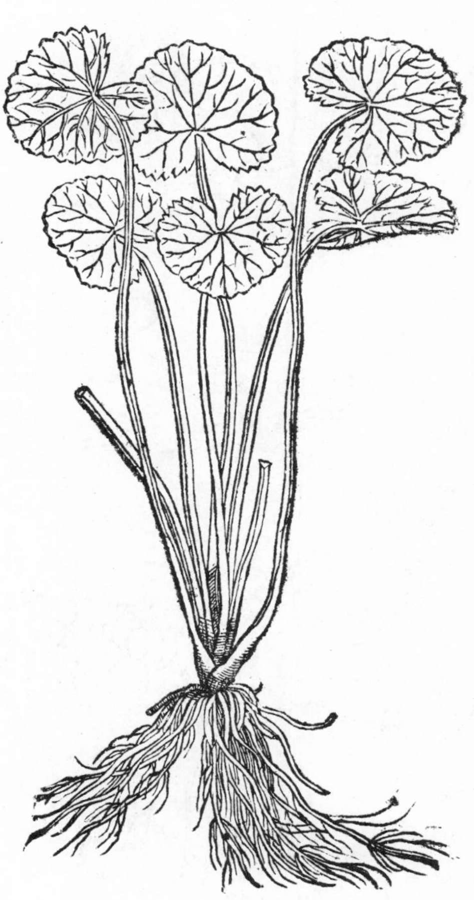

2 *Caltha paluſtris minor.*
The ſmall Marſh Marigold

2 The ſmaller Marſh Marigold hath many round leaues ſpred vpon the ground, of a darke greene colour: amongſt which riſe vp diuers branches, charged with the like leaues: the floures grow at the toppes of the branches, of a moſt ſhining yellow colour: the root is alſo like the former

3 The great Marſh Marigold with double floures is a ſtranger in England, his natiue Countrey ſhould ſeeme to be in the furtheſt part of Germanie, by the relation of a man of thoſe Countries that I haue had conference withall the which hee thus deſcribed: it hath (ſaith hee) leaues, roots, and ſtalkes like thoſe of our common ſort, and hath double floures like thoſe of the garden Marigold, wherein conſiſteth the difference

‡ *Camerarius* writes iuſt contrarie to that which our Authour here affirmes; for hee ſaith, *In Anglia ſua ſponte non ſolum plenis, ſed oderatis etiam floribus paſſim ſeſe offert.* But I feare that both our Authour

3 *Calthapalustris multiplex.*
Double floured Marsh Marigold.

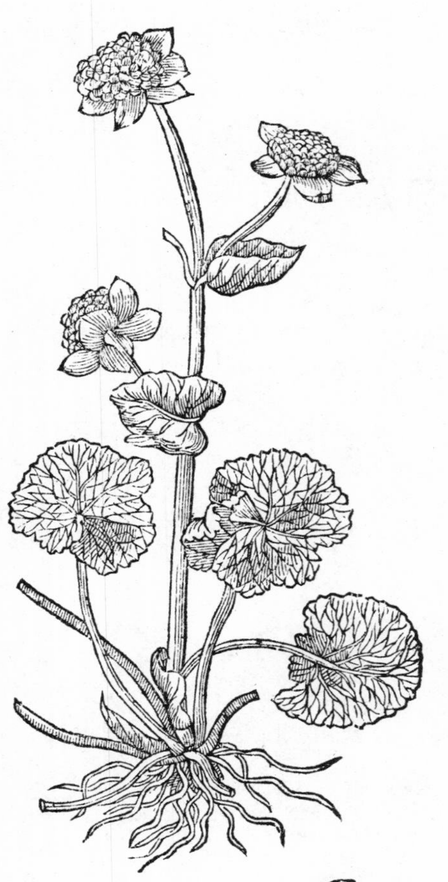

Authour and *Camerarius* were deceiued by trusting the report of some lying, or else ignorant persons, for I could neuer finde it growing wilde with double floures here, nor *Camerarius* there: yet I do not denie but by chance some one with double floures may be found both here & there, but this is not euery where. ‡

The Place.

They ioy in moist and marish grounds, and in watery medowes. ‡ I haue not found the double one wilde, but seene it preserued in diuers gardens for the beautie of the floure. ‡

¶ *The Time.*

They floure in the Spring when the Crowfoots doe, and oftentimes in Sommer: the leaues keepe their greenenesse all the Winter long.

¶ *The Names.*

Marsh Marigold is called of *Valerius Cordus*, *Caltha palustris*: of *Tabernamontanus*, *Populago*: but not properly: in English, Marsh Marigolds: in Cheshire and those parts it is called Bootes.

¶ *The Temperature and Vertues.*

Touching the faculties of these plants, wee haue nothing to say, either out of other mens writings, or our owne experiences

Chap. 295. *Of Frogge-bit.*

Morsus Ranæ.
Frogge-bit.

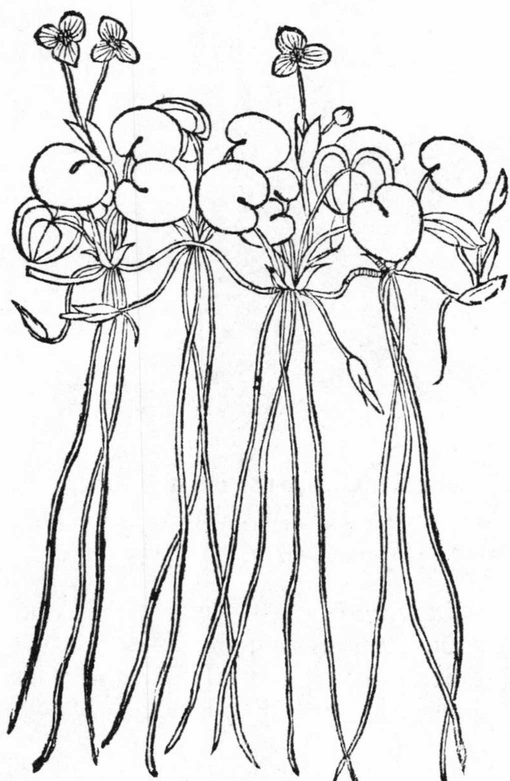

¶. *The Description.*

There floteth or swimmeth vpon the vpper parts of the water a small plant, which wee vsually call Frog-bit, hauing little round leaues, thicke and full of iuice, very like to the leaues of wall Peniwoort: the floures grow vpon long stems among the leaues, of a white colour, with a certaine yellow thrum in the middle, consisting of three leaues: in stead of roots it hath slender strings, which grow out of a short and small head, as it were, from whence the leaues spring, in the bottom of the water: from which head also come forth slopewise certaine strings, by which growing forth it multiplieth it selfe.

¶ *The Place.*

It is found swimming or floting almost in euery ditch, pond, poole, or standing water, in all the ditches about Saint George his fields, and in the ditches by the Thames side neere to Lambeth Marsh, where any that is disposed may see it.

¶ *The Time.*

It flourisheth and floureth most part of all the yeare.

¶ *The Names.*

It is called of some *Ranæ morsus*, and *Morsus Ranæ*, and *Nymphæa parua.*

¶ *The*

¶ *The Temperature and Vertues.*

A It is thought to be a kinde of Pond-weed (or rather of Water Lillie) and to haue the ſame faculties that belong vnto it.

CHAP. 296. *Of Water Lillie.*

¶ *The Deſcription.*

1 THe white water Lillie or *Nenuphar* hath great round leaues, in ſhape of a Buckler, thick, fat, and full of iuice, ſtanding vpon long round and ſmooth foot-ſtalkes, ful of a ſpungious ſubſtance; which leaues do ſwim or flote vpon the top of the water: vpon the end of each ſtalk groweth one floure onely, of colour white, conſiſting of many little long ſharpe pointed leaues, in the middeſt whereof bee many yellow threds: after the floure it bringeth forth a round head, in which lieth blackiſh glittering ſeed. The roots be thicke, full of knots, blacke without, white and ſpungie within, out of which groweth a multitude of ſtrings, by which it is faſtened in the bottome.

1 *Nymphæa alba.*
White Water Lillie.

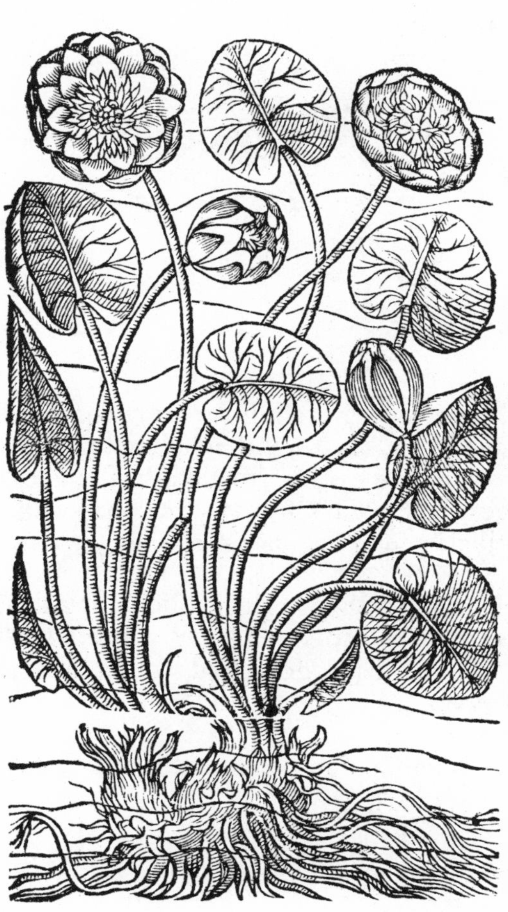

2 *Nymphæa lutea.*
Yellow Water Lillie.

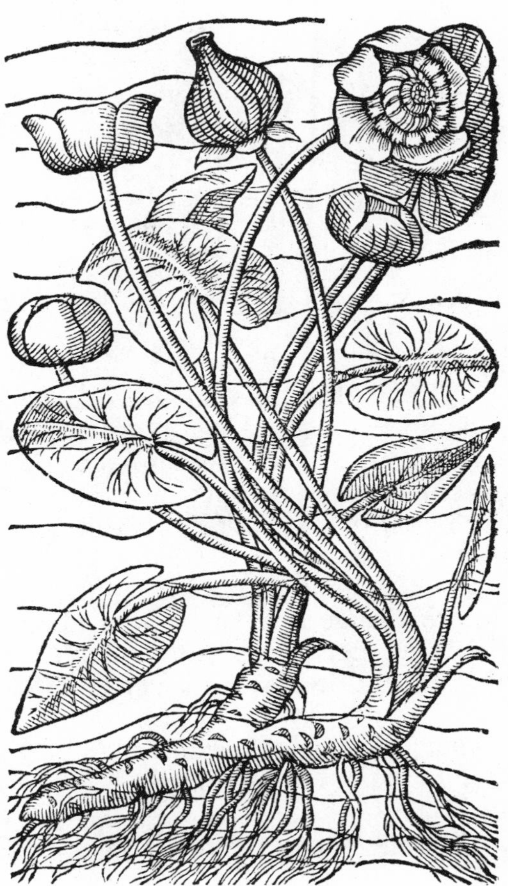

2 The leaues of the yellow water Lillie be like to the other, yet are they a little longer. The ſtalkes of the floures and leaues be like: the floures be yellow, conſiſting onely of fiue little ſhort leaues ſomething round; in the midſt of which groweth a ſmall round head, or button, ſharpe towards the point, compaſſed about with many yellow threds, in which, when it is ripe, lie alſo glittering ſeeds, greater than thoſe of the other, and leſſer than wheat cornes. The roots be thick, long, ſet with certaine dents, as it were white both within and without, of a ſpungious ſubſtance.

3 The ſmal white water Lillie floteth likewiſe vpon the water, hauing a ſingle root, with ſome few fibres faſtened thereto: from which riſeth vp many long, round, ſmooth, and ſoft foot-ſtalkes, ſome of which doe bring forth at the end faire broad round buckler leaues like vnto the precedent,

dent,but lesser: on the other foot-stalkes stand prettie white floures,consisting of fiue small leaues apeece, hauing a little yellow in the middle thereof.

3 *Nymphæa alba minor.*
The small white Water Lillie.

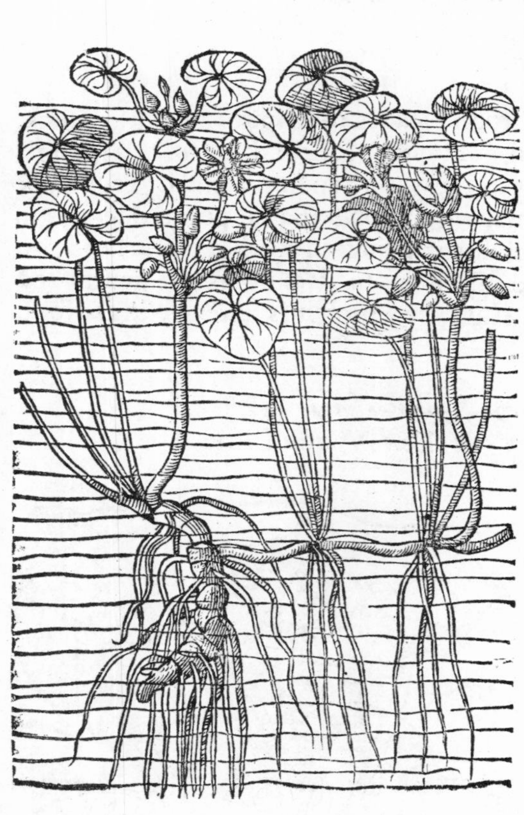

5 *Nymphæa lutea minima.*
Dwarfe Water Lillie.

4 The small yellow water Lillie hath a little threddie root,creeping in the bottome of the water, and dispersing it selfe far abroad: from which rise small tender stalkes, smooth and soft, whereon do grow little buckler leaues like the last described: likewise on the other small stalke standeth a tuft of many floures likewise floting vpon the water as the others do. ‡ This hath the floures larger than those of the next described, wherefore it may be fitly named ***Nymphæa lutea minor flore amplo.*** ‡

5 This dwarfe water Lillie differeth not from the other small yellow water Lillie,sauing that, that this kinde hath sharper pointed leaues, and the whole plant is altogether lesser, wherein lieth the difference. ‡ This hath the floures much lesse than those of the last described, wherefore it is fitly for distinction sake named *Nymphæa lutea minor flore paruo.* ‡

¶ *The Place.*

These herbes do grow in fennes,standing waters,broad ditches, and in brookes that run slowly, and sometimes in great riuers.

¶ *The Time.*

They floure and flourish most of the Sommer moneths.

¶ *The Names.*

Water Lillie is called in Greeke Νυμφαία: and in Latine also *Nymphæa*, so named because it loues to grow in waterie places,as ***Dioscorides*** saith: the Apothecaries call it *Nenuphar:* of *Apuleius*, *Mater Herculania*, *Alga palustris*, *Papauer palustre*, *Clavus veneris*, and ***Digitus veneris*** : ***Marcellus*** a very old writer reporteth, that it is called in Latine *Claua Herculis:* in French, *Badittin*: in high Dutch, **Wasser Mahem**: in low Dutch, **Plompen**: in English, Water Lillie, water Rose.

¶ *The Temperature.*

Both the root and seed of water Lillie haue a drying force without biting.

¶ *The Vertues.*

A Water Lillie with yellow floures stoppeth laskes, the ouerflowing of seed which commeth away by dreames or otherwise, and is good for them that haue the bloudie flix.

But

But water Lillie which hath the white floures is of greater force, inſomuch as it ſtaieth the whites:but both this and the other that hath the black root muſt be drunke in red wine:they haue alſo a ſcouring quality, therfore they both clenſe away the morphew,and be alſo good againſt the pilling away of the haire of the head; againſt the morphew they are ſteeped in water, and for the pilling away of the haire in Tarre: but for theſe things that is fitter which hath the black root,and for the other, that which hath the white root. B

Theopraſtus ſaith,that being ſtamped and laid vpon the wound, it is reported to ſtay the bleeding. C

The Phyſitions of our age do commend the floures of white *Nymphæa* againſt the infirmities of the head which come of a hot cauſe: and do certainely affirme, that the root of the yellow cureth hot diſeaſes of the kidnies and bladder, and is ſingular good againſt the running of the reines. D

The root and ſeed of the great water Lillie is very good againſt venery or fleſhly deſire, if one do drinke the decoction thereof, or vſe the ſeed or root in powder in his meates, for it dryeth vp the ſeed of generation, and ſo cauſeth a man to be chaſt, eſpecially vſed in broth with fleſh. E

The conſerue of the floures is good for the diſeaſes aforeſaid, and is good alſo againſt hot burning feuers. F

The floures being made into oile, as yee do make oile of roſes, doth coole and refrigerate, cauſing ſweate and quiet ſleepe, and putteth away all venereous dreames: the temples of the head and palmes of the hands and feet, and the breſt being annointed for the one, and the genitors vpon and about them for the other. G

The greene leaues of the great water Lillie, either the white or the yellow laid vpon the region of the backe in the ſmall, mightily ceaſe the inuoluntary flowing away of the ſeed called *Gonorrhæa*, or running of the raines, being two or three times a day remooued, and freſh applied thereto. H

CHAP. 297. *Of Pond-weed, or water Spike.*

1 *Potamogeiton latifolium.*
Broad leafed Pondweed.

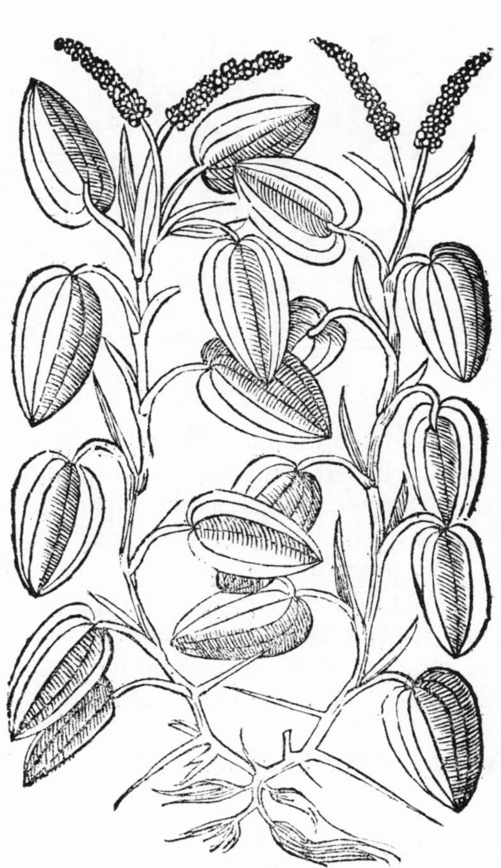

2 *Potamogeiton anguſtifolium.*
Narrow leafed Pondweed.

¶ *The Deſcription.*

1 POnd-weed hath little ſtalkes,ſlender, ſpreading like thoſe of the vine, and jointed: the leaues be long,ſmaller than the leaues of Plantaine,and harder,with manifeſt veines running alongſt them as in Plantains,which ſtanding vpon ſlender and long ſtems or footſtalkes, ſhew themſelues aboue the water, and lie flat along vpon the ſuperficiall or vpper part thereof, as do the leaues of the water Lillie: the floures grow in ſhort eares, and are of a light red purple colour, like thoſe of Red-ſhankes or Biſtort: the ſeed is hard.

‡ 2 This (whoſe figure was formerly vnfitly put by our Authour to the following deſcription) hath longer,narrower, and ſharper pointed leaues than thoſe of the laſt deſcribed, hauing the veines running from the middle rib to the ſides of the leaues,as in a willow leafe,which they ſomewhat reſemble; at the tops of the ſtalkes grow reddiſh ſpikes or eares like thoſe of the laſt deſcribed: the root is long,jointed, and fibrous. ‡

‡ 3 *Potamogeiton 3 Dodonæi.*
Small Pondweed.

‡ 4 *Potamogeiton longis acutis folijs.*
Long ſharpe leaued Pondweed.

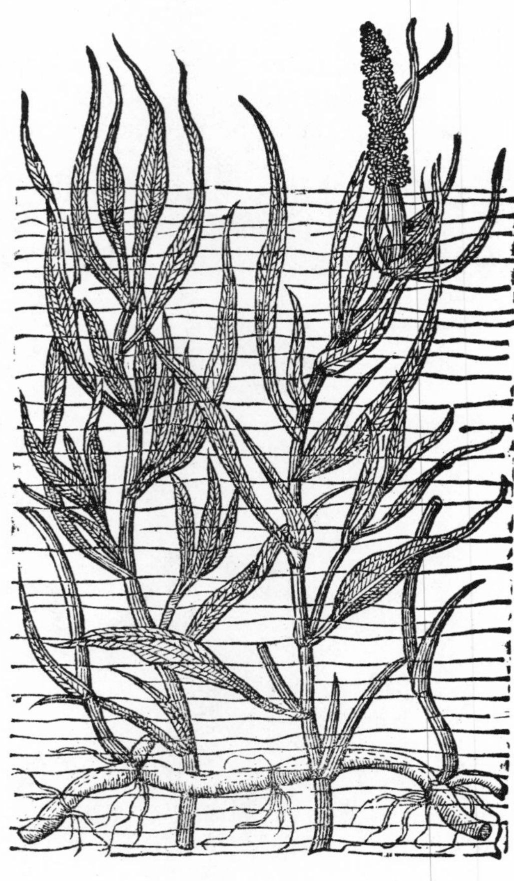

3 There is another Pondweed deſcribed thus; it ſhooteth forth into many ſlender and round ſtems,which are diſtributed into ſundry branches: his leaues are broad, long, and ſharpe pointed, yet much leſſe than the firſt kinde:out of the boſomes of the branches and leaues there ſpring certaine little ſtalkes which beare ſundry ſmall white moſſie floures, which doe turne into plaine and round ſeeds,like the common Tare or Vetch:his root is fibrous,throughly faſtened in the ground.

‡ 4 There is alſo another Pondweed, which hath whitiſh and jointed roots creeping in the bottome of the water, and ſending downe ſome fibres, but ſending vp ſlender jointed and long ſtalkes, ſmall below, and bigger aboue, hauing long narrow and very ſtiffe ſharpe pointed leaues. The floures grow in a reddiſh ſpike like thoſe of the firſt deſcribed. This is the *Potamogeiton altera* of *Dodonæus.* ‡

¶ *The Place.*

Theſe herbes do grow in ſtanding waters,pooles,ponds,and ditches,almoſt euery where.

¶ *The Time.*

They do floure in Iune and Iuly.

¶ *The Names.*

It is called of the Grecians, ποταμογείτων: in Latine, *Fontalis*, and *Spicata*: in high Dutch, **Zamkraut**: in low Dutch, **Fonteyncruyt**: in French, *Eſpi d'eaue*: in Engliſh, Pondweed, and water Spike.

¶ *The*

¶ *The Temperature.*

Pondweed, saith *Galen*, doth binde and coole, like as doth Knot-grasse, but his essence is thicker than that of Knot-grasse.

¶ *The Vertues.*

It is good against the itch, and consuming or eating Vlcers, as *Dioscorides* writeth. A

Also it is good being applied to the inflammation of the legges, wherein *Ignis sacer* hath gotten the superioritie. B

CHAP. 298. *Of Water Saligot, water Caltrops, or water Nuts.*

¶ *The Description.*

1 WAter Caltrops haue long slender stalkes, growing vp, and rising from the bottome of the water, and mounting aboue the same: the root is long, hauing here and there vnder the water certaine tassels full of small strings and threddie haires: the stem towards the top of the water, is very great in respect of that which is lower, the leaues are large and somewhat round, not vnlike those of the Poplar or Elme tree leaues, a little creuised or notched about the edges: amongst and vnder the leaues groweth the fruit, which is triangled, hard, sharpe pointed, and prickly: in shape like those hurtfull engines in the warres, cast in the passage of the enemie to annoy the feet of their horses, called Caltrops, whereof this tooke its name. within these heads or Nuts is contained a white kernell, in taste almost like the Chesnut, which is reported to be eaten greene, and being dried and ground to serue in stead of bread.

‡ There are two other plants which are found growing in many ponds and ditches of this kingdome, both about London and else-where, and I will here giue you the figures out of *Lobel* and *Clusius*, and their descriptions as they were sent me by Mr. *Goodyer*, who hath saued me the labour of describing them.

Tribulus aquaticus minor quercus floribus, Clus. p. 252.
Pusillum fontila pathum, Lobelij.

2 This water herbe bringeth forth from the root, thin, flat, knottie stalkes, of a reddish colour, two or three cubits long, or longer, according to the depth of the water (which when they are drie, are pliant and bowing) diuided towards the top into many parts or branches, bearing but one leafe at euery ioint, sometimes two inches long, and halfe an inch broad, thin, and as it were shining, so wrinckled and crompled by the sides that it seemeth to be torne, of a reddish greene colour: the foot-stalkes are something long and thicke, and rise vp from amongst those leaues, which alwaies grow two one opposit against another, in a contrarie manner to those that grow below on the stalk: neere the top of which foot-stalke groweth small grape like huskes, out of which spring very small reddish floures, like those of the Oke, euery floure hauing foure very small round topped leaues: after euery floure commeth commonly foure sharpe pointed graines growing together, containing within them a little white kernell. The lower part of the stalke hath at euery ioint small white threddie roots, somewhat long, whereby it taketh hold in the mudde, and draweth nourishment vnto it. The whole plant is commonly couered ouer with water. It floureth in Iune and the beginning of Iuly. I found it in the standing pooles or fish-ponds adioyning to a dissolued Abbey called Durford, which ponds diuide Hampshire and Sussex, and in other standing waters elswhere. This description was made vpon sight of the plant the 2. of Iune, 1622.

Tribulus aquaticus minor, muscatellæ floribus,

3 This hath not flat stalkes like the other, but round, kneed, and alwaies bearing two leaues at euery ioint, one opposite against another, greener, shorter and lesser than the other, sharpe pointed, not much wrinckled and crumpled by the edges. *Clusius* saith, that they are not at all crumpled. I neuer obserued any without crumples and wrinckles: the floures grow on short small foot-stalkes, of a whitish green colour, like those of *Muscatella Cordi*, called by *Gerard*, *Radix caua minima viridi flore: viz.* two floures at the top of euery foot-stalke, one opposite against another, euery floure containing foure small leaues: which two floures beeing past there come vp eight small huskes

1 *Tribulus aquaticus.*
Water Caltrops.

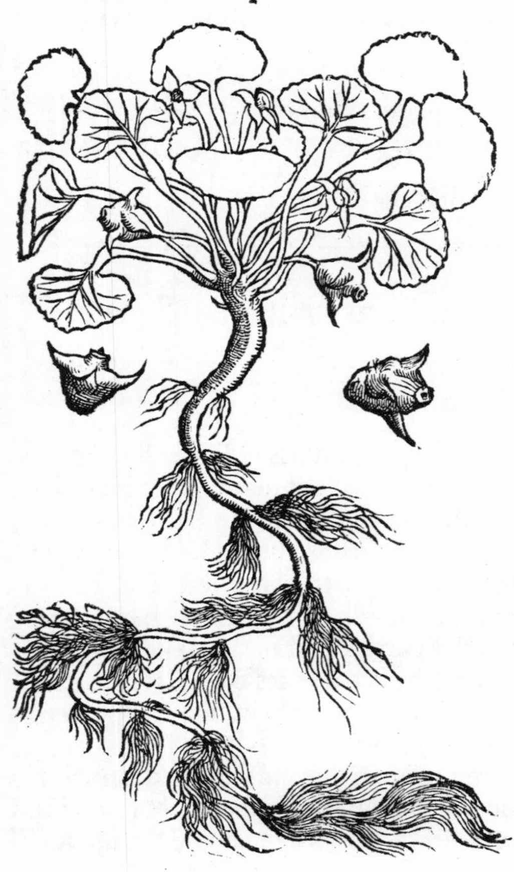

‡ 2 *Tribulus aquaticus minor quercus floribus.*
Small water Caltrops, or Frogs-lettuce.

‡ 3 *Tribulus aquaticus minor, Muscatellæ floribus.*
Small Frogs-Lettuce.

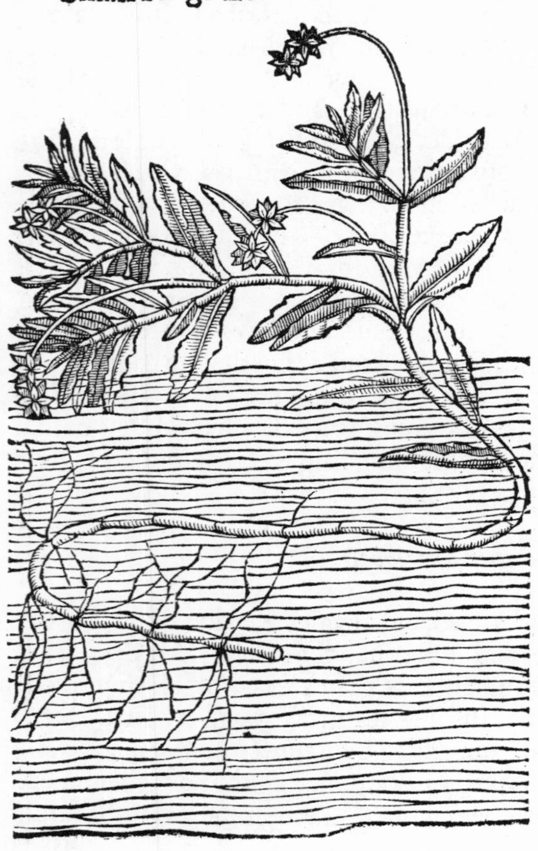

husks making six seueral waies a square of flours. The roots are like the former. This groweth abundantly in the riuer by Droxford in Hampshire. It floureth in Iune and Iuly when the other doth, and continueth couered ouer with water, greene, both winter and Sommer. *Iohn Goodyer.* ‡

¶ *The Place.*

Cordus saith that it groweth in Germany in myrie lakes, and in citie ditches that haue mud in them: in Brabant and in other places of the Low-countries, it is found oftentimes in standing waters, and springs: *Matthiolus* writeth, that it groweth not only in lakes of sweet water, but also in certaine ditches by the sea neere vnto Venice.

¶ *The Time.*

It flourisheth in Iune, Iuly, and August.

¶ *The Names.*

The Grecians call it τρίβολος ἔνυδρος: the Latins, *Tribulus aquatilis*, and *aquaticus*, and *Tribulus lacustris*: the Apothecaries, *Tribulus marinus*: in High Dutch, **Wasser nusz**: the Brabanders, **Water noten**: and of the likenesse of yron nailes, **Minckijsers**: the

the French men, *Macres*: in Engliſh it is named water Caltrops, Saligot, and Water-nuts: moſt do call the fruit of this Caltrops, *Caſtaneæ aquatiles*, or water Cheſ-nuts.

¶ *The Temperature.*

Water Caltrop is of a cold nature, it conſiſteth of a moiſt eſſence, which in this is more waterie than in the land Caltrops, wherein an earthie cold is predominant, as *Galen* ſaith.

¶ *The Vertues.*

A The herbe vſed in manner of a pultis, as *Dioſcorides* teacheth, is good againſt all inflammations or hot ſwellings: boiled with honie and water, it perfectly healeth cankers in the mouth, ſore gums, and the Almonds of the throat.

B The Thracians, ſaith *Plinie*, that dwell in Strymona, do fatten their horſes with the leaues of Saligot, and they themſelues do feed of the kernels, making very ſweet bread thereof, which bindeth the belly.

C The green nuts or fruit of *Tribulus aquaticus*, or Saligot, being drunke in wine, is good for them which are troubled with the ſtone and grauell.

D The ſame drunke in like manner, or laied outwardly to the place, helpeth thoſe that are bitten with any venomous beaſt, and reſiſteth all venome and poiſon.

E The leaues of Saligot be giuen againſt all inflammations and vlcers of the mouth, the putrifaction and corruption of the iawes, and againſt the Kings euill.

F A pouder made of the nuts is giuen to ſuch as piſſe bloud, and are troubled with grauell, and it doth bind the belly very much.

G ‡ The two leſſer water Caltrops here deſcribed are in my opinion much agreeable in temper to the great one, and are much fitter *Succidanea* for it then Aron, which ſome in the compoſition of *Vnguentum Agrippæ* haue appointed for it. ‡

CHAP. 299. *Of water Sengreene, or freſh water Soldier.*

Militaris Aizoides.
Freſh water Soldier.

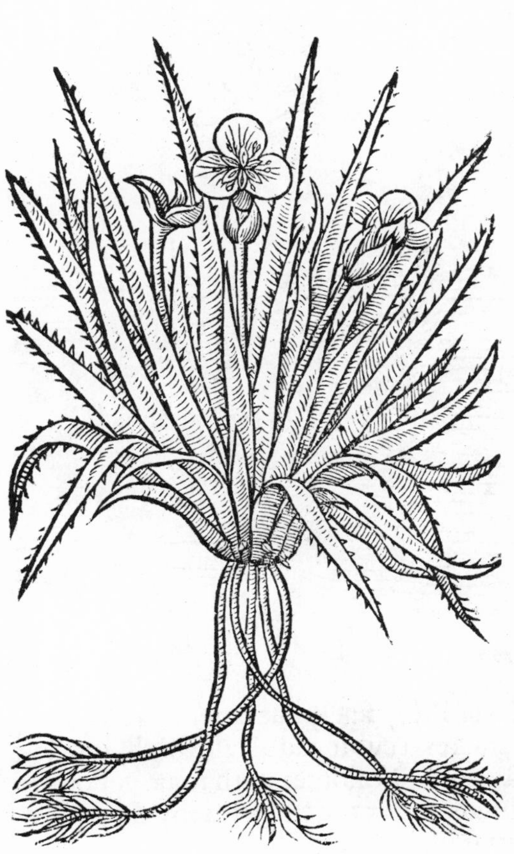

¶ *The Deſcription.*

FReſh water Soldier or water Houſleeke, hath leaues like thoſe of the herbe Aloe, or *Semper viuum*, but ſhorter and leſſer, ſet round about the edges with certaine ſtiffe and ſhort prickles: amongſt which commeth forth diuers caſes or huskes, verie like vnto crabbes clawes: out of which when they open grow white floures, conſiſting of three leaues, altogether like thoſe of Frogs-bit, hauing in the middle little yellowiſh threds: in ſtead of roots there be long ſtrings, round, white, verie like to great Harp-ſtrings, or to long wormes, which falling downe from a ſhort head that brought forth the leaues, go to the bottom of the water, and yet be they ſeldome there faſtened: there alſo grow from the ſame other ſtrings aſlope, by which the plant is multiplied after the manner of Frogs-bit.

¶ *The Place.*

‡ I found this growing plentifully in the ditches about Rotſey a ſmal village in Holderneſſe. And my friend M^r. *William Broad* obſerued it in the Fennes in Lincolne-ſhire. ‡ The leaues and floures grow vpon the top of the water, and the roots are ſent downe through the water to the mud.

¶ *The Time.*

It floures in Iune, and ſometimes in Auguſt.

¶ *The Names.*

It may be called *Sedum aquatile*, or water Sengreen, that is to say, of the likenesse of herbe Aloe, which is also called in Latine *Sedum* : of some, *Cancri chela*, or *Cancri forficula* : in English, VVater Housleeke, Knights Pondwoort; and of some, Knights water Sengreene, fresh water Soldier, or wading Pondweed: it seemeth to be *Stratiotes aquatilis*, or *Stratiotes potamios*, or Knights water VVoundwoort, which may also be named in Latine *Militaris aquatica*, and *Militaris Aizoides*, or Soldiers Yarrow; for it groweth in the water, and floteth vpon it, and if those strings which it sendeth to the bottome of the wat er be no roots, it also liueth without roots.

¶ *The Temperature.*

This herbe is of a cooling nature and temperament.

¶ *The Vertues.*

A This Housleeke staieth the bloud which commeth from the kidneies, it keepeth green wounds from being inflamed, and it is good against S. Anthonies fire and hot swellings, being applied vnto them : and is equall in the vertues aforesaid with the former.

Chap. 300. *Of Water Yarrow, and water Gillofloure.*

1 *Viola palustris.*
Water Violet.

‡ *Viola Palustris tenuifolia.*
The smaller leaued water violet.

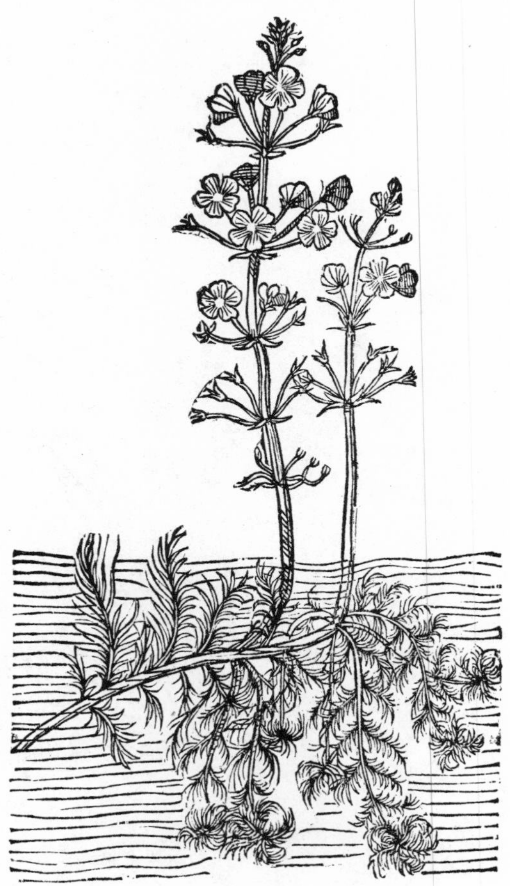

¶ *The Description.*

1 WAter Violet hath long and great iagged leaues, very finely cut or rent like Yarrow, but smaller: among which come vp small stalkes a cubit and a halfe high, bearing at the top small white floures like vnto stocke Gillofloures, with some yellownes in the middle. The roots are long and small like blacke threds, and at the end whereby they are fastened to the ground they are white, and shining like Chrystall.

‡ There is another varietie of this plant, which differs from it only in that the leaues are much smaller, as you may see them exprest in the figure. ‡

2 VVater

2 Water Milfoile, or water Yarrow hath long and large leaues deepely cut with many diuiſions like Fennell, but finelier iagged, ſwimming vpon the water. The root is ſingle, long, and round, which brings vp a right ſtraight and ſlender ſtalke, ſet in ſundry places with the like leaues, but ſmaller. The floures grow at the top of the ſtalke tuft faſhion, and like vnto the land Yarrow.

3 This water Milfoile differeth from all the kindes aforeſaid, hauing a root in the bottom of the water, made of many hairy ſtrings, which yeeldeth vp a naked ſlender ſtalke within the water, and the reſt of the ſtalke which floteth vpon the water diuideth it ſelfe into ſundry other branches and wings, which are bedaſht with fine ſmall iagged leaues like vnto Cammomill, or rather reſembling hairy taſſels or fringe, than leaues. From the boſomes whereof come forth ſmall and tender branches, euery branch bearing one floure like vnto water Crow-foot, white of colour, with a little yellow in the midſt: the whole plant reſembleth water Crow-foot in all things ſaue in the broad leaues.

† 4 There is another kinde of water violet very like the former, ſauing that his leaues are much longer, ſomewhat reſembling the leaues of Fennell, faſhioned like vnto wings, and the floures are ſomewhat ſmaller, yet white, with yellowneſſe in their middles, and ſhaped like thoſe of the laſt deſcribed. And the ſeed alſo growes like vnto that of the Water *Ranunculus*, laſt deſcribed.

5 There is alſo another kinde of water Milfoile, which hath leaues very like vnto water Violet, ſmaller, and not ſo many in number: the ſtalke is ſmall and tender, bearing yellow gaping floures faſhioned like a hood or the ſmall Snapdragon; which cauſed *Pena* to put vnto his name this additament *Galericulatum*, that is, hooded. The roots are ſmall and threddy, with ſome few knobs hanging thereat like the ſounds of fiſh.

2 *Millefolium aquaticum.*
Water Yarrow.

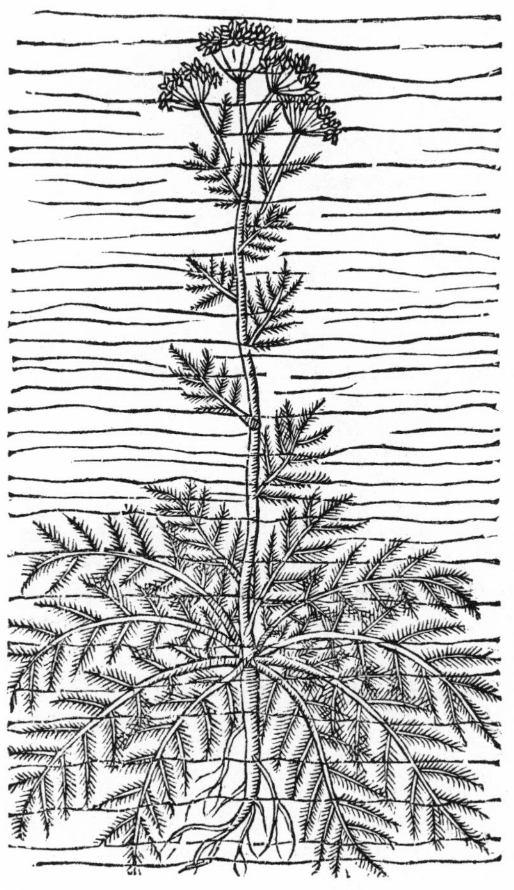

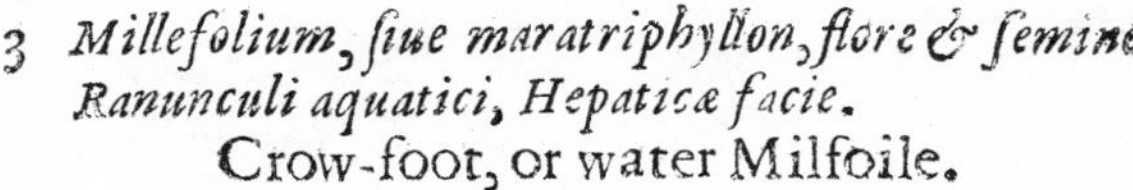

3 *Millefolium, ſiue maratriphyllon, flore & ſemine Ranunculi aquatici, Hepaticæ facie.*
Crow-foot, or water Milfoile.

‡ 6 To theſe may we adde a ſmall water Milfoile, ſet forth by *Cluſius*. It hath round greene ſtalkes ſet with many ioynts, whereout come at their lower ends many hairy fibres, whereby it taketh hold of the mud: the tops of theſe ſtems ſtand ſome handfull aboue the water, and at each ioynt ſtand fiue long finely winged leaues, very greene, and ſome inch long; which wax leſſe and leſſe,

lesse, as they stand higher or neerer the top of the stalke : and at each of these leaues about the top of the stem growes one small white floure consisting of six little leaues ioyned together, and not opening themselues : and these at length turne into little knobs, with foure little pointals standing out of them. *Clusius* calls this *Myriophyllon aquaticum minus*. ‡

‡ 4 *Millefolium tenuifolium*.
Fennell leaued water Milfoile.

‡ 5 *Millefolium palustre galericulatum*.
Hooded water Milfoile.

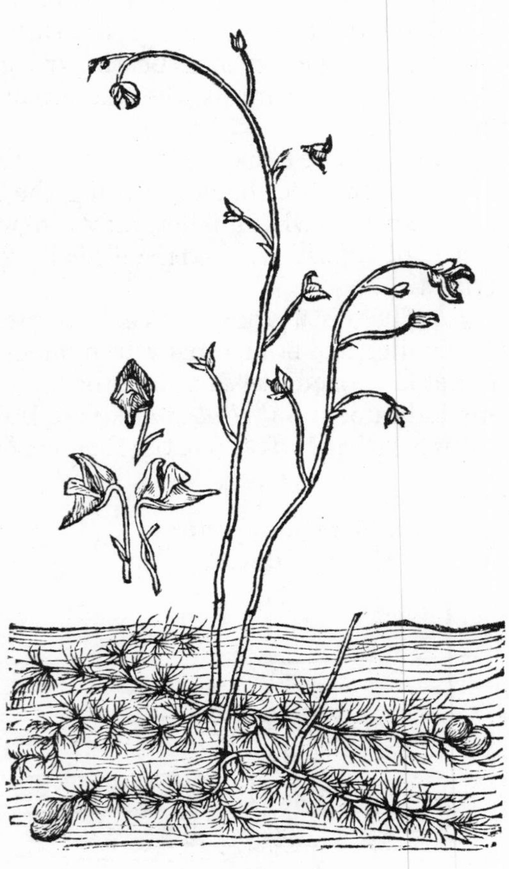

¶ *The Place.*

They be found in lakes and standing waters, or in waters that run slowly : I haue not found such plenty of it in any one place, as in the water ditches adioyning to Saint George his field neere London.

¶ *The Time.*

They floure for the most part in May and Iune.

¶ *The Names.*

The first is called in Dutch water **Violerian**, that is to say, *Viola aquatilis* : in English, Water Gillofloure, or water Violet : in French, *Gyroflees d' eaue* : *Matthiolus* makes this to be also *Myrophylli species*, or a kinde of Yarrow, although it doth not agree with the description thereof; for neither hath it one stalke onely, nor one single root, as *Myriophyllon* or Yarrow is described to haue; for the roots are full of strings, and it bringeth forth many stalkes.

The second is called in Greeke Μυριόφυλλον : in Latine, *Millefolium*, and *Myriophyllon*, and also *Supercilium Veneris* : in shops it is vnknowne. This Yarrow differeth from that of the land : the rest are sufficiently spoken of in their titles.

¶ *The Nature and Vertues.*

A Water Yarrow, as *Dioscorides* saith, is of a dry facultie; and by reason that it taketh away hot inflammations and swellings, it seemeth to be of a cold nature; for *Dioscorides* affirmeth, that water Yarrow is a remedie against inflammations in greene wounds, if with vineger it be applied greene or dry : and it is giuen inwardly with vineger and salt, to those that haue fallen from a high place.

B Water Gillofloure or water Violet is thought to be cold and dry, yet hath it no vse in physicke at all.

Chap.

CHAP. 301. *Of Ducks meate.*

Lens paluſtris.
Ducks meate.

¶ *The Deſcription.*

DVckes meate is as it were a certaine greene moſſe, with very little round leaues of the bigneſſe of Lentils: out of the midſt whereof on the nether ſide grow downe very fine threds like haires, which are to them in ſtead of roots: it hath neither ſtalke, floure, nor fruit.

¶ *The Place.*

It is found in pounds, lakes, city ditches, and in other ſtanding waters euery where.

¶ *The Time.*

The time of Ducks meate is knowne to all.

¶ *The Names.*

Duckes meate is called in Latine *Lens lacuſtris, Lens aquatilis*, and *Lens paluſtris*: of the Apothecaries it is named *Aquæ Lenticula*: in high-Dutch, **Meerlinſen**: in low-Dutch, **Waterlinſen**, and more vſually **Enden gruen**, that is to ſay, *Anatum herba*, Ducks herbe, becauſe Ducks doe feed thereon; whereupon alſo in Engliſh it is called Ducks meate: ſome terme it after the Greeke water Lentils; and of others it is named Graines. The Italians call it *Lent di palude*: in French, *Lentille d' eaue*: in Spaniſh, *Lenteias de agua.*

¶ *The Temperature.*

Galen ſheweth that it is cold and moiſt after a ſort in the ſecond degree.

¶ *The Vertues.*

Dioſcorides ſaith that it is a remedie againſt all manner of inflammations, Saint Anthonies fire, and hot Agues, if they be either applied alone, or elſe vſed with partched barley meale. It alſo knitteth ruptures in young children. A

Ducks meate mingled with fine wheaten floure and applied, preuaileth much againſt hot ſwellings, as Phlegmons, Eriſipelas, and the paines of the ioynts. B

The ſame doth helpe the fundament fallen downe in yong children. C

CHAP. 302. *Of Water Crow-foot.*

1 *Ranunculus aquatilis.*
Water Crow-foot.

¶ *The Description.*

1 Water Crow-foot hath slender branches trailing far abroad, whereupon grow leaues vnder the water most finely cut and iagged like those of Cammomill. Those aboue the water are somwhat round, indented about the edges, in forme not vnlike the smal tender leaues of the mallow, but lesser: among which do grow the floures, small, and white of colour, made of fine little leaues, with some yellownesse in the middle like the floures of the Straw-berry, and of a sweet smell: after which there come round rough and prickly knaps like those of the field Crow-foot. The roots be very small hairy strings.

‡ There is sometimes to be found a varietie of this, with the leaues lesse, and diuided into three parts after the manner of an Iuy leafe; and the floures are also much lesser, but white of colour, with a yellow bottome. I question whether this be not the *Ranunculus hederaceus Daleschampij, pag.* 1031. of the *hist. Lugd.* ‡

2 There is another plant growing in the water, of smal moment, yet not amisse to be remembred, called *Hederula aquatica*, or water Iuie: the which is very rare to finde; neuerthelesse I found it once in a ditch by Bermondsey house neere to London, and neuer elsewhere: it hath small threddy strings in stead of roots and stalkes, rising from the bottome of the water to the top; wherunto are fastned small leaues swimming or floting vpon the water, triangled or three cornered like to those of barren Iuie, or rather noble Liuerwort: barren of floures and seeds.

2 *Hederula aquatica.*
Water Iuie.

‡ 3 *Stellaria aquatica.*
Water Starwort.

3 There is likewise another herbe of small reckoning that floteth vpon the water, called *Stellaria aquatica*, or water Star-wort, which hath many small grassie stems like threds, comming from the bottome of the water vnto the vpper face of the same: whereupon do grow smal double floures of a greenish or herby colour. ‡ I take this *Stellaria* to be nothing else but a water Chickeweed, which growes almost in euery ditch, with two long narrow leaues at each ioynt, and halfe a dozen or more lying close together at the top of the water, in fashion of a starre: it may be seene in this shape in the end of Aprill and beginning of May: I haue not yet obserued either the floure or seed thereof. ‡

¶ The

¶ *The Place.*

Water Crow-foot groweth by ditches and ſhallow Springs, and in other moiſt and plaſhie places.

¶ *The Time.*

It floureth in Aprill and May, and ſometimes in Iune.

¶ *The Names.*

Water Crow-foot is called in Latine *Ranunculus aquatilis*, and *Polyanthemum aquatile*: in Engliſh, Water Crow-foot, and white water Crow-foot: moſt Apothecaries and Herbariſts do erroneouſly name it *Hepatica aquatica*, and *Hepatica alba*; and with greater error they mix it in medicines in ſtead of *Hepatica alba*, or graſſe of Parnaſſus. ‡ I know none that commit this great error here mentioned, neither haue I knowne either the one or the other euer vſed or appointed in medicine with vs in England, though *Dodonæus* (from whom our Author had this and moſt elſe) doe blame his countreymen for this miſtake and error. ‡

¶ *The Temperature and Vertues.*

Water Crow-foot is hot, and like to common Crow-foot.

Chap. 303. *Of Dragons.*

1 *Dracontium maius.*
Great Dragons.

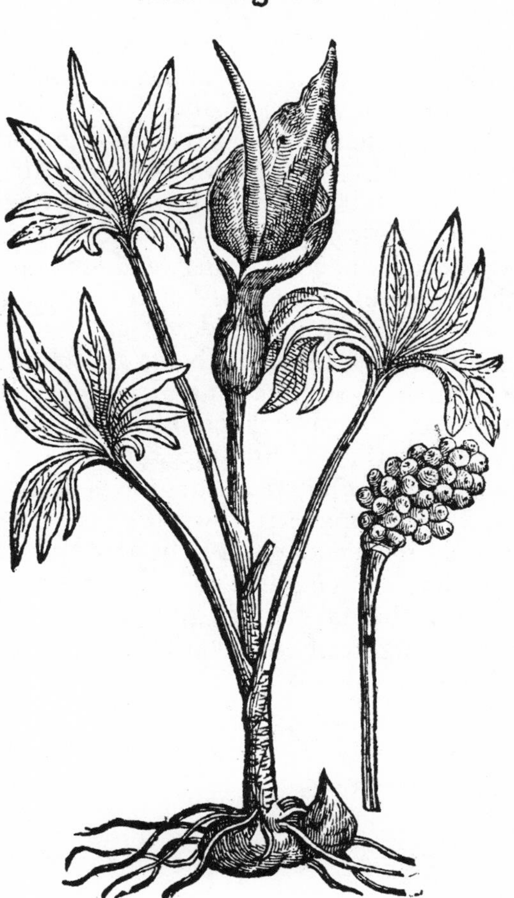

† 2 *Dracontium minus.*
Small Dragons.

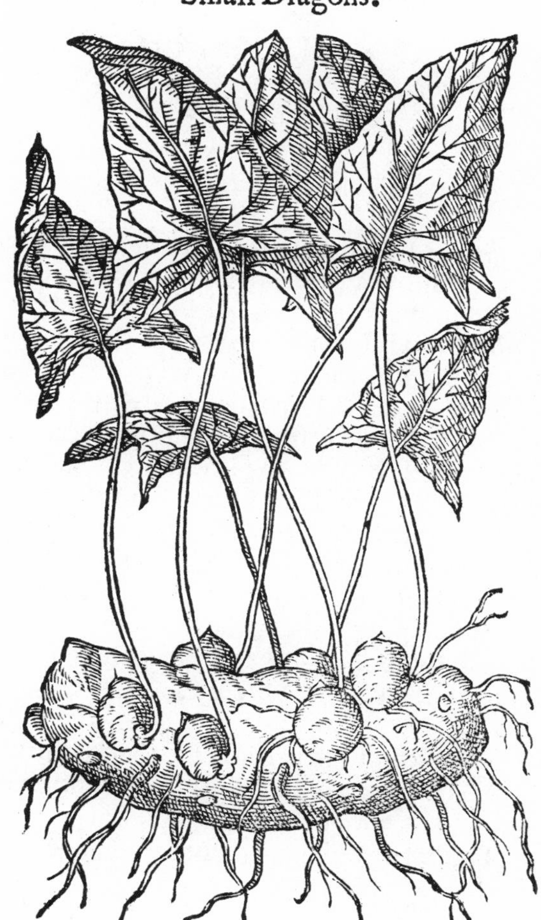

¶ *The Deſcription.*

1 The great Dragon riſeth vp with a ſtraight ſtalke a cubit and a halfe high or higher, thicke, round, ſmooth, ſprinkled with ſpots of diuers colours, like thoſe of the adder or ſnake: the leaues are great and wide, conſiſting of ſeuen or more ioyned together in order; euery one of which is long and narrow, much like to the leaues of Docke, ſmooth and ſlipperie: out of the top of the ſtalke groweth a long hoſe or huske greater than that of the Cuckow pintle, of a greeniſh colour without, and within crimſon, with his peſtell which is blackiſh, long, thicke, and pointed like a horne; the skin or filme whereof when the ſeed waxeth big, being ſtretched

1 *Dracunculus aquaticus.*
Water Dragons.

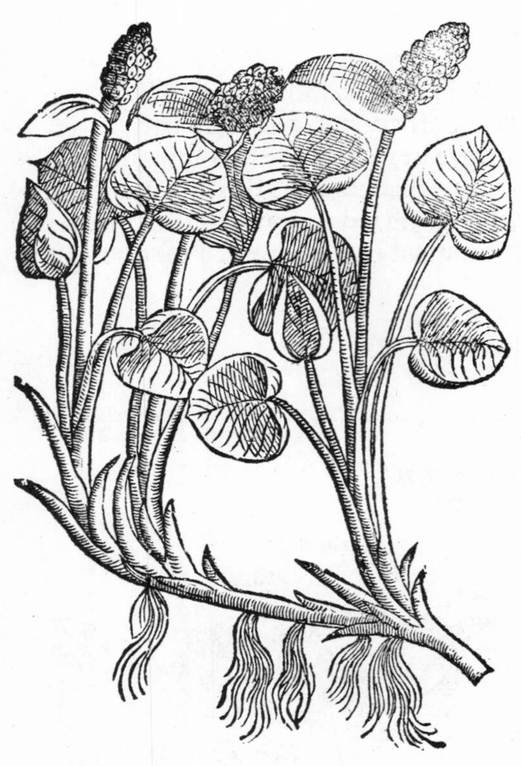

stretched or broken asunder, there appeareth the fruit, like to a bunch or cluster of grapes: the berries whereof at the first be greene, afterwards red and full of iuyce; in which is contained seed that is somewhat hard: The root continueth fresh, thicke, like to a knob, white, couered with a thin pilling, oftentimes of the bignesse of a meane apple, full of white little threds appendant thereunto.

2 The lesser Dragon is like *Aron* or wake Robin, in leaues, hose, or huske, pestell, and berries, yet are not the leaues sprinkled with blacke but with whitish spots, which perish not so soone as those of wake-Robin, but endure together with the berries euen vntil winter: these berries also be not of a deepe red, but of a colour enclining to Saffron. The root is not vnlike to the Cuckow-pint, hauing the forme of a bulbe, full of strings, with diuers rude shapes of new plants, whereby it greatly encreaseth.

‡ The figure which our Authour heere gaue by the title of *Dracuntium minus*, was no other than of *Aron*, which is described in the first place of the next chapter: neither is the description of any other plant, than of that sort thereof which hath leaues spotted either with white or blacke spots, though our Author say onely with white. I haue giuen you *Clusius* his figure of *Arum Byzantinum*, in stead of that which our Author gaue. ‡

3 The root of water Dragon is not round like a bulbe, but very long, creeping, and ioynted, and of meane bignesse; out of the ioynts whereof arise the stalkes of the leaues, which are round, smooth, and spongie within, and there grow downewards certaine white and slender strings. The fruit springs forth at the top vpon a short stalke, together with one of the leaues, being at the beginning couered with little white threds, which are in stead of the floures: after that it groweth into a bunch or cluster, at the first greene, and when it is ripe, red, lesser than that of Cuckow-pint, but not lesse biting: the leaues are broad, greenish, glib, and smooth, in fashion like those of Iuy, yet lesser than those of Cuckow-pint; and that thing whereunto the clustered fruit growes is also lesser, and in that part which is towards the fruit (that is to say the vpper part) is white.

4 The great Dragon of *Matthiolus* his description is a stranger not onely in England, but elsewhere for any thing that we can learne: my selfe haue diligently enquired of most strangers skilfull in plants, that haue resorted vnto me for conference sake, but no man can giue me any certaintie thereof; and therefore I thinke it amisse to giue you his figure or any description, for that I take it for a feigned picture.

¶ *The Place.*

The greater and the lesser Dragons are planted in gardens. The water Dragons grow in watery and marish places, for the most part in fenny and standing waters.

¶ *The Time.*

The berries of these plants are ripe in Autumne.

¶ *The Names.*

The Dragon is called in Greeke Δρακόντιον: in Latine, *Dracunculus*. The greater is named *Serpentaria maior*: of some, *Bisaria*, and *Colubrina*: *Cordus* calleth it *Dracunculus Polyphyllos*, and *Luph Crispum*: in high-Dutch, **Schlangenkraut**: in low-Dutch, **Speerwortele**: in French, *Serpentaire*: in Italian, *Dragontea*: in Spanish, *Taragontia*: in English, Dragons, and Dragon-wort. *Apuleius* calleth Dragon *Dracontea*, and setteth downe many strange names thereof, which whether they agree with the greater or the lesser, or both of them, he doth not expound; as *Pythonion*, *Anchomanes*, *Sauchromaton*, *Therion*, *Schœnos*, *Dorcadion*, *Typhonion*, *Theriophonon*, and *Eminion*. *Athenæus* sheweth, that Dragon is called *Aronia*, because it is like to *Aron*.

¶ *The Temperature.*

Dragon, as *Galen* saith, hath a certaine likenesse with *Aron* or wake-Robin, both in leaues, and also in root, yet more biting and more bitter than it, and therefore hotter, and of thinner parts: it is

also

alſo ſomething binding, which by reaſon that it is adjoined with the two former qualities, that is to ſay, biting and bitter, is is made in like manner a ſingular medicine of very great efficacy.

¶ *The Vertues.*

The root of Dragons doth clenſe and ſcoure all the entrailes, making thinne, eſpecially thicke and tough humours; and it is a ſingular remedy for vlcers that are hard to be cured, named in Greeke κακοήθη. A

It ſcoureth and clenſeth mightely, aſwell ſuch things as haue need of ſcouring, as alſo white and blacke morphew, being tempered with vineger. B

The leaues alſo by reaſon that they are of like qualitie are good for vlcers and greene wounds: and the leſſe dry they are, the fitter they be to heale, for the dryer ones are of a more ſharpe or biting quality than is conuenient for wounds.

The fruit is of greater operation than either the leaues or the root: and therefore it is thought to be of force to conſume and take away cankers and proud fleſh growing in the noſtrils, called in Greeke *Polypus*: alſo the juice doth clenſe away webs and ſpots in the eies. C

Furthermore, *Dioſcorides* writeth, that it is reported that they who haue rubbed the leaues or root vpon their hands, are not bitten of the viper. D

Pliny ſaith, that ſerpents will not come neere vnto him that beareth Dragons about him, and theſe things are read concerning both the Dragons, in the two chapters of *Dioſcorides*. E

Galen alſo hath made mention of Dragon in his booke of the faculties of nouriſhments, where he ſaith, that the root of Dragon being twice or thrice ſod, to the end it may loſe all his acrimony or ſharpeneſſe, is ſometimes giuen as Aron, or wake-Robin is, when it is needfull to expell the more forceable thicke and clammy humours that are troubleſome to the cheſt and lungs. F

And *Dioſcorides* writeth, that the root of the leſſer Dragon being both ſodde and roſt with honie, or taken of it ſelfe in meate, cauſeth the humours which ſticke faſt in the cheſt to be eaſily voided. G

The juice of the garden Dragons, as ſaith *Dioſcorides*, being dropped into the eies, doth clenſe them, and greatly amend the dimneſſe of the ſight. H

The diſtilled water hath vertue againſt the peſtilence or any peſtilentiall feuer or poiſon, being drunke bloud-warme with the beſt treacle or mithridate. I

The ſmell of the floures is hurtfull to women newly conceiued with child. K

CHAP. 304. *Of Cockow pint, or wake-Robin.*

¶ *The Deſcription.*

1 A*rum* or Cockow pint hath great, large, ſmooth, ſhining, ſharpe pointed leaues, beſpotted here and there with blackiſh ſpots, mixed with ſome blewneſſe: among which riſeth vp a ſtalke nine inches long, beſpeckled in many places with certaine purple ſpots. It beareth alſo a certaine long hoſe or hood, in proportion like the eare of an hare: in the middle of which hood commeth forth a peſtle or clapper of a darke murrie or pale purple colour: which being paſt, there ſucceedeth in place thereof a bunch or cluſter of berries in manner of a bunch of grapes, greene at the firſt, but after they be ripe of a yellowiſh red like corall, and full of pith, with ſome threddy additaments annexed thereto.

2 There is in Ægypt a kinde of *Arum* which alſo is to be ſeene in Africa, and in certaine places of Luſitania, about riuers and floods, which differeth from that which groweth in England and other parts of Europe. This plant is large and great, and the leaues thereof are greater than thoſe of the water Lillie: the root is thicke and tuberous, and toward the lower end thicker and broader, and may be eaten. It is reported to be without floure and ſeed, but the increaſe that it hath is by the fibres which runne and ſpread from the roots. ‡ This plant hath alſo peſtells and cluſters, of berryes as the common Aron, but ſomewhat different, the leaues are not cut into the ſtalke, but joined before the ſetting thereto: the root alſo is very large. Thoſe that deſire to ſee more of this plant, and the queſtion which ſome haue mooued, whither this be the *Colocaſia*, or *Faba Ægyptia* of the Antients? let them haue recourſe to the firſt chapter of *Fabius Columna* his *Minus cognitarum ſtripium pars altera*, and there they ſhall finde ſatisfaction. ‡

1 *Arum vulgare.*
Cockow pint.

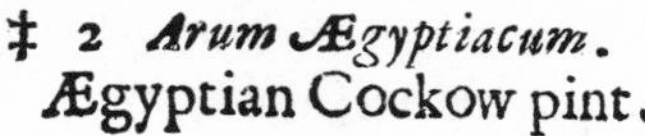
‡ 2 *Arum Ægyptiacum.*
Ægyptian Cockow pint.

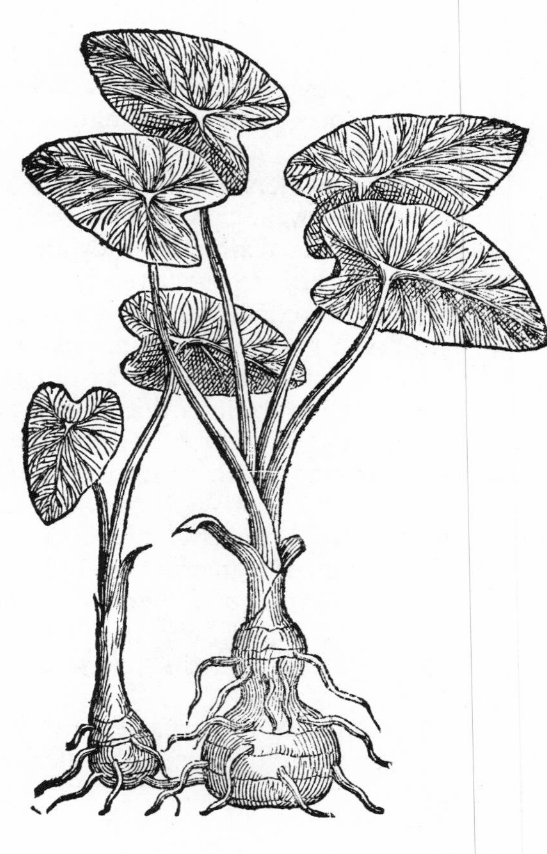

¶ *The Place.*

Cockow pint groweth in woods neere vnto ditches vnder hedges, euery where in ſhadowie places.

¶ *The Time.*

The leaues appeare preſently after winter: the peſtell ſheweth it ſelfe out of his huske or ſheath in Iune, whileſt the leaues are in withering: and when they are gone, the bunch or cluſter of berries becommeth ripe, which is in Iuly and Auguſt.

¶ *The Names.*

There groweth in Ægypt a kinde of Aron or Cuckow pint which is found alſo in Africa, and likewiſe in certaine places of Portingale neere vnto riuers and ſtreames, that differeth from thoſe of our countries growing, which the people of Caſtile call *Manta de nueſtra ſenora*: moſt would haue it to be called *Colocaſia*; but *Dioſcorides* ſaith that *Colocaſia* is the root of *Faba Ægyptia*, or the Beane of Egypt. ‡ *Fabius Columna* (in the place formerly alledged) prooues this not to be the true *Colocaſia*, and yet *Proſper Alpinus* ſince in his ſecond booke *de plantis exoticis, cap.* 17. and 18. labours to proue the contrary: let the curious haue recourſe to theſe, for it is too tedious for me in this place to inſiſt vpon it, being ſo large a point of controuerſie, which hath ſo much troubled all the late writers. ‡

The common Cuckow pint is called in Latine, *Arum*: in Greeke, ἄρον: in ſhops, *Iarus*, and *Barba-Aron*: of others, *Pes vituli*: of the Syrians, *Lupha*: of the men of Cyprus, *Colocaſia*, as we finde among the baſtard names. *Pliny* in his 24. booke, 16. chapter, doth witneſſe, that there is great difference betweene *Aron* and *Dracontium*, although there hath been ſome controuerſie about the ſame among the old writers, affirming them to be all one: in high Dutch it is called, **Pfaffen pint**: in Italian, *Gigora*: in Spaniſh, *Yaro*: in low Dutch, **Calfsuoet**: in French, *Pied d'veau*: in Engliſh, Cuckow pint, and Cuckow pintle, wake-Robin, Prieſts pintle, Aron; Calfes foot, and Rampe; and of ſome Stratchwoort.

¶ *The Temperature.*

The faculties of Cuckow pint doe differ according to the varietie of countries: for the root hereof, as *Galen* in his booke of the faculties of nouriſhments doth affirme, is ſharper and more biting in ſome countries than in others, almoſt as much as Dragons, contrariwiſe in Cyren a city in Africke, it is generally in all places hot and dry, at the leaſt in the firſt degree.

¶ *The*

¶ *The Vertues.*

If any man would haue thicke and tough humours which are gathered in the cheſt and lungs to be clenſed and voided out by coughing, then that Cuckowpint is beſt that biteth moſt.

It is eaten being ſodden in two or three waters, and freſh put to, whereby it may loſe his acrimonie; and being ſo eaten, they cut thicke humors meanely, but Dragons is better for the ſame purpoſe.

Dioſcorides ſheweth, that the leaues alſo are preſerued to be eaten, and that they muſt be eaten after they be dried and boyled; and writeth alſo, that the root hath a peculiar vertue againſt the gout, being laid on ſtamped with Cowes dung.

Beares after they haue lien in their dens forty dayes without any manner of ſuſtenance, but what they get with licking and ſucking their owne feet, do as ſoone as they come forth eate the herbe Cuckowpint, through the windie nature whereof the hungry gut is opened and made fit againe to receiue ſuſtenance: for by abſtaining from food ſo long a time, the gut is ſhrunke or drawne ſo cloſe together, that in a manner it is quite ſhut vp, as *Ariſtotle, Ælianus, Plutarch, Pliny*, and others do write.

The moſt pure and white ſtarch is made of the roots of Cuckowpint; but moſt hurtfull to the hands of the Laundreſſe that hath the handling of it, for it choppeth, bliſtereth, and maketh the hands rough and rugged, and withall ſmarting.

CHAP. 305. *Of Friers Cowle, or hooded Cuckowpint.*

1 *Ariſarum latifolium.*
Broad leaued Friers Cowle.

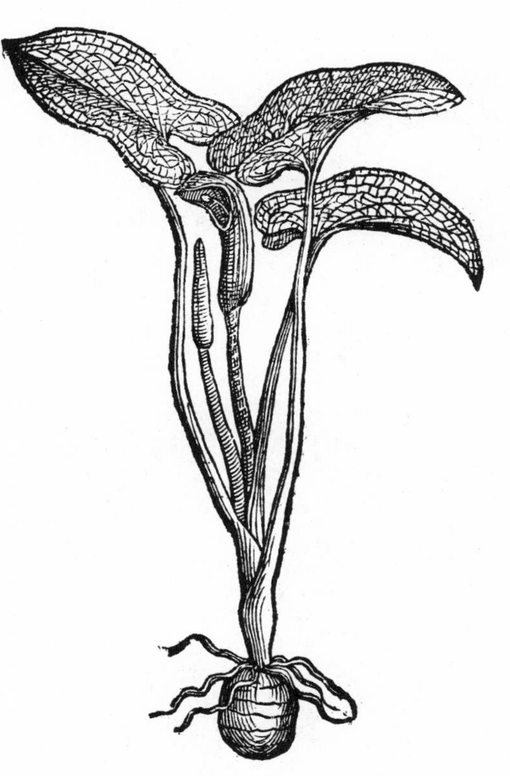

2 *Ariſarum anguſtifolium.*
Narrow leaued Friers Cowle.

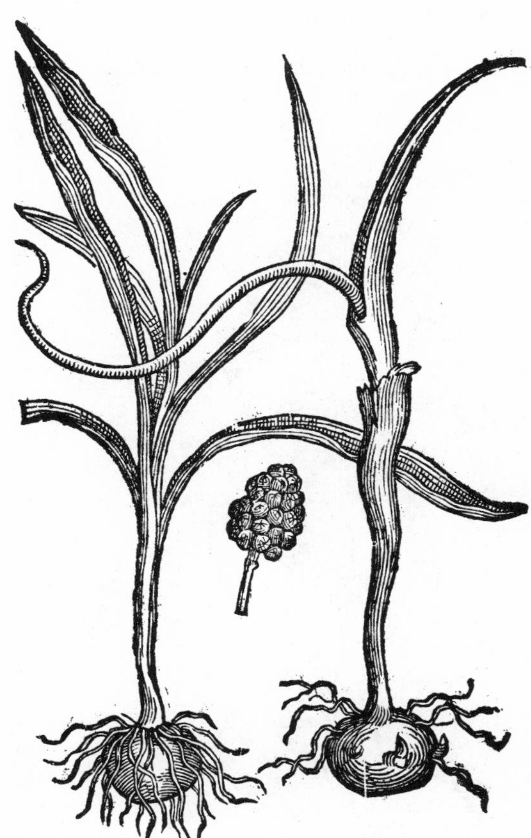

¶ *The Deſcription.*

1 Broad leaued Friers hood hath a leafe like Iuy, broad and ſharpe pointed, but far leſſe, approching neere to the forme of thoſe of Cuckowpint: the ſtalke thereof is ſmall and ſlender: the huske or hoſe is little; the peſtel ſmall, and of a blacke purpliſh colour; the cluſter when it is ripe is red; the kernels ſmall; the root white, hauing the forme of Aron or Cuckowpint, but leſſer, whereof doubtleſſe it is a kinde.

2 The second Friers hood hath many leaues, long and narrow, smooth and glittering: The huske or hose is narrow and long; the pestell that commeth forth of it is slender, in forme like a great earth worme, of a blackish purple colour, as hath also the inside of the hose, vpon which, hard to the ground, and sometimes a little within the ground, groweth a certaine bunch or cluster of berries, greene at the first, and afterwards red: the root is round and white like the others.

¶ *The Place.*

These plants are strangers in England, but common in Italy, and especially in Tuscane about Rome, and in Dalmatia, as *Aloisius Anguillara* witnesseth: notwithstanding I haue them in my Garden.

¶ *The Time.*

The floures and fruit of these come to perfection with those of Cuckowpint and Dragons.

¶ *The Names.*

Friers hood is called of *Dioscorides*, Ἀρίσαρον: in Latine, *Arisarum*: but *Pliny* calleth it Ἄρις, or *Aris*; for in his twenty fourth booke, *cap.* 16. he saith, That *Aris* which groweth in Egypt is like Aron or Cuckowpint: it may be called in English after the Latine name *Arisarum*; but in my opinion it may be more fitly called Friers hood, or Friers cowle, to which the floures seeme to be like; whereupon the Spaniards name it *Frailillos*, as *Daleschampius* noteth.

¶ *The Temperature.*

Friers-Cowle is like in power and facultie to the Cuckow-pint, yet is it more biting, as *Galen* saith.

¶ *The Vertues.*

A There is no great vse of these plants in physicke; but it is reported that they stay running or eating sores or vlcers: and likewise that there is made of the roots certaine compositions called in Greeke *Collyria*, good against fistula's: and being put into the secret part of any liuing thing, it rotteth the same, as *Dioscorides* writeth.

† That which was formerly figured and described in the third place, vnder the title of *Arisarum latifolium Matthioli*, was the same with that described by the name of *Dracontium minus*, in the precedent chapter, and therefore here omitted.

Chap. 306. *Of Astrabacca.*

1 *Asarum.*
Asarabacca.

2 *Asarina Matthioli.*
Italian Asarabacca.

¶ The

¶ *The Description.*

1 THe leaues of Asarabacca are smooth, of a deepe greene colour, rounder, broader, and tenderer than those of Iuy, and not cornered at all, not vnlike to those of Sow-bread: the floures lie close to the roots, hid vnder the leaues, standing vpon slender foot-stalkes, of an ill fauoured purple colour, like to the floures and husks of Henbane, but lesse, wherein are contained small seeds, cornered, and somewhat rough: the roots are many, small and slender, growing aslope vnder the vpper crust of the earth, one folded within another, of an vnpleasant taste, but of a most sweet and pleasing smell, hauing withall a kinde of biting qualitie.

2 This strange kinde of Asarabacca, which *Matthiolus* hath set forth creeping on the ground, in manner of our common Astrabacca, hath leaues somwhat rounder and rougher, sleightly indented about the edges, and set vpon long slender foot-stalkes: the floures grow hard vnto the ground like vnto those of Cammomill, but much lesser, of a mealy or dusty colour, and not without smel. The roots are long and slender, creeping vnder the vpper crust of the earth, of a sharpe taste, and bitter withall. ‡ This *Asarina* of *Matthiolus*, *Clusius* (whose opinion I here follow) hath iudged to be the *Tussilago Alpina* 2. of his description; wherefore I giue you his figure in stead of that of our Author, which had the floures exprest, which this wants. ‡

¶ *The Place.*

It delighteth to grow in shadowie places, and is very common in most gardens.

¶ *The Time.*

The herbe is alwaies greene; yet doth it in the Spring bring forth new leaues and floures.

¶ *The Names.*

It is called in Greeke Ἄσαρον, *Asarum*: in Latine, *Nardus rustica*: and of diuers, *Perpensa*: *Perpensa* is also *Baccharis* in *Pliny*, *lib.* 21. *cap.* 21. *Macer* saith, That *Asarum* is called *Vulgago*, in these words:

Est Asaron Græcè, Vulgago dicta Latinè.

This herbe, *Asaron* do the Grecians name;
Whereas the Latines *Vulgago* clepe the same.

It is found also amongst the bastard names, that it was called of the great learned Philosophers Αἷμα ἄρεως: that is, *Martis sanguis*, or the bloud of *Mars*: and of the French men *Baccar*; and thereupon it seemeth that the word *Asarabacca* came, which the Apothecaries vse, and likewise the common people: but there is another *Baccharis* differing from *Asarum*, yet notwithstanding *Crateuas* doth also call *Baccharis*, *Asarum*.

This confusion of both the names hath been the cause, that most could not sufficiently expound themselues concerning *Asarum* and *Baccharis*; and that many things haue beene written amisse in many copies of *Dioscorides*, in the chapter of *Asarum*: for when it is set downe in the Greek copies a sweet smelling garden herbe, it belongeth not to the description of this *Asarum*, but to that of *Baccharis*: for *Asarum* (as *Pliny* saith) is so called, because it is not put into garlands: and so by that meanes it came to passe, that oftentimes the descriptions of the old Writers were found corrupted and confused: which thing, as it is in this place manifest, so oftentimes it cannot so easily be marked in other places. Furthermore, *Asarum* is called in French *Cabaret*: in high-Dutch, **Haselwurtz**: in low-Dutch, **Mans ooren**: in English, Asarabacca, Fole-foot, and Hazel-wort.

¶ *The Temperature.*

The leaues of Asarabacca are hot and dry, with a purging qualitie adioyned thereunto, yet not without a certaine kinde of astriction or binding. The roots are also hot and dry, yet more than the leaues; they are of thin and subtill parts: they procure vrine, bring downe the desired sicknes, and are like in facultie, as *Galen* saith, to the roots of *Acorus*, but yet more forceable; and the roots of *Acorus* are also of a thinne essence, heating, attenuating, drying, and prouoking vrine, as he affirmeth: which things are happily performed by taking the roots of Asarabacca, either by themselues, or mixed with other things.

¶ *The Vertues.*

A The leaues draw forth by vomit, thicke phlegmaticke and cholericke humours, and withall moue the belly; and in this they are more forceable and of greater effect than the roots themselues.

B They are thought to keepe in hard swelling cankers that they encrease not, or come to exulceration, or creeping any farther, if they be outwardly applied vpon the same.

C The roots are good against the stoppings of the liuer, gall, and spleene, against wens and hard swellings, and agues of long continuance: but being taken in the greater quantitie, they purge flegme and choler not much lesse than the leaues (though *Galen* say no) by vomit especially, and also by siege.

D One dram of the pouder of the roots giuen to drinke in ale or wine, groſſely beaten, prouoketh vomit for the purpoſes aforeſaid; but being beaten into fine pouder, and ſo giuen, it purgeth very little by vomit, but worketh moſt by procuring much vrine; therefore the groſſer the pouder is, ſo much the better.

E But if the roots be infuſed or boyled, then muſt two, three, or foure drams be put to the infuſion; and of the leaues eight or nine be ſufficient: the iuyce of which ſtamped with ſome liquid thing, is to be giuen. The roots may be ſteeped in wine, but more effectually in whay or honied water, as *Meſues* teacheth.

F The ſame is good for them that are tormented with the Sciatica or gout in the huckle bones, for thoſe that haue the dropſie, and for ſuch alſo as are vexed with a quartaine ague, who are cured and made whole by vomiting.

Chap. 307. *Of Sea Binde-weed.*

1 *Soldanella marina.*
Sea Binde-weed.

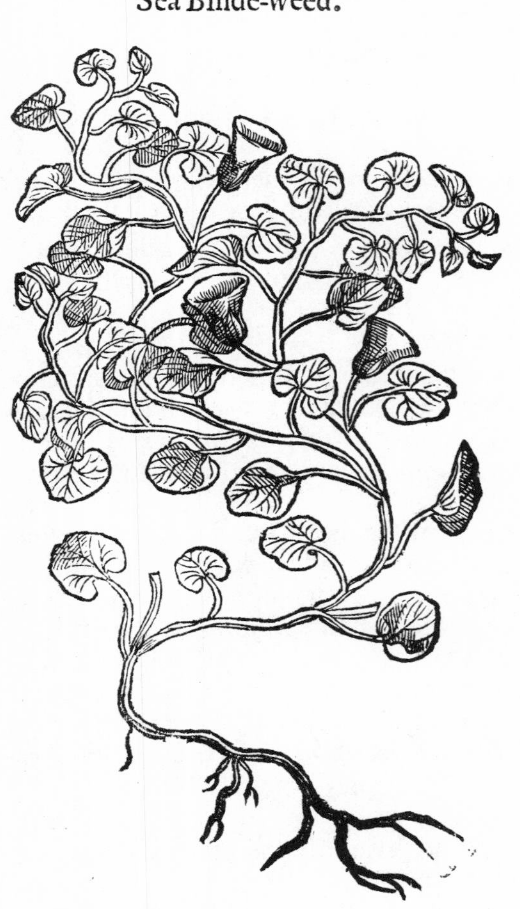

‡ 2 *Soldanella Alpina maior.*
Mountaine Binde-weed.

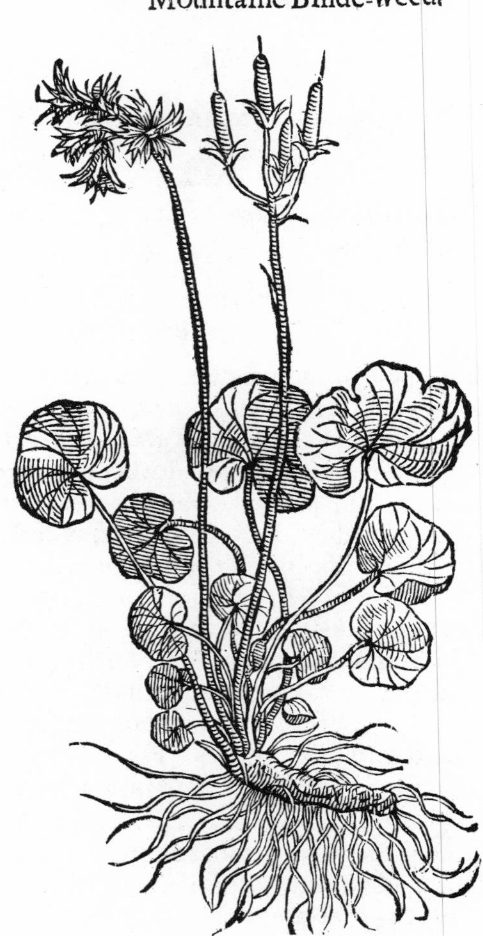

¶ *The Deſcription.*

1 *Soldanella* or Sea Binde-weed hath many ſmall branches, ſomwhat red, trailing vpon the ground, beſet with ſmall and round leaues, not much vnlike Aſarabacca, or the leaues of Ariſtolochia, but ſmaller; betwixt which leaues and the ſtalkes come forth floures formed like a bell, of a bright red incarnate colour, in euery reſpect anſwering the ſmall Binde-weed, whereof it is a kinde, albeit I haue here placed the ſame, for the reaſons rendred in my Proeme. The ſeed is blacke, and groweth in round huskes: the root is long and ſmall, thruſting it ſelfe far abroad, and into the earth like the other Binde-weeds.

2 *Soldanella* or mountaine Binde-weed hath many round leaues ſpred vpon the ground, not much vnlike the former, but rounder, and more full of veines, greener, of a bitter taſte like ſea Binde-weed: among which commeth forth a ſmall and tender ſtalke a handfull high, bearing at the top little floures like the ſmall Bell-floure, of a sky colour. The root is ſmall and threddy.

‡ 3 There

‡ 3 *Soldanella Alpina minor.*
Small Mountaine Bindweed.

‡ 3 There is of this kinde another hauing all the parts ſmaller, and the leaues redder and rounder: the floures alſo blew, and compoſed of one leafe diuided into fiue parts, and ſucceeded by a longiſh cod, round and ſharp pointed. ‡

¶ *The Place.*

The firſt grows plentifully by the Sea ſhore in moſt places of England, eſpecially neere to Lee in Eſſex, at Merſey in the ſame countie, in moſt places of the Iſle of Thanet, and Shepey, and in many places along the Northern coaſt.

The ſecond groweth vpon the mountains of Germanie, and the Alpes; it groweth vpon the mountains of VVales, not far from Cowmers Meare in North-Wales.

¶ *The Time.*

Theſe herbes do floure in Iune, and are gathered in Auguſt to be kept for medicine.

¶ *The Names.*

The firſt called *Soldanella* is of the Apothecaries and the Antients called *Marina Braſſica*, that is to ſay, Sea Colewoort: but what reaſon hath moued them ſo to doe I cannot conceiue, vnleſſe it be penurie and ſcarſitie of names, and becauſe they know not otherwiſe how to terme it: of this I am ſure, that this plant and *Braßica* are no more like than things which are moſt vnlike; for *Braſſica Marina* is the Sea Colewoort, which doth much reſemble the garden Cabbage or Cole, both in ſhape and in nature, as I haue in his due place expreſſed. A great fault and ouerſight therefore it hath been of the old writers and their ſucceſſors which haue continued the cuſtome of this error, not taking the paines to diſtinguiſh a Bindeweed from a Cole-woort. But to auoid controuerſies, the truth is, as I haue before ſhewed, that this *Soldanella* is a Bindeweed, and cannot be eſteemed for a *Braſſica*, that is a Colewoort. The later Herbariſts call it *Soldana*, and *Soldanella*: in Dutch, **Zeewind**, that is to ſay, *Conuoluulus Marinus*: of *Dioſcorides* κράμβη θαλασσία, (i) *Braßica Marina*: in Engliſh, Sea VVithwinde, Sea Bindweed, Sea-bels, Sea-coale, of ſome, Sea Fole-foot, and Scottiſh Scuruie-graſſe.

The ſecond is called *Soldanella montana*: in Engliſh, Mountaine Bindweed.

¶ *The Nature.*

Sea Bindeweed is hot and drie in the ſecond degree: the ſecond is bitter and very aſtringent.

¶ *The Vertues.*

Soldanella purgeth downe mightily all kinde of watriſh humours, and openeth the ſtoppings of the liuer, and is giuen with great profit againſt the dropſie: but it muſt be boiled with the broth of ſome fat meat or fleſh, and the broth drunke, or elſe the herbe taken in pouder worketh the like effect. A

Soldanella hurteth the ſtomack, and troubleth the weake and delicate bodies which doe receiue it in pouder, wherefore aduice muſt be taken to mix the ſaid pouder with Anniſe ſeeds, Cinnamon, ginger, and ſugar, which ſpices do correct his malignitie. B

Practitioners about Auſpurge and Rauiſpurge (cities of Germanie) do greatly boaſt that they haue done wonders with this herbe *Soldanella montana*; ſaying, that the leaues taken and emplaiſtred vpon the nauell and ſomewhat lower, draw forth water from their bellies that are hydroptike, that is, troubled with water or the dropſie: this effect it worketh in other parts without heating. C

It doth alſo wonderfully bring fleſh in wounds, and healeth them. D

Dioſcorides witneſſeth, that the whole herbe is an enemie to the ſtomacke, biting and extremely purging (both ſodden, and taken with meat) and bringeth troubleſome gripings thereunto, and doth oftentimes more hurt than good. E

‡ My friend M[r]. *Goodyer* hath told me, that in Hampſhire at Chicheſter and thereabout they make vſe of this for Scuruie-graſſe, and that not without great errour, as any that know the qualities may eaſily perceiue. F

Chap. 308. *Of the Grasse of Parnassus.*

† 1 *Gramen Parnassi.*
Grasse of Parnassus.

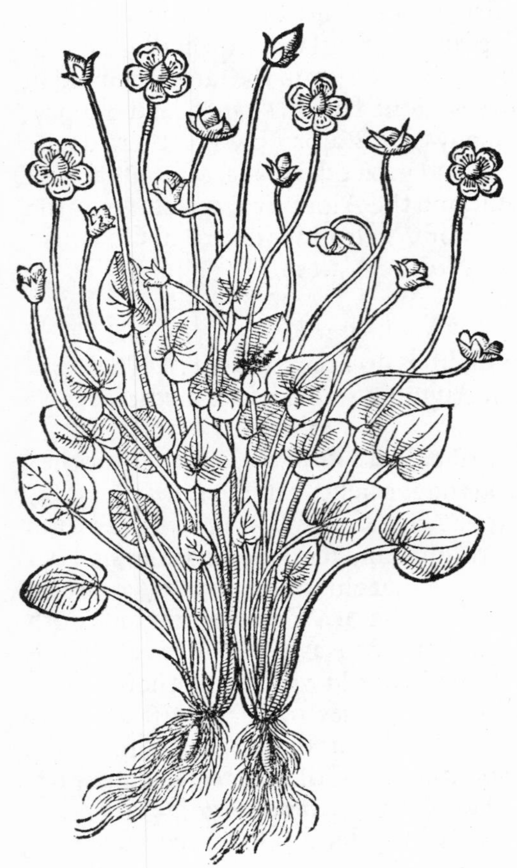

‡ 2 *Gramen Parnassi flore duplici.*
Grasse of Parnassus with double floures.

¶ *The Description.*

1 The Grasse of Parnassus hath small round leaues, very much differing from any kind of Grasse, much resembling the leaues of Iuie, or Asarabacca, but smaller, and not of so darke a colour: among these leaues spring vp small stalkes a foot high, bearing little white floures consisting of fiue round pointed leaues; which beeing falne and past, there come vp round knops or heads, wherein is contained a reddish seed. The root is somewhat thicke, with many strings annexed thereto.

2 The second kinde of *Gramen Parnassi* doth answer the former in each respect, sauing that the leaues are somewhat larger, and the floures double, otherwise verie like.

¶ *The Place.*

The first groweth very plentifully in Lansdall and Crauen, in the North parts of England; at Doncaster, and in Thornton fields in the same countrie: moreouer in the Moore neere to Linton, by Cambridge, at Hesset also in Suffolke, at a place named Drinkstone, in the medow called Butchers mead. ‡ Mr. *Goodyer* found it in the boggy ground below the red well of Wellingborough in Northampton shire: and Mr. *William Broad* obserued it to grow plentifully in the Castle fields of Berwicke vpon Tweed. ‡

The second is a stranger in England.

¶ *The Time.*

These herbes do floure in the end of Iuly, and their seed is ripe in the end of August.

¶ *The Names.*

Valerius Cordus hath among many that haue written of these herbes said something of them to good purpose, calling them by the name of *Hepatica alba* (whereof without controuersie they are kindes) in English, white Liuerwoort: although there is another plant called *Hepatica alba*, which for

for diſtinction ſake I haue thought good to Engliſh, Noble white Liuerwoort.

The ſecond may be called Noble white Liuerwoort with the double floure.

¶ *The Nature.*

The ſeed of Parnaſſus Graſſe, or white Liuer-woort, is drie, and of ſubtill parts.

¶ *The Vertues.*

The decoction of the leaues of Parnaſſus Graſſe drunken, doth drie and ſtrengthen the feeble and moiſt ſtomacke, ſtoppeth the bellie, and taketh away the deſire to vomite. A

The ſame boiled in wine or water, and drunken, eſpecially the ſeed thereof, prouoketh vrine, breaketh the ſtone, and driueth it forth. B

† The figure that was formerly in the firſt place of this Chapter was of *Vnifolium*, deſcribed before, *cap.* 90. *pag.* 409. that which was in the ſecond place belonged to the firſt deſcription.

CHAP. 309. *Of white Saxifrage, or Golden Saxifrage.*

¶ *The Deſcription.*

1 THe white Saxifrage hath round leaues ſpred vpon the ground, and ſomewhat iagged about the edges, not much vnlike the leaues of ground Iuie, but ſofter and ſmaller, and of a more faint yellowiſh greene: among which riſeth vp a round hairie ſtalke a cubit high, bearing at the top ſmall white floures, almoſt like Stockgillo floures: the root is compact of a number of blacke ſtrings, whereunto are faſtened very many ſmall reddiſh graines or round roots as bigge as pepper cornes, which are vſed in medicine, and are called *Semen Saxifragæ albæ*; that is, the ſeede of white Saxifrage, or Stone-breake, although (beſide theſe foreſaid round knobbes) it hath alſo ſmall ſeed contained in little huskes, following his floure as other herbes haue.

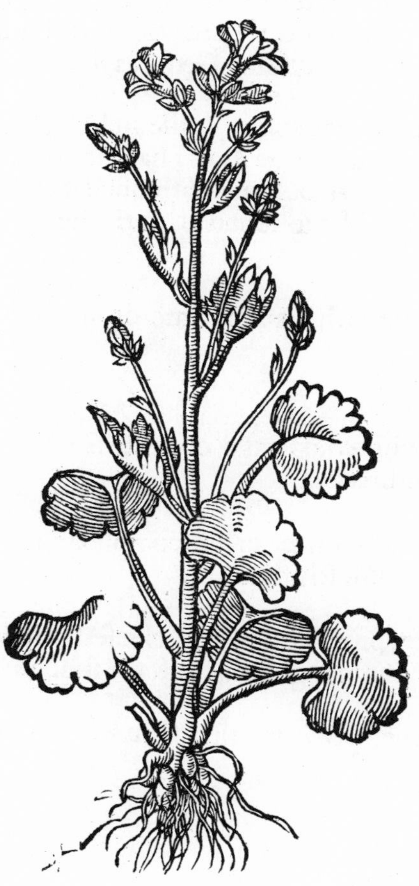

1 *Saxifraga alba.*
White Saxifrage.

2 *Saxifraga aurea.*
Golden Saxifrage.

Golden

‡ 3 *Saxifraga alba petræa.*
White Rocke Saxifrage.

2 Golden Saxifrage hath round compaſſed leaues,bluntly indented about the borders like the former, among which riſe vp ſtalkes a handfull high, at the top whereof grow two or three little leaues together: out of the middle of them ſpring ſmall floures of a golden color, after which come little husks,wherein is contained the red ſeed,not vnlike the former: the roote is tender, creeping in the ground with long threds or haires.

‡ 3. *Pona* hath ſet forth this plant by the name of *Saxifraga alba petræa*, and therefore I haue placed it here; though I thinke I might more fitly haue ranked him with *Paronychia rutaceo folio* formerly deſcribed. It hath a ſmall ſingle root from which ariſe diuers fat longiſh leaues,ſomewhat hairy,and diuided into three parts: amongſt thoſe riſes vp a round knottie ſtalke,roughiſh, and of a purpliſh colour,ſome halfe foot high,diuided into ſundry branches, which carry white floures, conſiſting of fiue leaues apiece, with ſome yellowiſh threds in their middles: theſe falling,there remaines a cup containing a very ſmall ſeed. It floures at the end of Iune in the ſhadowie places of the Alpes,whereas *Pona* firſt obſerued it. ‡

¶ *The Place.*

The white Saxifrage groweth plentifully in ſundrie places of England, and eſpecially in a field on the left hand of the high way, as you goe from the place of execution called Saint Thomas Waterings vnto Dedford by London. It groweth alſo in the great field by Iſlington called the Mantles: alſo in the greene places by the ſea ſide at Lee in Eſſex,among the ruſhes,and in ſundrie other places thereabout, and elſe where. ‡ It alſo growes in Saint Georges fields behinde Southwarke. ‡

The golden Saxifrage groweth in the moiſt and mariſh grounds about Bathe and Wels,alſo in the Moores by Boſton and Wisbich in Lincolnſhire: ‡ and Mr. *George Bowles* hath found it growing in diuers woods at Chiſſelhurſt in Kent: Mr. *Goodyer* alſo hath obſerued it abundantly on the ſhadowie moiſt rockes by Mapledurham in Hampſhire: and I haue found it in the like places in Yorkſhire. ‡

¶ *The Time.*

The white Saxifrage floureth in May and Iune: the herbe with his floure are no more ſeen vntill the next yeare.

The golden Saxifrage floureth in March and Aprill.

¶ *The Names.*

The firſt is called in Latine *Saxifraga Alba*: in Engliſh,white Saxifrage, or white Stone-breake: The ſecond is called Golden Saxifrage,or golden Stone-breake.

¶ *The Nature.*

The firſt of theſe,eſpecially the root and ſeed thereof,is of a warme or hot complexion.

Golden Saxifrage is of a cold nature,as the taſte doth manifeſtly declare.

¶ *The Vertues.*

A The root of white Saxifrage boiled in wine and drunken, prouoketh vrine, clenſeth the kidneis and bladder,breaketh the ſtone,and driueth it forth,and is ſingular good againſt the ſtrangurie,and all other griefes and imperfections in the reines.

B The vertues of golden Saxifrage are yet vnto vs vnknowne,notwitſtanding I am of this minde, that it is a ſingular wound herbe,equall with Sanicle.

CHAP. 310 *Of Sow-bread.*

¶ *The Deſcription.*

1 THe firſt being the common kinde of Sowbread, called in ſhops *Panis porcinus*, and *Arthanita*, hath many greene and round leaues like vnto Aſarabacca, ſauing that the vpper part of the leaues are mixed here and there confuſedly with white ſpots, and vnder the leaues next the ground of a purple colour: among which riſe vp little ſtemmes like vnto the ſtalks of violets, bearing at the top ſmall purple floures, which turne themſelues backward (beeing full blowne) like a Turks cap, or Tulepan, of a ſmall ſent or ſauour, or none at all: which being paſt there ſucceed little round knops or heads which containe ſlender browne ſeedes: theſe knoppes

1 *Cyclamen orbiculato folio.*
Round Sowbread.

2 *Cyclamen folio Hederæ.*
Iuie Sowbread.

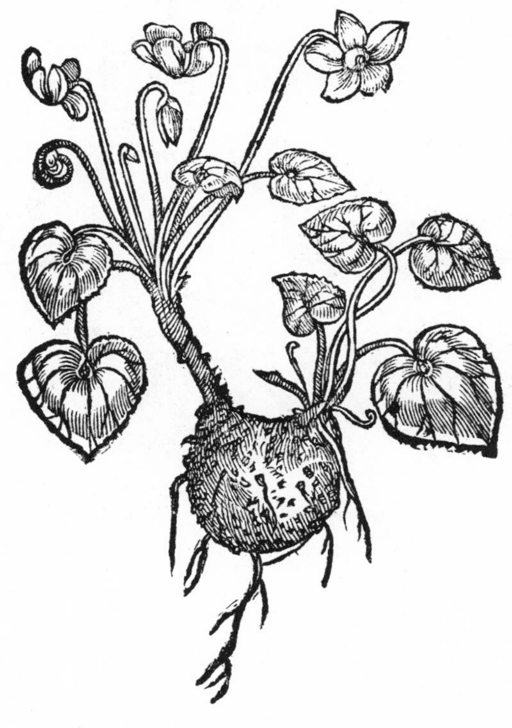

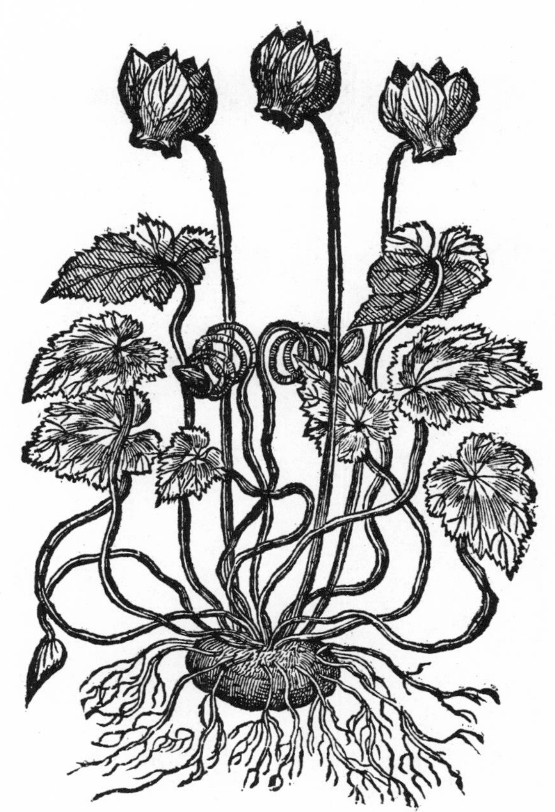

‡ 3 *Cyclamen Vernum.* Spring Sowbread.

are

are wrapped after a few daies in the small stalkes, as thred about a bottome, where it remaineth so defended from the iniurie of Winter close vpon the ground, couered also with the greene leaues aforesaid, by which meanes it is kept from the frost, euen from the time of his seeding, which is in September vntill Iune: at which time the leaues do fade away, the stalkes and seed remaining bare and naked, whereby it inioyeth the Sun (whereof it was long depriued) the sooner to bring them vnto maturitie: the root is round like a Turnep, blacke without and white within, with many smal strings annexed thereto.

‡ 4 *Cyclamen Vernum album.*
White floured Sowbread.

‡ 5 *An Cyclaminos altera, hederaceis folijs planta?*

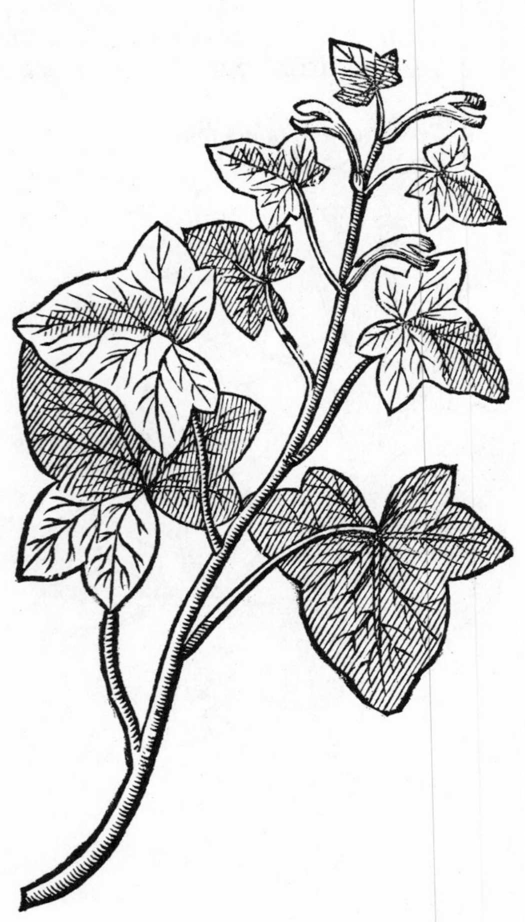

2 The second kinde of Sowbread, hath broad leaues spred vpon the ground, sharpe pointed, somewhat indented about the edges, of a darke greene colour, with some little lines or strakes of white on the vpper side, and of a darke reddish colour on that side next the ground: among which rise vp slender foot-stalks of two or three inches long: at the tops whereof stand such floures as the precedent, but of a sweeter smell, and more pleasant colour. The seed is also wrapped vp in the stalk for his further defence against the iniurie of winter. The root is somewhat greater, and of more vertue, as shall be declared.

3 There is a third kinde of Sowbread that hath round leaues without peaked corners, as the last before mentioned, yet somewhat snipt about the edges, and speckled with white about the brims of the leaues, and of a blackish colour in the middle: the floures are like to the rest, but of a deeper purple: the root also like, but smaller, and this commonly floures in the Spring.

‡ 4 This in leaues and roots is much like the last described, but the floures are smaller, snow white, and sweet smelling. There are diuers other varieties of these plants which I thinke it not necessarie for me to insist vpon: wherefore I referre the curious to the Garden of floures set forth by Mr. *Iohn Parkinson*, where they shall finde satisfaction. ‡

5 There is a plant which I haue set forth in this place that may very well be called into question, and his place also, considering that there hath been great contention about the same, and not fully determined on either part, which hath moued me to place him with those plants that most do resemble one another, both in shape and name: this plant hath greene cornered leaues like to Iuie, long

long and small gaping floures like the small Snapdragon : more hath not been said of this plant, either of stalke or root, but is left vnto the consideration of the learned.

‡ The plant which our Author here would acquaint you with, is that which *Lobel* figures with this title which I here giue, and saith it was gathered amongst other plants on the hils of Italy, but in what part or place, or how growing he knew not; and he onely questions whether it may not be the *Cyclaminos altera* of *Dioscorides, lib. 2. cap.* 195. ‡

¶ *The Place.*

Sow-bread groweth plentifully about Artoies and Vermandois in France, and in the Forest of Arden, and in Brabant : but the second groweth plentifully in many places of Italie.

It is reported vnto mee by men of good credit, that *Cyclamen* or Sow-bread groweth vpon the mountaines of Wales; on the hils of Lincolnshire, and in Somersetshire by the house of a gentleman called Mr. *Hales*; vpon a Fox-borough also not far from Mr. *Bamfields*, neere to a towne called Hardington. The first two kindes do grow in my garden, where they prosper well. ‡ I cannot learne that this growes wilde in England. ‡

¶ *The Time.*

Sow-bread floureth in September when the plant is without leafe, which doth afterwards spring vp, continuing greene all the Winter, couering and keeping warme the seede vntill Midsommer next, at what time the seed is ripe as aforesaid. The third floureth in the spring, for which cause it was called *Cyclamen vernum*: and so doth also the fourth.

¶ *The Names.*

Sow-bread is called in Greeke κυκλάμινος : in Latine, *Tuber terræ*, and *Terræ rapum* : of *Marcellus*, *Orbicularis* : of *Apuleius*, *Palalia*, *Rapum Porcinum*, and *Terræ malum* : in shops, *Cyclamen*, *Panis porcinus*, and *Arthanita* : in Italian, *Pan Porcino* : in Spanish, *Mazan de Puerco* : in High Dutch, **Schweinbrot** : in Low Dutch, **Uercking broot** : in French, *Pain de Porceau* : in English, Sow-bread. *Pliny* calleth the colour of this floure in Latine, *Colossinus color* : in English, Murrey colour.

¶ *The Nature.*

Sow-bread is hot and drie in the third degree.

¶ *The Vertues.*

The root of Sow-bread dried into pouder and taken inwardly in the quantitie of a dram and a halfe, with mead or honied water, purgeth downeward tough and grosse flegme, and other sharpe humours. A

The same taken in wine as aforesaid, is very profitable against all poison, and the bitings of venomous beasts, and to be outwardly applied to the hurt place. B

The pouder taken as aforesaid, cureth the iaundise and the stoppings of the liuer, taketh away the yellow colour of the bodie, if the patient after the taking hereof be caused to sweat. C

The leaues stamped with honie, and the iuice put into the eies, cleereth the sight, taketh away al spots and webs, pearle or haw, and all impediments of the sight, and is put into that excellent ointment called *Vnguentum Arthanitæ*. D

The root hanged about women in their extreame trauell with childe, causeth them to be deliuered incontinent, and taketh away much of their paine. E

The leaues put into the place hath the like effect, as my wife hath prooued sundrie times vpon diuers women, by my aduise and commandement, with good successe. F

The iuice of Sow-bread doth open the Hemorrhoids, and causeth them to flow beeing applied with wooll or flocks. G

It is mixed with medicines that consume or waste away knots, the Kings euill, and other hard swellings: moreouer it clenseth the head by the nostrils, it purgeth the belly being annointed therwith, and killeth the childe. It is a strong medicine to destroy the birth, being put vp as a pessarie. H

It scoureth the skin, and taketh away Sun-burning, and all blemishes of the face, pilling of the haire, and marks also that remaine after the small pocks and mesels : and giuen in wine to drinke, it maketh a man drunke. I

The decoction thereof serueth as a good and effectuall bath for members out of ioint, the gout, aud kibed heeles. K

The root being made hollow and filled with oile, closed with a little wax, and rosted in the hot embers, maketh an excellent ointment for the griefes last rehearsed. L

Being beaten and made vp into trochisches, or little flat cakes, it is reported to be a good amorous medicine to make one in loue if it be inwardly taken. M

¶ *The Danger.*

It is not good for women with childe to touch or take this herbe, or to come neere vnto it, or stride ouer the same where it groweth for the naturall attractiue vertue therein contained is such, that without controuersie they that attempt it in maner abouesaid, shall be deliuered before their

time:

time: which danger and inconuenience to auoid, I haue (about the place where it groweth in my garden) faſtened ſticks in the ground, and ſome other ſtickes I haue faſtened alſo croſſe-waies ouer them, leſt any woman ſhould by lamentable experiment finde my words to bee true, by their ſtepping ouer the ſame.

‡ I iudge our Author ſomething too womaniſh in this, that is, led more by vain opinion than by any reaſon or experience, to confirme this his aſſertion, which frequent experience ſhews to be vaine and friuolous, eſpecially for the touching, ſtriding ouer, or comming neere to this herbe. ‡

Chap. 311. *Of Birthwoorts.*

¶ *The Kindes.*

Birthwoort, as *Dioſcorides* writeth, is of three ſorts, long, round, and winding: *Plinie* hath added a fourth kinde called *Piſtolochia*, or little Birthwoort. The later writers haue ioined vnto them a fifth, named Saracens Birthwoort.

1 *Ariſtolochia longa.*
Long Birthwoort.

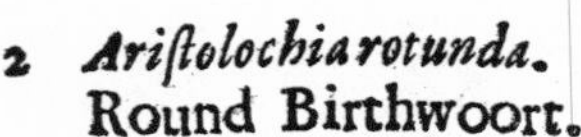
2 *Ariſtolochia rotunda.*
Round Birthwoort.

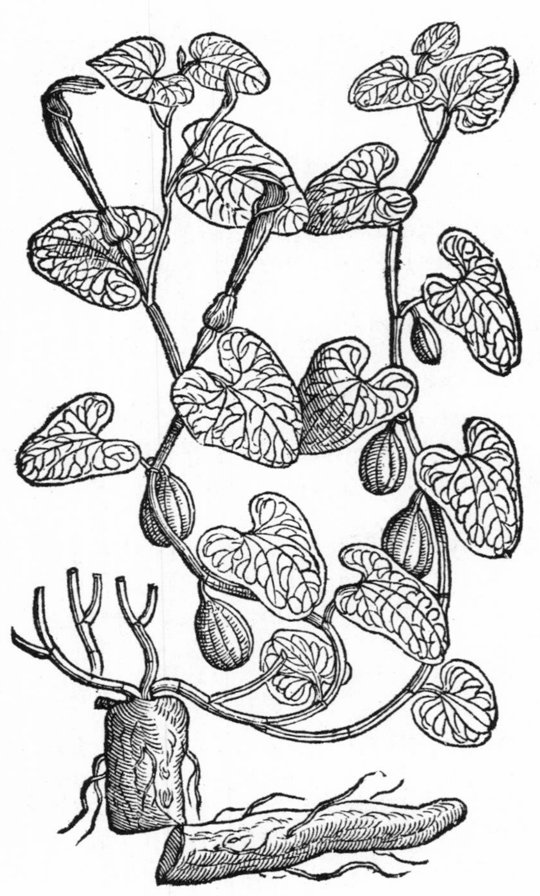

¶ *The Deſcription.*

1 Long Birthwoort hath many ſmall long ſlender ſtalkes creeping vpon the ground, tangling one with another very intricately, beſet with round leaues not much vnlike Sowbread or Iuie, but larger, of a light or ouerworne greene colour, and of a grieuous or lothſome ſmell and ſauour: among which come forth long hollow floures, not much vnlike the floures of Aron, but without any peſtell or clapper in the ſame; of a dark purple colour: after which do follow ſmall fruit like vnto little peares, containing triangled ſeeds of a blackiſh colour. The root is long, thicke, of the colour of box, of a ſtrong ſauour and bitter taſte.

2 The round Birthwoort in ſtalkes and leaues is like the firſt, but his leaues are rounder: the floures differ onely in this, that they be ſomewhat longer and narrower, and of a faint yellowiſh colour, but the ſmall flap or point of the floure that turneth backe againe, is of a darke or blacke purple

ple colour. The fruit is formed like a peare,ſharpe toward the top,more ribbed and fuller than the former : the root is round like vnto Sow-bread, in taſte and ſauour like the former.

3 *Ariſtolochia clematitis*. Climing Birthwoort.

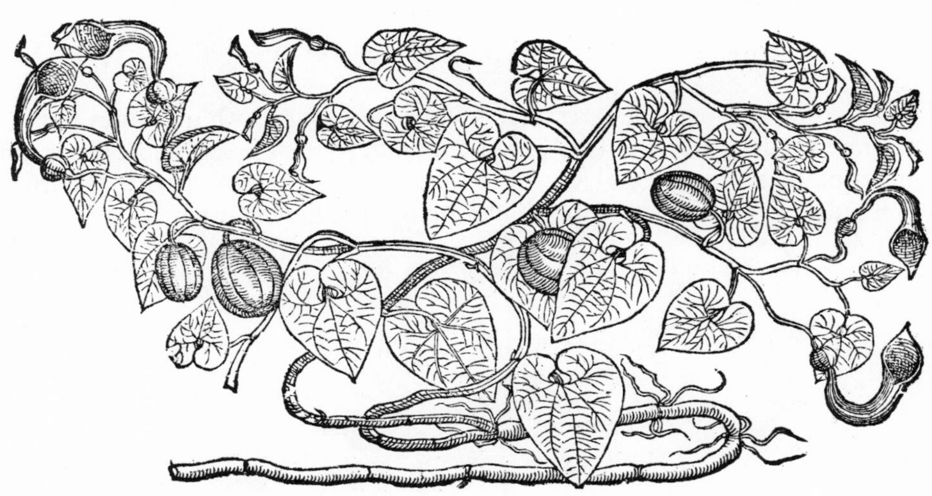

‡ 4 *Ariſtolochia Saracenica*.
Saracens Birthwoort.

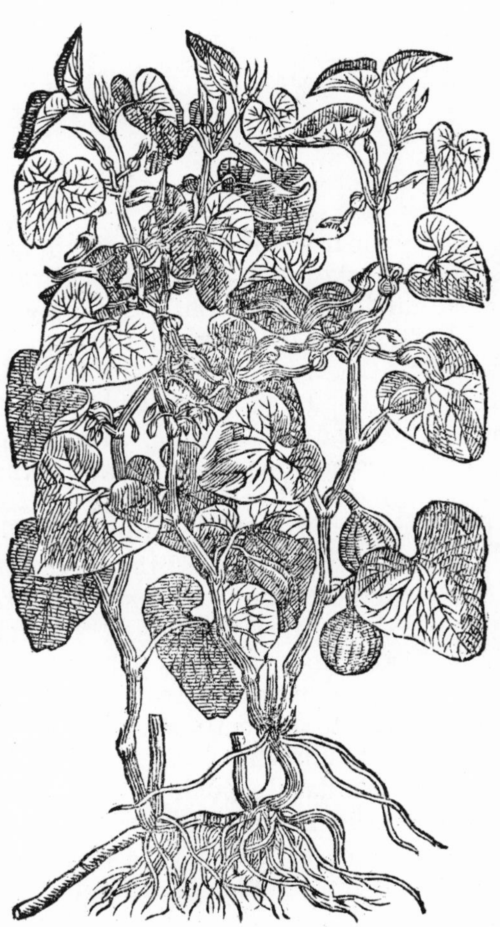

‡ 5 *Piſtolochia*.
Small Birthwoort.

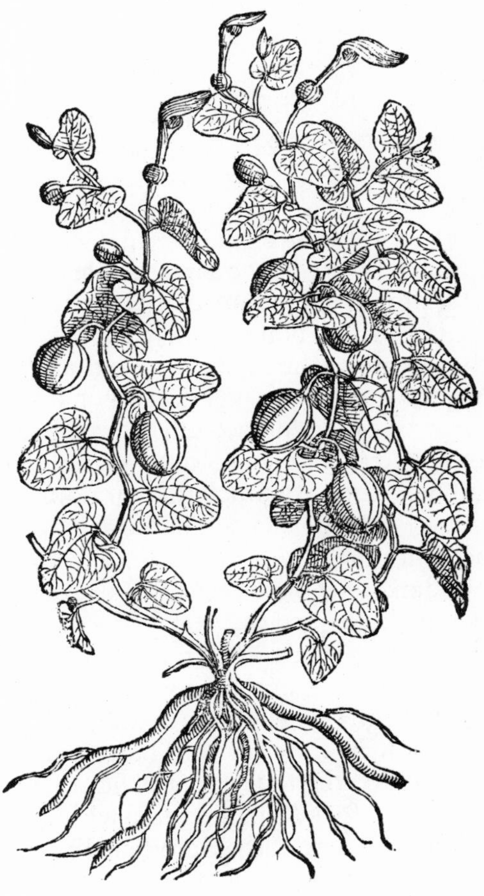

3 Climing Birthwoort taketh hold of any thing that is next vnto it,with his long and claſping ſtalks,which be oftentimes branched, and windeth it ſelfe like Bindweed : the ſtalks of the leaues are longer,whoſe leaues be ſmooth,broad,ſharpe pointed, as be thoſe of the others : the floures likewiſe hollow,long,yellow,or of a blackiſh purple colour : the fruit differeth not from that of the others : but the roots be ſlender and very long,ſometimes creeping on the top of the earth,and ſometimes growing deeper,being of like colour with the former ones.

4 There is a fourth kinde of Birthwoort reſembling the reſt in leaues and branched ſtalkes, yet higher

higher, and longer than either the long or the round: the leaues thereof be greater than those of *Asarabacca*; the floures hollow, long, and in one side hanging ouer, of a yellowish colour: the fruit is long and round like a peare, in which the seeds lie seuered, of forme three square, of an ill fauored blackish colour: the root is somewhat long, oftentimes of a mean thicknesse, yellow like to the colour of Box, not inferior in bitternesse either to the long or to the round Birthwoort: and sometimes these are found to be small and slender, and that is when they were but lately digged vp and gathered: for by the little parcels of the roots which are left, the young plants bring forth at the beginning tender and branched roots.

5 Small Birthwoort is like to the long and round Birthwoort both in stalkes and leaues, yet is it lesser and tenderer: the leaues thereof are broad, and like those of Iuie: the floure is long, hollow in the vpper part, and on the outside blackish: the fruit something round like the fruit of round Birthwoort: in stead of roots there grow forth a multitude of slender strings.

‡ 6 *Pistolochia Cretica siue Virginiana.* Virginian Snake-root.

‡ 6 *Clusius* figures and describes another smal *Pistolochia* by the name of *Pistolochia Cretica*, to which I thought good to adde the Epithite *Virginia* also, for that the much admired Snakeweed of Virginia seems no otherwise to differ from it than an inhabitant of Candy from one of the Virginians, which none I thinke will say to differ in *specie*. I will first giue *Clusius* his description, and then expresse the little varietie that I haue obserued in the plants that were brought from Virginia, and grew here with vs: it sends forth many slender stalks a foot long, more or lesse, and these are cornered or indented, crested, branched, tough, and bending towards the ground, or spred thereon, and of a darke green colour: vpon which without order grow leaues, neruous, and like those of the last described, yet much sharper pointed, and after a sort resembling the shape of those of *Smilax aspera*; but lesse, and of a darke and lasting greene colour, fastened to longish stalkes: out of whose bosomes grow long and hollow crooked floures, in shape like those of the long Birthwoort, but of a darker red on the outside, but somewhat yellowish within: and these are also fastened to pretty long stalks; and they are succeeded by fruit, not vnlike, yet lesse than that of the long Birthwort. This hath abundance of roots, like as the former, but much smaller, and more fibrous, and of a stronger smell. It floures in Iuly and August. Thus *Clusius* describes his, to which that Snakeweed that was brought from Virginia, and grew with M[r]. *Iohn Tradescant* at South-Lambeth, *An.* 1632. was agreeable in all points, but here and there one of the lower leaues were somwhat broader and rounder pointed than the rest: the floure was long, red, crooked, and a little hairie, and it did not open the top, or shew the inner side, which I iudge was by reason of the coldnesse and vnseasonablenesse of the later part of the Sommer when it floured: the stalks in the figure should haue been exprest more crooking or indenting, for they commonly grow so How hard it is to iudge of plants by one particle or facultie may very well appeare by this herbe I now treat of: for some by the similitude the root had with *Asarum*, and a vomiting qualitie which they attributed to it (which certainly is no other than accidentall) would forthwith pronounce and maintaine it an *Asarum*: some also refer it to other things, as to Primroses, *Vincitoxicum*, &c. Others more warily named it *Serpentaria Virginiana*, and *Radix Virginiana*, names as it were offering themselues and easily to be fitted and imposed vpon sundry things, but yet too generall, and therefore not fit any more to be vsed, seeing the true and specifick denomination is found. ‡

¶ *The Place.*

Pliny sheweth, that the Birthwoorts grow in fat and champion places, the fieldes of Spaine are full

full of theſe three long and round Birthwoorts: they are alſo found in Italie and Narbone or Languedock, a countrey in France. *Petrus Bellonius* writeth, that he found branched Birthwoort vpon Ida, a mountaine in Candie: *Carolus Cluſius* faith, that he found this ſame about Hiſpalis, and in many other places of Granado in Spain, among buſhes and brambles: they grow all in my garden.

¶ *The Time.*

They floure in May, Iune, and Iuly.

¶ *The Names.*

Birthwoort is called in Greeke ἀριστολοχία: in Latine likewiſe *Ariſtolochia*, becauſe it is ἄριστα ταῖς λόχοις: that is to ſay, good for women newly brought a bed, or deliuered with childe: in Engliſh, Birthwort, Hartwoort, and of ſome, Ariſtolochia.

The firſt is called *Ariſtolochia longa*, or long Birthwoort, of the forme of his root, and likewiſe *Ariſtolochia mas*, or male Birthwoort: the ſecond is thought to be *Fœmina* or female Birthwoort, & it is called *Rotunda Ariſtolochia*, or round Birthwoort: of diuers alſo *Terræ malum*, the Apple of the earth: yet *Cyclaminus* is alſo called *Terræ malum*, or the Apple of the earth.

¶ *The Temperature.*

All theſe Birthwoorts are of temperature hot and drie, and that in the third degree, hauing beſides a power to clenſe.

¶ *The Vertues.*

Dioſcorides writeth, that a dram weight of long Birthwoort drunke with wine and alſo applied, is A
good againſt ſerpents and deadly things: and that being drunke with myrrhe and pepper, it expelleth whatſoeuer is left in the matrix after the childe is deliuered, the floures alſo & dead children: and that being put vp in a peſſarie it performeth the ſame.

Round Birthwoort ſerueth for all theſe things, and alſo for the reſt of the other poiſons: it is B
likewiſe auaileable againſt the ſtuffing of the lungs, the hicket, the ſhakings or ſhiuerings of agues, hardneſſe of the milt or ſpleene, burſtings, cramps, and couvulſions, paines of the ſides if it be drunk with water.

It plucketh out thornes, ſplinters, and ſhiuers, and being mixed in plaiſters, or pulteſſes, it draws C
forth ſcales or bones, remoueth rottenneſſe or corruption, mundifieth and ſcoureth foule and filthy vlcers, and filleth them vp with new fleſh, if it be mixed with Ireos and honie.

Galen ſaith, that branched Birthwoort is of a more ſweet and pleaſant ſmell; and therefore is v- D
ſed in ointments, but it is weaker in operation than the former ones.

Birthwoort, as *Pliny* writeth, being drunk with water is a moſt excellent remedie for cramps and E
convulſions, bruiſes, and for ſuch as haue falne from high places.

It is good for them that are ſhort-winded, and are troubled with the falling ſickneſſe. F

The round *Ariſtolochia* doth beautifie, clenſe, and faſten the teeth, if they be often fretted or rub- G
bed with the pouder thereof.

‡ The root of the Virginian *Piſtolochia*, which is of a ſtrong and aromaticke ſent, is a ſingular & H
much vſed Antidote againſt the bite of the Rattle-ſnake, or rather Adder or Viper, whoſe bite is very deadly, and therfore by the prouidence of the Creator he hath vpon his taile a skinny dry ſubſtance parted into cels which containe ſome looſe, hard drie bodies that rattle in them (as if one ſhould put little ſtones or peaſe into a ſtiffe and very dry bladder) that ſo he may by this noiſe giue warning of his approch, the better to be auoided; but if any be bitten, they know, nor ſtand in need of no better antidote, than this root, which they chew, and apply to the wound, & alſo ſwallow ſome of it downe, by which means they quickly ouercome the malignitie of this poiſonous bite, which otherwiſe in a very ſhort time would proue deadly. Many alſo commend the vſe of this againſt the plague, ſmall pox, meaſels, and ſuch like maligne and contagious diſeaſes. ‡

CHAP. 312. *Of Violets.*

The Kindes.

THere might be deſcribed many kinds of floures vnder this name of violets, if their differences ſhould be more curiouſly looked into than is neceſſarie: for we might ioine hereunto the ſtock Gillofloures, the Wall floures, Dames Gillofloures, Marians violets, and likewiſe ſome of the bulbed floures, becauſe ſome of them by *Theophraſtus* are termed Violets. But this was not our charge, holding it ſufficient to diſtinguiſh and diuide them as neere as may be in kindred and neighbourhood; addreſſing my ſelfe vnto the Violets called the blacke or purple violets, or March Violets of the Garden, which haue a great prerogatiue aboue others, not onely becauſe the minde conceiueth a certaine pleaſure and recreation by ſmelling and handling of thoſe moſt odoriferous flours, but alſo for that very many by theſe Violets receiue ornament and comely grace: for there bee made of them Galands for the head, Noſe-gaies, and poeſies, which are delightfull to looke on, and pleaſant to ſmell to, ſpeaking nothing of their appropriate vertues; yea Gardens themſelues receiue by theſe the greateſt ornament of all, chiefeſt beautie and moſt gallant grace; and the recreation

creation of the minde which is taken hereby, cannot be but very good and honest: for they admonish and stir vp a man to that which is comely and honest; for floures through their beautie, variety of colour, and exquisite forme, do bring to a liberall and gentle manly minde, the remembrance of honestie, comelinesse, and all kindes of vertues. For it would be an vnseemely and filthie thing (as a certaine wise man saith) for him that doth looke vpon and handle faire and beautifull things, and who frequenteth and is conuersant in faire and beautifull places, to haue his minde not faire, but filthie and deformed.

¶ *The Description.*

1 The blacke or purple Violet doth forthwith bring from the root many leaues, broad, sleightly indented in the edges, rounder than the leaues of Iuie: among the midst wherof spring vp fine slender stems, and vpon euerie one a beautifull floure sweetly smelling, of a blew darkish purple, consisting of fiue little leaues, the lowest whereof is the greatest; and after them doe appeare little hanging cups or knaps, which, when they be ripe, do open and diuide themselues into three parts. The seed is small, long, and somewhat round withall. The root consisteth of many threddie strings.

1 *Viola nigra siue purpurea.*
The purple Garden Violet.

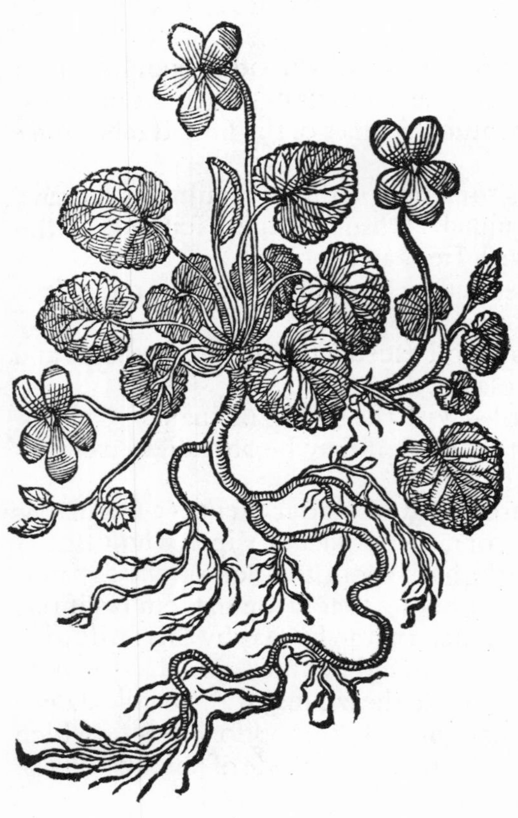

2 *Viola flore albo.*
The white Garden Violet.

2 The white garden Violet hath many milke white floures, in forme and figure like the precedent: the colour of whose floures especially setteth forth the difference.

3 The double garden violet hath leaues, creeping branches, and roots like the garden single violet; differing in that, that this sort of Violet bringeth forth most beautifull sweet double floures, and the other single.

4 The white double Violet likewise agreeth with the other of his kinde, and only differeth in the colour. For as the last described bringeth double blew or purple floures: contrariwise this plant beareth double white floures, which maketh the difference.

5 The yellow Violet is by nature one of the wilde Violets, for it groweth seldome any where but vpon most high and craggie mountains, from whence it hath bin diuers times brought into the garden, but it can hardly be brought to culture, or grow in the garden without great industrie. And by the relation of a Gentleman often remembred, called M^r. *Thomas Hesketh*, who found it growing

3 *Viola martia purpurea multiplex.*
The double garden purple Violet.

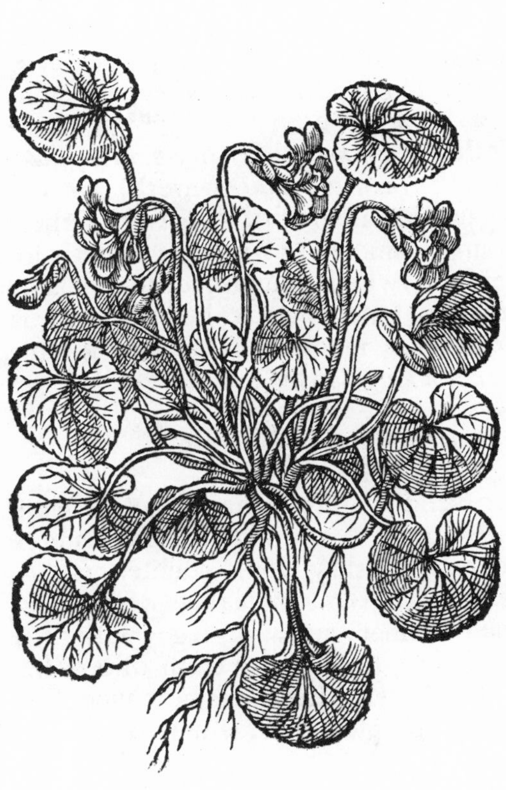

5 *Viola martia lutea.*
Yellow Violets.

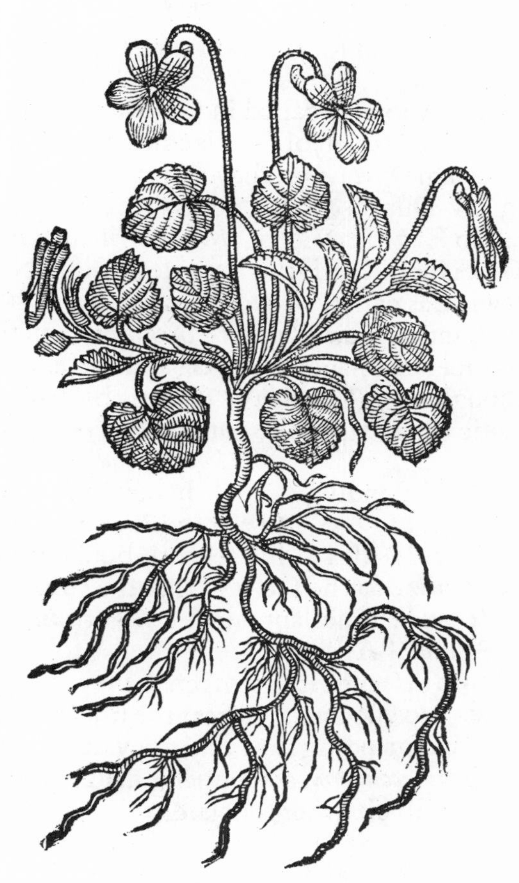

† 6 *Viola canina sylvestris.*
Dogs Violets, or wilde Violets.

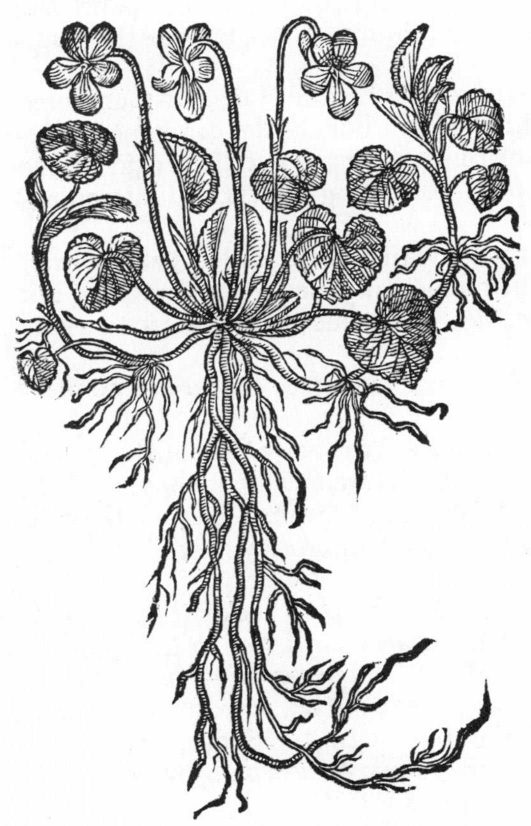

growing vpon the hills in Lancashire, neere vnto a village called Latham; and though he brought them into his garden, yet they withered and pined. The whole plant is described to be like vnto the field Violet, and differeth from it, in that this plant bringeth forth yellow floures, yet like in forme and figure, but without smell.

6 The wilde field Violet with round leaues riseth forth of the ground fro n a fibrous root, with long slender branches, whereupon do grow round smooth leaues. The floures grow at the top of the stalkes, of a light blew colour: ‡ and this growes commonly in Woods and such like places, and floures in Iuly and August. There is another varietie of this wilde Violet, which hath the leaues longer, narrower, and sharper pointed. And this was formerly figured and described in this place by our Author. ‡

7 There is found in Germanie about Noremberg and Strasborough, a kinde of Violet which is altogether a stranger in these parts. It hath (saith my Author) a thicke and tough root of a wooddy substance, from which riseth vp a stalke diuiding it selfe into diuers branches, of a wooddy substance; whereupon grow long iagged leaues like those of the Pansey. The floures grow at the top, compact of fiue leaues apiece, of a watchet colour.

¶ *The*

¶ *The Place.*

The Violet groweth in gardens almoſt euery where: the others which are ſtrangers haue beene touched in their deſcriptions.

¶ *The Time.*

The floures for the moſt part appeare in March, at the fartheſt in Aprill.

¶ *The Names.*

The Violet is called in Greeke Ἴον: of *Theophraſtus*, both Ἴον μέλαν, and μελάνιον: in Latine, *Nigra viola*, or blacke Violet, of the blackiſh purple colour of the floures. The Apothecaries keepe the Latine name *Viola*; but they call it *Herba Violaria*, and *Mater Violarum*: in high-Dutch, **Blau Viel**: in low-Dutch, **Violeten**: in French, *Violette de Mars*: in Italian, *Viola mammola*: in Spaniſh, *Violeta*: in Engliſh, Violet. *Nicander* in his Geoponickes beleeueth, (as *Hermolaus* ſheweth) that the Grecians did call it Ἴον, becauſe certaine Nymphs of Ionia gaue that floure firſt to *Iupiter*. Others ſay it was called Ἴον, becauſe when *Iupiter* had turned the young Damoſell *Io*, whom he tenderly loued, into a Cow, the earth brought forth this floure for her food: which being made for her ſake, receiued the name from her; and thereupon it is thought that the Latines alſo called it *Viola*, as though they ſhould ſay *Vitula*, by blotting out the letter *t*. *Seruius* reporteth, That for the ſame cauſe the Latines alſo name it *Vaccinium*, alledging the place of *Virgil* in his Bucolicks:

Alba liguſtra cadunt, vaccinia nigra leguntur.

Notwithſtanding *Virgil* in his tenth Eclog ſheweth, that *Vaccinium* and *Viola* do differ.

Ei nigræ violæ ſunt, & vaccinia nigra.

† *Vitruvius* alſo in his ſeuenth booke of Architecture or Building doth diſtinguiſh *Viola* from *Vaccinium*: for he ſheweth that the colour called *Sile Atticum*, or the Azure of Athens, is made *ex Viola*; and the gallant purple, *ex Vaccinio*. The Dyers, ſaith he, when they would counterfeit *Sile*, or Azure of Athens, put the dried Violets into a fat, kettle, or caldron, and boyle them with water; afterwards when it is tempered they poure it into a linnen ſtrainer, and wringing it with their hands, receiue into a mortar the liquor coloured with the Violets; and ſteeping earth of *Erethria* in it, and grinding the ſame, they make the Azure colour of Athens. After the ſame manner they temper *Vaccinium*, and putting milke vnto it, do make a gallant purple colour. But what *Vaccinia* are we will elſewhere declare.

¶ *The Temperature.*

The floures and leaues of the Violets are cold and moiſt.

¶ *The Vertues.*

A The floures are good for all inflammations, eſpecially of the ſides and lungs; they take away the hoarſeneſſe of the cheſt, the ruggedneſſe of the winde-pipe and iawes, allay the extream heate of the liuer, kidneyes, and bladder; mitigate the fierie heate of burning agues; temper the ſharpneſſe of choler, and take away thirſt.

B There is an oyle made of Violets, which is likewiſe cold and moiſt. The ſame being anointed vpon the teſticles, doth gently prouoke ſleepe which is hindred by a hot and dry diſtemper: mixed or laboured together in a woodden diſh with the yelke of an egge, it aſſwageth the pain of the fundament and hemorrhoides: it is likewiſe good to be put into cooling cliſters, and into pulteſſes that coole and eaſe paine.

C But let the oyle in which the Violets be ſteeped be either of vnripe oliues, called *Omphacinum*, or of ſweet Almonds, as *Meſues* ſaith, and the Violets themſelues muſt be freſh and moiſt: For being dry, and hauing loſt their moiſture, they doe not coole, but ſeeme to haue gotten a kinde of heate.

D The later Phyſitians do thinke it good to mix dry Violets with medicines that are to comfort and ſtrengthen the heart.

E The leaues of Violets inwardly taken do coole, moiſten, and make the belly ſoluble. Being outwardly applied, they mitigate all kinde of hot inflammations, both taken by themſelues, and alſo applied with Barley floure dried at the fire, after it hath lien ſoking in the water. They are likewiſe laid vpon a hot ſtomacke, and on burning eyes, as *Galen* witneſſeth. *Dioſcorides* writeth, that they be moreouer applied to the fundament that is fallen out.

F They may helpe the fundament that is fallen out, not as a binder keeping back the fundament, but as a ſuppler and a mollifier. Beſides, *Pliny* ſaith that Violets are as well vſed in garlands, as ſmelt vnto; and are good againſt ſurfeting, heauineſſe of the head; and being dried in water and drunke, they remoue the Squinancie or inward ſwellings of the throat. They cure the falling ſickneſſe, eſpecially in yong children, and the ſeed is good againſt the ſtinging of Scorpions.

G There is a ſyrrup made of Violets and Sugar, whereof three or foure ounces being taken at one time, ſoften the belly, and purge choler. The manner to make it is as followeth.

H Firſt make of clarified ſugar by boyling a ſimple ſyrrup of a good conſiſtence or meane thickeneſſe, whereunto put the floures cleane picked from all manner of filth, as alſo the white ends

nipped

nipped away, a quantitie according to the quantitie of the syrrup, to your owne discretion, wherein let them infuse or steepe foure and twenty houres, and set vpon a few warme embers; then strain it, and put more Violets into the same syrrup: thus do three or foure times, the oftner the better; then set them vpon a gentle fire to simper, but not to boyle in any wise: so haue you it simply made of a most perfect purple colour, and of the smell of the floures themselues. Some do adde thereto a little of the iuyce of the floures in the boyling, which maketh it of better force and vertue. Likewise some do put a little quantitie of the iuyce of Lymons in the boyling, which doth greatly encrease the beauty thereof, but nothing at all the vertue.

There is likewise made of Violets and sugar certain plates called Sugar Violet, or Violet tables, or Plate, which is most pleasant and wholesome, especially it comforteth the heart and the other inward parts. I

The decoction of Violets is good against hot feuers, and the inflammation of the liuer and all other inward parts: the like propertie hath the iuyce, syrrup, or conserue of the same. K

Syrrup of Violets is good against the inflammation of the lungs and brest, against the pleurisie and cough, against feuers and agues in yong children, especially if you put vnto an ounce of Syrrup eight or nine drops of oyle of Vitrioll, and mix it together, and giue it to the childe a spoonefull at once. L

The same giuen in manner aforesaid is of great efficacie in burning feuers and pestilent diseases, greatly cooling the inward parts: and it may seeme strange to some, that so sharpe a corrosiue as oyle of Vitriol should be giuen into the body; yet being delayed and giuen as aforesaid, sucking children may take it without any perill. M

The same taken as aforesaid cureth all inflammations of the throat, mouth, uvula, squinancie, and the falling euill in children. N

Sugar-Violet hath power to cease inflammations, roughnesse of the throat, and comforteth the heart, asswageth the paines of the head, and causeth sleepe. O

The leaues of Violets are vsed in cooling plaisters, oyles, and comfortable cataplasmes or pultesses; and are of greater efficacie among other herbes, as Mercurie, Mallowes, and such like, in clisters, for the purposes aforesaid. P

Chap. 313. *Of Hearts-ease, or Pansies.*

¶ *The Description.*

1 The Hearts-ease or Paunsie hath many round leaues at the first comming vp; afterward they grow somewhat longer, sleightly cut about the edges, trailing or creeping vpon the ground. The stalkes are weake and tender, whereupon do grow floures in forme and figure like the Violet, and for the most part of the same bignesse, of three sundry colours; whereof it tooke the syrname *Tricolor*, that is to say, purple, yellow, and white or blew: by reason of the beauty and brauerie of which colours they are very pleasing to the eye, for smell they haue little or none at all. The seed is contained in little knaps, of the bignesse of a Tare, which come forth after the floures be fallen, and do open of themselues when the seed is ripe. The root is nothing else but as it were a bundle of threddy strings.

2 The vpright Paunsie bringeth forth long leaues deepely cut in the edges, sharpe pointed, of a bleake or pale greene colour, set vpon slender vpright stalkes, cornered, ioynted, or kneed a foot high or higher; whereupon do grow very faire floures of three colours, *viz.* of purple, blew, and yellow, in shape like the common Hearts-ease, but greater and fairer: which colours are so excellently and orderly placed, that they bring great delectation to the beholders, though they haue little or no smell at all. For oftentimes it hapneth, that the vppermost floures are differing from those that grow vpon the middle of the plant, and those varie from the lowermost, as Nature list to dally with things of such beauty. The seed is like the precedent.

3 The wilde Paunsie differeth from that of the garden, in leaues, roots, and tender branches: the floures of this wilde one are of a bleake and pale colour, far inferiour in beauty to that of the garden, wherein consisteth the difference.

4 Stony Hearts-ease is a base and low plant: The leaues are rounder, and not so much cut about the edges as the others: The branches are weake and feeble, trailing vpon the ground: The floures are likewise of three colours, that is to say, white, blew, and yellow, void of smell. The root perisheth when it hath perfected his seed.

5 There is found in sundry places of England a wilde kinde hereof, bringing floures of a faint yellow colour, without mixture of any other colour, yet hauing a deeper yellow spot in the lowest leafe,

1 *Viola tricolor.*
Hearts-ease.

2 *Viola assurgens tricolor.*
Vpright Hearts-ease.

3 *Viola tricolor sylvestris.*
Wilde Paunsies.

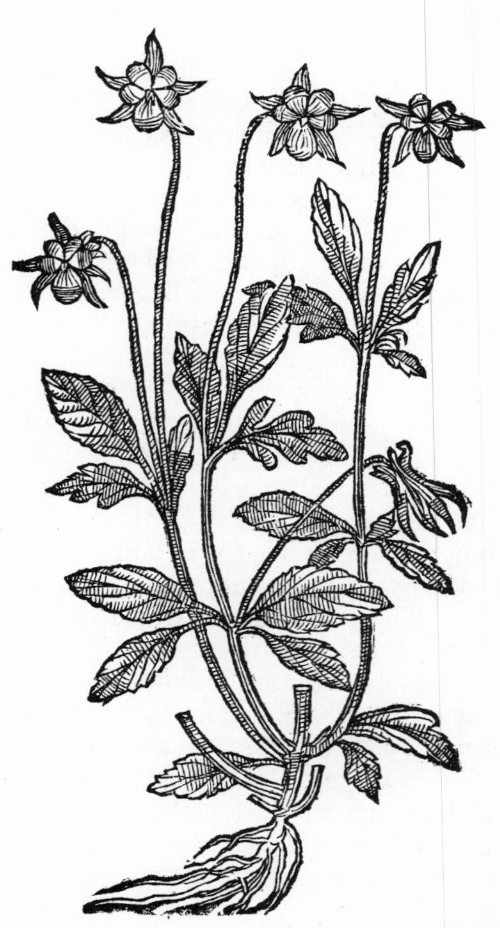
4 *Viola tricolor petræa.*
Stony Hearts-ease.

leafe with foure or fiue blackish purple lines, wherein it differeth from the other wilde kinde: and this hath beene taken of some yong Herbarists to be the yellow Violet.

¶ *The Place.*

The Hearts-ease groweth in fields in many places, and in gardens also, and that oftentimes of it selfe: it is more gallant and beautifull than any of the wilde ones.

Matthiolus reporteth, that the vpright Paunsie is found on mount Baldus in Italy. *Lobel* saith that it groweth in Languedocke in France, and on the tops of some hills in England; but as yet I haue not seene the same.

Those with yellow floures haue been found by a village in Lancashire called Latham, foure miles from Kyrckham, by M[r]. *Thomas Hesketh* before remembred.

¶ *The Time.*

They floure not onely in the Spring, but for the most part all Sommer thorow, euen vntill Autumne.

¶ *The Names.*

Hearts-ease is named in Latine *Viola tricolor*, or the three coloured Violet; and of diuers, *Iacea*; (yet there is another *Iacea* syrnamed *Nigra*: in English, Knap-weed, Bull-weed, and Matfellon of others, *Herba Trinitatis*, or herbe Trinitie, by reason of the triple colour of the floures: of some others, *Herba Clauellata*: in French, *Pensees*: by which name they became knowne to the Brabanders and others of the Low-countries that are next adioyning. It seemeth to be *Viola flammea*, which *Theophrastus* calleth φλόγα, which is also called φλόγιον: in English, Hearts-ease, Paunsies, Liue in idlenesse, Cull me to you, and Three faces in a hood.

The vpright Pansie is called not vnproperly *Viola assurgens*, or *Surrecta*, and withall *Tricolor*, that is to say, straight or vpright Violet three coloured: of some, *Viola arborescens*, or Tree Violet, for that in the multitude of branches and manner of growing it resembles a little tree.

¶ *The Temperature.*

It is of temperature obscurely cold, but more euidently moist, of a tough and slimie iuyce, like that of the Mallow; for which cause it moistneth and suppleth, but not so much as the Mallow doth.

¶ *The Vertues.*

It is good, as the later Physitions write, for such as are sicke of an ague, especially children and infants, whose convulsions and fits of the falling sicknesse it is thought to cure. A

It is commended against inflammations of the lungs and chest, and against scabs and itchings of the whole body, and healeth vlcers. B

The distilled water of the herbe or floures giuen to drinke for ten or more dayes together, three ounces in the morning, and the like quantitie at night, doth wonderfully ease the paines of the French disease, and cureth the same, if the Patient be caused to sweat sundry times, as *Costaeus* reporteth, in his booke *de natura Vniuers. stirp.* C

CHAP. 314. *Of Ground-Iuy, or Ale-hoofe.*

¶ *The Description.*

1 GRound Iuy is a low or base herbe; it creepeth and spreads vpon the ground hither and thither all about, with many stalkes of an vncertaine length, slender, and like those of the Vine, something cornered, and sometimes reddish: whereupon grow leaues something broad and round, wrinkled, hairy, nicked in the edges, for the most part two out of euerie ioynt: amongst which come forth the floures gaping like little hoods, not vnlike to those of Germander, of a purplish blew colour: the roots are very threddy: the whole plant is of a strong smell and bitter taste.

‡ 2 Vpon the rockie and mountainous places of Prouince and Daulphine growes this other kinde of Ale-hoofe, which hath leaues, stalkes, floures, and roots like in shape to those of the former, but the floures and leaues are of a light purple colour, and also larger and longer. This by *Lobel* is called *Asarina, siue Saxatilis hedera.* ‡

¶ *The Place.*

It is found as well in tilled as in vntilled places, but most commonly in obscure and darke places, vpon banks vnder hedges, and by the sides of houses.

¶ *The Time.*

It remaineth greene not onely in Sommer, but also in Winter at any time of the yeare: it floureth from Aprill till Sommer be far spent.

¶ The

1 *Hedera terreſtris.*
Ale-hoofe.

‡ 2 *Hedera ſaxatilis.*
Rocke Ale-hoofe.

¶ *The Names.*

It is commonly called *Hedera terreſtris* : in Greeke, χαμαικισσος : alſo *Corona terræ* : in high-Dutch, **Gundelreb** : in low-Dutch, **Onderhaue** : in French, *Lierre terreſtre* : *Hedera humilis* of ſome, and *Chamæciſſum* : in Engliſh, Ground-Iuy, Ale-hoofe, Gill go by ground, Tune-hoofe, and Cats-foot. ‡ Many queſtion whether this be the *Chamæciſſus* of the Antients : which controuerſie *Dodonæus* hath largely handled, *Pempt*. 3. *lib*. 3. *cap*. 4. ‡

¶ *The Temperature.*

Ground-Iuie is hot and dry, and becauſe it is bitter it ſcoureth, and remoueth ſtoppings out of the intrals.

¶ *The Vertues.*

A Ground-Iuy is commended againſt the humming noyſe and ringing ſound of the eares, being put into them, and for them that are hard of hearing.

B *Matthiolus* writeth, That the iuyce being tempered with Verdugreaſe, is good againſt fiſtulaes and hollow vlcers.

C *Dioſcorides* teacheth, That halfe a dram of the leaues being drunke in foure ounces and a halfe of faire water, for fourty or fifty dayes together, is a remedie againſt the Sciatica, or ache in the huckle bone.

D The ſame taken in like ſort ſix or ſeuen dayes doth alſo cure the yellow jaundice. *Galen* hath attributed (as we haue ſaid) all the vertue vnto the floures : Seeing the floures of Ground-Iuy (ſaith he) are very bitter, they remoue ſtoppings out of the liuer, and are giuen to them that are vexed with the Sciatica.

E Ground-Iuy, Celandine, and Daiſies, of each a like quantitie, ſtamped and ſtrained, and a little ſugar and roſe water put thereto, and dropped with a feather into the eyes, taketh away all manner of inflammation, ſpots, webs, itch, ſmarting, or any griefe whatſoeuer in the eyes, yea although the ſight were nigh hand gone : it is proued to be the beſt medicine in the world.

F The herbes ſtamped as aforeſaid, and mixed with a little ale and honey, and ſtrained, takes away the pinne and web, or any griefe out of the eyes of horſe or cow, or any other beaſt, being ſquirted into the ſame with a ſyringe, or I might haue ſaid the liquor iniected into the eyes with a ſyringe. But I liſt not to be ouer eloquent among Gentlewomen, to whom eſpecially my Works are moſt neceſſarie.

G The women of our Northerne parts, eſpecially about Wales and Cheſhire, do tunne the herbe Ale-hoof into their Ale, but the reaſon thereof I know not : notwithſtanding without all controuerſie

uersie it is most singular against the griefes aforesaid : being tunned vp in ale and drunke, it also purgeth the head from rhumaticke humors flowing from the braine.

Hedera terrestris boyled in water stayeth the termes ; and boyled in mutton broth it helps weake and aking backes. I

They haue vsed to put it into ointments against burning with fire, gunpouder, and such like. K

Hedera terrestris being bound in a bundle, or chopt as herbes for the pot, and eaten or drunke as thin broth stayeth the flux in women. L

CHAP. 315. *Of Iuy.*

¶ *The Kindes.*

THere be two kindes of Iuy, as *Theophrastus* witnesseth, reckoned among the number of those plants which haue need to be propped vp ; for they stand not of themselues, but are fastned to stone walls, trees, and such like, and yet notwithstanding both of a wooddy substance, and yet not to be placed among the trees, shrubs, or bushes, because of the affinitie they haue with climbing herbes ; as also agreeing in forme and figure with many other plants that climbe, and are indeed simply to be reckoned among the herbes that clamber vp. But if any will cauill, or charge me with my promise made in the beginning of this historie, where we made our diuision, namely, to place each plant as neere as may be in kindred and neighbourhood ; this promise I haue fulfilled, if the curious eye can be content to reade without rashnesse those plants following in order, and not onely this climbing Iuy that lifteth her selfe to the tops of trees, but also the other Iuy that creepeth vpon the ground.

Of the greater or the climing Iuy there are also many sorts ; but especially three, the white, the blacke, and that which is called *Hedera Helix*, or *Hedera sterilis*.

¶ *The Description.*

1 THe greater Iuy climbeth on trees, old buildings, and walls : the stalkes thereof are wooddy, and now and then so great as it seemes to become a tree ; from which it sendeth a multitude of little boughes or branches euery way, whereby as it were with arms it creepeth and wandereth far about : it also bringeth forth continually fine little roots, by which it fastneth it selfe and cleaueth wonderfull hard vpon trees, and vpon the smoothest stone walls : the leaues are smooth, shining especially on the vpper side, cornered with sharpe pointed corners. The floures are very small and mossie ; after which succeed bundles of black berries, euery one hauing a small sharpe pointall.

There is another sort of great Iuy that bringeth forth white fruit, which some call *Acharnicam irriguam* ; and also another lesser, the which hath blacke berries. This *Pliny* calleth *Selinitium*.

We also finde mentioned another sort hereof spred abroad, with a fruit of a yellow Saffron colour, called of diuers *Dionysias*, as *Dioscorides* writeth : others *Bacchica*, of which the Poets vsed to make garlands, as *Pliny* testifieth, *lib.* 16. *cap.* 34.

2 Barren Iuy is not much vnlike vnto the common Iuy aforesaid, sauing that his branches are both smaller and tenderer, not lifting or bearing it selfe vpward, but creeping along by the ground vnder moist and shadowie ditch bankes. The leaues are most commonly three square, cornered, of a blackish greene colour, which at the end of Sommer become brownish red vpon the lower side. The whole plant beareth neither floures nor fruit, but is altogether barren and fruitlesse.

‡ 3 There is kept for nouelties sake in diuers gardens a Virginian, by some (though vnfitly) termed a Vine, being indeed an Iuy. The stalkes of this grow to a great heighth, if they be planted nigh any thing that may sustaine or beare them vp : and they take first hold by certaine small tendrels, vpon what body soeuer they grow, whether stone, boords, bricke, yea glasse, and that so firmely, that oftentimes they will bring pieces with them if you plucke them off. The leaues are large, consisting of foure, fiue, or more particular leaues, each of them being long, and deeply notched about the edges, so that they somewhat resemble the leaues of the Chesnut tree : the floures grow clustering together after the manner of Iuy, but neuer with vs shew themselues open, so that we cannot iustly say any thing of their colour, or the fruit that succeeds them. It puts forth his leaues in April, and the stalkes with the rudiments of the floures are to be seene in August. It may as I said be fitly called *Hedera Virginiana*. ‡

¶ *The Place.*

Iuy groweth commonly about walls and trees ; the white Iuy groweth in Greece, and the barren Iuy groweth vpon the ground in ditch bankes and shadowie woods.

1 *Hedera corymbosa.*
Clymbing or berried Iuy.

2 *Hedera Helix.*
Barren or creeping Iuy.

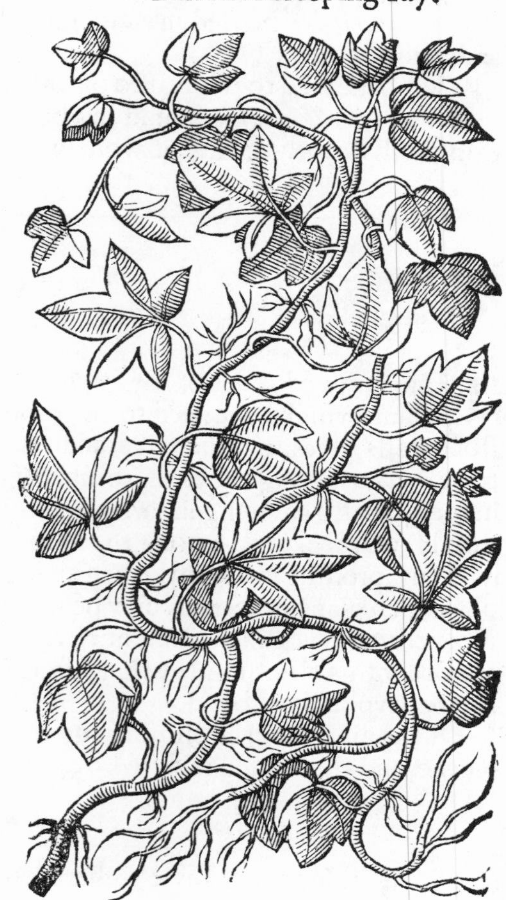

¶ *The Time.*

Iuy flourisheth in Autumne: the berries are ripe after the Winter Solstice.

¶ *The Names.*

Iuy is called in Latine *Hedera*: in Greeke, κιττὸς, and κισσὸς: in high-Dutch, **Epheu**: in low-Dutch, **Weyle**: in Spanish, *Yedra*: in French, *Liarre*.

The greater Iuy is called of *Theophrastus* ὀρθὸς κισσὸς: in Latine, *Hedera attollens*, or *Hedera assurgens*: *Gaza* interpreteth it *Hedera excelsa*. The later Herbarists would haue it to be *Hedera arborea*, or tree Iuy, because it groweth vpon trees, and *Hedera muralis*, which hangeth vpon walls.

Creeping or barren Iuy is called in Greeke ἐπίγειος κισσὸς: in English, Ground-Iuy: yet doth it much differ from *Hedera terrestris*, or Ground-Iuy before described: of some it is called *Clauicula*, *Hedera Helix*, and *Hedera sterilis*; and is that herbe wherein the Bore delighteth, according to *Iohannes Khuenius*.

¶ *The Temperature.*

Iuy, as *Galen* saith, is compounded of contrarie faculties; for it hath a certaine binding earthy and cold substance, and also a substance somewhat biting, which euen the very taste doth shew to be hot. Neither is it without a third facultie, as being of a certaine warme waterie substance, and that is if it be greene: for whilest it is in drying, this watery substance being earthy, cold, and binding consumeth away, and that which is hot and biting remaineth.

¶ *The Vertues.*

A The leaues of Iuy fresh and greene boyled in wine, do heale old vlcers, and perfectly cure those that haue a venomous and malicious quality ioyned with them; and are a remedy likewise against burnings and scaldings.

B Moreouer, the leaues boyled with vineger are good for such as haue bad spleens; but the floures or fruit are of more force, being very finely beaten and tempered with vineger, especially so vsed they are commended against burnings.

C The iuyce drawne or snift vp into the nose doth effectually purge the head, stayeth the running of the eares that hath beene of long continuance, and healeth old vlcers both in the eares and also in the nosthrils: but if it be too sharpe, it is to be mixed with oyle of Roses, or sallad oyle.

D The gum that is found vpon the trunke or body of the old stocke of Iuy, killeth nits and lice, and taketh away haire: it is of so hot a qualitie, as that it doth obscurely burne: it is as it were a certaine

certaine wateriſh liquor congealed of thoſe gummie drops. Thus farre *Galen*.

The very ſame almoſt hath *Dioſcorides*, but yet alſo ſomewhat more: for ouer and beſides hee E
ſaith, that fiue of the berries beaten ſmall, and made hot in a Pomegranat rinde, with oyle of roſes, and dropped into the contrarie eare, doth eaſe the tooth-ache; and that the berries make the haire blacke.

Iuy in our time is very ſeldome vſed, ſaue that the leaues are layd vpon little vlcers made in the F
thighes, legs, or other parts of the body, which are called Iſſues; for they draw humors and wateriſh ſubſtance to thoſe parts, and keepe them from hot ſwellings or inflammations, that is to ſay, the leaues newly gathered, and not as yet withered or dried.

Some likewiſe affirme that the berries are effectuall to procure vrine; and are giuen vnto thoſe G
that be troubled with the ſtone and diſeaſes of the kidneyes.

The leaues laid in ſteepe in water for a day and a nights ſpace, helpe ſore and ſmarting wate- H
riſh eyes, if they be bathed and waſhed with the water wherein they haue beene infuſed.

CHAP. 316. *Of rough Binde-weed.*

1 *Smilax Peruviana, Salſaparilla.*
Rough Binde-weed of Peru.

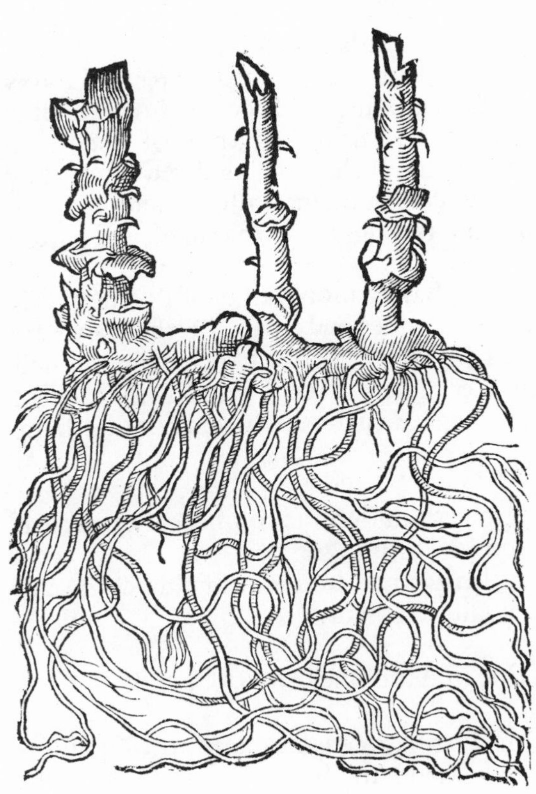

2 *Smilax aſpera.*

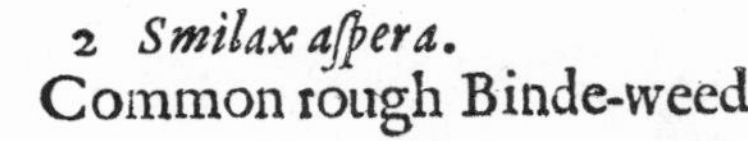

Common rough Binde-weed.

¶ *The Deſcription.*

1 ALthough we haue great plenty of the roots of this Binde-weed of Peru, which we vſually cally *Zarza*, or *Sarſa Parilla*, wherewith diuers griefes and maladies are cured, and that theſe roots are very well knowne to all; yet ſuch hath beene the careleſneſſe and ſmall prouidence of ſuch as haue trauelled into the Indies, that hitherto not any haue giuen vs inſtruction ſufficient, either concerning the leaues, floures, or fruit: onely *Monardus* ſaith, that it hath long roots deepe thruſt into the ground: which is as much as if a great learned man ſhould tell the ſimple, that our common carrion Crow were of a blacke colour. For who is ſo blinde that ſeeth the root it ſelfe, but can eaſily affirme the roots to be very long? Notwithſtanding, there is in the reports of ſuch as ſay they haue ſeene the plant it ſelfe growing, ſome contradiction or contrarietie: ſome report that it is a kind of Bind-weed, and eſpecially one of theſe rough Bindweeds: others,

3 *Smilax aspera Lusitanica.*
Rough Binde-weed of Portugall.

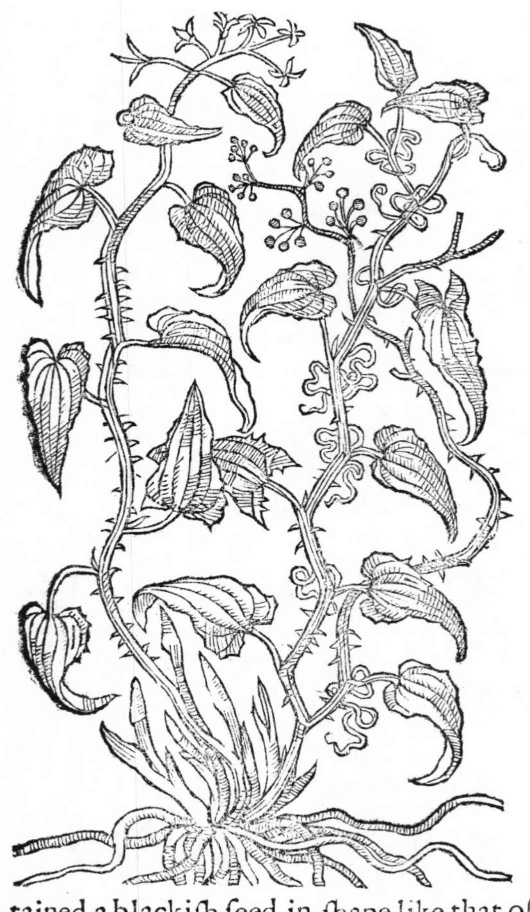

others, as one Mr. *White* an excellent painter, who carried very many people into Virginia (or after some Norembega) there to inhabit, at which time hee did see thereof great plentie, as he himselfe reported vnto me, with this bare description; It is (saith he) the root of a small shrubbie tree, or hedge tree, such as are those of our country called Haw-thorns, hauing leaues resembling those of Iuy, but the floures or fruit he remembreth not. ‡ It is most certaine, that *Sarsa parilla* is the root of the Americane *Smilax aspera*, both by consent of most Writers, and by the relation of such as haue seene it growing there. ‡

2 The common rough Binde-weed hath many branches set full of little sharpe prickles, with certaine clasping tendrels, wherewith it taketh hold vpon hedges, shrubs, and whatsoeuer standeth next vnto it, winding and clasping it selfe about from the bottom to the top; whereon are placed at euery ioint one leafe like that of Iuy, without corners, sharpe pointed, lesser and harder than those of smooth Binde-weed, oftentimes marked with little white spots, and garded or bordered about the edges with crooked prickles. The floures grow at the top of crooked stalks of a white colour, and sweet of smell. After commeth the fruit like those of the wilde Vine, greene at the first, and red when they be ripe, and of a biting taste; wherein is contained a blackish seed in shape like that of hempe. The root is long, somewhat hard, and parted into very many branches.

3 This rough Binde-weed, found for the most part in the barren mountaines of Portugal, differeth not from the precedent in stalkes and floures, but in the leaues and fruit; for the leaues are softer, and lesse prickly, and sometimes haue no prickles at all, and they are also oftentimes much narrower: the fruit or berry is not red but blacke when as it commeth to be ripe. The root hereof is one single root of a wooddy substance, with some fibres annexed thereto, wherein consisteth the difference.

¶ *The Place.*

Zarza parilla, or the prickly Binde-weed of America, groweth in Peru a prouince of America, in Virginia, and in diuers other places both in the East and West Indies.

The others grow in rough and vntilled places, about the hedges and borders of fields, on mountaines and vallies, in Italy, Languedock in France, Spaine, and Germany.

¶ *The Time.*

They floure and flourish in the Spring: their fruit is ripe in Autumne, or a little before.

¶ *The Names.*

It is named in Greeke σμίλαξ τραχεῖα. *Gaza* (*Theophrastus* his Translator) names it *Hedera Cilicia*; as likewise *Pliny*, who *lib.* 24. *cap.* 10 writeth, that it is also syrnamed *Nicophoron*. Of the Hetrurians, *Hedera spinosa*, and *Rubus ceruinus*: of the Castilians in Spaine, as *Lacuna* saith, *Zarza parilla*, as though they should say *Rubus viticula*, or Bramble little Vine. *Parra*, as *Matthiolus* interpreteth it, doth signifie a Vine; and *Parilla*, a small or little Vine.

Diuers affirme that the root (brought out of Peru a prouince in America) which the later Herbarists do call *Zarza*, is the root of this Bindeweed. *Garcias Lopius Lusitanus* granteth it to be like thereunto, but yet he doth not affirme that it is the same. Plants are oftentimes found to be like one another, which notwithstanding are proued not to be the same by some little difference; the diuers constitution of the weather and of the soile making the difference.

Zarzaparilla of Peru is a strange plant, and is brought vnto vs from the Countries of the new world called America; and such things as are brought from thence, although they also seeme and are like to those that grow in Europe, notwithstanding they doe often differ in vertue and operation: for the diuersitie of the soile and of the weather doth not only breed an alteration in the form, but

but doth moſt of all preuaile in making the vertues and qualities greater or leſſer. Such things as grow in hot places be of more force, and greater ſmell; and in cold, of leſſer. Some things that are deadly and pernitious, being remoued wax milde, and are made wholeſome: ſo in like manner, although *Zarza parilla* of Peru be like to rough Binde-weed, or to Spaniſh *Zarza parilla*, notwithſtanding by reaſon of the temperature of the weather, and alſo through the nature of the ſoile, it is of a great deale more force than that which groweth either in Spaine or in Africke.

The roots of *Zarza parilla* of Peru, which are brought alone without the plant, be long and ſlender, like to the leſſer roots of common liquorice, very many oftentimes hanging from one head, in which roots the middle ſtring is hardeſt. They haue little taſte, and ſo ſmall a ſmell that it is not to be perceiued. Theſe are reported to grow in Honduras a prouince of Peru. They had their name of the likeneſſe of rough Binde-weed, which among the inhabitants it keepeth; ſignifying in Spaniſh, a rough or prickly vine, as *Garcias Lopius* witneſſeth.

¶ *The Temperature.*

The roots are of temperature hot and dry, and of thin and ſubtill parts, inſomuch as their decoction doth very eaſily procure ſweat.

¶ *The Vertues.*

The roots are a remedie againſt long continuall paine of the ioynts and head, and againſt cold diſeaſes. They are good for all manner of infirmities wherein there is hope of cure by ſweating, ſo that there be no ague ioyned. A

The cure is perfected in few dayes, if the diſeaſe be not old or great; but if it be, it requireth a longer time of cure. The roots here meant are as I take it thoſe of *Zarza parilla*, whereof this *Smilax aſpera* or rough Binde-weed is holden for a kinde: notwithſtanding this of Spain and the other parts of Europe, though it be counted leſſe worth, yet is it commended of *Dioſcorides* and *Pliny* againſt poyſons. The leaues hereof, ſaith *Dioſcorides*, are a counterpoyſon againſt deadly medicines, whether they be drunke before or after. B

☞ † The ſecond and fourth were both formerly of one plant, I meane the hiſtorie; for the figure in the fourth place ſhould haue been in the third, and the figure in the third was the ſame with the ſecond, and ſhould haue been in the fourth place.

Chap. 317. *Of ſmooth or gentle Binde-weed.*

1 *Smilax lenis ſiue læuis maior.*
Great ſmooth Binde-weed.

2 *Smilax lenis minor.*
Small Binde-weed.

¶ *The Deſcription.*

1 IT is a ſtrange thing vnto me, that the name of *Smilax* ſhould be ſo largely extended, as that it ſhould be aſſigned to thoſe plants that come nothing neere the nature, and ſcarſly vnto any part of the forme of *Smilax* indeed. But we will leaue controuerſies to the further conſideration of ſuch as loue to dance in quag-mires, and come to this our common ſmooth *Smilax*, called and knowne by that name among vs, or rather more truly by the name of *Convolvulus maior*, or *Volubilis maior*: It beareth the long branches of a Vine, but tenderer, and for the length and great ſpreading therof it is very fit to make ſhadows in arbors: the leaues are ſmooth like Iuie, but ſomewhat bigger, and being broken are full of milke: amongſt which come forth great white and hollow floures like bells. The ſeed is three cornered, growing in ſmall huskes couered with a thin skin. The root is ſmall, white and long, like the great Dogs graſſe.

2 *Smilax lenis minor* is much like vnto the former in ſtalkes, leaues, floures, ſeed, and roots, ſauing that in all reſpects it is much ſmaller, and creepeth vpon the ground. The branches are ſmall and ſmooth: the little leaues tender and ſoft: the floures like vnto little bells, of a purple colour: the ſeed three cornered like vnto the others.

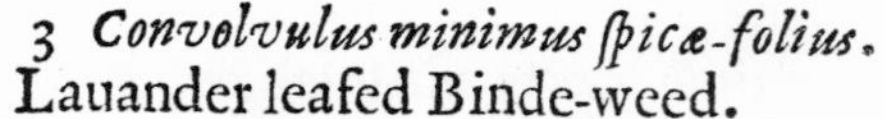

3 *Convolvulus minimus ſpicæ-folius.*
Lauander leafed Binde-weed.

† 4 *Convolvulus argenteus Altheæ folio.*
Siluer leafed Binde-weed.

3 This Bindweed *Pena* ſaith he neuer ſaw but in the brinks of quicke-ſets and Oliuets in Prouence, Sauoy, and Narbone; notwithſtanding I found it growing in the corne fields about great Dunmow in Eſſex, in ſuch abundance, that it doth much hurt vnto their corne. This kind of Bindweed or *Volubilis* is like vnto the ſmall Bindweed before mentioned, but it hath a finer floure, plaited or folded in the compaſſe of the bell very orderly, eſpecially before the Sun riſe (for after it opens it ſelfe the welts are not ſo much perceiued) and it is of a darke purple colour: the ſeed is not vnlike the reſt, cornered and flat, growing out of ſlender branches which ſtand vpright and thicke together, proceeding out of a woody white root. The leaues are long and narrow, reſembling *Linaria* both in colour and hairineſſe, in taſte drying, and ſomewhat heating.

‡ 4 The

5 *Volubilis nigra.*
Blacke Bindweed.

‡ 4 The stalkes and branches of this are some cubite long, slender, weake and hairy, so that they lie vpon the ground, if they haue nothing to sustaine them: vpon these without any order grow leaues, shaped like those of Iuy, or the marsh Mallow, but lesse, and couered ouer with a siluer-like downe or hairinesse, and diuided somewhat deep on the edges, sometimes also curled, and otherwhiles onely snipt about. The floure growes vpon long stalkes like as in other plants of this kinde, and consists of one folding leafe, like as that of the last mentioned, and it is either of a whitish purple, or else absolute purple colour: The root is small and creeping. It growes in many places of Spaine, and there floures in March and Aprill. *Clusius* calls this *Conuoluulus Altheæ folio*, and saith that the Portugals name it *Verdezilla*, and commend it as a thing most effectuall to heale wounds. Our Authour gaue the figure hereof (how fitly let the Reader iudge) by the name of *Papauer cornutum luteum minus*, making it a horned Poppy, as you may see in the former Edition, *Pag.* 294. ‡

† 5 This kinde of Bindweed hath a tough root full of threddie strings, from which rise vp immediatly diuers trailing branches, wherupon grow leaues like the common field Bindweed, or like those of Orach, of a black green colour, whereof it tooke his name: the floures are smal, and like those of Orach: the seed is black, three square, like, but lesse than that of Buck-wheat. The whole plant is not onely a hurtfull weed, but of an euill smell also, and too frequently found amongst corne. *Dodonæus* calls this *Convolvulum nigrum*: and *Helxine, Cissampelos*: *Tabernamontanus, Volubilis nigra*: and *Lobel, Helxine Cissampelos altera Atriplicis effigie.*

¶ *The Place.*

All these kindes of Bindweeds do grow very plentifully in most parts of England, ‡ The third and fourth excepted. ‡

¶ *The Time.*

They do floure from May to the end of August.

¶ *The Names.*

The great Bindweed is called in Greeke σμίλαξ λεία: in Latine, *Smilax Læuis*: of *Galen* and *Paulus Ægineta*, μίλαξ λεία: it is surnamed *Læuis* or smooth, because the stalkes and branches thereof haue no prickles at all. *Dolichus* called also *Smilax hortensis*, or Kidney beane, doth differ from this: and likewise *Smilax* the tree, which the Latines call *Taxus*: in English, the Yew tree. The later Herbarists do call this Bindweed *Volubilis maior, Campanella, Funis arborum, Convoluulus albus*, and *Smilax læuis maior*: in like manner *Pliny* in his 21. booke, 5. chapt. doth also name it *Conuoluulus*. It is thought to be *Ligustrum*, not the shrub priuet, but that which *Martial* in his first booke of Epigrams speaketh of, writing against *Procillus*.

The small Bindweed is called *Convoluulus minor*, and *Smilax læuis minor, Volubilis minor*: in high Dutch, **Windkraut**: in Low Dutch, **Wrange**: in French, *Liseron*: in Italian, *Vilucchio*: in Spanish, *Campanilla Yerua*: in English, Withwinde, Bindeweed, and Hedge-bels.

¶ *The Nature.*

These herbs are of an hot and dry temperature.

¶ *The Vertues.*

A The leaues of blacke Bindweed called *Helxine Cissampelos*, stamped and strained, and the iuice drunken, doth loose and open the bellie exceedingly.

B The leaues pound and laid to the grieued place, dissolueth, wasteth, and consumeth hard lumps and swellings, as *Galen* saith.

The

D The rest of the Bindweeds are not fit for medicine, but vnprofitable weeds, and hurtfull vnto each thing that groweth next vnto them.

† The description which our Author intended in the first place for *Volubilis nigra*, and took out of the 274. page of the *Aduersaria*, but so confusedly and imperfectly, neither agreeing with that he intended, I haue omitted as impertinent, and made his later, though also vnperfect description, somewhat more compleat and agreeable to the plant figured and intended.

Chap. 318. *Of Blew Bindweed.*

¶ *The Description.*

1 Blew Bindweed bringeth forth long, tender, and winding branches, by which it climeth vpon things that stand neere vnto it, and foldeth it selfe about them with many turnings and windings, wrapping it selfe against the Sun, contrary to all other things whatsoeuer, that with their clasping tendrels do embrace things that stand neere vnto them; whereupon doe grow broad cornered leaues very like vnto those of Iuie, something rough and hairy, of an ouerworne russet greene colour: among which come forth most pleasant floures bell fashion, somthing cornered as are those of the common Bindweed, of a most shining azure colour tending to purple: which being past, there succeed round knobbed seed vessels, wherein is contained long blackish seed of the bignesse of a Tare, and like vnto those of the great hedge Bindweed. The root is thready, and perisheth at the first approch of Winter.

1 *Convolvulus Cæruleus.*
Blew Binde-weed.

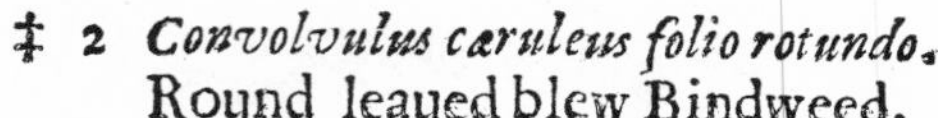

‡ 2 *Convolvulus cæruleus folio rotundo.*
Round leaued blew Bindweed.

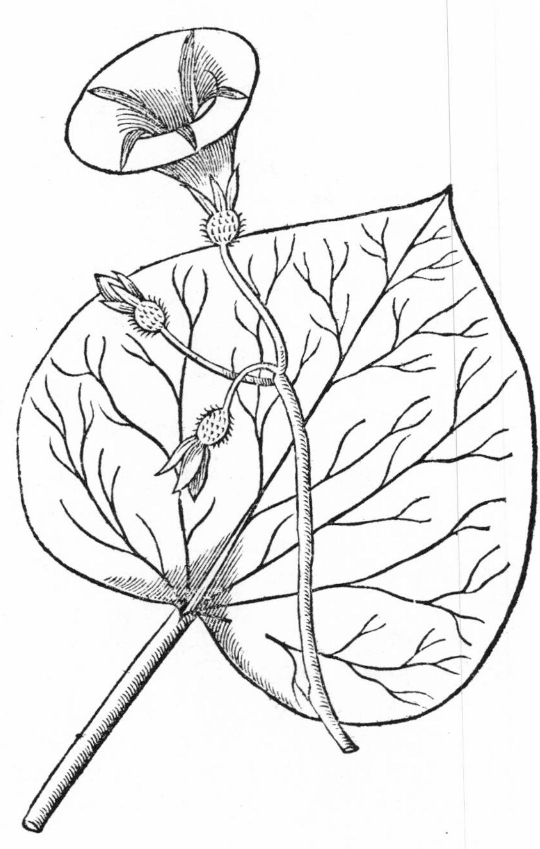

‡ 2 There are also kept in our gardens two other blew floured Bindweeds. The one a large and great plant, the other a lesser. The great sends vp many large and long winding branches, like those of the last described, and a little hairie: the leaues are large and roundish, ending in a sharpe point: the floures are as large as those of the great Bindweed, and in shape like them, but blew of colour, with fiue broad purplish veines equally distant each from other: and these floures commonly grow three neere together vpon three seuerall stalks some inch long, fastened to another stalke

‡ 3 *Convolvulus cæruleus minor, folio oblongo.*
Small blew Bindweed.

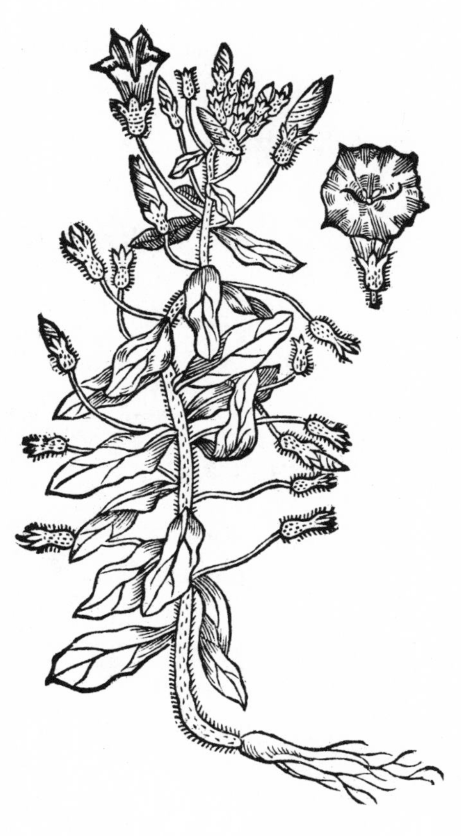

ſtalke ſome handfull long: the cup which holds the floures, and afterwards becomes the ſeed veſſell, is rough and hairie: the ſeed is blacke, and of the bigneſſe of a Tare: the root is ſtringie, and laſts no longer than to the perfecting of the ſeed. I haue onely giuen the figure of the leafe and floure largely expreſt, becauſe for the root and manner of growing it reſembles the laſt deſcribed.

3 This ſmall blew Bindweed ſendeth forth diuers long ſlender creeping hairie branches, lying flat vpon the ground, vnleſſe there be ſomething for it to reſt vpon: the leaues be longiſh and hairy, and out of their boſomes (almoſt from the bottome to the tops of the ſtalks) come ſmall foot-ſtalkes carrying beautifull floures of the bigneſſe and ſhape of the common ſmal Bindweed, but commonly of three colours; that is, white in the verie bottome, yellow in the middle, and a perfect azure at the top; and theſe twine themſelues vp, open and ſhut in fiue plaits like as moſt other floures of this kinde doe. The ſeed is contained in round knaps or heads, and is blacke and cornered: the root is ſmall, and periſhes euery yeare. *Bauhine* was the firſt that ſet this forth, and that by the name of *Convolvulus peregrinus cæruleus folio oblongo.* ‡

¶ *The Place.*

The ſeede of this rare plant was firſt brought from Syria and other remote places of the world, and is a ſtranger in theſe Northern parts; yet haue I brought vp and nouriſhed it in my garden vnto flouring, but the whole plant periſhed before it could perfect his ſeed.

¶ *The Time.*

The ſeed muſt be ſowne as Melons and Cucumbers are, and at the ſame time: it floured with me at the end of Auguſt.

¶ *The Names.*

It is called *Campana Lazula*, and *Lazura*: of the later Herbariſts *Campana Cærulea*, and alſo *Convolvulum Cæruleum*: it is thought to be the *Liguſtrum nigrum*; of which *Columella* in his tenth booke hath made mention.

Fer calathis violam, & nigro permiſta liguſtro
Balſama cum Caſſia nectens, &c.

In baskets bring thou Violets, and blew Bindweed withall,
But mixed with pleaſant Baulme, and Caſſia medicinall.

For if the greater ſmooth Withwinde, or Bindweed be *Liguſtrum*, then may this be not vnproperly called *Liguſtrum nigrum*: for a blew purple colour is oftentimes called blacke, as hath beene ſaid in the blacke Violet. But there be ſome that would haue this Bindweed to be *Granum nil Auicennæ*, of which he writeth in the 306. chapter; the which differeth from that *Nil* that is deſcribed in the 512. chapter. For this is *Iſatis Græcorum*, or the Grecian Woad: but that is a ſtrange plant, and is brought from India, as both *Auicen* and *Serapio* doe teſtifie: *Auicen* in this manner: what is *Granum Nil?* It is *Cartamum Indum*: and *Serapio* thus; *Habal Nil*, is *Granum Indicum*, in *cap.* 283. where the ſame is deſcribed in theſe words: [The plant thereof is like to the plant of *Leblab*, that is to ſay of *Convolvulus*, or Bindweed, taking hold of trees with his tender ſtalks: it hath both green branches and leaues, and there commeth out by euery leafe a purple floure, in faſhion of the Belfloures: and when the floure doth fall away, it yeeldeth a ſeed in ſmall cods (I read little heads) in in which are three graines, leſſer than the ſeedes of Staueſaker] to which deſcription this blew Bindweed is anſwerable.

There be alſo other ſorts of Bindweeds, which be referred to *Nil Auicennæ*; which no doubt may be kinds of *Nil*; for nothing gainſaith it why they ſhould not be ſo. Therefore to conclude, this beautifull Bindeweed, which we call *Convolvulus Cæruleus*, is called of the Arabians *Nil*: of *Serapio, Hab al Nil*: about Alepo and Tripolis in Syria the inhabitants call it *Haſmiſen*: the Italians, *Campana aZurea*: of the beautifull azured floures, and alſo *Fior de notte*, becauſe his beautie appeares moſt in the night.

¶ *The Temperature.*

Convolvulus Cæruleus, or *Nil*, as *Auicen* ſaith, is hot and drie in the firſt degree: but *Serapio* maketh it to be hot and drie in the third degree.

¶ *The Vertues.*

A It purgeth and voideth forth raw, thicke, flegmaticke, and melancholicke humours: it driues out all kinde of wormes, but it troubleth the belly, and cauſeth a readineſſe to vomit, as *Auicen* ſaith: it worketh ſlowly, as *Serapio* writeth; in whom more hereof may be found, but to little purpoſe, wherefore we thinke good to paſſe it ouer.

Chap. 319. *Of Scammonie, or purging Bindweed.*

1 *Scammonium Syriacum*
Syrian Scammonie.

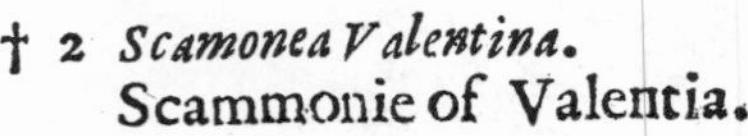
† 2 *Scamonea Valentina.*
Scammonie of Valentia.

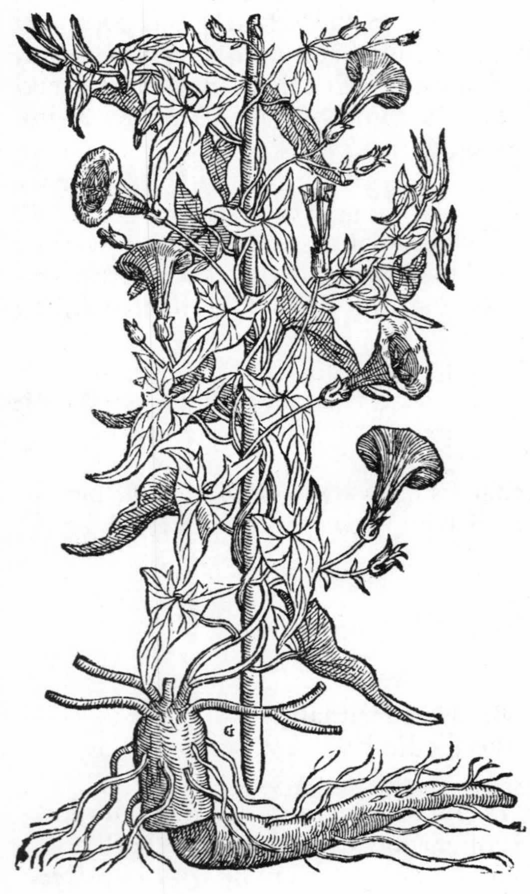

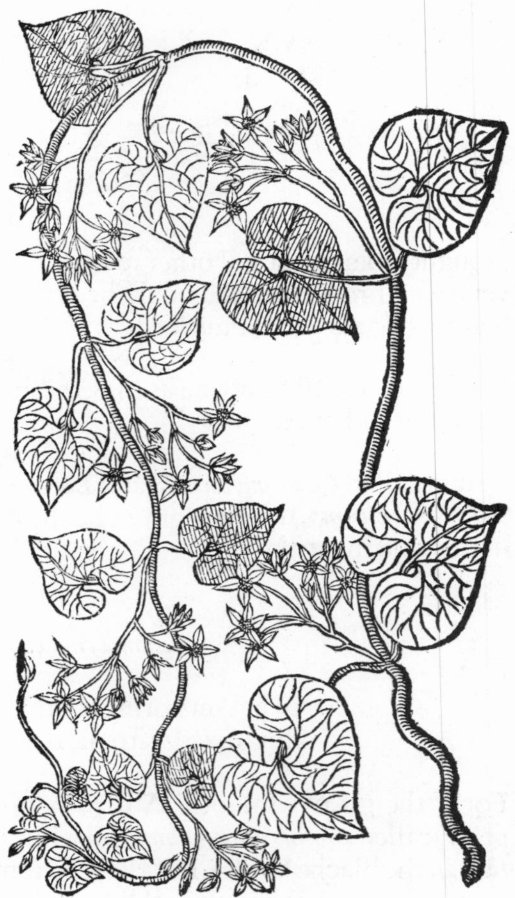

¶ *The Deſcription.*

1 SCammonie of Syria hath many ſtalks riſing from one root, which are long, ſlender, and like the claſping tendrels of the vine, by which it climeth and taketh hold of ſuch things as are next vnto it. The leaues bee broad, ſharpe pointed like thoſe of the ſmooth or hedge Bindweed: among which come forth very faire white floures tending to a buſh colour, bell faſhion. The root is long, thicke, and white within; out of which is gathered a iuice that beeing hardned

hardned,is greatly vsed in Physicke: for which consideration, there is not any plant growing vpon the earth,the knowledge whereof more concerneth a Physition,both for his shape and properties, than this Scammonie, which *Pena* calleth *Lactaria scansoriaque volvula*, that is, milkie and climbing Windweed, whereof it is a kinde; although for distinction sake I haue placed them as two seuerall kindes. And although this herbe be suspected, and halfe condemned of some learned men, yet there is not any other herbe to be found, whereof so small a quantitie will do so much good: neither could those which haue carped at it, and reproued this herbe, finde any simple in respect of his vertues to be put in his roome: and hereof insueth great blame to all practitioners, who haue not endeuoured to bee better acquainted with this herbe, chiefely to auoid the deceit of the craftie Drug-seller and Medicine-maker of this confected Scammonie, brought vs from farre places, rather to be called I feare infected Scammonie, or poisoned Scammonie, than confected. But to auoid the inconueniences hereof, by reason of the counterfeiting and ill mixing thereof: I haue therefore thought good to set downe what I haue taken out of the diligent, and no lesse learned obseruations of *Pena*, concerning this plant, *Anno* 1561, or 1562. *Vid. advers. pag.* 272.

‡ 3 *Scammonium Monspeliense.*
French Scammonie.

Sequinus Martinellus an Apothecarie of Venice, being a most diligent searcher of Simples; that he might haue the right Scammony of Antioch, trauelled into Syria, where from the citie of Alepo hee sent an 100. weight of the iuice of Scammonie of Antioch, prepared and hardned into a lumpe, at the making whereof he was present himselfe. This man sent also of the seeds thereof, which in all points answered the cornered seed of *Volubilis*; which being sowne in the beginning of the Spring at Padua and Venice, grew vp to the form of a braue & goodly *Convolvulus*, in leaues, floures, and shew so like vnto our *Cissampelos*, that a man would haue taken it for the same without controuersie, sauing that the root was great, and in bignesse equall to the great Brionie, as also in tendernesse. The outward bark of the root was of a dusky colour, and white within: the inner pith beeing taken forth seemeth in all mens iudgements to be the same and the best allowed *Turbith officinarum*: and yet it differeth from Turbith, in that, that it is more brittle, and will more easily bee broken, though the pith in Scammonie bee no lesse gummie and ful of milkie iuice, than Turbith. Further *Pena* reporteth, that afterward hee sent of this seed vnto Antwerpe, where it grew very brauely, the climing strings and branches growing vp to the height of fiue or six cubites, not differing from that which was sowne in Italie. Also *William Dries* of Antwerp, a most excellent Apothecary, did cut off the branches of his Antwerpian Scammonie from the root, and dried them, planted the seeds in his garden, and conferred the superfluous branched roots with the Turbith of Alexandria and could not find them to differ or disagree the one from the other in any point. But he that will know more concerning the making, difference, choice and vse of Scammony, let him reade *Pena* in his chapter of Scammonie, in the place formerly cited, where he shall finde many excellent secrets worthy the noting of those which would know how to vse such rare and excellent medicines.

2 Scammony of Valentia (whereof I haue plentie in my garden) is also a kinde of Bindweed, growing naturally by the sea side vpon the grauelly shore, by the mouth of the riuer Rhodanus, at the waters called *Aquas Marianas*, where the Apothecaries of Montpellier gather of it great plentie, who haue attempted to harden the milkie iuice thereof, to vse it in stead of Scammonie of Antioch. This plant bringeth forth many slender branches, which will climbe and very well run vpon a pole; as being supported therewith, and mounteth to the height of fiue or six cubits, climbing & ramping like the first kinde of Scammonie. The leaues are greene, smooth, plaine, and sharpe pointed,

pointed, which being broken do yeeld abundance of milke: the floures are white, small, and starre-fashion: the roots white and many, shooting forth sundry other roots, whereby it mightily increaseth.

† 3 This strange kinde of Scammonie, which *Clusius* maketh rightly to be *Periplocæ species*, hath very many long branches ramping and taking hold of such things as do grow neere vnto them, of a darkish ashe colour: whereupon do grow leaues sharp pointed, crooked at the setting on of the stalke like those of the blacke Bryonie, and likewise of an ashe colour, set together by couples: from the bosome whereof thrust forth small tender foot-stalkes, whereon are placed small white floures starre-fashion: the seeds are contained in long coddes, and are wrapped vp in downe, like as those of Swallow-wort. The root is very long, slender, and creeping, like that of the small Bindweed, so that if it once take in any ground, it can hardly be destroyed.

¶ *The Place.*

It doth grow in hot regions, in a fat soile, as in Misia, Syria, and other like countries of Asia; it is likewise found in the iland of Candia, as *Bellonius* witnesses; from whence I had some seeds, of which seed I receiued two plants that prospered exceeding well; the one whereof I bestowed vpon a learned Apothecarie of Colchester, which continueth to this day, bearing both floures and ripe seed. But an ignorant weeder of my garden plucked mine vp, and cast it away in my absence, in stead of a weede: by which mischance I am not able to write heereof so absolutely as I determined: it likewise groweth neere vnto the sea side about Tripolis in Syria, where the inhabitants doe call it *Meudheudi*.

¶ *The Time.*

It floured in my garden about S. *Iames* tide, as I remember; for when I went to Bristow Faire, I left it in floure; but at my returne it was destroyed as aforesaid.

¶ *The Names.*

The Greekes call it σκαμμωνία: the Latines, *Scammonium*, so naming not onely the plant it selfe, but also the hard and condensed iuice: of the Apothecaries, *Scammonea*; and when it is prepared, *Diagridium*: as though they should say, δακρύδιον: which signifieth a little teare: both the herbe and iuice are named *Scamony*: of *Rhasis*, *Coriziola*.

¶ *The Temperature.*

The iuice doth mightily purge by the stoole, and is the strongest purge whatsoeuer; for as *Oribasius* saith, it is in no part ouercome by those things which stir and moue the body. It worketh the same not vehemently by any hot qualitie, but by some other hid and secret propertie of the whole substance; for there is no extremitie of heat perceiued in it by taste: for with what liquor or thing soeuer it is mixed, it giueth vnto it no bitternesse, biting, or other vnpleasant taste at all, and therefore it is not to be accounted among the extreme hot medicines, but among those that are moderately hot and drie.

¶ *The Vertues.*

A It clenseth and draweth forth especially choler: also thinne and waterish humours, and oftentimes flegme, yet is it as *Paulus* teacheth more hurtfull to the stomacke than any other medicine.

B *Mesues* thinketh that it is not onely troublesome and hurtfull to the stomacke, but also that it shaueth the guts, gnawing and fretting the intrails; openeth the ends of the veins, and through the essence of his whole substance, it is an enemy to the heart, and to the rest of the inward parts: if it be vsed immoderately and in time not conuenient, it causeth swounings, vomitings, and ouerturnings of the stomacke, scouring, the bloudy flux and vlcers in the lower gut, which bring a continuall desire to the stoole.

C These mischiefes are preuented if the Scammonie be boiled in a Quince and mixed with the slime or mucilage of *Psillium*, called Fleawoort, the pap or pulp of Prunes, or other things that haue a slimie iuice, with a little Masticke added, or some other easie binding thing.

D *Plinie* affirmeth that the hurt thereof is taken away if Aloes be tempered with it: [Scammonie (saith he) ouerthroweth the stomack, purgeth choler, looseth the belly vnlesse two drams of Aloes be put vnto one scruple of it] which also *Oribasius* alloweth of in the first booke of his *Synopses*, and the seuenth booke of his medicinall Collections.

E The old Physitions were also wont to boile Scammonie in a Quince, and to giue the Quince to be eaten, hauing cast away the Scammonie: and this Quince so taken doth moue the belly without any hurt vnto the stomacke, as *Galen* in his first booke of the Faculties of Nourishments doth set downe, and likewise in his third booke of the Faculties of simple Medicines.

F The Apothecaries do vse Scammonie prepared in a Quince, which as we haue said they name *Diagridium*, and do mix it in diuers compositions.

They

They keepe vſually in their ſhops two compoſitions, or electuaries, the one of *Pſillium* or Flea- G
woort, ſet downe by *Meſue*: the other of Prunes fathered vpon *Nicolaus*, which were deuiſed for the tempering and correction of Scammony, and be commended for hot burning agues, and tertians, and for what diſeaſes ſoeuer that proceed of choler.

Galen hath taken Maſticke and *Bdellium* out of the pilles called *Cochiæ*, which alſo conteine in H
them a great and ſufficient quantitie of Scammonie, as we may reade in his firſt booke of medicines according to the places affected, which alſo we meane to touch in the chapter of Coloquintida, where we intend to intreat at large concerning maſticke, and other binding things, that are accuſtomed to be mixed for the correction of ſtrong and violent purgers.

The quantitie of Scammony, or of *Diagridium* it ſelfe, as *Meſue* writeth, is from fiue graines to I
ten or twelue: it may be kept as the ſame Author ſheweth, foure yeeres: *Pliny* iudgeth it to be after two yeeres little worth: it is to be vſed, ſaith he, when it is two yeeres old, and it is not good before, nor after. The mixing or otherwiſe the vſe thereof, more than is ſet downe, I thinke it not expedient to ſet forth in the Phyſicall vertues of Scammony, vpon the receipt whereof many times death inſueth: my reaſons are diuers, for that the ſame is very dangerous, either if too great a quantitie thereof be taken, or if it be giuen without correction; or taken at the hands of ſome runnagate phyſicke-monger, quackſaluer, old women-leaches, and ſuch like abuſers of Phyſicke, and deceiuers of people. The vſe of Scammony I commit to the learned, vnto whome it eſpecially and onely belongeth, who can very carefully and curiouſly vſe the ſame.

‡ The titles of the ſecond and third were formerly tranſpoſed, and both the figures belonged to the ſecond deſcription, which was of the *Scammonium Monſpelienſe* of the *Aduerſ.* being the ſame with the *Scammonea Valentina* of *Cluſius*.

CHAP. 320. *Of Briony, or the white Vine.*

¶ *The Kindes.*

There be two kindes of Bryony, the one white, the other blacke: of the white Briony as followeth.

Bryonia alba.
White Bryonie.

¶ *The Deſcription.*

White Briony bringeth forth diuers long and ſlender ſtalkes with many claſping tendrels like the Vine, wherewith it catcheth hold of thoſe things that are next vnto it. The leaues are broad, fiue cornered, and indented like thoſe of the Vine; but rougher, more hairie, and whiter of colour. The floures be ſmall and white, growing many together. The fruite conſiſteth in little cluſters, the berries whereof are at the firſt greene, and red when they be ripe. The roote is very greate, long, and thicke, growing deepe in the earth, of a white yellowiſh colour, extreame bitter, and altogether of an vnpleaſant taſte. The Queenes chiefe Surgion Mr. *William Godorous*, a very curious and learned gentleman ſhewed me a root hereof, that waied halfe an hundred weight, and of the bigneſſe of a child, of a yeere old.

¶ *The Place.*

Briony groweth almoſt euery where among pot-herbes, hedge-buſhes, and ſuch like places.

¶ *The Time.*

It floureth in May, and bringeth forth his grapes in Autumne.

¶ *The Names.*

Bryony is called in Greek, ἄμπελος λευκή: in Latine,

 Vitis

Vitis alba, or white Vine, and it is named, ἄμπελος, because it is not onely like the Vine in leaues, but also for that it bringeth forth his fruite made vp after the likenesse of a little cluster, although the berries stand not close together: it is called of *Pliny*, *Bryonia*, and *Madon*: of the Arabians, *Alphesera*: of *Matthaeus Sylvaticus*, *Viticella*: in the poore mans Treasure, *Rorastrum*: of *Apuleius*, *Apiastellum*, *vitis Taminia*, *Vitis alga*, and *Vitalba*: in high Dutch, **Stuchwurtz**: in low Dutch, **Brionie**: in English, Bryony, white Bryony, and tetter Berrie: in French, *Couleuree*: in Italian, *Zucca sylvatica*: in Spanish, *Nueza blanca*.

¶ *The Temperature.*

White Briony is in all parts hot and dry, exceeding the third degree, especially of heate, with an exceeding great force of clensing and scouring, by reason whereof it purgeth and draweth forth, not onely cholericke and flegmaticke humours, but also watrie.

¶ *The Vertues.*

A *Dioscorides* writeth that the first springs or sproutings being boiled and eaten, do purge by siege and vrine. *Galen* saith, that all men vse accustomably to eate of it in the spring time, and that it is a nourishment wholesome, by reason of the binding qualitie that it hath; which is to be vnderstood of those of the wilde Vine, called in Latine, *Tamus*; and not of the sproutings of this plant; for the sproutings of the first springs of white Bryony are nothing binding at all, but do mightily purge the belly, and torment the stomacke.

B *Dioscorides* also affirmeth, that the juice of the root being pressed out in the spring, and drunke with meade or honied water, purgeth flegme: and not onely the juice, but also the decoction of the root draweth forth flegme, choler, and waterish humours, and that very strongly; and it is withall oftentimes so troublesome to the stomacke, as it procureth vomite.

C This kinde of strong purgation is good for those that haue the dropsie, the falling sickenesse, and the dizzinesse and swimming of the braine and head, which hath continued long, and is hardly to be remooued: yet notwithstanding it is not dayly to be giuen (as *Dioscorides* admonisheth) to them that haue the falling sickenesse, for it will be troublesome enough to take it now and then: and it is (as we haue said) an exceeding strong medicine, purging with violence, and very forceable for mans nature.

D The root put vp in manner of a pessary bringeth forth the dead child and afterbirth: being boiled for a bath to sit in, it worketh the same effect.

E It scoureth the skin, and taketh away wrinckles, freckles, sunne burning, blacke marks, spots, and scars of the face, being tempered with the meale of vetches or Tares, or of Fenugreeke: or boiled in oile till it be consumed; it taketh away blacke and blew spots which come of stripes: it is good against Whitlowes: being stamped with wine and applied it breaketh biles, and small apostumes, it draweth forth splinters and broken bones, if it be stamped and laid thereto.

F The same is also fitly mixed with eating medicines, as *Dioscorides* writeth.

G The fruit is good against scabs and the leprie, if it be applied and annointed on, as the same Author affirmeth.

H *Galen* writeth, that it is profitable for Tanners to thicken their leather hides with.

I Furthermore, an electuary made of the roots and hony or sugar, is singular good for them that are short winded, troubled with an old cough, paine in the sides, and for such as are hurt and bursten inwardly: for it dissolueth and scatereth abroad congealed and clottered bloud.

K The root stamped with salt is good to be laid vpon filthy vlcers and scabbed legs. The fruite is likewise good to the same intent if it be applied in manner aforesaid.

L The root of Bryony and of wake-Robin stamped with some sulphur or brimstone, and made vp into a masse or lump and wrapped in a linnen clout, taketh away the morphew, freckles, and spots of the face, if it be rubbed with the same being dipped first in vineger.

Chap. 321. *Of blacke Brionie, or the wilde Vine.*

¶ *The Description.*

1 The black Bryony hath long flexible branches of a woodie substance, couered with a gaping or clouen barke growing very farre abroad, winding it selfe with his small tendrels about trees, hedges, and what else is next vnto it, like vnto the branches of the Vine. The leaues are like vnto those of Iuie or garden Nightshade, sharpe pointed, and of a shining greene colour: the floures are white, small, and mossie; which being past, there succeed little clusters of red berries

1 *Bryonia nigra.*
Blacke Bryonie.

berries, somwhat bigger than those of the small Raisons, or Ribes, which wee call Currans, or small Raisins. The root is very great and thick, oftentimes as bigge as a mans legge, blackish without, and verie clammie or slimie within; which being but scraped with a knife, or any other thing fit for that purpose, it seemeth to be a matter fit to spread vpon cloth or leather in manner of a plaister or Seare-cloth: which being so spread and vsed, it serueth to lay vpon many infirmities, and vnto verie excellent purposes, as shall bee declared in the proper place.

2 The wilde Blacke Bryonie resembleth the former, as well in slender Vinie stalkes as leaues; but clasping tendrels hath it none, neuerthelesse by reason of the infinite branches, and the tendernesse of the same, it taketh hold of those things that stand next vnto it, although easie to bee loosed, contrarie vnto the other of his kinde. The berries heereof are blacke of colour when they be ripe. The root also is blacke without, and within of a pale yellow colour like box. ‡ This which is here described is the *Bryonia nigra* of *Dodonæus*; But *Bauhine* calleth it *Bryonia Alba*; and saith it differeth from the common white Bryonie, onely in that the root is of a yellowish boxe colour on the inside, and the fruit or berries are blacke when as they come to ripenesse.

Bryonia nigra florens non fructum ferens.

3 This is altogether like the first described in roots, branches, and leaues; onely the foot-stalks whereon the floures grow are about eight or nine inches long: the floures are something greater, hauing neither before or after their flouring any berries or shew thereof; but the floures and foot-stalks do soone wither and fall away: this I haue heretofore, and now this Sommer, 1621, diligently obserued, because it hath not beene mentioned or obserued by any that I know. *Iohn Goodyer.* ‡

¶ *The Place.*

The first of these plants doth grow in hedges and bushes almost euery where.

The second groweth in Hessia, Saxonie, Westphalia, Pomerland, and Misnia, where white Bryonie doth not grow, as *Valerius Cordus* hath written, who saith that it growes vnder Hasell-trees, neere vnto a citie of Germanie called Argentine, or Strawsborough.

¶ *The Time.*

They spring in March, bring forth their floures in May, and their ripe fruit in September.

¶ *The Names.*

Blacke Bryonie is called in Greeke ἄμπελος ἀγρία: in Latine, *Bryonia nigra*: and *Vitis sylvestris*, or wilde Vine; notwithstanding it doth not a little differ from *Labrusca*, or *Vitis Vinefera sylvestris*, that is to say, from the wilde vine, which bringeth forth wine, which is likewise called *Ampelos agria*: Why both these were called by one name, *Pliny* was the cause, who could not sufficiently expound them in his 23. booke, first chapter; but confounded them, and made them all one, in which errour are also the Arabians.

This wilde Vine also is called in Latine, *Tamus*, and the fruit thereof *Vua Taminia*. *Pliny* nameth it also *Salicastrum*. *Ruellius* saith that in certaine shops it is called *Sigillum B. Mariæ*; it is also called *Cyclaminus altera* but not properly: in English, Blacke Bryonie, wilde Vine, and our Ladies-seale.

¶ *The Temperature.*

The roots of the wild Vine are hot and drie in the third degree: the fruit is of like temperature, but yet nor so forceable: both of them scoure and waste away.

¶ *The Vertues.*

A *Dioſcorides* ſaith, that the roots do purge wateriſh humours, and are good for ſuch as haue the dropſie; if they be boiled in wine, adding vnto the wine a little ſea water, and bee drunke in three ounces of faire freſh water: he ſaith furthermore, that the fruit or berries doth take away the Sunburne and other blemiſhes of the skin.

B The berries do not onely clenſe and remoue ſuch kinde of ſpots, but do alſo very quickly waſte and conſume away blacke and blew marks that come of bruiſes and drie beatings, which thing alſo the roots performe being laid vpon them.

C The young and tender ſproutings are kept in pickle, and reſerued to be eaten with meat as *Dioſcorides* teacheth. *Matthiolus* writeth that they are ſerued at mens tables alſo in our age in Tuſcanie: others report the like alſo to be done in Andaloſia, one of the kingdomes of Granado.

D It is ſaid that ſwine ſeeke after the roots hereof, which they dig vp and eat with no leſſe delight than they do the roots of *Cyclaminus*, or *Panis porcinus*, whereupon it was called *Cyclaminus altera*; or Sow-bread; if this reaſon ſtand for good, then may we in like manner ioine hereunto many other roots, and likewiſe call them *Cyclaminus altera*, or Sow-bread: for ſwine do not ſeeke after the roots of this onely, digge them vp and greedily deuoure them, but the roots of diuers other plants alſo, of which none are of the kindes of Sow-bread. It would therefore be a point of raſhneſſe to affirm *Tamus* or our Ladies-ſeale to be a kinde of Sowbread, becauſe the roots thereof are pleaſant meat to ſwine.

E The root ſpred vpon a piece of ſheepes leather, in manner of a plaiſter whileſt it is yet freſh and green, taketh away blacke or blew marks, all ſcars and deformitie of the skin, breaketh hard apoſtems, draweth forth ſplinters and broken bones, diſſolueth congealed bloud; and being laid on and vſed vpon the hip or huckle bones, ſhoulders, armes, or any other part where there is great pain and ache, it taketh it away in ſhort ſpace, and worketh very effectually.

† The figure that was formerly in the ſecond place of this chapter did no waies agree with the diſcription, for it was of the *Viorna* or Trauellors ioy (hereafter to be mentioned; which *Tabernamontanus*, (whoſe figures our Author made vſe of) calls *Vitis nigra ſecunda*.

CHAP. 322. *Of Bryonie of Mexico.*

¶ *The Deſcription.*

1 THat plant which is now called *Mechoacan*, or Bryonie of Mexico, commeth verie neere the kinds of Bindweeds, in leaues and trailing branches, but in roots like the Brionies; for there ſhooteth from the root thereof many long ſlender tendrels, which do infinitly graſpe and claſpe about ſuch things as grow or ſtand next vnto them: whereupon grow great broad leaues ſharpe pointed, of a darke greene colour, in ſhape like thoſe of our Ladies-ſeale, ſomwhat rough and hairie, and a little biting the tongue: among the leaues come forth the floures (as *Nicolaus Monardus* writeth) not vnlike thoſe of the Orenge tree, but rather of the golden Apple of loue, conſiſting of fiue ſmall leaues: out of the middeſt whereof commeth forth a little clapper or peſtell in manner of a round lumpe, as big as a Haſell nut; which being diuided with a thin skin, or membrane, that commeth through it, openeth into two parts, in each whereof are contained two ſeeds, as bigge as Peaſe, in colour blacke and ſhining. The root is thicke and long, verie like vnto the root of white Bryonie, whereof we make this a kinde, although in the taſte of the roots there is ſome difference: for the root of white Bryonie hath a bitter taſte, and this hath little or no taſte at all.

2 The Bryony, or *Mechoacan* of Peru groweth vp with many long trailing flexible branches, interlaced with diuers Vinie tendrels, which take hold of ſuch things as are next or neere vnto them, euen in ſuch manner of claſping and climing as doth the blacke Bryonie, or wine Vine, whereunto it is very like almoſt in each reſpect, ſauing that his moſſie floures do ſmell very ſweetly. The fruit as yet I haue not obſerued, by reaſon that the plant which doth grow in my garden did not perfect the ſame, by occaſion of the great rain and intemperate weather that hapned in *An.* 1596. but I am in good hope to ſee it in his perfection, & then we ſhall eaſily iudge whether it be that right *Mechoacan* that hath been brought from Mexico and other places of the Weſt Indies or no? The root by the figure ſhould ſeeme to anſwer that of the wilde Vine, but as yet thereof I cannot write certainly.

‡ 3 There is brought to vs and into vſe of late time the root of another plant, which ſeemes

to

1 Mechoacan.
Bryonie of Mexico.

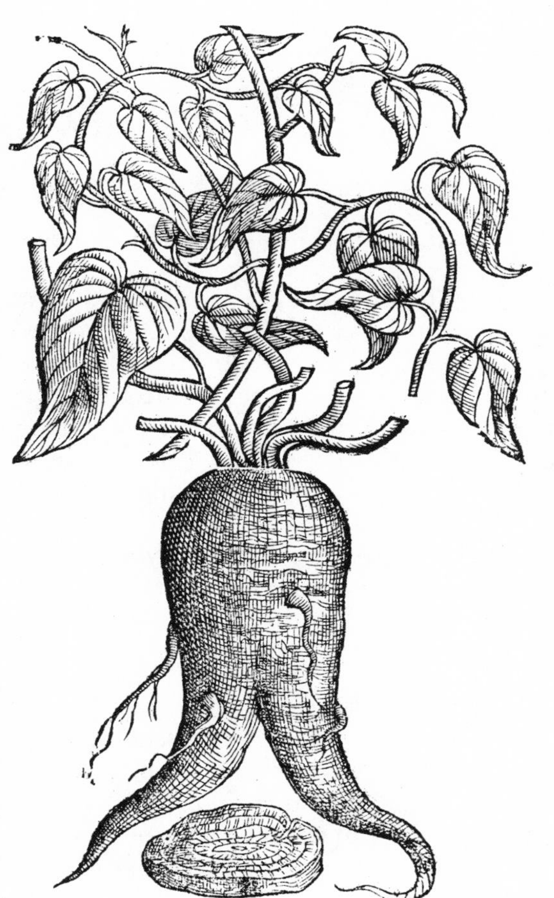

2 Mechoacan Peruviana.
Bryonie of Peru.

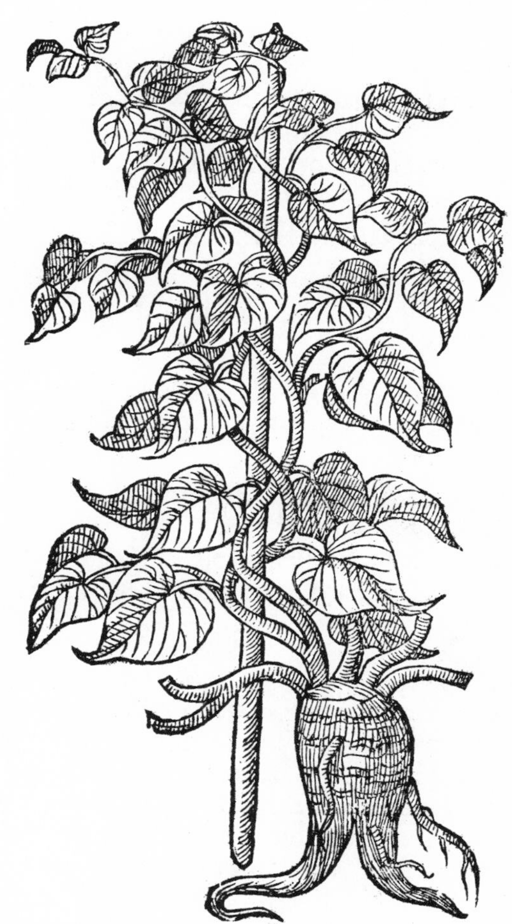

haue much affinitie with *Mechoacan*, and therefore *Bauhine* hath called it *Bryonia Mechoacan nigricans*, and thus deliuers the hiſtorie thereof. [It is a root like *Mechoacan*, but couered with a blackiſh barke, and reddiſh (or rather grayiſh) on the inſide: and cut into ſlices, it was brought ſome yeares agone out of India by the name of *Chelapa*, or *Gelapa*. It is called by thoſe of Alexandria and Marſeilles *Ialapium* or *Gelapum*: and of thoſe of Marſeilles it is thought the blacke or male *Mechoacan*: The taſte is not vngratefull, but gummy, and by reaſon of the much gummineſſe, put to the fire it quickly flames: it in facultie exceeds the common *Mechoacan*; for by reaſon of the great gummineſſe it more powerfully purgeth ſerous humours with a little griping, alſo it principally ſtrengthens the liuer and ſtomacke; wherefore it is ſafely giuen in the weight of ʒj. and performes the operation without nauſeouſneſſe It is vſually giuen in Succorie water, or ſome thin broth three houres before meat.] Thus much *Bauhine*, who ſaith it was firſt brought to theſe parts eleuen yeres before he ſet forth his *Prodromus*, ſo that was about 1611. It hath beene little vſed here till within this ten yeares. ‡

¶ *The Place.*

Some write that *Mechoacan* was firſt found in the Prouince of New Spaine, neere vnto the citie of Mexico or Mexican, whereof it tooke his name. It groweth likewiſe in a prouince of the Weſt Indies called *Nicaragua* and *Quito*, where it is thought the beſt doth grow.

¶ *The Names.*

It beareth his name as is ſaid, of the prouince in which it is found. Some take it to be *Bryonia ſpecies*, or to be a kinde of Bryonie: but ſeeing the root is nothing bitter, but rather without taſte, it hath little agreement with Bryonie; for the root of Bryonie is verie bitter. Diuers name it *Rha album*, or white Rubarbe, but vnproperly, being nothing like. It commeth neere vnto Scammony, and if I might yeeld my cenſure, it ſeemeth to be *Scammonium quoddam Americanum*, or a certain Scammonie of America. Scammonie creepeth, as wee haue ſayd, after the manner of Bindweed. The root is both white and thicke: the iuice hath but little taſte, as alſo hath this of *Mechoacan*: it is called in Engliſh, Mechoca and Mechocan, and may bee called Indian Bryonie.

¶ The

¶ *The Temperature.*

The root is of a meane temperature between hot and cold, but yet drie.

¶ *The Vertues.*

A It purgeth by siege, especially flegme, and then waterish humours. It is giuen from one ful dram weight to two, and that with wine, or with some distilled water (according as the disease requireth) or els in flesh broth.

B It is to be giuen with good effect to all, whose diseases proceed of flegme and cold humors. It is good against head-ache that hath continued long, old coughes, hardnesse of breathing, the colick, paine of the kidneies and ioints, the diseases of the reines and belly.

Chap. 323. *Of the Manured Vine.*

¶ *The Kindes.*

THe Vine may be accounted among those plants that haue need of staies and props, and cannot stand by themselues; it is held vp with poles and frames of wood, and by that meanes it spreadeth all about and climbeth aloft: it ioyneth it selfe vnto trees, or whatsoeuer standeth next vnto it.

Of Vines that bring forth wine, some be tame and husbanded; and others that be wilde: of tame Vines there are many that are greater, and likewise another sort that be lesser.

¶ *The Description.*

THe trunke or bodie of the Vine is great and thicke, very hard, couered with many barkes, and those full of cliffes or chinkes; from which grow forth branches, as it were armes, many waies spreading; out of which come forth iointed shoots and springs: and from the bosome of those ioints, leaues, and clasping tendrels; and likewise bunches or clusters filled ful of grapes: the leaues be broad, something round, fiue cornered, and somewhat indented about the edges; amongst which come forth many clasping tendrels, that take hold of such props or staies as do stand next vnto it. The grapes do differ both in colour and greatnesse, and also in many other things, the which to distinguish seuerally were impossible, considering the infinite sorts or kindes, and also those which are transplanted from one region or climate to another, do likewise alter both from the forme and taste they had before; in consideration whereof it shall be sufficient to set forth the figure of the manured grape, and speake somewhat of the rest.

There is found in Græcia and the parts of Morea, as *Pantalarea, Zante, Cephalonia,* and *Petras* (wherof some are Islands, and the other of the continent) a certaine Vine that hath a trunke or bodie of a wooddie substance, with a scaly or rugged bark, of a grayish colour, whereupon do grow faire broad leaues, sleightly indented about the edges, not vnlike vnto those of the Marsh-mallow: from the bosome whereof come forth many small clasping tendrels, and also tough and pliant foot-stalkes, whereon do grow verie faire bunches of grapes, of a watchet blewish colour: from the which fruit commeth forth long tender laces or strings, such as is found among Sauorie; whereupon wee call that plant which hath it laced Sauorie, not vnlike that that groweth among, and vpon Flax, which we call Dodder, or *Podagra lini*, whereof is made a blacke wine, which is called Greeke wine, yet of the taste of Sacke. The laced fruit of this Vine may be fitly termed *Vua barbata*, Laced or bearded grapes.

The plant that beareth those small Raisins which are commonly called Corans or Currans, or rather Raisins of Corinth, is not that plant which among the vulgar people is taken for Currans, being a shrubbe or bush that bringeth forth small clusters of berries, differing as much as may bee from Corans, hauing no affinitie with the Vine or any kinde thereof. The Vine that beareth small Raisins or Corans hath a bodie or stocke as other Vines haue, branches and tendrels likewise. The leaues are larger than any of the others, snipt about the edges like the teeth of a saw: among which come forth clusters of grapes, in forme like the other, but smaller, of a blewish colour; which being ripe are gathered and laid vpon hurdles, carpets, mats, and such like, in the Sun to drie: then are they carried to some house and laied vpon heapes, as we lay apples and corne in a garner, vntill the merchants do buy them: then do they put them into large Buts or other woodden vessels, and tread them downe with their bare feet, which they call Stiuing, and so are they brought into these parts for our vse. ‡ And they are commonly termed in Latine, *Vuæ Corinthiacæ*, and *Passulæ minores*. ‡

‡ There

Vitis Vinifera.
The manured Vine.

‡ There is also another which beareth exceeding faire grapes, whereof they make Raisins, whiter coloured, and much exceeding the bignesse of the common Raisin of the Sunne: yet that Grape whereof the Raisin of the Sun is made is a large one, and thought to be the *Vua Zibibi* of the Arabians; and it is that which *Tabernamontanus* figured vnder that name, who therein was followed by our Authour: but the figures being little to the purpose, I haue thought good to omit them. ‡

There is another kinde of Vine, which hath great leaues very broad, of an ouerworne colour; whereupon do grow great bunches of Grapes of a blewish colour: the pulpe or meate whereof sticketh or cleaueth so hard to the graines or little stones, that the one is not easily diuided from the other; resembling some starued or withered berrie that hath been blasted, whereof it was named *Duracina.*

There be some vines that bring forth grapes of a whitish or reddish yellow colour: others of a deepe red, both in the outward skinne, pulpe, and iuyce within.

There be others whose grapes are of a blew colour, or something red, yet is the iuyce like those of the former. These grapes do yeeld forth a white wine before they are put into the presse, and a reddish or paller Wine when they are trodden with the husks, and so left to macerate or ferment, with which if they remaine too long they yeeld forth a wine of a higher colour.

There be others which make a blacke and obscure red wine, whereof some bring bigger clusters, and consist of greater grapes; others of lesser: some grow more clustred and closer together, others looser: some haue but one stone, others more: some make a more austere or harsh wine; others a more sweet: of some the old wine is best; of diuers, the first yeares wine is most excellent: some bring forth fruit foure square, of which sorts or kindes we haue great plenty.

¶ *The Place.*

A fit soile for Vines, saith *Florentinus*, is euery blacke earth, which is not very close nor clammy, hauing some moisture; notwithstanding *Columella* saith that great regard is to be had what kinde or sort of Vine you would nourish, according to the nature of the countrey and soile.

A wise husbandman will commit to a fat and fruitfull soile a leane Vine, and of his own nature not too fruitfull: to a leane ground a fruitfull vine: to a close and compact earth a spreading vine, and that is full of matter to make branches of: to a loose and fruitfull soile a Vine of few branches. The same *Columella* saith, that the Vine delighteth not in dung, of what kinde soeuer it be; but fresh mould mixed with some shauings of horne is the best to be disposed about the roots, to cause fertilitie.

¶ *The Time.*

Columella saith, that the Vines must be pruned before the young branches bud forth. *Palladius* writeth, in Februarie: if they be pruned later they lose their nourishment with weeping.

¶ *The Names.*

The Vine is called in Greeke Ἄμπελος οἰνόφορος: as much to say in Latine as, *Vitis Vinifera*, or the Vine which beareth wine; and Ἄμπελος ἥμερος: that is, *Vitis mansuefacta, siue cultiua*, Tame or manured Vine. And it is called οἰνόφορος, that it may differ from both the Bryonies, the white and the blacke, and from *Tamus*, or our Ladies Seale, which be likewise named Ἄμπελοι. It is called *Vitis*, because *inuitatur ad uvas pariendas*. It is cherished to the intent to bring forth full clusters, as *Varro* saith.

Pliny maketh *Vua Zibeba, Alexandrina vitis*, or Vine of Alexandria, in his fourteenth booke, and third chapter, describing the same by those very words that *Theophrastus* doth. *Dioscorides* setteth it downe to be *altera species Vitis syluestris*, or a second kinde of wilde Vine; but wee had rather retaine it among the tame Vines. We may name it in English, Raisin Vine. The fruits hereof are

called

called in ſhops by the name of *Paſſularum de Corintho* : in Engliſh, Currans, or ſmall Raiſins.

Sylueſtris Vitis or wilde Vine is called in Greeke Ἄμπελος ἀγρία : and in Latine *Labruſca* ; as in *Virgils* Eclogs :

——— Adſpice vt antrum
Sylueſtris raris ſparſit labruſca racemis.

——— See how the wilde Vine
Bedecks the caue with ſparſed cluſters fine.

To this wilde Vine doth belong thoſe which *Pliny* in his ſixteenth booke, chapter 27. reporteth to be called *Trifera*, or that bring three ſundry fruits in one yeare, as *Inſanæ* and mad bearing Vines, becauſe in thoſe ſome cluſters are ripe and full growne, ſome in ſwelling, and others but flouring.

The fruit of the Vine is called in Greeke βότρυς, and σταφυλὴ : in Latine, *Racemus*, and *Vua*: in Engliſh, a bunch or cluſter of Grapes.

The cluſter of Grapes that hath been withered or dried in the Sun is named in Greeke σταφίς : in Latine, *Vua paſſa* : in ſhops, *Paſſula* : in Engliſh, Raiſins of the Sun.

The berry or Grape it ſelfe is called in Latine *Acinus*, and alſo *Granum*, as *Democritus* ſaith, ſpeaking of the berry.

The ſeeds or ſtones contained within the berries are called in Latine, *Vinacea*, and ſometimes *Nuclei* : in ſhops, *Arilli*, as though they ſhould ſay *Ariduli*, becauſe they are dry, and yeeld no iuyce ; notwithſtanding *Vinacea* are alſo taken in *Columella* for the droſſe or remnant of the Grapes after they be preſſed.

The ſtalke, which is in the middle of the cluſters, and vpon which the grapes do hang, is called of *Galen*, ϭότρυχος : of *Varro*, *Scapus uvarum*.

¶ *The Temperature and Vertues.*

A The tender and claſping branches of the Vine and the leaues do coole, and mightily bind. They ſtay bleeding in any part of the body : they are good againſt the laske, the bloudy flix, the heart-burne, heate of the ſtomacke, or readineſſe to vomit. It ſtayeth the luſting or longing of women with childe, though they be but outwardly applied, and alſo taken inwardly any manner of waies. They be moreouer a remedie for the inflammation of the mouth, and almonds of the throat, if they be gargled, or the mouth waſhed therewith.

B Of the ſame faculty be alſo the cluſters gathered before they be ripe ; and likewiſe the bunches of the wilde grape, which is accounted to be more effectuall againſt all thoſe infirmities.

C *Dioſcorides* ſaith, That the liquor which falleth from the body and branches being cut, and that ſometime is turned as it were into a gum (which driueth forth ſtones out of the kidnies and bladder, if the ſame be drunke in wine) healeth ring wormes, ſcabs, and leprie, but the place is firſt to be rubbed with Nitre. Being often anointed or layd on it taketh away ſuperfluous haires : but yet he ſaith that the ſame is beſt which iſſueth forth of the greene and ſmaller ſtickes, eſpecially that liquor which falleth away whileſt the branches are burning, which taketh away warts, if it be laid on them.

D The ſtones and other things remaining after the preſſing are good againſt the bloudy flix, the laske of long continuance, and for thoſe that are much ſubiect to vomiting.

E The aſhes made of the ſtickes and droſſe that remaine after the preſſing, being laid vpon the piles and hard ſwellings about the fundament, doe cure the ſame, being mixed with oile of Rue, or Herbe-grace and vineger, as the ſame Author affirmeth, it helpeth to ſtrengthen members out of ioynt, and ſuch as are bitten with any venomous beaſt, and eaſeth the paine of the ſpleene or milt, being applied in manner of a plaiſter.

F The later age do vſe to make a lie of the aſhes of Vine ſticks, in compoſitions of cauſticke and burning medicines, which ſerue in ſtead of an hot-iron : the one we call a potentiall cauterie, and the other actuall.

¶ *Of Grapes.*

G Of Grapes, thoſe that are eaten raw do trouble the belly, and fill the ſtomacke full of winde, eſpecially ſuch as are of a ſowre and auſtere taſte ; ſuch kindes of grapes doe very much hinder the concoction of the ſtomacke ; and while they are diſperſed through the liuer and veins they ingender cold and raw iuyce, which cannot eaſily be changed into good bloud.

H Sweet grapes and ſuch as are thorow ripe, are leſſe hurtfull ; their iuyce is hotter, and is eaſilier diſperſed. They alſo ſooner paſſe thorow the belly, eſpecially being moiſt, and moſt of all if the liquor with the pulpe be taken without the ſtones and skin, as *Galen* ſaith.

I The ſubſtance of the ſtones, although it be drier, and of a binding quality, doth deſcend thorow all

all the bowels, and is nothing changed : as alſo the skins, which are nothing at all altered in the body, or very little.

Thoſe grapes which haue a ſtrong taſte of wine are in a meane betweene ſoure and ſweet. A

Such grapes as haue little iuyce do nouriſh more, and thoſe leſſe that haue more iuyce : but B theſe do ſooner deſcend ; for the body receiueth more nouriſhment by the pulpe than by the iuice ; by the iuyce the belly is made more ſoluble.

Grapes haue the preheminence among the Autumne fruits, and nouriſh more than they all, but C yet not ſo much as figs : and they haue in them little ill iuyce, eſpecially when they be thorow ripe.

Grapes may be kept the whole yeare, being ordered after that manner as *Ioachimus Camerarius* D reporteth. You ſhall take (ſaith he) the meale of muſtard ſeed, and ſtrew in the bottome of any earthen pot well leaded ; whereupon you ſhall lay the faireſt bunches of the ripeſt grapes, the which you ſhall couer with more of the foreſaid meale, and lay vpon that another ſort of Grapes, ſo doing vntill the pot be full. Then ſhall you fill vp the pot to the brim with a kinde of ſweete Wine called Muſt. The pot being very cloſe couered ſhall be ſet into ſome Cellar or other cold place. The Grapes you may take forth at your pleaſure, waſhing them with faire water from the powder.

¶ *Of Raiſins.*

OF Raiſins moſt are ſweet ; ſome haue an auſtere or harſh taſte. Sweet Raiſins are hotter ; au- E ſtere colder : both of them do moderately binde, but the auſtere ſomewhat more, which doe more ſtrengthen the ſtomacke. The ſweet ones do neither ſlacken the ſtomacke, nor make the belly ſoluble, if they be taken with their ſtones, which are of a binding qualitie : otherwiſe the ſtones taken forth, they do make the belly looſe and ſoluble.

Raiſins do yeeld good nouriſhment to the body, they haue in them no ill iuyce at all, but doe F ingender ſomewhat a thicke iuyce, which notwithſtanding doth nouriſh the more.

There commeth of ſweet and fat Raiſins moſt plenty of nouriſhment : of which they are the G beſt that haue a thin skin.

There is in the ſweet ones a temperate and ſmoothing qualitie, with a power to clenſe mode- H rately. They are good for the cheſt, lungs, winde-pipe, kidneyes, bladder, and for the ſtomacke ; for they make ſmooth the roughneſſe of the winde-pipe, and are good againſt hoarſeneſſe, ſhortneſſe of breath, or difficultie of breathing : they ſerue to concoct the ſpittle, and to cauſe it to riſe more eaſily in any diſeaſe whatſoeuer of the cheſt, ſides, and lungs, and do mitigate the paine of the kidneyes and bladder, which hath ioyned with it heate and ſharpeneſſe of vrine : they dull and allay the malice of ſharpe and biting humors that hurt the mouth of the ſtomacke.

Moreouer, Raiſins are good for the liuer, as *Galen* writeth in his ſeuenth booke of medicines, I according to the places affected : for they be of force to concoct raw humors, and to reſtrain their malignitie, and they themſelues do hardly putrifie : beſides, they are properly and of their owne ſubſtance familiar to the intrals and cure any diſtemperature, and nouriſh much ; wherein they are chiefely to be commended, for Raiſins nouriſh, ſtrengthen, reſiſt putrifaction, and if there be any diſtemperature by reaſon of moiſture or coldneſſe, they helpe without any hurt, as the ſaid *Galen* affirmeth.

The old Phyſitians haue taught vs to take forth the ſtones, as we may ſee in diuers compoſiti- K ons of the antient writers ; as in that compoſition which is called in *Galen*, *Arteriaca Mithridatis*, which hath the ſeeds of the Raiſins taken forth : for ſeeing that Raiſins containe in them a thicke ſubſtance, they cannot eaſily paſſe through the veines, but are apt to breed obſtructions and ſtoppings of the intrals : which things happen the rather by reaſon of the ſeeds ; for they ſo much the harder paſſe through the body, and do quicklier and more eaſily cauſe obſtructions, in that they are more aſtringent or binding. Wherefore the ſeeds are to be taken out, for ſo ſhall the iuyce of the Raiſins more eaſily paſſe, and the ſooner be diſtributed through the intrals.

Dioſcorides reporteth, That Raiſins chewed with pepper draw flegme and water out of the head. L

Of Raiſins is made a pulteſſe good for the gout, rottings about the ioynts, gangrens, and mortified vlcers : being ſtamped with the herbe All-heale it quickly takes away the nailes that are looſe in the fingers or toes, being laid thereon.

¶ *Of Muſt.*

MVſt, called in Latine *Muſtum*, that is to ſay, the liquor newly iſſuing out of the grapes when M they be trodden or preſſed, doth fill the ſtomacke and intrals with winde ; it is hardly digeſted ; it is of a thicke iuyce, and if it do not ſpeedily paſſe through the body it becommeth more hurtfull.

hurtfull. It hath onely this one good thing in it (as *Galen* saith) that it maketh the body soluble.

A That which is sweetest and pressed out of ripe Grapes doth soonest passe through; but that which is made of soure and austere grapes is worst of all: it is more windy, it is hardly concocted, it ingendreth raw humors; and although it doth descend with a loosenesse of the belly, notwithstanding it oftentimes withall bringeth the collicke and paines of the stone: but if the belly be not mooued all things are the worse, and more troublesome; and it oftentimes brings an extreame laske, and the bloudy flix.

B That first part of the wine that commeth forth of it selfe before the Grapes be hard pressed, is answerable to the Grape it selfe, and doth quickly descend; but that which issues forth afterward, hauing some part of the nature of the stones, stalks, and skins, is much worse.

¶ *Of Cute.*

C Of Cute that is made of Must, which the Latines call *Sapa*, and *Defrutum*, is that liquor which we call in English Cute, which is made of the sweetest Must, by boyling it to a certain thicknesse, or boyling it to a third part, as *Columella* writeth.

D *Pliny* affirmeth, That *Sapa* and *Defrutum* do differ in the manner of the boyling; and that *Sapa* is made when the new wine is boyled away till onely a third part remaineth: and *Defrutum* till halfe be boyled.

E *Siræum*, (saith he in his fourteenth booke, *cap.* 17.) which others call ἕψημα, and we *Sapa*, a worke of wit, and not of nature, is made of new wine boyled to a third part; which being boiled to halfe we call *Defrutum*.

F *Palladius* ioyneth to these *Carænum*, which as he saith is made when a third part is boiled away, and two remaine.

G *Leontius* in his Geoponicks sheweth, that *Hepsema* must be made of eight parts of new wine, and an hundred of wine it selfe boyled to a third.

H *Galen* testifieth, that ἕψημα is new wine very much boyled. The later Physitians do call *Hepsema* or *Sapa* boyled wine.

I Cute or boyled wine is hot, yet not so hot as wine, but it is thicker; yet not so easily distributed or carried through the body, and it slowly descendeth by vrine, but by the belly oftentimes sooner: for it moderately maketh the same soluble.

K It nourisheth more, and filleth the body quickly; yet doth it by reason of his thicknesse sticke in the stomacke for a time, and is not so fit for the liuer or for the spleene. Cute also doth digest raw humors that sticke in the chest and lungs, and raiseth them vp speedily. It is therefore good for the cough and shortnesse of breath.

L The Vintners of the Low-countries (I will not say of London) doe make of Cute and Wine mixed in a certain proportion, a compound and counterfeit wine, which they sell for Candy wine, commonly called Malmsey.

M *Pliny lib.* 14. *cap.* 9. saith, that Cute was first deuised for a bastard hony.

¶ *Of Wine.*

N To speake of Wine, the iuyce of Grapes, which being newly pressed forth is called as we haue said *Mustum* or new wine: after the dregs and drosse are setled, and now it appeareth pure and cleere, it is called in Greeke οἶνος: in Latine *Vinum*: in English, Wine, and that not vnproperly. For certaine other iuyces, as of Apples, Pomegranats, Peares, Medlars, or Seruices, or such as otherwise made (for examples sake) of barley and Graine, be not at all simply called wines, but with the name of the thing added whereof they do consist. Hereupon is the wine which is pressed forth of the pomegranat berries named *Rhoites*, or wine of pomegranats: out of Quinces, *Cydonites*, or wine of Quinces: out of Peares, *Apyites*, or Perry: and that which is compounded of barley is called *Zythum*, or Barley wine: in English, Ale or Beere.

O And other certaine wines haue borrowed syrnames of the plants that haue beene steeped or infused in them; and yet all wines of the Vine, as Wormwood wine, Myrtle wine, and Hyssop wine, and these are all called artificiall wines.

P That is properly and simply called wine which is pressed out of the grapes of the vine, and is without any manner of mixture.

Q The kindes of wines are not of one nature, nor of one facultie or power, but of many differing one from another: for there is one difference thereof in taste, another in colour; the third is referred to the consistence or substance of the wine; the fourth consisteth in the vertue and strength thereof. *Galen* addeth that which is found in the smell, which belongs to the vertue and strength of the wine.

That

That may alſo be ioyned vnto them which reſpecteth the age: for by age wines become hotter and ſharper, and doe withall change oftentimes the colour, the ſubſtance,and the ſmell: for ſome wines are ſweet of taſte; others auſtere or ſomething harſh; diuers of a rough taſte, or altogether harſh; and moſt of them ſufficient ſharpe: there be likewiſe wines of a middle ſort, inclining to one or other qualitie. A

Wine is of colour either white or reddiſh, or of a blackiſh deepe red,which is called blacke,or of ſome middle colour betweene theſe. B

Some wine is of ſubſtance altogether thin; other ſome thicke and fat; and many alſo of a middle conſiſtence. C

One wine is of great ſtrength, and another is weake, which is called a wateriſh wine: a ful wine is called in Latine *Vinoſum*. There be alſo among theſe very many that be of a middle ſtrength. D

There is in all wines, be they neuer ſo weake, a certaine winie ſubſtance thin and hot. There be likewiſe waterie parts, and alſo diuers earthy: for wine is not ſimple,but (as *Galen* teſtifieth in his fourth booke of the faculties of medicines) conſiſteth of parts that haue diuers faculties. E

Of the ſundry mixture and proportion of theſe ſubſtances one with another there riſe diuers and ſundry faculties of the wine. F

That is the beſt and fulleſt wine in which the hot and winie parts do moſt of all abound: and the weakeſt is that wherein the waterie haue the preheminence. G

The earthy ſubſtance abounding in the mixture cauſeth the wine to be auſtere or ſomething harſh,as a crude or raw ſubſtance doth make it altogether harſh. The earthy ſubſtance being ſeuered falleth downe, and in continuance of time ſinketh to the bottome, and becomes the dregs or lees of the wine: yet it is not alwaies wholly ſeuered, but hath both the taſt and other qualities of this ſubſtance remaining in the wine. H

All wines haue their heate, partly from the proper nature and inward or originall heate of the vine, and partly from the Sun: for there is a double heate which ripeneth not only the grapes, but alſo all other fruits, as *Galen* teſtifieth; the one is proper and naturall to euery thing; the other is borrowed of the Sun: which if it be perceiued in any thing, it is vndoubtedly beſt and eſpecially in the ripening of grapes. I

For the heate which proceeds from the Sun concocteth the grapes and the iuyce of the grapes, and doth eſpecially ripen them,ſtirring vp and increaſing the inward and naturall heat of the wine, which otherwiſe is ſo ouerwhelmed with aboundance of raw and wateriſh parts, as it ſeemes to be dulled and almoſt without life. K

For vnleſſe wine had in it a proper and originall heate, the grapes could not be ſo concocted by the force of the Sun, as that the wine ſhould become hot; no leſſe than many other things naturally cold, which although they be ripened and made perfect by the heate of the Sun, do not for all that loſe their originall nature; as the fruits, iuyces, or ſeeds of Mandrake,Nightſhade,Hemlocke, Poppy, and of other ſuch like, which though they be made ripe, and brought to full perfection, yet ſtill retaine their owne cold qualitie. L

Wherefore ſeeing that wine through the heate of the Sunne is for the moſt part brought to his proper heate, and that the heate and force is not all alike in all regions and places of the earth; therefore by reaſon of the diuerſitie of regions and places, the wines are made not a little to differ in facultie. M

The ſtronger and fuller wine groweth in hot countries and places that lie to the Sun; the rawer and weaker in cold regions and prouinces that lie open to the North. N

The hotter the Sommer is the ſtronger is the wine; the leſſe hot or the moiſter it is.the leſſe ripe is the wine. Notwithſtanding not onely the manner of the weather and of the Sunne maketh the qualities of the wine to differ,but the natiue propertie of the ſoile alſo; for both the taſt and other qualities of the Wine are according to the manner of the Soile. And it is very well knowne, that not only the colour of the wine, but the taſte alſo dependeth vpon the diuerſity of the grapes. O

Wine (as *Galen* writeth) is hot in the ſecond degree, and that which is very old in the third;but new wine is hot in the firſt degree: which things are eſpecially to be vnderſtood concerning the meane betweene the ſtrongeſt and the weakeſt; for the fulleſt and mightieſt (being but *Horna*,that is as I take it of one yeare old) are for the moſt part hot in the ſecond degree. The weakeſt and the moſt wateriſh wines, although they be old, do ſeldome exceed the ſecond degree. P

The drineſſe is anſwerable to the heate in proportion,as *Galen* ſaith in his booke of Simples:but in his bookes of the gouernment of health he ſheweth, that wine doth not onely heate, but alſo moiſten our bodies, and that the ſame doth moiſten and nouriſh ſuch bodies as are extreme dry: and both theſe opinions be true. Q

For the faculties of wine are of one ſort as it is a medicine, and of another as it is a nouriſhment; which *Galen* in his booke of the faculties of nouriſhments doth plainly ſhew, affirming that thoſe qualities of the wine which *Hippocrates* writeth of in his booke of the manner of diet, be not R

as

as a nouriſhment, but rather as of a medicine. For wine as it is a medicine doth dry, eſpecially being outwardly applied; in which caſe, for that it doth not nouriſh the body at all, the drines doth more plainly appeare, and is more manifeſtly perceiued.

A Wine is a ſpeciall good medicine for an vlcer, by reaſon of his heate and moderate drying, as *Galen* teacheth in his fourth booke of the method of healing.

B *Hippocrates* writeth, That vlcers, what manner of ones ſoeuer they are, muſt not be moiſtned vnleſſe it be with wine: for that which is dry (as *Galen* addeth) commeth neerer to that which is whole, and the thing that is moiſt, to that which is not whole.

C It is manifeſt that Wine is in power or facultie dry, and not in act; for Wine actually is moiſt and liquid, and alſo cold: for the ſame cauſe it likewiſe quencheth thirſt, which is an appetite or deſire of cold and moiſt, and by this actuall moiſture (that we may ſo terme it) it is if it be inwardly taken, not a medicine, but a nouriſhment; for it nouriſheth, and through his moiſture maketh plenty of bloud; and by increaſing the nouriſhment it moiſtneth the body, vnleſſe peraduenture it be old and very ſtrong: for it is made ſharpe and biting by long lying, and ſuch kinde of Wine doth not onely heate, but alſo conſume and dry the body, for as much as it is not now a nouriſhment, but a medicine.

D That wine which is neither ſharpe by long lying, nor made medicinable, doth nouriſh and moiſten, ſeruing as it were to make plenty of nouriſhment and bloud, by reaſon that through his actuall moiſture it more moiſtneth by feeding, nouriſhing, and comforting, than it is able to dry by his power.

E Wine doth refreſh the inward and naturall heate, comforteth the ſtomacke, cauſeth it to haue an appetite to meate, moueth coucoction, and conueyeth the nouriſhment through all parts of the body, increaſeth ſtrength, inlargeth the body, maketh flegme thinne, bringeth forth by vrine cholericke and waterie humors, procureth ſweating, ingendreth pure bloud, maketh the body wel coloured, and turneth an ill colour into a better.

F It is good for ſuch as are in a conſumption by reaſon of ſome diſeaſe, and that haue need to haue their bodies nouriſhed and refreſhed (alwaies prouided they haue no feuer,) as *Galen* ſaith in his ſeuenth booke of the Method of curing. It reſtoreth ſtrength moſt of all other things, and that ſpeedily: It maketh a man merry and ioyfull: It putteth away feare, care, troubles of minde, and ſorrow: It moueth pleaſure and luſt of the body, and bringeth ſleepe gently.

G And theſe things proceed of the moderate vſe of wine: for immoderate drinking of wine doth altogether bring the contrarie. They that are drunke are diſtraughted in minde, become fooliſh, and oppreſſed with a drowſie ſleepineſſe, and be afterward taken with the Apoplexy, the gout, or altogether with other moſt grieuous diſeaſes; the braine, liuer, lungs, or ſome other of the intrals being corrupted with too often and ouermuch drinking of wine.

H Moreouer, wine is a remedy againſt taking of Hemlocke or green Coriander, the iuyce of black Poppy, Wolfs-bane, and Leopards-bane, Tode-ſtooles, and other cold poyſons, and alſo againſt the biting of ſerpents, and ſtings of venomous beaſts, that hurt and kill by cooling.

I Wine alſo is a remedie againſt the ouer-fulneſſe and ſtretching out of the ſides, windy ſwellings, the greene ſickneſſe, the dropſie, and generally all cold infirmities of the ſtomack, liuer, milt, and alſo of the matrix.

K But Wine which is of colour and ſubſtance like water, through ſhining bright, pure, of a thin ſubſtance, which is called white, is of all wines the weakeſt; and if the ſame ſhould be tempered with water it would beare very little: and hereupon *Hippocrates* calleth it ὀλιγόφορον, that is to ſay, bearing little water to delay it withall.

L This troubbleth the head and hurteth the ſinewes leſſe than others do, and is not vnpleaſant to the ſtomacke: it is eaſily and quickly diſperſed thorow all parts of the body: it is giuen with far leſſe danger than any other wine to thoſe that haue the Ague (except ſome inflammation or hot ſwelling be ſuſpected) and oftentimes with good ſucceſſe to ſuch as haue intermitting feuers; for as *Galen, lib.* 8. of his Method ſaith, it helpeth concoction, digeſteth humors that be halfe raw, procureth vrine and ſweat, and is good for thoſe that cannot ſleepe, and that be full of care and ſorrow, and for ſuch as are ouerwearied.

M Blacke wine, that is to ſay wine of a deepe red colour, is thicke, and hardly diſperſed, and doth not eaſily paſſe through the bladder: it quickly taketh hold of the braine, and makes a man drunk: it is harder of digeſtion: it remaineth longer in the body; it eaſily ſtoppeth the liuer and ſpleene; for the moſt part it bindes, notwithſtanding it nouriſheth more, and is more fit to ingender bloud: it filleth the body with fleſh ſooner than others do.

N That which is of a light crimſon red colour is for the moſt part more delightfull to the taſte, fitter for the ſtomacke; it is ſooner and eaſier diſperſed: it troubleth the head leſſe, it remains not ſo long vnder the ſhort ribs, and eaſilier deſcendeth to the bladder than blacke wine doth: it doth alſo

alſo make the belly coſtiue,if ſo be that it be not ripe. For ſuch crude and rough wines do oftentimes moleſt weake ſtomackes, and are troubleſome to the belly.

Reddiſh yellow wine ſeemeth to be in a meane betweene a thin and thicke ſubſtance: otherwiſe it is of all vines the hotteſt; aand ſuffereth moſt water to be mixed with it, as *Hippocrates* writeth. A

The old vine of this kinde, being of a thin ſubſtance and good ſmell, is a ſingular medicine for all thoſe that are much ſubject to ſwouning, although the cauſe thereof proceed of choler that hurteth the mouth of the ſtomacke, as *Galen* teſtifieth in the 12. booke of his method. B

Sweete wine the leſſe hot it is, the leſſe doth it trouble the head, and offend the minde; and it better paſſeth through the belly, making it oftentimes ſoluble: but it doth not ſo eaſily paſſe or deſcend by vrine. C

Againe, the thicker it is of ſubſtance, the harder and ſlowlier it paſſeth through: it is good for the lungs,and for thoſe that haue the cough. It ripeneth raw humours that ſticke in the cheſt,and cauſeth them to be eaſilier ſpit vp; but it is not ſo good for the liuer, whereunto it bringeth no ſmall hurt when either it is inflamed, or ſchirrous, or when it is ſtopped. It is alſo an enemy to the ſpleene,it ſticketh vnder the ſhort ribs,and is hurtfull to thoſe that are full of choler. For this kind of wine, eſpecially the thicker it is,is in them very ſpeedily turned into choler:and in others when it is well concocted, it increaſeth plenty of nouriſhment. D

Auſtere wine,or that which is ſomewhat harſh in taſt,nouriſheth not much; and if ſo be that it be thin and white, it is apt to prouoke vrine, it leſſe troubleth the head, it is not quickly digeſted, for which cauſe it is the more to be ſhunned,as *Galen* ſaith in his 12. booke of his method. E

That wine which is altogether harſh or rough in taſt, the leſſe ripe it is, the neerer it commeth to the qualities of Veriuice made of ſower grapes, being euidently binding. It ſtrengheneth a weake ſtomacke;it is good againſt the vnkindely luſting or longing of women with child; it ſtaieth the laske, but it ſticketh in the bowels: breedeth ſtoppings in the liuer and milt;it ſlowly deſcendeth by vrine, and ſomething troubleth the head. F

Old wine which is alſo made ſharpe by reaſon of age, is not onely troubleſome to the braine, but alſo hurteth the ſinewes: it is an enemy to the entrailes, and maketh the body leane. G

New wine,and wine of the firſt yeere,doth eaſily make the body to ſwell,and ingendreth winde, it cauſeth troubleſome dreames,eſpecially that which is not throughly refined, or thicke, or very ſweet: for ſuch do ſooner ſticke in the intrailes than others do. Other wines that are in a meane in colour,ſubſtance, taſte,or age,as they do decline in vertues and goodneſſe from the extreames; ſo alſo they be free from their faults and diſcommodities. They come neere in faculties to thoſe wines whereunto they be next,either in colour,taſte, or ſubſtance,or elſe in ſmell or in age. H

Wine is fitteſt for thoſe that be of nature cold and dry; and alſo for old men, as *Galen* ſheweth in his fifth booke of the gouernment of health: for it heateth all the members of their bodies, and purgeth away the watery part of the bloud,if their be any. I

The beſt wines are thoſe that be of a fat ſubſtance: for thoſe both increaſe bloud,and nouriſh the body;both which commodities they bring to old men, eſpecially at ſuch time as they haue no ſerous humour in their veines, and haue need of much nouriſhment. It happeneth that oftentimes there doth abound in their bodies a wateriſh excrement, and then ſtand they in moſt need of all of ſuch wines as do prouoke vrine. K

As wine is beſt for old men, ſo it is worſt for children: by reaſon that being drunke,it both moiſteneth and dryeth ouermuch, and alſo filleth the head with vapours, in thoſe who are of a moiſt and hot complection, or whoſe bodies are in a meane betweene the extreames, whom *Galen* in his booke of the gouernment of health doth perſuade, that they ſhould not ſo much as taſte of wine for a very long time:for neither is it good for them to haue their heads filled,nor to be made moiſt and hot, more than is ſufficient,becauſe they are already of ſuch a heate and moiſture, as if you ſhould but little increaſe either qualitie, they would forthwith fall into the extreme. L

And ſeeing that euery exceſſe is to be ſhunned,it is expedient moſt of all to ſhun this,by which not onely the body, but alſo the minde receiueth hurt. M

Wherefore we thinke,that wine is not fit for men that be already of full age,vnleſſe it be moderately taken,becauſe is carieth them headlong into fury and luſt, and troubleth and dulleth the reſonable part of the minde. N

¶ *Of the delaying, or tempering of Wine.*

IT was an ancient cuſtome, and of long continuance in old time, for wines to be mixed with water, as it is plaine and euident not onely by *Hippocrates*, but alſo by other old mens writings. Wine firſt began to be mixed with water for health and wholeſomeneſſe ſake: for as *Hippocrates* writeth in his booke of ancient Phyſicke, being ſimply and of it ſelfe much drunke, it ma- O keth a man in some sort weake and feeble: which thing *Ouid*, seemeth also to allow of writing thus.

Vt Venus eneruat vires, sic copia vini
Et tentat gressus, debilitatque pedes.

As Venery the vigour spends, so store of wine
Makes man to stagger, makes his strength decline.

A Moreouer, wine is the sweeter, hauing water poured into it, as *Athenæus* saith. *Homer* likewise commendeth that wine which is well and fitly allaied. *Philocorus* writeth (as *Athenæus* reporteth) that *Amphictyon* king of Athens was the first that allaied wine, as hauing learned the same of *Dionysius*: wherefore he saith, that those who in that manner drunke it remained in health, that before had their bodies feebled and ouerweakened with pure and vnmixed wine.

B The maner of mingling or tempering of wine was diuers: for sometimes to one part of wine, there were added two, and sometimes three or foure of water; or two parts of wine three of water: of a lesse delay was that which consisted of equall parts of wine and water.

C The old Comedians did thinke that this lesser mixture was sufficient to make men mad, among whom was *Mnesitheus*, whose words be extant in *Athenæus*.

D *Hippocrates* in the seuenth booke of his Aphorismes saith, that this manner of tempering of wine and water by equall parts bringeth as it were a light pleasant drunkennesse, and that it is a kinde of remedy against disquietnesse, yawnings, and shiuerings; and this mingling belongeth to the strongest wines.

E Such kinde of wines they might be which in times past the Scythians were reported of the old writers to drinke, who for this cause do call vnmixed wine the Scythians drinke. And they that drinke simple wine say, that they will *Scythizare*, or do as the Scythians do; as we may reade in the tenth booke of *Athenæus*.

F The Scythians, as *Hippocrates* and diuers other of the old writers affirme, be people of Germany beyond the floud Danubius, which is also called Ister: Rhene is a riuer of Scythia: and *Cyrus* hauing passed ouer Ister is reported to haue come into the borders of the Scythians.

G And in this our age all the people of Germany do drinke vnmixed wine, which groweth in their owne countrey, and likewise other people of the North parts, who make no scruple at all to drinke of the strongest wines without any mixture.

¶ *Of the liquor which is destilled out of wine, commonly called,* Aqua vitæ.

H There is drawne out of Wine a liquor, which in Latine is commonly called *Aqua vitæ*, or water of life, and also *Aqua ardens*, or burning water, which as distilled waters are drawne out of herbes and other things, is after the same manner distilled out of strong wine, that is to say, by certaine instruments made for this purpose, which are commonly called Limbeckes.

I This kinde of liquor is in colour and substance like vnto waters distilled out of herbes, and also resembleth cleere simple water in colour, but in facultie it farre differeth.

K It beareth the syrname of life, because that it serueth to preserue and prolong the life of man.

L It is called *Ardens*, burning, for that it is easily turned into a burning flame: for seeing it is not any other thing than the thinnest and strongest part of the wine, it being put to the flame of fire, is quickly burned.

M This liquor is very hot, and of most subtill and thin parts, hot and dry in the later end of the third degree, especially the purest spirits thereof: for the purer it is, the hotter it is, the dryer, and of thinner parts: which is made more pure by often distilling.

N This water distilled out of wine is good for all those that are made cold either by a long disease, or through age, as for old and impotent men: for it cherisheth and increaseth naturall heate, vpholdeth strength, repaireth and augmenteth the same: it prolongeth life, quickeneth all the senses, and doth not only preserue the memory, but also recouereth it when it is lost: it sharpeneth the sight.

O It is fit for those that are taken with the Catalepsie (which is a disease in the braine proceeding of drinesse and cold) and are subiect to dead sleepes, if there be no feuer joined; it serueth for the weakenesse trembling, and beating of the hart; it strengtheneth and heateth a feeble stomacke; it consumeth winde both in the stomacke, sides, and bowels; it maketh good concoction of meate, and is a singular remedy against cold poisons.

P It hath such force and power, in strengthening of the hart, and stirreth vp the instruments of the senses,

ſenſes; that it is moſt effectuall, not onely inwardly taken to the quantitie of a little ſpoonefull, but alſo outwardly applied: that is to ſay, ſet to the noſthrils, or laid vpon the temples of the head, and to the wreſts of the armes; and alſo to foment and bath ſundry hurts and griefes.

Being held in the mouth it helpeth the tooth-ache: it is alſo good againſt cold cramps and conuulſions, being chafed and rubbed therewith. A

Some are bold to giue it in quartaines before the fit, eſpecially after the height or prim of the diſeaſe. B

This water is to be giuen in wine with great iudgement and diſcretion; for ſeeing it is extreme hot, and of moſt ſubtill parts, and nothing elſe but the very ſpirit of the wine, it moſt ſpeedily peirceth through, and doth eaſily aſſault and hurt the braine. C

Therefore it may be giuen to ſuch as haue the apoplexie and falling ſickneſſe, the megrim, the headach of long continuance, the Vertigo, or giddineſſe proceeding through a cold cauſe: yet can it not be alwaies ſafely giuen, for vnleſſe the matter the efficient cauſe of the diſeaſe be ſmall, and the ſicke man of temperature very cold, it cannot be miniſtred without danger: for that it ſpredeth and diſperſeth the humours, it filleth or ſtuffeth the head, and maketh the ſicke man worſe: and if the humours be hot, as bloud is, it doth not a little increaſe inflammations alſo. D

This water is hurtfull to all that be of nature and complexion hot, and moſt of all to cholericke men: it is alſo offenſiue to the liuer, and likewiſe vnprofitable for the kidnies, being often and plentifully taken. E

If I ſhould take in hand to write of euery mixture, of each infuſion, of the ſundry colours, and euery other circumſtance that the vulgar people doe giue vnto this water, and their diuers vſe, I ſhould ſpend much time but to ſmall purpoſe.

¶ *Of Argall, Tartar, or wine Lees.*

The Lees of wine which is become hard like a cruſt, and ſticketh to the ſides of the veſſell, and wine casks, being dried, hard, ſound, and well compact, and which way be beaten into powder, is called in ſhops *Tartarum*: in Engliſh, Argall, and Tartar. F

Theſe Lees are vſed for many things; the ſiluer-Smiths poliſh their ſiluer herewith: the Diers vſe it: and it is profitable in medicine. G

It doth greatly dry and waſt away, as *Paulus Ægineta* ſaith: it hath withall a binding facultie, proceeding from the kinde of wine, of which it commeth. H

The ſame ſerueth for moiſt diſeaſes of the body: it is good for them that haue the greene ſicknes and the dropſie, eſpecially that kinde that lieth in the fleſh, called in Latine, *Leucophlagmatica*: being taken euery day faſting halfe a penny weight or a full penny weight (which is a dram and nine graines after the Romanes computation) doth not onely dry vp the wateriſh excrements, and voideth them by vrine, but it preuaileth much to clenſe the belly by ſiege. I

It would worke more effectually, if it were mixed either with hot ſpices, or with other things that breake winde, or elſe with diuretickes, which are medicines that prouoke vrine; likewiſe to be mixed with gentle purgers, as the ſicke mans caſe ſhall require. K

The ſame of it ſelfe, or tempered with oile of Myrtles, is a remedy againſt ſoft ſwellings, as *Dioſcorides* teacheth: it ſtaieth the laske, and vomiting, being applied outwardly vpon the region of the ſtomacke in a pultis; and if it be laid to the bottome of the belly and ſecret parts, it ſtoppeth the whites, waſteth away hot ſwellings of the kernels in the flankes, and other places, which be not yet exulcerated: it aſſwageth great breſts, and dryeth vp the milke, if it be annointed on with vineger. L

Theſe Lees are oftentimes burnt: if it become all white it is a ſigne of right and perfect burning, for till then it muſt be burned: being ſo burnt, the Grecians terme it, σφίκλην, as *Ægineta* ſaith: the Apothecaries call it, *Tartarum vſtum*, and *Tartarum calcinatum*: that it to ſay, burnt or calcined Tartar. M

It hath a very great cauſticke or burning qualitie: it clenſeth and throughly heateth, bindeth, eateth, and very much drieth, as *Dioſcorides* doth write: being mixed with Roſin, it maketh rough and ill nailes to fall away: *Paulus* ſaith, that it is mixed with cauſticks or burning medicines to increaſe their burning qualitie: it muſt be vſed whileſt it is new made, becauſe it quickly vaniſheth: for the Lees of wine burned, do ſoone relent or wax moiſt, and are ſpeedily reſolued into liquor. therefore he that would vſe it dry, muſt haue it put in a glaſſe, or glaſſed veſſell well ſtopped, and ſet in a hot and dry place. It melteth and is turned into liquor if it be hanged in a linnen bag in ſome place in a celler vnder the ground. N

The Apothecaries call this liquor that droppeth away from it, oile of Tartar. It retaineth a cauſticke and burning quality, and alſo a very dry facultie: it very ſoon taketh away leprie, ſcabs, tetters, and other filth and deformitie of the skin and face: with an equall quantitie of Roſe water O added, and as much Ceruse as is sufficient for a liniment, wherewith the blemished or spotted parts must be anointed ouer night.

¶ *The briefe summe of that hath been said of the Vine.*

A THe iuyce of the greene leaues, branches, and tendrels of the Vine drunken, is good for those that vomit and spit bloud, for the bloudy flix, and for women with childe that vomite ouermuch. The kernell within the grapes boyled in water and drunke hath the same effect.

B Wine moderately drunke profiteth much, and maketh good digestion, but it hurteth and distempereth them that drinke it seldome.

C White wine is good to be drunke before meate; it preserueth the body, and pierceth quickely into the bladder: but vpon a full stomacke it rather maketh oppilations or stoppings, because it doth swiftly driue downe meate before Nature hath of her selfe digested it.

D Claret wine doth greatly nourish and warme the body, and is wholesome with meate, especially vnto phlegmaticke people; but very vnwholesome for yong children, as *Galen* saith, because it heateth aboue nature, and hurteth the head.

E Red wine stops the belly, corrupteth the bloud, breedeth the stone, is hurtfull to old people, and good or profitable to few, saue to such as are troubled with the laske, bloudy flix, or any other loosenesse of the body.

F Sacke or Spanish wine hath beene vsed of a long time to be drunke after meate, to cause the meate the better to digest; but common experience hath found it to be more beneficiall to the stomacke to be drunke before meate.

G Likewise Malmsey, Muskadell, Bastard, and such like sweet wines haue been vsed before meat, to comfort the cold and weake stomacke, especially being taken fasting: but experience teacheth, that Sacke drunke in stead thereof is much better, and warmeth more effectually.

H Almighty God for the comfort of mankinde ordained Wine; but decreed withall, That it should be moderatly taken, for so it is wholsome and comfortable: but when measure is turned into excesse, it becommeth vnwholesome, and a poyson most venomous, relaxing the sinewes, bringing with it the palsey and falling sicknesse: to those of a middle age it bringeth hot feuers, frensie, and lecherie; it consumeth the liuer and other of the inward parts: besides, how little credence is to be giuen to drunkards it is euident; for though they be mighty men, yet it maketh them monsters, and worse than brute beasts. Finally in a word to conclude; this excessiue drinking of Wine dishonoreth Noblemen, beggereth the poore, and more haue beene destroyed by surfeiting therewith, than by the sword.

Chap. 324. *Of Hops.*

¶ *The Kindes.*

THere be two sorts of Hops: one the manured or the Garden Hop; the other wilde or of the hedge.

¶ *The Description.*

1 THe Hop doth liue and flourish by embracing and taking hold of poles, pearches, and other things vpon which it climeth. It bringeth forth very long stalkes, rough, and hairie; also rugged leaues broad like those of the Vine, or rather of Bryonie, but yet blacker, and with fewer dented diuisions: the floures hang downe by clusters from the tops of the branches, puffed vp, set as it were with scales like little canes, or scaled Pine apples, of a whitish colour tending to yellownesse, strong of smell: the roots are slender, and diuersly folded one within another.

2 The wilde Hop differeth not from the manured Hop in forme or fashion, but is altogether lesser, as well in the clusters of floures, as also in the franke shoots, and doth not bring forth such store of floures, wherein especially consisteth the difference.

¶ *The Place.*

The Hop ioyeth in a fat and fruitfull ground: it prospereth the better by manuring: also it groweth among briers and thornes about the borders of fields, I meane the wilde kinde.

¶ The

1 *Lupus salictarius.*
Hops.

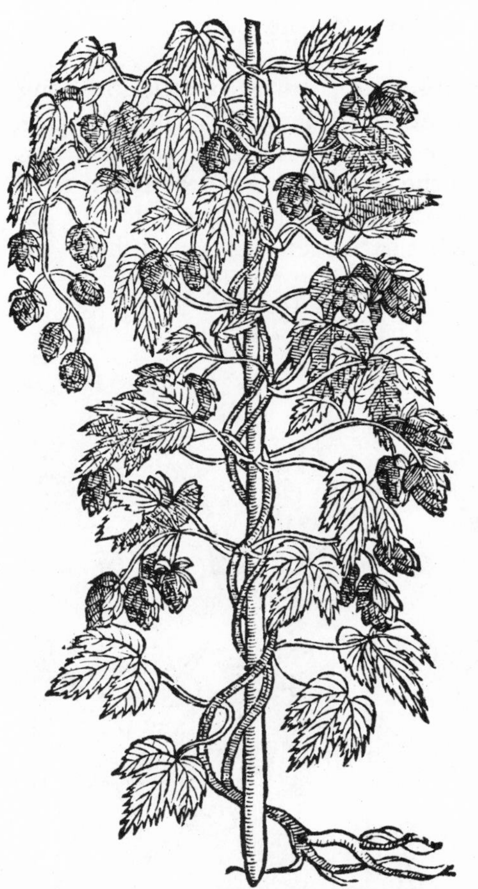

¶ *The Time.*

The floures of hops are gathered in August and September, and reserued to be vsed in beere: in the Spring time come forth new shoots or buds: in the Winter onely the roots remaine aliue.

¶ *The Names.*

It is called in shops and in all other places *Lupulus*: of some, *Lupus salictarius*, or *Lupulus salictarius*: in high-Dutch, **Hopffen**: in low-Dutch, **Hoppe**: in Spanish, *HombreZillos*: in French, *Houblon*: in English, Hops.

Pliny, lib. 21. *cap.* 15. maketh mention of Hops among the prickly plants.

¶ *The Temperature.*

The floures of the hop are hot and dry in the second degree: they fill and stuffe the head, and hurt the same with their strong smell. Of the same temperature also are the leaues themselues, which doe likewise open and clense.

¶ *The Vertues.*

A The buds or first sprouts which come forth in the Spring are vsed to be eaten in sallads; yet are they, as *Pliny* saith, more toothsome than nourishing, for they yeeld but very small nourishment: notwithstanding they be good for the intrals, both in opening and procuring of vrine, and likewise in keeping the body soluble.

B The leaues and little tender stalkes, and also the floures themselues remoue stoppings out of the liuer and spleene, purge by vrine, helpe the spleene, clense the bloud, and be profitable against long lingering Agues, scabs, and such like filth of the skin, if they be boyled in whay.

C The iuyce is of more force, and doth not onely remoue obstructions out of the intrals, but it is also thought to auoid choler and flegme by the stoole. It is written, that the same dropped into the eares taketh away the stench and corruption thereof.

D The floures are vsed to season Beere or Ale with, and too many do cause bitternesse thereof, and are ill for the head.

E The floures make bread light, and the lumpe to be sooner and easilier leauened, if the meale be tempered with liquor wherein they haue been boyled.

F The decoction of hops drunke openeth the stoppings of the liuer, the spleene, and kidneyes, and purgeth the bloud from all corrupt humors, causing the same to come forth with the vrine.

G The iuyce of Hops openeth the belly, and driueth forth yellow and cholericke humours, and purgeth the bloud from all filthinesse.

H The manifold vertues of Hops do manifestly argue the wholsomenesse of beere aboue ale; for the hops rather make it a physicall drinke to keepe the body in health, than an ordinary drinke for the quenching of our thirst.

CHAP. 325. *Of Trauellers-Ioy.*

¶ *The Description.*

1 THe plant which *Lobel* setteth forth vnder the title of *Viorna*, *Dodonæus* makes *Vitis alba*, but not properly; whose long wooddy and viny branches extend themselues very far, and into infinite numbers, decking with his clasping tendrels and white starre-like floures (being very sweet) all the bushes, hedges, and shrubs that are neere vnto it. It sends forth many branched stalkes, thicke, tough, full of shoots and clasping tendrels, wherewith it foldeth it selfe vpon the hedges, and taketh hold and climeth vpon euery thing that standeth neere vnto it.

it. The leaues are fastned for the most part by fiues vpon one rib or stem, two on either side, and one in the midst or point standing alone; which leaues are broad like those of Iuy, but not cornered at all: among which come forth clusters of white floures, and after them great tufts of flat seeds, each seed hauing a fine white plume like a feather fastned to it, which maketh in the winter a goodly shew, couering the hedges white all ouer with his feather-like tops. The root is long, tough, and thicke, with many strings fastned thereto.

2 *Clusius* hath set forth a kind of *Clematis*, calling it *Clematis Bætica*, hauing a maruellous long small branch full of ioynts, with many leaues indented about the edges like those of the peare tree, but stiffer and smaller, comming from euery ioynt; from whence also at each ioynt proceed two small clasping tendrels, as also the small foot-stalkes whereon the seeds do stand, growing in great tufted plumes or feathers, like vnto the precedent, whereof it is a kinde. The floures are not expressed in the figure, nor seene by the Author, and therefore what hath been said shall suffice.

1 *Viorna.*
The Trauellers Ioy.

2 *Clematis Bætica.*
The Spanish Trauellers Ioy.

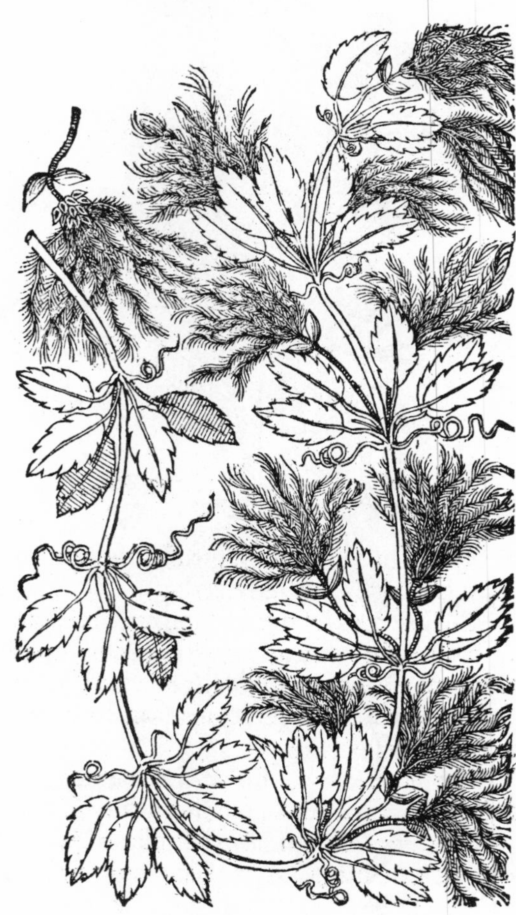

¶ *The Place.*

The Trauellers Ioy is found in the borders of fields among thornes and briers, almost in euerie hedge as you go from Grauesend to Canturbury in Kent; in many places of Essex, and in most of these Southerly parts about London, but not in the North of England that I can heare of.

The second is a stranger in these parts: yet haue I found it in the Isle of Wight, and in a wood by Waltham abbey.

¶ *The Time.*

The floures come forth in Iuly: the beauty thereof appeares in Nouember and December.

¶ *The Names.*

The first is called commonly *Viorna, quasi vias ornans*, of decking and adorning waies and hedges, where people trauel; and thereupon I haue named it the Trauellers Ioy: of *Fuchsius* it is called *Vitis nigra*: of *Dodonæus*, *Vitalba*: of *Matthiolus*, *Clematis altera*: of *Cordus*, *Vitis alba*: of *Dioscorides*, *Vitis syluestris*: of *Theophrastus*, *Atragene*: in Dutch, **Linen**: in French, as *Ruellius* writeth, *Viorne*.

¶ *The Temperature and Vertues.*

These plants haue no vse in physicke as yet found out, but are esteemed onely for pleasure, by reason of the goodly shadow which they make with their thicke bushing and clyming, as also for the beauty of the floures, and the pleasant sent or sauor of the same.

CHAP.

Chap. 326. *Of Ladies Bower, or Virgins Bower.*

¶ *The Description.*

1 That which *Lobel* describeth by the name *Clematis peregrina*, hath very long and slender stalks like the Vine, which are iointed, of a darke colour; it climeth aloft, and taketh hold with his crooked claspers vpon euery thing that standeth neere vnto it: it hath many leaues diuided into diuers parts; among which come the floures that hang vpon slender foot-stalkes, something like to those of Peruinckle, consisting onely of foure leaues, of a blew colour, and sometimes purple, with certaine threds in the middle: the seeds be flat, plaine, and sharpe pointed. The roots are slender, and spreading all about.

1 2 *Clematis peregrina Cærulea, siue rubra.*
Blew or red floured Ladies-bower.

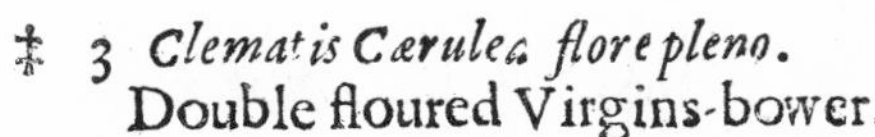

‡ 3 *Clematis Cærulea flore pleno.*
Double floured Virgins-bower.

2 The second differeth not from the other, in leaues, stalkes, hranches nor seed. The onely difference consisteth in that, that this plant bringeth forth red floures, and the other blew.

‡ 3 There is preserued also in some Gardens another sort of this *Clematis*, which in rootes, leaues, branches, and manner of growing differs not from the former: but the floure is much different, being composed of abundance of longish narrow leaues, growing thicke together, with foure broader or larger leaues lying vnder, or bearing them vp, and these leaues are of a darke blewish purple colour. *Clusius* calls this *Clematis altera flore pleno.* ‡

¶ *The Place.*

These plants delight to grow in Sunnie places: they prosper better in a fruitfull soile than in barren. They grow in my garden, where they flourish exceedingly.

¶ *The Time.*

They floure in Iuly and August, and perfect their seed in September.

¶ *The Names.*

Ladies Bower is called in Greeke κληματὶς: in Latine, *Ambuxum*: in English you may call it Ladies

dies bower,which I take from his aptneſſe in making of Arbors,Bowers, and ſhadie couertures in gardens.

¶ *The Temperature and Vertues.*

The facultie and the vſe of theſe in Phyſicke is not yet knowne.

Chap. 327. *Of purging Peruinckle.*

¶ *The Deſcription.*

1 Among theſe plants which are called *Clematides* theſe be alſo to be numbred,as hauing certaine affinitie,becauſe of the ſpreading,branching,and ſemblance of the Vine;and this is called *Flammula vrens*,by reaſon of his fierie and burning heate, becauſe that being laid vpon the skin,it burneth the place,and maketh a ſchar, euen as our common cauſtick or corraſiue medicines do. The leaues hereof anſwer both in colour and ſmoothneſſe, *Vinca,Peruinca*,or Peruinckle, growing vpon long clambring tender branches, like the other kindes of climbing plants.The floures are very white,ſtar-faſhion, and of an exceeding ſweet ſmell,much like vnto the ſmell of Hawthorne floures,but more pleaſant,and leſſe offenſiue to the head: hauing in the middle of the floures certaine ſmall chiues or threds. The root is tender, and diſperſeth it ſelfe far vnder the ground.

1 *Clematis vrens.*
Virgins Bower.

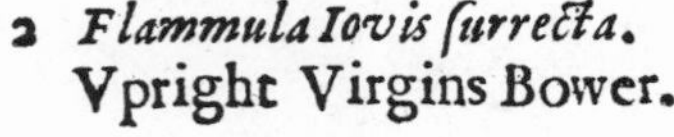
2 *Flammula Iouis ſurrecta.*
Vpright Virgins Bower.

2 Vpright Clamberer or Virgins Bower is alſo a kinde of *Clematis*, hauing long tough roots not vnlike to thoſe of Licorice;from which riſeth vp a ſtraight vpright ſtalke,of the height of three or foure cubits,ſet about with winged leaues, compoſed of diuers ſmall leaues, ſet vpon a middle rib,as are thoſe of the aſhe tree,or Valerian,but fewer in number:at the top of the ſtalks come forth ſmall white floures,very like the precedent,but not of ſo pleaſant a ſweet ſmell; after which come the ſeeds, flat and ſharpe pointed.

3 There

3 There is another *Clematis* of the kinde of the white *Clematis* or burning *Clematis*, which I haue recouered from seed, that hath been sent me from a curious and learned citizen of Strawsborough, which is like vnto the others in each respect, sauing that, that the floures heereof are very double, wherein consisteth the especiall difference.

4 Amongst the kindes of climbing or clambering plants, *Carolus Clusius*, and likewise *Lobel* haue numbred these two, which approch neere vnto them in leaues and floures, but are far different in clasping tendrels, or climbing otherwise, beeing low and base plants in respect of the others of their kinde. The first hath for his roots a bundle of tough tangling threddes, in number infinite, and thicke thrust together; from which rise vp many small stalkes, of a brownish colour, foure square, and of a wooddie substance: whereupon doe grow long leaues, of a biting taste, set together by couples, in shape like those of *Asclepias*, or silken Swallow-woort. The floures grow at the toppe of the stalkes, of a faire blew or skie colour, consisting of foure parts in manner of a crosse, hauing in the middle a bunched pointell, like vnto the head of field Poppie when it is young, of a whitish yellow colour, hauing little or no smell at all. The floures beeing past, then commeth the seed, such as is to be seen in the other kindes of *Clematis*. The whole plant dieth at the approch of Winter, and recouereth it selfe againe from the root, which indureth, whereby it greatly increaseth.

4 *Clematis Pannonica.*
Bush Bower.

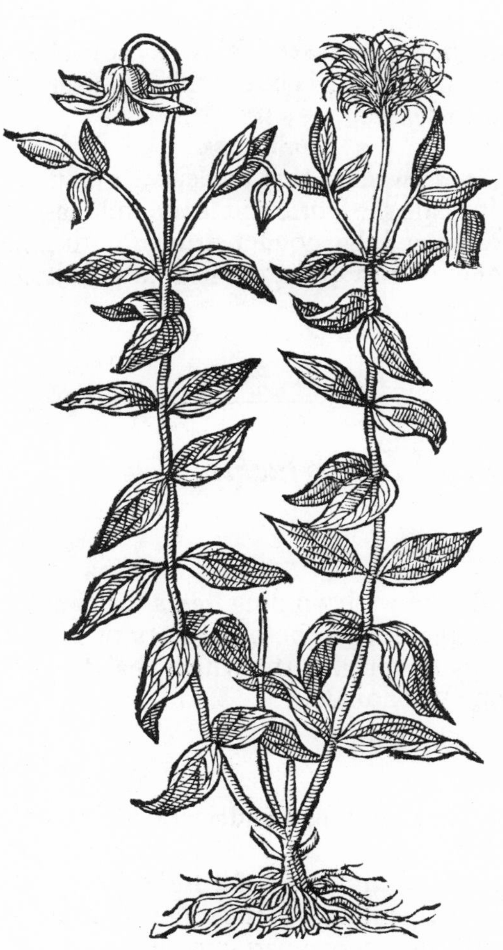

5 *Clematis maior Pannonica.*
Great Bush Bower.

5 The great Bush Bower differeth not from the former last described, but in greatnesse: which name of greatnesse setteth forth the difference.

‡ 6 Of these there is another, whose bending crested stalkes are some three cubites high, which send forth sundry small branches, set with leaues growing together by threes vpon short foot-stalkes, and they are like myrtle leaues, but bigger, more wrinckled, darke coloured, and snipt about the edges: the floure resembles a crosse, with foure sharpe pointed rough leaues of a whitish blew colour, which containe diuers small loose little leaues in their middles. The root is long and lasting. It growes vpon the rocky places of mount Baldus in Italy, where *Pona* found it, and he calls it *Clematis cruciata Alpina*. ‡

¶ The

‡ 6 *Clematis cruciata Alpina.*
Virgins Bower of the Alps.

A

¶ *The Place.*

Theſe plants do not grow wilde in England, that I can as yet learne; notwithſtanding I haue them all in my garden, where they flouriſh exceedingly.

¶ *The Time.*

Theſe plants do floure from Auguſt to the end of September.

¶ *The Names.*

There is not much more found of their names than is expreſſed in their ſeuerall titles, notwithſtanding there hath beene ſomewhat ſaid, as I thinke, by hearеſay, but nothing of certaintie: wherefore let that which is ſet downe ſuffice. We may in Engliſh call the firſt, Biting Clematis, or white Clematis, Biting Peruinkle or purging Peruinkle, Ladies Bower, and Virgins Bower.

¶ *The Temperature.*

The leafe hereof is biting, and doth mightily bliſter, being, as *Galen* ſaith, of a cauſticke or burning qualitie: it is hot in the beginning of the fourth degree.

¶ *The Vertues.*

Dioſcorides writeth, that the leaues being applied do heale the ſcurfe and lepry, and that the ſeed beaten, and the pouder drunke with faire water or with mead, purgeth flegme and choler by the ſtoole.

Chap. 328. *Of Wood-binde, or Hony-ſuckle.*

The Kindes.

There be diuers ſorts of Wood-bindes, ſome of them ſhrubs with winding ſtalks, that wrappe themſelues vnto ſuch things as are neere about them. Likewiſe there be other ſorts or kindes found out by the later Herbariſts, that clime not at all, but ſtand vpright, the which ſhall bee ſet forth among the ſhrubbie plants. And firſt of the common Woodbinde.

¶ *The Deſcription.*

1 Woodbinde or Honiſuckle climeth vp aloft, hauing long ſlender wooddie ſtalkes, parted into diuers branches: about which ſtand by certaine diſtances ſmooth leaue's, ſet together by couples one right againſt another; of a light greene colour aboue, vnderneath of a whitiſh greene. The floures ſhew themſelues in the topps of the branches many in number, long, white, ſweet of ſmell, hollow within; in one part ſtanding more out, with certaine threddes growing out of the middle. The fruit is like to little bunches of grapes, red when they be ripe, wherein is contained ſmall hard ſeed. The root is wooddie, and not without ſtrings.

2 This ſtrange kind of Woodbind hath leaues, ſtalks, and roots like vnto the common Woodbinde or Honiſuckle, ſauing that neere vnto the place where the floures come forth, the ſtalkes doe grow through the leaues, like vnto the herbe Thorow-wax, called *Perfoliata*; which leaues do reſemble little ſaucers: out of which broad round leaues proceed faire, beautifull, and well ſmelling floures, ſhining with a whitiſh purple colour, and ſomewhat daſht with yellow, by little and little ſtretched out like the noſe of an Elephant, garniſhed within with ſmall yellow chiues or threddes: and when the floures are in their flouriſhing, the leaues and floures do reſemble ſaucers filled with the

the floures of Woodbinde: many times it falleth out, that there is to be found three or foure saucers one aboue another, filled with floures, as the first, which hath caused it to be called double Hony-suckle, or Woodbinde.

1 *Periclymenum.*
Woodbinde or Honisuckles.

2 *Periclymenum perfoliatum.*
Italian Woodbinde.

¶ *The Place.*

The VVoodbinde groweth in woods and hedges, and vpon shrubbes and bushes, oftentimes winding it selfe so straight and hard about, that it leaueth his print vpon those things so wrapped.

The double Honisuckle groweth now in my garden, and many others likewise in great plenty, although not long since, very rare and hard to be found, except in the garden of some diligent Herbarists.

¶ *The Time.*

The leaues come forth betimes in the spring: the floures bud forth in May and Iune: the fruit is ripe in Autumne.

¶ *The Names.*

It is called in Greeke [illegible]: in Latine, *Volucrum maius*: of *Scribonius Largus*, *Syluæ mater*: in shops, *Caprifolium*, and *Matrisylua*: of some, *Lilium inter spinas*: in Italian, *Vincibosco*: in High Dutch, **Geysbladt**: in Low Dutch, **Gheytenbladt**, and **Mammekens Cruit**: in French, *Cheurefueille*: in Spanish, *Madreselua*: in English, VVoodbinde, Honisuckle, and Caprifoly.

¶ *The Temperature.*

There hath an errour in times past growne amongst a few, and now almost past recouerie to bee called againe, being growne an errour vniuersall, which errour is, how the decoction of the leaues of Honisuckles, or the distilled water of the floures, are rashly giuen for the inflammations of the mouth and throte, as though they were binding and cooling. But contrariwise Honisuckle is neither cold nor binding; but hot, and attenuating or making thinne. For as *Galen* saith, both the fruit of VVoodbinde, and also the leaues, do so much attenuate and heat, as if somewhat too much of them be drunke, they will cause the vrine to be as red as bloud, yet do they at the first onely prouoke vrine.

¶ The

¶ *The Vertues.*

A *Dioſcorides* writeth that the ripe ſeed gathered and dried in the ſhadow, and drunke vnto the quantitie of one dram weight, fortie daies together, doth waſte and conſume away the hardneſſe of the ſpleene, remoueth weariſomneſſe, helpeth the ſhortneſſe and difficultie of breathing, cureth the hicket, procureth bloudie vrine after the ſixt day, and cauſeth women to haue ſpeedie trauell in childe bearing.

B The leaues be of the ſame force: which being drunk thirty daies together, are reported to make men barren, and deſtroy their naturall ſeed.

C The floures ſteeped in oile and ſet in the Sun, is good to annoint the bodie that is benummed, and growne verie cold.

D The diſtilled water of the floures are giuen to be drunke with good ſucceſſe againſt the piſſing of bloud.

E A ſyrrup made of the floures is good to be drunke againſt the diſeaſes of the lungs and ſpleene that is ſtopped, being drunke with a little wine.

F Notwithſtanding the words of *Galen* (or rather of *Dodonæus*) it is certainely found by experience, that the water of Honiſuckles is good againſt the ſoreneſſe of the throte and uvula: and with the ſame leaues boiled, or the leaues and floures diſtilled, are made diuers good medicines againſt cankers, and ſore mouths, as well in children as elder people, and likewiſe for vlcerations and ſcaldings in the priuie parts of man or woman; if there be added to the decoction hereof ſome allome or Verdigreace, if the ſore require greater clenſing outwardly, prouided alwaies that there be no Verdigreace put into the water that muſt be iniected into the ſecret parts.

Chap. 329. *Of Iaſmine, or Gelſemine.*

1 *Iaſminum album.*
VVhite Geſſemine.

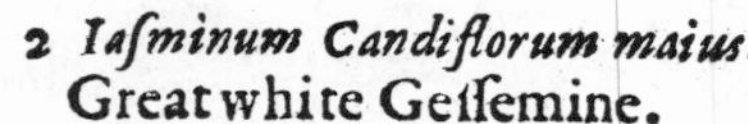

2 *Iaſminum Candiflorum maius.*
Great white Geſſemine.

¶ *The Deſcription.*

Iaſmine, or Gelſemine, is of the number of thoſe plants which haue need to be ſupported or propped vp, and yet notwithſtanding of it ſelfe claſpeth not or windeth his ſtalkes about

3 *Iaſminum luteum.*
Yellow Iaſmine.

bout ſuch things as ſtand neere vnto it, but onely leaneth and lyeth vpon thoſe things that are prepared to ſuſtaine it about arbors and banqueting houſes in gardens, by which it is held vp. The ſtalks therof are long,round,branched,iointed or kneed,and of a green colour, hauing within a white ſpungeous pith. The leaues ſtand vpon a middle ribbe, ſet together by couples like thoſe of the aſhe tree, but much ſmaller, of a deepe greene colour. The floures grow at the vppermoſt part of the branches, ſtanding in a ſmal tuft,far ſet one from another,ſweet in ſmel, of colour white. The ſeed is flat and broad like thoſe of Lupines,which ſeldome come to ripeneſſe. The root is tough and threddie.

2 *Lobel* reporteth that he ſaw in a garden at Bruxels, belonging to a reuerend perſon called M^r^. *Iohn Boiſot*, a kinde of Gelſemine, very much differing from our Iaſmine,which he nouriſhed in an earthen pot: it grew not aboue, ſaith he, to the height of a cubit, diuided into diuers branches,wherupon did grow leaues like thoſe of the common white Iaſmine, but blacker and rounder. The floures to the ſhew were moſt beautiful, in ſhape like thoſe of the common Iaſmine,but foure times bigger,gaping wide open, white on the vpper ſide,and of a bright red on the vnder ſide.

3 There is a kind hereof with yellow floures: but ſome doe deſcribe for the yellow Iaſmine, the ſhrubbie Trefoile, called of ſome *Trifolium fruticans*: and of others, *Polemonium*. But this yellow Iaſmine is one,and that is another plant,differing from the kindes of Iaſmine,as ſhall be declared in his proper place. The yellow Iaſmine differeth not from the common white Geſmine, in leaues,ſtalks,nor faſhion of the floures: the onely difference is,that this plant bringeth forth yellow floures,and the other white.

4 There is likewiſe another ſort that differeth not from the former in any reſpect,but in the colour of the floure, for this plant hath floures of a blew colour,and the others not ſo, wherein conſiſteth the difference.

¶ *The Place.*

Gelſemine is foſtered in gardens, and is vſed for arbors,and to couer banquetting houſes in gardens: it groweth not wilde in England that I can vnderſtand of,though M^r^. *Lyte* be of another opinion: the white Iaſmine is common in moſt places of England: the reſt are ſtrangers,and not ſeene in theſe parts as yet.

¶ *The Time.*

They bring forth their pleaſant floures in Iuly and Auguſt.

¶ *The Names.*

Among the Arabians *Serapio* was the firſt that named Geſſemine, *Zambach*: it is called *Iaſminum*, and *Ieſeminum*, and alſo *Geſſeminum*: in Engliſh, Iaſmine, Geſſemine, and Ieſſe.

There is in *Dioſcorides* a compoſition of oile of Iaſmine,which he ſaith is made in Perſia of the white floures of Violets,which Violets ſeeme to be none other than the floures of this Geſſemine: for *Dioſcorides* oftentimes hath reckoned faire and elegant floures amongſt the Violets; ſo that it muſt not ſeeme ſtrange that he calleth the floures of Geſſemine Violets,eſpecially ſeeing that the plant it ſelfe was vnknowne vnto him,as it is euident.

¶ *The Temperature.*

Geſſemine,and eſpecially the floures thereof be hot in the beginning of the ſecond degree, as *Serapio* reporteth out of *Meſue*.

¶ *The Vertues.*

The oile which is made of the floures hereof,waſteth away raw humors,and is good againſt cold rheumes; but in thoſe that are of a hot conſtitution it cauſeth head-ache, and the ouermuch ſmell thereof maketh the noſe to bleed,as the ſame Author affirmeth. It is vſed (as *Dioſcorides* writeth, A and after him *Aetius*) of the Persians in their banquets for pleasure sake: it is good to be annointed after bathes, in those bodies that haue need to be suppled and warmed, but by reason of smell it is not much vsed.

The leaues boiled in wine vntill they be soft, and made vp to the forme of a pultis, and applied, dissolue cold swellings, wens, hard lumps, and such like outgoings.

Chap. 330. *Of Peruinkle.*

¶ *The Description.*

1 PEruinkle hath slender and long branches trailing vpon the ground, taking hold here and there as it runneth; small like to rushes, with naked or bare spaces betweene ioint and ioint. The leaues are smooth, not vnlike to the Bay leafe, but lesser. The floures grow hard by the leaues, spreading wide open, composed of fiue small blew leaues.

We haue in our London gardens a kinde hereof bearing white floures, which maketh it to differ from the former.

1 *Vinca Peruinca minor.*
Peruinkle.

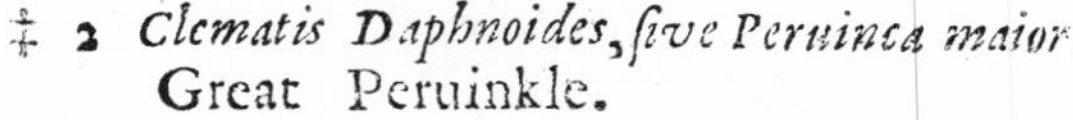

‡ 2 *Clematis Daphnoides, siue Peruinca maior.*
Great Peruinkle.

There is another with purple floures, doubling it selfe somewhat in the middle, with smaller leaues, wherein is the difference.

2 There is another sort, greater than any of the rest, which is called of some *Clematis Daphnoides*, of the similitude the leaues haue with those of the Bay. The leaues and floures are like those of the precedent, but altogether greater; wherein consisteth the difference.

¶ *The Place.*

They grow in most of our London gardens; they loue a moist and shadowie place: the branches remaine alwaies greene.

¶ *The Time.*

The floures of them do flourish in March, Aprill, and May, and oftentimes later.

¶ *The*

¶ *The Names.*

Peruinkle is called in Greeke κληματὶς δαφνοειδὴς: becauſe it bringeth forth ſtalkes, which creepe like thoſe of the Vine, and *Daphnoides* by reaſon that the leaues are like thoſe of the Bay, as aforeſaid. *Pliny* calleth it *Vinca Peruinca*, and *Chamædaphne*: notwithſtanding there is another *Chamædaphne*, of which in his place. The ſame Author likewiſe calleth it *Centunculus*: in High Dutch, **Jngrun**: in Low Dutch, **Uincooꝛte, maegden crupt**: in French, *Pucellage, Vauche & Peruauche*: in Italian, *Prouenca*: in Spaniſh, *Peruinqua*: in ſhoppes, *Clematis peruinca*: in Engliſh, Peruinkle, Peruinkle, and Periwinkle.

¶ *The Temperature.*

Peruinkle is ſomething hot, but within the ſecond degree, ſomething drie and aſtringent.

¶ *The Vertues.*

The leaues boiled in wine and drunken, ſtop the laske and bloudie flix. A

An handfull of the leaues ſtamped, and the iuice giuen to drinke in red wine, ſtoppeth the laske B
and bloudy flix, ſpitting of bloud, which neuer faileth in any bodie, either man or woman: it likewiſe ſtoppeth the inordinate courſe of the monethly ſickneſſe.

CHAP. 331. *Of Capers.*

¶ *The Kindes.*

THere be two ſorts of Capers eſpecially, one with broad leaues ſharpe pointed: the other with rounder leaues. The Brabanders haue alſo another ſort, called *Capparis fabago*, or Bean Capers.

1 *Capparis folio acuto.*
Sharpe leafed Capers.

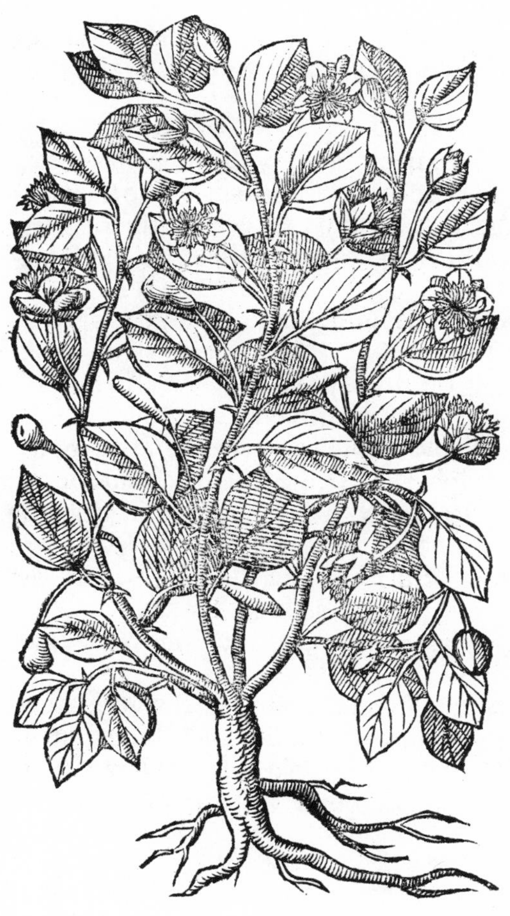

2 *Capparis rotundiore folio.*
Round leafed Capers.

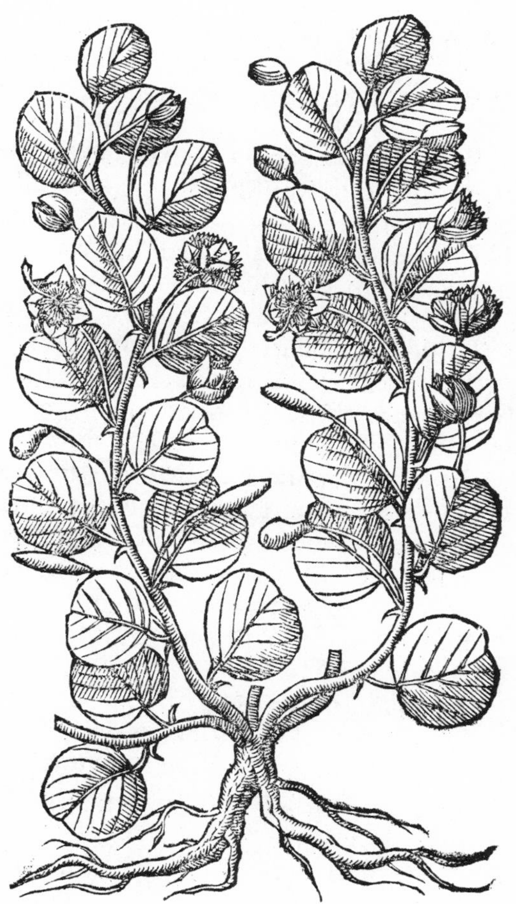

¶ *The Deſcription.*

1 THe Caper is a prickly ſhrub, the ſhoots or branches whereof be full of ſharpe prickly thornes, trailing vpon the ground if they bee not ſupported or propped vp: whereupon

doe grow leaues like those of the Quince-tree, but rounder: amongst the which come forth long slender foot-stalkes, whereon do grow round knoppes, which doe open or spread abroad into faire floures, after which commeth in place long fruit, like to an oliue, and of the same colour, wherein is contained flat rough seeds, of a duskie colour. The root is wooddie, and couered with a thick bark or rinde, which is much vsed in Physicke.

2 The second kinde of Caper is likewise a prickly plant, much like the bramble bush, hauing many slender branches set full of sharpe prickles. The whole plant traileth vpon the place where it groweth, beset with round blackish leaues disorderly placed, in shape like those of Astrabacca, but greater, approching to the forme of Fole-foot: among which commeth forth a small and tender naked twig, charged at the end with a small knap or bud, which openeth it selfe to a small starlike floure, of a pleasant sweet smell; in place whereof comes a small fruit, long and round like the Cornell berrie, of a browne colour. The root is long and wooddie, and couered with a thicke barke or rinde, which is likewise vsed in medicine.

¶ *The Place.*

The Caper groweth in Italy, Spaine, and other hot Regions without manuring, in a leane soyle, in rough places amongst rubbish, and vpon old walls, as *Dioscorides* reporteth.

Theophrastus writeth, that it is by nature wild, and refuseth to be husbanded, yet in these our daies diuers vse to cherish the same, and to set it in dry and stony places: my selfe at the impression hereof, planted some seeds in the bricke walls of my garden, which as yet do spring and grow green, the successe I expect.

¶ *The Time.*

The Caper floureth in Sommer, euen vntill Autumne. The knoppes of the floures before they open are those Capers or sauce that wee eat, which are gathered and preserued in pickle or Salt.

¶ *The Names.*

It is called in Greeke κάππαρις: and in Latine also *Capparis*: but properly *Cynosbatos*, or *Canirubus*: which is also taken for the wilde Rose; it is generally called Cappers in most languages: in English, Cappers, Caper, and Capers.

¶ *The Temperature.*

Capers, or the floures not yet fully growne, be of temperature hot, and of thinne parts; if they be eaten greene, they yeeld very little nourishment, and much lesse if they be salted. And therefore they be rather a sauce and medicine, than a meat.

¶ *The Vertues.*

A They stir vp an appetite to meat; they be good for a moist stomack, and stay the watering thereof, and clenseth away the flegme that cleaueth vnto it. They open the stoppings of the liuer and milt, with meat; they are good to be taken of those that haue a quartaine Ague, and ill spleenes. They are eaten boiled (the salt first washed off) with oile and vineger, as other sallads be, and sometimes are boiled with meat.

B The rinde or barke of the root consisteth of diuers faculties, it heateth, clenseth, purgeth, cutteth and digesteth, hauing withall a certaine binding qualitie.

C This barke is of a singular remedie for hard spleenes, being outwardly applied, and also inwardly taken, and the same boiled in vineger or oxymel, or being beaten and mixed with other simples: for after this manner it expelleth thicke and grosse humours, and conueieth away the same mixed with bloud, by vrine, and also by siege, whereby the milt or spleene is helped, and the paine of the huckle bones taken away: moreouer it bringeth downe the desired sicknesse, purgeth and draweth flegme out of the head, as *Galen* writeth.

D The same barke (as *Dioscorides* teacheth) doth clense old filthie sores, and scoureth away the thicke lips and crusts about the edges, and being chewed it taketh away the tooth-ache.

E Being stamped with vineger, it scoureth away tetters or Ring-wormes, hard swellings, and cures the Kings-euill.

F The barke of the roots of Capers is good against the hardnesse and stopping of the spleene, and profiteth much if it be giuen in drinke to such as haue the Sciatica, the Palsie, and those that are bursten or bruised by falling from some high place: it doth mightily prouoke vrine, insomuch that if it be vsed ouermuch, or giuen in too great a quantity, it procureth bloud to come with the vrine.

CHAP. 332. Of Beane Capers.

Capparis fabago.
Beane Capers.

¶ *The Deſcription.*

THis plant which the Germanes cal **Fabago**, and *Dodonæus* ſauoring of Dutch, calleth it in his laſt Edition *Capparis Fabago*, and properly: *Lobel* calleth it *Capparis Leguminoſa*: between which there is no great difference, who labour to refer this plant vnto the kindes of Capers, which is but a low and baſe herbe, and not a ſhrubbie buſh, as are the true Capers. It bringeth forth ſmooth ſtalks tender and branched, whereupon doe grow long thicke leaues, leſſer than thoſe of the true Capers, and not vnlike to the leaues of Purſlane, comming out of the branches by couples, of a light greene colour. The floures before they be opened are like to thoſe of the precedent, but when they be come to maturitie and full ripeneſſe they waxe white, with ſome yellow chiues in the midſt: which being paſt, there appeare long cods, wherein is contained ſmall flat ſeed. The root is tender, branching hither and thither.

¶ *The Place.*

It groweth of it ſelfe in corne fields of the low Countries, from whence I haue receiued ſeeds for my garden, where they flouriſh.

¶ *The Time.*

It floureth when the Caper doth.

¶ *The Names.*

It is called in Latine of the later Herbariſts *Capparis fabago*: of moſt, *Capparis Leguminoſa*: it is thought to be that herbe which *Auicen* deſcribeth in his 28. chapter, by the name of *Ardifrigi*: wee may content our ſelues that *Capparis fabago* retaine that name ſtill, and ſeeke for none other, vnleſſe it be for an Engliſh name, by which it may be called after the Latine, Beane Caper.

¶ *The Temperature and Vertues.*

Touching the faculties thereof we haue nothing left in writing worth the remembrance.

CHAP. 333. Of Swallow-wort.

¶ *The Deſcription.*

1 SWallow-wort with white floures hath diuers vpright branches of a browniſh colour, of the height of two cubits, beſet with leaues not vnlike to thoſe of *Dulcamara* or wooddie Night-ſhade, ſomewhat long, broad, ſharpe-pointed, of a blackiſh greene colour, and ſtrong fauour: among which come forth very many ſmall white floures ſtar-faſhion, hanging vpon little ſlender foot-ſtalkes: after which come in place thereof long ſharpe pointed cods, ſtuffed full of a moſt perfect white cotton reſembling ſilke, as well in ſhew as handling; (our London Gentlewomen haue named it Silken Ciſlie) among which is wrapped ſoft browniſh ſeed. The roots are very many, white, threddie, and of a ſtrong ſauour.

2 The ſecond kinde is oftentimes found with ſtalkes much longer, climing vpon props or ſuch things as ſtand neere vnto it, attaining to the height of fiue or ſix cubites, wrapping it ſelfe vpon them with many and ſundry foldings: the floures hereof are blacke: the leaues, cods, and roots be like thoſe of the former.

1 *Asclepias flore albo.*
White Swallow-woort.

2 *Asclepias flore nigro.*
Blacke Swallow-woort.

¶ *The Place.*

Both these kindes do grow in my garden, but not wilde in England; yet haue I heard it reported that it groweth in the fields about Northampton, but as yet I am not certaine of it.

¶ *The Time.*

They floure about Iune, in Autumne the downe hangeth out of the cods, and the seed falleth to the ground.

¶ *The Names.*

It is called of the later Herbarists *Vincetoxicum*: of *Ruellius*, *Hederalis*: in High Dutch, **Swaluwe wortele**, that is to say in Latine *Hirundinaria*: in English, Swallow-woort: of our Gentlewomen it is called Silken Cislie; *Æsculapius* (who is said to be the first inuenter of Physicke, whom therefore the Greekes and Gentiles honored as a God) called it after his owne name *Asclepias*, or *Æsculapius* herbe, for that he was the first that wrote thereof, and now it is called in shoppes *Hirundinaria*.

¶ *The Temperature.*

The roots of Swallow-woort are hot and dry; they are thought to be good against poison.

¶ *The Vertues.*

A *Dioscorides* writeth, that the roots of *Asclepias* or Swallow-woort boiled in wine, and the decoction drunke, are a remedie against the gripings of the belly, the stingings of Serpents, and against deadly poison, being one of the especiallest herbes against the same.

B The leaues boiled and applied in forme of a pultis, cure the euill sores of the paps or dugs, and matrix, that are hard to be cured.

Chap. 334. *Of Indian Swallow-woort.*

¶ *The Description*

There groweth in that part of Virginia, or Norembega, where our English men dwelled (intending there to erect a certaine Colonie) a kinde of *Asclepias*, or Swallow-woort, which the

the Sauages call *Wisanck* : there riseth vp from a single crooked root one vpright stalke a foot high, slender, and of a greenish colour: whereupon do grow faire broad leaues sharpe pointed, with many ribs or nerues running through the same like those of Ribwort or Plantaine, set together by couples at certaine distances. The floures come forth at the top of the stalks, which as yet are not obserued, by reason the man that brought the seeds and plants hereof did not regard them. After which, there come in place two cods (seldome more) sharpe pointed like those of our Swallow-wort, but greater, stuffed full of a most pure silke of a shining white colour: among which silke appeareth a small long tongue (which is the seed) resembling the tongue of a bird, or that of the herbe called Adders tongue. The cods are not only full of silke, but euerie nerue or sinew wherewith the leaues be ribbed are likewise most pure silke; and also the pilling of the stemmes, euen as flax is torne from his stalks. This considered, behold the iustice of God, that as he hath shut vp those people and nations in infidelity and nakednesse, so hath he not as yet giuen them vnderstanding to couer their nakednesse, nor matter wherewith to doe the same; notwithstanding the earth is couered ouer with this silke, which dayly they tread vnder their feet, which were sufficient to apparell many kingdomes, if they were carefullly manured and cherished.

Wisanck, siue Vincetoxicum Indianum.
Indian Swallow-wort.

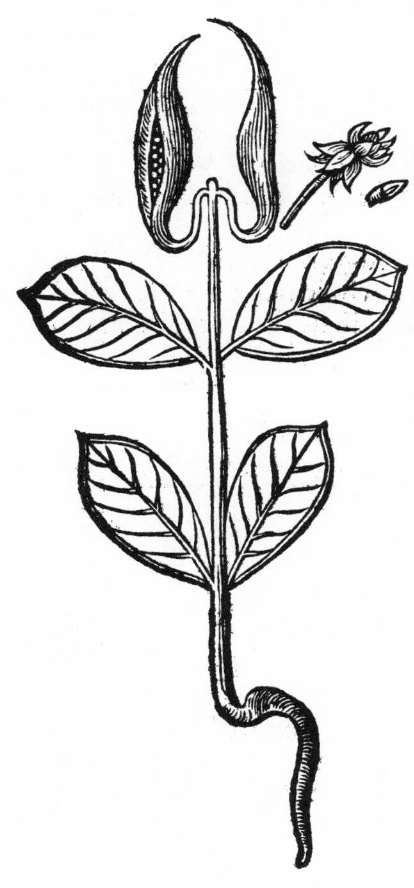

‡ *Apocynum Syriacum Clusij.*

‡ This Plant, which is kept in some gardens by the name of Virginia Silke Grasse, I take to be the same, or very like the *Beidelsar* of *Alpinus*; and the *Apocynum Syriacum* of *Clusius* : at Padua they call it *Esula Indica*, by reason of the hot milky iuyce. *Bauhinus* hath very vnfitly named it *Lapathum Ægyptiacum lactescens siliqua Asclepiadis*. But he is to be pardoned; for *Iohannes Carolus Rosenbergus*, *cap.16.p.46.* of his *Animad. & Exerc. Medicæ*, or *Rosa nobilis iatrica*, hath taken vpon him the credit and inuention of this absurd denomination: I may call it absurd, for that neither any way in shape or qualitie it resembles or participates any thing with a Docke. I haue giuen you the figure of our Author with his title, and that of *Clusius* with his: in the former the cods are only well exprest; in the later the leaues and floures reasonably well, but that they are too few in number, and set too far asunder. Vpon the sight of the growing and flouring plant I tooke this description: The root is long and creeping; the stalkes two or three cubits high, square, hollow, a finger thicke, and of a light greene colour, sending out towards the top some few branches: vpon this at

certaine

certaine ſpaces grow by couples leaues ſome halfe foot long, and three inches broad, darke greene on their vpper ſides, more whitiſh below, and full of large and eminent veines: at the top of the ſtalke and branches it carries moſt commonly an hundred or more floures, growing vpon foot-ſtalkes ſome inch long, all cloſe thruſt together after the manner of the Hyacinth of Peru at the firſt flouring: each floure is thus compoſed; firſt it hath fiue ſmall greene leaues bending backe, which ſerue for the cup: then hath it other fiue leaues foure times larger than the former, which bend backe and couer them; and theſe are greene on the vnder ſide, and of a pale colour with ſome redneſſe aboue: then are there fiue little graines (as I may ſo terme them) of a pleaſant red colour, and on their outſide like cornes of Millet, but hollow on their inſides, with a little thred or chiue comming forth of each of them: theſe fiue ingirt a ſmall head like a button, greeniſh vnderneath, and whitiſh aboue. I haue giuen you the figure of one floure by the ſide of our Authors figure. The leaues and ſtalkes of this plant are very full of a milky iuyce. ‡

¶ *The Place.*

It groweth, as before is rehearſed, in the countries of Norembega, now called Virginia by the honourable Knight Sir *Walter Raleigh,* who hath beſtowed great ſummes of money in the diſcouerie thereof; where are dwelling at this preſent Engliſh men.

¶ *The Time.*

It ſpringeth vp, floureth, and flouriſheth both Winter and Sommer, as do many or moſt of the plants of that countrey. ‡ It dies downe with vs euery Winter and comes vp in the Spring, and floures in Auguſt, but neuer bringeth forth the cods with vs, by reaſon of the coldneſſe of our Climate. ‡

¶ *The Names.*

The ſilke is vſed of the people of Pomeioc and other of the prouinces adioyning, being parts of Virginia, to couer the ſecret parts of maidens that neuer taſted man; as in other places they vſe a white kinde of moſſe Wiſanck: we haue thought *Aſclepias Virginiana,* or *Vincetoxicum Indianum* fit and proper names for it: in Engliſh, Virginia Swallow-wort, or the Silke-wort of Norembega.

¶ *The Nature and Vertues.*

A We finde nothing by report, or otherwiſe of our owne knowledge, of his phyſicall vertues, but onely report of the aboundance of moſt pure ſilke wherewith the whole plant is poſſeſſed.

B ‡ The leaues beaten either crude, or boyled in water, and applied as a pulteſſe, are good againſt ſwellings and paines proceeding of a cold cauſe.

C The milky iuyce, which is very hot, purges violently; and outwardly applied is good againſt tetters, to fetch haire off skins, if they be ſteeped in it, and the like. *Alpinus.* ‡

Chap. 335.

Of the Bombaſte or Cotton-Plant.

¶ *The Deſcription.*

THe Cotton buſh is a low and baſe Plant, hauing ſmall ſtalkes of a cubit high, and ſomtimes higher; diuided from the loweſt part to the top into ſundry ſmall branches, whereupon are ſet confuſedly or without order a few broad leaues, cut for the moſt part into three ſections, and ſometimes more, as Nature liſt to beſtow, ſomewhat indented about the edges, not vnlike to the leafe of the Vine, or rather the Veruaine Mallow, but leſſer, ſofter, and of a grayiſh colour: among which come forth the floures, ſtanding vpon ſlender foot-ſtalkes, the brimmes or edges whereof are of a yellow colour, the middle part purple: after which appeareth the fruit, round, and of the bigneſſe of a Tenniſe ball, wherein is thruſt together a great quantitie of fine white Cotton wooll; among which is wrapped vp blacke ſeed of the bigneſſe of peaſen, in ſhape like the trettles or dung of a cony. The fruit being come to maturitie or ripeneſſe, the huske or cod opens it ſelfe into foure parts or diuiſions, and caſteth forth his wooll and ſeed vpon the ground, if it be not gathered in his time and ſeaſon. The root is ſmall and ſingle, with few threds anexed thereto, and of a wooddy ſubſtance, as is all the reſt of the plant.

¶ *The Place.*

It groweth in India, in Arabia, Egypt, and in certaine Iſlands of the Mediterranean ſea, as Cyprus, Candy, Malta, Sicilia, and in other prouinces of the continent adiacent. It groweth about Tripolis and Aleppo in Syria, from whence the Factor of a worſhipfull merchant in London, Maſter *Nicholas Lete* before remembred, did ſend vnto his ſaid maſter diuers pounds weight of the ſeed; whereof ſome were committed to the earth at the impreſſion hereof, the ſucceſſe we leaue to the

Goßipium, ſiue Xylon.
The Cotton buſh.

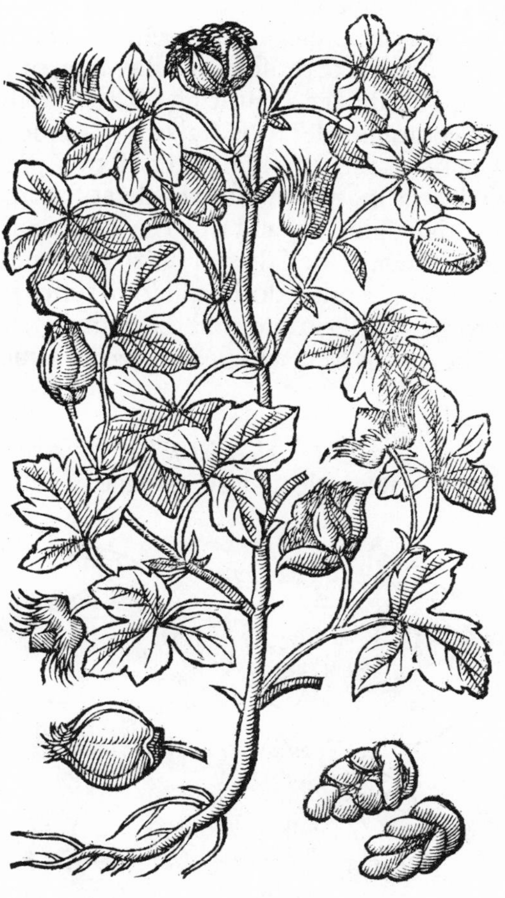

the Lord. Notwithſtanding my ſelfe 3 yeares paſt did ſow of the ſeed, which did grow verie frankly, but periſhed before it came to perfection, by reaſon of the cold froſts that ouertooke it in the time of flouring.

¶ *The Time.*

Cotton ſeed is ſowen in plowed fields in the Spring of the yeare, and reaped and cut down in harueſt, euen as corne with vs; and the ground muſt be tilled and ſowne new againe the next yeare, and vſed in ſuch ſort as we do the tillage for corne and grain: for it is a plant of one yere, and periſheth when it hath perfected his fruit, as many other plants do.

¶ *The Names.*

Cotton is called in Greeke ξύλον, and γοσσίπιον: in Latine, *Xylum*, and *Goßipium* after the Greeke: in ſhops, *Lanugo*, *Bombax*, and *Cotum*: in Italian, *Bombagia*: in Spaniſh, *Algodon*: in high-Dutch, **Baumwool**: in Engliſh and French, Cotton, Bombaſte and Bombace.

Theophraſtus hath made mention hereof in his fourth booke, *cap. 9.* but without a name; and he ſaith it is a tree in Tylus which beares wool. Neither is it any maruell if he took an vnknown ſhrub or plant, and that groweth in Countries farre off, for a tree: ſeeing alſo in this age (in which very many things come to be better knowne than in times paſt) the cotton or wooll hereof is called of the Germanes (as wee haue ſaid) **Baum wooll**, that is, Wooll of a tree, whereas indeed it is rather an herbe or ſmall ſhrub, and not to be numbred among trees.

Of this *Theophraſtus* writeth thus; It is reported that the ſame Iſland (*viz.* Tylus) doth bring forth many trees that beare wooll, which haue leaues like thoſe of the Vine, &c.

Pliny writing of the ſame, *lib. 19. cap. 1.* ſaith thus: The vpper part of Egypt toward Arabia bringeth forth a ſhrub which is called *Goßipion*, or *Xylon*, and therefore the linnen that is made of it is called *Xylina*. It is (ſaith he) the plant that beareth that wooll wherewith the garments are made which the Prieſts of Egypt do weare.

¶ *The Temperature.*

The ſeed of Cotton (according to the opinion of *Serapio*) is hot and moiſt: the wooll it ſelfe is hot and dry.

¶ *The Vertues.*

The ſeed of Cotton is good againſt the cough, and for them that are ſhort winded: it alſo ſtirreth vp the luſt of the body by increaſing naturall ſeed, wherefore it ſurpaſſeth. A

The oyle preſſed out of the ſeed taketh away freckles, ſpots, and other blemiſhes of the skin. B

The aſhes of the wooll burned ſtancheth the bleeding of wounds, vſed in reſtrictiue medicines, as Bole Armonicke, and is more reſtrictiue than Bole it ſelfe. C

To ſpeake of the commodities of the wooll of this plant were ſuperfluous, common experience and the dayly vſe and benefit we receiue by it ſhew them. So that it were impertinent to our hiſtorie to ſpeake of the making of Fuſtian, Bombaſies, and many other things that are made of the wooll thereof. D

CHAP. 336. *Of Dogs-bane.*

¶ *The Kindes.*

THere be two kindes of Dogs-banes: the one a clymbing or clambering plant; the other an vpright ſhrub.

¶ *The*

¶ *The Description.*

1 Dogs-bane riseth vp like vnto a small hedge bush, vpright and straight, vntill it haue attained to a certaine heighth; then doth it claspe and clime with his tender branches as do the Bindeweeds, taking hold vpon props or poles, or whatsoeuer standeth next vnto it: whereupon do grow faire broad leaues, sharpe pointed like those of the Bay tree, of a deepe greene colour. The floures come forth at the top of the stalkes, consisting of fiue small white leaues: which being past, there succeed long cods, set vpon a slender foot-stalke by couples, ioyning themselues together at the extreme point, and likewise at the stalke, making of two pieces knit together one entire cod; which cod is full of such downy matter and seed as that of *Asclepias*, but more in quantity by reason the cods are greater; which being dry and ripe, the silken cotton hangeth forth, and by little and little sheddeth, vntill the whole be fallen vpon the ground. The whole plant yeeldeth that yellow stinking milky iuyce that the other doth, and sometimes it is of a white colour, according to the climate where it groweth; for the more cold the country is, the whiter is the iuyce; and the more hot, the yellower. The root is long and single, with some threds anexed thereto.

1 *Periploca repens angustifolia.*
Climing Dogs-bane.

‡ 2 *Periploca latifolia.*
Broad leafed Dogs-bane.

2 There is another Dogs-bane that hath long and slender stalkes like those of the Vine, but of a browne reddish colour, wherewith it windeth it selfe about such things as stand neere vnto it, in manner of a Bindweed: whereupon are set leaues not vnlike to those of the Iuy, but not so much cornered, of a darke greene colour, and of a ranke smell being bruised betweene the fingers, yeelding forth a stinking yellow milky iuyce when it is so broken: amongst which come forth little white floures, standing scatteringly vpon little huskes: after the floures come long cods, very like vnto *Asclepias* or Swallow-wort, but greater, stuffed with the like soft downy silke; among which downe is wrapped vp flat blacke seed. The roots are many and threddy, creeping all about within the ground, budding forth new shoots in sundry places, whereby it greatly increaseth.

¶ *The Place.*

They grow naturally in Syria, and also in Italy, as *Matthiolus* reporteth: my louing friend *Iohn Robin* Herbarist in Paris did send me plants of both the kinds for my garden, where they floure and flourish; but whether they grow in France, or that he procured them from some other region, as yet I haue no certaine knowledge.

¶ The

¶ *The Time.*

They begin to bud forth their leaues in the beginning of May, and ſhew their floures in September.

¶ *The Names.*

Dogs-bane is called by the learned of our age *Periploca*: it is euident that they are to be referred to the *Apocynum* of *Dioſcorides*. The former of the two hath been likewiſe called κυνοκράμβη, and *Braſſica Canina*, or Dogs-Cole: notwithſtanding there is another Dogs Cole, which is a kind of wilde Mercurie. We may call the firſt Creeping Dogs-bane; and the other, Vpright or Syrian Dogs-bane.

¶ *The Temperature.*

Theſe plants are of the nature of that peſtilent or poyſonous herbe *Thora*, which being eaten of Dogs or any other liuing creature doth certainly kill them, except there be in readineſſe an Antidote or preſeruatiue againſt poyſon, and giuen, which by probabilitie is the herbe deſcribed in the former chapter, called *Vincetoxicum*; euen as *Anthora* is the Antidote and remedy againſt the poyſon of *Thora*; and *Herba Paris* againſt *Pardalianches*.

¶ *The Vertues.*

Dogs-bane is a deadly and dangerous plant, eſpecially to foure footed beaſts; for as *Dioſcorides* A
writeth, the leaues hereof being mixed with bread and giuen, killeth dogs, wolues, Foxes, and leopards, the vſe of their legs and huckle-bones being preſently taken from them, and death it ſelfe followeth incontinent, and therefore not to be vſed in medicine.

Chap. 337. *Of Solomons Seale.*

1 *Polygonatum.*
Solomons Seale.

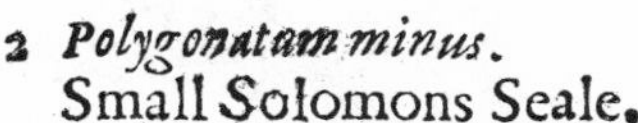

2 *Polygonatum minus.*
Small Solomons Seale.

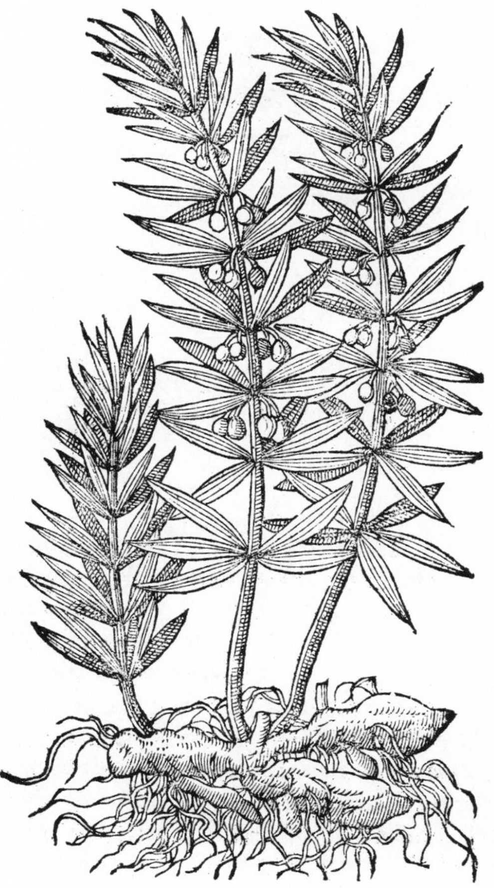

¶ *The Deſcription.*

1 The firſt kinde of Solomons Seale hath long round ſtalkes, ſet for the moſt part with long leaues ſomewhat furrowed and ribbed, not much vnlike Plantaine, but narrower, which for the moſt part ſtand all vpon one ſide of the ſtalk, and hath ſmal white floures reſembling the

the floures of Lilly conuall: on the other side when the floures be vaded there come forth round berries, which at the first are greene, and of a blacke colour tending to blewnesse, and when they be ripe be of the bignesse of Iuy berries, of a very sweet and pleasant taste. The root is white and thicke, full of knobs or ioynts, which in some places resemble the marke of a seale, whereof I think it tooke the name *Sigillum Solomonis*; and is sweet at the first, but afterward of a bitter taste, with some sharpenesse.

2 The second kinde of *Polygonatum* doth not much vary from the former, sauing in the leaues, which be narrower, and grow round about the stalke like a spur, in fashion like vnto Woodroofe or red Madder: among the leaues come forth floures like the former, but of a greener white colour: which being past, there succeed berries like the former, but of a reddish colour: which being past, there succeed berries like the former, but of a reddish colour: the roots are thick and knobby like the former, with some fibres anexed thereto.

3 *Polygonatum latifolium 2. Clusij.*
Sweet smelling Solomons Seale.

4 *Polygonatum ramosum.*
Branched Solomons Seale.

3 The third kinde of Solomons Seale, which *Carolus Clusius* found in the wooddy mountaines of Leitenberg, aboue Manderstorf, and in many other mountaines beyond the riuer Danubius, especially among the stones, hee sent to London to M^r. *Garth* a worshipfull Gentleman, and one that greatly delighteth in strange plants, who very louingly imparted the same vnto me. This plant hath stalkes very like vnto the common Solomons Seale, a foot high, beset with leaues vpon one side of the stalke like the first and common kinde, but larger, and more approching to the bignesse of the broad leafed Plantaine, the taste whereof is not very pleasant: from the bosome of which leaues come forth small well smelling greenish white floures not much vnlike the first: which being past, there follow seeds or berries that are at the first green, but afterward blacke, containing within the same berries a small seed as big as a Vetch, and as hard as a stone. The roots are like vnto the other of his kinde, yet not so thicke as the first.

4 The fourth kind according to my account, but the third of *Clusius* (which he found also in the mountaines aforesaid) groweth a foot high, but seldome a cubit, differing from all the others of his kinde; for his stalkes diuide themselues into sundry other branches, which are garnished with goodly leaues, larger and sharper pointed than any of the rest, which do embrace the stalks about after the manner of *Perfoliata* or Thorow-wax, yet very like vnto the kindes of Solomons Seale in shew,

ſhew, ſaue that they are ſomewhat hoarie vnderneath the leaues; which at the firſt are ſweete in taſte,but ſomewhat acride or biting towards the later end. From the backe part of the leaues ſhoot forth ſmall long tender and crooked ſtems,bearing at the end little gaping white floures not much vnlike *Lilium conuallium*, ſauouring like Hawthorne floures, ſpotted on the inner ſide with blacke ſpots : which being paſt, there come forth three cornered berries like the narrow leafed Solomons ſeale, greene at the firſt, and red when they be ripe,containing many white hard graines. The roots differ from all the other kindes,and are like vnto the crambling roots of *Thalictrum*,which the grauer hath omitted in the picture.

5 *Polygonatum anguſtifolium ramoſum.*
Narrow leaued Solomons ſeale.

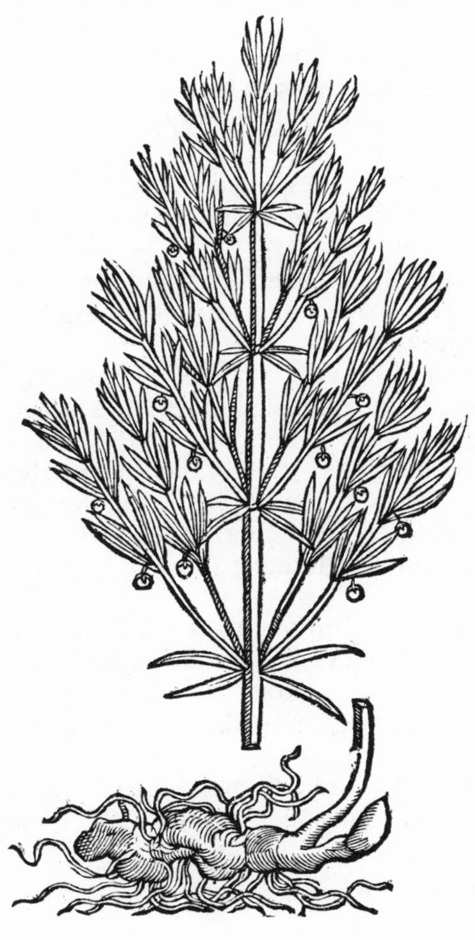

5 This rare ſort of Solomons Seale riſes vp from his tuberous or knobby root, with a ſtraight vpright ſtalke ioynted at certaine diſtances, leauing betweene each ioynt a bare and naked ſtalke, ſmooth, and of a greeniſh colour tending to yellownes; from the which ioynts thruſt forth diuers ſmal branches,with foure narrow leaues ſet about like a ſtar or the herbe Woodroofe : vpon which tender branches are ſet about the ſtalkes by certaine ſpaces long narrow leaues incloſing the ſame round about:among which leaues come forth ſmall whitiſh floures of little regard. The fruit is ſmall,and of a red colour,full of pulpe or meate; among which is contained a hard ſtony ſeed like that of the firſt Solomons ſeale.

‡ 6 There is kept in our gardens, and ſaid to be brought from ſome part of America another *Polygonatum*, which ſends vp a ſtalk ſome foot and more high, and it hath leaues long, neruous, and very greene and ſhining, growing one by another without any order vpon the ſtalke, which is ſomewhat creſted, crooked, and very greene; bearing at the very top thereof, aboue the higheſt leafe, vpon little foot-ſtalks, ſome eight or nine little white floures, conſiſting of ſix leaues apiece, which are ſucceeded by berries, as in the former. This floures in May, and is vulgarly named *Polygonatum Virginianum*, or Virginian Solomons ſeale. ‡

¶ *The Place.*

The firſt ſort of Solomons ſeale growes naturally wilde in Somerſet-ſhire, vpon the North ſide of a place called Mendip, in the pariſh of Shepton Mallet : alſo in Kent by a village called Crayford, vpon Rough or Row hill : alſo in Odiam parke in Hampſhire; in Bradfords wood, neere to a town in Wiltſhire foure miles from Bathe; in a wood neere to a village called Horſley, fiue miles from Gilford in Surrey, and in diuers other places.

That ſort of Solomons ſeale with broad leaues groweth in certaine woods in Yorkſhire called Clapdale woods, three miles from a village named Settle.

¶ *The Time.*

They ſpring vp in March, and ſhew their floures in May : the fruit is ripe in September.

¶ *The Names.*

Solomons ſeale is called in Greeke πολυγόνατον : in Latine likewiſe *Polygonatum*, of many Knees, for ſo the Greeke word doth import : in ſhops, *Sigillum Salomonis*, and *Scala cœli* : in Engliſh likewiſe Scala cœli, Solomons ſeale, and White-wort, or white root : in high-Dutch, Weiſſwurtz : in French, *Seau de Solomon* : of the Hetrurians, *Fraſinella*, and *Fraxinella*.

¶ *The Temperature.*

The roots of Solomons ſeale, as *Galen* ſaith, haue both a mixt facultie and qualitie alſo : For they haue (ſaith he) a certaine kinde of aſtriction or binding, and biting withall, and likewiſe a certaine loathſome bitterneſſe, as the ſame Author affirmeth : which is not to be found in thoſe that do grow in our climate.

¶ *The Vertues.*

A *Dioscorides* writeth, That the roots are excellent good for to seale or close vp greene wounds, being stamped and laid thereon; whereupon it was called *Sigillum Salomonis*, of the singular vertue that it hath in sealing or healing vp wounds, broken bones, and such like. Some haue thought it tooke the name *Sigillum* of the markes vpon the roots: but the first reason seemes to me more probable.

B The root of Solomons seale stamped while it is fresh and greene, and applied, taketh away in one night, or two at the most, any bruise, blacke or blew spots gotten by falls or womens wilfulnesse, in stumbling vpon their hasty husbands fists, or such like.

C *Galen* saith, that neither herbe nor root hereof is to be giuen inwardly: but note what experience hath found out, and of late dayes, especially among the vulgar sort of people in Hampshire, which *Galen*, *Dioscorides*, or any other that haue written of plants haue not so much as dreamed of; which is, That if any of what sex or age soeuer chance to haue any bones broken, in what part of their bodies soeuer; their refuge is to stampe the roots hereof, and giue it vnto the patient in ale to drinke: which sodoreth and glues together the bones in very short space, and very strangely, yea although the bones be but slenderly and vnhandsomely placed and wrapped vp. Moreouer, the said people do giue it in like manner vnto their cattell, if they chance to haue any bones broken, with good successe; which they do also stampe and apply outwardly in manner of a pultesse, as well vnto themselues as their cattell.

D The root stamped and applied in manner of a pultesse, and layd vpon members that haue been out of ioynt, and newly restored to their places, driueth away the paine, and knitteth the ioynt very firmely, and taketh away the inflammation, if there chance to be any.

E The same stamped, and the iuyce giuen to drinke with ale or white wine, as aforesaid, or the decoction thereof made in wine, helps any inward bruse, disperseth the congealed and clotted bloud in very short space.

F That which might be written of this herbe as touching the knitting of bones, and that truely, would seeme vnto some incredible; but common experience teacheth, that in the world there is not to be found another herbe comparable to it for the purposes aforesaid: and therefore in briefe, if it be for bruises inward the roots must be stamped, some ale or wine put thereto, strained, and giuen to drinke.

G It must be giuen in the same manner to knit broken bones, against bruises, black or blew marks gotten by stripes, falls, or such like; against inflammation, tumors or swellings that happen vnto members whose bones are broken, or members out of ioynt, after restauration: the roots are to be stamped small, and applied pultesse or plaisterwise, wherewith many great works haue beene performed beyond credit.

H *Matthiolus* teacheth, That a water is drawne out of the roots, wherewith the women of Italy vse to scoure their faces from sunne-burning, freckles, morphew, or any such deformities of the skinne.

† That which our Author formerly figured and described in the fifth place of this chapter, by the name of *Polygonatum acutum Clusii*, was that described by him in the fourth place; but the figure was not so well exprest.

Chap. 338. *Of Knee-holme, or Butchers broome.*

¶ *The Description.*

Knee-holme is a low wooddy plant, hauing diuers small branches, or rather stems, rising immediately from the ground, of the height of a foot; whereupon are set many leaues like vnto those of the Box tree, or rather of the Myrtle, but sharpe and pricking at the point. The fruit groweth vpon the middle rib of the leafe, greene at the first, and red as Corall when it is ripe, like those of *Asparagus*, but bigger. The roots are white, branched, of a meane thicknesse, and full of tough sprouting shoots thrusting forth in other places, whereby it greatly encreaseth.

¶ *The Place.*

It groweth plentifully in most places in England in rough and barren grounds, especially vpon Hampsted heath foure miles from London; in diuers places of Kent, Essex, and Barkshire, almost in euery copse and low wood.

¶ *The Time.*

The young and tender sprouts come forth at the first of the Spring, which are eaten in soms places

places, as the yong tender ſtalkes of Aſparagus and ſuch like herbes. The berries are ripe in Auguſt.

Ruſcus, ſiue Bruſcus.
Knee-holme, or Butchers broome.

¶ *The Names.*

It is called in Greeke ὀξυμυρσίνη, as though they ſhould ſay *Acuta Myrtus*, or pricking Myrtle; and *Myrtus ſylueſtris*, or wild Myrtle: in Latine, *Ruſcum*, or *Ruſcus*: in ſhops, *Bruſcus*: of diuers, *Scopa regia*, as teſtifieth *Marcellus Empericus* an old Writer: in high-Dutch, **Mueſedorn**: in low-Dutch, **Steckende palm**: in Italian, *Ruſco*, and *Pontogopi*: in Spaniſh, *Gilbarbeyra*: in Engliſh, Knee-holme, Knee-huluer, Butchers broome, and Petigree.

There be ſome (ſaith *Pliny*, *lib.* 25. *cap.* 13.) that call it *Oxymyrſine*.

Serapio, *cap.* 288. ſuppoſeth that *Myrtus Agria*, or wilde Myrtle, is the ſame that *Cubebæ* are: he alledgeth a reaſon, becauſe *Galen* hath not deſcribed *Myrtus Agria*, or Knee-holme; neither *Dioſcorides Cubebæ*. Which as it is a reaſon of no account, ſo is it alſo without truth: for *Galen* doth no where make mention of *Cubebæ*; and be it that he had, it ſhould not therefore follow that Knee-holme is *Cubebæ*. *Galen* ſpeaketh of *Carpeſium*, which *Auicen* in his 137 chapter maketh to be *Cubebæ*: and that *Carpeſium* doth much differ from Kneeholme, thoſe things do euidently declare which *Galen* hath left written hereof in his firſt book of of Counterpoyſons. *Carpeſium* (ſaith hee) is an herbe like in kinde to that which is called *Phu*, or Setwall, but of greater force, and more aromaticall or ſpicie. This groweth very plentifully in Sida a city of Pamphilia. Alſo he ſaith further, that ſome of the ſtickes of Carpeſium are like to thoſe of Cinnamon: there be two kinds thereof, one which is named *Laërtium*; and another that is called *Ponticum*. They both take their names of the mountaines on which they grow: but *Ponticum* is the better, which is put into medicines in which the herbe *Phu* ought to be put. For *Carpeſium*, as I haue ſaid, is like vnto *Phu*, or Setwall, yet is it ſtronger, and yeeldeth a certaine aromaticall qualitie both in taſte and ſmell. Thus far *Galen*. By which it plainly appeareth, that Knee-holme is not *Carpeſium*, that is to ſay, *Auicenna* his *Cubebæ*, as ſhall be further declared in the chapter of *Cubebæ*.

Herein *Serapio* was likewiſe deceiued, who ſuſpected it to be ſuch a like thing; ſaying, There be certaine fruits or graines called *Cubebæ*, not ſticks: yet do they neither agree with Knee-holm, neither yet were they knowne vnto *Galen*.

Iſaac in the ſecond booke of his Practiſe doth number it among the graines: and likewiſe *Haliabbas* in the ſecond booke of his Practiſe alſo, *num.* 162. The later Grecians, among whom is *Nicolaus Myrepſus*, call them *Cubebæ*.

¶ *The Temperature.*

The roots of Knee-holme, which be chiefely vſed, are of temperature hot, and meanly dry, with a thinneſſe of eſſence.

¶ *The Vertues.*

The decoction of the roots of Knee-holme made in wine and drunken, prouoketh vrine, breaketh the ſtone, driueth forth grauell and ſand, and eaſeth thoſe that make their water with great paine. A

Dioſcorides writeth the ſame things of the leaues and berries, which moreouer (ſaith hee) bring downe the deſired ſickneſſe, helpe the head-ache and the yellow jaundice. Ouer and beſides, the roots do ſerue to raiſe vp gently tough and groſſe flegm which ſticketh in the lungs and cheſt, and do concoct the ſame. B

CHAP. 339. *Of Horſe-tongue or Double-tongue.*

¶ *The Deſcription.*

1 Horſe-tongue ſendeth forth round ſtalkes of a ſpan long; wherupon are ſet long broad and ſharpe pointed leaues, but not pricking as are thoſe of Knee-holme, not vnlike to the leaues of the Bay tree, but leſſer; greater than thoſe of Knee-holm: out of the middle rib whereof commeth forth another leafe, ſharpe pointed alſo, but ſmall, and of the bigneſſe of the leafe of Knee-holme, reſembling a little tongue. From the boſome of which two leaues commeth forth a berry of the bigneſſe of a peaſe, of colour red when it is ripe, which is ſometimes in a manner all hid vnder the leafe. The root is white, long, and tough, and of a ſweet and pleaſant ſmell.

1 *Hippogloſſum mas.*
The male Horſe-tongue.

2 *Hippogloſſum fœmina.*
The female Horſe-tongue.

2 The female Horſe-tongue differeth not from the precedent but in ſtature and colour of the fruit: it riſeth vp (ſaith my Author) foure or fiue handfulls high: the berries come forth of the middle part of the greater leafe, and the ſetting on of the leſſer, of a feint yellowiſh red colour, wherein conſiſteth the difference. ‡ This is all one with the former. ‡

3 There is likewiſe another ſort of Double-tongue ſet forth by *Matthiolus*, which ſeemes vnto ſome not to differ from the firſt deſcribed or beſt known Horſe tongue, being in truth the ſelf ſame plant without any difference: notwithſtanding I haue ſet forth the figure, that it may appeare to be the ſame, or very little different, and that not to be diſtinguiſhed: but *Matthiolus* may not eſcape without reprehenſion, who knowing the vntrue tranſlation of *Ruellius*, would ſet forth ſo falſe a picture in his Commentaries.

‡ Our Author here, as in many other places, miſtakes himſelfe; for *Matthiolus* did not ſet forth that figure that our Author giues in this place, for *Hippogloſſum*, but by the title of *Laurus Alexandrina altera*: and it thus differs from the common Horſe-tongue; it hath ſhorter and rounder leaues, yet ſharpe pointed, and the berries are not couered with little leaues as in the other, neither haue they any apparant ſtalkes at all, but grow cloſe to the leaues, as you may ſee them expreſt in the figure. ‡

¶ The

3 *Hippoglossum Matthioli.*
Italian Horse-tongue.

¶ *The Place.*

They are found on the Alps of Liguria, and on the mountaines of Austria. *Bellonius* writeth, that they do grow very plentifully about the hil Athos.

The first of the Horse-tongues growes in my garden very plentifully.

¶ *The Time.*

That which groweth in my garden floured in the beginning of May: the fruit is ripe in the fall of the leafe.

¶ *The Names.*

Horse-tongue is called in Greek ἱππόγλωσσον: of the later Herbarists, *Bonifacia, Vvularia, Bislingua, Lingua Pagana,* and *Victoriola.* The same is also named δάφνη ἰδαῖα, of Ida a mountaine of Troy, which is called *Alexanders* Troy: of some, *Laurus Alexandrina,* or the Bay of Alexandria, and *Laurus Idæa.*

This *Hippoglossum Bonifacia* is called in high-Dutch, **Zapflinkraut:** in low-Dutch, **Tongheneblad:** in Spanish, *Lengua de Cauallo:* in English, Horse-tongue, Tongne-blade, Double-tongue, and Laurel of Alexandria.

¶ *The Temperature.*

Horse-tongue is euidently hot in the second degree, and dry in the first.

¶ *The Vertues.*

A The roots of Double-tongue boiled in wine, and the decoction drunke, helpeth the strangurie, prouoketh vrine, easeth women that haue hard trauell in childe-bearing. It expelleth the secondine or after-birth. The root beaten to pouder, whereof six drams giuen in sweet wine, doth helpe the diseases aforesaid: it bringeth downe the termes, as *Dioscorides* teacheth. The like writeth *Pliny* also: adding further, That it causeth women to haue speedy deliuerance, especially if halfe an ounce of the pouder of the root be giuen to drink in a draught of sweet wine.

B *Baptista Sardus* doth notably commend this herb for the diseases of the mother; by giuing, saith he, a little spoonfull of the pouder either of the herbe, the fruit, or of the root, to her that is troubled with the mother, she is thereby forthwith recouered. He also writeth, that the same is a singular good medicine for those that be bursten, if a spoonfull of the pouder of the root be drunke in the broth of flesh certaine dayes together.

CHAP. 340. *Of Cucumbers.*

¶ *The Kindes.*

THere be diuers sorts of Cucumber; some greater, others lesser; some of the Garden, some wilde; some of one fashion, and some of another, as shall be declared in the following chapters.

¶ *The Description.*

1 THe Cucumber creepes alongst vpon the ground all about, with long rough branches; whereupon do grow broad rough leaues vneuen about the edges: from the bosome whereof come forth crooked clasping tendrels like those of the Vine. The floures shoot forth betweene the stalkes and the leaues, set vpon tender foot-stalks composed of fiue small yellow leaues: which being past, the fruit succeedeth, long, cornered, rough, and set with certaine bumpes or risings, greene at the first, and yellow when they be ripe, wherein is contained a firme and sollid pulpe or substance transparent or thorow-shining, which together with the seed is eaten a little before they be fully ripe. The seeds be white, long, and flat.

1 *Cucumis vulgaris.*
Common Cucumber.

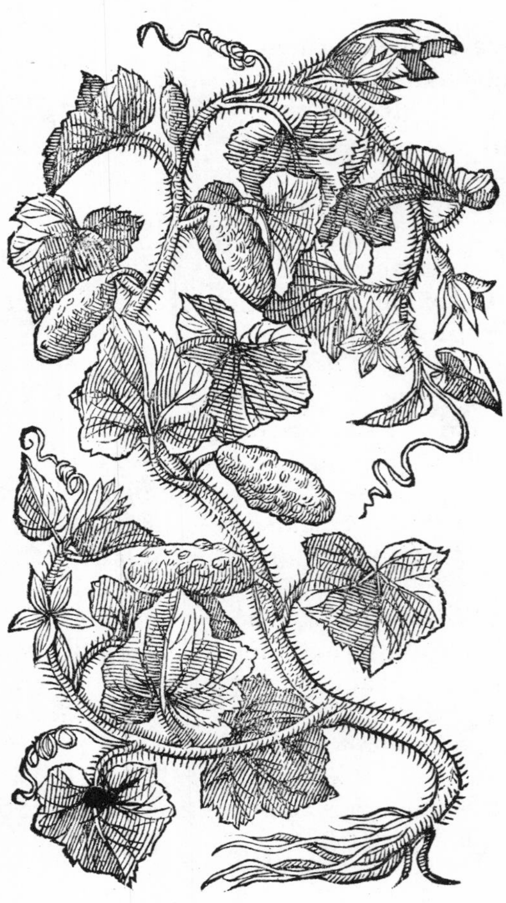

2 *Cucumis Anguina.*
Adders Cucumber.

4 *Cucumis ex Hispanica semine natus.*
Spanish Cucumber.

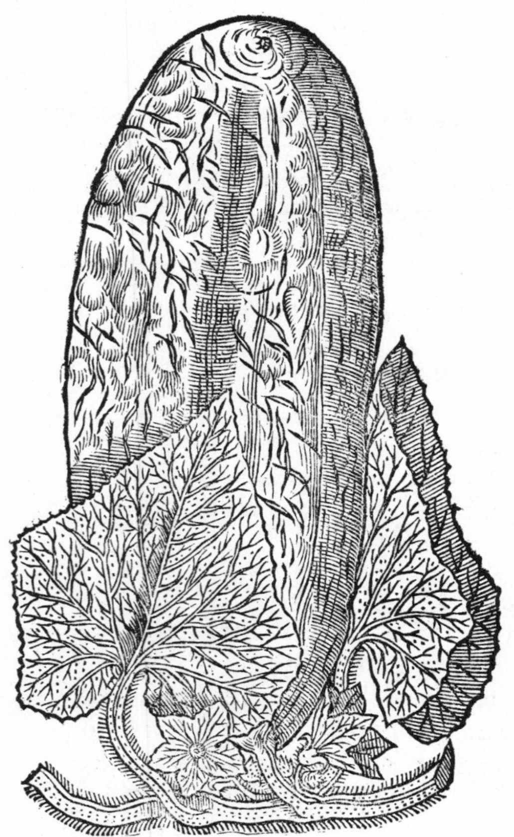

2 There be also certaine long cucumbers, which were first made (as is said) by art and manuring, which Nature afterwards did preserue: for at the first, when as the fruit is very little, it is put into some hollow cane, or other thing made of purpose, in which the cucumber groweth very long, by reason of that narrow hollownesse, which being filled vp, the cucumber encreaseth in length. The seeds of this kinde of cucumber being sowne bringeth forth not such as were before, but such as art hath framed; which of their own growth are found long, and oftentimes very crookedly turned: and thereupon they haue beene called *Anguini*, or long Cucumbers.

3 The peare fashioned Cucumber hath many trailing branches lying flat vpon the ground, rough and prickly; whereon doe stand at each ioynt one rough leafe, sharpe pointed, and of an ouerworn green colour; among which come forth clasping tendrels, and also slender foot-stalks, whereon do grow yellow starre-like floures. The fruit succeeds, shaped like a peare, as big as a great Warden. The root is threddy.

4 There hath bin not long since sent out of Spain some seeds of a rare & beautiful cucumber, into Strausburg a city in Germany, which there brought forth long trailing branches, rough & hairy, set with very large rough leaues sharp pointed, fashioned like vnto the leaues of the

the great Bur-docke,but more cut in or diuided : amongſt which come forth faire yellow floures growing nakedly vpon their tender foot-ſtalks : the which beeing paſt,the fruit commeth in place, of a foot in length,greene on the ſide toward the ground, yellow to the Sun ward, ſtraked with many ſpots and lines of diuers colours. The pulpe or meat is hard and faſt like that of our Pompion.

¶ *The Place.*

Theſe kindes of Cucumbers are planted in gardens in moſt countries of the world.

¶ *The Time.*

According to my promiſe heretofore made,I haue thought it good and conuenient in this place to ſet downe not onely the time of ſowing and ſetting of Cucumbers, Muske-melons, Citruls, Pompions, Gourds,and ſuch like, but alſo how to ſet or ſow all manner and kindes of other colde ſeeds,as alſo whatſoeuer ſtrange ſeeds are brought vnto vs from the Indies, or other hot Regions: *videl.*

Firſt of all in the middeſt of Aprill or ſomewhat ſooner(if the weather be any thing temperate) you ſhall cauſe to be made a bed or banke of hot and new horſe dung taken forth of the ſtable(and not from the dunghill) of an ell in breadth,and the like in depth or thickneſſe,of what length you pleaſe according to the quantitie of your ſeed:the which bank you ſhal couer with hoops or poles, that you may the more conueniently couer the whole bed or banke with Mats, old painted cloth, ſtraw or ſuch like,to keepe it from the iniurie of the cold froſtie nights, and not hurt the things planted in the bed : then ſhall you couer the bed all ouer with the moſt fertileſt earth finely ſifted, halfe a foot thick,wherein you ſhall ſet or ſow your ſeeds : that being done,caſt your ſtraw or other couerture ouer the ſame;and ſo let it reſt without looking vpon it, or taking away of your couering for the ſpace of ſeuen or eight daies at the moſt,for commonly in that ſpace they will thruſt themſelues vp nakedly forth of the ground:then muſt you caſt vpon them in the hotteſt time of the day ſome water that hath ſtood in the houſe or in the Sun a day before, becauſe the water ſo caſt vpon them newly taken forth of a well or pumpe,will ſo chill and coole them being brought and nouriſhed vp in ſuch a hot place,that preſently in one day you haue loſt all your labour ; I mean not onely your ſeed,but your banke alſo ; for in this ſpace the great heat of the dung is loſt and ſpent,keeping in memorie that euery night they muſt be couered and opened when the day is warmed with the Sun beames : this muſt be done from time to time vntill that the plants haue foure or ſix leaues a piece,and that the danger of the cold nights is paſt : then muſt they be replanted verie curiouſly, with the earth ſticking to the plant,as neere as may be vnto the moſt fruitfull place, and where the Sun hath moſt force in the garden;prouided that vpon the remouing of them you muſt couer them with ſome Docke leaues or wiſpes of ſtraw, propped vp with forked ſtickes,as well to keepe them from the cold of the night, as alſo the heat of the Sun : for they cannot whileſt they be young and newly planted,indure neither ouermuch cold nor ouermuch heat,vntill they are wel rooted in their new place or dwelling.

Oftentimes it falleth out that ſome ſeeds are more franker and forwarder than the reſt, which commonly do riſe vp very nakedly with long necks not vnlike to the ſtalke of a ſmall muſhrome,of a night old. This naked ſtalke muſt you couer with the like fine earth euen to the greene leaues,hauing regard to place your banke ſo that it may be defended from the North-windes.

Obſerue theſe inſtructions diligently,and then you ſhall not haue cauſe to complaine that your ſeeds were not good,nor of the intemperancie of the climat(by reaſon wherof you can get no fruit) although it were in the furtheſt parts of the North of Scotland.

¶ *The Names.*

The Cucumber is named generally *Cucumis* : in ſhops,*Cucumer* : and is taken for that which the Grecians call σικυς ἥμερος : in Latine,*Cucumis ſatiuus*,or garden Cucumber: in High Dutch,**Cucumen**: in Italian,*Concomero* : in Spaniſh, *Cogombro*: in French,*Concombre*: in Low Dutch,**Concommeren**: in Engliſh,Cowcumbers and Cucumbers.

¶ *The Temperature and Vertues.*

All the Cucumbers are of temperature cold and moiſt in the ſecond degree. They putrifie ſoon A in the ſtomacke,and yeeld vnto the body a cold and moiſt nouriſhment,and that very little,and the ſame not good.

Thoſe Cucumbers muſt be choſen which are green and not yet ripe : for when they are ripe and B yellow they be vnfit to be eaten.

The ſeed is cold,but nothing ſo much as the fruit. It openeth and clenſeth, prouoketh vrine, o- C peneth the ſtoppings of the liuer,helpeth the cheſt and lungs that are inflamed;and being ſtamped and outwardly applied in ſtead of a clenſer,it maketh the skin ſmooth and faire.

Cucumber(ſaith my Author) taken in meats,is good for the ſtomack and other parts troubled D with heat.It yeeldeth not any nouriſhment that is good,inſomuch as the vnmeaſurable vſe thereof filleth the veines with naughty cold humours.

The ſeed ſtamped and made into milke like as they do with Almonds,or ſtrained with milke or E ſweet

ſweet wine and drunke, looſeth the belly gently, and is excellent againſt the exulceration of the bladder.

F The fruit cut in pieces or chopped as herbes to the pot and boiled in a ſmall pipkin with a piece of mutton, being made into potage with Ote-meale, euen as herb potage are made, whereof a meſſe eaten to break-faſt, as much to dinner, and the like to ſupper; taken in this manner for the ſpace of three weekes together without intermiſſion, doth perfectly cure all manner of ſawce-flegme and copper faces, red and ſhining fierie noſes (as red as red Roſes) with pimples, pumples, rubies, and ſuch like precious faces.

G Prouided alwaies that during the time of curing you do vſe to waſh or bathe the face with this liquour following.

H Take a pinte of ſtrong white wine vineger, pouder of the roots of Ireos or Orrice three dragmes, ſearced or bolted into moſt fine duſt, Brimmeſtone in fine pouder halfe an ounce, Camphire two dragmes, ſtamped with two blanched Almondes, foure Oke Apples cut thorow the middle, and the iuice of foure Limons: put them all together in a ſtrong double glaſſe, ſhake them together very ſtrongly, ſetting the ſame in the Sunne for the ſpace of ten daies: with which let the face be waſhed and bathed daily, ſuffering it to drie of it ſelfe without wiping it away. This doth not only helpe firie faces, but alſo taketh away lentils, ſpots, morphew, Sun-burne, and all other deformities of the face.

† That which formerly was in the ſecond place by the name of *Cucumis Turcicus*, was the ſame with the fifth of the former Edition (now the fourth) and is therefore omitted.

Chap. 341. *Of Wilde Cucumber.*

Cucumis Aſininus.
Wilde Cucumber.

¶ *The Deſcription.*

The wilde Cucumber hath many fat hairie branches, very rough and full of iuice, creeping or trailing vpon the ground, wherupon are ſet very rough leaues, hairy, ſharp pointed, & of an ouerworne grayiſh greene colour: from the boſome of which come forth long tender footſtalkes: on the ends whereof doe grow ſmall floures compoſed of fiue ſmall leaues of a pale yellow colour: after which commeth forth the fruit, of the bignes of the ſmalleſt pullets egge, but ſomewhat longer, verie rough and hairy on the outſide, and of the colour and ſubſtance of the ſtalkes, wherein is contained very much water and ſmal hard blackiſh ſeeds alſo, of the bigneſſe of tares; which being come to maturitie and ripeneſſe, it caſteth or ſquirteth forth his water with the ſeeds, either of it owne accord, or being touched with the moſt tender or delicate hand neuer ſo gently, and oftentimes ſtriketh ſo hard againſt thoſe that touch it (eſpecially if it chance to hit againſt the face) that the place ſmarteth long after: whereupon of ſome it hath been called *Noli me tangere*, Touch me not. The root is thicke, white and long laſting.

The Place.

It is found in moſt of the hot countries among rubbiſh, grauell, & other vntilled places: it is planted in gardens in the Low-countries, and being once planted, ſaith *Dodonæus*, it eaſily commeth vp againe many yeares after (which is true:) and yet ſaith he further, that it doth not ſpring againe of the root, but of the ſeeds ſpirted or caſt about: which may likewiſe be true where he hath obſerued it, but in my garden it is otherwiſe, for as I ſaid before, the root is long laſting, and continueth from yeare to yeare.

¶ The

¶ *The Time.*

It ſpringeth vp in May, it floureth and is ripe in Autumne, and is to be gathered at the ſame time, to make that excellent compoſition called *Elaterium*.

¶ *The Names.*

It is called in Greeke σίκυς ἄγριος: in Latine, *Agreſtis*, and *Erraticus Cucumis*: in ſhoppes, *Cucumer aſininus*: in Italian, *Cocomero ſaluatico*: in Spaniſh, *Cogumbrillo amargo*: in Engliſh, wilde Cucumber, ſpirting Cucumbers, and touch me not: in French, *Concombres ſauvages*.

¶ *The Temperature.*

The leaues of wilde Cucumbers, roots and their rindes as they are bitter in taſte, ſo they be likewiſe hot and clenſing. The iuice is hot in the ſecond degree, as *Galen* witneſſeth, and of thin parts. It clenſeth and waſteth away.

¶ *The Vertues.*

A The iuice called *Elaterium* doth purge forth choler, flegme, and waterie humours, and that with force, and not onely by ſiege, but ſometimes alſo by vomit.

B The quantity that is to be taken at one time is from fiue grains to ten, according to the ſtrength of the patient.

C The iuice dried or hardened, and the quantitie of halfe a ſcruple taken, driueth forth by ſiege groſſe flegme, cholericke humours, and preuaileth mightily againſt the dropſie, and ſhortneſſe of breath.

D The ſame drawne vp into the noſthrils mixed with a little milk, taketh away the redneſſe of the eies.

E The iuice of the root doth alſo purge flegme, cholericke and wateriſh humours, and is good for the dropſie: but not of ſuch force as *Elaterium*, which is made of the iuice of the fruit: the making whereof I commend to the learned and curious Apothecaries: among which number Mr. *William Wright* in Bucklers Burie my louing friend hath taken more paines in curious compoſing of it, and hath more exactly performed the ſame than any other whatſoeuer that I haue had knowledge of.

CHAP. 342. *Of Citrull Cucumbers.*

1 *Citrullus officinarum.*
Citrull Cucumber.

‡ 2 *Citrullus minor.*
Small Citrull.

¶ *The*

¶ *The Description.*

1 THe Citrull Cucumber hath many long, flexible, and tender stalkes trailing vpon the ground,branched like vnto the Vine,set with certaine great leaues deeply cut,and very much iagged : among which come forth long clasping tendrels,and also tender footstalkes,on the ends whereof do grow floures of a gold yellow colour : the fruit is somewhat round,straked or ribbed with certaine deepe furrowes alongst the same,of a green colour aboue,and vnderneath on that side that lyeth vpon the ground something white : the outward skin whereof is very smooth ; the meat within is indifferent hard,more like to that of the Pompion than of the Cucumber or Muske melon : the pulpe wherein the seed lieth,is spungie, and of a slimie substance : the seed is long,flat, and greater than those of the Cucumbers : the shell or outward barke is blackish,sometimes of an ouerworne reddish colour. The fruit of the Citrull doth not so easily rot or putrifie as doth the Melon,which being gathered in a faire dry day may be kept a long time, especially being couered in a heape of wheat,as *Matthiolus* saith ; but according to my practise you may keepe them much longer and better in a heape of dry sand.

2 The second kinde of Citrull differeth not from the former, sauing that it is altogether lesser,and the leaues are not so deepely cut or iagged,wherein consisteth the difference.

¶ *The Place and Time.*

The Citrull prospereth best in hot Regions, as in Sicilia,Apulia,Calabria,and Syria, about Alepo and Tripolis. We haue many times sown the seeds,and diligently obserued the order prescribed in planting of Cucumbers.

¶ *The Names.*

The later Herbarists do call it *Anguria* : in shoppes,*Citrullus*,and *Cucumis Citrullus* : in English, Citruls,and Cucumber Citruls,and the seed is knowne by the name of *Semen Citrulli* : or Citrull seed. But if *Cucumis Citrullus* , be so called of the yellow colour of the Citron, then is the common Cucumber properly *Cucumis Citrullus* : which is knowne vnto all to be contrarie.

¶ *The Temperature and Vertues.*

A The meat or pulpe of Cucumer Citrull which is next vnto the bark is eaten raw,but more commonly boiled : it yeeldeth to the bodie little nourishment, and the same cold : it ingendreth a waterish bloud,mitigateth the extremity of heat of the inner parts,and tempereth the sharpnesse and feruent heat of choler: being raw and held in the mouth,it takes away the roughnesse of the tongue in Agues,and quencheth thirst.

B The seeds are of the like facultie with those of Cucumbers.

Chap. 343. *Of the wilde Citrull called* Colocynthis.

¶ *The Description.*

1 COloquintida hath beene taken of many to be a kinde of the wild Gourd,it lieth along creeping on the ground as doe the Cucumbers and Melons,comming neerest of all to that which in those daies of some Herbarists is called Citrull Cucumber:it bringeth forth vpon his long branches smal crooked tendrels like the Vine,and also very great broad leaues deepely cut or iagged : among which come forth small floures of a pale yellow colour; then commeth the fruit round as a bowle,couered with a thin rinde,of a yellow colour when it is ripe, which when it is pilled or pared off, the white pulpe or spungie substance appeareth full of seedes, of a white or else an ouerworne browne colour; the fruit so pared or pilled , is dryed for medicine; the which is most extreame bitter, and likewise the seede, and the whole plant it selfe in all his parts.

2 The second kinde of Coloquintida hath likewise many long branches and clasping tendrels, wherewith it taketh hold of such things as are neere vnto it. It bringeth forth the like leaues , but not so much iagged. The floures are small and yellow:the fruit is fashioned like a peare,and the other sort round,wherein the especiall difference consisteth.

¶ *The Place.*

Coloquintida is sowne and commeth to perfection in hot regions,but seldome or neuer in these Northerly and cold countries.

¶ The

1 *Colocynthis.*
The wilde Citrull or Coloquintida.

2 *Colocynthis pyriformis.*
Peare fashioned Coloquintida.

¶ *The Time.*

It is sowne in the Spring, and bringeth his fruit to perfection in August.

It hath beene diuers times deliuered vnto me for a truth, that they doe grow in the sands of the Mediterranean sea shore, or verie neere vnto it, wilde, for euery man to gather that list, especially on the coast of Barbarie, as also without the mouth of the Streights neere to *Sancta Crux* and other places adiacent; from whence diuers Surgions of London that haue trauelled thither for the curing of sicke and hurt men in the ship haue brought great quantities thereof at their returne.

¶ *The Names.*

It is vulgarly called *Coloquintida*: in Greeke κολοκυνθίς: the Latine translators for *Colocynthis* doe oftentimes set downe *Cucurbita syluestris*: notwithstanding there is a *Cucurbita syluestris* that differeth from *Colocynthis*, or Coloquintida: for *Cucurbita syluestris* is called in Greeke κολοκύνθα ἀγρία: or wilde Coloquintida, whereof shall be set forth a peculiar chapter next after the *Cucurbita* or Gourd: in English it is called Coloquintida, or Apple of Coloquintida.

¶ *The Temperature.*

Coloquintida as it is in his whole nature and in all his parts bitter, so is it likewise hot and drie in the later end of the second degree; and therefore it purgeth, clenseth, openeth and performeth all those things that most bitter things do: but that the strong qualitie which it hath to purge by the stoole, is, as *Galen* saith, of more force than the rest of his operations.

¶ *The Vertues.*

A Which operation of purging it worketh so violently, that it doth not onely draw forth flegme and choler maruellous speedily, and in very great quantitie: but oftentimes fetcheth forth bloud and bloudy excrements, by shauing the guts, and opening the ends of the meseraicall veines.

B So that therefore the same is not to bee vsed either rashly, or without some dangerous and extreme disease constraine thereunto: neither yet at all, vnlesse some tough and clammie thing bee mixed therewith, whereby the vehemencie thereof may be repressed, the hurtfull force dulled, and the same speedily passing through the belly, the guts be not fret or shaued. *Mesues* teacheth to mixe with it either Mastich, or gum Tragacanth.

C There be made of it Trochises, or little flat cakes, with Mastich, gum Arabick, Tragacanth and Bdellium,

Bdellium,of these,Mastich hath a manifest binding qualitie : but tough and clammie things are much better,which haue no astriction at all in them,or very little.

D For by such binding or astringent things,violent medicines being restrained and brideled, do afterward work their operation with more violence and trouble : but such as haue not binding things mixed with them do easilier worke,and with lesser paine, as be those pils which *Rhasis* in his ninth booke of *Almanzor* calleth *Illiacæ* : which are compounded of Coloquintida and Scamony, two of the strongest medicines that are ; and of a third called gum *Sagapene*, which through his clamminesse doth as it were daube the intrails and guts,and defend them from the harme that might haue come of either of them.

E The which composition,although it be wonderfull strong,and not to be vsed without very great necessitie vrge thereunto,doth notwithstanding easily purge, and without any great trouble, and with lesser torment than most of the mildest and gentlest medicins which haue Mastich and other things mixed with them that are astringent.

F And for this cause it is very like that *Galen* in his first booke of Medicines,according to the places affected,would not suffer Mastich and Bdellium to be in the pilles, which are surnamed *Cochiæ*: the which notwithstanding his Schoolemaster *Quintus* was also woont before to adde vnto the same.

G But Coloquintida is not onely good for purgations,in which it is a remedie for the dissinesse or the turning sicknesse,the megrim, continuall head-ache,the Apoplexie, the falling sicknesse, the stuffing of the lungs,the gnawings and gripings of the guts and intrailes,and other most dangerous diseases, but also it doth outwardly worke his operations, which are not altogether to be reiected.

H Common oile wherein the same is boiled,is good against the singing in the eares ,and deafenes: the same killeth and driueth forth all manner of wormes of the belly,and doth oftentimes prouoke to the stoole,if the nauell and bottome of the belly be therewith annointed.

I Being boiled in vineger, and the teeth washed therewith, it is a remedie for the tooth-ache, as *Mesues* teacheth.

K The seed is very profitable to keepe and preserue dead bodies with ; especially if Aloes and Myrrhe be mixed with it.

L The white pulpe or spungious pith taken in the weight of a scruple openeth the belly mightily,and purgeth grosse flegme,and cholericke humors.

M It hath the like force if it be boiled and laid to infuse in wine or ale,and giuen to drinke.

N Being taken after the same manner it profiteth the diseases before remembred,that is,the Apoplexie,falling sicknesse,giddinesse of the head,the collicke,loosenesse of sinewes,and places out of ioint,and all diseases proceeding of cold.

O For the same purposes it may be vsed in clisters.

P The same boiled in oile,and applied with cotton or wooll,taketh away the pain of the Hemorrhoides.

Q The decoction made in wine, and vsed as a fomentation or bathe, bringeth downe the desired sicknesse.

Chap. 344. *Of Muske-Melon,or Million.*

The Kindes.

There be diuers sorts of Melons found at this day, differing very notably in shape and proportion,as also in taste, according to the climate and countrie where they grow : but of the Antients there was onely one and no more,which is that *Melopepo* called of *Galen*, *Cucumis*,or *Galens* Cucumber : notwithstanding some haue comprehended the Muske-Melons vnder the kindes of Citruls,wherein they haue greatly erred : for doubtlesse the Muske-Melon is a kinde of Cucumber, according to the best approued Authors.

¶ *The Description.*

1 That which the later Herbarists do call Muske-Melons is like to the common Cucumber in stalks,lying flat vpon the ground,long,branched,and rough. The leaues be much alike,yet are they lesser,rounder,and not so cornered : the floures in like manner bee yellow:the fruit is bigger,at the first somwhat hairy,somthing long,now and then somwhat round; oftentimes greater,and many times lesser : the barke or rinde is of an ouerworne russet greene colour,

1 *Melo.*
The Muske Melon.

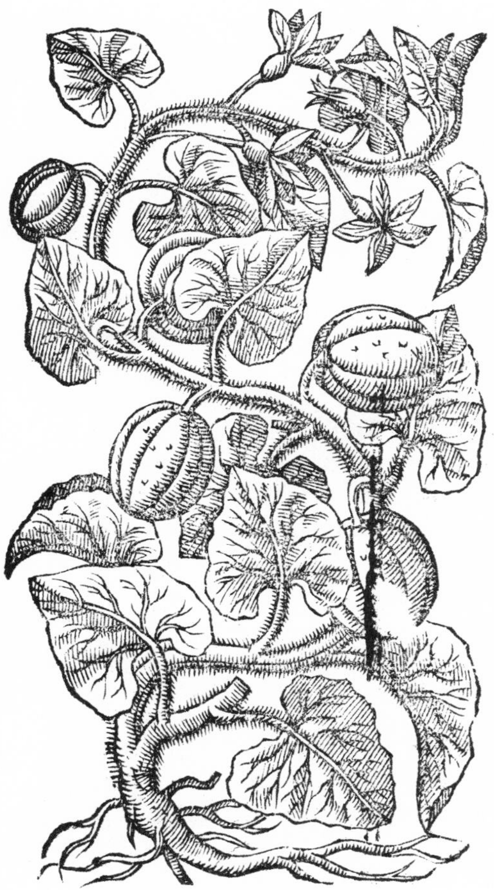

2 *Melo Saccharinus.*
Sugar melon.

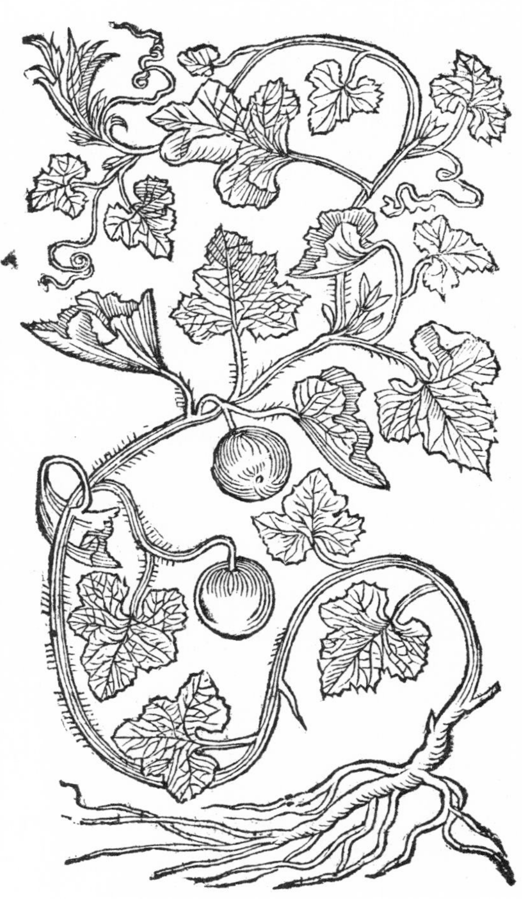

4 *Melo Hiſpanicus.*
Spaniſh Melons.

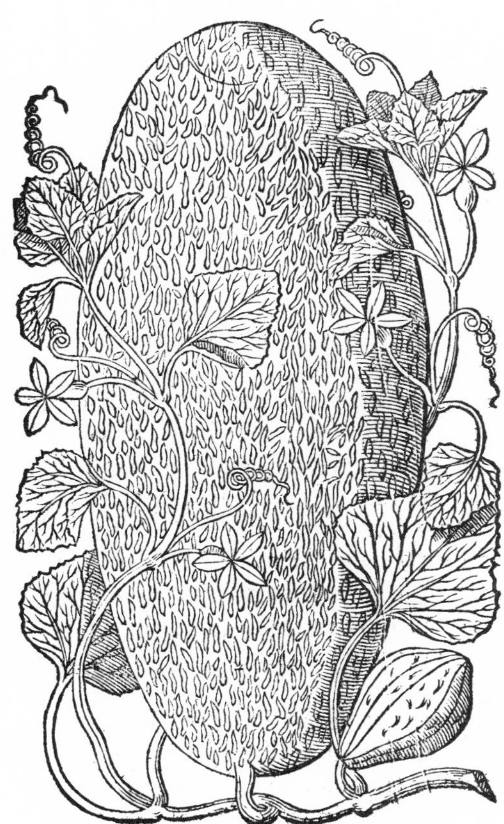

colour, ribbed and furrowed very deepely, hauing often chappes or chinkes, and a confuſed roughneſſe: the pulpe or inner ſubſtance which is to be eaten, is of a faint yellow colour. The middle part whereof is full of a ſlimie moiſture, among which is conteined the ſeed, like vnto thoſe of the Cucumber, but leſſer, and of a browner colour.

2 The ſugar Melon hath long trailing ſtalkes lying vpon the ground, whereon are ſet ſmall claſping tendrels like thoſe of the Vine, and alſo leaues like vnto the common Cucumber, but of a greener colour: the fruite commeth forth among thoſe leaues, ſtanding vpon ſlender footſtalkes, round as the fruite of *Coloquintida*, and of the ſame bigneſſe, of a moſt pleaſant taſte like Sugar, whereof it tooke the ſurname *Saccharatus*.

3 The Peare faſhioned Melon hath many long vinie branches, whereupon doe grow cornered leaues like thoſe of the Vine, and likewiſe great ſtore of long tendrels, claſping and taking hold of each thing that it toucheth: the fruite groweth vpon ſlender footſtalkes, faſhioned like vnto a Peare, of the bignes of a great Quince.

4 The Spaniſh Melon bringeth forth long trailing

trailing branches, whereon are set broad leaues slightly indented about the edges, not diuided at all, as are all the rest of the Melons. The fruite groweth neere vnto the stalke, like vnto the common Pompion, very long, not crested or furrowed at all, but spotted with very many such markes as are on the backeside of the Harts-tongue leafe. The pulpe or meate is not so pleasing in taste as the other.

¶ *The Place.*

They delight in hot regions, notwithstanding I haue seen at the Queenes house at Saint Iames very many of the first sort ripe, through the diligent and curious nourishing of them by a skilfull Gentleman the keeper of the said house, called Mr. *Fowle*, and in other places neere vnto the right Honorable the Lord of *Sussex* house, of Bermondsey by London, where from yeere to yeere there is very great plenty, especially if the weather be any thing temperate.

¶ *The Time.*

They are set or sowne in Aprill as I haue already shewne in the chapter of Cucumbers: their fruite is ripe in the end of August, and sometimes sooner.

¶ *The Names.*

The Muske Melon is called in Latine, *Melo*: in Italian, *Mellone*: in Spanish, *Melon*: in French, *Melons*: in High Dutch, **Melaun**: in low Dutch, **Meloenen**: in Greeke, μῆλον, which doth signifie an apple; and therefore this kinde of Cucumber is more truely called μηλοπέπων, or *Melopepon*: by reason that *Pepo* hath the smell of an apple, whereto the smell of this fruit is like; hauing withall the smell as it were of Muske: which for that cause are also named *Melones Muschatellini*, or Muske Melons.

¶ *The Temperature.*

The meate of the Muske Melon, is very cold and moist.

¶ *The Vertues.*

A It is harder of digestion than is any of the rest of Cucumbers: and if it remaine long in the stomacke is putrifieth, and is occasion of pestilent feuers: which thing also *Aëtius* witnesseth in the first booke of his *Tetrabibles*, writing that the vse of *Cucumeres*, or Cucumbers, breedeth pestilent feauers; for he also taketh *Cucumis* to be that which is commonly called a Melon: which is vsually eaten of the Italians and Spaniards rather to represse the rage of lust, than for any other Physicall vertue.

B The seed is of like operation with that of the former Cucumber.

Chap. 345. *Of Melons, or Pompions.*

¶ *The Kindes.*

There be found diuers kindes of Pompions which differ either in bignesse or forme: it shall be therefore sufficient to describe some one or two of them, and referre the rest vnto the view of the figures, which most liuely do expresse their differences; especially because this volume waxeth great, the description of no moment, and I hasten to an end.

¶ *The Description.*

1 The great Pompion bringeth forth thicke and rough prickly stalkes, which with their clasping tendrells take hold vpon such things as are neere vnto them, as poles, arbours, pales, and ledges, which vnlesse they were neere vnto them would creepe along vpon the ground; the leaues be wilde, and great, very rough, and cut with certaine deepe gashes, nicked also on the edges like a saw; the floures be very great like vnto a bell cup, of a yellow colour like gold, hauing fiue corners standing out like teeth: the fruite is great, thicke, round, set with thicke ribbes, like edges sticking forth. The pulpe or meate whereof which is next vnder the rinde is white, and of a meane hardnesse: the pith or substance in the middle is spungie, and slimie: the seed is great, broad, flat, something white, much greater than that of the Cucumber, otherwise not differing at all in forme. The colour of the barke or rinde is oftentimes of an obscure greene, sometimes gray. The rinde of the greene Pompion is harder, and as it were of a woody substance: the rinde of the gray is softer and tenderer.

2 The second kinde of Melons or Pompions is like vnto the former in stalkes and leaues, and also in clasping tendrels: but the gashes of the leaues are not so deepe, and the stalkes be tenderer: the floures are in like manner yellow, gaping, and cornered at the top, as be those of the former: but the fruite is somewhat rounder; sometimes greater, and many times lesser: and oftentimes

of

of a greene colour with an harder barke; now and then ſofter and whiter. The meat within is like the former: the ſeeds haue alſo the ſame forme, but they be ſomewhat leſſer.

1 *Pepo maximus oblongus.*
The great long Pompion.

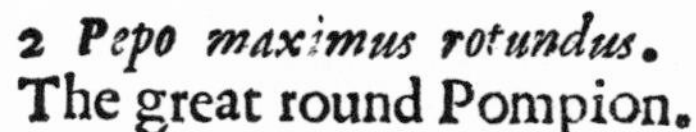
2 *Pepo maximus rotundus.*
The great round Pompion.

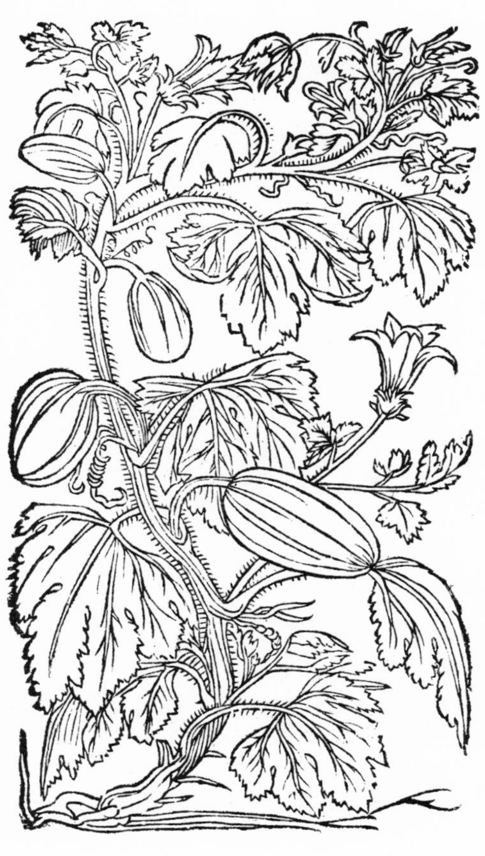

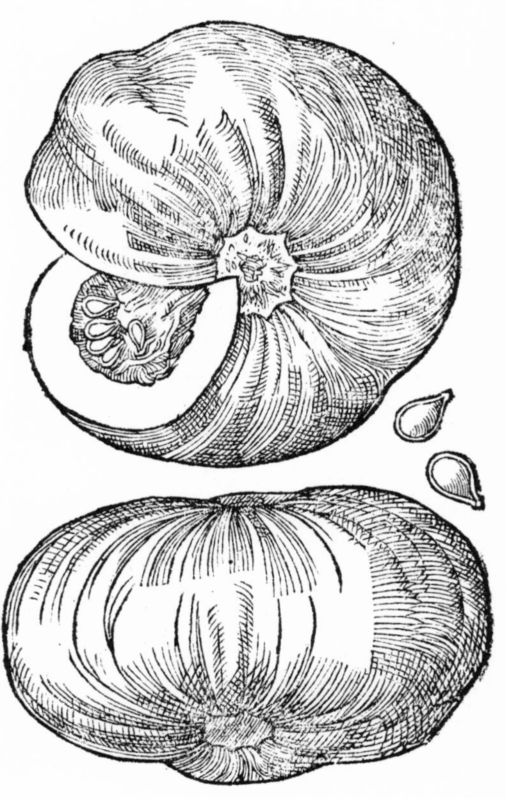

3 Of this kinde there is alſo another Pompion like vnto the former in rough ſtalkes, and in gaſhed and nicked leaues: the floure is alſo great and yellow, like thoſe of the others: the fruit is of a great bigneſſe, whoſe barke is full of little bunnies or hillie welts, as is the rinde of the Citron, which is in like manner yellow when it is ripe.

4 The fourth Pompion doth very much differ from the others in form: the ſtalks, leaues, and floures are like thoſe of the reſt: but the fruit is not long or round, but altogether broad, and in a manner flat like vnto a ſhield or buckler; thicker in the middle, thinner in the compaſſe, and curled or bumped in certaine places about the edges, like the rugged or vneuen barke of the Pomecitron; the which rinde is very ſoft, thin, and white: the meat within is meetely hard and dureable. The ſeed is greater than that of the common Cucumber, in forme and colour all one.

‡ *Macocks Virginiani, ſive Pepo Virginianus.*
The Virginian Macocke, or Pompion.

‡ This hath rough cornered ſtraked trailing branches proceeding from the root, eight or nine foot long, or longer, and thoſe againe diuided into other branches of a blackiſh greene colour, trailing, ſpreading, or running alongſt the earth, couering a great deale of ground, ſending forth broad cornered rough leaues, on great groſſe, long, rough, hairy foot-ſtalks, like and fully as big as the leaues of the common Pompion, with claſping tendrels and great broad ſhriueled yellow floures alſo like thoſe of the common Pompion: the fruit ſucceedeth, growing alongſt the ſtalkes, commonly not neere the root, but towards the vpper part or toppes of the branches, ſomewhat round, not extending in length, but flat like a bowle, but not ſo bigge as an ordinarie bowle, beeing ſeldome foure inches broad, and three inches long of a blackiſh greene colour when it is ripe. The ſubſtance or eatable part is of a yellowiſh white colour, containing in the middeſt a great deale of pulpe or ſoft matter, wherein the ſeed lyeth in certaine rowes alſo, like the common Pompion, but ſmaller. The root is made of many whitiſh branches, creeping far abroad in the earth, and periſh at the firſt approch of Winter.

3 *Pepo maximus compressus.* The great flat bottommed Pompion.

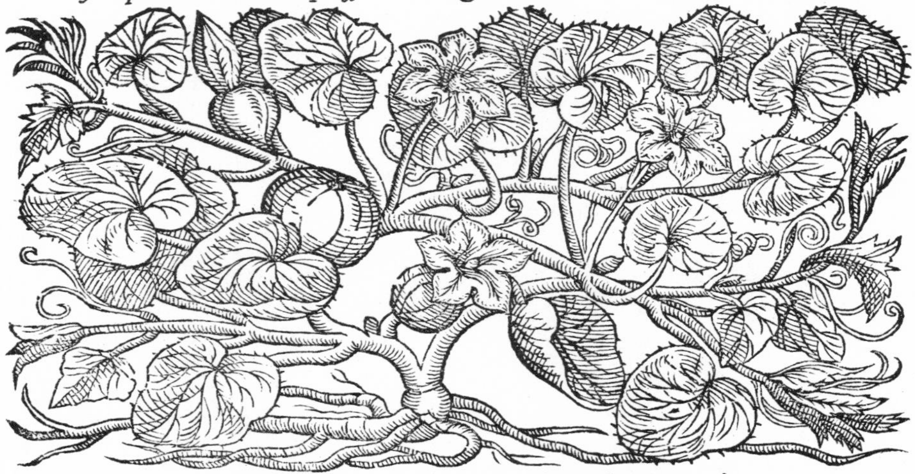

4 *Pepo maximus clypeatus.* The great buckler Pompion.

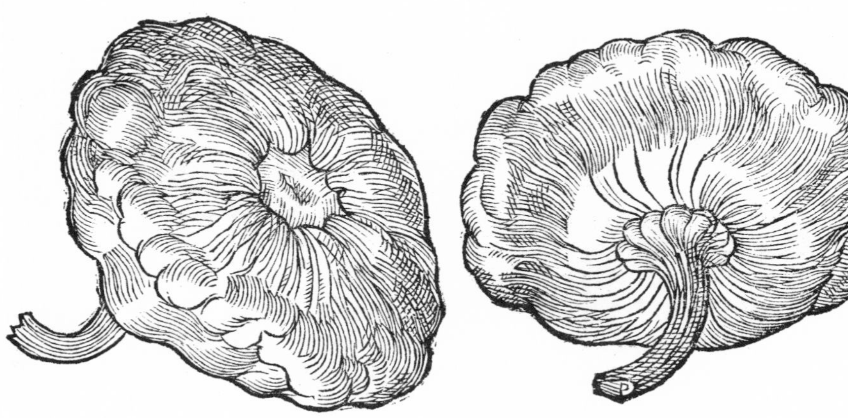

5 *Pepo Indicus minor rotundus.*
The small round Indian Pompion.

6 *Pepo Indicus angulosus.*
The cornered Indian Pompion.

Melones aquatici edules Virginiani.
The Virginian Water-Melon.

This Melon or Pompion is like and fully as bigge as the common Pompion, in spreading, running, creeping branches, leaues, floures, and clasping tendrels: the fruit is of a very blackish greene colour, and extendeth it selfe in length neere foure inches long, and three inches broad, no bigger nor longer than a great apple, and grow alongst the branches forth of the bosomes of the leaues not farre from the root euen to the toppes of the branches, containing a substance, pulpe, and flat seed, like the ordinary Pompion: the root is whitish, and dispersethit selfe verie farre abroad in the earth, and perisheth about the beginning of VVinter. October the tenth, 1621. *Iohn Goodyer*. ‡

¶ *The Place.*

All these Melons or Pompions be garden plants: they ioy best in a fruitfull soile, and are common in England; except the last described, which is as yet a stranger.

¶ *The Time.*

They are planted at the beginning of Aprill: they floure in August: the fruit is ripe in September.

¶ *The Names.*

The great Melon or Pompion is named in Greeke πέπων: in Latine likewise *Pepo*: The fruits of them all when they be ripe are called by a common name in Greeke, πέπονες: in English, Millions or Pompion. Whereupon certaine Physitions, saith *Galen*, haue contended, that this fruit ought to be called σικυοπέπων, that is to say in Latine, *Pepo Cucumeralis*, or Cucumber Pompion. *Pliny* in his ninth booke the fifth Chapter writeth, that *Cucumeres* when they exceed in greatnesse are named *Pepones*: it is called in High Dutch, **Pluker**: in Low Dutch, **Peponen**: in French, *Pompons*.

¶ *The Temperature and Vertues.*

All the Melons are of a cold nature, with plenty of moisture: they haue a certaine clensing qualitie, by meanes whereof they prouoke vrine, and do more speedily passe through the bodie than do either the Gourd, Citrull, or Cucumber, as *Galen* hath written A

The pulpe of the Pompion is neuer eaten raw, but boiled. For so it doth more easily descend, making the belly soluble. The nourishment which commeth hereof is little, thin, moist and cold, (bad, saith *Galen*) and that especially when it is not well digested: by reason whereof it maketh a man apt and readie to fall into the disease called the Cholericke Passion, and of some the Felonie. B

The seed clenseth more than the meat, it prouoketh vrine, and is good for those that are troubled with the stone of the kidnies. C

The fruit boiled in milke and buttered, is not onely a good wholesome meat for mans body, but being so prepared, is also a most physicall medicine for such as haue an hot stomacke, and the inward parts inflamed. D

The flesh or pulpe of the same sliced and fried in a pan with butter, is also a good and wholsome meat: but baked with apples in an ouen, it doth fil the body with flatuous or windie belchings, and is food vtterly vnwholesome for such as liue idlely; but vnto robustious and rustick people nothing hurteth that filleth the belly. E

CHAP. 346. *Of Wilde Pompions.*

¶ *The Description.*

1 AS there is a wilde sort of Cucumbers, of Melons, Citruls and Gourds, so likewise there be certaine wilde Pompions, that be so of their owne nature. These bring forth rough stalks, set with sharp thorny prickles. The leaues be likewise rough: the floures yellow as be those of the garden Melon, but euery part is lesser. The fruit is thicke, round, and sharp pointed, hauing a hard greene rinde. The pulpe or meat whereof, and the middle pith, with the seed are like those of the garden Pompion, but very bitter in taste.

2 The second is like vnto the former, but it is altogether lesser, wherein consisteth the difference.

1 *Pepo maior syluestris.*
The great wilde Pompion.

2 *Pepo minor syluestris.*
The small wilde Pompion.

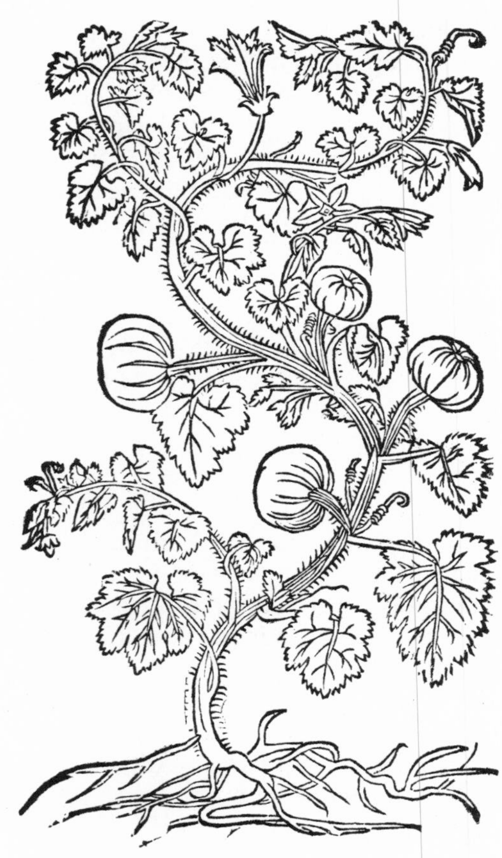

¶ *The Place.*

These Melons do grow wilde in Barbarie, Africa, and most parts of the East and West Indies. They grow not in these parts except they be sowne.

¶ *The Time.*

Their time of flouring and flourishing answereth that of the garden Pompion.

¶ *The Names.*

Although the Antient Physitions haue made no mention of these plants, yet the thing it selfe doth shew, that there be such, and ought to be called in Greeke πέπονες ἄγριοι in Latine, *Pepones syluesters:* in English, wilde Melons or Pompions.

¶ *The Temperature.*

Like as these wilde Melons be altogether of their owne nature very bitter, so be they also of temperature hot and drie, and that in the later end of the second degree. They haue likewise a clensing facultie, not inferior to the wilde Cucumbers.

¶ *The Vertues.*

A The wine, which when the pith and seed is taken forth, is powred into the rinde, and hath remained so long therein till such time as it becommeth bitter, doth purge the belly, and bringeth forth flegmaticke and cholerick humors. To be briefe, the iuice hereof is of the same operation that the wilde Cucumber is of; and being dried it may be vsed in stead of *Elaterium*, which is the dried iuice of the wilde Cucumber.

Chap. 347. *Of Gourds.*

¶ *The Kindes.*

There be diuers sorts of Gourds, some wilde, and others tame of the garden; some bringing forth fruit like vnto a bottle; others long, bigger at the end, keeping no certaine forme or fashion; some greater, others lesser. ‡ I will onely figure and describe two or three of the chiefest, and so passe ouer the rest, because each one vpon the first sight of them knowes to what kinde to referre them. ‡

¶ The

¶ *The Deſcription.*

1 THe Gourd bringeth forth very long ſtalkes as be thoſe of the Vine, cornered and parted into diuers branches, which with his claſping tendrels taketh hold and clymeth vpon ſuch things as ſtand neere vnto it: the leaues be very great, broad, and ſharpe pointed, almoſt as great as thoſe of the Clot-Burre, but ſofter, and ſomewhat couered as it were with a white freeſe, as be alſo the ſtalkes and branches, like thoſe of the mariſh Mallow: the floures be white, and grow forth from the boſome of the leaues: in their places come vp the fruit, which are not all of one faſhion, for oftentimes they haue the forme of flagons or bottles, with a great large belly and a ſmall necke. The Gourd (ſaith *Pliny, lib.* 19. *cap.* 5.) groweth into any forme or faſhion that you would haue it, either like vnto a wreathed Dragon, the leg of a man, or any other ſhape, according to the mould wherein it is put whileſt it is young: being ſuffered to clime vpon any Arbour where the fruit may hang, it hath beene ſeene to be nine foot long, by reaſon of his great weight which hath ſtretched it out to the length. The rinde when it is ripe is verie hard, wooddy, and of a yellow colour: the meate or inward pulpe is white; the ſeed long, flat, pointed at the top, broad below, with two peakes ſtanding out like hornes, white within, and ſweet in taſte.

2 The ſecond differeth not from the precedent in ſtalkes, leaues, or floures: the fruit hereof is for the moſt part faſhioned like a bottle or flagon, wherein eſpecially conſiſteth the difference.

1 *Cucurbita anguina.*
Snakes Gourd.

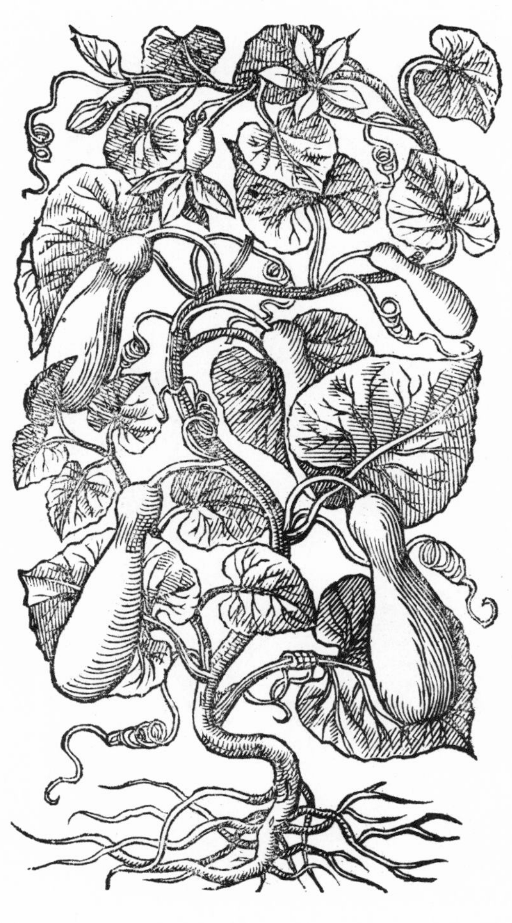

2 *Cucurbita lagenaria.*
Bottle Gourds.

¶ *The Place.*

The Gourds are cheriſhed in the gardens of theſe cold regions rather for pleaſure than for profit: in the hot countries where they come to ripeneſſe there are ſometimes eaten, but with ſmall delight; eſpecially they are kept for the rindes, wherein they put Turpentine, Oyle, Hony, and alſo ſerue them for pales to fetch water in, and many other the like vſes.

¶ *The Time.*

They are planted in a bed of horſe-dung in April, euen as we haue taught in the planting of cucumbers: they flouriſh in Iune and Iuly; the fruit is ripe in the end of Auguſt.

¶ *The Names.*

The Gourd is called in Greeke Κολόκυνθα ἥμερος: in Latine, *Cucurbita edulis, Cucurbita ſatiua*: of *Pliny,*

Cucurbita

Cucurbita Cameraria, becauſe it climeth vp, and is a couering for arbours and walking places, and banqueting houſes in gardens: he calleth the other which climeth not vp, but lyeth crawling on the ground, *Cucurbita plebeia*: in Italian, *Zucca*: in Spaniſh, *Calabazza*: in French, *Courge*: in high Dutch, Kurbs: in low-Dutch, Cauwoorden: in Engliſh, Gourds.

¶ *The Temperature.*

The meate or inner pulpe of the Gourd is of temperature cold and moiſt, and that in the ſecond degree.

¶ *The Vertues.*

A The iuyce being dropped into the eares with oyle of roſes is good for the paine thereof proceeding of a hot cauſe.

B The pulpe or meate mitigateth all hot ſwellings, if it be laid thereon in manner of a pultis, and being vſed in this manner it taketh away the head-ache and the inflammation of the eyes.

C The ſame Author affirmeth, that a long Gourd or elſe a Cucumber being laid in the cradle or bed by the young infant whileſt it is aſleepe and ſicke of an ague, it ſhall be very quickely made whole.

D The pulpe alſo is eaten ſodden, but becauſe it hath in it a wateriſh and thinne iuyce, it yeeldeth ſmall nouriſhment to the body, and the ſame cold and moiſt; but it eaſily paſſeth thorow, eſpecially being ſodden, which by reaſon of the ſlipperineſſe and moiſtneſſe alſo of his ſubſtance mollifieth the belly.

E But being baked in an ouen or fried in a pan it loſeth the moſt part of his naturall moiſture, and therefore it more ſlowly deſcendeth, and doth not mollifie the belly ſo ſoone.

F The ſeed allayeth the ſharpneſſe of vrine, and bringeth downe the ſame.

Chap. 348. *Of the wilde Gourd.*

1 *Cucurbita lagenaria ſylueſtris.*
Wilde Bottle Gourd.

2 *Cucurbita ſylueſtris fungiformis.*
Muſhrome wilde Gourd.

¶ *The*

¶ *The Deſcription*

1 There is beſides the former ones a certaine wilde Gourd: this is like the garden Gourd in clymbing ſtalkes, claſping tendrels, and ſoft leaues, and as it were downy; all and euerie one of which things being farre leſſe: this alſo clymbeth vpon Arbours and banquetting houſes: the fruit doth repreſent the great bellied Gourd, and thoſe that be like vnto bottles in forme, but in bigneſſe it is very farre inferiour; for it is ſmall, and ſcarſe ſo great as an ordinarie Quince, and may be held within the compaſſe of a mans hand: the outward rinde at the firſt is greene, afterwards it is as hard as wood, and of the colour thereof: the inner pulpe is moiſt, and very full of iuyce, in which lieth the ſeed. The whole is as bitter as Coloquintida, which hath made ſo many errors, one eſpecially, in taking the fruit Coloquintida for the wilde Gourd.

2 The ſecond wilde Gourd hath likewiſe many trailing branches and claſping tendrels, wherwith it taketh hold of ſuch things as be neere vnto it: the leaues be broad, deepely cut into diuers ſections, like thoſe of the Vine, ſoft and very downy, whereby it is eſpecially knowne to be one of the Gourds: the floures are very white, as are alſo thoſe of the Gourds. The fruit ſucceedeth, growing to a round forme, flat on the top like the head of a Muſhrome, whereof it tooke his ſyrname.

¶ *The Place.*

They grow of themſelues wilde in hot regions; they neuer come to perfection of ripeneſſe in theſe cold countries.

¶ *The Time.*

The time anſwereth thoſe of the garden.

¶ *The Names.*

The wilde Gourd is called in Greeke κολοκυνθα ἀγρια: in Latine, *Cucurbita ſylueſtris*, or wilde Gourd. *Pliny, lib.* 20. *cap.* 3. affirmeth, that the wilde Gourd is named of the Grecians, σόμφος, which is hollow, an inch thicke, not growing but among ſtones, the iuyce whereof being taken is very good for the ſtomacke. But the wilde Gourd is not that which is ſo deſcribed; for it is aboue an inch thicke, neither is it hollow, but full of iuyce, and by reaſon of the extreme bitterneſſe offenſiue to the ſtomacke.

Some alſo there be that take this for Coloquintida, but they are far deceiued; for Colocynthis is the wilde Citrull Cucumber, whereof we haue treated in the chapter of Citruls.

¶ *The Temperature.*

The wilde Gourd is as hot and dry as Coloquintida, that is to ſay, in the ſecond degree.

¶ *The Vertues.*

The wilde Gourd is extreme bitter, for which cauſe it openeth and ſcoureth the ſtopped paſſages of the body; it alſo purgeth downwards as do wilde Melons. A

Moreouer, the wine which hath continued all night in this Gourd likewiſe purgeth the belly mightily, and bringeth forth cholericke and flegmaticke humors. B

CHAP. 349. *Of Potato's.*

Siſarum Peruvianum, ſiue Batata Hiſpanorum.
Potatus, or Potato's.

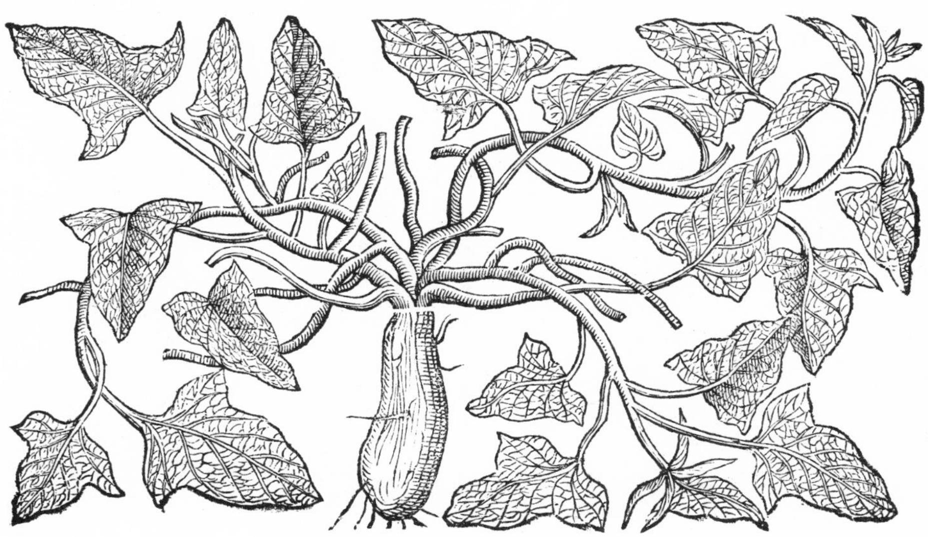

¶ *The Defcription.*

THis Plant (which is called of fome *Sifarum Peruvianum*, or Skyrrets of Peru) is generally of vs called Potatus, or Potatoes. It hath long rough flexible branches trailing vpon the ground like vnto Pompions; whereupon are fet greene three cornered leaues, very like vnto thofe of the wilde Cucumber. There is not any that haue written of this plant haue faid any thing of the floures: therefore I refer their defcription vnto thofe that fhall hereafter haue further knowledge of the fame. Yet haue I had in my garden diuers roots that haue flourifhed vnto the firft approch of Winter, and haue growne vnto a great length of branches, but they brought not forth any floures at all; whether becaufe the Winter caufed them to perifh before their time of flouring, or that they be of nature barren of floures, I am not certaine. The roots are many, thicke, and knobbie, like vnto the roots of Peionies, or rather of the white Afphodill, ioyned together at the top into one head, in maner of the Skyrrit, which being diuided into diuers parts and planted, do make a great increafe, efpecially if the greateft roots be cut into diuers goblets, and planted in good and fertile ground.

¶ *The Place.*

The Potatoes grow in India, Barbarie, Spaine, and other hot regions; of which I planted diuers roots (that I bought at the Exchange in London) in my garden, where they flourifhed vntil Winter, at which time they perifhed and rotted.

¶ *The Time.*

It flourifheth vnto the end of September: at the firft approch of great frofts the leaues together with the roots and ftalkes do perifh.

¶ *The Names.*

Clufius calleth it *Battata*, *Camotes*, *Amotes*, and *Ignames*: in Englifh, Potatoes, Potatus, and Potades.

¶ *The Temperature.*

The leaues of Potatoes are hot and dry, as may euidently appeare by the tafte. The roots are of a temperate qualitie.

¶ *The Vertues.*

A The Potato roots are among the Spaniards, Italians, Indians, and many other nations common and ordinarie meate; which no doubt are of mighty and nourifhing parts, and do ftrengthen and comfort nature; whofe nutriment is as it were a meane betweene flefh and fruit, but fomwhat windie; but being tofted in the embers they lofe much of their windineffe, efpecially being eaten fopped in wine.

B Of thefe roots may be made conferues no leffe toothfome, wholefome, and dainty than of the flefh of Quinces: and likewife thofe comfortable and delicate meats called in fhops *Morfelli*, *Placentulæ*, and diuers other fuch like.

C Thefe Roots may ferue as a ground or foundation whereon the cunning Confectioner or Sugar-Baker may worke and frame many comfortable delicate Conferues, and reftoratiue fweete meates.

D They are vfed to be eaten rofted in the afhes. Some when they be fo rofted infufe them and fop them in Wine; and others to giue them the greater grace in eating, doe boyle them with prunes, and fo eate them. And likewife others dreffe them (being firft rofted) with Oyle, Vineger, and falt, euerie man according to his owne tafte and liking. Notwithftanding howfoeuer they bee dreffed, they comfort, nourifh, and ftrengthen the body, procuring bodily luft, and that with greedineffe.

CHAP. 350. *Of Potatoes of Virginia.*

¶ *The Defcription.*

VIrginia Potato hath many hollow flexible branches trailing vpon the ground, three fquare, vneuen, knotted or kneed in fundry places at certaine diftances: from the which knots commeth forth one great leafe made of diuers leaues, fome fmaller, and others greater, fet together vpon a fat middle rib by couples, of a fwart greene colour tending to redneffe; the whole leafe refembling thofe of the Winter-Creffes, but much larger; in tafte at the firft like graffe, but afterward fharpe and nipping the tongue. From the bofome of which leaues come forth long round

round slender foot-stalkes, whereon do grow very faire & pleasant floures, made of one entire whole leafe, which is folded or plaited in such strange sort, that it seemeth to be a floure made of fiue sundry small leaues, which cannot easily be perceiued except the same be pulled open. The whole floure is of a light purple colour, striped downe the middle of euery fold or welt with a light shew of yellownesse, as if purple and yellow were mixed together. in the middle of the floure thrusteth forth a thicke flat pointall yellow as gold, with a small sharpe greene pricke or point in the middest thereof. The fruit succeedeth the floures, round as a ball, of the bignesse of a little Bullesse or wilde plum, greene at the first, and blacke when it is ripe; wherein is contained small white seed lesser than those of Mustard. The root is thicke, fat, and tuberous, not much differing either in shape, colour, or taste from the common Potatoes, sauing that the roots hereof are not so great nor long; some of them are as round as a ball, some ouall or egge-fashion; some longer, and others shorter: the which knobby roots are fastened vnto the stalkes with an infinite number of threddie strings.

Battata Virginiana, siue Virginianorum, & Pappus.
Virginian Potatoes.

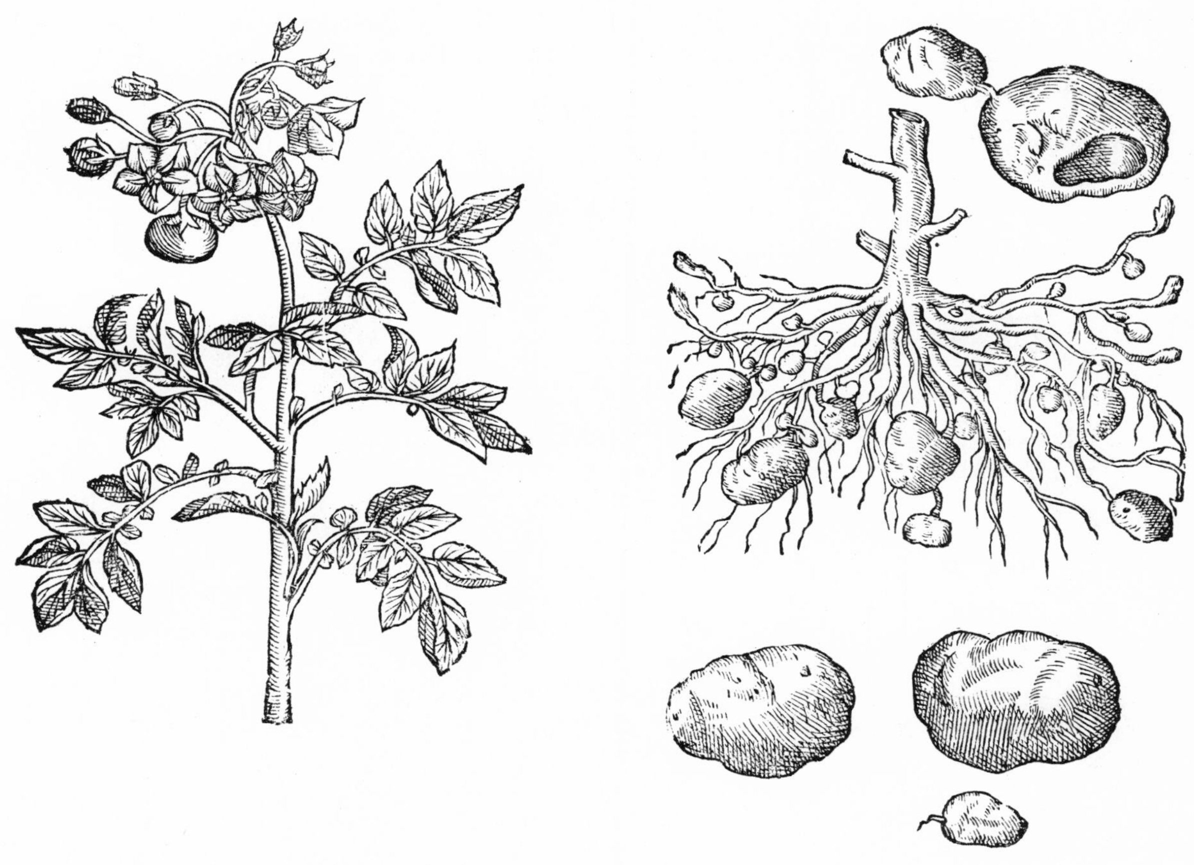

¶ *The Place.*

It groweth natnrally in America, where it was first discouered, as reports *C. Clusius*, since which time I haue receiued roots hereof from Virginia, otherwise called Norembega, which grow and prosper in my garden as in their owne natiue countrey.

¶ *The Time.*

The leaues thrust forth of the ground in the beginning of May: the floures bud forth in August. The fruit is ripe in September.

¶ *The Names.*

The Indians do call this plant *Pappus*, meaning the roots: by which name also the common Potatoes are called in those Indian countries. We haue the name proper vnto it mentioned in the title. Because it hath not onely the shape and proportion of Potatoes, but also the pleasant taste and vertues of the same, we may call it in English, Potatoes of America or Virginia.

‡ *Clusius* questions whether it be not the *Arachidna* of *Theophrastus*. *Bauhine* hath referred it to the Nightshades, and calleth it *Solanum tuberosum Esculentum*, and largely figures and describes it in his *Prodromus, pag.* 89. ‡

¶ *The*

¶ *The Temperature and Vertues.*

A The temperature and vertues be referred vnto the common Potatoes, being likewiſe a food, as alſo a meate for pleaſure, equall in goodneſſe and wholeſomeneſſe vnto the ſame, being either roſted in the embers, or boyled and eaten with oyle, vineger, and pepper, or dreſſed any other way by the hand of ſome cunning in cookerie.

B ‡ *Bauhine* ſaith, That he heard that the vſe of theſe toots was forbidden in Bourgondy (where they call them Indian Artichokes) for that they were perſuaded the too frequent vſe of them cauſed the leproſie. ‡

Chap. 351.

Of the Garden Mallow called Hollihocke.

¶ *The Kindes.*

There be diuers ſorts or kindes of Mallowes; ſome of the garden: there be alſo ſome of the Mariſh or ſea ſhore; others of the field, and both wilde. And firſt of the Garden Mallow or Hollihocke.

1 *Malua hortenſis.*
Single Garden Hollihocke.

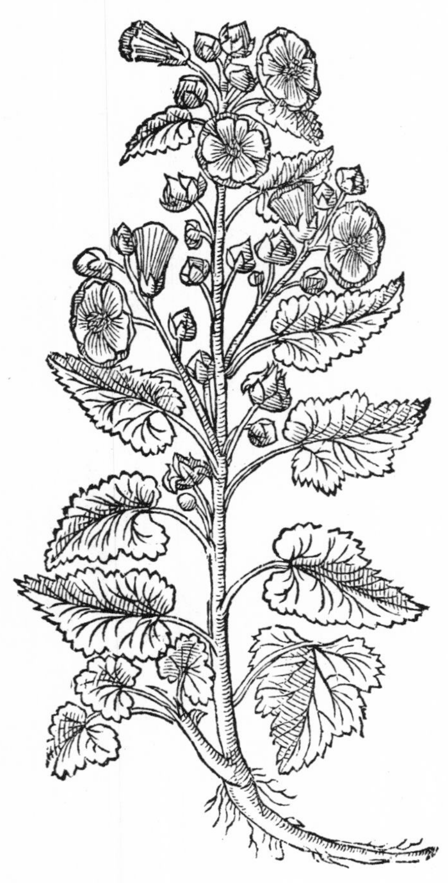

2 *Malua roſea ſimplex peregrina.*
Iagged ſtrange Hollihocke.

¶ *The Deſcription.*

1 The tame or garden Mallow bringeth forth broad round leaues of a whitiſh greene colour, rough, and greater than thoſe of the wilde Mallow. The ſtalke is ſtraight, of the height of foure or ſix cubits; whereon do grow vpon ſlender foot-ſtalks ſingle floures not much vnlike to the wilde Mallow, but greater, conſiſting only of fiue leaues, ſometimes white or red, now and then of a deepe purple colour, varying diuerſly, as Nature liſt to play with it: in their places groweth vp a round knop like a little cake, compact or made vp of a multitude of flat ſeeds like little cheeſes. The root is long, white, tough, eaſily bowed, and groweth deepe in the ground.

2 The

3 *Malua purpurea multiplex.*
Double purple Hollihocke.

2 The second being a strange kinde of Hollihocke hath likewise broad leaues, rough and hoarie, or of an ouerworne russet colour, cut into diuers sections euen to the middle ribbe, like those of Palma Christi. The floures are very single, but of a perfect red colour, wherein consisteth the greatest difference. ‡ And this may be called *Malua rosea simplex peregrina folio Ficus.* Iagged strange Hollihocke. ‡

3 The double Hollihocke with purple floures hath great broad leaues, confusedly indented about the edges, and likewise toothed like a saw. The stalke groweth to the height of foure or fiue cubits. The floures are double, and of a bright purple colour.

4 The Garden Hollihocke with double floures of the colour of scarlet, groweth to the height of fiue or six cubits, hauing many broad leaues cut about the edges. The stalke and root is like the precedent. ‡ This may be called *Multea hortensis rubra multiplex,* Double red Hollihockes, or Rose mallow. ‡

5 The tree mallow is likewise one of the Hollihockes; it bringeth forth a great stalke of the height of ten or twelue foot, growing to the forme of a small tree, whereon are placed diuers great broad leaues of a russet greene colour, not vnlike to those of the great Clot Burre Docke, deepely indented about the edges. The floures are very great and double as the greatest Rose, or double Peiony, of a deep red colour tending to blacknesse. The roote is great, thicke, aud of a wooddy substance, as is the rest of the plant. ‡ This may be called *Malua hortensis atrorubente multiplici flore.* ‡

¶ *The Place.*

These Hollihockes are sowne in gardens, almost euery where, and are in vaine sought else where.

¶ *The Time.*

The second yeere after they are sowne they bring forth their floures in Iuly and August, when the seed is ripe the stalke withereth, the root remaineth and sendeth forth new stalkes, leaues and floures, many yeres after.

¶ *The Names.*

The Hollihocke is called in Greeke, μολόχη: of diuers, *Rosa vltramarina,* or outlandish Rose, and *Rosa hyemalis,* or winter Rose. And this is that Rose which *Pliny* in his 21. book, 4. chapter writes to haue the stalke of a mallow, and the leaues of a pot-herbe, which they cal *Mosceuton:* in high Dutch, **Garten pappelen:** in low Dutch, **Winter Roosen:** in French, *Rose d'outre mer:* in English, Hollihocke, and Hockes.

¶ *The Temperature.*

The Hollihocke is meetely hot, and also moist, but not so much as the wilde Mallow: it hath likewise a clammie substance, which is more manifest in the seed and root, than in any other part.

¶ *The Vertues.*

The decoction of the floures, especially those of the red, doth stop the ouermuch flowing of the monthly courses, if they be boiled in red wine. A

The roots, leaues, and seeds serue for all those things for which the wilde Mallowes do, which are more commonly and familiarly vsed. B

Chap. 352. *Of the wilde Mallowes.*

¶ *The Description.*

1 The wilde Mallow hath broad leaues somewhat round and cornered, nickt about the edges, smooth, and greene of colour: among which rise vp many slender tough stalkes,

clad with the like leaues,but smaller. The floures grow vpon little footstalkes of a reddish colour mixed with purple strakes, consisting of fiue leaues, fashioned like a bell: after which commeth vp a knap or round button, like vnto a flat cake, compact of many small seeds. The root is white, tough, and full of a slimie juice, as is all the rest of the plant.

2 The dwarfe wilde Mallow creepeth vpon the ground: the stalkes are slender and weake, yet tough and flexible. The leaues be rounder, and more hoary than the other. The floures are small and of a white colour.

3 The crispe or curled Mallow, called of the vulgar sort French Mallowes, hath many small vpright stalkes, growing to the height of a cubit, and sometimes higher; whereon do grow broad leaues somewhat round and smooth, of a light greene color, plaited or curled about the brims like a ruffe. The floures be small and white. The root perisheth when it hath perfected his seed.

1 *Malua syluestris.*
The field Mallow.

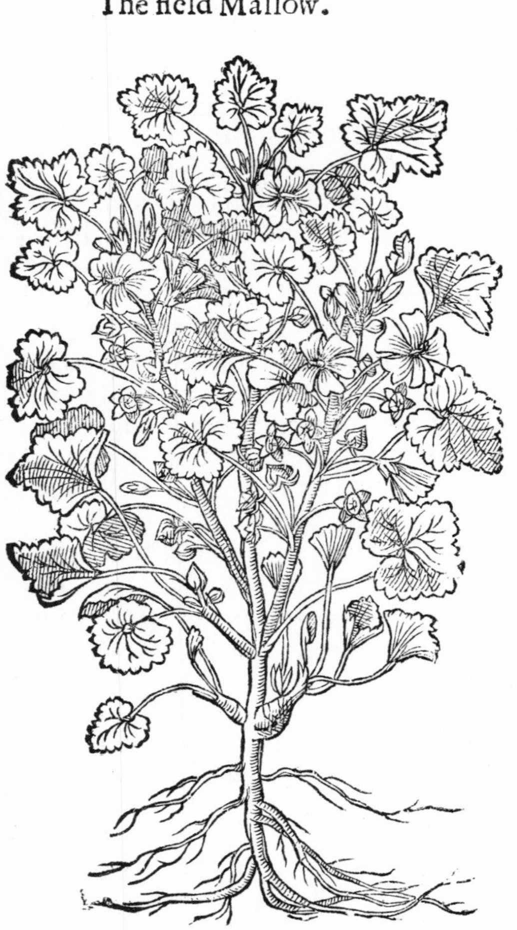

2 *Malua syluestis pumila.*
The wilde dwarfe Mallow.

4 The Veruaine Mallow hath many straight stalkes, whereon doe grow diuers leaues deepely cut and jagged euen to the middle rib, not vnlike to the leaues of Veruaine, whereof it tooke his name: among which come forth faire and pleasant floures like vnto those of the common Mallow in forme, but of a more bright red colour, mixed with stripes of purple, which setteth forth the beautie. The root is thicke, and continueth many yeeres. ‡ This is sometimes though more rarely found with white floures. ‡

‡ 5 This annuall Mallow, called by *Clusius*, *Malua trimestris*, is very like our common Mallow sending vp slender branched stalkes some three foot high; the bottome leaues are round, those on the stalkes more sharpe pointed, greene aboue, and whiter vnderneath, the floures consist of fiue leaues of a light carnation colour, the seed is like that of the ordinary mallow, but smaller; and such also is the root which perishes euery yeere as soon as the seed is ripe: it is sowne in some gardens, and growes wilde in Spaine. ‡

¶ *The Place.*

The two first mallowes grow in vntoiled places among pot-herbes, by high waies, and the borders of fields.

The French mallow is an excellent pot-herbe, for the which cause it is sowne in gardens, and is not to be found wilde that I know of.

3 *Malua criſpa.*
The French curled Mallow.

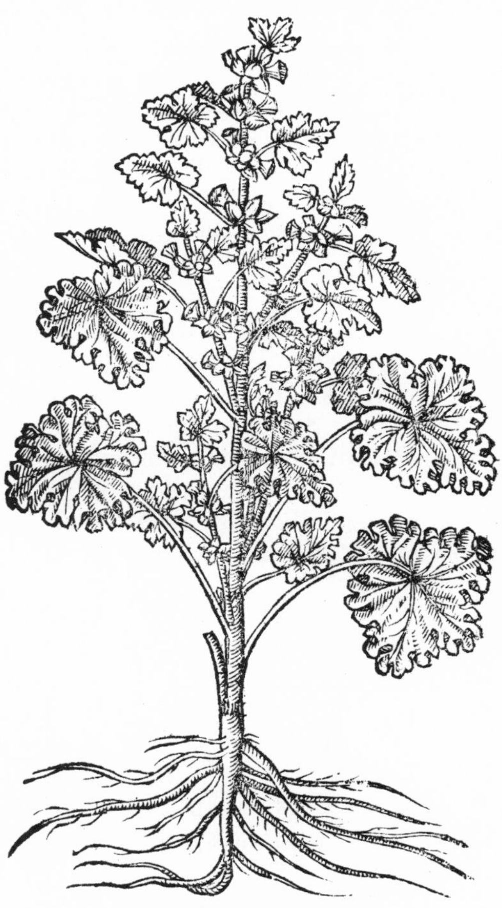

4 *Malua verbenaca.*
Veruaine Mallow.

‡ 5 *Malua æſtiua Hiſpanica.*
The Spaniſh Mallow.

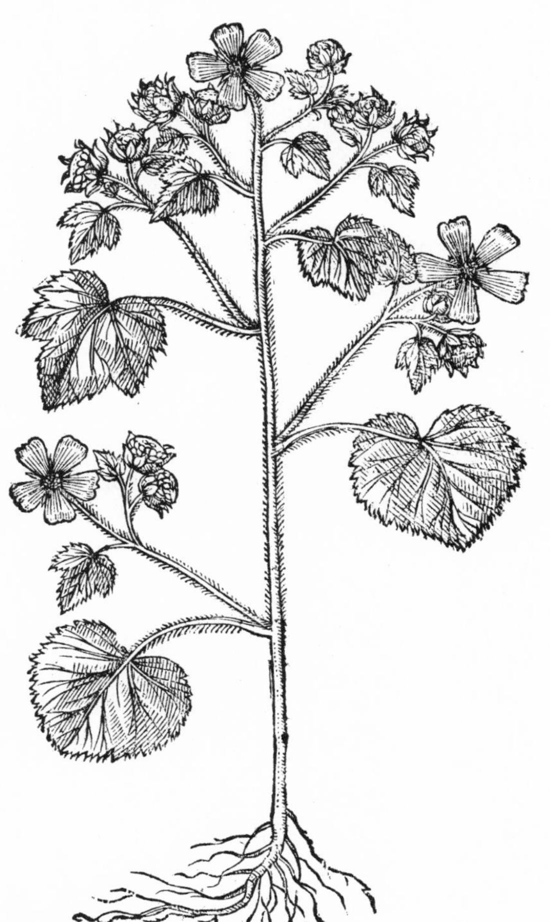

The Veruaine Mallow groweth not euerie where: it growes on the ditch ſides on the left hand of the place of execution by London, called Tyborn: alſo in a field neere vnto a village fourteene miles from London called Buſhey, on the backe-ſide of a Gentlemans houſe named M[r]. *Robert Wylbraham*: likewiſe amongſt the buſhes and hedges as you go from London to a bathing place called the Old Foord; and in the buſhes as you go to Hackny a village by London, in the cloſes next the town, and in diuers other places, as at Baſſingburne in Hartfordſhire, three miles from Roiſton.

‡ M[r]. *Goodyer* found the Veruain Mallow with white floures growing plentifully in a cloſe neere Maple-durham in Hampſhire, called Aldercrofts. ‡

¶ *The Time.*

Theſe wilde Mallowes do floure from Iune till Sommer be well ſpent: in the meane time their ſeed alſo waxeth ripe.

¶ *The Names.*

The wilde Mallow is called in Latine *Malua ſylueſtris*: in Greeke, Μαλάχη ἀγρία, or χερσαῖα: and ἄκοπος, as though they ſhould ſay a mitigator of paine: of ſome, *Oſiriaca*: in high-Dutch, **Pappelin**: in Low-Dutch, **Maluwe**, and **Keeſkens cruit**: in Engliſh, Mallow.

The Veruaine Mallow is called of *Dioſcorides, Alcea* : in Greeke, ἀλκέα : of ſome, *Herba Hungarica*, and *Herba Simeonis*, or Simons Mallow : in Engliſh, Veruaine Mallow, and iagged Mallow.

The name of this herbe *Malua* ſeemeth to come from the Hebrewes, who call it in their tongue מלוח *Malluach*, of the ſaltneſſe, becauſe the Mallow groweth in ſaltiſh and old ruinous places, as in dung-hills and ſuch like, which in moſt aboundant manner yeeldeth forth Salt-peter and ſuch like matter: for מלח *Melach* ſignifieth ſalt, as the Learned know. I am perſuaded that the Latine word *Malua* commeth from the Chaldee name *Mallucha*, the gutturall letter ח, *Ch*, being left out for good ſounds ſake : ſo that it were better in this word *Malüa* to reade *u* as a vowell, than as a conſonant : which words are vttered by the learned Doctor *Rabbi Dauid Kimhi*, and ſeeme to carrie a great ſhew of truth : in Engliſh it is called Mallow ; which name commeth as neere as may be to the Hebrew word.

¶ *The Temperature.*

The wilde Mallowes haue a certaine moderate and middle heate, and moiſtneſſe withall : the iuyce thereof is ſlimie, clammie, or gluing, the which are to be preferred before the garden Mallow or Hollihocke, as *Diphilus Siphinus* in *Athenæus* doth rightly thinke ; who plainely ſheweth, that the wilde Mallow is better than that of the garden : although ſome do prefer the Hollihocke, whereunto we may not conſent, neither yet yeeld vnto *Galen*, who is partly of that minde, yet ſtandeth he doubtfull : for the wilde Mallow without controuerſie is fitter to be eaten, and more pleaſant than thoſe of the garden, except the French Mallow, which is generally holden the wholſommeſt, and amongſt the pot-herbes not the leaſt commended by *Heſiod* : of whoſe opinion was *Horace*, writing in his ſecond Ode of his *Epodon*,

——— *& graui*
Maluæ ſalubres corpori.

The Mallow (ſaith *Galen*) doth nouriſh moderately, ingendreth groſſe bloud, keepeth the bodie ſoluble, and looſeth the belly that is bound. It eaſily deſcendeth, not onely becauſe it is moiſt, but alſo by reaſon it is ſlimy.

¶ *The Vertues.*

A The leaues of Mallowes are good againſt the ſtinging of Scorpions, Bees, Waſps, and ſuch like: and if a man be firſt anointed with the leaues ſtamped with a little oyle, he ſhal not be ſtung at all, *Dioſcorides* ſaith.

B The decoction of Mallowes with their roots drunken are good againſt all venome and poyſon, if it be incontinently taken after the poyſon, ſo that it be vomited vp againe.

C The leaues of Mallowes boyled till they be ſoft and applied, do mollifie tumors and hard ſwellings of the mother, if they do withall ſit ouer the fume thereof, and bathe themſelues therewith.

D The decoction vſed in cliſters is good againſt the roughneſſe and fretting of the guts, bladder, and fundament.

E The roots of the Veruaine Mallow do heale the bloudy flix and inward burſtings, being drunke with wine and water, as *Dioſcorides* and *Paulus Ægineta* teſtifie.

Chap. 353. *Of Marſh Mallow.*

¶ *The Deſcription.*

1 Marſh Mallow is alſo a certaine kinde of wilde Mallow : it hath broad leaues, ſmall toward the point, ſoft, white, and freeſed or cottoned, and ſleightly nicked about the edges : the ſtalkes be round and ſtraight, three or foure foot high, of a whitiſh gray colour ; whereon do grow floures like vnto thoſe of the wilde Mallowes, yet not red as they are, but commonly white, or of a very light purple colour out of the white : the knop or round button wherein the ſeeds lie is like that of the firſt wilde Mallow. The root is thicke, tough, white within, and containeth in it a clammy and ſlimy iuyce.

† 2 This ſtrange kinde of Mallow is holden amongſt the beſt writers to be a kinde of marſh Mallow : ſome excellent Herbariſts haue ſet it downe for *Sida Theophraſti*, wherto it doth not fully anſwer : it hath ſtalks two cubits high, wheron are ſet without order many broad leaues hoarie and whitiſh, not vnlike thoſe of the other marſh Mallow : the floures conſiſt of fiue leaues, and are larger than thoſe of the marſh Mallow, and of a purple colour tending to redneſſe : after which there come round bladders of a pale colour, in ſhape like the fruit or ſeeds of round *Ariſtochia*, or Birthwort, wherein is contained round blacke ſeed. The root is thicke and tough, much like that of the common Mallow.

3 This

1 *Althæa Ibiſcus.*
Marſh Mallow.

2 *Althæa paluſtris.*
Water Mallow.

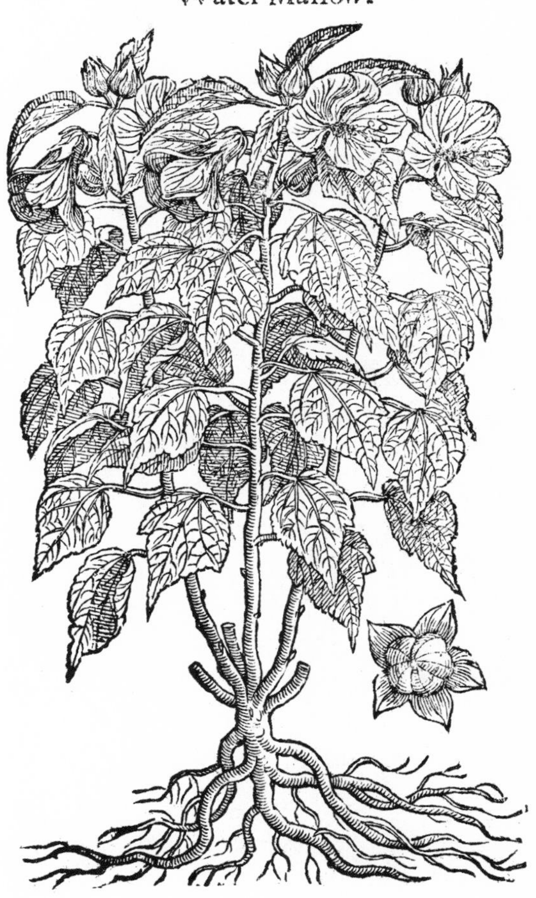

3 *Althæa Arboreſcens.*
Tree Mallow.

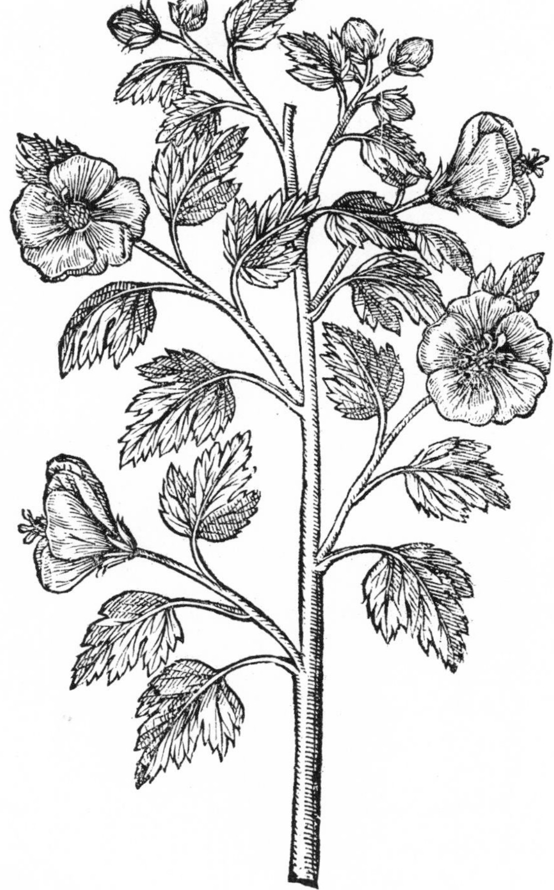

4 *Althæa frutex Cluſij.*
Shrubbed Mallow.

‡ 5 *Alcea fruticosa cannabina.*
Hempe-leaued Mallow.

3 This wilde Mallow is likewise referred vnto the kinds of marsh Mallow, called generally by the name of *Althæa*, which groweth to the form of a smal hedge tree, approching neerer to the substance or nature of wood than any of the other; wherewith the people of Olbia and Narbone in France doe make hedges, to seuer or diuide their gardens and vineyards (euen as we doe with quicke-sets of priuet or thorne) which continueth long: the stalke whereof groweth vpright, very high, comming neere to the Willow in wooddinesse and substance. The floures grow alongst the same, in fashion and colour of the common wild mallow.

4 The shrubby mallow riseth vp like vnto a hedge bush, and of a wooddy substance, diuiding it selfe into diuers tough and limber branches, couered with a barke of the colour of ashes; whereupon do grow round pointed leaues, somewhat nickt about the edges, very soft, not vnlike to those of the common marsh mallow, and of an ouerworne hoary colour. The floures grow at the top of the stalks, of a purple colour, consisting of fiue leaues, very like to the common wilde mallow, and the seed of the marsh mallow.

5 We haue another sort of mallow, called of *Pena, Alcea fruticosior pentaphylla*: it bringeth forth in my garden many twiggy branches, set vpon stiffe stalkes of the bignesse of a mans thumbe, growing to the height of ten or twelue foot: whereupon are set very many leaues deepely cut euen to the middle rib, like vnto the leaues of hempe: the floures and seeds are like vnto the common mallow: the root is exceeding great, thicke, and of a wooddy substance. ‡ *Clusius* calls this *Alcea fruticosa cannabino folio*: and it is with good reason thought to be the *Cannabis syluestris* described by *Dioscorides*, *lib.* 3. *cap.* 166. ‡

¶ *The Place.*

The common marsh mallow groweth very plentifully in the marshes both on the Kentish and Essex shore alongst the riuer of Thames, about Woolwych, Erith, Greenhyth, Grauesend, Tilburie, Lee, Colchester, Harwich, and in most salt marshes about London: being planted in gardens it prospereth well, and continueth long.

The second groweth in the moist and fenny places of Ferraria, betweene Padua in Italy, and the riuer Eridanus.

The others are strangers likewise in England: notwithstanding at the impression hereof I haue sowen some seeds of them in my garden, expecting the successe.

¶ *The Time.*

They floure and flourish in Iuly and August: the root springeth forth afresh euery yeare in the beginning of March, which are then to be gathered, or in September.

¶ *The Names.*

The common marsh mallow is called in Greeke ἀλθαία, and ἰβίσκος: the Latines retaine the names *Althæa* and *Ibiscus*: in shops, *Bismalua*, and *Maluauiscus*; as though they should say *Malua Ibiscus*: in high-Dutch, **Ibisch**: in low-Dutch, **Witte Maluwe**, and **Witten Hemst**: in Italian and Spanish, *Maluauisco*: in French, *Guimaulue*: in English, marsh mallow, moorish mallow, and white mallow.

The rest of the mallowes retaine the names expressed in their seuerall titles.

¶ *The Temperature.*

Marsh mallow is moderately hot, but drier than the other mallowes: the roots and seeds hereof are more dry, and of thinner parts, as *Galen* writeth; and likewise of a digesting, softning, or mollifying nature.

¶ *The*

¶ *The Vertues.*

A The leaues of Marſh Mallow are of the power to digeſt,mitigate paine,and to concoct:

B They be with good effect mixed with fomentations and pulteſſes againſt paines of the ſides, of the ſtone,and of the bladder; in a bath alſo they ſerue to take away any manner of paine.

C The decoction of the leaues drunke doth the ſame, which doth not only aſſwage paine which proceedeth of the ſtone,but alſo is very good to cauſe the ſame to deſcend more eaſily,and to paſſe forth.

D The roots and ſeeds are profitable for the ſame purpoſe: moreouer the decoction of the roots helpeth the bloudy flix, yet not by any binding qualitie, but by mitigating the gripings and frettings thereof: for they doe not binde at all, although *Galen* otherwiſe thought; but they cure the bloudy flix,by hauing things added vnto them,as the roots of Biſtort, Tormentill, the floures and rindes of Pomegranates and ſuch like.

E The mucilage or ſlimie iuice of the roots,is mixed very effectually with all oils,ointments,and plaiſters that ſlacken and mitigate paine.

F The roots boiled in wine,and the decoction giuen to drinke, expell the ſtone and grauell, helpe the bloudy flix,ſciatica,crampes, and conuulſions.

G The roots of Marſh Mallows, the leaues of common Mallowes,and the leaues of Violets,boiled in water vntill they be verie ſoft,and that little water that is left drained away, ſtamped in a ſtone morter,adding thereto a certaine quantitie of Fenugreeke,and Lineſeed in pouder; the root of the blacke Bryonie,and ſome good quantitie of Barrowes greaſe,ſtamped altogether to the forme of a pultis,and applied very warme,mollifie and ſoften Apoſtumes and hard ſwellings, ſwellings in the ioints,and ſores of the mother: it conſumeth all cold tumors,blaſtings,and windie outgrowings; it cureth the rifts of the fundament; it comforteth, defendeth, and preſerueth dangerous greene wounds from any manner of accidents that may happen thereto, it helpeth digeſtion in them, and bringeth old vlcers to maturation.

H The ſeeds dried and beaten into pouder and giuen to drinke, ſtoppeth the bloudy flix and laske, and all other iſſues of bloud.

CHAP. 354. *Of the yellow Lillie.*

Althæa Lutea.
Yellow Mallow.

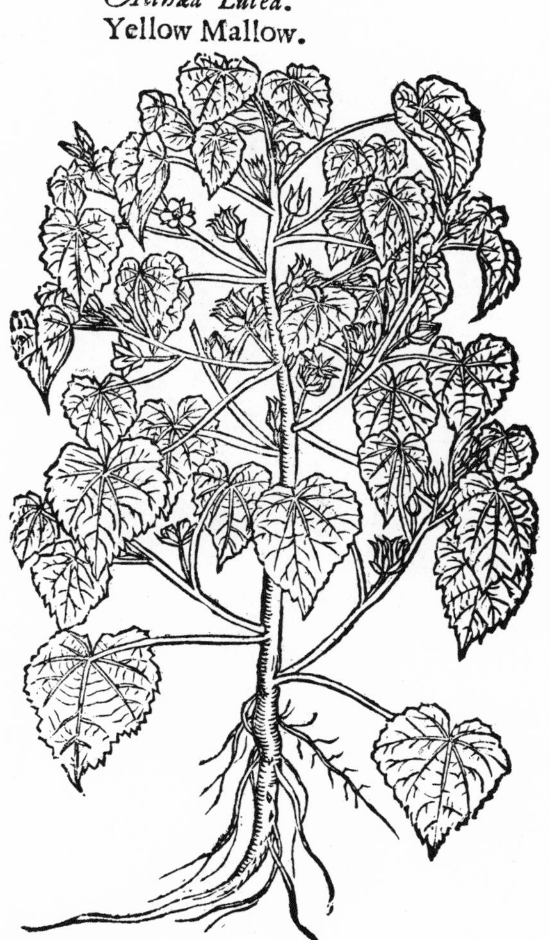

¶ *The Deſcription.*

THe yellow Mallow riſeth vp with a round ſtalke,ſomething hard or wooddie,three or foure cubits high, couered with broad leaues ſomething round, but ſharpe pointed, white, ſoft,ſet with very fine haires like to the leaues of gourds,hanging vpon long tender footſtalks: from the boſome of which leaues come forth yellow floures, not vnlike to thoſe of the common Mallow in forme: the knops or ſeed veſſels are blacke,crooked,or wrinckled, made vp of many ſmall cods, in which is black ſeed:the root is ſmall,and dieth when it hath perfected his ſeed.

¶ *The Place.*

The ſeed hereof is brought vnto vs from Spaine and Italy: we doe yearely ſow it in our gardens, the which ſeldom or neuer doth bring his ſeed to ripeneſſe: by reaſon whereof,we are to ſeeke for ſeeds againſt the next yeare.

¶ *The Time.*

It is ſowne in the midſt of Aprill, it brings forth his floures in September.

¶ *The Names.*

Some thinke this to be *Abutilon*: whereupon that agreeth which *Auicen* writeth, that it is like to the Gourd,that is to ſay in leafe, and to be named *Abutilon*, and *Arblutilon*: diuers take

take it to be that *Althæa* or Marſh Mallow, vnto which *Theophraſtus* in his ninth booke of the Hiſtorie of Plants doth attribute *Florem* μήλινον, or a yellow floure: for the floure of the common Marſh Mallow is not yellow, but white; yet may *Theophraſtus* his copie, which in diuers places is faultie, and hath many emptie and vnwritten places, be alſo faultie in this place; therefore it is hard to ſay, that this is *Theophraſtus* marſh Mallow, eſpecially ſeeing that *Theophraſtus* ſeemeth alſo to attribute vnto the root of Marſh Mallow ſo much ſlime, as that water may bee thickened therewith, which the roots of common Marſh Mallow can very well doe: but the root of *Abutilon* or yellow Mallow not at all: it may be called in Engliſh, yellow Mallow, and *Auicen* his Mallow.

¶ *The Temperature.*

The temperature of this Mallow is referred vnto the Tree-mallow.

¶ *The Vertues.*

A *Auicen* ſaith, that *Abutilon* or yellow Mallow, is held to be good for greene wounds, and doth preſently glew together, and perfectly cure the ſame.

B The ſeed drunke in wine preuaileth mightily againſt the ſtone.

C *Bernardus Paludanus* of Anchuſen reporteth, that the Turks do drinke the ſeed to prouoke ſleepe and reſt.

Chap. 355. *Of Venice Mallow, or Good-night at Noone.*

1 *Alcea Peregrina.*
Venice Mallow.

2 *Sabdariſa.*
Thornie Mallow.

¶ *The Deſcription.*

† 1 The Venice mallow riſeth vp with long, round, feeble ſtalkes, whereon are ſet vpon long ſlender foot-ſtalkes, broad iagged leaues, deepely cut euen to the middle rib: amongſt which come forth very pleaſant and beautifull floures, in ſhape like thoſe of the common mallow, ſomething

‡ 3 *Alcea Ægyptia.*
The Ægyptian Codded Mallow.

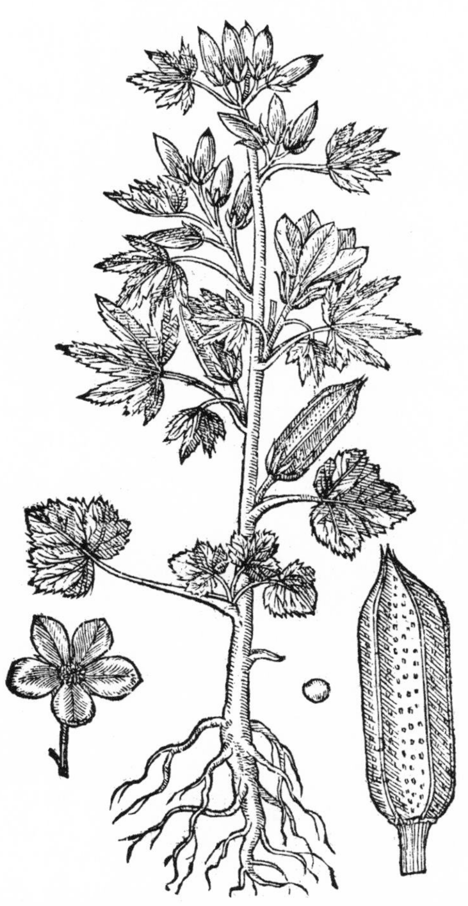

something white about the edges, but in the middle of a fine purple: in the middest of this floure standeth forth a knap or pestel, as yellow as gold: it openeth it selfe about eight of the clocke, and shutteth vp againe at noone, about twelue a clock when it hath receiued the beams of the Sun, for two or three houres, whereon it should seeme to reioice to look, and for whose departure, being then vpon the point of declension, it seemes to grieue, and so shuts vp the floures that were open, and neuer opens them againe; whereupon it might more properly be called *Malva horaria*, or the Mallow of an houre: and this *Columella* seemeth to call *Moloche*, in this verse;

——*Et Moloche, Prono sequitur quæ vertice solem.*

The seed is contained in thicke rough bladders, whereupon *Dodonæus* calleth it *Alcea Vesicaria:* within these bladders or seed vessels are contained blacke seed, not vnlike to those of *Nigella Romana.* The root is small and tender, & perisheth when the seed is ripe, and must be increased by new and yearely sowing of the seed, carefully reserued.

2 Thorn Mallow riseth vp with one vpright stalk of two cubits high, diuiding it selfe into diuers branches, whereupon are placed leaues deeply cut to the middle rib, and likewise snipt about the edges like a saw, in taste like Sorrel the floures for the most part thrust forth of the trunke or body of the small stalke, compact of fiue small leaues, of a yellowish colour; the middle part whereof is of a purple tending to rednesse: the husk or cod wherein the floure doth stand is set or armed with sharpe thornes: the root is small, single, and most impatient of our cold clymate, insomuch that when I had with great industrie nourished vp some plants from the seed, and kept them vnto the midst of May; notwithstanding one cold night chancing among many, hath destroied them all.

‡ 3 This also is a stranger cut leaued Mallow, which *Clusius* hath set forth by the name of *Alcea Ægyptia:* and *Prosper Alpinus* by the title of *Bammia:* the stalke is round, straight, green, some cubit and halfe high: vpon which without order grow leaues at the bottome of the stalk, like those of Mallow, cornered and snipt about the edges; but from the middle of the stalke to the top they are cut in with fiue deep gashes like as the leaues of the last described: the floures grow forth by the sides of the stalke, in forme and colour like those of the last mentioned, to wit, with fiue yellowish leaues: after these follow long thicke fiue cornered hairy and sharpe pointed seed vessels, containing a seed like *Orobus*, couered with a little downinesse: this growes in Egypt, where they eat the fruit thereof as we do Pease and Beanes: *Alpinus* attributes diuers vertues to this plant, agreeable to those of the common Marsh-mallow. ‡

¶ *The Place.*

The seeds hereof haue been brought out of Spaine and other hot countries. The first prospereth well in my garden from yeare to yeare.

¶ *The Time.*

They are to be sowne in the most fertill ground and sunnie places of the garden, in the beginning of May, or in the end of Aprill.

¶ *The Names.*

Their names haue beene sufficiently touched in their seueral descriptions. The first may be called in English, Venice-mallow, Good-night at noone, or the Mallow flouring but an houre: of *Matthiolus* it is called *Hypecoon*, or Rue Poppie; but vnproperly.

¶ *The Temperature and Vertues.*

There is a certaine clammie iuice in the leaues of the Venice Mallow, whereupon it is thought A to to come neere vnto the temperature of the common Mallow, and to be of a mollifying facultie: but his vſe in Phyſicke is not yet knowne, and therefore can there be no certaintie affirmed.

Chap. 356. *Of Cranes-bill.*

¶ *The Kindes.*

THere be many kindes of Cranes-bil, whereof two were known to *Dioſcorides*, one with the knobby root, the other with the Mallow leafe.

Geranium Columbinum.
Doues foot, or Cranes-bill.

¶ *The Deſcription.*

DOues-foot hath many hairy ſtalks, trailing or leaning toward the ground, of a browniſh color, ſomewhat kneed or iointed; wherupon do grow rough leaues of an ouerworne green colour, round, cut about the edges, and like vnto thoſe of the common Mallow: amongſt which come forth the floures of a bright purple color: after which is the ſeed, ſet together like the head and bil of a bird; wherupon it was called Cranes-bill, or Storks-bill, as are alſo all the other of his kinde. The root is ſlender, with ſome fibres annexed thereto.

‡ 2 There is another kinde of this with larger ſtalkes and leaues, alſo the leaues are more deeply cut in and diuided, and the floures are either of the ſame colour as thoſe of the common kinde, or elſe ſomewhat more whitiſh. This may be called *Geranium columbinum maius diſſectis foliis*, Great Doues foot.

3 To this kinde may alſo fitly be referred the *Geranium Saxatile* of *Thalius*: the root is ſmal and threddy, the leaues are ſmoother, redder, more bluntly cut about the edges, and tranſparent than thoſe of the firſt deſcribed, yet round, and otherwiſe like them: the floures are ſmall and red, and the bills like thoſe of the former. Maſter *Goodyer* found it growing plentifully on the bankes by the high way leading from Gilford towards London, neere vnto the Townes end. ‡

¶ *The Place.*

It is found neere to common high waies, deſart places, vntilled grounds, and ſpecially vpon mud walls almoſt euery where.

¶ *The Time.*

It ſpringeth vp in March and Aprill: floureth in May, and bringeth his ſeede to ripeneſſe in Iune.

¶ *The Names.*

It is commonly called in Latine, *Pes Columbinus*: in High Dutch, **Scarter kraut**: in Low Dutch, **Duyuen voet**: in French, *Pied de Pigeon*: hereupon it may be called *Geranium Columbinum*. in Engliſh, Doues-foot, and Pigeons-foot: of *Dioſcorides*, *Geranium alterum*: of ſome, *Pulmonia*, and *Gruina*.

¶ *The Temperature.*

Doues foot is cold and ſomewhat drie, with ſome aſtriction or binding, hauing power to ſoder or ioine together.

¶ *The Vertues.*

A It ſeemeth, ſaith my Author, to be good for greene and bleeding wounds, and aſſwageth inflammations or hot ſwellings.

The

The herbe and roots dried, beaten into most fine pouder, and giuen halfe a spoonfull fasting, and the like quantitie to bedwards in red wine, or old claret, for the space of one and twentie daies together, cureth miraculously ruptures or burstings, as my selfe haue often prooued, whereby I haue gotten crownes and credit: if the ruptures be in aged persons, it shall be needfull to adde thereto the powder of red snailes (those whithout shels) dried in an ouen, in number nine, which fortifi. the herbs in such sort, that it neuer faileth, although the rupture be great and of long continuance: it likewise profiteth much those that are wounded into the body, and the decoction of the herbe made in wine, preuaileth mightily in healing inward wounds, as my selfe haue likewise proued.

CHAP. 357. *Of Herbe Robert.*

Geranium Robertianum.
Herbe Robert.

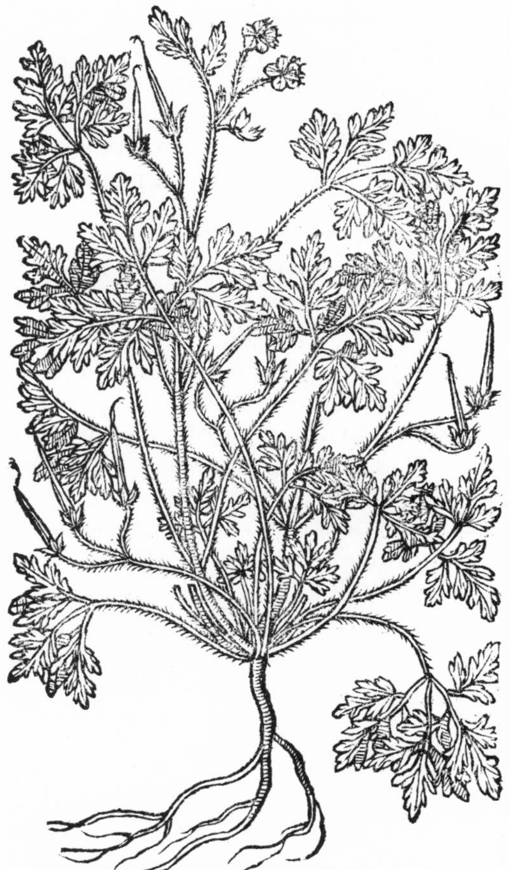

¶ *The Description.*

HErbe Robert bringeth forth slender weake and brittle stalks, somewhat hairie, and of a reddish colour, as are oftentimes the leaues also, which are iagged and deepely cut, like vnto those of Cheruile of a most loathsome stinking smell. The floures are of a most bright purple colour; which being past, there follow certaine smal heads, with sharpe beaks or bils like those of birds: the root is small and threddie.

¶ *The Place.*

Herbe Robert groweth vpon old walls, as wel those made of bricke and stone, as those of mud or earth: it groweth likewise among rubbish, in the bodies of trees that are cut downe, and in moist and shadowie ditch banks.

¶ *The Time.*

It floureth from Aprill till Sommer be almost spent: the herbe is green in winter also, and is hardly hurt with cold.

¶ *The Names.*

It is called in high Dutch, **Ruprechts kraut:** in low Dutch, **Robrechts kruit:** and thereupon it is named in Latine, *Ruberta*, and *Roberti herba: Ruellius* calleth it *Robertiana*; and we, *Robertianum:* of *Tabernamontanus*, *Rupertianum:* in English, Herbe Robert. Hee that conferreth this Cranes bill with *Dioscorides* his third *Sideritis* shall plainely perceiue, that they are both one, and that this is most apparently *Sideritis 3. Dioscoridis*; for *Dioscorides* setteth downe three *Sideritides*, one with the leafe of Horehound; the next with the leafe of Fearne; and the third groweth in walls and Vineyards: the natiue soile of Herbe Robert agree thereunto, and likewise the leaues, being like vnto Cheruile, and not vnlike to those of Corianders, according to *Dioscorides* description.

¶ *The Temperature.*

Herbe Robert is of temperature somewhat cold: and yet both scouring and somewhat binding, participating of mixt faculties.

¶ *The Vertues.*

A It is good for wounds and vlcers of the dugs & secret parts; it is thought to stanch bloud, which thing *Dioscorides* doth attribute to his third *Sideritis:* the vertue of this, saith he, is applied to heale vp bloudy wounds.

CHAP.

Chap. 358. *Of knobbed Cranes-bill.*

Geranium tuberosum.
Knobbie Cranes-bill.

¶ *The Description.*

This kinde of Cranes-bill hath many flexible branches, weake and tender, fat, and full of moisture, wheron are placed very great leaues cut into diuers small sections or diuisions, resembling the leaues of the tuberous *Anemone*, or Wind-floure, but somewhat greater, of an ouer-worn greenish colour: among which come forth long foot-stalks, whereon do grow faire floures, of a bright purple colour, and like vnto the smallest brier Rose in forme: which being past, there succeed such heads and beaks as the rest of the Cranes-bill haue: the root is thick, bumped or knobbed, which we call tuberous.

¶ *The Place.*

This kinde of Cranes-bill is a stranger in England, notwithstanding I haue it growing in my garden.

¶ *The Time.*

The time answereth the rest of the Cranes bills.

¶ *The Names.*

Cranes bill is called in Greeke γεράνιον: in Latine, *Gruinalis*, commonly *Rostrum Gruis*, or *Rostrum Ciconiæ*: of the likenesse of a Cranes-bill, or storkes-bill: of some, *Acus moscata*: but that name doth rather belong to another of this kind: it is also called *Acus Pastoris*: in Italian, *Rostro di grua*: in French, *Bec de Grue*: in Spanish, *Pico di Ciquena, pico del grou*: in High Dutch **Storckenschuable**: in Low Dutch, **Oiievaers beck**: in English, Storks-bill, Cranes-bill, Herons-bill, and Pincke-needle: this is also called for distinctions sake, *Geranium tuberosum*, and *Geranium bulbosum*: it is likewise *Geranium primum Dioscoridis* or *Dioscorides* his first Cranes-bill.

¶ *The Temperature.*

The roots of this Cranes-bill haue a little kinde of heat in them.

¶ *The Vertues.*

A *Dioscorides* saith that the roots may be eaten, and that a dram weight of them drunk in wine doth waste and consume away the windinesse of the Matrix.

B Also *Pliny* affirmeth, that the root hereof is singular good for such as after weaknesse craue to be restored to their former strength.

C The same Author affirmeth that the weight of a dram of it drunke in wine three times in a day, is excellent good against the Ptisicke, or consumption of the lungs.

Chap. 359. *Of Musked Cranes-bill.*

¶ *The Description.*

Mvsked Cranes-bil hath many weake and feeble branches trailing vpon the ground, whereon doe grow long leaues, made of many smaller leaues, set vpon a middle rib, snipt or cut about the edges, of a pleasant sweete smell, not vnlike to that of Muske: among which come forth the floures set vppon tender foote-stalkes, of a red colour, compact of fiue small leaues apiece: after which appeare small heads and pointed beakes or bills like the other kindes of Cranes bills. the root is small and threddy.

¶ The

Geranium moschatum.
Musked Cranes bill.

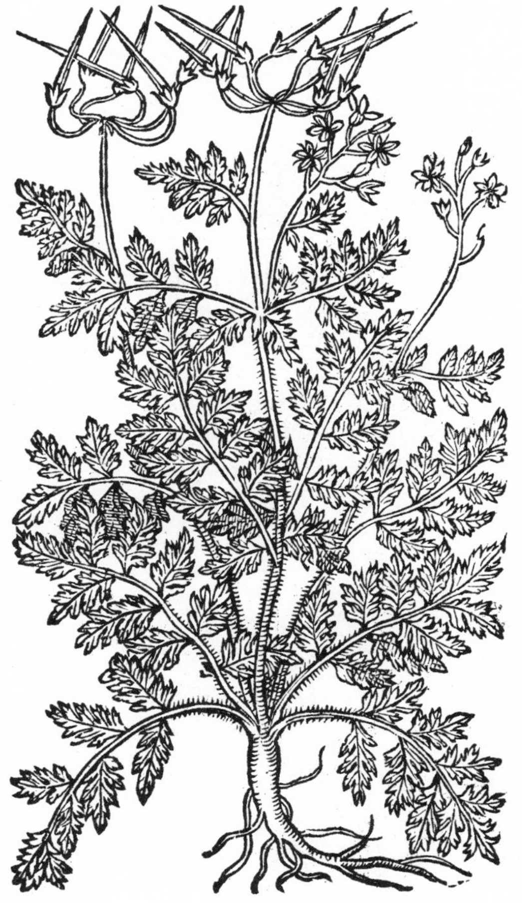

¶ *The Place.*

It is planted in Gardens for the sweet smell that the whole plant is possessed with, ‡ but if you rub the leaues and then smell to them, you shall finde them to haue a sent quite contrary to the former. ‡

¶ *The Time.*

It floureth and flourisheth all the sommer long.

¶ *The Names.*

It is called *Myrrhida Pliny Rostrum Ciconiæ, Arcus moschata*, in shops, and *Acus pastoris*, and likewise *Geranium moschatum*: in English, Musked Storkes bill, and Cranes bill, *Muschatum*, and of the vulgar sort Muschata, and also Pickneedle.

¶ *The Temperature.*

This Cranes bill hath not any of his faculties found out or knowne: yet it seemeth to be colde and a little dry, with some astriction or binding.

¶ *The Vertues.*

A The vertues are referred vnto those of Doues foot, and are thought of *Dioscorides* to be good for greene and bloudy wounds, and hot swellings that are newly begun.

Chap. 360. *Of Crow-foot Cranes-bill, or* Gratia Dei.

¶ *The Description.*

1 Crowfoot Cranes bill hath many long and tender branches tending to rednesse, set with great leaues deepely cut or jagged, in forme like those of the fielde Crowfoot, whereof it tooke his name; the floures are pretty large, and grow at the top of the stalkes vpon tender footstalkes, of a perfect blew colour: which being past, there succeed such heads, beakes, and bils as the other Cranes bils.

I haue in my garden another sort of this Cranes bill, bringing forth very faire white floures, which maketh it to differ from the precedent; in other respects there is no difference at all.

‡ 2 This which is the *Geranium* 2. *Batrachiodes minus* of *Clusius* hath large stalkes and leaues, and those very much diuided or cut in; the stalkes also are diuided into sundry branches, which vpon long footstalkes carry floures like in shape, but lesse than those of the formerly described, and not blew, but of a reddish purple colour, hauing ten threds and a pointall comming forth of the middle of the floure; the beakes or bils which are the seed stand vpright, and hang not downe their points as most others do. The root is large and liues many yeares.

3 The stalkes of this are stiffe, greene, and hairy, diuided at their tops into sundry branches which end in long footstalkes, vpon which grow floures commonly by couples, and they consist of fiue leaues apiece, and these of a darke red colour. The leaues are large, soft, and hairy, diuided into six or seuen parts, and snipt about the edges; the roots are large and lasting. It is kept with vs in gardens, and floures in May. *Clusius* calls it *Geranium* 1. *pullo flore*.

4 This also hath stalkes and leaues much like those of the last described, but somewhat lesse: the floures are as large as those of the last described, but of a more light red, and they are conteined in thicker and shorter cups, and succeeded by shorter seeds or bills, and are commonly of a sweet muske-like smell: The root is very long, red, and lasting. It floures in the middest of May, and is called

† 1 *Geranium Batrachioides.*
Crow-foot Cranes-bill.

2 *Geranium Batrachioides alterum.*
Small Crow-foot Cranes-bill.

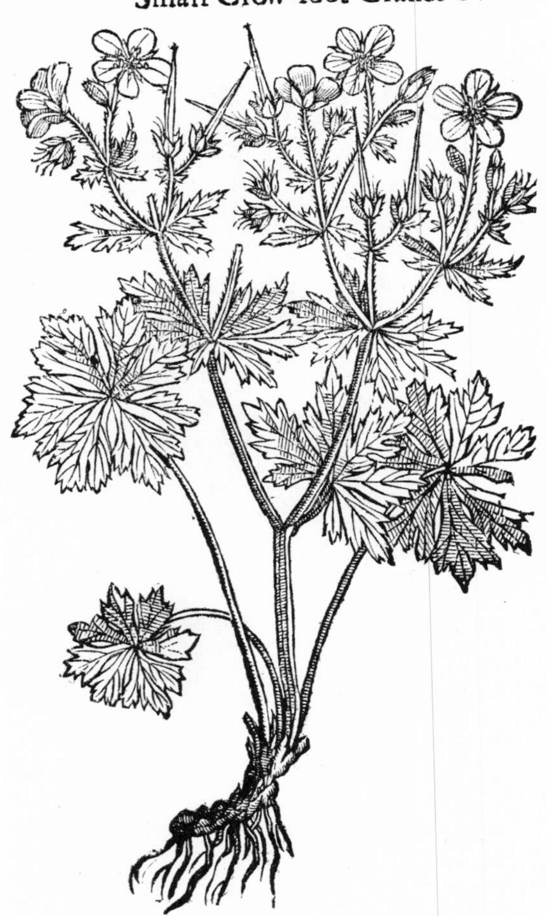

‡ 3 *Geranium Batrachioides pullo flore.*
Duskie Cranes-bill.

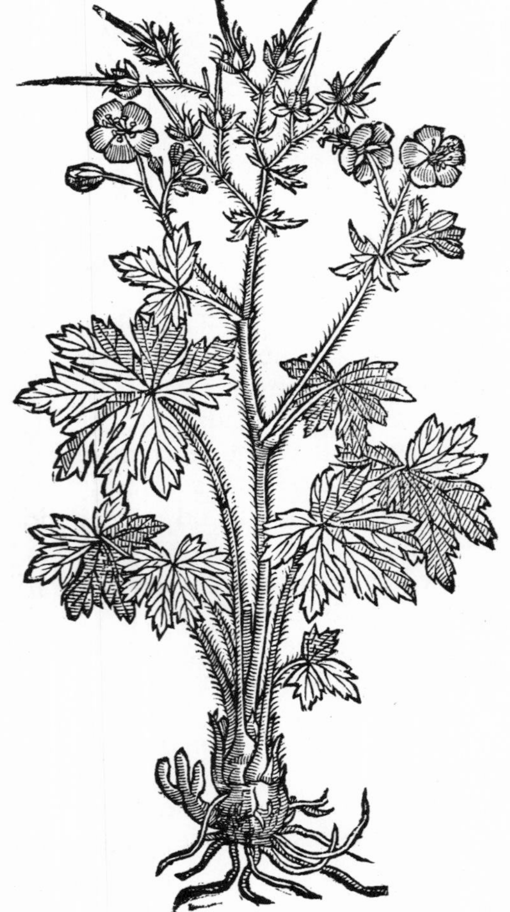

‡ 4 *Geranium Batrachioides longius radicatum.*
Long rooted Cranes-bill.

called by *Geſner*, *Geranium montanum*: by *Dodonæus*, *Geranium batrachioides alterum*: and by *Lobell*, *Geranium batrachioides longius radicatum*. ‡

¶ *The Place.*

Theſe Cranes bils are wilde of their owne nature, and grow in barren places, and in vallies rather than in mountaines; both of them do grow in my garden.

¶ *The Time.*

They floure, flouriſh, and grow greene moſt part of the Summer.

¶ *The Names.*

It is called in Greeke, βατραχοειδὴς, and *Geranium batrachioides*, which name it taketh from the likeneſſe of Crowfoot: of ſome it is called *Ranunculus cæruleus*, or blew Crowfoot: *Fuchſius* calleth it **Gottes gnad**, that is in Latine, *Gratia Dei*: in Engliſh alſo Gratia dei, blew Cranes bill, or Cranes bill with the blew floures, or blew Crowfoot Cranes bill.

¶ *The Temperature.*

The Temperature is referred to the other Cranes bils.

¶ *The Vertues.*

None of theſe plants are now in vſe in Phyſicke; yet *Fuchſius* ſaith, that Cranes bill with the blew floure is an excellent thing to heale wounds.

CHAP. 361. *Of Candy Cranes bill.*

1 *Geranium Creticum.*
Candy Cranes bill.

2 *Geranium Malacoides.*
Baſtard Candy Cranes bill.

¶ *The Deſcription.*

1 THe Cranes bill of Candie hath many long tender ſtalks, ſoft, and full of iuice: diuiding it ſelfe into diuers branches, whereon are ſet great broad leaues, cut, or jagged in diuers

sections or cuts: among which come forth flowers composed of fiue leaues apiece, of a blewish or watchet colour, in the middle part whereof come forth a few chiues, and a small pointell of a purplish colour: the head and beake is like to the rest of the Cranes bills, but greater: the root dieth when it hath perfected his seed.

2 This Cranes-bill, being a bastard kinde of the former, hath long slender branches growing to the height of two or three cubits, set about with very great leaues, not vnlike to those of Hollihocks, but somewhat lesser, of an ouerworne greene colour: among which rise vp little foot-stalks, on the ends whereof do grow small floures, lesser than those of the precedent, and of a murrey colour: the head and seeds are like also, but much lesser: the roots doe likewise die at the first approch of Winter.

¶ *The Place.*

These are strangers in England, except in the gardens of some Herbarists: they grow in my garden very plentifully.

¶ *The Time.*

The time answereth the rest of the Cranes-bils, yet doth that of Candie floure for the most part with me in May.

¶ *The Names.*

There is not more to be said of the names than hath been remembred in their seueral titles: they may be called in English, Cranes-bils, or Storkes-bils.

¶ *The Temperature.*

Their temperature answereth that of Doues-foot.

¶ *The Vertues.*

A Their faculties in working are equall to those of Doues-foot, and vsed for the same purposes, (& rightly) specially being vsed in wound drinks, for the which it doth far excel any of the Cranes bils, and is equall with any other herbe whatsoeuer for the same purpose.

CHAP. 362. *Of diuers wilde Cranes-bills.*

¶ *The Kindes.*

THere be diuers sorts or kindes of Cranes-bils which haue not been remembred of the antient, nor much spoken of by the later writers, all which I meane to comprehend vnder this chapter, making as it were of them a Chapter of wilde Cranes-bils, although some of them haue place in our London gardens, and that worthily, especially for the beautie of the floures: their names shall be expressed in their seueral titles, their natures and faculties are referred to the other Cranes-bils, or if you please to a further consideration.

¶ *The Description.*

1 SPotted Cranes-bill, or Storkes-bill, the which *Lobel* describeth in the title thus, *Geranium Fuscum flore liuido purpurante, & medio Candicante,* whose leaues are like vnto Crowfoot (beeing a kinde doubtlesse of Cranes bill, called *Gratia Dei*) of an ouerworne dustie colour, and of a strong sauour, yet not altogether vnpleasant: the stalkes are drie and brittle, at the tops whereof doe grow pleasant floures of a darke purple colour, the middle part of them tending to whitenesse: from the stile or pointel thereof, commeth forth a tuft of small purple hairy threds. The root is thick and very brittle, lifting it selfe forth of the ground, insomuch that many of the said roots lie aboue the ground naked without earth, euen as the roots of Floure-de-luces doe.

2 Of these wilde ones I haue another sort in my garden, which *Clusius* in his Pannonicke obseruations hath called *Geranium Hæmatoides,* or sanguine Cranes bill: and *Lobel, Geranium Gruinum,* or *Gruinale*: it hath many flexible branches creeping vpon the ground: the leaues are much like vnto Doues foot in forme, but cut euen to the middle rib: the floures are like those of the small wilde mallow, and of the same bignesse, of a perfect bright red colour, which if they be suffered to grow

1 *Geranium maculatum ſiue fuſcum.*
Spotted Cranes bill.

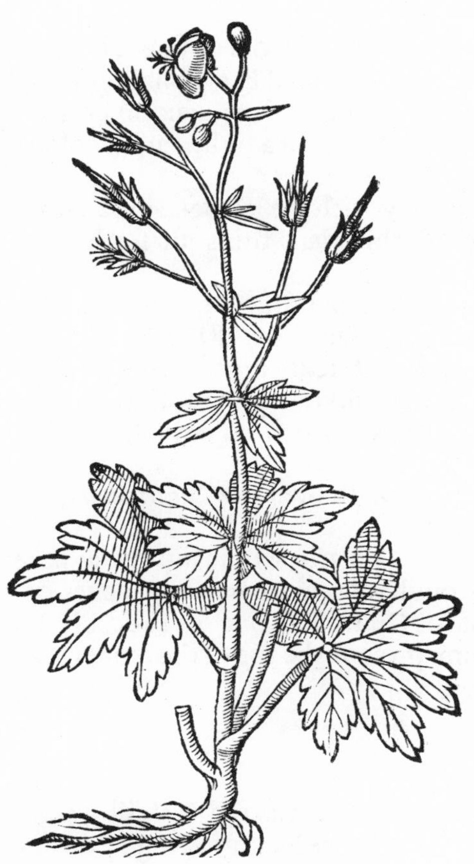

2 *Geranium ſanguinarium.*
Bloudy Cranes bill.

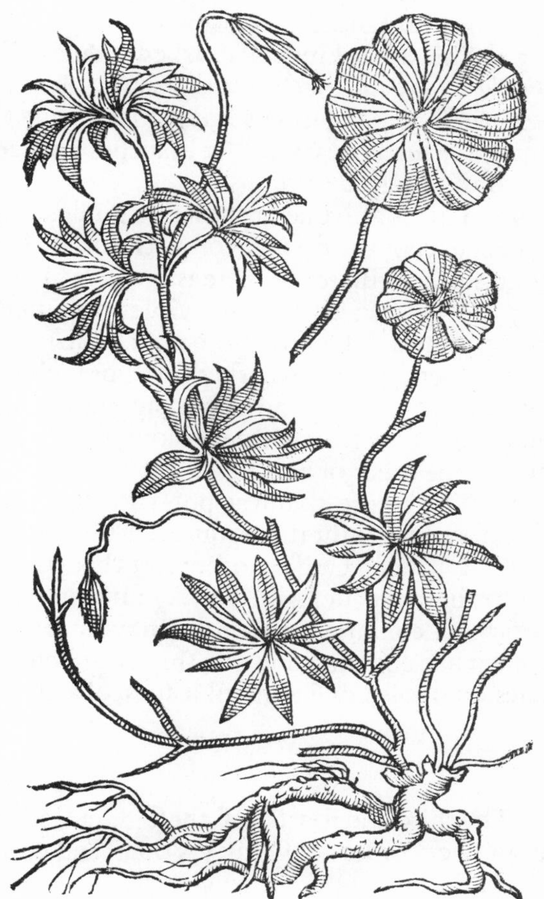

3 *Geranium Cicutæ folio inodorum.*
Vnſauorie fielde Cranes-bill.

5 *Geranium Violaceum.*
Violet Storkes-bill.

grow and stand vntill the next day, will be a murry colour; and if they stand vnto the third day, they will turne into a deep purple tending to blewnesse; their changing is such, that you shall finde at one time vpon one branch floures like in forme, but of diuers colours. The root is thicke, and of a wooddie substance.

3 This wilde kinde of musked Cranes bill, being altogether without sauour or smell, is called *Myrrhida inodorum*, or *Geranium arvense inodorum*, which hath many leaues spred flat vpon the ground, euery leafe made of diuers smaller leafes, and those cut or iagged about the edges, of no smel at all: amongst which rise vp slender branches, whereon doe grow small floures of a light purple colour: the root is long and fibrous.

4 This is also one of the wilde kindes of Cranes-bills, agreeing with the last described in each respect, except the floures, for as the other hath purple floures, so this plant bringeth forth white floures, other difference there is none at all.

5 The Cranes-bill with violet coloured floures, hath a thicke wooddie root, with some few strings annexed thereto: from which rise immediatly forth of the ground diuers stiffe stalks, which diuide themselues into other small branches, whereupon are set confusedly broad leaues, made of three leaues apiece, and those iagged or cut about the edges: the floures grow at the top of the branches of a perfect Violet colour, whereof it tooke his name: after which come such beakes or bils as the other of his kinde.

‡ The figure that was put vnto this Description is the same with *Geranium Robertianum*, and therefore I thought it not much amisse to put it here againe. ‡

6 I haue likewise another sort that was sent me from *Robinus* of Paris, whose figure was neuer set forth, neither described of any: it bringeth from a thicke tough root, with many branches of a brownish colour: wherupon do grow leaues not vnlik to those of Gratia Dei, but not so deeply cut, somewhat cornered, and of a shining greene colour: the floures grow at the top of the tender branches, composed of sixe small leaues, of a bright scarlet colour.

¶ *The Place.*

The third and fourth of these Cranes-bills growe of themselues about old VValls, and about the borders of fields, VVoods and copses; and most of the rest wee haue growing in our gardens.

¶ *The Time.*

Their time of flouring and seeding answereth the rest of the Cranes bills.

¶ *The Names.*

Their seuerall titles shall serue for their names, referring what might haue been said more to a further consideration.

¶ *The Nature and Vertues.*

A There hath not as yet any thing beene found either of their temperature or faculties, but may be referred vnto the other of their kinde.

‡ Chap. 363. *Of certaine other Cranes-bills.*

¶ *The Description.*

‡ 1 This which *Clusius* receiued from Doctour *Thomas Pennie* of London, and sets forth by the same title as you finde it here exprest, hath a root consisting of sundry long and small bulbes, and which is fibrous towards the top: the stalke is a cubit high, ioynted, and red neere vnto the roote, and about the ioints: out of each of these ioyntes come two leaues which are fastened vnto somewhat long foot-stalkes, and diuided into fiue parts, which also are snipt about the edges: out of each of which ioints by the setting on the foot-stalkes

‡ 1 *Geranium bulbosum Pennæi.*
Peanies bulbous Cranes bill.

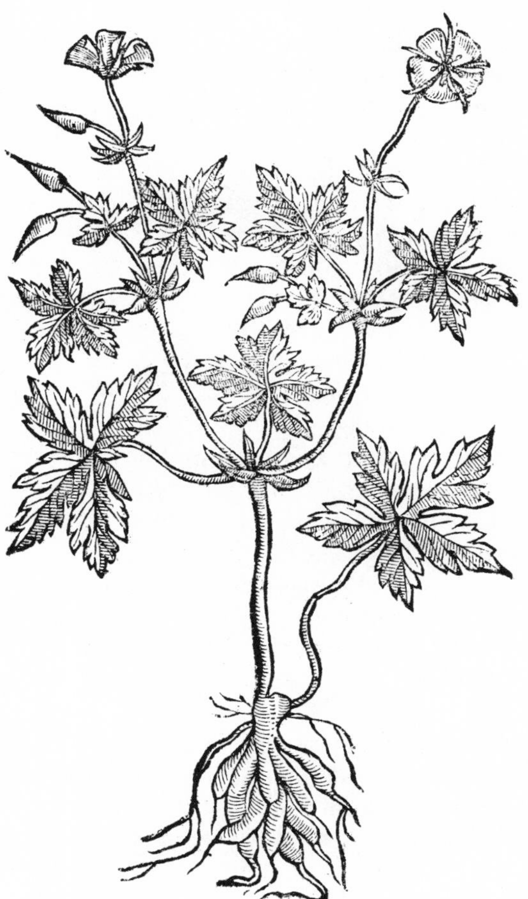

‡ 2 *Geranium nodosum, Plateau.*
Knotty Cranes bill.

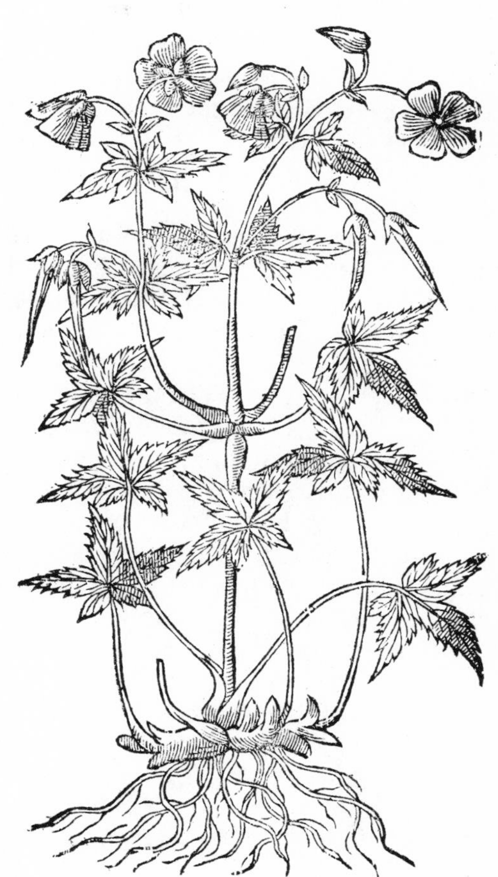

‡ 3 *Geranium argenteum Alpinum.*
Siluer leaued Mountaine Cranes bill.

foot-stalkes come forth fiue little sharpe pointed leaues: the floures grow by couples vpon the tops of the stalkes, and are of a reddish purple colour. It growes wilde in Denmark; whence Dr. *Turner* brought it, and bestowed it vpon Dr. *Penny* before mentioned.

2 This hath stalks some foot high, iointed, and of a purplish colour: vpon which grow leaues diuided into three parts; but those below are cut into fiue, and both the one and the other are snipt about the edges: the floures are composed of fiue reddish purple leaues of a pretty largenesse, with a reddish pointall in the middle; and falling, the seed follows, as in other plants of this kind: the root is knotty, and ioynted, with some fibres: it floures in May, and so continueth a great part of the Sommer after. *Clusius* calls this *Geranium* 5. *nodosum, Plateau*. This somtimes is found to carry tuberous excrescences vpon the stalkes, toward the later end of Sommer, whence *Plateau* distinguished it from the other, but afterwards found it to be the same: and *Clusius* also figures and describeth this later varietie by the name of *Geranium 6. tuberiferum Plateau.*

3 The root of this is some two handfuls long,

long, blacke without, and white within, and towards the top diuided into sundry parts; whence put forth leaues couered ouer with a fine siluer downe; and they are diuided into fiue parts, each of which againe is diuided into three others, and they are fastned to long slender and round foot-stalkes: the floures grow vpon foot-stalkes shorter than those of the leaues; the floures in colour and shape are like those of the Veruaine Mallow, but much lesse; and after it is vaded there followes a short bill, as in the other plants of this kinde. It floures in Iuly, and growes vpon the Alps, where *Pona* found it, and first set it forth by the name of *Geranium Alpinum longius radicatum*.

4 The stalkes of this pretty Cranes bill are some foot or better high, whereon grow leaues parted into fiue or six parts like those of the *Geranium fuscum*, but of a lighter greene colour: the floures are large, composed of fiue thin and soone fading leaues of a whitish colour, all ouer intermixt with fine veines of a reddish colour, which adde a great deale of beauty to the floure: for these veines are very small, and curiously dispersed ouer the leaues of the floure. It floures in Iune, and is preserued in diuers of our gardens; some cal it *Geran. Romanum striatum*: in the *Hortus Estetensis* it is set forth by the name of *Geranium Anglicum variegatum*. *Bauhine* calls it *Geranium batrachiodes flore variegato*. We may call it Variegated or striped Cranes bill.

5 There is of late brought into this kingdome, and to our knowledge, by the industry of Mr. *Iohn Tradescant*, another more rare and no lesse beautifull than any of the former; and he had it by the name of *Geranium Indicum noctù odoratum*: this hath not as yet beene written of by any that I know; therefore I will giue you the description thereof, but cannot as yet giue you the figure, because I omitted the taking thereof the last yeare, and it is not as yet come to his perfection. The leaues are larger, being almost a foot long, composed of sundry little leaues of an vnequal bignes, set vpon a thicke and stiffe middle rib; and these leaues are much diuided and cut in, so that the whole leafe somewhat resembles that of *Tanacetum inodorum*: and they are thicke, greene, and somewhat hairy: the stalke is thicke, and some cubit high; at the top of each branch, vpon foot-stalkes some inch long grow some eleuen or twelue floures, and each of these floures consisteth of fiue round pointed leaues of a yellowish colour, with a large blacke purple spot in the middle of each leafe, as if it were painted, which giues the floure a great deale of beauty, and it also hath a good smell. I did see it in floure about the end of Iuly, 1632. being the first time that it floured with the owner thereof. We may fitly call it Sweet Indian Storks bill, or painted Storks bill: and in Latine, *Geranium Indicum odoratum flore maculato*. ‡

Chap. 364. *Of Sanicle.*

Sanicula, siue Diapensia. Sanicle.

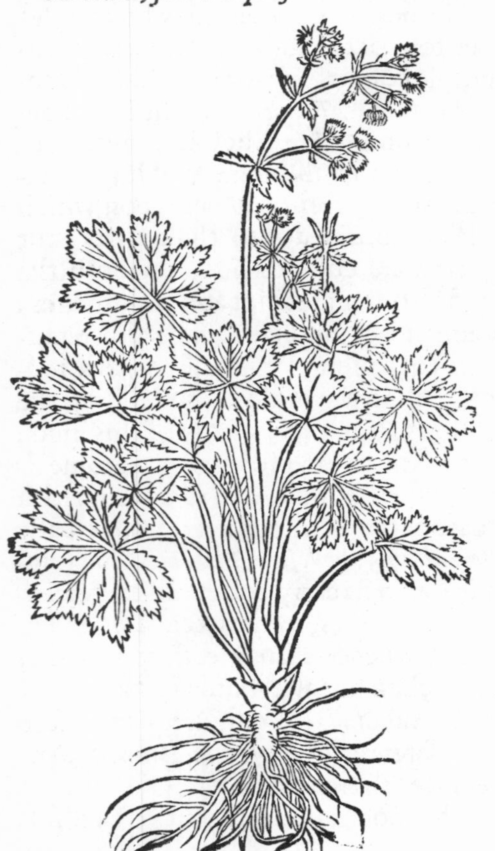

¶ *The Description.*

Sanicle hath leaues of a blackish greene colour, smooth and shining, somewhat round, diuided into fiue parts like those of the Vine, or rather those of the maple: among which rise vp slender stalkes of a browne colour, on the tops whereof stand white mossie floures: in their places come vp round seed, rough, cleauing to mens garments as they passe by, in manner of little burs: the root is blacke, and full of threddie strings.

¶ *The Place.*

It groweth in shadowie woods and copses almost euerie where: it ioyeth in a fat and fruitful moist soile.

¶ *The Time.*

It floureth in May and Iune: the seed is ripe in August: the leaues of the herbe are greene all the yeare, and are not hurt with the cold of Winter.

¶ *The Names.*

It is commonly called *Sanicula*; of diuers, *Diapensia*: in high and low Dutch, **Sanikel**: in French, *Sanicle*: in English, Sanickle, or Sanikel: it is so called, *à sanandis vulneribus*, or of healing of wounds, as *Ruellius* saith: there be also other

other Sanicles, ſo named of moſt Herbariſts, as that which is deſcribed by the name of *Dentaria*, or Coral-wort, and likewiſe *Auricula vrſi*, or Beares eare, which is a kind of Cowſlip; and likewiſe another ſet forth by the name of *Sanicula guttata*, whereof we haue entreated among the kindes of Beares eares.

¶ *The Temperature.*

Sanicle as it is in taſte bitter, with a certaine binding qualitie; ſo beſides that it clenſeth, and by the binding faculty ſtrengthneth, it is hot and dry, and that in the ſecond degree, and after ſome Authors, hot in the third degree, and aſtringent.

¶ *The Vertues.*

The iuyce being inwardly taken is good to heale wounds.

The decoction of it alſo made in wine or water is giuen againſt ſpitting of bloud, and the bloudie flix: alſo foule and filthy vlcers be cured by being bathed therewith. The herbe boyled in water, and applied in manner of a pulteſſe, doth diſſolue and waſte away cold ſwellings: it is vſed in potions which are called Vulnerarie potions, or wound drinkes, which maketh whole and ſound all inward wounds and outward hurts: it alſo helpeth the vlcerations of the kidnies, ruptures, or burſtings. A

CHAP. 365. *Of Ladies Mantle, or great Sanicle.*

Alchimilla.
Lyons foot, or Ladies mantle.

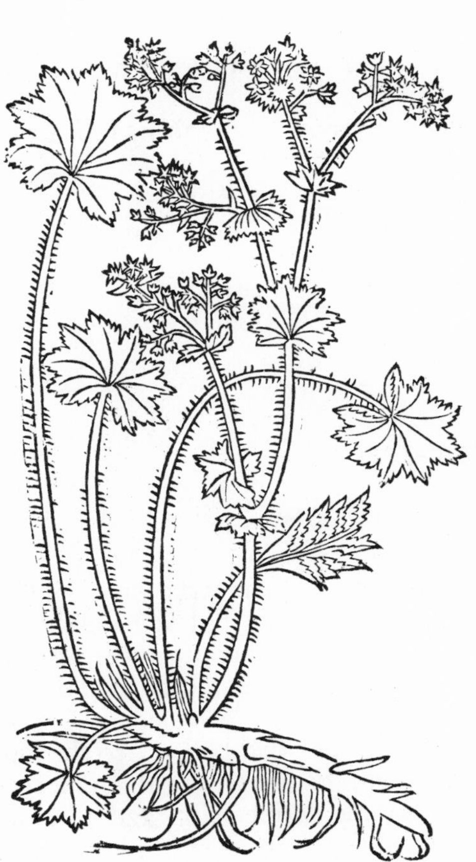

¶ *The Deſcription.*

Ladies mantle hath many round leaues, with fiue or ſix corners finely indented about the edges, which before they be opened are plaited and folded together, not vnlike to the leaues of Mallowes, but whiter, and more curled: among which riſe vp tender ſtalks ſet with the like leaues but much leſſer: on the tops whereof grow ſmall moſſie floures cluſtering thicke together, of a yellowiſh greene colour. The ſeed is ſmall and yellow, incloſed in greene husks. The root is thicke, and full of threddy ſtrings.

¶ *The Place.*

It groweth of it ſelfe wilde in diuers places, as in the towne paſtures of Andouer, and in many other places in Barkſhire and Hampſhire, in their paſtures and copſes, or low woods, and alſo vpon the banke of a mote that incloſeth a houſe in Buſhey called Bourn hall, fourteen miles from London, and in the high way from thence to Watford, a ſmall mile diſtant from it.

¶ *The Time.*

It floureth in May and Iune: it flouriſheth in Winter as well as in Sommer.

¶ *The Names.*

It is called of the later Herbariſts *Alchimilla*: and of moſt, *Stellaria*, *Pes Leonis*, *Pata Leonis*, and *Sanicula maior*: in high-Dutch, **Synnauw**, and **Vnser Frauwen mantel**: in French, *Pied de Lion*: in Engliſh, Ladies mantle, great Sanicle, Lyons foot, Lyons paw; and of ſome, Padelyon.

¶ *The Temperature.*

Ladies mantle is like in temperature to little Sanicle, yet is it more drying and more binding.

¶ *The Vertues.*

It is applied to wounds after the ſame manner that the ſmaller Sanicle is, being of like efficacie: it ſtoppeth bleeding, and alſo the ouermuch flowing of the natural ſickneſſe: it keeps downe maidens paps or dugs, and when they be too great or flaggy it maketh them leſſer or harder. A

CHAP.

Chap. 366. Of Neese-wort Sanicle.

Elleborine Alpina.
Neesewort Sanicle.

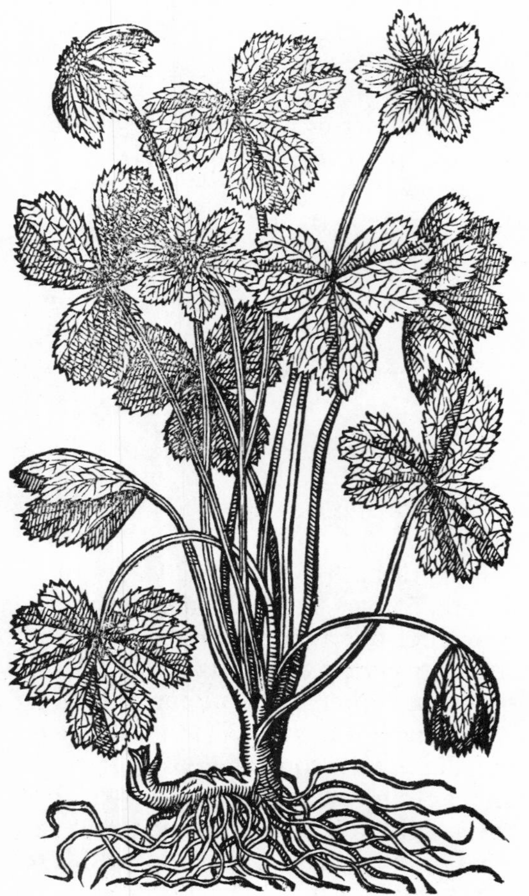

¶ *The Description.*

When I made mention of *Helleborus albus*, I did also set downe my censure concerning *Elleborine*, or *Epipactis*: but this *Elleborine* of the Alpes I put in this place, because it approcheth neerer vnto Sanicle and *Ranunculus*, as participating of both: it groweth in the mountaines and highest parts of the Alpish hills, and is a stranger as yet in our English gardens. The root is compact of many small twisted strings like black Hellebor: from thence arise small tender stalkes, smooth, and easie to bend; in whose tops grow leaues with fiue diuisions, somewhat nickt about the edges like vnto Sanicle: the floures consist of six leaues somewhat shining, in taste sharp, yet not vnpleasant. This is the plant which *Pena* found in the forrest of Esens, not farre from Iupiters mount, and sets forth by the name of *Alpina Elleborine Saniculæ & Ellebori nigri facie*.

¶ *The Nature and Vertues.*

I haue not as yet found any thing of his nature or vertues.

Chap. 367. Of Crow-feet.

¶ *The Kindes.*

There be diuers sorts or kinds of these pernitious herbes comprehended vnder the name of *Ranunculus*, or Crowfoot, whereof most are very dangerous to be taken into the body, and therefore they require a very exquisite moderation, with a most exact and due manner of tempering, not any of them are to be taken alone by themselues, because they are of most violent force, and therefore haue the greater need of correction.

The knowledge of these plants is as necessarie to the Physitian as of other herbes, to the end they may shun the same, as *Scribonius Largus* saith, and not take them ignorantly: or also, if necessitie at any time require, that they may vse them, and that with some deliberation and speciall choice, and with their proper correctiues. For these dangerous Simples are likewise many times of themselues beneficiall, and oftentimes profitable: for some of them are not so dangerous, but that they may in some sort, and oftentimes in fit and due season profit and do good, if temperature and moderation be vsed: of which there be foure kindes, as *Dioscorides* writeth; one with broad leaues, another that is downy, the third very small, and the fourth with a white floure: the later herbarists haue obserued also many moe: all these may be brought into two principall kindes, so that one be a garden or tame one, and the other wilde and of these some are common, and others rare, or forreigne. Moreouer, there is a difference both in the roots and in the leaues; for one hath a bumped or knobby root, another a long leafe as Speare-wort: and first of the wilde or field Crowfeet, referring the Reader vnto the end of the stocke and kindred of the same, for the temperature and vertues.

1 *Ranunculus pratenſis, etiamque hortenſis.*
Common Crow-foot.

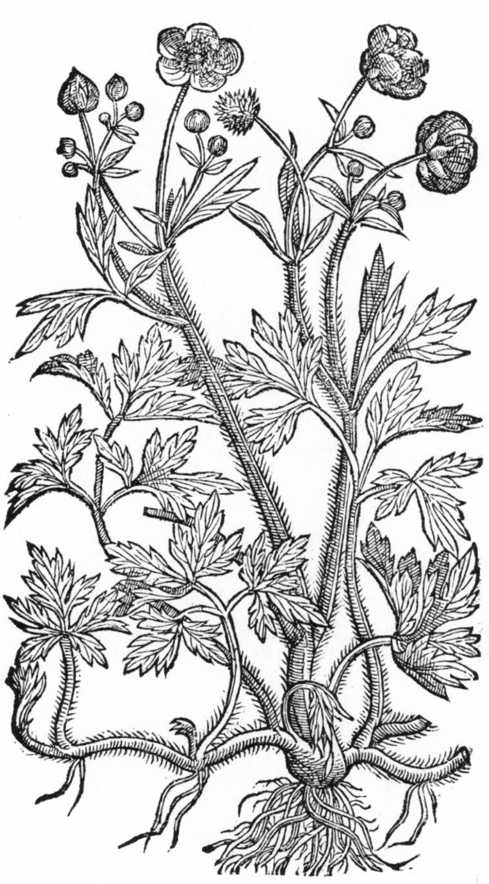

2 *Ranunculus ſurrectis cauliculis.*
Right Crow-foot.

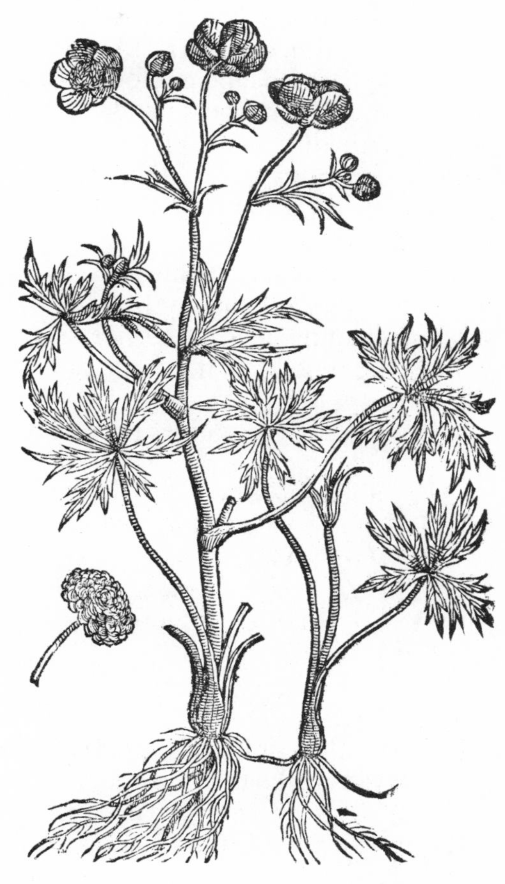

3 *Ranunculus aruorum.*
Crowfoot of the fallowed field.

4 *Ranunculus Alpinus albus.*
White mountaine Crow-foot.

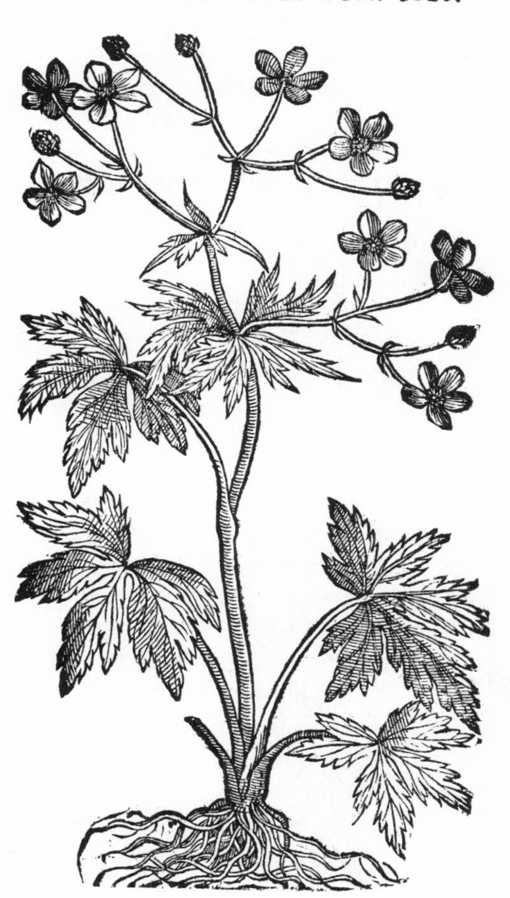

¶ *The*

¶ *The Description.*

1 THe common Crow-foot hath leaues diuided into many parts, commonly three, sometimes fiue, cut here and there in the edges, of a deepe greene colour, in which stand diuers white spots: the stalkes be round, something hairie, some of them bow downe toward the ground, and put forth many little roots, whereby it taketh hold of the ground as it traileth along: some of them stand vpright, a foot high or higher; on the tops whereof grow small floures with fiue leaues apiece, of a yellow glittering colour like gold: in the middle part of these floures stand certaine small threds of like colour: which being past, the seeds follow, made vp in a rough ball: the roots are white and threddy.

2 The second kinde of Crow-foot is like vnto the precedent, sauing that his leaues are fatter, thicker, and greener, and his small twiggy stalkes stand vpright, otherwise it is like: of which kind it chanced, that walking in the field next to the Theatre by London, in the company of a worshipfull merchant named Mr. *Nicolas Lete*, I found one of this kinde there with double floures, which before that time I had not seene.

¶ *The Place.*

They grow of themselues in pastures and medowes almost euery where.

¶ *The Time.*

They floure in May and many moneths after.

¶ *The Names.*

Crow-foot is called of *Lobel*, *Ranunculus pratensis*: of *Dodonæus*, *Ranunculus hortensis*, but vnproperly: of *Pliny*, *Polyanthemum*, which he saith diuers name *Batrachion*: in high-Dutch, **Schmalkbluom**: in low Dutch, **Boter bloemen**: in English, King Kob, Gold cups, Gold knobs, Crowfoot, and Butter-floures.

¶ *The Description.*

3 The third kinde of Crow-foot, called in Latine *Ranunculus aruorum*, because it growes commonly in fallow fields where corne hath beene lately sowne, and may be called Corne Crow-foot, hath for the most part an vpright stalke of a foot high, which diuides it selfe into other branches: whereon do grow fat thicke leaues very much cut or iagged, resembling the leaues of Sampire, but nothing so greene, but rather of an ouerworne colour. The floures grow at the top of the branches, compact of fiue small leaues of a faint yellow colour: after which come in place clusters of rough and sharpe pointed seeds. The root is small and threddy.

4 The fourth Crow-foot, which is called *Ranunculus Alpinus*, because those that haue first written thereof haue not found it elsewhere but vpon the Alpish mountains (notwithstanding it groweth in England plentifully wilde, especially in a wood called Hampsted Wood, and is planted in gardens) hath diuers great fat branches two cubits high, set with large leaues like the common Crow foot, but greater, of a deepe greene colour, much like to those of the yellow Aconite, called *Aconitum luteum Ponticum*. The floures consist of fiue white leaues, with small yellow chiues in the middle, smelling like the floures of May or Haw-thorne, but more pleasant. The roots are greater than any of the stocke of Crow-feet.

¶ *The Place and Time.*

Their place of growing is touched in their description: their time of flouring and seeding answereth the other of their kindes.

¶ *The Names.*

The white Crow-foot of the Alps and French mountaines is the fourth of *Dioscorides* his description; for he describeth his fourth to haue a white floure: more hath not bin said touching the names, yet *Tabern.* calls it *Batrachium album*: in English, white Crow-foot.

¶ *The Description.*

5 Among the wilde Crow-feet there is one that is syrnamed *Illyricus*, which brings forth slender stalks, round, and of a meane length: whereupon doe grow long narrow leaues cut into many long gashes, somthing white, and couered with a certaine downinesse: the floures be of a pale yellow colour: the root consisteth of many small bumpes as it were graines of corne, or little long bulbes growing close together like those of Pilewort. It is reported, that it was first brought out of Illyria into Italy, and from thence into the Low-Countries: notwithstanding we haue it growing very common in England. ‡ But only in gardens that I haue seene. ‡

6 The sixth kinde of Crow-foot, called *Ranunculus bulbosus*, or Onion rooted Crow-foot, and round rooted Crow-foot, hath a round knobby or onion-fashioned root, like vnto a small Turnep, and of the bignesse of a great Oliue: from the which rises vp many leaues spred vpon the ground, like those of the field Crow-foot, but smaller, and of an ouerworne greene colour: amongst which rise vp slender stalkes of the height of a foot: whereupon do grow floures of a feint yellow colour. ‡ This growes wilde in most places, and floures at the beginning of May. ‡

¶ The

¶ *The Place.*

It is alſo reported to be found not only in Illyria and Sclauonia, but alſo in the Iſland Sardinia, ſtanding in the Midland, or Mediteranian ſea.

¶ *The Names.*

This Illyrian Crowfoot is named in Greeke σέλινον ἄγριον, that is, *Apium ſylueſtre*, or wilde Smallage: alſo *Herba Sardoa:* it may be, ſaith my Author, that kinde of Crowfoot called *Apium riſus*, and γελωτοφυλλὶς; and this is thought to be that *Gelotophillis*, of which *Pliny* maketh mention in his 24. booke, 17. chap. which being drunke, ſaith he, with wine and myrrh, cauſeth a man to ſee diuers ſtrange ſights, and not to ceaſe laughing till he hath drunke Pine apple kernells with Pepper in wine of the Date tree, (I thinke he would haue ſaid vntill he be dead) becauſe the nature of laughing Crowfoot is thought to kill laughing, but without doubt the thing is cleane contrary; for it cauſeth ſuch conuulſions, cramps and wringings of the mouth and jawes, that it hath ſeemed to ſome that the partyes haue dyed laughing, whereas in truth they haue died in great torment.

5 *Ranunculus Illyricus.*
Crowfoot of Illyria.

6 *Ranunculus bulboſus.*
Roundrooted Crowfoot.

¶ *The Deſcription.*

7 The ſeuenth kinde of Crowfoot, called *Auricomus* of the golden lockes wherewith the floure is thrummed, hath for his root a great buſh of blackiſh hairy ſtrings; from which ſhoote forth ſmall jagged leaues, not much vnlike to Sanicle, but diuided onely into three parts, yet ſometimes into fiue; among which riſe vp branched ſtalkes of a foot high, whereon are placed the like leaues but ſmaller, ſet about the top of the ſtalkes, whereon do grow yellow floures, ſweet ſmelling, of which it hath been called *Ranunculus dulcis*, *Tragi*, or *Tragus* his ſweet Crowfoot. ‡ It growes in medowes and about the ſides of woods, and floures in Aprill. ‡

† 8 Frogge Crowfoot, called of *Pena*, *Aconitum Batrachioides:* of *Dodonæus*, *Batrachion Apulei*, is that formerly deſcribed in the fourth place, whereto this is much alike, but that the ſtalkes and leaues are larger, as alſo the floures, which are white: the root is tough and threddy.

9 The ninth Crowfoote hath many graſſie leaues, of a deepe greene tending to blewneſſe, ſomewhat long, narrow and ſmooth, very like vnto thoſe of the ſmall Biſtort, or Snakeweed: among

7 *Ranunculus auricomus.*
Golden-haired Crow-foot.

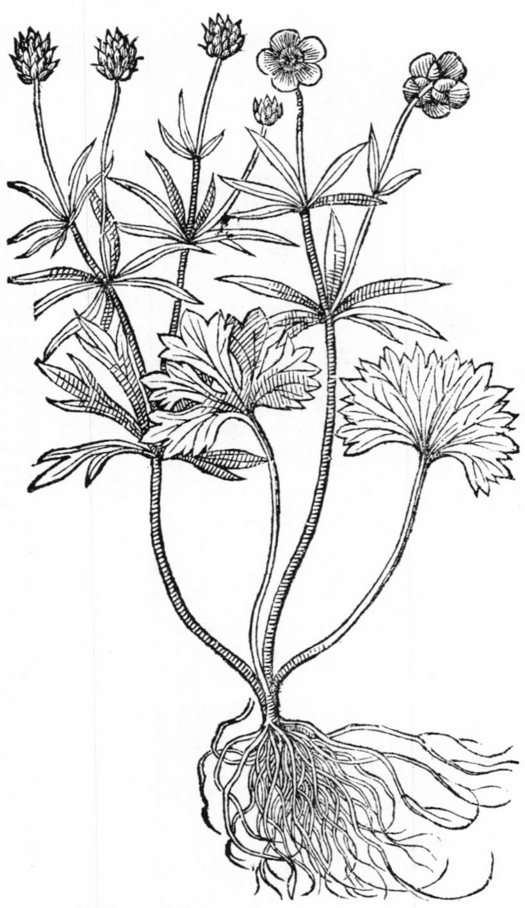

† 8 *Ranunculus Aconiti folio.*
Frog Crow-foot.

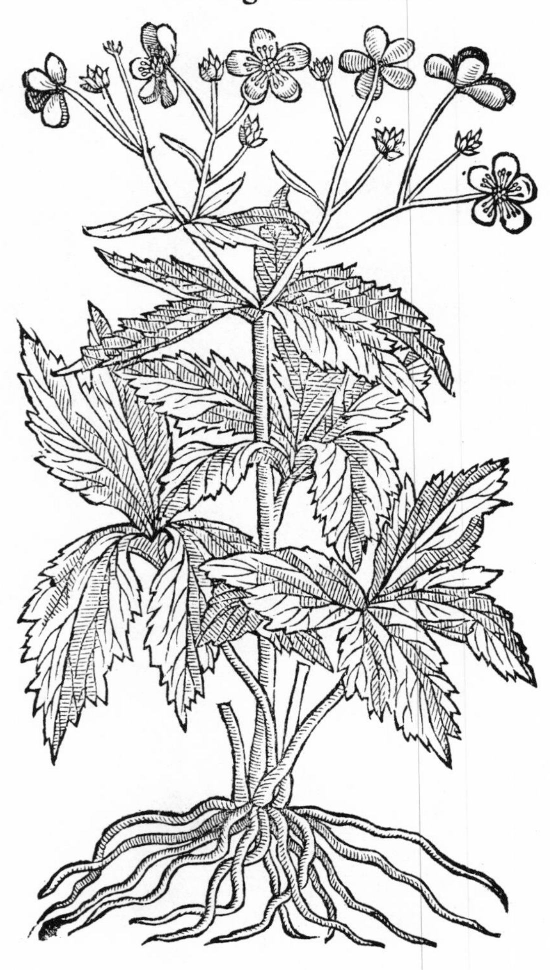

9 *Ranunculus gramineus Lobelij.*
Grassie Crow-foot.

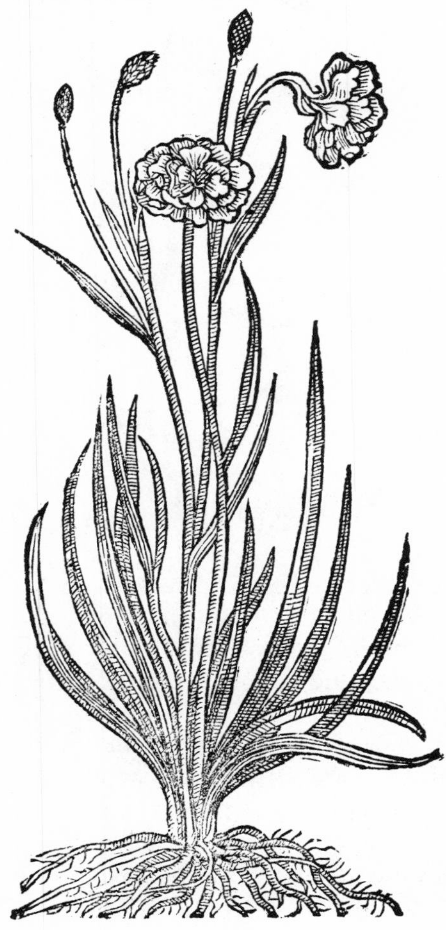

10 *Ranunculus Autumnalis Clusij.*
Winter Crow-foot.

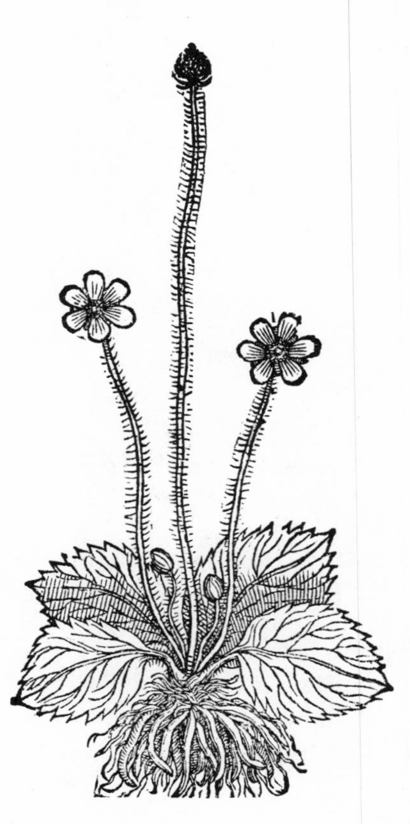

among which riſe vp ſlender ſtalkes, bearing at the top ſmall yellow floures like the other Crow-feet: the root is ſmall and threddy. ‡ There is a variety of this hauing double floures; and I haue giuen you the figure thereof in ſtead of the ſingle that was formerly in this place. ‡

10 The Autumne or Winter Crow-foot hath diuers broad leaues ſpred vpon the ground, ſnipt about the edges, of a bright ſhining greene colour on the vpper ſide, and hoary vnderneath, full of ribs or ſinewes as are thoſe of Plantaine, of an vnpleaſant taſte at the firſt, afterward nipping the tongue: among which leaues riſe vp ſundry tender foot-ſtalkes, on the tops whereof ſtand yellow floures conſiſting of ſix ſmall leaues apiece: after which ſucceed little knaps of ſeed like to a dry or withered ſtraw-berry. The root is compact of a number of limber roots, rudely thruſt together in manner of the Aſphodill.

11 The Portugall Crow-foot hath many thicke clogged roots faſtned vnto one head, very like thoſe of the yellow Aſphodill: from which riſe vp three leaues, ſeldome more, broad, thicke, and puffed vp in diuers places, as if it were a thing that were bliſtered, by meanes whereof it is very vneuen. From the middle of which leaues riſeth vp a naked ſtalke, thicke, fat, very tender, but yet fragile, or eaſie to breake: on the end whereof ſtandeth a faire ſingle yellow floure, hauing in the middle a naked rundle of a gold yellow tending to a Saffron colour.

11 *Ranunculus Luſitanicus Cluſij.*
Portugall Crow-foot.

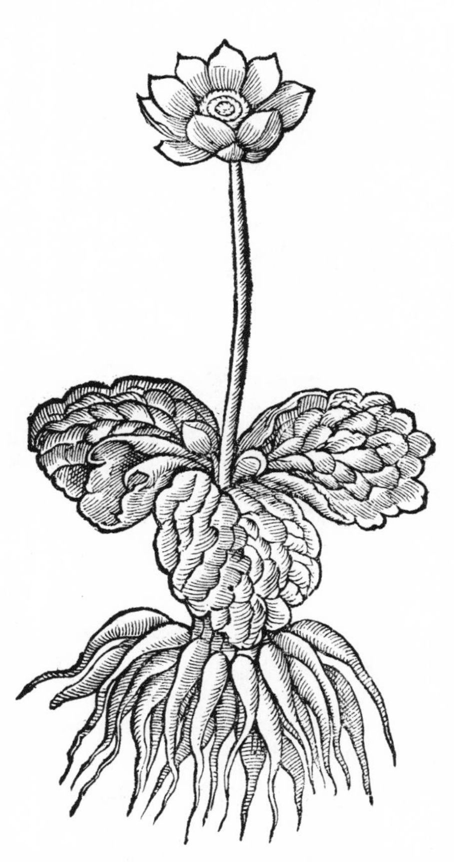

12 *Ranunculus globoſus.*
Locker Gowlons, or Globe Crowfoot.

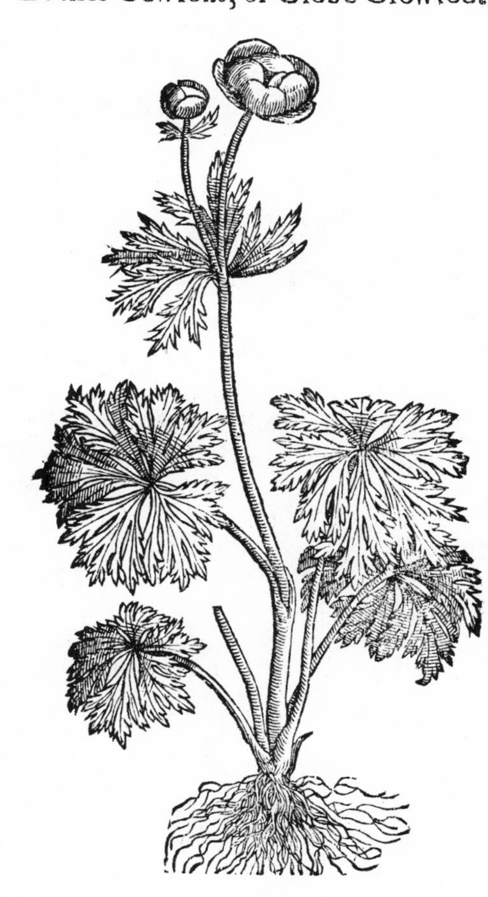

12 The Globe Crow-foot hath very many leaues deepely cut and iagged, of a bright greene colour like thoſe of the field Crow-foot: among which riſeth vp a ſtalke, diuided toward the top into other branches, furniſhed with the like leaues of thoſe next the ground, but ſmaller: on the tops of which branches grow very faire yellow floures, conſiſting of a few leaues folded or rolled vp together like a round ball or globe: whereupon it was called *Ranunculus globoſus*, or the Globe Crow-foot, or Globe floure: which being paſt, there ſucceed round knaps, wherein is blackiſh ſeed. The root is ſmall and threddy.

‡ 13 This hath large leaues like thoſe of the laſt deſcribed, but rough and hairy: the ſtalk is ſome foot high: the floures are pretty large, compoſed of fiue white ſharpiſh pointed leaues. It floures in Iuly, and growes in the Alps: it is the *Ranunculi montani 2. ſpecies altera* of *Cluſius*.

14 This other hath leaues not vnlike thoſe of the precedent, and ſuch ſtalkes alſo, but the floures conſiſt of 5 round leaues, purpliſh beneath; the edges of the vpper ſide are of a whitiſh purple, & the reſidue wholly white, with many yellow threds in the middle; it grows in the mountain

‡ 13 *Ranunculus hirsutus Alpinus flo. albo.*
Rough white floured mountaine Crow-foot.

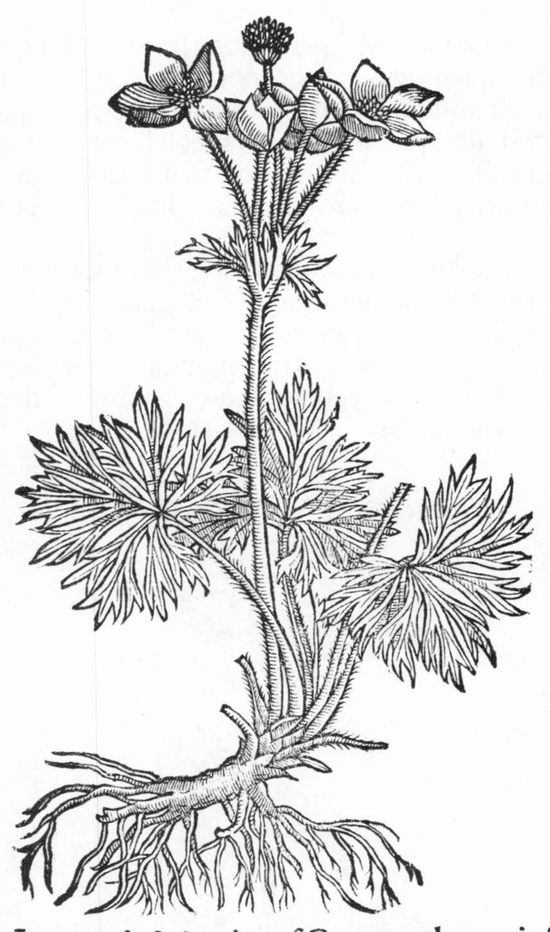

‡ 14 *Ranunculus montanus hirsutus purpureus.*
Rough purple floured mountain Crowfoot.

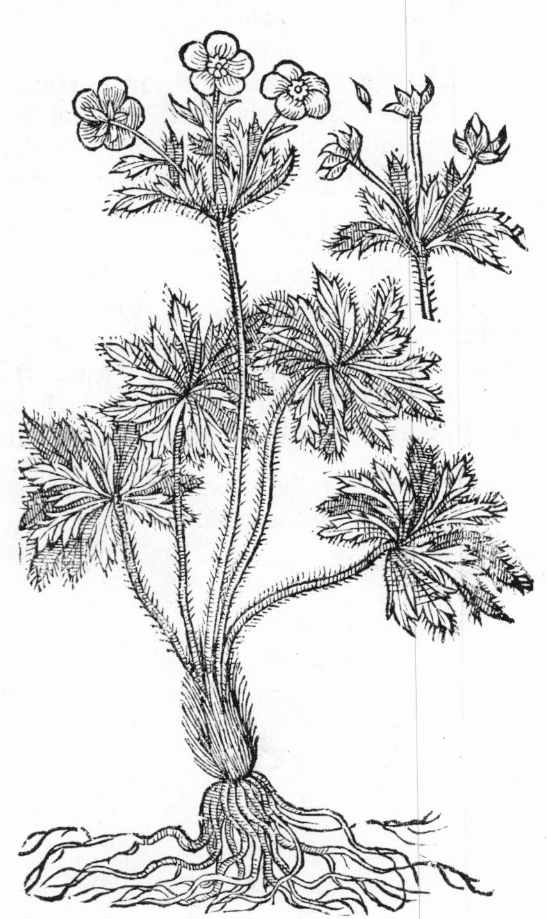

Iura, against the city of Geneua, whereas it floures in Iune, and ripens the seed in August. *Clusius* had the figure and description hereof from Dr. *Penny*, and he calls it *Ranunculus montanus* 3. ‡

¶ *The Place.*

The twelfth kind of Crowfoot groweth in most places of York-shire and Lancashire, and other bordering shires of the North countrey, almost in euery medow, but not found wilde in these Southerly or Westerly parts of England that I could euer vnderstand of.

¶ *The Time.*

It floureth in May and Iune: the seed is ripe in August.

¶ *The Names.*

The Globe floure is called generally *Ranunculus globosus* of some, *Flos Trollius*, and *Ranunculus Alpinus*: in English, Globe Crow-foot, Troll floures, and Lockron gowlons.

Chap. 368. *Of Double yellow and white Batchelors Buttons.*

¶ *The Description.*

1 The great double Crow-foot or Batchelors button hath many iagged leaues of a deepe greene colour: among which rise vp stalkes, whereon do grow faire yellow floures exceeding double, of a shining yellow colour, oftentimes thrusting forth of the middest of the said floures one other smaller floure: the root is round, or fashioned like a Turnep; the form whereof hath caused it to be called of some S. Anthonies Turnep, or Rape Crow-foot. The seed is wrapped in a cluster of rough knobs, as are most of the Crow-feet.

2 The double yellow Crow-foot hath leaues of a bright greene colour, with many weake branches trailing vpon the ground; whereon do grow very double yellow floures like vnto the precedent, but altogether lesser. The whole plant is likewise without any manifest difference, sauing that these floures do neuer bring forth any smaller floure out of the middle of the greater, as the other doth, and also hath no Turnep or knobby root at all, wherein consists the greatest difference.

3 The

† *Ranunculus maximus Anglicus.*
Double Crow-foot, or Batchelors buttons.

2 *Ranunculus dulcis multiplex.*
Double wilde Crow-foot.

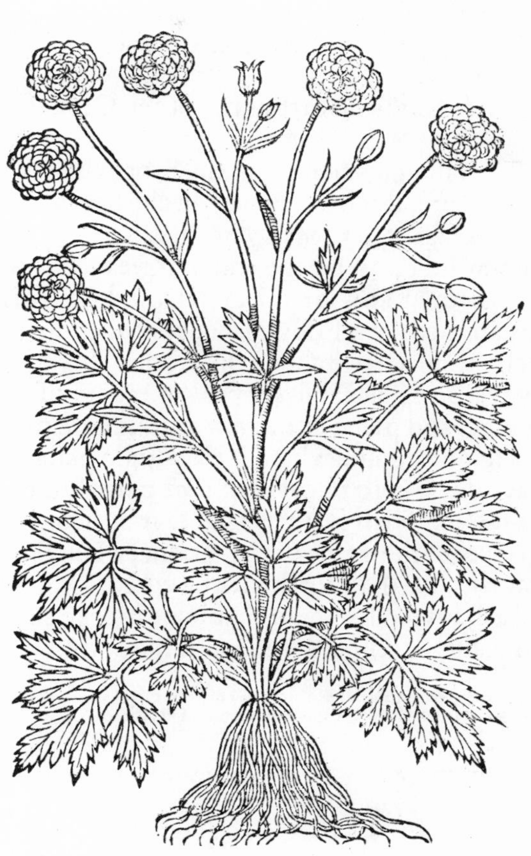

3 *Ranunculus albus multiflorus.*
Double white Crow-foot.

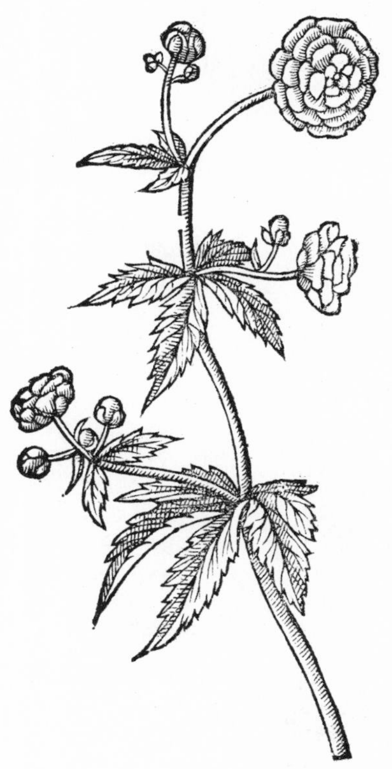

3 The white double Crow-foot hath many great leaues deeply cut with great gaſhes, and thoſe ſnipt about the edges. The ſtalks diuide themſelues into diuers brittle branches, on the tops whereof do grow very double floures as white as ſnow, and of the bigneſſe of our yellow Batchelors button. The root is tough, limber, and diſperſeth it ſelfe farre abroad, whereby it greatly increaſeth.

¶ *The Place.*

The firſt and third are planted in gardens for the beauty of the floures, and likewiſe the ſecond, which hath of late beene brought out of Lancaſhire vnto our London gardens, by a curious gentleman in the ſearching forth of Simples, M[r]. *Thomas Hesketh*, who found it growing wilde in the towne fields of a ſmal village called Hesketh, not farre from Latham in Lancaſhire.

¶ *The Time.*

They floure from the beginning of May to the end of Iune.

¶ *The Names.*

Dioſcorides hath made no mention hereof; but *Apuleius* hath ſeparated the firſt of theſe from the others, intreating of it apart, and naming it by a peculiar name *Batrachion*; whereupon it is alſo called *Apuleij Batrachion*, or *Apuleius* Crow-foot.

It is commonly called *Rapum D. Anthonij*, or Saint Anthonies Rape: it may be called in Engliſh, Rape Crow-foot: it is called generally about London, Batchelors buttons, and double Crow-foot: in

Dutch, S. Anthony Rapkin. ‡ These names and faculties properly belong to the *Ranunculus bulbosus*, described in the sixt place of the last chapter; and also to the first double one here described; for they vary little but in colour, and the singlenesse and doublenesse of their floures. ‡

The third is called of *Lobel, Ranunculus niueus polyanthos*: of *Tabern. Ranunculus albus multiflorus*: in English, Double white Crow-foot, or Batchelors buttons.

¶ *The Temperature.*

These plants do bite as the other Crow-feet do.

¶ *The Vertues.*

A The chiefest vertue is in the root, which being stamped with salt is good for those that haue a plague sore, if it be presently in the beginning tied to the thigh, in the middle between the groin or flanke and the knee: by meanes whereof the poyson and malignitie of the disease is drawn from the inward parts, by the emunctorie or clensing place of the flanke, into those outward parts of lesse account: for it exulcerateth and presently raiseth a blister, to what part of the body soeuer it is applied. And if it chance that the sore hapneth vnder the arme, then it is requisite to apply it to the arme a little aboue the elbow. My opinion is, that any of the Crow-feet will do the same: my reason is, because they all and euery of them do blister and cause paine, wheresoeuer they be applied, and paine doth draw vnto it selfe more paine; for the nature of paine is to resort vnto the weakest place, and where it may finde paine; and likewise the poyson and venomous qualitie of that disease is to resort vnto that painefull place.

B *Apuleius* saith further, That if it be hanged in a linnen cloath about the necke of him that is lunaticke, in the waine of the moone, when the signe shall be in the first degree of *Taurus* or *Scorpio*, that then he shall forthwith be cured. Moreouer, the herbe *Batrachion* stamped with vineger, root and all, is vsed for them that haue blacke skars or such like marks on their skins, it eats them out, and leaues a colour like that of the body.

† The figure that formerly was in the first place of this chapter was the double one mentioned in the second description of the foregoing chapter, where also you may finde a double floure exprest by the side of the figure.

Chap. 369. Of Turkie or Asian Crow-feet.

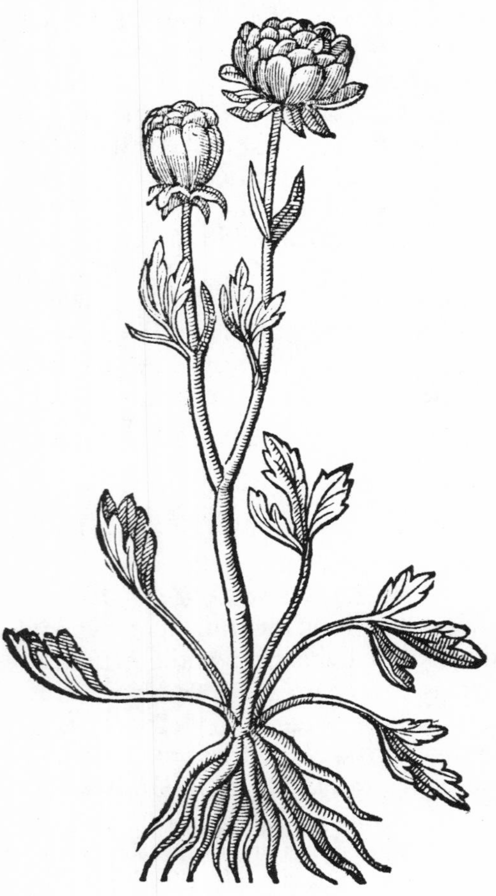

1 *Ranunculus sanguineus multiplex.*
The double red Crow-foot.

‡ 2 *Ranunculus Asiaticus flo. pleno miniato.*
The double Asian skarlet Crow-foot.

‡ 3 *Ranunculus Aſiaticus flore pleno prolifero.*
The double buttoned ſcarlet Aſian Crow foot.

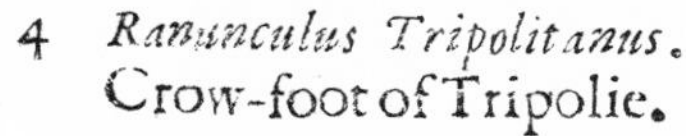

4 *Ranunculus Tripolitanus.*
Crow-foot of Tripolie.

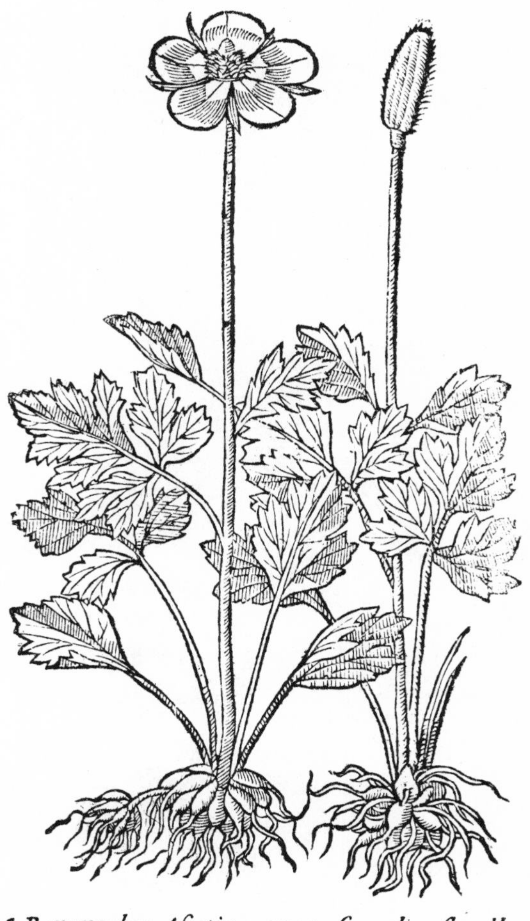

‡ 5 *Ranunculus grumoſa radice ramoſus.*
Branched red Aſian Crow-foot.

‡ 6 *Ranunculus Aſiaticus grumoſa radice flo. albo.*
White floured Aſian Crow-foot.

‡ 7 *Ranunculus Asiaticus grumosa radice flore flavo vario.*
Asian Crow-foot with yellow striped floures.

¶ *The Description.*

1 The double red Crow-foot hath a few leaues rising immediatly forth of the ground, cut in the edges with deepe gashes, somewhat hollow, and of a bright shining green colour. The stalk riseth vp to the height of a foot, smooth and very brittle, diuiding it selfe into other branches, sometimes two, seldome three: whereon do grow leaues confusedly, set without order: the floures grow at the tops of the stalks, very double, and of great beauty, of a perfect scarlet colour, tending to rednesse. The root is compact of many long tough roots, like those of the yellow Asphodill.

‡ 2 Of this kinde there is also another, or other the same better exprest; for *Clusius* the author of these neuer see the former, but makes it onely to differ, in that the floures are of a sanguine colour, and those of this of a kinde of scarlet, or red lead colour.

3 This differs nothing from the former, but that it sends vp another floure somewhat lesser, out of the middle of the first floure, which happens by the strength of the root, and goodnesse of the soile where it is planted. ‡

4 The Crow-foot of Tripolis or the single red *Ranunculus* hath leaues at the first comming vp like vnto those of Groundswell: among which riseth vp a stalke of the height of halfe a cubit, somwhat hairy, wheron grow broad leaues deeply cut, euen to the middle rib, like those of the common Crow-foot, but greener: the floure groweth at the top of the stalke, consisting of fiue leaues, on the outside of a darke ouerworne red colour, on the inside of a red lead colour, bright and shining, in shape like the wilde corne Poppie: the knop or stile in the middle which containeth the seede is garnished or bedeckt with very many small purple thrummes tending to blacknesse: the root is as it were a roundell of little bulbes or graines like those of the small Celandine or Pilewoort.

‡ 5 There be diuers other Asian Crow-feet which *Clusius* hath set forth, and which grow in the most part in the gardens of our prime Florists, and they differ little in their roots, stalkes, or leaues, but chiefely in the floures; wherefore I will onely briefely note their differences, not thinking it pertinent to stand vpon whole descriptions, vnlesse they were more necessary: this fift differs from the fourth in that the stalkes are diuided into sundry branches, which beare like, but lesse floures than those which stand vpon the main stalke: the colour of these differs not from that of the last described.

6 This is like the last described, but the floures are of a pure white colour, and sometimes haue a few streaks of red about their edges.

7 This in stalkes and manner of growing is like the precedent: the stalke seldome parting it selfe into branches; but on the top thereof it carries a faire floure consisting commonly of round topped leaues of a greenish yellow colour, with diuers red veines here and there dispersed and running alongst the leaues, with some purple thrums, and a head standing vp in the middle as in the former. ‡

¶ *The Place.*

The first groweth naturally in and about Constantinople, and in Asia on the further side of Bosphorus, from whence there hath been brought plants at diuers times, and by diuers persons, but they haue perished by reason of their long iourney, and want of skill of those bringers, that haue suffered them to lie in a box or such like so long, that when we haue receiued them they haue been as dry as ginger; notwithstanding *Clusius* saith he receiued a plant fresh and greene, the which a domestical theefe stole forth of his garden. My Lord and Master the right Honorable the Lord Treasurer

ſurer had diuers plants ſent him from thence which were drie before they came,as aforeſaid. The other groweth in Aleppo and Tripolis in Syria naturally,from whence we haue receiued plants for our gardens,where they flouriſh as in their owne countrey.

¶ *The Time.*

They bring forth their pleaſant floures in May and Iune the ſeed is ripe in Auguſt.

¶ *The Names.*

The firſt is called *Ranunculus Conſtantinopolitanus:* Of *Lobel*, *Ranunculus ſanguineus multiplex*, *Ranunculus Bizantinus*, *ſiue Aſiaticus:* in the Turkiſh tongue, *Torobolos*, *Catamer laile:* in Engliſh,the double red Ranunculus,or Crow-foot.

The fourth is called *Ranunculus Tripolitanus*,of the place from whence it was firſt brought into theſe parts: of the Turks,*Tarobolos Catamer*, without that addition *laile:* which is a proper word to all floures that are double.

¶ *The Temperature and Vertues.*

Their temperature and vertues are referred to the other Crow-feet, whereof they are thought to be kindes.

Chap. 370. *Of Speare-woort, or Bane-woort.*

¶ *The Deſcription.*

1 Speare-woort hath an hollow ſtalke full of knees or ioynts,whereon do grow long leaues, a little hairy,not vnlike thoſe of the willow,of a ſhining green colour: the floures are very large,and grow at the tops of the ſtalks,conſiſting of fiue leaues of a faire yellow colour,verie like to the field gold cup, or wilde Crow-foot: after which come round knops or ſeed veſſels,wherein is the ſeed: the root is contract of diuers bulbes or long clogs, mixed with an infinite number of hairy threds.

1 *Ranunculus flammeus maior.*
Great Speare-woort.

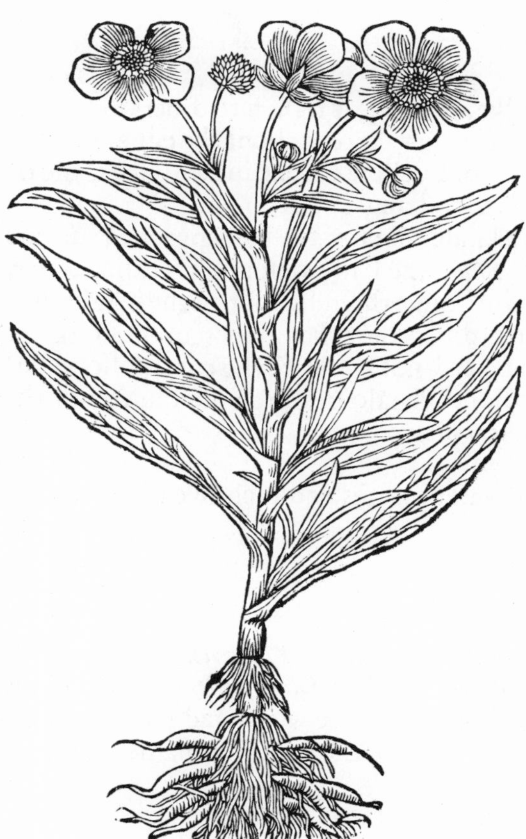

2 *Ranunculus flammeus minor.*
The leſſer Speare-woort.

2 The

2 The common Spearewoort being that which we haue called the leſſer, hath leaues, floures, and ſtalks like the precedent, but altogether leſſer: the roote conſiſteth of an infinite number of threddie ſtrings.

3 *Ranunculus flammeus ſerratus.*
Iagged Speare-woort.

4 *Ranunculus paluſtris rotundifolius.*
Mariſh Crow-foot, or Speare-worts.

3 Iagged Speare-woort hath a thicke fat hollow ſtalke, diuiding it ſelfe into diuers branches, whereon are ſet ſomtimes by couples two long leaues, ſharp pointed, & cut about the edges like the teeth of a ſaw. The floures grow at the top of the branches, of a yellow colour, in form like thoſe of the field Crowfoot: the root conſiſteth of a number of hairy ſtrings.

4 Marſh Crow-foot, or Speare-woort (whereof it is a kinde, taken of the beſt approued authors to be the true *Apium riſus*, though diuers thinke that *Pulſatilla* is the ſame: of ſome it is called *Apium hæmorrhoidarum*) riſeth forth of the mud or wateriſh mire from a threddie root, to the height of a cubit, ſometimes higher. The ſtalke diuideth it ſelfe into diuers branches, whereupon doe grow leaues deeply cut round about like thoſe of Doues-foot, and not vnlike to the cut Mallow, but ſomewhat greater, and of a moſt bright ſhining green colour: the floures grow at the top of the branches, of a yellow colour, like vnto the other water Crow-feet.

¶ *The Place.*

They grow in moiſt and dankiſh places, in brinkes or water courſes, and ſuch like places almoſt euery where.

¶ *The Time.*

They floure in May when other Crow-feet do.

¶ *The Names.*

Speare-woort is called of the later Herbariſts *Flammula*, and *Ranunculus Flammeus*; of *Cordus*, *Ranunculus* πλατύφυλλος, or broad leaued Crow-foot: of others, *Ranunculus longifolius*, or long leafed Crow foot: in Low Dutch, **Egelcoolen**: in Engliſh, Speare-Crowfoot, Speare-woort, and Banewoort, becauſe it is dangerous and deadly for ſheep; and that if they feed of the ſame it inflameth their liuers, fretteth and bliſtereth their guts and intrails.

¶ *The Temperature of all the Crowfeet.*

Speare-woort is like to the other Crow-feet in facultie, it is hot in the mouth or biting, it exulcerateth

cerateth and raiſeth bliſters, and being taken inwardly it killeth remedileſſe. Generally all the Crow-feet, as *Galen* ſaith, are of a very ſharpe or biting qualitie, inſomuch as they raiſe bliſters with paine: and they are hot and drie in the fourth degree.

¶ *The Vertues of all the Crowfeet.*

A The leaues or roots of Crowfeet ſtamped and applied vnto any part of the body, cauſeth the skin to ſwell and bliſter, and raiſeth vp wheales, bladders, cauſeth ſcars, cruſts, and ouglie vlcers: it is laid vpon cragged warts, corrupt nailes, and ſuch like excreſcences, to cauſe them to fall away.

B The leaues ſtamped and applied vnto any peſtilentiall or plague ſore, or carbuncle, ſtaieth the ſpreading nature of the ſame, and cauſeth the venomous or peſtilentiall matter to breath forth, by opening the parts and paſſages in the skin.

C It preuaileth much to draw a plague ſore from the inward parts, being of danger, vnto other remote places further from the heart, and other of the ſpirituall parts, as hath beene declared in the deſcription.

D Many do vſe to tie a little of the herbe ſtamped with ſalt vnto any of the fingers, againſt the pain of the teeth; which medicine ſeldome faileth; for it cauſeth greater paine in the finger than was in the tooth, by the meanes whereof, the greater paine taketh away the leſſer.

E Cunning beggers do vſe to ſtampe the leaues, and lay it vnto their legs and arms, which cauſeth ſuch filthy vlcers as we daily ſee (among ſuch wicked vagabonds) to moue the people the more to pittie.

F The kinde of Crowfoot of Illyria, being taken to be *Apium riſus* of ſome, yet others thinke *Aconitum Batrachioides* to be it. This plant ſpoileth the ſences and vnderſtanding, and draweth together the ſinewes and muſcles of the face in ſuch ſtrange manner, that thoſe who beholding ſuch as died by the taking hereof, haue ſuppoſed that they died laughing; ſo forceably hath it drawne and contracted the nerues and ſinewes, that their faces haue been drawne awry, as though they laughed, whereas contrariwiſe they haue died with great torment.

‡ Chap. 371. *Of diuers other Crowfeet.*

‡ 1 *Ranunculus Creticus latifolius.*
Broad leaued Candy Crowfoot.

‡ 2 *Ranunculus folio Plantaginis.*
Plantaine leaued Crowfoot.

¶ The

¶ *The Deſcription.*

‡ 1 THe roots of this are ſomwhat like thoſe of the Aſian *Ranunculus* : the leaues are verie large & roundiſh, of a light green colour, cut about the edges, & here and there deeply diuided: the ſtalke is thicke, round, and ſtiffe, diuided into two or three branches ; at the ſetting on of which grow longiſh leaues a little nickt about the end: the floures are of an indifferent bigneſſe, and conſiſt of fiue longiſh round pointed leaues, ſtanding a little each from other, ſo that the green points of the cups ſhew themſelues between them : there are yellow threds in the middle of theſe floures, which commonly ſhew themſelues in Februarie, or March. It is found only in ſome gardens, and *Cluſius* onely hath ſet it forth by the name we here giue you.

2 This alſo that came from the Pyrenæan hills is made a Denizen in our gardens: it hath a ſtalke ſome foot high, ſet with neruous leaues, like thoſe of Plantaine, but thinner, and of the colour of Woad, and they are ſomething broad at their ſetting on, and end in a ſharpe point : at the top of the ſtalke grow the floures; each conſiſting of fiue round ſlender pure white leaues, of a reaſonable bigneſſe, with yellowiſh threds and a little head in the middle : the root is white and fibrous. It floures about the beginning of May. *Cluſius* alſo ſet forth this by the title of *Ranunculus Pyrenæus albo flore*.

3 The ſame Author hath alſo giuen vs the knowledge of diuers other plants of this kinde, and this hee calls *Ranunculus montanus* 1. It hath many round leaues, here and there deeply cut in, and ſnipt about the edges, of a darke greene colour, and ſhining, pretty thicke, and of a very hot taſte : amongſt which riſes vp a ſlender, ſingle, and ſhort ſtalke, bearing a white floure made of fiue little leaues with a yellowiſh thrum in the middle: which falling, the ſeeds grow cluſtering together as in other plants of this kinde: the root is white and fibrous.

‡ 3 *Ranunculus montanus flo. minore.*
Mountain Crowfoot with the leſſer floure.

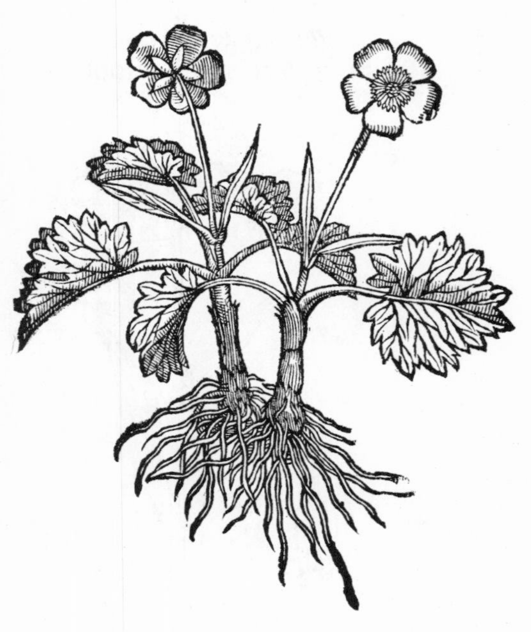

‡ 4 *Ranunculus montanus flore maiore.*
Mountain Crowfoot with the bigger floure.

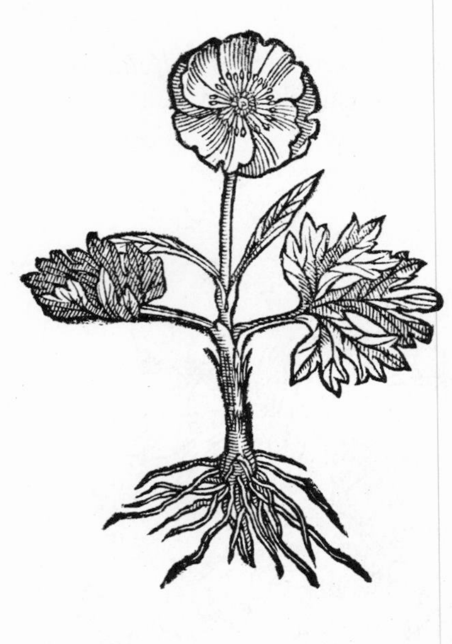

4 This alſo is nothing elſe but a varietie of the laſt deſcribed, and differs from it in that the floures are larger, and it is ſometimes found with them double. Both theſe grow on the tops of the Alpes, and there they floure as ſoone as the ſnow is melted away, which is vſually in Iune : but brought into gardens they floure very early, to wit, in Aprill.

5 The leaues of this are cut or diuided into many parts, like thoſe of Rue, but ſofter, & greener (whence *Cluſius* names it *Ranunculus Rutæ folio*) or not much vnlike thoſe of Coriander (whereupon

Pona

‡ 5 *Ranunculus præcox rutaceo folio.*
Rue leaued Crowfoot.

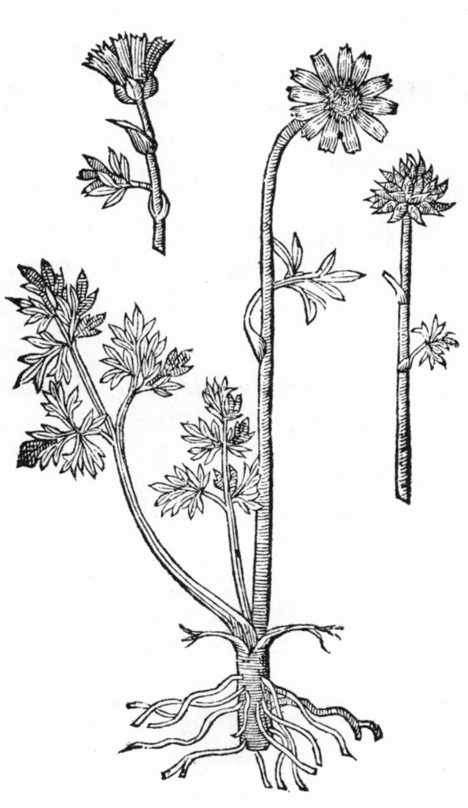

‡ 6 *Ranunculus Præcox Thalictri folio.*
Columbine Crowfoot.

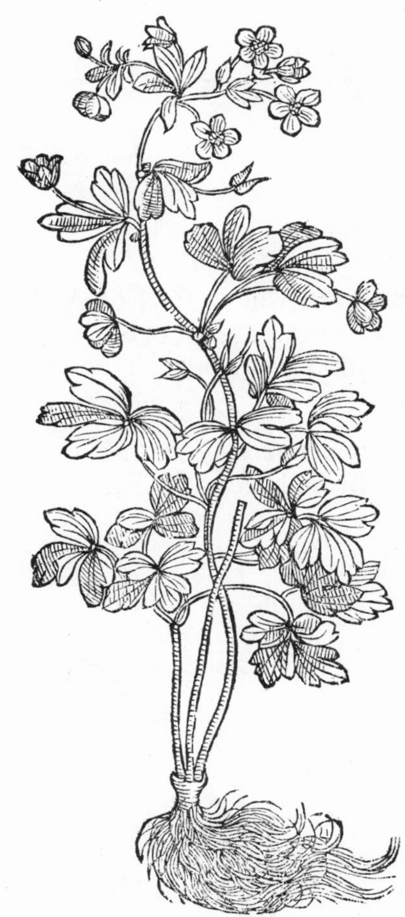

‡ 7 *Ranunculus parvus echinatus.*
Small rough headed Crowfoot.

Pona calls it *Ranunculus Coriandri folio*:) amongst or ratherbefore these comes vp a stalk some hand full high,bearing at the top thereof one floure of a reasonable bignesse: on the outside before it be throughly open of a pleasing red color, but white within,composed of twelue or more leaues.

6 This hath a stalke some foot high,small and reddish,whereon grow sundry leaues like those of the greater *Thalictrum*, or those of Columbines, but much lesse,and of a bitter taste: out of the bosomes of these leaues come the floures at each space one,white,and consisting of fiue leaues apiece: which falling, there succeed two or three little hornes containing a round reddish seed: the root is fibrous,white,very bitter, and creepes here and there,putting vp new shoots. It growes in diuers woods of Austria, and floures in Aprill, and the seed is ripe in May,or Iune. *Clusius* calls it *Ranunculus præcox* 2.*Thalictri folio*. It is the *Aquilegia minor Daleschampij* in the *Hist.Lugd.*

7 This which (as *Clusius* saith) some call the *Ranunculus* of *Apuleius*, hath also a fibrous root, with small leaues diuided into three parts, & cut about the edges, and they grow vpon short footstalkes; the stalkes are some two handfulls high, commonly leaning on the ground, and on them grow such leaues as the former: and out of their bosomes come little foot-stalks carrying floures of a pale yellow color,made of fiue leaues apiece, which follow there ſucced fiue or ſix ſharpe pointed rough cods, conteining ſeed almoſt like that of the former. ‡

CHAP. 372. *Of Woolfes-bane.*

¶ *The kindes.*

There be diuers ſorts of Wolfes-bane: whereof ſome bring forth flowers of a yellow colour; others of a blew, or tending to purple: among the yellow ones there are ſome greater, others leſſer; ſome with broader leaues, and others with narrower.

1 *Thora Valdenſis.*
Broad leafed VVolfes-bane.

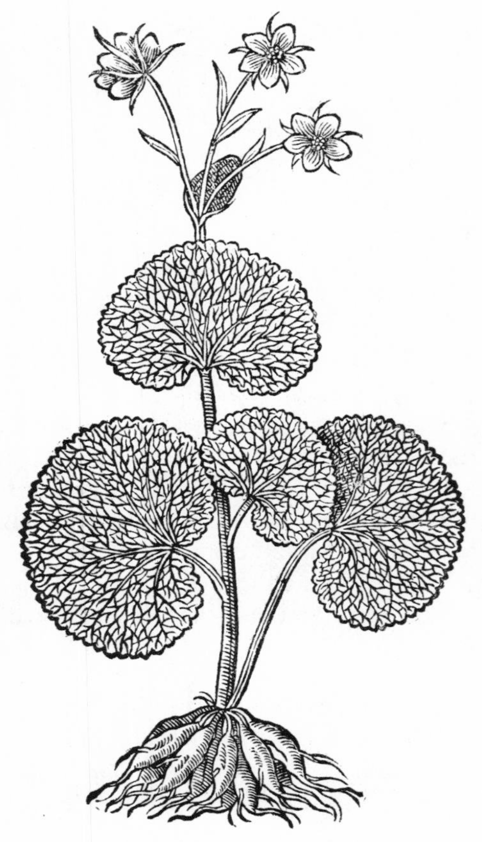

2 *Thora montis Baldi, ſiue Sabaudica.*
Mountaine VVolfes-bane.

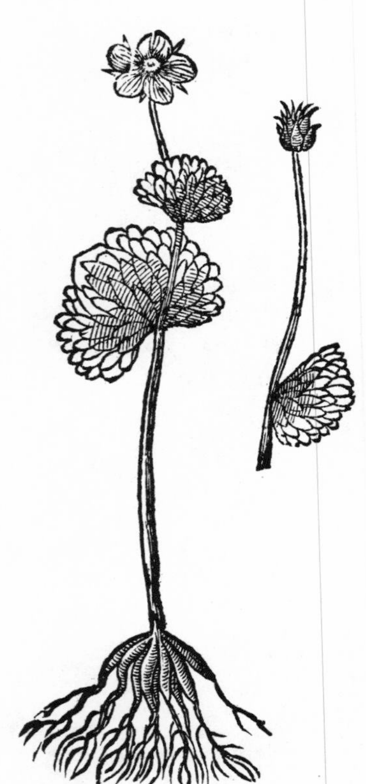

¶ *The Deſcription.*

1 THe firſt kinde of *Aconite*, of ſome called *Thora*, others adde thereto the place where it groweth in great abundance, which is the Alps, and call it *Thora Valdenſium*. This plant tooke his name of the Greek word φθόρα, ſignifying corruption, poiſon, or death, which are the certaine effects of this pernicious plant: for this they vſe very much in poiſons, and when they meane to infect their arrow heads, the more ſpeedily and deadly to diſpatch the wilde beaſts, which greatly annoy thoſe mountaines of the Alpes: to which purpoſe alſo it is brought into the Mart-townes neere vnto thoſe places to be ſold vnto the hunters, the iuice thereof beeing prepared by preſſing forth, and ſo kept in hornes and hoofes of beaſts, for the moſt ſpeedie poiſon of all the *Aconites*; for an arrow touched therewith, leaueth the wound vncureable (if it but onely fetch bloud where it entereth in) except that round about the wound the fleſh bee ſpeedily cut away in great quan-

argueth alſo that *Matthiolus* hath vnproperly called it *Pſeudoaconitum*, that is, falſe or baſtard Aconite; for without queſtion there is no worſe or more ſpeedie venome in the world, nor no Aconite or toxicall plant comparable hereunto. And yet let vs conſider the fatherly care and prouidence of God, who hath prouided a conquerour and triumpher ouer this plant ſo venomous, namely his *Antigoniſt*, *Antithora*, or to ſpeake in ſhorter and fewer ſyllables, *Anthora*, which is the very antidote or remedie againſt this kinde of Aconite. The ſtalke of this plant is ſmall and ruſhie, very ſmooth, two or three handfulls high: whereupon do grow two, three, or foure leaues, ſeldome more, which be ſomething hard, round, ſmooth, of a light greene colour tending to blewneſſe, like the colour of the leaues of Woad, nicked in the edges. The floures grow at the top of the ſtalkes, of a yellow colour, leſſer than thoſe of the field Crowfoot, otherwiſe alike: in the place therof groweth a knop or round head, wherein is the ſeed: the root conſiſteth of nine or ten ſlender clogs, with ſome ſmall fibers alſo, and they are faſtened together with little ſtrings vnto one head, like thoſe of the white Aſphodill.

2 Wolfes-bane of the mount Baldus hath one ſtalke, ſmooth and plaine, in the middle whereof come forth two leaues and no more, wherein it differeth from the other of the Valdens, hauing likewiſe three or foure ſharpe pointed leaues, narrow and ſomewhat iagged at the place where the ſtalke diuideth it ſelfe into ſmaller branches; whereon do grow ſmall yellow floures like the precedent, but much leſſer.

¶ *The Place.*

Theſe venomous plants doe grow on the Alpes, and the mountaines of Sauoy and Switzerland: the firſt grow plentifully in the countrey of the Valdens, who inhabite part of thoſe mountaines towards Italie. The other is found on Baldus, a mountaine of Italy. They are ſtrangers in England.

¶ *The Time.*

They floure in March and Aprill, their ſeed is ripe in Iune.

¶ *The Names.*

This kinde of Aconite or Wolfes-bane is called *Thora*, *Taura*, and *Tura*, it is ſurnamed *Valdenſis*, that it may differ from *Napellus*, or Monkes hood, which is likewiſe named *Thora*.

Auicen maketh mention of a certaine deadly herbe in his fourth booke, ſixt Fen. called *Farſiun*; it is hard to affirme this ſame to be *Thora Valdenſis*.

‡ *Geſner* iudges this to be the *Aconitum pardalianches* of *Dioſcorides*, and herein is followed by *Bauhine*. ‡

¶ *The Temperature and Vertues.*

The force of theſe Wolfes-banes, is moſt pernicious and poiſonſome, and (as it is reported) exceedeth the malice of *Napellus*, or any of the other Wolfes-banes, as we haue ſaid. A

They ſay that it is of ſuch force, that if a man eſpecially, and then next any foure footed beaſt be wounded with an arrow or other inſtrument dipped in the iuice hereof, they die within halfe an houre after remedileſſe. B

† There were formerly foure figures in this chapter, with as many deſcriptions, though the plants figured and deſcribed were but two, to which number they are now reduced. The two former, which were by the names of *Pthora Valdenſis mas* and *fœmina*, thus differed, the male had only two large round leaues, and the female foure. The other two being alſo of one plant are more deeply cut in vpon the top of the leaues, which are fewer and leſſer than thoſe of the former.

CHAP. 373. *Of Winter Wolfes-bane.*

¶ *The Deſcription.*

THis kinde of Aconite is called *Aconitum hyemale Belgarum*; of *Dodonæus*, *Aconitum luteum minus*: in Engliſh, VVolfes-bane, or ſmal yellow wolfes-bane, whoſe leaues come forth of the ground in the dead time of winter, many times bearing the ſnow vpon their heads of his leaues and floures, yea the colder the weather is, and the deeper that the ſnow is, the fairer and larger is the floure; and the warmer that the weather is, the leſſer is the floure, and worſe coloured: theſe leaues I ſay come forth of the ground immediatly from the root, with a naked, ſoft, and ſlender ſtem, deeply cut or iagged on the leaues, of an exceeding faire greene colour, in the midſt of which commeth forth a yellow floure, in ſhew or faſhion like vnto the common field Crow-foot: after which follow ſundry cods full of browne ſeeds like the other kindes of Aconites: the root is thicke, tuberous, and knottie, like to the kindes of Anemone.

Aconitum hyemale.
Winter Woolfes-bane.

A

¶ *The Place.*

It groweth vpon the mountaines of Germany: we haue great quantitie of it in our London gardens.

¶ *The Time.*

It floureth in Ianuarie; the seed is ripe in the end of March.

¶ *The Names.*

It is called *Aconitum hyemale,* or *Hibernum*, or winter Aconite: that it is a kinde of Aconite or Woolfs-bane, both the form of the leaues and cods, and also the dangerous faculties of the herbe it selfe do declare.

It is much like to *Aconitum Theophrasti*: which he describeth in his ninth booke, saying, it is a short herbe hauing no περιττὸν, or superfluous thing growing on it, and is without branches as this plant is: the root, saith he, is like to καρύῳ, or to a nut, or els to καρυκῆ, a dry fig, onely the leafe seemeth to make against it, which is nothing at all like to that of Succorie, which he compareth it vnto.

¶ *The Temperature and Vertues.*

This herbe is counted to be very dangerous and deadly, hot and drie in the fourth degree, as *Theoph.* in plaine words doth testifie concerning his owne Aconite; for which he saith that there was neuer found his Antidote or remedie: whereof *Athenæus* and *Theopompus* write, that this plant is the most poisonous herb of all others, which moued *Ouid* to say *Quæ quia nascuntur dura viuacia caute:* notwithstanding it is not without his peculiar vertues. *Ioachimus Camerarius* now liuing in Noremberg saith, the water dropped into the eies ceaseth the pain and burning: it is reported to preuaile mightily against the bitings of scorpions, and is of such force, that if the scorpion passe by where it groweth and touch the same, presently he becommeth dull, heauy, and sencelesse, and if the same scorpion by chance touch the white Hellebor, he is presently deliuered from his drowsinesse.

Chap. 374. *Of Mithridate Woolfes-bane.*

¶ *The Description.*

This plant called *Anthora*, being the antidote against the poison of *Thora*, *Aconite* or wolfes bane, hath slender hollow stalkes, very brittle, a cubit high, garnished with fine cut or iagged leaues, very like to *Nigella Romana*, or the common Larkes spurre, called *Consolida Regalis:* at the top of the stalkes doe grow faire flowers, fashioned like a little helmet, of an ouerworne yellow colour; after which come small blackish cods, wherein is conteined blacke shining seed like those of Onions: the root consisteth of diuers knobs or tuberous lumpes, of the bignesse of a mans thumbe.

¶ *The Place.*

This plant which in Greeke we may terme Αντιθορα: groweth abundantly in the Alps, called *Rhetici*, in Sauoy, and in Liguria. The Ligurians of Turin, and those that dwell neere the lake Lemane, haue found this herbe to be a present remedy against the deadly poison of the herb *Thora* and the rest of the Aconites, prouided that when it is brought into the garden there to be kept for Physicks vse, it must not be planted neere to any of the Aconites: for through his attractiue qualitie, it will draw

Anthora sive Aconitum salutiferum.
Wholsome Wolfes-bane.

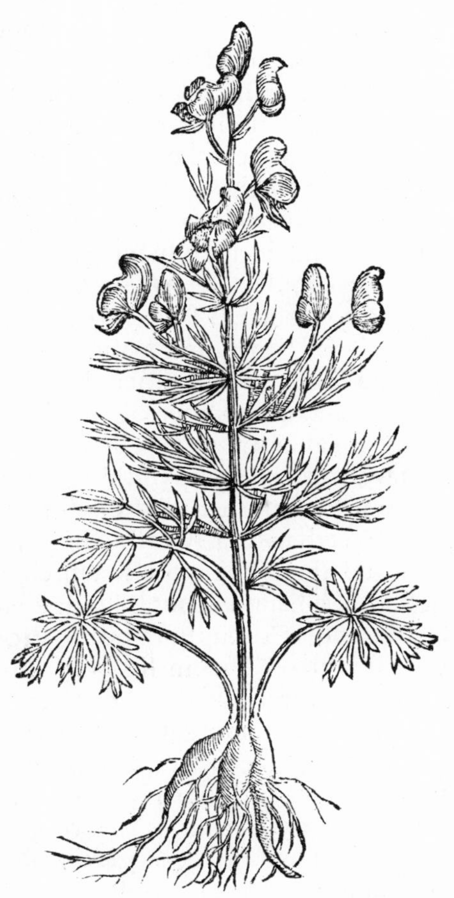

draw vnto it self the maligne and venomous poison of the Aconite, whereby it will become of the like qualitie, that is, to become poisonous likewise: but being kept far off, it retaineth his owne naturall qualitie still.

¶ *The Time.*

It floureth in August, the seed is ripe in the end of September.

¶ *The Names.*

The inhabitants of the lake of Geneua, & the Piemontoise do call it *Anthora*, and the common people *Anthoro*. *Auicen* calleth a certaine herbe which is like to Monks hood, as a remedy against the poison thereof, by the name of *Napellus Moysis*, in the 500 chap. of his second booke, and in the 745. chap. he saith, that *Zedoaria* doth grow with *Napellus* or Monkes hood, and that by reason of the neerenesse of the same, the force and strength thereof is dulled and made weaker, and that it is a treacle, that is, a counterpoison against the Viper, Monks hood, & all other poisons: and hereupon it followeth, that it is not onely *Napellus Moysis*, but allso *Zedoaria Auicennæ*: notwithstanding the Apothecaries do sell another *Zedoaria* differing from *Anthora*, which is a root of a longer forme, which not without cause is thought to be *Auicens* and *Serapio's Zerumbeth*, or *Zurumbeth*.

It is called *Anthora*, as though they should say *Antithora*, because it is an enemie to *Thora*, and a counterpoison to the same. *Thora* and *Anthora*, or *Tura* and *Antura*, seeme to be new words, but yet they are vsed in *Marcellus Empericus*, an old writer, who teacheth a medicine to be made of *Tura* and *Antura*, against the pin and web in the eies: in English, yellow Monks-hood, yellow Helmet floure, and Aconites Mithridate.

¶ *The Vertues.*

The root of *Anthora* is wonderfull bitter, it is an enemie to all poisons: it is good for purgations; for it voideth by the stoole both waterie and slimie humours, killeth and driueth forth all manner of wormes of the belly. A

Hugo Solerius saith, that the roots of *Anthora* do largely purge, not onely by the stoole, but also by vomite: and that the measure thereof is taken to the quantitie of *Faselus* (which is commonly called a beane) in broth or wine, and is giuen to strong bodies. B

Antonius Guanerius doth shew in his treatie of the plague, the second difference, the third chapter, that *Anthora* is of great force, yea and that against the plague: and the root is of like vertues, giuen with Dittanie, which I haue seene, saith he, by experience: and he further saith, it is an herbe that groweth hard by that herbe *Thora*, of which there is made a poison, wherewith they of Sauoy and those parts adiacent do enuenome their arrowes, the more speedily to kill the wilde Goats, and other wilde beasts of the Alpish mountaines. And this root *Anthora* is the *Bezoar* or counterpoison to that *Thora*, which is of so great a venome, as that it killeth all liuing creatures with his poisonsome qualitie: and thus much *Guanerius*. C

Simon Ianuensis hath also made mention of *Anthora*, and *Arnoldus Villanovanus* in his treatie of poisons: but their writings do declare that they did not well know *Anthora*. D

CHAP. 375. *Of yellow Wolfes-bane.*

¶ *The Description.*

THe yellow kinde of Wolfes-bane called *Aconitum luteum Ponticum*, or according to *Dodonæus Aconitum Lycoctonon luteum maius*: in English, yellow Wolfes-bane, whereof this our age hath found out sundry sorts not knowne to *Dioscorides*, although some of the sorts seeme to stand

Aconitum luteum Ponticum.
Yellow Woolfes bane.

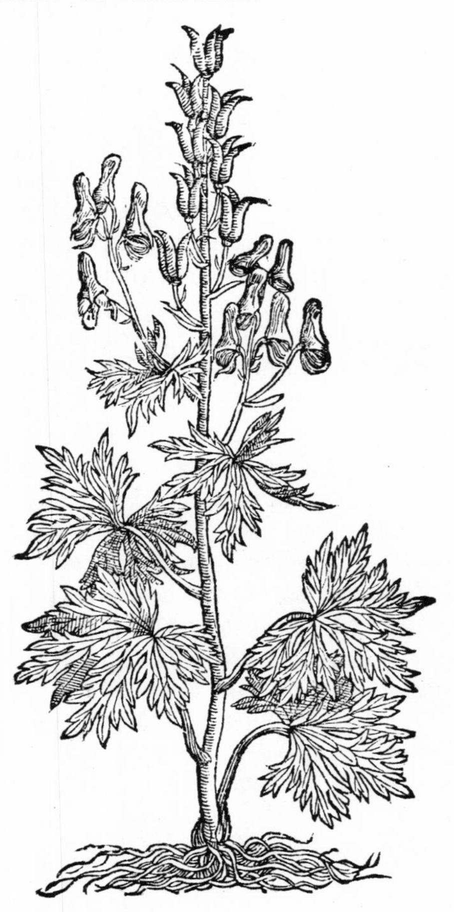

indifferent betweene the kindes of *Ranunculus, Helleborus*, and *Napellus*:) this yellow kinde I say hath large shining greene leaues fashioned like a vine, and of the same bignesse, deepely indented or cut, not much vnlike the leaues of *Geranium Fuscum*, or blacke Cranes-bill: the stalkes are bare or naked, not bearing his leaues vpon the same stalkes, one opposite against another, as in the other of his kinde: his stalkes grow vp to the height of three cubits, bearing very fine yellow floures, fantastically fashioned, and in such manner shaped, that I can very hardly describe them vnto you. They are somewhat like vnto the helmet Monkes hood, open and hollow at one end, firme and shut vp at the other: his roots are many, compact of a number of threddy or blacke strings, of an ouerworne yellow colour, spreading far abroad euery way, folding themselues one within another very confusedly. This plant groweth naturally in the darke hillie forrests, and shadowie woods, which are not trauelled nor haunted, but by wilde and sauage beasts, and is thought to be the strongest and next vnto *Thora* in his poisoning qualitie, of all the rest of the Aconites, or Woolfes banes; insomuch that if a few of the floures be chewed in the mouth, and spit forth againe presently, yet forthwith it burneth the jawes and tongue, causing them to swell, and making a certaine swimming or giddinesse in the head. This calleth to my remembrance an history of a certaine Gentleman dwelling in Lincolneshire, called *Mahewe*, the true report whereof my very good friend Mr. *Nicholas Belson*, sometimes fellow of Kings Colledge in Cambridge, hath deliuered vnto me: Mr. *Mahewe* dwelling in Boston, a student in Physicke, hauing occasion to ride through the Fennes of Lincolneshire, found a root that the hogs had turned vp, which seemed vnto him very strange and vnknowne, for that it was in the spring before the leaues were out,: this he tasted, and it so inflamed his mouth, tongue, and lips, that it caused them to swell very extreamely, so that before he could get to the towne of Boston he could not speake, and no doubt had lost his life if that the Lord God had not blessed those good remedies which presently he procured and vsed. I haue here thought good to expresse this history, for two especiall causes; the first is, that some industrious and diligent obseruer of nature may be prouoked to seeke forth that venemous plant, or some of his kindes: for I am certainely perswaded that it is either the *Thora Valdensium*, or *Aconitum luteum*, whereof this gentleman tasted, which two plants haue not at any time been thought to grow naturally in England: the other cause is, for that I would warne others to beware by that gentlemans harme. ‡ I am of opinion that this root which Mr. *Mahewe* tasted was of the *Ranunculus flammeus maior*, described in the first place of the 370. chapter aforegoing; for that growes plentifully in such places, and is of a very hot taste and hurtfull qualitie. ‡

¶ *The Place.*

The yellow Woolfes bane groweth in my garden, but not wilde in England, or in any other of these Northerly regions.

¶ *The Time.*

It floureth in the end of Iune, somewhat after the other Aconites.

¶ *The Names.*

This yellow Woolfes-bane is called of *Lobel*, *Aconitum luteum Ponticum*, or Ponticke Woolfesbane. There is mention made in *Dioscorides* his copies of three Woolfes-banes, of which the hunters vse one, and Physitions the other two. *Marcellus Virgilius* holdeth opinion that the vse of this plant is vtterly to be refused in medicine.

¶ *The Temperature and Vertues.*

A The facultie of this Aconite, as also of the other Woolfes-banes, is deadly to man, and likewise to all other liuing creatures.

It

It is vſed among the hunters which ſeek after wolues, the iuyce whereof they put into raw fleſh, which the wolues deuoure, and are killed. B

CHAP. 376.
Of other Wolfes-banes and Monkes-hoods.

¶ *The Deſcription.*

1 THis kinde of Wolfes-bane (called *Aconitum Lycoctonum*: and of *Dodonæus*, *Aconitum Lycoctonon flore Delphinij*, by reaſon of the ſhape and likenes that the floure hath with *Delphinium*, or Larkes-ſpur: and in Engliſh it is called blacke Wolfes-bane) hath many large leaues of a very deepe greene or ouerworne colour, very deepely cut or iagged: among which riſeth vp a ſtalke two cubits high; whereupon do grow floures faſhioned like a hood, of a very ill fauoured blewiſh colour, and the thrums or threds within the hood are blacke: the ſeed is alſo blacke and three cornered, growing in ſmall husks: the root is thicke and knobby.

† 1 *Aconitum lycoctonon flore Delphinij.*
Larks-heele Wolfes-bane.

† 2 *Aconitum lycoctonon cæruleum parvum.*
Small blew Wolfes-bane.

2 This kinde of Wolfes-bane, called *Lycoctonon cæruleum parvum, facie Napelli*: in Engliſh ſmall Wolfes-bane, or round Wolfes-bane, hath many ſlender brittle ſtalkes two cubits high, beſet with leaues very much iagged, and like vnto *Napellus*, called in Engliſh, Helmet-floure. The floures do grow at the top of the ſtalkes, of a blewiſh colour, faſhioned alſo like a hood, but wider open than any of the reſt: the cods and ſeed are like vnto the other: the root is round and ſmall, faſhioned like a Peare or ſmall Rape or Turnep: which moued the Germanes to call the ſame **Rapen-bloemen**, which is in Latine, *Flos rapaceus*: in Engliſh, Rape-floure.

3 This kinde of Wolfes-bane, called *Napellus verus*, in Engliſh, Helmet-floure, or the great Monkes-hood, beareth very faire and goodly blew floures in ſhape like an helmet; which are ſo beautifull, that a man would thinke they were of ſome excellent vertue, but *non eſt ſemper fides habenda fronti*. This plant is vniuerſally knowne in our London gardens and elſewhere; but naturally it

it groweth in the mountaines of Rhetia, and in sundry places of the Alps, where you shall find the grasse that groweth round it eaten vp with cattell, but no part of the herbe it selfe touched, except by certain flies, who in such aboundant measure swarme about the same that they couer the whole plant: and (which is very strange) although these flies do with great delight feed hereupon, yet of them there is confected an Antidote or most auailable medicine against the deadly bite of the spider called *Tarantala*, or any other venomous beast whatsoeuer; yea, an excellent remedie not onely against the Aconites, but all other poysons whatsoeuer. The medicine of the foresaid flies is thus made: Take of the flies which haue fed themselues as is aboue mentioned, in number twentie, of *Aristolochia rotunda*, and bole Armoniack, of each a dram.

4 There is a kinde of Wolfes-bane which *Dodonæus* reports he found in an old written Greeke booke in the Emperors Librarie at Vienna, vnder the the title of *Aconitum lycoctonum*, that answereth in all points vnto *Dioscorides* his description, except in the leaues. It hath leaues (saith hee) like vnto the Plane tree, but lesser, and more full of iags or diuisions; a slender stalke as Ferne, of a cubit high, bearing his seed in long cods: it hath blacke roots in shape like Creauises. Hereunto agreeth the Emperors picture in all things sauing in the leaues, which are not so large, nor so much diuided, but notched or toothed like the teeth of a saw.

3 *Napellus verus cæruleus.*
Blew Helmet-floure, or Monks-hood.

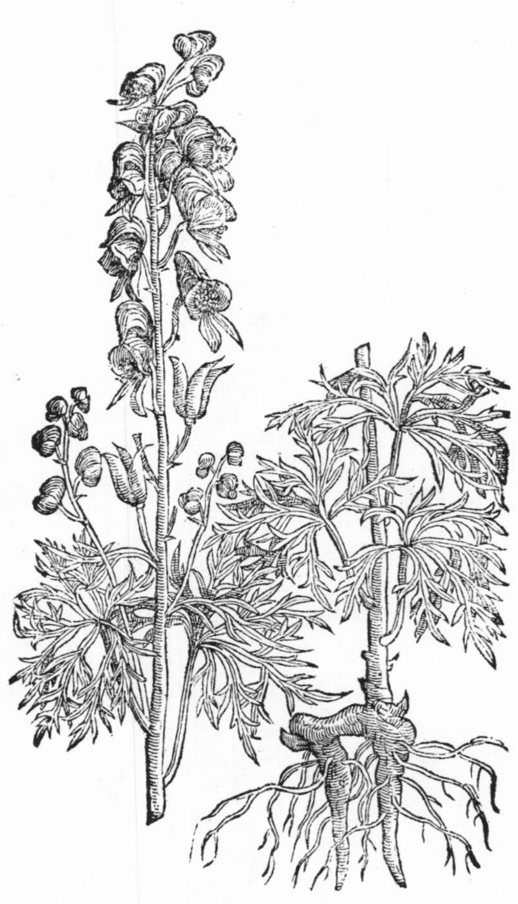

‡ 4 *Aconitum lycoctonum ex Cod. Cæsareo.*

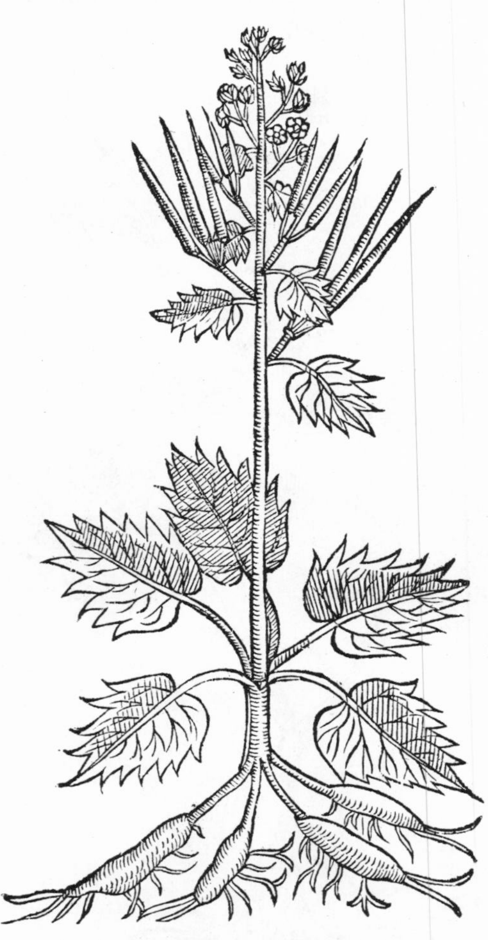

‡ 5 Besides these mentioned by our Author there are sundry other plants belonging to this pernitious Tribe, whose historie I will briefely runne ouer: The first of these is that which *Clusius* hath set forth by the name of *Aconitum lycoctonum flo. Delphinij Silesiacum*: it hath stalks some two or three cubits high, smooth and hollow, of a greenish purple colour, and couered with a certaine mealinesse: the leaues grow vpon long stalks, being rough, and fashioned like those of the yellow Wolfes bane, but of a blacker colour: the top of the stalke ends in a long spike of spurre-floures, which before they be open resemble locusts or little Lyzards, with their long and crooking tailes; but opening they shew fiue leaues, two on the sides, two below, and one aboue, which ends in a crooked taile or horne: all these leaues are wrinckled, and purple on their outsides, but smooth, and of an elegant blew within. After the floures are past succeed three square cods, as in other Aconites, wherein is contained an vnequall brownish wrinckled seed: the root is thicke, black, and tuberous. This growes naturally in some mountaines of Silesia, and floures in Iuly and August.

6 The

‡ 5 *Aconitum lycoct. hirsutum flo. Delphinij.*
Rough Larks-heele Wolfes-bane.

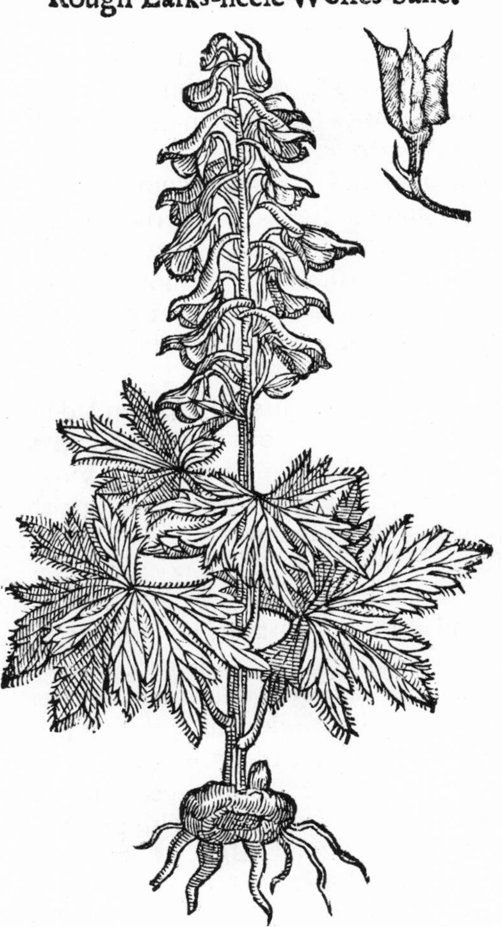

‡ 6 *Aconitum violaceum.*
Violet coloured Monks hood.

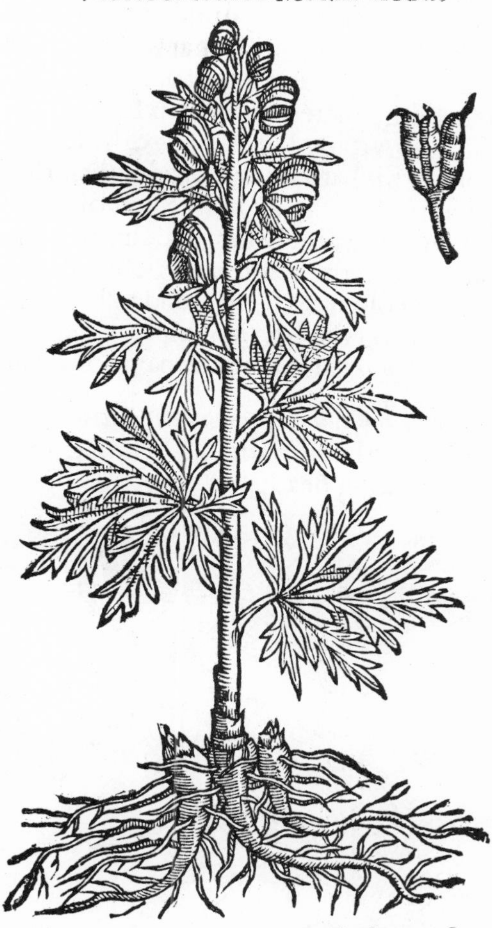

‡ 7 *Aconitum purpureum Neubergense.*
Purple Monks-hood of Newburg.

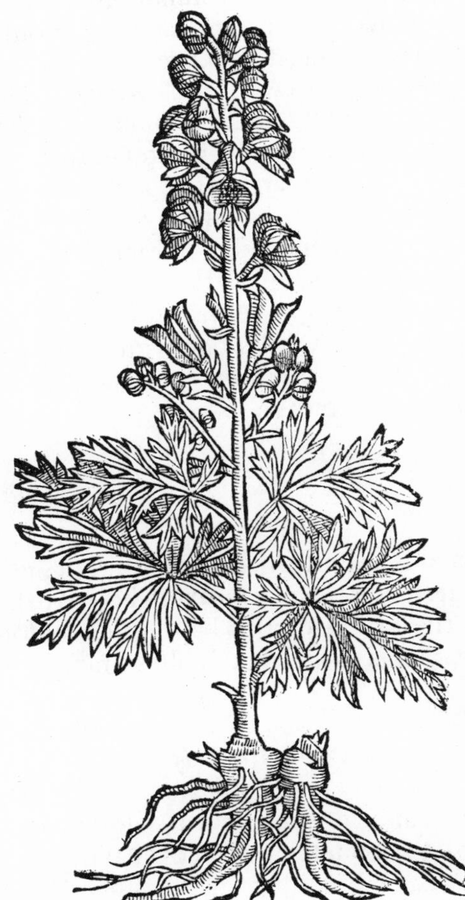

‡ 8 *Aconitum maximum Iudenbergense.*
Large floured Monks-hood.

6 The leaues of this are ſomwhat like, yet leſſe than thoſe of our common Monks-hood, blackiſh on the vpper ſide and ſhining. The ſtalke is ſome cubit and halfe high, firme, full of pith, ſmooth, and ſhining; diuided towards the top into ſome branches carrying few floures, like in forme to thoſe of the vulgar Monks-hood, of a moſt elegant and deepe violet colour: the ſeeds are like the former, and roots round, thicke, and ſhort, with many fibres. It growes vpon the hils nigh Saltsburg, where it floures in Iuly: but brought into gardens it floures ſooner than the reſt of this kinde, to wit in May. *Cluſius* calls this *Aconitum lycoctonum 4. Tauricum.*

7 This hath leaues broader than thoſe of our ordinarie Monks-hood, yet like them: the ſtalke is round, ſtraight, and firme, and of ſome three cubits height, and oft-times toward the top diuided into many branches, which carry their floures ſpike-faſhion, of a purple colour, abſolutely like thoſe of the common ſort, but that the thrummie matter in the middeſt of the floures is of a duskier colour. The root and reſt of the parts are like thoſe of the common kinde: it growes naturally vpon the Styrian Alpes, whereas it floures ſomewhat after the common kinde, to wit, in Iuly. *Cluſius* hath it by the name of *Aconitum lycoctonum 5. Neubergenſe.*

‡ 9 *Aconitum maximum nutante coma.*
Monkes-hood with the bending or nodding head.

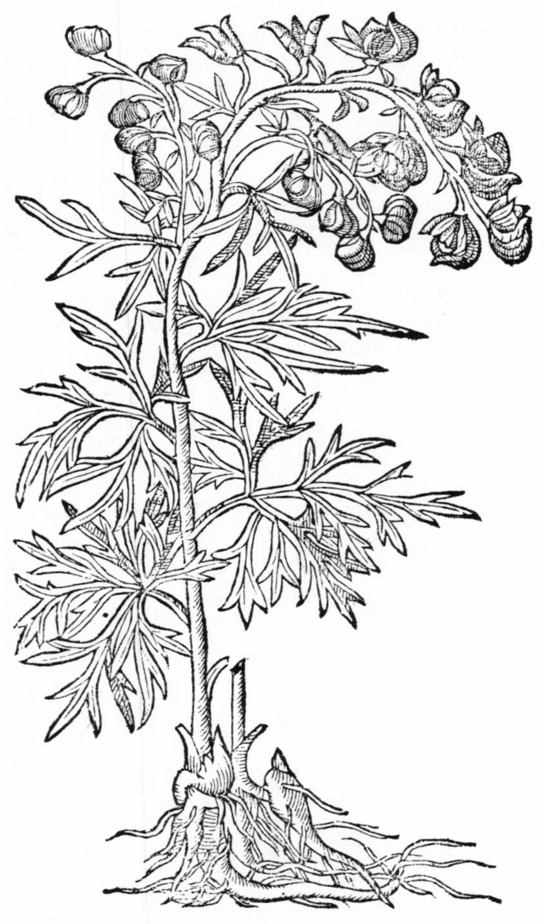

8 The leaues of this are alſo diuided into fiue parts, and ſnipt about the edges, and doe very much reſemble thoſe of the ſmal Wolfsbane deſcribed in the ſecond place, but that the leaues of that ſhine, when as theſe do not: the ſtalke is two cubits high, not very thicke, yet firme and ſtraight, of a greeniſh purple colour; and at the top carries fiue or ſix floures, the largeſt of all the Monks-hoods, conſiſting of foure leaues, as in the reſt of this kind, with a very large helmet ouer them, being ſometimes an inch long, of an elegant blewiſh purple color: the ſeed-veſſels, ſeeds, and roots are like the reſt of this kinde. This growes on Iudenberg, the higheſt hill of all Stiria, and floures in Auguſt; in gardens about the end of Iuly. *Cluſius* names it *Aconitum Lycoct. 9. Iudenbergenſe.*

9 This riſes vp to the height of three cubits, with a ſlender round ſtalke which is diuided into ſundry branches, and commonly hangs downe the head; whence *Cluſius* cals it *Aconitum lycoctonum 8. coma nutante.* The floures are like thoſe of the common Monkshood, but of ſomewhat a lighter purple colour. The leaues are larger and long, and much more cut in or diuided than any of the reſt. The roots, ſeeds, and other particles are not vnlike thoſe of the reſt of this kinde. ‡

¶ *The Place.*

Diuers of theſe Wolfs-banes grow in ſome gardens, except *Aconitum lycoctonon*, taken forth of the Emperors booke.

¶ *The Time.*

Theſe plants do floure from May vnto the end of Auguſt.

¶ *The Names.*

The firſt is *Lycoctoni ſpecies*, or a kinde of Wolfes-bane, and is as hurtfull as any of the reſt, and called of *Lobel, Aconitum flore Delphinij*, or Larke-ſpur Wolfes-bane. *Auicen* ſpeaketh hereof in his ſecond booke, and afterwards in his fourth booke, Fen. 6. the firſt Treatiſe: hauing his reaſons why and wherefore he hath ſeparated this from *Canach adip*, that is to ſay, the Wolfes ſtrangler, or the Wolfes bane.

The later and barbarous Herbariſts call the third Wolfes bane in Latine *Napellus*, of the figure and ſhape of the roots of *Napus*, or *Nauet*, or Nauew gentle: it is likewiſe *Aconiti lycoctoni ſpecies*, or a kinde of Wolfes-bane: alſo it may be called *Toxicum*; for *Toxicum* is a deadly medicine wherewith the Hunters poyſon their ſpeares, darts, and arrowes, that bring preſent death: ſo named of arrowes which the Barbarians call *Toxeumata*, and *Texa*. *Dioſcorides* ſetting downe the ſymptomes or or accidents cauſed by *Toxicum*, together with the remedies, reckoneth vp almoſt the verie ſame that *Auicen* doth concerning *Napellus*: notwithſtanding *Auicen* writes of *Napellus* and *Toxicum* ſeuerally; but not knowing what *Toxicum* is,as he himſelfe confeſſeth: ſo that it is not to be maruelled, that hauing written of *Napellus*, he ſhould afterward entreat againe of *Toxicum*.

¶ *The Nature and Vertues.*

All theſe plants are hot and dry in the fourth degree, and of a moſt venomous qualitie.

A The force and facultie of Wolfes-bane is deadly to man and all kindes of beaſts: the ſame was tried of late in Antwerpe, and is as yet freſh in memorie, by an euident experiment, but moſt lamentable; for when the leaues hereof were by certaine ignorant perſons ſerued vp in ſallads, all that did eate thereof were preſently taken with moſt cruell ſymptomes, and ſo died.

B The ſymptomes that follow thoſe that do eate of theſe deadly herbes are theſe; their lips and tongues ſwell forthwith, their eyes hang out, their thighes are ſtiffe, and their wits are taken from them, as *Auicen* writeth in his fourth booke. The force of this poyſon is ſuch, that if the points of darts or arrowes be touched with the ſame, it bringeth deadly hurt to thoſe that are wounded therewith.

C Againſt ſo deadly a poyſon *Auicen* reckoneth vp certaine remedies, which helpe after the poyſon is vomited vp; and among theſe he maketh mention of the Mouſe (as the copies euery where haue it) nouriſhed and fed vp with *Napellus*, which is altogether an enemie to the poyſonſome nature thereof, and deliuereth him that hath taken it from all perill and danger.

D *Antonius Guanerius* of Pauia, a famous phyſition in his age, in his treaty of poyſons is of opinion, that it is not a mouſe that *Auicen* ſpeakes of, but a fly: for he telleth of a certaine Philoſopher that did very carefully and diligently make ſearch after this Mouſe, and neither could find at any time either Mouſe, or the root of Wolfes-bane gnawne or bitten, as he had read; but in ſearching he found many flies feeding on the leaues, which the ſame Philoſopher tooke, and made of them an Antidote or counterpoyſon, which he found to be good and effectuall againſt other poyſons, but eſpecially the poyſon of Wolfes-bane.

E This compoſition conſiſteth of two ounces of *Terra lemnia*, as many of the berries of the Bay tree, and the like weight of Mithridate, 24 of the flies that haue taken their repaſt vpon Wolfes-bane, of honey and oyle Oliue a ſufficient quantitie.

F The ſame opinion that *Guanerius* is of, *Petrus Pena* and *Matthias de Lobel* doe alſo hold; who affirme, that there was neuer ſeene at any time any Mouſe feeding thereon, but that there be Flies which reſort vnto it by ſwarmes, and feed not onely vpon the floures, but on the herbe alſo.

¶ *The Danger.*

There hath beene little heretofore ſet downe concerning the vertues of the Aconites, but much might be ſaid of the hurts that haue come hereby, as the wofull experience of the lamentable example at Antwerpe, yet freſh in memorie, doth declare, as we haue ſaid.

† The figure that was in the firſt place formerly was of the *Aconitum luteum Ponticum*; and that in the ſecond place was of a *Napellus*.

Chap. 377. *Of blacke Hellebore.*

¶ *The Deſcription.*

1 The firſt kinde of blacke Hellebor *Dodonæus* ſetteth forth vnder this title *Veratrum nigrum*; and it may properly be called in Engliſh, blacke Hellebor, which is a name moſt fitly agreeing vnto the true and vndoubted blacke Hellebor, for the kindes and other ſorts hereof which hereafter follow are falſe and baſtard kindes thereof. This plant hath thicke and fat leaues of a deepe greene colour, the vpper part whereof is ſomewhat bluntly nicked or toothed, hauing ſundry diuiſions or cuts; in ſome leaues many, in others fewer, like vnto the female Peony, or *Smyrnium Creticum*. It beareth Roſe faſhioned floures vpon ſlender ſtems, growing immediately out of the ground an handfull high; ſomtimes very white, and oftentimes mixed with a little ſhew of purple: which being vaded, there ſucceed ſmall huskes full of blacke ſeeds: the roots are many, with long blacke ſtrings comming from one head.

2 The ſecond kinde of blacke Hellebor, called of *Pena*, *Helleboraſtrum*; and of *Dodonæus*, *Veratrum ſecundum* (in Engliſh, baſtard Hellebor) hath leaues muh like the former, but narrower and blacker, each leafe being much iagged or toothed about the edges like a ſaw. The ſtalkes grow to the height of a foot or more, diuiding themſelues into other branches toward the top; whereon do grow floures not much vnlike to the former in ſhew, ſaue that they are of a greeniſh herby colour. The roots are ſmall and threddy, but not ſo blacke as the former.

3 The

1 *Helleborus niger verus.*
The true blacke Hellebor.

2 *Helleborastrum.*
Wilde blacke Hellebor.

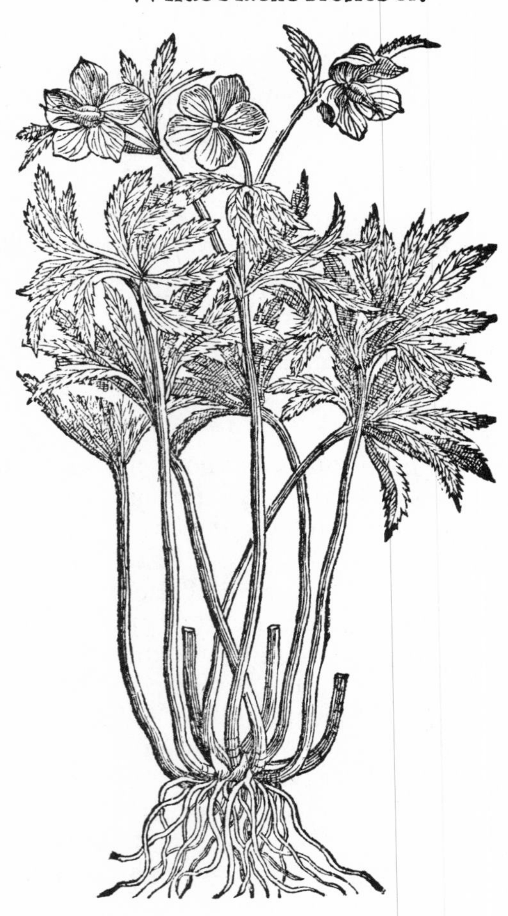

3 *Helleboraster maximus.*
The great Ox-heele.

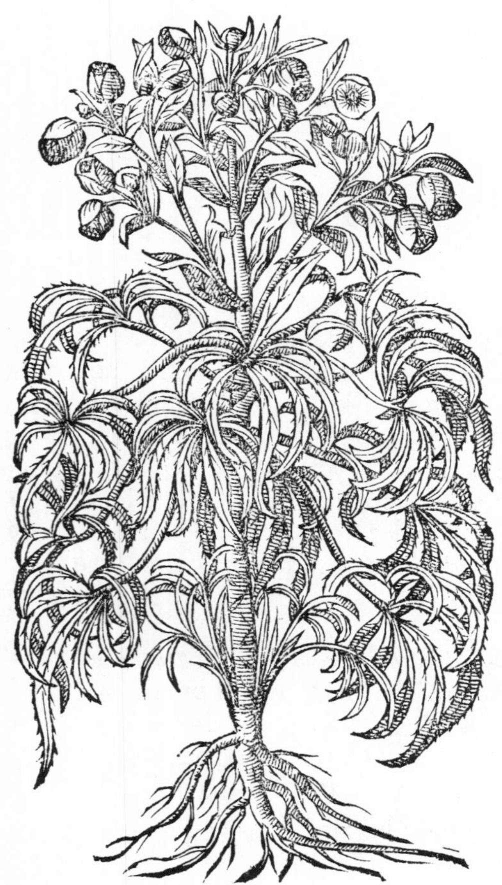

4 *Consiligo Ruellij, & Sesamoides magnum Cordi.*
Setter-wort, or Beare-foot.

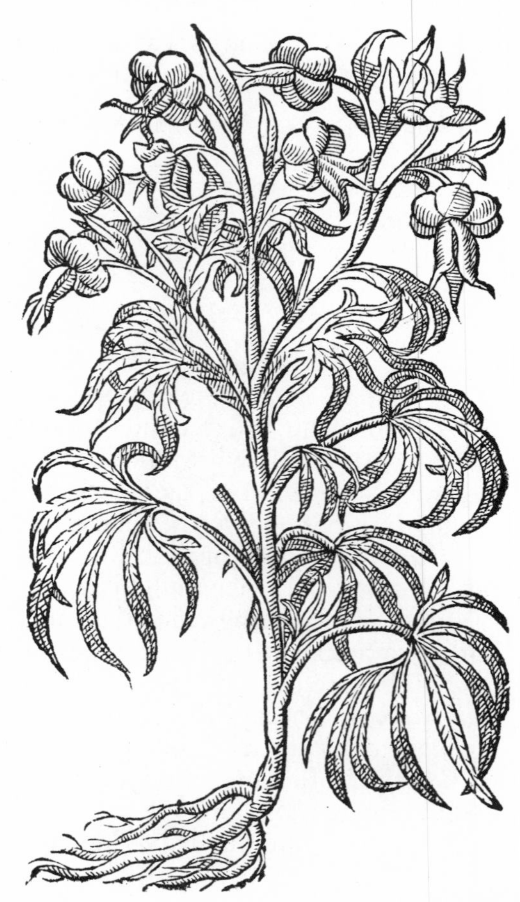

3 The third kinde of blacke Hellebor, called of *Pena*, *Helleboraster maximus*, with this addition, *flore & semine prægnans*, that is, full both of floures and seed, hath leaues somewhat like the former wilde Hellebor, saue that they be greater, more iagged, and deepely cut. The stalks grow vp to the height of two cubits, diuiding themselues at the top into sundry small branches, whereupon grow little round and bottle-like hollow greene floures; after which come forth seeds which come to perfect maturitie and ripenesse. The root consisteth of many small blacke strings, inuolued or wrapped one within another very intricately.

4 The fourth kinde of blacke Hellebor (called of *Pena* and *Lobel*, according to the description of *Cordus* and *Ruellius*, *Sesamoides magnum*, and *Consiligo*: in English, Ox-heele, or Settter-woort; which names are taken from his vertues in curing Oxen and such like cattell, as shall be shewed afterward in the names thereof) is so well knowne vnto the most sort of people by the name of Beare-foot, that I shall not haue cause to spend much time about the description. ‡ Indeed is was not much needfull for our Author to describe it, for it was the last thing he did; for both these two last are of one plant, both figures and descriptions; the former of these figures expressing it in floure, and the later in seed: but the former of our Author was with somewhat broader leaues, and the later with narrower. ‡

¶ *The Place.*

These Hellebors grow vpon rough and craggy mountains: the last growes wilde in many woods and shadowie places in England: we haue them all in our London gardens.

¶ *The Time.*

The first floureth about Christmasse, if the Winter be milde and warme: the others later:

¶ *The Names.*

It is agreed among the later writers, that these plants are *Veratra nigra*: in English, blacke Hellebor: in Greeke, ἐλλέβορος μέλας: in Italian, *Elleboro nero*: in Spanish, *Verde gambre negro*: of diuers, *Melampodium*, because it was first found by *Melampos*, who was first thought to purge therewith *Prætus* his mad daughters, and to restore them to health. *Dioscorides* writeth, that this man was a shepheard: others, a Sooth-sayer. In high Dutch it is called **Christwurtz**, that is, Christs herbe, or Christmasse herbe: in low Dutch, **Heylich Kerst cruyt**, and that because it floureth about the birth of our Lord Iesus Christ.

The third kinde was called of *Fuchsius*, *Pseudohelleborus*, and *Veratrum nigrum adulterinum*, which is in English, false or bastard blacke Hellebor. Most name it *Consiligo*, because the husbandmen of our time do herewith cure their cattell, no otherwise than the old Farriers or horse-leeches were wont to do, that is, they cut a slit or hole in the dew-lap, as they terme it (which is an emptie skin vnder the throat of the beast) wherein they put a piece of the root of Setterwort or Beare-foot, suffering it there to remaine for certaine dayes together: which manner of curing they do call Settering of their cattell, and is a manner of rowelling, as the said Horse-leeches doe their horses with horse haire twisted, or such like, and as in Surgerie we do vse with silke, which in stead of the word *Seton*, a certaine Physitian called it by the name Rowell; a word very vnproperly spoken of a learned man, because there would be some difference betwixt men and beasts. This manner of settering of cattell helpeth the disease of the lungs, the cough, and wheesing. Moreouer, in the time of pestilence or murraine, or any other diseases affecting cattell, they put the root into the place aforesaid, which draweth vnto it all the venomous matter, and voideth it forth at the wound. The which *Absyrtus* and *Hierocles* the Greeke Horse-leeches haue at large set downe. And it is called in English, Beare-foot, Setter-wort, and Setter-grasse.

The second is named in the German tongue, **Lowszkraut**, that is, *Pedicularis*, or Low sie grasse: for it is thought to destroy and kill lice, and not onely lice but sheepe and other cattell: and may be reckoned among the Beare-feet, as kindes thereof.

¶ *The Temperature.*

Blacke Hellebor, as *Galen* holdeth opinion, is hotter in taste than the white Hellebor: in like manner hot and dry in the third degree.

¶ *The Vertues.*

Black Hellebor purgeth downwards flegme, choler, and also melancholy especially, and all melancholy humors, yet not without trouble and difficultie: therfore it is not to be giuen but to robustious and strong bodies, as *Mesues* teacheth. A purgation of Hellebor is good for mad and furious men, for melancholy, dull, and heauy persons, for those that are troubled with the falling sicknes, for lepers, for them that are sicke of a quartane Ague, and briefely for all those that are troubled with blacke choler, and molested with melancholy. A

The manner of giuing it (meaning the first blacke Hellebor) saith *Actuarius* in his first booke, is three scruples, little more or lesse. B

It is giuen with wine of raisins or oxymel, but for pleasantnes sake some sweet and odoriferous seeds C seeds must be put vnto it: but if you would haue it stronger, adde thereunto a grain or two of Scamonie. Thus much *Actuarius*.

D The first of these kindes is best, then the second; the rest are of lesse force.

E The roots take away the morphew and blacke spots in the skin, tetters, ring-wormes, leprosies, and scabs.

F The root sodden in pottage with flesh, openeth the bellies of such as haue the dropsie.

G The root of bastard Hellebor, called among our English women Beare-foot, steeped in wine and drunken, looseth the belly euen as the true blacke Hellebor, and is good against all the diseases whereunto blacke Hellebor serueth, and killeth wormes in children.

H It doth his operation with more force and might, if it be made into pouder, and a dram thereof be receiued in wine.

I The same boyled in water with Rue and Agrimony, cureth the jaundice, and purgeth yellow superfluities by siege.

K The leaues of bastard Hellebor dried in an ouen, after the bread is drawne out, and the pouder thereof taken in a figge or raisin, or strawed vpon a piece of bread spred with honey and eaten, killeth wormes in children exceedingly.

Chap. 378. *Of Dioscorides his blacke Hellebor.*

Astrantia nigra, siue Veratrum nigrum Dioscoridis, Dod.
Blacke Master-worts, or *Dioscorides* his blacke Hellebor.

¶ *The Description.*

This kinde of blacke Hellebor, set forth by *Lobel* vnder the name of *Astrantia nigra*, agreeth very well in shape with the true *Astrantia*, which is called *Imperatoria*: neuertheles by the consent of *Dioscorides* and other Authors, who haue expressed this plant for a kinde of *Veratrum nigrum*, or blacke Hellebor, it hath many blackish green leaues parted or cut into foure or fiue deepe cuts, after the maner of the vine leafe very like vnto those of Sanicle, both in greennes of colour, and also in proportion. The stalke is euen, smooth, and plain: at the top wherof grow floures it little tufts or vmbels, set together like those of Scabious, of a whitish light greene colour, dashed ouer as it were with a little darke purple: after which come the seed like vnto *Carthamus* or bastard Saffron. The roots are many blackish threds knit to one head or master root.

¶ *The Place.*

Blacke Hellebor is found in the mountains of Germany, and in other vntilled and rough places: it prospereth in gardens.

Dioscorides writeth, That blacke Hellebor groweth likewise in rough and dry places: and that is the best which is taken from such like places; as that (saith hee) which is brought out of Anticyra a city in Greece. It groweth in my garden.

¶ *The Time.*

This blacke Hellebor flowreth not in Winter, but in the Sommer moneths. The herb is green all the yeare thorow.

¶ *The Names.*

It is called of the later Herbarists, *Astrantia nigra*: of others, *Sanicula fœmina*: notwithstanding it differeth much from *Astrantia*, an herbe which is also named *Imperatoria*, or Master-wort. The vulgar people call it Pellitorie of Spaine, but vntruly: it may be called blacke Master-wort, yet doubtlesse a kinde of Hellebor, as the purging facultie doth shew: for it is certaine, that diuers experienced physitians can witnesse, that the roots hereof do purge melancholy and other humors, and that they themſelues haue perfectly cured mad melancholy people being purged herewith. And that it hath a purging qualitie, *Conradus Geſnerus* doth likewiſe teſtifie in a certaine Epiſtle written to *Adolphus Occo*, in which he ſheweth, that *Aſtrantia nigra* is almoſt as ſtrong as white Hellebor, and that he himſelfe was the firſt that had experience of the purging facultie thereof by ſiege: which things confirme that it is *Dioſcorides* his blacke Hellebor.

Dioſcorides hath alſo attributed to this plant all thoſe names that are aſcribed to the other black Hellebors. He ſaith further, that the ſeed thereof in Anticyra is called *Seſamoides*, the which is vſed to purge with, if ſo be that the Text be true, and not corrupted. But it ſeemeth not to be altogether perfect; for if *Seſamoides*, as *Pliny* ſaith, and the word it ſelfe doth ſhew, hath his name of the likeneſſe of *Seſamum*, the ſeed of this blacke Hellebor ſhall vnproperly be called *Seſamoides*; being not like that of *Seſamum*, but of *Cnicus* or baſtard Saffron. By theſe proofes we may ſuſpect, that theſe words are brought into *Dioſcorides* from ſome other Author.

¶ *The Temperature and Vertues.*

A The faculties of this plant we haue already written to be by triall found like to thoſe of the other blacke Hellebor: notwithſtanding thoſe that are deſcribed in the former chapter are to be accounted of greater force.

† This whole Chapter (as moſt beſides) was out of *Dodonæus*, who, *Pempt*. 3. *lib*. 2. *cap*. 30. labours to proue this plant to be the true blacke Hellebor of *Dioſcorides*. There was alſo another deſcription thruſt by our Author into this chapter, being of the *Perſicaria ſiliquoſa* or *Noli me tangere* formerly deſcribed in the fourth place of the 114 Chap. *pag*. 446.

CHAP. 379. *Of Herbe Chriſtopher.*

Chriſtophoriana.
Herbe Chriſtopher.

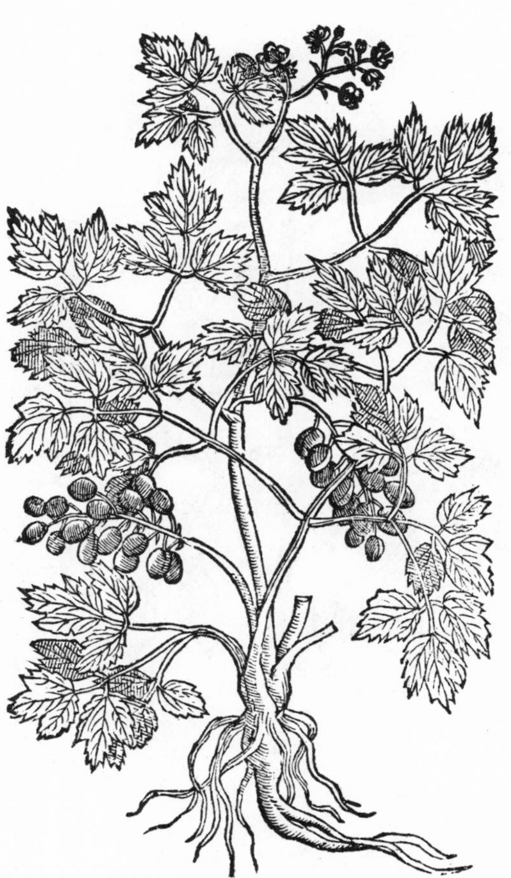

¶ *The Deſcription.*

ALthough Herbe Chriſtopher be none of the Binde-weeds, or of thoſe plants that haue need of ſupporting or vnderpropping, wherewith it may clime or rampe, yet becauſe it beareth grapes, or cluſters of berries, it might haue been numbred among the Ἄμπελοι, or thoſe that grow like Vines. It brings forth little tender ſtalkes a foot long, or not much longer; whereupon do grow ſundry leaues ſet vpon a tender foot-ſtalke, which do make one leafe ſomewhat iagged or cut about the edges, of a light greene colour: the floures grow at the top of the ſtalks, in ſpokie tufts conſiſting of four little white leaues apiece: which being paſt, the fruit ſucceeds, round, ſomwhat long, and blacke when it is ripe, hauing vpon one ſide a ſtreaked furrow or hollowneſſe growing neere together as doe the cluſters of grapes. The root is thicke, blacke without, and yellow within like Box, with many trailing ſtrings anexed therto, creeping far abroad in the earth, whereby it doth greatly increaſe, and laſteth long.

¶ *The Place.*

Herbe Chriſtopher groweth in the North parts of England, neere vnto the houſe of the right worſhipfull Sir *William Bowes*. I haue receiued plants thereof from *Robinus* of Paris, for my garden, where they flouriſh.

¶ *The Time.*

It floureth and flouriſheth in May and Iune, and the fruit is ripe in the end of Sommer.

¶ *The Names.*

It is called in our age *Christophoriana*, and *S. Christophori herba*: in English, Herbe Christopher: some there be that name it *Costus niger*: others had rather haue it *Aconitum bacciferum*: it hath no likenes at all nor affinitie with *Costus*, as the simplest may perceiue that do know both. But doubtlesse it is of the number of the Aconites, or Wolfs-banes, by reason of the deadly and pernicious qualitie that it hath, like vnto Wolfes-bane, or Leopards-bane.

¶ *The Temperature.*

The temperature of Herbe Christopher answereth those of the Aconites, as we haue said.

¶ *The Vertues.*

I finde little or nothing extant in the antient or later writers, of any one good propertie wherewith any part of this plant is possessed: therefore I wish those that loue new medicines to take heed that this be none of them, because it is thought to be of a venomous and deadly qualitie.

Chap. 380. *Of Peionie.*

¶ *The Kindes.*

There be three Peionies, one male, and two females, described by the Antients: the later writers haue found out foure more; one of the female kinde, called *Pæonia pumila*, or dwarfe Peonie; and another called *Pæonia promiscua siue neutra*, Bastard, Mis-begotten, or neither of both, but as it were a plant participating of the male and female; one double Peionie with white floures, and a fourth kinde bearing single white floures.

1 *Pæonia mas.*
Male Peionie.

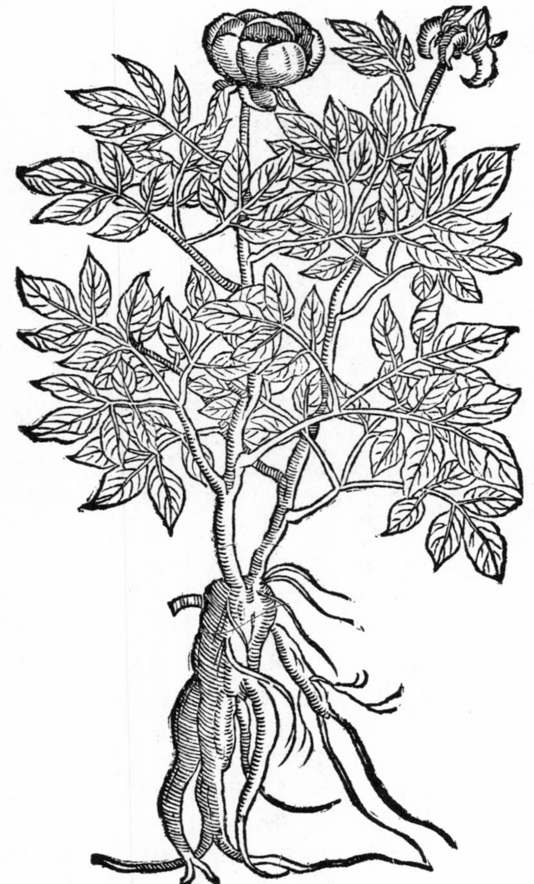

Pæonia mas cum semine.
Male Peionie in seed.

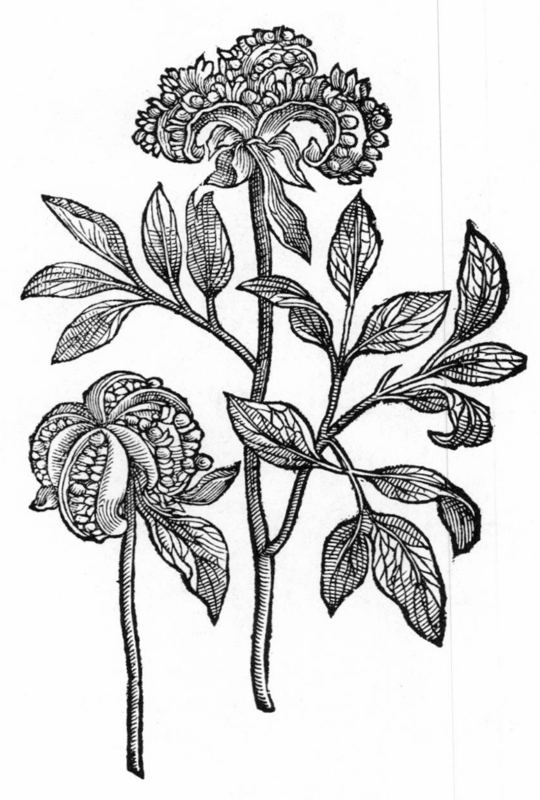

¶ *The Description.*

1 The first kinde of Peionie (being the male, called *Pæonia mas*: in English, Male Peiony) hath thicke red stalkes a cubit long: the leaues be great and large, consisting of diuers leaues growing or ioyned together vpon one slender stemme or rib, not much vnlike the leaues of the

the Wall-nut tree both in faſhion and greatneſſe: at the top of the ſtalkes grow faire large red floures very like roſes, hauing alſo in the midſt yellow threds or thrums like them in the roſe called *Anthera*: which being vaded and fallen away, there come in place three or foure great cods or huskes, which do open when they are ripe; the inner part of which cods is of a faire red colour, wherein is contained blacke ſhining and poliſhed ſeeds as big as a peaſe, and betweene euery black ſeed is couched a red or crimſon ſeed, which is barren and empty. The root is thicke, great, and tuberous, like vnto the common Peionie.

2 There is another kinde of Peionie, called of *Dodonæus, Pæonia fœmina prior*: of *Lobel, Pæonia fœmina*: in Engliſh, female Peonie, which is ſo well knowne vnto all that it needeth not any deſcription.

3 The third kinde of Peionie (which *Pena* ſetteth forth vnder the name *Pæonia fœmina polyanthos*: *Dodonæus, Pæonia fœmina multiplex*: in Engliſh, Double Peionie) hath leaues, roots, and floures like the common female Peionie, ſaue that his leaues are not ſo much iagged, and are of a lighter greene colour: the roots are thicker and more tuberous, and the floures much greater, exceeding double, of a very deep red colour, in faſhion very like the great double roſe of Prouince, but greater and more double.

2 *Pæonia fœmina.*
Female Peionie.

3 *Pæonia fœmina multiplex.*
Double red Peionie.

4 There is found another ſort of the double Peionie, not differing from the precedent in ſtalks, leaues, or roots: this plant bringeth forth white floures, wherein conſiſteth the difference.

5 There is another kinde of Peionie (called of *Dodonæus, Pæonia fœmina altera*: but of *Pena, Pæonia promiſcua, ſiue neutra*: in Engliſh, Maiden or Virgin Peiony) that is like to the common Peiony, ſauing that his leaues and floures are much leſſe, and the ſtalks ſhorter: it beareth red floures and ſeed alſo like the former.

6 We haue likewiſe in our London gardens another ſort bearing floures of a pale whitiſh colour, very ſingle, reſembling the female wilde Peiony, in other reſpects like the double white Peiony, but leſſer in all the parts thereof.

‡ 7 *Cluſius* by ſeed ſent him from Conſtantinople had two other varieties of ſingle Peionies; the one had the leaues red when they came out of the ground; and the floure of this was of a deep red colour: the other had them of a whitiſh greene, and the floures of this were ſomewhat larger, and of a lighter colour. In the leaues & other parts they reſembled the common double Peiony. ‡

4 *Pæonia fœmina polyanthos flore albo.*
The double white Peionie.

‡ 5 *Pæonia promiscua.*
Maiden Peionie.

‡ 6 *Pæonia fœmina pumila.*
Dwarfe female Peionie.

‡ 7 *Pæonia Byzantina.*
Turkish Peionie.

¶ *The Place.*

All the sorts of Peionies do grow in our London gardens, except that double Peiony with white floures, which we do expect from the Low-countries of Flanders.

The male Peionie groweth wilde vpon a cony berry in Betsome, being in the parish of Southfleet in Kent, two miles from Graues-end, and in the ground somtimes belonging to a farmer there called *Iohn Bradley*.

‡ I haue been told that our Author himselfe planted that Peionie there, and afterwards seemed to finde it there by accident: and I do beleeue it was so, because none before or since haue euer seen or heard of it growing wild since in any part of this Kingdome. ‡

¶ *The Time.*

They floure in May: the seed is ripe in Iuly.

¶ *The Names.*

The Peionie is called in Greeke παιωνία: in Latine also *Pæonia*, and *Dulcisida*: in shops, *Pionia*: in high Dutch, **Peonien blumen**: in low Dutch, **Mast bloemen**: in French, *Pinoine*: in Spanish, *Rosa del monte*: in English, Peionie: it hath also many bastard names, as *Rosa fatuina, Herba Casta*: of some, *Lunaris*, or *Lunaria Pæonia*: because it cureth those that haue the falling sicknesse, whom some men call *Lunaticos*, or Lunaticke. It is called *Idæus Dactylus*: which agreeth with the female Peionie; the knobbie roots of which be like to *Dactyli Idæi*, and *Dactyli Idæi* are certaine precious stones of the forme of a mans finger, growing in the Island of Candie: it is called of diuers *Aglaophotis*, or brightly shining, taking his name of the shining and glittering graines, which are of the colour of scarlet.

There be found two *Aglaophotides*, described by *Ælianus* in his 14. booke; one of the sea, in the 24. Chapter: the other of the earth, in the 27. chapter. That of the sea is a kinde of *Fucus*, or sea mosse, which groweth vpon high rocks, of the bignesse of Tamarisk, with the head of Poppy; which opening in the Sommer Solstice doth yeeld in the night time a certain fierie, and as it were sparkling brightnesse or light.

That of the earth, saith he, which by another name is called *Cynospastus*, lieth hid in the day time among other herbes, and is not knowne at all, and in the night time it is easily seene: for it shineth like a star, and glittereth with a fierie brightnesse.

And this *Aglaophotus* of the earth, or *Cynospastus*, is *Pæonia*; for *Apuleius* saith, that the seedes or graines of Peionie shine in the night time like a candle, and that plenty of it is in the night season found out and gathered by the shepheards. *Theophrastus* and *Pliny* do shew that Peionie is gathered in the night; which *Ælianus* also affirmeth concerning *Aglaophotus*.

This *Aglaophotus* of the earth, or *Cynospastus*, is called of *Iosephus* the writer of the Iewes warre, in his seuenth booke, 25. chapter, *Baaras*, of the place wherein it is found; which thing is plaine to him that conferreth those things which *Ælianus* hath written of *Aglaophotis* of the earth, or *Cynospastus*, with those which *Iosephus* hath set downe of *Baaras*: for *Ælianus* saith, that *Cynospastus* is not plucked vp without danger; and that it is reported how he that first touched it, not knowing the nature thereof, perished. Therefore a string must be fastned to it in the night, and a hungrie dog tied therto, who being allured by the smell of rosted flesh set towards him, may plucke it vp by the rootes. *Iosephus* also writeth, that *Baara* doth shine in the euening like the day star, and that they who come neere, and would plucke it vp, can hardly do it, except that either a womans vrine, or her menses be poured vpon it, and that so it may be pluckt vp at the length.

Moreouer, it is set downe by the said Author, as also by *Pliny* and *Theophrastus*, that of necessitie it must be gathered in the night; for if any man shall pluck off the fruit in the day time, being seene of the VVood pecker, he is in danger to loose his eies; and if he cut the root, it is a chance if his fundament fall not out. The like fabulous tale hath been set forth of Mandrake, the which I haue partly touched in the same chapter. But all these things be most vaine and friuolous: for the root of Peionie, as also the Mandrake, may be remoued at any time of the yeare, day or houre whatsoeuer.

But it is no maruell, that such kindes of trifles, and most superstitious and wicked ceremonies are found in the books of the most antient writers; for there were many things in their time verie vainly fegned and cogged in for ostentation sake, as by the Egyptians and other counterfeit mates, as *Pliny* doth truly testifie: an imitator of whom in times past, was one *Andreas* a Physition, who, as *Galen* saith, conueied into the art of Physick, lies and subtill delusions. For which cause *Galen* commanded his Schollers to refraine from the reading of him, and of all such like lying and deceitfull sycophants. It is reported that these herbes tooke the name of Peionie, or *Pæan*, of that excellent Physition of the same name, who first found out and taught the knowledge of this herbe vnto posteritie.

¶ The

¶ *The Temperature.*

The root of Peionie, as *Galen* saith, doth gently binde with a kinde of sweetnesse: and hath also ioined with it a certaine bitterish sharpnesse: it is in temperature not very hot, little more than meanly hot; but it is drie, and of subtill parts.

¶ *The Vertues.*

A *Dioscorides* writeth, that the root of the Male Peionie being dried, is giuen to women that be not well clensed after their deliuerie, being drunke in Mead or honied water to the quantitie of a bean; for it scowreth those plants, appeaseth the griping throwes and torments of the belly, and bringeth downe the desired sicknesse.

B *Galen* addeth, that it is good for those that haue the yellow iaundise, and pain in the kidnies and bladder, it clenseth the liuer and kidnies that are stopped.

C It is found by sure and euident experience made by *Galen*, that the fresh root tied about the necks of children, is an effectuall remedie against the falling sicknesse; but vnto those that are growne vp in more yeares, the root thereof must also be ministred inwardly.

D It is also giuen, saith *Pliny*, against the disease of the minde. The root of the male Peionie is preferred in this cure.

E Ten or twelue of the red berries or seeds drunke in wine that is something harsh or sower, and red, do stay the inordinate flux, and are good for the stone in the beginning.

F The blacke graines (that is the seed) to the number of fifteene taken in wine or mead, helpes the strangling and paines of the matrix or mother, and is a speciall remedie for those that are troubled in the night with the disease called *Ephialtes* or night Mare, which is as though a heauy burthen were laid vpon them, and they oppressed therewith, as if they were ouercome by their enemies, or ouerprest with some great weight or burthen; and they are also good against melancholicke dreames.

G Syrrup made of the floures of Peionie helpeth greatly the falling sicknesse: likewise the extraction of the roots doth the same.

Chap. 381. *Of toothed Violets or Corallwoorts.*

1 *Dentaria Bulbifera.*
Toothed Violet.

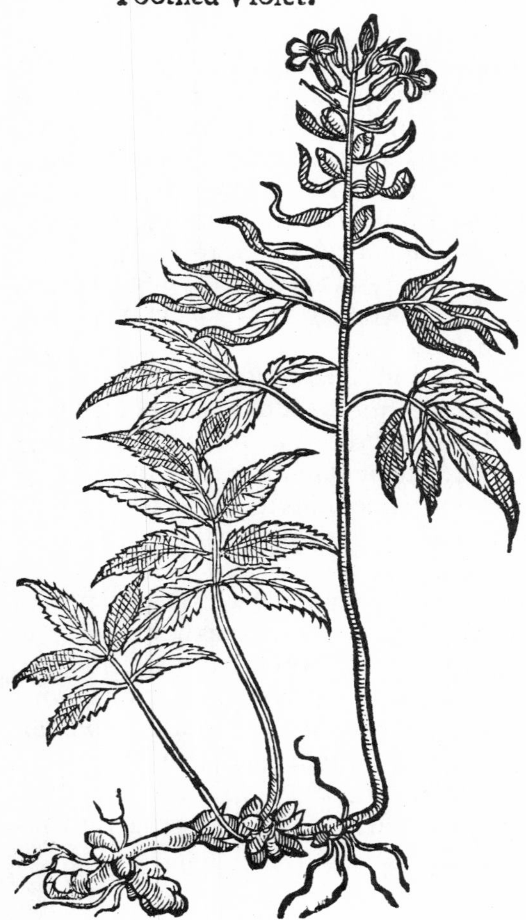

2 *Dentaria Coralloide radice, siue Dent. Enneaphyllos.*
The Corall toothed Violet.

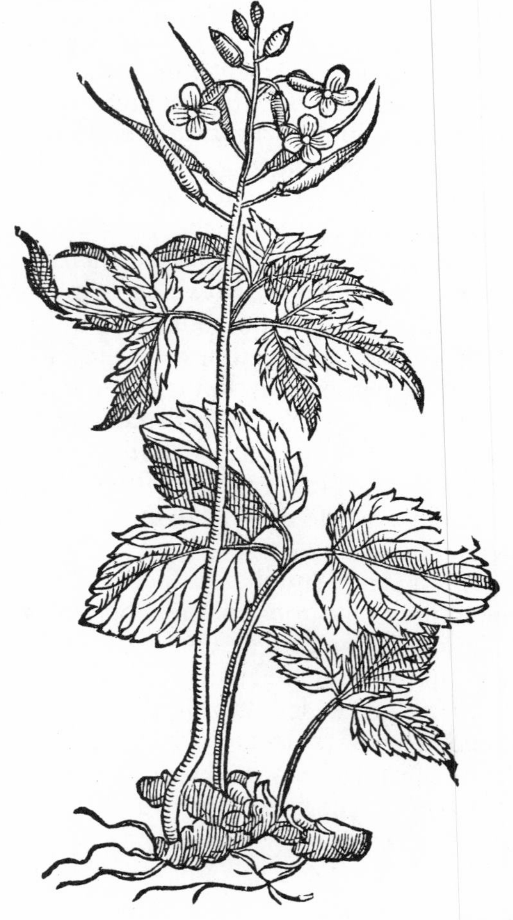

¶ *The*

3 *Dentaria Heptaphyllos Clusij.*
The seuen leafed toothed Violet.

4 *Dentaria Pentaphyllos Clusij.*
Fiue leafed toothed Violet.

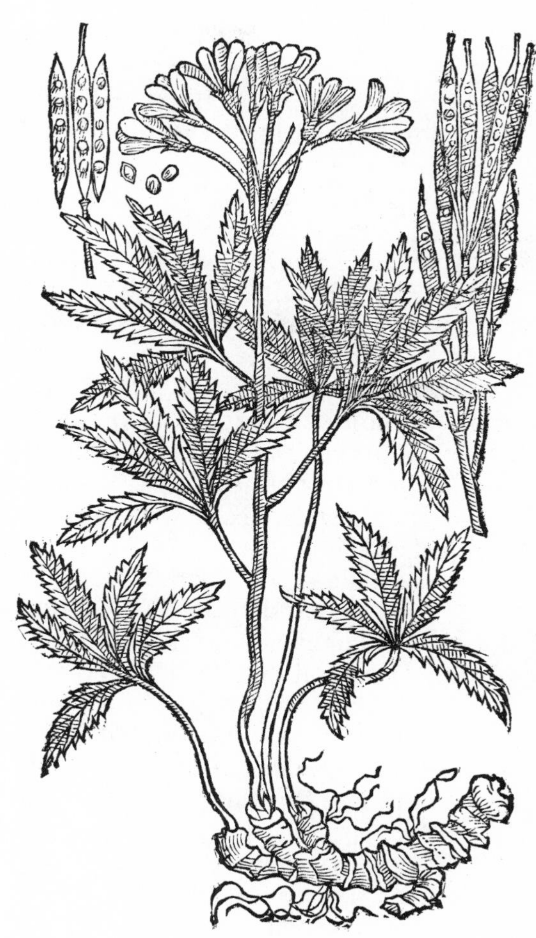

‡ 5 *Dentaria Pentaphyllos altera.*
The other fiue leaued Corall-wort.

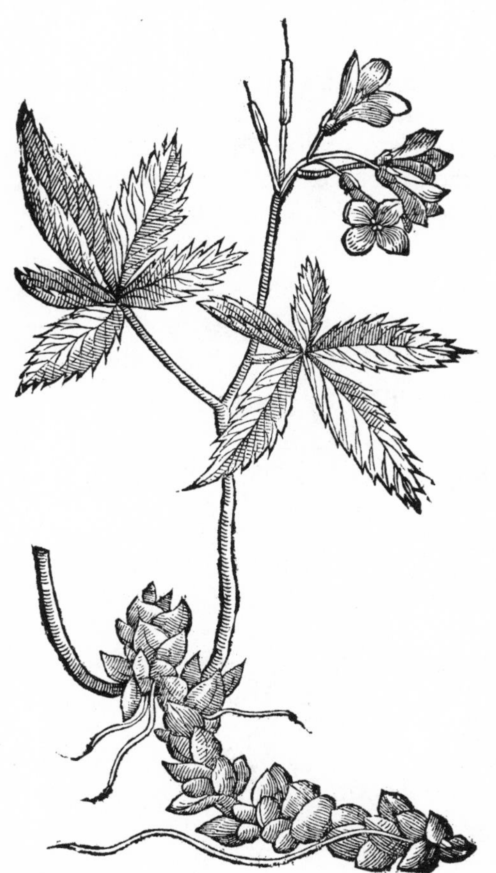

¶ *The Description.*

1 THe first kinde of *Dentaria* (called in Latine *Dentaria baccifera*: of *Dodonæus*, *Dentaria prior*: in English, Dogs tooth violet) hath a tuberous and knobbie root, toothed, or as it were kneed like vnto the crags of Corall, of an vnpleasant sauor, and somewhat sharp in taste: from which spring forth certaine small and slender stalkes a foot high, which haue leaues verie much cut or iagged, like vnto those of Hempe, of the forme and fashion of Ashen leaues: at the top of the stalkes doe grow small white floures, in shape like *Violæ matronales*, that is, Queenes Gillofloures, or rather like stocke-Gillofloures, of a white yellow colour, laid ouer with a light sprinkling of purple: among which come forth small knobs growing vpon the stalks among the leaues, such as are to be seen vpon the *Chimists Martagon*, which being ripe, do fall vpon the ground, whereof many other plants are ingendred.

2 The second kinde of Dogs-tooth violet bringeth forth small round stalks, firm and stiffe, a foot high, beset with leaues much broader, rounder, and greener than the former, bearing at the top many little floures consisting of foure small leaues, of a pale herbie colour; which beeing past, there succeed long and slender coddes somewhat

somewhat like the cods of Queenes Gillofloures, wherein is contained small blackish seed: the root is like the former, but not in euery respect much resembling Corall, yet white and tuberous notwithstanding.

3 The third kinde of Dogs-tooth Violet is called of *Clusius*, *Dentaria heptaphyllos*, that is, consisting of seuen leaues fastened vpon one rib, sinew, or small stem: of *Lobel* with this title, *Alabastrites altera*, or *Dentaria altera*: but *Cordus* calleth it *Coralloides altera*: in English, Corall violet; it hath stalkes, floures, and roots like vnto the first of his kinde, sauing that the floures are much fairer, and white of colour, and the roots haue a greater resemblance of Corall than the other.

4 The fourth kinde of Dogs-tooth violet, called in English Codded violet (which *Clusius* setteth forth vnder the title *Dentaria Matthioli Pentaphyllos*; which *Pena* doth also expresse vnder the title of *Nemoralis alpina Herbariorum Alabastrites*; *Cordus* calleth it *Coralloides*, and may very well bee called in English Cinkfoile violet) hath leaues so like the greater Cinkfoile, that it is hard to know one from another; therefore it might very well haue been reckoned among the herbes called *Pentaphylla*, that is, fiue leaued herbes. This plant groweth in the shadowie forrest about Turin, and the mountain Sauena called Calcaris, and by the Rhene not far from Basill. The stalks grow to the height of a cubit, beset with a tuft of floures at the top like to that of the first, but of a deeper purple colour: which being vaded, there succeed long and flat cods like vnto Rocket, or the great Celandine, wherein is contained a small seed. All the whole plant is of a hot and bitter taste. The roots are like vnto Corall, of a pale whitish colour: the leaues are rough and harsh in handling, and of a deep greene colour.

‡ 5 *Clusius* giues vs another varietie of *Dentaria pentaphyllos*, whose roots are more vneuen and knobby than the last described: the stalke is some foot high: the leaues fiue vpon a stalke, but not so rough, nor of so deep a greene as those of the former; yet the floures are of a deep purple colour, like those of the last described. ‡

¶ *The Place.*

They grow on diuers shadowie and darke hills. *Valerius Cordus* writeth, that they are found about the forest Hercinia, not far from Northusium, most plentifully, in a fat soile that hath quaries of stone in it. The first I haue in my garden.

¶ *The Time.*

They floure especially in Aprill and May: the seed commeth to perfection in the end of August.

¶ *The Names.*

The toothed Violet, or after some, Dogs-tooth violet, is commonly called *Dentaria*: of *Cordus*, *Coralloides*, of the root that is in forme like to Corall. *Matthiolus* placeth it *inter Solidagines & Symphyta*, among the Consounds and Comfries. Wee had rather call them *Violæ Dentariæ*, of the likenesse the floures haue with Stocke-gillofloures. They may be called in English, Toothed Violets, or Corall-woorts.

¶ *The Temperature and Vertues.*

A I haue read of few or no vertues contained in these herbes, sauing those which some women haue experienced to be in the first kinde thereof, and which *Matthiolus* ascribeth vnto *Pentaphylla dentaria* the fourth kinde, in the fourth booke of his Commentaries vpon *Dioscorides*, and in the chap. concerning *Symphytum*, where he saith that the root is vsed in drinkes which are made against *Enterocele* and inward wounds, but especially those wounds and hurts which haue entred into the hollownesse of the brest.

Chap. 382. *Of Cinkefoile, or fiue finger Grasse.*

¶ *The Description.*

1 The first kinde of Cinkfoile is so common and so vniuersally knowne, that I thinke it a needlesse trauell to stand about the description. ‡ It hath many long slender stalks, lying spred vpon the ground, out whereof grow leaues made of fiue longish snipt leaues fastened to one long foot-stalke: the floures also grow vpon the like foot-stalks, and are composed of fiue yellow leaues. The root is pretty large, of a reddish colour, and round; but dried, it becomes square. ‡

2 The second kinde of Cinkfoile or Quinquefoile hath round and smal stalks of a cubit high; the leaues are large, and very much iagged about the edges, very like the common Cinkfoile: the floures grow at the top of the stalks, in fashion like the common kind, but much greater, and of a pale or bleake yellow or else whitish colour: the root is blacke without, and full of strings annexed thereto, and of a wooddie substance.

† 3 The

Quinquefolium vulgare.
Common Cinkfoile.

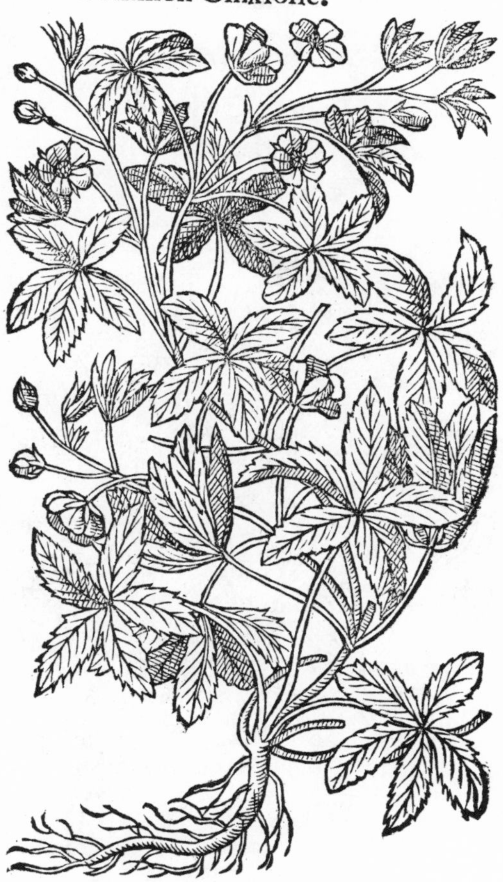

† 2 *Quinquefolium maius rectum.*
Great vpright Cinkfoile.

3 *Pentaphyllum purpureum.*
Purple Cinkfoile.

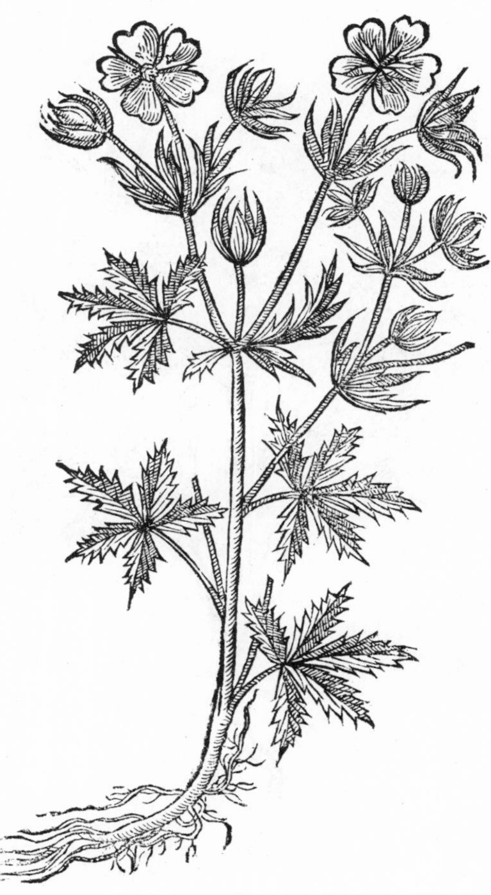

4 *Pentaphyllum rubrum palustre,*
Marsh Cinkfoile.

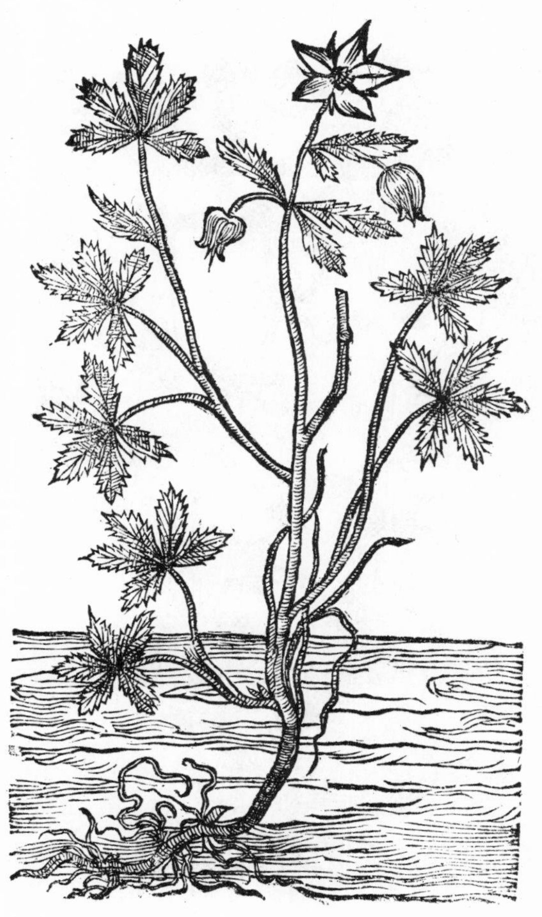

5 *Pentaphyllum petroſum, Heptaphyllum Cluſij.*
Stone Cinkfoile.

† 6 *Pentaphyllon ſupinum Potentillæ facie.*
Siluerweed Cinkfoile.

7 *Quinquefolium Tormentillæ facie.*
Wall Cinkfoile.

8 *Pentaphyllum Incanum.*
Hoarie Cinkfoile.

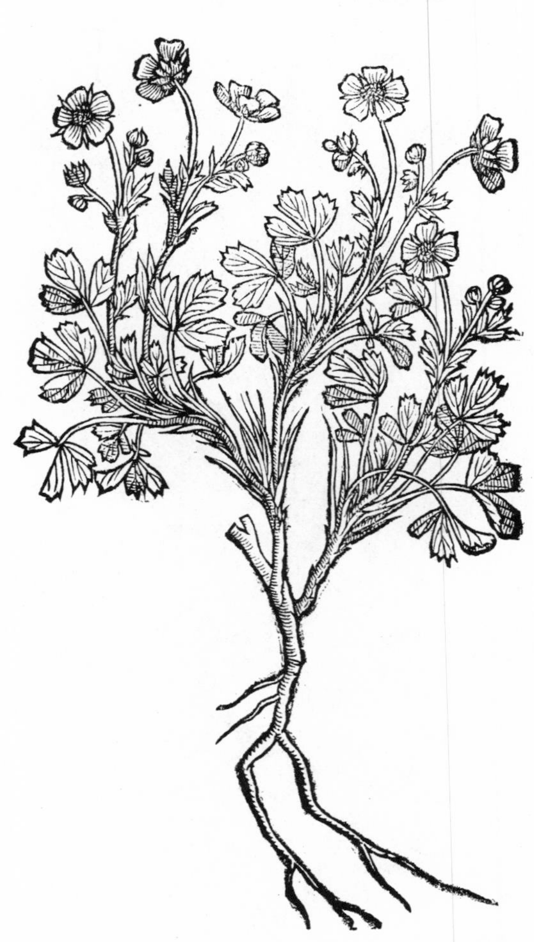

† 3 The third kinde of Cinkefoile hath leaues like thoſe of the laſt deſcribed, and his floures are of a purple colour; which being paſt, there ſucceedeth a round knop of ſeed like a Strawberry before it be ripe: the ſtalkes are creeping vpon the ground: the root is of a woody ſubſtance, full of blacke ſtrings appendant thereto. ‡ This differs not from the laſt deſcribed, but in the colour of the floures. ‡

4 The fourth kinde of Cinkefoile is very like vnto the other, eſpecially the great kinde: the ſtalkes are a cubit high, and of a reddiſh colour: the leaues conſiſt of fiue parts, ſomewhat ſnipt about the edges: the floures grow at the tops of the ſtalkes like vnto the other Cinkefoiles, ſauing that they be of a darke red colour: the root is of a wooddy ſubſtance, with ſome fibres or threddy ſtrings hanging thereat.

9 *Pentaphyllum incanum minus repens.*
Small hoary creeping Cinkefoile.

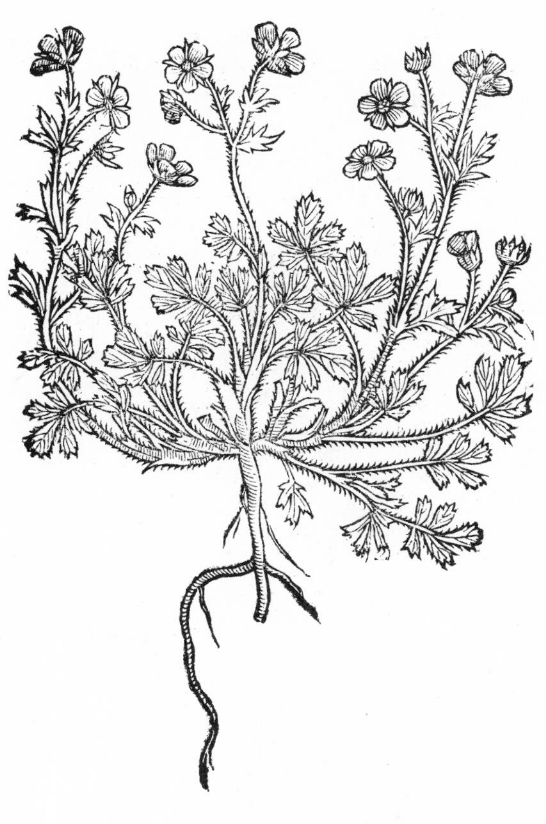

10 *Quinquefolium ſyluaticum majus flo. albo.*
Wood Cinkefoile, with white floures.

5 The fifth kinke of Cinkefoile groweth vpon the cold mountaines of Sauoy, and in the vallie of Auſtenſie, and in Narbone in France, and (if my memory faile not) I haue ſeen the ſame growing vpon Beeſton caſtle in Cheſhire: the leaues hereof are few, and thinne ſet, conſiſting of fiue parts like the other Cinkefoiles, oftentimes ſix or ſeuen ſet vpon one foot-ſtalke, not ſnipt about the edges as the other, but plaine and ſmooth; the leafe is of a ſhining white ſiluer colour, very ſoft and ſhining: the floures grow like ſtarres, vpon ſlender ſtalkes by tufts and bunches, of a white colour, and ſometimes purple, in faſhion like the floures of *Alchimilla*, or Ladies mantle: the root is thicke and full of ſtrings, and of a browne purple colour.

‡ 6 This plant, whoſe figure our Author formerly gaue for *Fragaria ſterilis*, & in his deſcription confounded with it, to auoid confuſion, I thinke fit to giue you here amongſt the Cinkefoiles, and in that place the *Fragaria ſterilis*, as moſt agreeable thereto. This ſeemes to challenge kindred of three ſeuerall plants, that is, Cinkefoile, Tormentill, and Siluer-weed, for it hath the vpper leaues, the yellow floures, creeping branches, and root of Cinkefoile, but the lower leaues are of a darke greene, and grow many vpon one middle rib like thoſe of Siluer-weed; the fruit is like an vnripe Strawberry. *Lobel* calls this *Pentaphyllum ſupinum Tormentilæ facie*: and *Tabernamontanus*, *Quinquefolium fragiferum repens.* ‡

7 The ſeuenth kinde of Cinkefoile, *Pena* that diligent ſearcher of Simples found in the Alpes of Rhetia, nere Clauena, and at the firſt ſight ſuppoſed it to be a kinde of *Tormentilla*, or *Pentaphyllum*,

taphyllum, saue that it had a more threddy root, rather like *Geranium*; it is of a darke colour outwardly, hauing some sweet smell, representing *Garyophyllata* in the sauor of his roots: in leaues and floures it resembles Cinkefoile and Tormentill, and in shape of his stalkes and roots *Auens* or *Garyophyllata*, participating of them all: notwithstanding it approcheth neerest vnto the Cinkefoiles, hauing stalkes a foot high, whereupon grow leaues diuided into fiue parts, and jagged round about the edges like the teeth of a saw, hauing the pale yellow floures of *Pentaphyllum* or *Tormentilla*; within which are little mossie or downy threddes, of the colour of saffron, but lesser than the common Auens.

8 The eighth kinde of Cinkefoile (according to the opinion of diuers learned men, who haue had the view thereof, and haue iudged it to be the true *Leucas* of *Dioscorides*, agreeable to *Dioscorides* his description) is all hoary, whereupon it tooke the addition *Incanum*. The stalkes are thicke, wooddy, and somewhat red, wrinckled also, and of a browne colour; which rise vnequall from the root, spreading themselues into many branches, shadowing the place where it groweth, beset with thicke and notched leaues like *Scordium*, or water Germander, which according to the iudgment of the learned is thought to be of no lesse force against poison than *Pentaphyllon*, or *Tormentilla*, being of an astringent and drying quality. Hereupon it may be that some trying the force hereof, haue yeelded it vp for *Leucas Dioscoridis*. This rare plant I neuer found growing naturally, but in the hollownesse of the peakish mountaines, and dry grauelly vallies.

‡ 11 *Quinquefolium syluaticum minus flo. albo.*
Small white floured wood Cinkefoile.

‡ 12 *Quinquefolium minus flo. aureo.*
Small golden floured Cinkefoile.

‡ 9 This hath the like creeping purple branches as the last described: the leaues are narrower, more hairy and deeper cut in; the floures are also of a more golden colour, in other respects they are alike. ‡

† 10 The wood Cinkefoile hath many leaues spred vpon the ground, consisting of fiue parts; among which rise vp other leaues, set vpon very tall foot-stalkes, and long in respect of those that did grow by the ground, and somewhat snipt about the ends, and not all alongst the edges. The floures grow vpon slender stalkes, consisting of fiue white leaues. The root is thicke, with diuers fibres comming from it.

‡ 11 This also from such a root as the last described sends forth many slender branches not creeping, but standing vpright, and set with little hoary leaues, snipt onely at the ends, like as

‡ 13 *Pentaphyllum fragiferum.*
Strawberry Cinkfoile.

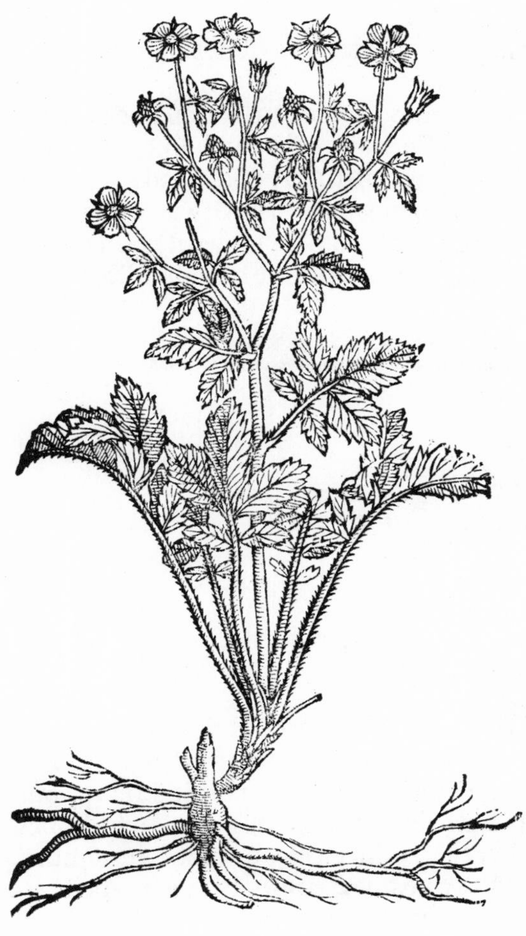

those of the last described: the tops of the branches carry pretty white floures like those of the last described, whereof it seems to be a kinde, yet lesse in each respect.

12 This from a blacke and fibrous root sends forth creeping branches, set with leaues like the common Cinkfoile, but lesse, somewhat hoary and shining; the stalks are some handfull high, and on their tops carry large floures in respect of the smalnesse of the plant, and these of a faire golden colour, with saffron coloured threds in their middle: the seedes grow after the manner of other Cinkfoiles: this floures in Iune, and it is *Clusius* his *Quinquefolium* 3. *aureo flore*. ‡

13 There is one of the mountain Cinkfoiles that hath diuers slender brittle stalks, rising immediatly out of the ground; whereupon are set by equall distances certain jagged leaues, not vnlike to the smallest leaues of Auens: the floures are white and grow at the top, hauing in them threds yellow of colour, and like to the other Cinkfoiles, but altogether lesser. The root is thicke, tough, and of a wooddie substance. ‡ The seedes grow clustering together like little Strawberries, whence *Clusius* calls it *Quinquefolium fragiferum*. ‡

¶ *The Place.*

They grow in low and moist medowes, vpon banks and by high waies sides: the second is onely to be found in gardens.

The third groweth in the woods of Sauena and Narbon, but not in England: The fourth groweth in a marsh ground adioining to the land called Bourne ponds, halfe a mile from Colchester; from whence I brought some plants for my garden, where they flourish and prosper well.

The fifth groweth vpon Beestone castle in Cheshire: the sixth vpon bricke and stone wals about London, especially vpon the bricke wall in Liuer-lane.

The place of the seuenth and eight is set forth in their descriptions.

¶ *The Time.*

These plants do floure from the beginning of May to the end of Iune.

¶ *The Names.*

Cinkfoile is called in Greeke πεντάφυλλον: in Latine, *Quinquefolium*: the Apothecaries vse the Greek name *Pentaphyllon*: and sometime the Latine name. There be very many bastard names, wherewith I will not trouble your eares: in High Dutch, **Funff fingerkraut**: in Low Dutch, **Uiiff Uinger kruijt**: in Italian, *Cinquefoglio*: in French, *Quinte fueille*: in Spanish, *Cinco en rama*: in English, Cinkfoile, Fiue finger Grasse, Fiue leaued grasse, and Sinkfield.

¶ *The Temperature.*

The roots of Cinkfoile, especially of the first, do vehemently drie, and that in the third degree, but without biting: for they haue very little apparant heat or sharpnesse.

¶ *The Vertues.*

The decoction of the roots of Cinkfoile drunke, cureth the bloudy flixe, and all other fluxes of the belly, and stancheth all excessiue bleeding. A

The iuice of the roots while they be yong and tender, is giuen to be drunke against the diseases of the liuer and lungs, and all poison. B

The same drunke in Mead or honied water, or wine wherein some pepper hath been mingled, cureth the tertian or quartaine feuers: and being drunken after the same manner for thirty daies together, it helpeth the falling sicknesse. C

The leaues vsed among herbes appropriate for the same purpose, cureth ruptures and burstings of the rim, and guts falling into the cods. D

 The

E The iuice of the leaues drunken doth cure the Iaundice, and comforteth the stomacke and liuer.

F The decoction of the roots held in the mouth doth mitigate the paine of the teeth, staieth putrifaction, and all putrified vlcers of the mouth, helpeth the inflammations of the almonds, throat, and the parts adioining, it staieth the laske, and helpeth the bloudy flix.

G The root boiled in vineger is good against the shingles, appeaseth the rage of fretting sores, and cankerous vlcers.

H It is reported, that foure branches hereof cureth quartaine agues, three tertians, and one branch quotidians: which things are most vaine and friuolous, as likewise many other such like, which are not onely found in *Dioscorides*, but also in other Authors, which we willingly withstand.

I *Ortolpho Morolto* a learned Physition, commended the leaues being boiled with water, and some *Lignum vitæ* added therto, against the falling sicknesse, if the patient be caused to sweat vpon the taking thereof. He likewise commendeth the extraction of the roots against the bloudy flix.

† Our Author formerly in his description, title, and place of growing mentioned that plant which he figured, and is yet kept in the second place; and in the first place he figured the common Cinke-foile, and made mention of it, yet without description in the second. That which formerly was in the sixth place, by the name of *Pentaphyllum lupinum*, was the same with that in the fifth place.

Chap. 383. *Of Setfoile, or Tormentill.*

Tormentilla.
Setfoile.

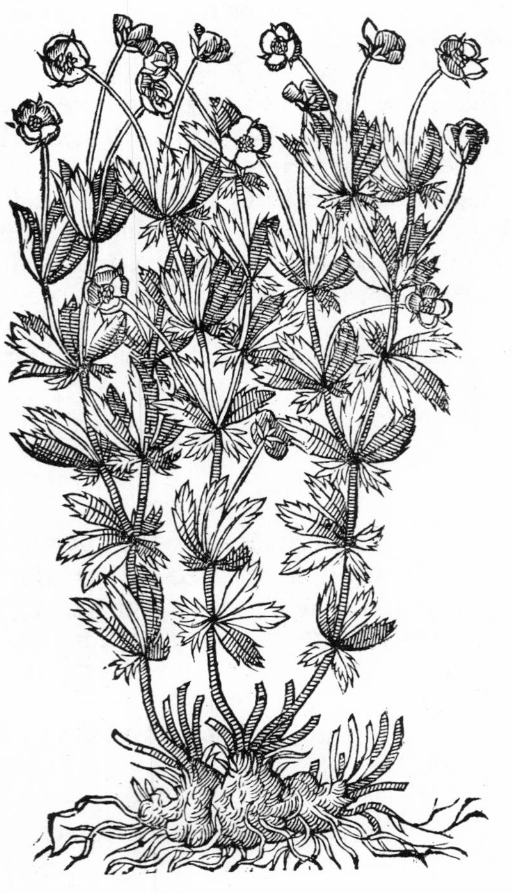

¶ *The Description.*

This herbe Tormentill or Setfoile is one of the Cinkfoiles, it brings forth many stalks slender, weake, scarse able to lift it selfe vp, but rather lieth downe vpon the ground: the leaues be lesser than Cinkefoile, but moe in number, somtimes fiue, but commonly seuen, whereupon it tooke his name Setfoile, which is seuen leaues, and those somewhat snipt about the edges: the floures grow on the toppes of slender stalkes, of a yellow colour, like those of the Cinkfoiles. The root is blacke without, reddish within, thicke, tuberous, or knobbie.

¶ *The Place.*

This plant loueth woods and shadowie places, and is likewise found in pastures lying open to the Sun, almost euery where.

¶ *The Names.*

It floureth from May, vnto the end of August.

¶ *The Names.*

It is called of the later Herbarists *Tormentilla*: some name it after the number of the leaues ἑπτάφυλλον, and *Septifolium*: in English, Setfoile and Tormentill: in high-Dutch, **Birkwurtz**: most take it to be *Chrysogonon*; whereof *Dioscorides* hath made a briefe description.

¶ *The Temperature.*

The root of Tormentill doth mightily dry, and that in the third degree, and is of thin parts: it hath in it very little heat, and is of a binding quality.

¶ *The Vertues.*

A Tormentill is not only of like vertue with Cinkefoile, but also of greater efficacie: it is much vsed against pestilent diseases: for it strongly resisteth putrifaction, and procureth sweate.

The

The leaues and roots boiled in wine, or the iuice thereof drunken prouoketh ſweat, and by that means driueth out all venome from the heart, expelleth poiſon, and preſerueth the bodie in time of peſtilence from the infection thereof, and all other infectious diſeaſes. B

The roots dried made into pouder and drunke in wine doth the ſame. C

The ſame pouder taken as aforeſaid, or in the water of a Smiths forge, or rather the water wherein hot ſteele hath been often quenched of purpoſe, cureth the laske and bloudy flix, yea although the patient haue adioined vnto his ſcouring a grieuous feuer. D

It ſtoppeth the ſpitting of bloud, piſſing of bloud, and all other iſſues of bloud, as well in men as women. E

The decoction of the leaues and rootes, or the iuice thereof drunke, is excellent good for all wounds, both outward and inward: it alſo openeth and healeth the ſtoppings of the liuer and lungs, and cureth the iaundice. F

The root beaten into pouder, tempered or kneaded with the white of an egge and eaten, ſtaieth the deſire to vomite, and is good againſt choler and melancholie. G

CHAP. 384. *Of wilde Tanſie or Siluer-weed.*

Argentina.
Siluerweed, or wilde Tanſie.

¶ *The Deſcription.*

Wilde Tanſie creepeth along vpon the ground with fine ſlender ſtalkes and claſping tendrels: the leaues are long made vp of many ſmall leaues, like vnto thoſe of the garden Tanſie, but leſſer; on the vpper ſide greene, and vnder very white. The floures be yellow, and ſtand vpon ſlender ſtems, as doe thoſe of Cinkfoile,

¶ *The Place.*

It groweth in moiſt places neere vnto high waies and running brookes euery where.

¶ *The Time.*

It floureth in Iune and Iuly.

¶ *The Names.*

The later Herbariſts do call it *Argentina*, of the ſiluer drops that are to be ſeene in the diſtilled water therof when it is put into a glaſſe, which you ſhall eaſily ſee rowling and tumbling vp and downe in the bottome; ‡ I iudge it rather ſo called of the fine ſhining Siluer coloured leaues. ‡ It is likewiſe called *Potentilla*: of diuers, *Agrimonia ſylueſtris*, *Anſerina*, and *Tanacetum ſylueſtre*: in High Dutch, **Genſerich**: in Low Dutch, **Ganſerick**: in French, *Argentine*: in Engliſh, Wilde Tanſie, and Siluerweed.

¶ *The Temperature.*

It is of temperature moderatly cold, and dry almoſt in the third degree, hauing withall a binding facultie.

¶ *The Vertues.*

Wilde Tanſie boiled in wine and drunk, ſtoppeth the laske and bloudy flix, and all other flux of bloud in man or woman. A

The ſame boiled in water and ſalt and drunke, diſſolueth clotted and congealed bloud in ſuch as are hurt or bruiſed with falling from ſome high place. B

The decoction hereof made in water, cureth the vlcers and cankers of the mouth, if ſome honie and allom be added thereto in the boiling. C

Wilde Tanſie hath many other good vertues, eſpecially againſt the ſtone, inward wounds, and wounds of the priuie or ſecret parts, and cloſeth vp all greene and freſh wounds. D

The

E The distilled water taketh away freckles, spots, pimples in the face and Sun-burning; but the herbe laid to infuse or steepe in white wine is far better: but the best of all is to steepe it in strong white wine vinegre, the face being often bathed or washed therewith.

CHAP. 385. *Of Auens, or Herbe Bennet.*

1 *Caryophyllata.*
Auens or herbe Bennet.

2 *Caryophyllata montana.*
Mountaine Auens.

¶ *The Description.*

1 THe common Auens hath leaues not vnlike to Agrimony, rough, blackish, and much clouen or deepely cut into diuers gashes: the stalke is round and hairy, a foot high, diuiding it selfe at the top into diuers branches, whereupon do grow yellow floures like those of Cinkefoile or wilde Tansie: which being past, there follow round rough reddish hairy heads or knops ful of seed, which being ripe wil hang vpon garments as the Burs doe. The root is thicke, reddish within, with certaine yellow strings fastened thereunto, smelling like vnto Cloues or like vnto the roots of Cyperus.

2 The Mountain Auens hath greater and thicker leaues than the precedent, rougher, and more hairie, not parted into three, but rather round, nicked on the edges: among which riseth vp slender stalkes, whereon doe grow little longish sharpe pointed leaues: on the toppe of each stalke doth grow

3 *Caryophyllata Alpina pentaphyllæa.*
Fiue leaued Auens.

‡ 4 *Caryophyllata montana purpurea.*
Red floured mountaine Auens.

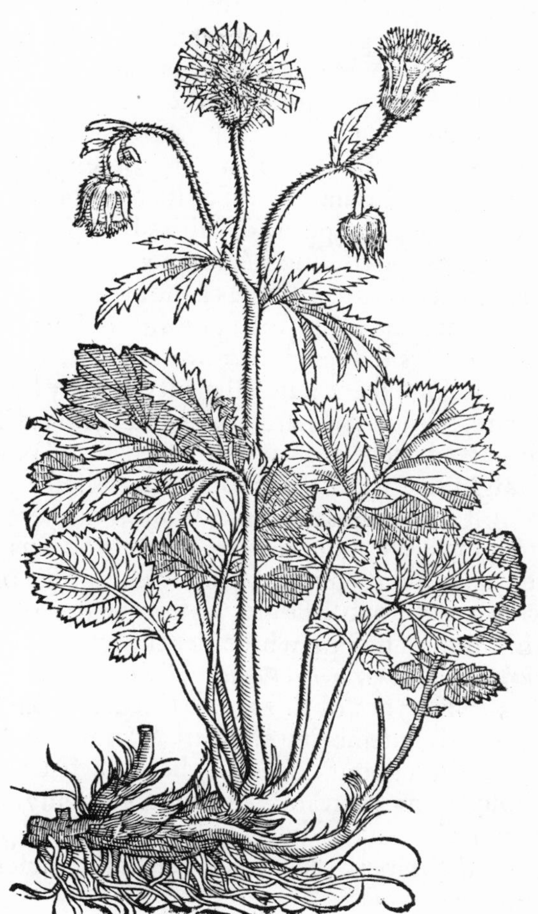

‡ 5 *Caryophyllata Alpina minima.*
Dwarfe mountaine Auens.

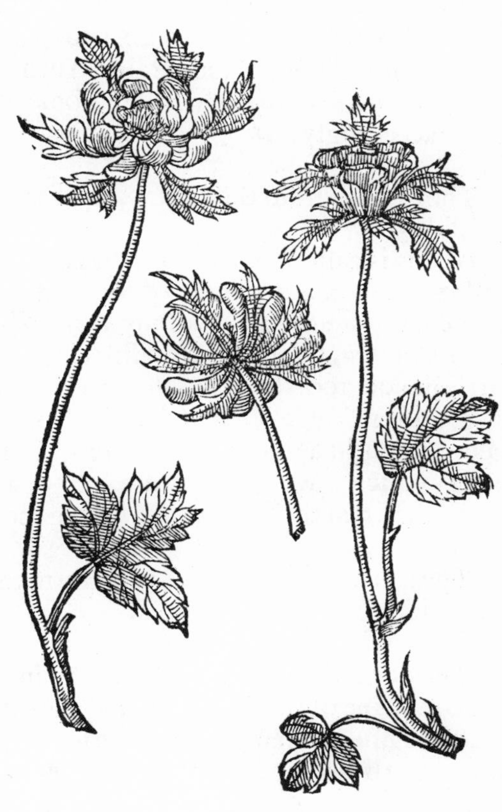

grow one floure greater than that of the former, which consisteth of fiue little leaues as yellow as gold: after which growes vp the seeds among long hairy threds. The root is long, growing aslope, somewhat thicke, with strings anexed thereto.

3 Fiue finger Auens hath many small leaues spred vpon the ground, diuided into fiue parts, somewhat snipt about the edges like Cinkefoile, whereof it tooke his name. Among which rise vp slender stalkes diuided at the top into diuers branches, whereon do grow small yellow floures like those of Cinkefoile: the root is composed of many tough strings of the smell of Cloues, which makes it a kind of Auens; otherwise doubtles it must of necessitie be one of the Cinkfoiles

‡ 4 This hath ioynted stringy roots some finger thick, from whence rise vp many large and hairy leaues, composed of diuers little leaues, with larger at the top, and these snipt about the edges like as the common Auens: amongst these leaues grow vp sundry stalkes some foot or better high, whereon grow floures hanging downe their heads, and the tops of the stalkes and cups of the floures are commonly of a purplish colour: the floures themselues are of a pretty red colour, and are of diuers shapes, and grow diuers wayes; which hath beene the reason that *Clusius* and other haue iudged them seuerall plants, as may be seene is *Clusius* his Workes, where he giues you the floures, which you here finde exprest, for a different kind. Now some of these floures, euen the greater part of them grow with fiue red round pointed leaues, which neuer lie faire open, but only stand straight out, the middle part being filled with a hairy matter and yellowish threds: other-some consist of seuen, eight, nine, or more leaues; and some againe lie wholly open, with greene leaues growing close vnder the cup of the floure, as you may see them represented in the figure; and some few now and then may be found composed of a great many little leaues thick thrust together, making a very double floure. After the floures are falne come such hairy heads as in other plants of this kinde, amongst which lies the seed. *Gesner* calls this *Geum rivale: Thalius, Caryophyllata maior purpurea: Camerarius, Caryophyllata aquatica: Clusius, Caryophyllata montana prima, & tertia.*

5 The root of this is also thicke, fibrous, and whitish; from which arise many leaues three fingers high, resembling those of Agrimonie, the little leaues standing directly opposite each against other, snipt about the edges, hairy, a little curld, and of a deepe greene colour: out of the midst of those, vpon a short stalk growes commonly on single floure of a gold-yellow colour, much like the mountaine Auens described in the second place. It floures at the beginning of Iuly, and groweth vpon the Alpes. *Pona* was the first that described it, and that by the name of *Caryophyllata Alpina omnium minima.* ‡

¶ *The Place.*

These kindes of Auens are found in high mountaines and thicke woods of the North parts of England: we haue them in our London gardens, where they flourish and encrease infinitely.

‡ The red floured mountaine Auens was found growing in Wales by my much honoured friend M[r]. *Thomas Glynn*, who sent some plants thereof to our Herbarists, in whose gardens it thriueth exceedingly. ‡

¶ *The Time.*

They floure from the beginning of May to the end of Iuly.

¶ *The Names.*

Auens is called *Caryophyllata*, so named of the smell of Cloues which is in the roots, and diuers call it *Sanamunda, Herba benedicta*, and *Nardus rustica*: in high-Dutch, **Benedicten wortz**: in French, *Galiot*: of the Wallons, *Gloria filia*: in English, Auens, and herbe Benet: it is thought to be *Geum Plinij*, which most do suspect, by reason he is so briefe. *Geum*, saith *Pliny, lib.26.cap.7.* hath little slender roots, blacke, and of a good smell.

The other kinde of Auens is called of the later Herbarists, *Caryophyllata montana*, Mountaine Auens: it might agtee with the description of *Baccharis*, if the floures were purple tending to whitenesse; which as we haue said are yellow, and likewise differ in that, that the roots of Auens smell of Cloues, and those of *Baccharis* haue the smell of Cinnamon.

¶ *The Temperature.*

The roots and leaues of Auens are manifestly dry, and something hot, with a kinde of scouring qualitie.

¶ *The Vertues.*

A The decoction of Auens made in wine is commended against cruditie or rawnesse of the stomacke, paine of the Collicke, and the biting of venomous beasts.

B The same is likewise a remedie for stitches and griefe in the side, for stopping of the liuer; it concocteth raw humours, scoureth away such things as cleaue to the intrals, wasteth and dissolueth winde, especially being boyled with wine: but if it be boyled with pottage or broth it is of great efficacie, and of all other pot-herbes is chiefe, not onely in physicall broths, but commonly to be vsed in all.

C The leaues and roots taken in this manner dissolue and consume clottered bloud in any inward part

part of the body; and therfore they are mixed with potions which are drunk of thoſe that are bruiſed, that are inwardly broken, or that haue fallen from ſome high place.

The roots taken vp in Autumne and dried, do keep garments from being eaten with moths, and F
make them to haue an excellent good odour, and ſerue for all the phyſicall purpoſes that Cinkefoiles do.

CHAP. 386. *Of Straw-berries.*

¶ The Kindes.

THere be diuers ſorts of Strawberries; one red, another white, a third ſort greene, and likewiſe a wilde Straw-berrie, which is altogether barren of fruit.

1 *Fragaria & Fraga.*
Red Straw-berries.

2 *Fragaria & Fraga ſubalba.*
White Straw-berries.

¶ The Deſcription.

1 THe Straw-berry hath leaues ſpred vpon the ground, ſomewhat ſnipt about the edges, three ſet together vpon one ſlender foot-ſtalke like the Trefoile, greene on the vpper ſide, and on the nether ſide more white: among which riſe vp ſlender ſtems, whereon do grow ſmall floures, conſiſting of fiue little white leaues, the middle part ſomwhat yellow, after which commeth the fruit, not vnlike to the Mulberrie, or rather the Raſpis, red of colour, hauing the taſte of wine, the inner pulpe or ſubſtance whereof is moiſt and white, in which is contained little ſeeds: the root is threddy, of long continuance, ſending forth many ſtrings, which diſperſe themſelues far abroad, whereby it greatly increaſeth.

2 Of theſe there is alſo a ſecond kinde, which is like to the former in ſtems, ſtrings, leaues, and floures. The fruit is ſomething greater, and of a whitiſh colour, wherein is the difference.

There is another ſort, which brings forth leaues, floures, and ſtrings like the other of his kinde. The fruit is green when it is ripe, tending to redneſſe vpon that ſide that lieth to the Sun, cleauing faſter

faster to the stemmes, and is of a sweeter taste, wherein onely consisteth the difference.

† 3 *Fragaria minime vesca, siue sterilis.*
Wilde or barren Straw-berry.

‡ There is also kept in our gardens (onely for varietie) another Strawberrie which in leaues and growing is like the common kinde; but the floure is greenish, and the fruit is harsh, rough, and prickely, being of a greenish colour, with some shew of rednesse. Mr. *Iohn Tradescant* hath told me that he was the first that tooke notice of this Straw-berry, and that in a womans garden at Plimouth, whose daughter had gathered and set the roots in her garden in stead of the common Straw-berry: but she finding the fruit not to answer her expectation, intended to throw it away: which labor he spared her, in taking it and bestowing it among the louers of such varieties, in whose gardens it is yet preserued. This may be called in Latine, *Fragaria fructu hispido*, The prickly Straw-berry. ‡

† 3 This wild Strawberry hath leaues like the other Straw-berry, but somewhat lesse, and softer, slightly indented about the edges, and of a light greene colour: among which rise vp slender stems bearing such floures as the common Straw-berries doe, but lesser, which doe wither away, leauing behinde a barren or chaffie head, in shape like a Straw-berrie, but of no worth or value: the root is like the others.

¶ *The Place.*

Straw-berries do grow vpon hills and vallies, likewise in woods and other such places that be somewhat shadowie: they prosper well in Gardens, the first euery where, the other two more rare, and are not to be found saue only in gardens.

‡ The barren one growes in diuers places, as vpon Blacke heath, in Greenwich parke, &c. ‡

¶ *The Time.*

The leaues continue greene all the yeare: in the Spring they spred further with their strings, and floure afterward: the berries are ripe in Iune and Iuly. ‡ The barren one floures in April and May, but neuer carries any berries. ‡

¶ *The Names.*

The fruit or berries are called in Latine by *Virgil* and *Ouid*, *Fraga*: neither haue they any other name commonly knowne: they are called in high-Dutch **Erdbeeren**: in low-Dutch, **Eertbesien**: in French, *Fraises*: in English, Strawberries.

¶ *The Temperature.*

The leaues and roots do coole and dry, with an astriction or binding quality: but the berries be cold and moist.

¶ *The Vertues.*

A The leaues boyled and applied in manner of a pultis taketh away the burning heate in wounds: the decoction thereof strengthneth the gummes, fastneth the teeth, and is good to be held in the mouth, both against the inflammation or burning heate thereof, and also of the almonds of the throat: they stay the ouermuch flowing of the bloudy flix, and other issues of bloud.

B The berries quench thirst, and do allay the inflammation or heate of the stomack: the nourishment which they yeeld is little, thin, and waterish, and if they happen to putrifie in the stomacke, their nourishment is naught.

C The distilled water drunke with white Wine is good against the passion of the heart, reuiuing the spirits, and making the heart merry.

D The distilled water is reported to scoure the face, to take away spots, and to make the face faire and smooth; and is likewise drunke with good successe against the stone in the kidnies.

E The leaues are good to be put into Lotions or washing waters, for the mouth and the priuie parts.

The

The ripe Straw-berries quench thirſt, coole heat of the ſtomack, and inflammation of the liuer, take away (if they be often vſed) the redneſſe and heate of the face. F

† That figure which formerly was in this place, and ſome part of the deſcription were (as I haue formerly noted) of the *Pentaphyllum ſupinum Potentillæ facie*, which you may finde deſcribed amongſt the Cinkfoiles in the ſixth place.

CHAP. 387. *Of Angelica.*

¶ *The Kindes.*

THere be diuers kindes of Angelica's; the garden Angelica, that of the water, and a third ſort wilde growing vpon the land.

1 *Angelica ſatiua.*
Garden Angelica.

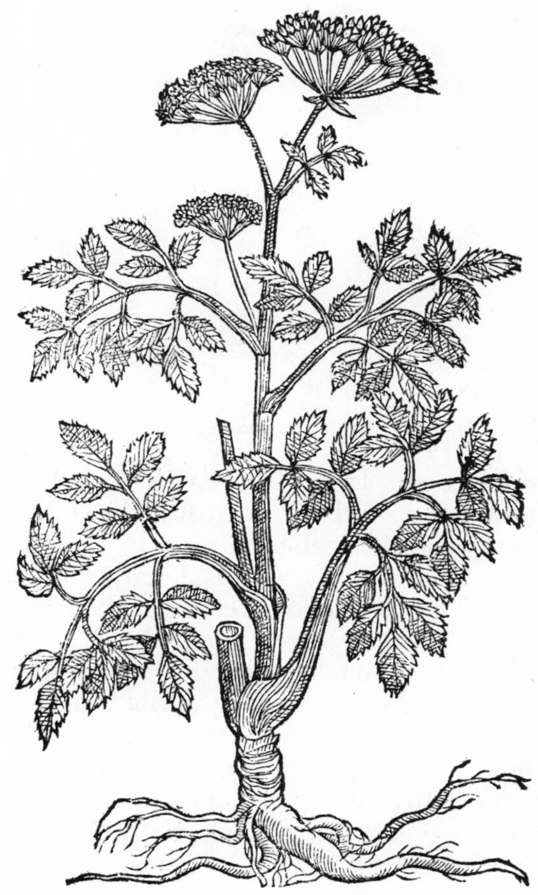

2 *Angelica ſylueſtris.*
Wilde Angelica.

¶ *The Deſcription.*

1 Concerning this plant Angelica there hath bin heretofore ſome contention and controuerſie; *Cordus* calling it *Smyrnium*: ſome later writers, *Coſtus niger*: but to auoid cauill, the controuerſie is ſoone decided, ſith it and no other doth aſſuredly retaine the name *Angelica*. It hath great broad leaues, diuided againe into other leaues, which are indented or ſnipt about much like to the vppermoſt leaues of *Sphondylium*, but lower, tenderer, greener, and of a ſtronger ſauor: among which leaues ſpring vp the ſtalkes, very great, thicke, and hollow, ſixe or ſeuen foot high, ioynted or kneed: from which ioynts proceed other armes or branches, at the top whereof grow tufts of whitiſh floures like Fennell or Dill: the root is thicke, great, and oilous, out of which iſſueth, if it be cut or broken, an oylie liquor: the whole plant, as well leaues, ſtalkes, as roots, are of a reaſonable pleaſant ſauour, not much vnlike *Petroleum*.

There is another kinde of true Angelica found in our Engliſh gardens (which I haue obſerued) being like vnto the former, ſauing that the roots of this kinde are more fragrant, and of a more aromaticke ſauor, and the leaues next the ground of a purpliſh red colour, and the whole plant leſſer.

2 The

‡ 3 *Archangelica.*
Great wilde Angelica.

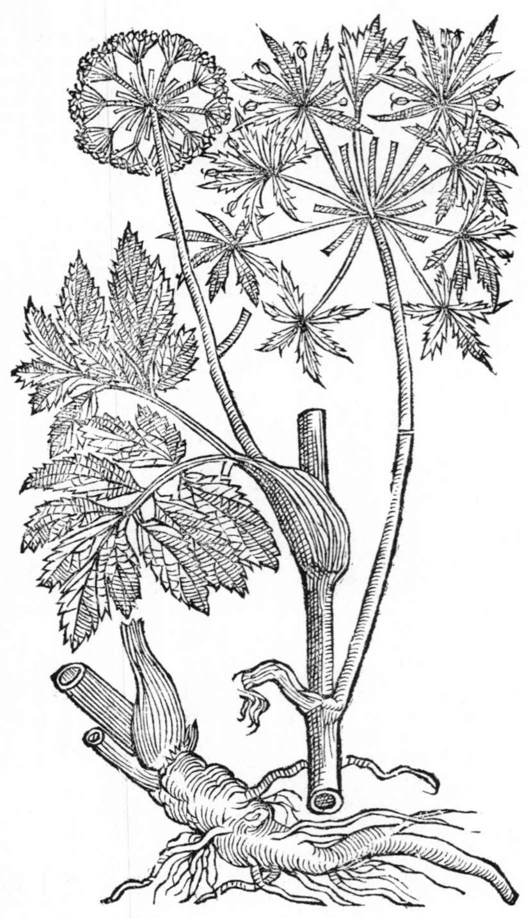

2 The wilde Angelica, which seldome growes in gardens, but is found to grow plentifully in water soken grounds and cold moist medowes, is like to that of the garden, saue that his leaues are not so deepely cut or iagged; they be also blacker and narrower: The stalkes are much slenderer and shorter, and the floures whiter: the root much smaller, and hath more threddy strings appendant thereunto, and is not so strong of sauour by a great deale.

3 *Matthiolus* and *Gesner* haue made mention of another kinde of Angelica, but we are very slenderly instructed by their insufficient descriptions: notwithstanding for our better knowledge and more certain assurance I must needs record that which my friend Mr. *Bredwell* related to me concerning his sight thereof, who found this plant growing by the more which compasseth the house of Mr. *Munke* of the parish of Iuer, two miles from Colbrook; and since that I haue seene the same in low fenny and marshy places of Essex, about Harwich. This plant hath leaues like vnto the garden Angelica, but smaller, and fewer in number, set vpon one rib a great stalke, grosse and thicke, whose ioynts and that small rib whereon the leafe growes are of a reddish colour, hauing many long branches comming forth of an husk or case, such as is in the common garden Parsnep: the floures doe grow at the top of the branches, and are of a white colour, and tuft fashion: which being past, there succeed broad long and thicke seeds, longer and thicker than garden Angelica: the root is great, thicke, white, of little sauour, with some strings appendant thereto.

‡ This of our Authors description seemes to agree with the *Archangelica* of *Lobel, Dodonæus*, and *Clusius*; wherefore I haue put their figure to it. ‡

¶ *The Place.*

The first is very common in our English gardens: in other places it growes wilde without planting; as in Norway, and in an Island of the North called Island, where it groweth very high. It is eaten of the inhabitants, the barke being pilled off, as we vnderstand by some that haue trauelled into Island, who were sometimes compelled to eate hereof for want of other food; and they report that it hath a good and pleasant taste to them that are hungry. It groweth likewise in diuers mountaines of Germanie, and especially of Bohemia.

¶ *The Time.*

They floure in Iuly and August, whose roots for the most part do perish after the seed is ripe: yet haue I with often cutting the plant kept it from seeding, by which meanes the root and plant haue continued sundry yeares together.

¶ *The Names.*

It is called of the later age *Angelica*: in high-Dutch, **Angelick**, **Brustwurtz**, or **des heilighen Geyst wurtzel**, that is, *Spiritus sancti radix*, the root of the holy Ghost, as witnesseth *Leonhartus Fuchsius*: in low-Dutch, **Angelika**: in French, *Angelic*: in English also Angelica.

It seemeth to be a kind of *Laserpitium*; for if it be compared with those things which *Theophrastus* at large hath written concerning *Silphium* or *Laserpitium*, in his sixth booke of the historie of plants, it shall appeare to be answerable thereunto. But whether wild *Angelica* be that which *Theophrastus* calleth *Magydaris*, that is to say, another kinde of *Laserpitium*, we leaue to be examined and considered of by the learned Physitians of our London Colledge.

¶ *The Temperature.*

Angelica, especially that of the garden, is hot and dry in the third degree; therefore it openeth, attenuateth or maketh thin, digesteth, and procureth sweat.

¶ *The*

¶ *The Vertues.*

The roots of garden Angelica is a singular remedy against poison, and against the plague, and all infections taken by euill and corrupt aire; if you do but take a peece of the root and hold it in your mouth, or chew the same between your teeth, it doth most certainely driue away the pestilentiall aire, yea although that corrupt aire haue possessed the hart, yet it driueth it out again by vrine and sweat, as Rue and Treacle, and such like *Antipharmaca* do. A

Angelica is an enemy to poisons: it cureth pestilent diseases if it by vsed in season: a dram weight of the pouder hereof is giuen with thin wine, or if the feuer be vehement, with the distilled water of *Carduus benedictus*, or of *Tormentill*, and with a little vineger, and by it selfe also, or with Treacle of Vipers added. B

It openeth the liuer and spleene: draweth downe the termes, driueth out or expelleth the secondine. C

The decoction of the root made in wine, is good against the cold shiuering of agues. D

It is reported that the root is auaileable against witchcraft and inchantments, if a man carry the same about them, as *Fuchsius* saith. E

It attenuateth and maketh thin, grosse and tough flegme: the root being vsed greene, and while it is full of juice, helpeth them that be asthmaticke, dissoluing and expectorating the stuffings therein, by cutting off and clensing the parts affected, reducing the body to health againe; but when it is dry it worketh not so effectually. F

It is a most singular medicine against surfeting and loathsomenesse to meate: it helpeth concoction in the stomacke, and is right beneficiall to the hart: it cureth the bitings of mad dogges, and all other venomous beasts. G

The wilde kindes are not of such force in working, albeit they haue the same vertues attributed vnto them. H

CHAP. 388. *Of Masterworts and herbe* Gerard.

1 *Imperatoria.*
Masterwoorts.

2 *Herba Gerardi.*
Herbe Gerard, or Aish-weed.

¶ *The Description.*

1 Imperatoria or Masterwoort hath great broad leaues not much vnlike wilde Angelica, but smaller, and of a deeper greene colour, in sauor like Angelica, and euery leafe diuided into sundry other little leaues: the tender knotted stalkes are of a reddish colour, bearing at the top round spokie tufts with white floures: the seed is like the seed of Dill: the root is thicke, knotty and tuberous, of a good sauour, and hot or biting vpon the tongue, which hath mooued the vnskilfull to call it Pellitory of Spaine, but very vnfitly and vntruely.

2 *Herba Gerardi*, which *Pena* doth also call *Imperatoria* and *Ostrutium*: the Germaines *Podagraria*, that is, Gout-woort: in English, herbe Gerard, or wilde Masterwoot, and in some places after *Lyte*, Ashweed, is very like the other in leaues, floures, and roots, sauing that they be smaller, growing vpon long stems: the roots tenderer, whiter, and not so thicke or tuberous. The whole plant is of a reasonable good sauour, but not so strong as Masterwoort.

¶ *The Place.*

Imperatoria groweth in darke woods and desarts; in my Garden and sundry others very plentifully.

Herbe Gerard groweth of it selfe in gardens without setting or sowing, and is so fruitfull in his increase, that where it hath once taken root, it will hardly be gotten out againe, spoiling and getting euery yeere more ground, to the annoying of better herbes.

¶ *The Time.*

They floure from the beginning of Iune to the beginning of August.

¶ *The Names.*

Imperatoria, or *Astrantia*, is called in English, Masterwoort, or bastard Pellitory of Spaine.

Herba Gerardi is called in English, Herb Gerard, Aishweed, and Goutwoort: in Latine also *Podagraria Germanica*.

¶ *The Nature.*

Imperatoria, especially the root, is hot and dry in the third degree. The wilde *Imperatoria*, or herbe Gerard, is almost of the same nature and quality, but not so strong.

¶ *The Vertues.*

A *Imperatoria* is not onely good against all poison, but also singular against all corrupt and naughty aire and infection of the pestilence, if it be drunken with wine.

B The roots and leaues stamped, dissolue and cure pestilentiall carbuncles and botches, and such other apostumations and swellings, being applied thereto.

C The root drunke in wine cureth the extreme and rigorous cold fits of agues, and is good against the dropsie, and prouoketh sweat.

D The same taken in manner aforesaid, comforteth and strengthneth the stomack, helpeth digestion, restoreth appetite, and dissolueth all ventosities or windinesse of the stomacke and other parts.

E It greatly helpeth such as haue taken great squats, bruses, or falls from some high place, dissoluing and scattering abroad congealed and clotted bloud within the body: the root with his leaues stamped and laid vpon the members infected, cureth the bitings of mad dogs, and of all other venomous beasts.

F Herbe Gerard with his roots stamped, and laid vpon members that are troubled or vexed with the gout, swageth the paine, and taketh away the swellings and inflammation thereof, which occasioned the Germaines to giue it the name *Podagraria*, because of his vertues in curing the gout.

G It cureth also the Hemorrhoids, if the fundament be bathed with the decoction of the leaues and roots, and the soft and tender sodden herbes laid thereon very hot.

H False Pellitory of Spaine attenuateth or maketh thin, digesteth, prouoketh sweate and vrine, concocteth grosse and colde humors, wasteth away windinesse of the entrailes, stomacke and matrix: it is good against the collicke and stone.

I One dram of the root in pouder giuen certaine daies together, is a remedy for them that haue the dropsie, and also for those that are troubled with conuulsions, cramps, and the falling sickenesse.

K Being giuen with wine before the fit come, it cureth the quartaine ague, and is a remedy against pestilent diseases.

L The same boiled in sharpe or sower wine, easeth the tooth-ach, if the mouth be washed therewith very hot.

M Being chewed it draweth forth water and flegme out of the mouth (which kinde of remedies in Latine are called *Apophlegmatismi*) and disburdeneth the braine of phlegmaticke humours, and are likewise vsed with good successe in apoplexies, drowsie sleepes, and other like infirmities.

Chap.

CHAP. 389.

Of Hercules Wound-wort, or All-heale.

¶ *The Kindes.*

PAnax is of ſundry kindes, as witneſſeth *Theophraſtus* in his ninth booke; one groweth in Syria, and likewiſe other three, that is to ſay, *Chironium*, *Heraclium*, and *Æſculapium*; or *Chirons* All-heale, *Hercules* All-heale, and *Æſculapius* All-heale. Beſides theſe there is one *Platyphyllon*, or broad leafed; ſo that in *Theophraſtus* there are ſix kindes of *Panax*: but *Dioſcorides* deſcribeth only three, *Horacleum*, *Aſclepium*, and *Chironium*: whereunto we haue added another ſort, whoſe vertues wee found out by meanes of a husbandman, and for that cauſe haue named it *Panax Coloni*, or Clownes wort.

1 *Panax Heracleum.*
Hercules All-heale.

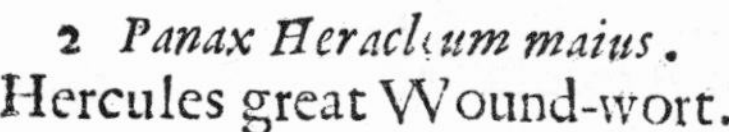

2 *Panax Heraclium maius.*
Hercules great Wound-wort.

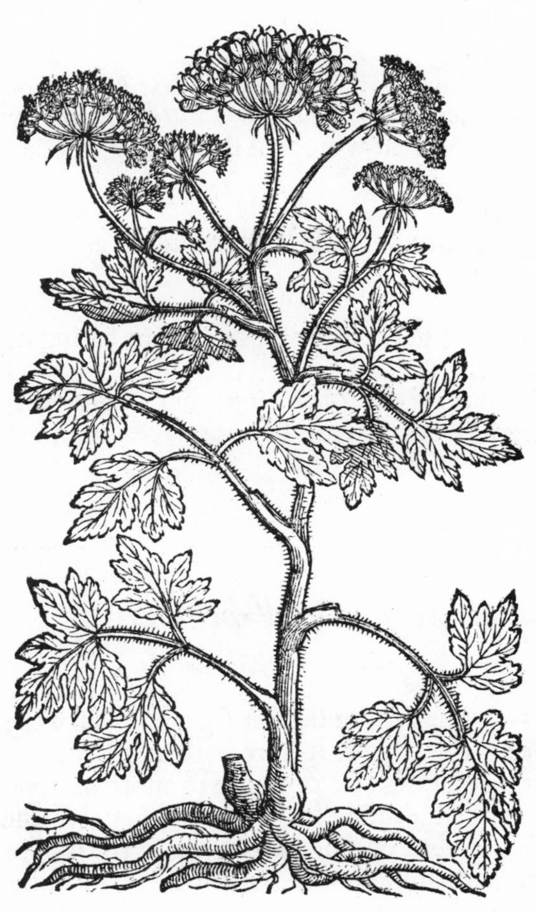

¶ *The Deſcription.*

1 HErcules All-heale or Wound-wort hath many broad leaues ſpread vpon the ground, very rough and hairy, of an ouerworne greene colour, and deepely cut into diuers ſections like thoſe of the Cow Parſenep, and not vnlike to the fig leaues: among the which riſeth vp a very ſtrong ſtalke couered ouer with a rough hairineſſe, of the height of foure or fiue cubits. Being wounded it yeeldeth forth a yellow gummie iuyce, as doth euery part of the plant, which is that precious gum called *Opopanax*: at the top of which ſtalks ſtand great tufts or vmbels of yellowiſh floures, ſet together in ſpoky rundles like thoſe of Dill, which turn into ſeed of a ſtraw colour, ſharpe and hot in taſte, and of a pleaſing ſauour: the root is very thicke, fat, and full of iuyce, and of a white colour.

2 The great Wound-wort, which the Venetians nouriſh in their gardens, hath great large leaues ſomewhat rough or hairie, conſiſting of diuers ſmall leaues ſet together vpon a middle rib, which make one entire leafe ioyned together in one, whereof each collaterall or ſide leafe is long and

and sharpe pointed: among which riseth vp a knotty stalke three or foure cubits high, diuiding it selfe into diuers branches; on the tops whereof do grow spokie tufts or rundles like the precedent, but the floures are commonly white: the seed is flat and plaine: the root long, thicke, and white, which being broken or wounded, yeeldeth forth liquor like that of the former, of a hot and biting taste.

¶ *The Place.*

These plants grow in Syria; the first of them also in my garden: but what *Panax* of Syria is, *Theophrastus* doth not expresse. *Pliny* in his 12 booke, Chap. 26. saith, that the leaues are round, and of a great compasse: but it is suspected that these are drawne from the description of Hercules Panax.

Broad leafed Panax is thought to be the great Centory: for *Pliny* witnesseth, that Panax which *Chiron* found out is syrnamed *Centaurium*, Centorie.

Matthiolus saith it growes of it selfe in the tops of the hills Apennini, in the Cape Argentaria, in the sea coasts of Siena, and it is cherished in the gardens of Italy: but he cannot affirme, That the liquor hereof is gathered in Italy; for the liquor *Opopanax* which is sold in Venice is brought, saith he, out of Alexandria a city in Egypt: it groweth also in Syria, Boœtia, and in Phocide, cities of Arcadia.

¶ *The Time.*

They floure and flourish from the first of May vnto the end of September.

¶ *The Names.*

That which is is called πάναξ in Greeke, is likewise named *Panax* in Latine: and that *Panax Heraclium* which *Dioscorides* setteth downe is called in Latine, *Panax Heraculanum*, or *Herculeum*, or Hercules Panax: it may be called in English, Hercules his Wound-wort or All-heale, or Opopanax wort, of the Greeke name.

¶ *The Temperature.*

The barke of the root of Hercules Wound-wort is hot and dry, yet lesse than the iuyce, as *Galen* teacheth.

¶ *The Vertues.*

A The seed beat to pouder and drunke in Wormwood wine is good against poyson, the biting of mad dogs, and the stinging of all manner of venomous beasts.

B The leafe or root stamped with honey, and brought to the forme of an Vnguent or Salue, cureth wounds and vlcers of great difficultie, and couereth bones that are bare or naked without flesh.

Chap. 390. *Of Clownes Wound-wort or All-heale.*

¶ *The Description.*

Clownes All-heale, or the Husbandmans Wound-wort, hath long slender square stalks of the height of two cubits, furrowed or chamfered along the same as it were with small gutters, and somewhat rough or hairy: wherupon are set by couples one opposite to another, long rough leaues somewhat narrow, bluntly indented about the edges like the teeth of a saw, of the forme of the leaues of Speare-mint, and of an ouerworne greene colour: at the top of the stalkes grow the floures spike fashion, of a purple colour mixed with some few spots of white, in forme like to little hoods. The root consisteth of many small threddy strings, whereunto are annexed or tied diuers knobby or tuberous lumpes, of a white colour tending to yellownesse: all the whole plant is of an vnpleasant sauour like *Stachys* or stinking Hore-hound. ‡ The root in the Winter time and the beginning of the Spring is somewhat knobby tuberous, and ioynted, which after the stalkes grow vp become flaccide and hollow, and so the old ones decay, and then it putteth forth new ones. ‡

¶ *The Place.*

It groweth in moist medowes by the sides of ditches, and likewise in fertile fields that are somwhat moist, almost euery where; especially in Kent about South-fleet, neere to Grauesend, and likewise in the medowes by Lambeth neere London.

¶ *The Time.*

It floureth in August, and bringeth his seed to perfection in the end of September.

¶ *The Names.*

That which hath bin said in the description shall suffice touching the names, as well in Latine as English.

‡ This

Panax Coloni.
Clownes All-heale.

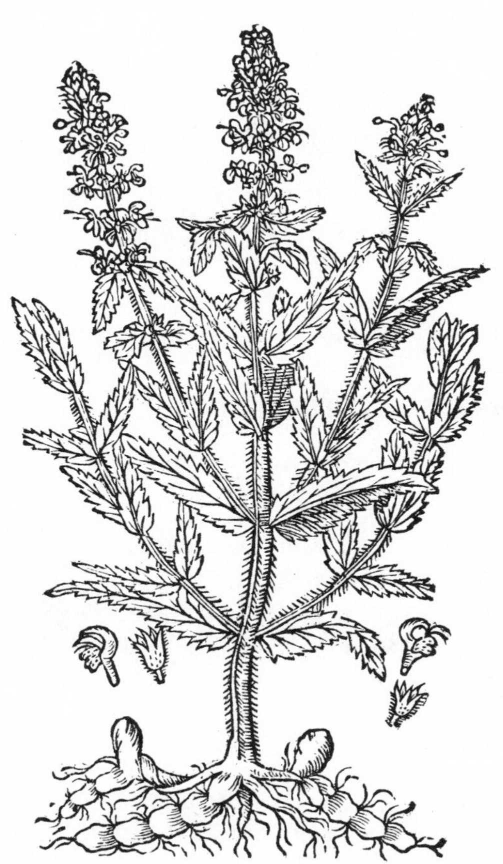

‡ This plant by *Gesner* was called *Stachys palustris*, and *Betonica fœtida*, and thought to be of the kinde of *Herba Iudaica*, or *Sideritis*; to which indeed I should, and *Thalius* hath referred it, calling it *Sideritis* 1. *grauis odoris*: *Cæsalpinus* calls it *Tertiola*; and giues this reason, *quod Tertianas sanet*, because it cures Tertians. *Tabernamontan*. called it *Stachys aquatica*, whose figure with a description our Authour in the former edition gaue, *pag*. 565. by the name of *Marrubium aquaticum acutum*; yet (as it seemeth) either not knowing, or forgetting what he had formerly done, he here againe setteth it forth as a new thing, vnder another title: but the former figure of *Tabern*. being in my iudgment the better, I haue here giuen you, with addition of the iointed tuberous roots as they are in Winter: yet by the Caruers fault they are not altogether so exquisitely exprest as I intended. ‡

¶ *The Temperature.*

This plant is hot in the second degree, and dry in the first.

¶ *The Vertues.*

A The leaues hereof stamped with *Axungia* or hogs grease, and applied vnto greene wounds in manner of a pultesse, healeth them in short time, and in such absolute manner, that it is hard for any that haue not had the experience thereof to beleeue: for being in Kent about a Patient, it chanced that a poore man in mowing of Peason did cut his leg with a sithe, wherein he made a wound to the bones, and withall very large and wide, and also with great effusion of bloud; the poore man crept vnto this herbe, which he bruised with his hands, and tied a great quantity of it vnto the wound with a piece of his shirt, which presently stanched the bleeding, and ceased the paine, insomuch that the poore man presently went to his dayes worke againe, and so did from day to day, without resting one day vntill he was perfectly whole, which was accomplished in a few dayes, by this herbe stamped with a little hogs grease, and so laid vpon in manner of a pultesse, which did as it were glew or soder the lips of the wound together, and heale it according to the first intention, as we terme it, that is, without drawing or bringing the wound to suppuration or matter; which was fully performed in seuen dayes, that would haue required forty dayes with balsam it selfe. I saw the wound, and offered to heale the same for charitie; which he refused, saying that I could not heale it so well as himselfe: a clownish answer I confesse, without any thankes for my good will; whereupon I haue named it Clownes Wound-wort, as aforesaid. Since which time my selfe haue cured many grieuous wounds, and some mortall, with the same herbe; one for example done vpon a Gentleman of Grayes Inne in Holborne, M[r]. *Edmund Cartwright*, who was thrust into the lungs, the wound entring in at the lower part of the *Thorax*, or the brest-blade, euen through that cartilaginous substance called *Mucronata Cartilago*, insomuch that from day to day the frothing and puffing of the lungs did spew forth of the wound such excrements as it was possessed of, besides the Gentleman was most dangerously vexed with a double quotidian feuer; whom by Gods permission I perfectly cured in very short time, and with this Clownes experiment, and some of my foreknowne helpes, which were as followeth.

B First I framed a slight vnguent hereof thus: I tooke foure handfulls of the herbe stamped, and put them into a pan, whereunto I added foure ounces of Barrowes grease, halfe a pinte of oyle Oliue, wax three ounces, which I boyled vnto the consumption of the iuyce (which is known when the stuffe doth not bubble at all) then did I straine it, putting it to the fire againe, adding thereto two ounces of Turpentine, the which I suffered to boyle a little, reseruing the same for my vse.

C The which I warmed in a sawcer, dipping therein small soft tents, which I put into the wound, defending the parts adioyning with a plaister of *Calcitheos*, relented with oyle of roses: which manner of dressing and preseruing I vsed euen vntill the wound was perfectly whole: notwithstanding once in a day I gaue him two spoonfulls of this decoction following.

D I tooke a quart of good Claret Wine, wherein I boyled an handfull of the leaues of *Solidago*

 Saracenica,

Saracenica, and Saracens Conſound, or foure ounces of honey, whereof I gaue him in the morning two ſpoonfulls to drinke in a ſmall draught of wine tempered with a little Sugar.

E In like manner I cured a Shoo-makers ſeruant in Holburne, who intended to deſtroy himſelfe for cauſes knowne vnto many now liuiug: but I deemed it better to couer the fault, than to put the ſame in print, which might moue ſuch a graceleſſe fellow to attempt the like: his attempt was thus; Firſt, he gaue himſelfe a moſt mortall wound in the throat, in ſuch ſort, that when I gaue him drinke it came forth at the wound, which likewiſe did blow out the candle: another deep and grieuous wound in the breſt with the ſaid dagger, and alſo two others in *Abdomine* or the nether belly, ſo that the *Zirbus* or fat, commonly called the caule, iſſued forth, with the guts likewiſe: the which mortall wounds, by Gods permiſſion, and the vertues of this herbe, I perfectly cured within twenty dayes: for the which the name of God be praiſed.

Chap. 391. *Of Magydare, or Laſer-wort.*

† *Laſerpitium.*
Laſer-wort.

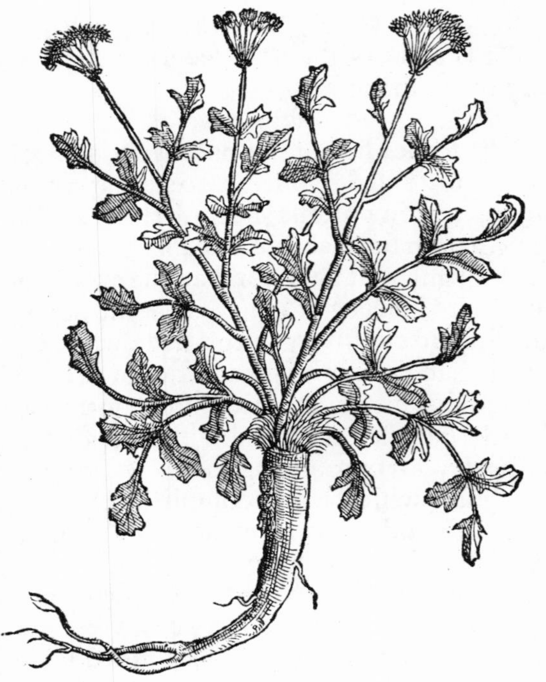

¶ *The Deſcription.*

It ſeemeth that neither *Dioſcorides* nor yet *Theophraſtus* haue euer ſeene *Laſerpitium*, *Sagapenum*, or any other of the gummiferous roots, but haue barely and nakedly ſet downe their iudgments vpon the ſame, either by heare-ſay, or by reading of other mens Workes. Now then ſeeing the old Writers be vnperfect herein, it behoooueth vs in this caſe to ſearch with more diligence the truth hereof; and the rather, for that very few haue ſet forth the true deſcription of that Plant which is called *Laſerpitium*, that is indeed the true *Laſerpitium*, from the roots whereof flow that ſap or liquor called *Laſer*. This plant, as *Pena* and *Lobel* themſelues ſay, was found out not far from the Iſles which *Dioſcorides* calls Stœchades, ouer againſt Maſſilia, among ſundry other rare plants. His ſtalke is great and thicke like Ferula, or Fennell gyant: The leaues are like vnto the common Smallage, and of an vnpleaſant ſauour. The floures grow at the top of the ſtalkes, tuft-faſhion like Ferula or Fennell: which being paſt, there ſucceed broad and flat ſeeds like Angelica, of a good ſauour, and of the colour of Box. The roots are many, comming from one head or chiefe root, and are couered ouer with a thicke and fat barke. Theſe roots and ſtalkes being ſcarified or cut, there floweth out of them a ſtrong liquor, which being dried is very medicinable, and is called *Laſer*.

¶ *The Place.*

There be ſundry ſorts of Laſer, flowing from the roots and ſtalkes of *Laſerpitium*, the goodneſſe or qualitie whereof varieth according to the countrey or clymate wherein the plant groweth. For the beſt groweth vpon the high mountaines of Cyrene and Africa, and is of a pleaſant ſmell: in Syria alſo, Media, Armenia, and Lybia; the liquour of which plant growing in theſe places is of a moſt ſtrong and deteſtable ſauour. *Lobel* reporteth, that *Iacobus Rainaudus* an Apothecarie of Maſſilia was the firſt that made it knowne, or brought the plants thereof to Montpellier in France, vnto the learned *Rondeletius*, who right well beholding the ſame, concluded, that of all the kindes of Ferula that he had euer ſeene, there was not any ſo anſwerable vnto the true *Laſerpitium* as this onely plant.

¶ *The Time.*

This Plant floureth in Montpellier about Midſommer.

¶ *The Names.*

It is called in Latine *Laserpitium*: in English, Laserwoort, and Magydare: the gum or liquour that issueth out of the same is called *Laser*, but that which is gathered from those plants that doe grow in Media and Syria, is called *Asa fœtida*.

¶ *The Nature.*

Laserpitium, especially the root, is hot and drie in the third degree: *Laser* is also hot and drie in the third degree, but it exceedeth much the heate of the leaues, stalkes, and rootes of *Laserpitium*.

¶ *The Vertues.*

The root of *Laserpitium* well pounded, or stamped with oile, scattereth clotted bloud, taketh away blacke and blew markes that come of bruises or stripes, cureth and dissolueth the Kings-euill, and all hard swellings and botches, the places being annointed or plaistered therewith. A

The same root made into a plaister with the oile of Ireos and wax, doth both asswage and cure the Sciatica, or gout of the hip or huckle bone. B

The same holden in the mouth and chewed, doth asswage the tooth-ache; for they are such roots as draw from the braine a great quantite of humors. C

The liquour or gum of *Laserpitium*, especially the *Laser* of Cyrene broken and dissolued in water and drunken, taketh away the hoarsenesse that commeth suddenly: and being supt vp with a reare egge, cureth the cough: and taken with some good broth or supping, is good against an old pleurisie. D

Laser cureth the iaundies and dropsie, taken with dried figs: also being taken in the quantitie of a scruple, with a little pepper and Myrrhe, is very good against shrinking of sinewes, and members out of ioint. E

The same taken with honie and vineger, or the syrrup of vineger, is very good against the falling sicknesse. F

It is good against the flux of the belly comming of the debilitie and weaknesse of the stomacke (called in Latine *Cœliacus morbus*) if it be taken with raisons of the Sun. G

It driueth away the shakings and shiuerings of agues, being drunke with wine, pepper, & white Frankincense. Also there is made an electuarie thereof called *Antidotus ex succo Cyrenaico*, which is a singular medicine against feuer quartaines. H

It is excellent against the bitings of all venomous beasts, and venomous shot of darts or arrowes, not onely taken inwardly, but also applied outwardly vpon wounds. I

It bringeth to maturation, and breaketh all pestilentiall imposthumes, botches and carbuncles, being applied thereto with Rue, Salt-peter, and honie: after the same manner it taketh away corns after they haue been scarified with a knife. K

Being laied to with Copperas and Verdigrease, it taketh away all superfluous outgrowings of the flesh, the Polypus that happeneth in the nose, and all scuruie manginesse. L

If it be applied with vineger, pepper and wine, it cureth the naughtie scurfe of the head, and falling off of the haire. M

The gum or liquour of *Laserpitium* which groweth in Armenia, Lybia, and sundry other places, is that stinking and lothsome gum called of the Arabian Physitions *Asa* and *Assa*, as also with vs in shoppes *Asa fœtida*: but the *Laserpitium* growing in Cyrene is the best, and of a reasonable pleasant smell, and is called *Laser* to distinguish and make difference betweene the two iuices; though *Asa fœtida* be good for all purposes aforesaid, yet is it not so good as *Laser* of Cyrene: it is good also to smell vnto, and to be applied vnto the nauels of women vexed with the choking, or rising of the mother. N

† That figure which formerly was in this place, was of the common Louage described in the following chapter.

CHAP. 392. *Of common Louage.*

¶ *The Description.*

ANtient writers haue added vnto this common kinde of Louage, a second sort, yet knowing that the plant so supposed is the true *Siler montanum*, and not *Leuisticum*, though others haue also deemed it *Laserpitium*. These two suppositions are easily answered, sith they bee sundrie kindes of plants, though they be very neere in shape and faculties one vnto another. This plant being

† *Leuisticum vulgare.*
Common Louage.

being our common garden Louage, hath large and broad leaues, almost like to smallage. The stalks are round, hollow and knottie, 3. cubits high, hauing spoky tufts, or bushy rundles; and at the top of the stalks of a yellow colour, a round, flat, and browne seed, like the seede of Angelica: the root is long and thicke, and bringeth forth euery yeare new stems.

¶ *The Place.*

The right *Leuisticum* or Louage groweth in sundry gardens, and not wild (as far as I know) in England.

¶ *The Time.*

Louage floureth most commonly in Iuly and August.

¶ *The Names.*

It is called in Latine *Leuisticum*: and by some, *Ligusticum*: of other some, *Siler montanum*, but not truly: in High Dutch, **Libstockel**: in French, *Liuische*: in Low Dutch, **Lauetse**: in English, Louage.

¶ *The Nature.*

This plant is hot and drie in the third degree.

¶ *The Vertues.*

A The roots of Louage are very good for all inward diseases, driuing away ventosities or windinesse, especially of the stomacke.

B The seed thereof warmeth the stomack, helpeth digestion; wherefore the people of Gennes in times past did vse it in their meates, as wee doe pepper, according to the testimonie of *Ant. Musa*.

C The distilled water of Louage cleareth the sight, and putteth away all spots, lentils, freckles, and rednesse of the face, if they be often washed therewith.

† The figure which was here was of the *Siler montanum*, or *Seseli Officinarum*.

Chap. 393. *Of Cow Parsnep.*

¶ *The Description.*

This plant *Sphondylium* groweth in all Countries, and is knowne by the name of wilde Parsnep or Sphondylium, whereunto it effectually answereth, both in his grieuous and ranke sauour, as also in the likenesse of the root, wereupon it was called *Sphondylium*; and of the Germanes, *Acanthus*, but vntruly: the leaues of this plant are long and large, not much vnlike the leaues of wilde Parsnep, or *Panax Heracleum*; deepely notched or cut about the edges like the teeth of a saw, and of an ouerworne greene colour. The floures grow in tufts or rundles, like vnto wilde Parsneps: the root is like to Henbane: this herbe in each part thereof hath an euill sauour, and differeth from the right *Acanthium*, not onely in faculties, but euen in all other things.

¶ *The Place.*

This plant groweth in fertile moist medowes, and feeding pastures, very commonly in all parts of England, or elsewhere, in such places as I haue trauelled.

¶ *The Time.*

Sphondylium floureth in Iune and Iuly.

¶ *The Names.*

It is called in Greek σφονδύλιον in Latine likewise *Sphondylium*: the in shops of High and Low Germanie

† *Sphondylium.*
Cow Parſnep.

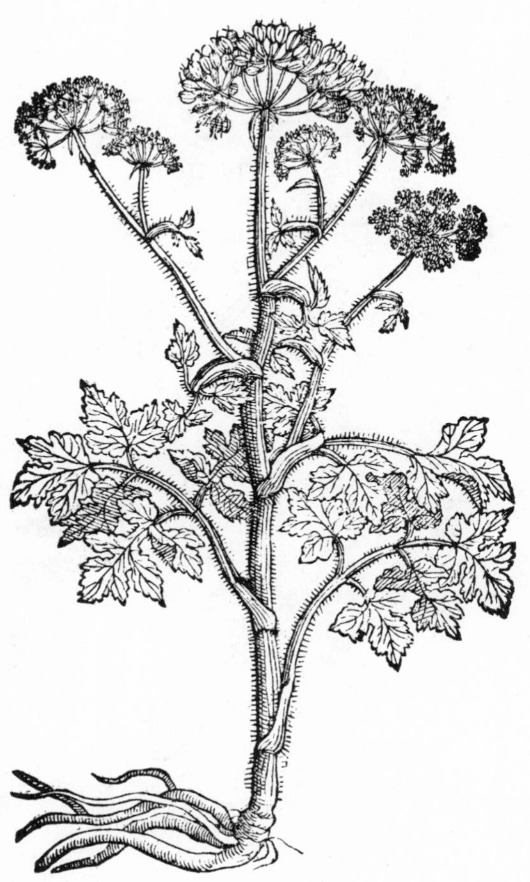

many *Branca vrſina*, who vnaduiſedly in times paſt haue vſed it in clyſters, in ſtead of Brancke Vrſine, and thereupon haue named it **Berneclaw**: in Engliſh, Cow Parſnep, medow Parſnep, and Madnep.

¶ *The Nature.*

Cow Parſnep is of a manifeſt warm complexion.

¶ *The Vertues.*

The leaues of this plant do conſume and diſſolue cold ſwellings if they be bruiſed and applied thereto. A

The people of Polonia and Lituania vſe to make drinke with the decoction of this herbe, and leuen or ſome other thing made of meale, which is vſed in ſtead of beere and other ordinarie drinke. B

The ſeede of Cow parſnep drunken, ſcoureth out flegmaticke matter through the guts, it healeth the iaundice, the falling ſickneſſe, the ſtrangling of the mother, and them that are ſhort winded. C

Alſo if a man be falne into a dead ſleepe, or a ſwoune, the fume of the ſeed will waken him againe. D

If a phrenticke or melancholicke mans head bee annointed with oile wherein the leaues and roots haue beene ſodden, it helpeth him very much, and ſuch as be troubled with the head-ache and the lethargie, or ſickneſſe called the forgetfull euill. E

† The figure formerly was of the *Paſtinata ſylueſtris*, or *Elaphoboſcum* of *Tabernamontanus*, and the figure that ſhould haue beene here was afterwards vnder the title of *Hippoſelinum*.

Chap. 394. *Of Herbe Frankincenſe.*

¶ *The Deſcription.*

1 There hath beene from the beginning diuers plants of ſundry kindes, which men haue termed by this glorious name *Libanotis*, onely in reſpect of the excellent and fragrant ſmell which they haue yeelded vnto the ſences of man, ſomewhat reſembling Frankincenſe. The ſent and ſmell *Dioſcorides* doth aſcribe to the root of this firſt kinde, which bringeth forth a long ſtalk with ioints like Fennell, whereon grow leaues almoſt like Cheruill or Hemlocks, ſauing that they be greater, broader, and thicker: at the top of the ſtalkes grow ſpokie taſſels bearing whitiſh floures, which do turne into ſweet ſmelling ſeed, ſomewhat flat, and almoſt like the ſeed of Angelica. The root is blacke without, and white within, hairie aboue, at the parting of the root and ſtalke like vnto *Meum* or *Peucedanum*, and ſauoreth like vnto Roſine, or Frankincenſe.

2 The ſecond kinde of *Libanotis* hath alſo a ſtraight ſtalke, full of knots and ioints: the leaues are like vnto Smallage: the floures grow in taſſels like vnto the former, and bring forth great, long and vneuen ſeed, of a ſharpe taſte: the root is like the former, and ſo is the whole plant very like, but leſſer.

3 The third kinde of *Libanotis* differeth ſomewhat from the others in forme and ſhape, yet it agreeth with them in ſmell, which in ſome ſort is like Frankinſence: the leaues are whiter, longer, and rougher than the leaues of Smallage: the ſtalks do grow to the height of two cubits, bearing at the top the ſpokie tufts of Dill, ſomewhat yellow: the root is like the former, but thicker, neither wanteth it hairie taſſels at the top of the root; which the others alſo haue, before rehearſed.

4 I can-

1 *Libanotis Theophrasti maior.*
Great herbe Frankinsence.

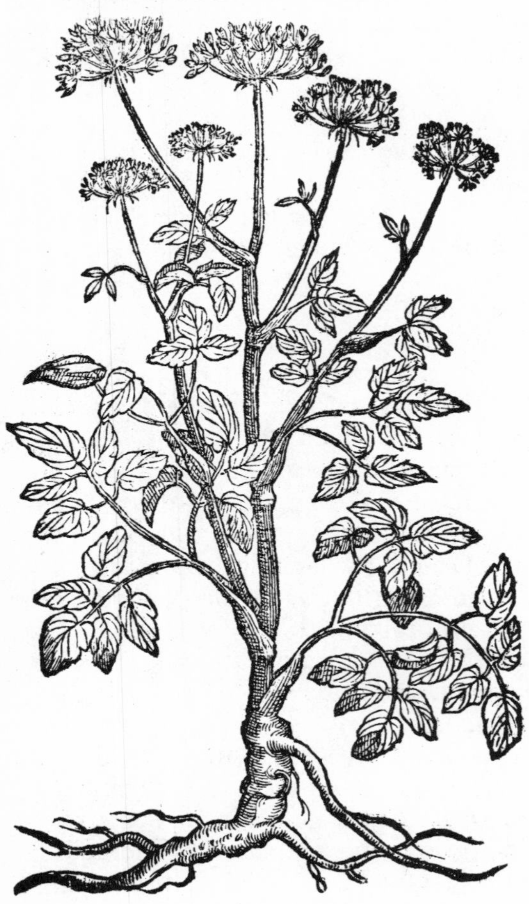

2 *Libanotis Theophrasti minor.*
Small herbe Frankinsence.

3 *Libanotis Theophrasti nigra.*
Blacke herbe Frankinsence.

4 *Libanotis Galeni, Cachrys verior.*
Rosemarie Frankincense.

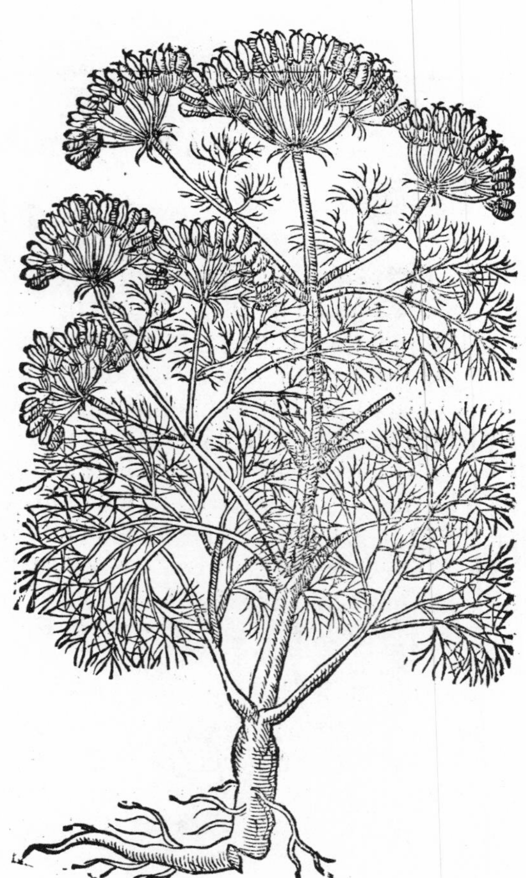

4 I cannot finde among all the plants called *Libanotides*, any one more agreeable to the true and right *Libanotis* of *Dioſcorides* than this herbe, which ariſeth vp to the height of fiue or ſix cubits with the cleere ſhining ſtalks of *Ferula*; diuiding it ſelfe from his knottie ioints into ſundry arms or branches, ſet full of leaues like Fennell, but thicker and bigger, and fatter than the leaues of *Cotula fœtida*, of a grayiſh greene colour, bearing at the top of the ſtalks the tufts of *Ferula*, or rather of Carrots, full of yellow floures: which being paſt there ſucceedeth long flat ſeed like the ſeed of the Aſh tree, ſmelling like Roſin, or Frankincenſe, which being chewed filleth the mouth with the taſt of Frankincenſe, but ſharper: all the reſt of the plant is tender, and ſomewhat hot, but not vnpleaſant: the plant is like vnto *Ferula*, and aboundeth with milke as *Ferula* doth, of a reaſonable good ſauour.

¶ *The Place.*

I haue the two laſt kindes growing in my garden; the firſt and ſecond grow vpon the high Deſerts and mountaines of Germanie.

¶ *The Time.*

Theſe herbs do floure in Iuly and Auguſt.

¶ *The Names.*

This herbe is called in Greeke Λιβανωτὶς, becauſe their roots do ſmell like incenſe, which is called in Greeke λίβανος: in Latine, *Roſmarinus*; the firſt may be Engliſhed great Frankincenſe Roſemarie; the ſecond ſmall Frankincenſe Roſemarie; Mr. *Lite* calleth the third in Engliſh, blacke Hart-root, the fourth white Hart-root: the ſeed is called *Cachrys* or *Canchrys*.

¶ *The Nature.*

Theſe herbes with their ſeeds and roots are hot and drie in the ſecond degree, and are of a digeſting, diſſoluing, and mundifying qualitie.

¶ *The Vertues.*

The leaues of *Libanotis* pounded, ſtoppe the fluxe of the Hemorrhoides or piles, and ſupple the ſwellings and inflammations of the fundament called *Condilomata*, concoct the ſwellings of the throat called *Strum*, and ripen botches that will hardly bee brought to ſuppuration or to ripeneſſe. A

The iuice of the leaues and roots mixed with honie, and put into the eies, doth quicken the ſight, and cleereth the dimneſſe of the ſame. B

The ſeed mingled with honie, doth ſcoure and clenſe rotten vlcers, and being applied vnto cold and hard ſwellings conſumeth and waſteth them. C

The leaues and roots boiled vntill they be ſoft, and mingled with the meale of Darnell and vineger, aſſwageth the paine of the gout, if they be applied thereto. D

Moreouer being receiued in wine and pepper, it helpeth the iaundice, and prouoketh ſweat, and being put into oile and vſed as an ointment, it cureth ruptures alſo. E

It purgeth the diſeaſe called in Greeke Ἀλφὸς: in Latine, *Vitiligo*, or *Impetigo*, that is, the white ſpottines of the skin, chaps, or rifts in the palms of the hands and ſoles of the feet, and by your patience couſin german to the ſcab of Naples, tranſported or tranſferred into France, and prettily well ſprinkled ouer our Northern coaſts. F

When the ſeed of *Libanotis* is put into receits, you muſt vnderſtand, that it is not meant of the ſeed of Cachris, becauſe it doth with his ſharpeneſſe exaſperate or make rough the gullet; for it hath a very heating qualitie, and doth drie very vehemently, yea this ſeed being taken inwardly, or the herbe it ſelfe, cauſeth to purge vpward and downeward very vehemently. G

CHAP. 395. *Of Corianders.*

¶ *The Deſcription.*

1 THe firſt or common kinde of Coriander is a very ſtinking herbe, ſmelling like the ſtinking worme called in Latine *Cimex*: it hath a round ſtalke full of branches, two foot long. The leaues are of a faint greene colour, very much cut or iagged: the leaues that grow loweſt, and ſpring firſt, are almoſt like the leaues of Cheruill or Parſley, but thoſe which come forth afterward, and grow vpon the ſtalks, are more iagged, almoſt like the leaues of Fumitorie, though a great deale ſmaller, tenderer, and more iagged. The floures are white, and do grow in round taſſels like vnto Dill. The ſeed is round, hollow within, and of a pleaſant ſent and ſauour when it is drie. The root is hard, and of a wooddie ſubſtance, which dieth when the fruit is ripe, and ſoweth it ſelfe from

from yeare to yeare, whereby it mightily increaseth.

1 *Coriandrum.*
Coriander.

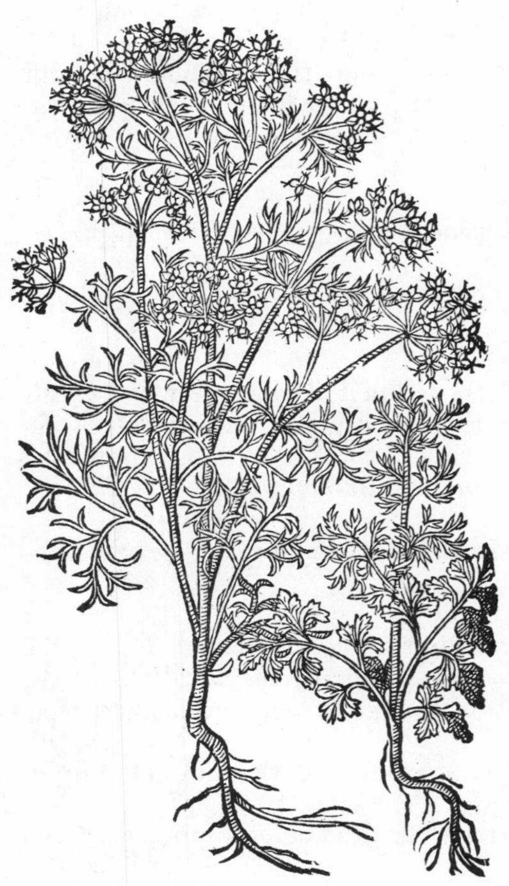

‡ 2 *Coriandrum alterum minus odorum.*
Bastard Coriander.

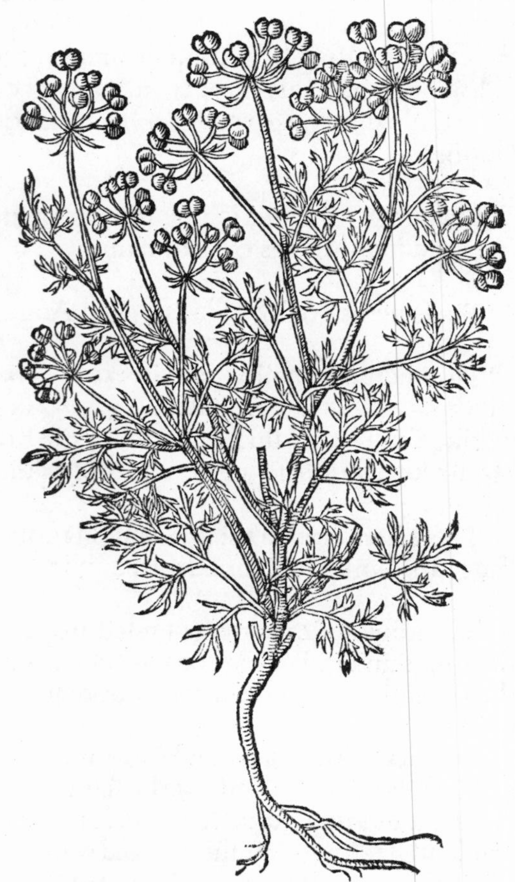

2 There is a second kinde of Coriander very like vnto the former, sauing that the bottome leaues and stalks are smaller: the fruit thereof is greater, and growing together by couples, it is not so pleasant of sauour nor taste, being a wilde kinde thereof, vnfit either for meat or medicine.

¶ *The Place.*

Coriander is sowne in fertile fields and gardens, and the first doth come of it selfe from time to time in my garden, though I neuer sowed the same but once.

¶ *The Time.*

They floure in Iune and Iuly, and deliuer their seed in the end of August.

¶ *The Names.*

The first is called in Latine *Coriandrum*: in English, Corianders. The second, *Coriandrum alterum*, wilde Corianders.

¶ *The Temperature.*

The greene and stinking leaues of Corianders are of complexion cold and dry, and very naught, vnwholesome and hurtfull to the body.

The drie and pleasant well sauouring seede is warme, and very conuenient to sundrie purposes.

¶ *The Vertues.*

A Coriander seed prepared and couered with sugar, as comfits, taken after meat closeth vp the mouth of the stomacke, staieth vomiting, and helpeth digestion.

B The same parched or rosted, or dried in an ouen, and drunk with wine, killeth and bringeth forth wormes, stoppeth the laske, and bloudy flix, and all other extraordinarie issues of bloud.

The manner how to prepare Coriander, both for meat and medicine.

C Take the seed well and sufficiently dried, whereupon poure some wine and vineger, and so leaue them to infuse or steepe foure and twentie houres, then take them forth and drie them, and keepe them for your vse.

D The greene leaues of Coriander boiled with the crums of bread or barly meale, consumeth all hot

hot swellings and inflammations : and with Beane meale dissolueth the Kings euill, wens, and hard lumpes.

The juice of the leaues mixed and laboured in a leaden mortar, with Ceruse, Litharge of siluer, vineger, and oile of Roses, cureth S. Anthonies fire, and taketh away all inflammations whatsoeuer. F

The juice of the greene Coriander leaues, taken in the quantitie of foure dragmes, killeth and poisoneth the body. G

The seeds of Coriander prepared with sugar, preuaile much against the gout, taken in some small quantitie before dinner vpon a fasting stomacke, and after dinner the like without drinking immediately after the same, or in three or foure houres. Also if the same be taken after supper it preuaileth the more, and hath more superiority ouer the disease. H

Also if it be taken with meate fasting, it causeth good digestion, and shutteth vp the stomacke, keepeth away fumes from rising vp out of the same : it taketh away the sounding in the eares, drieth vp the rheume, and easeth the squinancy. I

CHAP. 396. *Of Parsley.*

Apium hortense.
Garden Parsley.

¶ *The Description.*

1 THe leaues of garden Parsley are of a beautiful greene, consisting of many little ones fastned together, diuided most commonly into three parts, and also snipt round about the edges : the stalke is aboue one cubit high, slender, something chamfered, on the top whereof stand spoked rundles, bringing forth very fine little floures, and afterwards small seeds somewhat of a fiery taste: the root is long and white, and good to be eaten.

2 There is another garden Parsley in taste and vertue like vnto the precedent : the onely difference is, that this plant bringeth forth leaues very admirably crisped or curled like fannes of curled feathers, whence it is called *Apium crispum, siue multifidum* ; Curl'd Parsley.

‡ 3 There is also kept in some gardens another Parsley called *Apium siue Petroselinum Virginianum*, or Virginian Parsly; it hath leaues like the ordinary, but rounder, and of a yellowish greene colour, the stalkes are some three foot high, diuided into sundry branches whereon grow vmbels of whitish floures : the seeds are like, but larger than those of the common Parsley, and when they are ripe they commonly sow themselues, and the old roots die, and the young ones beare seed the second yeere after there sowing. ‡

¶ *The Place.*

It is sowne in beds in gardens ; it groweth both in hot and cold places, so that the ground be either by nature moist, or be oftentimes watered : for it prospereth in moist places, and is delighted with water, and therefore it naturally commeth vp neere to fountaines or springs : *Fuchsius* writeth that it is found growing of it selfe in diuers fenny grounds in Germany.

¶ *The Time.*

It may be sowne betime, but it slowly commeth vp ; it may oftentimes be cut and cropped : it bringeth forth his stalkes the second yeare : the seeds be ripe in Iuly or August.

¶ *The Names.*

Euery one of the Parsleyes is called in Greeke σέλινον: but this is named, σέλινον κηπαῖον, that is to say, *Apium hortense:* the Apothecaries and common Herbarists name it *Petroselinum :* in high Dutch,

 Petersilgen:

Petersilgen: in low Dutch, **Trimen Peterselie**: in French, *du Persil*: in Spanish, *Perexil Iuliuert*, and *Salsa*: in Italian, *Petrosello*: in English, Persele, Parsely, common Parsley, and garden Parsley. Yet is it not the true and right *Petroselinum* which groweth among rockes and stones, whereupon it tooke his name, and whereof the best is in Macedonia: therefore they are deceiued who thinke that garden Parsley doth not differ from stone Parsley, and that the onely difference is, for that Garden Parsley is of lesse force than the wilde; for wilde herbes are more strong in operation than those of the garden.

¶ *The Temperature.*

Garden Parsley is hot and dry, but the seed is more hot and dry, which is hot in the second degree, and dry almost in the third: the root is also of a moderate heate.

¶ *The Vertues.*

A The leaues are pleasant in sauces and broth, in which besides that they giue a pleasant taste, they be also singular good to take away stoppings, and to prouoke vrine: which thing the roots likewise do notably performe if they be boiled in broth: they be also delightfull to the taste, and agreeable to the stomacke.

B The seeds are more profitable for medicine; they make thinne, open, prouoke vrine, dissolue the stone, breake and waste away winde, are good for such as haue the dropsie, draw downe menses, bring away the birth, and after-birth: they be commended also against the cough, if they be mixed or boiled with medicines made for that purpose: lastly they resist poisons, and therefore are mixed with treacles.

C The roots or the seeds of any of them boiled in ale and drunken, cast forth strong venome or poison, but the seed is the strongest part of the herbe.

D They are also good to be put into clysters against the stone or torments of the guts.

CHAP. 367. Of water Parsley, or Smallage.

Eleoselinum, siue Paludapium.
Smallage.

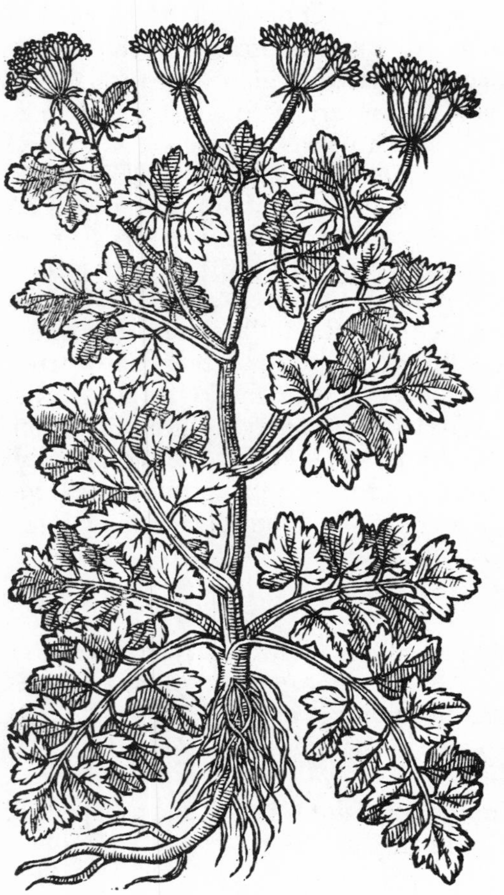

¶ *The Description.*

SMallage hath greene smooth and glittering leaues, cut into very many parcels, yet greater and broader than those of common Parsley: the stalkes be chamfered and diuided into branches, on the tops whereof stand little white floures; after which doe grow seeds something lesser than those of common Parsley: the roote is fastened with many strings.

¶ *The Place.*

This kinde of Parsley delighteth to grow in moist places, and is brought from thence into gardens. ‡ It growes wilde abundantly vpon the bankes in the salt marshes of Kent and Essex. ‡

¶ *The Time.*

It flourishes when the garden Parsley doth, and the stalke likewise commeth vp the next yeere after it is sowne, and then also it bringeth forth seeds which are ripe in Iuly and August.

¶ *The Names.*

It is called in Greeke ἐλεοσέλινον: of *Gaza*, *Paludapium*: in shops, *Apium*, absolutely without any addition: in Latine, *Palustre Apium*, and *Apium rusticum*: in high Dutch, **Epffich**: in low Dutch, **Eppe**, and of diuers **Iouffrouwmerck**: in Spanish and Italian, *Apio*: in French, *de L'ache*: in English, Smallage, Marsh Parsley, or water Parsley.

¶ *The Temperature.*

This Parsley is like in temperature and vertues to that of the garden, but it is both hotter and drier,

drier, and of more force in most things: this is seldome eaten, neither is it counted good for sauce, but it is very profitable for medicine.

¶ *The Vertues.*

A The juice thereof is good for many things, it clenseth, openeth, attenuateth or maketh thin; it remooueth obstructions, and prouoketh vrine, and therefore those syrrups which haue this mixed with them, as that which is called *Syrupus Bizantinus*, open the stoppings of the liuer and spleene, and are a remedy for long lasting agues, whether they be tertians or quartains, and all other which proceed both of a cold cause and also of obstructions or stoppings, and are very good against the yellow jaundise.

B The same juice doth perfectly cure the malicious and venomous vlcers of the mouth, and of the almonds of the throat with the decoction of Barly and *Mel Rosarum*, or hony of Roses added, if the parts be washed therewith: it likewise helpeth all outward vlcers and foule wounds: with hony it is profitable also for cankers exulcerated, for although it cannot cure them, yet it doth keep them from putrifaction, and preserueth them from stinking: the seed is good for those things for which that of the Garden Parsley is: yet is not the vse thereof so safe, for it hurteth those that are troubled with the falling sickenesse, as by euident proofes it is very well knowne.

C Smallage, as *Pliny* writeth, hath a peculiar vertue against the biting of venomous spiders.

D The juice of Smallage mixed with hony and beane floure, doth make an excellent mundificatiue for old vlcers and malignant sores, and staieth also the weeping of the cut or hurt sinewes in simple members, which are not very fatty or fleshie, and bringeth the same to perfect digestion.

E The leaues boiled in hogs grease, and made into the forme of a pultis, take away the paine of felons and whitlowes in the fingers, and ripen and heale them.

CHAP. 398. *Of Mountaine Parsley.*

† *Oreoselinum.*
Mountaine Parsley.

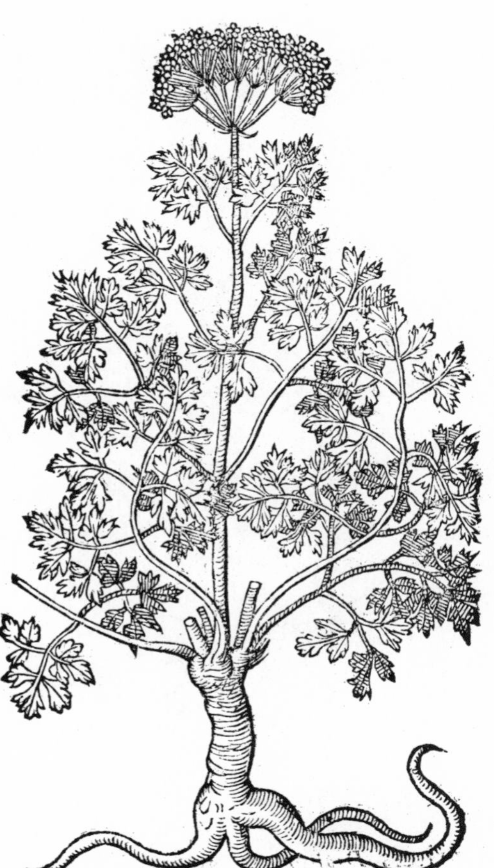

¶ *The Description.*

THe stalke of mountaine Parsley, as *Dioscorides* writeth, is a span high, growing from a slender root; vpon which are branches and little heads like those of Hemlock, yet much slenderer: on which stalkes do grow the seed, which is long, of a sharpe or biting taste, slender, and of a strong smell, like vnto Cumin: but we can not find that this kinde of Mountaine Parsley is knowne in our age: the leaues of this we here giue are like those of common parsley, but greater and broader, consisting of many slender footstalkes fastened vnto them; the stalke is short, the floures on the spoked tufts be white; the seed small: the root is white, and of a meane length or bignesse, in taste somewhat biting and bitterish, and of a sweet smell.

¶ *The Place.*

† *Dioscorides* writeth, that mountaine Parsley groweth vpon rockes and mountaines. And *Dodonæus* affirmeth that this herbe described growes on the hills which diuide Silesia from Morauia, called in times past the counntrey of the Marcomans: also it is said to be found on other mountaines and hills in the North parts of England.

¶ *The Names.*

The Grecians doe name it of the mountaines ὀρεοσέλινον, which the Latines also for that cause doe call *Apium Montanum*, and *Montapium*: in English, mountaine Parsley: in Latine, *Apium*: but *Dioscorides* maketh *Petroselinum* or stone Parsley to differ from mountaine Parsley; for saith he, we we must not be deceiued, taking mountaine Parsley to be that which groweth on rockes: for rocke Parsley is another plant, of some it is called, **Veelgutta:** in Latine, *Multibona,* (in English, much good:) for it is so named because it is good, and profitable for many things: and this is not altogether vnproperly termed *Oreoselinum*, or mountaine Parsley; for it groweth as we haue said on mountaines, and is not vnlike to stone Parsley: the seed is not like to that of Cumin, for if it were so: who would deny it to be *Oreoselinum*, or *Dioscorides*, his mountaine Parsley.

¶ *The Temperature and Vertues.*

A *Oreoselinum*, or mountaine Parsley is, as *Galen* saith, like in faculty vnto Smallage, but more effectuall; *Dioscorides* writeth that the seed and root being drunke in wine prouoke vrine, bring downe the menses, and that they are mixed with counterpoisons, diureticke medicines, and medicines that are hot.

B The root of *Veelgutta*, or much good, is also hot and dry, and that in the later end of the second degree, it maketh thin, it cutteth, openeth, prouoketh, breaketh the stone and expelleth it, openeth the stoppings of the liuer and spleene, and cureth the yellow jaundise: being chewed it helpeth the tooth-ach, and bringeth much water out of the mouth.

† This whole chapter was wholly taken from *Dodonæus Pempt.* 5. *lib.* 4. *cap.* 3. wherefore I haue giuen his figure, which was agreeable to the history, for the figure our Author here gaue, was of the *Selinum montanum pumilum*, farre different from this, as I shall hereafter shew you in the chap. of *Peucedanum*.

Chap. 399. *Of stone Parsley of Macedonie.*

† 1 *Petroselinum Macedonium, Fuchsij.*
Bastard stone Parsley.

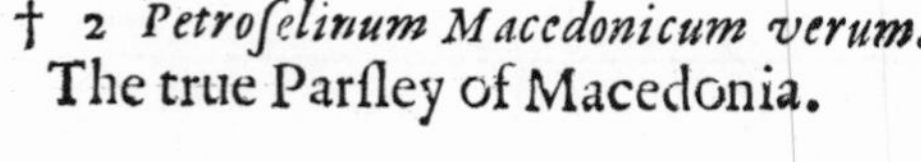

† 2 *Petroselinum Macedonicum verum.*
The true Parsley of Macedonia.

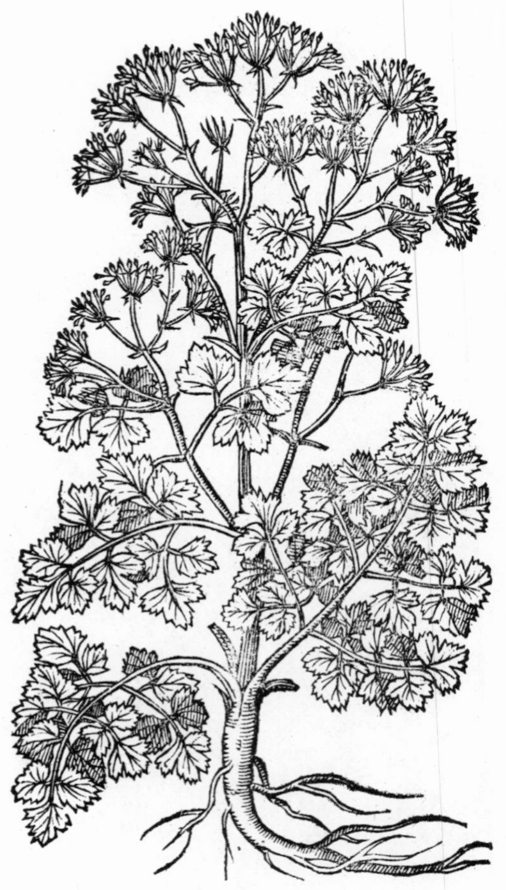

¶ *The Description.*

Of stone Parsley very little is written of the old writers, *Dioscorides* onely saith that this hath seed like to that of *Ameos*, but of a more pleasant smell, sharpe, aromaticall, or spiced: touching the forme of the leaues, the colour of the floures, and fashion of the root

root he writeth nothing at all: and *Pliny* is more briefe; as for *Theophraſtus* he doth not ſo much as name it, making mention onely of Parſley, Alexander, Smallage, and mountaine Parſley.

1 For ſtone Parſley *Leonhartus Fuchſius* hath ſet down a plant, hauing leaues not ſpred and cut after the manner of garden Parſley, but long, and ſnipped round about, made vp and faſtened to a rib or ſtem in the midſt, ſomething like, but yet not altogether, to the firſt leaues of the leſſer Saxifrage; the ſtalke is ſlender, and a cubit and a halfe high; the floures on the ſpokie tufts are white: the ſeed ſomething blacke, like to that of Ameos, and garden Parſley, very ſweet of ſmell, ſomething ſharpe or biting: the root is ſlender and full of ſtrings.

2 *Lobel* alſo in ſtead of the right ſtone Parſley deſcribeth another, which the Venetians call ſtone Parſley of Macedonia: this hath leaues like thoſe of garden Parſley, or rather of the Venetian Saxifrage which is the blacke herbe Frankincenſe formerly deſcribed: the ſtalke is a cubit high; the ſpokie tufts ſomething white: the ſeed ſmall, quickely vading (as he ſaith) inferiour to that of garden Parſley in temperature and vertues: but whether this be the true and right ſtone Parſley, he addeth, he is ignorant.

¶ *The Place.*

It groweth on craggy rocks, and among ſtones: but the beſt in Macedonia, whereupon it beareth the ſurname *Macedonicum*, of Macedonia.

¶ *The Time.*

It floureth in the ſommer moneths.

¶ *The Names.*

It is called in Greeke, πετροσέλινον, of the ſtony places where it groweth: in Latine, *Petrapium*, and *Petroſelinum Macedonicum*: in Engliſh, ſtone Parſley: the Apothecaries know it not: they are farre deceiued that would haue the herbe which *Fuchſius* pictureth to be *Amomum*: for *Amomum* differeth from this, as it is very plaine by the deſcription thereof in *Dioſcorides*: but we hold this for the true ſtone Parſley, till ſuch time as we may learne ſome other more like in leaues to the Parſlies, and in ſeed, ſuch as that of ſtone Parſley ought to be: and the very ſeed it ſelfe may cauſe vs to hold this opinion, being ſo agreeing to the deſcription as no herbe more; for it is ſharpe and biting, and of a ſweeter ſmell than is that of Ameos, and of a more ſpicy ſent; yet do not the leaues gainſay it, which though they haue not the perfect forme of other Parſleyes, yet notwithſtanding are not altogether vnlike. ‡ The firſt of theſe is thought by *Anguillara*, *Turner*, *Geſner*, *Cordus*, and others, to be the *Siſon* of *Dioſcorides*, and *Tragus* calls it, *Amomum Germanicum*, and the ſeeds in ſhops retaine the name of *Sem. Amomi*. The ſecond is thought by *Columna* to be the ſecond *Daucus* of *Dioſcorides*. ‡

¶ *The Temperature.*

The ſeed of ſtone Parſley which is moſt commonly vſed, is hot and dry, hauing withall a cutting quality.

¶ *The Vertues.*

It prouoketh vrine, and bringeth downe the floures: it is profitable againſt winde in the ſtomacke, and collicke gut, and gripings in the belly: for it is, as *Galen* ſaith, ἄφυσον, that is to ſay, a waſter or conſumer of winde: it is a remedy againſt paine in the ſides, kidnies, and bladder, it is alſo mixed in counterpoiſons: *Dioſcorides*. A

† The firſt figure that was formerly in this chapter ſhould haue been in the ſecond place, and that in the ſecond place was of Alexanders, and ſhould haue been put in the following chapter.

‡ Chap. 400. *Of Corne Parſley, or Hone-wort.*

¶ *The Deſcription.*

This herbe commeth vp at the firſt from ſeed like Parſley, with two ſmall long narrow leaues, the next that ſpring are two ſmall round ſmooth leaues nickt about the edges, and ſo for two or three couples of leaues of the next growth there are ſuch round leaues growing on a middle rib by couples, and one round one, alſo at the top; after as more leaues ſpring vp, ſo the faſhion of them alſo change, that is to ſay, euery leafe hath about eight or nine ſmall ſmooth greene leaues, growing on each ſide of a middle rib one oppoſite againſt another, and one growing by it ſelfe at the top, and are finely ſnipt or indented about the edges, in forme reſembling thoſe of *Sium odoratum Tragi*, but not ſo bigge, long, or at all browniſh; amongſt which riſe vp many ſmall round ſtraked ſtalkes or branches, about two foot long, now and then aboue twenty from one root, ſometimes growing vpright, ſometimes creeping not farre from the ground, joynted or kneed,

‡ *Selinum Sij folijs.*
Honewoort.

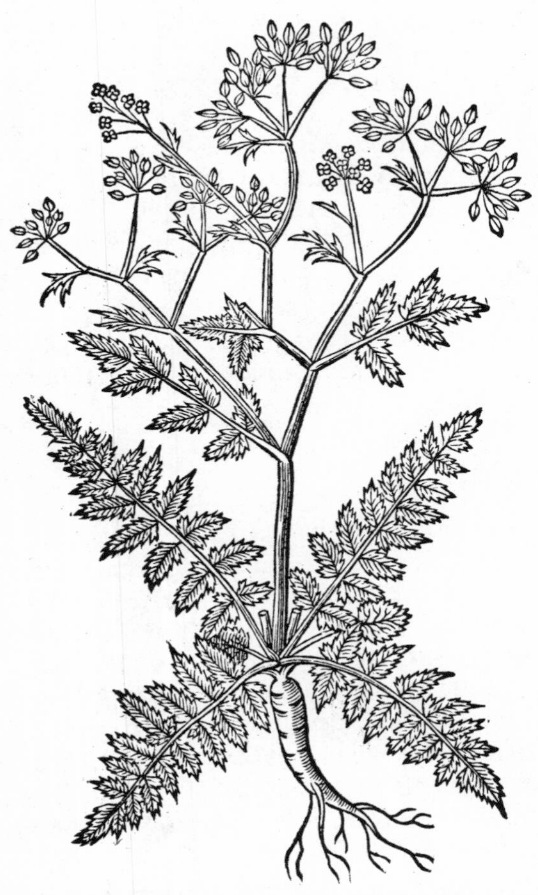

and diuiding themselues into very many branches, at euery joynt groweth one leafe smaller than the former, which together with the lowermost perish, so that there is seldome one greene leafe to be seen on this herbe when the seed is ripe; the floures are white, and grow most commonly at the tops of the branches, sometimes at most of the joynts euen from the earth, in vneuen or vnorderly vmbells, euery floure hauing fiue exceeding small leaues, flat, and broad at the toppe, and in the middle very small cheiues with purple tops, the whole floure not much exceeding the bignesse of a small pins head, which being past there commeth vp in the place of euery floure two small gray crooked straked seeds, like Parsley seeds, but bigger, in taste hot and aromaticall. The root is small and whitish, with many threds not so big as Parsley roots. It beginneth to floure about the beginning of Iuly, & so continues flouring a long time; part of the seed is ripe in August, and some scarse in the beginning of October, mean while some falleth, wherby it renueth it selfe, and groweth with flourishing greene leaues all the winter.

I tooke the descript on of this herbe the yeere, 1620. but obserued it long before, not knowing any name for it: first I refered it to *Sium*, calling it, *Sium terrestre*, and *Sium segetum & agrorum*; afterwards vpon sight of *Selinum peregrinum primum Clusii*, which in some respects resembleth this herbe, I named it *Selinum Sij folijs*; yet wanting an English name, at length about the yeere 1625. I saw Mistris *Vrsula Leigh* (then seruant to Mistris *Bilson* of Mapledurham in Hampshire, and now (5. *Marcij* 1632. wife to Master *William Mooring* Schoolemaster of Petersfield, a Towne neere the said Mapledurham) gather it in the wheate ershes about Mapledurham aforesaid (where in such like grounds it still groweth, especially in clay grounds) who told me it was called Honewort, and that her Mother mistris *Charitie Leigh* late of Brading in the Isle of Wight deceased, taught her to vse it after the manner heere expressed, for a swelling which shee had in her left cheeke, which for many yeeres would once a yere at the least arise there, and swelt with great heat, rednesse, and itching, vntil by the vse of this herbe it was perfectly cured, and rose no more nor swelled, being now (5. *Martij* 1632) about twenty yeeres since, only the scar remaineth to this day. This swelling her mother called by the name of a Hone, but asking whether such tumors werein the said Isle vsually called Hones she could not tell, by reason shee was brought from Brading aforesaid young, and not being aboue twelue yeeres old when shee vsed this medicine.

¶ *The Vertues.*

A Take one handfull of the greene leaues of this Honewort, and stampe them, put to it about halfe a pinte or more of beere, straine it, and drinke it, and so continue to drinke the like quantity euery morning fasting till the swelling doth abate, which with or in her was performed in the space of two weekes at the most. August, 18. 1620. *Iohn Goodyer:* ‡

Chap. 401. *Of Alexander.*

¶ *The Description.*

The leaues of Alexander are cut into many parcells like those of Smallage, but they be much greater and broader, smooth also, and of a deepe greene colour: the stalke is thicke, oftentimes a cubit high: the floures be white, and grow vpon spokie tufts: the seed is thicke, long,

long, blacke ſomething bitter, and of an aromaticall or ſpicy ſmell: the root is thicke, blacke without, white within, like to a little Radiſh, and is good to be eaten. out of which being broken or cut, there iſſueth forth a juice that quickely waxeth thicke, hauing in it a ſharpe bitterneſſe, like in taſte vnto Myrrhe: which thing alſo *Theophraſtus* hath noted, there iſſueth out of it, ſaith he, a juice like Myrrhe.

† *Hippoſelinum.*
Alexanders.

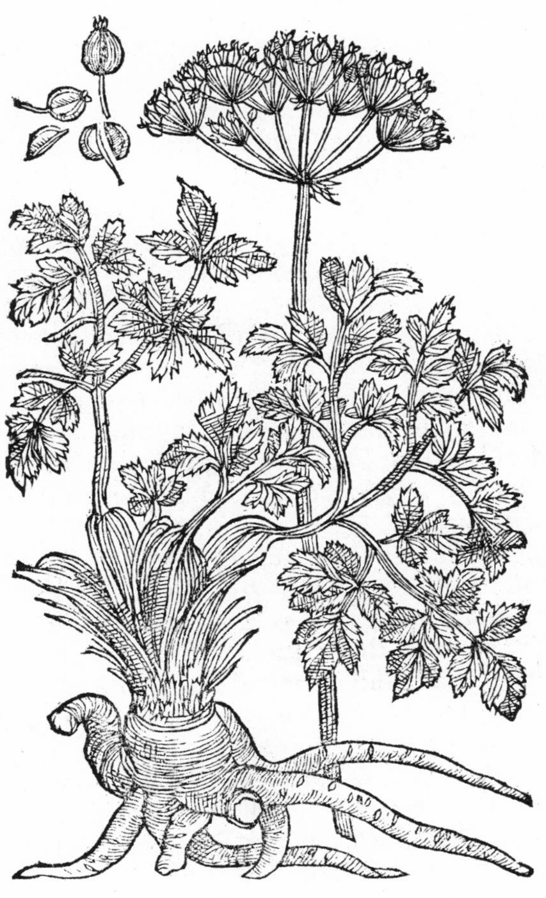

¶ *The Place.*

Alexanders or great Parſley groweth in moſt places of England.

¶ *The Time.*

The ſeed waxeth ripe the ſecond yeere, in the Moneth of Auguſt.

¶ *The Names.*

It is called in Greeke, of the greatneſſe wherein, it excelleth the other Parſleyes ἱπποσέλινον or Horſe Parſley; of *Gaza*, *Equapium*. it is alſo named *Olus atrum*, or the blacke pot-herbe; and of diuers *Syluestre Apium*, or wilde Parſley; of *Galen* and certaine others, σμύρνιον, by reaſon of the juice that iſſueth forth thereof, that is, as we haue ſaid, like vnto Myrrhe, which is called in Greeke σμύρνιον: there is alſo another *Smyrnium* of mount Aman, of which we do write in the 404. chapter: the Apothecaries cal it *Petroſelinum Macedonicum*: others, *Petroſelinum Alexandrinum*: the Germaines, **Groſz Epffich**: the Low Country-men, **Peterſelie van Macedonien**: in Spaniſh, *Perexil Macedonico*: the French, and Engliſhmen, Alexandre, Alexanders.

¶ *The Temperature.*

The ſeed & root of Alexanders, are no leſſe hot and dry than are thoſe of the Garden Parſley, they clenſe and make thinne, being hot and dry in the third degree.

¶ *The Vertues.*

A *Dioſcorides* ſaith, that the leaues and ſtalkes are boiled and eaten, and dreſſed alone by themſelues, or with fiſhes: that they are preſerued raw in pickle: that the root eaten both raw and ſod, is good for the ſtomacke: the root hereof is alſo in our age ſerued to the table raw for a ſallade herbe.

B The ſeeds bring downe the floures, expell the ſecondine, breake and conſume winde, prouoke vrine, and are good againſt the ſtrangury: the decoction alſo of the root doth the ſame, eſpecially if it be made with wine.

† The figure formerly here was of *Sphondylium*, and that belonging to this place was put in the foregoing chapter.

Chap. 402. *Of wilde Parſley.*

¶ *The Deſcription.*

This is like to the kindes of Parſleyes in the ſundry cuts of the leaues, and alſo in the bigneſſe for they be broad and cut into diuers parcels: the ſtalkes are round, chamfered, ſet with certaine joints, hollow within, a cubit high or higher, two or three comming forth together out of one root, and in the nether part many times of a darke reddiſh colour. The floures be white, and grow vpon ſpokie tufts: the ſeed is round, flat, like that of Dill: the root is white within, and diuided into many branches and ſtrings. This plant in what part ſoeuer it be cut or broken, yeeldeth forth a milky juice.

¶ The

† *Apium syluestre siue Thisselium.*
Wilde Parsley.

¶ *The Place.*

It is found by ponds sides in moist and dankish places, in ditches also, hauing in them standing waters, and oftentimes by old stockes of Alder trees.

‡ I haue not as yet obserued this plant growing wilde with vs. ‡

¶ *The Time.*

It floureth and bringeth forth seed in Iune and Iuly.

¶ *The Nature.*

The shops of the Low countries haue miscalled it in times past by the name of *Meum*, and vsed it for the right Mew, or Spiknel wort. The Germaines name it **Olsenich**: *Valerius Cordus*, *Olsenithium*: diuers in the Low-Countries call it **wilde Eppe**: that is to say in Latine, *Apium syluestre*, or wilde Parsley: and some, **water Eppe**: that is, *Hydroselinon*, or *Apium aquatile*, water Parsley: and oftentimes is it named, as we haue already written, *Eleoselinum*, and *Sium*. It may be more rightly termed in Latine, *Apium syluestre*, and in English, wilde Parsley.

Dioscorides hath made mention of wilde Parsley in the chapter of *Daucus* or wilde Carrot: and *Theophrastus* in his seuenth booke, where he maketh the Parsleyes to differ both in leaues and stalkes, and sheweth that some haue white stalkes, others purple, or else of sundry colours, and that there is also a certaine wilde Parsley; for he saith that those which haue the purple stalkes, and the stalkes of diuers colours, come neerest of all to the wilde Parsley. And therefore seeing that *Olsenichium*, or wilde Parsley, hath the lower part of the stalke of a purplish colour, and like in leaues to Parsley, which in times past we thought good rather to call *Apium syluestre*, or wilde Parsley, than to erre with the Apothecaries, and to take it for Mew. And after when we now know that it was held to be *Thysselium Plinij*, and that we could alledge nothing to the contrary, we also setled our selues to be of their opinion; and the rather, because the faculties are agreeable. *Thysselium*, saith *Pliny*, *lib.* 25. chapter 11. is not vnlike to Parsley: the root hereof purgeth flegme out of the head; which thing also the root of *Olsenichium* doth effectually performe, as we will forthwith declare. The name also is agreeable, for it seemeth to be called θυσσέλιον, because it extendeth it selfe, in Greeke, θύειν, thorow ἕλεος, or marish places.

¶ *The Temperature.*

The root hereof is hot and dry in the third degree.

¶ *The Vertues.*

A The root being chewed, bringeth by the mouth flegme out of the head, and is a remedy for the tooth-ach, and there is no doubt but that it also makes thin, cutteth and openeth, prouoketh vrine, and bringeth downe the floures, and doth likewise no lesse but more effectually performe those things that the rest of the Parsleyes do.

† The figure formerly put in this place was of the *Cerefolium syluestre* of *Tabernamontanus*, whose history I intend hereafter to giue you.

Chap. 403. *Of bastard Parsley.*

¶ *The Description.*

1 The first kinde of bastard Parsley is a rough hairy herbe, not much vnlike to Carrots: the leaues are like to those of Corianders, but parted into many small jagges: at the top of the branches do grow shadowie vmbels, or spokie rundles, consisting of many small white floures:

1 *Caucalis albis floribus.*
Baſtard Parſley with white floures.

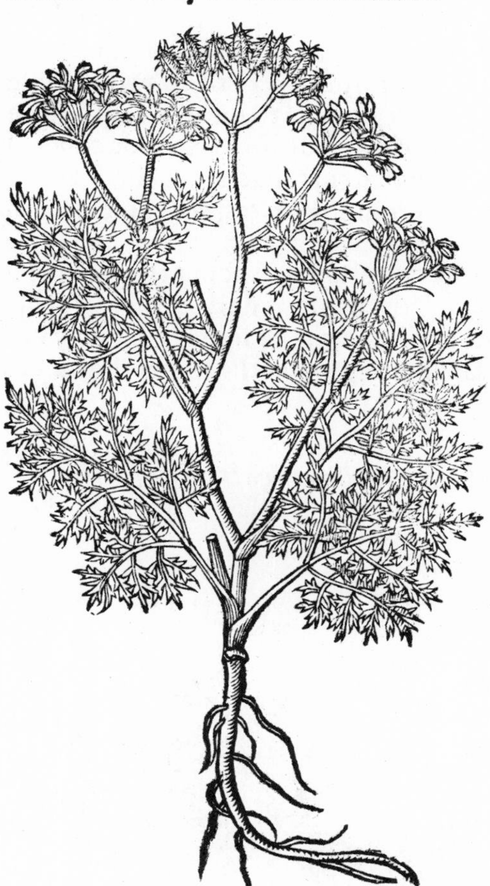

‡ 2 *Caucalis Apij folijs flore rubro.*
Baſtard Parſley with red floures.

‡ 3 *Caucalis Peucedanij folio.*
Hogs Parſley

‡ 4 *Caucalis maior Cluſ.*
Great rough Parſley.

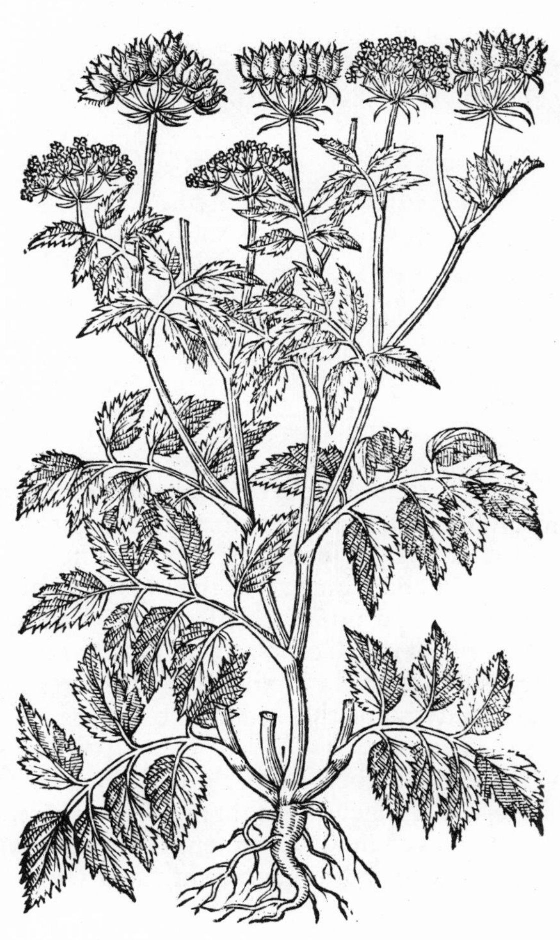

floures: the seed is long and rough, like the seed of Carrots, but greater: the root is straight and single, growing deepe into the ground, of a white colour, and in taste like the Parsnep.

2 There is another sort like vnto the former, sauing that the leaues thereof are broader, and the floures are of a reddish colour: there hath great controuersie risen about the true determination of *Caucalis*, because the Latine interpretation of *Dioscorides* is greatly suspected, conteining in it selfe much superfluous matter, not pertinent to the history: but wee deeme that this plant is the true *Caucalis*, the notes set downe declare it so to be: the floures, saith he, are reddish: the seeds couered with a rough huske set about with prickles, which cleaue vnto garments that it toucheth, as doe Burs; which roughnesse being pilled off, the seed appeares like vnto hulled Otes, not vnpleasant in taste, all which do shew it to be the same.

3 There is likewise another sort that hath a long single root, thrummed about the vpper end with many thrummy threds of a browne colour: from which riseth vp diuers stalkes full of joynts or knees, couered with a sheath or skinnie filme, like vnto that of *Meum*: the leaues are finely cut or jagged, resembling the leaues of our English Saxifrage: the floures grow at the top of the stalkes in spoky rundles like Fennell: the seed is small like that of Parsley.

‡ 5 *Caucalis minor flosculis rubentibus.*
Hedge Parsley.

‡ 6 *Caucalis nodosa echinato semine.*
Knotted Parsley.

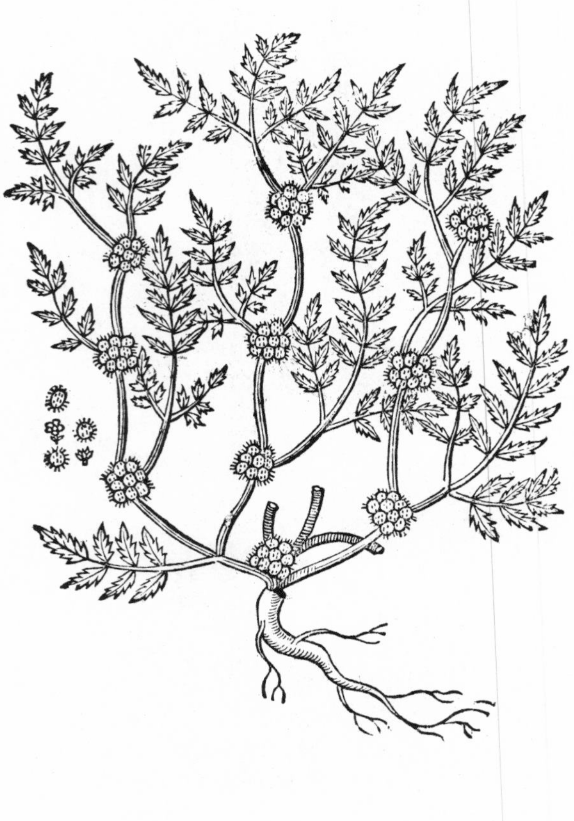

‡ 4 *Clusius* vnder the name of *Caucalis maior* hath described and figured this, which hath many crested straight stalkes some two cubits high or more, which are diuided into sundry branches, and at each joynt send forth large & winged leaues somewhat like those of Angelica, but rougher, and of a darker greene; at the tops of the branches grow vmbels of whitish floures, being of somewhat a purplish or flesh colour vnderneath; and these are succeeded by broad seed almost like those of the Cow-Parsnep, but that they are rougher, and forked at the top, and prickly: the root is white, hard and wooddy. It floures in Iune, ripens the seed in Iuly and August, and then the root dyes, and the seed must be sowne in September, and so it will come vp and continue greene all the winter.

5 Besides these formerly described there are two others growing wilde with vs; the first of these, which I haue thought good to call Hedge, or field Parsley, (because it growes about hedges, and in plowed fields very plentifully euery where) hath crested hollow stalkes growing vp to some cubit and halfe high, whereon stand winged leaues made of sundry little longish ones, set one against

against another, snipt about the edges, and ending in a long and sharpe pointed leafe: these leaues as also the stalkes are somewhat rough and harsh, and of a darke greene color: the floures are small and reddish, and grow in little vmbels, and are succeeded by longish little rough seed of somewhat a strong and aromaticke taste and smell. It is an annuall plant, and floures commonly in Iuly, and the seeds are ripe in August. *Cordus* and *Thalius* call it *Daucoides minus*; and *Bauhine*, *Caucalis semine aspeo flosculis subrubentibus*. There is a bigger and lesser variety or sort of this plant, for you shall find it growing to the height of two cubits, with leaues and all the vpper parts answerable, and you may againe obserue it not to exceed the height of halfe a foot.

6 This other, which *Bauhine* hath first set forth in writing by the name of *Caucalis nodosa echinato semine*, hath a white and long root, from which it sends vp sundry small crested and rough branches which commonly lie along vpon the ground, and they are commonly of an vnequall length, some a cubit long, other some scarse two handfulls: the leaues are small, rough, winged, and deeply jagged, and at the setting on of each leafe close to the stalkes vsually vpon very short foot-stalkes grow small little floures of colour white, or reddish, and made of fiue little leaues apeece: after these follow the seed, round, small and rough, and they grow close to the stalkes. It floures in Iune and Iuly, and growes wilde in sundry places, as in the fields, and vpon the bankes about S. Iames, and Pickadilla. *Fabius Columna* iudges it to be the true *Scandix* of the Antients. ‡

There is likewise one of these found in Spaine, called *Caucalis Hispanica*, like the first: but it is an annuall plant, which perishes at the first approach of winter, the which I haue sowne in my garden, but it perished before the seed was perfected.

¶ *The Place.*

These plants do grow naturally vpon rockes and stony grounds: we haue the first and the third in our pastures in most places of England: that with red floures is a stranger in England.

‡ I haue not heard that the third growes wilde with vs, but the second was found growing in the corne fields on the hilles about Bathe, by Mr. *Bowles*. ‡

¶ *The Time.*

They floure and flourish from May to the end of August.

¶ *The Names.*

Bastard Parsley is called in Greeke καυκαλίς: in Latine also *Caucalis*: of some, *Daucus syluestris*: among the bastard names of *Democritus*, [illegible]: in Latine, *Pes Gallinaceus*, *Pes Pulli*: the Egyptians name it *Seselis*: the country-men of Hetruria, *Petrosello saluatico*: in English, bastard Parsley, and Hennes foot.

¶ *The Temperature and Vertues.*

Dioscorides saith, that bastard Parsley is a pot-herbe which is eaten either raw or boiled, and prouoketh vrine. A

Pliny doth reckon it vp also among the pot-herbes: *Galen* addeth, that it is preserued in pickle for sallades in winter. B

The seed of bastard Parsley is euidently hot and dry, and that in the second degree: it prouoketh vrine, and bringeth downe the desired sickenesse: it dissolueth the stone, and driueth it forth. C

It taketh away the stoppings of the liuer, spleene, and kidnies: it cutteth and concocteth raw and flegmaticke humours: it comforteth a cold stomack, dissolueth winde, it quickneth the sight, and refresheth the heart, if it be taken fasting. D

Matthiolus in his Commentaries vpon *Dioscorides*, the second booke, attributeth vnto it many excellent vertues, to prouoke venery and bodily lust, and erection of the parts. E

† The figure which belonged to the third description in this chapter was formerly put for English Saxifrage.

Chap. 404. *Of Candy Alexanders.*

¶ *The Description.*

Dioscorides and *Pliny* haue reckoned *Smyrnium* among the kindes of Parsley, whose iudgements while this plant is young, and not growne vp to a stalke, may stand with very good reason, for that the young leaues next the ground are like to Parsley, but somewhat thicker and larger: among which riseth vp a stalke a cubit high, and somewhat more, garnished with round leaues, farre different from those next the ground, inclosing the stalke about like Thorow wax, or *Perfoliata*, which leaues are of a yellow colour, and do rather resemble the leaues of Fole-foot than Parsley: at

Smyrnium Creticum.
Candy Alexander.

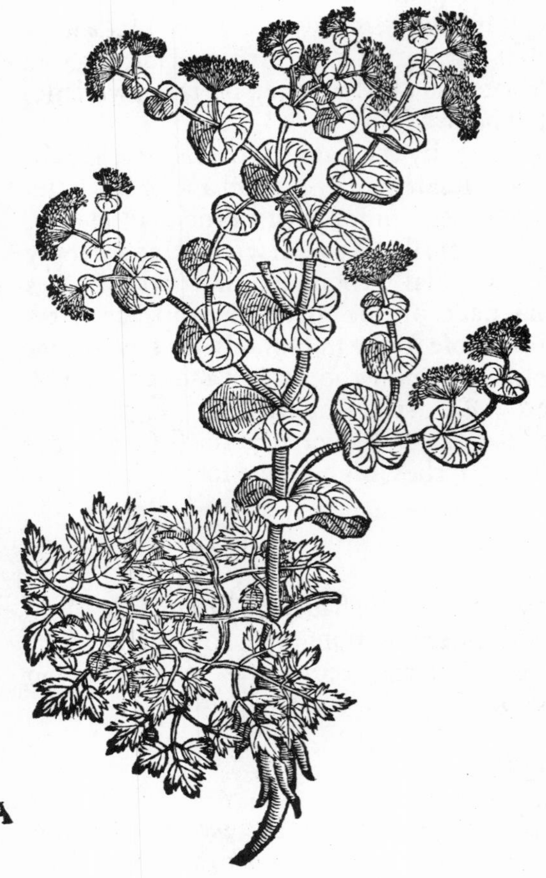

at the top of the stalkes doe grow round spokie tufts of a yellow color, after which commeth round and blacke seed like Coleworts, of a sharpe and bitter taste like Myrrhe: the root is white and thicke, contrary to the opinion of *Dodonæus*, who saith it is blacke without, but I speake that which I haue seen and prooued.

¶ *The Place.*

Smyrnium groweth naturally vpon the hils and mountaines of Candy, and in my garden also in great plenty: also vpon the mountain Amanus in Cilicia.

¶ *The Time.*

Smyrnium floureth in Iune, and the seed is ripe in August.

¶ *The Names.*

This plant is called in Latine, *Smyrnium*: in Greeke, σμύρνιον in Cilicia, *Petroselinum*, and as *Galen* testifieth, some haue called it, *Hipposelinum agreste*: in English, Candy Alexanders, or Thorow bored Parsley.

¶ *The Nature.*

Smyrnium is hot and dry in the third degree.

¶ *The Vertues.*

A The leaues of *Smyrnium* dissolue wens and hard swellings, dry vp vlcers and excoriations, and glew wounds together.

B The seeds are good against the stoppings of the spleene, kidnies, and bladder.

C Candy Alexanders hath force to digest and wast away hard swellings, in other things it is like to garden Parsley, and stone Parsley, and therefore we vse the seed heereof to prouoke the desired sickenesse, and vrine, and to helpe those that are stuffed in the lungs, as *Galen* writeth.

D The root is hot, so is the herbe and seed, which is good to be drunke against the biting of serpents: it is a remedy for the cough, and profitable for those that cannot take their breath vnlesse they do sit or stand vpright: it helpeth those that can hardly make their water: the seed is good against the infirmity of the spleene or milt, the kidnies and bladder: it is likewise a good medicine for those that haue the dropsie, as *Dioscorides* writeth.

Chap. 405. *Of Parsneps.*

¶ *The Description.*

1 The leaues of the tame or Garden Parsneps are broad, consisting of many small leaues fastened to one middle rib like those of the ash tree: the stalke is vpright, of the height of a man: the floures stand vpon spokie tufts, of colour yellow; after which commeth the seed flat and round, greater than those of Dil: the root is white, long, sweet, and good to be eaten.

2 The wilde Parsnep is like to that of the Garden, in leaues, stalke, tuft, yellow floures, flat and round seed, but altogether lesser: the root is small, hard, wooddy, and not fit to be eaten.

¶ *The Place.*

The garden Parsnep requireth a fat and loose earth, and that that is digged vp deepe.

The wilde Parsnep groweth in vntoiled places, especially in the salt marshes, vpon the bankes and borders of the same: the seed whereof being gathered and brought into the garden, and sowed

in

1 *Pastinaca latifolia sativa.*
Garden Parsneps.

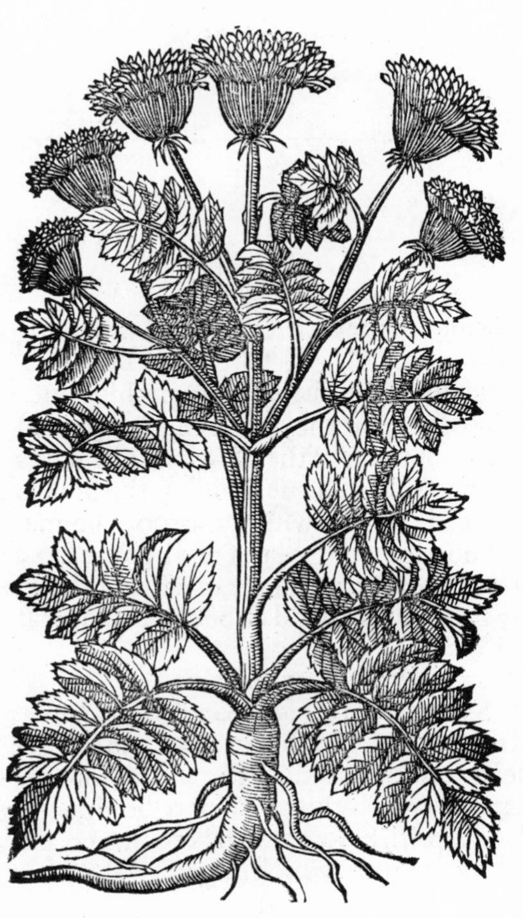

2 *Pastinaca latifolia sylvestris.*
Wilde Parsneps.

in fertill ground, do proue better roots, sweeter and greater than they that are sowne of seeds gathered from those of the garden.

They floure in Iuly and August, and seed the second yeare after they be sowne.

¶ *The Names.*

The Herbarists of our time do call the garden Parsneps σταφυλῖνος, and *Pastinaca*, and therefore wee haue surnamed it *Latifolia*, or broad leafed, that it may differ from the other garden Parsnep with narrow leaues, which is truly and properly called *Staphylinus*, that is, the garden Carrot. Some Physitions doubting, and not knowing to what herbe of the Antients it should be referred, haue fained the wilde kinde hereof to be *Panacis species*, or a kind of Alheale: diuers haue named it *Baucia*; others, *Branca Leonina*, but if you diligently marke and confer it with *Elaphoboscum* of *Dioscorides*, you shal hardly finde any difference at all: but the plant called at Montpelier *Pabulum Ceruinum*: in English, Harts fodder, supposed there to be the true *Elaphoboscum*, differeth much from the true notes thereof. Now *Baucia*, as *Iacobus Manlius* reporteth in *Luminari maiore*, is *Dioscorides*, and the old Writers *Pastinaca*, that is to say, *Tenuifolia*, or Carrot: but the old writers, and especially *Dioscorides* haue called this wilde Parsnep by the name of *Elaphoboscum*: and wee doe call them Parsneps and Mypes.

¶ *The Temperature.*

The Parsnep root is moderately hot, and more drie than moist.

¶ *The Vertues.*

The Parsneps nourish more than doe the Turneps or the Carrots, and the nourishment is somewhat thicker, but not faultie nor bad; notwithstanding they be somwhat windy: they passe through the bodie neither slowly nor speedily: they neither binde nor loose the belly: they prouoke vrine, and lust of the bodie: they be good for the stomacke, kidneies, bladder, and lungs. A

There is a good and pleasant food or bread made of the roots of Parsneps, as my friend Mr. *Plat* hath set forth in his booke of experiments, which I haue made no triall of, nor meane to do. B

The seed is hotter and drier euen vnto the second degree, it mooueth vrine, and consumeth winde. C

D It is reported, ſaith *Dioſcorides*, that Deare are preſerued from bitings of Serpents, by eating of the herbe *Elaphoboſcum*, or wilde Parſnep, wherupon the ſeed is giuen with wine againſt the bitings and ſtingings of Serpents.

† Both the figures that formerly were in this chapter were of the Garden Parſnep; the firſt being that of *Lobel*, and the ſecond that of *Tabernamontanus*: that hich ſhould haue beene in the ſecond place, was formerly put for *Sphondylium*.

Chap. 406. *Of Skirrets.*

Siſarum. Skirrets.

¶ *The Deſcription.*

THe leaues of the Skirret do likewiſe conſiſt of many ſmall leaues faſtened to one rib, euerie particular one whereof is ſomething nicked in the edges, but they are leſſer, greener, and ſmoother than thoſe of the Parſnep. The ſtalks be ſhort, and ſeldome a cubit high; the floures in the ſpoked tufts are white, the roots be many in number, growing out of one head an hand bredth long, moſt commonly not a finger thick, they are ſweet, white, good to be eaten, and moſt pleaſant in taſte.

¶ *The Place and Time.*

This Skirret is planted in Gardens, and eſpecially by the root, for the greater and thicker ones being taken away, the leſſer are put into the earth againe: which thing is beſt to be done in March or Aprill, before the ſtalks come vp, and at this time the roots which be gathered are eaten raw, or boiled.

¶ *The Names.*

This herbe is called in Latin *Siſarum*, and alſo in Greeke σίσαρον; the Latines do likewiſe call it *Siſer*; and diuers of the later Herbariſts, *Seruillum* or *Cheruillum*, or *Seruilla*. the Germans name it **Gierlin**: *Tragus*, **Zam garten Rapunkelen**: in the Low-countries, **Suycker wortelen**, that is to ſay, Sugar roots, and oftentimes **Serillen**: in Spaniſh, *Cherinia*: in Italian, *Siſaro*: in French, *Cheruy*: in Engliſh, Skirret and Skirwort. And this is that *Siſer* or Skirret which *Tiberius* the Emperour commanded to be conueied vnto him from Gelduba a caſtle about the riuer of Rhene, as *Pliny* reporteth in *lib.* 19. *cap.* 5. The Skirret is a medicinable herb, and is the ſame that the foreſaid Emperour did ſo much commend, inſomuch that he deſired the ſame to be brought vnto him euery yeare out of Germanie. It is not, as diuers ſuppoſe, *Serapio* his *Secacul*, of which he hath written in his 89. chapter: for *Secacul* is deſcribed by the leafe of *Iulben*, that is to ſay, of the peaſe, as *Matthiolus Syluaticus* expoundeth it: and it bringeth forth a black fruit of the bigneſſe of a Cichpeaſe, full of moiſture, and of a ſweet taſte, which is called *Granum Culcul*: But the Skirret hath not the leafe of the peaſe, neither doth it bring forth fruit like to the Ciche peaſe; whereupon it is manifeſt, that the Skirret doth very much differ from *Serapio* his *Secacul*: ſo farre is it from beeing the ſame.

¶ *The Nature and Vertues.*

A The roots of the Skirret be moderately hot and moiſt; they be eaſily concocted; they nouriſh meanly, and yeeld a reaſonable good iuice: but they are ſomething windie, by reaſon whereof they alſo prouoke luſt.

B They be eaten boiled, with vineger, ſalt, and a little oile, after the manner of a ſallad, and oftentimes they be fried in oile and butter, and alſo dreſſed after other faſhions, according to the skil of the cooke, and the taſte of the eater.

The

The women in Sueuia, saith *Hieronymus Heroldus*, prepare the roots hereof for their husbands, and know full well wherefore and why, &c. C

The iuice of the roots drunke with goats milke stoppeth the laske. The same drunke with wine putteth away windinesse out of the stomacke, and gripings of the belly, and helpeth the hicket or yeoxing. They stir vp appetite, and prouoke vrine. D

CHAP. 407. *Of Carrots.*

¶ *The Description.*

1 THe leaues of the garden Carrots are of a deepe greene colour, composed of many fine Fennell-like leaues, very notably cut or iagged; among which riseth vp a stalk straight and round, foure cubits high, somwhat hairie and hollow, hauing at the top round spoked tufts, in which do grow little white floures: in their places commeth the seed, rough and hairie, of a sweet smell when it is rubbed. The root is long, thicke and single, of a faire yellow colour, pleasant to be eaten, and very sweet in taste.

1 *Pastinaca sativa tenuifolia.*
Yellow Carrot.

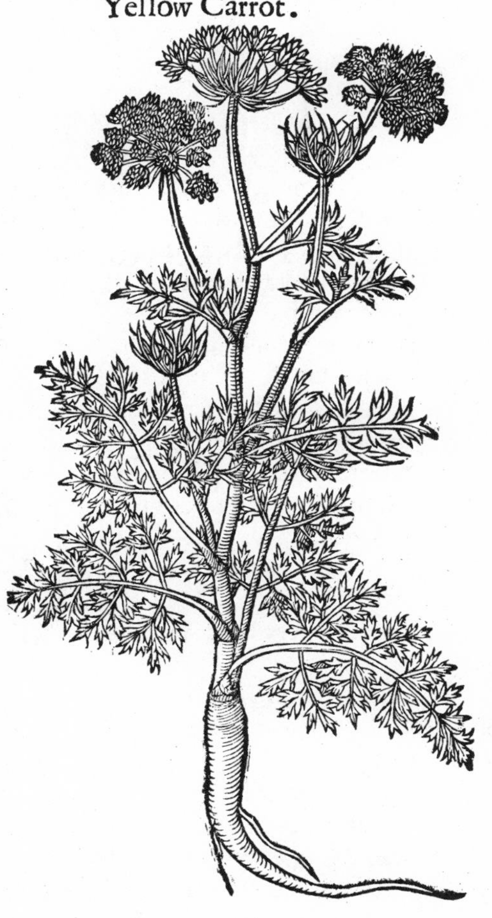

‡ 2 *Pastinaca satiua atro-rubens.*
Red Carrot.

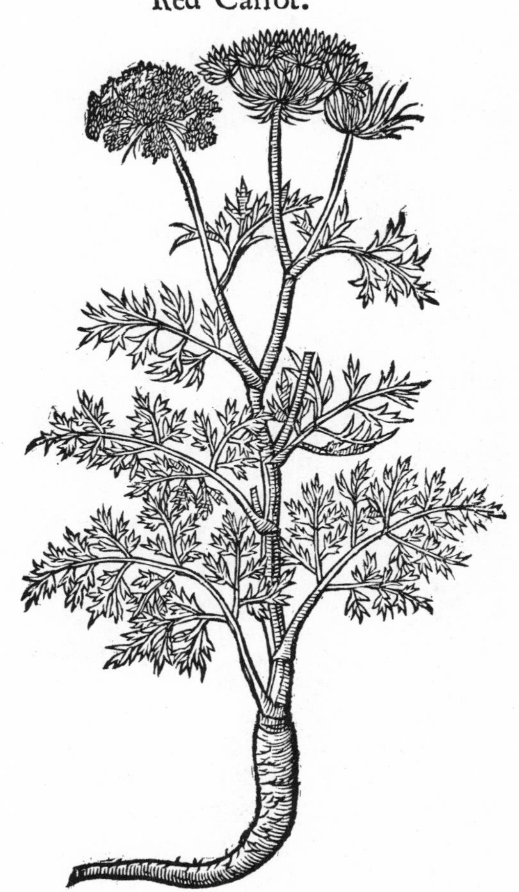

2 There is another kinde hereof like to the former in all parts, and differeth from it only in the colour of the root, which in this is not yellow, but of a blackish red colour.

¶ *The Place.*

These Carrots are sowne in the fields, and in gardens where other pot herbes are: they require a loose and well manured soile.

¶ *The Time.*

They are to be sowne in Aprill; they bring forth their floures and seed the yeare after they be sowne.

¶ *The Names.*

The Carrot is properly called in Greeke σταφυλῖνος, for that which we haue termed in Latine by the name

name of *Pastinaca latioris folij*, or the Garden Parsnep, is described of the old writers by another name: this Carrot is called in Latine likewise, *Pastinaca sativa*, but with this addition *tenuifolia*, that it may differ from the garden Parsnep with broad leaues, and white roots. *Theophrastus* in the ninth booke of his historie of plants nameth this *Staphylinus*, or Carrot, δαῦκος, and writeth that it groweth in Arcadia, and saith that the best is found in *Spartensi Achaia*, but doubtlesse he meant that *Daucus* which we call *Cretensis*, that may be numbred among the Carrots: *Galen* in his booke of the faculties of Simple medicines doth also make it to be *Daucus*, but yet not simply *Daucus*; for he addeth also *Staphilinus* or *Pastinaca*: in High Dutch it is called **Geel ruben**: in Low Dutch, **Geel Peen**, **Geel Pooten**, and **Geel Wortelen**: in French, *Carrotte*, and *Racine iaulne*: in Italian, *Pastinaca*: in Spanish, *Canahoria*: in English, Yellow Carrots: the other is called red Carrot, and blacke Carrot.

¶ *The Temperature and Vertues.*

A The root of the yellow Carrot is most commonly boiled with fat flesh and eaten: it is temperately hot and something moist. The nourishment which commeth thereof is not much, and not verie good: it is something windie, but not so much as be the Turneps, and doth not so soon as they passe through the bodie.

B The red Carrot is of like facultie with the yellow. The seed of them both is hot and drie, it breaketh and consumeth windinesse, prouoketh vrine, as doth that of the wilde Carrot.

Chap. 408. *Of Wilde Carrot.*

Pastinaca syluestris tenuifolia.
Wilde Carrot, or Bees-nest.

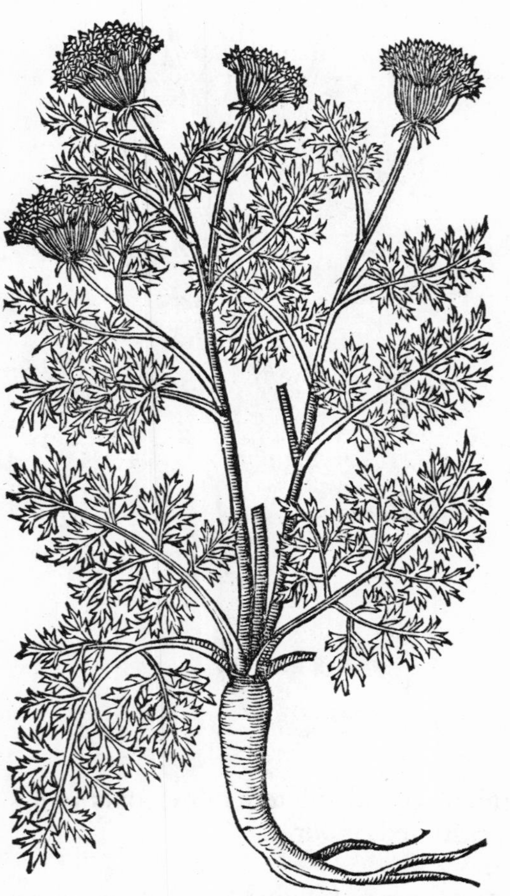

¶ *The Description.*

The leaues of the wilde Carrot are cut into diuers slender narrow parcels, very like vnto those of the garden Carrots, but they be somewhat whiter, and more hairie: the stalks be likewise hairie and somewhat rough: the floures are little, and stand vpon broad spoked tuftes, of a white color, of which tuft of floures the middlemost part is of a deep purple: the whole tuft is drawn together when the seed is ripe, resembling a birds nest; whereupon it hath been named of some Birds-nest: the root slender, and of a mean length.

¶ *The Place.*

It groweth of it selfe in vntoyled places, in fields, and in the borders thereof, almost euerie where.

¶ *The Time.*

It floures and flourishes in Iune and Iuly, the seed is ripe in August.

¶ *The Names.*

The wilde Carrot is called in Greeke σταφυλῖνος ἄγριος: in Latine, *Pastinaca syluestris tenuifolia*: in shops, *Daucus*: and it is vsed in stead of the true *Daucus*, and not amisse, nor vnprofitably: for *Galen* also in his time doth testifie that it was taken for *Daucus*, or bastard Parsly, and is without doubt *Dauci syluestris genus*, or a wilde kinde of bastard Parsly, so called of *Theophrastus*: in high Dutch it is named **Wild Pastenen**, **Vogel nest**: in Low Dutch, **Vogels nest**, and **Wilde Caroten Crookens cruyt**: in French, *Pastena de Sauvage*: in English, wilde Carrot, and after the Dutch, Birds-nest, and in some places Bees-nest.

Athenæus citing *Diphilus* for his Author, saith, that the Carrot is called φίλτρον, because it serueth for loue matters; and *Orpheus*, as *Pliny* writeth, said, that the vse hereof winneth loue: which things be bewritten of wilde Carrot,the root whereof is more effectuall than that of the garden,and containeth in it,as *Galen* ſaith,a certaine force to procure luſt.

¶ *The Temperature and Vertues.*

A The ſeed of this wild Carrot,and likewiſe the root is hot and drie in the ſecond degree,and doth withall open obſtructions.

B The root boiled and eaten,or boiled with wine, and the decoction drunke, prouoketh vrine, expelleth the ſtone,bringeth forth the birth; it alſo procureth bodily luſt.

C The ſeed drunke bringeth downe the deſired ſickneſſe, it is good for them that can hardly make water,it breaketh and diſſolueth winde,it remedieth the dropſie,it cureth the collick and ſtone,being drunke in wine.

D It is alſo good for the paſſions of the mother,and helpeth conception: it is good againſt the bitings of all manner of venomous beaſts: it is reported, ſaith *Dioſcorides*, that ſuch as haue firſt taken of it are not hurt by them.

Chap. 409. *Of Candie Carrots.*

Daucus Cretenſis verus.
Candie Carrots.

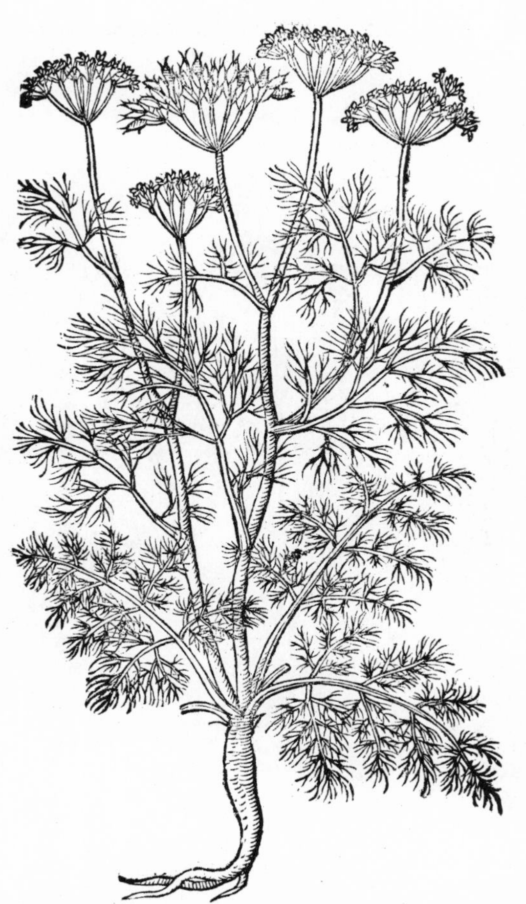

¶ *The Deſcription.*

This *Daucus Cretenſis*, being the true *Daucus* of *Dioſcorides*,doth not grow in Candy only,but is found vpon the mountaines of Germany, and vpon the hills and rockes of Iura about Geneua, from whence it hath beene ſent and conueied by one friendly Herbariſt vnto another, into ſundrie regions: it beareth leaues which are ſmall,and very finely iagged,reſembling either Fennel or wild Carrot: among which riſeth vp a ſtalke of a cubit high,hauing at the top white ſpokie tufts,and the floures of Dill:which being paſt,there come great plentie of long ſeed,well ſmelling, not vnlike the ſeed of Cumin, ſaue that it is whitiſh,with a certaine moſſineſſe, and a ſharpe taſte,and is in greater vſe than any part of the plant. The root alſo is right good in medicine,being leſſer than the root of a Parſnep, but hotter in taſte, and of a fragrant ſmell.

¶ *The Time.*

This floures in Iune and Iuly,his ſeed is ripe in Auguſt.

¶ *The Names.*

There is ſufficient ſpoken in the deſcription as touching the name.

¶ *The Nature.*

Theſe plants are hot and drie, eſpecially the ſeed of *Daucus Creticus*, which is hot and drie in the third degree: but the ſeed of the wilde Carrot is hot and drie in the ſecond degree.

¶ *The Vertues.*

A The ſeed of *Daucus* drunken is good againſt the ſtrangurie,and painfull making of water,it preuaileth againſt the grauell and ſtone,and prouoketh vrine.

B It aſſwageth the torments and gripings of the belly,diſſolueth windines,cureth the collick, and ripeneth an old cough.

C The ſame beeing taken in VVine, is verie good againſt the bitings of beaſts, and expelleth poiſon.

D The ſeed of *Daucus Creticus* is of great efficacie and vertue being put into Treacle, Mithridate, or any antidotes,againſt poiſon or peſtilence.

E The root thereof drunke in wine ſtoppeth the laske, and is alſo a ſoueraigne remedie againſt venome and poiſon.

Chap. 410. *Of ſtinking and deadly Carrots.*

¶ The Deſcription.

1 THe great ſtinking Carrot hath very great leaues, ſpread abroad like wings, reſembling thoſe of Fennell gyant (whereof ſome haue taken it to be a kinde, but vnproperly) of a bright greene colour, ſomewhat hairie: among which riſeth vp a ſtalk of the height of two cubits, and of the bigneſſe of a mans finger; hollow, and full of a ſpungious pith; whereupon are ſet at certaine ioints, leaues like thoſe next the ground, but ſmaller. The floures are yellow, ſtanding at the top of the ſtalkes in ſpokie rundles, like thoſe of Dill: after which commeth the ſeed, flat and broad like thoſe of the Parſnep, but much greater and broader. The root is thicke, garniſhed at the top with certaine capillaments or hairy threds, blacke without, white within, full of milkie iuice, of a moſt bitter, ſharpe, and lothſome taſte and ſmell, inſomuch that if a man do ſtand where the wind doth blow from the plant, the aire doth exulcerate and bliſter the face, and euery other bare or naked place that may be ſubiect to his venomous blaſt, and poiſonous qualitie.

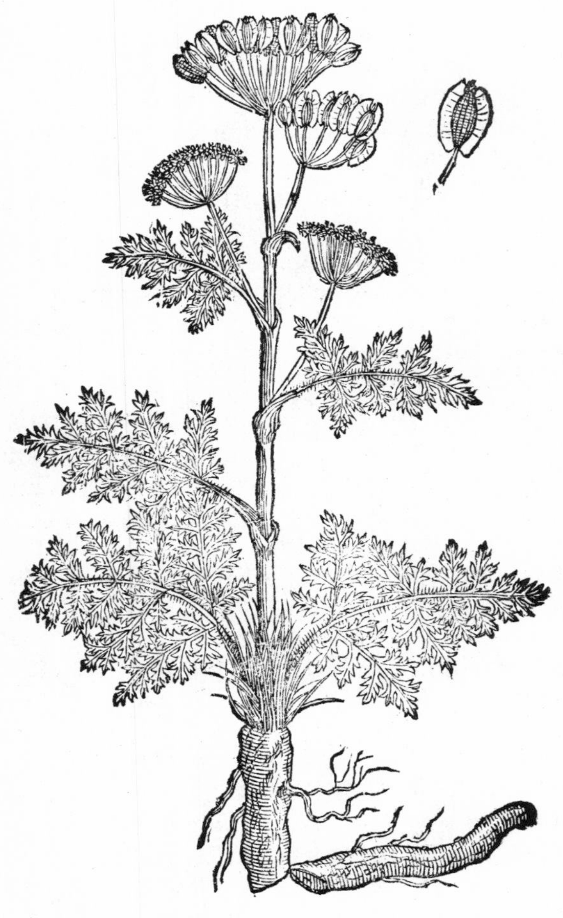

1 *Thapſia latifolia Cluſij.*
Stinking Carrots.

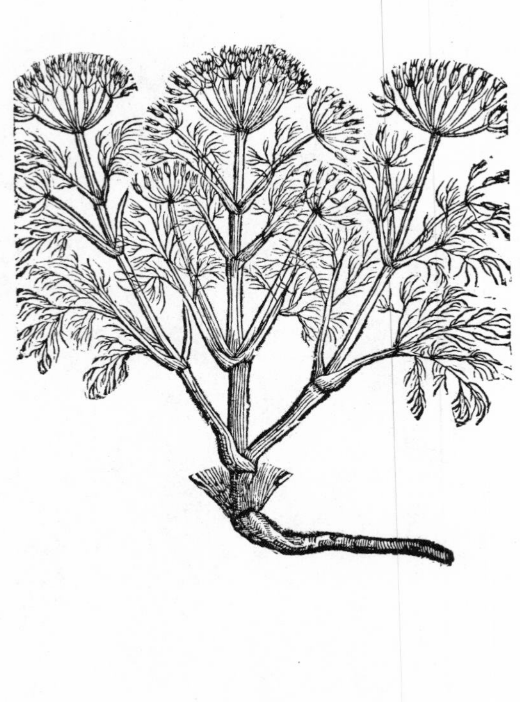

2 *Thapſia tenuifolia.*
Small leafed ſtinking Carrot.

2 This ſmall kind of ſtinking or deadly Carrot is like to the laſt deſcribed in each reſpect, ſauing that the leaues are thinner and more finely minced or iagged, wherein conſiſts the difference.

3 The common deadly Carrot is like vnto the precedent, ſauing that he doth more neerely reſemble the ſtalkes and leaues of the garden carrot, and is not garniſhed with the like buſh of haire about the top of the ſtalks: otherwiſe in ſeed, root, and euill ſmell, taſte and qualitie like.

¶ The Place.

Theſe pernicious plants delight in ſtonie hills and mountaines: they are ſtrangers in England.

¶ *The*

3 *Thapſia vulgaris.*
Deadly Carrots.

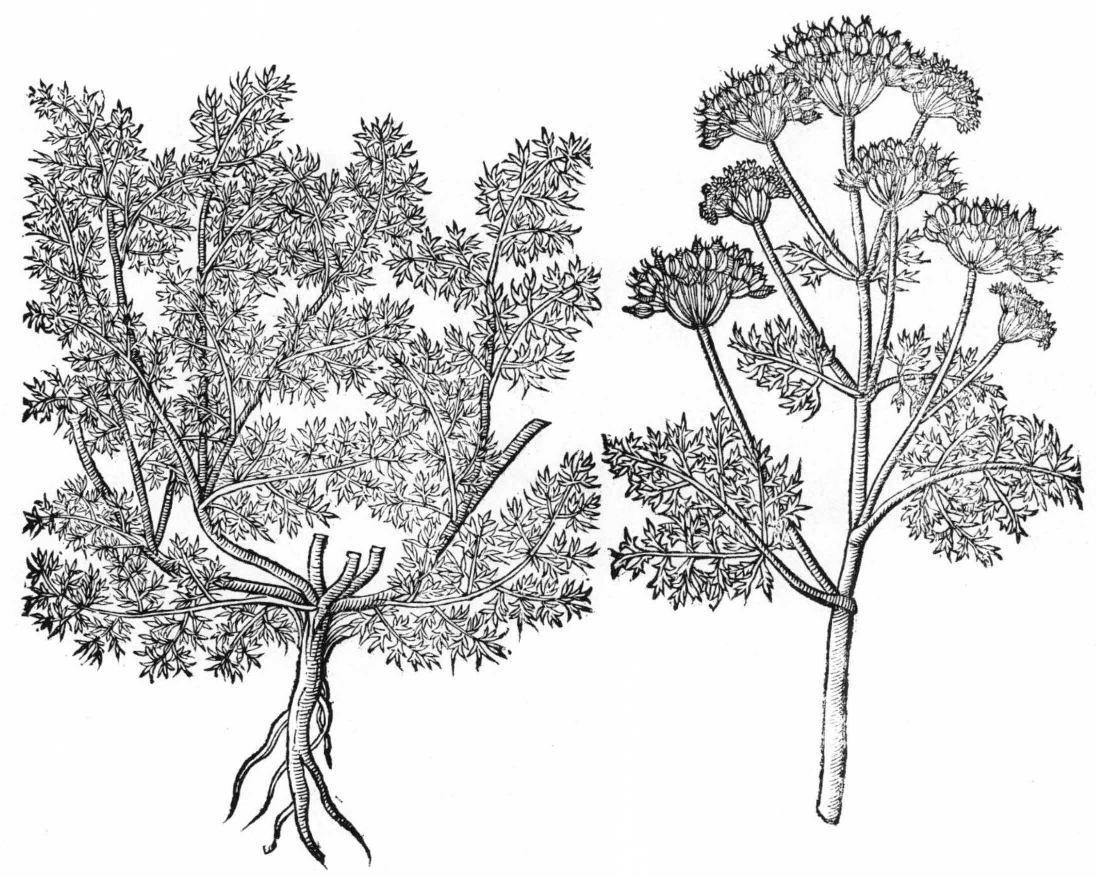

¶ *The Time.*

They floure in Auguſt,or ſomewhat after.

¶ *The Names.*

The French Phyſitians haue accepted the root of *Thapſia* for a kinde of Turbith,calling it *Turpetum Cineritium*; notwithſtanding vpon better conſideration they haue left the vſe thereof,eſpecially in purging, for it mightily hurteth the principall parts, and doth often cauſe cruell gripings in the guts and belly, with convulſions and cramps: neuertheleſſe the venomous qualitie may bee taken away with thoſe correctiues which are vſed in mitigating the extreme heate and virulent qualitie of *Sarcocolla, Hammoniacum,* and *Turpetum*: but where there be ſo many wholeſome Simples, and likewiſe compounds, they are not to be vſed.

Of ſome it is called *Turpetum Griſeum*: it is called *Thapſia*, as ſome thinke, of the Iſland Thapſus, where it was firſt found; or as we deeme, of the likeneſſe it hath with Carrots.

Of the people of Sicilia and Apulia it is called *Ferulacoli*, where it doth grow in great aboundance.

¶ *The Temperature and Vertues.*

The temperature and faculties in working haue been touched in the deſcription, and likewiſe in the names.

Chap. 411. *Of Fennell.*

¶ *The Deſcription.*

1 The firſt kinde of Fennell, called in Latine *Fœniculum*: in Greeke, Μάραθρον, is ſo well knowne amongſt vs, that it were but loſt labour to deſcribe the ſame.

2 The ſecond kinde of Fennell is likewiſe well knowne by the name of Sweet Fennell, ſo called becauſe the ſeeds thereof are in taſte ſweet like vnto Anniſe ſeeds, reſembling the common Fennell, ſauing that the leaues are larger and fatter, or more oleous: the ſeed greater and whiter, and the whole plant in each reſpect greater.

¶ The

Fœniculum vulgare.
Common Fennell.

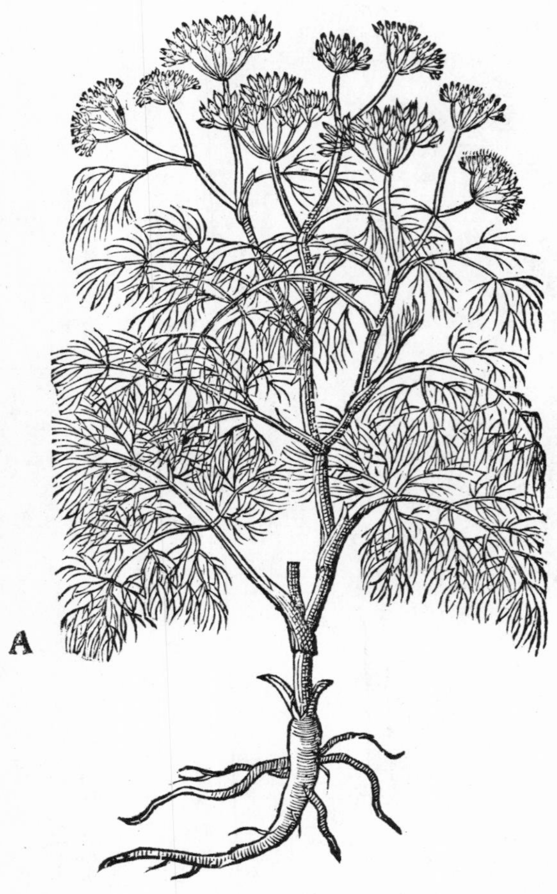

¶ *The Place.*

These herbs are set and sowne in gardens; but the second doth not prosper well in this Countrey: for being sowne of good and perfect seed, yet in the second yeare after his sowing, it will degenerate from the right kinde, and become common Fennell.

¶ *The Time.*

They floure in Iune and Iuly, and the seed is ripe in the end of August.

¶ *The Names.*

Fennell is called in Greeke Μάραθρον: in Latine, *Marathrum*, and *Fœniculum*: in high Dutch, **Fenckell**: in low-Dutch, **Venckell**: in Italian, *Finocchio*: in Spanish, *Hinoio*: in French, *Fenoil*: in English, Fennell, and Fenckell.

¶ *The Nature.*

The seed of Fennel is hot and dry in the third degree.

¶ *The Vertues.*

A The pouder of the seed of Fennell drunke for certaine dayes together fasting preserueth the eye-sight: whereof was written this Distichon following:

Fœniculus, Rosa, Verbena, Chelidonia, Ruta,
Ex his fit aqua quæ lumina reddit acuta.

Of Fennell, Roses, Veruain, Rue, and Celandine,
Is made a water good to cleere the sight of eine.

B The greene leaues of Fennell eaten, or the seed drunken made into a Ptisan, do fill womens brests with milke.

C The decoction of Fennell drunke easeth the paines of the kidnies, causeth one to auoid the stone, and prouoketh vrine.

D The roots are as effectuall, and not onely good for the intents aforesaid, but against the dropsie also, being boyled in wine and drunken.

E Fennell seed drunke asswageth the paine of the stomacke, and wambling of the same, or desire to vomit, and breaketh winde.

F The herbe, seed, and root of Fennell are very good for the lungs, the liuer, and the kidneyes, for it openeth the obstructions or stoppings of the same, and comforteth the inward parts.

G The seed and herbe of sweet Fennell is equall in vertues with Annise seed.

Chap. 412. *Of Dill.*

¶ *The Description.*

Dill hath a little stalke of a cubit high, round and ioynted; whereupon doe grow leaues verie finely cut, like to those of Fennell, but much smaller: the floures be little and yellow, standing in a spokie tuft or rundle: the seed is round, flat and thin: the whole plant is of a strong smell: the root is threddy.

¶ *The Place.*

It is sowne in gardens, and is also sometimes found wilde.

¶ *The*

Anethum.
Dill.

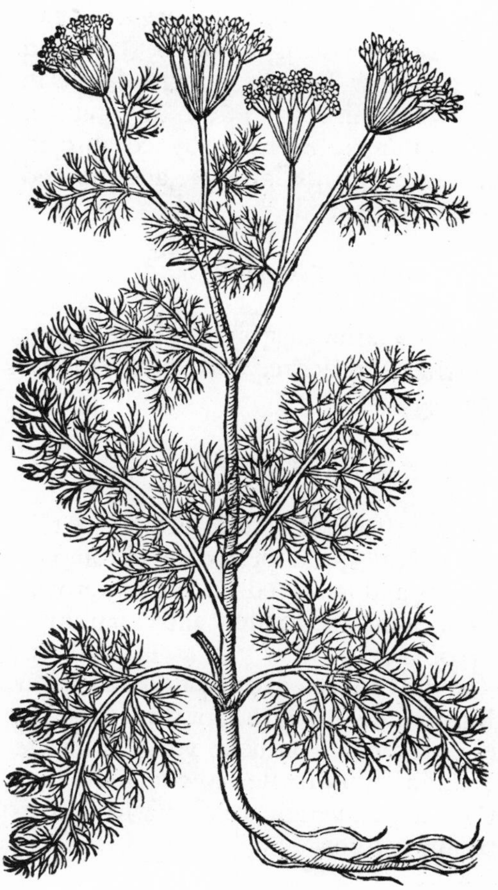

¶ *The Time.*

It bringeth forth floures and ſeed in Auguſt.

¶ *The Names.*

Dil is called in Greek Ἄνηθον: in Latine likewiſe *Anethum*, and *Anetum*: in high-Dutch, **Dyllen**: in low-Dutch, **Dille**: in Italian, *Anetho*: in Spaniſh, *Eneldo*: in French, *Anet*: in Engliſh, Dill, and Anet.

¶ *The Temperature.*

Dill, as *Galen* ſaith, is hot in the end of the ſecond degree, and dry in the beginning of the ſame, or in the end of the firſt degree.

¶ *The Vertues.*

A The decoction of the tops of dried Dil, and likewiſe of the ſeed, being drunke, ingendreth milke in the breſts of nurſes, allayeth gripings and windineſſe, prouoketh vrine, increaſeth ſeed, ſtayeth the yeox, hicket, or hicquet, as *Dioſcorides* teacheth.

B The ſeed likewiſe if it be ſmelled vnto ſtayeth the hicket, eſpecially if it be boyled in wine, but chiefely if it be boyled in Wormewood Wine, or Wine and a few branches of Worme-wood, and Roſe leaues, and the ſtomacke bathed therewith.

C *Galen* ſaith, that being burnt and layd vpon moiſt vlcers, it cureth them, eſpecially thoſe in the ſecret parts, and likewiſe thoſe *ſub Præputio*, though they be old and of long continuance.

D Common oyle, in which Dill is boyled or ſunned, as we do oyle of Roſes, doth digeſt, mitigate paine, procureth ſleepe, bringeth raw and vnconcocted humors to perfect digeſtion, and prouoketh bodily luſt.

E Dill is of great force or efficacie againſt the ſuffocation or ſtrangling of the mother, if the woman do receiue the fume thereof being boyled in wine, and put vnder a cloſe ſtoole or hollow ſeat fit for the purpoſe.

CHAP. 413. *Of Caruwaies.*

¶ *The Deſcription.*

CAruwaies haue an hollow ſtalke foure ſquare, of two cubits high, full of knots or ioynts; from which proceed ſundry other ſmall branches, ſet full of leaues very finely cut or iagged, like vnto thoſe of Carrots or Dill: at the top of the ſtalkes grow ſpokie white tufts like thoſe of Dill: after which commeth the ſeed, ſharpe in eating, yet of a pleaſant taſte: the root is like that of Parſley, often white, ſeldome yellow, and in taſte like vnto the Carrot.

¶ *The Place.*

It groweth almoſt euery where in Germanie and in Bohemia, in fat and fruitfull fields, and in medowes that are now and then ouer-run with water: it groweth alſo in Caria, as *Dioſcorides* ſheweth, from whence it tooke his name.

¶ *The Time.*

It floureth and ſeedeth from May to the end of Auguſt.

¶ The

Carum, siue Careum.
Caruwaies.

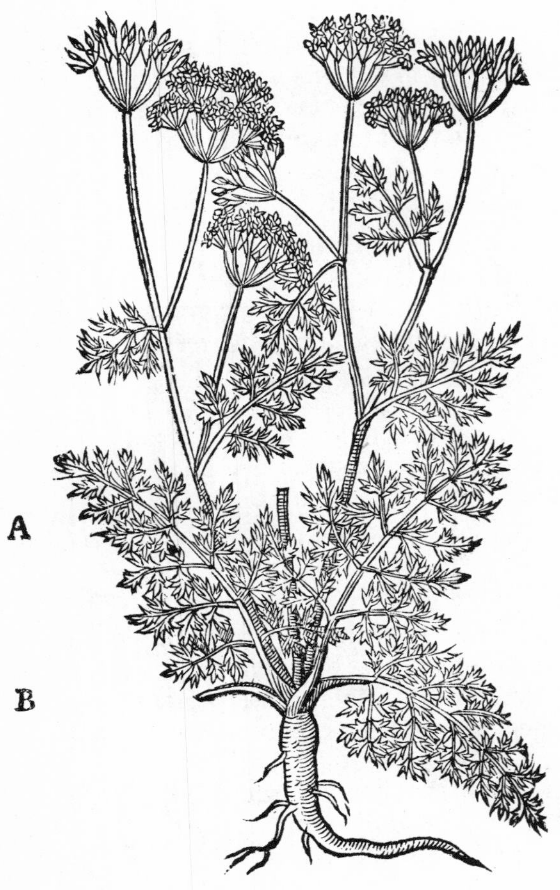

¶ *The Names.*

It is called in Greeke κάρος: in Latine, *Carum* and *Careum*: in shops, *Carui*. *Simeon Zethy* calleth it *Carnabadion*: in high-Dutch, **Kym**, and **Kymmel**: in low-Dutch, **Caruy saet**: in French, *du Caruy*: in Italian, *Caro*: in Spanish, *Carauea*, and an article being ioyned vnto it, *Alkarauea*: in English, Caruwaie, and the seed is called Caruwaie seed.

¶ *The Temperature.*

The seed of Caruwaies, as *Galen* saith, is hot and dry in the third degree, and hath a moderate biting qualitie.

¶ *The Vertues.*

A It consumeth winde, it is delightfull to the stomacke and taste, it helpeth concoction, prouoketh vrine, and is mixed with counterpoysons: the root may be sodden, and eaten as the Parsenep or Carrot is.

B The seeds confected, or made with sugar into Comfits, are very good for the stomacke, they helpe digestion, prouoke vrine, asswage and dissolue all windinesse: to conclude in a word, they are answerable to Anise seed in operation and vertues.

Chap. 414. *Of Annise.*

¶ *The Description.*

1 The stalke of Annise is round and hollow, diuided into diuers small branches, set with leaues next the ground somewhat broad and round: those that grow higher are more iagged, like those of yong Parsley, but whiter: on the top of the stalkes do stand spokie rundles or tufts of white floures, and afterward seed, which hath a pleasant taste as euerie one doth know.

‡ 2 This other Annise (whose vmbels *Clusius* had out of England from Master *Morgan* the Queenes Apothecarie, and *Iames Garret*; and which were brought from the Philippines by M^r^. *Tho Candish* in his voyage when he incompassed the world) is thus described by *Clusius*: The vmbels were large, no lesse than those of the Archangelica, made of diuers thicke stiffe foot-stalks, each whereof carried not double seed as the common Annise, but more, in a round head some inch ouer, made of cods set star-fashion, six, 8, or more, of a dusky colour, wrinkled, diuided into two equall parts, and open aboue: most of these huskes were empty, yet some of them contained one smooth shining ash-coloured seed, of the bignesse of that of *Orobus*; the taste and smell was the same with our common Anise seed, wherefore they which sent it to *Clusius* called it Anise: yet in the place where it grew it was called *Damor*; for M^r^. *Candish* had the name so written in the China characters, after their manner of writing. ‡

¶ *The Place.*

It groweth plentifully in Candy, Syria, Egypt, and other countries of the East. I haue often sowne it in my garden, where it hath brought forth his ripe seed when the yeare hath fallen out to be temperate.

¶ The

1 *Aniſum.*
Aniſe.

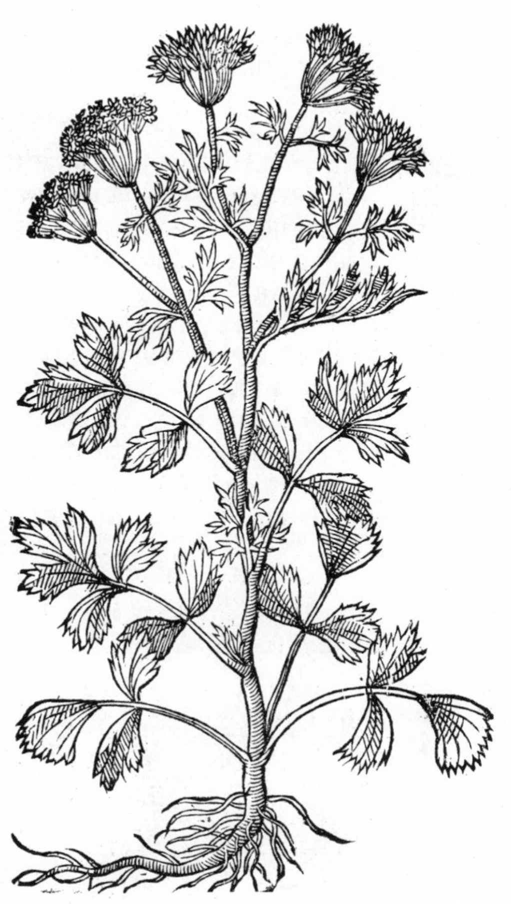

‡ 2 *Aniſum Indicum ſtellatum.*
Starry headed Aniſe.

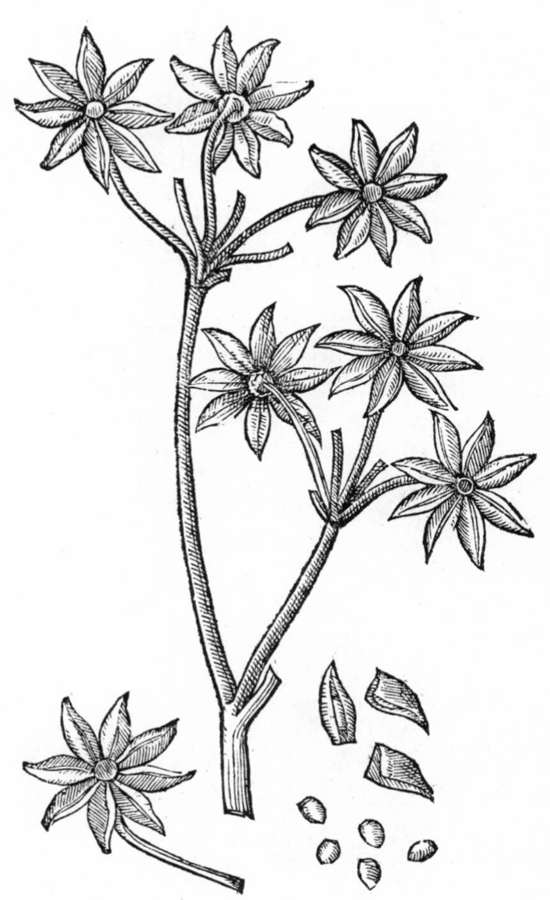

¶ *The Time.*

It is to be ſowne in theſe cold regions in the moneth of May: the ſeed is ripe in Auguſt.

¶ *The Names.*

It is called in Latine *Aniſum*: in Greeke, Ἄνισον: in high-Dutch, **Aniſz**: in low-Dutch, **Aniſſaet**: in Italian, *Aniſo*: in Spaniſh, *Matahalua*: in French, *Anis*: in Engliſh, Aniſe, and Anniſe ſeed.

¶ *The Temperature.*

Galen writeth, That the ſeed of Aniſe is hot and dry in the third degree: after others, it is hot in the ſecond degree, and much leſſe than dry in the ſecond degree; for it ingendreth milke, which it could not do if it were very dry, as *Galen* in his chapter of Fennell doth whether hee will or no declare and teſtifie; in that it doth ingender milke, his opinion is that it is not hot aboue the firſt degree: which thing alſo may be in Aniſe ſeed, both by this reaſon, and alſo becauſe it is ſweet. Therefore to conclude, Aniſe ſeed is dry in the firſt degree, and hot in the ſecond.

¶ *The Vertues.*

The ſeed waſteth and conſumeth winde, and is good againſt belchings and vpbraidings of the ſtomacke, allayeth gripings of the belly, prouoketh vrine gently, maketh aboundance of milke, and ſtirreth vp bodily luſt: it ſtayeth the laske, and alſo the white flux in women. A

Being chewed it makes the breath ſweet, and is good for them that are ſhort winded, and quencheth thirſt, and therefore it is fit for ſuch as haue the dropſie: it helpeth the yeoxing or hicket, both when it is drunken or eaten dry: the ſmell thereof doth alſo preuaile very much. B

The ſame being dried by the fire and taken with honey clenſeth the breſt very much from flegmaticke ſuperfluities: and if it be eaten with bitter almonds it doth helpe the old cough. C

It is to be giuen to yong children and infants to eate which are like to haue the falling ſicknes, or to ſuch as haue it by patrimonie or ſucceſſion. D

It taketh away the Squinancie or Quincie (that is, a ſwelling in the throat) being gargled with honey, vineger, and a little Hyſſop gently boiled together. E

CHAP.

CHAP. 414.

Of Bishops Weed, Herbe-William, or Ameos.

¶ *The Description.*

1 THe common Ameos, especially with vs here in England, hath round greene stalks, with diuers boughes and branches, and large long leaues, diuided into diuers other narrow long and small leaues, dented or snipt about the edges, hauing at the top of the stalke white floures in great spoky tufts, which bring forth a little sharpe and bitter seed: the root thereof is white and threddie.

2 This excellent and aromaticall Ameos of Candy hath tufts and leaues like *Daucus Creticus*, and a root like vnto the garden Carrot, of a yellow colour, and hot seed like *Origanum*, of an excellent spicie sauour or smell, growing in spoky tufts or roundles like *Carum*: it hath beene brought from Candy and Syria into Venice, and from Venice into France, Flanders, and England, where we haue often sowne it; but without doubt we haue beene beguiled therein by the deceitful drug-masters, who haue first boyled it, or vsed some other false and deceitfull deuice, to bring greater admiration vnto the Venice treacle, for the confection whereof this seed is a chiefe and most principall ingredient.

Ammi vulgare.
Common Bishops-weed.

† 2 *Ammi Creticum.*
Candy Bishops-weed.

3 There is another kinde of Ameos, which is an herbe very small and tender, hauing stalkes a foot and a halfe high, very small and tender, beset with leaues like vnto Dill, finely iagged, and somewhat slender; and at the top of the stalkes grow little tufts or spokie white rundles, which afterwards do turne into small gray seed, hot and sharpe in taste. The root is small and slender.

¶ *The Place.*

These plants do all grow in my garden, except *Ammi Creticum*, whereof hath beene sufficiently spoken in the description.

¶ *The*

‡ 3 *Ammi perpusillum.*
Small Bishops-weed.

¶ *The Time.*

They floure in Iune and Iuly, and yeeld their seed in the end of August.

¶ *The Names.*

The Grecians call it Ἄμμι: the Latines also *Ammi*: diuers call it *Cuminum Æthiopicum*: others, *Cuminum Regium*, or Comin Royall: in shops, *Ammios*, or *Ameos* in the Genitiue case: the Germanes, **Amey**: in English, Ameos, or Ammi: of some, Herbe-William, Bull-wort, and Bishops-weed.

¶ *The Temperature.*

The seed of Ameos is hot and dry in the later end of the third degree.

¶ *The Vertues.*

A It auaileth against gripings of the belly in making of vrine, against the bitings of serpents taken in wine, and also it bringeth downe the floures: being applied with honey it taketh away blacke and blew spots which come of stripes: the seed of *Sison* doth also the like, for it is hot and dry, and that in the third degree; likewise of thin parts, prouoking vrine, and bringing downe the desired sicknesse.

B The seed of Ameos is good to be drunken in wine against the biting of all manner of venomous beasts, and hath power against all maner of poyson & pestilent feuers, or the plague, and is vsed in the correcting of Cantharides, whereby those flies are made medicinable to be applied to the body without danger.

C Ameos brayed and mingled with honey scattereth congealed bloud, and putteth away blacke and blew markes which come by stripes or falls, if it be applied thereto in manner of a plaister.

† The figure which was formerly in the second place was of the *Hippomarathrum album* of *Tabernamontanus*.

CHAP. 416. *Of Cheruill.*

¶ *The Description.*

1 The leaues of Cheruill are slender, and diuersly cut, something hairy, of a whitish green: the stalks be short, slender, round, and hollow within, which at the first together with the leaues are of a whitish green, but tending to a red when the seeds are ripe: the floures be white, and grow vpon scattered tufts. The seed is long, narrow, slender, sharpe pointed: the root is full of strings.

‡ 2 There is found in Iune and Iuly, almost in euerie hedge, a certaine plant which *Tabernamont*. and *Bauhine* fitly cal *Chærophyllum*, or *Cerefolium syluestre*, and the figure was vnfitly giuen by our Author for *Thysselinum*: It hath a whitish wooddy root, from which arise round red and hairy stalkes some two cubits high, sometimes more, and oft times somewhat big and swolne about the ioynts, and they are not hollow but full of pith: toward the top it is diuided into sundry branches, which on their tops carry vmbels of small pure white little floures, which are succeeded by longish seeds. The leaues are vsually parted into three chiefe parts, and these againe subdiuided into fiue, and they are snipt about the edges, soft and hairy, of a darke greene or else reddish colour. It floureth in Iune and Iuly, and then ripens the seed. ‡

3 Great Cheruill hath large leaues deepely cut or iagged, in shew very like vnto Hemlocks, of a very good and pleasant smell and taste like vnto Cheruill, and something hairy, which hath caused vs to call it sweet Cheruill. Among these leaues riseth vp a stalke somwhat crested or furrowed, of the height of two cubits, at the top whereof grow spoky tufts or rundles with white

floures, which do turne into long browne crested and shining seed, one seed being as big as foure Fennell seeds, which being greene do taste like Anise seed. The root is great, thicke, and long, as big as *Enula Campana*, exceeding sweet in smell, and tasting like vnto Anise seeds.

1 *Cerefolium vulgare satiuum.*
Common Cheruill.

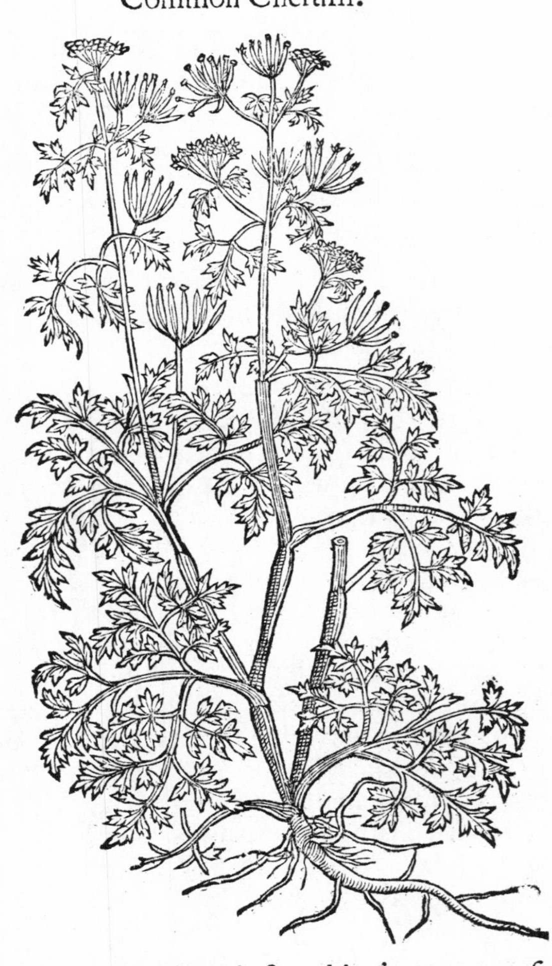

† 2 *Cerefolium syluestre.*
Hedge Cheruill.

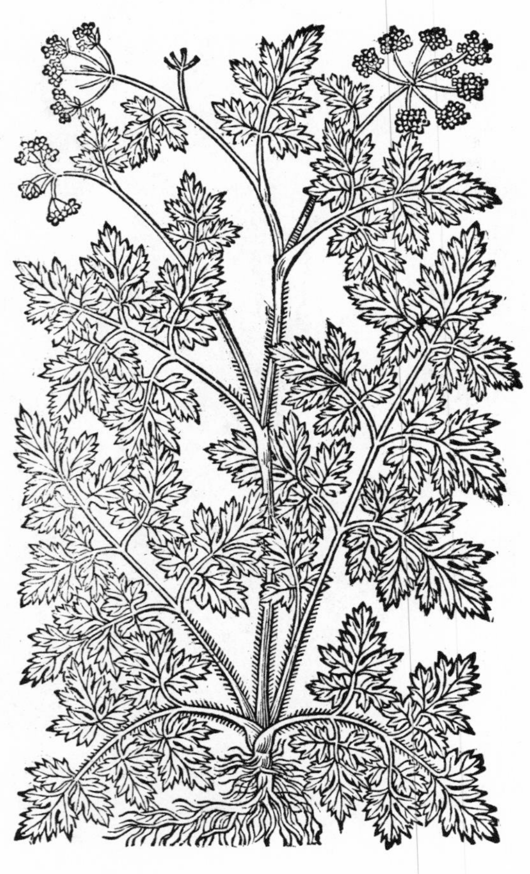

‡ 4 There is found in some parts of the Alps, as about Geneua and in other places, another *Myrrhis*, which in the leaues and vmbels is like that of the last described, but the whole plant is lesse; the seed is long, small, smooth, and shaped like an Oat, and in taste somewhat like that of the *Daucus Creticus*. *Lobel* hath this by the same name as we here giue it you.

5 About mud walls, high-wayes, and such places, here about London, and in diuers other places, is found growing a small plant, which in all things but the smell and height agrees with that referred to this kinde by *Fabius Columna*, and called *Myrrhis Æquicolorum noua*. The root hereof is small and white, perishing euery yeare when it hath perfected his seed: the stalks are slender, hollow, smooth, and not hairy, seldome exceeding the height of a cubit, or cubit and halfe; it is diuided into sundry branches, vpon the sides whereof against the setting on of the leaues, or out of their bosomes, grow forth the stalks, which carry vmbels of small white floures: after which follow the seeds, growing two together, and these longish, rough, round, and hairy, about the bignesse of Anise seeds. The leaues are small, and finely cut or diuided like those of Hemlock, but of a whitish colour, and hairy: it comes vp in March, floures in May, and ripens his seed in Iune. In Italy they eate the yong leaues in sallads, and call it wilde Cheruile: we may in English for distinctions sake call it small Hemlocke Cheruill.

6 To these we may fitly adde that plant which in the *hist. Lugd.* is called *Cicutaria alba*, and by *Camerarius*, *Cicutaria palustris*; for it floures at the same time with the last mentioned, and is found in floure and seed in May and Iune very frequently almost in all places; but afterwards his stalkes die downe, yet his roots liue, and the leaues are greene all the yeare. The root of this is very large, and diuided into sundry parts, white also and spungie, of a pleasing strong smell, with a hot and biting taste: the stalks grow vp in good ground to be some three cubits high, and they are hollow, ioynted, pretty thicke, greene, and much crested, sending forth of the bosomes of the leaues many branches, which vpon their tops carry vmbels composed of many white floures, each floure consisting of fiue little leaues, whereof the lowest is twice as big as the rest, the two side-ones lesse, and the vppermost the least of all. The leaues are large like those of *Myrrhis*, but of a dark green colour, and

and those that grow about the tops of the stalkes are commonly diuided into into three parts, and these subdiuided into sundry long sharpe pointed and snipt leaues like as in *Myrrhis*. The seeds grow two together, being longish, round, sharpe pointed, blacke, and shining. We may fitly terme this plant, wilde Cicely, for that it so much resembles the *Myrrhis* or garden Cicely, not onely in shape, but (if I be not deceiued) in vertues also. ‡

3 *Cerefolium magnum, siue Myrrhis.*
Great Cheruill or Myrrh.

‡ 4 *Myrrhis altera parua.*
Small sweet Cheruill.

¶ *The Place.*

The common Cheruill groweth in gardens with other pot-herbes: it prospers in a ground that is dunged and somewhat moist. The great sweet Cheruill groweth in my garden, and in the gardens of other men who haue been diligent in these matters.

¶ *The Time.*

These herbes do floure in May, and their seed is ripe in Iuly.

¶ *The Names.*

Cheruill is commonly called in Latine *Cerefolium*, and as diuers affirme, *Chærofolium*, with *o* in the second syllable. *Columella* nameth it *Chærephyllum*, and it is thought to be so called because it delighteth to grow with many leaues, or rather in that it causeth ioy and gladnes: in high-Dutch, **Korffelkraut**: in low-Dutch, **Keruell**: in Italian, *Cerefoglio*: in French, *Du Cerfueil*. in English, Cheruell, and Cheruill.

Myrrhis is also called *Myrrha*, taken from his pleasant sauour of Myrrh: of some, *Conila*, as it is found noted among the bastard names. It is also, by reason of the similitude it hath with Hemlocke, called by most late writers, *Cicutaria*. Of this, *Pliny* maketh mention, *lib. 24. cap. 16.* where he reporteth that it is called *Smyrrhiza*: in English it is called Cheruill, sweet Cheruill, or sweet Cicely.

¶ *The Temperature and Vertues.*

A Cheruill is held to be one of the pot-herbes, it is pleasant to the stomacke and taste: it is of a temperate heate and moderate drinesse, but nothing so much as the Parsleyes.

B It prouoketh vrine, especially being boyled in wine, and applied hot to the share or nethermost part of the belly, and the wine drunke in which it was boyled.

C It hath in it a certaine windinesse, by meanes whereof it procureth lust.

D It is vsed very much among the Dutch people in a kinde of Loblolly or hotch-pot which they do eate, called Warmus.

The

E The leaues of sweet Cheruill are exceeding good, wholesome, and pleasant, among other sallad herbs, giuing the taste of Anise seed vnto the rest.

F The root, saith *Galen*, is hot in the second degree, hauing a thinnesse of substance ioined with it.

G *Dioscorides* teacheth, that the root drunke in wine is a remedie against the bitings of the venomous spiders called in Latine *Phalangia*; and that it bringeth downe the menses and secondines; and being boyled and drunke it is good for such as haue the ptysick or consumption of the lungs.

H The seeds eaten as a sallad whilest they are yet greene, with oyle, vineger, and pepper, exceed all other sallads by many degrees, both in pleasantnesse of taste, sweetnesse of smell, and wholsomnesse for the cold and feeble stomacke.

I The roots are likewise most excellent in a sallad, if they be boyled and after dressed as the cunning Cooke knoweth how better than my selfe: notwithstanding I doe vse to eate them with oile and vineger, being first boyled; which is very good for old people that are dull and without courage; it reioyceth and comforteth the heart, and increaseth their lust and strength.

Chap. 417. *Of Shepheards needle or wilde Cheruill.*

¶ *The Description.*

1 *Scandix*, or *Pecten Veneris*, doth not much differ in the quantitie of the stalks, leaues, and floures, from Cheruill; but *Scandix* hath no such pleasant smell as Cheruill hath: the leaues be lesser, more finely cut, and of a browne greene colour: the floures grow at the top of the stalkes in small white tufts; after which come vp long seeds very like vnto pack-needdles, orderly set one by another like the great teeth of a combe, whereof it tooke the name *Pecten Veneris*, or Venus combe, or Venus needle: the root is white, a finger long.

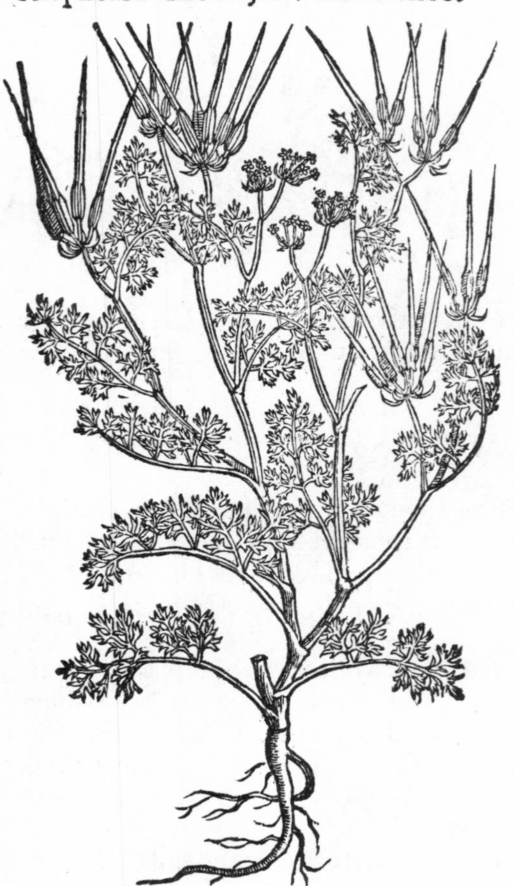

1 *Pecten Veneris, siue Scandix.*
Shepheards needle, or Venus combe.

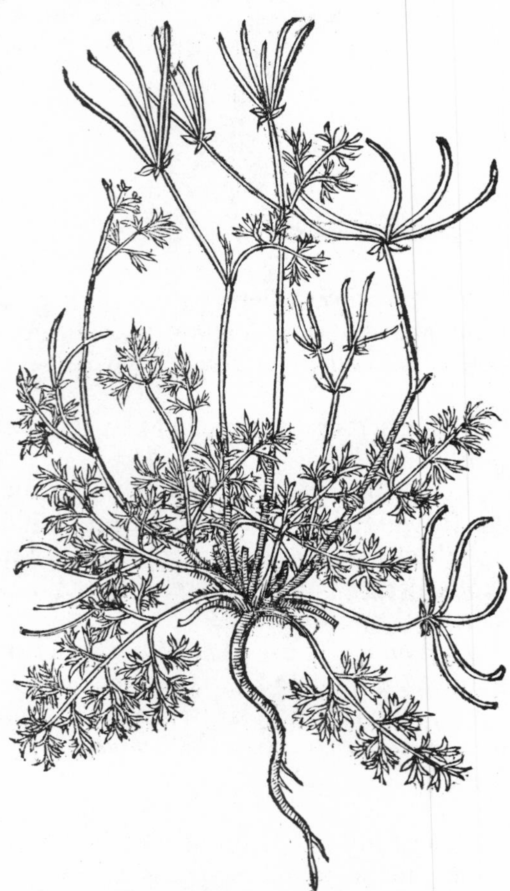

‡ 2 *Scandix minor, siue Anthriscus.*
Small Shepheards needle.

‡ 2 This from a slender long and whitish root sends vp many small leaues like those of the last described, but of a pleasing smell and taste something like that of the common Cheruill; amongst these leaues grow vp slender stalks a little hairy, diuided into short green and slender branches carrying little vmbels, consisting of fiue, six, seuen, or eight smal white floures, composed of fiue leaues apiece,

apiece, with a darke purplish chiue in the middle : the floures are succeeded by, or rather grow vpon long slender cods, which become some inch long, and resemble those of the last described. It floures in Iune, as *Clusius* affirmeth, who giues vs the history of it ; and he receiued it from *Honorius Bellus* out of Candy ; who writes, that in the Spring time it is much vsed in sallads, and desired, for that it much excites to Venery. He also thinks this plant to be the *Anthriscus* of *Pliny*, and by the same name *Clusius* sets it forth. *Columna* hath called it *Aniso-marathrum*, because the smell and taste is betweene that of Anise and Fennell. ‡

¶ *The Place.*

It groweth in most corne fields in England, especially among wheate and barley.

¶ *The Time.*

It floureth in May : the seed is ripe in August with corne.

¶ *The Names.*

The Latines call it *Scandix*, hauing borrowed that name of the Grecians, who call it σκάνδιξ : we finde among the bastard words, that the Romans did call it *Scanaria*, and *Acula*, of the seed that is like vnto a needle. *Ruellius* describeth it vnder the name *Pecten Veneris* . of others, *Acus Veneris*, and *Acus Pastoris*, or Shepheards Needle, wilde Cheruill, and Ladies combe : in high-Dutch, **Naelde Karnel** : This is that herbe (saith *Pliny, lib.22.cap.22.*) which *Aristophanes* obiected in sport to the Poet *Euripides*, that his mother was wont to sell no right pot-herbe but *Scandix*, or Shepheards needle, meaning, as I take it, *Visnaga*, wherewith the Spaniards doe picke their teeth when they haue eaten no meate at all except a few oranges or such a like trifle, called also *Scandix*.

¶ *The Temperature.*

Shepheards needle, saith *Galen*, is an herbe somewhat binding, and bitter in taste, insomuch that it is hot and dry either in the later end of the second degree, or in the beginning of the third.

¶ *The Vertues.*

Dioscorides saith it is eaten both raw and boyled, and that it is an wholesome pot-herbe among the Greekes ; but in these dayes it is of small estimation or value, and taken but for a wilde Wort, as appeareth by *Aristophanes* taunting of *Euripides*, as aforesaid. A

The decoction thereof is good for the bladder, kidneyes, and liuer ; but as I deeme hee meant Cheruill, when he set the same downe to be vsed in physicke. B

CHAP. 418. *Of Tooth-picke Cheruill.*

¶ *The Description.*

1 THe first of these Tooth-picke Cheruils beareth leaues like wilde Turneps, a round stalke furrowed, ioynted, blackish, and hairy, diuided into many branches, on the tops whereof grow spokie tufts, beset round about with many small leaues. The floures thereof are whitish : after commeth the the seed, which being once ripe do cluster and are drawne together, in a round thicke tuft like a small birds nest, as be those of the wilde Carrot ; whose seeds whoso toucheth, they will cleaue and sticke to his fingers, by reason of the glutinous or slimie matter they are possessed with. The root is small and whitish, bitter in taste, as is all the rest of the plant.

2 The Spanish Tooth-picke hath leaues, floures, and knobby stalkes like vnto wilde carrots, sauing that the leaues are somewhat finer, cut or iagged thicker, and tenderer, but not rough or hairy at all as is the former, of a bitter taste, and a reasonable good smell : among which rise vp bushie rundles or spokie tufts like those of the wilde Carrot or Birds nest, closely drawne together when the seed is ripe ; at what time also the sharpe needles are hardned, fit to make Tooth-pickes and such like, for which purpose they do very fitly serue.

¶ *The Place.*

Both of them grow in Syria, and most commonly in Cilicia : the later is to be found likewise in Spaine almost euery where ; and I haue it likewise in my garden in great plentie.

¶ *The Time.*

They floure in my garden about August, and deliuer their seed in October.

¶ *The Names.*

That which the Grecians call γιγγίδιον, the Latines do likewise name *Gingidium* : and it is called in Syria *Lepidium* : yet is there another *Lepidium*. It is reported among the bastard names to be called by the Romans, *Bisacutum* : of which name some shew remaines among the Syrians, who commonly call the later, *Gingidium*, *Visnaga* : this is named in English, Tooth-picke Cheruill.

1 *Gingidium latifolium.*
Broad Tooth-picke Cheruill.

2 *Gingidium Hifpanicum.*
Spanifh Tooth-picke Cheruill.

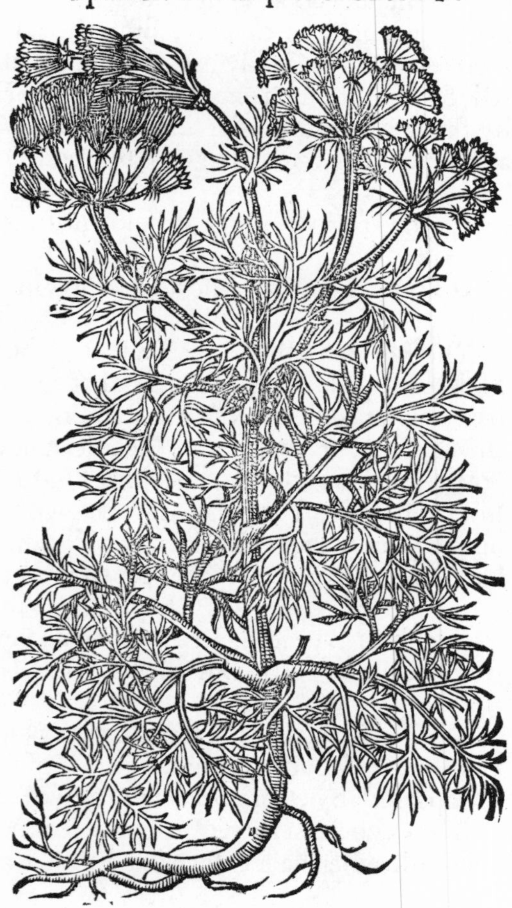

¶ *The Temperature and Vertues.*

A There is, faith *Galen*, great increafe of *Gingidium* in Syria, and it is eaten no otherwife than *Scandex* is with vs at Pergamum: it is, faith he, very wholefome for the ftomacke, whether it be eaten raw or boyled; notwithftanding it is euident that it is a medicine rather than a nourifhment. As it is bitter and binding, fo is it likewife of a temperate heate and drineffe. The heate is not very apparant, but it is found to be dry in the later end of the fecond degree, as alfo the faid Author alledgeth in his difcourfe of the faculties of fimple medicines.

B *Diofcorides* doth alfo write the fame: This pot-herbe (faith he) is eaten raw, fodden, and preferued with great good to the ftomacke; it prouoketh vrine, and the decoction thereof made with wine and drunke, is profitable to fcoure the bladder, prouoketh vrine, and is good againft the grauell and ftone.

C The hard quills whereon the feeds do grow are good to cleanfe the teeth and gums, and do eafily take away all filth and baggage fticking in them, without any hurt vnto the gums, as followeth after many other Tooth-picks, and they leaue a good fent or fauor in the mouth.

Chap. 419. *Of Mede-fweet, or Queene of the Medowes.*

¶ *The Defcription.*

1 THis herbe hath leaues like thofe of Agrimonie, confifting of diuers leaues fet vpon a middle rib like thofe of the Afh tree, euery fmall leafe fleightly fnipt about the edges, white on the inner fide, and on the vpper fide crumpled or wrinkled like vnto thofe of the Elme Tree; whereof it tooke the name *Vlmaria*, of the fimilitude or likeneffe that the leaues haue with the Elme leaues. The ftalke is three or foure foot high, rough, and very fragile or eafie to bee broken, of a reddifh purple colour: on the top whereof are very many little floures cluftering and growing together, of a white colour tending to yellowneffe, and of a pleafant fweete fmell,

1 *Regina prati.*
Queene of the Medow.

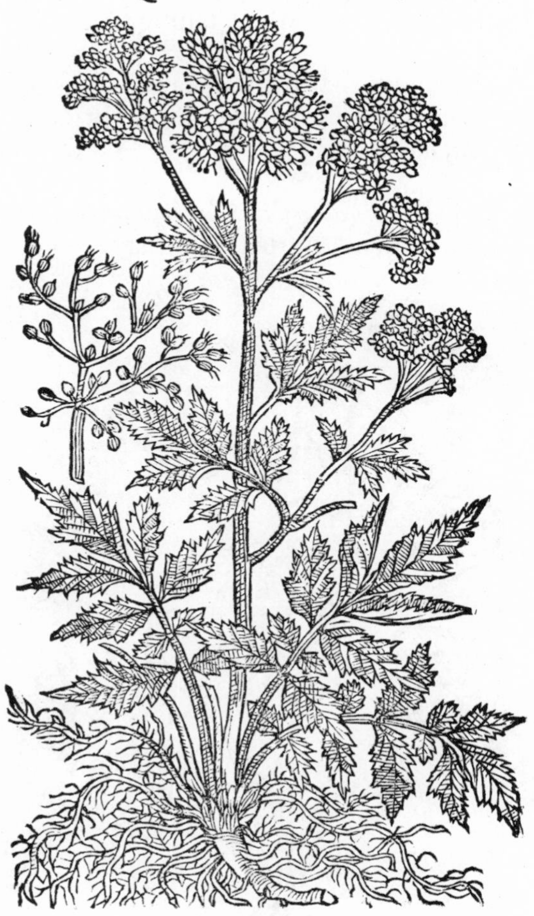

smel, as are the leaues likewise: after which come the seeds, small, crookedly turning or winding one with another, made into a fine little head. The root hath a sweet smel, spreding far abroad, blacke without, and of a darkish red colour within.

‡ 2 There is also another which by *Fuchsius, Tragus, Lonicerus, Gesner*, and others, is called *Barba Capri*: it hath large woodie rootes, leaues of the bignesse, and growing somewhat after the manner of the wild Angelica: the stalks are crested, and diuided into sundry branches, which carry long bending spikes or eares of white floures & seeds somewhat like those of the common kinde. This floures at the same time as the former, and I haue not yet heard of it wilde with vs, but onely seene it growing with Mr. *Tradescant*. ‡

¶ *The Place.*

It groweth in the brinkes of waterie ditches and riuers sides, and also in medowes: it liketh waterie and moist places, and groweth almost euery where.

¶ *The Time.*

It floureth and flourisheth in Iune, Iuly, and August.

¶ *The Names.*

It is called of the later age *Regina prati*, & *Barba Capri*: of some, *Vlmaria, à foliorum Vlmi similitudine*, from the likenesse it hath with the Elme tree leafe: in high Dutch, **Geisbart**. It is called *Barba Hirci*, which name belongeth to the plant which the Grecians do call *Tragopogon*: of *Anguillara, Potentilla maior*. It hath some likenesse with *Rhodora Plinij*, but yet we cannot affirme it to be the same. It is called in low Dutch **Reijnette**: in French, *Barbe de Cheure, Reine des Praiz*: in English, Meadesweet, Medow-sweet, and Queene of the medowes. *Camerarius* of Noremberg saith it is called of the Germanes his countrimen, **Wurme kraut**: because the roots, saith he, seem to be eaten with wormes. I rather suppose they call it so, because the antient hackny men and horsleaches do giue the decoction therof to their horses and asses, against the bots and wormes, for the which it is greatly commended.

¶ *The Temperature.*

Mede-sweet is cold and drie, with an euident binding qualitie adioined.

¶ *The Vertues.*

The root boiled, or made into pouder and drunke, helpeth the bloudy flix, staieth the laske, and all other fluxes of bloud in man or woman. A

It is reported, that the floures boiled in wine and drunke, do take away the fits of a quartaine ague, and make the heart merrie. B

The leaues and floures farre excell all other strowing herbes, for to decke vp houses, to straw in chambers, halls, and banqueting houses in the sommer time; for the smell thereof makes the heart merrie, delighteth the senses: neither doth it cause head-ache, or lothsomenesse to meat, as some other sweet smelling herbes do. C

The distilled water of the floures dropped into the eies, taketh away the burning and itching thereof, and cleareth the sight. D

CHAP. 420. *Of Burnet Saxifrage.*

¶ *The Description.*

1 THis great kinde of Pimpinell, or rather Saxifrage, hath great and long roots, fashioned like a Parsnep, of an hot and biting taste like Ginger: from which riseth vp an hollow stalke

ſtalke with ioints and knees two cubits high, beſet with large leaues, which do more neerely repreſent Smallage than Pimpernell, or rather the garden Parſnep. This plant conſiſteth of many ſmall leaues growing vpon one ſtem, ſnipt or dented about the edges like a ſaw: the floures do grow at the top of the ſtalkes in white round tufts: the ſeed is like the common Parſley, ſauing that it is hotter and biting vpon the tongue.

‡ There is a bigger and leſſer of this kinde, which differ little, but that the ſtalkes and veins of the leaues of the leſſer are of a purpliſh colour, and the root is hotter. Our Authour formerly gaue the figure of the leſſer in the ſecond place, in ſtead of that of *Bipinella*. ‡

1 *Pimpinella Saxifraga.*
Burnet Saxifrage.

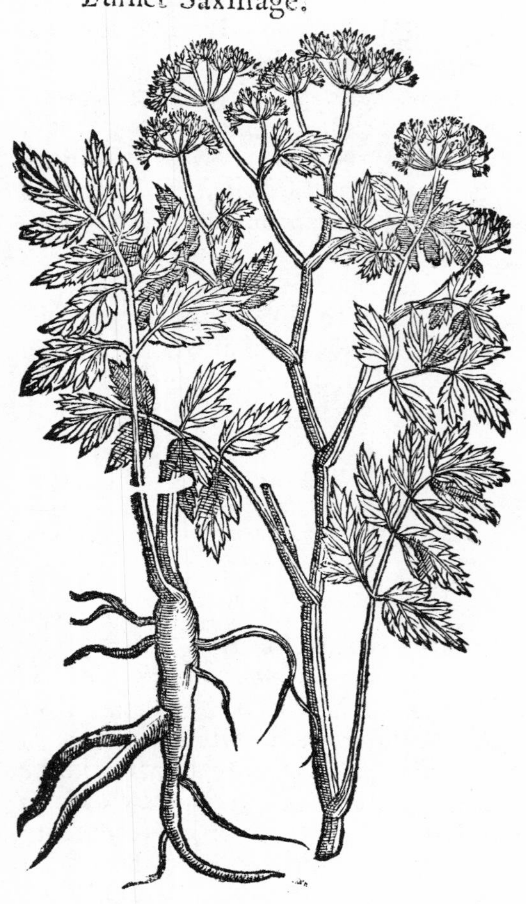

† 2 *Bipinella, ſive Saxifraga minor.*
Small Burnet Saxifrage.

2 *Bipinella* is likewiſe a kinde of Burnet or Pimpinell, vpon which *Pena* hath beſtowed this addition *Saxifraga minor*: vnder which name *Saxifraga* are comprehended diuers herbs of diuers kinds, and the one very vnlike to the other: but that kinde of Saxifrage which is called *Hircina*, which is rough or hairie Saxifrage, of others *Bipinella*, is beſt knowne, and the beſt of all the reſt, like vnto the ſmall Burnet, or common Parſley, ſauing that it is void of haires, as may appeare by the old Latine verſe,

Pimpinella habet pilos, Saxifraga non habet vllos.
Pimpinell hath haires ſome, but Saxifrage hath none.

Notwithſtanding, I haue found a kinde hereof growing in our paſtures adioining to London, the leaues whereof if you take and tenderly breake with your hands, you may draw forth ſmall threds, like the web of a ſpider, ſuch as you may draw from the leaues of Scabious. The ſtalke is hollow, diuiding it ſelfe from the ioints or knees, into ſundry other ſmall branches; at the top whereof doe grow ſmall tufts or ſpokie rundles, of a white colour: after which commeth the ſeed like to *Carui*, or Caruwaies, of a ſharpe taſte: the root is alſo ſharpe and hot in taſte.

¶ *The Place.*

Theſe plants do grow in drie paſtures and medowes in this countrey very plentifully.

¶ *The Time.*

They floure from Iune to the end of Auguſt.

¶ *The Names.*

That which *Fuchſius* calleth *Pimpinella maior*, *Dodonæus* termeth *Saxifragia maior*, which kinde of Saxifrage doth more abſolutely anſwer the true *Phellandrium* of *Pliny*, than any other plant whatſoeuer:

soeuer: wherein the Physitions of Paris haue been deceiued, calling or supposing the medow Rue to be the right *Phellandrium*, whereunto it is not like either in shape or facultie: for it is nothing so effectuall in breaking the stone, or prouoking of vrine, as either of these plants, especially *Pimpinella Hircina*, which is not so called, because it hath any rammish smell of a goat, but because practitioners haue vsed to feed goats with it, whose flesh and bloud is singular good against the stone, but we rather take it to be named *Hircina*, of *Hircinia sylua*, where it doth grow in great abundance, the sauour of the herbe not being vnpleasant, somewhat resembling the smell and taste of *Daucus, Ligustrum*, and *Pastinaca*: so to conclude, both these are called *Saxifragia*: the smaller is called of some *Petræfindula, Bipinella*, and *Bipenula*: of *Baptista Sardus*, and also of *Leonardus Fuchsius, Pimpinella maior*: wherefore diuers call it *Pimpinella Saxifraga*: for there is also another *Pimpinella*, called *Pimpinella Sanguisorba*: notwithstanding the verse before rehearsed sheweth a difference betweene *Pimpinella* and *Saxifraga*: in high Dutch, it is called **Biberneil**: in Low Dutch, **Bauenaert**: in English the greater may be called great Saxifrage, and the other small Saxifrage.

Bipinella is called *Saxifragia minor*: in English, Small Saxifrage, as *Pimpinella* is called great Saxifrage. ‡ *Columna* iudges it to be the *Tragium* of *Dioscorides*. ‡

¶ *The Nature.*

Saxifrage of both kindes, with their seed, leaues, and roots, are hot and drie in the third degree, and of thin and subtill parts.

¶ *The Vertues.*

The seed and root of Saxifrage drunken with wine, or the decoction thereof made with wine, causeth to pisse well, breaketh the stone in the kidnies and bladder, and is singular against the strangurie, and the stoppings of the kidnies and bladder: whereof it tooke the name *Saxifragia*, or breake stone. A

The iuice of the leaues of Saxifrage doth clense and take away all spots and freckles of the face, and leaueth a good colour. B

The distilled water thereof mingled with some vineger in the distillation, cleareth the sight, and taketh away all obscuritie and darknesse of the same. C

CHAP. 421. *Of Burnet.*

1 *Pimpinella hortensis.*
Garden Burnet.

2 *Pimpinella sylvestris.*
Wilde Burnet.

¶ *The*

¶ *The Kinds.*

Burnet of which we will intreat, doth differ from *Pimpinella*, which is also called *Saxifraga*. One of the Burnets is lesser, for the most part growing in gardens, notwithstanding it groweth in barren fieldes, where it is much smaller: the other greater, is altogither wilde.

¶ *The Description.*

1 Garden Burnet hath long leaues made vp together of a great many vpon one stem, euery one whereof is something round, nicked on the edges, somwhat hairie: among these riseth a stalke that is not altogether without leaues, something chamfered: vpon the tops whereof grow little round heads or knaps, which bring forth small floures of a browne purple colour, and after them cornered seeds, which are thrust vp together. The root is long: the whole plant doth smell something like a Melon, or Cucumber.

2 Wilde Burnet is greater in all parts, it hath wider and bigger leaues than those of the former: the stalke is longer, sometimes two cubits high: the knaps are greater, of a darke purple colour, and the seed is likewise cornered and greater: the root longer, but this Burnet hath no pleasant smell at all.

‡ 3 There is kept in some gardens another of this kinde, with very large leaues, stalkes, and heads, for the heads are some inch and halfe long, yet but slender considering the length, and the floures (as I remember) are of a whitish colour: in other respects it differs not from the precedent: it may fitly be called *Pimpinella sanguisorba hortensis maxima*, Great Garden Burnet. ‡

¶ *The Place.*

The small Pimpinell is commonly planted in gardens, notwithstanding it doth grow wilde vpon many barren heaths and pastures.

The great wilde Burnet groweth (as M^r. *Lyte* saith) in dry medowes about Viluord, and my selfe haue found it growing vpon the side of a causey which crosseth the one halfe of a field, whereof the one part is earable ground, and the other part medow, lying between Paddington and Lysson green neere vnto London, vpon the high way.

¶ *The Time.*

They floure from Iune, vnto the end of August.

¶ *The Names.*

The later herbarists doe call Burnet *Pimpinella sanguisorba*, that it may differ from the other, and yet it is called by seuerall names, *Sanguisorba*, and *Sanguinaria*: *Gesner* had rather it should be called *Peponella* of the smell of Melons or Pompions, to which it is like, as we haue said: of others it is named *Pimpinella*, or *Bipennula*: of most men, *Solbastrella*: in High Dutch, **Kolbleskraut, her Gots Bartlin, Blutkraut, megelkraut**: in French, *Pimpennelle, Sanguisorbe*: in English, Burnet. It agreeth *cum altera Dioscoridis Sideritide*; that is to say, with *Dioscorides* his second Iron-woort: the leafe (and especially that of the lesser sort) which we haue written to consist of many nicks in the edges of the leaues; and this may be the very same which *Pliny* in his 24 book, chapter 17. reporteth to be named in Persia, *Sissitiepterus*, because it made them merry; he also calleth the same *Protomedia*, and *Casigneta*, and likewise *Dionysionymphas*, for that it doth maruellously agree with wine; to which also this *Pimpinella* (as we haue said) doth giue a pleasant sent: neither is that repugnant, that *Pliny* in another place hath written, *De Sideritibus*, of the Iron-woorts; for it often falleth out that he intreateth of one and the selfe same plant in diuers places, vnder diuers names: which thing then hapneth sooner when the writers themselues do not well know the plant, as that *Pliny* did not well know *Sideritis* or Iron-woort, it is euen thereby manifest, because he setteth not downe his owne opinion hereof, but other mens.

¶ *The Temperature.*

Burnet, besides the drying and binding facultie that it hath, doth likewise meanly coole: and the lesser Burnet hath likewise withall a certaine superficiall, sleight, and temperate sent, which when it is put into the wine it doth leaue behind it: this is not in the dry herbe, in the iuice, nor in the decoction.

¶ *The Vertues.*

A Burnet is a singular good herb for wounds (which thing *Dioscorides* doth attribute to his second Ironwoort) and commended of a number: it stancheth bleeding, and therefore it was named *Sanguisorba*, as well inwardly taken, as outwardly applied.

B Either the iuice is giuen, or the decoction of the pouder of the drie leaues of the herbe, beeing bruised, it is outwardly applied, or else put among other externall medicines.

C It staieth the laske and bloudy flix: it is also most effectuall to stop the monthly course.

D The lesser Burnet is pleasant to be eaten in sallads, in which it is thought to make the heart merry and glad, as also being put into wine, to which it yeeldeth a certaine grace in the drinking.

The

The decoction of Pimpinell drunken, cureth the bloudy flix, the ſpitting of bloud, and all other fluxes of bloud in man or woman. F

The herbe and ſeed made into pouder, and drunke with wine, or water wherein iron hath beene quenched doth the like. G

The leaues of Pimpinell are very good to heale wounds, and are receiued in drinkes that are made for inward wounds. H

The leaues of Burnet ſteeped in wine and drunken, comfort the heart, and make it merry, and are good againſt the trembling and ſhaking thereof. I

CHAP. 422. *Of Engliſh Saxifrage.*

¶ *The Deſcription.*

1 THis kinde of Saxifrage our Engliſh women Phyſitions haue in great vſe, and is familiarly knowne vnto them, vouchſafing that name vnto it of his vertues againſt the ſtone: it hath the leaues of Fennel, but thicker and broader, very like vnto *Seſeli pratenſe, Monſpelienſium* (which addition *Pena* hath beſtowed vpon this our Engliſh Saxifrage) among which riſeth vp a ſtalke, of a cubit high or more, bearing at the top ſpokie rundles beſet with whitiſh yellow floures: the root is thicke, blacke without, and white within, and of a good ſauour.

† 1 *Saxifraga Anglicana facie Seſeli pratenſis.*
Engliſh Saxifrage.

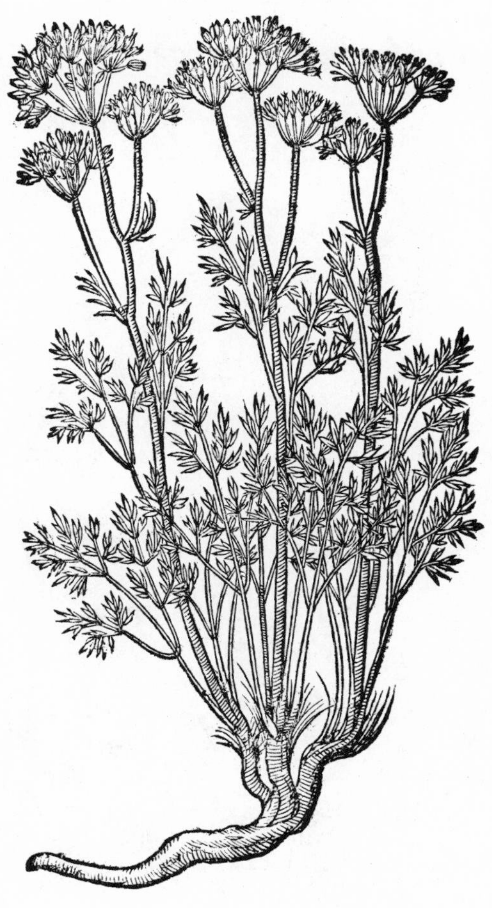

‡ 2 *Saxifraga Pannonica Cluſij.*
Auſtrian Saxifrage.

‡ 2 *Cluſius* hath ſet forth another plant not much different from this our common Saxifrage, and called it *Saxifraga Pannonica*, which I haue thought fit here to inſert: the leaus, ſaith he, are much ſhorter than thoſe of Hogs-Fennell, and ſomewhat like thoſe of Fumitorie: the ſtalkes are ſome foot high, ſlender, hauing ſome few ſmall leaues, and at the top carrying an vmbel of white floures: the root is not much vnlike that of Hogs-Fennel, but ſhorter and more acride; it is hairie at the top thereof

thereof, whence the ſtalkes and leaues come forth: it growes vpon ſome hils in Hungarie and Auſtria, and floures in Iuly. ‡

¶ *The Place.*

Saxifrage groweth in moſt fields and medowes euery where throughout this our kingdome of England.

¶ *The Time.*

It floureth from the beginning of May to the end of Auguſt.

¶ *The Names.*

Saxifraga Anglicana is called in our mother tongue Stone-breake or Engliſh Saxifrage: *Pena* and *Lobel* call it by this name *Saxifraga Anglicana*: for that it groweth more plentifully in England than in any other countrey.

¶ *The Nature.*

Stone-breake is hot and drie in the third degree.

¶ *The Vertues.*

A A decoction made with the ſeeds and roots of Saxifrage, breaketh the ſtone in the bladder and kidneies, helpeth the ſtrangurie, and cauſeth one to piſſe freely.

B The root of Stone-breake boiled in wine, and the decoction drunken, bringeth downe womens ſickneſſe, expelleth the ſecondine and dead childe.

C The root dried and made into pouder, and taken with ſugar, comforteth and warmeth the ſtomack, cureth the gnawings and griping paines of the belly.

D It helpeth the collicke, and driueth away ventoſities or windineſſe.

E Our Engliſh women vſe to put it in their running or rennet for cheeſe, eſpecially in Cheſhire (where I was borne) where the beſt cheeſe of this Land is made.

† I haue formerly Chap. 188. deliuered the hiſtory of the *Saxifraga maior* of *Matthiolus*, and *Saxifraga Antiquorum of Lobel*; not thinking that our Author had put their deſcriptions here amongſt the *Vmbelliferæ*, for if I had, I ſhould haue ſpared my labour there beſtowed, and haue giuen their figures here to the deſcriptions of our Author, which are now omitted. The figure formerly here was of the *Caucalis*, deſcribed in the third place of the 403 Chapter.

Chap. 423. *Of Siler Mountaine or baſtard Louage.*

† 1 *Siler montanum Officinarum.*
Baſtard Louage.

† 2 *Seſeli pratenſe Monſpelienſium.*
Horſe Fennell.

¶ *The*

¶ *The Deſcription.*

1 THe naturall plants of *Seſeli*, being now better knowne than in times paſt, eſpecially among our Apothecaries, is called by them *Siler montanum*, and *Seſcleos*: this plant they haue retained to very good purpoſe and conſideration; but the errour of the name hath cauſed diuers of our late writers to erre, and to ſuppoſe that *Siler montanum*, called in ſhops, *Seſeleos*, was no other than *Seſeli Maßilienſium* of *Dioſcorides*. But this plant containeth in his ſubſtance much more acrimony, ſharpeneſſe and efficacy in working, than any of the plants called *Seſelios*. It hath ſtalkes like Ferula, two cubits high. The root ſmelleth like *Liguſticum*: the leaues are very much cut or diuided, like the leaues of Fennell or *Seſeli Maßilienſe*, and broader than the leaues of *Peucedanum* At the top of the ſtalkes grow ſpoky tufts like Angelica, which bring forth a long and leafie ſeed like Cumine, of a pale colour; in taſte ſeeming as though it were condited with ſugar, but withall ſomewhat ſharpe, and ſharper than *Seſeli pratenſe*.

2 There is a ſecond kinde of *Siler* which *Pena* and *Lobel* ſet forth vnder the title of *Seſeli pratenſe Monſpelienſium*, which *Dodonæus* in his laſt edition calleth *Siler pratenſe alterum*, that is in ſhew very like the former. the ſtalkes thereof grow to the height of two cubits, but his leaues are ſomewhat broader and blacker: there are not ſo many leaues growing vpon the ſtalke, and they are leſſe diuided than the former, and are of little ſauour. The ſeed is ſmaller than the former, and ſauouring very little or nothing. The root is blacke without, and white within, diuiding it ſelfe into ſundry diuiſions.

¶ *The Place.*

It groweth of it ſelfe in Liguria, not far from Genua in the craggy mountaines, and in the gardens of diligent Herbariſts.

¶ *The Time.*

Theſe plants do floure from Iune to the end of Auguſt.

¶ *The Names.*

It is called commonly *Siler Montanum*: in French and Dutch by a corrupt name *Ser-Montain*. in diuers ſhops, *Seſeleos*, but vntruly: for it is not *Seſeli*, nor a kinde thereof: in Engliſh, Siler mountaine, after the Latine name, and baſtard Louage. ‡ The firſt is thought to be the *Liguſticum* of the Antients, and it is ſo called by *Matthiolus* and others. ‡

¶ *The Nature.*

This plant with his ſeed is hot and dry in the third degree.

¶ *The Vertues.*

The ſeeds of *Siler* drunke with Wormewood wine, or wine wherein Wormewood hath been ſodden, mooueth womens diſeaſes in great abundance: cureth the ſuffocation and ſtrangling of the matrix, and cauſeth it to returne vnto the naturall place againe. A

The root ſtamped with hony, and applied or put into old ſores, doth cure them and couer bare and naked bones with fleſh. B

Being drunke it prouoketh vrine, eaſeth the paines of the guts or entrailes proceeding of cruditie or rawneſſe, it helpeth concoction, conſumeth winde, and ſwelling of the ſtomacke. C

The root hath the ſame vertue or operation, but not ſo effectuall, as not being ſo hot and dry. D

† The figure which formerly was here was of the *Seſile Maſsilienſe* deſcribed in the next chapter in the fourth place, and that which belonged to this place was put for our common Louage. Alſo that figure which belonged to the ſecond deſcription was formerly vnder the title of *Fœniculum dulce*.

CHAP. 224. *Of Seſelios, or Harte-worts of Candy.*

¶ *The Deſcription.*

1 THis plant being the *Seſeli* of Candy, and in times paſt not elſewhere found, tooke his ſurname of that place where it was firſt found, but now adaies it is to be ſeen in the corne fields about Narbon in France, from whence I had ſeeds, which proſper well in my garden. This is but an annuall plant, and increaſeth from yeere to yeere by his owne ſowing. The leaues grow at the firſt euen with the ground, ſomewhat hairy, of an ouerworne greene colour, in ſhape much like vnto Cheruill, but thicker: among which riſeth vp an hairy rough ſtalke, of the height of a cubit, bearing at the top ſpokie tufts with white floures: which being vaded, there followeth round and flat ſeed, compaſſed and cunningly wreathed about the edges like a ring.

The seed is flat like the other, ioyned two together in one, as you may see in the seed of Ferula or Angelica, in shape like a round target, in taste like *Myrrhis*. *Matthiolus* did greatly mistake this plant.

2 There is a kinde of *Seseli Creticum*, called also *Tordylion*: and is very like vnto the former, sauing that his leaues are more like vnto common Parsneps than Cheruil, and the whole plant is bigger than the former.

1 *Seseli Creticum minus:*
Small Seseleos of Candie.

‡ 2 *Seseli Creticum maius.*
Great Seselios of Candie.

3 There is likewise a kinde of *Seseli* that hath a root as big as a mans arme, especially if the plant be old, but the new and young plants beare roots an inch thicke, with some knobs and tuberous sprouts, about the lower part, the root is thicke, rough, and couered ouer with a thicke barke, the substance whereof is first gummie, afterward sharpe, and as it were full of spattle; from the vpper part of the root proceed many knobs or thicke swelling roots, out of which there issueth great and large wings or branches of leaues, some whereof are notched and dented round about, growing vnto one side or rib of the leafe, standing also one opposite vnto another, of a darke and delaid green colour, and somewhat shining aboue, but vnderneath of a grayish or ashe colour: from amongst these leaues there ariseth a straked or guttered stalke, a cubit and a halfe high, sometimes an inch thicke, hauing many ioints or knees, and many branches growing about them, and vpon each ioint lesser branches of leaues. At the top of the stalkes, and vpper ends of the branches grow little cups or vmbels of white floures; which being vaded, there commeth in place a seed, which is very like *Siler montanum*. ‡ I take this here described to be the *Seseli montanum* 1. of *Clusius*, or *Ligusticum alterum Belgarum* of *Lobel*: and therefore I haue giuen you *Clusius* his figure in this place. ‡

There is also a kinde of *Seseli*, which *Pena* setteth forth for the first kinde of *Daucus*, wherof I take it to be a kinde, growing euery where in the pastures about London, that hath large leaues, growing for a time euen with the earth, and spred thereupon, and diuided into many parts, in manner almost like to the former for the most part in all things, in the round spokie tufts or vmbels, bearing stiffe and faire white floures in shape like them of Cinkefoile; in smell like *Sambucus* or Elder. When the floure is vaded, there commeth in place a yellow guttered seed, of a spicie and very hot taste. The root is thicke, and blacke without, which rotteth and perisheth in the ground (as wee may

may see in many gummie or Ferulous plants) after it hath seeded, neither will it floure before the second or third yeare after it is sowne. ‡ I am ignorant what our Author means by this description. ‡

‡ 3 *Seseli Creticum maius.*
Mountaine Seselios.

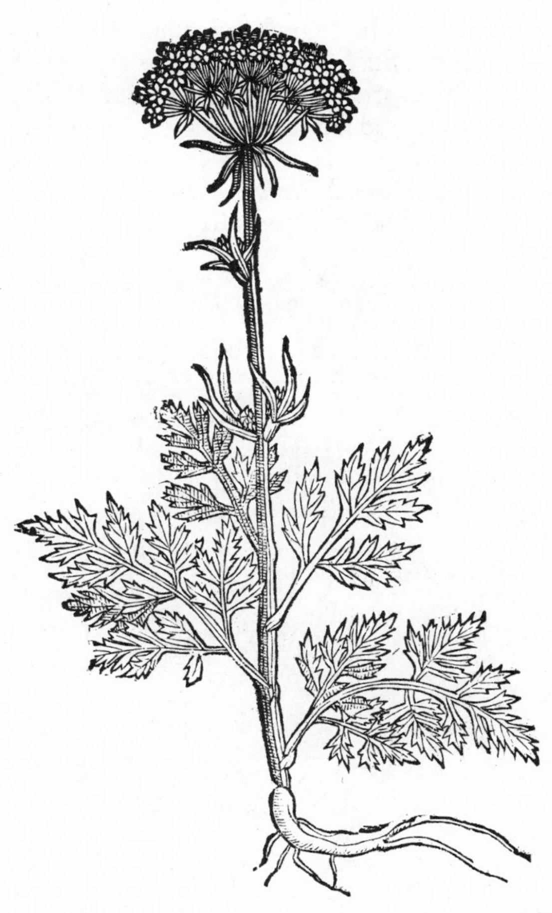

‡ 4 *Seseli Massiliense.*
Seselios of *Marseilles.*

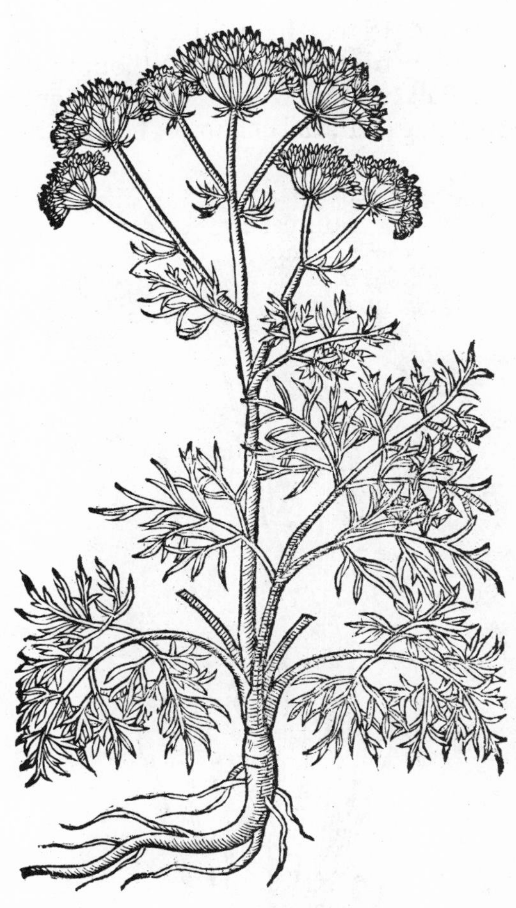

4 There is likewise a kinde of *Seseli* called *Seseli Massiliense*, which hath leaues very much clouen or cut, and finely iagged, very much like vnto the leaues of sweet Fennell, greater and thicker than the common Fennell. The stalke groweth to the height of three cubits, hauing knotty ioints, as it were knees; bearing at the top thereof tufts like vnto Dill, and seed somewhat long and cornered, of a sharpe and biting taste. The root is long and thicke like vnto great Saxifrage, of a pleasant smell, and sharpe in taste.

There is another *Seseli* of Massilia, which hath large and great leaues like vnto Ferula, and not much vnlike *Siler Montanum*: among which rise vp stalkes foure cubits high, bearing at the tops spokie tufts like vnto the last before rehearsed, of a good sauour. The root is like vnto the former in shape, substance, and sauour, but that it is greater.

¶ *The Place.*

These plants are strangers in England, notwithstanding I haue them in my garden.

¶ *The Time.*

They floure and flourish in September.

¶ *The Names.*

Their names haue been touched in their seuerall descriptions.

¶ *The Temperature and Vertues.*

It prouoketh vrine, and helpeth the strangurie, bringeth downe the sicknesse and dead birth: it helpeth the cough and shortnesse of breath, the suffocation of the mother, and helpeth the falling sicknesse. A

The seed drunke with wine concocteth raw humours, taketh away the griping and torments of the belly, and helpeth the ague, as *Dioscorides* saith. B

The iuice of the leaues is giuen to Goats and other cattell to drinke, that they may the sooner be deliuered of their young ones, as the same Author reporteth. C

CHAP. 425. *Of Spignell, Spicknell, or Mewe.*

¶ *The Description.*

1 Spignell hath stalkes rising vp to the height of a cubit and a halfe, beset with leaues resembling Fennell or Dill, but thicker, more bushie, and more finely iagged; and at the top of the stalkes do grow spokie tufts like vnto Dil. The roots are thick, and full of an oleous substance, smelling well, and chafing or heating the tongue, of a reasonable good sauour.

1 *Meum.*
Spignell.

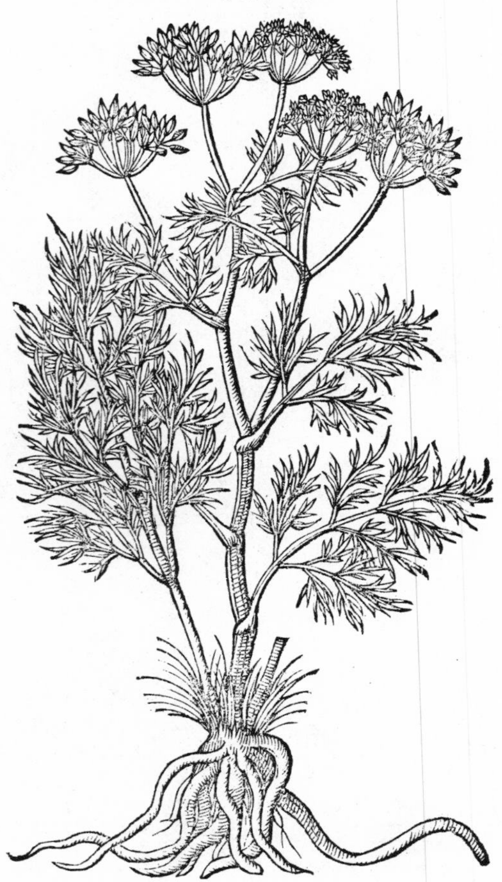

‡ 2 *Meum alterum Italicum.*
Italian Spignell.

2 There is a bastard kinde of Spignell like vnto the former, sauing that the leaues are not so finely cut or iagged: the floures are tufted more thicker than the former: the roots are many, thick, and full of sap.

¶ *The Place.*

Mew, or Meon, groweth in Westmerland, at a place called Round-twhat betwixt Aplebie and Kendall, in the parish of Orton.

Bastard Mewe, or *Meum*, groweth in the waste mountaines of Italie, and the Alps, and (as it hath been told me) vpon Saint Vincents rocke by Bristow, where I spent two daies to seeke it, but it was not my hap to find it, therefore I make some doubt of the truth thereof.

¶ *The Time.*

These herbes doe floure in Iune and Iuly, and yeeld their seed in August.

¶ *The Names.*

It is called of the Grecians μεῖον or μῆον: likewise of the Latines, *Meum*: of the Italians, *Meo*: in Apulia, as *Matthiolus* declareth, it is called *Imperatrix*: in diuers places of Spaine, *Sistra*: in others, *Pinello*: in High Dutch, **Beerewurtz**: in French, *Sistre*: *Ruellius* saith that it is named in France *Anethum tortuosum*, and *syluestre*, or writhed Dil, and wilde Dill: also it is called in English, Spignel, or Spicknell, of some Mew, and Bearewoort.

The second may be called bastard Spicknell.

¶ *The*

¶ *The Temperature.*

These herbes, especially the roots of right Meon, is hot in the third degree, and drie in the second.

¶ *The Vertues.*

A The roots of Meon, boiled in water and drunke, mightily open the stoppings of the kidnies and bladder, prouoke vrine and bodily lust, ease and helpe the strangurie, and consume all windinesse and belchings of the stomacke.

B The same taken with honie doth appease the griefe of the belly, and is exceeding good against all Catarrhes, rheumes, and aches of the iointes, as also any phlegme which falls vpon the Lungs.

C If the same be laied plaisterwise vpon the bellies of children, it maketh them to pisse well.

D They clense the entrails, and deliuer them of obstructions or stoppings: they prouoke vrine, driue forth the stone, and bring downe the floures: but if they be taken more than is requisite, they cause the head-ache; for seeing they haue in them more heat than drinesse, they carry to the head raw moisture and windie heat, as *Galen* saith.

CHAP. 426. *Of Horestrange, or Sulphurwoort.*

¶ *The Description.*

1 SVlphurwoort or Hogs-fennell hath a stiffe and hard stalke full of knees or knots, beset with leaues like vnto Fennell, but greater, comming neerer vnto Ferula, or rather like the leaues of wilde Pine-tree, and at the top of the stalkes round spokie tufts full of little yellow floures, which do turne into broad browne seed. The root is thicke and long: I haue digged vp roots thereof as big as a mans thigh, blacke without, and white within, of a strong and grieuous smell, and full of yellow sap or liquour, which quickly waxeth hard or dry, smelling not much vnlike brimstone, called *Sulphur*; which hath induced some to call it Sulphurwoort; hauing also at the top toward the vpper face of the earth, a certain bush of haire, of a browne colour, among which the leaues and stalkes do spring forth.

2 The second kinde of *Peucedanum* or Hogs-fennell is very like vnto the former, sauing that the leaues be like Ferula: the roots are nothing so great as the former, but all the rest of the plant doth far exceed the other in greatnesse.

3 There is another kinde of *Peucedanum* or Hogs Fennell, which *Pena* found vpon Saint Vincents rock by Bristow, whose picture he hath set forth in his *Aduersaria*, which that famous English Physition of late memorie, D. *Turner* found there also, supposing it to be the right and true *Peucedanum*, whereof no doubt it is a kinde: it groweth not aboue a foot high, and is in shape and leaues like the right *Peucedanum*, but they be shorter and lesser, growing somwhat like the writhed Fennell of Massilia, but the branches are more largely writhed, and the leaues are of the colour of the branches, which are of a pale greene colour. At the top of the branches grow small white tufts, hauing seed like Dill, but shorter and slenderer, of a good taste, somewhat sharpe. The root is thicker than the smalnesse of the herbe will well beare. Among the people about Bristow, and the rocke aforesaid, this hath been thought good to eat.

‡ The figure of this our Authour formerly gaue (yet vnfitly, it not agreeing with that description) for *Oreoselinum*: it may be he thought it the same with that of *Dodon.* his description, because he found it vnder the same title in *Tabernamontanus*. This is the *Selinum montanum pumilum* of *Clusius*; and the *Peucedani facie pusilla planta* of *Pena* and *Lobel*; wherfore *Bauhine* was mistaken in his *Pinax*, whereas he refers that of *Lobel* to his third *Peucedanum*: the root of this is black without, and white within, but short, yet at the top about the thicknesse of ones finger: the leaues are small and green, commonly diuided into fiue parts; and these againe subdiuided by threes: the stalke is some sixe inches or halfe a foot high, diuided into sundry branches, crested, broad, and at the toppes of the branches, euen when they first shoot vp, appeare little vmbels of white floures very small, and consisting of fiue leaues apiece. The seed is blacke, shining and round, two being ioined together, as in most vmbelliferous plants. It floures in May, and ripens the seede in Iuly: I receiued in Iuly 1632, some plants of this from Bristow, by the meanes of my oft mentioned friend Master *George Bowles*, who gathered it vpon Saint Vincents Rocke, whereas the Authours of the *Aduersaria* report it to grow. ‡

1 *Peucedanum.*
Sulphurwoort.

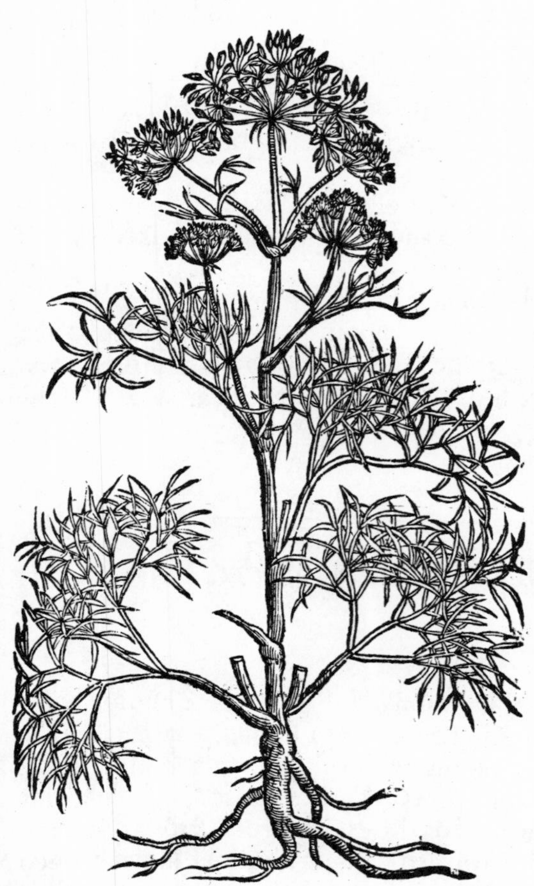

2 *Peucedanum maius.*
Great Sulphurwoort.

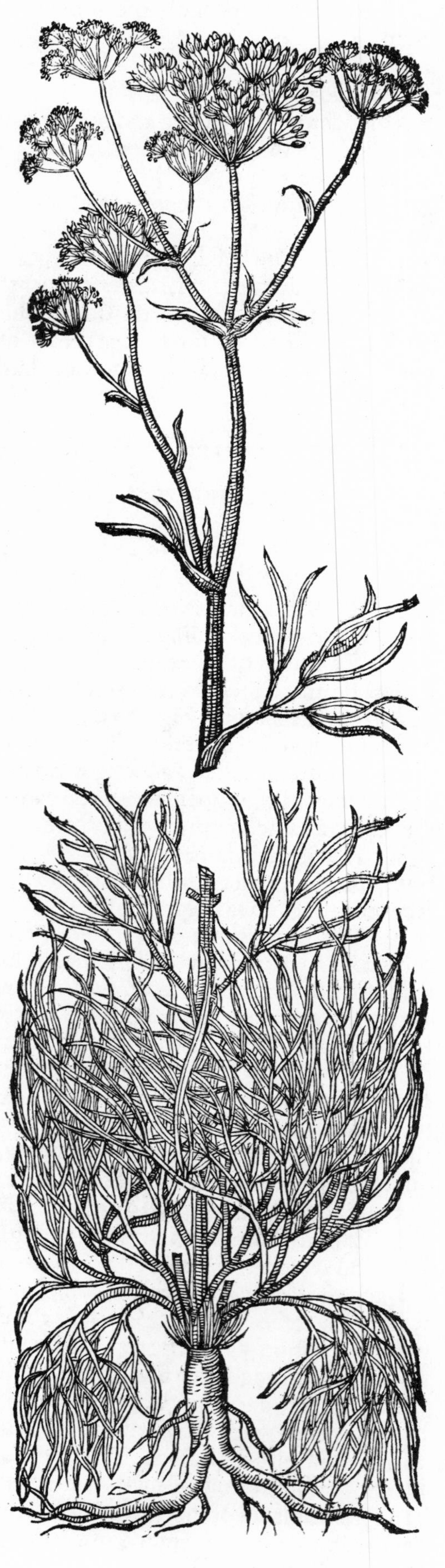

‡ 3 *Peucedanum pumilum.*
Dwarfe Hogs-Fennell.

¶ *The Place.*

The firſt kinde of *Peucedanum* or Hogs Fennell groweth very plentifully on the South ſide of a wood belonging to Waltham, at the Naſe in Eſſex by the high-way ſide; alſo at Whitſtable in Kent, in a medow neere to the ſea ſide, ſometime belonging to Sir *Henry Criſpe*, and adioyning to his houſe there. It groweth alſo in great plenty at Feuerſham in Kent, neere vnto the hauen vpon the bankes thereof, and in the medowes adioyning.

The ſecond kinde groweth vpon the ſea coaſts of Montpellier in France, and in the coaſts of Italy.

¶ *The Time.*

Theſe plants do floure in Iune, Iuly, and Auguſt.

¶ *The Names.*

The Grecians call it πευκέδανος: the Latines in like manner *Peucedanos*, or *Peucedanum*, and alſo *Pinaſtellum*: moſt of the ſhops, and likewiſe the common people name it *Fœniculum Porcinum*: of diuers, *Stataria*: of the Prophets, ἀγαθὸς δαίμων: that is to ſay, a good Angell or Ghoſt: in high-Dutch, **Harſtrang, Schweffel wurkel, Sewfenckel**: in Italian and French, *Peucedano*: in Spaniſh, *Herbatum*: in Engliſh, Hore-ſtrange, and Hore-ſtrong, Sow-Fennell, or Hogs Fennell, Sulphur-wort, or Brimſtone-wort. It is called *Peucedanum* and *Pinaſtellum*, of the Greeke and Latine words, πεύκη and *Pinus*.

¶ *The Temperature.*

Theſe herbes, eſpecially the yellow ſap of the root, is hot in the ſecond degree, and dry in the beginning of the third.

¶ *The Vertues.*

A The yellow ſap of the root of Hogs Fennell, or as they call it in ſome places of England, Hore-ſtrange, taken by it ſelfe, or with bitter almonds and Rue, is good againſt the ſhortneſſe of breath, it aſſwageth the griping paines of the belly, diſſolueth and driueth away ventoſitie or windineſſe of the ſtomacke; it waſteth the ſwelling of the milt or ſpleene, looſeth the belly gently, and purgeth by ſiege both flegme and choler.

B The ſame taken in manner aforeſaid prouoketh vrine, eaſeth the paine of the kidneyes and bladder, cauſeth eaſie deliuerance of childe, and expelleth the ſecondine, or after-birth, and the dead childe.

C The ſap or iuyce of the root mixed with oyle of Roſes, or Vineger, and applied, eaſeth the palſie, crampes, contraction or drawing together of ſinewes, and all old cold diſeaſes, eſpecially the Sciatica.

D It is vſed with good ſucceſſe againſt the rupture or burſtings in yong children, and is very good to be applied vnto the nauels of children that ſtand out ouer much.

E The decoction of the root drunke is of like vertue vnto the iuyce, but not altogether ſo effectuall againſt the foreſaid diſeaſes.

F The root dried and made into pouder doth mundifie and clenſe old ſtinking and corrupt ſores and vlcers, and healeth them: it alſo draweth forth the corrupt and rotten bones that hinder the ſame from healing, and likewiſe ſplinters and other things fixed in the fleſh.

G The ſaid pouder or iuyce of the root mixed with oyle of Roſes, cauſeth one to ſweat, if the body be anointed therewith, and therefore good to be put into the vnction or ointment for the French diſeaſe.

H The congealed liquor tempered with oyle of Roſes, and applied to the head after the manner of an ointment, is good for them that haue the Lethargie, that are franticke, that haue dizzineſſe in the head, that are troubled with the falling ſickneſſe, that haue the palſie, that are vexed with conuulſions and crampes, and generally it is a remedie for all infirmities of the ſinewes, with Vineger and oyle, as *Dioſcorides* teacheth.

I The ſame being ſmelt vnto reuiueth and calleth them again that be ſtrangled with the mother, and that lie in a dead ſleepe.

K Being taken in a reare egge it helpeth the cough and difficultie of breathing, gripings and windineſſe, which, as *Galen* addeth, proceedeth from the groſſeneſſe and clammineſſe of humors.

L It purgeth gently, it diminiſheth the ſpleene, by cutting, digeſting, and making thin humours that are thicke: it cauſeth eaſie trauell, and openeth the matrix.

M A ſmall piece of the root holden in the mouth is a preſent remedie againſt the ſuffocation of the mother.

CHAP. 427.
Of Herbe Ferula, or Fennell Gyant.

¶ *The Kindes.*

Dioscorides maketh mention of a *Ferula*, out of which is gathered the Gum *Sagapene*; and also he declareth, that the Gums *Galbanum* and *Ammoniacum* are liquors of this herb *Ferula*: but what difference there is in the liquors, according to the clymat or countrey where it groweth, he doth not set downe; for it may be that out of one kinde of *Ferula* sundry iuyces may be gathered, that is to say, according to the diuersitie of the countries where they grow, as we haue said: for as in Laser, the iuyce of Laserwort that groweth in Cyrene doth differ from that liquor which groweth in Media and Syria; so it is likely that the herbe *Ferula* doth bring forth in Media *Sagapenum*, in Cyrene *Ammoniacum*, and in Syria *Galbanum*. *Theophrastus* saith that the herbe *Ferula* is diuided into mo kindes, and he calleth one great, by the name of *Ferula*; and another little, by the name *Ferulago*.

1 *Ferula.*
Fennell Gyant.

‡ 2 *Ferulago.*
Small Fennell-Gyant.

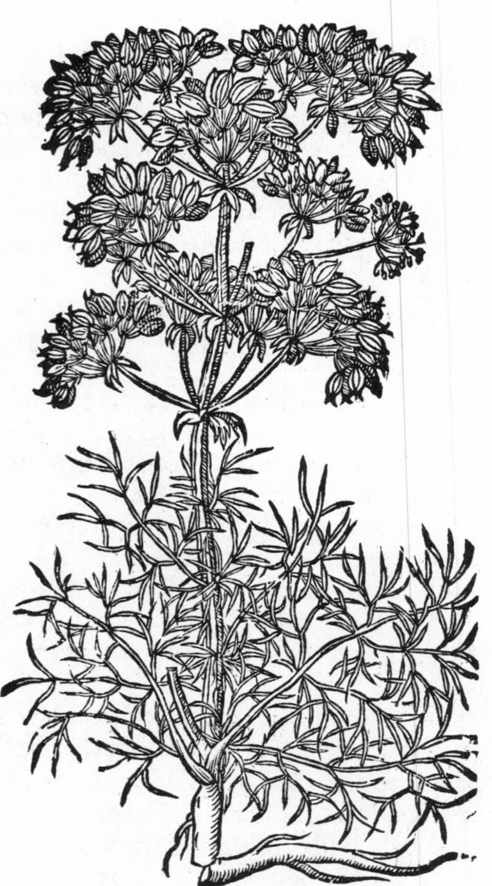

¶ *The Description.*

1 *Ferula*, or Fennell Gyant, hath very great and large leaues of a deepe greene colour, cut and iagged like those of Fennell, spreading themselues abroad like wings: amongst which riseth vp a great hollow stalke, somewhat reddish on that side which is next vnto the Sun, diuided into certaine spaces, with ioynts or knees like those of Hemlocks or Kexes, of the bignesse of a mans arme in the wrest, of the height of foure or fiue cubits where it groweth naturally, as in Italy, Greece, and other hot countries; notwithstanding it hath attained to the height of fourteene or fifteene foot in my garden, and likewise groweth fairer and greater than from whence it came, as it fareth with other plants that come hither from hot regions: as for example our great Artichoke, which first was brought out of Italy into England, is become (by reason of the great moisture which our countrey is subiect vnto) greater and better than those of Italy; insomuch

insomuch that diuers Italians haue sent for some plants of our Artichokes, deeming them to be of another kinde; neuerthelesse in Italy they are small and dry as they were before. Euen so it happeneth to this *Ferula*, as we haue said. This foresaid stalke diuideth it selfe toward the top into diuers other smaller branches, whereon are set the like leaues that grow next the ground, but much lesser. At the top of the branches at the first budding of the floures appeare certaine bundles inclosed in thin skins, like the yolke of an egge, which diuers call *Corculum Ferulæ*, or the little heart of *Ferula*; which being brought to maturitie, open themselues into a tuft or vmbel like that of Dil, of a yellowish colour: after which come the seed, in colour and fashion like those of the Parsnep, but longer and greater, alwaies growing two together, so closely ioyned, that it cannot be discerned to be more than one seed vntill they be diuided: the root is very thicke and great, full of a certaine gummie iuyce, that floweth forth, the root being bruised, broken, or cut; which being dried or hardned, is that gum which is called *Sagapenum*, and in some shops *Serapinum*.

‡ 3 *Panax Asclepium Ferulæ facie.*
Æsculapius his All-heale.

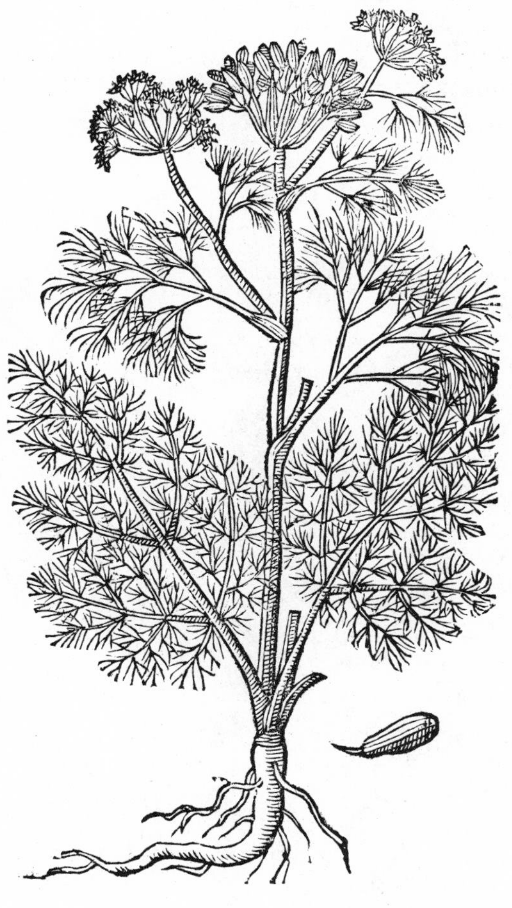

2 There is likewise another smaller *Ferula* like vnto the former in each respect, sauing that it is altogether lesse: the root likewise being wounded yeeldeth forth a sap or iuyce, which when it is hardned is called *Galbanum*: of the Assyrians, *Metopium*.

I haue likewise another sort sent mee from Paris, with this title *Ferula nigra*; which prospereth exceeding well in my garden, but difference I cannot finde any from the former, sa- that the leaues are of a more blacke or swart colour.

‡ 3 I know not where more fitly than in this place to giue you the historie of that *Ferula* or Ferulaceous plant that *Dodonæus*, *Lobel*, and others haue set downe vnder the name of *Panax Asclepium*. The stalke hereof is slender, a cubit high, crested and ioynted, and from these ioynts proceed leaues bigger than those of Fennell, and also rougher, and of a strong smell: at the tops of the branches grow vmbels of yellow floures: the seed is flattish, like that of the other *Ferula*: the root long, white, and of a strong smell. This growes naturally in Istria. ‡

¶ *The Place.*

These plants are not growing wilde in England; I haue them all in my garden.

¶ *The Time.*

They floure in Iune and Iuly; they perfect their seed in September; not long after, the stalke with his leaues perish: the root remaineth fresh and greene all Winter.

¶ *The Names.*

The first is called in Greeke Νάρθηξ: in Latine, *Ferula*: in Italian, *Ferola*: in Spanish, *Cananheia*: in English, Herbe Ferula, and Fennell Gyant.

¶ *The Temperature.*

These plants with their Gums are hot in the third degree, and dry in the second.

¶ *The Vertues.*

A The pith or marrow, called *Corculum Ferulæ*, as *Galen* teacheth, is of an astringent or binding qualitie, and therefore good for them that spit bloud, and that are troubled with the flix.

B *Dioscorides* saith, that being put into the nosthrils it stayeth bleeding, and is giuen in Wine to those that are bitten with Vipers.

C It is reported to be eaten in Apulia rosted in the embers, first wrapped in leaues or in old clouts, with pepper and salt; which, as they say, is a pleasant sweet food, that stirreth vp lust, as they report.

D The seed doth heate, and attenuate or make thinne: it is a remedie against cold fits of an Ague, by procuring sweat, being mixed with oyle, and the body anointed therewith.

E A dram of the iuyce of *Ferula* which beareth *Sagapenum*, purgeth by siege tough and slimie humors,

mors, and all groſſe flegme and choler, and is alſo good againſt all old and cold diſeaſes which are hard to be cured; it purgeth the brain, and is very good againſt all diſeaſes of the head, againſt
F the Apoplexie and Epilepſie.

Being taken in the ſame manner, it is good againſt crampes, palſies, ſhrinkings and paines of
G the ſinewes.

It is good againſt the ſhortneſſe of breath, the cold and long cough, the paine in the ſide and
H breſt, for it mundifieth and clenſeth the breſt from all cold flegme and rheumaticke humors.

Sagapenum infuſed or ſteeped in vineger all night, and ſpread vpon leather or cloath, ſcattereth, diſſolueth, and driueth away all hard and cold ſwellings, tumors, botches, and hard lumpes growing about the ioynts or elſewhere, and is excellent good to be put into or mingled with all oynt-
I ments or emplaiſters which are made to mollifie or ſoften.

The iuyce of *Ferula Galbanifera*, called *Galbanum*, drunke in wine with a little myrrh, is good againſt all venome or poyſon that hath beene taken inwardly, or ſhot into the body with venomous
K darts, quarrels, or arrowes.

It helps womens paineful trauel, if they do take therof in a cup of wine the quantitie of a bean.

L The perfume of *Galbanum* helpeth women that are grieued with the riſing of the mother, and is good for thoſe that haue the falling ſickneſſe.

M *Galbanum* ſoftneth, mollifieth, and draweth forth thornes, ſplinters, or broken bones, and conſumeth cold and flegmaticke humors, ſeruing in ſundry ointments and emplaiſters for the vſe of Surgerie, and hath the ſame phyſicall vertues that are attributed vnto *Sagapenum*.

Chap. 428. *Of Drop-wort, or Filipendula.*

1 *Filipendula.*
Drop-wort.

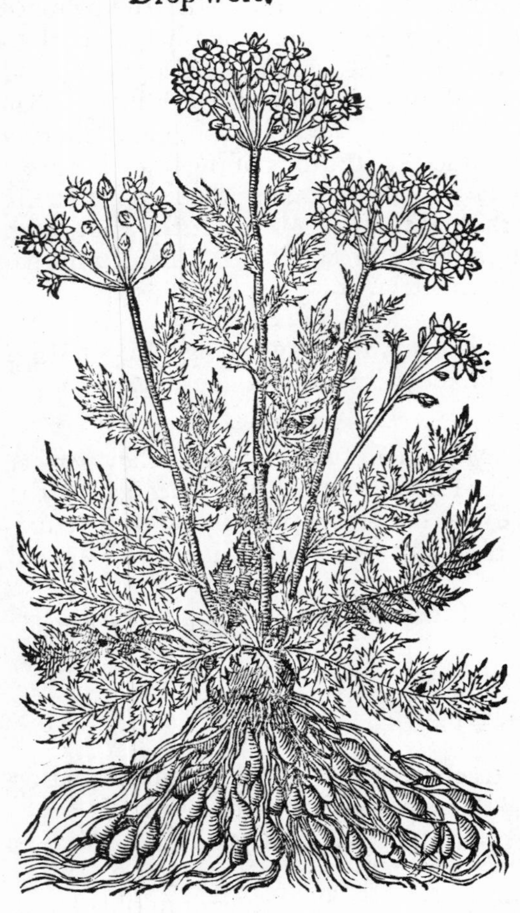

2 *Filipendula montana.*
Mountaine Drop-wort.

¶ *The Kindes.*

There be diuers ſorts of Drop-worts, ſome of the champion or fertill paſtures, ſome of more moiſt and dankiſh grounds, and ſome of the mountaine.

¶ *The*

¶ *The Deſcription.*

1 THe firſt kinde of Filipendula hath leaues growing and ſpred abroad like feathers, each leafe conſiſting of ſundry ſmall leaues dented or ſnipt round about the edges, growing to the ſtalke by a ſmall and ſlender ſtem: theſe leaues reſemble wilde Tanſie or Burnet, but that they be longer and thicker, ſet like feathers, as is aforeſaid: among theſe riſe vp ſtalkes a cubit and a halfe high, at the top whereof grow many faire white floures, each ſmall floure conſiſting of ſix ſlender leaues, like a little ſtar, buſhing together in a tuft like the floures of Medeſweet, of a ſoft ſweet ſmell: the ſeed is ſmall, and groweth together like a button: the roots are ſmall and blacke, whereupon depend many little knops or blacke pellets, much like the roots of the female Peonie, ſauing that they be a great deale ſmaller.

2 The ſecond kinde of Filipendula, called of *Pena* in his Obſeruations, *Oenanthe, ſiue Philipendula altera montana*, is neither at this day very well knowne, neither did the old writers heretofore once write or ſpeake of it: but *Pena* that painefull Herbariſt found it growing naturally in Narbone in France, neere vnto Veganium, on the top of the high hills called *Paradiſus Dei*, and neere vnto the mountaine Calcaris: this rare plant hath many knobby long roots, in ſhape like to *Aſphodelus luteus*, or rather like the roots of *Corruda*, or wilde Aſparagus; from which riſeth vp a ſtalke a foot high, and more, which is thicke, round, and chanelled, beſet full of leaues like thoſe of common Filipendula, but they be not ſo thicke ſet or winged, but more like vnto the leaues of a Thiſtle, conſiſting of ſundry ſmall leaues, in faſhion like to *Coronopus Ruellij*, that is, *Ruellius* his Bucks horne: round about the top of the ſtalke there groweth a very faire tuft of white floures, reſembling fine ſmall hoods, growing cloſe and thicke together like the floures of *Pedicularis*, that is, Red Rattle, called of *Carolus Cluſius*, *Alectorolophos*, whereof he maketh this plant a kinde, but in my iudgement and opinion it is rather like *Cynoſorchis*, a kinde of Satyrion.

3 *Filipendula anguſtifolia.*
Narrow leafed Drop-wort.

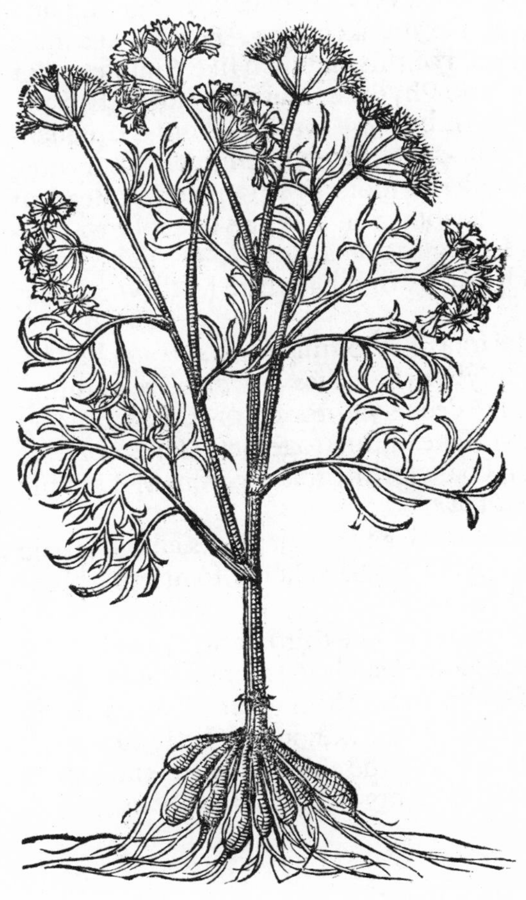

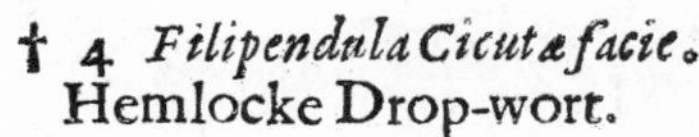
† 4 *Filipendula Cicutæ facie.*
Hemlocke Drop-wort.

3 There is another kinde of Filipendula ſet forth vnder the name of *Oenanthe*, that hath many tuberous and thicke roots like thoſe of Drop-wort, but white of colour, and euery one of thoſe knobs hath a certaine ſtring or fibre annexed thereto; from whence ariſeth a creſted ſtalk two foot high,

5 *Filipendula aquatica.*
Water Drop-wort.

high, diuiding it ſelf toward the top into ſundry armes or branches : from the hollow place or boſome of euery ioynt (out of which doe grow thoſe branches) the leaues doe alſo proceed, very much cut or iagged like Fennell : at the top of thoſe branches come forth ſpokie rundles of white floures faſhioned like ſtars.

† 4 The fourth kinde of Filipendula is as ſtrange a plant as the former, eſpecially with vs here in England, except in the waterie places and rilles in the North, where *Paludapium* or water Smallage groweth ; whereunto in leaues it is not vnlike, but more like *Ruta pratenſis* : it hath many large branches, a naughty ſauor, and in colour and ſhape like *Cicuta*, that is, Hemlocke. The ſtalkes are more than two cubits high, comming from a root which exceedingly multiplies it ſelfe into bulbes, like *Aſphodelus albus*. The ſmell of this plant is ſtrong and grieuous ; the taſte hot and biting, it being full of a iuyce, at firſt milky, but afterwards turning yellow. The ſpoky tufts or rundles growing at the top are like *Cicuta*, yea, it much reſembleth Hemlocke in propertie and qualities, and ſo doe they affirme that haue proued and ſeene the experience of it : for being eaten in ſallads it did well nigh poyſon thoſe which ate of it, making them giddie in their heads, waxing very pale, ſtaggering and reeling like drunken men. Beware and take good heed of this and ſuch like Simples ; for there is no Phyſitian that will giue it, becauſe there be many other excellent good Simples which God hath beſtowed vpon vs from the preuenting and curing of diſeaſes. ‡ Pernitious and not excuſable is the ignorance of ſome of our time, that haue bought and (as one may probably coniecture) vſed the roots of this plant in ſtead of thoſe of Peionie ; and I know they are dayly by the ignorant women in Cheape-ſide ſold to people more ignorant than themſelues, by the name of water Louage ; *Caueat Emptor*. The danger that may enſue by vſing them may be gathered by that which our Author hath here ſet downe, being taken out of the *Aduerſaria, pag. 326.* ‡

5 The fifth and laſt kinde of Filipendula, which is the fourth according to *Matthiolus* his account, hath leaues like water Smallage, which *Pliny* calleth *Sylaus*, the leaues very much reſembling thoſe of *Lauer Crateuæ* : among which riſeth vp a ſmall ſtalke deepely furrowed or creſted, bearing at the top thereof ſpokie or buſhy rundles of white floures thicke thruſt together. The roots are compact of very many filaments or threds, ; among which come forth a few tuberous or knobbie roots like vnto the ſecond.

¶ *The Place and Time.*

The firſt groweth plentifully vpon ſtonie rockes or mountaines, and rough places, and in fertile paſtures. I found great plenty thereof growing in a field adioyning to Sion houſe, ſomtime a Nunnerie, neere London, on the ſide of a medow called Sion Medow.

The ſecond hath been ſufficiently ſpoken of in the deſcription. The third groweth neere vnto brookes and riuers ſides. The fourth groweth betweene the plowed lands in the moiſt and wet furrowes of a field belonging to Batterſey by London. ‡ It alſo groweth in great aboundance in many places by the Thames ſide ; as amongſt the oyſiers againſt Yorke houſe, a little aboue the Horſe-ferrey, againſt Lambeth, &c. ‡ The fifth groweth neere the ſides of riuers and water-ſtreames, eſpecially neere the riuer of Thames or Tems, as in S. Georges fields, and about the Biſhop of Londons houſe at Fulham, and ſuch like places.

They floure from May to the end of Iune.

¶ *The Names.*

They are commonly called *Filipendulæ*. The firſt is called of *Nicolaus Myrepſus*, *Philipendula* : of ſome, *Saxifragra rubra*, and *Millefolium ſylueſtre* : of *Pliny*, *Molon* : in Italian and Spaniſh, *Filipendula* : in Engliſh, Filipendula and Drop-wort. Water Filipendula is called *Filipendula aquatica*, *Oenanthe aquatica*, and *Silaus Plinij*.

The

The fourth, whoſe leaues are like to Homlocks, is as ſome thinke called of *Co dus*, *Oiſenichium* in Engliſh Homlocke Filipendula.

¶ *The Nature.*

Theſe kindes of Filipendula are hot and dry in the third degree, opening and clenſing, and yet with a little aſtriction or binding. All the kindes of Oenanthes haue the ſame facultie, except the fourth, whoſe pernitious facultie we haue formerly touched.

¶ *The Vertues.*

The root of common Filipendula boiled in wine and drunken, is good againſt all paines of the bladder, cauſeth one to make water, and breaketh the ſtone. The like *Dioſcorides* hath written of Oenanthe; the root, ſaith he, is good for them that piſſe by drops. A

The powder of the roots of Filipendula often vſed in meate, will preſerue a man from the falling ſickeneſſe. B

† The figures that were formerly in the fourth and fifth places, were both of the plant deſcribed in the fifth place. I haue giuen you in the fourth place the figure which *Lobel* and others haue giuen for the plant there deſcribed, but it is not well expreſt, for the leaues are large like thoſe of Smallage, the ſtalke, branches and vmbells very large, and like thoſe of Hemlocke, but rather bigger.

CHAP. 429. *Of Homlocks, or herbe Bennet.*

¶ *The Deſcription.*

1 THe firſt kinde of Hemlocke hath a long ſtalke, fiue or ſix foot high, great and hollow, full of joints like the ſtalkes of Fennell, of an herby colour; poudered with ſmall red ſpots, almoſt like the ſtems of Dragons. The leaues are great, thicke, and ſmall cut or jagged like the leaues of Cheruill, but much greater, and of a very ſtrong and vnpleaſant ſauor. The floures are white, growing by tufts or ſpokie tops, which do change and turne into a white flat ſeed: the root is ſhort, and ſomewhat hollow within.

1 *Cicuta.*
Hemlocks.

2 The Apothecaries in times paſt not knowing the right *Seſeli Peloponnenſe*, haue erroniouſly taken this *Cicuta latifolia* for the ſame. The leaues whereof are broad, thicke, and like vnto *Cicutaria*, yet not the ſame; they called it *Seſeli Peloponnenſe cum folio Cicutæ*, the faculties whereof deny and refute that aſſertion and opinion, yea and the plant it ſelfe, which being touched, yeeldeth or breatheth out a moſt virulent or lothſome ſmell: theſe things ſufficiciently argue, that it is not a kinde of *Seſeli*, beſides the reaſons following: *Seſeli* hath a reaſonable good ſauour in the whole plant, the root is bare and ſingle, without fibres, like a Carrot; but *Cicuta* hath not onely a lothſome ſmell, but his roots are great, thicke, and knobby, like the roots of *Myrrhis*: the whole plant doth in a manner reſemble the leaues, ſtalkes and floures of *Myrrhis odorata*, whoſe ſmall white floures doe turne into long and crooked ſeeds, growing at the top of the branches three cubits high.

‡ 3 This in leaues, ſtalkes, and roots is larger than the laſt deſcribed, the ſtalkes equalling or exceeding the height of a man; the ſmell is ſtrange and greiuous, and in all the parts thereof it is like to the other plants of this kinde. *Lobel* figures it by the name of *Cicutaria maxima Brancionis*, and queſtions whither it be not *Thapſia tertia Salamanticenſium* of *Cluſius*; but *Cluſius* denies it ſo to be. ‡

2 *Cicuta latifolia fœtidissima.*
Broad leafed stinking Hemlocks.

‡ 3 *Cicuta latifolia altera.*
Gyant Hemlocke.

¶ *The Place.*

Common Hemlocke groweth plentifully about towne walls and villages in shadowie places, and fat soiles neere ditches.

The second groweth vpon mountaines and desart places, and is a stranger in England; yet I haue plants thereof in my garden.

¶ *The Time.*

They flourish and seed in September.

¶ *The Names.*

Homlocke is called in Greeke κώνειον: in Latine, *Cicuta* · in high Dutch, **Schirling:** in low Dutch, **Scheerlinck:** in Spanish, *Ceguta y Canaheia.* in French, *Cigue* in English, Hemlocke, Homlocke, Kexe, and herbe Bennet.

The second is called *Cicuta latifolia*, and *Cicutaria latifolia*, and *Seseli Peloponnense quorundam* · in English, great Hemlockes, and garden Homlocke.

¶ *The Temperature.*

Galen saith, that Homlocke is extreme cold in operation, euen in the fourth degree of coldnesse.

¶ The

¶ *The Vertues.*

A It is therefore a very raſh part to lay the leaues of Hemlocke to the ſtones of yong boyes or virgin ,breſts, and by that meanes to keepe thoſe parts from growing great ; for it doth not only eaſily cauſe thoſe members to pine away, but alſo hurteth the heart and liuer, being outwardly applied : then muſt it of neceſſitie hurt more being inwardly taken ; for it is one of the deadly poyſons which killeth by his cold qualitie, as *Dioſcorides* writeth, ſaying, Hemlocke is a very euill, dangerous, hurtfull, and poyſonous herbe, inſomuch that whoſoeuer taketh of it into his body dieth remedileſſe, except the party drinke ſome wine that is naturally hot, before the venome haue taken the heart, as *Pliny* ſaith : but being drunke with wine the poyſon is with greater ſpeed caried to the heart, by reaſon whereof it killeth preſently ; therefore not to be applied outwardly, much leſſe taken inwardly into the body.

B The great Hemlocke doubtleſſe is not poſſeſſed with any one good facultie, as appeareth by his lothſome ſmell, and other apparant ſignes, and therefore not to be vſed in phyſicke.

CHAP. 430. *Of wilde and water Hemlockes.*

¶ *The Deſcription.*

† 1 THis wilde kinde of Hemlocke hath a ſmall tough white root, from which ariſe vp diuers ſtiffe ſtalkes, hollow, ſomewhat reddiſh toward the Sun, ioynted or kneed at certaine diſtances : from which ioynts ſpring forth long leaues very greene, and finely minced or iagged like the common Cheruill or Parſley : the floures ſtand at the tops of the ſtalks in ſmall ſpokie vmbels, with little longiſh greene leaues about them : the ſeed followeth, like thoſe of Hemlocke, or as they grow together on the tops of the ſtalks they reſemble Coriander ſeeds, but leſſer : the whole plant is of a naughty ſmell.

† 1 *Cicutaria tenuifolia.*
Thin leafed wilde Hemlocke.

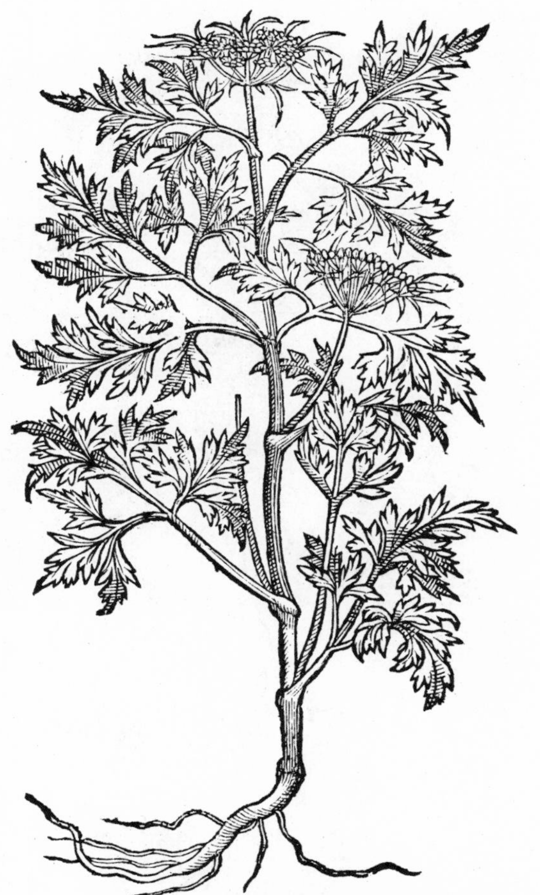

2 *Cicutaria paluſtris.*
Wilde water Hemlocke.

2 Water Hemlock, which *Lobel* calleth *Cicutaria paluſtris* : *Cluſius* and *Dodonæus*, *Phellandrium*, riſeth vp with a thicke fat and empty hollow ſtalke, full of knees or ioynts, creſted, chamfered, or furrowed,

furrowed, of a yellowish greene colour: the leaues shoot forth of the ioynts and branches, like vnto wilde Hemlocke, but much thicker, fatter, and oileous, very finely cut or iagged, almost like those of the smallest *Visnaga*, or Spanish Tooth-pickes: the floures stand at the top of the stalkes in small whitish tufts: the seed followeth, blackish, of the bignesse of Anise seed, and of a sweet sauour: the root is thicke and long, within the water, very soft and tender, with very many strings fastned thereto.

¶ *The Place.*

1 This growes among stones and rubbish, by the walls of cities and towns almost euery where.

The other groweth in the midst of water ditches and standing pooles and ponds, in most places of England: it groweth very plentifully in the ditches by a causey, as you goe from Redriffe to Detford neere London, and in many other places.

¶ *The Time.*

They floure and flourish in Iuly and August.

¶ *The Names.*

‡ 1 This is *Petroselini vitium* of *Tragus*; and *Dauci inutilis genus* of *Gesner*: *Thalius* calls it *Apium cicutarium*: *Lobel*, *Cicutaria fatua*: *Tabernamontanus*, *Petroselinum caninum*; which name we may fitly make English, and call it Dogs-parsley.

2 This is *Ligusticum syl.* & *Fœniculum syl.* of *Tragus*: *Cicutaria palustris* of *Lobel* and others: *Dodonæus* thinkes it *Plinies Phellandrion*; and *Cæsalpinus* iudges it his *Silaus*. ‡

¶ *The Temperature and Vertues.*

Their temperature and faculties are answerable to the common Hemlocke, which haue no vse in physicke, as we haue said.

† The figure formerly in the first place was of *Myrrhis*; the description I thinke was intended, yet not throughly agreeing with this I here giue you, wherefore I haue a little altered it.

Chap. 431.

Of Earth-nut, Earth Chest-nut, or Kipper-nut.

† 1 *Bulbocastanon minus.*
Small Earth-nut.

2 *Bulbocastanon maius.*
Great Earth-nut.

¶ *The Deſcription.*

1 EArth-nut or Kipper-nut, called after *Lobelius, Nucula terreſtris,* hath ſmall euen creſted ſtalkes a foot or ſomewhat more high: whereon do grow next the ground leaues like thoſe of Parſley, and thoſe that doe grow higher like vnto thoſe of Dill; the white floures doe ſtand on the top of the ſtalkes in ſpokie rundles, like the tops of Dill, which turne into ſmall ſeed, growing together by couples, of a very good ſmell, not vnlike to thoſe of Fennell, but much ſmaller: the root is round, knobbed, with certaine eminences or bunchings out; browne without, white within, of a firme and ſollid ſubſtance, and of a taſte like the Cheſſe-nut or Cheſt-nut, whereof it tooke his name.

2 There is alſo another Earth-nut that hath ſtalkes a foot high, whereon doe grow iagged leaues like thoſe of Engliſh Saxifrage, of a bright greene colour: the floures grow at the top of the branches, in ſmall ſpokie tufts conſiſting of little white floures: the root is like the other, bulbous faſhion, with ſome few ſtrings hanging at the bottome, of a good and pleaſant taſte. ‡ This differs from the former, in that the leaues are larger and greener: the root alſo is not ſo far within the ground, and it alſo ſends forth ſome leaues from the bulbe it ſelfe; whereas our common kind hath only the end of a ſmall root that carries the ſtem and leaues vpon it, faſtned vnto it as you ſee it expreſt in the former figure. ‡

¶ *The Place.*

Theſe herbes do grow in paſtures and corne fields almoſt euery where: there is a field adioining to High-gate, on the right ſide of the middle of the village, couered ouer with the ſame; and likewiſe in the next field vnto the conduit heads by Maribone, neere the way that leadeth to Paddington by London, and in diuers other places. ‡ I haue not yet obſerued the ſecond to grow wilde with vs. ‡

¶ *The Time.*

They floure in Iune and Iuly: the ſeed commeth to perfection afterward.

¶ *The Names.*

Alexander Trallianus hath made mention of Βολβοκάστανον, *Lib.* 7. reckoning it vp among thoſe kinds of meate or ſuſtenances which be good for ſuch as haue rotten lungs: of ſome it is called *Agriocaſtanon.*

Guinterius thought the word was corrupted, and that *Balanocaſtanon* ſhould be read: but this is as ſtrange a word as *Bolbocaſtanon,* which was deriued of the forme of a bulbe, and the taſt of a Cheſt-nut: of ſome, *Nucula terreſtris,* or the little Earth-nut: it is thought to be *Bunium Dioſcoridis* of ſome; but we thinke not ſo: of D^r. *Turner, Apios*; yet there is another *Apios,* being a kinde of *Tithymale*: of *Matthiolus, Oenanthe,* making it a kinde of Filipendula: in high-Dutch, **Eerdnuſz**: in low-Dutch, **Eertnoten**: the people of Sauoy call it *Fauerottes*: in Engliſh, Earth-nuts, Kipper-nuts, and Earth Cheſt-nuts.

¶ *The Temperature.*

The roots of Earth-nuts are moderately hot and dry, and alſo binding: but the ſeed is both hotter and drier.

¶ *The Vertues.*

A The ſeed openeth and prouoketh vrine, and ſo doth the root likewiſe.

B The root is good for thoſe that ſpit and piſſe bloud, if the root be eaten raw, or roſted in the embers.

C The Dutch people doe vſe to eate them boyled and buttered, as we doe Parſeneps and Carrots, which ſo eaten comfort the ſtomacke, and yeeld nouriſhment that is good for the bladder and kidneyes.

D There is a plaiſter made of the ſeeds hereof, whereof to write in this place were impertinent to our hiſtorie.

CHAP. 432. Of Cumin.

¶ *The Deſcription.*

THis garden Cumin is a low or baſe herbe of a foot high: the ſtalke diuideth it ſelfe into diuers ſmall branches, whereon doe grow little iagged leaues very finely cut into ſmall parcels, like thoſe of Fennell, but more finely cut, ſhorter and leſſer: the ſpoky tufts grow at the top of the branches and ſtalkes, of a red or purpliſh colour: after which come the ſeed, of a ſtrong or rancke ſmell, and a biting taſte: the root is ſlender, which periſheth when it hath ripened his ſeed.

Cuminum ſativum Dioſcoridis.
Garden Cumin.

¶ *The Place.*

Cumin is husbanded and ſown in Italy and Spain, and is very common in other hot countries, as in Æthiopia, Egypt, Cilicia, and all the leſſer Aſia.

It delights to grow eſpecially in putrified and hot ſoiles: I haue proued the ſeeds in my garden, where they haue brought forth ripe ſeed much fairer and greater than any that commeth from beyond the ſeas.

¶ *The Time.*

It is to be ſown in the middle of the ſpring; a ſhewre of raine preſently following doth much hinder the growth thereof, as *Ruellius* ſaith.

My ſelfe did ſow it in the midſt of May, which ſprung vp in ſix days after: and the ſeed was ripe in the end of Iuly.

¶ *The Names.*

It is called in Greeke κύμινον ἥμερον, that is, tame or garden Cumin, that it may differ from the wilde ones: it is named in Latine *Cuminum*: in ſhops, *Cyminum*: in high-Dutch, **Roomiſche kymmel**: in Italian, *Comino*: in Spaniſh, *Cominchos*: in French, *Comin* · in Engliſh, Cumin.

¶ *The Temperature.*

The ſeed of garden Cumin, as *Galen* ſaith, is hot and dry in the third degree: *Dioſcorides* ſaith that it hath in it alſo a binding qualitie.

¶ *The Vertues.*

A The ſeed of Cumin ſcattereth and breaketh all the windineſſe of the ſtomacke, belly, guts, and matrix: it is good againſt the griping torments, gnawing or fretting of the belly, not onely receiued inwardly by the mouth, but alſo in cliſters, and outwardly applied to the belly with wine and barley meale boyled together to the forme of a pultis.

B Being handled according to art, either in a cataplaſme, pultis, or plaiſter, or boyled in wine and ſo applied, it taketh away blaſtings, ſwellings of the cods or genitors: it conſumeth windie ſwellings in the ioynts, and ſuch like.

C Being taken in a ſupping broth it is good for the cheſt and for cold lungs, and ſuch as are oppreſſed with aboundance of raw humors.

D It ſtancheth bleeding at the noſe, being tempered with vineger and ſmelt vnto.

E Being quilted in a little bag with ſome ſmall quantitie of Bay ſalt, and made hot vpon a bedpan with fire or ſuch like, and ſprinkled with good wine vineger, and applied to the ſide very hot, it taketh away the ſtitch and paines thereof, and eaſeth the pleuriſie very much.

Chap. 433. Of wilde Cumin.

¶ *The Kindes.*

There be diuers plants differing very notably one from another in ſhape, and yet all comprehended vnder the title of wilde Cumin.

¶ *The Deſcription.*

1 The wilde Cumin hath ſmall white roots with ſome fibres thereto appendant; from the which ariſe ſundry little iagged leaues, conſiſting of many leſſer leaues, finely dented about the edges, in faſhion like the ſmalleſt leaues of wilde Parſnep: among which ſpringeth vp a ſlender bending ſtalke a foot high, like vnto *Pecten Veneris*, bearing at the top thereof white round

1 *Cuminum ſylueſtre.*
Wilde Cumin.

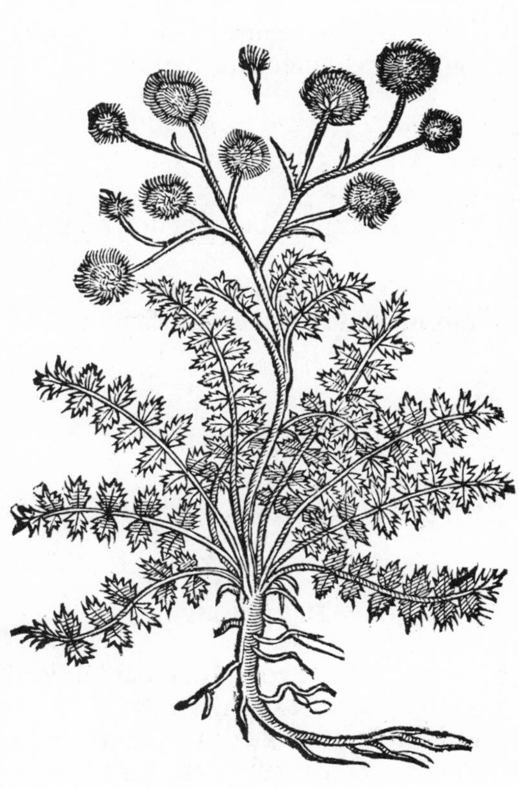

2 *Cuminum ſiliquoſum.*
Codded wilde Cumin.

3 *Cuminum Corniculatum, ſiue Hypecoum Cluſij.*
Horned wild Cumin.

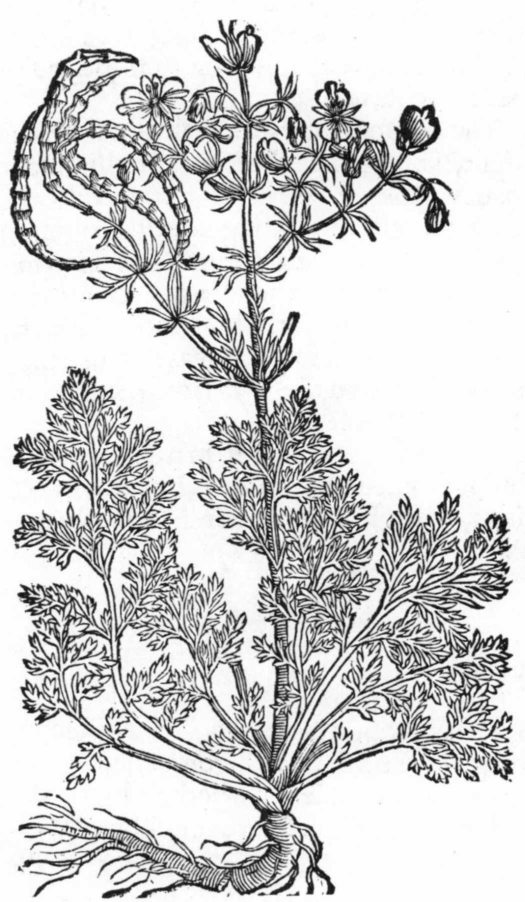

round and hairie buttons or knops, like *Arction*, as *Dioſcorides* hath right well obſerued: within which knoppes is contained a tender downie ſubſtance, among which is the ſeed, like the ſeed of *Dens Leonis*, but much leſſer.

2 The ſecond kinde of Cumin is verie like vnto the foreſaid wilde Cumin, ſaue that it beareth a number of horned or crooked cods, after the manner of *Scorpioides*, but thicker, and leſſe crooked, and the ſeedes within the cods are ſeuerally diſtinct and ſeparated one from another by equall partitions, in ſmall croſſes, yellow of colour, & ſomewhat long: the ſtalkes are little and tender, beſet with leaues much like vnto the ſmall leaues of *Carui*, or *Pecten Veneris*: and at the top of the ſtalks there do grow pretty yellow floures, like thoſe of great Celandine or Rocket, ſauing that they be ſomwhat leſſer.

† 3 The third kinde of Cumin is very like vnto the laſt before mentioned, but the leaues are much greater, more ſlender, & more finely cut or iagged, like the leaues of *Seſeli* of *Maßylia*: among which riſeth vp a ſtalke a cubit high or ſomewhat more, very ſmooth and whitiſh: at the top whereof ſpring forth fine yellow floures, not like the former, but conſiſting of ſix leaues apiece; whereof

whereof two are large, and edged with greene on the outside: the other foure are small ones, and grow two on a side betweene the two larger leaues: these floures being vaded, there succeed crooked cods, greater, and more full of knots or diuisions than the former, wherein is contained a small and flat yellow seed like *Galega*: the root is long, thicke, and single.

¶ *The Place.*

These wilde Cumins do grow in Lycia, and Galatia, a prouince of Asia, and in Carthage a citie of Spaine; seldome seene in these Northerne parts: notwithstanding at the impression hereof, the last did floure and flourish in my garden. ‡ These grow in Prouince in France, and in diuers parts of Spaine. ‡

¶ *The Time.*

They floure in August, and perfect their seed in September.

¶ *The Names.*

Their names haue been touched in their titles in as ample manner as hath been set down by any Author.

¶ *The Temperature and Vertues.*

Their temperature and vertues are referred to the garden Cumin; notwithstanding I cannot reade in any Author of their vse in Physicke.

Chap. 434. Of Flixweed.

1 *Sophia Chirurgorum.*
Flixweed.

¶ *The Description.*

1 Flixweed hath round and hard stalks, a cubit & a halfe high, wheron do grow leaues most finely cut and diuided into innumerable fine iags, like those of the sea Wormewood called *Seriphium*, or *Absinthium tenuifolium*, but much finer and smaller, drawing neere vnto the smallest leaues of Corianders, of an ouerworne greene colour: the floures grow alongst the tops of the spriggie branches, of a dark yellow colour: after which come long cods full of small red seeds: the root is long, straight, and of a wooddie substance.

2 The second sort differeth not from the precedent, sauing that the leaues of this plant are broader, wherein especially consisteth the difference; notwithstanding in mine opinion *Tabernamontanus* found this second sort growing in some fertill place, whereby the leaues did grow broader and greater, which moued him to make of this a second sort, whereas in truth they are both but one and the selfe same plant.

¶ *The Place.*

This Flixeweede groweth in most places of England, almost euery where in the ruins of old buildings, by high waies, and in filthie obscure base places.

¶ *The Time.*

It floureth and seedeth from Iune to the end of September.

¶ *The Names.*

Flixweed is called *Thalietrum*; and of some, *Thalictrum*, but vnproperly; for *Thalictrum* belongeth to English Rubarbe: the Paracelsians do vaunt and brag very much of an herbe called *Sophia*, adding thereto the surname *Paracelsi*, wherewith they imagine to do wonders, whether this be the same plant it is disputable, the controuersie not as yet decided; neuerthelesse we must be content

to

to accept of this for the true *Sophia*, vntil ſome diſciple or other of his do ſhew or ſet forth the plant wherewith their maſter *Paracelſus*, did ſuch great matters: in Engliſh we call it Flixweed, of his facultie againſt the flix.

¶ *The Temperature.*

Sophia drieth without any manifeſt ſharpeneſſe or heate.

¶ *The Vertues.*

The ſeed of *Sophia* or Flixweed drunke with wine, or Smithes water, ſtoppeth the bloudy flix, the laske, and all other iſſues of bloud. A

The herbe bruiſed or put into vnguents, cloſeth and healeth vlcers, or old ſores and wounds, as *Paracelſus* ſaith, and that becauſe it drieth without acrimonie or ſharpneſſe. B

Chap. 435. *Of the great Celandine, or Swallow-woort.*

¶ *The Deſcription.*

1 The great Celandine hath a tender brittle ſtalke, round, hairie, and full of branches, each whereof hath diuers knees or knottie ioints, ſet with leaues not vnlike to thoſe of Columbine, but tenderer and deeper cut or iagged, of a grayiſh greene vnder, and greene on the other ſide tending to blewneſſe: the floures do grow at the top of the ſtalkes, of a gold yellow colour, in ſhape like thoſe of the Wall-floure: after which come long cods, full of bleake or pale ſeeds: the whole plant is of a ſtrong ſmell, nothing pleaſant, and yeeldeth a thicke iuice of a milkie ſubſtance, of the colour of Saffron: the root is thicke and knobbie, with ſome threds annexed thereto, which being broken or bruiſed, yeeldeth a ſap or iuice of the colour of gold.

1 *Chelidonium maius.*
Great Celandine.

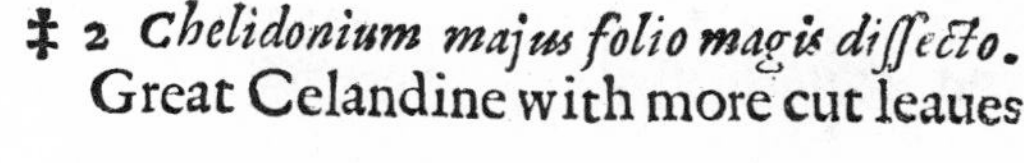

‡ 2 *Chelidonium majus folio magis diſſecto.*
Great Celandine with more cut leaues.

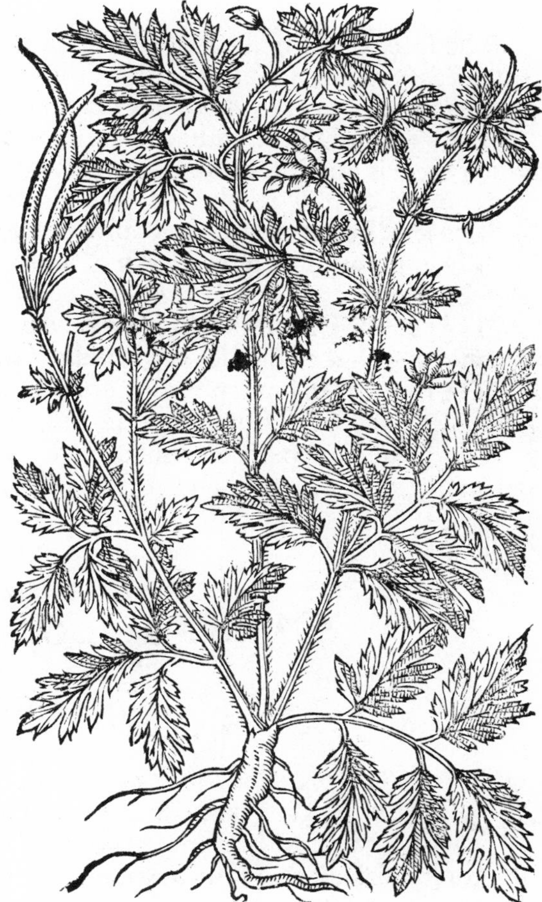

‡ 2 This other doth not in forme and magnitude differ from the former, but in the leaues, which are finelier cut and iagged, and ſomewhat in their ſhape reſemble an Oken leafe: the floures alſo

alſo are a little iagged or cut about the edges: and in theſe two particulars conſiſts the whole difference. *Cluſius* calls it *Chelidonium maius laciniato flore*; and *Bauhine*, *Chelidonium maius foliis quernis*. ‡

¶ *The Place.*

It groweth in vntilled places, by common way ſides, among briers and brambles, about old wals, and in the ſhade, rather than in the Sun.

¶ *The Time.*

It is greene all the yeare, it floureth from Aprill to a good part of Sommer, the coddes are perfected in the meane time.

¶ *The Names.*

It is called in Greeke χελιδόνιον: in Latine, *Chelidonium maius*, and *Hirundinaria maior*: amongſt the Apothecaries, *Chelidonia*: diuers miſcall it by the name *Celidonium*: it is named in Italian, *Celidonia*: in Spaniſh, *Celiduhenha*, *Yerua de las golundrinhas*: in high Dutch, **Groſz Scholwurtz**: in low Dutch **Stinkende Gouwe**: in French, *Eſclere*, or *Eſclayre*, and *Celidoine*: in Engliſh, Celandine, or great Celandine, Swallow-woort, and Tetterwoort.

It is called Celandine, not becauſe it then firſt ſpringeth at the comming in of the Swallows, or dieth when they goe away: for as we haue ſaid, it may be found all the yeare, but becauſe ſome hold opinion, that with this herbe the dammes reſtore ſight to their young ones when their eies be out: the which things are vaine and falſe; for *Cornelius Celſus* in his ſixth booke doth witneſſe, that when the ſight of the eies of diuers young birds be put forth by ſome outward meanes, it will after a time be reſtored of it ſelfe, and ſooneſt of all the ſight of the Swallow, wherupon (as the ſame Author ſaith) that the tale or fable grew, how thorow an herbe the dams reſtore that thing, which healeth of it ſelfe: the very ſame doth *Ariſtotle* alledge in the ſixt booke of the hiſtorie of Liuing creatures: [The eies of Swallowes (ſaith he) that are not fledge, if a man do pricke them out, do grow againe, and afterwards do perfectly recouer their ſight.]

¶ *The Temperature.*

The great Celandine is manifeſtly hot and drie, and that in the third degree, and withall ſcoures and clenſeth effectually.

¶ *The Vertues.*

A The iuice of the herbe is good to ſharpen the ſight, for it clenſeth and conſumeth away ſlimie things that cleaue about the ball of the eie, and hinder the ſight, and eſpecially being boiled with honie in a braſen veſſell, as *Dioſcorides* teacheth.

B The root cureth the yellow iaundiſe, which commeth of the ſtopping of the gall, eſpecially when there is no ague adioined with it, for it openeth and deliuereth the gall and liuer from ſtoppings.

C The root being chewed, is reported to be good againſt the tooth-ache.

D The iuice muſt be drawn forth in the beginning of Sommer, and dried in the Sunne, ſaith *Dioſcorides*.

E The root of Celandine boiled with Anniſe-ſeed in white wine, openeth the ſtoppings of the liuer, and cureth the iaundies very ſafely, as hath been often proued.

F The root cut in ſmall pieces is good to be giuen vnto Haukes againſt ſundry diſeaſes, whereunto they are ſubiect, as wormes, craie, and ſuch like.

G ‡ I haue by experience found (ſaith *Cluſius*) that the iuice of the great Celandine dropped into ſmall greene wounds of what ſort ſoeuer, wonderfully cures them. ‡

Chap. 436. *Of Coxcombe, or Yellow Rattle.*

¶ *The Deſcription.*

Criſta Galli, or *Criſta Gallinacea*, hath a ſtraight vpright ſtalke, ſet about with narrow leaues, ſnipt round about the edges: the floures grow at the top of the ſtems, of a yellow colour; after which come vp little flat pouches or purſes, couered ouer or contained within a little bladder, or flat skin, open before like the mouth of a fiſh, wherein is contained flat yellowiſh ſeed, which being ripe and drie, will make a noiſe or ratling when it is ſhaken or moued, of which propertie it tooke the name yellow Rattle.

¶ The

Crista Galli.
Yellow Rattle, or Coxcombe.

Pedicularis.
Lousewoort, or red Rattle.

¶ *The Place.*

It groweth in drie medowes and pastures, and is to them a great annoiance.

¶ *The Time.*

It floureth most part of the Sommer.

¶ *The Names.*

It is called in low Dutch **Ratelen**, and **Geele Ratelen**: commonly in Latine, *Crista Galli*, and *Gallinacea Crista*: in English, Coxcombe, Penie grasse, yellow or white Rattle: in High Dutch it is called **geel Rodel**: in French *Creste de Coc*: diuers take it to be the old writers *Alectorolophos*. ‡ Some thinke it to be the *Mimmulus*: or as others (& that more fitly) reade it, *Nummulus*, mentioned by *Pliny*, *lib*.18.*cap*.28. ‡

¶ *The Temperature and Vertues.*

A But what temperature or vertue this herbe is of, men haue not as yet beene carefull to know, seeing it is accounted vnprofitable.

CHAP. 437. *Of red Rattle, or Lousewoort.*

¶ *The Description.*

RED Rattle (of *Dodonæus* called *Fistularia*, and according to the opinion & censure of *Carolus Clusius*, *Pena* & others, the true *Alectorolophos*) hath very small, rent, or iagged leaues, of a browne red colour, and weake, small and tender stalkes, whereof some lie along trailing vpon the ground; within very moorish medowes they grow a cubit high and more, but in moist and wet heathes, and such like barren grounds not aboue an handful high: the floures grow round about the stalke, from the middest thereof euen to the top, and are of a brown red colour, in shape like the floures of dead Nettle: which being past, there succeed little flat pouches, wherin is contained flat and blackish seed, in shew very like vnto the former: the root is small, white, and tender.

¶ *The Place.*

It groweth in moist and moorish medowes, the herbe is not onely vnprofitable, but also hurtfull, and an infirmitie of the medowes.

¶ *The Time.*

It is found with his floures and stalkes in May and Iune.

¶ *The Names.*

It is called in Greeke φθίριον: in High Dutch, **Braun Rodel**: in Latine, *Pedicularis*, of the effect, because it filleth sheep and other cattel that feed in medowes where this groweth full of lice: diuers of the later Herbarists call it *Fistularia*: of some, *Crista Galli*: and diuers take it to be *Mimmulus herba*: in English, Rattle-grasse, Red Rattle-grasse, and Louse-woort.

¶ The

¶ *The Temperature.*

It is cold and drie and astringent.

¶ *The Vertues.*

A It is held to be good for Fistulaes and hollow vlcers, and to stay the ouermuch flowing of the menses, or any other flux of bloud, if it be boiled in red wine and drunke.

CHAP. 438. *Of Yarrow, or Nose-bleed.*

¶ *The Description.*

1 Common Yarrow hath very many stalkes comming vp a cubit high, round, and somewhat hard: about which stand long leaues, cut in the sides sundry wise, and as it were made vp of many small iagged leaues, euery one of which seeme to come neere to the slender leaues of Coriander: there stand at the top tufts or spoked rundles: the floures whereof are either white or purple, which being rubbed do yeeld a strong smell, but vnpleasant; the root sendeth downe many strings.

1 *Millefolium terrestre vulgare.*
Common Yarrow.

2 *Millefolium flore rubro.*
Red floured Yarrow.

2 The second kinde of Milfoile or Yarrow hath stalkes, leaues and roots like vnto the former, sauing that his spokie tufts are of an excellent faire red or crimson colour, and being a little rubbed in the hand, of a reasonable good sauour.

¶ *The Place.*

The first groweth euery where in drie pastures and medowes: red Milfoile groweth in a field by Sutton in Kent called Holly-Deane, from whence I brought those plants that do grow in my Garden; but it is not common euery where as the other is.

¶ *The Time.*

They floure from May to the end of October.

¶ *The*

¶ *The Names.*

Yarrow is called of the Latine Herbarists *Millefolium*: it is *Dioscorides* his *στρατιώτης*: in Latine, *Achillea*, and *Achillea sideritis*; which thing he may very plainely see that will compare with that description which *Dioscorides* hath set downe: this was found out, saith *Pliny* in his 25. booke, chap. 5. by *Achilles*, *Chirons* disciple, which for that cause is named *Achilleios*: of others, *Sideritis*: among vs, *Millefolium*: yet be there other *Sideritides*, and also another *Panaces Heracleion*, whereof we will intreat in another place: *Apuleius* setteth downe diuers names hereof, some of which are also found among the bastard names in *Dioscorides*: in Latine it is called *Militaris*, *Supercilium Veneris*, *Acrum*, or *Acorum syluaticum*: of the French-men, *Millefueille*: in high Dutch, Garben, Schafgras: in low Dutch, Geruwe: in Italian, *Millefoglio*: in Spanish, *Milhoyas yerua*: in English, Yarrow, Nose-bleed, common Yarrow, red Yarrow, and Milfoile.

¶ *The Temperature.*

Yarrow, as *Galen* saith, is not vnlike in temperature to the *Sideritides*, or Iron worts, that is to say, clensing, and meanely cold, but it most of all bindeth.

¶ *The Vertues.*

A The leaues of Yarrow doe close vp wounds, and keepe them from inflammation, or fiery swelling: it stancheth bloud in any part of the body, and it is likewise put into bathes for women to sit in: it stoppeth the laske, and being drunke it helpeth the bloudy flixe.

B Most men say that the leaues chewed, and especially greene, are a remedy for the tooth-ache.

C The leaues being put into the nose, do cause it to bleed, and ease the paine of the megrim.

D It cureth the inward excorations of the yard of a man, comming by reason of pollutions or extreme flowing of the seed, although the issue do cause inflammation and swelling of those secret parts, and though the spermaticke matter do come downe in great quantity, if the juice be injected with a syringe, or the decoction. This hath been prooued by a certain friend of mine, sometimes a Fellow of Kings Colledge in Cambridge, who lightly brused the leaues of common Yarrow, with Hogs-grease, and applied it warme vnto the priuie parts, and thereby did diuers times helpe himselfe, and others of his fellowes, when he was a student and a single man liuing in Cambridge.

E One dram in powder of the herbe giuen in wine, presently taketh away the paines of the colicke.

CHAP. 439. *Of yellow Yarrow, or Milfoile.*

1 *Millefolium luteum.*
Yellow Yarrow.

† 2 *Achillea, siue Millefolium nobile.*
Achilles Yarrow.

¶ *The Deſcription.*

1 YEllow Yarrow is a ſmall plant ſeldome aboue a ſpan high: the ſtalkes whereof are couered with long leaues, very finely cut in the edges like feathers in the wings of little birds: the tufts or ſpokie rundles bring forth yellow floures, of the ſame ſhape and forme of the common Yarrow: the root conſiſteth of threddy ſtrings.

2 Achilles Yarrow, or noble Milfoile, hath a thicke and tough root, with ſtrings faſtened thereto: from which immediately riſe vp diuers ſtalkes, very greene and creſted, whereupon doe grow long leaues compoſed of many ſmall jagges, cut euen to the middle rib: the floures ſtand on the top of the ſtalkes with ſpokie vmbels or tufts, of a whitiſh colour, and pleaſant ſmell.

¶ *The Place.*

Theſe kinds of Yarrow are ſeldome found: they grow in a fat and fruitfull ſoile, and ſometimes in medowes, and are ſtrangers in England.

¶ *The Time.*

They floure from May vntill Auguſt.

¶ *The Names.*

Dioſcorides deſcription doth ſufficiently declare, that this herbe is *Stratiotes Millefolium*: in Greeke, στρατιώτης χιλιόφυλλος: the height of the herbe ſheweth it, the forme of the leaues agree; there is ſome ambiguity or doubt in the colour of the floures, which *Dioſcorides* deſcribeth to be white, as the vulgar copies haue; but *Andreas Lacuna* addeth out of the old booke, of a yellow colour: it is named of the later age, *Millefolium minus*, or little Yarrow, and *Millefolium luteum*, yellow Yarrow, or Noſe-bleede: the Apothecaries and common people know it not.

¶ *The Temperature.*

Yarrow is meanely cold and ſomewhat binding.

¶ *The Vertues.*

A It is a principall herbe for all kinde of bleedings, and to heale vp new and old vlcers and greene wounds: there be ſome, ſaith *Galen*, that vſe it for fiſtulaes.

B This plant *Achillea* is thought to be the very ſame wherewith *Achilles* cured the wounds of his ſouldiers, as before in the former chapter.

† The plant here figured and deſcribed in the ſecond place, was alſo figured and deſcribed formerly in the fifth place of the 209. chapter of this booke, by the title of *Tanacetum minus album*, but the figure of *Lobels* which is put there being ſomewhat imperfect, I thought it not amiſſe here to giue that of *Dodonæus* which is ſomewhat more exquiſite, otherwiſe both the figure and hiſtory might in this place haue been omitted.

Chap. 440. *Of Valerian, or Setwall.*

¶ *The Deſcription.*

1 THe tame or garden Valerian hath his firſt leaues long, broad, ſmooth, green, and vndiuided; and the leaues vpon the ſtalkes greater, longer, and deepely gaſhed on either ſide, like the leaues of the greater Parſnep, but yet leſſer: the ſtalke is aboue a cubit high, ſmooth, and hollow, with certaine joints farre diſtant one from another: out of which joints grow forth a couple of leaues, and in the tops of the ſtalkes vpon ſpokie rundles ſtand floures heaped together, which are ſmall, opening themſelues out of a long little narrow necke, of colour whitiſh, and ſometimes withall of a light red: the root is an inch thicke, growing aſlope, faſtned on the vpper part of the earth by a multitude of ſtrings, the moſt part of it ſtanding out of the ground, of a pleaſant ſweet ſmell when it is broken.

2 The greater wilde Valerian hath leaues diuided and jagged, as thoſe of the former; thoſe about the ſtalke hereof are alſo ſmooth, hollow, and jointed, and aboue a cubit high: the floures ſtand on ſpokie rundles like to thoſe of the former, but of a light purple colour: the roots are ſlender, and full of ſtrings and ſmall threds, not altogether without ſmell.

3 The other wilde one is much like in forme to the garden Valerian, but farre leſſer: the firſt leaues thereof be vndiuided, the other are parted and cut in ſunder: the ſtalkes a ſpan long: the floures which ſtand on ſpokie rundles are like to thoſe of the others, of a light whitiſh purple colour: the roots be ſlender, growing aſlope, creeping, and full of fine ſmall threds, of little ſmell.

4 There is a ſmall Valerian growing vpon rockes and ſtony places, that is like vnto the laſt deſcribed, ſauing it is altogether leſſe. ‡ The ſtalk is ſome halfe foot high, and ſtrait, diuiding it ſelfe into branches toward the top, and that alwaies by couples: the bottome leaues are whole, the top leaues

1 *Valeriana hortensis.*
Garden Valerian, or Setwall.

2 *Valeriana maior sylvestris.*
Great wilde Valerian.

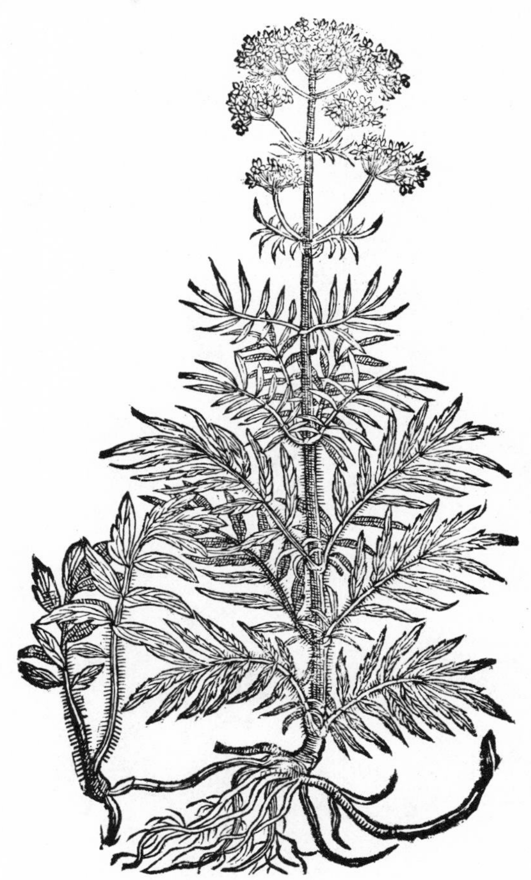

3 *Valeriana minor.*
Small Valerian.

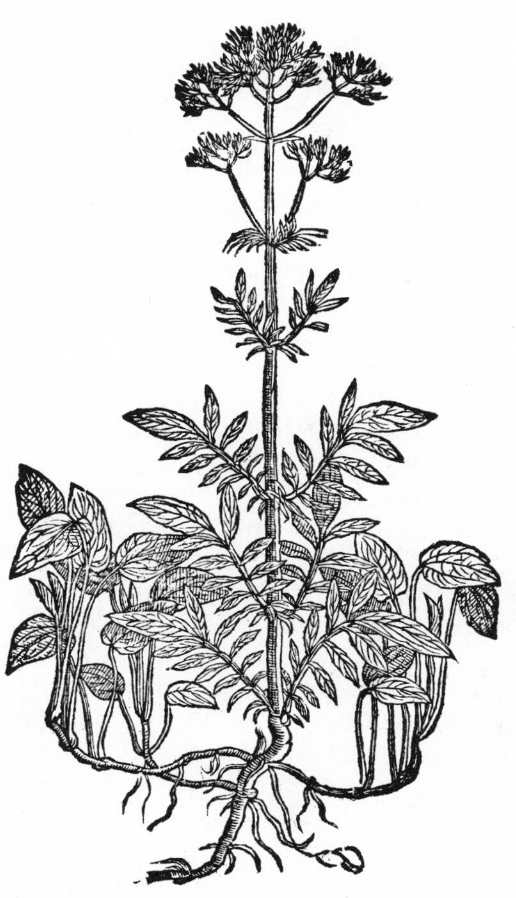

4 *Valeriana Petræa.*
Stone Valerian.

leaues much diuided, the floures are small, of a whitish purple colour, parted into fiue, and standing vpon round rough heads, which when the floures are falne, become star-fashioned, diuided into six parts: it floures in Iune, and is an annuall plant. ‡

5 *Valeriana Græca.*
Greekish Valerian.

‡ 6 *Valeriana Mexicana.*
Indian Valerian.

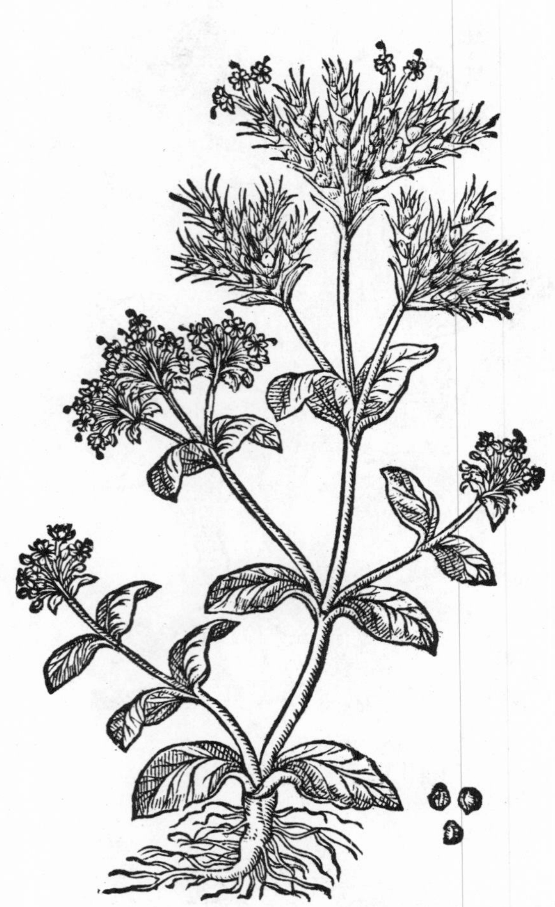

5 The fifth sort of Valerian hath diuers small hollow stalkes, a foot high and somewhat more, garnished with leaues like vnto those that do grow on the vpper part of the stalks of common Valerian, but smaller, cut or iagged almost to the middle rib: at the top of the stalkes doe grow the floures clustering together, of a blew colour, consisting of fiue leaues apiece, hauing in the middle thereof small white threds tipped with yellow: the seed is small, growing in little huskes or seed vessels: the root is nothing else but as it were all of threds.

6 I haue another sort of Valerian (the seed whereof was sent me from that reuerend Physition *Bernard Paludane*, vnder the title of *Valeriana Mexicana:*) hauing small tender stalkes trailing vpon the ground, very weake and brittle: whereupon doe grow smooth greenish leaues like those of Corne Sallade (which wee haue set forth amongst the Lettuce, vnder the title *Lactuca Agnina*, or Lambs Lettuce:) among the leaues come forth the floures clustering together, like vnto the great Valerian in forme, but of a deepe purple colour: the root is very small and threddie, which perisheth with the rest of the plant, when it hath brought his seed to maturitie or ripenesse, and must bee sowne anew the next yeare in May, and not before.

7 There is also another sort or kinde of Valerian called by the name *Phyteuma*, of the learned Physitions of Montpelier and others (set forth vnder the stocke or kindred of the Valerians, resembling the aforesaid Corn-sallad, which is called of some *Prolifera*, from the Greeke title *Phyteuma*; as if you should say, good to make conception, and to procure loue:) the lowest leaues are like those of the small Valerian, of a yellowish colour: the vpper leaues become more iagged: the stalks are an handfull high: on the tops whereof do stand small round spokie tufts of white floures; which being past, the seeds appeare like small round pearles, which being ripe, grow to be somewhat flat, hauing in the middle of each seed the print of an hole, as it were grauen or bored therein. The root is small and single, with some fibres annexed thereto.

‡ 8 This sends forth from a white and wooddie root many leaues spred vpon the ground, green, and

‡ 8 *Valeriana annua, Clus.*
Annuall Valerian.

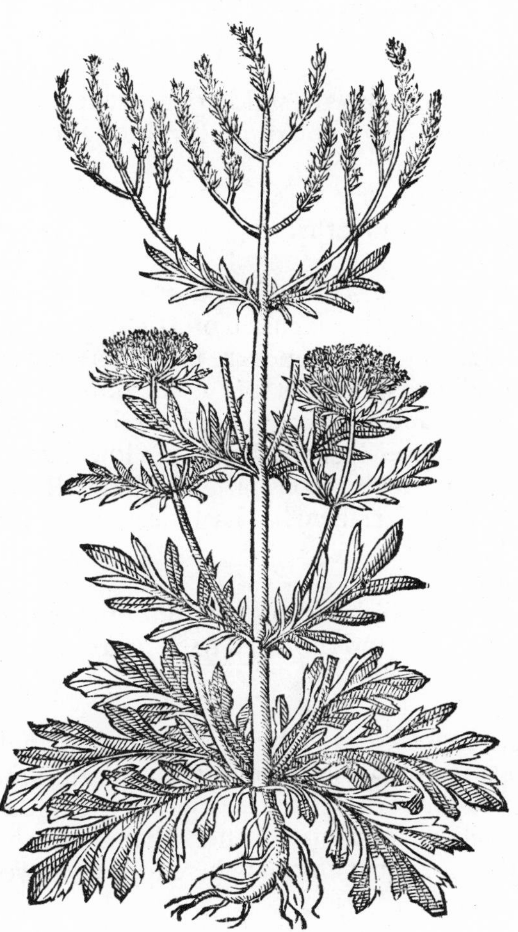

‡ 9 *Valeriana Alpina latifolia.*
Broad leaued Setwall of the Alps

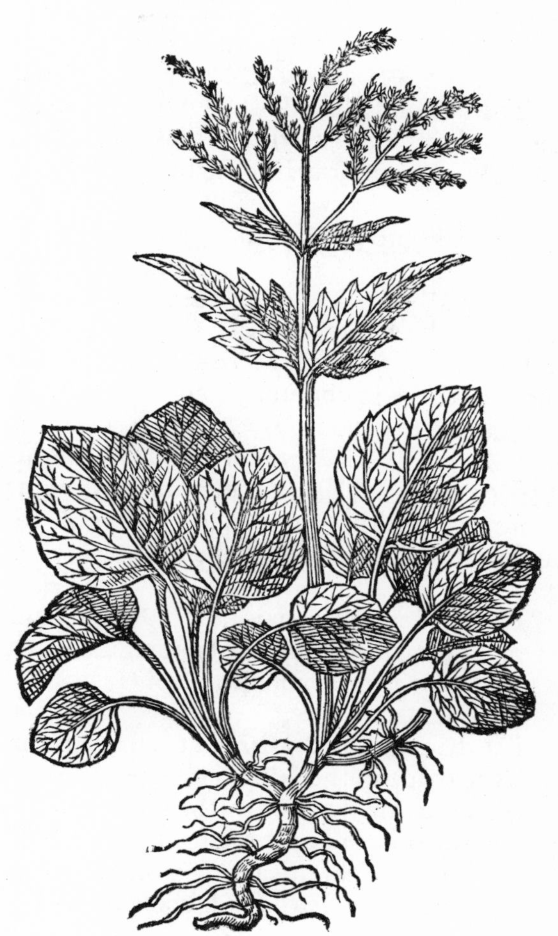

‡ 10 *Valeriana Alpina angustifolia,*
Small Alpine Setwall.

and not vnlike those of the Star-Thistle: among these rise vp some round hollow branched stalks two cubites high: at each ioint grow forth two leaues lesser, yet like the lower: at the tops of the branches grow the floures as it were in little vmbels, consisting of fiue leaues apiece; and these of a light red, or flesh colour: and then these as it were vmbels grow into longish branches bearing seed almost like, yet lesse than the red Valerian: it floures in Iuly, and perisheth when it hath ripened the seed. *Clusius* hath set this forth by the name of *Valeriana annua altera*. ‡

9 The same Author hath also giuen vs the historie of some other Plants of this kind; and this he cals *Valeriana syl. Alpina* 1 *latifol.* the stalk hereof is some foot high, round, greene, and crested: vpon which stand leaues longish, sharpe pointed, and cut in with two or three deepe gashes: but the bottome leaues are more round and larger, comming neere to these of *Trachelium*, yet lesser, slenderer, and bitter of taste: the floures which are white of colour, and the seed, are like those of the other Valerians: the root is small, creeping, fibrous, white and aromatick: it growes vpon the Alpes, and floures in Iune and Iuly.

10 This sendes forth leaues like those of the mountain Daisie: out of the midst of which riseth

riseth vp a stalk some foot high, iointed, and at the top diuided into little branches, carrying white floures like the other Valerians: the root is as aromaticke as that of the last mentioned; and grows in the chinkes of the Alpine rockes, where it floures in Iune and Iuly. *Clusius* hath it by the name of *Valeriana syluestris Alpin. 2. Saxatilis.* ‡

¶ *The Place.*

The first and likewise the Greeke Valerian are planted in gardens; the wilde ones are found in moist places hard to riuers sides, ditches, and waterie pits; yet the greater of these is brought into gardens where it flourisheth, but the lesser hardly prospereth.

¶ *The Time.*

These floure in May, Iune, and Iuly, and most of the Sommer moneths.

¶ *The Names.*

Generally the Valerians are called by one name, in Latine, *Valeriana*: in Greeke, φοῦ: in shoppes also *Phu*, which for the most part is meant by the garden Valerian, that is called of *Diosc.* νάρδος ἀγρία: in Latine, *Syluestris*, or *Rustica Nardus*: of *Pliny*, *Nardus Cretica*: which names are rather referred to those of the next chapter, although these be reckoned as wilde kindes thereof: of certaine in our age, *Marinella, Amantilla, Valentiana, Genicularis, Herba Benedicta*, and *Theriacaria*: in most shops, *Valeriana Domestica*: of *Theophrastus Paracelsus*, *Terdina*: in high Dutch, **Grosz baldrian**: in low Dutch, **Speercrupt, S. Joris crupt**, and **Valeriane**: in English, Valerian, Capons taile, and Setwall; but vnproperly, for that name belongeth to *Zedoaria*, which is not Valerian: what hath been set downe in the titles shall serue for the distinctions of the other kindes.

¶ *The Temperature.*

The garden Valerian is hot, as *Dioscorides* saith, but not much, neither the green root, but the dried ones; for the green is easily perceiued to haue very little heate, and the dried to be hotter, which is found by the taste and smell.

¶ *The Vertues.*

A The drie root, as *Dioscorides* teacheth, prouoketh vrine, bringeth downe the desired sicknesse, helpeth the paine in the sides, and is put into counterpoisons and medicines preseruatiue against the pestilence, as are treacles, mithridates, and such like: whereupon it hath been had (and is to this day among the poore people of our Northerne parts) in such veneration amongst them, that no broths, pottage, or physicall meats are worth any thing, if Setwall were not at an end: whereupon some woman Poet or other hath made these verses;

They that will haue their heale,
Must put Setwall in their keale.

B It is vsed generally in sleight cuts, wounds, and small hurts.

C The extraction of the roots giuen, is a most singular medicine against the difficultie of making water, and the yellow iaundies.

D Wilde Valerian is thought of the later Herbarists to be good for them that are bursten, for such as be troubled with the crampe and other convulsions, and also for all those that are bruised with falls.

E The leaues of these and also those of the garden, are good against vlcers and sorenes of the mouth and gums, if the decoction thereof be gargarized or held in the mouth.

F Some hold opinion that the roots of wilde Valerian dried and poudered, and a dramme weight thereof taken with wine, do purge vpward and downeward.

Chap. 441. *Of Mountaine Setwall, or Nardus.*

¶ *The Description.*

1 THe *Nardus* named *Celtica*, but now by some, *Ligustica Nardus*, flourisheth in high mountaines. The Vallesians in their mother tongue call it *Selliga*; whence *Gesner* thought it to be *Saliunca*; neither do I doubt, but that it is the same which *Virgil* speaketh of in these verses:

Puniceis humilis quantum Saliunca rosetis,
Iudicio nostro tantum tibi cedit Amintas.

For it is a very little herbe creeping on the ground, and afterward lifting vp it selfe with a stalke of a handfull high, whereupon from the lower part grow small thin leaues, first green, but afterwards somewhat yellowish: vpon the roughnesse of the root there are many scales, platted one vpon another; but vnder the root there are many browne strings and hairy threds, in smell like the roots of *Astrabacca*, or rather the wilde mountaine Valerian, whereof it seemes to be a kinde, in taste sharpe and bitter. The floures grow along the vpper branches, white or yellowish, and very small.

2 The

1 *Nardus Celtica.*
Celticke Spikenard.

3 *Hirculus.*
Vrine-wort.

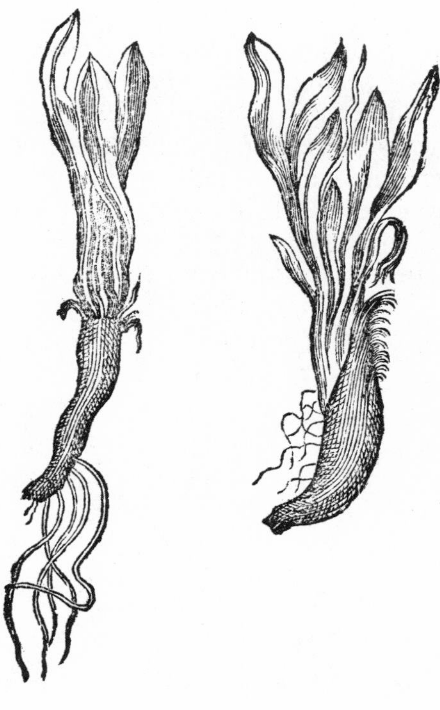

4 ‡ *Nardus montana germinans.*
Mountaine Nard at the first springing vp.

4 *Nardus montana.*
Mountaine Spikenard.

2 The second sort of Spikenard hath many threddy roots, from the which rise vp many scaly rough and thicke stalkes, hauing at the top certaine flat hoary leaues growing vpon smal and tender footstalkes. The whole plant is of a pleasant sweet smell.

3 *Hirculus* is a plant very rare, which as yet I neuer saw, notwithstanding we are greatly beholding to *Carolus Clusius* the father of forreine Simples, who finding this plant among many bunches or handfulls of mountaine Spikenard, hath made it knowne vnto posteritie, as he hath done many other rare plants, in

translating

tranflating of *Garcias* the Lufitanian Phyfitian, he fetteth it forth with a light defcription, faying, It is a bafe and low herbe two handfulls high, bringing forth leaues without any ftalkes at all, ‡ very hairy about the root, and blackifh, hauing no pleafant fent at all. The leaues chewed yeeld no aromaticke tafte, but are clammie, or vifcide; whereas the leaues of Celticke Narde are hot, with a little aftriction, and of a pleafant fmell and tafte. ‡

4 Mountaine Spikenard hath a great thicke knobbed root, fet here and there with fome tender fibres, of a pleafant fweet fmel; from the which come forth three or foure fmooth broad leaues, and likewife iagged leaues deepely cut euen to the middle rib: among which rife vp naked ftalks, garnifhed in the middle with a tuft of iagged leaues. The floures grow at the top of the ftalks, in an vmbel or tuft like thofe of the wilde Valerian in fhape and colour, and fuch alfo is the feed. ‡ I haue giuen you the figure of the root and whole leaues as they fhew themfelues when they firft appeare, as it was taken by *Clufius*. ‡

5 *Nardus Indica.*
Indian Spikenard.

6 *Nardus Narbonenfis.*
French Spikenard.

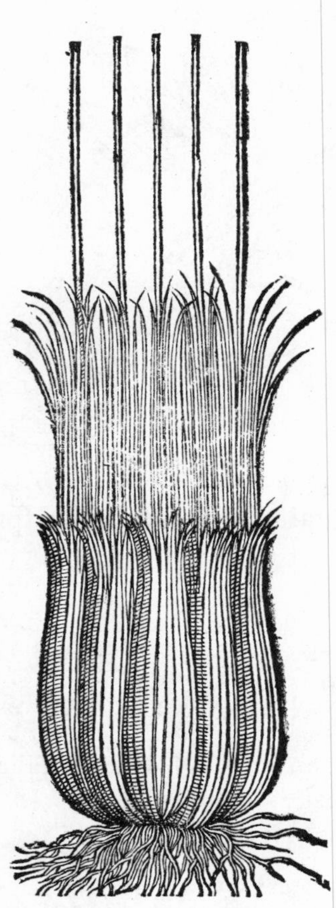

5 The Spikenard of India is a low plant, growing clofe vnto the ground, compofed of many rough browne hairy cloues, of a ftrong, yet not vnpleafant fmell. The root is fmall and threddie. ‡ It hath certainly ftalkes, floures, and feeds; but none of our Indian Writers or Trauellers haue as yet defcribed them. I haue feene little pieces of flender hollow ftalkes fome two inches long faftned to the roots that are brought to vs. ‡

† 6 This French Spikenard, being a baftard kinde, groweth clofe vpon the ground like the precedent, compact of fcaly rough leaues: in the middle whereof commeth forth a great bufh of round greene ftiffe and rufhy leaues: among the which fhoot vp diuers round ftalkes a cubit high, fet from the middle to the top with greenifh little cods, ftanding in chaffie huskes like thofe of Schœnanth. The root is fmall and threddy: the whole plant is altogether without fmell, which fheweth it to be a baftard kinde of Spikenard.

¶ *The Place.*

Thefe plants [the firft foure] are ftrangers in England, growing in great plentie vpon the mountaines of Iudenberg and Heluetia, on the rockes among the moffe, and in the mountains of Tiroll and Saltzburg.

The firft and fecond, if my memorie faile me not, do grow in a field in the North part of England

land, called Crag close, and in the foot of the mountaine called Ingleborow Fels. ‡ The fourth may be found in some gardens with vs. The fifth growes in the East Indies, in the prouinces of Mandou and Chito in the kingdome of Bengala and Decan. The last growes in Prouince in France, neere a little city called Gange. ‡

¶ *The Time.*

The leaues grow to withering in September, at which time they smell more pleasantly than when they flourished and were greene.

¶ *The Names.*

Nardus is called in Pannonia or Hungarie, of the countrey people, *Speick*: of some, *Bechi fu*; that is, the herbe of Vienna, because it doth grow there in great aboundance, from whence it is brought into other countries: of *Gesner*, *Saliunca*: in English, Celticke Spikenard: of the Valletians, *Selliga*, and *Nardus Celtica*.

¶ *The Temperature and Vertues.*

Celticke Narde mightily prouokes vrine, as recordeth *Rondeletius*; who trauelling through the desart countrey, chanced to lodge in a monasterie where was a Chanon that could not make his water, but was presently helped by the decoction of this herbe, through the aduice of the said *Rondeletius*. A

‡ The true Spikenard or Indian Nard hath a heating and drying facultie, being (according to *Galen*) hot in the first degree [yet the Greeke copy hath the third] and dry in the second. It is composed of a sufficiently astringent substance, and not much acride heate, and a certaine light bitternesse. Consisting of these faculties, according to reason, both inwardly and outwardly vsed it is conuenient for the liuer and stomacke. B

It prouoketh vrine, helps the gnawing paines of the stomacke, dries vp the defluxions that trouble the belly and intrals, as also those that molest the head and brest. C

It stayes the fluxes of the belly, and those of the wombe, being vsed in a pessarie, and in a bath it helpes the inflammation thereof. D

Drunke in cold water, it helpes the nauseousnesse, gnawings, and windinesse of the stomacke, the liuer, and the diseases of the kidneyes, and it is much vsed to be put into Antidotes. E

It is good to cause haire to grow on the eye lids of such as want it, and is good to be strewed vpon any part of the body that abounds with superfluous moisture, to dry it vp. F

The Celticke-Nard is good for all the forementioned vses, but of lesse efficacie, vnlesse in the prouoking of vrine. It is also much vsed in Antidotes. G

The mountaine Nard hath also the same faculties, but is much weaker than the former, and not in vse at this day that I know of. ‡

CHAP. 442. *Of Larkes heele or Larkes claw.*

¶ *The Description.*

1 THe garden Larks spur hath a round stem ful of branches, set with tender iagged leaues very like vnto the small Sothernwood: the floures grow alongst the stalks toward the tops of the branches, of a blew colour, consisting of fiue little leaues which grow together and make one hollow floure, hauing a taile or spur at the end turning in like the spurre of Tode-flax. After come the seed, very blacke, like those of Leekes: the root perisheth at the first approch of Winter.

2 The second Larks spur is like the precedent, but somewhat smaller in stalkes and leaues: the floures are also like in forme, but of a white colour, wherein especially is the difference. These floures are sometimes of a purple colour, sometimes white, murrey, carnation, and of sundry other colours, varying infinitely, according to the soile or countrey wherein they liue.

‡ 3 Larks spur with double floures hath leaues, stalkes, roots, and seeds like the other single kinde, but the floures of this are double; and hereof there are as many seuerall varieties as there be of the single kinde, to wit, white, red, blew, purple, blush, &c.

4 There is also another varietie of this plant, which hath taller stalkes and larger leaues than the common kinde: the floures also are more double and larger, with a lesser heele: this kind also yeeldeth vsually lesse seed than the former. The colour of the floure is as various as that of the former, being either blew, purple, white, red, or blush, and sometimes mixed of some of these. ‡

5 The wilde Larks spur hath most fine iagged leaues, cut and hackt into diuers parts, confusedly set vpon a small middle tendrell: among which grow the floures, in shape like the others, but a great

1 *Conſolida regalis ſatiua.*
Garden Larks heele.

2 *Crnſolida ſatiua flore albo vel rubro.*
White or red Larks ſpur.

‡ 3 *Conſolida regalis flore duplici.*
Double Larks ſpur.

‡ 4 *Conſolida reg. elatior flo. pleno.*
Great double Larks ſpur.

a great deale lesser, sometimes purple, otherwhiles white, and often of a mixt colour. The root is small and threddy.

5 *Consolida regalis syluestris.*
Wilde Larkes heele.

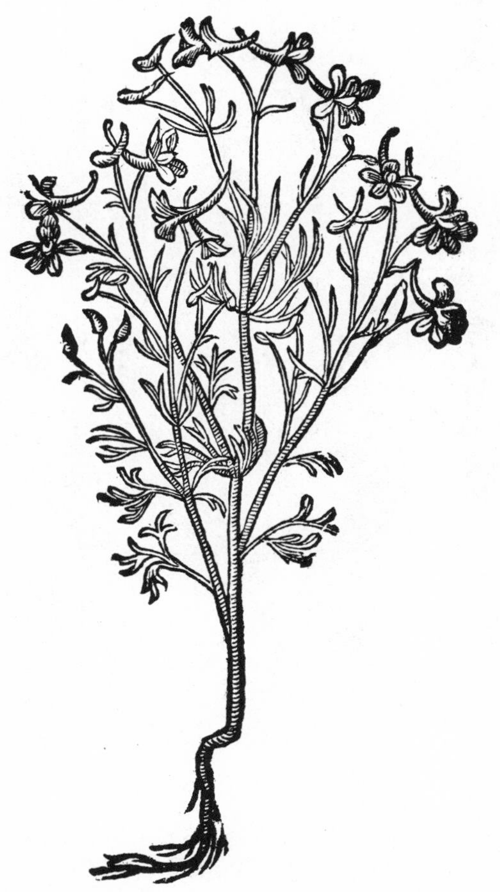

¶ *The Place.*

These plants are set and sowne in gardens: the last groweth wilde in corne fields, and where corn hath grown, ‡ but not with vs, that I haue yet obserued; though it be frequently found in such places in many parts of Germanie. ‡

¶ *The Time.*

They floure for the most part all Sommer long, from Iune to the end of August, and ofttimes after.

¶ *The Names.*

Larks heele is called *Flos Regius* of diuers, *Consolida regalis*; who make it one of the Consounds or Comfreyes. It is also thought to be the *Delphinium* which *Dioscorides* describes in his third booke; wherewith it may agree. It is reported by *Gerardus* of Veltwijcke, who remained Lieger with the great Turke from the Emperor *Charles* the fifth, That the said *Gerard* saw at Constantinople a copy which had in the chap. of *Delphinium*, not leaues but floures like Dolphines: for the floures, and especially before they be perfected, haue a certaine shew and likenesse of those Dolphines, which old pictures and armes of certain antient families haue expressed with a crooked and bending figure or shape; by which signe also the heauenly Dolphine is set forth. And it skilleth not, though the chapter of *Delphinium* be thought to be falsified and counterfeited; for although it be some other mans, and not of *Dioscorides*, it is notwithstanding some one of the old Writers, out of whom it is taken, and foisted into *Dioscorides* his bookes: of some it is called *Bucinus*, or *Bucinum*: in English, Larks spur, Larks heele, Larks toes, and Larks claw: in high-Dutch, **Ridder spooren**; that is, *Equitis calcar*, Knights spur: in Italian, *Sperone*: in French, *Pied d' alouette.*

¶ *The Temperature.*

These herbes are temperate and warme of nature.

¶ *The Vertues.*

A We finde little extant of the vertues of Larks heele, either in the antient or later writers, worth the noting, or to be credited; for it is set downe, that the seed of Larks spur drunken is good against the stingings of Scorpions; whose vertues are so forcible, that the herbe onely thrown before the Scorpion or any other venomous beast, causeth them to be without force or strength to hurt, insomuch that they cannot moue or stirre vntill the herbe be taken away: with many other such trifling toyes not worth the reading.

CHAP. 443. *Of Gith, or Nigella.*

¶ *The Kindes.*

THere be diuers sorts of Gith or Nigella, differing some in the colour of the floures, others in the doublenesse thereof, and in smell of the seed.

¶ *The Description.*

1 THe first kind of Nigella hath weake and brittle stalks of the height of a foot, full of branches, beset with leaues very much cut or iagged, resembling the leaues of Fumiterie, but much greener: the floures grow at the top of the branches, of a whitish blew colour, each floure being

1 *Melanthium.*
Garden Nigella.

2 *Melanthium ſylveſtre.*
Wilde Nigella.

3 *Melanthium Damaſcenum.*
Damaske Nigella.

‡ 4 *Melanthium Damaſcenum flo. pleno.*
Double floured Damaske Nigella.

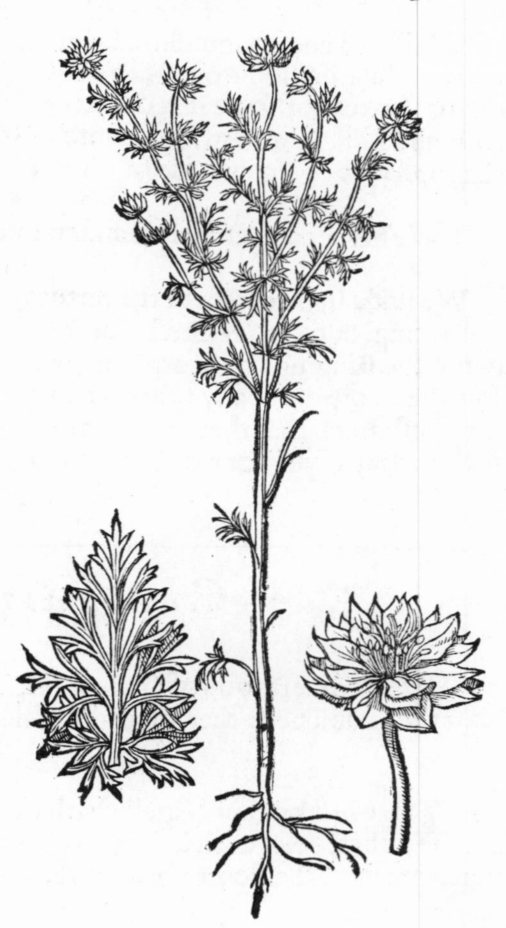

being parted into fiue ſmall leaues, ſtarre faſhion: the floures being vaded, there come vp ſmall knobs or heads, hauing at the end thereof fiue or ſix little ſharpe hornes or pointalls, and euery knob or head is diuided into ſundry ſmall cels or partitions, wherein the ſeed is conteined, which is of a blackiſh colour, very like vnto Onion ſeed, in taſte ſharpe, and of an excellent ſweet ſauour.

2 The wilde Nigella hath a ſtreaked ſtalke a foot or more high, beſet full of grayiſh eaues, very finely jagged, almoſt like the leaues of Dil: the floures are like the former, ſaue that they are blewer: the cods or knops are like the heads or huskes of Columbines, wherein is conteined the ſweet and pleaſant ſeed, like the former.

5 *Nigella flore albo multiplici.*
Damaske Nigella.

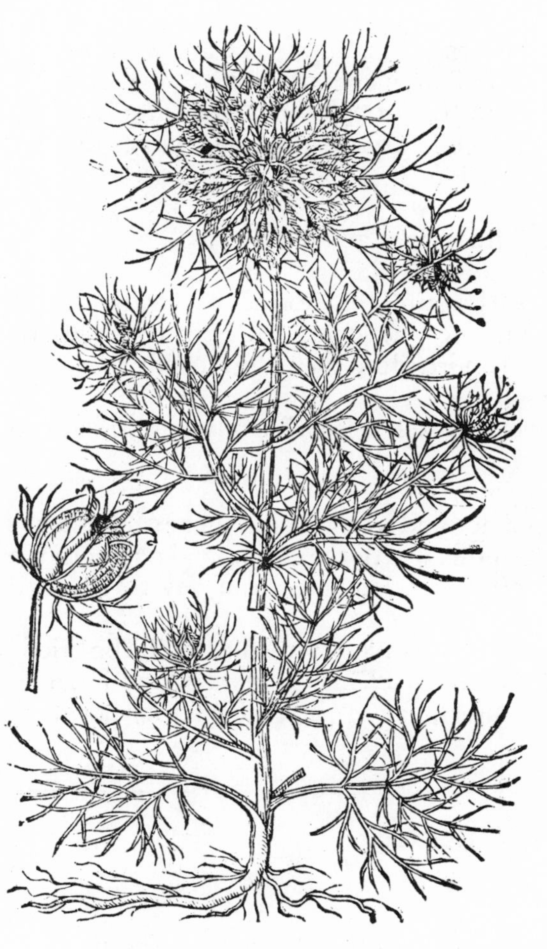

‡ 6 *Nigella Hiſpanica flore amplo.*
Great Spaniſh Nigella.

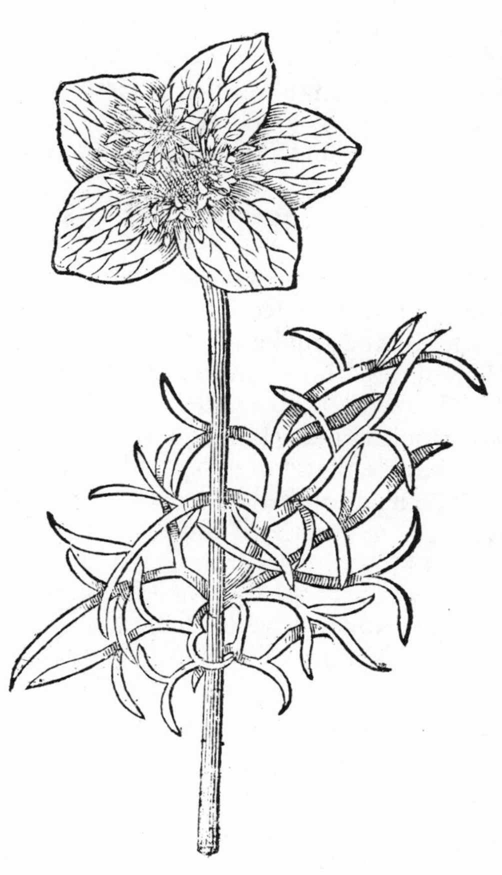

3 The third kinde of Nigella, which is both faire and pleaſant, called Damaske Nigella, is very like vnto the wilde Nigella in his ſmall cut and jagged leaues, but his ſtalke is longer: the floures are like the former, but greater, and euery floure hath fiue ſmall greene leaues vnder him, as it were to ſupport and beare him vp: which floures being gone, there ſucceed and follow knops and ſeed like the former, but without ſmell or ſauour.

‡ 4 This in the ſmalneſſe, and ſhape of the leaues and the manner of growing is like to the laſt deſcribed hauing ſmal leaues growing vnder the floure, which is not ſingle, as in the laſt deſcribed, but double, conſiſting of fiue or more rankes of little blewiſh leaues, which are ſucceeded by ſuch cornered heads as thoſe of the former, hauing in them a blacke ſeed without any manifeſt ſmell. ‡

5 The fifth kinde of Nigella hath many ſmall and ſlender ſtalkes, ſet full of ſlender and thin leaues deepely cut or jagged, of a faint yellowiſh greene colour: the floures grow at the top of the ſtalkes, of a whitiſh colour, and exceeding double: which being vaded, there ſucceed bowles or knobs, full of ſweet blacke ſeed like the former: ths root is ſmall and tender.

‡ 6 The root of this is ſlender, and yellowiſh; the ſtalke ſome cubit high, round, green, creſted, and toward the top diuided into ſundry branches, the leaues toward the bottome are ſomewhat ſmall cut, but ſomewhat larger vpon the ſtalkes. The floure is much larger than any of the former, compoſed of fiue leaues, of a light blew aboue, and ſomewhat whitiſh vnderneath, with

large

large veines running about them: in the middle stands vp the head, encompassed with blackish threds, and some 7. or 8. little gaping blewish floures at the bottomes of them; the leaues of the floures decaying the head becomes bigger, hauing at the tops thereof 6. 7. or 8. longish twined hornes growing, in a star fashion; the inside is parted into cels conteyning a yellowish green, or else blackish seed. It is set forth in the *Hortus Eystettensis* by the name of *Melanthium Hispanicum maius*; by Mr. *Parkinson* it is called *Nigella Hispanica flore simplici*; and *Bauhine* in his *Prodromus* hath it by the name of *Nigella latifolia flore maiore simplici cæruleo.* It is an annuall plant, and floures in Iuly; it is sometimes to be found in the gardens of our Florists. ‡

¶ *The Place.*

The tame are sowne in gardens: the wilde ones do grow of themselues among corne and other graine, in diuers countries beyond the seas.

¶ *The Time.*

The seed must be sowne in Aprill: it floureth in Iuly and August.

¶ *The Names.*

Gith is called in Greeke μελάνθιον: in Latine also *Melanthium*: in shops, *Nigella*, and *Nigella Romana*: of diuers, *Gith*, and *Salusandria*, and some among the former bastard names, *Papauer nigrum*: in high Dutch, **Swartzkymmich**: in low Dutch **Nardus saet**: in Italian, *Nigella*: in Spanish, *Axenuz*, *Alipiure*: in French, *Nielle odorante*: in English, Gith, and Nigella Romana, in Cambridgeshire, Bishops woort: and also *Diuæ Catharinæ flos*, Saint Katharines floure.

¶ *The Temperature.*

The seed of the garden Nigella is hot and dry in the third degree, and of thin parts.

¶ *The Vertues.*

A The seed of *Nigella Romana* drunke with wine, is a remedy against the shortnes of breath, dissolueth and putteth forth windinesse, prouoketh vrine, the menses, increaseth milke in the brests of nurses if it be drunke moderately; otherwise it is not onely hurtfull to them, but to any that take thereof too often, or in too great a quantity.

B The seed killeth and driueth forth wormes, whether it be taken with wine or water, or laid to the nauell in manner of a plaister.

C The oile that is drawne forth thereof hath the same property.

D The seed parched or dried at the fire, brought into pouder, and wrapped in a piece of fine lawne or sarcenet, cureth all murs, catarrhes, rheumes, and the pose, drieth the braine, and restoreth the sence of smelling vnto those which haue lost it, being often smelled vnto from day to day, and made warme at the fire when it is vsed.

E It takes away freckles, scurfs, and hard swellings, being laid on mixed with vineger. To be briefe, as *Galen* saith, it is a most excellent remedy where there is need of clensing, drying, and heating.

F It serueth well among other sweets to put into sweet waters, bagges, and odoriferous powders.

† The figures of the third and fourth of the former edition were transposed.

Chap. 444. *Of Cockle.*

¶ *The Description.*

Cockle is a common and hurtfull weed in our Corne, and very well knowne by the name of Cockle, which *Pena* calleth *Pseudomelanthium*, and *Nigellastrum*, by which names *Dodonæus* and *Fuchsius* do also terme it; *Mutonus* calleth it *Lolium*; and *Tragus* calleth it *Lychnoides segetum.* This plant hath straight, slender, and hairy stems, garnished with long hairy and grayish leaues, which grow together by couples, inclosing the stalke round about: the floures are of a purple colour, declining to rednesse, consisting of fiue small leaues, in proportion very like to wilde Campions; when the floures be vaded there follow round knobs or heads full of blackish seed, like vnto the seed of *Nigella*, but without any smell or sauour at all.

¶ *The Place and Time.*

The place of his growing, and time of his flouring, are better knowne then desired.

¶ *The Names.*

Cockle is called *Pseudomelanthium*, and *Nigellastrum*, wilde or bastard Nigella; of *Fuchsius*, *Lolium*: of Mouton, *Lychnoides segetum*: of *Tragus*, *Githago*: in high Dutch, **Kornegele**: in low Dutch, **Corne roosen**: in French, *Nielle des Bledz*: in English, Cockle, field Nigella, or wilde Nigella: in Italian, *Githone*; whereupon most Herbarists being mooued with the likenesse of the word, haue thought it to be the true Gith or *Melanthium*; but how farre they are deceiued it is better knowne, than needfull to be confuted: for it doth not onely differ in leaues from the true Gith, but also in other properties, and yet it is called Gith or *Melanthium*, and that is of the blackenes of the seed, yet not properly, but with a certain addition, that it may differ from the true *Melanthium*: for

Hippocrates

Pseudomelanthium.
Bastard Nigella, or Cockle.

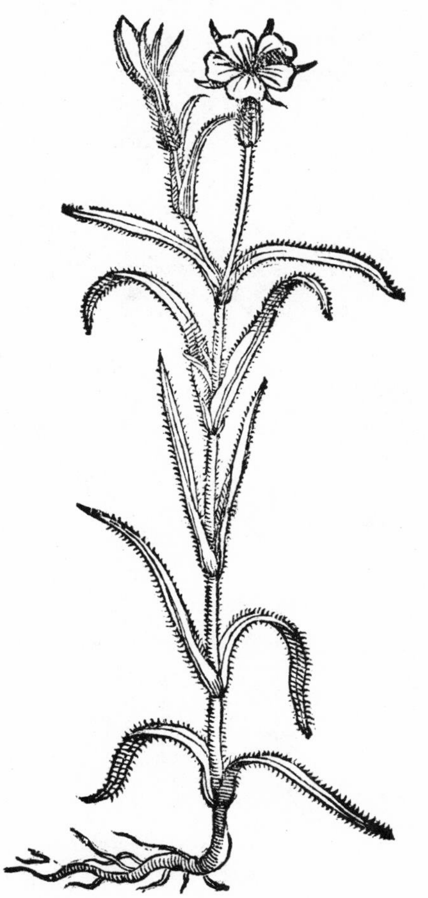

Hippocrates calleth it *Melanthium ex Tritico*, of wheate: *Octauius Horatianus* calleth that Gith which groweth among Corne: and for the same cause it is named of the learned of this our time *Nigellastrum*, *Gigatho*, and *Pseudomelanthium*: *Ruellius* saith it is called in French *Niele*, and *Flos Micancalus*.

¶ *The Temperature.*

The seed of Cockle is hot and dry in the later end of the second degree.

¶ *The Vertues.*

The seed made in a pessarie or mother suppositorie, with honey put vp, bringeth downe the desired sicknesse, as *Hippocrates* in his booke of womens diseases doth witnesse. A

Octauius Horatianus giueth the seed parched and beaten to pouder to be drunke against the yellow jaundice. B

Some ignorant people haue vsed the seed hereof for the seed of Darnell, to the great danger of those who haue receiued the same: what hurt it doth among corne, the spoyle vnto bread, as well in colour, taste, and vnwholesomnes, is better known than desired. C

CHAP. 445. *Of Fumitorie.*

¶ *The Kindes.*

THere be diuers herbes comprehended vnder the title of Fumitorie; some wilde, and others of the garden; some with bulbous or tuberous roots, and others with fibrous or threddy roots: and first of those whose roots are nothing but strings.

¶ *The Description.*

1 FVmitorie is a very tender little herbe: the stalkes thereof are slender, hauing as it were little knots or ioynts full of branches, that scarse grow vp from the ground without proppings, but for the most part they grow sidelong: the leaues round about are small, cut on the edges as those of Coriander, which as well as the stalkes are of a whitish greene: the floures be made vp in clusters at the tops of the small branches, of a red purple colour: then rise vp huskes, round and little, in which lieth the small seed: the root is slender, and groweth straight downe. ‡ This is also found with floures of a purple violet colour, and also somtimes with them white. ‡

2 The second kinde of Fumitorie hath many small long and tender branches, wherupon grow little leaues, commonly set together by threes or fiues, in colour and taste like vnto the former; hauing at the top of the branches many small clasping tendrels, with which it taketh hold vpon hedges, bushes, and whatsoeuer groweth next vnto it: the floures are small, and clustering together, of a white colour, with a little spot in their middles; after which succeed cods containing the seed: the root is single, and of a fingers length.

3 The third kinde of Fumitorie hath a very small root, consisting of diuers little strings; from which arise small and tender branches trailing here and there vpon the ground, beset with many small and tender leaues most finely cut and iagged, like the little leaues of Dill, of a deepe greene colour tending to blewnesse: the floures stand on the tops of the branches, in bunches or clusters thicke thrust together, like those of the medow Clauer, or three leafed grasse, of a most bright red colour, and very beautifull to behold: the root is very small and threddy.

1 *Fumaria purpurea.*
Common or purple Fumitory.

† 2 *Fumaria alba latifolia clauiculata.*
White broad leafed Fumitorie.

3 *Fumaria tenuifolia.*
Fine leafed Fumitorie.

4 *Fumaria lutea.*
Yellow Fumitorie.

4 The yellow Fumitorie hath many crambling thready roots, somewhat thicke, grosse, and fat, like those of *Asparagus*: from which rise diuers vpright stalkes a cubit high, diuiding themselues toward the top into other smaller branches; wheron are confusedly placed leaues like those of *Thalictrum*, or English Rubarb, but lesser and thinner: alongst the tops of the branches grow yellow floures, resembling those of Sage: which being past, there followeth small seed like vnto dust.

¶ *The Place.*

The Fumitories grow in corne fields among Barley and other graine; in vineyards, gardens, and such like manured places. I found the second and third growing in a corne field betweene a small village called Charleton and Greenwich.

¶ *The Time.*

Fumitorie is found with his floure in the beginning of May, and so continues to the end of sommer. When it is in floure is the best time to gather it to keepe dry, or to distill.

¶ *The Names.*

Fumitorie is called in Greeke Καπνὸς, and Κάπνιον, and often Καπνῖτις: in Latine, *Fumaria*: of *Pliny*, *Capnos*: in shops, *Fumus terræ*: in high-Dutch, **Erdtrauch**: in low-Dutch, **Grijsecom, Duyuen Kernel**: in Spanish, *Palomilha*: in French and English, Fumiterre.

¶ *The Temperature.*

Fumitorie is not hot, as some haue thought it to be, but cold and something dry; it openeth and clenseth by vrine.

¶ *The Vertues.*

It is good for all them that haue either scabs or any other filth growing on the skinne, and for them also that haue the French disease. A

It remoueth stoppings from the liuer and spleene: it purifieth the bloud, and is oft times good for them that haue a quartane ague. B

The decoction of the herbe is vsed to be giuen, or else the syrrup that is made of the iuyce: the distilled water thereof is also profitable against the purposes aforesaid. C

It is oftentimes boyled in whay, and in this manner it helpeth in the end of the Spring and in Sommer time those that are troubled with scabs. D

Paulus Ægineta saith that it plentifully prouoketh vrine, and taketh away the stoppings of the liuer, and feeblenesse thereof; that it strengthneth the stomacke, and maketh the belly soluble. E

Dioscorides affirmeth, that the iuyce of Fumitorie, of that which groweth among Barley, as *Ægineta* addeth, with gum Arabicke, doth take away vnprofitable haires that pricke the eyes, growing vpon the eye lids, the haires that pricke being first plucked away, for it will not suffer others to grow in their places. F

The decoction of Fumitorie drunken driueth forth by vrine and siege all hot cholericke burnt and hurtfull humors, and is a most singular digester of salt and pituitous humors. G

† There were formerly six figures and descriptions in this chapter whereof the two first figures were of the common Fumitorie, the one with purple, the other with white floures; and the two later were of the *Fumaria latifolia claiculata*, differing onely in the largenesse and smallnesse of the leafe. The description in the second place belonged to the *Fumaria claviculata*, which also was againe described in the fifth and sixth places, yet not to much purpose; wherefore I haue put the figure to the second, and omitted the other as superfluous.

CHAP. 446. *Of bulbous Fumitorie, or Hollow-root.*

¶ *The Description.*

1 THe leaues of great Hollow-root are iagged and cut in sunder, as be those of Coriander, of a light greenish colour, that is to say, like the gray colour of the leaues of Columbine, whereunto they be also in forme like, but lesser: the stalks be smooth, round, and slender, an handfull long; about which, on the vpper part stand little floures orderly placed, long, with a little horne at the end like the floures of Tode-flax, of a light red tending to a purple colour: the seed lieth in flat cods, very soft and greenish when it is ready to yeeld vp his black shining ripe seed: the root is bumped or bulbous, hollow within, and on the vpper part pressed down somewhat flat, couered ouer with a darke yellow skin or barke, with certaine strings fastned thereto, and of a bitter and austere taste.

2 The second is like vnto the first in each respect, sauing that it bringeth floures of a white colour, and the other not so.

3 The small purple Hollow-root hath roots, leaues, stalkes, floures, and seeds like the precedent, the especiall difference is, that this plant is somewhat lesse.

 4 The

4 The ſmall white Hollow-root likewiſe agreeth with the former in each reſpect, ſauing that this plant bringeth white floures, and the other not ſo.

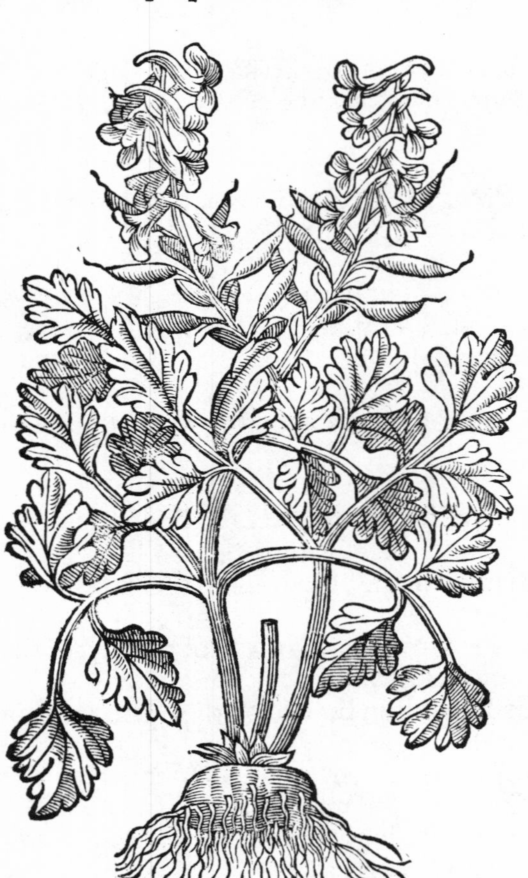

1 *Radix caua maior purpurea.*
Great purple Hollow-root.

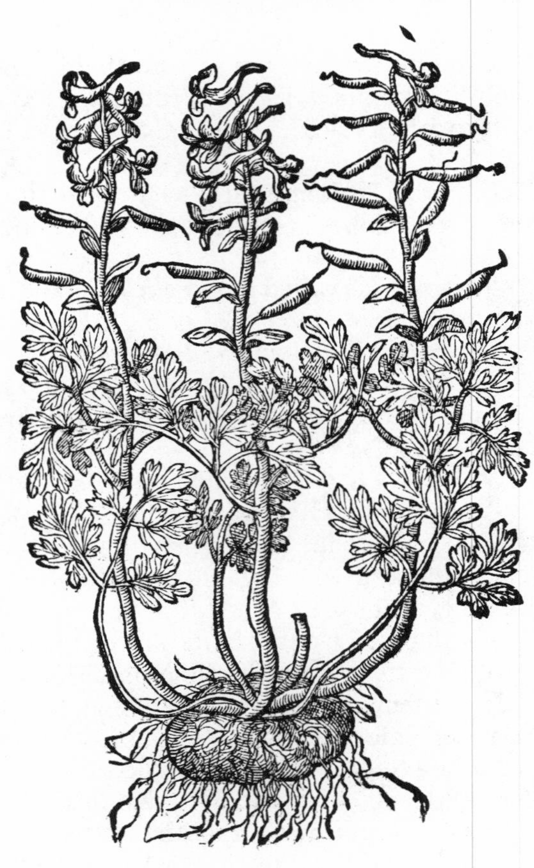

2 *Radix caua maior alba.*
Great white Hollow-root.

5 This kinde of Hollow-root is alſo like the laſt deſcribed, ſauing that the floures hereof are mixed with purple and white, which maketh it to differ from the others.

6 There is no difference in this, that can poſſibly be diſtinguiſhed, from the laſt deſcribed, ſauing that the floures hereof are of a mixt colour, white and purple, with ſome yellow in the hollowneſſe of the ſame, wherein conſiſteth the difference from the precedent.

7 This thin leafed Hollow-root hath likewiſe an hollow root, couered ouer with a yellow pilling, of the bigneſſe of a tenniſe ball: from which ſhoot vp leaues ſpred vpon the ground, very like vnto the leaues of Columbines, as well in forme as colour, but much thinner, more iagged, and altogether leſſer: among which riſe vp ſmall tender ſtalkes, weake and feeble, of an handfull high, bearing from the middle thereof to the top very fine floures, faſhioned vnto one piece of the Columbine floure, which reſembleth a little bird of a purple colour.

8 This other thin leafed Hollow-root is like the precedent, ſauing that this plant brings forth white floures tending to yellowneſſe, or as it were of the colour of the field Primroſe.

9 **Bunnykens holwortele**, as the Dutch men doe call it, hath many ſmall iagged leaues growing immediately from the ground, among which riſe vp very ſlender ſtalkes, whereon doe grow ſuch leaues as thoſe next the ground: on the top of the branches ſtand faire purple floures like vnto the others of his kinde, ſauing that the floures hereof are as it were ſmall birds, the bellies or lower parts whereof are of a white colour, wherein it differeth from all the reſt of the Hollow-roots.

10 The laſt and ſmall hollow-root is like the laſt deſcribed, ſauing that it is altogether leſſe, and the floures hereof are of a greene colour, not vnlike in ſhape to the floures of Cinkefoile. ‡ This plant, whoſe figure our Author here gaue with this ſmall deſcription, is that which from the ſmel of muske is called *Moſchatella*, by *Cordus* and others: it is the *Denticulata* of *Daleſchampius*: the *Fumaria bulboſa tuberoſa minima* of *Tabernamontanus*: and the *Ranunculus minimus ſeptentrionalium herbido muſcoſo flore* of *Lobel*. The root hereof is ſmall and toothed, or made of little bulbes reſembling teeth, and ending in white hairy fibres: it ſendeth vp diuers little branches ſome two

or

or three inches high: the leaues are somewhat like those of the yellow Fumitorie, or *Radix caua*, but much lesse: the floures grow clustering on the top of the stalke, commonly fiue or seuen together, each of them made of foure yellowish green leaues with some threds in them; it floures in Aprill, and is to be found in diuers places amongst bushes at that time, as in Kent about Chislehurst, especially in *Pits* his wood, and at the further end of Cray heath, on the left hand vnder a hedge among bryers and brambles, which is his proper seat. ‡

9 *Radix caua minor.*
Bunnikens Holwoort.

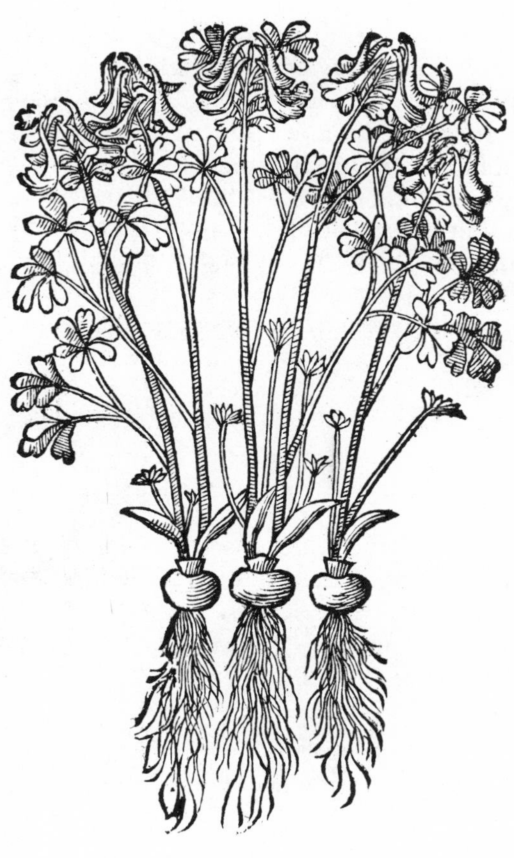

10 *Radix caua minima viridi flore.*
Small Bunnikens Holwoort.

¶ *The Place.*

These plants do grow about hedges, brambles, and in the borders of fields and vineyards, in low and fertile grounds, in Germanie and the Low-countries, neuerthelesse the two first, and also the two last described do grow in my garden.

¶ *The Time.*

These do floure in March, and their seed is ripe in Aprill: the leaues and stalkes are gon in May, and nothing remaining saue onely the roots, so little a while do they continue.

¶ *The Names.*

Hollow root is called in high Dutch **Holwurtz**: in low Dutch, **Hoolewortele**, that is, *Radix caua*: in English, Hollow root, and Holewoort: it is vsed in shops in steed of *Aristolochia*, or round Birthwoort; which errour is better knowne than needfull to be confuted: and likewise their errour is apparant, who rashly iudge it to be *Pistolochia*, or little Birthwoort. It should seem the old Writers knew it not; wherefore some of our later Authors haue made it *Leontopetali species*, or a kinde of Lions Turnep: others, *Eriphium*: and othersome, *Thesium*: most men, *Capnos Chelidonia*: it seemeth to agree with *Leontopetalon* in bulbed roots, and somewhat in leaues, but in no other respects, as may be perceiued by *Dioscorides* and *Plinies* description of *Leontopetalon*. And if *Eriphium* haue his name ἀπὸ τοῦ ἦρος, that is to say of the Spring, then this root may be not vnproperly *Eriphium*, and *Veris Planta*: or the Plant of the Spring: for it is euident that it appeareth and is greene in the Spring onely: some thinke it hath beene called *Eriphium, ab Hœdo*, or of the Goat: but this *Eriphion* is quite another plant, as both *Apuleius* writeth, and that booke also mentioneth which is attributed to *Galen*, and dedicated to *Paternianus*. In the booke which is dedicated to *Paternianus*, there be read these words; [*Eriphion* is an herbe which is found vpon high mountaines, it hath leaues like Smallage, a fine floure like the Violet, and a root as great as an Onion: it hath likewise other roots which

which ſend forth roots after roots. Whereby it is euident that this root whereof we intreat is not this kinde of *Eriphium*. Concerning *Theſium* the old Writers haue written but little: *Theophraſtus* ſaith, that the root thereof is bitter, and being ſtamped purgeth the belly. *Pliny* in his 21. booke, chap. 17. ſheweth, that the root which is called *Theſium* is like the bulbed plants, and is rough in taſte: *Athenæus* citing *Timachida* for an Authour, ſaith, that *Theſium* is called a floure, of which *Ariadnes* garland was made. Theſe things ſeeme well to agree with Hollow root; for it is bumped or bulbous, of taſte bitter and auſtere, or ſomething rough, which is alſo thought to purge: but what certaintie can be affirmed, ſeeing the old writers are ſo briefe? what manner of herbe *Capnos Chelidonia* is, which groweth by hedges, and hereupon is ſurnamed φραγμίτης, *Aetius* doth not expound, onely the name thereof is found in his ſecond Tetrab. the third booke, chap. 110. in *Martianus* his *Collyrium*, and in his Tetrab. 3. booke, 2. chap. among ſuch things as ſtrengthen the liuer. But if *Capnos Chelidonia* be that which *Pliny* in his 25. booke, chap. 13. doth call *Prima Capnos*, or the firſt *Capnos*, and commendeth it for the dimneſſe of the ſight, it is plain enough that *Radix caua*, or the Hollow root, is not *Capnos Chelidonia*: for *Plinies* firſt *Capnos* is branched, and foldeth it ſelfe vpon hedges: but Hollow root hath no ſuch branches growing on it, and is a low herbe, and is not held vp with props, nor needeth them. But if *Aetius* his *Capnos Chelidonia* be another herb differing from that of *Pliny* (which thing perchance was the cauſe why it ſhould bee ſurnamed *Chelidonia*) there is ſome reaſon why it ſhould be called *Capnos Chelidonia*; for it is ſomewhat like Fumitorie in leaues, though greater, and commeth vp at the firſt ſpring, which is about the time when the Swallowes do come in; neuertheleſſe it doth not follow, that it is true and right *Capnos Chelidonia*, for there be alſo other herbs comming vp at the ſame ſeaſon, and periſh in ſhort time after, which notwithſtanding are not called *Chelidonia*.

¶ *The Temperature.*

Hollow root is hot and drie, yet more drie than hot, that is to ſay, dry in the third degree, and hot in the ſecond; it bindeth, clenſeth, and ſomewhat waſteth.

¶ *The Vertues.*

A Hollow root is good againſt old and long laſting ſwellings of the Almonds in the throat, and of the iawes: it likewiſe preuaileth againſt the paines of the hemorrhoides, which are ſwolne and painefull, being mixed with the ointment of Poplar buds, called *Vng. Populeon.*

B It is reported that a dram weight hereof being taken inwardly, doth purge by ſiege, and draweth forth flegme.

† I haue reduced the eight figures which were formerly here put to the firſt 8. deſcriptions, being all of one and the ſame plant, to two, yet haue I left the deſcriptions, which in my opinion might haue been as well ſpared as the figures, for excepting the various colour of the floures there are but two diſtinct differences of the *Fumaria bulboſa maior*, the one hauing a hollowneſſe in the bottome of the root, and the other wanting it; and this which hath the ſollid root hath alſo the greene leaues betweene the floures cut in or diuided, the floures alſo are leſſe, more in number, and of an elegant red purple colour; and ſeldome found of any other colour, whereas the other varies much in the colour of the floures.

CHAP. 447. *Of Columbine.*

¶ *The Deſcription.*

1 THe blew Columbine hath leaues like the great Celandine, but ſomewhat rounder, indented on the edges, parted into diuers ſections, of a blewiſh greene colour, which being broken yeeld forth little iuice or none at all: the ſtalke is a cubit and a halfe high, ſlender, reddiſh, and ſleightly haired: the ſlender ſprigs whereof bring forth euerie one one floure with fiue little hollow hornes, as it were hanging forth, with ſmall leaues ſtanding vpright, of the ſhape of little birds. theſe floures are of colour ſomtimes blew, at other times of a red or purple, often white, or of mixt colors, which to diſtinguiſh ſeuerally would be to ſmal purpoſe, being things ſo familiarly knowne to all: after the floures grow vp cods, in which is contained little blacke and glittering ſeed: the roots are thicke, with ſome ſtrings thereto belonging, which continue manie yeares.

2 The ſecond doth not differ ſauing in the colour of the floures; for like as the others are deſcribed to be blew, ſo theſe are of a purple red, or horſe-fleſh colour, which maketh the difference.

3 The double Columbine hath ſtalks, leaues, and roots, like the former: the floures hereof are very double, that is to ſay, many of thoſe little floures (hauing the forme of birds) are thruſt one into the belly of another, ſometimes blew, often white, and other whiles of mixt colours, as nature liſt to play with her little ones, differing ſo infinitely, that to diſtinguiſh them apart would require

1 *Aquilegia cærulea.*
Blew Columbines.

2 *Aquileia rubra.*
Red Columbines.

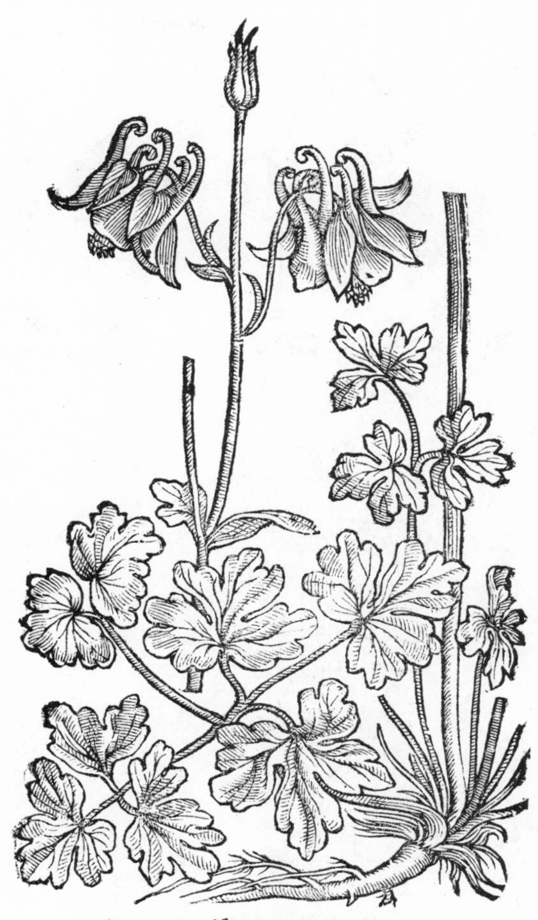

3 *Aquilina multiplex.*
Double Columbines.

‡ 4 *Aquilegia variegata.*
Variegated Columbine.

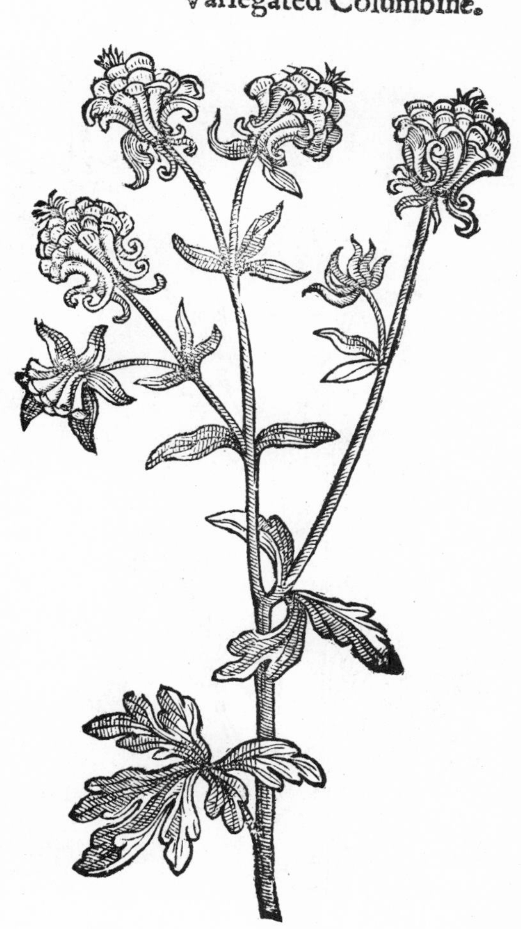

‡ 5 *Aquilegia flo. inverso rubro.*
Columbine with the inuerted red floure.

‡ 6 *Aquilegia flo. inverso albo.*
Inuerted Columbine with the white floure.

‡ 7 *Aquilegia flore roseo.*
Rose Columbine.

‡ 8 *Aquilegia degener.*
Degenerate Columbine.

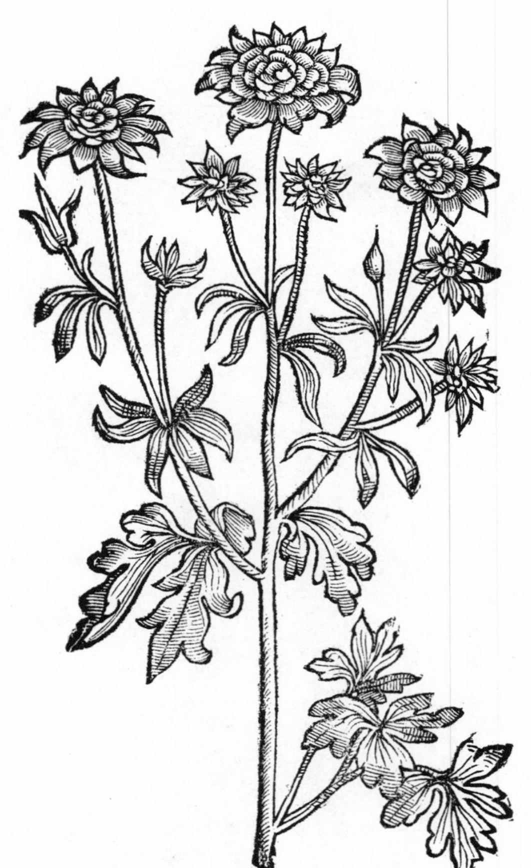

more time than were requisite to leese: and therefore it shall suffice what hath beene said for their descriptions.

‡ 4 There are also other varieties of this double kinde, which haue the floures of diuers or partie colours, as blew and white, and white and red variously marked or spotted.

5 This kinde hath the floures with their heeles or spurres turned outward or in the middle of the floure, whence it is called *Aquilina inuersa*: the floures of this are commonly reddish, or of a light or darke purple colour, and double.

6 This differs from the last in the colour of the floures which are white, yet double, and inuerted as the former.

7 The roots, leaues, and stalks of this are not vnlike those of the precedent, but the floure is much different in shape; for it hath no heels or spurs, but is made of sundrie long leaues lying flat open, being sometimes more single, and otherwhiles more double. The colour of the floure is either red, white, blew, or variously mixt of these as the former.

8 This though it be termed degenerate, is a kinde of it selfe, and it differs from the last described in that the vtmost leaues are the largest, and the colour thereof is commonly greene, or greene somewhat inclining to a purple. ‡

¶ *The Place.*

They are set and sowne in gardens for the beautie and variable colours of the floures.

¶ *The Time.*

They floure in May, Iune, and Iuly.

¶ *The Names.*

Columbine is called of the later Herbarists *Aquileia*, *Aquilina*, and *Aquilegia*: of *Costeus*, *Pothos*: of *Gesner*, *Leontostomum*: of *Daleschampius*, *Iouis flos*: of some, *Herba Leonis*, or the herbe wherein the Lion doth delight: in High Dutch, **Agley**: in Low Dutch, **Akeleyen**: in French, *Ancoiles*: in English, Columbine. ‡ *Fabius Columna* iudges it to be the *Isopyrum* described by *Dioscorides*. ‡

¶ *The Temperature.*

Columbines are thought to be temperate betweene heate and moisture.

¶ *The Vertues.*

A Notwithstanding what temperature or vertues Columbines haue is not yet sufficiently known; for they are vsed especially to decke the gardens of the curious, garlands, and houses: neuerthelesse *Tragus* writeth, that a dram weight of the seed, with halfe a scruple or ten graines of Saffron giuen in wine, is a good and effectuall medicine for the stopping of the liuer, and the yellow iaundise, but saith he, that who so hath taken it must be well couered with cloathes, and then sweat.

B Most in these daies following others by tradition, do vse to boile the leaues in milke against the sorenesse of the throat, falling and excoriation of the uvula: but the antient writers haue said nothing hereof: *Ruellius* reporteth, that the floures of Columbines are not vsed in medicine: yet some there be that do affirme they are good against the stopping of the liuer, which effect the leaues doe also performe.

C ‡ *Clusius* saith, that D[r]. *Francis Rapard* a Physition of Bruges in Flanders, told him that the seed of this common Columbine very finely beaten to pouder, and giuen in wine, was a singular medicine to be giuen to women to hasten and facilitate their labour, and if the first taking it were not sufficiently effectuall, that then they should repeat it againe. ‡

Chap. 448. *Of Wormewood.*

¶ *The Description.*

1 The first kind being our common and best knowne Wormwood, hath leaues of a grayish colour, very much cut or iagged, and very bitter: the stalkes are of a wooddie substance, two cubits high, and full of branches, alongst which doe grow little yellowish buttons, wherein is found small seed like the seed of Tansey, but smaller: the root is likewise of a wooddie substance, and full of fibres.

2 The second kinde of Wormwood bringeth forth slender stalkes about a foot high or somewhat more, garnished with leaues like the former, but whiter, much lesser, and cut or iagged into most fine and small cuts or diuisions: the floures are like the former, hanging vpon small stemmes with their heads downeward: the roots are whitish, small and many, crawling and crambling one ouer another, and thereby infinitely do increase, of sauour lesse pleasant than the common Wormwood.

wood. Some haue termed this plant *Absinthium santonicum*, but they had slender reason so to do: for if it was so called because it was imagined to grow in the Prouince of Saintoirge, it may very wel appeare to the contrarie; for in the Alpes of Galatia, a countrey in *Asia minor*, it groweth in great plenty, and therefore may rather be called *Galatium Sardonicum*, and not *Santonicum*: but leauing controuersies impertinent to the History, it is the Ponticke Wormwood of *Galens* description, and so holden of the learned *Paludane* (who for his singular knowledge in plants is worthy triple honor) and likewise many others.

1 *Absinthium latifolium siue Ponticum.*
Broad leafed Wormwood.

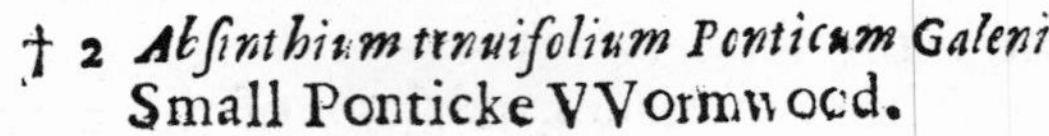

† 2 *Absinthium tenuifolium Ponticum Galeni.*
Small Ponticke VVormwood.

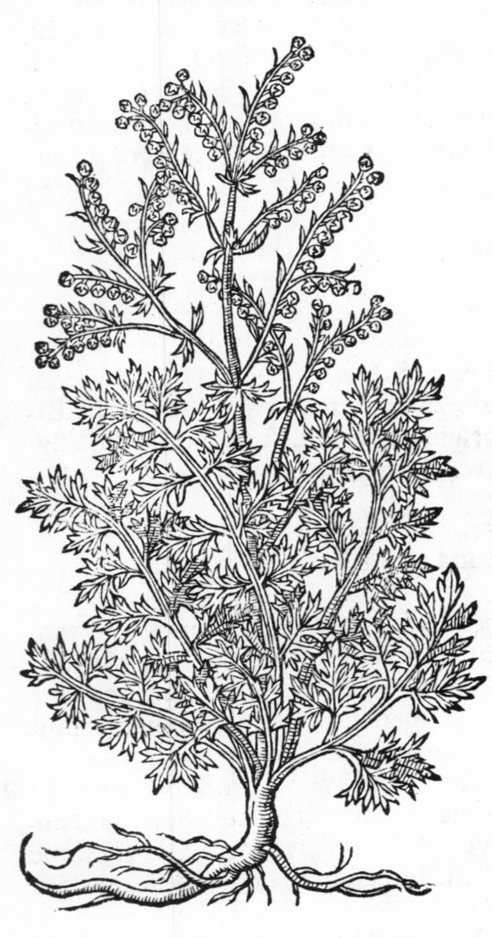

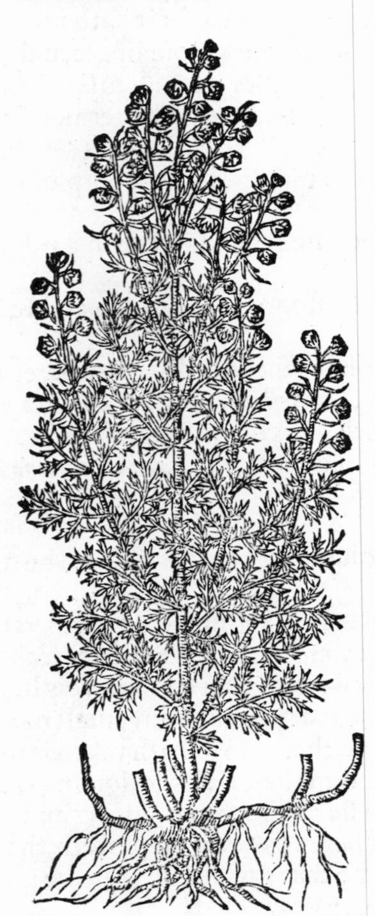

¶ *The Place.*

This broad leafed Wormewood delighteth to grow on rocks and mountaines, and in vntilled places; it groweth much vpon dry bankes, it is common euery where in all countries: the best, saith *Dioscorides*, is found in Pontus, Cappadocia, and on mount Taurus: *Pliny* writeth, that Ponticke Wormwood is better than that of Italie: *Ouid* in these words doth declare that Ponticke Wormwood is extreme bitter.

Turpia deformes gignunt Absinthia campi,
Terraque de fructu, quam sit amara docet.

Vntilled barren ground the lothsome Wormwood yeelds,
And knowne it's by the fruit how bitter are the fields.

And *Bellonius* in his first booke of Singularities, chap. 76. doth shew, that there is also a broad leafed Wormwood like vnto ours, growing in the Prouinces of Pontus, and is vsed in Constantinople by the Physitions there; it is likewise found in certain cold places of Switzerland, which by reason of the chilnesse of the aire riseth not vp, but creepeth vpon the ground, whereupon diuers cal it creeping Wormwood.

¶ *The Time.*

The little flours and seeds are perfected in Iuly and August, then may Wormwood be gathered and laied vp for profitable vses.

¶ *The Names.*

It is called in Greeke ἀψίνθιον: it is named of *Apuleius*, *Absinthium rusticum*, countrey Wormwood, or or pesants Wormewood: we haue named it *Absinthium latifolium*, broad leafed Wormewood, that it may differ from the rest: the Interpretors of the Arabians call the better sort, which *Dioscorides* nameth Ponticke Wormwood, *Romanum Absinthium*, Roman Wormwood: and after these, the barbarous Physitions of the later age: the Italians name Wormwood *Assenso*: the Spaniards, *Axenxios*, *Assensios*, most of them *Donzell*: the Portingales, *Alosna*: in high Dutch, **Weronnut, Wermut**: in French, *Aluyne*: in English, common VVormwood.

Victor Trincauilla, a singular Physition, in his practise tooke it for *Absinthium Ponticum*.

2 This is commonly called *Absinthium Romanum*: and in low Dutch, **Roomsche Alsene**: by which name it is knowne to very many Physitions and Apothecaries, who vse this in stead of Pontick wormwood: furthermore it hath a leafe and floure far lesse than the other wormwoods: likewise the smell of this is not onely pleasant, but it yeeldeth also a spicie sent, wheras all the rest haue a strong and lothsome smell: and this Ponticke Wormwood doth differ from that which *Dioscorides* commendeth: for *Dioscorides* his Pontick wormwood is accounted among them of the first kinde, or of broad leafed wormwood, which thing also *Galen* affirmeth in his sixt booke of the Faculties of medicines, in the chapter of Sothernwood. There be three kinds of Wormwood (saith he) wherof they vse to call one by the generall name, and that is especially Pontick: whereby it is manifest that *Galen* in this place hath referred Ponticke to no other than to the first wormwood; and therefore many not without cause maruell, that *Galen* hath written in his booke of the Method of curing, how Pontick wormwood is lesse in floure and leafe: many excuse him, and lay the fault vpon the corruption of the booke, and in his 9. booke of Method, the lesser they would haue the longer: therefore this wormwood with the lesser leafe is not the right Pontick wormewood, neither againe the Arabians Romane wormewood, who haue no other Romane than Ponticke of the Grecians. Also many beleeue that this is called *Santonicum*, but this is not to be sought for in Mysia, Thracia, or other countries Eastward, but in France beyond the Alps, if we may beleeue *Dioscorides* his copies there be that would haue it grow not beyond the Alps of Italy, but in Galatia a countrie in Asia, & in the region of the Sardines, which is in the lesser Asia, whereupon it was called in Greeke Σαρδόνιον, which was changed into the name *Santonicum* through the errour of the translators: *Dioscorides* his copies keep the word *Sardonium*, & *Galens* copies *Santonicum*, which came to posterity as it seemeth. iscalled in English, Romane Wormewood, garden or Cypres Wormewood, and French Wormwood.

¶ *The Temperature.*

Wormewood is of temperature hot and drie, hot in the second degree, and drie in the third: it is bitter and clensing, and likewise hath power to binde or strengthen.

¶ *The Vertues.*

It is very profitable to a weake stomacke that is troubled with choler, for it clenseth it through his bitternesse, purgeth by siege and vrine: by reason of the binding qualitie, it strengthneth and comforteth the stomacke, but helpeth nothing at all to remoue flegme contained in the stomacke, as *Galen* addeth. A

If it be taken before a surfeit it keepeth it off, and remoueth lothsomenesse, saith *Dioscorides*, and it helpeth not only before a surfeit, but also it quickly refresheth the stomack and belly after large eating and drinking. B

It is oftentimes a good remedie against long and lingring agues, especially tertians: for it doth not onely strengthen the stomacke and make an appetite to meat, but it yeeldeth strength to the liuer also, and riddeth it of obstructions or stoppings, clensing by vrine naughtie humours. C

Furthermore, Wormewood is excellent good for them that vomite bloud from the spleene, the which hapneth when the spleene being ouercharged and filled vp with grosse bloud doth vnburden it selfe, and then great plenty of bloud is oftentimes cast vp by vomite. It happeneth likewise that store of blacke and corrupt bloud mixed with excrements passeth downewards by the stoole, and it oftentimes hapneth that with violent and large vomiting the sicke man fainteth or swowneth, or when he is reuiued doth fall into a difficult and almost incureable tympanie, especially when the disease doth often happen; but from these dangers Wormewood can deliuer him, if when he is refreshed after vomite, and his strength any way recouered, he shall a good while vse it, in what manner soeuer he himselfe shall thinke good. D

Againe, Wormewood voideth away the wormes of the guts, not onely taken inwardly, but applied outwardly: it withstandeth all putrifactions; it is good against a stinking breath; it keepeth garments also from the Mothes; it driueth away gnats, the bodie being annointed with the oile thereof. E

Likewise it is singular good in pultesses and fomentations to binde and to drie. F

Besides all this *Dioscorides* declareth, that it is good also against windinesse and griping pains of the stomacke and belly, with Seseli and French Spikenard: the decoction cureth the yellow iaundies or the infusion, if it be drunke thrise a day some ten or twelue spoonfuls at a time. G

H It helpeth them that are strangled with eating of Mushroms, or toad stools, if it be drunk with vineger.

I And being taken with wine, it is good against the poison of *Ixia* (being a viscous matter proceeding from the thistle *Chamælion*) and of Hemlock, and against the biting of the shrew mouse, and of the Sea Dragon: it is applied to the squincie or inflammations of the throat with honie and niter, and with water to night wheales, and with heny to swartish markes that come vpon bruses.

K It is applied after the same manner to dim eies, and to mattering eares.

L *Ioachimus Camerarius* of Noremberg commendeth it greatly against the iaundice, giuing of the floures of Wormwood, Rosemarie, Sloes, of each a small quantitie, and a little saffron, boiled in wine, the body first being purged and prepared by the learned Physition.

† The figure which formerly was in the second place, was of a small wormwood, not different from the common kinde, but only in the smalnesse, and more aromatike taste; it growes on mountanous places, and *Gesner* calls it *Absinthium commune minus, vel Alpinum*: now our Authors description was intended for this, whose figure we haue giuen you, for it is the *Absinthium santonicum*, of some, as *Ruellius* and *Cesalpinus*; and the *Galatium Sardonium* of *Pena* and *Lobel*.

CHAP. 449. *Of Small leafed Wormewood.*

Absinthium tenuifolium Austriacum.
Austrian Wormewood.

¶ *The Description.*

Small leafed Wormwood bringeth forth very many little branches, slender, a span or a foot high, full of leaues, lesse by a great deale, and tenderer than the former, most finely and nicely minced: the floures like those of the former, hang vpon the little branches and sprigs: the roots are small, creeping ouertwhart, from whence do rise a great number of yong sprouts: this VVormwood also is somewhat white, and no lesse bitter than the broad leafed one, and hath not so ranke, or so vnpleasant a smell, but rather delightfull.

¶ *The Place.*

It grows plentifully in Mysia, Thracia, Hungarie, and Austria, and in other regions neere adioining: it is also found in Bohemia, and in many vntilled places of Germanie; it is a garden plant in the low Countries, and in England.

¶ *The Time.*

It bringeth forth floures and seed in Autumne: a little while after when winter commeth, the herbe withereth away, but the roote remaineth aliue, from which leaues and stalks do come againe in the spring.

¶ *The Names.*

‡ This *Lobel* calls *Absinthium Ponticum Tridentinum Herbariorum*: *Clusius*, *Absinthium tenuifolium Austriacum*: *Tabernamontanus*, *Absinthium Nabathæum Auicennæ*: wee may call it in English, small leaued Wormwood. ‡

¶ *The Temperature.*

Small leafed VVormwood is of facultie hot and drie, it is as bitter also as the broad leafed one, and of like facultie.

¶ *The Vertues.*

The faculties are referred vnto the common VVormwood.

CHAP.

CHAP. 450. *Of Sea Wormewood.*

¶ *The Description.*

1 THe white or common Sea VVormwood hath many leaues cut and diuided into infinite fine iags, like those of Sothernwood, of a white hoarie colour and strong smell, but not vnpleasant: among which rise vp tough hoarie stalks set with the like leaues, on the top wherof do grow smal yellowish floures: the root is tough, and creepeth far abroad, by means whereof it greatly increaseth.

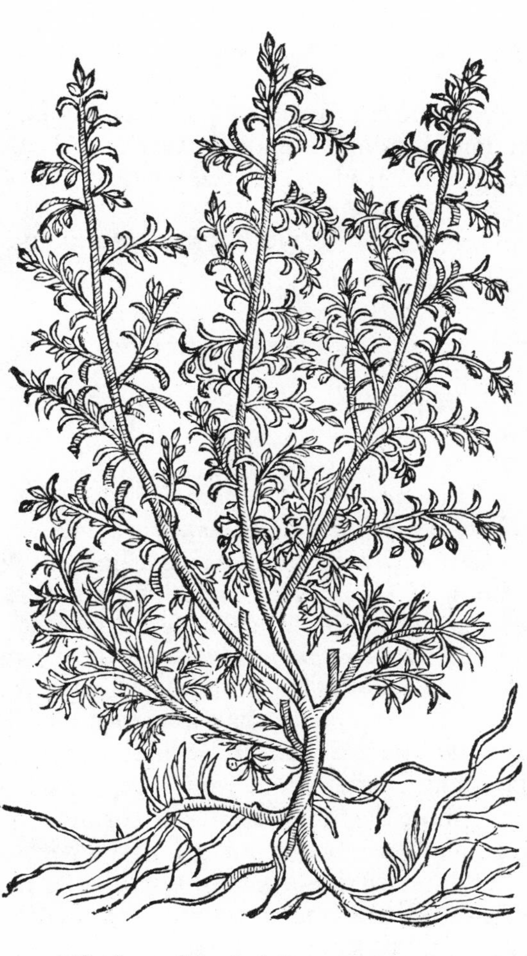

1 *Absinthium marinum album.*
VVhite Sea VVormwood.

2 *Absinthium marinum repens.*
Creeping Sea VVormwood.

2 The broad leafed Sea VVormwood hath very many soft leaues, growing close by the ground, of a darke swart colour, nothing so finely cut or iagged as the other of his kinde: the floures grow vpon the tops of the stalks, of a yellowish colour: the root is tough and creeping. ‡ This hath many weake slender branches commonly two foot long at their ful growth, red of colour, and creeping vpon the ground: the leaues are small, narrow, long and iagged, or parted towards their ends into sundry parcels: they are greene aboue, and grayish vnderneath: the toppes of the branches are set with many little stalkes, some inch long: which vpon short foot-stalkes comming out of the bosomes of little longish narrow leaues carry small round knops, like as in other plants of this kind: the taste is a little bitterish, and the smell not vnpleasant: this growes with Mr. *Parkinson* and others, and (as I remember) it was first sent ouer from the Isle of Rees by Mr. *Iohn Tradescant. Lobel* in his Obseruations mentions it by the name of *Absinth. Ponticum supinum Herbariorum*, and *Tabern.* sets it forth by the title of *Absinthium repens.* ‡

¶ *The Place.*

Thse VVormwoods do grow vpon the raised grounds in the salt marshes neere vnto the sea, in most places of England; which being brought into gardens doth there flourish as in his naturall place, and retaineth his smell, taste, and naturall qualitie, as hath beene often proued. ‡ I haue not heard

heard that the later growes wilde in any place with vs in England. ‡

¶ *The Time.*

These bring forth floures and seeds when the other Wormwoods doe. ‡ The later scarce seedes with vs, it floures so late in the yeare. ‡

¶ *The Names.*

Sea VVormwood is called in Greeke ἀψίνθιον θαλάσσιον: in Latine, *Absinthium marinum*, and likewise *Seriphium*: in Dutch, See Alsene: of diuers, *Santonicum*, as witnesseth *Dioscorides*: neuerthelesse there is another *Santonicum* differing from sea VVormwood: in English of some women of the countrey, Garden Cypresse.

¶ *The Temperature.*

Sea VVormwood is of nature hot and drie, but not so much as the common.

¶ *The Vertues.*

A *Dioscorides* affirmeth, that being taken of it selfe, or boiled with Rice, and eaten with hony, it killeth the small wormes of the guts, and gently looseth the belly, the which *Pliny* doth also affirme.

B The iuice of sea VVormwood drunke with wine resisteth poison, especially the poison of Hemlockes.

C The leaues stamped with figs, salt-peter, and the meale of Darnel, and applied to the belly, sides, or flankes, help the dropsie, and such as are spleenticke.

D The same is singular against all inflammations, and heat of the stomacke and liuer, exceeding all the kindes of VVormwood for the same purposes that common VVormwood serueth.

E It is reported by such as dwell neere the sea side, that the cattell which do feed where it groweth become fat and lusty very quickly.

F The herbe with his stalks laid in chests, presses, and ward-robes, keepeth clothes from moths and other vermine.

Chap. 451. Of Holy Wormewood.

Sementina.
Holie VVormewood.

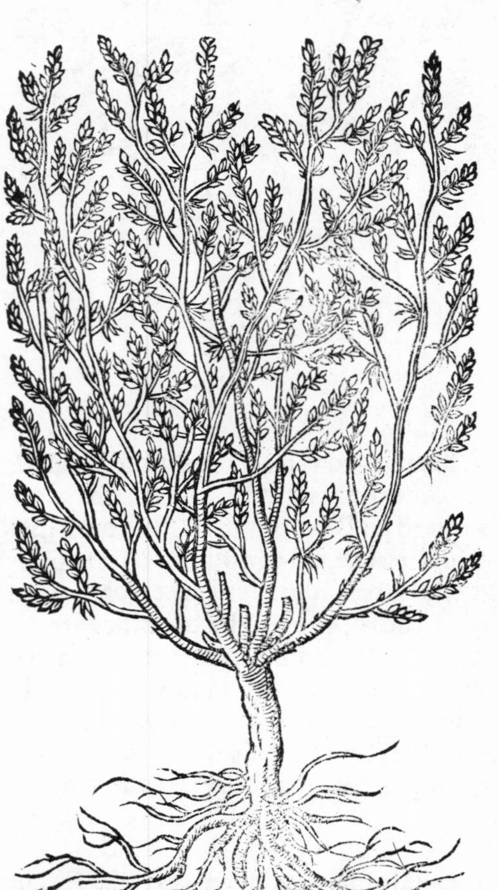

¶ *The Description.*

This Wormwood called *Sementina*, and *Semen sanctum*, which we haue Englished, Holy, is that kinde of Wormwood which beareth that seed which we haue in vse, called VVormeseed: in shops, *Semen Santolinum*: about which there hath been great controuersie amongst writers: some holding that the seed of *Santonicum Galatium* to be the true VVormseed: others deeming it to be that of *Romanum Absinthium*: it doth much resemble the first of the sea VVormwoods in shape and proportion: it riseth vp with a wooddie stalke, of the height of a cubite, diuided into diuers branches and wings; whereupon are set very small leaues: among which are placed clusters of seeds in such abundance, that to the first view it seemeth to be a plant consisting all of seed.

¶ *The Place.*

It is a forreine plant: the seeds being sowne in the gardens of hot regions doe prosper well; in these cold countries it will not grow at all. Neuertheles there is one or two companions about London, who haue reported vnto mee that they had great store of it growing in their gardens yearely, which they sold at a great price vnto our London Apothecaries, and gained much money thereby; one of the men dwelleth by the Bagge and Bottle neere London, whose name is *Cornewall*; into whose garden I was brought to see the thing that I would not beleeue; for being often told

told that there it did grow, I still persisted it was not true : but when I did behold this great quantitie of VVormwood, it was nothing else but common *Ameos*. How many Apothecaries haue been deceiued, how many they haue robbed of their money, and how many children haue been nothing the better for taking it, I refer it to the iudgement of the simplest, considering their owne report, to haue sold many hundreth pounds weight of it; the more to their shame be it spoken, and the lesse wit or skill in the Apothecaries : therefore haue I set downe this as a caueat vnto those that buy of these seeds, first to taste and trie the same before they giue it to their children, or commit it to any other vse. ‡ Certainely our Author was either misinformed, or the people of these times were very simple, for I dare boldly say there is not any Apothecary, or scarce any other so simple as to be thus deceiued now. ‡

¶ *The Time.*

It floureth and bringeth forth his seed in Iuly and August.

¶ *The Names.*

The French men call it *Barbotine* ; the Italians, *Semen Zena* · whereupon also the Latine name *Sementina* came : the seed is called euery where *Semen sanctum* · Holy-seed ; and *Semen contra Lumbricos* : in English, VVormseed; the herb it selfe is also called VVormseed, or wormseed-wort: some name it *Semen Zedoariæ*, Zedoarie seede, because it hath a smell somewhat resembling that of Zedoarie.

¶ *The Temperature.*

The seed is very bitter, and for that cause of nature hot and drie.

¶ *The Vertues.*

It is good against wormes of the belly and entrailes, taken any way, and better also if a little Rubarbe bee mixed withall, for so the wormes are not onely killed, but likewise they are driuen downe by the siege, which thing must alwaies be regarded. A

The seed mixed with a little *Aloe succotrina*, and brought to the forme of a plaister, and applied to the nauell of a childe doth the like. B

CHAP. 452. *Of forreine and Bastard Wormewoods.*

1 *Absinthium album.*
VVhite wormwood.

2 *Absinthium Ægyptium.*
VVormwood of Ægypt.

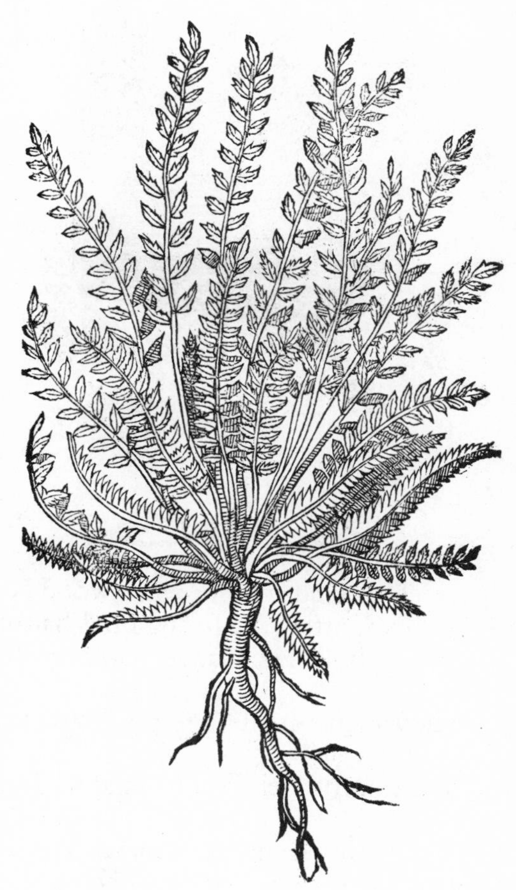

 ¶ The

¶ The Deſcription.

† 1 A*Bſinthium album* hath ſtraight and vpright ſtalkes, a foot high, beſet with broad leaues, but very deeply cut or clouen, in ſhew like vnto thoſe of the great Daiſy, but white of colour: at the top of the ſtalkes, out of ſcaly heads, as in an vmbell grow floures, compact of ſix ſmall white leaues: the root is long, with ſome fibres annexed vnto it.

2 This kinde of Wormwood *Geſner* and that learned Apothecarie *Valerandus Donraz*, called *Abſinthium Egyptium*: the leaues of this plant are very like to the leaues of *Trichomanes*, which is our common Maiden haire, of a white colour, euery ſmall leafe ſtanding one oppoſite againſt another, and of a ſtrong ſauour.

3 This VVormwood, which *Dodonæus* calleth *Abſinthium inodorum*, and *Inſipidum*, is very like vnto the ſea VVormwood, in his ſmall and tender leaues: the ſtalke beareth flowers alſo like vnto the foreſaid Sea-Wormwood, but it is of a ſad or deep colour, hauing neither bitter taſte, nor any ſauour at all; whereupon it was called, and that very fitly, *Abſinthium inodorum*, or *Abſinthium inſipidum*: in Engliſh, fooliſh, or vnſauory wormwood. ‡ *Dodonæus* ſaith not that his *Abſinthium inſipidum* is like the ſea wormwood, but that it is very like our common broad leaued VVormwood, and ſo indeed it is, and that ſo like, that it is hard to be diſcerned therefrom, but onely by the want of bitterneſſe and ſmell. ‡

3 *Abſinthium inodorum.*
Vnſauorie VVormwood.

4 *Abſinthium marinum, Abrotani fœminæ facie.*
Small Lauander Cotton.

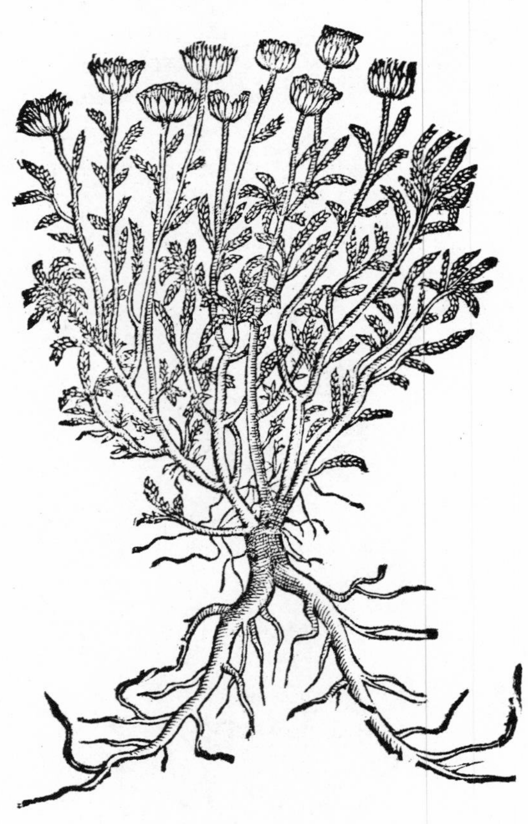

4 This kinde of Sea-wormwood is a ſhrubby and wooddie plant, in face and ſhew like to Lauander Cotton, of a ſtrong ſmel; hauing floures like thoſe of the common wormwood, at the firſt ſhew like thoſe of Lauander Cotton: the root is tough and wooddie.

¶ The Place.

Theſe plants are ſtrangers in England, yet we haue a few of them in Herbariſts gardens.

¶ The Time.

The time of their flouring and ſeeding is referred to the other wormwoods.

¶ The Names.

The white wormwood *Conradus Geſnerus* nameth *Seriphium fœmina*, and ſaith, that it is commonly called *Herba alba*, or white herbe: another had rather name it *Santonicum*; for as *Dioſcorides* ſaith, *Santonicum* is found in France beyond the Alpes, and beareth his name of the ſame country countrey where it groweth . but that part of Swisserland which belongeth to France is accounted of the Romans to be beyond the Alps ; and the prouince of Santon is far from it : for this is a part of Guines, scituate vpon the coast of the Ocean, beneath the floud Gerond Northward : therefore Santon Wormwood, if it haue his name from the Santons, groweth farre from the Alps : but if it grow neere adioyning to the Alps, then hath it not his name from the Santons.

¶ *The Temperature and Vertues.*

White Wormwood is hot and somewhat dry.

A Vnsauorie Wormwood, as it is without smell and taste, so is it scarse of any hot qualitie, much lesse hath it any scouring facultie. These are not vsed in physicke, where the others may be had, being as it were wilde or degenerate kindes of Wormwood ; some of them participating both of the forme and smell of other plants.

† The figure which was here formerly in the first place, by the name of *Absinthium arborescens*, is the first of the next chapter saue one, where you may see more thereof. The white Wormwood mentioned here in the Names, but no where else in the Chapter, is either the same with, or one very like our Sea Wormewood. Let such as are curious looke into *Camerarius* his *Hort. Med.* in the title of *Absinthium Santonicum* : and in *Dodonæus*, *Pempt.* 1. *lib.* 2. *cap.* 5. where the first description is of this Wormewood.

Chap. 453. *Of Mugwort.*

1 *Artemisia, mater Herbarum.*
Common Mugwort.

¶ *The Description.*

1 THe first kinde of Mug wort hath broad leaues, very much cut or clouen like the leaues of common Wormewood, but larger, of a darke greene colour aboue, and hoarie vnderneath : the stalkes are long and straight, and full of branches, whereon do grow small round buttons, which are the floures, smelling like Marierome when they wax ripe : the root is great, and of a wooddie substance.

2 The second kinde of Mugwort hath a great thicke and wooddy root, from whence arise sundry branches of a reddish colour, beset full of small and fine iagged leaues, verie like vnto sea Sothernwood : the seed groweth alongst the small twiggy branches, like vnto little berries, which fall not from their branches in a long time after they be ripe. ‡ I know not how this differeth from the former, but only in the colour of the stalk and floures, which are red or purplish ; whereas the former is more whitish. ‡

3 There is also another Mugwort, which hath many branches rising from a wooddie root, standing vpright in distances one from another, of an ashie colour, beset with leaues not much vnlike sea Purslane ; about the lower part of the stalkes, and toward the top of the branches they are narrower and lesser, and cut with great and deepe iagges, thicke in substance, and of a whitish colour, as all the rest of the plant is : it yeeldeth a pleasant smell like *Abrotanum marinum*, and in taste is somewhat saltish. the floures are many, and yellow : which being vaded, there followeth mossie seed like vnto that of the common Wormwood. ‡ The leaues of this plant are of two sorts ; for some of them are long and narrow, like those of Lauander (whence *Clusius* hath called it *Artemisia folio Lauendulæ*) other some are cut in or diuided almost to the middle rib ; as you may see it exprest apart in a figure by it selfe, which shewes both the whole, as also the diuided leaues.

¶ *The*

3 *Artemisia marina.*
Sea Mugwort.

‡ *Artemisiæ marinæ ramulus, folia integra & dissecta exprimens.*
A branch shewing the cut and vncut leaues.

¶ *The Place.*

The common Mugwort groweth wilde in sundry places about the borders of fields, about high waies, brooke sides, and such like places.

Sea Mugwort groweth about Rie and Winchelsea castle, and at Portsmouth by the Isle of Wight.

¶ *The Time.*

They floure in Iuly and August.

¶ *The Names.*

Mugwort is called in Greeke Ἀρτεμισία: and also in Latine *Artemisia*, which name it had of *Artemisia* Queene of Halicarnassus, and wife of noble *Mausolus* King of Caria, who adopted it for her owne herbe: before that it was called παρθένις, *Parthenis*, as *Pliny* writeth. *Apuleius* affirmeth that it was likewise called *Parthenion*; who hath very many names for it, and many of them are placed in *Dioscorides* among the bastard names: most of these agree with the right *Artemisia*, and diuers of them with other herbes, which now and then are numbred among the Mugworts: it is also called *Mater Herbarum*: in high-Dutch, **Beifusz**, and **Sant Johanus Gurtell**: in Spanish and Italian, *Artemisia*: in French, *Armoise*: in low-Dutch, **Bijuoet, Sint Jans kruyt**: in English, Mugwort, and common Mugwort.

¶ *The Temperature.*

Mugwort is hot and dry in the second degree, and somewhat astringent.

¶ *The Vertues.*

A *Pliny* saith that Mugwort doth properly cure womens diseases.

B *Dioscorides* writeth, that it bringeth downe the termes, the birth, and the after-birth.

C And that in like manner it helpeth the mother, and the paine of the matrix, to be boyled as bathes for women to sit in; and that being put vp with myrrh, it is of like force that the bath is of. And that the tender tops are boiled and drunk for the same infirmities; and that they are applied in manner of a pultesse to the share, to bring downe the monethly course.

D *Pliny* saith, that the traueller or wayfaring man that hath the herbe tied about him feeleth no wearisomnesse at all; and that he who hath it about him can be hurt by no poysonsome medicines, nor by any wilde beast, neither yet by the Sun it selfe; and also that it is drunke against *Opium*, or the

the iuyce of blacke Poppy. Many other fantaſticall deuices inuented by Poets are to be ſeene in the Works of the antient Writers, tending to witchcraft and ſorcerie, and the great diſhonour of God; wherefore I do of purpoſe omit them, as things vnworthie of my recording, or your reviewing.

Mugwort pound with oyle of ſweet almonds, and laid to the ſtomacke as a plaiſter, cureth all the paines and griefes of the ſame. B

It cureth the ſhakings of the ioynts, inclining to the palſie, and helpeth the contraction or drawing together of the nerues and ſinewes. C

† There were formerly two deſcriptions of the *Artemiſia marina*; wherefore I omitted the former, being the more vnperfect.

CHAP. 454. *Of Sothernwood.*

¶ *The Kindes.*

D*Ioſcorides* affirmeth that Sothernwood is of two kindes, the female and the male, which are euery where knowne by the names of the greater and of the leſſer: beſides theſe there is a third kinde, which is of a ſweeter ſmell, and leſſer than the others, and alſo others of a baſtard kinde.

† 1 *Abrotanum fœmina arboreſcens.*
Female Sothernwood.

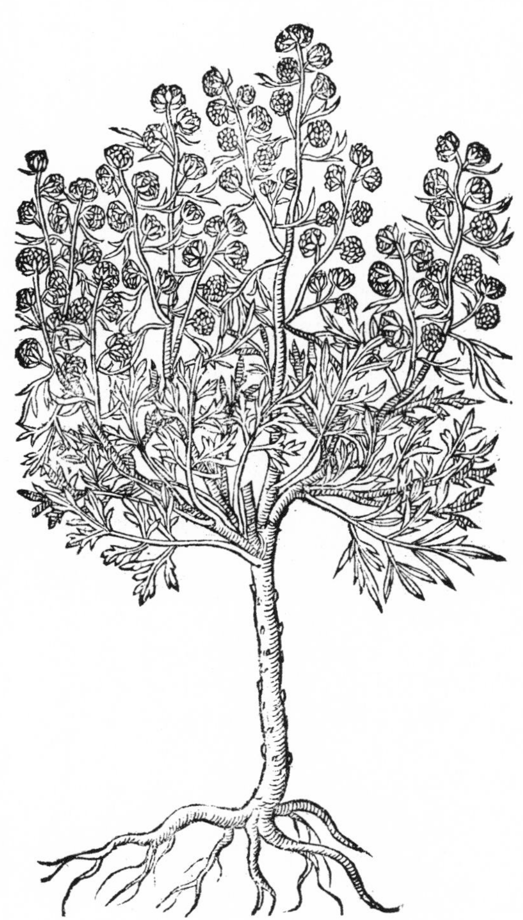

2 *Abrotanum mas.*
Male Sothernwood.

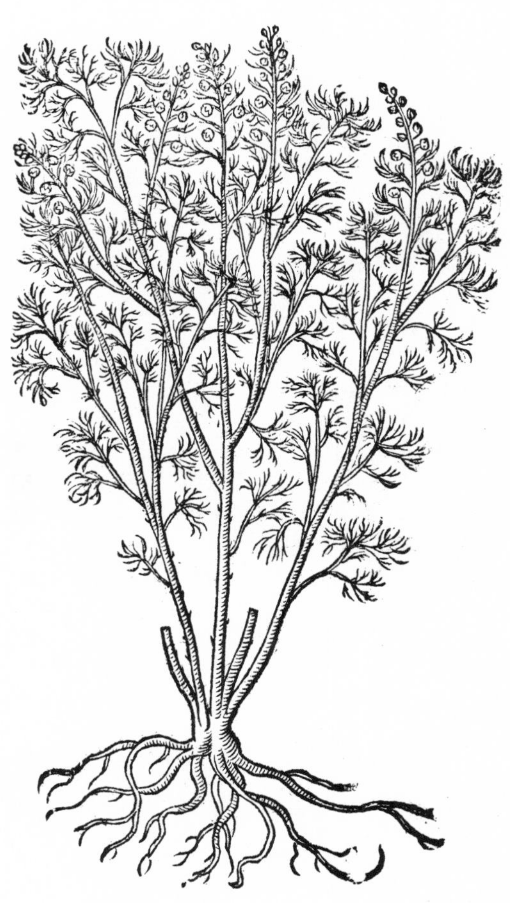

¶ *The Deſcription.*

1 THe greater Sothernwood by carefull manuring doth oftentimes grow vp in manner of a ſhrub, and commeth to be as high as a man, bringing forth ſtalkes an inch thicke, or more; out of which ſpring very many ſprigs or branches, ſet about with leaues diuerſly iagged and finely indented, ſomewhat white, and of a certaine ſtrong ſmell: in ſtead of floures, little ſmal cluſters of buttons do hang on the ſprigs, from the middle to the very top, of colour yellow, and at the length turne into ſeed. The root hath diuers ſtrings.

2 The

3 *Abrotanum humile.*
Dwarfe Sothernwood.

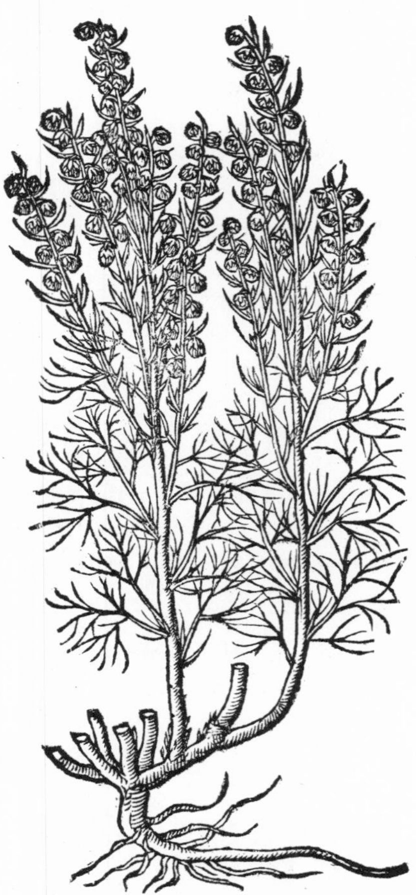

4 *Abrotanum inodorum.*
Vnſauorie Sothernwood.

5 *Abrotanum campeſtre.*
Wilde Sothernwood.

2 The leſſer Sothernwood groweth low, ful of little ſprigs of a woody ſubſtance: the leaues are long, and ſmaller than thoſe of the former, not ſo white: it beareth cluſtering buttons vpon the tops of the ſtalks: the root is made of many ſtrings.

3 The third kinde is alſo ſhorter: the leaues hereof are iagged and deeply cut after the maner of the greater Sothernwood, but they are not ſo white, yet more ſweet, wherein they are like vnto Lauander cotton. This kinde is very full of ſeed: the buttons ſtand alongſt on the ſprigs, euen to the very top, and be of a glittering yellow. The root is like to the reſt.

4 The vnſauorie Sothernwood groweth flat vpon the ground, with broad leaues deepely cut or iagged in the edges like thoſe of the common Mugwort: among which riſe vp weake and feeble ſtalkes trailing likewiſe vpon the ground, ſet confuſedly here and there with the like leaues that grow next the ground, of a grayiſh or hoary colour, altogether without ſmell. The floures grow alongſt the ſtalkes, of a yellowiſh colour, ſmall and chaffie: the root is tough and wooddy, with ſome ſtrings anexed thereto.

5 This wilde Sothernwood hath a great long thicke root, tough and wooddy, couered

ouer

ouer with a scaly barke like the scaly backe of an adder, and of the same colour: from which rise very many leaues like those of Fennell, of an ouerworne greene colour: among which grow small twiggy branches on the tops, and alongst the stalkes do grow small clustering floures of a yellow colour: the whole plant is of a darke colour, as well leaues as stalkes, and of a strong vnsauourie smell.

¶ *The Place.*

Theophrastus saith that Sothernwood delighteth to grow in places open to the Sun: *Dioscorides* affirmeth that it groweth in Cappadocia, and Galatia a countrey in Asia, and in Hierapolis a city in Syria: it is planted in gardens almost euery where: that of Sicilia and Galatia is most commended of *Pliny*.

¶ *The Time.*

The buttons of Sothernwood do flourish and be in their prime in August, and now and then in September.

¶ *The Names.*

It is called in Greeke Ἀβρότονον: the Latines and Apothecaries keepe the same name *Abrotanum*: the Italians and diuers Spaniards call it *Abrotano*: and other Spaniards, *Yerua lombriguera*: in high Dutch, **Stabwurtz**: in low-Dutch, **Aueroone**, and **Auercruijt**: the French, *Aurone*, and *Auroesme*: the English men, Sothernwood: it hath diuers bastard names in *Dioscorides*; the greater kinde is *Dioscorides* his *Fœmina*, or female Sothernwood; and *Pliny* his *Montanum*, or mountaine Sotherne-wood: the mountaine Sothernwood we take for the female, and the champion for the male. There be notwithstanding some that take Lauander Cotton to be the female Sothernwood; grounding thereupon, because it bringeth forth yellow floures in the top of the sprigs like cluster buttons: but if they had more diligently pondered *Dioscorides* his words, they would not haue been of this opinion: the lesser Sothernwood is *Mas*, the male, and is also *Plinies* champion Sothernwood; in Latine, *Campestre*. The third, as we haue said, is likewise the female, and is commonly called sweet Sothernwood, because it is of a sweeter sent than the rest. *Dioscorides* seemeth to call this kind *Siculum*, Sicilian Sothernwood.

¶ *The Temperature.*

Sothernwood is hot and dry in the end of the third degree: it hath also force to distribute and to rarifie.

¶ *The Vertues.*

The tops, floures, or seed boyled, and stamped raw with water and drunke, helpeth them that cannot take their breaths without holding their neckes straight vp and is a remedie for the cramp, and for sinewes shrunke and drawne together; for the sciatica also, and for them that can hardly make water; and it is good to bring downe the termes. A

It killeth wormes, and driueth them out: if it be drunke with wine it is a remedie against deadly poysons. B

Also it helpeth against the stinging of scorpions and field spiders, but it hurts the stomacke. C

Stamped and mixed with oyle it taketh away the shiuering cold that commeth by the ague fits, and it heateth the body if it be anointed therewith before the fits do come. D

If it be pouned with barley meale and laid to pushes it taketh them away. E

It is good for inflammations of the eyes, with the pulpe of a rosted Quince, or with crummes of bread, and applied pultis wise. F

The ashes of burnt Sothernwood, with some kinde of oyle that is of thin parts, as of *Palma Christi*, Radish oyle, oyle of sweet Marierome, or Organie, cureth the pilling of the haire off the head, and maketh the beard to grow quickly: being strewed about the bed, or a fume made of it vpon hot embers, it driueth away serpents: if but a branch be layd vnder the beds head they say it prouoketh venerie. G

The seed of Sothernwood made into pouder, or boyled in wine and drunke, is good against the difficultie and stopping of vrine; it expelleth, wasteth, consumeth, and digesteth all cold humors, tough slime and flegme, which do vsually stop the spleene, kidneyes, and bladder. H

Sothernwood drunke in wine is good against all venome and poyson. I

The leaues of Sothernwood boyled in water vntill they be soft, and stamped with barley meale and barrowes greafe vnto the forme of a plaister, dissolue and waste all cold tumors and swellings, being applied or laid thereto. K

† The description here in the first place is that of the *Abrotanum fœmina arborescens* of *Dodonæus*, being the very first in his *Pemptades*. The figure which our Author put thereto was of the Lauander Cotton, which should haue beene in the next chapter saue one: Now the figure that hee should haue put here was put two chapters before, by the name of *Absinthium arborescens*, by which name *Lobel* also calls it: but I haue thought it fitter to put it here, because here was the better description, and the plant is the better referred to this kinde.

CHAP.

Chap. 455.

Of Oke of Jeruſalem, and Oke of Cappadocia.

1 *Botrys.*
Oke of Ieruſalem.

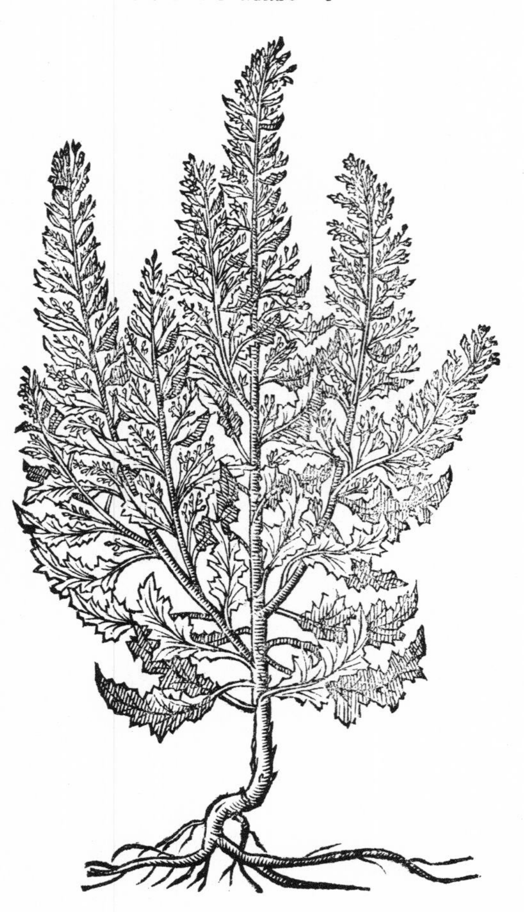

2 *Ambroſia.*
Oke of Cappadocia.

¶ *The Deſcription.*

1 Oke of Ieruſalem, or *Botrys*, hath ſundry ſmall ſtems a foot and a halfe high, diuiding themſelues into many ſmall branches, beſet with ſmall leaues deeply cut or iagged, very much reſembling the leafe of an Oke, which hath cauſed our Engliſh women to call it Oke of Ieruſalem; the vpper ſide of the leafe is of a deepe greene, and ſomewhat rough and hairy, but vnderneath it is of a darke reddiſh or purple colour: the ſeedie floures grow cluſtering about the branches, like the yong cluſters or blowings of the Vine: the root is ſmall and threddy: the whole herbe is of a pleaſant ſmell and ſauour, and of a feint yellowiſh colour, and the whole plant dieth when the ſeed is ripe.

2 The fragrant ſmell that this kind of *Ambroſia* or Oke of Cappadocia yeeldeth, hath moued the Poets to ſuppoſe that this herbe was meate and food for the gods: *Dioſcorides* ſaith it groweth three handfuls high: in my garden it groweth to the height of two cubits, yeelding many weake crooked and ſtreaked branches, diuiding themſelues into ſundrie other ſmall branches, hauing from the middeſt to the top thereof many moſſie yellowiſh floures not much vnlike common Wormwood, ſtandiug one before another in good order; and the whole plant is as it were couered ouer with bran or a mealy duſt: the floures do change into ſmall prickly cornered buttons, much like vnto *Tribulus terreſtris*; wherein is contained blacke round ſeed, not vnpleaſant in taſte and ſmell: the leaues are in ſhape like the leaues of Mugwort, but thinner and more tender: all the whole plant is hoary, and yeeldeth a pleaſant ſauor: the whole plant periſhed with me at the firſt approch of Winter.

¶ *The Place.*

Theſe plants are brought vnto vs from beyond the ſeas, eſpecially from Spaine and Italy.

¶ *The Time.*

They floure in Auguſt, and the ſeed is ripe in September.

¶ *The*

¶ *The Names.*

Oke of Ierusalem is called in Greeke Βότρυς : in Latine *Botrys* : In Italian, *Botri* : in Spanish, *Bien granada* : in high-Dutch, **Traukenkraut**, and **krottenkraut** : in French and low-Dutch, *Pyment* : in English, Oke of Ierusalem ; and of some, Oke of Paradise.

Oke of Cappadocia is called in Greeke Ἀμβροσία : in Latine, *Ambrosia* ; neither hath it any other knowne name. *Pliny* saith that *Ambrosia* is a wandering name, and is giuen vnto other herbes : for *Botrys* (Oke of Ierusalem, as we haue written) is of diuers also called *Ambrosia* : In English it is called Oke of Cappadocia.

¶ *The Temperature.*

These plants are hot and dry in the second degree, and consist of subtill parts.

¶ *The Vertues.*

These plants be good to be boyled in wine, and ministred vnto such as haue their brests stopt, and are short winded, and cannot easily draw their breath ; for they cut and waste grosse humours and tough flegme. The leaues are of the same force ; being made vp with sugar they commonly call it a conserue. A

It giueth a pleasant taste to flesh that is sodden with it, and eaten with the broth. B

It is dried and layd among garments, not onely to make them smell sweet, but also to preserue them from moths and other vermine ; which thing it doth also performe. C

There were formerly two more descriptions in this chapter, both which were made by looking vpon the figures in *Lobels Icons* ; the former being of his *Ambrosia spontanea strigosior*, which is nothing else but the *Coronopus Ruellij*, or Swines Cresses. The later was of his *Ambrosia tenuifolia*, which our Author in the last chapter set forth by the name of *Abrotanum campestre*.

CHAP. 456. *Of Lauander Cotton.*

† *Chamæcyparissus.*
Lauander Cotton.

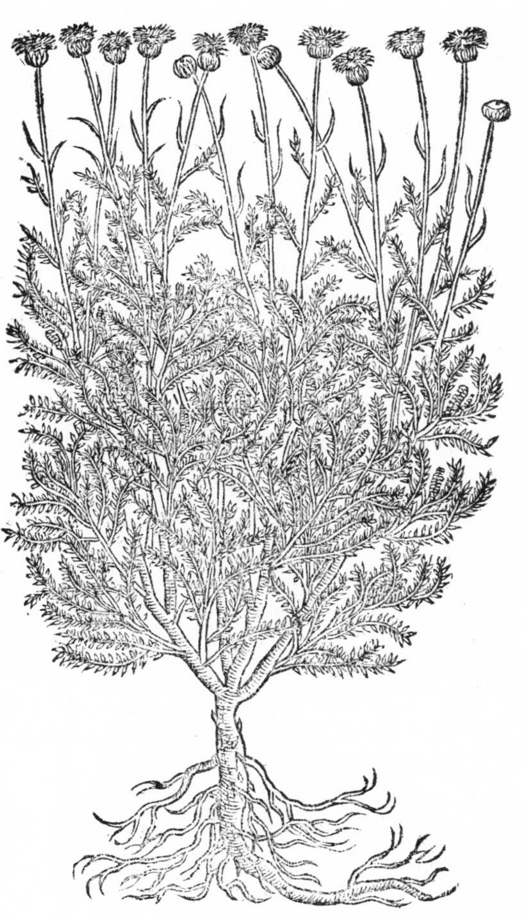

¶ *The Description.*

Lauander Cotton bringeth forth clustred buttons of a golden colour, and of a sweet smell, and is often vsed in garlands, and decking vp of gardens and houses. It hath a wooddy stocke, out of which grow forth branches like little boughes, slender, very many, a cubit long, set about with little leaues, long, narrow, purled, or crumpled ; on the tops of the branches stand vp floures, one alone on euery branch, made vp with short threds thrust close together, like to the floures of Tansie, and to the middle buttons of the floures of Cammomill, but yet something broader, of colour yellow, which be changed into seed of an obscure colour. The root is of a wooddy substance. The shrub it selfe is white both in branches and leaues, and hath a strong sweet smell.

‡ There are some varieties of this plant, which *Matthiolus*, *Lobel*, and others refer to *Abrotanum fœmina*, and so call it; and by the same name our Authour gaue the figure thereof in the last chapter saue one, though the description did not belong thereto, as I haue formerly noted. Another sort thereof our Authour, following *Tabernamontanus* and *Lobel*, set forth a little before by the name of *Absinthium marinum Abrotani fœminæ facie*, that *Dodonæus* calls *Santolina prima* ; and this here figured, *Santolina altera*. He also mentioneth three other differences thereof, which chiefely consist in the leaues ; for his third hath very short and small leaues like those of Heath ; whence *Bauhine* calls it *Abrotanum fœmina folijs Ericæ*. The fourth hath the leaues lesse toothed, and more like to Cypresse,

hence it is called in the *Aduerſ. Abrotanum peregrinum cupreſſi folijs*. The fifth hath not the ſtalkes growing vpright, but creeping: the leaues are toothed, more thicke and hoary than the reſt; in other reſpects alike. *Bauhine* calls it *Abrotanum fœmina repens caneſcens*. ‡

¶ *The Place.*

Lauander Cotton groweth in gardens almoſt euery where.

¶ *The Time.*

They floure in Iuly and Auguſt.

¶ *The Names.*

They are called by one name *Santolina*, or Lauander Cotton: of moſt, *Chamæcypariſſus*. But *Pliny* concerning *Chamæcypariſſus* is ſo ſhort and briefe, that by him their opinions can neither be reiected nor receiued.

They are doubtleſſe much deceiued that would haue Lauander Cotton to be *Abrotanum fœmina*, or the female Sothernwood: and likewiſe they are in the wrong who take it to be *Seriphium*, ſea Wormewood; and they who firſt ſet it abroch to be a kinde of Sothernwood we leaue to their errors; becauſe it is not abſolutely to be referred to one, but a plant participating of Wormewood and Sothernwood.

¶ *The Temperature.*

The ſeed of Lauander Cotton hath a bitter taſte, being hot and dry in the third degree.

¶ *The Vertues.*

A *Pliny* ſaith, That the herbe *Chamæcypariſſus* being drunke in wine is a good medicine againſt the poyſons of all ſerpents and venomous beaſts.

B It killeth wormes either giuen greene or dry, and the ſeed hath the ſame vertue againſt wormes, but auoideth them with greater force. It is thought to be equall with the vſuall worme-ſeed.

† The figure which formerly was in this place was of a kinde of Moſſe, which *Tragus* ſet forth by the name of *Sauina ſylueſtris*: *Turner* and *Tabernamontanus* called it *Chamæcypariſſus*. See more thereof in the Moſſes.

Chap. 457. *Of Sperage, or Aſparagus.*

1 *Aſparagus ſatiuus.*
Garden Sperage.

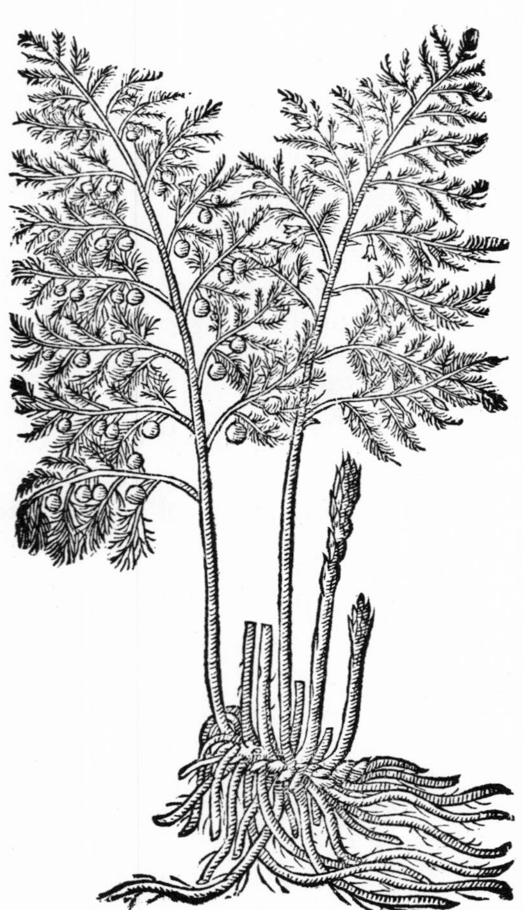

2 *Aſparagus petræus.*
Stone or mountaine Sperage.

¶ *The Deſcription.*

1 THe firſt being the manured or garden Sperage, hath at his firſt riſing out of the ground thicke tender ſhoots very ſoft and brittle, of the thickeneſſe of the greateſt ſwans quil, in taſte like vnto the greene beane, hauing at the top a certaine ſcaly ſoft bud, which in time groweth to a branch of the height of two cubits, diuided into diuers other ſmaller branches, whereon are ſet many little leaues like haires, more fine than the leaues of Dill: among which come forth ſmall moſſie yellowiſh floures, which yeeld forth the fruit, greene at the firſt, afterward red as Corall, of the bigneſſe of a ſmall peaſe; wherein is contained groſſe blackiſh ſeed exceeding hard, which is the cauſe that it lieth ſo long in the ground after the ſowing, before it do ſpring vp. The roots are many thicke ſoft and ſpongie ſtrings hanging downe from one head, and ſpred themſelues all about, whereby it greatly increaſeth.

2 We haue in our mariſh and low grounds neere vnto the ſea, a Sperage of this kinde, which differeth a little from that of the garden, and yet in kinde there is no difference at all, but only in manuring, by which all things or moſt things are made more beautifull, and larger. This may be called *Aſparagus paluſtris*, mariſh Sperage.

4 *Aſparagus ſylueſtris aculeatus.*
Wilde prickly Sperage.

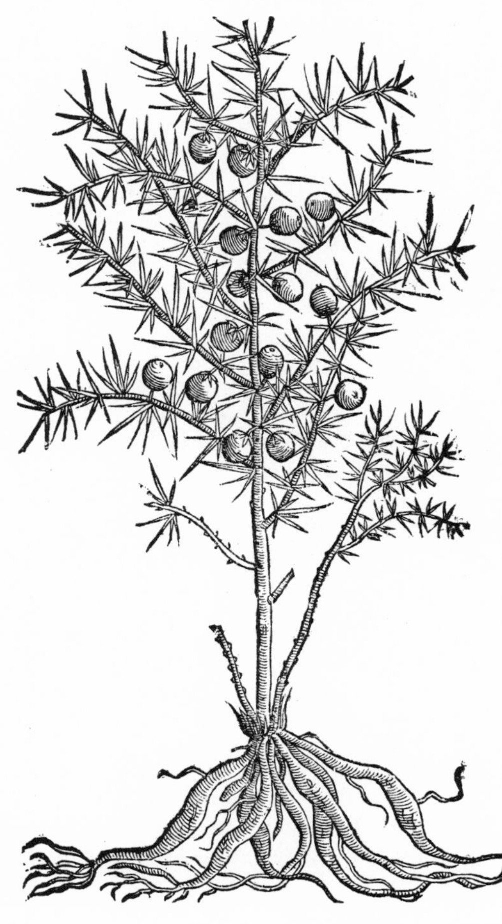

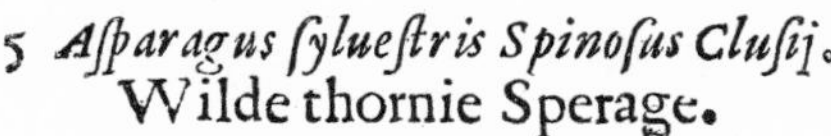

5 *Aſparagus ſylueſtris Spinoſus Cluſij.*
Wilde thornie Sperage.

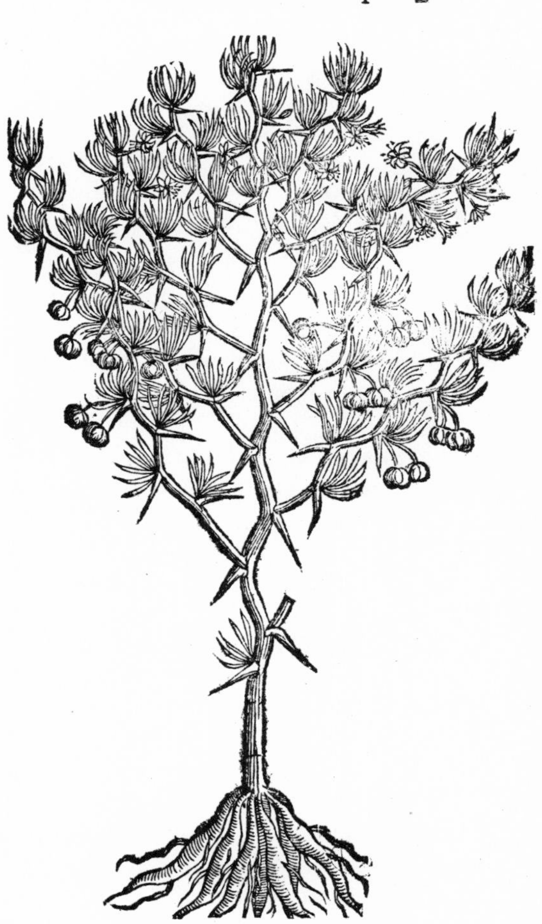

3 Stone or mountaine Sperage is one of the wilde ones, ſet forth vnder the title of *Corruda*; which *Lobel* calleth *Aſparagus petræus*; and *Galen, Myacanthinus*, that doth very well reſemble thoſe of the garden, in ſtalkes, roots, and branches, ſauing that thoſe fine hairy leaues which are in the garden Sperage be ſoft, blunt, and tender; and in this wilde Sperage, ſharpe hard and pricking thornes, though they be ſmall and ſlender: the root hereof is round, of the bigneſſe of a peaſe, and of a blacke colour: the roots are long, thicke, fat, and very many.

4 This fourth kinde differeth from the laſt deſcribed, being a wilde Sperage of Spaine and Hungarie: the plant is altogether ſet with ſharpe thornes (three or foure comming forth together) as are the branches of Whinnes, Goſe, or Furſen: the fruit is blacke when it is ripe, and full of a greeniſh pulpe, wherein lie hard and blacke ſeeds, ſometimes one, otherwhiles two in a berry: the roots are like the others, but greater and tougher.

 5 *Carolus*

‡ 6 *Drypis.*
Sperage Thistle.

5 *Carolus Clusius* describeth also a certain wilde Sperage with sharp prickles all alongst the stalkes, orderly placed at euery ioynt one, hard, stiffe, and whitish, the points of the thornes pointing downward: from the which ioynts also doe grow out a few long greene leaues fastned together, as also a little yellow floure, and one berry three cornered, and of a blacke colour, wherein is contained one black seed, seldome more: the roots are like the other.

6 *Drypis* being likewise a kinde hereof, hath long and small roots, creeping in the ground like Couch grasse; from which spring vp branches a cubit high, ful of knotty ioints: the leaues are small like vnto Iuniper, not much differing from *Corruda* or *Nepa*: the floures grow at the top of the stalke in spokie tufts or rundles, of a white colour, closely thrust together: the seed before it bee taken out of the huske is like vnto Rice; being taken out, like that of Melilot, of a saffron colour.

¶ *The Place.*

The first being our garden Asparagus groweth wild in Essex, in a medow adioining to a mill, beyond a village called Thorp; and also at Singleton not far from Carbie, and in the medows neere Moulton in Lincolnshire. Likewise it groweth in great plentie neere vnto Harwich, at a place called Bandamar lading, and at North Moulton in Holland, a part of Lincolnshire.

The wilde Sperages grow in Portugal and Biscay among stones, one of the which *Petrus Bellonius* doth make mention to grow in Candie, in his first booke of Singularities, *cap.* 18.

¶ *The Time.*

The bare naked tender shoots of Sperage spring vp in Aprill, at what time they are eaten in sallads; they floure in Iune and Iuly; the fruit is ripe in September.

¶ *The Names.*

The garden Sperage is called in Greeke Ἀσπάραγος: in Latine likewise *Asparagus*: in shops, *Sparagus*, and *Speragus*: in high-Dutch, **Spargen**: in low-Dutch, **Asparges**, and **Coralcruijt**; that is to say, *Herba Coralli*, or Corall-wort, of the red berries, which beare the colour of Corall: in Spanish, *Asparragos*: in Italian, *Asparago*: in English, Sperage, and likewise Asparagus, after the Latine name: in French, *Asperges*. It is named *Asparagus* of the excellencie, because *asparagi*, or the springs hereof are preferred before those of other plants whatsoeuer; for this Latine word *Asparagus* doth properly signifie the first spring or sprout of euerie plant, especially when it is tender, and before it do grow into a hard stalke, as are the buds, tendrels, or yong springs of wild Vine or hops, nnd such like.

Wilde Sperage is properly called in Greeke Μυάκανθα, which is as much to say as Mouse prickle, and Ἀσπάραγος πετραῖος, that is to say, *Petræus Asparagus*, or Stone Sperage: it is also named in Latine, *Asparagus syluestris*, and *Corruda*.

¶ *The Temperature.*

The roots of the garden Sperage, and also of the wilde, doe clense without manifest heate and drinesse.

¶ *The Vertues.*

A The first sprouts or naked tender shoots hereof be oftentimes sodden in flesh broth and eaten, or boyled in faire water, and seasoned with oyle, vineger, salt, and pepper, then are serued at mens tables for a sallad; they are pleasant to the taste, easily concocted, and gently loose the belly.

B They somewhat prouoke vrine, are good for the kidnies and bladder, but they yeeld vnto the body little nourishment, and the same moist, yet not faultie: they are thought to increase seed, and stir vp lust.

† The *Nepa* formerly mentioned in this chapter, but now omitted, was againe set forth by our Author amongst the Furses, where you may finde it.

Chap.

CHAP. 458.

Of Horſe-taile, or Shaue-graſſe.

¶ *The Deſcription.*

1 GReat Horſe-taile riſeth vp with a round ſtalke, hollow within like a Reed, a cubit high, compact as it were of many ſmall pieces, one put into the end of another, ſometimes of a reddiſh colour, very rough, and ſet at euery ioint with many ſtiffe ruſh-like leaues or rough briſtles, which maketh the whole plant to reſemble the taile of a horſe, whereof it tooke his name: on the top of the ſtalke do ſtand in ſtead of floures cluſtered and thicke Catkins, not vnlike to the firſt ſhoots of Sperage, which is called *Myacantha*: the root is ioynted, and creepeth in the ground.

2 This ſmall or naked Shaue-graſſe, wherewith Fletchers and Combe-makers do rub and poliſh their worke, riſeth out of the ground like the firſt ſhoots of Aſparagus, iointed or kneed by certaine diſtances like the precedent, but altogether without ſuch briſtly leaues, yet exceeding rough and cutting: the root groweth aſlope in the earth, like thoſe of the Couch-graſſe.

1 *Equiſetum maius.*
Great Horſe-taile.

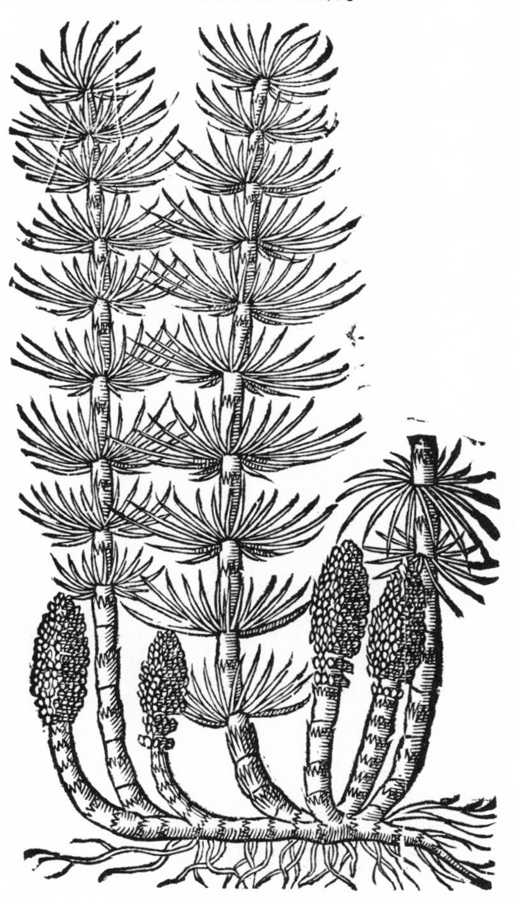

2 *Equiſetum nudum.*
Naked Horſe-taile.

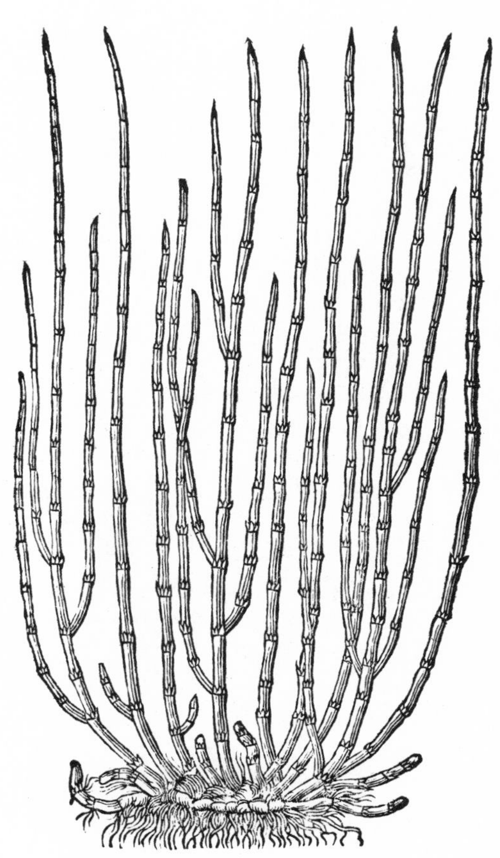

3 Horſe-taile which for the moſt part groweth among corne, and where corne hath been, hath a very ſlender root, and ſingle; from which riſe vp diuers iointed ſtalkes, whereon doe grow verie long rough narrow iointed leaues, like vnto the firſt deſcribed, but thicker and rougher, as is the reſt of the plant.

4 Water Horſe-taile, that growes by the brinks of riuers and running ſtreams, and often in the midſt of the water, hath a very long root, according to the depth of the water, groſſe, thicke, and iointed, with ſome threds anexed thereto: from which riſeth vp a great thick iointed ſtalke, whereon do grow long rough ruſhy leaues, pyramide or ſteeple faſhion. The whole plant is alſo tough, hard, and fit to ſhaue and rub wooden things as the other.

5 This kinde of Horſe-taile that growes in woods and ſhadowie places, hath a ſmall root, and ſingle, from which riſeth vp a rough chamfered ſtalke ioynted by certaine ſpaces, hauing at each ioynt two buſhes of rough briſtly leaues ſet one againſt another like the other of his kinde.

3 *Equiſetum ſegetale.*
Corne Horſe-taile.

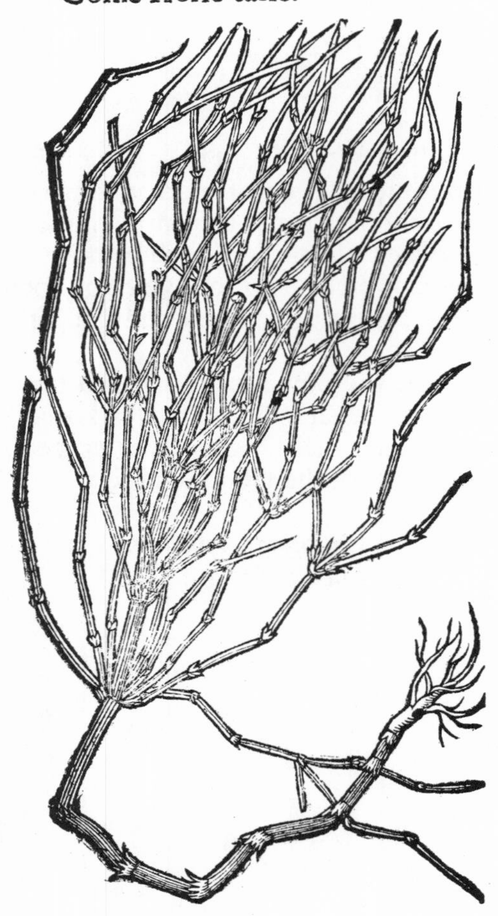

4 *Equiſetum paluſtre.*
Water Horſe-taile.

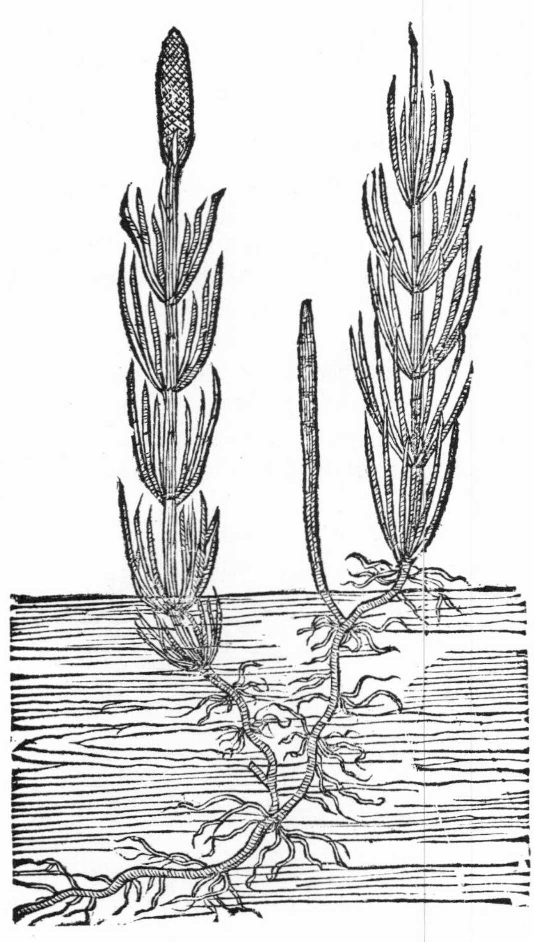

5 *Equiſetum ſyluaticum.*
Wood Horſe-taile.

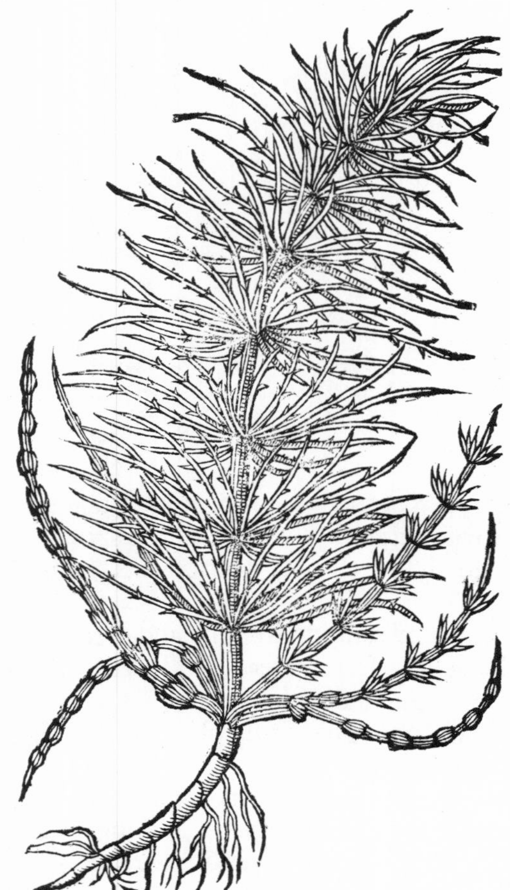

6 *Cauda equina fœmina.*
Female Horſe-taile.

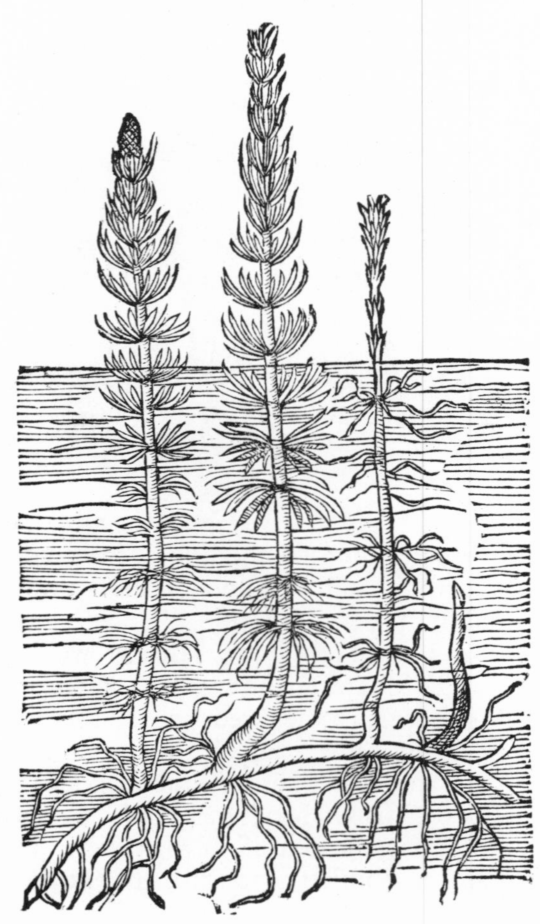

9 *Iuncaria Salmanticensis.*
Italian rushie Horse-taile.

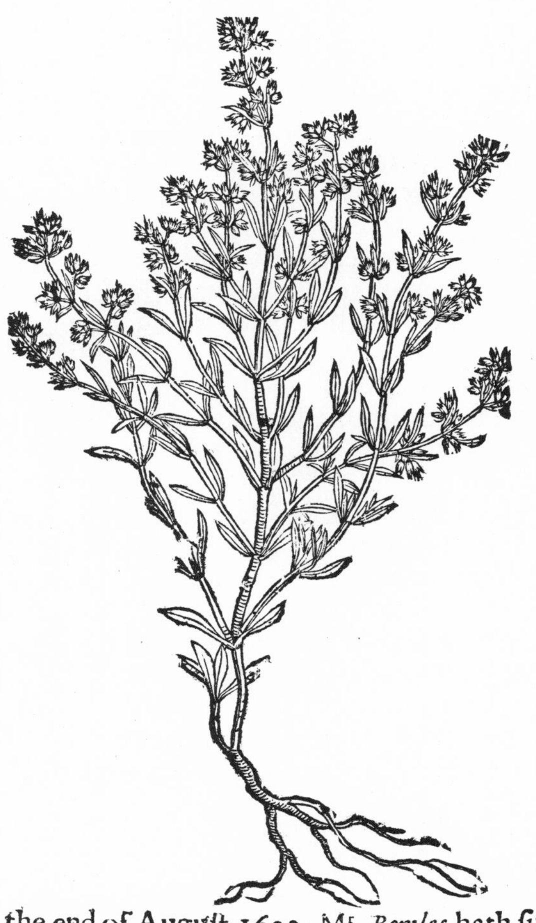

6 The female Horse taile groweth for the most part in waterish places, and by the brinks of small rills and pirling brookes; it hath a long root like that of Couch grasse, from which rise vp diuers hollow stalkes, set about at certaine distances with smal leaues in rundles like those of Woodroofe, altogether barren of seed and floure, whereof it was called by *Lobel*, *Polygonon fœmina semine Vidua*. ‡ This is sometimes found with tenne or more seedes at each ioynt; whence *Bauhine* hath called it *Equisetum palustre breuioribus folijs poly spermon*.‡

‡ 7 In some boggie places of this kingdome is found a rare and pretty *Hippuris* or Horse taile, which growes vp with many little branches, some two or three inches high, putting forth at each ioynt many little leaues, clustering close about the stalke, and set after the manner of other Horse-tailes: towards the tops of the branches the ioynts are very thicke: the colour of the whole plant is gray, a little inclining to green, very brittle, and as it were stony or grauelly like Coralline, and will crash vnder your feet, as if it were frozen; and if you chew it, you shall finde it all stonie or grauelly. My friend M[r]. *Leonard Buckner* was the first that found this plant, and brought it to me; he had it three miles beyond Oxford, a little on this side Euansham-ferry, in a bog vpon a common by the Beacon hill neere Cumner-wood, in the end of August, 1632. M[r]. *Bowles* hath since found it growing vpon a bog not far from Chisselhurst in Kent. I question whether this bee not the *Hippuris lacustris quædam folijs mansu arenosis* of *Gesner*: but if *Gesners* be that which *Bauhine* in his *Prodromus*, *pag*. 24. sets forth by the name of *Equisetum nudum minus variegatum*, then I iudge it not to be this of my description: for *Bauhines* differs from this in that it is without leaues, and ofttimes bigger: the stalks of his are hollow, these not so: this may be called *Hippuris Coralloides*, Horse-taile Coralline.

8 Towards the later end of the yeare, in diuers ditches, as in Saint Iames his Parke, in the ditches on the backe of Southwarke towards Saint Georges fields, &c. you may finde couered ouer with water a kinde of stinking Horse-taile: it growes sometimes a yard long, with many ioints and branches, and each ioint set with leaues, as in the other Horse-tailes, but they are somewhat iagged or diuided towards the tops. I take this to be the *Equisetum fœtidum sub aqua repens*, described in the fift place of *Bauhinus* his *Prodromus*: we may call it in English, Stinking water Horse-taile.‡

9 *Clusius* hath set forth a plant, that he referreth vnto the stocke of Horse-tailes, which he thus describeth: it hath many twiggie or rushie stalks, whereupon it was called *Iuncaria*: and may bee Englished, Rush-weed: the leaues grow vpon the branches like those of Flax: on the toppes of the stalks grow small chaffie floures of a whitish colour. The seed is small, and blacke of colour. The root is little and white: the whole plant is sweetish in taste.

10 *Dodonæus* setteth forth another Horse-taile, which he called climing Horse-taile, or horstaile of Olympus. There is (saith he) another plant like Horse-taile, but greater and higher. It riseth vp oftentimes with a stalke as big as a mans arme, diuided into many branches: out of which there grow long slender sprigs very full of ioints, like to the first Horse-taile. The floures stand about the ioints, of a mossie substance, small as are those of the Cornell tree; in places whereof grow vp red fruit full of sowre iuice, not vnlike to little Mulberries, in which is the seed. The root is hard and wooddie. This growes now and then to a great height, and sometimes lower. *Bellonius* writeth in his Singularities, that it hath been seene to be equall in height with the Plane tree: it commeth vp lower, neere to shorter and lesser trees or shrubs, yet doth it not fasten it selfe to the trees with any tendrels or clasping aglets; much lesse doth it winde it selfe about them, yet doth it delight to stand neere and close vnto them.

¶ The

¶ *The Place.*

The titles and deſcriptions ſhew the place of their growing: the laſt *Bellonius* reporteth to grow in diuers vallies of the mountaine Olympus, and not far from Raguſa a citie in Sclauonia.

¶ *The Time.*

They floure from Aprill to the end of Sommer.

¶ *The Names.*

Horſe-taile is called in Greeke ἵππουρις, *Hippuris*: in Latine, *Equiſetum* and *Equinalis*: of *Plinie* in his 15. booke, 28 chap. *Equiſetis*, of the likeneſſe of a horſe haire: of ſome, *Salix equina*: in ſhoppes, *Cauda equina*: in high Dutch, **Schaffthew**: in low Dutch, **Peertſteert**: in Italian, *Coda di Cauallo*: in Spaniſh, *Coda de mula*: in French, *Queue de cheual*: and *Caqueue*: in Engliſh, Horſe-taile, and Shaue-graſſe.

Shaue-graſſe is not without cauſe named *Aſprella*, of his ruggedneſſe, which is not vnknowne to women, who ſcoure their pewter and woodden things of the kitchin therewith: which the German women call **Kannenkraut**: and therefore ſome of our huſwiues do call it Pewterwoort. Of ſome the tenth is called *Ephedra, Anobaſis*, and *Caucon*.

¶ *The Temperature.*

Horſe-taile, as *Galen* ſaith, hath a binding facultie, with ſome bitterneſſe, and therefore it doth mightily dry, and that without biting.

¶ *The Vertues.*

A *Dioſcorides* ſaith, that Horſe-taile being ſtamped and laied to, doth perfectly cure wounds, yea though the ſinewes be cut in ſunder, as *Galen* addeth. It is of ſo great and ſo ſingular a vertue in healing of wounds, as that it is thought and reported for truth, to cure the wounds of the bladder, and other bowels, and helpeth ruptures or burſtings.

B The herbe drunke either with water or wine, is an excellent remedy againſt bleeding at the noſe, and other fluxes of bloud. It ſtaieth the ouermuch flowing of womens floures, the bloudy flix, and the other fluxes of the belly.

C The iuice of the herbe taken in the ſame manner can do the like, and more effectually.

D Horſe-taile with his roots boiled in wine, is very profitable for the vlcers of the kidnies & bladder, the cough and difficultie of breathing.

Chap. 459. Of Sea-Cluſter, or Sea Raiſon.

† 1 *Vua marina minor.*
Small Sea Grape.

¶ *The Deſcription.*

1 Small Sea Grape is not vnlike to horſtaile: it bringeth forth ſlender ſtalks, almoſt like ruſhes, ſet with many little ioints, ſuch as thoſe are of the Horſe-taile, and diuided into many wings and branches; the tops whereof are ſharpe pointed, ſomewhat hard and pricking: it is without leaues: the flours grow in cluſters out of the ioints, with little ſtems, they are ſmall and of a whitiſh green colour: the fruit conſiſteth of many little pearles, like to the vnripe berries of Raſpis, or Hind-berry: when it is ripe it is red with a ſaffron colour, in taſte ſweet and pleaſant: the ſeede or kernell is hard, three ſquare, ſharpe on euery ſide, in taſte binding: the root is iointed, long, and creeps aſlope: the plant it ſelfe alſo doth rather lie on the ground than ſtand vp: it groweth all full of ſmall ſtalkes and branches, caſting themſelues all abroad.

2 *Carolus Cluſius* hath ſet forth another ſort of ſea Grape, far different from the precedent; it riſeth vp to the height of a man, hauing manie branches of a wooddie ſubſtance, in form like to Spaniſh Broome, without any leaues at all: wherupon doe grow cluſters of floures vpon ſlender foot-ſtalks, of a yellowiſh moſſie or herby colour, like thoſe of the Cornell tree: after which come the fruit like vnto the mulberrie, of a reddiſh colour and ſower taſte, wherein lieth hid one or two ſeeds

feeds like those of Millet,blacke without,and white within: the root is hard, tough,and wooddie.

2 *Vua marina maior.*
Great shrubbie sea Grape.

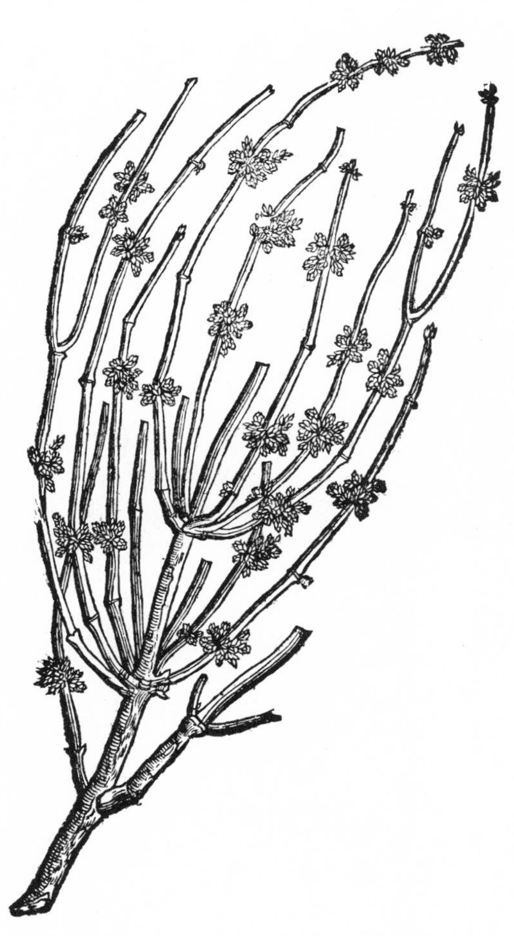

3 *Tragos Matthioli.*
Bastard Sea Grape.

3 *Tragon Matthioli*,or rather *Tragos improbus*,*Matthioli*, which he vnaduisedly called *Tragon*,is without controuersie nothing else but a kinde of *Kali*: this plant riseth vp out of the ground with stalks seldome a cubite high,diuided into sundry other grosse, thicke,and writhen branches, set,or armed with many pricking leaues, of the colour and shape of *Aizoon*,and somewhat thicke and fleshie: among which come forth such prickley burres,as are to be seen in *Tribulus terrestris*, as that it is hard for a man to touch any part thereof without pricking of the hands : the floures are of an herbie colour, bringing forth flat seed like vnto *Kali*: the root is slender,and spreadeth vnder the turfe of the earth: the whole plant is full of clammie iuice,not any thing astringent,but somewhat saltish,and of no singular vertue that is yet knowne :wherefore I may conclude,that this cannot be *Tragos Dioscoridis*,and the rather,for that this *Tragon* of *Matthiolus* is an herbe, and not a shrub, as I haue before spoken in *Vua marina*,neither beareth it any berries or graines like wheat,neither is it pleasant in taste and smell,or any thing astringent,all which are to be found in the right *Tragos* before expressed; which (as *Dioscorides* saith) is without leaues,neither is it thorney as *Tragus improbus Matthioli* is:this plant I haue found growing in the Isle of Shepey,in the tract leading to the house of Sir *Edward Hobby*,called Sherland.

¶ *The Place.*

It loueth to grow vpon dry banks and sandy places neere to the sea:it is found in Languedocke, not far from Montpelier,and in other places by the sea side,and is a stranger in England.

¶ *The Time.*

When it groweth of it selfe the fruit is ripe in Autumne,the plant it selfe remaineth long green, for all the cold in Winter.

¶ *The Names.*

It is called of the later Herbarists,*Vua marina*: in French, *Raisin de Mer*,of the pearled fruit, and the likenesse that it hath with the Raspis berrie,which is as it were a Raison or Grape,consisting of many little ones : it is named in Greeke τράγος, but it is not called *Tragus*,or *Traganos*,of a Goat (for so signifieth the Greeke word) or of his ranke and rammish smell,but because it bringeth forth fruit

fruit fit to be eaten,of the Verbe τρώγω,which signifieth to eat: it may be called Scorpion, because the sprigs thereof are sharpe pointed like to the Scorpions taile.

¶ *The Temperature.*

The berries or Raisons,and especially the seed that is in them haue a binding quality, as we haue said,and they are drie in the later end of the second degree.

¶ *The Vertues.*

A *Dioscorides* writeth,that the Raisons of sea Grape do stay the flix, and also the whites in women, when they much abound.

† Our Author as you see gaue the history of the lesser in the first place, but formerly the figure was in the third place, and another figure of the same in the second place,and the figure of the greater was in the first place.

Chap. 460. *Of Madder.*

¶ *The Kindes.*

There is but one kinde of Madder onely which is manured or set for vse,but if all those that are like vnto it in leaues and manner of growing were referred thereto,there should be many sorts: as Goose-grasse,soft Cliuer,our Ladies Bedstraw,Woodroofe,and Crosse-woort,all which are like to Madder in leaues,and therefore they be thought to be wilde kinds thereof.

1 *Rubia tinctorum.* Red Madder.

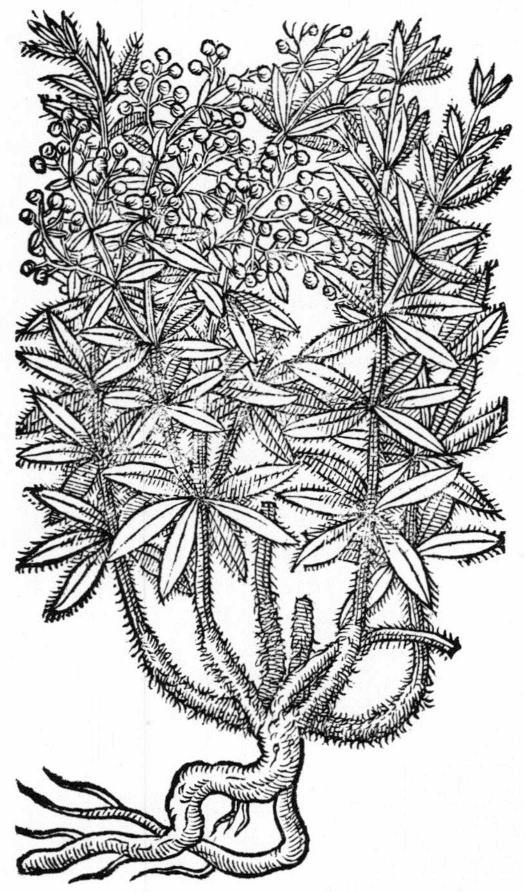

2 *Rubia syluestris.* Wilde Madder.

¶ *The Description.*

1 The garden or manured Madder hath long stalks or trailing branches dispersed farre abroad vpon the ground,square, rough, and full of ioints ; at euery ioint set round with greene rough leaues,in manner of a starre,or as those of Woodroofe: the floures grow at the toppe of

of the branches,of a faint yellow colour: after which come the ſeed,round,greene at the firſt,afterward red,and laſtly of a blacke colour: the root long,fat,full of ſubſtance,creepeth far abroad within the vpper cruſt of the earth, and is of a reddiſh colour when it is greene and freſh.

2 Wilde Madder is like in forme vnto that of the garden, but altogether ſmaller, and the leaues are not ſo rough,but ſmooth and ſhining: the floures are white: the root is very ſmall and tender,and oftentimes of a reddiſh colour.

3 *Rubia marina.*
Sea Madder.

‡ 4 *Rubia ſpicata Cretica.*
Small Candie Madder.

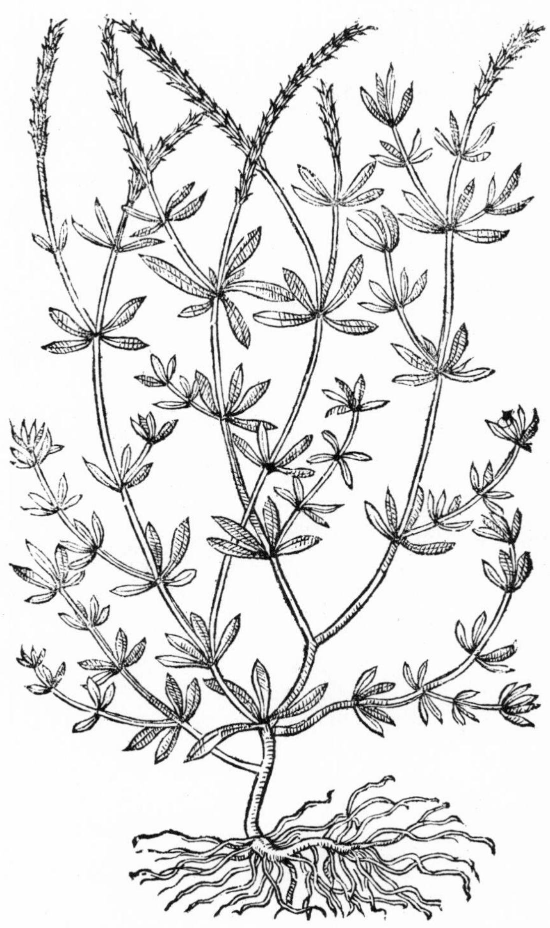

3 Sea Madder hath a root two foot long,with many dry threds hanging thereat,of a reddiſh colour like Alkanet,on the outſide of the ſame forme and bigneſſe, but within it of the colour of the ſcrapings of Iuniper,or Cedar wood,ſending forth diuers ſlender ſtalks round and ful of ioints: from which come forth ſmall thin leaues, ſtiffe and ſharpe pointed, ſomewhat hairie, in number commonly foure,ſtanding like a Burgonion croſſe; from the boſome of which come forth certain tufts of ſmaller leaues thruſt together vpon a heape: the floures grow at the top of the ſtalks,of a pale yellowiſh colour.

‡ *Rubia ſpicata Cretica Cluſij.*

‡ 4 This hath proceeding from the root many knottie foure ſquare rough little ſtalks, a foot high, diuided immediately from the root into many branches, hauing but one ſide branch growing forth of one ioint: about which ioints grow ſpred abroad foure or fiue, ſometimes ſixe narrow, ſhort, ſharpe pointed leaues,ſomewhat rough the toppes of the ſtalkes and branches are nothing but long ſmall foure ſquare ſpikes or eares,made of three leafed greene huskes: out of the top of each huske groweth a very ſmall greeniſh yellow floure,hauing foure exceeding ſmal leaues ſcarce to be ſeene: after which followeth in each huske one ſmall blackiſh ſeed, ſomewhat long, round on the one ſide,with a dent or hollowneſſe on the other. The root is ſmall, hard, woodie, crooked or ſcragged,with many little branches or threds,red without,and white within,and periſheth when the ſeeds are ripe. Iuly,19.1621.

*Synanchica Lug.p.*1185.

5 The root is crooked, blackish without, yellow vnderneath the skinne, white within that and wooddie; about fiue or six inches long, with many hairy strings: from the root arise many foure-square branches trailing vpon the ground, sometimes reddish towards the root: the leaues are small and sharpe pointed, like those of *Gallium*, and grow along the stalke, on certaine knees or ioints, foure or fiue together, sometimes fewer: from those ioints the stalk diuideth it selfe towards the top into many parts, whereon grow many floures, each floure hauing foure leaues, sometimes white, sometimes of a flesh colour, and euery leafe of these flesh coloured leaues is artificially straked in the middle, and neere the sides with three lines of a deeper red, of no pleasant smell: after which commeth the seed something round, growing two together like stones. It floureth all the Sommer long, and groweth in drie Chalkie grounds aboundantly. August 13. 1619. *Iohn Goodyer.* ‡

‡ 6 *Rubia minima.*
Dwarfe Madder.

‡ 6 *Lobel* thus describes this Dwarfe Madder: there is another (saith hee) which I gathered, growing vpon Saint Vincents rocks not farre from Bristow: the leaues are of the bignesse of those of Rupture-woort, sharpe pointed, and growing after the manner of those of Madder, vpon little creeping stalkes, some inch and halfe high, whereon grow yellowish small floures. The root is small, and of the colour of Corall. ‡

¶ *The Place.*

Madder is planted in gardens, and is verie common in most places of England. Master *George Bowles* found it growing wilde on Saint Vincents rocke; and out of the Cliffes of the rockes at Aberdovie in Merioneth shire.

The second groweth in moist medowes, in moorish grounds, and vnder bushes almost euery where.

3 This grows by the sea side in most places.

‡ The fourth growes onely in some few gardens with vs, but the fifth may bee found wilde in many places: I found it in great plenty on the hill beyond Chattam in the way to Canturburie. ‡

¶ *The Time.*

They flourish from May vnto the end of August: the roots are gathered and dried in Autumne, and sold to the vse of Diers and Medicine.

¶ *The Names.*

Madder is called in Greeke ἐρυθρόδανον; *Erythrodanum*: in Latine, *Rubia*, and *Rubeia*: in shops, *Rubia tinctorum*: *Paulus Ægineta* sheweth that it is named *Thapson* which the Diers vse, and the Romanes call it *Herba Rubia*: in Italian *Rubbia*, and *Robbia*: in Spanish, *Ruvia, Roya*, and *Granza*: in French, *Garance*: in high Dutch, **Rotte**: in low Dutch, **Mee**, and **Mee Crappen**: in English, Madder, and red Madder.

¶ *The Temperature.*

Of the temperature of Madder, it hath beene disputed among the learned, and as yet not censured, whether it doe binde or open; some say both; diuers diuersly deeme: a great Physition (I do not say the great learned) called me to account as touching the faculties heereof, although he had no commission so to doe, notwithstanding I was content to be examined vpon the point, what the nature of Madder was, because I haue written that it performeth contrary effects, as shall be shewed: the roots of Madder, which both the Physitions and diers doe vse, as they haue an obscure binding power

power and force; so be they likewise of nature and temperature cold and dry: they are withall of diuers thin parts, by reason whereof there colour doth easily pierce: yet haue they at the first a certain little sweetnes, with an harsh binding quality presently following it; which not onely we our selues haue obserued, but also *Auicen* the prince of Physitions, who in his 58. Chapter hath written, that the root of Madder hath a rough and harsh taste: now M^r. Doctor, whether it binde or open I haue answered, attending your censure: but if I haue erred, it is not with the multitude, but with those of the best and best learned.

¶ *The Vertues.*

The decoction of the roots of Madder is euery where commended for those that are bursten, brused, wounded, and that are fallen from high places. A

It stencheth bleeding, mitigateth inflammations, and helpeth those parts that be hurt and brused. B

For these causes they be mixed with potions, which the later Physitians call wound drinkes: in which there is such force and vertue, as *Matthiolus* also reporteth, that there is likewise great hope of curing of deadly wounds in the chest and intrails. C

Our opinion and judgement is confirmed by that most expert man, sometimes Physition of Louaine, *Iohannes Spiringus*, who in his *Rapsodes* hath noted, that the decoction of Madder giuen with *Triphera*, that great composition is singular good to stay the reds, the hemorrhoides and bloudy flixe, and the same approoued by diuers experiments: which confirmeth Madder to be of an astringent and binding qualitie. D

Of the same opinion as it seemeth is also *Eros Iulia* her freed man (commonly called *Trotula*) who in a composition against vntimely birth doth vse the same: for if he had thought that Madder were of such a qualitie as *Dioscorides* writeth it to be of, he would not in any wise haue added it to those medicines which are good against an vntimely birth. E

For *Dioscorides* reporteth, that the root of Madder doth plentifully prouoke vrine, and that grosse and thicke, and oftentimes bloud also, and it is so great an opener, that being but onely applied, it bringeth downe the menses, the birth, and after-birth: but the extreme rednes of the vrine deceiued him, that immediately followeth the taking of Madder, which rednesse came as he thought, from bloud mixed therewith, which notwithstanding commeth no otherwise then from the colour of the Madder. F

For the root hereof taken any maner of way doth by & by make the vrine extreme red: no otherwise than Rubarb doth make the same yellow, not changing in the meane time the substance thereof, nor making it thicker than it was before, which is to be vnderstood in those which are in perfect health, which thing doth rather shew that it doth not open, but binde, no otherwise than Rubarbe doth: for by reason of his binding quality the waterish humors do for a while keepe their colour. For colours mixed with binding things do longer remaine in the things coloured, and do not so soone vade: this thing they will know that gather colours out of the juices of floures and herbes, for with them they mixe allume, to the end that the colour may be retained and kept the longer, which otherwise would be quickely lost. By these things it manifestly appeareth that Madder doth nothing vehemently either clense or open, and that *Dioscorides* hath rashly attributed vnto it this kinde of qualitie, and after him *Galen* and the rest that followed, standing stiffely to his opinion. G

Pliny saith, that the stalkes with the leaues of Madder, are vsed against serpents. H

The root of Madder boiled in Meade or honied water, and drunken, openeth the stopping of the liuer, the milt and kidnies, and is good against the jaundise. I

The same taken in like maner prouoketh vrine vehemently, insomuch that the often vse thereof causeth one to pisse bloud, as some haue dreamed. K

Langius and other excellent Physitions haue experimented the same to amend the lothsome colour of the Kings-euill, and it helpeth the vlcers of the mouth, if vnto the decoction be added a little allume and hony of Roses. L

‡ 5 The fifth being the *Synanchica* of *Daleschampius*, dries without biting, and it is excellent against squinancies, either taken inwardly, or applied outwardly, for which cause they haue called it *Synanchica*, *Hist. Lugd.* ‡ M

CHAP. 461. *Of Goose-grasse, or Cliuers.*

¶ *The Description.*

1 *Aparine*, Cliuers or Goose-grasse, hath many small square branches, rough and sharpe, full of joints, beset at euery joint with small leaues star fashion, and like vnto small Madder:

the floures are very little and white, pearking on the tops of the ſprigs: the ſeeds are ſmall, round, a little hollow in the middeſt in maner of a nauell, ſet for the moſt part by couples: the roots ſlender and full of ſtrings: the whole plant is rough, and his ruggedneſſe taketh hold of mens veſtures and woollen garments as they paſſe by: being drawne along the tongue it fetcheth bloud: *Dioſcorides* reports, that the ſheepheards in ſtead of a Cullender do vſe it to take haires out of milke, if any remaine therein.

2 The great Gooſe-graſſe of *Pliny* is one of the Moone-worts of *Lobel*, it hath a very rough tender ſtalke, whereupon are ſet broad leaues ſomewhat long, like thoſe of Scorpion graſſe, or *Alyſſon Galeni*, *Galens* Moone-woort, very rough and hairy, which grow not about the joints, but three or foure together on one ſide of the ſtalke: the floures grow at the top of the branches, of a blew colour: after which commeth rough cleauing ſeeds, that do ſticke to mens garments which touch it: the root is ſmall and ſingle.

1 *Aparine.*
Gooſe-graſſe or Cleuers.

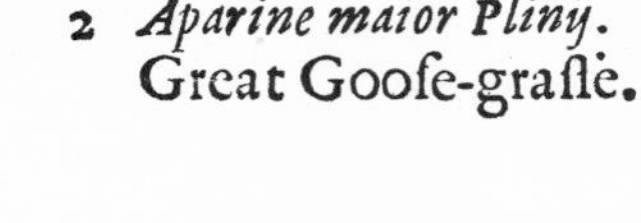

2 *Aparine maior Plinij.*
Great Gooſe-graſſe.

¶ *The Place.*

Gooſe-graſſe groweth neere the borders of fields, and oftentimes in the fields themſelues mixed with the corne alſo by common waies, ditches, hedges, and among thornes: *Theophraſtus* and *Galen* write, that it groweth among Lentles, and with hard embracing it doth choke it, and by that meanes is burdenſome and troubleſome vnto it.

¶ *The Time.*

It is found plentifully euery where in ſummer time.

¶ *The Names.*

It is named in Greeke ἀπαρίνη: *Apparine*: in Latine, *Lappa minor*, but not properly: *Pliny* affirmeth it to be *Lappaginis ſpeciem*: of ſome, *Philanthropos*, as though he ſhould ſay, a mans friend, becauſe it taketh hold of mens garments; of diuers alſo for the ſame cauſe, *Philadelphos*: in Italian, *Speronella*: in Spaniſh, *Preſera*, or *amor di Hortalano*: in high Dutch, **Kleeb kraut**: in French, *Reble, ou Grateron*: in low Dutch, **Kleefcruyt**: in Engliſh, Gooſe-ſhare, Gooſe-graſſe, Cleuer, or Clauer.

¶ *The Temperature.*

It is, as *Galen* ſaith, moderately hot and dry, and ſomewhat of thin parts.

¶ The

¶ *The Vertues.*

The iuice which is pressed out of the seeds, stalks, and leaues, as *Dioscorides* writeth, is a remedie for them that are bitten of the poisonsome spiders called in Latine *Phalangia*, and of vipers if it be drunke with wine. A

And the herbe stamped with swines grease wasteth away the kernels by the throte. B

Pliny teacheth that the leaues being applied do also stay the aboundance of bloud issuing out of wounds. C

Women do vsually make pottage of Cleuers with a little mutton and Otemeale, to cause lanknesse, and keep them from fatnesse. D

CHAP. 462. *Of Crosse-woort.*

¶ *The Description.*

1 Crosse-woort is a low and base herbe, of a pale greene colour, hauing many square feeble rough stalks full of ioints or knees, couered ouer with a soft downe: the leaues are little, short, & smal, alwaies foure growing together, and standing crossewise one right against another, making a right Burgunion crosse: toward the top of the stalke, and from the bosome of those leaues come forth very many small yellow floures, of a reasonable good sauour, each of which is also shaped like a Burgunion crosse: the roots are nothing else but a few small threds or fibres.

1 *Cruciata.*
Crosse-woort.

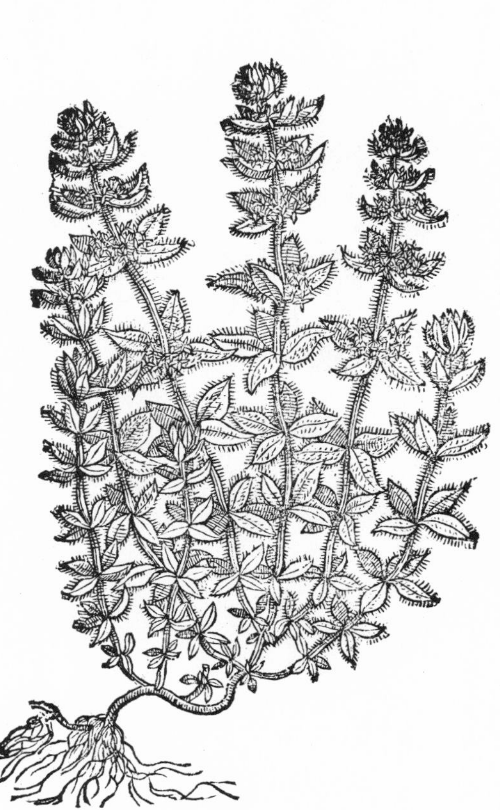

‡ 2 *Rubia Cruciata lævis.*
Crosse-woort Madder.

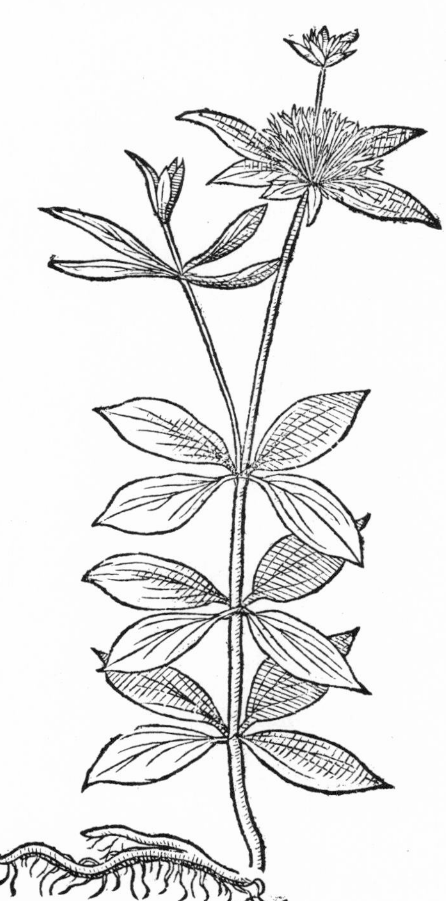

‡ 2 This in mine opinion may be placed here as fitly as any where els; for it hath the leaues standing crosse-waies foure at a ioint, somewhat like those of the largest Chickweed: the stalkes are betweene a foot and a halfe and two cubites high. The white Starre-fashioned floures stand in roundles about the tops of the stalks. It growes plentifully in Piemont, on the hills not farre from Turine. *Lobel* sets it forth by the name of *Rubia Lævis Taurinensium*. ‡

¶ *The Place.*

Cruciata, or Croſſe-woort, groweth in moiſt and fertile medowes; I found the ſame growing in the Churchyard of Hampſtead neere London, and in a paſture adioining thereto, by the mill: alſo it groweth in the Lane or high way beyond Charlton, a ſmall village by Greenwich, and in ſundry other places.

¶ *The Time.*

It floureth for the moſt part all Sommer long.

¶ *The Names.*

It is called *Cruciata*, and *Cruciatis*, of the placing of the leaues in manner of a Croſſe: in Engliſh, Croſſe-woort, or Golden Mugweet.

¶ *The Temperature.*

Croſſewoort ſeemeth to be of a binding and dry qualitie.

¶ *The Vertues.*

A Croſſewoort hath an excellent propertie to heale, ioine, and cloſe wounds together, yea it is very fit for them, whether they be inward or outward, if the ſaid herbe be boiled in wine and drunke.

B The decoction thereof is alſo miniſtred with good ſucceſſe to thoſe that are burſten: and ſo is the herbe, being boiled vntill it bee ſoft, and laied vpon the burſten place in manner of a pultis.

CHAP. 463. *Of Woodrooffe.*

1 *Aſperula.*
Woodrooffe.

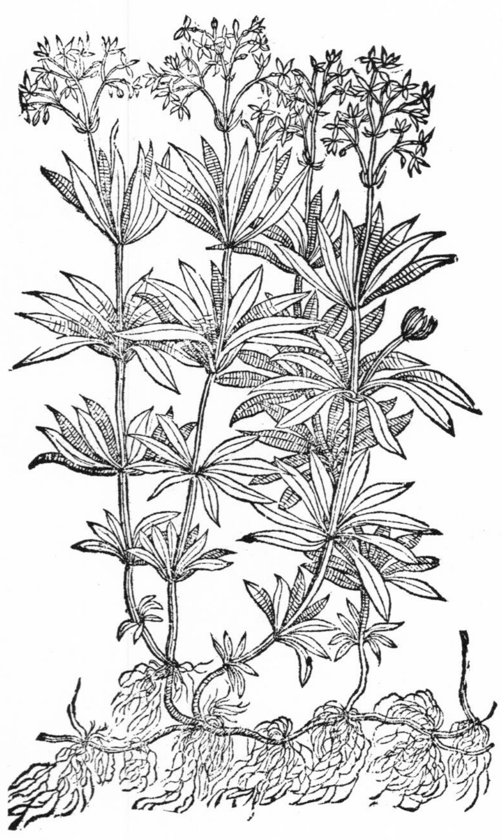

‡ 2 *Aſperula flore cæruleo.*
Blew Woodrooffe.

¶ *The Deſcription.*

1 WOodrooffe hath many ſquare ſtalkes full of ioints, and at euery knot or ioint ſeuen or eight long narrow leaues, ſet round about like a ſtar, or the rowell of a ſpurre: the floures grow at the top of the ſtems, of a white colour, and of a very ſweet ſmell, as is the reſt of the herbe, which being made vp into garlands or bundles, and hanged vp in houſes in the heat of Sommer, doth very wel attemper the aire, coole and make freſh the place, to the delight and comfort of ſuch as are therein.

2 There

‡ 3 *Saginæ Spergula.*
Spurrye.

2 There is another sort of Woodrooffe called *Asperula Cærulea*, or blew Woodrooffe; it is an herbe of a foot high, soft, hairy, and something branched, with leaues & stalks like those of white Woodrooffe the floures thereof are blew, standing vpon short stems on the tops of the stalks: the seed is small, round, and placed together by couples: the root is long, and of a red colour.

3 There is another herb called *Saginæ spergula*, or Spurry, which is sown in Brabant, Holland, and Flanders, of purpose to fatten cattel, and to cause them to giue much milke, and there called Spurrey, and Franke Spurrey: it is a base and low herbe, very tender, hauing many iointed stalks, whereupon do grow leaues set in round circles like those of Woodrooffe, but lesser and smoother, in forme like the rowell of a spur: at the top of the stalks do grow small white floures; after which come round seed like those of Turneps: the root is small and thred-die.

‡ 4 There are one or two plants more, which may fitly be here mentioned: the first of them is the *Spergula marina* of *Daleschampius*, which from a pretty large wooddy and roughish root sends vp iointed stalks some foot long: at each ioint come forth two long thick round leaues, and out of their bosomes other lesser leaues: the top of the stalks is diuided into sundry branches, bearing floures of a faint reddish colour, composed of fiue little leaues, with yellowish threds in the middle: after which follow cups or seed vessels, which open into foure parts, and containe a little flat reddish seed: it grows in the salt marshes about Dartford, and other such places; floures in Iuly and August, and in the meane space ripens the seed. We may call this in English, Sea Spurrie.

5 This other hath a large root, considering the smalnesse of the plant: from which arise many weak slender branches some three or foure inches long, somtimes more, lying commonly flat on the ground, hauing many knots or ioints: at each whereof vsually grow a couple of white scaly leaues, and out of their bosomes other small sharpe pointed little greene leaues: at the tops of the branches grow little red floures, succeeded by such, yet lesser heads than those of the former: it floures in Iuly and August, and growes in sandy grounds, as in Tuthill-fields nigh Westminster: the figure set forth in *Hist. Lugd. p.* 2179, by the title of *Chamæpeuce Plinij*; *Camphorata minor Daleschampij*, seems to be of this plant, but without the floure: *Bauhine* in his *Prodromus* describes it by the name of *Alsine Spergulæ facie*. This may be called Chickweed Spurrey, or small red Spurrey. ‡

¶ *The Place.*

White Woodroofe groweth vnder hedges, and in woods almost euery where: the second groweth in many places of Essex, and diuers other parts in sandy grounds. The third in Corne fields.

¶ *The Time.*

They floure in Iune and Iuly.

¶ *The Names.*

Most haue taken Woodrooffe to be *Pliny* his *Alyssos*, which as he saith, doth differ from *Erythrodanum*, or Garden Madder, in leaues onely, and lesser stalks: but such a one is not onely this, but also that with blew floures: for *Galen* doth attribute to *Alyssos*, a blew floure: notwithstanding *Galens* and *Plinies Alyssos* are thought to differ by *Galens* owne words, writing of *Alyssos* in his second booke of Counterpoisons, in *Antonius Cous* his composition, in this maner: *Alyssos* is an herb very like vnto Horehound, but rougher and fuller of prickles about the circles: it beareth a floure tending to blew.

Woodrooffe is named of diuers in Latine *Asperula odorata*, and of most men *Aspergula odorata*: of others, *Cordialis*, and *Stellaria*: in high Dutch, **Hertzfreydt**: in low Dutch, **Leuerkraut**: that is to

say,

ſay *Iecoraria*, or *Hepatica*, Liuerwoort : in French, *Muguet* : in Engliſh, Woodrooffe, Woodrowe, and Woodrowell.

¶ *The Temperature.*

Woodrooffe is of temperature ſomething like vnto our Ladies Bedſtraw, but not ſo ſtrong, being in a meane between heate and drineſſe.

¶ *The Vertues.*

A It is reported to be put into wine, to make a man merry, and to be good for the heart and liuer: it preuaileth in wounds, as *Cruciata*, and other vulnerarie herbes do.

Chap. 464. *Of Ladies Bedſtraw.*

¶ *The Kindes.*

THere be diuers of the herbes called Ladies Bedſtraw, or Cheeſe-renning; ſome greater, others leſſe; ſome with white floures, and ſome with yellow.

¶ *The Deſcription.*

1 LAdies Bedſtraw hath ſmall round euen ſtalkes, weake and tender, creeping hither and thither vpon the ground : whereon doe grow very fine leaues, cut into ſmall iags, finer than thoſe of Dill, ſet at certaine ſpaces, as thoſe of Woodrooffe : among which come forth floures of a yellow colour, in cluſters or bunches thicke thruſt together, of a ſtrong ſweet ſmel but not vnpleaſant: the root is ſmall and threddie.

1 *Gallium luteum.*
Yellow Ladies Bedſtraw.

2 *Gallium album.*
Ladies Bedſtraw with white floures.

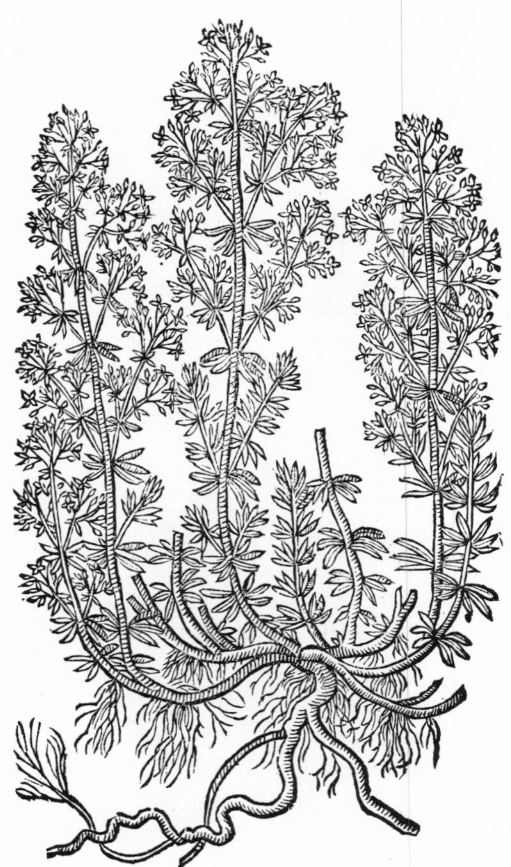

2 Ladies Bedſtraw with white floures is like vnto Cleauers or Gooſe-graſſe, in leaues, ſtalkes, and manner of growing, yet nothing at all rough, but ſmooth and ſoft: the floures be white, the ſeed round : the roots ſlender, creeping within the ground: the whole plant rampeth vpon buſhes, ſhrubs and all other ſuch things as ſtand neere vnto it: otherwiſe it cannot ſtand, but muſt reele and fall to the groud.

3 This

3 This small *Gallium*, or Ladies little red bed-strow, hath been taken for a kind of wild Madder; neuerthelesse it is a kinde of Ladies bed-strow, or cheese-renning, as appeareth both by his vertues in turning milke to cheese, as also by his forme, being in each respect like vnto yellow *Gallium*, and differs in the colour of the floures, which are of a dark red colour, with a yellow pointal in the middle, consisting of foure small leaues: the seed hereof was sent me from a Citisen of Strausburg in Germanie, and it hath not been seen in these parts before this time.

4 There is likewise another sort of *Gallium* for distinctions sake called *Mollugo*, which hath stalks that need not to be propped vp, but of it selfe standeth vpright, and is like vnto the common white *Gallium*, but that it hath a smoother leafe. The floures thereof be also white, and very small. The root is blackish.

† 3 *Gallium rubrum.*
Ladies Bed-strow with red floures.

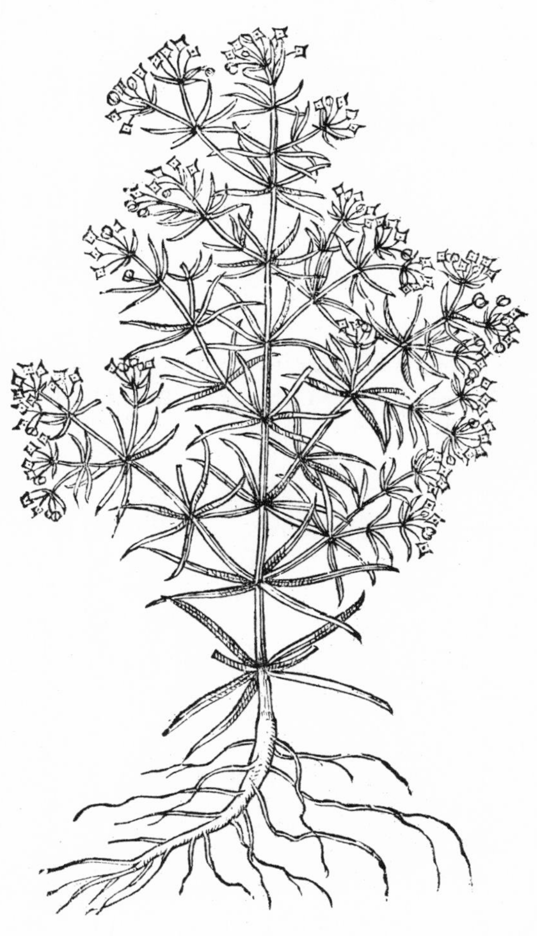

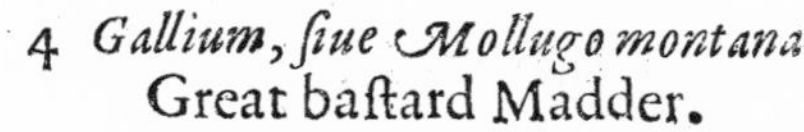

4 *Gallium, siue Mollugo montana.*
Great bastard Madder.

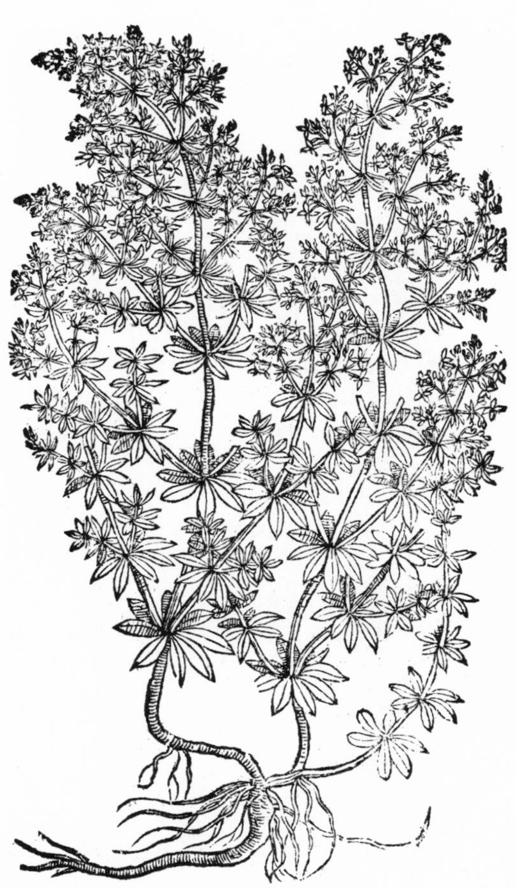

¶ *The Place.*

The first groweth vpon sunnie bankes neere the borders of fields, in fruitfull soiles almost euery where.

The second groweth in marish grounds and other moist places.

The third groweth vpon mountaines and hilly places, and is not yet found in England.

The fourth and last groweth in hedges among bushes in most places.

¶ *The Time.*

They floure most of the Sommer moneths.

¶ *The Names.*

The first is called in Greeke γάλιον: it hath that name of milke, called in Greeke γάλα, into which it is put as cheese-renning: in Latine likewise *Gallium*: in high-Dutch, **Megerkraut, Walstroo**: in low-Dutch, **Walstroo**: in French, *Petit Muguet*: in Italian, *Galio*: in Spanish, *Coaia leche yerua*: in English, our Ladies Bed-strow, Cheese-renning, Maids haire, and pety Mugwet.

The others are *Species Lappaginis*, or kindes of small Burres, so taken of the Antients: The last, of the softnesse and smoothnesse of the leaues, is commonly called *Mollugo*: diuers take it for a kinde of wilde Madder, naming it *Rubia syluestris*, or wilde Madder.

¶ *The Temperature.*

These herbes, especially that with yellow floures, is dry and something binding, as *Galen* saith.

¶ *The*

¶ *The Vertues.*

A The floures of yellow Maids haire, as *Dioſcorides* writeth, is vſed in ointments againſt burnings, and it ſtancheth bloud: it is put into the Cerote or Cere-cloath of Roſes: it is ſet a ſunning in a glaſſe, with Oyle Oliue, vntill it be white: it is good to anoint the wearied Traueller: the root thereof drunke in wine ſtirreth vp bodily luſt; and the floures ſmelled vnto worke the ſame effect.

B The herbe thereof is vſed for Rennet to make cheeſe, as *Matthiolus* reporteth, ſaying, That the people of Tuſcanie or Hetruria do vſe to turne their milke, that the Cheeſe which they make of Sheeps and Goats milke might be the ſweeter and more pleaſant in taſte, and alſo more wholſome, eſpecially to breake the ſtone, as it is reported.

C The people in Cheſhire, eſpecially about Namptwich, where the beſt cheeſe is made, do vſe it in their Rennet, eſteeming greatly of that cheeſe aboue other made without it.

D We finde nothing extant in the antient writers, of the vertues and faculties of the white kinde, but are as herbes neuer had in vſe either for phyſicke or Surgerie.

† The figure that was formerly in the third place was of the *Gallium album minus* of *Tabern* which commonly hath but two leaues at a ioynt, yet ſometimes it is found with three.

Chap. 465. *Of Ferne.*

¶ *The Kindes.*

There be diuers ſorts of Ferne, differing as well in forme as place of growing; whereof there be two ſorts according to the old writers, the male and the female; and theſe be properly called Ferne: the others haue their proper names, as ſhall be declared.

1 *Filix mas.*
Male Ferne.

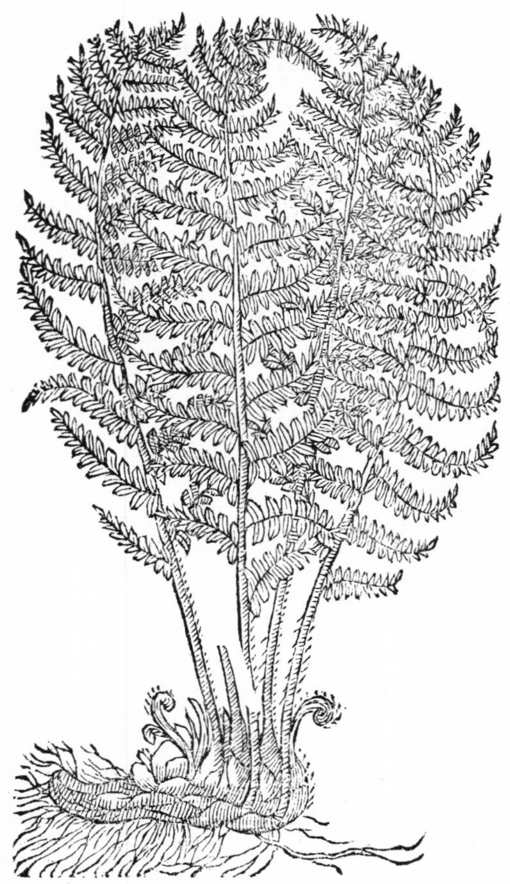

2 *Filix fœmina.*
Female Ferne, or Brakes.

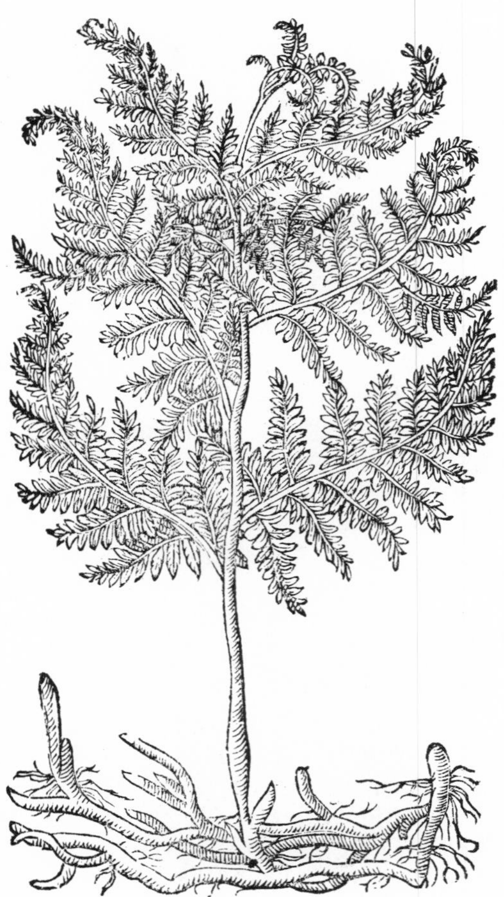

¶ The

¶ *The Deſcription.*

1 THe male Ferne bringeth forth preſently from the root broad leaues and rough, ſomewhat hard, eaſie to be broken, of a light greene colour, and ſtrong ſmell, more than a cubit long, ſpred abroad like wings, compounded as it were of a great number ſtanding vpon a middle rib, euerie one whereof is like a feather, nicked in the edges, and on the backeſide are ſprinkled as it were with a very fine earthy-coloured duſt or ſpots, which many raſhly haue taken for ſeed: the root conſiſteth of a number of tufts or threds, and is thicke and blacke, and is without ſtalke and ſeed, and altogether barren.

‡ *Filicis (vulgo) maris varietates & differentiæ.*
Differences of the male Ferne.

I haue obſerued foure ſorts of Ferne, by moſt writers eſteemed to be the male Ferne of *Dioſcorides*: by *Anguillara, Geſner, Cæſalpinus,* and *Cluſius*, accounted to be the female, and ſo indeed doe I thinke them to be, though I call them the male, with the multitude. If you looke on theſe Fernes according to their ſeuerall growths and ages, you may make many more ſorts of them than I haue done; which I am afraid hath beene the occaſion of deſcribing more ſorts than indeed there are in nature. Theſe deſcriptions I made by them when they were in their perfect growths.

1 *Filix mas ramoſa pinnulis dentatis.*

The roots are nothing but an aboundance of ſmall blacke hairy ſtrings, growing from the lower parts of the maine ſtalkes (for ſtalkes I will call them) where thoſe ſtalkes are ioyned together. At the beginning of the Spring you may perceiue the leaues to grow forth of their folding cluſters, couered with browniſh ſcales at the ſuperficies of the earth, very cloſely ioyned together: a young plant hath but a few leaues; an old one, ten, twelue, or more: each ſtalke at his lower end neere the ioyning to his fellowes, at his firſt appearing, before he is an inch long hauing ſome of thoſe blacke fibrous roots for his ſuſtenance. The leaues being at their full growth hath each of them a three-fold diuiſion, as hath that Ferne which is commonly called the female: the maine ſtalke, the ſide branches growing from him, and the nerues growing on thoſe ſide branches bearing the leaues: the maine ſtalke of that plant I deſcribe was fully foure foot long (but there are vſually from one foot to foure in length) full of thoſe browniſh ſcales, eſpecially toward the root, firme, one ſide flat, the reſt round, naked fully one and twenty inches, to the firſt paire of ſide branches. The ſide branches, the longeſt being the third paire from the root, were nine inches long, and ſhorter and ſhorter towards the top, in number about twenty paire; for the moſt part towards the root they grow by couples, almoſt oppoſite, the neerer the top the further from oppoſition: the nerues bearing the leaues, the longeſt were two inches and a quarter long, and ſo ſhorter and ſhorter toward the tops of the ſide branches; about twentie in number on each ſide of the longeſt ſide branch. The leaues grow for the moſt part by couples on the nerue, eight or nine paire on a nerue; each leafe being gaſhed by the ſides, the gaſhes ending with ſharpe points, of a deep green on the vpper ſide, on the vnder ſide paler, and each leafe hauing two rowes of duſty red ſcales, of a browne or blackiſh colour: toward the top of the maine ſtalke thoſe ſide branches change into nerues, bearing only the leaues. When the leaues are at their full growth, you may ſee in the middeſt of them at their roots the ſaid ſcaly folding cluſter; and as the old leaues with their blacke threddy roots wholly periſh, they ſpring vp; moſt yeares you may finde many of the old leaues greene all the Winter, eſpecially in warme places. This groweth plentifully in the boggy ſhadowie moores neere Durford Abbey in Suſſex, and alſo on the moiſt ſhadowie rockes by Mapledurham in Hampſhire, neere Peters-field; and I haue found it often on the dead putrified bodies and ſtems of old rotten okes, in the ſaid moores; neere the old plants I haue obſerued verie many ſmall yong plants growing, which came by the falling of the ſeed from thoſe duſty ſcales: for I beleeue all herbes haue ſeeds in themſelues to produce their kindes, *Gen.* 1. 11. & 12.

The three other haue but a twofold diuiſion, the many ſtalks and the nerues bearing the leaues. The roots of them all are blacke fibrous threds like the firſt, their maine ſtalks grow many thicke and cloſe together at the root, as the firſt doth: the difference is in the faſhion of their leaues, and manner of growing, and for diſtinctions ſake I haue thus called them:

2 *Filix mas non ramoſa pinnulis latis denſis minutim dentatis.*

The leaues are of a yellowiſh greene colour on both ſides, ſet very thicke and cloſe together on the

the nerue, that you cannot see betweene them, with maruellous small nickes by their sides, and on their round tops: each leafe hath also two rowes of dusty seed scales; the figures set forth by *Lobel, Tabern.* and *Gerard,* vnder the title of *Filix mas,* do well resemble this Ferne. This growes plentifully in most places in shadowie woods and copses.

3 *Filix mas non ramosa pinnulis angustis, raris, profunde dentatis.*

The leaues are of a deepe greene, not closely set together on the the nerue, but you may far off see betwixt them, deeply indented by the sides, ending with a point not altogether sharpe: each leafe hath also two rowes of dusty seed scales. I haue not seene any figure well resembling this plant. This groweth also in many places in the shade.

4 *Filix mas non ramosa pinnulis latis auriculatis spinosis.*

The leaues are of a deeper greene than either of the two last described, placed on the nerue not very close together, but that you may plainly see between them; each leafe (especially those next the stalke) hauing on that side farthest off the stalk a large eare or outgrowing ending, with a sharp pricke like a haire, as doth also the top of the leafe: some of the sides of the leaues are also nicked, ending with the like pricke or haire. Each leafe hath two rowes of dusty seed scales. This I take to be *Filix mas aculeata maior Bauhini.* Neither haue I seene any figure resembling this plant. It groweth abundantly on the shadowie moist rockes by Maple-durham neere Peters-field in Hampshire. *Iohn Goodyer.* Iuly 4. 1633. ‡

2 The female Ferne hath neither floures nor seed, but one only stalke, chamfered, something edged, hauing a pith within of diuers colours, the which being cut aslope, there appeareth a certain forme of a spred-Eagle: about this stand very many leaues which are winged, and like to the leaues of the male Ferne, but lesser: the root is long and blacke, and creepeth in the ground, being now and then an inch thicke, or somewhat thinner. This is also of a strong smell, as is the male.

¶ *The Place.*

Both the Fernes are delighted to grow in barren dry and desart places: and as *Horace* testifieth,

Neglectis vrenda Filix innascitur agris.

It comes not vp in manured and dunged places, for if it be dunged (as *Theophrastus, lib.8. cap.8.* reporteth) it withereth away.

The male ioyeth in open and champion places, on mountaines and stony grounds, as *Dioscorides* saith. ‡ It growes commonly in shadowie places vnder hedges. ‡

The female is often found about the borders of fields vnder thornes and in shadowie woods.

¶ *The Time.*

Both these Fernes wither away in winter: in the spring there grow forth new leaues, which continue greene all Sommer long.

¶ *The Names.*

The former is called in Greeke πτέρις: *Nicander* in his discourse of Treacle nameth it βλῆχνον: in Latine *Filix mas*: in Italian, *Felce*: in Spanish, *Helecho, Falguero,* and *Feyto*: in high-Dutch, **Waldt Farne**: in French, *Fougere,* or *Feuchiere masle*: in low-Dutch, **Waren Manneken**: in English, male Ferne.

The second kinde is called in Greeke θηλυπτέρις, that is, *Filix fœmina,* or female Ferne: in Latine, as *Dioscorides* noteth among the bastard names, *Lingua ceruina*: in high-Dutch, **Waldt Farn weiblin,** and **Grosz Farnkraut**: in low-Dutch, **Waren Wijfken**: in French, *Fougere femelle*: in English, Brake, common Ferne, and female Ferne.

¶ *The Temperature.*

Both the Fernes are hot, bitter, and dry, and something binding.

¶ *The Vertues.*

A The roots of the male Ferne being taken to the weight of halfe an ounce, drieueth forth long flat wormes out of the belly, as *Dioscorides* writeth, being drunke in Mede or honied water; and more effectually, if it be giuen with two scruples or two third parts of a dram of Scamonie, or of blacke Hellebor: they that will vse it, saith he, must first eate Garlicke. After the same manner, as *Galen* addeth, it killeth the childe in the mothers wombe. The root hereof is reported to be good for them that haue ill spleenes: and being stamped with swines grease and applied, it is a remedie against the pricking of the reed: for proofe hereof, *Dioscorides* saith the Ferne dieth if the Reed be planted about it; and contrariwise, that the Reed dieth if it be compassed with Ferne: which is vaine to thinke, that it hapneth by any antipathie or naturall hatred, and not by reason this Ferne prospereth not in moist places, nor the Reed in dry.

B The female Ferne is of like operation with the former, as *Galen* saith. *Dioscorides* reports, That this bringeth barrennesse, especially to women; and that it causeth women to be deliuered before their time: he addeth, that the pouder hereof finely beaten is laid vpon old vlcers, and healeth the

the galled neckes of oxen and other cattell: it is alſo reported, that the root of Ferne caſt into an hogſhead of wine keepeth it from ſouring.

The root of the male Ferne ſodden in Wine is good againſt the hardneſſe and ſtopping of the milt: and being boyled in water, ſtayeth the laske in yong children, if they be ſet ouer the decoction thereof to eaſe their bodies by a cloſe ſtoole. C

CHAP. 466. *Of Water-Ferne, or* Oſmund *the water-man.*

¶ The Deſcription.

WAter Ferne hath a great triangled ſtalke two cubits high, beſet vpon each ſide with large leaues ſpred abroad like wings, and dented or cut like Polypodie: theſe leaues are like the large leaues of the Aſh tree; for doubtleſſe when I firſt ſaw them afar off it cauſed me to wonder thereat, thinking that I had ſeene yong Aſhes growing vpon a bog; but beholding it a little neerer, I might eaſily diſtinguiſh it from the Aſh, by the browne rough and round graines that grew on the top of the branches, which yet are not the ſeed thereof, but are very like vnto the ſeed. The root is great and thicke, folded and couered ouer with many ſcales and interlacing roots, hauing in the middle of the great and hard woody part thereof ſome ſmall whiteneſſe, which hath beene called the heart of *Oſmund* the water-man.

Filix florida, ſiue Oſmunda Regalis.
Water Ferne, or Oſmund Royall.

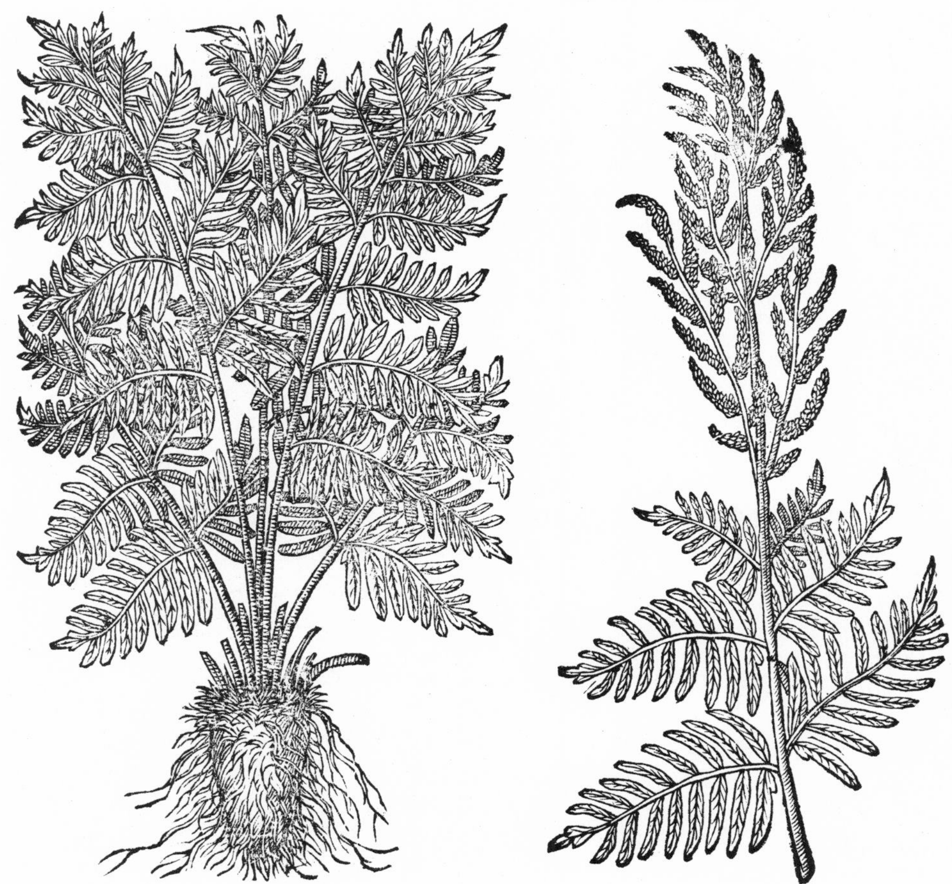

¶ The Place.

It groweth in the midſt of a bog at the further end of Hampſted heath from London, at the bottome of a hill adioyning to a ſmall cottage, and in diuers other places, as alſo vpon diuers bogges on a heath or common neere vnto Bruntwood in Eſſex, eſpecially neere vnto a place there that ſome haue digged, to the end to finde a neſt or mine of gold; but the birds were ouer fledge, and flowne away before their wings could be clipped. ‡ It did grow plentifully in both theſe places, but of late it is all deſtroyed in the former. ‡

¶ The

¶ *The Time.*

It flourisheth in Sommer, as the former Fernes : the leaues decay in Winter: the root continueth fresh and long lasting; which being brought into the garden prospereth as in his natiue soile, as my selfe haue proued.

¶ *The Names.*

It is called in Latine *Osmunda* : it is more truly named *Filix palustris*, or *aquatilis* : some terme it by the name of *Filicastrum* : most of the Alchimists call it *Lunaria maior* : *Valerius Cordus* nameth it *Filix latifolia* : it is named in high-Dutch, **Grosz Farn** : in low-Dutch, **Groot Varen, wilt Varen** : in English, Water-Ferne, Osmund the Water-man : of some, Saint Christophers herbe, and Osmund.

¶ *The Temperature.*

The root of this also is hot and dry, but lesse than they of the former ones.

¶ *The Vertues.*

A The root, and especially the heart or middle part thereof, boiled or else stamped, and taken with some kinde of liquor, is thought to be good for those that are wounded, dry-beaten, and bruised; that haue fallen from some high place : and for the same cause the Empericks do put it in decoctions, which the later Physitians do call wound-drinks : some take it to be so effectuall, and of so great a vertue, as that it can dissolue cluttered bloud remaining in any inward part of the body, and that it also can expell or driue it out by the wound.

B The tender sprigs thereof at their first comming forth are excellent good vnto the purposes aforesaid, and are good to be put into balmes, oyles, and consolidatiues, or healing plaisters, and into vnguents appropriate vnto wounds, punctures, and such like.

CHAP. 467. *Of Polypodie or wall-Ferne.*

1 *Polypodium.*
Wall Ferne, or Polypodie of the wall.

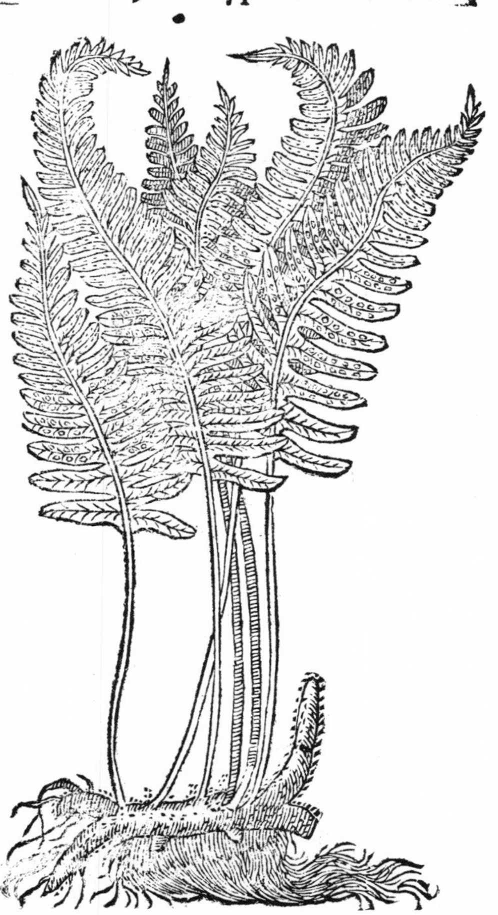

2 *Polypodium quercinum.*
Polypodie of the Oke.

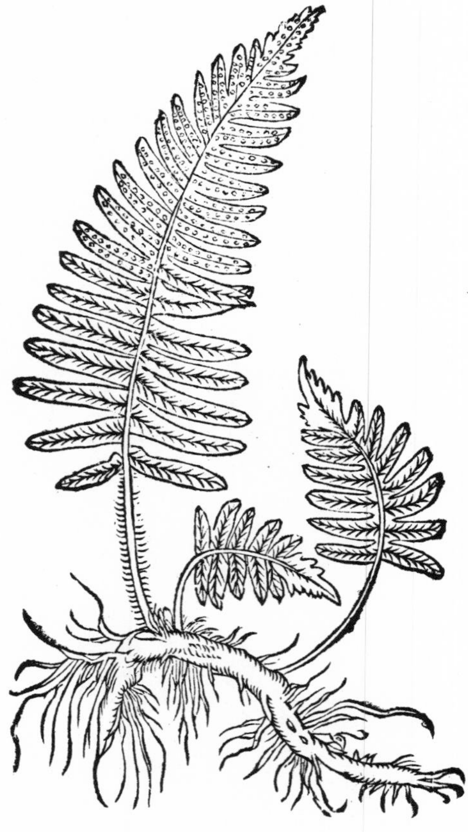

¶ *The*

‡ 3 *Polypodium Indicum.*
Indian Polypody.

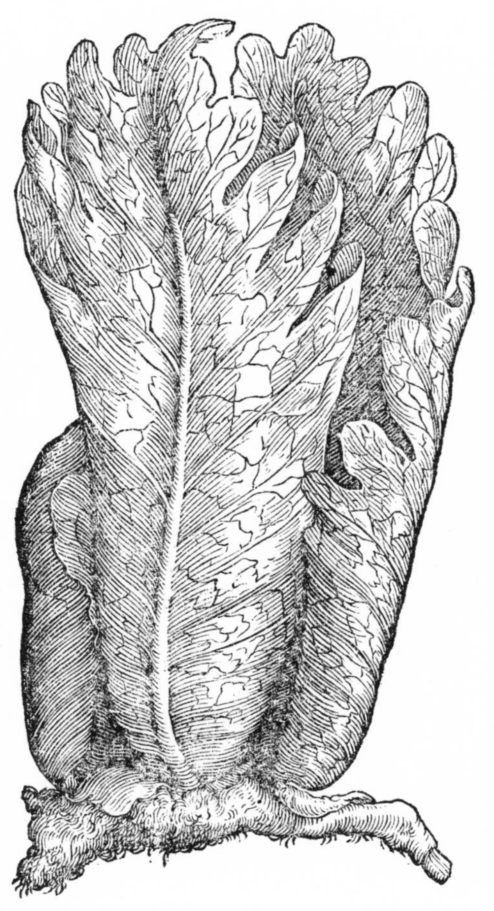

¶ *The Deſcription.*

1 THe leaues of Polypodie might be thought to be like thoſe of male Ferne, but that they are far leſſer, and not nicked at all in the edges: theſe do preſently ſpring vp from the roots, being cut on both the edges with many deepe gaſhes, euen hard to the middle rib; on the vpper ſide they are ſmooth, on the nether ſide they are lightly powdred as it were with duſty markes: the root is long, not a finger thick, creeping aſlope, on which are ſeen certaine little buttons like to thoſe pits and dents that appeare in the tailes of cuttle fiſhes: this hath in it a certaine ſweetneſſe, with a taſte ſomething harſh: this kinde of Ferne likewiſe wanteth not onely floures and ſeed, but ſtalkes alſo.

2 Polypodie of the Oke is much like vnto that of the wall, yet the leaues of it are more finely cut, ſmooth on the vpper ſide, of a pale green color, together with the ſtalkes and middle ribs; on the nether ſide rough like thoſe of Ferne: this Ferne alſo liueth without a ſtalke: it groweth without ſeed: the root hath many ſtrings faſtned to it, one folded within another, of a meane bigneſſe, and ſweet in taſte: it ſendeth forth heere and there new dodkins or ſprings, whereby it increaſeth.

‡ 3 *Cluſius* in his *Exotickes, lib.* 4. *cap.* 17. giues vs the Hiſtory of an Indian Ferne or Polypody found amongſt the papers of one D^r^. *Nicholas Colie* a Dutch Phyſitian, who died in his returne from the Eaſt-Indies. The root of it was ſix inches long, and almoſt one thicke, of the ſame ſhape and colour as the ordinarie one is: from this came vp three leaues, of which the third was leſſer than the other two; the two larger were eleuen inches long, and their breadth from the middle rib (which was very large) was on each ſide almoſt fiue inches; the edges were diuided almoſt like an Oken caſe: from the middle rib came other veines that ran to the ends of the diuiſions, and betweene theſe be ſmaller veines variouſly diuaricated and netted, which made the leafe ſhew prettily. The colour of it was like that of a dry oken leafe. Where D^r^. *Colie* gathered this it was vncertaine, for he had left nothing in writing. ‡

¶ *The Place.*

It groweth on the bodies of old rotten trees, and alſo vpon old walls, and the tops of houſes: it is likewiſe found among rubbiſh neere the borders of fields, eſpecially vnder trees and thornes, and now and then in woods: and in ſome places it groweth ranke and with a broader leafe, in others not ſo ranke, and with a narrower leafe.

That which groweth on the bodies of old Okes is preferred before the reſt; in ſtead of this moſt do vſe that which is found vnder the Okes, which for all that is not to be termed *Quercinum*, or Polypodie of the Oke.

¶ *The Time.*

Polypody is greene all the yeere long, and may be gathered at any time; it bringeth forth new leaues in the firſt ſpring.

¶ *The Names.*

The Grecians call it πολυπόδιον, of the holes of the fiſhes *Polypi*, appearing in the roots: it is called in Latine, *Polypodium*, after the Greeke name, and many times *Filicula*, as though they ſhould ſay *Parua Filix*, or little Ferne: the Italians name it *Polipodio*: the Spaniards, *Filipodio*, and *Polypodio*: in high Dutch, **Engelfuſz, Baumfarn, Dropffoourtz**: in low Dutch, **Boom varen**: in French, *Polypode*: and we of England, Polypodie: that which groweth vpon the wall we call Polypodie of the wall, and that on the Oke, Polypodie of the Oke.

¶ *The Temperature.*

Polypodie doth dry, but yet without biting as *Galen* writeth.

¶ *The Vertues.*

A *Dioscorides* writeth, that it is of power to purge and to draw forth choler and flegme. *Actuarius* addeth, that it likewise purgeth melancholy: other suppose it to be without any purging force at all, or else to haue very little: of the same minde is also *Iohannes Monardus*, who thinketh it purgeth very gently; which thing is confirmed by Experience, the mistris of things. For in very deed Polypody of it selfe doth not purge at all, but onely serueth a little to make the belly soluble, being boiled in the broth of an old cocke, with Beetes or Mallowes, or other like things that mooue to the stoole by their slipperines. *Ioannes Mesue* reckoneth vp Polypodie among those things that do especially dry and make thin: peraduenture he had respect to a certaine kinde of *Arthritis*, or ache in the joints: in which not one only part of the body, but many together most commonly are touched: for which it is very much commended by the Brabanders and other inhabitants about the riuer Rhene, and the Maze. In this kinde of disease the hands, the feet, and the joints of the knees and elbowes do swell. There is joined withall a feeblenesin mooung, through the extremity of the paine: sometimes the vpper parts are lesse grieued, and the lower more. The humors do also easily run from one place to another, and then settle. Against this disease the Geldres and Cleuelanders do vse the decoction of Polypodie, whereby they hope that the superfluous humours may be wasted and dried vp, and that not by and by, but in continuance of time: for they appoint that this decoction should be taken for certaine daies together.

B But this kinde of gout is sooner taken away either by bloud letting, or by purgations, or by both, and afterwards by sweate; neither is it hard to be cured if these generall remedies be vsed in time: for the humors do not remaine fixed in those joints, but are rather gathered together than settled about them.

C Therefore the body must out of hand be purged, and then that which remaineth is to be wasted and consumed away by such things as procure sweate.

D Furthermore, *Dioscorides* saith, that the root of Polypody is very good for members out of joint, and for chaps betweene the fingers.

E The root of Polypodie boiled with a little honie, water, and pepper, and the quantitie of an ounce giuen, emptieth the belly of cholericke and pituitous humours; some boile it in water and wine, and giue thereof to the quantitie of three ounces for some purposes with good successe.

Chap. 468. *Of Oke-Ferne.*

‡ Our Author here (as in many other places) knit knots, somewhat intricate to loose, for first he confounds in the names and nature the Polypody of the Oke, or lesser Polypodie with the Dryopteris, or Oke-Ferne; but that I haue now put backe to the former chapter, his fit place; then in the second place did he giue the Description of the Dryopteris of the *Aduers.* taken from thence, *pag.* 363. Then were the place, times, names, &c. taken out of the chapter of *Dryopteris Candida* of *Dodonæus*, being, *Pempt.* 3. *lib.* 5. *cap.* 4. But the figure was of the *Filicula fœmina petræa* 4. of *Tabernamontanus*. Now I will in this chapter giue you the *Dryopteris* of the *Aduersaria*, then that of *Dodonæus*, and thirdly that of *Tragus*; for I take them to be different; and this last to be that figured by our Author, out of *Tabernamontanus*. ‡

¶ *The Description.*

1 This kinde of Ferne called *Dryopteris*, or *Filix querna*, hath leaues like vnto the female Ferne before spoken of, but much lesser, smaller, and more finely cut or jagged, and is not aboue a foot high, being a very slender and delicate tender herbe. The leaues are so finely jagged that in shew they resemble feathers, set round about a small rib or sinew; the backe side being sprinckled, not with russet or browne markes or specks, as the other Fernes are, but as it were painted with white spots or markes, not standing out of the leaues in scales, as the spots in the male Ferne, but they are double in each leafe close vnto the middle rib or sinew. The root is long, browne, and somewhat hairy, very like vnto Polypody, but much slenderer, of a sharpe and causticke taste. ‡ *Rondeletius* affirmed that he found the vse of this deadly, being put into medicines in stead of Polypody by the ignorance of some Apothecaries in Dauphenye in France. M[r]. *Goodyer* hath sent me an acurate description together with a plant of this Ferne which I haue thought good here also to set forth. ‡

Dryopteris

‡ *Dryopteris Aduerſ.*
True Oke Ferne.

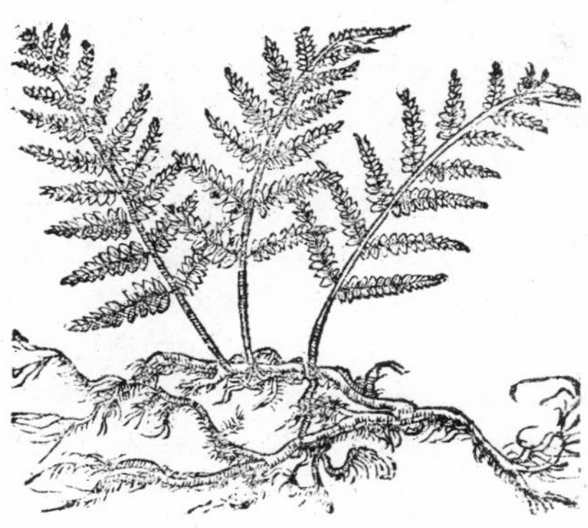

‡ 2 *Dryopteris alba Dod.*
White Oke-Ferne.

‡ 3 *Dryopteris Tragi.*
Tree Ferne.

Dryopteris Penæ & Lobelij.

The roots creepe in the ground or mire, neere the turfe or vpper part thereof, and fold amongſt themſelues, as the roots of *Polypodium* do, almoſt as big as a wheat ſtraw, and about fiue, ſix, or ſeuen inches long, cole blacke without, and white within, of a binding taſte inclining to ſweetneſſe, with an innumerable companie of ſmall blacke fibres like haires growing thereunto. The ſtalkes ſpring from the roots in ſeuerall places, in number variable, according to the length and encreaſe of the root; I haue ſeene ſmall plants haue but one or two, and ſome bigger plants haue fourteene or fifteene: they haue but a two-fold diuiſion, the ſtalke growing from the root, and the nerue bearing the leaues: the ſtalke is about fiue, ſix, or ſeuen inches long, no bigger that a bennet or ſmall graſſe ſtalke, one ſide flat, as are the male Fernes, the reſt round, ſmooth, and green. The firſt paire of nerues grow about three inches from the root, and ſo do all the reſt grow by couples, almoſt exactly one againſt another, in number about eight, nine, or ten couples, the longeſt ſeldome exceeding an inch in length. The leaues grow on thoſe nerues alſo by couples, eight or nine couples on a nerue, without any nickes or indentures, of a yellowiſh greene colour. This Ferne may be ſaid to be like *Polypodium* in his creeping root, like the male Ferne in his stalke, and like the female Ferne in his nerues and leaues. I could finde no seed-scales on the backesides of any of the leaues of this Ferne. Many yeares past I found this same in a very wet moore or bog, being the land of *Richard Austen*, called Whitrow Moore, where Peate is now digged, a mile from Peters-field in Hampshire; and this sixth of Iuly, 1633, I digged vp there many plants, and by them made this description. I neuer found it growing in any other place: the leaues perish at Winter, and grow vp againe very late in the Spring. *Iohn Goodyer*. Iuly 6. 1633.

2 *Dodonæus* thus describes his: *Dryopteris* (saith he) doth well resemble the male Ferne, but the leaues are much smaller, and more finely cut, smooth on the foresaid, and of a yellowish green together with the stalkes and middle nerues; on the backe it is rough as other Fernes, and also liueth without stalke or seed. The root consists of fibres intricately folded together, of an indifferent thicknesse, here and there putting vp new buds. This is the *Adianthum* of the *Aduers.* who affirme the vse thereof to be safe, and not pernitious and deleterie, as that of *Dryopteris*. It thus differs from the former; the leaues of this are not set directly one opposite to another, the diuisions of the leaues are larger and more diuided. The root is more threddy, and creepes not so much as that of the former.

3 This (which is *Clusius* his *Filix pumila saxatilis prima*, and which I take to be the *Dryopteris* or *Filix arborea* of *Tragus*) hath blacke slender long creeping roots, with few small hard hairy fibres fastned to them, of a very astringent taste: from these rise vp sundry stalkes a foot high, diuided into certaine branches of winged leaues, like to those of the female Ferne, but much lesse, tenderer and finer cut, and hauing many blackish spots on their lower sides. This differs from the two former, in that the leaues are branched, which is a chiefe difference; and *Bauhinus* did very well obserue it, if he had as well followed it, when he diuided *Filix* into *ramosa & nonramosa*. ‡

¶ *The Place.*

It is oftentimes found in sunny places, in the vallies of mountains and little hils, and in the tops of the trunks of trees in thicke woods.

¶ *The Time.*

The leaues hereof perish in Winter; in the Spring new come forth.

¶ *The Names.*

This is called in Greeke Δρυοπτερίς: in Latine, *Querna Filix*: *Oribasius* in his eleuenth book of physicall Collections calleth it *Bryopteris*, of the mosse with which it is found; for, as *Dioscorides* writeth, it groweth in the mosse of Okes. The Apothecaries in times past miscalled it by the name of *Adiantum*: but they did worse in putting it in compound medicines in stead of *Adiantum*. *Valerius Cordus* calleth it *Pteridion*: in low-Dutch, **Eijcken varen**: the Spaniards, *Helecho de Roble*: it is named in English, Oke-Ferne, Petty-Ferne; and it may most fitly be called Mosse-Ferne.

¶ *The Temperature and Vertues.*

A Oke-Ferne hath many tastes, it is sweet, biting, and bitter, it hath in the root a harsh or choking taste, and a mortifying qualitie, and therefore it taketh away haires. *Dioscorides* saith further, that Oke-Ferne stamped roots and all is a remedie to root vp haires, if it be applied to the body after sweating, the sweat being wiped away.

Chap. 469. *Of blacke Oke-Ferne.*

¶ *The Description.*

1 There is also a certaine other kinde of Ferne like to the former Oke-Ferne of *Dodonæus* his description, but the stalkes and ribs of the leaues are blackish, and the leaues of a deeper greene colour: this groweth out also immediately from the root, and is likewise diuersly, but not so finely indented: the root is made vp of many strings, not vnlike to the male Ferne, but much lesser.

2 The female blacke Ferne is like vnto the male, sauing his leaues are not so sharpe at the points, more white and broad than the male, wherein consisteth the difference.

¶ *The Place.*

They grow likewise vpon trees in shadowie woods, and now and then in shadowie sandy banks, and vnder hedges.

¶ The

1 *Onopteris mas.*
The male blacke Ferne.

¶ *The Time.*

They remaine greene all the yeare long, otherwise than Polypodie & Maidens haire do; yet do they not cease to bring forth new leaues in Summer: they are destitute of floures and seed, as is the former.

¶ *The Names.*

This is called of diuers of the later Herbarists, *Dryopteris nigra*, or blacke Oke-Ferne, of the likenesse that it hath with *Dryopteris*; which we haue called in English, Oke-Ferne, or mosse Ferne: of others, *Adiantum nigrum*, or blacke Maidens haire, that it may differ from the former, which is falsly called *Adiantum*. There are of the later Herbarists who would haue it to be *Lonchitis aspera*, or rough Spleen-wort; but what likenesse hath it with the leaues of *Scolopendrium*? none at all: therfore it is not *Lonchitis aspera*, much lesse *Adiantum Plinij*, which differeth not from *Adiantum Theophrasti*; for what he hath of *Adiantum*, the same he taketh out of *Theophrastus*: the right *Adiantum* we will describe hereafter. Notwithstanding blacke Oke-Ferne was vsed of diuers vnlearned Apothecaries of France and Germany for *Adiantum*, or Maiden-haire of Lumbardy: but these men did erre in doing so; yet not so much as they who take Polypodie of the Oke for the true Maiden-haire.

¶ *The Temperature and Vertues.*

The blacke Oke-Ferne hath no stipticke qualitie at all, but is like in facultie to *Trichomanes*, or English Maiden-haire.

CHAP. 470. *Of Harts-tongue.*

¶ *The Description.*

1 THe common kinde of Harts-tongue, called *Phyllitis*, that is to say, a plant consisting only of leaues, bearing neither stalke, floure, nor seed, resembling in shew a long tongue, whereof it hath been and is called in shops *Lingua ceruina*, that is, Harts tongue: these leaues are a foot long, smooth and plaine vpon one side, but vpon that side next the ground straked ouerthwart with certaine long rough markes like small wormes, hanging on the backside thereof. The root is blacke, hairy, and twisted, or so growing as though it were wound together.

2 The other kind of Ferne, called *Phyllitis multifida*, or *Laciniata*, that is, iagged Harts tongue, is very like vnto the former, sauing that the leaues thereof are cut or iagged like a mans hand, or the palme and browantles of a Deare, bearing neither stalke, floure, nor seed.

3 There is another kinde of Harts-tongue called *Hemionitis*, which hath bred some controuersie among writers: for some haue tooke it for a kinde of Harts-tongue, as it is indeed; others describe it as a proper plant by it selfe, called *Hemionitis*, of ἡμίονος, that is, *Mulus*, a Mule, because Mules do delight to feed thereon: it is barren in seeds, stalkes, and floures, and in shape it agreeth very well with our Harts-tongue: the roots are compact of many blackish haires: the leaues are spotted on the backside like the common Harts-tongue, and differ in that, that this *Hemionitis* in the base or lowest parts of the leaues is arched after the manner of a new Moone, or a forked arrow, the yongest and smallest leaues being like vnto the great Binde-weed, called *Volubilis*.

4 There is a kinde of Ferne called likewise *Hemionitis sterilis*, which is a very small and base herbe not aboue a finger high, hauing foure or fiue small leaues of the same substance and colour, spotted on the backe part, and in taste like Harts-tongue; but the leaues beare the shape of them of *Totabona*, or good *Henry*, which many of our Apothecaries do abusiuely take for Mercurie: The roots are very many, smooth, blacke, and threddie, bearing neither stalke, floure, nor seed: this plant

1 *Phyllitis.*
Harts-tongue.

2 *Phylliiis multifida.*
Finger Harts-tongue.

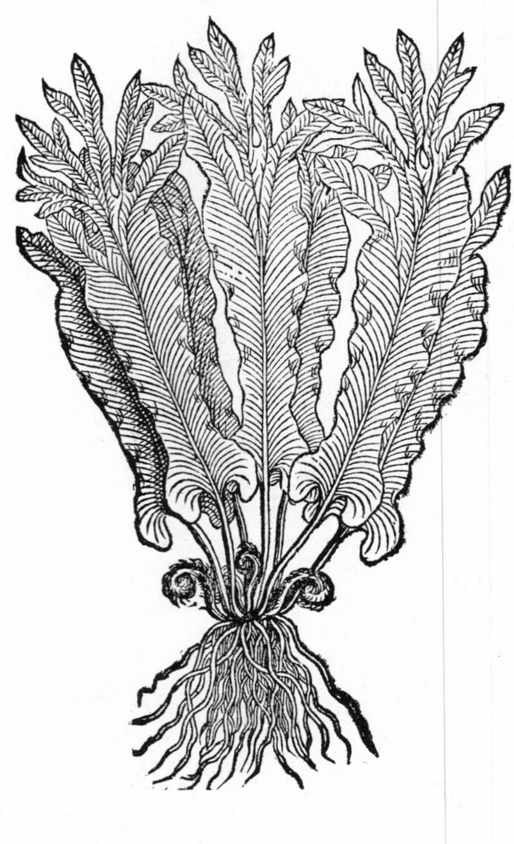

‡ 3 *Hemionitis maior.*
Mules Ferne, or Moone-Ferne.

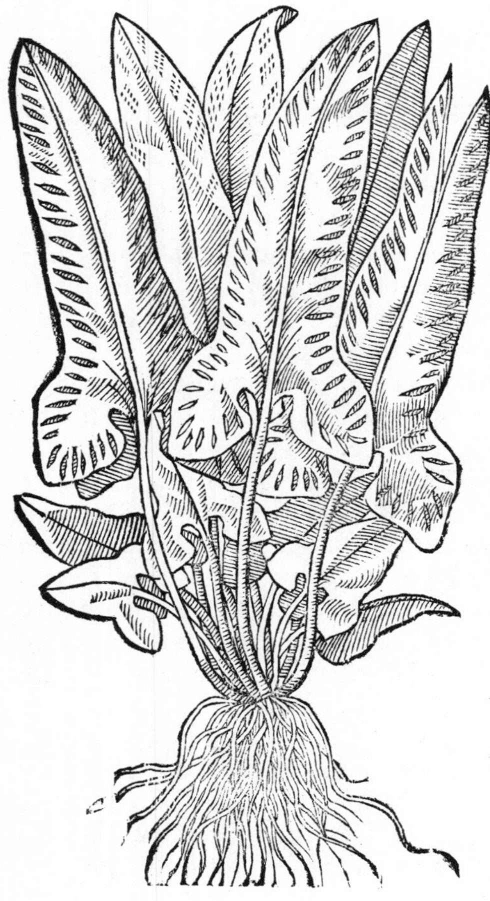

‡ 4 *Hemionitis minor.*
Small Moone-Ferne.

‡ 5 *Hemionitis perigrina.*
Handed Moone-Ferne.

my very good friend Mr. *Nicholas Belson* found in a grauelly lane in the way leading to Oxey parke neere vnto Watford, fifteene miles from London: it growes likewise on the stone walls of Hampton Court, in the garden of Mr. *Huggens*, keeper of the said house or pallace.

5 There is a kinde of Ferne called also *Hemionitis*, but with this addition *Peregrina*, that is very seldome found, and hath leaues very like to Harts-tongue, but that it is palmed or branched in the part next the ground, almost in manner of the second *Phyllitis*, at the top of the leaues; otherwise they resemble one another in nature and forme.

¶ *The Place.*

The common Harts-tongue groweth by the waies sides as you trauell from London to Excester in great plenty, in shadowie places, and moist stonie vallies and wels, and is much planted in gardens.

The second I found in the garden of Master *Cranwich* a Chirurgion dwelling at Much-dunmow in Essex, who gaue me a plant for my garden.

‡ Mr. *Goodyer* found it wilde in the banks of a lane neere Swaneling, not many miles from Southampton. ‡

It groweth vpon Ingleborough hils, and diuers other mountains of the North of England.

¶ *The Time.*

It is greene all the yeare long, yet lesse greene in winter: in Sommer it now and then bringeth forth new leaues.

¶ *The Names.*

It is called in Greeke φυλλῖτις: in Latine also *Phyllitis*: in shops, *Lingua ceruina*: and falsely *Scolopendria*, for it differeth much from the right *Scolopendria*, or Stone Ferne: it is called in high Dutch, Hirtzong: in low Dutch, Herstonge: in Spanish, *Lengua ceruina*: in French, *Langue de Cerf*: in English, Harts-tongue: of some, Stone Harts-tongue: *Apuleius* in his 83. Chapter nameth it *Radiolus*.

¶ *The Temperature.*

It is of a binding and drying facultie.

¶ *The Vertues.*

A This common Harts-tongue is commended against the laske and bloudy flix: *Dioscorides* teacheth, that being drunke in wine it is a remedie against the bitings of serpents.

B It doth open the hardnesse and stopping of the spleen and liuer, and all other griefes proceeding of oppilations or stoppings whatsoeuer.

CHAP. 471. *Of Spleene-woort, or Milt-waste.*

¶ *The Description.*

1 Spleene-woort being that kinde of Ferne called *Asplenium*, or *Ceterach*, and the true *Scolopendria*, hath leaues a span long, iagged or cut vpon both sides, euen hard to the middle ribbe; euery cut or incisure being as it were cut halfe round (whereby it is knowne from the rough Spleene-woort) not one cut right against another, but one besides the other, set in seuerall order, being slipperie and greene on the vpper side, soft and downie vnderneath; which when they be withered are folded vp together like a scrole, and hairie without, much like to the rough Beare-worme wherewith men bait their hookes to catch fish: the root is small, blacke, and rough, much platted or interlaced, hauing neither stalke, floure, nor seeds.

2 Rough

1 *Asplenium sive Ceterach.*
Spleenewoort or Miltwaste.

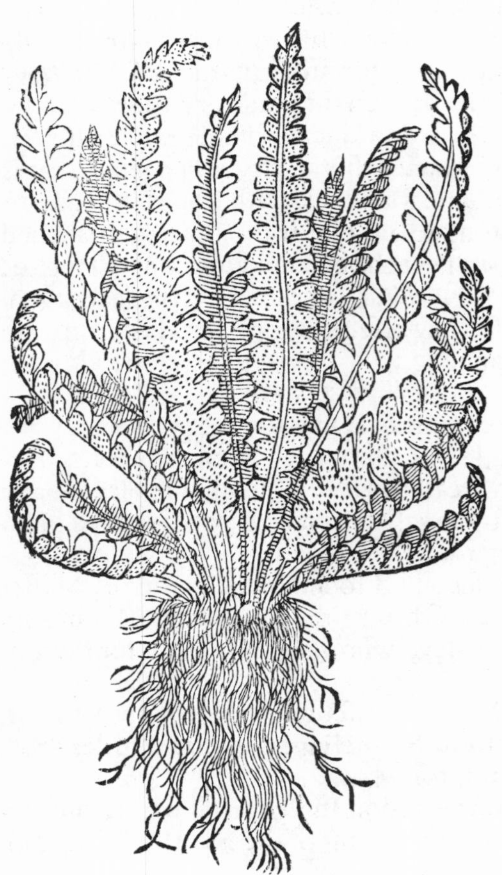

2 *Lonchitis aspera.*
Rough Spleenewoort.

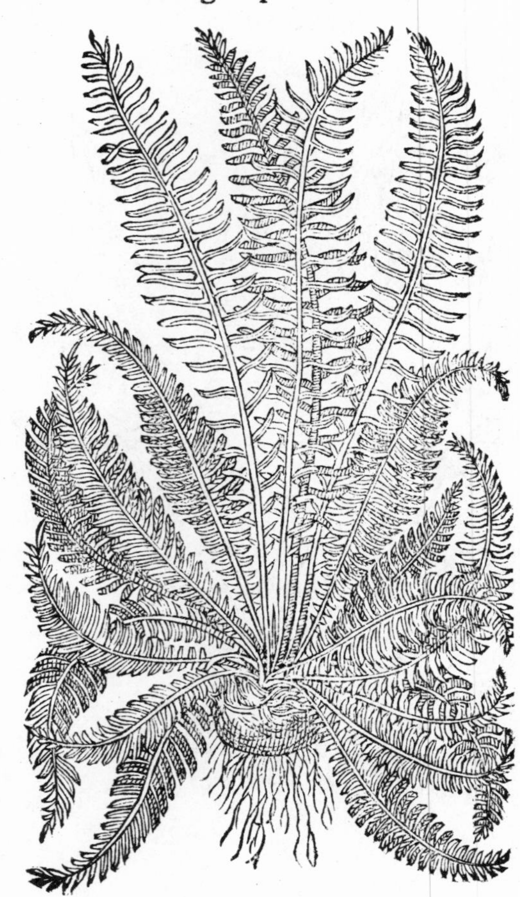

† 3 *Lonchitis aspera maior.*
Great rough Spleene-woort.

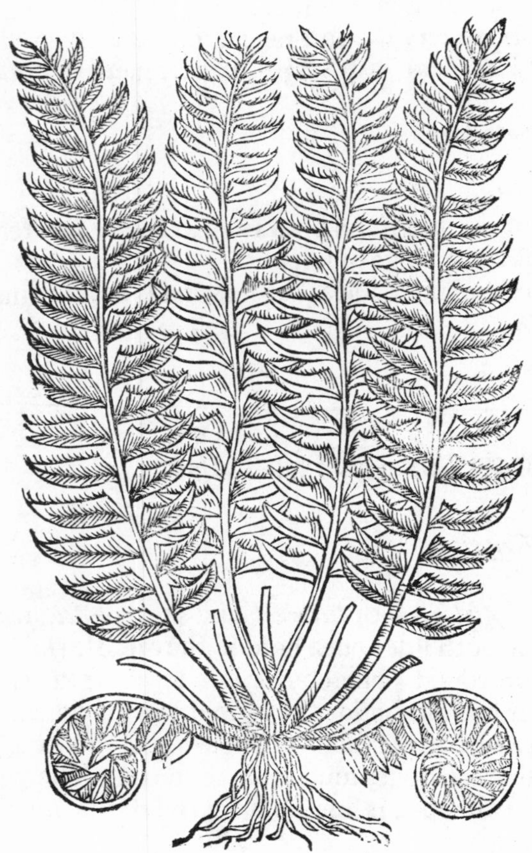

† 4 *Lonchitis Maranthæ.*
Bastard Spleene-woort.

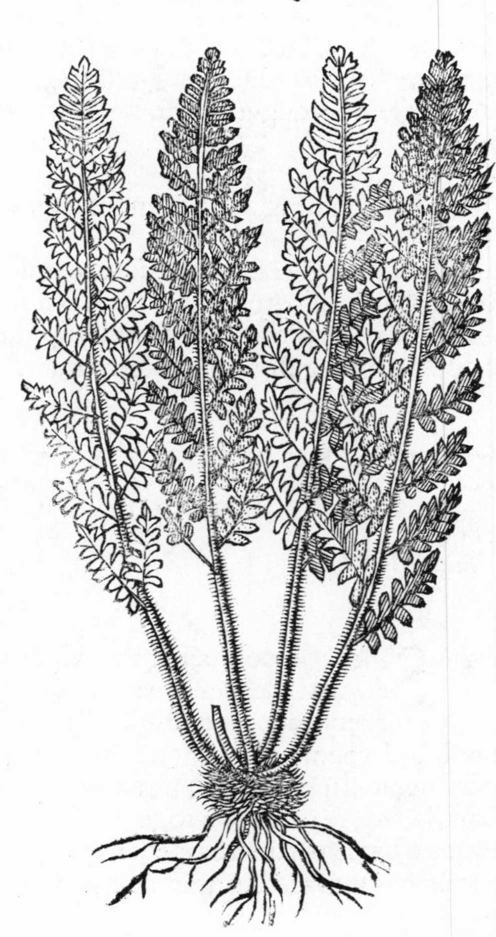

2 Rough Spleenewoort is partly like the other Fernes in shew, and beareth neither stalke nor seed, hauing narrow leaues a foot long, and somewhat longer, slashed on the edges euen to the middle rib, smooth on the vpper side, and of a swart greene colour; vnderneath rough, as is the leaues of Polypodie: the root is blacke, and set with a number of slender strings.

‡ 3 This greater Spleenwoort hath leaues like *Ceterach*, of a spanne long, somewhat resembling those of Polypodie, but that they are more diuided, snipt about the edges, and sharpe pointed: the root is fibrous and stringie. This growes on the rockes and mountainous places of Italy, and is the *Lonchitis aspera maior* of *Matthiolus* and others. ‡

4 This kinde of Spleenewoort is not onely barren of stalks and seeds, but also of those spots and marks wherewith the others are spotted: the leaues are few in number, growing pyramidis or steeplewise, great and broad below, and sharper toward the top by degrees: the root is thick, black, and bushie, as it were a Crowes nest.

¶ *The Place.*

Ceterach groweth vpon old stone walls and rockes, in darke and shadowie places throughout the West part of England; especially vpon the stone walls by Bristow, as you go to Saint Vincents Rocke, and likewise about Bathe, VVells, and Salisburie, where I haue seene great plentie thereof.

The rough Spleenwoort groweth vpon barren heaths, drie sandie bankes, and shadowie places in most parts of England, but especially on a heath by London called Hampstead heath, where it groweth in great abundance.

¶ *The Names.*

Spleene-woort or Milt-waste is called in Greeke, ἀσπλήνιον: in Latine likewise, *Asplenium*, and also *Scolopendria*: of *Gaza*, *Mula herba*: in shops, *Ceterach*: in high Dutch, **Steynfarn**: in low Dutch, **Steynuaren**, and **Miltcruyt**: in English, Spleenwoort, Miltwaste, Scaleferne, and Stoneferne: it is called *Asplenion*, because it is speciall good against the infirmities of the Spleene or Milt, and *Scolopendria*, of the likenesse that it hath with the Beare-worme, before remembred.

Rough Miltwaste is called of diuers of the later writers *Asplenium syluestre*, or wilde Spleenwoort: of some, *Asplenium magnum*, or great Spleene-woort: *Valerius Cordus* calleth it *Strutiopteris*: and *Dioscorides*, *Lonchitis aspera*, or rough Spleene-woort: in Latine according to the same Authour, *Longina*, and *Calabrina*: in English, rough Spleen-wort, or Miltwaste.

¶ *The Temperature.*

These plants are of thin parts, as *Galen* witnesseth, yet are they not hot, but in a meane.

¶ *The Vertues.*

Dioscorides teacheth, that the leaues boiled in wine and drunk by the space of forty daies, do take away infirmities of the spleen; help the strangurie, and yellow iaundice, cause the stone in the bladder to moulder nnd passe away, all which are performed by such things as be of thinne and subtill parts: he addeth likewise that they stay the hicket, or yeoxing, and also hinder conception, either inwardly taken, or hanged about the partie, and therefore, saith *Pliny*, Spleenewoort is not to be giuen to women, because it bringeth barrennesse. A

There be Empericks or blinde practitioners of this age, who teach, that with this herbe not onely the hardnesse and swelling of the Spleene, but all infirmities of the liuer also may be effectually, and in very short time remooued, insomuch that the sodden liuer of a beast is restored to his former constitution againe, that is, made like vnto a raw liuer, if it bee boyled againe with this herbe. B

But this is to be reckoned among the old wiues fables, and that also which *Dioscorides* telleth of, touching the gathering of Spleenewoort in the night, and other most vaine things, which are found here and there scattered in old books: from which most of the later Writers do not abstaine, who many times fill vp their pages with lies and friuolous toies, and by so doing do not a little deceiue yong students. C

† Formerly vnder the title of *Lonchitis Maranthae* was put the figure now in the third place, and the figure which should haue beene there, was in the third place of the next chapter, vnder the title of *Filicula petraea mas*.

CHAP.

CHAP. 472. *Of diuers small Fernes.*

¶ *The Description.*

1 THis small or dwarfe Ferne, which is seldome found except in the banks of stony fountaines, wells, and rockes bordering vpon riuers, is very like vnto the common Brakes in leaues, but altogether lesser: the root is composed of a bundle of blacke threddie strings.

2 The female, which is found likewise by running streames, wells, and fountaines, vpon rockes, and stonie places, is like the precedent, but is a great deale smaller, blacker of colour, fewer rootes, and shorter.

1 *Filicula fontana mas.*
The male fountaine Ferne.

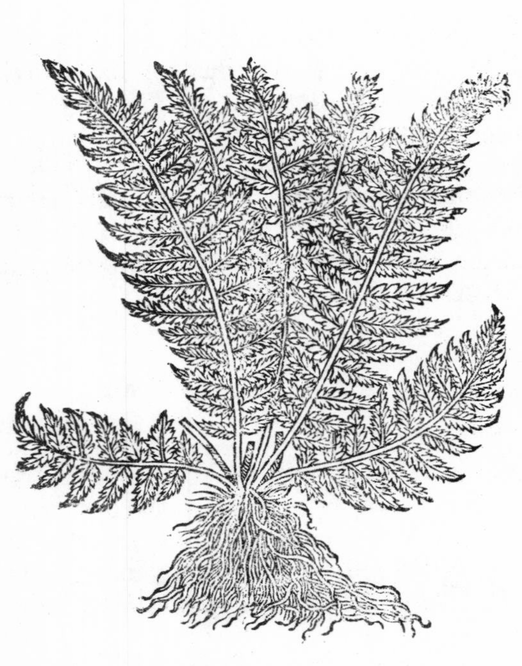

† 3 *Filicula petræa mas.*
The male dwarfe stone Ferne.

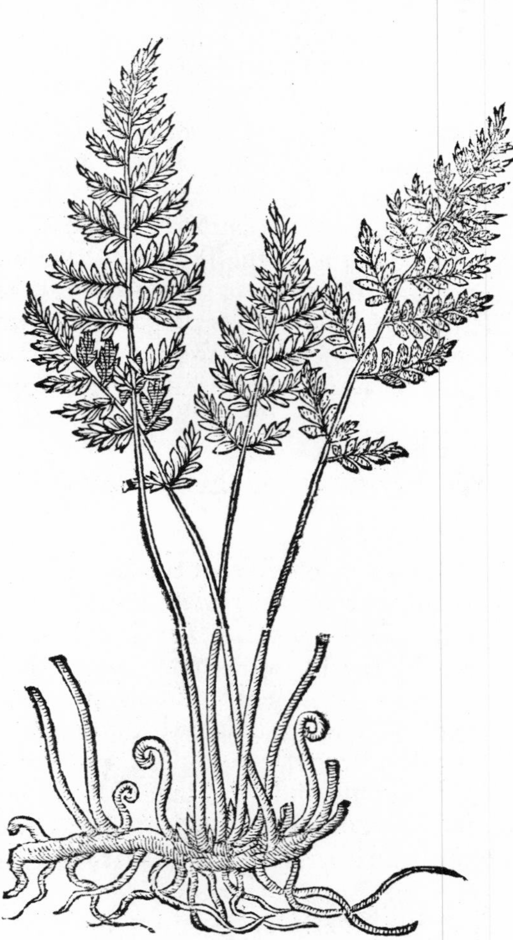

3 The male dwarfe Ferne that groweth vpon the stonie mountaines of the North and West parts of England, especially toward the sea, and also in the ioints of stone walls among the morter, hath small leaues deepely cut on both sides, like vnto *Ceterach* or Spleene-woort, barren both of seeds and stalks, as also of those spots or markes that are to be seene vpon the backe part of the other Fernes: the root creepeth along, set with some few hairie strings, resembling those of the Oke Ferne, called *Dryopteris.*

4 The female stone Ferne hath diuers long leaues rising from a threddy root, contrarie to that of the male, composed of many small leaues finely minced or cut like the teeth of a saw, of a whitish green colour, without any spots or marks at all, seeds or stalks, which groweth vnder shadowie rocks, and craggie mountaines in most places. ‡ From a small root composed of many blacke, hairie, and intricately folding strings, come vp many leaues two or three inches high, stiffe, thicke, darke greene, and shining: in the diuision, growth, position, shape and taste, it resembles the male Ferne, and hath also rustie spots on the backe: the middle ribbe and stalke is of a shining brownish

4 *Filicula petræa fœmina, sive Chamæfilix marina Anglica.*
The female dwarfe stone Ferne.

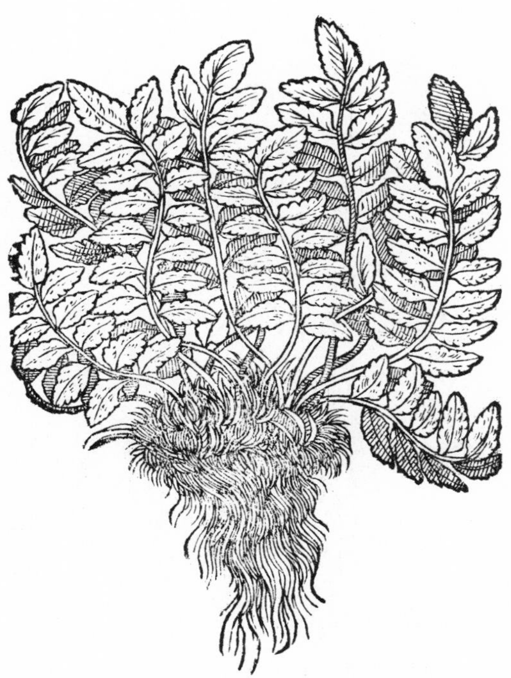

nish silken colour: it growes in the chinkes of the rockes by the Sea side in Cornewall.

¶ *The Place.*

The place is sufficiently touched in the description.

¶ *The Time.*

They flourish both Winter and Sommer, for when the leaues wither by reason of age, there arise young to supplie the place, so that they are not to be seene without greene and withered leaues both at once.

¶ *The Names.*

It sufficeth what hath bin said of the names in their seuerall titles: notwithstanding the last described we haue called *Chamæfilix marina Anglica*: which groweth vpon the rockie cliffe neere Harwich, as also at Douer, among the Sampire that there groweth.

¶ *The Temperature and Vertues.*

Their temperature and faculties in working A are referred vnto the kindes of blacke Oke Fernes, called *Dryopteris*, and *Onopteris*.

† It is hard to say what our Author in this chapter meant, by his figures and descriptions, wherefore I haue left his descriptions as I found them: the second figure which was very like the first I haue omited: for the third, which was of the *Lonchitis Maranthæ*, mentioned in the foregoing chapter, I haue put *Clusius* his figure of his *Filix saxatilis* 2. which growes in such places, and reasonable well fits our Authors description: in the fourth place I haue put *Lobbells Chamæfilix marina Anglica*, and his description, whch our Author, as I iudge, intended in that place to haue giuen vs.

CHAP. 473. *Of true Maiden-haire.*

¶ *The Kindes.*

THeophrastus and *Pliny* haue set downe two Maiden-haires, the blacke and the white, whereunto, may be added another called *Ruta muraria*, or wall Rue, equall to the others in facultie, whereof we will intreat.

¶ *The Description.*

1 *Capillus Veneris verus.*
True Maiden-haire.

1 WHoso will follow the variable opinions of writers concerning the Ferne called *Adianthum verum*, or *Capillus Veneris verus*, must of necessitie be brought into a labyrinth of doubts, considering the diuers opinions thereof: but this I know that Venus-haire, or Maiden-haire, is a low herb growing an hand high, smooth, of a darke crimson colour, and glittering withall: the leaues be smal, cut in sunder, and nicked in the edges something like those of Coriander, confusedly or without order placed, the middle rib whereof is of a blacke shining colour: the root consisteth of manie small threddie strings.

2 This Assyrian Maiden-haire is likewise a base or low herbe, hauing leaues, flat, smooth, and plaine, set vpon a blackish middle rib, like vnto that of the other Maiden-haire, cut or notched in the edges, nature keeping no certaine forme, but making one leafe of this fashion, and another far different from it: the root is tough and threddie.

3 This plant which we haue inserted among the Adianthes as a kinde thereof, may without errour so passe, which is in great request in Flanders and Germanie, where the practitioners in Physicke do vse the same in stead of *Capillus Veneris*, and with better successe than any of the Capillare herbs,

herbs,although *Matthiolus* and *Dioſcorides* himſelfe hath made this wall Rue to be a kinde of *Paronychia*,or Nailewoort:notwithſtanding the Germanes wil not leaue the vſe thereof,but receiue it as the true Adianth,eſteeming it equal,if not far better,than either *Ceterach*, *Capillus Veneris verus*, or *Tricomanes*,called alſo *Polytrichon* : it bringeth forth very many leaues, round and ſlender, cut into two or three parts,very hard in handling,ſmooth and greene on the outſide,of an ill fauoured dead colour vnderneath,ſet with little fine ſpots,which euidently ſheweth it to be a kinde of Ferne: the root is blacke and full of ſtrings.

2 *Capillus Veneris Syriaca.*
Aſſyrian Maiden-haire.

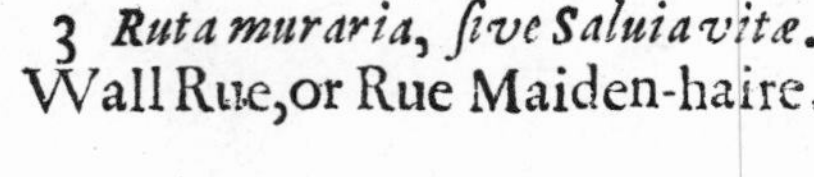

3 *Ruta muraria, ſive Saluia vitæ.*
Wall Rue,or Rue Maiden-haire.

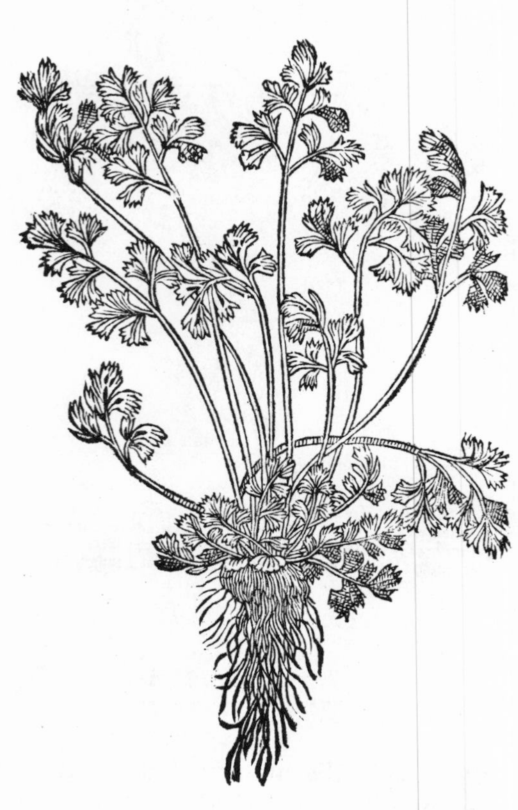

¶ *The Place.*

The right Maiden-haire groweth vpon walls, in ſtonie, ſhadowie,and moiſt places, neere vnto fountaines,and where water droppeth : it is a ſtranger in England: notwithſtanding I haue heard it reported by ſome of good credit, that it groweth in diuers places of the Weſt countrey of England.

The Aſſyrian Maiden-haire taketh his ſurname of his natiue countrey Aſſyria,it is a ſtranger in Europe.

Stone Rue groweth vpon old walls neere vnto waters,wells,and fountaines : I found it vpon the wall of the churchyard of Dartford in Kent, hard by the riuer ſide where people ride through, and alſo vpon the walls of the Churchyard of Sittingburne in the ſame Countie, in the middle of the towne hard by a great lake of water,and alſo vpon the Church walls of Railey in Eſſex, and diuers other places.

¶ *The Time.*

Theſe plants are greene both winter and ſommer,and yet haue neither floures nor ſeed.

¶ *The Names.*

Maiden-haire is called in Greek Ἀδίαντον : *Theophraſtus* and *Pliny* name it *Adiantum nigrum*,or black Maiden-haire : for they ſet downe two Maiden-haires, the blacke and the white, making this the blacke,and the Rue of the wall the white:it is called in Latine *Polytrichum*,*Callitrichum*, *Cincinalis*, *Terræ Capillus*,*Supercilium terræ*: of *Apuleius*,*Capillus Veneris*,*Capillaris*,*Crinita*: & of diuers,*Coriandrum putei* : the Italians keepe the name *Capillus Veneris* : in Engliſh, blacke Maiden haire,and Venus haire, and it may be called our Ladies haire.

It

It is called *Adianton* because the leafe, as *Theophrastus* saith, is neuer wet, for it casteth off water that falleth thereon, or being drowned or couered in water, it remaineth still as if it were dry, as *Pliny* likewise writeth; and is termed *Callitricon* and *Polytricon*, of the effect it hath in dying haire, and maketh it to grow thicke.

VVall Rue is commonly called in Latine, *Ruta muraria*, or *Ruta muralis*: of some, *Saluia vitæ*, but wherefore I know not, neither themselues, if they were liuing: of the Apothecaries of the Low-Countries *Capillus Veneris*, or Maiden haire, and they haue vsed it a long time for the right Maiden haire; it is that kinde of *Adiantum* which *Theophrastus* termed *Adiantum Candidum*, or white Maiden haire, for he maketh two, one blacke, and the other white, as we haue said. *Pliny* doth likewise set downe two kindes, one he calleth *Polytricon*; the other, *Tricomanes*, or English Maiden-haire, whereof we will intreate in the chapter following, which he hath falsely set downe for a kinde of *Adiantum*, for *Tricomanes* doth differ from *Adiantum*.

Some there be that thinke, Wall-Rue is *Paronychia Dioscoridis*, or *Dioscorides* his Whitlow wort, wherein they haue been greatly deceiued: it is called in high Dutch, **Mauerrauten**: in low Dutch, **Steencruyt**: in French, *Rue de muraille*: in English, Wall-Rue, and white Maiden-haire.

¶ *The Temperature and Vertues.*

A The true Maiden haire, as *Galen* testifieth, doth dry, make thin, waste away, and is in a meane betweene heate and coldnesse: *Mesues* sheweth that it consisteth of vnlike or disagreeing parts, and that some are watery and earthy, and the same binding, and another superficially hot and thinne: And that by this it taketh away obstructions or stoppings, maketh things thinne that are thicke, looseneth the belly, especially when it is fresh and greene: for as this part is thin, so is it quickly resolued, and that by reason of his binding and earthy parts: it stoppeth the belly, and stayeth the laske and other fluxes.

B Being drunke it breaketh the stone, and expelleth not onely the stones in the kidnies, but also those which sticke in the passages of the vrine.

C It raiseth vp grosse and slimie humors out of the chest and lungs, and also those which sticke in the conduits of the winde pipe, it breaketh and raiseth them out by spetting, if a loch or licking medicine be made thereof.

D Moreouer, it consumeth and wasteth away the Kings-euill, and other hard swellings, as the same Author affirmeth, and it maketh the haire of the head or beard to grow that is fallen and pilled off.

E *Dioscorides* reckoneth vp many vertues and operations of this Maiden-haire, which do not onely differ, but are also contrary one to another. Among others he saith, that the same stancheth bloud: and a little before, that it draweth away the secondines, and bringeth downe the desired sickenes. which words do confound one another with contrarieties; for whatsoeuer things do stanch bloud, the same do also stay the termes.

F He addeth also in the end, that it is sowne about sheepe-folds for the benefit of the sheepe, but what that benefit should be, he sheweth not.

G Besides, that it cannot be sowne, by reason it is without seed, it is euident, neither can it fitly be remooued. Therefore in this place it seemeth that many things are transposed from other places, and falsly added to this chapter: and peraduenture some things are brought hither out of discourse of *Cytisus*, or Milke Trefoile, whereof here to write were to small purpose.

H Wall-Rue is not much vnlike to blacke Maiden-haire in temperature and facultie.

I Wall-Rue is good for them that haue a cough, that are short winded, and that be troubled with stitches and paine in their sides.

K Being boiled, it causeth concoction of raw humors which sticke in the lungs; it taketh away the paine of the kidnies and bladder, it gently prouoketh vrine, and driueth forth stones.

L It is commended against ruptures in young children, and some affirme it to be excellent good, if the powder thereof be taken continually for forty daies together.

CHAP. 474. *Of English, or common Maiden-haire.*

¶ *The Description.*

1 English Maiden-haire hath long leaues of a darke green colour, consisting of very many small round leaues set vpon a middle rib, of a shining blacke colour, dashed on the nether side with small rough markes or speckes, of an ouerworne colour: the roots are small and threddy.

1 *Trichomanes mas.*
The Male English Maiden-haire.

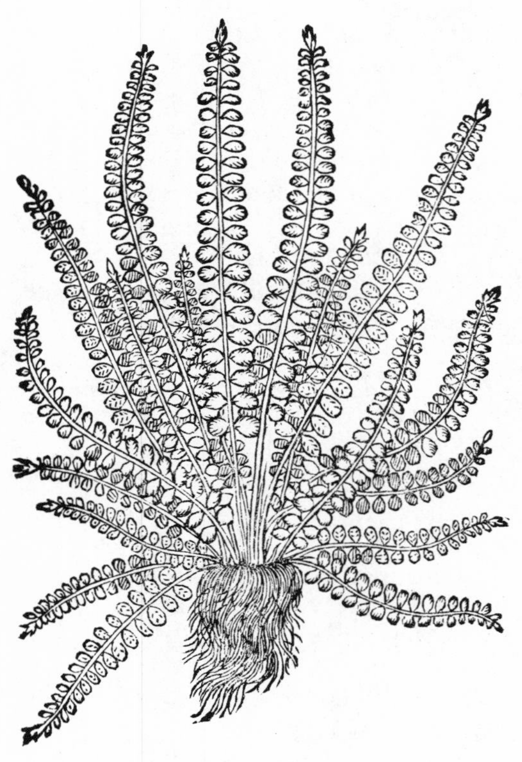

2 The female English Maiden haire is like vnto the precedent, sauing that it is lesser, and wanteth those spots or markes that are in the other, wherein consisteth the difference. ‡ Our Authors figure was of the *Trochomanes fœmina* of *Tabernamontanus*, which expresses a variety with branched leaues, and therein only was the difference. ‡

¶ *The Place.*

It growes for the most part nere vnto springs and brookes, and other moist places, vpon old stone walls and rockes: I found it growing in a shadowie sandie lane in Betsome, in the parish of Southfleet in Kent, vpon the ground whereas there was no stones or stony ground neere vnto it, which before that time I did neuer see; it groweth likewise vpon stone walls at her Majesties palace of Richmond, & in most stone wals of the West and North parts of England. ‡ Mr. *Goodyer* saith, that in Ianuary, 1624. he saw enough to lade an horse growing on the bancks in a lane, as he rode betweene Rake and Headly in Hampshire neere Wollmer Forrest. ‡

¶ *The Time.*

It continueth a long time, the coldnesse of winter doth it no harme, it is barren as the other Fernes are, whereof it is a kinde.

¶ *The Names.*

It is called in Greeke, τριχόμανες: in Latine, *Filicula*, as though we should say, *Parua Filex*, or little Ferne; also *Capillaris*: in shops, *Capillus Veneris*. *Apuleius* in his 51 chapter maketh it all one with *Callitrichon*: of some it is called *Polytrichon*: in English, common Maiden-haire.

¶ *The Temperature and Vertues.*

A These, as *Dioscorides* and *Galen* do write, haue all the faculties belonging to *Adiantum*, or blacke Maiden-haire.

B The decoction made in wine and drunke, helpeth them that are short winded, it helpeth the cough, ripeneth tough flegme, and auoideth it by spitting.

C The lie wherein it hath been sodden, or laid to infuse, is good to wash the head, causing the scurfe and scales to fall off, and haire to grow in places that are pild and bare.

Chap. 475. *Of Thistles.*

¶ *The Kindes.*

THe matter of the Thistles is diuers, some Thistles serue for nourishment, as the Artichoke without prickles, and the Artichoke with prickles; other for medicine, as the root of Carline which is good for many things; the blessed thistle also, otherwise called *Carduus benedictus*; Sea Huluer, and diuers others: some are poisonsome, as *Chamæleon niger*; one smooth, plaine, and without prickles, as the Thistle called Beares Breech, or *Acanthus satiuus*, whereof there is another with prickles, which we make the wilde, of the which two we intend to write in this chapter.

¶ *The Description.*

1 BEares breech of the garden hath broad leaues, smooth, somewhat blacke, gashed on both the edges, and set with many cuts and fine nickes: betweene which riseth vp in the midst a big stalke brauely deckt with floures, set in order from the middle vpward, of colour white, of forme long, which are armed as it were with two catkins, one higher, another lower: after them grow forth the huskes, in which is found broad seed: the roots be blacke without, and white within,

in, andfull of clammie iuice, and are diuided into many off-springs, which as they creepe far, so do they now and then bud forth and grow afresh: these roots are so full of life, that how little soeuer of them remaine, it oftentimes also bringeth forth the whole plant.

1 *Acanthus satiuus.*
Garden Beares-breech.

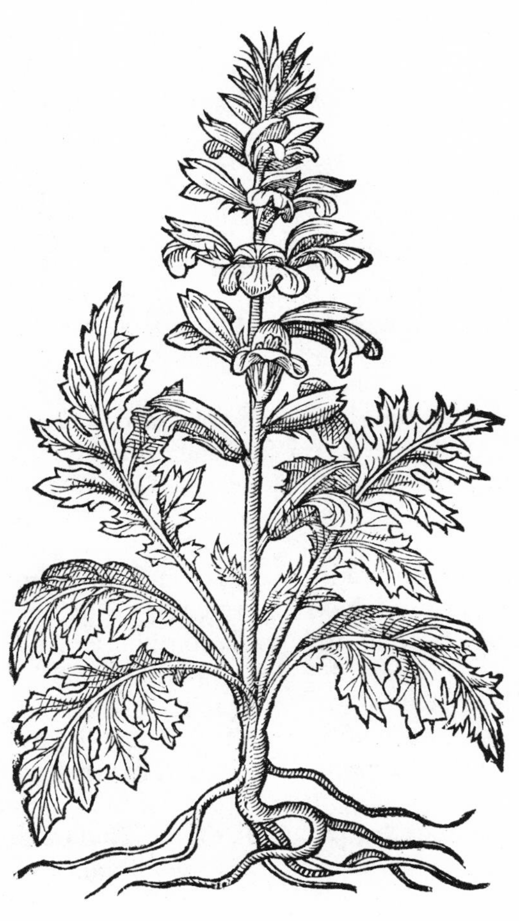

‡ 2 *Acanthus syl. aculeatus.*
Prickley Beares-breech.

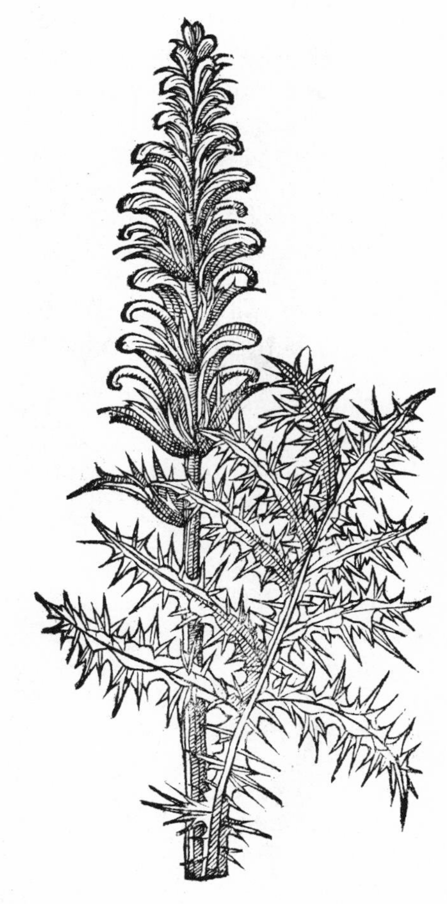

2 Wilde Beares-breech,called *Acanthus syluestris,Pena* setteth forth for *Chameleonta Monspeliensium*,and reporteth that he found it growing amongst the grauelly and moist places neere to the walls of Montpellier,and at the gate of Aegidia, betweene the fountaine and the brooke neere to the wall: this thistle is in stalke, floures, colour of leaues and seed like the first kinde, but shorter and lower,hauing large leaues,dented or iagged with many cuts and incisions, not onely in some few parts of the leaues,as some other Thistles,but very thickly dented or clouen,and hauing many sharpe,large, white and hard prickles about the sides of the diuisions and cuts,not very easie to be handled or touched without danger to the hand and fingers.

¶ *The Place.*

Dioscorides writeth,that garden Branke Vrsine groweth in moist and stonie places, and also in gardens: it were vnaduisedly done to seeke it in either of the Germaines any where, but in gardens onely; in my garden it doth grow very plentifully.

The wilde was found in certain places of Italy neere to the sea,by that notable learned man *Alfonsus Pancius*,Physition to the Duke of Ferrara,and professor of simples and Physick,and is a stranger in England.‡ I haue seene it growing in the garden of Mr. *Iohn Parkinson.* ‡

¶ *The Time.*

Both the Branke Vrsines do floure in the sommer season, the seed is ripe in Autumne: the root remaineth fresh; yet now and then it perisheth in winter in both the Germaines, if the weather be too cold: but in England the former seldome or neuer dieth.

¶ *The Names.*

It is called in Greeke ἄκανθος: the Latines keepe the same name *Acanthus*: yet doth *Acanthus* signifie generally all kinde of Thistles,and that is called *Acanthus* by the figure *Antonomasia*:the English name is Branke Vrsine, and Beares breech.

The tame or garden Branke Vrsine is named in Latine *Satiuus*,or *Hortensis Acanthus*: in Greeke, παιδέρως: and of *Galen*, *Oribasius*,and *Pliny*,μελάμφυλλος: *Pliny* also calleth this *Acanthus læuis*, or smooth

Branke Vrſine,and reporteth it to be a citie herbe,and to ſerue for arbors : ſome name it *Branca Vrſina* (others vſe to cal Cow-parſnep by the name of *Branca Vrſina*,but with the addition *Germanica*:) the Italians call it *Acantho*,and *Branca Orſina* : the Spaniards,*Yerua Giguante* : the Ingrauers of old time were wont to carue the leaues of this Branke Vrſine in pillers, and other works,and alſo vpon the eares of pots;as among others *Virgill* teſtifieth in the third Eclog of his Bucolicks:

Et nobis idem Alcimedon duo pocula fecit,
Et molli circum eſt anſas amplexus Acantha.

‡ I take *Virgils Acanthus* to be that which we now commonly call *Pyracantha*, as I ſhall hereafter ſhew when I come to treat thereof. ‡

The other Branke Vrſine is named in Greeke ἄγριος ἄκανθος:and in Latine,*Syluestris Acanthus*,or wilde Branke Vrſine, and they may be called properly *Acantha*, or *Spina*,a prickle ; by which name it is found called of moſt Herbariſts,*Acanthus* : yet there is alſo another *Acanthus* a thornie ſhrub:the liquour which iſſueth forth of it,as *Herodotus* and *Theophraſtus* affirme, is a gumme : for difference whereof peraduenture this kinde of *Acanthus* is named *Herbacantha*: There is likewiſe found among the baſtard names of *Acanthus* the word *Mamolaria*,and alſo *Crepula*,but it is not expreſſed to which of them,whether to the wilde or tame it ought to be referred.

¶ *The Temperature.*

Theleaues of the garden Branke Vrſine conſiſt in a meane as it were betweene hot and cold,being ſomwhat moiſt,with a mollifying and gentle digeſting facultie,as are thoſe of the Mallow,and therefore they are profitably boyled in clyſters,as well as Mallow leaues. The root, as *Galen* teacheth,is of a more drying qualitie.

¶ *The Vertues.*

A *Dioſcorides* ſaith,that the roots are a remedie for lims that are burnt with fire,and that haue been out of ioint,if they be laied thereunto: that being drunke they prouoke vrine, and ſtop the belly: that they helpe thoſe that be broken,and be troubled with the crampe, and be in a conſumption of the lungs.

B They are good for ſuch as haue the ptiſicke and ſpet bloud withal; for thoſe that haue faln from ſome high place,that are bruiſed and drie beaten,and that haue ouerſtrained themſelues, and they are as good as the roots of the greater Comfrey, whereunto they are verie like in ſubſtance,tough iuice,and qualitie.

C Of the ſame root is made an excellent plaiſter againſt the ache and numneſſe of the hands and feet.

D It is put into clyſters with good ſucceſſe againſt ſundry maladies.

Chap. 476. *Of the Cotton Thiſtle.*

¶ *The Deſcription.*

1 THe common Thiſtle,whereof the greateſt quantitie of down is gathered for diuers purpoſes,as well by the poore to ſtop pillowes, cuſhions, and beds for want of feathers,as alſo bought of the rich Vpholſters to mixe with the feathers and downe they do ſell,which deceit would be looked vnto : this Thiſtle hath great leaues,long and broad, gaſhed about the edges, and ſet with ſharp and ſtiffe prickles all alongſt the edges,couered all ouer with a ſoft cotton or downe: out from the middeſt whereof riſeth vp a long ſtalke aboue two cubits high ,cornered,and ſet with filmes, and alſo full of prickles : the heads are likewiſe cornered with prickles,and bring foorth floures conſiſting of many whitiſh threds:the ſeed which ſucceedeth them is wrapped vp in down; it is long,of a light crimſon colour, and leſſer than the ſeede of baſtard ſaffron : the root groweth deep in the ground,being white,hard, wooddie,and not without ſtrings.

2 The Illyrian cotton thiſtle hath a long naked root,beſet about the top with a fringe of many ſmall threds or iags:from which ariſeth a very large and tall ſtalke,higher than any man,rather like a tree than an annuall herbe or plant : this ſtalke is garniſhed with ſcroles of thinne leaues, from the bottome to the top,ſet full of moſt horrible ſharpe prickes, and ſo is the ſtalke and euerie part of the plant, ſo that it is impoſſible for man or beaſt to touch the ſame without great hurt or danger: his leaues are very great,far broader and longer than any other thiſtle whatſoeuer, couered with an hoarie cotten or downe like the former : the floures doe grow at the top of the ſtalkes, which

which is diuided into ſundry branches,and are of a purple colour,ſet or armed round about with the like,or rather ſharper thornes than the aforeſaid.

1 *Acanthium album.*
The white Cotton Thiſtle.

2 *Acanthium Illyricum purpureum.*
The purple Cotton Thiſtle.

¶ *The Place.*

Theſe Thiſtles grow by high waies ſides, and in ditches almoſt euery where.

¶ *The Time*

They floure from Iune vntill Auguſt, the ſecond yeare after they be ſown:and in the mean time the ſeed waxeth ripe,which being thorow ripe the herbe periſheth, as doe likewiſe moſt of the other Thiſtles,which liue no longer than till the ſeed be fully come to matutitie.

¶ *The Names.*

This Thiſtle is taken for that which is called in Greeke ἀκάνθιον, which *Dioſcorides* deſcribeth to haue leaues ſet with prickles round about the edges, and to be couered with a thin downe like a copweb,that may be gathered and ſpun to make garments of, like thoſe of ſilke : in high Dutch it is called **Weiſz wege Diſtill :** in Low Dutch, **Witte Wech Diſtel :** in French, *Chardon argentin :* in Engliſh, Cotton-Thiſtle,white Cotton-Thiſtle, wilde white Thiſtle, Argentine or the Siluer Thiſtle.

¶ *The Temperature and Vertues.*

Dioſcorides ſaith, that the leaues and roots hereof are a remedy for thoſe that haue their bodies A
drawne backwards; thereby *Galen* ſuppoſeth that theſe are of temperature hot.

Chap. 477. *Of our Ladies-Thiſtle.*

¶ *The Deſcription.*

THe leaues of our Ladies Thiſtle are as bigge as thoſe of white Cotton-Thiſtle:for the leaues thereof be great,broad,large,gaſhed in the edges,armed with a multitude of ſtiffe and ſharpe prickles,as are thoſe of Ote-Thiſtle,but they are without down,altogether ſlippery,of a light greene

Carduus Mariæ
Ladies Thistle.

green and speckled,with white and milky spots and lines drawne diuers waies: the stalk is high, and as big as a mans finger: the floures grow forth of heads full of prickles, being threds of a purple colour: the seed is wrapped in downe like that of Cotton Thistle: the root is long, thicke, and white.

¶ *The Place.*

It groweth vpon waste and common places by high waies,and by dung-hils almost euerie where.

¶ *The Time.*

It floureth and seedeth when Cotton Thistle doth.

¶ *The Names.*

It is called in Latine, *Carduus Lacteus*, and *Carduus Mariæ*; in high Dutch, **Unser Vrouwen Distell**: in French, *Chardon ne nostre dame*: in English,our Ladies Thistle: it may properly be called *Leucographus*,of the white spots and lines that are on the leaues: *Pliny* in his 27. booke, chap. 11. maketh mention of an herb called *Leucographis*,but what maner of one it is he hath not expressed; therefore it would be hard to affirme this to be the same that his *Leucographis* is; and this is thought to bee *Spina alba*, called in Greeke ἄκανθα λευκή,or white Thistle, Milk Thistle, and *Carduus Ramptarius*: of the Arabians, *Bedoard*,or *Bedeguar*,as *Matthæus Syluaticus* testifieth.

¶ *The Temperature and Vertues.*

The tender leaues of *Carduus Leucographus*, the prickles taken off, are sometimes vsed to bee eaten with other herbes.

A *Galen* writeth,that the roots of *Spina alba* do drie and moderately binde,that therefore it is good for those that be troubled with the lask and the bloudy flix, that it staieth bleedings,wasteth away cold swellings; easeth the paine of the teeth if they bee washed with the decoction thereof.

B The seed thereof is of a thin essence and hot facultie, therefore he saith that it is good for those that be troubled with cramps.

C *Dioscorides* affirmeth that the seeds being drunke are a remedie for infants that haue their sinews drawne together, and for those that be bitten of serpents: and that it is thought to driue away serpents,if it be but hanged about the necke.

Chap. 478. *Of the Globe Thistle.*

¶ *The Description.*

Globe Thistle hath a very long stalke,and leaues iagged,great,long,and broad,deeply gashed, strong of smell,somewhat greene on the vpper side,and on the nether side whiter and downy: the floures grow forth of a round head like a globe,which standeth on the tops of the stalkes; they are white and small, with blew threds in the midst: the seed is long, with haires of a meane length: the root is thicke and branched.

2 There is another Globe Thistle that hath lesser leaues, but more full of prickles,with round heads also: but there groweth out of them besides the floures,certaine long and stiffe prickles.

3 There is likewise another kinde resembling the first in forme and figure,but much lesser,and the floures thereof tend more to a blew.

4 There is also another Globe Thistle, which is the least,and hath the sharpest prickles of all the rest: the head is small; the floures whereof are white,like to those of the first.

5 There

1 *Carduus globoſus.*
The Globe-Thiſtle.

‡ 2 *Carduus globoſus acutus.*
Prickly headed Globe-Thiſtle.

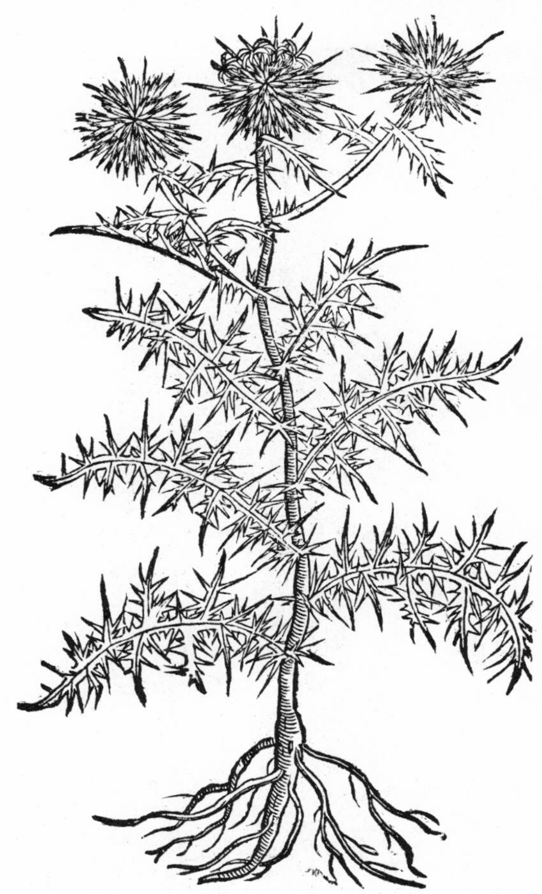

‡ 3 *Carduus globoſus minor.*
Small Globe-Thiſtle.

‡ 5 *Carduus globoſus capitulo latiore.*
Flat headed Globe-Thiſtle.

5 There is a certaine other kinde hereof, yet the head is not ſo round, that is to ſay, flatter and broader aboue; out of which ſpring blew floures: the ſtalke hereof is ſlender, and couered with a white thin downe: the leaues are long, gaſhed likewiſe on both ſides, and armed in euery corner with ſharpe prickles.

6 There is another called the Down-Thistle, which riseth vp with thicke and long stalks. The leaues thereof are iagged, set with prickles, white on the nether side: the heads be round and many in number, and are couered with a soft downe, and sharpe prickles standing forth on euerie side, being on the vpper part fraughted with purple floures all of strings: the seed is long, and shineth, as doth the seed of many of the Thistles.

‡ 6 *Carduus eriocephalus.*
Woolly headed Thistle.

¶ *The Place.*

They are sown in gardens, and do not grow in these countries that we can finde.

‡ I haue found the sixth by Pocklington and in other places of the Woldes in Yorkeshire. Mr. *Goodyer* also found it in Hampshire. ‡

¶ *The Names.*

They floure and flourish when the other Thistles do.

¶ *The Names.*

Fuchsius did at the first take it to be *Chamæleon niger*; but afterwards being better aduised, he named it *Spina peregrina*, and *Carduus globosus*. *Valerius Cordus* doth fitly call it *Sphærocephalus*: the same name doth also agree with the rest, for they haue a round head like a ball or globe. Most would haue the first to be that which *Matthiolus* setteth downe for *Spina alba*: this Thistle is called in English, Globe Thistle, and Ball-Thistle.

The downe or woolly headed Thistle is called in Latine, being destitute of another name, *Eriocephalus*, of the woolly head: in English, Downe Thistle, or woolly headed Thistle. It is thought of diuers to be that which *Bartholomæus Vrbeveteranus* and *Angelus Palea*, Franciscan Friers, report to be called *Corona Fratrum*, or Friers Crowne: but this Thistle doth far differ from that, as is euident by those things which they haue written concerning *Corona Fratrum*; which is thus: In the borders of the kingdome of Aragon towards the kingdome of Castile we finde another kind of Thistle, which groweth plentifully there, by common wayes, and in wheate fields, &c. *Vide Dod. Pempt. 5. lib. 5. cap. 5.*

¶ *The Temperature and Vertues.*

Concerning the temperature and vertues of these Thistles we can alledge nothing at all.

Chap. 479. *Of the Artichoke.*

¶ *The Kindes.*

There be three sorts of Artichokes, two tame or of the garden; and one wilde, which the Italian esteemeth greatly of, as the best to be eaten raw, which he calleth *Cardune*.

¶ *The Description.*

1 The leaues of the great Artichoke, called in Latine *Cinara*, are broad, great, long, set with deepe gashes in the edges, with a deepe channell or gutter alongst the middle, hauing no prickles at all, or very few, and they be of a greene ash colour: the stalke is aboue a cubit high, and bringeth forth on the top a fruit like a globe, resembling at the first a cone or Pine apple, that is to say, made vp of many scales, which is when the fruit is great or loosed of a greenish red colour within, and in the lower part full of substance and white; but when it opens it selfe there growes also

1 *Cinara maxima Anglica.*
The great red Artichoke.

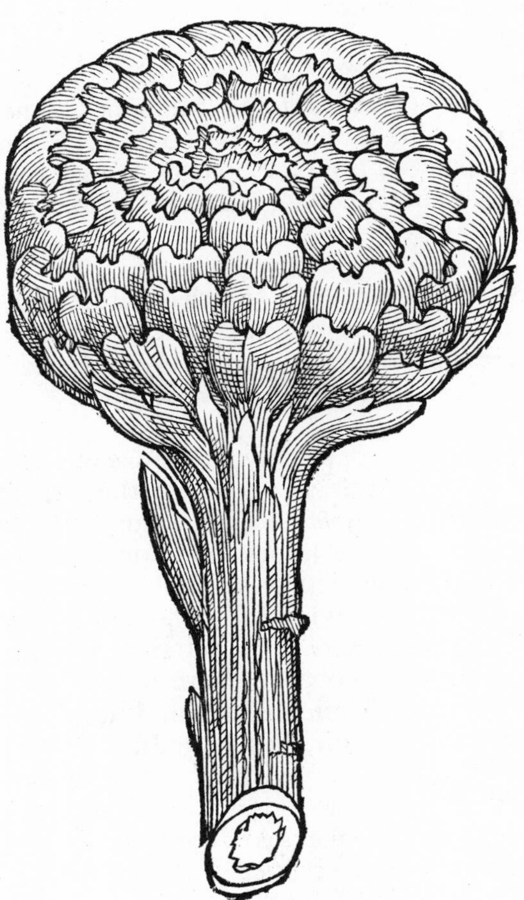

2 *Cinara maxima alba.*
The great white Artichoke.

3 *Cinara sylvestris.*
Wilde Artichoke.

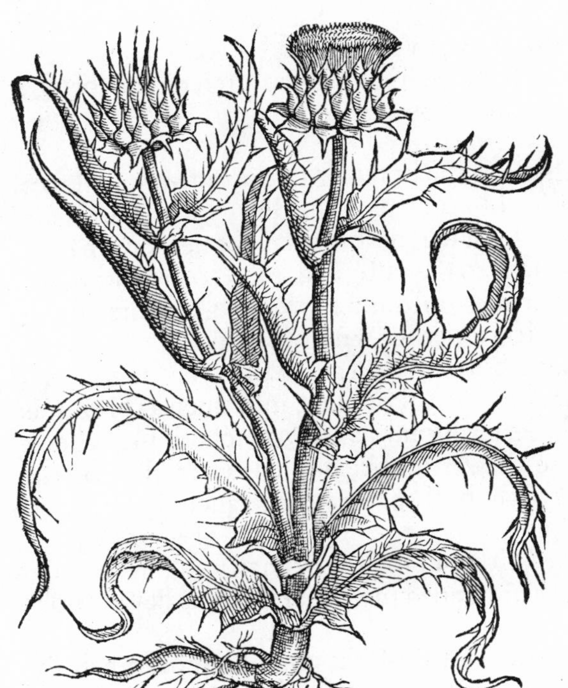

also vpon the cone a floure all of threds, of a gallant purple tending to a blew colour. The seed is long, greater and thicker than that of our Ladies thistle, lying vnder soft and downy haires which are contained within the fruit. The root is thicke, and of a meane length.

2 The second great Artichoke differeth from the former in the colour of the fruit, otherwise there is little difference, except the fruit hereof dilateth it selfe further abroad, and is not so closely compact together, which maketh the difference.

3 The prickly Artichoke, called in Latine *Carduus*, or *Spinosa Cinara*, differeth not from the former, saue that all the corners of the leaues hereof, and the stalkes of the cone or fruit, are armed with stiffe and sharp prickles, whereupon it beareth well the name of *Carduus*, or Thistle.

¶ *The Place.*

The Artichoke is to be planted in a fat and fruitfull soile: they do loue water and moist ground. They commit great error who cut away the side or superfluous leaues that grow by the sides, thinking thereby to increase the greatnesse of the fruit, when as in truth they depriue the root from much water by that meanes, which should nourish it to the feeding of the fruit; for if you marke the trough or hollow channell that is in euery leafe, it shall appeare very euidently, that the Creator

Creator in his secret wisedome did ordaine those furrowes, euen from the extreme point of the leafe to the ground where it is fastned to the root, for no other purpose but to guide and leade that water which falls far off, vnto the root; knowing that without such store of water the whole plant would wither, and the fruit pine away and come to nothing.

¶ *The Time.*

They are planted for the most part about the Kalends of Nouember, or somewhat sooner. The plant must be set and dunged with good store of ashes, for that kinde of dung is thought best for planting thereof. Euery yeare the slips must be torne or slipped off from the body of the root, and these are to be set in Aprill, which will beare fruit about August following, as *Columella, Palladius*, and common experience teacheth.

¶ *The Names.*

The Artichoke is called in Latine *Cinara*, of *Cinis*, Ashes, wherewith it loueth to be dunged. *Galen* calleth it in Greeke κύναρα, but with *k* and *v* in the first syllable: of some it is called *Cactos*: it is named in Italian, *Carcioffi, Archiocchi*: in Spanish, *Alcarrhofa*: in English, Artichoke: in French, *Artichaux*: in low-Dutch, **Artichoken**: whereupon diuers call it in Latine *Articocalus*, and *Articoca*: in high-Dutch, **Strobildorn**.

The other is named in Latine commonly not onely *Spinosa cinara*, or prickly Artichoke, but also of *Palladius, Carduus*: of the Italians, *Cardo*, and *Cardino*: of the Spaniards, *Cardos*: of the French men, *Chardons*: *Leonhartus Fuchsius* and most writers take it to be *Scolymus Dioscoridis*; but *Scolymus Dioscoridis* hath the leafe of Chameleon or *Spina alba*, with a stalke full of leaues, and a prickly head: but neither is *Cinara* the Artichoke which is without prickles, nor the Artichok with prickles any such kinde of herbe; for though the head hath prickles, yet the stalke is not full of leaues, but is many times without leaues, or else hath not past a leafe or two. *Cinara* doth better agree with that which *Theophrastus* and *Pliny* call κάκτος, *Cactus*, and yet it doth not bring forth stalkes from the root creeping alongst the ground: it hath broad leaues set with prickles; the middle ribs of the leaues, the skin pilled off, are good to be eaten, and likewise the fruit, the seed and down taken away; and that which is vnder is as tender as the braine of the Date tree: which things *Theophrastus* and *Pliny* report of *Cactus*. That which they write of the stalkes, sent forth immediately from the root vpon the ground, which are good to be eaten, is peraduenture the ribs of the leaues: euerie side taken away (as they be serued vp at the table) may be like a stalke, except euen in Sicilia, where they grew only in *Theophrastus* time. It bringeth forth both certaine stalks that lie on the ground, and another also standing straight vp; but afterwards being remoued and brought into Italy or England, it bringeth forth no more but one vpright: for the soile and clyme do much preuaile in altering of plants, as not onely *Theophrastus* teacheth, but also euen experience it selfe declareth: and of *Cactus, Theophrastus* writeth thus; κάκτος (*Cactus*) groweth onely in Sicilia: it bringeth forth presently from the root stalkes lying along vpon the ground, with a broad and prickly leafe: the stalkes being pilled are fit to be eaten, being somewhat bitter, which may be preserued in brine: it bringeth forth also another stalke, which is likewise good to be eaten.

¶ *The Temperature and Vertues.*

A The nailes, that is, the white and thicke parts which are in the bottome of the outward scales or flakes of the fruit of the Artichoke, and also the middle pulpe whereon the downy seed stands, are eaten both raw with pepper and salt, and commonly boyled with the broth of fat flesh, with pepper added, and are accounted a dainty dish, being pleasant to the taste, and good to procure bodily lust: so likewise the middle ribs of the leaues being made white and tender by good cherishing and looking to, are brought to the table as a great seruice together with other junkets: they are eaten with pepper and salt as be the raw Artichokes: yet both of them are of ill iuyce; for the Artichoke containeth plenty of cholericke iuyce, and hath an hard substance, insomuch as of this is ingendred melancholy iuyce, and of that a thin and cholerick bloud, as *Galen* teacheth in his book of the Faculties of nourishments. But it is best to eate the Artichoke boyled: the ribbes of the leaues are altogether of an hard substance: they yeeld to the body a raw and melancholy iuice, and containe in them great store of winde.

B It stayeth the inuoluntarie course of the naturall seed either in man or woman.

C Some write, that if the buds of yong Artichokes be first steeped in wine, and eaten, they prouoke vrine, and stir vp the lust of the body.

D I finde moreouer, that the root is good against the ranke smell of the arme-holes, if when the pith is taken away the same root be boyled in wine and drunke: for it sendeth forth plenty of stinking vrine, whereby the ranke and rammish sauor of the whole body is much amended.

Chap. 480. *Of Golden Thistles.*

¶ *The Description.*

1 The stalkes of Golden Thistle rise vp forthwith from the root, being many, round, and branched. The leaues are long, of a beautifull green, with deepe gashes on the edges, and set with most sharpe prickles: the floures come from the botome of the leaues, set in a scalie chaffie knap, very like to Succorie floures, but of colour as yellow as gold: in their places come vp broad flat and thin seeds, not great, nor wrapped in downe: the root is long, a finger thicke, sweet, soft, and good to be eaten, wherewith swine are much delighted: there issueth forth of this thistle in what part soeuer it is cut or broken, a iuyce as white as milke.

‡ There is some varietie of this Thistle; for it is found much larger about Montpelier than it is in Spaine, with longer branches, but fewer floures: the leaues also are spotted or streaked with white like as the milke Thistle: whence *Clusius*, whom I here follow, hath giuen two figures thereof; the former by the name of *Scolymus Theophrasti Hispanicus*; and the other by the title of *Scolymus Theophrasti Narbonensis*. This with white spots I saw growing this yere with Mr. *Tradescant* at South Lambeth. ‡

1 *Carduus Chrysanthemus Hispanicus.*
The Spanish golden Thistle.

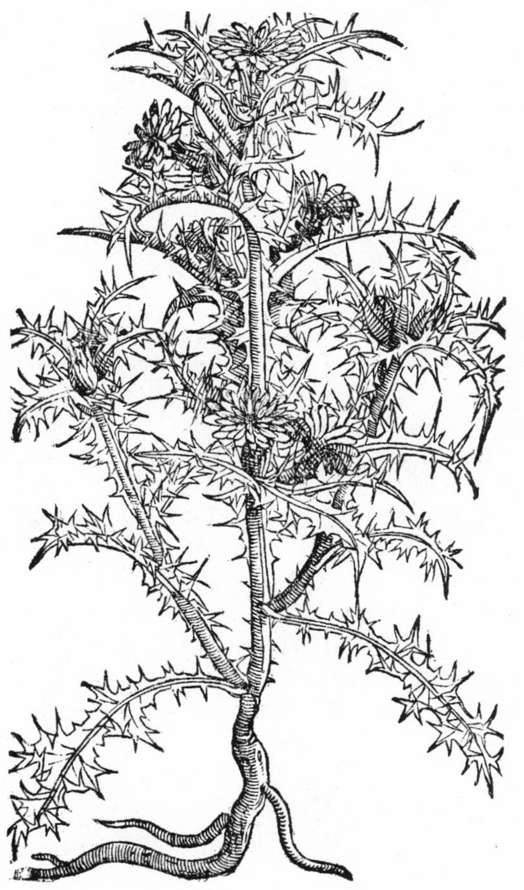

‡ *Carduus Chrysanthemus Narbonensis.*
The French golden Thistle.

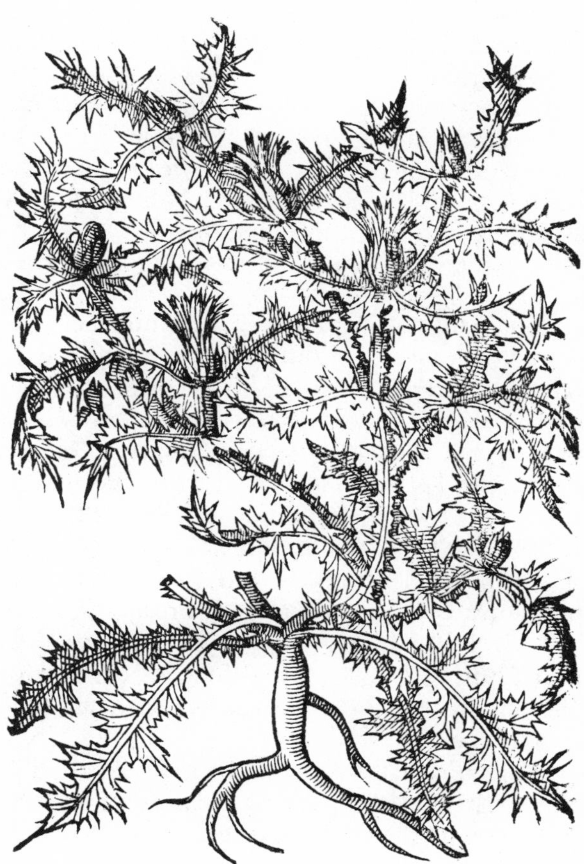

2 The golden Thistle of Peru, called in the West Indies, *Fique del Inferno*, a friend of mine brought it vnto me from an Island there called Saint Iohns Island, among other seeds. What reason the inhabitants there haue to call it so, it is vnto me vnknowne, vnlesse it be because of his fruit, which doth much resemble a fig in shape and bignesse, but so full of sharpe and venomous prickles, that whosoeuer had one of them in his throat, doubtlesse it would send him packing either to heauen or to hell. This plant hath a single wooddy root as big as a mans thumbe, but somwhat long: from which ariseth a brittle stalke full of ioynts or knees, diuiding it selfe into sundry other small branches, set full of leaues like vnto the milke Thistle, but much smaller, and straked with many white lines or streakes: and at the top of the stalks come forth faire and goodly yellow floures, very like vnto the sea Poppy, but more elegant, and of greater beauty, hauing in the midst thereof

thereof a small knop or boll, such as is in the middle of our wild Poppy, but full of sharpe thorns, and at the end thereof a staine or spot of a deepe purple: after the yellow floures be fallen, this foresaid knop groweth by degrees greater and greater, vntill it come to full maturitie, which openeth it selfe at the vpper end, shewing his seed, which is very blacke and round like the seeds of mustard. The whole plant and each part thereof doth yeeld verie great aboundance of milkie iuyce, which is of a golden colour, falling and issuing from any part thereof, if it be cut or bruised: the whole plant perisheth at the approch of Winter. The vertues hereof are yet vnknowne vnto me, wherefore I purpose not to set downe any thing thereof by way of coniecture, but shall, God willing, be ready to declare that which certaine knowledge and experience either of myne owne or others, shall make manifest vnto me.

¶ *The Place.*

The golden Thistle is sowne in gardens of the Low-Countries. *Petrus Bellonius* writes, That it groweth plentifully in Candy, and also in most places of Italy: *Clusius* reporteth that he found it in the fields of Spaine, and of the kingdome of Castile, and about Montpelier, with fewer branches, and of a higher growth.

The Indian Thistle groweth in Saint Iohns Island in the West Indies, and prospereth very well in my garden.

¶ *The Time.*

They floure from Iune to the end of August: the seed of the Indian golden Thistle must be sowne when it is ripe, but it doth not grow vp vntill May next after.

¶ *The Names.*

This Thistle is called in Latine *Carduus Chrysanthemus*: in Greeke of *Theophrastus*, Σκόλυμος; for those things which he writeth of *Scolymus* in his sixth and seuenth bookes doe wholly agree with this Thistle *Chrysanthemus*: which are these; *Scolymus*, doth floure in the Sommer solstice, brauely and a long time together; it hath a root that may be eaten both sod and raw, and when it is broken it yeeldeth a milky iuyce: *Gaza* nameth it *Carduus*. Of this *Pliny* also makes mention, *lib.21.ca.16.* *Scolymus*, saith he, differs from those kindes of Thistles, *viz. Acarna*, and *Atractilis*, because the root thereof may be eaten boyled. Againe, *Lib.22.Cap.22.* The East Countries vse it as a meate: and he calleth it by another name Λειμώνιον. Which thing also *Theophrastus* seemeth to affirme, in his sixt booke; for when he reckoneth vp herbes whose leaues are set with prickles, he addeth *Scolymus*, or *Limonia*.

Notwithstanding, *Pliny* maketh mention likewise of another *Scolymus*, which hee affirmeth to bring forth a purple floure, and betweene the middle of the prickes to wax white quickely, and to fall off with the winde; in his twentieth booke, *cap.2*. Which Thistle doubtlesse doth not agree with *Carduus Chrysanthemus*, that is, with *Theophrastus* his *Scolymus*, and with that which we mentioned before: so that there be in *Pliny* two *Scolymi*; one with a root that may be eaten, and another with a purple floure, turning into downe, and that speedily waxeth white. *Scolymus* is likewise described by *Dioscorides*; but this differs from *Scolymus Theophrasti*, and it is one of those which *Pliny* reckoneth vp, as we wil more at large declare hereafter. But let vs come againe to *Chrysanthemus*: This the inhabitants of Candy, keeping the marks of the old name, do call *Ascolymbros*: the Italians name it *Anconitani Rinci*: the Romans, *Spina borda*: the Spaniards, *Cardon lechar*: and of diuers it is also named *Glycyrrhizon*, that is to say, *dulcis Radix*, or sweet Root: it is called in English, golden Thistle: some would haue it to be that which *Vegetius* in *Arte Veterinaria* calls *Eryngium*: but they are deceiued; for that *Eryngium* whereof *Vegetius* writeth is *Eryngium marinum*, or sea Huluer, of which we will intreat.

The golden Thistle of India may be called *Carduus Chrysanthemus*, of his golden colour, adding thereto his natiue countrey *Indianus*, or *Peruanus*, or the golden Indian Thistle, or the golden Thistle of Peru: the seed came to my hands by the name *Fique del Inferno*: in Latine, *Ficus infernalis*, the infernall fig, or fig of hell.

¶ *The Temperature and Vertues.*

A The root and tender leaues of this *Scolymus*, which are sometimes eaten, are good for the stomacke, but they containe very little nourishment, and the same thinne and waterie, as *Galen* teacheth.

B *Pliny* saith, that the root hereof was commended by *Eratosthenes*, in the poore mans supper, and that it is reported also to prouoke vrine especially; to heale tetters and dry scurfe, being taken with vineger; and with Wine to stir vp fleshly lust, as *Hesiod* and *Alcæus* testifie; and to take away the stench of the arme-holes, if an ounce of the root, the pith picked out, be boyled in three parts of wine, till one part be wasted, and a good draught taken fasting after a bath, and likewise after meat: which

which later words *Dioſcorides* likewiſe hath concerning his *Scolymus* : out of whom *Pliny* is thought to haue borrowed theſe things.

† The plant our Author here deſcribes in the ſecond place, is that which I deſcribed and figured formerly, pag 401. by the name of *Papauer ſpinoſum*. I muſt confeſſe, I there ſhould haue omitted it, becauſe it is here ſet forth ſufficiently by our Author whereof indeed I had a little remembrance, and therefore at that time ſought his Index by all the names I could remember, but not making it a *Carduus*, I at that time miſſed thereof; but here finding it, I haue let the hiſtory ſtand as it was, and onely omitted the figure which you may finde before, and ſomething alſo in the hiſtory not here deliuered.

CHAP. 481. *Of white Carline Thiſtle of* Dioſcorides.

¶ *The Deſcription.*

1 THe leaues of Carline are very full of prickles, cut on both edges with a multitude of deepe gaſhes, and ſet along the corners with ſtiffe and very ſharpe prickles ; the middle ribs whereof are ſometimes red : the ſtalke is a ſpan high or higher, bringing forth for the moſt part onely one head or knap being full of prickles, on the outward circumference or compaſſe like the Vrchin huske of a cheſnut : and when this openeth at the top, there groweth forth a broad floure, made vp in the middle like a flat ball, of a great number of threds, which is compaſſed about with little long leaues, oftentimes ſomewhat white, very ſeldome red : the ſeed vnderneath is ſlender and narrow, the root is long, a finger thicke, ſomething blacke, ſo chinked as though it were ſplit in ſunder, ſweete of ſmell, and in taſte ſomewhat bitter.

‡ 1 *Carlina cauleſcens magno flore.*
Tall Carline Thiſtle.

2 *Carlina, ſeu Chameleon albus* Dioſcoridis.
The white Carline Thiſtle of *Dioſcorides* with the red floure.

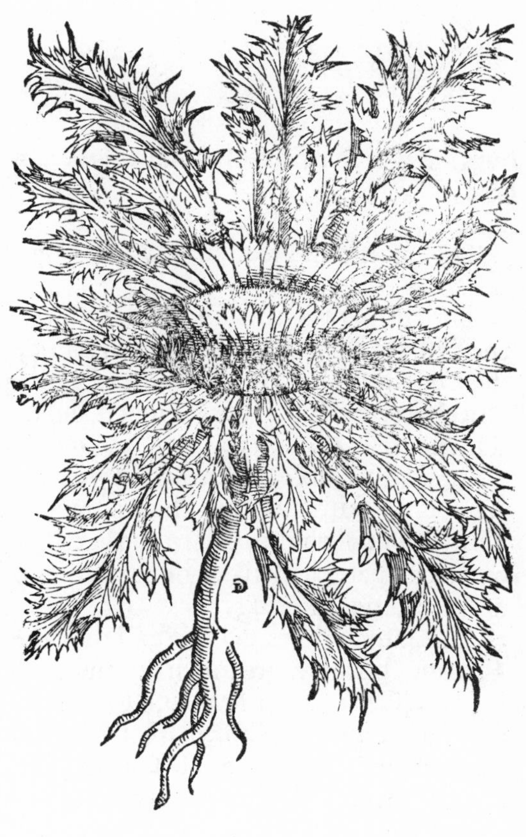

2 There is alſo another hereof without a ſtalke, with leaues alſo very full of prickles, like almoſt to thoſe of the other, lying flat on the ground on euery ſide : among which there groweth forth in the middle a round head or knap, ſet with prickles without after the ſame maner, but greater: the floure whereof in the middle is of ſtrings, and paled round about with red leaues, and ſometimes with white, in faire and calme weather the floures both of this and alſo of the other laie

 themſelues

‡ 3 *Carlina acaulos minor flore purp.*
Dwarfe Carline Thistle.

themselues wide open, and when the weather is foule and misty, are drawne close together: the root hereof is long, and sweet of smell, white, sound, not nicked or splitted as the other.

‡ 3 This small purple Carline Thistle hath a prety large root diuided oft times at the top into diuers branches, from which rise many green leaues lying spred vpon the ground, deeply cut and set with sharpe prickles; in the midst of these leaues come vp sometimes one, but otherwhiles more scaly heads, which carry a pretty large floure composed of many purple threds, like that of the Knapweed, but larger, and of a brighter colour; these heads grow vsually close to the leaues, yet sometimes they stand vpon stalkes three or foure inches high: when the floure is past they turne into downe, and are carried away with the winde: the seed is small and grayish. This growes vpon Blacke-Heath, vpon the chalky hills about Dartford, and in many such places. It floures in Iuly and August. *Tragus* calls it *Chamæleon albus, vel exiguus*; *Lobel, Carduus acaulis, Septentrionalium,* and *Chamæleon albus, Cordi*; *Clusius, Carlina minor purpureo flore,* and he saith in the opinion of some, it seemes not vnlike to the *Chamæleon* whereof *Theophrastus* makes mention, *lib. 6. cap. 3. Hist. plant.* ‡

¶ *The Place.*

They both grow vpon high mountaines in desart places, and oftentimes by high way sides: but that which bringeth forth a stalke groweth euery where in Germany, and is a stranger in England.

¶ *The Time.*

They floure and seed in Iuly and August, and many times later.

¶ *The Names.*

The former is called in Latine, *Carlina,* and *Cardopatium*; and of diuers, *Carolina,* of *Charlemaine* the first Romane Emperor of that name, whose armie (as it is reported) was in times past through the benefit of this root deliuered and preserued from the plague: it is called in high Dutch, **Eberwurtz**: in low Dutch, French, and other languages, as likewise in English, *Carline,* and Carline Thistle: it is *Dioscorides* his *Leucacantha* the strong and bitter roots shew the same; the faculties also are answerable, as forthwith we will declare: *Leucacantha* hath also the other names, but they are counterfeit, as among the Romanes *Gniacardus*; and among the Thuscans, *Spina alba,* or white Thistle, yet doth it differ from that Thistle which *Dioscorides* calleth *Spina alba,* of which he also writting apart, doth likewise attribute to both of them their owne proper faculties and operations and the same differing.

The later writers do also call the other *Carlina altera,* and *Carlina humilis,* or *minor,* low or little Carline: but they are much deceiued who go about to referre them both to the Chamæleons; for in Italy, Germany, or France, *Chamæleones,* the Chamæleons do neuer grow, as there is one witnesse for many, *Petrus Bellonius,* in his fift booke of Singularities, who sufficiently declareth what difference there is betweene the Carlines and the Chamæleons; which thing shall be made manifest by the description of the Chamæleons.

¶ *The Temperature and Vertues.*

A The root of Carline, which is chiefely vsed, is hot in the later end of the second degree, and dry in the third, with a thinnes of parts and substance; it procureth sweate, it driueth forth all kinde of wormes of the belly, it is an enemy to all maner of poisons, it doth not onely driue away infections of the plague, but also cureth the same, if it be drunke in time.

B Being chewed it helpeth the tooth-ache; it openeth the stoppings of the liuer and spleene.

C It prouoketh vrine, bringeth downe the menses, and cureth the dropsie.

D And it is giuen to those that haue been dry beaten, and fallen from some high place.

The like operations *Dioscorides* hath concerning *Leucacantha* : *Leucacantha* (saith he) hath a root like Cyperus, bitter and strong, which being chewed easeth the paine of the teeth : the decoction thereof with a draught of wine is a remedie against paines of the sides, and is good for those that haue the Sciatica or ache in the huckle bones, and for them that be troubled with the crampe. E

The iuyce also being drunke is of like vertues. F

CHAP. 482. *Of wilde Carline Thistle.*

¶ *The Description.*

1 THe great wilde Carline Thistle riseth vp with a stalke of a cubit high or higher, diuided into certaine branches : the leaues are long, and very full of prickles in the edges, like those of Carline : the floures grow also vpon a prickely head, being set with threds in the middest, and paled round about with a little yellowish leaues : the root is slender, and hath a twinging taste.

2 *Carolus Clusius* describeth a certaine other also of this kinde, with one onely stalke, slender, short, and not aboue a handfull high, with prickly leaues like those of the other, but lesser, both of them couered with a certaine hoary downe : the heads or knaps are for the most part two, they haue a pale downe in the midst, and leaues standing round about, being somewhat stiffe and yellow : the root is slender, and of a reddish yellow.

1 *Carlina syluestris maior.*
The great wilde Carline Thistle.

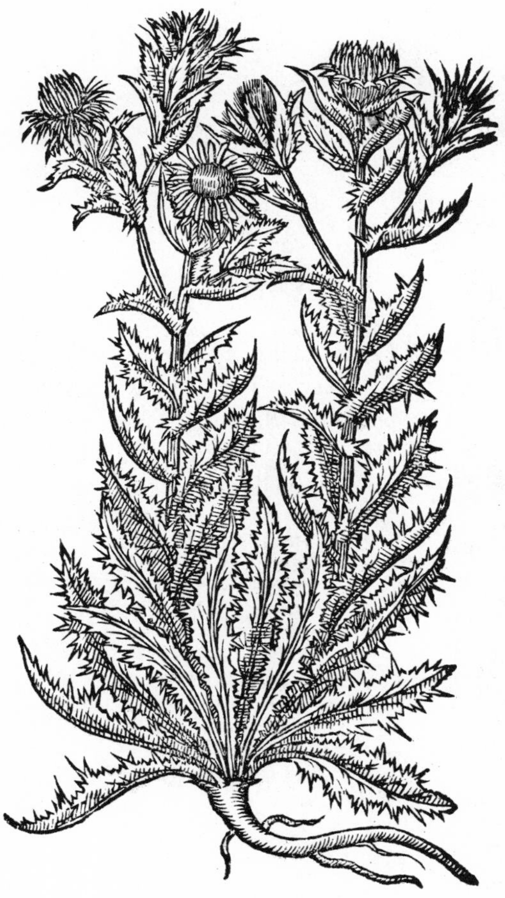

2 *Carlina syluestris minor.*
The little wilde Carline Thistle.

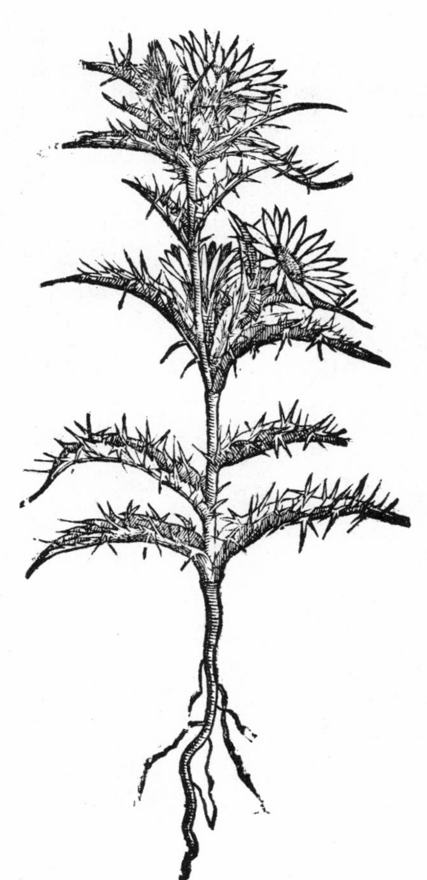

¶ *The Place.*

The great Carline is found in vntoiled and desart places, and oftentimes vpon hills. ‡ It groweth vpon Blacke Heath, and in many other places of Kent. ‡

The lesser Carline *Carolus Clusius* writeth that he found growing in dry stony and desart places, about Salmantica a city of Spaine.

¶ *The Time.*

They floure and flourish in Iune and Iuly.

¶ *The Names.*

It is commonly called in Latine, and that not vnfitly, *Carlina sylnestris*; for it is like to Carline in floures, and is not very vnlike in leaues. And that this is Ἄκορνα, it is so much the harder to affirme, by how much the briefer *Theophrastus* hath written hereof; for he saith that this is like bastard saffron, of a yellow colour and fat iuyce: and *Acorna* differs from *Acarna*; for *Acarna*, as *Hesychius* saith, is the Bay tree; but *Acorna* is a prickly plant.

¶ *The Temperature and Vertues.*

It is hot, especially in the root, the twinging taste thereof doth declare; but seeing it is of no vse, the other faculties be vnsearched out.

Chap. 483. *Of Chamæleon Thistle.*

¶ *The Kindes.*

There be two Chamæleons, and both blacke: the vertues of their roots do differ, and the roots themselues do differ in kinde, as *Theophrastus* declareth.

† 1 *Chamæleon niger.*
The blacke Chamæleon Thistle.

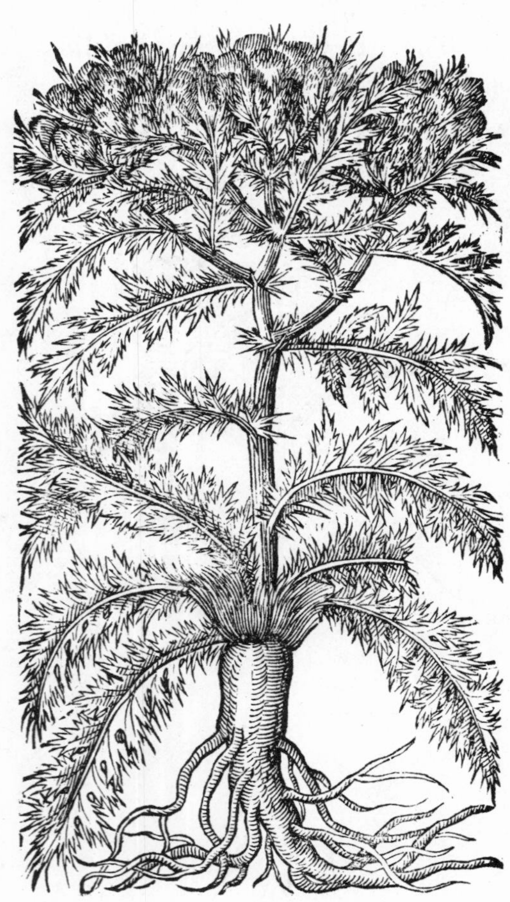

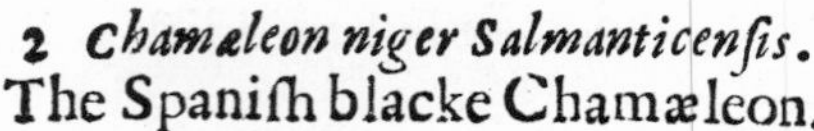

2 *Chamæleon niger Salmanticensis.*
The Spanish blacke Chamæleon.

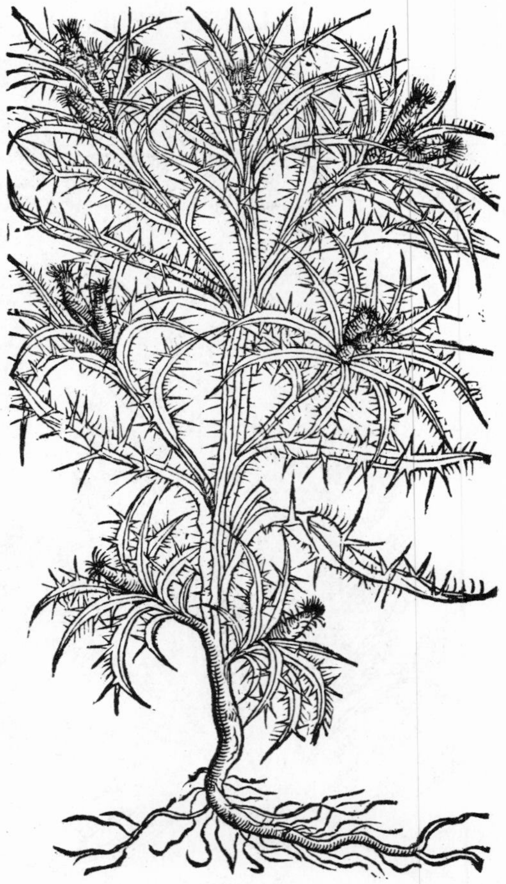

¶ *The Description.*

1 The leaues of blacke Chamæleon are lesser and slenderer than those of the prickely Artichoke, and sprinckled with red spots: the stalke is a cubit high, a finger thicke, and somewhat red: it beareth a tufted rundle, in which are slender prickely floures of a blew colour like the Hyacinth. The root is thicke, blacke without, of a close substance, sometimes eaten away, which being cut is of a yellowish colour within, and being chewed it bites the tongue.

2 This

2 This blacke Chamæleon hath many leaues, long and narrow, very full of prickles, of a light greene, in a manner white: the ſtalke is chamfered, a foot high, and diuided into branches, on the tops whereof ſtand purple floures growing forth of prickly heads: the root is blacke, and ſweet in taſte. This is deſcribed by *Cluſius* in his Spaniſh Obſeruations, by the name of *Chamæleon Salmanticenſis*, of the place wherein he found it: for he ſaith that this groweth plentifully in the territory of Salmantica a city in Spaine: but it is very manifeſt that this is not blacke Chamæleon neither doth *Cluſius* affirme it.

¶ *The Place.*

It is very common, ſaith *Bellonius*, in Lemnos, where it beareth a floure of ſo gallant a blew, as that it ſeemeth to contend with the skie in beautie; and that the floure of Blew-Bottle being of this colour, ſeemes in compariſon of it to be but pale. It groweth alſo in the fields neere Abydum, and hard by the riuers of Helleſpont, and in Heraclea in Thracia.

Chamæleon Salmanticenſis groweth plentifully in the territorie of Salmantica a city in Spaine.

¶ *The Time.*

They floure and flouriſh when the other Thiſtles do.

¶ *The Names.*

The blacke Chamæleon is called in Greeke χαμαιλέων μέλας: in Latine, *Chamæleon niger*: of the Romans, *Carduus niger*, and *Vernilago*: of ſome, *Crocodilion*: in Engliſh, the Chamæleon Thiſtle, or the Thiſtle that changeth it ſelfe into many ſhapes and colours.

¶ *The Temperature and Vertues.*

The root hereof, as *Galen* ſaith, containeth in it a deadly qualitie: it is alſo by *Nicander* numbred among the poyſonous herbes, in his booke of Treacles; by *Dioſcorides*, *lib. 6.* and by *Paulus Ægineta*: and therefore it is vſed only outwardly, as for ſcabs, morphewes, tetters, and to be briefe, for all ſuch things as ſtand in need of clenſing: moreouer, it is mixed with ſuch things as doe diſſolue and mollifie, as *Galen* ſaith. A

† The figure which was formerly in the firſt place did not agree with the hiſtorie (which was taken out of *Dodonæus*) though *Tabern.* gaue it for *Chamæleon Niger*; for it is the *Picnomos Cretæ, &c.* of *Lobel.* You ſhall finde it hereafter with the *Acarna Valerandi.*

CHAP. 484. *Of Sea Holly.*

¶ *The Kindes.*

D*Ioſcorides* maketh mention onely of one ſea Holly: *Pliny, lib. 22. cap. 7.* ſeemes to acknowledge two, one growing in rough places, another by the ſea ſide. The Phyſitians after them haue obſerued more.

¶ *The Deſcription.*

1 SEa Holly hath broad leaues almoſt like to Mallow leaues, but cornered in the edges, and ſet round about with hard prickles, fat, of a blewiſh white, and of an aromaticall or ſpicie taſte: the ſtalke is thicke, aboue a cubit high, now and then ſomewhat red below: it breaketh forth on the tops into prickly or round heads or knops, of the bigneſſe of a Wall-nut, held in for the moſt part with ſix prickely leaues, compaſſing the top of the ſtalke round about: which leaues as wel as the heads are of a gliſtring blew: the floures forth of the heads are likewiſe blew, with white threds in the midſt: the root is of the bigneſſe of a mans finger, very long, and ſo long, as that it cannot be all plucked vp, vnleſſe very ſeldome; ſet here and there with knots, and of taſte ſweet and pleaſant.

2 The leaues of the ſecond ſea Holly are diuerſly cut into ſundry parcels, being all ful of prickles alongſt the edges: the ſtalke is diuided into many branches, and bringeth forth prickly heads, but leſſer than thoſe of the other: from which there alſo grow forth blew floures, ſeldome yellow: there ſtand likewiſe vnder euery one of theſe, ſix rough and prickly leaues like thoſe of the other, but thinner and ſmaller: the root hereof is alſo long, blacke without, white within, a finger thicke, of taſte and ſmell like that of the other, as be alſo the leaues, which are likewiſe of an aromaticall or ſpicie taſte, which being new ſprung vp, and as yet tender, be alſo good to be eaten.

1 *Eryngium marinum.*
Sea Holly.

2 *Eryngium mediterraneum.*
Leuant sea Holly.

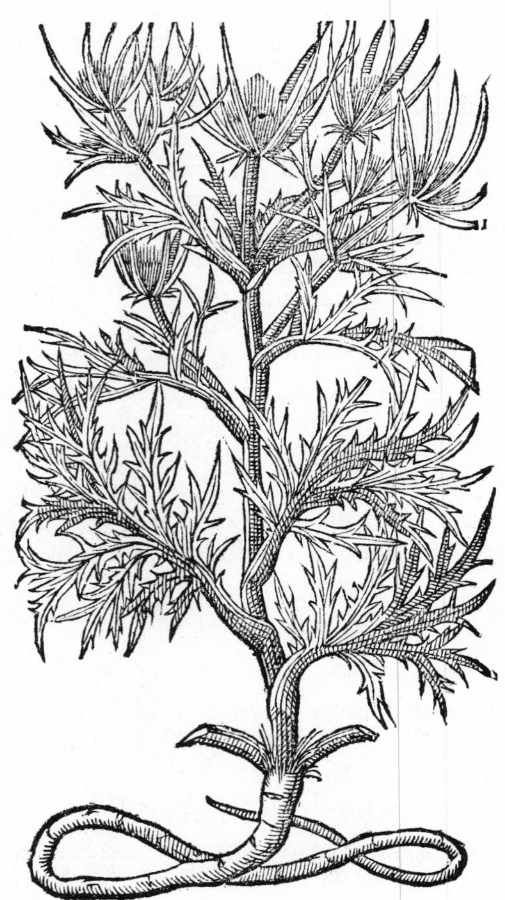

¶ *The Place.*

Eryngium marinum growes by the sea side vpon the baich and stony ground : I found it growing plentifully at Whitstable in Kent, at Rie and Winchelsea in Sussex, and in Essex at Landamer lading, at Harwich, and vpon Langtree point, on the other side of the water, from whence I haue brought plants for my garden.

Eryngium Campestre groweth vpon the shores of the Mediterranean sea, and in my garden likewise.

¶ *The Time.*

Both of them do floure after the Sommer solstice, and in Iuly.

¶ *The Names.*

This Thistle is called in Greeke Ἐρύγγιον : and likewise in Latine *Eryngium :* and of *Pliny* also *Erynge :* in shops, *Eringus ;* in English, Sea Holly, sea Holme, or sea Huluer.

The first is called in Latine *Eryngium marinum :* in low-Dutch euery where, **Cruys distil, Eindeloos, Meerwortele :** in English, sea Holly.

The second is named of *Pliny, lib.22.cap.8. Centum capita,* or hundred headed Thistle : in high-Dutch, **Manstrew, Brachendistell, Radendistel :** in Spanish, *Cardo corredor :* in Italian, *Eringio,* and *Iringo :* this is syrnamed *Campestre,* or Champion sea Holly, that it may differ from the other.

¶ *The Temperature.*

The roots of them both are hot, and that in a mean ; and a little dry also, with a thinnesse of substance, as *Galen* testifieth.

¶ *The Vertues.*

A The roots of sea Holly boyled in wine and drunken are good for them that are troubled with the Collicke, it breaketh the stone, expelleth grauell, and helpeth also the infirmities of the kidnies, prouoketh vrine, greatly opening the passages, being drunke fifteene dayes together.

B The roots themselues haue the same propertie if they be eaten, and are good for those that be liuer-sicke, and for such as are bitten with any venomous beast : they ease cramps, conuulsions, and the falling sicknesse, and bring downe the termes.

The

The roots condited or preſerued with ſugar, as hereafter followeth,are exceeding good to be gi- C
uen vnto old and aged people that are conſumed and withered with age,and which want naturall moiſture : they are alſo good for other ſorts of people that haue no delight or appetite to venerie, nouriſhing and reſtoring the aged,and amending the defects of nature in the younger.

¶ *The manner to condite Eryngos.*

Refine ſugar fit for the purpoſe,and take a pound of it,the white of an egge, and a pint of cleere D
water,boile them together and ſcum it,then let it boile vntill it be come to good ſtrong ſyrrup,and when it is boiled,as it cooleth, adde thereto a ſaucer full of Roſe-water,a ſpoone full of Cinnamon water,and a graine of Muske,which haue been infuſed together the night before,and now ſtrained; into which ſyrrup being more than halfe cold, put in your roots to ſoke and infuſe vntill the next day; your roots being ordered in manner hereafter following :

Theſe your roots being waſhed and picked, muſt be boiled in faire water by the ſpace of foure E
houres, vntill they be ſoft,then muſt they be pilled cleane,as ye pill parſneps,and the pith muſt bee drawne out at the end of the root; and if there be any whoſe pith cannot be drawne out at the end, then you muſt ſlit them, and ſo take out the pith : theſe you muſt alſo keepe from much handling, that they may be cleane, let them remaine in the ſyrrup till the next day, and then ſet them on the fire in a faire broad pan vntill they be verie hot,but let them not boile at all : let them there remaine ouer the fire an houre or more, remoouing them eaſily in the pan from one place to another with a woodden ſlice. This done, haue in a readineſſe great cap or royall papers, whereupon you muſt ſtraw ſome Sugar, vpon which lay your roots after that you haue taken them out of the pan. Theſe papers you muſt put into a Stoue,or hot houſe to harden; but if you haue not ſuch a place,lay them before a good fire. In this manner if you condite your roots, there is not any that can preſcribe you a better way. And thus may you condite any other root whatſoeuer, which will not onely bee exceeding delicate, but very wholeſome, and effectuall againſt the diſeaſes aboue named.

A certaine man affirmeth,ſaith *Aetius*,that by the continual vſe of Sea Holly,he neuer afterward F
voided any ſtone,when as before he was very often tormented with that diſeaſe.

It is drunke,ſaith *Dioſcorides*,with Carrot ſeed againſt very many infirmities,in the weight of a G
dramme.

The iuice of the leaues preſſed forth with wine is a remedie for thoſe that are troubled with the H
running of the reines.

They report that the herbe Sea Holly,if one Goat take it into her mouth,it cauſeth her firſt to I
ſtand ſtill,and afterwards the whole flocke, vntill ſuch time as the Shepheard take it forth of her mouth,as *Plutarch* writeth.

CHAP. 485. *Of baſtard Sea Hollies.*

¶ *The Deſcription.*

THis *Eryngium* which *Dodonæus* in his laſt edition calleth *Eryngium planum*; and *Pena* more fitly and truely,*Eryngium Alpinum cæruleum*,hath ſtalkes a cubite and a halfe high, hauing ſpaces betweene euery ioint : the lower leaues are greater and broader, and notched about the edges, but thoſe aboue are leſſer, compaſſing or enuironing each ioint ſtar-faſhion, beſet with prickles which are ſoft and tender,not much hurtful to the hands of ſuch as touch them;the knobs or heads are alſo prickley,and in colour blew. The root is bunchie or knottie,like that of *Helenium*, that is, Elecampane, blacke without, and white within, and like the Eringes in ſweetneſſe and taſte.

2 The ſecond baſtard Sea Holly, whoſe picture is ſet forth in *Dodonæus* his laſt Edition verie gallantly,being alſo a kind of Thiſtle,hath leaues like vnto the former Erynges, but broader next the rootes than thoſe which grow next the ſtalkes, ſomewhat long,greeniſh,ſoft,and not prickley, but lightly creuiſed or notched about the edges,greater than Quince leaues.The ſtalks grow more than a cubit high;on the tops whereof there hang downwards fiue or ſix knobs or heads, in colour and floures like the other; hauing three or foure whitiſh roots of a foot long.

3 The third kinde of baſtard *Eryngium* hath his firſt leaues (which grow next the ground) great,broad,and ſoft,growing as it were in a rundle about the root. The ſtalke is ſmall and ſlender, diuided into ſome branches,which beare many little leaues,turning or ſtanding many waies,which

be

1 *Eryngium cæruleum.*
Blew Sea Holly.

2 *Eryngium spurium primum Dodonæi.*
Bastard Sea Holly.

3 *Eryngium pumilum Clusij.*
Dwarfe Sea Holly.

4 *Eryngium Montanum.*
Mountaine Sea Holly.

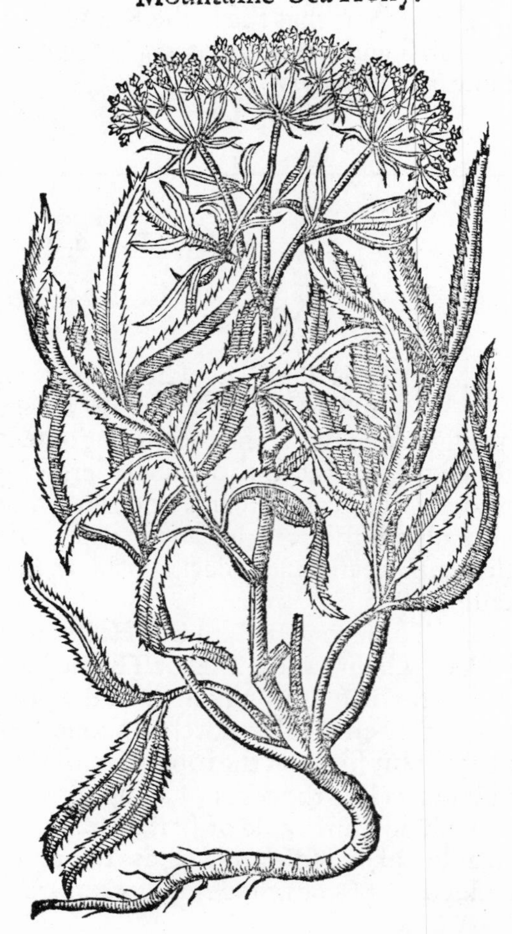

‡ 5 *Eryngium pusillum planum.*
Small smooth Sea Holly.

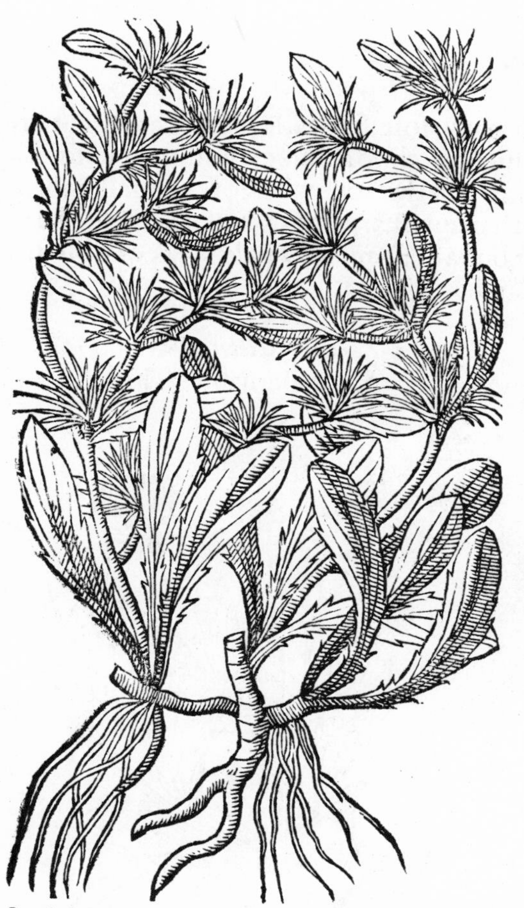

be also slender, prickly, and set about the stalks star-fashion. The knops or heads growing at the tops of the branches are round and prickly, bearing little blew floures and leaues, which compasse them about: the root is slender, and lasteth but one yeare.

4 The fourth kinde of bastard Sea Holly, which *Pena* calleth *Eryngium montanum recentiorum*, and is the fourth according to *Dodonæus* his account, is like to the Erynges, not in shape but in taste: this beareth a very small and slender stalke, of a meane height; whereupon doe grow three or foure leaues, & seldom fiue, made of diuers leaues set vpon a middle rib, narrow, long, hard, and of a darke greene colour, dented on both edges of the leafe like a saw: the stalke is a cubit high, iointed or kneed, and diuiding it selfe into many branches, on the tops whereof are round tufts or vmbels, wherin are contained the floures, and after they be vaded, the seedes, which are small, somewhat long, well smelling, and sharpe in taste: the root is white and long, not a finger thicke, in taste sweet, but afterwards somewhat sharpe, and in sent and sauour not vnpleasant: when the root is dried, it may be crumbled in pieces, and therefore quickly braied.

‡ 5 This is a low plant presently from the root diuided into sundry branches, slender, round & lying on the ground: at each ioint grow leaues without any certain order, broad toward their ends, and narrower at their setting on, snipt about their edges: those next the root were some inch broad, and two or more long, of a yellowish greene colour: the stalkes are parted into sundry branches, and at each ioint haue little leaues, and rough and greene heads, with blewish floures in them: the roots creepe, and are somewhat like those of *Asparagus*. This neither *Clusius* nor *Lobel* found wilde; but it grew in the garden of *Iohn Mouton* of Tourney, a learned Apothecarie, verie skilfull in the knowledge of plants: whereupon they both called it *Eryngium pusillum planum Moutoni*. ‡

¶ *The Place.*

These kindes of sea Holly are strangers in England: we haue the first and second in our London gardens.

¶ *The Time.*

They floure and flourish when the Thistles do.

¶ *The Names.*

These plants be *Eryngia spuria*, or bastard Sea Hollies, and are lately obserued: and therefore they haue no old names.

The first may bee called in Latine *Eryngium Borussicum*, or *Non spinosum*: Sea Hollie without prickles.

The second is called by *Matthiolus*, *Eryngium planum*, or flat Sea Holly: others had rather name it *Alpinum Eryngium*, or Sea Holly of the Alpes.

The third is rightly called *Eryngium pumilum*, little Sea Huluer.

Matthiolus maketh the fourth to be *Crithmum quartum*, or the fourth kinde of Sampier: and others, as *Dodonæus* and *Lobel*, haue made it a kinde of Sea Huluer.

¶ *The Temperature and Vertues.*

Touching the faculties hereof we haue nothing to set downe, seeing they haue as yet no vse in medicine, nor vsed to be eaten. But yet that they be hot, the very taste doth declare.

Chap.

Chap. 486. *Of Star-Thiſtle.*

¶ *The Deſcription.*

1 THe Star-Thiſtle, called *Carduus ſtellatus*, hath many ſoft frizled leaues, deepely cut or gaſht, altogether without prickles: among which riſeth vp a ſtalke, diuiding it ſelfe into many other branches, growing two foot high; on the tops whereof are ſmall knops or heads like the other Thiſtles, armed round about with many ſharpe prickles, faſhioned like a blaſing ſtar, which at the beginning are of a purple colour, but afterwards of a pale bleak or whitiſh colour: the ſeed is ſmall, flat, and round; the root is long, and browne without.

1 *Carduus ſtellatus.*
The Star-Thiſtle.

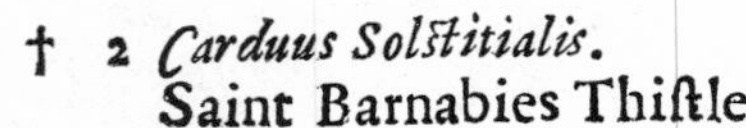
† 2 *Carduus Solſtitialis.*
Saint Barnabies Thiſtle.

2 Saint Barnabies Thiſtle is another kinde of Star-Thiſtle; notwithſtanding it hath prickles no where faue in the head onely, and the prickles of it ſtand forth in manner of a ſtar: the ſtalks are two cubits high, parted into diuers branches ſofter than are thoſe of ſtar-Thiſtle, which ſtalks haue velmes or thin skins cleauing vnto them all in length, by which they ſeeme to be foure-ſquare: the leaues are ſomewhat long, ſet with deep gaſhes on the edges: the floures are yellow, and conſiſt of threds: the ſeed is little; the root long and ſlender.

¶ *The Place.*

The two firſt do grow vpon barren places neere vnto cities and townes, almoſt euery where.

¶ *The Time.*

They floure and flouriſh eſpecially in Iuly and Auguſt.

¶ *The Names.*

The firſt is called in Latine, *Stellaria*; as alſo *Carduus Stellatus*, and likewiſe *Carduus Calcitrapa*; but they are deceiued, who take it to be *Eryngium*, or Sea-Holly, or any kinde thereof. *Matthiolus* ſaith, that it is called in Italian *Calcatrippa*: in high Dutch, **Wallen Diſtel**: in low Dutch, **Sterre Diſtell**: in French, *Chauſſe trappe*: in Engliſh, Star-Thiſtle.

S. Barnabies Thiſtle is called in Latine *Solſtitialis ſpina*, becauſe it floureth in the Sommer Solſtice

ſtice,as *Geſner* ſaith, or rather becauſe after the Solſtice the prickles thereof be ſharpeſt:of *Guillandinus*,*Eryngium*,but not properly,and *Stellaria Horatij Augerij*, who with good ſucceſſe gaue it againſt the ſtone,dropſies,greene ſickneſſe,and quotidian feuers.It is called in Engliſh as aboue ſaid,Saint Barnabies Thiſtle.

¶ *The Temperature.*

The Star-Thiſtle is of a hot nature.

¶ *The Vertues.*

A The ſeed is commended againſt the ſtrangurie: it is reported to driue forth the ſtone,if it bee drunke with wine.

B *Baptiſta Sardus* affirmeth,that the diſtilled water of this Thiſtle is a remedie for thoſe that are infected with the French Pox, and that the vſe of this is good for the liuer, that it taketh away the ſtoppings thereof.

C That it clenſeth the bloud from corrupt and putrified humours.

D That it is giuen with good ſucceſſe againſt intermitting feuers; whether they be quotidian or tertian.

E As touching the faculties of Saint Barnabies Thiſtle, which are as yet not found out, we haue nothing to write.

† There were formerly three figures and deſcriptions in this chapter,and all of them out of the 14. and 15.chapter.of the fifth booke, and fifth *Pemptas* of *Dodonæus*;but the firſt and ſecond figures were both of the firſt deſcribed,the third figure was of the *Acanthium peregrinum* of *Tabernamontanus*,which *Bauhine* knowes not what to make of,but I thinke it was drawne for,and(if the tuberous clogs of the roots were ſomewhat larger) might very well ſerue for the *Cirſium maximum Aſphodeli radice*,whoſe figure as I drew it from the plant I will hereafter giue you: the third deſcription was of the *Iacea maior lutea*, deſcribed in the third place of the 249. Chap.*pag.*727.

CHAP. 487 *Of Teaſels.*

¶ *The Kindes.*

OVr age hath ſet downe two kindes of Teaſels:the tame,and the wilde. Theſe differ not ſaue only in the husbanding; for all things that are planted and manured doe more flouriſh, and become for the moſt part fitter for mans vſe.

1 *Dipſacus ſatiuus.*
Garden Teaſell.

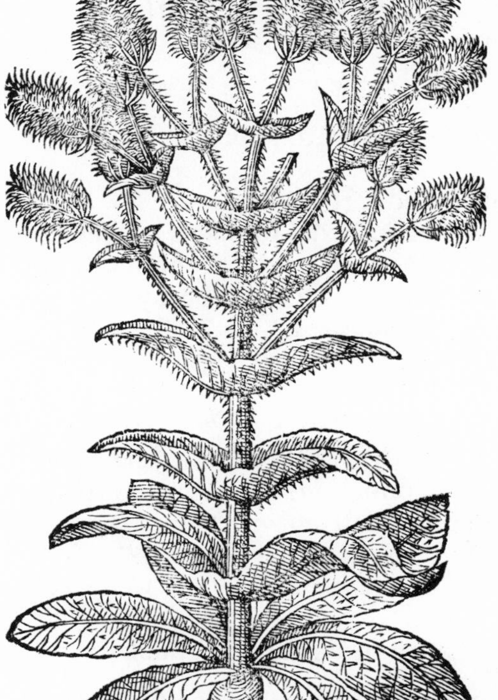

2 *Dipſacus ſylueſtris.*
Wilde Teaſell.

‡ 3 *Dipſacus minor, ſive Virga paſtoris.*
Sheepheards-rod.

¶ *The Deſcription.*

1 Garden Teaſel is alſo of the number of the Thiſtles; it bringeth forth a ſtalke that is ſtraight, very long, iointed, and ful of prickles: the leaues grow forth of the ioints by couples, not onely oppoſite or ſet one right againſt another, but alſo compaſſing the ſtalke about, and faſtened together; and ſo faſtened, that they hold dew and raine water in manner of a little baſon: theſe be long, of a light greene colour, and like to thoſe of Lettice, but full of prickles in the edges, and haue on the outſide all alongſt the ridge ſtiffer prickles: on the tops of the ſtalkes ſtand heads with ſharpe prickles like thoſe of the Hedge-hog, and crooking backward at the point like hookes: out of which heads grow little floures: The ſeed is like Fennell ſeed, and in taſte bitter: the heads wax white when they grow old, and there are found in the midſt of them when they are cut, certaine little magots: the root is white, and of a meane length.

2 The ſecond kinde of Teaſell which is alſo a kinde of Thiſtle, is very like vnto the former, but his leaues are ſmaller & narrower: his floures of a purple colour, and the hooks of the Teaſell nothing ſo hard or ſharpe as the other, nor good for any vſe in dreſſing of cloath.

3 There is another kinde of Teaſell, being a wilde kinde therof, and accounted among theſe Thiſtles, growing higher than the reſt of his kindes; but his knobbed heads are no bigger than a Nutmeg, in all other things elſe they are like to the other wilde kindes. ‡ This hath the lower leaues deeply cut in with one gaſh on each ſide at the bottome of the leafe, which little ears are omitted in the figure: the leaues alſo are leſſe than the former, and narrower at the ſetting on, and hold no water as the two former do: the whole plant is alſo much leſſe. ‡

¶ *The Place.*

The firſt called the tame Teaſell is ſowne in this countrey in gardens, to ſerue the vſe of Fullers and Clothworkers.

The ſecond kinde groweth in moiſt places by brookes, riuers, and ſuch like places.

The third I found growing in moiſt places in the high way leading from Braintree to Henningham caſtle in Eſſex, and not in any other place except here & there a plant vpon the high way from Much-Dunmow to London. ‡ I found it growing in great plentie at Edgecombe by Croyden, cloſe by the gate of the houſe of my much honoured friend Sir *Iohn Tunſtall.*

¶ *The Time.*

Theſe floure for the moſt part in Iune and Iuly.

¶ *The Names.*

Teaſell is called in Greeke δίψακος, and likewiſe in Latine, *Dipſacus*, *Labrum Veneris*, and *Carduus Veneris* it is termed *Labrum Veneris*, and *Lauer Lauacrum*, of the forme of the leaues made vp in faſhion of a baſon, which is neuer without water: they commonly call it *Virga paſtoris minor*, and *Carduus fullonum*: in high Dutch, **Karden Diſtell**: in low Dutch **Caerden**: in Spaniſh, *Cardencha*: and *Cardo Penteador*: in Italian, *Diſſaco*, and *Cardo* · in French, *Chardon de foullon*, *Verge à bergier*: in Engliſh, Teaſell, Carde Teaſell, and Venus baſon.

The third is thought to be *Galedragon Plinij*: of which he hath written in his 27. book the tenth Chapter.

¶ *The Temperature.*

A The rootes of theſe plants are drie in the ſecond degree, and haue a certaine clenſing facultie.

¶ *The*

¶ *The Vertues.*

There is small vse of Teasell in medicines : the heads (as we haue said) are vsed to dresse woollen cloth with. A

Dioscorides writeth, that the root being boiled in wine, & stamped till it is come to the substance of a salue, healeth chaps and fistulaes of the fundament, if it be applied thereunto; and that this medicine must be reserued in a box of copper, and that also it is reported to be good for all kindes of warts. B

It is needlesse here to alledge those things that are added touching the little wormes or magots found in the heads of the Teasell, and which are to be hanged about the necke, or to mention the like thing that *Pliny* reporteth of Galedragon : for they are nothing else but most vaine and trifling toies, as my selfe haue proued a little before the impression hereof, hauing a most grieuous ague, and of long continuance: notwithstanding Physicke charmes, these worms hanged about my neck, spiders put into a walnut shell, and diuers such foolish toies that I was constrained to take by fantasticke peoples procurement; notwithstanding I say, my helpe came from God himselfe, for these medicines and all other such things did me no good at all. C

† The figure which formerly was put into the second place, was of the *Dipsacus secundus* of *Tabernamontanus*, which differs from our common one, in that the leaues are deeply diuided, or cut in on their edges.

CHAP. 488. *Of Bastard Saffron.*

‡ 1 *Carthamus siue Cnicus.*
Bastard Saffron.

† 2 *Cnicus alter cæruleus.*
Blew floured Bastard Saffron.

¶ *The Description.*

1 CNicus, called also bastard Saffron, which may very wel be reckoned among the Thistles, riseth vp with a stalke of a cubite and a halfe high, straight, smooth, round, hard, and woody, & branched at the top : it is defended with long leaues, somthing broad, sharp pointed,

pointed, and with prickles in the edges: from the tops of the ſtalks ſtand out little heads or knops of the bigneſſe of an Oliue or bigger, ſet with many ſharpe pointed and prickly ſcales: out of which come forth floures like threds, cloſely compact, of a deepe yellow ſhining colour, drawing neere to the colour of Saffron: vnder them are long ſeeds, ſmooth, white, ſomewhat cornered, bigger than a Barly corne, the huske whereof is ſomething hard, the inner pulpe or ſubſtance is fat, white, ſweet in taſte: the root ſlender and vnprofitable.

2 There is alſo another kinde of Baſtard Saffron, that may very well be numbred amongſt the kindes of Thiſtles, and is very like vnto the former, ſauing that his flockie or threddie floures, are of a blew colour: the root is thicker, and the whole plant is altogether more ſharpe in prickles: the ſtalks alſo are more creſted and hairie.

¶ *The Place.*

It is ſowne in diuers places of Italy, Spaine, and France, both in gardens and in fields: *Pliny, lib. 25. cap. 15.* ſaith, that in the raigne of *Veſpaſian* this was not knowne in Italy; being in Egypt onely of good account, and that they vſed to make oile of it, and not meat.

¶ *The Time.*

The floures are perfected in Iuly and Auguſt: the root after the ſeed is ripe, the ſame yeare it is ſowne withereth away.

¶ *The Names.*

It is called in Greeke κνῖκος: in Latine alſo *Cnicus*, or *Cnecus*: in ſhops, *Cartamus*, or *Carthamum*: of diuers, *Crocus hortenſis*, and *Crocus Saracenicus*: in Italian, *Zaffarano Saracineſco*, and *Zaffarano ſaluatico*: in Spaniſh, *Alaſor*, and *Semente de papagaios*: in high Dutch, **wilden Zaffran**: in French, *Safran Sauuage*: in Engliſh, Baſtard Saffron of ſome, Mocke Saffron, and Saffron D'orte, as though you ſhould ſay Saffron *de horte*, or of the garden. *Theophraſtus* and *Pliny* call it *Cnecus vrbana*, and *ſatiua*, or tame and garden baſtard Saffron, that it may differ from *Atractilis*, which they make to be a kinde of *Cnicus ſylueſtris*, or wilde Baſtard Saffron, but rather a *ſpecies* of the Holy Thiſtle.

¶ *The Temperature.*

We vſe ſaith *Galen*, the ſeed onely for purgations: it is hot, and that in the firſt degree, as *Meſues* writeth.

¶ *The Vertues.*

A The iuice of the ſeed of baſtard Saffron bruiſed and ſtrained into honied water or the broth of a chicken, and drunke, prouoketh to the ſtoole, and purgeth by ſiege ſlimy flegme, and ſharp humors: Moreouer it is good againſt the collicke, and difficultie of taking breath, the cough, and ſtopping of the breſt, and is ſingular againſt the dropſie.

B The ſeed vſed as aforeſaid, and ſtrained into milke, cauſeth it to curdle and yeeld much cruds, and maketh it of great force to looſe and open the belly.

C The floures drunke with honied water open the liuer, and are good againſt the iaundice: and the floures are good to colour meat in ſtead of Saffron.

D The ſeed is very hurtfull to the ſtomacke, cauſing deſire to vomite, and is of hard ſlow digeſtion, remaining long in the ſtomacke and entrailes.

E Put to the ſame ſeed things comfortable to the ſtomacke, as Anniſe ſeed, Galingale, or Maſtick, Ginger, *Sal gemmæ*, and it ſhall not hurt the ſtomacke at all, and the operation thereof ſhall be the more quicke and ſpeedy.

F Of the inward pulpe or ſubſtance hereof is made a moſt famous and excellent compoſition to purge water with, commonly called *Diachartamon*, a moſt ſingular and effectual purgation for thoſe that haue the dropſie.

G The perfect deſcription hereof is extant in *Guido* the Surgion, in his firſt Doctrine, and the ſixt Tractate.

H We haue not read, or had in vſe that Baſtard Saffron with the blew floure, and therefore can ſay nothing of his vertues.

† The figure formerly was of the *Cnicus cæruleus*.

Chap. 489. *Of Wilde Baſtard Saffron.*

¶ *The Deſcription.*

1 Atractylis, otherwiſe called wilde Baſtard Saffron, bringeth forth a ſtraight and firme ſtalke, verie fragile or brittle, diuided at the toppe into certaine branches: it hath long,

long iagged leaues set with prickles:the heads on the tops of the branches are very ful of sharp prickles: out of which grow floures all of threds, like those of bastard Saffron, but they are of a light yellow colour, and sometimes purple: the seed is somewhat great, browne, and bitter, otherwise like that of bastard Saffron: the root is of a meane bignesse.

1 *Atractylis.*
Wilde Bastard Thistle.

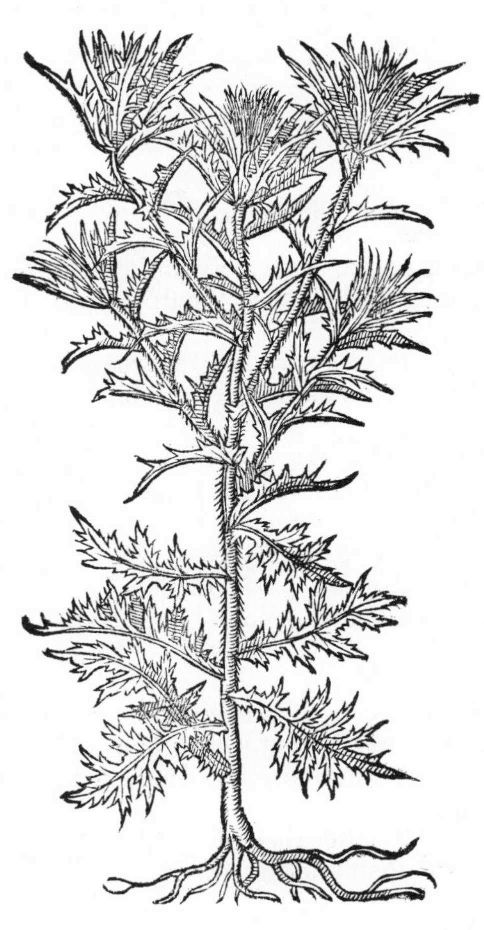

2 *Carduus Benedictus.*
The blessed Thistle.

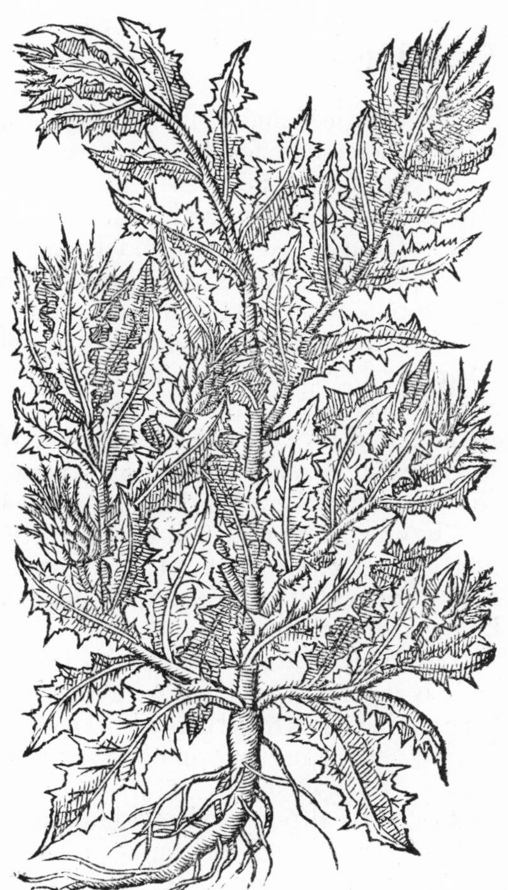

2 The stalkes of *Carduus Benedictus*, or Blessed Thistle, are round, rough, and pliable, and being parted into diuers branches, do lie flat on the ground: the leaues are iagged round about, and full of harmlesse prickles in the edges: the heads on the tops of the stalks are set with prickles, and inuironed with sharpe prickling leaues, out of which standeth a yellow floure: the seed is long, and set with haires at the top like a beard: the root is white, and parted into strings: the whole herb, leaues and stalks, and also the heads, are couered with a soft and thin downe.

¶ *The Place.*

Atractylis groweth in Candie, and in diuers prouinces and Islands of Greece, and also in Languedocke: and is an herbe growing in our English gardens.

Carduus Benedictus is found euery where in Lemnos, an Island of the Midland Sea, in Champion grounds, as *Petrus Bellonius* testifieth: it is diligently cherished in Gardens in these Northerne parts.

¶ *The Time.*

Atractylis is very late before it floureth and seedeth.

Carduus Benedictus floureth in Iuly and August, at which time it is especially to be gathered for Physicke matters.

¶ *The Names.*

Atractylis is called in Greeke Ἀτρακτυλὶς ἀγρία: of the Latins likewise, *Atractylis*, and *Cnicus syluestris*; and because women in the old time were wont to vse the stiffe stalk thereof *pro fuso aut colo*, for a spindle or a distaffe, it is named *Fusus agrestis*, and *Colus Rustica*; which thing *Petrus Bellonius* reporteth the women in Greece do also euen at this day; who call *Atractylis* by a corrupt name *Ardactyla*: diuers of the later herbarists name it *Syluestris Carthamus*: that is to say in low Dutch, **wilden Carthamus**: and in English, wilde Bastard Saffron: or Spindle Thistle.

Blessed Thistle is called in Latine euery where *Carduus Benedictus*, and in shops by a compound word,

word, *Cardo-benedictus*: it is moſt plaine, that it is *Species Atractylidis*, or a kind of wil de baſtard Saffron: it is called *Atractylis hirſutior*, hairie wilde baſtard Saffron: *Valerius Cordus* nameth it *Cnicus ſupinus*: it is called in high Dutch, **Beſeegnete diſtell, Kardo Benedict**: the later name whereof is knowne to the low Countrey men: in Spaniſh it is called *Cardo Sancto*: in French, *Chardon benoiſt*, or *beneiſt*. in the Iſle Lemnos, *Garderacantha*: in Engliſh, Bleſſed Thiſtle, but more commonly by the Latine name *Carduus Benedictus*.

¶ *The Temperature.*

Wilde baſtard Saffron doth drie and moderately digeſt, as *Galen* witneſſeth.

As *Carduus Benedictus* is bitter, ſo is it alſo hot and drie in the ſecond degree, and withall clenſing and opening.

¶ *The Vertues.*

A The tops, ſeed, and leaues of *Atractylis*, ſaith *Dioſcorides*, being beaten and drunk with pepper and wine, are a remedie for thoſe that are ſtung of the ſcorpion.

B Bleſſed Thiſtle taken in meat or drinke, is good for the ſwimming and giddineſſe of the head, it ſtrengthneth memorie, and is a ſingular remedie againſt deafeneſſe.

C The ſame boiled in wine and drunke hot, healeth the griping paines of the belly, killeth and expelleth wormes, cauſeth ſweat, prouoketh vrine, and driueth out grauel; clenſeth the ſtomack, and is very good againſt the Feuer quartaine.

D The iuice of the ſaid *Carduus* is ſingular good againſt all poiſon, as *Hierome Bocke* witneſſeth, in what ſort ſoeuer the medicine be taken; and helpeth the inflammation of the liuer, as reporteth *Ioachimus Camerarius* of Noremberg.

E The pouder of the leaues miniſtred in the quantitie of halfe a dram, is very good againſt the peſtilence, if it be receiued within 24. houres after the taking of the ſicknes, and the party ſweat vpon the ſame: the like vertue hath the wine, wherein the herbe hath been ſodden.

F The green herb pounded and laid to, is good againſt all hot ſwellings, as *Eryſipelas*, plague-ſores, and botches, eſpecially thoſe that proceed of the peſtilence, and is alſo good to be laied vpon the bitings of mad dogs, ſerpents, ſpiders, or any venomous beaſt whatſoeuer; and ſo is it likewiſe if it be inwardly taken.

G The diſtilled water thereof is of leſſe vertue.

H It is reported that it likewiſe cureth ſtubborne and rebellious vlcers, if the decoction be taken for certaine daies together; and likewiſe *Arnoldus de Villa noua* reporteth, that if it be ſtamped with Barrows greaſe to the form of an vnguent, adding thereto a little wheat floure, it doth the ſame, being applied twice a day.

I The herbe alſo is good being ſtamped and applied, and ſo is the iuice thereof.

K The extraction of the leaues drawne according to Art, is excellent good againſt the French diſeaſe, and quartaine agues, as reporteth the foreſaid *Camerarius*.

L The ſame Author reporteth, that the diſtilled water taken with the water of Louage, and Dodder, helpeth the ſauce-flegme face, if it be drunke for certaine daies together.

Chap. 490. *Of Thiſtle vpon Thiſtle, and diuers other Wilde Thiſtles.*

¶ *The Deſcription.*

1 Among all the Thornes and Thiſtles, this is moſt full of prickles; the ſtalks thereof are verie long, and ſeem to be cornered by reaſon of certaine thin skins growing to them, being ſent downe forth of the leaues: the leaues are ſet round about with many deep gaſhes, being very full of prickles as well as the ſtalks: the heads are very thicke ſet in euery place with ſtiffe prickles, and conſiſt of a multitude of ſcales; out of which grow purple floures, as they do out of other Thiſtles, ſeldome white: the root is almoſt ſtraight, but it groweth not deep.

2 To this alſo may be referred that which *Lobel* writeth to be named of the Italians *Leo*, and *Carduus ferox*, for it is ſo called of the wonderfull ſharpe and ſtiffe prickles, wherewith the whole plant aboundeth. the ſtalke thereof is ſhort, ſcarce a handfull high: the floure groweth forth of a prickly head, and is of a pale yellow colour, like that of wilde baſtard Saffron, and it is alſo inuironed and ſet round about on euery ſide with long hard thornes and prickles.

3 The third groweth ſeldome aboue a cubite or two foot high: it bringeth forth many round ſtalkes, parted into diuers branches; the leaues are like thoſe of white Cotton Thiſtle, but leſſer, and blacker, and not couered with downe or Cotton: vpon the tops of the ſtalks grow little heads like

† 1 *Polyacanthos.*
Thistle vpon Thistle.

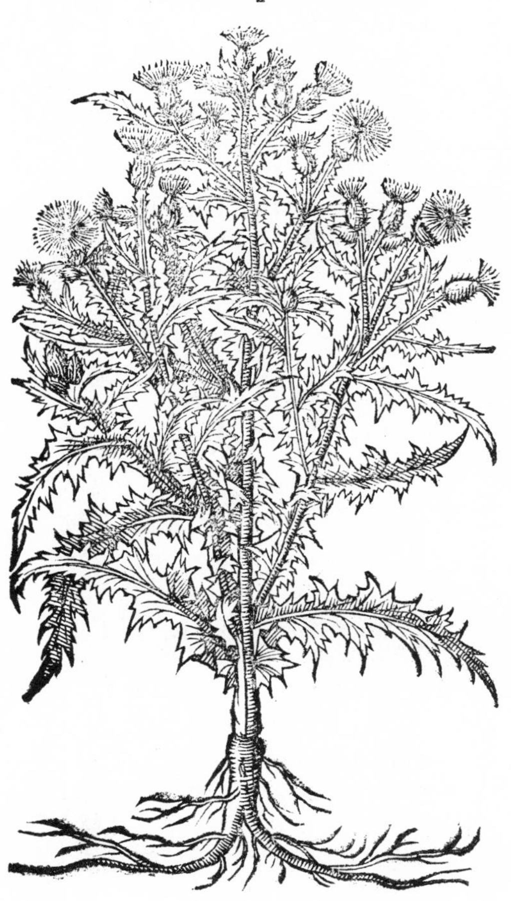

2 *Carduus ferox.*
The cruell Thistle.

† 3 *Carduus Asininus siue Onopyxos.*
The Asses Thistle, or Asses box.

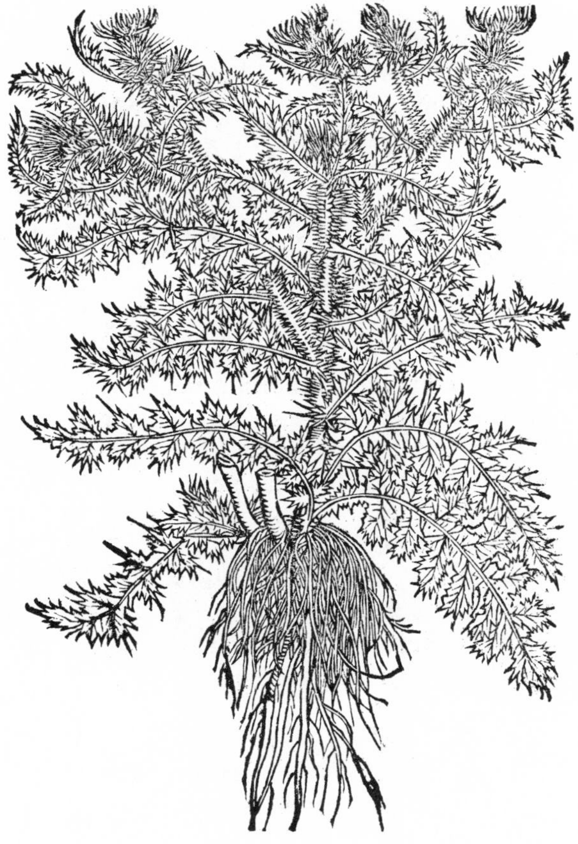

† 4 *Carduus vulgatissimus viarum.*
The Way Thistle.

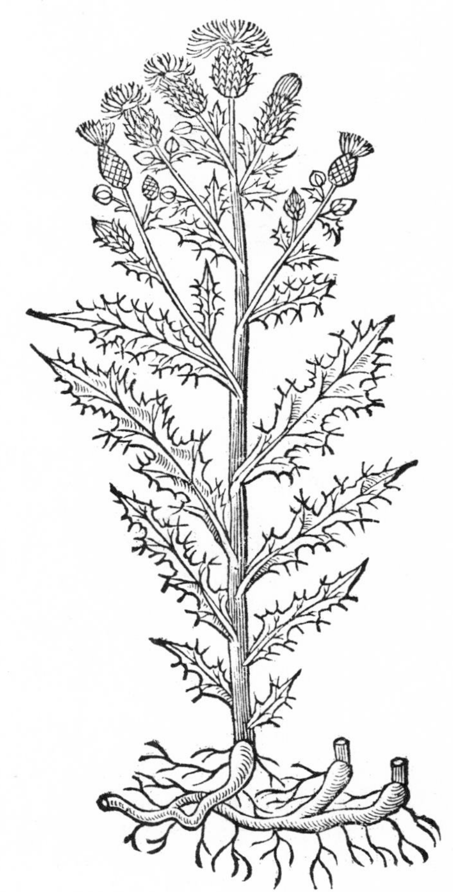

like Hedge-hogs; out of which ſpring gallant purple floures, that at length are turned into downe, leauing ſeedes behinde them like thoſe of the other Thiſtles : the root conſiſteth of many ſmall ſtrings.

4 The fourth riſeth vp with an higher ſtalke, now and then a yard long, round, and not ſo full of branches nor leaues, which are ſharpe and full of prickles, but leſſer and narrower : the heads be alſo leſſer, longer, and not ſo full of ſtiffe prickles : the floures are of a white colour, and vaniſh into downe : the root is blacke, and of a foot long.

5 This wilde Thiſtle which groweth in the fields about Cambridge, hath an vpright ſtalke, whereon do grow broad prickley leaues : the floures grow on the tops of the branches, conſiſting of a flockie downe, of a white colour tending to purple, of a moſt pleaſant ſweet ſmell, ſtriuing with the ſauour of muske : the root is ſmall, and periſheth at the approch of Winter. ‡ I had no figure directly fitting this ; wherefore I put that of *Dodonæus* his *Onopordon*, which may well ſerue for it, if the leaues were narrower, and more diuided. ‡

† 5 *Carduus Muſcatus.*
The musked Thiſtle.

6 *Carduus lanceatus.*
The Speare Thiſtle.

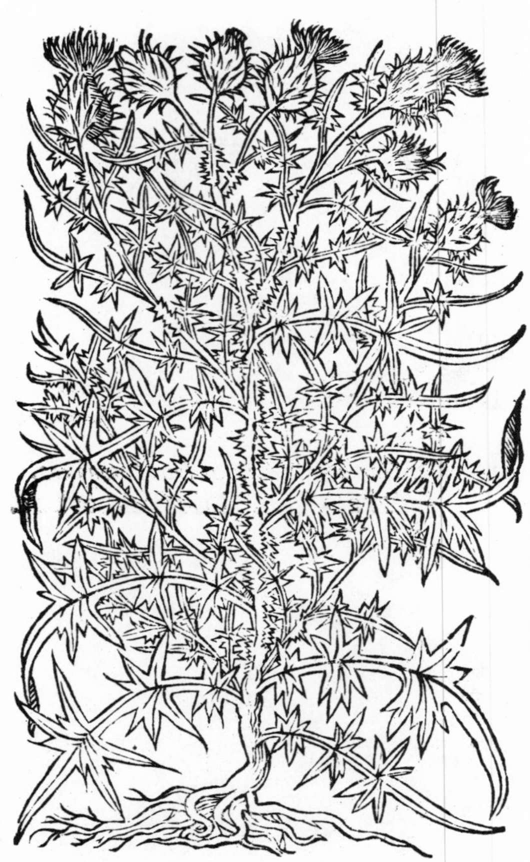

6 The Speare Thiſtle hath an vpright ſtalke, garniſhed with a skinnie membrane, full of moſt ſharpe prickles: whereon do grow very long leaues, diuided into diuers parts with ſharp prickles; the point of the leaues are as the point of a ſpeare, whereof it tooke his name: the floures grow on the tops of the branches, ſet in a ſcaly prickly head, like vnto the heads of Knapweed in forme, conſiſting of many threds of a purple colour : the root conſiſteth of many tough ſtrings.

7 *Theophraſtus* his fiſh Thiſtle called *Acarna*, which was brought from Illyria to Venice, by the learned *Valerandus Donrez*, deſcribed by *Theophraſtus*, hath horrible ſharpe yellow prickles, ſet vpon his greene indented leaues, which are couered on the backe ſide with an hoarie downe (as all the reſt of the plant) hauing a ſtalke of a cubit and a halfe high, and at the top certaine ſcaly knops containing yellow thrummie floures, armed or fenced with horrible ſharp prickles: the root is long and threddie.

8 The other kinde of fiſh Thiſtle, being alſo another *Acarna* of *Valerandus* deſcription, hath long and large leaues, ſet ful of ſharpe prickles, as though it were ſet full of pins: all the whole plant is couered with a certaine hoarineſſe, like the former : there ariſeth vp a ſtalke nine inches long, yea in ſome fertile grounds a cubite high, bearing the floure of *Carduus benedictus*, ſtanding thicke together, but leſſer.

‡ 9 This

7 *Acarna* Theophrasti.
Theophrastus his fish Thistle.

8 *Acarna Valerandi Donrez.*
Donrez his fish Thistle.

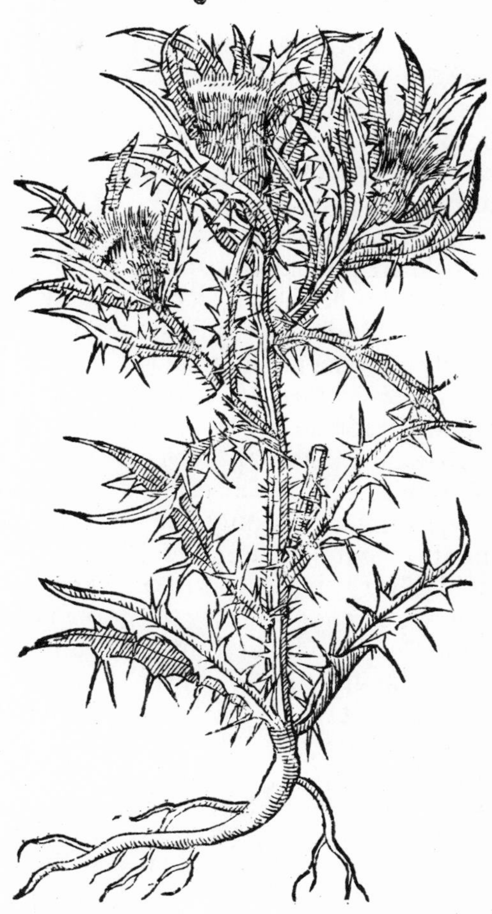

† 9 *Picnomos.*
The thicke or bush headed Thistle.

‡ 9 This Thistle in the opinion of *Bauhine*, whereto I much incline, is the same with the former. The root is small, the leaues long, welting the stalks at their setting on, and armed on the edges with sharpe prickles: the stalkes lie trailing on the ground like those of the star-Thistle, so set with prickles, that one knoweth not where to take hold thereof: it hath many closely compact vmbels, consisting of pale yellowish little floures like those of Groundswell: the seed is like that of *Carthamus*, smal and chaffie. *Pena* and *Lobel* call this *Picnomos Cretæ Salonensis*, of a place in Prouince where they first found it, called the Crau. being not farre from the city Salon. *Tabernamontanus* set it forth for *Chamæleon niger*, and our Author formerly gaue the figure hereof by the same title, though his historie belonged to another, as I haue formerly noted. ‡

¶ *The Time.*

The two first grow on diuers banks not farre from mount Apennine, and sometimes in Italy, but yet seldome.

The way Thistles grow euery where by highwaies sides and common paths in great plenty.

The places of the rest haue beene sufficiently spoken of in their descriptions.

¶ *The*

¶ *The Time.*

Theſe kindes of Thiſtles do floure from the beginning of Iune vntill the end of September.

¶ *The Names.*

Theſe Thiſtles comprehended in this preſent chapter are by one generall name called in Latine *Cardui ſylueſtres*, or wilde Thiſtles; and that which is the ſecond in order is named *Scolymus*: but not that *Scolymus* which *Theophr.* declareth to yeeld a milky iuyce (of which wee haue written before) but one of thoſe which *Pliny* in his twentieth booke, *cap.* 23. deſcribeth: of ſome they are taken for kindes of Chamæleon: their ſeuerall titles do ſet forth their ſeuerall Latine names, and alſo the Engliſh.

‡ There was formerly much confuſion in this chapter, both in the figures and hiſtorie, which I will here endeauour to amend, and giue as much light as I can, to the obſcuritie of our Authour and ſome others; to which end I haue made choice of the names as the fitteſt place.

1 This deſcription was taken out of *Dodonæus*, and the title alſo of *Onopordon* which was formerly put ouer the figure, and they belong to the Thiſtle our Author before deſcribed by the name of *Acanthium purp. Illyricum*, *cap.* 476. I haue therefore changed the title, yet let the deſcription ſtand, for it reaſonable well agrees with the figure which is of the *Carduus ſpinoſiſsimus vulgaris* of *Lobel*, and *Polyacantha Theophraſti* of *Tabern.* Of this Thiſtle I obſerue three kindes: the firſt is a Thiſtle ſome two cubits and a halfe high, with many ſlender ſtalkes and branches exceeding prickly, hauing commonly fiue prickly welts running alongſt the ſtalks: the leaues on the vpper ſides as alſo the ſtalkes are of a reaſonable freſh greene colour, but the vnderſide of the leafe is ſomwhat whitiſh: the heads conſiſt of ſundry hairy greene threds which looke like prickles, but they are weake, and not prickly: the floure is of the bigneſſe, and of the like colour and ſhape as the common Knapweed, yet ſomwhat brighter: it growes on ditch ſides, and floures in Iuly. This I take to be the *Aculeoſa Gaſæ* of the *Aduerſ. pag.* 374. but not that which *Lobel* figures for it in his *Icones.* This is that which *Tabernamontanus* figures for *Polyacantha*, and our Author gaue his figure in this place. The ſecond of theſe I take to be that which *Lobel* hath figured for *Polyacantha*, and *Dodonæus* for *Carduus ſyl.* 3. (which figure we here giue you) and in the *Hiſt. Lugd. pag.* 1473. it is both figured and deſcribed by the name of *Polyacanthos Theophraſti.* In the figure there is little difference: in the things themſelues this; the ſtalkes of this are as high as thoſe of the laſt, but ſlenderer, with fewer and ſtraighter branches, and commonly edged with foure large welts, which haue fewer, yet longer prickles than thoſe of the former: the leaues and ſtalkes of this are of a grayiſh or whitiſh colour: the heads are longiſh, but much ſmaller than thoſe of the former, and they ſeldom open or ſpred abroad their floures, but onely ſhew the tops of diuers reddiſh threds of a feint colour. This growes as frequently as the former, and commonly in the ſame places. The third, which I thinke may fitly be referred vnto theſe, growes on wet heaths and ſuch like places, hauing a ſtalke ſometimes foure or fiue cubits high, growing ſtraight vp, with few branches, and thoſe ſhort ones: the floures are of an indifferent bigneſſe, and commonly purple, yet ſometimes white. I thinke this may be the *Onopyxos alter Lugdunenſ.* or the *Carduus paluſtris* deſcribed in *Bauhinus* his *Prodromus*, *pag.* 156.

2 The ſecond, which is a ſtranger with vs, is the *Phœnix, Leo & Carduus ferox* of *Lobel* and *Dod. Bauhine* both refer'd it to *Acarna*, calling it *Acarna minor caule non folioſo.*

3 The third deſcription was alſo out of *Dodonæus*, being of his *Carduus ſylueſtris primus*, or the *Onopyxos Dodonæi* of the *Hiſt. Lugd.* The figures formerly both in the third and fourth place of this chapter were of the *Acanthium Illyricum* of *Lobel*, or the *Onopordon* of *Dodonæus*, formerly mentioned.

4 This deſcription alſo was out of *Dodonæus*, being of his *Carduus ſylueſtris alter*, agreeing in all things but the colour of the floures, which ſhould be purple. *Lobel* in his Obſeruations deſcribeth the ſame Thiſtle by the name of *Carduus vulgatiſsimus viarum*: but both he and *Dodon.* giue the figure of *Carlina ſylueſtris* for it: but neither the floures nor the heads of that agree with that deſcription. I iudge this to be the Thiſtle that *Fabius Columna* hath ſet forth for the *Ceanothos* of *Theophraſtus*; and *Tabern.* for *Carduus arvenſis*: and our Author, though vnfitly, gaue it in the next place for *Carduus muſcatus.*

5 The Muske-Thiſtle I haue ſeene growing about Deptford, and (as far as my memory ſerues me) it is very like to the third here deſcribed: it growes better than a cubit high, with reaſonable large leaues, and alſo heads which are a little ſoft or downy, large, with purple floures: the heads before the floures open ſmell ſtrong of muske. I haue found no mention of this but only in *Geſner*, *de Collectione in parte*, where he hath theſe words; *Carduus arvenſis maior purpureo flore (qui flore nondum nato Moſchum olebat) floret Iulio.* Our Author formerly gaue an vnfit figure for this, as I formerly noted.

There is ſufficient of the reſt in their titles and deſcriptions. ‡

¶ The

¶ *The Temperature and Vertues.*

Theſe wild Thiſtles (according to *Galen*) are hot and dry in the ſecond degree, and that through the propertie of their eſſence they driue forth ſtinking vrine, if the roots be boyled in Wine and drunke; and that they take away the ranke ſmell of the body and arme-holes. A

Dioſcorides ſaith, that the root of the common Thiſtle applied plaiſterwiſe correcteth the filthy ſmell of the arme holes and whole body. B

And that it workes the ſame effect if it be boyled in wine and drunke, and that it expelleth plentie of ſtinking vrine. C

The ſame Author affirmeth alſo, that the herbe being as yet greene and tender is vſed to be eaten among other herbes after the manner of Aſparagus. D

This being ſtamped before the floure appeareth, ſaith *Pliny*, and the iuyce preſſed forth, cauſeth haire to grow where it is pilled off, if the place be bathed with the iuyce. E

The root of any of the wilde Thiſtles being boyled in water and drunke, is reported to make them dry that drinke it. F

It ſtrengthneth the ſtomacke; and it is reported (if we beleeue it) that the ſame is alſo good for the matrix, that boyes may be ingendred: for ſo *Chereas* of Athens hath written, and *Glaucias*, who is thought to write moſt diligently of Thiſtles. G

This Thiſtle being chewed is good againſt ſtinking breath. Thus farre *Pliny*, in his twentieth booke, *cap.* 23. H

CHAP. 491. *Of the Melon or Hedge-hog Thiſtle.*

Melocarduus Echinatus Penæ & Lob.
The Hedge-hog Thiſtle.

¶ *The Deſcription.*

WHo can but maruel at the rare and ſingular workmanſhip which the Lord God almighty hath ſhewed in this Thiſtle, called by the name *Echino-Melocactos*, or *Melo-carduus Echinatus?* This knobby or bunchy maſſe or lump is ſtrangely compact and context together, containing in it ſundry ſhapes and formes, participating of a Pepon or Melon, and a Thiſtle, both being incorporate in one body; which is made after the forme of a cock of hay, broad and flat below, but ſharp toward the top, as big as a mans body from the belly vpward: on the outſide hereof are fourteene hard ribbes, deſcending from the crowne to the loweſt part, like the bunchy or out ſwelling rib of a Melon ſtanding out, and chanelled betweene: at the top or crowne of the plant iſſueth forth a fine ſilken cotton, wherewith it is full fraught; within which cotton or flockes lie hid certain ſmal ſheaths or cods, ſharpe at the point, and of a deep ſanguine colour, anſwering the cods of *Capſicum* or Indian Pepper, not in ſhew only, but in colour, but the cods are ſomewhat ſmaller. The furrowed or chanelled ribs on the outſide are garniſhed or rather armed with many prickly ſtars, ſtanding in a compaſſe like ſharpe crooked hornes or hookes, each ſtar conſiſting of ten or twelue pricks, wherewith the outward barke or pilling is garded, ſo that without hurt to the fingers it cannot be touched: this rinde is hard, thicke, and like vnto Aloes, of the colour of the Cucumber: the fleſh or inner pulpe is white, fat, wateriſh, of taſte ſoure, vnſauory, and cooling, much like vnto the meate of a raw Melon or Pompion. This plant groweth without leafe or ſtalke, as our Northerne Thiſtle doth, called *Carduus Acaulos*, and is bigger than the largeſt

Pompion:

Pompion: the roots are ſmall, ſpreading farre abroad in the ground, and conſiſting of blacke and tough twigs, which cannot endure the iniurie of our cold clymate.

¶ *The Place.*

This admirable Thiſtle groweth vpon the cliffes and grauelly grounds neere vnto the ſea ſide, in the Iſlands of the Weſt Indies, called S. *Margarets* and S. *Iohns* Iſle, neere vnto *Puerto rico*, or *Porto rico*, and other places in thoſe countries, by the relation of diuers trauellers that haue iournied into thoſe parts, who haue brought me the plant it ſelfe with his ſeed; the which would not grow in my garden by reaſon of the coldneſſe of the clymate.

¶ *The Time.*

It groweth, floureth, and flouriſheth all the yeare long, as do many other plants of thoſe Countries.

¶ *The Names.*

It is called *Carduus Echinatus*, *Melocarduus Echinatus*, and *Echino Melocactos*: In Engliſh, the Hedge-hog Thiſtle, or prickly Melon Thiſtle. ‡ Such as are curious may ſee more hereof in *Cluſius* his *Exoticks*, *lib*. 4. *cap*. 24. ‡

¶ *The Temperature and Vertues.*

There is not any thing extant ſet forth of the antient or of the later writers, neither by any that haue trauelled from the Indies themſelues: therefore we leaue it to a further conſideration.

Chap. 492. *Of the gummie Thiſtle, called Euphorbium.*

1 *Euphorbium.*
The poyſonous gum Thiſtle.

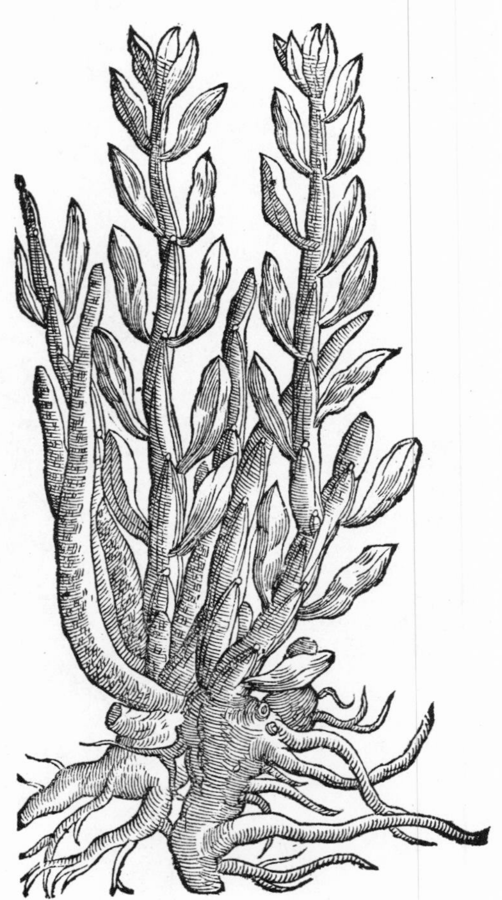

2 *Anteuphorbinm.*
The Antidote againſt the poyſonous Thiſtle.

¶ *The Deſcription.*

1 Euphorbium (whereout that liquor or gum called in ſhops *Euphorbium* is extracted) hath very great thicke groſſe and ſpreading roots, diſperſed far abroad in the ground: from which ariſe long and round leaues, almoſt like the fruit of a great Cucumber, a foot and a halfe long, ribbed, walled, and furrowed like vnto the Melon: theſe branched ribs are ſet or armed

armed for the most part with certaine prickles standing by couples, the point or sharpe end of one garding one way, and the point of another looking directly a cleane contrarie way : these prickes are often found in the gumme it selfe, which is brought vnto vs from Libya and other parts : the leaues hereof being planted in the ground will take root well, and bring forth great increase, which thing I haue proued true in my garden : it hath perished againe at the first approch of winter. The sap or liquor that is extracted out of this plant is of the colour and substance of the Creame of Milke ; it burneth the mouth extremely, and the dust or pouder doth very much annoy the head and the parts thereabout, causing great and vehement sneesing, and stuffing of all the pores.

2 This rare plant called *Anteuphorbium* hath a very thicke grosse and farre spreading root, very like vnto *Euphorbium* ; from which riseth vp many round greene and fleshie stalkes, whereupon do grow thicke leaues like Purslane, but longer, thicker, and fatter : the whole plant is full of cold and clammie moisture, which represseth the scorching force of *Euphorbium* ; and it wholly seemes at the first view to be a branch of greene Corall.

3 *Cereus Peruvianus spinosus Lobelij.*
The Torch-Thistle or thorny Euphorbium.

4 *Calamus Peruvianus spinosus Lobelij.*
The thorny Reed of Peru.

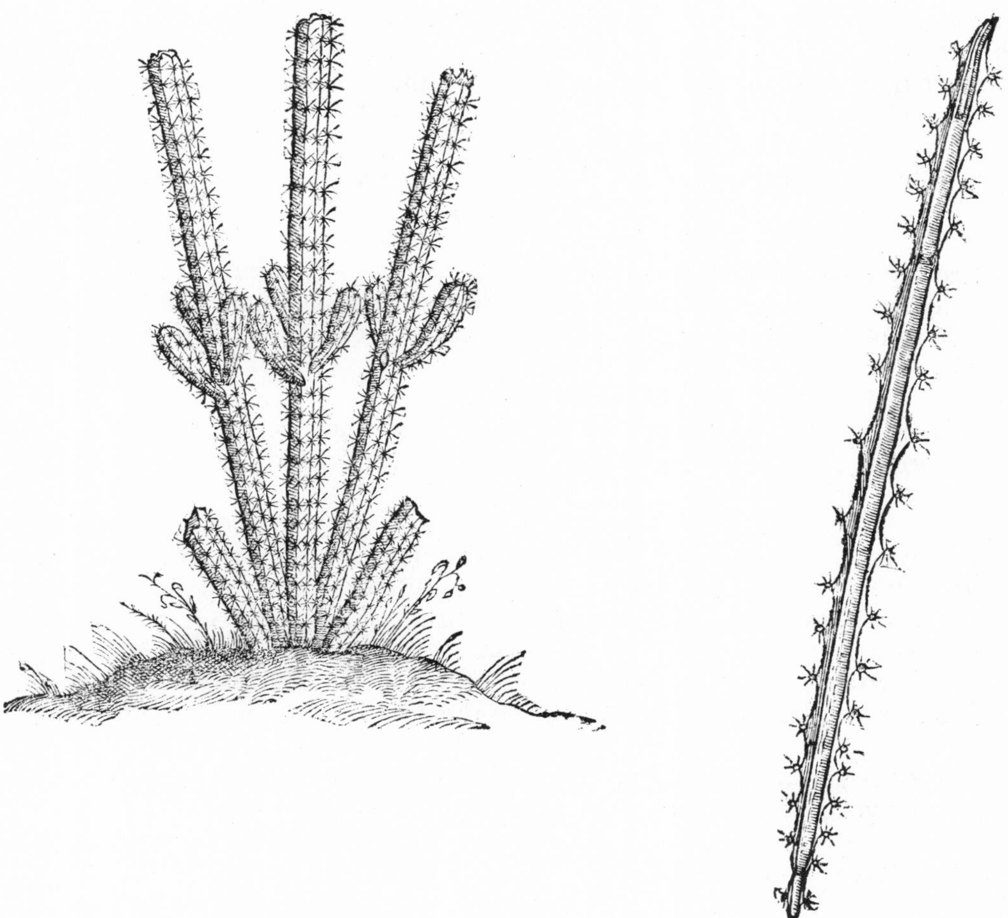

3 There is not among the strange and admirable plants of the world any one that giues more cause of maruell, or more moueth the minde to honor and laud the Creator, than this plant, which is called of the Indians in their mother tongue *Vragua*, which is as much to say, a torch, taper, or wax candle ; whereupon it hath been called in Latine by those that vnderstood the Indian tongue, *Cereus*, or a Torch. This admirable plant riseth vp to the height of a speare of twenty foot long, although the figure expresse not the same ; the reason is, the plant when the figure was drawn came to our view broken : it hath diuers bunches and vallies, euen as is to be seene in the sides of the Cucumber, that is, furrowed, guttered, or chamfered alongst the same, and as it were laid by a direct line, with a welt from one end vnto the other : vpon which welt or line do stand small star-like Thistles, sharpe as needles, and of the colour of those of the Melon Thistle, that is to say, of a browne colour : the trunke or body is of the bignesse of a mans arme, or a cable rope ; from the middle whereof thrust forth diuers knobby elbowes of the same substance, and armed with the like prickles that the body of the trunke is set withall : the whole plant is thicke, fat, and full of a fleshie substance, hauing much iuyce like that of Aloes, when it is hardned, and of a bitter taste : the floures

floures grow at the top or extreme point of the plant: after which follow fruit in shape like a fig, full of a red iuyce, which being touched staineth the hands of the colour of red leade: the taste is not vnpleasant.

4 There hath been brought from the Indies a prickly reed of the bignesse of a good big staf, of the length of six or eight foot, chamfered and furrowed, hauing vpon two sides growing vnto it an vneuen membrane or skinny substance, as it were a iag or welt set vpon the wing of a garment, and vpon the very point of euery cut or iagge armed with most sharpe prickles: the whole trunke is filled full of a spongeous substance, such as is in the hollownes of the brier or bramble; amongst the which is to be seene as it were the pillings of Onions, wherein are often found liuing things, that at the first seeme to be dead. The plant is strange, and brought dry from the Indies, therefore we cannot write so absolutely hereof as we desire; referring what more might be said to a further consideration or second edition.

¶ *The Place.*

These plants grow vpon Mount Atlas, in Libya, in most of the Islands of the Mediterranean sea, in all the coast of Barbarie, especially in S. Crux neere vnto the sea side, in a barren place there called by the English men Halfe Hanneken; which place is appointed for Merchants to confer of their businesse, euen as the Exchange in London is: from which place my friend Mr. *William Martin*, a right expert Surgeon, did procure me the plants of them for my garden, by his seruant that he sent thither as Surgeon of a ship. Since which time I haue receiued plants of diuers others that haue trauelled into other of those parts and coasts: notwithstanding they haue not endured the cold of our extreme Winter.

¶ *The Time.*

They put forth their leaues in the Spring time, and wither away at the approch of Winter.

¶ *The Names.*

It is called both in Greeke and Latine Εὐφόρβιον, *Euphorbium*: *Pliny* in one place putteth the herbe in the feminine gender, naming it *Euphorbia*: the iuyce is called also *Euphorbion*, and so it is likewise in shops: we are faine in English to vse the Latine word, and to call both the herbe and iuyce by the name of Euphorbium, for other name we haue none: it may be called in English, the Gum Thistle.

¶ *The Temperature.*

Euphorbium (that is to say, the congealed iuyce which we vse) is of a very hot, and, as *Galen* testifieth, caustick or burning facultie, and of thinne parts: it is also hot and dry in the fourth degree.

¶ *The Vertues.*

A An emplaister made with the gumme Euphorbium, and twelue times so much oyle, and a little wax, is very singular against all aches of the ioynts, lamenesse, palsies, crampes, and shrinking of sinewes, as *Galen, lib.4. de medicamentis secundum genera*, declareth at large, which to recite at this present would but trouble you ouermuch.

B Euphorbium mingled with oyle of Bay and Beares grease cureth the scurfe and scalds of the head, and pildnesse, causing the haire to grow againe, and other bare places, being anointed therewith.

C The same mingled with oyle, and applied to the temples of such are very sleepie, and troubled with the lethargie, doth awaken and quicken their spirits againe.

D If it be applied to the nuque or nape of the necke, it bringeth their speech againe that haue lost it by reason of the Apoplexie.

E Euphorbium mingled with vineger and applied taketh away all foule and ill fauoured spots, in what part of the body soeuer they be.

F Being mixed with oyle of Wall-floures, as *Mesues* saith, and with any other oyle or ointments, it quickly heateth such parts as are ouer cold.

G It is likewise a remedie against old paines in the huckle bones, called the Sciatica.

H *Aetius, Paulus, Actuarius*, and *Mesue* doe report, That if it be inwardly taken it purgeth by siege water and flegme; but withall it setteth on fire, scortcheth and fretteth, not onely the throat and mouth, but also the stomacke, liuer, and the rest of the intrals, and inflames the whole bodie.

I For that cause it must not be beaten smal, and it is to be tempered with such things as allay the heate and sharpenesse thereof, and that make glib and slipperie; of which things there must be such a quantitie, as that it may be sufficient to couer all ouer the superficiall or outward part thereof.

K But it is a hard thing so to couer and fold it vp, or to mix it, as that it will not burne or scortch. For though it be tempered with neuer so much oyle, if it be outwardly applied it raiseth blisters, especially in them that haue soft and tender flesh, and therfore it is better not to take it inwardly.

It is

It is troublesome to beate it, vnlesse the nostrils of him that beats it be carefully stopped and L
defended; for if it happen that the hot sharpnesse thereof do enter into the nose, it presently causeth itching, and moueth neesing, and after that, by reason of the extremitie of the heate, it draweth out aboundance of flegme and filth, and last of all bloud, not without great quantity of teares.

But against the hot sharpnesse of *Euphorbium*, it is reported that the inhabitants are remedied M
by a certaine herbe, which of the effect and contrarie faculties is named *Anteuphorbium*. This plant likewise is full of iuyce, which is nothing at all hot and sharpe but coole and slimy, allaying the heate and sharpnesse of *Euphorbium*. We haue not yet learned that the old writers haue set downe any thing touching this herbe; notwithstanding it seemeth to be a kinde of Orpine, which is the antidote or counterpoyson against the poyson and venome of *Euphorbium*.

‡ CHAP. 493. *Of soft Thistles, and Thistle gentle.*

‡ THere are certaine other plants by most writers referred to the Thistles; which being omitted by our Author, I haue thought fit here to giue you.

‡ *Cirsium maximum Asphodeli radice.*
Great soft bulbed Thistle.

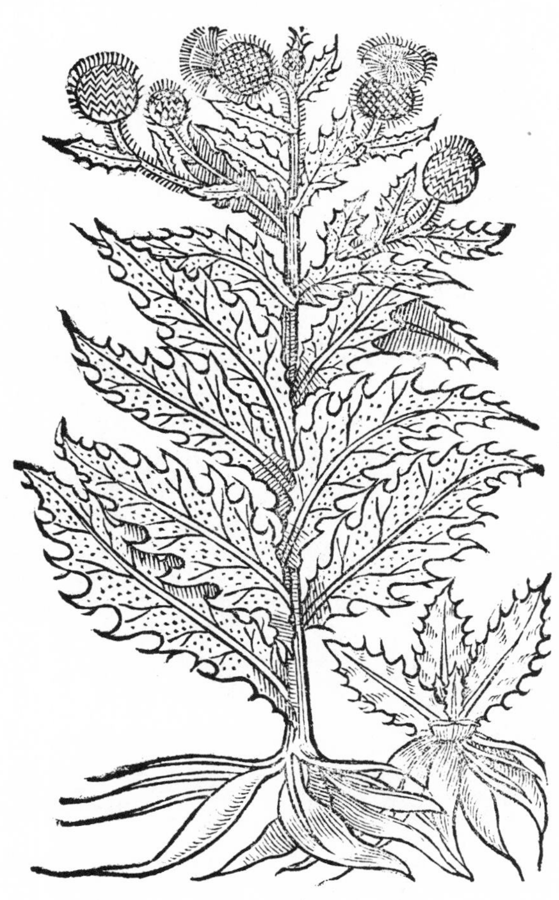

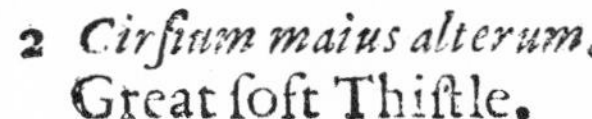

2 *Cirsium maius alterum.*
Great soft Thistle.

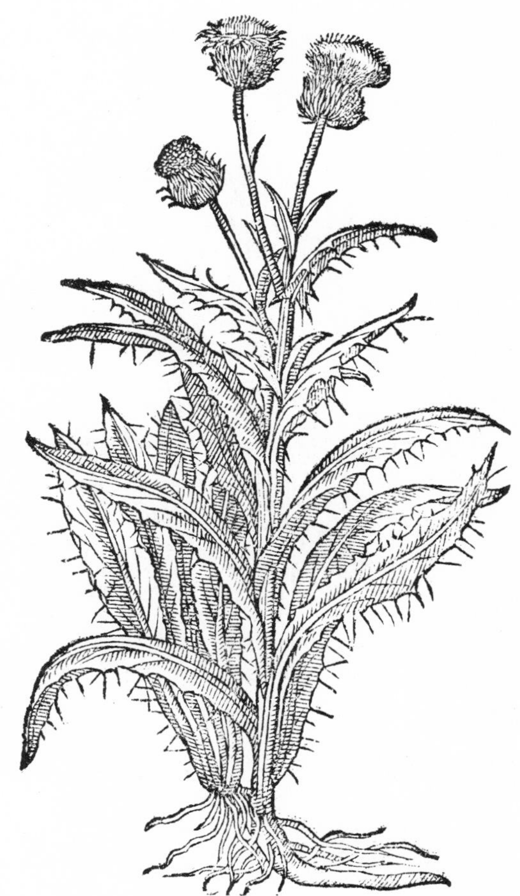

¶ *The Description.*

1 THe first and largest of these hath roots consisting of great longish bulbes like those of the Asphodill: from whence arise many large stalkes three or foure cubits high, crested and downy: the leaues are very long and large, iuycie, greenish, and cut about the edges, and set with soft prickles. At the tops of the stalkes and branches grow heads round and large, out whereof come floures consisting of aboundance of threds, of a purple colour, which flie away in downe. This growes wilde in the mountainous medowes and in some wet places of Austria. I haue seene it growing in the garden of M[r]. *Iohn Parkinson*, and with M[r]. *Tuggye*. It floures in Iuly. *Clusius* hath called it *Cirsium maximum mont. incano folio, bulbosa radice*. But he gaue no figure thereof, nor any else, vnlesse the *Acanthium peregrinum* in *Tabernamont*. (which our Author formerly

formerly, as I before noted, gaue by the name of *Solſtitialis lutea peregrina)* were intended for this plant, as I verily thinke it was. I haue giuen you a figure which I drew ſome yeares agoe by the plant it ſelfe.

2 The root of this is long, yet ſending forth of the ſides creeping fibres, but not bulbous: the leaues are like thoſe of the laſt mentioned, but leſſe, and armed with ſharpe prickles of a greeniſh colour, with the middle rib white: the heads ſometimes ſtand vpright, and otherwhiles hang downe; they are very prickly, and ſend forth floures conſiſting of many elegant purple threds. The ſtalkes are thicke, creſted and welted with the ſetting on of the leaues. This growes wilde vpon the ſea coaſts of Zeeland, Flanders, and Holland: it floures in Iune and Iuly: it is the *Cirſium tertium* of *Dodonæus*; and *Cirſium maius* of *Lobel*.

3 This whoſe root is fibrous and liuing, ſends forth leſſer, narrower, and ſofter leaues than thoſe of the former, not iagged or cut about their edges, nor hoary, yet ſet about with prickles: the ſtalkes are creſted: the heads are ſmaller, and grow three or foure together, carrying ſuch purple floures as the former. This is that which *Matthiolus, Geſner*, and others haue ſet forth for *Cirſium: Dodonæus*, for *Cirſium* 2. and *Cluſius* hath it for his *Cirſium quartum*, or *Montanum ſecundum*.

‡ 3 *Cirſium folijs non hirſutis.*
Soft ſmooth leaued Thiſtle.

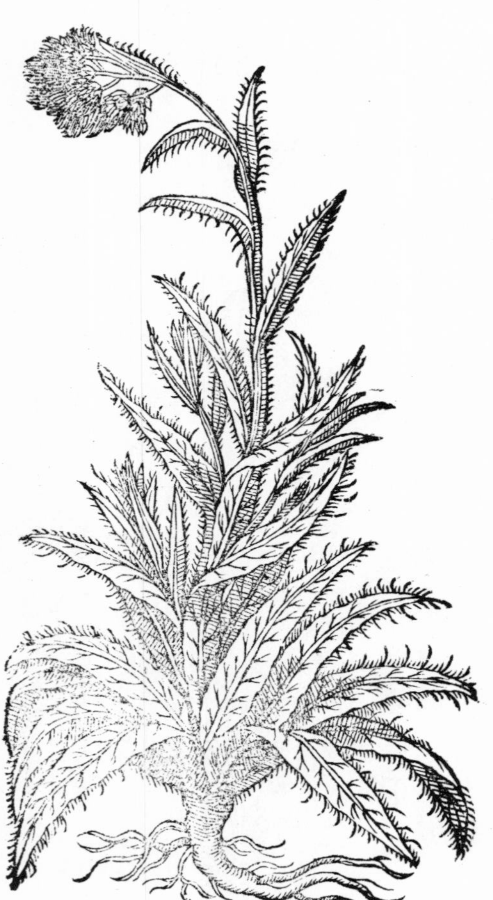

‡ 4 *Cirſium montanum capitulis paruis.*
Small Burre Thiſtle.

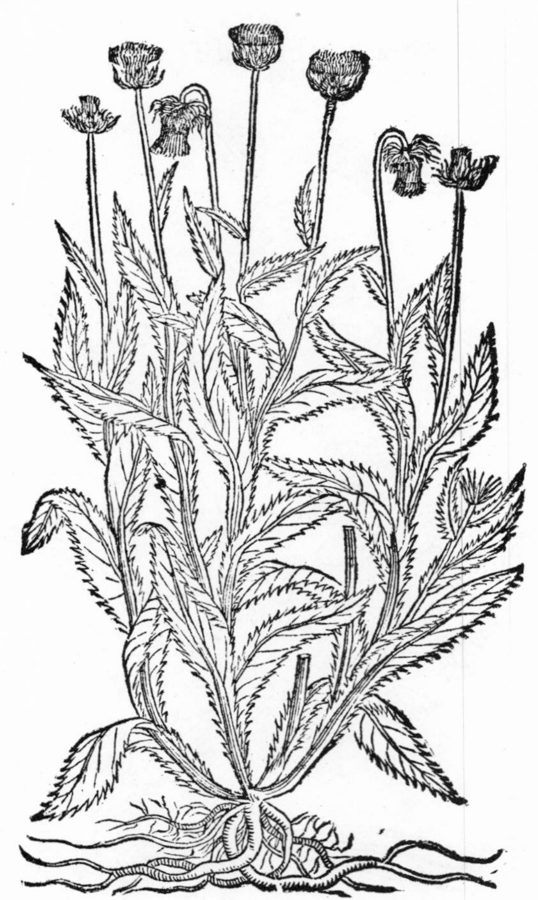

4 The leaues of this are ſomewhat like thoſe of the laſt deſcribed, but larger, and welting the ſtalkes further at their ſetting on: they are alſo ſet with prickles about the edges: the ſtalks are ſome two cubits high, diuided into ſundry long ſlender branches: on whoſe tops grow little rough prickly heads, which after the floures come to perfection doe hang downewards, and at the length turne into downe; amongſt which lies hid a ſmooth ſhining ſeed. This groweth wilde in diuers wooddy places of Hungarie and Auſtria. It is the *Cirſium* of *Dodonæus*; the *Cirſium* 2. or *Montanum* 1. of *Cluſius*; and *Cirſium alterum* of *Lobel*. It floures in Iune: the root is about the thickneſſe of ones little finger, fibrous alſo, and liuing.

5 This ſends vp long narrow leaues, hairy, and ſet about the edges with ſlender prickles: out of the middeſt of theſe leaues growes vp a ſtalke ſometimes a foot, otherwhiles a cubit high, ſlender, ſtiffe, and downy: vpon which grow leaues ſomewhat broad at their ſetting on, and there alſo a little nicked or cut in: this ſtalke ſometimes hath no branches, otherwhiles two or three long ſlender ones, at the tops whereof grow out of ſcaly heads ſuch floures as the common Knap-weed, which

‡ 5 *Cirsium montanum Anglicum.*
Single headed Thistle.

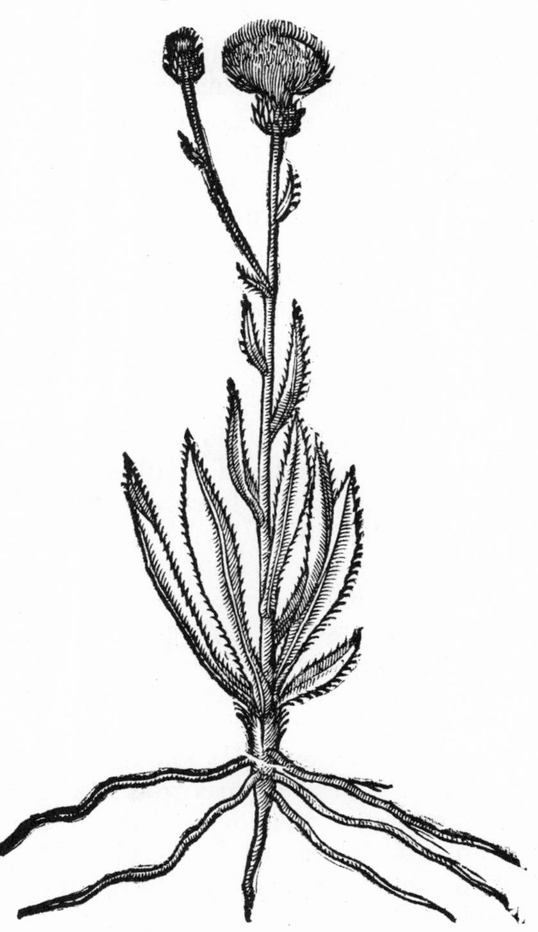

‡ *Cirsij Anglici alia Icon Pennei.*
Pennes figure of the same.

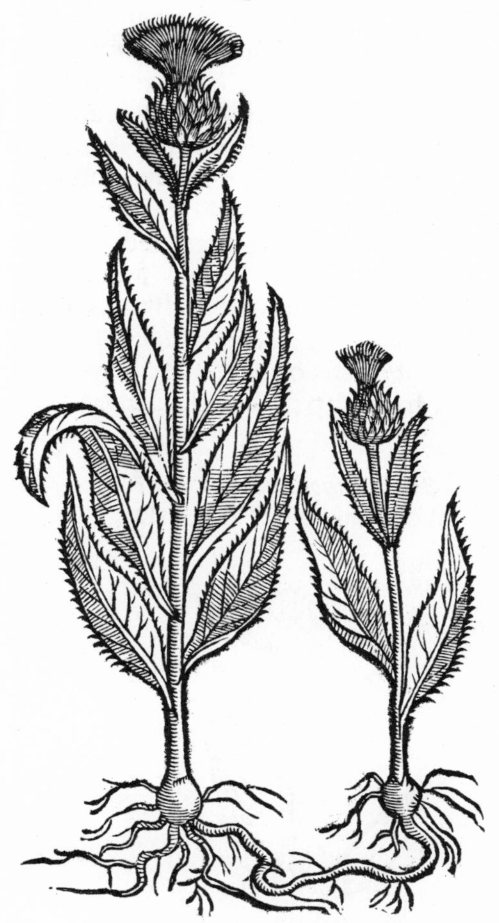

‡ *Cirsij Anglici Icon Lobelij.*
Lobels figure of the same.

‡ 6 *Carduus mollis folijs dissectis.*
Iagged leaued Thistle gentle.

which at length turne into downe; among which lies hid a ſmall ſhining ſeed like the other plants of this kind. The root is made of diuers thicke fibres, which run in the ground, and here and there put vp new heads. This plant wants no ſetting forth; for *Cluſius* giues vs the figure and hiſtorie thereof, firſt by the name of *Cirſium Pannonicum* 1. *pratenſe*; then he giues another hiſtorie thereof, with a worſer figure, (which he receiued of Dr. *Thomas Penny* of London) by the name of *Cirſium Anglicum* 2. *Lobel* alſo deſcribed it, and ſet it forth with a figure expreſſing the floure alreadie faded, by the name of *Cirſium Anglicum*. *Bauhine* in his *Pinax*, deceiued by theſe ſeuerall expreſſions, hath made three ſeuerall plants of this one; a fault frequent in many Writers of plants. *Cluſius* found it growing in the mountainous medowes alongſt the ſide of the Danow in Auſtria: *Penny*, in the medowes at the foot of Ingleborow hill in Yorke-ſhire: *Lobel*, in the medowes at a place called Acton in Glouceſter-ſhire. I found this onely once, and that was in a medow on this ſide Highgate, hauing beene abroad with the Companie of Apothecaries, and returning that way home, in the companie of Mr. *Iames Walſall*, *William Broad*, and ſome others. I haue giuen you both the figures of *Cluſius* his owne in the firſt place, and that of Dr. *Penny* in the ſecond, but the former is the better: I haue alſo giuen you that of *Lobel*.

‡ 7 *Carduus mollis folijs Lapathi.*
Docke leaued Thiſtle-gentle.

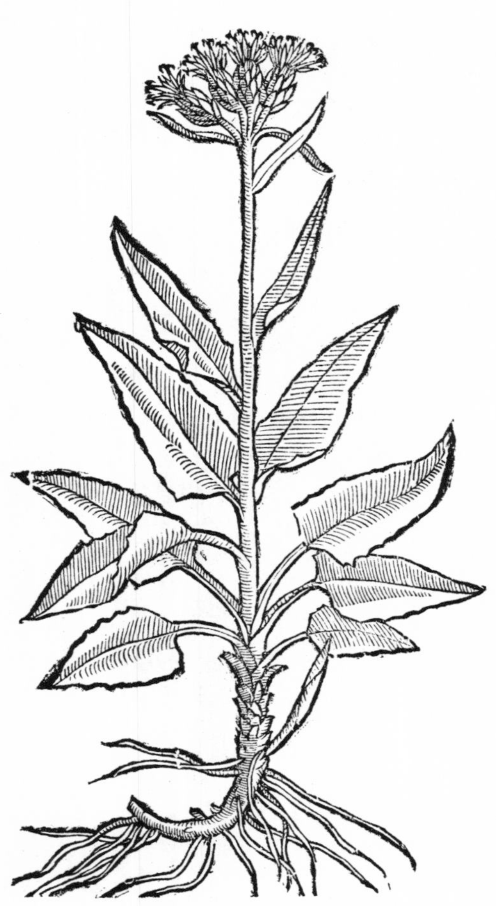

6 Theſe alſo *Cluſius* (whom I herein follow) addeth to the kindes of Thiſtles. This iagged leafed one, which he calleth *Carduus mollior primus*, hath many leaues at the root, both ſpred vpon the ground, and alſo ſtanding vpright; and they are couered with a white and ſoft downineſſe, yet greene on the vpper ſide: they are alſo much diuided or cut in euen to the middle rib, like to the ſofter or tenderer leaues of the Starre Thiſtle: they haue no prickles at all vpon them: out of the middeſt of theſe leaues grow vp one or two ſtalkes, round, creſted, purpliſh, hoarie, and ſome cubit or better high. The leaues that grow vpon the lower part of the ſtalke are diuided, thoſe aboue not ſo; the tops of the ſtalkes ſometimes, yet very ſeldome, are parted into branches, which cary ſcaly heads containing elegant floures made of many purple ſtrings. The floure decaying, there ſucceeds a cornered ſeed: the root ſomtimes equalls the thickeneſſe of ones finger, browniſh, long, and ſomewhat fibrous. It floures in May, and growes vpon the hilly places of Hungarie.

7 The ſtalke of this is ſome foot or better high, thicke, creſted, and ſomwhat hairy: the leaues about the root are ſomwhat large, and in ſhape like thoſe of *Bonus Henricus*, (abuſiuely called in Engliſh, Mercurie) ſomewhat ſinuated about the edges, and ſet with harmeleſſe prickles, greene aboue, and verie hoarie vnderneath, like the leaues of the white Poplar: thoſe that grow vpon the ſtalke are leſſer and narrower: out of whoſe boſomes towards the tops of the ſtalke grow out little branches which carry three, foure, or more little ſcaly heads like thoſe of the Blew-Bottle, or Knapweed, whereout grow threddy blewiſh purple floures: the ſeed is wrapped in downe, and not vnlike that of Blew-Bottle: the root is blacke, hard, and liuing, ſending forth ſhoots on the ſides. It growes vpon the higheſt Auſtrian Alpes, and floures in Iuly. *Cluſius* calls this *Carduus mollior Lapathi folio.*

¶ *The Temperature and Vertues.*

Theſe plants ſeeme by their taſte to be of a moderately heating and drying facultie, but none of them are vſed in medicine, nor haue their vertues ſet downe by any Author. ‡

Chap.

CHAP. 494.
Of three leafed Graſſe, or Medow Trefoile.

¶ *The Kindes.*

THere be diuers ſorts of three leafed Graſſes, ſome greater, others leſſer; ſome beare floures of one colour, ſome of another: ſome of the water, and others of the land: ſome of a ſweet ſmel, others ſtinking: and firſt of the common medow Trefoiles, called in Iriſh *Shamrockes*.

1 *Trifolium pratenſe.*
Medow Trefoile.

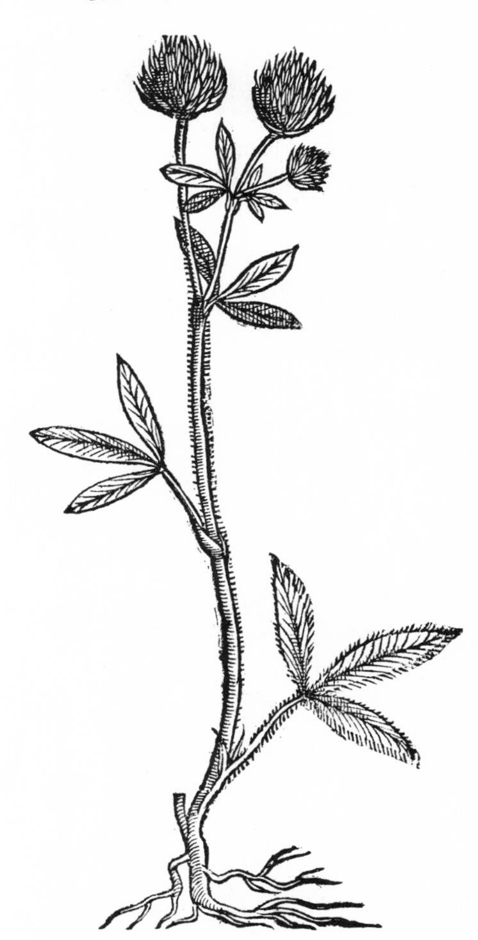
‡ 3 *Trifolium maius flore albo.*
Great white Trefoile.

¶ *The Deſcription.*

1 MEdow Trefoile bringeth forth ſtalkes a cubit long, round, and ſomething hairy, the greater part of which creepeth vpon the ground; whereon do grow leaues conſiſting of three ioyned together, one ſtanding a little from another, of which thoſe that are next the ground and roots are rounder, and they that grow on the vpper part longer, hauing for the moſt part in the midſt a white ſpot like a halfe moon. The floures grow on the tops of the ſtalkes in a tuft or ſmall Fox-taile eare, of a purple colour, and ſweet of taſte. The ſeed groweth in little huskes, round and blackiſh: the root is long, wooddy, and groweth deepe.

2 There is another of the field Trefoiles, differing from the precedent eſpecially in the colour of the floures; for as thoſe are of a bright purple, contrariwiſe theſe are very white, which maketh the difference. The leaues, floures, and all the whole plant is leſſe than the former.

3. 4. There is alſo a Trefoile of this kinde which is ſowne in fields of the Low-Countries, in Italy and diuers other places beyond the ſeas, that commeth vp ranker and higher than that which groweth in medowes, and is an excellent food for cattell, both to fatten them, and cauſe them to giue great ſtore of milke.

‡ Of this there is one more with white floures, which hath ſtalks ſome foot high, and narrow hairy leaues, with a root of the thickneſſe of ones little finger. This is *Cluſius* his *Trifolium maius primum.*

‡ 4 *Trifolium maius flo. purpureo.*
Great purple Trefoile.

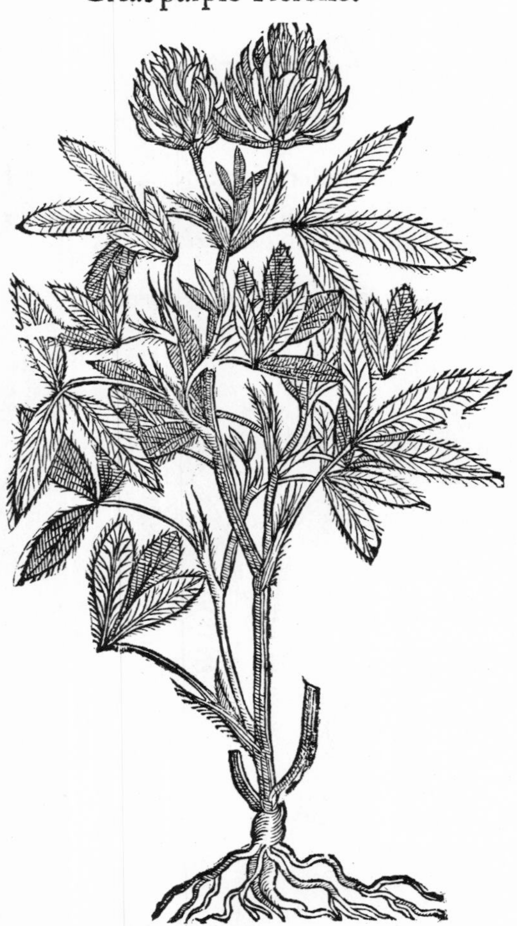

‡ 5 *Trifolium luteum Lupulinum.*
Hop Trefoile.

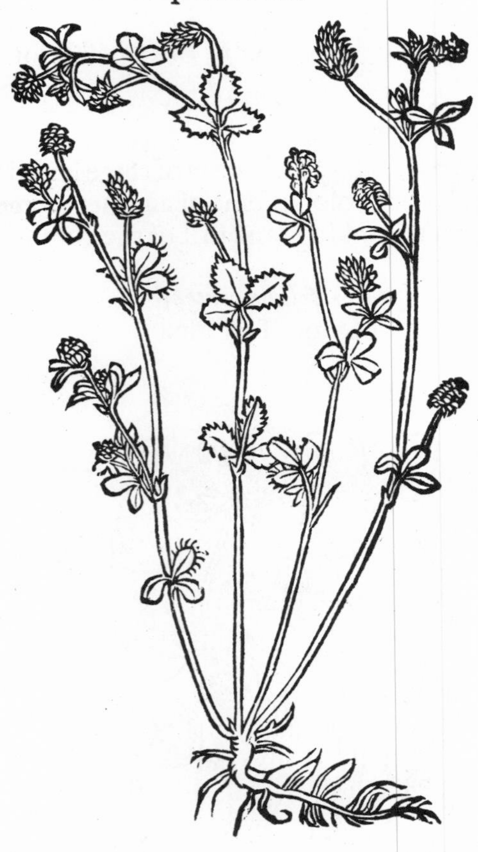

‡ 6 *Trifolium luteum minimum.*
Little yellow Trefoile.

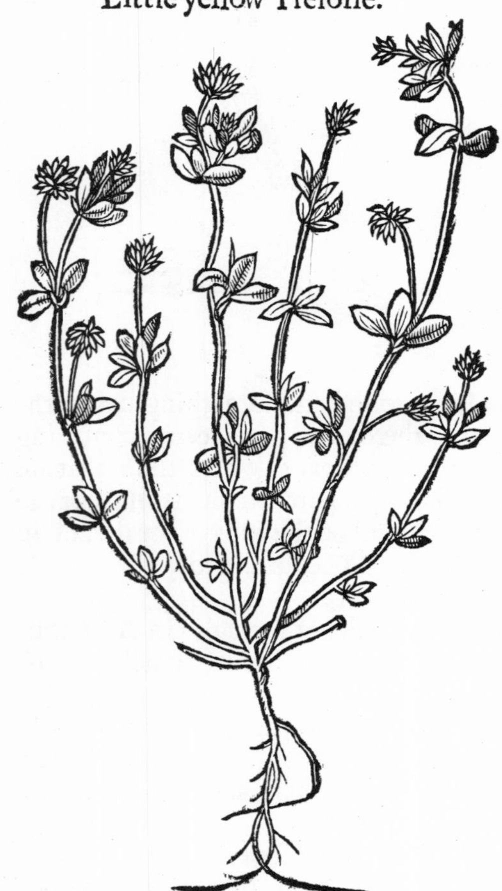

primum. The other hath stalkes some cubit high, with larger ioynts and leaues: the floure or head of floures is also larger, of an elegant red colour. This *Clusius* calls *Trifolium maius tertium*. ‡

5. 6. Likewise we haue in our fields a smaller Trefoile that bringeth forth yellow floures, a greater and a lesser, and diuers others also, differing from these in diuers notable points, the which to distinguish apart would greatly inlarge our volume, and yet to small purpose: therefore we leaue them to be distinguished by the curious, who may at the first view easily perceiue the difference, and also that they be of one stocke or kindred.

‡ The greater of these yellow Trefoiles hath prety large yellow heads, which afterward become of a brownish colour, and somewhat resemble a Hop: whence *Thalius* called it *Lupulus syluaticus*, or *Trifolium luteum alterum lupulinum*: *Dodonæus* cals it *Trifolium agrarium*. The leaues are small, and lightly nickt about the edges. The lesser hath smaller and far lesser yellow heads, which are succeeded by many little crooked clustring seeds: the leaues of this are small, and also snipt about the edges: both this & the other haue two little leaues close by the fastning of the foot-stalkes of the leaues to the main stalks; wherfore I refer them to the Medicks, and vsually cal this later, *Medica sem. racemoso*. It is the *Trifol. luteum minimum* of *Pena* and *Lobel*; and *Trifolium arvense* of *Tabern*. ‡

¶ The

¶ *The Place.*

Common Medow Trefoile groweth in medowes, fertile pastures, and waterish grounds. The others loue the like soile.

¶ *The Time.*

They floure from May to the end of Sommer.

¶ *The Names.*

Medow Trefoile is called in Latine *Trifolium pratense:* in High Dutch, **Wisenklee**: in low Dutch, **Claueren**: in French, *Treffle* and *Trainiere,* and *Visumarus,* as *Marcellus* an old writer testifieth: in English, Common Trefoile, Three-leafed grasse: of some, Suckles, and Honi-suckles, Cocksheads; and in Irish, *Shamrocks.*

¶ *The Temperature.*

The leaues and floures of Medow Trefoiles are cold and drie.

¶ *The Vertues.*

The decoction of three leaued Grasse made with honie, and vsed in a clyster, is good against the frettings and paines of the guts, and driueth forth tough and slimie humours that cleaue vnto the guts. A

The leaues boiled with a little barrowes grease, and vsed as a pultis, take away hot swellings and inflammations. B

Oxen and other cattell do feed of the herbe, and also calues and young lambs. The floures are acceptable to Bees. C

Pliny writeth, and setteth it downe for certaine, that the leaues hereof do tremble, and stand right vp against the comming of a storme or tempest. D

The medow Trefoile (especially that with the blacke halfe Moon vpon the leafe) stamped with a little honie, takes away the pin and web in the eies, ceaseth the paine and inflammation thereof, if it be strained and dropped therein. E

Chap. 495. *Of stinking Trefoile, or Treacle Clauer.*

Trifolium bituminosum,
Treacle Clauer.

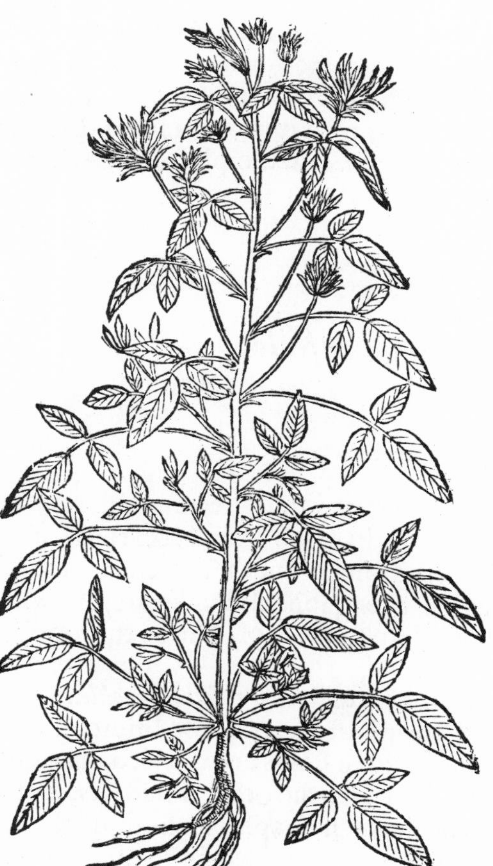

¶ *The Description.*

Treacle Clauer groweth vpright like a shrubbie plant, with stalkes of a cubit and a halfe high, whereupon do grow next the ground broad leaues, 3 ioined together, those vpon the stalkes are longer and narrower. The stalks are couered ouer with a rough euill coloured hairinesse: the leaues are of a dark black greene colour, and of a lothsome smell, like the pitch called *Bitumen Iudaicum,* whereof it took his name: the floures grow at the toppe of the stalks, of a darke purplish colour tending vnto blewnesse, in shape like those of Scabious: the seed is broad, rough, long, and sharpe pointed: the root is small and tender, and cannot indure the coldnesse of our winter, but perisheth at the first approch thereof.

¶ *The Place.*

It groweth naturally, saith *Hippocrates Hippiatros,* not *Cous,* in rough places, as *Ruellius* translateth it: in Germanie, France and England it neuer commeth vp of it selfe, but must be sown in gardens, as my selfe haue proued diuers times, and was constrained to sow it yearely, o else it would not come vp, neither of his owne sowing or otherwise.

¶ *The Time.*

It floureth not in my garden vntil the end of August.

¶ The

¶ *The Names.*

Nicander calleth this Trefoile [Greek]: in Latine, *Trifolium acutum*, or sharpe pointed Trefoile: of *Pliny*, *Trifolium odoratum*, but not properly; of others, *Trifolium Asphaltæum*, *siue Bituminosum*, or Stone Pitch Trefoile.

Auicen calleth it *Tarfilon*, and not *Handacocha: Auicen* doth comprehend *Dioscorides* his *Loti*, that is to say, *Lotus vrbana syluestris*, and *Ægyptia*; which *Dioscorides* confoundeth one with another in one chapter: in English it is called Clauer gentle, Pitch Trefoile, stinking Trefoile, & Treacle Clauer.

¶ *The Temperature.*

This Trefoile, called *Asphaltæum*, as *Galen* saith, is hot and drie, as *Bitumen* is, and that in the third degree.

¶ *The Vertues.*

A Being drunke, it taketh away the pain of the sides, which commeth by obstructions or stoppings, prouoketh vrine, and bringeth downe the desired sicknesse.

B *Hippocrates* writeth, that it doth not onely bring them downe, but likewise the birth, not onely inwardly taken, but also outwardly applied. If a woman, saith he, be not well clensed after her child bearing, giue her this Trefoile to drinke in white wine.

C *Dioscorides* saith, that the seeds and leaues being drunke in water, are a remedie for the pleurisie, difficultie of making water, the falling sicknesse, the dropsie when it first beginneth, and for those that are troubled with the mother: the quantity to be taken at once is three drams of the seeds, and foure of the leaues.

D The leaues drunke in Oxymel, or a syrrup of vineger made with honie, is good for those that are bitten with serpents.

E Some affirme that the decoction of the whole plant, root and leaues, taketh away paine from those whom serpents haue bitten, if they be washed therewith; but if any other man hauing an vlcer be washed with that water wherwith he was bathed that was bitten of the serpent, they say that he shall be troubled in the same manner that the stinged partie was.

F Some also giue with wine three leaues, or a small quantitie of the seeds in tertian agues, and in quartaine foure, as a sure remedie against the fits.

G The root also is put into antidotes or counterpoisons, saith *Dioscorides*: but other antient Physitions do not onely mix the root with them, but also the seed, as we may see in *Galen*, by a great many compositions in his 2. booke of Antidotes; that is to say, in the Treacles of *Ælius Gallus*, *Zeno Laudoceus*, *Claudius Apollonius*, *Eudemus*, *Heraclides*, *Dorothæus*, and *Heras*.

H The herbe stamped and applied vpon any enuenomed wound, or made with poisoned weapon, it draweth the poison from the depth most apparantly. But if it be applied vpon a wound where there is no venomous matter to work vpon, it doth no lesse infect that part, than if it had been bitten with some serpent or venomous beast: which wonderfull effect it doth not performe in respect of any vitious qualitie that it hath in it selfe, but because it doth not finde that venomous matter to work vpon, which it naturally draweth (as the Load-stone doth iron) wherupon it is constrained through his attractiue qualitie, to draw and gather together humours from far vnto the place, whereby the paine is greatly increased.

CHAP. 496. *Of diuers other Trefoiles.*

¶ *The Description.*

1 THree leaued grasse of America hath diuers crooked round stalks, leaning this way and that way, and diuided into diuers branches: whereon do grow leaues like those of the medow Trefoile, of a black greene colour, and of the smel of Pitch Trefoile, or Treacle Clauer: the floures grow at the top of the branches, made vp in a long spiked chaffie eare, of a white colour: after which commeth the seed, somewhat flat, almost like to those of Tares: the roots are long strings of a wooddie substance.

2 This three leafed grasse (which *Dodonæus* in his last Edition calleth *Trifolium cochleatum primum*: and *Lobel*, *Fœnum Burgundiacum*) hath diuers round vpright stalks, of a wooddie rough substance, yet not able of it selfe to stand without a prop or stay: which stalks are diuided into diuers small branches, whereupon do grow leaues ioined three together like the other Trefoiles, but of a darke swart greene colour: the floures grow at the top of the stalks in shape like those of the codded Trefoile, but of a darke purple colour: the seede followeth, contained in small wrinckled huskes

1 *Trifolium Americum.*
Trefoile of America.

† 2 *Trifolium Burgundiacum*
Burgundie Trefoile.

3 *Trifolium Salmanticum.* Portingale Trefoile.

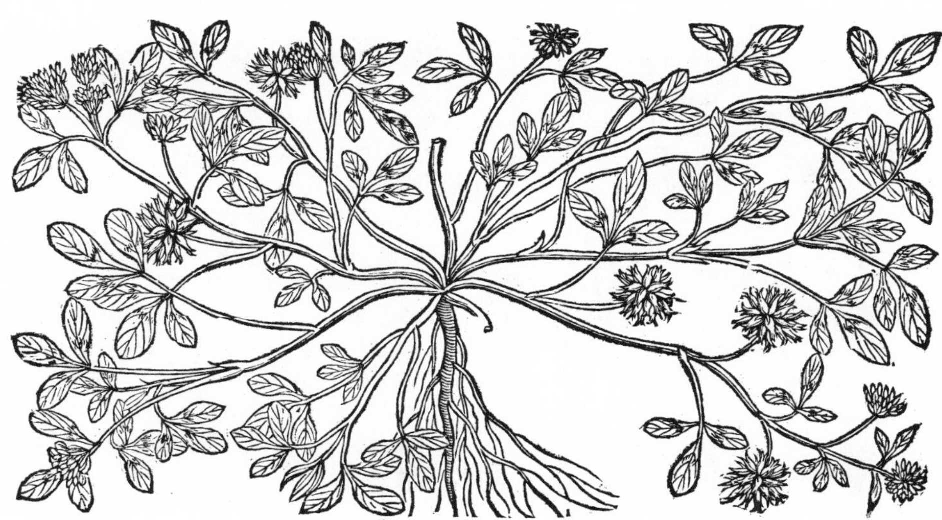

huskes turned round,after the manner of a water snaile:the root is thick,composed of diuers tough threddie strings,and lasteth long in my garden with great increase.

3 This three leaued grasse of Salmanca, a citie as I take it of Portingale, differeth not much from our field Trefoile : it hath many branches weake and tender,trailing vpon the ground, of two cubites and a halfe high : whereupon doe grow leaues set together by three vpon a stemme ; from the bosome whereof thrust forth tender foot-stalkes, whereon doe stand most fine floures of a bright red tending vnto purple : after which come the seed wrapped in small skinnes, of a red colour.

4 The Hart Trefoile hath very many flexible branches, set vpon a slender stalke, of the length of two or three foot, trailing hither and thither : whereupon doe grow leaues ioined together by three

4 *Trifolium cordatum.*
Heart Trefoile.

5 *Trifolium siliquosum minus.*
Small codded Trefoile.‡

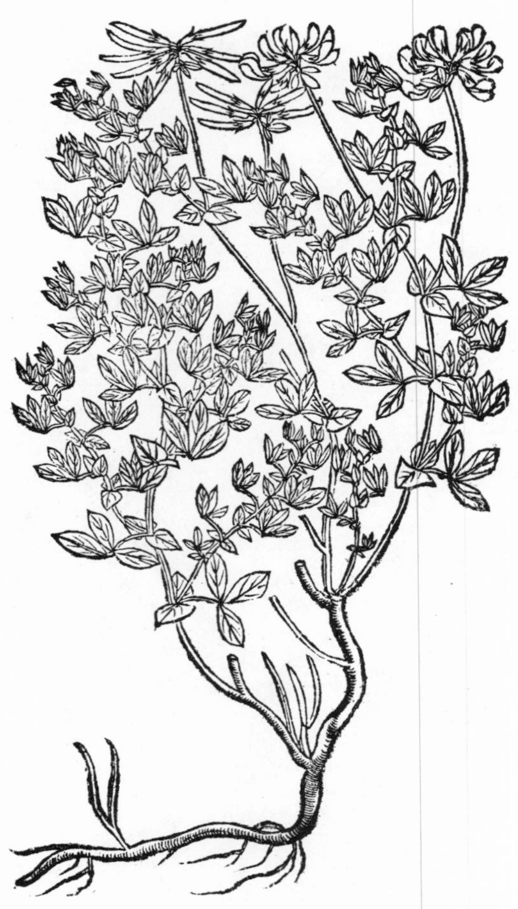

‡ 6 *Coronopus ex Codice Cæsareo.*
Crow-foot Trefoile.

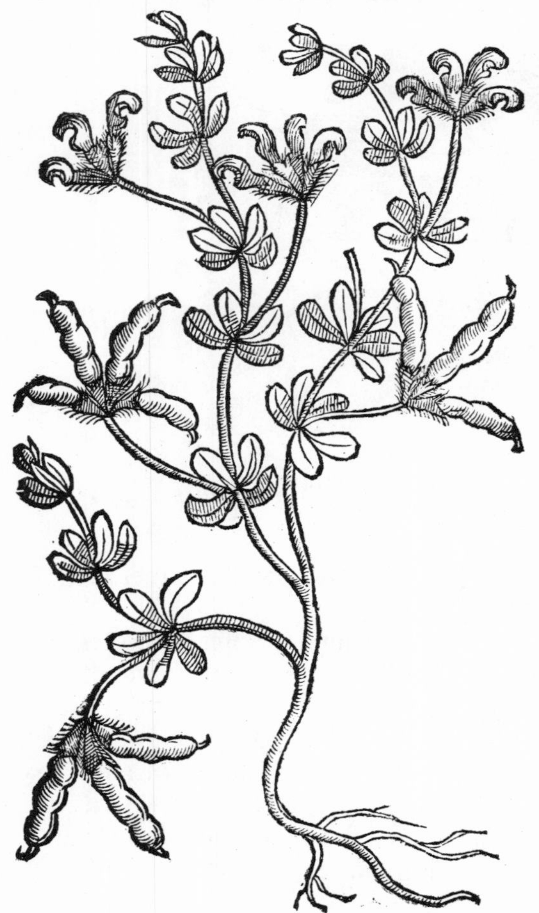

three on little slender foot-stalks, euery little leafe of the fashion of a heart, whereof it took his name: among which come forth scalie, or chaffie yellow floures: the root is thick and threddie. ‡ I take the plant which our Authour here figured and intended to describe vnto vs, to be of that *Medica* which *Camerarius* calls *Arabica*, which growes wilde in many places with vs, hauing the leaues a little dented in at the ends, so that they resemble the vulgar figure of a heart; and each leafe is marked with a blackish, or red spot: the floures be small and yellow: the seeds are contained in rough buttons, wound vp like the other Snaile Trefoiles, whereof it is a kinde. I haue giuen you the figure a little more exquisite, by the addition of the spots and cods. ‡

5 This kinde of three leaued grasse is a low herb, creeping vpon the ground: the leaues are like those of the common Trefoile, but lesser, and of a grayish greene colour: the floures are faire and yellow, fashioned like those of broome, but lesser: after come three or foure cods, wherein is contained round seed: the root is long and reddish. ‡ This is the *Trifolium corniculatum*, or *Melilotus Coronata* of *Lobel*: *Lotus pentaphyllos* of *Gesner*. ‡

This codded Trefoile is like vnto the last described in euery respect, sauing that this plant is altogether ‡ larger, hauing stalks a cubit bite and a halfe high: the leaues are alſo foure times as large,two roundiſh leaues growing by the ſtalke,and three longiſh ones growing vpon a ſhort foot-ſtalke comming forth betweene the two roundiſh leaues: both the ſtalke and leaues haue a little ſoft downineſſe or hairineſſe on them:the floures grow cluſtering together on the tops of the ſtalks,in ſhape,bigneſſe,and colour like that of the laſt deſcribed,but commonly more in number: they are alſo ſucceeded by ſuch cods as thoſe of the former.

6 The figure which *Dodonæus* hath ſet forth out of an old Manuſcript in the Emperors Library,being there figured for *Coronopus*, ſeemes to be of the laſt deſcribed,or ſome plant very like thereto,though the fiue leaues at each ioint be not put in ſuch order as they ſhould be, yet all the parts are well expreſt,according to the drawing of thoſe times,for you ſhall finde few antient expreſſions come ſo neere as this doth. ‡

7 There is a kinde of Clauer growing about Narbone in France, that hath many twiggie tough branches comming from a wooddy root, whereon are ſet leaues three together,after th maner of the other Trefoiles,ſomewhat long hairy,and of a hoarie or ouerworne greene colour. The floures are yellow,and grow at the tops of the branches like thoſe of Broome.

7 *Lotus incana,ſive Oxytriphyllon Scribonij Largi.*
Hoarie Clauer.

‡ 8 *Trifolium luteum ſiliqua cornuta.*
Yellow horned Trefoile.

‡ This ſends vp many branches from one root,ſome cubit or more long,commonly lying along vpon the ground,round,flexable,and diuided into ſundrie branches: the leaues ſtand together by threes,and are like thoſe of the true *Medica*, or Burgundie Trefoile, but much leſſe: the floures grow cluſtering together on the tops of the branches, like in ſhape to thoſe of the former; of a yellow colour,and not without ſmell: they are ſucceeded by ſuch,yet narrower crooked coddes,as the Burgundy Trefoile hath(but the Painter hath not wel expreſſed them:) in theſe cods are contained ſeeds like thoſe alſo of that Trefoile,and ſuch alſo is the root,which liues long,and much increaſes. It growes in Hungarie, Auſtria, and Morauia: it floures in Iune and Iuly: *Cluſius* calls it *Medica flore flavo: Tabernamontanus, Lens maior repens*: and *Tragus*, *Meliloti maioris ſpecies tertia: Bauhine* ſaith that about Nimes in Narbone it is found with floures either yellow white,greene,blew, purple, blacke,or mixed of blew and greene; and hee calleth it *Trifolium ſylueſtre luteum ſiliqua cornuta*; or *Medica frutescens*. ‡

¶ *The Place.*

The feuerall titles of most of these plants set forth their naturall place of growing: the rest grow in most fertile fields of England.

¶ *The Time.*

They floure and flourish most of the sommer moneths.

¶ *The Names.*

There is not much to be said as touching their names, more than hath beene set downe.

¶ *The Temperature and Vertues.*

The temperature and faculties of these Trefoiles are referred vnto the common medow Trefoiles.

‡ The figure formerly put in the second place was of the lesser yellow Trefoile described in the last chapter saue one.

Chap. 497. *Of the great Trefoiles, or winged Clauers.*

¶ *The Description.*

† 1 The great Hares foot being a kinde of Trefoile, hath a hard and wooddie root, full of blacke threddie strings: from whence arise diuers tough and feeble branches, whereupon do grow leaues, set together by threes, making the whole plant to resemble those of the Medow Trefoile: the floures grow at the top of the stalks, composed of a bunch of gray haires: among the which soft matter commeth forth small floures of a most bright purple colour, somwhat resembling the floures of the common medow Trefoile, but far greater. *Lobel* calls this *Lagopus maximus folio, & facie Trifolij pratensis*: *Dodonæus, Lagopus maior folio Trifolij.*

‡ 1 *Lagopus maximus.*
The great Hares foot Trefoile.

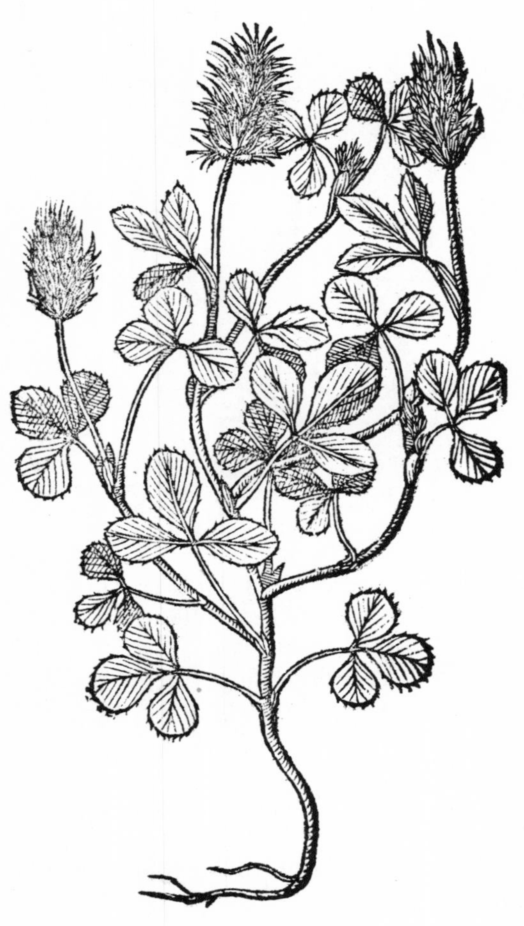

‡ 2 *Lagopus maior spica longiore.*
Great large headed Hares foot.

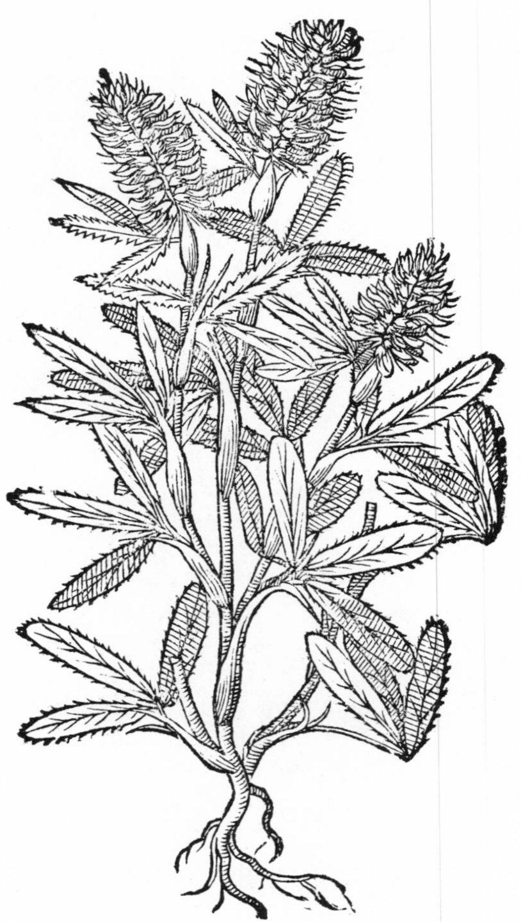

‡ 2 This

‡ 2 This elegant plant (which *Tragus* hath set forth for *Cytisus*, *Lobel* by the name of *Lagopus altera folio prinnato*, and *Clusius* for his *Trifolij maioris* 3. *altera species*) hath stalkes some foot and better high, whereon grow leaues set together by threes, long, hoary and lightly snipt about the edges, with elegant nerues or veines, running from the middle rib to the sides of the leaues, which are most conspicuous in hot Countries, and chiefly then when the leafe begins to decay. At the tops of the branches, in long and large heads grow the floures, of an elegant sanguine colour. This floures in May and Iune, and growes wilde vpon some mountaines of Hungary and Austria; I haue seene them, both this and the former, growing in the gardens of some of our Florists.

3 This other great kinde of Hares-foot sends forth one slender, yet stiffe stalke, whereon grow leaues whose foot-stalkes are large at the setting on, encompassing the stalkes: the leaues themselues grow by threes, long, narrow, and sharpe pointed, of a grayish colour like those of the common Hares-foot; the spike at the top is soft and downy, with little reddish floures amongst the whitish hairinesse. This growes wilde in Spaine: *Clusius* calls it *Lagopus angustifolius Hispanicus maior*.

There is another sort of this descrbed by *Lobel* and *Pena* in the *Aduers*. whose leaues are longer and narrower than this, the whole plant also is oft times lesser: they call it *Lagopus altera angustifolia*. ‡

‡ 3 *Lagopus angustifolius Hispanicus*.
Narrow leafed Spanish Hare-foot.

4 *Lagopodium*, *Pes leporis*.
Little Hares-foot Trefoile.

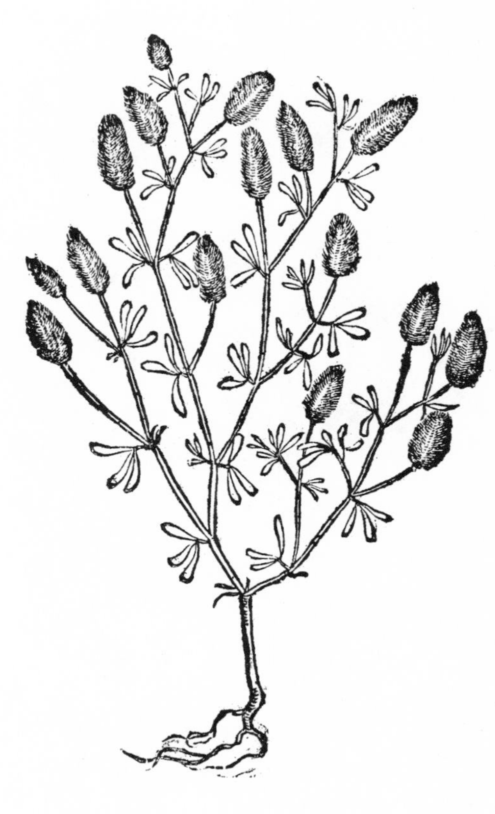

4 The small Hares-foot hath a round rough and hairy stalke, diuiding it selfe into diuers other branches; whereupon do grow small leaues, three joined together, like those of the small yellow Trefoile: the floures grow at the very point of the stalkes, consisting of a rough knap or bush of haires or downe, like that of *Alopecuros*, or Fox-taile, of a whitish colour tending to a light blush, with little white floures amongst the downinesse: the root is small and hard.

¶ *The Place.*

The first groweth in the fields of France and Spaine, and is a stranger in England; yet it groweth in my garden.

The small Hare-foot groweth among corne, especially among Barly, and likewise in barren pastures almost euery where.

¶ *The Time.*

They floure and flourish in Iune, Iuly, and August.

¶ *The Names.*

The great Hare-foot Trefoile is called of *Tragus*, *Cytiſus*: of *Cordus*, *Trifolium magnum*: of *Lobelius*, *Lagopum maximum*, and *Lagopodium*: in Greeke, λαγώπους: in Engliſh, the great Hares-foot.

The laſt, being the ſmalleſt of theſe kindes of Trefoiles, is called *Lagopus*, and *Pes Leporis*: in Dutch, **Hasen pootkens**: in high Dutch, **Hasen fusz**: in French, *Pied de lieure*: in Engliſh Hare-foot.

¶ *The Temperature and Vertues.*

A The temperature and faculties are referred vnto the other Trefoiles, whereof theſe are kindes: notwithſtanding *Dioſcorides* ſaith, that the ſmall Hares-foot doth binde and dry. It ſtoppeth, ſaith he, the laske, if it be drunke with red wine. But it muſt be giuen to ſuch as are feueriſh with water.

† Our Author in the firſt place formerly gaue the figure of *Tabern.* his *Lagopodium flore albo*, being only a variety of that plant: you ſhall hereafter finde it deſcribed by the name of *Anthyllis leguminoſa*; now he made the deſcription ſomewhat in the leaues to agree with the figure, though nothing almoſt with the truth of that he intended to deſcribe, for (as it is euident by the names) he intended to deſcribe both the firſt and ſecond (which are here now deſcribed) in the firſt place, for he hath confounded them both together in the names.

Chap. 498. *Of Water Trefoile, or Bucks Beanes.*

Trifolium paludoſum.
Marſh Trefoile.

¶ *The Deſcription.*

1 The great Marſh Trefoile hath thicke fat ſtalkes, weake and tender, full of a ſpungious pith, very ſmooth, and of a cubit long: whereon do grow leaues like to thoſe of the garden Beane, ſet vpon the ſtalkes three joined together like the other Trefoiles, ſmooth, ſhining, and of a deepe greene colour: among which toward the top of the ſtalkes ſtandeth a buſh of feather like floures of a white colour, daſht ouer ſlightly with a waſh of light carnation: after which the ſeed followeth, contained in ſmall buttons, or knobby huskes, of a browne yellowiſh colour like vnto Millet, and of a bitter taſte: the roots creepe diuers waies in the middle mariſh ground, being full of joints, white within, and full of pores, and ſpungie, bringing forth diuers by-ſhoots, ſtalkes, and leaues, by which meanes it is eaſily increaſed, and largely multiplied.

2 The ſecond differeth not from the precedent, ſauing it is altogether leſſer, wherein conſiſteth the difference, if there be any: for doubtleſſe I thinke it is the ſelfe ſame in each reſpect, and is made greater and leſſer, according to his place of growing, clymate, and countrey.

¶ *The Place.*

Theſe grow in marish and Fenny places, and vpon boggie grounds almoſt euery where.

¶ *The Time.*

They floure and flouriſh from Iune to the end of Auguſt.

¶ *The Names.*

Mariſh Trefoile is called in high Dutch, **Biberklee**, that is to ſay, *Caſtoris Trifolium*, or *Trifolium fibrinum* in low Dutch, of the likeneſſe that the leaues haue with the garden Beanes, **Bocxboonen**, that is to ſay, *Faſelus Hircinus*, or *Boona Hircina*: the later Herbariſts call it *Trifolium paluſtre*, and *Paludoſum*: of ſome, *Iſopyrum*: in Engliſh, marſh-Clauer, marſh-Trefoile, and Buckes-Beanes.

¶ *The Temperature and Vertues.*

A The ſeed of *Iſopyrum*, ſaith *Dioſcorides*, if it be taken with meade or honied water, is good againſt the cough and paine in the cheſt.

B It is alſo a remedy for thoſe that haue weake liuers and ſpet bloud, for as *Galen* ſaith it clenſeth and cutteth tough humours, hauing alſo adjoined with it an aſtringent or binding quality.

Chap.

CHAP. 499 *Of ſweet Trefoile, or garden Clauer.*

Trifolium odoratum.
Sweet Trefoile.

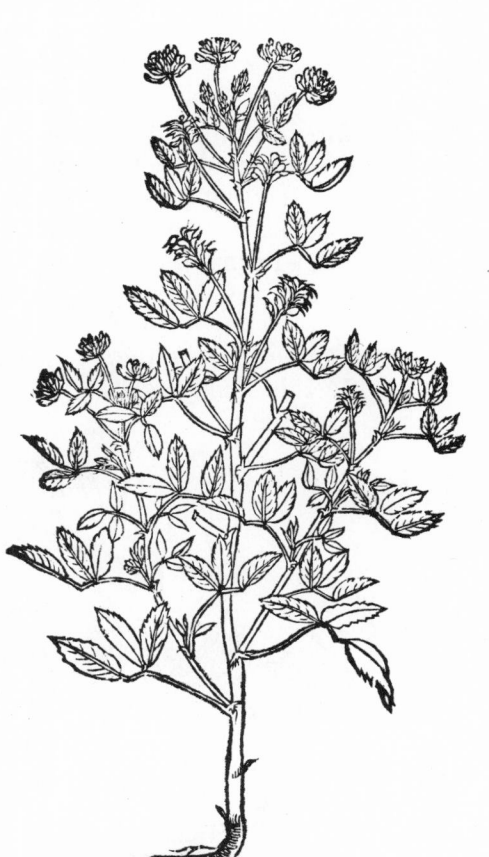

¶ *The Deſcription.*

SWeet Trefoile hath an vpright ſtalk, hollow, and of the height of two cubits, diuiding it ſelfe into diuers branches: whereon do grow leaues by three and three like to the other Trefoiles, ſleightly and ſuperficiouſly nicked in the edges: from the boſom wherof come the floures, euery one ſtanding on his owne ſingle foot-ſtalk, conſiſting of little chaffie husks, of a light or pale blewiſh colour: after which come vp little heads or knops, in which lieth the ſeed, of a whitiſh yellow colour, and leſſer than that of Fenu-greeke: the root hath diuers ſtrings: the whole plant is not onely of a whitiſh green colour, but alſo of a ſweet ſmell, and of a ſtrong aromaticall or ſpicie ſent, and more ſweet when it is dried: which ſmel in the gathered and dried plant doth likewiſe continue long: and in moiſt and rainie weather, it ſmelleth more than in hot and drie weather: and alſo when it is yet freſh and greene it loſeth and recouereth againe his ſmell ſeuen times a day; whereupon the old wiues in Germanie do call it **Sieuen gezeiten kraut**, that is, the herbe that changeth ſeuen times a day.

¶ *The Place.*

It is ſowne in gardens not onely beyond the ſeas, but in diuers gardens in England.

¶ *The Time.*

It is ſowne in May, it floureth in Iune and Iuly, and perfecteth his ſeed in the end of Auguſt, the ſame yere it is ſowne.

¶ *The Names.*

It is commonly called in Latine *Trifolium odoratum*: in high Dutch as we haue ſaid **Sieuen gezeiten**: in low Dutch, **Seuenghetijtcruijt**, that is to ſay, an herb of ſeuen times: it is called in Spaniſh, *Trebol real*: in French, *Treffle oderiferant*: in Engliſh, Sweet Trefoile, and garden Clauer: it ſeemeth to be *Lotus Vrbana*, or *ſatiua*, of which *Dioſcorides* writeth in his fourth booke: neuertheleſſe diuers Authors ſet downe Melilot, for *Lotus vrbana*, and *Trifolium odoratum*, but not properly. ‡ The Gardiners and herbe women in Cheapſide commonly call it, and know it by the name of Balſam, or garden Balſam. ‡

¶ *The Temperature.*

Galen ſaith, that ſweet Trefoile doth in a meane concoct and drie, and is in a meane and temperate facultie betweene hot and cold: the which faculties vndoubtedly are plainely perceiued in this ſweet Trefoile.

¶ *The Vertues.*

The iuice preſſed forth, ſaith *Dioſcorides*, with hony added thereto, clenſeth the vlcers of the eies, A
called in Latine *Argema*, and taketh away ſpots in the ſame, called *Albugines*; and remooueth ſuch things as doe hinder the ſight.

The oile wherein the floures are infuſed or ſteeped, doth perfectly cure greene wounds in very B
ſhort ſpace; it appeaſeth the paine of the gout, and all other aches, and is highly commended againſt ruptures, and burſtings in young children.

The iuice giuen in white wine cureth thoſe that haue fallen from ſome high place, auoideth C
congealed and clotted bloud, and alſo helpeth thoſe that do piſſe bloud, by meanes of ſome great bruiſe, as was prooued lately vpon a boy in Fanchurch ſtreet, whom a cart went ouer, where-

upon he did not onely piſſe bloud, but alſo it moſt wonderfully guſhed forth, both at his noſe and mouth.

D The dried herbe laied among garments keepeth them from Mothes and other vermine.

Chap. 500. *Of Fenugreeke.*

¶ *The Deſcription.*

1 Fenugreeke hath a long ſlender trailing ſtalke, greene, hollow within, and diuided into diuers ſmall branches: whereon do grow leaues like thoſe of the medow Trefoile, but rounder and leſſer, greene on the vpper ſide, on the lower ſide tending to an aſh colour: among which come ſmall white floures, after them likewiſe long ſlender narrow cods, in which do lie ſmall vneuen ſeeds, of a yellowiſh colour: which being dried, haue a ſtrong ſmell, yet not vnpleaſant: the root is ſmall, and periſheth when it hath perfected his ſeed.

1 *Fœnumgræcum.*
Fenugreeke.

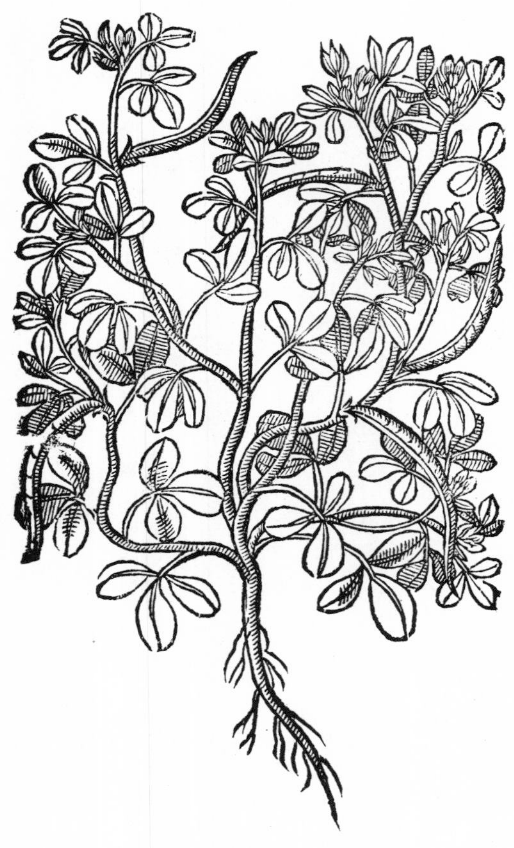

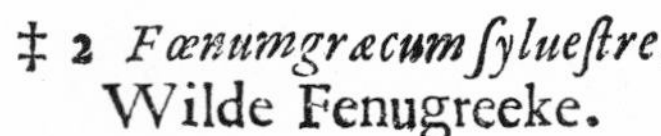

‡ 2 *Fœnumgræcum ſylueſtre.*
Wilde Fenugreeke.

2 There is a wilde kinde hereof ſeruing to little vſe, that hath ſmall round branches, full of knees or ioints: from each ioint proceedeth a ſmal tender footſtalk, whereon do grow three leaues and no more, ſomewhat ſnipt about the edges, like vnto thoſe of Burgundie Haie: from the boſoms whereof come forth ſmall yellow floures, which turne into little cods: the root is thicke, tough, and pliant.

¶ *The Place.*

Fenugreeke is ſowne in fields beyond the ſeas: in England wee ſow a ſmall quantitie thereof in our gardens.

¶ *The Time.*

It hath two ſeaſons of ſowing, according to *Columella*, of which one is in September, at what time it is ſowne that it may ſerue for fodder againſt winter; the other is in the end of Ianuarie, or the beginning of Februarie, notwithſtanding we may not ſow it vntill Aprill in England.

¶ *The*

¶ *The Names.*

It is called in Greeke τῆλις, or as it is found in *Pliny* his copies *Carphos*: in Latine, *Fœnum Græcum*: *Columella* saith that it is called *Siliqua*: in *Pliny* we read *Silicia*: in *Varro*, *Silicula*: in high Dutch, **Bockshorne**: in Italian, *Fiengreco*: in Spanish, *Alfornas*: in French, *Fenegrec*: and in English, Fenegreeke.

¶ *The Temperature and Vertues.*

A It is thought according to *Galen* in his booke of the Faculties of nourishments, that it is one of those simples which do manifestly heat, and that men do vse it for food, as they do Lupines; for it is taken with pickle to keep the body soluble, and for this purpose it is more agreeable than Lupines, seeing it hath nothing in his owne proper substance, that may hinder the working.

B The iuice of boiled Fenegreeke taken with honie is good to purge by the stoole all manner of corrupt humors that remaine in the guts, making soluble through his siminesse, and mitigating paine through his warmnesse.

C And because it hath in it a clensing or scouring facultie, it raiseth humors out of the chest: but there must be added vnto it no great quantitie of honie least the biting qualitie should abound.

D In old diseases of the chest without a feuer, fat dates are to be boiled with it, but when you haue mixed the same iuice pressed out with a great quantitie of hony, and haue againe boiled it on a soft fire to a mean thicknesse, then must you vse it long before meat.

E In his booke of the Faculties of simple medicines he saith, that Fenegreek is hot in the second degree, and dry in the first: therefore it doth kindle and make worse hot inflammations, but such as are lesse hot and more hard are thereby cured by being wasted and consumed away.

F The meale of Fenegreeke, as *Dioscorides* saith, is of force to mollifie and waste away: being boiled with mead and applied it taketh away inflammations, as well inward as outward

G The same being tempered or kneaded with niter and vineger, doth soften and waste away the hardnesse of the milt.

H It is good for women that haue either imposthume, vlcer, or stopping of the matrix, to bathe and sit in the decoction thereof.

I The iuice of the decoction pressed forth doth clense the haire, taketh away dandraffe, scoureth running sores of the head, called of the Græcians ἀχῶρας: being mingled with goose grease, and put vp in manner of a pessarie, or mother supposititorie, it doth open and mollifie all the parts, about the mother.

K Greene Fenegreeke bruised and pounded with vineger, is a remedie for weak and feeble parts, and that are without skin, vlcerated and raw.

L The decoction thereof is good against vlcers in the low gut, and foule stinking excrements of those that haue the bloudy flix.

M The oile which is pressed out thereof scoureth haires and scars in the priuie parts.

N The decoction of Fenegreeke seed, made in wine, and drunke with a little vineger, expelleth all euill humors in the stomacke and guts.

O The seed boiled in wine with dates and hony, vnto the form of a syrrup, doth mundifie and clense the breast, and easeth the paines thereof.

P The meale of Fenegreek boiled in mead or honied water, consumeth and dissolueth all cold hard imposthumes and swellings, and being mixed with the roots of Marsh Mallows and Linseed effecteth the same.

Q It is very good for women that haue any griefe or swelling in the matrix, or other lower parts, if they bathe those parts with the decoction thereof made in wine, or sit ouer it and sweat.

R It is good to wash the head with the decoction of the seed, for it taketh away the scurfe, scailes, nits, and all other such like imperfections.

CHAP. 501. *Of Horned Clauer, and blacke Clauer.*

¶ *The Description.*

1 THe horned Clauer, or codded Trefoile, groweth vp with many weake and slender stalks lying vpon the ground: about which are set white leaues, somewhat long, lesser, aud narrower than any of the other Trefoiles: the floures grow at the tops, of the fashion of those of Peason, of a shining yellow colour: after which come certain straight cods, bigger than those of Fenegreek, but blunter at their ends, in which are contained little round seed; the root is hard and woodie, and sendeth forth young springs euery yeare.

1 *Lotus trifolia corniculata.*
Horned or codded Clauer.

2 *Lotus quadrifolia.*
Foure leafed grasse.

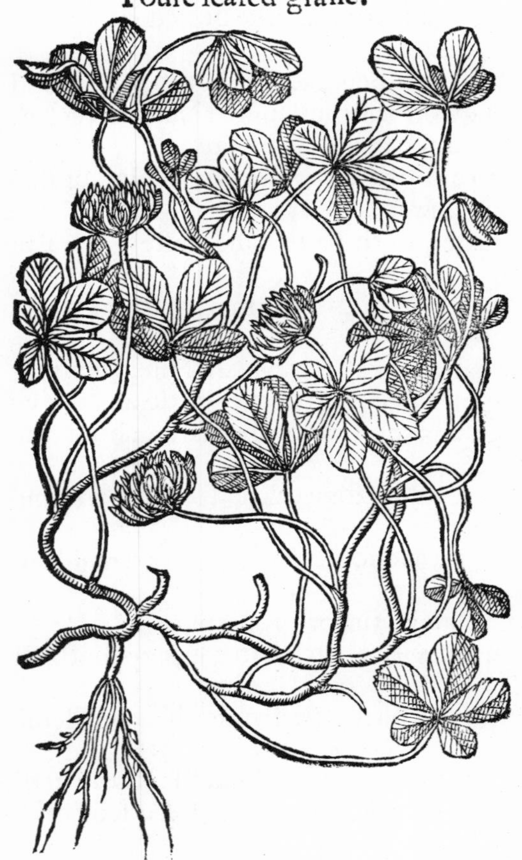

2 This kinde of three leafed grasse, or rather foure leafed Trefoile, hath leaues like vnto the common Trefoile, sauing that they bee lesser, and of a browne purplish colour, knowne by the name of Purple-wort, or Purple-grasse; whose floures are in shape like the medow Trefoile, but of a dustieo uerworn colour tending to whitenesse; the which doth oftentimes degenerate, sometime into three leaues, sometimes in fiue, and also into seuen, and yet the plant of his nature hath but foure leaues & no more. ‡ I do not thinke this to be the purple leaued Trefoile with the white floure, which is commonly called Purple-grasse; for I could neuer obserue it to haue more leaues than three vpon a stalke. ‡

‡ 3 The root of this is small and white, from which arise many weake hairie branches some cubit long: wheron grow soft hairy leaues three on one foot-stalke, with two little leaues at the root therof, & out of the bosoms of these vpon like footstalkes grow three lesser leaues, as also floures of the bignes and shape of those of a Vetch, but of a braue deep crimson veluet colour: after these are past come cods set with foure thinne welts or skins which make them seem foure square; whence *Camerarius* called it *Lotus pulcherrima tetragonolobus:* the seed is of an ash colour, somewhat lesse than a pease. It floures most of the Sommer moneths, and is for the prettinesse of the floure preserued in many Gardens by yearely sowing the seede, for it is an annuall plant. *Clusius* hath it by the name of *Lotus siliquosus rubello flore:* and hee saith the seeds were diuers times sent out of Italy by the name of *Sandalida.* It is also commonly called in Latine *Pisum quadratum.* ‡

¶ *The Place.*

The first groweth wilde in barren ditch bankes, pastures, and drie Mountaines.

‡ 3 *Lotus siliqua qaudrata.* Square crimson veluet pease.

The

The second groweth likewise in pastures and fields, but not so common as the other; and is planted in gardens.

¶ *The Time.*

They floure in Iuly and August.

¶ *The Names.*

The second is called *Lotus Trifolia*: in English, horned Clauer, or codded Trefoile.

The other is called *Lotus quadrifolia*, or foure leafed Grasse, or Purple-wort: of *Pena* and *Lobel*, *Quadrifolium phæum fuscum hortorum*.

¶ *The Temperature and Vertues.*

Their faculties in working are referred vnto the medow Trefoiles: notwithstanding it is reported, that the leaues of Purple-wort stamped, and the iuyce giuen to drinke, cureth young children of the disease called in English the Purples. A

Chap. 502. *Of Medicke Fodder, or snaile Clauer.*

¶ *The Description.*

1 This kinde of Trefoile, called *Medica*, hath many small and slender ramping branches, crawling and creeping along vpon the ground, set full of broad leaues slightly indented about the edges: the floures are very small, and of a pale yellow colour, which turne into round wrinkled knobs, like the water Snaile, or the fish called Periwinckle: wherein is contained flat seed fashioned like a little kidney, in colour yellow, in taste like a Vetch or pease: the root is small, and dieth when the seed is ripe: it growes in my garden, and is good to feed cattell fat.

1 *Trifolium Cochleatum.*
Medicke Fodder.

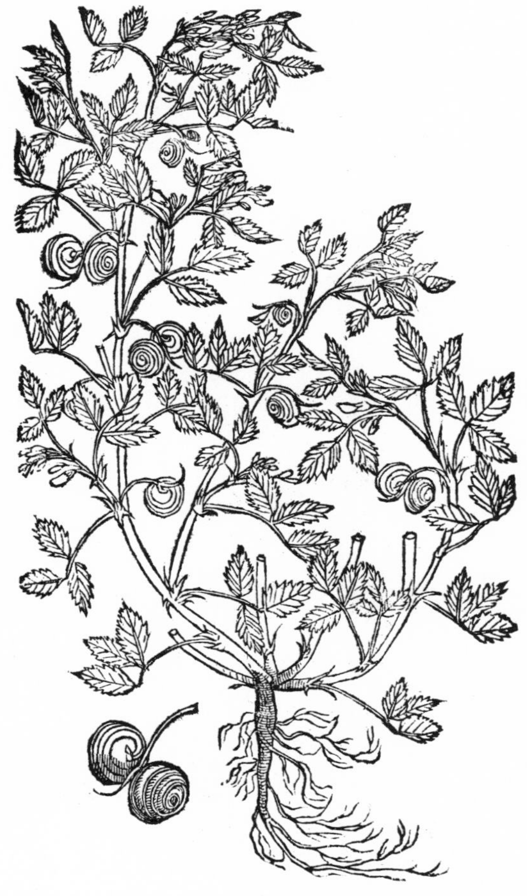

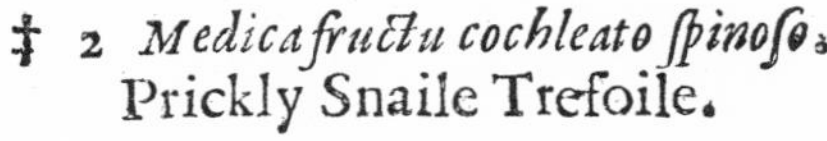

‡ 2 *Medica fructu cochleato spinoso.*
Prickly Snaile Trefoile.

‡ There are many varieties of these plants, and they chiefely consist in the fruit; for some are smooth and flat, as this first described: other some are rough and prickely, some with lesser, and other-

other ſome with bigger prickles; as alſo with them ſtanding diuers wayes, ſome are onely rough, and of thoſe ſome are as big as a ſmall nut, other ſome no bigger than a peaſe. I giue you here the deſcriptions of three rough ones, (as I receiued them from Mr. *Goodyer)* whereof the laſt is of the ſea, which, as you may ſee, our Author did but ſuperficially deſcribe.

2 *Medicæ maioris Bæticæ ſpecies prima, ſpinulis intortis.*

This hath foure ſquare reddiſh ſtreaked hairy trailing branches, like the ſmall Engliſh *Medica,* greater and longer, foure or fiue foot long: the leaues are alſo ſmooth, growing three together, neither ſharpe pointed, nor yet ſo broad at the top as the ſaid Engliſh Medica, but blunt topped, with a ſmall blacke ſpot in the midſt, not crooked: the floures are alſo yellow, three, foure, or fiue on a foot-ſtalke: after commeth a round writhed fruit fully as big as a haſell nut, with ſmall prickles not ſtanding fore-right, but lying flat on the fruit, finely wrapped, plaited, folded, or interlaced together, wherein lieth wrapped the ſeed in faſhion of a kidney, very like a kidney beane, but foure times ſmaller, and flatter, of a ſhining blacke colour without, like poliſhed Ieat; containing a white kernell within: the root is like the former, and periſheth alſo at Winter.

Medicæ maioris Bæticæ ſpinoſæ ſpecies altera.

The branches alſo creepe on the ground, and are ſtraked ſmooth foure ſquare, reddiſh here and there, three or foure foot long: the leaues are ſmooth, finely notched about the edges, ſharp pointed, without blacke ſpots, very like *Medica pericarpio plano:* the floures are ſmall and yellow like the other: the fruit is round, writhed or twined in alſo, fully as big as a haſell nut, ſomewhat cottonie or woolly, with ſhort ſharpe prickles: wherein lyeth alſo wrapped a ſhining blacke kidney-like ſeed, ſo like the laſt deſcribed, that they are not to be diſcerned apart: the root is alſo alike, and periſheth at Winter.

Medicæ marinæ ſpinoſæ ſpecies.

The branches of this are the leaſt and ſhorteſt of all the reſt, little exceeding a foot or two in length, and are foure ſquare, greene, ſomewhat hairie, and trailing on the ground: the leaues are like to thoſe of *Medica pericarpio plano,* not fully ſo ſharpe pointed, without blacke ſpots, ſoft, hairy, three on a foot-ſtalke: the floures grow alongſt the branches, on very ſmall foot-ſtalkes, forth of the boſomes of the leaues, (not altogether on or neere the tops of the branches) and are very ſmall and yellow, but one on a foot-ſtalke: after commeth ſmall round writhed fruit, no bigger than a peaſe, with very ſhort ſharpe prickles, wherein is contained yellowiſh ſeed of the faſhion of a kidney like the former, and is the hardeſt to be plucked forth of any of the reſt: the root is alſo whitiſh like the roots of the other, and alſo periſheth at Winter. Aug. 2. 1621. *Iohn Goodyer.* ‡

3 *Trifolium Cochleatum marinum.*
Medick Fodder of the ſea.

3 This kinde alſo of Trefoile, (called *Medica marina:* in Engliſh, ſea Trefoile, growing naturally by the ſea ſide about Weſtcheſter, and vpon the Mediterranean ſea coaſt, and about Venice) hath leaues very like vnto the common medow Trefoile, but thicker, and couered ouer with a flockie

flockie hoarinesse like *Gnaphalium*, after the manner of most of the sea herbes: the floures are yellow: the seeds wrinkled like the former, but in quantitie they be lesser.

¶ *The Place.*

The first is sowne in the fields of Germanie, Italy, and other countries, to feed their cattell, as we in England do Bucke-wheat: we haue a small quantitie thereof in our gardens, for pleasures sake

The third groweth neere vnto the sea side in diuers places.

¶ *The Time.*

Medica must be sowne in Aprill; it floureth in Iune and Iuly: the fruit is ripe in the end of August.

¶ *The Names.*

Medick fodder is called of some *Trifolium Cochleatum*, and *Medica*: in French, *L'herbe à Limasson*: in Greeke, Μηδική: in Spanish, *Mielguas*: of the Valentians and Catalons, *Alfafa*, by a word either barbarous or Arabicke: for the chiefe of the Arabian writers, *Auicen*, doth call *Medica*, *Cot*, *Alaseleti*, and *Alfasfasa*.

The other is called Sea Clauer, and Medick fodder of the sea.

¶ *The Temperature and Vertues.*

A Medick fodder is of temperature cold, for which cause it is applied greene to such inflammations and infirmities as haue need of cooling.

CHAP. 503. *Of Wood Sorrell, or Stubwort.*

1 *Oxys alba.*
White wood Sorrell.

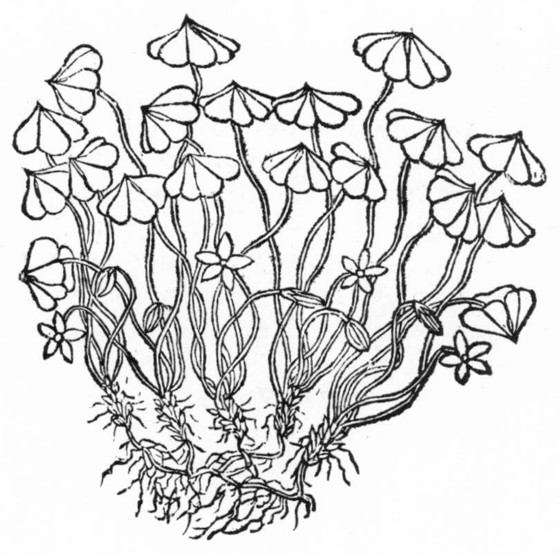

¶ *The Description.*

1 Oxys Pliniana, or *Trifolium acetosum*, being a kind of three leafed grasse, is a low and base herbe without stalk; the leaues immediately rising from the root vpon short stems at their first comming forth folded together, but afterward they do spred abroad, and are of a faire light greene colour, in number three, like the rest of the Trefoiles, but that each leafe, hath a deepe cleft or rift in the middle: amongst these leaues come vp small and weake tender stems, such as the leaues do grow vpon, which beare small star-like floures of a white colour, with some brightnes of carnation dasht ouer the same: the floure consisteth of fiue small leaues; after which come little round knaps or huskes full of yellowish seed: the root is very threddy, and of a reddish colour: the whole herbe is in taste like Sorrell, but much sharper and quicker, and maketh better greene sauce than any other herbe or Sorrell whatsoeuer.

‡ My oft mentioned friend Mr. *George Bowles* sent me some plants of this with very faire red floures, which he gathered in Aprill last, in a wood of Sir *Thomas Walsinghams* at Chisselhurst in Kent, called Stockwell wood, and in a little round wood thereto adioyning. ‡

2 The second kinde of *Oxys* or wood Sorrell is very like the former, sauing that the floures are of

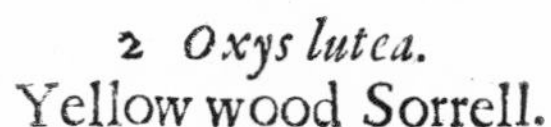
2 *Oxys lutea.*
Yellow wood Sorrell.

of a yellow colour, and yeeld for their ſeed veſſels ſmall and long horned cods; in other reſpects alike.

¶ *The Place.*

Theſe plants grow in woods and vnder buſhes, in ſandie and ſhadowie places in euery countrie.

‡ I haue not as yet found any of the yellow growing with vs. ‡

¶ *The Time.*

They floure from the beginning of Aprill vnto the end of May and midſt of Iune.

¶ *The Names.*

Wood Sorrell or Cuckow Sorrell is called in Latine *Trifolium acetoſum*: the Apothecaries and Herbariſts call it *Alleluya*, and *Panis Cuculi*, or Cuckowes meate, becauſe either the Cuckow feedeth thereon, or by reaſon when it ſpringeth forth and floureth the Cuckow ſingeth moſt, at which time alſo *Alleluya* was wont to be ſung in Churches. *Hieronymus Fracaſtorius* nameth it *Lujula*. *Alexander Benedictus* ſaith that it is called *Alimonia*: in high-Dutch, **Saurelklee**: in Low-Dutch, **Coeckcoecxbroot**: in French, *Pain de Cocu*: in Engliſh, wood Sorrel, wood Sower, Sower Trefoile, Stubwort, Alleluia, and Sorrell du Bois.

It is thought to be that which *Pliny, lib.* 27. *cap.* 12. calleth *Oxys*; writing thus: *Oxys* is three leafed, it is good for a feeble ſtomacke, and is alſo eaten of thoſe that are burſten. But *Galen* in his fourth booke of Simples ſaith, that *Oxys* is the ſame which *Oxalis* or Sorrell is: and *Oxys* is found in *Pliny* to be alſo *Iunci ſpecies*, or a kinde of Ruſh.

¶ *The Nature.*

Theſe herbes are cold and dry like Sorrell.

¶ *The Vertues.*

A Sorrell du Bois or wood Sorrell ſtamped and vſed for greene ſauce, is good for them that haue ſicke and feeble ſtomackes; for it ſtrengthneth the ſtomacke, procureth appetite, and of all Sorrel ſauces is the beſt, not onely in vertue, but alſo in the pleaſantneſſe of his taſte.

B It is a remedie againſt putrified and ſtinking vlcers of the mouth, it quencheth thirſt, and cooleth mightily an hot peſtilentiall feuer, eſpecially being made in a ſyrrup with ſugar.

Chap. 504. *Of noble Liuer-wort, or golden Trefoile.*

¶ *The Deſcription.*

1 Noble Liuerwort hath many leaues ſpred vpon the ground, three cornered, reſembling the three leaued graſſe, of a perfect graſſe greene colour on the vpper ſide, but grayiſh vnderneath: among which riſe vp diuers ſmall tender foot-ſtalkes of three inches long; on the ends whereof ſtands one ſmal ſingle blew floure, conſiſting of ſix little leaues, hauing in the middle a few white chiues: the ſeed is incloſed in little round knaps, of a whitiſh colour; which being ripe do ſtart forth of themſelues: the root is ſlender, compoſed of an infinite number of blacke ſtrings.

2 The ſecond is like vnto the precedent in leaues, roots, and ſeeds: the floures hereof are of a ſhining red colour, wherein conſiſteth the difference.

3 This

This strange three leaued Liuerwort differeth not from the former, sauing that this brings forth double blew floures tending to purple, and the others not so.

There is another in my garden with white floures, which in stalks and euery other respect is like the others.

1 *Hepaticum trifolium.*
Noble Liuerwort.

2 *Hepatica trifolia rubra.*
Noble red Liuerwort.

3 *Hepatica multiflora Lobelij.*
Noble Liuerwort with double floures.

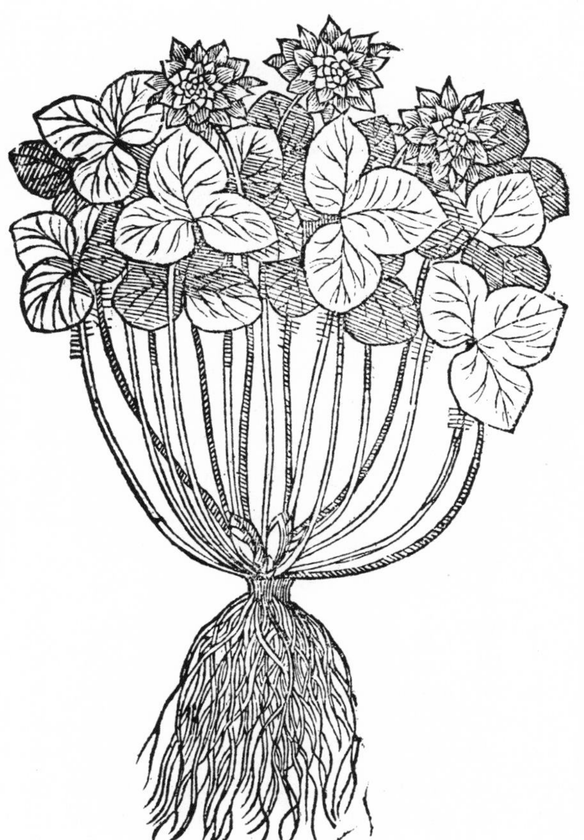

¶ *The Place.*

These pretty floures are found in most places of Germanie in shadowie woods among shrubs, and also by high-waies sides: in Italy likewise, and that not onely with the blew floures, but the same with double floures also, by the report of *Alphonsus Pancius* D. of Physicke in the Vniuersity of Ferrara, a man excellently well seen in the knowledge of Simples. They do all grow likewise in my garden, except that with double floures, which as yet is a stranger in England: ‡ it is now plentiful in many gardens. ‡

¶ *The Time.*

They floure in March and April, and perfect their seed in May.

¶ *The Names.*

Noble Liuerwort is called *Hepatica trifolia, Hepatica aurea, Trifolium aureum:* of *Baptista Sardus, Herba Trinitatis:* in high-Dutch, **Edel Leber kraut:** in low-Dutch, **Edel leuer cruijt:** in French, *Hepatique:* in English, Golden Trefoile, three leaued Liuerwort, noble Liuerwort, and herbe Trinitie.

¶ *The Temperature.*

These herbes are cold and drie, with an astringent or binding qualitie.

¶ *The*

¶ *The Vertues.*

A It is reported to be good against the weakenesse of the liuer which proceedeth of an hot cause; for it cooleth and strengthneth it not a little.

B *Baptista Sardus* commendeth it, and writeth that the chiefe vertue is in the root; if a spoonfull of the pouder thereof be giuen certaine dayes together with wine, or with some kinde of broth, it profiteth much against the disease called *Enterocele*.

Chap. 505. *Of Melilot, or plaister Clauer.*

¶ *The Description.*

1 THe first kinde of Melilot hath great plenty of small tough and twiggy branches, and stalkes full of ioynts or knees, in height two cubits, set full of leaues three together, like vnto Burgondie hay. The floures grow at the top of the stalke, of a pale yellow colour, standing thickly set and compact together, in order or rowes, very like the floures of *Securidaca altera*: which being vaded, there follow certaine crooked cods bending or turning vpward with a sharpe point, in fashion not much vnlike a Parrets bill, wherein is contained seed like Fenugreeke, but flatter and slenderer: the whole plant is of a reasonable good smell, much like vnto honey, and very full of iuyce: the root is very tough and pliant.

1 *Melilotus Syriaca odora.*
Assyrian Clauer.

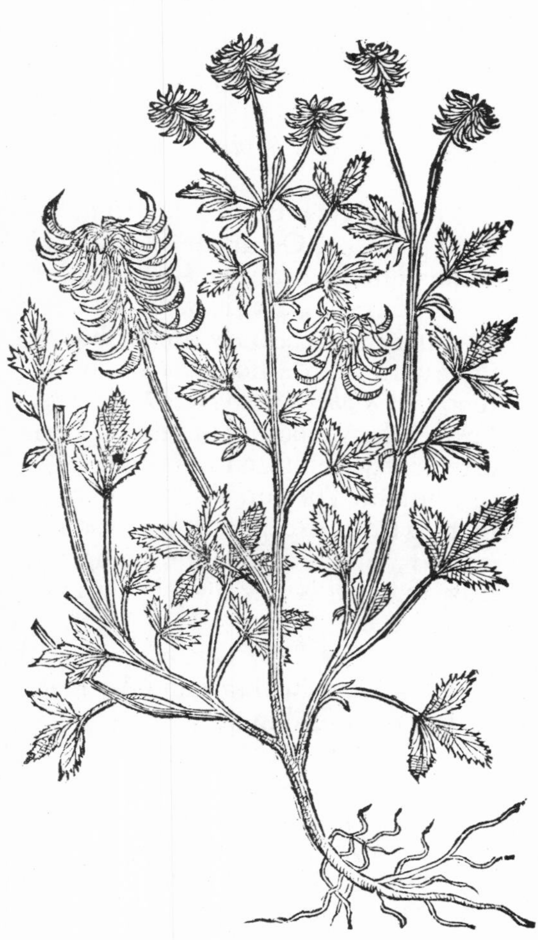

2 *Melilotus Italica & Patauina.*
Italian Clauer.

2 The second kinde of Melilot hath small and tender vpright stalkes, a cubit high, and somewhat more, of a reddish colour, set full of round leaues three together, not snipt about the edges like the other Trefoiles; and they are of a very deepe greene colour, thicke, fat, and full of iuyce. The floures grow alongst the tops of the stalkes, of a yellow colour, which turne into rough round seeds as big as a Tare, and of a pale colour. The whole plant hath also the sauour of honey, and perisheth when it hath borne his seed.

3 The

3 The third kind of Melilot hath round ſtalks and iagged leaues ſet round about,not much vnlike the leaues of Fenugreeke, alwaies three growing together like the Trefoiles, and oftentimes couered ouer with an hoarineſſe,as though meale had been ſtrewed vpon them. The floures be yellow and ſmall, growing thicke together in a tuft, which turne into little cods, wherein the ſeed is contained: the root is ſmall, tough, and pliant.

4 The fourth kinde of Melilot growes to the height of three cubits, ſet full of leaues like the common Melilot, and of the ſame ſauour: the floures grow alongſt the top of the ſtalks,of a white colour, which turne into ſmall ſoft huskes, wherein is contained little blackiſh ſeed: the root is alſo tough and pliant.

3 *Melilotus Coronata.*
Kings Clauer.

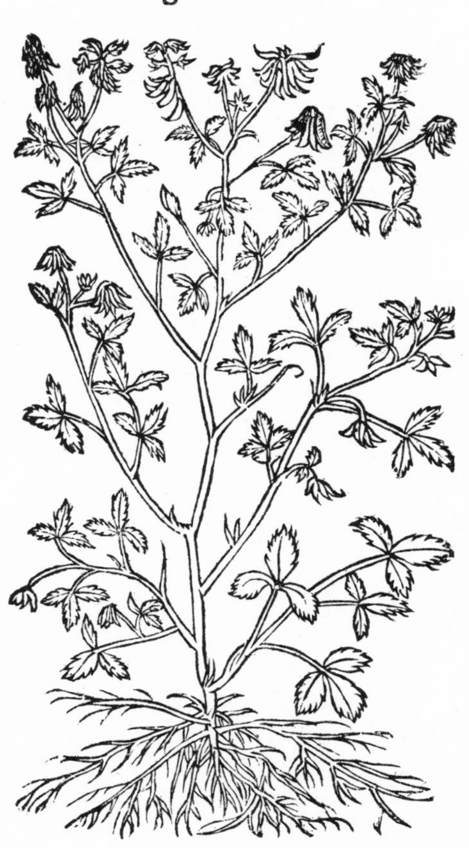

4 *Melilotus Germanica.*
Germane Clauer.

‡ Although our Author intended this laſt deſcription for our ordinarie Melilot,yet he made it of another which is three times larger, growing in ſome gardens (where it is onely ſowne) aboue two yards high, with white floures and many branches: the whole ſhape thereof is like the common kinde, as far as I remember. The common Melilot hath weake cornered greene ſtalkes ſome two foot and better high, whereon grow longiſh leaues ſnipt and oftentimes eaten about the edges, of a freſh greene colour: out of the boſomes of the leaues come little ſtalkes ſome handfull long, ſet thicke on their tops with little yellow floures hanging downe and turning vp again, each floure being compoſed of two little yellow leaues, whereof the vppermoſt turnes vp again,and the vndermoſt ſeemes to be parted into three. The floures paſt, there ſucceed little cods wherein is the ſeed. ‡

¶ *The Place.*

Theſe plants grow in my garden: the common Engliſh Melilot *Pena* ſetteth forth for *Melilotus Germanica*: but for certaintie no part of the world doth enioy ſo great part thereof as England, and eſpecially Eſſex; for I haue ſeene betweene Sudbury in Suffolke,and Clare in Eſſex,and from Clare to Heningham, and from thence to Ouendon, Bulmare, and Pedmarſh, very many acres of earable paſture overgrowne with the ſame; inſomuch that it doth not onely ſpoyle their land, but the corne alſo, as Cockle or Darnel, and as a weed that generally ſpreadeth ouer that corner of the Shire.

¶ *The Time.*

These herbes do floure in Iuly and August.

¶ *The Names.*

Plaister Clauer is called by the generall name, *Melilotus*, of some, *Trifolium odoratum*; yet there is another sweet Trefoile, as hath been declared. Some call it *Trifolium Equinum*, and *Caballinum*, or Horse-Trefoile, by reason it is good fodder for horses, who do greedily feed thereon: likewise *Trifolium Vrsinum*, or Beares Trefoile: of *Fuchsius*, *Saxifraga lutea*, and *Sertula Campana*: of *Cato*, *Serta Campana*, which most do name *Corona Regia*: in high Dutch, **Groote Steenclaueren**: of the Romanes and Hetrurians, *Tribolo*, as *Matthiolus* writeth: in English, Melilot, and Plaister-Clauer: in Yorkeshire, Harts-Clauer.

¶ *The Temperature.*

Melilote, saith *Galen*, hath more plenty of hot substance than cold (that is to say, hot and dry in the first degree) it hath also a certaine binding qualitie, besides a wasting and ripening facultie. *Dioscorides* sheweth, that Melilote is of a binding and mollifying qualitie, but the mollifying qualitie is not proper vnto it, but in as much as it wasteth away, and digesteth humors gathered in hot swellings, or otherwise: for so far doth it mollifie or supple that thing which is hard, which is not properly called mollifying, but digesting and wasting away by vapors: which kinde of quality the Grecians call διαφορητική.

¶ *The Vertues.*

A Melilote boiled in sweet wine vntill it be soft, if you adde thereto the yolke of a rosted egge, the meale of Fenegreeke and Linesеed, the roots of Marsh Mallowes and hogs greace stamped together, and vsed as a pultis or cataplasma, plaisterwise, doth asswage and soften all manner of swellings, especially about the matrix, fundament and genitories, being applied vnto those places hot.

B With the juice hereof, oile, wax, rosen and turpentine, is made a most soueraigne healing and drawing emplaster, called Melilote plaister, retaining both the colour and sauour of the herbe, being artificially made by a skilfull Surgion.

C The herbe boiled in wine and drunke prouoketh vrine, breaketh the stone, and asswageth the paine of the kidnies, bladder and belly, and ripeneth flegme, and causeth it to be easily cast forth.

D The juice thereof dropped into the eies cleereth the sight, consumeth, dissolueth, and cleane taketh away the web, pearle, and spot in the eies.

E Melilote alone with water healeth *Recentes melicerides*, a kinde of wens or rather apostems conteyning matter like honey; and also the running vlcers of the head, if it be laid to with chalke, wine and galls.

F It likewise mitigateth the paine of the eares, if the juice be dropped therein mixed with a little wine, and taketh away the paine of the head, which the Greekes call κεφαλαλγίαν, especially if the head be bathed therewith, and a little vineger and oile of Roses mixed amongst it.

‡ Chap. 506. *Of certaine other Trefoiles.*

‡ Those Trefoiles being omitted by our Author, I haue thought good to put into a chapter by themselues, though they haue little affinity with one another, the two last excepted.

¶ *The Description.*

1 The first of those in roots, stalkes, and manner of growing is like the Medicke or snaile Trefoiles formerly described: the leaues are hairie; the floures yellow and small: after which follow crooked flat cods, of an indifferent bredth, wherein is contained seeds made after the fashion of little kidnies; this the Italians, according to *Lobel*, call *Lunaria radiata*; in the *Hist. Ludg.* it is called *Medica syl. altera lunata*.

2 The root of this is long and thicke, couered with a yellowish rinde, and hauing a white sweet pith in the inside, couered with a hairinesse on the top, and sending forth sundry fibres: from this rise vp many weake long foot-stalkes, whereon grow leaues set together by threes, long, narrow, smooth, lightly nickt on the edges: amongst these rises vp commonly one stalke (yet sometimes two) smooth and naked, three or foure inches long; on the top thereof grow spike fashion, 8. or ten pretty large light purple floures, each of them being set in a cup diuided into 5. parts. This growes vpon diuers parts of the Alpes: and *Pena* in his *Mons Baldus* set it forth by the name of *Trifolium angustifolium Alpinum*. *Bauhinus* saith, the root hereof tasts like Liquorice, wherefore it may be called *Glycyrhiza Astragaloides*, or *Astragalus dulcis*: and he receiued it out of Spaine by the name of *Glycyrhiza*. He calls it in his *Prodromus*, *Trifolium Alpinum flore magno radice dulci*.

This

‡ 1 *Trifolium ſiliqua lunata.*
Moone Trefoile.

‡ 2 *Trifol. anguſtifol. Alpinum.*
Liquorice Trefoile.

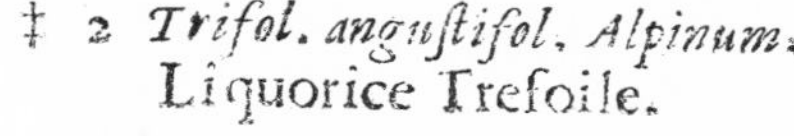

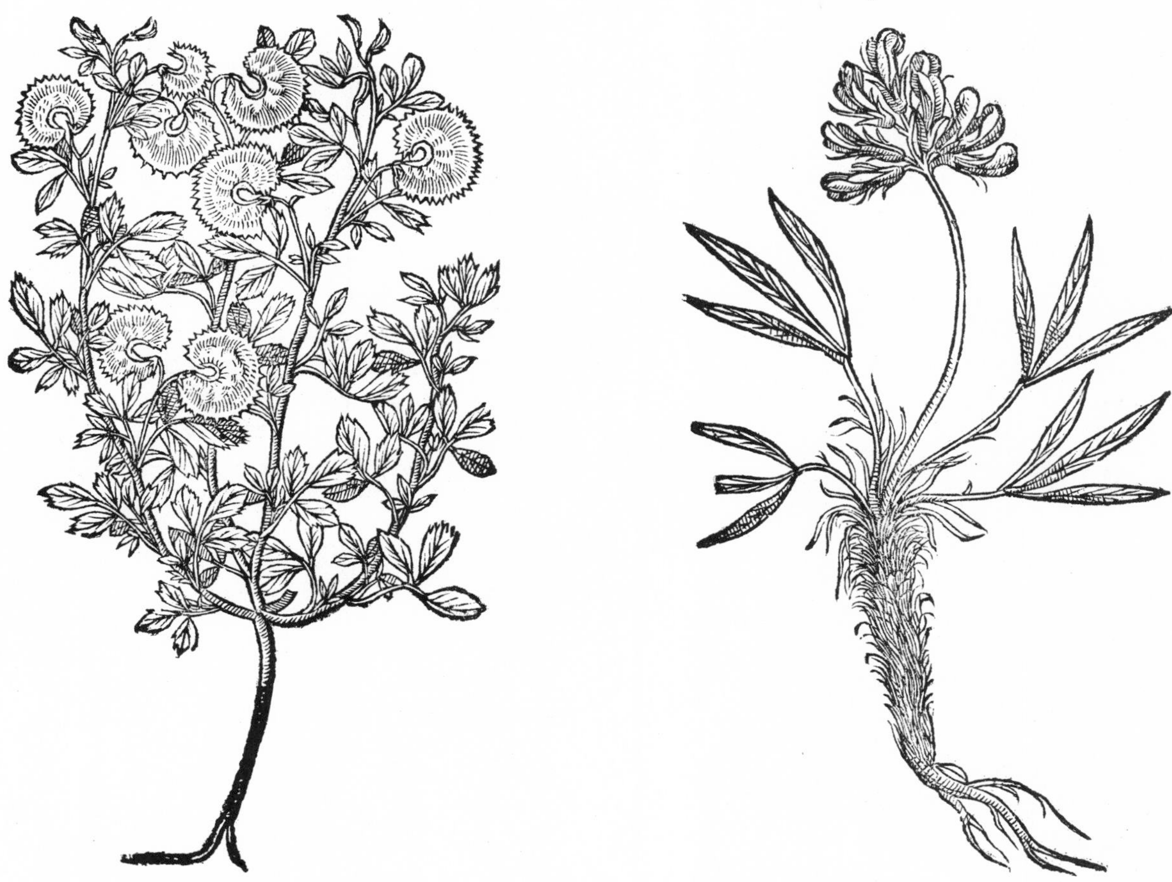

‡ 3 *Trifolium ſpinoſum Creticum.*
Prickly Trefoile.

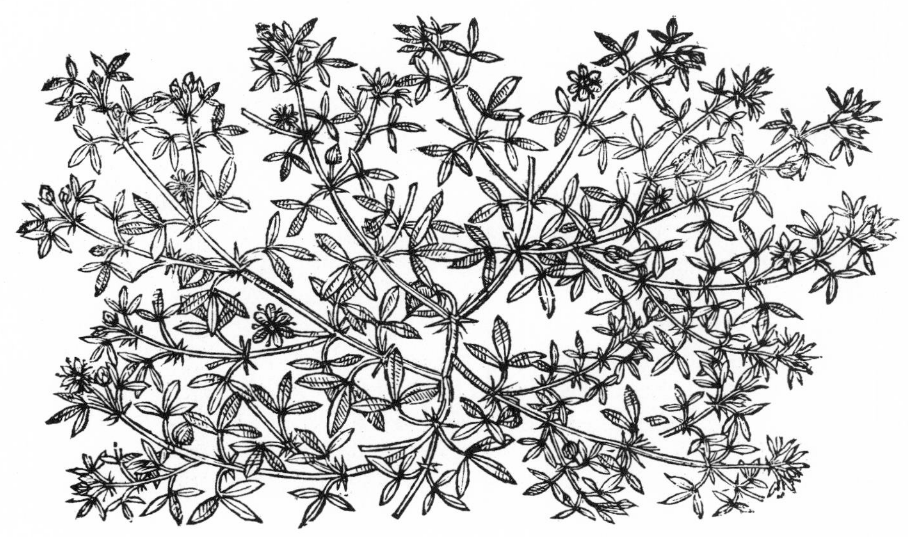

3 This thornie Trefoile hath a long threddy root, from which ariſe many ſhort branched ſtalkes ſome two handfulls high, cornered, and ſpred vpon the ground; the ioynts, which are many, are commonly red, and armed with foure ſharpe prickles, and out of each of them, vpon ſhort foot-ſtalkes grow two trifoile leaues, greene, longiſh, and ending in a little prickle: out of theſe ioynts alſo grow little foot-ſtalkes, which carry ſingle floures made of fiue little leaues, of the ſhape and colour of the little blew Bell-floure, with ten chiues in the middle tipt with yellow: after theſe follow fiue cornered ſharpe pointed heads, containing a ſingle flat red ſeed in each corner. *Cluſius* ſet forth this by the name of *Trifolium ſpinoſum Creticum*: the ſeed was ſent out of Candy by the name of ὀρβιτοχορτο: he queſtions whether it may not be the true *Tribulus terreſtris* of *Dioſcorides*.

4 The roots, ſtalkes, and leaues of this pretty Trefoile do not much differ from the common

 white

‡ 4 *Trifolium fragiferum.*
Straw-berry Trefoile.

white Trefoile, but there is some difference in the floures and seed; for the floures of this are small, grow thick together, & are of a whitish blush colour: after which follow heads made of little bladders or thinne skins, after such a manner as they resemble a Strawberrie or Raspas, and they are of a grayish colour, here and there marked with red: the stalkes seldome grow aboue three inches high. It growes in most salt marishes, as in Dartford salt marish, in those below Purfleet, and such like: it floures in Iuly and August. *Clusius* hath set it forth by the name of *Trifolium fragiferum Frisicum:* some had rather call it *Trifolium vesicarium*, Bladder Trefoile.

5 There are two other Trefoiles with which I thinke good to acquaint you, and those by the similitude of the cups, which containe the floures, and become the seed vessels, may be fitly called *Stellata*; and thus *Bauhine* calls the first *Trifolium stellatum*; whereto for distinctions sake I adde *hirsutum*, calling it *Trifol. stellatum hirsutum*, Rough starrie headed Trefoile: it hath a smal long white root, from which arise stalkes some foot high, round, slender, hairie, and reddish, hauing few leaues or branches: the leaues stand three on a stalke, as in other Trefoiles, smooth on the vpper side, and hairy below: the floures are small and red, like in shape to those of the common red Trefoile, but lesser; and they stand each of them in a cup reddish and rough below, and on the vpper part cut into fiue long sharpe leaues standing open as they commonly figure a starre: the floures fallen, these cups dilate themselues, and haue in the middle a longish transuerse whitish spot. I saw this flouring in May in the garden of Mr. *Tradescant*, who did first bring plants hereof from Fermentera a small Island in the Mediterranean sea.

6 This other (which for any thing that I know is not figured nor described by any) hath stalks sometimes a foot, otherwhiles little aboue an inch high, hairy, and diuided but into few branches: the leaues, which stand by threes, are fastned to long foot-stalkes, and they themselues are somewhat longish, hauing two little sharpe pointed leaues growing at the setting on of the foot-stalkes to the stalkes: they are greene of colour, and not snipt about the edges. The heads that grow on the tops of the stalkes are round, short, and greene, with small purple or else whitish floures like those of the common Trefoile, but lesser, standing in cups diuided into fiue parts, which when the floures are fallen become somewhat bigger, harsher, and more prickly, but open not themselues so much as those of the former: the seed is like that of Millet, but somwhat rounder. This floures in Iune, and the seed is ripe in Iuly. I first obserued it in Dartford salt marish, the tenth of Iune, 1633. I haue named this *Trifolium stellatum glabrum*, Smooth starrie headed Trefoile. ‡

¶ *The Temperature and Vertues.*

These, especially the three last, seeme to be of the same temper and vertue as the common Medow Trefoiles, but none of them are at this day vsed in Physicke, or knowne, vnlesse to some few. ‡

Chap. 597. *Of Pulse.*

¶ *The Kindes.*

There be diuers sorts of Pulse, as Beanes, Peason, Tares, Chiches, and such like, comprehended vnder this title Pulse: and first of the great Beane, or garden Beane.

¶ *The*

¶ *The Description.*

1 THe great Beane riseth vp with a foure square stalke, smooth, hollow, without ioynts, long and vpright, which when it is thicke sowne hath no need of propping, but when it is sowne alone by it selfe it soone falleth downe to the ground: it bringeth forth long leaues one standing from another, consisting of many growing vpon one rib or stem, euerie one whereof is somewhat fat, set with veines, slipperie, more long than round. The floures are eared, in forme long, in colour either white with blacke spots, or of a blackish purple: after them come vp long cods, thicke, full of substance, slenderer below, frized on the inside with a certaine white wooll as it were, or soft flockes; which before they be ripe are greene, and afterwards being dry they are blacke and somewhat hard, as be also the cods of broome, yet they be longer than those, and greater: in which are contained three, foure, or fiue Beanes, seldome more, long, broad, flat, like almost to a mans naile, great, and oftentimes to the weight of halfe a dram; for the most part white, now and then of a red purplish colour; which in their vpper part haue a long black nauell as it were, which is couered with a naile, the colour whereof is a light greene: the skin of the fruit or beane is closely compacted, the inner part being dry is hard and sound, and easily cleft in sunder; and it hath on the one side an euident beginning of sprouting, as haue also the little pease, great Pease, Ciches, and many other Pulses. The roots hereof are long, and fastned with many strings.

1 *Faba maior hortensis.*
The great garden Beane.

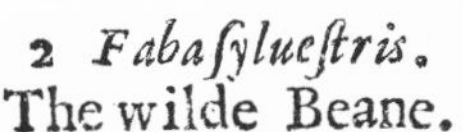
2 *Faba syluestris.*
The wilde Beane.

2 The second kinde of Beane (which *Pena* setteth forth vnder the title of *Syluestris Græcorum Faba*, and *Dodonæus*, *Bona syluestris*; which may be called in English Greeke Beanes) hath square hollow stalkes like the garden Beanes, but smaller. The leaues be also like the common Beane, sauing that the ends of the rib whereon those leaues do grow haue at the very end small tendrels or claspers, such as the pease leaues haue. The floures are in fashion like the former, but of a darke red colour: which being vaded, there succeed long cods which are blacke when they be ripe, within which is inclosed blacke seed as big as a Pease, of an vnpleasant taste and sauour.

 ‡ 3 The

‡ 3 The common Beane in stalkes, leaues, floures, and cods is like the former great garden Beane, but lesser in them all; yet the leaues are more, and grow thicker, and out of the bosomes of the leaues vpon little foot-stalkes grow the floures, commonly six in number, vpon one stalk, which are succeeded by so many cods, lesser and rounder than those of the former: the beans themselues are also lesse, and not so flat, but rounder, and somewhat longish: their colour are either whitish, yellowish, or else blacke. This is sowne in most places of this kingdome, in corne fields, and known both to man and beast. I much wonder our Author forgot to mention so common and vulgarly knowne a Pulse. It is the *Bona* or *Faselus minor* of *Dodonæus*; and the *Faba minor* of *Pena* and *Lobel*. ‡

¶ *The Place.*

The first Beane is sowne in fields and gardens euery where about London.

This blacke Beane is sowne in a few mens gardens who be delighted in varietie and study of herbes, whereof I haue great plenty in my garden.

¶ *The Time.*

They floure in Aprill and May, and that by parcels, and they be long in flouring: the fruit is ripe in Iuly and August.

¶ *The Names.*

The garden Beane is called in Latine *Faba*: in English, the garden Beane: the field Beane is of the same kinde and name, although the fertilitie of the soile hath amended and altered the fruit into a greater forme. ‡ The difference betweene the garden and field Beane is a specificke difference, and not an accidentall one caused by the soile, as euery one that knoweth them may well perceiue. ‡

The blacke Beane, whose figure we haue set forth in the second place, is called *Faba syluestris*: of some thought to be the true physicke Beane of the Antients; whereupon they haue named it *Faba Veterum*, and also *Faba Græcorum*, or the Greeke Beane. Some would haue the garden Beane to be the true *Phaseolus*, or Kidney Bean; of which number *Dodonæus* is chiefe, who hath so wrangled and ruffled among his relatiues, that all his antecedents must be cast out of dores: for his long and tedious tale of a tub we haue thought meet to commit to obliuion. It is called in Greeke πύανος, whereupon the Athenians feast dayes dedicated to *Apollo* were named πυανέψια, in which Beans and Pulses were sodden: in Latine it is also called *Faba fresa* or *fracta*, broken or bruised Beane.

‡ *Dodonæus* knew well what he did, as any that are either iudicious or learned may see, if they looke into the first chapter of the second booke of his fourth *Pemptas*. But our Authors words are too iniurious, especially being without cause, & against him, from whom he borrowed all that was good in this his booke, except the figures of *Tabernamontanus*. It may be Dr. *Priest* did not fit his translation in this place to our Authors capacitie; for *Dodonæus* did not affirme it to be the *Phaseolus*, but *Phaselus*, distinguishing betweene them. ‡

¶ *The Temperature and Vertues.*

A The Beane before it be ripe is cold and moist: being dry it hath power to bind and restraine, according to some Authors: further of the temperature and vertues out of *Galen*.

B The Beane (as *Galen* saith in his booke of the Faculties of nourishments) is windie meate, although it be neuer so much sodden and dressed any way.

C Beanes haue not a close and heauy substance, but a spongie and light, and this substance hath a scouring and clensing facultie; for it is plainly seene, that the meale of Beanes clenseth away the filth of the skin; by reason of which qualitie it passeth not slowly through the belly.

C And seeing the meale of Beanes is windie, the Beanes themselues if they be boyled whole and eaten are yet much more windie.

E If they be parched they lose their windinesse, but they are harder of digestion, and doe slowly descend, and yeeld vnto the body thicke or grosse nourishing iuyce; but if they be eaten green before they be ripe and dried, the same thing hapneth to them which is incident to all fruits that are eaten before they be fully ripe; that is to say, they giue vnto the body a moist kinde of nourishment, and therefore a nourishment more full of excrements, not onely in the inward parts, but also in the outward, and whole body thorow: therefore those kindes of Beans do lesse nourish, but they do more speedily passe thorow the belly, as the said Author in his booke of the Faculties of simple Medicines saith, that the Beane is moderately cold and dry.

F The pulpe or meate thereof doth somewhat clense, the skin doth a little binde.

G Therefore diuers Physitians haue giuen the whole Beane boyled with vineger and salt to those that were troubled with the bloudy flix, with laskes and vomitings.

H It raiseth flegme out of the chest and lungs: being outwardly applied it drieth without hurt the watery humors of the gout. We haue oftentimes vsed the same being boiled in water, and so mixed with swines grease.

We

We haue laid the meale therof with Oxymel, or ſyrrup of vineger, both vpon bruiſed and wounded ſinewes, and vpon the wounded parts of ſuch as haue been bitten or ſtung, to take away the fierie heat. I

It alſo maketh a good plaiſter and pultis for mens ſtones and womens paps: for theſe parts when they are inflamed, haue need of moderate cooling, eſpecially when the paps are inflamed through the cluttered and congealed milke contained in them. K

Alſo milke is dried vp with that pultis. L

The meale thereof (as *Dioſcorides* further addeth) being tempered with the meale of Fenugreek and hony, doth take away blacke and blew ſpots, which come by drie beatings, and waſteth away kernels vnder the eares. M

With Roſe leaues, Frankincenſe, and the white of an egge, it keepeth backe the watering of the eies; the pin and the web, and hard ſwellings. N

Being tempered with wine it healeth ſuffuſions, and ſtripes of the eies.

The Beane being chewed without the skin, is applied to the forehead againſt rheumes and falling downe of humours. O P

Being boiled in wine it taketh away the inflammation of the ſtones. Q

The skins of Beans applied to the place where the hairs were firſt plucked vp, wil not ſuffer them to grow big, but rather conſumeth their nouriſhment. R

Being applied with Barly meale parched and old oile, they waſte away the Kings euill. S

The decoction of them ſerueth to die woollen cloth withall. T

This Beane being diuided into two parts (the skin taken off) by which it was naturally ioined together, and applied, ſtancheth the bloud which doth too much iſſue forth after the biting of the horſeleach, if the one halfe be laied vpon the place. V

The blacke Beane is not vſed with vs at all, ſeeing, as we haue ſaid, it is rare, and ſowne onely in a few mens gardens, who be delighted in varietie and ſtudie of herbes. X

CHAP. 508. *Of Kidney Beane.*

¶ *The Kindes.*

THe ſtocke or kindred of the Kidney Bean are wonderfully many; the difference eſpecially conſiſteth in the colour of the fruit: there be other differences, wherof to write particularly would greatly ſtuffe our volume with ſuperfluous matter, conſidering that the ſimpleſt is able to diſtinguiſh apart the white Kidney Beane from the blacke, the red from the purple, and likewiſe thoſe of mixt colours from thoſe that are onely of one colour: as alſo great ones from little ones Wherefore it may pleaſe you to be content with the deſcription of ſome few, and the figures of the reſt, with their ſeuerall titles in Latine and Engliſh, referring their deſcriptions vnto a further conſideration, which otherwiſe would be an endleſſe labour, or at the leaſt needleſſe.

¶ *The Deſcription.*

1 THe firſt kinde of *Phaſeolus* or garden Smilax hath long and ſmall branches growing very high, taking hold with his claſping tendrels vpon poles and ſtickes, and whatſoeuer ſtandeth neere vnto him, as doth the hop or vine, which are ſo weake and tender, that without ſuch props or ſupporters they are not able to ſuſtaine themſelues, but will run ramping on the ground fruitleſſe: vpon the branches do grow broad leaues almoſt like Iuie, growing together by three, as in the common Trefoile or three leaued Graſſe: among which come the floures, that do vary and differ in their colours, according to the ſoile where they grow, ſometimes white, ſometimes red, and oftentimes of a pale colour: afterwards there come out long cods, whereof ſome are crooked, and ſome are ſtraight, and in thoſe the fruit is contained, ſmaller than the common Beane, ſomwhat flat, and faſhioned like a Kidney, which are of diuers colours, like vnto the floures: whereto for the moſt part theſe are like.

2 There is alſo another *Dolichus* or Kidney Beane, leſſer, ſhorter, and with ſmaller cods, whoſe floures and fruit are like in forme to the former Kidney Beanes, but much leſſer, and of a blacke colour.

3 There is likewiſe another ſtrange Kidney Beane, which doth alſo winde it ſelfe about poles and props neere adioining, that hath likewiſe three leaues hanging vpon one ſtem, as haue the other Kidney Beans, but euery one is much narrower and alſo blacker: the cods be ſhorter, plainer, and flatter, and containe fewer ſeeds.

1 *Phaseolus albus.*
White Kidney Beane.

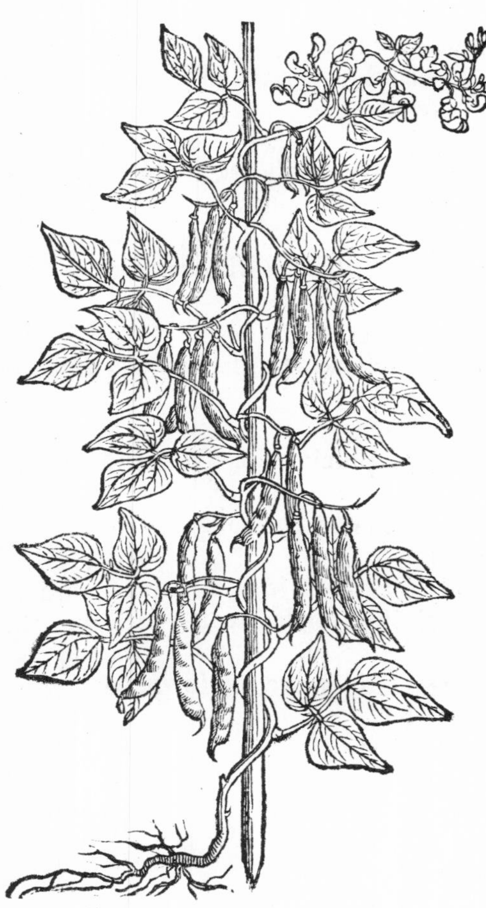

2 *Phaseolus niger.*
Blacke Kidney Beane.

3 *Smilax hortensis rubra.*
Red Kidney Beane.

4 *Smilax hortensis flava.*
Pale yellow Kidney Beane.

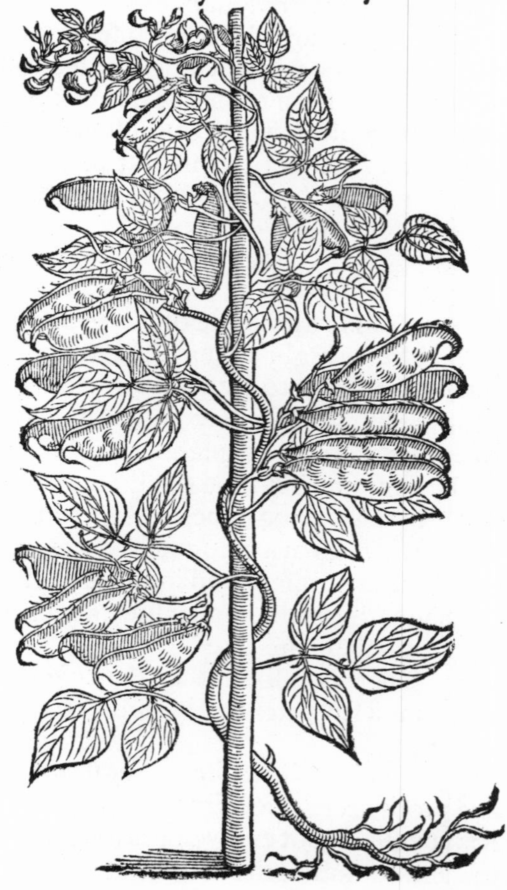

‡ 5 *Phaseolus peregrinus fructu minore albo.*
Indian Kidney Beane with a small white fruit.

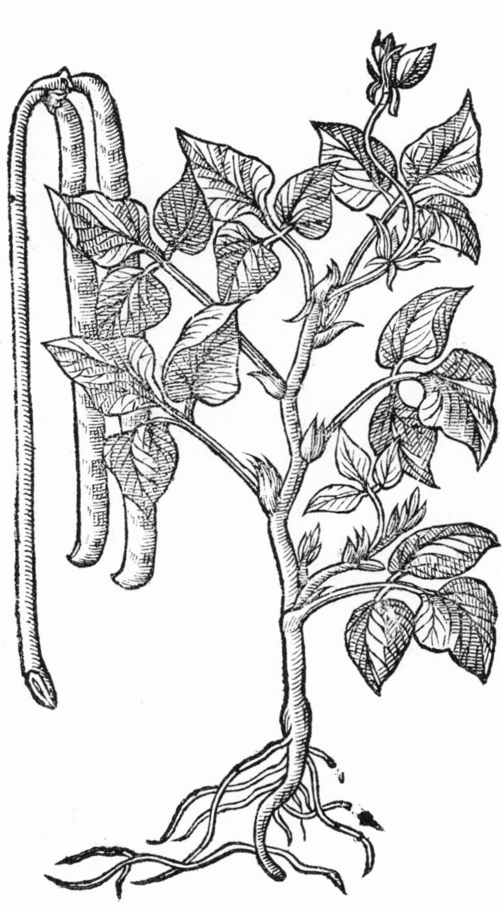

‡ 6 *Phaseolus peregrinus fructa minore frutescens.*
Indian Kidney Beane with a small red fruit.

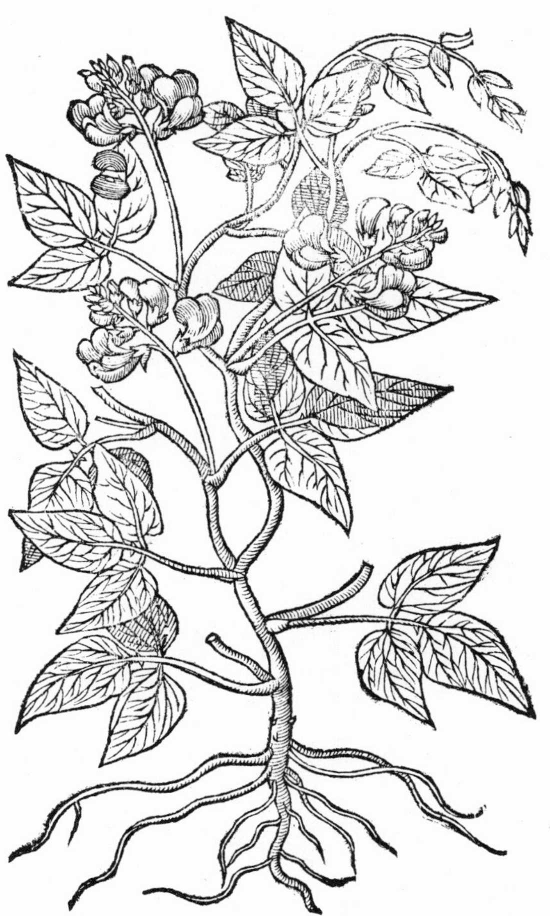

‡ 7 *Phaseolus peregrinus angustifolius.*
Narrow leafed Kidney Beane.

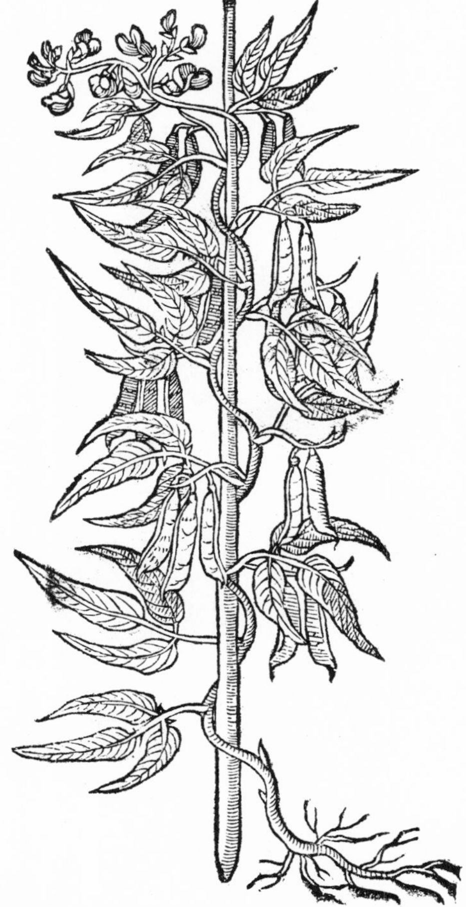

4 This Kidney Bean differeth not from the others, but onely in the colour of the fruit, which are of a pale yellow colour, wherein consisteth the difference.

‡ Besides the varieties of these Kidney Beans mentioned by our Author, there are diuers other reckoned vp by *Clusius*, which haue been brought out of the East and West Indies, and from some parts of Africa; I will only giue you the figures of two or three of them out of *Clusius*, with the colours of their floures and fruit.

5 The stalke of this is low and stiffe, the floures of a whitish yellow on the outside, and of a violet colour within: the fruit is snow white, with a blacke spot in the eye: This is *Phaseolus peregrinus* 4. of *Clusius*.

6 This hath leaues like the Marsh Trefoile, floures growing many together, in shape and magnitude like those of common Pease: the cods were narrow, and contained three or foure seeds, which were small, no bigger than the seeds of *Laburnum*; the Painter expressed two of them in the leafe next vnder the vppermost tuft of floures: this is *Clusius* his *Phaseolus peregrinus*. 5.

7 This growes high, winding about poles or other supporters: the leaues are narrower than the former: the fruit lesser and flatter, of a reddish colour. This is the *Phaseolus peregrinus* 6. of *Clusius*.

8 This windes about poles, and growes to a great

8 *Phaseolus Brasilianus.*
Kidney Beane of Brasile.

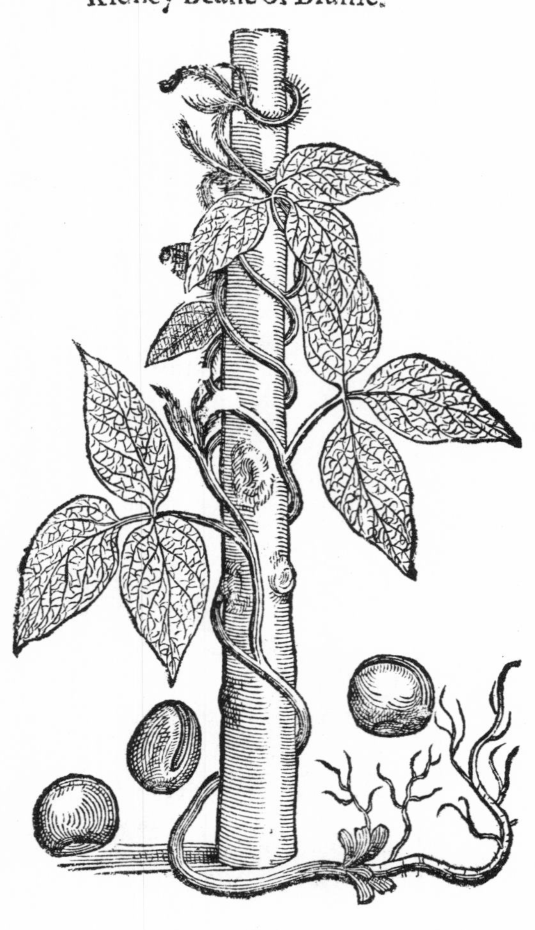

8 *Phaseoli Brasiliani ad vivum.*
The Brasile Kidney Bean in his ful bignesse.

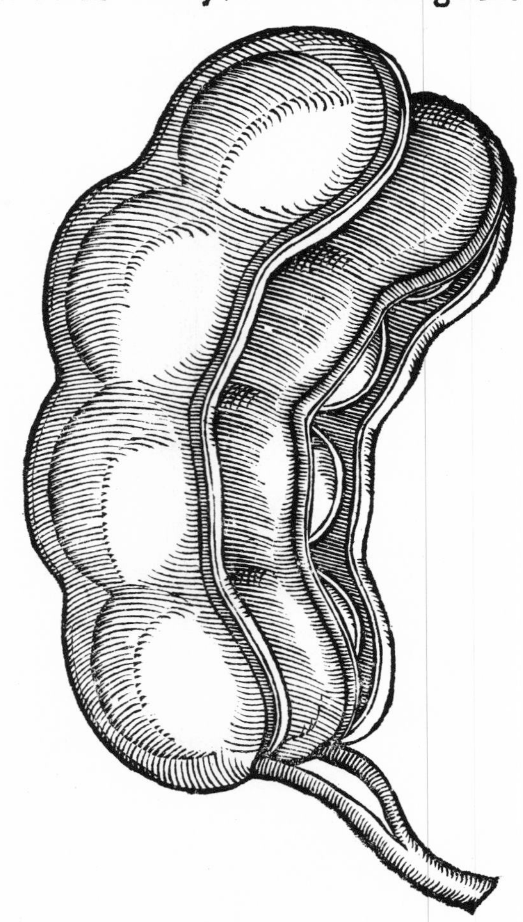

9 *Phaseolus Ægyptiacus.* The party coloured Beane of Egypt.

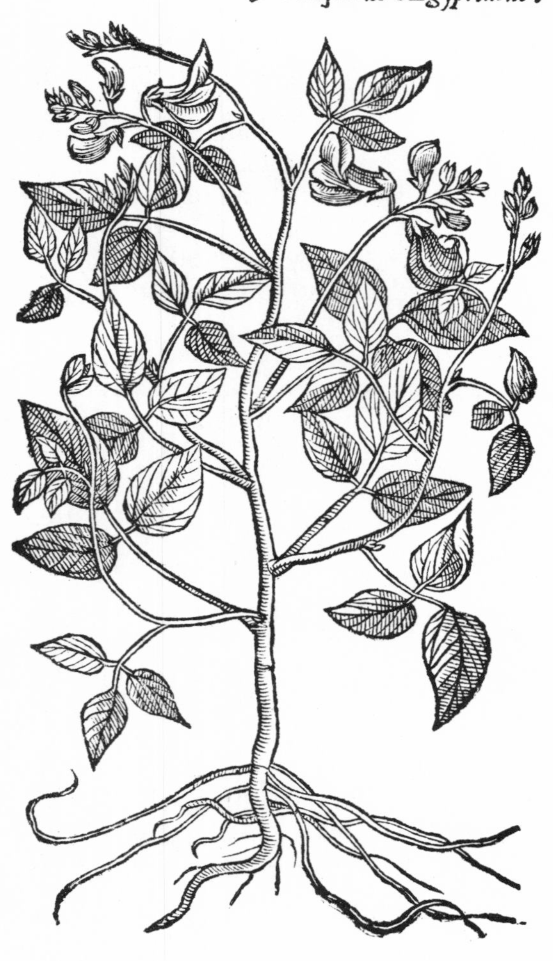

10 *Phaseoli Americi purgantes.*
Purging Kidney Bean of America.

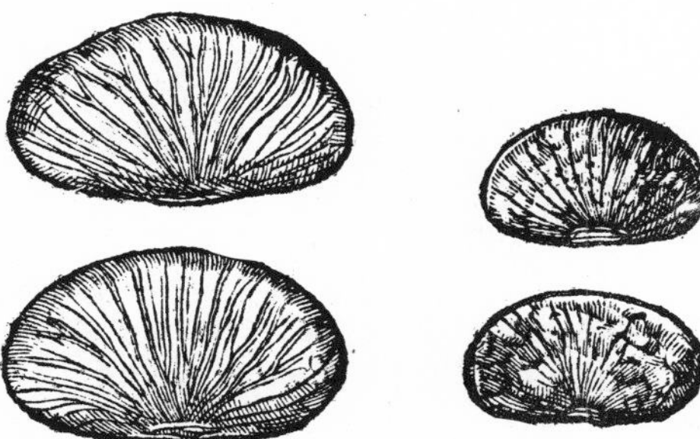

Phaseoli magni lati albi.

Phaseoli Brasiliani.

Phaseoli parui ex America delati.

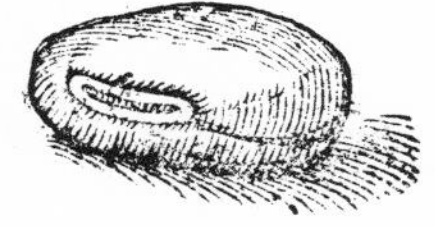

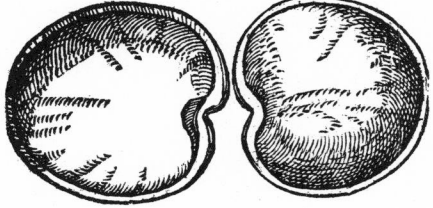

Phaseoli parui pallido albi ex America delati.

Phaseoli rubri.

Phaseoli rubri Indiani durissimi.

great height, with soft hairy leaues and large cods, wherein are contained seeds of diuers colours; sometimes they are red, otherwhiles of a whitish ash colour, sometimes wholly black, and otherwhiles spotted.

9 The Egyptian Beane is somewhat like the other Kidney Beanes in his growing: his fruit is of the bignesse of a small Hasell nut, blacke on one side, and of a golden yellow or Orenge colour on the other.

Besides these you finde here figured, and diuers others described by *Clusius*, I think it not amisse to mention two more. The first of these, which was procured by M^r^. *Tradescant*, and growes in our Gardens, is a large plant, not differing in manner of growth from the former Indian Kidney Beanes, but his floures are large, many, and of an elegant scarlet colour: whence it is vulgarly termed by our Flourists, the Scarlet Beane. The other I haue seene grow to a little height, but it would not indure; but the cods of it which were brought to vs were some three inches long, and couered with a hairie downe of a reddish colour, which put vpon the hands or skin in any part of the body would sting like a Nettle, and this was called the Stinging Beane: I thinke it came from some part of the East Indies. ‡

¶ The

¶ *The Place.*

Kidney Beanes doe easily and soone spring vp, and grow into a very great length, being sowne neere to long poles fastned hard by them, or hard by arbors or banqueting places, otherwise they lie flat on the ground, slowly come vp, hardly bring forth fruit, and become faultie and smitted, as *Theophrastus* writeth.

¶ *The Time.*

It is sowne in the Spring, especially in the midst of April, but not before: the fruit is ripe about the end of Sommer.

¶ *The Names.*

Hippocrates, Diocles, Theophrastus, and most of the other old Writers do call it δόλιχον: diuers of the bignesse of the seed do name it λοβὸν and λοβίον: in Latine, *Siliqua*: *Dioscorides* calleth it *Smilax*, because it climeth vp as *Smilax* doth, and taketh hold of props, staies, and shrubbes standing neere vnto it: others name it φασίολον, a Diminitiue deriued from φάσηλος: for φάσηλος and φασίολος are not one and the selfe same pulse called by diuers names, as some suppose, but sundry fruits one differing from the other, as *Galen* in his first booke of the Faculties of Nourishments doth sufficiently declare, where he intreateth of them both. For first he disputeth of *Phaseli* and *Ochri*, Beans, and Pease; then afterward others comming betweene, he writeth of *Dolichus*, which also is named *Phaseolus*: and though hee may be thought to doubt what manner of pulse that is which *Theophrastus* calleth *Dolichus*: notwithstanding he gathereth and concludeth that it is a fruit of a garden plant in Italy, and in Caria, growing in the fields, which is in forme longer than the Cichlings, and was commonly called in his time *Faseolus*. Of his opinion is *Paulus Ægineta*, writing of *Phaselus*, which hee nameth *Dolichus*, in the 79. chap. of his first booke. Moreouer, *Faselus* was in times past a common pulse in Italy and Rome, and *Dolichus* a strange pulse; for *Columella* and *Palladius*, writers of husbandry, haue made mention of the sowing of *Phaselus*: and *Virgil* calleth it *Vilis* in the first of his Georgicks: but concerning the sowing of *Dolichus* or Kidney Beane, none of the Latines haue written, by reason that the same was rare in Italy, and sowne onely in gardens, as *Galen* hath affirmed, naming it oftentimes a garden plant, and shewing that the same, as we haue said, is sowne in Caria; and likewise *Dioscorides* nameth it σμίλαξ κηπαία, that is to say *Smilax hortensis*, or garden Smilax, because it groweth in gardens: who also writing of this in another seuerall chapter, sheweth plainely, that *Smilax hortensis*, or *Dolichus* is another plant differing from *Faselus*, which he nameth *Phaseolus*.

For which causes it is not to be doubted, but that *Phaselus* with three syllables, differeth from *Faseolus* with foure syllables, no otherwise than *Cicer*, *Cicercula*, and *Cicera* differ, which notwithstanding be neere one to another in names: and it is not to be doubted but that they are deceiued, who thinke it to be one and the selfe same Pulse called by sundry names.

This plant is named in English, Kidney Beane, Sperage Beanes: of some, Faselles, or long Peason, French Beanes, garden Smilax, and Romane Beanes: in French, *Feues de Romme*: in Dutch, **Turckshoonen.**

¶ *The Temperature.*

Kidney Beanes, as *Dioscorides* teacheth, do more loose the belly than Peason; they are lesse windy, and nourish well, and no lesse than Peason, as *Diocles* saith: they be also without ingendring windinesse at all: the Arabian Physitions say that they are hot and moist of nature.

¶ *The Vertues.*

A The fruit and cods of Kidney Beanes boiled together before they be ripe, and buttered, and so eaten with their cods, are exceeding delicate meat, and do not ingender winde as the other Pulses doe.

B They doe also gently loose the belly, prouoke vrine, and ingender good bloud reasonably well; but if you eat them when they be ripe, they are neither toothsome nor wholsome. Therefore they are to be taken whilest they are yet greene and tender, which are first boiled vntill they be tender; then is the tib or sinew that doth run alongst the cod to be taken away; then must they be put into a stone pipkin, or some other vessell with butter, and set to the fire againe to stew, or boile gently: which meat is very wholsome, nourishing, and of a pleasant taste.

Chap. 509. *Of the flat Beane called Lupine.*

¶ *The Description.*

1 The tame or garden Lupine hath round hard stems, which of themselues do stand vpright without any succour, help or stay: the leaues consist of fiue, six, or seuen ioined together,

1 *Lupinus sativus.*
Garden Lupines.

2 *Lupinus flore luteo.*
Yellow Lupines.

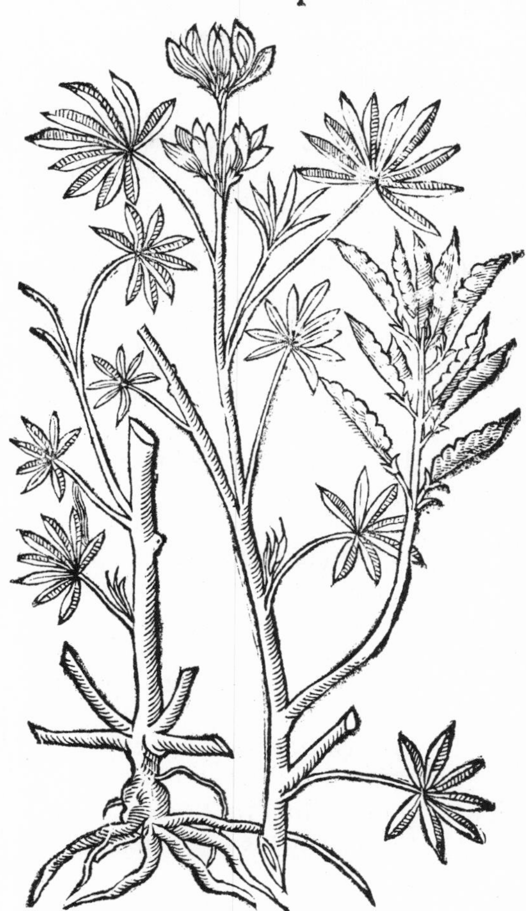

3 *Lupinus flore cæruleo.*
Blew Lupine.

‡ 4 *Lupinus maior flo. cæruleo.*
The great blew Lupine.

gether,like those of the Chast tree,green on the vpper side, and on the nether side white and downy;and in the euening about the setting of the Sun they hang flagging downwards as though they were withered:among these there commeth vp a tuft of floures of a pale or light blush colour,which turne into great rough cods, wherein is the fruit, which is flat and round like a cake,of a white colour,and bitter in taste : and where they cleaue vnto the cod, in that part they haue a certaine dent like a little nauell. This Lupine hath but one root,which is slender and wooddie, hauing hanging on it a few small threds like haires.

2 The yellow Lupine is like to the garden one in stalke and leaues,yet both of these lesser and shorter. It hath beautifull floures of an exceeding faire gold yellow colour,sweet of smell, made vp into an eare,of the colour of the yellow violet,and somewhat of the smell : the coddes are small, hard,somewhat hairy : the seeds be little,flat,round,in taste extreme bitter, of sundry colours , ill fauored,far lesser than the tame one.

3 The blew Lupines are longer than the yellow,and diuided into more wings and branches:the leaues be lesser and thinner : the floures small,and lesser than the yellow,of a blew colour:the seeds be also of diuers colours,bitter,and lesser than any of them all.

‡ 4 There is also another blew Lupine,whole leaues,stalks,floures, and cods are like,but larger than those of the first described : the floures are of colour blew,with some whitenesse here and there intermixt.‡

¶ *The Place and Time.*

They require(saith *Theophrastus*)a sandy and bad soile : they hardly come vp in tilled places,being of their owne nature wilde : they grow in my garden,and in other mens gardens about London. They are planted in Aprill, and bring forth their fruit at two or three sundrie times,as though it did floure often,and bring forth many crops:the first in May,the second in Iuly,the last in September,but it seldome commeth to ripenesse.

¶ *The Names.*

This pulse is named in Greeke θέρμος ἥμερος : in Latine,*Lupinus*,and *Lupinus satiuus* : in high Dutch, **Feigbonen** : in Italian,*Lupino domestico* . in Spanish,*Entramocos*: in the Brabanders language,**Uijch boonen**,and **Lupinen** : in French, *Lupins* : in English,Garden Lupine, tame Lupine, and of some after the German name Fig-beane.

¶ *The Temperature and Vertues.*

A The seed of the garden Lupine is πολύχρηστον, that is to say, much and often vsed,as *Galen* saith in his books of the Faculties of Nourishments:for the same being boiled and afterwards steeped in faire water,vntill such time as it doth altogether lose his naturall bitternes,and lastly being seasoned with a reasonable quantitie of salt,it is eaten with pickle. The Lupine is of an hard and earthy substance,wherefore it is necessarily of hard digestion, and containeth in it a thicke iuice; of which being not perfectly concocted in the veines,is ingendred a bloud or iuice which is properly called crude,or raw:but when it hath lost all his bitternes by preparing or dressing of it(as aforesaid)it is like τοῖς ἀποίοις,that is to say,to such things as are without relish,which is perceiued by the taste;& being so prepared, it is,as *Galen* writeth in his books of the Faculties of simple medicines, one of the emplaistickes or clammers.

B But whilest the naturall bitternesse doth as yet remaine, it hath power to clense and to consume or waste away; it killeth wormes in the belly,being both applied in manner of an ointment and giuen with hony to licke on,and also drunke with water and vineger.

C Moreouer,the decoction thereof inwardly taken,voideth the wormes; and likewise if it be sundry times outwardly vsed as a bath, it is a remedy against the morphew,sore heads,the small Pox, wilde scabs,gangrenes,venomous vlcers,partly by clensing, and partly by consuming and drying without biting;being taken with Rew and Pepper,that it may be the pleasanter, it scoureth the liuer and milt.

D It bringeth downe the menses, and expelleth the dead childe if it be layed to with myrrh and honie.

E Moreouer,the meale of Lupines doth waste or consume away without any biting qualitie,for it doth not onely take away blacke and blew spots that come of dry beatings,but also it cureth *Choeradas*,and *Phymata* : but then it is to be boiled either in vineger or oxymell, or else in water and vineger,and that according to the temperature of the grieued parties,and the diuersities of the diseases, *Quod ex vsu est eligendo*: and it also taketh away blew marks,and what thing soeuer else we haue said the decoction could do,all the same doth the meale likewise performe.

F These Lupines,as *Dioscorides* doth furthermore write, being boiled in raine water till they yeeld a certaine creame,are good to clense and beautifie the face.

G They cure the scabs in sheepe with the root of blacke Chameleon Thistle, if they be washed with the warme decoction.

The

The root boiled with water and drunke, prouoketh vrine.

The Lupines being made ſweet and pleaſant, mixed with vineger and drunk, take away the loth-ſomneſſe of the ſtomacke, and cauſe a good appetite to meat. H I

Lupines boiled in that ſtrong leigh which Barbars do vſe, and ſome Wormwood, Centorie, and bay ſalt added thereto, ſtay the running and ſpreading of a *Gangræna*, and thoſe parts that are depriued of their nouriſhment and begin to mortifie, and ſtaieth the ambulatiue nature of running and ſpreading vlcers, being applied thereto very hot, with ſtuphes of cloth or tow. K

CHAP. 510. *Of Peaſon.*

¶ *The Kindes.*

THere be diuers ſorts of Peaſon, differing very notably in many reſpects; ſome of the garden, and others of the field, and yet both counted tame: ſome with tough skinnes or membranes in the cods, and others haue none at all, whoſe cods are to be eaten with the Peaſe when they be young as thoſe of the young Kidney Beane: others carrying their fruit in the tops of the branches, are eſteemed and taken for Scottiſh Peaſon, which is not very common. There be diuers ſorts growing wild, as ſhall be declared.

1 *Piſum maius.*
Rownciuall Peaſe.

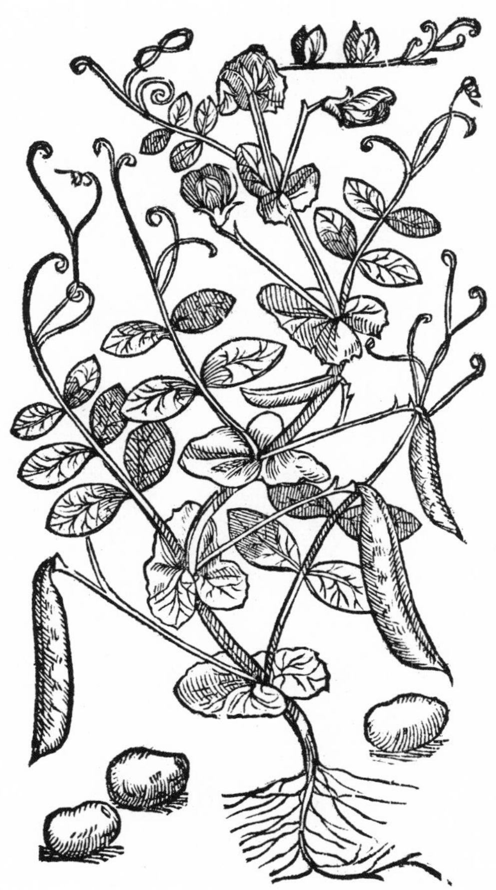

2 *Piſum minus.*
Garden and field Peaſe.

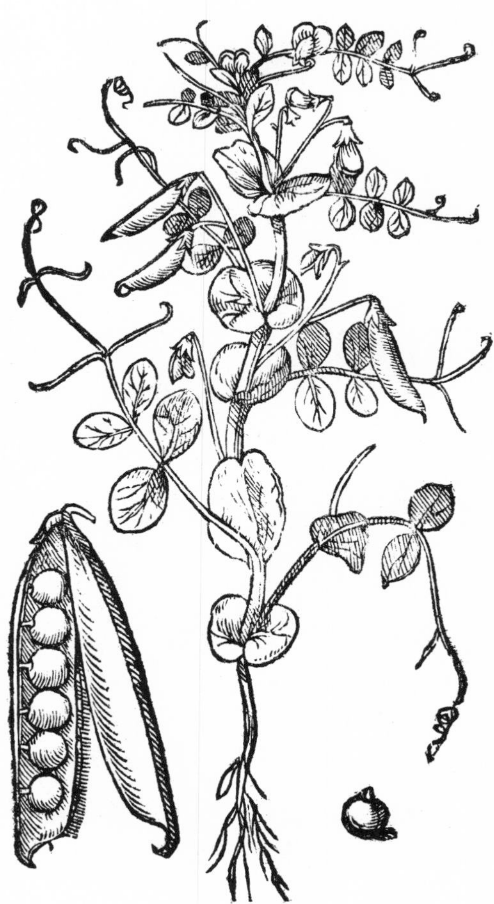

¶ *The Deſcription.*

1 THe great Peaſe hath long ſtalks, hollow, brickle, of a whitiſh green colour, branched, and ſpread vpon the ground, vnleſſe they be held vp with proppes ſet neere vnto them: the leafe thereof is wide and long, made vp of many little leaues which be ſmooth, white, growing vpon one little ſtalke or ſtem, and ſet one right againſt another: it hath alſo in the vpper part long claſping tendrels, wherewith it foldeth it ſelfe vpon props and ſtaies ſtanding next vnto it:

3 *Pisum umbellatum.*
Tufted or Scottish Pease.

4 *Pisum excorticatum.*
Pease without skins in the cod.

5 *Pisum sylvestre.*
Wilde Pease.

6 *Pisum perenne sylvestre.*
Euerlasting wilde Pease.

it: the floure is white and hath about the middle of it a purple spot: the cods be long, round *Cilindriforma*: in which are contained seeds greater than *Ochri*, or little Peason, which being drie are cornered, and that vnequall, of colour sometimes white and sometimes gray: the roots are small.

2 The field Pease is so very well knowne to all, that it were a needlesse labour to spend time about the description.

3 Tufted Pease are like vnto those of the field, or of the garden in each respect, the difference consisteth onely in that, that this plant carrieth his floures and fruit in the tops of the branches in a round tuft or vmbel, contrary to all other of his kinde, which bring forth their fruit in the midst, and alongst the stalks: the root is thicke and fibrous.

4 Pease without skins, in the cods differ not from the precedent, sauing that the cods hereof want that tough skinny membrane in the same, which the hogs cannot eat by reason of the toughnesse; whereas the other may be eaten cods and all the rest, euen as Kidney beanes are: which being so dressed are exceeding delicate meat.

5 The wilde Pease differeth not from the common field Pease in stalke and leaues, sauing that this wilde kinde is somewhat lesser: the floures are of a yellow colour, and the fruit is much lesser.

6 The Pease whose root neuer dies differeth not from the wilde Pease, onely his continuing without sowing, being once sowne or planted, setteth forth the difference.

¶ *The Place.*

Pease are set and sown in gardens, as also in the fields in all places of England. The tufted Pease are in reasonable plenty in the West part of Kent, about Sennocke or Seuenock; in other places not so common.

The wilde Pease do grow in pastures and earable fields in diuers places, specially about the field belonging vnto Bishops Hatfield in Hartfordshire.

¶ *The Time.*

They be sowne in the Spring time, like as be also other pulses, which are ripe in Summer: they prosper best in warme weather, and easily take harme by cold, especially when they floure.

¶ *The Names.*

The great Pease is called in Latine *Pisum Romanum*, or *Pisum maius*: in English, Roman Pease, or the greater Pease, also garden Pease: of some, Branch Pease, French Pease, and Rounciuals. *Theophrastus* and other old Writers do call it in Greeke πίσον: in Latine also *Pisum*: in low Dutch, **Roomsche erwiten**: in French, *des Pois*. The little Pease is called of the Apothecaries euery where *Pisum*, and *Pisum minus*: it is called in English, little Pease, or the common Pease.

¶ *The Temperature and Vertues.*

The Pease, as *Hippocrates* saith, is lesse windie than Beans, but it passeth sooner through the belly. A *Galen* writeth, that Peason are in their whole substance like vnto Beanes, and be eaten after the same manner that Beans are, notwithstanding they differ from them in these two things, both because they are not so windie as be the beanes, and also for that they haue not a clensing faculty, and therefore they do more slowly descend through the belly. They haue no effectuall qualitie manifest, and are in a meane between those things which are of good and bad iuice, that nourish much and little, that be windie and without winde, as *Galen* in his booke of the Faculties of Nourishments hath written of these and of beans.

CHAP. 511. *Of the tame or Garden Ciche.*

¶ *The Description.*

Garden Ciche bringeth forth round stalks, branched and somewhat hairy, leaning on the one side: the leaues are made of many little ones growing vpon one stem or rib, and set one right against another: of which euery one is small, broad, and nicked on the edges, lesser than the leaues of wilde Germander: the floures be smal, of colour either white, or of a reddish purple: after which come vp little short cods, puffed vp as it were with winde like little bladders, in which doe lie two or at the most three seeds cornered, small towards the end, with one sharp corner, not much vnlike to a Rams head, of colour either white, or of a reddish blacke purple; in which is plainly seen the place where they begin first to sprout. The root is slender, white and long: For as *Theophrastus* saith, the Ciche taketh deepest root of all the Pulses.

¶ *The Place.*

It is sowen in Italy, Spaine and France, euery where in the fields. It is sowen in our London gardens, but not common.

Cicer ſatiuum.
Garden Ciche.

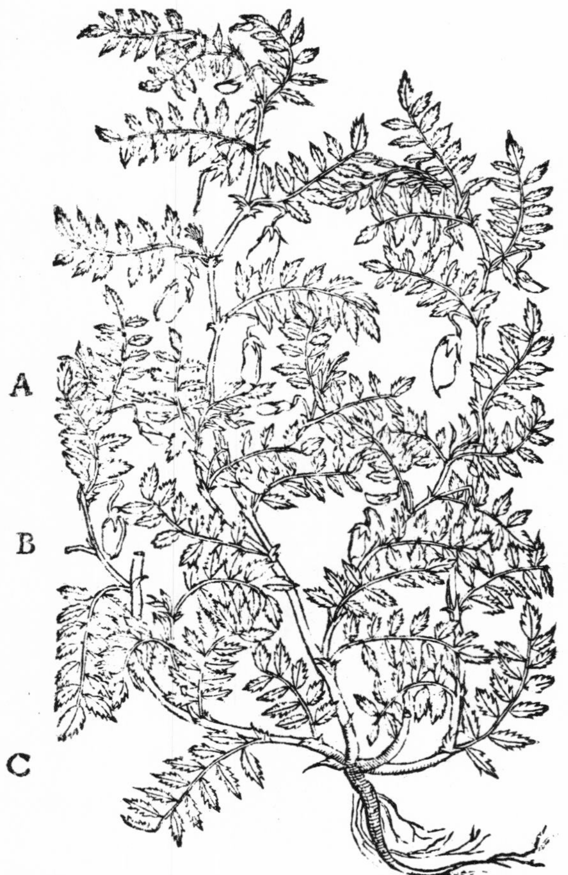

¶ *The Time.*

It is ſowne in Aprill,being firſt ſteeped in water a day before : the fruit is ripe in Auguſt.

¶ *The Names.*

It is called in Greeke ἐρέβινθος κριός: in Latine, *Cicer arietinum*,or Rams Ciches, & of the blackiſh purple colour, *Cicer nigrum* , *vel rubrum*, blacke or red Ciche: and the other is named *Candidum vel album Cicer* : or white Ciche: in Engliſh,Common Cich,or Ciches,red Cich, of ſome, Sheepes Ciche Peaſe, or Sheepes Ciche Peaſon.

¶ *The Temperature and Vertues.*

The Ciche,as *Galen* writeth in his booke of the Faculties of nouriſhments, is no leſſe windie than the true Bean,but it yeeldeth a ſtronger nouriſhment than that doth : it prouoketh luſt,and it is thought to ingender ſeed.

Some giue the ſame to ſtalion horſes.Moreouer,Ciches do ſcoure more than do the true Beanes : inſomuch as certaine of them do manifeſtly diminiſh or waſte away the ſtones in the Kidneyes : thoſe be the blacke and little Ciches called *Arietina*,or Rams Ciches, but it is better to drinke the broth of them ſodden in water.

Both the Rams Ciches, as *Dioſcorides* ſaith, the white and the blacke prouoke vrine, if the decoction therof be made with Roſemary and giuen vnto thoſe that haue either the Dropſie or yellow iaundice ; but they are hurtfull vnto the bladder and Kidneies that haue vlcers in them.

Chap. 512. *Of wilde Ciches.*

¶ *The Kindes.*

THe wilde Ciche is like to the tame(ſaith *Dioſcorides*)but it differeth in ſeed: the later writers haue ſet downe two kindes hereof,as ſhall be declared.

¶ *The Deſcription.*

1 THe firſt wilde Cich bringeth forth a great number of ſtalks branched, lying flat on the ground : about which be the leaues,conſiſting of many vpon one rib as do thoſe of the garden Ciche,but not nicked in the edges , more like to the leaues of Axcich : the floures come forth faſtned on ſmall ſtems,which grow cloſe to the ſtalks, of a pale yellow colour, and like vnto eares:in their places come vp little cods,in forme and bigneſſe of the fruit of garden Ciches,black and ſomething hairie,in which lieth the ſeed,that is ſmal,hard, flat,and glittering,in taſte like that of Kidney Beane : the root groweth deepe,faſtened with many ſtrings.

2 There is another kinde of wilde Cich that hath alſo a great number of ſtalks lying vpon the ground,about which ſtand ſoft leaues,ſomething hairy and white, conſiſting of three broad leaues ſtanding vpon a middle rib,the leaſt of which ſtand neereſt to the ſtem,and the greateſt at the very top : the floures come forth at the bottome of the leaues many together, of colour yellow ; after which grow ſmall long huskes,ſoft and hairy,in euery one whereof is a little cod, in which lie two ſeeds like little Cichlings.

¶ The

1 *Cicer ſylueſtre.*
The wilde Cich.

2 *Cicer ſylueſtre lat folium.*
Broad leafed wilde Cich.

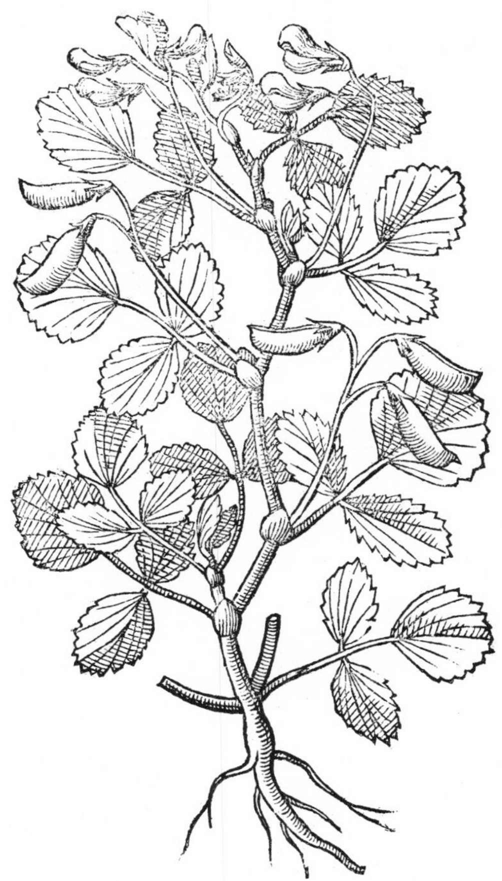

¶ *The Place.*

Theſe plants are ſowne in the parts beyond the ſeas for to feed their cattell with in winter, as we do tares, vetches, and ſuch other baſe pulſe.

¶ *The Time.*

The time anſwereth the Vetch or tare.

¶ *The Names.*

The wild Cich hath no other name in Latine but *Cicer ſylueſtre:* the later writers haue not found any name at all.

¶ *The Temperature and Vertues.*

Their temperature and vertues are referred to the garden Cich, as *Theophraſtus* affirmes; and *Galen* ſaith that the wilde Cich is in all things like vnto that of the garden, but in Phyſicks vſe more effectuall, by reaſon it is more hotter and drier, and alſo more biting and bitter.

CHAP. 513. *Of Lentils.*

¶ *The Deſcription.*

1 THe firſt Lentil growes vp with ſlender ſtalks, and leaues which be ſomwhat hard, growing aſlope from both ſides of the rib or middle ſtalke, narrow and many in number like thoſe of Tares, but narrower and leſſer: the floures be ſmall, tending ſomewhat towards a purple: the cods are little and broad: the ſeeds in theſe are in number three or foure, little, round, plaine, and flat: the roots are ſmall and threddy.

2 The ſecond kinde of Lentill hath ſmall tender and pliant branches a cubit high, wheron do grow leaues diuided or conſiſting of ſundry other ſmall leaues, like the wilde Vetch, ending at the middle rib with ſome claſping tendrels, wherewith it taketh hold of ſuch things as are neere vnto it: among theſe come forth little browniſh floures mixed with white, which turne into ſmall flat cods, containing little browne flat ſeed, and ſometimes white.

¶ *The*

1 *Lens maior.*
Great Lentils.

2 *Lens minor.*
Little Lentils.

¶ *The Place.*

Theſe Pulſes do grow in my garden; and it is reported vnto me by thoſe of good credit, that about Watford in Middleſex and other places of England the husbandmen do ſow them for their cattell, euen as others do Tares.

¶ *The Time.*

They both floure and wax ripe in Iuly and Auguſt.

¶ *The Names.*

They are called in Greeke φακὸς, or φακῆ: in Latine, *Lens*, and *Lenticula*: in high-Dutch, Linſen: in French, *Lentille*: in Italian, *Lentichia*: in Spaniſh, *Lenteia*: in Engliſh, Lentils.

¶ *The Temperature and Vertues.*

A Lentils, as *Galen* ſaith, are in a meane betweene hot and cold, yet are they dry in the ſecond degree: their skin is aſtringent or binding, and the meate or ſubſtance within is of a thicke and earthy iuyce, hauing a qualitie that is a little auſtere or ſomething harſh, much more the skin thereof; but the iuyce of them is quite contrarie to the binding qualitie; wherefore if a man ſhal boile them in faire water, and afterwards ſeaſon the water with ſalt and pickle, *aut cum ipſis oleo condiens*, and then take it, the ſame drinke doth looſe the belly.

B The firſt decoction of Lentils doth looſe the belly; but if they be boyled againe, and the firſt decoction caſt away, then doe they binde, and are good againſt the bloudy flixe or dangerous laskes.

C They do their operation more effectually in ſtopping or binding, if all or any of theſe following be boyled therewith, that is to ſay, red Beets, Myrtles, pils of Pomegranats, dried Roſes, Medlars, Seruice berries, vnripe Peares, Quinces, Plantaine leaues, Galls, or the berries of Sumach.

D The meale of Lentils mixed with honey doth mundifie and clenſe corrupt vlcers and rotten ſores, filling them with fleſh againe; and is moſt ſingular to be put into the common digeſtiues vſed among our London Surgeons for greene wounds.

E The Lentil hauing the skin or coat taken off, as it loſeth that ſtrong binding qualitie, and thoſe accidents that depend on the ſame, ſo doth it more nouriſh than if it had the skin on.

F It ingendreth thicke and naughty iuyce, and ſlowly paſſeth thorow the belly, yet doth it not ſtay the looſneſſe as that doth which hath his coat on; and therefore they that vſe to eat too much thereof

hereof do necessarily become Lepers, and are much subiect to cankers, for thicke and dry nourishments are apt to breed melancholy.

Therefore the Lentill is good food for them that through waterish humours be apt to fall into G
the dropsie, and it is a most dangerous food for dry and withered bodies; for which cause it bringeth dimnesse of sight, though the sight be perfect, through his excessiue drinesse, whereby the spirits of the sight be wasted; but it is good for them that are of a quitecontrarie constitution.

It is not good for those that want their termes; for it breedeth thicke bloud, and such as slowly H
passeth through the veines.

But it is singular good to stay the menses, as *Galen* in his booke of the faculties of nourishments I
affirmeth.

It causeth troublesome dreames (as *Dioscorides* doth moreouer write) it hurteth the head, si- K
newes, and lungs.

It is good to swallow downe thirty graines of Lentils shelled or taken from their husks, against L
the ouercasting of the stomacke.

Being boyled with parched barly meale and laid to, it asswageth the paine and ach of the gout. M

With honey it filleth vp hollow sores, it breaketh aschares, clenseth vlcers: being boyled in
wine it wasteth away wens and hard swellings of the throat. N

With a Quince, and Melilot, and oyle of Roses it helpeth the inflammation of the eyes and fun- O
dament; but in greater inflammations of the fundament, and great deep vlcers, it is boyled with the rinde of a pomegranat, dry Rose leaues, and honey.

And after the same maner against eating sores that are mortified, if sea water be added; it is also P
a remedie against pushes, the shingles, and the hot inflammation called S Anthonies fire, and for kibes, in such manner as we haue written: being boyled in sea water and applied, it helps womens brests in which the milke is cluttered, and cannot suffer too great aboundance of milke.

Chap. 514. *Of Cich or true Orobus.*

Orobus receptus Herbariorum.
The true Orobus.

¶ *The Description.*

This Pulse, which of most Herbarists is taken for the true Orobus, and called of some, bitter Fitch, is one of the Pulses whose tender branches traile vpon the ground, as *Theophrastus* saith, and whose long tender branches spred far abroad, whereon doe grow leaues like those of the field Vetch: among which grow white floures; after which come long cods, that appeare bunched on the outside against the place where the seeds do lie, which are small, round, russet of colour, and of a bitter taste: the root is small and single.

¶ *The Place.*

It prospereth best in a leane soile, according to *Columella*: it groweth in woods and copses in sundry places of Spaine and Italy, but here only in gardens.

¶ *The Time.*

This is sowne early and late, but if it be sowne in the spring it easily commeth vp, and is pleasant, and vnpleasant if it be sowne in the fall of the leafe.

¶ *The Names.*

This is called in Greeke ὄροβος the shops of Germanie haue kept the name *Orobus*: the Italians cal it *Macho*: the Spaniards, *Yeruo*, and *Yeruos*: in English it is called bitter Vetch, or bitter Fitch, and Orobus, after the Latine name. Of some Ers, after the French name.

¶ *The*

¶ *The Temperature and Vertues.*

A *Galen* in his first booke of the Faculties of nourishments saith, That men do altogether abstaine from the bitter Vetch, for it hath a very vnpleasant taste, and naughty iuyce, but Kine in Asia and in most other countries do eate thereof, being made sweet by steeping in water; notwithstanding men being compelled through necessitie of great famine, as *Hippocrates* also hath written, do oftentimes feed thereof; and we also dressing them after the manner of Lupines, vse the bitter Vetches with honey, as a medicine that purgeth thicke and grosse humors out of the chest and lungs.

B Moreouer, among the bitter Vetches the white are not so medicinable, but those which are neere to a yellow, or to the colour of Okar; and those that haue beene twice boyled, or sundrie times soked in water, lose their bitter and vnpleasant taste, and withall their clensing and cutting qualitie, so that there is onely left in them an earthy substance, which serues for nourishment, that drieth without any manifest bitternesse.

C And in his booke of the Faculties of simple medicines he saith, That bitter Vetch is dry in the later end of the second degree, and hot in the first: moreouer, by how much it is bitter, by so much it clenseth, cutteth, and remoueth stoppings: but if it be ouermuch vsed it bringeth forth bloud by vrine.

D *Dioscorides* writeth, that bitter Vetch causeth head-ache and heauy dulnesse, that it troubles the belly, and driueth forth bloud by vrine; notwithstanding being boyled it serueth to fatten Kine.

E There is made of the seed a meale fit to be vsed in medicine, after this maner: the full and white graines are chosen out, and being mixed together they are steeped in water, and suffered to lie till they be plumpe, and afterwards are parched till the skinne be broken; then are they ground, and searsed or shaken thorow a meale sieue, and the meale reserued.

F This looseth the belly, prouoketh vrine, maketh one well coloured: being ouermuch eaten or drunke it draweth bloud by the stoole, with gripings, and also by vrine.

G With honey it clenseth vlcers, taketh away freckles, sun-burnes, blacke spots in the skinne, and maketh the whole body faire and cleane.

H It stayeth running vlcers or hard swellings, and gangrens or mortified sores; it softeneth the hardnesse of womens breasts, it taketh away and breaketh eating vlcers, carbuncles, and sores of the head: being tempered with wine and applied it healeth the bitings of dogs, and also of venomous beasts.

I With vineger it is good against the stranguric, and mitigateth paine that commeth thereof.

K It is good for them that are not nourished after their meat, being parched and taken with hony in the quantitie of a nut.

L The decoction of the same helpeth the itch in the whole body, and taketh away kibes, if they be washed or bathed therewith.

M *Cicer* boyled in fountaine water with some *Orobus* doth asswage the swelling of the yard and priuie parts of man or woman, if they be washed or bathed in the decoction thereof; and the substance hereof may also be applied plaisterwise.

N It is also vsed for bathing and washing of vlcers and running sores, and is applied vnto the scurfe of the head with great profit.

Chap. 515. *Of the Vetch or Fetch.*

¶ *The Description.*

1 The Vetch hath slender and foure squared stalkes almost three foot long: the leaues be long, with clasping tendrels at the end made vp of many little leaues growing vpon one rib or middle stem; euery one whereof is greater, broader, and thicker than that of the Lentil: the floures are like to the floures of the garden beane, but of a blacke purple colour: the cods be broad, small, and in euery one are contained fiue or six graines, not round, but flat like those of the Lentil, of colour blacke, and of an vnpleasant taste.

‡ 2 There is another of this kinde which hath a creeping and liuing root, from which it sendeth forth crested stalkes some cubit and halfe high: the leaues are winged, commonly a dozen growing vpon one rib, which ends in a winding tendrel: each peculiar leafe is broader toward the bottome, and sharper towards the top, which ends not flat, but somewhat round. Out of the bosomes of the leaues towards the tops of the stalkes, on short foot-stalkes grow two, three or more pretty large pease-fashioned blewish purple floures, which are succeeded by such cods as the former, but somewhat lesser; which when they grow ripe become blacke, and fly open of themselues, and

‡ 1 *Vicia.*
Tare, Vetch, or Fetch.

† 2 *Vicia maxima dumetorum.*
Buſh Vetch.

‡ 3 *Vicia ſyl. flo. albo.*
White floured Vetch.

4 *Vicia ſylueſtris, ſiue Cracca maior.*
Strangle Tare, Tine, or wilde Fetch.

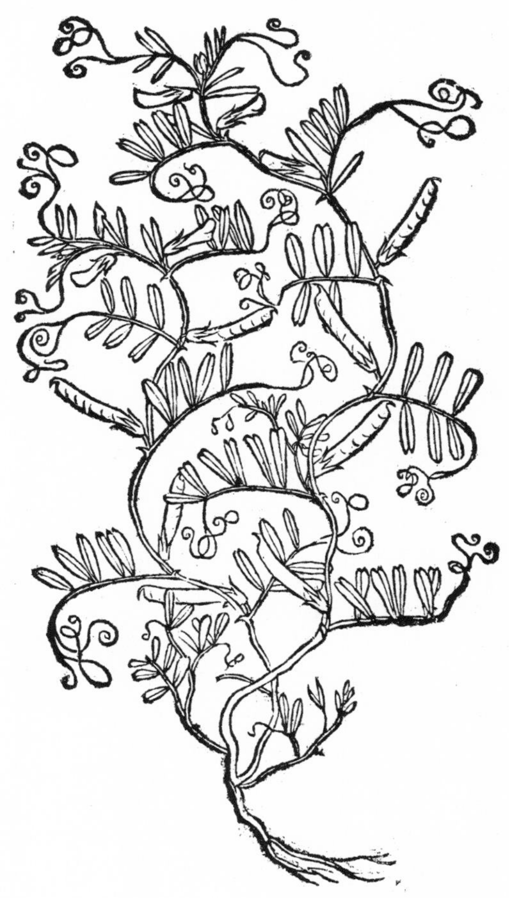

and ſo ſcatter their ſeed. This growes in many places wilde among buſhes, both here and in Germanie, as appeares by that name *Bauhine* thence giues it, calling it *Vicia maxima dumetorum*. *Tragus* makes it his *Vicia ſyl. altera*; and iudges it to be the *Aphace* of *Dioſcorides*; and he ſaith the Latines call it *Os mundi*: the high-Dutch, **S. Criſtoffels kraut**, and **Schwartz Linſen**. *Tabern.* calls it *Cracca maior*.

‡ 5 *Vicia ſyl ſiue Cracca minima.*
Small wilde Tare.

3 This alſo hath a laſting root, which ſendeth vp round creſted branches, a foot and ſomtimes a cubit high, whereon grow ſuch leaues as thoſe of the former, but more white and downie: the floures, which grow on ſhort foot-ſtalkes, out of the boſomes of the leaues, towards the top of the ſtalks, are of a whitiſh colour, with veines of a dusky colour, diuaricated ouer the vpper leafe: the cods are like thoſe of the common Fetch. *Cluſius* found this in ſome wilde places of Hungarie; it floured in May: he calls it *Vicia ſylueſtris albo flore*. ‡

4 Strangle Tare, called in ſome countries Tine, and of others wilde Vetch, is a ramping herbe like vnto the common Tare, ramping and clymbing among corne where it chanceth, that it plucketh it downe to the ground, and ouergroweth the ſame in ſuch ſort, that it ſpoileth and killeth not only wheat, but all other graine whatſoeuer: the herbe is better known than deſired, therefore theſe few lines ſhall ſuffice for the deſcription. ‡ This groweth pretty long, with many ſlender weake branches: the leaues are much ſmaller than the former, and end in claſping tendrels: the floures are of a purple colour, and commonly grow but one at a ioint, and they are ſucceeded by flat ſharpe pointed cods which containe ſome nine or ten ſeeds apiece.

5 This alſo growes a good height, with ſlenderer ſtalks than the former, which is diuided into ſundry branches: the leaues grow foure or ſix vpon foot-ſtalkes, ending alſo in claſping tendrels: the floures grow vpon pretty long but very ſlender foot-ſtalkes, ſometimes two or three, otherwhiles more, very ſmall, and of a whitiſh colour inclining to blewneſſe: which are ſucceeded by little ſhort flat cods, containing commonly foure or fiue little ſeeds of a blackiſh colour: this is the *Arachus, ſiue Cracca minima* of *Lobel*; but I queſtion whether it be that which *Bauhine* in his *Pinax* hath made the ſame with it, calling it *Vicia ſegetum cum ſiliquis plurimis hirſutis*: for that which I haue deſcribed, and which exactly agrees with this figure of *Lobel* and that deſcription in the *Aduerſ.* hath cods very ſmooth without any haireneſſe at all. This floures moſt part of Sommer, and growes in moſt places both in corne fields and medowes. ‡

¶ *The Place.*

The Tare is ſowne in any ground or ſoile whatſoeuer.

¶ *The Time.*

It floureth in May, and perfecteth his ſeed toward September.

¶ *The Names.*

It is called in Latine *Vicia, à vinciendo*, of binding or wrapping, as *Varro* noteth, becauſe, ſaith he, it hath likewiſe claſping tendrels ſuch as the vine hath, by which it crawles vpward vpon the ſtalks of the weeds which are next vnto it: of ſome, *Cracca*, and *Arachus*, and alſo *Aphaca*: it is called in high-Dutch, **Wicken**: in low-Dutch, **Uitſen**: in French, *Veſce*: in moſt ſhops it is falſely termed Ὄροβος, and *Eruum*; for *Eruum* doth much differ from *Vicia*: it is called in Engliſh, Vetch, or Fetch. The countrey men lay vp this Vetch with the ſeeds and whole plant, that it may be a fodder for their cattell.

¶ *The Temperature and Vertues.*

A Notwithſtanding I haue knowne, ſaith *Galen*, ſome, who in time of famin haue fed hereof, eſpecially in the ſpring, it being but greene; yet is it hard of digeſtion, and bindeth the belly.

Therefore

Therefore ſeeing it is of this kinde of nature, it is manifeſt that the nouriſhment which comes thereof hath in it no good iuyce at all, but ingendreth a thicke bloud, and apt to become melancholy.

† The figure of the common Fetch was formerly wanting, and in the ſtead thereof was put that of the other, deſcribed here in the ſecond place.

CHAP. 516. *Of Chichlings, Peaſe, and Tare euerlaſting.*

¶ *The Deſcription.*

1 THere is a Pulſe growing in our high and thicke woods, hauing a very thicke tough and wooddy root; from which riſe vp diuers long weake and feeble branches, conſiſting of a tough middle rib, edged on both ſides with a thin skinny membrane, ſmooth, and of a graſſe greene colour; whereon do grow at certaine diſtances ſmall flat ſtems, vpon which ſtand two broad leaues ioyning together at the bottome: from betwixt thoſe leaues come forth tough claſping tendrels which take hold of ſuch things as grow next vnto them: from the boſome of the ſtem, whereon the leaues do grow, ſhooteth forth a naked ſmooth foot-ſtalke, on which doe grow moſt beautifull floures like thoſe of the Peaſe, the middle part whereof is of a light red, tending to a red Purple in graine; the outward leaues are ſomewhat lighter, inclining to a bluſh colour: which being paſt, there ſucceed long round cods, wherein is contained ſeed of the bigneſſe of a Tare, but rounder, blackiſh without, and yellowiſh within, and of a bitter taſte.

‡ 5 *Lathyrus maior latifolius.*
Peaſe euerlaſting.

‡ 2 *Lathyrus anguſtifolius flore albo.*
White floured Chichelings.

† 2 Of which kinde there is likewiſe another like vnto the precedent in each reſpect, ſauing that the leaues hereof are narrower and longer, and therefore called of moſt which ſet forth the deſcription, *Lathyrus anguſtifolia*: the floures of this are white, and ſuch alſo is the colour of the fruit: the root is ſmall, and not laſting like that of the former.

‡ 3 The ſtalks, leaues, and floures of this are like thoſe of the precedent, but the floures are of a reddiſh purple colour: the cods are leſſer than thoſe of the former, and in them are contained

leſſer, harder, and rounder ſeeds, of a darke or blackiſh colour. This growes not wilde with vs, but is ſometimes ſowne in gardens, where it floures in Iune and Iuly.

4 This Egyptian differs not in ſhape from the reſt of his kinde, but the floures are of an elegant blew on the inſide, but of an aſh colour inclining to purple on the outſide: the cods grow vpon long foot-ſtalkes, and are a little winged or welted, and containe but two or three little cornered ſeeds ſpotted with blacke ſpots. This floures in Iune and Iuly; and the ſeed thereof was ſent to *Cluſius* from Conſtantinople, hauing been brought thither out of Egypt.

‡ 3 *Lathyrus anguſtifol. flo. purp.*
Purple floured Chichelings.

‡ 4 *Lathyrus Ægyptiacus.*
Egyptian Chichelings.

5 The ſtalkes of this are ſome two or three foot long, winged, weake, and lying on the ground vnleſſe they haue ſomewhat to ſupport them. Vpon theſe at certaine diſtances grow winged leaues with two little eares at their ſetting on to the ſtalke: theſe leaues conſiſt of ſix long and narrow greene leaues like thoſe of the other plants of this kinde; and theſe ſix leaues commonly ſtand vpright, by couples one againſt another; otherwhiles alternately: the footſtalke whereon theſe ſtand ends in claſping tendrels: the floures are in ſhape like the former, but the outer leafe is of a faire red or crimſon colour, and the inner leafe white: after the floures come the cods, containing ſome foure or fiue pretty large flat ſeeds, which ſwell out of the cods where they lie, which in the ſpaces betweene each ſeed are depreſt, like that of *Orobus*. This is only a garden plant with vs, and floures in Iune and Iuly, the ſeed is ripe in Auguſt. I haue for this giuen you *Lobels* figure of his *Lathyris anguſtiore graminco folio*; which may ſerue, if you but make the leaues and cods to agree with this deſcription. ‡

6 The yellow wilde Tare or Fetch hath diuers very ſmall ramping ſtalkes, tough, and leaning this way and that way, not able to ſtand of it ſelfe without the helpe of props or things that ſtand by it: the leaues are very thin and ſharpe pointed: the floures grow alongſt the leaues in faſhion of the peaſe floures, of a bright yellow colour: the roots are very ſmall, long, tough, and in number infinite, inſomuch that it is impoſſible to root it forth, being once gotten into the ground, vnleſſe the earth be digged vp with the roots, and both caſt into the riuer, or burned. Doubtleſſe it is the moſt pernicious and hurtfull weed of all others, vnto all manner of greene wholſome herbes or any wood whatſoeuer.

¶ The

‡ 5 *Lathyrus annuus siliquis Orobi.*
Party coloured Cicheling.

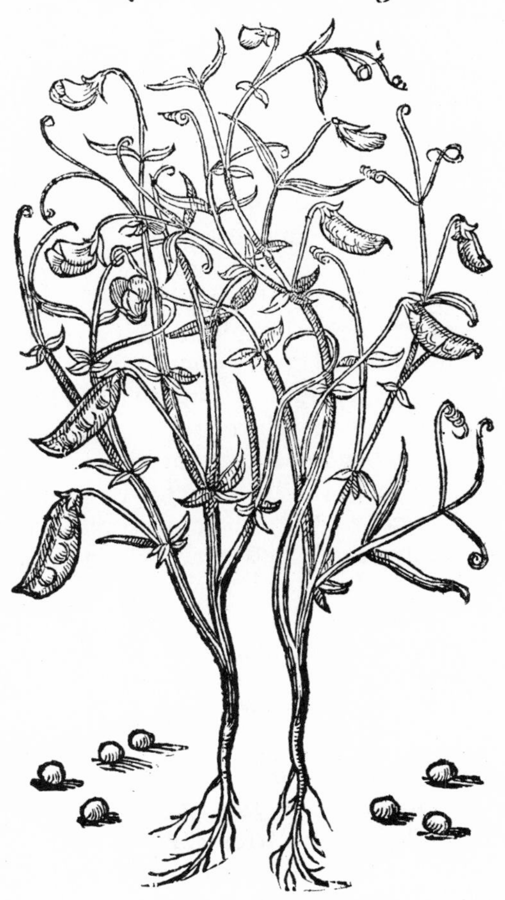

‡ 6 *Lathyrus sylvestris flo. luteo.*
Tare euerlasting.

¶ *The Place.*

The first growes in shadowie woods, and among bushes: there groweth great store thereof in Swainscombe wood, a mile and a halfe from Greenhithe in Kent, as you go to a village thereby called Betsome, and in diuers other places.

The sixth groweth in most grassie pastures, borders of fields, and among graine almost euerie where.

¶ *The Time.*

The time answereth the other Pulses.

¶ *The Names.*

The first is called *Lathyrus*, to make a difference betweene it and *Lathyris*, or Spurge: of *Matthiolus*, *Clymenum*: of *Cordus*, *Eruum satiuum*: of *Tragus*, *Pisum Græcorum*: in English, Pease euerlasting, great wilde Tare, and Cichling.

‡ The second is the *Evum album satiuum* of *Fuchsius*: *Lathyrus* or *Cicercula* of *Dodonæus*: *Lathyrus angustiore gramineo folio* of *Lobel*.

The third is the *Aracus siue Cicera* of *Dodonæus*: the *Lathyrus flore purpureo* of *Camerarius*.

The fourth by *Clusius* is called *Cicercula Ægyptiaca*: by *Camerarius*, *Aracus Hispanicus, siue Lathyrus Ægyptiacus*.

The fifth is not mentioned by any (that I remember) but M^r^. *Parkinson*, in his garden of floures, and that by the name I giue you it.

The sixth is the *Lathyrus sylvestris flo. luteis* of *Thalius*: *Legumen terræ glandibus simile* of *Dodonæus*: *Vicia* of *Tabernamontanus*: and it may be, the *Aracus flore luteo* of the *Aduers*. Howeuer, I haue put *Lobels* figure of *Aracus* for it, which well enough agrees with it. I vse for some resemblance it hath to *Aphaca* to call it *Aphacoides*. ‡

¶ *The Temperature and Vertues.*

The temperature and vertues are referred to the manured Tare or Vetch; notwithstanding they are not vsed for meate or medicine.

Chap. 517.
Of the oylie Pulse called Sesamum.

Sesamum, sive Sisamum.
The Oylie Graine.

¶ *The Description.*

† Sesamum hath a thicke and fat vpright stalke a cubit and a halfe high, garnished with leaues much like the Peach or Almond, but rougher, and cut in with somwhat deepe gashes on their sides: amongst these leaues come forth large white or else red floures, somewhat shaped like those of Foxgloues, which turne into round long crested cods, containing white flat oileous seed. *Theophrastus* affirmeth that there is a kinde thereof which is white, bearing only one root. No kinde of beast will eate this plant while it is greene, because of his bitternesse; but being withered and dried, the seed thereof becommeth sweet, and the cattell will feed on the whole plant.

¶ *The Place.*

It groweth both in Egypt and in India: *Sesama*, saith *Pliny*, came from the Indies; they make an oile of it. It is a stranger in England.

¶ *The Time.*

It is one of the Sommer grains, and is sowne before the rising of the seuen starres, as *Pliny* writeth; yet *Columella* saith, that *Sesamum* must be sowne after the Autumne Æquinoctial, against the Ides of October: they require for the most part a rotten soile, which the husbandmen of Campania do call a blacke mold.

¶ *The Names.*

The Grecians cal this grain Σήσαμον: the Latines also *Sesamum*, and *Sisamum*, and often in the fœminine gender *Sesama*: we are constrained for want of an English name to vse the Latine: it is vnknowne to the Apothecaries, especially the plant it selfe; but the seed and oyle thereof is to be found among them in other countries: we may call it Turky Millet.

¶ *The Temperature and Vertues.*

A According to some it is hot and dry in the first degree: the seed thereof, as *Galen* saith, is fat, and therefore being layd vp it commeth to be oylie very quickly; wherfore it speedily filleth and stuffeth vp those that feed thereof, and ouerthroweth the stomacke, and is slow of digestion, and yeeldeth to the body a fat nourishment: therefore it is manifest that it cannot strengthen the stomack, or any part thereof, as also no other kind of fat thing: and the iuice that commeth thereof is thick, and therefore it cannot speedily passe thorow the veines. Men do not greedily feed of it alone, but make cakes thereof with honey, which they call Σησαμίδας: it is also mixed with bread, and is of an hot temperature, for which cause it procureth thirst: and in his booke of the faculties of simple Medicines he saith, that *Sesamum* is not a little clammy and fat, and therefore it is an emplasticke, and a softner, and is moderately hot: the oile which commeth thereof is of like temperature, and so is the decoction of the herbe also.

B *Dioscorides* writeth, That *Sesamum* is an enemie to the stomacke, it causeth a stinking breath, if it remaine sticking betweene the teeth after it is chewed.

C It wasteth away grossenes of the sinewes; it is a remedie against bruises of the eares, inflammations, burnings and scaldings, pains of the ioynts, and biting of the poysonsom horned serpent called *Cerastes*. Being mixed with oile of Roses it takes away the head-ache which commeth of heate.

D Of the same force is the herbe boyled in wine, but it is especially good for the heate and paine of the eyes.

E Of the herb is made an oile vsed of the Egyptians, which as *Pliny* saith is good for the eares.

F It is a remedie against the sounding and ringing of the eares.

Chap. 518. *Of Hatchet Fetch.*

¶ *The Deſcription.*

1 THe firſt kinde of Hatchet Fetch hath many ſmall branches trailing here and there vpon the ground: vpon which grow ſmall leaues ſpred abroad like the leaues of the wilde Fetch; among which come forth cluſters of ſmall yellow floures, which fade away, and turne into little flat thin and browne cods, wherein is contained ſmall reddiſh ſeed of a bitter taſte.

2 The ſecond kinde of hatchet Fetch hath many round tough and flexible branches, trailing vpon the ground: whereupon do grow leaues like the former, but more like the leaues of Liquorice, and hauing the taſte of the Liquorice root; which hath giuen occaſion to ſome to deeme it a kinde of Liquorice: among theſe leaues come forth pale yellow floures, after which there ſucceed ſmall crooked cods, turning their points inwardly, one anſwering another like little hornes, containing ſmall flat ſeeds foure cornered, and faſhioned like a little wedge: the root is tough, of a woody ſubſtance, and doth continue fruitfull a very long time.

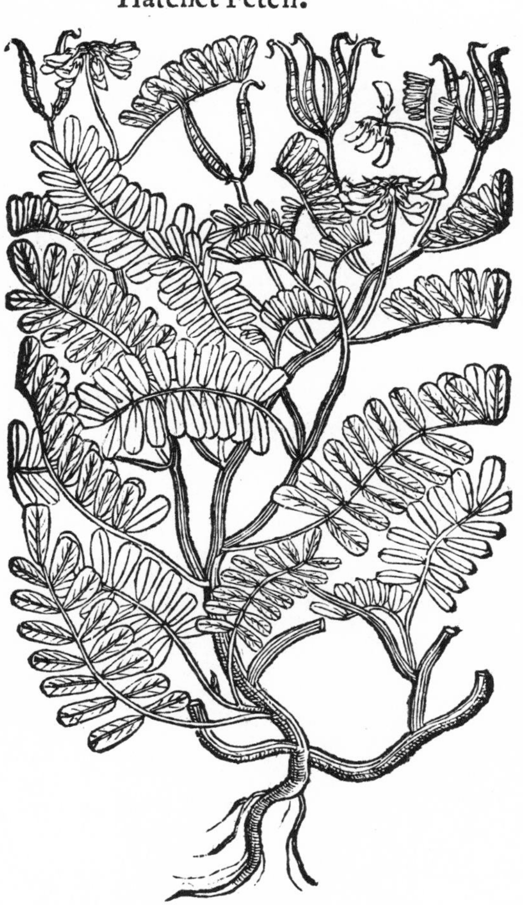

‡ 1 *Hedyſarum maius.*
Hatchet Fetch.

‡ 2 *Hedyſarum Glycyrhizatum.*
Liquorice hatchet Fetch.

3 There is another kind of *Securidaca* or hatchet Fetch, which hath branches, leaues, and roots like the laſt before remembred, and differeth in that, that the floures of this plant are mixed, and do vary into ſundry colours, being on the vpper part of a fleſh colour, and on the lower of a white or ſnowie colour, with a purple Storks bill in the middle: the leaues are in taſte bitter: the cods are ſmall like thoſe of Birds foot, and not much vnlike the cods of *Orobus*.

4 There is likewiſe another kinde of *Securidaca* or hatchet Fetch, which is dedicated vnto *Carolus Cluſius* by the aforenamed Dr. *Penny*, who found it in the North parts of England, hauing leaues, roots, and branches like vnto the former: but the floures of this are white, and mixed with ſome purple, and bitter alſo in taſte: the cods are like the claw of a crab, or (as *Cluſius* ſaith) like the knife which ſhoo-makers do vſe in Flanders; in which cods are contained ſmall reddiſh ſeed: this root alſo is of long continuance. ‡ *Cluſius* doth not ſay that Dr. *Penny* found this in the North of England, but in the territorie of Geneua not far from Pontetremile, amongſt the buſhes, and no where elſe. ‡

‡ 3 *Hedysarum maius siliquis articulatis.*
Hatchet Fetch with ioynted cods.

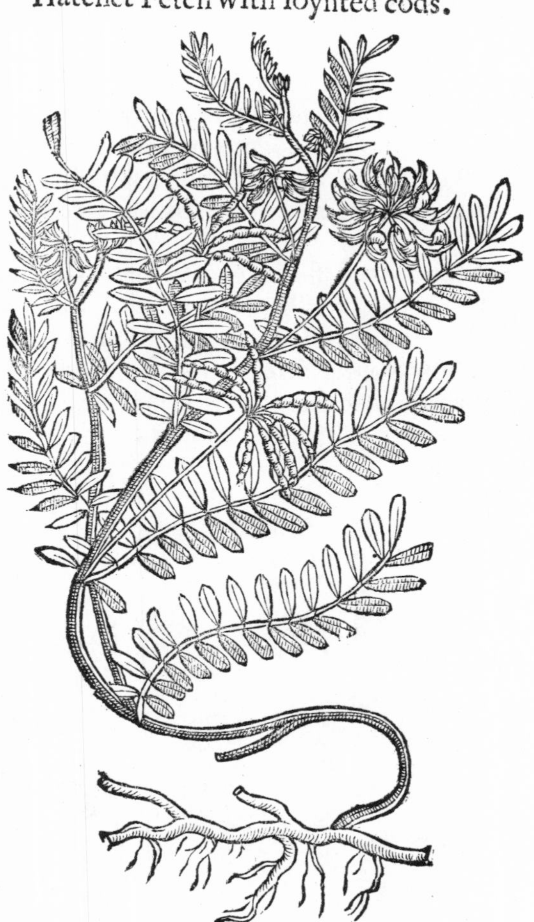

‡ 4 *Securidaca minor pallide cærulea.*
Small blew floured hatchet Fetch.

‡ 5 *Securidaca minor lutea.*
Small yellow hatchet Fetch.

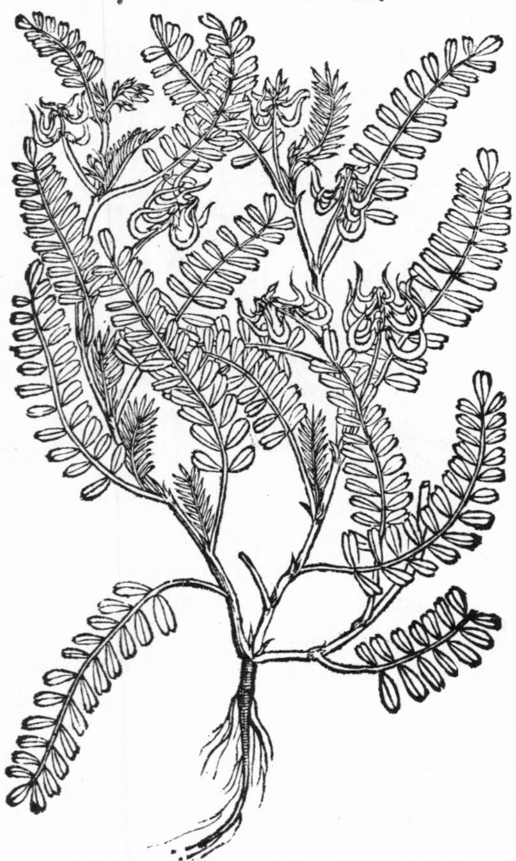

‡ 6 *Securidaca siliquis planis dentatis.*
Indented hatchet Fetch.

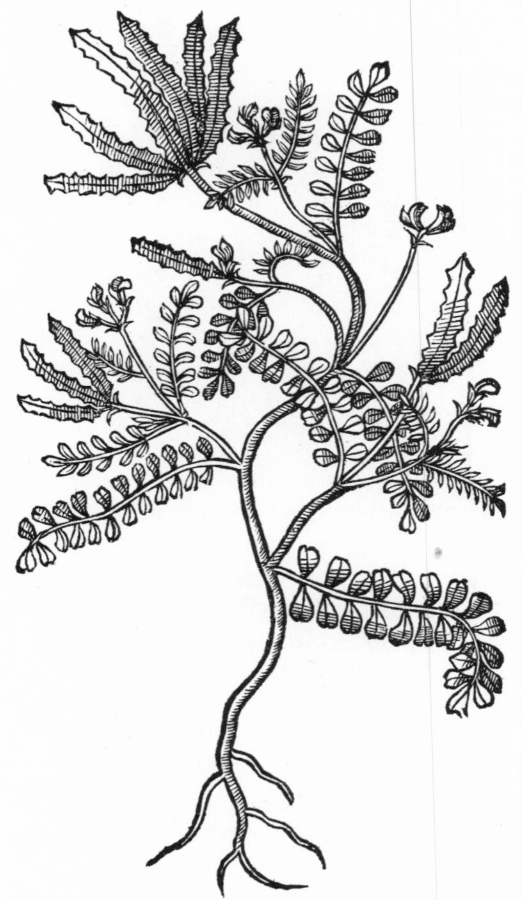

‡ 5 This in the ſtalks,leaues,colour, and ſhape of the floures is like,yet leſſe than the firſt deſcribed ; the cods are alſo ſmaller,leſſer,and more crooked : and herein onely conſiſts the chiefe difference,it is an annuall plant,and grows onely in ſome gardens. *Matthiolus,Lobel,Dodonæus*,and other,make this their *Hedyſarum*,or *Securidaca minor* .

6 This hath many creſted branches,whereon great winged leaues,that is,ſome twentie or more faſtened to one rib : the floures are like thoſe of the other plants of this kinde ; but the cods are of an inch long,flat,or indented or toothed on their ſides. But of what colour the floures and ſeeds are of it is not expreſt by *Cluſius*, who onely ſet this forth by a picture, and ſome pieces of a dried plant thereof,which he receiued from *Cortuſus*,by the name of *Scolopendria leguminoſa*, or *Hedyſarum peregrinum : Cortuſus* had it from *Honorius Bellus*,who obſerued it growing vpon the Rocks at Sebeṡico,a citie of Illyria. ‡

7 There is alſo another ſort of Hatchet Fetch,which hath very long and tough branches trailing vpon the ground,beſet with leaues like the former, but much greater : the floures do grow at the top of the branches,of a pale colour, and turne into rough round and flat cods, faſhioned like little bucklers : the root of this (as of the firſt) dieth at the firſt approch of winter, as ſoone as the ſeed is ripe : ‡ The ſtalks of this are ſtiffe and creſted, growing to the height of two cubits, with leaues as large as thoſe of Liquorice : the floures are of a faire bright red colour : the cods are made as it were of many rough buckler-like ſeeds,or rather ſeed veſſels wherin are contained ſmal brown ſeeds. ‡

‡ 7 *Hedyſarum Clypeatum.*
French Honyſuckle.

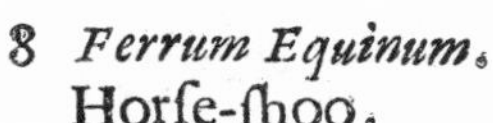

8 *Ferrum Equinum.*
Horſe-ſhoo.

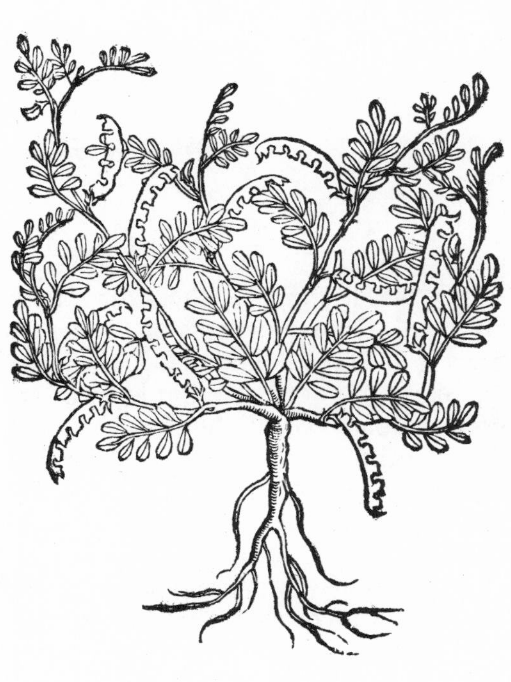

8 Horſe-ſhoo hath many ſtalks ſlender and lying vpon the ground : the leaues be thinne,and leſſer than thoſe of Axſeed : the floures along the ſtalks are little : after which come vp long cods ſomething broad,and a little bowing ; which haue vpon the one ſide deepe round and indented cuts,like after a ſort to an Horſe-ſhoo: the root is ſomewhat long.

¶ *The Place.*

Theſe plants do grow iu my garden: the ſecond kinde I found growing in Suffolke, in the highway on the right hand,as you goe from Sudbury to Corner Church, about an hundred paces from the end of the towne,as alſo in ſundry other places of the ſame countrey ; and in Eſſex about Dunmow,

mow,and in the townes called Clare and Hennyngham. ‡ Alſo it growes by Purfleet, about the foot of the hill whereon the Wind-mill ſtands; and in diuers parts of Kent. ‡

Horſe-ſhooe commeth vp in certaine vntilled and ſunnie places of Italy and Languedocke: it groweth likewiſe in my garden.

¶ *The Time.*

Theſe plants do floure in Iune,and their ſeed is ripe in Auguſt.

¶ *The Names.*

The Grecians name this,whether it be a pulſe or an infirmitie among corne, ἡδύσαρον: the Latines, of the forme of the ſeed, *Securidaca*, and *Hedyſarum*: in Engliſh, Axſeed, Axwoort, Ax-fitch, and Hatchet Fitch: it is vnknowne to the Apothecaries.

‡ The ſecond is the *Fœnugræcum ſylueſtre* of *Tragus* and *Dodonæus*: the *Glycyrhiza ſylueſtris* of *Geſner*;and the *Glaux vulgaris* of *Lobel*. ‡

Horſe-ſhooe is commonly called in Italian *Sferro de cauallo*: you may name it in Latine *Ferrum equinum*: in Engliſh,Horſe-ſhooe.

¶ *The Temperature.*

The ſeeds of theſe plants are hot and drie of complexion.

¶ *The Vertues.*

A Being drunke it is acceptable to the ſtomacke,and remoueth ſtoppings out of the intrailes, and of like vertue be the new leaues and tender crops of the whole plant.

B *Dioſcorides* ſheweth that it is alſo good for the ſtomacke being taken in drink,and is mixed with counterpoiſons.

C And it is thought to hinder conception, if it be applied with honie before the act.

D The ſeed of Axwoort openeth the ſtoppings of the liuer,the obſtruction of the ſpleen,and of all the inward parts.

E Horſe-ſhooe is bitter and like in nature to Axſeed.

† The figure which formerly was in the firſt place,agreed with the third deſcription: that which was in the ſecond place was of the *Hedyſarum minus*, of *Tabermontanus*,beeing a kinde of *Ferrum equinum*,which carries the cods many together on the tops of the branches,and growes in Germanie: whence *Bauhine* calls it *Ferrum equinum Germanicum ſiliquis in ſummitate*.

CHAP. 519. *Of Peaſe Earth-Nut.*

¶ *The Deſcription.*

1 THe Peaſe Earth nut commeth vp with ſlender and weake ſtalkes:the leaues be thin,and little,growing vpon ſlender ſtems,with claſping tendrels at the ends,with which it imbraceth and taketh hold of ſuch things as ſtand neere vnto it:the floures on the tops of the ſtalkes are like to thoſe of Peaſe,but leſſer,of a red purple colour, in ſmell not vnpleaſant: in their places come vp long cods,in which are three or foure round ſeeds:the roots be thicke,long,like after a ſort to acorns,but much greater,blacke without, gray within, in taſte like to the Cheſſe-nut: out of which beneath doth hang a long ſlender ſtring: there grow out of the ſame alſo other ſtrings, hard to the ſetting on of the ſtalk,vnto which creeping a ſlope do grow other kernelled roots whilſt the plant doth thus multiplie it ſelfe.

‡ 2 This with *Thalius* in his *Sylua Harcynia*, ſet forth by the name of *Aſtragalus ſyluaticus*,was by our Author taken for,and confounded with the *Terræglandes*,and therefore I haue put it with it, that the difference might the better appeare,which is not a little to ſuch as heedfully obſerue it: But our Author in this is to be pardoned,ſeeing Dr.*Turner*,a man more exquiſite in the knowledge of plants,and who had ſeene the true *Terræglandes* in Germany,miſtooke this for it, as may appeare by that little Tract of his of the names of plants in Latine and Engliſh,ſet forth *Anno*, 1548; for there he ſaith,[I haue ſeene this herbe of late in Come parke more aſtringent than it of Germany:] and indeed this growes there,and is much more aſtringent and wooddie than that of Germany,and no wiſe fit to be eaten. The root conſiſts of many blacke tuberous particles,here and there ſending forth fibers:from hence ariſe cornered ſtalks ſome foot high,ſmal below,& ſomwhat larger aboue: the leaues grow forth of the ſtalks,conſiſting ſometimes of two,& otherwhiles of 4. longiſh narrow leaues faſtned to one footſtalk, which at the ſetting on hath two little leaues or eares: forth of the boſomes of theſe leaues grow ſtalks ſome two inches long,each of which vſually carry a couple of Peaſe-faſhioned floures of a purple colour:which fading, vſually become blew: after theſe follow cods,ſtraight,roundand, blacke; and in each of them are commonly contained nine or ten white round

round seeds: it floures most part of Summer, and perfects the seed in Iuly and August. ‡

1 *Terræglandes.*
Pease Earth-nut.

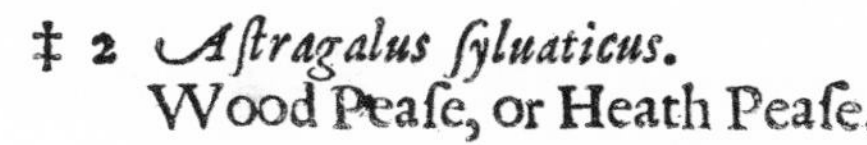
‡ 2 *Astragalus syluaticus.*
Wood Pease, or Heath Pease.

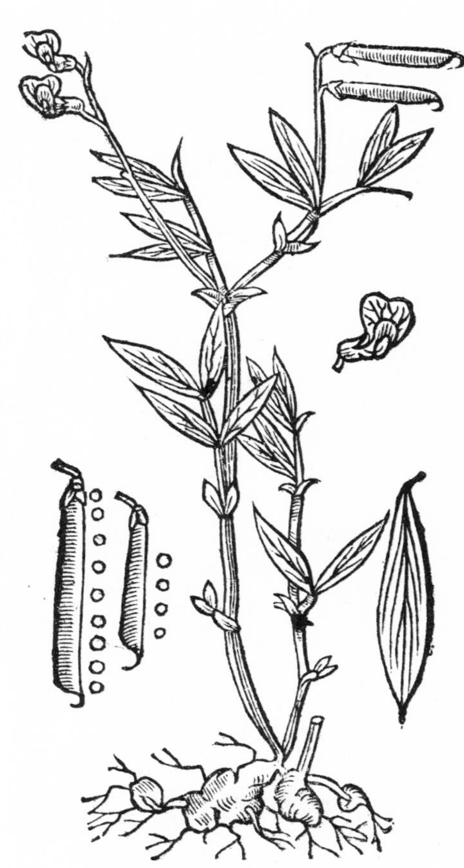

¶ *The Place.*

† 1 This groweth in corne fields, both with the corne it selfe, and also about the borders of fields among briers and brambles: it is found in diuers places of Germany, but not with vs that I can yet learne.

2 This is found in the woods and pastures of England, especially in Hampstead wood neere London: it groweth in Richmond Heath, and in Come parke likewise.

¶ *The Time.*

It floureth in Iune and Iuly, the nuts after haruest be digged vp and gathered.

¶ *The Names.*

It is called in high Dutch, **Erdnussen**: in low Dutch, **Eerdnoten, Eerdeeckelen**, and **Muysen metsteerten**, that is to say, tailed Mise, of the similitude or likenesse of domesticall mise, which the blacke, round, and long nuts, with a piece of the slender string hanging out behind, do represent: the later writers do call it in Latine *Terræ glandes* or *Terrestres glandes*: and in Greeke, χαμαιβάλανοι, *Chamæbalani*: in English, Pease Earth nut.

¶ *The Temperature and Vertues.*

The Nuts of this Pease being boiled and eaten, are hardlier digested than be either Turneps or A
Parsneps, yet do they nourish no lesse than the Parsneps: they are not so windie as they, they doe more slowly passe through the belly, by reason of their binding qualitie, and being eaten raw they be yet harder of digestion, and do hardlier and slowlier descend.

They be of temperature meanly hot, and somewhat drie, being withall not a little binding: wher- B
upon also they do not onely stay the fluxes of the belly, but also all issues of bloud, especially from the mother or bladder.

The root of Pease Earth-nut stoppeth the belly, and the inordinate course of womens sicknesse. C

Chap.

Chap. 520. *Of Milke Vetch.*

¶ *The Kindes.*

There be diuers ſorts of herbes contained vnder the title of *Aſtragalus*; whether I may, without breach of promiſe made in the beginning, inſert them among the *Legumina*, pulſes, or herbie plants, it is doubtfull: but ſeeing the matter is diſputable, I think it not amiſſe to ſuffer them thus to paſſe, vntill ſome other ſhall finde a place more conuenient and agreeing vnto them in neighbourhood.

¶ *The Deſcription.*

1 The firſt kinde of *Aſtragalus* hath reddiſh ſtalks, a cubit high, a finger thicke, ſomewhat creſted or furrowed, and couered ouer with an hairy moſſines; which diuide themſelues into ſundry ſmal branches, beſet with leaues conſiſting of ſundry little leaues ſet vpon a middle rib, like the wilde Vetch, placed on the ſmall pliant branches like feathers, which are likewiſe couered ouer with a woollie hoarineſſe; in taſte aſtringent at the firſt, but afterwards burning hot: among theſe leaues come forth many ſmall white floures, in faſhion like the floures of Lupines, which before their opening ſeeme to be ſomewhat yellow: the root is maruellous great and large, conſidering the ſmalneſſe of the plant; for ſometimes it groweth to the bigneſſe of a mans arme, keeping the ſame bigneſſe for the ſpace of a ſpan in length, and after diuideth it ſelfe into two or more forks or branches, blacke without, and wrinckled; white within, hard and wooddie, and in taſte vnpleaſant, which being dried becommeth harder than an horne.

1 *Aſtragalus Luſitanicus Cluſij.*
Portingale milke Vetch.

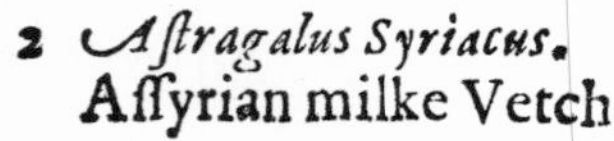
2 *Aſtragalus Syriacus.*
Aſſyrian milke Vetch.

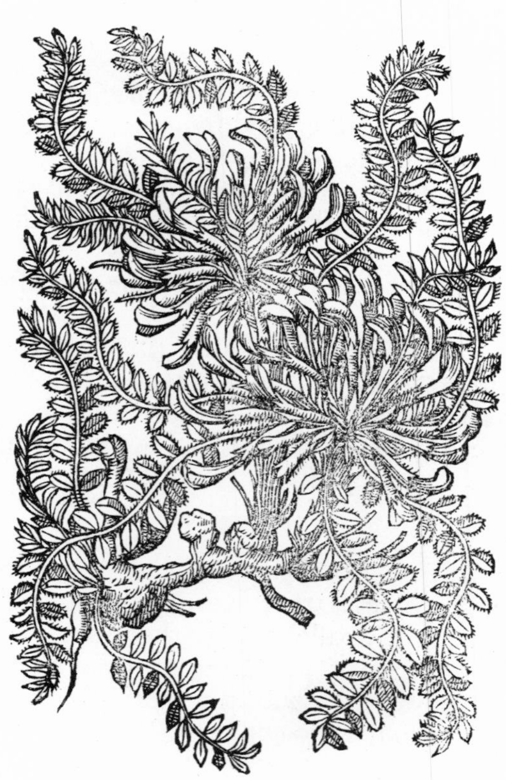

2 The ſecond kinde of *Aſtragalus* is a rare and gallant plant, and may well be termed *Planta Leguminoſa*, by reaſon that it is accounted for a kinde of *Aſtragalus*, reſembling the ſame in the ſimilitude of his ſtalkes and leaues, as alſo in the thickneſſe of his rootes, and the creeping and folding

thereof; and is garniſhed with a moſt thicke and pleaſant comlineſſe of his delectable red floures, growing vp together in great tufts, which are very ſeemly to behold.

3 There hath been ſome controuerſie about this third kinde, which I am not willing to proſecute or enter into: it may very well be *Aſtragalus* of *Matthiolus* his deſcription, or elſe his *Polygala*, which doth exceeding well reſemble the true *Aſtragalus*: his ſmall ſtalkes grow a foot high, beſet with leaues like *Cicer* or *Galega*, but that they are ſomewhat leſſer: among which come forth ſmall Peaſe like floures, of an Orange colour, very pleaſant in ſight: the root is tough and flexible, of a finger thicke.

‡ 3 *Aſtragalus Matthioli.*
Matthiolus his milke Vetch.

‡ 4 *Aſtragaloides.*
Baſtard Milke Vetch.

4 The fourth is called of *Mutonus* and other learned Herbariſts, *Aſtragaloides*, for that it reſembleth the true *Aſtragalus*, which groweth a cubit high, and in ſhew reſembleth Liquorice: the floures grow at the tops of the ſtalks, in ſhape like the Peaſe bloome, of a faire purple colour, which turne into ſmall blacke cods when they be ripe: the root is tough and very long, creeping vpon the vpper part of the earth, and of a woody ſubſtance.

The Place.

They grow amongſt ſtones, in open places, or as *Oribaſius* writeth, in places ſubiect to winds, and couered with ſnow: *Dioſcorides* copies do adde, in ſhadowie places: it groweth plentifully in Phenea a citie in Arcadia, as *Galen* and *Pliny* report: in *Dioſcorides* his copies there is read, in Memphis a citie of Arcadia; but Memphis is a citie of Egypt, and in Arcadia there is none of that name: ſome of them grow in my garden, and in ſundrie other places in England wilde; they grow in the medowes neere Cambridge, where the ſchollers vſe to ſport themſelues: they grow alſo in ſundrie places of Eſſex, as about Dunmow and Clare, and many other places of that countrey.

‡ I ſhould be glad to know which or how many of theſe our Authour heere affirmes to grow wilde in England; for as yet I haue not heard of, nor ſeene any of them wilde, nor in gardens with vs, except the laſt deſcribed, which growes in ſome few gardens. ‡

¶ *The Time.*

They floure in Iune and Iuly, and their ſeed is ripe in September.

¶ *The Names.*

Milke Vetch is called of *Matthiolus*, *Polygala*, but not properly: of most it is called *Astragalus*; in Spanish, *Garauancillos*: in the Portingales tongue, *Alphabeca*: in Dutch, **Cleyne Ciceren.**

A ¶ *The Temperature and Vertues.*

Astragalus, as *Galen* saith, hath astringent or binding roots, and therefore it is of the number of those simples that are not a little drying, for it glueth and healeth vp old vlcers, and staieth the flux of the belly, if they be boiled in wine and drunke: the same things also touching the vertues of *Astragalus Dioscorides* hath mentioned: the root, saith he, being drunke in wine staieth the laske, and prouoketh vrine, being dried and cast vpon old vlcers it cureth them: it likewise procureth great
B store of milke in cattell that do eat thereof, whence it tooke his name.

It stoppeth bleeding, but it is with much ado beaten, by reason of his hardnesse.

Chap. 521. *Of Kidney Vetch.*

¶ *The Description.*

1 Kidney Vetch hath a stalke of the height of a cubit, diuiding it selfe into other branches; whereon do grow long leaues, made of diuers leaues, like those of the Lentill, couered as it were with a soft white downinesse: the floures on the tops of the stalks of a yellow colour, verie many ioined together, as it were in a spokie rundle: after which grow vp little cods, in which is contained small seed: the root is slender, and of a wooddie substance. ‡ This is sometimes found with white floures: whereupon *Tabernamontanus* gaue two figures, calling the one *Lagopodium flore luteo*, and the other *Lagopodium flo. albo*. Our Author vnfitly gaue this later mentioned figure in the chapter of *Lagopus*, by the name of *Lagopum maximum*. ‡

1 *Anthyllis Leguminosa.*
Kidney Vetch.

2 *Stella leguminosa.*
Starry Kidney Vetch.

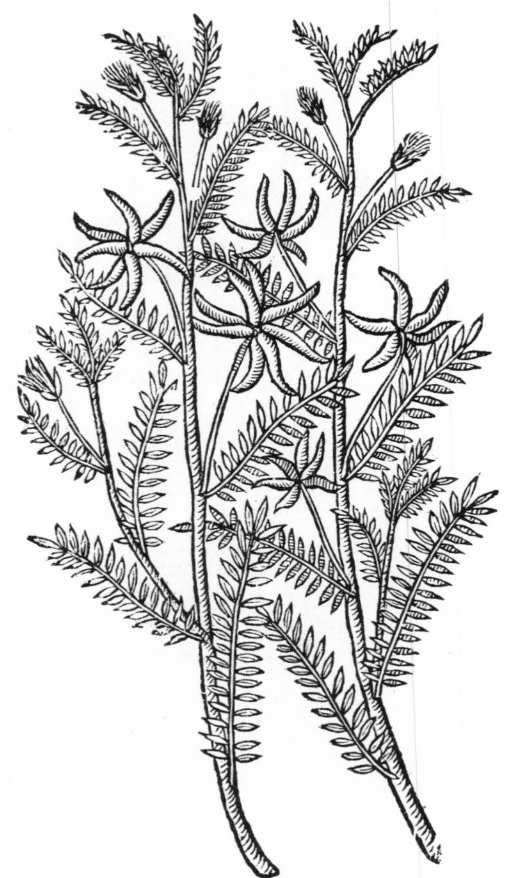

2 The Starry Kidney Vetch, called *Stella leguminosa*, or according to *Cortusus*, *Arcturo* hath many

many ſmall flexible tough branches, full of ſmall knots or knees, from each of which ſpringeth forth one long ſmall winged leafe, like birds foot, but bigger : from the boſome of thoſe leaues come forth little tender ſtems, on the ends whereof do grow ſmall whitiſh yellow floures, which are very ſlender, and ſoone vaded, like vnto them of Birds-foot : theſe floures turne into ſmall ſharpe pointed cods, ſtanding one diſtant from another, like the diuiſions of a ſtar, or as though it conſiſted of little hornes; wherein is contained ſmall yellowiſh ſeeds : the root is tough, and deeply growing in the ground.

3 There is another ſort of Kidney Vetch called Birds-foot, or *Ornithopodium*, which hath very many ſmall and tender branches, trailing here and there cloſe vpon the ground, ſet full of ſmall and ſoft leaues, of a whitiſh greene, in ſhape like the leaues of the wilde Vetch, but a great deale leſſer, and finer, almoſt like ſmall feathers : amongſt which the floures doe grow, that are very ſmall, yellowiſh, and ſometimes whitiſh; which being vaded there come in place thereof little crooked cods, fiue or ſix growing together, which in ſhew and ſhape are like vnto a ſmall birds foot, and each and euery cod reſembling a claw; in which are incloſed ſmall ſeed like that of Turneps.

‡ 3 *Ornithopodium maius.*
The great Birds-foot.

‡ 4 *Ornithopodium minus.*
Small Birds-foot.

‡ 5 *Scorpoides Leguminoſa.*
Small Horned pulſe.

4 There is alſo another kinde of *Ornithopodium*, or Birds-foot, called ſmall Birds-foot, which is very like vnto the firſt, but that it is much ſmaller: the branches or ſprigs grow not aboue a hand or halfe an hand in length, ſpreading themſelues vpon the ground with his ſmall leaues and branches, in maner of the leſſer *Arachus* : the floures are like vnto thoſe of the former, but very ſmall, and of a red colour.

‡ 5 This ſmall horned pulſe may fitly here take place : The root thereof conſiſts of many little fibres, from which ariſe two or three little ſlender ſtraight ſtalkes ſome handfull and halfe or foot high : at the tops of theſe grow little ſharpe pointed crooked hornes, rounder and ſlenderer than thoſe of Fenugreeke, turning their ends inwards like the tailes of Scorpions and ſo jointed; the floures are ſmall and yellow; the leaues little, and winged like thoſe of Birds-foot. *Pena* and *Lobel* found this amongſt the corne in the fields in Narbon in France, and they ſet it forth by the name as I haue here giuen you it. ‡

¶ *The Place.*

1. 3. 4. Theſe plants I found growing vpon Hampſtead Heath neere London, right againſt the

the Beacon, vpon the right hand as you go from London, neere vnto a grauell pit: they grow alſo vpon blacke Heath, in the high way leading from Greenwich to Charleton, within halfe a mile of the towne.

¶ *The Time.*

They floure from Iune to the middle of September.

¶ *The Names.*

‡ 1 This *Geſner* calls *Vulneraria ruſtica*: *Dodonæus*, *Lobel*, and *Cluſius*, call it *Anthyllis*, and *Anthyllis leguminoſa*. ‡

3. 4. I cannot finde any other name for theſe plants, but *Ornithopodium*: the firſt is called in Engliſh, great Birds-foot; the ſecond ſmall Birds-foot.

¶ *The Nature and Vertues.*

Theſe herbes are not vſed either in meate or medicine, that I know of as yet; but they are very good food for cattel, and procure good ſtore of milke, whereupon ſome haue taken them for kindes of *Polygala*.

Chap. 522. *Of Blacke milke Tare.*

Glaux Dioſcoridis.
Dioſcorides his milke Tare.

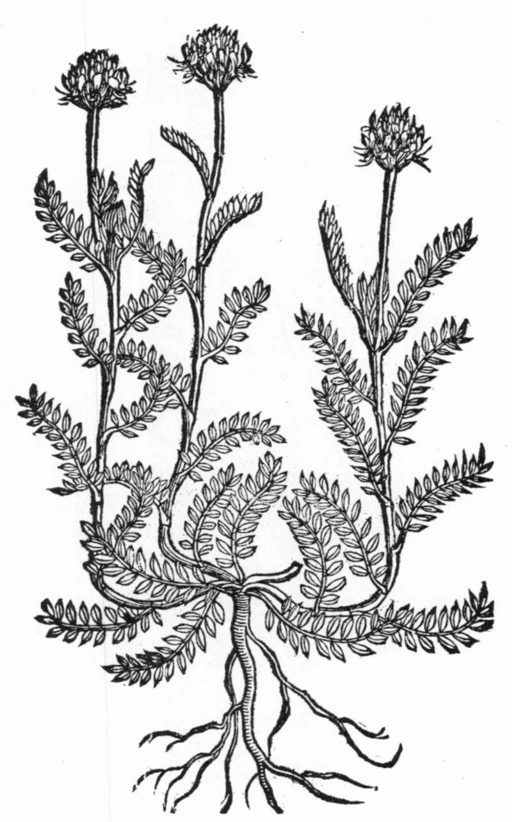

¶ *The Deſcription.*

The true *Glaux* of *Dioſcorides* hath very many tough and wooddy branches trailing vpon the ground, ſet full of ſmall winged leaues, in ſhape like the common *Glaux*, but a great deale ſmaller, reſembling the leaues of Tares, but rather like Birds-foot, of a very gray colour: amongſt which come forth knobby and ſcaly, or chaffie heads, very like the Medow Trefoile, of a faire purple colour: the root is exceeding long and wooddy, which the figure doth not expreſſe and ſet forth.

¶ *The Place.*

The true *Glaux* groweth vpon Barton hill, foure miles from Lewton in Bedfordſhire, vpon both the ſides of the declination of the hill.

¶ *The Time.*

Theſe plants do floure and flouriſh about Midſommer.

¶ *The Names.*

Theſe plants haue in times paſt been called *Glaux*, *i. folia habens glauca, ſiue pallentia*; that is, hauing skie coloured, or pale leaues.

Sithens that in times paſt, ſome haue counted *Glaux* among the kindes of *Polygala*, or Milkewoorts, we may therefore call this kinde of *Glaux*, blacke Milke-woort.

¶ *The Nature.*

Theſe herbes are dry in the ſecond degree.

¶ *The Vertues.*

A The ſeeds of the common *Glaux* are in vertue like the Lentils, but not ſo much aſtringent: they ſtop the flux of the belly, dry vp the moiſture of the ſtomacke, and ingender ſtore of milke.

† Our Author either not knowing, or forgetting what he had done, againe in this chapter, deſcribed the *Glaux Vulgaris*, whoſe hiſtory he gaue vs but foure chapters before, by the name of *Hedyſarum glycyrhizatum*; wherefore I haue omitted it here as not neceſſary.

Chap.

Chap. 523. Of red Fitchling, Medick Fitch, and Cockes-head.

¶ *The Description.*

1 The first kinde of *Onobrychis* hath many small and twiggie pliant branches, ramping and creeping through and about bushes, or whatsoeuer it groweth neere vnto: the leaues and all the rest of the pulse or plant is very like to the wilde Vetch or Tare: the floures grow at the top of small naked stalks, in shape like the pease bloome, but of a purple colour layed ouer with blew, which turne into small round prickly husks, that are nothing else but the seed.

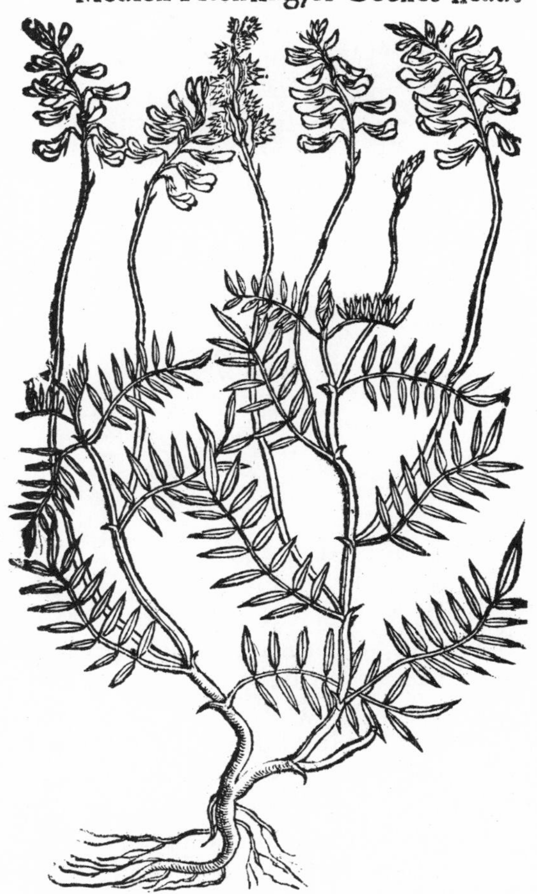

1 *Onobrychis, siue Caput Gallinaceum.*
Medick Fitchling, or Cockes-head.

2 *Onobrychis flore purpureo.*
Purple Cockes-head.

2 The second kind of Fitchling or Cocks-head, of *Clusius* his description, hath very many stalks, especially when it is growne to an old plant, round, hard, and leaning to the ground like the other pulses; and leaues very like *Galega*, or the wilde Vetch, of a bitter taste and lothsome sauour: among which come forth small round stems, at the ends whereof do grow floures spike fashion, three inches long, in shape like those of the great *Lagopus*, or medow Trefoile, but longer, of an excellent shining purple colour, but without smell: after which there follow small coddes, containing little hard and blacke seed, in taste like the Vetch. The root is great and long, hard, and of a wooddy substance, spreading it selfe far abroad, and growing very deep into the ground.

3 The third kinde of Fitchling or Cocks-head hath from a tough smal and wooddie root, many twiggie branches growing a cubit high, full of knots, ramping and creeping on the ground. The leaues are like the former, but smaller and shorter: among which come forth smal tender stemmes, whereupon do grow little floures like those of the Tare, but of a blew colour tending to purple: the floures being vaded, there come the small cods, which containe little blacke seed like a Kidney, of a blacke colour.

4 The fourth kinde of Fitchling hath firme green hard stalks a cubit and a halfe high, whereupon grow leaues like to the wilde Tare or *Galega*, but smaller and somewhat hairie, bitter and vnpleasant in taste, and in the end somewhat sharpe. At the top of the stalks come forth long spiked floures,

3 *Onobrychis* 2: *Clusij.*
Blew Medicke Fitch.

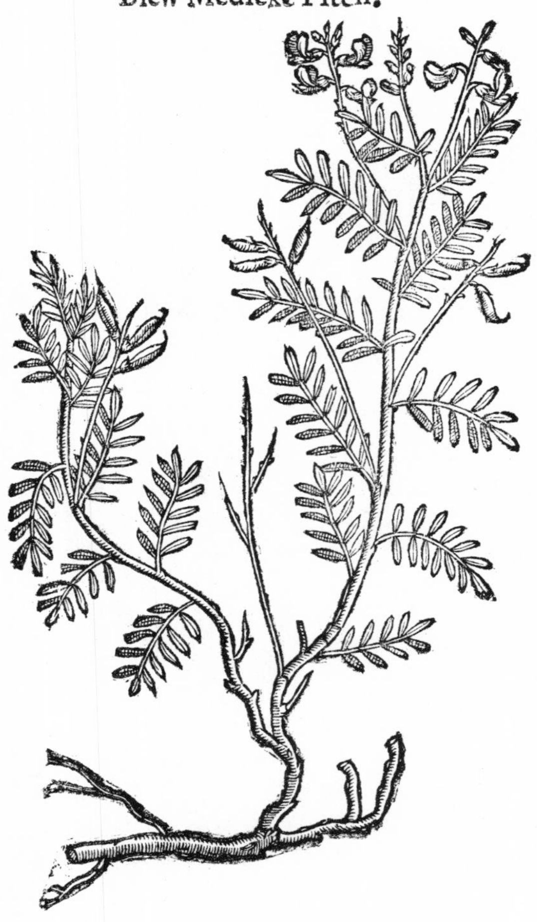

4 *Onobrychis* 3. *Clusij flore pallido.*
Pale coloured Medicke Fitch.

5 *Onobrychis montana* 4. *Clusij.*
Mountaine Medick Fetch.

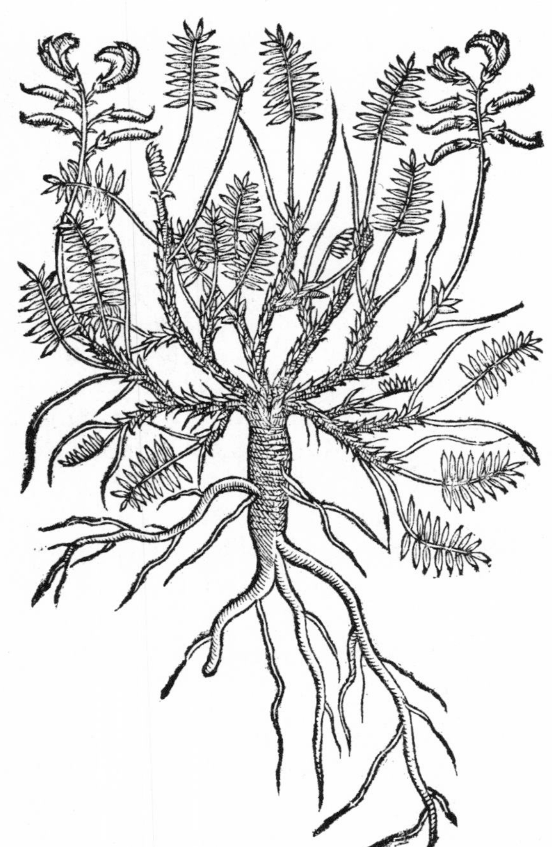

floures, of a pale colour, and in shape like those of the second kinde; which being vaded, there follow small bottle cods, wherein is contained little blacke seed like the seed of Fenegreek, but smaller. The root is thicke and hard, and of a wooddie substance, and lasting very long.

5 The fifth kinde of *Onobrychis* hath many grosse and wooddie stalks, proceeding immediatly from a thick, fat, and fleshie tough root: the vpper part of which are small, round, and pliant, garnished with little leaues like those of Lentils, or rather *Tragacantha*, somewhat soft, and couered ouer with a woollie hairinesse: amongst which come forth little long and naked stems, eight or nine inches long, whereon do grow many small floures of the fashion of the Vetch or Lentill, but of a blew colour tending to purple; and after them come smal cods, wherein the seed is contained.

¶ *The Place.*

The first and second grow vpon Barton hill, foure miles from Lewton in Bedfordshire, vpon both the sides of the hill: and likewise vpon the grassie balks between the lands of corn two miles from Cambridge, neere to a water mill towards London; & diuers other places by the way from London to Cambridge: the rest are strangers in England.

¶ *The Time.*

These plants do floure in Iuly, & their seed is ripe shortly after.

¶ *The*

¶ *The Names.*

It is ἀνώνυμος, or without a name among the later writers: the old and antient Physitions do call it ὀνοβρυχίς: for all those things that are found written in *Dioscorides* or *Pliny* concerning *Onobrychis*, doe especially agree hereunto. *Dioscorides* writeth thus; *Onobrychis* hath leaues like a Lentill, but longer; a stalk a span high; a crimson floure; a little root: it groweth in moist and vntilled places: and *Pliny* in like manner; *Onobrychis* hath the leaues of a Lentill, somwhat longer, a red floure, a small and slender root: it groweth about springs or fountaines of water.

All which things and euery particular are in this ἀνώνυμος, or namelesse herbe, as it is manifest: and therefore it is not to be doubted at all, but that the same is the *Onobrychis* of the old Writers: it may be called in English red Fetchling, or as some suppose Medick Fitch, or Cockes-head.

¶ *The Temperature.*

These herbs as *Galen* hath written in his books of the Faculties of simple Medicines, do rarifie or make thin and waste away.

¶ *The Vertues.*

A Therefore the leaues thereof when it is greene, being but as yet layed vpon hard swellings, waxen kernals, in manner of a salue, do waste and consume them away, but beeing dried and drunke in wine they cure the strangurie; and laied on with oile it procureth sweat.

B Which things also concerning *Onobrychis*, *Dioscorides* hath in these words set downe: the herbe stamped and applied wasteth away hard swellings of the kernels; but beeing drunke with wine it helpeth the strangurie, and rubbed on with oile it causeth sweatings.

CHAP. 524. *Of Bastard Dittanie.*

Fraxinella.
Bastard Dittanie.

¶ *The Description.*

Bastard Dittanie is a very rare and gallant plant, hauing many browne stalks, somwhat rough, diuided into sundry small branches, garnished with leaues like Liquorice, or rather like the leaues of the Ash tree, but blacker, thicker, and more ful of iuice, of an vnpleasant sauor: among which grow floures, consisting of fiue whitish leaues stripped with red, whereof one which groweth vndermost hangeth downe low; but the four which grow vppermost grow more stiffe and vpright: out of the midst of this floure commeth forth a tassell, which is like a beard, hanging also downwards, and somewhat turning vp at the lower end: which beeing vaded, there come in place foure huskes ioined together, much like the husks or coddes of Columbines, somewhat rough without, slimie to handle, and of a lothsome sauour, almost like the smell of a goat; whereupon some Herbarists haue called it *Tragium*: in the cods are contained small black shining seeds like Peonie seeds in colour: the roots are white, a finger thicke, one twisting or knotting within another, in tast somwhat bitter.

There is another kinde hereof growing in my garden, not very much differing: the leaues of the one are greater, greener, harder, and sharper pointed: of the other blacker, not so hard, nor so sharpe pointed: the floures also hereof be somthing more bright coloured, and of the other a little redder.

¶ *The Place.*

Bastard Dittany groweth wilde in the mountaines of Italy and Germanie, and I haue it growing in my garden.

¶ *The Time.*

It floureth in Iune and Iuly: the seed is ripe in the end of August.

¶ *The Names.*

The later Herbarists name it *Fraxinella:* most, χαμαιμέλιον, as though they should say *Humilis Fraxinus* or a low Ash: in English, bastard, or false Dittanie: the shops call it *Dictamnum*, and *Diptamum*, but not truly, and vse oftentimes the roots hereof in stead of the right Dittanie. That it is not the right Dittanie it is better knowne than needfull at all to be confuted; and it is as euident that the same is not *Dioscorides* his *Pseudodictamnum*, or bastard Dittanie: but it is plaine to be a kinde of *Tragium* of the old Writers, wherewith it seemeth to agree in shew, but not in substance.

‡ The root of this is onely vsed in shops, and there knowne by the name of *Radix Diptamni*, or *Dictamni*. ‡

¶ *The Temperature.*

The root of bastard Dittanie is hot and dry in the second degreee, it is of a wasting, attenuating, and opening facultie.

¶ *The Vertues.*

A It bringeth downe the menses, it also bringeth away the birth and after birth; it helpeth cold diseases of the matrix: and it is reported to be good for those that haue ill stomackes and are short winded.

B They also say, that it is profitable against the stingings and bitings of venomous serpents; against deadly poisons, against contagious and pestilent diseases, and that it is with good successe mixed with counterpoisons.

C The seed of Bastard Dittanie taken in the quantitie of a dram is good against the strangury, prouoketh vrine, breaketh the stone in the bladder, and driueth it forth.

D The like vertue hath the leaues and iuice taken after the same sort, and being applied outwardly, it draweth thornes and splinters out of the flesh.

E The root taken with a little Rubarb killeth and driueth forth wormes.

F *Dioscorides* reporteth, that the wilde Goats being stricken with darts or arrowes, will eat *Dictam*, and thereby cause them to fall out of their bodies; which is meant of the right *Dictam*, though *Dodonæus* reporteth that this plant will do the like (which I do not beleeue) ‡ nor *Dodonæus* affirme. ‡

Chap. 525. Of Land Caltrops.

Tribulus terrestris.
Land Caltrops.

¶ *The Description.*

Land Caltrops hath long branches full of ioints, spred abroad vpon the ground, garnished with many leaues set vpon a middle rib, after the manner of Fetches; amongst which grow little yellow branches, consisting of fiue smal leaues, like vnto the floures of Tormentill: I neuer saw the plant beare yellow, but white floures, agreeing with the description of *Dodonæus* in each respect, saue in the colour of the floures, which doe turne into small square fruit, rough, and full of prickles, wherein is a small kernell or seed: the root is white, and full of strings.

¶ *The Place.*

It groweth plentifully in Spain in the fields: it is hurtful to corne, but yet as *Pliny* saith, it is rather to be accounted among the diseases of corne, than among the plagues of the earth: it is also found in most places of Italy & France; I found it growing in a moist medow adioyning to the wood or Park of Sir *Francis Carew*, neere Croidon, not far from London, and not elsewhere; from whence I brought plants for my garden.

¶ The

¶ *The Time.*

It floureth in Iune and Iuly: the fruit is ripe in Auguſt.

¶ *The Names.*

It is called in Greeke τρίβολος: and in Latine *Tribulus*: and that it may differ from the other which groweth in the water, it is named τρίβολος χερσαῖος, or *Tribulus terreſtris*: it may be called in Engliſh, land Caltrops, of the likeneſſe which the fruit hath with Caltrops, that are inſtruments of Warrre caſt in the way to annoy the feet of the Enemies horſes, as is before remembred in the Water Saligot.

¶ *The Temperature and Vertues.*

In this land Caltrop there is an earthy and cold qualitie abounding, which is alſo binding, as *Galen* ſaith. A

The fruit thereof being drunke waſteth away ſtones in the kidneyes, by reaſon that it is of thin parts. B

Land Caltrops, ſaith *Dioſcorides*, being drunke to the quantitie of a French crowne weight, and ſo applied, cureth the bitings of the Viper. C

And if it be drunke in wine it is a remedie againſt poyſons: the decoction thereof ſprinckled about killeth fleas. D

‡ Chap. 526. *Of Spring or mountaine Peaſe or Vetches.*

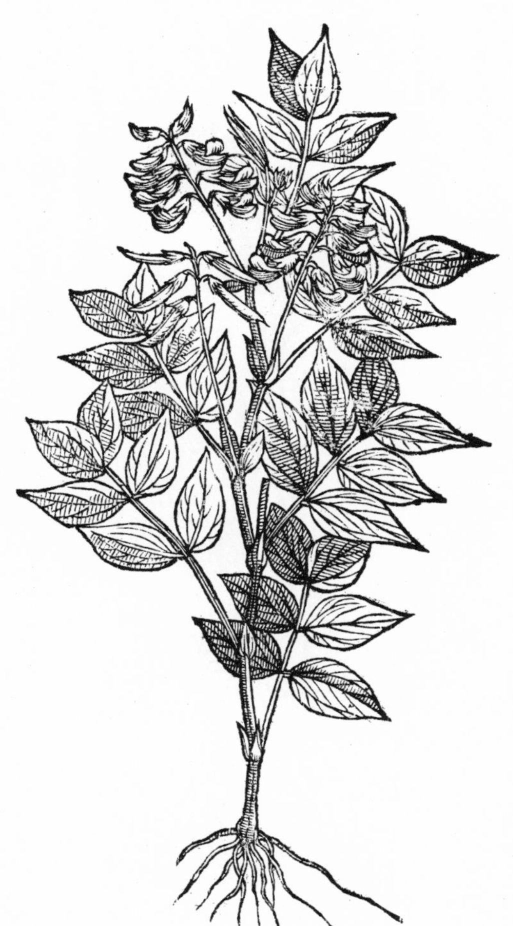
‡ 1 *Orobus Venetus.*
Venice Peaſe.

‡ 2 *Orobus ſyluaticus vernus.*
Spring Peaſe.

¶ *The Deſcription.*

‡ 1 This, which *Cluſius* calls *Orobus Venetus*, hath many cornered ſtalkes ſome foot long, whereon grow winged leaues, foure or ſix faſtned to one rib, ſtanding by couples one againſt another, without any odde leafe at the end: theſe leaues are of an indifferent largeneſſe, and

and of a light greene colour: the floures grow vpon long foot-stalks comming forth of the bosoms of the leaues, many together, hanging downe, small, yet shaped like those of other Pulses, and of a purple colour: after these follow cods almost like those of Fetches, but rounder, red when they be ripe, and containing in them a longish white seed: the root is hard and wooddy, running diuers wayes with many fibres, and liuing sundry yeares: this varies somtimes with yellower green leaues and white floures. It floures in May, and growes onely in some few gardens with vs.

2 The stalkes of this also are a foot or more high, stiffe, cornered, and green; on these do grow winged leaues six or eight on a rib, after the manner of those of the last described: each of these leaues hath three veines running alongst it: the floures in shape and manner of growing are like those of the former, but of a most elegant purple colour: which fading, they become blew. The floures are succeeded by such cods as the former, wherein are contained longish small variegated seed: which ripe, the cods fly open, and twine themselues round, as in most plants of this kinde: the root is blacke, hard, tuberous and wooddy, sending forth each yeare new shoots. This floures in April and May, and ripeneth the seed in Iune. This was found by *Clusius* in diuers mountainous wooddie places of Hungarie: he calls it *Orobus Pannonicus* 1.

‡ 3 *Orobus montanus flo. albo.*
White mountaine Pease.

‡ 4 *Orobus montanus angustifolius.*
Narrow leafed mountaine Pease.

3 This hath stalkes some cubit high, stiffe, straight, and crested; whereon by turnes are fastned winged leaues, consisting of foure sufficiently large and sharpe pointed leaues, whereto sometimes at the very end growes a fifth: the veines in these run from the middle rib towards their edges: their taste is first somewhat sourish, afterwards bitterish. The floures grow vpon short stalks comming forth of the bosomes of the leaues, fiue or six together, like those of the Fetch, but of colour white, with some little yellownesse on the two little leaues that turne vpwards. The cods are like those of the last described, and containe in them a brownish seed, larger than in any of the other kindes. This is an annuall plant, and perishes as soone as it hath perfected the seed. *Clusius* giues vs this by the name of *Orobus Pannonicus* 4. *Dodonæus* giues the same figure for his *Arachus latifolius*: and *Bauhine* affirmes this to be the *Galega montana*, in the *Hist. Lugd. pag.* 1139. But these seeme to be of two seuerall plants; for *Dodonæus* affirmes his to haue a liuing root, and such seemes also that in the *Hist. Lugd.* to be: yet *Clusius* saith expressely that his is an annuall, and floureth in

Aprill

Aprill and May, and groweth in some wooddy mountainous places of the kingdom of Hungarie.

4 This fourth hath straight firme cornered stalkes some foot or more high, whereupon grow leaues vsually foure on a foot-stalke, standing two against two, vpright, being commonly almost three inches long, at first of a sourish taste, but afterwards bitter: it hath no clauicles, because the stalkes need no supporters: the floures grow vpon long foot-stalkes, spike-fashion like those of Pease, but lesse, and white of colour: after these follow long blackish cods, full of a blacke or else spotted seed: the roots are about the length of ones little finger, fashioned like those of the Asphodill or lesser female Peionie, but lesser, blacke without, and white within. *Clusius* found this on the mountainous places nigh the baths of Baden, and in the like places in Hungarie: he calls it *Orobus Pannonicus* 3.

¶ *The Temperature and Vertues.*

These are not knowne nor vsed in physicke; yet if the third be the *Galega montana* of the *Historia Lugd.* then it is there said to be effectuall against poyson, the wormes, the falling sicknesse, and the Plague. ‡

‡ CHAP. 527. *Of some other Pulses.*

‡ 1 *Ochrus, siue Eruilia.*
Birds Pease.

‡ 2 *Eruum syluestre.*
Crimson grasse Fetch.

¶ *The Description.*

‡ 1 THe first of these hath cornered broad stalks like those of euerlasting Pease, and they are weake, and commonly lie vpon the ground, vnlesse they haue something to support them: the lower leaues are broad, and commonly welt the stalke at their setting on, and at the end of the first leafe do vsually grow out after an vnusuall manner, two, three, or more other prety large leaues more long than broad, and the middle rib of the first leafe runnes out beyond the setting on of the highest of the out-growing leaues, and then it ends in two or three clasping tendrels. Those leaues that grow the lowest vpon the stalkes haue commonly the fewest comming out of them.

them. The floures are like those of other Pulses, of colour white: the cods are some inch and halfe long, containing some halfe dozen darke yellow or blackish small Pease: these cods grow one at a ioynt, on short foot-stalkes comming forth of the bosomes of the leaues, and are welted on their broader side, which stands towards the maine stalke. This growes with vs only in gardens. *Dodonæus, Pena*, and *Lobel* call it *Ochrus sylvestris, siue Ervilia*.

2 The stalkes of this grow vp sometimes a cubit high, being very slender, diuided into branches, and set vnorderly with many grasse-like long narrow leaues: on the tops of the stalkes and branches, vpon pretty long foot-stalkes grow pretty pease-fashioned floures of a faire and pleasant crimson colour: which fallen, there follow cods, long, small, and round, wherein are nine, ten, or more round hard blacke shining graines: the root is small, with diuers fibres, but whether it die when the seed is perfected, or no, as yet I haue not obserued. This growes wilde in many places with vs, as in the pasture and medow grounds about Pancridge Church. *Lobel* and *Dodon.* call this *Ervum sylvestre*; and they both partly iudge it to be the first *Catanance* of *Dioscorides*, and by that name it is vsually called. It floures in Iune and Iuly, and the seed is ripe in August.

3 This also, though it be not frequently found, is no stranger with vs; for I haue found it in the corne fields about Dartford in Kent and some other places. It hath long slender ioynted creeping stalkes, diuided into sundry branches, whereon stand pretty greene three cornered leaues two at a ioynt, in shape and bignesse like those of the lesser Binde-weed. Out of the bosomes of these leaues at each ioynt comes a clasping tendrel, and commonly together with it a foot-stalke some inch or more long, bearing a pretty little pease-fashioned yellow floure, which is succeeded by a short flattish cod containing six or seuen little seeds. This floures in Iune, Iuly, and August, and so ripens the seed. It is by *Lobel* and others thought to be the *Aphace* of *Dioscorides, Galen*, and *Pliny*: and the *Pitine* of *Theophrastus*, by *Anguillara*.

‡ 3 *Aphaca.*
Small yellow Fetch.

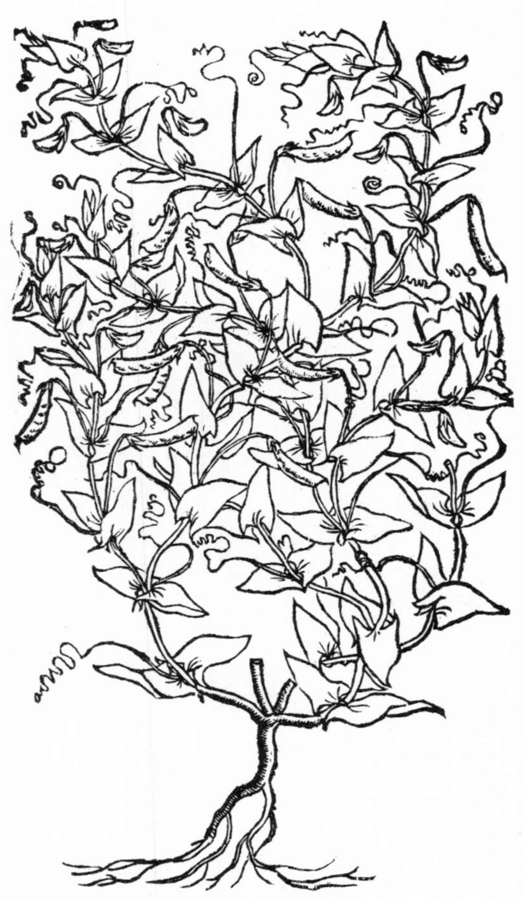

I finde mention in *Stowes* Chronicle, in *Anno* 1555, of a certaine Pulse or Pease, as they term it, wherewith the poore people at that time, there being a great dearth, were miraculously helped: he thus mentions it; In the moneth of August (saith he) in Suffolke, at a place by the sea side all of hard stone and pibble, called in those parts a shelfe, lying betweene the townes of Orford and Aldborough, where neither grew grasse, nor any earth was euer seene; it chanced in this barren place suddenly to spring vp without any tillage or sowing, great aboundance of Peason, whereof the poore gathered (as men iudged) aboue an hundred quarters, yet remained some ripe and some blossoming, as many as euer there were before: to the which place rode the Bishop of Norwich and the Lord *Willoughby*, with others in great number, who found nothing but hard rockie stone the space of three yards vnder the roots of these Peason: which roots were great and long, and very sweet.

Gesner also, *de Aquatilibus, lib.* 4. *pag.* 256. making mention, out of D^r. *Cajus* his letters, of the spotted English Whale, taken about that time at Lin in Norfolke, also thus mentions those pease: *Pisa* (saith he) *in littore nostro Britannico quod Orientem spectat, certo quodam in loco Suffolciæ, inter Alburnum & Ortfordium oppida, saxis insidentia (mirabile dictu) nulla terra circumfusa, autumnali tempore Anno* 1555, *sponte nata sunt, adeo magna copia, vt sufficerent vel millibus hominum.* These Pease, which by their great encrease did such good to the poore that yeare, without doubt grew there for many yeares before, but were not obserued till [*Magister artis, ingenique largitor Venter*]—— hunger made them take notice of them, and quickned their inuention, which commonly in our people is very dull, especially in finding out food of this nature.

My

My Worshipfull friend Dr. *Argent* hath told me, that many yeares ago he was in this place, and caused his man to pull away the beach with his hands, and follow the roots so long, vntill hee got some equall in length vnto his height, yet could come to no ends of them: hee brought these vp with him to London, and gaue them to Dr. *Lobel*, who was then liuing; and he caused them to be drawne, purposing to set them forth in that Worke which he intended to haue published, if God had spared him longer life. Now whether these Pease be truly so called, and be the same with the *Pisum syluestre Perenne*, or different; or whether they be rather of the stocke of the *Lathyrus maior*, or of some other Pulse here formerly described, I can affirme nothing of certaintie, because I haue seene no part of them, nor could gather by any that had, any certaintie of their shape or figure: yet would I not passe them ouer in silence, for that I hope this may come to be read by some who liue thereabout, that may by sending me the things themselues, giue me certaine knowledge of them; that so I may be made able, as I am alwaies willing, to impart it to others.

¶ *The Temperature and Vertues.*

I haue not haue not found any thing written of the faculties of the two first; but of *Aphace*, *Galen* saith it hath an astringent facultie like as the Lentill, and also is vsed to be eaten like as it, yet it is harder of concoction, but it dries more powerfully, and heates moderately. The seeds (saith hee) haue an astringent facultie; wherefore parched, broken, and boyled, they stay fluxes of the belly. We know (saith *Dodonæus*) by certaine experience, that the *Aphace* here described hath this astringent force and facultie. ‡ A

CHAP. 528. *Of bastard Rubarb.*

1 *Thalietrum, siue Thalictrum maius.*
Great bastard Rubarb.

Thalictrum minus.
Small bastard Rubarb.

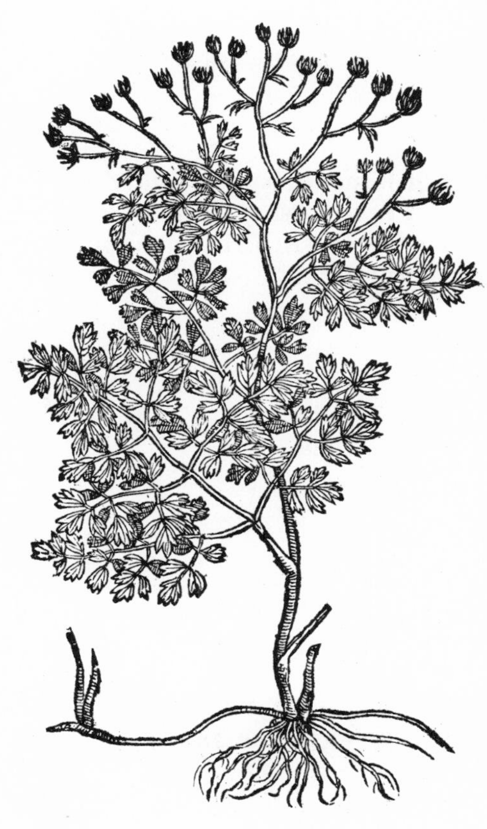

¶ *The Description.*

1 THe great *Thalietrum* or bastard Rubarb hath large leaues parted or diuided into diuers other small leaues, somewhat snipt about the edges, of a blacke or darke green colour: the

the ſtalkes are creſted or ſtreaked, of a purple colour, growing to the height of two cubits: at the top whereof grow many ſmall and hairy white floures, and after them come ſmall narrow huskes like little cods, foure or fiue growing together: the root is yellow, long, round, and knotty, diſperſing it ſelfe far abroad on the vpper cruſt of the earth.

2 The ſmall baſtard Rubarb is very like vnto the precedent, but that it is altogether leſſer: his ſtalkes are a ſpan or a foot long: his leaues be thin and tender; the root fine and ſlender: the little floures grow together in ſmall bundles or tufts, of a light yellow colour, almoſt white, and are of a grieuous ſauour.

‡ 3 There is kept in ſome gardens a plant of this kinde growing vp with large ſtalkes to the height of three cubits: the leaues are very like thoſe of Columbines: the floures are made of many white threds: it floures in Iune, and is called *Thalictrum maius Hiſpanicum*, Great Spaniſh Baſtard Rubarb. ‡

¶ *The Place.*

Theſe Plants doe grow alongſt the Ditch ſides leading from Kentiſh ſtreet vnto Saint Thomas a-Waterings (the place of Execution) on the right hand. They grow alſo vpon the bankes of the Thames, leading from Blacke-wall to Woolwich, neere London, and in ſundry other places alſo.

¶ *The Time.*

The floure for the moſt part in Iuly and Auguſt.

¶ *The Names.*

Diuers of the later Herbariſts do call it *Pigamum*, as though it were πήγανον, that is, Rue; whereupon moſt call it *Ruta paluſtris*, or Fen Rue: others, *Pſeudo-Rhabarbarum*, and *Rhabarbarum Monachorum*, by reaſon of the yellow colour of the root. But neither of their iudgements is greatly to be eſteemed of: they iudge better that would haue it to be *Thalietrum*, which *Dioſcorides* deſcribeth to haue leaues ſomething flatter than thoſe of Coriander; and the ſtalke like that of Rue, vpon which the leaues doe grow. *Pena* calleth it *Thalietrum*, *Thalictrum*, and *Ruta pratenſis*: in Engliſh, baſtard Rubarb, or Engliſh Rubarb: which names are taken of the colour, and taſte of the roots.

¶ *The Temperature.*

Theſe herbes are hot and dry of complexion.

¶ *The Vertues.*

A The leaues of baſtard Rubarb with other pot-herbes do ſomewhat moue the belly.

B The decoction of the root doth more effectually.

C *Dioſcorides* ſaith, that the leaues being ſtamped do perfectly cure old vlcers. *Galen* addeth, that they dry without biting.

Chap. 529. *Of Goats Rue.*

¶ *The Deſcription.*

Galega or Goats Rue hath round hard ſtalkes two cubits or more high, ſet full of leaues diſplayed or winged abroad; euerie leafe conſiſting of ſundrie ſmall leaues ſet vpon a ſlender rib, reſembling the leaues of the field Vetch or Tare, but greater and longer. The floures grow at the top of the ſtalke, cluſtering together after the manner of the wilde Vetch, of a light skie colour, which turne into long cods ſmall and round, wherein the ſeed is contained. The root is great, thicke, and of a white colour.

¶ *The Place.*

It groweth plentifully in Italy euery where in fat grounds and by riuers ſides: it groweth likewiſe in my garden.

¶ *The Time.*

It floureth in Iuly and Auguſt.

¶ *The Names.*

The Italians call it *Galega*, and *Ruta Capraria*: diuers name it corruptly *Gralega*: *Hieronymus Fracaſtorius*

Galega.
Goats Rue.

Fracaſtorius calleth it *Herba Galleca* : the Hetruſcians, *Lauaneſe*; and it is alſo called by diuers other names in ſundry places of Italy, as *Geſner* ſaith, as are *Caſtracane, Lauanna, Thorina*, or *Taurina, Martanica, Sarracena, Capragina, Herbaneſa, Fœnum græcum ſylueſtre*, and as *Braſauolus* witneſſeth, *Giarga*. It is named in Engliſh, Italian Fitch, and Goats Rue.

Some iudge that the old Phyſitions were wont to call it *Onobrychis*: others, *Glauce*: diuers would haue it to be *Polemonium*, but not ſo much *Petr. And Matthiolus* in his commentaries, as euery one of the deſcriptions mentioned by *Dioſcorides* do gainſay them; as alſo thoſe, who thinke that *Galega* is *Polygalon*, & that the name of *Galega* came of *Polygalon*, the very deſcription alſo of *Polygalon* is againſt them. for *Galega* is higher and greater than that it may be called a little ſhrub onely of an hand bredth high.

¶ *The Temperature.*

This plant is in a meane temperature betweene hot and cold.

¶ *The Vertues.*

Goats Rue is a ſingular herbe againſt all venome and poiſon, and againſt wormes, to kill and driue them forth, if the iuice be giuen to little children to drinke. A

It is of like vertue if it be fryed with Lineſeed oyle, and bound vpon the childes nauell. B

It is miniſtred vnto children which are poſſeſſed with the falling euill, a ſpoonefull euery morning in milke. C

Being boiled in vineger, and drunke with a little Treacle, it is very good againſt the infection of the plague, eſpecially if the medicine be taken within twelue houres. D

The herbe it ſelfe is eaten, being boiled with fleſh, as we vſe to eate Cabbage and other woorts, and likewiſe in ſallades, with oile, vineger and pepper, as we do eate boiled Spinage, and ſuch like; E Which is moſt excellent being ſo eaten, againſt all poiſon and peſtilence, or any venomous infirmitie whatſoeuer, and procureth ſweat.

It alſo helpeth the bitings and ſtingings of venomous beaſts, if either the iuice or the herbe ſtamped be laid vpon the wound. F

Halfe an ounce of the iuice inwardly taken is reported to helpe thoſe that are troubled with conuulſions, crampes, and all other the diſeaſes aforeſaid. G

The ſeedes do feed pullen exceedingly, and cauſe them to yeeld greater ſtore of egs than ordinary. H

‡ The iuice of the leaues, or the leaues themſelues bruiſed and applied to any part ſwollen by the ſting of a bee or waſpe, mitigate the paine, and are a preſent remedy, as Mr. *Cannon* a louer of Plants, and frind of mine, hath aſſured me he hath ſeen by frequent experience. ‡ I

CHAP. 530. *Of* Pliny *his Leadwoot.*

¶ *The Deſcription.*

DEntaria or *Dentillaria* hath offended in the ſuperlatiue degree, in that he hath hid himſelfe like a runnagate ſouldier, when the aſſault ſhould haue been giuen to the plant *Lepidium*, whereof doubtleſſe it is a kinde. But if the fault be mine, as without queſtion it is, I craue pardon for the ouerſight, and do intreate thee gentle reader to cenſure me with fauour, whereby I may more boldly inſert it in this place, rather than to leaue it vntouched. The learned of Narbone (eſpecially *Rondeletius*) haue not without good cauſe accounted this goodly plant for a kinde

Plumbago Pliny.
Leadwoort.

thereof, because the whole plant is of a biting taste, and a burning faculty, and that in such extremity, that it will raise blisters vpon a mans hand: for which cause some of the learned sort haue accounted it *Plinies Molybdæna,* or *Ægineta* his *Lepidium*: but the new Herbarists call it *Dentaria,* or *Dentillaria Rondeletij,* who made the like vse hereof, as he did of *Pyrethrum,* & such burning plants, to appease the immoderate pain of the tooth-ache and such like. This plant hath great thicke tough roots, of a wooddy substance, from whence spring vp long and tough stalkes two cubits high, confusedly garnished and beset with long leaues, in colour like Woad, of a sharpe and biting taste. The floures grow at the top of the stalkes of a purple colour; which being past, there succeed close glistering and hairy huskes, wherein is contained small blackish seed.

¶ *The Place.*

Pena reporteth that *Dentillaria* groweth about Rome, nigh the hedges and corne fields: it likewise groweth in my Garden in great plenty.

¶ *The Time.*

It floureth in Iuly and August.

¶ *The Names.*

Leadwoort is called *Molybdæna, Plumbago Plinij,* & *Dentillaria Rondeletij*: in Italian, *Crepanella,* the Romanes, *Herba S. Antonij*: in Illyria, *Cucurida*: in English, Leadwoort.

¶ *The Temperature.*

Dentillaria is of a caustick quality.

¶ *The Vertues.*

A It helpeth the tooth-ache, and that as some say if it be holden in the hand some small while.

Chap. 531. *Of Rue, or herbe Grace.*

¶ *The Description.*

1 Garden Rue or planted Rue, is a shrub full of branches, now and then a yard high, or higher: the stalkes whereof are couered with a whitish barke, the branches are more green: the leaues hereof consist of diuers parts, and be diuided into wings, about which are certaine little ones, of an odde number, something broad, more long than round, smooth and somewhat fat, of a gray colour, or greenish blew: the floures in the top of the branches are of a pale yellow, consisting of foure little leaues, something hollow: in the middle of which standeth vp a little head or button foure square, seldome fiue square, containing as many little coffers as it hath corners, being compassed about with diuers little yellow threds: out of which hang pretie fine tips of one colour; the seed groweth in the little coffers: the root is wooddy, and fastned with many strings: this Rue hath a very strong and ranke smell, and a biting taste.

2 The second being the wilde or mountaine Rue, called *Ruta syluestris,* is very like to garden Rue, in stalkes, leaues, floures, seed, colour, taste, and sauour, sauing that euery little leafe hath smaller cuts, and is much narrower: the whole plant dieth at the approch of winter, being an annuall plant, and must either stand till it do sow himselfe, or else must be sowne of others. ‡ This second is a variety of the garden Rue differing from the former onely in smallnesse. ‡

3 This plant is likewise a wilde kinde of Rue, and of all the rest the smallest, and yet more virulent, biting, and stinking than any of the rest: the whole plant is of a whitish pale greene, agreeing with the last before mentioned in each respect, saue in greatnesse, and in that the venomous fumes or vapors that come from this small wilde Rue are more noisome and hurtfull than the former. † The leaues lie spred vpon the ground, & are very finely cut and diuided: the whole plant is of such

1 *Ruta hortensis.*
Garden Rue.

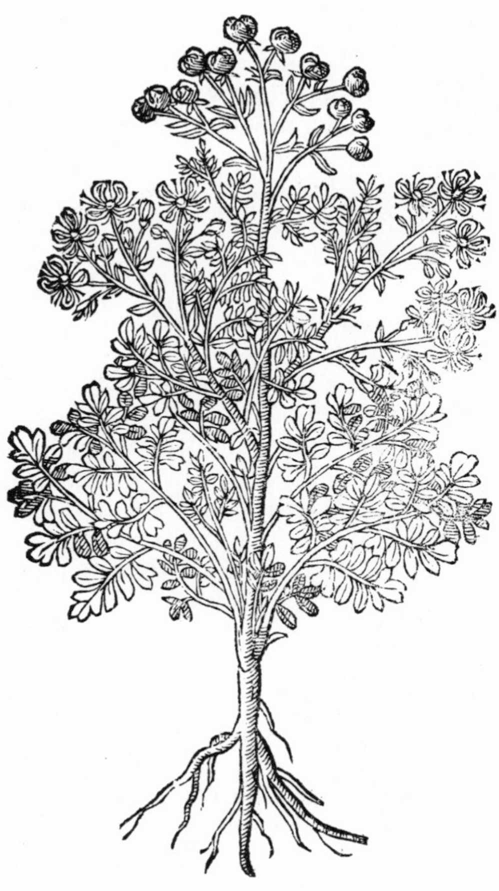

3 *Ruta sylvestris minima.*
The smallest wilde Rue.

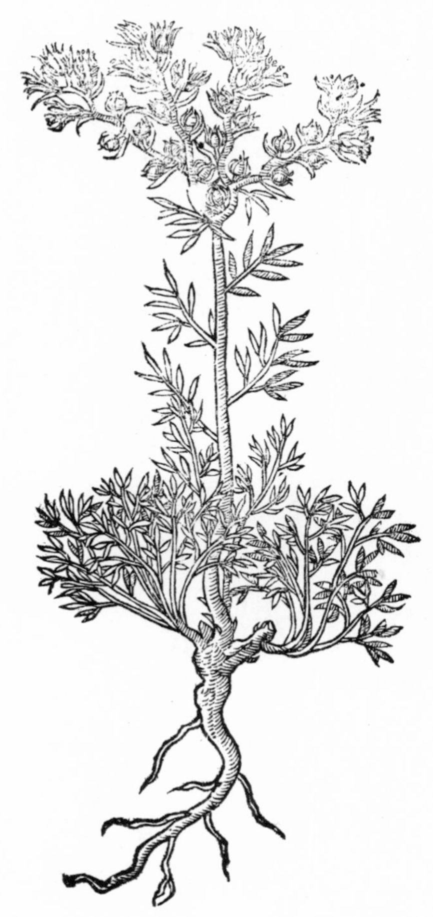

4 *Ruta montana.*
Mountaine Rue.

5 *Harmala.*
Wilde Rue with white floures.

‡ 6 *Ruta Canina.*
Dogs Rue.

such acrimonie, that *Clusius* saith he hath oftner than once obserued it to pierce through three paire of gloues to the hand of the gatherer; and if any one rub his face with his hand that hath newly gathered it, forthwith it will mightily inflame his face. He tells a history of a Dutch Student of Mompelier that went with him a simpling, who putting some of it betweene his hat and his head to keepe him the cooler, had by that meanes all his face presently inflamed and blistred wheresoeuer the sweat ran downe. ‡

4 There is another wilde Rue growing vpon the mountaines of Sauoy and other places adioyning, hauing a great thicke root, from which do arise great shoots or stalkes; wheron do grow leaues very thicke and fat, parted into diuers longish sections, otherwise resembling the leaues of the first described, of a strong and stinking smell: the floures grow on the tops of the stalkes, consisting of foure small yellow leaues: the seeds are like the other.

5 *Harmel* is one of the wilde Rues: it bringeth forth immediatly from the root diuers little stalks of a cubit high; whereupon do grow greene leaues diuersly cut into long pieces, longer and narrower than those of the wild strong smelling Rue: the floures be white, composed of fiue white leaues: the fruit is three square, bigger than that of the planted Rue, in which the seed lieth: the root is thick, long, and blackish: this Rue in hot countries hath a maruellous strong smell, in cold Countries not so.

‡ 6 This, which *Matthiolus* gaue for *Sideritis* 3. and *Lobel*, *Clusius*, and others for *Ruta canina*, hath many twiggy branches some cubit and halfe high; whereon grow leaues resembling those of the *Papauer Rhæas* or *Argemone*, lesser, thicker, and of a blackish greene: the floures are of a whitish purple colour, fashioned somewhat like those of *Antirrhinum*: the seed is small, and contained in such vessels as those of Rue, or rather those of *Blattaria*. The whole plant is of a strong and vngratefull smell: it growes in the hot and dry places about Narbon in France, Rauenna and Rome in Italy. ‡

¶ *The Place.*

Garden Rue ioyeth in sunny and open places: it prospereth in rough and brickie ground, and among ashes: it cannot in no wise away with dung.

The wilde are found on mountaines in hot countries, as in Cappadocia, Galatia, and in diuers prouinces of Italy and Spaine, and on the hills of Lancashire and Yorke.

Pliny saith that there is such friendship betweene it and the fig tree, that it prospers no where so well as vnder the fig tree. The best for physicks vse is that which groweth vnder the fig tree, as *Dioscorides* saith: the cause is alledged by *Plutarch* in the first booke of his *Symposiacks* or Feasts, for he saith it becommeth more sweet and milde in taste, by reason it taketh as it were some part of the sweetnes of the fig tree, whereby the ouer rancke quality of the Rue is allayd; vnlesse it be that the fig tree whilest it draweth nourishment vnto it selfe, it likewise draweth away the rancknesse of the Rue.

¶ *The Time.*

They floure in these cold countries in Iuly and August; in other countries sooner.

¶ *The Names.*

The first, which is *Hortensis Ruta*, garden Rue: in high-Dutch, **Rauten**: in low-Dutch, **Ruite**: the Italians and Apothecaries keepe the Latine name: in Spanish, *Aruda*: in French, *Rue de Iardin*: in English, Rue, and Herbe-Grace.

Wilde Rue is called in Greeke πήγανον, *Peganon*: in Latine, *Ruta sylvestris*, or wilde Rue: in Galatia and Cappadocia, μῶλυ: of diuers, *Harmala*: of the Arabians, *Harmel*: of the Syrians, *Besara*.

¶ *The Temperature.*

Rue is hot and dry in the later end of the third degree, and wild Rue in the fourth: it is of thin and ſubtill parts: it waſts and conſumes winde,it cutteth and digeſteth groſſe and tough humors.

¶ *The Vertues.*

Rue or Herbe-Grace prouokes vrine, brings downe the ſicknes,expels the dead child and after-birth,being inwardly taken,or the decoction drunke; and is good for the mother,if but ſmelled to. A

*Plin.lib.20.ca.*13.ſaith it opens the matrix,and brings it into the right place,if the belly all ouer and the ſhare(the breſt ſay the old falſe copies)be anointed therewith: mixed with hony it is a remedie againſt the inflammation and ſwelling of the ſtones,proceeding of long abſtinence from venerie, called of our Engliſh Mountebanks the Colts euill, if it be boyled with Barrowes greaſe, Bay leaues, and the pouders of Fenugreeke and Linſeed be added thereto, and applied pultis wiſe. B

It takes away crudity and rawneſſe of humors,and alſo windines and old paines of the ſtomack. C

Boiled with vineger it eaſeth paines,is good againſt the ſtitch of the ſide and cheſt,and ſhortnes of breath vpon a cold cauſe,and alſo againſt the paine in the ioynts and huckle bones. D

The oile of it ſerues for the purpoſes laſt recited: it takes away the collicke and pangs in the guts,not only in a cliſter, but alſo anointed vpon the places affected. But if this oile be made of the oile preſſed out of Lineſeed it will be ſo much the better,and of ſingular force to take away hard ſwellings of the ſpleene or milt. E

It is vſed with good ſucceſſe againſt the dropſie called in Greeke ὑποσάρκας, being applied to the belly in manner of a pultis. F

The herb a little boiled or ſcalded,and kept in pickle as Sampier,and eaten,quickens the ſight. G

The ſame applied with honey and the iuyce of Fennell is a remedie againſt dim eyes. H

The iuyce of Rue made hot in the rinde of a pomegranat and dropped into the eares,takes away the paine thereof. I

S. Anthonies fire is quenched therewith: it killeth the ſhingles,and running vlcers and ſores in the heads of yong children, if it be tempered with Ceruſe or white Lead,vineger,and oile of roſes, and made into the forme of *Nutritum* or *Triapharmacon*. K

Dioſcorides ſaith,that Rue put vp in the noſthrils ſtayeth bleeding. L

Of whoſe opinion *Pliny* alſo is; when notwithſtanding it is of power rather to procure bleeding through the ſharpe and biting qualitie that it hath. M

The leaues of Rue beaten and drunke with wine,are an antidote againſt poiſons,as *Pliny* ſaith. N

Dioſcorides writeth,that a twelue penny weight of the ſeed drunke in wine is a counterpoyſon againſt deadly medicines or the poyſon of Wolfs-bane, *Ixia*,Muſhroms,or Tode ſtooles,the biting of Serpents,ſtinging of Scorpions,ſpiders,bees,hornets,and waſps; and it is reported,that if a man be anointed with the iuyce of Rue theſe will not hurt him; and that the Serpent is driuen away at the ſmell thereof when it is burned, inſomuch that when the Weeſell is to fight with the Serpent, ſhe armeth her ſelfe by eating Rue againſt the might of the Serpent. O

The leaues of Rue eaten with the kernels of wallnuts or figs ſtamped together and made into a maſſe or paſte,is good againſt all euill aires, the peſtilence or plague,reſiſts poyſon and all venom. P

Rue boiled with Dil,Fennell ſeed,and ſome Sugar,in a ſufficient quantitie of wine, ſwageth the torments and griping paines of the belly, the paines in the ſides and breaſt, the difficulty of breathing,the cough,and ſtopping of the lungs, and helpeth ſuch as are declining to a dropſie. Q

The iuyce taken with Dill,as aforeſaid,helpeth the cold fits of agues,and alters their courſe: it helpeth the inflammation of the fundament, and paines of the gut called *Rectum inteſtinum*. R

The iuyce of Rue drunke with wine purgeth women after their deliuerance, driuing forth the ſecondine, the dead childe, and the vnnaturall birth. S

Rue vſed very often either in meate or drinke, quencheth and drieth vp the naturall ſeed of generation, and the milke of thoſe that giue ſucke. T

The oile wherein Rue hath beene boyled, and infuſed many dayes together in the Sun warmeth and chafeth all cold members if they be anointed therewith: alſo it prouoketh vrine if the region of the bladder be anointed therewith. V

If it be miniſtred in cliſters it expells windineſſe,and the torſion or gnawing paines of the guts. X

The leaues of garden Rue boiled in water and drunke,cauſeth one to make water,prouoketh the termes,and ſtoppeth the laske. Y

Ruta ſylueſtris or wilde Rue is much more vehement both in ſmell and operation, and therefore the more virulent or pernitious; for ſomtimes it fumeth out a vapor or aire ſo hurtfull that it ſcorcheth the face of him that looketh vpon it, raiſing vp bliſters, wheales, and other accidents; it venometh their hands that touch it, and will infect the face alſo, if it be touched with them before they be cleane waſhed; wherefore it is not to be admitted vnto meate or medicine. Z

The end of the ſecond Booke.

THE THIRD BOOKE OF THE HISTORIE OF PLANTS.

Containing the Deſcription, Place, Time, Names, Nature, and Vertues, of Trees, Shrubs, Buſhes, Fruit-bearing Plants, Roſins, Gums, Roſes, Heath, Moſſes: ſome Indian Plants, and other rare Plants not remembred in the Proeme to the firſt Booke. Alſo Muſhroms, Corall, and their ſeuerall kindes, &c.

The Proeme.

HAuing finiſhed the Treatiſe of Herbes and Plants in generall, vſed for meat, medicine, or ſweet ſmelling vſe, onely ſome few omitted for want of perfect inſtruction, and alſo being hindered by the ſlackeneſſe of the Cutters or Grauers of the thoſe; which wants we intend to ſupplie in this third and laſt part. The Tables as well generall as particular ſhall be ſet forth in the end of this preſent Volume.

CHAP. I. *Of Roſes.*

¶ *The Kindes.*

THe Plant of Roſes, though it be a ſhrub full of prickles, yet it had been more fit and conuenient to haue placed it with the moſt glorious floures of the world, than to inſert the ſame here among baſe and thornie ſhrubs: for the Roſe doth deſerue the chiefeſt and moſt principall place among all floures whatſoeuer; beeing not onely eſteemed for his beautie, vertues, and his fragrant and odoriferous ſmell; but alſo becauſe it is the honour and ornament of our Engliſh Scepter, as by the coniunction appeareth in the vniting of thoſe two moſt royall houſes of Lancaſter and Yorke. Which pleaſant floures deſerue the chiefeſt place in Crownes and garlands, as *Anacreon Thius* a moſt antient Greeke Poet (whom *Henricus Stephanus* hath tranſlated in a gallant Latine verſe) affirmes in thoſe verſes of a Roſe, beginning thus

Τὸ ῥόδον τὸ τῶν ἐρώτων, &c.

Roſa honos, decuſq; florum,
Roſa, cura, amorq; Veris.
Roſa, cælitum voluptas,
Roſeus puer Cytheres.
Caput implicat Corollis,
Charitum Choros frequentans

The

The Roſe is the honour and beautie of floures,
The Roſe is the care and loue of the Spring,
The Roſe is the pleaſure of th'heauenly powres:
The Boy of faire *Venus*, *Cytheras* darling,
Doth wrap in his head round with garlands of Roſe,
When to the dances of the Graces he goes.

Augerius Busbeckius ſpeaking of the eſtimation and honor of the Roſe, reporteth that the Turks can by no meanes indure to ſee the leaues of Roſes fall to the ground, becauſe that ſome of them haue dreamed, that the firſt or moſt antient Roſe did ſpring of the bloud of *Venus*; and others of the Mahumetans ſay, that it ſprang of the ſweat of *Mahumet*.

But there are many kindes of Roſes differing either in the bigneſſe of the floures, or the plant it ſelfe, roughneſſe or ſmoothneſſe, or in the multitude of the floures, or in the fewneſſe, or elſe in colour and ſmell: for diuers of them are high and tall, others ſhort and low; ſome haue fiue leaues, others very many. *Theophraſtus* telleth of a certaine Roſe growing about Philippi, with an hundred leaues, which the inhabitants brought forth of Pangæum, and planted it in Campania, as *Pliny* ſaith; which wee hold to be the Holland Roſe, that diuers call the Prouince Roſe, but not properly.

Moreouer, ſome be red, others white, and moſt of them or all, ſweetly ſmelling, eſpecially thoſe of the garden.

1 *Roſa alba.*
The White Roſe.

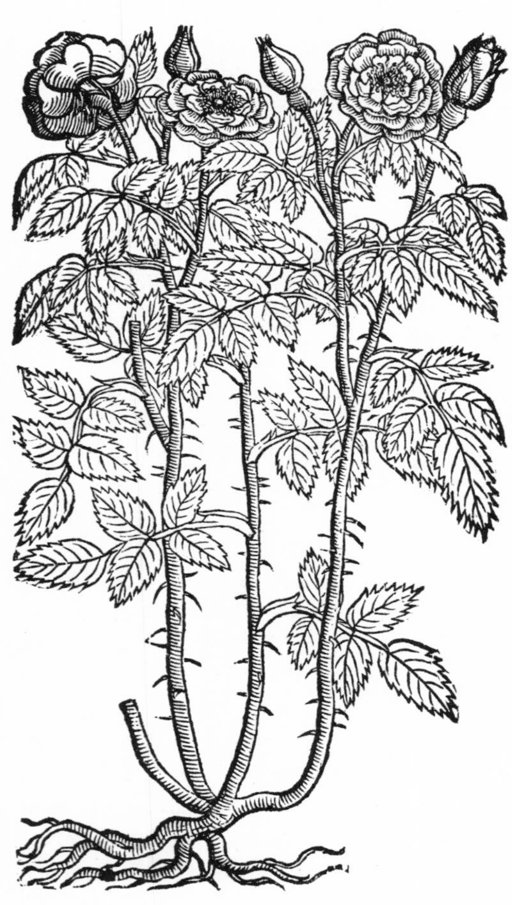

¶ *The Deſcription.*

1 If the curious could ſo be content, one general deſcription might ſerue to diſtinguiſh the whole ſtocke or kindred of the Roſes, beeing things ſo well knowne: notwithſtanding I thinke it not amiſſe to ſay ſomething of them ſeuerally, in hope to ſatisfie all. The white Roſe hath very long ſtalkes of a wooddie ſubſtance, ſet or armed with diuers ſharpe prickles: the branches whereof are likewiſe full of prickles, whereon doe grow leaues conſiſting of fiue leaues for the moſt part, ſet vpon a middle rib by couples; the old leafe ſtanding at the point of the ſame, and euery one of thoſe ſmall leaues ſomewhat ſnipt about the edges, ſomewhat rough, and of an ouerworne green colour: from the boſom wherof ſhoot forth long foot-ſtalks, whereon do grow very faire double floures, of a white colour, and very ſweet ſmell, hauing in the middle a few yellow threds or chiues; which being paſt there ſucceedeth a long fruit, greene at the firſt, but red when it is ripe, and ſtuffed with a downie choaking matter, wherein is contained ſeed as hard as ſtones. The root is long, tough, and of a wooddie ſubſtance.

2 The Red Roſe groweth very low in reſpect of the former: the ſtalkes are ſhorter, ſmoother, and browner of colour: the leaues are like, yet of a worſe duſtie colour: the floures grow on the tops of the branches, conſiſting of many leaues, of a perfect red colour: the fruit is likewiſe red when it is ripe: the root alſo wooddie.

3 The common Damaske Roſe in ſtature, prickley branches, and in other reſpects is like the white

2 *Rosa rubra.*
The red Rose.

3 *Rosa Provincialis, sive Damascena.*
The Prouince, or Damaske Rose.

5 *Rosa sine spinis.*
The Rose without prickles.

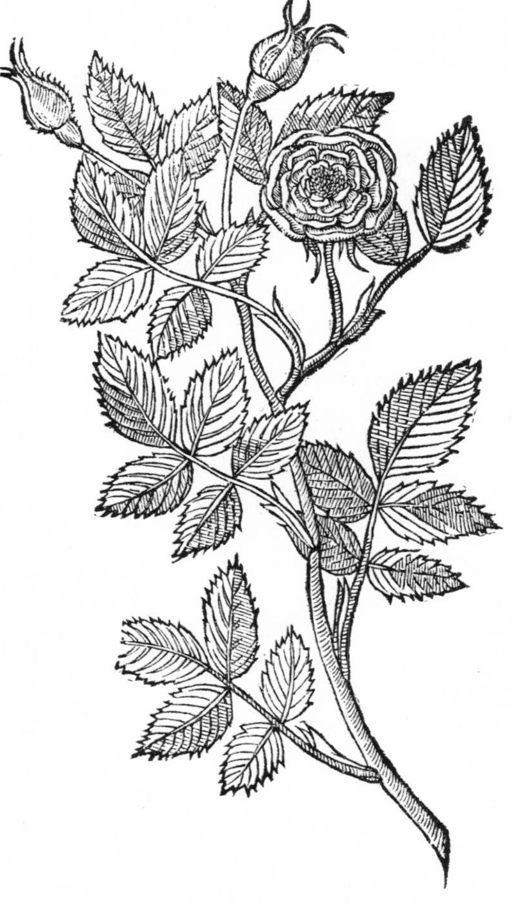

white Rose; the especiall difference consisteth in the colour and smell of the floures; for these are of a pale red colour, and of a more pleasant smell, and fitter for meate or medicine.

4 The *Rosa Provincialis minor*, or lesser Prouince Rose differeth not from the former, but is altogether lesser: the floures and fruit are like: the vse in physick also agreeth with the precedent.

5 The Rose without prickles hath many young shootes comming from the root, diuiding themselues into diuers branches, tough, and of a woody substance as are all the rest of the Roses, of the hight of two or three cubites, smooth and plaine without any roughnesse or prickles at all; whereon do grow leaues like those of the Holland Rose, of a shining deepe greene colour on the vpper side, vnderneath somewhat hoarie and hairy. The floures grow at the toppes of the branches, consisting of an infinite number of leaues, greater than those of the Damaske Rose, more double, and of a colour betweene the Red and Damaske Roses, of a most sweet smell. The fruit is round, red when it is ripe, and stuffed with the like flockes and seeds of those of the Damask Rose. The root is great, wooddie, and far spreading.

6 The

6 The Holland or Prouince Rose hath diuers shoots proceeding from a wooddie root, full of sharpe prickles, diuiding it selfe into diuers branches, whereon do grow leaues consisting of fiue leaues set vpon a rough middle rib, and those snipt about the edges: the floures grow on the tops of the branches, in shape and colour like the Damaske Rose, but greater and more double, insomuch that the yellow chiues in the middle are hard to be seene; of a reasonable good smell, but not ful so sweet as the common Damaske Rose: the fruit is like the other of his kinde.

6 *Rosa Hollandica, siue Batava.*
The great Holland Rose, commonly called the great Prouince Rose.

We haue in our London gardens one of the red Roses, whose floures are in quantitie and beauty equal with the former, but of greater estimation, of a perfect red colour, wherein especially it differeth from the Prouince Rose; in stalks, stature, and manner of growing it agreeth with our common red Rose.

¶ *Te Place.*

All these sorts of Roses we haue in our London gardens, except that Rose without prikles, which as yet is a stranger in England. The double white Rose doth grow wilde in many hedges of Lancashire in great abundance, euen as Briers do with vs in these Southerly parts, especially in a place of the countrey called Leyland, and in a place called Roughford, not far from Latham. Moreouer, in the said Leyland fields doth grow our garden Rose wilde, in the plowed fields among the corne in such abundance, that there may be gathered daily, during the time, many bushels of Roses, equall with the best garden Rose in each respect: the thing that giueth great cause of wonder is, that in a field in the place aforesaid, called Glouers field, euery yeare that the field is plowed for corne, that yeare the field will be spred ouer with Roses; and when it lyeth as they call it ley, and not plowed, then shall there be but few Roses to be gathered, by the relation of a curious Gentleman there dwelling, so often remembred in our Historie.

‡ I haue heard that the Roses which grow in such plenty in Glouers field, euery yeare the field is plowed, are no other than corne Rose, that is, red Poppies, howeuer our Author was informed. ‡

¶ *The Time.*

These floure from the end of May to the end of August, and diuers times after, by reason the tops and superfluous branches are cut away in the end of their flouring: & then do they sometimes flou re euen vntill October, and after.

¶ *The*

¶ *The Names.*

The Rose is called in Latine *Rosa*: in Greeke ῥόδον: and the plant it selfe ῥοδωνία: (which in Latine keepeth the same name that the floure hath) and it is called *Rodon* (as *Plutarch* saith) because it sendeth forth plenty of smell.

The middle part of the Roses, that is, the yellow chiues, or seeds and typs, is called *Anthos*, and *Flos Rosæ*, the floure of the Rose: in shops, *Anthera*, or the blowing of the rose.

The white parts of the leaues of the floure it selfe, by which they are fastened to the cups, be named *Vngues* or nails. That is called *Calix*, or the cup, which containeth and holdeth in together the yellow part and leaues of the floure.

Alabastri, are those parts of the cup which are deeply cut, & that compasse the floure close about before it be opened, which be in number fiue, two haue beards and two haue none, and the fift hath but halfe one: most do call them *Cortices Rosarum*, or the husks of the roses: the shoots of the plant of roses, *Strabo Gallus* in his little garden doth call *Viburna*.

The white Rose is called *Rosa alba*: in English, the white Rose: in high Dutch, **Weisz Roosen**: in low Dutch, **Witte Roosen**: in French, *Rose Blanche*: of *Plinie*, *Spineola Rosa*, or *Rosa Campana*.

The red Rose is called in Latine, *Rosa rubra*: the Frenchmen, *Rose Franche*, *Rose de Prouins*, a towne in Campaigne: of *Plinie*, *Trachinia*, or *Prænestina*.

The Damaske Rose is called of the Italians *Rosa incarnata*: in high Dutch, **Leibfarbige Roosen**: in low Dutch, **Prouencie Roose**: of some, *Rosa Prouincialis*, or Rose of Prouence: in French of some, *Melesia*: the Rose of Melaxo, a citie in Asia, from whence some haue thought it was first brought into those parts of Europe.

The great Rose, which is generally called the great Prouence rose, which the Dutch men cannot endure; for say they, it came first out of Holland, and therefore to be called the Holland Rose: but by all likelihood it came from the Damaske rose, as a kinde thereof, made better and fairer by art, which seemeth to agree with truth.

The rose without prickles is called in Latine, *Rosa sine spinis*, and may be called in English, the rose without thornes, or the rose of Austrich, because it was first brought from Vienna, the Metropolitan citie of Austrich, and giue nto that famous Herbarist *Carolus Clusius*.

¶ *The Temperature.*

The leaues of the floures of roses, because they doe consist of diuers parts, haue also diuers and sundry faculties: for there be in them certain that are earthy and binding, others moist and watery, and sundrie that are spirituall and airie parts, which notwithstanding are not all after one sort, for in one kinde these excell, in another those, all of them haue a predominant or ouerruling cold temperature, which is neerest to a meane, that is to say, of such as are cold in the first degree, moist, airie, and spirituall parts are predominant in the White roses, Damaske and Muske.

¶ *The Vertues.*

The distilled water of roses is good for the strengthning of the heart, & refreshing of the spirits, and likewise for all things that require a gentle cooling. A

The same being put into iunketting dishes, cakes, sauces, and many other pleasant things, giueth a fine and delectable taste. B

It mitigateth the paine of the eies proceeding of a hot cause, bringeth sleep, which also the fresh roses themselues prouoke through their sweet and pleasant smell. C

The iuice of these roses, especially of Damask, doth moue to the stoole, and maketh the belly soluble: but most effectually that of the Musk roses: next to them is the iuice of the Damask, which is more commonly vsed. D

The infusion of them doth the same, and also the syrrup made thereof, called in Latine *Drosatum*, or *Serapium*: the Apothecaries call it Syrrup of roses solutiue, which must be made of the infusion in which a great number of the leaues of these fresh roses are diuers and sundry times steeped. E

It is profitable to make the belly loose & soluble, when as either there is no need of other stronger purgation, or that it is not fit and expedient to vse it: for besides those excrements which stick to the bowels, or that in the first and neerest veines remaine raw, flegmaticke, and now and then cholericke, it purgeth no other excrements, vnlesse it be mixed with certaine other stronger medicines. F

This syrrup doth moisten and coole, and therefore it alayeth the extremitie of heat in hot burning feuers, mitigateth the inflammations of the intrails, and quencheth thirst: it is scarce good for a weake and moist stomacke, for it leaueth it more slacke and weake. G

Of like vertue also are the leaues of these preserued in Sugar, especially if they be onely bruised with the hands, and diligently tempered with Sugar, and so heat at the fire rather than boiled. H

¶ *The Temperature of Red Roses.*

There is in the red Roses, which are common euery where, and in the other that be of a deep purple, called Prouence roses, a more earthie substance, also a drying and binding qualitie, yet not I without certaine moiſture ioined, being in them when they are as yet freſh, which they loſe when they be dried: for this cauſe their iuice and infuſion doth alſo make the bodie ſoluble, yet not ſo much as of the others aforeſaid. Theſe roſes being dried and their moiſture gone, do bind and dry; and likewiſe coole, but leſſer than when they are freſh.

¶ *The Vertues.*

I They ſtrengthen the heart, and helpe the trembling and beating thereof.

K They giue ſtrength to the liuer, kidneies, and other weake intrails; they dry and comfort a weak ſtomacke that is flaſhie and moiſt; ſtay the whites and reds, ſtanch bleedings in any part of the body, ſtay ſweatings, binde and looſe, and moiſten the body.

L And they are put into all manner of counterpoiſons and other like medicines, whether they be to be outwardly applied or to be inwardly taken, to which they giue an effectuall binding, and certaine ſtrengthning qualitie.

M Honie of Roſes, or *Mel Roſarum*, called in Greeke ῥοδόμελι, which is made of them, is moſt excellent good for wounds, vlcers, iſſues, and generally for ſuch things as haue need to be clenſed and dried.

N The oile doth mitigate all kindes of heat, and will not ſuffer inflammations or hot ſwellings to riſe, and being riſen it doth at the firſt aſſwage them.

¶ *The Temperature and Vertues of the parts.*

O The floures or bloomings of Roſes, that is to ſay, the yellow haires and tips, do in like maner dry and binde, and that more effectually than of the leaues of the roſes themſelues: the ſame temperature the cups and beards be of; but ſeeing none of theſe haue any ſweet ſmell, they are not ſo profitable, nor ſo familiar or beneficiall to mans nature: notwithſtanding in fluxes at the ſea, it ſhall auaile the Chirurgion greatly, to carry ſtore thereof with him, which doth there preuaile much more than at the land.

P The ſame yellow called *Anthera*, ſtaieth not onely thoſe lasks and bloudy fluxes which do happen at the ſea, but thoſe at the land alſo, and likewiſe the white flux and red in women, if they bee dried, beaten to pouder, and two ſcruples thereof giuen in red wine, with a little powder of Ginger added thereto: and being at the ſea, for want of red wine you may vſe ſuch liquour as you can get in ſuch extremitie.

Q The little heads or buttons of the Roſes, as *Pliny* writeth, do alſo ſtanch bleeding, and ſtoppe the laske.

R The nailes or white ends of the leaues of the floures are good for watering eies.

S The iuice, infuſion, or decoction of Roſes, are to be reckoned among thoſe medicines which are ſoft, gentle, looſing, opening and purging gently the belly, which may be taken at all times and in all places, of euery kinde or ſex of people, both old and yong, without danger or perill.

T The ſyrrup made of the infuſion of Roſes, is a moſt ſingular & gentle looſing medicine, carrying downwards cholericke humors, opening the ſtoppings of the liuer, helping greatly the yellow iaundies, the trembling of the heart, & taking away the extreme heat in agues and burning feuers which is thus made:

V Take two pound of Roſes, the white ends cut away, put them to ſteepe or infuſe in ſix pintes of warme water in an open veſſell for the ſpace of twelue houres: then ſtraine them out, and put thereto the like quantitie of Roſes, and warme the water again, ſo let it ſtand the like time: do thus foure or fiue times; in the end adde vnto that liquor or infuſion, foure pound of fine ſugar in powder; then boyle it vnto the forme of a ſyrrup, vpon a gentle fire, continually ſtirring it vntill it be cold; then ſtraine it, and keepe it for your vſe, whereof may be taken in white wine, or other liquour, from one ounce vnto two.

X Syrrup of the iuice of Roſes is very profitable for the griefes aforeſaid, made in this manner:

Y Take Roſes, the white nailes cut away, what quantitie you pleaſe, ſtampe them, and ſtraine out the iuice, the which you ſhall put to the fire, adding thereto ſugar, according to the quantity of the iuice: boiling them on a gentle fire vnto a good conſiſtence.

Z Vnto theſe ſyrrups you may adde a few drops of oyle of Vitriol, which giueth it a moſt beautifull colour, and alſo helpeth the force in cooling hot and burning feuers and agues: you likewiſe may adde thereto a ſmall quantitie of the iuice of Limons, which doth the like.

A The conſerue of Roſes as well that which is crude and raw, as that which is made by ebullition or boiling, taken in the morning faſting, and laſt at night, ſtrengthneth the heart, and taketh away the ſhaking and trembling thereof, ſtrengthneth the liuer, kidneies, and other weake intrails, comforteth a weake ſtomacke that is moiſt and raw; ſtaieth the whites and reds in women, and in a word is the moſt familiar thing to be vſed for the purpoſes aforeſaid, and is thus made:

B Take the leaues of Roſes, the nails cut off, one pound, put them into a clean pan; then put thereto a pinte and a halfe of ſcalding water, ſtirring them together with a woodden ſlice, ſo let them ſtand

to

to macerate, close couered some two or three houres; then set them to the fire slowly to boyle, adding thereto three pounds of sugar in powder, letting them to simper together according to discretion, some houre or more; then keepe it for your vse.

The same made another way, but better by many degrees: take Roses at your pleasure, put them C
to boyle in faire water, hauing regard to the quantity; for if you haue many roses, you may take the more water; if fewer, the lesse water will serue: the which you shall boyle at the least three or foure houres, euen as you would boyle a piece of meat, vntill in the eating they be very tender, at which time the roses will lose their colour, that you would thinke your labour lost, and the thing spoyled. But proceed, for though the Roses haue lost their colour, the water hath gotten the tincture thereof; then shall you adde vnto one pound of Roses, foure pound of fine sugar in pure powder, and so according to the rest of the roses. Thus shall you let them boyle gently after the Sugar is put therto, continually stirring it with a woodden Spatula vntill it be cold, whereof one pound weight is worth six pound of the crude or raw conserue, as well for the vertues and goodnesse in taste, as also for the beautifull colour.

The making of the crude or raw conserue is very well knowne, as also Sugar roset, and diuers D
other pretty things made of roses and sugar, which are impertent vnto our historie, because I intend neither to make thereof an Apothecaries shop, nor a Sugar bakers storehouse, leauing the rest for our cunning confectioners.

CHAP. 2. *Of the Muske Roses.*

¶ *The Kindes.*

THere be diuers sorts of Roses planted in gardens, besides those written of in the former chapter, which are of most writers reckoned among the wilde roses, notwithstanding we thinke it conuenient to put them into a chapter betweene those of the garden and the brier roses, as indifferent whether to make them of the wilde roses, or of the tame, seeing we haue made them denizons in our gardens for diuers respects, and that worthily.

1 *Rosa Moschata simplici flore.*
The single Muske rose.

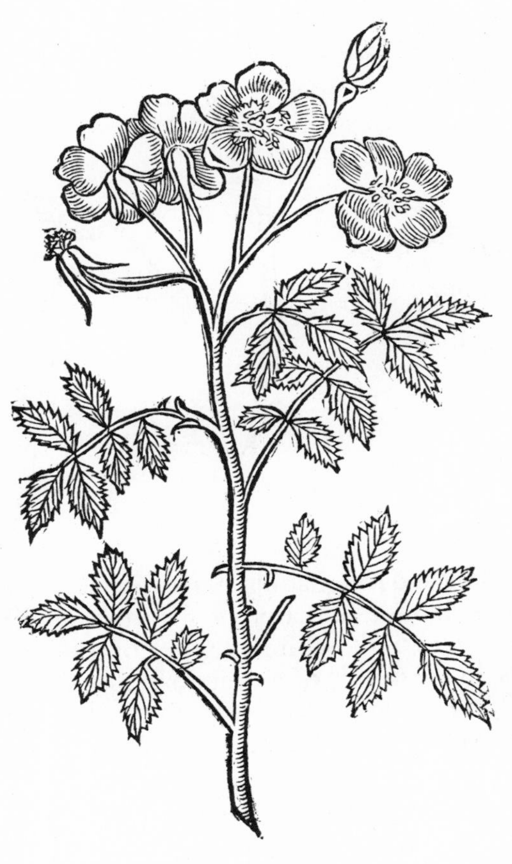

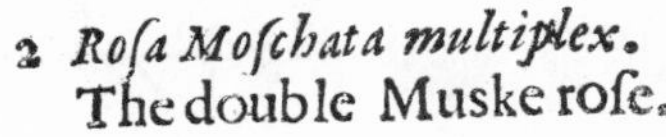

2 *Rosa Moschata multiplex.*
The double Muske rose.

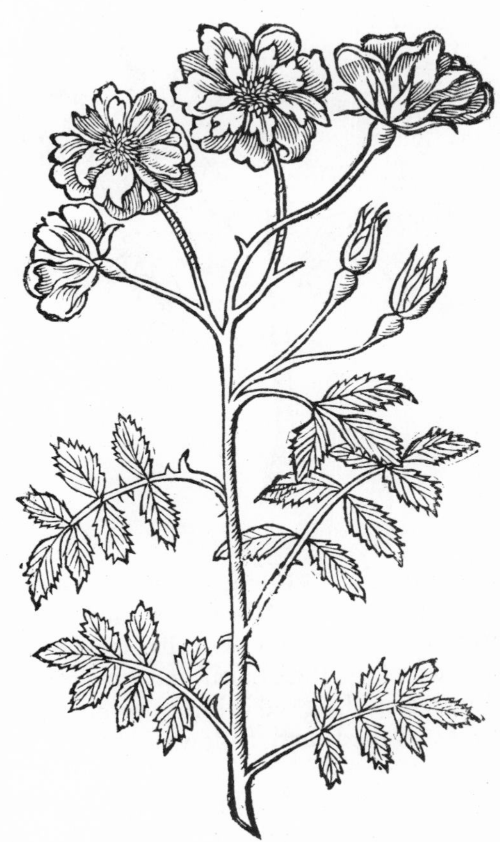

¶ *The Deſcription.*

1 THe ſingle Muske Roſe hath diuers long ſhoots of a greeniſh colour and wooddie ſubſtance, armed with very ſharpe prickles, diuiding it ſelfe into diuers branches: whereon do grow long leaues, ſmooth & ſhining, made of diuers leaues ſet vpon a middle rib, like the other roſes: the floures grow on the tops of the branches, of a white colour, and pleaſant ſweet ſmell, like that of Muske, whereof it tooke his name; hauing certain yellow ſeeds in the middle, as the reſt of the roſes haue: the fruit is red when it is ripe, and filled with ſuch chaffie flocks and ſeeds as thoſe of the other roſes: the root is tough and wooddie.

2 The double Muske roſe differeth not from the precedent in leaues, ſtalks, and roots, nor in the colour of the floures, or ſweetneſſe thereof, but onely in the doubleneſſe of the floures, wherein conſiſteth the difference.

3 Of theſe roſes we haue another in our London gardens, which of moſt is called the bluſh roſe; it floureth when the Damaske roſe doth: the floures hereof are very ſingle, greater than the other Muske roſes, and of a white colour, daſht ouer with a light waſh of carnation, which maketh that colour which wee call a bluſh colour: the proportion of the whole plant, as alſo the ſmell of the floures, are like the precedent.

3 *Roſæ Moſchatæ ſpecies maior.*
The great Muske roſe.

4 *Roſa Holoſericea.*
The veluet roſe.

4 The Veluet roſe groweth alwaies very low, like vnto the red roſe, hauing his branches couered with a certaine hairie or prickley matter, as fine as haires, yet not ſo ſharpe or ſtiffe that it will harme the moſt tender skin that is: the leaues are like the leaues of the white roſe: the floures grow at the top of the ſtalks, doubled with ſome yellow thrums in the midſt, of a deepe and blacke red colour, reſembling red crimſon veluet, whereupon ſome haue called it the Veluet roſe: when the floures be vaded, there follow red berries full of hard ſeeds, wrapped in a downe or woollineſſe like the others.

5 The yellow roſe which (as diuers do report) was by Art ſo coloured, and altered from his firſt eſtate, by graffing a wilde roſe vpon a Broome ſtalke; whereby (ſay they) it doth not onely change his colour, but his ſmell and force. But for my part I hauing found the contrarie by mine owne experience, cannot be induced to beleeue the report: for the roots and off-ſprings of this roſe haue brought

brought forth yellow roses, such as the maine stocke or mother bringeth out, which euent is not to be seen in all other plants that haue been graffed. Moreouer, the seeds of yellow roses haue brought forth yellow roses, such as the floure was from whence they were taken; which they would not do by any coniecturall reason, if that of themselues they were not a naturall kind of rose. Lastly, it were contrary to that true principle,

Naturæ sequitur semina quodque suæ: that is to say;

Euery seed and plant bringeth forth fruit like vnto it selfe, both in shape and nature: but leauing that errour, I will proceed to the description: the yellow rose hath browne and prickly stalks or shoots, fiue or six cubits high, garnished with many leaues, like vnto the Muske rose, of an excellent sweet smell, and more pleasant than the leaues of the Eglantine: the floures come forth among the leaues, and at the top of the branches of a faire gold yellow colour: the thrums in the middle, are also yellow: which being gone, there follow such knops or heads as the other roses do beare.

5 *Rosa lutea.*
The yellow rose.

‡ 6 *Rosa Lutea multiplex.*
The double yellow rose.

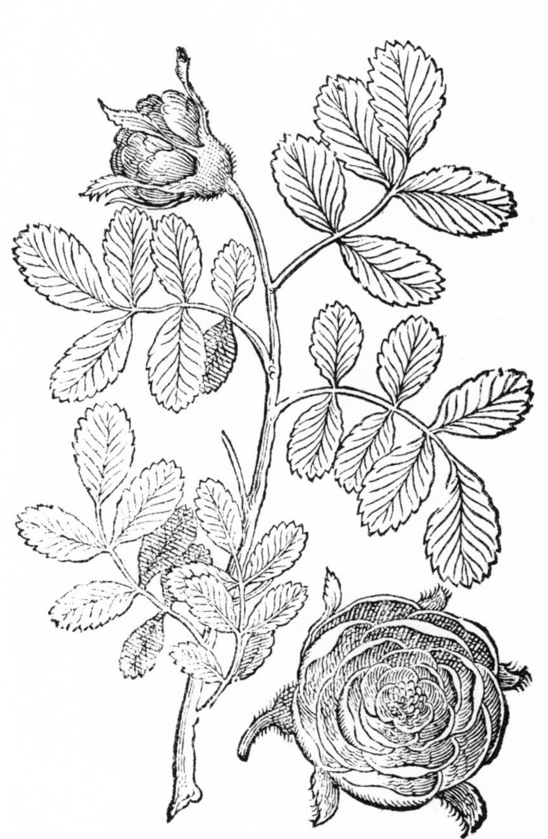

‡ 6 Of this kinde there is another more rare and set by, which in stalks, leaues, and other parts is not much different from the last described, onely the floure is very double, and it seldome fairly shewes it selfe about London, where it is kept in our chiefe gardens as a prime raritie. ‡

7 The Canell or Cinnamon rose, or the rose smelling like Cinnamon, hath shoots of a brown colour, foure cubits high, beset with thorny prickles, and leaues like vnto those of Eglantine, but smaller and greener, of the sauour or smell of Cinnamon, whereof it tooke his name, and not of the smell of his floures (as some haue deemed) which haue little or no sauour at all: the floures be exceeding double, and yellow in the middle, of a pale red colour, and sometimes of a carnation: the root is of a wooddie substance.

8 We haue in our London gardens another Cinnamon or Canell rose, not differing from the last described in any respect, but onely in the floures; for as the other hath very double floures, contrariwise these of this plant are verie single, wherein is the difference.

7 *Roſa Cinnamomea pleno flore.*
The double Cinnamon Roſe.

‡ 8 *Roſa Cinnamomea flore ſimplici.*
The ſingle Cinnamon Roſe.

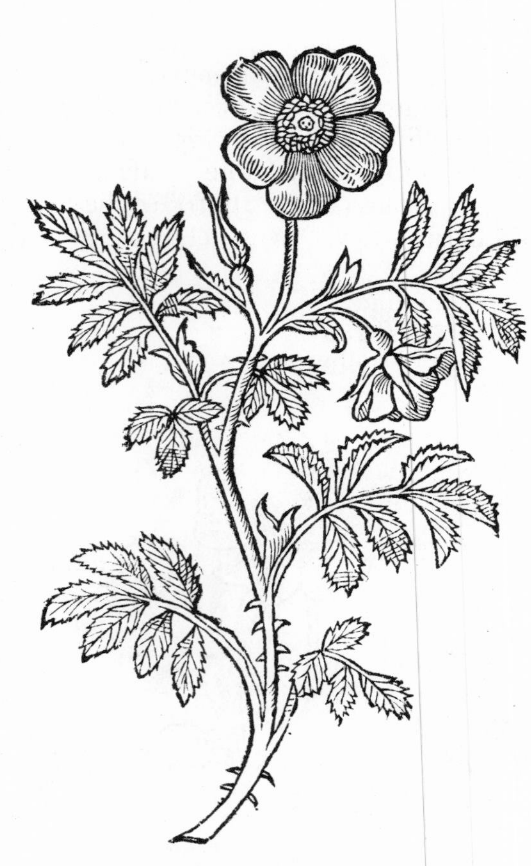

¶ *The Place.*

Theſe Roſes are planted in our London gardens, and elſewhere, but not found wilde in England.

¶ *The Time.*

The Muske Roſe floureth in Autumne, or the fall of the leafe: the reſt floure when the Damask and red Roſe do.

¶ *The Names.*

The firſt is called *Roſa Moſchata*, of the ſmell of Muske, as we haue ſaid: in Italian, *Roſa Moſchetta*: in French, *Roſes Muſquees*, or *Muſcadelles*: in Low Dutch, **Musket roosen**: in Engliſh, Musk Roſe: the Latine and Engliſh titles may ſerue for the reſt.

¶ *The Temperature.*

The Muske roſe is cold in the firſt degree, wherein airie and ſpiritual parts are predominant: the reſt are referred to the Brier roſe and Eglantine.

¶ *The Vertues.*

A Conſerue or ſyrrup made of the Muske roſe, in manner as before told in the Damaske and red roſes, doth purge very mightily wateriſh humors, yet ſafely, and without all danger, taken in the quantitie of an ounce in weight.

B The leaues of the floures eaten in the morning, in manner of a ſallad, with oile, vineger and pepper, or any other way according to the appetite and pleaſure of them that ſhall eat it, purge very notably the belly of wateriſh and cholericke humors, and that mightily, yet without all perill or paine at all, inſomuch as the ſimpleſt may vſe the quantitie, according to their owne fancie; for if they do deſire many ſtooles, or ſieges, they are to eat the greater quantity of the leaues; if fewer, the leſſe quantitie; as for example: the leaues of twelue or foureteene floures giue ſix or eight ſtooles, and ſo increaſing or diminiſhing the quantitie, more or fewer, as my ſelfe haue often proued.

C The white leaues ſtamped in a woodden diſh with apeece of Allum and the iuice ſtrained forth into ſome glaſed veſſell, dried in the ſhadow, and kept, is the moſt fine and pleaſant yellow colour that may be diuiſed, not only to limne or waſh pictures and Imagerie in books, but alſo to colour meates and ſauces, which notwithſtanding the Allum is very wholſome.

There

There is not any thing extant of the others, but are thought to be equall with the white Muske Rose, whereof they are taken and holden to be kindes.

CHAP. 3. *Of the wilde Roses.*

¶ *The Description.*

1 THe sweet Brier doth oftentimes grow higher than all the kindes of Roses; the shoots of it are hard, thicke, and wooddie; the leaues are glittering, and of a beautifull greene colour, of smell most pleasant: the Roses are little, fiue leaued, most commonly whitish, seldom tending to purple, of little or no smell at all: the fruit is long, of colour somewhat red, like a little oliue stone, and like the little heads or berries of the others, but lesser than those of the garden: in which is contained rough cotton, or hairie downe and seed, folded and wrapped vp in the same, which is small and hard: there be likewise found about the slender shoots hereof, round, soft, and hairie spunges, which we call Brier Balls, such as grow about the prickles of the Dog-rose.

1 *Rosa syluestris odora.* The Eglantine, or sweet Brier.

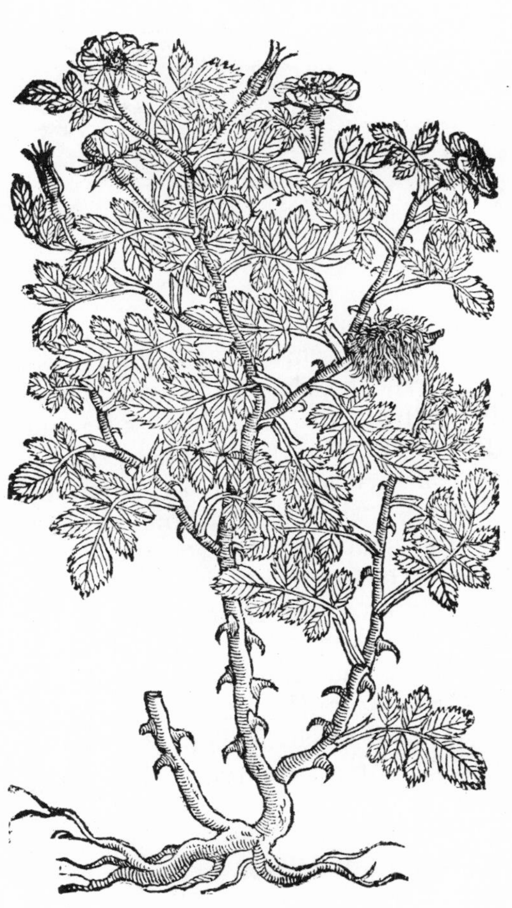

2 We haue in our London gardens another sweet Brier, hauing greater leaues, and much sweeter: the floures likewise are greater, and somewhat doubled, exceeding sweet of smell, wherein it differeth from the former.

3 The Brier Bush or Hep tree, is also called *Rosa canina*, which is a plant so common and well knowne, that it were to small purpose to vse many words in the description thereof: for euen children with great delight eat the berries thereof when they be ripe, make chaines and other prettie gewgawes of the fruit: cookes and gentlewomen make Tarts and such like dishes for pleasure thereof, and therefore this shall suffice for the description.

4 The Pimpinell rose is likewise one of the wilde ones, whose stalks shoot forth of the ground in many places, of the height of one or two cubits, of a browne colour, and armed with sharpe pric-

‡ 2 *Rosa syl. odora flore duplici.*
The double Eglantine.

3 *Rosa Canina inodora.*
The Brier Rose, or Hep tree.

4 *Rosa Pimpinellæ folio.*
The Pimpinell Rose.

kles, which diuide themselues toward the tops into diuers branches, whereon doe grow leaues consisting of diuers small ones, set vpon a middle rib like those of Burnet, which is called in Latine *Pimpinella*, whereupon it was called *Rosa Pimpinella*, the Burnet Rose. The floures grow at the tops of the branches, of a white colour, very single, and like vnto those of the Brier or Hep tree: after which come the fruit, blacke, contrarie to all the rest of the roses, round as an apple; whereupon some haue called it *Rosa Pomifera*, or the Rose bearing apples: wherein is contained seed, wrapped in chaffie or flockie matter, like that of the Brier: the root is tough and wooddie.

¶ *The Place.*

These wilde Roses do grow in the borders of fields and woods, in most parts of England. The last groweth very plentifully in a field as you go from a village in Essex, called Graies (vpon the brinke of the riuer Thames) vnto Horndon on the hill, insomuch that the field is full fraught therewith all ouer.

It groweth likewise in a pasture as you goe from a village hard by London called Knights bridge, vnto Fulham, a village thereby, and in many other places.

We haue them all except the Brier Bush in our London gardens, which we thinke vnworthy the place.

¶ *The*

¶ *The Time.*

They floure and flourish with the other Roses.

¶ *The Names.*

The Englantine Rose, which is *Cynorrhodi*, or *Caninæ Rosæ species*, a kinde of Dogs Rose: and *Rosa syluestris*, the wild Rose: in low-Dutch, **Eglantier:** in French, *Esglentine*; and as *Ruellius* testifies, *Eglenterium*: who also suspects it to be *Cynosbaton*, or *Canirubus*: of which *Dioscorides* hath written in these words; *Cynosbatus*, or *Canirubus*, which some call *Oxycantha*, is a shrub growing like a tree, full of prickles, with a white floure, long fruit like an oliue stone, red when it is ripe, and downie within: in English we call it Eglantine, or sweet Brier.

The spongie balls which are found vpon the branches are most aptly and properly called *Spongiolæ syluestris Rosæ*, the little sponges of the wilde Rose. The shops mistake it by the name of *Bedeguar*; for *Bedeguar* among the Arabians is a kinde of Thistle, which is called in Greeke Ἄκανθα λευκή: that is to say, *Spina alba* the white Thistle, not the white Thorne, though the word doe import so much.

The Brier or Hep tree is called *Syluestris Rosa*, the wilde Rose: in high-Dutch, **Wilden Rosen:** in French, *Roses sauuages*: *Pliny*, *lib*.8.*cap*.25. saith that it is *Rosa Canina*, Dogs Rose: of diuers, *Caninasentis*, or Dogs Thorne: in English, Brier bush, and Hep tree: the last hath been touched in the description.

¶ *The Temperature and Vertues.*

A The faculties of these wilde Roses are referred to the manured Rose, but not vsed in physicke where the other may be had: notwithstanding *Pliny* affirmeth, that the root of the Brier bush is a singular remedie found out by oracle, against the biting of a mad dog, which he sets downe in his eighth booke, chap.41.

B The same Author, *lib*.25.*cap*.2. affirmeth, that the little spongie Brier ball stamped with honey and ashes causeth haires to grow which are fallen through the disease called *Alopecia*, or the Foxes euill, in plaine termes the French pocks.

C *Fuchsius* affirmes, that the spongie excrescence or ball growing vpon the Brier are good against the stone and strangurie, if they be beaten to pouder and inwardly taken.

D They are good not as they be diureticks or prouokers of vrine, or as they are wearers away of the stone, but as certaine other binding medicines that strengthen the weake and feeble kidneyes; which do no more good to those that be subiect to the stone, than many of the diuretickes, especially of the stronger sort; for by too much vsing of diureticks or pissing medicines, it hapneth that the kidneyes are ouer-weakened, and often times too much heated, by which meanes not only the stones are not diminished, worne away, or driuen forth, but oftentimes are also increased and made more hard: for they separate and take away that which in the bloud is thin, waterie, and as it were wheyish; and the thicker part, the stronger sorts of diuretickes do draw together and make hard: and in like maner also others that are not so strong, by the ouermuch vsing of them, as *Galen*.*lib*.5. of the faculties of simple medicines reporteth.

E The fruit when it is ripe maketh most pleasant meats and banqueting dishes, as tarts and such like; the making whereof I commit to the cunning cooke, and teeth to eat them in the rich mans mouth.

CHAP. 4. *Of the Bramble or black-Berry bush.*

¶ *The Description.*

1 THe common Bramble bringeth forth slender branches, long, tough, easily bowed, tamping among hedges and whatsoeuer stands neere vnto it; armed with hard and sharpe prickles, whereon doe grow leaues consisting of many set vpon a rough middle rib, greene on the vpper side, and vnderneath somewhat white: on the tops of the stalks stand certaine floures, in shape like those of the Brier Rose, but lesser, of colour white, and sometimes washt ouer with a little purple: the fruit or berry is like that of the Mulberry, first red, blacke when it is ripe, in taste betweene sweet and soure, very soft, and full of grains: the root creepeth, and sendeth forth here and there yong springs.

‡ *Rubus repens fructu cæsio.*

‡ 2 This hath a round stalke set full of small crooked and very sharpe pricking thornes, and creepeth on hedges and low bushes of a great length, on the vpper side of a light red colour, and vnderneath greene, and taketh root with the tops of the trailing branches, whereby it doth mightily encrease:

encreaſe: the leaues grow without order, compoſed of three leaues, and ſometimes of fiue, or elſe the two lower leaues are diuided into two parts, as Hop leaues are now and then, of a light greene colour both aboue and vnderneath. The floures grow on the tops of the branches, *racematim*, many together, ſometimes white, ſometimes of a very light purple colour, euery floure containing fiue leaues, which are crompled or wrinkled, and do not grow plaine: the fruit followes, firſt green, and afterwards blew, euerie berry compoſed of one or two graines, ſeldome oboue foure or fiue growing together, about the bigneſſe of corans; wherein is contained a ſtony hard kernell or ſeed, and a iuyce of the colour of Claret wine, contrarie to the common *Rubus* or Bramble, whoſe leaues are white vnderneath: the berries being ripe are of a ſhining blacke colour, and euery berry containes vſually aboue forty graines cloſely compacted and thruſt together. The root is wooddy and laſting. This growes common enough in moſt places, and too common in ploughed fields. Sept. 6. 1619. *Iohn Goodyer*. ‡

3 The Raſpis or Framboiſe buſh hath leaues and branches not much vnlike the common Bramble, but not ſo rough nor prickly, and ſometimes without any prickles at all, hauing onely a rough hairineſſe about the ſtalkes: the fruit in ſhape and proportion is like thoſe of the Bramble, red when they be ripe, and couered ouer with a little downineſſe; in taſte not very pleaſant. The root creepeth far abroad, whereby it greatly encreaſeth. ‡ This growes either with prickles vpon the ſtalkes, or elſe without them: the fruit is vſually red, but ſometimes white of colour. ‡

1 *Rubus*.
The Bramble buſh.

2 *Rubus Idæus*.
The Raſpis buſh or Hinde-berry.

4 Stone Bramble ſeldome groweth aboue a foot high, hauing many ſmall flexible branches without prickles, trailing vpon the ground, couered with a reddiſh barke, and ſomwhat hairy: the leaues grow three together, ſet vpon tender naked foot-ſtalkes ſomewhat ſnipt about the edges: the floures grow at the end of the branches, conſiſting of foure ſmall white leaues like thoſe of the Cherry tree: after which come ſmall Grape-like fruit, conſiſting of one, two, or three large tranſparent berries, ſet together as thoſe of the common Bramble, of a red colour when they be ripe, and of a pleaſant taſte, but ſomewhat aſtringent. The roots creepe along in the ground very farre abroad, whereby it greatly increaſeth.

4 *Chamæmorus* (called in the North part of England, where they eſpecially doe grow, Knotberries, and Knought-berries) is likewiſe one of the Brambles, though without prickles: it brings forth

forth small weake branches or tender stems of a foot high; whereon do grow at certaine distances rough leaues in shape like those of the Mallow, not vnlike to the leaues of the Gooseberrie bush: on the top of each branch standeth one floure and no more, consisting of fiue small leaues of a dark purple colour: which being fallen, the fruit succeedeth, like vnto that of the Mulberrie, whereof it was called *Chamæmorus*, dwarfe Mulberry; at the first white and bitter, after red and somwhat pleasant: the root is long, something knotty; from which knots or ioynts thrust forth a few threddie strings. ‡ I take that plant to which our Author hereafter hath allotted a whole chapter, and called *Vaccinia nubis*, or Cloud-berries, to be the same with this, as I shall shew you more largely in that place. ‡

4 *Rubus Saxatilis.*
Stone blacke-Berry bush.

5 *Chamæmorus.*
Knot berry bush.

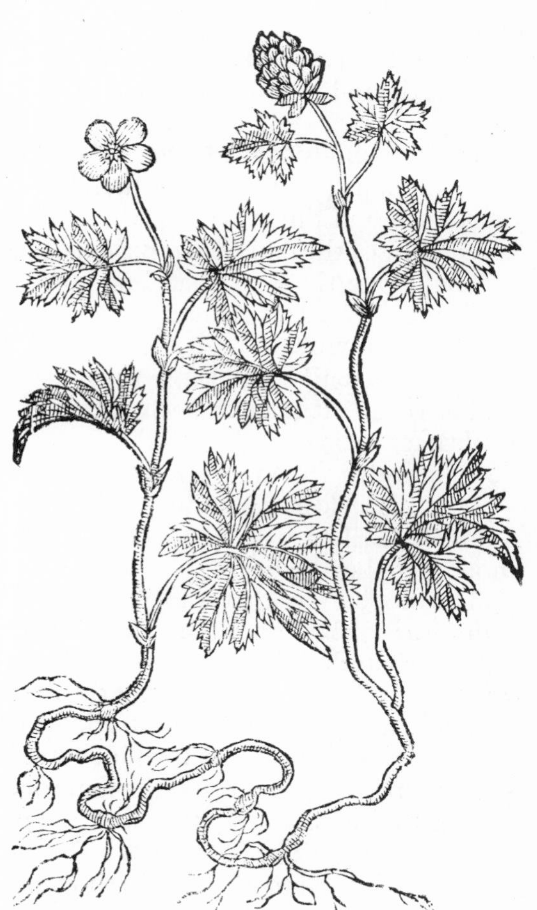

¶ *The Place.*

The Bramble groweth for the most part in euery hedge and bush.

The Raspis is planted in gardens: it groweth not wilde that I know of, except in the field by a village in Lancashire called Harwood, not far from Blackburne.

I found it among the bushes of a causey, neere vnto a village called Wisterson, where I went to schoole, two miles from the Nantwich in Cheshire.

The stone Bramble I haue found in diuers fields in the Isle of Thanet, hard by a village called Birchinton, neere Queakes house, sometimes Sir *Henry Crispes* dwelling place. ‡ I feare our Author mistooke that which is here added in the second place, for that which he figured and described in the third (now the fourth) which I know not yet to grow wilde with vs. ‡

Knot-berries do loue open snowie hills and mountaines; they grow plentifully vpon Ingleborow hils among the heath and ling, twelue miles from Lancashire, being thought to be the highest hill in England.

They grow vpon Stane-more betweene Yorkshire and Westmerland, and vpon other wet Fells and mountaines.

¶ *The Time.*

These floure in May and Iune with the Roses: their fruit is ripe in the end of August and September.

¶ The

¶ *The Names.*

The Bramble is called in Greeke βάτος: in French, *Rouges, Loi Duyts Brelmers*: in Latine, *Rubus*, and *Sentis*, and *Vepres*, as *Ouid* writeth in his first booke of Metamorpho sis.

Aut Lepori qui vepre latens hostilia cernit
Ora canum. ———

Or to th' Hare, that vnder Bramble closely lying, spies
The hostile mouthes of Dogs. ———

Of diuers it is called *Cynosbatus*, but not properly; for *Cynosbatus* is the wild Rose, as we haue written: in high-Dutch, **Bremen**: in low-Dutch, **Breemen**: in French, *Rouce*: in Italian, *Garza*: in English, Bramble bush, and Black-berry bush.

The fruit is named in Latine *Morum rubi*; and as *Fuchsius* thinketh, *Vacinium*, but not properly: in shops, *Mora Bati*: and in such shops as are more barbarous, *Mora Bassi*: in English, Blackeberries.

The Raspis is called in Greeke βάτος ἰδαῖα: in Latine, *Rubus Idæus*, of the mountaine Ida on which it groweth: in English, Raspis, Framboise, and Hinde-berry.

¶ *The Temperature and Vertues.*

A The yong buds or tender tops of the Bramble bush, the floures, the leaues, and the vnripe fruit, do very much dry and binde withall: being chewed they take away the heate and inflammation of the mouth, and almonds of the throat: they stay the bloudy flix, and other fluxes, and all maner of bleedings: of the same force is their decoction, with a little honey added.

B They heale the eyes that hang out, hard knots in the fundament, and stay the hemorrhoids, if the leaues be layd thereunto.

C The iuyce which is pressed out of the stalks, leaues, and vnripe berries, and made hard in the Sun, is more effectuall for all those things.

D The ripe fruit is sweet, and containeth in it much iuyce of a temperate heate, therefore it is not vnpleasant to be eaten.

E It hath also a certaine kinde of astriction or binding qualitie.

F It is likewise for that cause wholsome for the stomack, and if a man eat too largely therof, saith *Galen*, he shall haue the head-ache: but being dried whilest it is yet vnripe it bindeth and drieth more than the ripe fruit.

G The root besides that it is binding containeth in it much thin substance, by reason whereof it wasteth away the stone in the kidnies, saith *Galen*.

H *Pliny* writeth, that the berries and floures do prouoke vrine, and that the decoction of them in wine is a present remedie against the stone.

I The leaues of the Bramble boiled in water, with honey, allum, and a little white wine added thereto, make a most excellent lotion or washing water to heale the sores in the mouth, the priuie parts of man or woman, and the same decoction fastneth the teeth.

K The Raspis is thought to be like the Bramble in temperature and vertues, but not so much binding or drying. The Raspis, saith *Dioscorides*, performeth those things which the Bramble doth. The fruit is good to be giuen to those that haue weake and queasie stomacks.

Chap. 5. *Of Holly Roses, or Cistus.*

¶ *The Kindes.*

Cistus hath been taken of diuers to be a kinde of Rose: the old Writers haue made two sorts thereof, male and female; and likewise a third sort, which is called *Ledum*: the later Herbarists haue discouered diuers more, as shall be declared.

¶ *A generall Description, wherein all the sorts of Cistus are comprised.*

Cistus and his kindes are woody shrubs full of branches, of the height of two or three cubits: some haue broad leaues, others rough, vneuen, wrinkled, somewhat downy, and most like the leaues of Sage; although some haue the leaues of Rosemary, others the forme of those of the Poplar tree: the floures grow on the tops of the branches, like vnto the wild Rose, yet such as very quickly fade, perish, and fall away: those of the male are most of a reddish blew or purple colour; and of the female white: in their places come vp little heads or knops somwhat round, in which is contained small seed: the roots of them all are woody.

There groweth vp ſometimes vnder the ſhrub hard to the roots, a certaine excreſcence or hypociſt, which is thicke, fat, groſſe, full of iuyce, without leaues, wholly conſiſting of many little caſes or boxes, as do thoſe of Henbane or of the Pomegranat tree; of a yellowiſh red colour in one kinde, and in another white, and in certaine other greene or graſſie, as *Dioſcorides* ſaith.

¶ *The Deſcription.*

1 THe firſt kinde of *Ciſtus* groweth vp like a ſmall buſh or ſhrub, of a wooddy ſubſtance, three or foure cubits high. garniſhed with many ſmall and brittle branches, ſet full of crumpled or rugged leaues very like vnto Sage leaues: at the top of the branches come floures of a purple colour, in ſhape like vnto a ſingle Brier Roſe, hauing leaues ſomwhat wrinkled like a cloath new dried before it be ſmoothed, and in the midſt a few yellow chiues or thrums: the floures for the moſt part do periſh and fall away before noone, and neuer ceaſe flouring in ſuch maner from the moneth of May vnto the beginning of September, at which time the ſeed is ripe, being of a reddiſh colour, and is contained in an hard hairie huske not much vnlike the husk of Henbane.

1 *Ciſtus mas anguſtifolius.*
The male Holly Roſe.

2 *Ciſtus mas cum Hypociſtide.*
The male Holly Roſe with his excreſcence.

2 The ſecond ſort of *Ciſtus*, being another kind of the male *Ciſtus*, which *Pena* calls *Ciſtus mas cum Hypociſtide*, is like vnto the former, but that from the root of this kinde there commeth a certaine excreſcence or out-growing, which is ſometimes yellow, ſometimes greene, and ſometimes white; from which is drawne by an artificiall extraction a certaine iuyce called in ſhops *Hypociſtis.*

3 This kinde of *Ciſtus* hath many wooddy ſtalks diuided into diuers brittle branches of a ruſſet colour; whereon do grow rough leaues ſomewhat cut or toothed on the edges, and of an ouerworne colour: the floures grow on the tops of the branches, in forme of a Muske Roſe, but of an excellent bright purple colour: after which come round knops, wherein is contained ſmal reddiſh ſeed: the root is tough and wooddy,

4 This fourth ſort of *Ciſtus* hath diuers wooddy branches, whereon are ſet, thicke thruſt together, diuers ſmal leaues narrow like thoſe of Winter Sauorie, but of an ouerworne ruſſet colour: the root and floures are like the precedent.

3 *Cistus mas dentatus.*
Toothed or snipt male Cistus.

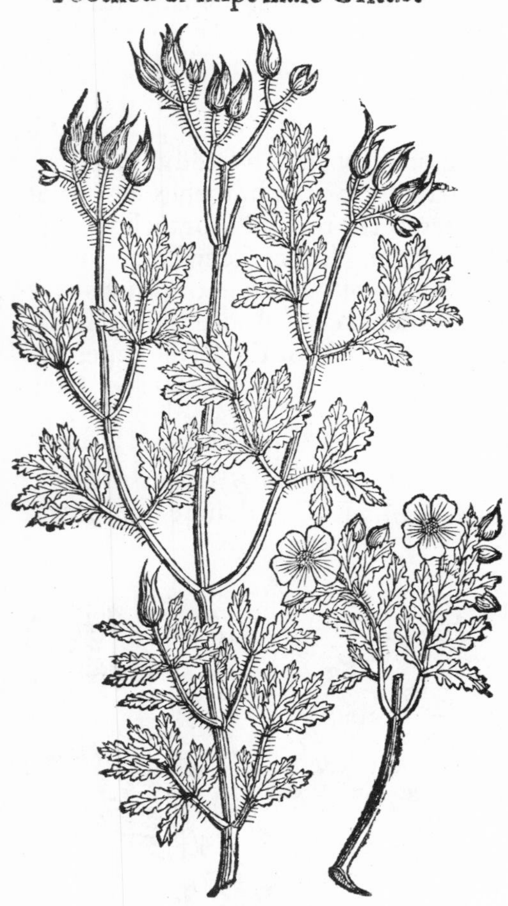

4 *Cistus mas tenuifolius.*
Thin leafed Cistus.

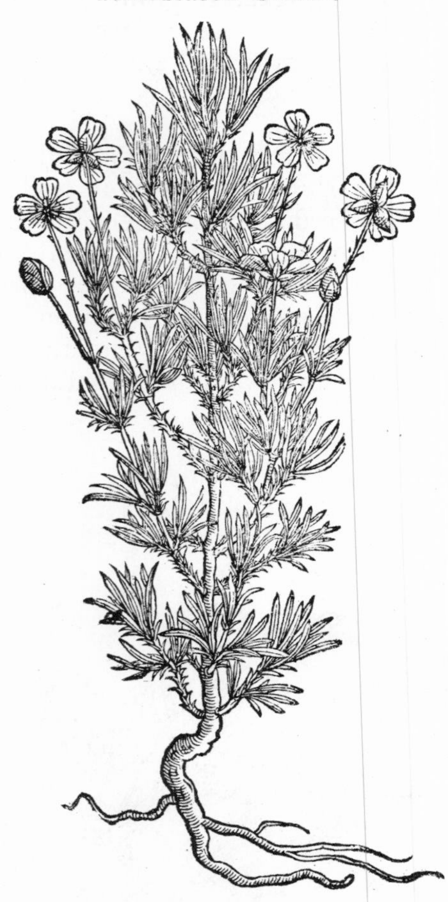

5 *Cistus fœmina.*
The female Cistus.

7 *Cistus folio Halimi.*
Cistus with leaues like Sea Purslane.

5 The firſt of the females is like vnto the male Ciſtus in each reſpect, ſauing that the floures hereof are of a white colour, with diuers yellow thrummes in the middle, and the others purple, wherein conſiſteth the difference.

6 The ſecond female of *Matthiolus* deſcription hath many hard and wooddie ſtalks, branched with diuers armes or wings: whereon are ſet by couples rough hoary and hairy leaues, of a darke ruſſet colour: among which come forth ſmall white floures like vnto thoſe of the Iaſmin: the root is tough and wooddy. ‡ This I iudge all one with the former, and therefore haue omitted the figure as impertinent, although our Authour followed it, making the floure ſo little in his deſcription. ‡

† 7 The ſeuenth ſort of Ciſtus groweth vp to the height of a ſmall hedge buſh, hauing diuers brittle branches full of pith: whereon are ſet leaues by couples, like thoſe of ſea Purſlane, that is to ſay, ſoft, hoary, and as it were couered ouer with a kinde of mealineſſe: the floures are yellow, and leſſe than thoſe of the former.

8 *Ciſtus folio Lauandulæ.*
Lauander leaued Ciſtus.

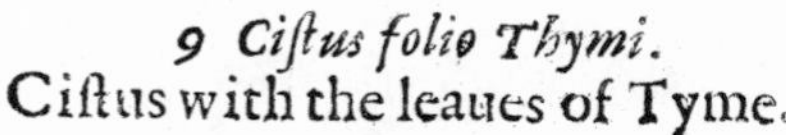
9 *Ciſtus folio Thymi.*
Ciſtus with the leaues of Tyme.

8 The eighth Ciſtus hath likewiſe ſhrubbie ſtalks in maner of a hedge tree, whereon do grow at certaine diſtances diuers leaues cloſe ioyned together at the ſtalke, like thoſe of the former, but ſomewhat lower and narrower: the floures we haue not expreſſed in the figure, by reaſon we haue no certaine knowledge of them.

9 This ninth Ciſtus is likewiſe a wooddy ſhrub ſome foot high: the ſtalks are very brittle, as are all the reſt of his kinde, whereon do grow very ſmall leaues like thoſe of Tyme: the floures are white, which maketh it one of the females.

10 The low or baſe Ciſtus with broad leaues, groweth like a ſmall ſhrub, of a wooddy ſubſtance: the leaues are many, of a darke greene colour: the floures are in forme like the other, but of a yellow colour: the roots are likewiſe wooddy.

11 This narrow leafed low Ciſtus hath diuers tough branches leaning to the ground, whereon do grow without order many ſmall narrow leaues ſomewhat long, of a gummy taſte at the firſt, afterwards bitter: the floures grow on the tops of the branches, of a yellow colour, conſiſting of fiue leaues, with certaine chiues in the middle; after which follow three ſquare cods or ſeed-veſſels: the root is tough and wooddy.

10 *Cistus humilis latifolius.*
Low Cistus with broad leaues.

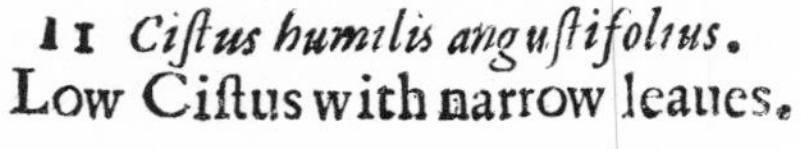

11 *Cistus humilis angustifolius.*
Low Cistus with narrow leaues.

12 *Cistus humilis Austriaca Clusij.*
Low Cistus of Austria.

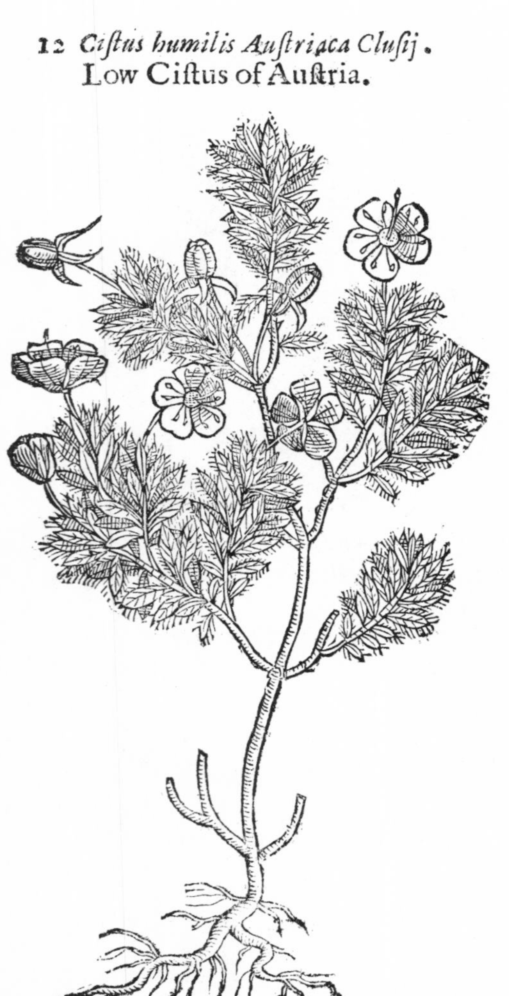

13 *Cistus humilis serpilli folio.*
Low Cistus with leaues like wilde Tyme.

12 The low or baſe Ciſtus of Auſtria groweth likewiſe leaning to the ground, hauing many wooddy branches very firme and tough, couered with a blackiſh barke; whereon do grow very many rough and hairy leaues in ſhape like thoſe of the ſmall myrtle, of a ſhining greene on the vpperſide, and of an aſtringent taſte: on euery branch ſtandeth one floure, ſeldome two, in forme like the other, but conſiſting of one leafe deeply diuided into fiue parts, and of a white colour tending to a fleſh colour.

13 This low ſort of Ciſtus hath many long tough branches trailing vpon the ground, of a reddiſh colour, whereon do grow ſmall leaues like thoſe of wilde Tyme, of a darke green colour, very thicke and fat, and ſomewhat hairy: the floures grow at the top of the branches, of a yellow gold colour, conſiſting of fiue ſmall leaues of a very ſweet ſmell. The root is thicke, hard, and wooddie.

14 This ſtrange and rare plant of *Lobels* obſeruation I haue thought meet to be inſerted amongſt the kindes of Ciſtus, as a friend of theirs, if not one of the kinde: it hath leaues like vnto the male Ciſtus (the firſt in this chapter deſcribed) but more hairy, bearing at the top of his branches a ſmall knop in ſhape like a rotten Strawberry, but not of the ſame ſubſtance; for it is compact of a ſcaly or chaffie matter ſuch as is in the middeſt of the Camomill floures, and of a ruſſet colour.

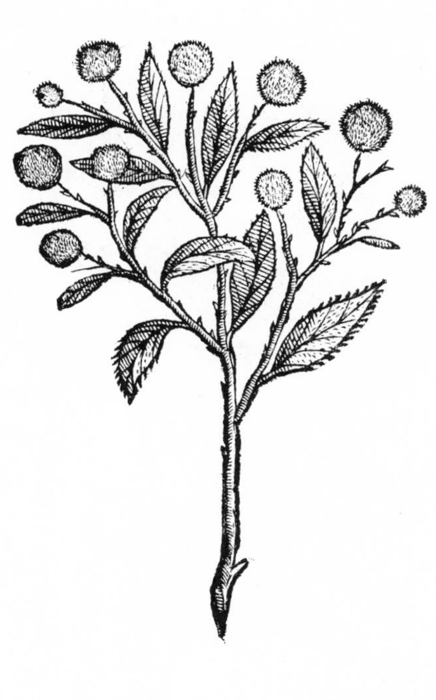

14 *Ciſtus exoticus Lobelij.*
Lobels ſtrange Ciſtus.

16 *Myrtociſtus Tho. Pennei Angli.*
D[r]. *Penny* his Ciſtus.

15 This adulterine or counterfeit or forged Ciſtus growes to the height of a hedge buſh: the branches are long or brittle, whereon do grow long leaues like thoſe of the Willow, of an ouerworne ruſſet colour: the floures are ſmall, conſiſting of fiue little yellow leaues: the whole plant being well viewed ſeemeth to be a Willow, but at the firſt ſight one of the Ciſtus; ſo that it is a plant participating of both: the root is wooddy. ‡ *Bauhine* iudges this (which our Author out of *Tabern.* figured and named *Ciſtus adulterinus*) to be the Ciſtus ſet forth in the eighth place of the next chapter ſaue one: but I rather iudge it to be of the *Ledum Sileſiacum* ſet forth in the eleuenth place of that chapter, and againe in the twelfth, where you may finde more thereof. ‡

16 This kinde of Ciſtus, which D[r]. *Penny* (a famous Phyſitian of London deceaſed) did gather vpon the Iſlands of Majorica or Majorca, and called it by the name μυρτοκίϛον, in Latine, *Myrtociſtus Balearica*, is a ſhrub growing to the height of three cubits, hauing a very rough barke, beſet round about with rough and ſcabbed warts; which bark wil of it ſelfe eaſily fall away from the old

old branches or boughes of the tree. The leaues of this tree are almoſt like them of *Myrtus*, very rough vnderneath like the branches aforeſaid; but the leaues that grow higher, and toward the top of the branches, are ſmooth, growing about the branches very thicke together, as in the other kindes of Ciſtus. The floures are yellow, growing on the top of the twigs, conſiſting of fiue long leaues full of many very long chiues within. When the floures be vaded, there followeth a verie long and fiue ſquare head or huske full of ſeed. The whole tree is very ſweet, out of which iſſueth a gum or roſine, or rather a thicke clammy and fat iuyce, ſuch as commeth forth of the kindes of *Ledum*.

17 This annual Ciſtus groweth vp from ſeed with one vpright ſtalke to the height of a cubit, oft times diuided into other ſmall branches; whereon grow rough leaues ſomwhat long, of a dark greene colour. The floures grow at the top of the ſtalks, conſiſting of 5 ſmall yellow leaues: which being paſt, there followeth a three ſquare ſeed veſſell full of ſmall reddiſh ſeed. The root is ſmall and wooddy, and periſheth when the ſeed is perfected.

17 *Ciſtus annuus.*
Ciſtus laſting one yeare.

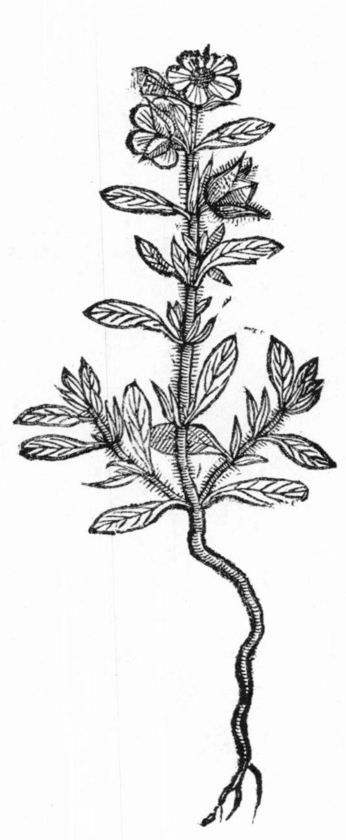

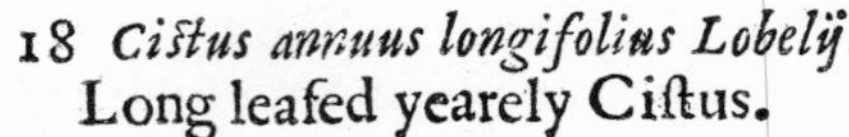
18 *Ciſtus annuus longifolius Lobelij.*
Long leafed yearely Ciſtus.

18 This other Ciſtus that laſteth but one yeare hath long ſtalks diuided into other branches of the height of two cubits; whereon do grow long rough leaues, ſet three together at certain diſtances, the middlemoſt whereof is longer than the other two: the floures grow on the ſides of the branches, like the female Ciſtus, of a white colour: the root is of a wooddy ſubſtance, as are all the reſt of his kinde.

‡ 19 This growes ſome foot high, with a ſquare rough greeniſh ſtalke, whereon by couples at certaine ſpaces ſtand little longiſh rough leaues, yet toward the top of the ſtalk they ſtand ſomtimes three together: vpon the top of the little branches grow floures like thoſe of the other Ciſtus, of colour yellow, with a fine ſanguine ſpot vpon each leafe of the floure. It groweth in ſome parts of France, as alſo on the Alps in Italy. *Cluſius* deſcribes it by the name of *Ciſtus annuus* 2. *Pona* in his *Mons Baldus* calls it *Ciſtus annuus flore guttato*.

20 This hath many ſlender branches whereon grow ſmall roundiſh leaues, hoarie, and ſomewhat like thoſe of Marjerome, ſomwhat leſſe, with the middle rib ſtanding out. The floures grow vpon the tops of the branches, and conſiſt of fiue white leaues, with a darke purple ſpot in the middle of each leafe: the threds in the middle of the floure are of a yellow colour: their ſeed-veſſels are

are of the bigneſſe of thoſe of flax, but three ſquare, containing a ſeed of the bigneſſe of that of Henbane. *Cluſius* found this in diuers parts of Spaine, and ſets it forth by the name of *Ciſtus folio Sampſuchi.* ‡

‡ 19 *Ciſtus annuus flore maculato.*
Spotted annuall Ciſtus.

‡ 20 *Ciſtus folio Sampſuchi.*
Marjerome leaued Ciſtus.

¶ *The Place.*

Holly Roſes grow in Italy, Spaine, and Languedoc, and in the countries bordering vpon the riuer Padus, in all Hetruria and Maſſiles, and in many other of the hotter prouinces of Europe, in dry and ſtony places, varying infinitely according to the diuerſitie of the regions where they doe grow; of which I haue two ſorts in my garden, the firſt, and the *Ciſtus annuus.*

¶ *The Time.*

They floure from May to September.

¶ *The Names.*

The Holly Roſe is called in Greeke κίσος, or κίσθος: in Latine alſo *Ciſtus,* and *Roſa ſyluatica:* of diuers, *Roſa Canina,* as *Scribonius Largus* writeth, but not properly: in Spaniſh, *Eſtepa:* of the Portugals, *Roſella:* in Engliſh, Holly Roſe, and Ciſtus, after the Greeke name. The fungous excreſcence growing at the root of Ciſtus, is called in Greeke ὑποκιστίς, becauſe it groweth vnder the ſhrub Ciſtus: it is alſo called *Limodoron:* ſome call it κύτινοι: among whom is *Paulus Ægincta,* who alſo doth not call that *Hypociſtis* which groweth vnder the ſhrub Ciſtus, but the iuyce hereof; whereupon might grow the word *Hypociſtis,* by which name the Apothecaries call this iuyce when it is hardned: of ſome it is called *Erithanon, Citinus,* and *Hypoquiſtidos.*

¶ *The Temperature.*

Ciſtus, as *Galen* ſaith, doth greatly dry, neere hand in the ſecond degree, and it is of that coldneſſe, that it hath withall a temperate heate: the leaues and the firſt buds being beaten do only dry and binde, in ſuch ſort as they may cloſe vp vlcers, and ioyne together greene wounds.

¶ *The Vertues.*

The floures are of moſt force, which being drunke with wine are good againſt the bloudy flix, A
weakeneſſe of the ſtomacke, fluxes, and ouerflowings of moiſt humors.

They cure putrified vlcers being applied in manner of a pultis: *Dioſcorides* teacheth that they B
are a remedie for eating vlcers, called in Greek Νομαὶ, being anointed therewith; and that they cure burnings, ſcaldings, and old vlcers.

 Hypociſtis

C *Hypociſtis* is much more binding: it is a ſure remedie for all infirmities that come of fluxes, as voiding of bloud, the whites, the laske, and the bloudy flix: but if it be requiſit to ſtrengthen that part which is ouerweakned with a ſuperfluous moiſture, it doth notably comfort and ſtrengthen the ſame.

D It is excellent to be mixed with fomentations that ſerue for the ſtomacke and liuer.

E It is put into the Treacle of Vipers, to the end it ſhould comfort and ſtrengthen weake bodies, as *Galen* writeth.

Chap. 6. *Of other Plants reckoned for dwarfe kindes of* Ciſtus.

1. 2. *Helianthemum Anglicum luteum vel album.*
English yellow or white dwarfe Ciſtus.

¶ *The Deſcription.*

1 The Engliſh dwarfe Ciſtus, called of *Lobel*, *Panax Chironium* (but there is another *Panax* of *Chirons* deſcription, which I hold to be the true and right *Panax*, notwithſtanding he hath inſerted it amongſt the kindes of Ciſtus, as being indifferent to ioyne with vs and others for the inſertion) is a low and baſe plant creeping vpon the ground, hauing many ſmal tough branches, of a browne colour; whereupon do grow little leaues ſet together by couples, thicke, fat, and ful of ſubſtance, and couered ouer with a ſoft downe: from the boſome whereof come forth other leſſer leaues: the floures before they be open are ſmall knops or buttons, of a browne colour mixed with yellow; and being open and ſpred abroad are like thoſe of the wild Tanſie, and of a yellow colour, with ſome yellower chiues in the middle: the root is thicke, and of a wooddy ſubſtance.

2 The ſecond is very like vnto the precedent, ſauing that the leaues are long, and doe not grow ſo thicke thruſt together, and are more woolly: the floures are greater, and of a white colour, wherein the eſpeciall difference conſiſteth. The root is like the former.

3 *Helianthemum luteum Germanicum.* The yellow dwarfe Ciſtus of Germanie.

3 There is found in Germanie, a certaine plant like to Ciſtus, and *Ledon*, but much leſſer, creeping vpon the ground, vnleſſe it be propped vp, hauing a multitude of twiggie branches, ſlender, and fine: whereupon do grow leaues leſſer than thoſe of Ledon or Ciſtus, very like to that of our Engliſh white dwarfe Ciſtus, of a full ſubſtance, ſleightly haired; wherein is contained a tough iuice: the floures are ſmall like little Roſes, or the wilde Tanſie, of a yellow colour: the roots be ſlender, wooddie, and ſomething red.

4 *Helianthemum album Germanicum.* The white dwarfe Ciſtus of Germanie.

5 *Helianthemum Sabaudicum.*
The dwarfe Ciſtus of Sauoy.

6 *Helianthemum anguſtifolium.*
Narrow leafed dwarfe Ciſtus.

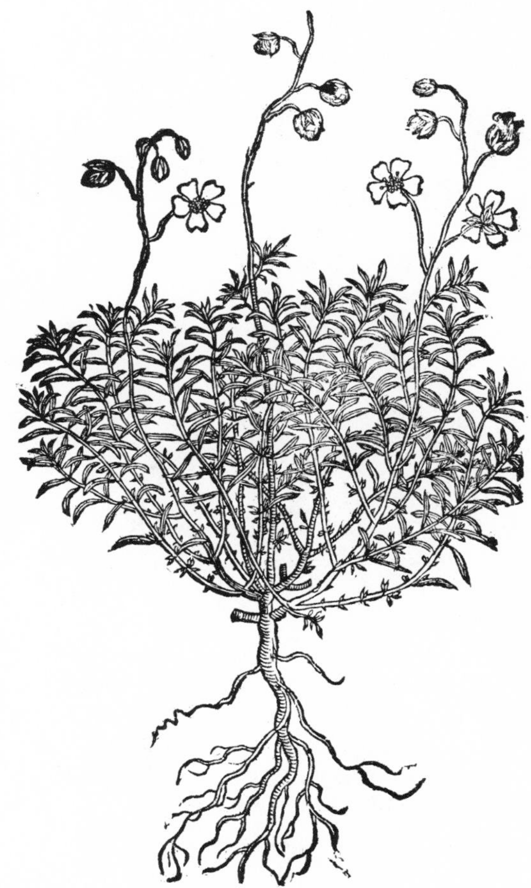

4 This differeth not from the laſt deſcribed, ſauing that the floures hereof are very white, and the others yellow, wherein they eſpecially differ.

5 The Dwarfe Ciſtus of Sauoy hath diuers tough branches, of a reddiſh colour, very tough and wooddy, diuided into diuers other branches: whereon are ſet ſmall leaues, foure together, by certain ſpaces; the floures grow at the top of the branches like thoſe of our yellow Dwarfe Ciſtus, of a yellow colour: the root is very wooddie.

6 This dwarfe Ciſtus with narrow leaues, hath very many ſmall flexible branches, of a browne colour, very ſmooth, and ramping vpon the ground; whereon do grow ſmall, long, narrow leaues, like thoſe of Time of Candie; from the boſome whereof come forth diuers other ſmaller leaues: the floures grow on the tops of the branches, of a bleak yellow colour: the root is likewiſe wooddy.

‡ 7 To theſe I may fitly adde two more: the firſt of theſe hath creeping ſtalks, ſome foot or two long, blackiſh, and diuided into ſundry ſmaller branches: the leaues grow thick and many together, ſet by couples (though the figure do not wel expreſſe ſo much:) theſe leaues are ſmal, of the bignes of thoſe of Time, thick, green aboue, and whitiſh vnderneath, and of a bitter taſt: at the ends of the branches grow two or foure floures neere together, very ſmall, compoſed of fiue little leaues, of a kinde of fleſh colour: to theſe ſucceed heads opening themſelues when they come to ripeneſſe into fiue parts, and containing a very ſmall ſeed: the root is hard and wooddie, ſending out certaine fibres: alſo the branches here and there put forth ſome fibres. This plant dryed hath a pretty pleaſing ſmell. This growes vpon the higheſt Auſtrian and Styrian alpes, and is ſet forth by *Cluſius* by the name of *Chamæciſtus ſeptimus*.

‡ 7 *Chamæciſtus ſerpillifolius.*
Tyme leaued dwarfe Ciſtus.

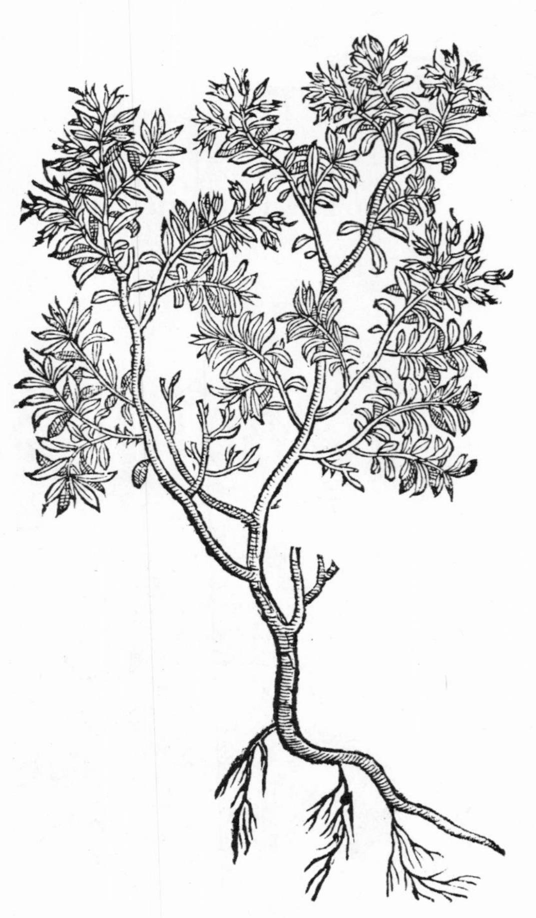

‡ 8 *Chamæciſtus Friſicus.*
Friſian Dwarfe Ciſtus.

8 The ſame Author alſo in his *Curæ poſteriores* giues vs the hiſtorie of this, which he receiued with ſome other rare plants from *Iohn Dortman*, a famous and learned Apothecarie of Groeningen: This little plant is in leafe and root almoſt like and neere of the ſame bigneſſe with the Celticke Nard, yet the ſtalks are vnlike, which are ſmall, ſet with a few longiſh leaues, and at the tops they carry fiue or ſix pretty floures like thoſe of Crowfeet, conſiſting of ſix leaues apiece, of a yellow colour, yet with ſome few ſpots of another colour, and theſe ſet in a double ring about the middle; after theſe follow heads or ſeed veſſels with forked tops, filled with a chaffie ſeed: the whole plant ſmells ſomewhat ſtrong. It growes together with *Gramen Pernaſſi* in rotten mooriſh places about a village in the county of Drent. *Dortman* called this *Hirculus Friſicus*: *Cluſius* addes, *qui Chamæciſti genus*. ‡

¶ The

¶ *The Place.*

Their seuerall titles haue touched their naturall countries: they grow in rough, drie, and sunnie places, in plaine fields and vpon mountaines.

Those of our English growing, I haue found in very many places, especially in Kent, vpon the chalkie bankes about Graues-end, Southfleet, and for the most part all the way from thence vnto Canturburie and Douer.

¶ *The Time.*

They floure from Iuly to the end of August.

¶ *The Names.*

Tragus calleth dwarfe Cistus in the high Dutch tongue, Heyden Ysope: in Latine, *Gratia Dei*: but there is another herbe called also of the later Herbarists *Gratia Dei*, which is *Gratiola*: *Valerius Cordus* nameth it *Helianthemum*, and *Solis flos*, or Sunne floure: of *Clusius*, *Chamæcistus*, or Dwarfe Cistus.

Pliny writeth, that *Helianthe* groweth in the champion countrey Temiscyra in Pontus, and in the mountaines of Cilicia neere to the sea: and he saith further, that the wise men of those countries, and the kings of Persia do annoint their bodies herewith, boiled with Lions fat, a little Saffron, and wine of Dates, that they may seeme faire and beautiful; and therefore haue they called it *Heliocaliden*, or the beautie of the Sun: *Matthiolus* saith, that *Helianthemum* is taken of some to be *Panaces Chironium*, or *Chirons* All-heale: but it is nothing likely, as we haue said.

¶ *The Temperature and Vertues.*

A The faculties and temperature are referred to the kindes of Cistus, for it healeth wounds, stancheth bloud, and stoppeth the spittings of bloud, the bloudie flixe, and all other issues of bloud.

B The same boiled in wine healeth vlcers in the mouth and priuie parts, if they be washed therewith: to be briefe, it ioineth together and strengthneth: which things doe plainely and euidently shew, that it is not onely like to Cistus and Ledon in forme, but in vertues and faculties also, and therefore it is manifest, that it is a certaine wilde kinde of Cistus and Ledon.

CHAP. 7. *Of Cistus Ledon, and Ladanum.*

¶ *The Kindes.*

THere be diuers sorts of Cistus, whereof that gummy matter is gathered, called in shops *Ladanum*, and *Labdanum*, but vnproperly.

¶ *The Description.*

1 CIstus Ledon is a shrub, growing to the height of a man, and sometimes higher; hauing many hard woddie branches, couered with a blackish bark: wherupon do grow leaues set together by couples, one right against another like vnto wings, of an inch broad, of a blacke swart greene on the vppersides, and whitish vnderneath: whereon is gathered a certain clammie transparent or through shining liquour, of a very hot sweet smell, which being gathered and hardned, is that which in shops is called *Labdanum*: the floures grow at the ends of the branches like little roses, consisting of fiue white leaues, euery one decked or beautified toward the bottome with pretty darke purplish spots tending to blacknesse, hauing in the middle very many yellow chiues, such as are in the middle of the Rose: after come the knaps or seed vessels, full of most smal reddish seed; the whole plant being dried, groweth somewhat whitish, and of a pleasant smell, the which it retaineth many yeares.

2 The second groweth likewise to the height of an hedge bush, the branches are long, and very fragile or easie to breake, whereon do grow leaues greener than any other of his kinde, yet vnderneath of a hoarie colour; growing toward winter to be somewhat reddish, of a sower and binding taste: the floures are like the precedent: the forme whereof the Grauer hath omitted, in other respects like the former.

3 The third sort of Cistus Ledon groweth vp to the height of a small hedge bush, hauing many twiggie branches; whereon do grow leaues like those of the Poplar tree, sharpe at the point, couered ouer with that clammie dew that the others are: the floures grow at the tops of the branches, of a white colour like the precedent.

4 The

1 *Cistus Ledon* 1.*Clusij*.
The first Cistus bringing *Ladanum*.

2 *Cistus ledon* 2.*Clusij*.
The second gum Cistus.

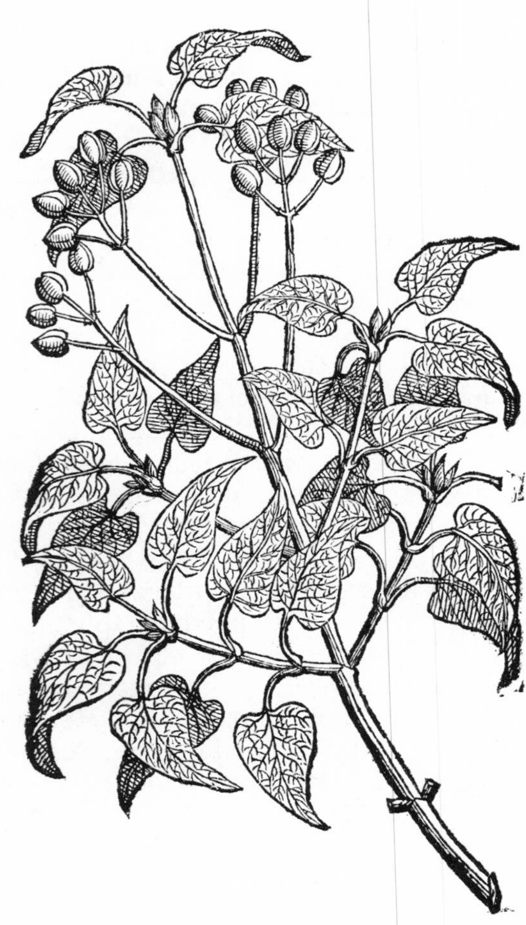

3 *Cistus ledon populea fronde.*
Cistus ledon with leaues like the Poplar.

4 *Cistus ledon* 4.*Clusij*.
Cistus ledon, the 4. of *Clusius*.

5 *Cistus Ledon* 5. *Clusij.*
The fift Cistus Ledon.

6 *Cistus Ledon* 6. *Clusij.*
The sixth Cistus Ledon.

7 *Cistus Ledon* 7. *Clusij.*
The 7. Cistus Ledon.

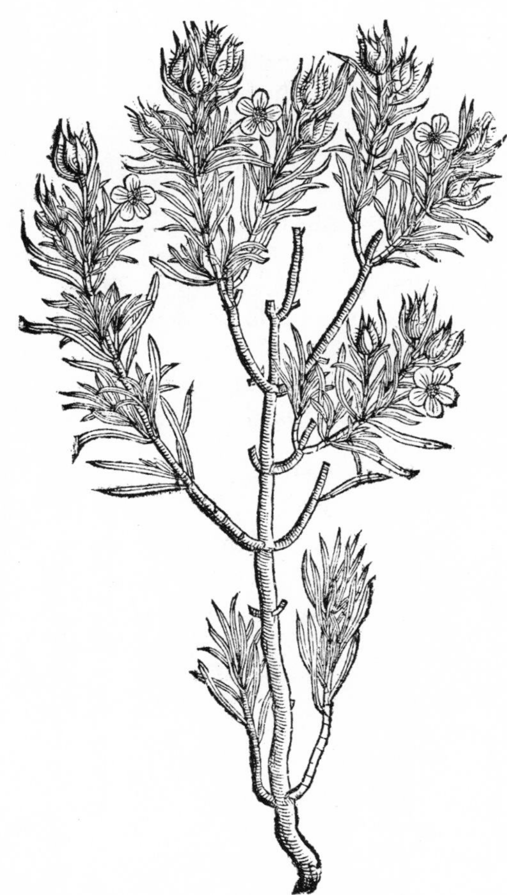

8 *Cistus Ledon cum Hypocistide Lobelij.*
The 8. Cistus Ledon, with his excrescence.

9 *Ciſtus Ledon* 10. *Cluſij.*
The 10. Ciſtus Ledon.

10 *Ciſtus Ledon Myrtifolium.*
Ciſtus Ledon with leaues like Myrtle.

11 *Ciſtus Ledum Sileſiacum.*
The Polonian Ciſtus Ledon.

4 The fourth of *Cluſius* deſcription groweth likewiſe to the height of a ſhrubby buſh, hauing many branches, flexible, hoarie, and hairie: the leaues are like the reſt of his kind, but ſofter, more hairy, of a ſwart green colour, daſht ouer with that dewie fatneſſe, not onely in the ſpring time, but in the heat of Sommer likewiſe: the floures are white, with yellow thrums in the middle: the reſt anſwereth the laſt deſcribed.

5 The fift groweth vp like a hedge buſh with many tough branches, whereon are ſet long rough leaues, hoarie vnderneath, ſomewhat daſht ouer with that fattie dew or humour that the reſt are poſſeſſed of: the floures are likewiſe of a white colour, with certaine yellow chiues in the middle: the root is wooddie.

† 6 The ſixth hath diuers ſmall branches couered with a blackiſh bark: the floures are ſet together at the tops of the branches by certaine ſpaces: they are yellow, and like the former in each reſpect.

7 The ſeuenth is a low ſhrub growing to the height of two cubits, hauing many branches couered with a barke of the colour of aſhes; whereon are confuſedly ſet diuers leaues at certaine diſtances, ſmall, narrow, like thoſe of winter Sauory, of an ouerworne ruſſet colour, very thick, fat, and glutinous: the floures are white, & differ not, nor the ſeed from the reſt.

8 The

8 The eighth groweth vp like a little hedge bush, hauing leaues like the common female Cistus, sauing that those of this plant are sprinckled ouer with that clammy moisture, and the other not so: the floures and seed are also like. From the root of this plant commeth such like excrescence called *Limodoron*, *Orobanche*, or *Hypocistis*, as there doth from the first male Cistus, wherein it differeth from all the rest vnder the name Ledon.

9 The ninth hath diuers brittle stalkes of an ash colour tending to a russet; whereon are set very many leaues like those of Thyme, of an ouerworne colour: the floures are white, with certaine yellow chiues in the middle, which the grauer hath omitted in the figure.

10 The tenth groweth vp like a small shrub, hauing brittle stalkes, couered with a blackish barke, and diuided into diuers branches; whereon are set vpon short truncheons or fat footstalkes, foure or fiue like those the Myrtle tree, of a strong smell: the floures are likewise of a white colour.

12 *Cistus Ledum Rorismarini folio.*
Cistus Ledon with leaues like Rosemarie.

13 *Cistus Ledum Matthioli.*
Cistus Ledon of *Matthiolus* description.

11. 12. The twelfth kinde of Cistus Ledon groweth vpright with a straight body or stocke, bringeth at the top many small twigs or rods of a cubit long, couered with a barke of the colour of ashes, which diuide themselues into other branches, of a purplish colour, beset with long and narrow leaues, not much vnlike to Rosemary, but longer; of a greene colour aboue, but vnderneath hauing as it were a long rib, made or compact of wooll or downe; of a sweet and pleasant smell, and somewhat sharpe in taste: on the tops of the branches grow knops or heads, compact as it were of many scales, of an iron or rustie colour: out of which commeth and proceedeth a cetaine round and long mane, or hairy panickled tuft of floures, with many long, tender, greene, and somewhat woolly stalkes or twigs growing vnto them, of a sweet sent and smell: the floures consist of fiue little white leaues, within which are contained ten white chiues with a long stile or pointal in the midst of the floure: when the floures be vaded, there succeed long knops or heads which are fiue cornered, in shape and bignesse like vnto the fruite and berries of *Cornus*; which being greene, are bespeckled with many siluer spots, but being ripe, are of a red colour; conteining within them a long yellow seed, which is so small and slender, that it is like to the dust or powder that falleth out of worme holes. ‡ This is the *Ledum Silesiacum* of *Clusius*; and the *Ledum Rorismarini folio* of *Tabernamontanus*: it is also the *Rosmarinum sylvestre* of *Matthiolus*; and *Chamæpeuce* of *Cordus*: and I am deceiued

deceiued if the figure which *Tabernamontanus* and our Author out of him gaue by the name of *Cistus adulterinus*, were not of this. ‡

13 Among the shrubby bushes comprehended vnder the title of *Cistus Ledum*, *Matthiolus* hath set forth one, whereof to write at large were impossible, considering the Author is so briefe, and of our selues we haue not any acquaintance with the plant it selfe: *Dioscorides* to helpe what may be, saith, that it is a shrub growing like vnto the stocke or kindred of the *Cisti*: from whose leaues is gathered a clammy dew which maketh that gummie matter that is in shops called *Lapdanum*: it groweth, saith he, in hot regions (but not with vs:) the Mauritanians call the juice or clammy matter, *Leden*, and *Laden*: of some, *Ladano* and *Odano*: in Spanish, *Xara* and further saith, it groweth in Arabia, where the bush is called *Chasus* thus much for the description. ‡ Our Author here seems to make *Dioscorides* to comment vpon *Matthiolus*, which shewes his learning, and how well he was exercised in reading or vnderstanding any thing written of Plants. Bbut of this enough; The plant here figured which *Matthiolus* iudges to be the true *Ledon*, or *Cistus Ladanifera* of *Dioscorides*, hath large stalkes and branches, whereon grow very thicke leaues, broad also and long, with the nerues running along it the leaues, the floure of this consists of fiue white leaues, and the seed is contained in a three cornered seed vessel. ‡

14 *Cistus Ledum Alpinum Clusij.*
The Mountaine Cistus.

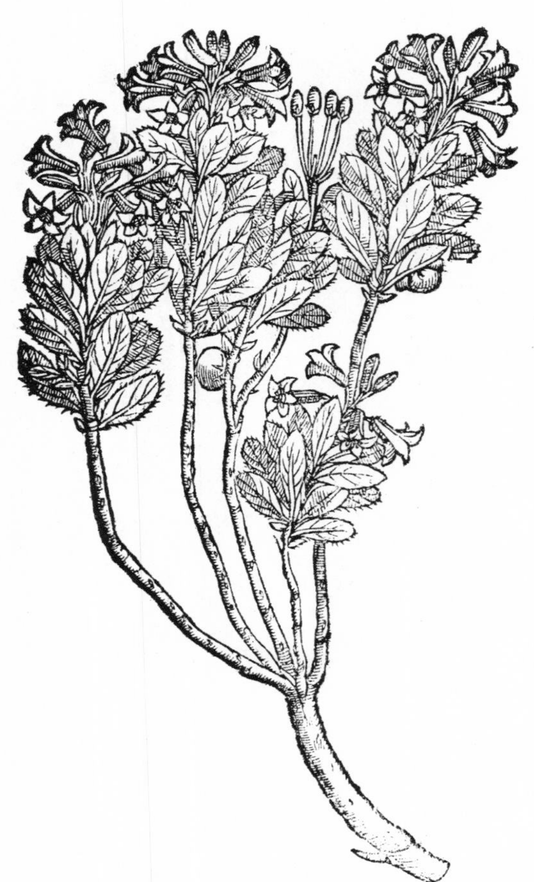

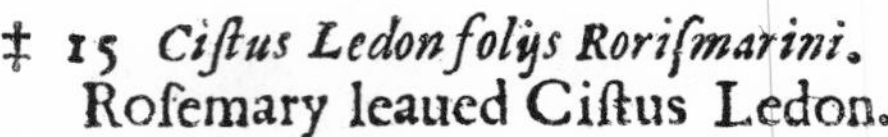

‡ 15 *Cistus Ledon folijs Rorismarini.*
Rosemary leaued Cistus Ledon.

14 The foureteenth Cistus, being one of those that do grow vpon the Alpish mountaines, which *Lobel* setteth downe to be *Balsamum alpinum* of *Gesner*: notwithstanding I thinke it not amisse to insert it in this place, hauing for my warrant that famous Herbarist *Carolus Clusius*: this plant is one of beautifullest, differing in very notable points, and yet resembleth them in the wooddy branches and leaues: it riseth vp hauing many weake branches leaning to the ground, yet of a wooddy substance, couered ouer with an ash coloured barke: the leaues are broad, and very rough, of a shining greene colour, and a binding taste: the floures grow at the tops of the branches like little bels, hanging downe their heads, diuided at the lips or brims into fiue diuisions, of a deepe red color on the out side, and dasht ouer here and there with some siluer spots; on the inside of a bright shining red colour, with certaine chiues in the middle, and of a very sweet smell, as is all the rest of the plant; after which come small heads or knaps, full of seed like dust, of a very strong smell, making the head of them to ake that smel thereto: the root is long, hard, and very woody: oftentimes there is found

found vpon the trunke or naked part of the ſtalks certaine excreſcences,or out-growings in manner of galls,of a fungous ſubſtance, like thoſe of Touchwood, white within, and red without, of an aſtringent or binding taſte.

‡ 15 This growes ſome cubite and better high,and hath long narrow glutinous leaues like in ſhape to thoſe of Roſemarie,ſet by couples,but not very thick : the branches whereon the floures do grow are ſlender, and the ſeed veſſels are diuided into fiue parts as in other plants of this kinde. This *Cluſius* found in Spaine,and ſets forth for his *Ledum nonum* ‡.

¶ *The Place.*

Ciſtus Ledon groweth in the Iſland of Candie, as *Bellonius* doth teſtifie,in vntilled places euery where : it is alſo found in Cyprus,as *Pliny* ſheweth,and likewiſe in many places of Spaine that lie open to the Sun : moreouer both the forme and bigneſſe of the leaues, and alſo of the plants themſelues,as well of thoſe that bring forth *Ladanum*,as the other Ciſtus,do varie in this wonderful maner,according to the diuerſitie of the places and countries where they grow : they are ſtrangers in theſe Northerly parts,being very impatient of our cold clymate.

¶ *The Time.*

They floure for the moſt part from May to the end of Auguſt:the clammie matter which falleth vpon the leaues,which is a liquid kinde of Roſen of a ſweet ſmell,is gathered in the Spring time as *Dioſcorides* ſaith : but as *Petrus Bellonius* affirmeth(being an eye witnes of the gathering) in the midſt of ſommer,and in the extreme heat of the Dog-daies,the which in our time not without great care and diligence,and as great labour,is gathered from the whole plant(with certain inſtruments made in manner of tooth pickes, or eare pickes,which in their tongue they call *Ergaſtiri*) and not gathered from the beards of Goats,as it is reported in the old fables of the lying Monks themſelues,called *Calohieros*,that is to ſay Greekiſh Monkes, who of very mockerie haue foiſted that fable among others extant in their workes.

‡ I thinke it not amiſſe for the better explanation of the matter here treated of,as alſo to ſhew you after what manner our Author in diuers places gaue the teſtimonies of ſundry Writers,and how well he vnderſtood them, here to ſet downe in Engliſh the words of *Bellonius* concerning the gathering of *Ladanum*,which are theſe. [The Greekes (ſaith he) for the gathering of *Ladanum*,prouide a peculiar inſtrument which in their vulgar tongue they terme *Ergaſtiri:* This is an inſtrument like to a Rake without teeth,to this are faſtened ſundry thongs cut out of a raw and vntanned hide; they gently rub theſe vpon the *Ladanum* bearing ſhrubs, that ſo the liquid moiſture concrete about the leaues may ſticke to them,which afterwards with kniues they ſhaue off theſe thongs in the heat of the day. Wherefore the labour of gathering *Ladanum* is exceeding great, yea intollerable,ſeeing they muſt of neceſſitie ſtay in the mountaines all the day long in the greateſt heat of the Dog-daies : neither vſually ſhall you finde any other who will take the paines to gather it, beſides, the *Calohieroi*,that is the Greeke Monkes. It is gathered no where in the whole Iſland of Candy in greater plenty,than at the foot of the mountaine Ida at a village called Cogualino, and at Milopotamo.‡]

¶ *The Names.*

The ſhrub it ſelfe is called in Greeke λῆδον, or λάδον : the Latines keep the name *Ledon* or *Ladon*,and is a kinde of *Ciſtus* or Hollie Roſes : the fat or clammie matter which is gathered from the leaues, is named *Ladanon* and *Ledanon*,according to the Greeke : the Apothecaries corruptly call it *Lapdanum* : *Dioſcorides* counteth that to be the beſt which is ſweet of ſmell, and ſomewhat greene, that eaſily waxeth ſoft, is fat, without ſand, and is not eaſily broken, but very full of Roſine or Gumme.

¶ *The Temperature.*

Ladanum,ſaith *Galen*,is hot in the later end of the firſt degree,hauing alſo a little aſtrictiue or binding qualitie;it is likewiſe of a thin ſubſtance,and therefore it ſofteneth, and withall doth moderately digeſt,and alſo concoct.

¶ *The Vertues.*

Ladanum hath a peculiar property againſt the infirmities of the mother,it keepeth haires from falling;for it waſteth away any ſetled or putrified humour that is at their roots. A

Dioſcorides ſaith,that *Ladanum* doth bind,heat, ſouple,& open,being tempered with wine,Myrrhe, and oile of Myrtles;it keepeth haires from falling,being annointed therewith ; or laied on mixed with wine,it maketh the markes or ſcars of wounds faire and well coloured. B

It taketh away the paine in the eares if it be powred or dropped therein,mixed with honied water,or with oile of Roſes. C

A fume made thereof draweth forth the afterbirth, and taketh away the hardneſſe of the matrix. D

E It is with good successe mixed with mollifying plaisters that mitigate paine.

F Being drunke with wine, it stoppeth the laske, and prouoketh vrine.

G There is made hereof diuers sorts of Pomanders, chaines, and bracelets, with other sweets mixed therewith.

Chap. 8. *Of Rosemarie.*

¶ *The Description.*

1 Rosemarie is a wooddie shrub, growing oftentimes to the height of three or foure cubits, especially when it is set by a wall: it consisteth of slender brittle branches, wheron do grow verie many long leaues, narrow, somewhat hard, of a quicke spicy taste, whitish vnderneath, and of a full greene colour aboue, or in the vpper side, with a pleasant sweet strong smell; among which come forth little floures of a whitish blew colour: the seed is blackish: the roots are tough and woody.

1 *Rosmarinum Coronarium.*
Garden Rosemarie.

2 *Rosmarinum syluestre.*
Wilde Rosemarie.

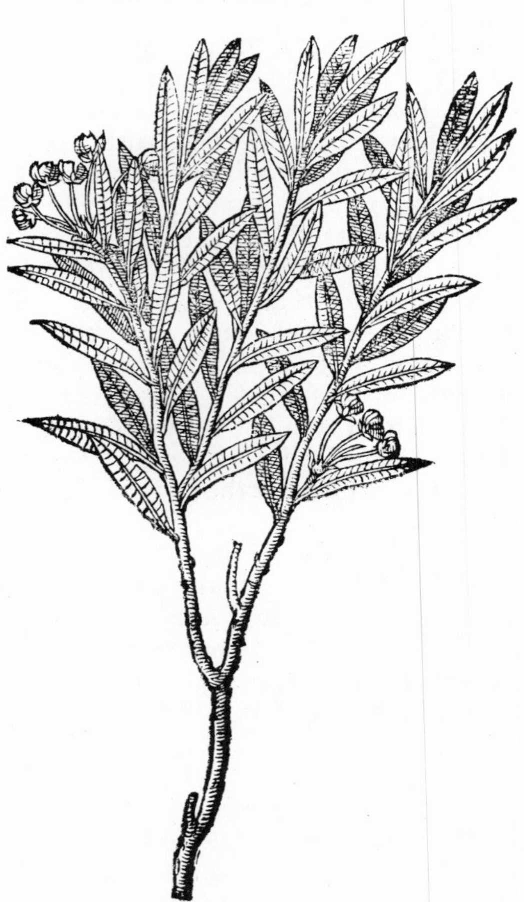

2 The wilde Rosemarie *Clusius* hath referred vnto the kindes of Cistus Ledon; we haue as a poore kinsman thereof inserted it in the next place, in kinred or neighbourhood at the least. This wilde Rosemarie is a small wooddie shrub, growing seldome aboue a foot high, hauing hard branches of a reddish colour, diuiding themselues into other smaller branches of a whitish color: wheron are placed without order diuers long leaues, greene aboue, and hoarie vnderneath, not vnlike to those of the dwarfe Willow, or the common Rosemarie, of a drie and astringent taste, of little smel or none at all: the floures stand on the tops of the branches, set vpon bare or naked footstalks, consisting of fiue small leaues of a reddish colour, somewhat shining; after which appeare little knaps full of small seed: the root is tough and wooddie.

3 This plant grows vp like an hedge shrub of a wooddie substance, to the height of two or three cubits;

3 *Caſia Poetica, Lobelij.*
The Poets Roſemarie or Gardrobe.

cubits; hauing many twiggie branches of a green colour: wherupon do grow narrow leaues like vnto *Linaria* or Toad-flax, of a bitter taſte; among which come forth ſmall moſſie floures, of a greeniſh yellow colour like thoſe of the Cornell tree, and of the ſmell of Roſemarie: which hath mouedme to placeit with the Roſemaries as a kinde thereof, not finding any other plant ſo neere vnto it in kinde and neighbourhood: after the floures be paſt, there ſucceed fruit like thoſe of the Myrtle tree, greene at the firſt, and of a ſhining red colour when they bee ripe, like Corall, or the berries of *Aſparagus*, ſoft and ſweet in taſte, leauing a certaine acrimonie or ſharpe taſte in the end: the ſtone within is hard as is the nut, wherein is contained a ſmal white kernel, ſweet in taſt: the root is of a wooddie ſubſtance: it floureth in the Sommer; the fruit is ripe in the end of October: the people of Granade, Montpelier, and of the kingdom of Valentia, doe vſe it in their preſſes and Wardrobes, whereupon they call it *Guardalobo*. ‡ This in *Cluſius* his time when he liued about Montpelier was called *Oſyris*; but afterwards they called it *Caſia*, thinking it that mentioned by the Poet *Virgil*; the which it cannot be, for it hath no ſweet ſmell. *Pena* and *Lobel* iudge it to be the *Caſia* of *Theophraſtus*, wherewith alſo it doth not well agree. ‡

¶ *The Place.*

Roſemarie groweth in France, Spaine, and in other hot countries; in woods, and in vntilled places: there is ſuch plentie thereof in Languedocke, that the inhabitants burne ſcarce any other fuell: they make hedges of it in the gardens of Italy and England, being a great ornament vnto the ſame: it groweth neither in the fields nor gardens of the Eaſterne cold countries; but is carefully and curiouſly kept in pots, ſet into the ſtoues and ſellers, againſt the iniuries of their cold Winters.

Wilde Roſemarie groweth in Lancaſhire in diuers places, eſpecially in a field called Little Reed, amongſt the Hurtleberries, neere vnto a ſmall village called Maudſley; there found by a learned Gentleman often remembred in our hiſtorie (and that worthily) Mr. *Thomas Hesketh*.

¶ *The Time.*

Roſemarie floureth twice a yeare, in the Spring, and after in Auguſt.

The wilde Roſemarie floureth in Iune and Iuly.

¶ *The Names.*

Roſemarie is called in Greeke λιβανωτὶς στεφανωματική: in Latine, *Roſmarinus Coronaria*: it is ſurnamed *Coronaria*, for difference ſake betweene it and the other *Libanotides*, which are reckoned for kindes of Roſemarie, and alſo becauſe women haue been accuſtomed to make crownes and garlands thereof: in Italian, *Roſmarino coronario*: in Spaniſh, *Romero*: in French and Dutch *Roſmarin*.

Wilde Roſemarie is called *Roſmarinus ſyluestris*: of *Cordus*, *Chamæpeuce*.

¶ *The Temperature.*

Roſemarie is hot and drie in the ſecond degree, and alſo of an aſtringent or binding quality, as being compounded of diuers parts, and taking more of the mixture of the earthy ſubſtance.

¶ *The Vertues.*

Roſemarie is giuen againſt all fluxes of bloud; it is alſo good, eſpecially the floures thereof, for A
all infirmities of the head and braine, proceeding of a cold and moiſt cauſe; for they dry the brain, quicken the ſences and memorie, and ſtrengthen the ſinewie parts.

Serapio witneſſeth, that Roſemarie is a remedie againſt the ſtuffing of the head, that commeth B
through coldneſſe of the braine, if a garland thereof be put about the head, whereof *Abin Meſuai* giueth teſtimonie.

Dioſcorides teacheth that it cureth him that hath the yellow iaundice, if it be boiled in water and C
drunk before exerciſe, & that after the taking therof the patient muſt bathe himſelfe & drink wine.

D The diſtilled water of the floures of Roſemarie being drunke at morning and euening firſt and laſt, taketh away the ſtench of the mouth and breath, and maketh it very ſweet, if there be added thereto, to ſteep or infuſe for certaine daies, a few Cloues, Mace, Cinnamon, and a little Anniſe ſeed.

E The Arabians and other Phyſitions ſucceeding, do write, that Roſemarie comforteth the brain the memorie, the inward ſenſes, and reſtoreth ſpeech vnto them that are poſſeſſed with the dumbe palſie, eſpecially the conſerue made of the floures and ſugar, or any other way confected with ſugar, being taken euery day faſting.

F The Arabians, as *Serapio* witneſſeth, giue theſe properties to Roſemarie: it heateth, ſay they, is of ſubtill parts, is good for the cold rheume which falleth from the braine, drieth away windines, prouoketh vrine, and openeth the ſtoppings of the liuer and milt.

G *Tragus* writeth, that Roſemarie is ſpice in the Germane Kitchins, and other cold countries. Further, he ſaith, that the wine boiled with Roſemarie, and taken of women troubled with the mother, or the whites, helpeth them, the rather if they faſt three or foure houres after.

H The floures made vp into plates with ſugar after the manner of Sugar Roſet and eaten, comfort the heart, and make it merry, quicken the ſpirits, and make them more liuely.

I The oile of Roſemarie chimically drawne, comforteth the cold, weake and feeble braine in moſt wonderfull maner.

K The people of Thuringia do vſe the wilde Roſemarie to prouoke the deſired ſickneſſe.

L Thoſe of Marchia vſe to put it into their drinke the ſooner to make their clients drunke, and alſo do put it into cheſts and preſſes among clothes, to preſerue them from mothes or other vermine.

† The vertues in the two laſt places properly belong to the *Roſmarinum ſylueſtre* of *Matthiolus*, which is the *Chamæpeuce* of *Cordus*, and is deſcribed in the 11. place of the foregoing Chapter, by the name of *Ciſtus Ledum Sileſiacum*.

Chap. 9. *Of Vpright Wood-binde.*

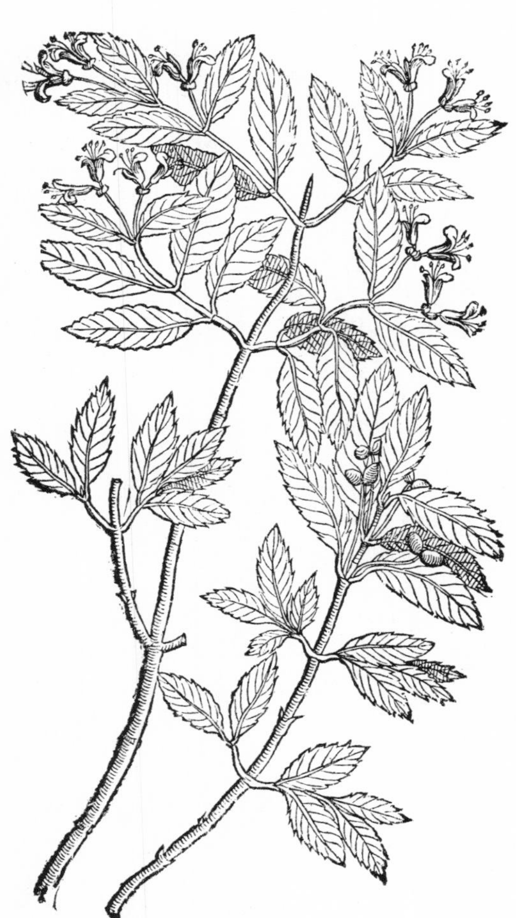

1 *Periclymenum rectum Sabaudicum.*
Sauoy Honiſuckles.

2 *Periclymenum rectum Germanicum:*
Germane Honiſuckles.

¶ *The*

¶ *The Deſcription.*

1 THis ſtrange kinde of Hony-ſuckle, found in the woods of Sauoy, repreſents vnto vs that ſhrub or hedge-buſh called *Cornus fœmina,* the Dog-berry tree, or Pricke-timber tree, hauing leaues and branches like the common Wood-binde, ſauing that this doth not clamber or clymbe as the others do, but contrariwiſe groweth vpright, without leaning to one ſide or other, like a ſmall tree or hedge-buſh: the floures grow vpon the tender ſprayes or twiggie branches, by couples, not vnlike in ſhape and colour to the common Wood-binde, but altogether leſſer, and of a white colour, hauing within the ſame many hairy chiues like the other of his kinde: after which come red berries ioyned together by couples: the root is tough and wooddy.

2 The ſtalkes of the ſecond be oftentimes of a meane thickneſſe, the wooddy ſubſtance ſomwhat whitiſh and ſoft: the branches be round, and couered with a whitiſh barke, notwithſtanding in the beginning when the ſprayes be yong they are ſomewhat reddiſh. The leaues be long, like thoſe of the common Hony-ſuckle, ſoft, and of a white greene: on the lower ſide they be whiter, and a little hairy: the floures be leſſer than any of the Wood-bindes, but yet of the ſame faſhion, and of a whitiſh colour, ioyned together by couples vpon ſeuerall ſlender foot-ſtalkes, like little wilde Cherries, of a red colour, the one leſſer oftentimes than the other.

3 *Periclymenum rectum fructu cæruleo.*
Vpright Wood-binde with blew berries.

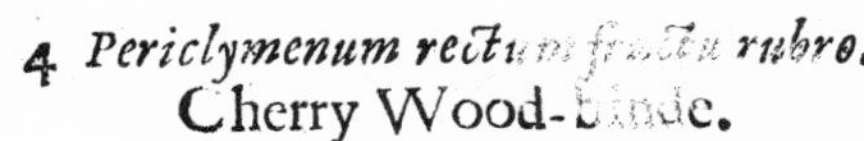
4 *Periclymenum rectum fructu rubro.*
Cherry Wood-binde.

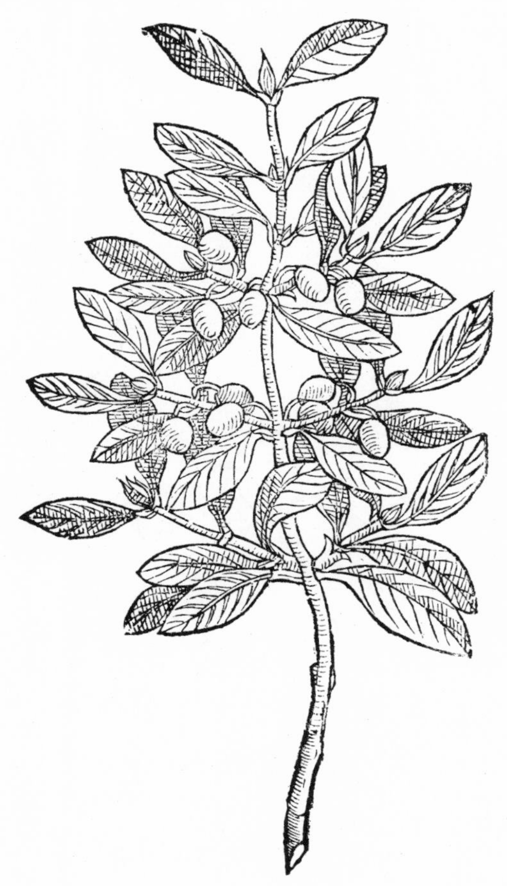

3 This ſtrange kinde of Wood-binde, which *Carolus Cluſius* hath ſet forth in his Pannonicke Obſeruations, riſeth vp oftentimes to the height of a man, euen as the former doth; which diuides it ſelfe into many branches, couered with a rough blacke barke, that choppeth and gapeth in ſundrie clefts as the barke of the Oke. The tender branches are of a whitiſh greene colour, couered with a woolly haírineſſe, or an ouerworne colour, whereupon do grow leaues ſet by couples one againſt the other, like vnto the common Wood-binde, of a drying bitter taſte: the floures grow by couples likewiſe, of a whitiſh colour. The fruit ſucceedeth, growing like little Cherries, each one on his owne foot-ſtalke, of a bright and ſhining blew colour; which being bruiſed, doe die the hands of a reddiſh colour, and they are of a ſharpe winie taſte, and containe in them many ſmall flat ſeeds. The root is wooddy, diſperſing it ſelfe far abroad.

4 This

4 This kind of vpright Wood-bind groweth vp likewise to the height of a man, and oftentimes more high, like to the last described, but altogether greater. The berries hereof are very blacke, wherein especially is the difference. ‡ The leaues of this are as large as Bay leaues, sharpe pointed, greene aboue, and whitish vnderneath, but not hairy, nor snipt about the edges: the floures grow by couples, of a whitish purple, or wholly purple: to these paires of floures there commonly succeeds but one berry, larger than any of the former, of the bignesse of a little cherry, and of the same colour, hauing two marks vpon the top therof, where the floures stood. ‡

Periclymeni 3. & 4. flores.
The floures of the third and fourth.

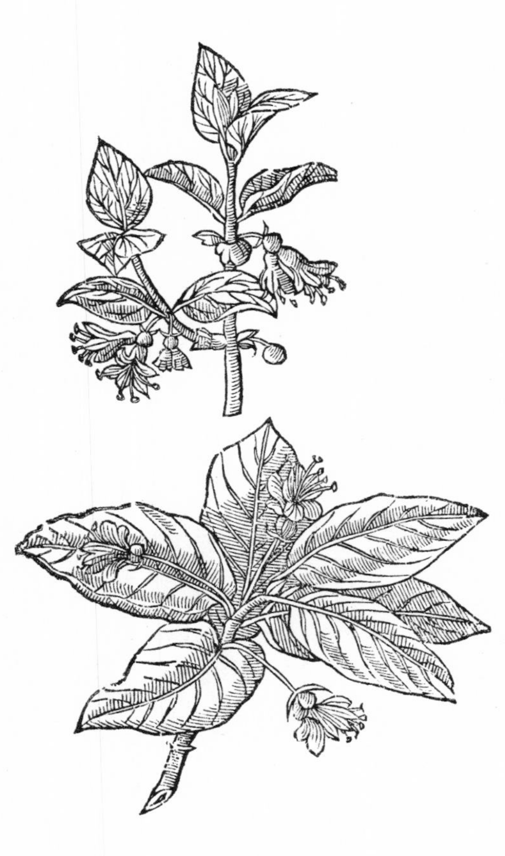

5 *Chamæpericlymenum.*
Dwarfe Hony-suckle.

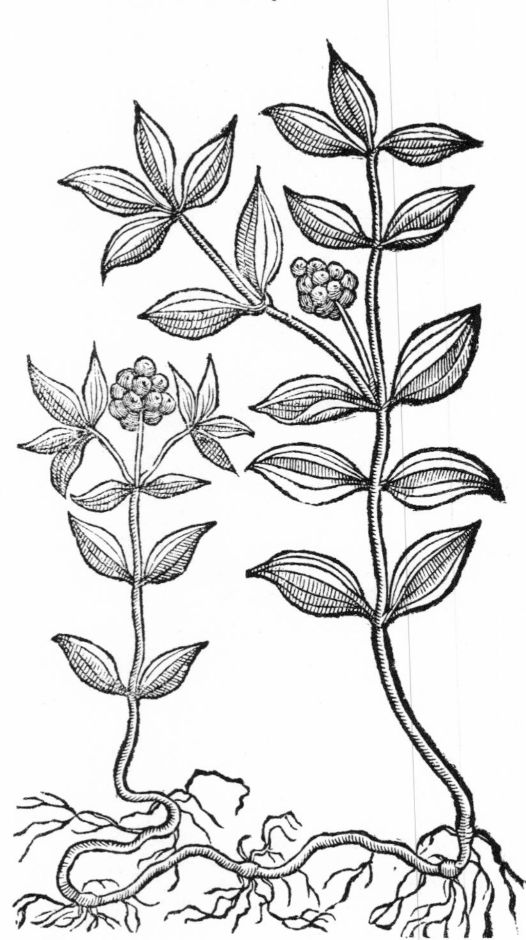

5 To the kindes of Wood-bindes this plant may likewise be referred, whose picture with this description was sent vnto *Clusius* long since by that learned Doctor in physicke *Thomas Penny* (of our London colledge of famous memorie:) it riseth vp with a stalke of a foot high; whereupon are set by couples faire broad leaues one right against another, ribbed with certaine nerues like those of Plantaine, sharpe pointed, and somewhat hollowed in the middle like Spoon-wort: from the bosome of which leaues come forth small floures, not seene or described by the Author: after which commeth forth a cluster of red berries, thrust hard together as those of Aaron or priests pint. The root is tough and very slender, creeping far abroad vnder the vpper crust of the earth, whereby it occupieth much ground.

¶ *The Place.*

These plants are strangers in England: they grow in the woods and mountaines of Switzerland, Germany, Sauoy, and other those parts tending to the East, East North-East, and East and by South.

I haue a plant of the first kinde in my garden: the rest as yet I haue not seene, and therefore cannot write so liberally thereof as I could wish.

‡ The dwarfe Hony-suckle growes in the maritime parts of Norway and Sweden, & the countries thereabout. ‡

¶ *The Time.*

They floure for the most part when the others do, that is to say in May and Iune, and their fruit is ripe in September.

¶ *The*

¶ *The Names.*

Vpright Wood-binde or Hony-suckle is called *Periclymenum stans*, and *Periclymenum rectum*, or vpright Wood-binde: of *Dodonæus*, *Xylosteum*: in high-Dutch, **Honds kirsen**, that is to say, *Canum Cerasa*, or Dog Cherries. The English names are expressed in their seuerall titles. It hath bin called *Chamæcerasus*, but not truly.

¶ *The Temperature and Vertues.*

Touching the temperature and vertues of these vpright Wood-bindes, we haue no experience at all our selues, neither haue we learned any thing of others.

CHAP. 10. *Of Sene.*

Sena folijs obtusis.
Italian Sene.

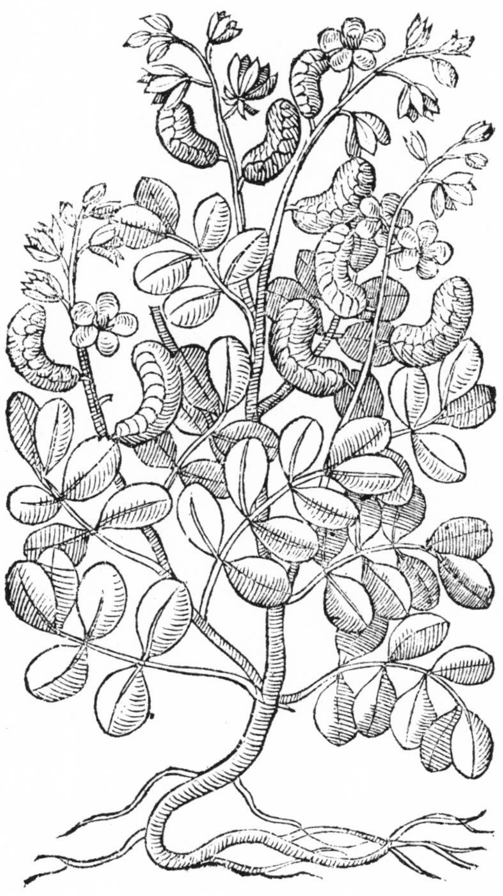

¶ *The Description.*

SEne bringeth forth stalks a cubit high, set with diuers branches: the leaues are long, winged, consisting of many small leaues like those of Liquorice, or of bastard Sene: the floures come forth of the bottom of the wings, of colour yellow, standing vpon slender footstalks; from which after the floures be gone hang forked cods, the same bowing inward like a halfe-moone, plain and flat, in which are contained seeds like to the seeds or kernells of grapes, of a blackish colour. The root is slender, long, and vnprofitable, which perisheth when the leaues are gathered for medicine, and the seeds be ripe, and must be sowne againe the next yeare, euen as we do corne.

There is another kinde of Sene growing in Italy, like the other in each respect, sauing that it is greater, and hath not that force in purging that the other hath.

¶ *The Place and Time.*

This is planted in Syria and Egypt, also in Italy, in Prouince in France, in Languedoc. It hardly groweth in high and low Germany, neither in England: it prospereth in hot Regions, and cannot away with cold; for that cause it is in Italy sowne in May, and continueth no longer than Autumne: the best is brought from Alexandria and Egypt. The Arabians were the first that found it out.

¶ *The Names.*

The Persians call it *Abalzemer*, as *Mesue* his copy teacheth: the Apothecaries *Sena*, by which name it was knowne to *Actuarius* the Grecian, and to the later Latines: it is called in English, Sene.

¶ *The Temperature.*

Sene is of a meane temperature, neither hot nor cold, yet inclining to heate, and dry almost in the third degree: it is of a purging facultie, and that by the stoole, in such sort as it is not much troublesome to mans nature, hauing withall a certaine binding qualitie, which it leaueth after the purging.

¶ *The Vertues.*

It voideth forth flegmaticke and cholericke humors, also grosse and melancholike, if it be helped with something tending to that end. A

It is a singular purging medicine in many diseases, fit for all ages and kindes. B

It purgeth without violence or hurt, especially if it be tempered with Anise seed or other like sweet smelling things added, or with gentle purgers or lenitiue medicines. It may be giuen in pouder, but commonly the infusion thereof is vsed. C

The

D The quantitie of the pouder is a dram weight, and in the infuſion, foure, fiue, or more. It may be mixed in any liquor.

E It is in the decoction or in the infuſion tempered with cold things in burning agues and other hot diſeaſes: in cold and long infirmities it is boyled with hot opening ſimples and ſuch like; or elſe it is ſteeped in wine, in which manner, as familiar to mans nature, it draweth forth gently by the ſtoole, almoſt without any kinde of paine, crude and raw humors.

F Moſt of the Arabians commend the cods, but our Phyſitions the leaues rather; for vnleſſe the cods be full ripe they ingender winde, and cauſe gripings in the belly. For they are oftentimes gathered before they be ripe, and otherwiſe eaſily fall away being ſhaken downe by the wind, by reaſon of their weake and ſlender ſtalks.

G Some alſo thinke that Sene is hurtfull to the ſtomacke, and weakneth the ſame, for which cauſe they ſay that Ginger or ſome ſweet kinde of ſpice is to be added, whereby the ſtomacke may be ſtrengthned. Likewiſe *Meſue* noteth that it is ſlow in operation, and therefore Salgem is to be mixed with it. Moreouer, Sene purgeth not ſo ſpeedily as ſtronger medicines do.

H Notwithſtanding it may be helped not only by Salgem, but alſo by other purging things mixed therewith, that is to ſay with ſimple medicines, as Rubarb, Agaricke, and others; and with compounds, as that which is called *Catholicon*, or the Electuarie *Diaphanicon*, or that which is made of the iuyce of Roſes, or ſome other, according as the condition or qualitie of the diſeaſe and of the ſicke man requireth.

I The leaues of Sene are a familiar purger to all people, but they are windie, and do binde the bodie afterwards, very much diſquieting the ſtomack with rumbling and belching: for the auoiding of which inconuenience there muſt be added Cinnamon, Ginger, Anniſe ſeed, and Fennell ſeed, Raiſins of the Sun, and ſuch like that do breake winde, which will the better help his purging qualitie.

K Sene doth better purge when it is infuſed or ſteeped, than when it is boyled: for doubtleſſe the more it is boiled the leſſe it purgeth, and the more windie it becommeth.

L Take Borage, Bugloſſe, Balme, Fumitorie, of each three drams, Sene of Alexandria very wel prepared and pouned, two ounces, ſtrow the pouder vpon the herbes and diſtill them: the water that commeth thereof reſerue to your vſe to purge thoſe that liue delicately, being miniſtred in white wine, with ſugar, in condited confections, and ſuch dainty waies, wherein delicate and fine people do greatly delight: you may alſo (as was ſaid before) adde hereunto according to the maladie, diuers purgers, as Agaricke, Mirobalans, &c.

M The pouder of Sene after it is well prepared two ounces, of the pouder of the root of Mechoacan foure drams, pouder of Ginger, Aniſe ſeeds, of each a little, a ſpoonfull of Aniſe ſeeds, but a very little Ginger, and a modicum or ſmall quantitie of *Salgemmæ*: this hath beene proued a verie fit and familiar medicine for all ages and ſexes. The patient may take one ſpoonful or two thereof faſting, either in pottage, ſome ſupping in drink, or white wine. This is right profitable to draw both flegme and melancholy from the breſt and other parts.

N The leaues of Sene and Cammomil are put in baths to waſh the head.

O Sene opens the inward parts of the body which are ſtopped, and is profitable againſt all griefes of the principall members of the body.

P Take Sene prepared according to art one ounce, Ginger halfa quarter of an ounce, twelue cloues, Fenell ſeed two drams, or in ſtead thereof Cinnamon and Tartar, of each halfe a dram, pouder all theſe; which done, take thereof in white wine one dram before ſupper, which doth maruellouſly purge the head.

Q Handle Sene in maner aboue ſpecified, then take halfe an ounce thereof, which don, adde thereto ſixty Raiſins of the Sunne with the ſtones pickt out, one ſpoonfull of Aniſe ſeeds braied, boile theſe in a quart of ale till one halfe be waſted, and while it is boiling put in your Sene: let it ſtand ſo till the morning, then ſtraine it, and put in a little Ginger: then take the one halfe of this potion and put thereunto two ſpoonfulls of ſyrrup of Roſes: drinke this together, I meane the one halfe of the medicine at one time, and if the patient canot abide the next day to receiue the other halfe, then let it be deferred vntil the third day after.

R Sene and Fumitorie (as *Raſis* affirmeth) do purge aduſt humors, and are excellent good againſt ſcabs, itch, and the ill affection of the body.

S If Sene be infuſed in whey, and then boyled a little, it becommeth good phyſicke againſt melancholy, clenſeth the braine and purgeth it, as alſo the heart, liuer, milt, and lungs, cauſeth a man to looke yong, ingendreth mirth, and taketh away ſorrow: it cleareth the ſight, ſtrengthneth hearing, and is very good againſt old feuers and diſeaſes ariſing of melancholy.

† There were formerly two figures in this chapter, which differed onely in that the firſt, which was the *Sena Orientalis*, had leſſer, narrower, and ſharper pointed leaues than the *Sena Italica*, which was the ſecond.

Chap.

Chap. 11. *Of bastard Sene.*

¶ *The Description.*

1 *Colutea* and Sene be so neere the one vnto the other in shape and shew, that the vnskilful Herbarists haue deemed *Colutea* to be the right Sene. This bastard Sene is a shrubby plant growing to the forme of a hedge bush or shrubby tree: his branches are straight, brittle, and wooddy; which being carelesly broken off, and as negligently prickt or stucke in the ground, will take root and prosper at what time of the yeare soeuer it be done; but slipt or cut, or planted in any curious sort whatsoeuer, among an hundred one will scarcely grow: these boughes or branches are beset with leaues like *Sena* or *Securidaca*, not much vnlike Liquorice: among which come forth faire broome-like yellow floures, which turne into small cods like the sownd of a fish or a little bladder, which will make a cracke being broken betweene the fingers: wherein are contained many blacke flat seeds of the bignesse of Tares, growing vpon a small rib or sinew within the cod: the root is hard, and of a wooddy substance.

1 *Colutea.*
Bastard Sene.

2 *Colutea Scorpioides.*
Bastard Sene with Scorpion cods.

2 Bastard Sene with Scorpion cods is a small wooddy shrub or bush, hauing leaues, branches, and floures like vnto the former bastard Sene, but lesse in each respect: when his small yellow floures are fallen there succeed little long crooked cods like the long cods or husks of *Matthiolus* his *Scorpioides*, whereof it tooke his name: the root is like the root of the Box tree, or rather resembling the roots of *Dulcamara* or Bitter-sweet, growing naturally in the shadowie woods of Valena in Narbone; whereof I haue a small plant in my garden, which may be called Scorpion Sene.

3 The low or dwarfe *Colutea* of *Clusius* description, hath a thicke wooddy root couered with a yellowish barke, with many fibres anexed thereto, which bringeth forth yearely new shoots, whereby it greatly encreaseth, of a cubit and a halfe high, smooth, and of a greene colour; whereon doe grow leaues composed of six or seuen leaues, and sometimes nine, set vpon a middle rib like those of the common kinde, of a stipticke taste, with some sharpnesse or biting: the floures grow vpon slender

3 *Colutea scorpioides humilis.*
Dwarfe bastard Sene.

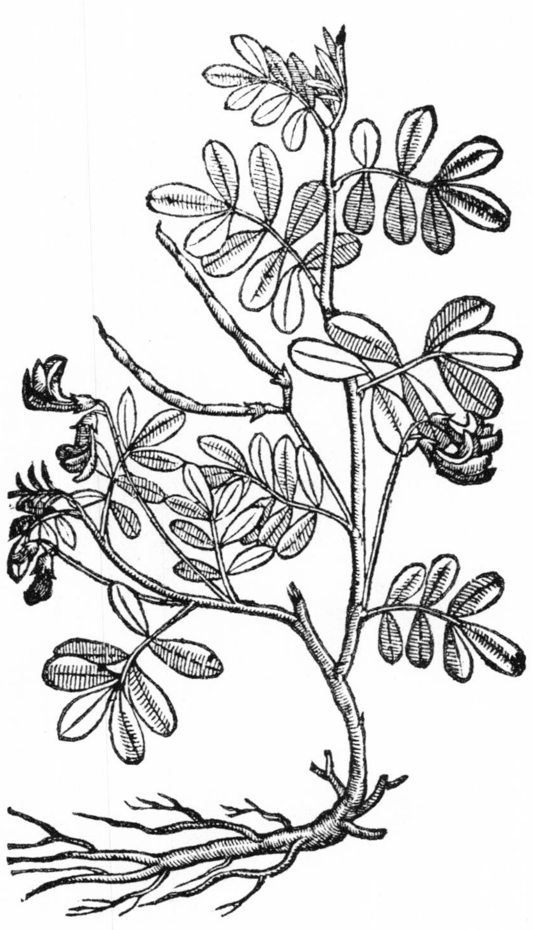

4 *Colutea scorpioides montana Clusij.*
Mountaine bastard Sene.

5 *Colutea minima, siue Coronilla.*
The smallest bastard Sene.

slender foot-stalkes, long and naked like those of the Pease, and of a yellow colour, of little or no smell at all, and yet that little nothing pleasant: after which come forth long cods, wherein is contained small seed like those of the Strangle Tare.

4 This mountaine bastard Sene hath stalks, leaues, and roots like the last described. The floures grow on the tops of the branches in maner of a crowne; whereupon some haue called it *Coronilla*: in shape like those of the pease, and of a yellow colour: the cods as yet we haue not seen, and therefore not expressed in the figure.

5 This small bastard Sene groweth like a small shrub creeping vpon the ground, halfe a cubit high, bringing forth many twiggie branches, in maner of those of the Spanish broome; wherupon do grow leaues like those of Lentils or the Strangle Tare, with many smal leaues set vpon a middle rib, somewhat fat or full of iuice, of the colour of the leaues of Rue or Herbe-grace, of an astringent and vnpleasant taste: the floures grow at the tops of the branches, of a yellow colour, in shape like those of the smallest broome: after which come little crooked cods like the clawes or toes of a bird, wherein is contained seed somwhat long, blacke, and of an vnsauorie taste: the root is long, hard, tough, and of a wooddy substance.

6 There

6 There is alſo found another ſort hereof, not much differing from the former, ſauing that this plant is greater in each reſpect, wherein eſpecially conſiſteth the difference.

¶ *The Place.*

Colutea or baſtard Sene groweth in diuers gardens, and commeth vp of ſeed; it quickly commeth to perfection, inſomuch that if a ſticke thereof be broken off and thruſt into the ground, it quickly taketh root, yea although it be done in the middle of ſummer, or at any other time, euen as the ſticks of Willow or Elder, as my ſelfe haue often prooued; the which bring forth floures and fruit the next yeere after.

The ſecond with Scorpion cods groweth likewiſe in my garden: the laſt doth grow in diuers barren chalky grounds of Kent towards Sittinbourne, Canturbury, and about Southfleet; I haue not ſeene them elſewhere: the reſt are ſtrangers in England.

¶ *The Time.*

They floure from May till ſummer be well ſpent, in the meane ſeaſon the cods bring forth ripe ſeed.

¶ *The Names.*

This ſhrub is called of *Theophraſtus* in Greeke κολουτέα with the diphthong ου in the ſecond ſillable: in Latine, as *Gaza* expoundeth it, *Coloutea* or *Colutea*: in high Dutch, **Welſch linſen**: in French, *Baguenaudier*: they are deceiued that thinke it to be *Sena*, or any kinde thereof, although we haue followed others in giuing it to name Baſtard Sene, which name is very vnproper to it: in low Dutch it is called **Sene boom**: and we may vſe the ſame name Sene tree, in Engliſh.

This *Calutea*, or baſtard Sene, doth differ from that plant κολυτέα with υ in the ſecond ſyllable, of which *Colytea*, *Theophraſtus* writeth in his third booke. ‡ The fifth is the *Polygala Valentina* of *Cluſius*. ‡

¶ *The Nature and Vertues.*

Theophraſtus, neither any other hath made mention of the temperature or faculties in working of theſe plants, more than that they are good to fatten cattell, eſpecially ſheepe. A

† There were formerly in the fifth and ſixth places here two figures no waies different, but that which was in the ſixth place was a little larger, and *Lobels* title which he puts in his *Icons*, ouer this was diuided betweene them: for as you ſee, *Colutea minima, ſiue Coronilla*, was ouer in the fifth; and *Colutea, ſiue Polygala Valentina Cluſij* was ouer the ſixth.

CHAP. 12. *Of Liquorice.*

¶ *The Deſcription.*

1 THe firſt kinde of Liquorice hath many wooddy branches, riſing vp to the height of two or three cubits, beſet with leaues of an ouerworne greene colour, conſiſting of many ſmall leaues ſet vpon a middle rib, like the leaues of *Colutea*, or the Maſtich tree, ſomewhat glutinous in handling: among which come ſmall knops growing vpon ſhort ſtems betwixt the leaues and the branches, cluſtering together, and making a round forme and ſhape: out of which grow ſmall blew floures, of the colour of an Engliſh Hyacinth; after which ſucceed round, rough, prickly heads, conſiſting of diuers rough or ſcaly huskes cloſely and thicke compact together; in which is contained a flat ſeed: the root is ſtraight, yellow within, and browne without: of a ſweet and pleaſant taſte.

2 The common and vſuall Liquorice hath ſtalkes and leaues very like the former, ſauing that his leaues are greener and greater, and the floures of a light ſhining blew colour: but the floures of this are ſucceeded by longiſh cods that grow not ſo thicke cluſtring together in round heads as the former, but ſpike faſhion, or rather like the wilde Vetch called *Onobrychis*, or *Galega* the cods are ſmall and flat like vnto the Tare: the roots are of a browniſh colour without, and yellow within like Box, and ſweeter in taſte than the former.

¶ *The Place.*

Theſe plants do grow in ſundry places of Germany wilde, and in France and Spaine, but they are planted in gardens in England, whereof I haue plenty in my garden: the poore people of the North parts of England do manure it with great diligence, wherby they obtain great plenty thereof, replanting the ſame once in three or foure yeares.

¶ *The Time.*

Liquorice floureth in July, and the ſeed is ripe in September.

1 *Glycyrrhiza Echinata* Dioscoridis.
Hedge-hogge Licorice.

‡ 2 *Glycyrrhiza vulgaris.*
Common Licorice.

¶ *The Names.*

The first is called in Greeke γλυκύρριζα: in Latine, *Dulcis radix*, or sweet Root: this Licorice is not knowne either to the Apothecaries or to the vulgar people: we call it in English, *Dioscorides* his Licorice.

It is most euident that the other is *Glycyrrhiza*, or Licorice: the Apothecaries call it by a corrupt word, *Liquiritia*: the Italians, *Regalitia*: the Spaniards, *Regeliza* and *Regalitia*: in high Dutch, **Suszholtz, Suszwurtzel**: in French, *Rigolisse*, *Raigalisse*, and *Reglisse*: in low Dutch, **Callissiehout, suethout**: in English, common Licorice: *Plin.* calleth it *Scythica herba*: it is named *Scythice* of the countrey Scythia, where it groweth.

¶ *The temperature.*

The Nature of *Dioscorides* his Licorice, as *Galen* saith, is familiar to the temperature of our bodies, and seeing it hath a certaine binding quality adioined, the temperature thereof so much as is hot and binding, is specially of a warme quality, comming neerest of all to a meane temperature; besides, for that it is also sweet, it is likewise meanely moist.

For as much as the root of the common Licorice is sweet, it is also temperately hot and moist; notwithstanding the barke thereof is something bitter and hot, but this must be scraped away; the fresh root when it is full of juice doth moisten more than the dry.

¶ *The Vertues.*

A The root of Licorice is good against the rough harshnesse of the throat and brest; it openeth the pipes of the lungs when they be stuffed or stopped, and ripeneth the cough, and bringeth forth flegme.

B The iuice of Licorice made according to Art, and hardned into a lumpe, which is called *Succus Liquiritiæ*, serueth well for the purposes aforesaid, being holden vnder the tongue, and there suffered to melt.

C Moreouer, with the juice of Licorice, Ginger, and other spices, there is made a certaine bread or cakes, called Ginger-bread, which is very good against the cough, and all the infirmities of the lungs and brest: which is cast into moulds, some of one fashion, and some of another.

D The iuice of Licorice is profitable against the heate of the stomacke, and of the mouth.

The

The same is drunke with wine of Raisons against the infirmities of the liuer and chest, scabs or sores of the bladder, and diseases of the kidneyes. E

Being melted vnder the tongue it quencheth thirst: it is good for greene wounds being layed thereupon, and for the stomacke if it be chewed. F

The decoction of the fresh roots serueth for the same purposes. G

But the dried root most finely poudered is a singular remedie for a pin and a web in the eye, if it be strewed thereupon. H

Dioscorides and *Pliny* also report, that Liquorice is good for the stomack and vlcers of the mouth, being cast vpon them. I

It is good against hoarsenesse, difficultie of breathing, inflammation of the lungs, the pleurisie, spitting of bloud or matter, consumption or rottennes of the lungs, all infirmities and ruggednes of the chest. K

It takes away inflammations, mitigateth and tempereth the sharpnesse and saltnes of humors, concocteth raw humors, and procureth easie spitting. L

The decoction is good for the kidnies and bladder that are exulcerated.

It cureth the strangurie, and generally all infirmities that proceed of sharpe, salt, and biting humors. M

These things concerning Liquorice hath also *Theophrastus*: *viz.* that with this and with cheese made of Mares milke the Scythians were reported to be able to liue eleuen or twelue dayes. N

The Scythian root is good for shortnesse of breath, for a dry cough, and generally for all infirmities of the chest. O

Moreouer, with honey it healeth vlcers, it also quencheth thirst if it be held in the mouth: for which cause they say that the Scythians do liue eleuen or twelue dayes with it and *Hippace*, which is cheese made of Mares milke, as *Hippocrates* witnesseth. P

Pliny in his twenty fifth booke, chap. 8. hath thought otherwise than truth, that *Hippace* is an herbe so called. Q

† Both the figures formerly were of the first described.

Chap. 13. *Of Milke Trefoile or Shrub Trefoile.*

¶ *The Kindes.*

There be diuers kindes or sorts of the shrubby Trefoile, the which might very well haue passed among the three leaued Grasses, had it not beene for my promise in the proeme of our first part, That in the last booke of our History the shrubbie or wooddy plants should be set forth, euerie one as neere as might be in kindred and neighbourhood.

¶ *The Description.*

† 1 The first kinde of *Cytisus* or shrubby Trefoile growes to the forme of a small shrub or wooddy bush two or three cubits high, branching into sundry small boughes or armes, set full of leaues like the small Trefoile, darke greene, and not hairie, three growing alwaies together: among these come forth smal yellow floures like them of French Broome, which doe turne into long and flat cods, containing small seed of a blackish colour.

2 The second kinde of *Cytisus* is likewise a small shrub, in shape after the manner of the former, but that the whole plant is altogether smaller, and the leaues rounder, set together by couples, and the small cods hairy at the ends, which sets forth the difference. ‡ The leaues of this are almost round, and grow three together close to the stalke: they are smooth, of a fresh greene, and the middlemost leafe of the three is the largest, and ends in a sharpe point: the floures are of the bignesse and colour of the *Trifolium corniculatum*: it floures in May. ‡

3 The root of this third kinde is single, from whence spring vp many smooth brittle stalks diuided into many wings and branches, whereon grow greene leaues smaller than those of medow Trefoile: the floures are yellow, lesser than Broome floures, otherwise very like, growing about the tops of the twiggie branches, diuided into spoky tufts: which being vaded, there follow thinne long narrow cods, lesser than those of the Broome, wherein is contained small blacke seed. The root is long, deeply growing into the ground, and sometimes waxeth crooked in the earth. ‡ This also hath smooth green leaues, and differs little (if any thing at all) from the first described, wherefore I thought it needlesse to giue a figure. Our Author called it *Cytisus siliquosus*, Codded shrub Trefoile, because one of the branches was fairely in the figure exprest with cods; I know no other reason, for all the *Cytisi* are codded as well as this. ‡

1 *Cytisus.*
The first shrub Tresoile.

2 *Cytisus.*
The second shrub Trefoile.

4 *Cytisus hirsutus.*
Hairy shrub Trefoile.

5 *Cytisus incanus.*
Hoary shrub Trefoile.

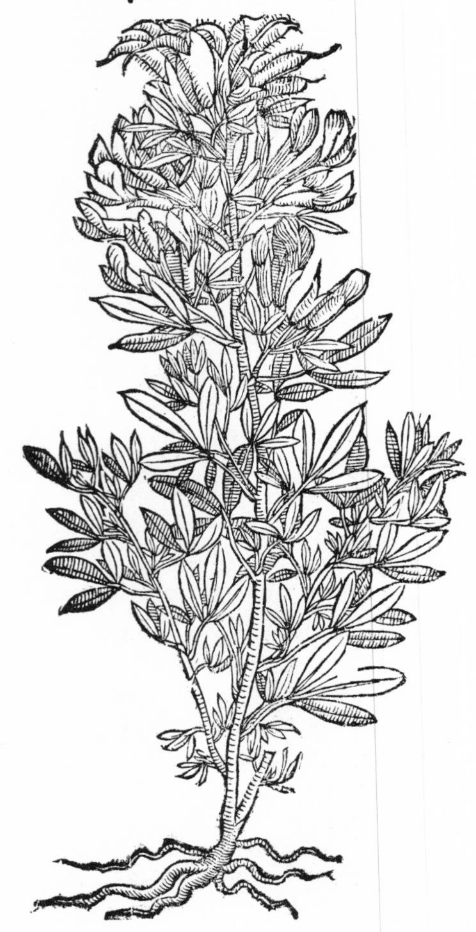

4 The fourth kinde of *Cytiſus* hath a great number of ſmall branches and ſtalkes like the former, but it is a lower plant, and more woolly; whoſe ſtalks and branches grow not very high, but yet very plentifully ſpred about the ſides of the plant: the leaues are greater than the former, but leſſer than thoſe of medow Trefoile: the floures grow cloſe together, as though they were bound vp or compact into one head or ſpokie tuft ſomewhat greater than the former: the cods are alſo greater, and more hairy: the root groweth very deepe into the ground, whereunto are adioyned a few fibres: it falleth out to be more hairy or woolly in one place than in another, and the more hairie and woolly that it is, the whiter it waxeth; for the roughneſſe bringeth it a certain whitiſh colour. ‡ The branches of this oft times lie along vpon the ground: the leaues are ſmooth and greene aboue, and hoarie vnderneath: the floures yellow, which fading ſometimes become orange coloured: the cods are round, and ſeeds browniſh. ‡

5 The fifth kinde of *Cytiſus* groweth to the height of a cubit or more, hauing many ſlender twiggy branches like Broome, ſtreaked and very hard: whereupon grow leaues very like Fenugreeke, yet all hoary, three together: from the boſome of which, or betweene the leaues and the ſtalkes, come forth yellow floures very like Broome, *Spartum*, or Peaſe, but ſmaller: the cods be like vnto Broome cods, of an aſh colour, but ſlenderer, rougher, and flatter; in the ſeueral cels or diuiſions whereof are contained bright ſhining ſeeds like the blacke ſeeds of Broome: all the whole plant is hoarie like *Rhamnus* or *Halymus*.

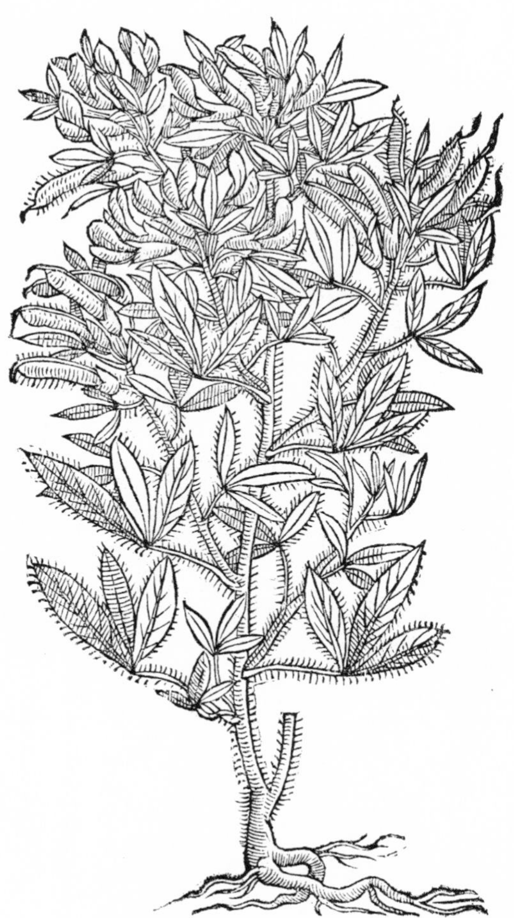

6 *Cytiſus Pinnatus*.
Winged ſhrub Trefoile.

7 *Cytiſus 7. Cornutus*.
The Horned ſhrub Trefoile.

6 The ſixth kinde of *Cytiſus* or buſh Trefoile groweth to the height of a tall man, with long ſtalkes couered ouer with a blackiſh barke, and a few boughes or branches, beſet or garniſhed with leaues like the common Trefoile, but ſmaller, growing alſo three together, whereof the middlemoſt of the three leaues is twice as long as the two ſide leaues; the vpper ſide whereof is green, and the lower ſide ſomwhat reddiſh and hairie: the floures grow along the ſtalks almoſt from the bottome to the top, of a golden yellow colour, faſhioned like the Broome floure, but greater than any of the reſt of his kinde, and of a reaſonable good ſauour: the ſeed hath the pulſie taſte of *Cicer*.

7 The ſeuenth kinde of *Cytiſus* hath many tough and hairy branches riſing from a wooddie root, foure or fiue cubits high, which are diuided into ſundry ſmaller branches beſet with leaues like the medow Trefoiles; among which come forth yellow floures like Broome, that turne into

crooked flat cods like a ſickle; wherein is contained the ſeed taſting like *Cicer* or *Legumen*. The whole plant is hoarie like *Rhamnus*, and being broken or bruiſed ſmelleth like Rocket.

8 This eighth kinde of *Cytiſus*, which *Pena* ſetteth forth, is doubtleſſe another kinde of *Cytiſus*, reſembling the former in leaues, floures, and cods, ſauing that the ſmall leaues (which are alwaies three together) are a little ſnipt about the edges: the whole plant is ſlenderer, ſofter, and greener, rather reſembling an herbe than a ſhrub: the root is ſmall and ſingle.

9 This baſtard or miſ-begotten ſhrub Trefoile, or baſtard *Cytiſus*, groweth vp like a ſhrub, but not of a wooddy ſubſtance, hauing tender ſtalks ſmooth and plaine: whereon do grow hairy leaues like the other, diuers ſet vpon one foot-ſtalke, contrarie to all the reſt: the floures grow along the ſtems like thoſe of the ſtocke Gillofloures, of a yellow colour: the root is tough and wooddy.

8 *Cytiſus* 8.
The eighth ſhrub Trefoile.

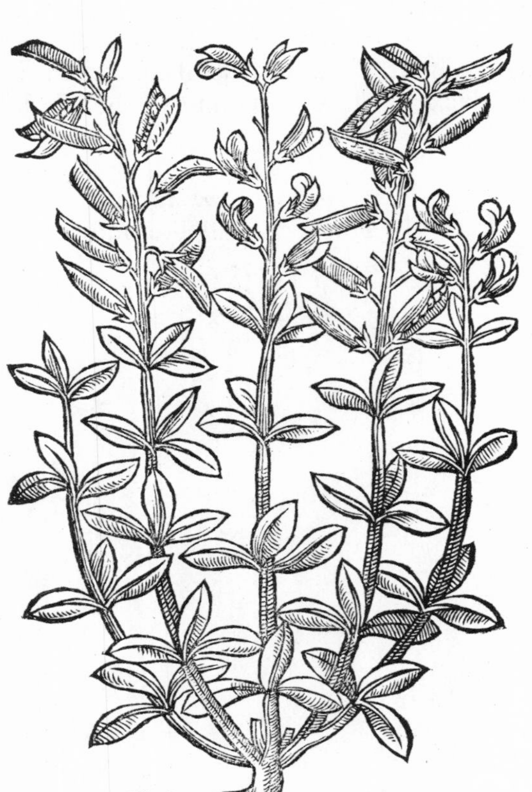

9 *Cytiſus adulterinus, ſiue Alyſſon fruticans.*
Baſtard ſhrub Trefoile.

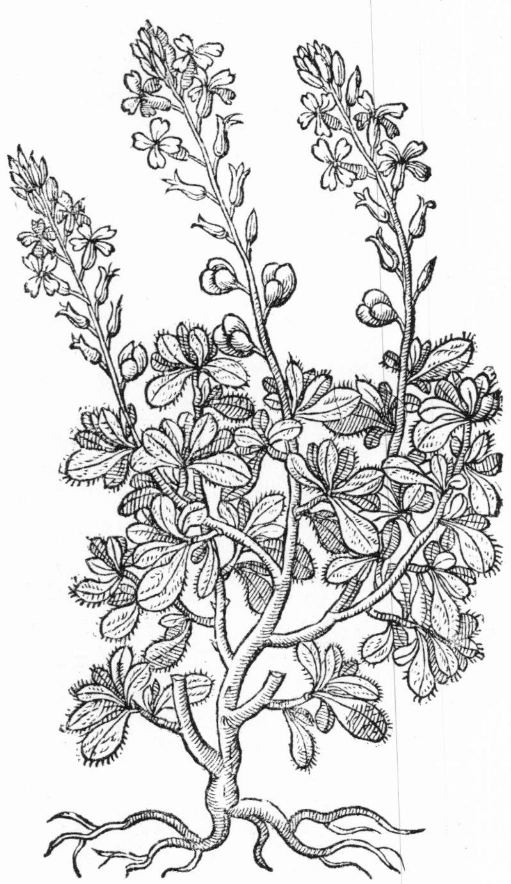

¶ *The Place.*

Theſe plants were firſt brought into Italy and Greece from one of the Iſles of Cyclades, called Cyntho or Cynthuſa, and ſince found in many places of France, as about Montpelier, Veganium, and other places: they are ſtrangers in England, though they grow very plentifully in Scotland, as it is reported; whereof I haue two ſorts in my garden, that is to ſay, *Cytiſus Maranthæ*, or the horned *Cytiſus*, and likewiſe one of the ſmalleſt, that is to ſay, the third in number. ‡ The ſecond groweth in the garden of Mr. *Iohn Tradeſcant*. ‡

¶ *The Time.*

Theſe plants floure for the moſt part in May, Iune, and Iuly, and ſome after: the ſeed is ripe in September.

¶ *The Names.*

The Grecians and Latines do call this ſhrub κύτισος, of Cynthuſa an Iſland before mentioned, in which place they are in great eſtimation for that they do ſo wonderfully feed cattell, and encreaſe milke in their dugs, nouriſh ſheepe and goats, which bring yong ones good for ſtore and increaſe. One Author doth call theſe plants in Greeke κύτισον, that is to ſay in Latine *Fœcundum fœnum*, fertile or fruitfull Hay, for that the kindes hereof cauſe milke to encreaſe, maketh good bloud and iuice, augmenteth ſtrength, and multiplieth the naturall ſeed of generation; they may be called in Engliſh, milke Trefoile, of the ſtore of milke which they encreaſe.

¶ *The*

¶ *The Temperature.*

The leaues of milke Trefoile do coole,as *Dioscorides* writeth; they asswage swellings in the beginning,if they be stamped and laid vnto them with bread: the decoction thereof drunke prouoketh vrine: *Galen* teacheth,that the leaues of Milke Trefoile haue a digesting or wasting qualitie mixed with a waterie and temperate facultie,as haue those of the Mallow.

¶ *The Vertues.*

Women,saith *Columella*, if they want milke must steepe dry milke Trefoile in faire water, and when it is throughly soked,they must the next day mix a quart or thereabouts of the same pressed or strained forth with a little wine,and so let it be giuen vnto them to drinke, and by that meanes they themselues shall receiue strength,and their children comfort by abundance of milke. A

Hippocrates reckoneth vp Milk-Trefoile among those things that encrease milke, in his booke of the Nature of women, and of womens diseases. B

Also *Aristomachus* of Athens in *Pliny*,commandeth to giue with wine the dry plant,and the same likewise boiled in water,to nurses to drinke when their milke is gone. C

Democritus and *Aristomachus* do promise that you shall want no Bees, if you haue milke Trefoile for them to feed on: for all writers with one consent do conclude (as *Galen* saith) that Bees doe gather of the floures of Milke Trefoile very great store of honie. D

Columella teacheth, that Milke Trefoile is notable good for hennes,Bees, Goats, Kine, and all kinde of Cattell,which quickely grow fat by eating thereof,and that it yeeldeth very great store of milke. E

The people of Betica and Valentia (where there is great store of *Cytisus*) doe vse it very much for the Silke Worms to hang their web vpon after they haue been well fed with the leaues of Mulberries. F

Milke Trefoile is likewise a maruellous remedie against the Sciatica, and all other kindes of gouts. G

† The description that formerly was in the first place belonged to that described and figured in the seuenth.

CHAP. 12. *Of Bastard Milke-Trefoiles.*

¶ *The Description.*

1 THis riseth vp with little stalks from the root,brittle, very many in number, parted into wings and branches,about which grow many leaues lesser than those of the medow Trefoile,of colour greene: the floures about the tops of the twigs be orderly placed in maner like ears, of colour yellow, lesser than those of broom,otherwise all alike:in their places grow vp slender cods long,narrow,and lesser than the cods of Broome: rough also and hairy; in which do lie little blackish seeds: the root is long,and groweth deepe,and oftentimes creepeth aslope.

2 The second kinde of bastard Milke-Trefoile is like vnto the former in plentifull stalkes and twigges,but that it is lower and more downie, neither doe the stalkes thereof stand vpright, but rather incline to the one side:the leaues also are somewhat greater, but yet lesser than those of the medow Trefoile, wholly white, and they neuer open themselues out,but keep alwaies folded with the middle rib standing out:the floures likewise be closelier ioined together, and compacted as it were into a little head,and be also something greater: the cods in like manner are a little bigger and hairy,and of a blackish purple or murrey: the root groweth deepe in the ground,being diuided into a few sprigs; it oftentimes happeneth to grow in one place more hairie or downie than in another: the more hairie and downie it is,the more white and hoarie it is; for the hairinesse doth also bring with it a certaine whitish colour.

3 The third kinde of bastard Milke Trefoile bringeth forth a companie of young shoots that are somewhat writhed and crooked,long leaues of a faire greene colour: the floures are closed together,long,white,or else galbineous,sweetly smelling,that is to say,hauing the smel of honie: the shrub it selfe is alwaies greene both Sommer and Winter. ‡ This growes some foot or better high, with slender hoarie branches,set with leaues three standing together vpon a very short stalke, and the middle leafe is as long againe as the other two; they are very white and hoarie, and the yellow floures grow out of the bosomes of the leaues all alongst the stalks. This is that mentioned in the vertues of the former chapter at *F* for the Silke wormes to worke vpon. ‡

4 The fourth shrub is likewise one of the wilde kinde,though in face and stature like the manured

1 *Pſeudocytiſus* 1.
The firſt baſtard ſhrub Trefoile.

2 *Pſeudocytiſus* 2.
The 2. baſtard ſhrub Trefoile.

3 *Cytiſus ſemper virens.*
The euer-greene ſhrub Trefoile.

4 *Pſeudocytiſus hirſutus.*
The hairie baſtard tree Trefoile.

nured *Cytisus*: It groweth vp like a small shrub or hedge bush to the height of two or three yards; on whose branches do grow three rough or hairie leaues, set vpon a slender foot-stalke, of a grasse greene colour aboue, with a reddish hairinesse below: the floures grow alongst the stalks from the middle to the toppe, of a bright shining yellow colour: the root is likewise wooddie.

¶ *The Place.*

These kindes of Milke Trefoiles are found in Morauia, so called in our age, which in times past was named *Marcomannorum prouincia*, and in the vpper Pannonia, otherwise called Austria, neere to high waies, and in the borders of fields; for they seeme after a sort to ioy in the shade. ‡ These grow (according to *Clusius*) in sundry parts of Spaine. ‡

¶ *The Time.*

They floure especially in Iune and Iuly.

¶ *The Names.*

It is euident enough that they are bastard kindes of Milke Trefoiles, and therefore they may be called and plainly termed *Pseudocytisi*, or bastard Milke Trefoiles, or *Cytisi syluestres*, that is to say, wilde Milke Trefoiles.

¶ *The Temperature and Vertues.*

What temperature these shrubs are of, or what vertues they haue we know not, neither haue wee as yet found out by our owne experience any thing, and therefore they may be referred to the other Milke-Trefoiles.

Chap. 15. *Of the venomous Tree Trefoile.*

† 1 *Dorycnium Monspeliensium.*
The venomous Trefoile of Montpelier.

2 *Dorycnium Hispanicum.*
The venomous Trefoile of Spain.

¶ *The Description.*

1 The venomous tree Trefoile of Montpelier hath many tough and pliant stalkes, two or three cubits high, diuided into sundry small twiggie braunches, beset with leaues three together

together, placed from ioint to ioint by spaces, somewhat hoarie, very like vnto the leaues of *Cytisus*, or Rue: among which come forth many small mossie white floures, tuft fashion, in small bundles like Nose-gaies, and very like the floures of the Oliue or Oke tree, which turne into small roundish bladders, as it were made of parchment: wherein is contained blacke seed like wilde *Lotus*, but in taste like the wilde tare: the whole plant is of an vnsauorie smell; the root is thicke, and of a wooddie substance.

2 The Spanish venomous Trefoile hath a wooddie stalke, rough and hoary, diuided into other small branches, whereon do grow leaues like the precedent: the floures grow on the tops of the branches, whereon do grow leaues like those of the Pease, and of a yellow, or rather greenish colour, wherein it differeth from the precedent.

¶ *The Place.*

These venomous Trefoiles grow in Narbone, on the barren and stonie craggie mountaines, at Frontignana, and about the sea coasts, and are strangers in England.

¶ *The Time.*

They flourish from May to the end of Iune.

¶ *The Names.*

Dorycnium, or Δορύκνιον, is that poisonous or venomous plant wherewith in times past they vsed to poison their arrow heads, or rather weapons, thereby to do the greater hurt vnto those whom they did assaile or pursue, whereupon it tooke his name: great controuersie hath been among Herbarists, what manner of plant *Dorycnium* should be; some saying one thing, and some another: which controuersies and sundry opinions are very well confuted by the true censure of *Rondeletius*, who hath for a definitiue sentence set downe the plant described for the true *Dorycnium*, and none other, which may be called in English, Venomous tree Trefoile. ‡ These plants do not sufficiently answer to the description of *Dioscorides*, neither can any one say certainly, that they are poisonous. ‡

¶ *The Temperature.*

Dorycnium is very cold, without moistning.

¶ *The Vertues.*

A Venomous Trefoile hath not one good qualitie that I can reade of, but it is a pestilent venomous plant, as hath been said in the description.

‡ The figures were formerly transposed.

Chap. 16. Of the shrub Trefoile called also Makebate.

Polemonium siue Trifolium fruticans.
Shrubby Trefoile, or yellow Iasmine.

¶ *The Description.*

This shrubby plant called *Polemonium*, hath many wooddie twigges, growing vnto the height of foure or fiue cubits, hauing smal twiggie branches, of a darke green colour, garnished with small leaues of a deepe greene colour, alwaies three ioined together vpon little footstalks, like the *Cytisus* bush, or the field Trefoile, but smaller: the floures be yellow, and round, diuided into fiue or six parts, not much vnlike the yellow Iasmine, which hath caused many to call it yellow Iasmine, euen vnto this day: when the floures be vaded, there succeed small round berries as big as a Pease, of a black purplish colour when they be ripe, which being broken will die or colour the fingers like Elder Berries: within these berries are contained a small flat seed, like vnto Lentils: the root is long and small, creeping hither and thither vnder the earth, putting forth new springs or shoots in sundry places, whereby it wonderfully increaseth.

¶ *The Place.*

It groweth plentifully in the countrey of Montpellier at New Castle vpon the drie hills, and hot banks of the Oliue fields, and in the stony fields and wood of Gramuntium: it growes in my garden, and in other Herbarists gardens in England.

¶ The

¶ *The Time.*

It floureth in Sommer: the seed is ripe in Autumne; the shrub it selfe is alwaies greene, and hath a lasting root.

¶ *The Names.*

Most do call it *Cytisus*, but we had rather name it *Trifolium fruticans*: for it doth not agree with *Cytisus* or Milk-Trefoile, as in the chapter before it is plaine enough by his description, vnlesse it be *Cytisus Marcelli*, or *Marcellus* his Milke-Trefoile, with which peraduenture it might be thought to haue some likenesse, if the floures which are yellow were white, or *galbineous*, that is to say, blew.

There be diuers also that take this Trefoile to be *Polemonium*, forasmuch as the leaues hereof seeme to be somewhat like those of common Rue, but *Polemonium* hath not the leafe of common Rue, otherwise called Herb-grace, but of the other, that is to say, of S. Iohns Rue: it is called in English, shrubby Trefoile, or Make-bait.

¶ *The Temperature.*

Polemonium is of temperature dry in the second degree, with some Acrimonie or sharpnesse.

¶ *The Vertues.*

This shrubby plant hath so many singular and excellent vertues contained in it, that some haue called it by the name *Chiliodunamis*, that is, hauing an hundred properties. A

It is very effectuall against the stinging of Scorpions, and (as some write) if a man hold it in his hand, he cannot be hurt with the biting of any venomous beast. B

Being taken in vineger it is very good for those that are spleneticke, and whose spleen or Milt is affected with oppilations or stoppings. C

If the root be taken in wine it helpeth against the bloudy flix, it prouoketh vrine being drunke with water, scoureth away grauell, and easeth the paine and ache called the Sciatica. D

Chap. 17. *Of Broome, and Broome Rape.*

1 *Genista.*
Broome.

2 *Rapum Genistæ, siue Orobanche.*
Broome Rape, or Orobanch.

¶ *The*

‡ *Orobanche Monſpeliaca flo. oblongis.*
Long floured Broome Rape.

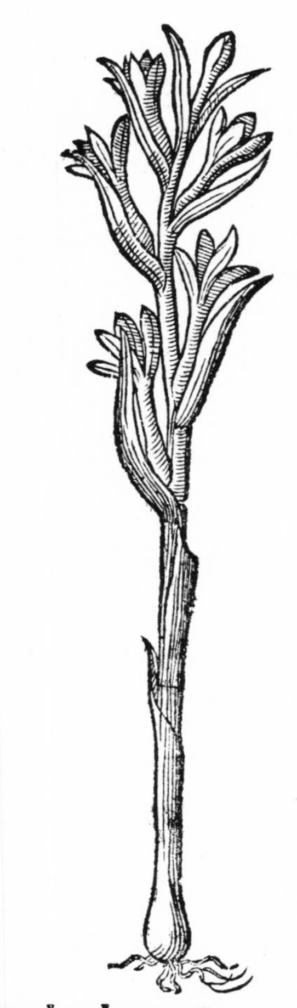

‡ *Orobanche flore maiore.*
Great floured Broome Rape.

‡ *Orobanche ramoſa.*
Branched Broome Rape.

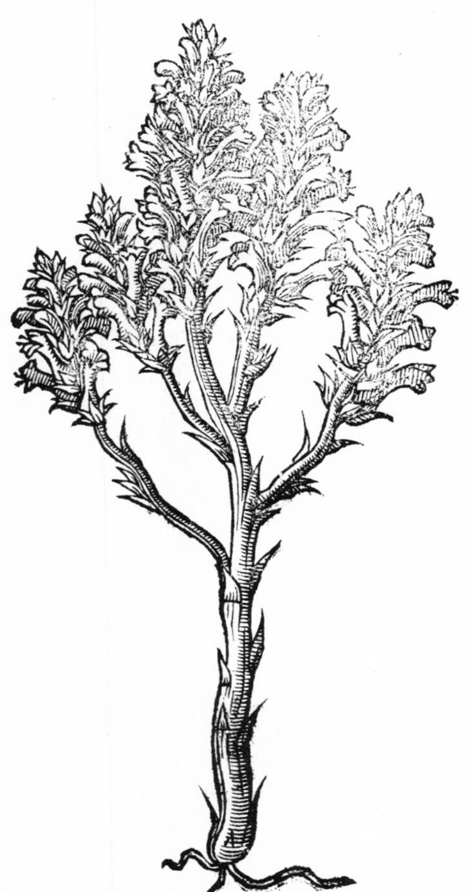

¶ *The Deſcription.*

1 Broome is a buſh or ſhrubby plant, it hath ſtalkes or rather woddie branches: from which do ſpring ſlender twigs, cornered, greene, tough, and that be eaſily bowed, many times diuided into ſmall branches: about which do grow little leaues of an obſcure green colour, and braue yellow floures; and at the length flat cods, which being ripe are blacke, as be thoſe of the common Vetch, in which do lie flat ſeeds, hard, ſomething browniſh, and leſſer than Lentils: the root is hard and wooddie, ſending forth diuers times another plant of the colour of an Oken leafe, in ſhape like vnto the baſtard Orchis, called Birds neſt, hauing a root like a Turnep or Rape, whereupon it is called *Rapum Geniſtæ*, or Broom Rape.

2 This is a certaine bulbed plant growing vnto the roots of broome, big below, and ſmaller aboue, couered with blackiſh ſcales, and of a yelowiſh pulpe within: from which doth riſe a ſtalke a ſpan long, hauing whitiſh floures about the top, like almoſt to thoſe of Dead Nettle: after which grow forth long, thicke, and round husks, in which are contained very many ſeeds, and good for nothing: the whole plant is of the colour of the Oken leafe.

‡ Of

3 *Genista Hispanica.*
Spanish Broome.

5 *Chamægenista Anglica.*
English Dwarfe Broome.

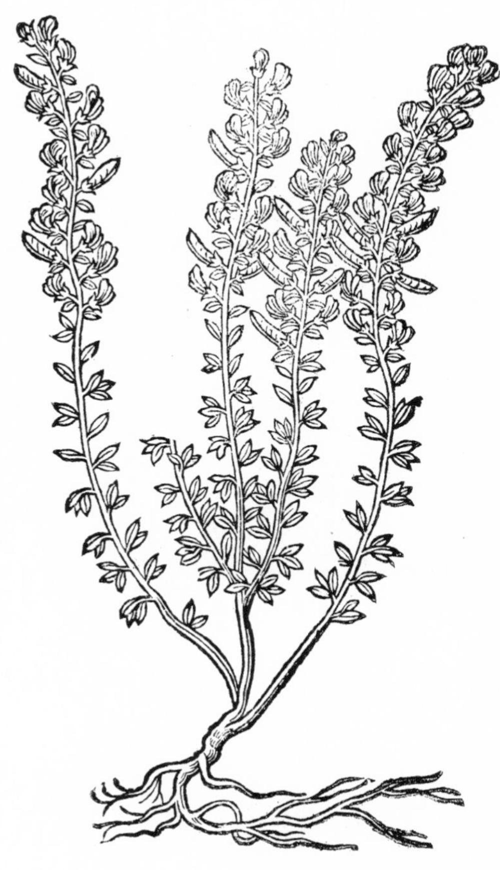

6 *Chamægenista Pannonica.*
Dwarfe broome of Hungarie.

‡ Of this *Orobanche* or Broome Rape there are some varieties obserued and set forth by *Lobel* and *Clusius*: the first of these varieties hath longer and smaller floures than the ordinarie. The second hath larger floures, and those of a blewish colour, and is sometimes found among corne. The third is parted towards the top into sundry branches; the floures of this are either blew, purplish, or else white, and it willingly growes among hempe. ‡

3 The Spanish Broome hath likewise wooddy stems, from whence grow vp slender pliant twigs, which be bare and naked without leaues, or at the least hauing but few small leaues, set here and there far distant one from another, with yellow floures not much vnlike the floures of common Broome, but greater, which turne into small long cods, wherein is conteined browne and flat seed: the roote is tough and wooddy.

4 Small leafed or thin leafed Broome hath many tough pliant shoots rising out of the ground, which grow into hard and tough stalks, which are diuided into diuers twiggy branches whereon doe grow very small thin leaues, of a whitish colour; whereupon some haue called it *Genista alba*, white Broome: the floures grow at the top of the stalkes, in shape like those of the common Broom, but of a white colour, wherein it specially differeth from the other Broomes.

5 English Dwarfe Broome hath many twiggy branches, very greene, tough, somewhat straked or cornered, leaning toward the ground: whereon do grow leaues set without order, sometimes two together, and often three or foure growing fast together, like vnto the common Broome, greene on the vpper side, hoary vnderneath, and of a bitter taste: among which leaues come forth yellow floures like those of common Broome, but lesser, of little or no smell at all: after which appeare small cods somewhat hairy, wherin is contained small seed: the root is tough and woody. ‡ *Bauhine* judges these two last described to be onely varieties of the common Broome; to whose opinion I do much incline, yet I haue let our Authrs description stand, together with the figure of this later, which seemingly expresses the greatest difference. ‡

6 The Dwarfe Broome of Hungary hath stalkes and yellow floures like those of the last described: the leaues hereof are different, they are longer, and more in number: the whole plant is altogether greater, wherein especially consisteth the difference.

¶ *The Place.*

The common Broome groweth almost euery where in dry pastures and low woods.

The Broome Rape is not to be found but where Broome doth grow; it groweth in a Broome field at the foot of Shooters hill next to London; vpon Hampstead Heath, and in diuers other places.

Spanish Broome groweth in diuers kingdomes of Spaine and Italy; we haue it in our London gardens.

The White Broome groweth likewise in Spaine and other hot regions; it is a stranger in England; of this *Titus Calphurnius* makes mention in his second Eclog of his Bucolicks, writing thus:

Cernis vt, ecce pater, quas tradidit Ornite vaccæ
Molle sub hirsuta latus explicuere genista.

See father, how the Kine stretch out their tender side
Vnder the hairy broome, that growes in fields so wide.

¶ *The Time.*

Broome floureth in the end of Aprill or May, and then the young buds of the floures are to be gathered, and laid in pickle or salt, which afterwards being washed or boyled, are vsed for sallads, as Capers be, and be eaten with no lesse delight: the cods and seeds be ripe in August; the Rape appeareth and is seene especially in the moneth of Iune.

The Spanish Broome doth floure sooner, and is longer in flouring.

¶ *The Names.*

This shrub is called in Latine, *Genista*, or as some would haue it *Genesta*: in Italian, *Genestra*: in Spanish likewise *Genestra*, or *Giestra*: in high Dutch, **Pfrimmen**: in low Dutch, **Brem**: in French, *Genest*: in English, Broome. ‡ The Spanish Broome by most writers is iudged to be the *Spartium* of *Dioscorides*. ‡

¶ *The Temperature and Vertues.*

A The twigs, floures, and seeds of Broome are hot and dry in the second degree: they are also of a thin essence, and are of force to clense and open, and especially the seed, which is dryer and not so full of superfluous moisture.

B The decoction of the twigs and tops of Broome doth clense and open the liuer, milt, and kidnies.

C It driueth away by the stoole watery humours, and therefore it is wholesome for them that haue the dropsie, especially being made with wine; but better for the other infirmities with water.

D The seed also is commended for the same purposes.

E There is also made of the ashes of the stalkes and branches dryed and burnt, a lie with thin white wine, as Rhenish wine, which is highly commended of diuers for the greene sickenesse and dropsie, and this doth mightily expell and driue forth thin and watery humors together with the vrine, and that by the bladder; but withall it doth by reason of his sharpe quality many times hurt and fret the intrailes.

F *Mesue* saith, that there is in the floures and branches a cutting moisture, but full of excrements, and therefore it causeth vomit: and that the plant doth in all his parts trouble, cut, attenuate, and violently purgeth by vomit and stoole, flegme and raw humours out of the ioints.

G But these things are not written of Broome, but of *Spartum*, which purgeth by vomit, after the manner of Hellebor, as both *Dioscorides* and *Pliny* do testifie.

H *Mesue* also addeth, that Broome doth breake the stone of the kidnies and bladder, and suffereth not the matter whereof the stone is made to lie long, or to become a stone.

I The young buds or little floures preserued in pickle, and eaten as a sallad, stir vp an appetite to meate and open the stoppings of the liuer and milt.

The same being fully blowne, stamped and mixed with swines grease, do ease the paine of the gout. L

And *Mesue* writeth, that this tempered with honie of Roses, or with an egge, doth consume away the Kings-euill. M

The Rape of the Broom or Broome Rape, being boyled in wine, is commended against the pains of the kidnies and bladder, prouoketh vrine, breaketh the stone, and expelleth it. N

The iuice pressed forth of Broom rape healeth green wounds, and clenseth old and filthy vlcers: the later Physitions do affirme that it is also good for old venomous and malicious vlcers. O

That worthy Prince of famous memorie *Henry* 8. King of England, was woont to drinke the distilled water of Broome floures, against surfets and diseases thereof arising. P

Sir *Thomas Fitzherbert* Knight, was woont to cure the blacke iaundice with this drinke onely. Q

Take as many handfuls (as you thinke good) of the dried leaues of Broom gathered and brayed to pouder in the moneth of May, then take vnto each handfull of the dried leaues, one spoonful and a halfe of the seed of Broom braied into pouder: mingle these together, and let the sicke drinke thereof each day a quantitie, first and last, vntill he finde some ease. The medicine must be continued and so long vsed, vntil it be quite extinguished: for it is a disease not very suddenly cured, but must by little and little be dealt withall. R

Orobanch or Broom rape sliced and put into oyle Oliue, to infuse or macerate in the same, as ye do Roses for oile of Roses, scoureth and putteth away all spots, lentils, freckles, pimples, wheals and pushes from the face, or any part of the body, being annointed therewith. S

Dioscorides writeth, that Orobanch may be eaten either raw or boiled, in manner as we vse to eat the sprigs or young shoots of *Asparagus*. T

The floures and seeds of Spanish Broome are good to be drunke with Mead or honied water in the quantitie of a dram, to cause one to vomite with great force and violence, euen as white Hellebor, or neesing pouder. V

If it be taken alone, it looseneth the belly, driueth forth great quantitie of waterie and filthie humours. X

Chap. 18. *Of base Broome or greening weed.*

¶ *The Description.*

1 This base kinde of Broom called Greene weed or Diers weed, hath many tough branches proceeding from a wooddie root: whereon do grow great store of leaues, of a deep green colour, somewhat long like those of Flax: the floures grow at the top of the branches not much vnlike the leaues of Broome, but smaller; of an exceeding faire yellow colour, which turne into small flat cods, wherein is contained a little flat seed.

2 *Carolus Clusius* setteth forth another kinde of Broome, which *Dodonæus* calleth *Genista tinctoria*, being another sort of Diers weed: it groweth like the Spanish Broome: vpon whose branches do grow long and small leaues like Flax, greene on the vpper side, and of an hoarie shining colour on the other. The floures grow at the top of the stalks, spike fashion, in forme and colour like the former: the roots are thicke and wooddie.

3 *Carolus Clusius* setteth forth two kindes of Broome. The first is a low and base plant, creeping and lying flat vpon the ground, whose long branches are nothing else, but as it were stalkes consisting of leaues thicke in the middest, and thinne about the edges, and as it were diuided with small nicks; at which place it beginneth to continue the same leafe to the end, and so from leafe to leafe, vntill it haue increased a great sort, all which doe as it were make one stalke; and hath none other leaues, sauing that in some of the nicks or diuisions there commeth forth a small leafe like a little eare. At the end of those flat and leafed stalks come forth the floures, much like the floures of the common Greening weed, but lesser, and of a yellow colour, which turne into small cods. The roots are very long, tough, and wooddie, ful of fibres, closing at the top of the root, from whence they proceed as from one body.

4 This kinde of Greenweed called of some *Chamæspartium*, hath a thicke wooddie root: from which rise vp diuers long leaues, consisting as it were of many pieces set together like a paire of Beads (as may better be perceiued by the figure, than expressed by words) greene on the vpper side, and whitish vnderneath, very tough, and as it were of a rushie substance: among which rise vp very small naked rushie stalkes; on the top whereof groweth an eare or spike of a chaffie matter, hauing here and there in the said eare diuers yellow floures like Broome, but very small or little.

5 The

1 *Genistella tinctoria.*
Greeneweed or Diers weed.

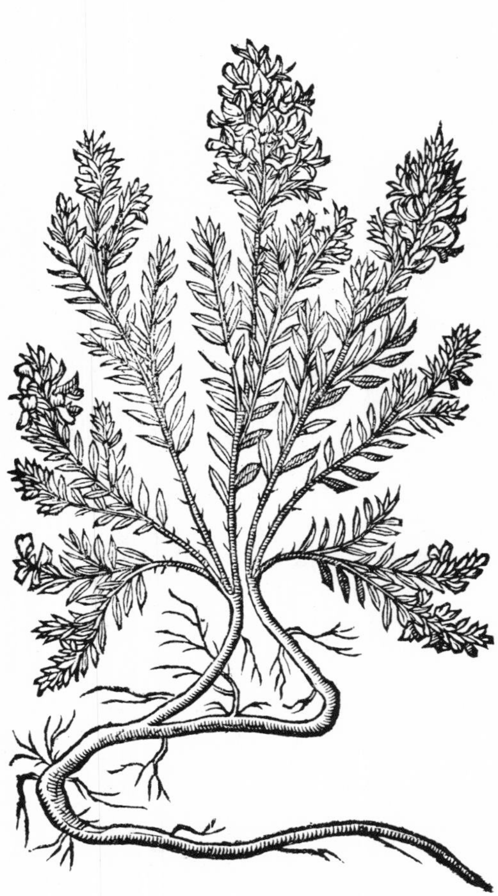

2 *Genistella infectoria.*
Wooddie Diers weed.

3 *Genistella pinnata.*
Winged Greeneweed.

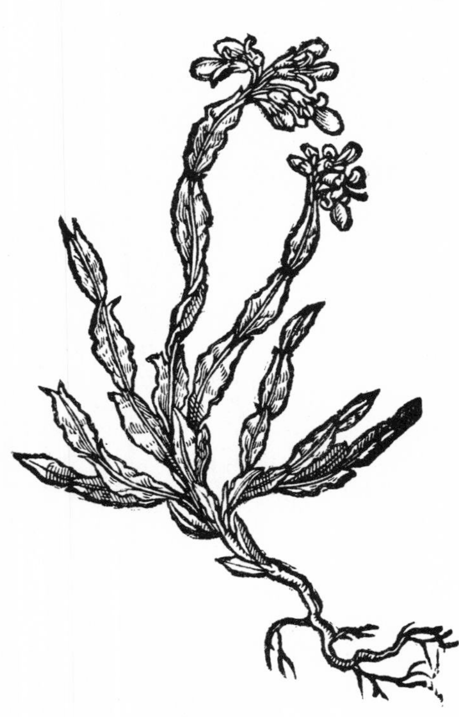

4 *Genistella globulata.*
Globe Greene weed.

5 The fiſt Greeneweed hath a wooddie tough root,with certaine ſtrings annexed thereto:from which riſe vp diuers long,flat leaues,tough,& very hard,conſiſting as it were of many little leaues, ſet one at the end of another, making of many one entire leafe,of a greene colour : amongſt which come forth diuers naked hard ſtalks ,very ſmall and ſtiffe, on the tops whereof ſtand ſpikie eares of yellow floures,like thoſe of Broome,in ſhape like that great three leafed graſſe, called *Lagopus*, or like the Fox-taile graſſe : after which come flat cods, wherein is incloſed ſmall ſeed like to Tares both in taſte and forme.

5 *Geniſtella Lagopoides maior.*
Hares foot Greeneweed.

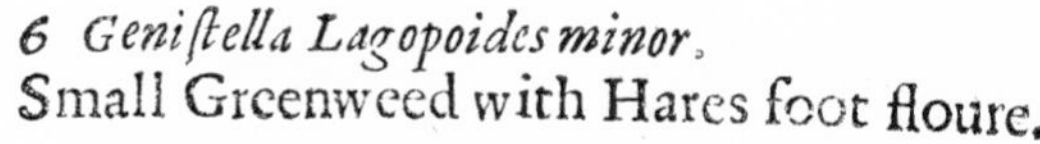

6 *Geniſtella Lagopoides minor.*
Small Greenweed with Hares foot floure.

6 This differeth not from the precedent in ſtalks,roots and leaues:the floures conſiſt of a flockie ſoft matter,not vnlike to the graſſie tuft of Foxtaile,reſembling the floure of *Lagopus*,or Hares-foot,but hauing ſmall yellow floures leſſer than the former,wherein it chiefely differeth from the other of his kinde.

¶ *The Place.*

The firſt being our common Diers-weed,groweth in moſt fertile paſtures and fields almoſt euery where. The reſt are ſtrangers in England.

¶ *The Time.*

They floure from the beginning of Iuly to the end of Auguſt.

¶ *The Names.*

The firſt of theſe Greenweeds is named of moſt Herbariſts *Flos Tinctorius*, but more rightly,*Geniſta Tinctoria*,of this *Pliny* hath made mention [The Greenweeds,ſaith he,do grow to dye cloths with] in his 18.booke 16.Chapter. It is called in high Dutch, **Ferblumen**, and **Ackerbrem** : in Italian,*Cerretta*,and *Coſaria*,as *Matthiolus* writeth in his chapter of *Lyſimachia*, or Looſe-ſtrife : in Engliſh,Diers Greening weed,baſe Broome,and Woodwaxen.

The reſt we refer to their ſeuerall titles.

¶ *The Temperature and Vertues.*

These plants are like vnto common Broome in bitterneſſe,and therefore are hot and drie in the A ſecond degree : they are likewiſe thought to be in vertues equall ; notwithſtanding their vſe is not ſo well knowne,and therefore not vſed at all where the other may be had: we ſhall not need to ſpeak of that vſe that Diers make thereof,being a matter impertinent to our Hiſtorie.

CHAP. 19. *Of Spanish base Broomes.*

¶ *The Description.*

‡ 1 THis growes to the height of a cubit, and is couered with a crested and rough barke, and diuided into many longish branches crested & green, which at their first springing vp haue some leaues vpon them, which fall away as soon as the plant comes to floure: from the sides of the branches come forth long foot-stalks whereon hang some small yellow floures, which are succeeded by short round yellowish red cods which commonly containe but one seed, seldome two, and these hard and blacke, and like a little Kidney, which when it is ripe will rattle in the cod being shaken. ‡

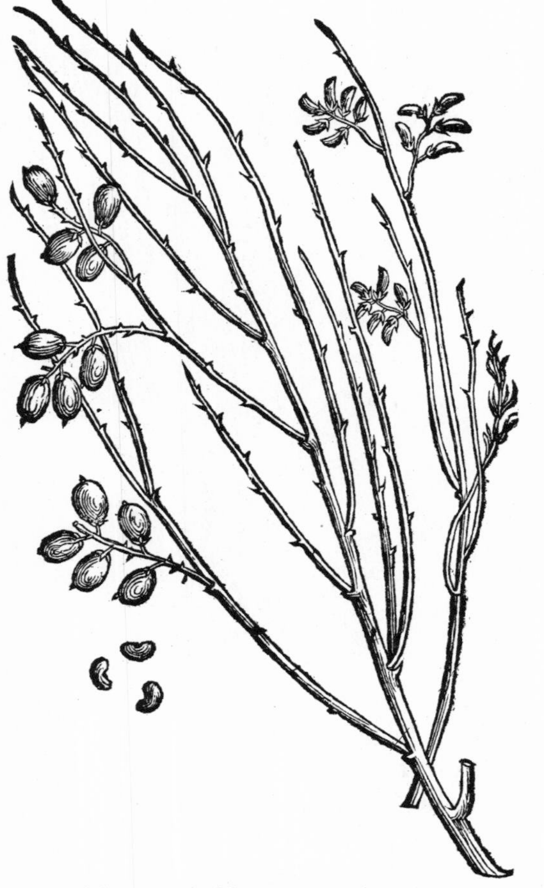

1 *Pseudospartum Hispanicum Aphyllum:* Spanish Broome without leaues.

2 *Pseudospartum album Aphyllum.* The white leafe-lesse Spanish broom.

2 This naked broome groweth vp to the height of a man: the stalk is rough, and void of leaues very greene and pliant, which diuideth it selfe into diuers twiggie branches, greene, and tough, like rushes: the floures grow all along the stalks like those of broome, but of a white colour, wherein it differeth from all the rest of his kinde.

¶ *The Place.*

These grow in the prouinces of Spaine, and are in one place higher and more bushie, and in an other lower.

¶ *The Time.*

‡ The first floures in May, and the second in Februarie. ‡

¶ *The Names.*

These base Spanish broomes may be referred to the true, which is called in Greeke σπάρτον: the Latines vse the same name, calling it sometimes *Spartum*, and *Spartium*: in Spanish, *Retama*: in English, Spanish broome, and bastard Spanish broome.

¶ *The Temperature and Vertues.*

A Both the seeds and iuice of the branches of these base broomes, wherewith they in Spaine and other hot regions do tie their vines, do mightily draw, as *Galen* writeth.

Dioſcorides ſaith, that the ſeeds and floures being drunke in the quantitie of a dram, with Mede or honied water, doth cauſe one to vomit ſtrongly, as the Hellebor or neeſing pouder doth, but yet without ieopardie or danger of life: the ſeed purgeth by ſtoole. B

The iuyce which is drawne from out of the branches ſteeped in water, being firſt bruiſed, is a remedie for thoſe that are tormented with the Sciatica, and for thoſe that be troubled with the Squincie, if a draught thereof be drunke in the morning; ſome vſe to ſteepe the branches in ſea water, and to giue the ſame in a cliſter, which purgeth forth bloudy and ſlimie excrements. C

† In this chapter formerly in the firſt place was againe figured and deſcribed the true *Spartium* or Spaniſh Broome; which I haue now omitted, becauſe it was figured and deſcribed in the laſt chapter ſaue one before. In the ſecond place was deſcribed that figured in the third: and in the third place was a deſcription to no purpoſe, which I therefore omitted, and as you ſee deſcribed anew and put in the firſt place that which formerly held the ſecond.

CHAP. 20.

Of Furzes, Gorſſe, Whin, or prickley Broome.

¶ *The Kindes.*

THere be diuers ſorts of prickly Broome, called in our Engliſh tongue by ſundry names, according to the ſpeech of the countrey people where they doe grow: in ſome places, Furzes; in others, Whins, Gorſſe, and of ſome, Prickly Broome.

† *Geniſta ſpinoſa vulgaris.*
Great Furze buſh.

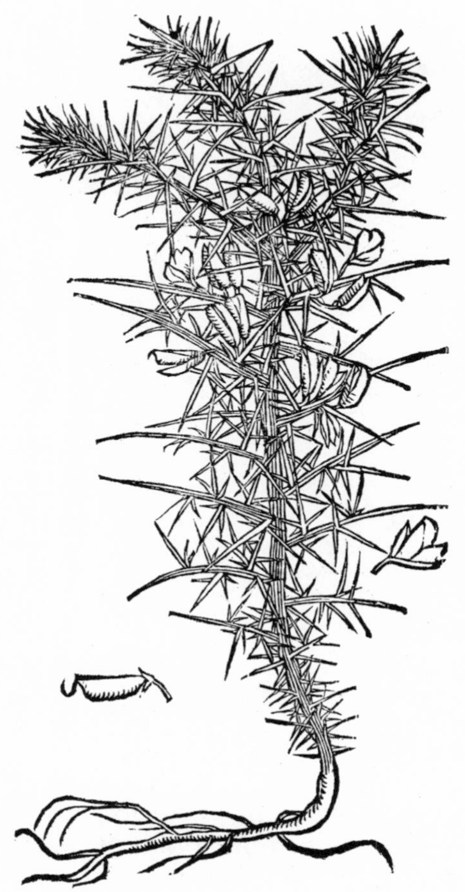

2 *Geniſta ſpinoſa minor.*
The ſmall Furze buſh.

¶ *The Deſcription.*

1 THe Furze buſh is a plant altogether a Thorne, fully armed with moſt ſharpe prickles, without any leaues at all except in the ſpring, and thoſe very few and little, and quickly falling away: it is a buſhy ſhrub, often riſing vp with many woody branches to the height of foure or fiue cubits, or higher, according to the nature and ſoile where they grow: the greateſt and higheſt that I did euer ſee do grow about Exceſter in the Weſt parts of England, where

where the great ſtalks are dearely bought for the better ſort of people,and the ſmall thorny ſpraies for the poorer ſort. From theſe thorny branches grow little floures like thoſe of Broome,and of a yellow colour,which in hot Regions vnder the extreme heate of the Sunne are of a very perfect red colour : in the colder countries of the Eaſt, as Danzicke, Brunſwicke, and Poland, there is not any branch hereof growing, except ſome few plants and ſeeds which my ſelfe haue ſent to Elbing otherwiſe called Meluin, where they are moſt curiouſly kept in their faireſt gardens, as alſo our common Broome, the which I haue ſent thither likewiſe, being firſt deſired by diuers earneſt letters : the cods follow the floures, which the Grauer hath omitted, as a German who had neuer ſeen the plant it ſelfe, but framed the figure by heare-ſay : the root is ſtrong, tough, and wooddy.

We haue in our barren grounds of the North parts of England another ſort of Furze, bringing forth the like prickley thornes that the others haue : the onely difference conſiſteth in the colour of the floures; for the others bring forth yellow floures,and thoſe of this plant are as white as ſnow.

† 2 To this may be ioyned another kinde of Furze which bringeth forth certaine branches that be ſome cubit high, ſtiffe, and ſet round about at the firſt with ſmall winged Lentill-like leaues and little harmeleſſe prickles, which after they haue been a yeare old, and the leaues gon,be armed onely with moſt hard ſharpe prickles, crooking or bending their points downwards. The floures hereof are of a pale yellow colour, leſſer than thoſe of Broome, yet of the ſame forme : the cods are ſmall, in which do lie little round reddiſh ſeeds : the root is tough and wooddy.

† 3 *Geniſta Spinoſa minor ſiliqua rotunda.*
Small round codded Furze.

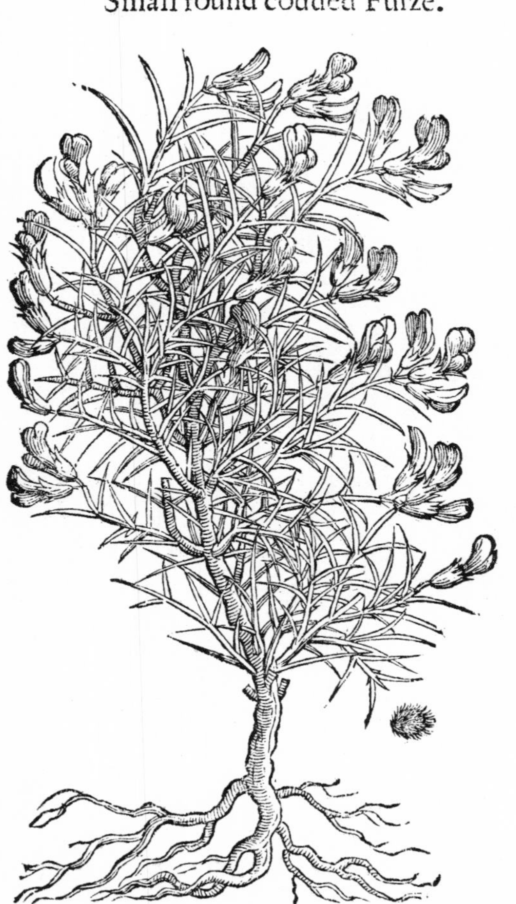

4 *Geniſtella aculeata.*
Needle Furze or petty Whin.

‡ Of this *Cluſius* reckons vp three varieties : the firſt growing ſome cubit high,with deepe yellow floures : the ſecond growes higher, and hath paler coloured floures : the third groweth to the height of the firſt,the floures alſo are yellow,the branches more prickly, and the leaues hairy ; and the figure I giue you is of this third varietie.

3 This ſeldome exceeds a foot in height, and it is on euerie ſide armed with ſharpe prickles, which grow not confuſedly, as in the common ſort, but keepe a certaine order, and ſtill grow forth by couples : they are of a lighter greene than thoſe of the common Furze : on the tops of each of the branches grow two or three yellow floures like thoſe of the former ; which are ſucceeded by little round rough hairy cods of the bigneſſe of Tares. This floures in March, and groweth in the way between Burdeaux and Bayone in France, and vpon the Pyrenean mountaines. *Cluſius* makes it his *Scorpius* 2. or ſecond ſort of Furze : *Lobel* calls it *Geniſta ſpartium ſpinoſum alterum.* ‡

4 This

4 This ſmall kinde of Furze (growing vpon Hampſtead heath neere London, and in diuers other barren grounds, where in manner nothing elſe wil grow) hath many weake and flexible branches of a wooddy ſubſtance: whereon do grow little leaues like thoſe of Tyme: among which are ſet in number infinite moſt ſharpe prickles, hurting like needles, whereof it tooke his name. The floures grow on the tops of the branches like thoſe of Broome, and of a pale yellow colour. The root is tough and wooddy.

‡ 5 This plant (ſaith *Cluſius*) is wholly new and elegant, ſome ſpan high, diuided into many branches, ſome ſpred vpon the ground, others ſtanding vpright, hauing plentifull ſtore of greene prickles: the floures in ſhape are like thoſe of Broome, but leſſe, and of a blewiſh purple colour, ſtanding in rough hairy whitiſh cups, two or three floures commonly growing neere together: ſometimes whileſt it floures it ſendeth forth little leaues, but not very often, and they are few, and like thoſe of the ſecond deſcribed, and quickly fall away, ſo that the whole plant ſeemes nothing but prickles, or like a hedge-hog when ſhe folds vp her ſelfe: the root is wooddy, and large for the proportion of the plant. It growes in the kingdome of Valentia in Spaine, where the Spaniards call it *Erizo*, that is, the Hedge-hog; and thence *Cluſius* alſo termed it *Erinacea*. It floureth in Aprill. ‡

5 *Geniſta ſpinoſa humilis.*
Dwarfe or low Furze.

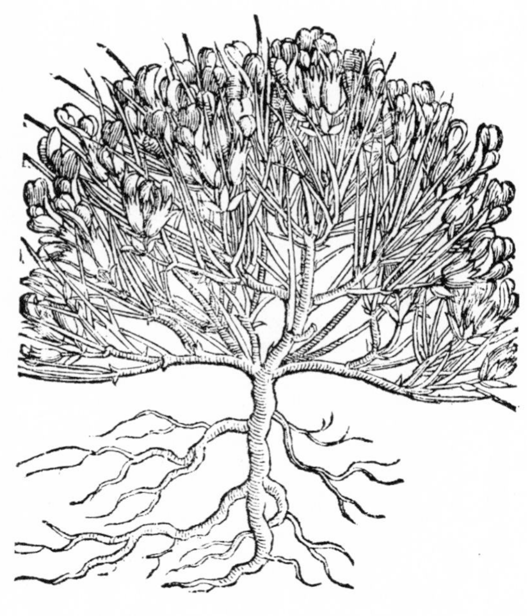

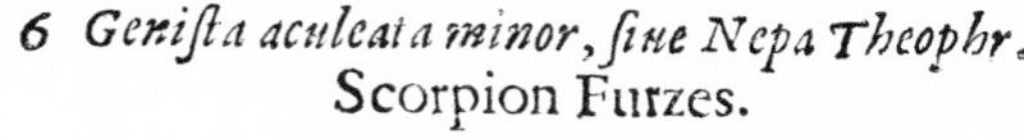

6 *Geniſta aculeata minor, ſiue Nepa Theophr.*
Scorpion Furzes.

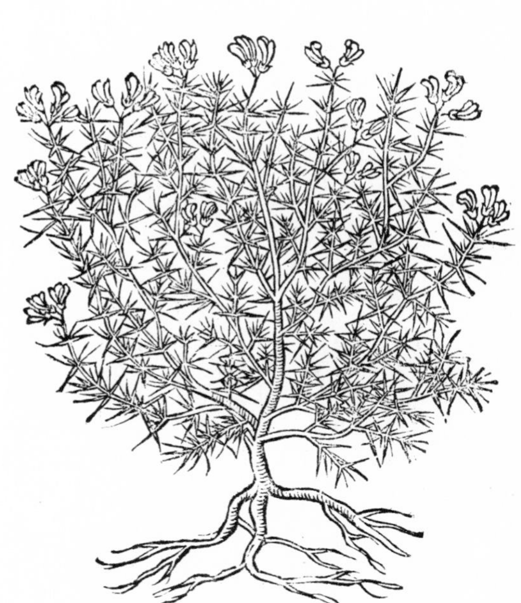

6 The ſmalleſt of all the Furzes is that of the Antients called *Nepa*, or Scorpion Furze, as the word *Nepa* ſeemeth to import: it is a ſtranger in England: it hath beene touched of the Antients in name onely: which fault they haue beene all and euerie of them to be complained of, being ſo briefe that nothing can be gathered from their deſcription: and therefore I refer what might hereof be ſaid to a further conſideration. ‡ This hath a thicke wooddy blacke root ſome halfe foot long, from whence ariſe many ſlender branches ſome foot high, which are ſet with many ſtiffe and ſharpe prickles, growing ſomewhat after the maner of the wilde prickly Sperage: the yong plants haue little leaues like thoſe of Tragacanth; the old ones none: the floures are ſmal, and come forth at the bottome of the prickles, and they are ſucceeded by broad cods wherein the ſeed is contained. It growes in diuers places of France and Spaine, and is thought to be the *Scorpius* of *Theophraſtus*, which *Gaza* tranſlates *Nepa*. ‡

¶ *The Place.*

The common ſort hereof are very well knowne to grow in paſtures and fields in moſt places of England. The reſt are likewiſe well knowne to thoſe that curiouſly obſerue the difference.

¶ *The Time.*

They floure from the beginning of May to the end of September.

¶ *The Names.*

Furze is commonly called *Geniſta ſpinoſa*: in high-Dutch, **Gaſpeldoren**: in Engliſh, Furze, Furzen buſhes, Whinne, Gorſſe, and Thorne-Broome.

This thorny Broome is taken for *Theophraſtus* his *Scorpius*, which *Gaza* nameth *Nepa*: the name *Scorpius* in *Pliny* is πολύσημον, that is to ſay, ſignifying many things, and common to certaine Plants: for

for beſides this *Scorpius* of which he hath made mention, *lib.* 25. *cap.* 5. ſetting downe *Theophraſtus* his words, where he maketh *Aconitum Thelyphonon* to be *Scorpius*, *lib.* 23. *cap.* 10. and likewiſe other plants vnder the ſame title, but vnproperly.

¶ *The Temperature and Vertues.*

A There is nothing written in *Theophraſtus* concerning the faculties of *Scorpius ſpinoſus*, or Furze: *Pliny* ſeemeth to attribute vnto it the ſame vertues that *Scorpioides* hath: notwithſtanding the later Writers do agree that it is hot and dry of complexion: the ſeeds are vſed in medicines againſt the ſtone, and ſtaying of the laske.

† This chapter hath vndergone a great alteration: as thus; the firſt, third, and fourth deſcriptions belonged to the third figure: the ſecond and fifth deſcription, to the fifth figure: and the firſt, ſecond, and fourth figures had no deſcriptions belonging to them. The figure that was in the firſt place is now in the third; the ſecond ſtill holds his place: the third is in the firſt, belonging thereto of right: and for handſomneſſe ſake I haue made the fourth and fifth change places. This *Nepa* alſo in the ſixth place was formerly mentioned by our Author (but now omitted) in the chapter of Aſparagus.

Chap. 21.

Of Cammocke Furze, Reſt-Harrow, or Petty Whinne.

¶ *The Kindes.*

There be diuers ſorts of Reſt-Harrow, which ſome haue inſerted among the ſmooth Broomes; others, among thoſe with prickles, whereof ſome haue purple floures and likewiſe ful of prickles; others, white floures, and ſharpe thornes: ſome alſo purple floures, others white, and alſo yellow, and euery of them void of prickles.

1 *Anonis, ſiue Reſta Bouis.*
Cammocke, or Reſt-Harrow.

3 *Anonis non ſpinoſa purpurea.*
Purple Reſt-Harrow without prickles.

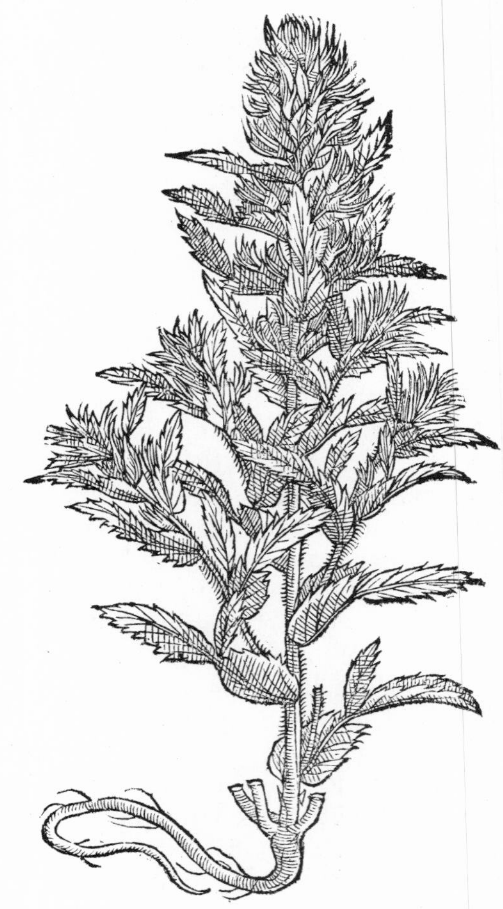

¶ *The Deſcription.*

1 Cammocke or ground Furze riſeth vp with ſtalkes a cubit high, and often higher, ſet with diuers ioynted branches, tough, pliable, and full of hard ſharpe thornes: among which do grow leaues in forme like thoſe of S. Iohns wort, or rather of the Lentill, of a deepe

deep green colour: from the bosome of which thorns and leaues come forth the floures, like those of Peason, of a purple colour: after which do come the cods, in which do lie flat seed: the root is long, and runneth far abroad, very tough, and hard to be torne in pieces with the plough, insomuch that the oxen can hardly passe forward, but are constrained to stand still; whereupon it was called Rest-Plough, or Rest-Harrow.

4 *Anonis, siue Spina lutea.*
Yellow Rest-Yarrow.

2 We haue in our London pastures, and likewise in other places, one of the Rest-Harrowes, not differing from the precedent in stalkes, leaues, or prickles: the onely difference is, that this plant bringeth forth white floures, and the others not so: whence we may call it *Anonis flore albo,* Cammocke with white floures.

3 Rest-Harrow without thornes hath a tough hoary rough stalke, diuided into other rough branches, whereon are set without order, long leaues sharpe pointed, sleightly cut about the edges, of an hoary colour, and somewhat hairy: from the bosome whereof commeth forth purple Pease-like floures of a reasonable good smell: the root is verie tough, long, and wooddy.

4 The yellow floured Cammock is a stranger in these parts, it is only found in the cold Easterne countries, for ought that I can learne; it differs not from the last described, sauing that the floures hereof are of a darke yellow colour, wherein it differeth from all the other of his kinde.

¶ *The Place.*

These grow in earable grounds in fertile pastures, and in the borders of fields, in a fat, fruitful, and long lasting soile: it is sooner found than desired of husbandmen, because the tough and wooddie roots are combersome vnto them, for that they stay the plough, and make the oxen stand.

¶ *The Time.*

They send forth new shoots in May: they be ful growne in Autumne, and then those that of nature are prickly be fullest of sharpe thornes: they floure in Iuly and August.

¶ *The Names.*

Cammocke is called in Greeke Ἀνωνὶς, or Ὀνωνὶς: and likewise in Latine *Anonis,* and *Ononis:* Of Herbarists commonly *Aresta Bouis,* and *Remora aratri,* because it maketh the Oxen whilest they be in plowing to rest or stand still: it is also called *Acutella,* of the stiffe and sharpe thorns which prick those that passe by: in French, *Areste beuf,* and *Boucrande.*

Crateuas nameth it *Ægipyrus:* in high-Dutch, **Stalkraut:** in low-Dutch, **Prangwortele:** in Italian, *Bonaga:* in Spanish, *Gattilhos:* in French, *Arreste beuf, Beuf & Bouerande:* in English, Cammocke, Rest-Harrow, Petty Whinne, and ground Furze.

¶ *The Temperature.*

The root of Cammocke is hot in the third degree, as *Galen* saith: it cutteth also and maketh thinne.

¶ *The Vertues.*

A The barke of the root drunke with Wine prouoketh vrine, breaketh the stone, and driueth it forth.

B The root boyled in water and vineger allayeth the paine of the teeth, if the mouth be often washed therewith hot.

C *Pliny* reporteth, that being boyled in Oxymel (or the syrrup made with honey and vineger) till the one halfe be wasted, it is giuen to those that haue the falling sicknesse. *Matthiolus* reporteth, that he knew a man cured of a rupture, by taking of the pouder of this root for many moneths together.

D The tender sprigs or crops of this shrub before the thornes come forth, are preserued in pickle, and be very pleasant sauce to be eaten with meat as sallad, as a *Dioscorides* teacheth.

CHAP.

Chap. 22.

Of Goofe-berrie, or Fea-berry Bufh.

¶ *The Kindes.*

THere be diuers forts of the Goofe-berries ; fome greater, others leffe : fome round, others long, and fome of a red colour : the figure of one fhall ferue for the reft.

‡ I will not much infift vpon diuerfities of fruits, becaufe my kinde friend Mr. *Iohn Parkinfon* hath fufficiently in his late Worke difcourfed vpon that fubiect ; onely becaufe I iudge many wil be defirous to know their names, and where to get them, I will briefely name the chiefe varieties our Kingdome affords ; and fuch as are defirous of them may finde them with Mr. *Iohn Millen* liuing in Old-ftreet.

The forts of Goofe-berries are thefe : the long greene, the great yellowifh, the blew, the great round red, the long red, and the prickly Goofe-berrie.

Vua Crifpa.
Goofe-berries.

¶ *The Defcription.*

THe Goofe-berry bufh is a fhrub of three or foure cubits high, fet thicke with moft fharpe prickles : it is likewife full of branches, flender, wooddy, and prickly : whereon doe grow round leaues cut with deepe gafhes into diuers parts like thofe of the Vine, of a very greene colour : the floures be very fmal, of a whitifh greene, with fome little purple dafhed here and there : the fruit is round, growing fcatteringly vpon the branches, greene at the firft, but waxing a little yellow through maturitie, full of a winie iuyce fomewhat fweet in tafte when they be ripe ; in which is contained hard feed of a whitifh colour : the root is woodie, and not without ftrings anexed thereto.

There is another whofe fruit is almoft as big as a fmall Chery, and very round in forme : as alfo another of the like bigneffe, of an inch in length, in tafte and fubftance agreeing with the common fort.

We haue alfo in our London gardens another fort altogether without prickles : whofe fruit is very fmal, leffer by much than the common kinde, but of a perfect red colour, wherein it differeth from the reft of his kinde.

¶ *The Place.*

Thefe plants do grow in our London gardens and elfewhere in great aboundance.

¶ *The Time.*

The leaues come forth in the beginning of Aprill or fooner : the fruit is ripe in Iune and Iuly.

¶ *The Names.*

This fhrub had no name among the old Writers, who as we deeme knew it not, or elfe efteemed it not : the later writers call it in Latine *Croffularia* : and oftentimes of the berries, *Vua Crifpa, Vua fpina, Vua fpinella*, and *Vua Crifpina* : in high-Dutch, **Kruſelbeer** : in low-Dutch, **Stekelbeſien** : in Spanifh, *Vua Crifpa*, or *Efpina* : in Italian, *Vua fpina* : in French, *Groifelles* : in Englifh, Goofe-berry, Goofe-berry bufh, and Fea-berry bufh in Chefhire, my natiue countrey.

¶ *The Temperature.*

The berries of this bufh before they be ripe are cold and dry, and that in the later end of the fecond degree, and alfo binding.

¶ *The Vertues.*

The fruit is vfed in diuers fauces for meate, as thofe that are skilfull in cookerie can better tel than my felfe.

They

They are vsed in broths in stead of Veriuice, which maketh the broth not onely pleasant to the B
taste, but is greatly profitable to such as are troubled with a hot burning ague.

They are diuersly eaten, but howsoeuer they be eaten they alwaies ingender raw and cold bloud: C
they nourish nothing or very little: they also stay the belly, and stench bleedings.

They stop the menses, or monethly sickenes, except they happen to be taken into a cold sto- D
mack, then do they not helpe, but rather clog or trouble the same by some manner of flix.

The ripe berries, as they are sweeter, so doe they also little or nothing binde, and are something E
hot, and yeeld a little more nourishment than those that be not ripe, and the same not crude or raw;
but these are seldome eaten or vsed as sauce.

The iuice of the greene Gooseberries cooleth all inflammations, *Erysipelas*, and Saint Antho- F
nies fire.

They prouoke appetite, and coole the vehement heate of the stomacke and liuer. G

The young and tender leaues eaten raw in a sallad, prouoke vrine, and driue forth the stone H
and grauell.

CHAP. 23. *Of Barberries.*

¶ *The Kindes.*

There de diuers sorts of Barberries, some greater, others lesser, and some without stones.

Spina acida, siue Oxyacantha.
The Barberry bush.

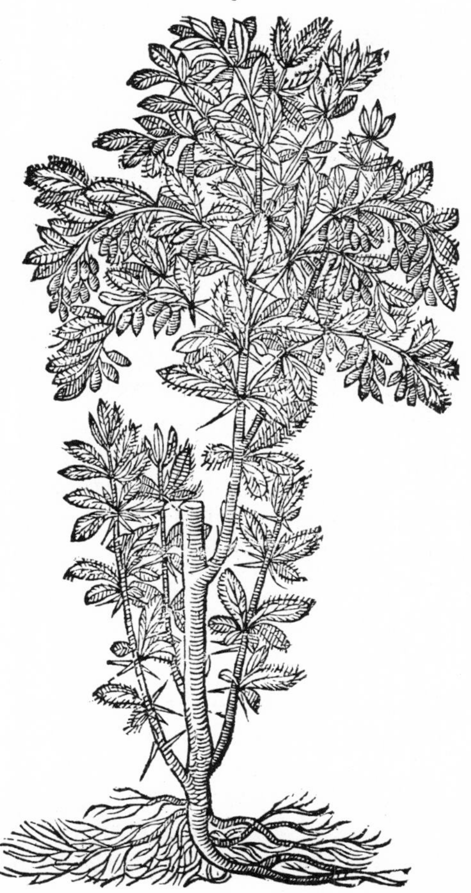

¶ *The Description.*

THe Barberry plant is an high shrub or bush, hauing many young straight shootes and branches, very full of white and prickly thornes; the rinde whereof is smooth and thin, the wood it selfe yellow: the leaues are long, very greene, slightly nicked about the edges, and of a sowre taste: the floures be yellow, standing in clusters vpon long stemmes: in their places come vp long berries, slender, red when they be ripe, with a little hard kernell or stone within; of a sowre and sharpe taste: the root is yellow, dispersethit selfe farre abroad, and is of a woody substance.

Wee haue in our London gardens another sort, whose fruite is like in forme and substance, but one berry is as big as three of the common kinde, wherein consisteth the difference.

We haue likewise another without any stone, the fruite is like the rest of the Barberries, both in substance and taste.

¶ *The Place.*

The Barberrie bush groweth of it selfe in vntoiled places and desart grounds, in woods, and the borders of fields, especially about a Gentlemans house called Mr. *Monke*, dwelling in a village called Iuer, two miles from Colebrooke, where most of the hedges are nothing else but Barberry bushes.

They are planted in gardens in most places of England.

¶ *The Time.*

The leaues spring forth in Aprill: the floures and fruite in September.

¶ *The Names.*

Galen calleth this thorne in Greeke, ὀξυάκανθα, who maketh it to differ from ὀξυάκανθος, in his booke of the Faculties of simple medicines: but more plainely in his booke of the Faculties of Nourishments; where he reckoneth vp the tender springs of Barberries among the tender shoots that are to be eaten, such as *Oxyacanthus* or the Hawthorne bringeth not forth, wherein he plainely made a difference, *Oxyacantha* the Barbery bush, and *Oxyacanthus* the Hawthorne tree.

Dioscorides hath not made mention of this Thorne; for that which he calleth *Oxyacantha* in the Fœminine gender, is *Galens Oxyacanthus* in the Masculine gender.

Auicen seemeth to containe both these shrubs vnder the name of *Amyrberis*, but we know they are neither of affinitie or neighbourhood, although they be both prickly.

The shrub it selfe is called in shops Barbaries, of the corrupted name *Amyrberis*, of the later writers *Crespinus*: in Italian, *Crespino* in Spanish, *Espino de maiuelas*: in high Dutch, **Paisselbeer**: in low Dutch, **Sauseboom**: in French, *Espine vinette*: and thereupon by a Latine name, *Spinuineta*, *Spina acida*, and *Oxyacantha Galeni*. ‡ In English, a Barbery bush, or Piprige Tree, according to D[r]. Turner. ‡

¶ The Temperature.

The leaues and berries of this thorne are cold and dry in the second degree: and as *Galen* also affirmeth, they are of thin parts, and haue a certaine cutting qualitie.

¶ The Vertues.

A The leaues are vsed of diuers to season meate with, and in stead of a sallad, as be those of Sorrell.

B The decoction thereof is good against hot burnings and cholericke agues: it allaieth the heate of the bloud, and tempereth the ouermuch heate of the liuer.

C The fruite or berries are good for the same things, and be also profitable for hot laskes, and for the bloudy flixe, and they stay all manner of superfluous bleedings.

D The greene leaues of the Barbery bush stamped, and made into sawce, as that made of Sorrell, called greene sauce, doth coole hot stomackes, and those that are vexed with hot burning agues, and procureth appetite.

E The conserue made of the fruite and sugar performeth all those things before remembred, but with better force and successe.

F The roots of the tree steeped for certaine daies together in strong lie, made with ashes of the ash-tree, and the haire often moistned therewith, maketh it yellow.

G ‡ The barke of the roots is also vsed in medicines for the iaundise, and that with good successe. ‡

Chap. 24. *Of the white Thorne, or Hawthorne Tree.*

¶ The Kindes.

There be two sorts of the white Thorn Trees described of the later writers, one very common in most parts of England: there is another very rare, and not found in Europe, except in some few rare gardens of Germanie; which differeth not from our common Hawthorne, sauing that the fruit here of is as yellow as Saffron: we haue in the West of England one growing at a place called Glastenburie, which bringeth forth his floures about Christmas, by the report of diuers of good credit, who haue seen the same; but my selfe haue not seen it; and therefore leaue it to be better examined.

¶ The Description.

1 The white Thorne is a great shrub growing oftentimes to the height of the Peare-tree the trunke or body is great: the boughes and branches hard and woody, set full of long sharpe thornes: the leaues be broad, cut with deepe gashes into diuers sections, smooth, and of a glistering greene colour: the floures grow vpon spokie rundles, of a pleasant sweet smell, sometimes white, and often dasht ouer with a light wash of purple; which hath moued some to thinke some difference in the plants: after which come the fruit, being round berries, green at the first, and red when they be ripe; wherein is found a soft sweet pulpe, and certaine whitish seed: the root groweth deepe in the ground, of a hard woody substance.

2 The second and third haue been touched in the first title, notwithstanding I haue thought it not vnfit to insert in this place a plant perticipating with the Hawthorne in floures and fruit, and with the Seruice tree in leaues, and not vnlike in fruit also.

Theophrastus hath set forth this tree vnder the name of *Aria*, which groweth vnto the forme of a small tree, delighting to grow in our shadowie woods of Cumberland and Westmerland, and many other places of the North country, where it is to be found in great quantitie: but seldome in Spaine,

Spaine, Italy, or any hot Region. This tree is garnished with many large branches beset with leaues like the Peare tree, or rather like the Aller leafe, of a darke greene colour aboue, and of a white colour vnderneath: among these leaues come forth tufts of white floures, very like vnto the Hawthorne floures, but bigger: after which succeed small red berries, like the berries of the Hawthorne, and in taste like the Neapolitan Medlar: the temperature and faculties whereof are not yet knowne.

1 *Oxyacanthus.*
The Haw-thorne tree.

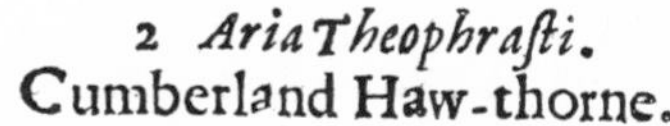

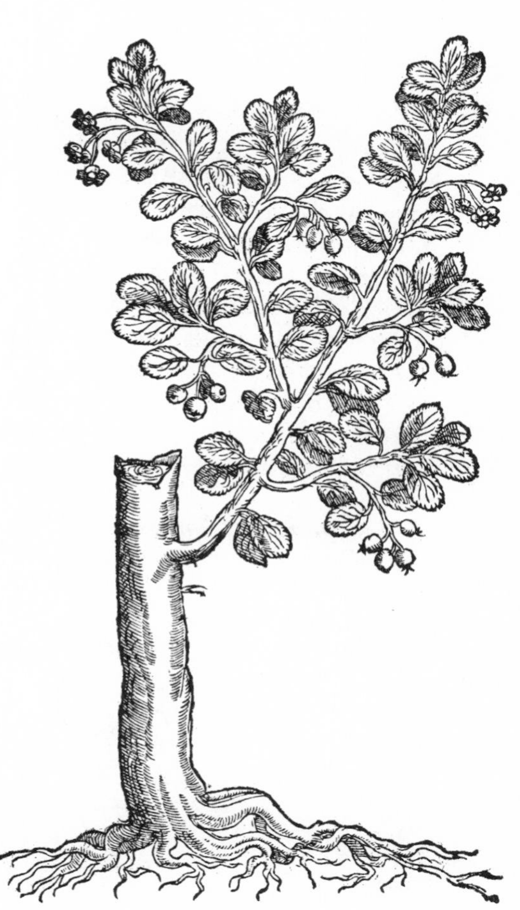

2 *Aria Theophrasti.*
Cumberland Haw-thorne.

¶ *The Place.*

The Haw-thorne groweth in woods and in hedges neere vnto high-waies almost euery where. The second is a stranger in England. The last groweth at Glastenbury Abbey, as it is credibly reported vnto me. ‡ The *Aria* groweth vpon Hampsted heath, and in many places of the West of England. ‡

¶ *The Time.*

The first and second floure in May; whereupon many do call the tree it selfe the May-bush, as a chiefe token of the comming in of May: the leaues come forth a little sooner: the fruit is ripe in the beginning of September, and is a food for birds in Winter.

¶ *The Names.*

Dioscorides describeth this shrub, and nameth it ὀξυάκανθα, in the fœminine gender: and *Galen* in his booke of the Faculties of simple medicines, ὀξυάκανθος, in the masculine gender: *Oxyacanthus*, saith he, is a tree, and is like to the wilde Peare tree in forme, and the vertues not vnlike, &c. Of *Oxyacantha*, *Dioscorides* writeth thus: It is a tree like to the wild Peare tree, very full of thorns, &c. *Serapio* calleth it *Amyrberis*: and some, saith *Dioscorides*, would haue it called πυρίνα, but the name *Pyrina* seemeth to belong to the yellow Haw-thorne: it is called in high-Dutch, **Hagdorn**: in low-Dutch, **Hagedoren**: in Italian, *Bagaia*: in Spanish, *Pirlitero*: in French, *Aub-espine*: in English, White-thorne, Haw-thorne tree; and of some Londoners, May-bush. ‡ This is not the *Oxyacantha* of the Greekes, but that which is called *Pyracantha*, as shall be shewed hereafter.

The second is thought to be the *Aria* of *Theophrastus*, and so *Lobel* and *Tabernamontanus* call it. Some, as *Bellonius*, *Gesner*, and *Clusius*, refer it to the *Sorbus*, and that not vnfitly: in some places of this kingdome they call it a white Beame tree. ‡

¶ *The*

¶ *The Temperature.*

The fruit of the Haw-thorne tree is very astringent.

¶ *The Vertues.*

A The Hawes or berries of the Haw-thorne tree, as *Dioscorides* writeth, do both stay the laske, the menses, and all other fluxes of bloud: some Authors write, that the stones beaten to pouder, and giuen to drinke are good against the stone.

Chap. 25. *Of Goats Thorne.*

¶ *The Description.*

1 THe first *Tragagantha* or Goats-thorne hath many branchie boughes and twigs, slender and pliant, so spred abroad vpon euerie side, that one plant doth sometimes occupie a great space or roome in compasse: the leaues are small, and in shape like Lentill leaues, whitish, and somewhat mossie or hairy, set in rowes one opposite against another: the floure is like the blossome of the Lentill, but much lesser, and of a whitish colour, and sometimes marked with purple lines or streaks: the seed is inclosed in small cods or husks, almost like vnto the wilde *Lotus* or horned Trefoile: the whole plant on euery side is set full of sharpe prickely thornes, hard, white, and strong: the roots run vnder the ground like Liquorice roots, yellow within, and blacke without, tough, limmer, and hard to breake; which being wounded in sundry places with some iron toole, and laid in the Sun at the highest and hottest time of Sommer, issueth forth a certain liquor, which being hardned by the Sun, is that gum which is called in shops *Tragacantha:* and of some, though barbarously *Dragagant.*

1 *Tragacantha, siue spina Hirci.*
Goats Thorne.

2 *Spina Hirci minor.*
Small Goats Thorne.

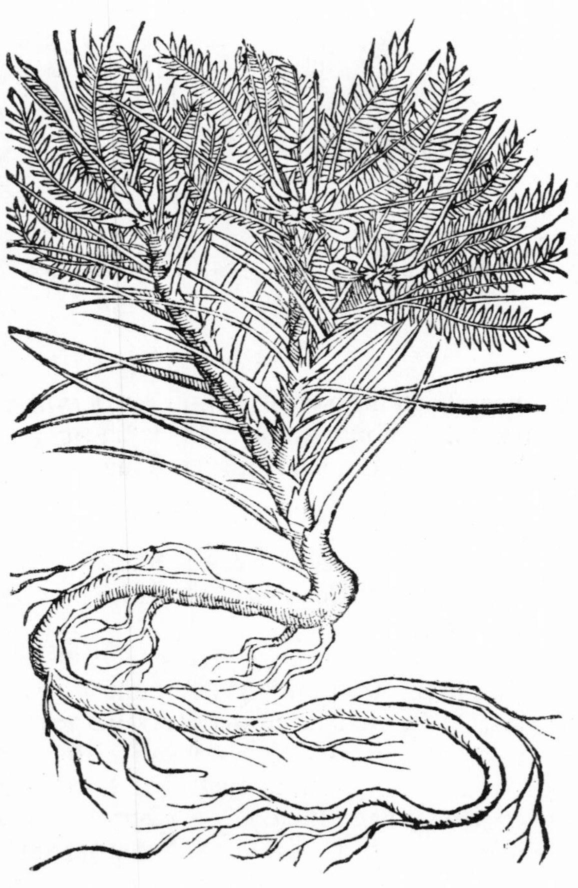

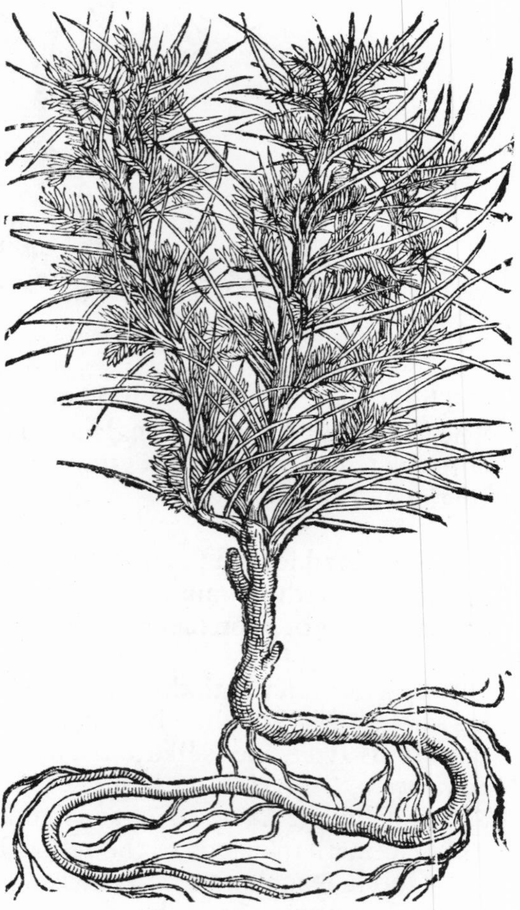

2 The second kinde of *Tragacantha* is a low and thicke shrub, hauing many shoots growing from one turfe: of a white or grayish colour, about a cubit high, stiffe and wooddy: the leaues are like the former, and garded with most stiffe pricks not very safely to be touched: among the thornie leaues come forth many floures in small tufts like *Genistella,* but that they are white: the cods

are many, ſtraight and thorny like *Geniſtella*, wherein are many ſmall white and three cornered ſeeds as big as muſtard ſeed. ‡ This differs from the former in that it is ſmaller, and loſeth the leaues euery Winter, when as the former keepes on the leaues vntill new ones come in the Spring. The middle rib of the winged leaues ends in a pricke, which by the falling of the leaues becommeth a long and naked thorne. I haue giuen you a more accurate figure hereof out of *Cluſius*, wherein the leaues, floures, cods, and ſeeds are all expreſſed apart. ‡

3 The Grecians haue called this plant Νευράδα, becauſe it is good for the ſinewes: it ſhould ſeeme it tooke the name *Poterion*, of *Potrix*, becauſe it loueth a watry or fenny ſoile: it hath ſmall branches, and leaues of *Tragacantha*, growing naturally in the tract of Piedmont in Italy: it ſpreadeth abroad like a ſhrub: the barke or rinde is blackiſh, and dry without great moiſture, very much writhed or wrinkled in and out as that of *Nepa* or *Corruda*: the ſharpe pricks ſtand not in order as *Tragacantha*, but confuſedly, and are finer and three times leſſer than thoſe of *Tragacantha*, growing much after the manner of *Aſtragalus*: but the particular leaues are greene aboue, and white below, ſhaped ſomewhat like Burnet: the ſeed is ſmall and red, like vnto Sumach, but leſſer.

‡ *Tragacanthæ minoris icon accuratior.*
A better figure of the Goats-thorne.

† 3 *Poterion Lob. ſiue Pimpinella ſpinoſa Camer.*
Burnet Goats-thorne.

¶ *The Place.*

Petrus Bellonius in his firſt booke of Singularities reports, that there is great plenty hereof growing in Candy vpon the tops of the mountaines. *Theophraſtus* ſaith that it was thought to grow no where but in Candy; but now it is certaine that it is found in Achaia, Peloponeſſus, and in Aſia: it doth alſo grow in Arcadia, which is thought not to be inferiour to that of Candy. It is thought by *Lobel* to grow in Languedock in France, whereof *Theophr.* hath written in his ninth booke, that the liquor or gum iſſueth out of it ſelfe, and that it is not needfull to haue the root broken or cut. The beſt is that, ſaith *Dioſcorides*, which is through-ſhining, thin, ſmooth, vnmixt, and ſweet of ſmel and taſte.

¶ *The Time.*

They floure and flouriſh in the Sommer moneth: I haue ſowne the ſeed of Poterion in Aprill, which I receiued from *Ioachimus Camerarius* of Noremberg, that grew in my garden two yeares together, and after periſhed by ſome miſchance.

 ¶ The

¶ *The Names.*

Goats-thorne is called in Greeke τραγάκανθα : of moſt Herbariſts likewiſe *Tragacantha* : we may cal it in Latine *Spina Hirci* : in French, *Barbe Renard* · and in Engliſh for want of a better name, Goats-Thorne : the liquor or gum that iſſueth forth of the roots beareth the name alſo of *Tragacantha* : it is called in ſhops *Gummi Tragacanthæ* ; and in a barbarous manner *Gummi Tragacanthi* : in Engliſh, Gum Dragagant.

¶ *The Temperature.*

This plant in each part thereof is of a drying facultie without biting. It doth conſolidate or glew together ſinewes that be cut : but the roots haue that facultie eſpecially, which are boyled in wine, and the decoction giuen vnto thoſe that haue any griefe or hurt in the ſinewes.

Gum Dragagant hath an emplaſticke qualitie, by reaſon whereof it dulleth or allayeth the ſharpneſſe of humors, and doth alſo ſomthing dry.

¶ *The Vertues.*

A The Gumme is ſingular good to be licked in with honey againſt the cough, roughneſſe of the throat, hoarſeneſſe, and all ſharpe and thin rheumes or diſtillations : being laid vnder the tongue it taketh away the roughneſſe thereof.

B Being drunke with Cute or the decoction of Liquorice it taketh away and allayeth the heat of the vrine : it is alſo vſed in medicines for the eyes.

C The greateſt part of thoſe artificiall beades, ſweet chaines, bracelets, and ſuch like pretty ſweet things of pleaſure are made hard and fit to be worne by mixing the gum hereof with other ſweets, being firſt ſteeped in Roſe water till it be ſoft.

† The figure which was in the third place was of the plant deſcribed in the ſecond which *Matthiolus* and *Tabern.* made their *Poterium*, but it agreed not with the deſcription which was taken out of the *Aduerſ.*

Chap. 26. *Of the Ægyptian Thorne.*

‡ 1 *Acacia Dioſcoridis.*
The Egyptian Thorne.

† 2 *Acacia altera trifolia.*
Thorny Trefoile.

¶ *The*

¶ *The Description.*

1 Dioscorides maketh mention of *Acacia*, whereof the first is the true and right *Acacia*, which is a shrub or hedge tree, but not growing right or straight vp as other small small trees do: his branches are wooddie, beset with many hard and long Thorns; about which grow the leaues, compact of many small leaues clustering about one side, as in the Lentill: the floures are whitish, the husks or cods be plaine and flat, yea very broad like vnto Lupines, especially on that side where the seed growes, which is contained sometimes in one part, and sometimes in two parts of the husk, growing together in a narrow necke: the seed is smooth and glistering. There is a blacke iuice taken out of these huskes, if they be dried in the shadow when they be ripe; but if when they are not ripe, then it is somewhat red: some do wring out a iuice out of the leaues and fruit: there floweth also a gum out of this tree, which is the gum of Arabia, called Gum Arabicke.

2 *Dioscorides* hauing described *Spina Acacia*, setteth downe a second kinde thereof, calling it *Acacia altera*, which hath the three leaues of Rue or *Cytisus*, and coddes like those of *Genistella*, but somewhat more blunt at the end, and thicke at the backe like a Rasor, and still groweth forward narrower and narrower, vntill it come to haue a sharpe edge: in these cods are contained three or foure flat seeds like *Genistella*, which before they wax ripe are yellow, but afterwards blacke: the whole plant groweth to the height of *Genista spinosa*, or Gorsse, both in shape, height, and resemblance, and not to the height of a tree, as *Matthiolus* would persuade vs, but full of sharpe Thornes like the former.

¶ *The Place.*

The true Acacia groweth in Egypt, Palestina, Lombardie, and Syria, as *Dioscorides* writeth: among the shrubs and trees that remaine alwaies greene, Acacia is noted for one by *Petrus Bellonius*, in his first booke of Singularities, chap. 44.

The other Acacia groweth in Cappadocia and Pontus, as *Dioscorides* writeth: it is also found in Corsica, and on diuers mountaines of Italy, and likewise vpon all the coast of Liguria and Lombardie, and vpon the Narboue coast of the Mediterranean sea.

¶ *The Time.*

These floure in May, and their fruit is ripe in the end of August.

¶ *The Names.*

The tree Acacia is named of the Græcians ἀκακία, yea euen in our time, and likewise of the Latins *Acacia*: it is also called *Ægyptia spina*: this strange thorne hath no English name that I can learn, and therefore it may keep still the Latine name Acacia; yet I haue named it the Egyptian thorne: the iuice is called also Acacia after the name of the plant: the Apothecaries of Germanie do vse in stead hereof, the iuice that is pressed forth of sloes or snags, which they therefore call *Acacia Germanica*: *Matthiolus* pictureth for Acacia the tree which the later Herbarists do call *Arbor Iudæ*, to which he hath vntruly added Thorns, that he might belie Acacia, and yet he hath not made it agree with *Dioscorides* his description.

They call this ἑτέρα ἀκακία: in Latine *Acacia altera*, or the other Acacia, and *Pontica Acacia*, or Ponticke Acacia.

¶ *The Nature.*

The iuice of Acacia, as *Galen* saith, consisteth not of one only substance, but is of substance both cold and earthie, to which also is coupled a certaine waterie essence, and it likewise hath thin and hot parts dispersed in it selfe: therefore it is dry in the third degree, and cold in the first if it be not washed; and in the second, if it be washed: for by washing it loseth his sharpe and biting quality and the hot parts.

¶ *The Vertues.*

A The iuice of Acacia stoppeth the laske, the inordinate course of womens termes, and mans inuoluntarie issue called *Gonorrhæa*, if it be drunke in red wine.

B It healeth the blastings and inflammations of the eies, and maketh the skin and palmes of the hands smooth after the healing of the *Serpigo*: it healeth the blisters and extreme heat in the mouth, and maketh the haires blacke that are washed therewith.

C It is good, saith *Dioscorides*, against S. Anthonies fire, the shingles, Chimetla, Pterygia, and whitlowes.

D The gum doth binde and somewhat coole: it hath also ioined vnto it an emplaistick quality, by which it dulleth or alayeth the sharpnesse of the medicines wherewith it is mixed. Being applied with the white and yolk of an egge, it suffereth not blisters to rise in burned or scalded parts. *Diosc.*

E The iuice of the other, saith *Dioscorides*, doth also binde, but it is not so effectuall nor so good in eie medicines.

† Our Author gaue but formerly one figure, which was that in the second place, and he would haue persuaded vs, that it was of the right *Acacia*, yet in his description he tells vs otherwaies.

CHAP.

Chap. 27. *Of box Thorne, and the iuice thereof called Lycium.*

¶ *The Description.*

1 BOx Thorne is a rare plant, in shape not vnlike the Box tree, whereof it hath beene reckoned for a wilde kinde, hauing many great branches set full of round and thicke leaues, very like that of the common Box tree: amongst which grow forth most sharpe pricking thornes: the floures grow among the leaues, which yeeld forth small blacke berries of a bitter tast, as big as a pepper corne: the iuice whereof is somewhat oilie, and of a reddish colour; which bitter iuice being set on fire, doth burne with a maruellous cracking and sparkling; the ashes thereof are of a red colour: it hath many wooddie roots growing aslope.

1 *Lycium, siue Pyxacantha.*
Box Thorne.

‡ 2 *Lycium Hispanicum.*
Spanish Box Thorne.

2 The other kinde of *Pyxacantha* or *Lycium*, groweth like vnto the common Priuet, hauing such like leaues, but somewhat narrower: the tops of the slender sprigs are furnished with prickles: the root is tough, and of a wooddie substance.

¶ *The Place.*

They grow in Cappadocia and Lycia, and in many other countries: it prospereth in rough places, it hath likewise been found in Languedoc, and Prouence in France: *Bellonius* writeth that hee found it in Palestina.

Matthiolus pictureth for Box Thorne, a plant with box leaues, with very many boughes, and certaine thornes standing among them: but the notable Herbarist *Anguillara* and others, hold opinion, that it is not the right; with whom we also do agree.

There is drawne out of the leaues and branches of box Thorn, or as *Pliny* saith, out of the boughs and roots being throughly boiled, a iuice, which is named *Lycium*.

Dioscorides saith, that the leaues and branches must be braied, and the infusion made many daies in in the decoction thereof, after which the feces or wooddie ſtuffe muſt be caſt away, and that which remaineth boiled againe till it become as thicke as honie: *Pliny* ſaith, that the roots and branches are very bitter, and for three daies together they muſt be boiled in a copper veſſell, and the wood and ſticks often taken out till the decoction be boiled to the thickneſſe of honie.

¶ *The Time.*

They floure in Februarie and March, and their fruit is ripe in September.

¶ *The Names.*

It is named in Greeke πυξάκανθα, which a man may call in Latine *Buxea ſpina*: and in Engliſh, Box Thorne: of ſome, Aſſes Box Tree, and prickley Boxe: it is alſo named *Lycium*, of the iuice which is boyled out of it: the iuice is properly called λύκιον, and retaineth in Latine the ſame name *Lycium*: it is termed in Engliſh Thorne box. But it ſeemeth to me, that the originall name *Lycium* is fitter, being a ſtrange thing, and knowne to very few: the Apothecaries know it not, who in ſtead thereof do vſe amiſſe the iuice of the fruit of Woodbinde, and that not without great errour, as we haue already written. ‡ It is vnknowne in our ſhops, neither is there any thing vſed for it, it being wholly out of vſe, wherefore our Author might here well haue ſpared *Dodonæus* his words. ‡

Dioſcorides teacheth to make a χύλισμα of Sumach which is good for thoſe things that *Lycium* is, and is vſed when *Lycium* is not to be had, and it is fit to be put in all medicines in ſtead thereof.

¶ *The Temperature.*

Lycium, or the iuice of Box Thorne, is as *Galen* teacheth, of a drying qualitie, and compounded of diuers kindes of ſubſtances, one of thinne parts digeſting and hot; another earthie and cold, by which it enioyeth his binding facultie: it is hot in a meane, and therefore it is vſed for ſeuerall purpoſes.

¶ *The Vertues.*

Lycium cleareth the ſight, ſaith *Dioſcorides*, it healeth the ſcuruie feſtred ſores of the eye lids, the itch, and old fluxes, or diſtillations of humors; it is a remedie for the running of the eares; for vlcers in the gummes, and almonds of the throat, and againſt the chappes or gallings of the lips and fundament.

† The figure which was in the 2. place, was of the *Lycium Italicum* of *Matthiolus* and others; but the deſcription and title better fitted this *Lycium Hiſpanicum* of *Lobel*, which therefore I put thereto. The figure alſo of the *Lycium Italicum* of *Matthiolus* our Author gaue againe in the next chapter ſaue two.

CHAP. 28. *Of Ramme or Harts Thorne.*

¶ *The Kindes.*

AFter the opinion of *Dioſcorides* there be three ſorts of *Rhamnus*, one with long, flat & ſoft leaues: the other with white leaues, and the third with round leaues, which are ſomewhat blackiſh; *Theophraſtus* and *Pliny* affirme that there are but two, the one white, and the other black, both which do beare Thornes: but by the labour and induſtry of the new and late writers there are found ſundry ſorts moe, all which and euery one of them are plants of a wooddie ſubſtance, hauing alſo many ſtraight twiggie and pliant branches, ſet with moſt ſharpe pricking thornes.

¶ *The Deſcription.*

1 THis is a ſhrubbe growing in the hedges, and bringing forth ſtraight branches and hard thornes, like to thoſe of the Hawthorne, with little leaues, long, ſomething fat and ſoft: and this hath that notable learned man *Cluſius* deſcribed more diligently in theſe words: the Ram is a ſhrub fit to make hedges of, with ſtraight branches, parting it ſelfe into many twigs, white, and ſet with ſtiffe and ſtrong thornes, hauing leaues, which for the moſt part grow by foures or fiues at the root of euery Thorne, long, ſomething fat, like to thoſe of the Oliue tree, ſomewhat white, but tender and full of iuice; which in Autumne doe ſometimes fall off, leauing new growing in their places: the floures in Autumne are ſomething long, whitiſh, diuided at the brims into fiue parts: in their places is left a ſeed, in ſhew as in *Gelſemine*: notwithſtanding it was neuer my chance to ſee the fruit: the root is thicke and diuerſly parted.

‡ I obſerued another (ſaith the ſame Author) almoſt like to the former, but lower, and diuided into more branches, with leſſer leaues, more thick and ſalt of taſte, and whiter alſo than the former: the floures are like, in all things but their colour, thoſe of the former, which in this are purple.

2 This hath more flexible ſtalks and branches, and theſe alſo ſet with thornes: the leaues are narrow, and not ſo thicke or fleſhie as thoſe of the former, yet remaine alwaies greene like as they do: the floures are ſmall and moſſie, of a greeniſh colour, growing thicke about the branches, and they are ſucceeded by a round fruit, yellowiſh when it is ripe, and remaining on the ſhrubbe all the Winter,

‡ 1 *Rhamnus* 1. *Clusij flo. albo.*
White floured Ram-thorne.

‡ *Rhamnus alter Clus. flore purpureo.*
Purple floured Ram-thorne.

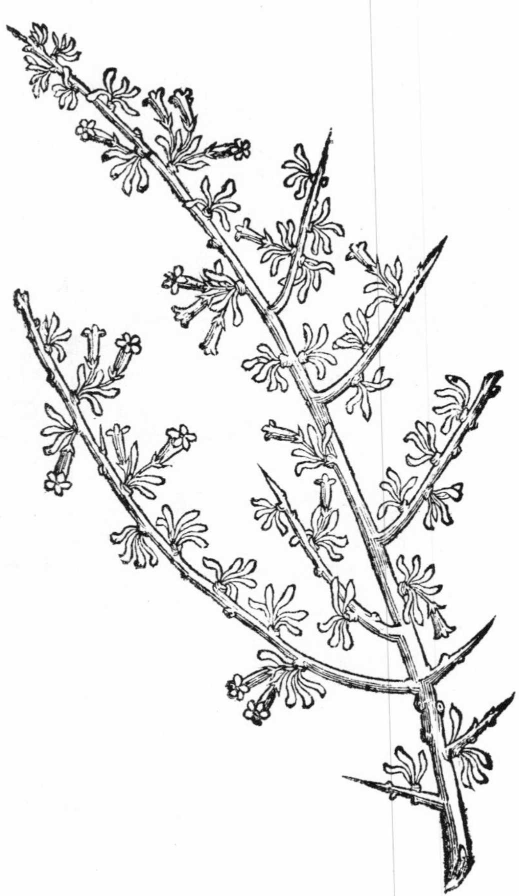

‡ 2 *Rhamnus* 2. *Clusij.*
Sallow-Thorne.

3 *Ramnus tertius Clusij.*
Ram or Harts-Thorne.

Winter: The whole shrubbe lookes as if it were sprinckled ouer with dust.

3 To these may be added another growing with many branches to the height of the Sloe tree or blacke Thorne, and these are couered with a blackish barke, and armed with long prickles: the leaues, as in the first, grow forth of certaine knots many together, long narrow, fleshie, greene, and continuing all the yeare: their taste is astringent, somewhat like that of Rhabarb: the floures shew themselues at the beginning of the Spring, of a greenish colour, growing thicke together, and neere the setting on of the leaues; in Summer it carries a blacke fruit almost like a Sloe, round, and harsh of taste.

¶ *The Place.*

The first of these growes in sundry places of Spaine, Portugall, and Prouince: the other varietie thereof *Clusius* saith he found but onely in one place, and that was neere the citie Horivela, called by the Antients *Orcellis*, by the riuer Segura, vpon the borders of the kingdome of Valentia: the second growes in many maritime places of Flanders and Holland, and in some vallies by riuers sides. The third growes in the vntilled places of the kingdome of Granado and Murcia. ‡

¶ *The Time.*

This Ram is euer greene together with his leaues: the fruit or berries remaine on the shrub, yea euen in Winter.

¶ *The Names.*

The Grecians call this thorne ῥάμνος: the Latines also *Rhamnus*: and of diuers it is also named περσεφόνιον, ἢ λευκάκανθα, that is *Spina alba*, or white Thorne, *Spina Cerualis*, or Harts-thorne, as we finde written among the bastard words. *Marcellus* nameth it *Spina salutaris*, and *Herba salutaris*, which hath, saith he, as it were a grape. It is called in Italian *Marruca* and *Rhamno*: in Spanish, *Scambrones*: in English, Ram, or Harts Thorne.

¶ *The Temperature.*

The Ram, saith *Galen*, doth drie and digest in the second degree, it cooleth in the later end of the first degree, and in the beginning of the second.

¶ *The Vertues.*

A The leaues, saith *Dioscorides*, are layed pultis wise vpon hot cholericke inflammations, and Saint Anthonies fire, but we must vse them whilest they be yet but tender, as *Galen* addeth.

B ‡ The leaues and buds or young shoots of the first, are eaten as sallads with oile, vineger, and salt, at Salamanca and other places of Castile, for they haue a certaine acrimonie and aciditie which are gratefull to the taste. A decoction of the fruit of the third is good to foment relaxed and weake or paralyticke members, and to ease the paine of the gout, as the Inhabitants of Granado told *Clusius*. ‡

† Our Author in this chapter gaue onely the figure of the third, and the description of the first, and the place of the second, with the names and faculties in generall.

CHAP. 29. *Of Christs Thorne.*

¶ *The Description.*

CHrists Thorn or Ram of Lybia, is a very tough and hard shrubby bush, growing vp sometimes vnto the height of a little tree, hauing very long and sharpe pricklie branches: but the thornes that grow about the leaues are lesser, and not so prickly as the former. The leaues are small, broad, and almost round, somewhat sharpe pointed, first of a darke greene colour, and then somwhat reddish. The floures grow in clusters at the top of the stalks, of a yellow colour: the husks wherein the seeds be contained, are flat and broad, very like vnto small bucklers as hard as wood, wherein are contained three or foure thin and flat seeds, like the seed of Line or Flax.

¶ *The Place.*

This Thorne groweth in Lybia; it is better esteemed of in the countrey of Cyrene than is their Lote tree, as *Pliny* affirmeth. Of this shrub *Diphilus Siphnius* in *Athenæus* in his foureteenth booke maketh mention, saying, that hee did verie often eat of the same in Alexandria that beautifull Citie.

Petrus Bellonius who trauelled ouer the Holly Land, saith, that this shrubbie thorne *Paliurus* was the

Paliurus.
Chriſts Thorne.

the thorne wherewith they crowned our Sauiour Chriſt: his reaſon for the proofe hereof is this, that in Iudæa there was not any thorne ſo common, ſo pliant, or ſo fit for to make a crown or garland of, nor any ſo full of cruell ſharpe prickles. It groweth throughout the whole countrey in ſuch aboundance, that it is their common fuell to burne; yea ſo common with them there, as our Gorſſe, Brakes, and broome is here with vs. *Ioſephus* in his firſt booke of Antiquities, and 11. chap. ſaith, that this Thorne hath the moſt ſharpe prickles of any other; and therefore that Chriſt might be the more tormented, the Iewes rather tooke this than any other. Of which I haue a ſmall tree growing in my garden, that I haue brought forth by ſowing of the ſeed.

The Time.

The leaues fall away and continue not alwaies green, as do thoſe of the Rams: it buddeth forth in the Spring, as *Pliny* teſtifieth.

¶ *The Names.*

This Thornie ſhrubbe is called in Greeke παλίουρος: the Latines and Italians retaine the ſame name *Paliurus*: for want of an Engliſh name, it may be termed Ramme of Lybia, or Chriſts Thorne: *Pliny* reporteth, that the ſeed is called *Zura*.

¶ *The Temperature.*

The leaues and root of Chriſts Thorne doe euidently binde and cut.

¶ *The Vertues.*

A By vertue of this cutting quality the ſeed doth weare away the ſtone, and cauſe tough and ſlimy humors to remoue out of the cheſt and lungs, as *Galen* ſaith.

B The decoction of the leaues and root of Chriſts Thiſtle, as *Dioſcorides* writeth, ſtoppeth the belly, prouoketh vrine, and is a remedy againſt poiſons, and the bitings of ſerpents.

C The root doth waſte and conſume away *Phymata*, and *Oedemata* if it be ſtamped and applied.

D The ſeed is good for the cough, and weareth away the ſtone in the bladder.

CHAP. 30. *Of Buck-Thorne, or laxatiue Rem.*

¶ *The Deſcription.*

1 BVck-thorne groweth in manner of a ſhrub or hedge tree; his trunke or body is often as big as a mans thigh; his wood or timber is yellow within, and his barke is of the colour of a Cheſtnut, almoſt like the bark of a Cherry tree. The branches are beſet with leaues that are ſomewhat round, and finely ſnipt about the edges like the leaues of the Crab or Wilding tree: among which come forth Thornes which are hard and prickly: the floures are white and ſmal, which being vaded there ſucceed little round berries, greene at the firſt, but afterwards black, wherof that excellent greene colour is made, which the Painters and Limners do call Sap-greene; but theſe berries before they be ripe do make a faire yellow colour, being ſteeped in vineger.

‡ 2 Beſides the common kinde, *Cluſius* mentions two other: the firſt of which hath branches ſome two cubits long, ſubdiuided into diuers others, couered with a ſmooth barke like that of the former, which, the vpper rinde being taken off, is of a yellowiſh greene colour, and bitteriſh taſte: the branches haue ſome few prickles vpon them, and commonly end in them: the leaues are almoſt like thoſe of the common kinde, but ſmaller, narrower, and ſomewhat reſembling thoſe of the blacke Thorn, hauing ſomewhat a drying taſte: the floure conſiſts of foure leaues of a yellowiſh greene

‡ 1 *Rhamnus solutivus.*
Buck-thorne.

‡ 2 *Rhamnus solutivus minor.*
Middle Buck-thorne.

† 3 *Rhamnus solutivus pumilus.*
Dwarfe Buck-thorne.

greene colour: the root is wooddie as in other shrubs: *Clusius* found this growing in the mountanous places of Austria, and calls it *Spina infectoria pumila.*

3 This other hath branches some cubite long, and of the thicknesse of ones little finger, or lesser, couered with a blacke and shriuelled barke: and towards the top diuided into little boughs, which are couered with a thin & smoother barke, and commonly end in a sharp thorn: the leaues much resemble those of the Slo-tree yet are they shorter and lesser, greene also, and snipt about the edges; first of an astringent, and afterwards of somewhat a bitterish taste; the floures which grow amongst the leaues are of an herby colour, and consist of foure leaues: the fruit is not much vnlike that of the former; but distinguished with two, & somtimes with three crests or dents, first green, and then black when it is ripe: the root is thicke, wooddie and hard. *Clusius* found this on the hill aboue the Bathes of Baden, hee calls it *Spina infectoria pumila 2.* This *Matthiolus* and others call *Lycium Italicum*: and our Author formerly gaue the figure of *Matthiolus* and *Tabernamontanus*, by the name of *Lycium Hispanicum*, and here againe another for his *Rhamnus solutivus*, which made mee to keepe it in this chapter, and omit it in the former, it being described in neither. ‡

¶ *The Place.*

Buck-thorne groweth neere the borders of fields, in hedges, woods, and in other vntoiled places: it delighteth to grow in riuers and in water ditches: it groweth in Kent in sundry places, as at Farningham vpon the cony burrowes belonging sometime to Mr. *Sibil*, as also vpon cony burrowes in Southfleet, especially in a small and narrow lane leading from the house of Mr. *William Swan* vnto Longfield downes, also in the hedge vpon the right hand at Dartford townes end towards London, and in many places more vpon the chalkie bankes and hedges.

¶ *The Time.*

It floureth in May, the berries be ripe in the fall of the leafe.

¶ *The Names.*

The later Herbarists call it in Latine *Rhamnus solutivus*, because it is set with thornes, like as the Ram, and beareth purging berries. *Matthiolus* nameth it *Spina infectoria*; *Valerius Cordus*, *Spina Cerui*, and diuers call it *Burgispina*. It is termed in high Dutch, **Creutzbeer weghdorn**: in Italian, *Spino Merlo*, *Spino Zerlino*, *Spino Ceruino*: in English, Laxatiue Ram, Way-thorne, and Buck-thorne: in low Dutch they call the fruit or berries **Rhijnbesien**, that is, as though you should say in Latine, *Bacca Rhenanæ*: in English, Rheinberries: in French, *Nerprun*.

¶ *The Temperature.*

The berries of this Thorne, as they be in taste bitter and binding, so be they also hot and dry in the second degree.

¶ *The Vertues.*

A The same do purge and void by the stoole thicke flegme, and also cholericke humors: they are giuen being beaten into pouder from one dram to a dram and a halfe: diuers do number the berries, who giue to strong bodies from fifteene to twenty or moe; but it is better to breake them and boile them in fat flesh broth without salt, and to giue the broth to drinke: for so they purge with lesser trouble and fewer gripings.

B There is pressed forth of the ripe berries a iuice, which being boyled with a little Allum is vsed of painters for a deep greene, which they do call Sap greene.

C The berries which be as yet vnripe, being dried and infused or steeped in water, do make a faire yellow colour; but if they be ripe they make a greene.

Chap. 31. *Of the Holme, Holly, or Huluer tree.*

Agrifolium.
The Holly tree.

¶ *The Description.*

The Holly is a shrubbie plant, notwithstanding it oftentimes growes to a tree of a reasonable bignesse: the boughes whereof are tough and flexible, couered with a smooth and green bark. The substance of the wood is hard and sound, and blackish or yellowish within, which doth also sinke in the water, as doth the Indian wood which is called *Guaiacum*: the leaues are of a beautifull green colour, smooth and glib, like almost the bay leaues, but lesser, and cornered in the edges with sharp prickles, which notwithstanding they want or haue few when the tree is old: the floures be white, and sweet of smell: the berries are round, of the bignesse of a little Pease, or not much greater, of colour red, of tast vnpleasant, with a white stone in the midst, which do not easily fall away, but hang on the boughes a long time: the root is wooddie.

There is made of the smooth barke of this tree or shrub, Birdlime, which the birders and country men do vse to take birds with: they pul off the barke, and make a ditch in the ground, specially in moist, boggy, or foggy earth, wherinto they put this bark, couering the ditch with boughes of trees, letting it remaine there till it be rotten and putrified, which will be done in the

the space of twelue daies or thereabout: which done, they take it forth, and beat in morters vntill it be come to the thicknesse and clamminesse of Lime: lastly, that they may cleare it from pieces of barke and other filthinesse, they do wash it very often: after which they adde vnto it a little oyle of nuts, and after that do put it vp in earthen vessells.

¶ *The Place.*

The Holly tree groweth plentifully in all countries. It groweth green both winter and sommer; the berries are ripe in September, and they do hang vpon the tree a long time after.

¶ *The Names.*

This tree or shrub is called in Latine *Agrifolium*: in Italian, *Agrifoglio*, and *Aguifoglio*: in Spanish, *Azebo*: in high Dutch, **Walddistell**, and of diuers **Steecpalmen**: in low Dutch, **Hulst**: in French, *Hous* and *Housson*: in English, Holly, Huluer, and Holme.

¶ *The Temperature.*

The berries of Holly are hot and drie, and of thin parts, and waste away winde.

¶ *The Vertues.*

They are good against the collicke: for ten or twelue being inwardly taken bring away by the stoole thicke flegmaticke humors, as we haue learned of them who oftentimes made triall thereof. A

The Birdlime which is made of the barke hereof is no lesse hurtfull than that of Misseltoe, for it is maruellous clammie, it glueth vp all the intrails, it shutteth and draweth together the guts and passages of the excrements, and by this meanes it bringeth destruction to man, not by any qualitie, but by his glewing substance. B

Holly beaten to pouder and drunke, is an experimented medicine against all the fluxes of the belly, as the dysenterie and such like. C

CHAP. 32. *Of the Oke.*

1 *Quercus vulgaris cum glande & musco suo.*
The Oke Tree with his Acornes and Mosse.

¶ *The Description.*

1 THe common Oke groweth to a great tree; the trunke or body whereof is couered ouer with a thicke rough barke full of chops or rifts: the armes or boughs are likewise great, dispersing themselues farre abroad: the leaues are bluntly indented about the edges, smooth, and of a shining greene colour, whereon is often found a most sweet dew and somewhat clammie, and also a fungous excrescence, which we call Oke Apples. The fruit is long, couered with a browne hard and tough pilling, set in a rough scaly cup or husk; there is often found vpon the body of the tree, and also vpon the branches, a certaine kind of long white mosse hanging downe from the same: and sometimes another wooddie plant, which we cal Misseltoe, being either an excrescence or outgrowing from the tree it selfe, or of the doung (as it is reported) of a bird that hath eaten a certaine berrie. ‡ Besides these there are about the roots of old Okes within the earth certaine other excrescences, which *Bauhine* and others haue called *Vuæ quercinæ*, because they commonly grow in clusters together, after the manner of Grapes and about their bignes, being sometimes round, & otherwhiles cornered, of a woody substance, hollow within; and somtimes of a purple, otherwhiles of a whitish colour on the outside: their taste is astringent, and vse singular in all Dysenteries

ries and fluxes of bloud, as *Encelius* affirmes, *Cap.* 51. *de Lapid. & Gen.* ‡

3 *Carolus Clusius* reporteth that hee found this base or low Oke not far from Lisbone, of the height of a cubite, which notwithstanding did also beare an acorne like that of our Oke tree, sauing that the cup is smoother, and the Acorne much bitterer, wherein it differeth from the rest of his kinde.

2 *Quercus vulgaris cum excrementis fungosis:*
The common Oke with his Apple or greene Gall.

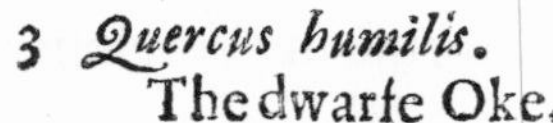

3 *Quercus humilis.*
The dwarfe Oke.

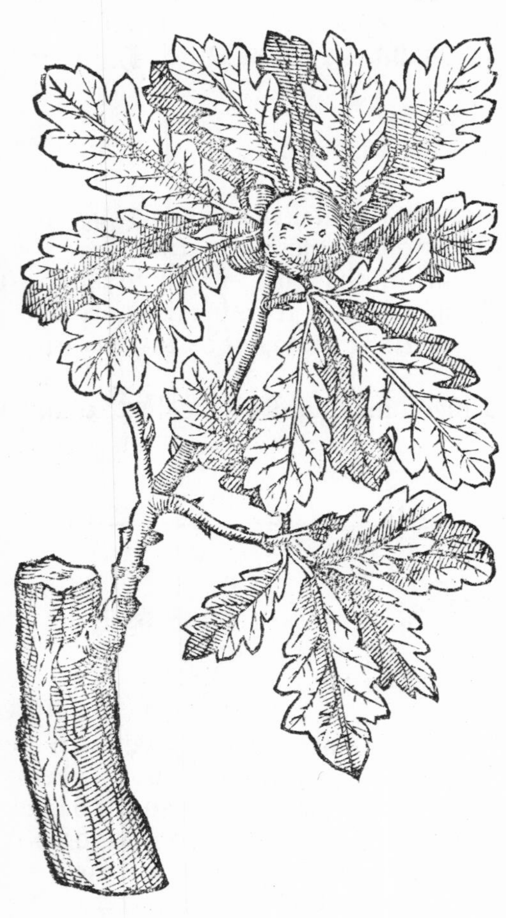

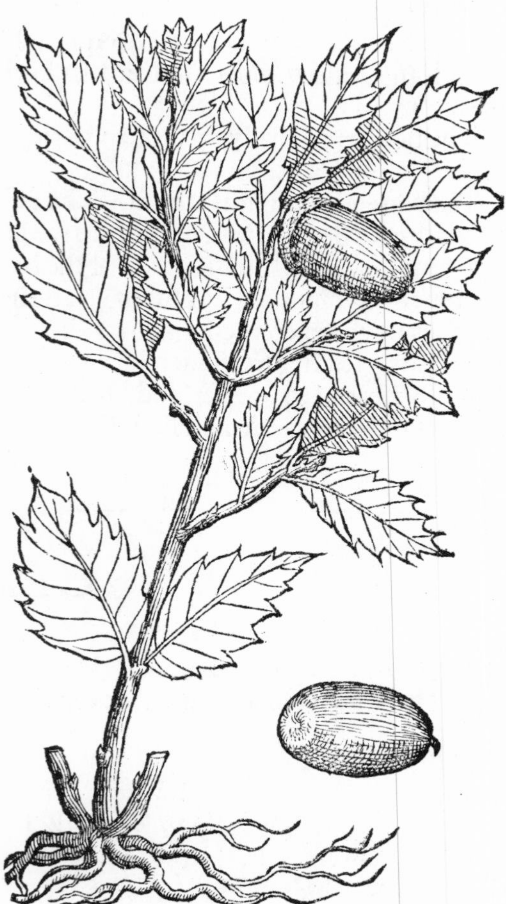

There is a wilde Oke which riseth vp oftentimes to a maruellous height, and reacheth very far with his armes and boughes, the body whereof is now and then of a mighty thicknesse, in compasse two or three fathoms: it sendeth forth great spreading armes, diuided into a multitude of boughs. The leaues are smooth, something hard, broad, long, gashed in the edges, greene on the vpper side: the Acornes are long, but shorter than those of the tamer Oke; euery one fastened in his owne cup, which is rough without: they are couered with a thin rinde or shell: the substance or kernell within is diuided into two parts, as are Beans, Pease, and Almonds: the bark of the yong Okes is smooth, glib, and good to thicken skins and hides with, but that of the old Okes is rugged, thicke, hard, and full of chops: the inner substance or heart of the wood is somthing yellow, hard and sound, and the older the harder: the white and outward part next to the barke doth easily rot, being subiect to the worme, especially if the tree be not felled in due time: some of the roots grow deepe into the earth, and othersome far abroad, by which it stiftely standeth.

¶ *The Place.*

The Oke doth scarcely refuse any ground; for it groweth in a drie and barren soile, yet doth it prosper better in a fruitfull ground: it groweth vpon hills and mountaines, and likewise in vallies: it commeth vp euery where in all parts of England, but is not so common in other of the South and hot regions.

¶ *The Time.*

The Oke doth cast his leaues for the most part about the end of Autumne: some keepe their leaues on, but dry all winter long, vntill they be thrust off by the new spring.

¶ *The Names.*

The Oke is called in Greeke δρῦς: in Latine, *Quercus*: of some, *Placida*, as *Gaza* translateth it. It may be called *Satiua*, *Vrbana*, or *Culta*, some also, *Emeros mudion*, and *Robur*. the Macedonians ἐτυμόδρυν, as as though you should say *Veriquercus*, as *Gaza* expoundeth it, or *Vere Quercus*, the true Oke. We may name it in English, the tamer Oke-tree: in French, *Chesne*: in Dutch, **Eyeken boom.**

The fruit is named in Greeke Βάλανος: in Latine, *Glans*: in high Dutch, **Eichel**: in low Dutch, **Eekel**: in Spanish, *Bellotus*: in Italian, *Chiande*: in English, Acorne and Mast.

The cup wherein the Acorne standeth is named in Greeke ὀμφακίς, as *Paulus Ægineta* in his third booke, 42 chapter testifieth, saying, *Omphacis* is the hollow thing out of which the Acorne groweth: in Latine, *Calix glandis*: in shops, *Cupula glandis*: in English, the Acorne cup.

¶ *The Temperature and Vertues.*

A The leaues, barke, Acorne cups, and the Acornes themselues, doe mightily binde and drie in the third degree, being somewhat cold withall.

B The best of them, saith *Galen*, is the thin skin which is vnder the barke of the tree, and that next, which lieth neerest to the pulpe, or inner substance of the Acorne; all these stay the whites, the reds, spitting of bloud and laskes: the decoction of these is giuen, or the pouder of them dried, for the purposes aforesaid.

C Acornes if they be eaten are hardly concocted, they yeeld no nourishment to mans body, but that which is grosse, raw, and cold.

D Swine are fatted herewith, and by feeding hereon haue their flesh hard and sound.

E The Acorns prouoke vrine, and are good against all venome and poison, but they are not of such a stopping and binding facultie as the leaues and barke.

F The Oke apples are good against all fluxes of bloud and lasks, in what manner soeuer they be taken, but the best way is to boile them in red wine, and being so prepared, they are good also against the excessiue moisture and swelling of the iawes and almonds or kernels of the throat

G The decoction of Oke apples staieth womens diseases, and causeth the mother that is falne downe to returne againe to the naturall place, if they doe sit ouer the said decoction being very hot.

H The same steeped in strong white wine vineger, with a little pouder of Brimstone, and the root of *Ireos* mingled together, and set in the Sun by the space of a moneth, maketh the haire blacke, consumeth proud and superfluous flesh, taketh away sun-burning, freckles, spots the morphew, with all deformities of the face, being washed therewith.

I The Oke Apples being broken in sunder about the time of their withering, do foreshew the sequell of the yeare, as the expert Kentish husbandmen haue obserued by the liuing things found in them: as if they finde an Ant, they foretell plenty of graine to insue: if a white worme like a Gentill or Magot, then they prognosticate murren of beasts and cattell; if a spider, then (say they) wee shall haue a pestilence or some such like sicknesse to follow amongst men: these things the learned also haue obserued and noted; for *Matthiolus* writing vpon *Dioscorides* saith, that before they haue an hole through them, they containe in them either a flie, a spider, or a worme: if a flie, then war insueth, if a creeping worme, than scarcitie of victuals, if a running spider, then followeth great sicknesse or mortalitie.

CHAP. 33. *Of the Scarlet Oke.*

¶ *The Kindes.*

Although *Theophrastus* hath made mention but of one of these Holme or Holly Okes onely, yet hath the later age set downe two kindes thereof; one bearing the scarlet grain, and the other onely the Acorn. which thing is not contrary to *Dioscorides* his opinion, for he intreateth of that which beareth the Acorne, in his first booke, among δρύες or the Okes; and the other hee describeth in his fourth booke, vnder the title κόκκος βαφική or *Coccus Baphice.*

¶ *The Description.*

The Oke which beareth the scarlet graine is a small tree, in manner of a hedge tree, of a meane bignesse, hauing many faire branches or boughes spread abroad: whereon are set leaues, green aboue, white vnderneath, snipt about the edges, and at euery corner one sharpe prickle, in manner of the smoother Holly: among which commeth sometimes, but not often, small Acornes, standing in little cups or husks, armed with prickes as sharpe as thornes, and of a bitter taste. Besides the Acornes, there is found cleauing vnto the wooddie branches, a certaine kinde of berries, or rather an excrescence, of the substance of the Oke Apple, and of the bignesse of a Pease, at the first white, and of the colour of ashes when they be ripe, in which are ingendred little Maggots, which seeme

Ilex Coccigera.
The Scarlet Oke.

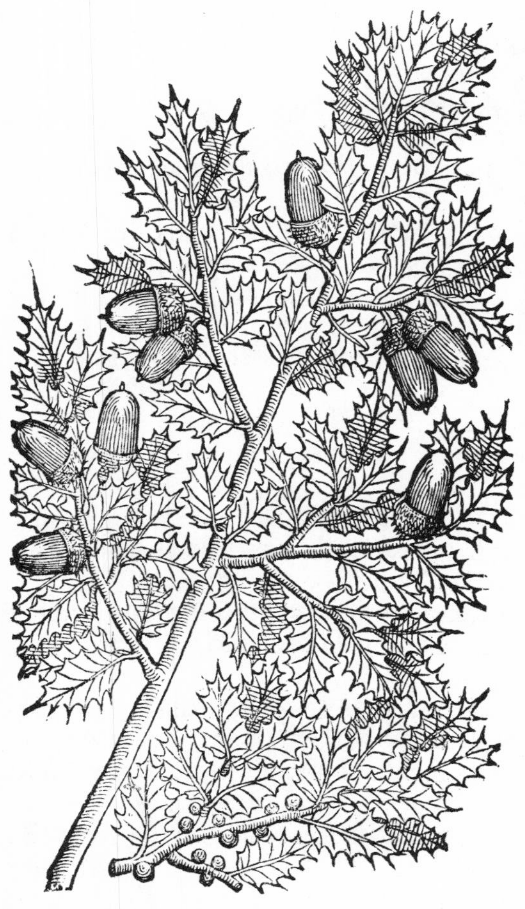

to be without life vntill they feele the heat of the sun, and then they creep, and seeke to flie away. But the people of the countrey (which make a gaine of them) doe watch the time of their flying, euen as we doe Bees, which they then take and put into a linnen bag, wherein they shake and boult them vp and downe vntil they be dead, which they make vp into great lumpes oftentimes, and likewise sell them to diers apart, euen as they were taken forth of the bag, whereof is made the most perfect Scarlet.

¶ *The Place.*

This Oke groweth in Languedocke, and in the countries thereabout, and also in Spain: but it beareth not the scarlet grain in all places, but in those especially, which lie towards the Midland sea, and which be subiect to the scorching heat of the Sun, as *Carolus Clusius* witnesseth; & not there alwaies, for when the tree waxeth old it growes to be barren. Then do the people cut and lop it downe, that after the young shoots haue attained to two or three yeares growth, it may become fruitfull againe.

Petrus Bellonius in his books of Singularities sheweth, that *Coccus Baphicus* or the Scarlet graine doth grow in the Holy land, and neere to the lake which is called the Sea of *Tiberias*, and that vpon little trees, whereby the inhabitants get great store of wealth, who seperat the husks from the pulpe or Magots, and sell this being made vp into balls or lumpes, much dearer than the emptie shels or husks.

Of this graine also *Pausanias* hath made mention in his tenth booke, and sheweth, that the tree which bringeth forth this graine is not great, and also groweth in Phocis, which is a countrey in Macedonia neere to the Boetians, not far from the mountain Parnassus.

Theophrastus writeth, that πρῖνος or the Scarlet Oke, is a great tree, and riseth vp to the height of the common Oke: amongst which writers there are some contrarietie. *Petrus Bellonius* reporteth it is a little tree, and *Theophrastus* a great one, which may chance according to the soyle and climate; for that vpon the stonie mountaines cannot grow to that greatnesse as those in the fertill grounds.

¶ *The Time.*

The little graines or berries which grow about the boughes begin to appeare especially in the Spring, when the Southwest windes do blow: the floures fall and are ripe in Iune, together with the Maggots growing in them, which receiuing life by the heat of the Sun, do forthwith flie away (in manner of a Moth or Butterflie) vnlesse by the care and diligence of the keepers, they be killed by much and often shaking them together, as aforesaid.

The tree or shrub hath his leaues alwaies greene: the Acornes be very late before they be ripe, seldome before new come vp in their place.

¶ *The Names.*

The Scarlet Oke is called in Greeke πρῖνος: in Latine *Ilex*: the later writers, *Ilex Coccigera*, or *Coccifera*; in Spanish, *Coscoia*: for want of a fit English name, we haue thought good to call it by the name of Scarlet Oke, or Scarlet Holme Oke: for *Ilex* is named of some in English, Holme, which signifieth Holly or Huluer. But this *Ilex*, as well as those that follow, might be called Holm Oke, Huluer Oke, or Holly Oke, for difference from the shrub or hedge tree *Agrifolium*, which is simply called Holme, Holly, and Huluer.

The graine or berrie that serueth to die with is properly called in Greeke κόκκος βαφική: in Latine, *Coccus infectoria*, or *Coccum infectorium*: *Pliny* also nameth it *Cusculium*: or as most men doe reade it, *Quisquilium*: the same Author saith, that it is likewise named *Scolecion*, or Maggot berrie.

The Arabians and the Apothecaries doe know it by the name of *Chesmes*, *Chermes*, and *Kermes*: They are deceiued who thinke that *Chesmes* doth differ from *Infectorium Coccum*: it is called in Italian, *Grano de tinctori*: in Spanish, *Grana de tintoreros*: in high Dutch, **Scharlachbeer**: in French, *Vermillon*,

Vermillon, and *Graine d'escarlate*: in English, after the Dutch, Scarlet Berry, or Scarlet graine, and after the Apothecaries word, *Coccus Baphicus*: the maggot within is that which is named Cutchonele, as most do deeme.

Theophrastus saith the Acorne or fruit hereof is called of diuers, Ἄκυλον, *Acylum*.

¶ *The Temperature and Vertues.*

This graine is astringent and somwhat bitter, and also dry without sharpnesse and biting, therefore, saith *Galen*, it is good for great wounds and sinewes that be hurt, if it be layd thereon: some temper it with Vineger; others with Oxymel or syrrup of vineger. A

It is commended and giuen by the later Physitians to stay the Menses: it is also counted among those Simple which be cordials, and good to strengthen the heart. Of this graine that noble and famous confection *Alkermes*, made by the Arabians, hath taken his name, which many doe highly commend against the infirmities of the heart: notwithstanding it was chiefly deuised in the beginning for purging of melancholy; which thing is plainly perceiued by the great quantitie of *Lapis Lazulus* added thereto: and therefore seeing that this stone hath in it a venomous quality, and likewise a property to purge melancholy, it canot of it selfe be good for the heart, but the other things be good, which be therefore added, that they might defend the heart from the hurts of this stone, and correct the malice thereof. B

This composition is commended against the trembling and shaking of the heart, and for swounings and melancholy passions, and sorrow proceeding of no euident cause: it is reported to recreate the minde, and to make a man merry and ioyfull. C

It is therefore good against melancholy diseases, vaine imaginations, sighings, griefe and sorrow without manifest cause, for that it purgeth away melancholy humors: after this maner it may be comfortable for the heart, and delightfull to the minde, in taking away the materiall cause of sorrow: neither can it otherwise strengthen a weake and feeble heart, vnlesse this stone called *Lapis Cyaneus* be quite left out. D

Therefore he that is purposed to vse this composition against beatings and throbbings of the heart, and swounings, and that not as a purging medicine, shall do well and wisely by leauing out the stone *Cyaneus*; for this being taken in a little weight or small quantitie, cannot purge at all, but may in the meane season trouble and torment the stomacke, and withall thorow his sharpe and venomous qualitie (if it be oftentimes taken) be very offensiue to the guts and intrailes, and by this meanes bring more harme than good. E

Moreouer, it is not necessarie, no nor expedient, that the bristle died with Cochenele, called *Chesmes*, as the Apothecaries terme it, should be added to this composition; for this bristle is not died without *Auripigmentum*, called also Orpiment, and other pernitious things ioyned therewith, whose poysonsome qualities are added to the iuyces together with the colour, if either the bristle or died silke be boyled in them. F

The berries of the Cochenele must be taken by themselues, which alone are sufficient to dy the iuices, and to impart vnto them their vertue: neither is it likewise needfull to boile the raw silke together with the graines, as most Physitians thinke: this may be left out, for it maketh nothing at all for the strengthning of the heart. G

CHAP. 34. *Of the great Skarlet Oke.*

¶ *The Description.*

THe great Skarlet Oke, or the great Holme Oke, groweth many times to the full height of a tree, sometimes as big as the Peare tree, with boughes far spreading like the Acorne or common Mast trees: the timber is firme and sound: the leaues are set with prickles round about the edges, like those of the former Skarlet Oke: the leaues when the tree waxeth old haue on them no prickles at all, but are somwhat bluntly cut or indented about the edges, greene on the vpper side, and gray vnderneath: the Acorne standeth in a prickly cup like our common Oke Acorne, which when it is ripe becommeth of a browne colour, with a white kernel within of taste not vnpleasant. There is found vpon the branches of this tree a certaine kinde of long hairy mosse of the colour of ashes, not vnlike to that of our English Oke. ‡ This tree is euer greene, and at the tops of the branches about the end of May, here in England, carrieth diuers long catkins of mossie yellow floures, which fall away, and are not succeeded by the acornes, for they grow out vpon other stalks. *Clusius* in the yeare 1581 obserued two trees; the one in a garden aboue the Bridge, and the other in the priuat garden at White-Hall, hauing lesser leaues than the former. The later of these is yet standing, and euery yeare beares small Acornes, which I could neuer obserue to come to any maturitie. ‡

¶ The

Ilex maior Glandifera.
The great Skarlet Oke.

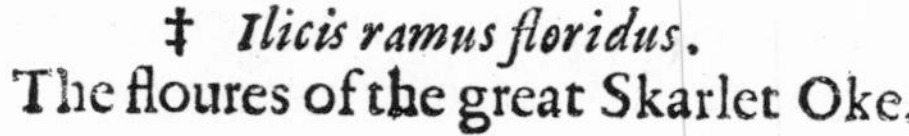
‡ *Ilicis ramus floridus.*
The floures of the great Skarlet Oke.

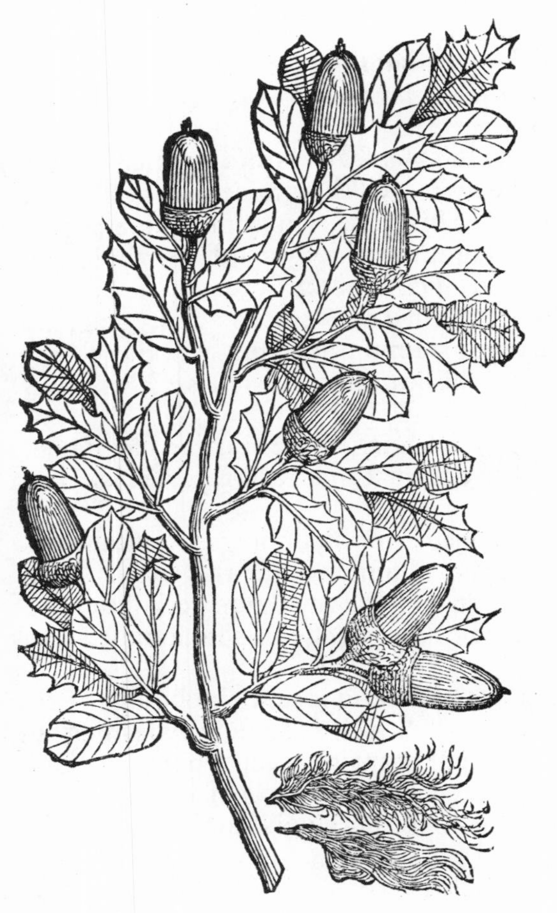

¶ *The Place.*

In diuers places there are great woods of these trees, hills also and vallies are beautified therewith: they grow plentifully in many countries of Spaine, and in Languedocke and Prouence in great plenty. It is likewise found in Italy. It beareth an Acorne greater, and of a larger size than doth the tame Oke; in some countries lesser and shorter: they are strangers in England, notwithstanding there is here and there a tree thereof, that hath been procured from beyond the seas: one groweth in her Maiesties Priuy Garden at White-Hall, neere to the gate that leadeth into the street, and in some other places here and there one.

¶ *The Time.*

It is greene at all times of the yeare: it is late before the Acornes be ripe. *Clusius* reporteth, that he saw the floures growing in clusters of a yellow colour in May.

¶ *The Names.*

This Oke is named in Greeke πρῖνος: in Latine, *Ilex:* in Spanish, *Enzina:* in Italian, *Elize:* in French, *Chesneuerd:* in English, Barren Skarlet Oke, or Holme Oke, and also of some, French or Spanish Oke.

The Spaniards call the fruit or Acorne *Bellota*, or *Abillota*. *Theophrastus* seemeth to call this tree not *Prinos*, but *Smilax*; for he maketh mention but of one *Ilex* onely, and that is of Scarlet Oke; and he sheweth that the Arcadians do not call the other *Ilex*, but *Smilax:* for the name *Smilax* is of many significations: there is *Smilax* among the Pulses, which is also called *Dolichus*, and *Phaseolus*; and *Smilax aspera*, and *Læuis*, amongst the Binde-weeds: likewise *Smilax* is taken of *Dioscorides* to be *Taxus*, the Yew tree. Of *Smilax*, *Theophrastus* writeth thus in his third booke: the inhabitants of Arcadia do call a certaine tree *Smilax*, being like vnto the Skarlet Oke: the leaues thereof be not set with such sharpe prickles, but tenderer and softer.

Of this *Smilax Pliny* also writeth, in his sixteenth booke, chap. 6. There be of *Ilex*, saith he, two kindes, *Ex ijs in Italia folio non multum ab oleis distant*, called of certain Grecians *Smilaces*, in the prouinces *Aquifolia:* in which words, in stead of Oliue trees may perchance be more truly placed *Suberis*, or the Corke tree; for this kinde of *Ilex* or *Smilax* is not reported of any of the old writers

to

to haue the leafe of the Oliue tree: but *Suber* in Greeke, called *Phellos*, or the Corke tree, hath a little leafe.

¶ *The Temperature and Vertues.*

The leaues of this Oke haue force to coole and repell or keepe backe, as haue the leaues of the Acornes or Mast trees: being stamped or beaten, and applied, they are good for soft swellings, and strengthen weake members. A

The barke of the root boiled in water vntill it be dissolued, and layd on all night, maketh the haire blacke, being first scoured with *Cimolia*, as *Dioscorides* saith. B

Clusius reporteth, that the Acorne is esteemed of, eaten, and brought into the market to be sold, in the city of Salamanca in Spaine, and in many other places of that countrey; and of this Acorne *Pliny* also hath peraduenture written, *lib.* 16. *cap.* 5. in these words: Moreouer, at this day in Spain the Acorne is serued for a second course. C

Chap. 35. *Of the great Holme-Oke.*

1 *Cerris maiore Glande.*
The Holme Oke with great Acornes.

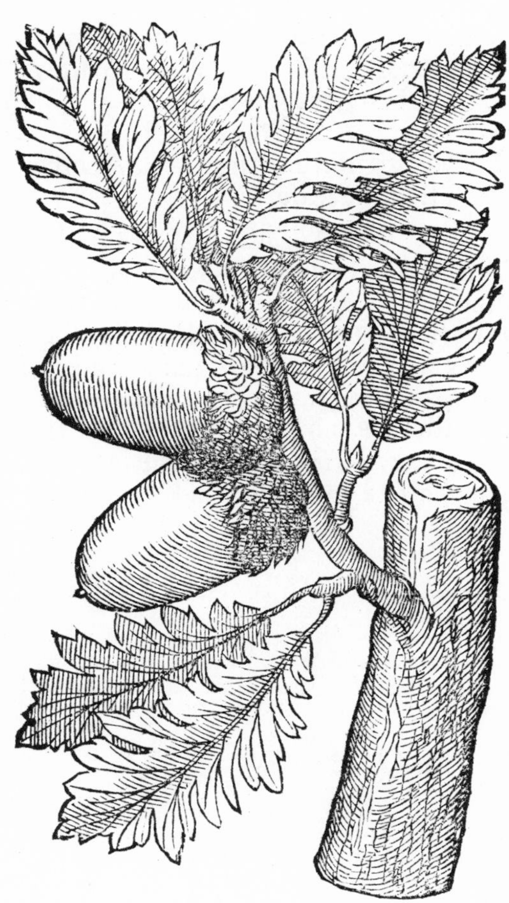

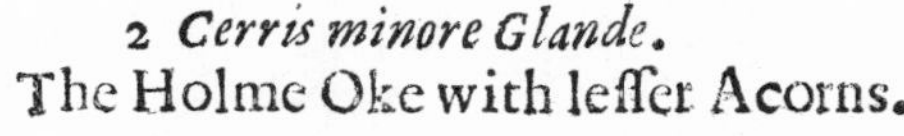

2 *Cerris minore Glande.*
The Holme Oke with lesser Acorns.

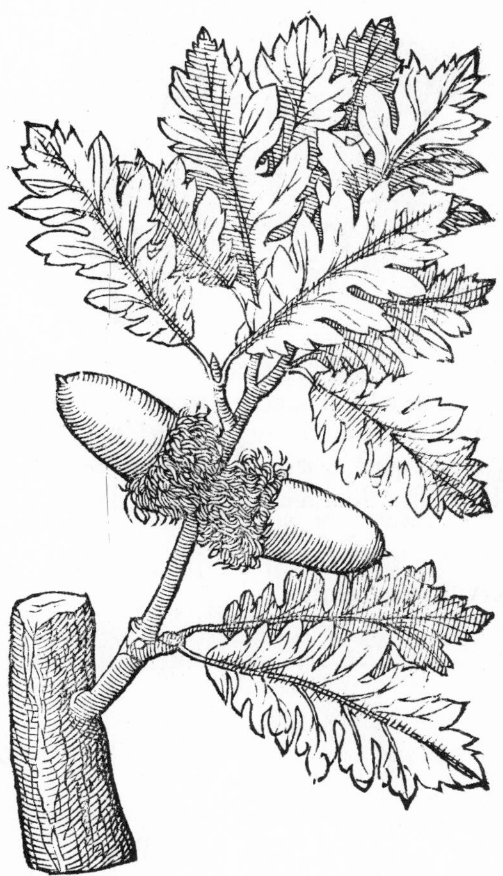

¶ *The Description.*

1 Among the wilder Okes this is not the least, for his comely proportion, although vnprofitable for timber, to make coles, carts, Wainscot, houses, or ships of: the fruit is not fit for any man or beast to eate, neither any propertie knowne for the vse of physicke or surgerie: it groweth vp to the height of a faire tree, the trunke or body is great, and verie faire to looke vpon: the wood or timber soft and spongie, scarce good to be burned: from which shooteth forth very comely branches dispersing themselues farre abroad; whereon are set for the most part by couples very faire leaues, greene aboue, and of an ouerworne russet colour vnderneath, cut or snipt about the edges very deepe: the Acorne groweth fast vnto the boughes, without any foot-stalke at all, being very like vnto our common Acorne, set in a rough and prickly cup like an Hedge hog or the Chesnut huske, of a harsh taste, and hollow within: this tree beareth or bringeth

‡ *Cerri minoris ramulus cum flore.*
A branch of the smaller Holme Oke with floures.

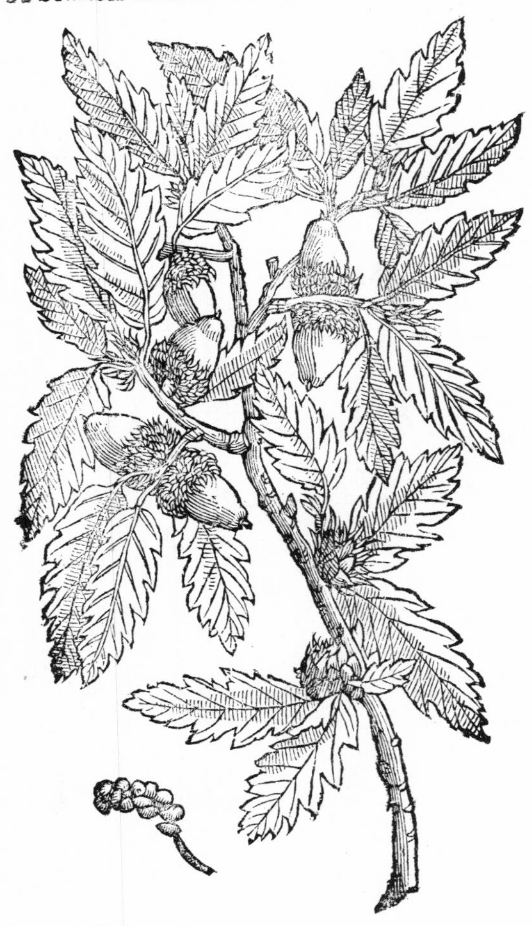

bringeth forth oft times a certain smooth kinde of Gall not altogether vnprofitable. This Oke likewise bringeth forth another kinde of excresence, which the Grauer hath omitted in the figure, which is called in Greeke φάσκον: *Gaza* nameth it *Penis*. This *Penis* or pricke is hollow, mossie, hanging downe halfe a yard long, like a long rag of linnen cloath.

2 The second is altogether like the first, sauing that this beareth smaller Acornes, and the whole tree is altogether lesse, wherein consisteth the difference.

‡ Both this & the former cary floures clustering vpon long stalkes, like as in the common Oke; but the fruit doth not succeed them, but grow forth in other places. ‡

¶ *The Place.*

This Oke groweth in vntoiled places, it is seldome times found, and that but in Woods onely: it is for the most part vnknowne in Italy, as *Pliny* reporteth.

¶ *The Time.*

They bring forth their fruit or Acornes in the fall of the leafe.

¶ *The Names.*

This Oke is called in Greeke Αἰγίλωψ: in Latine, *Cerrus*: yet doth *Pliny* make mention both of *Ægilops*, and also of *Cerrus*: Αἰγίλωψ is likewise one of the diseases of corne, called in Latine *Festuca*: in English, wilde Otes, and far differing from the tree *Ægilops*.

That which hangeth from the boughs, *Pliny, lib.* 16. *cap.* 8. calleth *Panus* onely: that acorne tree named *Ægilops* bringeth forth *Panos arentes*, withered prickes, couered with white mossie iags hanging downe, not only in the barke, but also from the boughes, halfe a yard in bignesse, bearing a sweet smell, as we haue said, among ointments.

¶ *The Temperature and Vertues.*

We finde nothing written of the faculties of this tree among the old Writers, neither of our owne experience.

Chap. 36. *Of the Corke Oke.*

¶ *The Description.*

1 The Corke tree is of a middle bignesse like vnto *Ilex*, or the barren skarlet Oke, but with a thicker body, and fewer boughes: the leaues be for the most part greater, broader, rounder, and more nicked in the edges: the barke of the tree is thicke, very rugged, and full of chinkes or crannies that cleaueth and diuideth it selfe into pieces, which vnlesse they be taken away in due time do giue place to another barke growing vnderneath, which when the old is remoued is maruellous red, as though it were painted with some colour: the Acorne standeth in a cup, which is great, bristled, rough, and full of prickles: this Acorne is also astringent or binding, more vnpleasant than the Holme Acorne, greater in one place, and lesse in another.

2 The Corke tree with narrow leaues groweth likewise to the height and bignesse of a great tree; the trunke or body whereof is couered with a rough and scabbed barke of an ouerworn blackish colour, which likewise cleaueth and casteth his coat when the inner barke groweth somwhat thicke: the branches are long, tough, and flexible, easie to be bowed any way, like those of the

Oziar;

Oziar ; whereupon do grow leaues like thoſe of the precedent, but longer, and little or nothing indented about the edges : the fruit groweth in ſmall cups as the Acornes doe : they are leſſer than thoſe of the other kinde, as is the reſt of the tree, wherein is the chiefeſt difference. ‡ This varies in the leafe, (as you may ſee in the figure) which in ſome is ſnipt about the edge, in other ſome not at all.

1, 2. *Suber latifolium & anguſtifolium.*
The Corke tree with broad and narrow leaues.

¶ *The Place.*

It groweth in the countrey of Aquitania, neere to the mountaines called Pyrenæi : it alſo groweth plentifully in the kingdomes of Spaine, differing ſomewhat from that of Aquitania, as *Cluſius* declareth : it is likewiſe found in Italy, and that in the territorie of Piſa, with a longer leafe, and ſharper pointed ; and about Rome with a broader, and cut in the edges like a ſaw, and rougher, as *Matthiolus* teſtifieth.

¶ *The Time.*

The leaues of the firſt are alwaies greene in Spaine and Italy, about the Pyrenæan mountaines they fall away in Winter.

¶ *The Names.*

This tree is called in Greeke φελλός : in Latine *Suber* : in French, *Liege* : in Italian, *Sugaro* : the ſame names do alſo belong to the barke : the Spaniards call the tree *Alcornoque* : the Engliſhmen, Corke tree ; and the barke, *Corcha de Alcornoque* ; whereupon the Low-countrey men and Engliſh men alſo do call it Corke ; and yet it is called in low-Dutch alſo **Wlothout.**

¶ *The Temperature and Vertues.*

This barke doth manifeſtly dry, with a binding facultie. A

Being beaten to pouder and taken in water it ſtancheth bleeding in any part of the body. The B
Corke which is taken out of wine veſſels, ſaith *Paulus*, being burnt, maketh aſhes which do mightily dry, and are mixed in compoſitions diuiſed againſt the bloudy flix.

Corke is alſo profitable for many things : it is vſed (ſaith *Pliny*) about the anchors of ſhips, Fi- C
ſhers nets, and to ſtop veſſels with ; and in Winter for womens ſhooes, which vſe remaines with vs euen to this day : fiſhermen hang this barke vpon the wings of their nets for feare of ſinking : and ſhoo-makers put it in ſhooes and pantofles for warmneſſe ſake.

CHAP.

Chap. 37. *Of the Gall tree.*

¶ *The Kindes.*

OF trees that bring forth Galls there be diuers ſorts, as may appeare by the diuers formes and ſorts of Galls ſet forth in this preſent chapter, which may ſerue for their ſeuerall diſtinctions, whereof ſome bring forth Acornes likewiſe, and ſome nothing but Galls: the figures of ſome few of the trees ſhall giue you ſufficient knowledge of the reſt: for all the Acorne or Maſt trees bring forth Galls, but thoſe trees whoſe figures we haue ſet forth do beare thoſe Galls fit for medicine, and to thicken skins with.

Dioſcorides and *Galen* make but two ſorts of Galls; the one little, yellow, full of holes, and more ſpongie in the inner part, both of them round, hauing the forme of a little ball, and the other ſmooth and euen on the out ſide: ſince, the later writers haue found moe, ſome hauing certain little knobs ſticking forth, like in forme to the Gall, which doth alſo cleaue and grow without ſtalke to the leafe. There is alſo found a certaine excreſcence of a light greene colour, ſpongie and waterie, in the middle whereof now and then is found a little flie or worme: which ſoft ball in hot countries doth oftentimes become hard, like the little ſmooth Gall, as *Theophraſtus* ſaith.

1 *Galla, ſiue Robur maius.*
The great Gall tree.

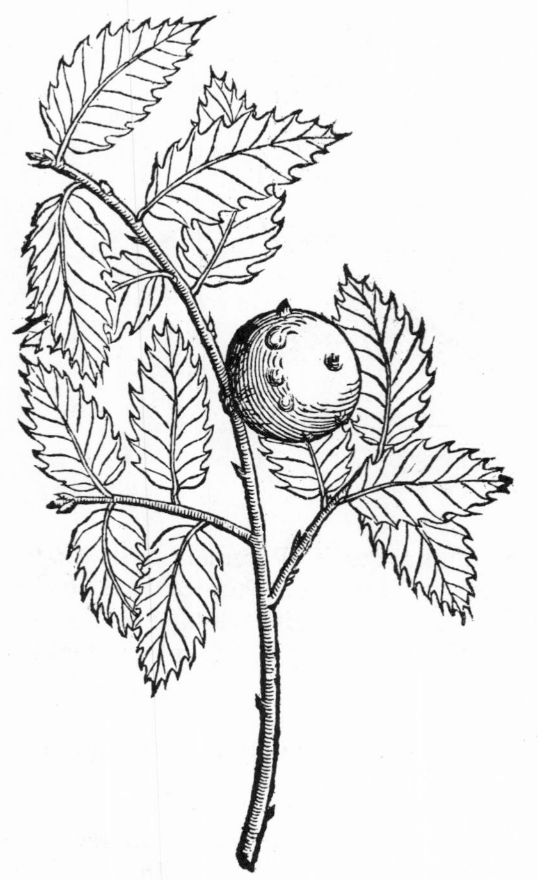

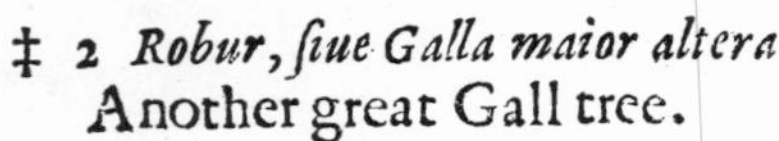

‡ 2 *Robur, ſiue Galla maior altera.*
Another great Gall tree.

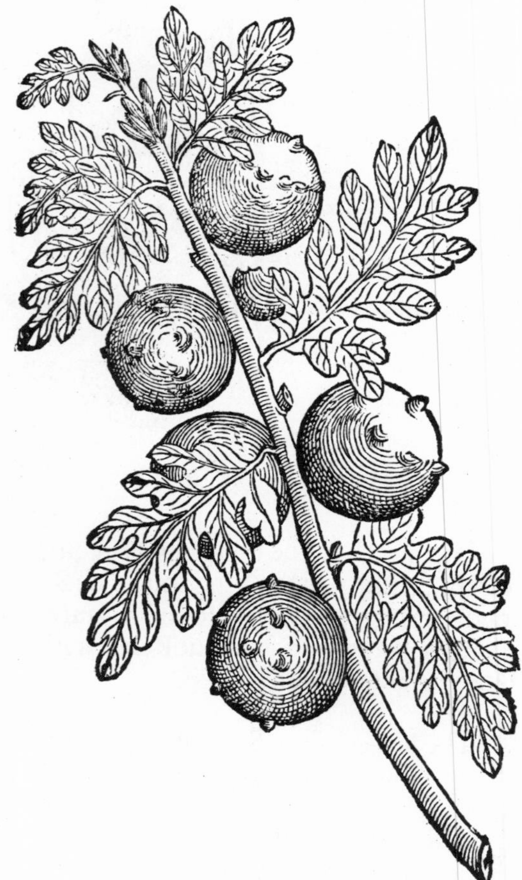

¶ *The Deſcription.*

1 THe Gall tree growes vp to a ſufficient height, hauing a very faire trunk or body, whereon are placed long twiggy branches bringing forth very faire leaues, broad, and nicked in the edges like the teeth of a ſaw: among which come forth acornes, although the figure expreſſe not the ſame, like thoſe of the Oke, and likewiſe a wooddy excreſcence, which we name the Gall, hauing certaine ſmall eminences or bunches on the out ſide, growing for the moſt part vpon the ſlender branches without ſtalks, and ſomtimes they grow at the ends thereof; which by the heate of the Sun are harder, greater, and more ſollid in one countrey than another, according to the ſoile and clymat.

‡ 2 This

3 *Galla minor.*
The little Gall tree.

‡ 2 This growes to the height of a tall man, hauing leaues deepely diuided on the edges like the Oke, and they are green aboue, but hairy and hoary below: it carries a great Gall of the bignesse of a little Apple, and that in great plenty, & without any order. This groweth in diuers parts of old Castile in Spain, and in all the mountainous woods about Vienna in Austria.‡

3 The lesser Gall tree differeth not from the former, sauing that it is altogether lesser: the fruit and Gall is likewise lesser, wherein especially consisteth the difference.

¶ *The Place.*

The Galls are found in Italy, Spaine, and Bohemia, and most of the hot regions.

¶ *The Time.*

The Gall, saith *Pliny*, appeareth or commeth forth when the Sun commeth out of the signe *Gemini*, and that generally in one night.

¶ *The Names.*

The Gall tree is called *Quercus, Robur*, and *Gallæ arbor*: the Gall is called in Greeke κηκὶς: the Apothecaries and Italians keepe the name *Galla* for the fruit: in high-Dutch, **Galopffel**: in low-Dutch, **Galnoten**: in Spanish, *Agalha Galha*, and *Bugalha*: in French, *Noix de Galle*: in English, Gaules, and Galls.

¶ *The Temperature and Vertues.*

A The Gall called *Omphacitis*, as *Galen* writeth, is dry in the third degree, and cold in the second: it is a very harsh medicine, it fastneth and draweth together faint and slacke parts, as the ouergrowings in the flesh: it repelleth and keepeth backe rheumes and such like fluxes, and doth effectually dry vp the same, especially when they haue a descent into the gums, almonds of the throat, and other places of the mouth.

B The other Gall doth dry and also binde; but so much lesser, by how much the harsh or choking qualitie is diminished: being boyled, beaten, and also applied in manner of a plaister is laid with good successe vpon the inflammations of the fundament, and falling downe thereof: it is boiled in water if there be need of a little astriction; and in wine, especially in austere wine, if more need require.

C Galls are very profitable against the Dysenterie and the Cœliacke passion, being drunk in wine, or the pouder thereof strewed vpon meats.

D Galls are vsed in dying and colouring of sundry things, and in making of inke.

E Last of all, burnt Galls doe receiue a further facultie, namely to stanch bloud, and are of thin parts, and of a greater vertue to dry than be those that are not burnt: they must be layd vpon hot burning coles vntill they come to be thorow white, and then they are to be quenched in Vineger and wine.

F Moreouer, Galls are good for those that are troubled with the bloudy flix and common laskes, being taken in wine or water, and also applied or vsed in meats: finally, these are to be vsed as oft as need requireth to dry and binde.

G Oke Apples are much of the nature of Galls, yet are they far inferiour to them, and of lesser force.

† Our Author out of *Tabernamontanus* gaue the figures of fourteene varieties of Galls; some being large, others small; some round, others longish, and other sorts diuersly cornered.

Chap. 38.
Of Miſſeltoe or Miſteltoe.

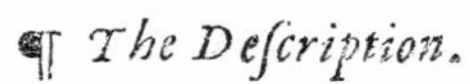

¶ *The Deſcription.*

1 *Viſcum.*
Miſſeltoe.

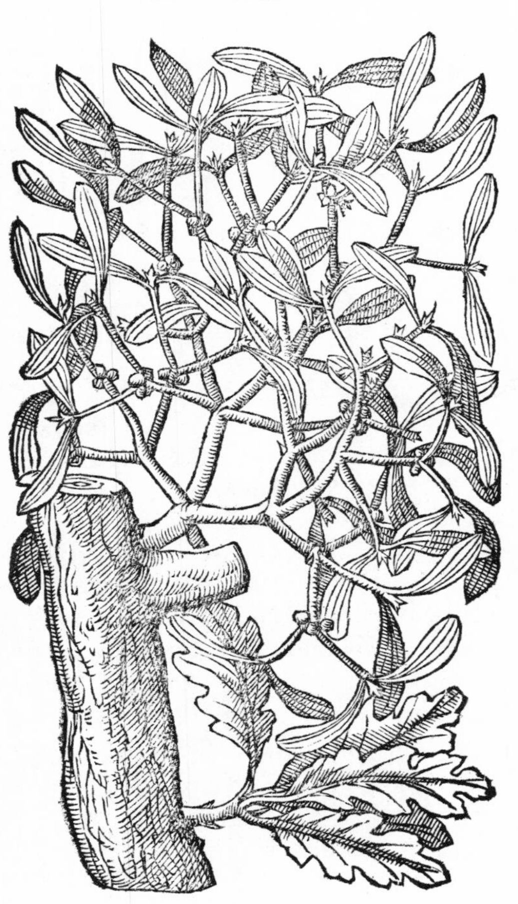

1 *Viſcum* or Miſſeltoe hath many ſlender branches ſpred ouerthwart one another, and wrapped and interlaced one within another: the bark whereof is of a light green or Popinjay colour: the leaues of this branching excreſcence be of a browne greene colour: the floures be ſmal and yellow: which being paſt, there appeare ſmall cluſters of white tranſlucent berries, which are ſo cleare that a man may ſee through them, and are full of clammy or viſcous moiſture, whereof the beſt Bird-lime is made, far exceeding that which is made of Holme or Holly barke: and within this berry is a ſmall blacke kernell or ſeed: this excreſence hath not any root, neither doth encreaſe himſelfe of his ſeed, as ſome haue ſuppoſed; but it rather commeth of a certaine moiſture and ſubſtance gathered together vpon the boughes and ioints of the trees, through the barke whereof this vaporous moiſture proceeding, bringeth forth the Miſſeltoe. Many haue diuerſly ſpoken hereof: ſome of the Learned haue ſet downe that it comes of the dung of the bird called a Thruſh, who hauing fed of the ſeeds thereof, as eating his owne bane, hath voided and left his dung vpon the tree, whereof was ingendred this berry, a moſt fit matter to make lime of to intrap and catch birds withall.

2 Indian Miſſeltoe groweth likewiſe vpon the branches of trees, running alongſt the ſame in manner of Polypodie: the ſtrings of the roots are like thoſe of Couch graſſe; from which riſe vp diuers ſtalks ſmooth and euen, ſet with ioints and knees at certain diſtances: toward the top comes forth one leafe ribbed like the Plantain leafe, whereon are marked certaine round eyes ſuch as are in the haft of a knife; from the boſome whereof commeth forth a chaffie branch, ſet with ſmall leaues which continue greene winter and Sommer.

2 *Viſcum Indicum Lobelij.*
Indian Miſſeltoe.

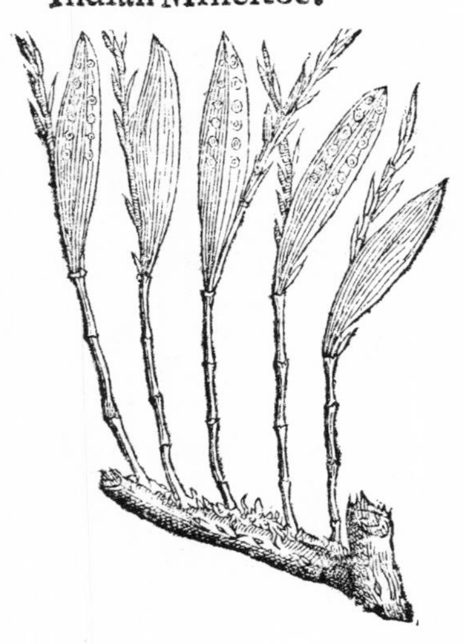

3 *Viſcum Peruvianum Lobelij.*
Miſſeltoe of Peru.

3 There is found alſo another plant growing vpon the boughes or branches of trees, in maner as our Miſſeltoe doth, and may very well be reckoned as a kinde thereof: the plant cleaueth vnto the branches, being ſet thereto as it were with the pillings of the ſea onion, of the bredth of a mans hand toward the bottome, and ſomewhat hollow: the tops whereof are very ſmall and ruſhy, hollow likewiſe, and of a purple colour: among which comes forth a branch like that of *Haſtula Regia*, or the Kings Speare, reſembling the buſh of Otes, couered with a white ſilke, ſuch as is to be found in *Aſclepias*, of a ſalt and nitrous taſte, and very vnpleaſant.

¶ *The Place.*

The firſt kinde of Miſſeltoe groweth vpon Okes and diuers other trees almoſt euery where; as for the other two they are ſtrangers in England.

¶ *The Time.*

Miſſeltoe is alwaies greene as well in Winter as in Sommer: the berries are ripe in Autumne; they remaine all Winter thorow, and are a food for diuers birds, as Thruſhes, Blacke-birds, and Ring-doues.

¶ *The Names.*

Miſſeltoe is called in Greeke ἰξὸς, and ἰξία: in Latine, *Viſcum*: in high-Dutch, **Miſtell**: in Low-Dutch, **Marentacken**: in Italian, *Viſchio*: in Spaniſh, *Liga*: in the Portugal language, *Viſgo*: in Engliſh, Miſſel, and Miſſeltoe.

The glue which is made of the berries of Miſſel is likewiſe called *Viſcum* and *Ixia*: in Engliſh, Bird-lime. *Ixia* is alſo called *Chamæleon albus*, by reaſon of the glue which is oftentimes found about the root thereof. This word is alſo aſcribed to *Chamæleon niger*, as we reade among the baſtard names. *Ixia* is likewiſe reckoned vp by *Dioſcorides, lib.6.* and by *Paulus Ægineta, lib.5.* among the poyſons: but what this poyſonſome and venomous *Ixia* is it is hard and doubtfull to declare: many would haue it to be *Chamæleon niger*: others, the glue or clammy ſubſtance which is made of the berries of Miſſel-toe; who do truly thinke that *Ixia* differeth from *Chamæleon niger* for *Paulus Ægineta, lib.5. cap.30.* in reckoning vp of ſimple poyſons hath firſt made mention of *Chamæleon niger*, then a little after of *Ixia*: and whileſt he doth particularly diſcourſe of euery one, he intreateth of *Chamæleon niger, cap.32.* and of *Ixia* (which hee alſo nameth *Vlophonon*) *cap.47.* and telleth of the dangerous and far differing accidents of them both. And *Dioſcorides* himſelfe, *lib.6.* where he ſetteth downe his iudgement of ſimple poyſons, intreateth firſt of *Chamæleon niger*, and then a little after of *Ixia*. Theſe things declare that *Chamæleon niger* doth differ from *Ixia*, which is reckoned among the poyſons. Moreouer, it can no where be read that *Chamæleon niger* beareth Bird-lime, or hath ſo glutinous and clammy a ſubſtance as that it ought to be called *Ixia*: therfore *Ixia*, as it is one of the poyſons, is the glue that is made of the berries of Miſſeltoe, which becauſe it is ſharpe and biting, inflameth and ſetteth the tongue on fire, and with his ſlimie and clammy ſubſtance doth ſo draw together, ſhut, and glue vp the guts, as that there is no paſſage for the excrements, which things are mentioned among the miſchiefes that *Ixia* bringeth.

‡ I can by no meanes approue of, or yeeld to this opinion here deliuered out of *Dodonæus* by our Author; which is, That the Bird-lime made of the berries of Miſſeltoe is poyſon; or that *Ixia* ſet forth by *Dioſcorides* and *Nicander* for a poyſon is meant of this: for this is manifeſtly treated of in *Dioſcorides, lib.3. cap.103.* by the name of ἰξὸς: when as the other is mentioned, *lib.6. cap.21.* by the name of ἰξίας. Alſo dayly experience ſhewes this plant to haue no maligne nor poiſonous, but rather a contrarie facultie, being frequently vſed in medicines againſt the Epilepſie. Such as would ſee more concerning *Ixia* or *Ixias*, let them haue recourſe to the firſt chapter of the firſt part of *Fabius Columna, de Stirpib. min. cognitis & rarioribus*, where they ſhall finde it largely treated of. ‡

¶ *The Temperature and Vertues.*

The leaues and berries of Miſſeltoe are hot and dry, and of ſubtill parts: the Bird-lime is hot and biting, and conſiſts of an airy and waterie ſubſtance, with ſome earthy qualitie; for according to the iudgement of *Galen*, his acrimony ouercommeth his bitterneſſe; for if it be vſed in outward applications it draweth humors from the deepeſt or moſt ſecret parts of the body, ſpreading and diſperſing them abroad, and digeſting them. A

It ripeneth ſwellings in the groine, hard ſwellings behinde the eares, and other impoſtumes, being tempered with roſin and a little quantitie of wax. B

With Frankincenſe it mollifieth old vlcers and malicious impoſtumes, being boyled with vnſlaked lime, or with *Gagate lapide*, or *Aſio*, and applied, it waſteth away the hardnes of the ſpleene. C

With Orpment or *Sandaraca* it taketh away foule ill fauoured nailes, being mixed with vnſlaked lime and wine lees it receiueth greater force. D

It hath been moſt credibly reported vnto me, that a few of the berries of Miſſeltoe bruiſed and ſtrained into Oyle, and drunken, hath preſently and forthwith rid a grieuous and ſore ſtitch. E

CHAP. 39. *Of the Cedar tree.*

¶ *The Kindes.*

THere be two Cedars, one great bearing Cones, the other small bearing berries like those of Iuniper.

Cedrus Libani.
The great Cedar tree of Libanus.

¶ *The Description.*

THe great Cedar is a very big and high tree, not onely exceeding all other resinous trees, and those which beare fruit like vnto it, but in his tallnesse and largenesse farre surmounting all other trees: the body or trunke thereof is commonly of a mighty bignesse, insomuch as foure men are not able to fathome it, as *Theophrastus* writeth: the barke of the lower part, which proceedeth out of the earth, to the first yong branches or shoots, is rough and harsh; the rest which is among the boughes is smooth and glib: the boughes grow forth almost from the bottome, and not farre from the ground, euen to the very top, waxing by degrees lesser and shorter still as they grow higher, the tree bearing the forme or shape of a Pyramide or sharpe pointed steeple: these compasse the body round about in maner of a circle, and are so orderly placed by degrees, as that a man may clymbe vp by them to the very top as by a ladder: the leaues be small and round like those of the Pine tree, but shorter, and not so sharp pointed; all the cones or clogs are far shorter and thicker than those of the Firre tree, compact of soft, not hard scales, which hang not downewards, but stand vpright vpon the boughes, whereunto also they are so strongly fastned, as they can hardly be plucked off without breaking of some part of the branches, as *Bellonius* writeth: the timber is extreme hard, and rotteth not, nor waxeth old: there is no wormes nor rottennesse can hurt or take the hard matter or heart of this wood, which is very odoriferous, and somewhat red: *Solomon* King of the Iewes did therefore build Gods Temple in Ierusalem of Cedar wood: the Gentiles were wont to make their Diuels or Images of this kinde of wood, that they might last the longer.

¶ *The Place.*

The Cedar trees grow vpon the snowie mountaines, as in Syria vpon mount Libanus, on which there remaine some euen to this day, saith *Bellonius*, planted as is thought by *Solomon* himselfe: they are likewise found on the mountaines Taurus and Amanus, in cold and stony places: the merchants of the factorie at Tripolis told me, that the Cedar tree groweth vpon the declining of the mountaine Libanus, neere vnto the Hermitage by the city Tripolis in Syria: they that dwel in Syria vse to make boats thereof for want of the Pine tree.

¶ *The Time.*

The Cedar tree remaineth alwaies greene, as other trees which beare such manner of fruit: the timber of the Cedar tree, and the images and other workes made thereof, seeme to sweat and send forth moisture in moist and rainy weather, as do likewise all that haue an oylie iuyce, as *Theophrastus* witnesseth.

¶ *The Names.*

This huge and mighty tree is called in Greeke κέδρος: in Latine likewise *Cedrus*: in English, Cedar, and Cedar tree. *Pliny, lib.* 24. *cap.* 5. nameth it *Cedrelate*, as though he should say, *Cedrus abies*, or *Cedrina abies*, Cedar Firre; both that it may differ from the little Cedar, and also because it is very like to the Firre tree.

The

The Rosin hath no proper name, but it may be syrnamed *Cedrina*, or Cedar Rosin.

The pitch which is drawne out of this is properly called κέδρια: yet *Pliny* writeth, that also the liquor of the Torch Pine is named *Cedrium*. The best, saith *Dioscorides*, is fat, thorow shining, and of a strong smell; which being poured out in drops vniteth it selfe together, and doth not remaine seuered.

¶ *The Temperature and Vertues.*

Cedar is of temperature hot and dry, with such an exquisite tenuitie and subtiltie of parts, that it seemeth to be hot and dry in the fourth degree, especially the Pitch or Rosin thereof. A

There issueth out of this tree a Rosin like vnto that which issueth out of the Fir tree, very sweet in smell, of a clammy or cleauing substance, the which if you chew in your teeth it will hardly be gotten forth againe, it cleaueth so fast: at the first it is liquid and white, but being dried in the sun it waxeth hard: if it be boiled in the fire an excellent pitch is made thereof, called Cedar Pitch. B

The Ægyptians were wont to coffin and embalme their dead in Cedar, and with Cedar pitch, although they vsed also other meanes, as *Herodotus* recordeth. C

The condited or embalmed body they call in shops *Mumia*, but very vnfitly; for *Mumia* among the Arabians is that which the Grecians call *Pissasphalton*, as appeareth by *Auicen, cap.* 474. and out of *Serapio, cap.* 393. D

He that interpreted and translated *Serapio* was the cause of this error, who translated and interpreted *Mumia* according to his owne fancie, and not after the sence and meaning of his Author *Serapio*, saying that this *Mumia* is a compsition made of Aloes and Myrrh mingled together with the moisture of mans body. E

The gum of Cedar is good to be put in medicines for the eyes, for being anointed therewith it cleareth the sight, and clenseth them from the Haw and from stripes. F

Cedar infused in vineger and put into the eares killeth the wormes therein, and being mingled with the decoction of Hyssop, appeaseth the sounding, ringing, and hissing of the eares. G

If it be washed or infused in vineger, and applied vnto the teeth, it easeth the tooth-ache. H

If it be put into the hollownesse of the teeth it breaketh them, and appeaseth the extreme griefe thereof. I

It preuaileth against *Angina's*, and the inflammation of the Tonsils, if a Gargarisme be made thereof. K

It is good to kill nits and lice and such like vermine: it cureth the biting of the serpent *Cerastes*, being layd on with salt. L

It is a remedie against the poyson of the sea Hare, if it be drunke with sweet wine. M

It is good also for Lepers: being put vp vnderneath it killeth all manner of worms, and draweth forth the birth, as *Dioscorides* writeth. N

CHAP. 40. *Of the Pitch tree.*

¶ *The Description.*

† 1 *Picea*, the tree that droppeth Pitch, called Pitch tree, groweth vp to be a tall, faire, and big tree, remaining alwaies greene like the Pine tree: the timber of it is more red than that of the Pine or Firre: it is set full of boughes not onely about the top, but much lower, and also beneath the middle part of the body, which many times hang downe, bending toward the ground: the leaues be narrow, not like those of the Pine tree, but shorter and narrower, and sharpe pointed like them, yet are they blacker, and withall couer the yong and tender twigs in manner of a circle, like those of the Firre tree; but being many, and thicke set, grow forth on all sides, and not onely one right against another, as in the Yew tree: the fruit is scaly, and like vnto the Pine apple, but smaller: the barke of the tree is somewhat blacke, tough and flexible, not brittle, as is the bark of the Firre tree: vnder which next to the wood is gathered a Rosin, which many times issueth forth, and is like to that of the Larch tree.

‡ 2 Of this sort (saith *Clusius*) there is found another that neuer growes high, but remaineth dwarfish, and it carries certaine little nugaments or catkins of the bignesse of a small nut, composed of scales lying one vpon another, but ending in a prickly leafe, which in time opening shew certaine emptie cauities or cels: from the tops of these sometimes grow forth branches set with many short and pricking little leaues: all the shrub hath shorter and paler coloured leaues than the former: I obserued neither fruit nor floure on this, neither know I whether it carry any. *Dalechampius* seemes to haue knowne this, and to haue called it *Pinus Tubulus* or *Tibulus*. ‡

1 *Picea maior.* The Pitch tree.	‡ 2 *Picea pumila.* The dwarfe Pitch tree.
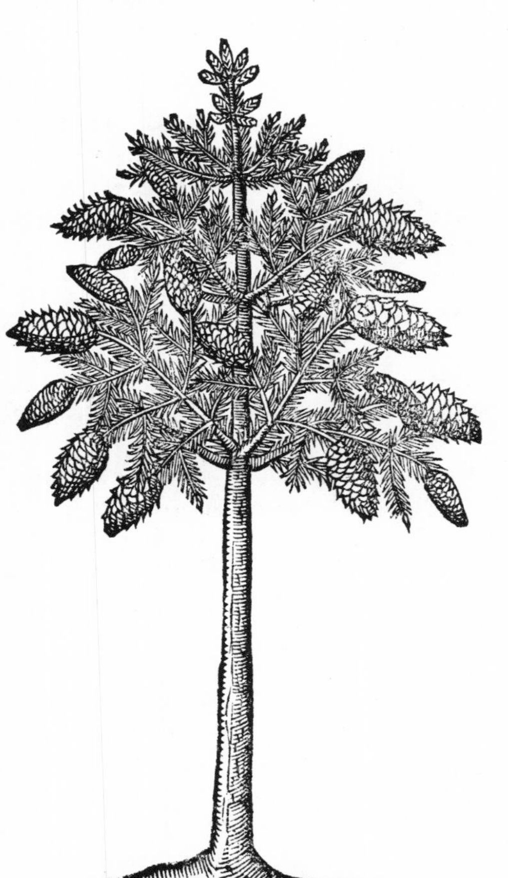	

¶ *The Place.*

The Pitch tree groweth in Greece, Italy, France, Germanie, and all the cold regions euen vnto Russia.

¶ *The Time.*

The fruit of the Pitch tree is ripe in the end of September.

¶ *The Names.*

The Grecians call this Cone tree πίτυς: the Latines, *Picea*, and not *Pinus*; for *Pinus* or the Pine tree, is the Grecians πεύκη, as shall be declared: that πίτυς is named in Latine *Picea, Scribonius Largus* testifieth, in his 201 Composition, writing after this manner; *Resinæ Petuinæ, id est, ex Picea arbore*, which signifies in English, of the Rosin of the tree *Pitys*, that is to say, of the Pitch tree. With him doth *Pliny* agree, *lib.16. cap.10.* where he translating *Theophrastus* his words concerning *Peuce* and *Pitys*, doth translate *Pitys*, *Picea*, although for *Peuce* he hath written *Larix*, as shall be declared. *Pliny* writeth thus; *Larix vstis radicibus non repullulat*: and the Larch tree doth not spring vp againe when the roots are burnt: the Pitch tree springeth vp againe, as it hapned in Lesbos, when the wood *Pyrthæus* was set on fire. Moreouer, the wormes *Pityocampæ* are scarce found in any tree but onely in the Pitch tree, as *Bellonius* testifieth: so that they are not rashly called *Pityocampæ*, or the wormes of the Pitch tree, although most Translators name them *Pinorum erucæ*, or the the wormes of the Pine trees: and therefore *Pitys* is firnamed by *Theophrastus*, φθειρόποιος, because wormes and magots are bred in it. But forsomuch as the name *Pitys* is common both to the tame Pine, and also to the Pitch tree, diuers of the late writers do for this cause suppose, that the Pitch tree is named by *Theophrastus*, πίτυς ἀγρία, or the wilde Pine tree. This *Picea* is named in high-Dutch, **Schwartz Tannebaum**, and **Rot Tannebaum**, and oftentimes also **Jozenholtz**; which name notwithstanding doth also agree with other plants: in English, Pitch tree: in low-Dutch, **Peck boom**.

¶ *The Temperature and Vertues.*

The leaues, barke, and fruit of the Pitch tree, are all of one nature, vertue, and operation, and of the same facultie with the Pine trees.

CHAP.

CHAP. 41. *Of the Pine Tree.*

¶ *The Kindes.*

THe Pine Tree is of two ſorts, according to *Theophraſtus*; the one ἥμερος, that is to ſay, tame, or of the garden; the other ἄγρια, or wilde: he ſaith that the Macedonians do adde a third, which is ἄκαρπος or barren, or without fruit, that vnto vs is vnknowne: the later writers haue found moe as ſhall be declared.

Pinus ſativa, ſive domeſtica.
The tame or manured Pine tree.

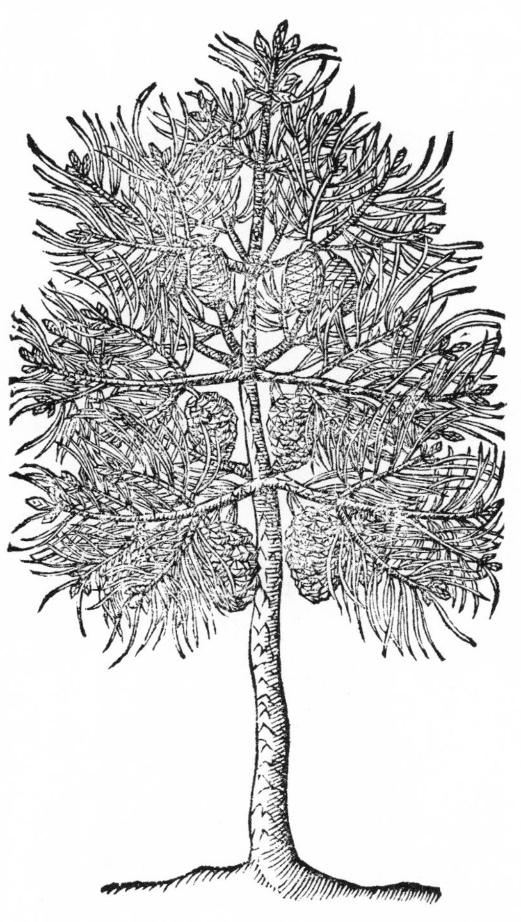

¶ *The Deſcription.*

THe Pine tree groweth high, and great in the trunk or bodie, which below is naked, but aboue it is clad with a multitude of boughes, which diuide themſelues into diuers branches, whereon are ſet ſmall leaues, verie ſtraight, narrow, ſomwhat hard and ſharp pointed: the wood or timber is hard, heauy, about the heart or middle ful of an oileous liquor, & of a reddiſh colour: the fruit or clogs are hard, great, and conſiſt of many ſound woody ſcales, vnder which are included certaine knobs, without ſhape, couered with a woodden ſhell, like ſmall nuts, wherein are white kernels, long, very ſweet, and couered with a thin skin or membrane, that eaſily is rubbed off with the fingers; which kernell is vſed in medicine.

¶ *The Place.*

This tree groweth of it ſelfe in many places of Italie, and eſpecially in the territorie of Rauenna, and in Languedock, about Marſiles, in Spain, & in other regions, as in the Eaſt countries: it is alſo cheriſhed in the gardens of pleaſure, as well in the Low-countries as England.

¶ *The Time.*

The Pine tree groweth greene both winter and Sommer: the fruit it commonly two yeres before it be ripe: wherfore it is not to be found without ripe fruit, and alſo others as yet verie ſmall, and not come to ripeneſſe.

¶ *The Names.*

It is called in Latine, *Pinus*, and *Pinus ſatiua, Vrbana*, or rather *Manſueta*: in Engliſh, tame and garden Pine: of the Macedonians and other Græcians, πεύκη ἥμερος: but the Arcadians name it πίτυς; for that which the Macedonians call πεύκη ἥμερος, the Arcadians name πίτυς, as *Theophraſtus* ſaith, and ſo doth the tame Pine in Arcadia, and about Elia change her name: and by this alteration of them it happens that the fruit or Nuts of the Pine tree found in the Cones or Apples, be named in Greeke by *Dioſcorides, Galen, Paulus*, and others, πιτυίδες, as though they ſhould terme it *Pityos fructus*, or the fruit of the Pine tree.

There is alſo another πίτυς: in Latine *Picea*, or the Pitch tree, which differeth much from the Pine tree: but *Pytis* of Arcadia differeth nothing from the Pine tree, as we haue ſaid.

The fruit or apples of theſe be called in Greeke κῶνοι, and in Latine *Coni* · notwithſtanding *Conos* is a common name to all the fruits of theſe kind of trees: they alſo be named in Latine, *Nuces pineæ*: by *Mneſitheus* in Greeke ὀστρακίδες, by *Diocles Caryſtius*, πιτυινα κάρυα, which be notwithſtanding the fruit or clogges of the tree that *Theophraſtus* nameth πεύκη, or the wilde Pine tree, as *Athenæus* ſaith. It is thought that the whole fruit is called by *Galen* in his 4. Commentarie vpon *Hipocrates* Bookes of Diet in ſharpe diſeaſes, *Strobilos*: yet in his 2. booke of the Faculties of Nouriſhments hee doth not call *Conos* or the apple by the name of *Strobilos*, but the nuts contained in it. And in like manner in his ſeuenth book of the Faculties of Simple medicines; the Pine Apple fruit, ſaith he, which they

they call *Coccalus*, and *Strobilus*, as we haue said before, that theſe are named in Greeke πιτύιδες. This apple is called in high Dutch, **Zyrbel:** in low Dutch, **Pijn appel:** in Engliſh, Pine apple, Clogge, and Cone.

¶ *The Temperature and Vertues.*

A The kernels of theſe nuts do concoct and moderately heate, being in a meane betweene cold and hot: it maketh the rough parts ſmooth; it is a remedy againſt an old cough, and long infirmities of the cheſt, being taken by it ſelfe or with hony, or elſe with ſome other licking thing.

B It cureth the Ptiſicke, and thoſe that pine and conſume away through the rottenneſſe of their lungs: it recouereth ſtrength; it nouriſheth and is reſtoratiue to the bodie.

It yeeldeth a thicke and good iuice, and nouriſheth much, yet is it not altogether eaſie of digeſtion, and therefore it is mixed with preſerues, or boyled with ſugar.

C The ſame is good for the ſtone in the Kidneies, and againſt frettings of the bladder, and ſcalding of the vrine, for it alayeth the ſharpneſſe, mitigateth paine, and gently prouoketh vrine: moreouer, it increaſeth milke and ſeed, and therefore it alſo prouoketh fleſhly luſt.

D The whole Cone or Apple being boyled with freſh Horehound, ſaith *Galen*, and afterwards boiled againe with a little hony til the decoction be come to the thickneſſe of hony, maketh an excellent medicine for the clenſing of the cheſt and lungs.

E The like thing hath *Dioſcorides*; the whole Cones, ſaith hee, which are newly gathered from the trees, broken aud boiled in ſweet wine are good for an old cough, and conſumption of the lungs, if a good draught of that liquour be drunke euery day.

F The ſcales of the Pine apple, with the barke of the tree, do ſtoppe the laske and the bloudy flixe, they prouoke vrine; and the decoction of the ſame hath the like propertie.

Chap. 42. *Of the Wilde Pine tree.*

1 *Pinus ſylueſtris.*
The wilde Pine tree.

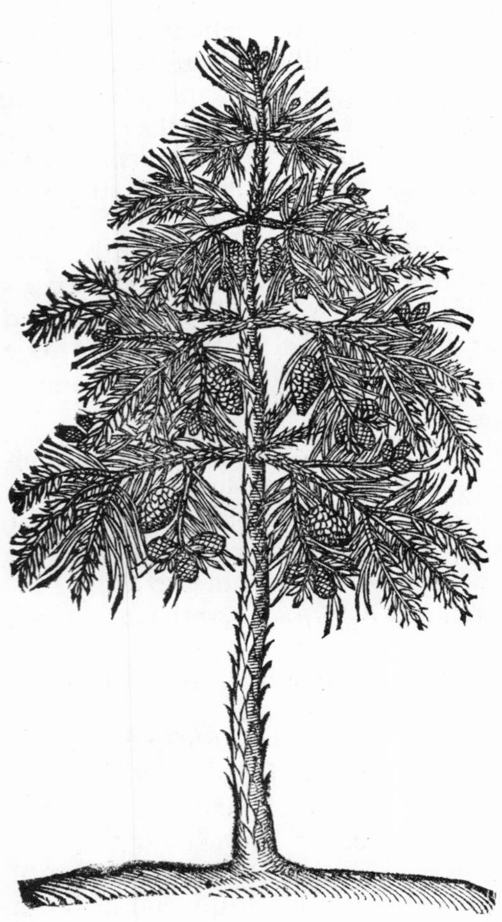

2 *Pinus ſylueſtris mugo.*
The low wilde Pine tree.

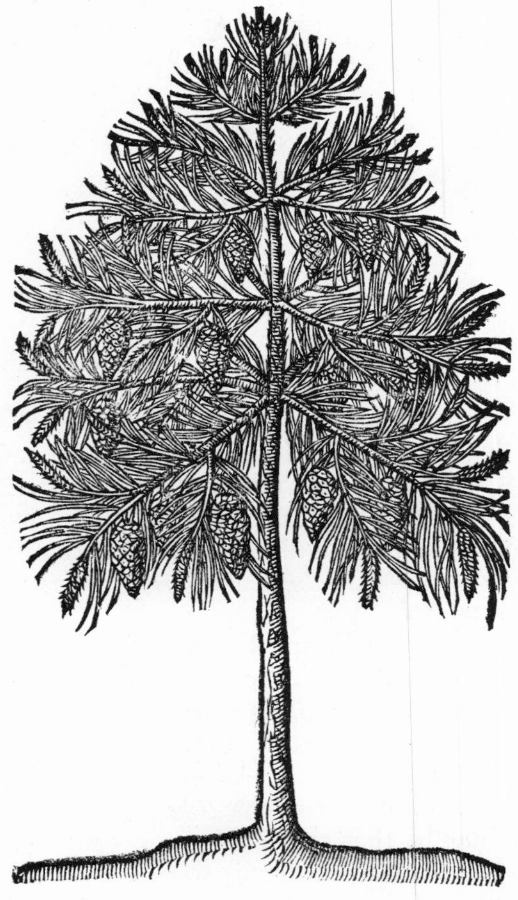

¶ *The*

¶ *The Description.*

THe first kinde of wilde Pine tree groweth very great,but not so high as the former, being the tame or manured Pine tree; the barke thereof is glib : the branches are spread abroad,beset with long sharpe pointed leaues : the fruit is somewhat like the tame Pine tree,with some Rosine therein, and sweet of smell, which doth easily open it selfe,and quickely falleth from the tree.

2 The second kinde of wilde Pine tree groweth not so high as the former, neither is the stem growing straight vp,but yet it bringeth forth many branches,long,slender,and so easie to be bent or bowed,that hereof they make Hoops for wine Hogs-heads and Tuns:the fruit of this pine is greater than the fruit of any of the other wilde Pines.

3 The third kinde of wilde Pine tree groweth straight vpright,and waxeth great and high,yet not so high as the other wilde kindes:the branches do grow like the pitch tree: the fruit is long and big,almost like the fruit of the said Pitch tree; wherein are contained smal triangled nuts,like the nuts of the Pine Apple tree,but smaller,& more brittle; in which is contained a kernell of a good taste,like the kernell of the tame Pine apple:the wood is beautiful,and sweet of smell,good to make tables and other workes of.

4 There is another wilde Pine of the mountaine, not differing from the precedent but in stature,growing for the most part like a hedge tree,wherein is the difference.

3 *Pinus sylvestris montana.*
The mountaine wilde Pine tree.

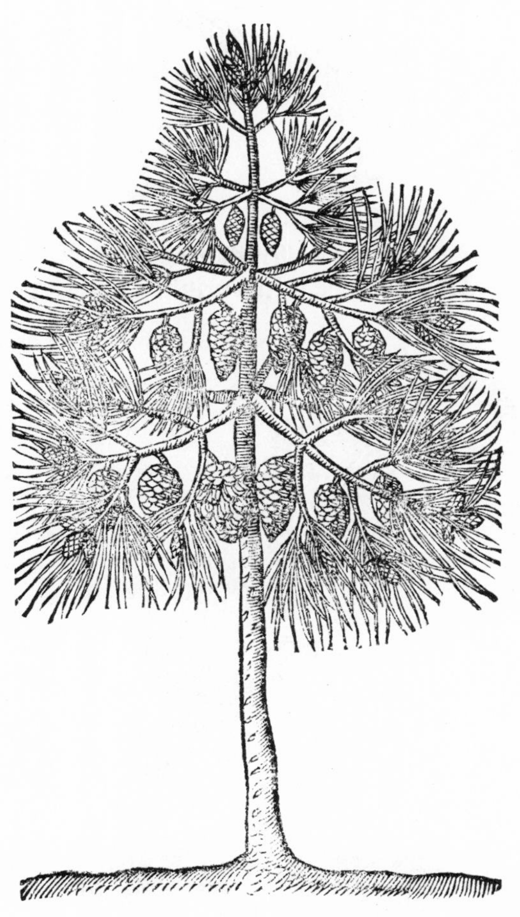

4 *Pinus montana minor.*
The smaller wilde Pine tree.

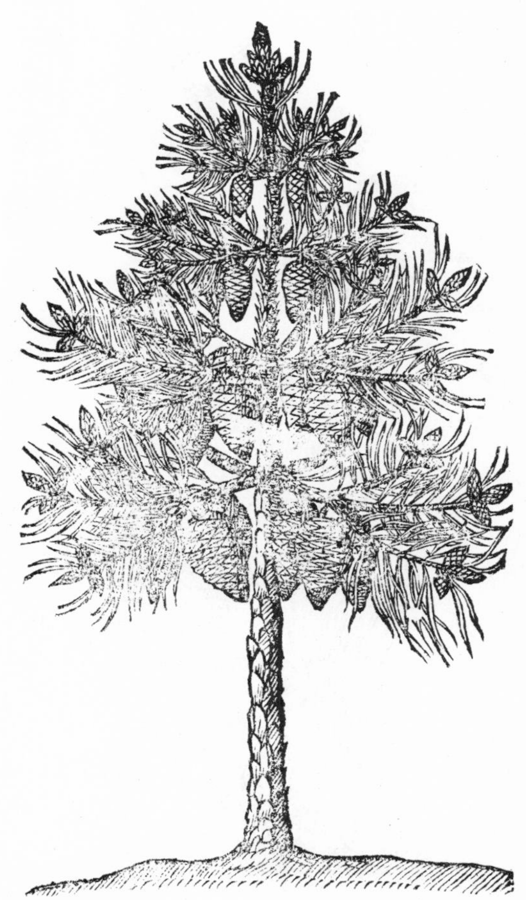

5 This kinde of Pine,called the sea Pine tree, groweth not aboue the height of two men, hauing leaues like the tame Pine tree,but shorter: the fruit is of the same forme, but longer somewhat fashioned like a Turnep : this tree yeeldeth very much Rosine. ‡ *Bauhine* iudges this all one with the third. ‡

6 The sixt kinde of wilde Pine being one of the Sea Pines,groweth like an hedge tree or shrub seldome exceeding the height of a man; with little leaues,like those of the Larch tree, but alwaies continuing with a very little cone,and fine small kernell.

7 The bastard wilde Pine tree groweth vp to a meane height ; the trunke or bodie, as also the branches

5 *Pinus maritima maior.*
The great Sea Pine tree.

6 *Pinus syluestris minor.*
The little Sea Pine Tree.

7 *Tæda sive pseudopinus.*
The bastard wilde Pine.

‡ 8 *Pinaster Austriacus.*
Dwarfe Pine with vpright Cones.

‡ 9 *Pinaster maritimus minor:*
Dwarfe Sea Pine.

ches & leaues are like vnto those of the manured Pine tree: the onely difference is, that some yeares it resembleth the Pine it selfe; and the other yeares as a wilde hedge tree, varying often, as nature listeth to play and sport her selfe amongst her delights, with other plants of lesse moment: the timber is soft, and not fit for building, but is of the substance of our Birch tree: the fruit is like those of the other wilde Pines, whereof this is a kinde.

‡ 8 This dwarfe Austrian Pine exceeds not the height of a man, but immediately from the root is diuided and spread abroad into tough, bending, pretty thicke branches, couered ouer with a rough barke: the leaues, as in the former, come two out of one hose, thicker, shorter, blunter pointed, and more greene than the former: the cones or clogs are but small, yet round, and compact, and hang not dowewards, but stand vpright: the root is tough and wooddie like other plants of this kinde: It growes on the Austrian and Styrian Alpes. *Clusius* sets it forth by the name of *Pinaster* 4. *Austriacus*.

9 This other Dwarfe is of the same height with the former, with such tough and bending branches, which are neither so thick nor clad with so rough a barke, nor so much spread. The leaues also are smaller, and not vnlike those of the Larix tree, but not so soft, nor falling euery yeare as they do. The cones are little and slender, the kernell small, blackish, and winged as the rest. *Clusius* found this onely in some few places of the kingdome of Murcia in Spaine, wherefore he calls it, *Pinaster* 3. *Hispanicus*. *Dodonæus* calls it *Pinus maritima minor*. ‡

¶ *The Place.*

These wilde Pines doe grow vpon the cold mountaines of Liuonia, Polonia, Noruegia, and Russia, especially vpon the Island called Holland within the Sownd, beyond Denmarke, and in the woods by Narua, vpon the Liefeland shore, and all the tract of the way, being a thousand Wersts, (each Werst containing three quarters of an English mile) from Narua vnto Moscouia, where I haue seene them grow in infinite numbers.

¶ *The Time.*

The fruit of these Pine trees is ripe in the end of September: out of all these issueth forth a white and sweet smelling Rosine: they are also changed into *Teda*, and out of these is boiled through the force of the fire, a blacke Pitch: the Pitch tree and the Larch tree be also sometimes changed into *Teda*; yet very seldome, for *Teda* is a proper and peculiar infirmitie of the wilde Pine tree. A tree is said to be changed into *Teda*, when not onely the heart of it, but also the rest of the substance is turned into fatnesse.

¶ *The Names.*

All these are called in Greeke πεῦκαι ἄγριαι: and in Latine *Syluestres Pini:* of *Pliny*, *Pinastri*: *Pinaster*, saith he in his 16. booke. 10 chapter, is nothing else but *Pinus syluestris*, or the wild Pine tree, of a lesser height, and ful of boughes from the middle, as the tame Pine tree in the top, (most of the copies haue falsely) of a maruellous height: they are far deceiued who thinke that the Pine tree is called in Greeke πίτυς, besides the tame Pine which notwithstanding is so called not of all men, but onely of the Arcadians (as we haue said before) πίτυς, all men do name the wilde πεύκη; and therefore *Teda*, or the Torch Pine, hereof is said to be in Latine not *Picea*, but *Pinea*, that is, not the Pitch-tree, but the Pine tree, as *Ouid* doth planly testifie in his Heroicall Epistles;

Vt

Vt vidi, vt perij, nec notis ignibus arsi,
Ardet vt ad magnos Pinea Teda deos.

Also in *Fastorum* 4.

Illic accendit geminas pro lampade Pinus:
Hinc Cereris sacris nunc quoque Teda datur.

The same doth *Virgill* also signifie in the seuenth of his Æneid.

Ipsa inter medias, flagrantem feruida Pinum
Sustinet.————————

Where in stead of *Flagrantem Pinum*, *Seruius* admonisheth vs to vnderstand *Teda Pinea*. *Catullus* also consenteth with them in the marriage song of *Iulia* and *Mallius*.

———— *Manu*
Pineam quate tedam.

And *Prudentius* in *Hymno Cerei Paschalis.*

Seu Pinus piceam fert alimoniam.

Moreouer, the herbe *Peucedanos*, or Horestrong, so named of the likenesse of πεύκης, is called also in Latine *Pinastellum*, of *Pinus* the Pine tree: all which things do euidently declare that πεύκη is called in Latine not *Picea*, but *Pinus*.

The first of these wilde kindes may be *Idæa Theophrasti*, or *Theophrastus* his Pine tree, growing on mount Ida, if the apple which is shorter were longer: for he nameth two kindes of wilde Pines, the one of mount Ida, and the other the Sea Pine with the round fruit: but we hold the contrarie, for the fruit or apple of the wilde mountaine Pine is shorter, and that of the Sea Pine longer. This may more truly be *Macedonum mas*, or the Macedonians male Pine, for they make two sorts of wilde Pines, the male and the female, and the male more writhed and harder to be wrought vpon, and the female more easie; but the wood of this is more writhed, and not so much in request for workes, as the other. and therefore it seemeth to be the male. This wilde Pine tree is called in high Dutch, **Hartzbaum**, and **Wilder Hartzbaum**: in Gallia Celtica, *Elvo Aleno*: and in Spanish, *Pino Carax*.

The second wild Pine tree is named commonly of the Italians *Tridentinis*, and *Ananiensibus*, *Cembro*, and *Cirmolo*, it seemeth to differ nothing at all from the Macedonians wilde female Pine, for the wood is easie to be wrought on, and serueth for diuers and sundry workes.

The third they call *Mugo*: this may be named not without cause χαμαιπεύκη, that is to say, *Humilis Pinus*, or Dwarfe Pine: yet doth it differ from *Chamæpeuce* the Herbe called in English, Ground Pine.

The fourth wilde Pine is named in Greeke παραλία πεύκη in Latine, *Maritima*, and *Marina Pinus*: in English, Sea Pine.

That which the Latines call *Teda*, is named in Greeke δᾷς, and δαδίον: in high Dutch, **Kynholtz**: it may be termed in English, Torch-pine.

Pliny is deceiued, in that he supposeth the Torch Pine to bee a tree by it selfe, and maketh it the sixth kinde of Cone-tree; as likewise he erreth in taking *Larix*, the Larch tree, for πεύκη, the Pine Tree. And as *Dioscorides* maketh so little difference as scarse any, betweene πεύκη and πίτυς, and supposeth them to be both of one kinde, so likewise he setteth downe faculties common to them both.

¶ *The Temperature and Vertues.*

A The barke of them both, saith he, doth binde; being beaten and applied it cureth Merigals, and also shallow vlcers and burnings if it be layed on with Litharge and fine Frankincense.

B With the Cerote of Myrtles it healeth vlcers in tender bodies: being beaten with Copperas it staieth tetters, and creeping vlcers: it draweth away the birth and after birth, if it be taken vnder in a fume: being drunke it stoppeth the belly, and prouoketh vrine.

C *Galen* hath almost the same things, but he saith, that the barke of the Pine tree is more temperate than that of the Pitch tree; the leaues stamped take away hot swellings and sores that come thereof.

D Being stamped and boyled in vineger, they asswage the paine of the teeth, if they be washed with this decoction hot: the same be also good for those that haue bad liuers, being drunke with water or mead.

E Of the same operation is likewise the barke of the pine nuts; but *Galen* affirmeth that the Cone or or apple, although it seeme to be like these is notwithstanding of lesser force, insomuch as it cannot effectually performe any of the aforesaid vertues, but hath in it a certaine biting qualitie, which hurteth.

The Torch Pine cut into small pieces and boiled in vineger, is a remedy likewise against the tooth-ache if the teeth be washed with the decoction. F

Of this there is made a profitable spather or slice to be vsed in making of compound plaisters and pessaries that ease paine. G

Of the smoke of this is made a blacke which serueth to make inke of, and for eating sores in the corners of eies, and against the falling away of the haire of the eie lids, and for watering and bleere eies, as *Dioscorides* teacheth. H

Of Rosins.

¶ *The Kindes.*

1 OVt of the Pine trees, especially of the wilde kinds, there issueth forth a liquid, whitish, and sweet smelling Rosin, and that many times by it selfe; but more plentifully either out of the cut and broken boughes, or forth of the body when the tree commeth to be a Torch Pine.

2 there issueth also forth of the crackes and chinkes of the barke, or out of the cut boughes, a certaine dry Rosin, and that forth of the Pine Tree or Firre Tree.

There is likewise found a certaine congealed Rosin vpon the cones or apples.

It is called in Latine, *Resina*: in Greeke, ῥητίνη: in high Dutch, **Hartz**: in low Dutch, **Herst**: in Italian, *Ragia*: in Spanish, *Resina*: in English, Rosin.

The first is named in Latine, *Liquida Resina*: in Greeke, ῥητίνη ὑγρά, and of diuers, αὐτόῤῥυτος, that it to say, issueth out of it selfe: of the Lacedemonians, πρωτόῤῥυτος, or *Primiflua*, the first flowing Rosin: and in Cicilia, καπνέλαιον, as *Galen* writeth in his third booke of medicines according to the kindes: in shops *Resina Pini*, or Rosin of the Pine tree, and common Rosin. It hapneth oftentimes through the negligent and carelesse gathering thereof, that certaine small pieces of wood, and little stones be found mixed with it: this kinde of Rosin *Galen* surnameth συγκεχυμένη, as though he should say, confused, which being melted and clensed from the drosse becommeth hard and brittle.

The like hapneth also to another liquid Rosin, which after it is melted, boiled, and cooled againe, is hard and brittle, and may likewise be beaten, ground, and searced; and this Rosin is named in Greeke φρυκτή: in Latine, *Fricta*, and many times *Colophonia*, in Greeke, κολοφωνία: which name is vsed among the Apothecaries, and may stand for an English name; for *Galen* in his third booke of Medicines according to their kindes saith, that it is called *Fricta*, and of some *Colophonia*. that, saith he, is the driest Rosin of all, which some call *Fricta*, others *Colophonia*: because in times past, as *Dioscorides* writeth it was fetched from *Colophon*, this being yellow or blacke in comparison of the rest, is white when it is beaten: *Pliny* in his 14 booke, 20. chapter.

The second Rosin is named in Greeke ῥητίνη ξηρά, specially that of the Pitch tree without fatnesse, and that soone waxeth dry, which *Galen* in his 6. booke of Medicines according to the kindes, calleth properly φύσημα πιτύινον: that which in Asia is made of the Pitch tree being very white, is called *Spagas*, as *Pliny* testifieth.

The third is called in Greeke ῥητίνη στροβιλίνη: the same is also named φύσημα στροβίλινον: this is vnknowne in shops. Yet there is to be sould a certaine dry Rosin, but the same is compounded of the Rosins of the Pine tree, of the cones or clogs, and of the Firre tree mixed altogether, which they call *Garipot*· this is vsed in perfumes in stead of Frankincense, from which notwithstanding it farre differeth.

¶ *The Temperature and Vertues.*

All the Rosins are hot and dry, but not all after one manner: for there is a difference among them: they which be sharper and more biting, are hotter, as that which commeth of the cones, being of Rosins the hottest, because it is also the sharpest: the Rosin of the Pitch Tree is not so much biting, and therefore not so hot: the Rosin of the Firre tree is in a meane between them both; the liquid Rosin of the Pine is moister, comming neere to the qualitie and facultie of the Larch Rosin. A

The Rosins which are burnt or dried, as *Dioscorides* testifieth, are profitable in plaisters, and compositions that ease wearisomenesse; for they do not onely supple or mollifie, but also by reason of the thinnesse of their parts and drynesse, they digest: therefore they both mollifie and wast away swellings, and through the same facultie they cure wearisomnesse, being vsed in compound medicines for that purpose. B

The liquid Rosins are very fitly mixed in ointments, commended for the healing vp of greene wounds, for they both bring to suppuration, and do also glue and vnite them together. C

D Moreouer, there is gathered out from the Rosins as from Frankencense, a congealed smoke, called in Latine *Fuligo*; in Greeke λιγνύς: and in English, Blacke, which serueth for medicines that beautifie the eie lids, and cure the fretting sores of the corners of the eies, and also watering eies, for it drieth without biting.

E There is made hereof saith *Dioscorides*, writing inke, but in our age not that which we write withall, but the same which serueth for Printers to print their bookes with, that is to say, of this blacke, or congealed smoke, and other things added.

Of Pitch and Tar.

The manner of drawing forth of Pitch.

Out of the fattest wood of the Pine tree changed into the Torch Pine, is drawne Pitch by force of fire. A place must be paued with stone, or some other hard matter, a little higher in the middle, about which there must also be made gutters, into which the liquor shall fall; then out from them other gutters are to be drawne, by which it may be receiued; being receiued, it is put into barrels. The place being thus prepared, the clouen wood of the Torch Pine must be set vpright; then must it be couered with a great number of Fir and Pitch boughes, and on euery part all about with much lome and earth: and great heed must be taken, least there be any cleft or chinke remaining, onely a whole left in the top of the furnace, thorow which the fire may be put in, and the flame and smoke may passe out: when the fire burneth the Pitch runneth forth, first the thin, and then the thicker.

This liquor is called in Greeke πίσσα: in Latine, *Pix*: in English, Pitch, and the moisture, euen the same that first runneth is named of *Plinie* in his 16. booke, 11. chapter, *Cedria*: There is boyled in Europe, saith he, from the Torch Pine a liquid Pitch vsed about ships, and seruing for many other purposes; the wood being clouen is burned with fire, and set round about the furnaces on euery side, after the manner of making Charcoles: the first liquor runneth thorow the gutter like water: (this in Syria is called *Cedrium*, which is of so great vertue, as in Ægypt the bodies of dead men are preserued, being all couered ouer with it) the liquor following being now thicker, is made Pitch. But *Dioscorides* writeth, that *Cedria* is gathered of the great Cedar tree, and nameth the liquor drawne out of the Torch tree by force of fire, πίσσα ὑγρά: this is, that which the Latines call *Pix liquida* the Italians, *Pece liquida*: in high Dutch, **Weich bach**: in low Dutch, **Teer**: in French, *Poix foudire*: in Spanish, *Pex liquida*: certaine Apothecaries, *Kitran*: and we in English, Tar.

And of this when it is boiled is made a harder Pitch: this is named in Greeke ξηρὰ πίσσα: in Latine, *Arida*, or *sicca Pix*: of diuers, παλίμπισσα: as though they should say, *Iterata Pix*, or Pitch iterated: because it is boiled the second time. A certaine kinde hereof being made clammie or glewing is named βοσκάς: in shops, *Pix nanalis*, or Ship Pitch: in high Dutch, **Bach**: in low Dutch, **Steenpeck**: in Italian, *Pece secca*: in French, *Poix seche*: in Spanish, *Pez seca*: in English, Stone Pitch.

¶ *The Temperature and Vertues.*

A Pitch is hot and dry, Tarre is hotter, and stone pitch more drying, as *Galen* writeth. Tar is good against inflammations of the almonds of the throte, and the uvula, and likewise the Squincie, being outwardly applied.

B It is a remedie for mattering eares with oile of Roses: it healeth the bitings of Serpents, if it be beaten with salt and applied.

C With an equall portion of wax it taketh away foule ilfauoured nailes, it wasteth away swellings of the kernels, and hard swellings of the mother and fundament.

D With barly meale and a boies vrine it consumeth χοιράδας, or the Kings euill: it staieth eating vlcers, if it be laid vnto them with brimstone, and the barke of the Pitch Tree, or with branne.

E If it be mixed with fine Frankincense, and a cerote made thereof, it healeth chops of the fundament and feet.

F Stone Pitch doth mollifie and soften hard swellings: it ripens and maketh matter, and wasteth away hard swellings and inflammations of kernels: it filleth vp hollow vlcers, and is fitly mixed with wound medicines.

G What vertue Tarre hath when it is inwardly taken we may reade in *Dioscorides* and *Galen*, but we set downe nothing thereof, for that no man in our age will easily vouchsafe the taking.

H There is also made of Pitch a congealed smoke or blacke, which serueth for the same purposes, as that of the Rosins doth.

Chap.

CHAP. 43. *Of the Firre or Deale Tree.*

¶ *The Description.*

1 THe Firre tree groweth very high and great, hauing his leaues euer greene; his trunke or body smooth, euen and straight, without ioints or knots, vntill it hath gotten branches; which are many and very faire, beset with leaues, not much vnlike the leaues of the Ewe tree, but smaller: among which come forth floures vpon the taller trees, growing at the bottomes of the leaues like little catkins, as you may see them exprest in a branch apart by themselues: the fruit is like vnto the Pine Apple, but smaller and narrower, hanging downe as the Pine Apple: the timber hereof excelleth all other timber for the masting of ships, posts, rails, deale boords, and sundry other purposes.

1 *Abies.*
The Firre tree.

‡ 2 *Abies mas.*
The male Firre tree.

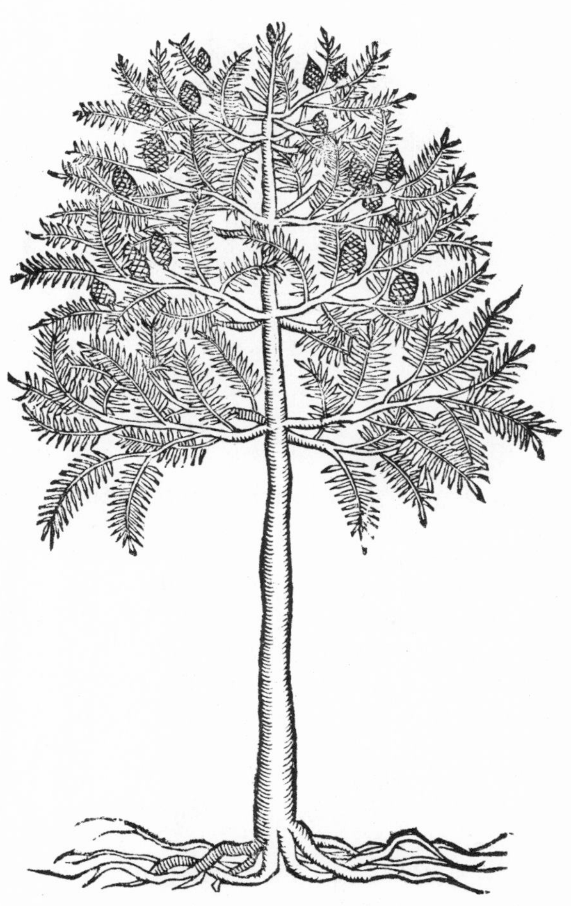

2 There is another kinde of Firre tree, which is likewise a very high and tall tree, and higher than the Pine: the body of it is straight without knots below, waxing smaller and smaller euen to the very top: about which it sendeth forth boughes, foure together out of one and the selfe same part of the body, placed one against another, in manner of a crosse, growing forth of the foure sides of the body, and obseruing the same order euen to the very top: out of these boughes grow others also, but by two and two, one placed right against another, out of the sides, which bend downwards when the other beare vpwards: the leaues compasse the boughes round about, and the branches thereof: they be long, round, and blunt pointed, narrower, and much whiter than those of the Pitch tree, that is to say, of a light greene, and in a manner of a white colour: the cones or clogs be long, and longer than any others of the cone trees, they consist of a multitude of soft scales, they hang downe from the end of the twigs, and doe not easily fall downe, but remaine on the tree a very long time: the kernels in these are small, not greater than the kernels of the Cherrie stone, with a thinne skin growing on the one side, very like almost to the wings of Bees, or great Flies: the timber or substance of the wood is white, and clad with many coats, like the head of an Onion.

‡ *Abietis ramus cum julis.*
A branch with Catkins or floures.

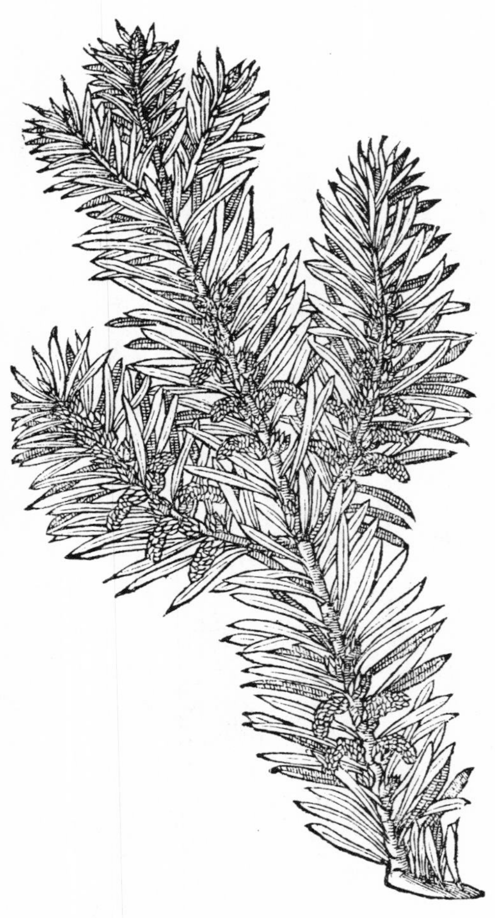

¶ *The Place.*

The Firre trees grow vpon high mountains, in many woods of Germany and Bohemia, in which it continueth alwaies greene; it is found alſo on hils in Italy, France, & other countries; it commeth downe oftentimes into the vallies: they are found likewiſe in Pruſe, Pomerania, Liefeland, Ruſſia, & eſpecially in Norway, where I haue ſeene the goodlieſt trees in the world of this kinde, growing vpon the rockie and craggie mountaines, almoſt without any earth about them, or any other thing, ſauing a little moſſe about the roots, which thruſt themſelues here and there into the chinkes and cranies of the rockes, and therefore are eaſily caſt downe with any extreme gale of winde. I haue ſeen theſe trees growing in Cheſhire, Staffordſhire, and Lancaſhire, where they grew in great plenty, as is reported, before *Noahs* floud: but then being ouerturned and ouerwhelmed haue lien ſince in the moſſes and waterie mooriſh grounds very freſh and ſound vntill this day, & ſo full of a reſinous ſubſtance, that they burne like a Torch or Linke, and the inhabitants of thoſe countries do call it Fir-wood, and Fire-wood vnto this day: out of this tree iſſueth the roſin called *Thus*, in Engliſh, Frankinſence: but from the young Fir trees proceedeth an excellent cleare and liquid Roſin, in taſte like to the peelings or outward rinde of the Pomecitron.

¶ *The Time.*

The time of the Fir tree agreeth with the Pine trees.

¶ *The Names.*

The tree is called in Latine *Abies*: in Greeke, ἐλάτη: amongſt the Græcians of our time the ſame name remaineth whole and vncorrupt: it is called in high Dutch, **Weiſz Thannen**, and **Weiſz Thannen baum**: in Low Dutch, **Witte Dennen boom**, or **Abel=boom**, and **Maſt=boom**: in Italian, *Abete*: in Spaniſh, *Abeto*: in Engliſh, Firre-tree, Maſt-tree, and Deale-tree. The firſt is called in rench, *du Sap*, or *Sapin*: the other is *Suiffe*.

The liquid roſin which is taken forth of the barke of the young Firre-trees, is called in Greeke δάκρυον τῆς ἐλάτης: in Latine, *Lachryma abietis*, and *Lachrvma abiegna*: in the ſhops of Germany, as alſo of England, *Terebinthina Veneta*, or Venice Turpentine: in Italian, *Lagrimo*: diuers do thinke that *Dioſcorides* calleth it ἐλαιώδης ῥητίνη, *Oleaſa Reſina*, or oile Roſin; but oile Roſin is the ſame that *Pix liquida*, or Tar is.

Arida Abietum Reſina, or drie Roſin of the Fir trees, is rightly called in Greeke ῥητίνη ἐλατίνη, and in Latine, *Abiegna Reſina*: it hath a ſweet ſmell, and is oftentimes vſed among other perfumes in ſtead of Frankincenſe.

¶ *The Temperature.*

The barke, fruit, and gums of the Fir tree, are of the nature of the Pitch tree and his gums.

¶ *The Vertues.*

A The liquid Roſin of the Fir tree called Turpentine, looſeth the belly, driueth forth hot cholerick humours, clenſeth and mundifieth the kidnies, prouoketh vrine, and driueth forth the ſtone and grauell.

B The ſame taken with Sugar and the pouder of Nutmegs, cureth the ſtrangurie, ſtaieth the Gonorrhœa or the inuoluntary iſſue of mans nature, called the running of the rains, and the white flux in women.

C It is very profitable for all green and freſh wounds, eſpecially the wounds of the head: for it healeth and clenſeth mightily, eſpecially if it be waſhed in Plantaine water, and afterward in Roſe water, the yolke of an egge put thereto, with the pouders of *Olibanum* and Maſticke finely ſearced, adding thereto a little Saffron.

CHAP.

Chap. 44. *Of the Larch Tree.*

¶ *The Description.*

1 THe Larch is a tree of no ſmall height,with a body growing ſtraight vp:the bark wherof in the nether part beneath the boughes is thicke, rugged,and full of chinks;which being cut in ſunder is red within,and in the other part aboue ſmooth, ſlipperie, ſomething white without : it bringeth forth many boughes diuided into other leſſer branches,which be tough and pliable. The leaues are ſmall,and cut into many iags, growing in cluſters thicke together like taſſels,which fall away at the approch of Winter : the floures or rather the firſt ſhewes of the cones or fruit be round,and grow out of the tendereſt boughes, being at the length of a braue red purple colour : the cones be ſmall, and like almoſt in bigneſſe to thoſe of the Cypreſſe tree, but longer,and made vp of a multitude of thin ſcales like leaues : vnder which lie ſmall ſeeds, hauing a thin velme growing on them very like to the wings of Bees and waſps :the ſubſtance of the wood is very hard,of colour,eſpecially that in the midſt,ſomewhat red, and very profitable for workes of long continuance.

1 *Laricis ramulus.*
A branch of the Larch tree.

2 *Larix cum Agarico ſuo.*
The Larch tree with his Agarick.

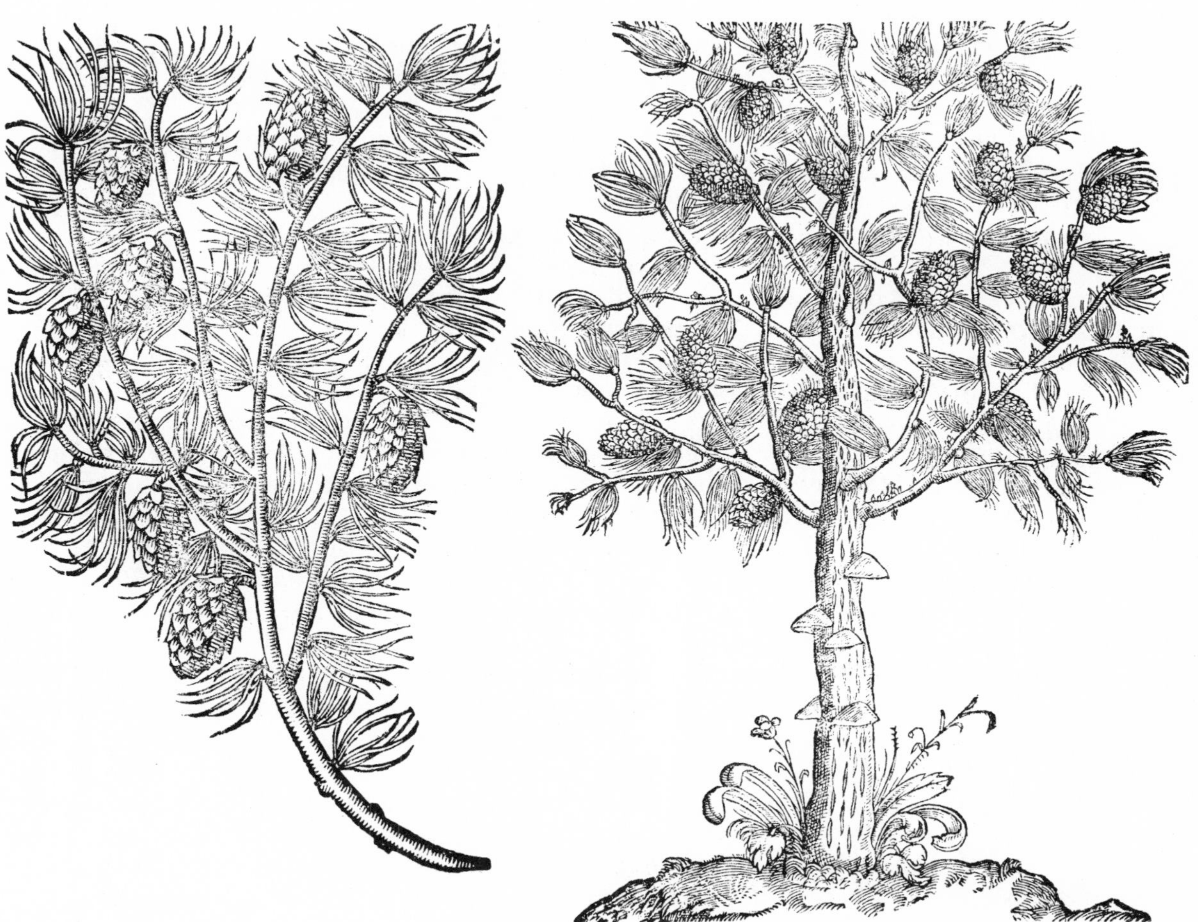

It is not true that the wood of the Larch tree tree cannot be ſet on fire,as *Vitruvius* reporteth of the caſtle made of Larch wood,which *Cæſar* beſieged;for it burneth in chimneies,and is turned into coles,which are very profitable for Smithes, as *Matthiolus* writeth.

There is alſo gathered of the Larch tree a liquid Roſin, very like in colour and ſubſtance to the whiter hony,as that of Athens or of Spaine,which notwithſtanding iſſueth not forth of it ſelfe,but runneth out of the ſtocke of the tree,when it hath been bored euen to the heart with a great and long auger and wimble.

Galen writeth,that there be after a ſort two kindes hereof,in his 4. booke of Medicines, according to the kinds, one like vnto Turpentine,the other more ſharper than this, hotter,more liquid of of a ſtronger ſmell,and in taſte bitterer and hotter : but the later is thought not to be the Roſine of the Larch,but of the Fir-tree, which *Galen* becauſe it is after a ſort like in ſubſtance,might haue taken for that of the Larch tree.

There groweth alſo vpon the Larch tree a kinde of Muſhrum or excreſcence,not ſuch as is vpon other trees,but whiter,ſofter,more looſe and ſpungie than any other of the Muſhrums, and good for medicine,which beareth the name of *Agaricus*,or Agaricke : I find that *Pliny* ſuppoſeth all the Maſticke trees,and thoſe that beare Galls, do bring forth this *Agaricum*: wherein he was ſomewhat deceiued,and eſpecially in that he took *Glandifera* for *Conifera*,that is,thoſe trees which beare maſt or Acornes,for the Pine apple trees : but among all the trees that beare *Agaricus*, the Larch is the chiefe,and bringeth moſt plenty of Agarick.

¶ *The Place.*

The Larch tree groweth not in Greece,or in Macedon, but chiefely vpon the Alpes of Italy,not far from Trent, hard by the riuers *Benacus* and *Padus*; and alſo in other places of the ſame mountaines : it is likewiſe found on hils in Morauia, which in times paſt was called the countrey of the Marcomans : *Fuchſius* writeth,that it groweth alſo in Sileſia : others,in Luſatia, in the borders of Poland : it alſo groweth plentifully in the woods of Gallia Ceſalpina.

Pliny hath ſaid ſomewhat hereof,contradicting the writings of others,in his 16 book,8 chapter, where he ſaith,that ſpecially the Acorne trees of France do beare Agaricke,and not only the acorn trees,but the Cone trees alſo;among which,ſaith he,the Larch tree is the chiefe that bringeth forth Agaricke,and that not onely in Gallia,which now is called France, but rather in Lumbardy and Piemont in Italy,where there be whole woods of Larch trees,although they be found in ſome ſmal quantitie in other countries.

The beſt Agarick is that which is whiteſt, very looſe and ſpungie, which may eaſily be broken, and is light,and in the firſt taſte ſweet,hard,and well compact : that which is heauy, blackiſh, and containing in it little threds as it were of ſinewes,is counted pernicious and deadly.

¶ *The Time.*

Of all the Cone trees onely the Larch tree is found to be without leaues in the Winter : in the Spring grow freſh leaues out of the ſame knobs, from which the former did fall. The cones are to be gathered before winter,ſo ſoone as the leaues are gone : but after the ſcales are looſed and opened,the ſeeds drop away : the Roſine muſt be gathered in the Sommer moneths.

¶ *The Names.*

This tree is called in Greeke λάριξ : in Latine alſo *Larix*,in Italian and Spaniſh, *Larice* : in high Dutch,**Lerchenbaum** : in low Dutch **Lorkenboom** : in French,*Meleſe* : in Engliſh,Larch tree,and of ſome Larix tree.

The liquid Roſin is named by *Galen* alſo λάριξ : the Latines call it *Reſina Larigna*, or *Reſina Laricea*, Larch Roſin : the Italians,*Larga* : the Apothecaries,*Terebinthina*,or Turpentine,and it is ſold and alſo mixed in medicines in ſtead thereof : neither is that a thing newly done; for *Galen* likewiſe in his time reporteth,that the Druggers ſold the Larch Roſine in ſtead of Turpentine : and this may bee done without errour; for *Galen* himſelfe in one place vſeth Larch Roſin for Turpentine;and in another, Turpentine for Larch Roſine,in his booke of medicines according to the kindes.

The Agaricke is alſo called in Greeke ἀγάρικον and ἀγάρικος : in Latine, *Agaricum* and *Agaricus*,and ſo likewiſe in ſhops : the Italians,Spaniards, and other nations do imitate the Greeke word ; and in Engliſh we call it Agaricke.

¶ *The Temperature and Vertues.*

A The leaues,barke,fruit and kernell,are of temperature like vnto the Pine,but not ſo ſtrong.

B The Larch Roſin is of a moiſter temperature than all the reſt of the Roſines,and is withall without ſharpneſſe or biting, much like to the right Turpentine, and is fitly mixed with medicines which perfectly cure vlcers and greene wounds.

C All Roſins,ſaith *Galen*,that haue this kinde of moiſture and clammineſſe ioined with them,do as it were binde together and vnite dry medicines,and becauſe they haue no euident biting qualitie, they doe moiſten the vlcers nothing at all : therefore diuers haue very well mixed with ſuch compound medicines either Turpentine Roſin,or Larch Roſin:thus far *Galen*. Moreouer,Larch Roſin performeth all ſuch things that the Turpentine Roſin doth, vnto which, as we haue ſaid, it is much like in temperature,which thing likewiſe *Galen* himſelfe affirmeth.

D Agaricke is hot in the firſt degree and dry in the ſecond,according to the old writers. It cutteth, maketh thin,clenſeth,taketh away obſtructions or ſtoppings of the intrailes, and purgeth alſo by ſtoole.

E Agaricke cureth the yellow iaundice proceeding of obſtructions,and is a ſure remedie for cold ſhakings,which are cauſed of thicke and cold humors.

F The ſame being inwardly taken and outwardly applied, is good for thoſe that are bit of venomous beaſts which hurt with their cold poiſon.

It prouoketh vrine, and bringeth downe the menſes : it maketh the body well co loured, driueth forth wormes, cureth agues, eſpecially quotidians and wandring feuers,and others that are of long continuance, if it be mixed with fit things that ſerue for the diſeaſe : and theſe things it performes by drawing forth and purging away groſſe, cold, and flegmaticke humors, which cauſe the diſeaſes. G

From a dram weight,or a dram and a halfe, to two,it is giuen at once in ſubſtance or in pouder: the weight of it in an infuſion or decoction is from two drams to fiue. H

But it purgeth ſlowly, and doth ſomewhat trouble the ſtomacke; and therefore it is appointed that Ginger ſhould be mixed with it, or wilde Carrot ſeed, or Louage ſeed, or Sal gem,in Latine, *Sal foſſilis*. I

Galen, as *Meſue* reporteth, gaue it with wine wherein Ginger was infuſed : ſome vſe to giue it with Oxymel, otherwiſe called ſyrrup of vineger, which is the ſafeſt way of all. K

Agaricke is good againſt the paines and ſwimming in the head,or the falling Euill,being taken with ſyrrup of vineger. L

It is good againſt the ſhortneſſe of breath, called *Aſthma*, the inueterate cough of the lungs, the ptyſicke, conſumption, and thoſe that ſpet bloud : it comforteth the weake and feeble ſtomacke, cauſeth good digeſtion, and is good againſt wormes. M

Chap. 45. *Of the Cypreſſe tree.*

Cupreſſus ſatiua & ſylueſtris.
The Garden and wild Cypreſſe tree.

¶ *The Deſcription.*

THe tame or manured Cypreſſe tree hath a long thicke and ſtraight body; whereupon many ſlender branches do grow, which do not ſpred abroad like the branches of other trees, but grow vp alongſt the body, yet not touching the top : they grow after the faſhion of a ſteeple,broad below, and narrow toward the top : the ſubſtance of the wood is hard, ſound,well compact, ſweet of ſmell, and ſomewhat yellow, almoſt like the yellow Saunders,but not altogether ſo yellow,neither

ther doth it rot nor wax old, nor cleaueth or choppeth it ſelf. The leaues are long, round like thoſe of Tamariske, but fuller of ſubſtance. The fruit or nuts do hang vpon the boughes, being in manner like to thoſe of the Larch tree, but yet thicker and more cloſely compact: which being ripe do of themſelues part in ſunder, and then falleth the ſeed, which is ſhaken out with the winde: the ſame is ſmall, flat, very thin, of a ſwart ill fauoured colour, which is pleaſant to Ants or Piſmires, and ſerueth them for food.

Of this diuers make two kindes, the female and the male; the female barren, and the male fruitfull. *Theophraſtus* reporteth, that diuers affirme the male to come of the female. The Cypreſſe yeelds forth a certaine liquid Roſin, like in ſubſtance to that of the Larch tree, but in taſte maruellous ſharpe and biting.

The wilde Cypreſſe, as *Theophraſtus* writeth, is an high tree, and alwaies greene, ſo like to the other Cypreſſe, as it ſeemeth to be the ſame both in boughes, body, leaues, and fruit, rather than a certaine wilde Cypreſſe: the matter or ſubſtance of the wood is ſound, of a ſweet ſmell, like that of the Cedar tree, which rotteth not: there is nothing ſo criſped as the root, and therefore they vſe to make precious and coſtly workes thereof.

‡ I know no difference betweene the wilde and tame Cypreſſe of our Author, but in the handſomneſſe of their growth, which is helped ſomewhat by art. ‡

¶ *The Place.*

The tame and manured Cypreſſe groweth in hot countries, as in Candy, Lycia, Rhodes, and alſo in the territorie of Cyrene: it is reported to be likewiſe found on the hills belonging to Mount Ida, and on the hills called *Leuci*, that is to ſay white, the tops whereof be alwaies couered with ſnow. *Bellonius* denieth it to be found vpon the tops of theſe hills, but in the bottoms on the rough parts and ridges of the hills: it groweth likewiſe in diuers places of England where it hath beene planted, as at Sion a place neere London, ſometime a houſe of Nunnes: it groweth alſo at Greenwich, and at other places, and likewiſe at Hampſted in the garden of M^r^. *Wade*, one of the Clerkes of her Maieſties priuy Councell.

The wilde kinde of Cypreſſe tree groweth hard by *Ammons* Temple, and in other parts of the countrey of Cyrene vpon the tops of mountaines, and in extreme cold countries. *Bellonius* affirmeth, that there is found a certaine wilde Cypreſſe alſo in Candy, which is not ſo high as other Cypreſſe trees, nor groweth ſharpe toward the top, but is lower, and hath his boughes ſpred flat, round about in compaſſe: he ſaith the body thereof is alſo thicke: but whether this be *Thya*, of which *Theophraſtus* and *Pliny* make mention, we leaue it to conſideration.

¶ *The Time.*

The tame Cypres tree is alwaies greene; the fruit may be gathered thrice a yeare, in Ianuarie, May, and September, and therefore it is ſyrnamed *Trifera*.

The wilde Cypres tree is late, and very long before it buddeth.

¶ *The Names.*

The tame Cypres is called in Greeke, κυπάρισσος, or κυπάριττος: in Latine, *Cupreſſus*: in ſhops, *Cypreſſus*: in Italian, *Cypreſſo*: in French and Spaniſh, *Cipres*: in high-Dutch, **Cipꝛeſſenbaum**: in low-Dutch, **Cypꝛeſſe boom**: in Engliſh, Cypres, and Cypres tree.

The fruit is named in Greeke, σφαίρια τῆς κυπαρίσσου: in Latine, *Pilulæ Cupreſſi*, *Nuces Cupreſſi*, and *Galbuli*: in ſhops, *Nuces Cypreſſi*: in Engliſh, Cypres nuts or clogs. This tree in times paſt was dedicated to *Pluto*, and was ſaid to be deadly; whereupon it is thought that the ſhadow thereof is vnfortunate.

The wilde Cypres tree is called in Greeke, θύα or θῦον, and θύον: from this doth differ θυεία, being a name not of a plant, but of a mortar in which dry things are beaten: *Thya*, as *Pliny* writeth, *lib.* 13. *cap.* 16. was well knowne to *Homer*: he ſheweth that this is burned among the ſweet ſmells, which *Circe* was much delighted withall, whom he would haue to be taken for a goddeſſe, to their blame that call ſweet and odoriferous ſmells, euen all of them, by that name; becauſe he doth eſpecially make mention withall in one verſe, of *Cedrus* and *Thya*: the copies haue falſly *Larix*, or Larch tree, in which it is manifeſt that he ſpake onely of trees: the verſe is extant in the fift booke of *Odyſſes*, where he mentioneth, that *Mercurie* by *Iupiters* commandement went to *Calypſus* den, and that he did ſmell the burnt trees *Thya* and *Cedrus* a great way off.

Theophraſtus attributeth great honor to this tree, ſhewing that the roofs of old Temples became famous by reaſon of that wood, and that the timber thereof, of which the rafters are made is euerlaſting, and it is not hurt there by rotting, cobweb, nor any other infirmitie or corruption.

¶ *The Temperature.*

The fruit and leaues of the Cypres are dry in the third degree, and aſtringent.

¶ *The Vertues.*

A The Cypres nuts being ſtamped and drunken in wine, as *Dioſcorides* writeth, ſtoppeth the laske and bloudy flix; it is good againſt the ſpitting of bloud and all other iſſues of bloud.

They

They glue and heale vp great vlcers in hard bodies: they ſafely and without harme ſoke vp and conſume the hid and ſecret moiſture lying deepe and in the bottome of weake and moiſt infirmities. B

The leaues and nuts are good to cure the rupture, to take away the *Polypus*, being an excreſcence growing in the noſe. C

Some do vſe the ſame againſt carbuncles and eating ſores, mixing them with parched Barley meale. D

The leaues of Cypres boyled in ſweet wine or Mede, helpes the ſtrangurie and difficulty of making water. E

It is reported, that the ſmoke of the leaues doth driue away gnats, and that the clogs do ſo likewiſe. F

The ſhauings of the wood laid among garments preſerueth them from the moths: the roſin killeth Moths, little wormes, and magots. G

† Our Author in this chapter hath put together two chapters of *Dodonæus*; the one of Cypreſſe, the other of *Thya*, out of *Theophraſtus* and others. *Vid. Pempt. 6. lib. 5. cap. 7 & 8.*

CHAP. 46. *Of the Tree of Life.*

Arbor Vitæ.
The Tree of Life.

¶ *The Deſcription.*

THe tree Tree of Life groweth to the height of a ſmall tree, the barke being of a darke reddiſh colour: the timber very hard, the branches ſpreading themſelues abroad, hanging down toward the ground by reaſon of the weakeneſſe of the twiggie branches ſurcharged with very oileous and ponderous leaues, caſting, and ſpreading themſelues like the feathers of a wing, reſembling thoſe of the Sauine tree, but thicker, broader, and more ful of gummie or oileous ſubſtance: which being rubbed in the hands do yeeld an aromatick, ſpicie, or gummie ſauor, very pleaſant and comfortable: amongſt the leaues come forth ſmall yellowiſh floures, which in my garden fall away without any fruit: but as it hath beene reported by thoſe that haue ſeene the ſame, there followeth a fruit in hot regions, much like vnto the fruit of the Cypres tree, but ſmaller, compact of little and thinne ſcales cloſely pact one vpon another, which my ſelfe haue not yet ſeene. The branches of this tree laid downe in the earth wil very eaſily take root, euen like the Woodbinde or ſome ſuch plant; which I haue often proued, and thereby haue greatly multiplied theſe trees.

¶ *The Place.*

This tree groweth not wilde in England, but it groweth in my garden very plentifully.

¶ *The Time.*

It endureth the cold of our Northerne clymat, yet doth it loſe his gallant greenes in the winter moneths: it floureth in my garden about May.

¶ *The Names.*

Theophraſtus and *Pliny*, as ſome thinke, haue called this ſweet and aromatical tree *Thuia*, or *Thya*: ſome call it *Cedrus Lycia*: the new writers do terme it *Arbor vitæ*: in Engliſh, the tree of life, I doe not meane that whereof mention is made, *Gen. 3. 22.*

¶ *The Temperature.*

Both the leaues and boughes be hot and dry.

¶ *The Vertues.*

Among the plants of the New-found land, this Tree, which *Theophraſtus* calls *Thuia*, or *Thua*, is is the most principal,and best agreeing vnto the nature of man,as an excellent cordial,and of a very pleasant smell.

Chap. 47. *Of the Yew tree.*

Taxus.
The Yew tree.

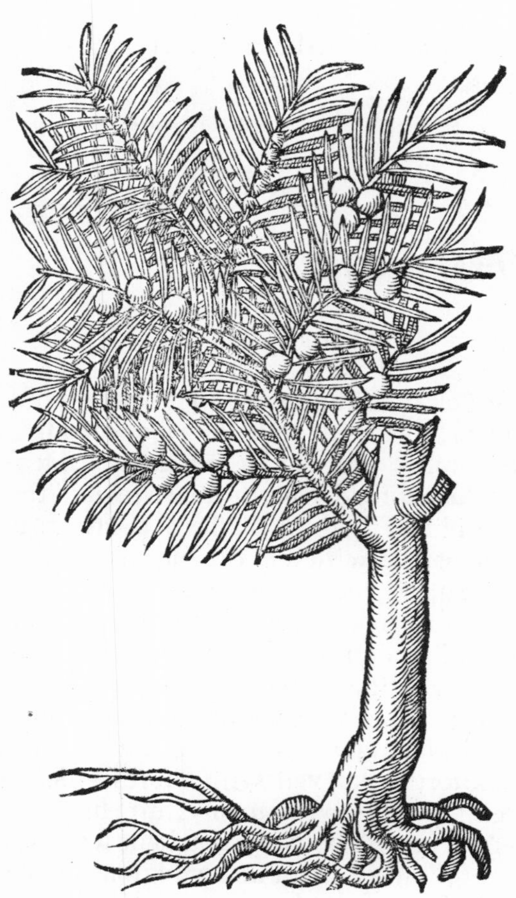

¶ *The Description.*

‡ IN stead of the description and place mentioned by our Author(which were not amisse) giue me leaue to present you with one much more accurate, sent me by Mr. *John Goodyer*.

Taxus glandifera bacciferáque.
The Yew bearing Acornes and berries.

THe Yew tree that beareth Acornes and berries is a great high tree remaining alwaies greene, and hath vsually an huge trunke or body as big as the Oke, coucred ouer with a scabbed or scaly barke, often pilling or falling off, and a yong smooth barke appearing vnderneath ; the timber hereof is somewhat red, neere as hard as Box, vniuersally couered next the barke with a thicke white sap like that of the Oke, and hath many big limmes diuided into many smal spreading branches : the leaues be about an inch long, narrow like the leaues of Rosemary, but smooth, and of a darker greene colour, growing all alongst the little twigs or branches close together, seldome one opposite against another, often hauing at the ends of the twigs little branches composed of many leaues like the former, but shorter and broader, closely compact or ioyned together : amongst the leaues are to be seene at all times of the yeare, small slender buds somewhat long, but neuer any floures ; which at the very beginning of the Spring grow bigger and bigger, till they are of the fashion of little Acornes, with a white kernell within : after they are of this forme, then groweth vp from the bottomes of the Acornes a reddish matter, making beautiful reddish berries more long than round, smooth on the out side, very clammie within, and of a sweet taste, couering all the Acorne, onely leauing a little hole at the top, where the top of the Acorne is to be seene : these fallen, or deuoured by birds, leaue behinde them a little whitish huske made of a few scales, appearing like a little floure, which peraduenture may deceiue some, taking it to be so indeed : it seemes this tree, if it were not hindred by cold weather, would alwaies haue Acornes and berries on him, for he hath alwaies little buds, which so soone as the Spring yeelds but a reasonable heate, they grow into the forme of Acornes : about the beginning of August, seldome before, you shall finde them turned into ripe berries, and from that time till Christmasse, or a little after, you may see on him both Acornes and red berries.

Taxus tantum florens.
The Yew which only floures.

The Yew which onely beareth floures and no berries, is like the other in trunke, timber, barke, and leaues , but at the beginning of Nouember, or before, this tree doth beginne to be very thicke set or fraught on the lower side or part of the twigs or little branches, with small round buds, verie neere as big, and of the colour of Radish seed, and do so continue all the Winter, till about the beginning or middle of Februarie, when they open at the top, sending forth one small sharpe pointall, little longer than the huske, diuided into many parts, or garnished towards the top with many small

ſmall duſty things like floures, of the colour of the husks; and if you ſhall beate or throw ſtones into this tree about the end of Februarie, or a good ſpace after, there will proceed and fly from theſe floures an aboundance of duſtie ſmoke. Theſe duſty floures continue on the trees till about harueſt, and then ſome and ſome fall away, and ſhortly after the round buds come vp as aforeſaid.

¶ *The Place.*

Theſe trees are both very common in England: in Hampſhire there is good plentie of them growing wilde on the chalkie hills, and in Church-yards where they haue been planted.

¶ *The Time.*

The time is expreſſed in their deſcriptions. Dec. 19. 1621. *Iohn Goodyer.* ‡

¶ *The Names.*

This tree is named by *Dioſcorides*, Σμίλαξ: by *Theophraſtus*, Μίλος: but *Nicander* in his book of Counterpoyſons, Σμίλος: *Galen* doth alſo call it Κάκτος: it is named in Latine *Taxus*: in high-Dutch, **Eybenbaum**: in low-Dutch, **Ibenboom**: in Italian, *Taſſo*: in Spaniſh, *Toxo*, and *Taxo*: in French, *Yf*: in Engliſh, Ewe, or Yew tree: in the vnlearned ſhops of Germany, if any of them remaine, it is called *Tamariſcus*; where in times paſt they were wont not without great error, to mix the bark hereof in compound medicines, in ſtead of the Tamariske barke.

¶ *The Temperature.*

The Yew tree, as *Galen* reporteth, is of a venomous qualitie, and againſt mans nature. *Dioſcorides* writeth, and generally all that heretofore haue dealt in the facultie of Herbariſme, that the Yew tree is very venomous to be taken inwardly, and that if any doe ſleepe vnder the ſhadow thereof it cauſeth ſickneſſe and oftentimes death. Moreouer, they ſay that the fruit thereof being eaten is not onely dangerous and deadly vnto man, but if birds do eat thereof, it cauſeth them to caſt their feathers, and many times to die. All which I dare boldly affirme is altogether vntrue: for when I was yong and went to ſchoole, diuers of my ſchoole-fellowes and likewiſe my ſelfe did eat our fils of the berries of this tree, and haue not only ſlept vnder the ſhadow thereof, but among the branches alſo, without any hurt at all, and that not one time, but many times. *Theophraſtus* ſaith, That Λόφουρα, *animalia*, *Gaza* tranſlates them *Iumenta*, or labouring beaſts, do die, if they do eat of the leaues; but ſuch cattell as chew their cud receiue no hurt at all thereby.

Nicander in his book of Counterpoiſons doth reckon the Yew tree among the venomous plants, ſetting downe alſo a remedie, and that in theſe words, as *Gorræus* hath tranſlated them.

Parce venenata Taxo, quæ ſurgit in Oeta
Abietibus ſimilis, letthoque abſumit acerbo
Ni præter morem pleno cratere meraca
Fundere vina pares, cum primum ſentiet æger
Arctari obſtructas fauces animæque canalem.

‡ Shun th' poys'nous Yew, the which on Oeta growes,
Like to the Firre, it cauſes bitter death,
Vnleſſe beſides thy vſe pure wine that flowes
From empty'd cups, thou drinke, when as thy breath
Begins to faile, and paſſage ot thy life
Growes ſtrait. ———

Pena and *Lobel* alſo obſerued that which our Author here affirmes, and dayly experience ſhewes it to be true, that the Yew tree in England is not poyſonous: yet diuers affirme, that in Prouince in France, and in moſt hot countries, it hath ſuch a maligne qualitie, that it is not ſafe to ſleepe or long to reſt vnder the ſhadow thereof. ‡

CHAP. 48. *Of the Iuniper tree.*

¶ *The Kindes.*

AMong the Iuniper trees one is leſſer, another greater, being a ſtrange and forreine tree: one of theſe bringeth forth a floure and no fruit; the other fruit and no floures.

¶ *The Deſcription.*

1 THe common Iuniper tree groweth in ſome parts of Kent vnto the ſtature and bignes of a faire great tree, but moſt commonly it growes very low like vnto ground Furres: this tree

1 *Iuniperus.*
The Iuniper tree.

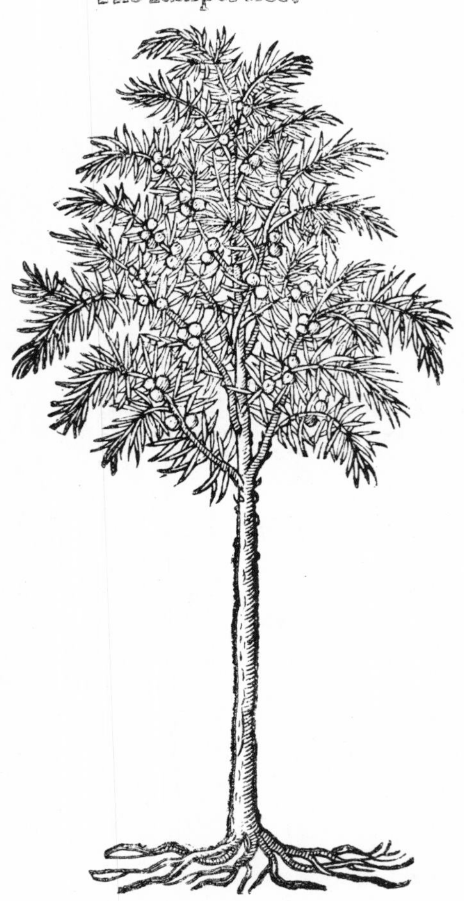

2 *Iuniperus maxima.*
The great Iuniper tree.

‡ 3 *Iuniperus Alpina minor.*
Small Iuniper of the Alps.

tree hath a thin bark or rinde, which in hot regions will chop and rend it ſelf into many cranies or pieces: out of which rifts iſſueth a certaine gum or liquour much like vnto Frankincenſe: the leaues are very ſmall, narrow, and hard, and ſomwhat prickly, growing euer green along the branches, thicke together: amongſt which come forth round and ſmall berries, greene at the firſt, but afterward blacke declining to blewneſſe, of a good ſauor, and ſweet in taſte, which do wax ſomwhat bitter after they be dry and withered.

2 The great Iuniper tree comes now and then to the height of the Cypres tree, with a greater and harder leafe, and alſo with a fruit as big as Oliue berries, as *Bellonius* writeth, of an exceeding faire blew colour, and of an excellent ſweet ſauor.

‡ 3 This exceeds not the height of a cubit, but growes low, and as it were creeps vpon the ground, and conſiſts of ſundry thicker and ſhorter branches than the common kind, tough alſo, writhen, and hard to breake; 3 leaues alwaies growing at equall diſtances, as in the common, but yet broader, ſhorter, and thicker, neither leſſe pricking than they, of a whitiſh greene colour on the inſide, and green without, incompaſſe the tender branches. *Cluſius*, who giues vs this figure and hiſtorie, obſerued not the floure, but the fruit is like that of the ordinarie,

nary,but yet somewhat longer; It growes vpon the Austrian Alpes,and ripens the fruite in August and September. ‡

¶ *The Place.*

The common Iuniper tree is found in very many places,especially in the South parts of England. *Bellonius* reporteth,that the greater groweth vpon mount Taurus : *Aloisius Anguillara* writeth, that it is found on the sea shores of the *Ligurian* and *Adriaticke* sea and in *Illyricum*, bringing forth great berries: and others say that it growes in Prouence of France: it commeth vp for the most part in rough places and neere to the sea, as *Dioscorides* noteth.

¶ *The Time.*

The Iuniper tree floureth in May; the floure whereof is nothing else but as it were a little yellowish dust or powder strowed vpon the boughes. The fruit is ripe in September,and is seldome found either winter or Sommer without ripe and vnripe berries, and all at one time.

¶ *The Names.*

The Iuniper tree is called in Greeke ἄρκευθος: the Apothecaries keepe the Latine name *Iuniperus*: the Arabians call it *Archonas* and *Archenas*: the Italians, *Ginepro*: in high Dutch, **Wechholter**: in Spanish, *Enebro*, *Ginebro*, and *Zimbro*. the French men and base Almaines *Geneue*: in English, Iuniper tree.

The lesser is named in Greeke ἀρκευθὶς: in Latine, *Iuniperus*. The great Iuniper Tree is called as some thinke in Greeke κυπάρισσος ἀγρία: in Latine (by *Lobel*) *Iuniperus maximus Illyricus cærulea bacca*, by reason of the colour of the berries, and may be called in English, blew Iuniper.

The berries are called *Grana Iuniperi*: in Greeke, ἀρκευθίς, although the Tree it selfe also is oftentimes called by the same name ἀρκευθὶς: it is termed in high Dutch, **Krametbeer, Weckholterbeer**: in low Dutch, **Geneverbessen**: in Spanish, *Neurinas*: in English, Iuniper berries.

The gum of the Iuniper tree is vsually called of the Apothecaries *Vernix*: in Latine, *Lachryma Iuniperi*: *Serapio* nameth it *Sandarax* and *Sandaracha*; but there is another *Sandaracha* among the Grecians,being a kinde of Orpment, which growes in the same minerals wherein Orpment doth, and this doth farre differ from *Vernix*,or the Iuniper gum. *Pliny* in his 11.booke,7.chapter maketh mention also of another *Sandaracha*, which is called *Erithrce* and *Cerinthus*: this is the meate of Bees whilest they be about their worke.

¶ *The Temperature.*

Iuniper is hot and dry, and that in the third degree, as *Galen* teacheth; the berries are also hot, but not altogether so drie: the gum is hot and dry in the first degree,as the Arabians write.

¶ *The Vertues.*

The fruite of the Iuniper tree doth clense the liuer and kidnies, as *Galen* testifieth: it also maketh thin clammie and grosse humors: it is vsed in counterpoysons and other wholesome medicines: being ouer largely taken it causeth gripings and gnawings in the stomacke, and maketh the head hot: it neither bindeth nor looseth the belly: it prouoketh vrine. A

Dioscorides reporteth, that this being drunke is a remedy against the infirmitie of the chest, coughes, windines, gripings and poisons,and that the same is good for those that be troubled with cramps, burstings, and with the disease called the mother. B

It is most certaine that the decoction of these berries is singular good against an old cough,and against that with which children are now and then extremely troubled, called the Chin cough, in which they vse to rise vp raw,tough and clammy humors,that haue many times bloud mixed with them. C

Diuers in Bohemia do take in stead of other drinke, the water wherein those berries haue been steeped, who liue in wonderfull good health. D

This is also drunke against poisons and pestilent feuers,and it is not vnpleasant in the drinking: when the first water is almost spent, the vessell is againe filled vp with fresh. E

The smoke of the leaues and wood driueth away serpents, and all infection and corruption of the aire,which bring the plague,or such like contagious diseases: the iuice of the leaues is laid on with wine, and also drunke against the bitings of the viper. F

The ashes of the burned barke, being applied with water, take away scurffe and filth of the skinne. G

The powder of the wood being inwardly taken, is pernicious and deadly, as *Dioscorides* vulgar copies do affirme; but the true copies vtterly deny it,neither do any of the old writers affirme it. H

The fume and smoke of the gum doth stay flegmaticke humors that distill out of the head, and stoppeth the rheume: the gum doth stay raw and flegmaticke humors that sticke in the stomacke and guts,if it be inwardly taken,and also drunke. I

It killeth all maner of wormes in the belly, it staieth the menses, and hemorrhodes: it is commended also against spitting of bloud; it dryeth hollow vlcers, and filleth them with flesh, if it be cast thereon: being mixed with oile of Roses, it healeth chops in the hands and feet. K

L There is made of this and of oile of Lineſeed, mixed together, a liquor called Verniſh, which is vſed to beautifie pictures and painted tables with, and to make iron gliſter, and to defend it from the ruſt.

Chap. 49. *Of the prickly Cedar, or Cedar Iuniper.*

¶ *The Kindes.*

THe prickly Cedar tree is like to Iuniper, and is called the ſmall or little Cedar, for difference from the great and tall Cedar, which bringeth Cones; and of this there are two kindes, as *Theophraſtus* and *Pliny* do teſtifie, that is to ſay, one of Lycia, and another crimſon.

¶ *The Deſcription.*

1 THe Crimſon or prickly Cedar ſeemeth to be very like to the Iuniper tree in body and boughes, which are writhed, knotty, and parted into very many wings: the ſubſtance of the wood is red, and ſweet of ſmell like that of the Cypreſſe; the tree is couered ouer with a rugged barke: the leaues be narrow and ſharpe pointed, harder than thoſe of Iuniper, ſharper and more pricking, and ſtanding thinner vpon the branches: the fruite or berry is ſometimes as big as a haſell nut, or, as *Theophraſtus* ſaith, of the bigneſſe of Myrtle berries, and being ripe it is of a reddiſh yellow, or crimſon colour, ſweet of ſmell, and ſo pleaſant in taſte, as euen the countrey-men now and than do eate of the ſame with bread.

1 *Oxycedrus Phœnicia.*
Crimſon prickly Cedar.

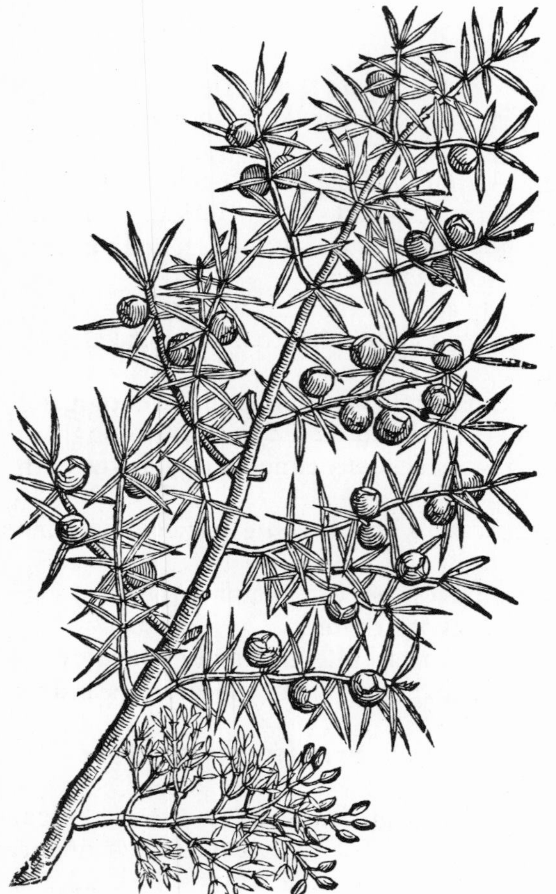

3 *Oxycedrus Lycia.*
Rough Lycian Cedar.

2 The other low Cedar which growes in Lycia is not ſo high as the former, hauing likewiſe a writhed body as big as a mans arme, full of boughes; the barke is rough, yellowiſh without, and red within: the leaues ſtand thicker, like at the firſt to thoſe of Iuniper, but yet ſomewhat ſhorter, and in the third or fourth yeere thicker, long and round withall, comming neere to the leaues of

‡ 3 *Cedrus Lycia altera.*
The other Lycian Cedar.

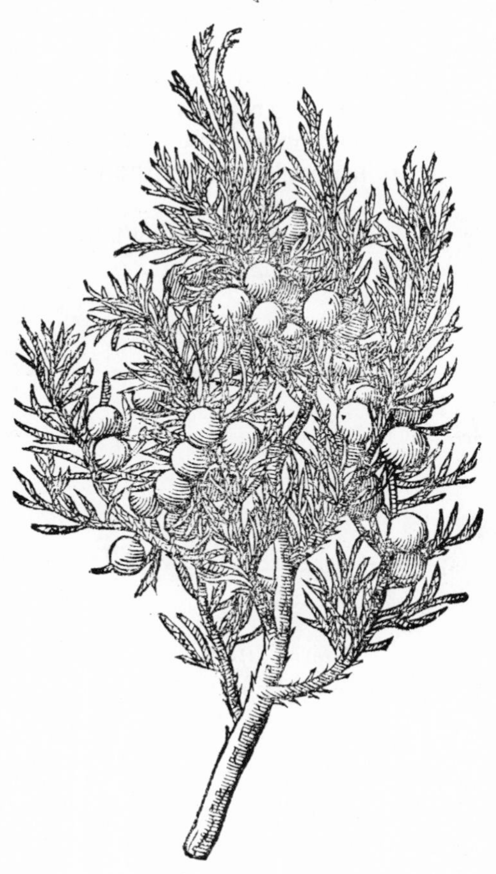

of the Cypres tree, or of the ſecond Sauine, that is, blunt, and not pricking at all, which being bruiſed betweene the fingers do yeeld a very pleaſant ſmel: ſo doth one and the ſelfe ſame plant bring forth below ſharpe and prickly leaues, and aboue thick and blunt ones, as that notable learned Herbariſt *Cluſius* hath moſt diligently obſerued: the fruit or berry is round like that of Iuniper, of colour yellow when it is ripe, inclining to a red, in taſte ſomwhat bitter, but ſweet of ſmell.

‡ 3 This alſo hath Cypreſſe-like leaues, not vnlike thoſe of the laſt deſcribed, yet ſomwhat thicker and broader: the fruit is alſo much larger, being as big as Haſell nuts, and of a red or skarlet colour; whence *Lobel* calleth it *Cedrus Phœnicia altera.* ‡

¶ *The Place.*

The prickely Cedar with the crimſon colour commeth vp higher and greater in certaine places of Italy, Spaine, and Aſia, and in other Countries; for that which grows on mount Garganus in Apulia is much higher and broader than thoſe that grow elſewhere, and bringeth forth greater berries, of the bigneſſe of an haſell nut, and ſweeter, as that moſt diligent writer *Bellonius* reporteth. *Carolus Cluſius* ſheweth, that the prickely Cedar and the Iuniper tree be of ſo great a growth in diuers places of Spaine, as he hath obſerued, as that the body of them is as thicke as a man.

The Lycian Cedar is found in Prouence of France, not far from Maſſilia, and groweth in a great part of Greece, in Illyricum and Epirum.

¶ *The Time.*

Both of them are alwaies greene, and in Winter alſo full of fruit, by reaſon that they continually bring forth berries, as when the old do fall new come in their places: in the ſpring grow vp new buds and beginnings of berries: in Autumne they wax ripe the ſecond yeare, as doe the berries of Iuniper.

¶ *The Names.*

They are called in Latine, *Minores*, and *Humiles Cedri*, little and low Cedars, for difference from the tall and great Cedar which beareth Cones.

The former is named in Greeke, Ὀξύκεδρος, and Κέδρος φοινικὴ in Latine, *Oxycedrus*, and *Cedrus Punica*: in Engliſh, Prickly Cedar, and Crimſon Cedar: *Pliny* ſyrnameth it *Phœnicea*, of the crimſon colour of the fruit: the Spaniards call this alſo *Enebro*, as *Cluſius* teſtifieth, euen by the ſame name which they giue to the Iuniper: wherein likewiſe they are thought to imitate diuers of the old Writers, who haue not by names diſtinguiſhed the Iuniper from the Cedar, but haue, as *Theophraſtus* noteth, called them *Cedros*, Cedar trees; yet with an addition, Ὀξύκεδρος, or prickly Cedar.

The other with the blunt leafe is named by *Theophraſtus*, Λύκια Κέδρος: of *Pliny* alſo, *Lycia Cedrus*: in Prouince of France, *Morucine*: diuers name this *Sabina*, and vſe it in ſtead of Sauine, which they want; as the Apothecaries of *Epidaurus*, and in diuers cities of Greece, and alſo in Illyricum and Epirum, as *Bellonius* teſtifieth. Some would haue it to be Θύα, *Thya*; but *Thya*, according to *Theophraſtus*, is like, not onely in body, leaues, and boughes, but in fruit alſo, to the Cypreſſe tree, but the fruit of this is nothing like to the Cypreſſe Cones.

The fruit of this Cedar is named by *Theophraſtus*, Κέδρις, *Cedris*: notwithſtanding *Cedrus*, as hee himſelfe doth alſo teſtifie (*Gaza* nameth it *Credula*) is a certaine little ſhrub which neuer groweth to a tree.

The gum or liquor which iſſueth forth of the prickly Cedar is alſo called *Vernix*, and is ſold in ſtead thereof.

¶ *The Temperature and Vertues.*

A The little Cedar, as *Galen* writeth, is hot and dry in a manner in the third degree: the matter or ſubſtance thereof is ſweet of ſmell, like that of Iuniper, and is vſed for perfumes and odoriferous ſmells together with the leaues.

The

B The berries or fruit of the low Cedar haue the faculties not so strong, as the same Author testifieth, insomuch as that they may also be eaten, yet if they be taken too plentifully, they cause head-ache, and breed heate and gnawings in the stomacke. Yet there is a difference between these two Cedar berries; for the crimson ones are not so hot and dry, by reason they are sweeter and pleasanter to the taste, and therefore they are better to be eaten, and do also yeeld vnto the body a kind of nourishment: but the berries of that of *Lycia* are biting, hotter and drier also than those of Iuniper, from which they differ especially in the biting qualitie, they bring no nourishment at all, and though a man eate neuer so few of them he shall feele gnawings in his stomacke, and paine in his head.

C The Peasants do feed thereon rather to satisfie their hunger, than for any delight they haue in the taste, or the physicall vertues thereof; albeit they be good against the strangurie, and prouoke vrine.

Chap. 50. *Of Sauin.*

¶ *The Kindes.*

There be two kindes of Sauin; one like in leafe to Tamariske, the other to the Cypresse tree; whereof the one beareth berries, the other is barren.

1 *Sabina sterilis.*
Barren Sauin.

2 *Sabina baccifera.*
Sauin bearing berries.

¶ *The Description.*

1 The first Sauin, which is the common kind, and best of all knowne in this country, groweth in manner of a low shrub or tree: the stem or trunke whereof is somtimes as big as a mans arme, diuiding it selfe into many branches set full of small leaues like vnto Cypres, or Tamariske, but thicker, and more sharpe or prickely, remaining greene Winter and Sommer, in smell ranke or very strong, barren both of floures and fruit.

2 The

‡ 3 *Sabina baccata altera.*
The lesser berry-bearing Sauin.

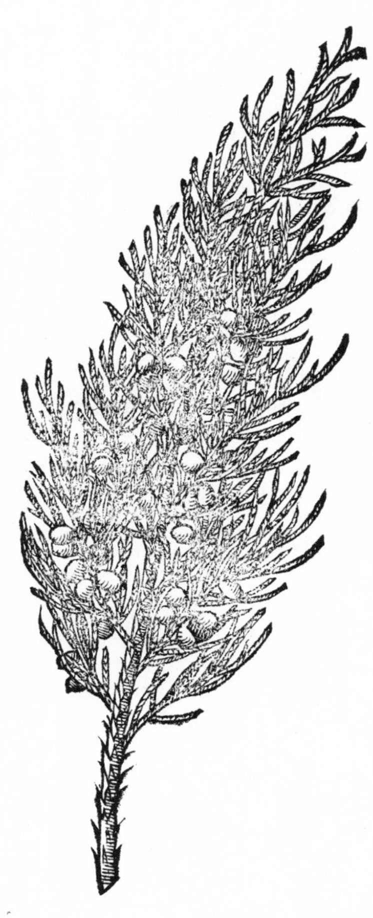

2 The other Sauin is an high tree, as *Bellonius* saith, as tall as the Almond tree, and much like to the tame Cypresse tree: the bodie is writhed, thicke, and sometimes of so great a compasse as that it cannot be fathomed; the substance of the wood is red within, as is that of the Iuniper, and of the prickely Cedar: the barke is not very thicke, and it is of a yellowish red: the leaues are of a maruellous gallant greene colour, like to those of the Cypres tree, yet thicker or more in number; in taste bitter, of a spicie smell, and like Rosin: the boughes are broader, and thicke set as it were with wings, like those of the Pitch tree and of the Yew tree: on which grow a great number of berries, very round like those of the little Cedars, which at the first are green, but when they be ripe they are of a blackish blew. Out of the root hereof issueth oftentimes a rosin, which being hard is like to that of the Iuniper tree, and doth also crumble in the chewing.

‡ 3 There is another, which differs from the last described onely in that the leaues are smaller and lesse pricking than those of the former, as also the branches lesser: *Lobel* calls this *Sauina baccata altera.* ‡

¶ *The Place.*

Both of them grow vpon hills in woods, and in other like vntoiled places, as in Candy, Mysia, and elsewhere. *P. Bellonius* reporteth that he found them both vpon the tops of the mountaines Taurus, Amanus, and Olympus.

The first is planted in our English gardens almost euery where: the second is planted both by the seed and by the slip: the slips must be set in a ground that is meanly moist and shadowie, till they haue taken root: the shrubs which grow of these decline toward the one side, retaining still the nature of the bough: but that Sauin which is planted by the seed groweth more vpright; this in continuance of time bringeth forth seeds, and the other for the most part remaines barren: both these grow in my garden.

¶ *The Time.*

They both continue alwaies greene: the one is found to be loden with ripe fruit commonly in Winter, but it hath fruit at all times; for before the old berries fall, new are come vp.

¶ *The Names.*

Sauine is called in Greeke Βράθυς, or Βράθυ: in Latine, *Sabina.*

The first is commonly called in the Apothecaries shops by the name *Sauina:* of diuers, *Sauimera:* the Italians and Spaniards keepe the Latine name: it is called in high-Dutch, **Siben baum:** in low-Dutch, **Sauelboom:** in French, *Sauenier:* in English, common Sauine, or garden Sauine.

Some name the other *Cupressus Cretica,* or Cypres of Candy, as *Pliny* saith, *lib.* 12. *cap.* 17. making mention of a tree called *Bruta:* some there are that take this to be *altera Sabina,* or the second Sauin, and to be read *Bruta* for Βράθυ, *Brathu,* by altering of the vowels. For it is described by *Plin. li.* 12. *cap.* 17. to be like the Cypresse tree, in these words; They seeke in the mountaine Elimæi the tree *Bruta,* being like to the broad Cypres tree, hauing white boughes, yeelding a sweet smell when it is set on fire; whereof mention is made with a miracle, in the stories of *Claudius Cæsar.* It is reported that the Parthians do vse the leaues in drinks; that the smell is very like to that of the Cypres tree, and that the smoke thereof is a remedie against other woods. It groweth beyond Pasitigris, neere vnto the towne Sittaca, on mount Zagrus. Thus far *Pliny.*

The mountaines Elimæi are described by *Strabo* in the countrey of the Assyrians, next after the mountaine Sagrus aboue the Babylonians; by *Ptolomæus* not far from the Persian gulfe: therefore it is hard to say that *Bruta* is *Sabina altera,* or the second Sauine, seeing that so great a distance of the place may vndoubtedly cause a difference, and that it is not largely but briefely described. It seemeth that *Thya* mentioned by *Theophrastus* is more like vnto Sauine: but yet forasmuch as *Thya* is like in fruit to the Cypres tree, and not to the fruit or berries of the little Cedars, it is also verie manifest

manifeſt, that the ſecond Sauine is not *Thya*, neither *Vitæ arbor*, ſo called of the later Herbariſts: it is likewiſe named by *Lobel*, *Sabina genuina baccifera, atrocærulea*, that is, the true Sauine that beareth berries of a blackiſh blew colour.

¶ *The Temperature.*

The leaues of Sauine, which are moſt vſed in medicine, are hot and dry in the third degree, and of ſubtill parts, as *Galen* ſaith.

¶ *The Vertues.*

A The leaues of Sauin boyled in Wine and drunke prouoke vrine, bring downe the menſes with force, draw away the after-birth, expell the dead childe, and kill the quicke: it hath the like vertue receiued vnder in a perfume.

B The leaues ſtamped with honey and applied, cure vlcers, ſtay ſpreading and creeping vlcers, ſcoure and take away all ſpots and freckles from the face or body of man or woman.

C The leaues boyled in oyle Oliue, and kept therein, kill the wormes in children, if you anoint their bellies therewith: and the leaues poudered and giuen in milke or Muſcadell do the ſame.

D The leaues dried and beate into fine pouder, and ſtrewed vpon thoſe kindes of excreſcences *ſub præputio*, called Caroles, and ſuch like, gotten by dealing with vncleane women, take them away perfectly, curing and healing them: but if they be inueterate and old, and haue been much tampered withall, it ſhall be neceſſarie to adde vnto the ſame a ſmall quantitie of *Auripigmentum* in fine pouder, and vſe it with diſcretion, becauſe the force of the medicine is greatly increaſed thereby, and made more corroſiue.

Chap. 51. *Of Tamariske.*

1 *Tamariſcus Narbonenſis.*
French Tamariske.

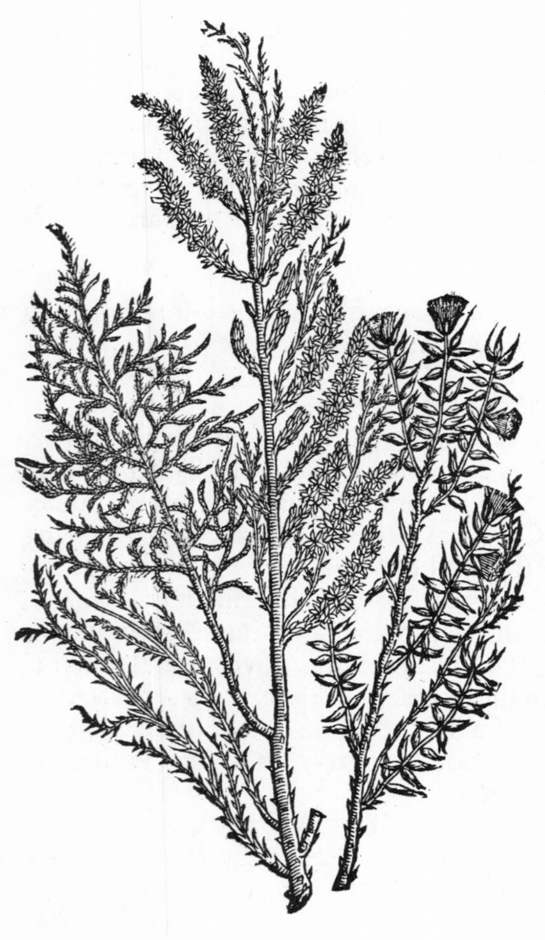

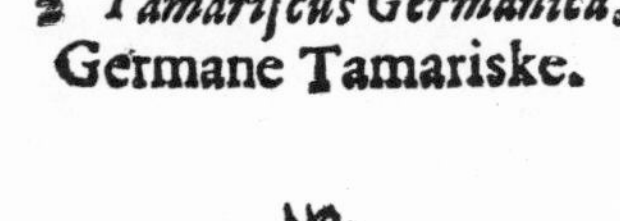

2 *Tamariſcus Germanica.*
Germane Tamariske.

¶ *The*

¶ *The Deſcription.*

1 THe firſt kinde of Tamariske groweth like a ſmall hedge tree, couered with a reddiſh barke, hauing many branches ſet and bedeckt with leaues, much like vnto Heath: among which come forth ſmall moſſie white floures declining to purple, which turne into a pappous or downie ſeed, that flieth away with the winde, as that of Willow doth: the root is wooddie as the roots of other ſhrubs be, and groweth diuers waies.

2 The Germane Tamariske hath many wooddie branches or ſhoots riſing from the root, with a white bark, hauing his leaues thicker and groſſer than the former, and not ſo finely iagged or cut: The floures are reddiſh, and larger than the former, growing not vpon foot-ſtalkes, many thick cluſtering together, as thoſe of the former, but each a pretty diſtance from another on the tops of the branches ſpike faſhion, and begin to floure below: which do turne into ſeed, that is likewiſe carried away with the winde.

¶ *The Place.*

Tamariske groweth by running ſtreames, and many times by riuers that breake forth, and not ſeldome about fenny grounds, commonly in a grauelly ſoile, for it beſt proſpereth in moiſt and ſtony places: it is found in Germany, Vindelicia, Italy, Spaine, and alſo in Greece.

The Tamarisks do alſo grow in Egypt and Syria, as *Dioſcorides* writeth, and likewiſe in Tylus an Iſland in Arabia, as *Theophraſtus* noteth: the wood wherof, ſaith he, is not weak as with vs in Greece, but ſtrong like πρίνινον, or timber, or any other ſtrong thing: this Tamariske *Dioſcorides* doth call ἥμερος, that is to ſay, tame, or planted; and ſaith that it bringeth forth fruit very like to Galls, in taſte rough and binding.

Petrus Bellonius in his ſecond booke of ſingularities reporteth, that hee ſaw in Egypt very high Tamarisks and great like other trees, and that ſometimes in moiſt places by riuers ſides, and many times alſo in dry and grauelly grounds where no other trees did grow, which now and then did beare hanging on the boughes ſuch a multitude of Galls, that the inhabitants call *Chermaſel*, as being ouer loden, they were ready to breake. Both theſe grow and proſper well in gardens with vs here in England.

¶ *The Time.*

Theſe trees or ſhrubs floure in May, and in the later end of Auguſt, their ſeed is carried away with the wind.

¶ *The Names.*

They are called in Greeke μυρίκη: and in Latine alſo *Myrica*, and *Tamarix*: in ſhops, *Tamariſcus*: of *Octauius Horatianus*, *Murica*: *Dioſcorides* maketh that which groweth in Greece and Italy to be ἀγρία, or wilde Tamariske: it is named in high Dutch **Tamariſchen holtz**, and **Portz**: in low Dutch, **Ibenboom, Tamariſchboome**: in Italian, *Tamarigio*: in Spaniſh, *Tamarguira*, and *Tamariz*: in French, *Tamaris*: in Engliſh, Tamariske.

¶ *The Temperature and Vertues.*

Tamariske hath a clenſing and cutting facultie with a manifeſt drying; it is alſo ſomewhat aſtringent or binding, and by reaſon of theſe qualities it is very good for an hard ſpleen, being boyled with vineger or wine, either the root or leaues, or tender branches, as *Galen* writeth. A

Moreouer *Dioſcorides* teacheth, that the decoction of the leaues made with wine, doth waſte the ſpleene, and that the ſame is good againſt the tooth-ache, if the mouth be waſhed therewith: that it bringeth downe the Menſes, if the patient ſit therein; that it killeth lice and nits, if the parts be bathed therewith. B

The aſhes of burnt Tamariske hath a drying facultie, and greatly ſcouring withall, and a little binding. C

The floures and downie ſeed of the greater Tamariske doth greatly binde, inſomuch as it commeth very neere to the Gall named *Galla Omphacitis*, but that the roughneſſe of taſte is more euident in the Gall; the which floures are of an vnequall temperature, for there is ioined to the nature therof a great thinneſſe of parts, and clenſing facultie, which the Gall hath not, as *Galen* writeth. D

Theſe floures we fitly vſe (ſaith *Dioſcor.*) in ſtead of Gall, in medicines for the eies and mouth: E

It is good to ſtanch bloud, and to ſtay the laske and womens whites, it helpeth the yellow iaundice, and alſo cureth thoſe that are bit of the venomous ſpider called *Phalangium*; the barke ſerueth for the ſame purpoſes. F

The leaues and wood of Tamariske haue great power and vertue againſt the hardneſſe and ſtopping of the ſpleene, eſpecially the leaues being boiled in water, and the decoction drunke, or elſe infuſed in a ſmall veſſell of Ale or Beere, and continually drunke: and if it bee drunke forth of a cup or diſh made of the wood or timber of Tamariske, is of greater efficacie. G

CHAP.

Chap. 52. *Of Heath, Hather, or Linge.*

¶ *The Kindes.*

There be diuers ſorts of Heath, ſome greater, ſome leſſer; ſome with broad leaues, and ſome narrower: ſome bringing forth berries, and others nothing but floures.

¶ *The Deſcription.*

1 The common Heath is a low plant, but yet wooddie and ſhrubby, ſcarce a cubit high: it bringeth forth many branches, whereupon do grow ſundry little leaues ſomewhat hard and rough, very like to thoſe of Tamariske, or the Cypreſſe tree: the floures are orderly placed alongſt the branches, ſmall, ſoft, and of a light red colour tending to purple: the root is alſo wooddie, and creepeth vnder the vpper cruſt of the earth: and this is the Heath which the Antients tooke to be the right and true Heath.

1 *Erica vulgaris, ſive Pumila.*
Common or dwarfe Heath.

‡ *Erica vulgaris hirſuta.*
Rough leaued Heath.

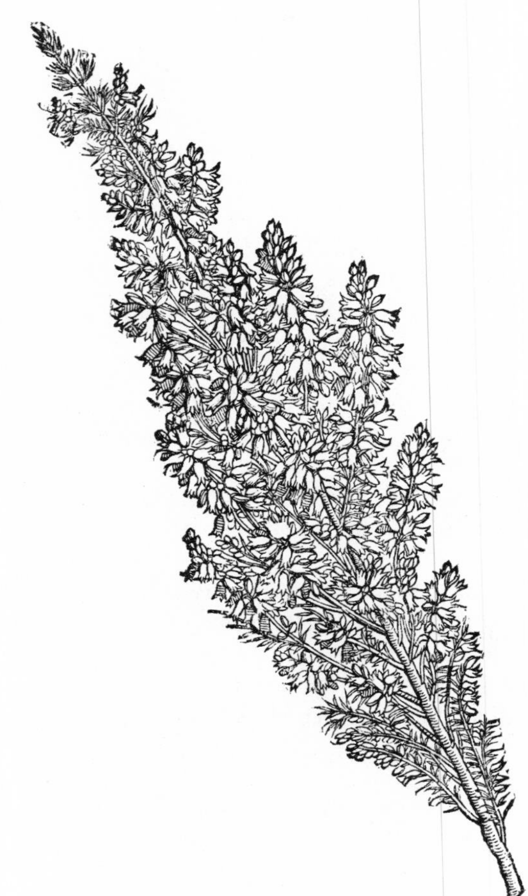

There is another Heath which differeth not from the precedent, ſauing that this plant bringeth forth floures as white as ſnow, wherein conſiſteth the difference: wherefore we may call it *Erica pumila alba*, Dwarfe Heath with white floures.

2 The great Heath, (which *Carolus Cluſius* at his being in England found in the barren grounds about Windſor, which in his Spaniſh trauels he maketh the firſt kinde): groweth to the height of two cubits, ſeldome higher, full of branches, couered with a blackiſh barke: whereon are ſet in very good order by couples, ſmall, rough, ſquare leaues finer than thoſe of Tamariske or Cypreſſe. The floures incloſe the little twiggie branches round about at certaine diſtances, from the lower part to the top faſhioned like little bottles, conſiſting of foure parts, of a ſhining purple colour, very beautifull to behold, and the rather to be eſteemed becauſe it floureth twiſe in the yeare: the root is likewiſe wooddie.

‡ 3 This

† 3 *Erica maior flore albo Clusij.*
The great Heath with white floures.

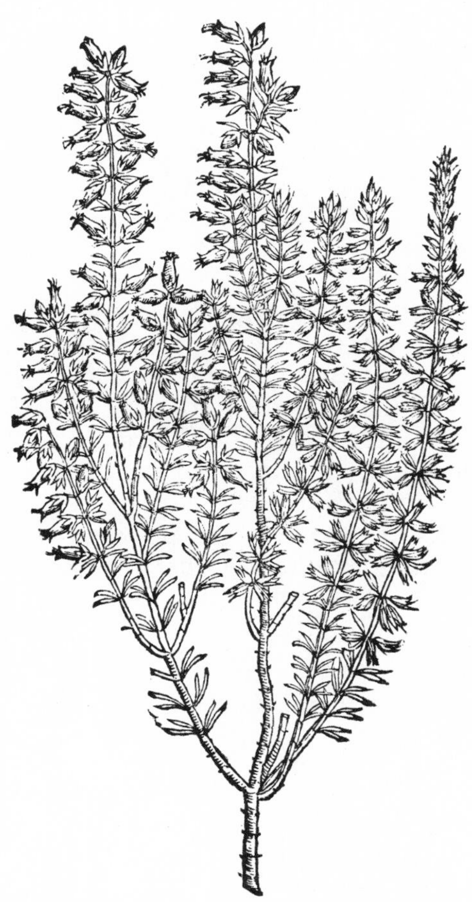

4 *Erica maior flore purpureo.*
Great Heath with purple floures.

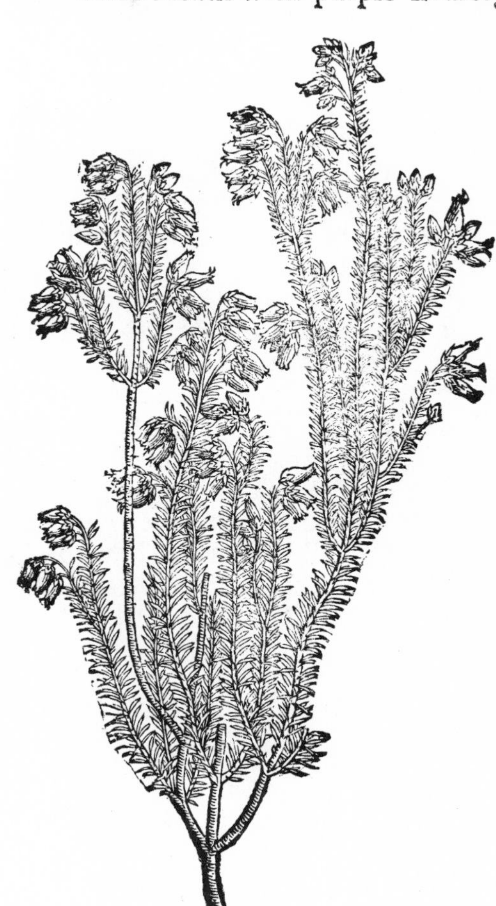

† 5 *Erica cruciata.*
Crossed Heath.

6 *Erica Pyramidalis.*
Steeple Heath.

‡ 3 This, saith *Clusius*, which is the largest that I haue seene, sometimes exceeds the height of a man, very shrubby, hauing a hard and blackish red wood: the leaues are small and short, growing about the branches by foures, of a very astringent taste: it hath plentiful store of floures growing all alongst the branches, so that somtimes the larger branches haue floures for a foot in length: this floure is hollow and longish, well smelling, white and beautifull. It growes betweene Lisbone and the Vniuersity of Conimbrica in Portugal where it floures in Nouember, December, and Ianuarie. ‡

† 4 Of this kind there is another sort with whitish purple floures, more frequently found than the other sort, which floures are somwhat greater than the former, but in forme like, and flouring at the same time. ‡ The leaues also are hairy, and grow commonly by foures: the hollow floures grow clustering together at the very tops of the branches, and are to be found in Iuly and August; it growes on diuers heathy places of this kingdome. ‡

5 Crossed Heath groweth to the height of a cubit and a halfe, full of branches, commonly lying along vpon the ground, of a swart darke colour: whereon do grow small leaues, set at certaine spaces by two vpon one side, and two on the other, opposite, one answering another, euen as doe the leaues of Crosse-wort. The floures in like manner stand alongst the branches Crosse fashion, of a darke ouerworne greenish colour. The root is likewise woody, as is all the rest of the plant.

6 This Steeple Heath hath likewise many wooddy braunches, garnished with small leaues which easily fall off from the dryed stalks; among which come forth diuers little mossie greenish floures of small moment. The whole bush for the most part groweth round together like a little cocke of hay, broad at the lower part, and sharp aboue like a Pyramide or steeple, whereof it tooke his name.

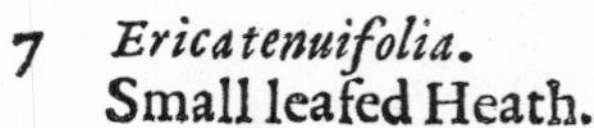

7 *Erica tenuifolia.*
Small leafed Heath.

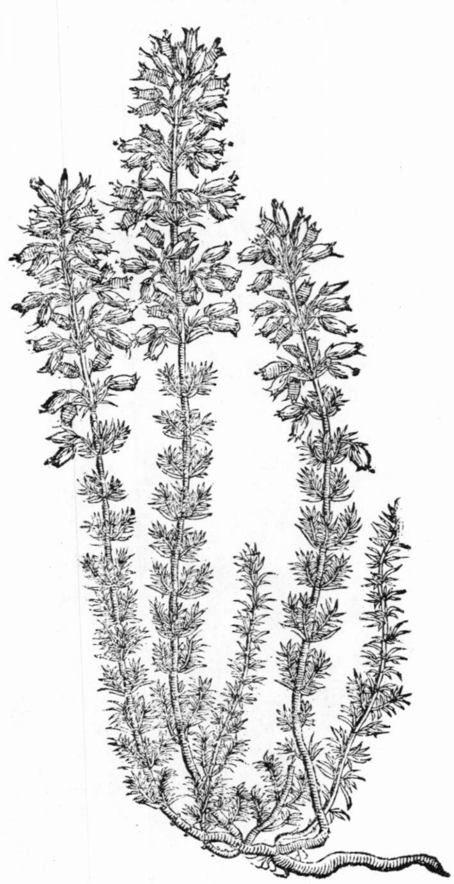

8 *Erica tenuifolia caliculata.*
Challice Heath.

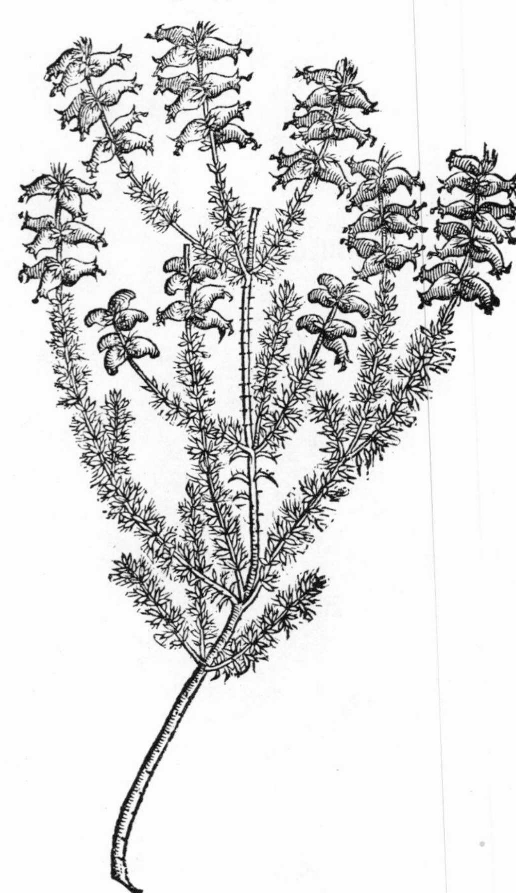

7 This small or thinne leafed Heath is also a low and base shrub, hauing many small and slender shoots comming from the root, of a reddish browne colour; whereupon doe grow verie manie small leaues, not vnlike to them of common Time, but much smaller and tenderer: the floures grow in tufts at certaine spaces, of a purple colour. The root is long and of a wooddie substance. ‡ The branches of this are commonly whitish, the leaues very green: the floures are smallest at both ends and biggest in the middest, hollow, and of a faire purple colour, which doth not easily decay: it floures

floures most part of Summer, and growes in many Heathie grounds. ‡

8 Challice Heath hath also many wooddy branches growing from the roots, slender, of a reddish browne colour, a foot and a halfe high, garnished with very little leaues, lesser than those of Time: the floures grow on the tops and vpper parts of the branches, and be in number, fiue, six or moe, hanging downewards, in fashion long, hollow within like a little tunnell or open cup or challice, of a light purplish colour: the root creepeth and putteth forth in diuers places new springs or shoots.

9 The Heath that bringeth forth berries hath many weake and slender branches of a reddish colour, which trailing vpon the ground do take hold thereof in sundry places, whereby it mightily increaseth: the leaues are somewhat broad, of a thicke and fleshie substance, in taste somthing drying at the first, but afterwards somewhat sharpe and biting the tongue: among which come forth small floures of an herbie colour: which being vaded there succeed small round berries, that at the first are greene, and afterward blacke, being as big as those of Iuniper; wherein is contained purple iuice like that of the Mulberry: within those berries are contained also small three cornered grains: the root is hard, and of a wooddy substance. ‡ I found this growing in great plenty in Yorkshire on the tops of the hills by Gisbrough, between it and Rosemary-topin (a round hill so called) and some of the people thereabouts told me they called the fruit Crake berries. This is the same that *Matthiolus* calls *Erica Baccifera*: and it is the *Erica Coris folio* 11. of *Clusius*. ‡.

‡ 9 *Erica baccifera procumbens.*
Heath bearing Berries.

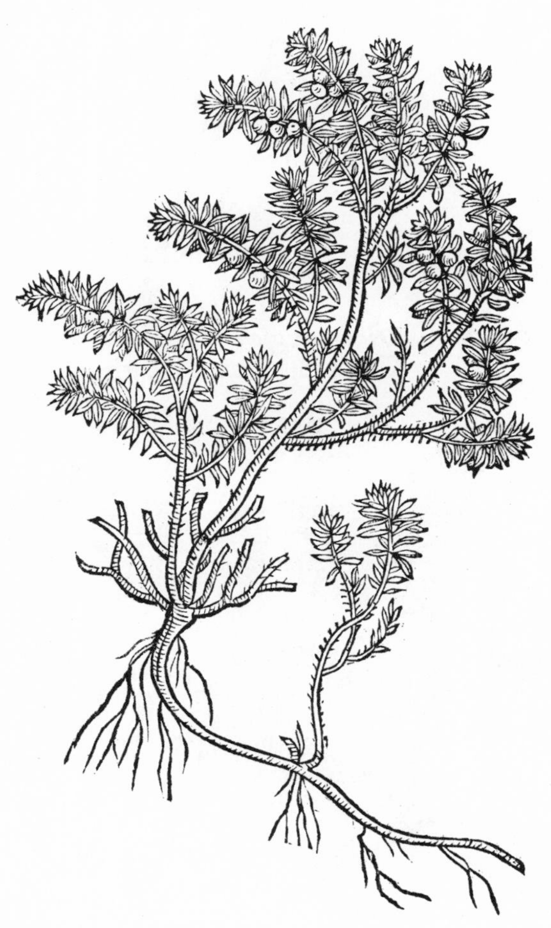

10 *Erica baccifera tenuifolia.*
Small leafed Heath with Berries.

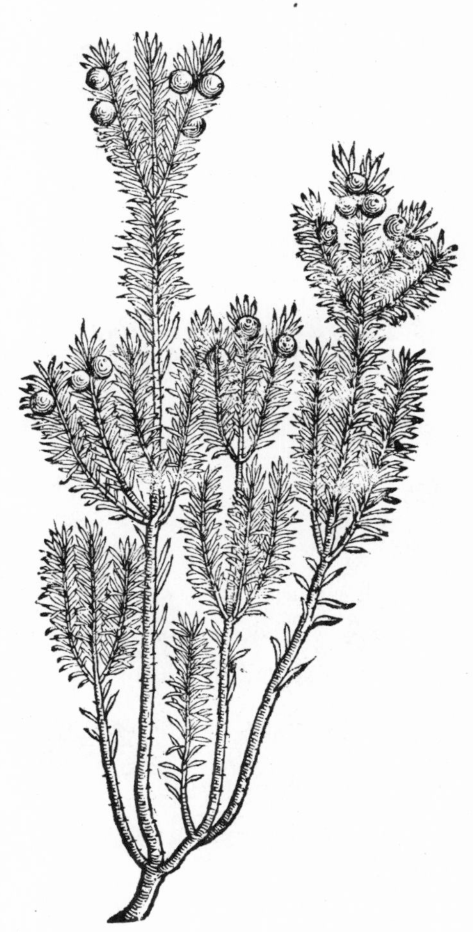

‡ 10 This which our Authour figured as you seee in the tenth place (putting the description of the former thereto) hath brittle branches growing some cubit high, couered with a barke blacker than the rest: the leaues are like those of the former, but blacker and smaller, growing about the stalks by threes, of a hottish taste with some astriction. In September and October it carries a fruit on the tops of the branches different from the rest, for it is very beautifull, white, transparent, resembling dusky and vneuen pearles in forme and colour, succulent also, and of an acide taste, commonly containing three little seeds in each berry: in Nouember this fruit becomes dry, and falls away of it selfe. *Clusius* onely obserued this in Portugall, and at the first sight a far off tooke the white berries to haue been graines of Manna. He calls it *Erica Coris folio*. 10.

11 I remember (saith *Dodonæus*) that I obserued another Heath which grew low; yet sent forth many

‡ 11 *Erica pumila, 3. Dod.*
Dodonæus his Dwarfe Heath.

‡ 12 *Erica ternis per intervalla ramis.*
Heath with three branches at a ioint.

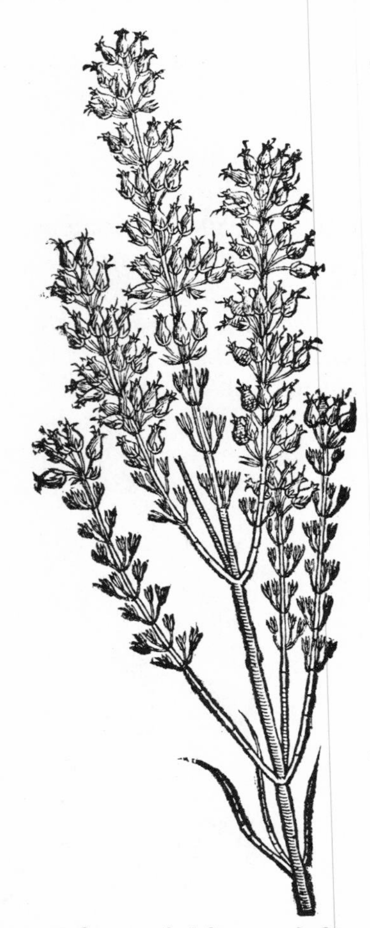

‡ 13 *Erica perigrina Lobelij.*
Lobels strange Heath.

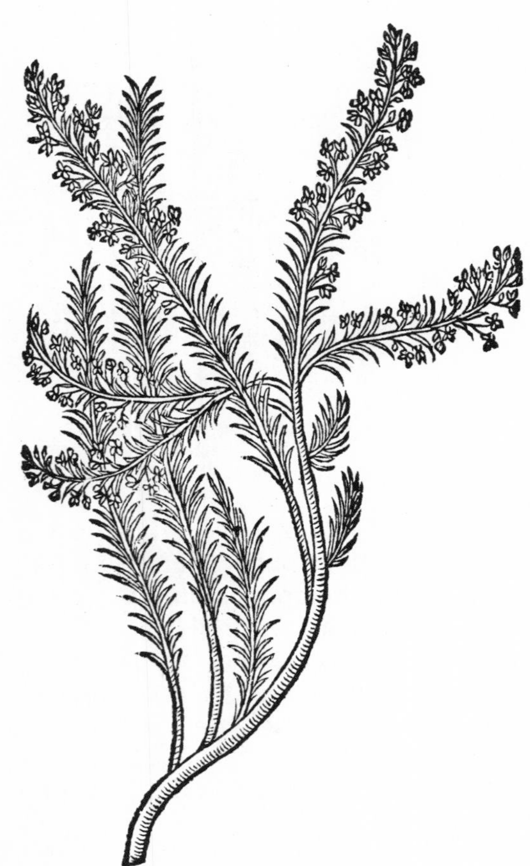

‡ 14 *Erica Coris folio 7. Clusij.*
Creeping Dutch Heath.

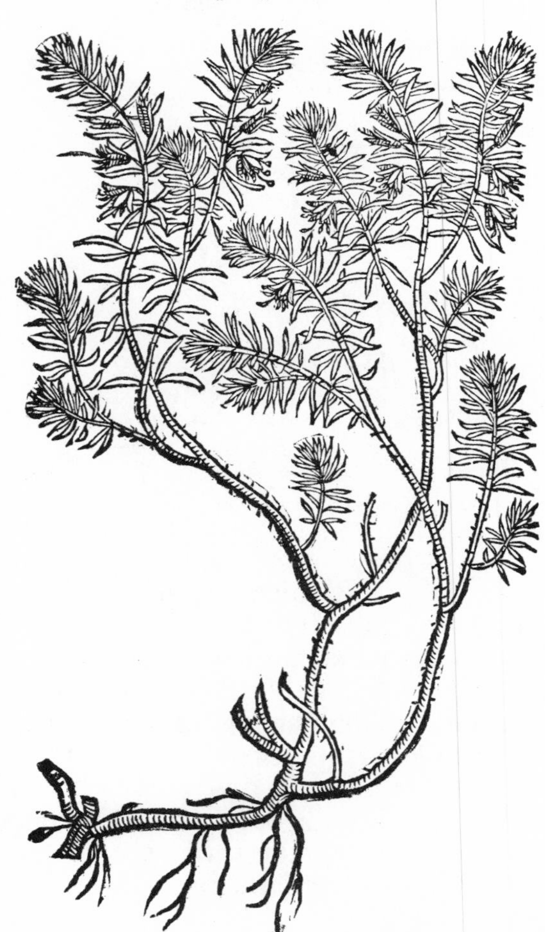

‡ 15 *Erica Coris folio. 9. Cluſij.*
Small Auſtrian Heath.

many wooddy and twiggy branches,hauing vpon them little narrow and longiſh leaues; on theſe ſtalkes ſpike faſhion to the tops of them, yet but on one ſide, grow elegant redde floures, pointed with blacke. This growes in that tract of Germany which leads from Bohemia to Noremberg on dry and vntilled places, and neere woods. It floures in Aprill.

12 This ſhrubby Heath is commonly ſome cubit high, hauing ſlender branches which come out of the maine ſtemmes commonly three together; and the leaues alſo grow in the ſame order; the tops of the branches are adorned with many floures of a darke purple colour, hollow, round, biggeſt below,and ſtanding vpon long footſtalks. *Cluſius* found this growing in the vntilled places of Portingale aboue Lisbone,where it floured in December; he calls it *Erica Coris folio*, 5.

13 Beſides all theſe (ſaith *Lobel*, hauing firſt treated of diuers plants of this kinde) there is a certaine rarer ſpecies growing like the reſt after the manner of a ſhrub in pots, in the Garden of M[r].*Iohn Brancion*: the leafe is long,and the purple floures, which as far as I remember conſiſted of foure little leaues apiece,grow on the tops of the branches. I know not whence it was brought,and therfore for the rarity I call it *Erica peregrina*,that is,Strange, or Forreine Heath.

14 This hath many round blackiſh purple branches ſome foot or cubit high,lying oft times along vpon the ground : theſe are beſet with many narrow little leaues, almoſt like thoſe of the third deſcribed, yet ſomewhat longer,commonly growing foure, yet ſometimes fiue together, of an aſtringent taſte; the little floures grow on the top of the branches, longiſh, hollow, and of a light purple colour, comming out of foure little leaues almoſt of the ſame colour; when theſe are ripe and dryed they containe a blackiſh and ſmall ſeed; the root is hard, wooddy, and runnes diuers waies; the weake branches alſo that lie vpon the ground now and then take root againe. *Cluſius* found this growing plentifully in diuers mountanous places of Germany where it floured in Iune, and Iuly.

15 The weake ſtalkes of this are ſome foot high,which are ſet with many ſmall greene leaues growing commonly together by threes; the tops of the branches are deckt with little hollow and longiſh floures diuided at their ends into foure parts, of a fleſh colour,together with the foure little leaues out of which they grow, hauing eight blackiſh little threds in them, with a purpliſh pointall in the middle. The ſeed is blacke and ſmall; the root wooddy as in other plants of this kinde.*Cluſius* found this in ſome mountanous woods of Auſtria, where it floured in Aprill and May. ‡

¶ *The Place.*

Heath groweth vpon dry mountaines which are hungry and barren, as vpon Hampſteed Heath neere London,where all the ſorts do grow, except that with the white floures,and that which beareth berries. ‡ There are not aboue three or foure ſorts that I could euer obſerue to grow there. ‡

Heath with the white floures groweth vpon the downes neere vnto Graueſend.

Heath which beareth berries groweth in the North parts of England, namely, at a place called Crosby Rauenſwaith, and in Crag cloſe alſo in the ſame countrey: from whence I haue receiued the red berries by the gift of a learned Gentleman called M[r]. *Iames Thwaites*.

¶ *The Time.*

Theſe kindes or ſorts of Heath do for the moſt part floure all the Sommer, euen vntill the laſt of September.

¶ *The Names.*

Heath is called in Greeke, ἐρείκη: in Latine alſo *Erica*: diuers do falſly name it *Myrica*: in high and low Dutch, Heiden: in Italian, *Erica*: in Spaniſh, *Breſo Quirro*: in French, *Bruyre*: in Engliſh, Heath, Hather, and Linge.

¶ *The Temperature.*

Heath hath, as *Galen* ſaith, a digeſting facultie, conſuming by vapors: the floures and leaues are to be vſed.

¶ *The Vertues.*

A The tender tops and floures, ſaith *Dioſcorides*, are good to be laid vpon the bitings and ſtingings of any venomous beaſt: of theſe floures the Bees do gather bad hony.

B The barke and leaues of Heath may be vſed for, and in the ſame cauſes that Tamariske is vſed.

† The figure which our Author gaue in the ninth place by the name of *Erica baccifera latifolia* I take to be the *Vitis Idæa*, 2. of *Cluſius* (which you ſhall finde in his due place) and in ſtead thereof I haue giuen you our ordinary berry-bearing Heath.

Chap. 53. *Of Heath of Ierico.*

1 *Roſa Hiericontea maior.*
The Heath Roſe of Ierico.

¶ *The Deſcription.*

1 This kinde of Heath which of the later writers hath been called by the name *Roſa Hiericontea*; the coiner ſpoiled the name in the mint, for of all plants that haue bin written of, there is not any more vnlike vnto the Roſe, or any kinde thereof than this plant: what moued them thereto I know not: but thus much of my owne knowledge, it hath neither ſhape, nature, nor facultie agreeing with any Roſe; the which doubtleſſe is a kinde of Heath, as the barren ſoile, and that among Heath, doth euidently ſhew, as alſo the Heathie matter wherewith the whole plant is poſſeſſed, agreeing with the kindes of Heath in very notable points. It riſeth vp out of the ground, of the height of four inches, or an hand breadth, compact or made of ſundry hard ſtickes, (which are the ſtalkes) claſping or ſhutting it ſelfe together into a round forme, intricately weauing it ſelfe one ſticke ouerthwart another, like a little net: vpon which wooddy ſtickes do grow leaus not vnlike to thoſe of the Oliue tree, which maketh the whole plant of a round forme, and hollow within; among the leaues on the inſide grow ſmall moſſie floures, of a whitiſh herbie colour, which

2 *Roſa Hiericontea ſiccata.* The Heath Roſe of Ierico dried.

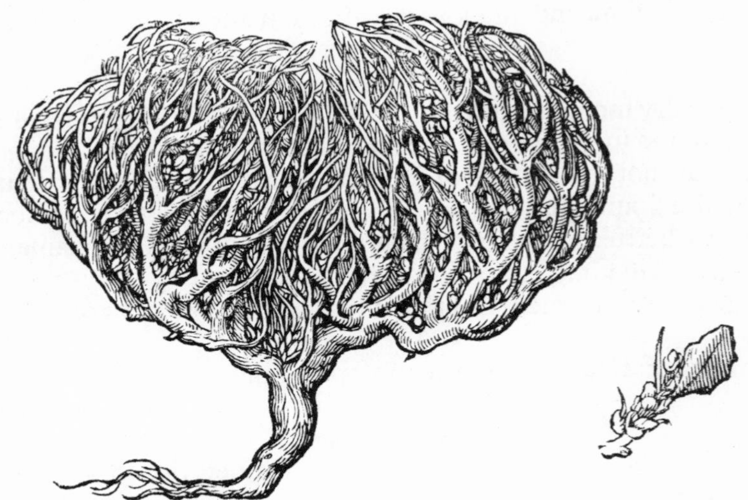

turne

turn into little seed, like the seed of Rocket, but lesser: the whole plant is of the substance of heath, and wooddie.

2 The second figure setteth forth the dried plant, as it is brought vnto vs from beyond the seas: which being set into a dish of warme water, for halfe an houre, openeth it selfe in forme, as when it did grow, and taken forth vntill it be drie, returneth shut vp againe as before.

¶ *The Place.*

It groweth in the barren grounds of France, and other hot regions, among the Heath and such like plants: it is a stranger in England, yet dried we haue them in great plenty. ‡ I haue not read nor heard that this growes wilde in France; but *Bellonius* saith it growes in Arabia *deserta*: *Bauhine* saith it easily grew and flourished many yeares in his garden at *Basill*. ‡

¶ *The Time.*

The seed being sowne in our cold climate, is sowne in Aprill; it perisheth when it is sprung vp, and bringeth neither floures nor seed.

¶ *The Names.*

This kinde of Heath is called *Rosa Hiericontea*, or *de Hiericho*, the Rose of Ierico: of some, the Rose of Ierusalem, and also *Rosa Mariæ*: in English, the Heath Rose.

¶ *The Temperature and Vertues.*

A There is not any of the antient nor later writers that haue set downe any certaintie of this plant as touching the temperature and faculties, but onely a bare picture with a slender description.

CHAP. 54. *Of the Chaste Tree.*

1 *Vitex, siue Agnus Castus.*
The Chaste tree.

‡ 2 *Vitex latiore serrato folio.*
Chaste tree with cut leaues.

¶ *The Deſcription.*

1 Vitex, or the Chaſte tree, groweth after the manner of a buſhie ſhrub or hedge tree, hauing many twiggie branches, very pliant and eaſie to be bent without breaking, like to the willow: the leaues are for the moſt part diuided into fiue or ſeuen ſections or diuiſions, much like the leaues of Hemp, whereof each part is long and narrow, very like vnto the willow leafe, but ſmaller: the floures do grow at the vppermoſt parts of the branches, like vnto ſpikie eares, cluſtering together about the branches, of a light purple or blew colour, and very ſweet ſmel: the fruit is ſmall and round, like vnto the graines or cornes of pepper.

‡ 2 *Lobel* mentions another varietie hereof that differs from the former onely in that it hath broader leaues, and theſe alſo ſnipt about the edges. ‡

¶ *The Place.*

Vitex groweth naturally in Italy, and other hot regions, by water courſes and running ſtreames: I haue it growing in my garden.

¶ *The Time.*

Vitex beginneth to recouer his laſt leaues in May, and the floures come forth in Auguſt.

¶ *The Names.*

† The Grecians call this ſhrub ἄγνος, and λύγος: *Agnos* (i.) *Caſtus*, Chaſte: becauſe, ſaith *Pliny* in his 24. booke, 9. Chapter, the Athenian Matrons in their feaſt called *Theſmophoria* dedicated to the honour of *Ceres*, deſirous to keepe themſelues chaſte, doe lay the leaues in their beds vnder them: the Latines name it *Vitex*, and of diuers it is termed, as wee finde among the baſtard and counterfeit names, ἄγονος: in Latine, *Salix marina*, or *Salix Amerina*, and *Piper Agreſte*: in high Dutch, **Schaffmulle, Keuſchbaum**: in low Dutch, and alſo of the Apothecaries, *Agnus Caſtus*: the Italians, *Vitice, Agno Caſto*: in Spaniſh, *Gattile caſto*: in Engliſh, Chaſte tree, Hempe tree, and of diuers *Agnus caſtus*. ‡ The name *Agnus Caſtus* comes by confounding the Greeke name *Agnos* with *Caſtus*, the Latine interpretation thereof. ‡

¶ *The Temperature.*

The leaues and fruit of *Agnus caſtus* are hot and drie in the third degree: they are of very thin parts, and waſte or conſume winde.

The Vertues.

A *Agnus Caſtus* is a ſingular medicine and remedie for ſuch as would willingly liue chaſte, for it withſtandeth all vncleanneſſe, or deſire to the fleſh, conſuming and drying vp the ſeed of generation, in what ſort ſoeuer it be taken, whether in pouder onely, or the decoction drunke, or whether the leaues be carried about the body; for which cauſe it was called *Caſtus*; that is to ſay, chaſte, cleane, and pure.

B The ſeed of *Agnus Caſtus* drunken, driueth away, and diſſolueth all windineſſe of the ſtomacke, openeth and cureth the ſtoppings of the liuer and ſpleen, and in the beginning of dropſies, it is good to be drunke in wine in the quantitie of a dram.

C The leaues ſtamped with butter, diſſolue and aſſwage the ſwellings of the genitories and cods, being applied thereto.

D The decoction of the herbe and ſeed is good againſt pain and inflammations about the matrix, if women be cauſed to ſit and bathe their priuy parts therein: the ſeed being drunke with Pennyroiall bringeth downe the menſes, as it doth alſo both in a fume and in a peſſary: in a Pultis it cureth the head-ache, the Phrenticke, and thoſe that haue the Lethargie are woont to be bathed herewith, oile and vineger being added thereto.

E The leaues vſed in a fume, and alſo ſtrowed, driue away ſerpents; and beeing layed on doe cure their bitings.

F The ſeed laied on with water doth heale the clifts or rifts of the fundament; with the leaues, it is a remedie for lims out of ioint, and for wounds.

G It is reported that if ſuch as iourney or trauell do carry with them a branch or rod of *Agnus Caſtus* in their hand, it will keep them from Merry-galls, and wearineſſe: *Dioſc.*

CHAP. 55. *Of the Willow Tree.*

¶ *The Deſcription.*

1 The common Willow is an high tree, with a body of a meane thickneſſe, and riſeth vp as high as other trees doe if it be not topped in the beginning, ſoone after it is planted; the barke

barke thereof is ſmooth, tough, and flexible: the wood is white, tough, and hard to be broken: the leaues are long, leſſer and narrower than thoſe of the Peach tree, ſomewhat greene on the vpper ſide and ſlipperie, and on the nether ſide ſofter and whiter: the boughes be couered either with a purple, or elſe with a white barke: the catkins which grow on the toppes of the branches come firſt of all forth, being long and moſſie, and quickly turne into white and ſoft downe, that is carried away with the winde.

1 *Salix.*
The common Willow.

2 *Salix aquatica.*
The Oziar or water Willow.

2 The leſſer bringeth forth of the head, which ſtandeth ſomewhat out, ſlender wands or twigs, with a reddiſh or greene barke, good to make bas. ets and ſuch like workes of: it is planted by the twigs or rods being thruſt into the earth, the vpper part whereof when they are growne vp, is cut off, ſo that which is called the head increaſeth vnder them, from whence the ſlender twigs doe grow, which being oftentimes cut, the head waxeth greater: many times alſo the long rods or wands of the higher Withy trees be lopped off and thruſt into the ground for plants, but deeper, and aboue mans height: of which do grow great rods, profitable for many things, and commonly for bands, wherewith tubs and casks are bound.

3 The Sallow tree or Goats Willow, groweth to a tree of a meane bigneſſe: the trunke or body is ſoft and hollow timber, couered with a whitiſh rough barke: the branches are ſet with leaues ſomewhat rough, greene aboue, and hoarie vnderneath: among which come forth round catkins, or aglets that turne into downe, which is carried away with the winde.

4 This other Sallow tree differeth not from the precedent, but in this one point, that is to ſay, the leaues are greater and longer, and euery part of the tree larger, wherein is the difference. ‡ Both thoſe laſt deſcribed haue little roundiſh leaues like little eares growing at the bottoms of the footſtalkes of the bigger leaues, whereby they may bee diſtinguiſhed from all other Plants of this kinde. ‡

5 The Roſe Willow groweth vp likewiſe to the height and bigneſſe of a ſhrubby tree, the body whereof is couered with a ſcabbed rough barke: the branches are many, whereupon do grow very many twigs of a reddiſh colour, garniſhed with ſmall long leaues, ſomewhat whitiſh: amongſt which come forth little floures, or rather a multiplication of leaues, ioined together in forme of a

3 *Salix Caprea rotundi folia.*
The Goat round leafed Willow.

4 *Salix Caprea latifolia.*
The Goat broad leafed Sallow.

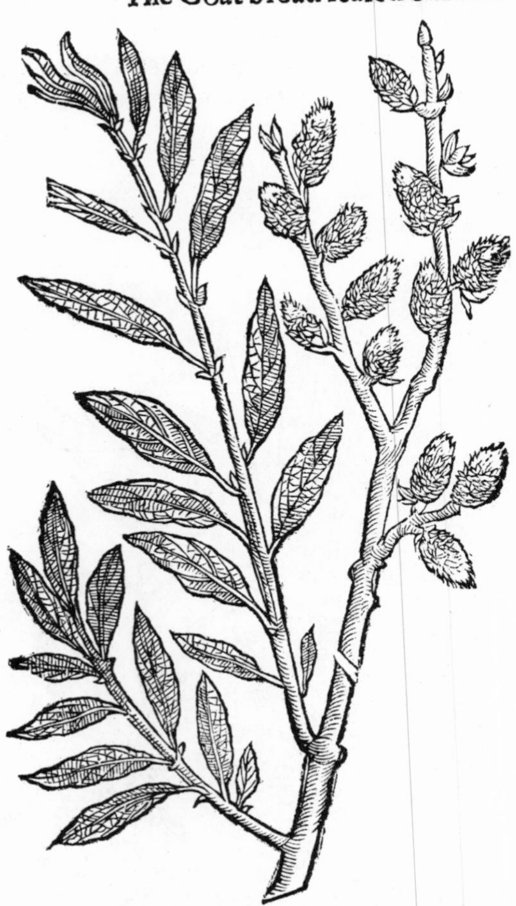

5 *Salix Roſea Anglica.*
The Engliſh Roſe Willow.

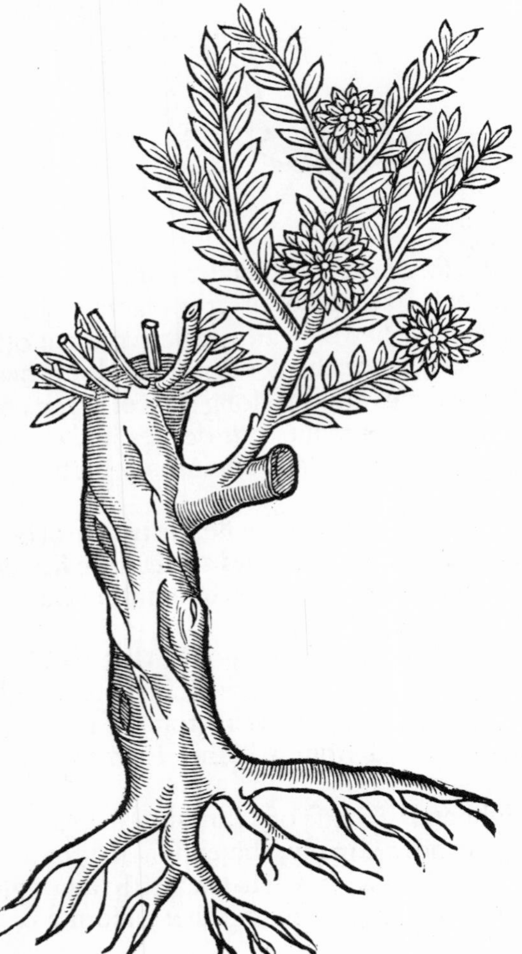

Roſe, of a greeniſh white colour, which doe not only make a gallant ſhew, but alſo yeeld a moſt cooling aire in the heat of Sommer, being ſet vp in houſes, for the decking of the ſame.

6 The low or baſe Willow groweth but low, & leaneth weakly vpon the ground, hauing many ſmall and narrow leaues, ſet vpon limber and pliant branches, of a darke or blackiſh greene colour: amongſt which come forth long ſlender ſtems full of moſſie floures, which turne into a light downie ſubſtance that flieth away with the winde.

7 The dwarfe Willow hath very ſmall and ſlender branches, ſeldome times aboue a foot, but neuer a cubit high, couered with a duskiſh barke, with very little and narrow leaues, of a greene colour aboue, and on the vpper ſide, but vnderneath of a hory or ouer-worne greeniſh colour, in bigneſſe and faſhion of the leaues of garden Flax: among which come forth little duskiſh floures, which doe turne into downe that is carried away with the winde: the root is ſmall and threddy, of the bigneſſe of a finger, and of a blackiſh colour.

8 There is another kinde of willow like to the former, and differeth from it in that, the leaues of this kinde are ſmaller and narrower, as big as the leaues of Myrtle, hauing ſmall knobbie floures of a duskiſh colour,

6 *Salix humilis.*
The low Willow.

7 *Chamæitea, siue Salix pumila.*
The dwarfe Willow.

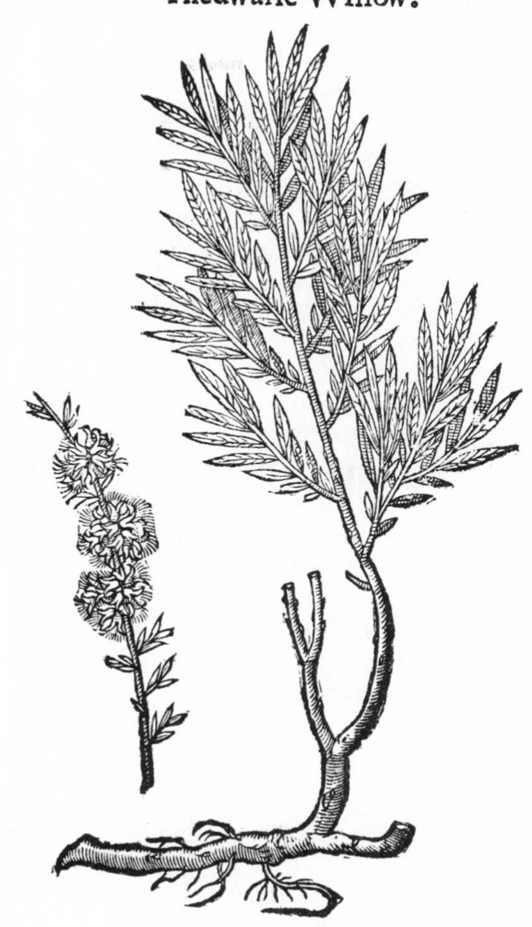

‡ 8 *Salix humilis repens.*
Creeping dwarfe Willow.

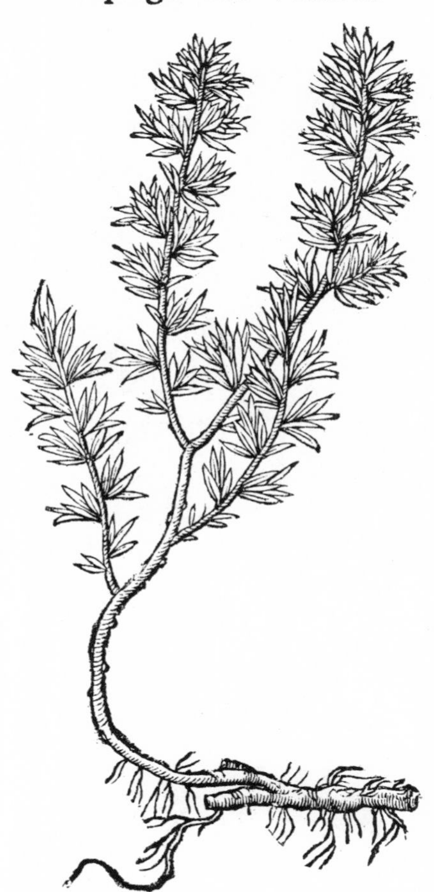

colour, which turne into downe that flyeth away with the winde: the root is small and limber, not growing deep, but running along vpon the vpper crust of the earth.

¶ *The Place.*

These Willowes grow in diuers places of England: the Rose-Willow groweth plentifully in Cambridge shire, by the riuers and ditches there in Cambridge towne they grow aboundantly about the places called Paradise and Hell-mouth, in the way from Cambridge to Grandchester: I found the dwarfe Willowes growing neere to a bog or marish ground at the further end of Hampsted heath vpon the declining of the hill, in the ditch that incloseth a small Cottage there, not halfe a furlong from the said house or cottage.

¶ *The Time.*

The willowes do floure at the beginning of the Spring.

¶ *The Names.*

The Willow tree is called in Greeke Ἰτέα: in Latine, *Salix*: in high-Dutch, **Weyden**: in low-Durch, **Wilgen**: in Italian, *Salice*, *Salcio*: in French, *Saux*: in Spanish, *Salgueiro*, *Salzer*, and *Sauz*: in English, Sallow, Withie, and Willow.

The

The greater is called in Latine *Salix perticalis*, common Withy, Willow, and Sallow, eſpecially that which being often lopped ſendeth out from one head many boughs: the kinde hereof with the red barke is called of *Theophraſtus*, blacke Withy; and the other, white: *Pliny* calleth the black *Græca*, or Greeke Withie (the red, being the Greeke Withy, ſaith he, is eaſie to be cleft) and the whiter, *Amerina*.

Theophraſtus writeth, that the Arcadians do call the leſſer Ἑλίκη, not Ἰτέα: *Pliny* alſo nameth this *Helice*: both of them do make this to be *Salicis tertia ſpecies*, the third kinde of Sallow: the ſame is likewiſe called in Latine, *Salix pumila, Salix viminalis, Gallica Salix*; and by *Columella*, *Sabina*, which he ſaith that many do terme *Amerina*: in high-Dutch, **Kleyn Weyden**: in low-Dutch, **Wijmen**: in Engliſh, Oſier, ſmall Withy, Twig Withy: *Petrus Creſcentius* nameth it *Vincus*.

The Temperature.

The leaues, floures, ſeed, and barke of Willowes are cold and dry in the ſecond degree, and aſtringent.

¶ *The Vertues.*

A The leaues and barke of Withy or Willowes do ſtay the ſpitting of bloud, and all other fluxes of bloud whatſoeuer in man or woman, if the ſaid leaues and barke be boiled in wine and drunke.

B The greene boughes with the leaues may very well be brought into chambers and ſet about the beds of thoſe that be ſicke of feuers, for they do mightily coole the heate of the aire, which thing is a wonderfull refreſhing to the ſicke Patients.

C The barke hath like vertues: *Dioſcorides* writeth, that this being burnt to aſhes, and ſteeped in vineger, takes away cornes and other like riſings in the feet and toes: diuers, ſaith *Galen*, doe ſlit the barke whileſt the Withy is in flouring, and gather a certain iuice, with which they vſe to take away things that hinder the ſight, and this is when they are conſtrained to vſe a clenſing medicine of thin and ſubtill parts.

Chap. 56. *Of the Oliue Tree.*

1 *Olea ſatiua.*
The manured Oliue tree.

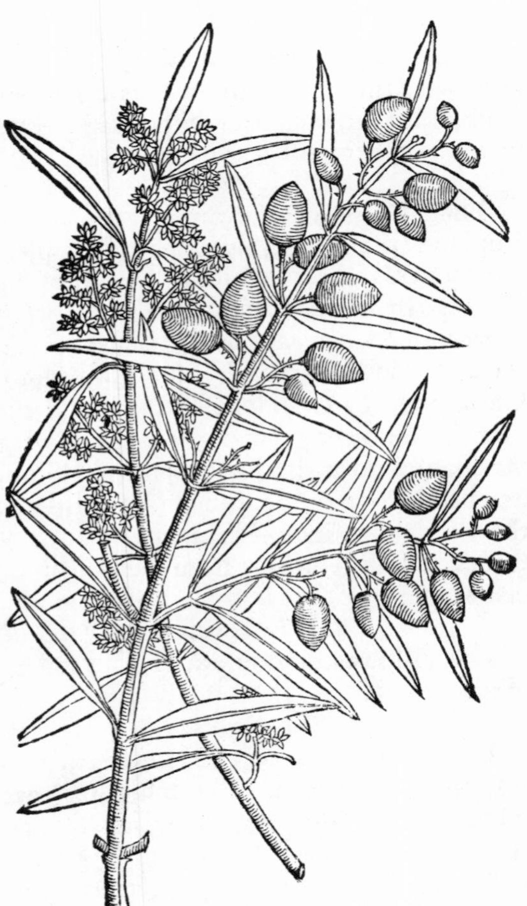

2 *Olea ſylueſtris.*
The wilde Oliue tree.

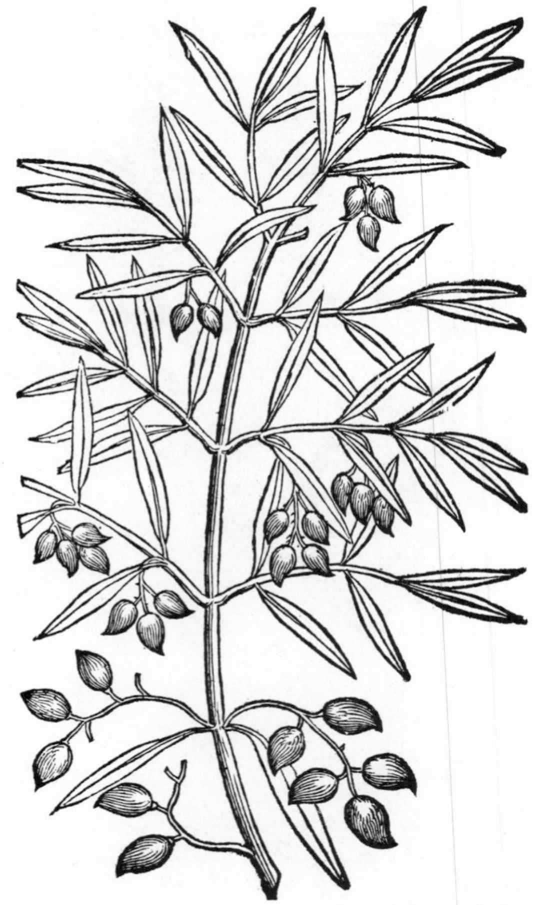

¶ *The*

¶ *The Description.*

1 THe tame or manured Oliue tree groweth high and great with many branches, full of long narrow leaues not much vnlike the leaues of Willowes, but narrower and smaller: the floures be white and very small, growing vpon clusters or bunches: the fruit is long and round, wherein is an hard stone: from which fruit is pressed that liquor which we call oyle Oliue.

2 The wilde Oliue is like vnto the tame or garden Oliue tree, sauing that the leaues are something smaller: among which sometimes do grow many prickely thornes: the fruit hereof is lesser than of the former, and moe in number, which do seldome come to maturitie or ripenes in somuch that the oile which is made of those berries continueth euer green, and is called Oile Omphacine, or oile of vnripe Oliues.

¶ *The Place.*

Both the tame and the wilde Oliue trees grow in very many places of Italy, France, and Spaine, and also in the Islands adioyning: they are reported to loue the sea coasts; for most do thinke, as *Columella* writeth, that aboue sixty miles from the sea they either dy, or else bring forth no fruit: but the best, and they that do yeeld the most pleasant Oyle are those that grow in the Island called Candy.

¶ *The Time.*

All the Oliue trees floure in the moneth of Iune: the fruit is gathered in Nouember or December: when they be a little dried and begin to wrinckle they are put into the presse, and out of them is sqeezed oile, with water added in the pressing: the Oliues which are to be preserued in salt and pickle must be gathered before they be ripe, and whilest they are greene.

¶ *The Names.*

The tame or garden Oliue tree is called in Greeke Ἐλαία, and Ἐλαία ἥμερος: in Latine, *Olea satiua*, and *Vrbana*: in high-Dutch, **Oelbaum:** in low-Dutch, **Olijfboome:** in Italian, *Oliuo domestico*: in French, *Oliuier*: in Spanish, *Oliuo*, and *Oliuera*: in English, Oliue tree.

The berry is called *Oliua*: in Greeke also Ἐλαία: in Spanish, *Azeytuna*: in French, Dutch, and English, Oliue.

Oliues preserued in brine or pickle are called *Colymbades*.

The wilde Oliue tree is named in Greeke, Ἀγριελαία: in Latine, *Olea syluestris*, *Oleaster*, *Cotinus*, *Olea Æthiopica*: in Dutch, **Wald Oelbaum**: in Italian, *Oliuo saluatico*: in Spanish, *Azebuche*, *Azambulheyro*: in French, *Oliuier sauuage*: in English, wilde Oliue tree.

¶ *The Temperature and Vertues.*

The Oliues which be so ripe as that either they fall off themselues, or be ready to fall, which are A
named in Greeke, Δρυπετεῖς, be moderatly hot and moist, yet being eaten they yeeld to the body little nourishment.

The vnripe oliues are dry and binding. B

Those that are preserued in pickle, called *Colymbades*, do dry vp the ouermuch moisture of the C
stomacke, they remoue the loathing of meate, stirre vp an appetite; but there is no nourishment at all that is to be looked for in them, much lesse good nourishment.

The branches, leaues, and tender buds of the Oliue tree do coole, dry, and binde, and especial- D
ly of the wilde Oliue; for they be of greater force than those of the tame: therefore by reason they be milder they are better for eye medicines, which haue need of binding things to be mixed with them.

The same do stay S. Anthonies fire, the shingles, epinyctides, night wheales, carbuncles, and ea- E
ting vlcers: being laid on with honey they take away eschares, clense foule and filthy vlcers, and quench the heate of hot swellings, and be good for kernels in the flanke: they heale & skin wounds in the head, and being chewed they are a remedie for vlcers in the mouth.

The iuyce and decoction also are of the same effect: moreouer, the iuice doth stay all maner of F
bleedings, and also the whites.

The iuice is pressed forth of the stamped leaues, with wine added thereto (which is better) or G
with water, and being dried in the Sun it is made vp into little cakes like perfumes.

The sweat or oyle which issueth forth of the wood whilest it is in burning healeth tetters, scurfs H
and scabs, if they be anointed therewith.

The same which is pressed forth of the vnripe Oliues is as cold as it is binding. I

The old oile which is made of sweet and ripe Oliues, being kept long, doth withall become hot- K
ter, and is of greater force to digest or waste away; and that oile which was made of the vnripe Oliue, being old, doth as yet retaine some part of his former astriction, and is of a mixt faculty, that is to say, partly binding, and partly digesting; for it hath got this digesting or consuming faculty by age, and the other propertie of binding of his owne nature.

The

L The oile of ripe Oliues mollifieth and aſſwageth paine, diſſolueth tumors or ſwellings, is good for the ſtiffeneſſe of the ioints, and againſt cramps, eſpecially being mingled according to art, with good and wholeſome herbes appropriate vnto thoſe diſeaſes and griefes, as *Hypericon*, Cammomill, Dill, Lillies, Roſes, and many others, which do fortifie and increaſe his vertues.

M The oile of vnripe Oliues, called *Omphacinum oleum*, doth ſtay, repreſſe, and driue away the beginning of tumors and inflammations, cooling the heate of burning vlcers and exulcerations.

Chap. 57. *Of Priuet or Prim Print.*

Liguſtrum.
Priuet, or Prim Print.

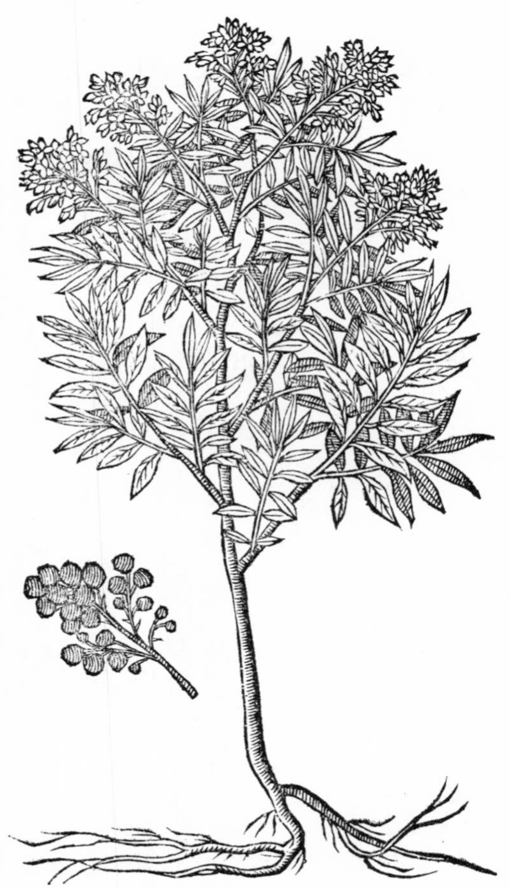

¶ *The Deſcription.*

Priuet is a ſhrub growing like a hedge tree, the branches and twigs wherof be ſtraight, and couered with ſoft gliſtring leaues of a deepe green colour, like thoſe of Peruincle, but yet longer, greater alſo than the leaues of the Oliue tree: the floures be white, ſweet of ſmell, very little, growing in cluſters; which being vaded there ſucceed cluſters of berries, at the firſt greene, and when they be ripe blacke like a little cluſter of grapes, which yeeld a purple iuice: the root groweth euery way aſlope.

¶ *The Place.*

The common Priuet groweth naturally in euery wood, and in the hedge rowes of our London gardens: it is not found in the countrey of Polonia and other parts adiacent.

¶ *The Time.*

It floureth in the end of May, or in Iune: the berries are ripe in Autumne or about Winter, which now and then continue all the Winter long; but in the meane time the leaues fall away, and in the Spring new come vp in their places.

¶ *The Names.*

It is called in Latine, *Liguſtrum*: in Italian at this day, *Guiſtrico*, by a corrupt word drawne from *Liguſtrum*: it is the Grecians φιλλύρεα, and in no wiſe κύπρος: for Cyprus is a ſhrub that groweth naturally in the Eaſt, and Priuet in the Weſt. They be very like one vnto another, as the deſcriptions doe declare, but yet in this they differ, as witneſſeth *Bellonius*, becauſe the leaues of Priuet do fall away in winter, and the leaues of Cyprus are alwaies greene: moreouer, the leaues of Cyprus do make the haire red, as *Dioſcorides* ſaith, and (as *Bellonius* reporteth) do giue a yellow colour: but the leaues of Priuet haue no vſe at all in dying. And therefore *Pliny*, *lib.* 24. *cap.* 10. was deceiued, in that he iudged Priuet to be the ſelfe ſame tree which Cyprus is in the Eaſt: which thing notwithſtanding he did not write as hee himſelfe thought, but as other men ſuppoſe; for, *lib.* 12. *cap.* 14. he writeth thus: Some (ſaith he) affirme this, *viz.* Cyprus, to be that tree which is called in Italy, *Liguſtrum*; and that *Liguſtrum* or Priuet is that plant which the Grecians call φιλλύρεα, the deſcription doth declare.

Phillyria, ſaith *Dioſcorides*, is a tree like in bigneſſe to Cyprus, with leaues blacker and broader than thoſe of the Oliue tree: it hath fruit like to that of the Maſtick tree, blacke, ſomething ſweet, ſtanding in cluſters, and ſuch a tree for all the world is Priuet, as we haue before declared.

Serapio the Arabian, *cap.* 44. doth call Priuet *Mahaleb*. There is alſo another *Mahaleb*, which is a graine or ſeed of which *Auicen* maketh mention, *cap.* 478. that it doth by his warme and comfortable heate diſſolue and aſſwage paine. *Serapio* ſeemeth to intreat of them both, and to containe diuers of the *Mahaleb* vnder the title of one chapter: it is named in high-Dutch, **Beinholtzlein, Mundtholtz, Rhein oder Schulweiden**: in low-Dutch, **Keelcruyt, Monthout**: in French, *Troeſne*: in Engliſh, Priuet, Primprint, and Print.

Some

Some there be that would haue the berries to be called *Vaccinia*, and *Vaccinium* to be that of which *Vitruvius* hath made mention in his ſeuenth booke of Architecture or the art of building, chap. 14. of purple colours: after the ſame manner, ſaith he, they temper *Vaccinium*, and putting milke vnto it do make a gallant purple: in ſuch breuitie of the old writers what can be certainely determined.

¶ *The Temperature.*

The leaues and fruit of Priuet are cold, dry, and aſtringent.

¶ *The Vertues.*

The leaues of Priuet do cure the ſwellings, apoſtumations, and vlcers of the mouth or throat, A
being gargariſed with the iuyce or decoction thereof, and therefore they be excellent good to be put into lotions, to waſh the ſecret parts, and the ſcaldings with women, cankers and ſores in childrens mouthes.

CHAP. 58. *Of Mocke-Priuet.*

1 *Phillyrea anguſtifolia.*
Narrow leaued Mock-Priuet.

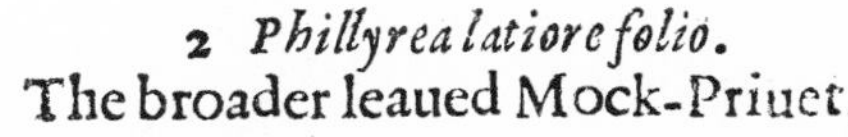
2 *Phillyrea latiore folio.*
The broader leaued Mock-Priuet.

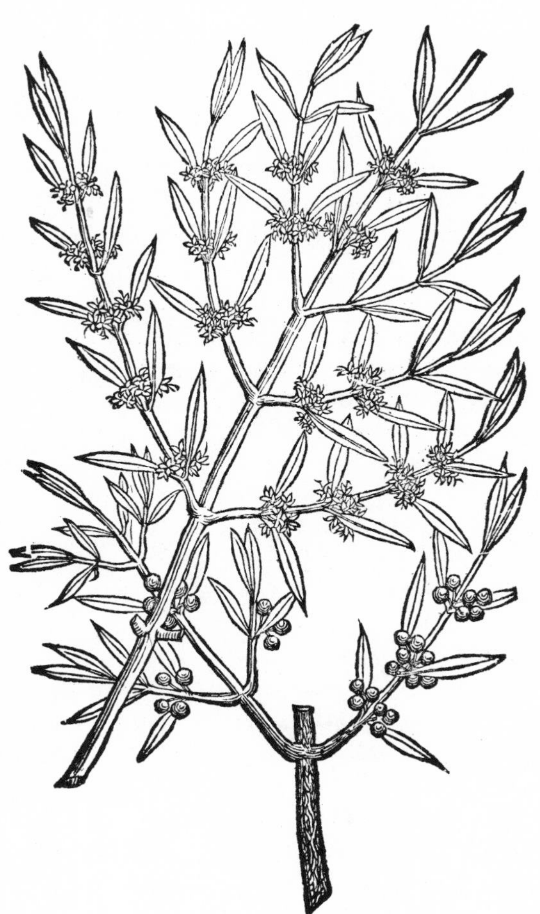

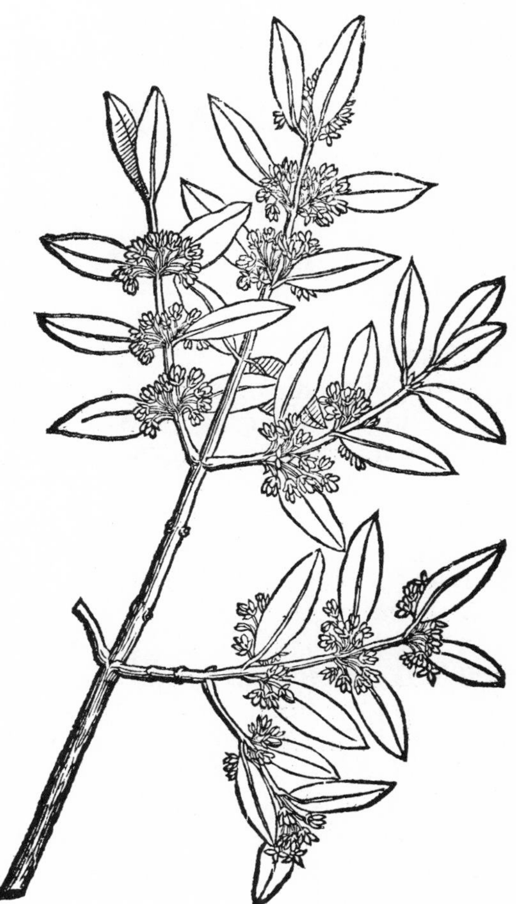

¶ *The Deſcription.*

1 Cyprus is a kinde of Priuet, and is called *Phillyrea*, which name all the ſorts or kindes thereof do retaine, though for diſtinctions ſake they paſſe vnder ſundry titles. This plant groweth like an hedge tree, ſometimes as big as a Pomegranat tree, beſet with ſlender twiggy boughes which are garniſhed with leaues growing by couples, very like the leaues of the Oliue tree, but broader, ſofter, and of a greene colour: from the boſomes of theſe leaues come forth great bunches of ſmall white floures, of a pleaſant ſweet ſmell: which being vaded, there ſucceed cluſters of blacke berries very like the berries of the Alder tree.

2 The ſecond Cyprus, called alſo *Phillyrea latifolia*, is very like the former in body, branches, leaues,

3 *Phillyrea serrata 2. Clusij.*
The second toothed Priuet of *Clusius.*

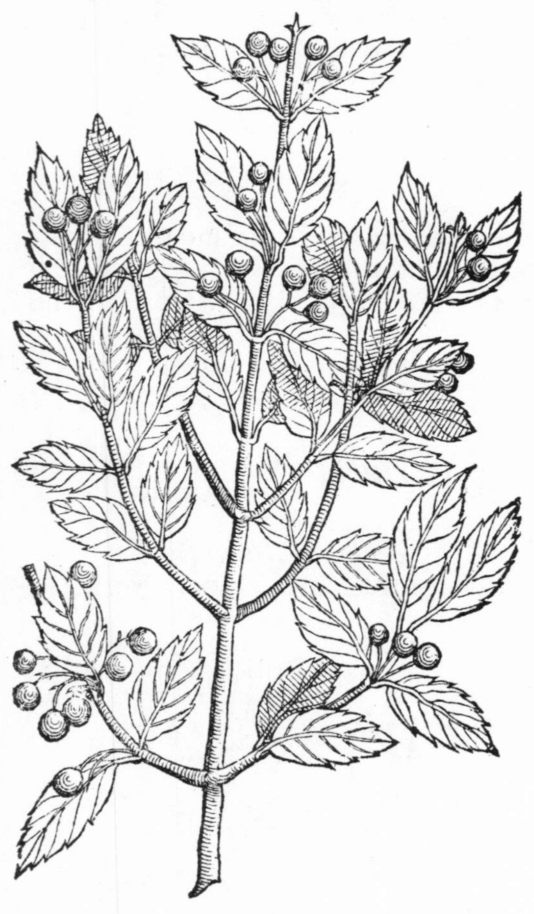

leaues, floures, and fruit; and the difference is this, that the leaues of this plant are broader, but in facultie they are like.

3 This kinde of Priuet riseth vp like an hedge bush, of the height of fiue or six cubits: the branches are long, fragile or brittle, couered with a whitish barke; whereon are set leaues somwhat broad, iagged on the edges like the teeth of a saw, and of a deep green colour: among which come forth the floures, which neither my Author nor my selfe haue seene: the berries grow vpon small footstalks, for the most part three together, being round, and of the bignesse of pepper graines, or Myrtle berries, of a blacke colour when they be ripe.

¶ *The Place.*

These plants do grow in Syria neere the city Ascalon, and were found by our industrious *Pena* in the mountaines neere Narbone and Montpelier in France: the which I planted in the garden at Barn-Elmes neere London, belonging to the right Honourable the Earle of Essex: I haue them growing in my garden likewise.

¶ *The Time.*

The leaues shoot forth in the first of the Spring: the floures shew themselues in May and Iune: the fruit is ripe in September.

¶ *The Names.*

This Priuet is called in Greeke, κύπρος, and in Latine also *Cyprus*; and may be named in English, Easterlin Priuet, and Mocke-Priuet, for the reason following: they are deceiued who taking *Pliny* for their Author, do thinke that it is *Ligustrum,* or our Westerne Priuet, as wee haue shewed in the former chap. it is the Arabians *Alcanna,* or *Henne:* and it is also called of the Turks *Henne* euen at this present time.

¶ *The Temperature.*

The leaues of these kindes of Priuet haue a binding qualitie, as *Dioscorides* writeth.

¶ *The Vertues.*

A Being chewed in the mouth they heale the vlcers thereof, and are a remedie against inflammations or hot swellings.

B The decoction thereof is good against burnings and scaldings.

C The same being stamped and steeped in the iuice of Mullen and laid on, do make the haire red, as *Dioscorides* noteth. *Bellonius* writeth, that not only the haire, but also the nether parts of mans body and nailes likewise are coloured and died herewith, which is counted an ornament among the Turks.

D The floures being moistned in vineger and applied to the temples asswageth head-ache.

E There is also made of these an oile called *Oleum Cyprinum,* sweet of smell, and good to heate and supple the sinewes.

Chap. 59. *Of bastard Priuet.*

¶ *The Description.*

1 This shrubby tree, called *Macaleb,* or *Mahaleb,* is also one of the Priuets: it riseth vp like vnto a small hedge tree, not vnlike vnto the Damson or Bullesse tree, hauing many vpright stalks and spreading branches: whereon do grow leaues not vnlike those of the *Phillyrea* of *Clusius* description: amongst which come forth mossie floures of a white colour, and of a perfect sweet

ſweet ſmell, growing in cluſters, many hanging vpon one ſtem, which the Grauer hath omitted: after which come the berries, greene at the firſt, and blacke when they be ripe, with a little hard ſtone within, in which lieth a kernell.

2 *Geſner* and *Matthiolus* haue ſet forth another *Macaleb*, being alſo another baſtard Priuet. It groweth to a ſmall hedge tree, hauing many greene branches ſet with round leaues like thoſe of the Elme tree, ſomwhat ſnipt about the edges: the floures are like thoſe of the precedent: The fruit, or rather the kernell thereof, is as hard as a beade of Corall, ſomewhat round, and of a ſhining blacke colour; which the cunning French Perfumers do bore thorow, making thereof bracelets, chaines, and ſuch like trifling toyes, which they ſend into England, ſmeared ouer with ſome odde ſweet compound or other, and they are here ſold vnto our curious Ladies and Gentlewomen for rare and ſtrange Pomanders, for great ſummes of money.

1 *Phillyrea arbor, verior Macaleb.*
Baſtard Priuet.

2 *Macaleb Geſneri.*
Corall Priuet.

¶ *The Place.*

Theſe trees grow in diuers places of France, as about Tholouſe, and ſundry other places: they are ſtrangers in England.

¶ *The Time.*

The floures bud forth in the Spring: the fruit is ripe in Nouember and December.

¶ *The Names.*

This baſtard Priuet is that tree which diuers ſuſpect to be that *Mahaleb* or *Macaleb* of which *Auicen* writeth, *cap.* 478. and which alſo *Serapio* ſpeaketh of out of *Meſue*: but it is an hard thing to affirme any certaintie thereby, ſeeing that *Auicen* hath deſcribed it without markes: notwithſtanding this is taken to be the ſame of moſt writers, and thoſe of the beſt: we may call it in Engliſh, baſtard Priuet, or Corall, or Pomander Priuet, being without doubt a kinde thereof.

¶ *The Temperature and Vertues.*

A Concerning this baſtard Priuet we haue learned as yet no vſe thereof in Phyſicke. The kernels which are found in the ſtones or fruit, as they be like in taſte to thoſe of Cherries, ſo be they alſo anſwerable to them in temperature; for they are of a temperat heate, and do gently prouoke vrine, and be therefore good for the ſtone: more we haue not to write than hath beene ſpoken in the deſcription.

CHAP. 60. *Of the fruitlesse Priuet.*

¶ *The Description.*

1 THis shrubby bush, called of *Pliny* and *Carolus Clusius, Alaternus*, groweth vp to a small hedge tree, in forme like vnto a bastard Priuet; but the leaues are more like those of *Ilex*, or the French Oke, yet stiffer and rounder than those of *Macaleb*: amongst which come forth tufts of greenish yellow floures like those of the Lentiske tree: vnder and among the leaues come forth the berries, like those of *Laurus Tinus*, in which are contained two kernels like to the Acines or stones of the Grape.

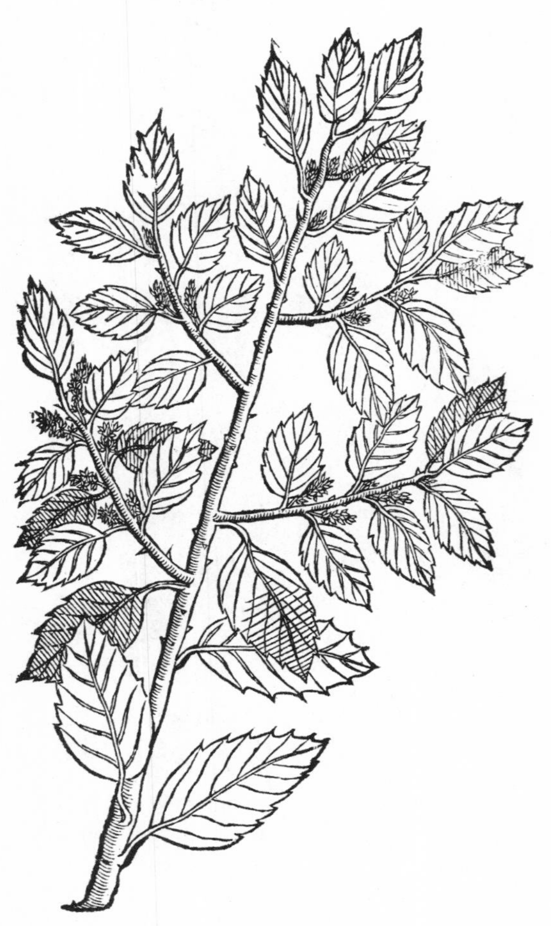

1 *Alaternus Plinij.*
Fruitlesse Priuet.

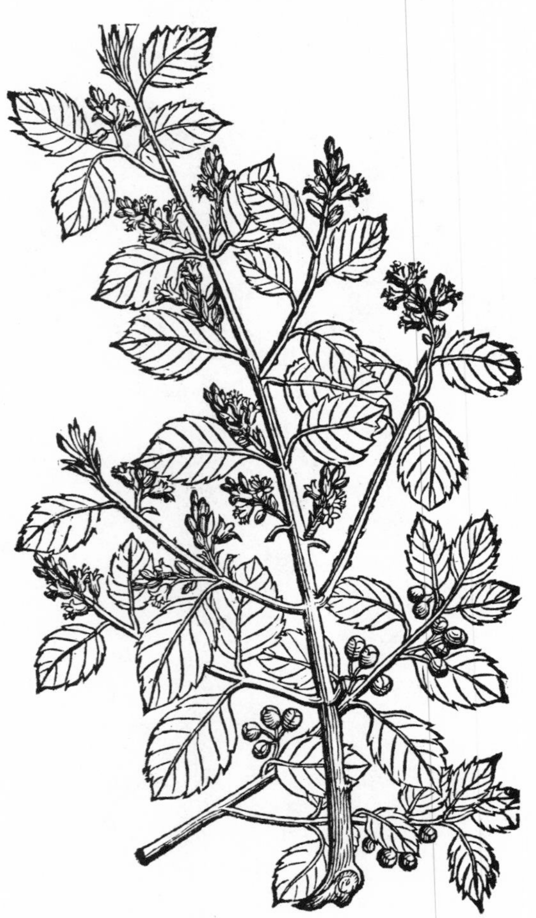

2 *Alaternus humilior.*
The lower fruitlesse Priuet.

2 The second kinde of *Alaternus* is likewise a fruitlesse kinde of Priuet, hauing narrow leaues somewhat snipt about the edges: from the bosomes whereof come forth small herby coloured floures; which being vaded, there succeedeth the fruit, whereof *Auicen* speaketh, calling it by the name *Fagaras*, being a fruit in bignesse and forme like those in shops called *Cocculus-indi*, and may be the same for any thing that hath been written to the contrarie. This fruit hangeth as it were in a darke ash-coloured skin or huske, which incloseth a slender stiffe shell like the shell of a nut, couered with a thin or blacke filme, whether it be the fruit of this plant it is not censured; notwithstanding you shall finde the figure hereof among the Indian fruits, by the name *Fagaras*.

‡ This hath shorter branches and rounder leaues than the former: the floures are larger and greener; to which succeed fruit clustering together, first greene, then red, and afterwards blacke, and consisting of three kernells: it floures in Februarie and the beginning of March, and growes in sundry places of Spaine. The fruit of this is not the *Fagaras*, neither doth the *Fagaras* mentioned by our Author any way agree with the *Cocculus Indi* of the shops, as shall be shewed hereafter in their fitplces. ‡

¶ *The Place.*

These plants do grow in the shadowie woods of France, and are strangers in England.

¶ *The*

¶ *The Time.*

The time anſwereth the reſt of the Priuets.

¶ *The Names.*

Alaternus of *Pliny* is the ſame *Phillyrea* which *Theophraſtus* hath written of by the name *Philyca*, and *Bellonius* alſo, *lib.* 1. *cap.* 42. of his Singularities, and the people of Candy call it *Elæprinon*: the Portugals, *Caſca*: in French, *Dalader*, and *Sangin blanc*: in Engliſh, barren or fruitleſſe Priuet: notwithſtanding ſome haue thought it to beare fruit, which at this day is called *Fagaras*: with vs, *Cocculus-Indi*, as we haue ſaid. ‡ I can by no meanes approue of the Engliſh name here giuen by our Author; but iudge the name of Euer-greene Priuet, (giuen it by Mr. *Parkinſon*) to be much more fitting to the thing. ‡

¶ *The Temperature and Vertues.*

Whether the plant be vſed in medicine I cannot as yet learne: the fiſhermen of Portugall do vſe to ſeethe the barke thereof in water, with the which decoction they colour their nets of a reddiſh colour, being very fit for that purpoſe: the wood alſo is vſed by Dyers to dye a darke blacke withall. A

CHAP. 61. *Of the white and blew Pipe-Priuet.*

1 *Syringa alba.*
White Pipe.

2 *Syringa cærulea.*
Blew Pipe.

¶ *The Deſcription.*

1 THe white Pipe groweth like an hedge tree, or buſhy ſhrub: from the root whereof ariſe many ſhoots, which in ſhort time grow to be equall with the old ſtocke, whereby in little time it increaſeth to infinite numbers, like the common Engliſh Prim or Priuet, whereof doubtleſſe it is a kinde, if we conſider euery circumſtance: the branches are couered with a rugged gray barke: the timber is white, with ſome pith or ſpongie matter in the middle like Elder, but leſſer in quantitie. Theſe little branches are garniſhed with ſmall crumpled leaues of the ſhape and bigneſſe of Peare tree leaues, and very like in forme: among which come forth the

‡ 3 *Syringa Arabica.*
Arabian Pipe.

4 *Balanus Myrepsica, siue Glans vnguentaria.*
The Oylie Acorne.

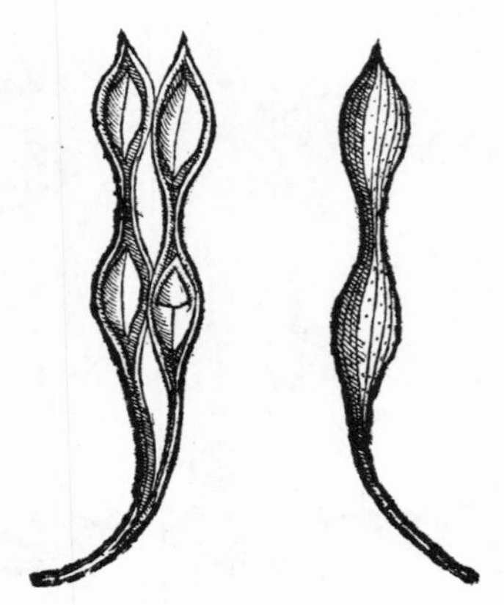

the floures, growing in tufts, compact of foure small leaues of a white colour, and of a pleasant sweet smell; but in my iudgement they are too sweet, troubling and molesting the head in very strange manner. I once gathered the floures and laid them in my Chamber window, which smelled more strongly after they had lien together a few houres, with such an vnacquainted sauor, that they awaked me out of my sleepe, so that I could not take any rest till I had cast them out of my chamber. When the floures be vaded then followeth the fruit, which is small, curled, and as it were compact of many little folds, broad towards the vpper part, and narrow towards the stalk, and black when it is ripe, wherin is contained a slender and long seed. The root hereof spreadeth it selfe abroad in the ground, after the manner of the roots of such shrubbie trees.

2 The blew Pipe groweth likewise in maner of a smal hedge tree, with many shoots rising from the root like the former, as our common Priuet doth, whereof it is a kinde. The branches haue some small quantitie of pith in the middle of the wood, and are couered with a darke blacke greenish barke or rinde. The leaues are exceeding greene, and crumpled or turned vp like the brimmes of an hat, in shape very like vnto the leaues of the Poplar tree: among which come the floures, of an exceeding faire blew colour, compact of many small floures in the forme of a bunch of grapes: each floure is in shew like those of *Valeriana rubra Dodonæi*, consisting of foure parts like a little star, of an exceeding sweet sauor or smel, but not so strong as the former. When these floures be gone, there succeed flat cods, and somewhat long, which being ripe are of a light colour, with a thinne membrane or filme in the midst, wherein are seeds almost foure square, narrow and ruddy.

‡ 3 This (which *Clusius* setteth forth by the name of *Iasminum Arabicum*, or *Syringa Arabica*) groweth some two or three cubits high, diuided into many slender branches, whereon by couples at each ioint stand leaues like those of the first described, but thinner, and not snipt about the edges: on the tops of the branches grow the floures, wholly white, consisting of nine, ten, or twelue leaues set in two rankes: these floures are very sweet, hauing a sent as it were compounded of the Spanish Iasmine, and Orange floures. It is a tender plant, and may be graffed vpon the common Iasmine, whereon it thriues well, and floures most part of the Sommer. It groweth plentifully in Egypt; and *Prosper Alpinus* is thought to mention this by the name of *Sambac Arabum, siue Gelseminum Arabicum*. ‡

4 *Glans vnguentaria*, or the oylie Acorne, is the fruit of a tree like Tamariske, of the bignesse of an Hasell Nut, out of the kernell whereof, no otherwise than out of bitter Almonds, is pressed an oylie iuyce which is vsed in pretious Oyntments, as *Dioscorides* affirmeth: neither is it in our time wholly reiected; for the oyle of this fruit mixed with sweet odours serueth to perfume gloues

gloues and diuers other things; and is vulgarly knowne by the name of Oyle of Ben.

¶ *The Place.*

1. 2. These trees grow not wilde in England, but I haue them growing in my garden in very great plenty.

¶ *The Time.*

They floure in Aprill and May, but as yet they haue not borne any fruit in my garden, though in Italy and Spaine their fruit is ripe in September.

¶ *The Names.*

The later Physitians call the first *Syringa*, or rather σύριγξ: that is to say, a Pipe, because the stalks and branches thereof, when the pith is taken out, are hollow like a pipe: it is also many times syrnamed *Candida*, or white, or *Syringa candido flore*, or Pipe with a white floure, because it should differ from *Lillach*, which is sometimes named *Syringa cœrulea*, or blew Pipe: in English, White Pipe.

Blew Pipe the later Physitians, as we haue said, do name *Lillach*, or *Lilac*: of some, *Syringa cœrulea*, or blew Pipe: most do expound the word *Lillach*, and call it *Ben*: *Serapio's* and the Arabians *Ben* is *Glans vnguentaria*, which the Grecians name βάλανος μυρεψική, from which *Lillach* doth very much differ: among other differences it is very apparant, that *Lillach* bringeth forth no Nut, howsoeuer *Matthiolus* doth falsly picture it with one; for it hath only a little cod, the seed whereof hath in it no oile at all. The figure of the *Balanus Myrepsica* we haue thought good to insert in this chapter, for want of a more conuenient roome.

¶ *The Temperature and Vertues.*

Concerning the vse and faculties of these shrubs neither we our selues haue found out any thing nor learned ought of others. A

‡ The *Balanus Myrepsica* taken in the quantitie of a dram, causeth vomit; drunk with *Hydromel* it purges by stoole, but is hurtfull to the stomacke. B

The oile pressed out of this fruit, which is vsually termed oyle of Ben, as it hath no good or pleasing smell, so hath it no ill sent, neither doth it become rancide by age, which is the reason that it is much vsed by perfumers. C

The oile smoothes the skin, softens and dissolues hardnesse, and conduces to the cure of all cold affects of the sinewes; and it is good for the paine and noise in the eares, being mixed with Goosegrease, and so dropped in warme in a small quantitie. ‡ D

Chap. 62. *Of Widow-Waile, or Spurge Oliue.*

¶ *The Description.*

Widow-waile is a small shrub about two cubits high. The stalke is of a wooddy substance, branched with many small twigs, full of little leaues like Priuet, but smaller and blacker, on the ends whereof grow small pale yellow floures: which being past, there succeedeth a three cornered berrie like the Tithymales, for which cause it was called *Tricoccos*, that is, three berried *Chamelæa*: these berries are greene at the first, red afterward, and browne when they be withered, and containe in them an oylie fatnesse like that of the Oliue, being of an hot and biting taste, and that doe burne the mouth, as do both the leaues and rinde. The root is hard and woody.

¶ *The Place.*

It is found in most vntilled grounds of Italy and Languedoc in France, in rough and desart places. I haue it growing in my garden.

¶ *The Time.*

It is alwaies greene: the seed is ripe in Autumne.

¶ *The Names.*

The Grecians call it χαμελαία, as though they should say, low or short Oliue tree: the Latines, *Oleago*, and *Oleastellus*, and likewise *Citocacium*: it is also named of diuers, *Oliuella*, as *Matthiolus Syluaticus* saith: it is called in English, Widow-Waile, *quia facit viduas*.

The fruit is named of diuers, Κόκκος κνίδειος: in Latine, *Coccus cnidicus*: but he is deceiued, saith *Dioscorides*, that nameth the fruit of Spurge-Oliue, *Coccus Cnidicus*: *Auicen* and *Serapio* call *Chamelæa*, or Spurge Oliue, *Mezereon*: vnder which name notwithstanding they haue also contained both the Chamæleons or Carlines; and so haue they confounded *Chamelæa* or Spurge Oliue with the Carlines, and likewise *Thymælea*, or Spurge flax.

Chamelæa Arabum Tricoccos.
Widow-Waile.

Chamelæa Germanica, siue Mezereon.
Spurge Flax, or the dwarfe Bay.

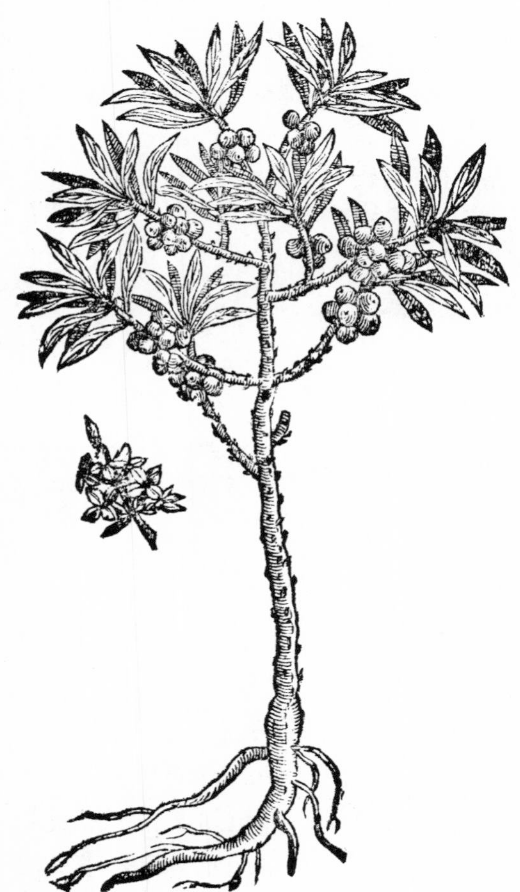

¶ *The Temperature.*

Both the leaues and fruit of Spurge-Oliue, as we haue said, are of a burning and extreme hot temperature.

¶ *The Vertues.*

The leaues, saith *Dioscorides*, purge both flegme and choler, especially taken in pills, so that two parts of Wormewood be mixed with one of Spurge Oliue, and made vp into pils with Mede or honied water. They melt not in the belly, but as many as be taken are voided whole.

Mesue likewise hath a description of pills of the leaues of *Mezereon*, that is, *Chamelæa*, or Spurge-Oliue (yet *Syluius* expoundeth it *Thymelæa*, or Spurge-Flax) but in stead of Wormwood he taketh the outward substance of the yellow Mirobalans and Cepula Mirobalans, and maketh them vp with Tereniabin, that is to say, with Manna and soure Dates, which they call Tamarinds, dissolued in Endiue water; and appointeth the same leaues to be first tempered with very strong vineger, and to be dried.

These pills are commended against the Dropsie, for they draw forth watery humours, but are violent to nature; therfore we must vse them as little as may be. Moreouer, *Dioscorides* addeth, that the leaues of Spurge Oliue beaten with hony do clense filthy or crusted vlcers.

CHAP. 63.

Of Germane Oliue Spurge.

¶ *The Description.*

The dwarfe Bay tree, called of Dutch men *Mezereon*, is a smal shrub two cubits high: the branches be tough, limber, & easie to bend, very soft to be cut; whereon grow long leaues like those of Priuet, but thicker and fatter. The floures appeare before the leaues, oft times in Ianuarie, clustring together about the stalks at certain distances, of a whitish colour tending to purple, and of a most fragrant and pleasant sweet smel: after come the small berries, green at the first, but being ripe, of a shining red colour, and afterward wax of a dark black colour, of a very hot and burning taste, inflaming the mouth and throat, being tasted, with danger of choking. The root is wooddy.

¶ *The Place and Time.*

This plant grows naturally in the moist and shadowy woods of most of the East countries, especially about Meluin in Poland, from whence I haue had great plenty thereof for my garden, where they floure in the first of the Spring, and ripen their fruit in August.

¶ *The Names.*

It is vsually called in high-Dutch, **Zeilant, Zeidelbast, Lenszkraut**, and **Kellerhals**: the Apothecaries

Apothecaries of our countrey name it *Mezereon*, but we had rather call it *Chamelæa Germanica*: in English, Dutch Mezereon, or it may be called Germane Oliue Spurge. We haue heard, that diuers Italians do name the fruit thereof *Piper Montanum*, Mountaine Pepper. Some say that *Laureola* or Spurge Laurell is this plant, but there is another *Laureola*, of which we will hereafter treat: but by what name it is called of the old writers, and whether they knew it or no, it is hard to tell. It is thought to be *Cneoron album Theophrasti*, but by reason of his breuitie, we can affirme no certainty.

There is, saith he, two kindes of *Cneoron*, the white and the blacke, the white hath a leafe, long, like in forme to Spurge Oliue: the black is ful of substance like Mirtle; the low one is more white, the same is with smell, and the blacke without smell. The root of both which groweth deepe, is great: the branches be many, thicke, wooddie, immediatly growing out of the earth, or little aboue the earth, tough: wherefore they vse these to binde with, as with Oziars. They bud and floure when the Autumne Equinoctiall is past, and a long time after. Thus much *Theophrastus*.

The Germane Spurge Oliue is not much vnlike to the Oliue tree in leafe: the floure is sweet of smell: the buds whereof, as we haue written, come forth after Autumne: the branches are wooddy and pliable: the root long growing deepe: all which shew that it hath great likenesse and affinity with *Cneoron*, if it be not the very same.

¶ *The Temperature.*

This plant is likewise in all parts extreme hot: the fruit, the leaues, and the rinde are very sharpe and biting: they bite the tongue, and set the throte on fire.

¶ *The Vertues.*

The leaues of Mezereon do purge downeward, flegme, choler, and waterish humours with great violence. A

Also if a drunkard do eat one graine or berry of this plant, hee cannot be allured to drinke any drinke at that time; such will be the heat of his mouth and choking in the throat. B

This plant is very dangerous to be taken into the body, & in nature like to the Sea Tithymale, leauing (if it be chewed) such an heat and burning in the throat, that it is hard to be quenched. C

The shops of Germany and of the Low-countries do when need require vse the leaues hereof in stead of Spurge Oliue, which may be done without errour; for this Germane Spurge Oliue is like in vertue and operation to the other, therefore it may be vsed in stead therof, and prepared after the like and selfe same manner. D

Chap. 64. *Of Spurge Flax.*

1 *Thymelæa.*
Spurge Flax, or mountaine Widow waile.

¶ *The Description.*

Spurge Flax bringeth forth many slender branched sprigs aboue a cubite high, couered round with long and narrow leaues like those of flax, narrower & lesser than the leaues of Spurge Oliue. The floures are white, small, standing on the vpper parts of the sprigs: the fruit is round, greene at the first, but red when it is ripe, like almost to the round berries of the Hawthorne, in which is a white kernel couered with a blacke skinne, very hot and burning the mouth like Mezereon: the root is hard and wooddie.

¶ *The Place.*

It groweth in rough mountains, and in vntoiled places in hot regions. It groweth in my garden.

¶ *The Time.*

It is greene at any time of the yeare, but the fruit is perfected in Autumne.

¶ *The Names.*

The Grecians call it θυμέλαια: the Syrians, as *Dioscorides* witnesseth, *Apolinon*: diuers also *Chamelæa*, but not properly: but as *Dioscorides* saith, the leafe is properly called *Cneoron*, & the fruit *Coccos Cnidios*; notwithstanding those which *Theophrastus* calleth *Cneora* seem to differ from *Thymelæa*, or Spurge Flax, vnlesse *Nigrum Cneoron* be *Thymelæa* for *Theophrastus* saith that

that there be two kindes of *Cneoron*; the one white, the other blacke: this may be called in English, Spurge Flax, or mountaine Widow Wayle: the seed of *Thymelæa* is called in shops, *Granum Gnidium*.

¶ *The Temperature.*

Spurge Flax is naturally both in leaues and fruit extreme hot, biting, and of a burning qualitie.

¶ *The Vertues.*

A The graines or berries, as *Dioscorides* saith, purge by siege choler, flegme and water, if twenty graines of the inner part be drunke, but t burneth the mouth and throat, wherefore it is to be giuen with fine floure or Barly meale, or in Raisons, or couered with clarified hony, that it may be swallowed.

B The same being stamped with Niter and vineger, serueth to annoint those with, which can hardly sweat.

C The leaues must be gathered about haruest, and being dried in the shade, they are to be layed vp and reserued.

D They that would giue them must beat them, and take forth the strings: the quantity of two ounces and two drams put into wine tempered with water, purgeth and draweth forth watery humors: but they purge more gently if they be boiled with Lentils, and mixed with pot-herbes chopped.

E The same leaues beaten to pouder and made vp into trochisces or flat cakes, with the iuice of sower grapes are reserued for vse.

F The herbe is an enemy to the stomacke, which also destroyeth the birth if it be applied.

† Our Author formerly following *Tabernamontanus* gaue two figures and descriptions in this Chapter, but being both of one thing I omitted the worser figure and description.

Chap. 65. *Of Spurge Laurell.*

Laureola florens.
Laurell, or Spurge Laurell flouring.

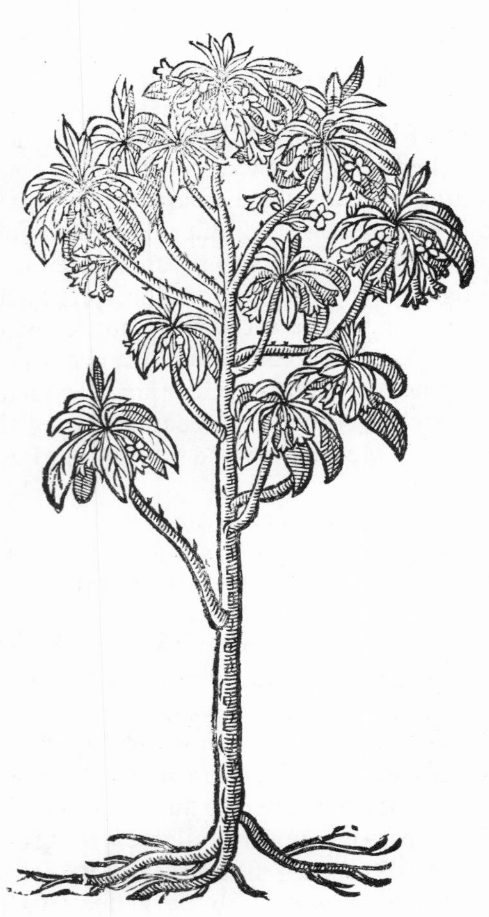

Laureola cum fructu.
Laurell with his fruit.

¶ The

¶ *The Deſcription.*

SPurge Laurell is a ſhrub of a cubit high, oftentimes alſo of two, and ſpreadeth with many little boughes, which are tough and lithy, and couered with a thicke rinde. The leaues be long, broad, groſſe, ſmooth, blackiſh greene, ſhining, like the leaues of Laurell, but leſſer, thicker, and without ſmell, very many at the top, cluſtering together. The floures be long, hollow, of a whitiſh greene, hanging beneath and among the leaues: the berries when they be ripe are blacke, with a hard kernell within, which is a little longer than the ſeed of Hempe: the pulpe or inner ſubſtance is white: the root wooddie, tough, long, and diuerſly parted, growing deepe: the leaues, fruit and barke, as wel of the root as of the little boughes, doe with their ſharpneſſe and burning qualitie bite and ſet on fire the tongue and throat.

¶ *The Place.*

It is found on mountaines, in vntilled, rough, ſhadowie, and wooddie places, as by the lake of Lozanna or Geneua, and in many places neere the riuer of Rhene and of the Maze. ‡ It growes abundantly alſo in the woods in the moſt parts of England. ‡

¶ *The Time.*

The floures bud very ſoon, a little after the Autume Equinoctiall: they are full blown in Winter, or in the firſt Spring: the fruit is ripe in May and Iune: the plant is alwaies greene, and indureth the cold ſtormes of winter.

¶ *The Names.*

It is called in Greeke δαφνοειδὲς, of the likeneſſe it hath with the leaues of the Laurell or Bay tree: in Latine likewiſe *Daphnoides*: the later Latiniſts for the ſame cauſe name it *Laureola*, as though they ſhould ſay *Minor Laurus*, or little Laurell. it is called χαμαιδάφνη, and πέπλιον, notwithſtanding there is another *Chamædaphne*, and another *Peplion*. This ſhrub is commonly called in Engliſh, Spurge Laurell; of diuers, Laurell or Lowry.

Some ſay that the Italians name the berries hereof *Piper montanum*, or Mountaine Pepper, as alſo the berries of Dutch Mezereon: others affirme them to bee called in High Dutch alſo, **Zeilant.**

It may be *Theophraſtus* his *Cneoron*: for it is much like to a Mirtle in leafe, it is alſo a branched plant, tough and pliable, hauing a deep root, without ſmell, with a blacke fruit.

¶ *The Temperature.*

It is like in temperature and facultie to the Germane Spurge Oliue, throughout the whole ſubſtance biting and extreme hot.

¶ *The Vertues.*

The drie or greene leaues of Spurge Laurell, ſaith *Dioſcorides*, purgeth by ſiege flegmaticke humors; it procureth vomite and bringeth downe the menſes, and being chewed it draweth water out of the head. **A**

It likewiſe cauſeth neezing; moreouer, fifteene graines of the ſeed thereof drunke, are a purgation. **B**

CHAP. 66. *Of Roſe Bay, or Oleander.*

¶ *The Deſcription.*

1 ROſe Bay is a ſmall ſhrub of a gallant ſhew like the Bay tree, bearing leaues, thicker, greater, longer and rougher than the leaues of the Almond tree: the floures be of a faire red colour, diuided into fiue leaues, not much vnlike a little Roſe: the cod or fruit is long, like *Aſclepias*, or *Vincetoxicum*, and full of ſuch white downe, among which the ſeed lieth hidden: the root is long, ſmooth, and wooddie.

2 The ſecond kinde of Roſe bay, is like the firſt, & differeth in that, that this plant hath white floures; but in other reſpects it is very like.

The

1 *Nerium, siue Oleander.*
The Rose Bay.

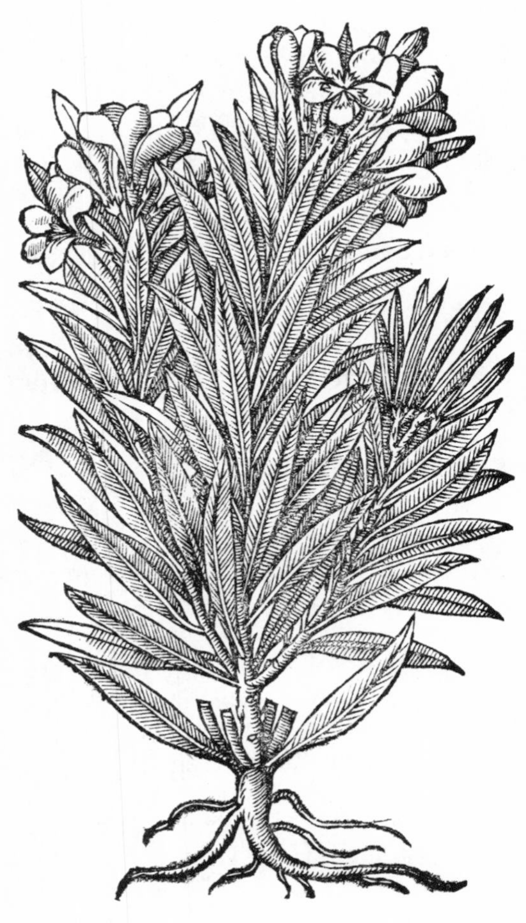

2 *Nerium flore albo.*
The Rose Bay with white floures

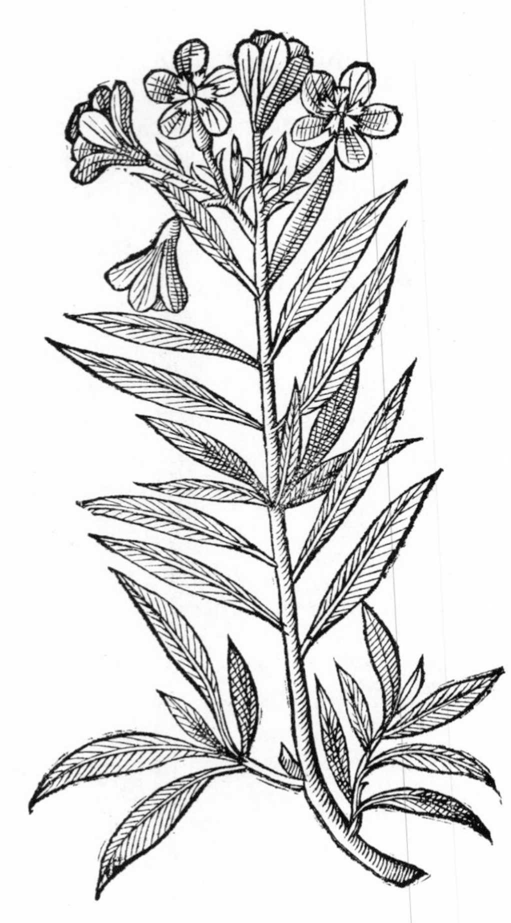

¶ *The Place.*

These grow in Italy and other hot regions, by riuers and the Sea side: I haue them growing in my garden.

¶ *The Time.*

In my garden they floure in Iuly and August: the cods be ripe afterwards.

¶ *The Names.*

This plant is named in Greeke Νήριον, by *Nicander*, Νήρεις: in Latine likewise *Nerion*, and also *Rhododendron*, and *Rhododaphne*, that is to say, *Rosea arbor*, and *Rosea Laurus*: in shops, *Oleander*: in Italian, *Oleandro*: in Spanish, *Adelfa*, *Eloendro*, and *Alendro*: in French, *Rosagine*: in English, Rose tree, Rose Bay, Rose Bay tree and Oleander.

¶ *The Temperature and Vertues.*

A This tree being outwardly applied, as *Galen* saith, hath a digesting facultie: but if it be inwardly taken it is deadly and poisonsome, not only to men, but also to most kindes of beasts.

B The floures and leaues kill dogs, asses, mules, and very many of other foure footed beasts: but if men drinke them in wine they are a remedy against the bitings of Serpents, and the rather if Rue be added.

C The weaker sort of cattell, as sheep and goats, if they drinke the water wherein the leaues haue been steeped, are sure to die.

Chap. 67. Of dwarfe Rose Bay.

¶ *The Description.*

Dwarfe *Nerium*, or Rose Bay, hath leaues which for the most part are alwaies green, rough, and small, of a pale yellow colour like Box, far lesser than Oleander: the whole plant is of a shrubbie stature, leaning this way and that way, as not able to stand vpright without helpe; his branches are couered and set full of small floures, of a shining scarlet or crimson colour; growing vpon the

1 *Chamærhododendros Alpigena.*
Dwarfe Roſe Bay.

Laurus.
The Bay tree.

the hils as ye go from Trent to Verona, which in Iune and Iuly are as it were couered with a ſcarlet coloured carpet, of an odoriferous ſauor, and delectable aſpect, which being fallen there commeth ſeed and faire berries like *Aſparagus*.

¶ *The Place.*

The place and time are expreſſed in the deſcription.

¶ *The Names.*

This may be called in Engliſh, Dwarfe Roſe Bay of the Alps. I find not any thing extant of the vertues, ſo that I am conſtrained to leaue the reſt vnto your owne diſcretion.

† The other plant our Author formerly deſcribed in this chapter in the 2. place by the name of *Chamærododendros montana*, I haue here omitted, becauſe he ſet it forth before by the name of *Ciſtus Ledum Sileſiacum*, giuing 2. figures and one deſcription, in the 11. and 12. places of the 8. chap. of this 3. Booke.

CHAP. 68. *Of the Bay or Laurell tree.*

¶ *The Deſcription.*

1 THe Bay or Laurell tree commeth oftentimes to the height of a tree of a mean bigneſſe; it is full of boughes, couered with a greene barke: the leaues thereof are long, broad, hard, of colour greene, ſweetly ſmelling, and in taſte ſomwhat bitter: the floures alongſt the boughes and leaues are of a greene colour: the berries are more long than round, and be couered with a black rind or pill: the kernell within is clouen into two parts, like that of the Peach and Almond, and other ſuch, of a browne yellowiſh colour, ſweet of ſmell, in taſte ſomewhat bitter, with a little ſharpe or biting qualitie.

2 There is alſo a certaine other kinde hereof more like to a ſhrub, ſending forth out of the roots many offſprings, which notwithſtanding groweth not ſo high as the former, and the barkes of the boughes be ſomewhat red: the leaues be alſo tenderer, and not ſo hard: in other things not vnlike.

Theſe two Bay trees *Dioſcorides* was not ignorant of; for he ſaith, that the one is narrow leafed, and the other broader leafed, or rather harder leafed which is more like.

¶ *The Place.*

The Laurell or bay tree groweth naturally in

ly in Spaine and ſuch hot regions;we plant and ſet it in gardens, defending it from cold at the beginning of March eſpecially.

I haue not ſeene any one tree thereof growing in Denmarke,Sweuia,Poland, Liuonia, or Ruſſia, or in any of thoſe cold countries where I haue trauelled.

¶ *The Time.*

The Bay tree groweth greene winter and Sommer : it floureth in the Spring,and the black fruit is ripe in October.

¶ *The Names.*

This tree is called in Greeke δάφνη: in Latine,*Laurus* : in Italian, *Lauro* · in high Dutch, **Loorbeerbaum** : in low Dutch,**Laurus boome** : in French,*Laurier* : in Spaniſh,*Laurel*,*Lorel*,and *Loureiro* : in Engliſh,Laurell,or Bay tree.

The fruit is named in Greeke δαφνίδες : in Latine,*Lauri baccæ* : in high Dutch,**Looerbeeren** : in low Dutch,**Bakeleer** : in Spaniſh,*Vayas* : in Engliſh, Bay berries.

The Poets faine that it tooke his name of *Daphne*, *Lado* his daughter, with whom *Apollo* fell in loue.

¶ *The Temperature and Vertues.*

A The Berries and leaues of the Bay tree,ſaith *Galen*,are hot and very drie,and yet the berries more than the leaues.

B The barke is not biting and hot,but more bitter,and it hath alſo a certaine aſtrictiue or binding qualitie.

C Bay Berries with Hony or Cute,are good in a licking medicine, ſaith *Dioſcorides*, againſt the pthiſicke or Conſumption of the lungs,difficulty of breathing,and all kinde of fluxes or rheumes about the cheſt.

D Bay Berries taken in wine,are good againſt the bitings and ſtingings of any venomous beaſt,and againſt all venome and poiſon : they clenſe away the morphew : the iuice preſſed out hereof is a remedy for paine of the eares,and deafeneſſe,if it be dropped in with old wine and oile of Roſes : this is alſo mixed with ointments that are good againſt weariſomneſſe, and that heate and diſcuſſe or waſte away humors.

E Bay berries are put into Mithridate,Treacle, and ſuch like medicines that are made to refreſh ſuch people as are growne ſluggiſh and dull by meanes of taking opiate medicines, or ſuch as haue any venomous or poiſoned quality in them.

F They are good alſo againſt cramps and drawing together of ſinewes.

G We in our time do not vſe the berries for the infirmities of the lungs,or cheſt, but miniſter them againſt the diſeaſes of the ſtomacke,liuer,ſpleene, and bladder : they warme a cold ſtomacke,cauſe concoction of raw humours,ſtirre vp a decaied appetite, take away the loathing of meat, open the ſtopping of the liuer and ſpleene,prouoke vrine, bring down the menſes,and driue forth the ſecondine.

H The oile preſſed out of theſe,or drawne forth by decoction, doth in ſhort time take away ſcabs and ſuch like filth of the skin.

I It cureth them that are beaten blacke and blew,and that be bruiſed by ſquats and falls,it remoueth blacke and blew ſpots and congealed bloud,and digeſteth and waſteth away the humors gathered about the grieued part.

K *Dioſcorides* ſaith,that the leaues are good for the diſeaſes of the mother and bladder, if a bath be made thereof to bathe and ſit in : that the greene leaues do gently binde, that being applied, they are good againſt the ſtingings of waſpes and Bees;that with Barly meale parched and bread, they aſſwage all kinde of inflammations, and that being taken in drinke they mitigate the paine of the ſtomacke,but procure vomite.

L The Berries of the Bay tree ſtamped with a little Scammonie and Saffron, and laboured in a mortar with vineger and oile of Roſes to the forme of a liniment, and applied to the temples and forepart of the head,do greatly ceaſe the paine of the Megrim.

M It is reported that common drunkards were accuſtomed to eat in the morning faſting two leaues thereof againſt drunkenneſſe.

N The later Phyſitions doe oftentimes vſe to boyle the leaues of Laurell with diuers meats,eſpecially fiſhes,and by ſo doing there happeneth no deſire of vomiting : but the meat ſeaſoned herewith becommeth more ſauory and better for the ſtomacke.

O The barke of the root of the Bay tree,as *Galen* writeth,drunken in wine prouoketh vrine, breakes the ſtone,and driueth forth grauell : it openeth the ſtoppings of the liuer, the ſpleene,and all other ſtoppings of the inward parts : which thing alſo *Dioſcorides* affirmeth, who likewiſe addeth that it killeth the childe in the mothers wombe.

It

It helpeth the dropſie and the iaundiſe, and procureth vnto women their deſired ſickneſſe. N

‡ Our Author here alſo gaue the two figures of *Tabernamontanus*; the firſt by the name of *Laurus mas*, or the male Bay tree, and the other by the name of *Laurus fœmina*, the female Bay: the difference in the figures was little or none, wherefore I haue made one ſerue.

CHAP. 69. *Of the Wilde Bay tree.*

¶ *The Deſcription.*

1 Laurus Tinus, or the wilde Bay tree, groweth like a ſhrub or hedge buſh, hauing many tough and pliant branches, ſet full of leaues very like to the Bay leaues, but ſmaller and more crumpled, of a deepe and ſhining greene colour: among which come forth tufts of whitiſh floures, turning at the edges into a light purple: after which follow ſmall berries of a blew colour, containing a few graines or ſeeds like the ſtones or ſeeds of grapes: the leaues and all the parts of the plant are altogether without ſmell or ſauour.

1 *Laurus Tinus.*
The wilde Bay tree.

2 *Laurus Tinus Luſitanica.*
The Portingale wilde Bay tree.

2 *Tinus Luſitanica* groweth verie like to *Cornus Fœmina*, or the Dog-berry tree, but the branches be thicker, and more ſtiffe, couered with a reddiſh barke mixed with greene: the leaues are like the former, but larger, hauing many ſinewes or vaines running through the ſame like as in the leaues of Sage: the floures hereof grow in tufts like the precedent, but they are of colour more declining to purple: the ſmall branches are likewiſe of a purple colour: the leaues haue no ſmell at all, either good or bad: the berries are ſmaller than the former, of a blew colour declining to blackneſſe.

¶ *The Place.*

The wilde Bay groweth plentifully in euery field of Italy, Spain, and other regions, which differ according to the nature and ſcituation of thoſe countries: they grow in my garden and proſper very well.

¶ *The Time.*

The wilde Laurell is euer greene, and may oftentimes be seene most part of the winter, and the beginning of the spring, with the floures and ripe berries growing both at one season.

¶ *The Names.*

It is called in Latine *Tinus*, and *Laurus syluestris*: in Greeke, δάφνη ἀγρία: *Cato* nameth it *Laurus syluatica*: in Italian, *Lauro syluatico*: in Spanish, *Vua de Perro*, otherwise *Follado*; and of diuers, *Durillo*: in English wilde Bay.

¶ *The Temperature and Vertues.*

Pliny nor any other of the Antients haue touched the faculties of this wilde Bay, neither haue we any vnderstanding thereof by the later writers, or by our owne experience.

Chap. 70. *Of the Box Tree.*

Buxus.
The Box tree.

¶ *The Description.*

The great Box is a faire tree, bearing a great body or trunke: the wood or timber is yellow and very hard, and fit for sundry workes, hauing many boughes and hard branches, beset with sundry small hard green leaues, both winter and Sommer like the Bay tree: the floures are very little, growing among the leaues, of a greene colour: which being vaded there succeed small blacke shining berries, of the bignes of the seeds of Corianders, which are inclosed in round greenish huskes, hauing three feet or legs like a brasse or boiling pot: the root is likewise yellow, and harder than the timber, but of greater beauty, and more fit for dagger haftes, boxes, and such like vses, whereto the trunke or body serueth, than to make medicines; though foolish empericks and women leaches, do minister it against the Apoplexie and such diseases: Turners and Cutlers, if I mistake not the matter, do call this wood Dudgeon, wherwith they make Dudgeon hafted daggers.

There is also a certaine other kinde hereof, growing low, and not aboue halfe a yard high, but it spreadeth all abroad: the branches hereof are many and very slender: the leaues bee round, and of a light greene.

¶ *The Place.*

Buxus, or the Box tree groweth vpon sundry waste and barren hils in England, and in diuers gardens.

¶ *The Time.*

The Box tree groweth greene winter and Sommer: it floureth in Februarie and March, and the seed is ripe in September.

¶ *The Names.*

The Grecians call it πύξος: in Latine, *Buxus*: in high Dutch, **Buchszbaum**: in low Dutch, **Bux-boom**: in Italian, *Bosso*: in English, Box tree.

The lesser may be called χαμαιπύξος: and in Latine, *Humi Buxus*, or *Humilis Buxus*: in English, dwarf Box, or ground Box, and it is commonly called Dutch Box.

¶ *The Temperature and Vertues.*

A The leaues of the Box tree are hot, drie, and astringent, of an euill and lothsome smell, not vsed in medicine, but onely as I said before in the description.

CHAP.

CHAP. 71. *Of the Myrtle Tree.*

¶ *The Description.*

1 THe first and greatest *Myrtus* is a small tree,growing to the height of a man,hauing many faire and pliant branches, couered with a browne barke, and set full of leaues much like vnto the Laurell or Bay leafe,but thinner and smaller,somewhat resembling the leaues of Peruincle, which being bruised do yeeld forth a most fragrant smell,not much inferiour vnto the smell of Cloues,as all the rest of the kindes do : among these leaues come forth small white floures, in shape like the floures of the Cherry tree,but much smaller,and of a pleasant sauour,which do turn into small berries,greene at the first, and afterwards blacke.

1 *Myrtus Laurea maxima.*
The Myrtle tree.

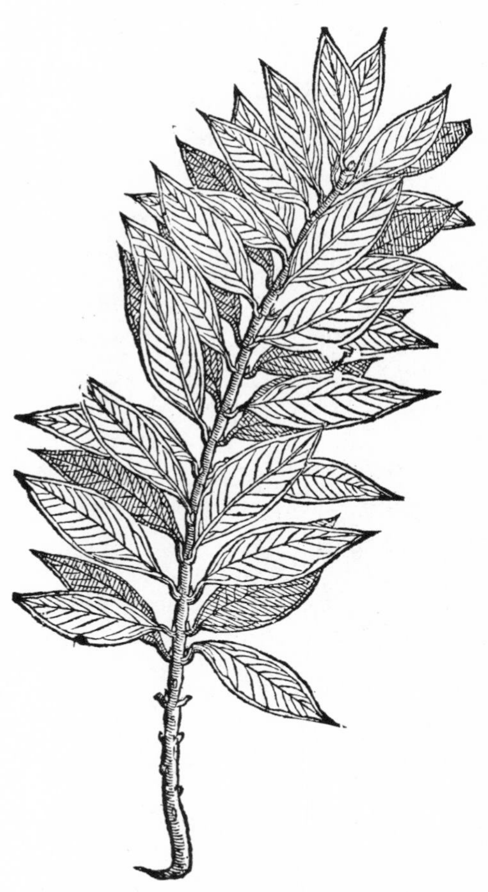

‡ 2 *Myrtus Bætica latifolia.*
Great Spanish Myrtle.

2 There is also another kind of *Myrtus* called *Myrtus Bætica latifolia*, according to *Clusius Myrtus Laurea*,that hath leaues also like Bay leaues, growing by couples vpon his pleasant greene branches, in a double row on both sides of the stalkes, of a light greene colour, and somewhat thicker than the former,in sent and smell sweet : the floures and fruit are not much differing from the first kinde.

3 There is likewise another kinde of *Myrtus* called *Exotica*, that is strange and not common : it groweth vpright vnto the height of a man like vnto the last before mentioned,but that it is replenished with greater plenty of leaues,which do fold in themselues hollow and almost double,broader pointed,and keeping no order in their growing,but one thrusting within another,and as it were crossing one another confusedly; in all other points agreeing with the precedent.

4 There is another sort like vnto the former in floures and branches, but the leaues are smooth, flat and plaine,and not crumpled or folded at all, they are also much smaller than any of the former. The fruit is in shape like the other, but that it is of a white colour, whereas the fruit of the other is blacke.

5 There is also another kinde of Myrtle , called *Myrtus minor*,or noble Myrtle,as being the

 chiefe

‡ 3 *Myrtus exotica.*
Strange Myrtle.

‡ 4 *Myrtus fructu albo.*
Myrtle with white berries.

‡ 5 *Myrtus minor.*
The little Myrtle.

‡ 6 *Myrtus Bætica sylvestris.*
Wilde Spanish Myrtle.

chiefe of all the rest (although most common and best knowne) and it groweth like a little shrub or hedge bush, very like vnto the former, but much smaller: the leaues are smal and narrow, very much in shape resembling the leaues of Masticke Time called *Marum*, but of a fresher greene colour: the floures be white, nothing differing from the former sauing in greatnesse, and that sometimes they are more double.

‡ 6 This growes not very high, neither is it so shrubby as the former: the branches are small and brittle: the leaues are of a middle bignesse, sharpe pointed, standing by couples in two rowes, seldome in foure as the former, they are blackish also and wel smelling. the floure is like that of the rest: the fruit is round, growing vpon long stalks out of the bosomes of the leaues, first greene, then whitish, lastly blacke, of a winy and pleasant taste with some astriction. This growes wilde in diuers places of Portugall, where *Clusius* found it flouring in October: he calls it *Myrtus Bætica syluestris*. ‡

¶ The Place.

These kindes of Myrtles grow naturally vpon the wooddy hills and fertill fields of Italy and Spain. ‡ The two last are nourished in the garden of Mistresse *Tuggy* in Westminster, and in some other gardens. ‡

¶ The Time.

Where they ioy to grow of themselues they floure when the Roses do: the fruit is ripe in Autumne: in England they neuer beare any fruit.

¶ The Names.

It is called in Greeke μυρσίνη: in Latine, *Myrtus*: in the Arabicke tongue, *Alas*: in Italian, *Myrto* in Spanish, *Arrayhan*: in the Portingale language, *Murta*, and *Murtella*: other Nations doe almost keepe the Latine name, as in English it is called Myrtle, or Myrtle tree.

Among the Myrtles that which hath the fine little leafe is surnamed of *Pliny*, *Tarentina*; & that which is so thicke and full of leaues is *Exotica*, strange or forreine. *Nigra Myrtus* is that which hath the blacke berries: *Candida*, which hath the white berries, and the leaues of this also are of a lighter greene: *Satiua*, or the tame planted one is cherished in gardens and orchards: *Syluestris*, or the wild Myrtle is that which groweth of it selfe; the berries of this are oftentimes lesser, and of the other, greater. *Pliny* doth also set downe other kindes; as *Patritia*, *Plebeia*, and *Coniugalis*: but what manner of ones they are he doth not declare: he also placeth among the Myrtles, *Oxymyrsine*, or Kneeholm, which notwithstanding is none of the Myrtles, but a thornie shrub.

Pliny in his 14. book, 16. chap. saith, that the wine which is made of the wilde Myrtle tree is called *Myrtidanum*, if the copie be true. For *Dioscorides* and likewise *Sotion* in his Geoponikes report, that wine is made of Myrtle berries when they be thorow ripe, but this is called *Vinum Myrteum*, or *Myrtites*, Myrtle wine.

Moreouer, there is also a wine made of the berries and leaues of Myrtle stamped and steeped in Must, or wine new pressed from the grape, which is called, as *Dioscorides* saith, *Myrsinite vinum*, or wine of Myrtles.

The Myrtle tree was in times past consecrated to *Venus*. *Pliny* in his 15. booke, 29. chapter, saith thus, There was an old Alter belonging to *Venus*, which they now call *Murtia*.

¶ The Temperature and Vertues.

The Myrtle consisteth of contrary substances, a cold earthinesse bearing the preheminence; it hath also a certaine subtill heat, therefore, as *Galen* saith, it drieth notably. A

The leaues, fruit, buds, and iuice do binde, both outwardly applied and inwardly taken: they stay the spitting of bloud, and all other issues thereof: they stop both the whites and reds in women, if they sit in a bath made therewith: after which manner and by fomenting also they stay the superfluous course of the hemorrhoides. B

They are a remedy for laskes, and for the bloudy flix, they quench the fiery heat of the eies, if they be laid on with parched Barly meale. C

They be also with good successe outwardly applied to all inflammations newly beginning, and also to new paine vpon some fall, stroke or straine. D

They are wholsome for a moist and watery stomacke: the fruit and leaues dried prouoke vrine: for the greene leaues containe in them a ceartaine superfluous and hurtfull moisture. E

It is good with the decoction herof made with wine, to bathe lims that are out of ioint, and burstings that are hard to be cured, and vlcers also of the outward parts: it helpeth spreading tetters, scoureth away the dandrafe and sores of the head, maketh the haires blacke, and keepeth them from F from shedding; withstandeth drunkennesse, if it be taken fasting, and preuaileth against poison, and the bitings of any venomous beast.

G There is drawne out of the green berries thereof a iuice, which is dried and reserued for the foresaid vses.

H There is likewise pressed out of the leaues a iuice, by adding vnto them either old wine or raine water, which must be vsed when it is new made, for being once drie it putrifieth, and as *Dioscorides* saith, loseth his vertues.

Chap. 72. *Of sweet Willow or Gaule.*

Myrtus Brabantica, siue Elæagnus Cordi.
Gaule, sweet willow, or Dutch Myrtle tree.

¶ *The Description.*

Gaule is a low and little shrub or wooddy plant, hauing many brown & hard branches: whereupon doe grow leaues somewhat long, hard, thicke, and oileous, of an hot sauour or smell somewhat like *Myrtus*: among the branches come forth other little ones, wherupon do grow many spokie eares or tufts, full of small floures, and after them succeed great store of square seeds clustering together, of a strong and bitter taste. The root is hard, and of a wooddie substance.

¶ *The Place.*

This Gaule groweth plentifully in sundry places of England, as in the Ile of Ely, & in the Fennie countries thereabouts, wherof there is such store in that countrey, that they make fagots of it and sheaues, which they call Gaule sheaues, to burne and heat their ouens. It groweth also by Colebrooke, and in sundry other places.

¶ *The Time.*

The Gaule floureth in May and Iune, and the seed is ripe in August.

¶ *The Names.*

This tree is called of diuers in Latine, *Myrtus Brabantica*, and *Pseudomyrsine*, and *Cordus* calleth it *Elæagnus*, *Chamæleagnus* and *Myrtus Brabantica*. *Elæagnus* is described by *Theophrastus* to be a shrubbie plant like vnto the Chaste tree, with a soft and downie leafe, and with the floure of the Poplar tree: and that which we haue described is no such plant. It hath no name among the old writers for ought we know, vnlesse it be *Rhus syluestris Plinij*, or *Pliny* his wilde Sumach, of which hee hath written in his 24. book, 11 chap. [There is, saith he, a wilde herbe with short stalkes, which is an enemy to poison, and a killer of mothes.] It is called in low Dutch, **Gagel**: in English, Gaule.

¶ *The Temperature.*

Gaule or the wilde Myrtle, especially the seed, is hot and drie in the third degree: the leaues be hot and drie, but not so much.

¶ *The Vertues.*

A The fruit is troublesome to the brain; being put into beere or aile whilest it is in boiling (which many vse to do) it maketh the same heady, fit to make a man quickly drunke.

B The whole shrub, fruit and all, being laied among clothes, keepeth them from moths and worms.

CHAP. 73. *Of Worts or Wortle berries.*

¶ *The Kindes.*

V*Accinia*, or Worts, of which we treat in this place, differ from Violets, neither are they eſteemed for their floures but berries: of theſe Worts there be diuers ſorts found out by the later Writers.

1 *Vaccinia nigra.*
Blacke Worts or Wortle berries.

2 *Vaccinia rubra.*
Red Worts or Wortle berries.

¶ *The Deſcription.*

1 V*Accinia nigra*, the blacke Wortle or Hurtle, is a baſe and low ſhrub or wooddy plant, bringing forth many branches of a cubit high, ſet full of ſmall leaues of a dark greene colour, not much vnlike the leaues of Box or the Myrtle tree: amongſt which come forth little hollow floures turning into ſmall berries, greene at the firſt, afterward red, and at the laſt of a blacke colour, and full of a pleaſant and ſweet iuyce: in which doe lie diuers little thinne whitiſh ſeeds: theſe berries do colour the mouth and lips of thoſe that eate them, with a black colour: the root is wooddy, ſlender, and now and then creeping.

2 *Vaccinia rubra*, or red Wortle, is like the former in the manner of growing, but that the leaues are greater and harder, almoſt like the leaues of the Box tree, abiding greene all the Winter long: among which come forth ſmall carnation floures, long and round, growing in cluſters at the top of the branches: after which ſucceed ſmall berries, in ſhew and bigneſſe like the former, but that they are of an excellent red colour, and full of iuyce, of ſo orient and beautifull a purple to limne withall, that Indian *Lacca* is not to be compared thereunto, eſpecially when this iuyce is prepared and dreſſed with Allom according to art, as my ſelfe haue proued by experience: the taſt is rough and aſtringent: the root is of a wooddy ſubſtance.

3 *Vaccinia alba*, or the white Wortle, is like vnto the former, both in ſtalks and leaues, but the berries are of a white colour, wherein conſiſteth the difference.

‡ The figure which our Author here giues in the third place hath need of a better deſcription, for

3 *Vaccinia alba.*
The white Worts or Wortle berries.

4 *Vaccinia Pannonica, ſiue Vitis Idæa.*
Hungarie Wortle berries.

5 *Vaccinia Vrſi, ſiue Vua Vrſi apud Cluſium.*
Beare Wortle berries.

† 6 *Vitis Idæa folijs ſubrotundis maior.*
Great round leaued Wortle berries.

for the difference is not onely in the colour of the berries. This differs from the former in forme and bignesse, for it sends forth many stalkes from the root, and these three, foure, or fiue cubits high, thicke, and diuided into sundry branches, couered for the most part with a blackish barke: at the beginning of the Spring from the buds at the sides of the branches it sends forth leaues all horie and hairy vnderneath, and greene aboue: from the midst of these, vpon little foot-stalkes stand clustering together many little floures, consisting of fiue white leaues apiece without smell; and then the leaues by little and little vnfold themselues and cast off their downinesse, and become snipt about the edges. The fruit that succeeds the floures is round, blacke, somewhat like, but bigger than a Haw, full of iuyce of a very sweet taste; wherein lies ten or more longish smooth blackish seeds. It growes vpon the Austrian and Stirian Alps, where the fruit is ripe in August. *Clusius* calls it *Vitis Idæa* 3. *Pena* and *Lobel, Amelancher: Gesner* by diuers names, as *Myrtomalus, Petromelis, Pyrus ceruinus, &c.* ‡

4 *Carolus Clusius* in his Pannonicke Obseruations hath set downe another of the Wortle berries, vnder the name of *Vitis Idæa,* which differeth from the other Wortle berries, not onely in stature, but in leaues and fruit also. ‡ The leaues are long, narrow, sharpe pointed, full of veines, a little hairy, and lightly snipt about the edges, greener aboue than below: the fruit growes from the tops of the branches of the former yeare, hanging vpon long foot-stalkes, and being as big as little Cherries, first greene, then red, and lastly blacke, full of iuyce, and that of no vnpleasant taste, containing no kernels, but flat white seeds commonly fiue in number: the stalkes are weake, and commonly lie vpon the ground: *Clusius* found it vpon the Austrian mountaine Snealben, with the fruit partly ripe, and partly vnripe, in August. It is his *Vitis Idæa* 1. ‡

5 The same Author also setteth forth another of the Wortle berries, vnder the title of *Vua Vrsi,* which is likewise a shrubby plant, hauing many feeble branches, whereon grow long leaues blunt at the points, and of an ouerworn green colour: among which, at the tops of the stalks come forth clusters of bottle-like floures of an herby colour: the fruit followeth, growing likewise in clusters, green at the first, and blacke when they be ripe: the root is of a wooddy substance. ‡ This is alwaies greene, and the floures are of a whitish purple colour. ‡

6 ‡ This differs from the second, in that the leaues are thinner, more full of veines, and whiter vnderneath: the floure is like the common kind, whitish purple, hollow, and diuided into fiue parts: the fruit also is blacke, and like that of the first described. This growes on diuers mountainous places of Germany, where *Clusius* obserued it, who made it his *Vitis Idæa* 2. ‡

¶ *The Place.*

These plants prosper best in a lean barren soile, and in vntoiled wooddy places: they are now and then found on high hils subiect to the winde, and vpon mountaines: they grow plentifully in both the Germanies, Bohemia, and in diuers places of France and England; namely in Middlesex on Hampsted heath, and in the woods thereto adioyning, and also vpon the hills in Cheshire called Broxen hills, neere Beeston castle, seuen miles from the Nantwich; and in the wood by Highgate called Finchley wood, and in diuers other places.

The red Wortle berry groweth in Westmerland at a place called Crosby Rauenswaith, where also doth grow the Wortle with the white berry, and in Lancashire also vpon Pendle hills.

‡ I haue seene none of these but only the first described, growing vpon Hampsted heath. The white formerly mentioned in the third description, and here againe in the place, seems only a varietie of the second hauing white berries, as far as I can gather by our Author; for it is most certaine, that it is not that which he figured, and I haue described in the third place. ‡

¶ *The Time.*

The Wortle berries do floure in May, and their fruit is ripe in Iune.

¶ *The Names.*

VVortle berries is called in high-Dutch, **Heydelbeeren:** in low-Dutch, **Crakebesien**, because they make a certaine cracke whilest they be broken betweene the teeth: of diuers, **Hauerbesien:** the French men, *Airelle,* or *Aurelle,* as *Iohannes de Choul* writeth: and we in England, VVorts, VVhortle berries, Blacke-berries, Bill-berries, and Bull-berries, and in some places, VVin-berries.

Most of the shops of Germany do call them *Myrtilli,* but properly *Myrtilli* are the fruit of the Myrtle tree, as the Apothecaries name them at this day. This plant hath no name for ought wee can learne, either among the Greekes or antient Latines; for whereas most doe take it to be *Vitis Idæa,* or the Corinth tree, which *Pliny* syrnameth *Alexandrina,* it is vntrue; for *Vitis Idæa* is not only like to the common Vine, but is also a kinde of Vine: and *Theophrastus,* who hath made mention hereof doth call it, without an Epethete, Ἄμπελος, simply, as a little after we wil declare; which without doubt he would not haue done if he had found it to differ from the common Vine: For what things soeuer receiue a name of some plant, the same are expressed with some Epethit added to be known to differ from others· as *Laurus Alexandrina, Vitis alba, Vitis nigra, Vitis syluestris,* and such like.

Moreouer, those things which haue borrowed a name from some plant are like thereunto, if not wholly,

wholly, yet either in leafe or fruit, or in ſome other thing. *Vitis alba & nigra,* that is, the white and blacke Bryonies, haue leaues and claſping tendrels as hath the common Vine, and clyme alſo after the ſame manner: *Vitis ſylueſtris*, or the wilde Vine, hath ſuch like ſtalks as the Vine hath, and bringeth forth fruit like to the little Grapes. *Laurus Alexandrina*, and *Chamedaphne*, and alſo *Daphnoides*, are like in leaues to the Laurell tree: *Sycomorus* is like in fruit to the Fig tree, and in leaues to the Mulberry tree: *Chamædrys* hath the leafe of an Oke; *Peucedanus* of the Pine tree: ſo of others which haue taken their names from ſome other: but this low ſhrub is not like the Vine either in any part, or in any other thing.

This *Vitis Idæa* groweth not on the vppermoſt and ſnowie parts of mount Ida (as ſome would haue it, but about Ida, euen the hill Ida, not of Candy, but of Troas in the leſſer Aſia, which *Ptolomie* in his fifth booke of Geographie, chap. 3. doth call *Alexandri Troas*, or *Alexander* his Troy: whereupon it is alſo aduiſedly named of *Pliny, lib.* 14. *cap.* 3. *Vitis Alexandrina,* no otherwiſe than *Alexandrina Laurus* is ſaid of *Theophraſtus* to grow there: *Laurus*, ſyrnamed *Alexandrina*, and *Ficus quædam*, or a certaine Fig tree, and Ἄμπελος, that is to ſay the Vine, are reported, ſaith he, to grow properly about Ida. Like vnto this Vine are thoſe which *Philoſtratus* in the life of *Apollonius* reporteth to grow in Mæonia, and Lydia, ſcituated not far from Troy, comparing them to thoſe vines which grow in India beyond Caucaſus: The Vines there, ſaith he, be very ſmall, like as be thoſe that do grow in Mæonia and Lydia, yet is the wine which is preſſed out of them of a maruellous pleaſant taſte.

This Vine which growes neere to mount Ida is reported to be like a ſhrub, with little twigs and branches of the length of a cubit, about which are grapes growing aſlope, blacke, of the bignes of a beane, ſweet, hauing within a certaine winie ſubſtance, ſoft: the leafe of this is round, vncut, and little.

This is deſcribed by *Pliny, lib.* 14, *cap.* 3. almoſt in the ſelfe ſame words: It is called, ſaith he, *Alexandrina vitis*, and groweth neere vnto Phalacra: it is ſhort, with branches a cubit long, with a blacke grape of the bignes of the Latines Beane, with a ſoft pulpe and very little, with very ſweet cluſters growing aſlope, and a little round leafe without cuts.

And with this deſcription the little ſhrub which the Apothecaries of Germany do call *Myrtillus* doth nothing at all agree, as it is very manifeſt; for it is low, ſcarce a cubit high, with a few ſhort branches not growing to a cubit in length: it doth not bring forth cluſters or bunches, nor yet fruit like vnto grapes, but berries like thoſe of the Yew tree, not ſweet, but ſomewhat ſoure and aſtringent; in which alſo there are many little white flat ſeeds: the leafe is not round, but more long than round, not like to that of the Vine, but of the Box tree. Moreouer, it is thought that this is not found in Italy, Greece, or in the leſſer Aſia, for that *Matthiolus* affirmeth the ſame to grow no where but in Germanie and Bohemia; ſo far is it from being called or accounted to be *Vitis Idæa* or *Alexandrina*.

The fruit of this may be thought not without cauſe to be named *Vaccinia,* ſith they are berries; for they may be termed of *Baccæ*, berries, *Vaccinia,* as though they ſhould be called *Baccinia*. Yet this letteth not that there may be alſo other *Vaccinia's*: for *Vaccinia* is πολύσημος *dictio*, or a word of diuers ſignifications. *Virgil* in the firſt booke of his Bucolicks, *Eclog.* 10. affirmeth, that the written Hyacinth is named of the Latines, *Vaccinium*, tranſlating into Latine *Theocritus* his verſe which is taken out of his tenth Eidyl.

Καὶ τὸ ἴον μέλαν ἐστὶ, καὶ ἁ γραπτὰ ὑάκινθος.

Virgil:

Et nigræ Violæ, ſunt & Vaccinia nigra.

Vitruvius, lib. 7. of his Architecture doth alſo diſtinguiſh *Vaccinium* from the Violet, and ſheweth, that of it is made a gallant purple; which ſeeing that the written Hyacinth cannot do, it muſt needs be that this *Vaccinium* is another thing than the Hyacinth is, becauſe it ſerues to giue a purple dye.

Pliny alſo, *lib.* 16. *cap.* 18. hath made mention of *Vaccinia*, which are vſed to dye bond-ſlaues garments with, and to giue them a purple colour.

But whether theſe be our *Vaccinia* or Whortle berries it is hard to affirme, eſpecially ſeeing that *Pliny* reckoneth vp *Vaccinia* amongſt thoſe plants which grow in waterie places; but ours grow on high places vpon mountaines ſubiect to windes, neither is it certainly knowne to grow in Italy. Howſoeuer it is, theſe our Whortles may be called *Vaccinia*, and do agree with *Plinies* and *Vitruvius* his *Vaccinia*, becauſe garments and linnen cloath may take from theſe a purple die.

The red Whortle berries haue their name from the blacke Whortles, to which they be in form very like, and are called in Latine, *Vaccinia rubra:* in high-Dutch, **Rooter Heidelbeere**: in low-Dutch, **Roode Crakebeſſen**: the French men, *Aurelles Rouges:* they be named in Engliſh Red,

Red Worts, or red Wortle berries. *Conradus Geſnerus* hath called this plant *Vitis Idæa rubris acinis:* but the growing of the berries doth ſhew, that this doth farre leſſe agree with *Vitis Idæa*, than the blacke; for they do not hang vpon the ſides of the branches as do the black (which deceiued them that thought it to be *Vitis Idæa*) but from the tops of the ſprigs in cluſters.

As concerning the names of the other they are touched in their ſeuerall deſcriptions.

¶ *The Temperature.*

Theſe *Vaccinia* or Wortle berries are cold euen in the later end of the ſecond degree, and dry alſo, with a manifeſt aſtriction or binding qualitie.

Red Wortle berries are cold and dry, and alſo binding.

¶ *The Vertues.*

The iuyce of the blacke Wortle berries is boyled till it become thicke, and is prepared or kept A
by adding hony and ſugar vnto it: the Apothecaries call it *Rob*, which is preferred in all things before the raw berries themſelues; for many times whileſt they be eaten or taken raw they are offenſiue to a weake and cold ſtomacke, and ſo far are they from binding the belly, or ſtaying the laske, as that they alſo trouble the ſame through their cold and raw qualitie, which thing the boiled iuyce called *Rob* doth not any whit at all.

They be good for an hot ſtomacke, they quench thirſt, they mitigate and allay the heate of hot B
burning agues, they ſtop the belly, ſtay vomiting, cure the bloudy flix proceeding of choler, and helpe the felonie, or the purging of choler vpwards and downwards.

The people of Cheſhire do eate the blacke wortles in creame and milke, as in theſe South parts C
we eate Strawberries, which ſtop and binde the belly, putting away alſo the deſire to vomit.

The red Wortle is not of ſuch a pleaſant taſte as the blacke, and therefore not ſo much vſed to D
be eaten; but (as I ſaid before) they make the faireſt carnation colour in the world.

CHAP. 74.

Of the Mariſh Worts or Fenne-Berries.

Vacciniapaluſtria.
Mariſh Worts.

¶ *The Deſcription.*

THe Mariſh Wortle berries grow vpon the bogs in mariſh or mooriſh grounds, creeping thereupon like vnto wilde Time, hauing many ſmall limmer and tender ſtalkes layd almoſt flat vpon the ground, beſet with ſmal narrow leaues faſhioned almoſt like the leaues of Thyme, but leſſer: among which come forth little berries like vnto the common blacke Wortle berrie in ſhape, but ſomewhat longer, ſometimes all red, and ſometimes ſpotted or ſpecked with red ſpots of a deeper colour: in taſte rough and aſtringent.

¶ *The Place.*

The Mariſh Wortle growes vpon bogs and ſuch like wateriſh and fenny places, eſpecially in Cheſhire and Staffordſhire, where I haue found it in great plenty.

¶ *The Time.*

The Berries are ripe about the end of Iuly, and in Auguſt.

¶ *The Names.*

They are called in high-Dutch, **Moſzbeeren, Veenbeſſen:** that is to ſay, Fen-Grapes, or Fen-Berries, and Mariſh-worts, or Mariſh-Berries. *Valerius Cordus* nameth them *Oxycoccon:* wee haue called them *Vacciniapaluſtria*, or Mariſh Wortle berries, of the likeneſſe they haue to the other berries: ſome alſo call them Moſſe-Berries, or Moore-berries.

¶ *The Temperature.*

Theſe Wortle berries are cold and dry, hauing withall a certain thinneſſe of parts and ſubſtance, with a certaine binding qualitie adioyned.

¶ The

¶ *The Vertues.*

A They take away the heate of burning agues, and alſo the drought, they quench the furious heate of choler, they ſtay vomiting, reſtore an appetite to meate which was loſt by reaſon of cholericke and corrupt humors, and are good againſt the peſtilent diſeaſes.

B The iuice of theſe alſo is boyled till it be thicke, with ſugar added that it may be kept, which is good for all things that the berries are, yea and far better.

† I haue brought this Chapter and the next following from the place they formerly held, and ſeated them here amongſt the reſt of their kindred.

Chap. 75. *Of Cloud-berry.*

Vaccinia Nubis.
Cloud-berries.

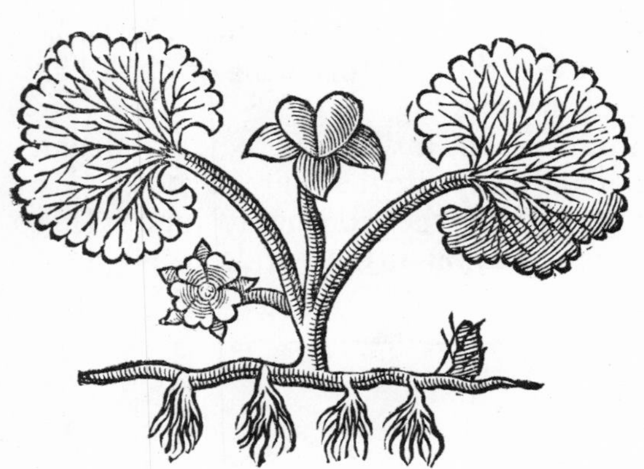

¶ *The Deſcription.*

THe Cloud-berrie hath many ſmall threddy roots, creeping farre abroad vnder the vpper cruſt of the earth, and alſo the moſſe, like vnto Couch-graſſe, of an ouerworn reddiſh colour, ſet here and there with ſmal tufts of hairy ſtrings: from which riſe vp two ſmall ſtalks, hard, tough, and of a wooddy ſubſtance (neuer more nor leſſe) on which doe ſtand the leaues like thoſe of the wilde Mallow, and of the ſame colour, full of ſmall nerues or ſinewes running in each part of the ſame: between the leaues commeth vp a ſtalke likewiſe of a wooddy ſubſtance, whereon doth grow a ſmall floure conſiſting of fiue leaues, of an herby or yellowiſh green colour like thoſe of the wilde Auens. After commeth the fruit, greene at the firſt, after yellow, and the ſides next the Sun red when they be ripe; in forme almoſt like vnto a little heart, made as it were of two, but is no more but one, open aboue, and cloſed together in the bottom, of a harſh or ſharpe taſte, wherein is contained three or foure little white ſeeds.

¶ *The Place.*

This plant groweth naturally vpon the tops of two high mountaines (among the moſſie places) one in Yorkſhire called Ingleborough, the other in Lancaſhire called Pendle, two of the higheſt mountaines in all England, where the clouds are lower than the tops of the ſame all Winter long, whereupon the people of the countrey haue called them Cloud-berries, found there by a curious gentleman in the knowledge of plants, called Mr. *Hesketh*, often remembred.

¶ *The Time.*

The leaues ſpring vp in May, at which time it floureth: the fruit is ripe in Iuly.

¶ *The Temperature.*

The fruit is cold and dry, and very aſtringent.

¶ *The Vertues.*

A The fruit quencheth thirſt, cooleth the ſtomacke, and allayeth inflammations, being eaten as Worts are, or the decoction made and drunke.

† My friend M. *Pimble* of Maribone receiued a plant hereof out of Lancaſhire: and by the ſhape of the leafe I could not iudge it to differ from the *Chamæmorus* formerly deſcribed, *pag.* 1273. neither doe the deſcriptions much differ in any materiall point: the figures differ more; but I iudge this a very imperfect one.

Chap. 76. *Of ſhrub Heart-Wort of Æthiopia.*

¶ *The Deſcription.*

THis kind of Seſely, being the Æthiopian Seſely, hath blackiſh ſtalks of a wooddy ſubſtance: this plant diuideth it ſelfe into ſundry other armes or branches, which are beſet with thicke fat and oileous leaues, faſhioned ſomewhat like the Wood-binde leaues, but thicker, and more

Seseli Æthiopicum frutex.
Shrub Sesely, or Hart-woort of Ethiopia.

more gummie, approching very neere vnto the leaues of Oleander both in shape and substance, being of a deepe or darke green colour, and of a very good sauour and smell, and continueth greene in my garden both winter and Sommer, like the Bay or Laurell. The floures do grow at the tops of the branches in yellow rundles like the floures of Dill; which being past, there succeedeth a darke or duskie seed resembling the seed of Fennell, and of a bitter taste. The root is thicke and of a wooddy substance.

¶ *The Place.*

It is found both in stony places, and on the sea coasts not farre from Marsilles, and likewise in other places of Languedocke: it also groweth in Ethiopia, in the darke and desart woods: it groweth in my garden.

¶ *The Time.*

It flourisheth, floureth and seedeth in Iuly and August.

¶ *The Names.*

The Grecians call it Αἰθιοπικὸν σέσελι: the Latines likewise *Æthiopicum Seseli*: the Ægyptians, κύονος φρίκη: that is, Dogs horrour: in English, Sesely of Erhiopia, or Ethiopian Hartwoort.

¶ *The Temperature and Vertues.*

Sesely of Ethiopia is thought to haue the same faculties that the Sesely of Marsilles hath, whereunto I refer it. A

CHAP. 77. *Of the Elder tree.*

¶ *The Kindes.*

THere be diuers sorts of Elders, some of the land, and some of the water or marish grounds; some with very jagged leaues, and others with double floures, as shall be declared.

¶ *The Description.*

1 THe common Elder groweth vp now and then to the bignesse of a meane tree, casting his boughes all about, and oftentimes remaineth a shrub: the body is almost all wooddie, hauing very little pith within; but the boughes and especially the young ones which be iointed, are full of pith within, and haue but little wood without: the barke of the body and great armes is rugged and full of chinks, and of an ill fauoured wan colour like ashes: that of the boughes is not very smooth, but in colour almost like; and that is the outward barke, for there is another vnder it neerer to the wood, of colour greene: the substance of the wood is sound, somewhat yellow, and that may be easily cleft: the leaues consist of fiue or six particular ones fastened to one rib, like those of the Walnut tree, but euery particular one is lesser, nicked in the edges, and of a ranke and stinking smell. The floures grow on spokie rundles, which be thin and scattered, of a white colour and sweet smell: after them grow vp little berries, greene at the first, afterwards blacke, whereout is pressed a purple juice, which being boiled with Allom and such like things, doth serue very well for the Painters vse, as also to colour vineger: the seeds in these are a little flat, and somewhat long. There groweth oftentimes vpon the bodies of those old trees or shrubs a certaine excrescence called *Auricula Iudæ*, or Iewes eare, which is soft, blackish, couered with a skin, somewhat like now and then to a mans eare, which being plucked off and dryed,

 shrinketh

shrinketh together and becommeth hard. This Elder groweth euery where, and is the common Elder.

2 There is another also which is rare and strange, for the berries of it are not blacke, but white: this is like in leaues to the former.

1 *Sambucus.*
The common Elder tree.

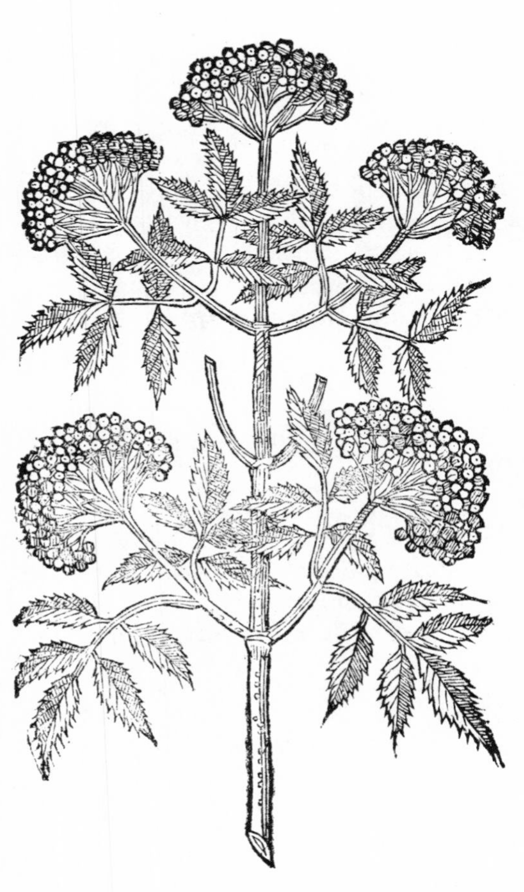

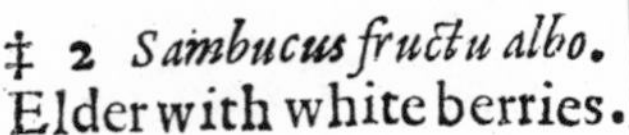

‡ 2 *Sambucus fructu albo.*
Elder with white berries.

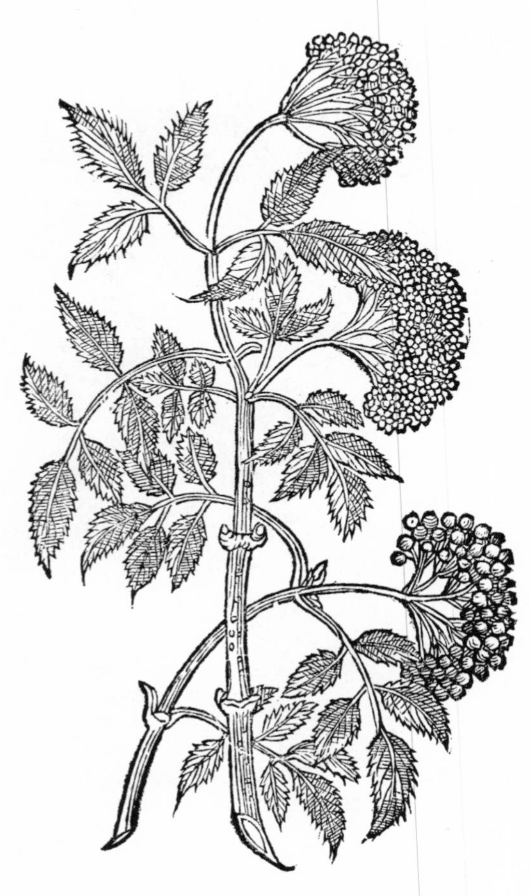

3 The jagged Elder tree groweth like the common Elder in body, branches, shootes, pith, floures, fruit, and stinking smell, and differeth onely in the fashion of the leaues, which doth so much disguise the tree, and put it out of knowledge, that no man would take it for a kinde of Elder, vntill he hath smelt thereunto, which will quickely shew from whence he is descended: for these strange Elder leaues are very much jagged, rent or cut euen vnto the middle rib. From the trunke of this tree as from others of the same kinde, proceedeth a certaine fleshie excrescence like vnto the eare of a man, especially from those trees that are very old.

4 This kinde of Elder hath floures which are white, but the berries redde, and both are not contained in spokie rundles, but in clusters, and grow after the manner of a cluster of grapes: in leaues and other things it resembleth the common Elder, saue that now and then it groweth higher.

¶ *The Place.*

The common Elder groweth euery where: it is planted about conie-burrowes for the shadow of the Conies; but that with the white berries is rare: the other kindes grow in like places; but that with the clustered fruit groweth vpon mountaines; that with the jagged leaues groweth in my garden.

¶ *The Time.*

These kindes of Elders do floure in Aprill and May, and their fruit is ripe in September.

¶ *The Names.*

This tree is called in Greeke, ἀκτῆ: in Latine and of the Apothecaries, *Sambucus*: of *Guillielmus Salicetus*, *Beza*: in high Dutch, **Holunder, Holder**: in low Dutch, **Vlier**: in Italian, *Sambuco*: in French, *Hus* and *Suin*: in Spanish, *Sauco*, *Sauch*, *Sambugueyro*: in English, Elder, and Elder tree: that with the white berries diuers would haue to be called *Sambucus syluestris*, or wilde Elder, but *Matthiolus* calleth it *Montana*, or mountaine Elder.

¶ *The*

3 *Sambucus laciniatis folijs.*
The iagged Elder tree.

4 *Sambucus racemoſa, vel Ceruina.*
Harts Elder, or Cluſter Elder.

¶ *The Temperature and Vertues.*

Galen attributeth the like facultie to Elder that he doth to Danewoort, and ſaith that it is of a drying qualitie, gluing, and moderatly digeſting: and it hath not only theſe faculties, but others alſo; for the barke, leaues, firſt buds, floures, and fruit of Elder, do not only dry, but alſo heate, and haue withall a purging qualitie, but not without trouble and hurt to the ſtomacke. A

The leaues and tender crops of common Elder taken in ſome broth or pottage open the belly, purging both ſlimie flegme and cholericke humors: the middle barke is of the ſame nature, but ſtronger, and purgeth the ſaid humors more violently. B

The ſeeds contained within the berries dried are good for ſuch as haue the dropſie, and ſuch as are too fat, and would faine be leaner, if they be taken in a morning to the quantity of a dram with wine for a certaine ſpace. C

The leaues of Elder boiled in water vntill they be very ſoft, and when they are almoſt boiled enough a little oile of ſweet Almonds added thereto, or a little Lineſeed oile; then taken forth and laid vpon a red cloath, or a piece of ſcarlet, and applied to the hemorrhoides or Piles as hot as can be ſuffered, and ſo let to remaine vpon the part affected, vntill it be ſomewhat cold, hauing the like in a readineſſe, applying one after another vpon the diſeaſed part, by the ſpace of an houre or more, and in the end ſome bound to the place, and the patient put warme a bed; it hath not as yet failed at the firſt dreſſing to cure the ſaid diſeaſe; but if the Patient be dreſſed twice it muſt needs doe good if the firſt faile. D

The greene leaues pounded with Deeres ſuet or Bulls tallow are good to be laid to hot ſwellings and tumors, and doth aſſwage the paine of the gout. E

The inner and greene barke doth more forcibly purge: it draweth forth choler and waterie humors; for which cauſe it is good for thoſe that haue the dropſie, being ſtamped, and the liquor preſſed out and drunke with wine or whay. F

Of like operation are alſo the freſh floures mixed with ſome kinde of meat, as fried with egges, they likewiſe trouble the belly and moue to the ſtoole: being dried they loſe as well their purging qualitie as their moiſture, and retaine the digeſting and attenuating qualitie. G

H The vinegar in which the dried floures are ſteeped are wholſome for the ſtomacke: being vſed with meate it ſtirreth vp an appetite, it cutteth and attenuateth or maketh thin groſſe and raw humors.

I The facultie of the ſeed is ſomewhat gentler than that of the other parts: it alſo moueth the belly, and draweth forth waterie humors, being beaten to pouder, and giuen to a dram weight: being new gathered, ſteeped in vineger, and afterwards dried, it is taken, and that effectually, in the like weight of the dried lees of wine, and with a few Aniſe ſeeds, for ſo it worketh without any maner of trouble, and helpeth thoſe that haue the dropſie. But it muſt be giuen for certaine daies together in a little wine, to thoſe that haue need thereof.

K The gelly of the Elder, otherwiſe called Iewes eare, hath a binding and drying qualitie: the infuſion thereof, in which it hath bin ſteeped a few houres, taketh away inflammations of the mouth, and almonds of the throat in the beginning, if the mouth and throat be waſhed therewith, and doth in like manner helpe the uvula.

L *Dioſcorides* ſaith, that the tender and greene leaues of the Elder tree, with barley meale parched, do remoue hot ſwellings, and are good for thoſe that are burnt or ſcalded, and for ſuch as be bitten with a mad dog, and that they glew and heale vp hollow vlcers.

M The pith of the young boughes is without qualitie: This being dried, and ſomewhat preſſed or quaſhed together, is good to lay vpon the narrow orifices or holes of fiſtula's and iſſues, if it be put therein.

Chap. 78. *Of Mariſh or Water Elder.*

1 *Sambucus aquatilis, ſiue paluſtris.*
Mariſh or water Elder.

2 *Sambucus Roſea.*
The Roſe Elder.

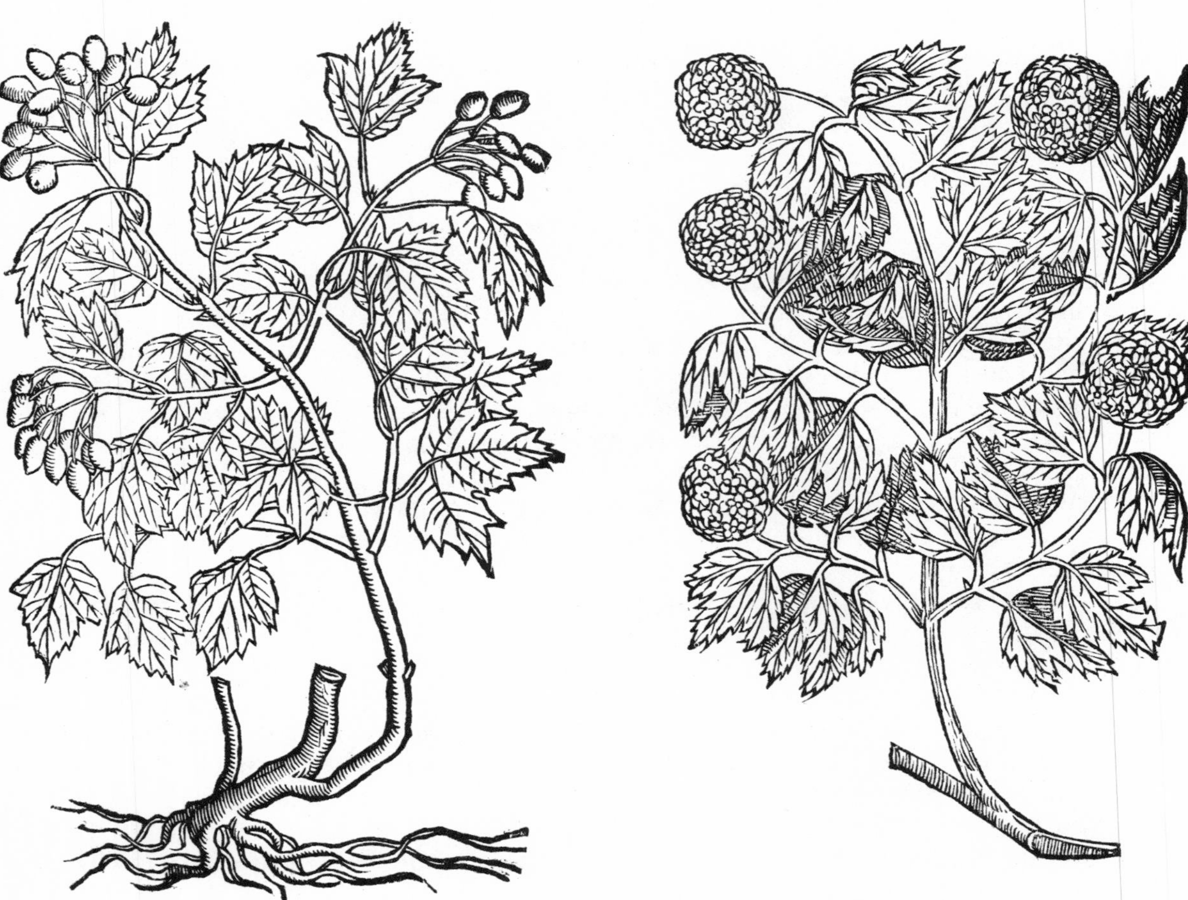

¶ *The Deſcription.*

1 Mariſh Elder is not like to the common Elder in leaues, but in boughes: it groweth after the manner of a little tree: the boughes are couered with a barke of an ill fauoured Aſh colour, as be thoſe of the common Elder: they are ſet with ioints by certaine

certaine diſtances, and haue in them great plenty of white pith, therefore they haue leſſe wood, which is white and brittle: the leaues be broad, cornered, like almoſt to Vine leaues, but leſſer and ſofter: among which come forth ſpoked rundles which bring forth little floures, the vttermoſt whereof alongſt the borders be greater, of a gallant white colour, euery little one conſiſting of fiue leaues: the other in the midſt and within the borders be ſmaller, and it floures by degrees, and the whole tuft is of a moſt ſweet ſmell: after which come the fruit or berries, that are round like thoſe of the common Elder, but greater, and of a ſhining red colour, and blacke when they be withered.

2 *Sambucus Roſea*, or the Elder Roſe groweth like an hedge tree, hauing many knotty branches or ſhoots comming from the root, full of pith like the common Elder: the leaues are like the vine leaues; among which come forth goodly floures of a white colour, ſprinkled and daſhed here and there with a light and thin Carnation colour, and do grow thicke and cloſely compact together, in quantitie and bulke of a mans hand, or rather bigger, of great beauty, and ſauoring like the floures of the Haw-thorne: but in my garden there groweth not any fruit vpon this tree, nor in any other place, for ought that I can vnderſtand.

3 This kinde is likewiſe an hedge tree, very like vnto the former in ſtalks and branches, which are iointed and knotted by diſtances, and it is full of white pith: the leaues be likewiſe cornered: the floures hereof grow not out of ſpoky rundles, but ſtand in a round thicke and globed tuft, in bigneſſe alſo and faſhion like to the former, ſauing that they tend to a deeper purple colour, wherin only the difference conſiſts.

¶ *The Place.*

Sambucus paluſtris, the water Elder, growes by running ſtreames and water courſes, and in hedges by moiſt ditch ſides.

The Roſe Elder groweth in Gardens, and the floures are there doubled by Art, as it is ſuppoſed.

¶ *The Time.*

Theſe kindes of Elders do floure in Aprill and May, and the fruit of the water Elder is ripe in September.

¶ *The Names.*

The water Elder is called in Latine, *Sambucus aquatica*, and *Sambucus paluſtris*: it is called *Opulus*, and *Platanus*, and alſo *Chamæplatanus*, or the dwarfe Plane tree, but not properly: *Valerius Cordus* maketh it to be *Lycoſtaphylos*: the Saxons, ſaith *Geſner*, do call it *Vua Lupina*; from whence *Cordus* inuented the name λυκοσταφυ'λος: it is named in high-Dutch, **Walt holder**, and **Hirſch holder**: in low Dutch, **Swelcken**, and **Swelckenhout**: of certaine French men, *Obiere*: in Engliſh, Mariſh Elder, and Whitten tree, Ople tree, and dwarfe Plane tree.

The Roſe Elder is called in Latine, *Sambucus Roſea*, and *Sambucus aquatica*, being doubtles a kind of the former water Elder, the floures being doubled by art, as we haue ſaid: it is called in Dutch, **Ghelderſche Rooſe**: in Engliſh, Gelders Roſe, and Roſe Elder.

¶ *The Temperature and Vertues.*

Concerning the faculties of theſe Elders, and the berries of the Water Elder, there is nothing found in any writer, neither can we ſet downe any thing hereof of our owne knowledge.

Chap. 79.

Of Dane-Wort, Wall-Wort, or Dwarfe Elder.

¶ *The Deſcription.*

Dane-wort, as it is not a ſhrub, neither is it altogether an herby plant, but as it were a Plant participating of both, being doubtles one of the Elders, as may appeare both by the leaues, floures, and fruit, as alſo by the ſmell and taſte.

Wall-wort is very like vnto Elder in leaues, ſpoky tufts, and fruit, but it hath not a wooddie ſtalke; it bringeth forth only greene ſtalks, which wither away in Winter: theſe are edged, and full of ioynts, like to the yong branches and ſhoots of Elder: the leaues grow by couples, with diſtances, wide, and conſiſt of many ſmall leaues which ſtand vpon a thicke ribbed ſtalke, of which euery one is long, broad, and cut in the edges like a ſaw, wider and greater than the leaues of the common Elder tree: at the top of the ſtalkes there grow tufts of white floures tipt with red, with fiue little chiues in them pointed with blacke, which turne into blacke berries like the Elder, in the which be little long ſeed: the root is tough, and of a good and reaſonable length, better for Phyſicks vſe than the leaues of Elder.

Ebulus, siue Sambucus humilis.
Dane-wort, or dwarfe Elder.

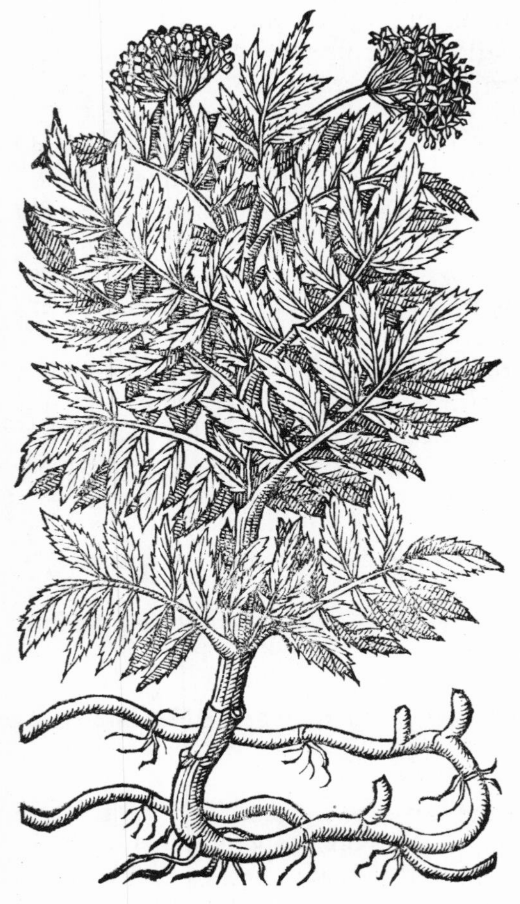

¶ *The Place.*

Dane-wort growes in vntoiled places neere common waies, and in the borders of fields: it groweth plentifully in the lane at Kilburne Abbey by London: also in a field by S. Ioans neere Dartford in Kent: and also in the highway at old Branford townes end next London, and in many other places.

¶ *The Time.*

The floures are perfected in Sommer, and the berries in Autumne.

¶ *The Names.*

It is named in Greeke, χαμαιάκτη, that is, *humilis Sambucus*, or low Elder: it is called in Latine, *Ebulus*, and *Ebulum*: in high-Dutch, **Attich**: in low-Dutch, **Hadich**: in Italian, *Ebulo*: in French, *Hieble* in Spanish, *Yezgos*: in English, Wall-wort, Dane-wort, and dwarfe Elder.

¶ *The Temperature.*

Wall-wort is of temperature hot and drie in the third degree, and of a singular qualitie, which *Galen* doth attribute vnto it, to wast and consume; and also it hath a strange and speciall facultie to purge by the stoole: the roots be of greatest force, the leaues haue the chiefest strength to digest and consume.

¶ *The Vertues.*

A The roots of Wall-wort boiled in wine and drunken are good against the dropsie, for they purge downwards watery humors.

B The leaues do consume and waste away hard swellings if they be applied pultis-wise, or in a fomentation or bath.

C *Dioscorides* saith, that the roots of Wall-wort doe soften and open the matrix, and also correct the infirmities thereof, if they be boiled for a bath to sit in; and dissolue the swellings and paines of the belly.

D The iuice of the root of Dane-wort doth make the haire blacke.

E The yong and tender leafe quencheth hot inflammations, being applied with Barly meale: it is with good successe laid vpon burnings, scaldings, and vpon the bitings of mad dogs; and with Bulls tallow or Goats suet it is a remedie for the gout.

F The seed of Wall-wort drunke in the quantitie of a dram is the most excellent purger of waterie humors in the world, and therefore most singular against the dropsie.

G If one scruple of the seed be bruised and taken with syrrup of Roses and a little Secke, it cureth the dropsie, and easeth the gout, mightily purging downwards waterish humors, being once taken in the weeke.

Chap. 80. *Of Beane Trefoile.*

¶ *The Description.*

1 The first kinde of *Anagyris* or *Laburnum* groweth like vnto a small tree, garnished with many small branches like the shoots of Oziars, set full of pale greene leaues, alwaies three together, like the *Lotus* or medow Trefoile, or rather like the leaues of *Vitex*, or the *Cytisus* bush: among which come forth many tufts of floures of a yellow colour, not much vnlike the floures of Broome: when these floures be gone there succeed small flat cods, wherein are contained seeds like Galega or the Cytisus bush: the whole plant hath little or no sauour at all: the root is soft and gentle, yet of a woody substance.

2 Stinking

2 Stinking Trefoile is a ſhrub like to a little tree, riſing vp to the height of ſix or eight cubits, or ſometimes higher: it ſendeth forth of the ſtalks very many ſlender branches; the barke whereof is of a deep greene colour: the leaues ſtand alwaies three together, like thoſe of *Lotus* or medow Trefoile, yet of a lighter greene on the vpper ſide: the floures be long, as yellow as gold, very like to thoſe of Broome, two or three alſo ioined together: after them come vp broad cods, wherein do lie hard fruit like Kidney Beanes, but leſſer; at the firſt white, afterwards tending to a purple, and laſt of all of a blackiſh blew: the leaues and floures hereof haue a filthy ſmell, like thoſe of the ſtinking Gladdon, and ſo ranke withall, as euen the paſſers by are annoied therewith.

1 *Anagyris.*
Beane Trefoile.

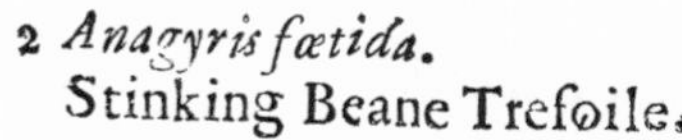
2 *Anagyris fœtida.*
Stinking Beane Trefoile.

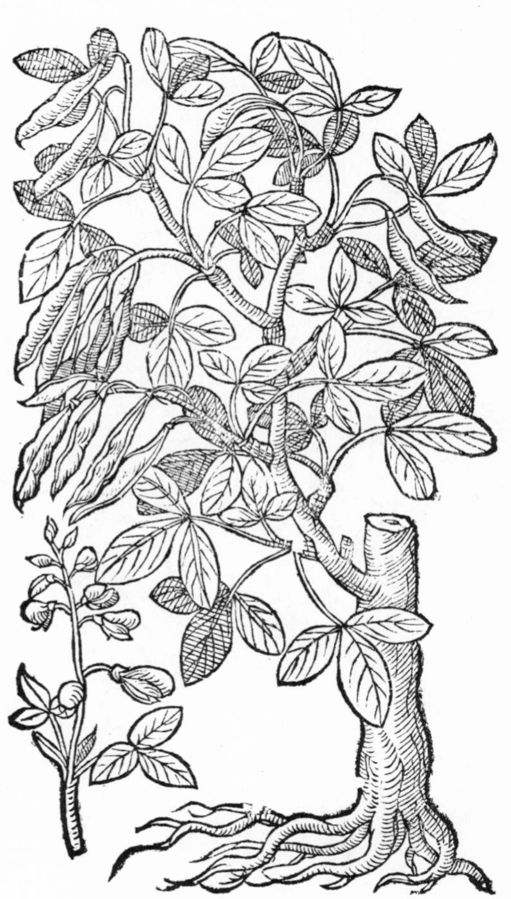

‡ Of *Anagyris* there are foure kindes, two with ſtinking leaues; the one with longiſh leaues, the other with rounder.

Two other whoſe leaues do not ſtinke; the one of theſe hath ſometimes foure or fiue leaues on one ſtalke, and the leaues are long and large. The other hath them leſſer and narrower. ‡

¶ *The Place.*

Theſe grow of themſelues in moſt places of Languedocke and Spaine, and in other countries alſo by high waies ſides, as in the Iſle of Candy, as *Bellonius* writeth: the firſt I haue in my garden; the other is a ſtranger in England. ‡ Maſter *Trad ſcant* hath two ſorts hereof in his garden. ‡

¶ *The Time.*

They floure in Iune, and the ſeed is ripe in September.

¶ *The Names.*

The Beane Trefoile is called in Greeke ἀνάγυρις which name remaineth vncorrupt in Candy euen to this day: in Latine alſo *Anagyris*, and *Laburnum*: of the people of Anagni in Italy named *Eghelo*, which is referred vnto *Laburnum*, of which *Pliny* writeth in his 16. booke, 18. chapter. In Engliſh, Beane Trefoile, or the Peaſcod tree.

¶ *The Temperature.*

Beane Trefoile, as *Galen* writeth, hath a hot and digeſting faculty.

¶ *The*

¶ *The Vertues.*

A The tender leaues, saith *Dioscorides*, being stamped and layed vpon cold swellings, do waste away the same.

B They are drunke with Cute in the weight of a dram against the stuffing of the lungs, and doe bring downe the menses, the birth, and the afterbirth.

C They cure the head-ache being drunke with wine; the iuice of the root digesteth and ripeneth, if the seed be eaten it procureth vomite, which thing, as *Matthiolus* writeth, the seed not onely of stinking Beane Trefoile doth effect, but that also of the other likewise.

Chap. 81. *Of Iudas Tree.*

Arbor Iudæ.
Iudas Tree.

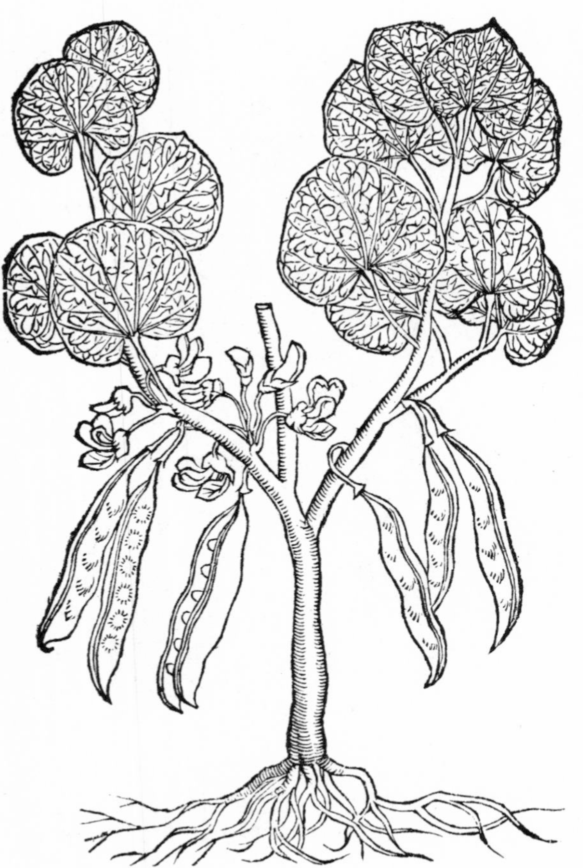

¶ *The Description.*

IVdas tree is likewise one of the hedge plants: it groweth vp vnto a tree of a reasonable bignesse, couered with a dark coloured barke, whereon doe grow many twiggie tough branches of a brown colour, garnished with round leaues, like those of round Birthwoort, or Sowbread, but harder, and of a deeper greene colour: among which come forth small floures like those of Peason, of a purple colour, mixed with red, which turn into long flat cods, pressed hard together, of a tawny or wan colour, wherein is contained small flat seeds, like the Lentill, or rather like the seed of Medica, fashioned like a little kidney: the root is great and wooddie.

¶ *The Place.*

This shrub is found in diuers prouinces of Spaine, in hedges, and among briers & brambles: the mountaines of Italy, and the fields of Languedocke are not without this shrub; it groweth in my garden.

¶ *The Time.*

The floures come forth in the Spring, and before the leaues: the fruit or cods be ripe in Sommer.

¶ *The Names.*

It is commonly named in Latine *Arbor Iudæ*: some haue called it *Sycomorus*, or Sycomore tree, and that because the floures and cods hang downe from the bigger branches: but the right Sycomore tree is like the Fig-tree in fruit, & in leaues to the Mulberrie tree, wherupon it is so named. Others take it to be κερκίς: of which *Theophrastus* writeth thus, *Cercis* bringeth forth fruit in a cod; which words are all so few, as that of this no certaintie can be gathered, for there be more shrubs that bring forth fruit in cods. The French men call it *Guainier*, as though they should say, *Vaginula*: or a little sheath: most of the Spaniards do name it *Algorouo loco*, that is, *Siliqua syluestris* or *fatua*, wilde or foolish cod: others, *Arbol d' amor*, for the brauenesse sake: it may be called in English, *Iudas* tree, for that it is thought to be that whereon *Iudas* did hang himselfe, and not vpon the Elder tree, as it is vulgarly said.

¶ *The Temperature and Vertues.*

The temperature and vertues of this shrub are vnknowne, and not found out: for whereas *Matthiolus* maketh this to be *Acacia*, by adding falsely thornes vnto it, it is but a surmise.

Chap.

CHAP. 82. *Of the Carob tree, or Saint Iohns Bread.*

¶ *The Description.*

THe Carob tree is alſo one of thoſe that beare cods; it is a tree of a middle bigneſſe, very full of boughes: the leaues long, and conſiſt of many ſet together vpon one middle rib, like thoſe of the Aſh, but euery particular one of them is broader, harder and rounder: the fruit or long cods in ſome places are a foot in length, in other places ſhorter by halfe, an inch broad, ſmooth, & thick; in which do lie flat and broad ſeeds: the cods themſelues are of a ſweet taſte, and are eaten of diuers, but not before they be gathered and dried; for being as yet green, though ripe, they are vnpleaſant to be eaten by reaſon of their ill fauoured taſte.

Ceratia ſiliqua, ſiue Ceratonia.
The Carob tree.

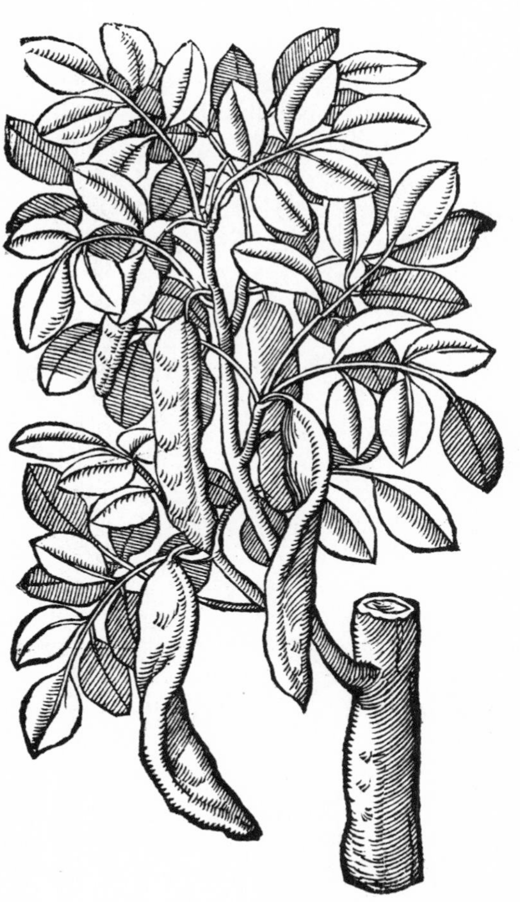

¶ *The Place.*

This groweth in Apulia, a Prouince of the kingdome of Naples, and alſo in diuers vntoiled places in Spaine: it is likewiſe found in India and other countries Eaſtward, where the cods are ſo full of ſweet iuice as that it is vſed to preſerue Ginger and other fruites, as *Matthiolus* ſheweth. *Strabo lib.* 15. ſaith, that *Ariſtobulus* reporteth how there is a tree in India of no great bignes, which hath great cods, ten inches long, full of hony; *Quas qui ederent non facile ſeruari*; which thing peraduenture is onely to be vnderſtood of the greene cods, & thoſe that are not yet dry: it is very wel known in the coaſts of Nicea and Liguria in Italy, as alſo in all the tracts and coaſts of the Weſt Indies, and Virginia. It groweth alſo in ſundry places of Paleſtine, where there is ſuch plenty of it, that it is left vnto ſwine and other wilde beaſts to feed vpon, as our Acornes and Beech maſt. Moreouer, both young and old feed thereon for pleaſure, and ſome haue eaten thereof to ſupply and help the neceſſary nouriſhment of their bodies. This of ſome is called Saint *Iohns* bread, and thought to be that which is tranſlated Locuſts, whereon S. *Iohn* did feed when he was in the wilderneſſe, beſides the wilde hony whereof he did alſo eat; but there is ſmall certainty of this: but moſt certaine that the people of that countrey doe feed vpon theſe cods, in Greeke called κεράτια: in Latine, *Siliquæ*: but Saint *Iohns* food is called in Greeke ἀκρίδες: which word is often vſed in the Reuelation written by Saint *Iohn*, and tranſlated Locuſts. Now wee muſt alſo remember that this Greeke word hath two ſeuerall interpretations or ſignifications, for taken in the good part, it ſignifieth a kinde of creeping creature, or flie, which hoppeth or skippeth vp and down, as doth the graſhopper; of which kinde of creatures it was lawfull to eat, *Leuit.* 11. 22. and *Mat.* 3. 4. It ſignifieth alſo thoſe Locuſts which came out of the ſmoke of the bottomles pit, mentioned *Apoc.* 9. *v.* 3. 4. &c. which were like vnto horſes prepared for battell. The Hebrew word which the Engliſh tranſlators haue turned Graſhoppers, *Tremelius* dares not giue the name *Locuſt* vnto it, but calleth it by the Hebrew name *Arbis*, after the letters and Hebrew name, ſaying thus in the note vpon the 22. verſe of the 11. chapter of *Leuit.* Theſe kindes of creeping things neither the Hebrews nor the Hiſtoriographers, nor our ſelues do know what they meane: wherefore we ſtill retaine the Hebrew words, for all the foure kindes thereof: but it is certaine that the Eaſt countrey Graſhoppers and Locuſts were ſometimes vſed in meat, as *Math.* 3. 4. and *Marc.* 1. 6. *Plin lib.* 11. *Natur. Hiſtor. cap.* 26. and 29.

Thus

Thus far *Tremelius* and *Iunius*. By that which hath been ſaid it appeareth what S. *Iohn* the Baptiſt fed of,vnder the title Locuſts : and that it is nothing like vnto this fruit *Ceratia ſiliqua* : I rather take the husks or ſhells of the fruit of this tree to be the cods or husks whereof the prodigall childe would haue fed,but none gaue them vnto him,though the ſwine had their fill thereof. Theſe cods being drie are very like beane cods,as I haue often ſeen. I haue ſowne the ſeeds in my garden,where they haue proſpered exceeding well.

‡ There is no doubt but the Κεράτια or *Siliquæ* mentioned in Saint *Lukes* Goſpel, Chap. 15. *v.* 16. were the cods or fruit of this tree. I cannot beleeue that either the fruit of this or the Locuſts, were the Ἀκρίδες, mentioned in the third chapter of Saint *Mat. v.* 4. But I am of the opinion of the Greeke Father *Iſodore Peluſiota*, who, *lib.* 1. *Epiſt.* 132. hath theſe words, Αἱ ἀκρίδες, ἃς Ἰωάννης ἐτρέφετο, οὐ ζῶα εἰσιν, ὡς τινες οἴονται ἀμαθῶς, κανθάροις ἐοικότα, μὴ γένοιτο, ἀλλ' ἀκρέμονες βοτανῶν ἢ φυτῶν : οὔτε ἡ πόα τις ἐστι πάλιν τὸ μέλι ἄγριον, ἀλλὰ μέλι ὄρειον ὑπὸ μελισσῶν ἀγρίων γινόμενον, &c. That is : The *Acrides* which *Iohn* fed vpon are not liuing creatures like to Beetles, as ſome vnlearnedly ſuppoſe, farre be it from vs ſo to thinke ; but they are the tender buds of herbes and plants or trees ; neither on the other ſide is the *Meli agrion* any herbe ſo called, but mountaine hony gathered by wilde Bees, &c. ‡

¶ *The Time.*

The Carob tree bringeth forth fruit in the beginning of the Spring, which is not ripe till Autumne.

¶ *The Names.*

The Carob tree is called κερατωνία : in Latine likewiſe, *Ceratonia* : in Spaniſh, *Garouo* : in Engliſh, Carob tree ; and of ſome, Beane tree, and Saint *Iohns* Bread : the fruit or cod is named κεράτιον : in Latine *Siliqua*, or *Siliqua dulcis* : in diuers ſhops, *Xylocaracta* : in other ſhops in Italy, *Carobe*, or *Carobole* : of the Apothecaries of Apulia, *Salequa* : it is called in Spaniſh, *Alfarobas*, or *Algarovas* : and without an article *Garouas* : in high Dutch, **S. Johans brot** : that is to ſay, *Sancti Iohan. panis*, or S. *Iohns* Bread, neither is it knowne by any other name in the Low-countries : Some call it in Engliſh, Carob.

¶ *The Temperature.*

The Carob tree is drie and aſtringent, as is alſo the fruit, and containeth in it a certaine ſweetnes as *Galen* ſaith.

¶ *The Vertues.*

A The fruit of the Carob Tree, beeing eat when it is greene, doth gently looſe the belly ; but beeing dry it is hard of digeſtion, and ſtoppeth the belly, it prouoketh vrine, it is good for the ſtomacke, and nouriſheth well, and much better than when it is greene and freſh.

Chap. 83. *Of* Caſsia Fiſtula, *or Pudding Pipe.*

¶ *The Deſcription.*

Caſſia purgatrix, or *Caſſia fiſtula*, groweth vp to be a faire tree, with a tough barke like leather, of the colour of Box, whereupon ſome haue ſuppoſed it to take the Greeke name Κάσυς, in Latine, *Coriaceus* : the armes and branches of this are ſmall and limber, beſet with many goodly leaues, like thoſe of the Wall-nut tree : among which come forth ſmall floures of a yellow colour, compact or conſiſting of ſix little leaues, like the floures of *Chelidonium minus*, or Pile-woort : after theſe be vaded, there ſucceed goodly blacke round, long cods, whereof ſome are two foot long, and of a woody ſubſtance : in theſe coddes are contained a blacke pulpe, very ſweet and ſoft, of a pleaſant taſte, and ſeruing to many vſes in Phyſicke, in which pulpe lieth the ſeed couched in little cels or partitions : this ſeed is flat and browniſh, not vnlike the ſeed of *Ceratia Siliqua*, and in other reſpects very like vnto it alſo.

¶ *The Place.*

This tree groweth much in Egypt, eſpecially about Memphis and Alexandria, and moſt parts of Barbarie, and is a ſtranger in theſe parts of Europe.

¶ *The Time.*

The Caſſia tree groweth green winter and ſommer : it ſheddeth his old leaues when new are come, by meanes whereof it is neuer void of leaues : it floureth early in the ſpring, and the fruit is ripe in Autumne.

¶ The

Cassia fistula.
Pudding Pipe tree.

¶ *The Names.*

This tree was vnknowne to the old writers, or so little accounted of, as that they haue made no mention of it at all: the Arabians were the first that esteemed of it, by reason they knew the vse of the pulpe which is found in the Pipes: and after them the later Grecians, as *Actuarius* & other of his time, by whom it was named κασία μέλαινα that is to say in Latine, *Casia nigra.* The fruit thereof, saith *Actuarius* in his fift booke, is like a long pipe, hauing within it a thicke humour or moisture, which is not congealed all alike thorow the pipe, but is separated and diuided with many partitions, being thin wooddy skins. The Apothecaries call it *Casia fistula*, and with a double ss *Cassia fistula*: it is called in English after the Apothecaries word, Cassia fistula, and may also be Englished, Pudding Pipe, because the cod or Pipe is like a pudding: but the old Cassia fistula, or σύριγξ in Greeke, is that sweet and odoriferous barke that is rolled together, after the manner of a long and round pipe, now named of the Apothecaries *Cassia lignea*, which is a kinde of Cinamon.

¶ *The Temperature.*

The pulpe of this pipe which is chiefely in request, is moist in the later end of the first degree, and little more than temperatly hot.

The Vertues.

The pulpe of *Casia fistula* extracted with violet water, is a most sweet and pleasant medicine, and may be giuen without danger to all weak people of what age and sex soeuer they be, yea it may be ministred to women with childe, for it gently purgeth cholericke humours and slimie flegme, if it be taken in the weight of an ounce. A

Cassia is good for such as be vexed with hot agues, pleurisies, iaundice, or any other inflammation of the liuer, being taken as afore is shewed. B

Cassia is good for the reines and kidneies, driueth forth grauell and the stone, especially if it bee mingled with the decoction of Parsley, and Fennell roots, and drunke. C

It purgeth and purifieth the bloud, making it more cleane than before, breaking therewith the acrimonie and sharpnesse of the mixture of bloud and choler together. D

It dissolueth all phlegmons and inflammations of the brest, lungs, and the rough artery called *Trachea arteria*, easing those parts exceeding well. E

Cassia abateth the vehemencie of thirst in agues, or any hot disease whatsoeuer, especially if it be taken with the iuice of *Intybum*, *Cichoreum*, or *Solanum*, depured according to Art: it abateth also the intemperate heat of the reines, if it be receiued with diureticke simples, or with the decoction of Licorice onely, and will not suffer the stone to grow in such persons as do receiue and vse this medicine. F

The best *Cassia* for your vse is to be taken out of the most ful, most heauy, & fairest cods, or canes, and those which do shine without, and are full of soft pulpe within; that pulpe which is newly taken forth is better than that which is kept in boxes, by what Art soeuer. G

Cassia being outwardly applied, taketh away the roughnesse of the skin, and being laid vpon hot swellings, it bringeth them to suppuration. H

Many singular compounded medicines are made with this *Cassia*, which here to recite belongs not to my purpose or history. I

CHAP.

CHAP. 84. *Of the Lentiske, or Maſticke tree.*

¶ *The Deſcription.*

Lentiſcus.
The Maſticke tree.

¶ *The Deſcription.*

THe Maſtick tree groweth commonly like a ſhrub without any great body, riſing vp with many ſprings and ſhoots like the Haſell; and oftentimes it is of the height and bigneſſe of a meane tree: the boughes thereof are tough, and flexible; the barke is of a yellowiſh red colour, pliable likewiſe, and hard to be broken: there ſtand vpon one rib for the moſt part 8 leaues, ſet vpon a middle rib, much like to the leaues of Licorice, but harder, of a deepe greene colour, and oftentimes ſomewhat red in the brims, as alſo hauing diuers vains running along of a red colour, and ſomthing ſtrong of ſmel: the floures be moſſie, and grow in cluſters vpon long ſtems: after them come vp the berries, of the bigneſſe of Vetches, greene at the firſt, afterwards of a purple colour, and laſt of all, black, fat, and oily, with a hard black ſtone within; the kernel wher-of is white, of which alſo is made oile, as *Dioſcorides* witneſſeth: it bringeth forth likewiſe cods beſides the fruit (which may be rather termed an excreſcence, than a cod) writhed like a horn; in which lieth at the firſt a liquour, and afterwards when this waxeth ſtale, little liuing things like vnto gnats, as in the Turpentine hornes, and in the folded leaues of the Elm tree. There commeth forth of the Maſtick tree a Roſin, but dry, called Maſticke.

¶ *The Place.*

The Maſticke tree groweth in many regions, as in Syria, Candy, Italy, Languedocke, and in moſt Prouinces of Spaine: but the chiefeſt is in Chios an Iland in Greece, in which it is diligently and ſpecially looked vnto, and that for the Maſticke ſake, which is there gathered from the husbanded Maſticke trees by the inhabitants euery yeare moſt carefully, and is ſent from thence into all parts of the world.

¶ *The Time.*

The floures be in their pride in the ſpring time, and the berries in Autumne: the Maſtick muſt be gathered about the time when the Grapes be.

¶ *The Names.*

This tree is named in Greeke σχῖνος: in Latine, *Lentiſcus*: in Italian, *Lentiſque*: in Spaniſh, *Mata*, and *Arcoyra*: in Engliſh, Maſticke tree; and of ſome, Lentiske tree.

The Roſin is called in Greeke ῥητίνη σχινίνη, and μαστίχη: in Latine, *Lentiſcina Reſina*, and likewiſe, *Maſtiche*: in ſhops, *Maſtix*: in Italian, *Maſtice*: in high and low Dutch and French alſo, *Maſtic*: in Spaniſh, *Almaſtiga*, *Maſtech*, and *Almecega*: in Engliſh, Maſticke.

Cluſius writeth, that the Spaniards call the oile that is preſſed out of the berries, *Azeyte de Mata*.

¶ *The Temperature.*

The leaues, barke, and gum of the Maſticke tree are of a meane and temperate heate, and are drie in the ſecond degree, and ſomewhat aſtringent.

¶ *The Vertues.*

A The leaues and barke of the Maſticke tree ſtoppe the laske, the bloudy flixe, the ſpitting of bloud,

bloud, the piſſing of bloud, and all other fluxes of bloud: they are alſo good againſt the falling ſickeneſſe, the falling downe of the mother, and comming forth of the fundament.

The gum Maſticke hath the ſame vertue, if it be relented in wine and giuen to be drunke. B

Maſticke chewed in the mouth is good for the ſtomacke, ſtaieth vomiting, increaſeth appetite, comforteth the braines, ſtaieth the falling downe of the rheumes and watery humors, and maketh a ſweet breath. C

The ſame infuſed in Roſe water is excellent to waſh the mouth withall, to faſten looſe teeth, and to comfort the iawes. D

The ſame ſpred vpon a piece of leather or veluet, and laid plaiſterwiſe vpon the temples, ſtaieth the rheume from falling into the iawes and teeth, and eaſeth the paines thereof. E

It preuaileth much againſt vlcers and wounds, being put into digeſtiues and healing Vnguents. F

It draweth flegme forth of the head gently and without trouble. G

It is alſo vſed in waters which ſerue to clenſe and make faire the face with. H

The decoction of this filleth vp hollow vlcers with fleſh if they be bathed therewith. I

It knitteth broken bones, ſtaieth eating vlcers, and prouoketh vrine. K

CHAP. 85. *Of the Turpentine Tree.*

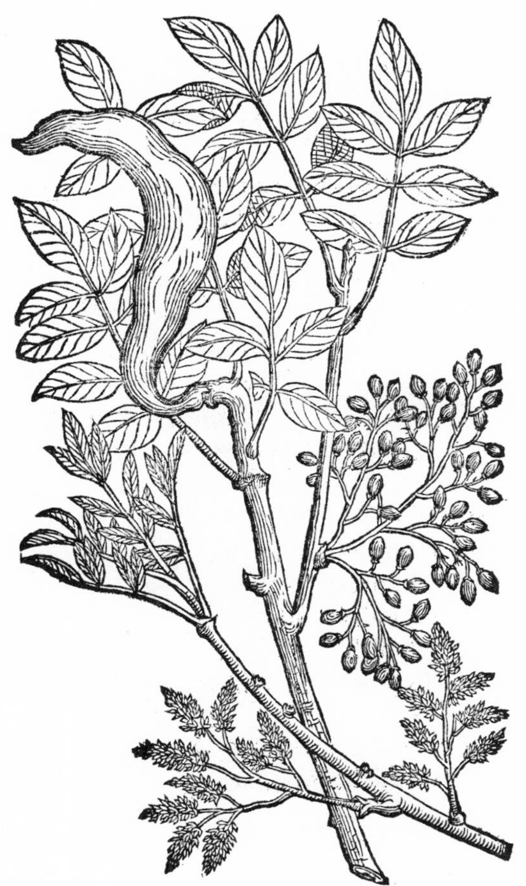

1 *Terebinthus.*
The Turpentine tree.

2 *Terebinthus latifolia.*
The broad leafed Turpentine tree.

¶ *The Deſcription.*

1 THe firſt Turpentine Tree groweth to the height of a tall and faire tree, hauing many long boughes or branches, diſperſed abroad, beſet with long leaues, conſiſting of ſundry other ſmall leaues, each whereof reſembleth the Bay leafe, growing one againſt another vpon a little ſtem or middle rib, like vnto the leaues of the Aſh tree: the floures be ſmall & reddiſh, growing vpon cluſters or bunches that turne into round berries, which at their beginning are greene, afterwards reddiſh, but being ripe wax blacke, or of a darke blew colour, clammie, full of fat and

and oilous in substance, and of a pleasant sauour: this plant beareth an empty cod, or crooked horne somewhat reddish, wherein are found small flies, wormes or gnats, bred and ingendred of a certaine humorous matter, which cleaueth to the inner sides of the said cods or hornes, which wormes haue no physicall vse at all. The right Turpentine issueth out of the branches of these trees, if you do cut or wound them, the which is faire and cleere, and better than that which is gathered from the barke of the Firre tree.

2 The second kinde of Turpentine tree is very like vnto the former, but that it groweth not so great: yet the leaues are greater and broader, and of the same fashion, but very like to the leaues of the Pistacia tree. The berries are first of a scarlet colour, and when they be ripe of a skie colour. The great horned cods are sharpe pointed, and somewhat cornered, consisting as it were of the substance of gristles. And out of those bladders being broken, do creepe and come small flies or gnats, bred of a fuliginous excrement, and ingendred in those bladders. The tree doth also yeeld his Turpentine by dropping like the former.

¶ *The Place.*

These trees grow, as *Dioscorides* saith, in Iurie, Syria, Cyprus, Africke, and in the Islands called *Cyclades*. *Bellonius* reporteth that there are found great store of them in Syria, and Cilicia, and are brought from thence to Damascus to be sold. *Clusius* saith, that it growes of it selfe in Languedocke, and in very many places of Portingale and Spaine, but for the most part like a shrub, and without bearing Turpentiue.

Theophrastus writeth, that it groweth about the hill Ida, and in Macedonia, short, in manner of a shrub, and writhed; and in Damascus and Syria great, in manner of a small tree: he also setteth downe a certaine male Turpentine tree, and a female: the male, saith he, is barren, and the female fruitfull. And of these he maketh the one with a berry red at the first, of the bignesse of a Lentill, which cannot come to ripenes; and the other with the fruit greene at the first, afterwards somewhat of a yellowish red, and in the end blacke, waxing ripe in the spring, of the bignesse of the Grecians Beane, and rosenny.

He also writeth of a certaine Indian Turpentine tree, that is to say, a tree like in boughes and leaues to the right Turpentine tree, but differing in fruit, which is like vnto Almonds.

¶ *The Time.*

The floures of the Turpentine tree come forth in the spring together with the new buds: the berries are ripe in September and October, in the time of Grape gathering. The hornes appeare about the same time.

¶ *The Names.*

This tree is called in Greeke τέρμινθος, and also many times τερέβινθος: in Latine, *Terebinthus*: in Italian, *Terebintho*: in Spanish, *Cornicabra*: in French, *Terebinte*: in English, Turpentine tree: the Arabians call it *Botin*, and with an article *Albotin*.

The Rosin is surnamed τερμινθίνη in Latine, *Terebinthina*: in high Dutch, Termintijn: in English, Turpentine, and right Turpentine: in the Arabian language *Albotin*, who name the fruit *Granum viride*, or greene berries.

¶ *The Temperature and Vertues.*

A The barke, leaues, and fruit of the Turpentine tree do somewhat binde, they are hot in the second degree, and being greene they dry moderately, but when they are dryed they dry in the second degeee; and the fruit approacheth more neere to those that be dry in the third degree, and also hotter. This is fit to be eaten, as *Dioscorides* saith, but it hurteth the stomacke.

B It prouoketh vrine, helpeth those that haue bad spleenes, and is drunke in wine against the bitings of the poysonsome spiders called *Phalangia*.

C The Rosin of the Turpentine tree excelleth all other Rosins, according to *Dioscorides* his opinion: but *Galen* writeth, that the Rosin of the masticke tree beareth the preheminence, and then the Turpentine.

D This Rosin hath also an astringent or binding facultie, and yet not so much as masticke; but it hath withall a certaine bitternesse ioyned, by reason whereof it digesteth more than that of the Masticke tree: thorow the same qualitie there is likewise in it so great a clensing, as also it healeth scabs, in his 8. booke of the faculties of simple medicines; but in his booke of medicines according to the kindes, he maketh that of the Turpentine tree to be much like the Rosins of the Larch tree, which he affirmeth to be moister than all the rest, and to be without both sharpnesse and biting.

E The fruit of Turpentine prouoketh vrine and stirreth vp fleshly lust.

F The Rosine of this tree, which is the right Turpentine, looseth the belly, openeth the stoppings of the liuer and spleene, prouoketh vrine, and driueth forth grauell, being taken the quantitie of two or three Beanes.

The

The like quantitie waſhed in water diuers times vntill it be white,then muſt be put thereto the like quantity of the yolk of an egge,and laboured togetheradding thereto by little and little(continually ſtirring it)a ſmall draught of poſſit drinke made of white wine, and giuen to drink in the morning faſting,ithelpeth moſt ſpeedily the Gonorrhæa,or running of the reines, commonly at the firſt time, but the medicine neuer faileth at the ſecond time of the taking of it,which giues ſtooles from foure to eight,according to the age and ſtrength of the patient. G

Chap. 86. *Of the Frankincenſe tree.*

¶ *The Deſcription.*

THe tree from which Frankincenſe floweth is but low,and hath leaues like the Maſtick tree; yet ſome are of opinion that the leafe is like the leafe of a Peare tree, and of a graſſie colour: the rinde is like that of the Bay tree,whereof there are two kindes: the one groweth in mountains and rockie places, the other in the plaine: but thoſe in the plaines are much worſe than thoſe of the mountaines: the gum hereof is alſo blacker, fitter to mingle with Pitch, and ſuch other ſtuffe to trim ſhips,than for other vſes.

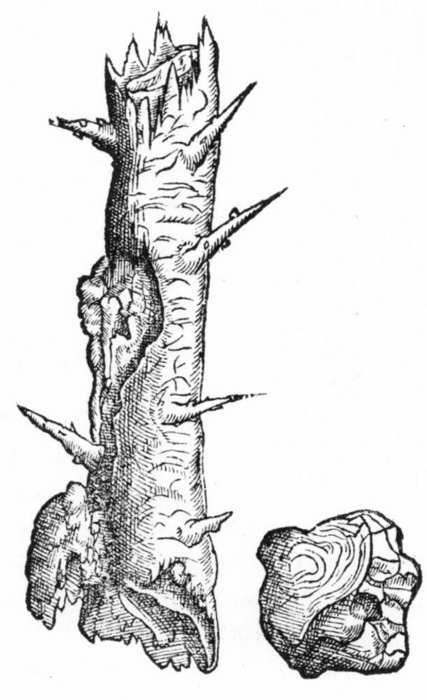

Arbor Thurifera.
The Frankincenſe tree.

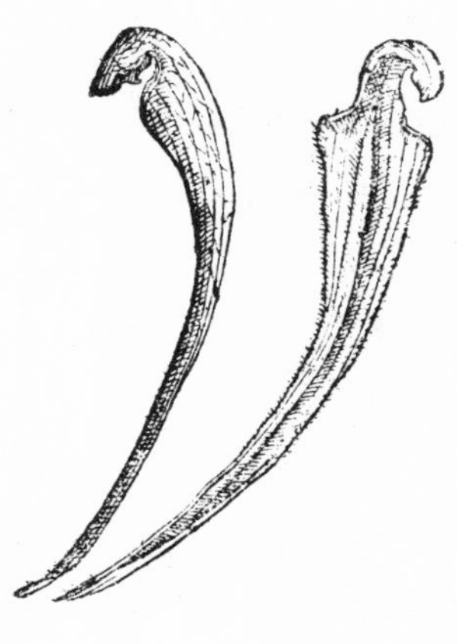

Thuris Limpidifolium Lobelij.
The ſuppoſed leafe of the Frankincenſe tree.

Theuet in his Coſmographie ſaith,that the Frankincenſe tree doth reſemble a gummie or roſiny Pine tree,which yeeldeth a iuice that in time groweth hard, and is called *Thus*, Frankincenſe, in whom is found ſometime certaine ſmall graines like vnto grauell, which they call the Manna of Frankincenſe.

Of this there is in Arabia two other ſorts,the one,the gum wherof is gathered in the Dog daies when the Sun is in Leo,which is white,pure, cleare,and ſhining. *Pena* writeth that he hath ſeene the cleare Frankincenſe called *Limpidum*, and yeelding a very ſweet ſmell when it is burnt, but the leafe hath been ſeldome ſeene; which the Phyſition *Lаunanus* gaue to *Pena* and *Lobel*, together with ſome pieces of the Roſine,which he had of certaine mariners, but he could affirme nothing of certaintie whether it were the leafe of the Frankincenſe,or of ſome other Pine tree, yeelding the like iuice or gum. It is,ſaith he (which doth ſeldom happen in other leaues) from the lower part or foot of the leafe,to the vpper end,as it were doubled,conſiſting of two thin rindes or coats,with a ſheath a ſpan and a halfe long,at the top gaping open like a hood or fooles coxcombe,and as it were couered with a helmet,which is a thing ſeldome ſeene in a leafe,but is proper to the floures of *Napellus*, or *Lonchitis*, as writers affirme; the other is gathered in the spring, which is reddish, worser than the other in price or value, because it is not so well concocted in the heat of the Sunne. The Arabians wound this tree with a knife, that the liquour may flow out more abundantly, whereof some trees yeeld threescore pounds of Frankinsence.

¶ *The Place.*

Dioscorides saith it groweth in Arabia, and especially in that quarter which is called Thurifera, the best in that countrey is called *Stagonias*, and is round, and if it be broken, is fat within, and when it is burned doth quickly yeeld a smel: next to it in goodnes is that which groweth in Smilo, lesser than the other, and more yellow.

¶ *The Time.*

The time is already declared in the description.

¶ *The Names.*

It is called in Greeke λίβανος: in Latine, *Thus*: in Italian, *Incenso*: in Dutch, **Weyrauch**: in Spanish, *Encenso*: in French, *Enceus*: in English, Frankincense, and Incense: in the Arabian tongue, *Louan*, and of some few, *Condur*. ‡ The Rosin carries the same name; but in shops it is called *Olibanum*, of the Greeke name and article put before it. ‡

The Temperature and Vertues.

It hath, as *Dioscorides* saith, a power to heate and binde.

A It drieueth away the dimnesse of the eye-sight, filleth vp hollow vlcers, it closes raw wounds, stai-
B eth all corruptions of bloud, although it fall from the head.

Galen writeth thus of it; *Thus* doth heate in the second degree, and drie in the first, and hath some
C small astriction, but in the white there is a manifest astriction; the rinde doth manifestly binde and dry exceedingly, and that most certainly in the second degree, for it is of more grosser parts than Frankincense, and not so sharpe, by reason whereof it is much vsed in spitting of bloud, swellings in the mouth, the collicke passion, the flux in the belly rising from the stomacke, and bloudy flixes.

D The fume or smoke of it hath a more drier and hotter quality than the Frankincense it selfe, being dry in the third degree.

E It doth also clense and fill vp the vlcers in the eies, like vnto Myrrhe: thus far *Galen*.

F *Dioscorides* saith, that if it be drunk by a man in health, it drieueth him into a frensie: but there are few Greekes of his minde.

Auicen reporteth that it doth helpe and strengthen the wit and vnderstanding, but the often ta-
G king of it will breed the head-ache, and if too much of it be drunke with wine it killeth.

Chap. 87. *Of Fisticke Nuts.*

Pistacia. The Fisticke Nut.

¶ *The Description.*

The tree which beareth Fisticke Nuts is like to the Turpentine tree: the leaues hereof be greater than those of the Masticke tree, but set after the same maner, and in like order that they are, being of a faint yellow colour out of a green; the fruit or Nuts do hang by their stalks in clusters, being greater than the Nuts of Pine Apples, and much lesser than Almonds: the husks without is of a grayish colour sometimes reddish, the shell brickle and white; the substance of the kernell greene; the taste sweet, pleasant to be eaten, and something sweet of smell.

¶ *The Place.*

Fisticke Nuts grow in Persia, Arabia, Syria, and in India; now they are made free Denizons in Italy, as in Naples and in other Prouinces there.

¶ *The Time.*

This tree doth floure in May, and the fruit is ripe in September.

¶ *The Names.*

This Nut is called in Greeke πιστάκιον in *Athenaeus*: *Nicander Colophonius* in his booke of Treacles nameth it ψιττάκιον: *Possidonius* nameth it βιστάκιον: others, φιστάκιον: the Latines obseruing the same termes, haue named it *Pistacion*, *Bistacion*, or *Phistacion*:

ſtacion: the Apothecaries, *Fiſtici*: the Spaniards, *Alhocigos*, and *Fiſticos*: in Italian, *Piſtacchi*: in English, Fiſticke Nut.

¶ *The Temperature and Vertues.*

The kernels of the Fiſticke Nuts are oftentimes eaten as be thoſe of the Pine Apples; they be A
of temperature hot and moiſt; they are not ſo eaſily concocted, but much eaſier than common nuts: the iuice is good, yet ſomewhat thicke; they yeeld to the body no ſmall nouriſhment, they nouriſh bodies that are conſumed: they recouer ſtrength.

They are good for thoſe that haue the phthiſicke, or rotting away of the lungs. B

They concoct, ripen, and clenſe forth raw humours that cleaue to the lights and cheſt. C

They open the ſtoppings of the liuer, and be good for the infirmities of the kidneies; they alſo D
remoue out of the kidneies ſand and grauell; and aſſwage their paine: they are alſo good for vlcers.

The kernels of Fiſticke nuts condited, or made into comfits, with ſugar, and eaten, doe procure E
bodily luſt, vnſtop the lungs and the breſt, are good againſt the ſhortneſſe of breath, and are an excellent preſeruatiue medicine being miniſtred in wine againſt the bitings of all manner of wilde beaſts.

Chap. 88. *Of the Bladder Nut.*

Nux veſicaria.
The Bladder Nut.

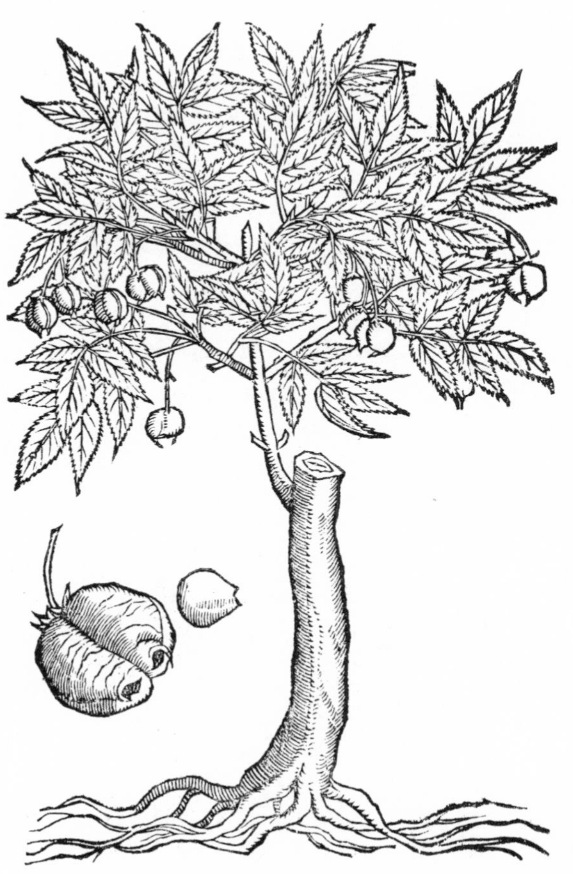

¶ *The Deſcription.*

This is a low tree, hauing diuers young ſprings growing forth of the root: the ſubſtance of the wood is white, very hard & ſound; the barke is of a light greene: the leaues conſiſt of fiue little ones, which be nicked in the edges like thoſe of the Elder, but leſſer, not ſo greene nor ranke of ſmell. It hath the pleaſant whitiſh floures of Bryonie or *Labruſca*, both in ſmell and ſhape, which turne into ſmal cornered bladders of winter Cherries, called Alkakengie, but of an ouerworne greeniſh colour: in theſe bladders are contained two little nuts, and ſometimes no more but one, leſſer than the Haſell nut, but greater than the Ram Cich, with a wooddie ſhel and ſomewhat red: the kernell within is ſomething green; in taſte at the firſt ſweet, but afterwards lothſome, and ready to prouoke vomit.

¶ *The Place.*

It groweth in Italy, Germany and France, it groweth likewiſe at the houſe of ſir *Walter Culpepper* neere Flimmewell in the Weild of Kent, as alſo in the Frier-yard without Saint Paules gate in Stamford, and about Spalding Abbey, and in the garden of the right honourable the Lord Treaſurer my very good Lord and Maſter, and by his houſe in the Strand. It groweth alſo in my garden, and in the garden hedges of ſir *Francis Carew* neere Croydon, ſeuen miles from London.

¶ *The Time.*

This tree floureth in May, the Nuts be ripe in Auguſt and September.

¶ *The Names.*

It is commonly called in high Dutch, **Pimpernuſz**, which ſignifieth in low Dutch **Pimpernoten**: diuers call it in Latine *Piſtacium Germanicum*: we thinke it beſt to call it *Nux veſicaria*. *Matthiolus* in his Epiſtles doth iudge the Turks *Coulcoul* and *Hebulben* to agree with this: *Gulielmus Quacelbenus* affirmeth, Coulcoul to be vſed of diuers in Conſtantinople for a daintie, eſpecially when they be new brought out of Egypt. This plant hath no old name, vnleſſe it be *Staphylodendron Plinij*:

 for

forwhich it is taken of the later writers:and *Pliny* hath written of it in his 16.book, 16.chap. There is also (saith he) beyond the Alpes a tree, the timber whereof is very like to that of white Maple, and is called *Staphylodendron*, it beareth cods,and in those kernels,hauing the taste of the Hasel nut. It is called in English,S.Anthonies nuts,wilde Pistacia,or Bladder nuts:the Italians call it *Pistachio Saluaticke*: the French men call it *Baguenaudes a patre nostres*,for that the Friers do vse to make beads of the nuts.

¶ *The Temperature and Vertues.*

A These nuts are moist and ful of superfluous raw humours,and therefore they easily procure a readinesse to vomite,and trouble the stomacke,by reason that withall they be somewhat binding,and therefore they be not to be eaten.

B They haue as yet no vse in medicine,yet notwithstanding some haue attributed vnto them some vertues in prouoking of Venerie.

Chap. 89. *Of the Hasell tree.*

¶ *The Description.*

1 The Hasell tree groweth like a shrub or small tree,parted into boughes without ioints, tough and pliable:the leaues are broad,greater and fuller of wrinckles than those of the Alder tree,cut in the edges like a saw,of colour greene,and on the backside more white,the bark is thin: the root is thicke,strong,and growing deep; in stead of floures hang downe catkins, aglets, or blowings, slender,and well compact: after which come the Nuts standing in a tough cup of a greene colour, and iagged at the vpper end, like almost vnto the beards in Roses. The shell is smooth and wooddie: the kernel within consisteth of a white,hard,and sound pulpe, and is couered with a thin skin,oftentimes red,most commonly white;this kernell is sweet and pleasant vnto the taste.

1 *Nux Auellana,siue Corylus.*
The Filberd Nut.

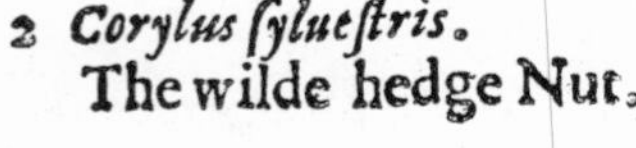

2 *Corylus syluestris.*
The wilde hedge Nut.

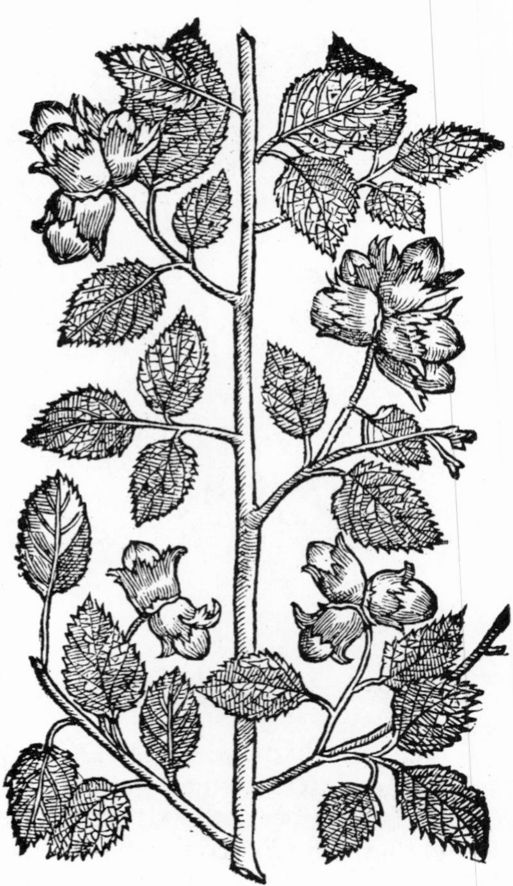

2 *Corylus syluestris* is our hedge Nut or Hasell Nut tree, which is very well knowne, and therefore needeth not any description: whereof there are also sundry sorts, some great, some little, some rathe ripe, some later, as also one that is manured in our gardens, which is very great, bigger than any Filberd, and yet a kinde of Hedge nut: this then that hath beene said shall suffice for Hedge-Nuts.

‡ 3 The small Turky Nut tree growes but low, and the leaues grow without order, vpon the twigs, they are in shape like those of the former, but somewhat longer: the chiefe difference consists in the fruit, which is small, and like an Hasell Nut, but shorter: the huske, wherein somtimes one, otherwhiles more Nuts are contained, is very large, tough, and hard, diuided both aboue and below into a great many iags, which on euery side couer and hold in the Nuts, and these cups are very rough without, but smooth on the inside. *Clusius* first set this forth (hauing receiued it from Constantinople) by the name of *Auellana pumila Byzantina*. ‡

3 *Auellana pumila Byzantina cum suo fructu.*
The Filberd Nut of Constantinople.

¶ *The Place.*

The Hasell trees do commonly grow in Woods and in dankish vntoiled places: they are also set in Orchards, the Nuts whereof are better, and of a sweeter taste, and be most commonly red within.

¶ *The Time.*

The catkins or aglets come forth very timely, before winter be fully past, and fall away in March or Aprill, so soone as the leaues come forth: the Nuts be ripe in August.

¶ *The Names.*

This shrub is called in Latine, *Corylus*: in Greeke, κάρυα Ποντικά, that is, *Nux Pontica*, or Ponticke Nut: in high-Dutch, **Hasel strauck**: in low-Dutch, **Haseleer**: in English, Hasel tree, and Filberd tree; but the Filberd tree is properly that which groweth in gardens and Orchards, and whose fruit is commonly wholly couered ouer with the huske, and the shell is thinner.

The Nut is named in Latine, *Nux Pontica, tenuis Nux, parua Nux*: it is also called *Nux Praenestina, Nux Heracleotica*, and commonly *Nux auellana*, by which name it is vsually knowne to the Apothecaries: in high-Dutch, **Hasel Nusz**: in low-Dutch, **Hasel Noten**: in Italian, *Nocciuole, Auellane, Nocelle*: in French, *Noisettes, & Noisilles*: in Spanish, *Auellanas*: in English, Hasell nut, and Filberd.

These

These Nuts that haue their skinnes red are the garden and planted Nuts, and the right Ponrick Nuts or Filberds: they are called in high-Dutch, **Rhurnusz**, and **Rotnusz**: in low-Dutch, **Roode Hasel Noten**: in English, Filberds, and red Filberds.

The other Nuts which be white are iudged to be wilde.

¶ *The Temperature and Vertues.*

A Hasell Nuts newly gathered, and not as yet dry, containe in them a certaine superfluous moisture, by reason whereof they are windie: not onely the new gathered Nuts, but the dry also, be very hard of digestion; for they are of an earthy and cold essence, and of an hard and sound substance, for which cause also they very slowly passe thorow the belly, therefore they are troublesome and clogging to the stomacke, cause head-ache, especially when they be eaten in too great a quantitie.

B The kernells of Nuts made into milke like Almonds do mightily bind the belly, and are good for the laske and the bloudy flix.

C The same doth coole exceedingly in hot feuers and burning agues.

D The catkins are cold and dry, and likewise binding: they also stay the lask.

E ‡ The kernels of Nuts rather cause than cure the bloudy flix and lasks, wherefore they are not to be vsed in such diseases. ‡

Chap. 90. *Of the Wall-nut tree.*

Nux Iuglans.
The Walnut tree.

¶ *The Description.*

This is a great tree with a thicke and tall body: the barke is somewhat greene, and tending to the colour of ashes, and oftentimes full of clefts: the boughes spread themselues far abroad: the leaues consist of fiue or six fastned to one rib, like those of the Ash tree, and with one standing on the top, which be broader and longer than the particular leaues of the Ash, smooth also, and of a strong smell: the catkins or aglets come forth before the Nuts: these Nuts do grow hard to the stalke of the leaues, by couples, or by three & three; which at the first when they be yet but tender haue a sweet smel, and be couered with a green huske: vnder that is a wooddy shell in which the kernell is contained, being couered with a thin skin, parted almost into foure parts with a woody skin as it were: the inner pulp whereof is white, sweet and pleasant to the tast; and that is when it is new gathered, for after it is dry it becommeth oily and ranck.

¶ *The Place.*

The Walnut tree groweth in fields neere common high-waves, in a fat and fruitfull ground, and in orchards: it prospereth on high fruitfull bankes, it loueth not to grow in waterie places.

¶ *The Time.*

The leaues together with the catkins come forth in the Spring: the Nuts are gathered in August.

¶ *The Names.*

The tree is called in Greeke, Κάρυα: in Latine, *Nux*, which name doth signifie both the tree and the fruit: in high Dutch, **Nuszbaum**: in low-Dutch, **Nooteboome**, and **Nootelaer**: in French, *Neisier*: in Spanish, *Nogueyra*: in English, Walnut tree, and of some, Walsh nut tree. The Nut is called in Greeke, Κάρυον βασιλικόν, that is to say, *Nux Regia*, or the Kingly Nut: it is likewise named

Nux

Nux Inglans, as though you should say *Iouis glans*, Iupiters Acorne; or *Iuvans glans*, the helping Acorne: and of diuers, *Persica Nux*, or the Persian Nut: in high-Dutch, Welsch Nusz, and Baumnusz: in low-Dutch, Okernoten, Walsch Noten: In Italian, *Noci*: in French, *Noix*: in Spanish, *Nuezes*, and *Nous*: in English, Walnut; and of some, Walsh nut.

¶ *The Temperature and Vertues.*

The fresh kernels of the nuts newly gathered are pleasant to the taste: they are a little cold, and haue no small moisture, which is not perfectly concocted: they be hard of digestion, and nourish little: they slowly descend. A

The dry nuts are hot and dry, and those more which become oily and ranke: these be very hurtfull to the stomacke, and besides that they be hardly concocted, they increase choler, cause headache, and be hurtfull for the chest, aud for those that be troubled with the cough. B

Dry Nuts taken fasting with a fig and a little Rue withstand poyson, preuent and preserue the body from the infection of the plague, and being plentifully eaten they driue wormes forth of the belly. C

The greene and tender Nuts boiled in Sugar and eaten as Suckad, are a most pleasant and delectable meate, comfort the stomacke, and expell poyson. D

The oile of Walnuts made in such manner as oile of Almonds, maketh smooth the hands and face, and taketh away scales or scurfe, blacke and blew marks that come of stripes or bruises. E

Milke made of the kernels, as Almond milke is made, cooleth and pleaseth the appetite of the languishing sicke body. F

With onions, salt, and hony, they are good against the biting of a mad dog or man, if they be laid vpon the wound. G

Being both eaten, and also applied, they heale in short time, as *Dioscorides* saith, Gangrens, Carbuncles, ægilops, and the pilling away of the haire: this also is effectually done by the oile that is pressed out of them, which is of thin parts, digesting and heating. H

The outward greene huske of the Nuts hath a notable binding facultie. I

Galen deuised and taught to make of the iuyce thereof a medicine for the mouth, singular good against all inflammations thereof. K

The leaues and first buds haue a certaine binding qualitie, as the same Authour sheweth; yet there doth abound in them an hot and dry temperature. L

Some of the later Physitions vse these for baths and lotions for the body, in which they haue a force to digest and also to procure sweat. M

CHAP. 91. *Of the Chestnut tree.*

¶ *The Description.*

1 THe Chestnut tree is a very great an high tree: it casteth forth very many boughes: the body is thicke, and sometimes of so great a compasse as that two men can hardly fathom it: the timber or substance of the wood is sound and durable: the leaues be great, rough, wrinkled, nicked in the edges, and greater than the particular leaues of the Walnut tree. The blowings or catkins be slender, long, and greene: the fruit is inclosed in round a rough and prickly huske like to an hedge-hog or Vrchin, which opening it selfe doth let fall the ripe fruit or Nut. This nut is not round, but flat on the one side, smooth, and sharpe pointed: it is couered with a hard shell, which is tough and very smooth, of a darke browne colour: the meate or inner substance of the nut is hard and white, and couered with a thin skin which is vnder the shell.

2 The Horse Chestnut groweth likewise to be a very great tree, spreading his great and large armes or branches far abroad, by which meanes it maketh a very good coole shadow. These branches are garnished with many beautifull leaues, cut or diuided into fiue, six, or seuen sections or diuisions, like to the Cinkfoile, or rather like the leaues of *Ricinus*, but bigger. The floures grow at the top of the stalks, consisting of foure small leaues like the Cherry blossome, which turne into round rough prickly heads like the former, but more sharpe and harder: the nuts are also rounder. ‡ The floures of this, saith *Clusius* (whose figure of them I here giue you) come out of the bosom of the leafe which is the vppermost of the branch, and they are many in number growing vpon prettie long foot-stalkes, consisting each of them of foure white leaues of no great bignesse; the two vppermost are a little larger than the rest, hauing round purple spots in their middles: out of the middle of the floure come forth many yellowish threds with golden pendants. The fruit is contained in a prickly huske that opens in three parts, and it is rounder and not so sharpe pointed as the

1 *Caſtanea.*
Cheſtnut tree.

2 *Caſtanea Equina cum flore.*
Horſe Cheſtnut tree in floure.

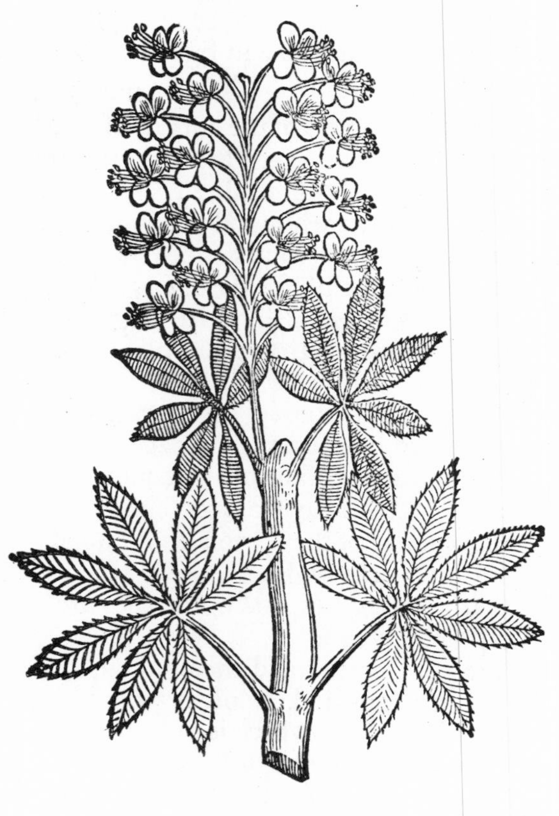

Caſtaneæ Equinæ fructus.

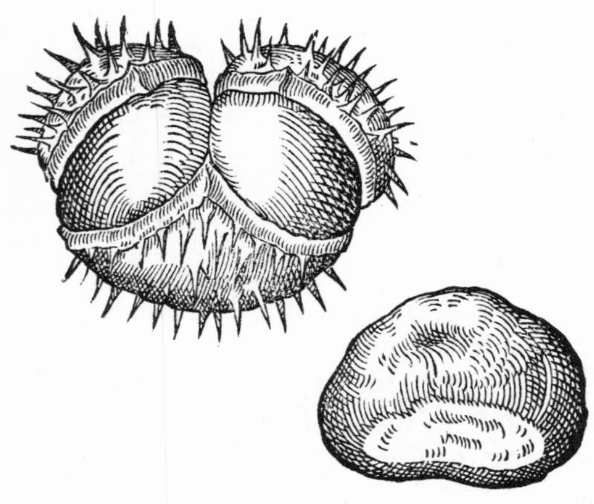

‡ 3 *Caſtaneæ Peruanæ fructus.*

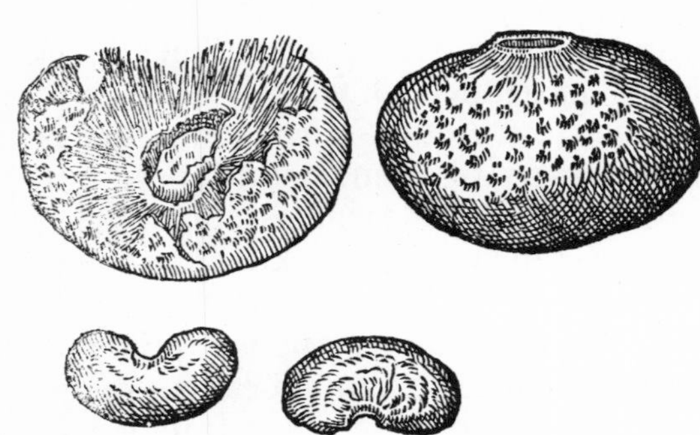

the ordinary Cheſtnut, neither vnder the vtter coat hath it any peeling within as the other hath, neither is it of ſo good a taſte. ‡

‡ 3 This Americane Cheſtnut is almoſt round, but that it is a little flatted on the ſides, eſpecially whereas it is faſtned to the ſtalke : the vtter coat is ſufficiently thicke, yet brittle, and as it were fungous, of a browniſh yellow colour : vnder this are aboundance of ſmall yet ſtiffe prickles, faſt ſticking to the ſhell that containes the kernell : the ſhell it ſelfe is browniſh, not thick, but tough and hard to breake, ſmooth and ſhining on the inſide, wherein is contained a kernel of the bigneſſe and colour of an hares kidney, white within, and ſweet in taſte like an almond or the common Cheſtnut. *Cluſius* cals this *Caſtanea Peruana*, or Cheſtnut of Peru; and hee ſaith hee had it from the famous Geographer *Abraham Ortelius*, who had it ſent him by *Benedictus Arias Montanus*. The figure is expreſt vnder that of the Horſe Cheſtnut. ‡

¶ *The Place.*

The firſt growes on mountaines and ſhadowie places, and many times in the vallies: they loue a ſoft and blacke ſoile. There be ſundry woods of Cheſtnuts in England, as a mile

mile and a halfe from Feuersham in Kent, and in sundry other places: in some countries they be greater and pleasanter: in others smaller, and of worse taste.

The Horse Chestnut groweth in Italy, and in sundry places of the East countries. ‡ It is now growing with Mr. *Tradescant* at South Lambeth. ‡

¶ *The Time.*

The blowings or aglets come forth with the leaues in Aprill; but the Nuts later, and be not ripe till Autumne.

¶ *The Names.*

The Chestnut tree beares the name of the Nut both in Greeke and Latine: in high-Dutch, **Kestenbaum**, and **Kastanibaum**: in low-Dutch, **Castaniboom**: in French, *Castaignier*: in English, Chestnut tree.

The Nut is called in Greeke, Κάστανον: in Latine, *Castanea, Iouis glans, Sardinia glans*: in high-Dutch **Kesten**: in low-Dutch, **Castanien**: in Italian, *Castagne*: in French, *Chastaigne*: in Spanish, *Morones, Castanas*: in English, Chestnut: the greater Nuts be named of the Italians, *Marroni*: of the French men and of diuers base Almaines, *Marons*.

The Horse Chestnut is called in Latine, *Equina Castanea*: in English, Horse Chestnut, for that the people of the East countries do with the fruit thereof cure their horses of the cough, shortnesse of breath, and such like diseases.

¶ *The Temperature and Vertues.*

Our common Chestnuts are very dry and binding, and be neither hot nor cold, but in a mean be- A
tweene both: yet haue they in them a certaine windinesse, and by reason of this, vnlesse the shell
be first cut, they skip suddenly with a cracke out of the fire whilest they be rosting.

Of all the Acornes, saith *Galen*, the Chestnuts are the chiefest, and doe onely of all the wilde B
fruits yeeld to the body commendable nourishment; but they slowly descend, they be hardly concocted, they make a thicke bloud, and ingender winde: they also stay the belly, especially if they
be eaten raw.

Being boiled or rosted they are not of so hard digestion, they more easily descend, and are lesse C
windy, yet they also make the body costiue.

Some affirme, that of raw Chestnuts dried, and afterwards turned into meale, there is made a D
kinde of bread: yet it must needs be, that this should be dry and brittle, hardly concocted, and verie slow in passing thorow the belly; but this bread may be good against the laske and bloudy flix.

An Electuarie of the meale of Chestnuts and hony is very good against the cough and spitting E
of bloud.

The barke of the Chestnut tree boiled in wine and drunke, stops the laske, the bloudy flix, and F
all other issues of bloud.

Chap. 92. *Of the Beech tree.*

¶ *The Description.*

The Beech is an high tree, with boughes spreading oftentimes in manner of a circle, and with a thicke body hauing many armes: the barke is smooth: the timber is white, hard, and verie profitable: the leaues be smooth, thin, broad, and lesser than those of the blacke Poplar: the catkins or blowings be also lesser and shorter than those of the Birch tree, and yellow: the fruit or Mast is contained in a huske or cup that is prickly, and rough bristled, yet not so much as that of the Chestnut: which fruit being taken forth of the shells or vrchin husks, be couered with a soft and smooth skin like in colour and smoothnesse to the Chestnuts, but they be much lesser, and of another forme, that is to say, triangled or three cornered: the kernell within is sweet, with a certaine astriction or binding qualitie: the roots be few, and grow not deepe, and little lower than vnder the turfe.

¶ *The Place.*

The Beech tree loueth a plaine and open countrey, and groweth very plentifully in many Forrests and desart places of Sussex, Kent, and sundry other countries.

¶ *The Time.*

The Beech floureth in Aprill and May, and the fruit is ripe in September, at what time the Deere do eate the same very greedily, as greatly delighting therein; which hath caused forresters and huntsmen to call it Buck-mast.

¶ The.

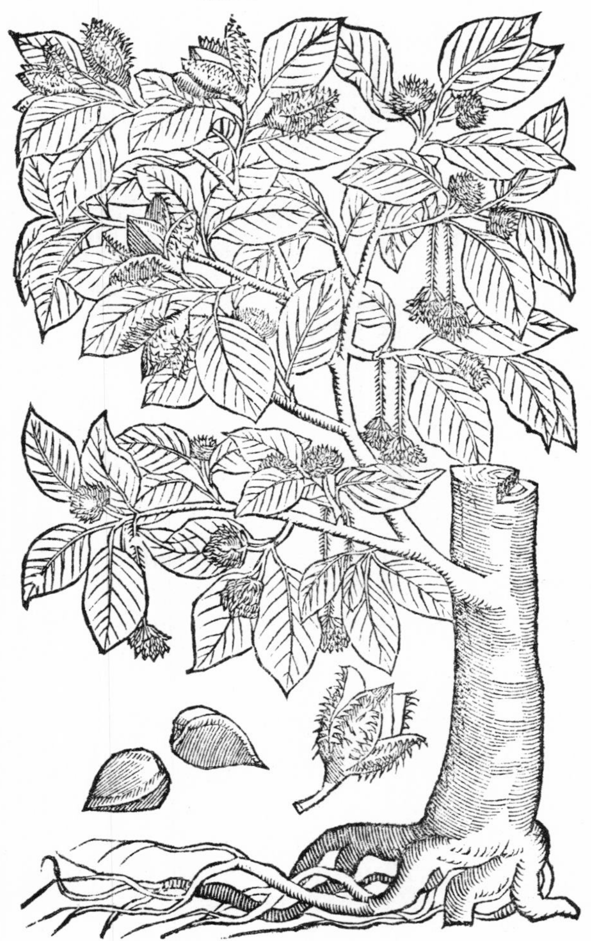
Fagus.
The Beech.

¶ *The Names.*

The tree is called in Greeke, φηγὸς : in Latine, *Fagus* : in high-Dutch, **Buchbaum**, or **Buch** : in low-Dutch, **Bukenboom** : in Italian, *Faggi* : in Spanish, *Haia*, *Faia*, and *Fax* : in French, *Fau*, or *Hestre* : in English, Beech tree, Beech-mast, and Buck-mast.

The fruit is called in Latine, *Nuces Fagi* : in Greeke, βάλανοι τῆς φηγοῦ : in low-Dutch, **Bueken nootkens** : in French, *Faine* : in English, Beech-mast. *Dioscorides* reckons the Beech among the Acorne trees; and yet is the mast nothing at all like to an Acorne. Of *Theophrastus* it is called *Oxya* : of *Gaza*, *Sciscina*.

Pliny also makes mention of this tree, but vnder the name of *Ostrya* (if so be in stead of *Ostrya* we must not reade *Oxya*) *lib.* 13. *ca.* 21. It bringeth forth (saith he, meaning Greece) the tree *Ostrys*, which they likewise call *Ostrya*, growing alone among waterie stones, like to the Ash tree in barke and boughes, with leaues like those of the Peare tree, but somewhat longer and thicker, and with wrinkled cuts which runne quite thorow, with a seed like in colour to a Chestnut, and not vnto barley : the wood is hard and firme, which being brought into the house there followes hard trauell of childe, and miserable deaths, as it is reported; and therefore it is to be forborne, and not vsed as fire wood, if *Plinies* copies be not corrupted.

¶ *The Temperature.*

The leaues of Beech do coole : the kernell of the Nut is somwhat moist.

¶ *The Vertues.*

A The leaues of Beech are very profitably applied vnto hot swellings, blisters, and excoriations; and being chewed they are good for chapped lips, and paine of the gums.

B The kernels or mast within are reported to ease the paine of the kidneyes proceeding of the stone, if they be eaten, and to cause the grauell and sand the easier to come forth. With these, mice and Squirrels are greatly delighted, who do mightily encrease by feeding thereon : Swine also be fatned herewith, and certaine other beasts : also Deere do feed thereon very greedily : they be likewise pleasant to Thrushes and Pigeons.

C *Petrus Crescentius* writeth, That the ashes of the wood is good to make glasse with.

D The water that is found in the hollownesse of Beeches cureth the naughty scurfe, tetters, and scabs of men, horses, kine, and sheepe, if they be washed therewith.

Chap. 93. *Of the Almond tree.*

¶ *The Description.*

The Almond tree is like to the Peach tree, yet is it higher, bigger, of longer continuance : the leaues be very long, sharpe pointed, snipt about the edges like those of the Peach tree : the floures be alike : the fruit is also like a peach, hauing on one side a cleft, with a soft skin without, and couered with a thin cotton; but vnder this there is none, or very little pulp, which is hard like a gristle not eaten : the nut or stone within is longer than that of the peach, not so rugged, but smooth; in which is contained the kernel, in taste sweet, and many times bitter : the root of the tree groweth deepe : the gum which soketh out hereof is like that of the peach tree.

‡ There are diuers sorts of Almonds, differing in largenes and taste : we commonly haue three or foure sorts brought to vs, a large sweet Almond, vulgarly termed a Iordan almond; and a lesser, called a Valence Almond : a bitter Almond of the bignesse of the Valence almond, and somtimes another bitter one lesse than it. ‡

¶ *The*

Amygdalus.
The Almond tree.

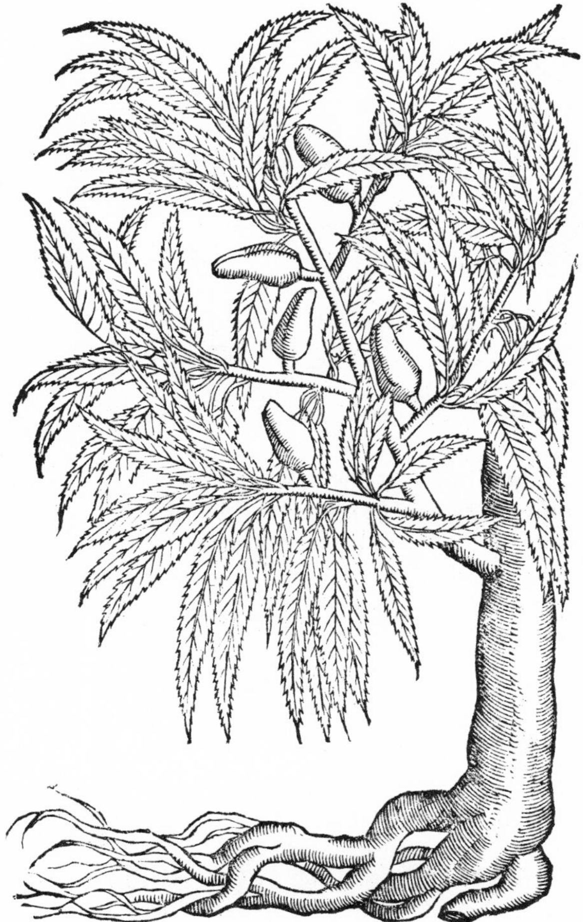

¶ *The Place.*

The natural place of the Almond is in the hot regions, yet we haue them in our London gardens and orchards in great plenty.

¶ *The Time.*

The Almond floureth betimes with the Peach: the fruit is ripe in Auguſt.

¶ *The Names.*

The tree is called is Greek, Ἀμυγδαλῆ: in Latine, *Amygdalus*: in French, *Amandier*: in Engliſh, Almond tree.

The fruit is called in Greeke, Ἀμύγδαλον: in Latine, *Amygdalum*: in ſhops, *Amygdala*: in high-Dutch, **Mandel**: in low-Dutch, **Amandelen**: in Italian, *Mandole*: in Spaniſh, *Almendras, Amelles*, and *Amendoas*: in French, *Amandes*: in Engliſh, Almond.

¶ *The Temperature and Vertues.*

Sweet Almonds when they be dry be mo- A
deratly hot; but the bitter ones are hot and dry in the ſecond degree. There is in both of them a certaine fat and oily ſubſtance, which is drawne out by preſſing.

Sweet Almonds being new gathered are B
pleaſant to the taſte, they yeeld ſome kind of nouriſhment, but the ſame groſſe and earthy, and groſſer than thoſe that be dry, and not as yet withered. Theſe do likewiſe ſlowly deſcend, eſpecially being eaten without their skins; for euen as the huskes or branny parts of corne doe ſerue to driue downe the groſſe excrements of the belly, ſo doe likewiſe the skins or husks of the almonds: therefore thoſe that be blanched do ſo ſlowly deſcend, as that they do withall binde the belly; whereupon they are giuen with good ſucceſſe to thoſe that haue the laske or the bloudy flix.

There is drawne out of ſweet Almonds, with liquor added, a white iuice like milke, which ouer C
and beſides that it nouriſheth, and is good for thoſe that are troubled with the laske and bloudie flix, it is profitable for thoſe that haue the pleuriſie and ſpit vp filthy matter, as *Alexander Trallianus* witneſſeth: for there is likewiſe in the Almonds an opening and concocting qualitie, with a certaine clenſing faculty, by which they are medicinable to the cheſt and lungs, or lights, and ſerue for the raiſing vp of flegme and rotten humors.

Almonds taken before meate do ſtop the belly, and nouriſh but little; notwithſtanding many D
excellent meates and medicines are therewith made for ſundry griefes, yea very delicat and wholſome meates, as Almond butter, creame of Almonds, marchpane, and ſuch like, which dry and ſtay the belly more than the extracted iuyce or milke; and they are alſo as good for the cheſt and lungs.

They do ſerue alſo to make the Phyſicall Barley water, and Barley Creame, which are giuen E
in hot Feuers, as alſo for other ſicke and feeble perſons, for their further refreſhing and nouriſhments.

The oile which is newly preſſed out of the ſweet Almonds is a mitigater of paine and all maner F
of aches. It is giuen to thoſe that haue the pleuriſie, being firſt let bloud; but eſpecially to thoſe that are troubled with the ſtone of the kidnies, it ſlackens the paſſages of the vrine, and maketh them glib or ſlipperie, and more ready to ſuffer the ſtone to haue free paſſage: it maketh the belly ſoluble, and therefore it is likewiſe vſed for the collicke.

It is good for women that are newly deliuered; for it quickly remoueth the throwes which re- G
maine after their deliuery.

The oile of Almonds makes ſmooth the hands and face of delicat perſons, and clenſeth the skin H
from all ſpots, pimples, and lentils.

Bitter Almonds doe make thinne and open, they remoue ſtoppings out of the liuer and ſpleene, I
therfore they be good againſt paine in the ſides: they make the body ſoluble, prouoke vrine, bring

downe the menses, helpe the strangurie, and clense forth of the chest and lungs clammie humors: if they be mixed with some kinde of looch or medicine to licke on: with starch they stay the spitting of bloud.

L And it is reported that fiue or six being taken fasting do keepe a man from being drunke.

M These also clense and take away spots and blemishes in the face, and in other parts of the body; they mundifie and make cleane foule eating vlcers.

N With hony they are laid vpon the biting of mad dogs; being applied to the temples with vineger or oile of Roses, they take away the head-ache, as *Dioscorides* writeth.

O They are also good against the cough and shortnesse of winde.

P They are likewise good for those that spit bloud, if they be taken with the fine floure of *Amylum*.

Q There is also pressed out of these an oile which prouoketh vrine, but especially if a few scorpions be drowned, and steeped therein.

R With oile it it singular good for those that haue the stone, and cannot easily make water but with extremitie of paine, if the share and place betweene the cods and fundament be anoynted therewith.

S *Dioscorides* saith, that the gum doth heate and binde, which qualities notwithstanding are not perceiued in it.

T It helpeth them that spit bloud, not by a binding facultie, but thorow the clamminesse of his substance, and that is by closing vp of the passages and pores, and so may it also cure old coughes, and mitigate extreame paines that proceed of the stone, and especially take away the sharpenesse of vrine, if it be drunke with Bastard, or with any other sweet potion, as with the decoction of Licorice, or of Raisons of the sunne. The same doth likewise kill tetters in the outward parts of the bodie (as *Dioscorides* addeth) if it be dissolued in vineger.

Chap. 94. *Of the Peach tree.*

¶ *The Kindes.*

‡ There are diuers sorts of Peaches besides the foure here set forth by our Author, but the trees do not much differ in shape, but the difference chiefely consists in the fruit, whereof I will giue you the names of the choice ones, and such as are to be had from my friend M^r^. *Millen* in Old-street, which are these; two sorts of Nutmeg Peaches; The Queenes Peach; The Newington Peach; The grand Carnation Peach; The Carnation Peach; The Blacke Peach; the Melocotone; the White; The Romane; The Alberza; The Island Peach; Peach du Troy. These are all good ones. He hath also of that kinde of Peach which some call *Nucipersica* or Nectorins, these following kindes; the Roman red, the best of fruits; the bastard Red; the little dainty green; the yellow; the white; the russet, which is not so good as the rest. Those that would see any fuller discourse of these may haue recourse to the late worke of M^r^. *Iohn Perkinson*, where they may finde more varieties, and more largely handled, and therefore not necessarie for me in this place to insist vpon them. ‡

¶ *The Description.*

1 The Peach tree is a tree of no great bignesse: it sendeth forth diuers boughes, which be so brittle, as oftentimes they are broken with the weight of the fruit or with the winde. The leaues be long, nicked in the edges, like almost to those of the Walnut tree, and in taste bitter: the floures be of a light purple colour. The fruit or Peaches be round, and haue as it were a chinke or cleft on the one side; they are couered with a soft and thin downe or hairie cotton, being white without, and of a pleasant taste; in the middle whereof is a rough or rugged stone, wherein is contained a kernell like vnto the Almond; the meate about the stone is of a white colour. The root is tough and yellowish.

2 The red Peach tree is likewise a tree of no great bignesse: it also sendeth forth diuers boughes or branches, which be very brittle. The leaues be long, and nicked in the edges like to the precedent. The floures be also like vnto the former; the fruite or Peaches be round, of a red colour on the outside; the meate likewise about the stone is of a gallant red colour. These kindes of Peaches are very like to wine in taste, and therefore maruellous pleasant.

3 *Persica præcocia*, or the d'auant Peach tree is like vnto the former, but his leaues are greater and larger. The fruit or Peaches be of a russet colour on the one side, and on the other side next vnto the sun of a red colour, but much greater than the red Peach: the stones whereof are like vnto the former: the pulpe or meate within is of a golden yellow colour, and of a pleasant taste.

4 *Persica*

Persica alba.
The white Peach.

4 *Persica lutea*, or the yellow Peach tree, is like vnto the former in leaues and floures: his fruit is of a yellow colour on the outside, and likewise on the inside, harder than the rest in the middle of the Peach is a wooddy, hard, and rough stone, full of crests and gutters, in which doth lie a kernell much like to that of the Almond, and with such a like skin: the substance within is white, and in taste something bitter. The fruit hereof is of greatest pleasure, and best taste of all the other of his kinde; although there be found at this day diuers other sorts that are of very good taste, not remembred of the ancient, or set downe by the later writers, whereof to speake particularly, would not be greatly to our pretended purpose, considering we hasten to an end.

‡ 5 There is also kept in some of our choise gardens a kind of Peach which hath a very double and beautiful floure, but it is seldome succeeded by any fruit; they call this, *Persica flore pleno*, The double blossomed Peach. ‡

¶ *The Place.*

They are set and planted in gardens and vineyards: I haue them all in my garden, with many other sorts.

¶ *The Time.*

The Peach tree soone commeth vp: it beares fruit the third or fourth yeere after it is planted, and it soone decaieth, and is not of long continuance; it floureth in Aprill, or a little while after that the leaues appeare, and hath his fruit ripe in September.

¶ *The Names.*

The Peach tree is called in Greeke, μηλέα περσική: in Latine, *Malus Persica*, and *Persica*: in high Dutch, **Persichboum**: in low Dutch, **Perse boom**: in French, *Perscher*: in English, Peach tree.

The fruit, as *Galen* testifieth, is named μῆλον περσικόν, and περσικὸν also without any addition: in Latine, *Malum Persicum*, and *Persicum*: in high Dutch, **Pfersing**: in low Dutch, **Persen**: in Italy, *Pesche*: in Spanish, *Pexegos*: in French, *Pesches*: in English, Peach.

¶ *The Temperature and Vertues.*

A Peaches be cold and moist, and that in the second degree; they haue a juice and also a substance that doth easily putrifie, which yeeldeth no nourishment, but bringeth hurt, especially if they be eaten after other meates; for then they cause the other meates to putrifie. But they are lesse hurtfull if they be taken first; for by reason that they are moist and slippery, they easily and quickly descend; and by making the belly slippery, they cause other meates to slip downe the sooner.

B The kernels of the Peaches be hot and dry, they open and clense; they are good for the stoppings of the liuer and spleene.

C Peaches before they be ripe do stop the laske, but being ripe they loose the belly, and ingender naughty humors, for they are soone corrupted in the stomacke.

D The leaues of the Peach tree do open the stopping of the liuer, and do gently loosen the belly: and being applied plaisterwise vnto the nauell of young children, they kill the wormes, and driue them forth.

E The same leaues boiled in milke, do kill the wormes in children very speedily.

F The same being dried, and cast vpon greene wounds, cure them.

G The floures of the Peach tree infused in warme water for the space of ten or twelue houres, and strained, and more floures put to the said liquor to infuse after the same manner, and so iterated six or eight times, and strained again, then as much sugar as it will require added to the same liquor and boiled vnto the consistence or thicknes of a syrrup, and two spoonefulls hereof taken, doth so singularly well purge the belly, that there is neither Rubarbe, Agaricke, nor any other purger comparable vnto it; for this purgeth downe waterish humors mightily, and yet without griefe or trouble, either to the stomacke, or lower parts of the body.

H The kernell within the Peach stone stamped small, and boiled with Vineger vntill it be brought to the forme of an ointment, is good to restore and bring again the haire of such as be troubled with the *Alopecia*.

I There is drawne forth of the kernels of Peaches, with Peniroyall water, a iuice like vnto milke, which is good for those that haue the Apoplexy: if the same be oftentimes held in the mouth it draweth forth water and recouereth the speech.

K The gum is of a meane temperature, but the substance thereof is tough and clammy, by reason whereof it dulleth the sharpnes of thin humors: it serueth in a looch or licking medicine for those that be troubled with the cough, and haue rotten lungs, and stoppeth the spitting and raising vp of bloud, and also stayeth other fluxes.

Chap. 95.

Of the Aprecocke or Abrecocke tree.

1 *Armeniaca malus maior.*
The greater Aprecocke tree.

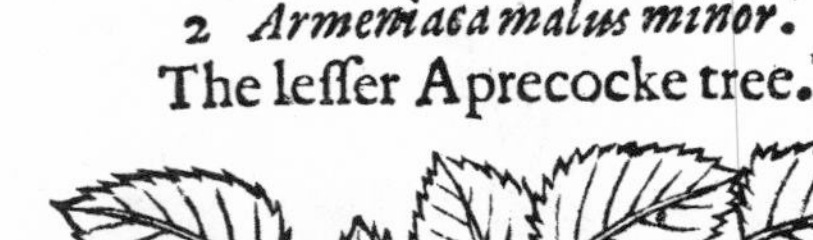

2 *Armeniaca malus minor.*
The lesser Aprecocke tree.

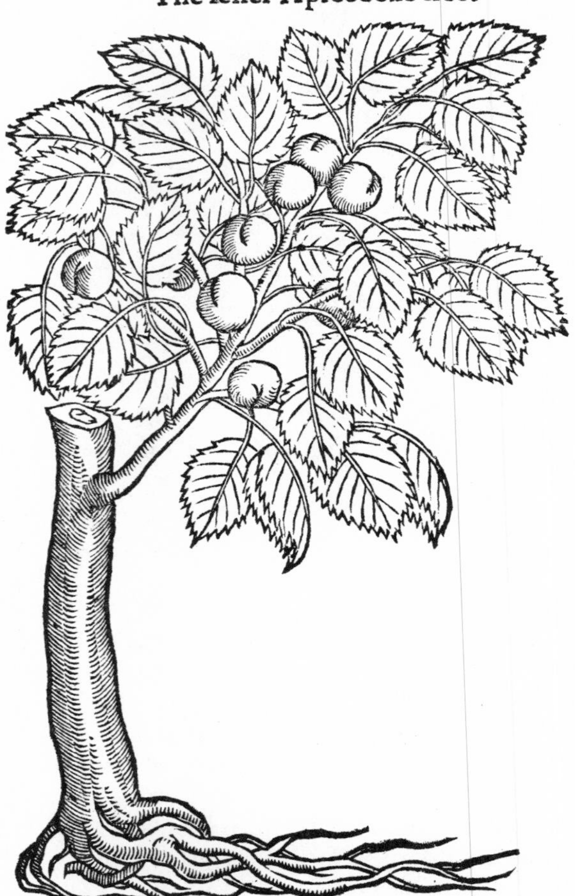

¶ *The Description.*

1 This tree is greater than the Peach tree, and hath a bigger body, it lasteth longer, especially if it be grafted or inoculated: the leaues hereof are broad, and sharpe pointed, like those of blacke Poplar, but lesser, and comming more neere to the leaues of birch, nicked in the edges: the floures are somewhat white: the fruit round like a peach, yellow within and without, in which doth lie a browne stone, nothing rough at all as is that of the Peach, shorter also, and lesser, in which is included a sweet kernell.

2 We haue another sort of Aprecocke, whose trunk or body is equall with the other in greatnesse, it is like also in leaues and brittle branches: his time of flouring, flourishing, and manner of growing accordeth: the only point wherein they differ is, that this tree bringeth forth lesse fruit, and not so good in taste, in euery other respect it is like.

‡ Of this also Mr. *Parkinson* hath set forth diuers varieties; and my forementioned friend Mr. *Millen* hath these fiue sorts; the common, the long and great, the Muske, the Barbary, and the early Aprecocke. ‡

¶ *The*

¶ *The Place.*

These trees do grow in my garden, and now adaies in many other gentlemens gardens throughout all England.

¶ *The Time.*

They floure and flourish in Aprill, and their fruit is ripe in Iuly.

¶ *The Names.*

This tree is called in Greeke, Μηλέα Ἀρμενιακή: in Latine, *Malus Armeniaca*: in English, Abrecocke tree, and Aprecocke tree.

The fruit is named Μῆλον Ἀρμενιακὸν, and of diuers πραικόκκιον, or Βερικόκκιον, which be words corrupted from the Latine; for *Præcox* in Latine is diuers times called *Præcoquum*: it is named *Malum Armeniacum*, and commonly *Armeniacum*: it is called in high-Dutch, **Molletten Mollelin, S. Johans Pfersing**: in low-Dutch, **Vroege Persen, Auant Persen**: in Italian, *Armeniache, Bacoche, Grisomele, Moniache*: in French, *Abricoz*: in Spanish, *Aluarcoques, Aluarchigas*, and *Albercocs*: in English, Abrecocke, and of some, Aprecocke, and Aprecox.

Galen seemeth to make a difference betweene *Præcocia* and *Armeniaca*, in his booke of the Faculties of nourishments, preferring *Præcocia* before *Armeniaca*; yet he doth confesse that both of them be called *Armeniaca*: others pronounce them *Armenia* with foure syllables. And in his booke of the faculties of simple medicines he affirmeth, that both the fruit and the tree are called πραικόκκιον: diuers of the later Physitions do between these also make a difference, saying, that the greater ones and those that are grafted be *Armeniaca* (which the French men call *Auant Perses*) and the lesser *Præcocia*: in French, *Abricoz*.

¶ *The Temperature and Vertues.*

Aprecocks are cold and moist in the second degree, but yet not so moist as Peaches, for which cause they do not so soone or easily putrifie, and they are also more wholesome for the stomacke, and pleasant to the taste; yet do they likewise putrifie, and yeeld but little nourishment, and the same cold, moist, and full of excrements: being taken after meate they corrupt and putrifie in the stomacke; being first eaten before other meate they easily descend, and cause the other meates to passe downe the sooner, like as also the Peaches do. A

The kernell with in the same is sweet, and nothing at all like in facultie to that of the Peach. B

The vertues of the leaues of this tree are not as yet found out. C

Chap. 96. *Of Pomegranat tree.*

¶ *The Kindes.*

As there be sundry sorts of Apples, Peares, Plums, and such like fruits; so there are two sorts of Pomegranates, the garden and the wilde, and a third sort which is barren and fruitles: the fruit of the garden Pomegranat is of three sorts; one hauing a soure iuyce or liquour; another hauing a very sweet and pleasant liquor, and the third the taste of Wine: of the wilde also there be two sorts, and the difference betweene them is no more than betwixt crabs and weildings, which are both wilde kindes of Apples: therefore the description of the garden Pomegranat shal suffice for the rest.

¶ *The Description.*

1 The manured Pomegranat tree groweth vp to the height of a hedge tree, being seuen or eight cubits high, hauing many pliant and twiggy branches, very limber, tough, and of a browne colour: whereon are set very many leaues in shape like those of the Priuet, but more like those of the Myrtle tree, of a bright shining greene colour tending to yellownesse: amongst which there stand certaine sharpe thornes confusedly set, and likewise hollow floures like vnto the hedge Rose, indented on the edges like a starre, of a Carnation colour, and very single: after which commeth the fruit, couered with a hard barke, of an ouerworne purplish colour, full of graines and kernels, which after they be ripe are of a gallant crimson colour, and ful of iuyce, which differeth in taste according to the soile, clymat, and countrey where they grow; some be sweet, others soure, and the third are in a middle betweene them both, hauing the taste of Wine.

1 *Malus Granata, siue Punica.*
The Pomegranat tree.

2 *Malus Punica syluestris.*
The wilde Pomegranat.

Balaustia, siue Pleni flores Gran. syl.
The double floures of wild Pomegranat.

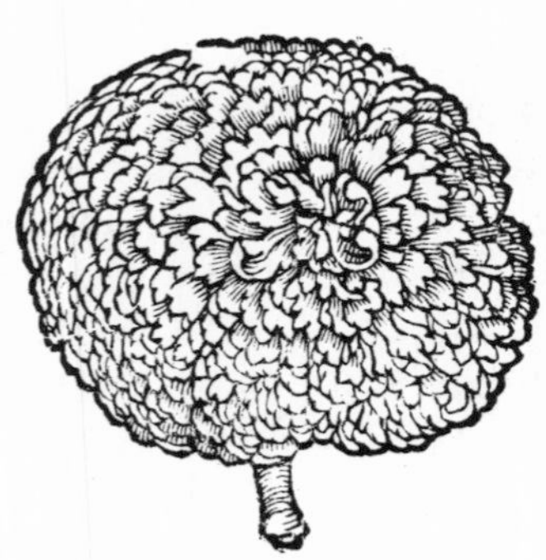

† 2 The wilde Pomegranat tree is like the other in leaues and twiggy branches, but it is more prickly and horrid: of this there are two sorts, the one hauing such floures & fruit as the tame Pomgranat; the other bearing floures very double, as may appeare by the figure, which wither and fall away, leauing no fruit behind them, as the double floured Cherry doth, and diuers other herbes and trees also; & it is altogether barren of fruit: of this *Dioscorides* makes sundry sorts, differing in colour: one is white, saith he; another yellowish red, and a third sort of the colour of the Rose: this with red floures is best knowne among the Apothecaries.

¶ *The Place.*

Pomegranats grow in hot countries toward the South, in Italy, Spaine, and chiefely in the kingdome of Granado, which is thought to be so named of the great multitude of Pomgranats, which be commonly called *Granata*: they grow in a number of places also without manuring: yet being manured they prosper better; for in gardens, vineyards, orchards, and other like husbanded grounds they come vp more cheerefully: I haue recouered diuers yong trees hereof, by sowing of the seed or grains, of the height of three or foure cubits, attending Gods leisure for floures and fruit.

¶ *The Time.*

The Pomegranate floureth in the moneths of May and Iune: the fruit is ripe in the end of August.

¶ *The*

¶ *The Names.*

The Pomegranate tree is called in Latine, *Malus Punica*: in Greeke, of the Athenians, ῥόα, and ῥιὰ, as *Galen* ſaith: in Engliſh, Pomegranate tree: the fruit is alſo named ῥόα, or ῥοιά: in Latine, *Malum Punicum*: in ſhops, *Malum*, or *Pomum Granatum*: in high Dutch, **Granatopffel**: in low Dutch, **Granatappel**: in Italian, *Melagrano*, and *Pomo Granato*: in Spaniſh, *Granadas*, and *Romanas*: in French, *Pommes Granades*: in Engliſh, Pomegranate.

The floure of the fruitfull Pomegranate tree is called of the Grecians, κύτινος: which is notwithſtanding properly the cup of the floure: the Latines name it alſo *Cytinus*.

The floure of the wilde and barren Pomegranate tree is called βαλαύστιον: the Apothecaries doe likewiſe terme it *Balauſtium*.

The pill or rinde of the Pomegranate which is ſo much in vſe, is named in Greeke σίδιον: in Latine, *Malicorium*, and *Sidium*: in ſhops it is called *Cortex granatorum*, or Pomegranate Pill.

¶ *The Temperature and Vertues.*

A The iuicie grains of the Pomegranate are good to be eaten, hauing in them a meetly good iuice: they are wholeſome for the ſtomacke, but they all containe in them a thin and ſmall nouriſhment, or none at all.

B The ſweet ones be not ſo cold as the reſt, but they eaſily cauſe hot ſwellings to ariſe, and they are not ſo much commended for agues.

C The ſoure ones, and eſpecially if they be withall ſomething harſh, doe euidently coole, dry, and ſomething binde.

D They are good for the heart burne, they repreſſe and ſtay the ouermuch vomiting of choler, called the Felonie: they are a remedie againſt the bloudy flixe, aptneſſe to vomite, and vomite it ſelfe.

E There is made of the iuice of theſe ſoure Pomegranats a ſyrrup, which ſerueth for the ſame purpoſes, and is alſo many times very profitable againſt the longing of women with childe, vnleſſe the coldneſſe of the ſtomacke be a hinderance thereunto.

F The ſeeds of the graines, and eſpecially of the ſower Pomegranat, being dried, do likewiſe coole and binde.

G They ſtop the flix, ſtay vomiting, and ſtanch the ſpitting vp of bloud, they ſtrengthen the ſtomacke.

H Of the ſame effect be the floures, both of the tame and wilde Pomegranate tree, being like to the ſeeds in temperature and vertues.

I They faſten the teeth, and ſtrengthen the gums, if the ſame be waſhed therewith.

K They are good againſt burſtings that come by falling downe of the guts, if they be vſed in plaiſters and applied.

L The rinde or pill is not onely like in facultie to the ſeeds, and both the ſorts of floures, but alſo more auaileable; for it cooleth and bindeth more forceably; it bringeth downe the hot ſwellings of the almonds in the throat, being vſed in a gargariſme or a lotion for the throat, and it is a ſingular remedy for all things that need cooling and binding.

M *Dioſcorides* writeth, that there is alſo gathered a iuice out of both thoſe ſorts of floures, which is very like in facultie and vertue to *Hypociſtis*, as the ſame Author affirmeth.

N The bloſſomes of the tame and wilde Pomegranates, as alſo the rinde or ſhell thereof made into pouder, and drunke in red wine, or boyled in red wine, and the decoction drunke, is good againſt the bloudy flix, and all other iſſues of bloud; yea it is good for women to ſit ouer, & bathe themſelues in the decoction hereof: theſe foreſaid bloſſomes and ſhels are good alſo to put into reſtraining pouders, for the ſtanching of bloud in wounds.

O The ſeeds or ſtones of Pomegranats dried in the Sun, and beaten to pouder, are of like operation with the floures: they ſtop the laske and all iſſues of bloud in man or woman, being taken in the manner aforeſaid.

Chap. 97. *Of the Quince Tree.*

¶ *The Kindes.*

Columella maketh three kindes of Quinces, *Struthia*, *Chryſomeliana*, and *Muſtela*, but what manner ones they be hee doth not declare, notwithſtanding wee finde diuers ſorts differing as well in forme,

forme, as taſte and ſubſtance of the fruit, wherof ſome haue much core and many kernels, and others fewer.

Malus Cotonea.
The Quince tree.

¶ *The Deſcription.*

THe Quince tree is not great, but groweth low, and many times in maner of a ſhrub: it is couered with a rugged barke, which hath on it now and then certaine ſcales: it ſpreadeth his boughes in compaſſe like other trees; about which ſtand leaues ſomewhat round, like thoſe of the common Apple tree, greene and ſmooth aboue, and vnderneath ſoft and white: the floures be of a white purple colour: the fruit is like an apple, ſaue that many times it hath certaine embowed and ſwelling diuiſions: it differeth in faſhion and bigneſſe; for ſome Quinces are leſſer and round truſt vp together at the top with wrinkles, others longer and greater: the third ſort be after a middle manner betweene both; they are all of them ſet with a thin cotton or freeze, and be of the colour of gold, and hurtfull to the head by reaſon of their ſtrong ſmell; they all likewiſe haue a kinde of choking taſte: the pulp within is yellow, and the ſeed blackiſh, lying in hard skins, as doe the kernels of other apples.

¶ *The Place.*

The Quince tree groweth in gardens and orchards, and is planted oftentimes in hedges and fences belonging to gardens & vineyards: it delighteth to grow on plaine and euen grounds, and ſomewhat moiſt withall.

¶ *The Time.*

Theſe apples be ripe in the fall of the leafe, and chiefely in October.

¶ *The Names.*

The tree is called in Greeke μηλέα κυδώνια: in Latine, *Malus Cotonea*: in Engliſh, Quince tree.

The fruit is named μῆλον κυδώνιον: *Malum Cotoneum, Pomum Cydonium*, and many times, *Cydonium*, without any addition; by which name it is made known to the Apothecaries: it is called in high Dutch, **Quitten, Quittenopfell**, or **Kuttenopffel**: in low Dutch, **Queappel**: in Italian, *Mele cotogne*: in Spaniſh, *Codoyons, Membrilhos*, and *Marmellos*: in French, *Pomme de coing*: in Engliſh, Quince.

¶ *The Temperature and Vertues.*

A Quinces be cold and dry in the ſecond degree, and alſo very much binding, eſpecially when they be raw: they haue likewiſe in them a certaine ſuperfluous and excrementall moiſture, which will not ſuffer them to lie long without rotting: they are ſeldom eaten raw: being roſted or baked they be more pleaſant.

B They ſtrengthen the ſtomacke, ſtay vomiting, ſtop lasks, and alſo the bloudy flix.

C They are good for thoſe that ſpit vp bloud, or that vomit bloud; and for women alſo that haue too great plenty of their monethly courſes.

D *Simeon Sethi* writeth, that the woman with childe, which eateth many Quinces during the time of her breeding, ſhall bring forth wiſe children, and of good vnderſtanding.

E The Marmalade, or Cotiniate, made of Quinces and ſugar, is good and profitable for the ſtrengthening of the ſtomacke, that it may retaine and keepe the meat therein vntill it be perfectly digeſted: it likewiſe ſtayeth all kindes of fluxes, both of the belly and other parts, and alſo of bloud: which Cotiniate is made in this manner:

F Take faire Quinces, pare them, cut them in pieces, and caſt away the core, then put vnto euery pound of Quinces a pound of ſugar, and to euery pound of ſugar a pinte of water: theſe muſt bee boiled together ouer a ſtill fire till they be very ſoft, then let it be ſtrained or rather rubbed through a ſtrainer, or an hairy ſieue, which is better, and then ſet it ouer the fire to boile againe, vntill it be ſtiffe,

ſtiffe, and ſo box it vp, and as it cooleth put thereto a little Roſe water, and a few graines of Muske, well mingled together, which will giue a goodly taſte vnto the Cotiniat. This is the way to make Marmalade:

Take whole Quinces and boile them in water vntill they be as ſoft as a ſcalded codling or apple, G
then pill off the skin, and cut off the fleſh, and ſtampe it in a ſtone morter; then ſtraine it as you did the Cotiniate; afterward put it into a pan to drie, but not to ſeeth at all: and vnto euery pound of the fleſh of Quinces, put three quarters of a pound of ſugar, and in the cooling you may put in roſe water and a little Muske, as was ſaid before.

There is boiled with Quinces oile which therefore is called in Greeke *Melinon*, or oile of Quin- H
ces, which we vſe, ſaith *Dioſcorides*, ſo oft as we haue need of a binding thing.

The ſeed of Quinces tempered with water, doth make a muſcilage, or a thing like iellv, which I
being held in the mouth, is maruellous good to take away the roughneſſe of the tongue in hot burning feuers.

The ſame is good to be layed vpon burnings or ſcaldings, and to be put into cliſters againſt the K
bloudy flix; for it eaſeth the paine of the guts, and alaieth the ſharpneſſe of biting humors.

Many other excellent, dainty and wholeſome confections are to be made of Quinces, as ielly of L
Quinces, and ſuch odde conceits, which for breuitie ſake I do now let paſſe.

Chap. 98. *Of the Medlar Tree.*

¶ *The Kindes.*

There are diuers ſorts of Medlars, ſome greater, others leſſer: ſome ſweet, and others of a more harſh taſte: ſome with much core, and many great ſtony kernels, others fewer: and likewiſe one of Naples called *Aronia*.

1 *Meſpilus ſatiua.*
The manured Medlar.

‡ 2 *Meſpilus ſativa altera.*
The other Garden Medlar.

The

¶ *The Deſcription.*

1 THe manured Medlar tree is not great, the body whereof is writhed, the boughes hard, not eaſie to be broken: the leaues be longer, yet narrower than thoſe of the apple tree, darke, greene aboue, and ſomewhat whiter and hairy below: the floures are white and great, hauing fiue leaues a piece: the fruit is ſmall, round, and hath a broad compaſſed nauell or crowne at the top: the pulpe or meat is at the firſt white, and ſo harſh or choking, that it cannot be eaten before it become ſoft, in which are contained fiue ſeeds or ſtones, which be flat and hard.

‡ 2 There is another which differeth from the laſt deſcribed, in that the leaues are longer and narrower, the ſtocke hath no prickles vpon it: the fruit alſo is larger and better taſted: in other reſpects it is like to the laſt deſcribed. This is the *Meſpilus fructu preſtantiore* of *Tragus*, and *Meſpilus Domeſtica* of *Lobel*. ‡

3 The Neapolitane Medlar tree groweth to the height and greatneſſe of an Apple tree, hauing many tough and hard boughes or branches, ſet with ſharp thornes like the white Thorne, or Hawthorne: the leaues are very much cut or iagged like the Hawthorne leaues, but greater, and more like Smallage or Parſley, which leaues before they fal from the tree do wax red: among theſe leaues come forth great tufts of floures of a pale herby colour: which being paſt, there ſucceed ſmall long fruit leſſer than the ſmalleſt Medlar, which at the firſt are hard, and greene of colour, but when they be ripe, they are both ſoft and red, of a ſweet and pleaſant taſte: wherein is contained three ſmall hard ſtones, as in the former, which be the kernels or ſeeds thereof.

3 *Meſpilus Aronia.*
The Neapolitane Medlar.

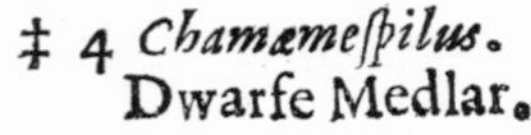

‡ 4 *Chamæmeſpilus.*
Dwarfe Medlar.

4 There is a dwarfe kinde of Medlar growing naturally vpon the Alpes, and hils of Narbone, and on the rocks of Mount Baldus nigh Verona, which hath been by ſome of the beſt learned eſteemed for a kinde of Medlar: others, whoſe iudgements cannot ſtand with truth or probability, haue ſuppoſed it to be *Euonymus*, of the Alpes: this dwarfe Medlar groweth like a ſmall hedge tree, of four or fiue cubits high, bearing many ſmal twiggie wands or crops, beſet with many ſlender leaues green aboue, and of a skie colour vnderneath, in ſhew like to a dwarfe Apple tree, but the fruit is very

very like the Haw, or fruit of the white Thorne, and of a red colour. ‡ The floures come forth in the Spring three or foure together, hollow, and of an herbie colour, it growes in diuers places of the Alpes: it is the *Chamæmespilum* of the *Aduers.* and the *Chamæmespilus Gesneri*, of *Clusius*. ‡

¶ *The Place.*

The Medlar trees do grow in Orchards, and oftentimes in hedges among Briars and Brambles; being grafted in a white Thorne it prospereth wonderfull well, and bringeth forth fruit twise or thrise bigger than those that are not grafted at all, almost as great as little apples: we haue diuers sorts of them in our Orchards.

¶ *The Time.*

It is very late before Medlars be ripe, which is in the end of October, but the floures come forth timely enough.

¶ *The Names.*

The first is called in Greeke by *Theophrastus* μεσπίλη: in Latine, *Mespilus*: in high Dutch, **Nespelbaum**: in low Dutch, **Mispelboome**: in French, *Nefflier*: in English, Medlar tree.

The Apple or fruit is named in Greeke, μέσπιλον: in Latine likewise, *Mespilum*: in high Dutch, **Nespel**, in low Dutch, **Mispele**: in Italian, *Nespolo*: in French, *Nefle*: in Spanish, *Nesperas*: in English, Medlar.

Dioscorides affirmeth, that this Medlar tree is called ἐπιμηλίς, and of diuers, *Sitanion*: *Galen* also in his booke of the faculties of simple medicines nameth this *Epimelis*, which is called, as he saith, by the countrey men in Italy, *Vnedo*, and groweth plentifully in Calabria; for vnder the name of *Mespilus*, or Medlar tree, he meaneth no other than *Tricoccus*, which is also named *Aronia*.

The Neapolitane Medlar tree is called in Greeke μεσπίλεια and μέσπιλη: *Galen* calleth it *Epimelis*.

The fruit hereof is called *Tricoccos*, of the three graines or stones that it hath: they of Naples call it *Azarolo*: and we may name it in English, three graine Medlar, or Neapolitane Medlar, or Medlar of Naples.

¶ *The Temperature.*

The Medlars are cold, drie, and astringent; the leaues are of the same nature: the dwarfe Medlar is dry, sharpe, and astringent.

¶ *The Vertues.*

Medlars do stop the belly, especially when they be greene and hard, for after that they haue been kept a while, so that they become soft and tender, they doe not binde or stop so much, but are then more fit to be eaten. A

The fruit of the three grain Medlar, is eaten both raw and boiled, and is more wholesome for the stomacke. B

These Medlars be oftentimes preserued with sugar or hony: and being so prepared they are pleasant and delightfull to the taste. C

Moreouer, they are singular good for women with childe: for they strengthen the stomacke, and stay the lothsomnesse thereof. D

The stones or kernels of the Medlars, made into pouder and drunke, doe breake the stone, expell grauell, and procure vrine. E

CHAP. 99. *Of the Peare tree.*

¶ *The Description.*

TO write of Peares and Apples in particular, would require a particular volume: the stocke or kindred of Peares are not to be numbred: euery country hath his peculiar fruit: my selfe knows one curious in graffing and planting of fruits, who hath in one piece of ground, at the point of three score sundry sorts of Peares, and those exceeding good, not doubting but if his minde had been to seeke after multitudes, he might haue gotten together the like number of those of worse kinds: besides the diuersities of those that be wilde, experience sheweth sundry sorts: and therefore I thinke it not amisse to set downe the figures of some few with their seuerall titles, as well in Latine as English, and one generall description for that, that might be said of many, which to describe apart, were to send an owle to Athens, or to number those things that are without number.

‡ Our Author in this chapter gaue eight figures with seuerall titles to them, so I pluckt a peare from each tree, and put his title to it, but not in the same order that he obserued, for hee made the Katherine peare tree the seuenth, which I haue now made the first, because the figure expresses the whole tree. ‡

¶ The

¶ *The generall description.*

THe Peare tree is for the most part higher than the Apple tree, hauing boughes not spread abroad, but growing vp in height: the body is many times great: the timber or wood it selfe is very tractable or easie to be wrought vpon, exceeding fit to make moulds or prints to be grauen on, of colour tending to yellownesse: the leafe is somewhat broad, finely nicked in the edges, greene aboue, and somewhat whiter vnderneath: the floures are white: the Peares, that is to say, the fruit, are for the most part long, and in forme like a Top; but in greatnes, colour, forme, and tast very much differing among themselues; they be also couered with skins or coats of sundry colours: the pulpe or meat differeth, as well in colour as tast: there is contained in them kernels, blacke when they be ripe: the root groweth strait downe with some braunches running aslope.

Pirus superba, siue Katherina.
The Katherine Peare tree.

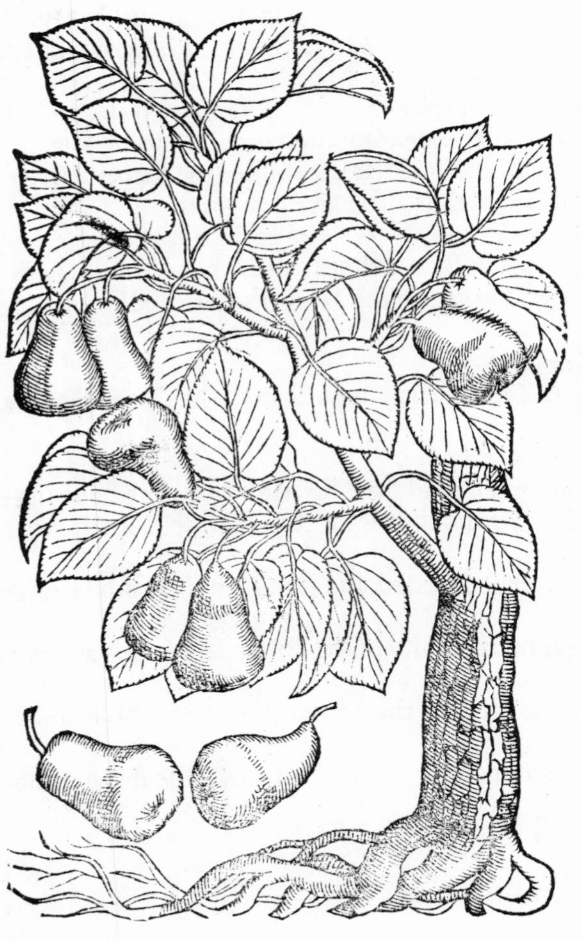

1 *Pyra Præcocia.* The Ienneting Peare.
2 *Pyra Iacobæa.* Saint Iames Peare.
3 *Pyrum regale.* The Peare royall.

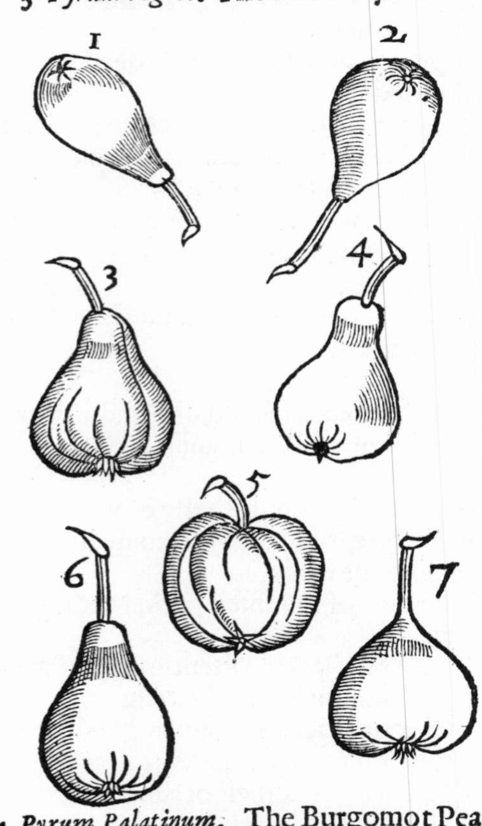

4 *Pyrum Palatinum.* The Burgomot Peare.
5 *Pyrum Cydonium.* The Quince peare.
6 *Pyrum Episcopatum.* The Bishops peare.
7 *Pyrum hyemale.* The Winter peare.

¶ *The Place.*

The tame Peare trees are planted in Orchards, as be the apple trees, and by grafting, though vpon wilde stockes, come much varietie of good and pleasant fruits. All these before specified, and many sorts more, and those most rare and good, are growing in the ground of Master *Richard Pointer*, a most cunning and curious graffer and planter of all manner of rare fruits, dwelling in a small village neere London called Twicknam; and also in the ground of an excellent graffer and painfull planter, Mr. *Henry Banbury*, of Touthill street neere Westminster, and likewise in the ground of a diligent and most affectionate louer of plants Mr. *Warner* neere Horsey downe by London, and in diuers other grounds about London. ‡ Most of the best peares are at this time to be had with Mr. *Iohn Millen* in Old-street, in whose nursery are ro be found the choisest fruits this kingdom yeelds. ‡

¶ *The Time.*

The floures do for the most part come forth in Aprill, the leaues afterwards: all peares are not ripe at one time: some be ripe in Iuly, others in August, and diuers in September and later.

¶ *The Names.*

The tame or Orchard peare tree is called in Greeke ἄπιος, or with a double ππ ἄππιος: in Latine, *Pyrus vrbana*, *urbana*, or *Cultiua* : of *Tarentinus* in his Geoponikes ἐπιμήλιος : in high Dutch, Bijrbaum, in low Dutch, Peerboom : in French, *Porrier*.

The Peare or fruit it selfe is called in Greeke ἄπιον : in Latine, *Pyrum* : in high Dutch, Biern : in low Dutch, Peere : in Italian, *Pere* : in French, *Poyre* : in Spanish, *Peras* : in English, Peare.

¶ *The Temperature and Vertues.*

Leauing the diuers and sundry surnames of Peares, let vs come to the faculties which the Physitions ought to know; which also varie according to the differences of their tastes : for some Peares are sweet, diuers fat and vnctuous, others soure, and most are harsh, especially the wilde peares, and some consist of diuers mixtures of tastes, and some hauing no taste at all, but as it were a waterish taste. A

All Pears are cold, and all haue a binding qualitie and an earthie substance: but the Choke pears and those that are harsh be more earthie, and the sweet ones lesse : which substance is so full of superfluous moisture in some, as that they cannot be eaten raw. All manner of Peares doe binde and stop the belly, especially the Choke and harsh ones, which are good to be eaten of those that haue the laske and the bloudy flix. B

The harsh and austere Peares may with good successe be laied vpon hot swellings in the beginning, as may be the leaues of the tree, which do both binde and coole. C

Wine made of the iuice of peares called in English, Perry, is soluble, purgeth those that are not accustomed to drinke thereof, especially when it is new; notwithstanding it is as wholsome a drink being taken in small quantitie as wine; it comforteth and warmeth the stomacke, and causeth good digestion. D

CHAP. 100. *Of the wilde Peare tree.*

¶ *The Kindes.*

AS there be sundry kindes of the manured Peares, so are there sundry wilde; wherof to write apart were to small purpose: therefore one description with their seuerall titles shall be sufficient for their distinctions.

Pyrum strangulatorium maius.
The great Choke peare.

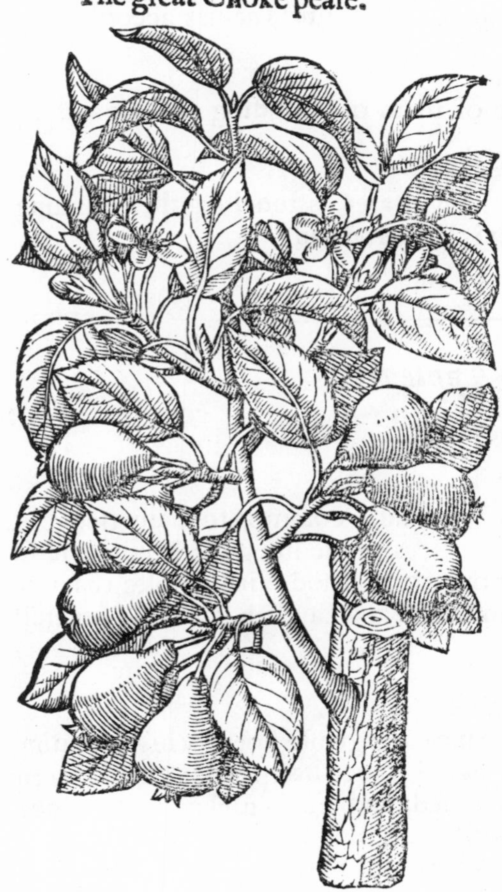

¶ *The generall Description.*

THe wilde Peare tree grows likewise great, vpright, full of branches, for the most part Pyramides like, or of the fashion of a steeple, not spread abroad as is the Apple or Crab tree: the timber of the trunke or body of the tree is very firme and sollid, and likewise smooth, a wood very fit to make diuers sorts of instruments of, as also the hafts of sundry tooles to worke withal; and likewise serueth to be cut into many kindes of moulds, not only such prints as these figures are made of, but also many sorts of pretty toies, for coifes, brest-plates, and such like, vsed among our English gentlewomen: the branches are smooth, couered with a blackish barke, very fragile or easie to break, whereon do grow leaues, in some greater, in other lesser: the floures are like those of the manured Pear-tree, yet some whiter than others: the fruit differ not in shape, yet some greater than others; but in taste they differ among themselues in diuers points, some are sharpe, soure, and of an austere taste; some more pleasant, others harsh and bitter, and some of such a choking taste, that they are not to be eaten of hogs & wild beasts, much lesse of men : they also differ in colour, euery circumstance whereof to distinguish apart would greatly enlarge our volume, and bring to the Reader small profit or commoditie.

 ¶ The

1 *Pyrum ſtrangulatorium maius.*
The great Choke peare.

2 *Pyrum ſtrangulatorium minus.*
The ſmall Choke peare.

3 *Pyrus ſylueſtris.*
The wilde hedge Peare tree.

4 *Pyrus ſylueſtris minima.*
The wilde Crab peare tree.

5 *Pyrus pedicularia.*
The Lowſie wilde peare.

6 *pyrus Coruina.*
The Crow peare tree.

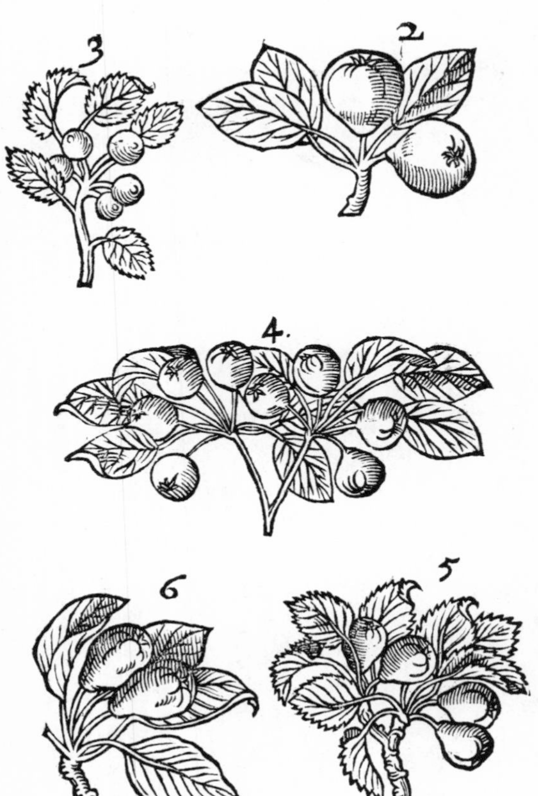

¶ *The Place.*

The wilde peares grow of themſelues without manuring in moſt places, as woods, or in the borders of fields, and neere to high waies.

¶ *The Time.*

The time of wilde peares anſwereth the tame or manured peare, notwithſtanding for the moſt part they are not ripe much before Winter.

¶ *The Names.*

The wilde peare tree is called in Latine, *Pyrus ſylueſtris* and *P[illegible]aſter*: in Greeke, ἀχρὰς: by which name both the fruit and tree are knowne. Peares haue diuers ſyrnames among the antient Writers, and ſpecially in *Pliny*, in his 15. booke, 15. chapter, none of which are knowne to the later Writers (or not deſired:) euery citie or euery countrey haue names of themſelues, and peares haue alſo diuers names according to to the places.

The Temperature.

All peares are of a cold temperature, and the moſt part of them of a binding qualitie and an earthie ſubſtance.

¶ *The Vertues.*

The vertues of the wilde peares are referred vnto the garden peares as touching their binding facultie, but are not to be eaten, becauſe their nouriſhment is little and bad.

Chap. 101. *Of the Apple tree.*

¶ *The Kindes.*

THe Latine name *Malus* reacheth far among the old Writers, and is common to many trees, but we will briefely firſt intreat of *Mali*, properly called Apple trees, whoſe ſtocke or kindred is ſo infinite, that we haue thought it not amiſſe, to vſe the ſame order or method with Apples that wee haue done with peares; that is, to giue them ſeuerall titles in Latine and Engliſh, and one generall deſcription for the whole.

¶ *The Deſcription.*

THe Apple tree hath a body or truncke commonly of a meane bigneſſe, not very high, hauing long armes or branches, and the ſame diſordered: the barke ſomewhat plaine, and not verie rugged: the leaues bee alſo broad, more long than round, and finely nicked in the edges. The floures are whitiſh tending vnto a bluſh colour. The fruit or Apples doe differ in greatneſſe, forme, colour, and taſte; ſome couered with a red skinne, others yellow or greene, varying infinitely

nitely according to the soyle and climate, some very great, some little, and many of a middle sort; some are sweet of taste, or something soure; most be of a middle taste betweene sweet and soure, the which to distinguish I thinke it impossible; notwithstanding I heare of one that intendeth to write a peculiar volume of Apples, and the vse of them; yet when he hath done what hee can doe, hee hath done nothing touching their seuerall kindes to distinguish them. This that hath beene said shall suffice for our Historie.

‡ Our Author gaue foure figures more out of *Tabernamontanus*, with these titles. 3. *Malum reginale*, the Queening or Queene of Apples. 5 *Platomela siue Pyra æstiua*: The Sommer Pearemaine. 6 *Platurchapia siue Pyra hyemalia*: the Winter Pearemaine. ‡

1 *Malus Carbonaria.*
The Pome Water tree.

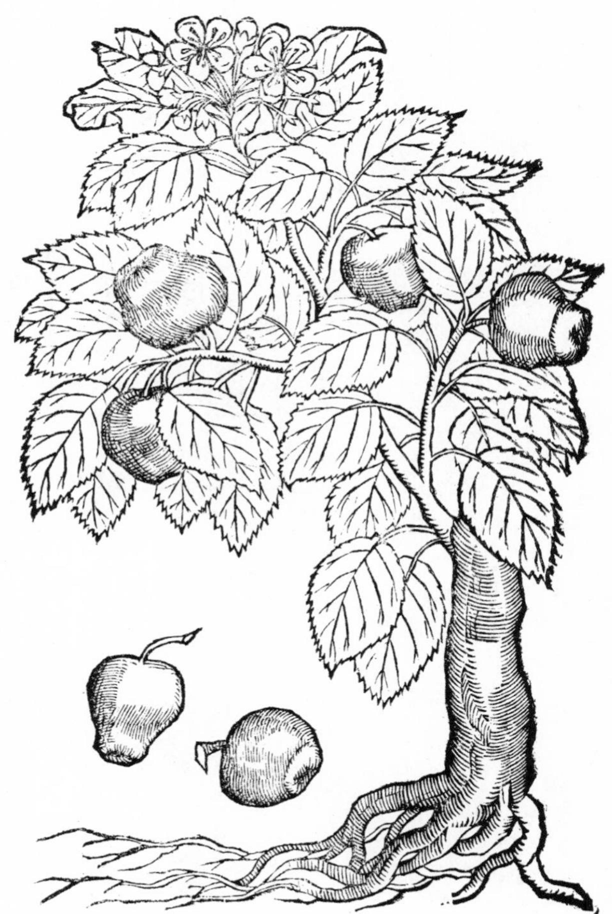

2 *Malus Carbonaria longo fructu.*
The Bakers ditch Apple tree.

¶ *The Place.*

The tame and graffed Apple trees are planted and set in gardens and orchards made for that purpose: they delight to grow in good and fertile grounds: Kent doth abound with apples of most sorts. But I haue seene in the pastures and hedge-rows about the grounds of a worshipful gentleman dwelling two miles from Hereford called Master *Roger Bodnome*, so many trees of all sorts, that the seruants drinke for the most part no other drinke but that which is made of Apples; The quantity is such, that by the report of the Gentleman himselfe, the Parson hath for tithe many hogsheads of Syder. The hogs are fed with the fallings of them, which are so many, that they make choise of those Apples they do eat, who will not taste of any but of the best. An example doubtles to be followed of Gentlemen that haue land and liuing: but enuie saith, the poore wil break down our hedges and we shall haue the least part of the fruit) but forward in the name of God, graffe, set, plant and nourish vp trees in euery corner of your grounds, the labour is small, the cost is nothing, the commoditie is great, your selues shall haue plenty, the poore shall haue somewhat in time of want to relieue their necessitie, and God shall reward your good mindes and diligence.

¶ *The Time.*

They bloom about the end of Aprill, or in the beginning of May. The forward apples be ripe about the Calends of Iuly, others in September.

¶ *The Names.*

The Apple tree is called in Greeke μηλέα: in Latine, *Malus* and *Pomus*: in high Dutch, **Opffelbaum**: in low Dutch, **Appelboom**: in French, *Pommier*: in English, Apple-tree.

The Grecians name the fruit μῆλον: the Latines, *Malum* or *Pomum*: in high Dutch, **Opfell**: in low Dutch, **Appel**: in French and Spanish, *Mansanas*: in English, Apple.

¶ *The Temperature.*

All Apples be of temperature cold and moist, and haue ioined with them a certaine excrementall or superfluous moisture: but as they be not all of like coldnesse, so neither haue they like quantitie of superfluous moisture. They are soonest rotten that haue greatest store of moisture, and they may be longer kept in which there is lesse store: for the abundance of excrementall moisture is the cause why they rot.

Sweet Apples are not so cold and moist, which being rosted or boyled, or otherwise kept, retaine or keepe the soundnesse of their pulpe.

They yeeld more nourishment, and not so moist a nourishment as do the other Apples, and doe not so easily passe through the belly.

Soure Apples are colder and also moister: the substance or pulpe of these when they be boiled doth run abroad, and retaineth not his soundnesse: they yeeld a lesser nourishment, and the same raw and cold.

They do easily and speedily passe through the belly, and therefore they do mollifie the belly, especially being taken before meat.

Harsh or Austere Apples being vnripe, are cold; they ingender grosse bloud, and great store of winde, and often bring the Collicke.

Those Apples which be of a middle taste containe in them oftentimes two or three sorts of tasts, and yet do they retaine the faculties of the other.

¶ *The Vertues.*

A Rosted Apples are alwaies better than the raw, the harm whereof is both mended by the fire, and may also be corrected by adding vnto them seeds or spices.

B Apples be good for an hot stomacke: those that are austere or somewhat harsh doe strengthen a weake and feeble stomacke proceeding of heat.

C Apples are also good for all inflammations or hot swellings, but especially for such as are in their beginning, if the same be outwardly applied.

D The iuice of Apples which be sweet and of a middle taste, is mixed in compositions of diuers medicines, and also for the tempering of melancholy humours, and likewise to mend the qualities of medicines that are dry: as are *Serapium ex pomis Regis Saporis*, *Confectio Alkermes*, and such like compositions.

E There is likewise made an ointment with the pulpe of Apples and Swines grease and Rose water, which is vsed to beautifie the face, and to take away the roughnesse of the skin, which is called in shops *Pomatum*: of the Apples whereof it is made.

F The pulpe of the rosted apples, in number foure or fiue, according to the greatnesse of the Apples, especially of the Pome-water, mixed in a wine quart of faire water, laboured together vntill it come to be as apples and Ale, which wee call Lambes Wooll, and the whole quart drunke last at night, within the space of an houre, doth in one night cure those that pisse by droppes with great anguish and dolour; the strangurie, and all other diseases proceeding of the difficultie of making water; but in twise taking it, it neuer faileth in any: oftentimes there happeneth with the foresaid diseases the Gonorrhæa, or running of the Raines, which it likewise healeth in those persons, but not generally in all; which my selfe haue often proued, and gained thereby both crownes and credit.

G The leaues of the tree do coole and binde, and be also counted good for inflammations, in the beginning.

H Apples cut in pieces, and distilled with a quantitie of Camphere and butter-milke, take away the markes and scarres gotten by the small pockes, being washed therewith when they grow vnto their state and ripenesse: prouided that you giue vnto the patient a little milk and Saffron, or milk and mithridate to drinke, to expell to the extreme parts that venome which may lie hid, and as yet not seene.

Chap.

CHAP. 102 *Of the Wilding or Crab tree.*

¶ The Kindes.

LIke as there be diuers manured Apples, ſo are there ſundry wilde Apples, or Crabs, whereof to write apart were to ſmall purpoſe,and therefore one deſcription ſhall ſuffice for the reſt.

Malus ſylueſtris.
The wilding or Crab tree.

¶ The generall Deſcription.

THere be diuers wilde Apple trees not huſbanded,that is to ſay,not grafted;the fruit whereof is harſh and binding: for by grafting both Apples and Peares become more milde and pleaſant. The crab or wilding tree growes oftentimes to a reaſonable greatneſſe, equall with the Apple tree: the wood is hard, firme, and ſollid; the barke rough; the branches or boughes many,the floures and fruit like thoſe of the apple tree,ſome red, others white,ſome greater,others leſſer: the difference is known to all,therefore it ſhall ſuffice what hath been ſaid for their ſeuerall diſtinctions: we haue in our London gardens a dwarfe kinde of ſweet Apple,called *Chamæmalus*, the dwarfe apple tree, or Paradiſe apple, which beareth apples very timely without grafting.

‡ Our Author here alſo (out of *Tabernamontanus*)gaue foure figures, whereof I onely retaine the beſt,with their ſeueral titles. *Malus ſylueſtris rubens.* The great wilding or red Crab tree: 2 *Malus ſylueſtris alba.* The white wilding or Crab tree: 3 *Malus ſylueſtris minor.* The ſmaller Crab tree: 4 *Malus duracina ſylueſtris.* The choking leane Crab-tree.‡

¶ The Place.

The Crab tree groweth wilde in woods and hedge rowes almoſt euery where.

¶ The Time.

The time anſwereth thoſe of the garden.

¶ The Names.

Their titles doth ſet forth their names in Latine and Engliſh.

¶ The Temperature.

Of the temperature of wilde apples hath beene ſufficiently ſpoken in the former Chapter.

¶ The Vertues.

The iuice of wilde Apples or crabs taketh away the heate of burnings,ſcaldings, and all inflammations: and being laid on in ſhort time after it is ſcalded,it keepeth it from bliſtering. A

The iuice of crabs or Veriuice is aſtringent or binding,and hath withall an abſterſiue or clenſing qualitie, beeing mixed with hard yeeſt of Ale or Beere, and applied in manner of a cold ointment,that is,ſpread vpon a cloth firſt wet in the Veriuice and wrung out, and then laid to, taketh away the heat of Saint Anthonies fire,all inflammations whatſoeuer,healeth ſcab'd legs, burnings and ſcaldings whereſoeuer it be. B

Chap. 103. *Of the Citron, Limon, Orange, and Assyrian Apple trees.*

¶ *The Kindes.*

THe Citron tree is of kindred with the Limon tree, the Orange is of the same house or stocke, and the Assyrian Apple tree claimeth a place as neerest in kinred and neighbourhood: wherefore I intend to comprehend them all in this one chapter.

¶ *The Description.*

1 THe Citron tree is not very great, hauing many boughes or branches, tough and pliable, couered with a greene barke: whereon do grow greene leaues, long, somewhat broad, very smooth, and sweet of smell like those of the Bay tree: among which come forth here and there certaine prickles, set far in sunder: from the bosome whereof come forth smal floures, consisting of fiue little leaues, of a white colour tending to purple, with certaine threds like haires growing in the middle: the fruit is long, greater many times than the Cucumber, often lesser, and not much

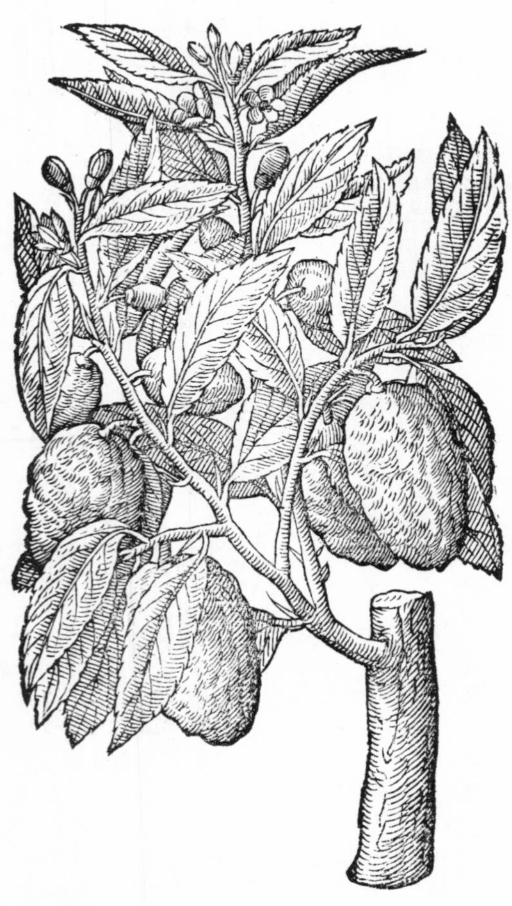

1 *Malus medica.*
The Pome Citron tree.

2 *Malus Limonia.*
The Limon tree.

greater than the Limon: the barke or rinde is of a light golden colour, set with diuers knobs or bumps, and of a very pleasant smell: the pulpe or substance next vnto it is thick, white, hard, hauing a kinde of aromaticall or spicie smell, almost without any taste at all: the softer pulpe within that is not so firme or solid, but more spungie, and full of a sower iuice, in which the seed lieth hid, greater and thicker than a graine of Barley.

2 The Limon tree is like vnto the Pome Citron tree in growth, thorny branches, and leaues of a pleasant

a pleaſant ſweet ſmell, like thoſe of the Bay-tree: the floures hereof are whiter than thoſe of the Citron tree, and of a moſt ſweet ſmell: the fruit is long and thicke, leſſer than the Pome Citron: the rinde is yellow, ſomewhat bitter in taſte, and ſweet of ſmell: the pulpe is white, more in quantitie than that of the Citron, reſpecting the bignes; in the middle part whereof is conteined more ſoft ſpungie pulpe, and fuller of ſoure juice: the ſeeds are like thoſe of the Pome Citron.

3 The Orenge tree groweth vp to the height of a ſmall Peare tree, hauing many thornie boughes or branches, like thoſe of the Citron tree: the leaues are alſo like thoſe of the Bay-tree, ‡ but that they differ in this, that at the lower end next the ſtalke there is a leſſer leafe made almoſt after the vulgar figure of an heart, whereon the bigger leafe doth ſtand, or is faſtned: ‡ & they are of a ſweet ſmell: the floures are white, of a moſt pleaſant ſweet ſmell alſo: the fruit is round like a ball, euery circumſtance belonging to the forme is very well knowne to all; the taſte is ſoure, ſometimes ſweet, and often of a taſte betweene both: the ſeeds are like thoſe of the Limon.

3 *Malus arantia.*
The Orange tree.

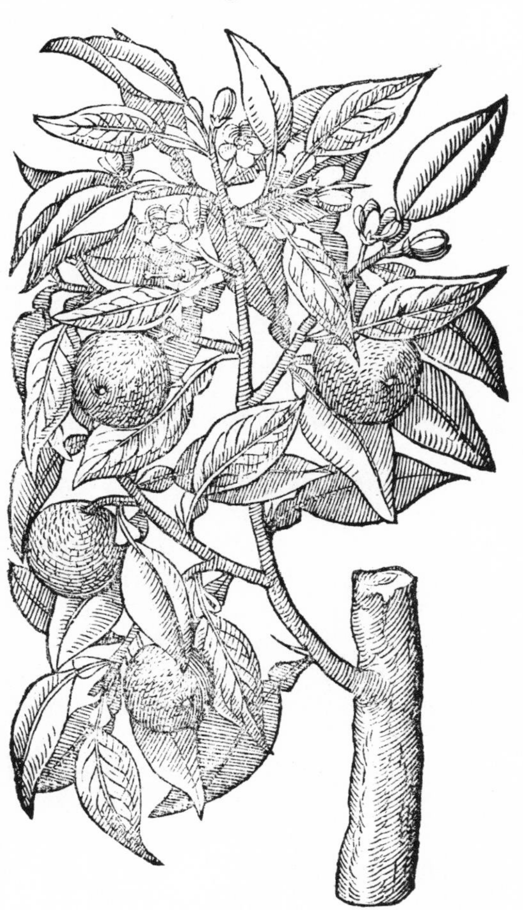

4 *Malus Aſſyria.*
The Aſſyrian Apple tree.

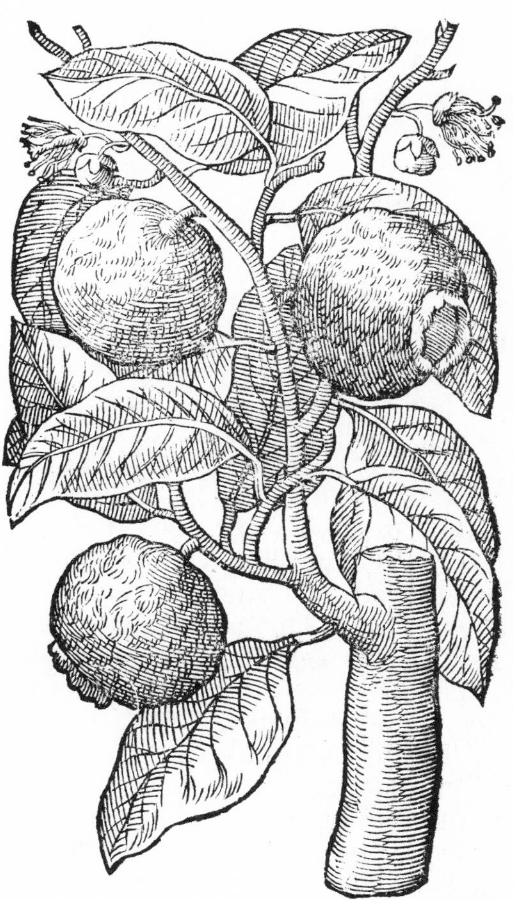

4 The Aſſyrian Apple tree is like vnto the Orange tree: the branches are like: the leaues are greater: the floures are like thoſe of the Citron tree: the fruit is round, three times as big as the Orange: the barke or peeling is thicke, rough, and of a pale yellow colout, wherein appeare often as it were ſmall clifts or crackes: the pulpe or inner ſubſtance is full of iuice, in taſte ſharpe, as that of the Limon, but not ſo pleaſant: the ſeeds are like thoſe of the Citron.

¶ *The Place.*

The Citron, Limon, and Orange trees do grow eſpecially on the ſea coaſts of Italy, and on the Iſlands of the Adriaticke Turrhene, and alſo Ægæan Seas, & likewiſe on the maine land, neer vnto meeres and great lakes: there is alſo great ſtore of them in Spaine, but in places eſpecially ioining to the ſea, or not farre off: they are alſo found in certaine prouinces of France which lie vpon the midland ſea. They were firſt brought out of Media, as not onely *Plinie* writeth, but alſo the Poet *Virgil* affirmeth in the ſecond book of his Georgickes, writing of the Citron tree after this maner:

Media fert triſtes ſuccos, tardumque ſaporem
Felicis mali, quo non præſentius vllum,
Pocula ſi quando ſæuæ infecere nouercæ,

Miſcueruntque

Miscueruntque herbas, & non innoxia verba,
Auxilium venit, ac membris agit atra venena.
Ipsa ingens arbos, faciesque simillima Lauro;
Et, si non alium late iactaret odorem,
Laurus erit; folia haud vllis labentia ventis;
Flos apprime tenax. Animas & olentia Medi
Ora fouent illo, & senibus medicantur anhelis.

The Countrey Media beareth iuyces sad,
And dulling tastes of happy Citron fruit,
Than which, no helpe more present can be had,
If any time stepmothers worse than brute
haue poyson'd pots, and mingled herbs of sute
With hurtfull charmes: this Citron fruit doth chase
Blacke venome from the body in euery place.
The tree it selfe in growth is large and big,
And very like in shew to th' Laurell tree;
And would be thought a Laurell, leafe and twig,
But that the smell it casts doth disagree:
The floure it holds as fast as floure may be:
Therewith the Medes a remedie do finde
For stinking breaths and mouthes a cure most kinde,
And helpe old men which hardly fetch their winde.

¶ *The Time.*

These trees be alwaies greene, and do, as *Pliny* saith, beare fruit at all times of the yere, some falling off, others waxing ripe, and others newly comming forth.

¶ *The Names.*

The first is called in Greeke, Μηλέα μηδική: in Latine, *Malus Medica*, and *Malus Citria*: in English, Citron tree, and Pomecitron tree.

The fruit is named in Greeke, Μῆλον μηδικὸν: in Latine, *Malum Medicum*, and *Malum Citrium*: and *Citromalum*. *Æmilyanus* in *Athenæus* sheweth, that *Iuba* King of Mauritania hath made mention of the Citron, who saith that this Apple is named among them, *Malum Hespericum*: *Galen* denieth it to be called any longer *Malum Medicum*, but *Citrium*; and saith, that they who call it *Medicum* do it to the end that no man should vnderstand what they say: the Apothecaries call these apples *Citrones*: in high-Dutch, **Citrin opffell**, **Citrinaten**: in low-Dutch, **Citroenen**: in Italian, *Citroni*, and *Cedri*: in Spanish, *Cidras*: in French, *Citrons*: in English, Citron Apple, and Citron.

The second kinde of Citron is called in Latine, *Limonium Malum*; in shops, *Limones*. in French, *Limons*: in low-Dutch, **Limonen**: in English, Limon, and Lemon.

The third is named in Latine, *Malnm anarantium* or *Anerantium*: and of some *Aurantium*: of others, *Aurengium*, of the yellow colour of gold: some would haue them called *Arantia*, of *Arantium*, a towne in Achaia or Arania, of a countrey bearing that name in Persia: it is termed in Italian *Arancio*: in high-Dutch, **Pomeranken**: in low-Dutch, **Araengie Appelen**: in French, *Pommes d Orenges*: in Spanish, *Naransas*: in English, Orenges.

The fourth is named of diuers, *Pomum Assyrium*, or the Citron of Assyria, and may be Englished Adams Apple, after the Italian name; and among the vulgar sort of Italians, *Lomie*, of whom it is also called *Pomum Adami*, or Adams Apple; and that came by the opinion of the common rude people, who thinke it to be the same Apple which *Adam* did eate of in Paradise, when he transgressed Gods commandment; whereupon also the prints of the biting appeare therein, as they say: but others say that this is not the Apple, but that which the Arabians do call *Musa* or *Mosa*, whereof *Auicen*, *cap*. 395. maketh mention: for diuers of the Iewes take this for that through which by eating, *Adam* offended, as *Andrew Theuet* sheweth.

¶ *The Temperature and Vertues.*

A All these fruits consist of vnlike parts, and much differing in facultie.

B The rindes are sweet of smell, bitter, hot, and dry.

C The white pulpe is cold, and containeth in it a grosse iuice, especially the Citron.

D The inner substance or pap is soure, as of the Citrons and Limons, cold and dry, with thinnesse of parts.

E The seed because it is bitter is hot and dry.

F The rinde of the Pomecitron is good against all poysons, for which cause it is put into treacles and such like confections.

G It is good to be eaten against a stinking breath, for it maketh the breath sweet; and being so taken it comforteth the cold stomacke exceedingly.

H The white, sound, and hard pulpe is now and then eaten, but very hardly concocted, and ingendreth a grosse, cold, and phlegmaticke iuyce; but being condite with sugar, it is both pleasant in taste, and easie to be digested, more nourishing, and lesse apt to obstruction and binding or stopping.

I *Galen* reporteth, that the inner iuice of the Pomecitron was not wont to be eaten, but it is now vsed for sauce; and being often vsed, it represseth choler which is in the stomacke, and procures appetite: it is excellent good also to be giuen in vehement and burning feuers, and against all pestilent and venomous or infectious diseases: it comforteth the heart, cooleth the inward parts, cutteth, diuideth, and maketh thin, grosse, tough, and slimy humors.

K Of this foresaid sharpe iuice there is a syrrup ptepared, which is called in shops, *Syrupus de Acetositate Citri*, very good against the foresaid infirmities.

L Such a sirrup is also prepared of the sharpe iuice of Limons, of the same quality and operation, so that in stead of the one, the other will serue very well.

M A dozen of Orenges cut in slices and put into a gallon of water, adding thereto an ounce of Mercurie sublimate, and boiled to the consumption of the halfe, cureth the itch and manginesse of the body.

N Men in old time (as *Theophrastus* writeth in his fourth booke) did not eate Citrons, but were contented with the smell, and to lay them amongst cloathes, to preserue them from Moths.

O As often as need required they vsed them against deadly poysons; for which thing they were especially commended euen by *Virgils* verses, which we haue before alledged.

P *Athenæus*, *lib*. 3. hath extant a story of some that for certaine notorious offences were condemned to be destroyed of Serpents, who were preserued and kept in health and safetie by the eating of Citrons.

Q The distilled water of the whole Limons, rinde and all, drawne out by a glasse Still, takes away tetters and blemishes of the skin, and maketh the face faire and smooth.

R The same being drunke prouoketh vrine, dissolueth the stone, breaketh and expelleth it.

S The rinde of Orenges is much like in facultie to that of the Citrons and Limons, yet it is so much the more hot as it is more biting and bitter.

T The inner substance or soure pap which is full of iuice is of like facultie, or not much inferiour to the facultie of the pap of Citrons or Limons; but the sweet pap doth not much coole or drie, but doth temperatly heate and moisten, being pleasant to the taste: it also nourisheth more than doth the soure pap, but the same nourishment is thin and little; and that which is of a middle tast, hauing the smacke of wine, is after a middle sort more cold than sweet, and lesser cold than soure: the sweet and odoriferous floures of orenges be vsed of the perfumers in their sweet smelling ointments.

V Two ounces of the iuice of Limons, mixed with the like quantitie of the spirit of wine, or the best *Aqua vitæ* (but the spirit of wine rectified is much better) and drunk at the first approch of the fit of an ague, taketh away the shaking presently: the medicine seldome faileth at the second time of the taking thereof perfectly to cure the same; but neuer at the third time, prouided that the Patient be couered warme in a bed, and caused to sweat.

X There is also distilled out of them in a glasse still, a water of a maruellous sweet smell, which being inwardly taken in the weight of an ounce and a halfe, moueth sweat, and healeth the ague.

Y The seed of all these doth kill wormes in the belly, and driueth them forth: it doth also mightily resist poyson, and is good for the stinging of scorpions, if it be inwardly taken.

Z Those which be called Adams Apples are thought to be like in faculties to the soure iuyce, especially of the Limons, but yet they be not so effectuall.

Chap. 104. *Of the Cornell tree.*

¶ *The Description.*

The tame Cornell tree groweth somtime of the height and bignesse of a smal tree, with a great number of springs: it is couered with a rugged barke: the wood or timber is very hard and dry, without any great quantity of sap therein: the leaues are like vnto the Dog berry leaues, crumpled rugged, and of an ouerworne colour: the floures grow in small bunches before any leaues do appeare, of colour yellow, and of no great value (they are so small) in shew like the floures of the Oliue

Cornus mas.
The male Cornel tree.

Oliue tree:which being vaded,there come small long berries, which at the first bee greene, and red when they be ripe; of an austere and harsh taste, with a certaine sourenesse: within this berry is a small stone, exceeding hard, white within like that of the Oliue, wherunto it is like both in the fashion and oftentimes in the bignesse of the fruit.

¶ *The Place.*

This groweth in most places of Germanie without manuring: it growes not wild in England. But yet there be sundry trees of them growing in the gardens of such as loue rare and dainty plants,whereof I haue a tree or two in my garden.

¶ *The Time.*

The tame Cornell tree floureth sometime in February, & commonly in March, and afterwards the leaues come forth as an vntimely birth: the berries or fruit are ripe in August.

¶ *The Names.*

The Grecians call it Κρανία: the Latines, *Cornus*: in high-Dutch, **Cornelbaum:** in low-Dutch, **Cornoele boom:** the Italians, *Corniolo*: in French, *Cornillier*: in Spanish, *Cornizolos*: in English, the Cornell tree, and the Cornelia tree; of some, long Cherrie tree.

The fruit is named in Latine, *Cornum*: in high-Dutch, **Cornell:** in low-Dutch, **Cornoele:** in Italian, *Cornole*: in English, Cornel berries and Cornelian Cherries.

This is *Cornus mas Theophrasti*, or *Theophrastus* his male Cornell tree; for he setteth downe two sorts of the Cornell trees, the male and the female: he maketh the wood of the male to be sound, as in this Cornell tree; which we both for this cause and for others also haue made to be the male. The female is that which is commonly called *Virga sanguinea*, or Dogs berry tree, and *Cornus syluestris*, or the wilde Cornell tree, of which we will treat in the next Chapter following.

¶ *The Temperature and Vertues.*

A The fruit of the Cornell tree hath a very harsh or choking taste: it cooleth, drieth, and bindeth, yet may it also be eaten, as it is oftentimes.

B It is a remedie against the laske and bloudy flix, it is hurtfull to a cold stomacke, and increaseth the rawnesse thereof: the leaues and tender crops of the tree are likewise of an harsh and choking taste, and do mightily dry.

C They heale greene wounds that are great and deepe, especially in hard bodies, but they are not so good for small wounds and tender bodies, as *Galen* writeth.

Chap. 105.

Of the female Cornell or Dog-Berry tree.

¶ *The Description.*

That which the Italians call *Virga sanguinea*, or the bloudy Rod, is like to the Cornel tree, yet it groweth not into a tree, but remaineth a shrub: the yong branches thereof are iointed, and be of an obscure red purple: they haue within a white spongie pith like that of Elder but the old stalks are hard and stiffe, the substance of the which is also white, and answerable to those of the Cornell tree: the leaues are also like, the middle rib whereof as also the brittle foot-stalkes are somewhat reddish: at the top whereof stand white floures in spoky rundles, which turne into berries,

Cornus fæmina.
The Dog-berry tree.

greene at the first, and of a shining black colour when they be ripe, in taste vnpleasant, and not cared for of the birds.

¶ *The Place.*

This shrub groweth in hedges and bushes in euery countrey of England.

¶ *The Time.*

The floures come forth in the Spring in the moneth of Aprill: the berries are ripe in Autumne.

¶ *The Names.*

The Italians do commonly call it *Sanguino*, and *Sanguinello*: *Petrus Crescentius* termes it *Sanguinus*; and *Matthiolus*, *Virga sanguinea*: *Pliny*, *lib.*24. *cap.*10. hath written a little of *Virga Sanguinea*: Neither is *Virga Sanguinea*, saith hee, counted more happy, the inner barke whereof doth breake open the scarres which they before haue healed. It is an hard thing, or peraduenture a rash part, to affirme by these few words, that *Pliny* his *Virga Sanguinea* is the same that the Italian *Sanguino* is. This is called in high-Dutch, **Hartriegel**: in low-Dutch, **Wilde Cornoelle**, that is to say, *Cornus syluestris*, or wilde Cornell tree: and in French, *Cornellier sauuage*: in English, Hounds tree, Hounds berry, Dogs berry tree, Pricke-Timber: in the North countrey they call it Gaten tree, or Gater tree; the berries whereof seem to be those which *Chaucer* calleth Gater berries: *Valerius Cordus* nameth it ψευδοκρανία, that is to say, *Falsa* or *Spuria Cornus*, false or bastard Cornell tree: this seemeth also to be *Theophrastus* his θηλυκρανία, or *Cornus fœmina*, female Cornell tree. This hath little branches hauing pith within, neither be they hard nor sound, like those of the male: the fruit is ἄβρωτος, that is, not fit to be eaten, and a late fruit which is not ripe till after the Autumne Æquinoctiall; and such is the wilde Cornell tree or Gater tree, the yong and tender branches whereof be red, and haue (as wee haue written) a pith within: the fruit or berries be vnpleasant, and require a long time before they can be ripe.

¶ *The Temperature.*

The berries hereof are of vnlike parts; for they haue some hot, bitter, and clensing, and very many cold, dry, harsh, and binding, yet they haue no vse in medicine.

¶ *The Vertues.*

Matthiolus writeth, that out of the berries first boiled, and afterwards pressed, there issueth an A
oyle which the Anagnian countrey people do vse in lamps: but it is not certaine, nor very like, that the barke of this wilde Cornell tree hath that operation which *Pliny* reporteth of *Virga Sanguinea*; for he saith, as we haue already set downe, that the inner barke thereof doth break and lay open the scars which they before haue healed.

CHAP. 106. *Of Spindle tree or Pricke-wood.*

¶ *The Description.*

1 PRickewood is no high shrub, of the bignesse of the Pomegranat tree: it spreadeth farre with his branches: the old stalks haue their barke somewhat white; the new and those that be lately growne be greene, and foure square: the substance of the wood is hard, and mixed with a light yellow: the leaues be long, broad, slender, and soft: the floures be white, many standing vpon one foot-stalke, like almost to a spoked rundle: the fruit is foure square, red, and containing foure white seeds, euery one whereof is couered with a yellow coat, which being taken off giueth a yellow die.

2 This

1 *Euonymus Theophrasti.*
English Prick-timber tree.

2 *Euonymus latifolius.*
Broad leafed Spindle tree.

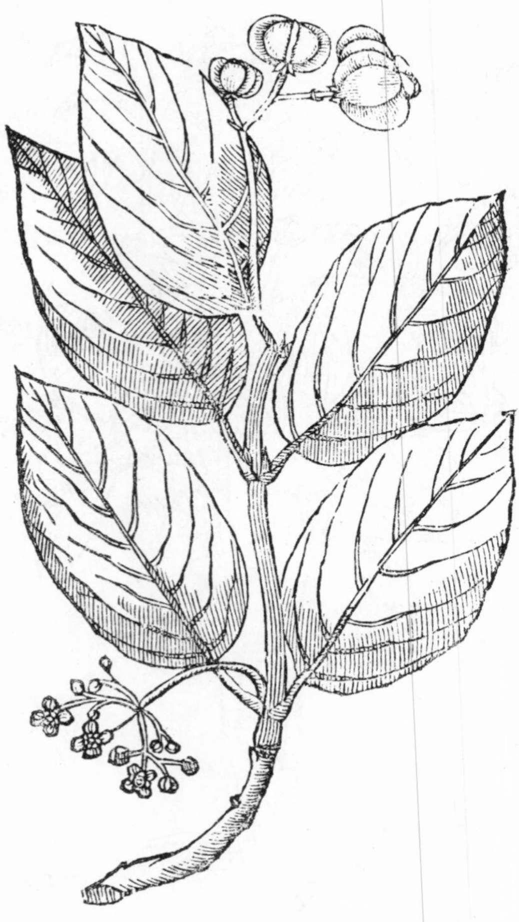

3 *Euonymus Pannonicus.*
Hungarie Spindle tree.

2 This other sort of *Euonymus* groweth to the forme of an hedge tree, of a meane bignesse; the trunke or body whereof is of the thicknesse of a mans leg, couered with a rough or scabbed barke of an ouerworn russet colour. The branches thereof are many, slender, and very euen, couered with a greene barke whilest they be yet young and tender; they are also very brittle, with some pith in the middle like that of the Elder. The leaues are few in number, full of nerues or sinues dispersed like those of Plantaine, in shape like those of the Pomecitron tree, of a lothsome smell and bitter taste: amongst which come forth slender footstalks very long and naked, whereon do grow small floures consisting of foure small leaues like those of the Cherrie tree, but lesser, of a white colour tending to a blush, with some yellownesse in the middle: after commeth the fruit, which is larger than the former, and as it were winged, parted commonly into foure, yet somtimes into fiue parts; and opening when it is ripe, it sheweth the white graines filled with a yellow pulpe. The root is tough and wooddy, dispersing it selfe farre abroad vnder the vpper crust of the earth.

3 The

3 The ſame Author ſetteth forth another ſort which he found in the mountaines of Morauia and Hungary, hauing a trunke or ſtocke of the height of three or foure cubits, couered with a bark greene at the firſt, afterward ſprinkled ouer with many blacke ſpots: the boughes are diuided toward the top into diuers ſmall branches, very brittle and eaſie to breake, whereon are placed leaues by couples alſo, one oppoſite to another, ſomewhat ſnipt about the edges, in ſhape like thoſe of the great Myrtle, of an aſtringent taſte at the beginning, after ſomewhat hot and bitter: amongſt which come forth ſmall floures ſtanding vpon long naked foot-ſtalkes, conſiſting of foure little leaues of a bright ſhining purple colour, hauing in the middle ſome few ſpots of yellow: after commeth the fruit, foure cornered, not vnlike to the common kinde, of a ſpongious ſubſtance, and a gold yellow colour: wherein is contained not red berries like the other, but blacke, very like to thoſe of *Fraxinella*, of a ſhining blacke colour like vnto burniſhed horne; which are deuoured of birds when they be ripe, and the rather becauſe they fall of themſelues out of their huskes, otherwiſe the bitterneſſe of the husks would take away the delight.

¶ *The Place.*

The firſt commeth vp in vntoiled places, and among ſhrubs, vpon rough bankes and heapes of earth. it ſerueth alſo oftentimes for hedges in fields, growing amongſt Brambles and ſuch other Thornes.

The other ſorts *Carolus Cluſius* found in a wood of Hungarie beyond the riuer Drauus, and alſo vpon the mountaines of Morauia and other places adiacent.

¶ *The Time.*

The floures appeare in Aprill: the fruit is ripe in the end of Auguſt, or in the moneth of September.

¶ *The Names.*

Theophraſtus calleth this ſhrub Εὐώνυμος, and deſcribeth it in his third booke of the Hiſtorie of Plants: diuers alſo falſly reade it *Anonymos*: *Petrus Creſcentius* calleth it *Fuſanum*, becauſe ſpindles be made of the wood hereof; and for that cauſe it is called in high-Dutch, **Spindelbaum**, yet moſt of them **Hanhodlin**: in low Dutch, **Papenhout**: in Italian, *Fuſano*: in French, *Fuſin*, and *Bonnet de preſtre*: in Engliſh, Spindle tree, Prick wood, and Prick-timber.

¶ *The Temperature and Vertues.*

This ſhrub is hurtfull to all things, as *Theophraſtus* writeth, and namely to Goats: hee ſaith the A
fruit hereof killeth; ſo doth the leaues and fruit deſtroy Goats eſpecially, vnleſſe they ſcoure as wel vpwards as downwards: if three or foure of theſe fruits be giuen to a man they purge both by vomit and ſtoole.

CHAP. 107. *Of the blacke Aller tree.*

¶ *The Deſcription.*

THe blacke Aller tree bringeth forth from the root ſtraight ſtalkes diuided into diuers branches: the outward barke whereof is blacke, and that next to the wood yellow, and giueth a colour as yellow as Saffron: the ſubſtance of the wood is white and brittle, with a reddiſh pith in the midſt: the leaues be like thoſe of the Alder tree, or of the Cherry tree, yet blacker, and a little rounder: the floures be ſomewhat white: the fruit are round berries, in which appeare a certaine rift or chinke, as though two were ioined together, at the firſt greene, afterwards red, and laſt of all blacke: in this there be two little ſtones: the root runneth along in the earth.

¶ *The Place.*

The Aller tree groweth in moiſt woods and copſes: I found great plenty of it in a wood a mile from Iſlington, in the way from thence toward a ſmall village called Harnſey, lying vpon the right hand of the way; and in the woods at Hampſted neere London, and in moſt woods in the parts about London.

¶ *The Time.*

The leaues and floures appeare in the beginning of the Spring; and the berries in Autumne.

¶ *The Names.*

This ſhrub is called *Alnus nigra*, or blacke Alder: and by others, *Frangula*: *Petrus Creſcentius* nameth it *Auornus*: in low-Dutch, **Sparkenhout**, and oftentimes **Pijlhout**, becauſe boies make for themſelues arrowes hereof: in high-Dutch, **Faulbaum**: it is called in Engliſh, blacke Aller tree; and of diuers Butchers pricke tree.

Alnus nigra, siue Frangula.
The blacke Aller tree.

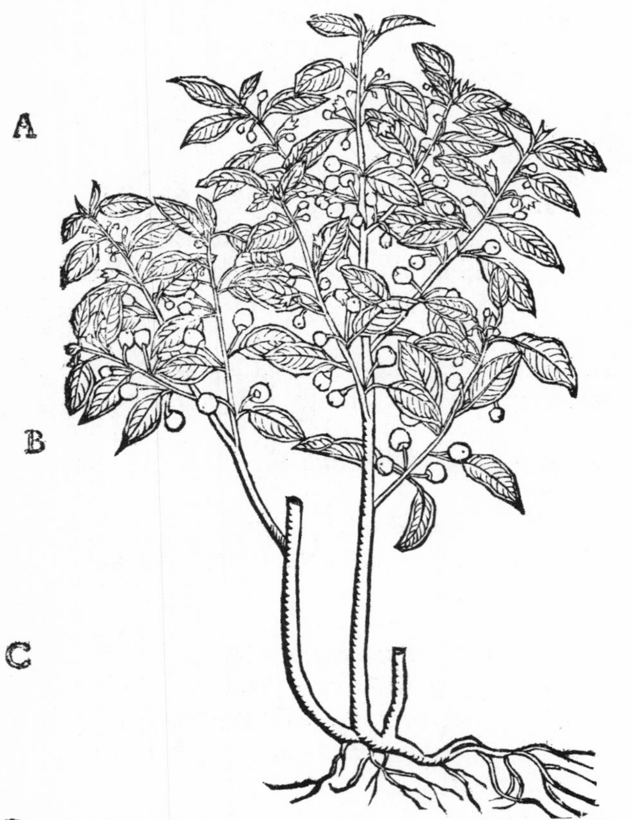

¶ *The Temperature.*

The inner barke of the blacke Aller tree is of a purging and dry qualitie.

¶ *The Vertues.*

A The inner barke hereof is vsed of diuers country men, who drink the infusion thereof when they would be purged: it purgeth thicke phlegmaticke humors, and also cholericke, and not only by the stoole, but many times also by vomit, not without great trouble and paine to the stomacke: it is therefore a medicine more fit for clownes than for ciuill people, and rather for those that feed grossely, than for dainty people.

B There be others who affirme that the dried barke is more gentle, and causeth lesser paine: for the greene bark (say they) which is not yet dried containeth in it a certaine superfluous moisture which causeth gripings and vomitings, and troubles the stomacke.

C The same barke being boiled in wine or vineger makes a lotion for the tooth ache; and is commended against scabs and filthinesse of the skin.

D The leaues are reported to be good fodder for cattell, especially for kine, and to cause them to yeeld good store of milke.

Chap. 108. *Of the Seruice tree.*

¶ *The Description.*

1 The Seruice tree groweth to the height and bignesse of a great tree, charged with many great armes or boughes which are set with sundry small branches, garnished with many great leaues somewhat long like those of the Ash: the floures are white, and stand in clusters, which turne into small browne berries somewhat long, which are not good to be eaten vntill they haue lien a while, and vntill they be soft like the Medlar, whereto it is like in taste and operation.

2 The common Seruice tree groweth likewise to the height of a great tree, with a straight body of a brownish colour, full of branches, set with large displayed leaues like the Maple or the White-Thorne, sauing that they are broader and longer: the floures are white, and grow in tufts; which being fallen, there come in place thereof small round berries, browne vpon one side and reddish toward the Sun, of an vnpleasant taste in respect of the former: in which are contained little blackish kernels.

¶ *The Place.*

These trees are found in woods and groues in most places of England: there be many smal trees thereof in a little wood a mile beyond Islington from London: in Kent it groweth in great aboundance, especially about Southfleet and Grauesend. ‡ The later of these I haue seene growing wilde in diuers places, but not the former in any place as yet. ‡

¶ *The Time.*

They floure in March, and their fruit is ripe in September.

¶ *The Names.*

The first is called in Greeke, ὄη, and ὄυα: in Latine, *Sorbus*: in high-Dutch, **Sperwerbaum**: in low-Dutch, **Sorbedboom**: in French, *Cormier*: in English, Seruice tree, and of some after the Latines, Sorbe tree.

The

1 *Sorbus.*
The Seruice tree.

2 *Sorbus torminalis.*
Common Seruice tree.

The common Seruice tree is named of *Pliny*, *Sorbus torminalis* : in high-Dutch, **Aressel**, **Eschrosel**, and **Wilder Sperwerbaum** : in Engliſh, Common Seruice tree.

The berries or fruit of the Seruice tree is called Ὄον, or Ὄυον : in Latine, *Sorbum* : in high Dutch, **Speierling**, **Sporopffel** : in low-Dutch, **Sorben** : in Italian, *Sorbe*, and *Sorbole* : in French, *Corme* : in Spaniſh, *Seruas*, and *Sorbas* : in Engliſh, Seruice ; of ſome, Sorbe Apple.

¶ *The Temperature and Vertues.*

Seruice berries are cold and binding, and much more when they be hard, than when they are milde and ſoft : in ſome places they are quickly ſoft, either hanged in a place which is not altogether cold, or laid in hay or chaffe : thoſe Seruices are eaten when the belly is too ſoluble, for they ſtay the ſame ; and if they yeeld any nouriſhment at all, the ſame is very little, groſſe, and cold and A
therefore it is not expedient to eate of theſe or other-like fruits, nor to vſe them otherwiſe than in medicines.

Theſe do ſtay all manner of fluxes of the belly, and likewiſe the bloudy flixe; as alſo vomiting: they ſtanch bleeding if they be cut and dried in the ſunne before they be ripe, and ſo reſerued for vſe : theſe we may vſe diuers waies according to the manner of the greife and grieued part.

CHAP. 109. *Of the Aſh tree.*

¶ *The Deſcription.*

1 THe Aſh alſo is an high and tall tree : it riſeth vp with a ſtraight body, now and then of no ſmall bigneſſe, now and then of a middle ſize, and is couered with a ſmooth bark : the wood is white, ſmooth, hard, and ſomewhat rough grained : the tender branches hereof and ſuch as be new growne vp are ſet with certaine ioints, and haue within a white and ſpongie pith : but the old boughes are wooddy throughout, and be without either ioints or much pith : the leaues are long and winged, conſiſting of many ſtanding by couples, one right againſt another vpon one rib or ſtalke, the vpermoſt of all excepted, which ſtandeth alone ; of which euery particular one is long, broad, like to a Bay leafe, but ſofter, and of a lighter greene, without any ſweet

ſmell, and nicked round about the edges: out of the yonger ſort of the boughes, hard to the ſetting on of the leaues, grow forth hanging together many long narrow and flat cods, as it were like almoſt to diuers birds tongues, where the ſeed is perfected, which is of a bitter taſte: the roots be many, and grow deepe in the ground.

Fraxinus.
The Aſh tree.

¶ *The Place.*

The Aſh doth better proſper in moiſt places, as about the borders of Medowes and Riuer ſides, than in dry grounds.

¶ *The Time.*

The leaues and keyes come forth in Aprill and May, yet is not the ſeed ripe before the fall of the leafe.

¶ *The Names.*

This tree is called in Greeke, Μελία, and of diuers, μέλεα: in Latine, *Fraxinus*: in high-Dutch, **Eſchernbaum**, **Eſchernholtz**, and **Steyneſchern**: in low-Dutch, **Eſſchen**, and **Eſſchenboom**: in Italian, *Fraßino*: in French, *Freſne*: in Spaniſh, *Freſno*, *Fraxino*, and *Freixo*: in Engliſh, Aſh tree.

The fruit like vnto cods is called of the Apothecaries, *Lingua Auis*, and *Lingua Paſſerina*: it may be named in Greeke, Ὀρνιθόγλωσσον: yet ſome would haue it called *Orneogloſſum*: others make *Ornus* or the wilde Aſh to be called *Orneogloſſum*: it is termed in Engliſh, Aſh keyes, and of ſome, Kite-keyes.

¶ *The Temperature and Vertues.*

A The leaues and bark of the Aſh tree are dry and moderatly hot: the ſeed is hot and dry in the ſecond degree.

B The iuice of the leaues or the leaues themſelues being applied, or taken with wine, cure the bitings of vipers, as *Dioſcorides* ſaith.

C The leaues of this tree are of ſo great vertue againſt ſerpents, as that they dare not ſo much as touch the morning and euening ſhadowes of the tree, but ſhun them afar off, as *Pliny* reports, *li.16. cap.13.* He alſo affirmeth, that the ſerpent being penned in with boughes layd round about, will ſooner run into the fire, if any be there, than come neere the boughes of the Aſh: and that the Aſh doth floure before the Serpents appeare, and doth not caſt his leaues before they be gon againe.

D We write (ſaith he) vpon experience, that if the ſerpent be ſet within the circle of a fire and the boughes, the ſerpent will ſooner run into the fire than into the boughes. It is a wonderfull courteſie in nature, that the Aſh ſhould floure before theſe ſerpents appeare, and not caſt his leaues before they be gon againe.

E Both of them, that is to ſay the leaues and the barke, are reported to ſtop the belly: and being boiled with vineger and water, do ſtay vomiting, if they be laid vpon the ſtomacke.

F The leaues and barke of the Aſh tree boiled in wine and drunk, do open the ſtoppings of the liuer and ſpleene, and do greatly comfort them.

G Three or foure leaues of the Aſh tree taken in wine each morning from time to time, doe make thoſe leane that are fat, and keepeth them from feeding which do begin to wax fat.

H The ſeed or Kite-keyes of the Aſh tree prouoke vrine, increaſe naturall ſeed, and ſtirre vp bodily luſt, eſpecially being poudred with nutmegs and drunke.

I The wood is profitable for many things, being exalted by *Homers* commendations, and *Achilles* ſpeare, as *Pliny* writeth. The ſhauings or ſmall pieces thereof being drunke are ſaid to be pernicious and deadly, as *Dioſcorides* affirmeth.

K The Lee which is made with the Aſhes of the barke cureth the white ſcurfe, and ſuch other like roughneſſe of the skin, as *Pliny* teſtifieth.

CHAP.

CHAP. 110.

Of the wilde Aſh, otherwiſe called Quicke-Beame or Quicken tree.

Sorbus ſylueſtris, ſiue Fraxinus Bubula.
The Quicken tree, wilde Aſh, or wilde Seruice tree.

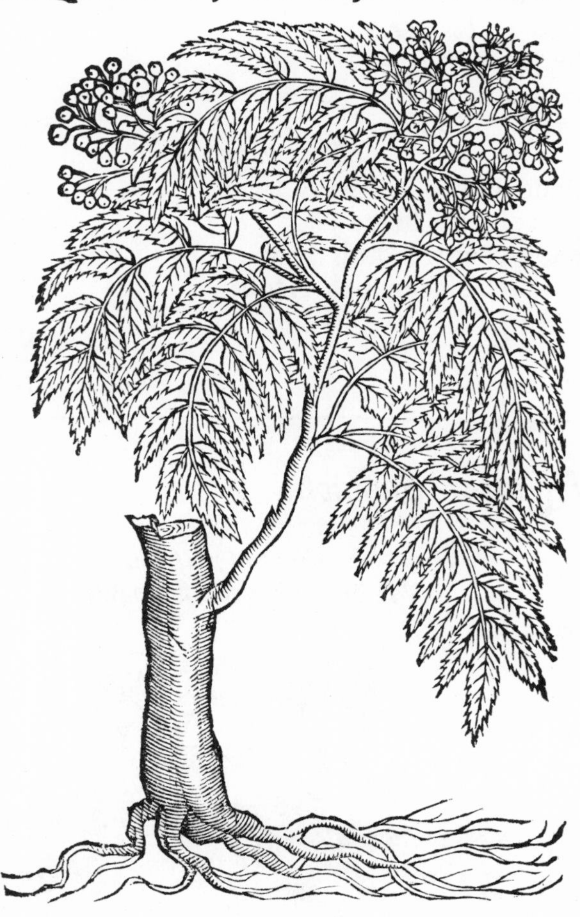

¶ *The Deſcription.*

THe wilde Aſh or Quicken Tree *Pena* ſetteth forth for the wilde Seruice: this tree groweth ſeldome or neuer to the ſtature and height of the Aſh tree, notwithſtanding it growes to the bignes of a large tree: the leaues be great and long, and ſcarcely be diſcerned from the leaues of the Seruice tree: the floures bee white, and ſweet of ſmell, and grow in tufts, which do turne into round berries, greene at the firſt, but when they be ripe of a deepe red colour, and of an vnpleaſant taſte: the branches are as full of iuice as the Oſiar, which is the cauſe that boyes doe make Pipes of the barke thereof as they doe of Willowes.

¶ *The Place.*

The wilde Aſh or Quicken tree groweth on high mountaines, and in thicke high woods in moſt places of England, eſpecially about Namptwich in Cheſhire, in the Weilds of Kent, in Suſſex and diuers other places.

¶ *The Time.*

The wild Aſh floures in May, and the berries are ripe in September.

¶ *The Names.*

The Latines call this tree *Ornus*, and oftentimes *Sylueſtris Fraxinus*, or wilde Aſh: and it is alſo *Fraxini ſpecies*, or a kinde of Aſh; for the Grecians (as not only *Pliny* writeth, but alſo *Theophraſtus*) hath made two kindes of Aſh, the one high and tall, the other lower: the high and tall one is *Fraxinis vulgaris*, or the common Aſh; and the lower *Ornus*, which alſo is named Ὀρεινὴ μελία, or *Montana Fraxinus*, mountaine Aſh; as the other, πεδινὴ, or field Aſh; which is alſo named Βουμελία, or as *Gaza* tranſlateth it, *Bubula Fraxinus*, but more truly *Magna Fraxinus*, or great Aſh; for the ſyllable Βοῦ is a ſigne of bigneſſe: this *Ornus* or great Aſh is named in high-Dutch, **Malbaum** in low-Dutch, **Hauereſchen**, or **Quereſchen**, of diuers, **Qualſter**: in French, *Freſne ſauuage*: in Engliſh, Wilde Aſh, Quicken tree, Quick-beame tree, and Wicken tree. *Matthiolus* makes this to be *Sorbus ſylueſtris*, or wilde Seruice tree.

¶ *The Temperature and Vertues.*

A Touching the faculties of the leaues, barke, or berries, as there is nothing found among the old, ſo is there nothing noted among the later writers: but *Pliny* ſeemeth to make this wilde Aſh like in faculties to the common Aſh; for *lib. 16. cap. 13.* where he writes of both the Aſhes, hee ſaith, that the common Aſh is *Criſpa*, and the mountaine Aſh *Spiſſa*: and forthwith he addeth this: The Grecians write, that the leaues of them do kill cattell, and yet hurt not thoſe that chew their cud; which the old writers haue noted of the Yew tree, and not of the Aſh tree. *Pliny* was deceiued by the neereneſſe of the words μίλος and μελία: μίλος is the Yew tree, and μελία the Aſh tree: ſo that hee hath falſly attributed that deadly facultie to the Aſh tree, which doth belong to the Yew tree.

B The leaues of the wilde Aſh tree boiled in wine are good againſt the paine in the ſides, and the ſtopping of the liuer, and aſſwage the bellies of thoſe that haue the tympanie and dropſie.

C *Benedictus Curtius Symphoryanus* is deceiued in the hiſtorie of *Ornus*, when he thinketh out of *Virgils Georgicks*, that *Ornus* hath the floure of the Peare tree; for out of *Virgils* verſes no ſuch thing at all

all can be gathered: for he in intreateth not of the forms of trees, but of the graffing of diuers into others, vnlike and differing in nature; as of the graffing of the Nut tree into the Strawberry tree; the Apple into the Plane tree, the Beech into the Chestnut tree; the Peare into the wilde Ash or Quick-beame tree, the Oke into the Elme tree: and in this respect hee writeth, that the Plane tree bringeth forth an Apple, the Beech tree a Chestnut; the wilde Ash tree bringeth forth the white floure of the Peare tree, as is most manifest out of *Virgils* owne words, after this manner, in the second booke of his Georgicks:

Inseritur vero ex fœtu nucis Arbutus horrida,
Et steriles Platani malos gessere valentes,
Castaneæ Fagos: Ornus incanuit albo
Flore Pyri, glandémque sues fregere sub Vlmis.

The Tree Strawb'ry on Walnuts stocke doth grow,
And barren Planes faire Apples oft haue borne;
Chestnuts, Beech-Mast; the Quicken tree doth shew
The Peares white floure; and swine oft times th' Acorn
Haue gathered vnder Elmes. ——

Chap. III. *Of Coriars Sumuch.*

1 *Rhus Coriaria.*
Coriar Sumach.

2 *Rhus Myrtifolia.*
Wilde or Myrtle Sumach.

¶ *The Description.*

1 Coriars Sumach groweth vp vnto the height of a hedge tree, after the manner of the Elder tree, bigger than *Dioscorides* reporteth it to be, or others, who affirme that *Rhus* groweth two cubits high: whose errors are the greater: but this *Rhus* is so like to the Seruice

Seruice tree in ſhape and manner of growing, that it is hard to know one from the other; but that the leaues are ſoft and hairy, hauing a red ſinew or rib thorow the midſt of the leafe: the floures grow with the leaues vpon long ſtems cluſtering together like cats taile, or the catkins of the nut tree, but greater, and of a whitiſh green colour: after which come cluſters of round berries, growing in bunches like grapes.

2 *Pliny* his Sumach, or the Sumach of *Plinies* deſcription, groweth like a ſmall hedge tree, hauing many ſlender twiggie branches, garniſhed with little leaues like *Myrtus*, or rather like the leaues of the *Iuiube* tree; among which come forth ſlender moſſie floures, of no great account or value, which bring forth ſmall ſeeds, incloſed within a cornered caſe or huske, faſhioned like a ſpoon: the trunke or body of both theſe kindes of Sumach being wounded with ſome iron Inſtrument, yeeldeth a gum or liquour.

¶ *The Place.*

Sumach groweth, as *Dioſcorides* ſaith, in ſtony places: it is found in diuers mountaines & woods in Spaine, and in many places on the mount Apennine in Italy, and alſo neere vnto Pontus. *Archigenes* in *Galen*, in the 8. book of medicines according to the places affected, ſheweth, that it groweth in Syria, making choice of that of Syria.

¶ *The Time.*

The floures of Sumach come forth in Iuly, the ſeed with the berries are ripe in Autumne.

¶ *The Names.*

This is called in Greeke ῥοῦς: *Rhus*, ſaith *Pliny*, hath no Latine name; yet *Gaza* after the ſignification of the Greeke word, faineth a name, calling it *Fluida*: the Arabians name it *Sumach*: the Italians, *Sumacho*: the Spaniards, *Sumagre*: in low Dutch, by contracting of the word they cal it **Smack** or **Sumach**: in Engliſh, Sumach, Coriars Sumach, and Leather Sumach: the leaues of the ſhrub be called ῥοῦς βυρσοδεψικὴ: in Latine, *Rhus coriaria*, or *Rhoe*.

The ſeed is named *Eruthros*: and ῥοῦς ἐπὶ τὰ ὄψα: in Latine, *Rhus Culinaria*, and *Rhus obſoniorum*: in Engliſh, Meat Sumach, and Sauce Sumach.

¶ *The Temperature.*

The fruit, leaues, and ſeed hereof do very much binde, they alſo coole and drie: drie they are in the third degree, and cold in the ſecond, as *Galen* teacheth.

¶ *The Vertues.*

A The leaues of Sumach boyled in wine and drunken, do ſtop the laske, the inordinate courſe of womens ſickneſſes, and all other inordinate iſſues of bloud.

B The ſeed of Sumach eaten in ſauces with meat, ſtoppeth all manner of fluxes of the belly, the bloudy flix, and all other iſſues, eſpecially the white iſſues of women.

C The decoction of the leaues maketh haires blacke, and is put into ſtooles to fume vpward into the bodies of thoſe that haue the Dyſenterie, and is to be giuen them alſo to drinke.

D The leaues made into an ointment or plaiſter with hony and vineger, ſtaith the ſpreading nature of *Gangrænes* and *Pterygia*.

E The drie leaues ſodden in water vntill the decoction be as thicke as hony, yeeld forth a certaine oilineſſe, which performeth all the effects of *Licium*.

F The ſeed is no leſſe effectuall to be ſtrowed in pouder vpon their meats which are *Cœliaci* or *Dyſenterici*.

G The ſeedes pouned, mixed with honie and the powder of Oken coles, healeth the Hemorrhoides.

H There iſſueth out of the ſhrub a gum, which being put into the hollowneſſe of the teeth, taketh away the paine, as *Dioſcorides* writeth.

CHAP. 112. *Of red Sumach.*

¶ *The Deſcription.*

1 THeſe two figures are of one and the ſelfe ſame plant; the firſt ſheweth the ſhrub being in floure: the other when it is full floured with the fruit growne to ripeneſſe, notwithſtanding ſome haue deemed them to be of two kindes, wherein they were deceiued.

† This excellent and moſt beautifull plant *Coggygria* (beeing reputed of the Italians and the Venetians for a kind of *Rhus* or Sumach, becauſe it is vſed for the ſame purpoſes whereto *Rhus* ſerueth.

ueth and therein doth farre excell it) is an hedge plant growing not aboue the height of foure or fiue cubits,hauing tough and pliant ſtalks and twiggie branches like to Oziers, of a brown colour. The leaues be round,thick,and ſtiffe like the leaues of *Capparis*,in colour and ſauor of *Piſtacia* leaues, or *Terebinthus*;among which ariſeth a ſmall vpright ſprig,bearing many ſmal cluſtering little greeniſh yellow floures,vpon long and red ſtalks. After which follow ſmall reddiſh Lentill-like ſeeds that carry at the tops a moſt fine woolly or flockie tuft, criſped and curled like a curious wrought ſilken fleece,which curleth and foldeth it ſelfe abroad like a large buſh of haires.

1 *Coggygria Theophraſti.* & *Cotinus Coriarius Pliny.*
Venice Sumach. or Red Sumach.

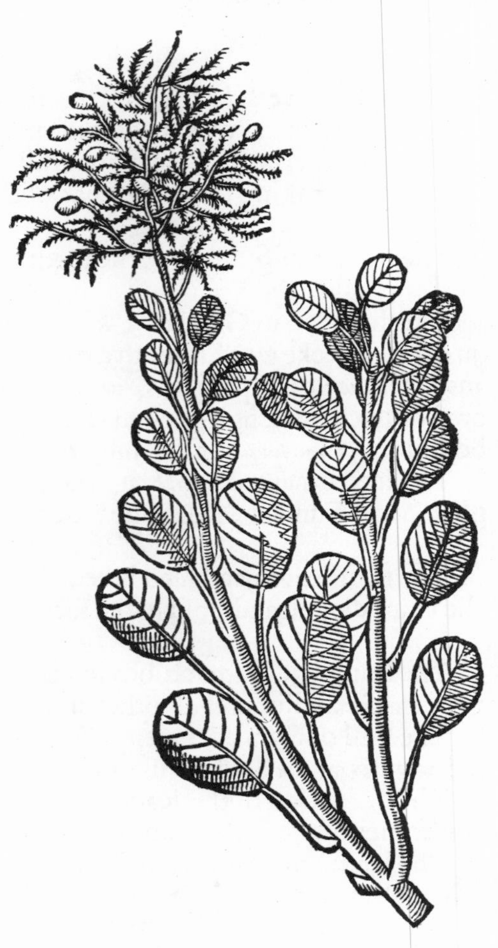

¶ *The Place.*

Coggygria groweth in Orleans neere Auignion, and in diuers places of Italy, vpon the Alpes of Styria,and many other places. It groweth on moſt of the hils of France, in the high woods of the vpper Pannonia or Auſtria,and alſo of Hungaria and Bohemia.

¶ *The Time.*

They floure and flouriſh for the moſt part in Iuly.

¶ *The Names.*

The firſt is called *Coggygria*, and *Coccygria* in Engliſh,Venice Sumach,or Silken Sumach;of *Pliny*,*Cotinus*,in his 16.booke,18.chapter. There is,ſaith he,on mount Apennine a ſhrub which is called *Cotinus ad lineamenta modo Conchylij colore inſignis*,and yet *Cotinus* is *Oleaſter*, or *Olea ſylueſtris*, the wilde Oliue tree,from which this ſhrub doth much differ;and therfore it may rightly be called *Cotinus Coriaria*. Diuers would haue named it *Scotinus*,which name is not found in any of the old writers. The Pannonians do call it *Farblauff*: it is alſo thought that this ſhrub is *Coggygria Plinij*, of which in his 13.booke,22.chapter he writeth in theſe words:*Coggygria* is alſo like to *Vnedo* in leafe, not ſo great;it hath a property to looſe the fruit with downe,which thing happeneth vnto no other tree.

¶ *The Temperature.*

The leaues and ſlender branches together with the ſeeds are very much binding, cold and drie as the other kindes of Sumach are.

¶ *The Vertues.*

A The leaues of *Coggygria*,or Silken Sumach,are ſold in the markets of Spaine and Italy for great ſummes

summes of money, vnto those that dresse Spanish skinnes, for which purpose they are verie excellent.

The root of *Cotinus*, as *Anguillara* noteth, serueth to die with, giuing to wooll and cloth a reddish colour, which *Pliny* knew, shewing that this shrub (that is to say the root) is *ad lineamenta modo Conchylij colore insignis*. B

CHAP. 113. *Of the Alder Tree.*

¶ *The Description.*

1 THe Alder tree or Aller, is a great high tree hauing many brittle branches, the barke is of a browne colour, the wood or timber is not hard, and yet it will last and indure verie long vnder the water, yea longer than any other timber whatsoeuer: wherefore in the fenny and soft marish grounds they do vse to make piles and posts thereof, for the strengthening of the walls and such like. This timber doth also serue very well to make troughes to conuey water in stead of pipes of Lead. The leaues of this tree are in shape somewhat like the Hasell, but they are blacker & more wrinckled, very clammie to handle, as though they were sprinckled with honie. The blossome or floures are like the aglets of the Birch tree: which being vaded, there followeth a scaly fruit closely growing together, as big as a Pigeons egge, which toward Autumne doth open, and the seed falleth out and is lost.

1 *Alnus.*
The Alder tree.

‡ 2 *Alnus hirsuta.*
Rough leaued Alder.

‡ 2 *Clusius* and *Bauhine* haue obserued another kinde of this which differs from the ordinary, in that it hath larger and more cut leaues, and these not shining aboue, but hoary vnderneath: the catkins, as also the rough heads are not so large as those of the former: the barke also is whiter. *Clusius* makes it his *Alnus altera*: and *Bauhine*, his *Alnus hirsutus*, or *folio incano*. ‡

The

¶ *The Place.*

The Aller or Alder tree delighteth to grow in low and moist waterish places.

¶ *The Time.*

The Aller bringeth forth new leaues in Aprill, the fruit whereof is ripe in September.

¶ *The Names.*

This tree is called in Greeke κλήθρα: in Latine, *Alnus*: *Petrus Crescentius* nameth it *Amedanus*: it is called in high Dutch, **Erlenbaum** and **Ellernbaum**: in low Dutch, **Elsen** and **Elsen-boom**: in Italian, *Alno*. in French, *Aulne*: in English, Alder and Aller.

¶ *The Temperature.*

The leaues and barke of the Alder tree are cold, drie, and astringent.

¶ *The Vertues.*

A The leaues of Alder are much vsed against hot swellings, vlcers, and all inward inflammations, especially of the Almonds and kernels of the throat.

B The barke is much vsed of poore countrey Diers, for the dying of course cloth, cappes, hose, and such like into a blacke colour, whereunto it serueth very well.

CHAP. 114. *Of the Birch tree.*

Betula.
The Birch tree.

¶ *The Description.*

THe common Birch tree waxeth likewise a great tree, hauing many boughes beset with many small rods or twigs, very limber and pliant: the barke of the young twigs and branches is plaine, smooth, and full of sap, in colour like the Chestnut; but the rind of the body or trunk is hard without, white, rough, and vneuen, full of chinkes or creuises: vnder which is found another fine barke, plaine, smooth, and as thin as paper which heretofore was vsed in stead of paper, to write vpon, before the making of paper was knowne: in Russia and these cold regions it serueth in stead of tiles and slate to couer their houses withall: this tree beareth for his floures certaine aglets like the Hasell tree, but smaller, wherein the seed is contained.

¶ *The Place.*

This common Birch tree grows in woods, fenny grounds, and mountaines, in most places of England.

¶ *The Time.*

The catkins or aglets do first appeare, and then the leaues, in Aprill or a little later.

¶ *The Names.*

Theophrastus calleth this tree in Greeke, σημύδα: diuers, σημυς: others σήμυος: it is named in Latine, *Betula*: diuers also write it with a double *ll Betulla*, as some of *Plinies* copies haue it: it is called in high Dutch, **Birkenbaum**: in low Dutch, **Berckenboom**: in Italian, *Betula*: by them of Trent, *Bedallo*: in French, *Bouleau*: in English, Birch tree.

¶ *The Temperature and Vertues.*

Concerning the medicinable vse of the Birch tree, or his parts, there is nothing extant either in the old or new writers.

This tree, saith *Pliny* in his 16-booke, 18. chapter, *Mirabili candore & tenuitate terribilis magistratuum virgis*: for in times past the Magistrates roddes were made heereof: and in our time also the Schoolemasters and parents do terrifie their children with rods made of Birch.

It serueth well to the decking vp of houses, and banquetting roomes, for places of pleasure, and beautifying of streets in the crosse or gang weeke, and such like.

CHAP.

CHAP. 115. *Of the Hornebeame, or Hard beame Tree.*

Betulus, siue Carpinus.
The Hornebeame tree.

¶ *The Description.*

BEtulus, or the Hornebeam tree grows great, and very like vnto the Elme, or Wich Hasel tree, hauing a great body: the wood or timber whereof is better for arrowes and shafts, pulleies for mills, and such like deuises, than Elme or Wich Hazell; for in time it waxeth so hard, that the toughnesse and hardnesse of it may be rather compared vnto horn than vnto wood, and therefore it was called Hornebeame, or Hardbeame: the leaues hereof are like the Elme, sauing that they be tenderer: among those hang certaine triangled things, vpon which be found knaps, or little heads of the bignesse of Ciches, in which is contained the fruit or seed: the root is strong and thicke.

¶ *The Place.*

Betulus or the Hornebeame tree growes plentifully in Northamptonshire, also in Kent by Grauesend, where it is commonly taken for a kinde of Elme.

¶ *The Time.*

This tree doth spring in Aprill, and the seed is ripe in September.

¶ *The Names.*

The Hornebeam tree is called in Greek ζυγία, which is as if you should say *Coniugalis*, or belonging to the yoke, because it serueth well to make ζυγία of, in Latine, *Iuga*, yokes wherewith oxen are yoked together, which are also euen at this time made thereof, as witnesseth *Benedictus Curtius Symphorianus*, and our selues haue sufficient knowledge thereof in our owne country; and therefore it may be Englished Yoke Elme. It is called of some, *Carpinus* and *Zugia*: it is also called *Betulus*, as if it were a kinde of Birch, but my selfe better like that it should be one of the Elmes: in high Dutch, Hayne: in French, *Carne*: in Italian, *Carpino*: in English, Hornebeame, Hardbeame, Yoke Elme, and in some places, Witch hasell.

¶ *The Temperature and Vertues.*

This tree is not vsed in medicine, the vertues are not expressed of the Antients, neither haue wee any certaine experiments of our owne knowledge more than hath beene said for the vse of Husbandrie. A

CHAP. 116. *Of the Elme tree.*

‡ OVr Author onely described two Elmes, and those not so accurately but that I thinke I shall giue the Reader content, in exchanging them for better receiued from M^r^. *Goodyer*; which are these.

Vlmus vulgatissima folio lato scabro. The common Elme.

1 THis Elme is a very great high tree, the barke of the young trees, and boughes of the Elder, which are vsually lopped or shred, is smooth and very tough, and wil strip or pil from the wood a great length without breaking: the bark of the body of the old trees as the trees grow in bignesse, teares or rents, which makes it very rough. The innermost wood of the tree is of reddish yellow

yellow or brownish colour, and curled, and after it is drie, very tough, hard to cleaue or rent, whereof aues of Carts are most commonly made: the wood next the barke, which is called the sap, is white. Before the leaues come forth the floures appeare, about the end of March, which grow on the twigs or branches, closely compacted or thrust together, and are like to the chiues growing in the middle of most floures, of a reddish colour: after which come flat seed, more long than broad, not much vnlike the garden Arach seed in forme and bignesse, and doe for the most part fall away before or shortly after the leaues spring forth, and some hang on a great part of the Sommer: the leaues grow on the twigges, of a darke greene colour; the middle size whereof are two inches broad, and three inches long, some are longer and broader, some narrower and shorter, rough or harsh in handling on both sides, nickt or indented about the edges, and many times crumpled, hauing a nerue in the middle, and many smaller nerues growing from him: the leafe on one side of the nerue is alwaies longer than on the other. On these leaues oftentimes grow blisters or small bladders, in which at the spring are little wormes, about the bignesse of Bed-fleas. This Elme is common in all parts of England, where I haue trauelled.

Vlmus minor folio angusto scabro.
The Narrow leaued Elme.

2 This tree is like the other, but much lesser and lower, the leaues are vsually about two inches and a halfe long, and an inch or an inch and a quarter broad, nickt or indented about the edges, and hath one side longer than the other, as the first hath, and are also harsh or rough on both sides, the

1 *Vlmus vulgatiss. folio lato scabro.*
The common Elme tree.

‡ 2 *Vlmus minor folio augusto scabro.*
The narrow leaued Elme.

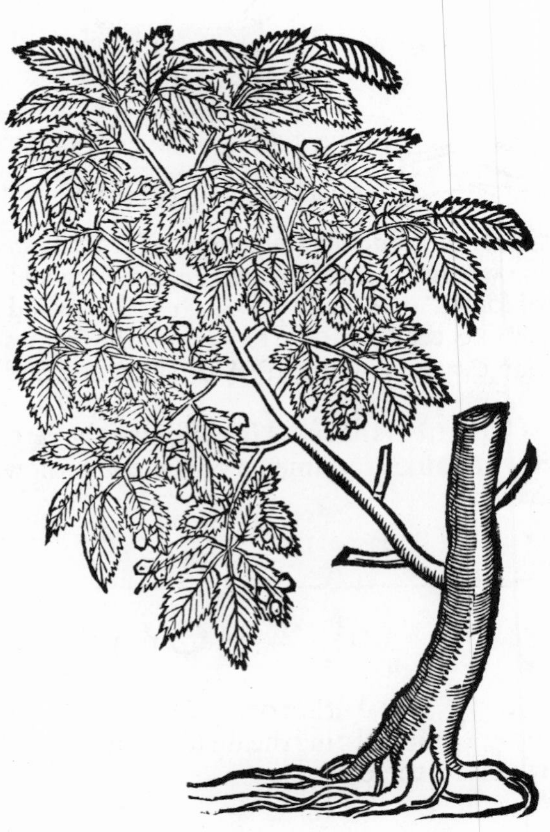

barke or rinde will also strip as the first doth: hitherto I haue not obserued either the floures or seed, or blisters on the leaues, nor haue I had any sight of the timber, or heard of any vse thereof. This kinde I haue seene growing but once, and that in the hedges by the high way as I rode betweene Christ Church and Limmington in the New Forrest in Hampshire, about

about the middle of September 1624. from whence I brought ſome ſmall plants of it, not a foot in length, which now, 1633. are riſen vp ten or twelue foot high, and grow with me by the firſt kinde, but are eaſily to be diſcerned apart, by any that will looke on both.

‡ 3 *Vlmus folio latiſsimo ſcabro.*
Witch Haſell, or the broadeſt leaued Elme.

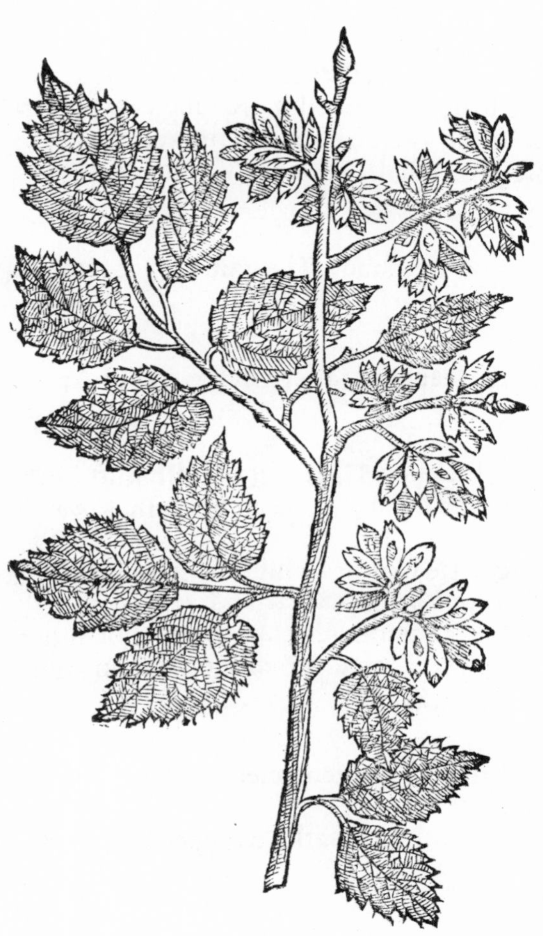

4 *Vlmus folio glabro.*
Witch Elme, or ſmooth leaued Elme.

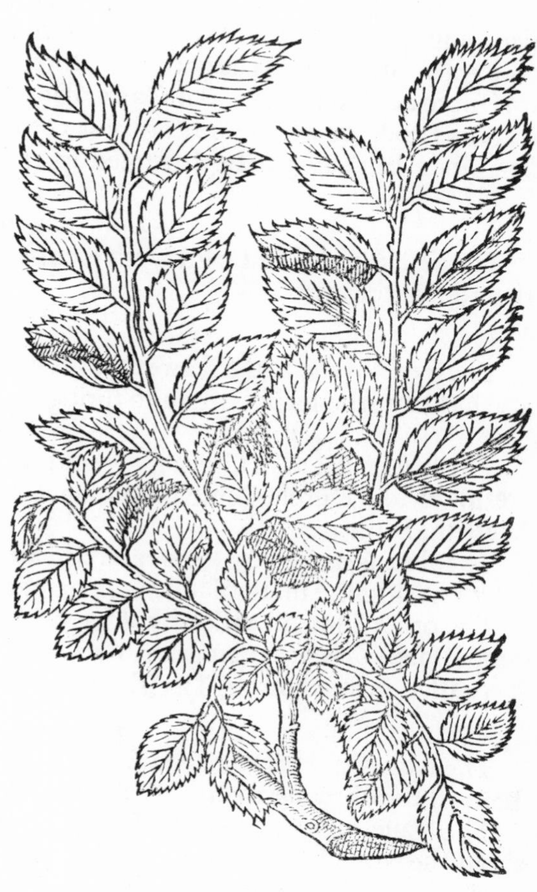

Vlmus folio latiſsimo ſcabro.
Witch Haſell, or the broadeſt leaued Elme.

3 This groweth to be a very great tree, and alſo very high, eſpecially when he groweth in woods amongſt other trees: the barke on the outſide is blacker than that of the firſt, and is alſo very tough, ſo that when there is plenty of ſap it will ſtrip or peele from the wood of the boughes from the one end to the other, a dozen foot in length or more, without breaking, whereof are often made cords or ropes: the timber hereof is in colour neere like the firſt; it is nothing ſo firme or ſtrong for naues of Carts as the fruit is, but will more eaſily cleaue; this timber is alſo coue-red with a white ſappe next the barke: the branches or young boughes are groſſer and bigger, and do ſpread themſelues broader, and hang more downewards than thoſe of the firſt; the floures are nothing but chiues, very like thoſe of the firſt kind: the ſeed is alſo like, but ſomething bigger: the leaues are much broader and longer than any of the kindes of Elme, vſually three or foure inches broad, and fiue or ſix inches long, alſo rough or harſh in handling on both ſides, ſnipt or indented about the edges, neere reſembling the leaues of the Haſell: the one ſide of the leaues are alſo moſt commonly longer than the other, alſo on the leaues of this Elme are ſometimes bliſters or bladders like thoſe on the firſt kinde. This proſpereth and naturally groweth in any ſoile moiſt or dry, on high hills, and in low vallies in good plenty in moſt places in Hampſhire, wher it is commonly called VVitch Haſell. Old men affirme, that when long boughes were in great vſe, there were very many made of the wood of this tree, for which purpoſe it is mentioned in the ſtatutes of England by the name of VVitch Haſell, as 8. *El.* 10. This hath little affinitie with *Carpinus*, which in Eſſex is called VVitch Haſell.

Vlmus folio glabro.
VVitch Elme, or smooth leauen Elme.

4 This kinde is in bignesse and height like the first, the boughes grow as those of the VVitch Hasell doe, that is hanged more downewards than those of the common Elme, the barke is blacker than that of the first kinde, it will also peele from the boughes: the floures are like the first, and so are the seeds: the leaues in forme are like those of the first kinde, but are smooth in handling on both sides. My worthy friend and excellent Herbarist of happy memorie Mr. *William Coys* of Stubbers in the parish of Northokington in Essex told me, that the wood of this kinde was more desired for naues of Carts than the wood of the first. I obserued it growing very plentifully as I rode between Rumford and the said Stubbers, in the yeere 1620. intermixed with the first kinde, but easily to be discerned apart, and is in those parts vsually called VVitch Elme. ‡

¶ *The Place.*

The first kinde of Elme groweth plentifully in all places of England. The rest are set forth in their descriptions.

¶ *The Time.*

The seeds of the Elme sheweth it selfe first, and before the leaues; it falleth in the end of Aprill, at what time the leaues begin to spring.

¶ *The Names.*

The first is called in Greeke, πτελέα: in Latine, *Vlmus*: in high Dutch, **Rust holtz, Rustbaum, Vlmbaum**: in low Dutch, **Olmen**: in French, *Orme*, and *Omeau*: in Italian, *Olmo*. in Spanish, *Vlmo*: in English, Elme tree.

The seed is named by *Plinie* and *Columella*, *Samera*. The little wormes which are found with the liquor within the small bladders be named in Greeke, κώνωπες: in Latine, *Culices*, and *Muliones*.

The other Elme is called by *Theophrastus*, ὀρειπτελέα; which *Gaza* translateth *Montiulmus* or mountaine Elme. *Columella* nameth it *Vernacula*, or *Nostras Vlmus*, that is to say, *Italica*, or Italian Elme: it is called in low Dutch, **Herseleer**, and in some places, **Heerenteer**.

¶ *The Temperature and Vertues.*

A The leaues and barke of the Elme be moderately hot, with an euident clensing facultie; they haue in the chewing a certaine clammie and glewing qualitie.

B The leaues of Elme glew and heale vp greene wounds, so doth the barke wrapped and swadled about the wound like a band.

C The leaues being stamped with vineger do take away scurffe.

D *Dioscorides* writeth, that one ounce weight of the thicker barke drunke with wine or water purgeth flegme.

E The decoction of Elme leaues, as also of the barke or root, healeth broken bones very speedily, if they be fomented or bathed therewith.

F The liquor that is found in the blisters doth beautifie the face, and scoureth away all spots, freckles, pimples, spreading tetters, and such like, being applied thereto.

G It healeth greene wounds, and cureth ruptures newly made, being laid on with Spleenwoort and the trusse closely set vnto it.

Chap. 117. *Of the Line or Linden Tree.*

¶ *The Description.*

1 The female Line or Linden tree waxeth very great and thicke, spreading forth his branches wide and farre abroad, being a tree which yeeldeth a most pleasant shadow, vnder and within whose boughes may be made braue sommer houses and banqueting arbors, because the more that it is surcharged with weight of timber and such like, the better it doth flourish. The barke is brownish, very smooth, and plaine on the outside, but that which is next to the timber is white, moist and tough, seruing very well for ropes, trases, and halters. The timber is whitish, plaine and without knots, yea very soft and gentle in the cutting or handling. Better gunpouder is made of the coles of this wood than of VVillow coles. The leaues are greene, smooth,

ſmooth, ſhining and large,ſomewhat ſnipt or toothed about the edges: the floures are little, whitiſh,of a good ſauour,and very many in number, growing cluſtering together from out of the middle of the leafe: out of which proceedeth a ſmall whitiſh long narrow leafe: after the floures ſucceed cornered ſharpe pointed Nuts,of the bigneſſe of Haſell Nuts. This tree ſeemeth to be a kinde of Elme,and the people of Eſſex about Heningham (wheras great plenty groweth by the way ſides) do call it broad leafed Elme.

1 *Tilia fæmina.*
The female Line tree.

2 *Tilia mas.*
The male Line tree.

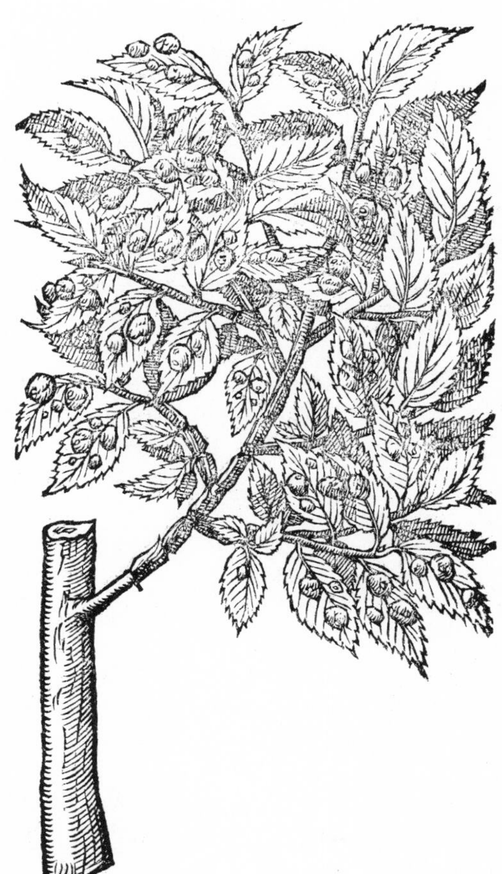

2 The male *Tilia* or Line tree groweth alſo very great and thicke,ſpreading it ſelfe far abroad like the other Linden tree: his bark is very tough and pliant,and ſerueth to make cords and halters of. The timber of this tree is much harder,more knottie,and more yellow than the timber of the other,not much differing from the timber of the Elme tree: the leaues hereof are not much vnlike Iuy leaues, not very greene,ſomewhat ſnipt about the edges: from the middle whereof come forth cluſters of little white floures like the former:which being vaded, there ſucceed ſmall round pellets,growing cluſtering together,like Iuy berries,within which is contained a little round blackiſh ſeed,which falleth out when the berry is ripe.

¶ *The Place.*

The female Linden tree groweth in ſome woods in Northampton ſhire, alſo neere Colcheſter, and in many places alongſt the high way leading from London to Henningham,in the countie of Eſſex.

The male Linden tree groweth in my Lord Treaſurers garden at the Strand,and in ſundry other places,as at Barn-elmes,and in a garden at Saint Katherines neere London. ‡ The female growes in the places here named,but I haue not yet obſerued the male. ‡

¶ *The Time*

Theſe trees floure in May,and their fruit is ripe in Auguſt.

¶ *The Names.*

The Linden tree is called in Greeke φιλύρα: in Latine,*Tilia:* in high Dutch,**Linden**,and **Lindenbaum**: in low Dutch,**Linde**,and **Lindenboom**: the Italians,*Tilia:* the Spaniards,*Teia:* in French, *Tilet* and *Tilieul:* in Engliſh, Linden tree,and Line tree.

 ¶ The

¶ *The Temperature.*

The barke and leaues of the Linden or Line tree, are of a temperate heat, somewhat drying and astringent.

¶ *The Vertues.*

A The leaues of *Tilia* boiled in Smithes water with a piece of Allom and a little honey, cure the sores in childrens mouthes.

B The leaues boiled vntill they be tender, and pouned very small with hogs grease, and the pouder of Fenugreeke and Lineseed, take away hot swellings and bring impostumes to maturation, being applied thereto very hot.

C The floures are commended by diuers against paine of the head proceeding of a cold cause, against dissinesse, the Apoplexie, and also the falling sicknesse, and not onely the floures, but the distilled water thereof.

D The leaues of the Linden (saith *Theophrastus*) are very sweet, and be a fodder for most kinde of cattle: the fruit can be eaten of none.

Chap. 118. *Of the Maple tree.*

‡ 1 *Acer maius.* The great Maple.

† 2 *Acer minus.* The lesser Maple.

¶ *The Description.*

THe great Maple is a beautifull and high tree, with a barke of a meane smoothnesse: the substance of the wood is tender and easie to worke on; it sendeth forth on euery side very many goodly boughes and branches, which make an excellent shadow against the heate of the Sun; vpon which

which are great, broad, and cornered leaues, much like to those of the Vine, hanging by long reddish stalks: the floures hang by clusters, of a whitish greene colour; after them commeth vp long fruit fastened together by couples, one right against another, with kernels bumping out neere to the place in which they are combined: in all the other parts flat and thin like vnto parchment, or resembling the innermost wings of grashoppers: the kernels be white and little.

2 There is a small Maple which doth oftentimes come to the bignes of a tree, but most commonly it groweth low after the maner of a shrub: the barke of the young shoots hereof is likewise smooth; the substance of the wood is white, and easie to be wrought on: the leaues are cornered like those of the former, slippery, and fastened with a reddish stalke, but much lesser, very like in bignes, and smoothnes to the leafe of Sanicle, but that the cuts are deeper: the floures be as those of the former, greene, yet not growing in clusters, but vpon spoked roundles: the fruit standeth by two and two vpon a stem or foot-stalke.

¶ *The Place.*

The small or hedge Maple groweth almost euery where in hedges and low woods.

The great Maple is a stranger in England, only it groweth in the walkes and places of pleasure of noble men, where it especially is planted for the shadow sake, and vnder the name of Sycomore tree.

¶ *The Time.*

These trees floure about the end of March, and their fruit is ripe in September.

¶ *The Names.*

This tree is called in Greeke σφένδαμνος: in Latine, *Acer*: in English, Maple, or Maple tree.

The great Maple is called in high Dutch, **Ahorne**, and **Waldescherne**: the French men, *Grand Erable*, and *Plasne* abusiuely, and this is thought to be properly called σφένδαμνος: but they are far deceiued that take this for *Platanus*, or the Plane tree, being drawne into this errour by the neerenesse of the French word; for the Plane tree doth much differ from this. ‡ This is now commonly (yet not rightly) called the Sycomore tree. And seeing vse will haue it so, I thinke it were not vnfit to call it the bastard Sycomore. ‡

The other is called in Latine, *Acer minor*: in high Dutch, **Masholder**: in low Dutch, **Booghout**: in French, *Erable*: in English, small Maple, and common Maple.

¶ *The Temperature and Vertues.*

A What vse the Maple hath in medicine we finde nothing written of the Grecians, but *Pliny* in his 14. booke, 8. chapter affirmeth, that the root pouned and applied, is a singular remedy for the paine of the liuer. *Serenus Sammonicus* writeth, that it is drunke with wine against the paines of the side.

Si latus immeritum morbo tentatur acuto,
Accensum tinges lapidem stridentibus vndis.
Hinc bibis: aut Aceris radicem tundir, & vna.
Cum vino capis: hoc præsens medicamen habetur.

Thy harmelesse side if sharpe disease inuade,
In hissing water quench a heated stone:
This drinke. Or Maple root in pouder made,
Take off in wine, a present med'cine knowne.

CHAP. 119. *Of the Poplar tree.*

¶ *The Kindes.*

THere be diuers trees vnder the title of Poplar, yet differing very notably, as shall be declared in the descriptions, whereof one is the white, another the blacke, and a third sort set downe by *Pliny*, which is the Aspe, named by him *Lybica*; and by *Theophrastus*, *Kerkis*: likewise there is another of America, or of the Indies, which is not to be found in these regions of Europe.

¶ *The Deſcription.*

1 THe white Poplar tree commeth ſoone to perfection,and groweth high in ſhort time,full of boughes at the top : the barke of the body is ſmooth,and that of the boughes is likewiſe white withall : the wood is white,eaſie to be cleft : the leaues are broad, deeply gaſhed,& cornered like almoſt to thoſe of the Vine,but much leſſer,ſmooth on the vpper ſide, glib,and ſomwhat greene ; and on the nether ſide white and woolly : the catkins are long, downy, at the firſt of a purpliſh colour : the roots ſpread many waies,lying vnder the turfe,and not growing deepe, and therefore it happeneth that theſe trees be oftentimes blowne downe with the winde.

1 *Populus alba.*
The white Poplar tree.

2 *Populus nigra.*
The blacke Poplar tree.

2 The black Poplar tree is as high as the white,and now and then higher, oftentimes fuller of boughes,and with a thicker body : the barke thereof is likewiſe ſmooth, but the ſubſtance of the wood is harder,yellower,and not ſo white, fuller of veines,and not ſo eaſily cleft:the leaues be ſomwhat long,and broad below towards the ſtem,ſharp at the point, and a little ſnipt about the edges, neither white nor woolly,like the leaues of the former, but of a pleaſant greene colour : amongſt which come forth long aglets or catkins,which do turne into cluſters : the buds which ſhew themſelues before the leaues ſpring out,are of a reaſonable good ſauour,of the which is made that profitable ointment called *Vnguentum Populeon.*

3 The third kinde of Poplar is alſo a great tree : the barke and ſubſtance of the wood is ſomewhat like that of the former : this tree is garniſhed with many brittle and tender branches, ſet full of leaues,in a manner round,much blacker and harder than the blacke Poplar, hanging vpon long and ſlender ſtems,which are for the moſt part ſtil wauering,and make a great noiſe by being beaten one to another,yea though the weather be calme,and ſcarce any winde blowing;and it is knowne by the name of the Aſpen tree : the roots hereof are ſtronger, and grow deeper into the ground than thoſe of the white Poplar.

4 This ſtrange Poplar,which ſome do call *Populus rotundifolia*,in Engliſh,the round leafed Poplar of India,waxeth a great tree,bedect with many goodly twiggie branches,tough and limmer like the

3 *Populus Libyca.*
The Aſpen tree.

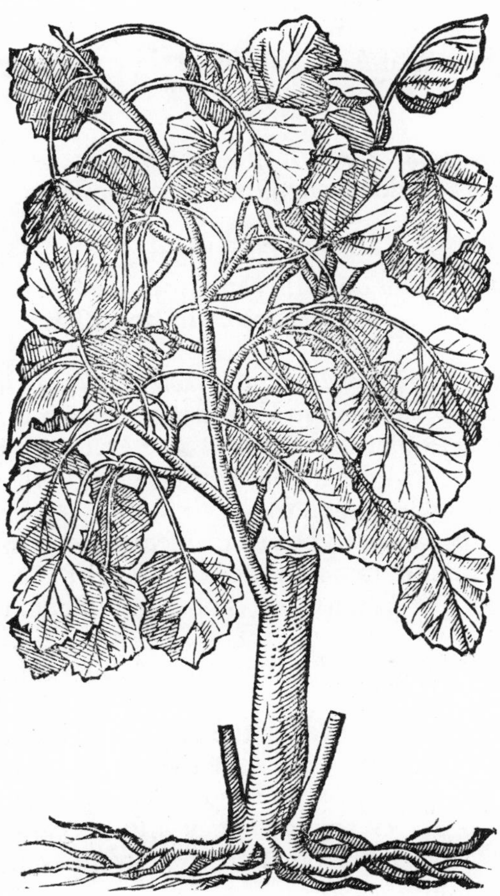

4 *Populus Americana.*
The Indian Poplar tree.

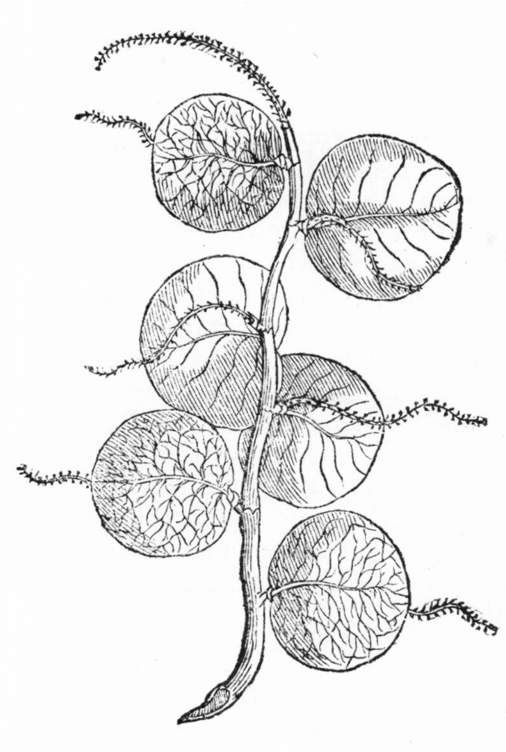

‡ 5 *Populus alba folijs minoribus.*
The leſſer leaued white Poplar.

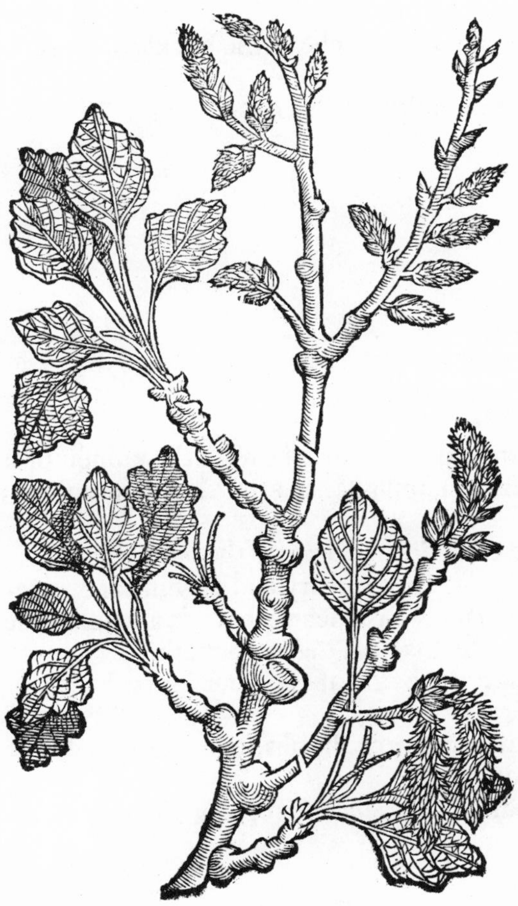

the Willow, full of ioints where the leaues do grow, of a perfect roundneſſe, ſaue where it cleaueth or groweth to the ſtalk: from the boſoms or corners of theſe leaues come forth ſmall aglets, like vnto our Poplar, but ſmaller: the leafe is thick, and very like the leaues of *Arbor Iudæ,* but broader, of an aſtringent taſte, ſomewhat heating the mouth, and ſaltiſh.

5 There is alſo another ſort of Poplar which groweth likewiſe vnto a great tree, the branches whereof are knotty and bunched forth as though it were full of ſcabs or ſores: the leaues come forth in tufts moſt commonly at the end of the boughes, not cut or iagged, but reſembling the leaues of that *Atriplex* called *Pes Anſerinus*; in colour like the former, but the aglets are not ſo cloſely packed together, otherwiſe it is like.

¶ *The Place.*

Theſe trees doe grow in low moiſt places, as in medowes neere vnto ditches, ſtanding waters and riuers.

The firſt kinde of white Poplar groweth not very common in England, but in ſome places here and there a tree: I found many both ſmall & great growing in a low medow turning

turning vp a lane at the farther end of a village called Black-wall, from London; and in Eſſex at a place called Ouenden, and in diuers other places.

The Indian Poplar groweth in moſt parts of the Iſlands of the Weſt Indies.

¶ *The Time.*

Theſe trees do bud forth in the end of March and beginning of Aprill, at which time the buds muſt be gathered to ſerue for *Vnguentum Populeon.*

¶ *The Names.*

The white Poplar is called in Greeke, Λευκή: in Latine, *Populus alba*: of diuers, *Farfarus*, as of *Plautus* in his Comedie *Penulus*, as you may ſee by his words ſet downe in the chap. of Colts-foot, *pag.* 813.

It is called in high-Dutch, **Poppelbaum, Weiſz Alberbaum**: in low-Dutch, **Abeel**, of his horie or aged colour, and alſo **Abeelboome**; which the Grammarians doe falſly interpret *Abies*, the Firre tree: in Italian, *Popolo nero*: in French, *Peuplier blanc*, *Aubel*, *Obel*, or *Aubeau*: in Engliſh, white Poplar tree, and Abeell, after the Dutch name.

The ſecond is called in Greeke, Αἴγειρος: in Latine, *Populus nigra*: by *Petrus Creſcentius*, *Albarus*: in high-Dutch, **Aſpen**: in low-Dutch, **Populier**: in Italian, *Popolo nero*: in French, *Peuplier noir*: in Spaniſh, *Alamo nigailho*: in Engliſh, Poplar tree, blacke Poplar, and Pepler. The firſt or new ſprung buds whereof are called of the Apothecaries, *Oculi Populi*, Poplar buds: others chuſe rather to call it *Gemma Populi*: ſome of the Grecians name it Σπίρμα: whereupon they grounded their error, who raſhly ſuppoſed that thoſe roſenny or clammy buds are not to be put or vſed in the compoſition of the ointment bearing the name of the Poplar, and commonly called in Engliſh, Popilion and Pompillion, but the berries that grow in cluſters, in which there is no clammineſſe at all.

They are alſo as far deceiued, who giuing credit to Poets fables, do beleeue that Amber commeth of the clammy roſin falling into the riuer Poo.

The third is called of diuers, *Populus tremula*, which word is borrowed of the French men, who name it *Tremble*: it alſo receiued a name amongſt the low Countrey men, from the noiſe and ratling of the leaues, *viz.* **Rateeler**: this is that which is named of *Pliny*, *Libyca*: and by *Theophr.* Κερκίς, which *Gaza* calleth *Populus montana* in Engliſh, Aſpe, and Aſpen tree, and may alſo be called Tremble, after the French name, conſidering it is the matter whereof womens tongues were made, (as the Poets and ſome others report) which ſeldom ceaſe wagging.

¶ *The Temperature and Vertues.*

A The white Poplar hath a clenſing facultie, ſaith *Galen*, and a mixt temperature, conſiſting of a waterie warme eſſence, and alſo a thin earthy ſubſtance.

B The barke, as *Dioſcorides* writeth, to the weight of an ounce (or as others ſay, and that more truly, of little more than a dram) is a good remedie for the Sciatica or ache in the huckle bones, and for the ſtrangurie.

C That this barke is good for the Sciatica, *Serenus Sammonicus* doth alſo write:

Sæpius occultus victa coxendice morbus
Perfurit, & greſſus diro languore moratur:
Populus alba dabit medicos de cortice potus.

An hidden diſeaſe doth oft rage and raine,
The hip ouercome and vex with the paine,
It makes with vile aking one tread ſlow and ſhrinke;
The barke of white Poplar is helpe had in drunke.

D The ſame barke is alſo reported to make a woman barren, if it be drunke with the kidney of a Mule, which thing the leaues likewiſe are thought to perform, being taken after the floures or reds be ended.

E The warme iuice of the leaues being dropped into the eares doth take away the paine thereof.

F The roſin or clammy ſubſtance of the blacke Poplar buds is hot and dry, and of thin parts, attenuating and mollifying: it is alſo fitly mixed *acopis & malagmatis*: the leaues haue in a manner the like operation for all theſe things, yet weaker, and not ſo effectuall, as *Galen* teacheth.

G The leaues and yong buds of blacke Poplar doe aſſwage the paine of the gout in the hands or feet, being made into an ointment with May butter.

H The ointment made of the buds is good againſt all inflammations, bruſes, ſquats, falls, and ſuch like: this ointment is very well knowne to the Apothecaries.

I *Paulus Ægineta* teacheth to make an oile alſo hereof, called *Ægyrinum*, or oile of blacke Poplar.

CHAP.

CHAP. 120. *Of the Plane tree.*

Platanus.
The Plane tree.

¶ *The Deſcription.*

THe Plane is a great tree, hauing very long and farre ſpreading boughes caſting a wonderfull broad ſhadow, by reaſon whereof it was highly commended and eſteemed of among the old Romans: the leaues are cornered like thoſe of *Palma Chriſti*, greater than Vine leaues, and hanging vpon little red foot-ſtalkes: the floures are ſmall and moſſie, and of a pale yellowiſh colour: the fruit is round like a ball, rugged, and ſomewhat hairy; but in Aſia more hairy and greater, almoſt as big as a Walnut: the root is great, diſperſing it ſelfe far abroad.

¶ *The Place.*

The Plane tree delighteth to grow by ſprings or riuers: *Pliny* reports that they were wont to be cheriſhed with wine: they grew afterward (ſaith he) to be of ſo great honour (meaning the Plane trees) as that they were cheriſhed and watered with wine: and it is found by experience that the ſame is very comfortable to the roots, and wee haue alreadie taught, that trees deſire to drinke wine. This tree is ſtrange in Italy, it is no where ſeene in Germany, nor in the low-Countries: in Aſia it groweth plentifully: it is found alſo in Candy, growing in vallies, and neere vnto the hill Athos, as *Petrus Bellonius* in his Singularities doth declare: it groweth in many places of Greece, and is found planted in ſome places of Italy, for pleaſure rather than for profit. My ſeruant *William Marſhall* (whom I ſent into the Mediterranean ſea as Surgeon vnto the Hercules of London) found diuers trees hereof growing in Lepanto, hard by the ſea ſide, at the entrance into the towne, a port of Morea, being a part of Greece, and from thence brought one of thoſe rough button , being the fruit thereof. ‡ There are one or two yong ones at this time growing with M^r^. *Tradeſcant.* ‡

The Plane trees caſt their leaues in Winter, as *Bellonius* teſtifieth, and therefore it is no maruel that they keepe away the Sun in Sommer, and not at all in Winter: there is, ſaith *Pliny*, no greater commendation of the tree, than that it keepeth away the Sunne in Sommer, and entertaineth it in Winter.

¶ *The Names.*

This tree is called in Greeke, πλάτανος: and likewiſe in Latine *Platanus*: it beareth his name of the bredth: the French mens *Plaſne* doth far differ from this, which is a kind of Maple: this tree is named in Engliſh, Plane tree.

¶ *The Temperature and Vertues.*

The Plane tree is of a cold and moiſt eſſence, as *Galen* ſaith: the greene leaues are good to be laid vpon hot ſwellings and inflammations in the beginning. A

Being boiled in wine they are a remedie for the running and the watering of the eyes, if they be applied. B

The barke and balls do dry: the barke boiled in vineger helpeth the tooth-ache,

The fruit of the Plane tree drunke with wine helpeth the bitings of mad dogs and ſerpents, and mixed with hogs greaſe it maketh a good ointment againſt burning and ſcalding. C

The burned barke doth mightily dry, and ſcoureth withall; it remoueth the white ſcurfe, and cureth moiſt vlcers. D

The

F The dust or downe, saith *Galen*, that lieth on the leaues of the tree is to be taken heed of, for if it be drawne in with the breath, it is offensiue to the winde-pipe by his extreme drinesse, and making the same rough, and hurting the voice, as it doth also the sight and hearing, if it fall into the eyes or eares. *Dioscorides* doth not attribute this to the dust or downe of the leaues onely, but also to that of the balls.

Chap. 121. *Of the Wayfaring Tree.*

Lantana, siue Viburnum.
The Wayfaringtree.

¶ *The Description.*

The Wayfaring mans tree growes vp to the height of an hedge tree, of a mean bignes: the trunke or body thereof is couered with a russet barke: the branches are long, tough, and easie to be bowed, and hard to be broken, as are those of the Willow, couered with a soft whitish barke, whereon are broad leaues thicke and rough, sleightly indented about the edges, of a white colour, and somewhat hairy whilest they be fresh and green; but when they begin to wither and fall away, they are reddish, and set together by couples one opposit to another. The floures are white, and grow in clusters: after which come clusters of fruit of the bignesse of a pease, somewhat flat on both sides, at the first greene, after red, and blacke when they be ripe: the root disperseth it selfe far abroad vnder the vpper crust of the earth.

¶ *The Place.*

This tree groweth in most hedges in rough and stony places, vpon hils and low woods, especially in the chalky grounds of Kent about Cobham, Southfleet, and Grauesend, and in all the tract to Canturbury.

¶ *The Time.*

The floures appeare in Sommer: the berries are ripe in the end of Autumne, and new leaues come forth in the Spring.

This hedge tree is called *Viurna* of *Ruellius*: in French, *Viorne*, and *Viorna*: in Italian, *Lantana*: it is reputed for the tree *Viburnum*, of which *Virgil* maketh mention in the first Eclog, where hee commendeth the city Rome for the loftinesse and statelinesse thereof, aboue other Cities, saying, that as the tall Cypres trees do shew themselues aboue the low and shrubby Viorn, so doth Rome aboue other cities lift vp her head very high; in these verses:

Verum hæc tantùm alias inter caput extulit vrbes,
Quantum lenta solent inter viburna cupressi.

But this all other cities so excels,
As Cypresse, which 'mongst bending Viornes dwels.

‡ I iudge *Viburnum* not to be a name to any particular plant, but a generall name to all low and bending shrubs; amongst which this here described may take place as one. I enquired of a countrey man in Essex, if he knew any name of this: he answered, it was called the Cotton tree, by reason of the softnesse of the leaues. ‡

¶ *The Temperature.*

The leaues and berries of Lantana are cold and dry, and of a binding qualitie.

¶ *The*

¶ *The Vertues.*

The decoction of the leaues of Lantana is very good to be gargled in the mouth against al swellings and inflammations thereof, against the scuruie and other diseases of the gums, and fastneth loose teeth. A

The same boiled in lee doth make the haires blacke if they be bathed or washed therewith, and suffered to dry of it selfe. B

The berries are of the like facultie, the pouder whereof when they be dried stay the laske, all issues of bloud, and also the whites. C

It is reported, that the barke of the root of the tree buried a certaine time in the earth, and afterwards boiled and stamped according to art, maketh good Bird-lime for Fowlers to catch Birds with. D

CHAP. 122. *Of the Beade tree.*

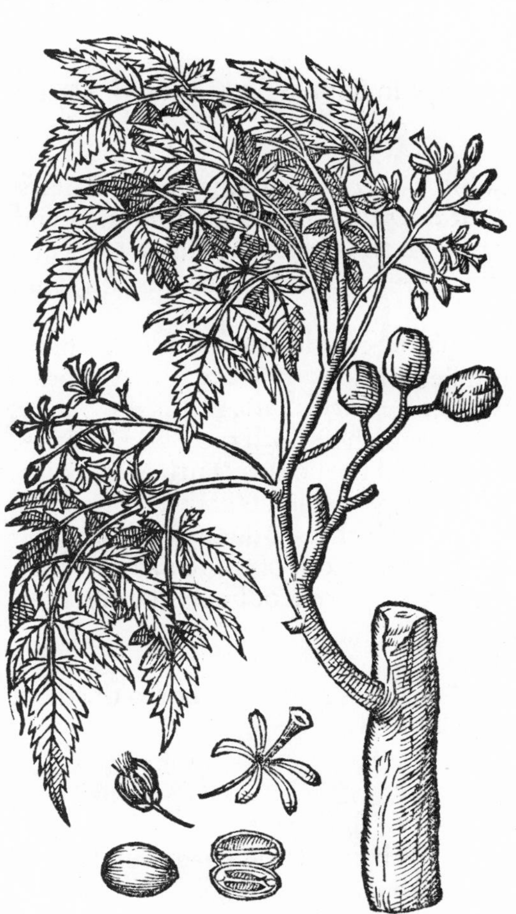

1 *Zizypha candida.*
The Beade tree.

‡ 2 *Zizypha Cappadocica.*
The Beade tree of Cappadocia.

¶ *The Description.*

1 THis tree was called *Zizypha candida* by the Herbarists of Montpellier; and by the Venetians and Italians, *Sycomorus*, but vntruly: the Portugals haue termed it *Arbor Paradiso*: all which and each whereof haue erred together, both in respect of the fruit and of the whole tree: some haue called it *Zizypha*, though in facultie it is nothing like; for the taste of this fruit is very vnpleasant, virulent, and bitter. But deciding all controuersies, this is the tree which *Auicen* calleth *Azederach*, which is very great, charged with many large armes, that are garnished with twiggie branches, set full of great leaues consisting of sundry small leaues, one growing right opposite to another like the leaues of the Ash tree or Wicken tree, but more deepely cut about the edges like the teeth of a saw: among which come the floures, consisting of fiue small blew leaues layd abroad in manner of a starre: from the middest whereof groweth forth a small hollow cup resembling

resembling a Chalice: after which succeedeth the fruit, couered with a brownish yellow shel, very like vnto the fruit of Iuiubes (whereof *Dodonæus* in his last edition maketh it a kinde) of a rancke, bitter, and vnpleasant taste, with a six cornered stone within, which being drawne on a string, serueth to make Beades of, for want of other things.

2 *Zizyphus Cappadocica* groweth not so great as the former, but is of a meane stature, and full of boughes: the barke is smooth and euen, and that which groweth vpon the trunke and great boughes is of a shining scarlet colour: out of these great armes or boughes grow slender twigges, white and soft, which are set full of whitish leaues, but more white on the contrarie or backe part, and are like to the leaues of Willow, but narrower and whiter: amongst these leaues come forth small hollow yellowish floures, growing at the ioints of the branches, most commonly three together, and of a pleasant sauour, with some few threds or chiues in the middle thereof. After which succeedeth the fruit, of the bignesse and fashion of the smallest Oliue, white both within and without, wherein is contained a small stone which yeeldeth a kernell of a pleasant taste, and very sweet.

¶ *The Place.*

Matthiolus writeth, that *Zizyphus candida* is found in the cloisters of many monasteries in Italy; *Lobel* saith that it groweth in many places in Venice and Narbon; and it is wont now of late to be planted and cherished in the goodliest orchards of all the low-Countries.

Zizyphus Cappadocica groweth likewise in many places of Italy, and specially in Spaine: it is also cherished in gardens both in Germany and in the low-Countries. ‡ It groweth also here in the garden of Mr. *Iohn Parkinson*. ‡

¶ *The Time.*

These trees floure in Iune in Italy and Spain; their fruit is ripe in September; but in Germany and the low-Countries there doth no fruit follow the floures.

¶ *The Names.*

Zizyphus candida Auicen calleth *Azederach*, or as diuers read it, *Azederaeth*: and they name it, saith he, in Rechi, *Arbor Mirobalanorum*, or the Mirobalane tree, but not properly, and in Tabrasten, and Kien, and Thihich. The later writers are far deceiued in taking it to be the Sycomore tree; and they as much, that would haue it to be the Lote or Nettle tree: it may be named in English, Bead tree, for the cause before alledged.

The other is *altera species Zizyphi*, or the second kinde of Iuiube tree, which *Columella* in his ninth booke and fourth chap. doth call *Zizyphus alba*, or white Iuiube tree, for difference from the other that is syrnamed *Rutila*, or glittering red. *Pliny* calleth this *Zizyphus Cappadocica*, in his 21 booke, ninth chapter, where he entreateth of the honour of Garlands, of which he saith there be two sorts, whereof some be made of floures, and others of leaues: I would call the floures (saith he) brooms, for of those is gathered a yellow floure, and *Rhododendron*, also *Zizypha*, which is called *Cappadocica*. The floures of these are sweet of smell, and like to Oliue floures. Neither doth *Columella* or *Pliny* vnaduisedly take this for *Zizyphus*, for both the leaues and floures grow out of the tender and yong sprung twigs, as they likewise do out of the former: the floures are very sweet of smel, and cast their sauor far abroad: the fruit also is like that of the former.

¶ *The Temperature.*

Auicen writing and intreating of *Azadaraeth*, saith, that the floures thereof be hot in the third degree, and dry in the end of the first.

Zizyphus Cappadocica is cold and dry of complexion.

¶ *The Vertues.*

A The floures of *Zizyphus*, or *Azadaraeth* open the obstructions of the braine.

B The distilled water thereof killeth nits and lice, preserueth the haire of the head from falling, especially being mixed with white wine, and the head bathed with it.

C The fruit is very hurtfull to the chest, and a troublesome enemie to the stomacke; it is dangerous, and peraduenture deadly.

D Moreouer, it is reported, that the decoction of the barke and of Fumitorie, with Mirobalans added, is good for agues proceeding of flegme.

E The iuice of the vppermost leaues with honey is a remedie against poison.

F The like also hath *Rhasis*: the Beade tree, saith he, is hot and dry: it is good for stoppings of the head, it maketh the haire long; yet is the fruit thereof very offensiue to the stomacke, and oftentimes found to be pernitious and deadly.

G *Matthiolus* writeth, that the leaues and wood bringeth death euen vnto beasts, and that the poyson thereof is resisted by the same remedies that *Oleander* is.

H *Zizyphus Cappadocica* preuaileth against the diseases aforesaid, but the decoction thereof is verie good for those whose water scaldeth them with the continuall issuing thereof, as also for such as haue the running of the reines and the exulcerations of the bladder and priuy parts.

A looch

A looch or licking medicine made thereof or the syrrup, is excellent good against spitting of bloud proceeding of the distillations of sharpe or salt humors. G

† The figure that formerly was in the second place,was of the narrow leaued kinde of *Guajacum Patavinum*, which you shall finde in the second place of the next chapter saue one.

CHAP. 123. *Of the Lote,or Nettle tree.*

Lotus arbor.
The Nettle tree.

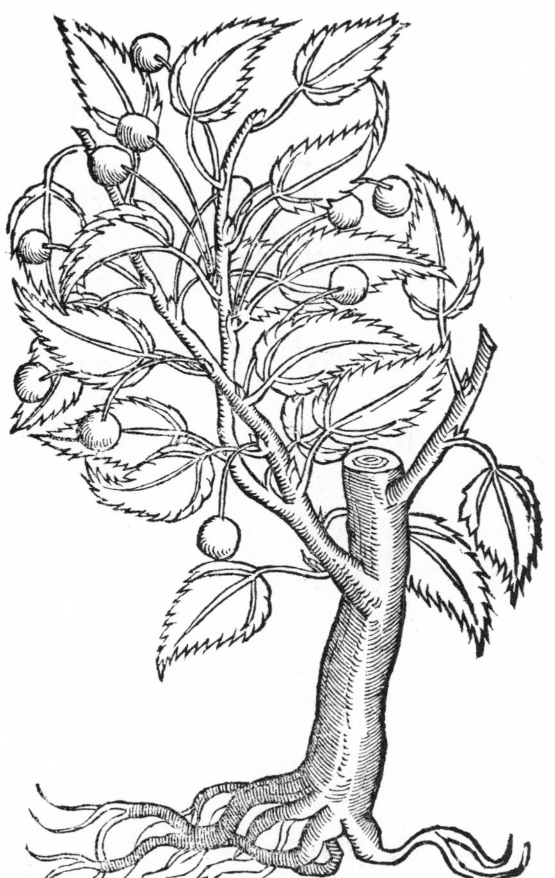

¶ *The Description.*

THe Lote whereof we write is a tree as big as a Peare tree,or bigger and higher: the body and armes are very thicke;the barke whereof is smooth,of a gallant green colour tending to blewnesse: the boughes are long, and spread themselues all about: the leaues be like those of the Nettle, sharpe pointed, and nicked in the edges like a saw, and dasht here and there with stripes of a yellowish white colour: the berries be round,and hang vpon long stalkes like Cherries, of a yellowish white colour at the first, and afterwards red, but when they be ripe they be somewhat blacke.

¶ *The Place.*

This is a rare and strange tree in both the Germanies: it was brought out of Italy, where there is found store thereof, as *Matthiolus* testifieth: I haue a small tree thereof in my garden. There is likewise a tree thereof in the garden vnder London wall, sometime belonging to Mr. *Gray*, an Apothecary of London; and another great tree in a garden neere Coleman street in London, being the garden of the Queenes Apothecarie at the impression hereof, called Mr. *Hugh Morgan*, a curious conseruer of rare simples. The Lote tree doth also grow in Africke, but it somewhat differeth from the Italian Lote in fruit, as *Pliny* in plaine words doth shew in his thirteenth booke, seuenteenth chapter. That part of Africke, saith he, that lieth towards vs, bringeth forth the famous Lote tree,which they call *Celtis*, and the same well knowne in Italy, but altered by the soile: it is as big as the Peare tree, although *Nepos Cornelius* reporteth it to be shorter: the leaues are full of fine cuts, otherwise they be thought to be like those of the Holme tree. There be many differences,but the same are made especially by the fruit: the fruit is as big as a Beane, and of the colour of Saffron,but before it is thorow ripe, it changeth his color as doth the Grape. It growes thicke among the boughes,after the manner of the Myrtle,not as in Italy,after the manner of the Cherry; the fruit of it is there so sweet, as it hath also giuen a name to that countrie and land, too hospitable to strangers, and forgetfull of their owne countrey.

It is reported that they are troubled with no diseases of the belly that eate it. The better is that which hath no kernell, which in the other kinde is stony: there is also pressed out of it a wine, like to a sweet wine; which the same *Nepos* denieth to endure aboue ten daies, and the berries stamped with *Alica* are reserued in vessels for food. Moreouer we haue heard say, that armies haue been fed therewith, as they haue passed too and fro thorow Africke. The colour of the wood is blacke: they vse to make flutes and pipes of it: the root serueth for kniues hafts, and other short workes: this is there the nature of the tree: thus farre *Pliny*. In the same place he saith, that this renowmed tree doth grow about Syrtes and Nasamonæ: and in his 5. booke, 7. chapter he sheweth that there is not far from the lesser Syrtis, the Island Menynx, surnamed *Lotophagitis*, of the plenty of Lote trees.

Strabo in his 17. booke affirmeth, that not onely *Menynx*, but also the lesser *Syrtis* is said to be *Lotophagitis*: first, saith he, lieth Syrtis a certaine long Island by the name Cercinna, and another lesser, called Circinnitis; next to this is the lesser Syrtis, which they call Lotophagitis Syrtis: the compasse of this gulfe is almost 1600. furlongs; the bredth of the mouth 600. By both the capes there be Islands ioined to the maine land, that is, Circinna and Menynx, of like bignesse: they thinke that Menynx is the countrey of the Lotophagi, or those that feed of the Lote trees, of which countrie *Homer* maketh mention, and there are certaine monuments to be seen, and *Vlysses* Altar, and the fruit is selfe, for there be in it great plenty of Lote trees, whose fruit is wonderful sweet: thus saith *Strabo*.

This Lote is also described by *Theophrastus*; in his fourth booke he saith, that there be very many kindes, which be seuered by the fruit: the fruit is of the bignes of a beane, which when it waxeth ripe doth alter his colour as grapes do: the fruit of which the Lotophagi do eate is sweet, pleasant, harmeles, and wholesome for the belly, but that is pleasanter which is without kernels, and of this they make their wine.

This Lote tree, as the same Author affirmeth, is by nature euerlasting: as for example, the Lote trees whereof *Pliny* hath written in his 16. booke, 44. chapter. At Rome, saith he, the Lote tree in *Lucinas* court, how much elder it was than the church of the citie, built in the yeere which was without magistrates, 469. it is vncertaine: there is no doubt but that it was elder, because *Lucina* bare the name of that *Lucus* or groue. This is now about 450. yeeres old. That is elder which is surnamed *Capillata*, or hairie; because the haire of the vestall virgins was brought vnto it: but the other Lote tree in *Vulcans* church, which *Romulus* built by the victory of tenths is taken to be as old as the citie, as *Massurius* witnesseth.

¶ *The Time.*

They lose their leaues at the first approch of winter; and recouer them againe in Aprill: the fruit is ripe in September.

¶ *The Names.*

This tree is called in Greeke, λωτός: in Latine by *Pliny*, *Celtis*: in Italian, *Perlaro*. by those of Trent, *Bagolaro*: and in English, Lote tree, and Nettle tree.

¶ *The Temperature and Vertues.*

A The Lote tree is not greatly binding as *Galen* saith, but of thin parts, and of a drying nature.

B The decoction of the wood beaten small, being either drunke or vsed clisterwise, is a remedy for the bloudy flix; and for the whites and reds.

C It stoppeth the laske, and maketh the haire yellow, and as *Galen* addeth, keepeth haires from falling.

D The shiuers or small pieces thereof, as the same Author alleageth, are boiled sometimes in water, sometimes in wine, as need shall require.

Chap. 124.

Of Italian wood of Life, or Pocke wood, vulgarly called Lignum vitæ.

¶ *The Description.*

1 Italian *Lignum vitæ*, or Wood of Life, groweth to a faire and beautifull tree, hauing a straight and vpright body, couered ouer with a smooth and darke greene barke, yeelding forth many twiggy branches, set forth of goodly leaues, like those of the Peare tree, but of greater beautie, and somewhat broader: among which commeth forth the fruit, growing close to the branches, almost without stalkes: this fruit is round, and at the first greene, but blacke when it is ripe, as big as Cherries, of an excellent sweet taste when it is dried: but this is not the Indian *Lignum sanctum*, or *Guaiacum*, whereof our bowles and physicall drinkes be made, but it is a bastard kind therof, first planted in the common garden at Padua, by the learned *Fallopius*, who supposed it to be the right *Guaiacum*.

‡ 2 The leaues of this are longer and narrower than the former, but firme also and neruous like as they are; the fruit is in shape like Sebestens, but much lesse, of a blewish colour when it is ripe, with many little stones within; the taste hereof is not vnpleasant. *Matthiolus* calls this *Pseudolotus*; and *Tabernamontanus*, *Lotus Africana*: whose figure our Author in the last chapter saue one gaue vnfitly for the *Zizyphus Cappadocica*. ‡

¶ *The Place.*

Guaiacum Patauinum groweth plentifully about Lugdunum, or Lions in France: I planted it in the

the garden of Barne-Elmes neere London two trees: beſides, there groweth another in the garden of M[r]. *Gray* an Apothecarie of London, and in my garden likewiſe.

1 *Guaiacum Patauinum latifolium.*
Broad leafed Italian Wood of life.

2 *Guaiacum Patauinum anguſtifol.*
Narrow leafed Italian Guaiacum.

¶ *The Time.*

It floureth in May, and the fruit is ripe in September.

¶ *The Names.*

Guaiacum Patauinum hath been reputed for the *Lotus* of *Theophraſtus*: in Engliſh it is called the baſtard Meuynwood.

‡ This hath no affinitie with the true Indian Guajacum which is frequently vſed in medicine. ‡

¶ *The Temperature and Vertues.*

‡ The fruit of this is thought to be of the ſame temper and qualitie with that of the Nettle-tree. ‡

Chap. 125. *Of the Strawberry tree.*

¶ *The Deſcription.*

THe Strawberry tree groweth for the moſt part low, very like in bigneſſe to the Quince tree (whereunto *Dioſcorides* compareth it.) The body is couered with a reddiſh barke, both rough and ſcaly: the boughes ſtand thicke on the top, ſomewhat reddiſh: the leaues bee broad, long, and ſmooth, like thoſe of Bayes, ſomewhat nicked in the edges, and of a pale greene colour: the floures grow in cluſters, being hollow and white, and now and then on the one ſide ſomewhat of a purple colour: in their places come forth certaine berries hanging downe vpon little long ſtems like vnto Strawberries, but greater, without a ſtone within, but onely with little ſeeds, at the firſt greene, and when they be ripe they are of a gallant red colour, in taſte ſomewhat harſh, and in a manner without any reliſh; of which Thruſhes and Black-birds do feed in Winter.

Arbutus.
The Strawberry tree.

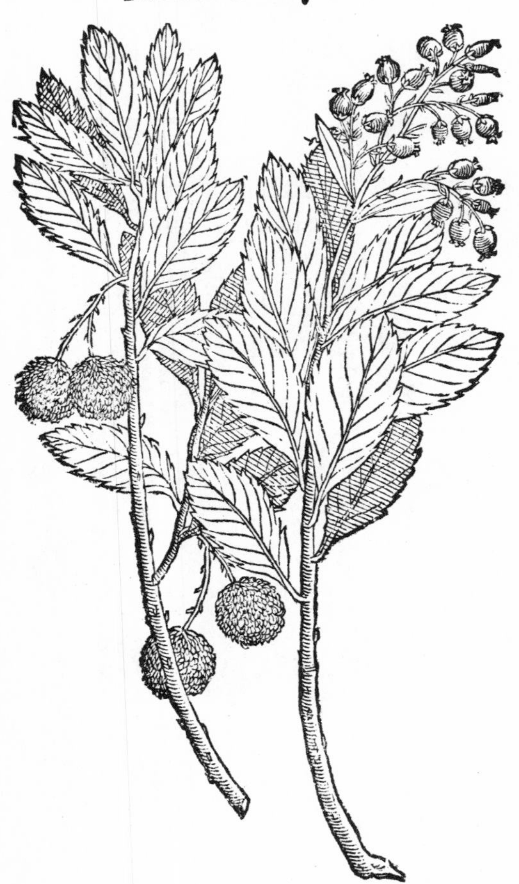

¶ *The Place.*

The Strawberry tree groweth in most Countries of Greece, in Candy, Italy, and Spaine, also in the vallies of the mountaine Athos, where, being in other places but little, they become great huge trees, as *P. Bellonius* writeth. *Iuba* also reporteth, that there be in Arabia of them fifty cubits high. They grow only in some few gardens with vs.

¶ *The Time.*

The Strawberry tree floureth in Iuly and August, and the fruit is ripe in September, after it hath remained vpon the tree by the space of an whole yeare.

¶ *The Names.*

This tree is called in Greek, κόμαρος: in Latine, *Arbutus*: in English, Strawberry tree, and of some, Arbute tree.

The fruit is named in Greeke, μιμαίκυλον, or as others reade it, μιμαίκυλον: in Latine, *Memæcylum*, and *Arbutus*; and *Pliny* calleth it *Vnedo*: Ground Strawberries (saith he) haue one body, and *Vnedo*, much like vnto them, another body, which onely in apple is like to the fruit of the earth: The Italians call this Strawberry *Albatro*: the Spaniards, *Madrono*, *Medronheyro*, and *Medronho*: in French, *Arboutes*, *Arbous*: It may be termed in English, Tree Strawberry.

¶ *The Temperature and Vertues.*

A The fruit of the Strawberry tree is of a cold temperature, hurting the stomacke, and causing head-ache; wherefore no wholesome food, though it be eaten in some places by the poorer sort of people.

Chap. 126. *Of the Plum tree.*

¶ *The Kindes.*

TO write of Plums particularly would require a peculiar volume, and yet the end not to be attained vnto, nor the stocke or kindred perfectly knowne, neither to be distinguished apart: the number of the sorts or kindes are not knowne to any one countrey: euery Clymat hath his owne fruit, far different from that of other countries: my selfe haue three score sorts in my garden, and all strange and rare: there be in other places many more common, and yet yearely commeth to our hands others not before knowne, therefore a few figures shall serue for the rest. ‡ Let such as require a larger historie of these varieties haue recourse to the oft mentioned Worke of Mr. *Parkinson*: and such as desire the things themselues may finde most of the best with Mr. *Iohn Millen* in Old street. ‡

¶ *The Description.*

1 THe Plum or Damson tree is of a meane bignesse: it is couered with a smooth barke: the branches are long, whereon do grow broad leaues, more long than round, nicked in the edges: the floures are white: the Plums do differ in colour, fashion, and bignes, they all consist of pulpe and skin, and also of kernell, which is shut vp in a shell or stone. Some Plums are of a blackish blew, of which some be longer, others rounder, others of the colour of yellow wax, diuers of a crimson red, greater for the most part than the rest. There be also green Plums, and withall very long, of a sweet and pleasant taste: moreouer, the pulpe or meate of some is drier, and easilier separated from the stone: of other-some it is moister, and cleaueth faster: our common Damson is knowne to all, and therefore not to be stood vpon.

2 The

1 *Prunus Domestica.*
The Damſon tree.

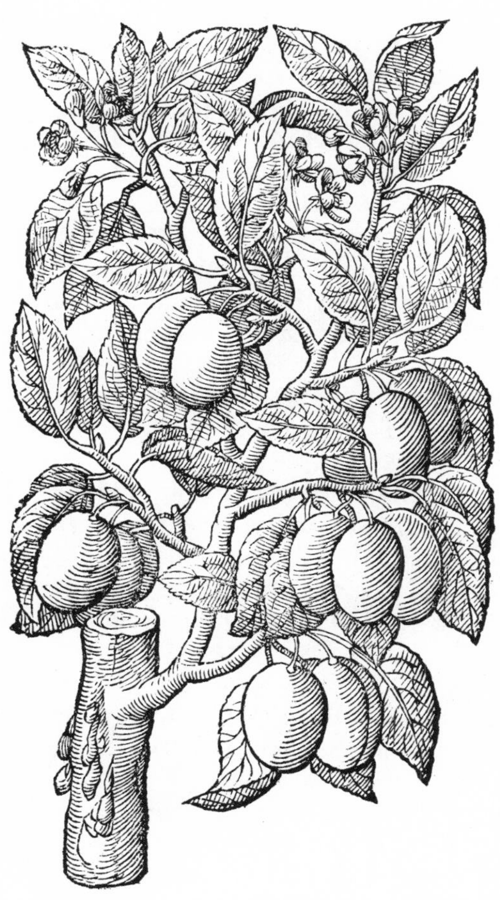

2 *Prunus Mirobalana.*
The Mirobalane Plum tree.

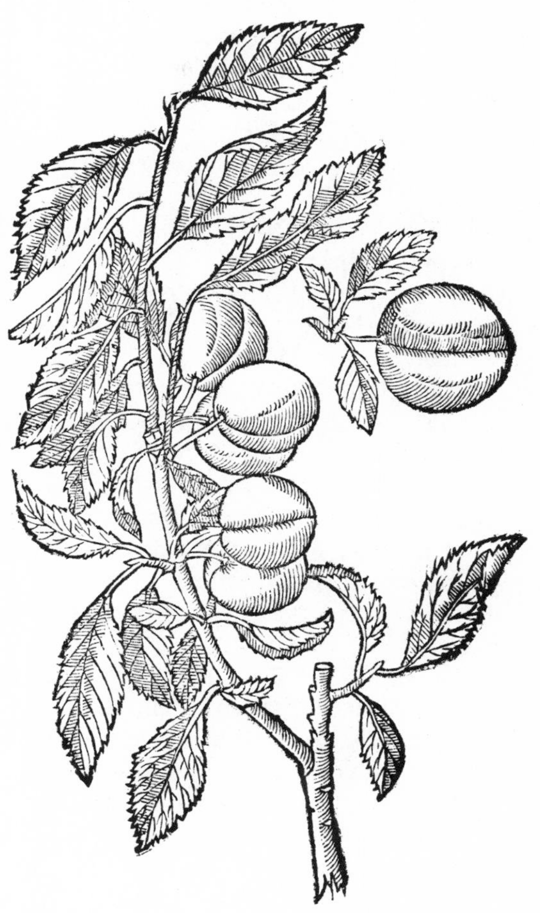

3 *Prunus Amygdalina.*
The Almond Plum tree.

5 *Prunus ſylueſtris.*
The Sloe tree.

2 The Mirobalan Plum tree groweth to the height of a great tree, charged with many great armes or boughes, which diuide themselues into small twiggy branches, by means whereof it yeeldeth a goodly and pleasant shadow: the trunke or body is couered with a finer and thinner barke than any of the other Plum trees: the leaues do somewhat resemble those of the Cherry tree, they are very tender, indented about the edges: the floures be white: the fruit is round, hanging vpon long foot-stalkes pleasant to behold, greene in the beginning, red when it is almost ripe, and being full ripe it glistereth like purple mixed with blacke: the flesh or meate is full of iuice pleasant in taste: the stone is small, or of a meane bignesse: the tree bringeth forth plenty of fruit euery other yeare.

3 The Almond Plum groweth vp to the height of a tree of a meane bignesse: the branches are long, smooth, and euen: the leaues are broad, somthing long, and ribbed in diuers places, with small nerues running through the same: the floures are white, sprinkled with a little dash of purple scarcely to be perceiued: the fruit is long, hauing a cleft downe the middle, of a browne red colour, and of a pleasant taste.

4 The Damascen Plum tree groweth likewise to a meane height, the branches very brittle; the leaues of a deepe green colour: the fruit is round, of a blewish blacke colour: the stone is like vnto that of the Cherry, wherein it differeth from all other Plums.

5 The Bullesse and the Sloe tree are wilde kindes of Plums, which do vary in their kind, euen as the greater and manured Plums do. Of the Bullesse, some are greater and of better taste than others. Sloes are some of one taste, and some of others, more sharpe; some greater, and others lesser; the which to distinguish with long descriptions were to small purpose, considering they be all and euery of them knowne euen vnto the simplest: therefore this shall suffice for their seueral descriptions.

¶ *The Place.*

The Plum trees grow in all knowne countries of the world: they require a loose ground, they also receiue a difference from the regions where they grow, not only of the forme or fashion, but especially of the faculties, as we will forthwith declare.

The Plum trees are also many times graffed into trees of other kindes, and being so ingraffed, they *faciem parentis, succum adoptionis, vt Plinius dicit, exhibent.*

The greatest varietie of these rare Plums are to be found in the grounds of M^r. *Vincent Pointer* of Twicknam, before remembred in the Chap. of Apples: although my selfe am not without some, and those rare and delicate.

The wilde Plums grow in most hedges through England.

¶ *The Time.*

The common and garden Plum trees do bloome in April: the leaues come forth presently with them: the fruit is ripe in Sommer, some sooner, some later.

¶ *The Names.*

The Plumme tree is called in Greeke, Κοκκυμηλέα: in Latine, *Prunus*: in high-Dutch, **Pflaumenbaum**: in low-Dutch, **Pruymen**: in Spanish, *Ciruelo*: in French, *Prunier*: in English, Plum tree.

The fruit is called in Greeke, Κοκκύμηλον: in Latine, *Prunum*: in high-Dutch, **Pflaumen**: in low-Dutch, **Pruymen**: in Italian and French, *Prune*: in Spanish, *Prunas*: in English, Prune, and Plum. These haue also names from the regions and countries where they grow.

The old Writers haue called those that grow in Syria neere vnto Damascus, *Damascena Pruna*: in English, Damsons, or Damaske Prunes: and those that grow in Spain, *Hispanica*, Spanish Prunes or Plums. So in our age we vse to call those that grow in Hungarie, *Hungarica*, or *Pannonica*, Plums of Hungarie: some, *Gallica Pruna*, or French Prunes, of the country of France. *Clearcus Peripateticus* saith, that they of Rhodes and Sicilia do call the Damaske Prunes *Brabula*.

¶ *The Temperature and Vertues.*

A Plummes that be ripe and new gathered from the tree, what sort soeuer they are of, do moisten and coole, and yeeld vnto the body very little nourishment, and the same nothing good at all: for as Plummes do very quickly rot, so is also the iuice of them apt to putrifie in the body, and likewise to cause the meate to putrifie which is taken with them: onely they are good for those that would keepe their bodies soluble and coole; for by their moisture and slipperinesse they do mollifie the belly.

B Dried Plums, commonly called Prunes, are wholsomer, and more pleasant to the stomack, they yeeld more nourishment, and better, and such as cannot easily putrifie. It is reported, saith *Galen* in his booke of the faculties of Nourishments, that the best doe grow in Damascus a city of Syria; and next to those, they that grow in Spaine: but these doe nothing at all binde, yet diuers of the Damaske Damson Prunes very much; for Damaske Damson Prunes are more astringent, but they of Spaine be sweeter. *Dioscorides* saith, that Damaske Prunes dried do stay the belly; but *Galen* affirmeth, in his books of the faculties of simple medicines, that they do manifestly loose the belly, yet

yet lesser than they that bee brought out of Spaine; being boiled with Mead or honied water, which hath a good quantitie of honey in it, they loose the belly very much (as the same Authour saith) although a man take them alone by themselues, and much more if the Mead be supped after them. We most commend those of Hungarie being long and sweet; yet more those of Morauia the chiefe and principall citie in times past of the Prouince of the Marcomans: for these after they be dried, that the waterie humour may be consumed away, be most pleasant to the taste, and do easily without any trouble so mollifie the belly, as that in that respect they go beyond Cassia and Manna, as *Thomas Iordanus* affirmeth.

The leaues of the Plum tree are good against the swelling of the Vuula, the throat, gums, & kernels vnder the throat and iawes; they stop the rheume and falling downe of humors, if the decoction thereof be made in wine, and gargled in the mouth and throat. C

The gumme which commeth out of the Plum-tree doth glew and fasten together, as *Dioscorides* saith. D

Being drunke in wine it wasteth away the stone, and healeth Lichens in infants and young children; if it be layed on with vineger, it worketh the same effects that the gum of the Peach and cherrie tree doth. E

The wilde Plums do stay and binde the belly, and so do the vnripe plummes of what sort soeuer, whiles they are sharpe and sower, for then are they astringent. F

The iuice of Sloes doth stop the belly, the lask and bloudy flix, the inordinat course of womens termes, and all other issues of bloud in man or woman, and may very well be vsed in stead of Acatia, which is a thornie tree growing in Ægypt, very hard to be gotten, and of a deere price, and therfore the better for wantons; albeit our Plums of this countrey are equall vnto it in vertues. G

CHAP. 127. *Of Sebesten, or the Assyrian Plum.*

Sebestenæ, Myxa, siue Myxara.
Assyrian Plums.

¶ *The Description.*

Sebestines are also a kinde of Plums: the tree whereof is not vnlike to the Plum tree, sauing it groweth lower than the most of the manured Plum trees; the leaues be harder and rounder; the floures grow at the tops of the branches consisting of fiue small white leaues, with pale yellowish threds in the middle, like those of the Plum tree: after followeth the fruit like to little Plummes, fastened in little skinny cups, which when they be ripe are of a greenish black colour, wherein is contained a small hard stone. The fruit is sweet in taste, the pulpe or meat is very tough and clammie.

¶ *The Place.*

The Sebesten trees grow plentifully in Syria and Egypt, they were in times past forreine and strange in Italy, now they grow almost in euery garden, being first brought thither in *Plinie* his time. Now do the Sebesten trees, saith he, in his 15. booke, 18. chapter, begin to grow in Rome, among the Seruice trees.

¶ *The Time.*

The time answereth the common Plums.

¶ *The Names.*

Pliny calleth the tree *Myxa*, it may bee suspected that this is the tree which *Matron Paradus* in his Atticke banket in *Athenæus* doth call ἐμμυξεῖς but we cannot certainely affirme it, and especially because diuers haue diuersly deemed thereof. The berry or fruit is named *Myxon*,

Myxon and *Myxarion*, neither haue the Latines any other name. The Arabians and the Apothecaries do call it *Sebesten*: which is also made an English name: we may call it the Assyrian Plum.

¶ *The Temperature and Vertues.*

A Sebestens be very temperately cold and moist, and haue a thicke and clammie substance; therefore they nourish more than most fruits do, but withall they easily stop the intrailes, and stuffe vp the narrow passages, and breed inflammations.

B They take away the ruggednesse of the throat and lungs, and also quench thirst, being taken in a looch or licking medicine, or prepared any other kinde of way, or else taken by themselues.

C The weight of ten drams, or of an ounce and a halfe of the pap or pulpe hereof being inwardly taken, doth loose the belly.

D There is also made of this fruit a purging Electuarie, but such an one as quickly mouldeth, and therefore it is not to be vsed but when it is new made.

Chap. 128. *Of the Indian Plums, or Mirobalans.*

¶ *The Kindes.*

There be diuers kinds of Mirobalans, as *Chebulæ*, *Belliricæ*, *Emblicæ*, &c. They likewise grow vpon diuers trees, and in countries far distant one from another, and *Garcias* the Portugall Physition is of opinion, that the fiue kindes grow vpon fiue diuers trees.

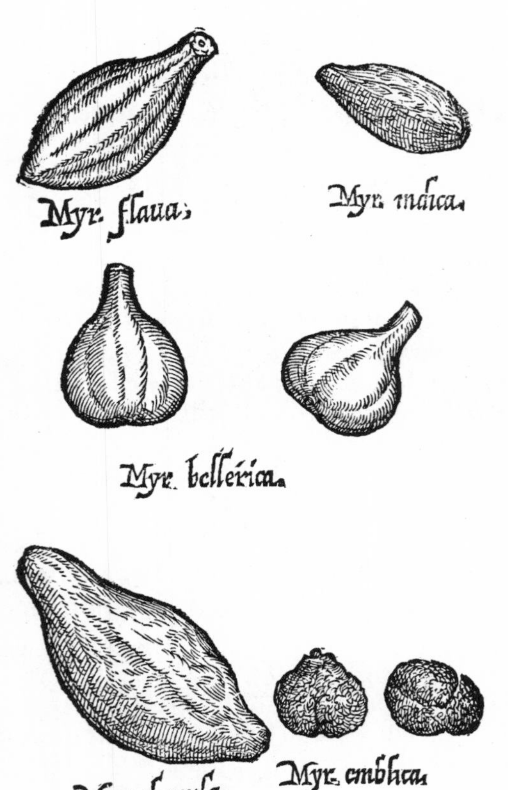

¶ *The Description.*

1 The first of the Mirobalan trees, called *Chebulæ*, is a shrubbie tree altogether wilde (which the Indians doe call *Aretca*:) in stature not vnlike to the Plum tree; the branches are many, and grow thicke together, whereon are set leaues like those of the Peach tree. The fruit is greater than any of the rest, somwhat long, fashioned like a peare.

2 This second kinde of Mirobalan, called *Flaua*, or *Citrina*, which some do call *Aritiqui*, but the common people of India, *Arare*, groweth vpon a tree of meane stature, hauing many boughes standing finely in order, and set full of leaues like vnto the Seruice tree.

3 The third kinde of Mirobalans, called *Emblicæ*, the Indians doe call *Amiale*, which grow vpon a tree of mean stature, like the former, but the leaues are very much iagged, in shape like the leaues of Ferne, but that they be somewhat thicker: the Indians do not put the fruit hereof vnto physicall vses, but occupie it for the thickening and tanning of their leather in stead of *Rhus*, or Coriars Sumach, as also to make inke and bletch for other purposes.

4 *Mirobalani Belliricæ*, called of the Sauages *Gotni*, and *Guti*, groweth vp to a meane stature, garnished with leaues like vnto Laurell or the Bay tree, but somewat lesser, thinner, and of a pale greene colour.

5 The fift kinde of Mirobalans is called *Indicæ*, which the Indians do call *Rezannale*; it groweth vpon a tree of meane stature, or rather vpon a shrub or hedge plant, bearing leaues like the Willow, and a fruit eight square. There is a fift kinde, the tree whereof is not mentioned in Authors.

¶ *The Place and Time.*

The last foure kindes of Mirobalans do grow in the kingdome of Cambaia: they grow likewise in Goa, Batecala, Malanor, and Dabul: the *Kebula* in Bisnager, Decan, Guzarate, and Bengala, & many other places of the East Indies. The time agreeth with other fruits in those countries.

¶ *The Names.*

Those which we haue said to be yellow, the inhabitants of those countries where they grow doe call

call them *Arare*;thoſe that be blacke they call *ReZennale*;the *Bellericæ*,*Gotim*;the *Chebulæ*, *Aretca*;the *Emblicæ* are called *Aretiqui*.

¶ *The Temperature.*

All the kindes of Mirobalans are in taſte aſtringent and ſharpe like vnto the vnripe *Sorbus* or Seruice berries,and therefore they are of complexion cold and drie.

¶ *The Vertues.*

The Indians vſe them rather to bind than purge,but if they do vſe them for a purge, they vſe the decoction of them,and vſe them much conſerued in ſugar,and eſpecially the *Chebulæ*; the yellow and blacke be good that way likewiſe. A

The yellow and *Bellericæ* taken before meat,are good againſt a laske,or weake ſtomacke,as *Garcias* writeth. B

The yellow and blacke,or *Indicæ*,and the *Chebulæ*,purge lightly, if two or three drams be taken, and draw ſuperfluous humors from the head. C

The yellow,as ſome write,purge choler,*Chebulæ* flegme, *Indicæ* melancholie, and ſtrengthen the inward parts,but roſted in the embers,or otherwiſe waſted,they drie more than they purge. D

There are two ſorts eſpecially brought into theſe parts of the world conſerued, the *Chebulæ*, and of them the beſt are ſomewhat long like a ſmall Limon,with a hard rinde and black pith,of the taſt of a conſerued Wall-nut;and the *Bellericæ*, which are round and leſſer,and tenderer in eating. E

Lobel writeth,that of them the *Emblicæ* do meanly coole,ſome do drie in the firſt degree, they purge the ſtomacke of rotten flegme,they comfort the braine,the ſinewes, the heart,and liuer, procure appetite,ſtay vomite,and coole the heat of choler, helpe the vnderſtanding, quench thirſt, and the heate of the intrailes:the greateſt and heauieſt be the beſt. F

They purge beſt,and with leſſer paine if they be laid in water in the Sun vntill they ſwell,& ſod on a ſoft fire,& after they haue ſod and be cold,preſerued in foure times ſo much white honey,put to them. G

Garcias found the diſtilled water to be right profitable againſt the French diſeaſe,and ſuch like infections. H

The *Bellericæ* are alſo of a milde operation,and do comfort,and are cold in the firſt degree, and drie in the ſecond:the others come neere to the *Emblicæ* in operation. I

† I haue in this chapter contented my ſelfe with the expreſſing of the fruits out of *Cluſius* and *Lobel*, and omitted the figures of the three Mirabalan trees, which our Author gaue vs out of *Tabernamontanus*; becauſe I iudge them rather drawne by fancy than by the things themſelues.

Chap. 129. *Of the Iuiube tree.*

Iuiube Arabum, ſiue Ziziphus Dodonæi.
The Iuiube tree.

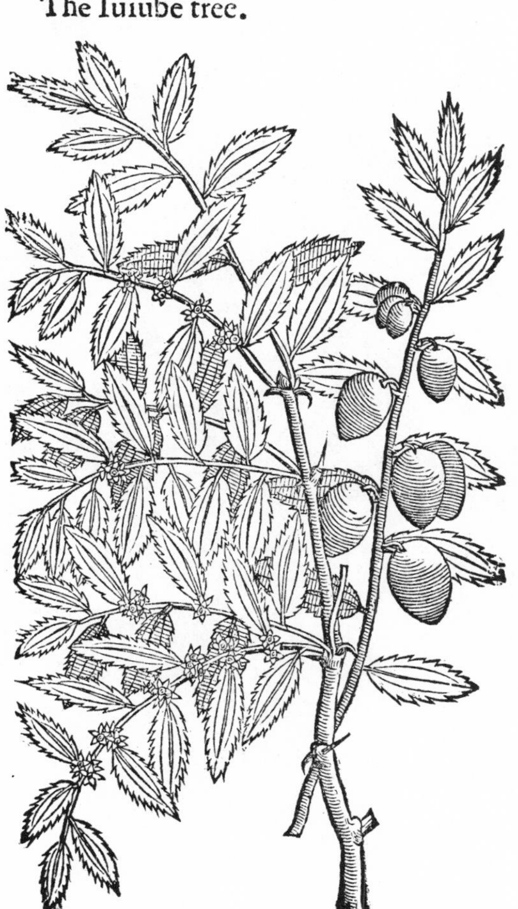

¶ *The Deſcription.*

The Iuiube tree is not much leſſer than *Ziziphus candida*,hauing a wreathed trunke or body,and a rough barke full of rifts or cranies, and ſtiffe branches,beſet with ſtrong and hard prickles; from whence grow out many long twigs,or little ſtalkes, halfe a foot or more in length,in ſhew like Ruſhes,limmer, and eaſily bowing themſelues, and very ſlender like the twigges of Spartum: about which come forth leaues one aboue another,which are ſomewhat long,not very great, but hard and tough like to the leaues of *Peruinca* or Peruinckle; & among theſe leaues come forth pale and moſſie little floures: after which ſucceed long red well taſted ſweet berries as big as Oliues (of a meane quantity) or little Prunes,or ſmal Plums,wherin there are hard round ſtones, or in which a ſmall kernell is contained.

¶ *The Place.*

There be now at this day Iuiube trees growing in very many places of Italy,which in times paſt were newly brought thither out of Syria, and that about *Pliny* his time, as he himſelfe hath written in his 17.book,10.chap.

¶ *The Time.*

It floureth in Aprill,at which time the ſeeds or ſtones are to be ſet or ſowne for increaſe.

¶ The

¶ *The Names.*

This tree is called in Greeke ζίζυφος and ζιζιφος with *Iota* in the second syllable: in Latine likewise, *Zizyphus*, and of *Petrus Crescentius*, *Zezulus* : in English, Iuiube tree.

The fruit or Plums are named in Greeke ζίζυφα, ζιζιφα: *Galen* calleth them σηρικά, as *Auicen* plainely sheweth in his 369. chapter, intreating of the Iuiube, in which be set downe those things that are mentioned concerning *Serica* in *Galens* books of the faculties of Nourishments: in Latine likewise *Zizypha* and *Serica* : in shops, *Iuiubæ* : in English, Iuiubes.

¶ *The Temperature.*

Iuiubes are temperate in heate and moisture.

¶ *The Vertues.*

A The fruit of the Iuiube tree eaten is of hard digestion, and nourisheth very little; but being taken in syrrups, electuaries, and such like confections, it appeaseth and mollifieth the roughnesse of the throat, the brest and lungs, and is good against the cough, but exceeding good for the reines of the backe, and kidneies and bladder.

Chap. 130. *Of the Cherrie Tree.*

¶ *The Kindes.*

THe antient Herbarists haue set down foure kindes of Cherrie trees, the first is great and wilde, the second tame or of the garden: the third, whose fruit is soure: the fourth is that which is called in Latine *Chamæcerasus*, or the dwarfe Cherrie tree. The later writers haue found diuers sorts more, some bringing forth great fruit, others lesser, some with white fruit, some with blacke, others of the colour of blacke bloud, varying infinitely according to the climate and countrey where they grow.

1 *Cerasus vulgaris.*
The common English Cherrie tree.

3 *Cerasus Hispanica.*
The Spanish Cherrie tree.

¶ The

¶ *The Deſcription.*

1 THe Engliſh Cherrie tree groweth to an high and great tree, the body whereof is of a meane bigneſſe, which is parted aboue into very many boughes, with a barke ſomewhat ſmooth, and of a browne crimſon colour, tough and pliable: the ſubſtance or timber is alſo browne in the middle, and the outward part is ſomewhat white. The leaues be great, broad, long, ſet with veines or nerues, and ſleightly nicked about the edges: the floures are white, of a mean bignes, conſiſting of fiue leaues, and hauing certaine threds in the middle, of the like colour: the Cherries be round, hanging vpon long ſtems or foot-ſtalks, with a ſtone in the midſt which is couered with a pulpe or ſoft meat; the kernell thereof is not vnpleaſant to the taſte, though ſomewhat bitter.

2 The Flanders Cherry tree differeth not from our Engliſh Cherrie tree in ſtature or forme of leaues or floures; the difference conſiſteth in this, that this tree bringeth forth his fruit ſooner, and greater than the other: wherefore it may be called in Latine, *Ceraſus precox, ſiue Belgica.*

5 *Ceraſus Serotina.*
Late ripe Cherrie tree.

6 *Ceraſus vno pediculo plura.*
The Cluſter Cherrie tree.

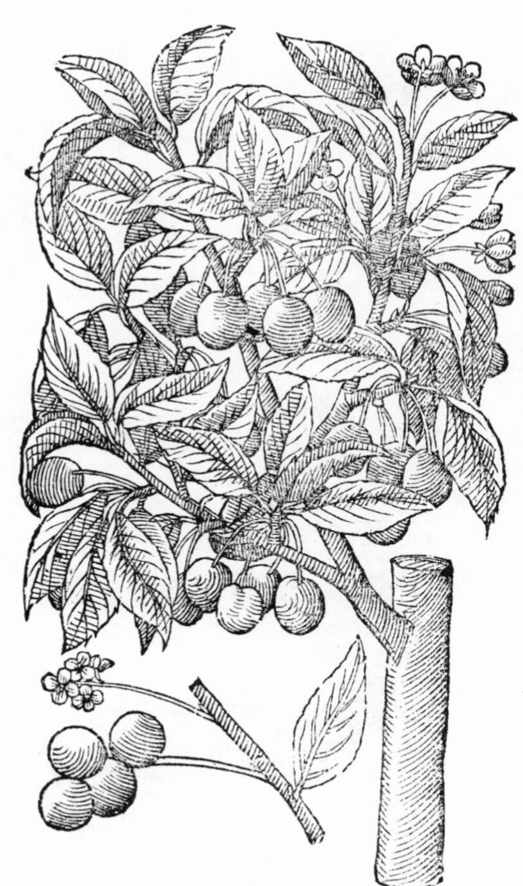

3 The Spaniſh Cherrie tree groweth vp to the height of our common Cherrie tree: the wood or timber is ſoft and looſe, couered with a whitiſh ſcalie barke: the branches are knottie, greater & fuller of ſubſtance than any other Cherry tree: the leaues are likewiſe greater and longer than any of the reſt, in ſhape like thoſe of the Cheſtnut tree: the floures are like the others in forme, but whiter of colour: the fruit is greater and longer than any, white for the moſt part all ouer, except thoſe that ſtand in the hotteſt place where the Sun hath ſome reflexion againſt a wall: they are alſo white within, and of a pleaſant taſte.

4 The Gaſcoine Cherrie tree groweth very like to the Spaniſh Cherry tree in ſtature, floures, and leaues: it differeth in that it bringeth forth very great Cherries, long, ſharpe pointed, with a certaine hollowneſſe vpon one ſide, and ſpotted here and there with certaine prickles of purple colour as ſmall as ſand: the taſte is moſt pleaſant, and excelleth in beautie.

5 The late ripe Cherry tree groweth vp like vnto our wilde Engliſh Cherry tree, with the like leaues,

7 *Cerasus multiflora fructus edens.*
The double floured Cherry tree bearing fruit.

8 *Cerasus multiflora pauciores fructus edens.*
The doule floured barren Cherry tree.

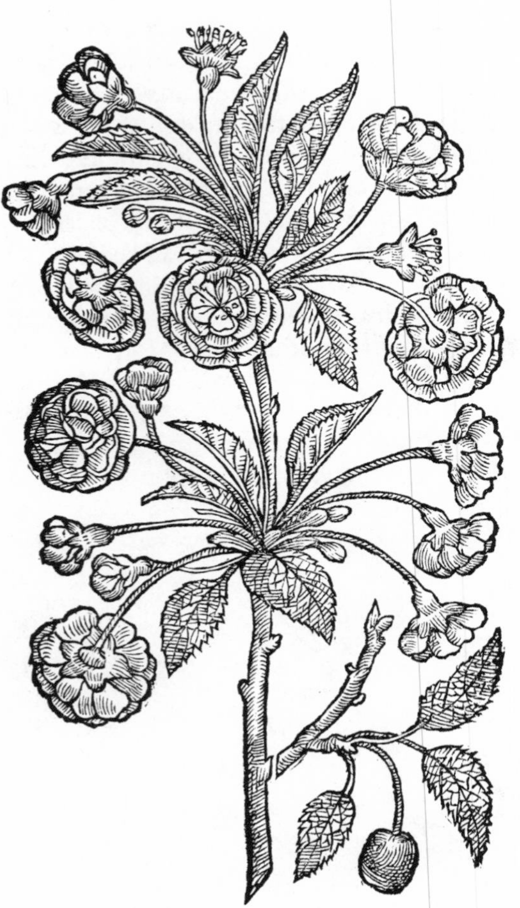

9 *Cerasus auium nigra & racemosa.*
Birds Cherry, and blacke Grape Cherry tree.

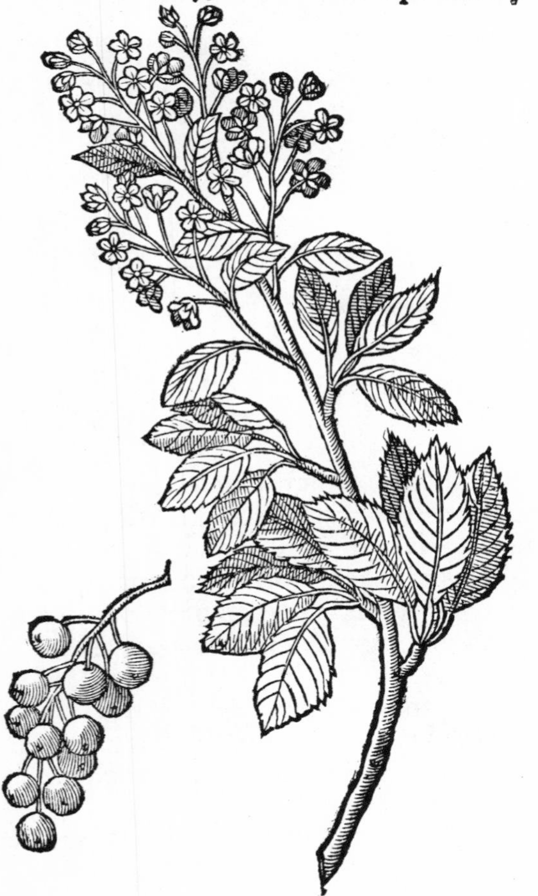

10 *Cerasus racemosa rubra.*
Red Grape Cherry tree.

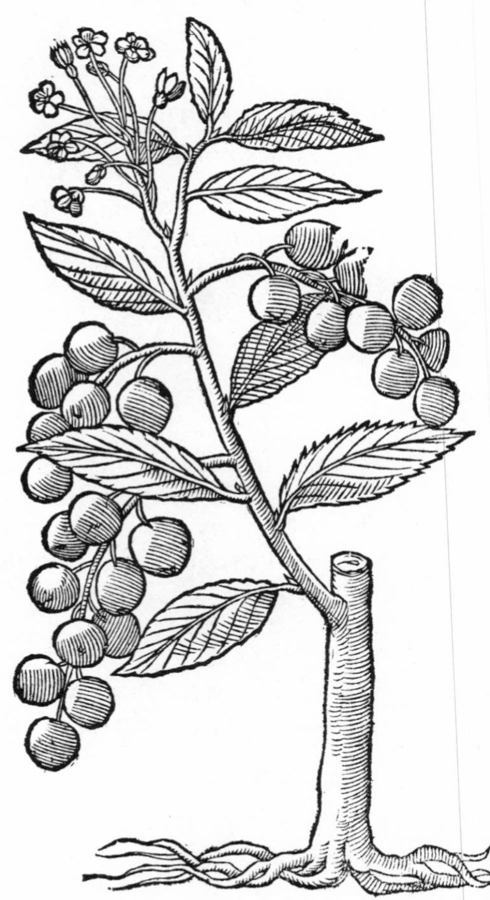

leaues,branches,and floures, fauing that they are fometimes once doubled : the fruit is fmall, round, and of a darke bloudy colour when they be ripe, which the French-men gather with their ftalkes,and hang them vp in their houfes in bunches or handfulls againft winter, which the Phyfitians do giue vnto their patients in hot and burning feuers, being firft fteeped in a little warme water,that caufeth them to fwell and plumpe vp as full and frefh as when they did grow vpon the tree.

6 The Clufter Cherry-tree differeth not from the laftdefcribed either in leaues,branches,or ftature : the floures are alfo like,but neuer commeth any one of them to be double. The fruit is round,red when they be ripe,and many growing vpon one ftem or footftalke in clufters, like as the Grapes do. The tafte is not vnpleafant,although fomewhat foure.

7 This Cherrie-tree with double floures growes vp vnto a fmall tree, not vnlike to the common Cherrie tree in each refpect,fauing that the flours are fomewhat doubled,that is to fay, three or foure times double ; after which commeth fruit(though in fmall quantitie)like the other common Cherrie.

8 The double floured Cherrie-tree growes vp like vnto an hedge bufh,but not fo great nor high as any of the others ; the leaues and branches differ not from the reft of the Cherrie-trees. The floures hereof are exceeding double, as are the floures of Marigolds, but of a white colour, and fmelling fomewhat like the Hawthorne floures ; after which come feldome or neuer any fruit, although fome Authors haue faid that it beareth fometimes fruit, which my felfe haue not at any time feen ; notwithftanding the tree hath growne in my garden many yeeres, and that in an excellent good place by a bricke wall,where it hath the reflection of the South funne, fit for a tree that is not willing to beare fruit in our cold climat.

11 *Cerafus nigra.*
The common blacke Cherry-tree

12 *Chamæcerafus.*
The dwarfe Cherry-tree.

9 The Birds Cherry-tree,or the blacke Cherry-tree, that bringeth forth very much fruit vpon one branch (which better may be vnderftood by fight of the figure, than by words) fpringeth vp like an hedge tree of fmall ftature, it groweth in the wilde woods of Kent,and are there vfed for ftockes to graft other Cherries vpon, of better tafte,and more profit, as efpecially thofe called the Flanders Cherries:this wilde tree growes very plentifully in the North of England,efpecially at a place called Heggdale,neer vnto Rofgill in Weftmerland, and in diuers other places about Crofbie Rauenfwaith, and there called Hegberrie-tree : it groweth likewife in Martome Parke, foure

 miles

miles from Blackeburne, and in Harward neere thereunto; in Lancashire almost in euery hedge: the leaues and branches differ not from those of the wilde Cherry-tree: the floures grow alongst the small branches, consisting of fiue small white leaues, with some greenish and yellow thrums in the middle: after which come the fruit, greene at the first, blacke when they be ripe, and of the bignesse of Sloes; of an harsh and vnpleasant taste.

10 The other birds Cherry-tree differeth not from the former in any respect, but in the colour of the berries; for as they are blacke; so on the contrary, these are red when they be ripe, wherein they differ.

11 The common blacke Cherry-tree growes vp in some places to a great stature: there is no difference betweene it and our common Cherry-tree, sauing that the fruit hereof is very little in respect of other Cherries, and of a blacke colour.

12 The dwarfe Cherry-tree groweth very seldome to the height of three cubits: the trunke or body small, couered with a darke coloured blacke: whereupon do grow very limber and pliant twiggy branches: the leaues are very small, not much vnlike to those of the Priuite bush: the floures are small and white: after which come Cherries of a deepe red colour when they be ripe, of taste somewhat sharpe, but not greatly vnpleasant: the branches laid downe in the earth, quickly take root, whereby it is greatly increased.

My selfe with diuers others haue sundry other sorts in our gardens, one called the Hart Cherry, the greater and the lesser; one of a great bignesse, and most pleasant in taste, which we call *Luke Wardes* Cherry, because he was the first that brought the same out of Italy; another we haue called the Naples Cherry, because it was first brought, into these parts from Naples: the fruit is very great, sharpe pointed, somewhat like a mans heart in shape, of a pleasant taste, and of a deepe blackish colour when it is ripe, as it were of the colour of dried bloud.

We haue another that bringeth forth Cherries also very great, bigger than any Flanders Cherrie, of the colour of Iet, or burnished horne, and of a most pleasant taste, as witnesseth Mr. *Bull,* the Queenes Maiesties Clockemaker, who did taste of the fruit (the tree bearing onely one Cherry, which he did eate; but my selfe neuer tasted of it) at the impression hereof. We haue also another, called the Agriot Cherry, of a reasonable good taste. Another we haue with fruit of a dun colour, tending to a watchet. We haue one of the dwarffe Cherries, that bringeth forth fruit as great as most of our Flanders Cherries, whereas the common sort hath very small Cherries, and those of an harsh taste. These and many sorts more we haue in our London gardens, whereof to write particularly would greatly enlarge our volume, and to small purpose: therefore what hath been said shall suffice. ‡ I must here (as I haue formerly done, in Peares, Apples and other such fruites) refer you to my two friends Mr. *Iohn Parkinson,* and Mr. *Iohn Millen,* the one to furnish you with the history, and the other with the things themselues, if you desire them. ‡

¶ *The Time.*

The Cherrie-trees bloome in Aprill; some bring forth their fruit sooner; some later: the red Cherries be alwaies better than the blacke of their owne kinde.

¶ *The Names.*

The Cherry-tree is called in Greeke, κέρασος: and also in Latine, *Cerasus*: in high-Dutch, **Kirschenbaum**: in low-Dutch, **Kersenboome,** and **Crieckenboom**: in French, *Cerisier*: in English, Cherry-tree.

The fruit or Cherries be called in Greeke, κεράσια, and κεράσια: and in Latine likewise, *Cerasa*: in English, Cherries: the Latine and English names in their seuerall titles shall suffice for the rest that might be said.

¶ *The Temperature and Vertues.*

A The best and principall Cherries be those that are somewhat sower: those little sweet ones which be wilde and soonest ripe be the worst: they containe bad juice, they very soon putrifie, and do ingender ill bloud, by reason whereof they do not onely breed wormes in the belly, but troublesome agues, and often pestilent feuers: and therefore in well gouerned common wealths it is carefully prouided, that they should not be sold in the markets in the plague time.

B Spanish Cherries are like to these in faculties, but they do not so soone putrifie: they be likewise cold, and the iuice they make is not good.

C The Flanders or Kentish Cherries that are through ripe, haue a better juice but watery, cold and moist: they quench thirst, they are good for an hot stomacke, and profitable for those that haue the ague: they easily descend and make the body soluble: they nourish nothing at all.

D The late ripe Cherries which the French-men keep dried against winter, and are by them called *Morelle*, and we after the same name call them Morell Cherries, are dry, and do somewhat binde; these being dried are pleasant to the taste, and wholesome for the stomacke, like as Prunes be, and do stop the belly.

Generally

Generally all the kindes of Cherries are cold and moiſt of temperature, although ſome more cold and moiſt than others : the which being eaten before meat doe ſoften the belly very gently, they are vnwholſome either vnto moiſt and rheumaticke bodies, or for vnhealthie and cold ſtomackes. E

The common blacke Cherries do ſtrengthen the ſtomack, and are wholeſomer than the red Cherries, the which being dried do ſtop the laske. F

The diſtilled water of Cherries is good for thoſe that are troubled with heate and inflammations in their ſtomackes, and preuaileth againſt the falling ſickneſſe giuen mixed with wine. G

Many excellent Tarts and other pleaſant meats are made with Cherries, ſugar, and other delicat ſpices, whereof to write were to ſmall purpoſe. H

The gum of the Cherrie tree taken with wine and water, is reported to helpe the ſtone; it may do good by making the paſſages ſlippery, and by tempering & alaying the ſharpneſſe of the humors; and in this maner it is a remedy alſo for an old cough. *Dioſcorides* addeth, that it maketh one well coloured, cleareth the ſight, and cauſeth a good appetite to meat. I

CHAP. 131. *Of the Mulberrie tree.*

1 *Morus.*
The Mulberrie tree.

2 *Morus alba.*
The white Mulberrie tree.

¶ *The Deſcription.*

1 THe common Mulberie tree is high, and ful of boughes: the body wherof is many times great, the barke rugged; & that of the root yellow: the leaues are broad and ſharp pointed, ſomething hard, and nicked on the edges; in ſtead of floures, are blowings or catkins, which are downie: the fruit is long, made vp of a number of little graines, like vnto a blacke-Berrie, but thicker, longer, and much greater, at the firſt greene, and when it is ripe blacke, yet is the iuice (whereof it is full) red: the root is parted many waies.

 2 The

2 The white Mulberrie tree groweth vntill it be come vnto a great and goodly stature, almost as big as the former: the leaues are rounder, not so sharpe pointed, nor so deeply snipt about the edges, yet sometimes sinuated or deeply cut in on the sides, the fruit is like the former, but that it is white and somewhat more tasting like wine.

¶ *The Place.*

The Mulberry trees grow plentifully in Italy and other hot regions, where they doe maintaine great woods and groues of them, that there Silke wormes may feed thereon. The Mulberry tree is fitly set by the slip; it may also be grafted or inoculated into many trees, being grafted in a white Poplar, it bringeth forth white Mulberies, as *Beritius* in his Geoponickes reporteth. These grow in sundry gardens in England.

¶ *The Time.*

Of all the trees in the Orchard the Mulberry doth last bloome, and not before the cold weather is gone in May (therefore the old Writers were wont to call it the wisest tree) at which time the Silke wormes do seeme to reuiue, as hauing then wherewith to feed and nourish themselues, which all the winter before do lie like small graines or seeds, or rather like the dunging of a flesh flie vpon a glasse, or some such thing, as knowing their proper time both to performe their duties for which they were created, and also when they may haue wherewith to maintaine and preserue their owne bodies, vnto their businesse aforesaid.

The berries are ripe in August and September. *Hegesander* in *Athenæus* affirmeth, that the Mulberie trees in his time did not bring forth fruit in twentie years together, and that so great a plague of the gout then raigned and raged so generally, as not onely men, but boies, wenches, eunuchs, and women were troubled with that disease.

¶ *The Names.*

This tree is named in Greeke μορία, and συκαμινέα: in Latine, *Morus*: in shops, *Morus Celsi*: in high Dutch, **Maulbeerbaum**: in low Dutch, **Moerbesie boom**: in French, *Meurier*: in English, Mulberry tree.

The fruit is called μόρον, and συκάμινον: in Latine, *Morum*: in shops, *Morum Celsi*: in high Dutch, **Moerbesie**: in Italian, *Moro*: in French, *Meure*: in Spanish, *Moras* and *Mores*: in English, Mulberry.

¶ *The Temperature and Vertues.*

A Mulberries being gathered before they be ripe, are cold and dry almost in the third degree, and do mightily binde; being dried they are good for the laske and bloudy flix, the pouder is vsed in meat, and is drunke with wine and water.

B They stay bleedings, and also the reds; they are good against inflammations or hot swellings of the mouth and iawes, and for other inflammations newly beginning.

C The ripe and new gathered Mulberries are likewise cold and be ful of iuice, which hath the taste of wine, and is something drying, and not without a binding qualitie: and therefore it is also mixed with medicines for the mouth, and such as helpe the hot swellings of the mouth, and almonds of the throat; for which infirmities it is singular good.

D Of the iuice of the ripe berries is made a confection with sugar, called *Diamorum*: that is, after the manner of a syrrup, which is exceeding good for the vlcers and hot swellings of the tongue, throat, and almonds, or Vuula of the throat, or any other malady arising in those parts.

E These Mulberries taken in meat, and also before meat, do very speedily passe through the belly, by reason of the moisture and slipperinesse of their substance, and make a passage for other meats, as *Galen* saith.

F They are good to quench thirst, they stir vp an appetite to meat, they are not hurtfull to the stomacke, but they nourish the body very little, being taken in the second place, or after meat, for although they be lesse hurtfull than other like fruits, yet are they corrupted and putrified, vnlesse they speedily descend.

G The barke of the root is bitter, hot and drie, and hath a scouring facultie: the decoction hereof doth open the stoppings of the liuer and spleen, it purgeth the belly, and driueth forth wormes.

H The same bark being steeped in vineger helpeth the tooth ache: of the same effect is also the decoction of the leaues and barke, saith *Dioscorides*, who sheweth that about haruest time there issueth out of the root a iuice, which the next day after is found to be hard, and that the same is very good against the tooth-ache; that it wasteth away *Phyma*, and purgeth the belly.

I *Galen* saith, that there is in the leaues and first buds of this tree a certaine middle facultie, both to binde and scoure.

CHAP. 132. *Of the Sycomore tree.*

Sycomorus.
The Sycomore tree.

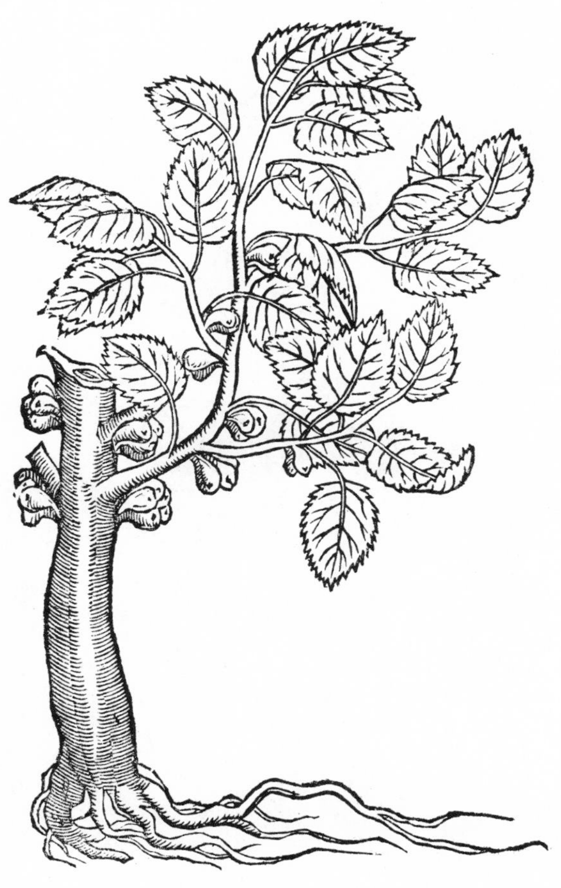

¶ *The Description.*

THe Sycomore tree is of no small height,being very like to the mulberie tree in bignesse & shew, as also in leafe: the fruit is as great as a Fig, and of the same fashion, very like in iuice and taste to the wilde Fig, but sweeter, and without any grains or seeds within,which groweth not forth of the tender boughes, but out of the body and great old armes very fruitfully: this tree hath in it plenty of milkie iuice,which so soon as any part is broken or cut, doth issue forth.

¶ *The Place.*

It groweth,as *Dioscorides* writeth, very plentifully in Caria and Rhodes, and in sundry places of Egypt, as at the great Cayre or Alkaire, and in places that doe not bring forth much wheat,in which it is an helpe,and sufficeth in stead of bread & corne when there is scarsitie of victuals. *Galen* writeth, that he saw a plant of the Sycomore tree like to the wilde Fig tree, fruit and all.

¶ *The Time.*

It bringeth forth fruit three or foure times in one yeare,and oftner if it be scraped with an iron knife, or other like instrument.

¶ *The Names.*

This tree is called in Greeke,συκόμορος, of the Fig tree and the Mulbery tree: in Latine, *Sycomorus*: *Cornelius Celsus* nameth it backward *Morosycos*: the Egyptians of our time do call it *Ficus Pharaonis*,or *Pharao* his Fig tree,as witnesseth *Bellonius*: and it is likewise termed *Ficus Ægyptia*, Egyptian Fig tree,and also *Morus Ægyptia*,or Egyptian Mulberrie tree. We cal it English, Sycomore tree after the Greek and Latine,and also Mulberry Fig tree,which is the right Sycomore tree,and not the great Maple,as we haue said in the chapter of the Maple.

The fruit is named in Greeke *Sycomoron*, and in Italian, *Sycomero* and *Fico d'Egitto*.

¶ *The Temperature and Vertues.*

A The fruit of the Sycomore tree hath no sharpnesse in it at all,as *Galen* saith. It is somwhat sweet in taste,and is of temperature moist after a sort,and cold as be Mulberries.

B It is good,saith *Dioscorides*, for the belly;but it is ἄτροφος, that is, without any nourishment, and troublesome to the stomacke.

C There issueth forth of the barke of this tree in the beginning of the Spring, before the fruit appeareth,a liquour,which being taken vp with a spunge,or a little wooll, is dried, made vp into fine cakes,and kept in gallie pots: this mollifieth,closeth wounds together, and dissolueth grosse humours.

D It is both inwardly taken and outwardly applied against the bitings of serpents,hardnesse of the milt or spleene,and paine of the stomacke proceeding of a cold cause: this liquor doth very quickly putrifie.

Chap. 133. *Of the Figtree.*

¶ *The Description.*

1 The garden Fig tree becommeth a tree of a meane stature,hauing many branches full of white pith within,like Elderne pith,and large leaues of a darke greene colour, diuided into sundry sections or diuisions. The fruit commeth out of the branches without any floure at all that euer I could perceiue,which fruit is in shape like vnto Peares,of colour either whitish,or somewhat red or of a deep blew,full of small graines within,of a sweet and pleasant taste; which beeing broken before it be ripe,doth yeeld most white milk,like vnto the kindes of Spurge, and the leaues also beeing broken doe yeeld the like liquour;but when the Figges be ripe,the iuice thereof is like honie.

1 *Ficus.*
The Fig tree.

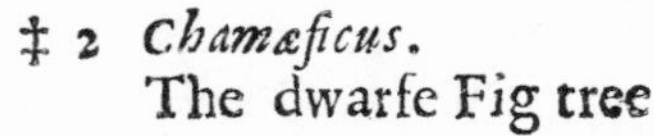

‡ 2 *Chamæficus.*
The dwarfe Fig tree.

2 The dwarfe Fig-tree is like vnto the former in leaues and fruit, but it neuer groweth aboue the height of a man,and hath many small shoots comming from the roots, whereby it greatly increaseth.

There is also another wilde kinde,whose fruit is neuer ripe;*Theophrastus* nameth it *Erineos*;*Pliny Caprificus.*

¶ *The Place.*

The Fig trees do grow plentifully in Spain and Italy,and many other countries,as in England; where they beare fruit,but it neuer commeth to kindely maturitie,except the tree be planted vnder an hot wall,whereto neither North, nor Northeast windes can come.

¶ *The Time.*

The dwarfe Fig tree groweth in my garden, and bringeth forth ripe and very great fruit in the moneth of August,of which Figs sundry persons haue eaten at pleasure.

In England the Fig trees put not forth their leaues vntill the end of May, where oftentimes the fruit commeth forth before the leaues appeare.

¶ The

¶ *The Names.*

The Fig tree is called in Greeke, συκῆ, and of diuers, for difference sake betweene it and the wild Fig tree, συκῆ ἥμερος: in Latine, *Ficus*, and *Ficus satiua*, and *Vrbana*: in high-Dutch, **Feygenbaum**: in low-Dutch, **Vijgheboom**: in French, *Figuier*: in Italian, *Fico*: in Spanish, *Higuera*. in English, Fig tree.

The fruit is named in Greeke, σῦκον: in Latine, *Ficus*: and the vnripe fruit, ὄλυνθος: in Latine, *Grossus*: that which is dried is called in Greeke, ἰσχάς: in Latine, *Carica*: in high-Dutch, **Feygen**: in low-Dutch, **Vijghen**: in French, *Figues*: in Italian, *Fichi*: in Spanish, *Higos*: in English, Fig: the little seeds which are found in them are named by *Galen*, κεγχραμίδες, *Cechramides*.

¶ *The Temperature.*

The greene Figs new gathered are somewhat warme and moist: the dry and ripe Figs are hot almost in the third degree, and withall sharpe and biting.

The leaues also haue some sharpnesse, with an opening power, but not so strong as the iuice.

¶ *The Vertues.*

A The dry Figs do nourish better than the greene or new Figs; notwithstanding they ingender not very good bloud, for such people as do feed much thereon doe become lowsie.

B Figs be good for the throat and lungs, they mitigate the cough, and are good for them that be short winded: they ripen flegme, causing the same to be easily spet out, especially when they be sodden with Hyssop, and the decoction drunke.

C Figges stamped with Salt, Rew, and the kernels of Nuts withstand all poyson and corruption of the aire. The King of Pontus, called *Mithridates*, vsed this preseruatiue against all venom and poyson.

D Figs stamped and made into the forme of a plaister with wheat meale, the pouder of Fenugreek, and Lineseed, and the roots of marish Mallowes, applied warme, do soften and ripen impostumes, phlegmons, all hot and angry swellings and tumors behinde the eares: and if you adde thereto the roots of Lillies, it ripeneth and breaketh Venerious impostumes that come in the flanke, which impostume is called *Bubo*, by reason of his lurking in such secret places: in plaine English termes they are called botches.

E Figs boiled in Wormwood wine with some Barly meale are very good to be applied as an implaister vpon the bellies of such as haue the dropsie.

F Dry Figges haue power to soften, consume, and make thinne, and may be vsed both outwardly and inwardly, whether it be to ripen or soften impostumes, or to scatter, dissolue, and consume them.

G The leaues of the Fig tree do waste and consume the Kings Euill, or swelling kernells in the throat, and do mollifie, waste, and consume all other tumors, being finely pouned and laid thereon: but after my practise, being boiled with the roots of marish Mallowes vntill they be soft, and so incorporated together, and applied in forme of a plaister.

H The milky iuyce either of the figs or leaues is good against all roughnesse of the skinne, lepries, spreading sores, tetters, small pockes, measels, pushes, wheales, freckles, lentiles, and all other spots, scuruinesse, and deformitie of the body and face, being mixed with Barley meale and applied: it doth also take away warts and such like excrescences, if it be mingled with some fattie or greasie thing.

I The milke doth also cure the tooth-ache, if a little lint or cotton be wet therein, and put into the hollownesse of the tooth.

K It openeth the veines of the hemorrhoids, and looseneth the belly, being applied to the fundament.

L Figs stamped with the pouder of Fenugreeke, and vineger, and applied plaisterwise, doe ease the intollerable paine of the hot gout, especially the gout of the feet.

M The milke thereof put into the wound proceeding of the biting of a mad dog, or any other venomous beast, preserueth the parts adioyning, taketh away the paine presently, and cureth the hurt.

N The greene and ripe Figs are good for those that be troubled with the stone of the kidneyes, for they make the conduits slipperie, and open them, and do also somewhat clense: whereupon after the eating of the same, it hapneth that much grauell and sand is conueyed forth.

O Dry or barrell Figs, called in Latine *Caricæ*, are a remedie for the belly, the cough, and for old infirmities of the chest and lungs: they scoure the kidnies, and clense forth the sand, they mitigate the paine of the bladder, and cause women with child to haue the easier deliuerance, if they feed thereof for certaine dayes together before their time.

P *Dioscorides* saith, that the white liquor of the Fig tree, and iuice of the leaues, do curdle milke as rennet doth, and dissolue the milke that is cluttered in the stomacke, as doth vineger.

Q It bringeth downe the menses, if it be applied with the yolke of an egge, or with yellow wax.

CHAP.

Chap. 134. *Of the prickly Indian Fig tree.*

Ficus Indica.
The Indian Fig tree.

Fructus.
The fruit.

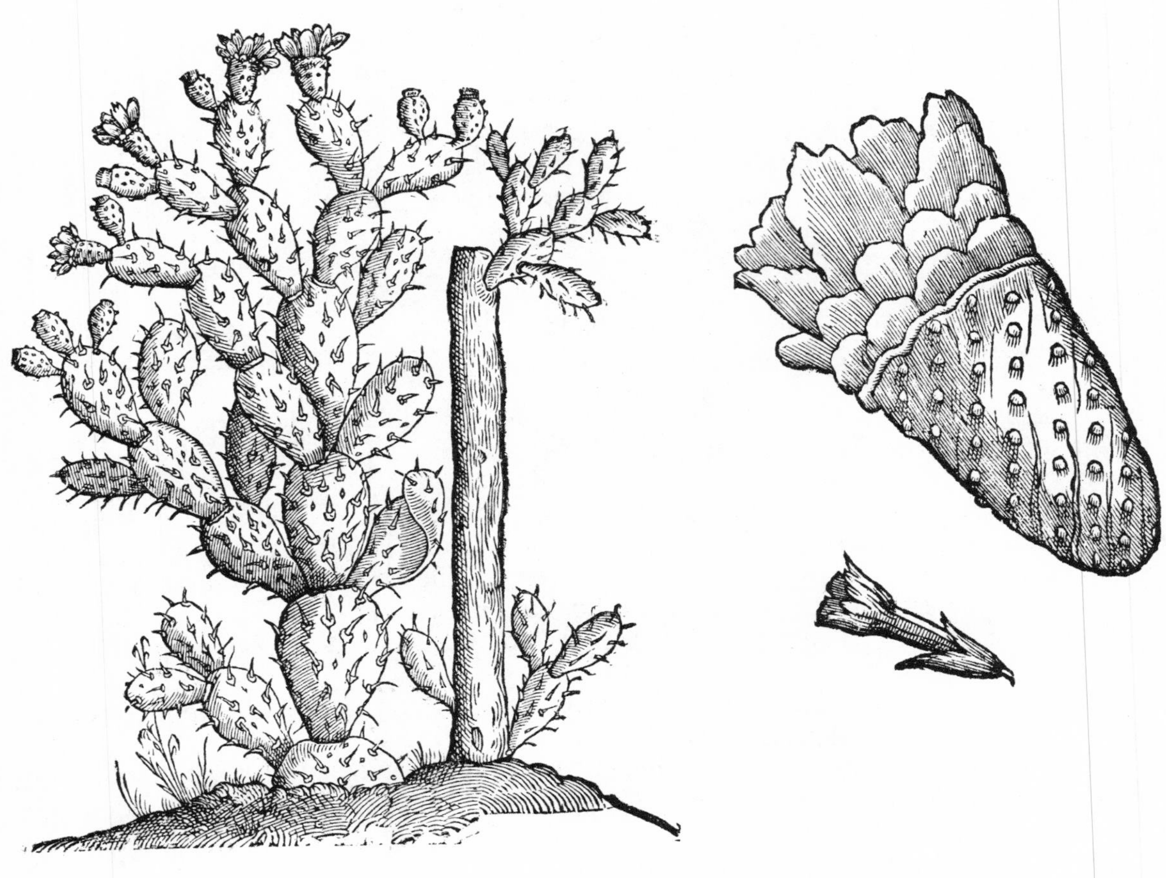

¶ *The Description.*

This strange and admirable plant, called *Ficus Indica*, seemes to be no other thing than a multiplication of leaues, that is, a tree made of leaues, without body or boughes; for the leafe set in the ground doth in short space take root, and bringeth out of it selfe other leaues, from which do grow others one after another, till such time as they come to the height of a tree, hauing also in the meane season boughes as it were comming from those leaues, sometimes more, otherwhiles fewer, as Nature list to bestow, adding leafe vnto leafe, whereby it occupieth a great piece of ground: these leaues are long and broad, as thicke as a mans thumbe, of a deepe greene colour, set full of long, slender, sharpe, and whitish prickles: on the tops of which leaues come forth long floures not vnlike to those of the manured Pomegrenat tree, of a yellow colour: after which commeth the fruit like vnto the common Fig, narrow below, and bigger aboue, of a greene colour, and stuffed full of a red pulpe and iuice, staining the hands of them that touch it, as do the Mulberries, with a bloudy or sanguine colour: the top of which Figs are inuironed with certaine scaly leaues like a crowne, wherein are also contained small graines that are the seeds: the which being sowne, do bring forth plants round bodied, like vnto the trunke of other trees, with leaues placed thereon like the other; which being set in the ground bring forth trees of leaues, as we haue shewed.

‡ Vpon this plant in some parts of the West Indies grow certain excrescences, which in continuance of time turn into Insects; and these out-growings are that high prized Cochenele wherwith they dye colours in graine. ‡

¶ *The Place.*

This plant groweth in all the tract of the East and West Indies, and also in the countrey Norembega, now called Virginia, from whence it hath beene brought into Italy, Spaine, England, and other countries: in Italy it sometimes beareth fruit, but more often in Spaine, and neuer as yet in England, although I haue bestowed great pains and cost in keeping it from the iniury of our cold clymat.

It groweth

It groweth also at S. Crux and other places of Barbary, and also in an Island of the Mediterranean sea, called Zante, about a day and nights sailing with a meane winde from Petrasse a port in Morea, where my seruant *William Marshall* (before remembred) did see not only great store of those trees made of leaues, but also diuers other round bodied plants of a woody substance: from whence he brought me diuers plants thereof in tubs of earth, very fresh and greene, which flourished in my garden at the impression hereof.

¶ *The Time.*

These plants do grow greene and fresh both Winter and Sommer, by the relation of my foresaid seruant: notwithstanding they must be very carefully kept in these countries from the extremitie of Winter.

¶ *The Names.*

This is thought to be the plant called of *Pliny*, *Opuntium*; whereof he hath written, *lib.* 21. *ca.* 17. in this manner: About Opuns is the herbe *Opuntia*, to mans taste sweet, and it is to be maruelled, that the root should be made of the leaues, and that it should so grow. Opuns is a city neere vnto Phocis in Greece, as *Pausanias*, *Strabo*, and *Pliny* testifie: but it is commonly called in Latine, *Ficus Indica*: of the Indians, *Tune*, and *Tunas*, and also *Anapallus*, as testifieth *Bellonius*: in English, Indian Fig tree.

There is a certaine other described for the Indian Fig tree, by *Theophrastus*, *lib.* 4. which *Pliny*, *lib.* 12. *cap.* 5. doth eloquently expresse almost in the same words, but turned into Latine, whereof we intend to speake in the next chapter.

¶ *The Temperature and Vertues.*

We haue no certaine instruction from the Antients, of the temperature or faculty of this plant, or of the fruit thereof: neither haue we any thing whereof to write of our owne knowledge, more than that we haue heard reported of such as haue eaten liberally of the fruit hereof, that it changed their vrine to the colour of bloud; who at the first sight thereof stood in great doubt of their life, thinking it had been bloud, whereas it proued afterwards by experience to be nothing but the tincture or colour the vrine had taken from the iuice of the fruit, and that without all hurt or griefe at all. A

It is reported of some, that the iuice of the fruit is excellent good against vlcers of long continuance. B

‡ Cochenele is giuen alone, and mixed with other things, in maligne diseases, as pestilent feuers and the like, but with what successe I know not. ‡ C

CHAP. 135. *Of the arched Indian Fig tree.*

¶ *The Description.*

THis rare and admirable tree is very great, straight, and couered with a yellowish bark tending to tawny: the boughes and branches are many, very long, tough, and flexible, growing very long in short space, as do the twigs of Oziars, and those so long and weake, that the ends thereof hang downe and touch the ground, where they take root and grow in such sort, that those twigs become great trees: and these being growne vp vnto the like greatnesse, doe cast their branches or twiggy tendrels vnto the earth, where they likewise take hold and root; by meanes whereof it commeth to passe, that of one tree is made a great wood or desart of trees, which the Indians do vse for couerture against the extreme heate of the Sun, wherewith they are grieuously vexed: some likewise vse them for pleasure, cutting downe by a direct line a long walke, or as it were a vault, through the thickest part, from which also they cut certaine loope-holes or windowes in some places, to the end to receiue thereby the fresh coole aire that entreth thereat, as also for light, that they may see their cattell that feed thereby, to auoid any danger that might happen vnto them either by the enemie or wilde beasts: from which vault or close walke doth rebound such an admirable echo or answering voice, if one of them speake vnto another aloud, that it doth resound or answer againe foure or fiue times, according to the height of the voice, to which it doth answer, and that so plainly, that it cannot be knowne from the voice it selfe: the first or mother of this wood or desart of trees is hard to be knowne from the children, but by the greatnesse of the body, which three men can scarsely fathom about: vpon the branches whereof grow leaues hard and wrinckled, in shape like those of the Quince tree, greene aboue, and of a whitish hoary colour vnderneath, whereupon the Elephants delight to feed: among which leaues come forth the fruit, of the bignes of a mans thumbe, in shape like a small Fig, but of a sanguine or bloudy colour and of a sweet tast, but

but not so pleasant as the Figs of Spaine; notwithstanding they are good to be eaten, and withall very wholsome.

Arbor ex Goa, siue Indica.
The arched Indian Fig tree.

¶ *The Place.*

This wondrous tree groweth in diuers places of the East Indies, especially neere vnto Goa, and also in Malaca: it is a stranger in most parts of the world.

¶ *The Time.*

This tree keepeth his leaues green winter and Sommer.

¶ *The Names.*

This tree is called of those that haue trauelled, *Ficus Indica*, the Indian Fig; and *Arbor Goa*, of the place where it groweth in greatest plenty: we may call it in English, the arched Fig tree.

‡ Such as desire to see more of this Fig tree, may haue recourse to *Clusius* his *Exoticks, lib.1.cap.1.* where he shewes it was mentioned by diuers antient Writers, as *Q. Curtius, lib.9. Plin. lib.12.ca.5. Strabo, lib.5.* and *Theophr. Hist.Plant. lib.4.cap.5.* by the name of *Ficus Indica.* ‡

¶ *The Temperature and Vertues.*

We haue nothing to write of the temperature or vertues of this tree, of our owne knowledge: neither haue wee receiued from others more, than that the fruit hereof is generally eaten, and that without any hurt at all, but rather good, and also nourishing.

Chap. 136.

Of Adams Apple tree, or the West-Indian Plantaine.

¶ *The Description.*

WHether this plant may be reckoned for a tree properly, or for an herby Plant, it is disputable, considering the soft and herby substance whereof it is made; that is to say, when it hath attained to the height of six or seuen cubits, and of the bignesse of a mans thigh, notwithstanding it may be cut downe with one stroke of a sword, or two or three cuts with a knife, euen with as much ease as the root of a Radish or Carrot of the like bignesse: from a thicke fat threddy root rise immediately diuers great leaues, of the length of three cubits and a halfe, sometimes more, according to the soile where it groweth, and of a cubit and more broad, of bignes sufficient to wrap a childe in of two yeares old, in shape like those of Mandrake, of an ouerworn green colour, hauing a broad rib running thorow the middle thereof: which leaues, whether by reason of the extreme hot scorching Sun, or of their owne nature, in September are so dry and withered, that there is nothing thereof left or to be seene but onely the middle rib. From the middest of these leaues riseth vp a thicke trunke, whereon doth grow the like leaues, which the people do cut off, as also those next the ground, by meanes whereof it riseth vp to the height of a tree, which otherwise would remaine a low and base plant. This manner of cutting they vse from time to time, vntill it come to a certaine height, aboue the reach of the Elephant, which greedily seeketh after the fruit. In the middest of the top among the leaues commeth forth a soft and fungous stumpe, whereon do grow diuers apples in forme like a small Cucumber, and of the same bignesse, couered with

with a thin rinde like that of the Fig, of a yellow colour when they be ripe: the pulpe or substance of the meate is like that of the Pompion, without either seeds, stones, or kernels, in tast not greatly perceiued at the first, but presently after it pleaseth, and entiseth a man to eat liberally thereof, by a certaine entising sweetnes it yeelds: in which fruit, if it be cut according to the length (saith myne Author) oblique, transuerse, or any other way whatsoeuer, may be seen the shape and forme of a crosse, with a man fastned thereto. My selfe haue seene the fruit, and cut it in pieces, which was brought me from Aleppo in pickle; the crosse I might perceiue, as the forme of a spred-Egle in the root of Ferne; but the man I leaue to be sought for by those that haue better eyes and iudgment than my selfe.

Musa Serapionis.
Adams Apple tree.

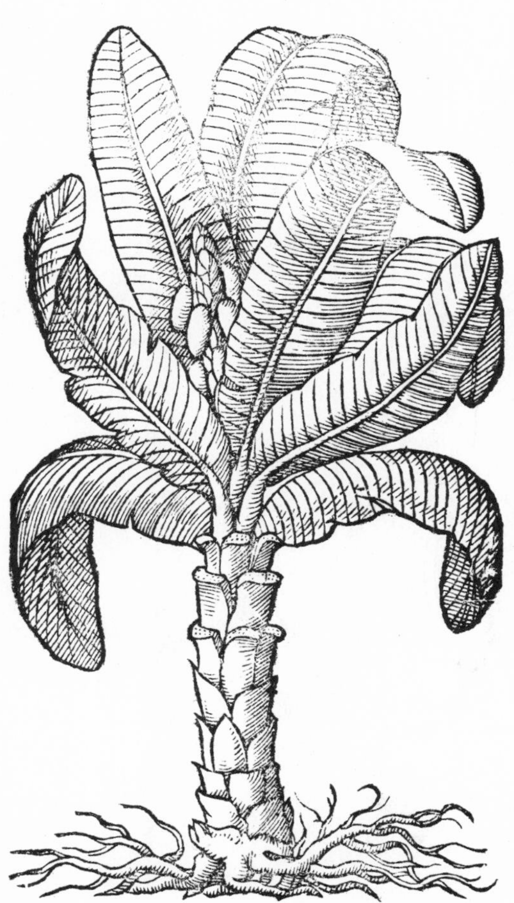

Musæ Fructus.
Adams Apple.

‡ Aprill 10. 1633. my much honored friend D. *Argent* (now President of the Colledge of Physitions of London) gaue me a plant he receiued from the Bermuda's: the length of the stalke was some two foot; the thicknesse thereof some seuen inches about, being crested, and full of a soft pith, so that one might easily with a knife cut it asunder. It was crooked a little, or indented, so that each two or three inches space it put forth a knot of some halfe inch thicknesse, and some inch in length, which incompassed it more than halfe about; and vpon each of these ioints or knots, in two rankes one aboue another, grew the fruit, some twenty, nineteene, eighteene, &c. more or lesse, at each knot: for the branch I had, contained nine knots or diuisions, and vpon the lowest knot grew twenty, and vpon the vppermost fifteene. The fruit which I receiued was not ripe, but greene, each of them was about the bignesse of a large Beane; the length of them some fiue inches, and the bredth some inch and halfe: they all hang their heads downewards, haue rough or vneuen ends, and are fiue cornered; and if you turne the vpper side downward, they somewhat resemble a boat, as you may see by one of them exprest by it selfe: the huske is as thicke as a Beanes, and will easily shell off it: the pulpe is white and soft: the stalke whereby it is fastned to the knot is verie short, and almost as thicke as ones little finger. This stalke with the fruit thereon I hanged vp in my shop, were it became ripe about the beginning of May, and lasted vntil Iune: the pulp or meat was very soft and tender, and it did eate somewhat like a Muske-Melon. I haue giuen you the figure of the whole branch, with the fruit thereon, which I drew as soone as I receiued it, and it is marked with this figure 1. The figure 2. sheweth the shape of one particular fruit, with the lower side

ſide vpwards. 3. The ſame cut through the middle long wayes. 4. The ſame cut ſide wayes. I haue been told (but how certaine it is I know not) that the floures which precede the fruit are bell-faſhioned, and of a blew colour. I could obſerue no ſeed in the fruit; it may be it was becauſe it had been cut from the ſtocke ſo long before it came to maturitie. This Plant is found in many places of Aſia, Africke, and America, eſpecially in the hot regions: you may find frequent mention of it amongſt the ſea voyages to the Eaſt and Weſt Indies, by the name of Plantaines, or *Platanus, Bannanas, Bonnanas, Bouanas, Dauanas, Poco, &c.* ſome (as our Author hath ſaid) haue iudged it the forbidden fruit; other-ſome, the Grapes brought to *Moſes* out of the Holy-land. ‡

Muſæ fructus exactior Icon.
An exacter figure of the Plantaine fruit.

¶ *The Place.*

This admirable tree groweth in Egypt, Cyprus, and Syria, neere vnto a chiefe city there called Alep, which we call Aleppo; and alſo by Tripolis, not far from thence: it groweth alſo in Canara, Decan, Guzarate, and Bengala, places of the Eaſt Indies.

¶ *The Time.*

From the root of this tree ſhooteth forth yong ſprings or ſhoots, which the people take vp and plant for the increaſe in the Spring of the yeare. The leaues wither away in September, as is aboue ſaid.

¶ *The Names.*

It is called *Muſa* by ſuch as trauell to Aleppo: by the Arabians, *Muſa Maum*: in Syria, *Moſe*: The Grecians and Chriſtians which inhabit Syria, and the Iewes alſo, ſuppoſe it to be that tree of whoſe fruit *Adam* did taſte; which others thinke to be a ridiculous fable: of *Pliny, Opuntia.*

It is called in the Eaſt Indies (as at Malauar where it alſo groweth) *Palan*: in Malayo, *Pican*: and in that part of Africa which we call Ginny, *Bananas*: in Engliſh, Adams Apple tree.

¶ *The Temperature.*

Dioſcorides and *Serapio* iudge, that it heateth in the end of the firſt degree, and moiſtneth in the end of the ſame.

¶ *The Vertues.*

A The fruit hereof yeeldeth but little nouriſhment: it is good for the heate of the breaſt, lungs, and bladder: it ſtoppeth the liuer, and hurteth the ſtomacke if too much of it be eaten, and procureth

cureth looſeneſſe in the belly: whereupon it is requiſit for ſuch as are of a cold conſtitution, in the eating thereof to put vnto it a little Ginger or other ſpice.

It is alſo good for the reines, or kidnies, and to prouoke vrine: it nouriſheth the childe in the mothers wombe, and ſtirreth to generation. B

CHAP. 137. *Of the Date tree.*

Palma.
The Date tree.

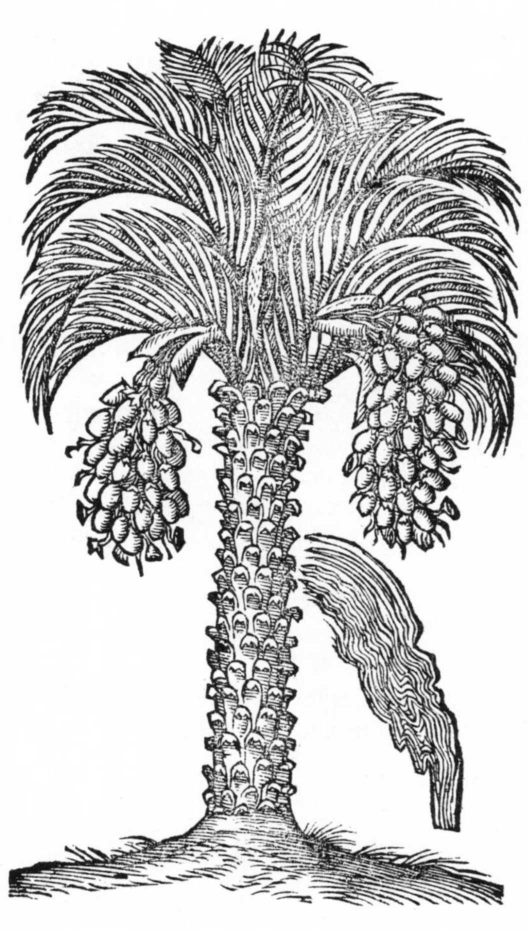

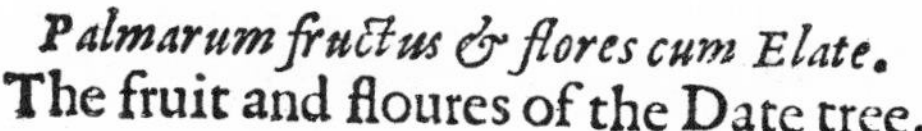

Palmarum fructus & flores cum Elate.
The fruit and floures of the Date tree.

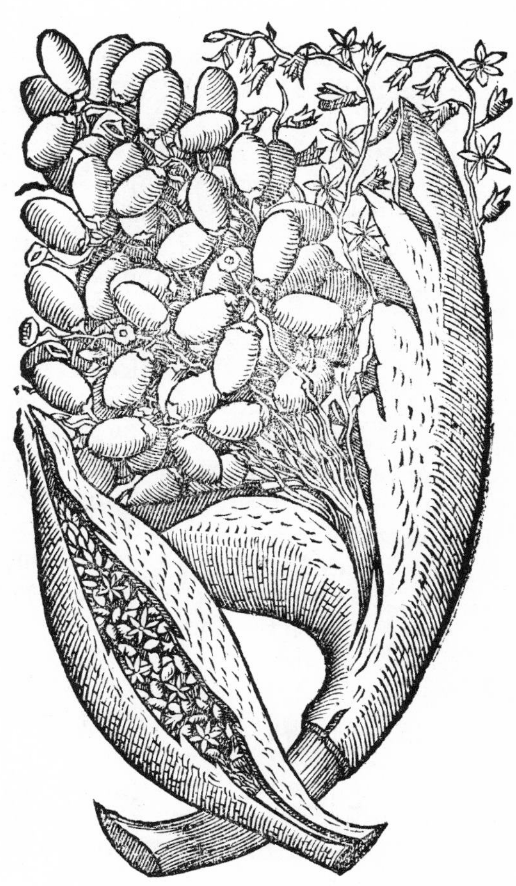

¶ *The Deſcription.*

THe Date tree groweth very great and high: the body or trunke thereof is thicke, and couered with a ſcaly rugged barke, cauſed by the falling away of the leaues: the boughes grow onely on the top, conſiſting of leaues ſet vpon a wooddy middle rib like thoſe of Reeds or Flags: the inner part of which rib or ſtalke is ſoft, light, hollow, and ſpongie. Among the leaues come forth the floures included in a long skinny membrane, as it were a ſheath or hoſe, like that which couereth the Floure de-Luce before it be blowne, which being opened of it ſelfe, white floures ſtart forth, ſtanding vpon ſhort and ſlender foot-ſtalkes, which are faſtened with certaine ſmall filaments or threddy ſtrings like vnto little branches: after which ſpring out from the ſame branches the fruit or Dates, which be in faſhion long and round, in taſte ſweet, and many times ſomewhat harſh, of a yellowiſh red colour; wherein is contained a long hard ſtone, which is in ſtead of kernell and ſeed; the which I haue planted many times in my Garden, and haue growne to the height of three foot: but the firſt froſt hath nipped them in ſuch ſort, that ſoone after they periſhed, notwithſtanding my induſtrie by couering them, or what elſe I could doe for their ſuccour.

¶ *The Place.*

The Date trees grow plentifully in Africa and Egypt; but thoſe which are in Paleſtina and

Syria,

Syria be the best: they grow likewise in most places of the East and West Indies, where there be diuers sorts, as well wilde, as tame or manured.

¶ *The Time.*

The Date tree is alwaies green, and floureth in the Spring time: the fruit is ripe in September, and being then gathered they are dried in the Sunne, that they may be the better both transported into other countries far distant, as also preserued from rotting at home.

¶ *The Names.*

The tree is called in Greeke, φοῖνιξ: in Latine, *Palma*: in English, Date tree.

The fruit is named in Greeke, βάλανος φοινίκων: that is to say, *Glans Palmarum*, or the fruit of the Date trees: and by one word, φοινικοβάλανος: in Latine, *Palmula*: in shops, *Dactylus*: in high-Dutch, **Datteln**: in low-Dutch, **Dadelen**: in Italian, *Dattoli*: in French, *Dattes*: in Spanish, *Tamaras*, and *Datiles*: in English, Date.

The cod or sheath wherein the floures and Dates are wrapped, is called ἐλάτη: and of some, βίρακος.

¶ *The Temperature and Vertues.*

A All manner of Dates whatsoeuer are hard of digestion, and cause head-ache: the worser sort be those that be dry and binding, as the Egyptian Dates; but the soft, moist, and sweet ones are lesse hurtfull.

B The bloud which is ingendred of Dates in mans body is altogether grosse, and somewhat clammy: by these the liuer is very quickly stopped, especially being inflamed and troubled with some hard swelling: so is the spleene likewise.

C The Dates which grow in colder regions, when they cannot come to perfect ripenesse, if they be eaten too plentifully, do fill the body full of raw humors, ingender winde, and oft times cause the leprosie.

D The drier sorts of Dates, as *Dioscorides* saith, be good for those that spet bloud, for such as haue bad stomacks, and for those also that be troubled with the bloudy flix.

E The best Dates, called in Latine *Caryotæ*, are good for the roughnesse of the throat and lungs.

F There is made hereof both by the cunning Confectioners and Cookes, diuers excellent cordiall, comfortable, and nourishing medicines, and that procure lust of the body very mightily.

G They do also refresh and restore such vnto strength as are entring into a consumption, for they strengthen the feeblenesse of the liuer and spleene, being made into conuenient broths, and physicall medicines directed by a learned Physitian.

H Dry Dates do stop the belly, and stay vomiting, and the wambling of womens stomackes that are with childe, if they be either eaten in meates or otherwise, or stamped and applied vnto the stomacke as a pectorall plaister.

I The ashes of the Date stones haue a binding qualitie, and emplastick facultie, they heale pushes in the eyes, *Staphylomata*, and falling away of the haire of the eye lids, being applied together with Spikenard: with wine it keepeth proud flesh from growing in wounds.

K The boughes and leaues do euidently binde, but especially the hose, that is to say, the sheath or case of the floures: and therefore it is good to vse these so oft as there is need of binding.

L The leaues and branches of the Date tree do heale greene wounds and vlcers, refresh and coole hot inflammations.

M *Galen* in his booke of Medicines according to the kindes mentioneth a composition called *Diapalma*, which is to be stirred with the bough of a Date tree in stead of a spature or a thing to stirre with, for no other cause than that it may receiue thereby some kinde of astriction or binding force.

Chap. 138. *Of the wilde Date trees.*

¶ *The Description.*

T*Heophrastus* maketh this plant to be a kinde of Date tree, but low and of small growth, seldome attaining aboue the height of a cubit: on the top whereof shoot forth for the most part long leaues like those of the Date tree, but lesser and shorter; from the sides whereof breakes forth a bush of threddy strings: among which riseth vp small branches garnished with clusters of white floures, in which before they be opened are to be seene vnperfect shapes of leaues, closely compassed about with an innumerable sort of thin skinny hulls; which rude shapes with the floures are serued vp and eaten at the second course among other iunkets, with a little salt and pepper, being pleasant to the taste. ‡ The stalke is about the thicknes of ones little

1 *Palmites, siue Chamærriphes.*
The little wilde Date tree.

2 *Palmapinus, siue Palma conifera.*
The wilde Date tree bearing cones.

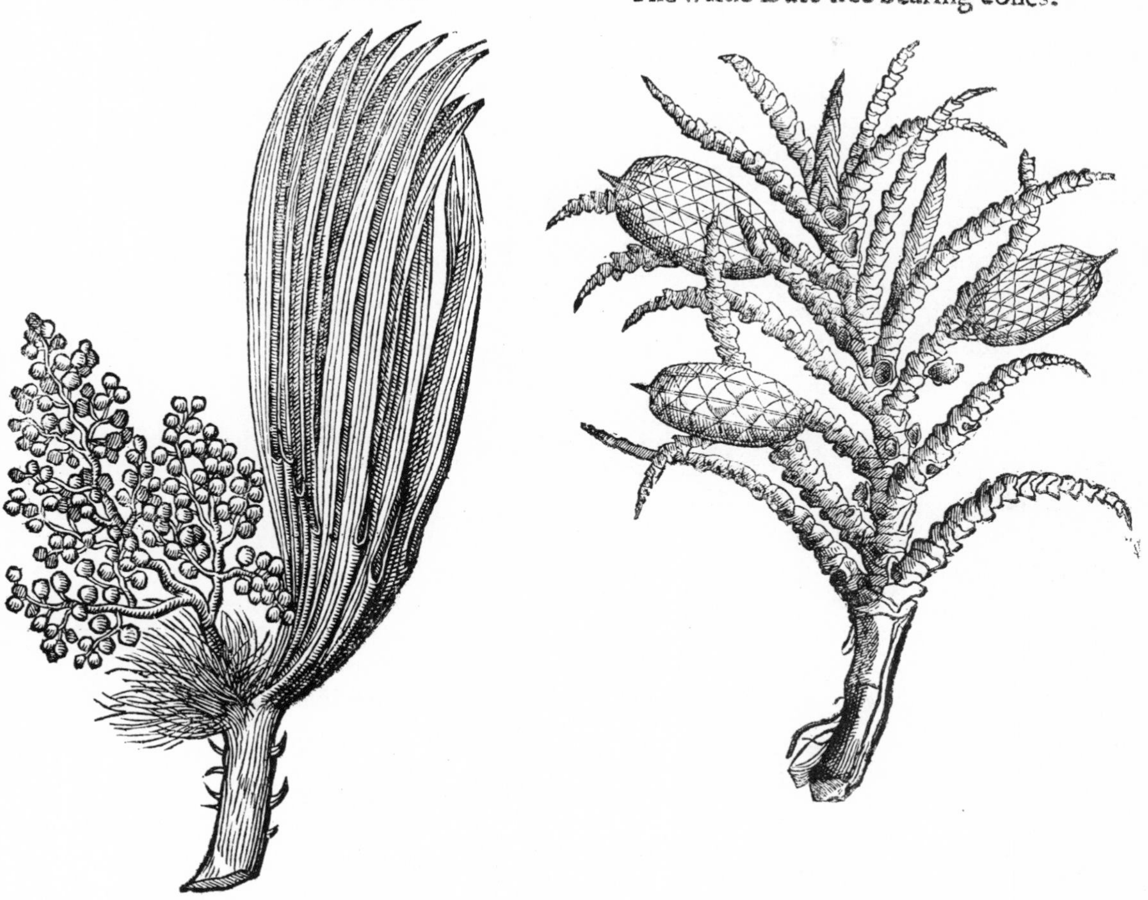

‡ *Fructus Palmapini.*
The fruit of the Cone-Date.

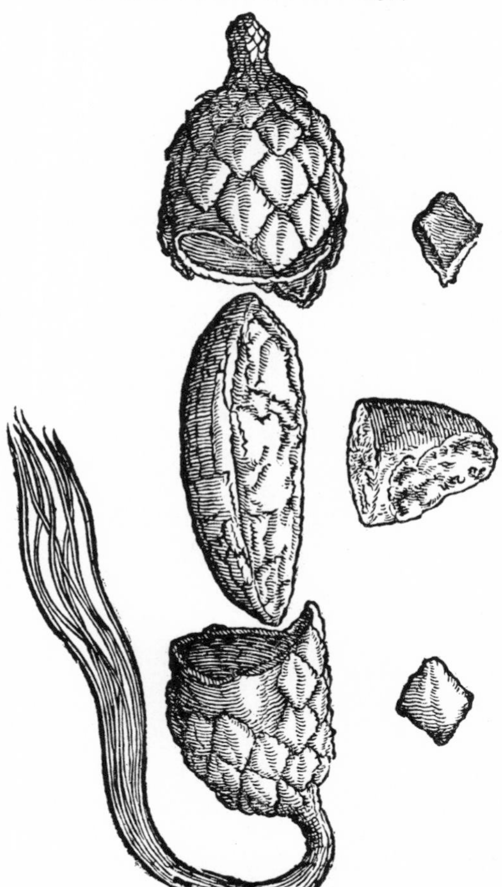

little finger, here and there set with a few crooked pricks: the leaues within some handfull or two of the stalke are cut vp and made into little besomes, which are sold in many glasse shops here in London. ‡

2 The wilde Date tree that brings forth cones or key-clogs, is of most trauellers into the Indies thought to be barren of Dates, except sometimes it yeeldeth forth some small berries like vnto Dates, but dry, and nothing worth. This tree groweth to the height and bignesse of a low tree; the trunke or body whereof is soft, of a fungous or pithy substance, vnfit for building, as is the manured Date tree: the branch it selfe was brought vnto vs from the Indies, dry & void of leaues, wherefore we must describe the leaues by report of the bringer. The branches (saith my Author) are couered ouer with long flaggie leaues, hanging downe of a great length like those of the Date tree: the branches are also couered with a scaly or scabbed barke, verie rough, one scale or plate lying ouer another, as tiles vpon a house: the fruit growes at the end of the branches, not vnlike a great Pine Apple cone, couered ouer with a skinne like the Indian Nut: wherein is contained a shel, within which shell lieth hid an acorn or long

kernell of an inch long, and sometimes longer, very hard to be broken, in taste like the Chestnut; which the sauage people do grate and stampe to pouder to make them bread.

¶ *The Place.*

Theophrastus saith the first growes in Candy, but much more plentifully in Cilicia, and are now found in certaine places of Italy by the sea side, and also in diuers parts of Spaine.

The other hath been found by trauellers into the West Indies, from whence haue bin brought the naked branches with the fruit.

¶ *The Time.*

The time answereth that of the manured Date tree.

¶ *The Names.*

The little Date tree or wilde Date tree is named of *Theophrastus*, χαμαιῤῥιφὴς: in Naples, *Cephaglione*. in Latine commonly *Palmites*. That which is found in the midst of the yong springs, and is vsed to be eaten in banquets, is called in Greeke, ἐγκέφαλος φοίνικος: in Latine, *Palmæ cerebrum*, the brain of the Date tree.

¶ *The Temperature and Vertues.*

A *Galen* supposeth that the brain of the Date tree consisteth of sundry parts, that is to say, of a certaine wateric and warme substance, and of an earthy and cold; therefore it is moist and cold, with a certaine astriction or binding qualitie.

B Being taken as a meat it ingendreth raw humors and winde, and therefore it is good to be eaten with pepper and salt.

Chap. 139. *Of the drunken Date tree.*

Areca, siue Fausel.
The drunken Date tree.

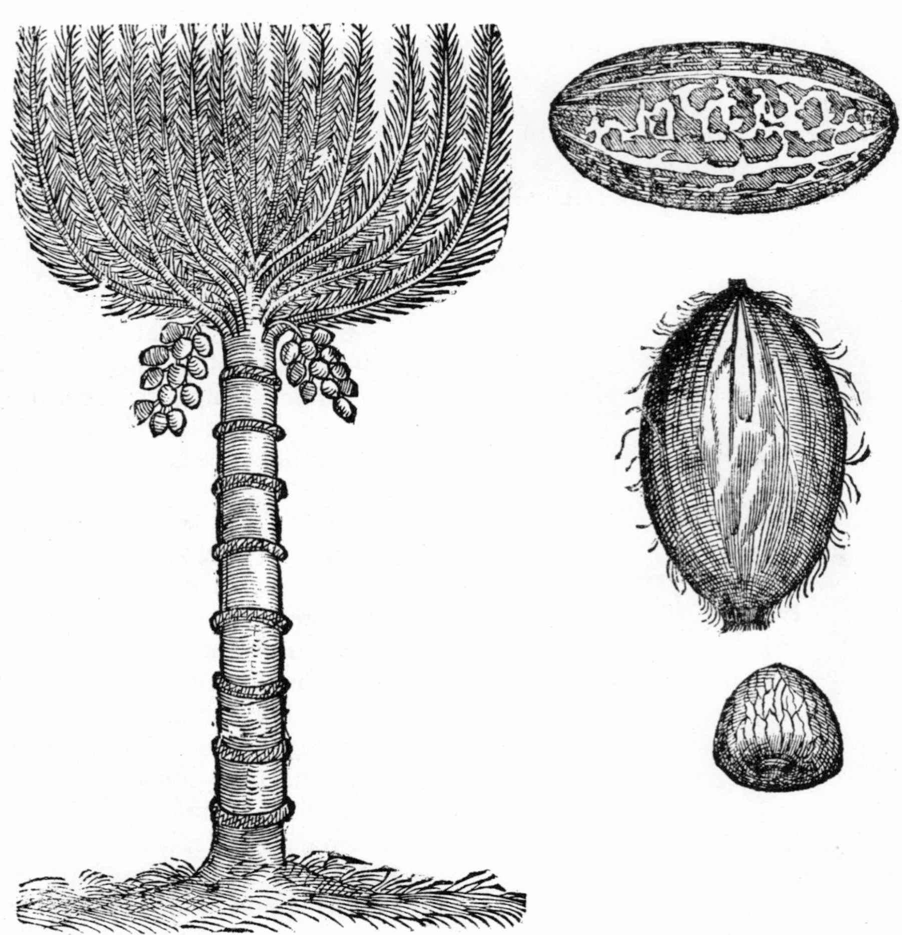

¶ *The Description.*

The drunken Date tree, which *Carolus Clusius* calleth *Fausel*, is an Indian tree of a great bignes, the timber whereof is very soft and spongious, exceeding smooth and plaine vnto a great height,

height, not possible to be climbed vp; and therefore the Indians for their easier ascending vp, at some distances do tie round about the tree certaine wyths or ropes made of the barkes of trees, as may be perceiued by the figure, whereby very easily they go vp and downe to gather the fruit at their pleasure. The top of the tree is diuided into sundry branches, in substance like to the great cane; whereupon do grow faire flaggie leaues like those of the Palme or Date tree, whereof doubtlesse this is a wilde kinde: from the bottome of which branches commeth forth fruit in long bunches like traces of Onions, couered with a soft pulpe like vnto the Wall-nut, rough, and verie full of haire of a yellowish colour, and like the dried Date when it is ripe: within which huske is contained fruit like vnto the Nutmeg, but greater, very hard, and striped ouer with red and white veines, or sinues.

¶ *The Place, Time, and Names.*

This Date tree, which the Arabians call *Faufel*, that is by interpretation, *Auellana Indica*, the Indian Nut or Filberd, *Auicen* and *Serapio* call *Filfel*, and *Fufel*. It groweth in the East Indies in diuers and sundry places, as in Malauar, where vulgarly it is called *Pac*; and of the Nobles and Gentlemen, *Areca*: which name is vsed amongst the Portugals which dwell in those Indies: in Guzarate and Decan it is called *Cupare*: in Zeilan, *Poaz*: in Malaca, *Pinan*: in Cochin, *Chacani*: in English, the drunken Date tree, which name we haue coined from his qualitie, because the fruit maketh those drunke that eate thereof.

¶ *The Temperature.*

It is cold and dry in the second degree.

¶ *The Vertues.*

The fruit of *Areca* before it be ripe is reckoned amongst the stupefactiue or astonishing medicines; for whosoeuer eateth thereof waxeth drunke, because it doth exceedingly amase and astonish the senses. A

When the Indians are vexed with some intolerable ache or paine, or must of necessitie endure some great torment or torture, then do they take of this fruit, whereby the rigor of that pain which otherwise they should feele, is very much mitigated. B

The iuice of the fruit of *Areca* doth strengthen the gums, fasten the teeth, comfort the stomack, stay vomiting and loosenesse of the belly: it doth also purge the body from congealed or clotted bloud gathered within the same. C

CHAP. 140. *Of the Indian Nut tree.*

¶ *The Description.*

1 THe Grecians haue not known, but the Arabians haue mentioned this Indian Nut tree, the body whereof is very great, smooth and plaine, void of boughes or branches, of a great height; wherefore the Indians do wrap ropes about the body thereof, as they doe vpon the tree last described, for their more ease in gathering the fruit: the timber whereof is verie spongie within, but hard without, a matter fit to make their Canoos and boats of: on the top of the tree grow the leaues like those of the Date tree, but broad, and sharpe at the point as thornes, whereof they vse to make needles, bodkins, and such like instruments, wherewith they sow the sailes of their ships, and do such like businesse: among these leaues come forth clusters of floures like those of the Chestnut tree, which turne into great fruit of a round forme, and somwhat sharp at one end; in that end next vnto the tree is one hole, somtimes two bored through: this Nut or fruit is wrapped in a couerture, consisting of a substance not vnlike to hempe before it be beaten soft: there is also a finer and gentler stuffe next vnto the shell, like vnto Flax before it be made soft: in the middle whereof is contained a great Nut couered with a very hard shell, of a browne colour before it be polished, afterward of a blacke shining colour like burnished horne: next vnto the shell vpon the inside there cleaueth a white cornelly substance firme and sollid, of the colour and taste of a blanched Almond: within the cauitie or hollownes thereof is contained a most delectable liquor like vnto milke, and of a most pleasant taste.

2 We haue no certaine knowledge from those that haue trauelled into the Indies, of the tree which beareth this little Indian Nut; neither haue we any thing of our owne knowledge, more, than that we see by experience that the fruit hereof is lesser, wherein consisteth the difference.

‡ The other, expressed in the same table with the former, by the name of *Mehenbethene*, *Clusius* receiued it by the same name from *Cortusus* of Padua: yet it doth not (as hee saith) well agree with the description; and he rather approues of their opinion who refer it to the *Nux vnguentaria*; of

1 *Nux Indica arbor.*
The Indian Nut tree.

Nux Indica.
The Indian Nut.

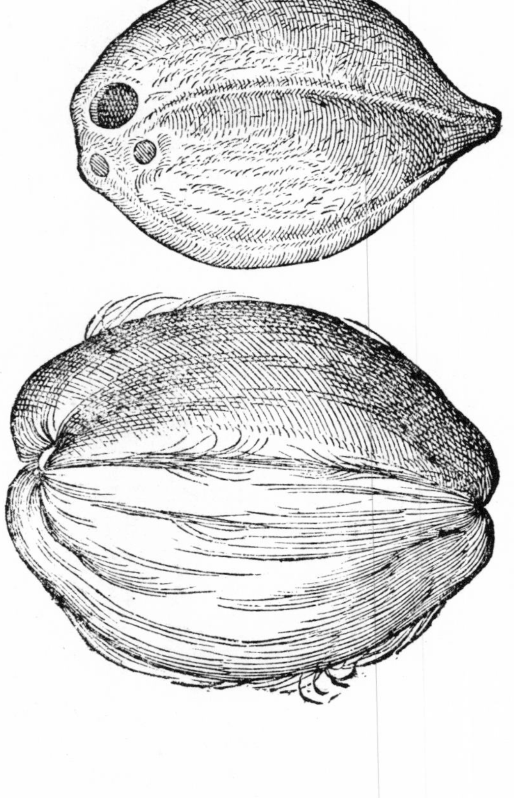

2 *Nucula Indica.*
The little Indian Nut.

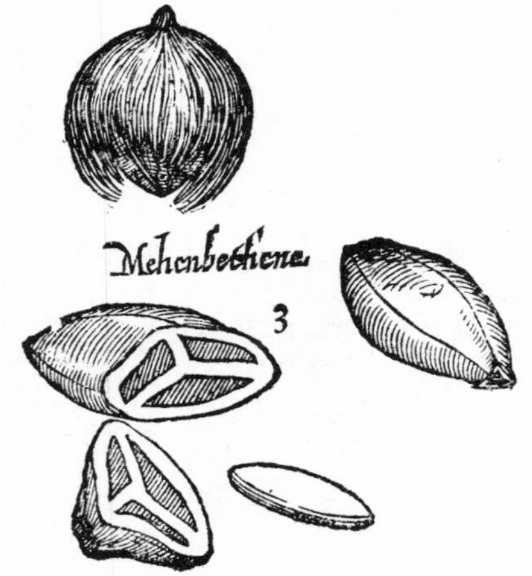

or *Ben*. It is some inch long, of a triangular figure, with a hard and wooddy shel: which broken, shewes three cells or partitions, in each whereof is contained a long kernell white and sweet. ‡

¶ *The Place.*

This Indian Nut groweth in some places of Africa, and in the East Indies, and in all the Islands of the West Indies, especially in Hispaniola, Cuba, and Saint Iohns Island, and also vpon the continent by Carthagena, Nombre de Dios, and Panama, and in Virginia, otherwise called Norembega, part of the same continent, for the most part neere vnto the sea side, and in moist places, but it is seldome found in the vplandish countries.

¶ *The Time.*

It groweth greene Winter and Sommer.

¶ *The Names.*

The fruit is called in Latine, *Nux Indica*: of the Indians, *Cocus*: of the Portugals that dwell in the East Indies, *Cocco*, taken from the end, wherein are three holes representing the head of a Monkie: *Serapio* and *Rhasis* do call this tree *Iaralnare*, *id est*, *Arborem Nuciferam*, the tree bearing Nuts: of *Auicen*, *Glauci al hend*: of the vulgar people, *Maro*, and the fruit *Narel*; which name *Narel* is common among the Persians and Arabians: it is called in Malauar, *Tengamaran*: the ripe fruit, *Tenga*; and the greene fruit, *Eleri*: in Goa it is called *Lanhan*: in Malaio, *Triccan*: and the Nut, *Nihor*.

The distilled liquor is called *Sula*; and the oile that is made thereof, *Copra*.

¶ *The Temperature.*

It is of a meane temper betwixt hot and cold.

¶ *The*

¶ *The Vertues and vse.*

The Indians do vse to cut the twigs and tender branches toward the euening, at the ends whereof they haue bottle gourds, hollow canes, and such like things, fit to receiue the water that droppeth from the branches thereof, which pleasant liquor they drinke in stead of wine; from the which is drawne a strong and comfortable Aqua Vitæ, which they vse in time of need against all manner of sicknesses; of the branches and boughes they make their houses; of the trunk or body of the tree, ships and boates; of the hempon the outward part of the fruit, they make ropes and cables; and of the finer stuffe, sailes for their ships. A

Likewise they make of the shell of the Nut, cups to drinke in, which we likewise vse in England, garnished with siluer for the same purposes. The kernell serueth them for bread and meat; the milkie iuice doth serue to coole and refresh their wearied spirits: out of the kernel when it is stamped, is pressed a most precious oile, not onely good for meat, but also for medicine, wherewith they annoint their feeble lims after their tedious trauell, by meanes whereof the ache and paine is mitigated, and other infirmities quite taken away proceeding of other causes. B

CHAP. 141. *Of the Dragon Tree.*

1 *Draco arbor.*
The Dragon tree.

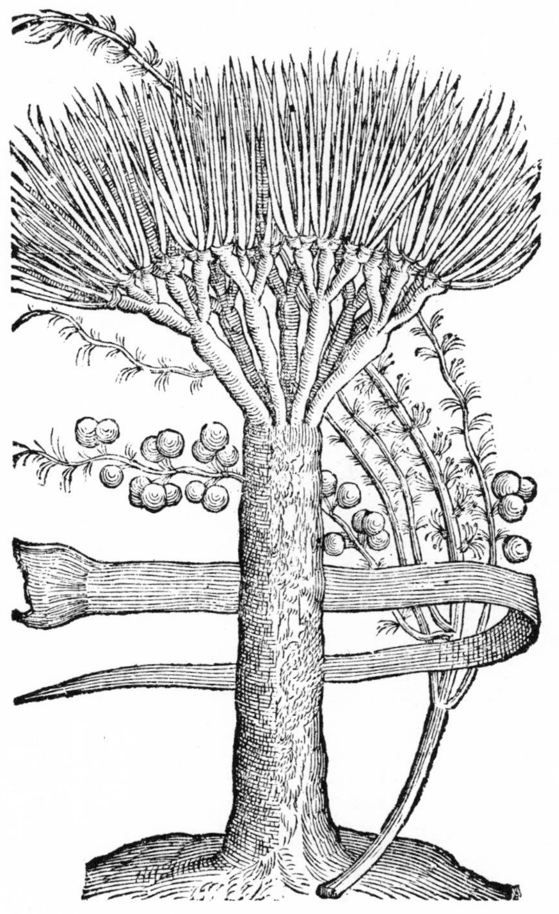

Draconis fructus.
The Dragon tree fruit.

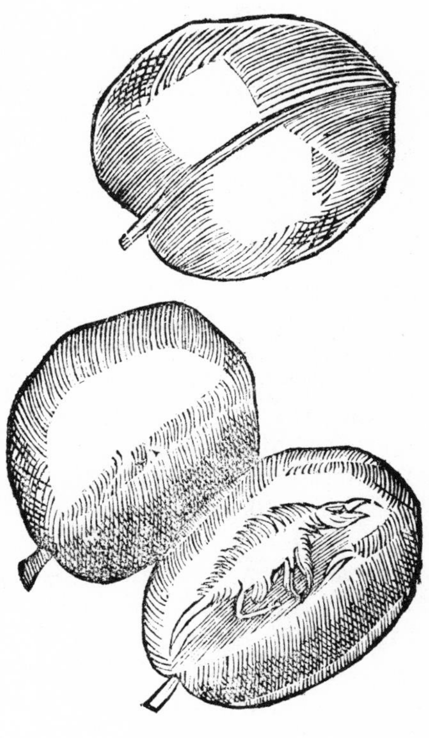

¶ *The Description.*

THis strange and admirable tree groweth very great, resembling the Pine tree, by reason it doth alwaies flourish, and hath his boughes or branches of equal length and bignesse, which are bare and naked, of eight or nine cubits long, and of the bignesse of a mans arme: from the ends of which do shoot out leaues of a cubit and a halfe long, and full two inches broad, somewhat thicke, and raised vp in the middle, then thinner and thinner like a two edged sword: among which come forth little mossie floures, of small moment, and turne into berries, of the bignesse of Cherries, of a yellowish

lowish colour, round, light and bitter, couered with a threefold skin or filme, wherein is to be seene, as *Monardus* and diuers other report, the forme of a Dragon, hauing a long necke or gaping mouth; the ridge or backe armed with sharpe prickles, like the Porcupine; it hath also a long taile, & foure feet, very easie to be discerned: the figure of it we haue set forth vnto you according to the greatnesse thereof, because our words and meaning may be the better vnderstood, and also the leafe of the tree in his full bignesse, because it is impossible to be expressed in the figure: the trunke or body of the tree is couered with a rough barke, very thin, and easie to be opened or wounded with any small toole or instrument; which being so wounded in the Dog daies, bruised or bored, doth yeeld forth drops of a thicke red liquour, which of the name of the tree are called Dragons teares, or *Sanguis Draconis*, Dragons bloud: diuers haue doubted whether the liquour or gummie iuice were all one with *Cinnabaris* of *Dioscorides* (not meaning that *Cinaber* made of Quicksiluer) but the receiued opinion is, they differ not, by reason their qualitie and temperature worke the like effect.

¶ *The Place.*

This tree groweth in an Island which the Portugals call Madera, and in one of the Canarie Islands, called *Insula Portus Sancti*; and as it seemeth it was first brought out of Africke, although some are of a contrary opinion, and say, that it was first brought from Carthagena, in America, by the Bishop of the same Prouince.

¶ *The Time.*

The time of his growing we haue touched in the description, where wee said that it flourisheth and groweth greene all the yeare.

¶ *The Names.*

The names haue beene sufficiently spoken of in the description and in their seuerall titles.

¶ *The Temperature and Vertues.*

A ‡ The *Sanguis Draconis* which is thought to proceed from this tree hath an astringent faculty, and is with good successe vsed in the ouermuch flowing of the courses, in fluxes, Dysenteries, spitting of bloud, fastening loose teeth, and such other affects which require astriction.

B Smiths also vse it to varnish ouer their workes to giue them a sanguine colour, and keep them from rust. ‡

Chap. 142. *Of the Sassafras or Ague tree.*

¶ *The Description.*

The Sassafras tree growes very great much like to the Pine tree: the trunk or body is straight, smooth, and void of boughs, of a great height: it is couered with a twofold grosse rind, the vppermost of the colour of ashes, that next the wood of a tawny colour: on the top come forth many goodly branches like those of the Palm tree, whereon grow green leaues somwhat like those of the fig tree, of a sweet smell when they be greene, but much sweeter when they be dry, declining to the smell of fennel, with much sweetnesse in tast: they are green Winter and Summer, neither bearing fruit nor floures, but is altogether barren as it is said: the roots are grosse, conformable to the greatnesse of the tree, of a tawny colour, dispersing themselues far abroad vnder the vpper crust of the earth, by means whereof they are often cast down with mean blasts of wind. ‡ The wood of the tree is very strong, hard, and brittle, it hath not so strong & pleasant a smell as that of the root, neither is it in such vse. The leaues are of two sorts, some long and smooth, and not snipt about the edges; othersome, and those chiefely on the end of the branches, are deepely gashed in, as it were diuided into three seuerall parts. I haue giuen the figure of a branch taken from a little tree, which grew in the garden of M[r] *Wilmot* at Bow; who died some few yeares ago.

¶ *The Place.*

This tree groweth in most parts of the West Indies, specially about the cape of Florida, Wingandico, and Virginia, otherwise named Norembega.

¶ *The Time.*

It flourisheth and keepeth green Winter and Summer.

¶ The

Sassafras.
The Sassafras tree.

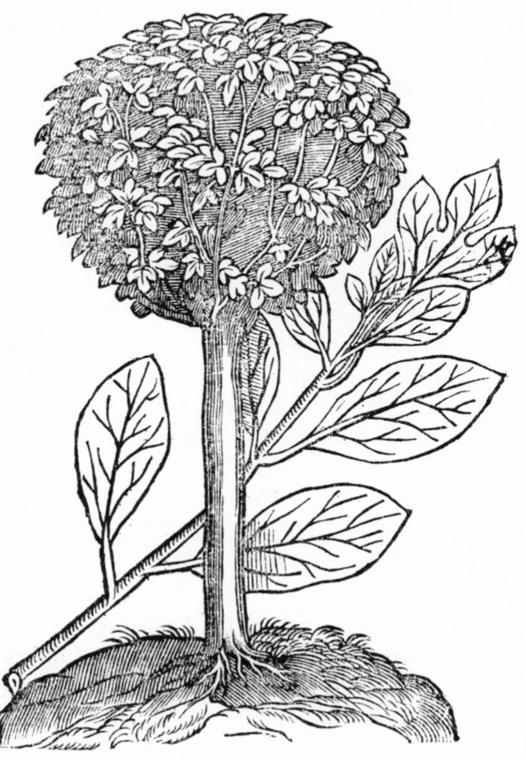

¶ *The Names.*

The Spaniards and French men haue named this tree, *Sassafras*: the Indians in their tongue, *Pauame*: for want of an English name we are contented to call it the Ague tree, of his vertue in healing the Ague.

¶ *The Temperature.*

The boughes and branches hereof are hot & dry in the second degree; the rinde is hotter, for that it entreth into the third degree of heate and drinesse, as is manifestly perceiued in the decoction.

¶ *The Vertues.*

The best of all the tree is the root, and that worketh the best effect, the which hath the rinde cleauing very fast to the inner part, and is of colour tawnie, and much more sweet of smell than all the tree and his branches. A

The rinde tasteth of a more sweet smell than the tree; and the water being sod with the root is of greater and better effects than any other part of the tree, and is of a more sweet smell, and therefore the Spaniards vse it, for that it worketh better and greater effects. B

It is a tree that groweth neere vnto the sea, and in temperate places that haue not much drouth, nor moisture. There be mountaines growing full of them, and they cast forth a most sweet smell, so that at the beginning when they saw them first, they thought they had been trees of Cinnamon, & in part they were not deceiued: for that the rinde of this tree hath as sweet a smell as Cinamon hath, and doth imitate it in colour and sharpnesse of taste, and pleasantnesse of smell: and so the water that is made of it is of a most sweet smell and taste, as the Cinamon is, and procureth the same works and effects as Cinamon doth. C

The wood hereof cut in smal pieces and boiled in water, to the colour of Claret wine, and drunk for certaine daies together, helpeth the dropsie, remoueth oppilation or stopping of the liuer, cureth quotidian and tertian agues, and long feuers. D

The root of Sassafras hath power to comfort the liuer, and to free from oppilations, to comfort the weake and feeble stomacke, to cause good appetite, to consume windinesse, the chiefest cause of cruditie and indigestion, stay vomiting, and make sweet a stinking breath. E

It prouoketh vrine, remoueth the impediments that doe cause barrennesse, and maketh women apt to conceiue. F

CHAP. 143. *Of the Storax tree.*

¶ *The Description.*

THe Storax tree groweth to the height and bignesse of the Quince tree: the trunke or bodie is couered with a barke or rinde like vnto the Birch tree: the branches are small and limmer, whereon do grow leaues like those of the Quince tree, greenish aboue, and whitish vnderneath: among which come forth white floures, like those of the Orange tree, of an vnpleasant smell: after commeth the fruit or berries, standing vpon long and slender footstalks, couered ouer with a little woollinesse, of the bignesse of a bladder nut, and of the same colour; wherein is contained small seed,

Styrax arbor.
The Storax tree.

ſeed, whereunto alſo cleaue certaine gummie teares,bearing the name of the tree, and which iſſue from the trunk or body when it is wounded.

¶ *The Place.*

This tree groweth in diuers places of France, Italy and Spaine, where it bringeth forth little or no gum at all: it groweth in Iudæa,Pamphylia,Syria,Piſidia,Sidon, and many other places of Iurie or Paleſtine, as alſo in diuers Iſlands in the Mediterranean ſea,namely Cyprus, Candy, Zant,and other places, where it bringeth forth his gummy liquour in full perfection of ſweetneſſe, and alſo in great plenty,where it is gathered and put into great Canes or Reeds, whereof as ſome deeme it took the name *Calamita*; others deeme of the leaues of Reeds wherein they wrap it: hereof I haue two ſmall trees in my garden, the which I raiſed of ſeed.

¶ *The Time.*

It floureth in May, and the fruit is ripe in September.

¶ *The Names.*

This tree, as may be gathered by ſome, was called *Styrax*, by reaſon of that gum or liquour which droppeth out of the ſame, being like vnto the hollow pipes of Iſe,that hang at the eaues of houſes in Winter,called Styria,or of the Canes or the leaues of Reeds ſpoken of before: in Latine, *Storax Calamitæ*: in Engliſh, Storax, which is kept in Canes or the leaues of Reeds: there floweth from ſome of theſe trees a certain gummie liquor,which neuer groweth naturally hard, but remaineth alwaies thinne,which is called liquid Styrax,or Storax.

¶ *The Temperature.*

The gum of this tree is of an heating,mollifying,and concocting qualitie.

¶ *The Vertues.*

A It helpeth the cough,the falling downe of rheumes and humours into the cheſt, and hoarſneſſe of the voice: it alſo helpeth the noiſe and ſounding of the eares, preuaileth againſt *Strumas*,or the Kings euill,nodes on the nerues,and hard ſwellings proceeding of a cold cauſe: it preuaileth alſo againſt all cold poiſons,as Hemlocks and ſuch like.

B Of this gum there are made ſundry excellent perfumes, pomanders, ſweet waters, ſweet bags, and ſweet waſhing balls, and diuers other ſweet chaines & bracelets, whereof to write were impertinent to this hiſtorie.

CHAP.144. *Of the Sorrowfull tree or Indian Mourner.*

¶ *The Deſcription.*

Arbor triſtis,the ſad or ſorrowfull tree waxeth as big as an Oliue tree,garniſhed with many goodly branches,ſet full of leaues like thoſe of the Plum tree: among which come forth moſt odoriferous and ſweet ſmelling floures,whoſe ſtalkes are of the colour of Saffron, which flouriſh and ſhew themſelues onely in the night time,and in the day time looke withered and with a mourning cheere: the leaues alſo at that time ſhrinke in themſelues together,much like a tender plant that is froſt bitten,very ſadly lumping, lowring, and hanging downe the head, as though it loathed the light,and could not abide the heate of the Sun. I ſhould but in vain loſe labour in repeating a fooliſh fanſie of the Poeticall Indians,who would make fooles beleeue, that this tree was once a faire daughter of a great Lord or King,and that the Sun was in loue with her, with other toies which I omit.

Arbor tristis.
The sorrowfull tree.

omit. ‡ The floures are white, somewhat like those of Iasmine, but more double, and they are of a very sweet smell: there succeed them many little cods, containing some six seeds a piece somewhat like those of *Stramonium*. ‡

¶ *The Place, Time, and Names.*

This tree groweth in the East Indies, especially in Goa, and Malayo: in Goa it is called *Parizataco*: in Malayo, *Singadi*: in Decan, *Pul*: of the Arabians, *Guart*: and of the Persians and Turkes, *Gul*: in English, the Sad or Sorrowfull tree, or the Indian mourner. The time is specified in the description.

¶ *The Temperature and Vertues.*

We haue no ceartaine knowledge of the temperature hereof, neuerthelesse we read that the Indians do colour their brothes and meates with the stalkes of the floures hereof in stead of Saffron, or whatsoeuer that they desire to haue of a yellow colour. A

It is reported, that if a linnen cloth be steeped in the distilled water of the floures; and the eyes bathed and washed therewith, helpeth the itching and paine therof, and staieth the humours that fall downe to the same. B

There is made of the splinters of the wood certaine tooth-picks, and many pretty toies for pleasure. C

CHAP. 145. *Of the Balsam tree.*

¶ *The Kindes.*

THere be diuers sorts of trees from which do flow Balsames, very different one from another, not onely in forme, but also in fruit, liquour, and place of growing; the which to distinguish would require more time and trauell than either our sm. ll time wil affoord, or riches for our maintenance to discouer the same in their naturall countries: which otherwise by report to set downe certaine matter by incertainties, would discredit the Author, and no profit shall arise thereby to the Reader: notwithstanding we wil set downe so much as we haue found in the workes of some trauellers, which best agree with the truth of the historie.

¶ *The Description.*

1 THere be diuers trees growing in the Indies, whose fruits are called by the name of the fruit of the Balsam tree: among the rest this whose figure we haue set forth vnto your view, we our selues haue seene and handled; and therefore the better able to describe it. It is a fruit very crooked, and hollowed like the palme of an hand, two inches long, halfe an inch thicke, couered with a thicke smooth rinde, of the colour of a drie Oken leafe; wherein is contained a kernell (of the same length and thicknesse, apt to fil the said shell or rinde) of the substance of an Almond; of the colour of ashes, fat, and oilie; of a good smell, and very vnpleasant in taste.

2 The wood we haue dry brought vnto vs from the Indies for our vse in Physicke (a small description may serue for a dry sticke) neuerthelesse wee haue other fruits brought from the Indies, whose figures are not set forth, by reason they are not so well knowne as desired; whereof one is of the bignes of a Wal-nut, somewhat broad on the vpper side, with a rough or rugged shell, vneuen, blacke of colour, and full of a white kernell, with much iuice in it; of a pleasant taste and smell, like the oile of Mace: the whole fruit is exceeding light, in respect of the quantitie or bignesse, euen as

it

it were a piece of Corke, which notwithstanding sinketh to the bottome when it falleth into the water, like as doth a stone.

1 *Balsami fructus.*
The fruit of the Balsam tree.

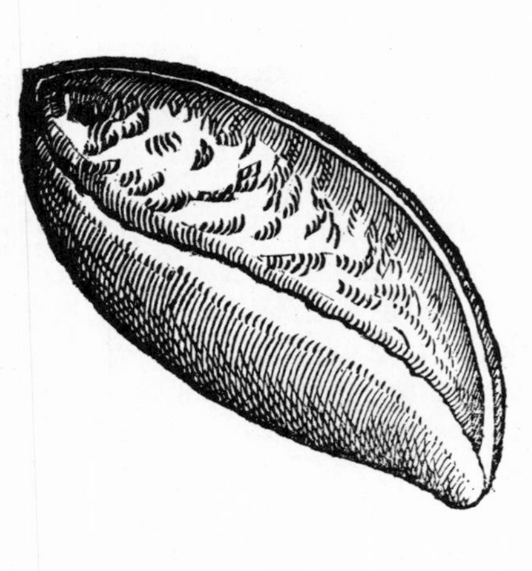

‡ 3 *Balsamum Alpini cum Carpobalsamo.*
The Balsam tree with the fruit.

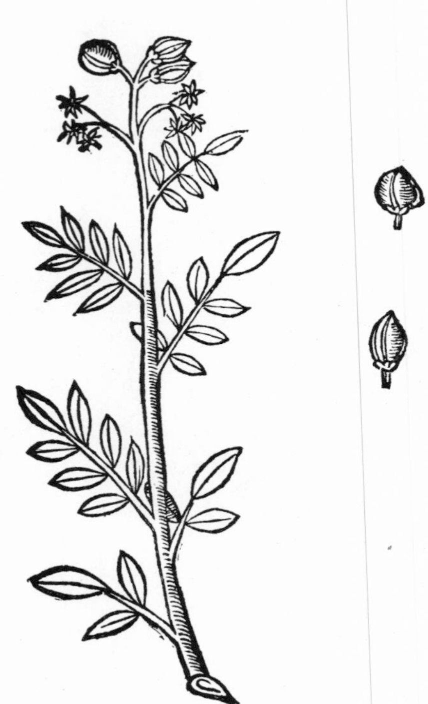

3 This tree, saith *Garcias*, that beareth the fruit *Carpobalsamum*, is also one of the Balsam trees: it groweth to the height and bignesse of the Pomegranate tree, garnished with very many branches: whereon do grow leaues like those of Rue, but of colour whiter, alwaies growing greene: amongst which come forth floures, whereof we haue no certaintie: after which commeth forth fruit like that of the Turpentine tree, which in shoppes is called *Carpobalsamum*, of a pleasant smell; but the liquour which floweth from the wounded tree is much sweeter: which liquour of some is called *Opobalsamum*.

‡ *Prosper Alpinus* hath writ a large Dialog e of the Balsam of the Antients, and also figured and deliuered the historie thereof in his booke *De Plant. Ægypti, cap.* 14. whether I refer the curious I haue presented you with a slip from his tree, and the *Carpobalsamum* set forth by our Author, which seemes to be of the same plant. The leaues of this are like to those of *Lentiscus*, alwaies greene, and winged, growing three, fiue, or seuen fastened to one foot-stalke; the wood is gummie, reddish, and well smelling: the floures are small and white like those of Acatia, growing vsually three nigh together: the fruit is of the shape and bignesse of that of the Turpentine tree, containing yellow and well smelling seeds, filled with a yellowish moisture like honey, their taste is bitterish, & somwhat biting the tongue. ‡

Of these Balsam trees there is yet another sort: the fruit whereof is as it were a kernell without a shell, couered with a thin skin straked with many veines, of a browne colour: the meat is firm and solid, like the kernell of the Indian Nut, of a white colour, and without smell, but of a grateful tast; and it is thought to be hot in the first degree, or in the beginning of the second.

There be diuers sorts more, which might be omitted because of tediousnesse: neuerthelesse I wil trouble you with two speciall trees worthy the noting: there is, saith my Author, in America a great tree of monstrous hugenesse, beset with leaues and boughes euen to the ground; the trunke wherof is couered with a twofold bark, the one thick like vnto Corke, & another thin next to the tree: from betweene which barks doth flow (the vpper barke being wounded) a white Balsam like vnto teares

or

or drops, of a most sweet sauour, and singular effects, for one drop of this which thus distilleth out of the tree, is worth a pound of that which is made by decoction: the fruit hereof is small in respect of the others; it seldome exceedeth the bignes of a Pease, of a bitter taste, inclosed in a narrow huske, of the length of a finger, something thin, and of a white colour; which the Indians do vse against head-ache: which fruit of most is that we haue before described, called *Carpobalsamum*.

It is also written, that in the Island called *Hispaniola*; there groweth a small tree, of the height of two men, without the industry of man, hauing stalkes or stems of the colour of ashes; whereon do grow greene leaues, sharpe at both ends, but more greene on the vpper side than on the lower; hauing a middle rib somewhat thicke and standing out; the foot-stalkes whereon they grow are somewhat reddish: among which leaues commeth fruit growing by clusters, as long as a mans hand, fingers and all: the stones or graines in the fruit be few, and greene; but growing to rednesse more and more as the fruit waxeth ripe. From the which is gathered a juice after this manner: they take the young shootes and buds of the tree, and also the clusters of the fruit, which they bruise, and boile in water to the thickenesse of hony, which being strained, they keepe it for their vses.

They vse it against wounds and vlcers; it stoppeth and stancheth the bloud; maketh them cleane; bringeth vp the flesh, and healeth them mightily, and with better successe than true Balsame. The branches of the tree being cut, do cast forth by drops a certaine cleare water, more worth than *Aqua vitæ*, most wholesome against wounds, and all other diseases proceeding from cold causes, if it be drunken some few daies together.

¶ *The Place.*

These trees grow in diuers parts of the world, some in Ægypt, and most of those countries adiacent: there groweth of them in the East and West Indies; as trauellers in those parts report.

¶ *The Time.*

These trees for the most part keepe greene winter and Sommer.

¶ *The Names.*

Balsame is called in Greeke, Βάλσαμον: in Latine also *Balsamum*: of the Arabians *Balseni*, *Balesina*, and *Belsan* in Italian, *Balsamo*: in French, *Baume*.

The liquor that floweth out of the tree when it is wounded, is called *Opobalsamum*: the wood *Xylobalsamum*: the fruit *Carpobalsamum*: and the liquor which naturally floweth from the tree in Ægypt *Balsamum*.

¶ *The Temperature.*

Balsame is hot and dry in the second degree, with astriction.

¶ *The Vertues.*

Naturall Balsame taken in a morning fasting, with a little Rose water or wine, to the quantitie
of fiue or six drops, helpeth those that be asthmatike, or short of winde: it preuaileth against the A
paines of the bladder, and stomacke, and comforteth the same mightily; and also amendeth a stinking breath; & takes away the shaking fits of the quotidian egue, if it be taken two or three times.

It helpeth consumptions, clenseth the barren wombe, especially being annointed vpon a pessary, B
or mother suppositorie, and vsed.

The stomacke being annointed therewith, digestion is helped thereby; it also preserueth the
stomacke from obstructions and windinesse; it helpeth the hardnesse of the spleene; easeth the C
griefes of the reines and belly, proceeding of cold causes.

It also taketh away all manner of aches, proceeding of cold causes, if they be annointed there- D
with; but more speedily, if a linnen cloth be wet therein, and laid thereon: vsed in the same manner, it dissolueth hard tumors, called *œdemata*; and strengthneth the weake members.

The same refresheth the braine, and comforteth the parts adioining, it helpeth the palsie, con- E
vulsions, and all griefes of the sinewes, if they be annoitned therewith.

The maruellous effects that it worketh in new and greene wound, were heere too long to set F
downe, and also superfluous; considering the skilfull Chirurgion whom it most concerneth, doth know the vse thereof, and as for the beggerly Quacksaluers, Runnagates, and knauish Mountibanks, we are not willing to instruct them in things so far aboue their reach, capacitie and worthinesse.

CHAP. 146. Of a kinde of Balme, or Balsame Tree.

¶ *The Description.*

This tree which the people of the Indies do call *Molli*, groweth to the bignesse of a great tree, hauing a trunke or body of a darke greene colour, sprinkled ouer with many ash coloured ſpots: the branches are many, and of very great beautie; whereupon do grow leaues not vnlike to thoſe of the Aſh-tree, conſiſting of many ſmall leaues, ſet vpon a middle rib; growing narrower euer towards the point, euery particular one jagged on the ſides like the teeth of a ſaw; which being plucked from the ſtem, yeeldeth forth a milkie juice, tough and clammie, ſauouring like the bruiſed leaues of Fenell, and as it ſeemeth in taſte ſomewhat aſtringent: the floures grow in cluſters vpon the twiggie branches, like thoſe of the Vine a little before the grapes be formed: after followeth the fruit or berries, ſomewhat greater than Pepper cornes, of an oilie ſubſtance, greene at the firſt, and of a darke reddiſh colour when they be ripe. ‡ The firſt of the figures was taken from a tree, only of three yeeres growth, but the latter from a tree come to his full growth, as it is affirmed in *Cluſius* his *Cur. Poſter.* It differs only in that the leaues of the old trees are not at all ſnipt or diuided on the edges. ‡

1 *Molli, ſiue Molly Cluſij, & Lobelij.*
The Balſame tree of *Cluſius* and *Lobels* deſcription.

‡ 2 *Molle arboris adultæ ramus.*
A branch of the old tree of Molle.

¶ *The Place.*

This tree, ſaith a learned Phyſition called *Ioh. Fragoſus*, doth grow in the King of Spaine his garden at Madryll, which was the firſt that euer he did ſee: ſince which time, *Iohn Ferdinando* Secretary vnto the foreſaid king did ſhew vnto the ſaid *Fragoſus* in his owne garden a tree ſo large, and of ſuch beautie, that he was neuer ſatisfied with looking on it, and meditating vpon the vertues thereof. Which words I haue receiued from the hands of a famous learned man, called Mr. *Lancelot Browne*, Dr. in Phyſicke, and Phyſition to the Queenes Maieſtie, at the impreſſion hereof; faithfully tranſlated out of the Spaniſh tongue, without adding or taking any thing away.

They grow plentifully in the vales and low grounds of Peru, as all affirme that haue trauelled to the VVeſt Indies; as alſo thoſe that haue deſcribed the ſingularities thereof. My ſelfe with diuers others, as namely Mr. *Nicholas Lete*, a worſhipfull Merchant of the Citie of London; and alſo a moſt skilfull Apothecary, Mr. *Iames Garret*, who haue receiued ſeeds hereof from the right Honorable the Lord Hunſdon, Lord high Chamberlaine of England, worthy of triple honour for his care in getting, as alſo for his curious keeping rare and ſtrange things brought from the fartheſt parts of the world; which ſeedes we haue ſowne in our gardens, where they haue brought forth plants of a foot high; and alſo their beautifull leaues: notwithſtanding our care, diligence, and

and induſtry, they haue periſhed at the firſt approch of winter, as not being able by reaſon of their tenderneſſe to indure the cold of our Winter blaſts.

¶ *The Time.*

As touching the time of his flouriſhing, and bringing his fruit to maturitie, we haue as yet no certaine knowledge, but is thought to be greene both VVinter and Sommer.

¶ *The Names.*

This moſt notable tree is called by the Indian name *Molle*: of ſome, *Molly*, and *Muelle*, taken from his tender ſoftneſſe, as ſome haue deemed: it may be called the Fennell tree, or one of the Balme, or Balſam trees.

¶ *The Temperature.*

This tree is thought to be of an aſtringent or binding qualitie; whereby it appeares beſides the hot temperature it hath, to be compounded of diuers other faculties.

¶ *The Vertues.*

The Indians vſe to ſeeth the fruit or berries hereof in water, and by a ſpeciall skill they haue in the boiling, do make a moſt wholeſome wine or drinke, as alſo a kind of vineger, and ſometimes hony; which are very ſtrange effects, theſe three things being ſo contrary in taſte. A

The leaues boiled, and the decoction drunke, helpeth them of any diſeaſe proceeding of a cold cauſe. B

The gum which iſſueth from the tree, being white like vnto Manna, diſſolued in milke, taketh away the web of the eies, and cleareth the ſight, being wiped ouer with it. C

The barke of this tree boiled, and the legs that be ſwolne and full of paine, bathed and waſhed with the decoction diuers times, taketh away both infirmities in ſhort ſpace. D

This tree is of ſuch eſtimation among the Indians, that they worſhip it as a god, according vnto their ſauage rites and ceremonies: much like as *Pliny* reporteth of *Homers Moly*, the moſt renowned of all plants, which they had in old time in ſuch eſtimation and reuerence, that as it is recorded, the gods gaue it the name of *Moly*, and ſo writeth *Ouid*: E

Pacifer huic dederat florem Cyllenius album,
Moly vocant Superi, nigra radice tenetur.

If any be deſirous to ſee more hereof, they may reade a learned diſcourſe of it ſet forth in the Latine tongue, by the learned *Lobel*, who hath at large written the hiſtorie thereof, dedicated vnto the right Honourable, the Lord Chamberlaine, at the Impreſſion hereof; faithfully ouerſeene and examined by the learned Phyſition before remembred, M^r. Doctor *Browne*, and his cenſure vpon the ſame. ‡ Together with *Lobels* reply, who iudged this plant (and not without good reaſon) to be a kinde of the true Balſam of the Antients, and not much different from that ſet forth by *Proſper Alpinus*, whereof I haue made mention in the foregoing chapter. ‡ F

Chap. 147. *Of the Canell, or Cinnamon tree.*

¶ *The Deſcription.*

1 THe tree which hath the Cinnamon for his barke is of the ſtature of an Oliue tree: hauing a body as thick as a mans thigh, from which the Cinnamon is taken; but that taken from the ſmaller branches is much better: which branches or boughes are many, and very ſtraight; whereon do grow beautifull leaues, in ſhape like thoſe of the Orenge tree, and of the colour of the Bay leafe (not as it hath been reported) like vnto the leaues of flags or floure de-Luce: among theſe pleaſant leaues and branches come forth many faire white floures, which turne into round blacke fruit or berries, of the bigneſſe of an Haſell Nut, or the Oliue berry, and of a blacke colour; out of which is preſſed an oile, that hath no ſmell at all vntill it be rubbed and chafed betweene the hands: the trunke or body with the greater armes or boughes of the tree are couered with a double or twofold barke, like that of *Suber*, the Corke tree: the innermoſt whereof is the true and pleaſant Cinnamon, which is taken from the tree, and caſt vpon the ground in the heate of the Sunne; through the heate

Canellæ folium, Bacillus, & Cortex.
The leafe, barke, and trunke of the Cinnamon tree.

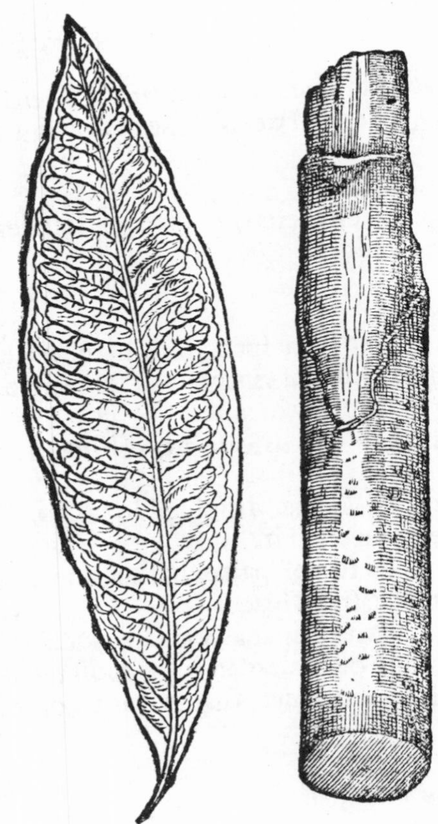

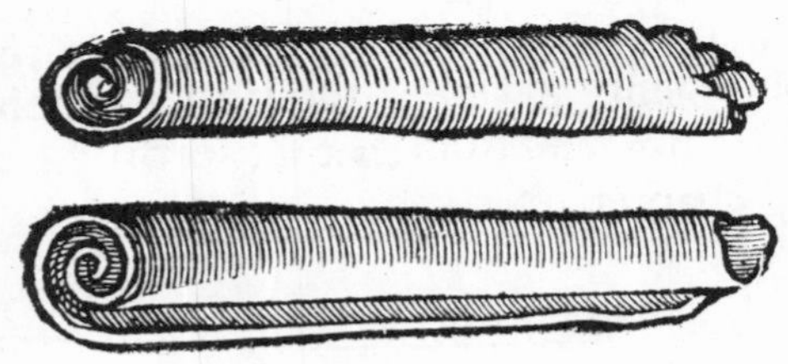

thereof it turneth and foldeth it selfe round together, as wee daily see by viewing the thing it selfe: this tree being thus peeled, recouereth a new barke in the space of three yeares, and is then ready to be disbarked as afore. That Cinnamon which is of a pale colour hath not been well dried in the Sunne: that of a faire browne colour is best; & that which is blackish, hath been too much dried, and also hath taken some wet in the time of drying.

‡ 2 Besides the Cinnamon vulgarly knowne and vsed, there is another sort which also is commonly receiued for the *Cassia* of *Dioscorides* and the Antients. Now this differs from the former in that it is of a redder colour, of a more hard, sollid, and compact substance, commonly also thicker, & if you chew it, more clammy and viscous: the taste and smell are much like Cinnamon, yet not altogether so strong as that of the best Cinnamon. There is much controuersie in late Writers concerning both the true Cinnamon, and Cassia of the Antients: the which I haue not time nor space here to mention, much lesse to insist vpon: I haue obserued that both the Cinnamon and Cassia that we haue are couered ouer with a rough grayish barke, like that of an Oke or other such tree, which is cleane scraped off, and taken away before it be brought to vs. ‡

¶ *The Place.*

The chiefest places where the trees doe grow that beare Cinnamon, are Zeilan and Malauar: but those of Zeilan are the best: they grow in other of the Molucca Ilands, as Iaoa, or Iaua, the greater and the lesse, and also in Mindanoa, for the most part vpon mountaines.

¶ *The Time.*

The Cinnamon tree groweth green winter and Sommer, as do all the other trees of the Moluccaes, and East Indies for the most part: the boughes whereof are cut off at seasonable times, by the expresse commandement of the King of the Country; and not before he haue appointed the time.

There hath beene some controuersie among writers concerning the tree whose bark is *Cassia*, and that tree that beareth Cinnamon, making them both one tree: but that opinion is not to be receiued: for there is a great difference betweene them, as there is betwixt an Oke, and a Chestnut tree; for the tree whose barke is *Cassia*, is doubtlesse a bastard kinde of *Canell*, or Cinnamon: in shew it is very like, but in sweetnesse of smell and other circumstances belonging to Cinnamon, farre inferiour.

¶ *The Names.*

Cinnamon is called in Italian *Canella*: in Spanish, *Canola*: in French, *Canelle*: in high Dutch, **Zimmet roezlin**: the Grecians, κιννάμωμον: the Latines likewise *Cinnamomum*: the Arabians, *Darseni*, and as some say, *Querfaa*, others, *Querfe*: in Zeilan, *Cuurde*: in the Island Iaua they name it *Cameaa*: in Ormus, *Darchini* (i.) *lignum Chinense*, the wood of China: in Malauar, *Cais mains*, which in their tongue signifieth *Dulce lignum*, Sweet wood: in English, Cinnamome, Cinnamon, and Canell. The other is called *Cassia*, and *Cassia lignea*.

¶ *The Temperature and Vertues.*

Dioscorides writeth, that Cinnamon hath power to warme, and is of thinne parts: it is also drie and

and aſtringent, it prouoketh vrine, cleareth the eies, and maketh ſweet breath.

The decoction bringeth downe the menſes, preuaileth againſt the bitings of venomous beaſts, the inflammation of the inteſtines and reines. B

The diſtilled water hereof is profitable to many, and for diuers infirmities, it comforteth the weake, cold, and feeble ſtomacke, eaſeth the paines and frettings of the guts and intralles proceeding of cold cauſes, it amendeth the euill colour of the face, maketh ſweet breath, & giueth a moſt pleaſant taſte vnto diuers ſorts of meats, and maketh the ſame not onely more pleaſant, but alſo more wholeſome for any bodies of what conſtitution ſoeuer they be, notwithſtanding the binding qualitie. C

The oile drawne chimically preuaileth againſt the paines of the breſt, comforteth the ſtomacke, breaketh windineſſe, cauſeth good digeſtion, and being mixed with ſome honie, taketh away ſpots from the face, being annointed therewith. D

The diſtilled water of the floures of the tree, as *Garcias* the Luſitanian Phyſition writeth, excelleth far in ſweetneſſe all other waters whatſoeuer, which is profitable for ſuch things as the barke it ſelfe is. E

Out of the berries of this tree is drawn by expreſſion, as out of the berries of the Oliue tree, a certaine oyle, or rather a kinde of fat like butter, without any ſmell at all, except it bee made warme, and then it ſmelleth as the Cinnamon doth, and is much vſed againſt the coldneſſe of the ſinewes all paines of the ioints, and alſo the paines and diſtemperature of the ſtomacke and breaſt. F

To write as the worthineſſe of the ſubiect requireth, would aske more time than we haue to beſtow vpon any one plant; therefore theſe few ſhall ſuffice, knowing that the thing is of great vſe among many, and knowne to moſt. G

‡ *Caſſia* vſed in a larger quantitie ſerueth well for the ſame purpoſes which Cinnamon doth. ‡ H

CHAP. 142. *Of Gum Lacke and his rotten tree.*

Lacca cum ſuis bacillis.
Gum Lacke with his ſtaffe or ſticke.

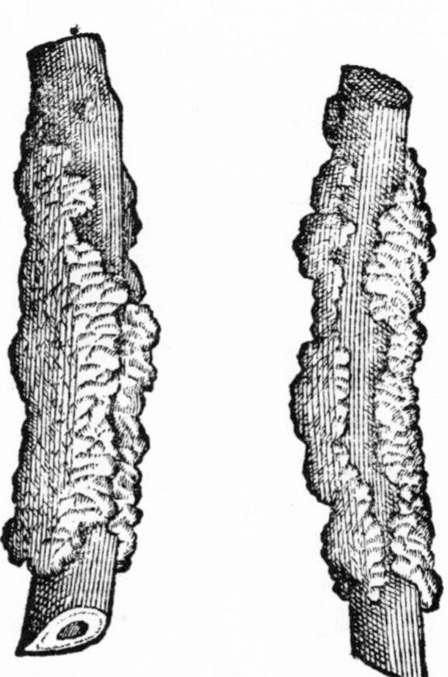

¶ *The Deſcription.*

THe tree that bringeth forth that excremental ſubſtance called *Lacca*, both in the ſhops of Europ and elſewhere, is called of the Arabians, Perſians and Turkes, *Loc Sumutri*, as who ſhould ſay, *Lacca* of Sumutra: ſome which haue ſo termed it, haue thought that the firſt plentie thereof came from Sumutra, but herein they haue erred, for the abundant ſtore thereof came from Pegu, where the inhabitants therof do cal it *Lac*, & others of the ſame Prouince, *Trec*: the hiſtory of which tree, according to that famous Herbariſt *Cluſius* is as followeth. [There is in the countrey of Pegu and Malabar, a great tree, whoſe leaues are like them of the Plum tree, hauing many ſmall twiggie branches; when the trunk or body of the tree waxeth old, it rotteth in ſundry places, wherein do breed certain great Ants or Piſmires, which continually work and labour in the time of Harueſt and Sommer, againſt the penurie of Winter: ſuch is the diligence of thoſe Ants, or ſuch is the nature of the tree wherein they harbour, or both, that they prouide for their winter food, a lumpe or maſſe of ſubſtance, which is of a crimſon colour, ſo beautifull and ſo faire, as in the whole World the like is not ſeene, which ſerueth not onely to phyſicall vſes but is a perfect and coſtly colour for Painters, called by vs, Indian Lack. The Piſmires (as I ſaid) worke out this colour, by ſucking the ſubſtance or matter of Lacca from the tree, as Bees do make honie and wax, by ſucking the matter thereof from all herbes, trees, and floures, and the inhabitants of that country, do as diligently seek for this Lacca, as we in England and other countries seeke in the woods for honie; which Lacca after they haue found, they take from the tree, and dry it into a lump; among which sometimes there come ouer some sticks and pieces of the tree with the wings of the Ants, which haue fallen among it, as we daily see.

‡ The Indian Lacke or Lake which is the rich colour vsed by Painters, is none of that which is vsed in shops, nor here figured or described by *Clusius*, wherefore our Author was much mistaken in that he here confounds together things so different; for this is of a resinous substance, and a faint red colour, and wholly vnfit for Painters, but vsed alone and in composition to make the best hard sealing wax. The other seemes to be an artificiall thing, and is of an exquisite crimson colour, but of what it is, or how made, I haue not as yet found any thing that carries any probabilitie of truth. ‡

¶ *The Place.*

The tree which beareth Lacca groweth in Zeilan and Malauar, and in other parts of the East Indies.

¶ *The Time.*

Of the time we haue no certaine knowledge.

¶ *The Names.*

Indian Lacke is called in shops *Lacca*: in Italian, *Lachetta*: *Auicen* calleth it *Luch*: *Paulus* and *Dioscorides*, as some haue thought, *Cancamum*: the other names are expressed in the description.

¶ *The Temperature and Vertues.*

A Lack or Lacca is hot in the second degree, it comforteth the heart and liuer, openeth obstructions, expelleth vrine, and preuaileth against the dropsie.

B There is an artificiall Lack made of the scrapings of Brasill and Saffron, which is vsed of Painters, and not to be vsed in Physicke as the other naturall Lacca.

Chap. 149 *Of the Indian leafe.*

Tamalapatra.
The Indian leafe.

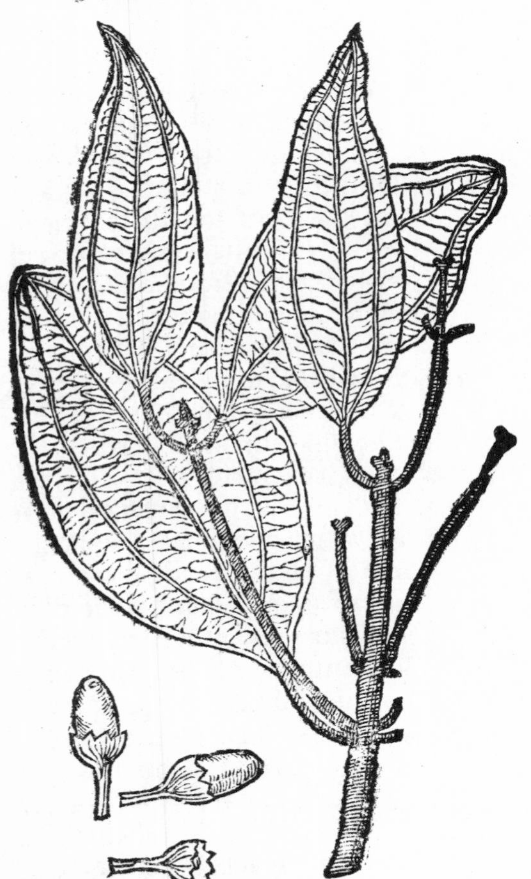

¶ *The Description.*

Tamalapatra, or the Indian leafe grows vpon a great tree like the Orenge tree, with like leaues also, but broader, a little sharp pointed, of a greene glistering colour, and three small ribs running through each leafe, after the manner of Ribwort, wherby it is easie to be known: it smelleth somewhat like vnto Cloues, but not so strong as Spikenard or Mace (as some haue deemed) nor yet of so subtill and quick a sent as Cinnamon. There was sent or added vnto this figure by *Cortusus* a certaine fruit like vnto a small Acorn, with this inscription, *Fructus Canellæ*, the fruit of the Canell tree, which may be doubted of, considering the description of the forenamed tree holden generally of most to be perfect.

¶ *The Place.*

The Indian leafe groweth not fleeting vpon the water like vnto *Lens palustris*, as *Dioscorides* and *Pliny* do set downe, (though learned and painfull writers) but is the leafe of a great tree, a branch whereof wee haue set forth vnto your view, which groweth in Arabia and Cambaya, far from the water side.

¶ *The Time.*

Of the time we haue no certain knowledge, but it is supposed to be green winter and sommer.

¶ The

¶ *The Names.*

Tamalapatra is called of the Indians in their mother tongue, especially of the Arabians, *Cadegi Indi*, or *La degi Indi*, that is, *Folium Indicum*, or *Indum*, the Indian leafe: but the Mauritanians doe call it *Tembul*. The Latines and Grecians following some of the Arabians, haue called it *Malabathrum*.

¶ *The Temperature and Vertues.*

The Indian leafe is hot and dry in the second degree, agreeing with Nardus in temperature, or as others report with Mace: it prouoketh vrine mightily, warmeth and comforteth the stomacke, and helpeth digestion. A

It preuaileth against the pin and web in the eyes, the inflamed and waterie eyes, and all other infirmities of the same. B

It is laid among cloathes, as well to keepe them from moths and other vermine, as also to giue vnto them a sweet smell. C

CHAP. 150. *Of the Cloue tree.*

Caryophylli veri Clusij.
The true forme of the Cloue tree.

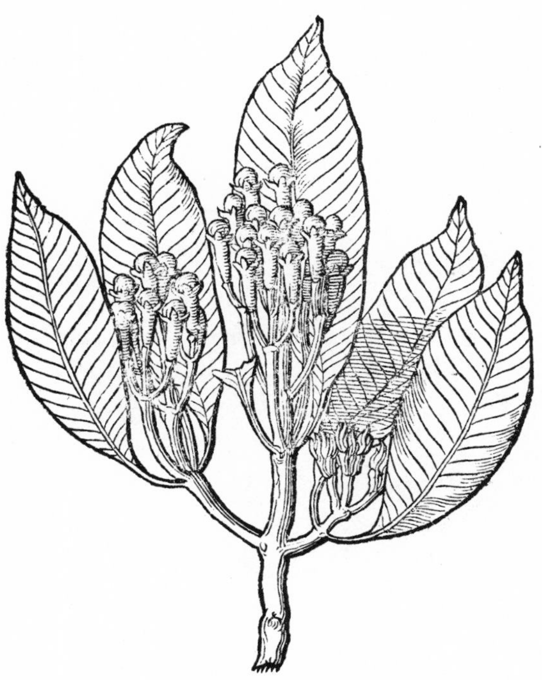

¶ *The Description.*

THe Cloue tree groweth great in forme like vnto the Bay tree, the trunke or bodie whereof is couered with a russet barke: the branches are many, long, and very brittle, whereupon do grow leaues like those of the Bay tree, but somewhat narrower: amongst which come the floures, white at the first, after of a greenish colour, waxing of a darke red colour in the end: which floures are the very cloues when they grow hard: after when they be dried in the Sunne they become of that dusky black colour which we dayly see, wherein they continue. For those that wee haue in estimation are beaten downe to the ground before they be ripe, and are suffered there to lie vpon the ground vntill they bee dried throughly, where there is neither grasse, weeds, nor any other herbes growing to hinder the same, by reason the tree draweth vnto it selfe for his nourishment all the moisture of the earth a great circuit round about, so that nothing can there grow for want of moisture, and therfore the more conuenient for the drying of the Cloues. Contrariwise, that grosse kinde of Cloues which hath beene supposed to be the male, are nothing else than fruit of the same tree tarrying there vntill it fall downe of it selfe vnto the ground, where by reason of his long lying, and meeting with some raine in the mean season, it loseth the quick taste that the others haue. Some haue called these *Fusti*, whereof we may English them Fusses. Some affirme that the floures hereof surpasse all other floures in sweetnesse when they are greene; and hold the opinion, that the hardned floures are not the Cloues themselues, (as wee haue written) but thinke them rather to be the seat or huske wherein the floures doe grow: the greater number hold the former opinion. And further, that the trees are increased without labour, graffing, planting, or other industrie, but by the falling of the fruit, which beare fruit within eight yeares after they be risen vp, and so continue bearing for an hundred yeares together, as the inhabitants of that countrey do affirme.

¶ *The Place.*

The Cloue tree groweth in some few places of the Molucca Islands, as in Zeilan, Iaua the greater and the lesse, and in diuers other places.

¶ *The*

¶ *The Time.*

The Cloues are gathered from the fifteenth of September vnto the end of Februarie, not with hands, as we gather Apples, Cherries, and such like fruit, but by beating the tree, as Wall-nuts are gotten, as we haue written in the description.

¶ *The Names.*

The fruit hereof was vnknowne to the antient Grecians: of the later writers called Καρυόφυλλον: in Latine also *Caryophyllus*, and *Clavus*: in French, *Clou de Gyrofle*: the Mauritanians, *Charhumfel*: in Italian, *Carofano*: in high-Dutch, **Naegel**: in Spanish, *Clauo de especia*: of the Indians, *Calafur*: in the Molucca's, *Changue*: of the Pandets, *Arumfel*, and *Charumfel*: in English, Cloue tree, & Cloues.

¶ *The Temperature.*

Cloues are hot and dry in the third degree.

¶ *The Vertues.*

A Cloues strengthen the stomacke, liuer, and heart, helpe digestion, and prouoke vrine.

B The Portugall women that dwell in the East Indies draw from the Cloues when they bee yet greene, a certaine liquor by distillation, of a most fragrant smell, which comforteth the heart, and is of all cordials the most effectuall.

C Cloues stop the belly: the oile or water thereof dropped into the eyes, sharpens the sight, and clenseth away the cloud or web in the same.

D The weight of foure drams of the pouder of Cloues taken in milke procureth the act of generation.

E There is extracted from the Cloues a certaine oile or rather thicke butter of a yellow colour; which being chafed in the hands smelleth like the Cloues themselues, wherewith the Indians do cure their wounds and other hurts, as we do with Balsam.

F The vse of Cloues, not onely in meat and medicine, but also in sweet pouder and such like, is sufficiently knowne: therefore this shall suffice.

‡ There were formerly three figures in this chapter: wherefore I omitted two as impertinent.

CHAP. 151. *Of the Nutmeg tree.*

1 *Nux Muscata rotunda, siue fœmina.*
The round or female Nutmeg.

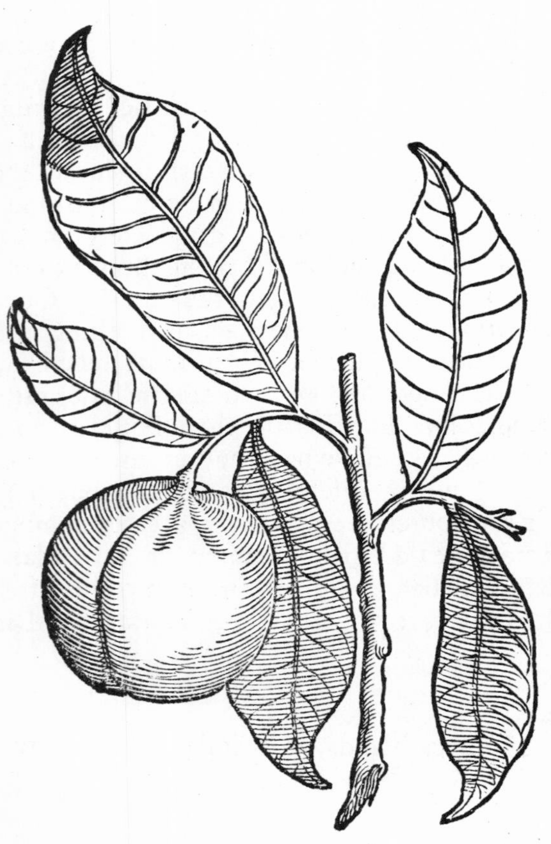

2 *Nux Myristica oblonga, siue mas.*
The longish or male Nutmeg.

Nux Moſchata,cum ſua Maci.
The Nutmeg with his Mace about him.

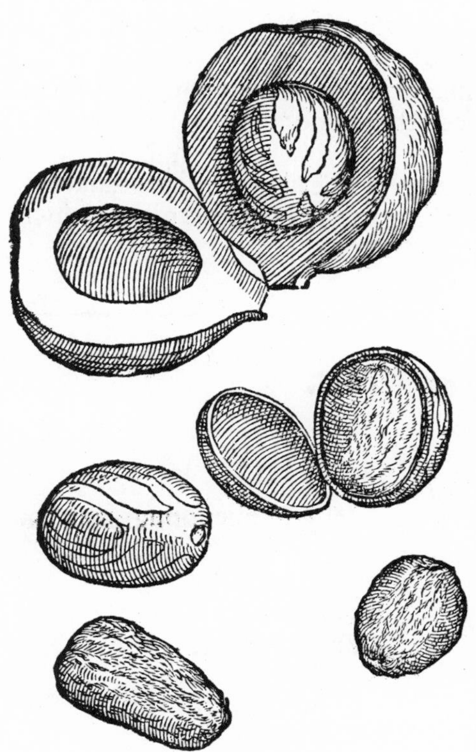

¶ *The Deſcription.*

1 THe tree that beareth the Nutmeg and the Mace is in forme like to the Peare tree, but the leaues of it are like thoſe of the Bay or Orenge tree, alwaies greene on the vpper ſide, and more whitiſh vnderneath; among which come forth the Nut and Mace as it were the floures. The Nut appeareth firſt, compaſſed about with the Mace, as it were in the middle of a ſingle roſe, which in proceſſe of time doth wrap and incloſe the Nut round on euery ſide: after commeth a huske like that of the Wall-nut,but of an harder ſubſtance, which incloſeth the Nut with his Mace as the Wall-nut husk doth couer the Nut, which in time of ripeneſſe doth cleaue of it ſelfe as the Wall-nut huske doth, and ſheweth his Mace, which then is of a perfect crimſon colour, and maketh a moſt goodly ſhew, eſpecially when the tree is well laden with fruit: after the Nut becommeth dry, the Mace likewiſe gapeth and forſaketh the Nut, euen as the firſt huske or couerture, and leaues it bare and naked, as we all do know; at which time it getteth to it ſelfe a kinde of darke yellow colour, and loſeth that braue crimſon dye which it had at the firſt.

‡ 2 The tree which carrieth the male Nutmeg (according to *Cluſius*) thus differs from the laſt deſcribed: the leaues are like thoſe of the former in ſhape, but much bigger, being ſometimes a foot long, and three or foure inches broad; their common length is ſeuen or eight inches,and bredth two and a halfe: they are of a whitiſh colour vnderneath, and greene and ſhining aboue. The Nuts alſo grow at the very ends of the branches, ſometimes two or three together, and not onely one, as in the common kinde. The Nut it ſelfe is alſo larger and longer: the Mace that incompaſſes it is of a more elegant colour,but not ſo ſtrong as that of the former.

I can ſcarſe beleeue our Authors aſſertion in the foregoing deſcription, that the Nut appeareth firſt, compaſſed about with the Mace as it were in the middeſt of a ſingle Roſe, &c. But I rather thinke they all come forth together, the Nutmeg, Mace, the greene outward huske and all, iuſt as we ſee Wall-nuts do, and onely open themſelues when they come to full maturitie. In the third figure you may ſee expreſt the whole manner of the growing of the Nutmeg, together with both the ſorts of Nutmegs taken forth of their ſhells. ‡

¶ *The Place.*

The Nutmeg tree groweth in the Indies, in an Iſland eſpecially called Banda,and in the Iſlands of Molucca, and in Zeilan, though not ſo good as the firſt.

¶ *The Time.*

The fruit is gathered in September in great aboundance, all things being common in thoſe countries.

¶ *The Names.*

The Nutmeg tree is called of the Grecians, κάρυον μυριστικὸν: of the Latines, *Nux Moſchata*, and *Nux Myriſtica*: in Italian, *Noce Moſcada*: in Spaniſh, *Nuez de eſcetie*: in French, *Noix Muſcade*: in high-Dutch, **Moſchat Nuſz**: of the Arabians, *Leuzbane*, or *Gianziban*: of the countrey people where they grow, *Palla*: The Maces, *Bunapalla*. In Decan the Nut is called *Iapatri*, and the Maces, *Iaifol*: of *Auicen*, *Iauſiband*, (i.) *Nux Bandenſis*. The Maces he calleth *Beſbaſe*: in Engliſh, Nutmeg.

¶ *The Temperature.*

The Nutmeg, as the Mauritanians write, is hot and dry in the ſecond degree complete, and ſomwhat aſtringent.

¶ *The Vertues.*

Nutmegs cauſe a ſweet breath, and amend thoſe that do ſtink, if they be much chewed and holden in the mouth. A

The

B The Nutmeg is good against freckles in the face, quickneth the sight, strengthens the belly and feeble liuer; it taketh away the swelling in the spleene, stayeth the laske, breaketh winde, and is good against all cold diseases in the body.

C Nutmegs bruised and boiled in Aqua vitæ vntill they haue wasted and consumed the moisture, adding thereto of *Rhodomel* (that is, honey of Roses) gently boiling them, being strained to the forme of a syrrup, cure all paines proceeding of windie aud cold causes, if three spoonfulls be giuen fasting for certaine dayes together.

D The same bruised and boyled in strong white wine vntill three parts be sodden away, with the roots of Mother-wort added thereto in the boyling, and strained: this liquor drunke with some sugar cureth all gripings of the belly proceeding of windinesse.

E As touching the choice, there is not any so simple but knoweth that the heauiest, fattest, and fullest of iuice are the best, which may easily be found out by pricking the same with a pinne or such like.

Chap. 152. *Of the Pepper Plant.*

¶ *The Kindes.*

There be diuers sorts of Pepper, that is to say, white, blacke, and long Pepper, one greater and longer than the other; and also a kinde of Ethiopian Pepper.

1 *Piper nigrum.*
Blacke Pepper.

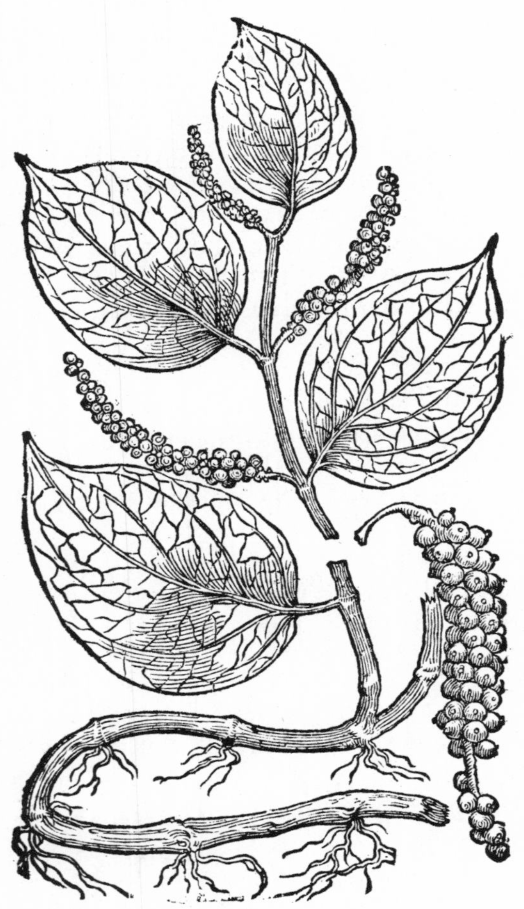

2 *Piper album.*
White Pepper.

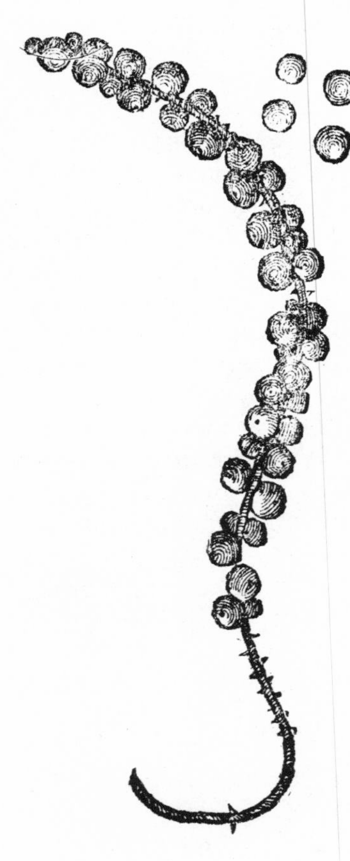

¶ *The Description.*

1 The Plant that beareth the blacke Pepper groweth vp like a Vine among bushes and brambles where it naturally groweth; but where it is manured it is sowne at the bottome of the tree *Fausel* and the Date trees, whereon it taketh hold, and clymbeth vp euen to the top, as doth the Vine, ramping and taking hold with his clasping tendrels of any other thing

thing it meeteth withall. The leaues are few in number, ‡ growing at each ioint one, first on one side of the stalke, then on the other, like in shape to the long vndiuided leaues of Iuy, but thinner, sharpe pointed, and sometimes so broad, that they are foure inches ouer, but most commonly two inches broad, and foure long, hauing alwaies fiue pretty large nerues running alongst them. The fruit grow clustering together vpon long stalks, which come forth at the ioints against the leaues, as you may see in the figure: the root (as one may coniecture) is creeping; for the branches that lie on the ground do at their ioints put forth new fibres or roots. We are beholden to *Clusius* for this exact figure and description, which he made by certaine branches which were brought home by the Hollanders from the East Indies. The curious may see more hereof in his Exotickes and notes vpon *Garcias*. ‡

† 3 *Piper longum.*
Long Pepper.

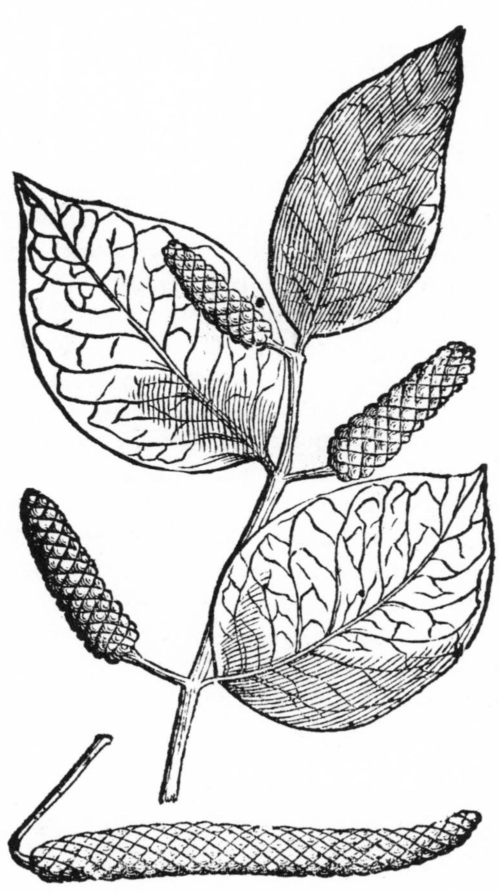

4 *Piper Æthiopicum, siue Vita longa.*
Pepper of Ethiopia.

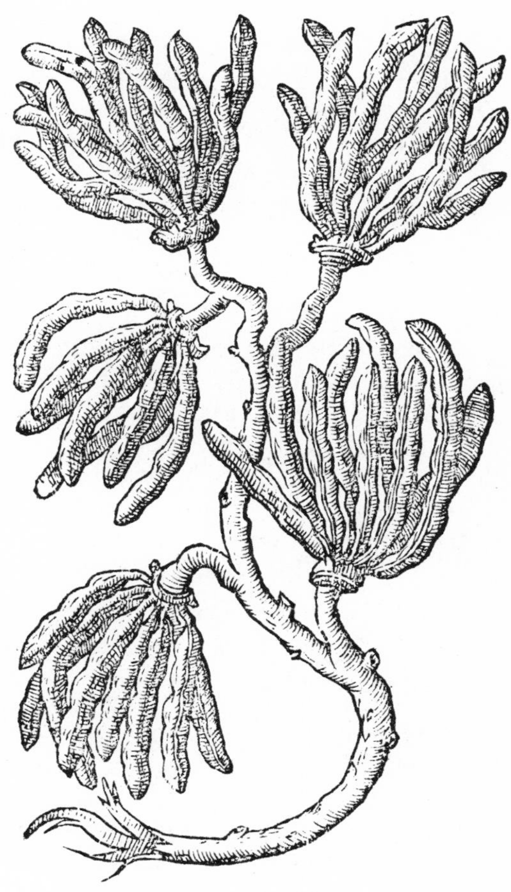

2 The Plant that brings white Pepper is not to be distinguished from the other plant, but only by the colour of the fruit, no more than a Vine that beareth blacke Grapes, from that which bringeth white: and of some it is thought, that the selfe same plant doth sometimes change it selfe from black to white, as diuers other plants do. ‡ Neither *Clusius*, nor any other else that I haue yet met with, haue deliuered vs any thing of certaine, of the plant whereon white Pepper growes: *Clusius* only hath giuen vs the manner how it growes vpon the stalkes, as you may see it here exprest ‡.

There is also another kinde of Pepper, seldome brought into these parts of Europe, called *Piper Canarium*: it is hollow within, light, and empty, but good to draw flegme from the head, to helpe the tooth-ache and cholericke affects.

3 The tree that beareth long Pepper hath no similitude at all with the plant that brings black and white Pepper: some haue deemed them to grow all on one tree, which is not consonant to truth, for they grow in countries far distant one from another, and also that countrey where there is blacke Pepper hath not any of the long Pepper; and therefore *Galen* following *Dioscorides*, were together both ouerseen in this point. This tree, saith *Monardes*, is not great, yet of a wooddy substance, dispersing here and there his clasping tendrels, wherewith it taketh hold of other trees and such other things as do grow neere vnto it. The branches are many and twiggie, whereon growes the fruit, consisting of many graines growing vpon a slender foot-stalke, thrust or compact close together,

‡ 5 *Piper Caudatum.*
Tailed Pepper.

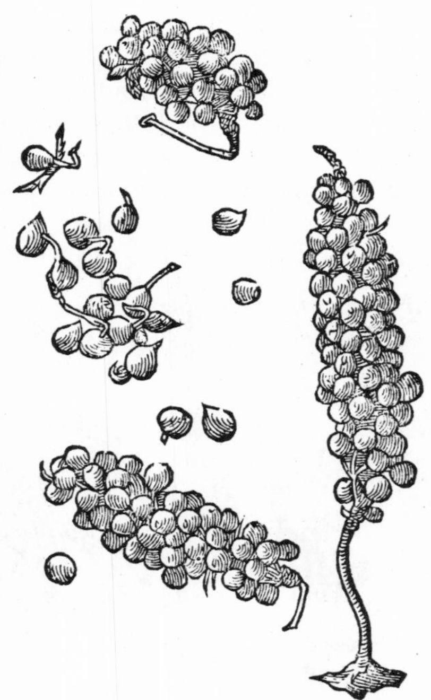

together, greene at the firſt, and afterward blackiſh; in taſte ſharper and hotter than common blacke Pepper, yet ſweeter, and of better taſte. ‡ For this figure alſo I acknowledge my ſelfe beholden to the learned and diligent *Cluſius*, who cauſed it to be drawne from a branch of ſome foot in length, that he receiued from D^r. *Lambert Hortenſius*, who brought it from the Indies. The order of growing of the leaues, and fruit is like that of the blacke; but the joints ſtand ſomewhat thicker together, the leafe alſo doth little differ from that of the blacke, onely it is thinner, of a lighter greene, and (as *Cluſius* thought) hath a ſhorter foot-ſtalke, the veines or nerues alſo were leſſe imminent, more in number, and run from the middle rib to the ſides, rather than alongſt the leafe. ‡

4 This other kinde of Pepper brought vnto vs from Æthiopia, called of the country where it groweth, *Piper Æthiopicum*: in ſhops, *Amomum*, and alſo *Longa Vita*. It groweth vpon a ſmall tree, in manner of an hedge buſh, whereupon grow long cods in bunches, a finger long, of a browne colour, vneuen, and bunched or puft vp in diuers places, diuided into fiue or ſix lockers or cels, each whereof containeth a round ſeed ſomewhat long, leſſer than the ſeeds of Pæony, in taſte like common Pepper, or *Cardamomum*, whoſe facultie and temperature it is thought to haue, whereof we hold it a kinde.

5 Another kinde of Pepper is ſometimes brought, which the Spaniards do call *Pimenta de rabo*, that is, Pepper with a taile: it is like vnto Cubebes, round, full, ſomewhat rough, blacke of colour, and of a ſharpe quicke taſte, like the common Pepper, of a good ſmell: it groweth by cluſters vpon ſmall ſtems or ſtalkes, which ſome haue vnaduiſedly taken for *Amomum*. The King of Portingal forbad this kinde of Pepper to be brought ouer, for feare leaſt the right Pepper ſhould be the leſſe eſteemed, and ſo himſelfe hindered in the ſale thereof.

¶ *The Place.*

Blacke and white Pepper grow in the kingdome of Malauar, and that very good; in Malaca alſo, but not ſo good; and alſo in the Iſlands Sunde and Cude: there is great ſtore growing in the kingdome of China, and ſome in Cananor, but not much.

Pepper of Æthiopia groweth in America, in all the tract of the country where Nata and Carthago are ſituated. The reſt hath been ſpoken of in their ſeuerall deſcriptions. The white Pepper is not ſo common as the blacke, and is vſed there in ſtead of ſalt.

¶ *The Time.*

The plant riſeth vp in the firſt of the ſpring; the fruit is gathered in Auguſt.

¶ *The Names.*

The Grecians, who had beſt knowledge of Pepper, do call it πίπερι: the Latines, *Piper*: the Arabians, *Fulfel* and *Fulful*: in Italian, *Pepe*: in Spaniſh, *Pimenta*: in French, *Poiure*. in high-Dutch, Pfeffer: in Engliſh, Pepper.

That of Æthiopia is called, *Piper Æthiopicum*, *Amomum*, *Vita longa*, and of ſome, *Cardamomum*, whereof we hold it to be a kinde. I receiued a branch hereof at the hands of a learned Phyſition of London, called M^r. *Steuen Bredwell*, with his fruit alſo.

¶ *The Temperature.*

The Arabians and Perſian Phyſitians iudge, that Pepper is hot in the third degree.

But the Indian Phyſitians which for the moſt part are Emperickes, hold that Pepper is cold, as almoſt all other ſpice, which are hot indeed: the long Pepper is hot alſo in the third degree, and as we haue ſaid, is thought to be the beſt of all the kindes.

¶ *The Vertues.*

A *Dioſcorides* and others agreeing with him, affirme, that Pepper reſiſteth poiſon, and is good to be put in medicaments for the eies.

All

All Pepper heateth, prouoketh vrine, digeſteth, draweth, diſperſeth, and clenſeth the dimneſſe of the ſight as *Dioſcorides* noteth. B

† I haue omitted in this chapter *Matthiolus* his counterfeit figure, which was formerly here.

CHAP. 153. *Of baſtard Pepper, called Betle, or Betre.*

Betle, ſiue Betre.
Baſtard Pepper.

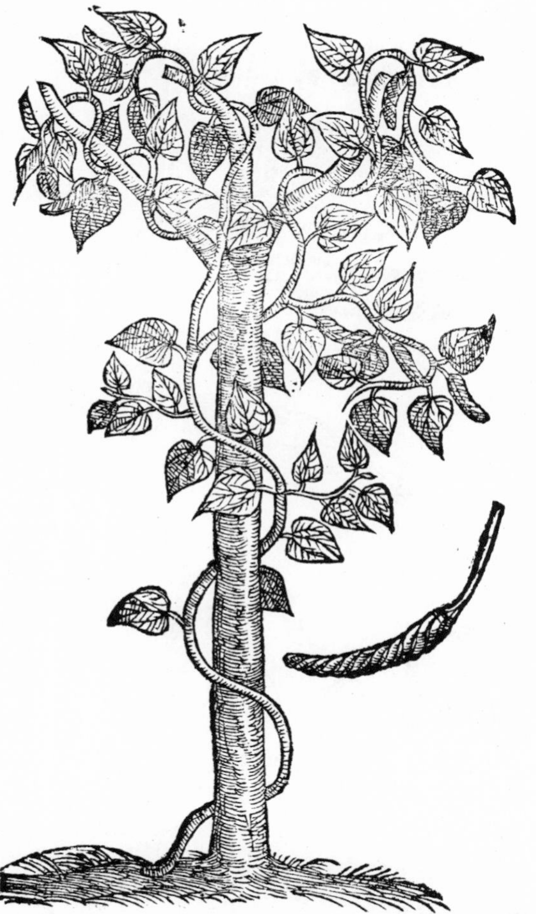

¶ *The Deſcription.*

THis plant climeth and rampeth vpon trees, buſhes, or whatſoeuer elſe it meeteth withall, like vnto the Vine, or the blacke Pepper, whereof ſome hold it for a kinde. The leaues are like thoſe of the greater Bindeweed, but ſomewhat longer, of a duſtie colour, with diuers veines or ribs running through the ſame. The fruit groweth among the leaues, very crookedly writhed, in ſhape like the taile of a Lyzard, of the taſte of Pepper, yet very pleaſant to the palate.

¶ *The Place.*

It groweth among the Date trees, and *Areca*, in moſt of the Molucca Iſlands, eſpecially in the marriſh grounds.

¶ *The Time.*

The time anſwereth that of Pepper.

¶ *The Names.*

This hath been taken for the Indian leafe, but not properly: of moſt it is called *Tembul*, and *Tambul* · in Malauar *Betre*: in Decan, Guzarat, and Canam it is called *Pam*: in Molaio, *Siri*.

¶ *The Vſe and Temperature.*

The leaues chewed in the mouth are of a bitter taſte, whereupon (ſaith *Garcias*) they put thereto ſome Areca and with the lime made o oyſter ſhels, whereunto they alſo adde ſome Amber Griece, *Lignum Aloes*, and ſuch like, which they ſtampe together, making it into a paſte, which they role vp into round balls, keepe dry for their vſe, and carry the ſame in their mouthes vntill by little and little it is conſumed; as when we carry ſugar-Candy in our mouthes, or the iuice of Licorice; which is not onely vnto the ſeely Indians meate, but alſo drinke in their tedious trauels, refreſhing their wearied ſpirits; and helping memory: which is eſteemed among the Empericke Phyſitions, to be hot and dry in the ſecond degree. ‡ *Garcias* doth not affirme that the Indians eate it for meate, or in want of drinke, but that they eate it after meate, and that to giue the breath a pleaſant ſent, which they count a great grace, ſo that if an inferiour perſon that hath not chewed Betre, or ſome ſuch thing, come to ſpeak with any great man, he holds his hand before his mouth leſt his breath ſhould offend him. ‡

CHAP. 154. *Of Graines, or Graines of Paradiſe.*

¶ *The Kindes.*

THere be diuers ſorts of Graines, ſome long, others Peare faſhion; ſome greater; and others leſſer.

¶ *The Description.*

† The first figure hereof setteth forth vnto your view the cod wherein the hot spice lieth, which we call Graines: in shops, *Grana Paradisi*: it groweth, by the report of the Learned, vpon a low herby plant: the leaues are some foure inches long, and three broad, with somewhat a thicke middle rib, from which run transuerse fibres; they much in shape resemble those of Cloues. The fruit is like a great cod or huske, in shape like a Fig when it groweth vpon the tree, but of colour russet, thrust full of small seeds or graines of a darke reddish colour (as the Figure sheweth which is diuided) of an exceeding hot taste.

Cardamomi genera.
The kindes or sorts of Graines.

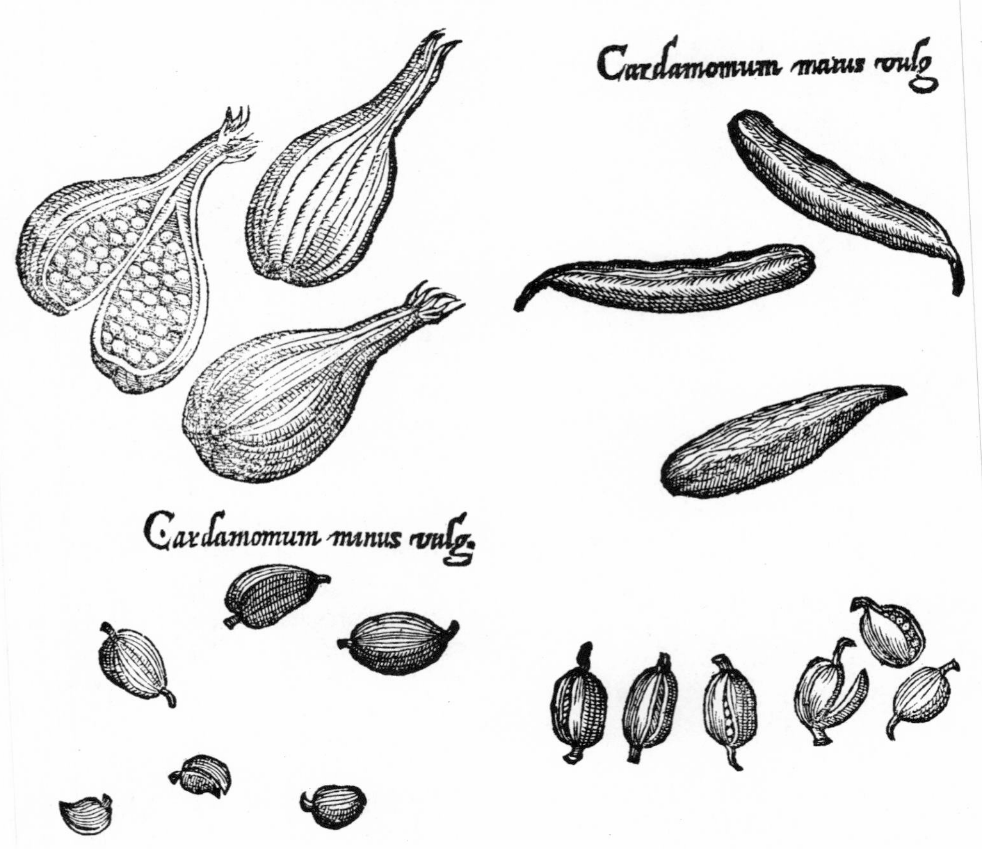

The other sorts may be distinguished by the sight of the picture, considering the onely difference consists in forme.

¶ *The Place.*

Graines grow in Ginny, and the Cardamones in all the East Indies, from the port of Calecute vnto Cananor; it groweth in Malauar, in Ioa, and in diuers other places.

¶ *The Time.*

They spring vp in May, being sowne of seed, and bring their fruit to ripenesse in September.

¶ *The Names.*

Graines are called in Greeke, καρδάμωμον: in Latine, *Cardamomum*: of the Arabians, *Corrumeni*: of diuers Gentiles, *Dore*: of *Auicen*, *Saccolaa quebir* (i.) *magnum*: the other, *Saccolaa ceguer* (i.) *minus*. It is called in Malauar, *Etremelli*: in Zeilan, *Ençal*: in Bengala, Guzarat, and Decan, *Hil*, and *Eluchi*. The first and largest sort are called of some, *Mileguetta*, and *Milegetta*: in English, Grains, and Graines of Paradice.

¶ *The Temperature.*

Auicen writeth, that *Saccolaa*, *Cardamomum*, or *Grana Paradisi*, are hot and dry in the third degree, with astriction.

¶ *The Vertues.*

A The Graines chewed in the mouth draw forth from the head and stomacke waterish and pituitous humors.

B They also comfort and warme the weake, cold, and feeble stomacke, helpe the ague, and rid the shaking fits, being drunke with Sacke.

CHAP.

Chap. 155. *Of Yucca or Iucca.*

Yucca, siue Iucca Peruana.
The root whereof the bread Casaua or Cazaua is made.

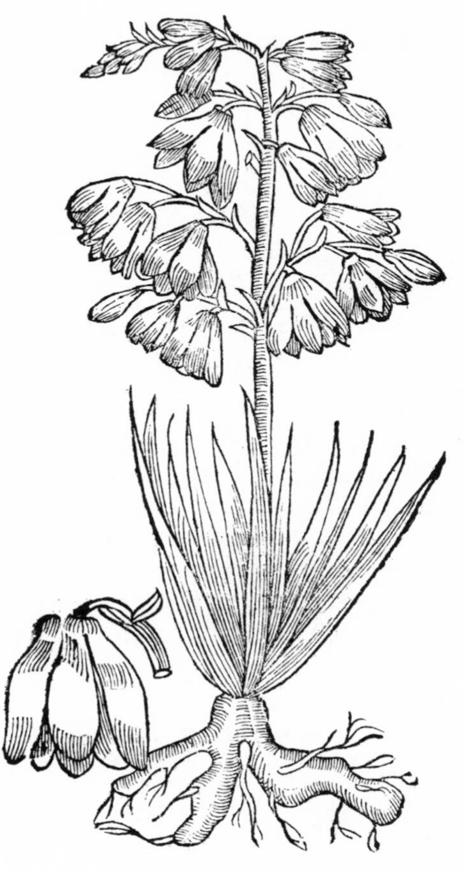

¶ *The Description.*

THe Plant of whose root the Indian bread called Cazaua is made, is a low herbe consisting onely of leaues and roots; it hath neither stalke, floures, nor fruit, that I can vnderstand of others, or by experience of the plant it selfe, which hath growne in my garden foure yeares together, and yet doth grow and prosper exceedingly; neuerthelesse without stalke, fruit, or floures, as aforesaid. It hath a very great root, thicke and tuberous, and verie knobby, full of iuice somewhat sweet in taste, but of a pernicious qualitie, as saith my Author: from which root riseth vp immediately forth of the ground very many leaues ioyned vnto the head of the root in a round circle; the which are long, of the length of a cubit, hollowed like a gutter or trough, verie smooth, and of a greene colour, like that of Woade: the edges of the leaues are sharpe like the edge of a knife, and of a browne colour: the point of the leafe is a prick as sharp as a needle, which hurteth those that vnaduisedly passe by it: the leafe with aduised eye viewed is like vnto a little wherry, or such like boat: they are also very tough, hard to break, and not easie to cut, except the knife be very sharpe.

‡ *Lobel* in the second part of his *Aduersaria* largely describes and figures this plant; and there hee affirmes hee wrot a description (the which he there sets downe) for our Author; but he did not follow it, and therefore committed these errours: First, in that hee saith it is the root whereof Cazaua bread was made; when as *Lobel* in his description said he thought it to be *Alia species à Yucca Indica ex qua panis communis fit.* Secondly, in that he set downe the place out of the *Historia Lugd.* (who tooke it out of *Theuet*) endeauouring by that meanes to confound it with that there mentioned, when as he had his from Mr. *Edwards* his man. And thirdly, (for which indeed he was most blame-worthy, and wherein he most shewed his weakenesse) for that hee doth confound it with the *Manihot* or true *Yuicca,* which all affirme to haue a leafe like that of hemp, parted into seuen or more diuisions: and also in that he puts it to the *Arachidna* of *Theophrastus*, when as he denies it both floure and fruit; yet within some few yeares after our Author had set forth this Worke it floured in his garden.

This some yeares puts forth a pretty stiffe round stalke some three cubits high, diuided into diuers vnequall branches carrying many pretty large floures, shaped somewhat like those of *Fritillaria*, but that they are narrower at their bottomes: the leaues of the floure are six: the colour on the inside white, but on the out side of an ouerworne reddish colour from the stalke to the middest of the leafe; so that it is a floure of no great beautie, yet to be esteemed for the raritie. I saw it once floure in the garden of Mr. *Wilmot* at Bow, but neuer since, though it hath been kept for many yeares in sundry other gardens, as with Mr. *Parkinson* and Mr. *Tuggy*. This was first written of by our Author; and since by *Lobel* and Mr. *Parkinson*, who keepe the same name, as also *Bauhine*, who to distinguish it from the other calls it *Yucca folijs Aloes*. ‡

¶ *The Place.*

This plant groweth in all the tract of the Indies, from the Magellane straights vnto the cape of Florida, and in most of the Islands of the Canibals, and others adioyning, from whence I had that plant brought me that groweth in my garden, by a seruant of a learned and skilfull Apothecary of Excester, named Mr. *Tho. Edwards*.

¶ *The Time.*

It keepeth greene both Winter and Sommer in my garden, without any couerture at all, notwithstanding the iniurie of our cold clymat.

¶ *The Names.*

It is reported vnto me by Trauellers, that the Indians do call it in some parts, *Manihot*, but generally *Yucca* and *Iucca*: it is thought to be the plant called of *Theophrastus*, *Arachidna*; and of *Pliny*, *Aracidna*.

¶ *The Temperature.*

This plant is hot and dry in the first degree, which is meant by the feces or drosse, when the poisonous iuice is pressed or strained forth, and is also dry in the middle of the second degree.

Chap. 156.

Of the fruit Anacardium, and Caious, or Caiocus.

¶ *The Description.*

The antient writers haue been very briefe in the historie of *Anacardium*: the Grecians haue touched it by the name of Ἀνακάρδιον, taking the name from the likenesse it hath of an heart both in shape and colour; called of the Portugals that inhabit the East Indies, *Faua de Malaqua*, the bean of Malaca; for being greene, and as it hangeth on the tree, it resembleth a Beane, sauing that it is much bigger: but when they be dry they are of a shining blackish colour, containing between the outward rinde and the kernell (which is like an Almond) a certaine oile of a sharpe causticke or burning qualitie, called *Mel Acardinum*, although the kernell is vsed in meates and sauces, as we do Oliues and such like, to procure appetite.

Anacardium.
The Beane of Malaca.

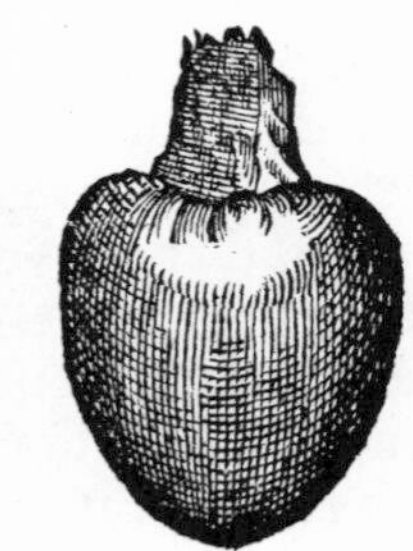

Caious.
The kidney Beane of Malaca.

The other fruit groweth vpon a tree of the bignesse of a Peare tree: the leaues are much like to those of the Oliue tree, but thicker and fatter, of a feint greene colour: the floures are white, consisting of many small leaues much like the floures of the Cherry tree, but much doubled, without smell: after commeth the fruit (according to *Clusius*, of the forme and magnitude of a goose egge, full of iuice; in the end whereof is a nut) in shape like an Hares kidney, hauing two rindes, between which is contained a most hot and sharp oile like that of *Anacardium*, whereof it is a kind.

The Beane or kernell it selfe is no lesse pleasant and wholsome in eating, than the *Pistacia*, or Fisticke nut, whereof the Indians do eate with great delight, affirming that it prouoketh Venerie, wherein is their chiefest felicitie. The fruit is contained in long cods like those of Beans, but greater: neere vnto which cods commeth forth an excrescence like vnto an apple, very yellow, of a good smell, spongious within, and full of iuice, without any seeds, stones, or graines at all, somewhat sweet in taste, at the one end narrower than the other, Peare fashion, or like a little bottle, which hath bin reputed of some for the fruit, but not rightly; for it is rather an excrescence, as is the oke Apple.

¶ *The Place.*

The first growes in most parts of the East Indies, especially in Cananor, Calecute, Cambaya, and Decan. The later in Brasile.

¶ The

¶ *The Time.*

These trees floure and flourish Winter and Sommer.

¶ *The Names.*

Their names haue been touched in their descriptions. The first is called *Anacardium*, of the likenesse it hath with an heart: of the Arabians, *Balador*: of the Indians, *Bibo*.

The second is called *Caious*, and is thus written, *Caiöüs*, and *Caius*: of some, *Caiocus*.

¶ *The Temperature and Vertues.*

A The oile of the fruit is hot and dry in the fourth degree, it hath also a caussticke or corrosiue qualitie: it taketh away warts, breaketh apostumes, preuaileth against leprie, *alopecia*, and easeth the paine of the teeth, being put into the hollownesse thereof.

B The people of Malauar do vse the said oile mingled with chalke, to marke their cloathes or any other thing they desire to be coloured or marked, as we do vse chalke, okar, and red marking stones, but their colour will not be taken forth againe by any manner of art whatsoeuer.

C They also giue the kernell steeped in whay to them that be asthmaticke or short winded; and when the fruit is yet green they sticke the same so steeped against the wormes.

D The Indians for their pleasure will giue the fruit vpon a thorne or some other sharpe thing, and hold it in the flame of a candle, or any other flame, which there will burne with such crackings, lightnings, and withall yeeld so many strange colours, that it is great pleasure to the beholders which haue not seene the like before.

CHAP. 157.

Of Indian Morrice Bells, and diuers other Indian Fruits.

† 1 *Ahouay Theueti.*
Indian Morrice Bels.

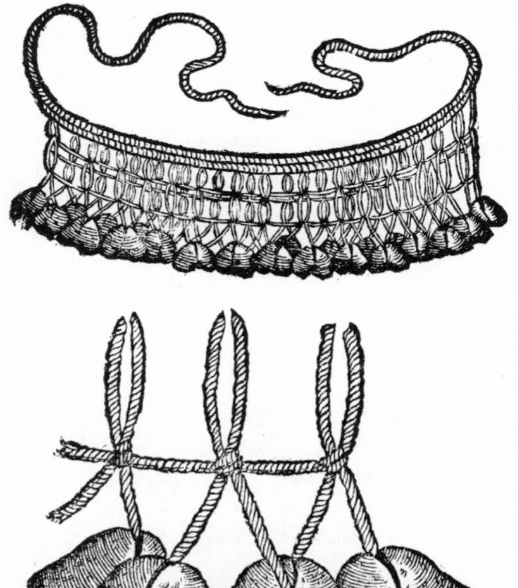

† 2 *Fructus Higuero.*
Indian Morosco bels.

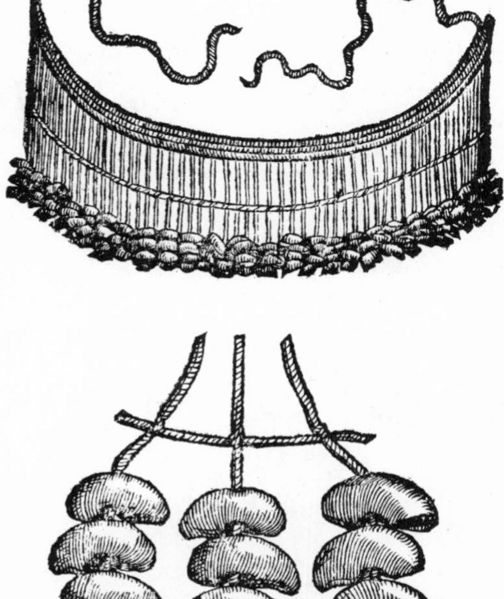

¶ *The Description.*

THis fruit groweth vpon a great tree of the bignesse of a Peare tree, full of branches, garnished with many leaues which are alwaies greene, three or foure fingers long, and in bredth two: when the branches are cut oft there issueth a milky iuice not inferiour to the fruit in his venomous

 qualitie;

qualitie. The trunke or body is couered with a grayish barke: the timber is white and soft, not fit to make fire of, much lesse for any other vse; for being cut and put to the fire to burne, it yeeldeth forth such a loathsome and horrible stinke, that neither man nor beast are able to endure it: wherefore the Indians haue no vse thereof, but onely of the fruit, which in shape is like the Greeke letter Δ, of the bignesse of a Chestnut, and couered with a most hard shell, wherein is contained a kernel of a most venomous and poysonsome qualitie, wherewith the men being angry with their wiues, do poyson them, and likewise the women their husbands: they likewise vse to dip or anoint and invenome their arrowes therewith, the more speedily to dispatch their enemies. Which kernell they take forth with some conuenient instrument, leauing the shell as whole as may be, not touching the kernell with their hands because of its venomous qualitie, which would spoile their hands, and sometimes take away their life also. In which shells they put some little stones, and tye them vpon strings (as you may perceiue by the figure) which they dry in the Sunne, and after tye them about their legs, as we do bells, to set forth their dances, and Morosco Matachina's, wherein they take great pleasure, by reason they thinke themselues to excell in those kindes of dances. Which ratling sound doth much delight them, because it setteth forth the distinction of sounds, for they tune them and mix them with great ones and little ones, in such sort as we doe chimes or bells.

2 There is also another sort hereof, differing onely in forme; they are of the like venomous qualitie, and vsed for the same purpose. ‡ The fruit of *Higuerro* is like that of a gourd in pulpe, and it may be eaten: the shape of the fruit is round, whereas the former is three cornered. ‡

¶ *The Place.*

These do grow in most parts of the West Indies, especially in some of the Islands of the Canibals, who vse them in their dances more than any of the other Indians. ‡ You may see these vpon strings as they are here figured, amongst many other varieties, with M[r]. *Iohn Tradescant* at South Lambeth. ‡

¶ *The Time.*

We haue no certaine knowledge of the time of flouring or bringing the fruit to maturitie.

¶ *The Names and Vse.*

We haue sufficiently spoken of the names and vse hereof, therefore what hath beene said may suffice.

† The figures were transposed.

Chap. 158. *Of the vomiting and purging Nuts.*

1 *Nuces vomicæ.*
Vomiting Nuts.

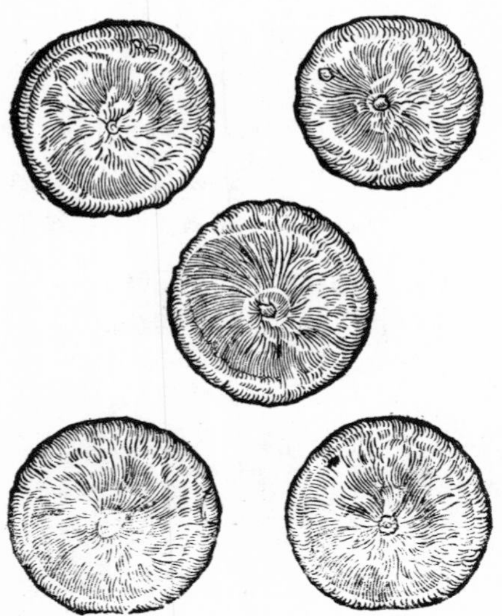

1 *Nuces purgantes.*
Purging Nuts.

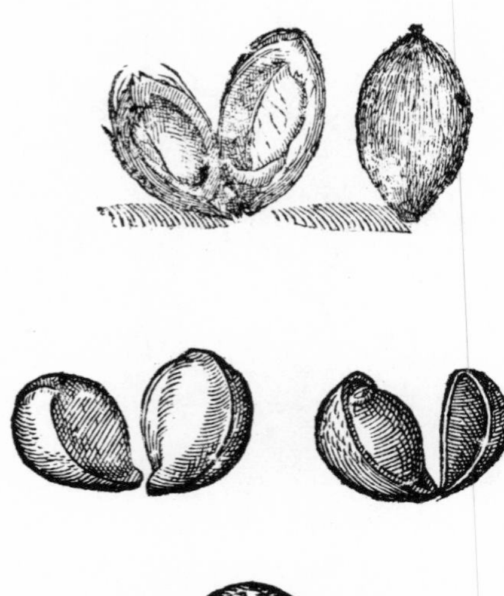

¶ *The*

¶ *The Description.*

1 AVicen and *Serapio* make *Nux vomica*, and *Nux Methel*, to be one, whereabout there hath been much cauelling; yet the case is plaine, if the text be true, that the Thorne Apple is *Nux Methel* Of the tree that beareth the fruit that is called in shops *Nux vomica*, and *Nux Methel*, we haue no certaine knowledge: some are of opinion, that the fruit is the root of an herbe, and not the nut of a tree: and therefore since the case among the learned resteth doubtful, we leaue the rest that might be said to a further consideration. The fruit is round, flat, like a little cake, of a russet ouerworne colour, fat and firme, in taste sweet, and of such an oily substance, that it is not possible to stampe it in a mortar to powder; but when it is to be vsed, it must be grated or scraped with some instrument for that purpose.

2 There be certaine Nuts brought from the Indies, called purging Nuts, of their qualitie in purging grosse and filthie humors, for want of good instruction from those that haue trauelled the Indies, we can write nothing of the tree it selfe: the Nut is somewhat long, ouall, or in shape like an egge, of a browne colour: within the shell is contained a kernell, in taste sweet, and of a purging facultie.

¶ *The Place and Time.*

These Nuts do grow in the desarts of Arabia, and in some places of the East Indies: we haue no certaine knowledge of their springing, or time of maturitie.

¶ *The Names.*

Auicen affirmeth the vomiting Nut to be of a poisonous qualitie, cold in the fourth degree, hauing a stupifying nature, and bringeth deadly sleepe.

¶ *The Vertues.*

Of the Physicall vertues of the vomitting Nuts we thinke it not necessarie to write, because the danger is great, and not to be giuen inwardly, but mixed with other compositions, and that very curiously by the hands of a faithfull Apothecarie. A

The pouder of the Nut mixed with some flesh, and cast vnto crowes and other rauenous fowles, doth kill and so dull their sences at the least, that you may take them with your hands. B

They make also an excellent sallet, mixed with some meat or butter, and laied in the garden where cats vse to scrape to burie their excrements, spoyling both the herbes and also seeds new sowne. C

CHAP. 159. *Of diuers sorts of Indian fruits.*

¶ *The Kindes.*

THese fruits are of diuers sorts and kinds, wherof we haue little knowledge, more than the fruits themselues, with the names of some of them: therefore it shall suffice to set forth vnto your view the forme onely, leauing vnto Time, and those that shall succeed, to write of them at large, which in time may know that, that in this time of infancie is vnknowne.

‡ OVr Authour formerly in this Chapter set forth diuers figures of Indian fruits, and amongst the rest *Beritinus*, *Cacao*, *Cocci Orientales*, *Buna*, *Fagaras*, *Cububæ*, *&c.* but he gaue but onely three descriptions, and these either false or to no purpose; wherefore I haue omitted them, and in this chapter giuen you most of these fruits which were formerly figured therin, together with an addition of sundry other out of *Clusius* his Exotickes, whose figures I haue made vse of, and here giuen you all those which came to my hands though nothing so many as are set forth in his Exotickes; neither, if I should haue had the figures, would the shortnesse of my time nor bignesse of the booke (being already growne to so large a volume) suffer mee to haue inserted them; therefore take in good part those I here giue, together with the briefe histories of them.

¶ *The Description.*

1 THe first and one of the best knowne of these fruits, are the *Cubibæ*, called of the Arabian Physitions *Cubibe* and *Quabeb*; but of the vulgar *Quabebchini*: in Iaoa where they plentifully grow, *Cumuc*: the other Indians, (the Malayans excepted) call them *Cubas sini*, not for that they grow in China, but because the Chinois vse to buy them in Iaoa and Sunda, and so carry them to the other ports of India. The plant which carries this fruit hath leaues like

1 *Cubibæ.* Cubibs.
2 *Cocci Orientales. Cocculus Indi.*

Cubibe

Cocci.

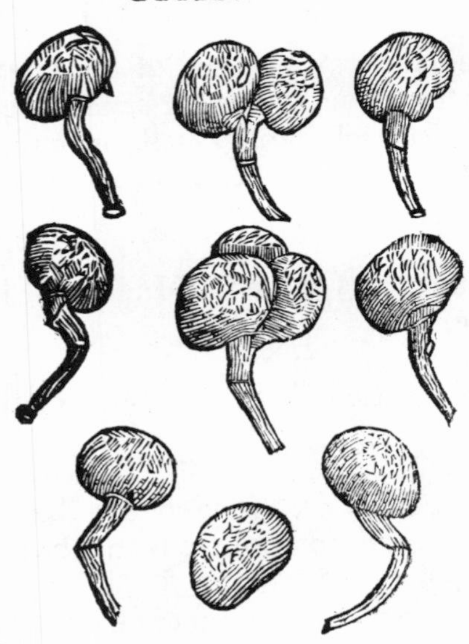

3 *Fagara.*

4 *Mungo.*

5 *Buna.*

‡ 6 *Amomum verum.*

‡ 7 *Amomum spurium.*

‡ 8 *Amomis.*

9 *Beritinus.*

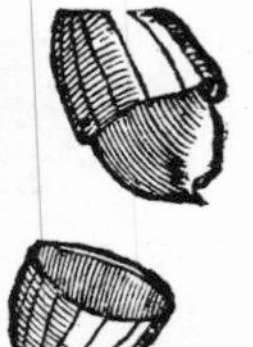

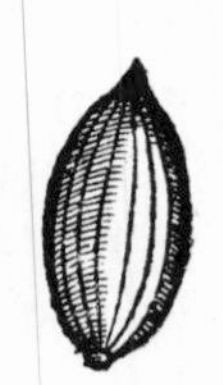

‡ 10 *Nuces insanæ.* Mad Nuts.

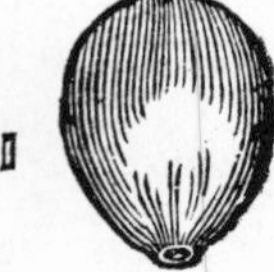

like those of pepper, but narrower, and it also windes about trees like as Iuy or Pepper doth: the fruit hangs in clusters, like as those we call red Currans, and not close thrust together in bunches, as grapes: the fruit or berries are of the bignesse of Pepper cornes, wrinckled, and of a brownish colour: they are of a hot and biting aromaticke taste, and oft times hollow within, but if theybe not hollow, then haue they a pretty reddish smooth round seed vnder their rough vtter huske; each of these berries commonly hath a piece of his foot-stalke adhering to it. It is reported that the Natiues where it growes first gently boile or scald these berries before they sell them, thar so none els may haue them, by sowing the seeds. Some haue thought these to haue beene the *Carpesium* of the Antients; and other-some haue iudged them the seeds of *Agnus Castus*, but both these opinions are erronious.

These are hot and dry in the beginning of the third degree; wherefore they are good against the cold and moist affects of the stomacke and flatulencies: they helpe to clense the breast of tough and thicke humours; they are good for the spleene, for hoarsnesse and cold affects of the wombe, chewed with Masticke, they draw much flegmaticke matter from the head, they heat and comfort the braine. The Indians vse them macerated in wine to excite venerie. A

2 The Plant which carries this fruit is vnknowne, but the berrie is well knowne in shoppes by the name of *Cocculus Indicus* some call them *Cocci Orientales*: others, *Cocculæ Orientales*: some, as *Cordus* for one, thinke them the fruit of *Solanum furiosum*: others iudge them the fruit of a Tithymale, or of a *Clematis*. These berries are of the bignesse of Bay berries commonly round, and growing but one vpon a stalke; yet sometimes they are a little cornered, and grow two or three clustering together: their outer coat or shell is hard, rough, and of a brownish duskie colour: their inner substance is very oily, of a bitter taste.

They are vsed with good successe to kill lice in childrens heads, being made into pouder and so strowed amongst the haire. They haue also another faculty which our Author formerly set downe in the chapter of *Alaternus* (where he confounded these with *Fagaras*) in these words, which I haue there omitted, to insert here; B

In England we vse the fruit called *Cocculus Indi* in pouder mixed with flower, hony, and crummes of bread to catch fish with, it being a numming, soporiferous, or sleeping medicine, causeth rhe fish to turne vp their bellies, as being sencelesse for a time. C

3 *Fagara* is a fruit of the bignesse of a Chich-pease, couered with a thin coat of a blackish ash colour, vnderwhich outer coat is a slender shell containing a sollid kernell, involued in a thin and blacke filme. The whole fruit both in magnitude, forme, and colour is so like the *Cocculus Indus* last described, that at the first sight one would take it to be the same. *Auicen* mentions this in his 266. Chap. after this manner. What is *Fagara?* It is a fruit like a Chich, hauing the seed of *Mahaleb*, and in the hollownesse is a blacke kernell as in *Schebedenegi*, and it is brought out of Sofale.

He places it amongst those that heate and dry in the third degree, and commends it against the coldnesse of the stomacke and liuer, it helps concoction, and bindes the belly.

4 This which *Clusius* thinkes to be *Mungo* (which is vsed in the East Indies about Guzarat and Decan for prouender for horses) is a small fruit of the bignesse of Pepper, crested, very like Coriander seed, but that it is bigger and blacke, it is of a hot taste.

5 *Buna* is a fruit of the bignesse of *Fagara*, or somewhat bigger or longer, of a blackish ash colour, couered with a thin skin, furrowed on both sides longwise, whereby it is easily diuided into two parts, which containe each a kernell longish and flat vpon one side, of a yellowish colour, and acide taste. They say that in Alexandria they make a certaine very cooling drinke hereof. *Rauwolfius* in his iournal seemes to describe this fruit by the name of *Bunnu*: and by the appellation, forme, and faculties; he thinkes it may be the *Buncho* of *Auicen*, and *Buncha* of *Rhasis*, to *Almansor*. *Clusius*.

6 This is a kind of Cardamome: and by diuers it is thought to be the true *Amomum* of the Antients, and to this purpose *Nicholas Marogna*, a Physition of Verona hath written a treatise which is set forth at the later end of *Pona's* description of Mount *Baldus*, to which I refer the curious: these cods or berries (whether you please to call them) grow thicke clustering together, they are round, and commonly of the bignesse of a cherry: the outer skin is tough, smoother, whiter, and lesse crested than that of the Cardamome: within this filme lye the seeds clustering together, yet with a thin filme parted into three, the particular seeds are cornered, somewhat smoother and larger than those of Cardamomes, but of the same aromaticke taste, and of a browne colour. Their temperature and faculties may be referred to those of Cardamomes.

7. 8. This with the next ensuing are by *Clusius* set forth by the names I here giue you them, though (as he saith) neither of them agree with the *Amomum* of *Dioscorides*, they were only branches set thicke with leaues, hauing neither any obseruable smell or taste: they were sent to the learned and diligent Apothecarie *Walarandus Donrez* of Lyons, from Ormuz the famous Mart & port town in the Persian Bay.

9 Those that accompanied the renowned Sir *Francis Drake* in his voyage about the World, light

11 *Cacao.* Small Cocoes.

‡ 14 *Guavobanus.* Tree Melon.

12 *Cucciophera.* Quince Dates.

‡ 13 *Baruce, Arara, Orukoria, Cropiot.*

‡ 15 *Ananas?* The Pinia, or Pine Thistle.

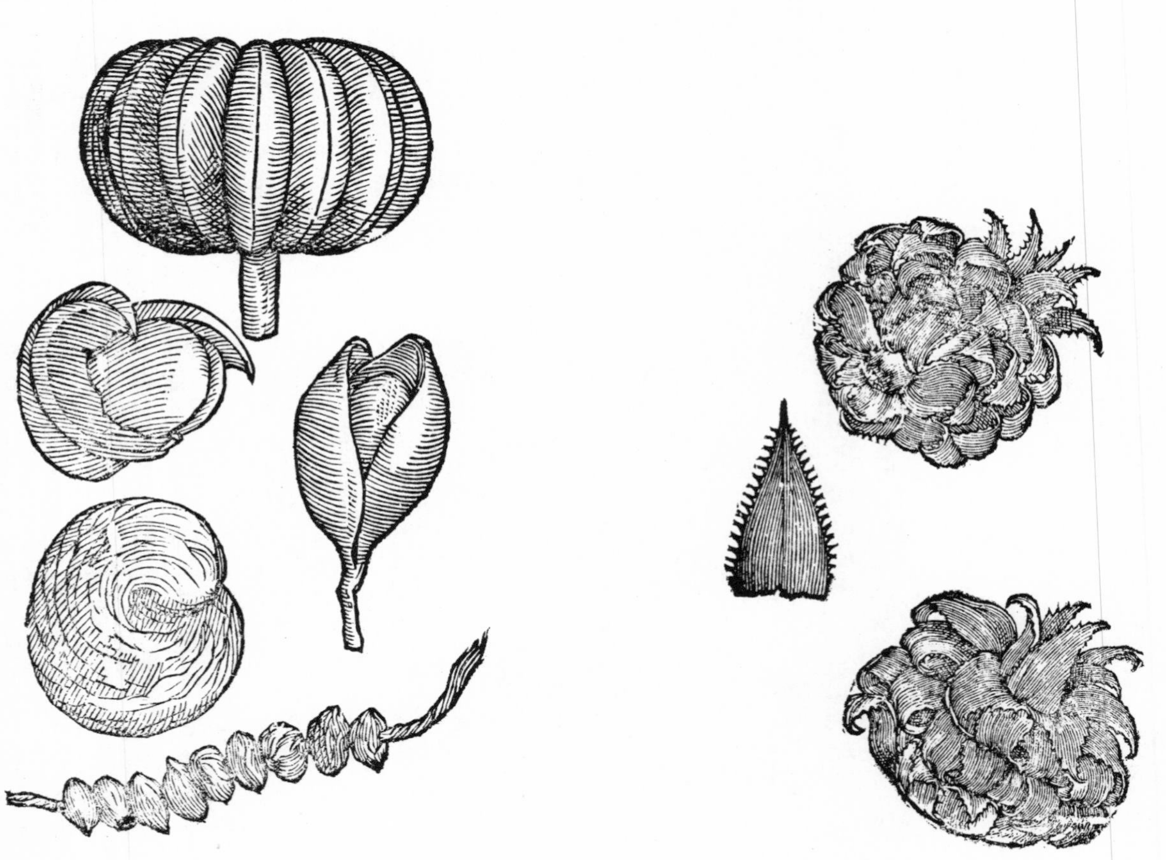

light vpon a certaine desert Island, wherein grew many very tall trees, and looking for something amongst these to refresh themselues, amongst others they obserued some bigger than Okes, hauing leaues like those of the Bay tree, thicke and shining, not snipt about the edges; their fruit was longish like to the small Acornes of the Ilex or Holme Oke, but without any cup; yet couered with a thin shell of an ash colour, and somtimes blacke, hauing within it a longish white kernell wrapped in a thin peeling, being without any manifest taste; They when they found it, though much opprest with hunger, yet durst not taste thereof, least it should haue been poisonous: but afterwards comming to the Island Beretina, not far from this, they found it to abound with these trees, & learned that their fruit was not poisonous, but might be eaten. Wherupon afterwards they in want of other victuals, boiled some as they do Pease, and ground others into floure, wherewith they made puddings. They found this tree also in the Moluccoes.

10 The first expressed in this table is the mad Plum, or as *Clusius* had rather terme it the Mad Nut; for he calls them Κάρυα μανικά, or *Insanæ Nuces*. The Hollanders finding them in their return from the East Indies, and eating the kernels, were for a time distracted, and that variously, according to the particular temperature of each that ate of them; as you may see in *Clusius Exot. lib. 2. Cap. 26.* This was round, little more than two inches about, with a shell not thicke, but sufficiently strong, brownish on the out side, and not smooth, but on the inside of a yellowish colour and smooth, containing a membranous stone or kernell couered with a black pulp, in form and bignesse not much vnlike a Bullas or Sloe, hauing a large white spot on the lower part whereas it was fastened to the stalke: vnder the pulpe lay the kernell, somewhat hard, and of an ash colour: the foot-stalke was short and commonly carried but one fruit, yet sometimes they obserued two growing together: the tree wheron this fruit grew was of the bignesse of a Cherry tree, hauing long and narrow leaues like those of the Peach tree: the other fruit figured in the 2. place was of a brownish yellow colour, somwhat bigger, but not vnlike a small Nut, and inch long, and somwhat more about, smaller below, and bigger aboue, and as it were parted into foure, being very hard and sollid. Of this see more in the fourteenth place.

11 The *Cacoa* is a fruit well knowne in diuers parts of America; for they in some places vse it in stead of money, and to make a drinke, of which, though bitter, they highly esteeme: the trees which beare them are but small, hauing long and narrow leaues, and will onely grow well in places shadowed from the Sun. The fruit is like an Almond taken out of his husk, and it is couered with a thin blacke skin, wherein is contained a kernell obliquely diuided into two or three parts, brownish, and distinguished with ash coloured veines, of an astringent and vngratefull taste.

12 This which *Clusius* had from *Cortusus*, for the fruit of *Bdellium*, is thought to be the *Cuci* of *Pliny*, and is the *Cuciophera* of *Matthiolus*, and by that name our Author had it in this Chapter. The whole fruit is of the bignesse of a Quince, and of the same colour, with a sweet and fibrous flesh, vnder which is a nut of the bignesse of a large Walnut or somewhat more, almost of a triangular form, bigger below, and smaller aboue, well smelling, of a darke ash colour, with a very hard shell, which broken there is therein contained a hard kernell of the colour and hardnesse of marble, hauing a hollownesse in the middle, as much as may containe a Hasell Nut.

13 In this table are foure seuerall fruits described by *Clusius Exot. lib. 2. c. 21.* The first is called *Baruce*, and is said to grow vpon a high tree in Guyana called Hura: it consisted of many Nuts of some inch long, strongly fastened or knit together, each hauing a hard wooddy shell, falling into two parts, containing a round and smooth kernell couered with an ash coloured filme.

They say the natiues there vse this fruit to purge and vomite. A

The second called Arara growes in Kaiana, but how, it is not knowne: it was some inch long, couered with a skin sufficiently hard and blacke, fastened to a long and rugged stalke that seemed to haue carried more than one fruit: the kernell is blacke, and of the bignesse of a wilde Oliue.

The natiues vse the decoction hereof to wash maligne vlcers, and they say the kernell will loose the belly. B

The third named *Orukoria* is the fruit of a tree in Wiapock, called *Iuruwa*, they vse this to cure their wounds, dropping the iuice of the fruit into them. This fruit is flat almost an inch broad, and two long, but writhen like the cod of the true *Cytisus*, but much bigger, very wrinckled, of an ash colour, containing a smooth seed.

The fourth called *Cropiot* is a small and shriuelled fruit, not much vnlike the particular ioints of the Æthiopian pepper.

The sauages vse to take it mixed amongst their Tabaco to asswage head-ache: there were diuers of them put vpon a string (as you may see in the figure) the better to dry them. C

14 This which by *Clusius* & *Lobel* is thought to be the *Guanabanus* mentioned by *Scaliger Exerc. 281. part. 6.* is a thicke fruit some foot and halfe long, couered with a thicke and hard rinde, freezed ouer with a soft downinesse, like as a Quince is, but of a greenish colour, with some veines, or rather furrowes running alongst it, as in Melons: the lower end is somewhat sharp: at the vpper end it is fastened

fastened to the boughes, with a firme, hard, and fibrous stalke: this fruit containes a whitish pulpe, which the Ethiopians vse in burning feuers to quench the thirst, for it hath a pleasant tartnesse: this dried becomes friable, so that it may be brought into pouder with ones fingers, yet retaineth its aciditie: in this pulp lye seeds like little Kidneis, or the seeds of the true *Anagyris*, of a black shining colour, with some fibres comming out of their middles: these sowne brought forth a plant hauing leaues like the Bay tree, but it dyed at the approch of Winter. *Clus.*

15 *Ananas Pinias*, or Pine Thistle is a plant hauing leaues like the *Aizoon aquaticum*, or water

‡ 16 *Fabæ Ægyptiæ affinis.*

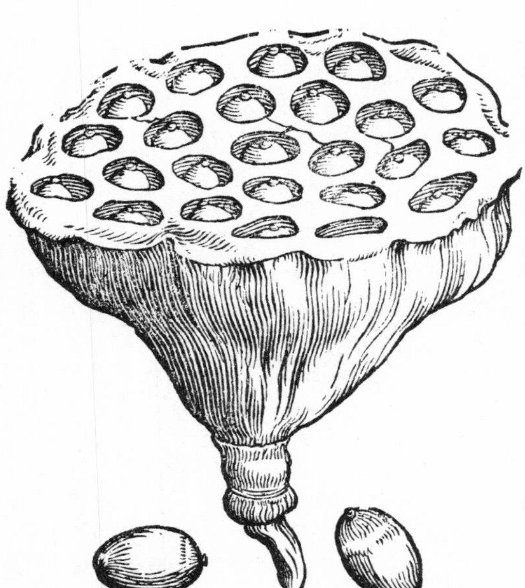

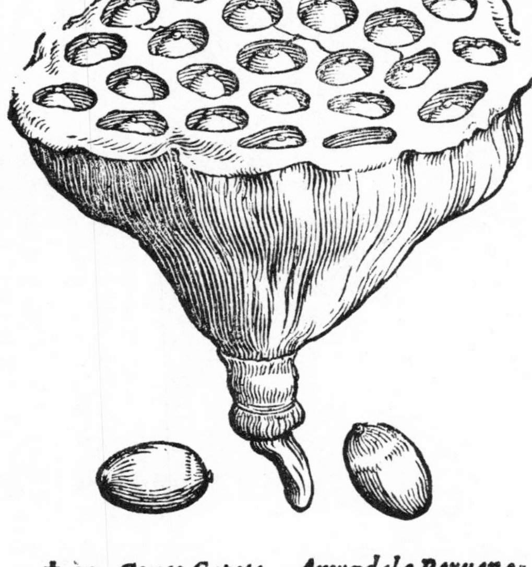

‡ 19 *Fructus tetragonus.* The square Coco.

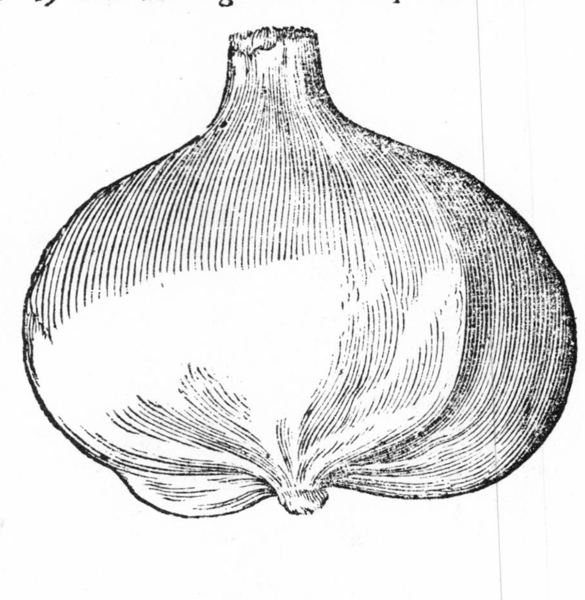

‡ 17 *Coxco Cypote. Amygdalæ Peruanæ.* Almonds of Peru.

11

‡ 18 *Buenas Noches.*

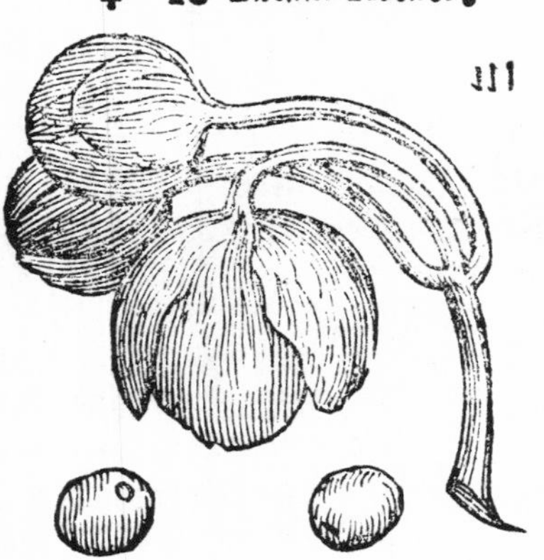

‡ 20 *Arboris laniferæ siliqua.* A cod of the wooll-bearing tree.

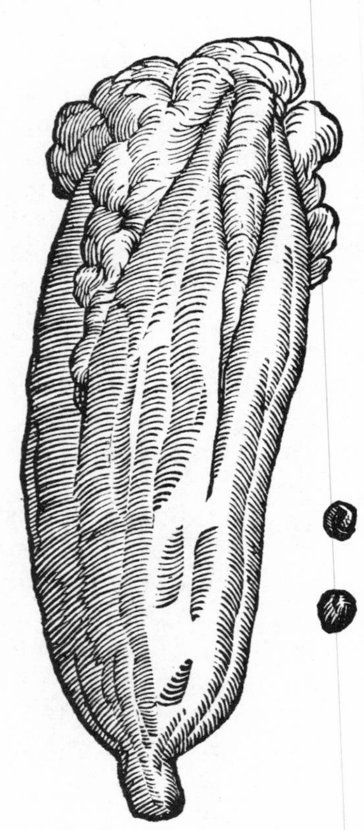

Sengreene,

Sengreene, somewhat sharpe and prickly about the edges: the stalke is round, carrying at the top therof one fruit of a yellowish colour when it is ripe, of the bignes of a Melon, couered with a scale-like rinde: the smell is gratefull, somewhat like that of the Malocotone: at the top of the fruit, and sometimes below it come forth such buds as you see here presented in the figure, which they set in the ground and preserue the kind by in stead of seed: the meat of this fruit is sweet & very pleasant of taste, & yeelds good nourishment; there are certaine small fibres in the meat thereof, which though they do not offend the mouth, yet hurt they the gums of such as too frequently feed thereon.

16 The forme of this is somewhat strange, for it is like a large Poppy head cut off nigh the top: the substance thereof was membranous and wrinckled, of a brownish colour, very smooth: the circumference at the top is about nine inches, and so it growes smaller and smaller euen to the stalke, which seems to haue carried a floure whereto this fruit succeeded: the top of the fruit was euen, and in it were orderly placed 24. cauities, in each whereof was contained a little Nut like an Acorn, almost an inch long, and as much thick; the vpper part was of a brownish colour, & the kernel within was rank and all mouldy. *Clusius* could learne neither whence this came, nor how it grew, but with a great deale of probability thinks it may be that which the Antients described by the name of *Faba Ægyptia*.

17 The former of these two *Clusius* receiued by the name of *Coxco Cypote*, that is the Nut Cypote: It is of a dusky browne colour, smooth, and shining, but on the lower part of an ash colour, rough, which the Painter did not well expresse in drawing the figure. The 2. hee receiued by the name of *Almendras del Peru*, (i.) Almonds of Peru: the shell was like in colour and substance to that of an almond, and the kernell not vnlike neither in substance nor taste: yet the forme of the shell was different, for it was triangular, with a backe standing vp, and two sharp sides, and these very rough.

18 This was the fruit of a large kind of *Convolvulus* which the Spaniards called *Buenas noches*, or Goodnight, because the floures vse to fade as soone as night came. The seeds were of a footy colour as big as large Pease, being three of them contained in a skinny three cornered head. You may see more hereof in *Clusius, Exot. lib. 2. cap. 18.*

19 This is the figure of a square fruit which *Clusius* coniectures to haue been some kind of Indian Nut or Coco: it was couered with a smooth rinde, was seuen inches long, and a foot and halfe about, being foure inches and a halfe from square to square.

20 About Bantam in the East Indies growes a tall tree sending forth many branches, which are set thicke with leaues long and narrow, bigger than those of Rosemary: it carries cods six inches long, and fiue about, couered with a thin skin, wrinckled and sharp pointed, which open themselues from below into fiue parts, and are full of a soft woolly or Cottony matter, wherewith they stuffe cushions, pillowes and the like, and also spin some for certaine vses: amongst the downe lye blacke seeds like those of Cotton, but lesse, and not fastened to the downe.

21 This which *Clus* calls *Palma saccifera*, or the Bag Date, because it carries the figure of an Hippocras bag, was found in a desart Island in the Antlantick ocean, by certaine Dutch mariners who obserued whole woods thereof: these bags were some of them 22. or more inches long, and some seuen inches broad in the broadest place, strongly woue with threds crossing one another, of a brownish yellow colour. These sachels (as they report who cut them from the tree) were filled with fruit of the bignesse of a Walnut huske and all: within these were others, as round as if they had bin torned, and so hard that you could scarce breake them with a hammer: in the midst of these were white kernels, tasting at the first somewhat like pulse, but afterwards bitter like a Lupine.

22 The tree which carries this rough cod is very large, as I haue been told by diuers: some who saw it in Persia, & others that obserued it in Mauritius Island. *Clusius* also notes that they haue bin brought from diuers places: the cod is some three inches long, and some two inches broad, of a duskie red colour, and all rough and prickly: in these cods are contained one, two or more round nuts or seeds of a grayish ash colour, hauing a little spot on one side, where they are fastened to the cod they are exceeding hard, and difficult to breake, but broken they shew a white kernel very bitter and vnpleasant of taste. I haue seen very many and haue some of these, and some haue offered to sel them for East Indian *Beazor*, whereto they haue some small resemblance, though nothing in facultie like them (if I may credit report, which I had rather do than make tryall) for I haue been told by some that they are poisonous; and by others, that they strongly procure vomit.

23 The long cod expressed in this figure is called in the East Indies (as *Clusius* was told) *Kaye baka*, it was round, the thicknesse of ones little finger, and six inches long: the rinde was thick, black, hard and wrinckled, and it contained a hard pulpe of a sowrish taste, which they affirm was eatable.

The other was a cod of some inch and halfe long, and some inch broad, membranous, rough, and of a brownish colour, sharp pointed, and opening into two parts, and distinguished with a thin film into foure cels, wherein were contained scarlet Peare fashioned little berries, hauing golden spots especially in the middles. This growes in Brasile, and as *Clusius* was informed was called *Daburi*.

24 In the second place of the tenth figure and description in this chapter you may finde the

‡ 21 *Palma saccifera.*
The Sachell Date.

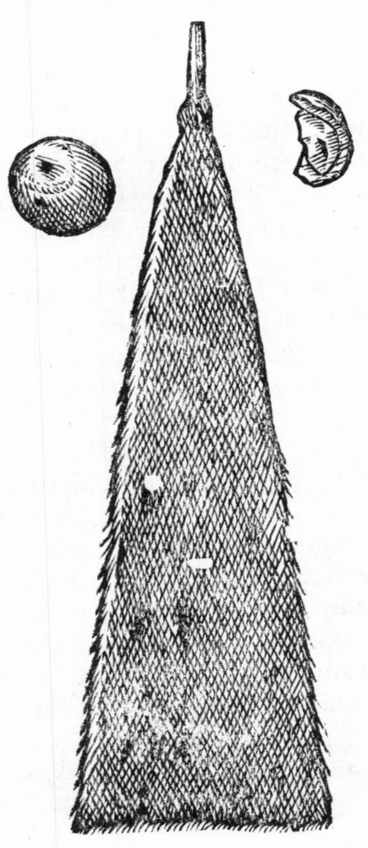

‡ 22 *Lobus Echinatus.*
Beazor Nuts.

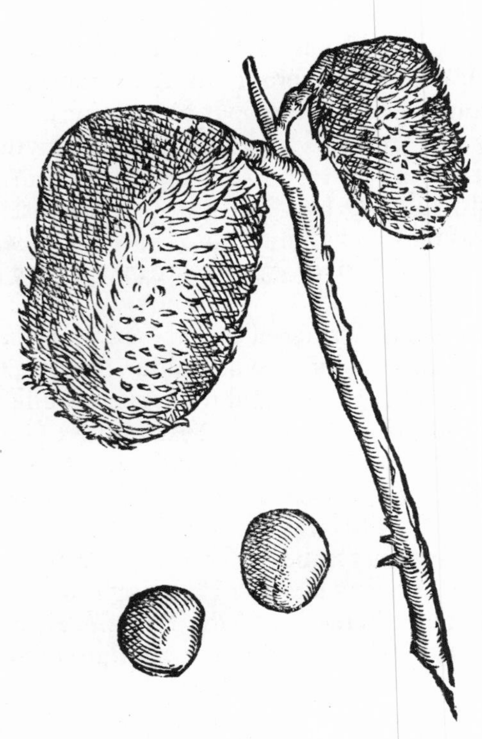

‡ 23 *Kaie baka.*
Daburi.

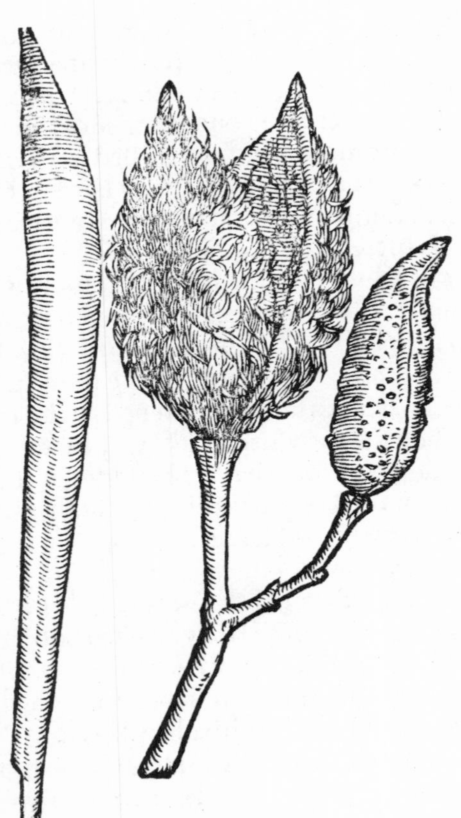

‡ 24 *Nucula Indica racemosa.*
The Indian, or rather Ginny Nut.

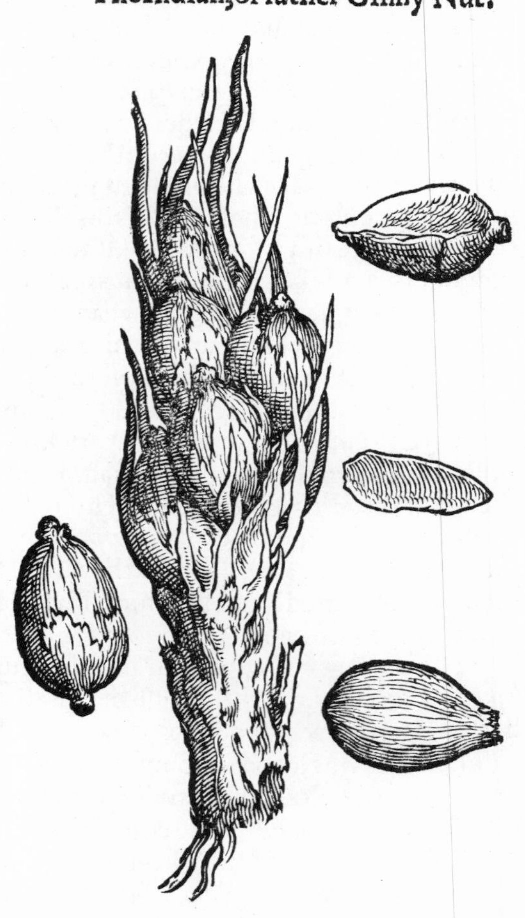

‡ 25 *Fructus ſquamoſi.*
Scalie fruits.

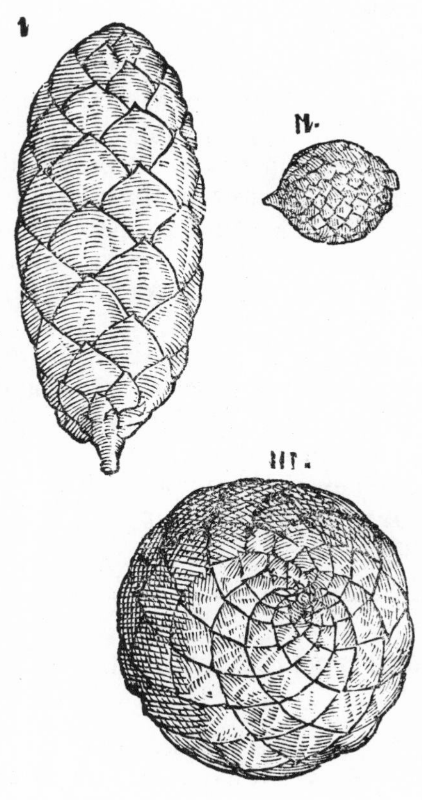

‡ 26 *Fructus alij Exotici.*
Other ſtrange fruits.

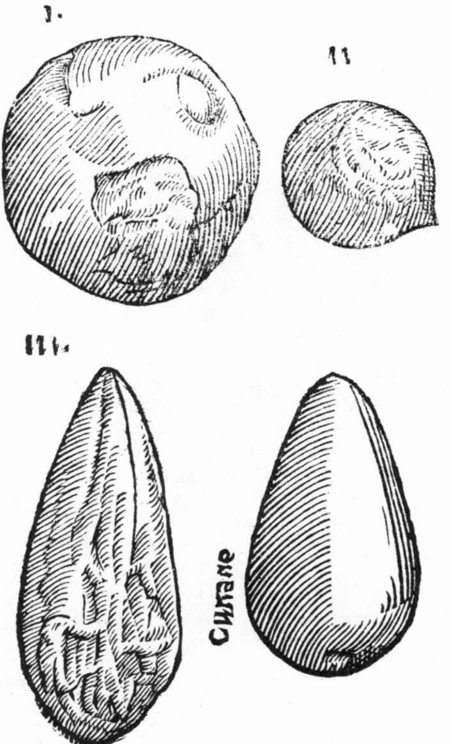

‡ 26 *Fructus alij Exotici.*
Other ſtrange fruits.

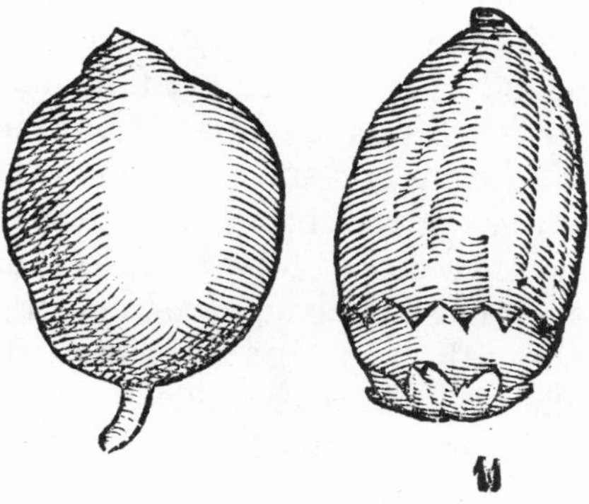

ſingle Nut here figured, deſcribed, & ſet forth; but this figure better expreſſes the manner of growing therof: for firſt it preſents to the view the nuts in their outer husks growing cloſe together, then the ſingle nuts in and out of their husks, and laſtly their kernell: the ſhell of this nut containes in it a certaine oilie ſubſtance, ſomwhat reſembling the oile of ſweet almonds: the tree whereof this nut is the fruit growes in Ginny, and is much vſed by the people there, for they preſſe a liquour forth of the leaues, or elſe boile them in water, & this ſerues them in ſtead of wine & beare, or at leaſt for a common drink, of the fruit they make bread of a very ſweet and pleaſant taſte.

25 Theſe ſcaily fruits are ſet forth by *Cluſius, Exot. l. 2 c. 3.* The firſt was three inches long and two inches about, and had in it a longiſh hard ſollid kernell, with many veines diſperſed ouer it, and ſuch kernels are ſomtimes poliſhed, whereby they become white, and then their blacke veines make a fine ſhew, which hath giuen occaſion to ſom e impoſtors to put them to ſaile for rare and precious ſtones.

The ſecond was ſmall, round and ſcaily, and the ſcailes turned their points downwards towards the ſtalke.

The third was alſo ſcaily, of the bigneſſe of a Walnut in his huske, with the ſcailes very orderly placed, and of a browniſh colour: it had a kernel which ratled in it when it was ſhaken.

26 The firſt of the two in the former table was brought from Ginny, it was of the bigneſſe and ſhape of a plum, two inches long, and one and halfe broad, of a thicke fungous ſubſtance, ſomewhat wrinckled, and blackiſh on the outſide, and within containing a certaine whitiſh inſiped friable pulpe, wherein lay a few ſmall ſeeds.

The ſecond was ſome inch and halfe long, an inch thicke, couered with an aſh coloured skin, compoſed within of many fibres almoſt like the huske of the Nut Fauſell, at the lower end it ſtood in a double cup, and it was ſharp pointed at the vpper end: in this skin was contained a kernell, or rather nut, blacke, hard, and very wrinckled not much vnlike to that of Fausell, whereto I refer it as a kinde thereof. These two are treated of by *Clusius, Exot. lib. 2. c. 23.*

The first of the second table (wherein are contained foure figures) was of a round forme, yet a little flat on one side, distinguished vnder the blacke and shining coat wherewith it was couered, with furrowes running euery way, not vnlike to the Nut Fausell taken forth of his couer: the inner pulpe was hard and whitish, first of a salt, and then of an astringent taste.

The second of these was an inch long, but rather the kernell of a fruit, than a fruit it selfe; it was round except at the one end, and all ouer knobby (though the picture expresse not so much) there was also some shew of a triangular forme at each end.

The third was two inches and a halfe long, and in the broadest part some inch and more broad: it was somewhat crooked, the backe high and rising, the top narrow, and the lower part sharp pointed, of an ash colour, with thicke and eminent nerues running alongst the back from the top to the lower part, exprest with such art, as if they had been done by some curious hand: it seemed to haue bin couered ouer with another rinde, but it was worne off by the beating of the waues of the sea vpon the shore.

The fruit *Cunane* figured in the fourth place of this table, was two inches long, and an inch broad at the head, and so smaller by little & little, with a back standing out, smooth, black, and shining, hauing three holes at the top, one aboue, & two below: they said it grew vpon a smal tree called Morremor, and was yet vnripe, but when it was ripe it would be as big again, and that the natiues where it grew (which was as I take it about Wiapock) rost it vpon the coles, and eate it against the headache. *Clusius* sets forth these foure in his *Exot. l. 2. c. 22.* he describes *Cunane cap. 21.* ‡

Chap. 160. *Of Sun-Dew, Youth woort, Ros Solis.*

1 *Ros Solis folio rotundo.*
Sun-Dew with round leaues.

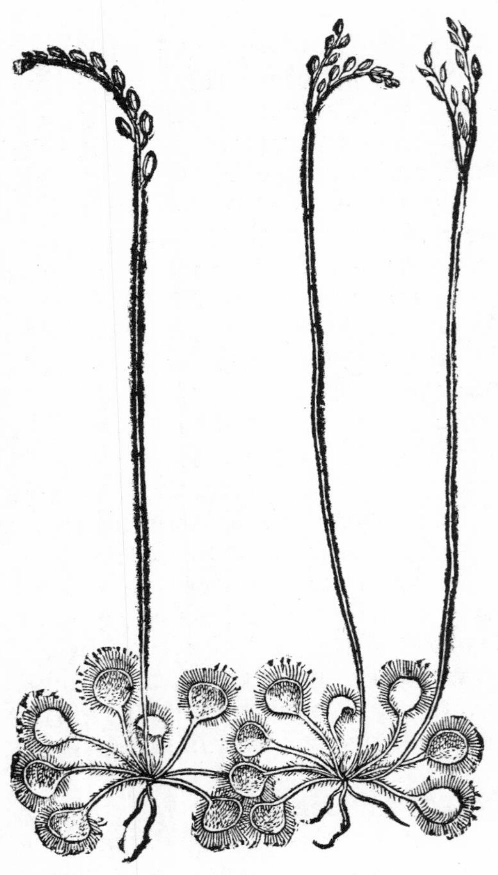

2 *Ros Solis folio oblongo.*
Sun-Dew with longish leaues.

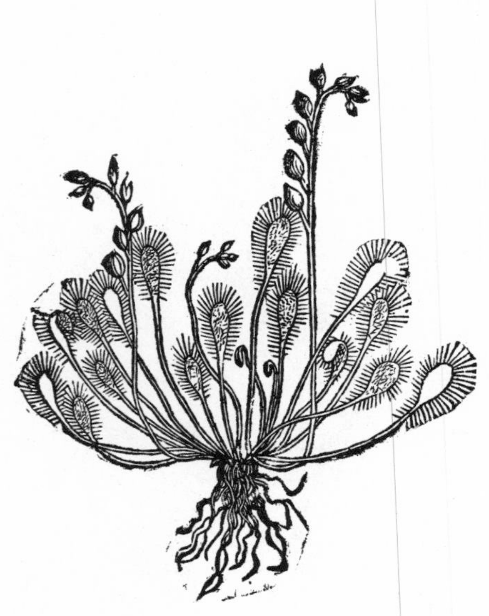

¶ The

¶ *The Description.*

1 SVn-Dew is a little herb, and groweth very low, it hath a few leaues standing vpon slender stems, very small, something round, a little hollow, and like an eare picker, hairy and reddish as be also the stems, hauing dew and moisture vpon them at the driest time of the yeare, and when the Sun shineth hottest euen at high noone; and a moneth after there spring vp little stalks, a hand breadth high, on which stand small whitish floures: the roots are very slender, and like vnto haires.

2 The second kinde is like vnto the former, in stalks and floures, but larger, and the leaues are longer, and not so round, wherein consisteth the difference.

¶ *The Place.*

They grow in desart, sandie and sunny places, but yet waterie, and seldome other-where than among the white marish mosse which groweth on the ground and also vpon bogs.

¶ *The Time.*

Sun-Dew flourisheth in Sommer, it floureth in May or Iune: it is to be gathered when the weather is most dry and calme. The distilled water hereof that is drawne forth with a glasse still, is of a glittering yellow colour like gold, and coloureth siluer put therein like gold.

¶ *The Names.*

It is called in Latine, *Ros Solis*: of diuers, *Rorella*: it is named of other, *Salsi Rosa*, of the dew which hangeth vpon it, when the Sun is at the hottest: it is called in high Dutch, Sondaw, and Suidaw: in low Dutch, Loopichecruit, which in English signifieth Lustwoort, because sheepe and other cattell, if they do but onely taste of it, are prouoked to lust. It is called in English, Sun-Dew, Ros Solis, Youth-woort: in the Northern parts, Red Rot, because it rotteth sheepe; and in Yorkeshire, Moore grasse.

¶ *The Temperature.*

It is a searing or causticke herbe, and very much biting, being hot and drie in the fourth degree.

¶ *The Vertues.*

The leaues being stamped with salt do exulcerate and raise blisters, to what part of the body soeuer they be applied. A

The later Physitions haue thought this herbe to be a rare and singular remedie for all those that be in a consumption of the lungs, and especially the distilled water thereof: for as the herbe doth keep and hold fast the moisture and dew, and so fast, that the extreme drying heate of the Sun cannot consume and waste away the same: so likewise men thought that herewith the naturall and radical humidity in mens bodies is preserued and cherished. But the vse therof doth otherwise teach, and reason sheweth the contrarie: for seeing it is an extreme biting herb, and that the distilled water is not altogether without this biting qualitie, it cannot be taken with safety: for it hath also bin obserued, that they haue sooner perished that vsed the distilled water hereof, than those that abstained from it, and haue followed the right and ordinarie course of diet. B

Cattell of the female kinde are stirred vp to lust by eating euen of a small quantity: which thing hath greatly increased their vain opinion, without sence or reason; for it doth not moue nor prouoke cattell to lust, for that it increaseth the substance of the seed, but because through his sharp and biting qualitie it stirreth vp a desire to lust, which before was dulled, and as it were asleepe. C

It strengthneth and nourisheth the body, especially if it be distilled with wine, and that liquour made thereof which the common people do call Rosa Solis. D

If any be desirous to haue the said drinke effectuall for the purposes aforesaid, let them lay the leaues of Rosa Solis in the spirit of wine, adding thereto Cinnamon, Cloues, Maces, Ginger, Nutmegs, Sugar, and a few graines of Muske, suffering it so to stand in a glasse close stopt from the aire, and set in the Sun by the space of ten daies, then straine the same, and keep it for your vse. E

Chap. 161. *Of Mosse of trees.*

¶ *The Description.*

TRee Mosse hath certaine things like haires, made vp as it were of a multitude of slender leaues, now and then all to be iagged, hackt, and finely carued, twisted and interlaced one in another, which cleaue fast to the barkes of trees, hanging downe from the bodies: one of this kinde is more

 slender

Muscus quernus.
The Mosse of the Oke & of other trees.

slender and thin, another more thicke, another shorter, another longer; all of them for the most being of a whitish colour, yet oftentimes there is a certaine one also which is blacke, but lesser and thinner: the most commendable of them all, as *Pliny* saith, be those that are whitish, then the reddish, and lastly such as be blacke.

¶ *The Place.*

This Mosse is found on the Oke tree, the white and blacke Poplar tree, the Oliue tree, the Birch tree, the Apple tree, the Peare tree, the Pine tree, the wilde Pine tree, the Pitch tree, the Firre tree, the Cedar tree, the Larch tree, & on a great sort of other trees. The best, as *Dioscorides* saith, is that of the Cedar tree, the next of the Poplar, in which kinde the white and the sweet smelling Mosse is the chiefest; the blackish sort is of no account. *Matthiolus* writeth, that in Italy that Mosse is sweet which groweth on the Pine tree, the Pitch tree, the Fir tree, & the Larch tree, and the sweetest that of the Larch tree.

¶ *The Time.*

Mosse vpon the trees continueth all the yeare long.

¶ *The Names.*

It is called of the Grecians βρύον of the Latins, *Muscus*: the Arabians and some Apothecaries in other countries call it *Vsnea*: in high Dutch, **Mosz**: in low Dutch, **Mosch**: the French men, *Lu Mousch*: the Italians, *Musgo*: in Spanish, *Musco de los arbores*: in English, Mosse, tree Mosse, or Mosse of trees.

¶ *The Temperature.*

Mosse is somewhat cold and binding, which notwithstanding is more and lesse according vnto the nature and facultie of that tree on which it groweth, and especially of his barke: for it taketh vnto it selfe and also retaineth a certaine propertie of that barke, as of his breeder of which hee is ingendred: therefore the Mosse which commeth of the Oke doth coole and very much binde, besides his owne and proper facultie, it receiueth also the extreme binding quality of the Oke barke it selfe.

The Mosse which commeth of the Cedar tree, the Pine tree, the Pitch tree, the Fir tree, the Larch tree, and generally all the Rosine trees are binding and do moreouer digest and soften.

¶ *The Vertues.*

Serapio saith, that the wine in which Mosse hath been steeped certain daies, bringeth sound sleep,
A strengtheneth the stomacke, staieth vomiting, and stoppeth the belly.

Dioscorides writeth, that the decoction of Mosse is good for women to sit in, that are troubled
B with the whites; it is mixed with the oile of Ben, and with oiles to thicken them withall.

It is fit to be vsed in compositions which serue for sweet perfumes, and that take away wearisom-
C nesse; for which things that is best of all which is most sweet of smell.

CHAP. 162. *Of ground Mosse.*

¶ *The Kindes.*

There groweth also on the superficiall or vppermost part of the earth diuers Mosses, as also vpon rocks and stony places, and marish grounds, differing in forme not a little.

¶ *The Description.*

1 The common Mosse groweth vpon the earth, and the bottome of old and antient trees, but specially vpon such as grow in shadowie woods, and alsoat the bottom of shadowy hedges,

1 *Muscus terrestris vulgaris.*
Common ground Mosse.

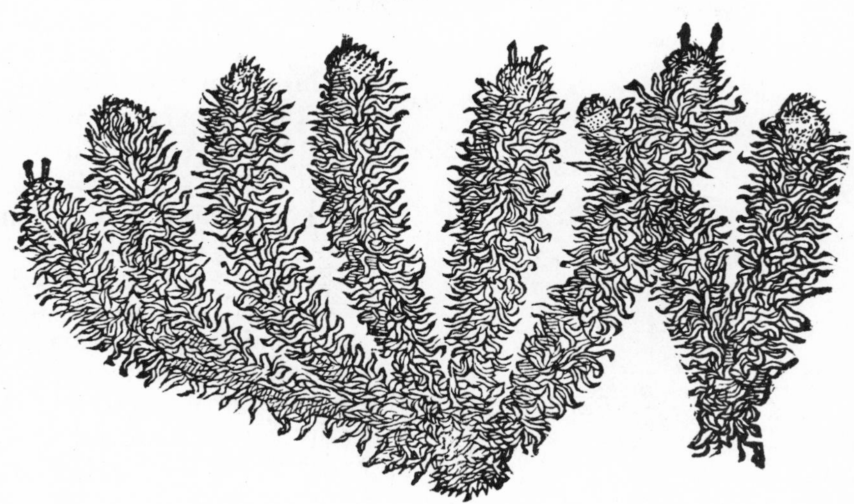

2 *Muscus terrestris scoparius.*
Beesome ground Mosse.

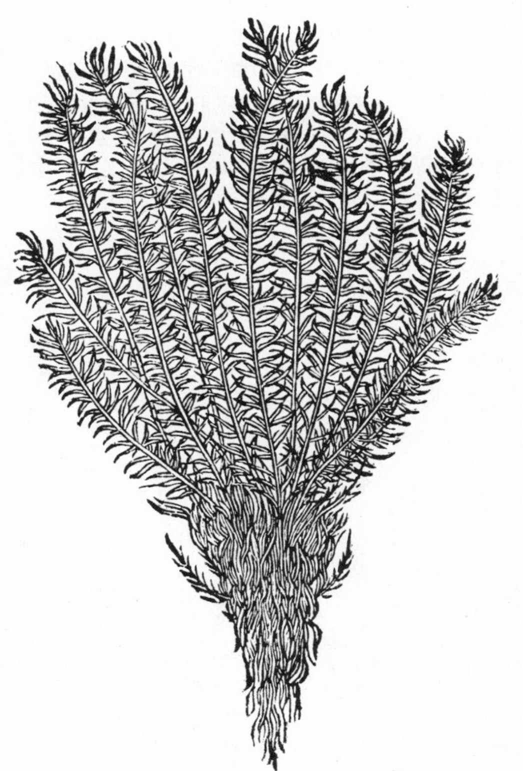

3. 4. *Muscus capillaris, siue Adianthum aureum maius & minus.*
Goldilockes or golden Maiden-haire the bigger and lesse.

hedges and ditches, and such like places: it is very well knowne by the softnesse and length thereof, being a mosse most common, and therefore needeth not any further description.

2 Beesome Mosse, which seldome or neuer is found but in bogs and marish places, yet sometimes haue I found it in shadowie dry ditches, where the Sun neuer sheweth his face: it groweth vp halfe a cubit high, euery particular leafe consisting of an innumerable sort of hairy threds set vpon a middle rib, of a shining blacke colour like that of Maiden-haire, or the Capillare Mosse *Adianthum aureum* whereof it is a kinde.

3 This kinde of Mosse, called *Muscus capillaris*, is seldome found but vpon bogs and moorish places, and also in some shadowie dry ditches where the Sun doth not come. I found it in great aboundance in a shadowie ditch vpon the left hand neere vnto a gate that leadeth from Hampsted heath,

5 *Muscus ramosus floridus.*
Flouring branched Mosse.

heath toward Highgate; which place I haue shewed vnto diuers expert Surgeons of London, in our wandering abroad for our farther knowledge in Simples. This kind of Mosse, the stalkes thereof are not aboue one handfull high, couered with short haires standing very thicke together, of an obscure yellow green colour; out of which stalkes spring vp sometimes very fine naked stems, somewhat blacke, vpon the tops of which hang as it were little graines like wheat cornes. The roots are very slender and maruellous fine.

‡ Of this *Adianthum aureum* there are three kindes, different onely in magnitude, and that the two bigger haue many hairie threds vpon their branches, when as the least hath onely three or foure close to the root; and this is the least of plants that I euer yet saw grow. ‡

4 Of this there is also another kinde altogether lesser and lower. This kind of mosse groweth in moist places also, commonly in old mossie and rotten trees, likewise vpon rocks, and oftentimes in the chinks and crannies of stone walls.

† 5 There is oftentimes found vpon old Okes and Beeches, and such like ouer-grown trees, a kinde of Mosse hauing many slender branches, which diuide themselues into other lesser branches; whereon are placed confusedly very many small threds like haires, of a greenish ash colour: vpon the ends of the tender branches sometimes there commeth forth a floure in shape like vnto a little buckler or hollow Mushrom, of a whitish colour tending to yellownes, and garnished with the like leaues of those vpon the lower branches.

6 *Muscus Pyxidatus.*
Cup or Chalice Mosse.

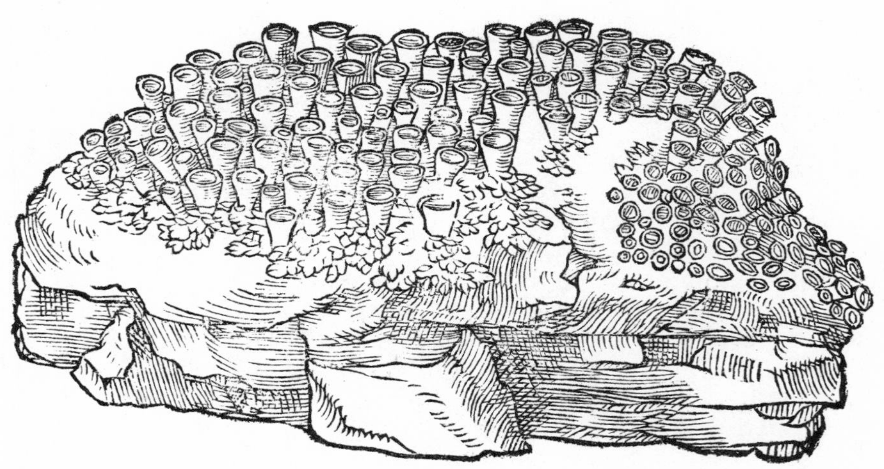

6 Of this Mosse there is another kinde, which *Lobel* in his Dutch Herbal hath set forth vnder the title of *Muscus Pyxidatus*, which I haue Englished, Cup Mosse or Chalice Mosse: it groweth in the most barren dry and grauelly ditch bankes, creeping flat vpon the ground like vnto Liuerwort, but of a yellowish white colour: among which leaues start vp here and there certaine little things fashioned like a little cup called a Beaker or Chalice, and of the same colour and substance of the lower leaues, which vndoubtedly may be taken for the floures: the pouder of which Mosse giuen to children in any liquor for certaine dayes together, is a most certaine remedie against that perillous malady called the Chin-cough

7 There is likewise found in the shadowie places of high mountaines, and at the foot of old and

and rotten trees, a certaine kinde of Moſſe in face and ſhew not vnlike to that kinde of Oke Ferne called *Dryopteris*. It creepeth vpon the ground, hauing diuers long branches, conſiſting of many ſmall leaues, euery particular leafe made vp of ſundry little leaues, ſet vpon a middle rib one oppoſite to another.

7 *Muſcus Filicinus.*
Moſſe Ferne.

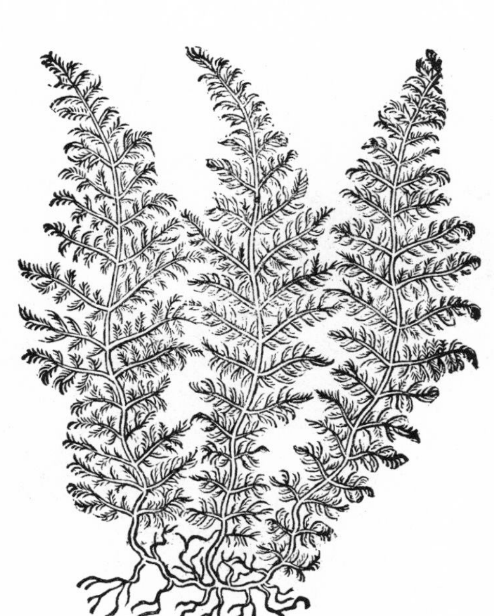

8 *Muſcus corniculatus.*
Horned or knagged Moſſe.

9 *Muſcus denticulatus.*
Toothed Moſſe.

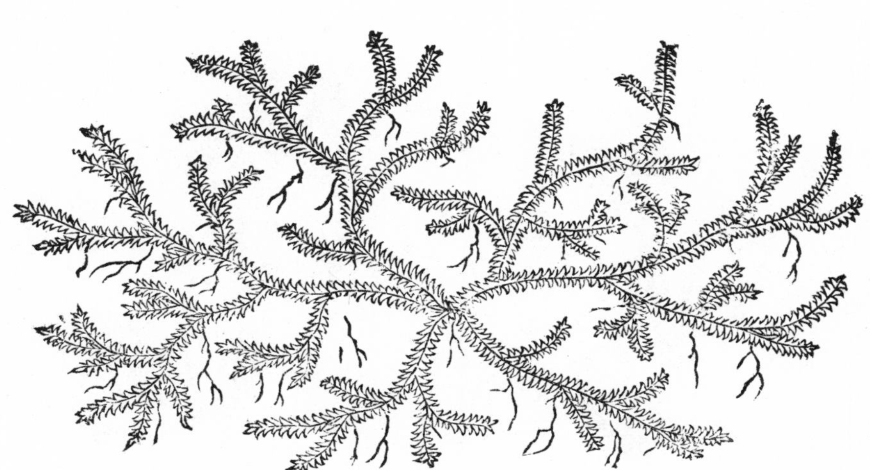

8 There is found vpon the tops of our moſt barren mountaines, but eſpecially were ſea Coles are accuſtomed to be digged, ſtone to make iron of and alſo where ore is gotten for tinne and lead, a certaine ſmall plant: it riſeth forth of the ground with many bare and naked branches, diuiding themſelues at the top into ſundry knags like the forked hornes of a Deere, euery part whereof is of an ouerworne whitiſh colour.

‡ Our Author formerly gaue another figure and deſcription of this plant, by the name of *Holoſteum petræum*, which I haue omitted, thinking this the better. *Tragus, Lonicerus,* and *Bauhine* referre this to the Fernes; and the laſt of them calleth it *Filix ſaxatilis corniculata*: *Pena* and *Lobel* made it their *Holoſtium alterum*: *Thalius* calls it *Adianthum acroſchiſton, ſeu furcatum.* ‡

9 There is found creeping vpon the ground a certaine kinde of Moſſe at the bottom of Heath and Ling, and ſuch like buſhes growing vpon barren mountaines, conſiſting as it were of ſcales made vp into a long rope or cord, diſperſing it ſelfe far abroad into ſundry branches, thruſting out here

here and there certain roots like threds, which take hold vpon the vpper cruſt of the earth, whereby it is ſent and diſperſed far abroad : the whole plant is of a yellowiſh greene colour.

10 This other kinde of Moſſe is found in the like places : it alſo diſperſeth it ſelfe far abroad, and is altogether leſſer than the precedent, wherein conſiſts the difference.

10 *Muſcus minor denticulatus.*
Little toothed Moſſe.

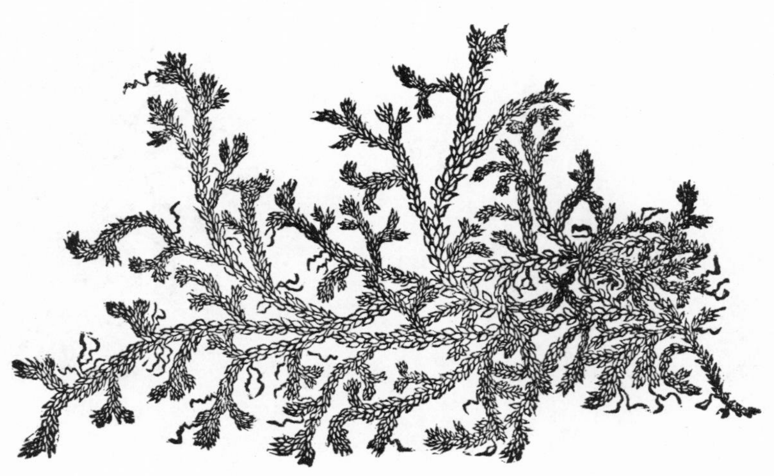

11 *Muſcus clauatus, ſiue Lycopodium.*
Club Moſſe, or Wolfe claw Moſſe.

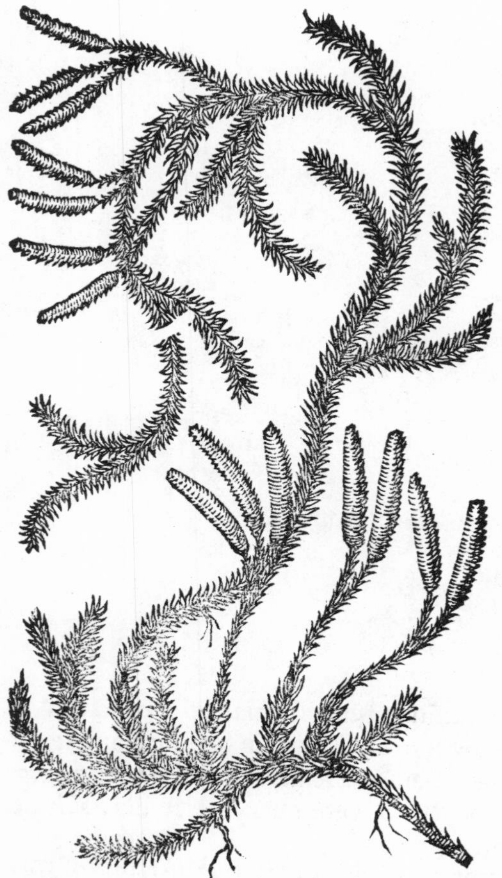

† 12 *Muſcus clauatus folijs Cypreſſi.*
Heath Cypres.

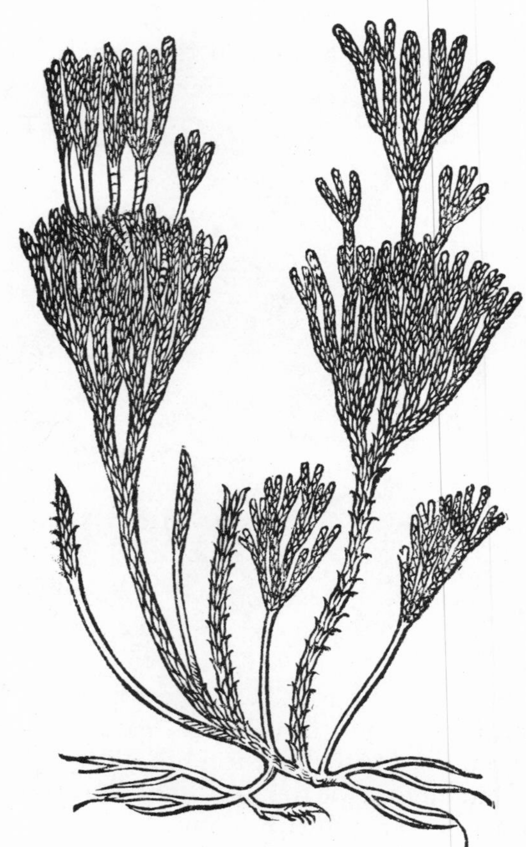

11 There is likewiſe another kinde of Moſſe, which I haue not elſewhere found than vpon Hampſted heath, neere vnto a little cottage, growing cloſe vpon the ground amongſt buſhes and brakes, which I haue ſhewed vnto diuers Surgeons of London, that haue walked thither with me for their further knowledge in Simples, who haue gathered this kinde of Moſſe, wherof ſome haue made

13 *Muſcus ex cranio humano.*
Moſſe growing vpon the skull of a man.

made them hat-bands, girdles, and alſo bands to tye ſuch things as they had before gathered, for the which purpoſe it moſt fitly ſerued; ſome pieces whereof are ſix or eight foot long, conſiſting as it were of many hairie leaues ſet vpon a tough ſtring, very cloſe couched and compact together, from which is alſo ſent forth certaine other branches like the firſt: in ſundry places there be ſent down fine little ſtrings, which ſerue in ſtead of roots, wherewith it is faſtened to the vpper part of the earth, and taketh hold likewiſe vpon ſuch things as grow next vnto it. There ſpring alſo from the branches bare and naked ſtalkes, on which grow certaine eares as it were like the catkins or blowings of the Haſell tree, in ſhape like a little club or the reede Mace, ſauing that it is much leſſer, and of a yellowiſh white colour, very well reſembling the claw of a Wolfe, whereof it tooke his name; which knobby katkins are altogether barren, and bring forth neither ſeed nor floure.

‡ 14 *Muſcus parvus ſtellaris.*
Small Heath Moſſe.

‡ 12 This, whoſe figure in the former edition was by our Author vnfitly put for Lauander Cotton (hauing more regard to the title of the figure in *Tabernamontanus*, than to ſee whether it were that which he there deſcribed) is no other than a kinde of *Muſcus clauatus*, or Club-Moſſe. It is thought to be the *Selago* mentioned by *Pliny, lib.* 25. *cap.* 11. *Tragus* and ſome others call it *Sauina ſylueſtris*: *Turner* and *Tabernamontanus*, *Chamæcypariſſus*: but *Bauhine* the moſt fitly nameth it *Muſcus clauatus folijs Cypreſſi*: and *Turner* not vnfitly in Engliſh, Heath Cypreſſe. This is a low plant, and keepes greene Winter and Sommer: the leaues are like thoſe of Cypreſſe, bitter in taſte, but without ſmell: it carries ſuch eares or catkins as the former, and thoſe of a yellowiſh colour: it is found growing in diuers wooddy mountainous places of Germanie, where they call it **Wald Seuenbaum**, or wilde Sauine. ‡

13 This kinde of Moſſe is found vpon the skulls or bare ſcalps of men and women, lying long in charnell houſes or other places, where the bones of men and women are kept together: it groweth very thicke, white, like vnto the ſhort moſſe vpon the trunkes of old Okes: it is thought to be a ſingular remedie againſt the falling Euill and the Chin-cough in children, if it be poudered, and then giuen in ſweet wine for certaine daies together.

‡ 14 Vpon diuers heathy places in the moneth of May is to be found growing a little ſhort Moſſe not much in ſhape different from the firſt deſcribed, but much leſſe, and parted at the top into ſtar-faſhioned heads. *Lobel* calls this, *Muſcus in Ericetis prouenieus*. ‡

¶ The

¶ *The Place.*

Their feuerall defcriptions fet forth their naturall places of growing.

¶ *The Time.*

They flourifh efpecially in the Sommer moneths.

¶ *The Names.*

Goldilocke is called in high-Dutch, **Widertodt, golden Wedertodt, Jung Frauwen har:** in low-Dutch, **Gulden Wederdoot:** *Fuchfius* nameth it *Polytrichon Apuleij*, or *Apuleius* his Maiden-haire; neuertheleffe *Apuleius* Maiden-haire is nothing elfe but *Diofcorides* his *Trichomanes*, called Englifh Maiden-haire; and for that caufe wee had rather it fhould be termed *Mufcus capillaris*, or hairy Moffe. This is called in Englifh, Goldilockes: it might alfo be termed Golden Moffe, or Hairy Moffe.

Wolfes claw is called of diuers Herbarifts in our age, *Mufcus terreftris*: in high-Dutch, **Beerlap, Surtelkraut, Seilkraut:** in low-Dutch, **Wolfs clauwen**; whereupon wee firft named it *Lycopodium*, and *Pes Lupi*: in Englifh, Wolfes foot, or Wolfes claw, and likewife Club-Moffe. Moft fhops of Germanie in former times did falfly terme it *Spica celtica*: but they did worfe, and were very much too blame, that vfed it in compound medicines in ftead of *Spica celtica*, or French Spikenard: as touching the reft, they are fufficiently fpoken of in their defcriptions.

¶ *The Temperature.*

The Moffes of the earth are dry and aftringent, of a binding qualitie, without any heate or cold.

Goldilocks and the Wolfes clawes are temperate in heate and cold.

¶ *The Vertues.*

A The Arabian Phyfitians do put Moffe amongft their cordiall medicines, as fortifying the ftomacke, to ftay vomit, and to ftop the laske.

B Moffe boiled in Wine and drunke ftoppeth the fpitting of bloud, piffing of bloud, the termes, and bloudy flix.

C Moffe made into pouder is good to ftanch the bleeding of greene and frefh wounds, and is a great helpe vnto the cure of the fame.

D Wolfes claw prouoketh vrine, and as *Hieronymus Tragus* reporteth, wafteth the ftone, and driueth it forth.

E Being ftamped and boyled in wine and applied, it mitigateth the paine of the gout.

F Floting wine, which is now become flimie, is reftored to his former goodneffe, if it be hanged in the veffell, as the fame Author teftifieth.

† The figure formerly in the firft place was of the *Mufcus Montanus* of *Tabern.* being a fmall kinde of *Mufcus denticulatus*. The fifth and fixth were both of one; and fo of the two defcriptions I haue made one more accurate, and referued the better figure.

Chap. 163. *Of Liuerwort.*

¶ *The Defcription.*

1 Liuerwort is alfo a kinde of Moffe which fpreadeth it felfe abroad vpon the ground, hauing many vneuen or crumpled leaues lying one ouer another, as the fcales of Fifhes do, greene aboue, and browne vnderneath: amongft thefe grow vp fmall fhort ftalkes, fpred at the top like a blafing ftarre, and certaine fine little threds are fent downe, by which it cleaueth and fticketh faft vpon ftones, and vpon the ground, by which it liueth and flourifheth.

2 The fecond kinde of Liuerwort differeth not but in ftature, being altogether leffe, and more fmooth or euen: the floures on the tops of the flender ftems are not fo much laid open like a ftar; but the efpeciall difference confifteth in one chiefe point, that is to fay, this kinde being planted in a pot, and fet in a garden aboue the ground, notwithftanding it fpitteth or cafteth round about the place great ftore of the fame fruit, where neuer any did grow before.

‡ Of this fort which is fmall, and oftentimes found growing in moift gardens among Beares-eares, and fuch plants, when they are kept in pots, there are two varieties, one hauing little ftalkes fome inch long, with a ftarre-fafhioned head at the top: the other hath the like tender ftalke, and a round head at the top thereof. ‡

3 This is found vpon rockes and ftony places, as well neere vnto the fea, as further into the land: it groweth flat vpon the ftones, and creepeth not far abroad as the ground Liuerwort doth, it only refteth it felfe in fpots and tufts fet here and there, of a dufty ruffet colour aboue, and blackifh vnderneath: among the crumpled leaues rife vp diuers fmall ftems, whereupon do grow little ftar-like floures of the colour of the leaues: it is often found at the bottom of high trees growing vpon high

1 *Hepatica terrestris.*
Ground Liuer-wort.

2 *Hepatica stellata & vmbellata.*
Small Liuer-wort with starry and round heads.

3 *Hepatica petræa.*
Stone Liuerwort.

high mountaines, especiall in shadowie places.

¶ *The Place.*

This is often found in shadowy and moist places, on rocks and great stones layd by the highway, and in other common paths where the Sun beams do seldome come, and where no traueller frequenteth.

¶ *The Time.*

It brings forth his blasing stars and leaues oftentimes in Iune and Iuly.

¶ *The Names.*

It is called of the Grecians, Λειχὴν: of the Latines, *Lichen*: and of some, Βρύον, that is to say, *Muscus*, or Mosse, as *Dioscorides* witnesseth: it is named in shops *Hepatica*, yet there are also many other herbes named *Hepaticæ*, or Liuer-worts, for difference whereof this may fitly be called *Hepatica petræa*, or Stone Liuer-wort, hauing taken that name from the Germanes, who call this Liuerwort, **Steyn Leberkraut**: and in low-Dutch, **Steen Leuercruyt**: in English, Liuerwort.

¶ *The Temperature.*

This Stone Liuerwort is of temperature cold and dry, and somewhat binding.

¶ *The Vertues.*

It is singular good against the inflammations of the liuer, hot and sharpe agues, and tertians which proceed of choler.

Dioſcorides teacheth, that Liuer-wort being applied to the place ſtancheth bleeding, takes away all inflammations, and that it is good for a tetter or ring-worme, called in Greeke, Λειχὴν : and that it is a remedie for them that haue the yellow iaundice, euen that which commeth by the inflammation of the liuer; and that furthermore it quencheth the inflammations of the tongue.

Chap. 164.

Of Lung-wort, or wood Liuer-wort, and Oiſter-greene.

1 *Lichen arborum.*
Tree Lung-wort.

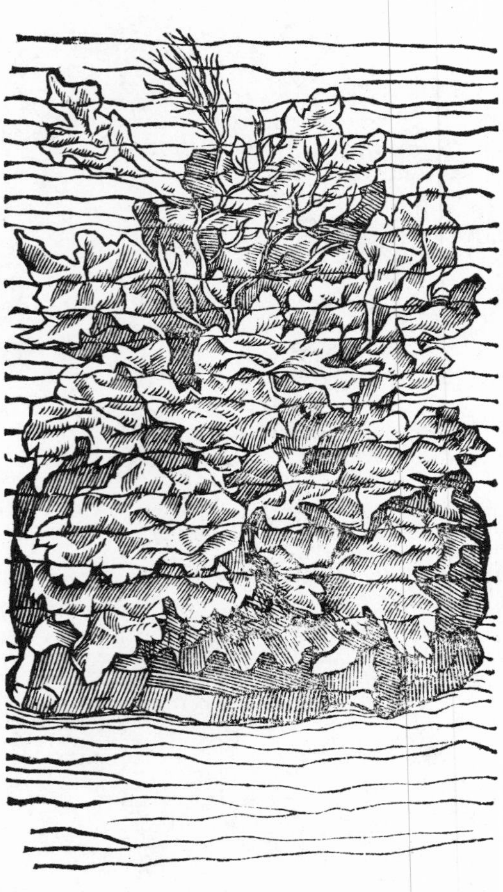

2 *Lichen marinus.*
Sea Lung-wort, or Oiſter-greene.

¶ *The Deſcription.*

1 TO Liuerwort there is ioyned Lung-wort, which is alſo another kinde of Moſſe, drier, broader, of a larger ſize, and ſet with ſcales: the leaues hereof are greater, and diuerſly folded one in another, not ſo ſmooth, but more wrinckled, rough and thicke almoſt like a Fell or hide, and tough withall: on the vpper ſide whitiſh, and on the nether ſide blackiſh or duſty, it ſeemeth to be after a ſort like to lungs or lights.

2 This kinde of ſea Moſſe is an herby matter much like vnto Liuer-wort, altogether without ſtalke or ſtem, bearing many greene leaues, very vneuen or crumpled, and full of wrinkles, and ſomwhat broad, not much differing from leaues of criſpe or curled Lettuce: this groweth vpon rockes within the bowels of the ſea, but eſpecially among oiſters, and in greater plenty among thoſe Oiſters which are called Wall-fleet Oiſters: it is very well knowne euen to the poore Oiſter-women which carry Oiſters to ſell vp and downe, who are greatly deſirous of the ſaid moſſe for the decking and beautifying of their Oiſters, to make them ſell the better. This moſſe they doe call Oiſter-greene.

‡ 3 The branches of this elegant plant are ſome handfull or better high, ſpred abroad on euerie ſide, and only conſiſting of ſundry ſingle roundiſh leaues, whereto are faſtned ſomtimes one,

ſometimes

ſometimes two or more ſuch leaues ; ſo that the whole plant conſiſts of branches made vp of ſuch round leaues, faſtned together by diuers little & very ſmal threds: the lower leaues which ſtick faſt to the rockes are of a browniſh colour, the other of a whitiſh or a light greene colour, ſmooth and ſhining. This growes vpon rockes in diuers parts of the Mediterranean. *Cluſius* ſetteth it forth by the name of *Lichen Marinus* ; and he receiued it from *Imperato* by the name of *Sertuloria* : and *Cortuſus* had it from *Corſica*, by the title of *Corallina latifolia*; and he called it *Opuntia marina*, hauing reference to that mentioned by *Theophraſt. lib. 1. cap. 12. Hiſt. Plant.* ‡

‡ 3 *Lichen marinus rotundifolius.* Round leaued Oiſter-weed.

4 *Quercus marina.*
Sea Oke or Wracke.

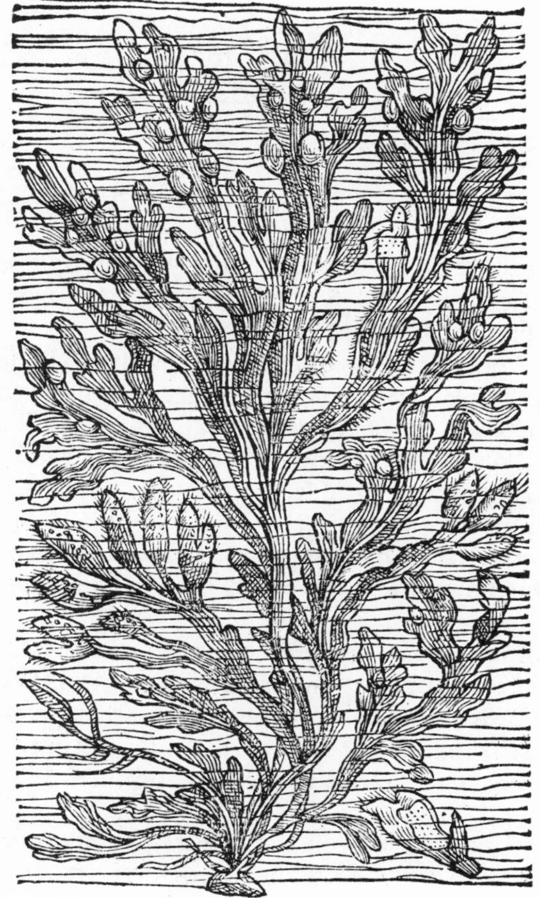

‡ 4 *Quercus marinæ varietas.*
A varietie of the ſea Oke or Wrack.

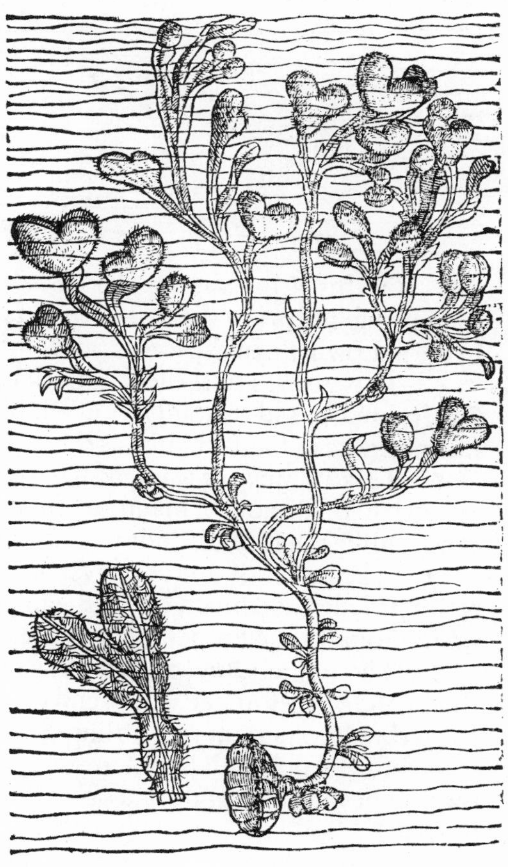

4 There is alſo another ſort of ſea Weed found vpon the drowned rockes, which are naked and bare of water at euery tyde. This ſea Weed groweth vnto the rocke, faſtned vnto the ſame at one end,

end, being a soft herby plant, very slipperie, insomuch that it is a hard matter to stand vpon it without falling: it rampeth far abroad, and here and there is set with certaine puft vp tubercles or bladders, full of winde, which giueth a cracke when it is broken: the leafe it selfe doth somewhat resemble the Oken leafe, whereof it tooke his name *Quercus marina*, the sea Oke: of some, Wracke, and Crow Gall. His vse in physicke hath not beene set forth, and therefore this bare description may suffice.

‡ 5 *Quercus marina secunda.*
Sea Thongs.

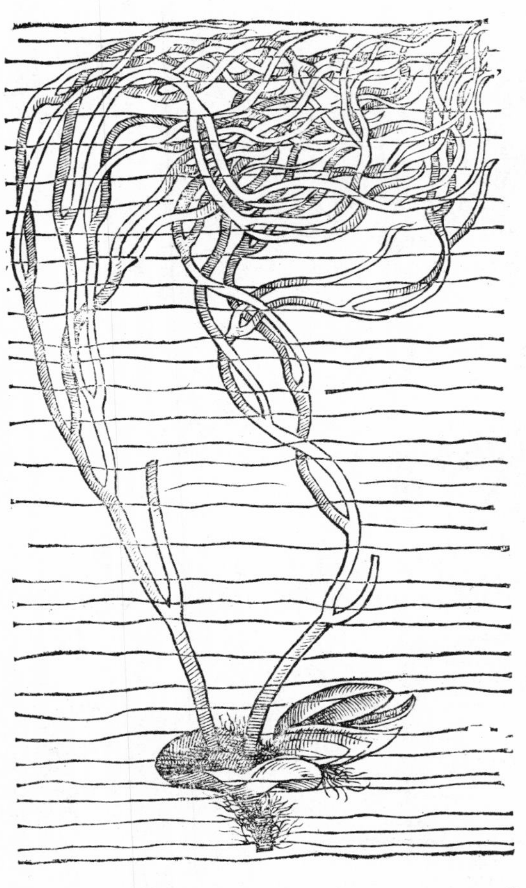

‡ 6 *Quercus marina tertia.*
The third sea Wracke.

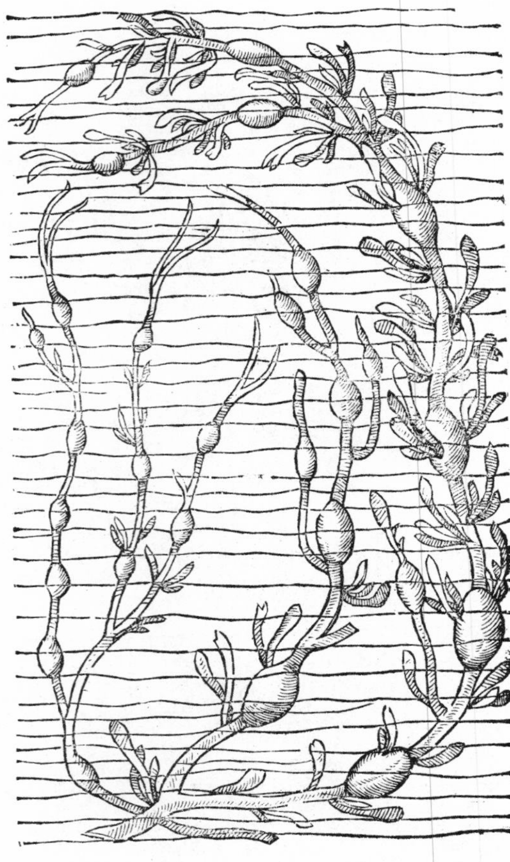

‡ Of this *Quercus marina*, or *Fucus*, there are diuers sorts, whereof I will giue you the figures and a briefe historie: the first of these is onely a varietie of the last described, differing there-from in the narrownesse of the leaues, and largenesse of the swolne bladders.

5 This growes to the length of fiue or six foot, is smooth and membranous, being some halfe fingers bredth, and variously diuided, like wet parchment or leather cut into thongs: this hath no swolne knots or bladders like as the former; and is the *Fucus marinus secundus* of *Dodonæus*.

6 This Wracke or sea weed hath long and flat stalkes like the former, but the stalks are thicke set with swolne knots or bladders, out of which sometimes grow little leaues, in other respects it is not vnlike the former kindes. *Dodonæus* makes this his *Fucus marinus* 3.

7 The leaues of this other Wracke, which *Dodonæus* makes his *Fucus marinus quartus*, are narrower, smaller, and much diuided; and this hath either none or very few of those swollen bladders which some of the former kindes haue.

8 This, which *Lobel* calleth *Alga marina*, hath iointed blacke branched creeping roots of the thicknesse of ones finger, which end as it were in diuers eares, or hairy awnes, composed of whitish hairy threds somewhat resembling Spikenard: from the tops of those eares forth leaues, long, narrow, soft, and grasse-like, first greene, but white when they are dry. It growes in the sea as the former. They vse it in Italy and other hot countries to packe vp glasses with, to keepe them from breaking.

9 Of this Tribe are diuers other plants; but I will onely giue you the history of two more, which I first obserued the last yeare, going in company with diuers London Apothecaries to finde Simples, as farre as Margate in the Island of Tenet; and whose figures (not before extant that I know of) I first gaue in my Iournall or enumeration of such plants as we there and in other places found. The first of these by reason of his various growth is by *Bauhine* in his *Prodromus* distinguished

guiſhed into two, and deſcribed in the ſecond and third places. The third he calls *Fucus longiſſimo, latiſſimo, craſſoque folio*, and this is marked with the figure 1. The ſecond he calls *Fucus arboreus polyſchides*; and this you may ſee marked with the figure 2. This ſea Weed (as I haue ſaid) hath a various face, for ſometimes from a fibrous root, which commonly groweth to a pibble ſtone, or faſtened to a rocke, it ſendeth forth a round ſtalke ſeldome ſo thicke as ones little finger, and about ſome halfe foot in length, at the top whereof growes out a ſingle leafe, ſometimes an ell long, and then it is about the bredth of ones hand, and it ends in a ſharpe point, ſo that it very well reſembles a two edged ſword. Sometimes from the ſame root come forth two ſuch faſhioned leaues, but then commonly they are leſſer. Otherwhiles at the top of the ſtalke it diuides it ſelfe into eight, nine, ten, twelue, more or fewer parts, and that iuſt at the top of the ſtalke, and theſe neuer come to that length that the ſingle leaues do. Now this I iudge to be the *Fucus polyſchides* of *Bauhine*. That theſe two are not ſeuerall kindes I am certaine; for I haue marked both theſe varieties from one and the ſame root, as you may ſee them here expreſt in the figure. At Margate where they grow they call them ſea Girdles, and that name well befits the ſingle one; and the diuided one they may call Sea Hangers, for if you do hang the tops downewards, they doe reaſonable well reſemble the old faſhioned ſword-hangers. Thus much for their ſhape: now for their colour, which is not the ſame in all; for ſome are more greene, and theſe can ſcarce be dried; other-ſome are whitiſh, and theſe do quickly dry, and then both in colour and ſubſtance are ſo like parchment, that ſuch as know them not would at the firſt view take them to be nothing elſe. This is of a glutinous ſubſtance, and a little ſaltiſh taſte, and diuers haue told me they are good meate, being boiled tender, and ſo eaten with butter, vineger, and pepper.

‡ 7 *Quercus marina quarta.*
Iagged Sea Wracke.

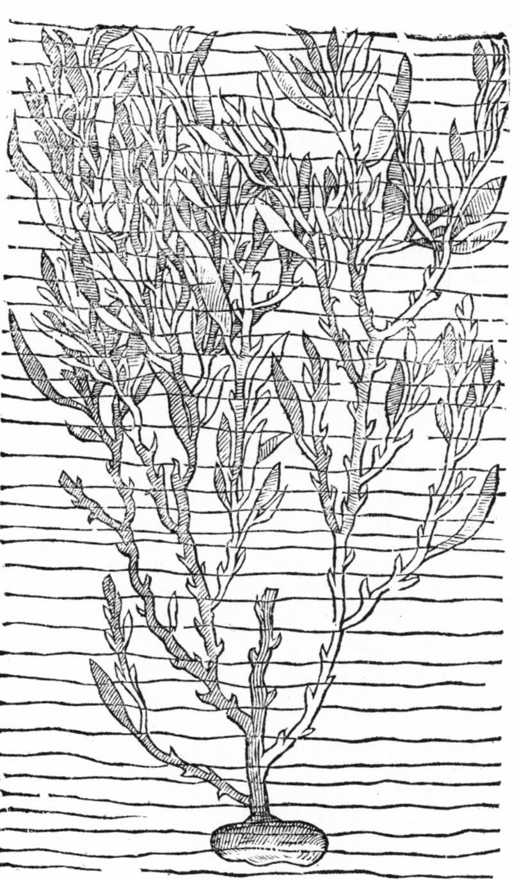

‡ 8 *Alga.*
Graſſe Wracke.

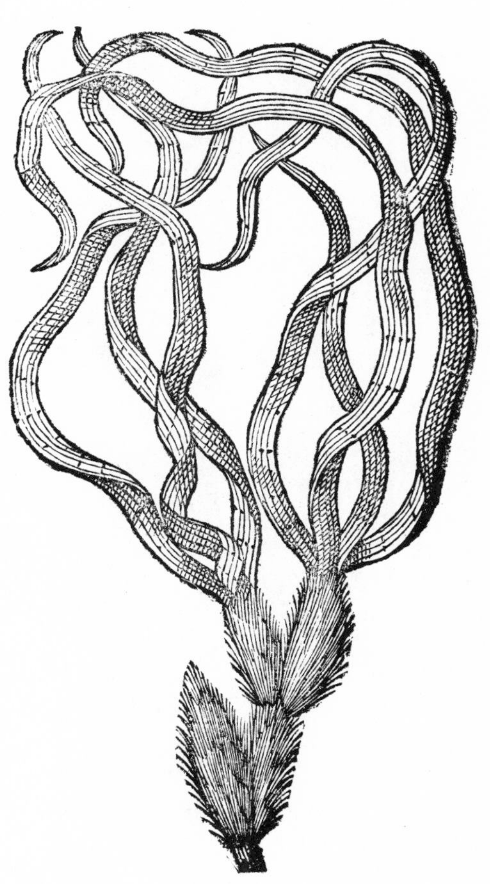

10 This which I giue you in the tenth place is not figured or deſcribed by any that as yet I haue met with; wherefore I gaue the figure and deſcription in the forementioned Iournall, which I will here repeate. This is a very ſucculent and fungous plant, of the thickneſſe of ones thumbe; it is of a darke yellowiſh colour, and buncheth forth on euerie ſide with many vnequall tuberoſities or knots: whereupon Mr. *Thomas Hickes* being in our companie did fitly name it Sea ragged Staffe. We did not obſerue it growing, but found one or two plants thereof ſome foot long apiece.

‡ 9 *Fucus phasganoides & polyschides.*
Sea Girdle and Hangers.

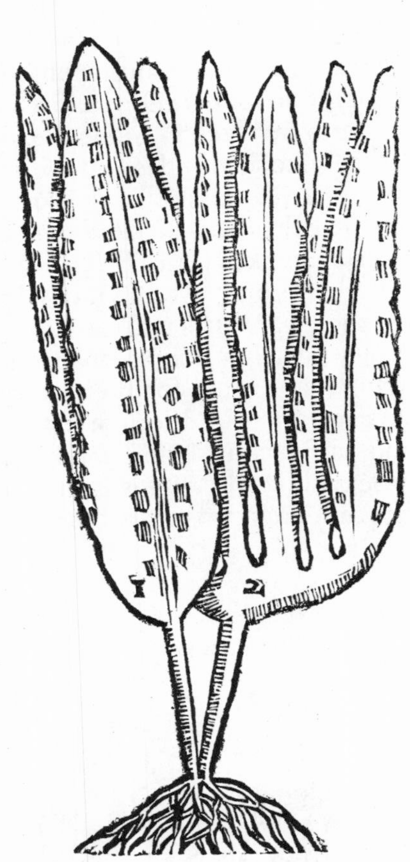

‡ 10 *Fucus spongiosus nodosus.*
Sea ragged Staffe.

‡ 11 *Conferua.* Hairy Riuerweed.

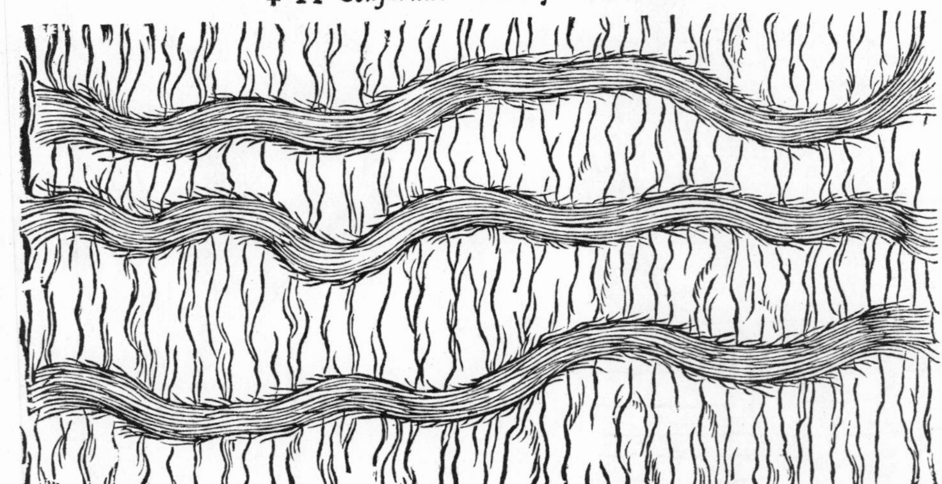

11 In some slow running waters is to be found this long greene hairy weed, which is thought to be the *Conferua* of *Pliny* : it is made vp onely of long hairy greene threds, thicke thrummed together without any particular shape or fashion, but only following the current of the streame. ‡

¶ *The Place.*

It groweth vpon the bodies of old Okes, Beech, and other wilde trees, in darke & thick woods: it is oftentimes found growing vpon rocks, and also in other shadowie places.

¶ *The Time and Names.*

It flourisheth especially in the Sommer moneths.

It taketh his name *Pulmonaria* of the likenesse of the forme which it hath with lungs or lights, called in Latine *Pulmones*, of some, *Lichen*: it is called in high-Dutch, **Lungenkraut**: in low-Dutch **Longhencruyt**: in French, *Herbe à Poulmon*: in English, Lung-wort, and wood Liuerwort.

¶ *The Temperature.*

This seemeth to be cold and dry.

¶ *The*

¶ *The Vertues.*

It is reported that shepheards and certaine horseleeches do with good successe giue the pouder hereof with salt vnto their sheepe and other cattell which be troubled with the cough, and be broken winded. A

Lungwoort is much commended of the learned Physitions of our time against the diseases of the lungs, especially for the inflammations and vlcers of the same, being brought into pouder, and drunke with water. B

It is likewise commended for bloudy and greene wounds, and for vlcers in the secret parts, and also to stay the reds. C

Moreouer, it stoppeth the bloudy flix, and other flixes and scourings, either vpwards or downewards, especially if they proceed of choler: it stayeth vomiting, as men say, and it also stoppeth the belly.

Oister greene fried with egges and made into a tansie & eaten, is a singular remedy for to strengthen the weaknesse of the backe. D

CHAP. 165. *Of Sea Mosse, or Coralline.*

¶ *The Kindes.*

THere be diuers sorts of Mosse, growing as well within the bowels of the sea, as vpon the rocks, distinguished vnder sundry titles.

1 *Muscus marinus, siue Corallina alba.*
White Coralline, or sea Mosse.

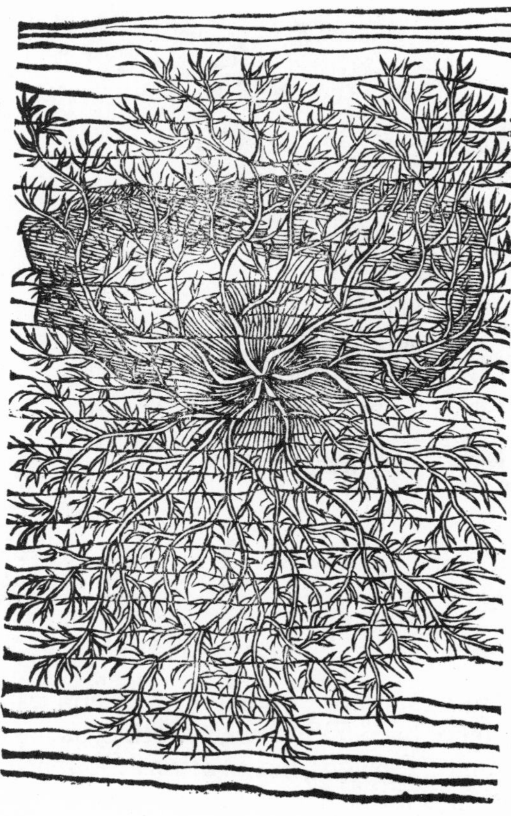

† 2 *Muscus marinus albidus.*
White sea Mosse.

3 *Corallina Anglica.*
English Coralline.

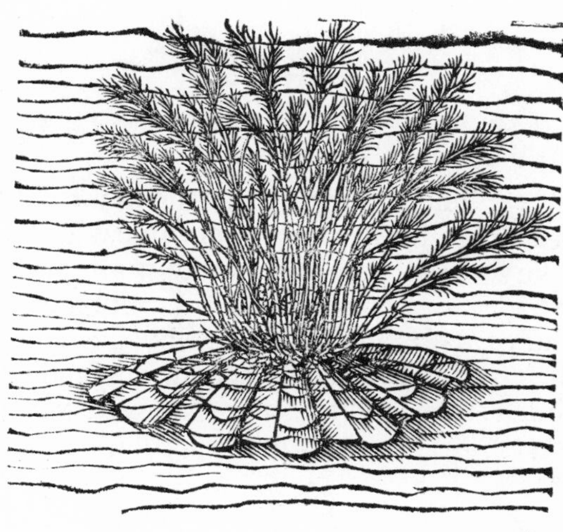

¶ *The*

¶ *The Deſcription.*

1 This kinde of Sea Moſſe hath many ſmall ſtalkes finely couered or ſet ouer with ſmall leaues, very much cut or iagged, euen like the leaues of Dill, but hard, and of a ſtonie ſubſtance.

2 The ſecond is much like vnto the former, yet not ſtony, but more finely cut, and growing more vpright, branching it ſelfe into many diuiſions at the top, growing very thicke together, and in great quantitie, out of a piece of ſtone, which is faſhioned like an hat or ſmall ſtonie head, wherby it is faſtened vnto the rocks.

3 This third kinde of ſea moſſe is very well knowne in ſhops by the name *Corallina*; it yeeldeth forth a great number of ſhoots, in ſhap much like vnto Corall; being full of ſmall branches diſperſed here and there, diuerſly varying his colour, according to the place where it is found, beeing in ſome places red, in otherſome yellow, and of an herby colour; in ſome gray, or of an aſh colour, and in otherſome very white.

4 The fourth kinde of Sea Moſſe is ſomewhat like the former, but ſmaller, and not ſo plentiful where it groweth, proſpering alwaies vpon ſhels, as of Oyſters, Muſcles, and Scallops, as alſo vpon rolling ſtones, in the bottome of the water, which haue tumbled downe from the high cliffes and rocks, notwithſtanding the old prouerbe, that rolling ſtones neuer gather Moſſe.

4 *Corallina minima.*
The ſmalleſt Coralline.

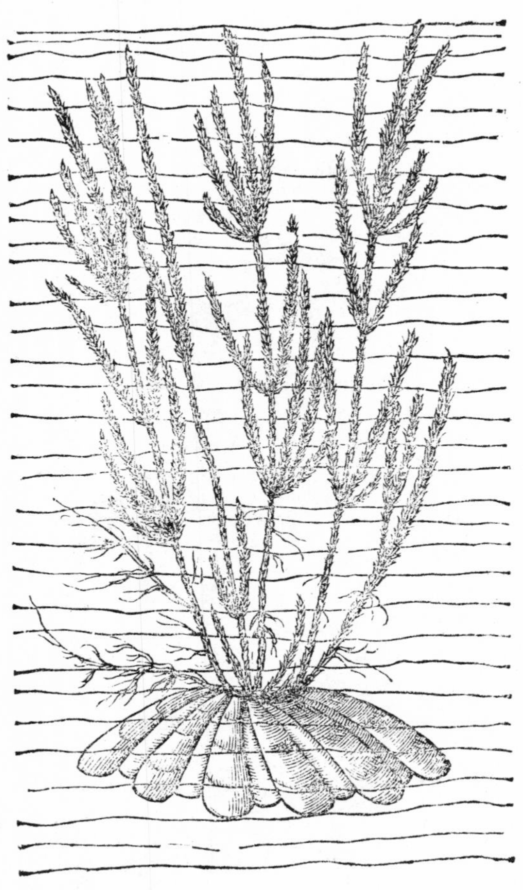

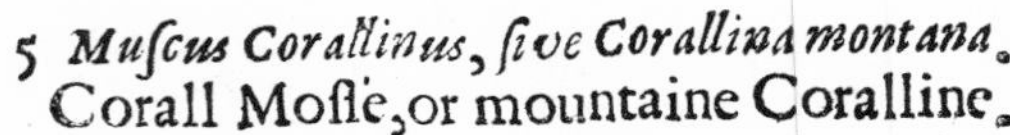
5 *Muſcus Corallinus, ſiue Corallina montana.*
Corall Moſſe, or mountaine Coralline.

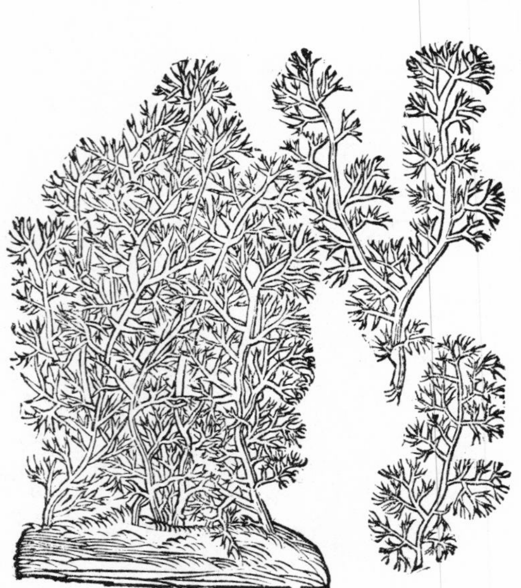

5 There is found vpon the rocks and mountaines of France, bordering vpon the Mediterranean ſea, a certaine kinde of Coralline, which in theſe parts hath not been found: it groweth in manner like vnto a branch of Corall, but altogether leſſer, of a ſhining red colour, and of a ſtony ſubſtance.

‡ I know not what our Author meant by this deſcription, but the plant which here is figured out of *Tabernamontanus* (and by the ſame title he hath it) is of a Moſſe growing vpon Hampſtead heath, and moſt ſuch places in England: it growes vp ſome two or three inches high, and is diuided into very many little branches ending in little threddy chiues: all the branches are hollow, and of a very light whitedry ſubſtance, which makes it ſomewhat to reſemble Coralline, yet is it not ſtony at all. ‡

6 There is alſo found vpon the rocks neere vnto Narbone in France, and not far from the ſea, a kinde

6 *Fucus marinus tenuifolius.*
Fenell Coralline, or Fenell Mosse.

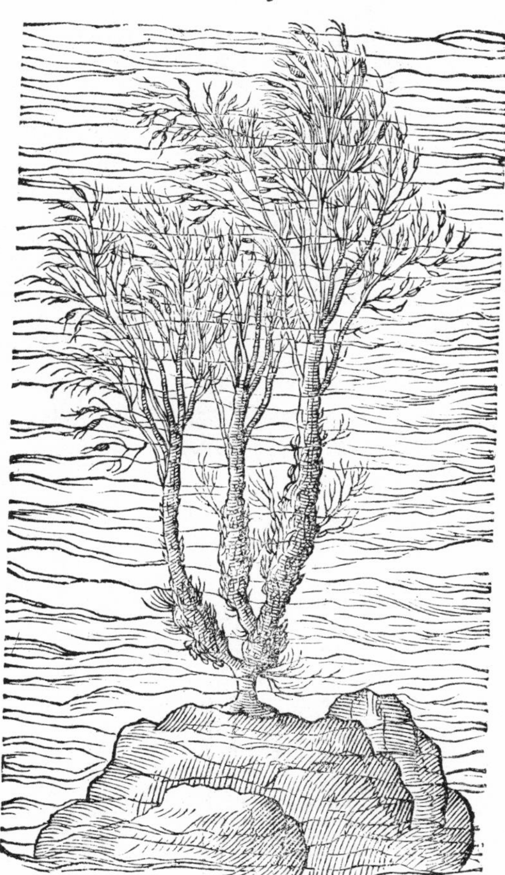

‡ 7 *Fucus ferulaceus.*
Sea Fenell.

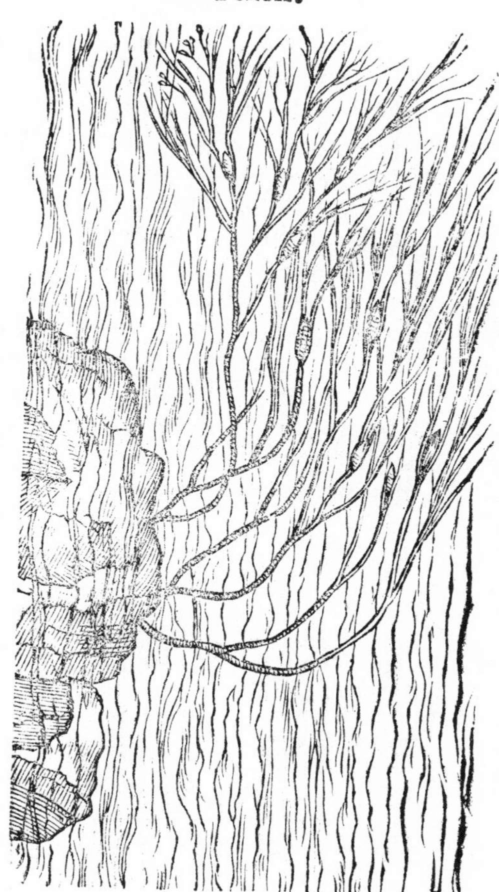

‡ 8 *Fucus tenuifolius alter.*
Bulbous sea Fennell.

‡ 9 *Muscus marinus Clusius.*
Branched Sea Mosse.

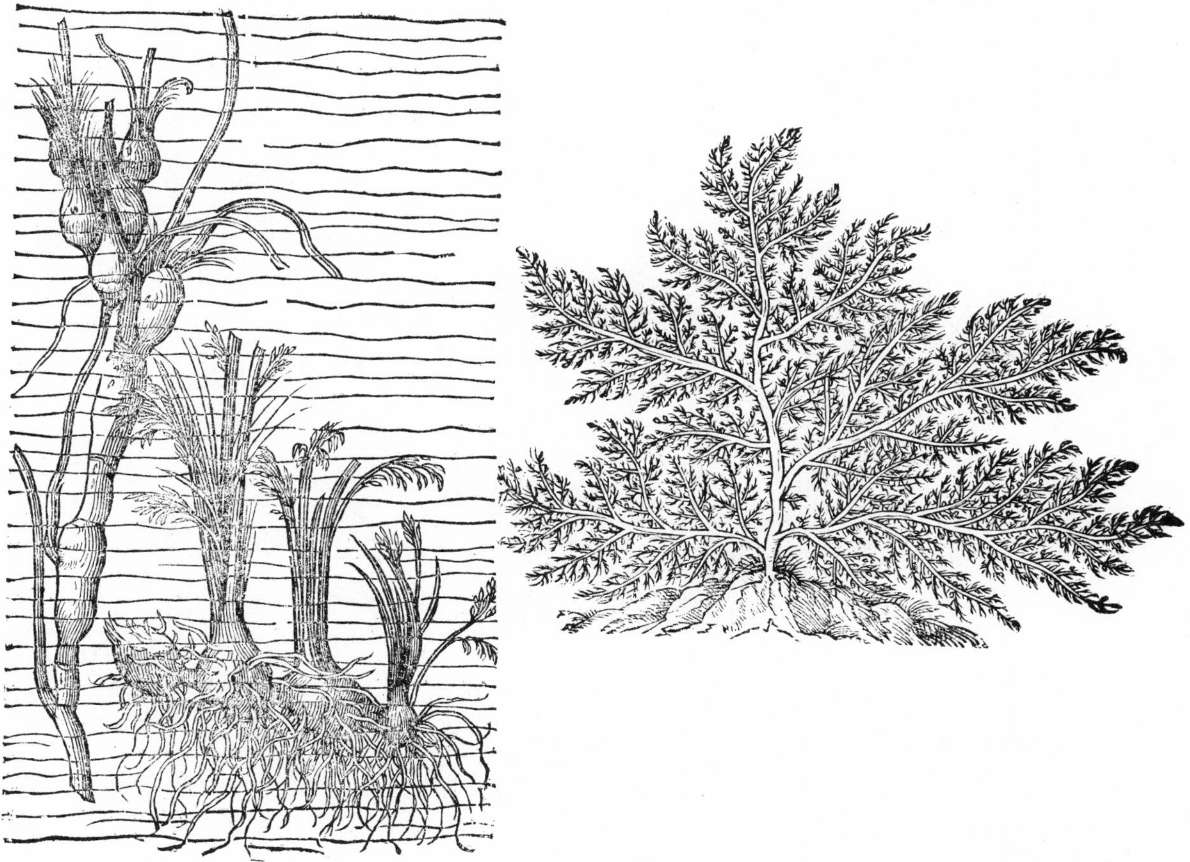

kinde of Coralline. it groweth vp to the forme of a ſmall ſhrub, branched diuerſly; whereon doe grow ſmall graſſe-like leaues, very finely cut or iagged, like vnto Fennel; yet are they of a ſtony ſubſtance, as are the reſt of the Corallines; of a darke ruſſet colour.

‡ 7 This growes alſo in the like places, hauing many ſmall long Fennell-like diuided leaues vpon ſtalks ſome foot long, with ſome ſwelling eminences here & there ſet in the diuiſions of the leaues: this is by *Lobel* called by the name I here giue you it.

8 This alſo hath fine cut leaues like thoſe of Fennell, but much leſſe & ſhorter, of a faire green colour: theſe grow vp from round tuberous roots, which together with the fibres they ſend forth are of a blackiſh colour: the ſtalks alſo are tuberous and ſwolne, as in other plants of this kind. It growes in the ſea with the former. *Dodonæus* calls this *Fucus marinus virens tenuifolius*.

9 This kinde of ſea Moſſe growes ſome foure or more inches long, diuided into many branches, which are ſubdiuided into ſmaller, ſet with leaues finely iagged, like thoſe of Cammomill; at firſt ſoft, flexible, and tranſparent, greene below, and purpliſh aboue; being dried, it becommeth rough and fragile, like as Coralline. It growes in the Mediterranean ſea.

10 This Sea Moſſe is a low little excreſcence, hauing ſomewhat broad cut leaues growing many from one root: in the whole face it reſembles the moſſe that grows vpon the branches of Okes and other trees, and is alſo white and very like it, but much more brittle. This by *Dodonæus* is called *Muſcus Marinus tertius*.

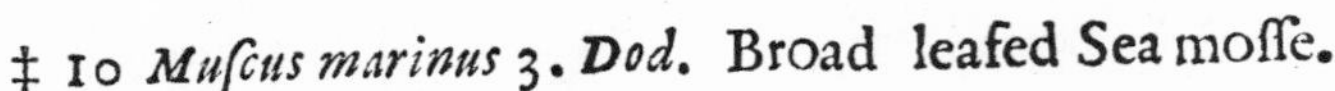

‡ 10 *Muſcus marinus* 3. *Dod.* Broad leafed Sea moſſe.

‡ 11 *Abies marina Belgica, Cluſ. Cluſius* his Sea Firr.

11 Vpon the rocks and ſhels of ſea fiſhes are to be found diuers ſmall plants, hauing reſemblance to others that grow vpon the land; and *Cluſius* ſaith, vpon the coaſt of the Low countries he obſerued one which very much reſembled the Fir-tree, hauing branches growing orderly on both ſides, but thoſe very brittle and ſmall, ſeldome exceeding a handfull in height, and couered as it were with many ſmall ſcales. He obſerued others that reſembled Cypreſſe trees, and other branches that reſembled Tameriske or heath. ‡

¶ *The*

¶ *The Place.*

Theſe Moſſes grow in the ſea vpon the rocks, and are oftentimes found vpon Oiſter ſhels, Muskell ſhells, and vpon ſtones: I found very great plenty thereof vnder Reculuers and Margate, in the Iſle of Thanet, and in other places alongſt the ſands from thence vnto Douer.

¶ *The Time.*

The time anſwereth the other Moſſes, and are found at all times of the yeare.

¶ *The Names.*

Sea Moſſe is called in Greeke Βρύϊ θαλάσσιον: in Latine, *Muſcus marinus*: of the Apothecaries, Italians, and French men, *Corallina*: in Spaniſh, *Malbarquiana yerua*: in high Dutch, **Meermoſz**: in low Dutch, **Zee Moſch**: in Engliſh, Sea Moſſe, and of many Corallina, after the Apothecaries word, and it may be called Corall Moſſe. The titles diſtinguiſh the other kindes.

¶ *The Temperature.*

Corallina conſiſteth, as *Galen* ſaith, of an earthie and wateriſh eſſence, both of them cold: for by his taſte it bindeth, and being applied to any hot infirmitie, it alſo euidently cooleth: the earthie eſſence of this Moſſe hath in it alſo a certaine ſaltneſſe, by reaſon whereof likewiſe it drieth mightily.

¶ *The Vertues.*

A *Dioſcorides* commendeth it to be good for the gout which hath need to be cooled.

B The later Phyſitions haue found by experience, that it killeth wormes in the belly; it is giuen to this purpoſe to children in the weight of a dram or thereabouts.

C That which cleaueth to Corall, and is of a reddiſh colour, is of ſome preferred and taken for the beſt: they count that which is whitiſh, to be the worſer. Notwithſtanding in the French Ocean, the Britain, the low countrey, or elſe in the Germane ocean ſea, there is ſcarce found any other than the whitiſh Coralline, which the nations neere adioyning do effectually vſe.

CHAP. 166. Of Corall.

1 *Corallium rubrum.*
Red Corall.

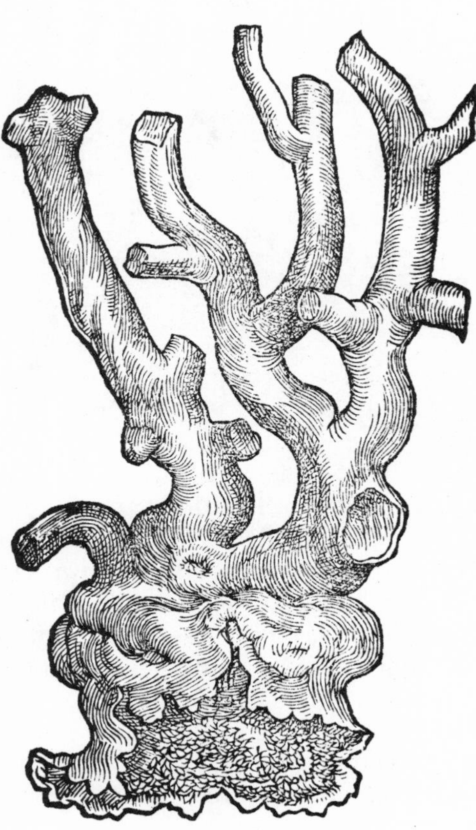

2 *Corallium nigrum, ſiue Antipathes.*
Blacke Corall.

¶ *The*

3 *Corallium album.* White Corrall.

4 *Corallium album alterum.*
The other white or yellow Corrall.

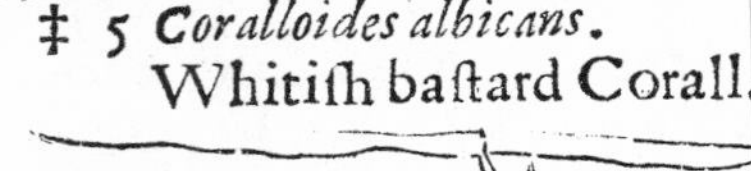

‡ 5 *Coralloides albicans.*
Whitish bastard Corall.

¶ *The Description.*

1 ALthough Corrall be a matter or substance, euen as hard as stones; yet I thinke it not amisse to place and insert it here next vnto the mosses, and the rather for that the kindes thereof do shew themselues, as well in the maner of their growing, as in their place and forme, like vnto the Mosses. This later age wherein we liue, hath found moe kindes hereof than euer were knowne or mentioned among the old writers. Some of these Corrals grow in the likenesse of a shrub, or stony matter; others in a straight forme, with crags and ioints, such as we see by experience: the which for that they are so well knowne, and in such request for Physicke, I will not stand to describe; only this remember, that there is some Corrall of a pale yellow colour, as there be some red, and some white.

2 The blacke Corrall groweth vpon the rocks neere to the sea about Massilia, in manner of the former; herein differing from it, in that this is of a shining blacke colour, and very smooth, growing vp rather like a tree, than like a shrub

3 The white Corrall is like to the former, growing vpon the rocks neere the sea, and in the West parts of England, about Saint Michaels mount; but the branches hereof are smaller, and more brittle, finelier dispersed into a number of branches, of a white colour.

4 The

‡ 6 *Coralloides rubens.*
Reddish bastard Corall.

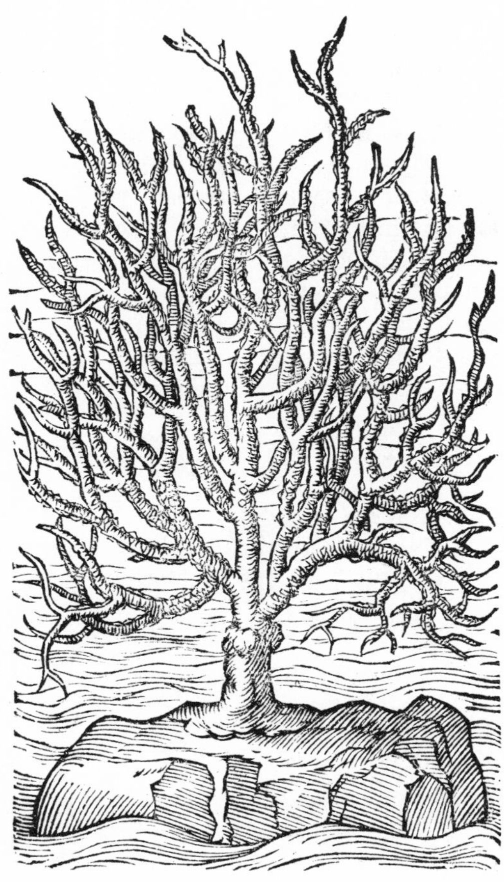

7 *Spongia marina alba.*
White Spunge.

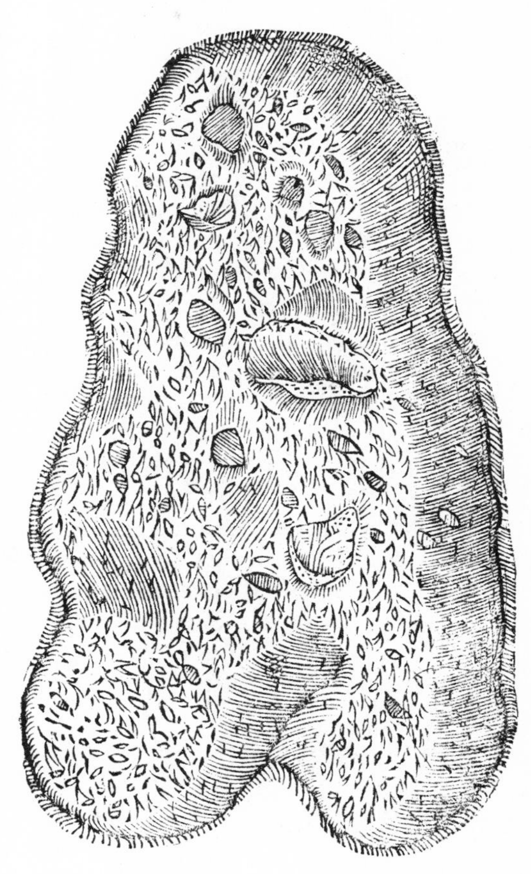

‡ 8 *Spongia infundibuli forma.*
Funnell fashioned Spunge.

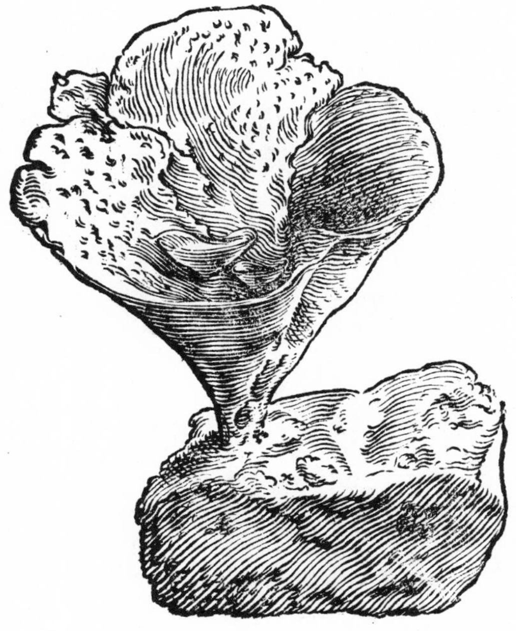

‡ 9 *Spongia ramosa.*
Branched Spunge.

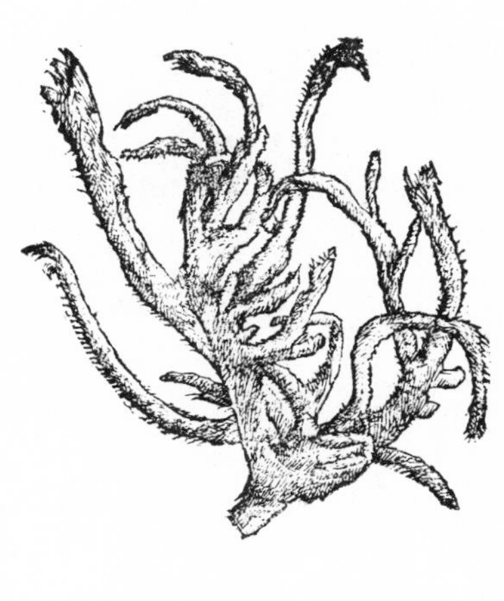

4 The fourth and last groweth also vpon the Westerne rocks of the sea, and in the place afore-named, and varieth his colour, sometimes waxing white, sometimes yellow, and sometimes red.

‡ 5 This growes vp with many branches some two or three handfulls high; the inner part is a hard wooddy substance, which is couered ouer with a white and hard stony matter, so that it much resembleth white Corall, but that it is neither so thicke, hard, nor smooth, but is rough and

and bends easily without breaking, which Corall will not do. *Lobel* calls this *Corallina alba*, it growes in the Mediteranian sea, and vpon the Coasts of Spaine.

6 This in all respects is like the last described, the colour excepted, which is a darke red, and therefore better resembles the red Corall. *Clusius* refers both these to the *Quercus marina* mentioned by *Theophrastus, Hist. plant. lib. 7. cap. 4.* ‡

7 There is found growing vpon the rockes neere vnto the sea, a certaine matter wrought together, of the fome or froth of the sea, which we call spunges, after the Latine name, which may very fitly be inserted among the sea Mosses, whereof to write at large would greatly increase our volume, and little profit the reader, considering we hasten to an end, and also that the vse is so well knowne vnto all: therefore these few lines may serue vntill a further consideration, or a second Edition. ‡ Spunges are not like the *Alcyonium*, that is, an accidentall matter wrought together of the froth of the sea, as our Author affirmes, but rather of a nobler nature than plants, for they are said to haue sence, and to contract themselues at the approach of ones hand that comes to cut them vp, or for feare of any other harme-threatning obiect, and therefore by most writers they are referred to the ζωόφυτα: which some render *Plantanimalia*, that is, such as are neither absolute plants, nor liuing creatures, but participate of both: they grow of diuers shapes and colours vpon the Rockes in the Mediterranian, as also in the Archipelago, or Ægean sea.

8 *Clusius* obserued one yet adhering to the stone whereon it grew, which in shape resembled a funnell, but in substance was like another Spunge.

9 There is also to be found vpon our English coast a small kinde of spunge cast vp by the sea, and this is also of different shapes and colour, for the shape it is alwaies diuided into sundry branches, but that after a different manner; and the colour is oft times brownish, and otherwhiles gray or white. *Lobel* makes it *Conferuæ marinæ genus.* ‡

¶ *The Place.*

The place of their growing is sufficiently spoken of in their seuerall descriptions.

¶ *The Time.*

The time answereth the other kindes of sea Mosses.

¶ *The Names.*

Corallium rubrum is called in English, red Corrall. *Corallium nigrum*, blacke Corrall. *Corallium album*, white Corrall.

¶ *The Temperature.*

Corrall bindeth, and meanely cooleth: it clerseth the scars and spots of the eies, and is very effectuall against the issues of bloud, and easeth the difficultie of making water.

¶ *The Vertues.*

A Corrall drunke in wine or water, preserueth from the spleene; and some hang it about the neckes of such as haue the falling sickenesse, and it is giuen in drinke for the same purpose.

B It is a soueraigne remedy to drie, to stop, and stay all issues of bloud whatsoeuer in man or woman, and the dysentery.

C Burned Corrall drieth more than when it is vnburned, and being giuen to drinke in water, it helpeth the gripings of the belly, and the griefes of the stone in the bladder.

D Corrall drunke in wine prouoketh sleepe: but if the patient haue an ague, then it is with better successe ministred in water, for the Corrall cooleth, and the water moistneth the body, by reason whereof it restraineth the burning heate in agues, and represseth the vapours that hinder sleepe.

Chap. 167. *Of Mushrumes, or Toadstooles.*

¶ *The Kindes.*

Some Mushrumes grow forth of the earth; other vpon the bodies of old trees, which differ altogether in kindes. Many wantons that dwell neere the sea, and haue fish at will, are very desirous for change of diet to feed vpon the birds of the mountaines; and such as dwell vpon the hills or champion grounds, do longe after sea fish; many that haue plenty of both, do hunger after the earthie excrescences, called Mushrumes: whereof some are very venomous and full of poison, others not so noisome; and neither of them very wholesome meate; wherefore for the auoiding of the venomous qualitie of the one, and that the other which is lesse venomous may be discerned from it, I haue thought good to set forth their pictures with their names and places of growth. ‡ Because the booke is already grown too voluminous, I will only giue you the figures of such as my Author hath here mentioned, with some few others, but not trouble you with any more history, yet distinguish betweene such as are eatable, and those that be poysonous, or at least not to be eaten; for the first figured amongst the poysonous ones, is that we call Iewes-eare, which hath no poysonous facultie in it. *Clusius* (all whose figures I could haue here giuen you) hath written a peculiar tract of these bastard plants, or excrescences, where such as desire it may finde them sufficiently discoursed of. ‡

¶ The

1 *Fungi vulgatissimi esculenti.* Common Mushrums to be eaten.

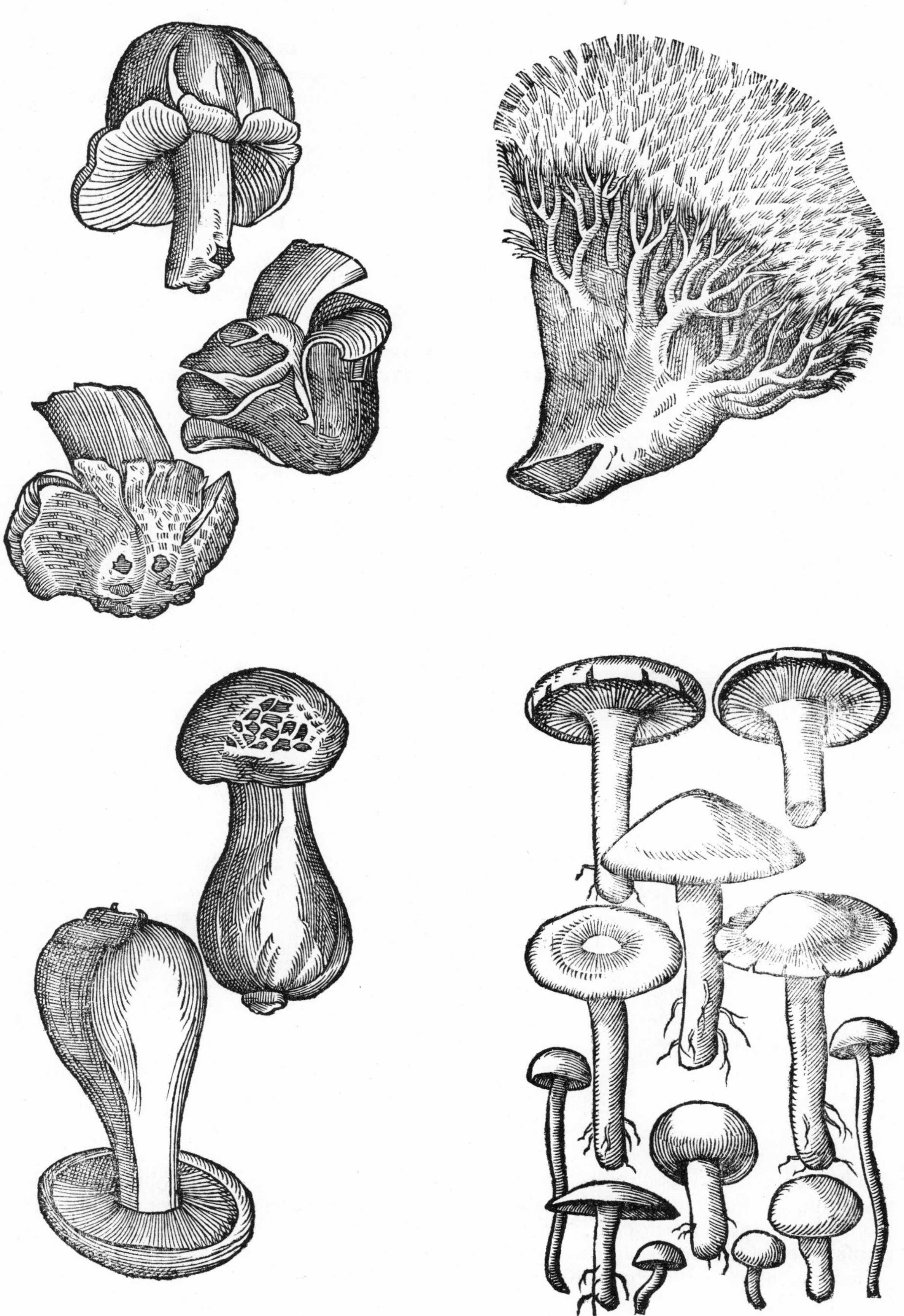

¶ *The Description.*

1 GRound Mushrums grow vp in one night, standing vpon a thicke and round stalke, like vnto a broad hat or buckler, of a very white colour vntil it begin to wither, at what time it loseth his faire white, declining to yellownesse: the lower side is somewhat hollow, set or decked with fine gutters, drawne along from the middle centre to the circumference or round edge of the brim.

2 All Mushroms are without pith, rib, or veine: they differ not a little in bignesse and colour; some are great, and like a broad brimmed hat; others smaller, about the bignesse of a siluer coine called a doller: most of them are red vnderneath; some more, some lesse; others little or nothing red at all: the vpper side which beareth out, is either pale or whitish, or else of an ill fauored colour like ashes (they commonly call it Ash colour) or else it seemeth to be somewhat yellow.

There is another kinde of Mushrums called *Fungi parui lethales galericulati*: in English, deadly Mushrums, which are fashioned like vnto an hood, and are most venomous and full of poison.

There is a kinde of Mushrum called *Fungus Clypeiformis lethalis*, that is also a deadly Mushrum, fashioned like a little buckler.

There is another kinde of Mushrum, which is also most venomous and full of poison, bearing also the shape of a buckler, being called *Fungus venenatus Clypeiformis*: in English, the stinking venomous Mushrom.

2 *Fungi lethales, aut saltem non esculenti.*
Poyson Mushrums, or at the least such as are not vulgarly eaten.

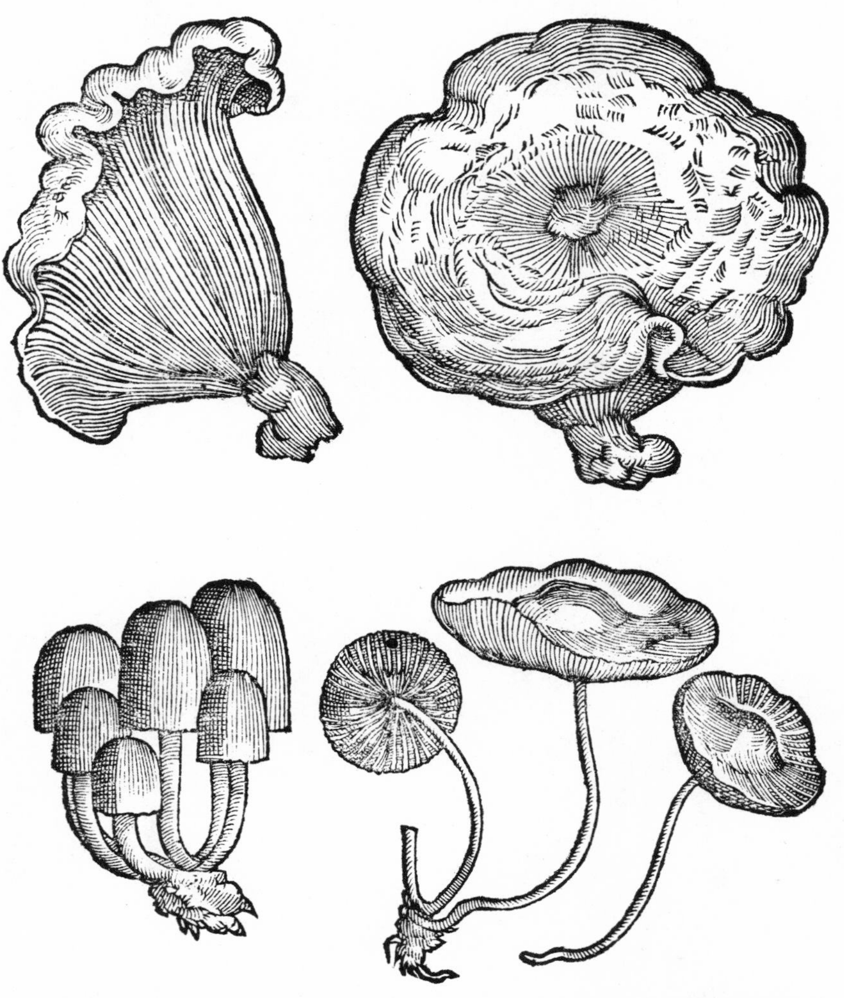

The Mushrums or Toodstooles which grow vpon the trunkes or bodies of old trees, very much resembling *Auricula Iudæ*, that is Iewes eare, do in continuance of time grow vnto the substance of wood, which the Foulers do call Touchwood, and are for the most halfe circuled or halfe round, whose vpper part is somewhat plaine, and sometime a little hollow, but the lower part is plaited or pursed together. This kinde of Mushrum the Grecians do call αἰγυρίτης, and is full of venome or poison as the former, especially those which grow vpon the Ilex, Oliue, and Oke trees.

There is likewise a kinde of Mushrum called *Fungus Fauaginosus*, growing vp in moist and shadowie woods, which is also venomous, hauing a thicke and tuberous stalke, an handfull high, of a duskish colour; the top whereof is compact of many small diuisions, like vnto the hony combe.

There

Fungus ſambucinus, ſive Auricula Iudæ. Iewes eares.

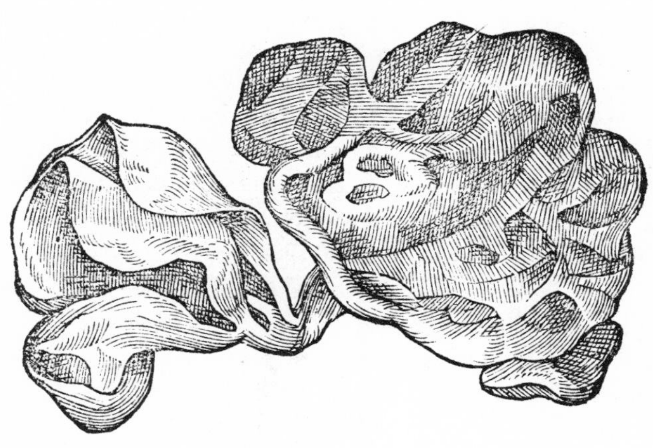

Fungi lethales, ſiue non eſculenti. Poyſonous Muſhrums.

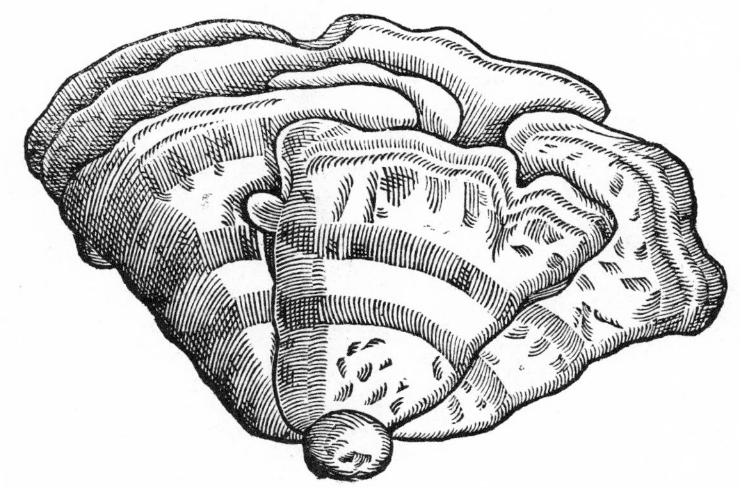

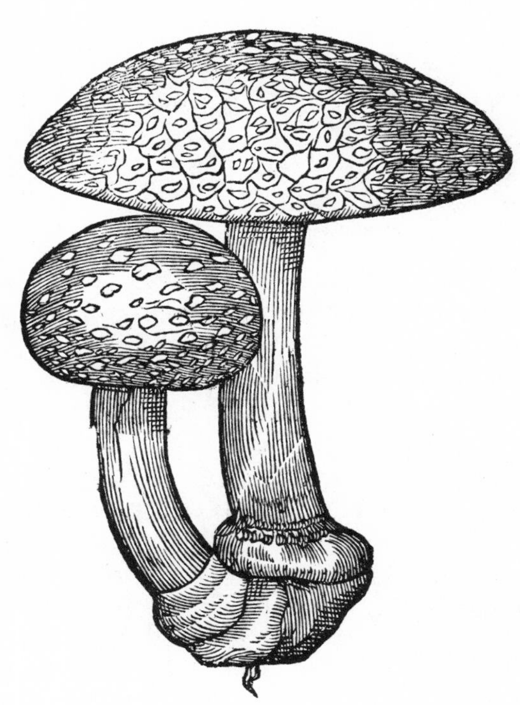

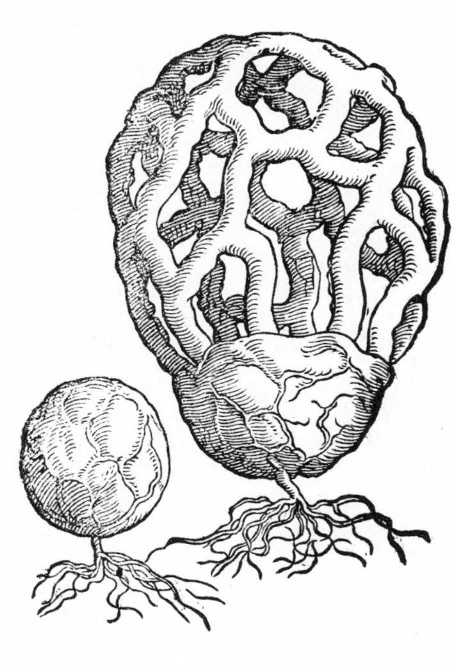

There is alſo found another, ſet forth vnder the title *Fungus virilis penis arecti forma*, which wee Engliſh, Pricke Muſhrum, taken from his forme.

3 *Fungus orbicularis*, or *Lupi crepitus*, ſome do call it *Lucernarum fungus*: in Engliſh, Fuſſe balls, Pucke Fuſſe, and Bulfiſts, with which in ſome places of England they vſe to kill or ſmolder their Bees, when they would driue the Hiues, and bereaue the poore Bees of their meat, houſes, and liues: theſe are alſo vſed in ſome places where neighbours dwell far aſunder, to carry and reſerue fire from place to place, whereof it tooke the name, *Lucernarum Fungus*: in forme they are very round, ſticking and cleauing vnto the ground, without any ſtalks or ſtems; at the firſt white, but afterward of a duskiſh colour, hauing no hole or breach in them, whereby a man may ſee into them, which being troden vpon do breath forth a moſt thin and fine pouder, like vnto ſmoke, very noiſome and hurtfull vnto the eies, cauſing a kinde of blindneſſe, which is called Poor-blinde, or Sand-blinde.

Fungi lethales, ſive non eſculenti. Poiſonous Muſhrums.

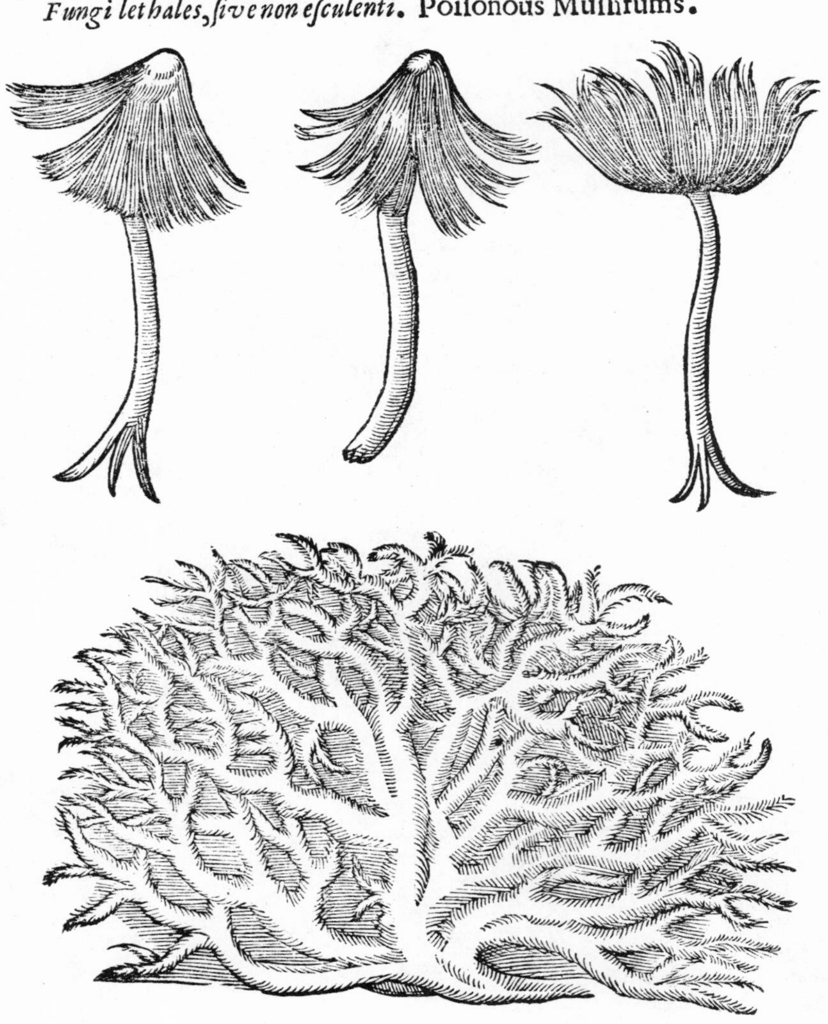

There is another kinde of *Fungus*, or Muſhrum, which groweth in moiſt medowes, and by ditch ſides, fiue or ſix inches high, couered ouer with a skin like a piece of ſheepes leather, of a ruſſet colour; which being taken away there appeareth a long and white ſtumpe, in forme not much vnlike to an handle, mentioned in the title, or like vnto the white and tender ſtalke of Aron, but greater: this kinde is alſo full of venome and poiſon.

There is likewiſe a kinde of Muſhrum, with a certaine round excreſcence, growing within the earth, vnder the vpper cruſt or face of the ſame, in dry and grauelly grounds in Pannonia and the Prouinces adioiningwhich do cauſe the ground to ſwel, and be full of hils like Mole-hils. The people where they grow, are conſtrained to dig them vp and caſt them abroad like as we do Mole-hils. ſpoiling their grounds, as Mole-hils are hurtfull vnto our ſoile: theſe haue neither ſtalks, leaues, fibres nor ſtrings annexed or faſtened vnto them, and for the moſt part are of a reddiſh colour, but within of a whitiſh yellow: the Grecians haue called this tuberous excreſcence, *Idna*, and the Latines *Tubera*: the Spaniards do call them *Turmas de tierra*: in Engliſh wee may call them Spaniſh Fuſſe bals.

¶ *The Place.*

Muſhrums come vp about the roots of trees, in graſſie places of medowes, and Ley Land newly turned

Fungus fauiginosus.
Hony-comb'd Mushrome.

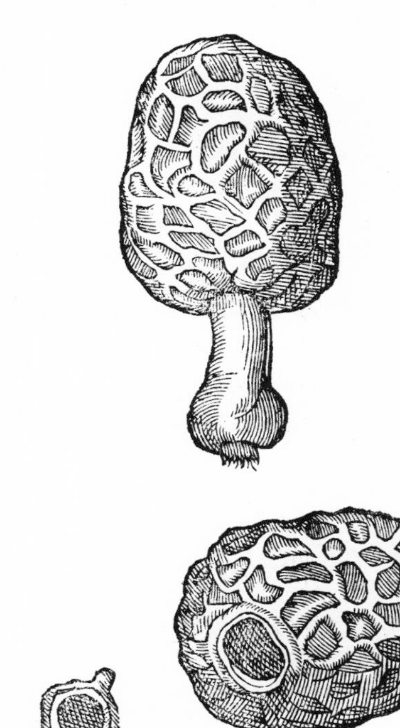

Fungus Virilis Penis effigie.
Pricke Mushrom.

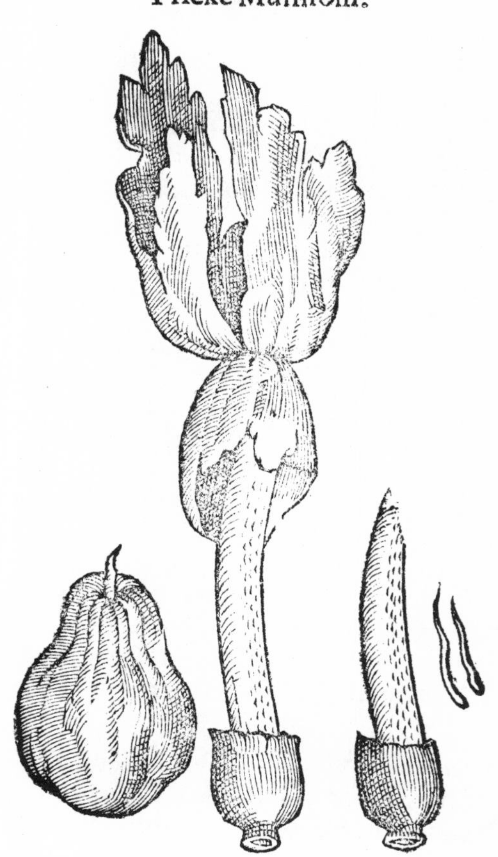

Tubera terræ.
Fusse-balls, or Puckfists.

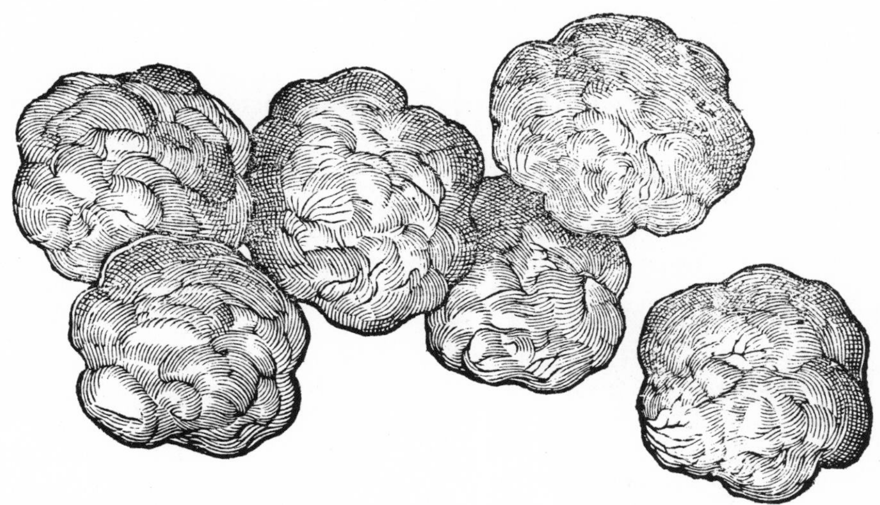

turned; in woods also where the ground is sandy, but yet dankish: they grow likewise out of wood, forth of the rotten bodies of trees, but they are vnprofitable and nothing worth. Poisonsome Mushroms, as *Dioscorides* saith, groweth where old rusty iron lieth, or rotten clouts, or neere to serpents dens, or roots of trees that bring forth venomous fruit. Diuers esteeme those for the best which grow in medowes, and vpon mountaines and hilly places, as *Horace* saith, *lib. ser.* 2. *satyr.* 4.

——— pratensibus optima fungis
Natura est, alijs malè creditur.

The medow Mushroms are in kinde the best.
It is ill trusting any of the rest.

¶ *The Time.*

Diuers come vp in Aprill, and last not till May, for they flourish but whilest Aprill continues: others grow later, about August; yet all of them after raine, and therefore they are found one yere sooner, and another later. Mushroms, saith *Pliny*, grow in shoures of raine: they come of the slime of trees, as the same Author affirmeth.

¶ *The Names.*

They are called in Latine, *Fungi*: in Greeke, μύκητες: in Italian, *Fonghi*: in Spanish, *Hungos*, *Cugumenos*: in French, *Campinion*, which word the low-Countrey men also vse, and call them **Campernoellen**: in high-Dutch, **Schwemme**, **Pfifferling**: in English, Mushroms, Toad-stooles, and Paddock-stooles.

The Mushroms that come vp in Aprill are called in Latine of some, *Spongiolæ*: of the Italians, *Prignoli*: and in high-Dutch, **Morchel**.

They that are of a light red are called of some *Boleti*, among the later ones which rise and fall away in seuen dayes. The white, or those which be somewhat yellow, are called in Latine, *Suilli*: which the later Physitions name *Porcini*, or Swine Mushroms. *Suilli*, saith *Pliny*, are dried, being hanged vpon rushes, which are thrust through them. The dry ones are in our age also eaten in Bohemia and Austria: they that grow by the roots of Poplar trees are called of the Latines, *Populnei*, Poplar Mushroms.

Puffes-fists are commonly called in Latine, *Lupi crepitus*, or Wolfes fists: in Italian, *Vescie de Lupo*: in English, Puffes-fists, and Fusse-balls in the North. *Pliny* nameth them *Pezicæ*, as though he should say, flat.

Tree Mushroms be called in Greeke, μυκῆτες: in Latine, *Fungi arborum*, and *Fungi arborei*: in English, tree Mushroms, or Touch-wood: in high-Dutch also **Schwemme**. They are all thought to be poisonsome, being inwardly taken. *Nicander* writeth, that the Mushroms of the Oliue tree, the Ilex tree, and of the Oke tree bring death.

¶ *The Temperature and Vertues.*

A *Galen* affirmes, that they are all very cold and moist, and therefore do approch vnto a venomous and murthering facultie, and ingender a clammy, pituitous, and cold nutriment if they be eaten. To conclude, few of them are good to be eaten, and most of them do suffocate and strangle the eater. Therefore I giue my aduice vnto those that loue such strange and new fangled meates, to beware of licking honey among thornes, lest the sweetnesse of the one do not counteruaile the sharpnesse and pricking of the other.

B Fusse-balls are no way eaten: the pouder of them doth dry without biting: it is fitly applied to merigalls, kibed heeles, and such like.

C In diuers parts of England where people dwell farre from neighbours, they carry them kindled with fire, which lasteth long: whereupon they were called *Lucernarum Fungi*.

D The dust or pouder hereof is very dangerous for the eyes, for it hath been often seen, that diuers haue been pore-blinde euer after, when some small quantitie thereof hath been blowne into their eyes.

E The countrey people do vse to kill or smother Bees with these Fusse-balls, being set on fire, for the which purpose it fitly serueth.

F ‡ The fungous excrescence of the Elder, commonly called a Iewes eare, is much vsed against the inflammations and all other sorenesses of the throat, being boiled in milke, steeped in beere, vineger, or any other conuenient liquor. ‡

CHAP. 168.

Of great Tooth-wort, or Clownes Lung-wort.

¶ *The Description.*

1 THere is often found among the Mushroms a certaine kinde of excrescence consisting of a jelly or soft substance, like that of the Mushroms, and therefore it may the more fitly be here inserted: it riseth forth of the ground in forme like vnto *Orobanche*, or the Broome-Rape, and also in substance, hauing a tender, thicke, tuberous, or mis-shapen body, consisting as it were of scales like teeth (whereof it tooke his name) of a dusty shining colour tending to purple. The stalke riseth vp in the middle, garnished with little gaping hollow floures like those of Satyrion; on the outside of an ouerworne whitish colour: the whole plant resembleth a rude forme of that jelly,

gellie, or slim ie matter, found in the fields, which we call the falling of stars: the root is small and tender.

2 There is also another sort hereof found, not differing from the precedent: the chiefe difference consisteth in that, that this plant is altogether lesser; ‡ and hath a root diuersly diuaricated like Corall, white of colour, full of juice, and without any fibres annexed thereto ‡; in other respects like.

1 *Dentaria maior Mathioli.*
Great Toothwoort, or Lungwoort.

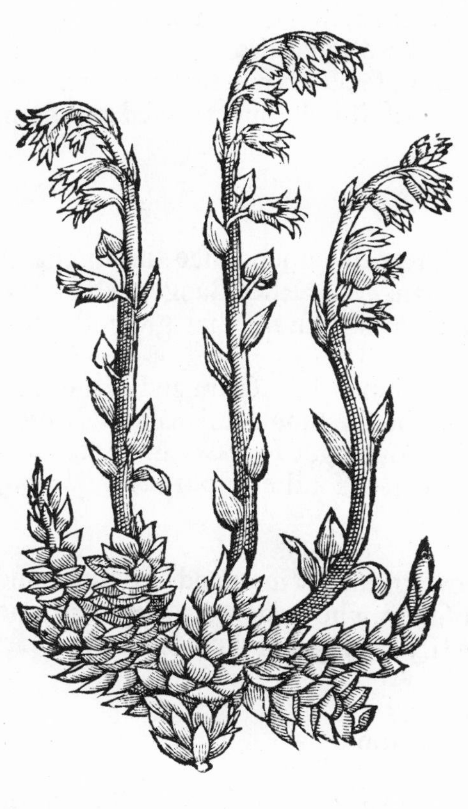

2 *Dentaria minor.*
Little Lungwoort.

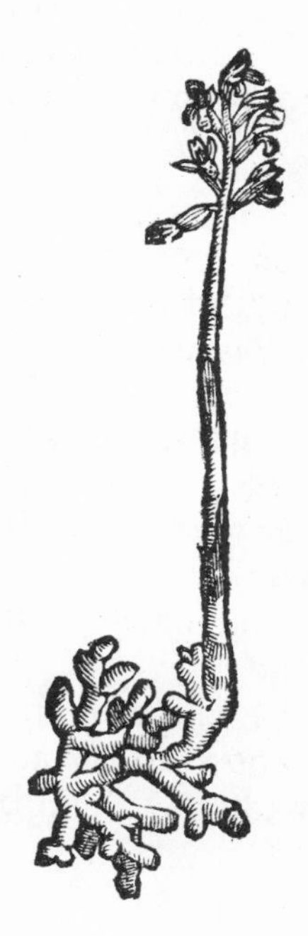

¶ *The Place.*

These plants do grow at the bottome of Elme trees, and such like, in shadowie places: I found it growing in a lane called East-lane, vpon the right hand as ye go from Maidstone in Kent vnto Cockes Heath, halfe a mile from the towne; and in other places thereabout: it doth also grow in the fields about Croidon, especially about a place called Groutes, being the land of a worshipfull Gentleman called M^r. *Garth*: and also in a wood in Kent neere Cravfoot, called Rowe, or Roughhill: it groweth likewise neere Harwood in Lancashire, a mile from Whanley, in a wood called Talbot banke.

¶ *The Time.*

They flourish in May and Iune.

¶ *The Names.*

There is not any other name extant, more than is set forth in the description.

¶ *The Temperature and Vertues.*

There is nothing extant of the faculties hereof, either of the ancient or later writers: neither haue we any thing of our owne experience; onely our countrie women do call it Lungwoort, and do vse it against the cough, and all other imperfections of the lungs: but what benefit they reape thereby I know not; neither can any of iudgement giue me further instruction thereof.

CHAP. 166. *Of Saunders.*

¶ *The Kindes.*

THe ancient Greekes haue not knowne the sorts of Saunders: *Garcias* and others describe three, *Album*, *Rubrum*, and *Pallidum*: which in shops is called *Citrinum*.

¶ The

¶ *The Description.*

1 THe Saunders tree groweth to the bignesse of the Walnut-tree, garnished with many goodly branches; whereon are set leaues like those of the Lentiske tree, alwaies greene; among which come forth very faire floures, of a blew colour tending to blacknesse; after commeth the fruit of the bignesse of a Cherry, greene at the first, and blacke when it is ripe; without taste, and ready to fall downe with euery little blast of winde: the timber or wood is of a white colour, and a very pleasant smell.

2 There is likewise another which groweth very great, the floures and fruit agree with the other of his kinde: the wood is of a yellowish colour, wherein consisteth the difference.

‡ 3 The third sort which wee call Red-Saunders is a very hard and sollid wood, hauing little or no smell, the colour thereof is very red, it groweth not in those places where the other grow, neither is the forme of the tree described by any that I know of, it is frequently vsed to colour sauces, and for such like vses. ‡

¶ *The Place.*

The white and yellow Saunders grow naturally, and that in great aboundance, in an Island called Timor, and also in the East-Indies beyond the riuer Sanges or rather Ganges, which the Indians call *Hanga*, and also about Iaua, where it is of better odour than any that groweth elsewhere.

The red Saunders growes within the riuer Ganges, especially about Tanasarim, and in the marrish grounds about Charamandell: *Auicen, Serapio* and most of the Mauritanians call it by a corrupt name, *Sandal:* in Timor, Malaca, and in places neere adioyning, *Chandama* in Decan and Guzarate, *Sercanda:* in Latine, *Sandalum* and *Santalum*, adding thereto for the colour *album, flauum,* or *Citrinum,* and *rubrum*, that is, white, yellow, and red Saunders.

¶ *The Time.*

These trees which are the white and yellow Saunders grow greene Winter and Sommer, and are not one knowne from another, but by the Indians themselues, who haue taken very certaine notes and markes of them, because they may the more speedily distinguish them when the Mart commeth.

¶ *The Names.*

Their names haue been sufficiently spoken of in their descriptions.

¶ *The Temperature.*

† Yellow and white Saunders are hot in the third degree, and dry in the second. The redde Saunders are not so hot. †

¶ *The Vertues.*

A The Indians do vse the decoction made in water, against hot burning agues, and the ouermuch flowing of the menses, *Erisipelas*, the gout, and all inflammations, especially if it be mixed with the iuice of Nightshade, Housleeke, or Purslane.

B The white Saunders mixed with Rose water, and the temples bathed therewith, ceaseth the paine of the megrim, and keepeth backe the flowing of humours to the eies.

C *Auicen* affirmeth it to be good for all passions of the hart, and maketh it glad and merry, and therefore good to be put in collises, iellies, and all delicate meates which are made to strengthen and reuiue the spirits.

D ‡ Red Saunders haue an astrictiue and strengthning facultie, but are not cordiall as the other two, they are vsed in diuers medicines and meates both for their facultie and pleasing red colour which they giue to them. ‡

Chap. 170. *Of Stony wood, or wood made Stones.*

¶ *The Description.*

AMong the wonders of England this is one of great admiration, and contrarie vnto mans reason and capacitie, that there should be a kinde of wood alterable into the hardnesse of a stone called Stonie wood, or rather a kinde of water, which hardneth wood and other things, into the nature and matter of stones. But we know that the workes of God are wonderfull,

Lignum Lapideum, ſiue in Lapides conuerſum.
Stonie wood, or wood made ſtones.

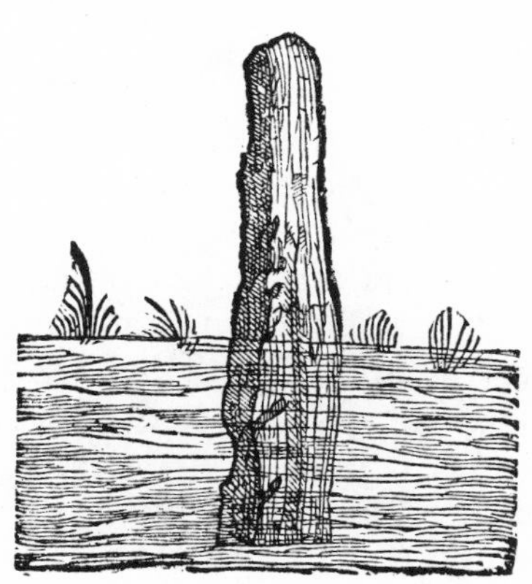

wonderfull, if we doe but narrowly ſearch the leaſt of them, which we dayly behold; much more if we turne our eyes vpon thoſe that are ſeldome ſeene, and knowne but of a few, and that of ſuch as haue painfully trauelled in the ſecrets of Nature. This ſtrange alteration of nature is to be ſeene in ſundry parts of England & Wales, through the qualities of ſome waters and earth, which change ſuch things into ſtone as do fall therein, or which are of purpoſe for triall put into them. In the North part of England there is a Well neere vnto Knaesborough, which will change any thing into ſtone, whether it be wood, timber, leaues of trees, moſſe, leather gloues, or ſuch like. There be diuers places in Bedfordſhire, Warwickſhire, and Wales, where there is ground of that qualitie, that if a ſtake be driuen into it, that part of the ſtake which is within the ground will be a firme and hard ſtone, and all that which is aboue the ground retaineth his former ſubſtance and nature. Alſo my ſelfe being at Rougby (about ſuch time as our fantaſticke people did with great concourſe and multitudes repaire and run headlong vnto the ſacred Wells of *Newnam Regis*, in the edge of Warwickſhire, as vnto the water of life, which could cure all diſeaſes) I went from thence vnto theſe Wells, where I found growing ouer the ſame a faire Aſh tree, whoſe boughes did hang ouer the ſpring of water, whereof ſome that were ſeare and rotten, and ſome that of purpoſe were broken off, fell into the water, and were all turned into ſtones. Of theſe boughes or parts of the tree I brought into London, which when I had broken in pieces, therein might be ſeene, that the pith and all the reſt was turned into ſtones; yea many buds and flourings of the tree falling into the ſaid water, were alſo turned into hard ſtones, ſtill retaining the ſame ſhape and faſhion that they were of before they were in the water. I doubt not but if this water were proued about the hardning of ſome Confections Phyſicall, for the preſeruation of them, or other ſpecial ends, it would offer greater occaſion of admiration for the health and benefit of mankinde, than it doth about ſuch things as already haue been experimented, tending to very little purpoſe.

CHAP. 171.

Of the Gooſe tree, Barnacle tree, or the tree bearing Geeſe.

Britannicæ Conchæ anatiferæ.
The breed of Barnacles.

¶ *The*

¶ *The Description.*

Hauing trauelled from the Grasses growing in the bottome of the fenny waters, the Woods, and mountaines, euen vnto Libanus it selfe; and also the sea, and bowels of the same, wee are arriued at the end of our Historie; thinking it not impertinent to the conclusion of the same, to end with one of the maruells of this land (we may say of the world.) The historie whereof to set forth according to the worthinesse and raritie thereof, would not only require a large and peculiar volume, but also a deeper search into the bowels of nature, than my intended purpose wil suffer me to wade into, my sufficiencie also considered; leauing the historie thereof rough hewen, vnto some excellent men, learned in the secrets of nature, to be both fined and refined: in the mean space take it as it falleth out, the naked and bare truth, though vnpolished. There are found in the North parts of Scotland and the Islands adiacent, called Orchades, certain trees whereon do grow certaine shells of a white colour tending to russet, wherein are contained little liuing creatures: which shells in time of maturitie do open, and out of them grow those little liuing things, which falling into the water do become fowles, which we call Barnakles; in the North of England, brant Geese; and in Lancashire, tree Geese: but the other that do fall vpon the land perish and come to nothing. Thus much by the writings of others, and also from the mouths of people of those parts, which may very well accord with truth.

But what our eyes haue seene, and hands haue touched we shall declare. There is a small Island in Lancashire called the Pile of Foulders, wherein are found the broken pieces of old and bruised ships, some whereof haue been cast thither by shipwracke, and also the trunks and bodies with the branches of old and rotten trees, cast vp there likewise; whereon is found a certaine spume or froth that in time breedeth vnto certaine shels, in shape like those of the Muskle, but sharper pointed, and of a whitish colour: wherein is contained a thing in forme like a lace of silke finely wouen as it were together, of a whitish colour, one end whereof is fastned vnto the inside of the shell, euen as the fish of Oisters and Muskles are: the other end is made fast vnto the belly of a rude masse or lumpe, which in time commeth to the shape and forme of a Bird: when it is perfectly formed the shell gapeth open, and the first thing that appeareth is the foresaid lace or string; next come the legs of the bird hanging out, and as it groweth greater it openeth the shell by degrees, til at length it is all come forth, and hangeth onely by the bill: in short space after it commeth to full maturitie, and falleth into the sea, where it gathereth feathers, and groweth to a fowle bigger than a Mallard, and lesser than a Goose, hauing blacke legs and bill or beake, and feathers blacke and white, spotted in such manner as is our Mag-Pie, called in some places a Pie-Annet, which the people of Lancashire call by no other name than a tree Goose: which place aforesaid, and all those parts adioyning do so much abound therewith, that one of the best is bought for three pence. For the truth hereof, if any doubt, may it please them to repaire vnto me, and I shall satisfie them by the testimonie of good witnesses.

Moreouer, it should seeme that there is another sort hereof; the historie of which is true, and of mine owne knowledge: for trauelling vpon the shore of our English coast betweene Douer and Rumney, I found the trunke of an old rotten tree, which (with some helpe that I procured by Fishermens wiues that were there attending their husbands returne from the sea) we drew out of the water vpon dry land: vpon this rotten tree I found growing many thousands of long crimson bladders, in shape like vnto puddings newly filled, before they be sodden, which were very cleere and shining; at the nether end whereof did grow a shell fish, fashioned somewhat like a small Muskle, but much whiter, resembling a shell fish that groweth vpon the rockes about Garnsey and Garsey, called a Lympit: many of these shells I brought with me to London, which after I had opened I found in them liuing things without forme or shape; in others which were neerer come to ripenes I found liuing things that were very naked, in shape like a Bird: in others, the Birds couered with soft downe, the shell halfe open, and the Bird ready to fall out, which no doubt were the Fowles called Barnakles. I dare not absolutely auouch euery circumstance of the first part of this history, concerning the tree that beareth those buds aforesaid, but will leaue it to a further consideration; howbeit that which I haue seene with mine eyes, and handled with mine hands, I dare confidently auouch, and boldly put downe for veritie. Now if any will obiect, that this tree which I saw might be one of those before mentioned, which either by the waues of the sea or some violent wind had been ouerturned, as many other trees are; or that any trees falling into those seas about the Orchades, will of themselues beare the like fowles, by reason of those seas and waters, these being so probable coniectures, and likely to be true, I may not without preiudice gainesay, or indeauour to confute.

‡ The Barnakle, whose fabulous breed my Author here sets downe, and diuers others haue also

alſo deliuered, were found by ſome Hollanders to haue another originall, and that by egges, as other Birds haue: for they in their third voyage to finde out the North-Eaſt paſſage to China, and the Molucco's, about the eightieth degree and eleuen minutes of Northerly latitude, found two little Iſlands, in the one of which they found aboundance of theſe Geeſe ſitting vpon their egges, of which they got one Gooſe, and tooke away ſixty egges, &c. *Vide Pontani, Rerum & vrb. Amſtelodam. Hiſt. lib. 2. cap. 22.* Now the ſhells out of which theſe birds were thought to fly, are a kinde of *Balanus marinus*; and thus *Fabius Columna*, in the end of his *Phytobaſanos*, writing *piſcium aliquot hiſtoria*, iudiciouſly proues: to whoſe opinion I wholly ſubſcribe, and to it I refer the Curious. His aſſeueration is this: *Conchas vulgò Anatiferas, non eſſe fructus terreſtres, neque ex ijs Anates oriri; ſed Balani marinæ ſpeciem.* I could haue ſaid ſomthing more hereof, but thus much I thinke may ſerue, together with that which *Fabius Columna* hath written vpon this point. ‡

¶ *The Place.*

The borders and rotten plankes whereon are found theſe ſhels wherein is bred the Barnakle, are taken vp in a ſmall Iſland adioyning to Lancaſhire, halfe a mile from the maine land, called the Pile of Foulders.

¶ *The Time.*

They ſpawne as it were in March and Aprill; the Geeſe are formed in May and Iune, and come to fulneſſe of feathers in the moneth after.

And thus hauing through Gods aſſiſtance diſcourſed ſomewhat at large of Graſſes, Herbes, Shrubs, Trees, and Moſſes, and certaine Excreſcences of the earth, with other things moe, incident to the hiſtorie thereof, we conclude and end our preſent Volume, with this wonder of England. For the which Gods name be euer honored and praiſed.

FINIS.

AN APPENDIX OR ADDITION OF certaine Plants omitted in the former Historie.

The Preface.

HAuing run through the Historie of Plants gathered by Mr. *Gerrard*, and much enlarged the same both by the addition of many Figures and histories of Plants not formerly contained in it, and by the amending and encreasing the historie of sundry of those which before were therein treated of; I finde that I haue forgotten diuers which I intended to haue added in their fitting places: the occasion hereof hath beene, my many businesses, the troublesomenesse, and aboue all, the great expectation and hast of the Worke, whereby I was forced to performe this task within the compasse of a yeare. Now being constant to my first resolution, I here haue, as time would giue me leaue, and my memorie serue, made a briefe collection and addition (though without method) of such as offered themselues vnto me; and without doubt there are sundrie others which are as fitting to be added as those; and I should not haue been wanting, if time would haue permitted me to haue entred into further consideration of them. In the meane time take in good part those that I haue here presented to your view.

Chap. 1. *Of the Maracoc or Passion-floure.*

¶ *The Description.*

THis Plant, which the Spaniards in the West Indies call *Granadilla*, because the fruit somewhat resembles a Pomegranat, which in their tongue they term *Granadas*, is the same which the Virginians call *Maracoc*. The Spanish Friers for some imaginarie resemblances in the floure, first called it *Flos Passionis*, the Passion floure, and in a counterfeit figure, by adding what was wanting, they made it as it were an Epitome of our Sauiours Passion: thus superstitious persons *semper sibi somnium fingunt*. *Bauhine* desirous to refer it to some stock or kindred of formerly knowne plants, giues it the name of *Clematis trifolia*: yet the floures and fruit pronounce it not properly belonging to their Tribe; but *Clematis* being a certaine genericke name to all wooddy winding plants, this as a species may come vnder the denomination, though little in other respects participating with them. The roots of this are long, somewhat like, yet thicker than those of *Sarsaparilla*, running vp and downe, and putting vp their heads in seuerall places: from these roots rise vp many long winding round stalkes, which grow two, three, foure, or more yards high, according to the heate and seasonablenesse of the yeare and soile whereas they are planted: vpon these stalkes grow many leaues diuided into three parts, sharpe pointed, and snipt about the edges: commonly out of the bosomes of each of the vppermost leaues there groweth a clasping tendrell and a floure: the floure growes vpon a little foot-stalke some two inches long, and is of a longish cornered forme, with fiue little crooked hornes at the top, before such time as it open it selfe; but opened, this longish head diuides it selfe into ten parts, and sustaines the leaues of the floure, which are very many, long, sharpe pointed, narrow, and orderly spred open one by another, some lying straight, others crooked: these leaues are of colour whitish, but thicke spotted with a Peach colour, and towards the bottome it hath a ring of a perfect Peach colour, and aboue and beneath it a white circle, which giue a great grace to the floure; in the middest whereof rises an vmbrane, which parts it selfe into foure or fiue crooked spotted hornes, with broadish heads: from

 the

the midst of these rises another roundish head which carries three nailes or hornes, biggest aboue, and smallest at their lower end: this floure with vs is neuer succeeded by any fruit, but in the West Indies, whereas it naturally growes, it beares a fruit, when it is ripe of the bignesse and colour of Pomegranats, but it wants such a ring or crown about the top as they haue; the rinde also is much thinner and tenderer, the pulpe is whitish, and without taste, but the liquor is somwhat tart: they open them as they do egges, and the liquor is supped off with great delight, both by the Indians and Spaniards, (as *Monardus* witnesseth) neither if they sup off many of them shall they find their stomack opprest, but rather their bellies are gently loosned. In this fruit are contained many seeds somwhat like Peare kernells, but more cornered and rough.

Clematis trifolia, siue Flos Passionis.
The Maracoc or Passion-floure.

This growes wilde in most of the hot countries of America, from whence it hath been brought into our English gardens, where it growes very well, but floures only in some few places, and in hot and seasonable yeares: it is in good plenty growing with Mistresse *Tuggy* at Westminster, where I haue some yeares seene it beare a great many floures.

Chap. 2. *Of Ribes or red Currans.*

¶ *The Description.*

1 THe plant which carries the fruit which we commonly terme red Currans, is a shrubbie bush of the bignesse of a Gooseberry bush, but without prickles: the wood is soft and white, with a pretty large pith in the middle: it is couered with a double barke, the vndermost, being the thicker, is greene, and the vppermost, which sometimes chaps and pills off, is of a brownish colour,

colour, and ſmooth: the barke of the yongeſt ſhoots is whitiſh and rough: the leaues, which grow vpon footſtalkes ſome two inches long, are ſomewhat like Vine leaues, but ſmaller by much, and leſſe cornered, being cut into three, and ſometimes, but ſeldomer, into fiue parts, ſomwhat thicke, with many veines running ouer them, greener aboue than they are below: out of the branches in Spring time grow ſtalkes hanging downe ſome ſix inches in length, carrying many little greeniſh floures, which are ſucceeded by little red berries, cleare and ſmooth, of the bigneſſe of the Whortle berries, of a pleaſant tart taſte. Of this kinde there is another, onely different from this in the fruit, which is twice ſo big as that of the common kind.

2 The buſh which beares the white Currans is commonly ſtraighter and bigger than the former: the leaues are leſſer, the floures whiter, and ſo alſo is the fruit, being cleare and tranſparent, with a little blackiſh rough end.

1 *Ribes vulgaris fructu rubro.*
Red Currans.

2 *Ribes fructu albo.*
White Currans.

3 Beſides theſe there is another, which differs little from the former in ſhape, yet grows ſomwhat higher, and hath leſſer leaues: the floures are of a purpliſh green colour, and are ſucceeded by fruit as big againe as the ordinary red, but of a ſtinking and ſomewhat loathing ſauour: the leaues alſo are not without this ſtinking ſmell.

¶ *The Place, Time, and Names.*

None of theſe grow wild with vs, but they are to be found plentifully growing in many gardens, eſpecially the two former, the red and the white.

The leaues and floures come forth in the Spring, and the fruit is ripe about Midſommer.

This plant is thought to haue been vnknowne to the antient Greekes: ſome thinke it the *Ribes* of the Arabian *Serapio*. *Fuchſius, Matthiolus*, and ſome other deny it; notwithſtanding *Dodonæus* affirmes it: neither is the controuerſie eaſily to be decided, becauſe the Author is briefe in the deſcription thereof, neither haue we his words but by the hand of a barbarous Tranſlator. Howeuer the ſhops of late time take it (the faculties conſenting thereto) for the true Ribes, and of the fruit hereof prepare their *Rob de Ribes. Dodonæus* calls it *Ribeſium, groſſularia rubra, & Groſſularia tranſmarina*; and they are diſtinguiſhed into three ſorts, *Rubra, Alba, Nigra Ribeſia*, red, white, and blacke Currans: the Germans call them S. Johans traubell, or traublin, and S. Johans Beerlin: the Dutch, Beſkins ouer Zee: the Italians, *Vuetta roſſa*: the French, *Groiſſeles, Groiſelles d'outre mer*: the Bohemians, Jahodi S. Jana: the Engliſh, Red Currans: yet muſt they not be confounded with those Currans which are brought from Zant, and the continent adioyning thereto, and which are vulgarly sold by our Grocers; for they are the fruit of a small Vine, and differ much from these.

The Temperature and Vertues.

A The berries of red Currans, as also of the white, are cold and dry in the end of the second degree, and haue some astriction, together with tenuitie of parts.

B They extinguish and mitigate feuerish heates, represse choler, temper the ouer-hot bloud, resist putrefaction, quench thirst, helpe the deiection of the appetite, stay cholericke vomitings and scourings, and helpe the Dysenterie proceeding of an hot cause.

C The iuice of these boiled to the height of honey, either with or without sugar (which is called *Rob de Ribes*) hath the same qualities, and conduces to the same purposes.

Chap. 3. *Of Parsley Breake-stone, and bastard Rupturewort.*

1 *Percepier Anglorum Lob.*
Parsley Breake-stone.

2 *Polygonũ Herniariæ facie.*
Bastard Rupture-wort.

¶ *The Description.*

1 I Thought it was not altogether inconuenient to couple these two Plants together in one Chapter; first, because they are of one stature; and secondly, taken out of one and the same History of Plants, to wit, the *Aduersaria* of *Pena* and *Lobel*.

The first of these, which the Authors of the *Aduersaria* set forth by the name of *Percepier*, (and rather assert, than affirme to be the *Scandix* of the Antients) is by *Tabernamontanus* called *Scandix minor*: and by *Fabius Columna*, *Alchimilla montana minima*: it hath a small woody yellowish fibrous root, from which rise vp one, two, or more little stalks, seldome exceeding the height of an handfull, and these are round and hairy, and vpon them grow little roundish leaues, like the tender leaues of Cheruill, but hairy, and of a whitish green colour, fastned to the stalkes with short foot-stalkes, and hauing little eares at their setting on: the floures are small, greene, and fiue cornered, many clustering together at the setting on of the leaues: the seed is small, smooth, and yellowish: the stalks of this plant grow sometimes vpright, and otherwhiles they lean on the ground: it is to be found vpon diuers dry and barren grounds, as in Hide Parke, Tuthill fields, &c. It floures in May, and ripens the seed in Iune and Iuly. It seemes by the Authors of the *Aduersaria*, that in the West countrey about Bristow they call this Herbe Percepier; but our herbe women in Cheapside know it by the name of Parsley Breakestone.

D This is hot and dry, and of subtil parts: it vehemently and speedily moues vrine, and by some is kept in pickle, and eaten as a sallad.

E The distilled water is also commended to be effectuall to moue vrine, and clense the kidnies of grauell.

2 The historie of this, by the forementioned Authors, *Aduers. pag.* 404. is thus set forth vnder this title, *Polygonium Herniariæ folijs & facie, perampla radice Astragalitidi*: Neither (say they) ought this to be despised by such as are studious of the knowledge of Plants; for it is very little knowne, being a very small herbe lying along vpon the ground, and almost ouerwhelmed or couered with the grasse, hauing little branches very full of ioints: the little leaues and seeds are whitish, and very like those of *Herniaria*, or Rupture-wort: the whole plant is white, hauing a very small and mossie floure: the root is larger than the smalnesse of the plant seemeth to require, hard, branched, diuersly turning and winding, and therefore hard to be plucked vp: the taste is dry and hottish. It growes vpon a large Plaine in Prouince, betweene the cities Arles and Selon. Thus much *Pena* and *Lobel*. I am deceiued, if some few yeares agone I was not shewed this plant, gathered in some part of this kingdome, but where, I am not able to affirme.

Chap. 4. *Of Heath Spurge and Rocke Roſe*

¶ *The Deſcription.*

1 These Plants by right ſhould haue followed the hiſtorie of *Thymelæa*, for in ſhape and facultie they are not much vnlike it. The firſt is a low ſhrub, ſending from one root many branches of ſome cubit long, and theſe bending, flexible, and couered with an outer blackiſh barke, which comprehends another within, tough, and which may be diuided into fine threds: the leaues are like thoſe of *Chamælea*, yet leſſer, ſhorter and thicker, a little rough alſo, and growing about the branches in a certaine order: if you chew them they are gummie, bitter at the firſt, and afterwards hot and biting: the floures grow amongſt the leaues, longiſh, yellowiſh, and diuided at the end into foure little leaues: the fruit is ſaid to be like that of *Thymælea*, but of a blackiſh colour, the root is thicke and wooddie. It growes frequently in the kingdome of Granado and Valentia in Spaine, it floures in March and Aprill. The Herbariſts there terme it *Sanamunda*, and the common people, *Mierda-cruz*, by reaſon of the purging facultie.

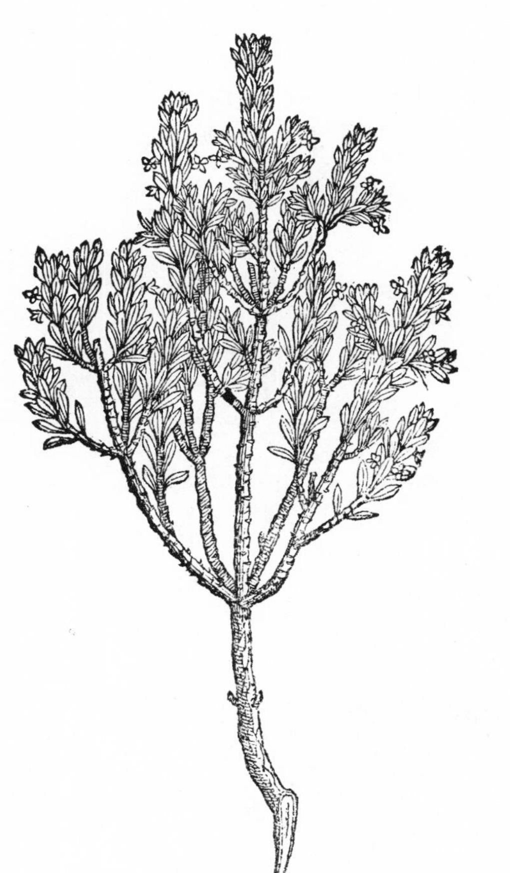

1 *Sanamunda* 1. *Cluſ.* Heath Spurge.

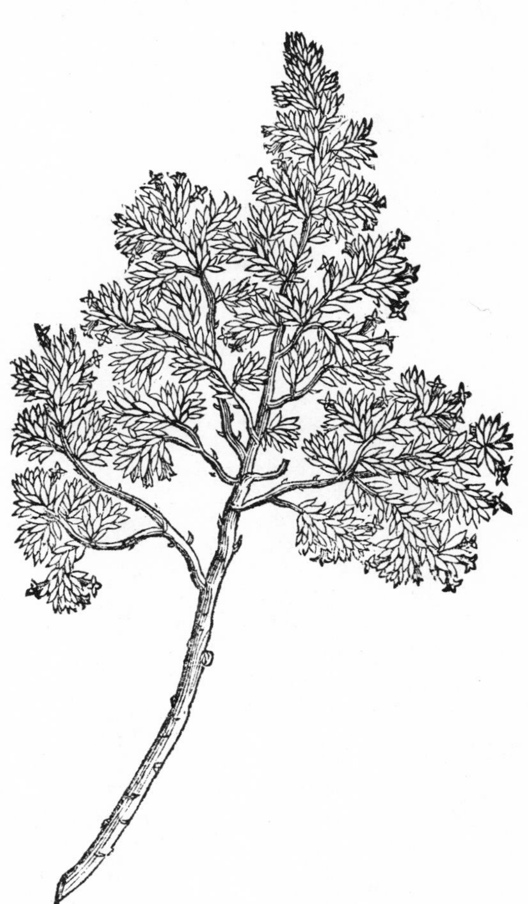

2 *Sanamunda* 2. *Cluſ.* The ſecond Heath Spurge.

2 The other is a ſhrub ſome cubit high, hauing tough flexible branches couered with a denſe and thick barke, which, the outward rinde being taken away, ouer all the plant, but chiefely next the root, may be drawn into threds like Flax or Hemp: the vpper branches are ſet with thick, ſhort, fat, rough ſharp pointed leaues, of ſomwhat a ſaltiſh taſte at the firſt, afterwards of a hot & biting taſte: the floures are many, little and yellow: the root is thicke and wooddie like as that of the former: this growes vpon the ſea coaſt of Spaine, and on the mountaines nigh Granado, where they call it *Sanamunda*, and the common people about Gibraltar call it Burhalaga, and they only vſe it to heat their ouens with. It floures in Februarie. *Anguillara* called this, *Empetron: Cæſalpinus, Cneoron*, and in the *Hiſtoria Lugd.* it is the *Cneoron nigrum Myconi: Seſamoides minus: Dalechampij*, and *Phacoides, Oribaſij quibuſdam*.

3 This

3 This is bigger than either of the two former, hauing whiter and more flexible branches, whose barke is vnmeasurably tough and hard to breake : the vpper branches are many, and those very downie, and hanging downe their heads, set thicke with little leaues like Stone-crop, and of the like hot or burning facultie : the floures are like those of the former; sometimes greenish, otherwhiles yellow : *Clusius* did not obserue the fruit, but saith, it floured at the same time with the former, and grew in all the sea coast, from the Straits of Gibralter, to the Pyrenæan mountaines. *Alfonsus Pantius* called this *Cneoron* : *Lobel* and *Tabernamontanus* call it *Erica Alexandrina*.

3 *Sanamunda* 3. *Clus.*
The third Heath Spurge.

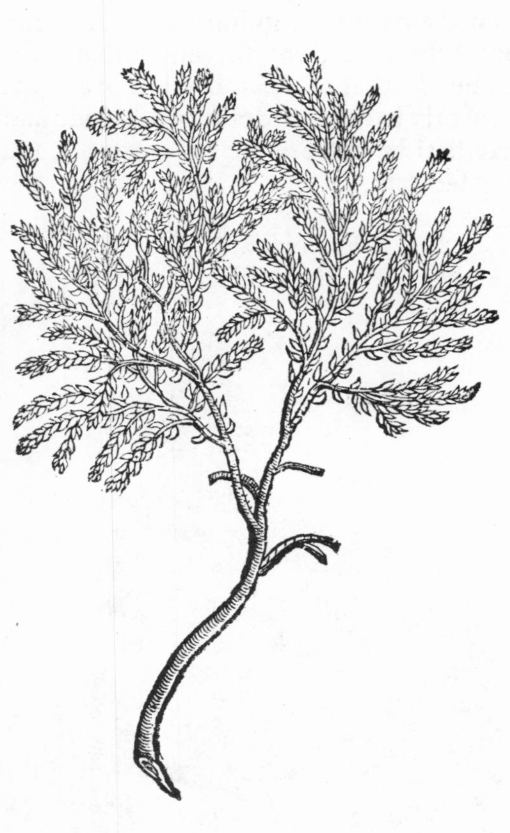

4 *Cneoron Matthioli.*
Rocke Rose.

4 This also may not vnfitly bee ioined to the former, for it hath many tender flexible tough branches commonly leaning or lying along vpon the ground, vponwhich without order grow leaues greeene, skinny, and like those of the true *Thymelæa*; at first of an vngratefull, and afterwards of a bitter taste, yet hauing none or very little acrimony (as far as may be perceiued by their taste :) the floures grow vpon the tops of the branches six seuen or more together, consisting of foure little leaues of a reddish purple colour, very beautifull and well smelling, yet offending the head if they be long smelt vnto : these are succeeded by small berries, of colour white, containing a round seed, couered with an ash coloured skin. The root is long, of the thickenesse of ones little finger, sometimes blackish, yet most commonly yellowish, tough, and smallest at the top where the branches come forth. It floures in Aprill and May, and ripens the fruit in Iune : it floures sometimes thrice in the yeare, and ripens the fruit twise; for *Clusius* affirmes that twise in one yeare he gathered ripe berries from one and the same plant. It growes plentifully vpon the mountainous places of Austria about Vienna ; whither the countrey women bring the floures to the market in great plenty to sell them to deck vp houses : it grows also in the dry medowes by Frankford on the Mœne, where there is obserued a variety with white floures. *Matthiolus* would haue this to be the *Cneoron album* of *Theophrastus* : *Cordus* calls it *Thym elæa minor* : it is the *Cneoron alterum Matthioli*, and *Oleander syl. Auicennæ Myconi*, in the *Hist. Lugd.* The Germans call it **Stein Roselin** : and wee may call it Rocke Rose, or dwarfe Oleander.

5 This plant by *Bauhine* is called *Cneorum album folio oleæ argenteo molli* : and by *Dalechampius*, *Cneorum album*, which hath been the reason I haue put it here, although *Cæsalpinus*, *Imperatus*, and *Plateau*,

5 *Cneorum album folys argenteis.*
White Rocke Roſe.

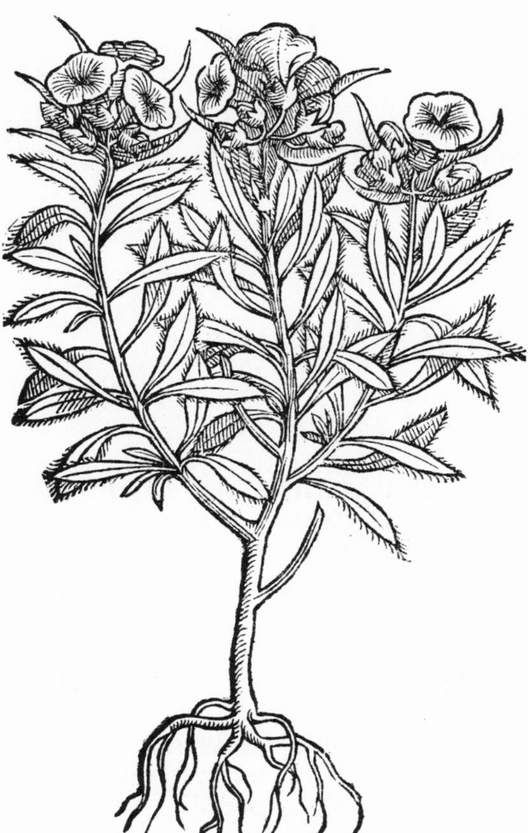

Chamæbuxus flore Coluteæ.
Baſtard dwarfe box.

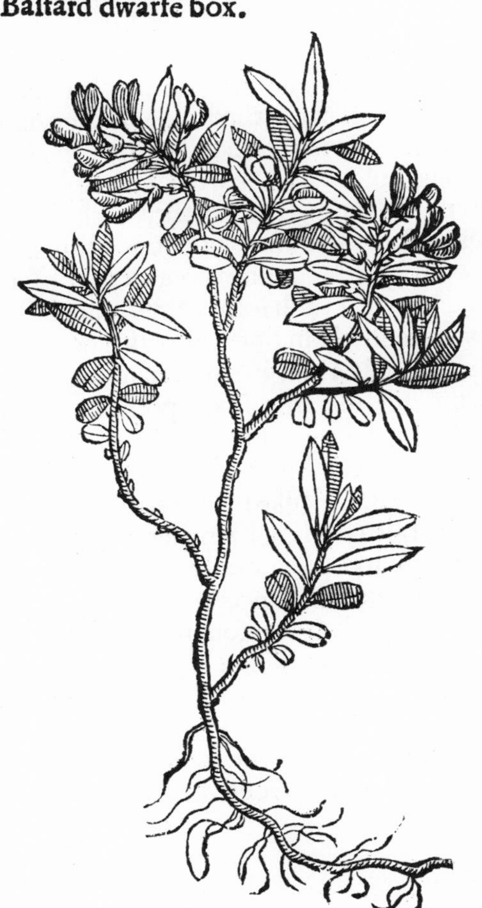

teau,who ſent it to *Cluſius*, would haue it to be and cal it *Dorycnium:* It is a ſhrubby herb ſending from one root many ſingle ſtalkes ſome halfe cubit or better high: the leaues which grow vpon the ſtalkes without order, are like thoſe of the Oliue, but ſomewhat narrower, and couered ouer with a ſoft ſiluer-like downineſſe: at the top of the ſtalks grow many floures cluſtering together, of the ſhape of thoſe of the leſſer Bindeweed, but white of colour. This growes wilde in ſome parts of Sicily, whence *Cæſalpinus* calls it *Dorychnium ex Sicilia.*

¶ *The Temperature and Vertues.*

A The three firſt are very hot, and two firſt haue a ſtrong purging facultie, for taken in the weight of a dram with the decoction of Cicers they mightily purge by ſtoole, both flegme, choller, and alſo wateriſh humours, and they are often vſed for this purpoſe by the Countrey people in ſome parts of Spaine.

The faculties of the reſt are not knowne. nor written of by any as yet.

CHAP. 5. *Of Baſtard dwarfe Box.*

¶ *The Deſcription.*

THis which *Cluſius* for want of a name calls *nonymos flore Coluteæ* *Geſner* called *Chamæbuxus* to which *Bauhine* addes *flore Coluteæ*; and *Beſler* in his *hortus Eyſtettenſis*, agreeable to the name I haue giuen it in Engliſh, calls it *Pſeudochamæbuxus*. It is a ſmall plant hauing many creping wooddy tough roots, here and there ſending forth ſmall fibers; from theſe ariſe many tough bending branches ſome ſpan long, hauing thicke ſharpe pointed greene leaues, almoſt like thoſe of Boxe, and theſe grow vpon the ſtalks without any order, and when you firſt chew them they are of an vngratefull taſte, afterwards bitter and hot; at the tops of the branches, do come forth amongſt the leaues three or foure longiſh floures, for the moſt part without ſmell; yet in ſome places they ſmell ſweet, like as ſome of the Narciſſes; they conſiſt of three leaues apiece; two whereof are white, and ſpread abroad as wings, a whitiſh little hood couering their lower ends: the third is wrapt vp in forme of a pipe, with the end hollow & crooked, and this

this is of a yellow colour, which by age oft times becomes wholly red: after those floures succeed cods, broad and flat, little lesse than those of the broad leaued *Thlaspi*, and greene of colour, rough, and in each of these cods are commonly contained a couple of seeds, of the bignes of little Chichlings, of a blackish ash colour, rough, and resembling a little dug.

This is sometimes found to vary, hauing the two winged leaues yellow or red, and the middle one yellow.

¶ *The Place.*

It floures in Aprill and May, and ripens the seed in Iune; it growes vpon most of the Austrian and Stirian Alpes, and in diuers places of Hungarie. It is neither vsed in Physicke, nor the faculties thereof in medicine knowne.

Chap. 6. *Of Winged Bindweed, or Quamoclit*

Quamoclit, siue Conuoluulus Pennatus.
Winged Windeweed.

¶ *The Description.*

The first that writ of and described this plant was *Cæsalpinus*, & that by the name of *Gelsiminum rubrum alterum*: after him *Camerarius* gaue a description and figure therof in his *Hortus Medicus*, by the name of *Quamoclit*: and after him *Fabius Columna* both figured and described it more accurately, whose description is put to the figure of it (we here giue) in *Clus.* his *Curæ posteriores*. It is so tender a plant that it will not come to any perfection with vs, vnlesse in extraordinary hot yeres, and by other artificiall helps, wherefore I will borrow the description thereof out of *Fabius Columna*. This exoticke plant, saith he, cannot more fitly be referred to any kinde, than to the family of the *Convoluuli*, or Bindweeds, for in the nature and whole habit it is almost like them, excepting the shape of the winged leaues: it is stored with lesse milk: the flours are long, hollow, but parted into fiue at the top, of a pleasing red colour, with streaked lines or folds, standing vpon long stalkes one or two together comming out of the bosomes of the leaues at each ioint of the branches, and they haue in them fiue yellowish pointalls; then succeeds a longish fruit standing in a scaly cup, ending in a sharp pointall, and couered with a tough skin, as that of the common *Convoluulus*, but lesser, hauing within it foure longish blacke hard seedes, of a biting taste. The leaues grow alternately out of the ioints of the purple winding branches, being winged and finely diuided, twise as small as the common *Rheseda*, of a darke greene colour, but the young ones are yellowish, first hauing a few diuisions, but afterwards more, till they come to haue thirteen on a side, and one at the top: but the lower ones are oft times forked: by reason of the great plenty of leaues and flouring stalks or branches, winding themselues about artificiall hoops, crossings, or other fashioned workes of Reeds, or the like, set for winding herbs to clime vpon, it much delights the eie of the beholder, and is therefore kept in pots in gardens of pleasure. The seed sowne in the beginning of the Spring growes vp in Iune, and the first leaues resemble the winged fruit of the Maple: it floures in the end of August, and ripens the seed in the end of September.

Chap.

Chap. 7. *Of the ſenſitiue Herbe.*

Herba mimoſa.
The ſenſitiue herbe.

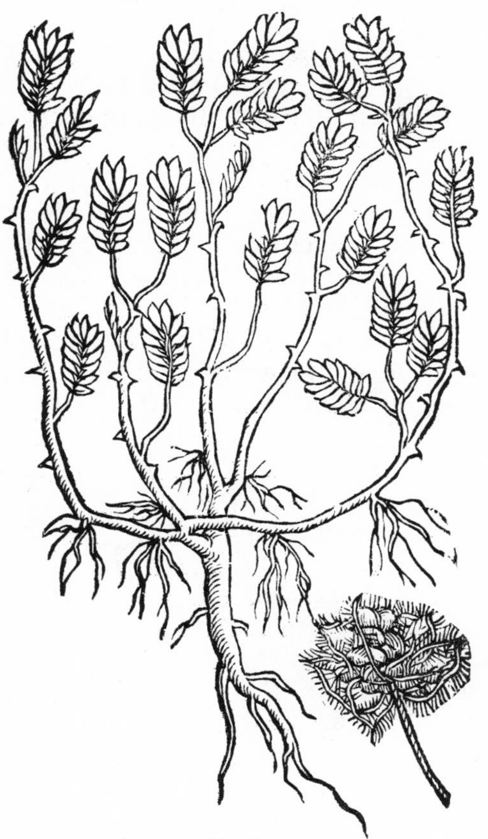

Eius exactior icon.
A perfect figure thereof.

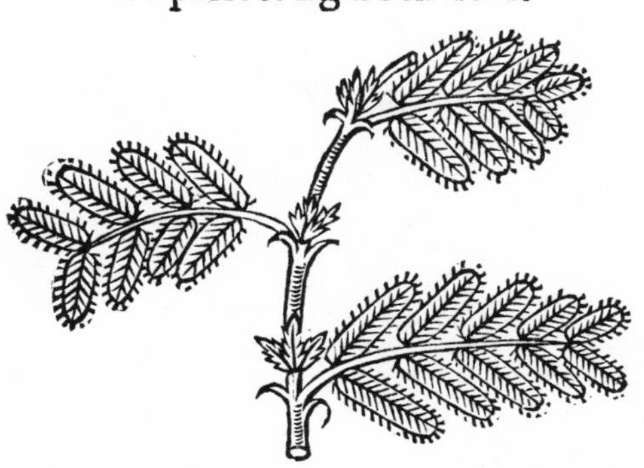

¶ *The Deſcription.*

THis which I here call the ſenſitiue herbe, is that which *Chriſtopher a Coſta* ſets forth by the name of *Herba mimoſa*, or the Mocking herbe, becauſe when one puts his hand thereto it forthwith ſeemes to wither and hang downe the leaues; but when you take it away againe it recouers the priſtine greeneſſe and vigor. I wil here giue you that which *Acoſta* writes thereof, & the figure & hiſtorie which *Cluſius* giues in his notes vpon him; and alſo another figure better expreſſing the leaues and manner of growing. There is found (ſaith *Acoſta*) in ſome Gardens another plant ſome fiue handfuls long, reſting vpon the neighbouring ſhrubs or walls, hauing a ſlender ſtalke of a freſh greene colour, not very round, ſet at certaine ſpaces with ſmall and pricking thornes: the leaues are not vnlike the former, [That is, the *Herba viua*, which in condition is little different from this] being ſomewhat leſſer than thoſe of the female Ferne. It loues to grow in moiſt and ſtony places, and is called *Herba mimoſa*, for the reaſon formerly giuen. The nature hereof is much different from that of *Arbor triſtis* for euery night at Sun-ſet it as it were withers and dries, ſo that one would thinke it were dead, but at Sun-riſe it recouers the former vigor, and by how much the Sun growes hotter, by ſo much it becomes the greener, and all the day it turnes the leaues to the Sun.

A This plant hath the ſmell and taſte of Liquorice, and the leaues are commonly eaten by the Indians againſt the cough, to clenſe the cheſt, & cleare the voice: it is alſo thought good againſt the paines of the kidneies, and to heale greene wounds. Thus much *Acoſta*.

Now, ſaith *Cluſius*, the leaues of many plants, eſpecially pulſes, vſe to contract or ſhrinke vp their leaues in the night time. Now I receiued a dry plant, which was ſent to me by the name of *Herba mimoſa*, by *Iames Garret* in the end of October, 1599, which he writ he had of the right Honourable the Earle of Cumberland, who returning from Saint *Iohn de Puerto rico* in the Weſt Indies, brought it put in a pot with ſome earth, but could not preſerue it aliue. But I cauſed the figure of that dried plant to be expreſſed as well as it might, ſo to fit it to the deſcription following, made alſo by the dried plant. This plant which was wholly drie and without leaues had a ſingle root, and that not thick, but hard and wooddy, with few fibres, from whence aroſe three or foure ſhort ſtalks, which ſtraight diuided themſelues into ſlender branches, which ſpread themſelues round about vpon the ground, at each ioint putting forth many long and ſlender fibres, like as in the branches of the common Woodbinde, which lye vpon the ground: theſe branches were a cubit long, and ſometimes more, round, tough, with ſome prickles, broader at their ſetting on, as you may ſee in the common bramble, yet leſſer, fewer, & leſſe firme; theſe againe were diuided into other more ſlender branches ſet with many little prickles, out of whoſe ioints betwixt two little leaues grew forth foot-ſtalks, bedeckt with their little leaues, which were many, ſet in order, with other to anſwer to them on the other ſide, but hauing no ſingle leafe at the end: they were tender & green, not vnlike the little leaues of *Acacia*, & theſe (at their firſt comming out) couered with a thin whitiſh hairines, as I gathered by a little branch retaining the foot-ſtalke and leaues thereon (which he ſent with the former) and it had alſo ſome fibres comming forth thereof. He alſo added to the former two little heads, which growing vpon the ſame plant, he writ he receiued of the

forementioned

forementioned right Honorable Earle, with some branches yet retaining the leaues. These little heads consisted of many slender, narrow, and as it were prickly little leaues; amongst which lay hid round seeds, smooth, blacke, and somewhat swoln in the middle : the floures I saw not, neither know I whether they were brought with the rest : but whether the leaues of this plant being green, & yet growing on the ground, do wither at the approch of ones hand, as *Christopher A Costa* writes, and for that cause imposes the name thereon, they best know who haue seene the greene and yet growing plant : for the faculties you may haue recourse to that which *A Costa* hath set downe. Thus much out of *Clusius*.

Novemb. 7. 1632. I being with M^r. *Iob Best* at the Trinity house in Ratcliffe; among other varieties, he shewed me a dry plant hereof, which I heedfully obserued, and carefully opening out some of the fairest leaues, which (as also the whole plant besides) were carelesly dried, I found the leaues grew vsually some dozen or more on a foot-stalke, iust as many on one side as on the other; & they were couered ouer with a little downines, which standing out on their edges made them look as if they had bin snipt about the edges, which they were not : also I found at euery ioint two little hooked prickles, & not two little leaues or appendices at the setting on of the foot-stalks, but three or foure little leaues, as the rudiment of a yong branch, comming forth at the bosom of each foot-stalk: the longest branch (as far as I remember) was not aboue a span long; I then drew as perfect a figure as I could of the perfectest branch therof, drawing as neere as I could the leaues to their ful bignesse, the which I here present you withall. There are two figures formerly extant, the one this of *Clusius*, which I here giue you, and the other in the 18. booke, & 144 chap. of the *Hist. Lug* which is out of *A Costa*, and this seeems to be so far different from that of *Clusius*, that *Bauhine* in his *Pinax* saith, *Clusius notis suis in Acostam diuersam plane figuram proposuit, herbam minosam nominans* : but he did not wel consider it, for if he had, he might haue found these so much different, thus far to agree; they both make the branches prickly & weak : the leaues many on one rib, one opposite to another without an odde one at the end : but *Clusius* figures the leaues so close together, that they seem but one leafe, and *Acosta* makes them too far a sunder, and both of them make them too sharp pointed; *Clus.* made his be taken from a dried plant, and *Acosta* I iudg made his by the Idæa thereof which he had in his memorie, and after this manner, if my iudgement faile me not, are most of the figures in him exprest: but of this enough, if not too much.

Chap. 8. *Of the Staffe tree, and euer-greene Priuet.*

1 *Celastrus Theophrasti.*
The staffe tree.

2 *Phillyrea* 1. *Clus.*
Clusius his 1. Mocke-Priuet.

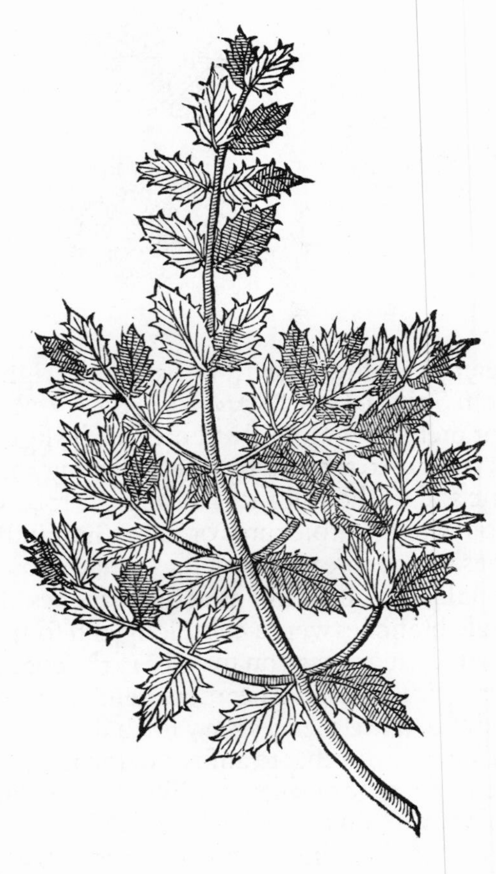

¶ *The Description.*

1 THe history and figure of this tree are set forth in *Clusius* his *Curæ poster.* and there it is asserted to be the κήλαστρος, or κήλαστρον of *Theophrastus*; for by diuers places in *Theophrastus* there collected, it is euident, that his *Celastus* was euer greene, grew vpon very high and cold mountaines, yet might be transplanted into plaine and milder places, that it floured exceeding late and could not perfect the fruit by reason of the nigh approch of winter, and that it was fit for no other vse but to make staues on for old men.

Now this tree growes but to a small height, hauing a firme and hard body, diuiding it selfe at the top into sundry branches, which being yonge are couered with a greene barke, but waxing old with a brownish one; it hath many leaues, growing alwaies one against another, and thicke together, of a deepe shining greene aboue, and lighter vnderneath, keeping their verdure both Winter and Sommer: they are of the bignesse of those of *Alaternus,* not snipt about the edges, but onely a little nickt, when they are yet yong; at the top of the tenderest branches among the leaues, vpon footstalkes of some inch long, grow fiue or six little floures consisting commonly of fiue little leaues of a yellowish greene colour, and these shew themselues in the end of Autumne, or the beginning of Winter, and also in the beginning of the Spring; but if the Sommer be cold and moist it shewes the buds of the floures in October; the fruit growes on a short stalke and is a berry of the bignesse of the Myrtle, first green, then red, of the colour of that of *Asparagus,* and lastly blacke when it is withered: the stone within the berry is little, and as it were three cornered, conteining a kernell couered with a yellow filme. Where this growes wilde I know not, but it was first taken notice of in the publike Garden at the Vniuersitie of Leyden, from whence it was brought into some few gardens of this Kingdome.

2 The first *Phyllyria* of *Clusius,* may fitly be refer'd to the rest of the same tribe and name described formerly in the 59. chapter of the the third booke. It growes somewhat taller than the Scarlet Oke, and hath branches of the thicknesse of ones thumbe or somewhat more, and those couered with a greene barke marked with whitish spots; the leaues somewhat resemble those of the Scarlet Oke, but greater, greener, thicker, somewhat prickley about the edges, of an astringent taste, but not vngratefull. The floure thereof *Clusius* did not see, the fruit is a little blacke berry, hanging downe out from the bosome of the leaues, and conteining a kernell or stone therein. It growes wilde in many wilde places of Portugale, where they call it Azebo.

The temperature and vertues are refer'd to those set downe in the formerly mentioned chapter.

CHAP. 9. *Of Mocke-Willow.*

Speiræa Theophrasti, Clus.
Mocke-Willow.

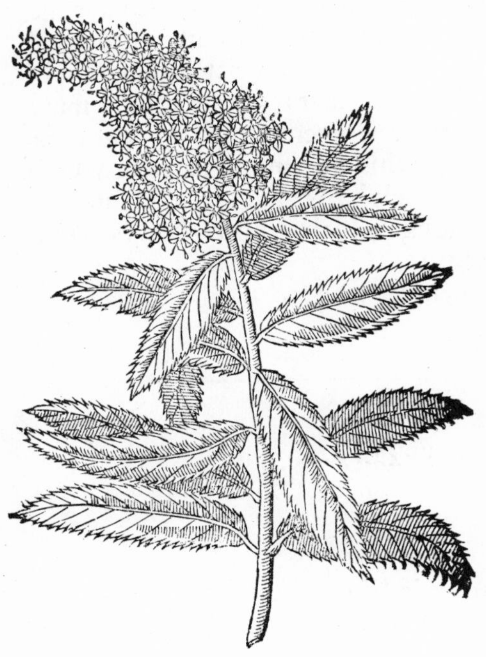

¶ *The Description.*

THis Willow leaued shrub, which *Clusius* coniectures may be refer'd to the *Speiræa* mentioned by *Theophrastus, lib. 1. cap. 23. hist. plant.* I haue named in English, Mocke-Willow, how fitly I know not; but if any will impose a fitter name I shall be well pleased therewith; but to the thing it selfe. It is a shrub, (saith *Clusius*) some two cubits high, hauing slender branches or twigs couered ouer with a reddish barke, whereon grow many leaues without order, long, narrow, like those of the Willow, snipt about the edges, of a light green aboue, and of a blewish greene vnderneath, of a drying taste conjoyned with some bitternes. The tops of the branches for some fingers length carry thicke spikes of small floures clustering together, and consisting of fiue leaues apiece, out of whose middle come forth many little threds of a whitish red or flesh colour, together with the floure, hauing no peculiar

 smell,

smell, but such as is in the floure of the Oliue tree; these floures fading there succeed small fiue cornered heads, which comming to full maturitie containe a small and yellowish dusty seed: it floures in Iuly, and ripens the seed in the end of August. *Clusius* had this plant from *Fredericke Sebizius* Physition to the Duke of Briga, and that from Briga in Silesia, and he (as I said) refers it to the Σπειραία of *Theophrastus*, which he reckons amongst the shrubs that carry spike fashioned floures.

This is not vsed in medicine, nor the Temperature and faculties thereof as yet knowne.

Chap. 10. *Of the Strawberry-Bay.*

Adrachne Theophrasti.
The Strawberry-Bay.

¶ *The Description.*

THe figure and history of this were sent by *Honorius Bellus* out of Candy to *Clusius*, from whom I haue it. It is that which *Theophrastus* calls *Adrachne* or (as most of the printed bookes haue it) *Andrachne*, but the former seemes the righter, and is the better liked by *Pliny, lib.*1 ; *cap.*22. At this day in Candy where it plentifully growes, it is called *Adracla*. It is rather a shrub than a tree, delighting in rockie and mountanous places, and keeping greene VVinter and Sommer, hauing leaues so like those of Bayes, that they are distinguishable only by the smell, which these are destitute of. The barke of the bole and all the branches is so smooth, red and shining, that they shew like branches of Corall, this barke crackes or breakes off in Sommer, and pills off in thinne fleakes, at which time it is neither red nor shining but in a meane betweene yellow and ash-colour. It hath floures twice in a yeere like as the *Arbutus*, or Strawberry tree, and that so like it, that you can scarse know the one from the other; yet this differs from it in that it growes onely in the mountaines, hath not the leaues jagged, neither a rough barke; the wood hereof is very hard, and so brittle that it will not bend, and they vse it to burne and to make whorles for their womens spindles. *Theophrastus* reckons vp this tree amongst those which die not when their barkes are taken off, and are alwaies greene, and retaine their leaues at their tops all winter long: which to be so *Honorius Bellus* obserued. *Bellonius* also obserued this tree in many places of Syria.

The fruit in Temperature, as in shape, is like that of the Stawberry-tree.

Chap. 11. *Of the Cherry-Bay.*

¶ *The Description.*

THe Cherry-bay is one of the euergreen trees: it rises vp to an indifferent height, and is diuided into sundry branches, couered ouer with a swart green barke: that of the yonger shoots is wholly greene.

green, the leaues alternately ingirt the branches, & they are long, ſmooth, thick, green, and ſhining, ſnipt alſo lightly about the edges: when the tree is growne to ſome height, at the tops of the branches amongſt the leaues of the former yeares growth, vpon a ſprig of ſome fingers length, it puts forth a great many little white floures, conſiſting of fiue leaues a piece, with many little chiues in them: theſe floures quickly fall away, and the fruit that ſucceeds them is a berry of an ovall figure, of the bigneſſe of a large Cherry or Damſon, and of the ſame colour, and of a ſweet and pleaſant

Laurocerafi flos.
The Cherry-bay in floure.

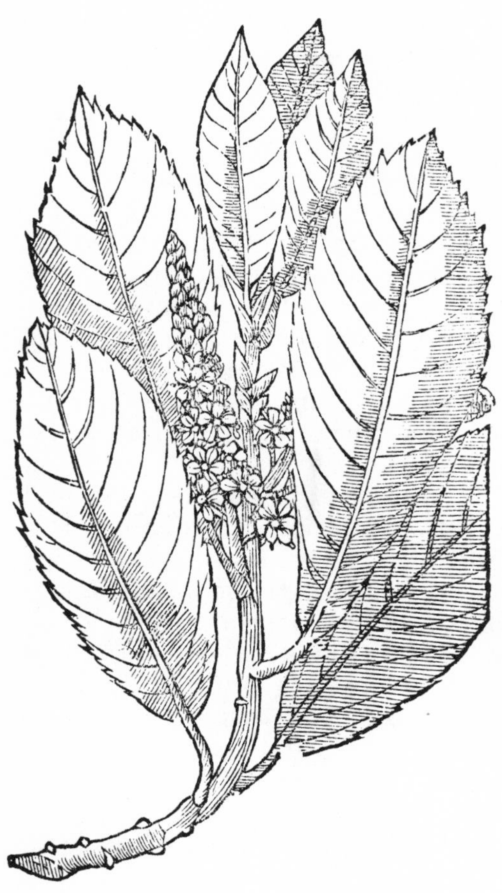

Laurocerafi fructus.
The Cherry-bay with the fruit.

taſte, with a ſtone in it like to a Cherry ſtone. This floures in May, and ripens the fruit in Auguſt or September: it was firſt ſent to *Cluſius* from Conſtantinople, and that by the name of *Trabiſon curmaſi.* 1. *Trapezuntina dactylus*, the Date of Trapeſon; but it hath no affinitie with the Date. *Dalechampius* refers it to the ſecond *Lotus* mentioned by *Theophraſtus, hiſt plant. lib.* 4. *cap.* 4. but therewith it doth not agree. *Cluſius* and moſt ſince, cal it fitly *Laurocerafus*, or *Ceraſus folio Laurino*. It is now got into many of our choiſe Engliſh gardens, where it is well reſpected for the beauty of the leaues and their laſting or continuall greeneneſſe.

The fruit hereof is good to be eaten, but what phyſicall vertues the tree or leaues thereof haue, it is not yet knowne.

Chap. 12. *Of the Euer-greene Thorne.*

This plant which *Lobel* and ſome other late writers haue called by the name of *Pyracantha*, is the *Oxyacantha* mentioned by *Theophraſtus, lib.* 1. *cap.* 15. *lib.* 3. *cap.* 4. *hiſt. plant.* among the euer green trees, and I thinke rather this than our white Thorn to be the *Oxyacantha* of *Dioſcorides, lib.* 1. *c.* 123. and certainely it was no other than this Thorne which *Virgil* makes mention of by the name of *Acanthus, lib.* 2. *Georg.* in theſe words, *Et baccas ſemper frondentis Acanthi.* That is, And the berries of the Eire-greene Thorn.

Oxyacantha Theophrasti.
The Euer-greene Thorne.

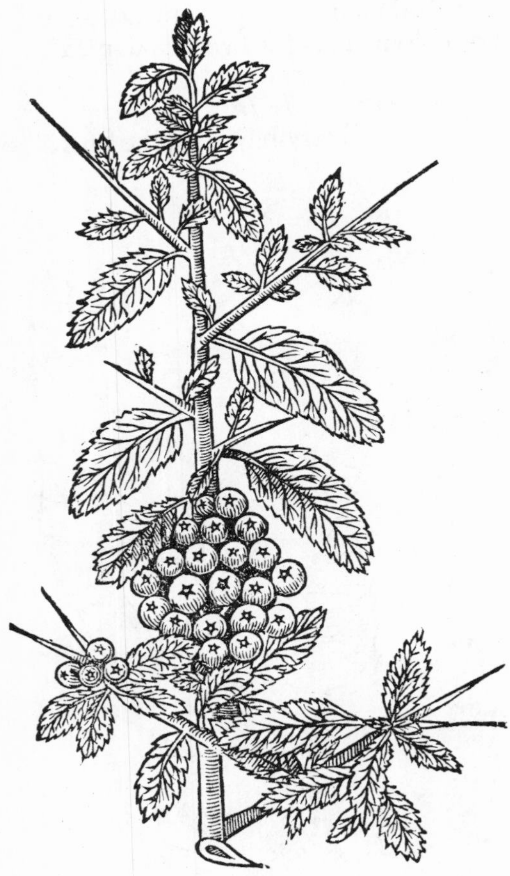

¶ *The Description.*

THis growes vp like a bush, vnlesse you keepe it with pruning, and then it will in time grow to the height of a smal tree, as the Hawthorne, whereto it is of affinitie, for the wood is white and hard, like it, and couered ouer with the like barke; but the leaues are somwhat like those of the Damson tree, longish, sharp pointed, and snipt about the edges: & they grow alongst the branches, without any order, yet sometimes they keep this maner of growing: at each knot, where commonly there is a sharpe prickle, growes out one of the larger leaues, which may be some inch and halfe long, and some three quarters of an inch broad: then vpon the prickle, and at the comming out therof are three or foure, more or lesse, much smaller leaues: now these leaues are of a faire and shining green aboue, but paler vnderneath, and they keep on al the yeare: At the ends, and oft times in the middles of the branches come forth clusters or vmbels of little whitish blush coloured floures, consisting of fiue leaues apiece, with some little chiues in their middles: then follow clusters of berries, in shape, taste, and bignesse like those of the Hawthorne, and of the same, but much more orient and pleasing colour, and containing in them the like seed: now these berries hang long vpon the tree, & make a gallant shew amongst the greene leaues, and chiefely then, when as the Autumne blasts haue depriued other trees of their wonted verdure. This floures in May and Iune, and ripens the fruit in September and October: it growes wilde in sundry places of Italy, and Prouince in France, but is kept in gardens with vs, where it is held in good esteeme for his euer greenesse and pliablenesse to any worke or forme you desire to impose vpon him.

The fruit haue the same faculties that are formerly attributed to Hawes, in the foregoing booke, *pag.* 1328. and therefore I will not here repeat them.

CHAP. 13. *Of the Ægyptian Nap, or great Iuiubes tree.*

¶ *The Description.*

THis tree, which for his leaues and manner of growing I thinke may fitly be referred to the Iuiubes tree, is of two sorts; that is, the one prickly, and the other not prickly, in other respects they are both alike, so that one figure and historie may serue for them both; which I will giue you out of *Clusius*, who receiued this figure together with a description thereof from *Honorius Bellus*, and also added therto that which *Prosper Alpin.* hath written of it in his 5. chap. *de Plant. Ægypt.* It grows to the height of an indifferent Peare-tree, and the bodie and branches thereof are couered with a whitish ash coloured barke: the leaues are like those of the Iuiubes tree, two inches long, and one broad, with three nerues running alongst them; of a deepe shining greene aboue, and more whitish vnderneath: and they grow alternately vpon the branches: and at their comming forth grow tufts of little white floures hanging vpon single long foot-stalks: after these followes the fruit like vnto a small Apple, of the bignesse for the most part of a large Cherry, and sometimes as big as a VValnut, of a sweet taste, containing therein a kernell or stone like that of an Oliue. It beares fruit twise a yeare, for it hath ripe fruit both in the Spring and fall; yet the vernall fruit seldom comes to good, by

Oenoplia non ſpinoſa.
The great Iuiubes tree.

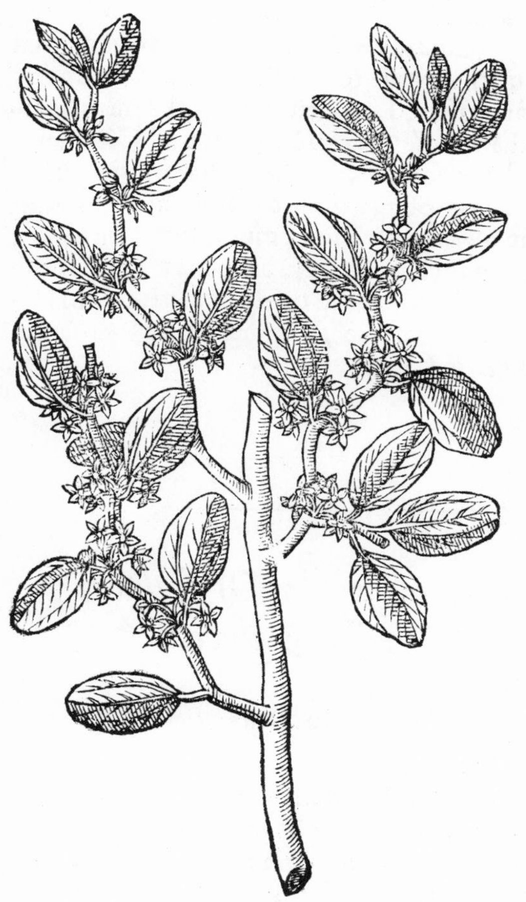

by reaſon of the too much moiſture of the ſeaſon, which cauſes it to become worme-eaten. The Thorny kinde is deſcribed by *Alpinus*, who rightly iudges it the *Connarus* of *Athenæus*, but the figure he giues is not very accurate. That which wants prickles growes (as well as the prickly one) in Ægypt and Syria, as alſo in the city Rhetimo in Candy, whither it was brought out of Syria.

The hiſtorie of both theſe trees is in *Serapio* by the name of *Sadar*: but he, according to his cuſtome confounds it with the *Lotus* of *Dioſcorides*, from which it very much differs. *Bellonius* in his ſecond booke, and 79. chap. of his Obſeruations, reckons vp *Napeca* amongſt the trees that are alwaies greene: which is true, in thoſe that grow in Egypt and Syria; but falſe in ſuch as grow in Candy. That tree in Ægypt and Syria is called *Nep*, or *Nap*. *Alpinus* calls it *Paliurus Athenæi*, or *Nabca Ægyptiorum*, thinking it (as I formerly ſaid) the *Connarus* mentioned in the 14. booke of *Athenæus* his Deipnoſophiſts.

¶ *The Vertues out of Alpinus.*

A The fruit is of a cold and dry facultie, and the vnripe ones are frequently vſed to ſtrengthen the ſtomacke, and ſtop lasks: the iuice of them being for this purpoſe either taken by the mouth, or injected by clyſter: of the ſame fruit dried and macerated in water is made an infuſion profitable againſt the relaxation and vlceration of the guts.

B The decoction or infuſion of the ripe dried fruit, is of a very frequent vſe againſt all peſtilent feuers: for they affirme that this fruit hath a wonderfull efficacie againſt venenate qualities, and putrifaction, and that it powerfully ſtrengthens the heart.

C Alſo the iuice of the perfectly ripe fruit is very good to purge choler forth of the ſtomacke and firſt veines: and they willingly vſe an infuſion made of them in all putride feuers to mitigate their heate or burning.

CHAP. 14. *Of the Perſian Plum.*

¶ *The Deſcription.*

1 THis tree is thought by *Cluſius* (to whom I am beholden for the hiſtorie and figure) to be the *Perſea arbor* mentioned by *Pliny* and *Plutarch*, but he ſomewhat doubts whither it be that which is mentioned by *Theophraſtus*. *Dioſcorides* alſo, *Galen* and *Strabo* make mention of the *Perſea arbor*, and they all make it a tree alwaies greene, hauing a longiſh fruit ſhut vp in the ſhell and coat of an Almond: with which how this agrees you may ſee by this deſcription of *Cluſius*.

This tree (ſaith he) is like to a Peare tree, ſpreading it ſelfe far abroad, and being alwaies green, hauing branches of a yellowiſh green colour. The leaues are like thoſe of the broadeſt leaued Baytree, greene aboue, and of a grayiſh colour vnderneath, firm, hauing ſome nerues running obliquely, of a good taſte and ſmell, yet biting the tongue with a little aſtriction. The floures are like thoſe of the Bay, growing many thicke together, and conſiſt of ſix ſmall whitiſh yellow leaues. The fruit at the firſt is like a Plum, and afterwards it becomes Peare faſhioned, of a blacke colour, and pleaſant taſte: it hath in it a heart faſhioned kernell, in taſte not vnlike a Cheſnut, or ſweet Almond. I found it flouring in the Spring, and I vnderſtood the fruit was ripe in Autumne, by the relation of Sig.

Persea arbor.
The Persian Plum.

Iohn Placa, Physition and Professor of Valentia, who shewed me the tree growing in the garden of a Monasterie a mile from Valentia, brought thither, as they say, out of America, and he said they called it *Mamay*: but the Spaniards who haue described America giue this name to another tree. But diuers yeares after, I vnderstood by the most learned *Simon de Tovar*, a Physition of Ciuil, who hath the same tree in his garden, with other exoticke plants, that it is not called *Mamay*, but *Aguacate*. Thus much out of *Clusius*; where such as are desirous, may finde more largely handled the question, whither this be the *Persea* of the Antients or no? *Rariorum plan. Hist. l.* 1. *c.* 2.

Cotonaster Gesneri.
Gesners wilde Quince

CHAP. 15. *Of Gesners wilde Quince.*

¶ *The Description.*

THe shrub which I here figure out of *Clusius*, is thought both by him and others, to be the *Cotonastrum* or *Cidonago*, mentioned by *Gesner* in his Epistles, *lib.* 3. *pag.* 88. It hath branches some cubit long, tough, and bare of leaues in their lower parts, couered with a blacke barke: and towards the tops of the branches grow leaues somewhat like those of Quinces: of a darke greene aboue, and whitish vnderneath, snipt about the edges: at the tops of the branches grow vsually many floures, consisting of fiue purplish coloured leaues a piece, with some threddes in their middles: these decaying, vnder them grow vp red dry berries without any pulp or iuice, each of them containing foure triangular seeds. *Clusius* found this flouring in Iune vpon the tops of the Austrian Alpes, and he questions whether it were not this which *Bellonius* found in the mountains of Candy, and called *Agriomælea*, *lib.* 1. *cap.* 17. This is not vsed in Physicke, nor the faculties thereof knowne.

CHAP.

Chap. 16. *Of Tamarindes.*

Tamarindus.
The Tamarinde.

Tamarindi filiqua.
The cod of the Tamarinde.

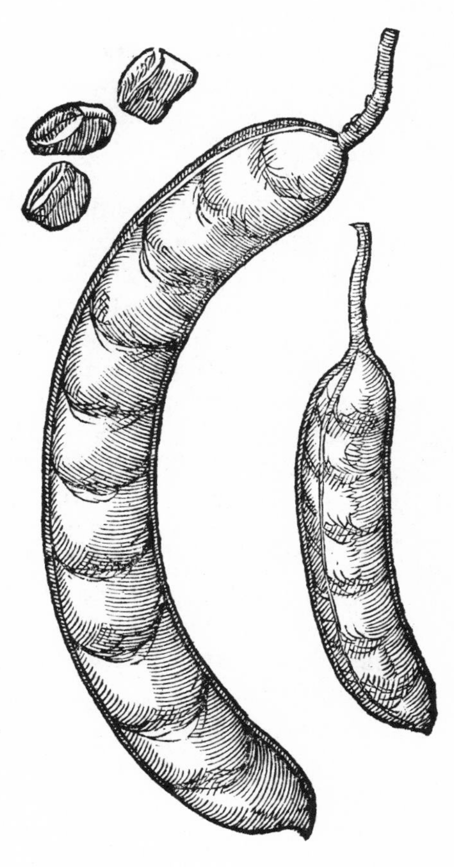

¶ *The Description.*

Tamarinds, which at this day are a medicine frequently vsed, and vulgarly knowne in shops, were not knowne to the antient Greekes, but to some of the later, as *Actuarius*, and that by the name of *Oxyphœnicæ*, that is, soure Dates, drawne as it may seeme from the Arabicke appellation, *Tamarindi*, that is, Indian Date: but this name is vnproper, neither tree nor fruit being of any affinitie with the Date, vnlesse the Arabicke *Tamar* be a word vsed in composition for fruits of many kindes, as the Greeke Μῆλον, the Latine *Malum*, and Apple with vs in English; for we call the Cone of the Pine, and excrescence of the Oke leafe, by the name of Pine Apple, and Oke Apple. But howsoeuer it be, it is no matter for the name, whether it be proper or no, if so be that it serue to distinguish the thing from others, and we know what is denoted by it. In Malauar they call it *Puti*: in Guzarat, *Ambili*, by which name it is knowne in most parts of the East Indies. This tree is thus described by *Prosper Alpinus, de Plant. Ægypti, cap.* 10. The Tamarind (saith he) is a tree of the bignesse of a Plum tree, with many boughes and leaues like those of the Myrtle, many standing vpon one rib [one against another, with a single one at the end:] it carrieth white floures very like those of the Orange tree: out of whose middle comes forth foure white and very slender threds: after these come thicke and large cods, at first greene, but when they are ripe of an ash colour; and within these are contained thicke, hard, brownish, cornered seeds, and a blacke acide pulpe. These trees grow in some few gardens of Egypt, whither they haue bin brought out of Arabia and Ethiopia. This plant hath this strange qualitie that the leaues alwaies follow the Sun, and when it sets they all contract themselues, and open out themselues againe at the rising thereof; and there is obserued to be such force in this motion, that they closely shut vp and hold their cods (if any be on the tree) and then at the rising of the Sun they forgoe them againe. But I haue obserued this folding vp of the leaues to be common to diuers other Egyptian plants, as *Acatia, Abrus, Absus*, and *Sesban*. Thus much out of *Alpinus*.

The

The figure I here giue in the first place, out of *Lobel*, is of a plant some six moneths old, arisen of a seed . and such by sowing of seeds I haue seene growing in the garden of my deceased friend Mr. *Tuggy*, but they still died at the first approch of Winter. The other figure expresses the cods, and some of the seeds apart, taken forth of the cods : now the cods are neuer brought whole to vs, but the vtter rindes are taken off, and the strings or nerues that runne alongst the cods : the pulpe and seeds in it are close thrust together, and so are brought to vs in pots and such like vessels.

¶ *The Temperature and Vertues.*

A The fruit or pulpe of Tamarindes is cold and dry in the third degree : it is of good vse in cholericke diseases, as burning Feuers, Tertians, and the like : it is a lenitiue and very gently purging medicine and therefore vsed to be put into medicines seruing to that purpose.

B They vse (saith *Alpinus*) the leaues of Tamarindes to kill wormes in young children ; and also their infusion or decoction to loosen the belly : the leaues are acide, and not vnpleasant vnto the taste.

C The Arabians preserue the small and yet greene cods of this tree, as also the ripe ones, either with sugar, or the honey boiled out of the fruit of the Carob tree : they also mix the pulpe with sugar, which trauellers carry with them in their iournies through the desart places of Africk, wherewith they being dry or ouerheated, may quench their thirst, coole and refresh themselues, and also euacuate many hot humors by stoole.

D In pestilent and all other burning putrid feuers they drinke the water with sugar, wherein a good quantitie of Tamarinds haue been infused ; for it is a drinke very pleasant to such as are thirsty by reason of too much heate, for it powerfully cooles and quenches thirst.

E They are also vsed in all putrid feuers caused by cholericke and adust humors, and also against the hot distempers and inflammations of the liuer and reines, and withall against the Gonorrhæa.

F Some also commend them against obstructions, the dropsie, iaundice, and the hot distempers of the Spleene : they conduce also to the cure of the itch, scab, leprosie, tetters, and all such vlcerations of the skin which proceed of adust humors.

G They are not good for such as haue cold stomacks, vnlesse their coldnesse be corrected by putting to them Mace, Anise seeds, Squinanth, or such like.

Chap. 17.

Of the Mamoera, the Male and Female.

¶ *The Description.*

THe historie of these two trees, together with the figures I here giue you, are in the *Curæ Posteriores* of *Clusius*, from whence I will take as much as concernes their history, and briefely here giue it you.

That of the Poet (saith he) is most true, *Non omnis fert omnia tellus* : for I thinke there is no prouince to be found, which produces not some peculiar plant not growing in other regions, as they can testifie who haue trauelled ouer forrein countries, especially if they haue applied themselues to the obseruation of plants. Amongst such I thinke I may reckon that honest and courteous man *Iohn Van Vsele*, who returning out of that part of America called Brasile, shewed me in the yeare 1607. a booke, wherein he in liuely colours had exprest some plants and liuing creatures : for as he told me, when he purposed to trauell he learned to paint, that so he might expresse in colours, for his memorie and delight after he was returned home, such singularities as he should obserue abroad. Now amongst those which hee in that booke had expressed, I obserued two very singular, and of a strange nature, whose figures without any difficultie he bestowed vpon me, as also the following historie.

These two trees, whose figures you see here exprest, are of the same kinde, and differ only in sex ; for the one of them, to wit the male, is barren, and only carries floures, without any fruit ; but the female onely fruit, and that without floure : yet they say they are so louing, and of such a nature, that if they be set far asunder, and the female haue not a male neere her, shee becomes barren, and beares no fruit : of which nature they also say the Palme is.

Now the bole or trunke of that tree which beares the fruit is about two foot thicke, and it groweth some nine foot high before it begin to beare fruit ; but when it hath acquired a iust magnitude, then shall you see the vpper part of the tree laden with fruit, and that it will be as it were thicke

girt

girt about therewith for ſome nine foot high more: the fruit is round and globe-faſhioned, of the ſhape and magnitude of a ſmall gourd, hauing when it is ripe a yellowiſh pulpe, which the inhabitants vſe to eate to looſen their bellies: this fruit contains many kernels of the bignes of a ſmal peaſe blacke and ſhining, of no vſe that he could learne, but which were caſt away as vnneceſſary: the leaues come forth amongſt the fruit, growing vpon long foot-ſtalkes, and they in ſhape much reſemble the Plane tree or great Maple.

Mamoera mas.
The male Dug tree.

Mamoera fœmina.
The female Dug tree.

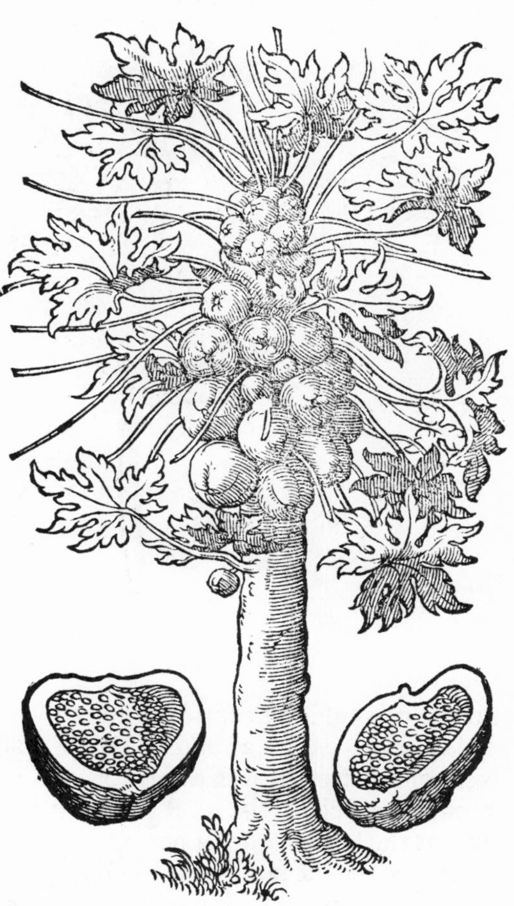

What name the Braſilians giue it he could not tell, but of the Portugals that dwelt there it was called *Mamoera*, and the fruit *Mamaon*, of the ſimilitude I thinke they haue with dugs, which by the Spaniards are called *Mamas* and *Tetas*.

There is no difference in the forme of the trunke or leaues of the male and female, but the male only carries floures hanging downe, cluſtering together vpon long ſtalks like to the floures of Elder, but of a whitiſh yellow colour, and theſe vnprofitable, as they affirme.

Both theſe trees grow in that part of America wherein is ſcituate the famous Bay called by the Portugals, *Baya de todos los ſanctos*, lying about thirteene degrees diſtant from the Equator towards the Antarticke pole.

Chap. 18.

Of the Cloue-Berry Tree.

¶ *The Deſcription.*

I Muſt alſo abſtract the hiſtorie of this out of the Works of the learned and diligent *Cluſius*, who ſets it forth in his Exoticks, *lib.* 1. *cap.* 17. in the next chapter after Cloues.

I put (ſaith he) the deſcription of this fruit next after the hiſtorie of Cloues, both for the affinitie

Amomum quorundam, fortè Garyophyllon Plinij.
The Cloue-berry tree.

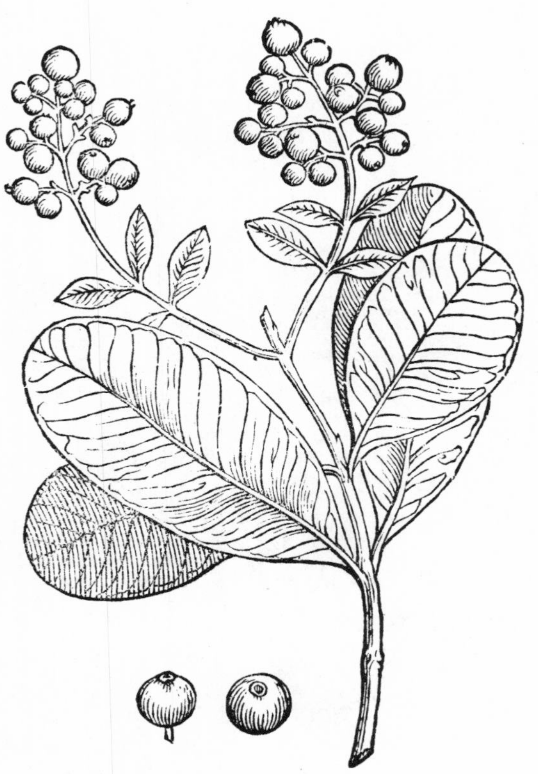

affinitie of smell it hath with Cloues, as also for another cause, which I will shew hereafter. *Iames Garret* in the yeare 1601 sent me from London this round fruit, commonly bigger than Pepper cornes, yet some lesse, wrinkled, of a brownish colour, sufficiently fragile; which opened, I found contained a seed round, black, which might be diuided into two parts, of no lesse aromaticke taste and smell than the fruit it selfe, and in some sort resembling that of Cloues: it growes in bunches or clusters, as I coniectured by many berries which yet kept their stalks, & two or three which stucke to one little stalke: to these were added leaues of one form, but of much different bignes, for some of them were seuen inches long, and three broad; some onely fiue inches long, and two and a half broad; others did not exceed 3 inches in length, and these were not two inches broad; and some a so were much lesse and narrower than these, especially those that were found mixed with the berries, differing according to the place in the boughes or branches which they possest. I obserued none among them which had snipt leaues, but smooth, with many small veines running obliquely from the middle rib to the sides, with their points now narrower, otherwhiles broader, and roundish: they were of a brownish ash colour, of a sufficient acride taste: the branches which were added to the rest were slender, quadrangular, couered with a barke of an ash colour, and those were they of a yeares growth; for those that were of an after growth were brownish, and they had yet remaining the prints where the leaues had growne, which for the most part were one against another, and these also were of an acride taste, as well as the leaues, and of no vngratefull smell.

I receiued the same fruit some yeares before, but without the stalks, and with this question propounded by him which sent it, *An Amomum?* And certainly the faculties of this fruit are not very much vnlike those which *Dioscorides* attributes to his *Amomum*; for it hath an heating astrictiue and drying facultie, and I thinke it may performe those things whereto *Dioscorides, Lib.* 1. *Cap.* 14. saith his is good; yet this wanteth some notes which he giues vnto his, as the leaues of Bryonie, &c.

But I more diligently considering this Exoticke fruit, finde some prime notes which do much moue me (for I will ingenuously professe what I thinke) to iudge it the *Garyophyllon* of *Pliny*; for he, *Hist. Nat. lib.* 12. *cap.* 7. after he hath treated of Pepper addes these words: [There is besides in the Indies a thing like to the Pepper corne, which is called *Garyophyllon*, but more great and fragil: they affirme it growes in an Indian groue; it is brought ouer for the smels sake.] Though this description be briefe and succinct, neither containes any faculties of the fruit it selfe, yet it hath manifest notes, which, compared with those which the fruit I here giue you possesse, you shal find them very like; as comparing them to Pepper cornes, yet bigger and more fragile, as for the most part these berries are: their smell is also very pleasing, and comming very neere to that of Cloues, and for the smells sake only they were brought ouer in *Plinies* time. I found, this fruit being chewed made the breath to smell well: and it is credible, that it would be good for many other purposes, if triall were made.

CHAP.

Chap. 19. Of Guaiacum, or Indian Pock-wood.

Guaiaci arboris ramulus.
A branch of the Guaiacum tree.

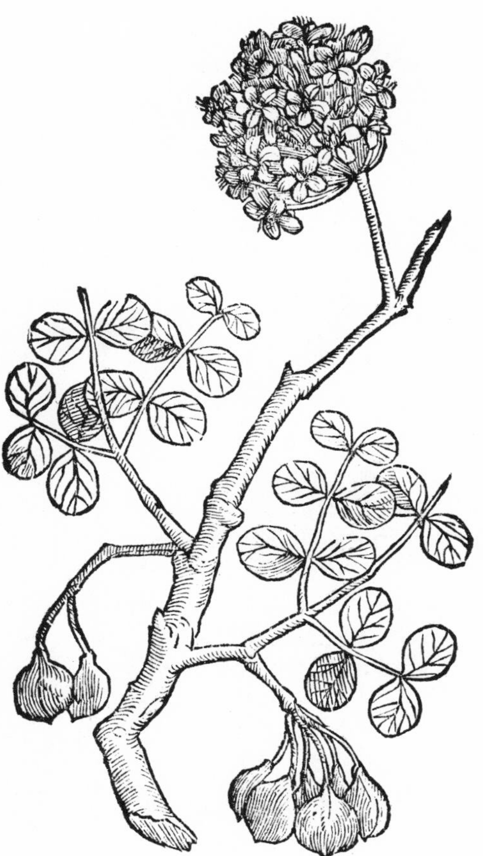

¶ *The Description.*

Gvaiacum, which some call *Lignum Sanctum*: others, *Lignum vitæ*, is a well kown wood, though of a tree vnknown, or at least not certainly knowne; for this figure which I here giue you out of *Clusius*, was gotten, and the historie framed as you shall heare by his own words, taken out of his *scholia* vpon the 21 Chapter of *Monardus*. About the beginning (saith he) of the yeare 1601. I receiued from *Peter Garret* a branch of a foot long, which he writ was giuen him by a certaine Surgeon lately returned from America, for a branch of the tree Guaiacum: which if it be a branch of the true *Guaiacum*, then hath *Nicolas Monardus* sleightly enough set downe the historie of this tree. I thus described this branch which was sent me.

This branch was a foot long, very writhen, and distinguished with many knots, scarse at the lower end equalling the thicknesse of a writing pen or goose quil, hauing an hard and yellowish wood, and a wrinkled barke of an ash colour: at the vpper end it was diuided into slender branches, whereof some yet retained their leaues, and other some the floures and the rudiment of the fruit: the leaues, or more truly the wings or foot-stalkes of the leaues grew vpon slender branches one against another, each winged leafe hauing foure or sixe little leaues, alwaies growing by couples one against other, as in the Masticke tree; and these were thickish, round, and distinguished with many veines, which by reason of their drinesse (as I obserued) would easily fall off, leauing the footstalks naked, and onely retaining the markes whereas the leaues had beene. In the knots of the vpper branches there grew as it were swellings, out of which together grew six, eight, ten, or more slender foot-stalkes, some inch long, each carrying a floure not great, consisting of six little leaues (but whether white, yellow, or blew, I could not by reason of the drinesse iudge:) out of the middle of the floure grew many little threds, and in some the rudiment of the fruit began to appeare, hauing two cels, almost shaped like the seed-vessell of the common Shepheards purse.

Thus much *Clusius*, who afterwards receiued the fruit from two or three, but the most perfect from the learned Apothecarie *Iohn Pona* of Verona: they are commonly parted into two parts or cels, yet he obserued one with three: he found longish stones in them almost like those of *Euonymus*, and they consisted of a very hard and hairy substance like to that of the Date stones, containing a smooth kernel of a yellowish colour.

Now will I giue you the descriptions of *Monardus*: then, what I haue obserued my selfe of this wood, which I must confesse is very little, yet which may giue some light to the ignorant. Of this wood (saith *Monardus*) many haue written many waies, saying that it is either Ebonie, or a kinde of Box, or calling it by some other names. But as it is a new kinde of tree, not found in these regions, or any other of the whole world described by the Antients, but only those of late discouered; so this shall be a new tree to vs: howeuer it be, it is a large tree of the bignes of the Ilex, ful of branches, hauing a great matrix or blackish pith, the substance of the wood being harder than Ebonie: the barke is thicke, gummie or fat, and when the wood is dry falleth easily off: the leaues are smal and hard: the floure yellow: the which is followed by a round sollid fruit, containing in it seeds like those of the Medlar.

It growes plentifully in the Isles of *Sancto Domingo*.

Another

Another kinde of this was afterwards found in the Island of S. Iohn de Puerto rico, neere to the former: it is also like the last described, but altogether lesse, and almost without matrix or pith, smelling stronger, and being bitterer than the former; which being left, this is now in vse, and of the wondrous effects it is called *Lignum sanctum*; neither without desert, being (experience giuing testimonie) it excells the other: yet both their faculties are admirable in curing the French disease, and therefore the water or decoction of both of them are drunke, either mixed together, or seuerally, both for the cure of the forementioned disease, as also against diuers other affects. Thus much for *Monardus* his description.

The wood which is now in vse with vs is of a large tree, whose wood is very heauy, sollid, and fit to turne into bowles or the like, and all that I haue yet seene hath been wholly without matrix or pith, and commonly it is of a darke brownish colour, somewhat inclining to yellow, hauing a ring of white ingirting it next to the barke; I haue obserued a tree whose diametre hath been two foot and a quarter, to haue had as little or lesse of this white wood as one whose diameter was thirteene inches; and this which was thirteene inches had only a white circle about it of one inch in bredth: I thinke the yonger the tree is, the bigger the white circle is: the best wood is dense, heauy, brownish, leauing a quicke and biting taste in the decoction, as also his smell and colour. The barke of this wood is also dense and heauy, of a hard substance and yellowish colour within, but rough and greenish, or else grayish without, and of somewhat a bitterish taste. Thus much for the description of the wood and his barke. Now let me say somewhat briefely of the temperature and qualities.

The Temperature and Vertues.

A It is iudged to be hot and dry in the second degree: it hath a drying, attenuating, dissoluing, and clensing facultie, as also to moue sweat, and resist contagion and putrefaction.

B The decoction of the barke or wood of Guajacum, made either alone or with other ingredients, as shall be thought most fit for the temper and age of the Patient, is of singular vse in the cure of the French Poxes, and it is the most antient and powerfull antidote that is yet known against that disease. I forbeare to specifie any particular medicine made thereof, because they are wel enough knowne to all to whom this knowledge belongs, and they are aboundantly set downe by all those that haue treated of that disease.

C It also conduceth to the cure of the dropsie, Asthma, Epilepsie, the diseases of the bladder and reines, paines of the ioints, flatulences, crudities, and lastly all chronicall diseases proceeding from cold and moist causes: for it oftentimes workes singular effects whereas other medicines little preuaile.

D It doth also open the obstructions of the liuer and spleene, warmes and comforts the stomacke and all the intrals, and helps to free them of any grosse viscous matter which may be apt to breed diseases in them.

Chap. 20.

Of the Guayaua, or Orange-Bay.

¶ The Description.

Simon de Touar sent *Clusius* a branch of the tree which the Spaniards call *Guayauas*, from which he drew this figure, and thus describes it. This branch (saith *Clusius*) whose vpper part together with the fruit I caused to be drawne, was some foot long, foure square, alternately set with leaues growing by couples, being foure inches long, and one and a halfe or two broad, of the forme of Bay leaues, very firme, hauing a swelling rib running alongst the lower side, with veins running obliquely from thence to the sides, of an ash or grayish colour beneath, but smooth aboue, with the veines lesse appearing; which broken, though old, yet retained the smell of Bay leaues, and also after some sort the taste: the fruit was smooth, yet shriueled, because peraduenture it was vnripe, of the bignesse of a small apple, longish, blackish on the out side like a ripe plum, but within full of a reddish pulpe, of an acide taste; and in the middle were many whitish seeds of the bignesse of Millet, or those that are in Figs.

Nicolas Monardus (as he is turned into Latine by *Clusius*) thus giues vs the historie of *Guayauas*, in his sixty fourth Chapter. It is a tree, saith he, of an indifferent bignesse, and hath spreading branches, the leafe of the Bay, and a white floure, like that of the Orange, yet somewhat bigger,

Guayavæ arboris ramus.
The Orange-Bay.

and well smelling; it easily growes, whereso-euer it be sowne, and so spreds and creepes that it is accounted as a weed, for it spoiles the grasse of many pastures, with the too much spreading as brambles do; the fruit is like to our apples, of the bignesse of those the Spaniards call *Camuesas*, green at the first, and of a golden colour when they be ripe, with their inner pulpe white, and sometimes red; diuided: it hath foure cells, wherein lie the seeds, like those of the Medlers, very hard, of a brownish colour, wholly stony, without kernell and taste.

A The fruit is vsually eaten, the rinde being first taken off; it is pleasing to the palate, wholesome and easie of concoction; being greene it is good in fluxes of the belly, for it powerfully bindes; and ouer, or throughly ripe it looseth the belly; but betweene both, that it is neither too greene, nor ouer-ripe, if rosted, it is good both for sound and sicke; for so handled it is wholesommer, and of a more pleasing taste; that also is the better which is gathered from domesticke and husbanded trees. The Indians profitably bathe their swolne legges in the decoction of the leaues; and by the same they free the spleene from obstruction. The fruit seemes to be cold, wherefore they giue it rosted to such as are in feuers. It growes commonly in all the VVest Indies. Thus much *Monardus*.

CHA. 21. *Of the Corall tree.*

¶ *The Description.*

THe same last mentioned *Simon de Touar* a learned and prime Physition of Ciuill sent *Clusius* three or foure branches of this tree, from whence he framed this history and figure. He writ (saith *Clus.*) that this tree grew in his garden, sprung vp of seeds sent from America, which had the name of Corall imposed on them, by reason the floures were like Corall, but he did not set downe there shape; writing onely this in his letter: That he had two little shrubs, which had borne floures, and that the greater of them bore also cods full of large beanes, but in the extreme Winter, which they had the yeere before, he lost not onely that tree, and others sprung vp of Indian seed, but also many other plants. Now seeing that this tree carries coddes, I coniecture the floures were in forme not vnlike to those of Pease, or of the tree called *Arbor Iudæ*, but of another colour, to wit, red like Corall, especially seeing that in the catalogue of his garden which hee sent me the yeere before, he had writ thus [*Arbor Indica dicta Coral, ob eius florem similem Corallo, &c.* that is, An Indian tree called Corrall, by reason of the floure like to Corrall, whose leaues are very like those of the *Arbor Iudæ*, but this hath thornes, which that wants.] And verily the branches which he sent (for he writ he sent the branches with the leaues, but the tree brought out some twice or thrice as bigge) had leaues not much vnlike those of *Arbor Iudæ*, but fastened to a shorter footstalke and growing one against another, with a single one at the end of the branch, which was here and there set with sharpe and crooked prickles; but whether these branches are onely the stalkes of the leaues, or perfect branches, I doubt, because all that hee sent had three leaues apiece; I could easily persuade my selfe, that they were onely leaues, seeing the vpper part ended in one leafe; and the lower end of one among the rest, yet shewed the place where it seemed it grew to the bough. But I affirme nothing, seeing there was none whereof I could inquire, by

Coral arboris ramus.
A branch of the Corall tree.

reason of his death who sent them me, which hapned shortly after; yet I haue made the forme of the leaues with the manner as I coniectured they grow, to be delineated in the figure which I here giue you. Whether *Matthiolus* in his last edition of his Commentaries vpon *Dioscorides* would haue exprest this, by the Icon of his first *Acacia*, which is prickly, and hath leaues resembling those of *Arbor Iudæ*, I know not; but if he would haue expressed this tree, the painter did not well play his part.

After that *Clusius* had set forth thus much of this tree in his *Hist rariorum plant.* the learned Dr. *Castaneda* a Physition also of Ciuill certified me, saith he, that the floures of this tree grow thicke together at the tops of the branches, ten, twelue, or more hanging vpon short foot-stalkes, growing out of the same place: whose figure he also sent, but so rudely drawne, that I could not thereby haue come to any knowledge of the floures, but that he therewith sent me two dried floures, by which I partly gathered their form. Now these flours were very narrow, 2. inches long or more, consisting of three leaues, the vppermost of which much exceeded the 2. narrow ones on the sides both in length and breadth, and it was doubled; but before the floure was opened it better resembled a horne or cod, than a floure, and the lower end of it stood in a short green cup, in the middest of the floure vnder the vpper leafe that was folded, but open at the top; there came forth a smooth pointall, diuided at the top into nine parts or threds, whose ends of what colour they were, as also the threds, I know not, because I could not gather by the dry floure, whose colour was quite decayed, and the picture it self expressed no separation of the leaues in the floure, no forme of threds, but onely the floures shut, and resembling rather cods than floures, and those of a deepe red colour. But if I could haue seen them fresher, I should haue been able to haue giuen a more exact description: wherefore let the reader take in good parr that which I haue here performed. Thus much *Clusius*.

Chap. 22. *Of the sea Lentill.*

¶ *The Description.*

SOme call this *Vua marina*, and others haue thought it the *Lenticula marina* of *Serapio*, but they are deceiued, for his *Lenticula marina* described in his 245. chapter, is nothing else than the *Muscus marinus* or *Bryon thalaßion*, described by *Dioscorides, lib.* 4. *cap.* 99. as any that compares these two places together may plainely see.

1 The former of these hath many winding stalkes, whereon grow short branches set thick with narrow leaues like those of Beluidere, or Besome flax, and among these grow many skinny, hollow, empty round berries of the bignesse and shape of Lentills, whence it takes the name: this growes in diuers places of the Mediterranian and Adriaticke seas.

2 This differs little from the former, but that the leaues are broader, shorter, and snipt about the edges. But this being in probabilitie the Sargazo of *Acosta*, you shall here what he saies thereof. In that famous and no lesse to be feared nauigation del Sergazo (for so they which saile into the Indies call all that space of the Ocean from the 18. to the 34. degree of Northerly latitude) is seen a deepe and spatious sea couered with an herbe called Sarguazo, being a span long, wrapped with the tender branches as it were into balls, hauing narrow and tender leaues some halfe inch long, much

1 *Lenticula marina anguſtifolia.*
Narrow leaued Sea Lentill.

2 *Lenticula marina ſerratis folys.*
Cut leaued Sea Lentill.

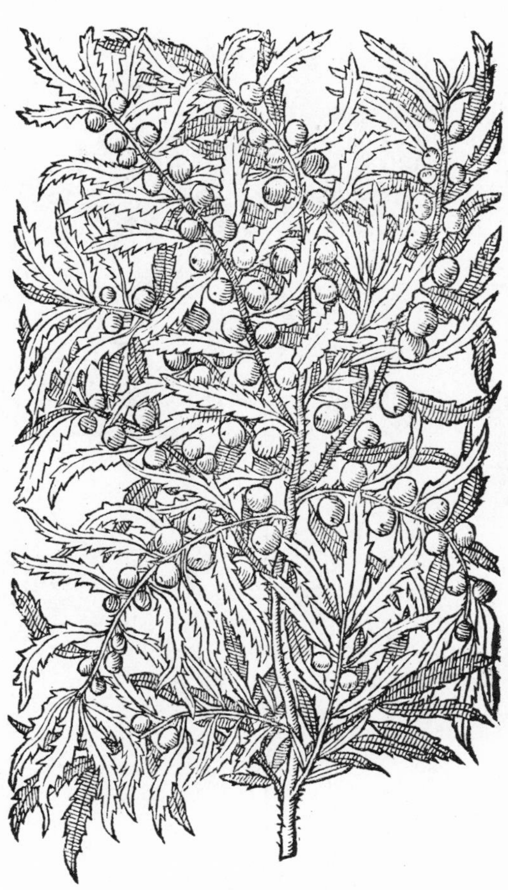

much ſnipt about the edges,of colour reddiſh,of taſte inſipide,or without any ſenſible biting, but what is rather drawne from the ſalt water, than naturally inherent in the plant. At the ſetting on of each leafe growes a ſeed round like a pepper corne, of a whitiſh colour, and ſometimes of white and red mixed, very tender when as it is firſt drawne forth of the water, but hard when it is dried, but by reaſon of the thinneſſe very fragile, and full of ſalt water: there is no root to be obſerued in this plant, but only the marks of the breaking off appeares; and it is likely it growes in the deepe and ſandy bottome of the ſea, and hath ſmall roots; yet ſome are of opinion that this herb is plucked vp and carried away by the rapide courſe of waters that fall out of many Iſlands into the Ocean. Now the Maſter of the ſhip wherein I was did ſtiffely maintaine this opinion; and in the ſailing here we were becalmed; but as far as euer wee could ſee wee ſaw the ſea wholly couered with this plant, and ſending down ſome yong Sailers which ſhould driue the weeds from the ſhip, and clenſe the water, we plainly ſaw round heapes thereof riſe vp from the bottom of the ſea where by ſounding we could finde no bottome.

This plant pickled with ſalt and vineger hath the ſame taſt as Sampier, and may be vſed in ſtead thereof, and alſo eaten by ſuch as ſaile, in place of Capers. I willed it ſhould be giuen newly taken forth of the ſea, to Goats which we carried in the ſhip, and they fed vpon it greedily. A

I found no faculties thereof; but one of the Sailers troubled with a difficultie of making water, casting out ſand and groſſe humors, ate thereof by chance both raw and boiled, onely for that the taſte thereof pleaſed him: after a few dayes hee told to me that he found great good by the eating thereof, and he tooke ſome of it with him, that ſo he might vſe it when he came aſhore. Hitherto *A Coſta.* B

CHAP. 23. *Of the Sea Feather.*

Myriophyllum marinum.
The Sea Feather.

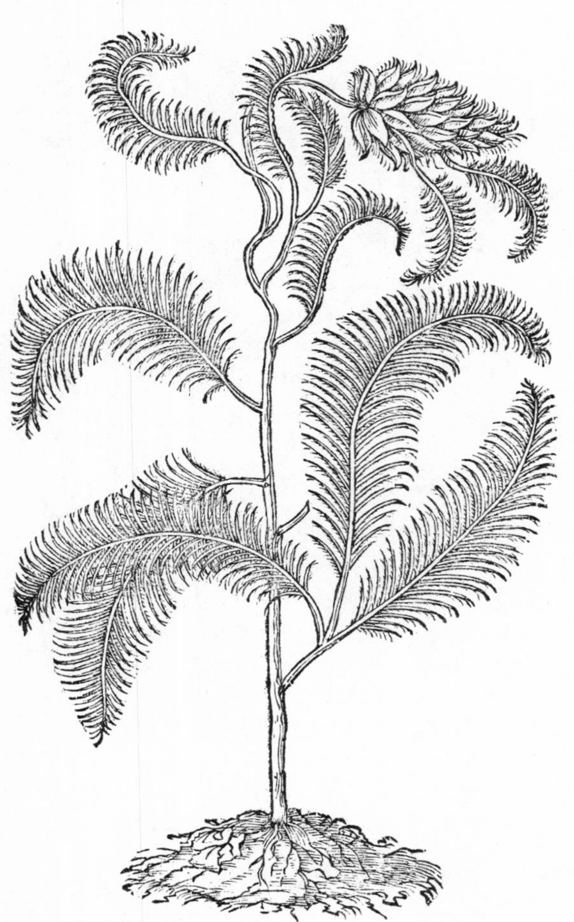

¶ *The Description.*

THis elegant plant, which *Clusius* receiued from *Cortusus* by the name of *Myriophylum Pelagicum*, is thus described by him: As much (saith hee) as I could coniecture by the picture, this was some cubit high, hauing a straight stalke, sufficiently slender, diuided into many branches, or rather branched leaues, almost like those of Ferne, but far finer, bending their tops like the branches of the Palme, of a yellowish colour: the top of the stalk adorned with lesser leaues, orded in certaine scales or cloues framed into a head; which are found to containe no other seed than tender plants already formed, in shape like to the old one: which falling, sinke to the bottome of the sea, and there take root and grow, and so become of the same magnitude as the old one from whence they came. The stalke is fastned with most slender and more than capillarie fibres, in stead of a root, not vpon rocks and Oister shells, as most other sea plants are, but vpon sand or mud in the bottome of the sea: this stalke when it is drie is no lesse brittle than glasse or Coralline; but greene and yet growing it is as tough and flexible as *Spartum* or Matweed.

¶ *The Place.*

It groweth in the deepest streames of the Illyrian sea, whence the Fishermen draw it forth with hooks and other instruments which they call Sperne. The whole plant, though dried, retains the faculties.

¶ *The Names.*

The Italian Fishermen call it *Penachio delle Ninfe*, and *Palma de Nettuno*: some also, *Scettro di Nettuno*.

¶ *The Vertues.*

A They say it is good against the virulent bites of the Sea serpents, and the venomous stings or prickes of Fishes

B Applied to small greene wounds it cures them in the space of 24 houres.

C *Cortusus* writ, that he had made triall thereof for the killing and voiding of wormes, and that he found it to be of no lesse efficacie than any Coralline, and that giuen in lesse quantitie.

CHAP. 24. *Of the Sea Fan.*

¶ *The Description.*

THis elegant shrub groweth vpon the rockes of the sea (where it is sometimes couered with the water) in diuers places; for it hath been brought both from the East and West Indies, and as I haue been informed it is to be found in great plenty vpon the rocks at the Burmuda Isles. *Clusius* calls

Frutex marinus reticulatus.
Sea Fan.

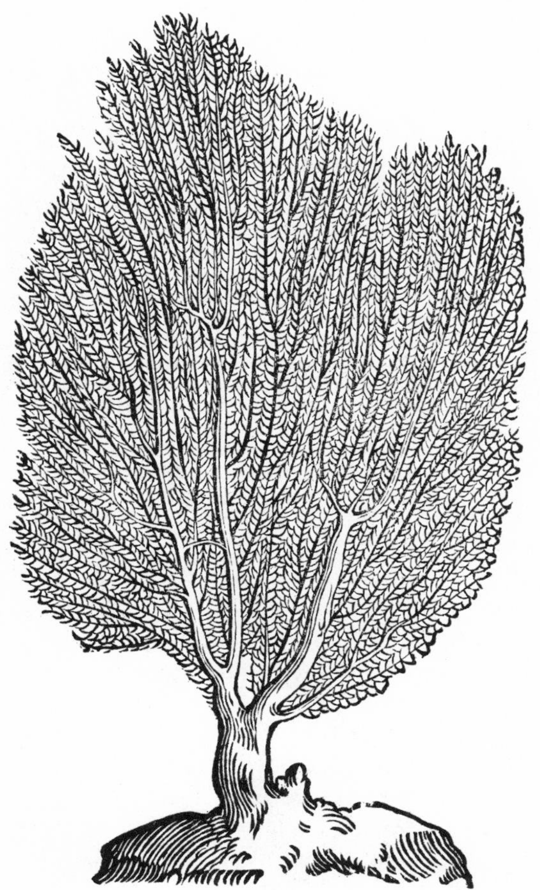

calls it *Frutex Marinus elegantiſsimus*, and thinkes it may be referred to the *Palma Marina* of *Theophraſtus*. *Bauhine* hath referred it to the *Corallina's*, calling it *Corallina cortice reticulato maculoſo purpuraſcente*. It growes vp ſomtimes to the height of three foot, hauing a ſtalke ſome handfull or two high before it part into branches: then is it diuided into three, foure, or more branches, which are ſubdiuided into infinite other leſſer ſtrings, which are finely interwouen and ioyned together as if they were netted, yet leauing ſometimes bigger, otherwhiles leſſer holes: and theſe twiggy branches become ſmaller and ſmaller, the farther they are from the root, and end as it were in ſmal threds: theſe branches grow not vp on euerie ſide, as in other plants, but flat one beſides another, ſo that the whole plant reſembles a fan, or a cabbage leafe eaten full of holes; yet ſomtimes vpon the ſides come forth other ſuch fanne-like branches, ſome bigger, ſome leſſe, ſometimes one or two, otherwhiles more. The inner ſubſtance of this Sea-Fan is a blackiſh tough, and hard wood, and it is all couered ouer with a rough Coral-like ſtony matter, of a reddiſh or purpliſh colour, and this you may with your naile or a knife ſcrape off from the ſmooth and blacke wood.

I know no vſe of this, but it is kept for the beauty and raritie thereof, by many louers of ſuch curioſities, amongſt which for the rareneſſe of the ſtructure this may hold a prime place.

Chap. 25.

Of China, and Baſtard China.

¶ *The Deſcription.*

THis root which is brought from the remoteſt parts of the world, and is in frequent vſe with vs, hath not been knowne in Europe little aboue foureſcore and ten yeares: for *Garcias ab Orta* the Portugall Phyſition writes, That he came to the firſt knowledge thereof in the Eaſt Indies, in the yeare 1535, and that by this meanes, as he relates it: It hapned (ſaith he) that about that time a merchant in the Iſle *Diu* told the noble gentleman S[r]. *Mart. Alfonſo de Souſa* my Patron, by what meanes he was cured of the French Poxes, which was by a certaine root brought from China; whoſe faculties he much extolled, becauſe ſuch as vſed it needed not obſerue ſo ſtrict a diet as was requiſit in the vſe of Guajacum, but ſhould onely abſtaine from Beefe, Porke, Fiſh, and crude fruits; but in China they do not abſtaine from fiſh, for they are there great gluttons. When the report of this root was diulged abroad, euery man wonderfully deſired to ſee and vſe it, becauſe they did not well like of the ſtrict dyet they were forced to obſerue in the vſe of Guajacum. Beſides, the inhabitants of theſe countries, by reaſon of their idle life are much giuen to gluttony. About this time the China ſhips arriue at Malaca, bringing a ſmall quantitie of this root for their owne vſe. But this little was ſought for with ſuch earneſtneſſe, that they gaue an exceſſiue rate for it; but afterwards the Chinois bringing a greater quantitie, the price fell, and it was ſold verie cheape. From this time Guajacum began to be out of vſe, and baniſhed the Indies, as a Spaniard that would famiſh the Natiues. Thus much *Garcias* concerning the firſt vſe thereof in the Eaſt Indies.

1 The China now in vſe is a root of the largeneſſe of that of the ordinarie Flag,or *Iris paluſtris*, and not much in ſhape vnlike thereto, but that it wants the rings or circles that are imprinted in the other : the outer coat or skin of this root is thin, ſometimes ſmooth, otherwhile rugged, of a browniſh red colour, and not to be ſeparated from the ſubſtance of the root,which is of an indifferent firmeneſſe,being not ſo hard as wood, but more ſollid than moſt roots which are not of ſhrubs or trees : the colour is ſometimes white, with ſome very ſmall mixture of redneſſe ; otherwhiles it hath a greater mixture of red, and ſome are more red than white : it is almoſt without taſt,yet that it hath is dry, without any bitterneſſe or acrimonie at all. The beſt is that which is indifferently ponderous, new, firme, not worme-eaten, nor rotten, and which hath a good and freſh colour, and that either white, or much inclining thereto. The plant whoſe root this is (if we may beleeue *Chriſtopher A Coſta*) hath many ſmall prickly and flexible branches, not vnlike the *Smilax aſpera*, or the prickly Binde-weed : the biggeſt of theſe exceedeth not the thickeneſſe of ones little finger. The leaues are of the bigneſſe of thoſe of the broad leaued Plantaine : the roots are as large as ones hand, ſometimes leſſe, ſollid, heauy, white, and alſo ſometimes red, and many oft times growing together.

1 *China vulgaris Officinarum.*
True China.

2 *Pſeudo-China.*
Baſtard China.

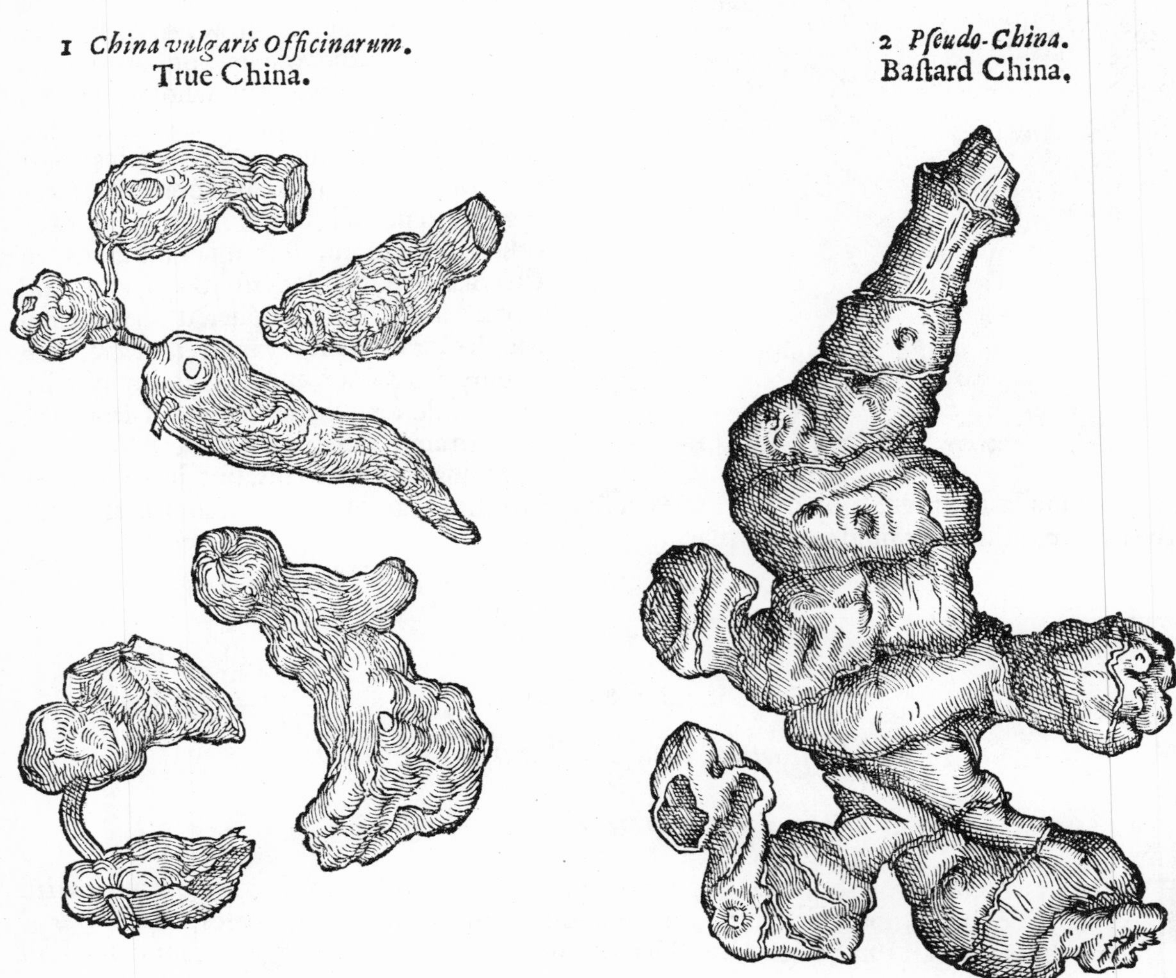

It groweth aboundantly in the territorie of China, and is alſo found in Malabar, Cochin, Cranganor, Coulan, Tanor, and other places.

The Chinois call it *Lampatan* : in Decan they call it *Lampatos* : in Canarin, *Bouti* : the Arabians, Perſians, and Turks terme it *Choph-China*.

2 This other root, whoſe figure you ſee here expreſt, was ſent from London to *Cluſius* in the yeare 1591, by *Iames Garret*, being brought out of Wingandecaow, or Virginia, with this inſcription, *Chinæ ſpecies*, A kinde of China. *Cluſius* cauſed this figure thereof to be drawne, and thus deſcribeth it. This root (ſaith hee) was very knotty, and formed with out-growings, or bunches ſtanding out, of a reddiſh colour, and it yet retained at the top ſome part of the ſtalke, being ſomwhat like vnto that of *Smilax aſpera*, or common rough Binde-weed, hard, wooddy, and full of veines, as the ſtalks of *Smilax aſpera* : the ſubſtance of the root was alſo reddiſh, as the root of the common Flagge, at the firſt of a ſaltiſh taſte, it being old, (for ſo it was when I receiued it) and

and then drying. Now I iudge this the ſame that the writer of the Virginian Hiſtorie mentions in his chapter of roots, and ſaith, it was brought into England for China, though the Natiues knew no vſe thereof: but they vſe another root very like China, which they call *Tſinaw*, of which beeing cut, beaten, and preſſed out with water, they draw a iuice wherewith they make their bread. Thus much *Cluſius*, to whoſe words I thinke it not amiſſe to adde that which M^r^. *Thomas Hariot* (who was the writer of the Virginian hiſtorie, here mentioned by *Cluſius*) hath ſet downe concerning this thing.

Tſinaw (ſaith he) is a kinde of root much like vnto that which in England is called the China root, brought from the Eaſt Indies. And we know not any thing to the contrarie but that it may be of the ſame kinde. Theſe roots grow many together in great cluſters, and doe bring forth a Brier ſtalk, but the leafe in ſhape is far vnlike: which being ſupported by the trees it groweth neereſt vnto, wil reach or clime to the top of the higheſt. From theſe roots whileſt they be new or freſh, being chopt into ſmall pieces and ſtampt, is ſtrained with water a iuice that maketh bread, and alſo beeing boiled, a very good ſpoonemeat in manner of a gelly, and is much better in taſte, if it be tempered with oyle. This *Tſinaw* is not of that ſort which by ſome was cauſed to be brought into England for the China root; for it was diſcouered ſince, and is in vſe as is aforeſaid; but that which was brought hither is not yet knowne, neither by vs, nor by the inhabitants, to ſerue for any vſe or purpoſe, although the roots in ſhape are very like. Thus much *Hariot*.

¶ *The Temperature and Vertues.*

China is thought to be moderately hot and drie: the decoction thereof made alone or with other things, as the diſeaſe and Symptomes ſhal require, is much commended by *Garcias*, for to cure the French pox, but chiefely that diſeaſe which is of ſome ſtanding: yet by moſt it is iudged leſſe powerfull than *Guajacum*, or *Sarſaparilla*. A

It attenuates, moues ſweat, and dries, and therefore reſiſts putrifaction: it ſtrengthens the liuer, helpes the dropſie, cures maligne vlcers, ſcabbes, and lepry. It is alſo commended in Conſumptions. B

The decoction of this root, ſaith *Garcias*, beſides the diſeaſes which haue communitie with the Poxe, conduces to the cure of the Palſie, Gout, Sciatica, ſchirrous and œdematous tumours. It alſo helps the Kings-euill. It cureth the weakeneſſe of the ſtomacke, the inueterate head-ache, the ſtone and vlceration of the bladder; for many by the vſe of the decoction hereof haue beene cured, which formerly receiued help by no medicine. C

CHAP. 26. *Of Coſtus.*

¶ *The Deſcription.*

THis ſimple medicine was briefely deſcribed by *Dioſcorides*, who mentions three kindes thereof, but what part of a plant, whether root, wood, or fruit, he hath not expreſt: but one may probablely coniecture it is a root, for that he writes toward the end of the Chapter where he treats thereof, *lib*.1.*cap*.15. that it is adulterated by mixing therewith the roots of *Helenium commagenum*; now a root cannot well be adulterated but with another. Alſo *Pliny*, *lib*.12.*cap*.12. calls it a root; but neither any of the antient or moderne Writers haue deliniated the plant, whoſe root ſhould be this *Coſtus*. *Dioſcorides* makes three ſorts, as I haue ſaid: the Arabian being the beſt; which was white, light, ſtrong, and well ſmelling: the Indian, which was large, light, and blacke: the Syrian, which was heauie, of the colour of Box, and ſtrong ſmelling. Now *Pliny* makes two kindes, the blacke, and the white, which he ſaith is the better; ſo I iudge his blacke to be the Indian of *Dioſcorides*, and his white, the Arabian. Much agreeable to theſe (but whether the ſame or no, I do not determine) are the two roots whoſe figures I here preſent to your view, and they are called by the names of *Coſtus dulcis* (I thinke they ſhould haue ſaid *odoratus*) and *Coſtus amarus*.

1 The firſt of theſe, which rather from the ſmell, than taſte, is called ſweet, is a pretty large root, light, white, and well ſmelling, hauing the ſmell of Orris, or a violet, but ſomewhat more quick and piercing, eſpecially if the root be freſh, and not too old: it is oft times diuided at the top into two, three, or more parts, from whence ſeuerall ſtalks haue growne, and you ſhall ſomtimes obſerue vpon ſome of them pieces of theſe ſtalks ſome two or three inches long, of the thickeneſſe of ones little

little finger, crested, and filled with a soft pith, like as the stalks of Elder, or more like those of the Bur-docke: the taste of the root is bitter, with some acrimonie, which also *Dioscorides* requires in his, for he saith, the taste should be biting and hot; thus much for the first, being *Costus dulcis* of the shoppes.

1 *Costus Indicus sive odoratus.*
Indian or sweet smelling Costus.

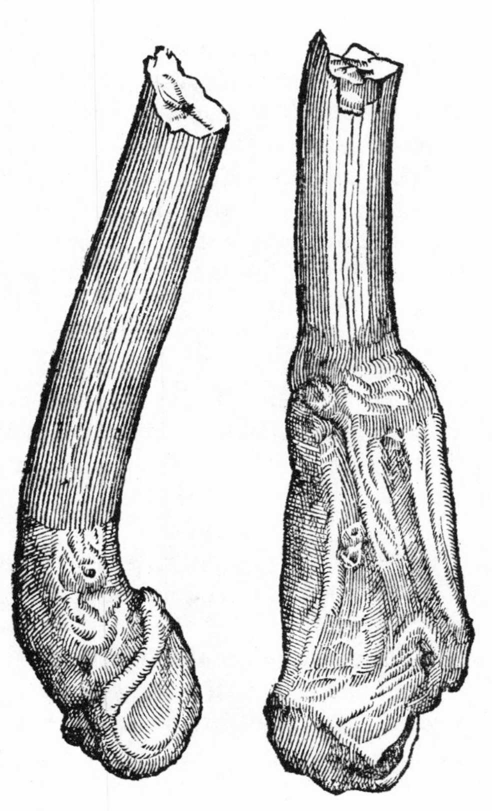

2 *Costus Officinarum Lobelij.*
Bitter Costus.

2 The second, which is the *Costus amarus*, and it may be the Indian of *Dioscorides*, and *Niger* of *Pliny*, is a root blacke both within and without, light, yet very dense. It seemes to be of some large root, for that it is brought ouer cut into large pieces, of the bignesse of ones finger, sometimes bigger sometimes lesse, which it seemes is for the more conuenient drying thereof, for a large root, vnlesse it be cut into pieces can scarcely be wel dried: the taste of this is bitter, somewhat clammy and ingrate: the smell is little or none.

There are some other roots which haue been set forth by late writers for *Costus*, but because they are neither in vse, knowne here with vs, nor more agreeable to the descriptions of the Antients, I hastening to an end, am willing to passe them ouer in silence.

¶ *The Temperature and Vertues out of the Antients.*

A It hath a heating and attenuating facultie, and therefore was vsed in oile to annoint the bodie against the cold fits of Agues, the Sciatica, and when it was needfull to draw any thing to the superficies of the body.

B It is also conuenient to moue vrine, to procure the termes, to help strains, conuulsions, or cramps and paines in the sides; and by reason of the bitternesse it kills wormes.

C It is good to be drunke against the bite of the viper: against paines of the chest, and windinesse of the stomacke taken in Wine with Worme-wood: and it is vsed to be put into sundrie Antidotes.

CHAP.

Chap. 27. *Of Drakes root, or Contra-yerua.*

¶ *The Deſcription.*

THat root which of late is knowne in ſome ſhops by the Spaniſh name *Contra-yerua*, is the ſame which *Cluſius* hath ſet forth by the title of *Drakena radix* : wherefore I will giue you the hiſtorie of *Cluſius*, and thereto adde that which *Monardus* writes of the *Contra-yerua*. For though *Bauhine*, and the Author of the *Hiſtoria Lugdunenſis* ſeeme to make theſe different, yet I finde that both *Cluſius* his figure and hiſtorie exactly agree with the roots ſent vs from Spaine by that title, wherefore I ſhall make them one, till ſome ſhall ſhew me how they differ: and *Cluſius* ſeemes to be of this minde alſo, who deſired but the degree of heate which *Monardus* giues theſe, and that is but the ſecond degree : now theſe haue no taſte at the firſt, vntill you haue chewed them a pretty while, and then you ſhall finde a manifeſt heate and acrimonie in them, which *Cluſius* did alſo obſerue in his.

In the yeare (ſaith *Cluſius*) 1581. the generous Knight Sir *Francis Drake* gaue me at London certain roots, with three or foure Peruvian Beazor ſtones, which in the Autumne before (hauing finiſhed his voyage, wherein paſſing the Straights of Magellan, he had encompaſſed the World) he had brought with him, affirming them to be of high eſteeme amongſt the Peruvians : now for his ſake that beſtowed theſe roots vpon me, I haue giuen them the title *Drakena radix*, or *Drakes* root, and haue made them to be expreſſed in a table, as you may here ſee them.

1 *Drakena radix.*
Contra-yerua.

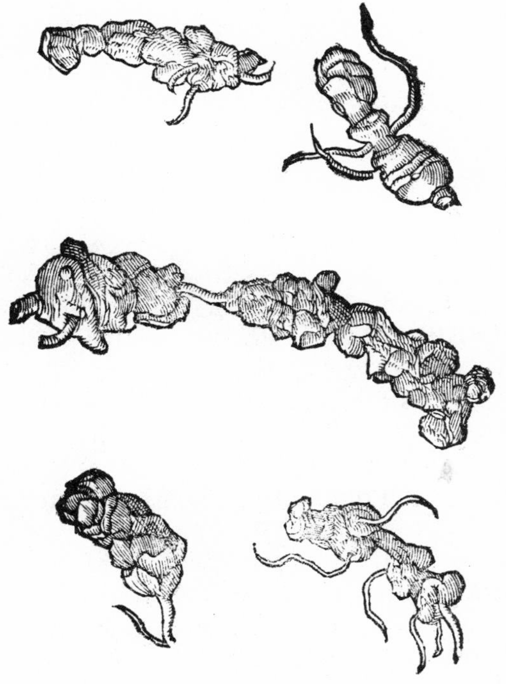

2 *Radix Drakenæ affinis.*
Another ſort of Contra-yerua.

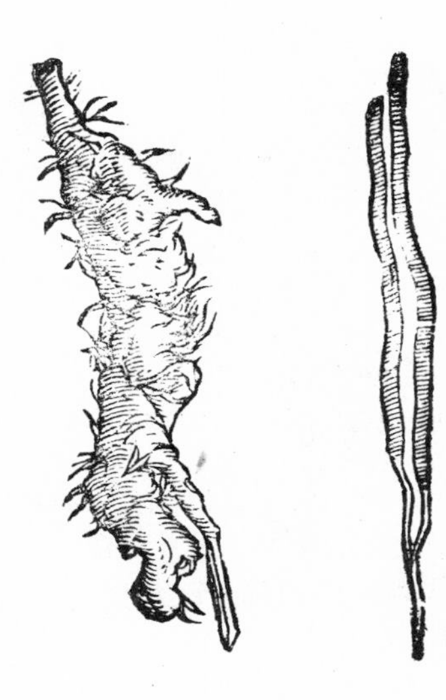

Theſe roots were for the moſt part ſome halfe inch thick, longiſh, now and then bunching out into knots and vnequall heads, and their tops looked as if they were compoſed of thicke ſcales, almoſt like thoſe of the *Dentaria enneaphyllos* ; blackiſh without, wrinckled, and hard, becauſe dried: their inner part was white ; they had ſlender fibres here and there growing out of them, and ſome more thicke and large, hard alſo and tough, at which hung other knots: I obſerued no manifeſt ſmel they had, but found them to haue a taſte ſomewhat aſtringent, & drying the tongue at the firſt ; but being long chewed, they left a quicke and pleaſing acrimonie in the mouth.

It ſeemed to haue great affinitie with the *Radix S. Helenæ*, whereof *Nic. Monardus* ſpeakes in his booke of the Simple Medicines brought from the Weſt Indies: but ſeeing *N. Eliot* (who accompanied

nied S[r].*Fran.Drake* in that voyage,said,that the Spaniards in Peru had them in great request; and they could not easily be got of them,and that he had learned by them, that the leaues were present poison, but the root an antidote,and that not only against the same poison,but also against other; and that it strengthned the heart and vitall faculties,if it were beaten to pouder, and taken in the morning in a little wine; and giuen in water,it mitigated the heat of Feuers. By reason of these faculties it should much agree with the *Radix Contra-yerua,* whereof *Monardus* writes in the same booke: yet in these I required the aromaticke taste and degree of heate, which he attributes vnto these roots. Thus much *Clus.*

A From Charcis a Prouince of Peru, saith *Monard.* are brought certaine roots very like the roots of *Iris*, but lesse, and hauing the smell of Fig leaues. The Spaniards that liue in the Indies call them *Contra-yerua*, as if you should say an Antidote against poison; because the pouder of them taken in white Wine is a most present remedy against all poison of what kinde soeuer it be (only sublimate excepted, whose malignitie is onely extinguished by the drinking of milke) it causes them to bee cast vp by vomite, or euacuated by sweat. They also say that Philtres or amorous potions are cast forth by drinking this pouder. It also killeth wormes in the belly. The root chewed hath a certain aromaticke taste ioined with acrimony; wherefore it seemes hot in the second degree. Thus farre *Monardus*.

2 *Clusius Exot. l.* 4.*c.*11. being the next after *Drakena radix*, describes this root, whose figure I giue you in the 2. place, & that by the same title as it is here set forth. These roots, saith he, seemed somwhat like the *Drakena radix* which were found in the great ship which brought backe the Viceroy from the East Indies, and was taken by the English: for they were tuberous, and as much as one may gather by their forme, crept vpon the surface of the earth, hauing vpon them many haires and fibres, and being of a sooty colour, yet somewhat inclining to yellow, dying the spittle in chewing them and being bitter: they as yet retained foot-stalks of the leaues, but of what fashion they were no man can easily guesse. But it was likely they were of great vse among the Indians, seeing that the Vice-roy brought them together with other precious medicines growing in the East Indies. *Iames Garret* sent this to *Clusius* with the little plant dryed, whose figure you see exprest by it.

Chap. 28. *Of* Lignum Aloes.

Lignum Aloes vulgare.

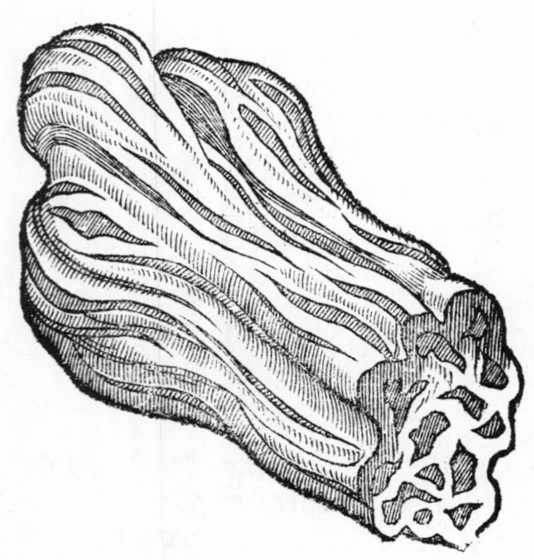

¶ *The Description.*

IT is a question whether the *Agallochum* descri-bed in the 21.*c.l.* 1 of *Dioscorides* be the same which the later Greeks and shops at this time call *Xyloaloe*, or *Lignum Aloes*, many make them the same: others, to whose opinion I adhere, make them differerent, yet haue, not the later, shew what *Agallochum* should bee, which I notwithstanding will do; and though I doe not now giue you my arguments, yet I will point at the things, & shew positiuely my opinions of them.

The first and best of these is that which some call *Calumbart*: others, *Calumba*, or *Calambec*: this is of high esteem in the Indies, & seldom found but amongst the Princes, and persons of great qualitie; for it is sold oft times for the weight in gold; I haue not seen any therof but in beads, it seemes to be a whiter wood than the ordinary, of a finer graine, not so subiect to rot, and of a more fragrant smell, and but light.

The second sort, which is vsually brought ouer, and called in shops by the name of *Lignum Aloes*, is also a precious and odoriferous wood, especially burnt: the stickes of this are commonly knotty & vnsightly: some parts of them being white, soft, and doted: othersome, dense, blackish,

blackish, or rather intermixt with blacke and white veines, but much more blacke than white, and this put to the fire will sweat out an oily moisture, and burnt, yeeld a most fragrant odour. This I take to be the true *Xyloaloe* of the late Greekes; and the *Agalugen* of *Auicen*; and that they call *Palo d' Agula* in the Indies.

The third is a wood of much lesse price than the former: and I coniecture it might well be substituted for *Thus*: and this I take to be the *Agallochum* of *Dioscorides*; the *Lignum Aloes syluestre* of *Garcias*; and *Agula braua* of *Linscoten*. It is a firme and sollid wood, somewhat like that of the Cedar, not subiect to rot or decay: the colour thereof is blackish, especially on the out-side; but on the in-side it is oft times brownish and speckled, containing also in it an oilie substance, and yeelding a sweet and pleasing smell when it is burnt, but not like that of the two former: the taste also of this is bitterer than that of the former: and the wood (though dense and sollid) may be easily cleft long-waies; it is also a farre handsomer and more sightly wood than the former, hauing not many knots in it.

Garcias ab Orta thus describes the tree that is the *Lignum Aloes* (I iudge it's that I haue set forth in the second place:) it is (saith he) like an Oliue tree, sometimes larger: the fruit or floure I could not yet see, by reason of the difficulties and dangers which are to be vndergone in the accurate obseruation of this tree (Tigers frequently there seeking their prey.) I had the branches with the leaues brought me from Malaca. Now they say that the wood new cut downe hath no fragrant odour, nor till it be dried: neither the smell to be diffused ouer the whole matter of the wood, but in the heart of the tree; for the barke is thicke, and the matter of the wood without smell. Yet may I not denie, but the barke and wood putrifying that oilie and fat moisture, may betake it selfe to the heart of the tree, and make it the more odoriferous: but there is no need of putrifaction to get a smell to the *Lignum Aloes*: for there are sundry so expert and skilfull in the knowledge thereof, that they will iudge of that which is new cut downe, whither it will be odoriferous or no. For in all sorts of wood some are better than othersome: thus much out of *Garcias*; where such as are desirous may see more vpon this subiect.

¶ *The Temperature and Vertues.*

It is of temperature moderately hot and dry, and also of somewhat subtill parts. Chewed it makes the breath smell sweet, and burnt it is a rich perfume. A

Taken inwardly it is good to helpe the stomack that is too cold and moist, as also the weak liuer. B

It is commended likewise in dysenteries and pleurisies: and put also into diuers Cordiall medicines and Antidotes as a prime ingredient. C

Chap. 29. *Of Gedwar.*

1 *Gedwar aut, Geiduar.*

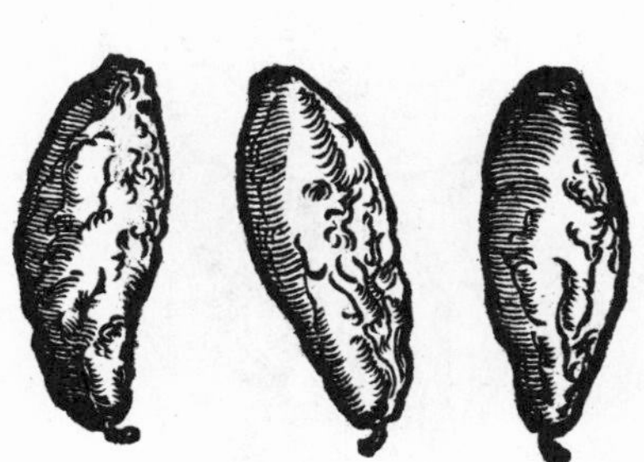

2 *Zedoariæ exactior icon.* A better figure of Zedoary.

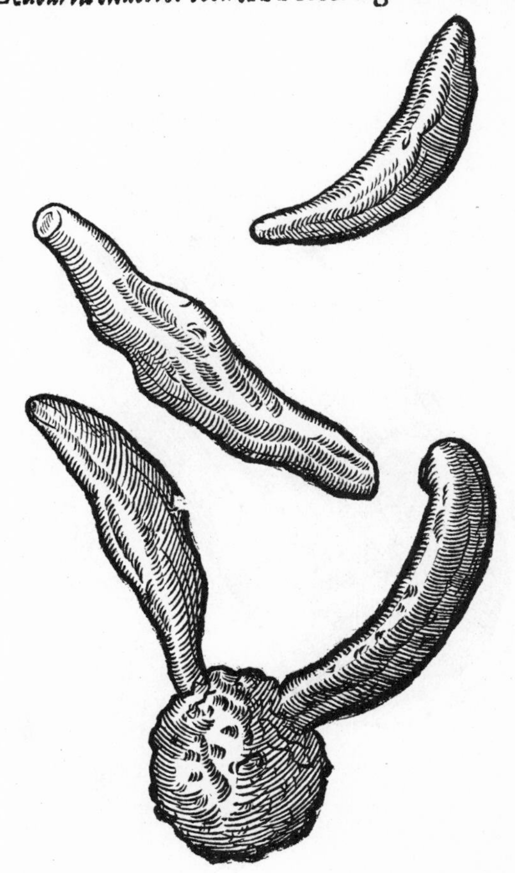

¶ *The Description.*

IN the Chapter of Zedoarie (which I made the 28. of the first booke) I might fitly haue giuen you this historie of Gedwar, which is thought to be that described by *Auicen, lib.* 2. *c.* 734. and a kinde of Zedoarie: *Garcias* saith, Gedwar is at a high rate, and not easily to be found, vnlesse with the Indian Mountibanks and juglers, which they call *Iogues*, which goe vp and downe the countrey like Rogues, and of these the Kings and Noblemen buy *Geiduar*: it is good for many things, but chiefely against poisons, and the bites and stings of venomous creatures. Now *Clusius* in his *Auctarium* at the end thereof giues this figure, with the following historie.

1 Because *Garcias*, saith he, *cap.* 42. *l.* 1. *Aromatum hist.* treating of Zedoarie writes, that *Auicen* calls it Gedwar; and saith that it is of the magnitude of an Acorne, and almost of the same shape, I in my notes at the end of that chapter affirmed that it was not knowne in Europe, and hard to be knowne. But in the yeare 1605, *Iohn Pona* sent me from Verona together with other things two roots written on by the name of *Gedwar verum*. They were not much vnlike a longish Acorne, or (that I may more truly compare them) the smaller bulbs of an Asphodil, or *Anthora*: the one of them was whole and not perished: the other rotten and broken, yet both of them very hard and sollid, of an ash colour without, but yellowish within, which tasted, seemed to possesse a heating facultie and acrimonie.

But although I can affirm nothing of certaintie of this root, yet I made the figure of the wholler of them to be exprest in a table, that so the forme might be conceiued in ones minde more easily, than by a naked description. Let the Studious thanke *Pona* for the knowledge hereof. Thus much *Clusius*.

2 In the 28 chapter of the first booke I gaue the figure of Zodoarie out of *Clusius*, hauing not at that time this figure of *Lobel*, which presents to your view both the long and the round, with the manner how they grow together, being not seuerall roots, but parts of one and the same.

Chap. 30, *Of Rose-wood.*

Aspalathus albicans torulo citreo.
White Rose-wood.

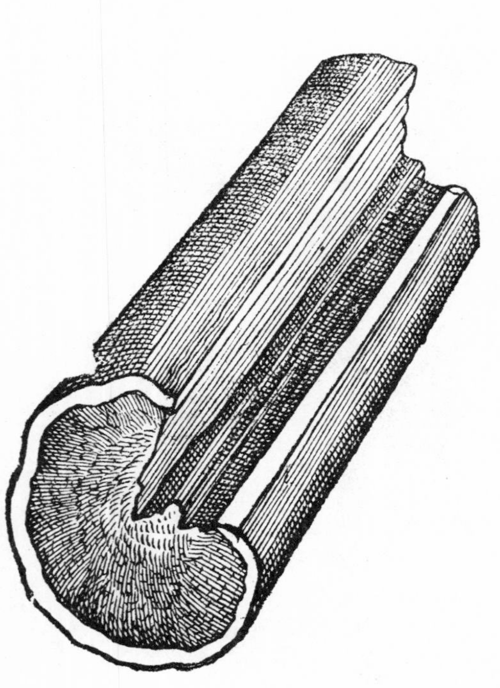

Aspalathus rubens.
Reddish Rose-wood.

¶ *The*

¶ *The Description.*

BOth these as also some other woods are referred to the *Aspalathus* described by *Dioscorides, l.1. c.19.* But the later of these I take to be the better of the two sorts there mentioned. The first of them is whitish without, hauing a yellowish or citrine coloured round in the middle: the taste is hottish, and smell somewhat like that of a white-Rose.

The other hath also a small ring of white, next the thicke and rugged barke, and the inner wood is of a reddish colour, very dense, sollid and firme, as also indifferent heauy: the smell of this is also like that of a Rose, whence they vulgarly call it *Lignum Rhodium*, Rose-wood, rather than from Rhodes the place where the later of them is said to grow.

¶ *The Faculties out of* Dioscorides.

A It hath a heating facultie with astriction, whence the decoction thereof made in wine is conuenient to wash the vlcers of the mouth, and the eating vlcers of the priuities and such vnclean sores as the *Ozæna* (a stinking vlcer in the nose so called.)

B Put vp in a pessarie it drawes forth the childe, the decoction thereof stayes the loosenesse of the belly, and drunke it helpes the casting vp of bloud, the difficultie of making water, and windinesse.

AT the end of this Appendix I haue thought good to giue you diuers descriptions of Plants, which I receiued from my often mentioned friend Mr. *Goodyer*, which also were omitted in their fitting places, partly through haste, and partly for that I receiued some of them after the printing of those chapters wherein of right they should haue been inserted. They are most of them of rare and not written of plants, wherefore more gratefull to the curious.

Hieracium stellatum Boelij.

THis plant is in round, hairy, straked, branched stalks, and long, rough, blunt indented leaues like to *Hieracium falcatum*, but scarce a foot high: the floures are also yellow three times smaller: which past, there succeed long crooked slender sharpe pointed cods or huskes, neere an inch long, spreading abroad, star-fashion, wherein a long seed is contained: this hath no heads or woolly down like any of the rest, but onely the said crooked coddes which doe at the first spread abroad. The root is small, threddie, full of milkie iuice, as is also the whole plant and it perisheth when the seed is ripe.

Hieracium medio nigrum flore maiore Boely.

This hath at the first spreading vpon the ground many long, narrow, green, smooth leaues bluntly indented about the edges, like those of *Hieracium falcatum*, but smaller: amongst which rise vp three, foure, or more, small, smooth, straked round stalks, diuided into other branches, which grow longer than the stalks themselues leaning or trayling neere the ground: the floures grow on the tops of the stalks, but one together, composed of many pale yellow leaues, the middle of each floure being of a blackish purple colour.

Hieracium medio nigrum flore minore Boelij.

This is altogether like the last before described in stalkes and leaues: the floures are also of a blackish purple in the middle, but they are three times smaller.

Hieracium lanosum

There groweth from one root three, foure or more round vpright soft cottonie stalks, of a reasonable bignesse, two foot high, diuided into many branches, especially neere the top, whereon groweth at each diuision one broad sharpe pointed leafe, diuided into corners, and very much crumpled, and also very soft cottonie and woolly, as is the whole plant: the floures are small, double, of a pale yellow colour, very like those of *Pilosella repens*, growing clustering very many together at the tops of the stalkes and branches, forth of small round soft cottonie heads: these foure plants grew from

 seed

feed which I receiued from M^r^. *Coys*,1620.and I made these descriptions by the Plants the 22. of August, 1621.

Blitum spinosum : est Beta Cretica semine aculeato Bauhini Matth. pag. 371.

This sendeth forth from one root many round greene strailing, ioynted, small branches, about a foot long : the leaues are of a light greene colour, and grow at euery ioint one, somewhat like the leaues of great Sorrell, but they are round topped without barbes or eares below, or any manifest taste or smell, very like the leaues of Beets, but much smaller : the floures grow clustering together about the ioints, and at the tops of the branches small and greenish, each floure containing fiue or six very small blunt topped leaues, and a few dustie chiues in the middle : which past, there commeth great prickly shriuelled seed, growing euen close to the root, and vpwards on the ioints, each seed hauing three sharpe prickes at the top growing side-waies, which indeed may be more properly called the huske; which huske in the in-side is of a darke reddish colour, and containeth one seed in forme like the seed of *Flos Adonis*, round at the lower end, and cornered towards the top, and sharp pointed, couered ouer with a darke yellowish skin; which skin pulled away, the kernell appeareth yellow on the out side, and exceeding white within, and will with a light touch fall into very small pouder like meale.

Gerany Bæticæ species Boelij.

This hath at the beginning many broad leaues, indented about the edges, somwhat diuided, like those of *Geranium Creticum*, but of a lighter greene colour, and smaller : amongst which grow vp many round hairy kneed trailing branches, diuided into many other branches, bearing leaues like the former, but smaller, and no more diuided. The floures are smal like those of *Geranum Moschatum*, but of a deeper reddish colour, each floure hauing fiue small round topped leaues: after followeth small long hairie seed, growing at the lower end of a sharpe pointed beak like that of *Geranium Moschatum*: the whole plant perisheth when the seed is ripe.

Boelius a Low-countrey-man gathered the seeds hereof in Bætica a part of Spaine, and imparted them to M^r^. *William Coys*, a man very skilfull in the knowledge of Simples, who hath gotten plants thereof, and of infinite other strange herbes, and friendly gaue me seeds hereof, and of many other, *Anno*, 1620.

Antirrhinum minus flore Linariæ luteum inscriptum.

This hath at the first many very small, round, smooth branches from one root, trayling on the ground, about foure or fiue inches long, set with many small greene short sharp pointed leaues, like those of *Serpillum*, but that these are longer, smooth, and three or foure growing opposite one against another: amongst which rise vp fiue or six, sometimes ten or twelue vpright round smooth little stalks a cubit high, diuided into branches bearing small long smooth greene leaues, growing without order, as narrow as the vpper leaues of *Oenanthe Angustifolia*: at the toppes of the stalks and branches grow clustering together fiue six or more small yellow floures, flouring vpwards, leauing a long spike of very small huskes, each huske hauing a small line or chinke as though two huskes were ioined together, the one side of the huske being a little longer than the other, wherein is contained exceeding small blackish seed. The root is very short, small, and white, with a few threds, and perisheth at winter.

This plant is not written of that I can finde. I receiued seed thereof from M^r^. *William Coys* often remembred.

Linaria minor æstiua.

The stalkes are round, smooth, of a whitish greene colour, a foot high, weake, not able to stand vpright : whereon grow long narrow sharpe pointed leaues, most commonly bending or turning downewards. The floures grow in spikes at the toppes of the branches, yet not very neere together, and are verie small and yellow, with a small tayle : the seed of this plant is small, flat, and of a blackish gray colour, inclosed in small round huskes, and you shall commonly haue at one time floures and ripe seed all on a stalke. The whole plant is like to the common *Linaria*, but that it is a great deale lesser, and the floures are six times as small, and perish at Winter. I also receiued seeds thereof from M^r^. *William Coys*.

Scorp.

Scorpioides multiflorus Boëlij.

This Plant is in creeping branches and leaues like the common *Scorpioides Bupleuri folio*: the floures are also alike, but a little bigger, and grow foure or fiue together on one foot-stalke: the cods are rougher, and very much turned round, or folded one within another: in all things else alike.

Scorpioides siliqua crassa Boëlij.

This is also like the other in creeping branches and leaues: the flouree are something bigger than any of the rest, and grow not aboue one or two together on a foot-stalke: the cods are crooked, without any rough haire, yet finely checkquered, and seuen times bigger than any of the rest, fully as big as a great Palmer-worme, wherein is the difference: the seed is almost round, yet extending somewhat in length, almost as big as small field Peason, of a browne or yellowish colour. This also perisheth when the seed is ripe. *Sept.* 1. 1621.

Silibum minus flore nutante Boëlij.

This Thistle is in stalkes and leaues much smaller than our Ladies Thistle, that is to say, The stalkes are round, straked, somewhat woolly, with narrow skinny prickly edges three or foure foot high, diuided into many branches, whereon grow long leaues, deeply diuided, full of white milke-like streakes and sharpe prickles by the edges: the floures grow on the top of the stalkes and branches full of small heads, commonly turning downewards, of the bignesse of an Oliue, set with very small slender sharpe prickes, containing nothing but small purple chiues, spreading abroad like those of *Iacea*, with some blewish chiues in the middle: the seed followeth, inclosed in downe, and is small and grayish like the seed of other Thistles, but it is as clammy as Bird-lime. The whole plant perisheth at Winter, and reneweth it selfe by the falling of the seed. I finde not this written of. It was first gathered by *Boelius* in Spaine, and imparted vnto Mr *William Coys*, who friendly gaue me seeds thereof.

Aracus major Bæticus Boelij.

It hath small weake foure square streaked trailing branches, two foot high, lesser, but like those of Fetches; whereon grow many leaues without order, and euery seuerall leafe is composed of six seuen, or more small sharp pointed leaues, like those of Lentils, set on each side of a middle rib, which middle rib endeth with clasping tendrels: the floures grow forth of the bosomes of the leaues, but one in a place, almost without any foot-stalkes at all, like those of Vetches, but of a whitish colour, with purple streakes, and of a deeper colour tending to purple towards the nailes of the vpper couering leaues: after which follow the cods, which are little aboue an inch long, not fully so big as those of the wilde beane, almost round, and very hairy: wherein is contained about foure peason, seldome round, most commonly somewhat flat, and sometimes cornered, of a blackish colour, neere as big as field peason, and of the taste of Fetches: the whole herbe perisheth when the seed is ripe. This plant *Boelius* sent to Mr *William Coys*, who hath carefully preserued the same kind euer since, and friendly imparted seeds to me in *Anno* 1620.

Legumen pallidum Vlissiponense, Nonij Brandonij.

This plant is very like, both in stalkes, leaues, and cods, to *Aracus major Bæticus*, but the floures of this are of a pale yellow or Primrose colour, and the whole herbe smaller, and nothing so hairy. It perisheth also when the seed is ripe. I receiued the seeds likewise from Mr *Coys*.

Vicia Indica fructu albo. Pisum Indicum Gerardo.

This Vetch differeth not in any thing at all, either in stalkes, leaues, cods, fashion of the floures, or colour thereof, from our common manured Vetch, but that it groweth higher, and the fruit is bigger and rounder, and of a very cleare white colour, more like to Peason than Vetches. Mr *Gerrard* was wont to call this Vetch by the name of *Pisum Indicum*, or Indian Pease, gotten by him after the publishing of his Herball, as Mr *Coys* reported to me. But the said Mr *Coys* hath in my judgement more properly named it *Vicia fructu albo*: which name I thought most fit to call it by, onely adding *Indica* to it, from whence it is reported to haue been gotten. *Iuly*, 30. 1621.

Astragalus marinus Lusitanicus Boely.

This hath fiue, six, or more round straked reddish hairy stalkes or branches, of a reasonable bignesse, proceeding from one root, sometimes creeping or leaning neere the ground, and sometimes standing vpright, a cubit high, with many greene leaues, set by certaine distances, out of order like those of *Glaux vulgaris*, but lesser, euery leafe being composed of fourteene or more round topped

ped leaues, a little hairy by the edges, set on each side of a long middle rib, which is about nine or ten inches in length, without tendrels: the floures grow forth of the bosomes of the leaues, neere the tops of the stalkes, on long round streaked hairy foot-stalkes, of a very pale yellow colour, like those of *Securidaca minor*, but bigger, growing close together in short spikes, which turne into spikes of the length of two or three inches, containing many smal three cornered cods about an inch long, growing close together like those of *Glaux vulgaris*, each cod containing two rowes of small flat foure cornered seeds, three or foure in each row, of a darke yellowish or leadish colour, like to those of *Securidaca minor*, but three or foure times as big, of little taste: the root is small, slender, white, with a few threds, and groweth downe right, and perisheth when the seed is ripe. I first gathered seeds of this plant in the garden of my good friend Mr *Iohn Parkinson* an Apothecary of London, *Anno*, 1616.

Faba veterum serratis folijs Boelij.

This is like the other wilde Beane in stalks, floures, cods, fruit, and clasping tendrels, but it differeth from it in that the leaues hereof (especially those that grow neere the tops of the stalkes) are notched and indented about the edges like the teeth of a saw. The root also perisheth when the seed is ripe. The seeds of this wilde Beane were gathered by *Boelius* a Low-country man, in Bætica a part of Spaine, and by him sent to Mr *William Coys*, who carefully preserued them, and also imparted seeds thereof to me, in *Anno* 1620. *Iuly* 31. 1621.

Pisum maculatum Boelij.

They are like to the small common field Peason in stalkes, leaues, and cods; the difference is, the floures are commonly smaller, and of a whitish green colour: the peason are of a darke gray colour, spotted with blacke spots in shew like to blacke Veluet; in taste they are also like, but somewhat harsher. These peason I gathered in the garden of Mr *Iohn Parkinson*, a skilfull Apothecary of London; and they were first brought out of Spaine by *Boelius* a Low-country man.

Lathyrus æstivus flore luteo. *Iuly*, 28. 1621

This is like *Lathyris latiore folio Lobelij*, in stalkes, leaues, and branches, but smaller: the stalks are two or three foot long, made flat with two skins, with two exceeding small leaues growing on the stalkes, one opposite against another: betweene which spring vp flat foot-stalkes, an inch long, bearing two exceeding narrow sharpe pointed leaues, three inches long: betweene which grow the tendrels, diuided into many parts at the top, and taking hold therewith: the floures are small, and grow forth of the bosomes of the leaues, on each foot-stalke one floure, wholly yellow, with purple strakes. After each floure followeth a smooth cod, almost round, two inches long, wherein is contained seuen round Peason, somewhat rough, but after a curious manner, of the bignesse and taste of field Peason, and of a darke sand colour.

Lathyrus æstivus Bæticus flore cæruleo Boelij.

This is also like *Lathyris latiore folio Lobelij*, but smaller, yet greater than that with yellow floures, hauing also adjoyning to the flat stalkes, two eared sharpe pointed leaues, and also two other slender sharpe pointed leaues, about foure inches long, growing on a flat foot-stalke betweene them, an inch and an halfe long, and one tendrel betweene them diuided into two or three parts: the flours are large, and grow on long slender foure-square foot-stalkes, from the bosomes of the leaues, on each foot-stalke one: the vpper great couering leafe being of a light blew, and the lower smaller leaues of a deeper blew: which past, there come vp short flat cods, with two filmes, edges, or skins on the vpper side, like those of *Eruilia Lobelij*, containing within, four or fiue great flat cornered Peason, bigger than field Peason, of a darke sand colour.

Lathyrus æstivus edulis Bæticus flore albo Boelij.

This is in flat skinny stalkes, leaues, foot-stalkes, and cods, with two skins on the vpper side, and all things else like the said *Lathyrus* with blew floures, only the floures of this are milke white: the fruit is also like.

Lrthyrus æstivus flore miniato.

This is also in skinnie flat stalkes and leaues like the said *Lathyris latiore folio*, but far smaller, not three foot high: it hath also small sharp pointed leaues growing by couples on the stalke, betweene which grow two leaues, about three inches long, on a flat foot-stalk halfe an inch long: also between those leaues grow the tendrels: the floures are coloured like red lead, but not so bright, growing on smooth

smooth short foot-stalks one on a foot-stalke: after which follow cods very like those of the common field peason, but lesser, an inch and a halfe long, containing foure, fiue, or six cornered Peason, of a sand colour, or darke obscure yellow, as big as common field peason, and of the same taste.

Lathyrus palustris Lusitanicus Boelij

Hath also flat skinny stalks like the said *Lathyrus latiore folio*, but the paire of leaues which grow on the stalke are exceeding small as are those of *Lathyrus flore luteo*, and are indeed scarce worthy to be called leaues: the other paire of leaues are about two inches long, aboue halfe an inch broad, and grow from betweene those small leaues, on flat foot-stalkes, an inch long: betweene which leaues also grow the tendrels: the floures grow on foot-stalks which are fiue inches long, commonly two on a foot-stalke, the great vpper couering leaues being of a bright red colour, and the vnder leaues are somewhat paler: after commeth flat cods, containing seuen or eight small round peason, no bigger than a Pepper corne, gray and blacke, spotted before they are ripe, and when they are fully ripe of a blacke colour, in taste like common Peason: the stalkes, leaues, foot-stalkes and cods are somewhat hairy and rough.

Lathyrus æstivus dumetorum Bæticus Boelij

Hath also flat skinny stalkes like the said *Lathyrus latiore folio*, but smaller, and in the manner of the growing of the leaues altogether contrary. This hath also two small sharp pointed leaues, adjoyning to the stalke: betweene which groweth forth a flat middle rib with tendrels at the top hauing on each side (not one against another) commonly three blunt topped leaues, sometimes three on the one side, and two on the other, and sometimes but foure in all about an inch and a halfe long, the floures grow on foot-stalks, about two or three inches long, each foot-stalke vsually bearing two floures, the great couering leafe being of a bright red colour; and the two vnder leaues of a blewish purple colour: after which follow smooth cods, aboue two inches long, containing fiue six or seuen smooth Peason, of a browne Chestnut colour, not round but somewhat flat, more long than broad, especially those next both the ends of the cod, of the bignesse and tast of common field Peason.

Iuniperus sterilis.

This shrub is in the manner of growing altogether like the Iuniper tree that beareth berries, only the vpper part of the leaues of the youngest and tenderest bowes and branches are of a more reddish greene colour: the flours grow forth of the bosoms of the leaues, of a yellowish colour, which neuer exceed three in one row, the number also of each row of leaues: each floure is like to a small bud, more long than round, neuer growing to the length of a quarter of an inch, being nothing else but very small short crudely chiues, very thicke and close thrust together, fastened to a very small middle stem, in the end turning into small dust, which flieth away with the winde, not much vnlike that of *Taxus sterilis*: on this shrub is neuer found any fruit. 15. *May*. 1621.

WHen the last sheets of this worke were on the Presse, I receiued a letter from Mr *Roger Bradshaghe*, wherein he sent me inclosed a note concerning some plants mentioned by our Author which I haue thought fitting here to impart to the Reader: he writ not then who it was that writ it, but since hath certified me that it was one Mr *Iohn Redman* a skilfull Herbarist, to whom, though vnknown, I giue thankes, for his desire to manifest the truth and satisfie our doubts in these particulars.

BEcause you write that *Gerards* Herball is vpon a review, I haue thought good to put you in mind what I haue obserued touching some plans which by him are affirmed to grow in our Northern parts: first the plant called *Pyrola*, which he saith groweth in Lansdale, I haue made search for it the space of twenty yeares, but no such is to be heard of.

Sea Campion with a red floure was told him groweth in Lancashire: no such hath euer beene seene by such as dwell neere where they should grow.

White Fox-gloues grow naturally in Lansdale, saith he, it is very rare to see one in Lansdale.

Garden Rose he writes groweth about Leiland in Glouers field wilde: I haue learned the truth from those to whom this Glouers field did belong, and I finde no such thing, onely aboundance of red wilde poppie, which the people call Corne-Rose, is there seene.

White Whortles, as he saith, grow at Crosbie in Westmerland, and vpon Wendle hill in Lancashire: I haue sought Crosbie very diligently for this Plant and others, which are said to grow there, but none could I finde, nor can I here of any of the countrey people in these parts, who dayly are labouring vpon the mountaines where the Whortle berries abound, that any wihte ones haue

beene

beene seene, sauing that those which *Gerard* calls red Whortles, and they are of a very pale white greene vntill they be full ripe, so as when the ripe ones looke red, the vnripe ones looke white.

Cloud-berry assuredly is no other than Knout-berry.

Heskets Primrose groweth in Clap-dale. If Mr *Hesket* found it there it was some extraordinary luxurious floure, for now I am well assured no such is there to be seene, but it is only cherished in our gardens.

Gerard saith many of these Northerne plants do grow in Crag-close. In the North euery towne and village neere vnto any craggie ground both with vs and in Westmerland haue closes so called, whereby *Gerards* Crag-close is kept close from our knowledge.

Chamæmorus, sex Vaccinia nubis. Knot, or Knout-berry, or Cloud-berry.

THis Knot, Knout, or Cloud-berry (for by all these names it is knowne by vs in the North, and taketh these names from the high mountaines whereon it groweth, and is perhaps, as *Gerard* saith, one of the brambles, though without any prickles) hath roots as small as packe-thred, which creepe far abroad vnder the ground, of an ouerworne red colour, here and there thrusting more fastly into the mossie hillockes tufts of small threddy strings, and at certaine joynts putting vp small stalks rather tough than wooddy, halfe a foot high, somthing reddish below, on which do grow two or three leaues of a reasonable sad greene colour, with foot-stalks an inch long, one aboue another without order: the highest is but little, and seldome will spread open; they are something rugged, crispie, full of nerues in euery part, notched about the edges, and with some foure gashes a little deeper than the rest, wherby the whole leafe is lightly diuided into fiue portions. On the top of the stalke commeth one floure consisting of foure, sometimes of fiue leaues apiece, very white and tender, and rather crumpled than plaine, with some few short yellow threds in the midst: it standeth in a little greene huske of fiue leaues, out of which when the floure fades, commeth the fruit, composed of diuers graines like that of the bramble, as of eight, ten, or twelue, sometimes of fewer, and perhaps through some mischance but of three or two, so joyned, as they make some resemblance of a heart, from whence (it may be) hath growne that errour in *Gerard* of diuiding this plant into two kindes: the fruit is first whitish greene, after becommeth yellow, and reddish on that side next the Sun.

It groweth naturally in a blacke moist earth or mosse, whereof the countrey maketh a fewell wee call Turfe, and that vpon the tops of wet fells and mountaines among the Heath, mosse, and brake: as about Ingleborough in the West part of Yorke-shire, on Graygreth a high fell on the edge of Lancashire, on Stainmor such a like place in Westmerland, and other such like high places.

The leaues come forth in May, and in the beginning of Iune the floures: the fruit is not ripe till late in Iuly.

The berries haue a harsh and something vnpleasant taste.

THis Worke was begun to be printed before such time as we receiued all the figures from beyond the Seas, which was the occasion I omitted these following in their fitting places: but thinking it not fit to omit them wholly, hauing them by me, I will giue you them with their titles, and the reference to the places whereto they belong.

* In August last whiles this worke was in the Presse, and drawing to an end, I and Mr *William Broad* were at Chissel-hurst with my oft mentioned friend Mr *George Bowles*, and going ouer the heath there I obserued this small *Spartum* whose figure I here giue, and whereof you shall find mention, in the place noted vnder the title of the figure; but it is not there described, for that I had not seen it, nor could finde the description therof in any Author, but in Dutch, which I neither had, nor vnderstood. Now this little Matweed hath some small creeping stringy roots: on which grow somwhat thicke heads, consisting of three or foure leaues, as it were wrapt together in one skin, biggest below, and so growing smaller vpwards, as in *Schænanth*, vntill they grow vp to the height of halfe an inch, then these rushie greene leaues (whereof the longest scarce exceeds two inches) breake out of these whitish skins wherein they are wrapped, and lie along vpon the ground, and amongst these growes vp a small grassie stalke, some hand full or better high, bending backe the top, which carries two rowes of small chaffie seeds. It is in the perfection about the beginning of August.

FINIS.

Cyperus Indicus, siue Curcuma.
Turmericke.
Pag.33. Lib.1. Cap.27.

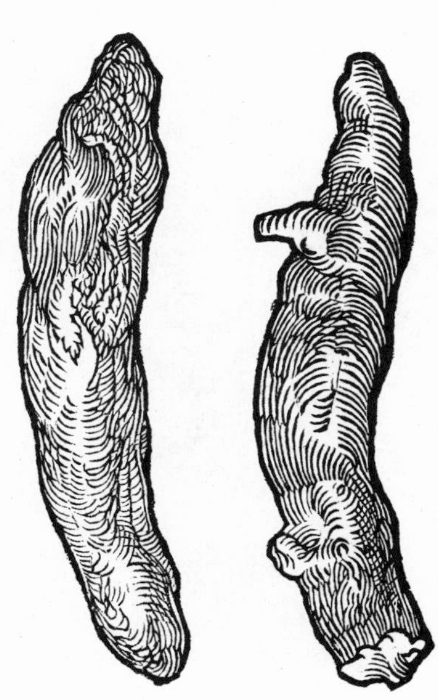

Iuncus minor capitulis Equiseti.
Club-Rush.
Pag.35. Lib.1. Cap.29. the fifth.

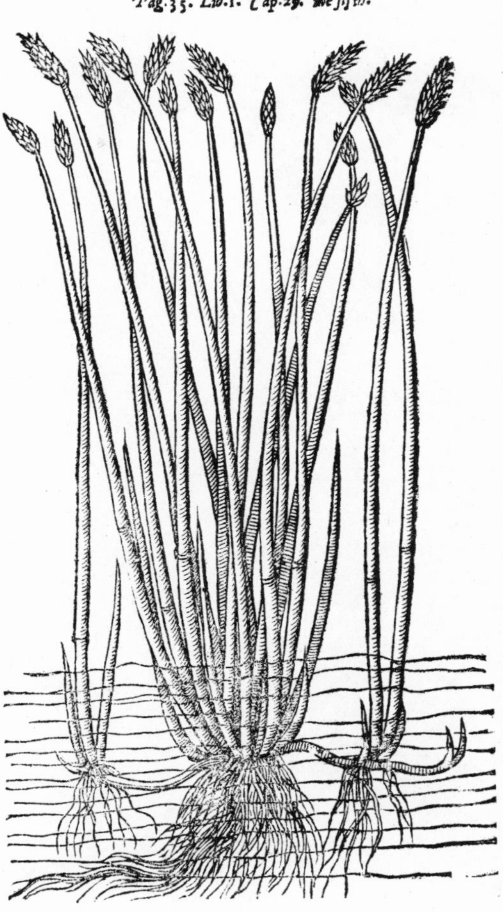

* *Spartum nostras parvum Lobelij.*
Heath Mat-weed.
Pag.41. lib.1. Cap.34. the fifth.

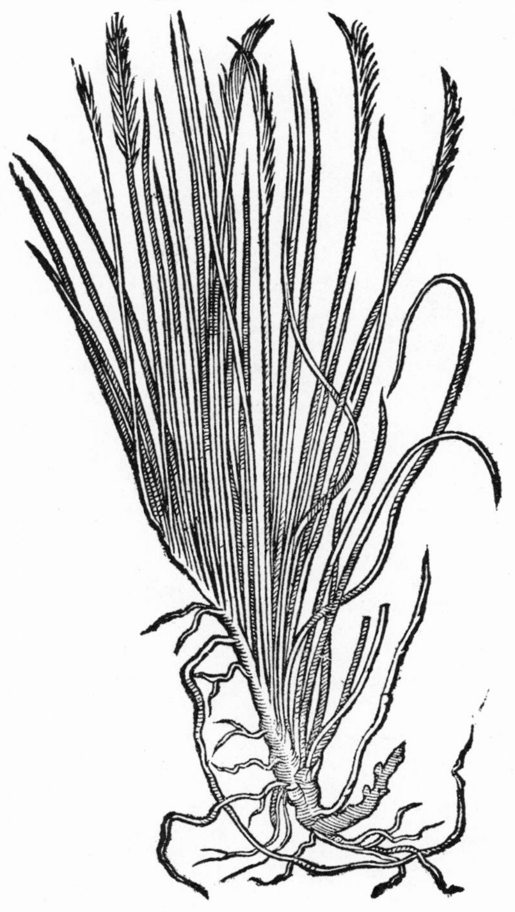

Schœnanthi flores.
The floures of Camels Hay.
Pag.43. lib.1. cap.35. the first.

INDEX LATINVS STIRPIVM IN HOC *opere descriptarum necnon nomina quædam Græca,* Arabica, Barbara, &c.

Aphaca.

B

Buna.

D

Daucus

Frute.

Pulicaris.

I

K

L

M

Molle.

T. Tabacum

Herba Lassulata, id est Balsamita maior.
Herba Pinnula, id est, Hyoscyamus.
Herba Turca, i. Herniaria.
Herba Hungarica Dodon. i. Alcea.
Herba Simeonis Dodon. id est, Alcea.
Herba Vrbana, i. Acanthus.
Herba Tunica Gordonij, id est Ocymastrum.
Herba Tunica Dodon, id est, Caryophyllata.
Herba Gallica Fracastorij, i. Galega.
Herba Rutinalis, i. Sphondylium.
Herba Sardoa, id est Ranunculus aquaticus,
Herba Sacra, i. Tabaco.
Herba Sacra Agrippæ, i. Melissa.
Hermodactylus Dodon. id est Colchicum.
Hermodact. Italorum, i. Iris tuberosa Lobel.
Hesperis Clusij, i. Leucoium marinum Lobelij.
Hippia, i. Alsine.
Hirundinaria, i. Asclepias.
Hortus Veneris, i. Cotyledon.
Horminum Tridentinum, id est Colus Iouis.
Humadh, i. Lapathum.
Hunen, i. Iuiube.
Huniure, i. Vrtica.
Hydroselinum, i. Paludapium.
Hydroselinum Camerarij, id est Lauer maius.
Hyoscyamus Peruvianus, i. Tabaco.
Hippoglossum, Bonifacia, id est Laurus Alex.
Hyosyris Plinij, i. Iacea nigra.
Hyophthalmon, i. Aster Atticus.
Hypecoon Dodon, id est Cuminum syluestre.
Hypecoon Clusii, i. Alcea Veneta.
Hipposelinon, i. Olus atrum.

I

Iarus, id est, Arum.
Iackaiak, i. Anemone.
Iasione, i. Campanula.
Iasin, i. Enula.
Iauatri, i. Nux Moscata.
Ibiga, i. Chamæpitys.
Iezar Serapionis, i. Pastinaca.
Imperatrix, i. Meum.
Inula Rustica Scribonij Largi, i. Consolida maior.
Inguinalis, i. Aster atticus.
Intybus, i. Cichorium.
Iouis Faba, i. Hyoscyamus.
Iouis Glans, i, Castanea.
Iouis Flos, i. Lychnis.
Iouis Arbor, i. Quercus.
Iorgir, i. Eruca.
Irio, i. Erysimum.
Iuncus quadratus Celsi, i. Cyperus.
Iua Muscata, i. Chamæpitys.
Iufacti, i. Sambucus.
Iusqiamus, i. Hyoscyamus.
Ixopus Cordi, i. Chondrilla.

K

Kanturion, i. Centaureum.
Kanz, i. Amygdalus.
Kaper, i. Capparis.
Kauroch. i. Chelidonium maius.
Kebikengi, i. Ranunculus.
Keiri, i. Leucoium.
Kemetri, i. Pyrus.
Kemum, i. Cuminum.
Kenne, i. Ligustrum.
Keruagh, i. Ricinus.
Kerugha, i. Ricinus.
Kermes, i. Coccus infectoria.
Kulb, i. Milium solis.
Kusbera Auerroij, id est Coriandrum.
Kusbor, i. Coriandrum.

L

Labrusca, i. Bryonia nigra.
Labrum Veneris, i. Dipsacus.
Laburnum, i. Anagyris.
Lactaria, i. Tithymalus.
Lactuca leporina, i. Sonchus.
Lactucella, i. Sonchus.
Lanata Cordi, i. Aria Theoph.
Lancea Christi, id est, Lingua Serpentina.
Lantana, i. Viburnum.
Lanaria, i. Radicula.
Lanceola, i. Quinqueneruia.
Laudata Nobilium, i. Veronica.
Lathyris, i. Cataputia.
Lathyrus, i. Pisum syluestre.
Lauer Lauacrum, i, Dipsacus.
Laurus Alexandrina, id est, Hippoglossum.
Laurus rosea, i. Oleander.
Laurus syluestris, id est, Laurus Tinus.
Laurentina Mathioli, i. Bugula.
Leo Columellæ, i. Aquilegia.
Leontostomium Gesneri, id est, Aquilegia.
Leo Herba Dodon. i. Aquilegia.
Lepidium Plinij, i. Piperitis.
Lesen Arthaur, i. Buglossum.
Leucacantha, i. Carlina.
Leucanthemum, i. Chamæmelum.
Libadion Plinij, i. Centaureum.
Libanium Apulei, i. Borago.
Limodoron Dodon, i. Orobanche.
Lingua auis, i. Fraxini semen.
Lingua Pagana, i. Hippoglossum.
Liliago Cordi, i. Phalangium Lobel.
Lisen, i. Plantago.
Lotus Vrbana, i. Trifolium odoratum Lobelij.
Longina, i Lonchitis.
Lichen, i. Hepatica officinarum.
Lunaria Arthritica Gesneri, i. Auricula Vrsi.
Luciola, i. Lingua serpentina.
Lunaria Græca, i. Bolbonac.
Lunaria maior Dioscor. id est, Alysson.
Luph Cordi, i. dracunculus.
Luiula, i. Trifolium Acetosum.
Lycostaphylos Cordi, id est, Sambucus aquatica,
Lycopsis, i. Buglossum syluestre.
Lycopersicum, i. Poma Amoris.

M

Machla, id est, Palma.
Madon Plinij, id est, Bryonia alba.
Mahaleb Auicennæ, id est, Pseudoligustrum.
Mahaleb, i. species Phillyreæ.
Magydaris Theoph. i. Laserpitium.
Malacocissos, id est, Hedera Terrestris.
Malinathalla Theop, i. Mala insana vel potius, Cyperus Esculentus.
Malacissus Cassani Bassi, i. Caltha palustris.
Maluauiscus, i. Ibiscus.
Manus Martis, i. Quinquefolium.
Marana, i. Stramonia.
Marathrum, i. Fœniculum.
Maru herba Dodon. id est, Cerinthe Plinij.
Marinella, i. Phu magnum.
Marmarites, i. Fumaria.
Marmorella, i. Agrimonia.
Mastastes, i. Laserpitium.
Maston Plinij, i. Scabiosa.
Mater Herbarum, i. Artemisia.
Materfilon, i. Iacea nigra.
Matrisaluia, i. Horminum.
Matrisylua, i. Periclymenum.
Maurohebræ Caput, id est, Antirrhinum.
Medium Dioscor. id est, Viola Mariana.
Medium Lobelij, i. Iris maritima Narbonensis.
Melochia, i. Corcorus.
Melampodium, i. Helleborus niger
Mel frugum Dioclis, i. Panicum.
Melampyrum, id est, Triticum Vaccinum.
Melaspermum, i. Nigella.
Melich Arab. id est, Trifolium fruticans.
Meleagris Flos, i. Frittillaria.
Melanthium, i. Nigella.
Melospinum, i. Pomum Spinosum.
Memiran Andr. Bellunensis, i. Chelid. maius.
Memiresin Auicen. idem.
Meud Hendi Arabibus, id est, Scamonea.
Memitha Arabibus, id est, Papauer Cornutum.
Memæsylum, i Arbutus.
Menogenion, i. Pæonia.
Mentha Saracenica, id est Balsamita maior.
Meu, id est Meum.
Memiren Serapionis, i. Chelidonium minus.
Methel, i. Stramonia.
Merzenius, i. Maiorana.
Mescatremsir, id est, Dictamnum.

Nominum quorundam interpretatio.

Mille grana, i. Herniaria.
Menianthe Theop. id est Trifolium paluſtre.
Militaris, i. Millefolium.
Miha, i. Styrax.
Millemorbia, i. Scrophularia.
Mixa, i. Sebeſten.
Molochia Serapionis, id est, Corcoros Matthioli.
Molybdena, id est, Dentillaria Rondeletij.
Momordica, i Balſamita mas.
Morghani Syriaca, id eſt, Fabago Belgarum.
Mochus Dodon. id eſt, Orobus Lobel.
Morella, i. Solanum Hortenſe.
Mula Herba Gazæ, i. Ceterach.
Multibona, i. Petroſelinum.
Mumeiz, i. Sycomorus.
Muralia Plin. i. Helxine.
Myophononon, i. Doronicum.
Myrtus ſylueſtris, i. Ruſcus.
Myrica, i. Tamariſcus.
Myriophyllum, i. Viola aquatilis.

N

NAbatnaho, id eſt, Mentha.
Nanochach, i. Ammi.
Nard & Naron Arab, i. Roſa.
Nardus Cretica, i. Phu magnum.
Nardus Ruſtica Plinij, i. id eſt, Conyza vel potius Aſarum.
Narf. i. Naſturtium.
Nargol, i. Palma.
Naſturtium hibernum, i. Barbarea.
Nenuphar, i. Nymphæa.
Neottia, i. Nidus auis.
Nepa Gazæ, i. Geniſta ſpinoſa.
Nerium, i. Oleander.
Nicophoron Plinij, i. Smilax aſpera.
Nicoſiana, i. Tabaco.
Nigellaſtrum, i. Pſeudomelanthium.
Nilofer, i. Nymphæa.
Nil Auicennæ, id eſt, Conuolvulus Cæruleus.
Nola Culinaria, i. Anemone.
Noli me tangere, i. Impatiens herba.
Noli me tangere, i. Cucumis ſylueſtris.
Nux Metel, i. Stramonia Fuchſij.
Nux Veſicaria, id eſt, Staphylodendron.
Nymphea minima, i. Morſus Ranæ.

O

OCulus Chriſti, id eſt, Horminum ſyl.
Odontis, id eſt. Dentillaria Rondeletij.
Olualidia, i. Chamamælum.
Olea Bohemica, i. Ziziphus alba.
Oleagnos, i. Chamelæa.
Oleaſtellum, i. Chamelæa.
Olus Iudiacum, i. Corcoros.
Olus album Dodon. i. Valeriana Campeſtris, vel Lactuca agnina.
Onagra Veterum, i. Chamænerium.
Onitis Plinij, i. Origanum.
Ononis, i. Reſta Bouis.
Onobrychis, id eſt, Caput Gallinaceum.
Onobrychis Belgarum, i. Campanula Aruenſis.
Onoſma, id eſt, Bugloſſum ſylueſtre.
Onopordon; id eſt, Acanthium Illyricum.
Ordelion Nicandri, i. Tordylion.
Ophris, i. Bifolium.
Ophiogloſſum, id eſt, Lingua ſerpentis.
Opuntia Plinii, i. Ficus Indica.
Opſago, i. Solanum ſomniferum.
Orbicularis, i. Cyclamen.
Orvala, i. Horminum.
Oreoſelinum, i. Petroſelinum.
Ornus, i. Fraxinus Bubula.
Orontium, i. antirrhinum.
Oſtria Cordi, i. Ornus Tragi.
Oſteocollon, i. Conſolida maior.
Oſtrutium, i. Imperatoria.
Oſyris, i. Linaria.
Othonna, i. Flos Africanus.
Oxyacantha, i. Berberis.
Oxyacanthus i. Spina appendix, vel pyracantha.
Oxys. i. Trifolium Acetoſum.
Oxymyrſine, i. Ruſcus.
Oxycoccus Cordi, id eſt, Vaccinia paluſtris.

P

PAlma Chriſti, id eſt, Ricinus.
Palalia, i. Cyclamen.
Pæderota, i. Acanthus.
Panis Cuculi, id eſt, Trifolium Acetoſum.
Pancratium, i. Squilla.
Panis porcinus, i. Cyclamen.
Papauer Spumeum, i. Ben album.
Paronychia Dioſcor. id eſt. Ruta Muraria.
Paſſerina, Ruellii, id eſt, Morſus Gallinæ.
Pedicularis herba, i. Staphiſagria.
Peduncularia Marcelli, id eſt, Staphiſagria.
Peganon, i. Ruta ſylueſtris.
Pentadactylon, i. Ricinus.
Peponella Geſneri, id eſt, Pimpinella.
Perlaro, i. Lotus arbor.
Perforata, i. Hypericon.
Perdicion, i. Helxine.
Periſterion, i. Scabioſa minima.
Perſonata, i. Bardana.
Pezicæ Plinii, ſunt fungi ſpecies.
Pes auis, i. Ornithopodium.
Pes Leonis, i. Alchimilla.
Pes vituli, i. Arum.
Pes Leporinus, i. Lagopus.
Petrum Americæ, i. Tabaco.
Petilius Flos, i. Flos africanus.
Pharnaceum, i. Coſtus Spurius.
Phaſganon Theop. i. Gladiolus.
Phalangitis, i. Phalangium.
Phellos, i. Suber.
Phellandrium, i. Cicutaria paluſtris.
Phellandrium Guillandini, i. Angelica.
Phœnix, i. Lolium.
Philomedium, i. Chelidonium maius.
Phileterium, i. Ben album.
Phleos, i. Sagittaria.
Phthirion, i. Pedicularis.
Phylateria, i. Polemonium.
Phillyrea Dodon. i. Liguſtrum.
Phyllon Theophraſti, i. Mercurialis.
Philanthropos, i. Aparine.
Picnacomon Anguill, i. Rheſeda.
Pimpinella ſpinoſa Camerarii, i. Poterion Lobel.
Pinaſtella, i. Peucedanum.
Piper aquaticum, i. Hydropiper.
Piper Calecuthium, Indum, Braſilianum, i. Capſicum.
Piper agreſte, i. Vitex.
Piſtatia ſylueſtris, id eſt Nux Veſicaria.
Piſtana, i. Sagittaria.
Planta leonis, i. Alchimilla.
Pneumonanthe Lobelii, i. Viola Calathiana Dodonæi.
Podagraria Germanica, id eſt Herba Gerardi.
Polytricum, i. Capillus Veneris.
Polytricum Fuchſii, id eſt Muſcus capillaris.
Polygonatum, id eſt, Sigillum Salomonis.
Polygonoides Dioſcoridis, id eſt, Vinca peruinca.
Polyanthemum, i. Ranunculus aquaticus.
Pologonum, i. Centumnodia.
Populago, i. Tuſſilago, vel Caltha paluſtris.
Potentilla maior, i. Vlmaria.
Pothos Coſtei, i. Aquilegia.
Pothos Theophraſti, i. Aquilegia.
Proſerpina herba, i. Chamomelum.
Protomedia, i Pimpinella.
Pſeudorchis, i. Bifolium.
Pſeudobunium, i. Barbaræa.
Pſeudocapſicum, i. Strichnodendron.
Pyrethrum ſylueſtre, i. Ptarmica.
Pteridion Cordi, i. Dryopteris Tragi.
Puſtech, i. Piſtacia.
Pulicaria, i. Conyza.

Q

Quemi, id eſt, Nigella.

R

RAdix Naronica, id eſt, Iris.
Ramel, i. Ciſtus.
Rapum terræ, i. Cyclamen.
Raginigi, i. Fœniculum.
Raledialemen Haliabbi, id eſt, Fumaria.
Rigina prati, i. Vlmaria.
Roſa fatuina, i. Pæonia.
Roſa Iunonis, i. Lilium.
Roraſtrum, i. Bryonia.
Rorella, i. Ros ſolis.
Rotula ſolis, i. Chamæleum.
Rhododaphne, i. Oleander.
Rhododendron, i. Oleander.
Rhuſelinum Apulei, i. Ranunculus.

Rima Maria, i. Alliaria.
Rincus Marinus, i. Crithmum.
Rubus ceruinus, i. Smilax aspera.
Rumex, id est Lapathum.
Ruta capraria, id est Galega.
Ruta palustris, id est Thalietrum.

S

Sabeteregi, id est Fumaria.
Sabaler, i. Satureia.
Sadeb, id est Ruta.
Sacra herba Agrippæ, i. Saluia.
Saffargel, i. Malus Cydonia.
Safarheramon, i. Sparganium.
Salicaria, i. Lysimachia.
Saliunca Gesneri, i. Nardus Celtica.
Salsirora, i. Ros solis.
Salicastrum Plin. i. Amara dulcis.
Salix Amerina, i. Salix humilis.
Saliuaris, i. Pyrethrum.
Saluia vitæ, i. Ruta muraria.
Saluia agrestis, id est Scordium alterum.
Salvia Romana, i. Balsamita maior.
Salusandria, i. Nigella.
Samalum Plin. i. Pulsatilla.
Samolum Plin. i. Anagallis Aquatica.
Sampsuchum, i. Amaracus.
Sanguis Herculis, id est Helleborus albus.
Sanguinaria, i. Cornu cerui.
Sanamunda. i. Caryophyllata quibusdã.
Sarax, i. Filix.
Sardinia glans, i. Castanea.
Sauch, i. Malus Persica.
Saxifragia lutea Fuchsii, id est Melilotus.
Saxifragia rubra, i. Philipendula.
Sagitta, i. Sagittaria.
Scammonea tenuis, i. Helxine Cissampelos.
Scandix, i. Pecten Veneris.
Scarlea, i. Horminum.
Scaunix Auerr. i. Nigella.
Scissima Gazæ, i. Fagus.
Schehedenegi, i. Cannabis.
Scheiteregi, i. Fumaria.
Scoparia, i Osyris.
Scolopendria, i. Lingua ceruina.
Scorodonia, i. scordium alterum, vel salvia agrestis.
Scorpio Theophrasti, i Genista spinosa.
Scolymos Dioscor. i. Cinara.
Scilla, i. Squilla.
Scuck Syriaca, i. Papaver Rhœas.
Secacul Monardi, i. Sigillum Salomonis.
Selago Plinij id est Sauina syluestris Tragi.
Seliem, i. Rapum.
Seligonion, i. Pæonia.
Selanion, i. Crocus vernus.
Selliga, i. Nardus Celtica.
Seminalis, i. Equisetum.
Sedum maius, i. sempervivum.
Seneffigi, i. Viola martia.
Serpentaria, i. Dracunculus.
Sertula Campana, i. Melilotus.
Serapias mas, i. Orchis fœmina Tragi.
Seygar, i. Nux moscata.
Sida Theoph. i. Althæa palustris.
Sideritis tertia Matth. i. Ruta canina Monspeliensium.
Sideritis, i. Marrubium aquaticum.
Sicilianiа Camerarij, i. Androsæmum Dodonæi.
Siger Indi, id est Palma.
Siringa cærulea Dodon. id est Lilac Matthioli.
Siliqua dulcis, i. Ceratia siliqua.
Silicula Varronis, i. Fœnugrecum.
Siliquastrum Plinij, i. Capsicum.
Sigillum Mariæ, i. Bryonia nigra.
Sin, id est Ficus.
Sinasbarium, i. Mentha aquatica.
Sinapi Persicum, i. Thlaspi.
Siser. id est Sisarum.
Silaus Plin. i. Thysselium.
Sison Syriacum, i. Ammi.
Sissitiepteris Plin. i. Pimpinella.
Siler Plin. i. Alnus nigra.
Sithim, i. Larix.
Smilax levis, id est Convoluulus maior flo. albo.
Smyrhiza Plin. i. Myrrhis.
Sorbus aucuparia, id est Fraxinus bubula.
Sorbus Alpina Gesn. i. Aria Theophrasti.
Sorbus syluestris, id est Fraxinus bubula.
Solanum rubrum, i. Capsicum.
Solanum lignosum Plinij, id est Amaradulcis.
Solanum tetraphyllum, id est Herba Paris.
Solanum vesicarium, i. Alkakengi.
Solatrum, i. Solanum hortense.
Solbastrella, i. Pimpinella.
Sosibio Theoph. i. Anemone.
Sparganion Matthioli, id est Platanaria.
Spina acuta, i. Oxyacanthus.
Spina acida, i. Oxyacantha.
Spina hirci, i. Tragacantha.
Spina infectoria, id est, Rhamnus solutivus.
Spina Iudaica, i. Paliurus.
Spiræa Theoph. i. Viburnum.
Sponsa solis, id est Ros solis.
Sphacelus Dodon. i. Scordium alterum Lobelii.
Splyte, i. radix cava.
Spicata, i. Potamogeiton.
Staphylodendron Plin. i. Nux vesicaria.
Statice Dalescamp. i. Caryophyllus marinus Lobelii.
Stataria, i. Peucedanum.
Stellaria Horat. Augerii, i. Carduus stellatus.
Struthiopteris Cordi, i. Lonchitis.
Struthium, i. Saponaria.
Strumaria Galeni, i. Lappa minor.
Strangulatoria Auicennæ, id est Doronicum.
Sucaram, i. Cicuta.
Succisa, i. Morsus Diaboli.
Surum Auicennæ, i. Nigella.
Symphytum, i. Consolida maior.
Symphoniaca, i. Hyoscyamus.
Supercilium Veneris, i. Viola aquatilis.
Supercilium terræ, id est, Capillus Veneris.
Sus, i. Liquiritia.

T

Tagetes Indica, id est Flos Africanus.
Tabaleb, i. Lens palustris.
Tamecnemum Cordi, i. Vaccaria.
Tarifilon Auicennæ. i. Trifolium bituminosum.
Tatula Clusii, i. Stramonia.
Tatoula Turcis, i. Pomum spinosum.
Tamus. Dodon. i. Bryonia nigra.
Taraxacon, i. Dens Leonis.
Tarfa, i. Tamariscus.
Teda arbor, i. Pinus syluestris.
Terzola, Baptistæ Sardi, i. Eupatorium cannabinum.
Tetrahit, i. herba Iudaica.
Terdina Paracelsi, i. Phu magnum.
Terpentaria, i. Betonica Aquatica.
Teliphano, i. Doronicum.
Thina, i. Larix.
Thut, i. Morus.
Thuia Theophrasti, i. Arbor vitæ.
Thysselium, i. Apium syluestre.
Thymbra, i. Satureia..
Tornsol bobo, i. Heliotropium.
Topiaria, i. Acanthus.
Trapezuntica Dactylus, id est Laurocerasus.
Tragium, i. Fraxinella.
Tragium Germanicum, i. Atriplex olida
Tremula, i. Populus Lybica.
Trifolium fibrinum, id est Trifolium palustre.
Trifolium cochleatum, i. Medica.
Trifolium fruticans, i. Polemonium.
Trifolium Asphaltites, i. Trifolium bituminosum
Tuber terræ, i. Cyclamen.
Turbith, i. Thapsia.
Turbith Auiceunæ, i. Tripolium.
Typhium Theophrast. i. Tussilago.

V

Vesicaria peregrina, i. Pisum cordatum.
Veelgutta, Dod. i. Petroselinum.
Veratrum, i. Helleborus.
Veratrum, nig. Dios. i. Astrantia nigra.
Verbascula, i. Primula veris.
Verdelhel Haliab, i. Ranunculus.
Victoriola, i. Hippoglossum.
Vitis alba, i. Bryonia.
Vitis Idæa, i. Vaccinia.
Virga sanguinea Matthioli, i Cornus fœmina.
Virga pastoris, i. Dipsacus.
Vitalis, i. Crassula.
Vitalba, i. Viorna.
Viticella, i. Momordica.

Vincetoxicum.

Nominum quorundam interpretatio.

Vincetoxicum, i. Aſclepias.
Viola nigra, i. Viola martia.
Viola flammea, i. Viola tricolor.
Viperaria, i. Scorzonera.
Viſnaga, i. Gingidium.
Vmbilicus Veneris, i. Cotyledon.
Vnedo Plin. i. Arbutus.
Vngula caballina, i. Tuſſilago.
Vlticana, i. Solanum ſomniferum.
Vlpicum Columellæ, i. Allium.
Vrinaria, i. dens leonis.
Vſnea, i. Muſcus.
Vua lupina Marcelli, i. Sambucus aquatica.
Vua taminia, i. Bryonia nigra.
Vua lupina, i. Herba Paris.
Vua verſa, i. Herba Paris.
Vua vulpis, i. Solanum hortenſe.
Vvularia, i. Hippogloſſum.
Vvularia, i. Laurus Alexandrina.
Vvularia Dodonæi, i. Trachelium.
Vulvaria, i. Atriplex olida.
Vulgago Maceri, i. Aſarum.
Vncata Caya, i. Stramonia.

X

Xaier, i. Alniriem Libanotis.
Xanium, i. Melanthium.
Xylon, i. Goſſipium.
Xylocaracta, i. Ceratia ſiliqua.
Xyphium, i. Gladiolus.

Y

Yebet, i. Anetum.

Z

Zahara Auicennæ, id eſt Anthyllis Lobel.
Zaiton, i. Olea.
Zarund, i. Ariſtolochia.
Zarza parilla, i. Sarſa parilla.
Zerumbeth, i. Zedoaria.
Zizania, i. Lolium,
Zinziber caninum, i. Capſicum.
Ziziphus, i. Iuiubæ.
Ziziphus alba, i. Elæagnus Matth.
Ziziphus alba Camerarii, i. Olea Bohemica.

A Table of ſuch Engliſh names as are attributed to the Herbes, *Shrubs, and Trees mentioned in this Hiſtorie.*

A

B

Blew

A Table of English names.

Cotton

A Table of English Names.

Foredit

A Table of English names.

M

N

O

P

A Table of English Names.

Sallow

A Table of English Names.

T

A Table of English Names.

A Supplement or Appendix vnto the generall Table, and to the *Table of English Names, gathered out of antient written* and printed Copies, and from the mouthes of plaine and ſimple country people.

A

Anet is Dill.
Amee, Ameos.
Argentill, Percepier.
Ache, Smallage.
Alliaria, in written copies Cardiaca.

B

Baldmoine, Gentian.
Baldmonie, Meum.
Baldwein, Gentian.
Belweed, Iacea nigra.
Biſhops worts, Betony.
Birds neſt, wilde Parſnep.
Birds tongue, Stitchwort
Bigold, Chryſanthemum ſegetum.
Blew ball. Blew bottle.
Bolts, Ranunculus globoſus.
Bow-wood, Knapweed.
Browne begle, Bugle.
Bruſewort Conſolida minor.
Brotherwort. Puliol mountaine.
Brydewort, Vlmaria.
Bright, Cheledonia.
Brokeleake, water Dragons.
Bruſewort, Sopewort.
Bucks beans, Trifolium paludoſum.
Buckram, Aron.

C

Cardiacke, Alliaria.
Carſes, Creſſes.
Catmint, Nepta.
Cencleffe, Daffodill.
Chaffeweed, Cottonweed.
Cheruell or Chenerell was called (though vntruly) Apium riſus.
Churles Treacle, Allium.
Churchwort, Penny-royall.
Ciderage, Arſmart.
Clithe, the Burre docke.
Clitheren, Gooſe graſſe or Cliuers.
Clite, Lappa.
Cloue tongue, Elleborus niger.
Cocks foot, Columbine.
Cocke foot, Cheledonia maior.
Cow fat, Cow Baſil.
Criſtaldre, the leſſer Centory.
Croneberries, Vaccinia paluſtris.
Crowbell, yellow Daffodil.
Crow berries, Erica baccifera.
Crowfoot is Orchis, in Lincolneſhire and Yorkſhire.
Crow ſope, Sopewort.
Crow leeke, Hyacinthus Anglicus.
Cropweed, Iacea nigra.
Culuerwort, Columbine.
Culrage, Arſmart.
Cutberdole and Cutbertill is Branke vrſine.

D

Dilnote, Cyclamen,
Donninethel. wilde Hempe.
Dragons female, water Dragons.
Dropwort, Filipendula.
Duncedowne, Catstaile.
Dwale is Nightſhade.

E

Edderwort, Dracontium.
Etleber. Alliaria.
Elfedocke, Enula Campana.
Earth gall, great Centory, or rather ſmall.
Euerferne is wall Ferne.
Exan, Croſwort, yet not our Cruciata.

F

Fane, white Floure deluce.
Fauerell, Cepea.
Field Cypreſſe is Chamæpitys.
Fieldwort. Felwort or Gentian,
Filewort, Filago minor.
Fleadocke, Petaſites.
Fleawort, Pſyllium.
Forget me not, Chamæpitys.
Forebitten more. Diuels bit.
Fauerole, water Dragons.
Franke, Spurry.
Freiſer is the herbe that beareth Straw-berries, Strawberrier.

G

Gallingal meke is Ariſtot. rotunda.
Gaten tree or Gater tree is Dogs berry tree.
Gandergoſſes is Zeekes.
Geckdor, Aparine.
Good King Harry English Mercury.
Gooſechite, Agrimony.
Gooſegraſſe was ſometime called Argentina.
Gooſe bill. Aparine.
Garden Ginger Piperitis.
Glond, Cow Baſil
Grace of God, S. Iohns wort,
Greene Muſtard, Dittander.
Groundwill, Groundſwell.
Ground needle, Geranium muſcatum.
Ground Enel, Venus combe.

H

Hairene, Cliuers.
Hammerwort, Pellitory of the wall.
Hardhow, Marygolds.
Hares eye. Lychnis ſylueſtris.
Harebell, Crow leeke.
Herbe Iuy. Chamæpitys.
Henbell, Henbane.
Heihow. Hedera terreſtris.
Herbe Bennet. Hemlocke.
Herbe Peter, Cowſlip.
Herba martis, Martagon.
Hertelowre, Chamædryos.
Hertwort, Fraxinus.
Hilwort, Puliol mountaine.
Hippia maior. common Pimpernel.
Holy rope, wilde Hempe,
Houndberry, Solanum.
Horewort, Filago.
Horſechire, Germander.
Horſeheale, Elecampane.
Horſe thiſtle, wilde Lettuce.
Honeſty, Bulbonack.
Horſelder. Enula campana.
Horſemint, Watermint,
Hulwort Polium.
Hyndheele, Ambroſia.

I

Iacea alba, wilde or White Tanſy.
Imbrecke, Houſleeke
Ioane ſiluer pin double Poppy.

K

Kandlegoſts, Gooſegraſſe.
Kings crowne, Melilotus.
King cob or King cup is Crowfoot.
Kiſſe me ere I riſe Panſies.
Kidneywort, Nauelwort.

L

Lungwort, Helleborus albus.
Little Wale is Gromell.
Lichwort is Pellitory of the wall.
Longwort, Pellitory of Spaine.
Lilly leeke, Moly.
Lilly riall, Pennyroyall.
Lodewort, water Crowfoot.
Louſewort, Staphiſacre.
Luſtwort is Sundew.
Lyngwort, Helleborus albus.

M

Mans Motherwort Palma Chriſti.
May bloſſomes, Conual Lillies.
Mawroll. White Horehound.
Mauthen or Mathes, Cotula fœtida.
March. Smallage.
March beetle, Cats taile.
Meedles Arage
Merecrop. Pimpernel.
Morel Nightſhade
Mouſepeaſe, Orobus.
Mugwet, Woodrooſe.

N

Neele. Lollium.
Neſpite, Calamint.
Nep, Cats mint.
Noſebleed, Yarrow.

O

Orual, Orpin.
Oxan Cruciata.
Oxtongue, Lingua boui.

P

Pagle, Stitchwort.
Palme de Dieu, *Palma Chriſti.*
Papwort, Mercury.
Paſtell, Woade.
Pedelion Helleborus niger.
Peters ſtaffe Tapſus barbatus.
Peuterwort, Horſetaile.
Pimentary, Baulme.
Powkneedle, Storks bill.
Primroſe, Liguſtrum.
Pygle, Gramen Leucanthemum.

A Table of obſolete and leſſe vſed English Names.

R

Rams foot is water Crowfoot.
Red knees is Hydropiper.
Robin in the hoſe is Lychnis ſylveſtris.
Rods gold is Marigold.

S

Scabwort is Enula Campana.
Sea Docke is Branke vrſine.
Seggrom is Ragwort.
Selfe heale was ſometimes called Pimpernel.
Sheep killing is Cotyledon aquatica,
Sleepewort is Lettuce.
Staggerwort and Stauerwort is Iacobea.
Stanmarch is Aliſander.
Standelwelks is Satyrion.
S. Maries ſeed is Sow thiſtle ſeed.
Small honeſty is Pinks.
Somerwort is Ariſtolochia.
Stike pile is Storks bill.
Stedfaſt is Palma Chriſti.
Stobwort is Oxys.
Sparrow tongue is Knot graſſe.
Stonnord and Stonehore is Stonecrop.
Stubwort is Wood Sorrell.
Swines graſſe is Knot graſſe.
Swine Carſe is Knot graſſe.
Swichen is Groundſwell.
Sowdwort is Columbine.

T

Talewort is wilde Borage.
Tanke is wilde Parſnep.
Tetterwort is great Celandine.
Toothwort is Shepheards purſe.
Tutſane is Clymenum Italorum.

W

Walwort is Ebulus, which was ſometime called Filipendula.
Warence is Madder.
Warmot is Wormwood.
Waywort is Pimpernell.
Waybread is Plantago.
Waywort is Hippia maior.
Waterwort is Maidenhaire.
Weythernoy is Feuerfew.
White Bothen is great Daiſy.
Wilde Sauager is Cockle.
Wilde Nardus is Aſarum.
White Golds is great Daiſy.
Wood march is Sanicle.
Wood ſower is Oxys,
Woodbroney is Fraxinus.
Woodnep is Ameos.
Wolfes thiſtle is Chamæleon.
Wyneberry is Vaccinea.
Wymot is Ibiſcus.
Wit is Hyoſcyamus luteus.

Y

Yron head is Knapweed.

Z

Zeekes was counted Satyrion minor, and is that which Lobel calleth Serapias fœmina pratenſis.

A Catalogue

A Catalogue of the Brittish Names of Plants, sent me by Master Robert Dauyes of Guissaney in Flint-Shire.

A

A Net. Dill.
Aurddanadl. Red Archangell Nettles.
Aûrvanadl. vide *Hwb yr ychen.*

B

B *Anadyl.* Broome.
Banatlos. Furze.
Berw yr Frengie. Cresses.
Berw yr dwr. water Cresses.
Bedwen. a Birch tree.
Biattus. Beets.
Blaen yr Twrch. Mercury.
Blaen y gwayw. Spearewort.
Bleidd dug. Wolfes bane.
Brialbu Mair. Cowslips.
Brwynen. a Rush.
Bylwg. Cockle, or field *Nigella.*
Bust yl y Ddayar. Centorie.

C

C *arn yr ebol.* Folefoot.
Cas gan gythrel. Veruaine.
Cacamweej. Burre.
Caliwlyn y mél. Agrimonie.
Cancwlwm. Knot grasse.
Camamill. Camomil.
Ceirch. Oats.
Cennin. Leekes.
Cennin Pedr. Daffodill.
Gedor y wrach. Horsetaile.
Cegid. Hemlocke.
Celynen. Holly.
Chwerwlys yr kithin. Wood Sage.
Clust yr ewic. Laurell.
Clofslops. Gillofloures.
Clustiev yr Derw. vide *Galladr.*
Clust llygoden. Mouse eare.
Claiarlys y dwr. Brookelime.
Coed Ceri. Seruice tree.
Cowarch. Hempe.
Cower y llaeth, Caliwlyn y mél,
Coed kirin. Plum trees.
Corsen. a Poole reed.
Cribe y Bleiddiev. v. *Cacamweei.*
Craith vnnos. Prunel or Selfe heale.
Crafanke y vrân. Crowfoot.
Cribe san Fraid. Betony.
Cynglennydd. white Mullen.
Cynfon y Celioc. Setwell.

D

D *Ail y gwaed.* Penny royall.
Danadl. Nettles.
Danadlen wenn. White Archangell Nettle.
Danit y llew. Dandeleon.
Danadlen ddall. dead Nettle.

E

E *Bolgarn yr ardd.* Assarabacca.
Efrev. Darnell.
Eiddew. Iuy.
Eiddew y ddayar. }
Eidral. } ground Iuy.
Eithin yr ieir, v. *Hwb yr ychen.*
Erienlys. S Iohns wort.
Erbin. Calamint.
Eulun persli. bastard Parsley.

F

F *A* Beanes.
Fenich y Cwn. wild Cammomil.
Fenich Fenell.
Fettes Fitches.

G

G *Alladr.* Lungwort like Liuerwort.
Garllec. Garlicke.
Glesyn y Coed. Bugle.
Gladyn. Gladiol or Corne Flag.
Geleudrem, v *Llysie Ewfras.*
Gold Mair, Marigold.
Gruc. v. *Banatlos.*
Grayanllys y dwr. Brooke lime.
Gwlydd. small Chickweed.
Gwlydd Mair. Pimpernell.
Gwenynddail Gwenynoc. Balme.
Gwyddfyd. Woodbind or Honisuckle
Gwden y Coed. Smooth Bindewood.
Gwallt gwener. Venus haire.
Gwallt y forwyn. Maiden haire.
Gwayw yr Brenhin. Daffodil.
Gwenith. Wheat.
Gwinwydden. Vine.

H

H *Ad y gramandi.* Gromel.
Haidd Barly.
Hesc melfedoc. Water Torch, or *Typha palust.*
Hoccys. Mallowes.
Hoccys y gors Marish Mallowes.
Hwb yr ychen. Camock, or rest harrow

LL

L *Laeth bron Mair.* Sage of Ierusalem.
Llaulys. Stauesacre.
Llawenllys. Borage.
Llewic ychwannen, v. *y Benfelen.*
Llewic yr iâr. Henbane.
Llewpard dûg. Aconitum.
Llysie Ivan. Mugwort.
Llysie llwydion, v. *Llsie Ivan.*
Llysie llewelyn. Pauls Betony.
Llysie y wennol. Celandine.
Llym y llygaid, v. *Llysie y wennol.*
Llysie Effras. Eyebright.
Llysie yr Crymman, v. *Gwlydd Mair.*
Llysie lliw, vide Dyars weed.
Llysie pen tû Housleeke.
Llysie yr gwaedlin. Yarrow or Milfoile
Llysie Mair, vide *Gold mair.*
Llysie Amor. Floure gentle.
Llygaid y Dydd. Daisies.
Llysie yr pwdin, v. *Dail y gwaed.*
Llysie yr gâth, v *Erbin.*
Llysie y Blaidd, v. *Bleid dûg.*
Llysie y moch. Nightshade.
Llysie y Cribev. Teasell.
Llysie Simion, v. *Cas gan gythrel.*
Llysie yr Cyrph. Periwinckle.
Llysie Eva. }
Lyriaid y mor. } Sea banke horne.
Llysie yr meddyglyn. wilde Carret.
Llwyfen. Elme tree.
Llwynlys. Scuruy grasse.

M

M *Afon* Raspis.
Marchalan. Elecampane.
March rhedyn y derw. Polypody, Oke Ferne
Maip. Turneps.

March

March yſgal y gerddi, Artichoke.
Meſys. Strawberries.
Menig ellyllion. Fox gloues.
Meirw. Iuniper tree.
Meillionen y meirch. Right Trefoile.
Mintas. Mints.
Moron. Parſneps.
Moron y maes. wilde Parſneps.
Mwg y ddayar. Fumetory.
Mwſogl. Moſſe.
Mynawyd y bigail. Storks bill.

N

Nyddoes. Spinage.

O

Onnen. an Aſh tree.

P

PAwen yr Arth. Beares breech.
Padere Mair. Croſſewort.
Perſli y dwr. water Parſley.
Perſli Frengic. Smallage.
Phion ffrwyth, v. *Menic yellyllion*.
Pidni y goc. Aron, or Cuckow pint.
Poerlys, v. *y laſlys*.
Poplys. a Poplar.
Pwrſ. y Bigail. Shepheards purſe.
Pys y Ceirw. Tares.

R

RHedyn. Ferne.
Rhedegat y derw. v. *Galladr*.
Rhûg. Rie.
Rhoſyn. a Roſe.

S

SAeds gwylle, v. *Chwerwlys*.
Siwdrmwt. Sothernwood.
Siacckedy melnydd, v. *Cynffon llwynoc*.
Sirian. Cherries.
Snoden Fair. Engliſh Galingall.
Sowdl y Crydd. v. *Blaen yr yiwrch*.
Suran y gôc wood Sorrell.
Suran. Sorrell.
Syſi, v. *Meſys*.

T

TAfod y ki. Dogs tongue.
Tafod y neidr. Adders tongue.
Tafod yr hydd. Harts tongue.
Tafol. a Docke.
Tafol Mair. Biſtort.
Tagaradr, v. *Hwb yr ychen*.
Tafod yr edn. Birds tongue.
Tafod yr ych. Bugloſſ.
Telephin. Orpin.
Tormaen, Filipendula.
Tryw, v. *Caliwlyn y mêl*.
Troed y glomen. Columbine.
Triacl y tylodion. Tormentilla.
Troed y dryw. Parſley Breakſtone, or ſmal Saxifrage.
Triacl y Cymro. Germander.
Troed yr hedydd. Larke heele.

W

Wilffraev, v. *Llyſie yr gwaedlin*
Winniwn. Onions.

Y

YBewfelen. Fleabane.
Y benlas wenn, v. *Claſrlys*.
Y bengaled. red Scabious.
Y benlas. Blewbottle, or Cornfloure.
Y bengoch. Horehound.
Y Claſrlyes Scabious.
Y Dorfagl. medow three leafed graſſe.
Y Droedrydd. Herbe Robert.
Y Drwynſawr. *Caliwlyn y mêl*.
Y Ddwy genniog. herb Twopence, or Moneywort.
Y Dorllwyd. wild tanſy or Siluerweed
Y dew bannoc, v. *Cynffon Llwynoc*.
Y Dinboeth. Arſmart.
Y Ddayarlys. Peony.
Y Doddedigc wenn. Pilewort.
Y fendigedi. Tutſan or Parke leaues.
Y Fabgoll. Poppy.
Y fiolud. Violet.
Y fylfen. *Y fronwys*. ſmall Celandine.
Y feidioc lâs, v. *Llyſſie Ivan*.
Y fyddarlys. Prickmadam.
Y fyddygyn, v. *Craith vn nos*.
Y fyw fyth. *Llyſſiev pentſi*.
Y gauri goch. v. *Bwſtl y Ddayar*.
Y gynga, v. *Llyſie yr bidl*.
Y gloria. wilde Roſe, or Spargwort.
Y gâs wenwyn. Diuels bit.
Y gyfog. a kinde of Spurge.
Y glaiarlys. *Y grevlvs*, Groundſwell.
Y gyſgadvr. Nightſhade or Morell.
Y gingroen. Todeflax.
Y llew gwynn dôf. Garden Orach.
Y llew gwynn gwyllt, wilde Orach.
Y lliwlys, v. *Llyſiev lliw*.
Y llwynhidydd. Ribwort.
Y llindro. Doder.
Y llyſiewyn bendigedic. Valerian.
Y lleûadlys. Lunaria.
Y Môr gelyn. Sea Holly.
Y Mûrlys. Pellitory of the wall.
Y Papi coch, v. red Poppy, or corne Roſe.

Yr Eſcarlys	*Hir*	Ariſtolochia,	long.
	geon	or Birthwort,	round.
	bychan	or Hartwort,	ſmall.

Yr Alaw. Water Lilly.
Yr hên lydan, i. *ffordd*. Waybread.
Yr Rhût. Rue, or herbe Grace.
Yr vchelfa. Miſſeltoe.
Yr yſcallen Fraith. our Ladies thiſtle.
Yr yſcallen Fendigedic. Card. Benedict.
Yr holliach. Clownes wort.
Yſcall drain gwynn Carline Thiſtle.
Yſcall. wilde Thiſtles.
Yſcall y moch. Sow thiſtle.
Yſcol fair. Peters wall or ſquare, S. Iohns wort.
Yſcaw. Elder trees.
Yſcaw Mair. Walwort.
Yſpaddaden. White thorne.
Yſniab. Muſtard.
Y wermod. Wormwood.
Y wermod wenn. Feuerfew.
Y winwydden wenn. white Bryonie.
Y winwydden ddû. blacke Bryony.
Y wilffrae. *Llyſie yr gwaedlin*.
Y wennwlydd. Great Thickweed.

A TABLE, WHEREIN IS CONTAINED THE NATVRE AND VERTVES OF ALL THE Herbes, Trees, and Plants, described in *this present Herbal.*

A

B

The Table of Vertues.

C

To

D

E

F

G

H

I

M. A

M

N

O

The Table of Vertues.

For

To

The Table of Vertues.

Y

Errata.

Errata.

I would wiſh the courteous Reader to take notice and amend theſe faults eſcaped in the printing, and to pardon other ſuch literall faults as he may perhaps here and there obſerue.

Faults in Figures tranſpoſed.

Pag. 48. The two figures of *Phalangium ramoſum* & *Phalangium non ramoſum* are put one for another.
Pag. 50. The two figures are put one for another. And likewiſe in *Pag.* 808. the two firſt figures are tranſpoſed.

Faults in Words and Marks.

Pag.9.lin.1. *elegaſis*, reade *elegans*. p.31,l.32, *Cyriacus*, r. *Syriacus*. p.84,l.22, *longiſſimo*, r. *longiſſima*. p.186,l.1. for 79, r.101. p.242, title, *Lepidium annum*, r. *annuum*. p.228,l.15, *abortinum*, r. *abortivum*. 229,l.14, *arbortirum*, r. *abortivum*. p.245,l.1. Wilde, reade white. p.256,l.1. in the title, adde the figure 2. p.282,l.17. *Itybus*, r. *Intybus*. p,289, l.4. *Verracarium*, r. *Verrucarium*. p.494. l.43. *Anticarbinum*, r. *Antirrbinum*. p.604,l.7, hath been abſurd from, r. had been abſurd, for. p.848,l.15. *Virginia*, r. *Virginiana*. p.929,l.21. *Multea*, r. *Malva*. p.935,l.28. Lilly, r, Mallow. p,941,l.13, *Arcus*, r. *Acus*. p.1011, l.25. *Strum*, r. *Strumæ*. p.1016,l.19. *Macedonium*, r. *Macedonicum*. p.1051,l.4. *Seſeli Creticum*, r. *Seſeli montanum*. p.1133,l.37, Oken caſe, r Oken leafe. p.1323,l.7, Reſt-Yarrow, r. Reſt-Harrow. p.1401, l.50 & 51. *Cnidicus*, r. *Cnidius*. p.1424,l.17. vpon, r. open. p.1524,l.40, a pleaſant, r. pleaſant a. p.1628, l.39. them, r. it.

Pag. 169, lin. vlt. put ‡. p.184,l. penult. † put ‡. p.257,l.16 & 20, put ‡.‡. pag.203, l.18, put ‡. pag. 261,l.13. put ‡. and l.17, put ‡. p.264,l.5 & 12, for † † put ‡ ‡. p.287,l.6, for † † put ‡ ‡. p.303,l.12, put ‡. p.1143,l.2, put ‡. p.1339,l.8, put ‡.